Biology

Eighth Edition

Jonathan B. Losos
Harvard University

Kenneth A. Mason
Purdue University

Susan R. Singer
Carleton College

based on the work of

Peter H. Raven
Director, Missouri Botanical Gardens;
Engelmann Professor of Botany
Washington University

George B. Johnson
Professor Emeritus of Biology
Washington University

Boston   Burr Ridge, IL   Dubuque, IA   New York   San Francisco   St. Louis
Bangkok   Bogotá   Caracas   Kuala Lumpur   Lisbon   London   Madrid   Mexico City
Milan   Montreal   New Delhi   Santiago   Seoul   Singapore   Sydney   Taipei   Toronto

Higher Education

BIOLOGY, EIGHTH EDITION

Published by McGraw-Hill, a business unit of The McGraw-Hill Companies, Inc., 1221 Avenue of the Americas, New York, NY 10020.

 This book is printed on recycled, acid-free paper containing 10% postconsumer waste.

1 2 3 4 5 6 7 8 9 0 DOW/DOW 0 9 8 7 6

ISBN 978–0–07–296581–0
MHID 0–07–296581–9

Publisher: *Janice Roerig-Blong*
Executive Editor: *Patrick E. Reidy*
Senior Developmental Editor: *Anne L. Winch*
Director of Marketing: *Chad E. Grall*
Lead Project Manager: *Peggy J. Selle*
Senior Production Supervisor: *Kara Kudronowicz*
Senior Media Project Manager: *Jodi K. Banowetz*
Senior Media Producer: *Eric A. Weber*
Senior Coordinator of Freelance Design: *Michelle D. Whitaker*
Cover/Interior Designer: *Christopher Reese*
Senior Photo Research Coordinator: *Lori Hancock*
Photo Research: *Pronk &Associates*
Supplement Producer: *Melissa M. Leick*
Art Development/Art Studio: *Electronic Publishing Services Inc., NYC*
Compositor: *Electronic Publishing Services Inc., NYC*
Typeface: *10/12 Janson*
Printer: *R. R. Donnelley Willard, OH*

(USE) Cover Image: *DNA: ©Doug Struthers/Getty Images; Female Mountain Gorilla: ©Daniel J. Cox/Getty Images; Dragonfly: ©Siede Preis/Getty Images; Maple Leaf: C Squared Studios/Getty Images; Knotweed Pollen: ©David Scharf/Getty Images*

The credits section for this book begins on page C-1 and is considered an extension of the copyright page.

Library of Congress Cataloging-in-Publication Data

Biology / Peter H. Raven ... [et al.]. -- 8th ed.
   p. cm.
Includes index.
ISBN 978–0–07–296581–0 --- ISBN 0–07–296581–9 (alk. paper)
   1. Biology--Textbooks. I. Raven, Peter H.
QH308.2.R38 2008
570--dc22
                    2006032857
www.mhhe.com

brief contents

about the authors...

Susan Singer is the Laurence McKinley Gould Professor of the Natural Sciences in the department of biology at Carleton College in Northfield, Minnesota, where she has taught introductory biology, plant biology, genetics, plant development, and developmental genetics for 20 years. Her research interests are focused on the development and evolution of flowering plants. Singer has authored numerous scientific publications on plant development, contributed chapters to developmental biology texts, and is actively involved with the education efforts of several professional societies. She received the American Society of Plant Biology's Excellence in Teaching Award, serves on the National Academies Board on Science Education, and chaired the National Research Council study committee that produced *America's Lab Report.*

Jonathan Losos is a Monique and Philip Lehner Professor for the Study of Latin America in the Department of Organismic and Evolutionary Biology and Curator of Herpetology at the Museum of Comparative Zoology at Harvard University. Losos's research has focused on studying patterns of adaptive radiation and evolutionary diversification in lizards. The recipient of several awards, including the prestigious Theodosius Dobzhanksy and David Starr Jordan Prizes for outstanding young evolutionary biologists, Losos has published more than 100 scientific articles.

Kenneth Mason is a Continuing Lecturer at Purdue University where he is responsible for the largest introductory biology course on campus. He is currently collaborating with faculty in the College of Engineering on an innovative new course supported by the National Science Foundation that combines biology, chemistry, and physics. Prior to Purdue, he was an Assistant Professor at the University of Kansas, where he did research on the genetics of pigmentation in amphibians, publishing both original work and reviews on the topic. While at KU he taught a variety of courses and wrote a lab manual used in the genetics laboratory course.

preface

Everywhere you look in this edition you will see a new Raven & Johnson's *Biology*. From the new author team, to the new design and art program and the completely revised content, *Biology* has undergone a transformation. This "new" text maintains the clear, accessible, and engaging writing style of past editions and the pervasive emphasis on evolution and scientific inquiry that have made this a leading textbook for students majoring in biology. Now we have coupled this approach to a modern integration of the exciting new research in molecular biology and genomics to offer our readers the most significant and important revision of our text since the first edition.

Our New Author Team

Perhaps the greatest change to this edition is the change in authoring responsibilities. Two of us, Jonathan Losos (Harvard University) and Susan Singer (Carleton College), became new co-authors in the seventh edition, and now we've taken on full authoring responsibilities for the eighth edition as we welcome Kenneth Mason of Purdue University to our team. Ken served as a key contributor in the genetics unit for the seventh edition. The quality of the work he provided, his expertise in genetics, and his experience in teaching majors in biology at the university level made him a natural choice to join our new author team.

We're excited about the opportunity to take what was already a high-quality textbook and move it forward in a significant way for a new generation of students. All of us have extensive experience teaching undergraduate biology and we've used this knowledge as a guide in producing a text that is up-to-date, beautifully illustrated, and pedagogically sound for the student. We've also worked to make it even easier to use and more closely integrated with its media support materials to provide instructors with an excellent complement to their teaching.

Our New Visual Program

Even a casual look through the pages of the eighth edition will show the care that went into developing the art and photo program and the related page design. A brand new visual program is rare in a revision, and our team found the opportunity to reevaluate the effectiveness of the artwork and photos to be an exciting challenge. Our goal has been to provide a clear, consistent, accurate art program that is easy-to-follow and beautifully three-dimensional.

To prepare for the revision, a variety of specialists reviewed the seventh edition art and photo program to assess its instructional effectiveness and presentational value. We worked closely with the artists and medical illustrators to produce sketches that effectively convey the chapter's most important concepts or that provide students with a particularly thought-provoking or dynamic example. A separate panel of reviewers then evaluated each newly rendered figure for pedagogical value and accuracy.

Our text-paging team then worked in conjunction with the artists to create innovative page spreads where the visuals and textual content function together in a well-coordinated and closely integrated manner. For complex processes, figures use numbered text boxes to lead the student step-by-step through the figure. For others, where the whole is more important than the pieces, the figure is not interrupted by text but explained thoroughly in the legend. Multilevel figures take students from a macro to micro view using "blow-out" arrows. Where figures can be put "inline" within the text this has been done to intimately connect the art with the text narrative and to allow the artwork to visually enliven the page.

Our Modern Content and Approach

One goal that unites our author team is the attempt to bring students an exciting and up-to-date view of modern biology. The extensive nature of this revision has produced exceptionally current content throughout. Likewise, rather than pasting paragraphs of new material into selected chapters, we have carefully worked together to reconsider our text's outline and coverage to provide a more consistent approach to concepts so that the reader is not buried in detail in one chapter and left wondering how something works in another.

Two new chapters in the evolution unit perfectly illustrate the importance of this goal. Chapter 24: *Genome Evolution* describes comparative genomics and explains how exciting new discoveries resulting from the genome sequencing of so many different species is revolutionizing our understanding of evolution. Chapter 25: *Evolution of Development* follows with a discussion of how changes in genes can produce changes in development patterns, which result in new characteristics and sometimes in speciation.

We've expanded our coverage of patterns of inheritance into chapter 12 on heredity and Mendelian principles and chapter 13 on chromosomal theory of inheritance. We've created a new chapter 23: *Systematics and the Phylogenetic Revolution* and we now provide chapter 26: *The Tree of Life* to offer a broad overview as a way of introduction to the unit.

Another individual example of the current content is the coverage of the newly discovered fossil that is transitional between fish and amphibians, *Tiktaalik*, which was announced in mid-2006 (see figure 35.14b). Similarly, our discussion of the state of the environment is based on up-to-the-moment data on population trends, global temperatures, and CO_2 levels.

The physiology unit has been reorganized and includes a new introductory chapter 43: *The Animal Body and Principles of Regulation*. This is designed to introduce the tissues and organ systems covered in later chapters and provide an understanding of control systems and their associated feedback mechanisms.

Our Consistent Themes

It is important to have consistent themes that organize and unify a text. We met extensively to discuss our approaches to teaching biology and to design the most effective text possible. A number of themes are used throughout the book to unify the broad-ranging material that makes up modern biology. This begins with the primary goal of this textbook to provide a comprehensive understanding of evolutionary theory and the scientific basis for this view. We use an experimental framework combining both historical and contemporary research examples to help students appreciate the progressive and integrated nature of science.

Biology Is Based Upon an Understanding of Evolution

When Peter Raven and George Johnson began work on *Biology* in 1982 they set out to write a text that presented biology the way they taught in their classrooms: as the product of evolution. Much as all biology "only makes sense in the light of evolution," this text is enhanced by a consistent evolutionary theme that is woven throughout the text, and we have enhanced this theme in the eighth edition.

The enhanced evolutionary thread can be found in obvious examples such as the two new chapters on molecular evolution, but can also be found throughout the text. As each section considers the current state of knowledge, the "what" of biological phenomenon, they also consider how each system may have arisen by evolution, the "where it came from" of biological phenomenon.

Our approach allows evolution to be dealt with in the context in which it is relevant. The material throughout this book is considered not only in terms of present structure and function, but how that structure and function may have arisen via evolution by natural selection.

Biology Uses the Methods of Scientific Inquiry

Another unifying theme within the text is that knowledge arises from experimental work that moves us progressively forward. The use of historical and experimental approaches throughout allow the student to not only see where the field is now, but more importantly, how we arrived here. The incredible expansion of knowledge in biology has created challenges for authors to decide what content to keep, and to what level an introductory text should strive. We have tried to keep as much historical context as possible and to provide this within an experimental framework consistently throughout the text.

Rather than interrupting the text with an experimental box, we describe experiments in the context of the concepts being provided. This keeps experimental approaches relevant to the story being told. Data are provided throughout the text and figures to illustrate how we have arrived at our present view of the various topics that make up the different sections of the book. Students are also provided with "Inquiry Questions" to stimulate thought about the material throughout the book. The questions often involve data that are presented in figures, but are not limited to this approach, also leading the student to question the material in the text as well.

Biology Is an Integrative Science

The explosion of molecular information has reverberated throughout all areas of biological study. Scientists are increasingly able to describe complicated processes in terms of the interaction of specific molecules, and this knowledge of life at the molecular level has illuminated relationships that were previously unknown. Using this cutting-edge information, we have made great strides to more strongly connect the different areas of biology in this edition.

One example of this integration concerns the structure and function of biological molecules, an emphasis of modern biology. This revision brings that focus to the entire book using this as a theme to weave together the different aspects of content material with a modern perspective. Given the enormous amount of information that has accumulated in recent years, this provides a necessary thread that integrates these new perspectives into the fabric of the traditional biology text.

Likewise, all current biology texts have added a genomics chapter, and our text was one of the first to do this. This chapter has been updated, but, more importantly, the results from the analysis of genomes and the proteomes that they encode have been added throughout the book wherever this information is relevant. This allows a more modern perspective throughout the book rather than limiting it to a few chapters. Examples, for instance, can be found in the diversity chapters, where classification of some organisms were updated based on new findings revealed by molecular techniques.

This systems approach to biology also shows up at the level of chapter organization. We introduce genomes in the genetics section in the context of learning about DNA and genomics. We then come back to this topic with an entire chapter at the end of the evolution unit where we look at the evolution of genomes, followed by a chapter on the evolution of development, which leads into our unit on the diversity of organisms.

Similarly, we introduce the topic of development with a chapter in the genetics section, return to it in the evolution unit, and have dedicated chapters in both the plant and animal units. This layering of concepts is important as we believe that students best understand evolution, development, physiology, and ecology when they can reflect on the connections between the microscopic and macroscopic levels of organization.

Our Enhanced Readability and Learning System

Biology has always been considered a user-friendly text, but the sheer volume of information in a major's biology text demands that authors do everything possible to make the content clear and well-organized to aid the student. In this eighth edition we have taken steps to help our readers through careful scrutiny of the narrative and thorough redesign of our pedagogical learning system.

Telling the Story of Biology

We had the benefit of an excellent developmental copyeditor who worked with us on each revised chapter prior to turning the manuscript over to production. The copyeditor focused on improving the use of headings to organize the content, improving the clarity of the writing, making the writing consistent between chapters, consistently identifying and defining the key terms, and eliminating redundant material within and among chapters. This process removed the clutter that can accumulate over several editions and ensured that each chapter reads smoothly from start to finish.

Providing a Learning System

Part of what makes *Biology* such an easy book to learn from is its consistent pedagogical framework, and we have strengthened this learning system in the eighth edition. Each chapter opens with an outline consisting of the numbered headings and the supporting headings. Throughout the chapter the interior design works with the content to remind the students of where they are in the chapter.

Each concept within a chapter is clearly demarcated at beginning and end. The declarative main headings and sentence-style supporting headings provide an excellent overview of each concept to be covered. Short interim summaries following each main concept within the narrative remind students of the most critical information they need to take away from that reading.

Extensive figure legends in conjunction with the new visual program provide a "visual outline" of all major ideas in every chapter. Great care was taken during the paging process to ensure that with few exceptions figures are displayed on the same pages where the narrative detail supporting those figures is provided.

We've also created an extensive set of summary tables throughout the text to make study and review as easy and efficient as possible. Content reviews at the end of the chapter recapitulate important content within the same conceptual framework provided in the opening outline. We've also developed new Challenge Questions to promote active learning and higher level critical thinking.

Our Commitment to You

We are united in our excitement about this opportunity to take a textbook that was already accurate and innovative and move it forward in a meaningful way. We believe our research and teaching experience provides us with the tools to couple the most significant modern research findings and scientific approach to an engaging pedagogical presentation.

In working with reviewers, contributors, focus group participants, students, and our colleagues around the globe, we have been impressed with the level of energy, thoughtfulness, and dedication faculty and students bring to the study of biology. It is a privilege to serve you through the pages of this textbook and its media support program.

Please let us know how we can serve you better by writing us at ravenbiology@mcgraw-hill.com. Best wishes to you in your own classroom. We look forward to hearing from you.

Jonathan Losos
Ken Mason
Susan Singer

ACKNOWLEDGMENTS

A revision of this scope relies on the talents and efforts of many people working behind the scenes and we have benefited greatly from their assistance.

Jody Larson was our developmental copyeditor who labored many hours and provided countless suggestions for improving the organization and clarity of the text. Her contributions had a tremendous impact on the quality of the final product.

We were fortunate to have Electronic Publishing Services on our side for the overhaul of the *Biology* art program. Kim Moss, Jen Christiansen, Martin Huber, Eliza Jewett, Patty O'Connell, and the rest of the team did a fantastic job in developing art and photo concepts and page spreads based on our manuscript and the ideas exchanged in our development meetings. Our close collaboration resulted in a text that is pedagogically effective as well as more beautiful than any biology text on the market. This was a giant undertaking and the staff at EPS handled it all with professionalism, skill, and good humor.

We were fortunate in our McGraw-Hill book team led by Patrick Reidy, executive editor, Anne Winch, senior developmental editor, Chad Grall, marketing director for life sciences, Peggy Selle, lead project manager, Michelle Whitaker, senior freelance design coordinator, Linda Davoli, production copyeditor, and many more people behind the scenes.

During this revision we have had the support of spouses and children, who have seen less of us than they might have liked because of the pressures of getting this revision completed. They have adapted to the many hours this book draws us away from them, and, even more than us, look forward to its completion.

As with every edition, acknowledgments would not be complete without thanking the generations of students who have used the many editions of this text. They have taught us as least as much as we have taught them, and their questions and suggestions continue to improve the text and supplementary materials.

Finally, we need to thank our reviewers and contributors. Instructors from across the country are continually invited to share their knowledge and experience with us through reviews and focus groups. The feedback we received shaped this edition, resulting in new chapters, reorganization of the table of contents, and expanded coverage in key areas. Several faculty members were asked to provide preliminary drafts of chapters to ensure that the content was as up to date and accurate as possible, and still others were asked to provide chapter outlines and assessment questions. All of these people took time out of their already busy lives to help us build a better edition of *Biology* for the next generation of introductory biology students, and they have our heartfelt thanks.

Preliminary Draft Revision Contributors

Brian Bagatto
University of Akron

Nancy Maroushek Boury
Iowa State University

Julia Emerson
Amherst College

T. H. Frazzetta
University of Illinois, Urbana-Champaign

Douglas Gaffin
University of Oklahoma

Gonzalo Giribet
Harvard University

Richard Hill
Michigan State University

Duncan S. MacKenzie
Texas A&M University

Elizabeth A. Weiss
University of Texas, Austin

End-of-Chapter Pedagogy and Inquiry Contributors

Arthur Buikema
Virginia Polytechnic Institute

Merri Lynn Casem
California State University-Fullerton

Mark Lyford
University of Wyoming

Peter Niewiarowski
University of Akron

Thomas Pitzer
Florida International University

Laurel Roberts
University of Pittsburgh

Michael Windelspecht
Appalachian State University

Reviewers and Accuracy Checkers

Barbara J. Abraham
Hampton University

Richard Adler
University of Michigan, Dearborn

Sylvester Allred
Northern Arizona University

Steven M. Aquilani
Delaware County Community College

Jonathan W. Armbruster
Auburn University

Gregory A. Armstrong
The Ohio State University

Jorge E. Arriagada
St. Cloud State University

David K. Asch
Youngstown State University

Brian Bagatto
University of Akron

Garen Baghdasarian
Santa Monica College

Anita Davelos Baines
The University of Texas, Pan American

Ronald A. Balsamo Jr.
Villanova University

Michael Bartlett
Portland State University

Vernon W. Bauer
Francis Marion University

James E. Baxter
Ohlone College

George W. Benz
Middle Tennessee State University

Gerald K. Bergtrom
University of Wisconsin, Milwaukee

Arlene G. Billock
University of Louisiana, Lafayette

Catherine S. Black
Idaho State University

Michael W. Black
California Polytechnic State University

Robert O. Blanchard
University of New Hampshire

Andrew R. Blaustein
Oregon State University

Mary A. Bober
Santa Monica College

Nancy Maroushek Boury
Iowa State University

M. Deane Bowers
University of Colorado

Scott A. Bowling
Auburn University

Benita A. Brink
Adams State College

Anne Bullerjahn
Owens Community College

Ray D. Burkett
Southwest Tennessee Community College

Helaine Burstein
Ohio University

Scott Burt
Truman State University

Carol T. Burton
Bellevue Community College

Jennifer Carr Burtwistle
Northeast Community College

Jorge Busciglio
University of California, Irvine

Pat Calie
Eastern Kentucky University

Christy A. Carello
The Metropolitan State College of Denver

Michael Carey
University of Scranton

Jeff Carmichael
University of North Dakota

Michael J. Carlisle
Trinity Valley Community College

John H. Caruso
University of New Orleans

Thomas T. Chen
University of Connecticut

Cynthia Church
The Metropolitan State College of Denver

Linda T. Collins
University of Tennessee, Chattanooga

Scott T. Cooper
University of Wisconsin, La Crosse

Joe R. Cowles
Virginia Tech

Nigel M. Crawford
University of California, San Diego

James Crowder
Brookdale Community College

Karen A. Curto
University of Pittsburgh

Bela Dadhich
Delaware County Community College

Lydia B. Daniels
University of Pittsburgh

Terry Davin
Penn Valley Community College

Joseph S. Davis
University of Florida

Neta Dean
Stony Brook University

Kevin W. Dees
Wharton County Junior College

D. Michael Denbow
Virginia Tech

Donald Deters
Bowling Green State University

Hudson DeYoe
University of Texas, Pan American

Randy DiDomenico
University of Colorado

Nd Dikeocha
College of the Mainland

Robert S. Dill
Bergen Community College

Diane M. Dixon
Southeastern Oklahoma State University

Kevin Dixon
University of Illinois

John S. Doctor
Duquesne University

Ernest F. DuBrul
University of Toledo

Charles Duggins Jr.
University of South Carolina

Richard P. Elinson
Duquesne University

Johnny El-Rady
University of South Florida

Frederick B. Essig
University of South Florida

David H. Evans
University of Florida

Guy E. Farish
Adams State College

Daphne G. Fautin
University of Kansas

Bruce E. Felgenhauer
University of Louisiana, Lafayette

Carolyn J. Ferguson
Kansas State University

Teresa G. Fischer
Indian River Community College

Irwin Forseth
University of Maryland

Gail Fraizer
Kent State University

Barbara A. Frase
Bradley University

Sylvia Fromherz
University of Northern Colorado

Phillip E. Funk
DePaul University

Caitlin R. Gabor
Texas State University, San Marcos

Purti P. Gadkari
Wharton County Junior College

John R. Geiser
Western Michigan University

Frank S. Gilliam
Marshall University

Miriam S. Golbert
College of the Canyons

Scott A. Gordon
University of Southern Indiana

John S. Graham
Bowling Green State University

David A. Gray
California State University, Northridge

William F. Hanna
Massasoit Community College

Kyle E. Harms
Louisiana State University

Kerry D. Heafner
University of Louisiana, Monroe

Susan E. Hengeveld
Indiana University

Charles Henry
University of Connecticut, Storrs

Peter Heywood
Brown University

Juliana G. Hinton
McNeese State University

Margaret L. Horton
University of North Carolina, Greensboro

James Horwitz
Palm Beach Community College

Laura A. Houston
Montgomery College

Feng Sheng Hu
University of Illinois

Allen N. Hunt
Elizabethtown Community and Technical College

David C. Jarrell
University of Mary Washington

Jennifer L. Jeffery
Wharton County Junior College

William Jeffery
University of Maryland, College Park

Lee Johnson
The Ohio State University

Craig T. Jordan
The University of Texas, San Antonio

Ronald L. Jones
Eastern Kentucky University

Robyn Jordan
University of Louisiana, Monroe

Walter S. Judd
University of Florida

David Julian
University of Florida

Daniel Kainer
Montgomery College

Ronald C. Kaltreider
York College of Pennsylvania

Thomas C. Kane
University of Cincinnati

Donald A. Kangas
Truman State University

William J. Katembe
Delta State University

Steven J. Kaye
Red Rocks Community College

Stephen R. Kelso
University of Illinois, Chicago

Nancy S. Kirkpatrick
Lake Superior State University

John Z. Kiss
Miami University

John C. Krenetsky
The Metropolitan State College of Denver

Karin E. Krieger
University of Wisconsin, Green Bay

David T. Kurjiaka
University of Arizona

Arlene T. Larson
University of Colorado, Denver

Peter Lavrentyev
University of Akron

Laura G. Leff
Kent State University

Michael R. Lentz
University of North Florida

Harvey Liftin
Broward Community College

Yue J. Lin
St. John's University

Amy Litt
New York Botanical Garden

Christopher R. Little
The University of Texas, Pan American

James Long
Boise State University

James O. Luken
Coastal Carolina University

Dennis J. Lye
Northern Kentucky University

P. T. Magee
University of Minnesota, Minneapolis

Richard Malkin
University of California, Berkeley

Mark D. Mamrack
Wright State University

Kathleen A. Marrs
Indiana University Purdue University, Indianapolis

Diane L. Marshall
University of New Mexico

Paul B. Martin
St. Philip's College

Peter J. Martinat
Xavier University, Los Angeles

Joel Maruniak
University of Missouri

Patricia Matthews
Grand Valley State University

Robin G. Maxwell
The University of North Carolina, Greensboro

Brenda S. McAdory
Tennessee State University

Nael A. McCarty
Georgia Institute of Technology

Brock R. McMillan
Minnesota State University, Mankato

Kay McMurry
The University of Texas, Austin

Elizabeth McPartlan
De Anza College

Brad Mehrtens
University of Illinois, Urbana-Champaign

Michael Meighan
University of California, Berkeley

Douglas Meikle
Miami University

Allen F. Mensinger
University of Minnesota, Duluth

Wayne B. Merkley
Drake University

Catherine E. Merovich
West Virginia University

Frank J. Messina
Utah State University

Brian T. Miller
Middle Tennessee State University

Sarah L. Milton
Florida Atlantic University

Subhash Minocha
University of New Hampshire

Hector C. Miranda Jr.
Texas Southern University

Patricia Mire
University of Louisiana, Lafayette

Robert W. Morris
Widener University

Satyanarayana Swamy Mruthinti
State University of West Georgia

Richard L. Myers
Southwest Missouri State University

Monica Marquez Nelson
Joliet Junior College

Jacalyn S. Newman
University of Pittsburgh

Harry Nickla
Creighton University

Richard A. Niesenbaum
Muhlenberg College

Kris M. Norenberg
Xavier University, Louisiana

Deborah A. O'Dell
University of Mary Washington

Sharman D. O'Neill
University of California, Davis

Cynthia P. Paul
University of Michigan, Dearborn

John S. Peters
College of Charleston

Jay Phelan
University of California, Los Angeles

Gregory W. Phillips
Blinn College

Thomas R. Pitzer
Florida International University

Gregory J. Podgorski
Utah State University

Alan Prather
Michigan State University

Mitch Price
The Pennsylvania State University

Carl Quertermus
State University of West Georgia

Shana Rapoport
California State University, Northridge

Kim Raun
Wharton County Junior College

Robert S. Rawding
Gannon University

Jill D. Reid
Virginia Commonwealth University

Linda R. Richardson
Blinn College

Robin K. Richardson
Winona State University

Carolyn Roberson
Roane State Community College

Kenneth R. Robinson
Purdue University

Kenneth H. Roux
Florida State University

Charles L. Rutherford
Virginia Tech University

Margaret Saha
College of William and Mary

Thomas Sasek
University of Louisiana, Monroe

Bruce M. Saul
Augusta State University

Deemah N. Schirf
The University of Texas, San Antonio

Christopher J. Schneider
Boston University

Timothy E. Shannon
Francis Marion University

Rebecca Sheller
Southwestern University

Mark A. Sheridan
North Dakota State University

Richard Showman
University of South Carolina

Michéle Shuster
New Mexico State University

William Simcik
Tomball College, a North Harris Community College

Rebecca B. Simmons
University of North Dakota

Phillip Snider Jr.
Gadsden State Community College

Thomas E. Snowden
Florida Memorial College

Dianne Snyder
Augusta State University

Farah Sogo
Orange Coast College

Nancy G. Solomon
Miami University

Kathryn H. Sorensen
American River College

Kevin N. Sorensen
Snow College

Bruce Stallsmith
University of Alabama, Huntsville

Patricia Steinke
San Jacinto College

Jacqueline J. Stevens
Jackson State University

John W. Stiller
East Carolina University

Antony Stretton
University of Wisconsin, Madison

Brett W. Strong
Palm Beach Community College

Gregory W. Stunz
Texas A&M University, Corpus Christi

Cynthia A. Surmacz
Bloomsburg University

Yves S. H. Tan
Cabrillo College

Sharon Thoma
University of Wisconsin, Madison

Anne M. S. Tokazewski
Burlington County College

Marty Tracey
Florida International University

Terry M. Trier
Grand Valley State University

Marsha R. Turell
Houston Community College

Linda Tyson
Santa Fe Community College

Rani Vajravelu
University of Central Florida

Jim Van Brunt
Rogue Community College

Judith B. Varelas
University of Northern Colorado

Neal J. Voelz
St. Cloud State University

Janice Voltzow
University of Scranton

Jyoti R. Wagle
Houston Community College System, Central

Charles Walcott
Cornell University

Randall Walikonis
University of Connecticut

Eileen Walsh
Westchester Community College

Steven A. Wasserman
University of California, San Diego

R. Douglas Watson
University of Alabama, Birmingham

Cindy Martinez Wedig
University of Texas, Pan American

Richard Weinstein
Southern New Hampshire University

Elizabeth A. Weiss
University of Texas, Austin

William R. Wellnitz
Augusta State University

Jonathan F. Wendel
Iowa State University

Sue Simon Westendorf
Ohio University

Vernon Lee Wiersema
Houston Community College, Southwest

Judy Williams
Southeastern Oklahoma State University

Lawrence R. Williams
University of Houston

Robert Winning
Eastern Michigan University

C. B. Wolfe
The University of North Carolina, Charlotte

Clarence C. Wolfe
Northern Virginia Community College

Eric Vivien Wong
University of Louisville

Gene K. Wong
Quinnipiac University

Denise Woodward
The Pennsylvania State University

Richard P. Wunderlin
University of South Florida

Douglas A. Wymer
The University of West Alabama

Lan Xu
South Dakota State University

H. Randall Yoder
Lamar University

Kathryn G. Zeiler
Red Rocks Community College

Scott D. Zimmerman
Missouri State University

Henry G. Zot
University of West Georgia

International Reviewers

Mari L. Acevedo
University of Puerto Rico, Arecibo

Heather Addy
University of Calgary

Heather E. Allison
University of Liverpool

David Backhouse
University of New England

Andrew Bendall
University of Guelph

Tony Bradshaw
Oxford Brookes University

D. Bruce Campbell
Okanagan College

Clara E. Carrasco
University of Puerto Rico, Ponce

Ian Cock
Griffith University

Margaret Cooley
University of New South Wales

R. S. Currah
University of Alberta

Logan Donaldson
York University

Theo Elzenga
University of Groningen

Neil Haave
University of Alberta, Augustana

Louise M. Hafner
QUT

Clare Hasenkampf
University of Toronto, Scarborough

Annika F. M. Haywood
Memorial University of Newfoundland

Rong-Nan Huang
National Central University

William Huddleston
University of Calgary

Wendy J. Keenleyside
University of Guelph

Chris Kennedy
Simon Fraser University

Alex Law
Nanyang Technical University, Singapore

Richard C. Leegood
University of Sheffield

R. W. Longair
University of Calgary

Thomas H. MacRae
Dalhousie University

Rolf W. Matthewes
Simon Fraser University

R. Ian Menz
Flinders University

Todd C. Nickle
Mount Royal College

Kirsten Poling
University of Windsor

Jim Provan
Queen's University Belfast

Roberto Quinlan
York University

Elsa I. Colón Reyes
University of Puerto Rico, Aguadilla Campus

Richard Roy
McGill University

Liliane Schoofs
Katholicke Universiteit Leuren

Joan Sharp
Simon Fraser University

Julie Smit
University of Windsor

Nguan Soon Tan
Nanyang Technological University

Fleur Tiver
University of South Australia

Llinil Torres-Ojeda
University of Puerto Rico, Aguadilla Campus

Han A. B. Wösten
University of Utrecht

H. H. Yeoh
National University of Singapore

Art Review Panel

David K. Asch
Youngstown State University

Karl J. Aufderheide
Texas A&M University

Brian Bagatto
University of Akron

Andrew R. Blaustein
Oregon State University

Mark Browning
Purdue University

Jeff Carmichael
University of North Dakota

Wes Colgan III
Pikes Peak Community College

Karen A. Curto
University of Pittsburgh

Donald Deters
Bowling Green State University

Ernest F. DuBrul
University of Toledo

Ralph P. Eckerlin
Northern Virginia Community College

Frederick B. Essig
University of South Florida

Sharon Eversman
Montana State University, Bozeman

Barbara A. Frase
Bradley University

John R. Geiser
Western Michigan University

John Graham
Bowling Green State University

Susan E. Hengeveld
Indiana University

David Julian
University of Florida

Pamela J. Lanford
University of Maryland, College Park

James B. Ludden
College of DuPage

Patricia Mire
University of Louisiana, Lafayette

Janice Moore
Colorado State University

Jacalyn S. Newman
University of Pittsburgh

Robert Newman
University of North Dakota

Nicole S. Obert
University of Illinois, Urbana-Champaign

David G. Oppenheimer
University of Florida

Ellen Ott-Reeves
Blinn College, Bryan

Laurel Bridges Roberts
University of Pittsburgh

Deemah N. Schirf
The University of Texas, San Antonio

Mark A. Sheridan
North Dakota State University

Richard Showman
University of South Carolina

Phillip Snider Jr.
Gadsden State Community College

Nancy G. Solomon
Miami University

David Tam
University of North Texas

Marty Tracey
Florida International University

Michael J. Wade
Indiana University

Jyoti R. Wagle
Houston Community College System, Central

Andy Wang
The University of Iowa

Cindy Martinez Wedig
University of Texas, Pan American

C. B. Wolfe
The University of North Carolina, Charlotte

Biology Symposium Attendees

Every year McGraw-Hill conducts several General Biology Symposia, which are attended by instructors from across the country. These events are an opportunity for editors from McGraw-Hill to gather information about the needs and challenges of instructors teaching the major's biology course. It also offers a forum for the attendees to exchange ideas and experiences with colleagues they might not have otherwise met. The feedback we have received has been invaluable, and has contributed to the development of *Biology* and its supplements.

2006

Michael Bell
Richland College

Scott Bowling
Auburn University

Peter Busher
Boston University

Allison Cleveland
University of South Florida, Tampa

Sehoya Cotner
University of Minnesota

Kathyrn Dickson
California State College, Fullerton

Cathy Donald-Whitney
Collin County Community College

Stanley Faeth
Arizona State University

Karen Gerhart
University of California, Davis

William Glider
University of Nebraska, Lincoln

Stan Guffey
The University of Tennessee

Bernard Hauser
University of Florida, Gainesville

Mark Hens
University of North Carolina, Greensboro

Jim Hickey
Miami University of Ohio, Oxford

Sherry Krayesky
University of Louisiana, Lafayette

Brenda Leady
University of Toledo

Michael Meighan
University of California, Berkeley

Comer Patterson
Texas A&M University

Debra Pires
University of California, Los Angeles

Robert Simons
University of California, Los Angeles

Steven D. Skopik
University of Delaware

Ashok Upadhyaya
University of South Florida, Tampa

Anthony Uzwiak
Rutgers University

Dave Williams
Valencia Community College, East Campus

Jay Zimmerman
St. John's University

2005

Donald Buckley
Quinnipiac University

Arthur Buikema
Virginia Polytechnic Institute

Anne Bullerjahn
Owens Community College

Garry Davies
University of Alaska-Anchorage

Marilyn Hart
Minnesota State University

Daniel Flisser
Camden County College

Elizabeth Godrick
Boston University

Miriam Golbert
College of the Canyons

Sherry Harrel
Eastern Kentucky University

William Hoese
California State University, Fullerton

Margaret Horton
University of North Carolina, Greensboro

Carol Hurney
James Madison University

James Luken
Coastal Carolina University

Mark Lyford
University of Wyoming

Gail McKenzie
Jefferson State Junior College

Melissa Michael
University of Illinois, Urbana-Champaign

Subhash C. Minocha
University of New Hampshire

Leonore Neary
Joliet Junior College

K. Sata Sathasivan
University of Texas, Austin

David Senseman
University of Texas, San Antonio

Sukanya Subramanian
Collin County Community College

Randall Terry
Lamar University

Sharon Thoma
University of Wisconsin, Madison

William Tyler
Indian River Community College

2004

Jonathan Akin
Northwestern State University of Louisiana

David Asch
Youngstown State University

Diane Bassham
Iowa State University

Donald Buckley
Quinnipiac University

Ruth Buskirk
University of Texas, Austin

Charles Creutz
University of Toledo

Lydia Daniels
University of Pittsburgh

Laura DiCaprio
Ohio University

Michael Dini
Texas Tech University

John Doctor
Duquesne University

Ernest DuBrul
University of Toledo

John Elam
Florida State University

Samuel Hammer
Boston University

Marilyn Hart
Minnesota State University

Marc Hirrel
University of Central Arkansas

Carol Johnson
Texas A&M University

Dan Krane
Wright State University

Karin Krieger
University of Wisconsin, Green Bay

Josephine Kurdziel
University of Michigan

Martha Lundell
University of Texas, San Antonio

Roberta Maxwell
University of North Carolina, Greensboro

John Merrill
Michigan State University

Melissa Michael
University of Illinois, Urbana-Champaign

Peter Niewarowski
University of Akron

Ronald Patterson
Michigan State University

Peggy Pollak
Northern Arizona University

Uwe Pott
University of Wisconsin, Green Bay

Mitch Price
Pennsylvania State University

Steven Runge
University of Central Arkansas

Thomas Shafer
*University of North Carolina
Wilmington*

Richard Showman
University of South Carolina

Michèle Shuster
New Mexico State University

Dessie Underwood
California State University, Long Beach

Mike Wade
Indiana University

Elizabeth Willott
University of Arizona

Carl Wolfe
University of North Carolina, Charlotte

Majors Biology Media Focus Group

Russell Borski
North Carolina State University

Mark Decker
University of Minnesota

John Merrill
Michigan State University

Melissa Michael
University of Illinois, Urbana-Champaign

Randall Phillis
University of Massachusetts, Amherst

Mitch Price
Pennsylvania State University

Vivid and Instructional New Art Program for Visual Learners

The *Biology* author team collaborated with a team of medical and scientific illustrators to create the new visual program for the eighth edition. Focusing on consistency, accuracy, and pedagogical value, the team created an art program that is intimately connected with the text narrative. The resulting realistic, 3-D illustrations will stimulate student interest and help instructors teach difficult concepts.

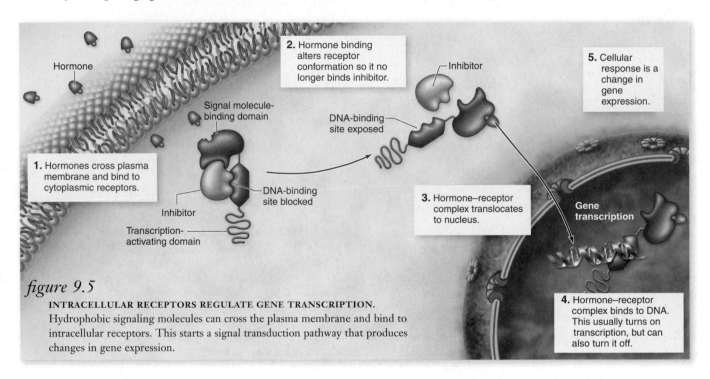

2. Hormone binding alters receptor conformation so it no longer binds inhibitor.

5. Cellular response is a change in gene expression.

Hormone

Inhibitor

Signal molecule-binding domain

DNA-binding site exposed

1. Hormones cross plasma membrane and bind to cytoplasmic receptors.

DNA-binding site blocked

Inhibitor

Transcription-activating domain

3. Hormone–receptor complex translocates to nucleus.

Gene transcription

4. Hormone–receptor complex binds to DNA. This usually turns on transcription, but can also turn it off.

figure 9.5

INTRACELLULAR RECEPTORS REGULATE GENE TRANSCRIPTION.
Hydrophobic signaling molecules can cross the plasma membrane and bind to intracellular receptors. This starts a signal transduction pathway that produces changes in gene expression.

For complex processes, figures use numbered text boxes to lead the student step-by-step through the figure.

1. A cell frozen in medium is cracked with a knife blade.

2. The cell often fractures through the interior, hydrophobic area of the lipid bilayer, splitting the plasma membrane into two layers.

3. The plasma membrane separates such that proteins and other embedded membrane structures remain within one or the other layers of the membrane.

4. The exposed membrane is coated with platinum, which forms a replica of the membrane. The underlying membrane is dissolved away, and the replica is then viewed with electron microscopy.

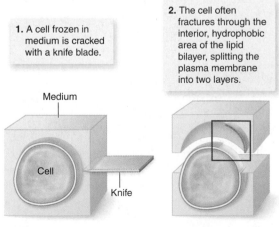

Medium

Cell

Knife

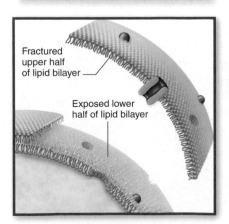

Fractured upper half of lipid bilayer

Exposed lower half of lipid bilayer

0.15 µm

Exposed lower half of lipid bilayer

External surface of plasma membrane

figure 5.3

VIEWING A PLASMA MEMBRANE WITH FREEZE-FRACTURE MICROSCOPY.

Multilevel figures

take students from a macro to micro view using "blow-out" arrows to help students put concepts into context.

Illustrations are paired with high-quality LM, SEM, and TEM photomicrographs to provide students with real-life examples of cellular structures.

Whenever possible, a measurement bar is provided with a micrograph to provide students with an appreciation of the scale of biological structures.

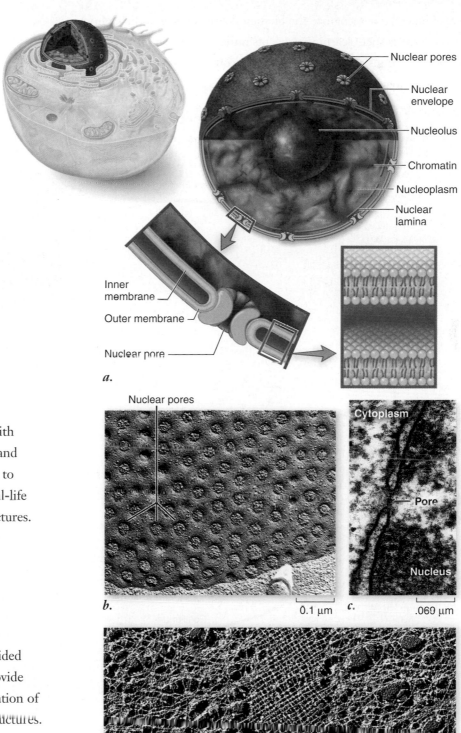

Nuclear pores

Nuclear envelope

Nucleolus

Chromatin

Nucleoplasm

Nuclear lamina

Inner membrane

Outer membrane

Nuclear pore

a.

Nuclear pores

b.　0.1 µm

Cytoplasm

Pore

Nucleus

c.　.060 µm

d.　1 µm

Consistent color coding

means that students immediately recognize the biological structures used throughout the book. Their study time is spent learning concepts rather than orienting themselves to figure conventions. In some figures, color coding is also used to give the student visual cues to how information is related.

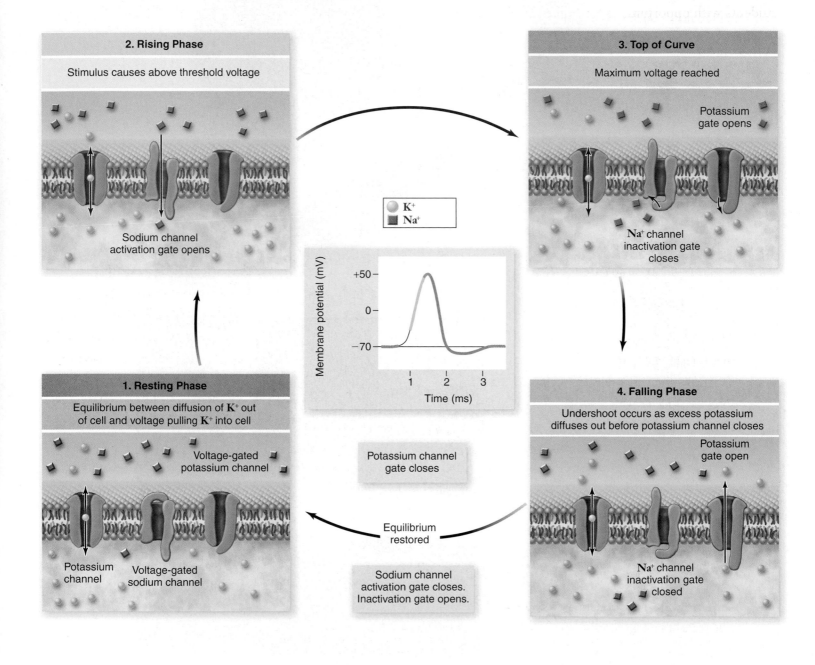

2. Rising Phase

Stimulus causes above threshold voltage

Sodium channel activation gate opens

3. Top of Curve

Maximum voltage reached

Potassium gate opens

Na⁺ channel inactivation gate closes

K⁺
Na⁺

1. Resting Phase

Equilibrium between diffusion of K⁺ out of cell and voltage pulling K⁺ into cell

Voltage-gated potassium channel

Potassium channel   Voltage-gated sodium channel

Membrane potential (mV)

+50
0
−70

1   2   3

Time (ms)

Potassium channel gate closes

Equilibrium restored

Sodium channel activation gate closes. Inactivation gate opens.

4. Falling Phase

Undershoot occurs as excess potassium diffuses out before potassium channel closes

Potassium gate open

Na⁺ channel inactivation gate closed

Consistent Pedagogical Aids to Promote Learning

Each chapter in the eighth edition is structured using the same set of pedagogical devices, which enables the student to develop a consistent learning strategy. These tools work together to provide a clear content hierarchy, break content into smaller, more accessible chunks, repeat important concepts, and provide students with opportunities for higher level thought.

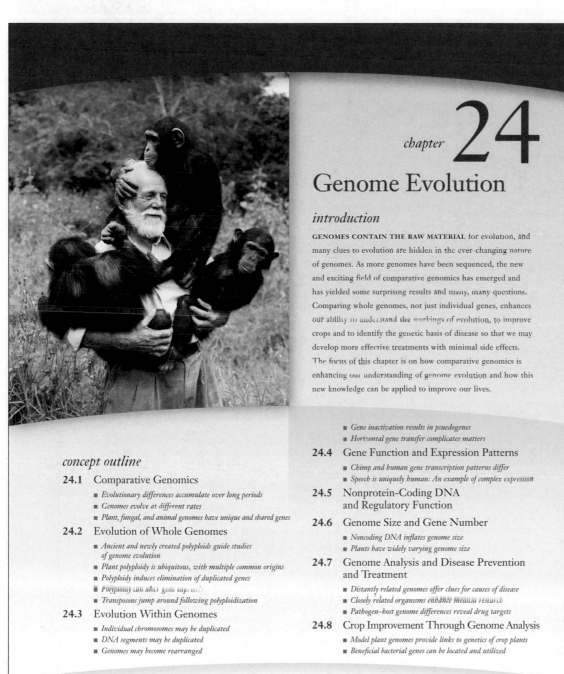

chapter 24

Genome Evolution

introduction

GENOMES CONTAIN THE RAW MATERIAL for evolution, and many clues to evolution are hidden in the ever-changing nature of genomes. As more genomes have been sequenced, the new and exciting field of comparative genomics has emerged and has yielded some surprising results and many, many questions. Comparing whole genomes, not just individual genes, enhances our ability to understand the workings of evolution, to improve crops and to identify the genetic basis of disease so that we may develop more effective treatments with minimal side effects. The focus of this chapter is on how comparative genomics is enhancing our understanding of genome evolution and how this new knowledge can be applied to improve our lives.

concept outline

24.1 Comparative Genomics
- *Evolutionary differences accumulate over long periods*
- *Genomes evolve at different rates*
- *Plant, fungal, and animal genomes have unique and shared genes*

24.2 Evolution of Whole Genomes
- *Ancient and newly created polyploids guide studies of genome evolution*
- *Plant polyploidy is ubiquitous, with multiple common origins*
- *Polyploidy induces elimination of duplicated genes*
- *Polyploidy can alter gene imprinting*
- *Transposons jump around following polyploidization*

24.3 Evolution Within Genomes
- *Individual chromosomes may be duplicated*
- *DNA segments may be duplicated*
- *Genomes may become rearranged*

- *Gene inactivation results in psuedogenes*
- *Horizontal gene transfer complicates matters*

24.4 Gene Function and Expression Patterns
- *Chimp and human gene transcription patterns differ*
- *Speech is uniquely human: An example of complex expression*

24.5 Nonprotein-Coding DNA and Regulatory Function

24.6 Genome Size and Gene Number
- *Noncoding DNA inflates genome size*
- *Plants have widely varying genome size*

24.7 Genome Analysis and Disease Prevention and Treatment
- *Distantly related genomes offer clues for causes of disease*
- *Closely related organisms enhance medical research*
- *Pathogen–host genome differences reveal drug targets*

24.8 Crop Improvement Through Genome Analysis
- *Model plant genomes provide links to genetics of crop plants*
- *Beneficial bacterial genes can be located and utilized*

471

Chapter openers include an outline comprised of the chapter headings, which provides a consistent framework for the student. Declarative, numbered main headings and sentence-style supporting headings result in a cogent overview of the content to be covered.

1. Interim summaries

review key points from the section so students can easily identify the take-away message.

2. Numbered main headings

clearly identify the start of a new concept section.

3. Inquiry questions

challenge students to think about what they are reading at a more sophisticated level.

are shown in figure 55.11. Oysters produce vast numbers of offspring, only a few of which live to reproduce. However, once they become established and grow into reproductive individuals, their mortality rate is extremely low (type III survivorship curve). Note that in this type of curve, survival and mortality rates are inversely related. Thus, the rapid decrease in the proportion of oysters surviving indicates that few individuals survive, thus producing a high mortality rate. In contrast, the relatively flat line at older ages indicates high survival and low mortality.

In hydra, animals related to jellyfish, individuals are equally likely to die at any age. The result is a straight survivorship curve (type II).

Finally, mortality rates in humans, as in many other animals and in protists, rise steeply later in life (type I survivorship curve).

Of course, these descriptions are just generalizations, and many organisms show more complicated patterns. Examination of the data for *P. annua*, for example, reveals that it is most similar to a type II survivorship curve (figure 55.12).

1 The growth rate of a population is a sensitive function of its age structure. The age structure of a population and the manner in which mortality and birthrates vary among different age cohorts determine whether a population will increase or decrease in size.

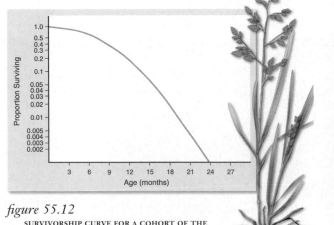

figure 55.12

SURVIVORSHIP CURVE FOR A COHORT OF THE MEADOW GRASS *POA ANNUA*. After several months of age, mortality increases at a constant rate through time.

? inquiry

Suppose you wanted to keep meadow grass in your room as a houseplant. Suppose, too, that you wanted to buy an individual plant that was likely to live as long as possible. What age plant would you buy? How might the shape of the survivorship curve affect your answer?

3

55.4 **2** ## Life History and the Cost of Reproduction

Natural selection favors traits that maximize the number of surviving offspring left in the next generation. Two factors affect this quantity: how long an individual lives, and how many young it produces each year.

Why doesn't every organism reproduce immediately after its own birth, produce large families of offspring, care for them intensively, and perform these functions repeatedly throughout a long life, while outcompeting others, escaping predators, and capturing food with ease? The answer is that no one organism can do all of this, simply because not enough resources are available. Consequently, organisms allocate resources either to current reproduction or to increasing their prospects of surviving and reproducing at later life stages.

The complete life cycle of an organism constitutes its **life history.** All life histories involve significant trade-offs. Because resources are limited, a change that increases reproduction may decrease survival and reduce future reproduction. As one example, a Douglas fir tree that produces more cones increases its current reproductive success—but it also grows more slowly. Because the number of cones produced is a function of how large a tree is, this diminished growth will decrease the number of cones it can produce in the fu-

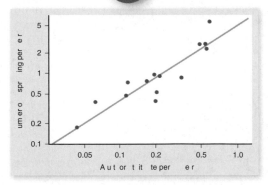

figure 55.13

REPRODUCTION HAS A PRICE. Data from many bird species indicate that increased fecundity in birds correlates with higher mortality, ranging from the albatross (lowest) to the sparrow (highest). Birds that raise more offspring per year have a higher probability of dying during that year.

1154 *part* VIII *ecology and behavior*

1

TABLE 15.2	Differences Between Prokaryotic and Eukaryotic Gene Expression	
Characteristic	**Prokaryotes**	**Eukaryotes**
Introns	No introns, although some archaeal genes possess them.	Most genes contain introns.
Number of genes in mRNA	Several genes may be transcribed into a single mRNA molecule. Often these have related functions and form an operon. This coordinates regulation of biochemical pathways.	Only one gene per mRNA molecule; regulation of pathways accomplished in other ways.
Site of transcription and translation	No membrane-bounded nucleus, transcription and translation are coupled.	Transcription in nucleus; mRNA moves out of nucleus for translation.
Initiation of translation	Begins at AUG codon preceded by special sequence that binds the ribosome.	Begins at AUG codon preceded by the 5′ cap (methylated GTP) that binds the ribosome.
Modification of mRNA after transcription	None; translation begins before transcription is completed.	A number of modifications while the mRNA is in the nucleus: Introns are removed and exons are spliced together; a 5′ cap is added; a poly-A tail is added.

1. Summary Tables are used extensively to help students study and review the chapter content. Illustrations are added in some cases to further aid students in recall.

2. Self Test Questions are a mixture of knowledge and comprehension questions that test a student's basic understanding of the main concepts from the chapter.

3. Challenge Questions are application and analysis questions that measure a student's ability to use terms and concepts learned from the chapter in new situations.

4. Concept Review summarizes the main concepts from the chapter and their supporting ideas. Key figures are cited to alert students to particularly relevant illustrations.

---

review questions

2

SELF TEST

1. Which of the following statements is NOT part of cell theory?
   a. All organisms are composed of one or more cells.
   b. Cells come from other cells by division.
   c. Cells are the smallest living things.
   d. Eukaryotic cells have evolved from prokaryotic cells.
2. The most important factor that limits the size of a cell is—
   a. the amount of proteins and organelles that can be made by a cell
   b. the rate of diffusion
   c. the surface-area-to-volume ratio of the cell
   d. the amount of DNA in the cell
3. What type of microscope would you use to examine the surface details of a cell?
   a. Compound light microscope
   b. Transmission electron microscope
   c. Scanning electron microscope
   d. Confocal microscope
4. All cells have all of the following except—
   a. Plasma membrane
   b. Genetic material
   c. Cytoplasm
   d. Cell wall
5. Eukaryotic cells are more complex than prokaryotic cells. Which of the following would you NOT find in a prokaryotic cell?
   a. Cell wall
   b. Plasma membrane
   c. Nucleus
   d. Ribosomes
6. The difference between a gram-positive and gram-negative bacteria is—
   a. the thickness of the peptidoglycan cell wall
   b. the type of polysaccharide present in the cell wall
   c. the type and amount of protein in the cell wall

11. Proteins can move from the Golgi apparatus to—
    a. the extracellular fluid
    b. transport vesicles
    c. lysosomes
    d. all of the above
12. Lysosomes function to—
    a. carry proteins to the surface of the cell
    b. add short-chain carbohydrates to make glycoproteins
    c. break down organelles, proteins, and nucleic acids
    d. remove electrons and hydrogen atoms from hydrogen peroxide
13. What do chloroplasts and mitochondria have in common?
    a. Both are present in animal cells.
    b. Both have an outer membrane and an elaborate inner membrane.
    c. Both are present in all eukaryotic cells.
    d. Both organelles function to produce glucose.
14. Eukaryotic cells are composed of three types of cytoskeletal filaments. How are these three filaments similar?
    a. They contribute to the shape of the cell.
    b. They are all made of the same type of protein.
    c. They are all the same size and shape.
    d. They are all equally dynamic and flexible.
15. Animal cells connect to the extracellular matrix through—
    a. glycoproteins
    b. fibronectins
    c. integrins
    d. collagen

CHALLENGE QUESTIONS

3

1. Eukaryotic cells are typically larger than prokaryotic cells (refer to figure 4.2). How might the difference in the cellular structure of eukaryotic versus a prokaryotic cell help to explain this observation?
2. The smooth endoplasmic reticulum is the site of synthesis of the phospholipids that make up all the membranes of a cell—especially the plasma membrane. Use the diagram of an animal cell (figure 4.6) to trace a pathway that would carry a phospholipid molecule from the SER to the plasma membrane. What endomembrane compartments would the phospholipids travel through? How can a phospholipid molecule move between membrane compartments?
3. Use the information provided in table 4.3 to develop a set of predictions about the properties of mitochondria and chloroplasts if these organelles were once free-living prokaryotic cells. How do your predictions match with the evidence for endosymbiosis?
4. In evolutionary theory, homologous traits are those with a similar structure and function derived from a common ancestor. Analogous traits represent adaptations to a similar environment, but from distantly related organisms. Consider the structure and function of the flagella found on eukaryotic and prokaryotic cells. Are the flagella an example of a homologous or analogous trait? Defend your answer.
5. The protist, *Giardia lamblia*, is the organism associated with waterborne diarrheal diseases. *Giardia* is an unusual eukaryote because it seems to lack mitochondria. Explain the existence of a mitochondria-less eukaryote in the context of the endosymbiotic theory.

---

concept review

4

4.1 Cell Theory

Modern cell theory states that organisms are composed of one or more cells. Cells are the smallest unit of life and arise from preexisting cells.
- Cell size is constrained by the effective distance of diffusion within a cell, from the surface to the interior of the cell.
- As a cell increases in size the surface area increases as a square function and the volume increases as a cubic function.
- Large cells deal with the diffusion problem by having more than one nucleus or by becoming flattened or elongated.
- The visualization of cells and their components is facilitated by microscopes and staining cell structures.
- All cells have DNA, a cytoplasm, a plasma membrane, and ribosomes.

4.2 Prokaryotic Cells (figure 4.3)

Prokaryotic cells do not have a nucleus or an internal membrane system, and they lack membrane-bounded organelles.
- The plasma membrane is surrounded by a rigid cell wall that maintains shape and helps maintain osmotic balance.
- The plasma membrane in some prokaryotes is infolded and provides similar functions to eukaryotic internal membranes.
- Bacteria have a cell wall made up of peptidoglycan. Archaeal cell walls have different architecture.
- Archaeal plasma membranes differ from bacteria and eukaryotes.
- The plasma membrane in some archaea is a monolayer composed of saturated lipids attached to glycerol at each end.
- Structurally Archaea resemble prokaryotes, but functionally they more closely resemble eukaryotes.
- Prokaryotic flagella rotate because of proton transfer.

4.3 Eukaryotic Cells (figures 4.6 and 4.7)

Eukaryotic cells have a membrane-bounded nucleus, an endomembrane system, and many different organelles.
- The nucleus contains genetic information.
- The nuclear envelope consists of two phospholipid bilayers; the outer layer is contiguous with the ER.
- The inside of the nuclear envelope is covered with nuclear lamins, which maintain the shape of the nucleus.
- Nuclear pores allow exchange of small molecules between the nucleoplasm and the cytoplasm.
- DNA is organized with proteins into chromatin.
- The nucleolus is a region of the nucleoplasm where rRNA is transcribed and ribosomes are assembled.
- Ribosomes are composed of RNA and protein and use information in mRNA to direct the synthesis of proteins.

4.4 The Endomembrane System

The endomembrane system forms compartments and vesicles and provides channels to carry molecules and surfaces for synthesis of macromolecules.
- The endoplasmic reticulum (ER) creates channels and passages within the cytoplasm (figure 4.11).
- The interior compartment of the ER is called the cisternal space, or lumen.
- The rough endoplasmic reticulum (RER) has ribosomes on the surface and is composed mainly of flattened sacs. RER is involved in protein synthesis and modification.
- The smooth endoplasmic reticulum (SER) lacks ribosomes

and is composed more of tubules. SER is involved in synthesis of carbohydrates and lipids and in detoxification.
- The Golgi apparatus receives vesicles from the ER on the *cis* face, modifies and packages macromolecules, and transports them in vesicles formed on the *trans* face (figure 4.13).
- Lysosomes are vesicles containing enzymes that break down macromolecules located in food vacuoles and recycle the components of old organelles (figure 4.14).
- Microbodies contain enzymes and grow by incorporating lipids and proteins before they divide.
- Peroxisomes contain enzymes that catalyze oxidation reactions, resulting in the formation of hydrogen peroxide.
- Plants have many specialized vacuoles. The conspicuous central vacuole, surrounded by the tonoplast membrane, is used for storage, maintaining water balance, and growth.

4.5 Mitochondria and Chloroplasts: Cellular Generators

Mitochondria and chloroplasts have a double-membrane structure, contain their own DNA, can synthesize proteins, can divide, and are involved in energy metabolism.
- Mitochondria produce ATP using energy-contained macromolecules (figure 4.17).
- *The inner membrane is extensively folded into layers called cristae.*
- *The intermembrane space is a compartment between the inner and outer membrane.*
- *The mitochondrial matrix is a compartment consisting of the fluid within the inner membrane.*
- Chloroplasts use light to generate ATP and sugars (figure 4.18).
- *In addition to a double membrane, chloroplasts also have stacked membranes called grana that contain vesicles called thylakoids.*
- *The fluid stroma surrounds the thylakoids.*
- Evidence indicates that mitochondria and chloroplasts arose via endosymbiosis.

4.6 The Cytoskeleton

The cytoskeleton is composed of three different fibers that support cell shape and anchors organelles and enzymes (figure 4.20).
- Actin filaments, or microfilaments, are long, thin polymers responsible for cell movement, cytoplasmic division, and formation of cellular extensions.
- Microtubules are hollow structures that are used in cell movement and movement of materials within a cell.
- Intermediate filaments are stable structures that serve a wide variety of functions.
- Paired centrioles, located in the centrosome, help assemble the nuclear division apparatus of animal cells (figure 4.21).
- Molecular motors move vesicles along microtubules.

4.7 Extracellular Structures and Cell Movement

Extracellular structures provide protection, support, strength, and cell recognition.
- Plants have cell walls composed of cellulose fibers. Fungi have cell walls composed of chitin.
- Animals have a complex extracellular matrix.
- Cell crawling occurs as actin polymerization forces the cell membrane forward while myosin pulls the cell forward.
- Eukaryotic flagella have a 9 + 2 structure and arise from a basal body.
- Cilia are shorter and more numerous flagella.

Designed to help students maximize their learning experience in biology—we offer the following options to students:

ARIS

ARIS (Assessment, Review, and Instruction System) is an electronic study system that offers students a digital portal of knowledge.

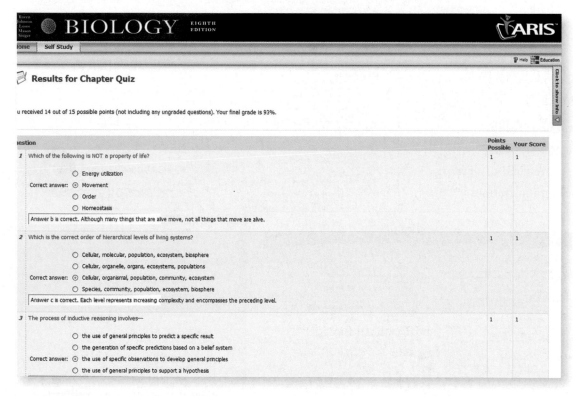

Students can readily access a variety of **digital learning objects,** which include

- chapter level quizzing
- pretests
- animations
- videos
- flashcards
- answers to Inquiry Questions
- answers to all end-of-chapter questions
- MP3 and MP4 downloads of selected content
- learning outcomes and assessment capability woven around key content

Student Study Guide

Helping students focus their time and energy on important concepts, the study guide offers students a variety of tools:

1. **Tips for Mastering Key Concepts**—provides an overview of the chapter, summarizes key points, and gives helpful hints on topics to consider.
2. **Map of Understanding**—illustrates relationships between the major concepts in the chapter.
3. **Key Terms**—listed by the section of the chapter in which they occur.

4. **Learning by Experience**—a variety of activities designed to help students actually use the knowledge acquired from the chapter.
5. **Exercising Your Knowledge**—approximately 30 multiple-choice questions per chapter.
6. **Assessing Your Knowledge**—provides answers to the chapter questions, including a guide to which questions test comprehension of the concepts and which test for knowledge of detail. Students are directed to the appropriate text sections for review.

Content Delivery Flexibility

Biology is available in many formats in addition to the traditional textbook to give instructors and students more choices when deciding on the format of their biology text. Choices include:

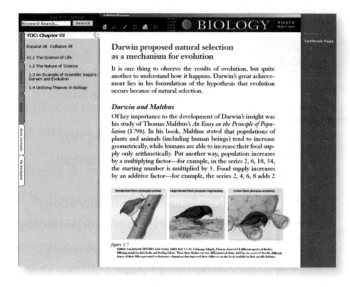

eBook

The entire text is available electronically through the ARIS website. This electronic text offers not only the text in a digital format but includes embedded links to figures, tables, animations, and videos to make full use of the digital tools available and further enhance student understanding.

Color Custom by Chapter

For even more flexibility, we offer the *Biology* text in a full-color, custom version that allows instructors to pick the chapters they want included. Students pay for only what the instructor chooses.

Volumes

The complete text has been split into three natural segments to allow instructors more flexibility and students more purchasing options.

Volume 1—Units 1 (Chemistry), **2** (Cell Biology), and **3** (Genetics)

Volume 2—Units 6 (Plant Biology) and **7** (Animal Biology)

Volume 3—Units 4 (Evolution), **5** (Diversity), and **8** (Ecology)

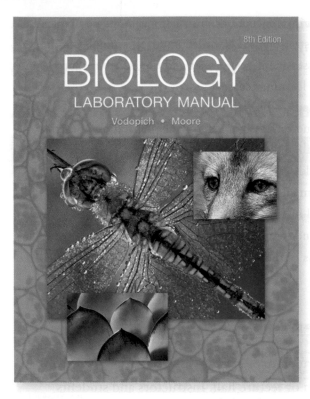

Biology Laboratory Manual, Eighth Edition
Darrell S. Vodopich, *Baylor University*
Randy Moore, *University of Minnesota-Minneapolis*
0-07-299522-X / 2008

This laboratory manual is designed for an introductory majors' biology course with a broad survey of basic laboratory techniques. The experiments and procedures are simple, safe, easy to perform, and especially appropriate for large classes. Few experiments require a second class meeting to complete the procedure. Each exercise includes many photographs, traditional topics, and experiments that help students learn about life. Procedures within each exercise are numerous and discrete so that an exercise can be tailored to the needs of the students, the style of the instructor, and the facilities available.

New to this edition:

* **Clearer Information on Safety**—A new safety icon will be used throughout the text, replacing the current "CAUTION" that appears in red font.
* *Laboratory Safety Rules* table has been added to the *Welcome* chapter.

* **Process of Science Exercises**—A new exercise on *Process of Science* now appears in chapter 1.
* **New Figures and Tables**—More than 70 tables and figures have been extended with additions, revisions, or replacements.

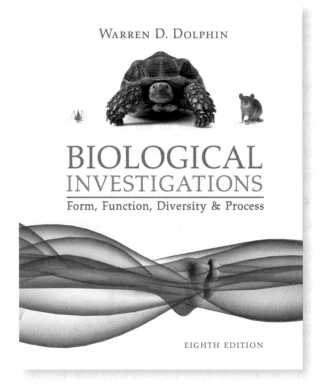

Biological Investigations Lab Manual, Eighth Edition
Warren D. Dolphin, *Iowa State University*
0-07-299287-5 / 2008

This independent lab manual can be used for a one- or two-semester majors' level general biology lab and can be used with any majors' level general biology textbook. The labs are investigative and ask students to use more critical thinking and hands-on learning. The author emphasizes investigative, quantitative, and comparative approaches to studying the life sciences.

New to this edition:

* **Revised Art Program**—More than 90 figures have been revised or replaced throughout the textbook.
* **How to Evaluate Web Material**—Information on evaluating material gathered on the web has been added to Lab Topic 1.

contents

Part VII Animal Form and Function

Biology

chapter

1

The Science of Biology

introduction

YOU ARE ABOUT TO EMBARK ON A JOURNEY—a journey of discovery about the nature of life. Nearly 180 years ago, a young English naturalist named Charles Darwin set sail on a similar journey on board H.M.S. *Beagle*; a replica of this ship is pictured here. What Darwin learned on his five-year voyage led directly to his development of the theory of evolution by natural selection, a theory that has become the core of the science of biology. Darwin's voyage seems a fitting place to begin our exploration of biology: the scientific study of living organisms and how they have evolved. Before we begin, however, let's take a moment to think about what biology is and why it's important.

concept outline

1.1 The Science of Life

- *Biology unifies much of natural science*
- *Life defies simple definition*
- *Living systems show hierarchical organization*

1.2 The Nature of Science

- *Much of science is descriptive*
- *Science uses both deductive and inductive reasoning*
- *Hypothesis-driven science makes and tests predictions*
- *Reductionism breaks larger systems into their component parts*
- *Biologists construct models to explain living systems*
- *The nature of scientific theories*
- *Research can be basic or applied*

1.3 An Example of Scientific Inquiry: Darwin and Evolution

- *The idea of evolution existed prior to Darwin*
- *Darwin observed differences in related organisms*
- *Darwin proposed natural selection as a mechanism for evolution*
- *Testing the predictions of natural selection*

1.4 Unifying Themes in Biology

- *Cell theory describes the organization of living systems*
- *The molecular basis of inheritance explains the continuity of life*
- *The relationship between structure and function underlies living systems*
- *The diversity of life arises by evolutionary change*
- *Evolutionary conservation explains the unity of living systems*
- *Cells are information-processing systems*
- *Emergent properties arise from the organization of life*

The Science of Life

This is the most exciting time to be studying biology in the history of the field. The amount of data available about the natural world has exploded in the last 25 years, and we are now in a position to ask and answer questions that previously were only dreamed of.

We have determined the entire sequence of the human genome, and are in the process of sequencing the genomes of other species at an ever increasing pace. We are closing in on a description of the molecular workings of the cell in unprecedented detail, and we are in the process of finally unveiling the mystery of how a single cell can give rise to the complex organization seen in multicellular organisms. With robotics, advanced imaging, and analytical techniques, we have tools available that were formerly the stuff of science fiction.

In this text, we attempt to provide a view of the science of biology as it is practiced now, while still retaining some sense of how we got to this exciting position. In this introductory chapter, we examine the nature of biology and the nature of science in general to begin to put into context the information presented in the rest of the text.

Biology unifies much of natural science

The study of biology does much to unify the information gained from all of the natural sciences. Biological systems are the most complex chemical systems that we know of on Earth, and their many functions are both determined and constrained by the principles of chemistry and physics. Put another way, there are no new laws of nature to be gleaned from the study of biology—but that study does illuminate and illustrate the workings of those natural laws.

The intricate chemical workings of cells are based on all that we have learned from the study of chemistry, and every level of biological organization is governed by the nature of energy transactions learned from the study of thermodynamics. Biological systems do not represent any new forms of matter, and yet they are the most complex organization of matter known. The complexity of living systems is made possible by a constant source of energy: the Sun. The conversion of this energy source into organic molecules by photosynthesis can be understood using the principles of chemistry and physics.

As modern scientists take on more difficult problems, the nature of how we do science is changing as well. Science is becoming more interdisciplinary, combining the expertise of a variety of scientists in exciting new fields such as nanotechnology. Biology is at the heart of this multidisciplinary approach because biological problems often require many different approaches to arrive at solutions.

Life defies simple definition

In its broadest sense, biology is the study of living things—*the science of life*. Living things come in an astounding variety of shapes and forms, and biologists study life in many different ways. They live with gorillas, collect fossils, and listen to whales. They read the messages encoded in the long molecules of heredity and count how many times a hummingbird's wings beat each second.

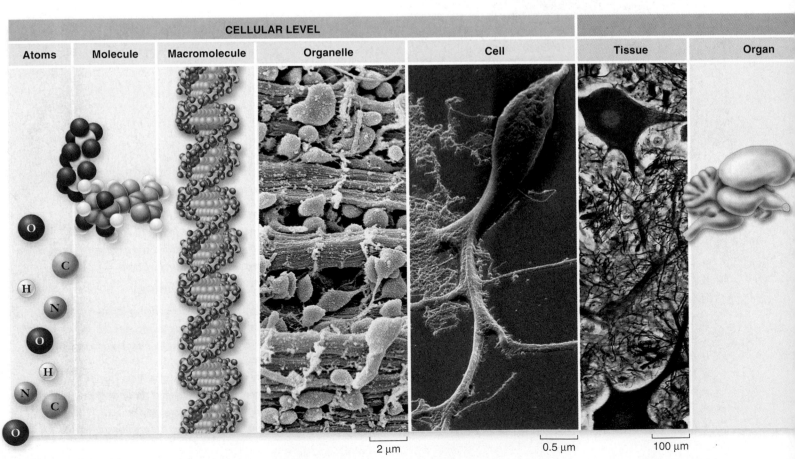

CELLULAR LEVEL						
Atoms	Molecule	Macromolecule	Organelle	Cell	Tissue	Organ

2 μm   0.5 μm   100 μm

What makes something "alive"? Anyone could deduce that a galloping horse is alive and a car is not, but why? We cannot say, "If it moves, it's alive," because a car can move, and gelatin can wiggle in a bowl. They certainly are not alive. Although we cannot define life with a single simple sentence, we can come up with a series of seven characteristics shared by living systems:

- **Cellular organization.** All organisms consist of one or more cells. Often too tiny to see, cells carry out the basic activities of living. Each cell is bounded by a membrane that separates it from its surroundings.
- **Ordered complexity.** All living things are both complex and highly ordered. Your body is composed of many different kinds of cells, each containing many complex molecular structures. Many nonliving things may also be complex, but they do not exhibit this degree of ordered complexity.
- **Sensitivity.** All organisms respond to stimuli. Plants grow toward a source of light, and the pupils of your eyes dilate when you walk into a dark room.
- **Growth, development, and reproduction.** All organisms are capable of growing and reproducing, and they all possess hereditary molecules that are passed to their offspring, ensuring that the offspring are of the same species.
- **Energy utilization.** All organisms take in energy and use it to perform many kinds of work. Every muscle in your body is powered with energy you obtain from the food you eat.
- **Homeostasis.** All organisms maintain relatively constant internal conditions that are different from their environment, a process called **homeostasis.**
- **Evolutionary adaptation.** All organisms interact with other organisms and the nonliving environment in ways that influence their survival, and as a consequence, organisms evolve adaptations to their environments.

Living systems show hierarchical organization

The organization of the biological world is hierarchical—that is, each level builds on the level below it.

1. **The Cellular Level.** At the cellular level (figure 1.1), **atoms,** the fundamental elements of matter, are joined together into clusters called **molecules.** Complex biological molecules are assembled into tiny structures called **organelles** within membrane-bounded units we call **cells.** The cell is the basic unit of life. Many independent organisms are composed only of single cells. Bacteria are single cells, for example. All animals and plants, as well as most fungi and algae, are multicellular—composed of more than one cell.
2. **The Organismal Level.** Cells in complex multicellular organisms exhibit three levels of organization. The most basic level is that of **tissues,** which are groups of similar

figure 1.1

HIERARCHICAL ORGANIZATION OF LIVING SYSTEMS. Life is highly organized from the simplest atoms to complex multicellular organisms. Along this hierarchy of structure, atoms form molecules that are used to form organelles, which in turn form the functional subsystems within cells. Cells are organized into tissue, and then into organs and organ systems such as the nervous system pictured. This extends beyond individual organisms to populations, communities, ecosystems and finally the entire biosphere.

ORGANISMAL LEVEL

| Organ system | Organism | Population | Species | Community | Ecosystem | Biosphere |

POPULATIONAL LEVEL

cells that act as a functional unit. Tissues, in turn, are grouped into **organs,** body structures composed of several different tissues that act as a structural and functional unit. Your brain is an organ composed of nerve cells and a variety of associated tissues that form protective coverings and contribute blood. At the third level of organization, organs are grouped into **organ systems.** The nervous system, for example, consists of sensory organs, the brain and spinal cord, and neurons that convey signals.

3. **The Populational Level.** Individual organisms can be categorized into several hierarchical levels within the living world. The most basic of these is the **population,** a group of organisms of the same species living in the same place. All populations of a particular kind of organism together form a **species,** its members similar in appearance and able to interbreed. At a higher level of biological organization, a **biological community** consists of all the populations of different species living together in one place.

4. **Ecosystem Level.** At the highest tier of biological organization, a biological community and the physical habitat within which it lives together constitute an ecological system, or **ecosystem.** For example, the soil, water, and atmosphere of a mountain ecosystem interact with the biological community of a mountain meadow in many important ways.

5. **Emergent Properties at Every Level.** At each higher level in the living hierarchy, novel properties emerge. These **emergent properties** result from the way in which components interact, and they often cannot be deduced just from looking at the parts themselves. Examining individual cells, for example, gives little clue of what a whole animal is like. You, as a human, have the same array of cell types as does a giraffe. It is because the living world exhibits many emergent properties that it is difficult to define "life."

6. **The Biosphere.** The entire planet can be thought of as an ecosystem that we call the biosphere.

The previous descriptions of the common features and organization of living systems begins to get at the nature of what it is to be alive. The rest of this book illustrates and expands on these basic ideas to try to provide a more complete account of living systems.

> Biology is a unifying science that uses the knowledge gained from other natural sciences to study living systems. There is no simple definition of life, but living systems share a number of properties that together describe life. Living systems are also organized hierarchically, with new properties emerging that may be greater than the sum of the parts.

1.2 The Nature of Science

Much like life itself, the nature of science defies simple description. For many years scientists have written about the "scientific method" as though there is a single way of doing science. This oversimplification has contributed to confusion on the part of nonscientists about the nature of science.

At its core, science is concerned with understanding the nature of the world around us by using observation and reasoning. To begin with, we assume that natural forces acting now have always acted, that the fundamental nature of the universe has not changed since its inception, and that it is not changing now. A number of complementary approaches allow understanding of natural phenomena—there is no one "right way."

Scientists also attempt to be as objective as possible in the interpretation of the data and observations they have collected. Because scientists themselves are human, this is not completely possible; because science is a collective endeavor subject to scrutiny, however, it is self-correcting. Results from one person are verified by others, and if the results cannot be repeated, they are rejected.

Much of science is descriptive

The classic vision of the scientific method is that observations lead to hypotheses that in turn make experimentally testable predictions. In this way, we dispassionately evaluate new ideas to arrive at an increasingly accurate view of nature. We discuss this way of doing science later in this chapter, but it is important to understand that much of science is purely descriptive: In order to understand anything, the first step is to describe it

completely. Much of biology is concerned with arriving at an increasingly accurate description of nature.

The study of biodiversity is an example of descriptive science that has implications for other aspects of biology in addition to social implications. Efforts are currently underway to classify all life on the Earth. This ambitious project is purely descriptive, but it will lead to a much greater understanding of biodiversity as well as the impact of our species on biodiversity.

One of the most important accomplishments of molecular biology at the dawn of the 21st century was the completion of the sequence of the human genome. Many new hypotheses about human biology will be generated by this knowledge, and many experiments will be needed to test these hypotheses, but the determination of the sequence itself was descriptive science.

Science uses both deductive and inductive reasoning

The study of logic recognizes two opposite ways of arriving at logical conclusions: deductive reasoning and inductive reasoning. Science makes use of both of these methods, although induction is the primary way of reasoning in hypothesis-driven science.

Deductive reasoning

Deductive reasoning applies general principles to predict specific results. Over 2200 years ago, the Greek scientist Eratosthenes used Euclidean geometry and deductive reasoning to accurately estimate the circumference of the Earth (figure 1.2). Deductive reasoning is the reasoning of mathematics and

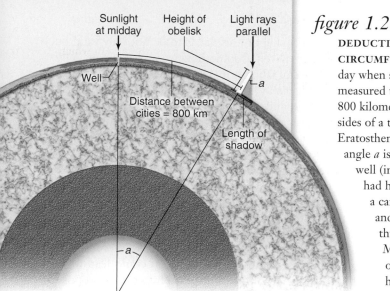

Sunlight at midday | Height of obelisk | Light rays parallel

Well

Distance between cities = 800 km

Length of shadow

a

a

figure 1.2

DEDUCTIVE REASONING: HOW ERATOSTHENES ESTIMATED THE CIRCUMFERENCE OF THE EARTH USING DEDUCTIVE REASONING. 1. On a day when sunlight shone straight down a deep well at Syene in Egypt, Eratosthenes measured the length of the shadow cast by a tall obelisk in the city of Alexandria, about 800 kilometers (km) away. **2.** The shadow's length and the obelisk's height formed two sides of a triangle. Using the recently developed principles of Euclidean geometry, Eratosthenes calculated the angle, *a*, to be 7° and 12′, exactly $\frac{1}{50}$ of a circle (360°). **3.** If angle *a* is $\frac{1}{50}$ of a circle, then the distance between the obelisk (in Alexandria) and the well (in Syene) must be equal to $\frac{1}{50}$ the circumference of the Earth. **4.** Eratosthenes had heard that it was a 50-day camel trip from Alexandria to Syene. Assuming that a camel travels about 18.5 km per day, he estimated the distance between obelisk and well as 925 km (using different units of measure, of course). **5.** Eratosthenes thus deduced the circumference of the Earth to be 50 × 925 = 46,250 km. Modern measurements put the distance from the well to the obelisk at just over 800 km. Employing a distance of 800 km, Eratosthenes's value would have been 50 × 800 = 40,000 km. The actual circumference is 40,075 km.

philosophy, and it is used to test the validity of general ideas in all branches of knowledge. For example, if all mammals by definition have hair, and you find an animal that does not have hair, then you may conclude that this animal is not a mammal. A biologist uses deductive reasoning to infer the species of a specimen from its characteristics.

Inductive reasoning

In **inductive reasoning,** the logic flows in the opposite direction, from the specific to the general. Inductive reasoning uses specific observations to construct general scientific principles. For example, if poodles have hair, and terriers have hair, and every dog that you observe has hair, then you may conclude that all dogs have hair. Inductive reasoning leads to generalizations that can then be tested. Inductive reasoning first became important to science in the 1600s in Europe, when Francis Bacon, Isaac Newton, and others began to use the results of particular experiments to infer general principles about how the world operates.

An example from modern biology is the action of homeobox genes in development. Studies in the fruit fly, *Drosophila melanogaster*, identified genes that could cause dramatic changes in developmental fate, such as a leg appearing in the place of an antenna. When the genes themselves were isolated and their DNA sequence determined, it was found that similar genes were found in many animals, including humans. This led to the general idea that the homeobox genes act as switches to control developmental fate.

Hypothesis-driven science makes and tests predictions

Scientists establish which general principles are true from among the many that might be true by systematically testing alternative proposals. If these proposals prove inconsistent with experimental observations, they are rejected as untrue. Figure 1.3 illustrates the process.

After making careful observations, scientists construct a **hypothesis,** which is a suggested explanation that accounts for those observations. A hypothesis is a proposition that might be

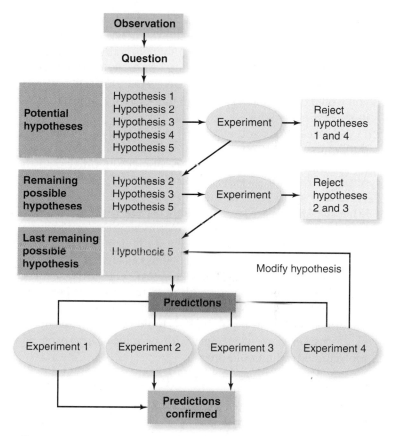

figure 1.3

HOW SCIENCE IS DONE. This diagram illustrates how scientific investigations proceed. First, scientists make observations that raise a particular question. They develop a number of potential explanations (hypotheses) to answer the question. Next, they carry out experiments in an attempt to eliminate one or more of these hypotheses. Then, predictions are made based on the remaining hypotheses, and further experiments are carried out to test these predictions. The process can also be iterative. As experimental results are performed, the information can be used to modify the original hypothesis to fit each new observation.

true. Those hypotheses that have not yet been disproved are retained. They are useful because they fit the known facts, but they are always subject to future rejection if, in the light of new information, they are found to be incorrect.

This process can also be *iterative*, that is, a hypothesis can be changed and refined with new data. For instance, geneticists Beadle and Tatum studied the nature of genetic information to arrive at their "one-gene/one-enzyme" hypothesis (chapter 15). This hypothesis states that a gene represents the genetic information necessary to make a single enzyme. As investigators learned more about the molecular nature of genetic information, the hypothesis was refined to "one-gene/one-polypeptide" because enzymes can be made up of more than one polypeptide. With still more information about the nature of genetic information, investigators found that a single gene can specify more than one polypeptide, and the hypothesis was refined again.

Testing hypotheses

We call the test of a hypothesis an **experiment.** Suppose that a room appears dark to you. To understand why it appears dark, you propose several hypotheses. The first might be, "There is no light in the room because the light switch is turned off." An alternative hypothesis might be, "There is no light in the room because the lightbulb is burned out." And yet another hypothesis might be, "I am going blind." To evaluate these hypotheses, you would conduct an experiment designed to eliminate one or more of the hypotheses.

For example, you might test your hypotheses by flipping the light switch. If you do so and the room is still dark, you have disproved the first hypothesis: Something other than the setting of the light switch must be the reason for the darkness. Note that a test such as this does not prove that any of the other hypotheses are true; it merely demonstrates that the one being tested is not. A successful experiment is one in which one or more of the alternative hypotheses is demonstrated to be inconsistent with the results and is thus rejected.

As you proceed through this text, you will encounter many hypotheses that have withstood the test of experiment. Many will continue to do so; others will be revised as new observations are made by biologists. Biology, like all science, is in a constant state of change, with new ideas appearing and replacing or refining old ones.

Establishing controls

Often scientists are interested in learning about processes that are influenced by many factors, or **variables.** To evaluate alternative hypotheses about one variable, all other variables must be kept constant. This is done by carrying out two experiments in parallel: a test experiment and a control experiment. In the **test experiment,** one variable is altered in a known way to test a particular hypothesis. In the **control experiment,** that variable is left unaltered. In all other respects the two experiments are identical, so any difference in the outcomes of the two experiments must result from the influence of the variable that was changed.

Much of the challenge of experimental science lies in designing control experiments that isolate a particular variable from other factors that might influence a process.

Using predictions

A successful scientific hypothesis needs to be not only valid but also useful—it needs to tell us something we want to know. A hypothesis is most useful when it makes predictions because those predictions provide a way to test the validity of the hypothesis. If an experiment produces results inconsistent with the predictions, the hypothesis must be rejected or modified. In contrast, if the predictions are supported by experimental testing, the hypothesis is supported. The more experimentally supported predictions a hypothesis makes, the more valid the hypothesis is.

As an example, in the early history of microbiology it was known that nutrient broth left sitting exposed to air becomes contaminated. There were two hypothesis proposed to explain this observation: spontaneous generation and the germ hypothesis. Spontaneous generation held that there was an inherent property in organic molecules that could lead to the spontaneous generation of life. The germ hypothesis proposed that preexisting microorganisms that were present in air could contaminate the nutrient broth.

These competing hypotheses were tested by a number of experiments that involved filtering air and boiling the broth to kill any contaminating germs. The definitive experiment was performed by Louis Pasteur, who constructed flasks with curved necks that could be exposed to air, but that would trap any contaminating germs. When such flasks were boiled to sterilize them, they remained sterile, but if the curved neck was broken off, they became contaminated (figure 1.4).

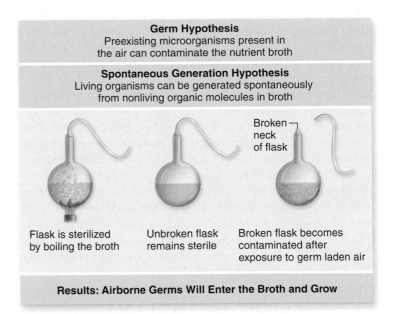

Germ Hypothesis
Preexisting microorganisms present in the air can contaminate the nutrient broth

Spontaneous Generation Hypothesis
Living organisms can be generated spontaneously from nonliving organic molecules in broth

Broken neck of flask

Flask is sterilized by boiling the broth

Unbroken flask remains sterile

Broken flask becomes contaminated after exposure to germ laden air

Results: Airborne Germs Will Enter the Broth and Grow

figure 1.4

EXPERIMENT TO TEST SPONTANEOUS GENERATION VS. GERM HYPOTHESIS. Pasteur built swan-necked flasks to prevent airborne contamination. When the flask is heated, it kills any germs in the flask. The flask will remain sterile unless the neck is broken, in which case it becomes contaminated. Spontaneous generation predicts growth should occur in either flask, while germ theory predicts growth only when the sterile flask is exposed to air.

This result was predicted by the germ hypothesis—that when the sterile flask is exposed to air, airborne germs will arrive in the broth and grow. The spontaneous generation hypothesis predicted no difference in results with exposure to air. This experiment disproved the hypothesis of spontaneous generation and supported the hypothesis of airborne germs under the conditions tested.

Reductionism breaks larger systems into their component parts

Scientists often use the philosophical approach of **reductionism** to understand a complex system by reducing it to its working parts. Reductionism has been the general approach of biochemistry, which has been enormously successful at unraveling the complexity of cellular metabolism by concentrating on individual pathways and specific enzymes. By analyzing all of the pathways and their components, scientists now have an overall picture of the metabolism of cells.

Reductionism has limits when applied to living systems, however—one of which is that enzymes do not always behave exactly the same in isolation as they do in their normal cellular context. A larger problem is that the complex interworking of many networked functions leads to emergent properties that cannot be predicted based on the workings of the parts. Biologists are just beginning to come to grips with this problem and to think about ways of dealing with the whole as well as the workings of the parts. The emerging field of systems biology is aimed toward this different approach.

Biologists construct models to explain living systems

Biologists construct models in many different ways for a variety of uses. Geneticists construct models of interacting networks of proteins that control gene expression, often even drawing cartoon figures to represent that which we cannot see. Population biologists build models of how evolutionary change occurs. Cell biologists build models of signal transduction pathways and the events leading from an external signal to internal events. Structural biologists build actual models of the structure of proteins and macromolecular complexes in cells.

Models provide a way to organize how we think about a problem. Models can also get us closer to the larger picture and away from the extreme reductionist approach. The working parts are provided by the reductionist analysis, but the model shows how they fit together. Often these models suggest other experiments that can be performed to refine or test the model.

As researchers gain more knowledge about the actual flow of molecules in living systems, more sophisticated kinetic models can be used to apply information about isolated enzymes to their cellular context. In systems biology, this modeling is being applied on a large scale to regulatory networks during development, and even to modeling an entire bacterial cell.

The nature of scientific theories

Scientists use the word **theory** in two main ways. The first meaning of *theory* is a proposed explanation for some natural phenomenon, often based on some general principle. Thus, we speak of the principle first proposed by Newton as the "theory of gravity." Such theories often bring together concepts that were previously thought to be unrelated.

The second meaning of *theory* is the body of interconnected concepts, supported by scientific reasoning and experimental evidence, that explains the facts in some area of study. Such a theory provides an indispensable framework for organizing a body of knowledge. For example, quantum theory in physics brings together a set of ideas about the nature of the universe, explains experimental facts, and serves as a guide to further questions and experiments.

To a scientist, theories are the solid ground of science, expressing ideas of which we are most certain. In contrast, to the general public, the word *theory* usually implies the opposite—a *lack* of knowledge, or a guess. Not surprisingly, this difference often results in confusion. In this text, *theory* will always be used in its scientific sense, in reference to an accepted general principle or body of knowledge.

Some critics outside of science attempt to discredit evolution by saying it is "just a theory." The hypothesis that evolution has occurred, however, is an accepted scientific fact—it is supported by overwhelming evidence. Modern evolutionary theory is a complex body of ideas, the importance of which spreads far beyond explaining evolution. Its ramifications permeate all areas of biology, and it provides the conceptual framework that unifies biology as a science. Again, the key is how well a hypothesis fits the observations. Evolutionary theory fits the observations very well.

Research can be basic or applied

In the past it was fashionable to speak of the "scientific method" as consisting of an orderly sequence of logical, either/or steps. Each step would reject one of two mutually incompatible alternatives, as though trial-and-error testing would inevitably lead a researcher through the maze of uncertainty that always impedes scientific progress. If this were the case, a computer would make a good scientist. But science is not done this way.

As the British philosopher Karl Popper has pointed out, successful scientists without exception design their experiments with a pretty fair idea of how the results are going to come out. They have what Popper calls an "imaginative preconception" of what the truth might be. Because insight and imagination play such a large role in scientific progress, some scientists are better at science than others—just as Bob Dylan stands out among songwriters or Claude Monet stands out among Impressionist painters.

Some scientists perform *basic research*, which is intended to extend the boundaries of what we know. These individuals typically work at universities, and their research is usually supported by grants from various agencies and foundations.

The information generated by basic research contributes to the growing body of scientific knowledge, and it provides the scientific foundation utilized by *applied research*. Scientists who conduct applied research are often employed in some kind of industry. Their work may involve the manufacture of food additives, the creation of new drugs, or the testing of environmental quality.

Research results are written up and submitted for publication in scientific journals, where the experiments and conclusions are reviewed by other scientists. This process of careful evaluation, called *peer review*, lies at the heart of modern science. It helps to ensure that faulty research or false claims are not given the authority of scientific fact. It also provides other scientists with a starting point for testing the reproducibility of experimental results. Results that cannot be reproduced are not taken seriously for long.

> Science uses many methods to arrive at an understanding of the natural world. Science can be descriptive, amassing observations to gain an increasingly accurate account of the world. Both deductive reasoning and inductive reasoning are used in science. Hypothesis-driven science builds hypotheses based on observation. When a hypothesis has been extensively tested, it becomes an accepted theory. Theories are coherent explanations of observed data at present, but they may be modified to fit new data.

1.3 An Example of Scientific Inquiry: Darwin and Evolution

Darwin's theory of evolution explains and describes how organisms on Earth have changed over time and acquired a diversity of new forms. This famous theory provides a good example of how a scientist develops a hypothesis and how a scientific theory grows and wins acceptance.

Charles Robert Darwin (1809–1882; figure 1.5) was an English naturalist who, after 30 years of study and observation, wrote one of the most famous and influential books of all time. This book, *On the Origin of Species by Means of Natural Selection*, created a sensation when it was published, and the ideas Darwin expressed in it have played a central role in the development of human thought ever since.

The idea of evolution existed prior to Darwin

In Darwin's time, most people believed that the different kinds of organisms and their individual structures resulted from direct actions of a Creator (and to this day, many people still believe this). Species were thought to have been specially created and to be unchangeable, or immutable, over the course of time.

In contrast to these ideas, a number of earlier naturalists and philosophers had presented the view that living things must have changed during the history of life on Earth. That is, **evolution** has occurred, and living things are now different from how they began. Darwin's contribution was a concept he called *natural selection*, which he proposed as a coherent, logical explanation for this process, and he brought his ideas to wide public attention.

Darwin observed differences in related organisms

The story of Darwin and his theory begins in 1831, when he was 22 years old. He was part of a five-year navigational mapping expedition around the coasts of South America (figure 1.6), aboard H.M.S. *Beagle*. During this long voyage, Darwin had the chance to study a wide variety of plants and animals on continents and islands and in distant seas. Darwin observed a number of phenomena that were of central importance to him in reaching his ultimate conclusion.

Repeatedly, Darwin saw that the characteristics of similar species varied somewhat from place to place. These geographical patterns suggested to him that lineages change gradually as species migrate from one area to another. On the Galápagos Islands, 960 km (600 miles) off the coast of Ecuador, Darwin encountered a variety of different finches on the various islands. The 14 species, although related, differed slightly in appearance, particularly in their beaks (figure 1.7).

Darwin thought it was reasonable to assume that all these birds had descended from a common ancestor arriving from the South American mainland several million years ago. Eating different foods on different islands, the finches' beaks had changed during their descent—"descent with modification," or evolution. (These finches are discussed in more detail in chapters 21 and 22.)

In a more general sense, Darwin was struck by the fact that the plants and animals on these relatively young volcanic islands resembled those on the nearby coast of South America. If each one of these plants and animals had been created independently and simply placed on the Galápagos Islands, why didn't they resemble the plants and animals of islands with similar climates—

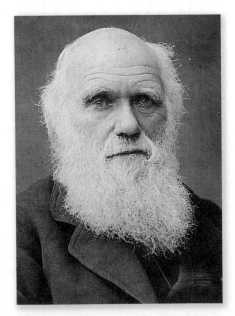

figure 1.5

CHARLES DARWIN. This newly rediscovered photograph taken in 1881, the year before Darwin died, appears to be the last ever taken of the great biologist.

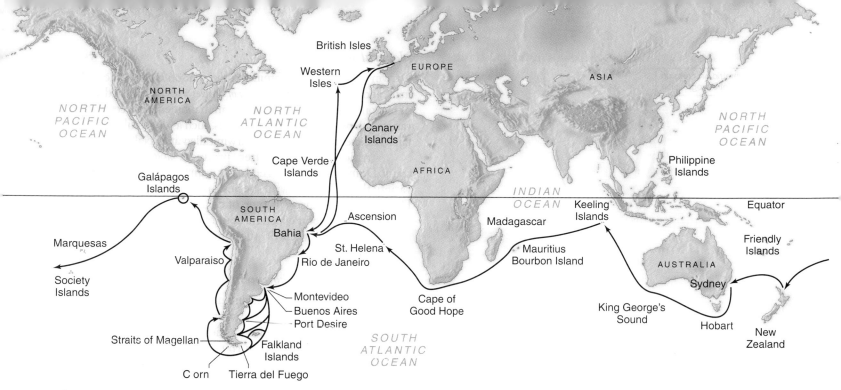

figure 1.6

THE FIVE-YEAR VOYAGE OF H.M.S. *BEAGLE*. Most of the time was spent exploring the coasts and coastal islands of South America, such as the Galápagos Islands. Darwin's studies of the animals of the Galápagos Islands played a key role in his eventual development of the concept of evolution by means of natural selection.

such as those off the coast of Africa, for example? Why did they resemble those of the adjacent South American coast instead?

Darwin proposed natural selection as a mechanism for evolution

It is one thing to observe the results of evolution, but quite another to understand how it happens. Darwin's great achievement lies in his formulation of the hypothesis that evolution occurs because of natural selection.

Darwin and Malthus

Of key importance to the development of Darwin's insight was his study of Thomas Malthus's *An Essay on the Principle of Population* (1798). In his book, Malthus stated that populations of plants and animals (including human beings) tend to increase geometrically, while humans are able to increase their food supply only arithmetically. Put another way, population increases by a multiplying factor—for example, in the series 2, 6, 18, 54, the starting number is multiplied by 3. Food supply increases by an additive factor—for example, the series 2, 4, 6, 8 adds 2

Woodpecker Finch (*Cactospiza pallida*)

Large Ground Finch (*Geospiza magnirostris*)

Cactus Finch (*Geospiza scandens*)

figure 1.7

THREE GALÁPAGOS FINCHES AND WHAT THEY EAT. On the Galápagos Islands, Darwin observed 14 different species of finches differing mainly in their beaks and feeding habits. These three finches eat very different food items, and Darwin surmised that the different shapes of their bills represented evolutionary adaptations that improved their ability to eat the foods available in their specific habitats.

chapter **1** *the science of biology*

be laid out with confidence by tracing the origins of particular nucleotide changes in the gene sequence. The pattern of descent obtained is called a **phylogenetic tree.** It represents the evolutionary history of the gene, its "family tree." Molecular phylogenetic trees agree well with those derived from the fossil record, which is strong direct evidence of evolution. The pattern of accumulating DNA changes represents, in a real sense, the footprints of evolutionary history.

Darwin's theory of evolution by natural selection presents an example of the process of science. Darwin observed differences in related organisms and proposed the hypothesis of natural selection to explain these differences. The predictions generated by natural selection have been tested and continue to be tested by analysis of the fossil record, genetics, comparative anatomy, and even the DNA of living organisms.

1.4 Unifying Themes in Biology

The study of biology encompasses a large number of different subdisciplines, ranging from biochemistry to ecology. In all of these, however, unifying themes can be identified. Among these are cell theory, the molecular basis of inheritance, the relationship between structure and function, evolution, and the emergence of novel properties.

Cell theory describes the organization of living systems

As was stated at the beginning of this chapter, all organisms are composed of cells, life's basic units (figure 1.11). Cells were discovered by Robert Hooke in England in 1665, using one of the first microscopes, one that magnified 30 times. Not long after that, the Dutch scientist Anton van Leeuwenhoek used microscopes capable of magnifying 300 times and discovered an amazing world of single-celled life in a drop of pond water.

In 1839, the German biologists Matthias Schleiden and Theodor Schwann, summarizing a large number of observations by themselves and others, concluded that all living organisms consist of cells. Their conclusion has come to be known as the **cell theory.** Later, biologists added the idea that all cells come from preexisting cells. The cell theory, one of the basic ideas in biology, is the foundation for understanding the reproduction and growth of all organisms.

The molecular basis of inheritance explains the continuity of life

Even the simplest cell is incredibly complex—more intricate than any computer. The information that specifies what a cell is like—its detailed plan—is encoded in **deoxyribonucleic acid (DNA),** a long, cablelike molecule. Each DNA molecule is formed from two long chains of building blocks, called nucleotides, wound around each other (figure 1.12). Four different nucleotides are found in DNA, and the sequence in which they occur encodes the cell's information. Specific sequences of several hundred to many thousand nucleotides make up a **gene,** a discrete unit of information.

The continuity of life from one generation to the next—heredity—depends on the faithful copying of a cell's DNA into daughter cells. The entire set of DNA instructions that specifies a cell is called its *genome.* The sequence of the human genome, 3 billion nucleotides long, was decoded in rough draft form in 2001, a triumph of scientific investigation.

a.   60 µm

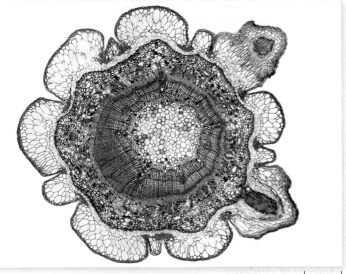

b.   568 µm

figure 1.11

LIFE IN A DROP OF POND WATER. All organisms are composed of cells. Some organisms, including the protists, shown in part (*a*) are single-celled. Others, such as the plant shown in cross section in part (*b*) consist of many cells.

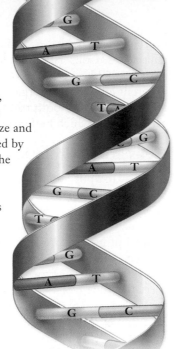

figure 1.12

GENES ARE MADE OF DNA. Winding around each other like the rails of a spiral staircase, the two strands of DNA make a double helix. Because of their size and shape, the nucleotide represented by the letter A can only pair with the nucleotide represented by the letter T, and likewise, the letter G with the letter C. This means that whatever the sequence is on one strand, the sequence on the other strand will be complementary. From each strand, the other can be easily assembled.

The relationship between structure and function underlies living systems

One of the unifying themes of molecular biology is the relationship between structure and function. Function in molecules, and larger macromolecular complexes, is dependent on their structure.

Although this observation may seem trivial, it has far-reaching implications. We study the structure of molecules and macromolecular complexes to learn about their function. When we know the function of a particular structure, we can infer the function of similar structures found in different contexts, such as in different organisms.

Biologists study both aspects, looking for the relationships between structure and function. On the one hand, this allows similar structures to be used to infer possible similar functions. On the other hand, this knowledge also gives clues as to what kinds of structures may be involved in a process if we know about the functionality.

For example, suppose that we know the structure of a human cell's surface receptor for insulin, the hormone that controls uptake of glucose. We then find a similar molecule in the membrane of a cell from a different species—perhaps even a very different organism, such as a worm. We might conclude that this membrane molecule acts as a receptor for an insulin-like molecule produced by the worm. In this way, we might be able to discern the evolutionary relationship between glucose uptake in worms and in humans.

The diversity of life arises by evolutionary change

The unity of life that we see in certain key characteristics shared by many related life-forms contrasts with the incredible diversity of living things in the varied environments of Earth. The underlying unity of biochemistry and genetics argues that all life has evolved from the same origin event. The diversity of life arises by evolutionary change leading to the present biodiversity we see.

Biologists divide life's great diversity into three great groups, called domains: Bacteria, Archaea, and Eukarya. The domains Bacteria and Archaea are composed of single-celled organisms with little internal structure (termed *prokaryotes*), and the domain Eukarya is made up of organisms composed of a complex, organized cell or multiple complex cells (termed *eukaryotes*).

Within Eukarya are four main groups called kingdoms (figure 1.13). Kingdom Protista consists of all the unicellular eukaryotes except yeasts (which are fungi), as well as the multicellular algae. Because of the great diversity among the protists, many biologists feel kingdom Protista should be split into several kingdoms.

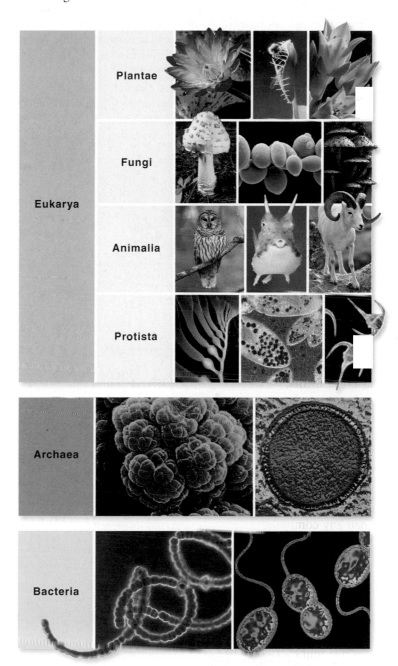

figure 1.13

THE DIVERSITY OF LIFE. Biologists categorize all living things into three overarching groups called domains: Bacteria, Archaea, and Eukarya. Domain Eukarya is composed of four kingdoms: Plantae, Fungi, Animalia, and Protista.

Kingdom Plantae consists of organisms that have cell walls of cellulose and obtain energy by photosynthesis. Organisms in the kingdom Fungi have cell walls of chitin and obtain energy by secreting digestive enzymes and then absorbing the products they release from the external environment. Kingdom Animalia contains organisms that lack cell walls and obtain energy by first ingesting other organisms and then digesting them internally.

Evolutionary conservation explains the unity of living systems

Biologists agree that all organisms alive today have descended from some simple cellular creature that arose about 3.5 BYA. Some of the characteristics of that earliest organism have been preserved. The storage of hereditary information in DNA, for example, is common to all living things.

The retention of these conserved characteristics in a long line of descent usually reflects that they have a fundamental role in the biology of the organism—one not easily changed once adopted. A good example is provided by the homeodomain

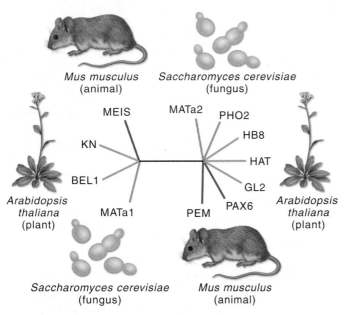

figure 1.14

TREE OF HOMEODOMAIN PROTEINS. Homeodomain proteins are found in fungi (*brown*), plants (*green*), and animals (*blue*). Based on their sequence similarities, these 11 different homeodomain proteins (uppercase letters at the ends of branches) fall into two groups, with representatives from each kingdom in each group. That means, for example, the mouse homeodomain protein PAX6 is more closely related to fungal and flowering plant proteins, such as PHO2 and GL2, than it is to the mouse protein MEIS.

proteins, which play a critical role in early development in eukaryotes. Conserved characteristics can be seen in approximately 1850 homeodomain proteins, distributed among three different kingdoms of organisms (figure 1.14). The homeodomain proteins are powerful developmental tools that evolved early, and for which no better alternative has arisen.

Cells are information-processing systems

One way to think about cells is as highly complex nanomachines that process information. The information stored in DNA is used to direct the synthesis of cellular components, and the particular set of components can differ from cell to cell. The control of gene expression allows differentiation of cell types in time and space, leading to changes over developmental time into different tissue types—even though all cells in an organism carry the same genetic information.

Cells also process information that they receive about the environment. Cells sense their environment through proteins in their membranes, and this information is transmitted across the membrane to elaborate signal-transduction chemical pathways that can change the functioning of a cell.

This ability of cells to sense and respond to their environment is critical to the function of tissues and organs in multicellular organisms. A multicellular organism can regulate its internal environment, maintaining constant temperature, pH, and concentrations of vital ions. This homeostasis is possible because of elaborate signaling networks that coordinate the activities of different cells in different tissues.

Emergent properties arise from the organization of life

As mentioned earlier, the hierarchical organization of life leads to emergent properties. The idea that the whole is greater than the sum of its parts is true of biological systems. At present, these emergent properties are not predictable, but they are observable. As biologists gain a greater understanding of the organization of biological systems, the problem of emerging properties becomes one of the most interesting challenges they face. The new science of systems biology is aimed at solving this problem, and it represents one of the most exciting areas of future research.

Biology as a science is broad and complex, but some unifying themes help to organize this complexity. Cells are the basic unit of life, and they are information-processing machines. The structures of molecules, macromolecular complexes, and even higher levels of organization are related to their functions. The diversity of life can be classified and organized based on similar features, and evolutionary conservation indicates important functions. The organization of living systems leads to emergent properties that cannot be predicted at present.

1.1 The Science of Life

Biological systems are complex chemical systems, and their functions are both determined and constrained by the principles of chemistry and physics.

■ The study of biological systems is interdisciplinary because solutions require many different approaches to solve a problem.

■ Although life is difficult to define, living systems have seven characteristics in common. All organisms:

• *are composed of one or more cells.*

• *are complex and highly ordered.*

• *respond to stimuli.*

• *are capable of growth, reproduction, and transmission of genetic information to their offspring.*

• *need energy to accomplish many kinds of work.*

• *maintain relatively constant internal conditions independent of the environment by a process called homeostasis.*

• *evolve adaptations to their environment.*

■ The organization of living systems is hierarchical, progressing from atoms to the biosphere.

■ At each higher level of organization emergent properties arise that are greater than the sum of the parts.

1.2 The Nature of Science

At its core, science is concerned with understanding the nature of the world by using observation and reasoning.

■ Much of science is concerned with an increasingly accurate description of nature.

■ There are two ways to arrive at a logical conclusion.

• *Deductive reasoning applies general principles to predict specific results.*

• *Inductive reasoning uses specific observations to construct general scientific principles.*

■ Hypothesis-driven science makes and tests predictions.

• *A hypothesis is constructed from careful observations.*

• *Hypotheses are iteratively changed and refined as new data are generated.*

• *A scientific experiment is a test of a hypothesis.*

• *Experiments involve a test in which a variable is manipulated and a control in which the variable is not manipulated.*

• *Hypotheses are rejected if they make a prediction that cannot be verified experimentally.*

• *A hypothesis can be supported by experiments but never proved.*

■ Scientists use reductionism to study components of a larger system. This has limits because parts may act differently when they are isolated from the larger system.

■ Biologists use models to organize how we think about a problem.

■ Scientists use the word *theory* in two main ways: as a proposed explanation for some natural phenomenon and as a body of concepts that explains facts in an area of study.

■ Scientists engage in basic and applied research.

1.3 An Example of Scientific Inquiry

Darwin's theory of evolution is a good example of how a scientist develops a hypothesis and how a scientific theory grows and gains acceptance.

■ Darwin proposed natural selection as a coherent, logical explanation for how life changed during the history of the Earth.

■ Darwin observed variation in similar species from place to place.

■ Sparked by Malthus's ideas, Darwin noted that species produce many offspring, but only a limited number survive and reproduce.

■ Darwin observed that the traits of offspring can be changed by artificial selection.

■ Darwin proposed that individuals that possess traits that increase survival and reproductive success will become more numerous in populations.

■ This is called natural selection, or to use Darwin's language, descent with modification.

■ Wallace independently came to the same conclusions from his own studies.

■ Natural selection has been tested using data from many fields.

• *The fossil record shows intermediate links between major groups of organisms.*

• *The age of the Earth, argued to be young in Darwin's time, has been shown by studying rates of radioactive decay to be 4.5 billion years.*

• *The research of Mendel and others provided evidence that traits can be inherited as discrete units.*

• *Comparative anatomy provides evidence of evolution from the study of homologous structures.*

• *Molecular data from the study of DNA and proteins provide evidence for changes over time.*

• *Phylogenetic trees using molecular data support the organismal relationships observed in the fossil record.*

■ Taken together, the preceding facts strongly support evolution by natural selection.

■ No data to conclusively disprove evolution has been found since Darwin.

1.4 Unifying Themes in Biology

Unifying themes in biology embrace the many complex subdisciplines of biology. They are:

■ Cell theory describes the basic unit of life and is the foundation of understanding growth and reproduction in all organisms.

■ Hereditary information encoded in genes and found in the molecule DNA is passed on from one generation to the next.

■ The structure and function of organic molecules are interdependent.

■ The diversity of life along with the underlying similarities in biochemistry and genetics support the contention that all life evolved from a single source.

■ Evolution is conservative, and all living organisms share characteristics found in original life forms because they serve an important function.

■ Cells can sense and respond to environmental changes through proteins located on their cell membranes.

■ At each higher level of hierarchical organization new emergent properties arise that could not have been predicted from lower levels of organization.

SELF TEST

1. Which of the following is NOT a property of life?
   a. Energy utilization
   b. Movement
   c. Order
   d. Homeostasis
2. Which is the correct order of hierarchical levels of living systems?
   a. Cellular, molecular, population, ecosystem, biosphere
   b. Cellular, organelle, organs, ecosystems, populations
   c. Cellular, organismal, population, community, ecosystem
   d. Species, community, population, ecosystem, biosphere
3. The process of inductive reasoning involves—
   a. the use of general principles to predict a specific result
   b. the generation of specific predictions based on a belief system
   c. the use of specific observations to develop general principles
   d. the use of general principles to support a hypothesis
4. A hypothesis in biology is best described as—
   a. a possible explanation of an observation
   b. an observation that supports a theory
   c. a general principle that explains some aspect of life
   d. an unchanging statement that correctly predicts some aspect of life
5. What is the significance of Pasteur's experiment to test the germ hypothesis?
   a. It proved that heat can sterilize a broth.
   b. It demonstrated that cells can arise spontaneously.
   c. It demonstrated that some cells are germs.
   d. It demonstrated that cells can only arise from other cells.
6. Which of the following is NOT an example of reductionism?
   a. Analysis of an isolated enzyme's function in an experimental assay
   b. Investigation of the effect of a hormone on cell growth in a petri dish
   c. Observation of the change in gene expression in response to specific stimulus
   d. An evaluation of the overall behavior of a cell
7. A scientific theory is—
   a. a guess about how things work in the world
   b. a statement of how the world works that is supported by experimental data
   c. a belief held by many scientists
   d. both a and c
8. How is the process of natural selection different from that of artificial selection?
   a. Natural selection produces more variation.
   b. Natural selection makes an individual better adapted.
   c. Artificial selection is a result of human intervention.
   d. Artificial selection results in better adaptations.
9. How does the fossil record help support the theory of evolution by natural selection?
   a. It demonstrates that simple organisms predate more complex organisms.
   b. It provides evidence of change in the form of organisms over time.
   c. It shows that diversity existed millions of years ago.
   d. Both a and b.
10. The theory of evolution by natural selection is a good example of how science proceeds because—
    a. It rationalizes a large body of observations.
    b. It makes predictions that have been tested by a variety of approaches.
    c. It represents Darwin's belief of how life has changed over time.
    d. Both a and b.
11. How does the field of molecular genetics help support the concept of evolution?
    a. Comparisons of genes demonstrate a relationship between all living things.
    b. Different organisms have different genomes.
    c. Sequencing allows for the identification of unique genes.
    d. The number of genes in an organism increase with the complexity of the organism.
12. The cell theory states—
    a. Cells are small.
    b. Cells are highly organized.
    c. There is only one basic type of cell.
    d. All living things are made up of cells.
13. The molecule DNA is important to biological systems because—
    a. It can be replicated.
    b. It encodes the information for making a new individual.
    c. It forms a complex, double-helical structure.
    d. Nucleotides form genes.
14. In which domain of life would you find only single-celled organisms?
    a. Eukarya
    b. Bacteria
    c. Archaea
    d. Both b and c
15. Evolutionary conservation occurs when a characteristic is—
    a. important to the life of the organism
    b. not influenced by evolution
    c. reduced to its least complex form
    d. found in more primitive organisms

CHALLENGE QUESTIONS

1. Exobiology is the study of life on other planets. In recent years, scientists have sent various spacecraft out into the galaxy in search for extraterrestrial life. Assuming that all life shares common properties, what should exobiologists be looking for as they explore other worlds?
2. The classic experiment by Pasteur (see figure 1.4) tested the hypothesis that cells arise from other cells. In this experiment cell growth was measured following sterilization of broth in a swan-neck flask or in a flask with a broken neck.
   a. Which variables were kept the same in these two experiments?
   b. How does the shape of the flask affect the experiment?
   c. Predict the outcome of each experiment based on the two hypotheses.
   d. Some bacteria (germs) are capable of producing heat-resistant spores that protect the cell and allow it to continue to grow after the environment cools. How would the outcome of this experiment have been affected if spore-forming bacteria were present in the broth?

The Nature of Molecules

introduction

ABOUT 12.5 BILLION YEARS AGO, an enormous explosion likely marked the beginning of the universe. With this explosion began a process of star building and planetary formation that eventually led to the formation of Earth, about 4.5 billion years ago. Around 3.5 billion years ago, life began on Earth and started to diversify. To understand the nature of life on Earth, we first need to understand the nature of matter that forms the building blocks of all life.

Starting with the earliest speculations about the world around us, the most basic question has always been, "What is it made of?" The ancient Greeks recognized that larger things may be built of smaller parts. This concept was not put on solid experimental ground until the early 20th century, when physicists began trying to break atoms apart. From those humble beginnings to the huge particle accelerators used today, the picture that emerges of the atomic world is fundamentally different from that of the macroscopic world around us.

To understand how living systems are assembled, we must first understand a little about atomic structure, about how atoms can be linked together by chemical bonds to make molecules, and about the ways in which these small molecules are joined together to make larger molecules, until finally we arrive at the structure of a cell. Our study of life on Earth therefore begins with physics and chemistry. For many of you, this chapter will be a review of material encountered in other courses.

concept outline

The Nature of Atoms

Any substance in the universe that has mass and occupies space is defined as **matter.** All matter is composed of extremely small particles called **atoms.** Because of their size, atoms are difficult to study. Not until early in the last century did scientists carry out the first experiments revealing the physical nature of atoms.

Atomic structure includes a central nucleus and orbiting electrons

Objects as small as atoms can be "seen" only indirectly, by using complex technology such as tunneling microscopy (figure 2.1). We now know a great deal about the complexities of atomic structure, but the simple view put forth in 1913 by the Danish physicist Niels Bohr provides a good starting point for understanding atomic theory. Bohr proposed that every atom possesses an orbiting cloud of tiny subatomic particles called **electrons** whizzing around a core, like the planets of a miniature solar system. At the center of each atom is a small, very dense nucleus formed of two other kinds of subatomic particles: **protons** and **neutrons** (figure 2.2).

Atomic number and the elements

Within the nucleus, the cluster of protons and neutrons is held together by a force that works only over short, subatomic distances. Each proton carries a positive (+) charge, and each neutron has no charge. Electrons carry a negative (−) charge. Typically, an atom has one electron for each proton and is, thus, electrically neutral. Different atoms are defined by the number of protons, a quantity called the **atomic number.** The chemical behavior of an atom is due to the number and configuration of electrons, as we will see later in this chapter. Atoms with the same atomic number (that is, the same number of protons) have the same chemical properties and are said to belong to the same element. Formally speaking, an **element** is any substance that cannot be broken down to any other substance by ordinary chemical means.

figure 2.1

SCANNING TUNNELING MICROSCOPE IMAGE. The scanning tunneling microscope is a nonoptical form of imaging. A probe ending in a single atom is passed over the surface of the material to be imaged. Quantum effects between probe and surface allow imaging. This image shows a lattice of oxygen atoms (shown in dark blue) on a rhodium crystal (shown in light blue).

Hydrogen	Oxygen
1 Proton 1 Electron	8 Protons 8 Neutrons 8 Electrons

a.

b.

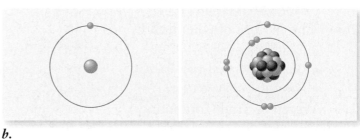

proton (positive charge) electron (negative charge) neutron (no charge)

figure 2.2

BASIC STRUCTURE OF ATOMS. All atoms have a nucleus consisting of protons and neutrons, except hydrogen, the smallest atom, which usually has only one proton and no neutrons in its nucleus. Oxygen, for example, typically has eight protons and eight neutrons in its nucleus. In the simple "Bohr model" of atoms pictured here, electrons spin around the nucleus at a relatively far distance. *a.* Atoms are depicted as a nucleus with a cloud of electrons. The cloud of electrons is not shown to scale. *b.* The electrons are shown in discrete energy levels. These are described in greater detail in the text and the next two figures.

Atomic mass

The terms *mass* and *weight* are often used interchangeably, but they have slightly different meanings. *Mass* refers to the amount of a substance, but *weight* refers to the force gravity exerts on a substance. An object has the same mass whether it is on the Earth or the Moon, but its weight will be greater on the Earth because the Earth's gravitational force is greater than the Moon's. The **atomic mass** of an atom is equal to the sum of the masses of its protons and neutrons. Atoms that occur naturally on Earth contain from 1 to 92 protons and up to 146 neutrons.

The mass of atoms and subatomic particles is measured in units called *daltons.* To give you an idea of just how small these units are, note that it takes 602 million million billion (6.02×10^{23}) daltons to make 1 gram. A proton weighs approximately 1 dalton (actually 1.007 daltons), as does a neutron (1.009 daltons). In contrast, electrons weigh only 1/1840 of a dalton, so their contribution to the overall mass of an atom is negligible.

Electrons

The positive charges in the nucleus of an atom are neutralized or counterbalanced by negatively charged electrons, which are located in regions called **orbitals** that lie at varying distances around the nucleus. Atoms with the same number of protons and electrons are electrically neutral, having no net charge, and are called **neutral atoms.**

Electrons are maintained in their orbitals by their attraction to the positively charged nucleus. Sometimes other forces overcome this attraction, and an atom loses one or more electrons. In other cases, atoms gain additional electrons. Atoms in which the number of electrons does not equal the number of protons are known as **ions,** and they are charged particles. An atom having more protons than electrons has a net positive charge and is called a **cation.** For example, an atom of sodium (Na) that has lost one electron becomes a sodium ion (Na^+), with a charge of +1. An atom having fewer protons than electrons carries a net negative charge and is called an **anion.** A chlorine atom (Cl) that has gained one electron becomes a chloride ion (Cl^-), with a charge of −1.

Isotopes

Although all atoms of an element have the same number of protons, they may not all have the same number of neutrons. Atoms of a single element that possess different numbers of neutrons are called **isotopes** of that element.

Most elements in nature exist as mixtures of different isotopes. Carbon (C), for example, has three isotopes, all containing six protons (figure 2.3). Over 99% of the carbon found in nature exists as an isotope that also contains six neutrons. Because the total mass of this isotope is 12 daltons (6 from protons plus 6 from neutrons), it is referred to as carbon-12 and is symbolized ^{12}C. Most of the rest of the naturally occurring carbon is carbon-13, an isotope with seven neutrons. The rarest carbon isotope is carbon-14, with eight neutrons. Unlike the other two isotopes, carbon-14 is unstable: Its nucleus tends to break up into elements with lower atomic numbers. This nuclear breakup, which emits a significant amount of energy, is called *radioactive decay*, and isotopes that decay in this fashion are **radioactive isotopes.**

Some radioactive isotopes are more unstable than others, and therefore they decay more readily. For any given isotope, however, the rate of decay is constant. The decay time is usually expressed as the **half-life,** the time it takes for one-half of the atoms in a sample to decay. Carbon-14, for example, often used in carbon dating of fossils and other materials, has a half-life of 5730 years. A sample of carbon containing 1 gram of carbon-14 today would contain 0.5 gram of carbon-14 after 5730 years, 0.25 gram 11,460 years from now, 0.125 gram 17,190 years from now, and so on. By determining the ratios of the different isotopes of carbon and other elements in biological samples and in rocks, scientists are able to accurately determine when these materials formed.

Radioactivity has many useful applications in modern biology. Radioactive isotopes are one way to label, or "tag," a specific molecule and then follow its fate, either in a chemical reaction or in living cells and tissue. The downside, however, is that the energetic subatomic particles emitted by radioactive substances have the potential to severely damage living cells, producing genetic mutations and, at high doses, cell death. Consequently, exposure to radiation is carefully controlled and regulated. Scientists who work with radioactivity follow strict handling protocols and wear radiation-sensitive badges to monitor their exposure over time to help ensure a safe level of exposure.

Electrons determine the chemical behavior of atoms

As mentioned earlier, the key to the chemical behavior of an atom lies in the number and arrangement of its electrons in their orbitals. The Bohr model of the atom shows individual electrons as following discrete circular orbits around a central nucleus; however, such a simple picture does not reflect reality. Modern physics indicates that it is not possible to precisely locate the position of any

Carbon-12	Carbon-13	Carbon-14
6 Protons 6 Neutrons 6 Electrons	6 Protons 7 Neutrons 6 Electrons	6 Protons 8 Neutrons 6 Electrons

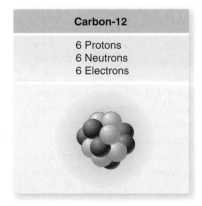 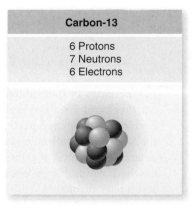 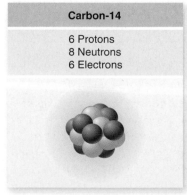

figure 2.3

THE THREE MOST ABUNDANT ISOTOPES OF CARBON. Isotopes of a particular element have different numbers of neutrons.

individual electron at any given time. In fact, an electron could be anywhere, from close to the nucleus to infinitely far away from it.

A particular electron, however, is more likely to be located in some areas than in others. An orbital is defined as the area around a nucleus where an electron is most likely to be found. These orbitals represent probability distributions for electrons, that is, regions with a high probability of containing an electron. Some electron orbitals near the nucleus are spherical (*s* orbitals), while others are dumbbell-shaped (*p* orbitals) (figure 2.4). Still other orbitals, more distant from the nucleus, may have different shapes. Regardless of its shape, no orbital may contain more than two electrons.

Almost all of the volume of an atom is empty space, because the electrons are on average very distant from the nucleus, relative to its size. If the nucleus of an atom were the size of a golf ball, the orbit of the nearest electron would be a mile away. Consequently, the nuclei of two atoms never come close enough in nature to interact with each other. It is for this reason that an atom's electrons, not its protons or neutrons, determine its chemical behavior, and it also explains why the isotopes of an element, all of which have the same arrangement of electrons, behave the same way chemically.

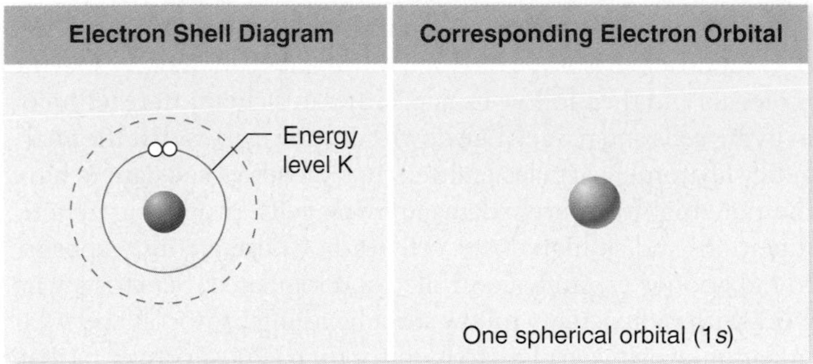

Atoms contain discrete energy levels

Because electrons are attracted to the positively charged nucleus, it takes work to keep them in their orbitals, just as it takes work to hold a grapefruit in your hand against the pull of gravity. The formal definition of energy, as you probably recall, is the ability to do work.

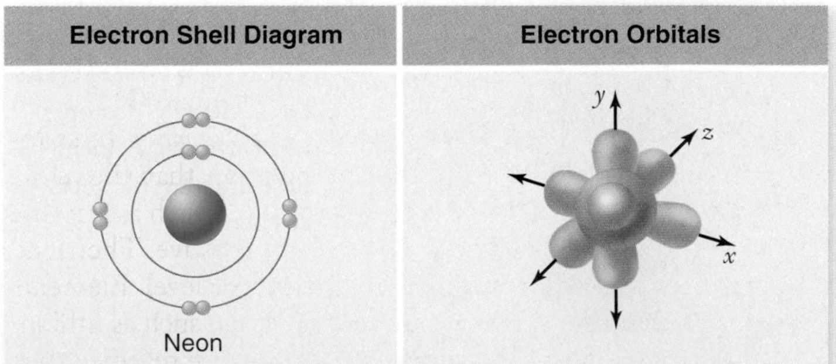

The grapefruit held above the ground is said to possess *potential energy* because of its position; if you release it, the grapefruit falls, and its potential energy is reduced. Conversely, if you carried the grapefruit to the top of a building, you would increase its potential energy. Electrons also have a potential energy that is related to their position. To oppose the attraction of the nucleus and move the electron to a more distant orbital requires an input of energy, which results in an electron with greater potential energy. Chlorophyll captures energy from light during photosynthesis in this way, as you'll see in chapter 8—light energy excites electrons in the chlorophyll molecule. Moving an electron closer to the nucleus has the opposite effect: Energy is released, usually as radiant energy (heat or light), and the electron ends up with less potential energy (figure 2.5).

One of the initially surprising aspects of atomic structure is that electrons within the atom have discrete **energy levels.** These discrete levels correspond to quanta (sing., quantum), which means specific amount of energy. To use the grapefruit analogy again, it is as though a grapefruit could only be raised to particular floors of a building. Every atom exhibits a ladder of potential energy values, a discrete set of orbitals at particular energetic "distances" from the nucleus.

Because the amount of energy an electron possesses is related to its distance from the nucleus, electrons that are the same distance from the nucleus have the same energy, even if they occupy different orbitals. Such electrons are said to occupy the same energy level. The energy levels are denoted

Electron Shell Diagram	Corresponding Electron Orbital
Energy level K	One spherical orbital (1*s*)

a.

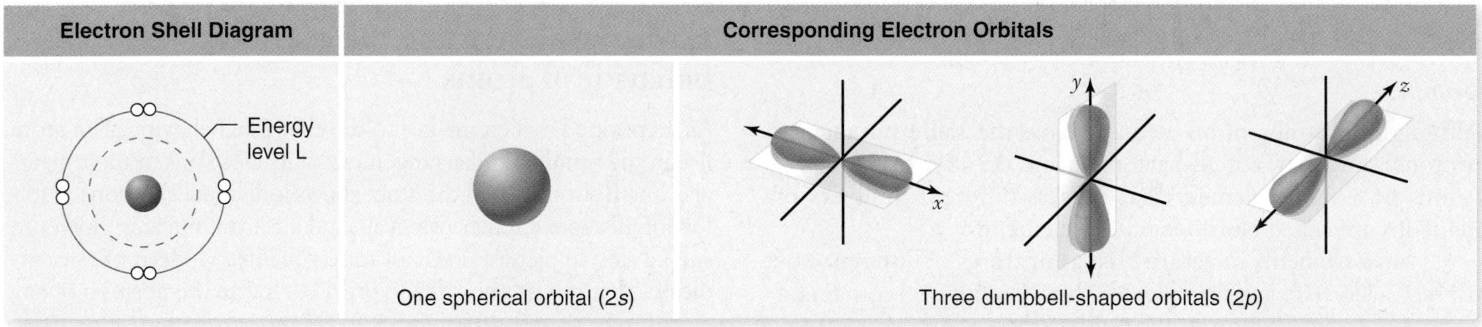

Electron Shell Diagram	Corresponding Electron Orbitals
Energy level L	One spherical orbital (2*s*) Three dumbbell-shaped orbitals (2*p*)

b.

Electron Shell Diagram	Electron Orbitals
Neon	

c.

figure 2.4

ELECTRON ORBITALS. *a.* The lowest energy level or electron shell—the one nearest the nucleus—is level K. It is occupied by a single *s* orbital, referred to as 1*s*. *b.* The next highest energy level, L, is occupied by four orbitals: one *s* orbital (referred to as the 2*s* orbital) and three *p* orbitals (each referred to as a 2*p* orbital). Each orbital holds two paired electrons with opposite spin. Thus, the K level is populated by two electrons, and the L level is populated by a total of eight electrons. *c.* The neon atom shown has the L and K energy levels completely filled with electrons and is thus unreactive.

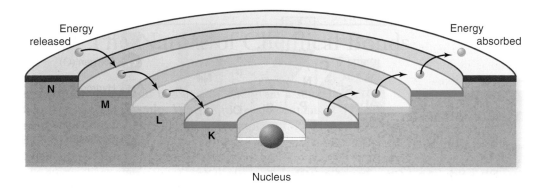

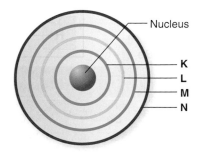

figure 2.5

ATOMIC ENERGY LEVELS. Electrons have energy of position. When an atom absorbs energy, an electron moves to a higher energy level, farther from the nucleus. When an electron falls to lower energy levels, closer to the nucleus, energy is released. The first two energy levels are the same as shown in the previous figure.

with letters K, L, M, and so on (figure 2.5). Be careful not to confuse energy levels, which are drawn as rings to indicate an electron's *energy*, with orbitals, which have a variety of three-dimensional shapes and indicate an electron's most likely *location*. Electron orbitals are arranged so that as they are filled, this successively fills each energy level. This filling of orbitals and energy levels is what is responsible for the chemical reactivity of elements.

During some chemical reactions, electrons are transferred from one atom to another. In such reactions, the loss of an electron is called **oxidation,** and the gain of an electron is called **reduction.**

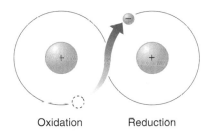

Oxidation   Reduction

Notice that when an electron is transferred in this way, it keeps its energy of position. In organisms, chemical energy is stored in high-energy electrons that are transferred from one atom to another in reactions involving oxidation and reduction (described in chapter 7). Because oxidation and reduction can be coupled, such that one atom or molecule is oxidized while another is reduced in the same reaction, we call these *redox reactions.*

An atom consists of a nucleus of protons and neutrons surrounded by a cloud of electrons. The number of electrons largely determines the chemical behavior of an atom. Atoms that have the same number of protons but different numbers of neutrons are called isotopes. Isotopes of an atom differ in atomic mass but have similar chemical properties. Electrons are localized about a nucleus in regions called orbitals. No orbital can contain more than two electrons, but many orbitals may be the same distance from the nucleus and, thus, contain electrons of the same energy.

2.2 Elements Found in Living Systems

Ninety elements occur naturally, each with a different number of protons and a different arrangement of electrons. When the nineteenth-century Russian chemist Dmitri Mendeleev arranged the known elements in a table according to their atomic number, he discovered one of the great generalizations of science: The elements exhibited a pattern of chemical properties that repeated itself in groups of eight. This periodically repeating pattern lent the table its name: the periodic table of elements (figure 2.6).

The periodic table displays elements according to atomic number and properties

The eight-element periodicity that Mendeleev found is based on the interactions of the electrons in the outermost energy level of the different elements. These electrons are called **valence elec-**

trons, and their interactions are the basis for the elements' differing chemical properties. For most of the atoms important to life, the outermost energy level can contain no more than eight electrons; the chemical behavior of an element reflects how many of the eight positions are filled. Elements possessing all eight electrons in their outer energy level (two for helium) are **inert,** or nonreactive. These elements, which include helium (He), neon (Ne), argon (Ar), and so on, are termed the *noble gases.* In sharp contrast, elements with seven electrons (one fewer than the maximum number of eight) in their outer energy level, such as fluorine (F), chlorine (Cl), and bromine (Br), are highly reactive. They tend to gain the extra electron needed to fill the energy level. Elements with only one electron in their outer energy level, such as lithium (Li), sodium (Na), and potassium (K), are also very reactive; they tend to lose the single electron in their outer level.

and an unfilled outer energy level; for these reasons, the hydrogen atom is unstable. However, when two hydrogen atoms are in close association, each atom's electron is attracted to both nuclei. In effect, the nuclei are able to share their electrons. The result is a diatomic (two-atom) molecule of hydrogen gas.

The molecule formed by the two hydrogen atoms is stable for three reasons:

1. **It has no net charge.** The diatomic molecule formed as a result of this sharing of electrons is not charged because it still contains two protons and two electrons.
2. **The octet rule is satisfied.** Each of the two hydrogen atoms can be considered to have two orbiting electrons in its outer energy level. This state satisfies the octet rule, because each shared electron orbits both nuclei and is included in the outer energy level of *both* atoms.
3. **It has no unpaired electrons.** The bond between the two atoms also pairs the two free electrons.

Unlike ionic bonds, covalent bonds are formed between two individual atoms, giving rise to true, discrete molecules.

The strength of covalent bonds

The strength of a covalent bond depends on the number of shared electrons. Thus **double bonds,** which satisfy the octet rule by allowing two atoms to share two pairs of electrons, are stronger than **single bonds,** in which only one electron pair is shared. In practical terms, more energy is required to break a double bond than a single bond. The strongest covalent bonds are **triple bonds,** such as those that link the two nitrogen atoms of nitrogen gas molecules (N_2).

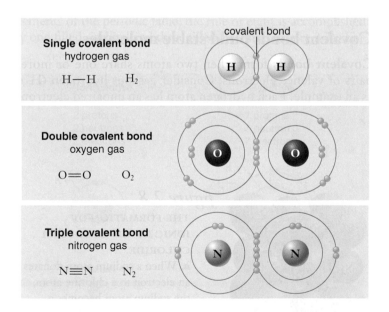

Covalent bonds are represented in chemical formulas as lines connecting atomic symbols, where each line between two bonded atoms represents the sharing of one pair of electrons. The *structural formulas* of hydrogen gas and oxygen gas are H—H and O=O, respectively, and their *molecular formulas* are H_2 and O_2. The structural formula for N_2 is N≡N.

Molecules with several covalent bonds

A vast number of biological compounds are composed of more than two atoms. An atom that requires two, three, or four additional electrons to fill its outer energy level completely may acquire them by sharing its electrons with two or more other atoms.

For example, the carbon atom (C) contains six electrons, four of which are in its outer energy level and are unpaired. To satisfy the octet rule, a carbon atom must form four covalent bonds. Because four covalent bonds may form in many ways, carbon atoms are found in many different kinds of molecules. CO_2 (carbon dioxide), CH_4 (methane), and C_2H_5OH (ethanol) are just a few examples.

Polar and nonpolar covalent bonds

Atoms differ in their affinity for electrons, a property called **electronegativity.** In general, electronegativity increases left to right in a row of the periodic table and decreases down the columns of the table. Thus the elements in the upper-right corner have the highest electronegativity.

For bonds between identical atoms, for example, H_2 or O_2, the affinity for electrons is obviously the same, and the electrons are equally shared. Such bonds are termed **nonpolar,** and the resulting compounds are also referred to as nonpolar.

For atoms that differ greatly in electronegativity, electrons are not equally shared. The shared electrons are more likely to be found near the atom with greater electronegativity, and less likely to be located near the atom of lower electronegativity. In this case, although the molecule is still electrically neutral (same number of protons as electrons), the distribution of charge is not uniform. This unequal distribution results in regions of partial negative charge near the more electronegative atom, and regions of partial positive charge near the less electronegative atom. Such bonds are termed **polar covalent bonds,** and the molecules polar molecules. When drawing polar molecules, these partial charges are usually symbolized by the lowercase Greek letter delta (δ). The partial charge seen in a polar covalent bond is relatively small—far less than the unit charge of an ion. For biological molecules, we can predict polarity of bonds by knowing the relative electronegativity of a small number of important atoms (table 2.2). Notice that although C and H differ slightly in electronegativity, this small difference is negligible, and C—H bonds are considered nonpolar.

The importance of polar and nonpolar molecules will be explored later as it is an important feature of the chemistry of water. Water (H_2O) is a polar molecule with electrons more concentrated around the oxygen atom.

TABLE 2.2	Relative Electronegativities of Some Important Atoms
Atom	**Electronegativity**
O	3.5
N	3.0
C	2.5
H	2.1

Chemical reactions alter bonds

The formation and breaking of chemical bonds, which is the essence of chemistry, is termed a *chemical reaction*. All chemical reactions involve the shifting of atoms from one molecule or ionic compound to another, without any change in the number or identity of the atoms. For convenience, we refer to the original molecules before the reaction starts as *reactants*, and the molecules resulting from the chemical reaction as *products*. For example:

$$6H_2O + 6CO_2 \longrightarrow C_6H_{12}O_6 + 6O_2$$

$$\underset{reactants}{} \longrightarrow \underset{products}{}$$

You may recognize this reaction as a simplified form of the photosynthesis reaction, in which water and carbon dioxide are combined to produce glucose and oxygen. Most animal life ultimately depends on this reaction, which takes place in plants. (Photosynthetic reactions will be discussed in detail in chapter 8.)

The extent to which chemical reactions occur is influenced by three important factors:

1. **Temperature.** Heating the reactants increases the rate of a reaction because the reactants collide with one another more often. (Care must be taken that the temperature is not so high that it destroys the molecules altogether.)
2. **Concentration of reactants and products.** Reactions proceed more quickly when more reactants are available, allowing more frequent collisions. An accumulation of products typically slows the reaction and, in reversible reactions, may speed the reaction in the reverse direction.
3. **Catalysts.** A catalyst is a substance that increases the rate of a reaction. It doesn't alter the reaction's equilibrium between reactants and products, but it does shorten the time needed to reach equilibrium, often dramatically. In living systems, proteins called enzymes catalyze almost every chemical reaction.

Many reactions in nature are reversible, meaning that the products may themselves be reactants, and the reaction proceeds in reverse. We can write the preceding reaction in the reverse order:

$$C_6H_{12}O_6 + 6O_2 \longrightarrow 6H_2O + 6CO_2$$

$$\underset{reactants}{} \longrightarrow \underset{products}{}$$

This reaction is a simplified version of the oxidation of glucose by cellular respiration, in which glucose is broken down into water and carbon dioxide in the presence of oxygen. Virtually all organisms carry out forms of glucose oxidation; details are covered later, in chapter 7.

An ionic bond is an attraction between ions of opposite charge in an ionic compound. Such bonds exist between an ion and all of the oppositely charged ions in its immediate vicinity. A covalent bond is a stable chemical bond formed when two atoms share one or more pairs of electrons. In polar covalent bonds, unequal electron sharing results in an imbalance of charge, called polarity, between the bonded atoms. Chemical reactions make and break bonds, combining reactants to form products.

2.4 Water: A Vital Compound

Of all the molecules that are common on Earth, only water exists as a liquid at the relatively low temperatures that prevail on the Earth's surface. Three-fourths of the Earth is covered by liquid water (figure 2.9). When life was beginning, water provided a medium in which other molecules could move around and interact, without being held in place by strong covalent or ionic bonds. Life evolved in water for 2 billion years before spreading to land. And even today, life is inextricably tied to water. About two-thirds of any organism's body is composed of water, and all organisms require a water-rich environment, either internal or external, for growth and reproduction. It is no accident that tropical rain forests are bursting with life, while dry deserts appear almost lifeless except when water becomes temporarily plentiful, such as after a rainstorm.

a. Solid

b. Liquid

c. Gas

figure 2.9

WATER TAKES MANY FORMS. *a.* When water cools below 0°C, it forms beautiful crystals, familiar to us as snow and ice. *b.* Ice turns to liquid when the temperature is above 0°C. *c.* Liquid water becomes steam when the temperature rises above 100°C, as seen in this hot spring at Yellowstone National Park.

Water's structure facilitates hydrogen bonding

Water has a simple molecular structure, consisting of an oxygen atom bound to two hydrogen atoms by two single covalent bonds (figure 2.10). The resulting molecule is stable: It satisfies the octet rule, has no unpaired electrons, and carries no net electrical charge.

The single most outstanding chemical property of water is its ability to form weak chemical associations, called **hydrogen bonds.** These bonds form between the partially negative O atoms and the partially positive H atoms of two water molecules. While these bonds have only 5 to 10% of the strength of covalent bonds, they are responsible for much of the chemical organization of living systems.

The electronegativity of O is much greater than that of H (see table 2.2), and so the bonds between these atoms are polar. In fact, water is a highly polar molecule, and *the polarity of water underlies water's chemistry and the chemistry of life.*

If we consider the shape of a water molecule, we can see that water's two covalent bonds have a partial charge at each end: δ^- at the oxygen end and δ^+ at the hydrogen end. The most stable arrangement of these charges is a *tetrahedron,* in which the two negative and two positive charges are approximately equidistant from one another. The oxygen atom lies at the center of the tetrahedron, the hydrogen atoms occupy two of the apexes, and the partial negative charges occupy the other two apexes (figure 2.10*b*). The bond angle between the two covalent oxygen–hydrogen bonds is 104.5°. This value is slightly less than the bond angle of a regular tetrahedron, which would be 109.5°. In water, the partial negative charges occupy more space than the partial positive regions, and therefore the oxygen–hydrogen bond angle is slightly compressed.

Water molecules are cohesive

The polarity of water allows water molecules to be attracted to one another: that is, water is **cohesive.** Each water molecule is attracted at its oxygen end, which is δ^-, to the hydrogen end, of other molecules, which is δ^+. The attraction produces hydrogen bonds among water molecules (figure 2.11). Each hydrogen bond is individually very weak and transient, lasting on average only a hundred-billionth (10^{-11}) of a second. The cumulative effects of large numbers of these bonds, however, can be enormous. Water forms an abundance of hydrogen bonds, which are responsible for many of its important physical properties (table 2.3).

Water's cohesion is responsible for its being a liquid, not a gas, at moderate temperatures. The cohesion of liquid water is also responsible for its **surface tension.** Small insects can walk on water (figure 2.12) because at the air–water interface, all the surface water molecules are hydrogen-bonded to molecules below them.

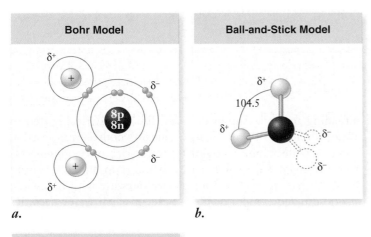

Bohr Model

Ball-and-Stick Model

104.5

a.

b.

Space-Filling Model

c.

figure 2.10

WATER HAS A SIMPLE MOLECULAR STRUCTURE.
a. Each water molecule is composed of one oxygen atom and two hydrogen atoms. The oxygen atom shares one electron with each hydrogen atom. *b.* The greater electronegativity of the oxygen atom makes the water molecule polar: Water carries two partial negative charges (δ^-) near the oxygen atom and two partial positive charges (δ^+), one on each hydrogen atom. *c.* Space-filling model shows what the molecule would look like if we could see it.

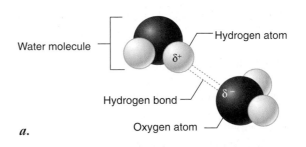

Water molecule — Hydrogen atom

Hydrogen bond

Oxygen atom

a.

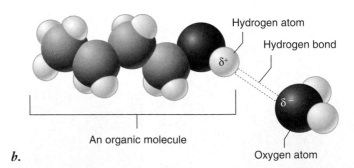

Hydrogen atom

Hydrogen bond

An organic molecule

b.

Oxygen atom

figure 2.11

STRUCTURE OF A HYDROGEN BOND. *a.* Hydrogen bond between two water molecules. *b.* Hydrogen bond between an organic molecule (*n*-butanol) and water. H in *n*-butanol forms a hydrogen bond with oxygen in water. This kind of hydrogen bond is possible any time H is bound to a more electronegative atom (see table 2.2).

TABLE 2.3	The Properties of Water	
Property	**Explanation**	**Example of Benefit to Life**
Cohesion	Hydrogen bonds hold water molecules together.	Leaves pull water upward from the roots; seeds swell and germinate.
High specific heat	Hydrogen bonds absorb heat when they break and release heat when they form, minimizing temperature changes.	Water stabilizes the temperature of organisms and the environment.
High heat of vaporization	Many hydrogen bonds must be broken for water to evaporate.	Evaporation of water cools body surfaces.
Lower density of ice	Water molecules in an ice crystal are spaced relatively far apart because of hydrogen bonding.	Because ice is less dense than water, lakes do not freeze solid, allowing fish and other life in lakes to survive the winter.
Solubility	Polar water molecules are attracted to ions and polar compounds, making them soluble.	Many kinds of molecules can move freely in cells, permitting a diverse array of chemical reactions.

Water molecules are adhesive

The polarity of water causes it to be attracted to other polar molecules as well. This attraction for other polar substances is called **adhesion.** Water is adhesive to any substance with which it can form hydrogen bonds. This property explains why substances containing polar molecules get "wet" when they are immersed in water, but those that are composed of nonpolar molecules (such as oils) do not.

The attraction of water to substances having surface electrical charges is responsible for capillary action: If a glass tube with a narrow diameter is lowered into a beaker of water, water will rise in the tube above the level of the water in the beaker, because the ad-hesion of water to the glass surface, drawing it upward, is stronger than the force of gravity, drawing it down. The narrower the tube, the greater the electrostatic forces between the water and the glass, and the higher the water rises (figure 2.13).

The chemistry of life is water chemistry. Water can form hydrogen bonds with itself and with other polar molecules because of its polar characteristics. Hydrogen bonding makes water cohesive; the molecules stick to each other. The cohesive nature of water is responsible for its high surface tension. Water molecules are also adhesive: they stick to other polar molecules. This property is responsible for the phenomenon of capillary action.

figure 2.12

COHESION. Some insects, such as this water strider, literally walk on water. In this photograph you can see how the insect's feet dimple the water as its weight bears down on the surface. Because the surface tension of the water is greater than the force that one foot brings to bear, the strider glides atop the surface of the water rather than sinking. The high surface tension of water is due to hydrogen bonding between water molecules.

figure 2.13

ADHESION. Capillary action causes the water within a narrow tube to rise above the surrounding water; the adhesion of the water to the glass surface, which draws water upward, is stronger than the force of gravity, which tends to draw it down. The narrower the tube, the greater the surface area available for adhesion for a given volume of water, and the higher the water rises in the tube.

Water moderates temperature through two properties: its high specific heat and its high heat of vaporization. Water also has the unusual property of being less dense in its solid form, ice, than as a liquid. In addition, water acts as a solvent for polar molecules and exerts an organizing effect on nonpolar molecules. Water can also dissociate to form ions. All these properties result from its polar nature.

Water's high specific heat helps maintain temperature

The temperature of any substance is a measure of how rapidly its individual molecules are moving. In the case of water, a large input of thermal energy is required to break the many hydrogen bonds that keep individual water molecules from moving about. Therefore, water is said to have a high **specific heat,** which is defined as the amount of heat that must be absorbed or lost by 1 gram of a substance to change its temperature by 1 degree Celsius (°C). Specific heat measures the extent to which a substance resists changing its temperature when it absorbs or loses heat. Because polar substances tend to form hydrogen bonds, the more polar a substance is, the higher is its specific heat. The specific heat of water (1 calorie/gram/°C) is twice that of most carbon compounds and nine times that of iron. Only ammonia, which is more polar than water and forms very strong hydrogen bonds, has a higher specific heat than water (1.23 calories/gram/°C). Still, only 20% of the hydrogen bonds are broken as water heats from 0° to 100°C.

Because of its high specific heat, water heats up more slowly than almost any other compound and holds its temperature longer when heat is no longer applied. This characteristic enables organisms, which have a high water content, to maintain a relatively constant internal temperature. The heat generated by the chemical reactions inside cells would destroy the cells if not for the absorption of this heat by the water within them.

Water's high heat of vaporization facilitates cooling

The **heat of vaporization** is defined as the amount of energy required to change 1 gram of a substance from a liquid to a gas. A considerable amount of heat energy (586 calories) is required to accomplish this change in water. Because the transition of water from a liquid to a gas requires the input of energy to break its many hydrogen bonds, the evaporation of water from a surface causes cooling of that surface. Many organisms dispose of excess body heat by evaporative cooling, for example, through sweating in humans and many other vertebrates.

Solid water is less dense than liquid water

At low temperatures, water molecules are locked into a crystal-like lattice of hydrogen bonds, forming solid ice (see figure 2.9). Interestingly, ice is less dense than liquid water because the hydrogen bonds in ice space the water molecules relatively far apart. This unusual feature enables icebergs to float. If water did not

have this property, nearly all bodies of water would be ice, with only the shallow surface melting annually. The buoyancy of ice is important ecologically because it means bodies of water freeze from the top down and not the bottom up. Liquid water beneath the surface of ice that covers most lakes in the winter allows fish and other animals to overwinter without being frozen.

The solvent properties of water help move ions and polar molecules

Water molecules gather closely around any substance that bears an electrical charge, whether that substance carries a full charge (ion) or a charge separation (polar molecule). For example, sucrose (table sugar) is composed of molecules that contain polar hydroxyl (OH) groups. A sugar crystal dissolves rapidly in water because water molecules can form hydrogen bonds with individual hydroxyl groups of the sucrose molecules. Therefore, sucrose is said to be *soluble* in water. Water is termed the *solvent*, and sugar is called the *solute*. Every time a sucrose molecule dissociates, or breaks away, from a solid sugar crystal, water molecules surround it in a cloud, forming a **hydration shell** that prevents it from associating with other sucrose molecules. Hydration shells also form around ions such as Na^+ and Cl^- (figure 2.14).

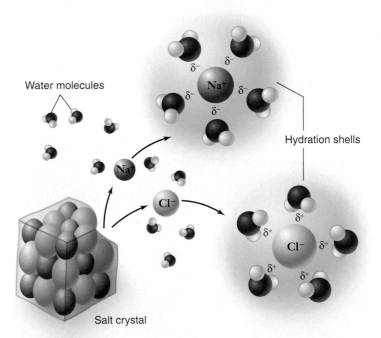

Water molecules

Hydration shells

Salt crystal

figure 2.14

WHY SALT DISSOLVES IN WATER. When a crystal of table salt dissolves in water, individual Na^+ and Cl^- ions break away from the salt lattice and become surrounded by water molecules. Water molecules orient around Cl^- ions so that their partial positive poles face toward the negative Cl^- ion; water molecules surrounding Na^+ ions orient in the opposite way, with their partial negative poles facing the positive Na^+ ion. Surrounded by hydration shells, Na^+ and Cl^- ions never reenter the salt lattice.

Water organizes nonpolar molecules

Water molecules always tend to form the maximum possible number of hydrogen bonds. When nonpolar molecules such as oils, which do not form hydrogen bonds, are placed in water, the water molecules act to exclude them. The nonpolar molecules are forced into association with one another, thus minimizing their disruption of the hydrogen bonding of water. In effect, they shrink from contact with water, and for this reason they are referred to as **hydrophobic** (Greek *hydros*, "water," and *phobos*, "fearing"). In contrast, polar molecules, which readily form hydrogen bonds with water, are said to be **hydrophilic** ("water-loving").

The tendency of nonpolar molecules to aggregate in water is known as **hydrophobic exclusion.** By forcing the hydrophobic portions of molecules together, water causes these molecules to assume particular shapes. This property can also affect the structure of proteins, DNA, and biological membranes. In fact, the interaction of nonpolar molecules and water is critical to living systems.

Water can form ions

The covalent bonds of a water molecule sometimes break spontaneously. In pure water at 25°C, only 1 out of every 550 million water molecules undergoes this process. When it happens, a proton (hydrogen atom nucleus) dissociates from the molecule. Because the dissociated proton lacks the negatively charged electron it was sharing, its positive charge is no longer counterbalanced, and it becomes a hydrogen ion, H^+. The rest of the dissociated water molecule, which has retained the shared electron from the covalent bond, is negatively charged and forms a hydroxide ion, OH^-. This process of spontaneous ion formation is called *ionization*:

$$H_2O \longrightarrow OH^- + H^+$$

water   *hydroxide ion*   *hydrogen ion (proton)*

At 25°C, a liter of water contains one ten-millionth (or 10^{-7}) mole of H^+ ions. A **mole** is defined as the weight of a substance in grams that corresponds to the atomic masses of all of the atoms in a molecule of that substance. In the case of H^+, the atomic mass is 1, and a mole of H^+ ions would weigh 1 gram. One mole of any substance always contains 6.02×10^{23} molecules of the substance. Therefore, the **molar concentration** of hydrogen ions in pure water, represented as $[H^+]$, is 10^{-7} moles/liter. (In reality, the hydrogen ion usually associates with another water molecule to form a hydronium ion, H_3O^+.)

Water does not change temperature rapidly because of its high specific heat. In living systems, high water content maintains a near-constant temperature. Water's high heat of vaporization allows cooling by evaporation. Because polar water molecules cling to one another, it takes considerable energy to separate them. Water also clings to other polar molecules, causing them to be soluble in a water solution, but water tends to exclude nonpolar molecules. Water dissociates to form ions. The hydrogen ion concentration of pure water is 10^{-7} moles per liter.

2.6 Acids and Bases

The concentration of hydrogen ions, and concurrently of hydroxide ions, in a solution is described by the terms *acidity* and *basicity*. Pure water, having an $[H^+]$ of 10^{-7} mole/liter, is considered to be neutral, that is, neither acidic nor basic. Recall that for every H^+ ion formed when water dissociates, an OH^- ion is also formed, meaning that the dissociation of water produces H^+ and OH^- in equal amounts.

The pH scale measures hydrogen ion concentration

The **pH scale** (figure 2.15) is a more convenient way to express the hydrogen ion concentration of a solution. This scale defines **pH,** which stands for "partial hydrogen," as the negative logarithm of the hydrogen ion concentration in the solution:

$$pH = -\log [H^+]$$

Because the logarithm of the hydrogen ion concentration is simply the exponent of the molar concentration of H^+, the pH equals the exponent times -1. For water, therefore, an $[H^+]$ of 10^{-7} moles/liter corresponds to a pH value of 7. This is the neutral point—a balance between H^+ and OH^-—on the pH scale. This balance occurs because the dissociation of water leads to equal amounts of H^+ and OH^-.

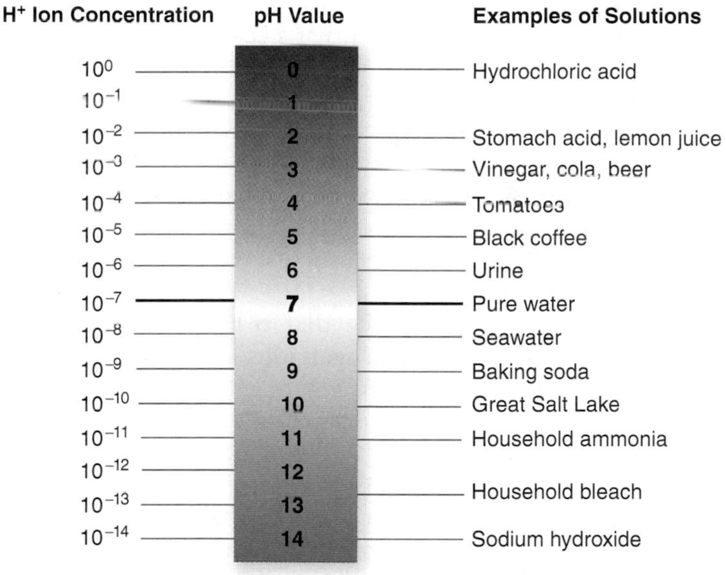

H^+ Ion Concentration	pH Value	Examples of Solutions
10^0	0	Hydrochloric acid
10^{-1}	1	
10^{-2}	2	Stomach acid, lemon juice
10^{-3}	3	Vinegar, cola, beer
10^{-4}	4	Tomatoes
10^{-5}	5	Black coffee
10^{-6}	6	Urine
10^{-7}	7	Pure water
10^{-8}	8	Seawater
10^{-9}	9	Baking soda
10^{-10}	10	Great Salt Lake
10^{-11}	11	Household ammonia
10^{-12}	12	Household bleach
10^{-13}	13	
10^{-14}	14	Sodium hydroxide

figure 2.15

THE pH SCALE. The pH value of a solution indicates its concentration of hydrogen ions. Solutions with a pH less than 7 are acidic, whereas those with a pH greater than 7 are basic. The scale is logarithmic, so that a pH change of 1 means a 10-fold change in the concentration of hydrogen ions. Thus, lemon juice is 100 times more acidic than tomato juice, and seawater is 10 times more basic than pure water, which has a pH of 7.

Note that, because the pH scale is *logarithmic*, a difference of 1 on the scale represents a 10-fold change in hydrogen ion concentration. A solution with a pH of 4 therefore has 10 times the [H⁺] of a solution with a pH of 5 and 100 times the [H⁺] concentration of a solution with a pH of 6.

Acids

Any substance that dissociates in water to increase the concentration of H⁺ ions (and lower the pH) is called an **acid.** The stronger an acid is, the more H⁺ ions it produces and the lower its pH. For example, hydrochloric acid (HCl), which is abundant in your stomach, ionizes completely in water. A dilution of 10^{-1} moles/liter of HCl dissociates to form 10^{-1} moles/liter of H⁺ ions, giving the solution a pH of 1. The pH of champagne, which bubbles because of the carbonic acid dissolved in it, is about 2.

Bases

A substance that combines with H⁺ ions when dissolved in water, and thus lowers the [H⁺], is called a **base.** Therefore, basic (or alkaline) solutions have pH values above 7. Very strong bases, such as sodium hydroxide (NaOH), have pH values of 12 or more. Many common cleaning substances, such as ammonia and bleach, accomplish their action because of their high pH.

Buffers help stabilize pH

The pH inside almost all living cells, and in the fluid surrounding cells in multicellular organisms, is fairly close to 7. Most of the enzymes in living systems are extremely sensitive to pH; often even a small change in pH will alter their shape, thereby disrupting their activities and rendering them useless. For this reason, it is important that a cell maintain a constant pH level.

But the chemical reactions of life constantly produce acids and bases within cells. Furthermore, many animals eat substances that are acidic or basic. Cola drinks, for example, are moderately strong (although dilute) acidic solutions. Despite such variations in the concentrations of H⁺ and OH⁻, the pH of an organism is kept at a relatively constant level by buffers (figure 2.16).

A **buffer** is a substance that resists changes in pH. Buffers act by releasing hydrogen ions when a base is added and absorbing hydrogen ions when acid is added, with the overall effect of keeping hydrogen ion concentration relatively constant.

Within organisms, most buffers consist of pairs of substances, one an acid and the other a base. The key buffer in human blood is an acid–base pair consisting of carbonic acid (acid) and bicarbonate (base). These two substances interact in a pair of reversible reactions. First, carbon dioxide (CO₂) and H₂O join to form carbonic acid (H₂CO₃), which in a second reaction dissociates to yield bicarbonate ion (HCO₃⁻) and H⁺.

If some acid or other substance adds H⁺ ions to the blood, the HCO₃⁻ ions act as a base and remove the excess H⁺ ions by forming H₂CO₃. Similarly, if a basic substance removes H⁺ ions from the blood, H₂CO₃ dissociates, releasing

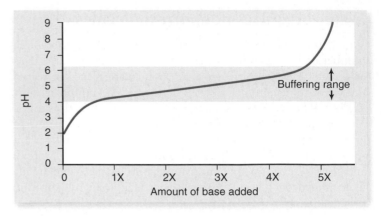

figure 2.16

BUFFERS MINIMIZE CHANGES IN pH. Adding a base to a solution neutralizes some of the acid present, and so raises the pH. Thus, as the curve moves to the right, reflecting more and more base, it also rises to higher pH values. A buffer makes the curve rise or fall very slowly over a portion of the pH scale, called the "buffering range" of that buffer.

inquiry

? *For this buffer, adding base raises pH more rapidly below pH 4 than above it. What might account for this behavior?*

more H⁺ ions into the blood. The forward and reverse reactions that interconvert H₂CO₃ and HCO₃⁻ thus stabilize the blood's pH.

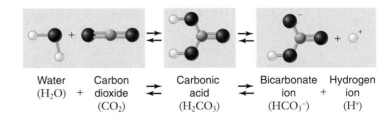

| Water (H₂O) | + | Carbon dioxide (CO₂) | ⇌ | Carbonic acid (H₂CO₃) | ⇌ | Bicarbonate ion (HCO₃⁻) | + | Hydrogen ion (H⁺) |

The reaction of carbon dioxide and water to form carbonic acid is a crucial one because it permits carbon, essential to life, to enter water from the air. The Earth's oceans are rich in carbon because of the reaction of carbon dioxide with water.

In a condition called blood acidosis, human blood, which normally has a pH of about 7.4, drops to a pH of about 7.1 or below. This condition is fatal if not treated immediately. The reverse condition, blood alkalosis, involves an increase in blood pH of a similar magnitude and is just as serious.

The pH of a solution is the negative logarithm of the H⁺ ion concentration in the solution. Thus, low pH values indicate high H⁺ concentrations (acidic solutions), and high pH values indicate low H⁺ concentrations (basic solutions). Even small changes in pH can be harmful to life. Buffer systems in organisms, such as the carbon dioxide/bicarbonate system in humans, help to maintain pH within a narrow range.

2.1 The Nature of Atoms

All matter is composed of atoms (figure 2.2).

■ Atoms are composed of a nucleus with positively charged protons and neutral neutrons surrounded by one or more orbitals containing negatively charged electrons.

■ To be electrically neutral an atom must have the same number of protons as electrons.

■ Atoms that gain or lose electrons are called ions.

■ If an atom loses electrons it has a positive charge and is called a cation. If the atom gains an electron it has a negative charge and is called an anion.

■ Each element is defined by its atomic number, which is the number of protons found in the nucleus.

■ Atomic mass is the sum of the mass of protons and neutrons in an atom.

An atom is called an isotope if the number of neutrons exceeds the number of protons.

■ Isotopes are different forms of the same element that have different numbers of neutrons and thus different atomic mass.

■ Radioactive elements are unstable and break up into smaller number elements.

■ The rate of decay for any radioactive element is constant.

■ Decay is expressed as a half-life, the amount of time for 50% of atoms to decay.

Electrons determine the behavior of atoms.

■ The potential energy of electrons increases as distance from the nucleus increases. Excited electrons can temporarily move to a higher energy level and increase their potential energy.

■ The loss of electrons from an atom is called oxidation.

■ The gain of electrons is called reduction.

■ Electrons can be transferred from one atom to another in coupled redox reactions (see figure in left column, page 21).

2.2 Elements Found in Living Systems

The periodic table is based on the interactions of valence electrons.

■ There are 90 naturally occurring elements in the Earth's crust.

■ Twelve of these elements are found in living organisms in greater than trace amounts.

■ Elements with filled outermost electron orbitals are inert. These are found in the last column of the periodic table.

2.3 The Nature of Chemical Bonds

Molecules contain two or more similar atoms joined by chemical bonds. Compounds contain two or more different elements.

■ Ionic bonds occur when two different ions with opposite electrical charges are attracted to each other. Ionic bonds can be very strong, but not as strong as a covalent bond (figure 2.8b).

■ Covalent bonds occur when one or more pairs of electrons are shared between two atoms. One atom may form covalent bonds with several other atoms.

■ Covalent bonds are the strongest and they are responsible for the stability of organic molecules.

■ Electronegativity is the affinity of an atom for electrons. It increases across rows and decreases down columns of the periodic table.

■ Nonpolar covalent bonds involve equal sharing of electrons between atoms.

■ Polar covalent bonds involve unequal sharing of electrons between atoms. This occurs between atoms with large differences in electronegativity.

■ Chemical reactions make, break, or otherwise alter bonds. Temperature, pH, and catalysts will affect reaction rates.

2.4 Water: A Vital Compound

Life can be understood through the chemistry of water (figure 2.11a).

■ Hydrogen bonds are weak interactions between a partially positive H in one molecule and a partially negative oxygen in another molecule.

■ Cohesion is the tendency of water molecules to adhere to one another due to hydrogen bonding.

■ Adhesion occurs when water molecules adhere to other polar molecules.

2.5 Properties of Water

Water has several properties because it is polar.

■ Water has a high specific heat because it takes a considerable amount of energy to disrupt hydrogen bonds. The large amount of water in organisms helps them maintain body temperature.

■ Water has a high heat of vaporization, which is used for cooling. It takes a considerable amount of heat to break enough hydrogen bonds for liquid water to become a gas.

■ The hydrogen bonds between water molecules in the solid phase are spaced farther apart than in the liquid phase. As a result ice floats.

■ Water is a good solvent for polar substances and ions. Water will tend to exclude nonpolar substances.

■ Molecules or portions of molecules that are polar are hydrophilic. These substances will be attracted to water.

■ Molecules that are nonpolar are hydrophobic molecules. These substances will be repelled by water.

■ Because of hydrophobic exclusion, nonpolar molecules or their components will tend to aggregate or form particular shapes. This can influence the shape of biological molecules.

2.6 Acids and Bases (figure 2.15)

Water forms two ions when covalent bonds break.

■ The hydrogen ion (H^+) is positively charged, and the hydroxide ion (OH^-) is negatively charged.

■ The relationship between H^+ and OH^- is expressed as pH. This is defined as the negative logarithm of H^+ ion concentration.

■ The pH scale is logarithmic and a difference of 1 on a pH scale means a 10-fold change in concentration of hydrogen ions.

■ If the concentration of hydrogen ions is greater than that of hydroxide ions, the solution is acidic and the pH is below 7 units. If the concentration of hydrogen ions is less than that of the hydroxide ions, the solution is basic and the pH is above 7 units.

SELF TEST

1. The property that distinguishes one atom (carbon for example) from another atom (oxygen for example) is—
   a. The number of electrons
   b. The number of protons
   c. The number of neutrons
   d. The combined number of protons and neutrons
2. If an atom has one valence (outer energy level) electron, it will most likely form—
   a. One polar, covalent bond
   b. Two nonpolar, covalent bonds
   c. Two covalent bonds
   d. An ionic bond
3. An atom with a net positive charge must have—
   a. More protons than neutrons
   b. More protons than electrons
   c. More electrons than neutrons
   d. More electrons than protons
4. The isotopes C^{12} and C^{14} differ in—
   a. The number of neutrons
   b. The number of protons
   c. The number of electrons
   d. Both b and c
5. An atom with more electrons than protons is called—
   a. An element
   b. An isotope
   c. A cation
   d. An anion
6. Which of the following is NOT a property of the elements most commonly found in living organisms?
   a. The elements have a low atomic mass
   b. The elements have an atomic number less than 21
   c. The elements possess eight electrons in their outer energy level
   d. The elements are lacking one or more electrons from their outer energy level
7. Which of the following atoms would you predict could be a cation?
   a. Fluorine (F)
   b. Helium (He)
   c. Potassium (K)
   d. Boron (B)
8. Refer to the element pictured. How many covalent bonds could this atom form?

   a. Two
   b. Three
   c. Four
   d. None
9. Refer to the element pictured. How many covalent bonds could this atom form?

   a. Two
   b. Three
   c. Four
   d. None
10. An ionic bond is held together by
    a. shared valence electrons
    b. attractions between ions of the same charge
    c. charge attractions between valence electrons
    d. attractions between ions of opposite charge
11. How do polar covalent bonds differ from nonpolar covalent bonds?
    a. In a polar covalent bond the electrons are shared equally between the atoms.
    b. In a nonpolar covalent bond there is a charge attraction between the atomic nuclei.
    c. There is a large difference in electronegativity of the atoms in a nonpolar bond.
    d. There is a large difference in electronegativity of the atoms in a polar bond.
12. A hydrogen bond can form—
    a. when hydrogen is part of a polar covalent bond
    b. only in water
    c. between any large electronegative atoms like oxygen
    d. when two atoms of hydrogen share an electron
13. Which of the following properties of water is NOT a consequence of its ability to form hydrogen bonds?
    a. Cohesiveness
    b. High specific heat
    c. Ability to function as a solvent
    d. Neutral pH
14. A substance with a high concentration of hydrogen ions is—
    a. called a base
    b. called an acid
    c. has a high pH
    d. both b and c

CHALLENGE QUESTIONS

1. Elements that form ions are important for a range of biological processes. You have learned about the cations, sodium (Na^+), calcium (Ca^{2+}) and potassium (K^+) in this chapter. Use your knowledge of the definition of a cation to identify other examples from the periodic table.
2. A popular theme in science fiction literature has been the idea of silicon-based life-forms in contrast to our carbon-based life. Evaluate the possibility of silicon-based life based on the chemical structure and potential for chemical bonding of a silicon atom.
3. Recent efforts by NASA to search for signs of life on Mars have focused on the search for evidence of liquid water in the planet's history rather than looking directly for biological organisms (living or fossilized). Use your knowledge of the influence of water on life on Earth to construct an argument justifying this approach.

 **Do you need additional review?** *Visit www.ravenbiology.com for practice quizzes, animations, videos, and activities designed to help you master the material in this chapter.*

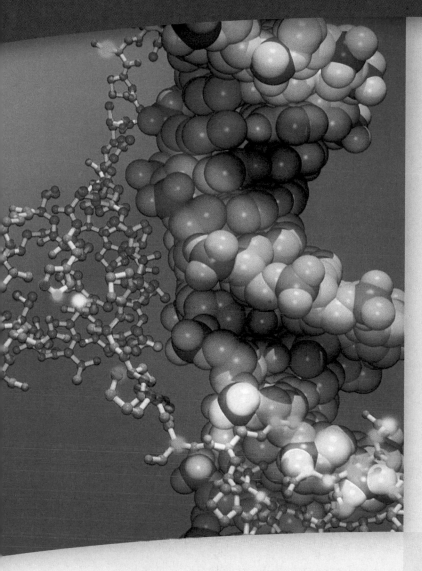

chapter 3

The Chemical Building Blocks of Life

introduction

A CUP OF WATER CONTAINS more molecules than there are stars in the sky. But many molecules are much larger than water molecules; they consist of thousands of atoms, forming hundreds of molecules that are linked together into long chains. These enormous assemblies, which are almost always synthesized by living things, are **macromolecules.** As you may know, biological macromolecules can be divided into four categories: *carbohydrates, nucleic acids, proteins,* and *lipids,* and they are the basic chemical building blocks from which all organisms are assembled.

Biological macromolecules all involve carbon-containing compounds, so we begin the discussion with a brief summary of carbon and its chemistry. The study of the chemistry of carbon, because of its biological significance, is known as *organic chemistry.*

concept outline

3.1 Carbon: The Framework of Biological Molecules

In chapter 2, we reviewed the basics of chemistry. No new laws of chemistry are found in biological systems, and biological systems do not violate the laws of chemistry. Thus, chemistry forms the basis of living systems.

The framework of biological molecules consists predominantly of carbon atoms bonded to other carbon atoms or to atoms of oxygen, nitrogen, sulfur, or hydrogen. Because carbon atoms can form up to four covalent bonds, molecules containing carbon can form straight chains, branches, or even rings, balls, and coils.

Molecules consisting only of carbon and hydrogen are called **hydrocarbons.** Because carbon–hydrogen covalent bonds store considerable energy, hydrocarbons make good fuels. Gasoline, for example, is rich in hydrocarbons, and propane gas, another hydrocarbon, consists of a chain of three carbon atoms, with eight hydrogen atoms bound to it. The chemical formula for propane is C_3H_8, and its structural formula is shown as

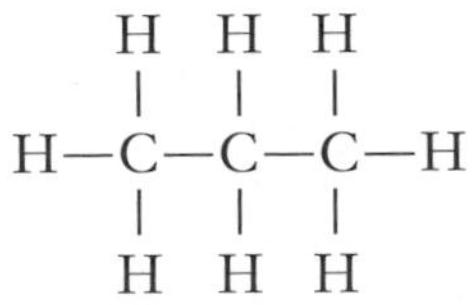

$$
\begin{array}{ccccccc}
 & H & & H & & H & \\
 & | & & | & & | & \\
H & - C & - & C & - & C & - H \\
 & | & & | & & | & \\
 & H & & H & & H &
\end{array}
\qquad \text{Propane structural formula}
$$

Theoretically speaking, there is no limit to the length of a chain of carbon atoms. As described in the rest of this chapter, the four main types of biological molecules often consist of huge chains of carbon-containing compounds.

Functional groups account for differences in molecular properties

Carbon and hydrogen atoms both have very similar electronegativities, so electrons in C—C and C—H bonds are evenly distributed, with no significant differences in charge over the molecular surface. For this reason, hydrocarbons are nonpolar. Most biological molecules that are produced by cells, however, also contain other atoms. Because these other atoms frequently have different electronegativities, molecules containing them exhibit regions of partial positive or negative charge, and so are polar. These molecules can be thought of as a C—H core to which specific molecular groups, called **functional groups,** are attached. For example, an attached —OH group is a functional group called a *hydroxyl group.*

Functional groups have definite chemical properties that they retain no matter where they occur. The hydroxyl and carbonyl (C=O) groups, for example, are both polar because of the electronegativity of the oxygen atoms (as described in chapter 2). Other functional groups include the acidic carboxyl (COOH) and phosphate (PO_4) groups, and the basic amino (NH_2) group. Many of these functional groups can also participate in hydrogen bonding. Hydrogen bond donors and acceptors can be predicted based on the electronegativities given previously in table 2.2. Figure 3.1 illustrates these biologically important functional groups and lists the macromolecules in which they are found.

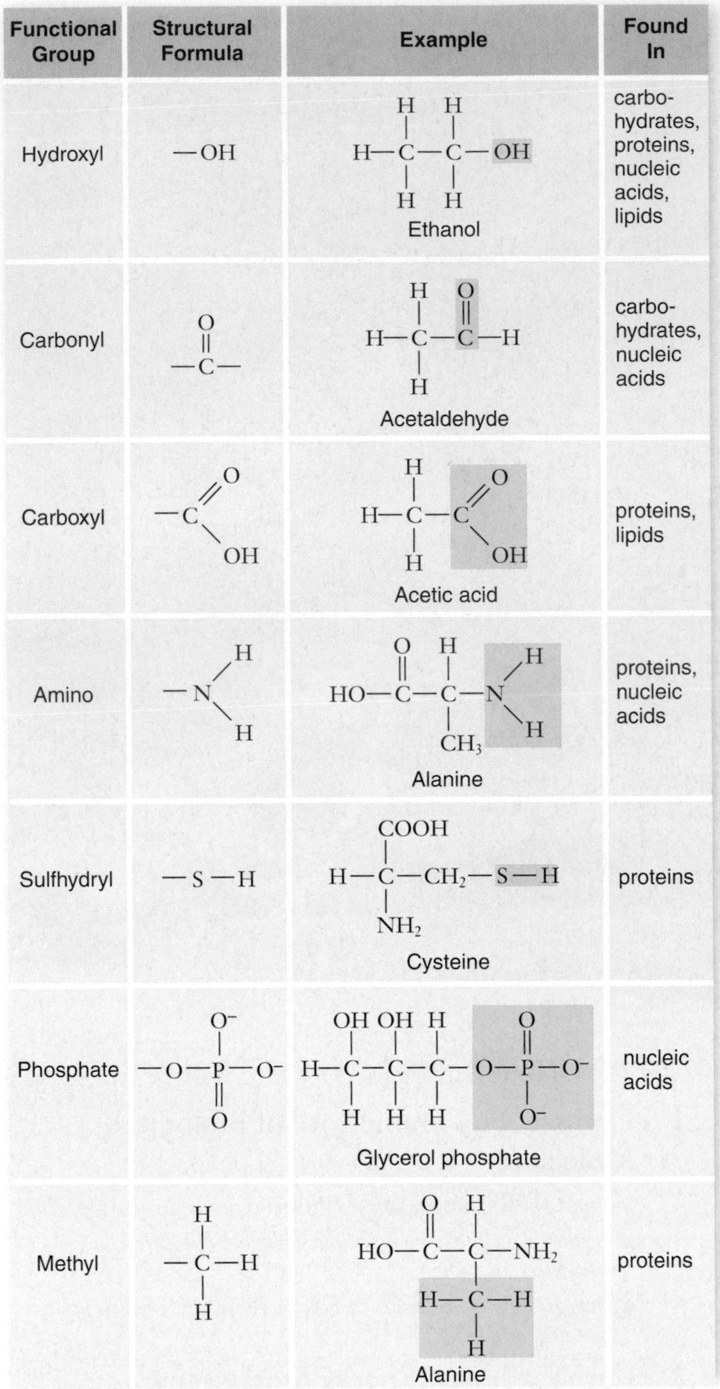

Functional Group	Structural Formula	Example	Found In
Hydroxyl	—OH	Ethanol	carbohydrates, proteins, nucleic acids, lipids
Carbonyl	$\overset{O}{\underset{\parallel}{-C-}}$	Acetaldehyde	carbohydrates, nucleic acids
Carboxyl	$-C\overset{O}{_{\diagdown OH}}$	Acetic acid	proteins, lipids
Amino	$-N\overset{H}{_{\diagdown H}}$	Alanine	proteins, nucleic acids
Sulfhydryl	—S—H	Cysteine	proteins
Phosphate	$-O-P\overset{O^-}{\underset{O}{-O^-}}$	Glycerol phosphate	nucleic acids
Methyl	$-\overset{H}{\underset{H}{C}}-H$	Alanine	proteins

figure 3.1

THE PRIMARY FUNCTIONAL CHEMICAL GROUPS. These groups tend to act as units during chemical reactions and confer specific chemical properties on the molecules that possess them. Amino groups, for example, make a molecule more basic, whereas carboxyl groups make a molecule more acidic. These functional groups are also not limited to the examples in the "Found In" column but are widely distributed in biological molecules.

TABLE 3.1	Macromolecules		
Macromolecule	Subunit	Function	Example
CARBOHYDRATES			
Starch, glycogen	Glucose	Energy storage	Potatoes
Cellulose	Glucose	Plant cell walls	Paper; strings of celery
Chitin	Modified glucose	Structural support	Crab shells
NUCLEIC ACIDS			
DNA	Nucleotides	Encodes genes	Chromosomes
RNA	Nucleotides	Needed for gene expression	Messenger RNA
PROTEINS			
Functional	Amino acids	Catalysis; transport	Hemoglobin
Structural	Amino acids	Support	Hair; silk
LIPIDS			
Fats	Glycerol and three fatty acids	Energy storage	Butter; corn oil; soap
Phospholipids	Glycerol, two fatty acids, phosphate, and polar R groups	Cell membranes	Phosphatidylcholine
Prostaglandins	Five-carbon rings with two nonpolar tails	Chemical messengers	Prostaglandin E (PGE)
Steroids	Four fused carbon rings	Membranes; hormones	Cholesterol; estrogen
Terpenes	Long carbon chains	Pigments; structural support	Carotene; rubber

Isomers have the same molecular formulas but different structures

Organic molecules having the same structural formula can exist in different forms called **isomers.** If the differences exist in the actual structure of the carbon skeleton, we call them *structural isomers.* Later you will see that glucose and fructose are structural isomers of $C_6H_{12}O_6$. A more subtle form of isomer is called a *stereoisomer,* and these molecules have the same carbon skeleton but differ in how the groups attached to this skeleton are arranged.

Enzymes in biological systems usually recognize only a single, specific stereoisomer. A subcategory of stereoisomers, called *enantiomers,* are actually mirror images of each other. A molecule that has mirror-image versions is called a **chiral** molecule. When carbon is bound to four different molecules, this inherent asymmetry exists (figure 3.2).

Chiral compounds are characterized by their effect on polarized light. Polarized light has a single plane, and chiral molecules rotate this plane either to the left or to the right. We therefore call the two chiral forms D for *dextrorotatory* and L for *levorotatory.* Living systems tend to produce only a single enantiomer of the two possible forms; for example, in most organisms we find primarily D-sugars and L-amino acids.

Biological macromolecules include carbohydrates, nucleic acids, proteins, and lipids

As mentioned at the beginning, biological macromolecules are traditionally grouped into carbohydrates, nucleic acids, proteins, and lipids (table 3.1). In many cases, these macromolecules are polymers. A **polymer** is a long molecule built by linking together a large number of small, similar chemical subunits called **monomers,** like railroad cars coupled to form a train. The nature of a polymer is determined by the monomers used to build the polymer. For example, complex carbohydrates such as starch are polymers of simple ring-shaped sugars; nucleic acids (DNA and RNA) are polymers of nucleotides (figure 3.3); proteins are polymers of amino acids, and fats are polymers of fatty acids (see figure 3.3). These long chains are built via chemical reactions termed *dehydration reactions* and are broken down by *hydrolysis reactions.*

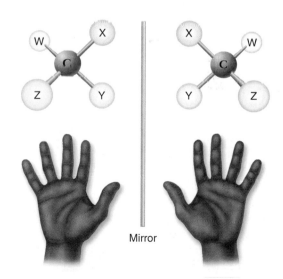

Mirror

figure 3.2

CHIRAL MOLECULES. When carbon is bound to four different groups, the resulting molecule is said to be chiral. This kind of molecule can have stereoisomers that are mirror images. The two molecules shown have the same four groups but cannot be superimposed, much like your two hands. These types of stereoisomers are called *enantiomers.*

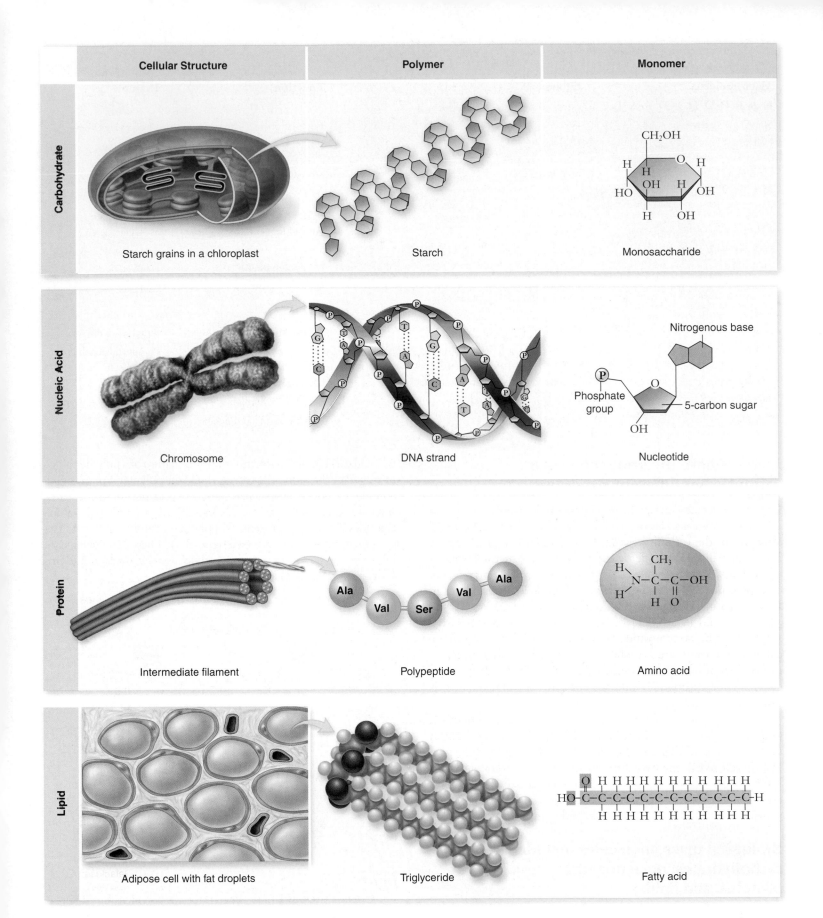

	Cellular Structure	Polymer	Monomer
Carbohydrate	Starch grains in a chloroplast	Starch	Monosaccharide
Nucleic Acid	Chromosome	DNA strand	Nucleotide
Protein	Intermediate filament	Polypeptide	Amino acid
Lipid	Adipose cell with fat droplets	Triglyceride	Fatty acid

figure 3.3

POLYMER MACROMOLECULES. The four major biological macromolecules are shown. Carbohydrates, nucleic acids, and proteins all form polymers and are shown with the monomers used to make them. Lipids do not fit this simple monomer–polymer relationship, however, because they are constructed from glycerol and fatty acids. All four types of macromolecule are also shown in their cellular context.

The dehydration reaction

Despite the differences between monomers of the major polymers, the basic chemistry of their synthesis is similar: To form a covalent bond between two monomers, an —OH group is removed from one monomer, and a hydrogen atom (H) is removed from the other (figure 3.4*a*). For example, this simple chemistry is the same for linking amino acids together to make a protein or assembling glucose units together to make starch. This reaction is also used to link fatty acids to glycerol in lipids. This chemical reaction is called condensation or a **dehydration reaction,** because the removal of —OH and —H is the same as the removal of a molecule of water (H_2O). For every subunit that is added to a macromolecule, one water molecule is removed. These and other biochemical reactions require that the reacting substances are held close together and that the correct chemical bonds are stressed and broken. This process of positioning and stressing, termed *catalysis,* is carried out in cells by enzymes.

The hydrolysis reaction

Cells disassemble macromolecules into their constituent subunits by performing reactions that are essentially the reverse of dehydration—a molecule of water is added instead of removed (figure 3.4*b*). In this process, which is called **hydrolysis,** a hydrogen atom

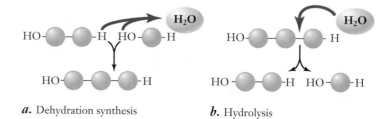

a. Dehydration synthesis   **b.** Hydrolysis

figure 3.4

MAKING AND BREAKING MACROMOLECULES.
a. Biological macromolecules are polymers formed by linking monomers together by dehydration synthesis. This process releases a water molecule for every bond formed. **b.** Breaking the bond between subunits involves a process called hydrolysis, which reverses the loss of a water molecule by dehydration.

is attached to one subunit and a hydroxyl group to the other, breaking a specific covalent bond in the macromolecule.

> Living systems are made up of four main types of macromolecule. Macromolecules are polymers, which consist of long chains of similar subunits that are joined by dehydration reactions and are broken down by hydrolysis reactions.

3.2 Carbohydrates: Energy Storage and Structural Molecules

Carbohydrates are a loosely defined group of molecules that contains carbon, hydrogen, and oxygen in the molar ratio 1:2:1. Their empirical formula (which lists the number of atoms in the molecule with subscripts) is $(CH_2O)_n$, where n is the number of carbon atoms. Because they contain many carbon–hydrogen (C—H) bonds, which release energy when oxidation occurs, carbohydrates are well suited for energy storage. Sugars are among the most important energy-storage molecules, and they exist in several different forms.

Monosaccharides are simple sugars

The simplest of the carbohydrates are the **monosaccharides** (Greek *mono,* "single," and Latin *saccharum,* "sugar"). Simple sugars may contain as few as three carbon atoms, but those that play the central role in energy storage have six (figure 3.5). The empirical formula of six-carbon sugars is:

$$C_6H_{12}O_6 \qquad \text{or} \qquad (CH_2O)_6$$

Six-carbon sugars can exist in a straight-chain form, but in an aqueous environment they almost always form rings.

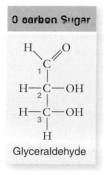

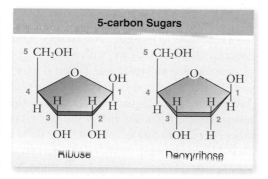

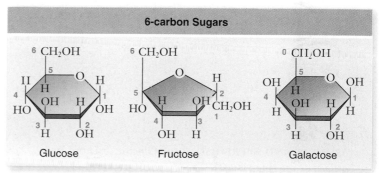

figure 3.5

MONOSACCHARIDES. Monosaccharides, or simple sugars, can contain as few as three carbon atoms and are often used as building blocks to form larger molecules. The five-carbon sugars ribose and deoxyribose are components of nucleic acids (see figure 3.14). The six-carbon sugar glucose is a component of large energy-storage molecules. The blue numbers refer to the carbon atoms; monosaccharides are conventionally numbered from the more oxidized end.

figure 3.6

STRUCTURE OF THE GLUCOSE MOLECULE. Glucose is a linear, six-carbon molecule that forms a six-membered ring in solution. Ring closure occurs such that two forms can result: α-glucose and β-glucose. These structures differ only in the position of the OH bound to carbon 1. The structure of the ring can be represented in many ways; the ones shown here are the most common, with the carbons conventionally numbered (in *blue*) so that the forms can be compared easily. The bold lines represent portions of the molecule that are projecting out of the page toward you.

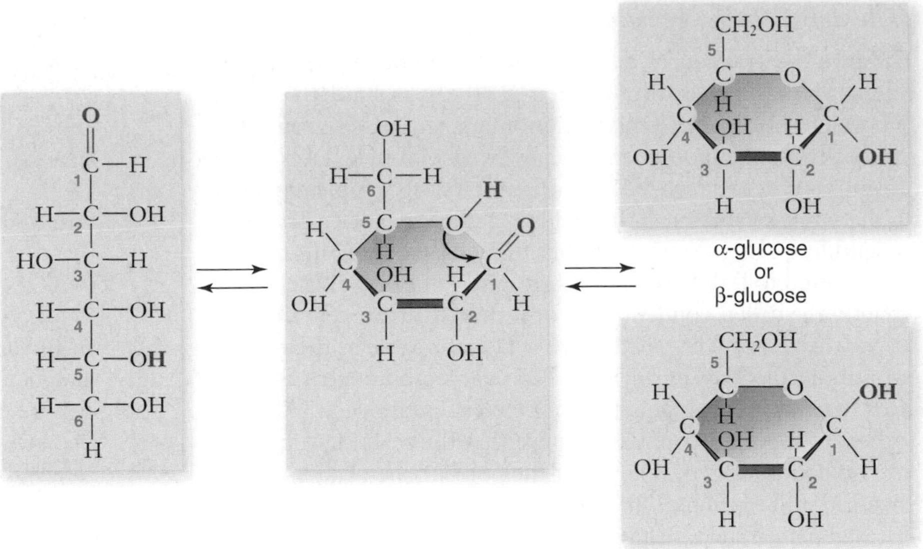

α-glucose
or
β-glucose

The most important of the six-carbon monosaccharides for energy storage is glucose, which you first encountered in the examples of chemical reactions in chapter 2. Glucose has seven energy-storing C—H bonds (figure 3.6). Depending on the orientation of the carbonyl group (C=O) when the ring is closed, glucose can exist in two different forms: alpha (α) or beta (β).

Sugar isomers have structural differences

Glucose is not the only sugar with the formula $C_6H_{12}O_6$. Both structural isomers and stereoisomers of this simple six-carbon skeleton exist in nature. Fructose is a structural isomer that differs in the positioning of the carbonyl carbon (C=O); galactose is a stereoisomer that differs in the position of OH and H groups relative to the ring (figure 3.7). These differences often account for substantial functional differences between the isomers. Your

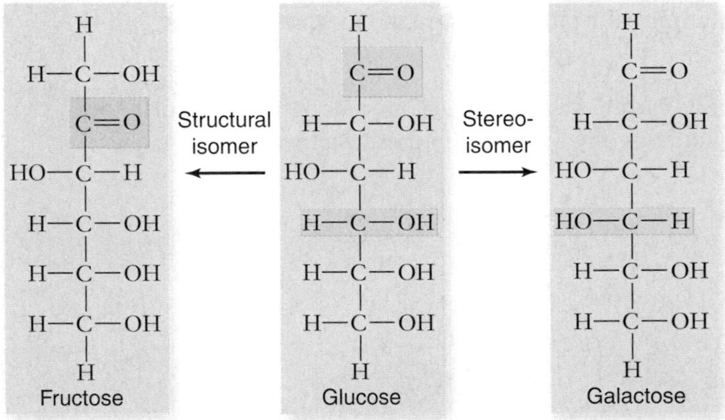

figure 3.7

ISOMERS AND STEREOISOMERS. Glucose, fructose, and galactose are isomers with the empirical formula $C_6H_{12}O_6$. A structural isomer of glucose, such as fructose, has identical chemical groups bonded to different carbon atoms. Notice that this results in a five-membered ring in solution (see figure 3.5). A stereoisomer of glucose, such as galactose, has identical chemical groups bonded to the same carbon atoms but in different orientations (the —OH at carbon 4).

taste buds can discern them: Fructose tastes much sweeter than glucose, despite the fact that both sugars have identical chemical composition. Enzymes that act on different sugars can distinguish both structural and stereoisomers of this basic six-carbon skeleton. The different stereoisomers of glucose are also important in the polymers that can be made using glucose as a monomer, as you will see later in this chapter.

Disaccharides serve as transport molecules in plants and provide nutrition in animals

Most organisms transport sugars within their bodies. In humans, the glucose that circulates in the blood does so as a simple monosaccharide. In plants and many other organisms, however, glucose is converted into a transport form before it is moved from place to place within the organism. In such a form, it is less readily metabolized during transport.

Transport forms of sugars are commonly made by linking two monosaccharides together to form a **disaccharide** (Greek *di*, "two"). Disaccharides serve as effective reservoirs of glucose because the normal glucose-utilizing enzymes of the organism cannot break the bond linking the two monosaccharide subunits. Enzymes that can do so are typically present only in the tissue where the glucose is to be used.

Transport forms differ depending on which monosaccharides are linked to form the disaccharide. Glucose forms transport disaccharides with itself and with many other monosaccharides, including fructose and galactose. When glucose forms a disaccharide with the structural isomer fructose, the resulting disaccharide is *sucrose*, or table sugar (figure 3.8a). Sucrose is the form in which most plants transport glucose and is the sugar that most humans and other animals eat. Sugarcane and sugar beets are rich in sucrose.

When glucose is linked to the stereoisomer galactose, the resulting disaccharide is *lactose*, or milk sugar. Many mammals supply energy to their young in the form of lactose. Adults have greatly reduced levels of lactase, the enzyme required to cleave lactose into its two monosaccharide components, and thus they cannot metabolize lactose efficiently. Most of the energy that is channeled into lactose production is therefore reserved for offspring. For this reason lactose as an energy source is primarily for offspring in mammals.

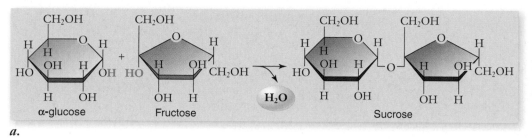 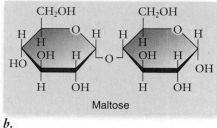

a.
b.

figure 3.8

HOW DISACCHARIDES FORM. Some disaccharides are used to transport glucose from one part of an organism's body to another; one example is sucrose *(a)*, which is found in sugarcane. Other disaccharides, such as maltose *(b)*, in grain are used for storage.

Polysaccharides provide energy storage and structural components

Polysaccharides are longer polymers made up of monosaccharides that have been joined through dehydration synthesis. **Starch,** a storage polysaccharide, consists entirely of α-glucose molecules linked in long chains. **Cellulose,** a structural polysaccharide, also consists of glucose molecules linked in chains, but these molecules are β-glucose. Because starch is built from α-glucose we call the linkages α linkages; cellulose has β linkages.

Starches and Glycogen

Organisms store the metabolic energy contained in monosaccharides by converting them into disaccharides, such as *maltose* (figure 3.8*b*), which are then linked together into the insoluble polysaccharides called *starches*. These polysaccharides differ mainly in the way in which the polymers branch.

The starch with the simplest structure is *amylose*, which is composed of many hundreds of α-glucose molecules linked together in long, unbranched chains. Each linkage occurs between the number 1 carbon of one glucose molecule and the number 4 carbon of another, making them α-1,4 linkages (figure 3.9*a*). The long chains of amylose tend to coil up in water, a property that renders amylose insoluble. Potato starch is about 20% amylose (figure 3.9*b*).

Most plant starch, including the remaining 80% of potato starch, is a somewhat more complicated variant of amylose called *amylopectin*. Pectins are branched polysaccharides with the branches occurring due to bonds between the 1 carbon of one molecule and the 6 carbon of another (α-1,6 linkages). These short amylose branches consist of 20 to 30 glucose subunits (figure 3.9*b*).

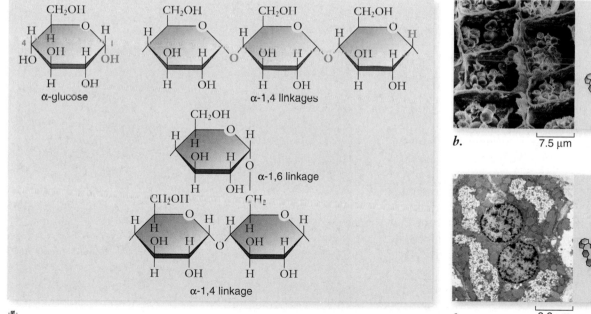

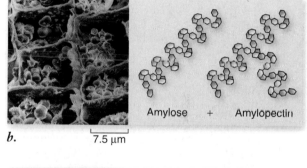

b. 7.5 µm

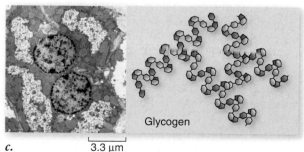

c. 3.3 µm

figure 3.9

POLYMERS OF GLUCOSE: STARCH AND GLYCOGEN.
a. Starch chains consist of polymers of α-glucose subunits joined by α-1,4 glycosidic linkages. These chains can be branched by forming similar α-1,6 glycosidic bonds. These storage polymers then differ primarily in their degree of branching. *b.* Starch is found in plants and is composed of amylose and amylopectin, which are unbranched and branched, respectively. The branched form is insoluble and forms starch granules in plant cells. *c.* Glycogen is found in animal cells and is highly branched and also insoluble, forming glycogen granules.

figure 3.10

POLYMERS OF GLUCOSE: CELLULOSE. Starch chains consist of α-glucose subunits, and cellulose chains consist of β-glucose subunits. *a.* Thus the bonds between adjacent glucose molecules in cellulose are β-1,4 glycosidic linkages. *b.* Cellulose is unbranched and forms long fibers. Cellulose fibers can be very strong and are quite resistant to metabolic breakdown, which is one reason wood is such a good building material.

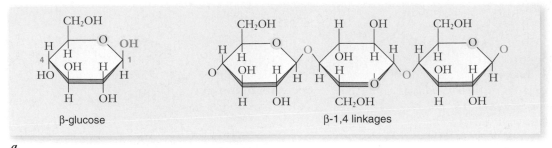

β-glucose

β-1,4 linkages

a.

b.

500 µm

The comparable molecule to starch in animals is **glycogen.** Like amylopectin, glycogen is an insoluble polysaccharide containing branched amylose chains. Glycogen has a much greater average chain length and more branches than plant starch (figure 3.9c).

Cellulose

Although some chains of sugars store energy, others serve as structural material for cells. For two glucose molecules to link together, the glucose subunits must be of the same form. *Cellulose* is a polymer of β-glucose (figure 3.10). The bonds between adjacent glucose molecules still exist between the 1 carbon of the first glucose and the 4 carbon of the next glucose, but these are β-1,4 linkages.

A chain of glucose molecules consisting of all β-glucose has properties that are very different from those of starch. These long, unbranched β-linked chains make tough fibers. Cellulose is the chief component of plant cell walls (figure 3.10). It is chemically similar to amylose, with one important difference: The starch-hydrolyzing enzymes that occur in most organisms cannot break the bond between two β-glucose units because they only recognize α linkages.

Because cellulose cannot be broken down readily by most creatures, it works well as a biological structural material. Those few animals able to break down cellulose find it a rich source of energy. Certain vertebrates, such as cows, can digest cellulose by means of bacteria and protists harbored in their digestive tracts that provide the necessary enzymes.

Chitin

Chitin, the structural material found in arthropods and many fungi, is a modified form of cellulose where an N-acetyl group replaces a hydroxyl in each glucose unit. When cross-linked by proteins, it forms a tough, resistant surface material that serves as the hard exoskeleton of insects and crustaceans (figure 3.11; see chapter 33). Few organisms are able to digest chitin and use it as a major source of nutrition, but most possess a chitinase enzyme, probably to protect against fungi.

Sugars are among the most important energy-storage molecules in organisms. Monosaccharides have three to six carbon atoms; the six-carbon monosaccharides typically have a ring form. The structural differences between sugar isomers can confer substantial functional differences upon the molecules. Disaccharides consist of two linked monosaccharides. Starches are polymers of α-glucose. Most starches are branched, rendering the polymer insoluble. Structural carbohydrates such as cellulose in plants are chains of sugars such as β-glucose that are not easily digested.

figure 3.11

CHITIN. Chitin is the principal structural element in the external skeletons of many invertebrates, such as this lobster.

Nucleic Acids: Information Molecules

The biochemical activity of a cell depends on production of a large number of proteins, each with a specific sequence. The information necessary to produce the correct proteins is passed through generations of organisms, even though the protein molecules themselves are not.

Nucleic acids are the information-carrying devices of cells, just as disks contain the information that computers use, blueprints carry the information that builders use, and road maps display the information that travelers use. Two main varieties of nucleic acids are **deoxyribonucleic acid (DNA;** figure 3.12) and **ribonucleic acid (RNA).**

The way DNA encodes the genetic information used to assemble proteins (as discussed in detail in chapter 14) is similar to the way the letters on a page encode information. Unique among macromolecules, nucleic acids are able to serve as templates to produce precise copies of themselves. This characteristic allows genetic information to be preserved during cell division and during the reproduction of organisms. DNA, found primarily in the nuclear region of cells, contains the genetic information necessary to build specific organisms.

Cells use RNA to read the cell's DNA-encoded information and to direct the synthesis of proteins. RNA is similar to DNA in structure and consists of transcribed copies of portions of the DNA. These transcripts serve as blueprints specifying the amino acid sequences of proteins. This process will be described in detail in chapter 15.

Nucleic acids are nucleotide polymers

Nucleic acids are long polymers of repeating subunits called **nucleotides.** Each nucleotide consists of three components: a pentose or five-carbon sugar (ribose in RNA and deoxyribose in DNA); a phosphate ($-PO_4$) group; and an organic nitrogenous (nitrogen-containing) base (figure 3.13). When a nucleic acid polymer forms, the phosphate group of one nucleotide binds to the hydroxyl group from the pentose sugar of another, releasing water and forming a phosphodiester bond by a dehydration reaction. A **nucleic acid,** then, is simply a chain of five-carbon sugars linked together by phosphodiester bonds with a nitrogenous base protruding from each sugar (figure 3.14). These chains of nucleotides, *polynucleotides,* have different ends: a phosphate on one end and an $-OH$ from a sugar on the other end. We conventionally refer to these ends as 5′ ("five-prime," $-PO_4$) and 3′ ("three-prime," $-OH$) taken from the numbering on carbons of the sugar (figure 3.14).

Two types of nitrogenous bases occur in nucleotides. The first type, *purines,* are large, double-ring molecules found in both DNA and RNA; the two types of purines are adenine (A) and guanine (G). The second type, *pyrimidines,* are smaller, single-ring molecules; they include cytosine (C, in both DNA and RNA), thymine (T, in DNA only), and uracil (U, in RNA only).

DNA carries the genetic code

Organisms use sequences of nucleotides in DNA to encode the information specifying the amino acid sequences of their proteins. This method of encoding information is very similar to the way in which sequences of letters encode information in a

a. ⌞ 2 nm ⌟   *b.*

figure 3.12

IMAGES OF DNA. *a.* A scanning-tunneling micrograph of DNA (false color; 2,000,000) showing approximately three turns of the DNA double helix. ***b.*** A space-filling model for comparison to the image of actual DNA in *(a).*

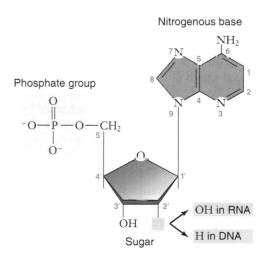

figure 3.13

STRUCTURE OF A NUCLEOTIDE. The nucleotide subunits of DNA and RNA are made up of three elements: a five-carbon sugar (ribose or deoxyribose), an organic nitrogenous base (adenine is shown here), and a phosphate group. Notice that the numbers on the sugar all are given as "primes" (1′, 2′, etc.) to distinguish them from the numbering on the rings of the bases.

It is therefore not surprising that changing a single amino acid can drastically alter the structure, and thus the function of a protein. The sickle cell version of hemoglobin (HbS), for example, is a change of a single glutamic acid for a valine in the β-globin chain that causes the protein to aggregate into clumps. Note that this exchanges a charged amino acid for a nonpolar one on the surface of the protein, leading the protein to become sticky and form aggregates. Another variant of hemoglobin called HbE, actually the most common in human populations, causes a change from glutamic acid to lysine at a different site in the β-globin chain. In this case the structural change is not as dramatic, but it still impairs function, resulting in multiple types of anemia and thalassemia. More than 700 structural variants of hemoglobin are known, with up to 7% of the world's population being carriers for forms that are clinically relevant.

Quaternary structure: subunit arrangements

When two or more polypeptide chains associate to form a functional protein, the individual chains are referred to as subunits of the protein. The arrangement of these subunits is termed the **quaternary structure** of the protein. In proteins composed of subunits, the interfaces where the subunits contact one another are often nonpolar, and they play a key role in transmitting information between the subunits about individual subunit activities.

As mentioned earlier, the protein hemoglobin is composed of two α-chain subunits and two β-chain subunits. So each α- and β-globin chain has a primary structure consisting of a specific sequence of amino acids. This then assumes a characteristic secondary structure consisting of α helices and β sheets that are then arranged into a specific tertiary structure for each α- and β-globin subunit. Lastly, these subunits are then arranged into their final quaternary structure that, in this case, is the final structure of the protein. For proteins that consist of only a single peptide chain, the enzyme lysozyme for example, the tertiary structure is the final structure of the protein.

Motifs and domains are additional structural characteristics

To directly determine the sequence of amino acids in a protein is a laborious task. Although the process has been automated, it remains slow and difficult.

The ability to sequence DNA changed this situation rather suddenly. Sequencing DNA was a much simpler process, and even before it was automated, the number of known sequences rose quickly. With the advent of automation, the known sequences increased even more dramatically, and today the entire sequence of hundreds of bacterial genomes and more than a dozen animal genomes, including that of humans, has been determined. Because the DNA sequence is directly related to amino acid sequence in proteins, biologists now have a large database of protein sequences to compare and analyze. This new information has also stimulated thought about the logic of the genetic code and whether underlying patterns exist in protein structure. Our view of protein structure has evolved with this new information. Researchers still view the four-part hierarchical structure as being important, but two new terms have entered the biologist's vocabulary: motif and domain.

Motifs

As biologists discovered the 3-D structure of proteins (an even more laborious task than determining the sequence) they noticed similarities between otherwise dissimilar proteins. These similar structures are called **motifs,** or sometimes "supersecondary structure." The term *motif* is borrowed from the arts and refers to a recurring thematic element in music or design.

One very common protein motif is the β-α-β motif, which creates a fold or crease; the so-called "Rossmann fold" at the core of nucleotide-binding sites in a wide variety of proteins. A second motif that occurs in many proteins is the β barrel, a β sheet folded around to form a tube. A third type of motif, the helix-turn-helix, consistes of two α helices separated by a bend. This motif is important because many proteins use it to bind to the DNA double helix (figure 3.23; see also chapter 16).

Motifs indicate a logic to structure that investigators still do not understand. Do they simply represent a reuse by evolution of something that already works, or do they represent an optimal solution to a problem, such as how to bind a nucleotide? One way to think about it is that if amino acids are letters in the language of proteins, then motifs represent repeated words or phrases. Motifs have been very useful in the analysis of the proliferation of known proteins. Databases of protein motifs are now maintained, and these can be used to search new unknown proteins for known motifs. This process can shed light on the function of an unknown protein.

Domains

Domains of proteins are functional units within a larger structure. They can be thought of as substructure within the tertiary structure of a protein (figure 3.23). To continue the metaphor: Amino acids are letters in the protein language, motifs are words or phrases, and domains are paragraphs.

Most proteins are made up of multiple domains that perform different aspects of the protein's function. In many cases, these domains can be physically separated. For example, transcription factors (discussed in chapter 16) are enzymes that bind to DNA and initiate its transcription. If the DNA-binding region is exchanged with a different transcription factor, using molecular biology techniques, then the specificity of the factor for DNA can be changed without changing its ability to stimulate transcription. Such "domain-swapping" experiments have been performed with many transcription factors, and they indicate, among other things, that the DNA-binding and activation domains are functionally separate.

These functional domains of proteins may also help the protein to fold into its proper shape. As a polypeptide chain folds, the domains take their proper shape, each more or less independently of the others. This action can be demonstrated experimentally by artificially producing the fragment of a polypeptide that forms the domain in the intact protein, and showing that the fragment folds to form the same structure as it exhibits in the intact protein. A single polypeptide chain connects the domains of a protein, like a rope tied into several adjacent knots.

Domains can also correspond to the structure of the genes that encode them. Later, in chapter 15, you will see that genes

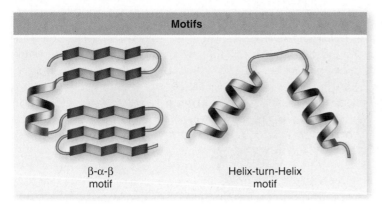

Motifs

β-α-β
motif

Helix-turn-Helix
motif

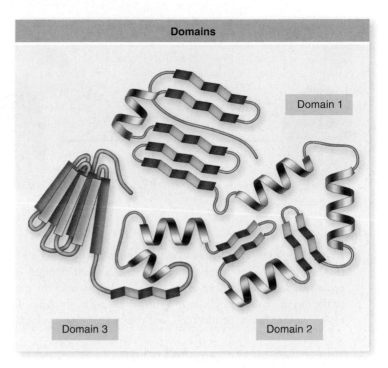

Domains

Domain 1

Domain 3

Domain 2

figure 3.23

MOTIFS AND DOMAINS. The elements of secondary structure can combine, fold, or crease to form motifs. These motifs are found in different proteins and can be used to predict function. Proteins also are made of larger domains, which are functionally distinct parts of a protein. The arrangement of these domains in space is the tertiary structure of a protein.

in eukaryotes are often in pieces within the genome, and these pieces, called *exons*, sometimes encode the functional domains of a protein. This finding led to the idea of evolution acting by shuffling protein-encoding domains.

The process of folding relies on chaperone proteins

Until recently, investigators thought that newly made proteins fold spontaneously as hydrophobic interactions with water shove nonpolar amino acids into the protein interior. We now know this view is too simple. Protein chains can fold in so many different ways that trial and error would simply take too long. In addition, as the open chain folds its way toward its final form, nonpolar "sticky" interior portions are exposed during intermediate stages. If these intermediate forms are placed in a test tube in an environment identical to that inside a cell, they stick to other, unwanted protein partners, forming a gluey mess.

How do cells avoid having their proteins clump into a mass? A vital clue came in studies of unusual mutations that prevent viruses from replicating in bacterial cells—it turns out that the virus proteins produced inside the cells could not fold properly. Further study revealed that normal cells contain **chaperone proteins,** which help other proteins to fold correctly.

Molecular biologists have now identified many proteins that act as molecular chaperones. This class of proteins has multiple subclasses, and representatives have been found in essentially every organism that has been examined. Furthermore, these proteins seem in all cases to be essential for viability as well, illustrating their fundamental importance. Many are heat shock proteins, produced in greatly increased amounts when cells are exposed to elevated temperature; high temperatures cause proteins to unfold, and heat shock chaperone proteins help the cell's proteins to refold properly.

One class of these proteins, called chaperonins, has been extensively studied. In the bacterium *Escherichia coli* (*E. coli*),

one example is the GroE chaperonin, an essential protein. In mutants in which the GroE chaperonin is inactivated, fully 30% of the bacterial proteins fail to fold properly. Chaperonins associate to form a large macromolecular complex that resembles a cylindrical container. Proteins can move into the container, and the container itself can change its conformation considerably (figure 3.24). Experiments have shown that an improperly folded protein can enter the chaperonin and be refolded. The details of how this is accomplished are not clear, but seem to involve changes in the hydrophobicity of the interior of the chamber.

The flexibility of the structure of chaperonins is amazing. We tend to think of proteins as being fixed structures, but this is clearly not the case for chaperonins and this flexibility is necessary for their function. It also illustrates that even domains that may be very widely separated in a very large protein are still functionally connected. The folding process within a chaperonin utilizes the hydrolysis of ATP to power these changes in structure necessary for function. This entire process can occur in a cyclic manner until the appropriate structure is achieved. Cells use these chaperonins both to accomplish the original folding of some proteins and to restore the structure of incorrectly folded ones.

Some diseases may result from improper folding

Chaperone protein deficiencies may be implicated in certain diseases in which key proteins are improperly folded. Cystic fibrosis is a hereditary disorder in which a mutation disables a vital protein that moves ions across cell membranes. In at least some cases, this protein appears to have the correct amino acid sequence, but it fails to fold into its final, functional form. Researchers have also speculated that chaperone deficiency may be a cause of the protein clumping in brain cells that produces the amyloid plaques characteristic of Alzheimer disease.

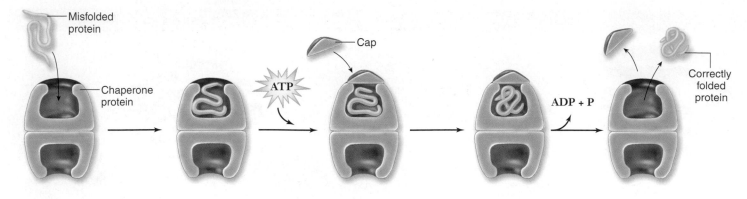

figure 3.24

HOW ONE TYPE OF CHAPERONE PROTEIN WORKS. This barrel-shaped chaperonin is from the GroE family of chaperone proteins. It is composed of two identical rings each with seven identical subunits, each of which has three distinct domains. An incorrectly folded protein enters one chamber of the barrel, and a cap seals the chamber. Energy from the hydrolysis of ATP allows for structural alterations to the chamber, changing it from hydrophobic to hydrophilic. This change allows the protein to refold. After a short time, the protein is ejected, either folded or unfolded, and the cycle can repeat itself.

Denaturation inactivates proteins

If a protein's environment is altered, the protein may change its shape or even unfold completely. This process is called **denaturation** (figure 3.25). Proteins can be denatured when the pH, temperature, or ionic concentration of the surrounding solution is changed.

When proteins are denatured, they are usually rendered biologically inactive. This action is particularly significant in the case of enzymes. Because practically every chemical reaction in a living organism is catalyzed by a specific enzyme, it is vital that a cell's enzymes remain functional.

Denaturation of proteins is involved in the traditional methods of salt curing and pickling: Prior to the general availability of refrigerators and freezers, the only practical way to keep microorganisms from growing in food was to keep the food in a solution containing a high concentration of salt or vinegar, which denatured the enzymes of most microorganisms and prevented them from growing on the food.

Most enzymes function within a very narrow range of physical parameters. Blood-borne enzymes that course through a human body at a pH of about 7.4 would rapidly become denatured in the highly acidic environment of the stomach. Conversely, the protein-degrading enzymes that function at a pH of 2 or less in the stomach would be denatured in the relatively basic pH of the blood. Similarly, organisms that live near oceanic hydrothermal vents have enzymes that work well at the temperature of this extreme environment (over 100°C). They cannot survive in cooler waters, because their enzymes do not function properly at lower temperatures. Any given organism usually has a tolerance range of pH, temperature, and salt concentration. Within that range, its enzymes maintain the proper shape to carry out their biological functions.

When a protein's normal environment is reestablished after denaturation, a small protein may spontaneously refold into its natural shape, driven by the interactions between its nonpolar amino acids and water (figure 3.26). This process is termed **renaturation,** and it was first established for the enzyme ribonuclease (RNase). The renaturation of RNase led to the doctrine that primary structure determines tertiary structure. Larger proteins can rarely refold spontaneously, however,

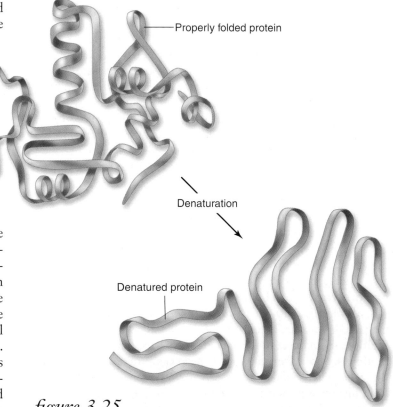

figure 3.25

PROTEIN DENATURATION. Changes in a protein's environment, such as variations in temperature or pH, can cause a protein to unfold and lose its shape in a process called denaturation. In this denatured state, proteins are biologically inactive.

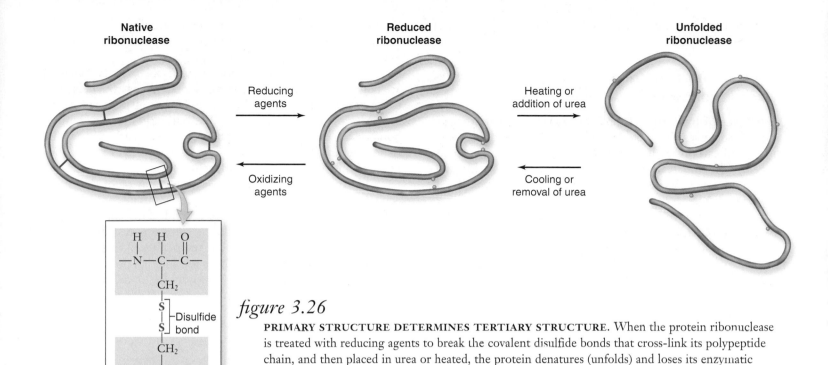

Native ribonuclease → Reducing agents → **Reduced ribonuclease** → Heating or addition of urea → **Unfolded ribonuclease**

Oxidizing agents

Cooling or removal of urea

Disulfide bond

figure 3.26

PRIMARY STRUCTURE DETERMINES TERTIARY STRUCTURE. When the protein ribonuclease is treated with reducing agents to break the covalent disulfide bonds that cross-link its polypeptide chain, and then placed in urea or heated, the protein denatures (unfolds) and loses its enzymatic activity. Upon cooling or removal of urea and treatment with oxidizing agents, it refolds and regains its enzymatic activity, demonstrating that for this protein, no information except the amino acid sequence of the protein is required for proper folding. This is not true of all proteins.

because of the complex nature of their final shape, so this simple dictum needs to be qualified.

The fact that some proteins can spontaneously renature implies that tertiary structure is strongly influenced by primary structure. In an extreme example, the *E. coli* ribosome can be taken apart and put back together experimentally. Although this process requires temperature and ion concentration shifts, it indicates an amazing degree of self-assembly. That complex structures can arise by self-assembly is a key idea in the study of modern biology.

It is important to distinguish denaturation from **dissociation.** For proteins with quaternary structure, the subunits may be dissociated without losing their individual tertiary structure. For example, the four subunits of hemoglobin may dissociate into four individual molecules (two α-globin and two β-globin) without denaturation of the folded globin

proteins, and they will readily reassume their four-subunit quaternary structure.

Proteins are a diverse class of macromolecule that perform many different functions. Proteins are made up of 20 different kinds of amino acids. The amino acids fall into five chemical classes, each with different properties that determine the nature of the resulting protein. Protein structure can be viewed at four levels: (1) the amino acid sequence, or primary structure; (2) coils and sheets, called secondary structure; (3) the three-dimensional shape, called tertiary structure; (4) individual polypeptide subunits associated in a quaternary structure. Different proteins often have similar substructures called motifs and can be broken down into functional domains. Proteins have a narrow range of conditions in which they fold properly; outside that range, proteins tend to unfold.

3.5 Lipids: Hydrophobic Molecules

Lipids are a somewhat loosely defined group of molecules with one main chemical characteristic: They are insoluble in water. Storage fats such as animal fat are one kind of lipid. Oils such as olive oil, corn oil, and coconut oil are also lipids, as are waxes such as beeswax and earwax.

Lipids have a very high proportion of nonpolar carbon–hydrogen (C—H) bonds, and so long-chain lipids cannot fold up like a protein to sequester their nonpolar portions away from

the surrounding aqueous environment. Instead, when they are placed in water, many lipid molecules spontaneously cluster together and expose what polar (hydrophilic) groups they have to the surrounding water, while sequestering the nonpolar (hydrophobic) parts of the molecules together within the cluster. You may have noticed this effect when you add oil to a pan containing water, and the oil beads up into cohesive drops on the water's surface. This spontaneous assembly of lipids is of paramount importance to cells, as it underlies the structure of cellular membranes.

Fats consist of complex polymers of fatty acids attached to glycerol

Many lipids are built from a simple skeleton made up of two main kinds of molecules: fatty acids and glycerol. Fatty acids are long-chain hydrocarbons with a carboxylic acid (COOH) at one end. Glycerol is a three-carbon polyalcohol (three —OH groups). Many lipid molecules consists of a glycerol molecule with three fatty acids attached, one to each carbon of the glycerol backbone. Because it contains three fatty acids, a fat molecule is commonly called a **triglyceride** (the more accurate chemical name is *triacylglycerol*). This basic structure is depicted in figure 3.27. The three fatty acids of a triglyceride need not be identical, and often they differ markedly from one another. The hydrocarbon chains of

fatty acids vary in length; the most common are even-numbered chains of 14 to 20 carbons. The many C—H bonds of fats serve as a form of long-term energy storage.

If all of the internal carbon atoms in the fatty acid chains are bonded to at least two hydrogen atoms, the fatty acid is said to be **saturated,** which refers to its having all the hydrogen atoms it can have (see figure 3.27). A fatty acid that has double bonds between one or more pairs of successive carbon atoms is said to be **unsaturated.** Those fatty acids with more than one double bond are termed **polyunsaturated.**

Having double bonds changes the behavior of the molecule because no free rotation can occur about a C=C double bond as it can with a C—C single bond. This characteristic mainly affects melting point: that is, whether the fatty acid is a solid fat or a liquid oil at room temperature. Fats containing polyunsaturated fatty acids have low melting points because their fatty acid chains bend at the double bonds, preventing

Structural Formula	Structural Formula
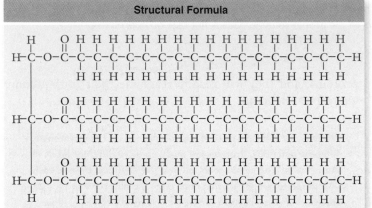	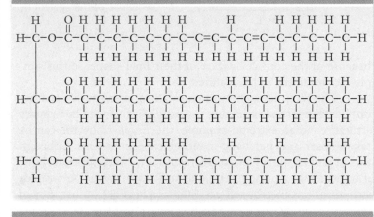

Space-Filling Model	Space-Filling Model
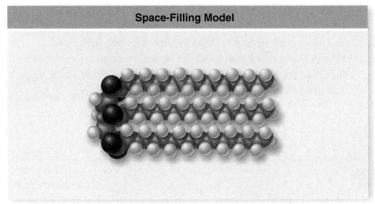	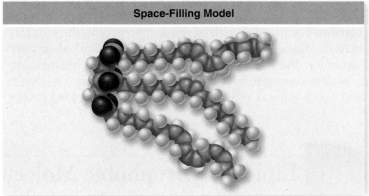

a.   *b.*

figure 3.27

SATURATED AND UNSATURATED FATS. *a.* A saturated fat is composed of triglycerides that contain three saturated fatty acids, those with no double bonds and, thus, a maximum number of hydrogen atoms bonded to the carbon chain. Many animal fats are saturated. *b.* Unsaturated fat is composed of triglycerides, contains three unsaturated fatty acids, that is, they have one or more double bonds and, thus, fewer than the maximum number of hydrogen atoms bonded to the carbon chain. This example includes both a monounsaturated and two polyunsaturated fatty acids. Plant fats are typically unsaturated. The many kinks of the double bonds prevent the triglyceride from closely aligning, producing liquid oils at room temperature.

the fat molecules from aligning closely with one another. Most saturated fats, such as animal fat or those in butter, are solid at room temperature.

Placed in water, triglycerides spontaneously associate together, forming fat globules that can be very large relative to the size of the individual molecules. Because fats are insoluble in water, they can be deposited at specific locations within an organism, such as in vesicles of adipose tissue.

Organisms contain many other kinds of lipids besides fats (figure 3.28). *Terpenes* are long-chain lipids that are components of many biologically important pigments, such as chlorophyll and the visual pigment retinal. Rubber is also a terpene. *Steroids*, another class of lipid, are composed of four carbon rings. Most animal cell membranes contain the steroid cholesterol. Other steroids, such as testosterone and estrogen, function as hormones in multicellular animals. *Prostaglandins* are a group of about 20 lipids that are modified fatty acids, with two nonpolar "tails" attached to a five-carbon ring. Prostaglandins act as local chemical messengers in many vertebrate tissues. Later chapters explore the effects of some of these complex fatty acids.

Fats are excellent energy-storage molecules

Most fats contain over 40 carbon atoms. The ratio of energy-storing C—H bonds in fats is more than twice that of carbohydrates (see section 3.2), making fats much more efficient molecules for storing chemical energy. On average, fats yield about 9 kilocalories (kcal) of chemical energy per gram, as compared with about 4 kcal per gram for carbohydrates.

Most fats produced by animals are saturated (except some fish oils), whereas most plant fats are unsaturated (see figure 3.27). The exceptions are the tropical plant oils (palm oil and coconut oil), which are saturated despite their fluidity

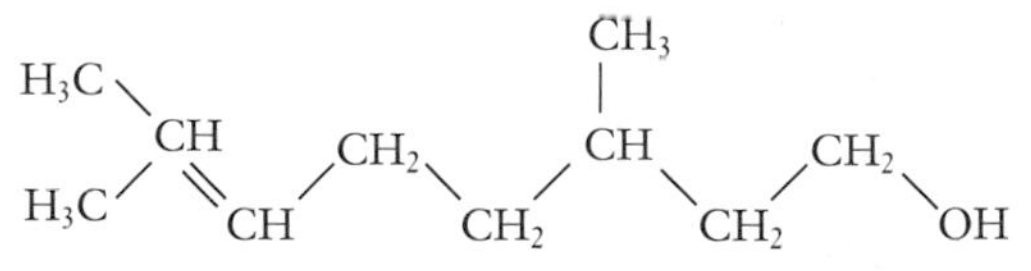

a. Terpene (citronellol)

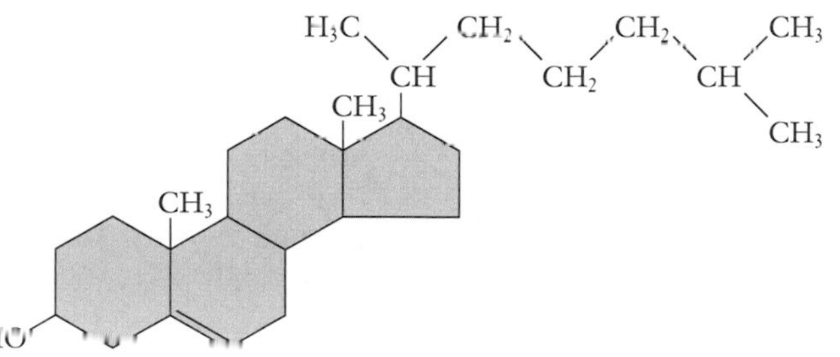

b. Steroid (cholesterol)

figure 3.28

OTHER KINDS OF LIPIDS. ***a.*** Terpenes are found in biological pigments, such as chlorophyll and retinal, and ***(b)*** steroids play important roles in membranes and as the basis for a class of hormones involved in chemical signaling.

at room temperature. An oil may be converted into a solid fat by chemically adding hydrogen. Peanut butter sold in stores is usually artificially hydrogenated to make the peanut fats solidify, preventing them from separating out as oils while the jar sits on the store shelf. However, artificially hydrogenating unsaturated fats seems to eliminate the health advantage they have over saturated fats. The hydrogenation reaction produces *trans*-fatty acids that appear to increase the level of cholesterol carried in blood. Therefore, it is currently thought that margarine, made from hydrogenated corn oil, is no better for your health than butter.

When an organism consumes excess carbohydrate, it is converted into starch, glycogen, or fats reserved for future use. The reason that many humans in developed countries gain weight as they grow older is that the amount of energy they need decreases with age, but their intake of food does not. Thus, an increasing proportion of the carbohydrates they ingest is available to be converted into fat.

A diet rich in fats is one of several factors that are thought to contribute to heart disease, particularly atherosclerosis, a condition in which deposits of fatty substances called plaque adhere to the lining of blood vessels, blocking the flow of blood. Fragments of plaque, breaking off from a deposit and clogging arteries to the brain, are a major cause of strokes.

Phospholipids form membranes

Complex lipid molecules called **phospholipids** are among the most important molecules of the cell because they form the core of all biological membranes. An individual phospholipid can be thought of as a substituted triglyceride, that is, a triglyceride with a phosphate replacing one of the fatty acids. The basic structure of a phospholipid includes three kinds of subunits:

1. *Glycerol*, a three-carbon alcohol, with each carbon bearing a hydroxyl group. Glycerol forms the backbone of the phospholipid molecule.
2. *Fatty acids*, long chains of —CH_2 groups (hydrocarbon chains) ending in a carboxyl (—COOH) group. Two fatty acids are attached to the glycerol backbone in a phospholipid molecule.
3. *A phosphate group* (—PO_4^{2-}) attached to one end of the glycerol. The charged phosphate group usually has a charged organic molecule linked to it, such as choline, ethanolamine, or the amino acid serine.

The phospholipid molecule can be thought of as having a polar "head" at one end (the phosphate group) and two long, very nonpolar "tails" at the other (figure 3.29). This structure is essential for these molecules' function, although it first appears paradoxical. Why would a molecule need to be soluble in water, but also not soluble in water? The formation of a membrane shows the unique properties of such a structure.

In water, the nonpolar tails of nearby lipid molecules aggregate away from the water, forming spherical *micelles*, with the tails facing inward (figure 3.30*a*). This is actually how detergent molecules work to make grease soluble in water. The grease is soluble within the nonpolar interior of the micelle and the polar surface of the micelle is soluble in water. With phospholipids, a

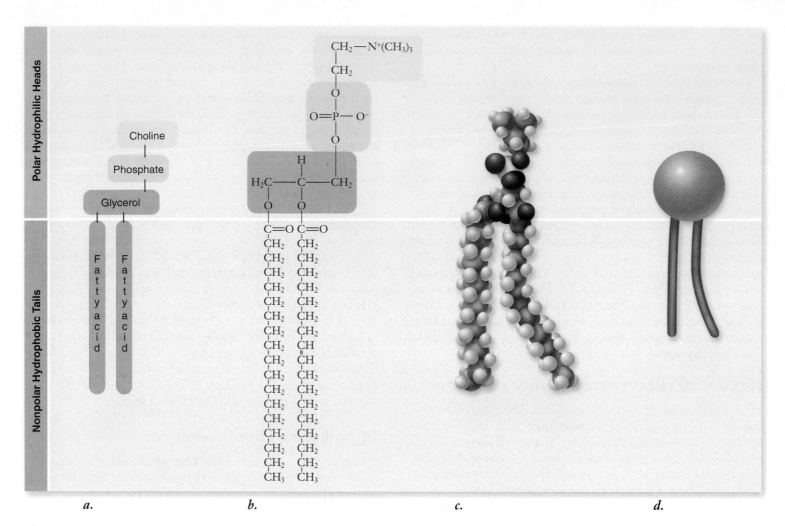

figure 3.29

PHOSPHOLIPIDS. The phospholipid phosphatidylcholine is shown as **(a)** a schematic, **(b)** a formula, **(c)** a space-filling model, and **(d)** an icon used in depictions of biological membranes.

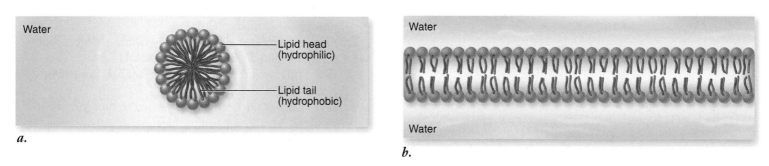

figure 3.30

LIPIDS SPONTANEOUSLY FORM MICELLES OR LIPID BILAYERS IN WATER. In an aqueous environment, lipid molecules orient so that their polar (hydrophilic) heads are in the polar medium, water, and their nonpolar (hydrophobic) tails are held away from the water. **a.** Droplets called micelles can form, or **(b)** phospholipid molecules can arrange themselves into two layers; in both structures, the hydrophilic heads extend outward and the hydrophobic tails inward. This second example is called a phospholipid bilayer.

more complex structure forms in which two layers of molecules line up, with the hydrophobic tails of each layer pointing toward one another, or inward, leaving the hydrophilic heads oriented outward, forming a bilayer (figure 3.30b). Lipid bilayers are the basic framework of biological membranes, discussed in detail in chapter 5.

Triglycerides are made of fatty acids and glycerol. Cells also contain a variety of different lipids that play important roles in cell metabolism. Because the C—H bonds in lipids are nonpolar, they are not water-soluble, and aggregate together in water. This kind of aggregation by phospholipids forms biological membranes.

3.1 Carbon: The Framework of Biological Molecules

Carbon forms the backbone of all biological molecules. Carbon can be arranged into chains and rings and is used in combination with other atoms to form biological molecules.

- Carbon can form four covalent bonds.
- Hydrocarbons consist of carbon and hydrogen, and their bonds store considerable energy.
- Biological molecules are built using functional groups that confer specific chemical properties.
- Carbon and hydrogen have similar electronegativity so C—H bonds are not polar.
- Oxygen and nitrogen have greater electronegativity than carbon and hydrogen leading to polar bonds.
- Isomers are molecules that have the same formula but different structure.
- Structural isomers differ in actual structure, whereas stereoisomers differ in how the structural groups are attached.
- Enantiomers are stereoisomers that are mirror images of each other and they rotate the plane of polarized light.
- Polymers are long chains of similar chemical subunits or monomers.
- Most important biological macromolecules are polymers.
- Biological polymers are formed by elimination of water, or a dehydration reaction.
- Biological polymers can be broken down by adding water, or a hydrolysis reaction.

3.2 Carbohydrates: Energy Storage and Structural Molecules

The empirical formula of a carbohydrate is $(CH_2O)_n$. Carbohydrates are used for energy storage and as structural molecules.

- Simple sugars contain three to six carbon atoms and exist as structural isomers and stereoisomers.
- Monosaccharides contain one subunit, disaccharides contain two and polysaccharides contain more than two.
- Glucose is used to make three important polymers: starch, glycogen, and cellulose.
- Starch and glycogen are branched polymers of α-glucose made by plants and animals, respectively, used for energy storage.
- Cellulose is an unbranched polymer of β-glucose made by plants for cell walls.

3.3 Nucleic Acids: Information Molecules

Nucleic acids are polymers formed by phosphodiester bonds between nucleotides. Nucleic acid molecules are used for information storage.

- DNA and RNA are polymers composed of nucleotide monomers.
- DNA uses the sugar deoxyribose, and RNA uses the sugar ribose.
- Nucleic acid polymers contain four different nucleotides. In DNA these include the bases adenine, guanine, cytosine, and thymine. In RNA, thymine is replaced by uracil.
- DNA exists as a double helix, but RNA is usually a folded single chain.

- DNA is stabilized by hydrogen bonds between bases. These form specific base pairs: adenine with thymine and guanine with cytosine.
- DNA encodes the information for amino acid sequences in proteins using four different nucleotides.
- RNA is made by copying DNA and is used to make proteins.
- Adenosine triphosphate (ATP) is a nucleotide that is also used to provide energy in cells.
- The nucleotides NAD^+ and FAD are used to transport electrons.

3.4 Proteins: Molecules with Diverse Structures and Functions

Proteins are a structurally diverse category of molecules. Proteins are made from amino acid monomers. Proteins carry out a diverse array of functions.

- Most enzymes are proteins that act as catalysts for metabolic reactions.
- Proteins defend our bodies, transport gases and ions, provide structure, contract and provide motion, receive information, regulate cell activities, and store bound ions.
- Proteins are linear polymers of 20 different kinds of amino acids.
- Amino acids are joined by peptide bonds to make polypeptides.
- The 20 common amino acids are characterized by R groups that may be polar, nonpolar, or charged.
- Protein structure is defined by the following hierarchy: primary, secondary, tertiary, and quaternary.
- Primary structure refers to the amino acid sequence. Secondary structure is based on hydrogen-bonding patterns that can create helices or planar sheets. Tertiary structure refers to the three-dimensional folded shape, and quaternary structure is formed by the association of two or more polypeptides.
- Motifs are similar structures found in dissimilar proteins. Domains are functional subunits within a tertiary structure.
- Chaperonins assist in the folding of proteins. Chaperone deficiencies may cause disease.
- Denaturation refers to an unfolding of tertiary structure. Disassociation refers to separation of quaternary subunits with no changes to their tertiary structure.

3.5 Lipids: Hydrophobic Molecules

Lipids are composed of fatty acids and glycerol and are insoluble in water. Lipids are long-term energy storage molecules. Phospholipids form the basis for biological membranes.

- Fatty acids can be saturated or unsaturated.
- Saturated fatty acids contain the maximum number of hydrogen atoms.
- Unsaturated fatty acids contain one or more double bonds between carbon atoms.
- Phospholipids contain two fatty acids and one phosphate attached to glycerol. The phosphate head is hydrophilic, and the lipid tail is hydrophobic.
- Membranes composed of phospholipids have the hydrophilic heads of each layer directed to the outside of the membrane, and the hydrophobic tails directed toward the center.

SELF TEST

1. How is a polymer formed from multiple monomers?
   a. From the growth of the chain of carbon atoms
   b. By the removal of an —OH group and a hydrogen atom
   c. By the addition of an —OH group and a hydrogen atom
   d. Through hydrogen bonding
2. Why are carbohydrates important molecules for energy storage?
   a. The C—H bonds found in carbohydrates store energy
   b. The double bonds between carbon and oxygen are very strong
   c. The electronegativity of the oxygen atoms means that a carbohydrate is made up of many polar bonds
   d. They can form ring structures in the aqueous environment of a cell
3. Plant cells store energy in the form of _____, and animal cells store energy in the form of _____
   a. fructose; glucose
   b. disaccharides; monosaccharides
   c. cellulose; chitin
   d. starch; glycogen
4. Which carbohydrate would you find as part of a molecule of RNA?
   a. Galactose
   b. Deoxyribose
   c. Ribose
   d. Glucose
5. What makes cellulose different from starch?
   a. Starch is produced by plant cells, and cellulose is produced by animal cells.
   b. Cellulose forms long filaments, and starch is highly branched.
   c. Starch is insoluble, and cellulose is soluble.
   d. All of the above.
6. A molecule of DNA or RNA is a polymer of—
   a. monosaccharides
   b. nucleotides
   c. amino acids
   d. fatty acids
7. What chemical bond is responsible for linking amino acids together to form a protein?
   a. Phosphodiester
   b. β-1,4 Linkage
   c. Peptide
   d. Hydrogen
8. The double helix structure of a molecule of DNA is stabilized by—
   a. phosphodiester bonds
   b. peptide bonds
   c. an α helix
   d. hydrogen bonds
9. Which of the following is NOT a difference between DNA and RNA?
   a. Deoxyribose sugar versus ribose sugar
   b. Thymine versus uracil
   c. Double-stranded versus single-stranded
   d. Phosphodiester versus hydrogen bonds
10. What monomers make up a protein?
    a. Monosaccharides
    b. Nucleotides
    c. Amino acids
    d. Fatty acids
11. Which part of an amino acid has the greatest influence on the overall structure of a protein?
    a. The (—NH₂) amino group
    b. The R-group
    c. The (—COOH) carboxyl group
    d. Both a and c
12. A mutation that alters a single amino acid within a protein can alter—
    a. the primary level of protein structure
    b. the secondary level of protein structure
    c. the tertiary level of protein structure
    d. all of the above
13. Which of these factors contributes to the diversity of protein form and function in the cell?
    a. Quaternary interactions between peptides
    b. Formation of α helices and β-pleated sheets
    c. The linear sequence of amino acids that makes up the polymer
    d. All of the above
14. What chemical property of lipids accounts for their insolubility in water?
    a. The length of the carbon chain
    b. The large number of nonpolar C—H bonds
    c. The branching of saturated fatty acids
    d. The C=C bonds found in unsaturated fatty acids
15. The spontaneous formation of a lipid bilayer in an aqueous environment occurs because—
    a. The polar head groups of the phospholipids can interact with water.
    b. The long fatty acid tails of the phospholipids can interact with water.
    c. The fatty acid tails of the phospholipids are hydrophobic.
    d. Both a and c.

CHALLENGE QUESTIONS

1. Spider webs are made of "silk," which is a long, fibrous protein. The threads you see in a web are actually composed of many individual proteins. One important structural motif within spider's silk protein is a "β-crystal". β-Crystals are regions where the β-pleated sheets from the multiple individual protein fibers stack one upon the other. What chemical bonds are required for the formation of the β-crystals? What level of protein structure is responsible for the formation of the silk you see in the web? Predict how the presence of the β-crystal motif would influence the physical properties of the silk protein.
2. How do the four biological macromolecules differ from one another? Refer to the diagram of the monomer structure in figure 3.3 and summarize what "clues" you use to distinguish between these important molecules.
3. Hydrogen bonds play an important role in stabilizing and organizing biological macromolecules. Consider the four macromolecules discussed in this chapter. Describe three examples where hydrogen bond formation affects the form or function of the macromolecule.
4. The cells of your body are distinct even though they all contain the same genetic information. Use the information in table 3.2 to develop an explanation for the diversity of specialized cellular structure and function found within your body.

ARIS™ **Do you need additional review?** *Visit* www.ravenbiology.com *for practice quizzes, animations, videos, and activities designed to help you master the material in this chapter.*

chapter 4

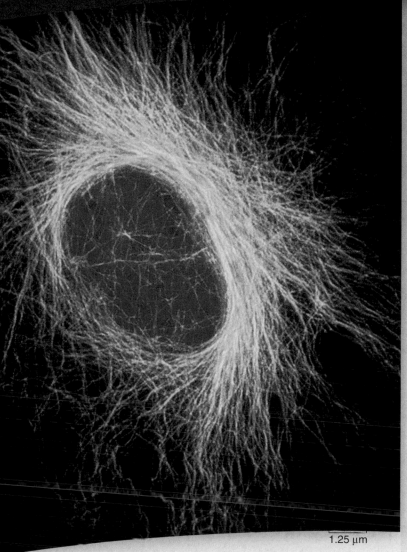

1.25 μm

Cell Structure

introduction

ALL ORGANISMS ARE COMPOSED OF CELLS. The gossamer wing of a butterfly is a thin sheet of cells and so is the glistening outer layer of your eyes. The hamburger or tomato you eat is composed of cells, and its contents soon become part of your cells. Some organisms consist of a single cell too small to see with the unaided eye, while others, such as humans, are composed of many specialized cells, such as the fibroblast cell shown in the striking fluorescence micrograph on this page. Cells are so much a part of life that we cannot imagine an organism that is not cellular in nature. In this chapter, we take a close look at the internal structure of cells. In chapters 5 to 10, we will focus on cells in action—how they communicate with their environment, grow, and reproduce.

Cell Theory

A general characteristic of cells is their microscopic size. Although there are exceptions, a typical eukaryotic cell is 10 to 100 micrometers (μm) (10 to 100 millionths of a meter) in diameter, most prokaryotic cells are only 1 to 10 μm in diameter.

Because cells are so small, their discovery did not occur until the invention of the microscope in the seventeenth century. Robert Hooke was the first to observe cells in 1665, naming the shapes he saw in cork *cellulae* (Latin, "small rooms"). This comes down to us as *cells*. Another early microscopist, Anton van Leeuwenhoek first observed live cells, which he termed tiny "animalcules." After these early efforts, a century and a half passed before biologists fully recognized the importance of cells. In 1838, botanist Matthias Schleiden stated that all plants "are aggregates of fully individualized, independent, separate beings, namely the cells themselves." In 1839, Theodor Schwann reported that all animal tissues also consist of individual cells. Thus was born cell theory.

Cell theory is the unifying foundation of cell biology

The cell theory was proposed to explain the observation that all organisms are composed of cells. While it sounds simple, it is a far-reaching statement about the organization of life.

In its modern form, **cell theory** includes the following three principles:

1. All organisms are composed of one or more cells, and the life processes of metabolism and heredity occur within these cells.
2. Cells are the smallest living things, the basic units of organization of all organisms.
3. Cells arise only by division of a previously existing cell.

Although life likely evolved spontaneously in the environment of early Earth, biologists have concluded that no additional cells are originating spontaneously at present. Rather, life on Earth represents a continuous line of descent from those early cells.

Cell size is limited

Most cells are relatively small for reasons related to the diffusion of substances into and out of cells. The rate of diffusion is affected by a number of variables, including surface area available for diffusion, temperature, concentration gradient of diffusing substance, and the distance over which diffusion must occur. As the size of a cell increases, the length of time for diffusion from the membrane to the interior of the cell increases as well. Larger cells need to synthesize more macromolecules, have correspondingly higher energy requirements, and produce a greater quantity of waste. Molecules used for energy and biosynthesis must be transported through the membrane. Any metabolic waste produced must be removed, also passing through the membrane. The rate at which this transport occurs depends on both the distance to the membrane, as well as

the area of membrane available. For this reason, an organism made up of many relatively small cells has an advantage over one composed of fewer, larger cells.

The advantage of small cell size is readily apparent in terms of the **surface area-to-volume ratio.** As a cell's size increases, its volume increases much more rapidly than its surface area. For a spherical cell, the surface area is proportional to the square of the radius, whereas the volume is proportional to the cube of the radius. Thus, if two cells differ by a factor of 10 in radius, the larger cell will have 10^2, or 100 times, the surface area, but 10^3, or 1000 times, the volume of the smaller cell (figure 4.1).

The cell surface provides the only opportunity for interaction with the environment, because all substances enter and exit a cell via this surface. The membrane surrounding the cell plays a key role in controlling cell function, and because small cells have more surface area per unit of volume than large ones, the control is more effective when cells are relatively small.

Although most cells are small, some quite large cells do exist, and these have apparently overcome the surface-area-to-volume problem by one or more adaptive mechanisms. For example, some cells, such as skeletal muscle cells, have more than one nucleus, allowing genetic information to be spread around a large cell. Some other large cells, such as neurons, are long and skinny, so that any given point in the cytoplasm is close to the plasma membrane, and thus diffusion between the inside and outside of the cell can still be rapid.

Microscopes allow visualization of cells and components

Other than egg cells, not many cells are visible to the naked eye (figure 4.2). Most are less than 50 μm in diameter, far smaller than the period at the end of this sentence. So, to visualize cells we

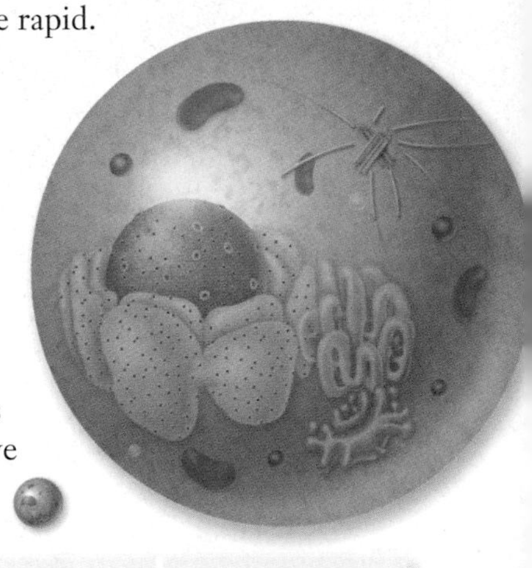

Cell radius (*r*)	1 unit	10 unit
Surface area ($4\pi r^2$)	12.57 unit^2	1257 unit^2
Volume ($\frac{4}{3}\pi r^3$)	4.189 unit^3	4189 unit^3
Surface Area / Volume	3	0.3

figure 4.1

SURFACE-AREA-TO-VOLUME RATIO. As a cell gets larger, its volume increases at a faster rate than its surface area. If the cell radius increases by 10 times, the surface area increases by 100 times, but the volume increases by 1000 times. A cell's surface area must be large enough to meet the metabolic needs of its volume.

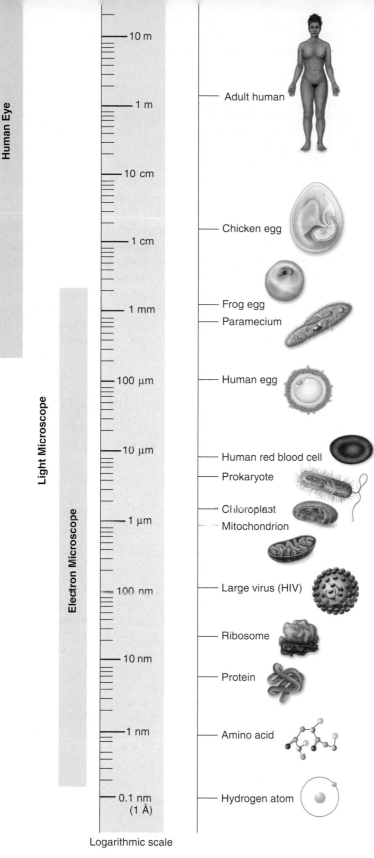

Human Eye

Light Microscope

Electron Microscope

— 10 m — Adult human

— 1 m

— 10 cm

— 1 cm — Chicken egg

— 1 mm — Frog egg
— Paramecium

— 100 μm — Human egg

— 10 μm — Human red blood cell
— Prokaryote

— 1 μm — Chloroplast
— Mitochondrion

— 100 nm — Large virus (HIV)

— Ribosome

— 10 nm — Protein

— 1 nm — Amino acid

— 0.1 nm — Hydrogen atom
(1 Å)

Logarithmic scale

figure 4.2

THE SIZE OF CELLS AND THEIR CONTENTS. Most cells are microscopic in size, although vertebrate eggs are typically large enough to be seen with the unaided eye. Prokaryotic cells are generally 1 to 10 μm across.
$1 \text{ m} = 10^2 \text{ cm} = 10^3 \text{ mm} = 10^6 \text{ μm} = 10^9 \text{ nm}$

---

need the aid of technology. The development of microscopes and their refinement has allowed us to explore cells.

The resolution problem

How do we study cells if they are too small to see? The key is to understand why we can't see them. The reason we can't see such small objects is the limited resolution of the human eye. **Resolution** is defined as the minimum distance two points can be apart and still be distinguished as two separated points. When two objects are closer together than about 100-μm, the light reflected from each strikes the same photoreceptor cell at the rear of the eye. Only when the objects are farther than 100 μm apart will the light from each strike different cells, allowing your eye to resolve them as two objects rather than one.

Types of microscopes

One way to increase resolution is to increase magnification so that small objects appear larger. The first microscopists used glass lenses to magnify small cells and cause them to appear larger than the 100-μm limit imposed by the human eye. The glass lens adds increased focusing power. Because the glass lens makes the object appear closer, the image on the back of the eye is bigger than it would be without the lens.

Modern **light microscopes,** which operate with visible light, use two magnifying lenses (and a variety of correcting lenses) to achieve very high magnification and clarity (table 4.1). The first lens focuses the image of the object on the second lens, which magnifies it again and focuses it on the back of the eye. Microscopes that magnify in stages using several lenses are called **compound microscopes.** They can resolve structures that are separated by at least 200 nanometers (nm).

Light microscopes, even compound ones, are not powerful enough to resolve many of the structures within cells. For example, a cell membrane is only 5 nm thick. Why not just add another magnifying stage to the microscope to increase its resolving power? The reason is that when two objects are closer than a few hundred nanometers, the light beams reflecting from the two images start to overlap each other. The only way two light beams can get closer together and still be resolved is if their wavelengths are shorter. One way to avoid overlap is by using a beam of electrons rather than a beam of light. Electrons have a much shorter wavelength, and an **electron microscope,** employing electron beams, has 1000 times the resolving power of a light microscope. **Transmission electron microscopes,** so called because the electrons used to visualize the specimens are transmitted through the material, are capable of resolving objects only 0.2 nm apart—just twice the diameter of a hydrogen atom!

A second kind of electron microscope, the **scanning electron microscope,** beams the electrons onto the surface of the specimen. The electrons reflected back from the surface, together with other electrons that the specimen itself emits as a result of the bombardment, are amplified and transmitted to a screen, where the image can be viewed and photographed. Scanning electron microscopy yields striking three-dimensional images, and it has improved our understanding of many biological and physical phenomena (see table 4.1).

TABLE 4.1	Microscopes

LIGHT MICROSCOPES

Bright-field microscope:
Light is simply transmitted through a specimen, giving little contrast. Staining specimens improves contrast but requires that cells be fixed (not alive), which can cause distortion or alteration of components.

28.4 µm

Dark-field microscope:
Light is directed at an angle toward the specimen; a condenser lens transmits only light reflected off the specimen. The field is dark, and the specimen is light against this dark background.

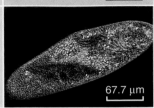
67.7 µm

Phase-contrast microscope:
Components of the microscope bring light waves out of phase, which produces differences in contrast and brightness when the light waves recombine.

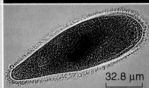
32.8 µm

Differential-interference–contrast microscope:
Out-of-phase light waves to produce differences in contrast are combined with two beams of light traveling close together, which create even more contrast, especially at the edges of structures.

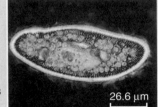
26.6 µm

Fluorescence microscope:
Fluorescent stains absorb light at one wavelength, then emit it at another. Filters transmit only the emitted light.

10.2 µm

Confocal microscope:
Light from a laser is focused to a point and scanned across the fluorescently stained specimen in two directions. Clear images of one plane of the specimen are produced, while other planes of the specimen are excluded and do not blur the image. Multiple planes can be used to reconstruct a 3-D image.

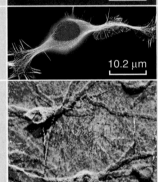
25.0 µm

ELECTRON MICROSCOPES

Transmission electron microscope:
A beam of electrons is passed through the specimen. Electrons that pass through are used to expose film. Areas of the specimen that scatter electrons appear dark. False coloring enhances the image.

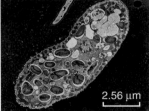
2.56 µm

Scanning electron microscope:
An electron beam is scanned across the surface of the specimen, and electrons are knocked off the surface. Thus, the surface topography of the specimen determines the contrast and the content of the image. False coloring enhances the image.

6.76 µm

Using stains to view cell structure

Although resolution remains a physical limit, we can improve the images we see by altering the sample. Certain chemical stains increase the contrast between different cellular components. Structures within the cell will absorb or exclude the stain, producing contrast that aids resolution.

Staining techniques have become even more powerful with the use of stains that bind to specific types of molecules. This method employs antibodies that bind, for example, to a particular protein. This process, called *immunohistochemistry*, uses antibodies generated in animals such as rabbits or mice. When these animals are injected with specific proteins, they produce antibodies that bind to the injected protein, and the antibodies can be purified from their blood. These purified antibodies can then be chemically bonded to enzymes, to stains, or to fluorescent molecules. When cells are incubated in a solution containing the antibodies, the antibodies bind to cellular structures that contain the target molecule and can be seen with light microscopy. This approach has been used extensively in the analysis of cell structure and function.

All cells exhibit basic structural similarities

The general plan of cellular organization varies in the cells of different organisms, but despite these modifications, all cells resemble one another in certain fundamental ways. Before we begin a detailed examination of cell structure, let's first summarize four major features all cells have in common: a nucleoid or nucleus where genetic material is located, cytoplasm, *ribosomes* to synthesize proteins, and a plasma membrane.

Centrally located genetic material

Every cell contains DNA, the hereditary molecule. In **prokaryotes,** the simplest organisms, most of the genetic material lies in a single circular molecule of DNA. It typically resides near the center of the cell in an area called the **nucleoid,** but this area is not segregated from the rest of the cell's interior by membranes.

By contrast, the DNA of eukaryotes, which are more complex organisms, is contained in the nucleus, which is surrounded by a double membrane structure called the nuclear envelope. In both types of organisms, the DNA contains the genes that code for the proteins synthesized by the cell. (Details of nucleus structure are described later in the chapter.)

The cytoplasm

A semifluid matrix called the **cytoplasm** fills the interior of the cell. The cytoplasm contains all of the sugars, amino acids, and proteins the cell uses to carry out its everyday activities. Although it is an aqueous medium, cytoplasm is more like jello than water due to the high concentration of proteins and other macromolecules. In eukaryotic cells, in addition to the nucleus, the cytoplasm also contains specialized membrane-bounded compartments called **organelles.** The part of the cytoplasm that contains organic molecules and ions in solution is called the **cytosol** to distinguish it from the larger organelles suspended in this fluid.

The plasma membrane

The **plasma membrane** encloses a cell and separates its contents from its surroundings. The plasma membrane is a phospholipid bilayer about 5 to 10 nm (5 to 10 billionths of a meter) thick, with proteins embedded in it. Viewed in cross section with the electron microscope, such membranes appear as two dark lines separated by a lighter area. This distinctive appearance arises from the tail-to-tail packing of the phospholipid molecules that make up the membrane (see chapter 5).

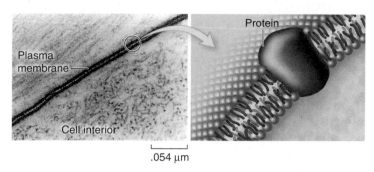

Plasma membrane

Cell interior

Protein

.054 μm

The proteins of the plasma membrane are in large part responsible for a cell's ability to interact with the environment. *Transport proteins* help molecules and ions move across the plasma membrane, either from the environment to the interior of the cell or vice versa. *Receptor proteins* induce changes within the cell when they come in contact with specific molecules in the environment, such as hormones, or with molecules on the surface of neighboring cells. These molecules can function as *markers* that identify the cell as a particular type. This interaction between cell surface molecules is especially important in multicellular organisms, whose cells must be able to recognize one another as they form tissues.

We'll examine the structure and function of cell membranes more thoroughly in chapter 5.

All organisms are cells or aggregates of cells. All cells arise from preexisting cells. Multicellular organisms usually consist of many small cells rather than a few large ones because of the limits of diffusion of small molecules into the cell.

Most cells and their components are so small they can only be viewed using microscopes that use lenses to focus beams of reflected light or electrons. Different kinds of microscopes and different staining procedures are used, depending on what subcellular component is being resolved.

All cells are bounded by a plasma membrane and filled with a semifluid substance called cytoplasm. The genetic material is found in the central portion of the cell and, in eukaryotic cells, is contained in a membrane-bounded structure called a nucleus.

4.2 Prokaryotic Cells

When cells were visualized with microscopes, two basic cellular architectures were recognized: eukaryotic and prokaryotic. These refer to the presence or absence, respectively, of a membrane-bounded nucleus that contains genetic material. We have already mentioned that in addition to lacking a nucleus, prokaryotic cells do not have an internal membrane system or numerous membrane-bounded organelles.

Prokaryote Cells Have Relatively Simple Organization

Prokaryotes are the simplest organisms. Prokaryotic cells are small, consisting of cytoplasm surrounded by a plasma membrane and encased within a rigid **cell wall,** with no distinct interior compartments (figure 4.3). A prokaryotic cell is like a one-room cabin in which eating, sleeping, and watching TV all occur.

Prokaryotes are very important in the ecology of living organisms. Some harvest light by photosynthesis, others break down dead organisms and recycle their components, cause disease, and have uses

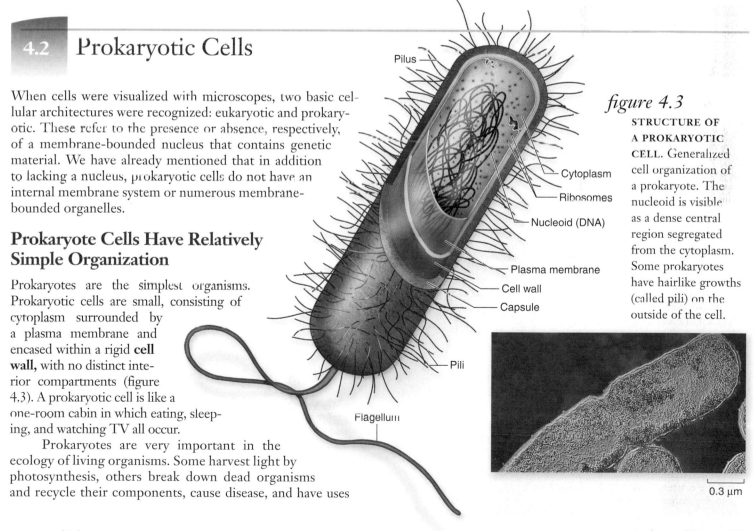

Pilus

Cytoplasm

Ribosomes

Nucleoid (DNA)

Plasma membrane

Cell wall

Capsule

Pili

Flagellum

figure 4.3

STRUCTURE OF A PROKARYOTIC CELL. Generalized cell organization of a prokaryote. The nucleoid is visible as a dense central region segregated from the cytoplasm. Some prokaryotes have hairlike growths (called pili) on the outside of the cell.

0.3 μm

in many important industrial processes. There are two main domains of prokaryotes: archaea and bacteria. Chapter 28 covers prokaryotic diversity in more detail.

Although prokaryotic cells do contain complex structures like **ribosomes,** which carry out protein synthesis, most have no membrane-bounded organelles characteristic of eukaryotic cells. Prokaryotes also lack the elaborate cytoskeleton found in eukaryotes, although they appear to have molecules related to actin, which is found in microfilaments (discussed later in the chapter). These actinlike proteins form supporting fibrils near the surface of the cell, but the cytoplasm of a prokaryotic cell appears to be one unit with no internal support structure. Consequently, the strength of the cell comes primarily from its rigid cell wall (see figure 4.3).

The plasma membrane of a prokaryotic cell carries out some of the functions organelles perform in eukaryotic cells. For example, some photosynthetic bacteria, such as the cyanobacterium *Prochloron* (figure 4.4), have an extensively folded plasma membrane, with the folds extending into the cell's interior. These membrane folds contain the bacterial pigments connected with photosynthesis. In eukaryotic plant cells, photosynthetic pigments are found in the inner membrane of the chloroplast.

Because a prokaryotic cell contains no membrane-bounded organelles, the DNA, enzymes, and other cytoplasmic constituents have access to all parts of the cell. Reactions are not compartmentalized as they are in eukaryotic cells, and the whole prokaryote operates as a single unit.

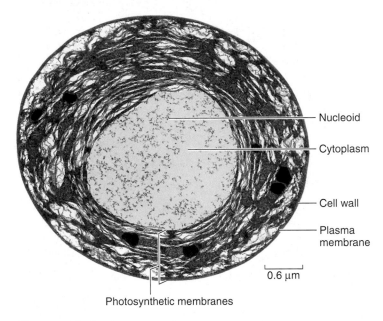

figure 4.4

ELECTRON MICROGRAPH OF A PHOTOSYNTHETIC BACTERIAL CELL. Extensive folded photosynthetic membranes are shown in green in this false color electron micrograph of a *Prochloron* cell.

 inquiry

If you were designing a cell to be as large as possible, what modifications would facilitate large size?

Bacterial cell walls consist of peptidoglycan

Most bacterial cells are encased by a strong **cell wall.** This cell wall is composed of *peptidoglycan*, which consists of a carbohydrate matrix (polymers of sugars) that is cross-linked by short polypeptide units. Cell walls protect the cell, maintain its shape, and prevent excessive uptake or loss of water. Plants, fungi, and most protists also have cell walls of a different chemical structure that lack peptidoglycan, discussed in later chapters.

With the exception of the class Mollicutes, commonly called mycoplasma, which lack a cell wall, bacteria can be classified into two types based on differences in their cell walls that can be detected by the Gram staining procedure. **Gram-positive** bacteria have a thick, single-layered peptidoglycan cell wall that retains a violet dye from the Gram stain, causing the stained cells to appear purple under a microscope. **Gram-negative** bacteria have evolved a more complex, multilayered cell wall that does not retain the purple dye. These bacteria appear red after staining due to a second background dye.

The susceptibility of bacteria to antibiotics often depends on the structure of their cell walls. The drugs penicillin and vancomycin, for example, interfere with the ability of bacteria to cross-link the peptides in their peptidoglycan cell wall. Like removing all the nails from a wooden house, this destroys the integrity of the matrix, which can no longer prevent water from rushing in and swelling the cell to bursting.

Some bacteria also secrete a jellylike protective capsule of polysaccharide around the cell. Many disease-causing bacteria have such a capsule, which enables them to adhere to teeth, skin, food—or to practically any surface that will support their growth.

Archaea lack peptidoglycan

We are still learning about the physiology and structure of archaea. Many of these organisms are difficult to culture in the laboratory, and so this group has not yet been studied in detail. More is known about their genetic makeup than about any other feature.

The cell walls of archaea have various chemical compositions, including polysaccharides and proteins, and possibly even inorganic components. A common feature distinguishing archaea from bacteria is the nature of their membrane lipids. The chemical structure of archaeal lipids is distinctly different from that of lipids in bacteria and can include saturated hydrocarbons that are covalently attached to glycerol at both ends, such that their membrane is a monolayer. These features seem to confer greater thermal stability on archaeal membranes, although the tradeoff seems to be an inability to alter the degree of saturation of the hydrocarbons—meaning that archaea with this characteristic lose the ability to adapt to changing environmental temperatures.

The cellular machinery that replicates DNA and synthesized proteins in archaea is more closely related to eukaryotic systems than to bacterial systems. Even though they share a

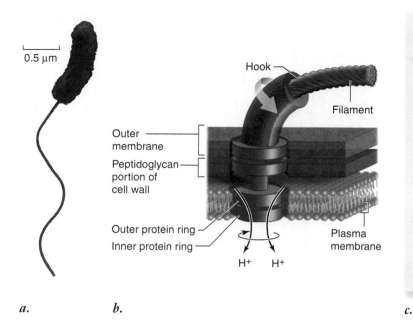

a. b.

c.

figure 4.5

SOME PROKARYOTES MOVE BY ROTATING THEIR FLAGELLA. *a.* The photograph shows *Vibrio cholerae*, the microbe that causes the serious disease cholera. *b.* The bacterial flagellum is a complex structure. The motor proteins, powered by a proton gradient, are anchored in the plasma membrane. Two rings are found in the cell wall. The motor proteins cause the entire structure to rotate. *c.* As the flagellum rotates it creates a spiral wave down the structure. This powers the cell forward.

similar overall cellular architecture with prokaryotes, archaea appear to be more closely related to eukaryotes.

Some prokaryotes move by means of rotating flagella

Flagella (singular, *flagellum*) are long, threadlike structures protruding from the surface of a cell that are used in locomotion. Prokaryotic flagella are protein fibers that extend out from the cell. There may be one or more per cell, or none, depending on the species. Bacteria can swim at speeds of up to 70 cell lengths per second by rotating their flagella like screws (figure 4.5). The rotary motor uses the energy stored in a gradient that transfers protons across the plasma membrane to power the movement of the flagellum. Interestingly, the same principle, in which a proton gradient powers the rotation of a molecule, is

used in eukaryotic mitochondria and chloroplasts by an enzyme that synthesizes ATP (see chapter 7).

Prokaryotes are small cells that lack complex interior organization. The two domains of prokaryotes are archaea and bacteria. Bacteria are encased by a cell wall composed of peptidoglycan, and archaea have cell walls made from a variety of carbohydrate and peptides.

Bacteria can be separated into gram positive and gram negative based on staining that detects differences in cell wall architecture. Archaea have unusual membrane lipids.

Same prokaryotes are motile, propelled by external flagella that rotate.

4.3   Eukaryotic Cells

Eukaryotic cells (figures 4.6 and 4.7) are far more complex than prokaryotic cells. The hallmark of the eukaryotic cell is compartmentalization, which is achieved by an extensive **endomembrane system** that weaves through the cell interior and by numerous *organelles*. Organelles are the membrane-bounded structures that form compartments within which multiple biochemical processes can proceed simultaneously and independently.

Plant cells often have a large, membrane-bounded sac called a **central vacuole**, which stores proteins, pigments, and waste materials. Both plant and animal cells contain **vesicles**, smaller sacs that store and transport a variety of materials. Inside the nucleus, the DNA is wound tightly around proteins and packaged into compact units called **chromosomes.**

All eukaryotic cells are supported by an internal protein scaffold, the **cytoskeleton.** Although the cells of animals and some protists lack cell walls, the cells of fungi, plants, and many protists have strong cell walls composed of cellulose or chitin fibers embedded in a matrix of other polysaccharides and proteins. In the remainder of this chapter, we will examine the internal components of eukaryotic cells in more detail.

The nucleus acts as the information center

The largest and most easily seen organelle within a eukaryotic cell is the **nucleus** (Latin, "kernel" or "nut"), first described by the Scottish botanist Robert Brown in 1831. Nuclei are

figure 4.6

STRUCTURE OF AN ANIMAL CELL. In this generalized diagram of an animal cell, the plasma membrane encases the cell, which contains the cytoskeleton and various cell organelles and interior structures suspended in a semifluid matrix called the cytoplasm. Some kinds of animal cells possess fingerlike projections called microvilli. Other types of eukaryotic cells—for example, many protist cells—may possess flagella, which aid in movement, or cilia, which can have many different functions.

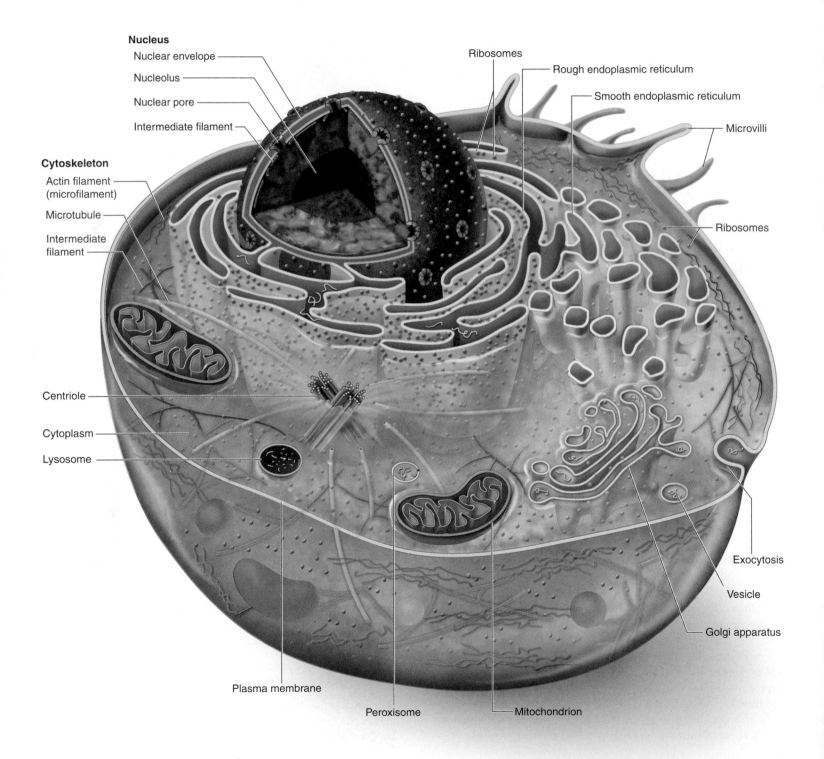

Nucleus
Nuclear envelope
Nucleolus
Nuclear pore
Intermediate filament

Cytoskeleton
Actin filament (microfilament)
Microtubule
Intermediate filament

Centriole
Cytoplasm
Lysosome

Plasma membrane
Peroxisome

Ribosomes
Rough endoplasmic reticulum
Smooth endoplasmic reticulum
Microvilli
Ribosomes

Exocytosis
Vesicle
Golgi apparatus
Mitochondrion

figure 4.7

STRUCTURE OF A PLANT CELL. Most mature plant cells contain a large central vacuole, which occupies a major portion of the internal volume of the cell, and organelles called chloroplasts, within which photosynthesis takes place. The cells of plants, fungi, and some protists have cell walls, although the composition of the walls varies among the groups. Plant cells have cytoplasmic connections to one another through openings in the cell wall called plasmodesmata. Flagella occur in sperm of a few plant species, but are otherwise absent from plant and fungal cells. Centrioles are also absent.

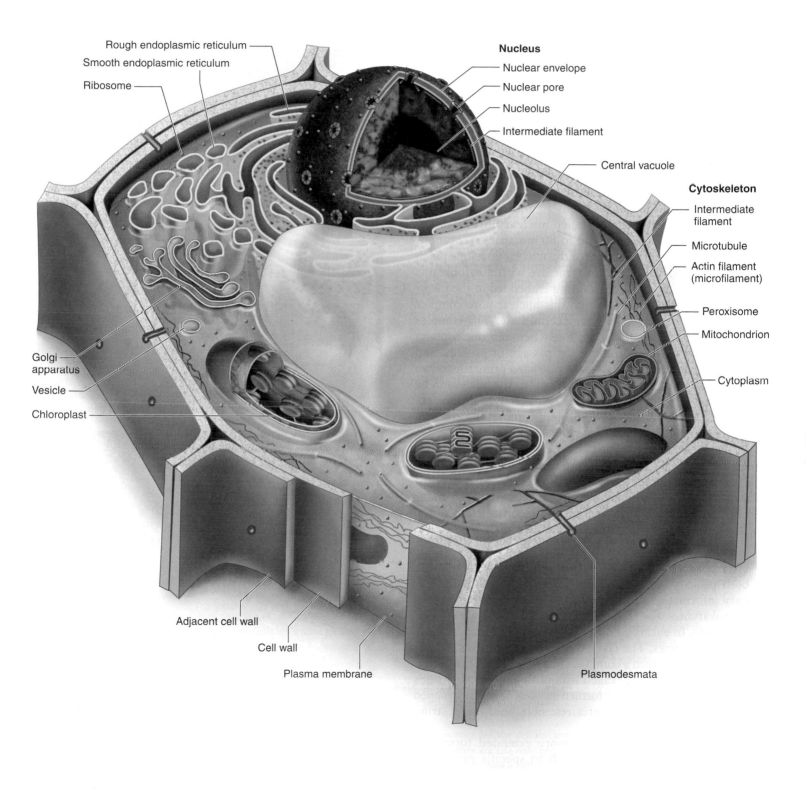

The Endomembrane System

The interior of a eukaryotic cell is packed with membranes so thin that they are invisible under the low resolving power of light microscopes. This endomembrane system fills the cell, dividing it into compartments, channeling the passage of molecules through the interior of the cell, and providing surfaces for the synthesis of lipids and some proteins. The presence of these membranes in eukaryotic cells constitutes one of the most fundamental distinctions between eukaryotes and prokaryotes.

The largest of the internal membranes is called the **endoplasmic reticulum (ER)**. *Endoplasmic* means "within the cytoplasm," and *reticulum* is Latin for "a little net." Like the plasma membrane, the ER is composed of a phospholipid bilayer embedded with proteins. It weaves in sheets through the interior of the cell, creating a series of channels between its folds (figure 4.11). Of the many compartments in eukaryotic cells, the two largest are the inner region of the ER, called the **cisternal space** or **lumen,** and the region exterior to it, the cytosol, which is the fluid component of the cytoplasm containing dissolved organic molecules such as proteins and ions.

The rough ER is a site of protein synthesis

The **rough ER (RER)** gets its name from its surface appearance, which is pebbly instead of smooth due to the presence of ribosomes. The RER is not easily visible with a light microscope, but it can be seen using the electron microscope. It appears to be composed of flattened sacs, the surfaces of which are bumpy with ribosomes (see figure 4.11).

The proteins synthesized on the surface of the RER are destined to be exported from the cell, sent to lysosomes or vacuoles (described in a later section), or embedded in the plasma membrane. These proteins enter the cisternal space as a first step in the pathway that will sort proteins to their eventual destinations. This pathway also involves vesicles and the Golgi

apparatus, described later. The sequence of the protein being synthesized determines whether the ribosome will become associated with the ER or remain a cytoplasmic ribosome.

In the ER, newly synthesized proteins can be modified by the addition of short-chain carbohydrates to form **glycoproteins.** Those proteins destined for secretion are then kept separate from other products and are later packaged into vesicles. The ER also manufactures membranes by producing membrane proteins and phospholipid molecules. The membrane proteins are inserted into the ER's own membrane, which can then expand and pinch off in the form of vesicles to be transferred to other locations.

The smooth ER has multiple roles

Regions of the ER with relatively few bound ribosomes are referred to as **smooth ER (SER).** The SER appears more like a network of tubules than the flattened sacs of the RER. The membranes of the SER contain many embedded enzymes. Enzymes anchored within the ER, for example, catalyze the synthesis of a variety of carbohydrates and lipids. Steroid hormones are synthesized in the SER as well. The majority of membrane lipids are assembled in the SER and then sent to whatever parts of the cell need membrane components.

The SER is used to store Ca^{2+} in cells. This keeps the cytoplasmic level low, allowing Ca^{2+} to be used as a signaling molecule. In muscle cells, for example, Ca^{2+} is used to trigger muscle contraction. In other cells, Ca^{2+} release from SER stores is involved in diverse signalling pathways.

The ratio of SER to RER depends on a cell's function. In multicellular animals such as ourselves, great variation exists in this ratio. Cells that carry out extensive lipid synthesis, such as those in the testes, intestine, and brain have abundant SER. Cells that synthesize proteins that are secreted, such as antibodies, have much more extensive RER.

Another role of the SER is the modification of foreign substances to make them less toxic. In the liver, the enzymes of the SER carry out this detoxification. This action can include neu-

figure 4.11

THE ENDOPLASMIC RETICULUM. Rough ER (RER), blue in the drawing, is composed more of flattened sacs and forms a compartment throughout the cytoplasm. Ribosomes associated with the cytoplasmic face of the RER extrude newly made proteins into the interior, or lumen. The smooth ER (SER), green in the drawing, is a more tubelike structure connected to the RER. The micrograph has been colored to match the drawing.

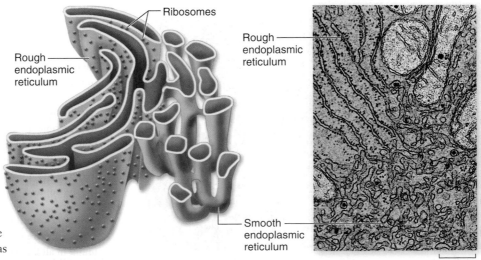

Ribosomes

Rough endoplasmic reticulum

Rough endoplasmic reticulum

Smooth endoplasmic reticulum

0.08 μm

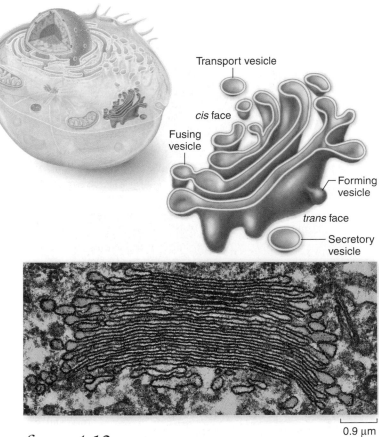

figure 4.12

THE GOLGI APPARATUS. The Golgi apparatus is a smooth, concave, membranous structure. It receives material for processing in transport vesicles on the *cis* face and sends the material packaged in transport or secretory vesicles off the *trans* face. The substance in a vesicle could be for export out of the cell or for distribution to another region within the same cell.

tralizing substances that we have taken for a therapeutic reason, such as penicillin. Thus relatively high doses are prescribed for some drugs to offset our body's efforts to remove them. Liver cells have extensive SER as well as enzymes that can process a variety of substances by chemically modifying them.

The Golgi apparatus sorts and packages proteins

Flattened stacks of membranes called **Golgi bodies** can be found within the endomembrane system, often interconnected with one another. These structures are named for Camillo Golgi, the nineteenth-century Italian physician who first identified them. The number of Golgi bodies a cell contains ranges from 1 or a few in protists, to 20 or more in animal cells and several hundred in plant cells. They are especially abundant in glandular cells, which manufacture and secrete substances. Collectively, the Golgi bodies are referred to as the **Golgi apparatus** (figure 4.12).

The Golgi apparatus functions in the collection, packaging, and distribution of molecules synthesized at one location and utilized at another within the cell or even outside of it. A Golgi body has a front and a back, with distinctly different membrane compositions at these opposite ends. The front, or receiving end, is called the *cis* face, and it is usually located near ER. Materials move to the *cis* face in transport vesicles that bud off the ER. These vesicles

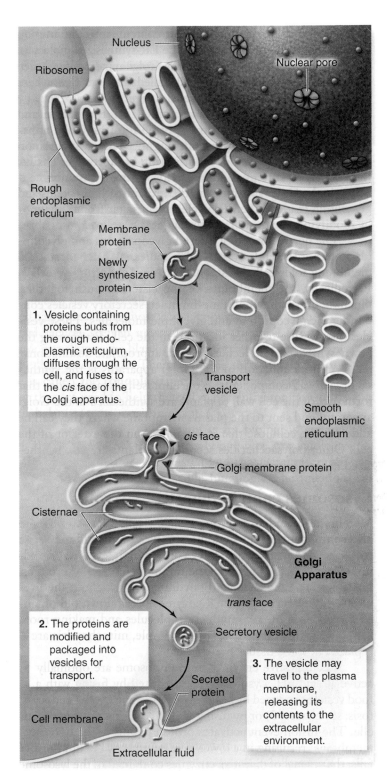

figure 4.13

PROTEIN TRANSPORT THROUGH THE ENDOMEMBRANE SYSTEM. Proteins synthesized by ribosomes on the RER are translocated into the internal compartment of the ER. These proteins may be used at a distant location in the cell or secreted. They are transported within vesicles that bud off the rough ER. These transport vesicles travel to the *cis* face of the Golgi apparatus. There they can be modified and packaged into vesicles that bud off the *trans* face of the Golgi apparatus. Vesicles leaving the *trans* face transport proteins to other locations in the cell, or fuse with the plasma membrane, releasing their contents to the extracellular environment.

Mitochondria and chloroplasts arose by endosymbiosis

Symbiosis is a close relationship between organisms of different species that live together. As noted in chapter 29, the theory of **endosymbiosis** proposes that some of today's eukaryotic organelles evolved by a symbiosis arising between two cells that were each free-living. One cell, a prokaryote, was engulfed by and became part of another cell, which was the precursor to modern eukaryotes (figure 4.19).

According to the endosymbiont theory, the engulfed prokaryotes provided their hosts with certain advantages associated with their special metabolic abilities. Two key eukaryotic organelles are believed to be the descendants of these endosymbiotic prokaryotes: mitochondria, which are thought to have originated as bacteria capable of carrying out oxidative metabolism, and chloroplasts, which apparently arose from photosynthetic bacteria.

The endosymbiont theory is supported by a wealth of evidence, summarized as follows:

- Both mitochondria and chloroplasts are surrounded by two membranes; the inner membrane may have evolved from the plasma membrane of the engulfed prokaryote; the outer membrane is probably derived from the plasma membrane or endoplasmic reticulum of the host cell.
- Mitochondria are about the same size as most prokaryotes, and the cristae formed by their inner membranes resemble the folded membranes in various groups of bacteria.
- Mitochondrial ribosomes are also similar to prokaryotic ribosomes in size and structure.
- Both mitochondria and chloroplasts contain circular molecules of DNA similar to those in prokaryotes.
- The genomes of mitochondria and chloroplasts show similarities to the genomes of α-proteobacteria and cyanobacteria, respectively.
- Finally, mitochondria divide by simple fission, splitting in two just as prokaryotic cells do, and they apparently replicate and partition their DNA in much the same way as prokaryotes do.

Mitochondria and chloroplasts are both involved in energy conversion. Both mitochondria and chloroplasts have their own DNA, which contains specific genes related to some of their functions, but both depend on nuclear genes for other functions.

Much evidence exists that both mitochondria and chloroplasts arose via endosymbiosis.

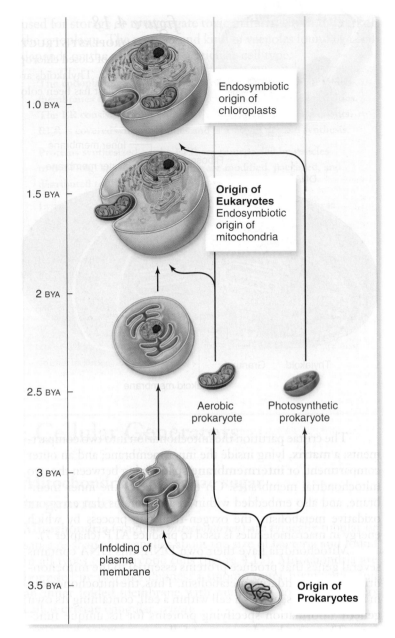

figure 4.19

PROPOSED ENDOSYMBIOTIC ORIGIN OF EUKARYOTIC CELLS. Both mitochondria and chloroplasts are thought to have arisen by endosymbiosis where a free-living cell is taken up but not digested. The cell that engulfed the future mitochondria and chloroplasts is thought to have first acquired a nuclear envelope and endomembrane system from infolding of the plasma membrane. The proposed series of events leading to modern eukaryotic cells is shown next to a rough time line of the history of Earth.

4.6   The Cytoskeleton

The cytoplasm of all eukaryotic cells is crisscrossed by a network of protein fibers that supports the shape of the cell and anchors organelles to fixed locations. This network, called the cytoskeleton, is a dynamic system, constantly forming and disassembling.

Individual fibers consist of polymers of identical protein subunits that attract one another and spontaneously assemble into long chains. Fibers disassemble in the same way, as one subunit after another breaks away from one end of the chain.

Three types of fibers compose the cytoskeleton

Eukaryotic cells may contain the following three types of cytoskeletal fibers, each formed from a different kind of subunit: (1) actin filaments, sometimes called microfilaments, (2) microtubules, and (3) intermediate filaments.

Actin filaments (microfilaments)

Actin filaments are long fibers about 7 nm in diameter. Each filament is composed of two protein chains loosely twined together like two strands of pearls (figure 4.20). Each "pearl," or subunit, on the chains is the globular protein **actin.** Actin filaments exhibit polarity in that they have plus (+) and minus (–) ends. These designate the direction of growth of the filaments. Actin molecules spontaneously form these filaments, even in a test tube.

Cells regulate the rate of actin polymerization through other proteins that act as switches, turning on polymerization when appropriate. Actin filaments are responsible for cellular movements such as contraction, crawling, "pinching" during division, and formation of cellular extensions.

Microtubules

Microtubules, the largest of the cytoskeletal elements, are hollow tubes about 25 nm in diameter, each composed of a ring of 13 protein protofilaments (see figure 4.20). Globular proteins consisting of dimers of α- and β-*tubulin* subunits polymerize to form the 13 protofilaments. The protofilaments are arrayed side by side around a central core, giving the microtubule its characteristic tube shape.

In many cells, microtubules form from nucleation centers near the center of the cell and radiate toward the periphery. They are in a constant state of flux, continually polymerizing and depolymerizing. The average half-life of a microtubule ranges from as long as 10 minutes in a nondividing animal cell to as short as 20 seconds in a dividing animal cell. The ends of the microtubule are designated as plus (+) (away from the nucleation center) or minus (–) (toward the nucleation center).

Along with facilitating cellular movement, microtubules provide organization to the cytoplasm and are responsible for moving materials within the cell itself, as described shortly.

Intermediate filaments

The most durable element of the cytoskeleton in animal cells is a system of tough, fibrous protein molecules twined together in an overlapping arrangement (see figure 4.20). These **intermediate filaments** are characteristically 8 to 10 nm in diameter, intermediate in size between actin filaments and microtubules. Once formed, intermediate filaments are stable and usually do not break down.

Intermediate filaments constitute a heterogeneous group of cytoskeletal fibers. The most common type, composed of protein subunits called *vimentin*, provides structural stability for many kinds of cells. *Keratin*, another class of intermediate filament, is found in epithelial cells (cells that line organs and body cavities) and associated structures such as hair and fingernails. The intermediate filaments of nerve cells are called *neurofilaments*.

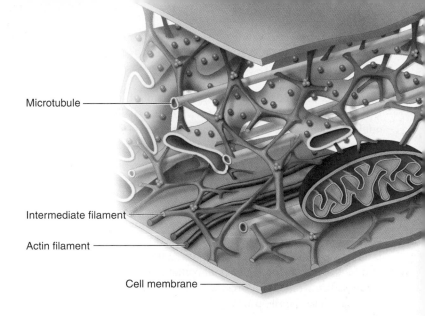

Microtubule

Intermediate filament

Actin filament

Cell membrane

a. Actin filaments

b. Microtubules

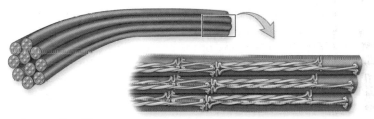

c. Intermediate filament

figure 4.20

MOLECULES THAT MAKE UP THE CYTOSKELETON. *a.* *Actin filaments:* Actin filaments, also called *microfilaments,* are made of two strands of the globular protein actin twisted together. They often are found in bundles, or a branching network. Actin filaments in many cells are concentrated below the plasma membrane in bundles known as stress fibers, which may have a contractile function. ***b.*** *Microtubules:* Microtubules are composed of α and β tubulin protein subunits arranged side by side to form a tube. Microtubules are comparatively stiff cytoskeletal elements with many functions in the cell including intracellular transport and the separation of chromosomes during mitosis. ***c.*** *Intermediate filaments:* Intermediate filaments are composed of overlapping staggered tetramers of protein. These tetramers are then bundled into cables. This molecular arrangement allows for a ropelike structure that imparts tremendous mechanical strength to the cell.

Centrosomes are microtubule-organizing centers

Centrioles are barrel-shaped organelles found in the cells of animals and most protists. They occur in pairs, usually located at right angles to each other near the nuclear membranes (figure 4.21); the region surrounding the pair in almost all animal cells is referred to as a **centrosome.** Surrounding the centrioles in the centrosome is the **pericentriolar material,** which contains ring-shaped structures composed of tubulin. The pericentriolar material can nucleate the assembly of microtubules in animal cells. Structures with this function are called **microtubule-organizing centers.** The centrosome is also responsible for the reorganization of microtubules that occurs during cell division. The centrosomes of plants and fungi lack centrioles, but still contain microtubule-organizing centers. You will learn more about the actions of the centrosomes when we describe the process of cell division in chapter 10.

The cytoskeleton helps move materials within cells

Actin filaments and microtubules often orchestrate their activities to affect cellular processes. For example, during cell reproduction (see chapter 10), newly replicated chromosomes move to opposite sides of a dividing cell because they are attached to shortening microtubules. Then, in animal cells, a belt of actin pinches the cell in two by contracting like a purse string.

Muscle cells also use actin filaments sliding relative to filaments of the motor protein myosin to contract. The fluttering of an eyelash, the flight of an eagle, and the awkward crawling of a baby all depend on these cytoskeletal movements within muscle cells.

Not only is the cytoskeleton responsible for the cell's shape and movement, but it also provides a scaffold that holds certain enzymes and other macromolecules in defined areas of the cytoplasm. For example, many of the enzymes involved in cell metabolism bind to actin filaments; so do ribosomes. By moving and anchoring particular enzymes near one another, the cytoskeleton, like the endoplasmic reticulum, helps organize the cell's activities.

Molecular motors

All eukaryotic cells must move materials from one place to another in the cytoplasm. One

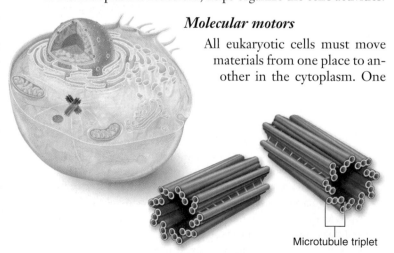

figure 4.21

CENTRIOLES. Each centriole is composed of nine triplets of microtubules. Centrioles are not found in plant cells. In animal cells they help to organize microtubules.

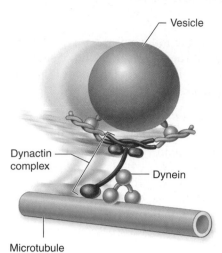

figure 4.22

MOLECULAR MOTORS. Vesicles can be transported along microtubules using motor proteins that use ATP to generate force. The vesicles are attached to motor proteins by connector molecules, such as the dynactin complex shown here. The motor protein dynein moves the connected vesicle along microtubules.

way that cells do this is by using the channels of the endoplasmic reticulum as an intracellular highway. Material can also be moved using vesicles loaded with cargo that can move along the cytoskeleton like a railroad track. For example, in a nerve cell with an axon that may extend far from the cell body, vesicles can be moved along tracks of microtubules from the cell body to the end of the axon.

Four components are required: (1) a vesicle or organelle that is to be transported, (2) a motor protein that provides the energy-driven motion, (3) a connector molecule that connects the vesicle to the motor molecule, and (4) microtubules on which the vesicle will ride like a train on a rail (figure 4.22).

The direction that a vesicle is moved depends on the identity of the motor protein involved, and the fact that microtubules are organized with their plus ends toward the periphery of the cell. In one case, a protein called kinectin binds vesicles to the motor protein **kinesin.** Kinesin uses ATP to power its movement toward the cell periphery, dragging the vesicle with it as it travels along the microtubule toward the plus end. As nature's tiniest motors, these proteins literally pull the transport vesicles along the microtubular tracks. Another set of vesicle proteins, called the dynactin complex, binds vesicles to the motor protein **dynein** (see figure 4.22), which directs movement in the opposite direction along microtubules toward the minus end, inward toward the cell's center. (Dynein is also involved in the movement of eukaryotic flagella, as discussed later.) The destination of a particular transport vesicle and its content is thus determined by the nature of the linking protein embedded within the vesicle's membrane.

The major eukaryotic cell structures and their respective functions are summarized in table 4.2.

The three principal fibers of the cytoskeleton are actin filaments (microfilaments), microtubules, and intermediate filaments. These fibers interact to modulate cell shape and permit cell movement and act to move materials within the cytoplasm.

Material is also moved in large cells using vesicles and molecular motors. The motor proteins move the vesicles along tracks of microtubules.

TABLE 4.2 Eukaryotic Cell Structures and Their Functions

Structure		Description	Function
Plasma membrane		Phospholipid bilayer with embedded proteins	Regulates what passes into and out of cell; cell-to-cell recognition; connection and adhesion; cell communication
Nucleus		Structure (usually spherical) that contains chromosomes and is surrounded by double membrane	Instructions for protein synthesis and cell reproduction; contains genetic information
Chromosomes		Long threads of DNA that form a complex with protein	Contain hereditary information used to direct synthesis of proteins
Nucleolus		Site of genes for rRNA synthesis	Synthesis of rRNA and ribosome assembly
Ribosomes		Small, complex assemblies of protein and RNA, often bound to ER	Sites of protein synthesis
Endoplasmic reticulum (ER)		Network of internal membranes	Intracellular compartment forms transport vesicles; participates in lipid synthesis and synthesis of membrane or secreted proteins
Golgi apparatus		Stacks of flattened vesicles	Packages proteins for export from cell; forms secretory vesicles
Lysosomes		Vesicles derived from Golgi apparatus that contain hydrolytic digestive enzymes	Digest worn-out organelles and cell debris; digest material taken up by endocytosis
Microbodies		Vesicles that are formed from incorporation of lipids and proteins and that contain oxidative and other enzymes	Isolate particular chemical activities from rest of cell
Mitochondria		Bacteria-like elements with double membrane	"Power plants" of the cell; sites of oxidative metabolism
Chloroplasts		Bacteria-like elements with membranes containing chlorophyll, a photosynthetic pigment	Sites of photosynthesis
Cytoskeleton		Network of protein filaments	Structural support; cell movement; movement of vesicles within cells
Flagella (cilia)		Cellular extensions with 9 + 2 arrangement of pairs of microtubles	Motility or moving fluids over surfaces
Cell wall		Outer layer of cellulose or chitin; or absent	Protection; support

Extracellular Structures and Cell Movement

Essentially all cell motion is tied to the movement of actin filaments, microtubules, or both. Intermediate filaments act as intracellular tendons, preventing excessive stretching of cells, and actin filaments play a major role in determining the shape of cells. Because actin filaments can form and dissolve so readily, they enable some cells to change shape quickly.

Some cells crawl

The arrangement of actin filaments within the cell cytoplasm allows cells to crawl, literally! Crawling is a significant cellular phenomenon, essential to such diverse processes as inflammation, clotting, wound healing, and the spread of cancer. White blood cells in particular exhibit this ability. Produced in the bone marrow, these cells are released into the circulatory system and then eventually crawl out of venules and into the tissues to destroy potential pathogens.

At the leading edge of a crawling cell, actin filaments rapidly polymerize, and their extension forces the edge of the cell forward. This extended region is stabilized when microtubules polymerize into the newly formed region. Forward movement of the cell overall is then achieved through the action of the protein **myosin,** which is best known for its role in muscle contraction. Myosin motors along the actin filaments contract, pulling the contents of the cell toward the newly extended front edge.

Overall crawling of the cell takes place when these steps occur continuously, with a leading edge extending and stabilizing, and then motors contracting to pull the remaining cell contents along. Receptors on the cell surface can detect molecules outside the cell and stimulate extension in specific directions, allowing cells to move toward particular targets.

Flagella and cilia aid movement

Earlier in this chapter, we described the structure of prokaryotic flagella. Eukaryotic cells have a completely different kind of flagellum, consisting of a circle of nine microtubule pairs surrounding two central microtubules; this arrangement is referred to as the **9 + 2 structure** (figure 4.23).

As pairs of microtubules move past each other using arms composed of the motor protein dynein, the eukaryotic flagellum *undulates,* rather than rotates. When examined carefully, each flagellum proves to be an outward projection of the cell's interior, containing cytoplasm and enclosed by the plasma membrane. The microtubules of the flagellum are derived from a **basal body,** situated just below the point where the flagellum protrudes from the surface of the cell.

The flagellum's complex microtubular apparatus evolved early in the history of eukaryotes. Although today the cells of many multicellular and some unicellular eukaryotes no longer possess flagella and are nonmotile, an organization similar to the 9 + 2 arrangement of microtubules can still be found within them, in structures called **cilia** (singular, *cilium*). Cilia are short cellular projections that are often organized in rows. They are more numerous than flagella on the cell surface, but have the same internal structure.

figure 4.23

FLAGELLA AND CILIA. A eukaryotic flagellum originates directly from a basal body. The flagellum has two microtubules in its core connected by radial spokes to an outer ring of nine paired microtubules with dynein arms (9 + 2 structure). The basal body consists of nine microtubule triplets connected by short protein segments. The structure of cilia is similar to that of flagella, but cilia are usually shorter.

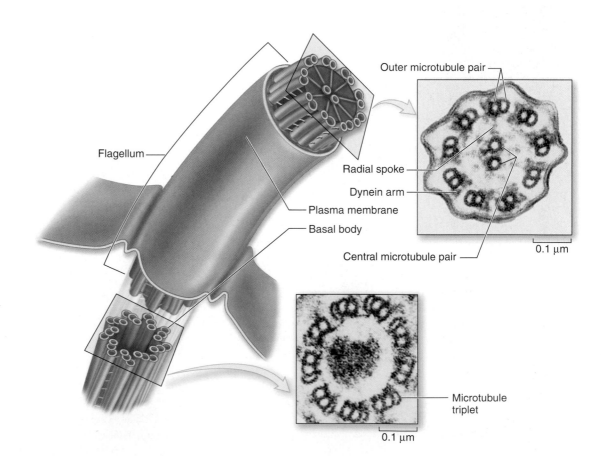

Flagellum

Plasma membrane

Basal body

Outer microtubule pair

Radial spoke

Dynein arm

Central microtubule pair

0.1 µm

Microtubule triplet

0.1 µm

In many multicellular organisms, cilia carry out tasks far removed from their original function of propelling cells through water. In several kinds of vertebrate tissues, for example, the beating of rows of cilia move water over the tissue surface. The sensory cells of the vertebrate ear also contain conventional cilia surrounded by actin-based stereocilia; sound waves bend these structures and provide the initial sensory input for hearing. Thus, the 9 + 2 structure of flagella and cilia appears to be a fundamental component of eukaryotic cells (figure 4.24).

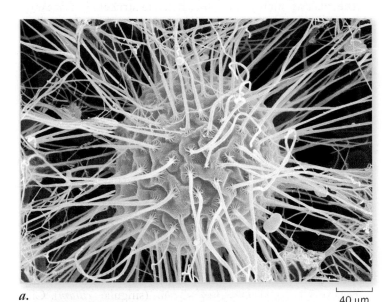

a.

40 μm

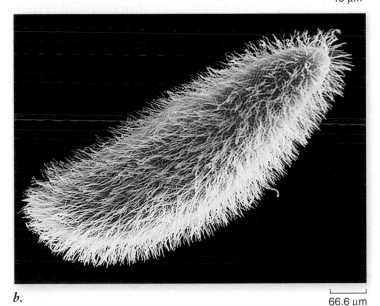

b.

66.6 μm

figure 4.24

FLAGELLA AND CILIA. *a.* A flagellated green alga with numerous flagella that allow it to move through the water. *b.* Paramecium are covered with many cilia, which beat in unison to move the cell. The cilia can also be used to move fluid into the mouth to ingest material.

inquiry

?   *The passageways of human trachea (the path of airflow into and out of the lungs) are known to be lined with ciliated cells. What function could these cilia perform?*

Plant Cell Walls Provide Protection and Support

The cells of plants, fungi, and many types of protists have cell walls, which protect and support the cells. The cell walls of these eukaryotes are chemically and structurally different from prokaryotic cell walls. In plants and protists, the cell walls are composed of fibers of the polysaccharide cellulose, whereas in fungi, the cell walls are composed of chitin.

In plants, **primary walls** are laid down when the cell is still growing, and between the walls of adjacent cells is a sticky substance called the **middle lamella**, which glues the cells together (figure 4.25). Some plant cells produce strong **secondary walls,** which are deposited inside the primary walls of fully expanded cells.

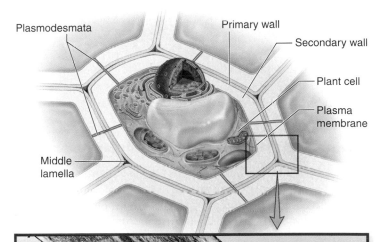

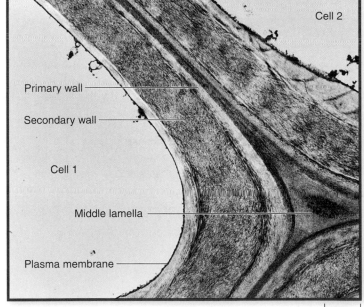

0.4 μm

figure 4.25

CELL WALLS IN PLANTS. Plant cell walls are thick, strong, and rigid. Primary cell walls are laid down when the cell is young. Thicker secondary cell walls may be added later when the cell is fully grown.

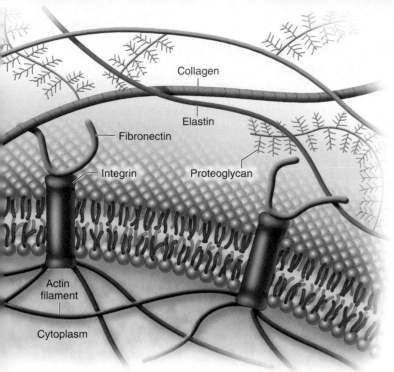

Collagen
Elastin
Fibronectin
Integrin
Proteoglycan
Actin filament
Cytoplasm

figure 4.26

THE EXTRACELLULAR MATRIX. Animal cells are surrounded by an extracellular matrix composed of various glycoproteins that give the cells support, strength, and resilience.

Animal cells secrete an extracellular matrix

Animal cells lack the cell walls that encase plants, fungi, and most protists. Instead, animal cells secrete an elaborate mixture of glycoproteins into the space around them, forming the **extracellular matrix (ECM)** (figure 4.26). The fibrous protein collagen, the same protein found in cartilage, tendons, and ligaments may be abundant in the ECM. Strong fibers of collagen and another fibrous protein, elastin, are embedded within a complex web of other glycoproteins, called proteoglycans, that form a protective layer over the cell surface.

The ECM of some cells is attached to the plasma membrane by a third kind of glycoprotein, **fibronectin.** Fibronectin molecules bind not only to ECM glycoproteins but also to proteins called **integrins,** which are an integral part of the plasma membrane. Integrins extend into the cytoplasm, where they are attached to the microfilaments and intermediate filaments of the cytoskeleton. Linking ECM and cytoskeleton, integrins allow the ECM to influence cell behavior in important ways, altering gene expression and cell migration patterns by a combination of mechanical and chemical signaling pathways. In this way, the ECM can help coordinate the behavior of all the cells in a particular tissue.

Table 4.3 compares and reviews the features of three types of cells.

Cell movement involves proteins. These can either be internal in the case of crawling cells that use actin and myosin or external in the case of cells powered by cilia or flagella.

Eukaryotic cilia and flagella are composed of bundles of microtubules in a 9 + 2 array. They undulate rather than rotate.

Plant cells have a cellulose-based cell wall.

In animal cells, which lack a cell wall, the cytoskeleton is linked by integrin proteins to a web of glycoproteins called the extracellular matrix.

TABLE 4.3	A Comparison of Prokaryotic, Animal, and Plant Cells		
	Prokaryote	**Animal**	**Plant**
EXTERIOR STRUCTURES			
Cell wall	Present (protein-polysaccharide)	Absent	Present (cellulose)
Cell membrane	Present	Present	Present
Flagella/cilia	Flagella may be present	May be present (9 + 2 structure)	Absent except in sperm of a few species (9 + 2 structure)
INTERIOR STRUCTURES			
ER	Absent	Usually present	Usually present
Ribosomes	Present	Present	Present
Microtubules	Absent	Present	Present
Centrioles	Absent	Present	Absent
Golgi apparatus	Absent	Present	Present
Nucleus	Absent	Present	Present
Mitochondria	Absent	Present	Present
Chloroplasts	Absent	Absent	Present
Chromosomes	A single circle of DNA	Multiple; DNA–protein complex	Multiple; DNA–protein complex
Lysosomes	Absent	Usually present	Present
Vacuoles	Absent	Absent or small	Usually a large single vacuole

4.1 Cell Theory

Modern cell theory states that organisms are composed of one or more cells. Cells are the smallest unit of life and arise from preexisting cells.

- Cell size is constrained by the effective distance of diffusion within a cell, from the surface to the interior of the cell.

- As a cell increases in size the surface area increases as a square function and the volume increases as a cubic function.

- Large cells deal with the diffusion problem by having more than one nucleus or by becoming flattened or elongated.

- The visualization of cells and their components is facilitated by microscopes and staining cell structures.

- All cells have DNA, a cytoplasm, a plasma membrane, and ribosomes.

4.2 Prokaryotic Cells (figure 4.3)

Prokaryotic cells do not have a nucleus or an internal membrane system, and they lack membrane-bounded organelles.

- The plasma membrane is surrounded by a rigid cell wall that maintains shape and helps maintain osmotic balance.

- The plasma membrane in some prokaryotes is infolded and provides similar functions to eukaryotic internal membranes.

- Bacteria have a cell wall made up of peptidoglycan. Archaeal cell walls have different architecture.

- Archaeal plasma membranes differ from bacteria and eukaryotes.

- The plasma membrane in some archaea is a monolayer composed of saturated lipids attached to glycerol at each end.

- Structurally Archaea resemble prokaryotes, but functionally they more closely resemble eukaryotes.

- Prokaryotic flagella rotate because of proton transfer.

4.3 Eukaryotic Cells (figures 4.6 and 4.7)

Eukaryotic cells have a membrane-bounded nucleus, an endomembrane system, and many different organelles.

- The nucleus contains genetic information.

- The nuclear envelope consists of two phospholipid bilayers; the outer layer is contiguous with the ER.

- The inside of the nuclear envelope is covered with nuclear lamins, which maintain the shape of the nucleus.

- Nuclear pores allow exchange of small molecules between the nucleoplasm and the cytoplasm.

- DNA is organized with proteins into chromatin.

- The nucleolus is a region of the nucleoplasm where rRNA is transcribed and ribosomes are assembled.

- Ribosomes are composed of RNA and protein and use information in mRNA to direct the synthesis of proteins.

4.4 The Endomembrane System

The endomembrane system forms compartments and vesicles and provides channels to carry molecules and surfaces for synthesis of macromolecules.

- The endoplasmic reticulum (ER) creates channels and passages within the cytoplasm (figure 4.11).

- The interior compartment of the ER is called the cisternal space, or lumen.

- The rough endoplasmic reticulum (RER) has ribosomes on the surface and is composed mainly of flattened sacs. RER is involved in protein synthesis and modification.

- The smooth endoplasmic reticulum (SER) lacks ribosomes

and is composed more of tubules. SER is involved in synthesis of carbohydrates and lipids and in detoxification.

- The Golgi apparatus receives vesicles from the ER on the *cis* face, modifies and packages macromolecules, and transports them in vesicles formed on the *trans* face (figure 4.13).

- Lysosomes are vesicles containing enzymes that break down macromolecules located in food vacuoles and recycle the components of old organelles (figure 4.14).

- Microbodies contain enzymes and grow by incorporating lipids and proteins before they divide.

- Peroxisomes contain enzymes that catalyze oxidation reactions, resulting in the formation of hydrogen peroxide.

- Plants have many specialized vacuoles. The conspicuous central vacuole, surrounded by the tonoplast membrane, is used for storage, maintaining water balance, and growth.

4.5 Mitochondria and Chloroplasts: Cellular Generators

Mitochondria and chloroplasts have a double-membrane structure, contain their own DNA, can synthesize proteins, can divide, and are involved in energy metabolism.

- Mitochondria produce ATP using energy-contained macromolecules (figure 4.17).

- *The inner membrane is extensively folded into layers called cristae.*

- *The intermembrane space is a compartment between the inner and outer membrane.*

- *The mitochondrial matrix is a compartment consisting of the fluid within the inner membrane.*

- Chloroplasts use light to generate ATP and sugars (figure 4.18).

- *In addition to a double membrane, chloroplasts also have stacked membranes called grana that contain vesicles called thylakoids.*

- *The fluid stroma surrounds the thylakoids.*

- Evidence indicates that mitochondria and chloroplasts arose via endosymbiosis.

4.6 The Cytoskeleton

The cytoskeleton is composed of three different fibers that support cell shape and anchors organelles and enzymes (figure 4.20).

- Actin filaments, or microfilaments, are long, thin polymers responsible for cell movement, cytoplasmic division, and formation of cellular extensions.

- Microtubules are hollow structures that are used in cell movement and movement of materials within a cell.

- Intermediate filaments are stable structures that serve a wide variety of functions.

- Paired centrioles, located in the centrosome, help assemble the nuclear division apparatus of animal cells (figure 4.21).

- Molecular motors move vesicles along microtubules.

4.7 Extracellular Structures and Cell Movement

Extracellular structures provide protection, support, strength, and cell recognition.

- Plants have cell walls composed of cellulose fibers. Fungi have cell walls composed of chitin.

- Animals have a complex extracellular matrix.

- Cell crawling occurs as actin polymerization forces the cell membrane forward while myosin pulls the cell forward.

- Eukaryotic flagella have a 9 + 2 structure and arise from a basal body.

- Cilia are shorter and more numerous flagella.

SELF TEST

1. Which of the following statements is NOT part of the cell theory?
   a. All organisms are composed of one or more cells.
   b. Cells come from other cells by division.
   c. Cells are the smallest living things.
   d. Eukaryotic cells have evolved from prokaryotic cells.
2. The most important factor that limits the size of a cell is—
   a. the amount of proteins and organelles that can be made by a cell
   b. the rate of diffusion
   c. the surface-area-to-volume ratio of the cell
   d. the amount of DNA in the cell
3. What type of microscope would you use to examine the surface details of a cell?
   a. Compound light microscope
   b. Transmission electron microscope
   c. Scanning electron microscope
   d. Confocal microscope
4. All cells have all of the following except—
   a. Plasma membrane
   b. Genetic material
   c. Cytoplasm
   d. Cell wall
5. Eukaryotic cells are more complex than prokaryotic cells. Which of the following would you NOT find in a prokaryotic cell?
   a. Cell wall
   b. Plasma membrane
   c. Nucleus
   d. Ribosomes
6. The difference between a gram-positive and gram-negative bacteria is—
   a. the thickness of the peptidoglycan cell wall
   b. the type of polysaccharide present in the cell wall
   c. the type and amount of protein in the cell wall
   d. the layers of cellulose in the cell wall
7. Which of the following is not true of bacterial flagella?
   a. Bacterial flagella rotate producing a spiral wave.
   b. Bacterial flagella are anchored to a basal body.
   c. Bacterial flagella are powered by a proton gradient.
   d. Bacterial flagella are composed of microtubules.
8. All eukaryotic cells possess each of the following except—
   a. Mitochondria
   b. Cell wall
   c. Cytoskeleton
   d. Nucleus
9. Ribosomal RNAs are manufactured in which specific area of a eukaryotic cell?
   a. The nucleus
   b. The cytoplasm
   c. The nucleolus
   d. The chromatin
10. Which of these organelles is NOT associated with the production of proteins in a cell?
    a. Ribosomes
    b. Smooth endoplasmic reticulum (SER)
    c. Rough endoplasmic reticulum (RER)
    d. Golgi apparatus
11. Proteins can move from the Golgi apparatus to—
    a. the extracellular fluid
    b. transport vesicles
    c. lysosomes
    d. all of the above
12. Lysosomes function to—
    a. carry proteins to the surface of the cell
    b. add short-chain carbohydrates to make glycoproteins
    c. break down organelles, proteins, and nucleic acids
    d. remove electrons and hydrogen atoms from hydrogen peroxide
13. What do chloroplasts and mitochondria have in common?
    a. Both are present in animal cells.
    b. Both have an outer membrane and an elaborate inner membrane.
    c. Both are present in all eukaryotic cells.
    d. Both organelles function to produce glucose.
14. Eukaryotic cells are composed of three types of cytoskeletal filaments. How are these three filaments similar?
    a. They contribute to the shape of the cell.
    b. They are all made of the same type of protein.
    c. They are all the same size and shape.
    d. They are all equally dynamic and flexible.
15. Animal cells connect to the extracellular matrix through—
    a. glycoproteins
    b. fibronectins
    c. integrins
    d. collagen

CHALLENGE QUESTIONS

1. Eukaryotic cells are typically larger than prokaryotic cells (refer to figure 4.2). How might the difference in the cellular structure of eukaryotic versus a prokaryotic cell help to explain this observation?
2. The smooth endoplasmic reticulum is the site of synthesis of the phospholipids that make up all the membranes of a cell—especially the plasma membrane. Use the diagram of an animal cell (figure 4.6) to trace a pathway that would carry a phospholipid molecule from the SER to the plasma membrane. What endomembrane compartments would the phospholipids travel through? How can a phospholipid molecule move between membrane compartments?
3. Use the information provided in table 4.3 to develop a set of predictions about the properties of mitochondria and chloroplasts if these organelles were once free-living prokaryotic cells. How do your predictions match with the evidence for endosymbiosis?
4. In evolutionary theory, homologous traits are those with a similar structure and function derived from a common ancestor. Analogous traits represent adaptations to a similar environment, but from distantly related organisms. Consider the structure and function of the flagella found on eukaryotic and prokaryotic cells. Are the flagella an example of a homologous or analogous trait? Defend your answer.
5. The protist, *Giardia lamblia*, is the organism associated with water-borne diarrheal diseases. *Giardia* is an unusual eukaryote because it seems to lack mitochondria. Explain the existence of a mitochondria-less eukaryote in the context of the endosymbiotic theory.

Do you need additional review? *Visit www.ravenbiology.com for practice quizzes, animations, videos, and activities designed to help you master the material in this chapter.*

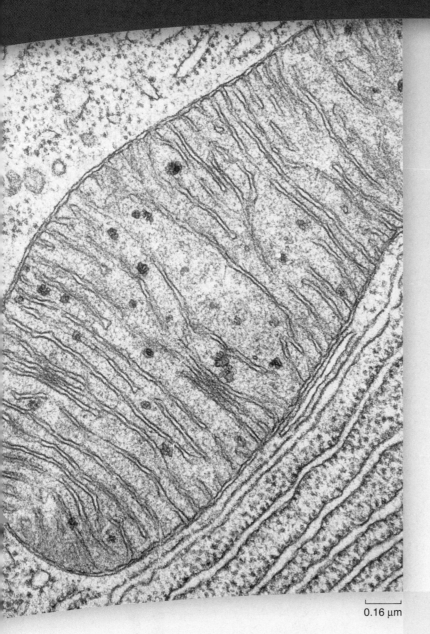

0.16 μm

chapter 5

Membranes

introduction

AMONG A CELL'S MOST IMPORTANT ACTIVITIES are its interactions with the environment, a give-and-take that never ceases. Without it, life could not persist. Living cells are encased within a lipid membrane through which few water-soluble substances can pass; but at the same time, the membrane contains protein passageways that permit specific substances to move into and out of the cell and allow the cell to exchange information with its environment. Eukaryotic cells also contain internal membranes like those of the mitochondrion and endoplasmic reticulum pictured here. We call the delicate skin of lipids with embedded protein molecules that encase the cell a **plasma membrane.** This chapter examines the structure and function of this remarkable membrane.

concept outline

5.1 The Structure of Membranes

- *The fluid mosaic model shows proteins embedded in a fluid lipid bilayer*
- *Cellular membranes consist of four component groups*
- *Electron microscopy has provided structural evidence*

5.2 Phospholipids: The Membrane's Foundation

- *Phospholipids spontaneously form bilayers*
- *The phospholipid bilayer is fluid*
- *Membrane fluidity can change*

5.3 Proteins: Multifunctional Components

- *Proteins and protein complexes perform key functions*
- *Structural features of membrane proteins*

5.4 Passive Transport Across Membranes

- *Transport can occur by simple diffusion*
- *Proteins allow membrane diffusion to be selective*
- *Osmosis is the movement of water across membranes*

5.5 Active Transport Across Membranes

- *Active transport uses energy to move materials against a concentration gradient*
- *The sodium–potassium pump runs directly on ATP*
- *Coupled transport uses ATP indirectly*

5.6 Bulk Transport by Endocytosis and Exocytosis

- *Bulk material enters the cell in vesicles*
- *Material can leave the cell by exocytosis*

5.1 The Structure of Membranes

The membranes that encase all living cells are sheets of lipid only two molecules thick; more than 10,000 of these sheets piled on one another would just equal the thickness of this sheet of paper. Biologists established the components of membranes—not only lipids, but also proteins and other molecules—through biochemical assays, but the nature of the membrane structure remained elusive.

We begin by considering the theories that have been advanced about membrane structure. We then look at the individual components of membranes more closely.

The fluid mosaic model shows proteins embedded in a fluid lipid bilayer

The lipid layer that forms the foundation of a cell's membranes is a bilayer formed of **phospholipids** (figure 5.1). For many years, biologists thought that the protein components of the cell membrane covered the inner and outer surfaces of the phospholipid bilayer like a coat of paint. An early model por-trayed the membrane as a sandwich; a phospholipid bilayer between two layers of globular protein. This model, however, was not consistent with what researchers were learning in the 1960s about the structure of membrane proteins.

Unlike most proteins found within cells, membrane proteins are not very soluble in water. These **globular proteins** possess long stretches of nonpolar hydrophobic amino acids. If such proteins indeed coated the surface of the lipid bilayer, then their nonpolar portions would separate the polar portions of the phospholipids from water, causing the bilayer to dissolve! This was clearly not the case, so the model required modification.

In 1972, S. Jonathan Singer and Garth J. Nicolson revised the model in a simple but profound way: They proposed that the globular proteins are *inserted* into the lipid bilayer, with their nonpolar segments in contact with the nonpolar interior of the bilayer and their polar portions protruding out from the membrane surface. In this model, called the **fluid mosaic model,** a mosaic of proteins floats in or on the fluid lipid bilayer like boats on a pond (figure 5.2).

figure 5.1

DIFFERENT VIEWS OF PHOSPHOLIPID STRUCTURE. Phospholipids are composed of glycerol (*pink*) linked to two fatty acids and a phosphate group. The phosphate group (*yellow*) can have additional molecules attached, such as the positively charged choline (*green*) shown. Phosphatidylcholine is a common component of membranes, it is shown in (*a*) with its chemical formula, (*b*) as a space-filling model and, (*c*) as the icon that is used in most of the figures in this chapter. The phosphate portion of the molecule is hydrophilic, and the fatty acid tails are hydrophobic. This allows them to associate into bilayers, with the hydrophobic tails in the middle, in water.

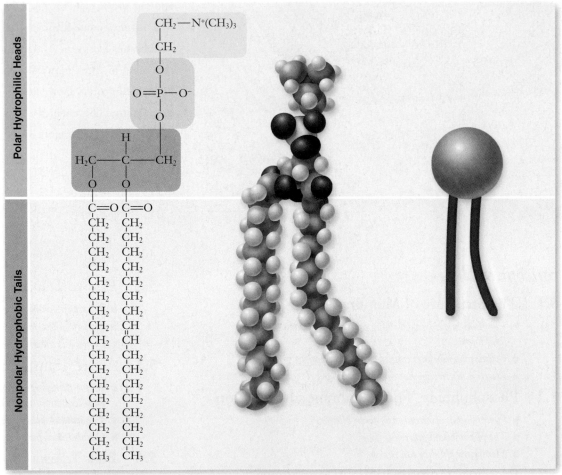

a. Formula   *b.* Space-filling model   *c.* Icon

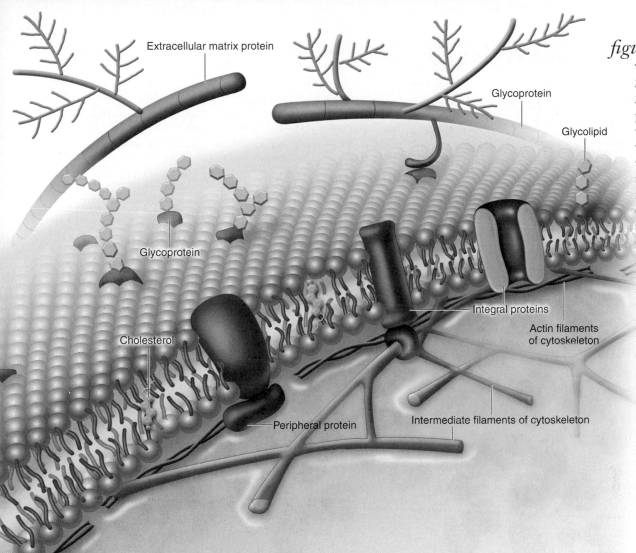

Extracellular matrix protein

Glycoprotein

Glycolipid

Glycoprotein

Integral proteins

Actin filaments
of cytoskeleton

Cholesterol

Peripheral protein

Intermediate filaments of cytoskeleton

figure 5.2

THE FLUID MOSAIC MODEL OF CELL MEMBRANES. Integral proteins protrude through the plasma membrane, with nonpolar regions that tether them to the membrane's nonpolar interior. Carbohydrate chains are often bound to the extracellular portion of these proteins forming glycoproteins. Peripheral membrane proteins are associated with the surface of the membrane. Membrane phospholipids can be modified by the addition of carbohydrates to form glycolipids. In addition, a variety of associated proteins are found inside and outside the cell. Inside the cell, actin filaments and intermediate filaments interact with membrane proteins. Outside the cell, many animal cells have an elaborate extracellular matrix composed primarily of glycoproteins.

Cellular membranes donsist of four component groups

A eukaryotic cell contains many membranes. Although they are not all identical, they share the same fundamental architecture. Cell membranes are assembled from four components (table 5.1):

1. **Phospholipid bilayer.** Every cell membrane is composed of phospholipids in a bilayer. The other components of the membrane are embedded within the bilayer, which provides a flexible matrix and, at the same time, imposes a barrier to permeability. Animal cell membranes also contain cholesterol, a steroid with a polar hydroxyl group (—OH).

2. **Transmembrane proteins.** A major component of every membrane is a collection of proteins that float in the lipid bilayer. These proteins provide passageways that allow substances and information to cross the membrane. Many membrane proteins are not fixed in position; they can move about, just as the phospholipid molecules do. Some membranes are crowded with proteins, but in others, the proteins are more sparsely distributed. Because these proteins are embedded in the membrane structure they are also called **integral membrane proteins.**

3. **Interior protein network.** Membranes are structurally supported by intracellular proteins that reinforce the membrane's shape. For example, a red blood cell has a characteristic biconcave shape because a scaffold made of a protein called spectrin links proteins in the plasma membrane with actin filaments in the cell's cytoskeleton.

   Membranes use networks of other proteins to control the lateral movements of some key membrane proteins, anchoring them to specific sites. Proteins that are associated with the membrane but not part of its structure are usually called **peripheral membrane proteins.**

4. **Cell surface markers.** As you learned in the preceding chapter, membrane sections assemble in the endoplasmic reticulum, transfer to the Golgi apparatus, and then are transported to the plasma membrane. The ER adds chains of sugar molecules to membrane proteins and lipids, converting them into **glycoproteins** and **glycolipids.** Different cell types exhibit different varieties of these glycoproteins and glycolipids, which act as cell identity markers, on their surfaces.

TABLE 5.1	Components of the Cell Membrane			
Component	**Composition**	**Function**	**How It Works**	**Example**
Phospholipid bilayer	Phospholipid molecules	Provides permeability barrier, matrix for proteins	Excludes water-soluble molecules from nonpolar interior of bilayer and cell	Bilayer of cell is impermeable to large water-soluble molecules, such as glucose
Transmembrane proteins	Carriers	Actively or passively transport molecules across membrane	Move specific molecules through the membrane in a series of conformational changes	Glycophorin carrier for sugar transport; sodium–potassium pump
	Channels	Passively transport molecules across membrane	Create a selective tunnel that acts as a passage through membrane	Sodium and potassium channels in nerve, heart, and muscle cells
	Receptors	Transmit information into cell	Signal molecules bind to cell surface portion of the receptor protein; this alters the portion of the receptor protein within the cell, inducing activity	Specific receptors bind peptide hormones and neurotransmitters
Interior protein network	Spectrins	Determine shape of cell	Form supporting scaffold beneath membrane, anchored to both membrane and cytoskeleton	Red blood cell
	Clathrins	Anchor certain proteins to specific sites, especially on the exterior plasma membrane in receptor-mediated endocytosis	Proteins line coated pits and facilitate binding to specific molecules	Localization of low-density lipoprotein receptor within coated pits
Cell surface markers	Glycoproteins	"Self" recognition	Create a protein/carbohydrate chain shape characteristic of individual	Major histocompatibility complex protein recognized by immune system
	Glycolipid	Tissue recognition	Create a lipid/carbohydrate chain shape characteristic of tissue	A, B, O blood group markers

Originally, it was believed that because of its fluidity, the plasma membrane was uniform, with lipids and proteins free to diffuse rapidly in the plane of the membrane. However, in the last decade evidence has accumulated suggesting the plasma membrane is not homogeneous and contains microdomains with distinct lipid and protein composition. One type of microdomain, the **lipid raft,** is heavily enriched in cholesterol, which fills space between the phospholipids, making them more tightly packed than the surrounding membrane.

Lipid rafts appear to be involved in many important biological processes, including signal reception and cell movement. The structural proteins of replicating HIV viruses are targeted to lipid raft regions of the plasma membrane during virus assembly within infected cells.

Electron microscopy has provided structural evidence

Electron microscopy allows biologists to examine the delicate, filmy structure of a cell membrane. We discussed two types of electron microscopes in chapter 4: the transmission electron microscope (TEM) and the scanning electron microscope (SEM). Both provide illuminating views of membrane structure.

When examining cell membranes with electron microscopy, specimens must be prepared for viewing. In one method of preparing a specimen, the tissue of choice is embedded in a hard epoxy matrix. The epoxy block is then cut with a microtome, a machine with a very sharp blade that makes incredibly thin, transparent "epoxy shavings" less than 1 μm thick that peel away from the block of tissue.

These shavings are placed on a grid, and a beam of electrons is directed through the grid with the TEM. At the high magnification an electron microscope provides, resolution is good enough to reveal the double layers of a membrane. False color can be added to the micrograph to enhance detail.

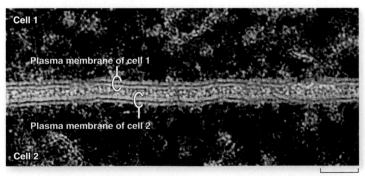

.038 μm

Freeze-fracturing a specimen is another way to visualize the inside of the membrane (figure 5.3). The tissue is embedded in a medium and quick frozen with liquid nitrogen. The frozen tissue is then "tapped" with a knife,

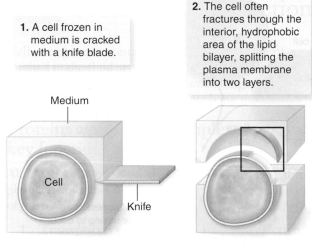

1. A cell frozen in medium is cracked with a knife blade.

Medium

Cell

Knife

2. The cell often fractures through the interior, hydrophobic area of the lipid bilayer, splitting the plasma membrane into two layers.

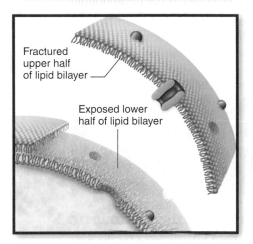

3. The plasma membrane separates such that proteins and other embedded membrane structures remain within one or the other layers of the membrane.

Fractured upper half of lipid bilayer

Exposed lower half of lipid bilayer

4. The exposed membrane is coated with platinum, which forms a replica of the membrane. The underlying membrane is dissolved away, and the replica is then viewed with electron microscopy.

0.15 μm

Exposed lower half of lipid bilayer

External surface of plasma membrane

figure 5.3

VIEWING A PLASMA MEMBRANE WITH FREEZE-FRACTURE MICROSCOPY.

causing a crack between the phospholipid layers of membranes. Proteins, carbohydrates, pits, pores, channels, or any other structure affiliated with the membrane will pull apart (whole, usually) and stick with one or the other side of the split membrane.

Next, a very thin coating of platinum is evaporated onto the fractured surface, forming a replica or "cast" of the surface. After the topography of the membrane has been preserved in the cast, the actual tissue is dissolved away, and the cast is examined with electron microscopy, creating a strikingly different view of the membrane.

Cellular membranes contain four components: (1) a phospholipid bilayer, (2) transmembrane proteins, (3) an internal protein network providing structural support, and (4) cell surface markers composed of glycoproteins and glycolipids.

From the results of electron microscopy and molecular biology techniques, we have gained a much clearer view of the structure of membranes and the interactions of membrane components. The modern view of the membrane is called the fluid mosaic model. This refers to the fluid nature of the membrane and that it is a mosaic of phospholipids and proteins floating in the phospholipid bilayer.

5.2 Phospholipids: The Membrane's Foundation

Like the fat molecules (triglycerides) described in chapter 3, a phospholipid has a backbone derived from the three-carbon polyalcohol *glycerol*. Attached to this backbone are one to three fatty acids, long chains of carbon atoms ending in a carboxyl (—COOH) group. A triglyceride molecule has three such chains, one attached to each carbon in the backbone; because these chains are nonpolar, they do not form hydrogen bonds with water, and triglycerides are not water-soluble.

A phospholipid, by contrast, has only two fatty acid chains attached to its backbone. The third carbon of the glycerol carries a phosphate group, thus *phospho*lipid. An additional polar organic molecule is often added to the phosphate group as well.

From this simple molecular framework, a large variety of lipids can be constructed by varying the polar organic group attached to the phosphate and the fatty acid chains attached to the glycerol. Mammalian membranes, for example, contain hundreds of chemically distinct species of lipids.

Phospholipids spontaneously form bilayers

The phosphate groups are charged, and other molecules attached to them are polar or charged. This creates a huge change in the molecule's physical properties compared with a triglyceride. The strongly polar phosphate end is hydrophilic, or "water-loving," while the fatty acid end is strongly nonpolar and hydrophobic, or "water-fearing." The two nonpolar fatty acids extend in one direction, roughly parallel to each other, and the polar phosphate group points in the other direction. To represent this structure, phospholipids are often diagrammed as a polar head with two dangling nonpolar tails, as in figure 5.1*c*.

What happens when a collection of phospholipid molecules is placed in water? The polar water molecules repel the long, nonpolar tails of the phospholipids while seeking partners for hydrogen bonding. Because of the polar nature of the water molecules, the nonpolar tails of the phospholipids end

The anchoring of proteins in the bilayer

Many membrane proteins are attached to the surface of the membrane by special molecules that associate strongly with phospholipids. Like a ship tied to a floating dock, these anchored proteins (peripheral proteins) are free to move about on the surface of the membrane tethered to a phospholipid. The anchoring molecules are modified lipids that have (1) nonpolar regions that insert into the internal portion of the lipid bilayer and (2) chemical bonding domains that link directly to proteins.

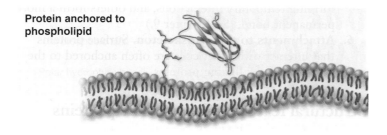

Protein anchored to phospholipid

In contrast, other proteins actually span the lipid bilayer (integral membrane proteins). The part of the protein that extends through the lipid bilayer, in contact with the nonpolar interior, consists of nonpolar amino acid helices or β-pleated sheets (see chapter 3). Because water avoids nonpolar amino acids, these portions of the protein are held within the interior of the lipid bilayer. The polar ends protrude from both sides of the membrane. Any movement of the protein out of the membrane, in either direction, brings the nonpolar regions of the protein into contact with water, which "shoves" the protein back into the interior. These forces prevent the transmembrane proteins from simply popping out of the membrane and floating away.

Transmembrane domains

Cell membranes contain a variety of different transmembrane proteins, which differ in the way they traverse the lipid bilayer. The primary difference lies in the number of times that the protein crosses the membrane. Each membrane spanning region is called a **transmembrane domain.** These domains are composed of hydrophobic amino acids usually arranged into α-helices (figure 5.6).

Proteins need only a single transmembrane domain to be anchored in the membrane, but they often have more than one such domain. An example of a protein with a single transmembrane domain is the linking protein that attaches the spectrin network of the cytoskeleton to the interior of the plasma membrane.

Biologists classify some types of receptors based on the number of transmembrane domains they have, such as G protein–coupled signal receptors with seven membrane-spanning domains (chapter 9). These receptors respond to external molecules, such as epinephrine, and initiate a cascade of events inside the cell.

Another example is bacteriorhodopsin, one of the key transmembrane proteins that carries out photosynthesis in halophilic (salt-loving) archaea. It contains seven nonpolar helical segments that traverse the membrane, forming a structure within the membrane through which protons pass during the light-driven pumping of protons (figure 5.7).

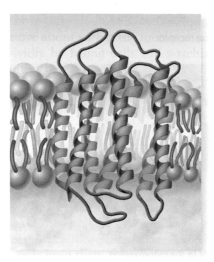 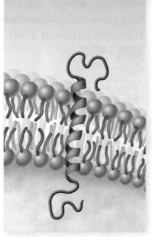

a.   *b.*

figure 5.6

TRANSMEMBRANE DOMAINS. Integral membrane proteins have at least one hydrophobic transmembrane domain (shown in blue) to anchor them in the membrane. *a.* Receptor protein with seven transmembrane domains. *b.* Protein with single transmembrane domain.

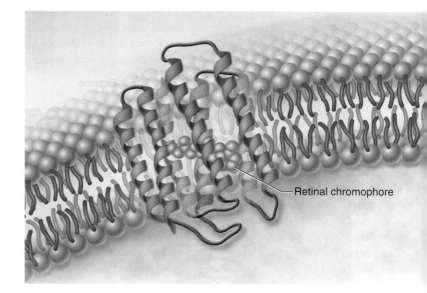

Retinal chromophore

figure 5.7

BACTERIORHODOPSIN. This transmembrane protein mediates photosynthesis in the archaean *Halobacterium salinarium.* The protein traverses the membrane seven times with hydrophobic helical strands that are within the hydrophobic center of the lipid bilayer. The helical regions form a structure across the bilayer through which protons are pumped by the retinal chromophore (*green*) using energy from light.

Pores

Some transmembrane proteins have extensive nonpolar regions with secondary configurations of β-pleated sheets instead of α-helices (chapter 3). The β-sheets form a characteristic motif, folding back and forth in a cylinder so the sheets come to be arranged like a pipe through the membrane. This forms a polar environment in the interior of the β-sheets spanning the membrane. This so-called *β-barrel*, open on both ends, is a common feature of the porin class of proteins that are found within the outer membrane of some bacteria, where they allow molecules to pass through the membrane (figure 5.8).

Proteins in the membrane have a variety of functions and confer the main differences between membranes of different cells. These functions include transport, enzymatic functions, reception of extracellular signals, cell-to-cell interactions, and cell identity markers.

Proteins that are embedded in the membrane have one or more hydrophobic regions, called transmembrane domains, that anchor them in the membrane.

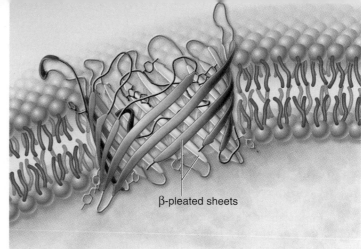

β-pleated sheets

figure 5.8

A PORE PROTEIN. The bacterial transmembrane protein porin creates large open tunnels called pores in the outer membrane of a bacterium. Sixteen strands of β-pleated sheets run antiparallel to one another, creating a so-called β-barrel in the bacterial outer cell membrane. The tunnel allows water and other materials to pass through the membrane.

inquiry

? *Based only on amino acid sequence, how would you recognize an integral membrane protein?*

5.4 Passive Transport Across Membranes

Many substances can move in and out of the cell without the cell's having to expend energy. This type of movement is termed **passive transport.** Some ions and molecules can pass through the membrane fairly easily and do so because of a **concentration gradient**—a difference between the concentration on the inside of the membrane and that on the outside. Some substances also move in response to a gradient, but do so through specific channels formed by proteins in the membrane.

Transport can occur by simple diffusion

Molecules and ions dissolved in water are in constant random motion. This random motion causes a net movement of these substances from regions of high concentration to regions of lower concentration, a process called **diffusion** (figure 5.9).

Net movement driven by diffusion will continue until the concentration is the same in all regions. Consider adding a drop of colored ink to a bowl of water; what happens to the ink? Over time the ink becomes dispersed throughout the solution. This is due to diffusion of the ink molecules. In the context of cells, we are usually concerned with differences in concentration of molecules across the plasma membrane. We need to consider the relative concentrations both inside and outside the cell, as well as how readily a molecule can cross the membrane.

The major barrier to crossing a biological membrane is the hydrophobic interior that repels polar molecules but not nonpolar molecules. If a concentration difference exists for a nonpolar molecule, it will move across the membrane until the concentration is equal on both sides. At this point, movement

a.   b.   c.   d.

figure 5.9

DIFFUSION. If a drop of colored ink is dropped into a beaker of water (*a*) its molecules dissolve (*b*) and diffuse (*c*). Eventually, diffusion results in an even distribution of ink molecules throughout the water (*d*).

in both directions still occurs, but there is no net movement in either direction. This includes molecules like O_2 and nonpolar organic molecules such as steroid hormones.

The plasma membrane has limited permeability to small polar molecules and very limited permeability to larger polar molecules and ions. The movement of water, one of the most important polar molecules, is discussed in its own section later on.

Proteins allow membrane diffusion to be selective

Many important molecules required by cells cannot easily cross the plasma membrane. These molecules can still enter the cell by diffusion through specific channel proteins or carrier proteins embedded in the plasma membrane, provided there is a higher concentration of the molecule outside the cell than inside. **Channel proteins** have a hydrophilic interior that provides an aqueous channel through which polar molecules can pass when the channel is open. **Carrier proteins,** in contrast to channels, bind specifically to the molecule they assist, much like an enzyme binds to its substrate. These channels and carriers are usually selective for one type of molecule, and thus the cell membrane is said to be **selectively permeable.**

Diffusion of ions through channels

You saw in chapter 2 that atoms with an unequal number of protons and electrons have electric charge and are called ions. Those that carry positive charge are called *cations* and those that carry negative charge are called *anions.*

Because of their charge, ions interact well with polar molecules such as water, but are repelled by nonpolar molecules such as the interior of the plasma membrane. Therefore, ions cannot move between the cytoplasm of a cell and the extracellular fluid without the assistance of membrane transport proteins.

Ion channels possess a hydrated interior that spans the membrane. Ions can diffuse through the channel in either direction, depending on their relative concentration across the membrane (figure 5.10). Some channel proteins can be opened or closed in response to a stimulus. These channels are called **gated channels** and depending on the nature of the channel, the stimulus can be either chemical or electrical.

Three conditions determine the direction of net movement of the ions: (1) their relative concentrations on either side of the membrane, (2) the voltage difference across the membrane and for gated channels, (3) the state of the gate (open or closed). A voltage difference is an electrical potential difference across the membrane called a **membrane potential.** Changes in membrane potential form the basis for transmission of signals in the nervous system and some other tissues. (We discuss this topic in detail in chapter 44.) Each type of channel is specific for a particular ion, such as calcium (Ca^{2+}), sodium (Na^+), potassium (K^+), or chloride (Cl^-), or in some cases, for more than one cation or anion. Ion channels play an essential role in signaling by the nervous system.

Carrier proteins and facilitated diffusion

Carrier proteins can help transport both ions and other solutes, such as some sugars and amino acids, across the membrane. Transport through a carrier is still a form of diffusion and therefore requires a concentration difference across the membrane (figure 5.11). Because this process is facilitated by a carrier protein, it is often called **facilitated diffusion.**

Carriers must bind to the molecule they transport, so the relationship between concentration and rate of transport differs from that due to simple diffusion. As concentration increases, transport by simple diffusion shows a linear increase in rate of transport. But when a carrier protein is involved, a concentration increase means that more of the carriers are bound to the transported molecule. At high enough concentrations all carriers will be occupied, and the

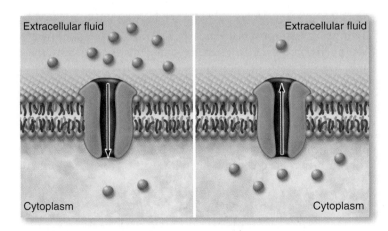

figure 5.10

ION CHANNELS. The movement of ions through a channel is shown. In the panel on the left the concentration is higher outside the cell, so the ions move into the cell. In the panel on the right the situation is reversed. In both cases, transport will continue until the concentration is equal on both sides of the membrane. At this point, ions continue to cross the membrane in both directions, but there is no net movement in either direction.

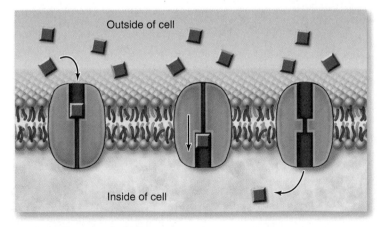

figure 5.11

FACILITATED DIFFUSION IS A CARRIER-MEDIATED TRANSPORT PROCESS. Molecules bind to a carrier protein on the extracellular side of the cell and pass through the plasma membrane via a conformational change in the carrier protein. This will only occur when there is a higher concentration outside the cell.

rate of transport will be constant. This means that the carrier exhibits **saturation.**

This situation is somewhat like that of a stadium (the cell) where a crowd must pass through turnstiles to enter. If there are unoccupied turnstiles, you can go right through, but when all are occupied, you must wait. When ticket holders are passing through the gates at maximum speed, the rate at which they enter cannot increase, no matter how many are waiting outside.

Facilitated diffusion provides the cell with a mechanism to prevent the buildup of unwanted molecules within the cell or to take up needed molecules that may be present outside the cell in high concentrations. Facilitated diffusion has three essential characteristics:

1. **It is specific.** A given carrier transports only certain molecules or ions.
2. **It is passive.** The direction of net movement is determined by the relative concentrations of the transported substance inside and outside the cell. The direction is always from high concentration to low concentration.
3. **It saturates.** If all relevant protein carriers are in use, increases in the concentration gradient do not increase the rate of movement.

Facilitated diffusion in red blood cells

Several examples of facilitated diffusion can be found in the plasma membrane of vertebrate red blood cells (RBCs). One RBC carrier protein, for example, transports a different molecule in each direction: chloride ion (Cl^-) in one direction and bicarbonate ion (HCO_3^-) in the opposite direction. As you will learn in chapter 49, this carrier is important in the uptake and release of carbon dioxide.

The glucose transporter is a second vital facilitated diffusion carrier in RBCs. Red blood cells keep their internal concentration of glucose low through a chemical trick: They immediately add a phosphate group to any entering glucose molecule, converting it to a highly charged glucose phosphate that can no longer bind to the glucose transporter, and therefore cannot pass back across the membrane. This maintains a steep concentration gradient for unphosphorylated glucose, favoring its entry into the cell.

The glucose transporter that assists the entry of glucose into the cell does not appear to form a channel in the membrane. Instead, this transmembrane protein appears to bind to a glucose molecule and then to flip its shape, dragging the glucose through the bilayer and releasing it on the inside of the plasma membrane. After it releases the glucose, the transporter reverts to its original shape and is then available to bind the next glucose molecule that comes along outside the cell.

Osmosis is the movement of water across membranes

The cytoplasm of a cell contains ions and molecules, such as sugars and amino acids, dissolved in water. The mixture of these substances and water is called an *aqueous solution.* Water is termed the **solvent,** and the substances dissolved in the water are **solutes.** Both water and solutes tend to diffuse from regions of high concentration to ones of low concentration; that is, they diffuse down their concentration gradients.

When two regions are separated by a membrane, what happens depends on whether the solutes can pass freely through that membrane. Most solutes, including ions and sugars, are not lipid-soluble and, therefore, are unable to cross the lipid bilayer. The concentration gradient of these solutes can lead to the movement of water.

Osmosis

Water molecules interact with dissolved solutes by forming hydration shells around the charged solute molecules. When a membrane separates two solutions having different concentrations of solutes, different concentrations of *free* water molecules exist on the two sides of the membrane. The side with higher solute concentration has tied up more water molecules in hydration shells and thus has fewer free water molecules.

As a consequence of this difference, free water molecules move down their concentration gradient, toward the higher solute concentration. This net diffusion of water across a membrane toward a higher solute concentration is called **osmosis** (figure 5.12).

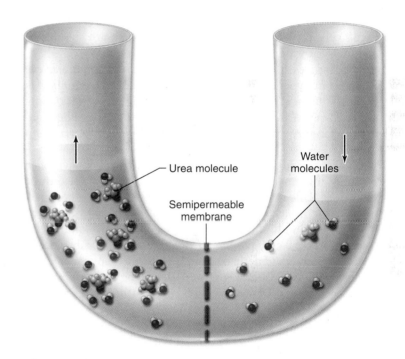

figure 5.12

OSMOSIS. Concentration differences in charged or polar molecules that cannot cross a semipermeable membrane will result in movement of water, which can cross the membrane. Water molecules form hydrogen bonds with charged or polar molecules forming a hydration shell around them in solution. A higher concentration of polar molecules (urea) shown on the left side of the membrane leads to water molecules gathering around each urea molecule. These water molecules are no longer free to diffuse across the membrane. The polar solute has reduced the concentration of free water molecules, creating a gradient. This causes a net movement of water by diffusion from right to left in the U-tube, raising the level on the left and lowering the level on the right.

The concentration of *all* solutes in a solution determines the **osmotic concentration** of the solution. If two solutions have unequal osmotic concentrations, the solution with the higher concentration is **hypertonic** (Greek *hyper,* "more than"), and the solution with the lower concentration is **hypotonic** (Greek *hypo,* "less than"). When two solutions have the same osmotic concentration, the solutions are **isotonic** (Greek *iso,* "the same"). The terms *hyperosmotic, hypoosmotic,* and *isosmotic* are also used to describe these conditions.

A cell in any environment can be thought of as a plasma membrane separating two solutions: the cytoplasm and the extracellular fluid. The direction and extent of any diffusion of water across the plasma membrane is determined by comparing the osmotic strength of these solutions. Put another way, water diffuses out of a cell in a hypertonic solution (that is, the cytoplasm of the cell is hypotonic, compared with the extracellular fluid). This loss of water causes the cell to shrink until the osmotic concentrations of the cytoplasm and the extracellular fluid become equal.

Aquaporins: water channels

The transport of water across the membrane is complex. Studies on artificial membranes show that water, despite its polarity, can cross the membrane, but this flow is limited. Water flow in living cells is facilitated by **aquaporins,** which are specialized channels for water.

A simple experiment demonstrates this. If an amphibian egg is placed in hypotonic spring water (the solute concentration in the cell is higher than that of the surrounding water), it does not swell. If aquaporin mRNA is then injected into the egg, the channel proteins are expressed and appear in the egg's plasma membrane. Water will now diffuse into the egg, causing it to swell.

More than 11 different kinds of aquaporins have been found in mammals. These fall into two general classes: those that are specific for only water, and those that allow other small hydrophilic molecules, such as glycerol or urea, to cross the membrane as well. This latter class explains how some membranes allow the easy passage of small hydrophilic substances.

The human genetic disease, hereditary (nephrogenic) diabetes insipidus (NDI), has been shown to be caused by a nonfunctional aquaporin protein. This disease causes the excretion of large volumes of dilute urine, illustrating the importance of aquaporins to our physiology.

Osmotic pressure

What happens to a cell in a hypotonic solution? (That is, the cell's cytoplasm is hypertonic relative to the extracellular fluid.) In this situation, water diffuses into the cell from the extracellular fluid, causing the cell to swell. The pressure of the cytoplasm pushing out against the cell membrane, or **hydrostatic pressure,** increases. The amount of water that enters the cell depends on the difference in solute concentration between the cell and the extracellular fluid. This is measured as **osmotic pressure,** defined as the force needed to stop osmotic flow.

If the membrane is strong enough, the cell reaches an equilibrium, at which the osmotic pressure, which tends to drive water into the cell, is exactly counterbalanced by the hy-

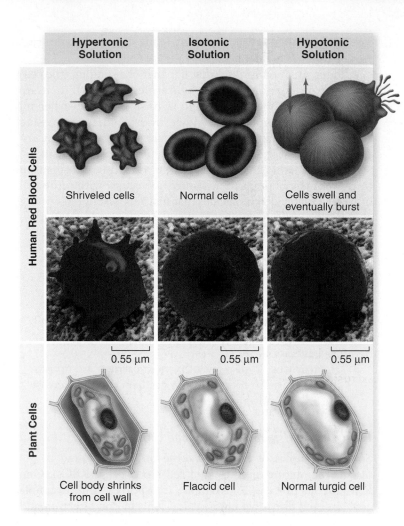

figure 5.13

HOW SOLUTES CREATE OSMOTIC PRESSURE. In a hypertonic solution, water moves out of the cell causing the cell to shrivel. In an isotonic solution, water diffuses into and out of the cell at the same rate, with no change in cell size. In a hypotonic solution, water moves into the cell. Direction and amount of water movement is shown with blue arrows (top). As water enters the cell from a hypotonic solution, pressure is applied to the plasma membrane until the cell ruptures. Water enters the cell due to osmotic pressure from the higher solute concentration in the cell. Osmotic pressure is measured as the force needed to stop osmosis. The strong cell wall of plant cells can withstand the hydrostatic pressure to keep the cell from rupturing. This is not the case with animal cells.

drostatic pressure, which tends to drive water back out of the cell. However, a plasma membrane by itself cannot withstand large internal pressures, and an isolated cell under such conditions would burst like an overinflated balloon (figure 5.13).

Accordingly, it is important for animal cells, which only have plasma membranes, to maintain osmotic balance. In contrast, the cells of prokaryotes, fungi, plants, and many protists are surrounded by strong cell walls, which can withstand high internal pressures without bursting.

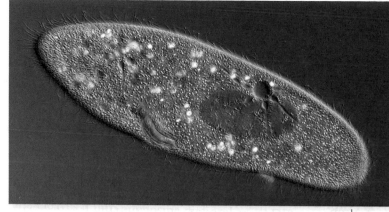

figure 5.14

CONTRACTILE VACUOLE. A micrograph of *Paramecium caudatum* with a large contractile vacuole visible near the center of the cell. The vacuole can contract to expel water. This helps the cell maintain osmotic balance in hypotonic solutions where water enters the cell by osmosis.

25 μm

Maintaining osmotic balance

Organisms have developed many strategies for solving the dilemma posed by being hypertonic to their environment and therefore having a steady influx of water by osmosis.

Extrusion. Some single-celled eukaryotes, such as the protist *Paramecium*, use organelles called contractile vacuoles to remove water. Each vacuole collects water from various parts of the cytoplasm and transports it to the central part of the vacuole, near the cell surface. The vacuole possesses a small pore that opens to the outside of the cell. By contracting rhythmically, the vacuole pumps out (extrudes) through this pore the water that is continuously drawn into the cell by osmotic forces (figure 5.14).

Isosmotic Regulation. Some organisms that live in the ocean adjust their internal concentration of solutes to match that of the surrounding seawater. Because they are isosmotic with respect to their environment, no net flow of water occurs into or out of these cells.

Many terrestrial animals solve the problem in a similar way, by circulating a fluid through their bodies that bathes cells in an isotonic solution. The blood in your body, for example, contains a high concentration of the protein albumin, which elevates the solute concentration of the blood to match that of your cells' cytoplasm.

Turgor. Most plant cells are hypertonic to their immediate environment, containing a high concentration of solutes in their central vacuoles. The resulting internal hydrostatic pressure, known as **turgor pressure,** presses the plasma membrane firmly against the interior of the cell wall, making the cell rigid. Most green plants depend on turgor pressure to maintain their shape, and thus they wilt when they lack sufficient water.

Passive transport involves diffusion, which requires a concentration gradient. Diffusion will only cause net movement of molecules until equilibrium is reached, when concentrations on both sides are equal.

Diffusion can occur directly through the membrane for hydrophobic molecules, or through channels for ions. Facilitated diffusion occurs through carrier proteins that bind to the molecule being transported. The selective nature of both channel proteins and carrier proteins gives the membrane selective permeability.

Water is transported in response to concentration differences inside and outside the cell of solutes that cannot cross the membrane. This process is called osmosis. The transport of water occurs through the membrane but is aided by channel proteins called aquaporins.

5.5 Active Transport Across Membranes

While diffusion, facilitated diffusion, and osmosis are passive transport processes that move materials down their concentration gradients, cells can also move substances across a cell membrane *up* their concentration gradients. This process requires the expenditure of energy, typically from ATP, and is therefore called **active transport.**

Active transport uses energy to move materials against a concentration gradient

Like facilitated diffusion, active transport involves highly selective protein carriers within the membrane that bind to the transported substance, which could be an ion or a simple

Stage One: Glycolysis The first stage is a 10-reaction biochemical pathway called *glycolysis* that produces ATP by substrate-level phosphorylation. The enzymes that catalyze the glycolytic reactions are in the cytoplasm of the cell, not associated with any membrane or organelle.

For each glucose molecule, two ATP molecules are used up early in the pathway, and four ATP molecules are produced by substrate-level phosphorylation. The net yield is two ATP molecules for each molecule of glucose catabolized. In addition, four electrons are harvested from the chemical bonds of glucose and carried by NADH for oxidative phosphorylation. Glycolysis yields two energy-rich **pyruvate** molecules for each glucose entering the pathway. This remaining energy can be harvested in later stages.

Stage Two: Pyruvate Oxidation In the second stage, pyruvate is converted into carbon dioxide and a two-carbon molecule called **acetyl-CoA.** For each molecule of pyruvate converted, one molecule of NAD^+ is reduced to NADH, again to carry electrons that can be used to make ATP. Remember that two pyruvate molecules result from each glucose.

Stage Three: The Krebs Cycle The third stage introduces acetyl-CoA into a cycle of nine reactions called the **Krebs cycle,** named after the German biochemist Hans Krebs, who discovered it. The Krebs cycle is also called the *citric acid cycle*, for the citric acid, or citrate, formed in its first step, and less commonly, the *tricarboxylic acid cycle*, because citrate has three carboxyl groups.

For each turn of the Krebs cycle, one ATP molecule is produced by substrate-level phosphorylation, and a large number of electrons are removed by the reduction of NAD^+ to NADH and FAD to $FADH_2$. Each glucose provides two acetyl-CoA to the Krebs cycle allowing two turns.

Stage Four: Electron Transport Chain and Chemiosmosis In the fourth stage, energetic electrons carried by NADH are transferred to a series of electron carriers that progressively extract the electrons' energy and use it to pump protons across a membrane.

The proton gradient created by electron transport is used by ATP synthase to produce ATP. This utilization of a proton gradient to drive the synthesis of ATP is called **chemiosmosis** and is the basis for oxidative phosphorylation.

Pyruvate oxidation, the reactions of the Krebs cycle, and ATP production by electron transport chains occur within many forms of prokaryotes and inside the mitochondria of all eukaryotes. Figure 7.5 provides an overview of the complete process of aerobic respiration beginning with glycolysis.

> The oxidation of glucose can be broken down into stages. These include glycolysis, which produces pyruvate, the oxidation of pyruvate, and the Krebs cycle. The electrons derived from oxidation reactions are used in the electron transport chain to produce a proton gradient that can be used by the enzyme ATP synthase to make ATP in the process called chemiosmosis.

7.3 Glycolysis: Splitting Glucose

Glucose molecules can be dismantled in many ways, but primitive organisms evolved a glucose-catabolizing process that releases enough free energy to drive the synthesis of ATP in enzyme-coupled reactions. Glycolysis occurs in the cytoplasm and converts glucose into two 3-carbon molecules of pyruvate (figure 7.6). For each molecule of glucose that passes through this transformation, the cell nets two ATP molecules.

Priming changes glucose into an easily cleaved form

The first half of glycolysis consists of five sequential reactions that convert one molecule of glucose into two molecules of the 3-carbon compound **glyceraldehyde 3-phosphate (G3P).** These reactions require the expenditure of ATP, so they are an endergonic process.

Step A: Glucose priming Three reactions "prime" glucose by changing it into a compound that can be cleaved readily into two 3-carbon phosphorylated molecules. Two of these reactions transfer a phosphate from ATP, so this step requires the cell to use two ATP molecules.

Step B: Cleavage and rearrangement In the first of the remaining pair of reactions, the 6-carbon product of step A is split into two 3-carbon molecules. One is G3P, and the other is then converted to G3P by the second reaction (figure 7.7).

ATP is synthesized by substrate-level phosphorylation

In the second half of glycolysis, five more reactions convert G3P into pyruvate in an energy-yielding process that generates ATP.

Step C: Oxidation Two electrons (and one proton) are transferred from G3P to NAD^+, forming NADH. A molecule of P_i is also added to G3P to produce 1,3-bisphosphoglycerate. The phosphate incorporated will later be transferred to ADP by substrate-level phosphorylation to allow a net yield of ATP.

Step D: ATP generation Four reactions convert 1,3-bisphosphoglycerate into pyruvate. This process generates two ATP molecules per G3P (see figures 7.4 and 7.7) produced in Step B.

mol
carr
type
port
two
two
used

of an
subst
tions
enabl
the e:

direct
to mc

5. Bindin
depho

figure

THE
(Na⁺)
K⁺ are
Na⁺ a:

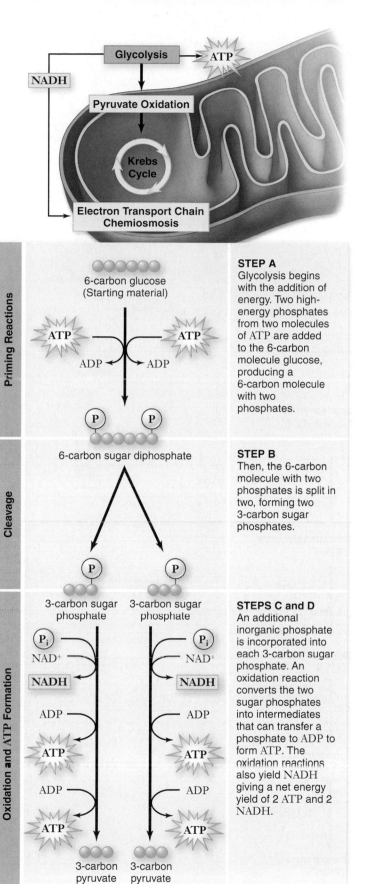

figure 7.6
HOW GLYCOLYSIS WORKS.

Priming Reactions

6-carbon glucose
(Starting material)

ATP → ADP
ATP → ADP

STEP A
Glycolysis begins with the addition of energy. Two high-energy phosphates from two molecules of ATP are added to the 6-carbon molecule glucose, producing a 6-carbon molecule with two phosphates.

P   P
6-carbon sugar diphosphate

Cleavage

STEP B
Then, the 6-carbon molecule with two phosphates is split in two, forming two 3-carbon sugar phosphates.

P   P
3-carbon sugar phosphate   3-carbon sugar phosphate

Oxidation and ATP Formation

P_i P_i
NAD^+ NAD^+
NADH   NADH
ADP   ADP
ATP   ATP
ADP   ADP
ATP   ATP

STEPS C and D
An additional inorganic phosphate is incorporated into each 3-carbon sugar phosphate. An oxidation reaction converts the two sugar phosphates into intermediates that can transfer a phosphate to ADP to form ATP. The oxidation reactions also yield NADH giving a net energy yield of 2 ATP and 2 NADH.

3-carbon pyruvate   3-carbon pyruvate

Because each glucose molecule is split into two G3P molecules, the overall reaction sequence has a net yield of two molecules of ATP, as well as two molecules of NADH and two of pyruvate:

4 ATP (2 ATP for each of the 2 G3P molecules in step D)
− 2 ATP (used in the two reactions in step A)

2 ATP (net yield for entire process)

The hydrolysis of one molecule of ATP yields a ΔG of −7.3 kcal/mol under standard conditions. Thus cells harvest a maximum of 14.6 kcal of energy per mole of glucose from glycolysis.

A brief history of glycolysis

Although far from ideal in terms of the amount of energy it releases, glycolysis does generate ATP. For more than a billion years during the anaerobic first stages of life on Earth, glycolysis was the primary way heterotrophic organisms generated ATP from organic molecules.

Like many biochemical pathways, glycolysis is believed to have evolved backward, with the last steps in the process being the most ancient. Thus, the second half of glycolysis, the ATP-yielding breakdown of G3P, may have been the original process. The synthesis of G3P from glucose would have appeared later, perhaps when alternative sources of G3P were depleted.

Why does glycolysis take place in modern organisms, since its energy yield in the absence of oxygen is comparatively little? The answer is that evolution is an incremental process: Change occurs by improving on past successes. In catabolic metabolism, glycolysis satisfied the one essential evolutionary criterion—it was an improvement. Cells that could not carry out glycolysis were at a competitive disadvantage, and only cells capable of glycolysis survived. Later improvements in catabolic metabolism built on this success. Metabolism evolved as one layer of reactions added to another. Nearly every present-day organism carries out glycolysis, as a metabolic memory of its evolutionary past.

The last section of this chapter discusses the evolution of metabolism in more detail.

NADH must be recycled to continue respiration

Inspect for a moment the net reaction of the glycolytic sequence:

$$\text{glucose} + 2\ ADP + 2\ P_i + 2\ NAD^+ \longrightarrow 2\ \text{pyruvate} + 2\ ATP + 2\ NADH + 2\ H^+ + 2\ H_2O$$

You can see that three changes occur in glycolysis: (1) Glucose is converted into two molecules of pyruvate; (2) two molecules of ADP are converted into ATP via substrate-level phosphorylation; and (3) two molecules of NAD^+ are reduced to NADH. This leaves the cell with two problems: extracting the energy that remains in the two pyruvate molecules, and regenerating NAD^+ to be able to continue glycolysis.

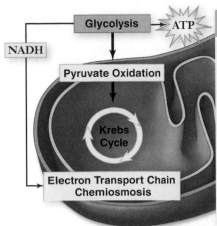

Glycolysis → ATP

NADH

Pyruvate Oxidation

Krebs Cycle

Electron Transport Chain Chemiosmosis

1. Phosphorylation of glucose by ATP.

2–3. Rearrangement, followed by a second ATP phosphorylation.

4–5. The 6-carbon molecule is split into two 3-carbon molecules—one G3P, another that is converted into G3P in another reaction.

6. Oxidation followed by phosphorylation produces two NADH molecules and two molecules of BPG, each with one high-energy phosphate bond.

7. Removal of high-energy phosphate by two ADP molecules produces two ATP molecules and leaves two 3PG molecules.

8–9. Removal of water yields two PEP molecules, each with a high-energy phosphate bond.

10. Removal of high-energy phosphate by two ADP molecules produces two ATP molecules and two pyruvate molecules.

figure 7.7

THE GLYCOLYTIC PATHWAY. The first five reactions convert a molecule of glucose into two molecules of G3P. The second five reactions convert G3P into pyruvate.

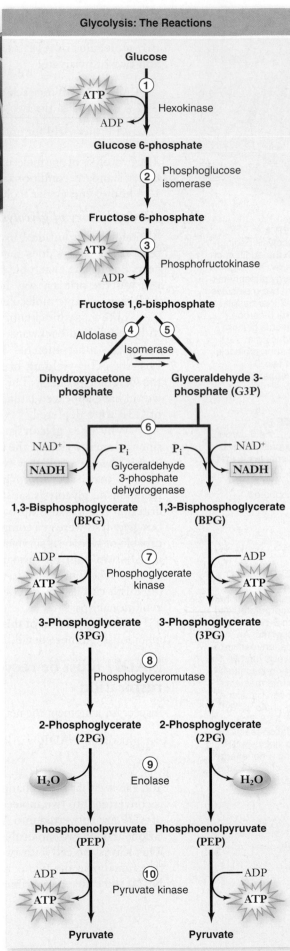

Glycolysis: The Reactions

Glucose

① Hexokinase
ATP → ADP

Glucose 6-phosphate

② Phosphoglucose isomerase

Fructose 6-phosphate

③ Phosphofructokinase
ATP → ADP

Fructose 1,6-bisphosphate

④ Aldolase   ⑤
Isomerase

Dihydroxyacetone phosphate   Glyceraldehyde 3-phosphate (G3P)

⑥ Glyceraldehyde 3-phosphate dehydrogenase
NAD⁺ P_i → NADH P_i NAD⁺ → NADH

1,3-Bisphosphoglycerate (BPG)   1,3-Bisphosphoglycerate (BPG)

⑦ Phosphoglycerate kinase
ADP → ATP   ADP → ATP

3-Phosphoglycerate (3PG)   3-Phosphoglycerate (3PG)

⑧ Phosphoglyceromutase

2-Phosphoglycerate (2PG)   2-Phosphoglycerate (2PG)

⑨ Enolase
H₂O   H₂O

Phosphoenolpyruvate (PEP)   Phosphoenolpyruvate (PEP)

⑩ Pyruvate kinase
ADP → ATP   ADP → ATP

Pyruvate   Pyruvate

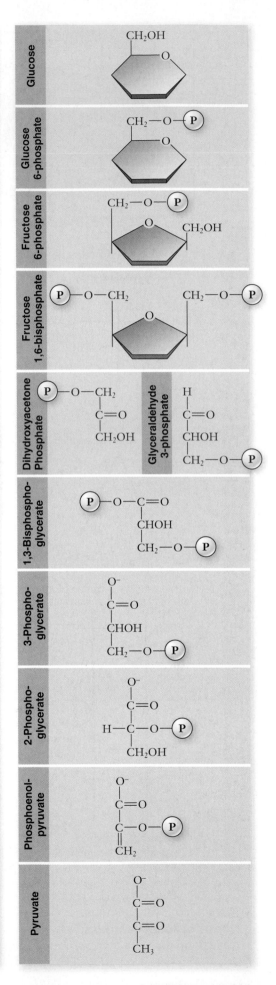

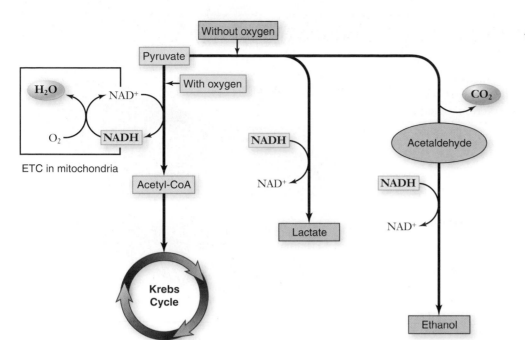

figure 7.8

THE FATE OF PYRUVATE AND NADH PRODUCED BY GLYCOLYSIS. In the presence of oxygen, NADH is oxidized by the electron transport chain (ETC) in mitochondria using oxygen as the final electron acceptor. This regenerates NAD^+ allowing glycolysis to continue. The pyruvate produced by glycolysis is oxidized to acetyl-CoA, which enters the Krebs cycle. In the absence of oxygen, pyruvate is instead reduced, oxidizing NADH and regenerating NAD^+ to allow glycolysis to continue. Direct reduction of pyruvate, as in muscle cells, produces lactate. In yeast, carbon dioxide is first removed from pyruvate producing acetaldehyde, which is then reduced to ethanol.

Recycling NADH

As long as food molecules that can be converted into glucose are available, a cell can continually churn out ATP to drive its activities. In doing so, however, it accumulates NADH and depletes the pool of NAD^+ molecules. A cell does not contain a large amount of NAD^+, and for glycolysis to continue, NADH must be recycled into NAD^+. Some molecule other than NAD^+ must ultimately accept the electrons taken from G3P and be reduced. Two processes can carry out this key task (figure 7.8):

1. **Aerobic respiration.** Oxygen is an excellent electron acceptor. Through a series of electron transfers, electrons taken from G3P can be donated to oxygen, forming water. This process occurs in the mitochondria of eukaryotic cells in the presence of oxygen. Because air is rich in oxygen, this process is also referred to as *aerobic metabolism*. A significant amount of ATP is also produced.
2. **Fermentation.** When oxygen is unavailable, an organic molecule, such as acetaldehyde in wine fermentation, can accept electrons instead (figure 7.9). This reaction plays an important role in the metabolism of most organisms, even those capable of aerobic respiration.

The fate of pyruvate

The fate of the pyruvate that is produced by glycolysis depends on which of these two processes takes place. The aerobic respiration path starts with the oxidation of pyruvate to produce acetyl-CoA, which is then further oxidized in a series of reactions called the Krebs cycle. The fermentation path, by contrast, uses the reduction of all or part of pyruvate to oxidize NADH back to NAD^+. We examine aerobic respiration next; fermentation is described in detail in a later section.

figure 7.9

HOW WINE IS MADE. The conversion of pyruvate to ethanol takes place naturally in grapes left to ferment on vines, as well as in fermentation vats of crushed grapes. Yeasts carry out the process to continue glycolysis under anaerobic conditions. When their conversion increases the ethanol concentration to about 12%, the toxic effects of the alcohol kill the yeast cells. What is left is wine.

Glycolysis splits the 6-carbon molecule glucose into two 3-carbon molecules of pyruvate. This process requires first using two ATP molecules in "priming" reactions eventually producing four molecules of ATP per glucose for a net yield of two ATP. The oxidation reactions of glycolysis require NAD^+ and produce NADH. This NAD^+ must be regenerated either by oxidation in the electron transport chain using O_2, or by using an organic molecule in a fermentation reaction.

7.4   The Oxidation of Pyruvate to Produce Acetyl-CoA

In the presence of oxygen, the oxidation of glucose that begins in glycolysis continues where glycolysis leaves off—with pyruvate. In eukaryotic organisms, the extraction of additional energy from pyruvate takes place exclusively inside mitochondria. In prokaryotes similar reactions take place in the cytoplasm and at the plasma membrane.

The cell harvests pyruvate's considerable energy in two steps. First, pyruvate is oxidized to produce a two-carbon compound and CO_2, with the electrons transferred to NAD^+ to produce NADH. Next, the two-carbon compound is oxidized to CO_2 by the reactions of the Krebs cycle.

Pyruvate is oxidized in a "decarboxylation" reaction that cleaves off one of pyruvate's three carbons. This carbon departs as CO_2 (figure 7.10). The remaining two-carbon compound, called an acetyl group, is then attached to coenzyme A; this entire molecule is called *acetyl-CoA*. A pair of electrons and one associated proton is transferred to the electron carrier NAD^+, reducing it to NADH, with a second proton donated to the solution.

The reaction involves three intermediate stages, and it is catalyzed within mitochondria by a *multienzyme complex*. As chapter 6 noted, a multienzyme complex organizes a series of enzymatic steps so that the chemical intermediates do not diffuse away or undergo other reactions. Within the complex, component polypeptides pass the substrates from one enzyme to the next, without releasing them. *Pyruvate dehydrogenase*, the complex of enzymes that removes CO_2 from pyruvate, is one of the largest enzymes known; it contains 60 subunits! The reaction can be summarized as:

$$\text{pyruvate} + NAD^+ + \text{CoA} \longrightarrow \text{acetyl-CoA} +$$
$$NADH + CO_2 + H^+$$

The molecule of NADH produced is used later to produce ATP. The acetyl group is fed into the Krebs cycle, with the CoA being recycled for another oxidation of pyruvate. The Krebs cycle then completes the oxidation of the original carbons from glucose.

> Pyruvate is oxidized in the mitochondria to produce acetyl-CoA and CO_2. This reaction is a link between glycolysis and the reactions of the Krebs cycle as acetyl-CoA is used by the Krebs cycle.

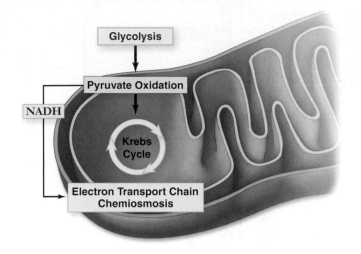

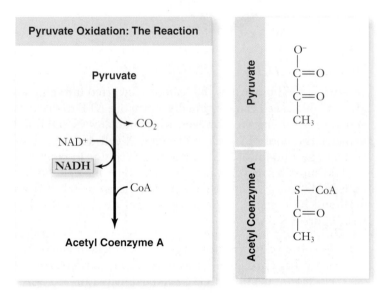

figure 7.10

THE OXIDATION OF PYRUVATE. This complex reaction uses NAD^+ to accept electrons reducing it to NADH. The product, acetyl-CoA, feeds the acetyl unit into the Krebs cycle, and the CoA is recycled for another oxidation of pyruvate. NADH provides energetic electrons for the electron transport chain.

7.5   The Krebs Cycle

In this third stage, the acetyl group from pyruvate is oxidized in a series of nine reactions called the *Krebs cycle*. These reactions occur in the matrix of mitochondria.

In this cycle, the two-carbon acetyl group of acetyl-CoA combines with a four-carbon molecule called oxaloacetate. The resulting six-carbon molecule, citrate, then goes through a several-step sequence of electron-yielding oxidation reactions, during which two CO_2 molecules split off, restoring oxaloacetate. The regenerated oxaloacetate is used to bind to another acetyl group for the next round of the cycle.

In each turn of the cycle, a new acetyl group is added and two carbons are lost as two CO_2 molecules, and more electrons are transferred to electron carriers. These electrons are then used by the electron transport chain to drive *proton pumps* that generate ATP.

The Krebs cycle has three segments: An overview

The nine reactions of the Krebs cycle can be grouped into three overall segments. These are described in the following sections and summarized in figure 7.11.

Segment A: Acetyl-CoA plus oxaloacetate This reaction produces the 6-carbon citrate molecule.

Segment B: Citrate rearrangement and decarboxylation Five more steps, which have been simplified in figure 7.11, reduce citrate to a 5-carbon intermediate and then to 4-carbon succinate. During these reactions, two NADH and one ATP are produced.

Segment C: Regeneration of oxaloacetate Succinate undergoes three additional reactions, also simplified in the figure, to become oxaloacetate. During these reactions, one NADH is produced; in addition, a molecule of **flavin adenine dinucleotide (FAD)**, another cofactor, becomes reduced to FADH$_2$.

The specifics of each reaction are described next.

The Krebs cycle is geared to extract electrons and synthesize one ATP

Figure 7.12 summarizes the sequence of the Krebs cycle reactions. A 2-carbon group from acetyl-CoA enters the cycle at the beginning, and two CO$_2$ molecules, one ATP, and four pairs of electrons are produced.

Reaction 1: Condensation Citrate is formed from acetyl-CoA and oxaloacetate. This condensation reaction is irreversible, committing the 2-carbon acetyl group to the Krebs cycle. The reaction is inhibited when the cell's ATP concentration is high and stimulated when it is low. The result is that when the cell possesses ample amounts of ATP, the Krebs cycle shuts down, and acetyl-CoA is channeled into fat synthesis.

Reactions 2 and 3: Isomerization Before the oxidation reactions can begin, the hydroxyl (—OH) group of citrate must be repositioned. This rearrangement is done in two steps: First, a water molecule is removed from one carbon; then water is added to a different carbon. As a result, an —H group and an —OH group change positions. The product is an isomer

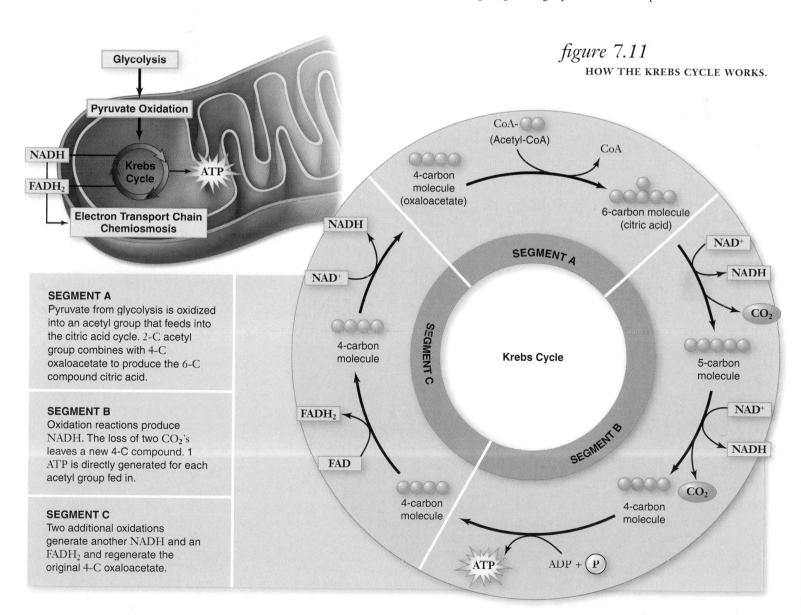

figure 7.11

HOW THE KREBS CYCLE WORKS.

SEGMENT A
Pyruvate from glycolysis is oxidized into an acetyl group that feeds into the citric acid cycle. 2-C acetyl group combines with 4-C oxaloacetate to produce the 6-C compound citric acid.

SEGMENT B
Oxidation reactions produce NADH. The loss of two CO$_2$'s leaves a new 4-C compound. 1 ATP is directly generated for each acetyl group fed in.

SEGMENT C
Two additional oxidations generate another NADH and an FADH$_2$ and regenerate the original 4-C oxaloacetate.

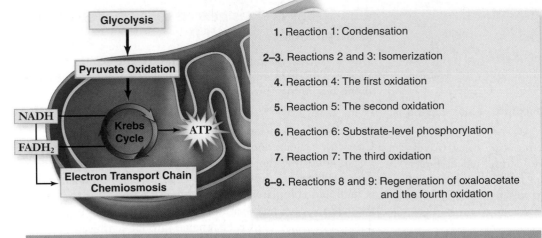

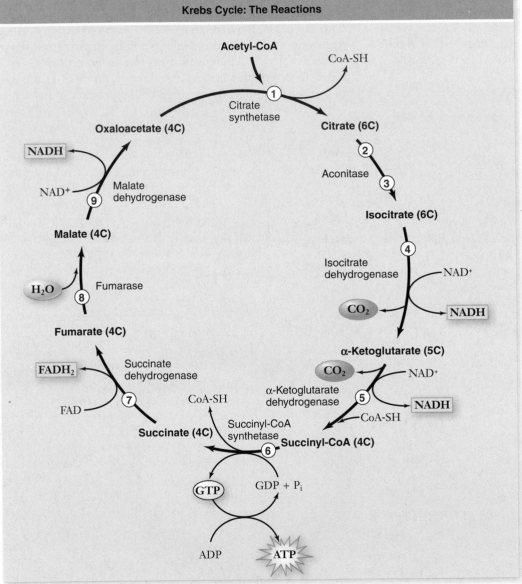

1. Reaction 1: Condensation

2–3. Reactions 2 and 3: Isomerization

4. Reaction 4: The first oxidation

5. Reaction 5: The second oxidation

6. Reaction 6: Substrate-level phosphorylation

7. Reaction 7: The third oxidation

8–9. Reactions 8 and 9: Regeneration of oxaloacetate and the fourth oxidation

figure 7.12

THE KREBS CYCLE. This series of reactions takes place within the matrix of the mitochondrion. For the complete breakdown of a molecule of glucose, the two molecules of acetyl-CoA produced by glycolysis and pyruvate oxidation each have to make a trip around the Krebs cycle. Follow the different carbons through the cycle, and notice the changes that occur in the carbon skeletons of the molecules and where oxidation reactions take place as they proceed through the cycle.

of citrate called *isocitrate*. This rearrangement facilitates the subsequent reactions.

Reaction 4: The First Oxidation In the first energy-yielding step of the cycle, isocitrate undergoes an oxidative decarboxylation reaction. First, isocitrate is oxidized, yielding a pair of electrons that reduce a molecule of NAD^+ to NADH. Then the oxidized intermediate is decarboxylated; the central carboxyl group splits off to form CO_2, yielding a 5-carbon molecule called *α-ketoglutarate*.

Reaction 5: The Second Oxidation Next, α-ketoglutarate is decarboxylated by a multienzyme complex similar to pyruvate dehydrogenase. The succinyl group left after the removal of CO_2 joins to coenzyme A, forming *succinyl-CoA*. In the process, two electrons are extracted, and they reduce another molecule of NAD^+ to NADH.

Reaction 6: Substrate-Level Phosphorylation The linkage between the 4-carbon succinyl group and CoA is a high-energy bond. In a coupled reaction similar to those that take place in glycolysis, this bond is cleaved, and the energy released drives the phosphorylation of guanosine diphosphate (GDP), forming guanosine triphosphate (GTP). GTP can transfer a phosphate to ADP converting it into ATP. The 4-carbon molecule that remains is called *succinate*.

Reaction 7: The Third Oxidation Next, succinate is oxidized to *fumarate* by an enzyme located in the inner mitochondrial membrane. The free-energy change in this reaction is not large enough to reduce NAD^+. Instead, FAD is the electron acceptor. Unlike NAD^+, FAD is not free to diffuse within the mitochondrion; it is tightly associated with its enzyme in the inner mitochondrial membrane. Its reduced form, $FADH_2$, can only contribute electrons to the electron transport chain in the membrane.

Reactions 8 and 9: Regeneration of Oxaloacetate In the final two reactions of the cycle, a water molecule is added to fumarate, forming *malate*. Malate is then oxidized, yielding a 4-carbon molecule of *oxaloacetate* and two electrons that reduce a molecule of NAD^+ to NADH. Oxaloacetate, the molecule that began the cycle, is now free to combine with another 2-carbon acetyl group from acetyl-CoA and reinitiate the cycle.

Glucose becomes CO_2 and potential energy

In the process of aerobic respiration, glucose is entirely consumed. The six-carbon glucose molecule is cleaved into a pair of 3-carbon pyruvate molecules during glycolysis. One of the carbons of each pyruvate is then lost as CO_2 in the conversion of pyruvate to acetyl-CoA. The two other carbons from acetyl-CoA are lost as CO_2 during the oxidations of the Krebs cycle.

All that is left to mark the passing of a glucose molecule into six CO_2 molecules is its energy, some of which is preserved in four ATP molecules and in the reduced state of 12 electron carriers. Ten of these carriers are NADH molecules; the other two are $FADH_2$.

Following the electrons in the reactions reveals the direction of transfer

As you examine the changes in electrical charge in the reactions that oxidize glucose, a good strategy for keeping the transfers clear is always to *follow the electrons*. For example, in glycolysis, an enzyme extracts two hydrogens—that is, two electrons and two protons—from glucose and transfers both electrons and one of the protons to NAD^+. The other proton is released as a hydrogen ion, H^+, into the surrounding solution. This transfer converts NAD^+ into NADH; that is, two negative electrons ($2e^-$) and one positive proton (H^+) are added to one positively charged NAD^+ to form NADH, which is electrically neutral.

As mentioned earlier, energy captured by NADH is not harvested all at once. The two electrons carried by NADH are passed along the electron transport chain, which consists of a series of electron carriers, mostly proteins, embedded within the inner membranes of mitochondria.

NADH delivers electrons to the beginning of the electron transport chain, and oxygen captures them at the end. The oxygen then joins with hydrogen ions to form water. At each step in the chain, the electrons move to a slightly more electronegative carrier, and their positions shift slightly. Thus, the electrons move *down* an energy gradient.

The entire process of electron transfer releases a total of 53 kcal/mol (222 kJ/mol) under standard conditions. The transfer of electrons along this chain allows the energy to be extracted gradually. Next, we will discuss how this energy is put to work to drive the production of ATP.

> The Krebs cycle completes the oxidation of glucose begun with glycolysis. Units of the 2-carbon molecule acetyl-CoA are added to the 2-carbon molecule oxaloacetate to produce citric acid (another name for the cycle is the citric acid cycle). The cycle then uses a series of oxidation, decarboxylation, and rearrangement reactions to return to oxaloacetate. This process produces NADH and $FADH_2$, which provide electrons and protons for the electron transport chain. It also produces one ATP per turn of the cycle.

7.6 The Electron Transport Chain and Chemiosmosis

The NADH and $FADH_2$ molecules formed during aerobic respiration each contain a pair of electrons that were gained when NAD^+ and FAD were reduced. The NADH molecules carry their electrons to the inner mitochondrial membrane, where they transfer the electrons to a series of membrane-associated proteins collectively called the *electron transport chain (ETC)*.

The electron transport chain produces a proton gradient

The first of the proteins to receive the electrons is a complex, membrane-embedded enzyme called **NADH dehydrogenase**. A carrier called *ubiquinone* then passes the electrons to a protein–cytochrome complex called the *bc₁ complex*. Each complex in the chain operates as a proton pump, driving a proton out across the membrane into the intermembrane space (figure 7.13*a*).

The electrons are then carried by another carrier, *cytochrome c*, to the cytochrome oxidase complex. This complex uses four electrons to reduce a molecule of oxygen. Each oxygen then combines with two protons to form water:

$$O_2 + 4 H^+ + 4 e^- \longrightarrow 2 H_2O$$

In contrast to NADH, which contributes its electrons to NADH dehydrogenase, FADH₂, which is located in the inner mitochondrial membrane, feeds its electrons to ubiquinone, which is also in the membrane. Electrons from FADH₂ thus "skip" the first step in the electron transport chain.

The plentiful availability of a strong electron acceptor, oxygen, is what makes oxidative respiration possible. As you'll see in chapter 8, the electron transport chain used in aerobic respiration is similar to, and may well have evolved from, the chain employed in photosynthesis.

The gradient forms as electrons move through electron carriers

Respiration takes place within the mitochondria present in virtually all eukaryotic cells. The internal compartment, or matrix, of a mitochondrion contains the enzymes that carry out the reactions of the Krebs cycle. As mentioned earlier, protons (H⁺) are produced when electrons are transferred to NAD⁺. As the electrons harvested by oxidative respiration are passed along the electron transport chain, the energy they release transports protons out of the matrix and into the outer compartment called the intermembrane space.

Three transmembrane complexes of the electron transport chain in the inner mitochondrial membrane actually accomplish the proton transport (see figure 7.13*a*). The flow of highly energetic electrons induces a change in the shape of pump proteins, which causes them to transport protons across the membrane. The electrons contributed by NADH activate all three of these proton pumps, whereas those contributed by FADH₂ activate only two because of where they enter the chain. In this way a proton gradient is formed between the intermembrane space and the matrix.

Chemiosmosis utilizes the electrochemical gradient to produce ATP

The internal negativity of the matrix with respect to the intermembrane space attracts the positively charged protons and induces them to reenter the matrix. The higher outer concentration of protons also tends to drive protons back in by diffusion, but because membranes are relatively impermeable to ions, this process occurs only very slowly. Most of the protons that reenter the matrix instead pass through ATP synthase, an

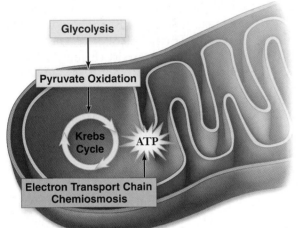

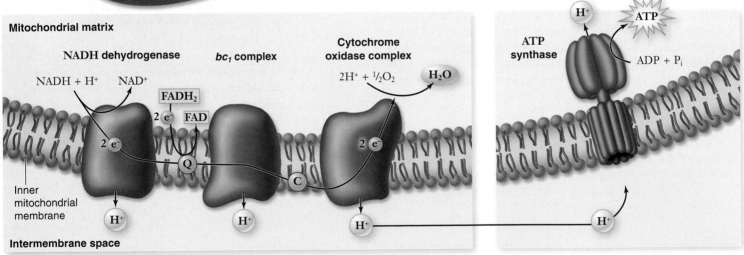

a. The electron transport chain

b. Chemiosmosis

figure 7.13

THE ELECTRON TRANSPORT CHAIN AND CHEMIOSMOSIS. *a.* High-energy electrons harvested from catabolized molecules are transported by mobile electron carriers (ubiquinone, marked Q, and cytochrome *c*, marked C) between three complexes of membrane proteins. These three complexes use portions of the electrons' energy to pump protons out of the matrix and into the intermembrane space. The electrons are finally used to reduce oxygen forming water. *b.* This creates a concentration gradient of protons across the inner membrane. This electrochemical gradient is a form of potential energy that can be used by ATP synthase. This enzyme couples the reentry of protons to the phosphorylation of ADP to form ATP.

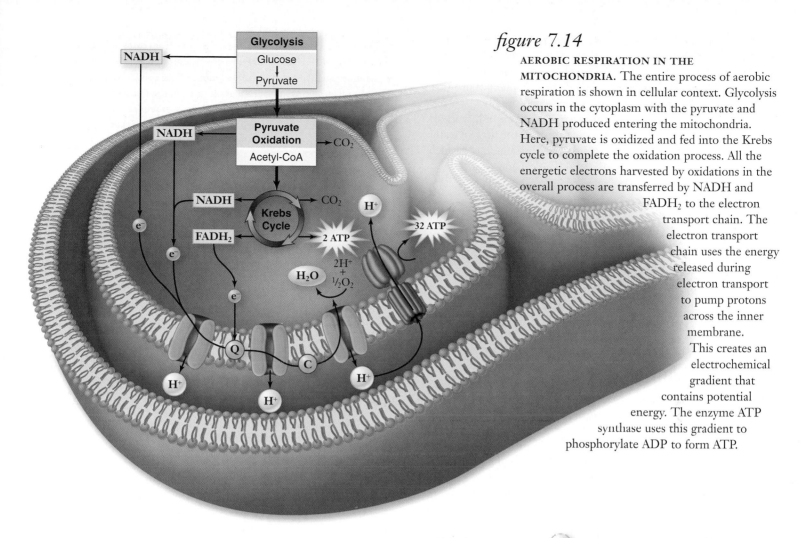

figure 7.14

AEROBIC RESPIRATION IN THE MITOCHONDRIA. The entire process of aerobic respiration is shown in cellular context. Glycolysis occurs in the cytoplasm with the pyruvate and NADH produced entering the mitochondria. Here, pyruvate is oxidized and fed into the Krebs cycle to complete the oxidation process. All the energetic electrons harvested by oxidations in the overall process are transferred by NADH and FADH₂ to the electron transport chain. The electron transport chain uses the energy released during electron transport to pump protons across the inner membrane. This creates an electrochemical gradient that contains potential energy. The enzyme ATP synthase uses this gradient to phosphorylate ADP to form ATP.

enzyme that uses the energy of the gradient to catalyze the synthesis of ATP from ADP and P$_i$. Because the chemical formation of ATP is driven by a diffusion force similar to osmosis, this process is referred to as *chemiosmosis* (figure 7.13*b*). The newly formed ATP is transported by facilitated diffusion to the many places in the cell where enzymes require energy to drive endergonic reactions.

The energy released by the reactions of cellular respiration ultimately drives the proton pumps that produce the proton gradient. The proton gradient provides the energy required for the synthesis of ATP. Figure 7.14 summarizes the overall process.

ATP synthase is a molecular rotary motor

ATP synthase uses a fascinating molecular mechanism to perform ATP synthesis (figure 7.15). Structurally, the enzyme has a membrane-bound portion and a narrow stalk that connects the membrane portion to a knoblike catalytic portion. This complex can be dissociated into two subportions: the F$_0$ membrane-bound complex, and the F$_1$ complex composed of the stalk and a knob, or head domain.

The F$_1$ complex has enzymatic activity. The F$_0$ complex contains a channel through which protons move across the membrane down their concentration gradient. As they do so, their movement causes part of the F$_0$ complex and the stalk to rotate relative to the knob. The mechanical energy of this

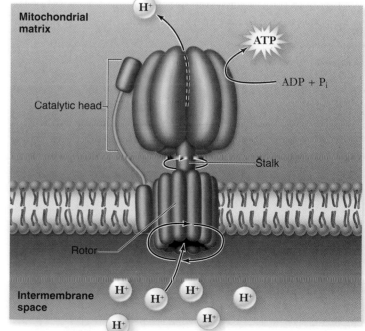

figure 7.15

THE ATP ROTARY ENGINE. Protons move across the membrane down their concentration gradient. The energy released causes the rotor and stalk structures to rotate. This mechanical energy alters the conformation of the ATP synthase enzyme to catalyze the formation of ATP.

rotation is used to change the conformation of the catalytic domain in the F_1 complex.

Thus, the synthesis of ATP is achieved by a tiny rotary motor, the rotation of which is driven directly by a gradient of protons. The process is like that of a water mill, in which the flow of water due to gravity causes a millwheel to turn and accomplish work or produce energy. The flow of protons is similar to the flow of water that causes the wheel to turn. Of course, ATP synthase is a much more complex motor, and it produces a chemical end-product.

The electron transport chain uses electrons from oxidation reactions carried by NADH and FADH$_2$ to create a proton gradient across the inner membrane of the mitochondria. The protein complexes of the electron transport chain are located in the inner membrane and use the energy from electron transfer to pump protons across the membrane, creating an electrochemical gradient. The enzyme ATP synthase can then use this gradient of protons to drive the endergonic reaction of phosphorylating ADP to ATP.

7.7 Energy Yield of Aerobic Respiration

How much metabolic energy in the form of ATP does a cell gain from aerobic breakdown of glucose? Knowing the steps involved in the process, we can calculate the theoretical yield of ATP and compare it with the actual yield.

The theoretical yield for eukaryotes is 36 ATP per glucose molecule

The chemiosmotic model suggests that one ATP molecule is generated for each proton pump activated by the electron transport chain. Because the electrons from NADH activate three pumps and those from FADH$_2$ activate two, we would expect each molecule of NADH and FADH$_2$ to generate three and two ATP molecules, respectively.

In doing this accounting, remember that everything downstream of glycolysis must be multiplied by 2 because two pyruvates are produced per molecule of glucose. A total of 10 NADH molecules is generated by respiration: 2 from glycolysis, 2 from the oxidation of pyruvate (1×2), and another 6 from the Krebs cycle (3×2). Also, two FADH$_2$ are produced (1×2). Finally, two ATP are generated directly by glycolysis and an-

other two ATP from the Krebs cycle (1×2). This gives a total of $10 \times 3 = 30$ ATP from NADH, plus $2 \times 2 = 4$ ATP from FADH$_2$, plus four ATP, for a total of 38 ATP (figure 7.16).

This number is accurate for bacteria, but it does not hold for eukaryotes because the NADH produced in the cytoplasm by glycolysis needs to be transported into the mitochondria by active transport, which costs 1 ATP per NADH transported. This reduces the predicted yield for eukaryotes to 36 ATP.

The actual yield for eukaryotes is 30 ATP per glucose molecule

The amount of ATP actually produced in a eukaryotic cell during aerobic respiration is somewhat lower than 36, for two reasons. First, the inner mitochondrial membrane is somewhat "leaky" to protons, allowing some of them to reenter the matrix without passing through ATP synthase. Second, mitochondria often use the proton gradient generated by chemiosmosis for purposes other than ATP synthesis (such as transporting pyruvate into the matrix).

figure 7.16

THEORETICAL ATP YIELD. The theoretical yield of ATP harvested from glucose by aerobic respiration totals 38 molecules. In eukaryotes this is reduced to 36 as the NADH generated by glycolysis in the cytoplasm has to be actively transported into the mitochondria costing the cell 1 ATP per NADH transported.

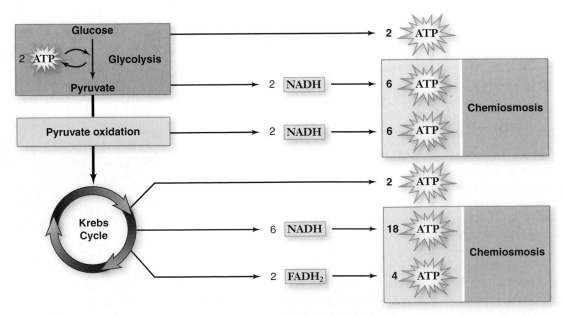

Total net ATP yield = **38**
(36 in eukaryotes)

Consequently, the actual measured values of ATP generated by NADH and FADH₂ are closer to 2.5 for each NADH, and 1.5 for each FADH₂. With these corrections, the overall harvest of ATP from a molecule of glucose in a eukaryotic cell is calculated as: 4 ATP from substrate-level phosphorylation + 25 ATP from NADH (2.5 × 10) + 3 ATP from FADH₂ (1.5 × 2) − 2 ATP for transport of glycolytic NADH = 30 molecules of ATP.

We mentioned earlier that the catabolism of glucose by aerobic respiration, in contrast to that by glycolysis alone, has a large energy yield. Aerobic respiration in a eukaryotic cell harvests about (7.3 × 30)/686 = 32% of the energy available in glucose. (By comparison, a typical car converts only about 25% of the energy in gasoline into useful energy.)

The higher yield of aerobic respiration was one of the key factors that fostered the evolution of heterotrophs. As this mechanism for producing ATP evolved, nonphotosynthetic organisms could more successfully base their metabolism on the exclusive use of molecules derived from other organisms. As long as some organisms captured energy by photosynthesis, others could exist solely by feeding on them.

> Passage of electrons down the electron transport chain produces roughly 3 ATP per NADH. This can result in a maximum of 38 ATP for all of the NADH generated by the complete oxidation of glucose, plus the ATP generated by substrate-level phosphorylation. NADH generated in the cytoplasm only lead to 2 ATP/NADH due to the cost of transporting the NADH into the mitochondria, leading to a total of 36 ATP for the mitochondria per glucose

7.8 Regulation of Aerobic Respiration

When cells possess plentiful amounts of ATP, the key reactions of glycolysis, the Krebs cycle, and fatty acid breakdown are inhibited, slowing ATP production. The regulation of these biochemical pathways by the level of ATP is an example of feedback inhibition. Conversely, when ATP levels in the cell are low, ADP levels are high, and ADP activates enzymes in the pathways of carbohydrate catabolism to stimulate the production of more ATP.

Control of glucose catabolism occurs at two key points in the catabolic pathway, namely at a point in glycolysis and at the beginning of the Krebs cycle (figure 7.17). The control point in glycolysis is the enzyme phosphofructokinase, which catalyzes the conversion of fructose phosphate to fructose bisphosphate. This is the first reaction of glycolysis that is not readily reversible, committing the substrate to the glycolytic sequence. ATP itself is an allosteric inhibitor (chapter 6) of phosphofructokinase, as is the Krebs cycle intermediate citrate. High levels of both ATP and citrate inhibit phosphofructokinase. Thus, under conditions when ATP is in excess, or when the Krebs cycle is producing citrate faster than it is being consumed, glycolysis is slowed.

The main control point in the oxidation of pyruvate occurs at the committing step in the Krebs cycle with the enzyme pyruvate dehydrogenase, which converts pyruvate to acetyl-CoA. This enzyme is inhibited by high levels of NADH, a key product of the Krebs cycle.

Another control point in the Krebs cycle is the enzyme citrate synthetase, which catalyzes the first reaction, the conversion of oxaloacetate and acetyl-CoA into citrate. High levels of ATP inhibit citrate synthetase (as well as phosphofructokinase, pyruvate dehydrogenase, and two other Krebs cycle enzymes), slowing down the entire catabolic pathway.

> Respiration is controlled by levels of ATP in the cell and levels of key intermediates in the process. The control point for glycolysis is the enzyme phosphofructokinase.

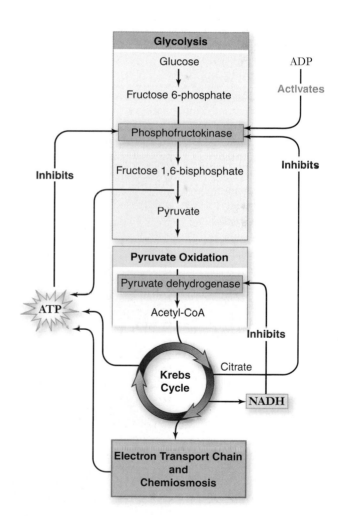

figure 7.17

CONTROL OF GLUCOSE CATABOLISM. The relative levels of ADP and ATP and key intermediates NADH and citrate control the catabolic pathway at two key points: the committing reactions of glycolysis and the Krebs cycle.

Oxidation Without O₂

In the presence of oxygen, cells can use oxygen to produce a large amount of ATP. But even when no oxygen is present to accept electrons, some organisms can still respire *anaerobically*, using inorganic molecules as final electron acceptors for an electron transport chain.

For example, many prokaryotes use sulfur, nitrate, carbon dioxide, or even inorganic metals as the final electron acceptor in place of oxygen (figure 7.18). The free energy released by using these other molecules as final electron acceptors is not as great as that using oxygen because they have a lower affinity for electrons. Less total ATP is produced, but the process is still respiration and not fermentation.

Methanogens use carbon dioxide

Among the heterotrophs that practice anaerobic respiration are primitive Archaea such as thermophiles and methanogens. Methanogens use carbon dioxide (CO_2) as the electron acceptor, reducing CO_2 to CH_4 (methane). The hydrogens are derived from organic molecules produced by other organisms. Methanogens are found in diverse environments, including soil and the digestive systems of ruminants like cows.

Sulfur bacteria use sulfate

Evidence of a second anaerobic respiratory process among primitive bacteria is seen in a group of rocks about 2.7 BYA, known as the Woman River iron formation. Organic material in these rocks is enriched for the light isotope of sulfur, ^{32}S, relative to the heavier isotope, ^{34}S. No known geochemical process produces such enrichment, but biological sulfur reduction does, in a process still carried out today by certain primitive prokaryotes.

In this sulfate respiration, the prokaryotes derive energy from the reduction of inorganic sulfates (SO_4) to hydrogen sulfide (H_2S). The hydrogen atoms are obtained from organic molecules other organisms produce. These prokaryotes thus are similar to methanogens, but they use SO_4 as the oxidizing (that is, electron-accepting) agent in place of CO_2.

The early sulfate reducers set the stage for the evolution of photosynthesis, creating an environment rich in H_2S. As dis-

figure 7.18

SULFUR-RESPIRING PROKARYOTE.
a. The micrograph shows the archaeal species *Thermoproteus tenax*. This organism can use elemental sulfur as a final electron acceptor for anaerobic respiration. *b.* *Thermoproteus* is often found in sulfur containing hot springs such as the Norris Geyser Basin in Yellowstone National Park shown here.

a.

0.625 μm

b.

cussed in chapter 8, the first form of photosynthesis obtained hydrogens from H_2S using the energy of sunlight.

Fermentation uses organic compounds as electron acceptors

In the absence of oxygen, cells that cannot utilize an alternative electron acceptor for respiration must rely exclusively on glycolysis to produce ATP. Under these conditions, the electrons generated by glycolysis are donated to organic molecules in a process called *fermentation*. This process recycles NAD^+, the electron acceptor that allows glycolysis to proceed.

Bacteria carry out more than a dozen kinds of fermentation reactions, often using pyruvate or a derivative of pyruvate to accept the electrons from NADH. Organic molecules other than pyruvate and its derivatives can be used as well; the important point is that the process regenerates NAD^+:

organic molecule + NADH $\longrightarrow$ reduced organic molecule + NAD^+

Often the reduced organic compound is an organic acid—such as acetic acid, butyric acid, propionic acid, or lactic acid—or an alcohol.

Ethanol fermentation

Eukaryotic cells are capable of only a few types of fermentation. In one type, which occurs in yeast, the molecule that accepts electrons from NADH is derived from pyruvate, the end-product of glycolysis.

Yeast enzymes remove a terminal CO_2 group from pyruvate through decarboxylation, producing a 2-carbon molecule called acetaldehyde. The CO_2 released causes bread made with yeast to rise; bread made without yeast (unleavened bread) does not rise. The acetaldehyde accepts a pair of electrons from NADH, producing NAD^+ and ethanol (ethyl alcohol) (figure 7.19).

This particular type of fermentation is of great interest to humans, because it is the source of the ethanol in wine and beer. Ethanol is a by-product of fermentation that is actually toxic to yeast; as it approaches a concentration of about 12%, it begins to kill the yeast. That explains why naturally fermented wine contains only about 12% ethanol.

Lactic acid fermentation

Most animal cells regenerate NAD^+ without decarboxylation. Muscle cells, for example, use the enzyme lactate dehydrogenase to transfer electrons from NADH back to the pyruvate that is produced by glycolysis. This reaction converts pyruvate into lactic acid and regenerates NAD^+ from NADH (see figure 7.19). It therefore closes the metabolic circle, allowing glycolysis to continue as long as glucose is available.

Circulating blood removes excess lactate, the ionized form of lactic acid, from muscles, but when removal cannot keep pace with production, the accumulating lactic acid interferes with muscle function and contributes to muscle fatigue.

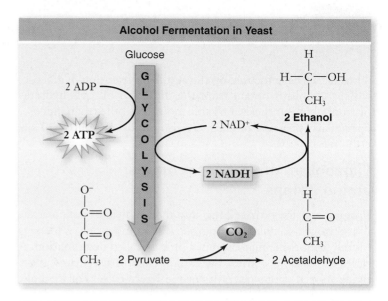

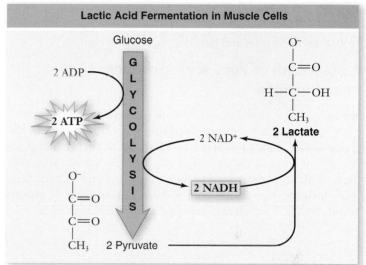

figure 7.19

FERMENTATION. Yeasts carry out the conversion of pyruvate to ethanol. Muscle cells convert pyruvate into lactate, which is less toxic than ethanol. In each case, the reduction of a metabolite of glucose has oxidized NADH back to NAD^+ to allow glycolysis to continue under anaerobic conditions.

O_2 is the electron acceptor for aerobic respiration. Due to oxygen's high affinity for electrons this results in the highest yield of ATP, but this is not the only form of respiration found in living systems. Nitrate, sulfur, and CO_2 are among other terminal electron acceptors used in anaerobic respiration. Organic molecules can also be used in fermentation reactions but allow only a partial oxidation of glucose via glycolysis. Fermentation reactions produce a variety of compounds, including ethanol in yeast and lactic acid in humans.

Catabolism of Proteins and Fats

Thus far we have focused on the aerobic respiration of glucose, which organisms obtain from the digestion of carbohydrates or from photosynthesis. Organic molecules other than glucose, particularly proteins and fats, are also important sources of energy (figure 7.20).

Catabolism of proteins removes amino groups

Proteins are first broken down into their individual amino acids. The nitrogen-containing side group (the amino group) is then removed from each amino acid in a process called **deamination**. A series of reactions convert the carbon chain that remains into a molecule that enters glycolysis or the Krebs cycle. For example, alanine is converted into pyruvate, glutamate into α-ketoglutarate (figure 7.21), and aspartate into oxaloacetate. The reactions of glycolysis and the Krebs cycle then extract the high-energy electrons from these molecules and put them to work making ATP.

Catabolism of fatty acids produces acetyl groups

Fats are broken down into fatty acids plus glycerol. Long-chain fatty acids typically have an even number of carbons, and the many C—H bonds provide a rich harvest of energy. Fatty acids are oxidized in the matrix of the mitochondrion. Enzymes remove the 2-carbon acetyl groups from the end of each fatty acid until the entire fatty acid is converted into acetyl groups (figure 7.22). Each acetyl group is combined with coenzyme

A to form acetyl-CoA. This process is known as **β-oxidation.** This process is oxygen-dependent, which explains why aerobic exercise burns fat, but anaerobic exercise does not.

How much ATP does the catabolism of fatty acids produce? Let's compare a hypothetical 6-carbon fatty acid with the six-carbon glucose molecule, which we've said yields about 30 molecules of ATP in a eukaryotic cell. Two rounds of β-oxidation would convert the fatty acid into three molecules of acetyl-CoA. Each round requires one molecule of ATP to prime the process, but it also produces one molecule of NADH and one of $FADH_2$. These molecules together yield four molecules of ATP (assuming 2.5 ATPs per NADH, and 1.5 ATPs per $FADH_2$).

The oxidation of each acetyl-CoA in the Krebs cycle ultimately produces an additional 10 molecules of ATP. Overall, then, the ATP yield of a 6-carbon fatty acid would be approximately: 8 (from two rounds of β-oxidation) − 2 (for priming those two rounds) + 30 (from oxidizing the three acetyl-CoAs) = 36 molecules of ATP. Therefore, the respiration of a 6-carbon fatty acid yields 20% more ATP than the respiration of glucose.

Moreover, a fatty acid of that size would weigh less than two-thirds as much as glucose, so a gram of fatty acid contains more than twice as many kilocalories as a gram of glucose. You can see from this fact why fat is a storage molecule for excess energy in many types of animals. If excess energy were stored instead as carbohydrate, as it is in plants, animal bodies would have to be much bulkier.

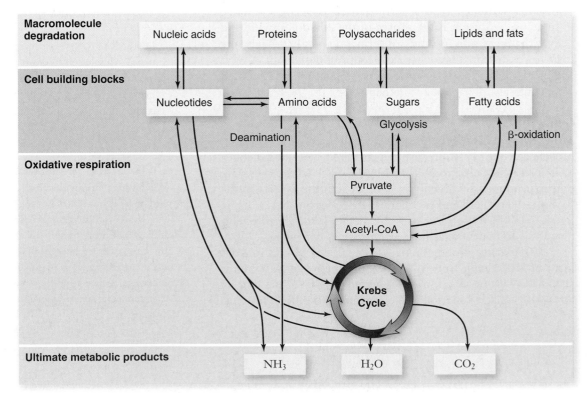

figure 7.20

HOW CELLS EXTRACT CHEMICAL ENERGY. All eukaryotes and many prokaryotes extract energy from organic molecules by oxidizing them. The first stage of this process, breaking down macromolecules into their constituent parts, yields little energy. The second stage, oxidative or aerobic respiration, extracts energy, primarily in the form of high-energy electrons, and produces water and carbon dioxide. Key intermediates in these energy pathways are also used for biosynthetic pathways, shown by reverse arrows.

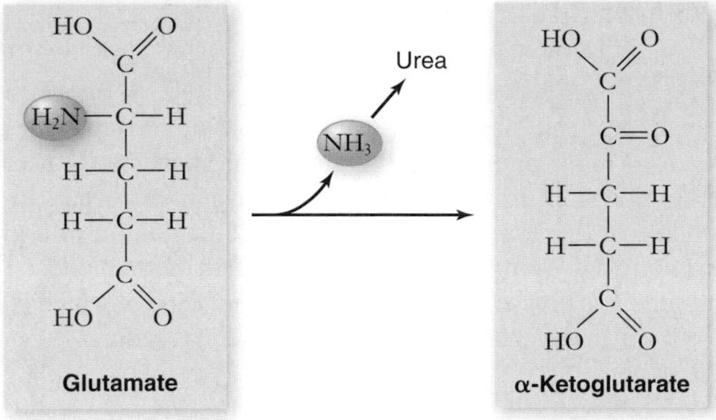

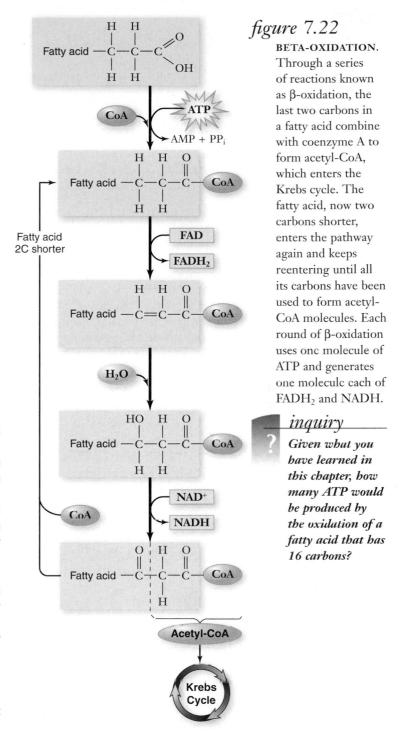

figure 7.21

DEAMINATION. After proteins are broken down into their amino acid constituents, the amino groups are removed from the amino acids to form molecules that participate in glycolysis and the Krebs cycle. For example, the amino acid glutamate becomes α-ketoglutarate, a Krebs cycle intermediate, when it loses its amino group.

A small number of key intermediates connect metabolic pathways

Oxidation pathways of food molecules are interrelated in that a small number of key intermediates, such as pyruvate and acetyl-CoA, link the breakdown from different starting points. These key intermediates allow the interconversion of different types of molecules, such as sugars and amino acids (see figure 7.20).

Cells can make glucose, amino acids, and fats, as well as getting them from external sources, and they use reactions similar to those that break down these substances. In many cases, the reverse pathways even share enzymes if the free-energy changes are small. For example, gluconeogenesis, the process of making new glucose, uses all but three enzymes of the glycolytic pathway. Thus, much of glycolysis runs forward or backward, depending on the concentrations of the intermediates—with only three key steps having different enzymes for forward and reverse directions.

Acetyl-CoA has many roles

Many different metabolic processes generate acetyl-CoA. Not only does the oxidation of pyruvate produce it, but the metabolic breakdown of proteins, fats, and other lipids also generates acetyl-CoA. Indeed, almost all molecules catabolized for energy are converted into acetyl-CoA.

Acetyl-CoA has a role in anabolic metabolism as well. Units of two carbons derived from acetyl-CoA are used to build up the hydrocarbon chains in fatty acids. Acetyl-CoA produced from a variety of sources can therefore be channeled into fatty acid synthesis or into ATP production, depending on the organism's energy requirements. Which of these two options is taken depends on the level of ATP in the cell.

When ATP levels are high, the oxidative pathway is inhibited, and acetyl-CoA is channeled into fatty acid synthesis. This explains why many animals (humans included) develop fat reserves when they consume more food than their activities

figure 7.22

BETA-OXIDATION. Through a series of reactions known as β-oxidation, the last two carbons in a fatty acid combine with coenzyme A to form acetyl-CoA, which enters the Krebs cycle. The fatty acid, now two carbons shorter, enters the pathway again and keeps reentering until all its carbons have been used to form acetyl-CoA molecules. Each round of β-oxidation uses one molecule of ATP and generates one molecule each of $FADH_2$ and NADH.

inquiry

? *Given what you have learned in this chapter, how many ATP would be produced by the oxidation of a fatty acid that has 16 carbons?*

require. Alternatively, when ATP levels are low, the oxidative pathway is stimulated, and acetyl-CoA flows into energy-producing oxidative metabolism.

Fats are a major energy storage molecule that can be broken down into units of acetyl-CoA by β-oxidation and fed into the Krebs cycle. The major metabolic pathways are connected by a number of key intermediates. This allows many processes to be used to either build up (anabolism) or break down (catabolism) the major biological macromolecules and allows interconversion of different types of molecules.

Evolution of Metabolism

We talk about cellular respiration as a continuous series of stages, but it is important to note that these stages evolved over time, and metabolism has changed a great deal in that time. Both anabolic processes and catabolic processes evolved in concert with each other. We do not know the details of this biochemical evolution, or the order of appearance of these processes. Therefore the following timeline is based on the available geochemical evidence and represents a hypothesis rather than a strict timeline.

The earliest life forms degraded carbon-based molecules present in the environment

The most primitive forms of life are thought to have obtained chemical energy by degrading, or breaking down, organic molecules that were abiotically produced, that is, carbon-containing molecules formed by inorganic processes on the early Earth.

The first major event in the evolution of metabolism was the origin of the ability to harness chemical bond energy. At an early stage, organisms began to store this energy in the bonds of ATP.

The evolution of glycolysis also occurred early

The second major event in the evolution of metabolism was glycolysis, the initial breakdown of glucose. As proteins evolved diverse catalytic functions, it became possible to capture a larger fraction of the chemical bond energy in organic molecules by breaking chemical bonds in a series of steps.

Glycolysis undoubtedly evolved early in the history of life on Earth, because this biochemical pathway has been retained by all living organisms. It is a chemical process that does not appear to have changed for well over 2 billion years.

Anaerobic photosynthesis allowed the capture of light energy

The third major event in the evolution of metabolism was anaerobic photosynthesis. Early in the history of life, a different way of generating ATP evolved in some organisms. Instead of obtaining energy for ATP synthesis by reshuffling chemical bonds, as in glycolysis, these organisms developed the ability to use light to pump protons out of their cells, and to use the resulting proton gradient to power the production of ATP through chemiosmosis.

Photosynthesis evolved in the absence of oxygen and works well without it. Dissolved H_2S, present in the oceans of the early Earth beneath an atmosphere free of oxygen gas, served as a ready source of hydrogen atoms for building organic molecules. Free sulfur was produced as a by-product of this reaction.

Oxygen-forming photosynthesis used a different source of hydrogen

The substitution of H_2O for H_2S in photosynthesis was the fourth major event in the history of metabolism. Oxygen-forming photosynthesis employs H_2O rather than H_2S as a source of hydrogen atoms and their associated electrons. Because it garners its electrons from reduced oxygen rather than from reduced sulfur, it generates oxygen gas rather than free sulfur.

More than 2 BYA, small cells capable of carrying out this oxygen-forming photosynthesis, such as cyanobacteria, became the dominant forms of life on Earth. Oxygen gas began to accumulate in the atmosphere. This was the beginning of a great transition that changed conditions on Earth permanently. Our atmosphere is now 20.9% oxygen, every molecule of which is derived from an oxygen-forming photosynthetic reaction.

Nitrogen fixation provided new organic nitrogen

Nitrogen is available from dead organic matter, and from chemical reactions that generated the original organic molecules. For life to expand, a new source of nitrogen was needed. Nitrogen fixation was the fifth major step in the evolution of metabolism. Proteins and nucleic acids cannot be synthesized from the products of photosynthesis because both of these biologically critical molecules contain nitrogen. Obtaining nitrogen atoms from N_2 gas, a process called *nitrogen fixation*, requires breaking an $N \equiv N$ triple bond.

This important reaction evolved in the hydrogen-rich atmosphere of the early Earth, where no oxygen was present. Oxygen acts as a poison to nitrogen fixation, which today occurs only in oxygen-free environments or in oxygen-free compartments within certain prokaryotes.

Aerobic respiration utilized oxygen

Aerobic respiration is the sixth and final event in the history of metabolism. Aerobic respiration employs the same kind of proton pumps as photosynthesis and is thought to have evolved as a modification of the basic photosynthetic machinery.

Biologists think that the ability to carry out photosynthesis without H_2S first evolved among purple nonsulfur bacteria, which obtain their hydrogens from organic compounds instead. It was perhaps inevitable that among the descendants of these respiring photosynthetic bacteria, some would eventually do without photosynthesis entirely, subsisting only on the energy and electrons derived from the breakdown of organic molecules. The mitochondria within all eukaryotic cells are thought to be descendants of these bacteria.

The complex process of aerobic metabolism developed over geological time, as natural selection favored organisms with more efficient methods of obtaining energy from organic molecules. The process of photosynthesis, as you have seen in this concluding section, has also developed over time, and the rise of photosynthesis changed life on Earth forever. The next chapter explores photosynthesis in detail.

Although the evolution of metabolism is not known in detail, major milestones can be recognized. These include the evolution of metabolic pathways that allow extraction of energy from organic compounds, the evolution of photosynthesis, and the evolution of nitrogen fixation. Photosynthesis began as an anoxygenic process that later evolved to produce oxygen, thus allowing the evolution of aerobic metabolism.

7.1 Overview of Respiration

Respiration occurs when carbohydrates and oxygen are converted to carbon dioxide, water, and energy.

- Autotrophs convert energy from sunlight to organic molecules.
- Heterotrophs use organic compounds made by autotrophs.
- Energy-rich molecules are degraded by oxidation reactions.
- Electron carriers can be reversibly oxidized and reduced.
- Energy released from redox reactions is used to make ATP.
- NAD^+ is an important electron carrier that can act as a coenzyme.
- Aerobic respiration uses oxygen as the final electron acceptor.
- Oxidizing food molecules in stages is more efficient than one step.

7.2 The Oxidation of Glucose: A Summary

Cells make ATP from the oxidation of glucose by two fundamentally different mechanisms.

- Substrate-level phosphorylation transfers a phosphate from a phosphate-bearing intermediate directly to ADP (figure 7.4).
- In oxidative phosphorylation ATP is generated by the enzyme ATP synthase, which is powered by a proton gradient.

7.3 Glycolysis: Splitting Glucose (figure 7.6)

Glycolysis is a series of chemical reactions that occur in the cell cytoplasm. Glucose yields 2 pyruvate, 2 NADH and 2 ATP.

- Priming reactions add two phosphates to glucose.
- This 6-carbon diphosphate is cleaved into two 3-carbon molecules of glyceraldehyde-3-phosphate (G3P).
- Oxidation of G3P transfers electrons to NAD^+ yielding NADH.
- The final product is two molecules of pyruvate.
- Glycolysis produces a net of 2 ATP, 2 NADH, and 2 pyruvate.
- NADH must be recycled into NAD^+ to continue glycolysis.
- In the presence of oxygen NADH is oxidized during respiration.

7.4 The Oxidation of Pyruvate to Produce Acetyl-CoA

Pyruvate from glycolysis is transported into the mitochondria where it is oxidized, and the product is fed into the Krebs cycle.

- Pyruvate oxidation results in 1 CO_2, 1 NADH, and 1-acetyl-CoA per pyruvate.
- Acetyl-CoA enters the Krebs cycle as two-carbon acetyl units.

7.5 The Krebs Cycle

Each acetyl compound that enters the Krebs cycle yields 2 CO_2, 1 ATP, 3 NADH and 1 $FADH_2$.

- An acetyl group combines with oxaloacetate producing citrate.
- Citrate is oxidized, removing CO_2 and generating NADH.

7.6 The Electron Transport Chain and Chemiosmosis (figure 7.13)

The electron transport chain is located on the inner membrane of mitochondria. It produces a proton gradient used in ATP synthesis.

- NADH is oxidized to NADH by NADH dehydrogenase.

- The electrons are transferred sequentially through 3 complexes to cytochrome oxidase, where electrons join with H^+ and oxygen.
- As the electrons move down the electron transport chain, three protons are pumped into the intermembrane space.
- This provides sufficient energy to produce 3 ATP.
- Protons diffuse back into the mitochondrial matrix through the ATP synthase channel, which phosphorylates ADP to ATP.
- As each proton passes through ATP synthase the energy causes the rotor and rod to rotate, altering the conformation of ATP synthase and catalyzing the formation of one ATP (figure 7.15).
- $FADH_2$ transfers electrons to ubiquinone. Only two protons are transported into the intermembrane space and 2 ATP are produced.

7.7 Energy Yield of Aerobic Respiration (figure 7.16)

Aerobic respiration theoretically yields 38 ATP per molecule of glucose.

- Eukaryotes yield 36 ATP per glucose molecule because it costs ATP to transport NADH formed during glycolysis into mitochondria.

7.8 Regulation of Aerobic Respiration (figure 7.17)

Glucose catabolism is controlled by the concentration of ATP molecules and products of the Krebs cycle.

- High ATP concentrations inhibit phosphofructokinase, the third enzyme in glycolysis; low ATP levels activate this enzyme.
- High concentrations of NADH inhibit pyruvate dehydrogenase.

7.9 Oxidation Without O_2 (figure 7.8)

In the absence of oxygen another final electron acceptor is necessary for respiration. For normally aerobic organisms, in the absence of oxygen, ATP can only be produced by glycolysis.

- In many prokaryotes inorganic molecules are used as final electron acceptors for an electron transport chain.
- The regeneration of NAD^+ by the oxidation of NADH and reduction of an organic molecule is called fermentation.
- In yeast, pyruvate is decarboxylated, then reduced to ethanol as NADH is oxidized to NAD^+.
- In animals, pyruvate is reduced to lactate as NADH is oxidized.

7.10 Catabolism of Proteins and Fats (figure 7.20)

Proteins, fats, and nucleic acids are built up and broken down through key intermediates

- Nucleic acids are metabolized through the Krebs cycle.
- Amino acids are deaminated before they are metabolized.
- The fatty acids are converted to acetyl-CoA by β-oxidation.
- With high ATP, acetyl-CoA is converted into fatty acids.

7.11 Evolution of Metabolism

Major milestones are recognized in the evolution of metabolism, the order of events is hypothetical.

- Five major metabolic processes evolved before atmospheric oxygen was present.
- *Early life-forms metabolized organic molecules that were abiotically produced and began to store energy as ATP.*
- *Glycolysis evolved incrementally.*
- *Early photosynthesis used H_2S to make organic molecules from CO_2.*
- *The substitution of H_2O for H_2S resulted in the formation of oxygen.*
- *Nitrogen fixation made N available.*

8.1 Overview of Photosynthesis

Life is powered by sunshine. The energy used by most living cells comes ultimately from the Sun, captured by plants, algae, and bacteria through the process of photosynthesis.

The diversity of life is only possible because our planet is awash in energy streaming Earthward from the Sun. Each day, the radiant energy that reaches Earth equals about 1 million Hiroshima-sized atomic bombs. Photosynthesis captures about 1% of this huge supply of energy (an amount equal to 10,000 Hiroshima bombs) and uses it to provide the energy that drives all life.

Photosynthesis combines CO_2 and H_2O, producing glucose and O_2

Photosynthesis occurs in a wide variety of organisms, and it comes in different forms. These include a form of photosynthesis that does not produce oxygen (anoxygenic) and a form that does (oxygenic). Anoxygenic photosynthesis is found in four different bacterial groups: purple bacteria, green sulfur bacteria, green nonsulfur bacteria, and heliobacteria. Oxygenic photosynthesis is found in cyanobacteria, seven groups of algae, and essentially all land plants. These two types of photosynthesis share similarities in the types of pigments used to trap light energy, but they differ in the arrangement and action of these pigments.

In the case of plants, photosynthesis takes place primarily in the leaves. Figure 8.1 illustrates the levels of organization in a plant leaf. As you learned in chapter 4, the cells of plant leaves contain organelles called chloroplasts,

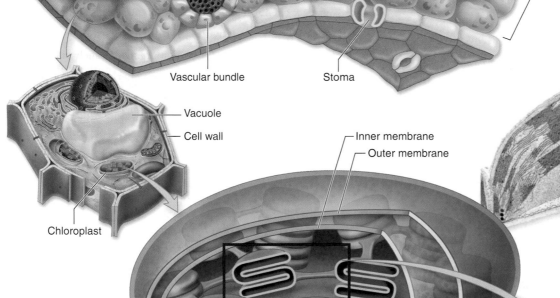

Cuticle

Epidermis

Mesophyll

Vascular bundle

Stoma

Vacuole

Cell wall

Inner membrane

Outer membrane

1.58 μm

Chloroplast

Thylakoid membrane   Thylakoid   Stroma

Granum

figure 8.1

JOURNEY INTO A LEAF. A plant leaf possesses a thick layer of cells (the mesophyll) rich in chloroplasts. The chloroplast consists of two membranes: an outer membrane enclosing the entire organelle and an inner one organized into flattened structures called thylakoid disks. The flattened thylakoids in the chloroplast are stacked into columns called grana. The rest of the interior is filled with a semifluid substance called stroma.

which carry out the photosynthetic process. No other structure in a plant cell is able to carry out photosynthesis (figure 8.2). Photosynthesis takes place in three stages:

1. capturing energy from sunlight;
2. using the energy to make ATP and to reduce the compound NADP$^+$, an electron carrier, to NADPH; and
3. using the ATP and NADPH to power the synthesis of organic molecules from CO_2 in the air.

The first two stages require light and are commonly called the **light-dependent reactions.**

The third stage, the formation of organic molecules from CO_2, is called **carbon fixation,** and this process takes place via a cyclic series of reactions. As long as ATP and NADPH are available, the carbon fixation reactions can occur either in the presence or in the absence of light, and so these reactions are also called the **light-independent reactions.**

The following simple equation summarizes the overall process of photosynthesis:

$$6\ CO_2 + 12\ H_2O + \text{light} \longrightarrow C_6H_{12}O_6 + 6\ H_2O + 6\ O_2$$

carbon   water   glucose   water   oxygen
dioxide

You may notice that this equation is the reverse of the reaction for respiration. In respiration, glucose is oxidized to CO_2 using O_2 as an electron acceptor. In photosynthesis, CO_2 is reduced to glucose using electrons gained from the oxidation of water. The oxidation of H_2O and the reduction of CO_2 requires energy that is provided by light. Although this statement is an oversimplification, it provides a useful "global perspective."

In plants, photosynthesis takes place in chloroplasts

In the preceding chapter, you saw that a mitochondrion's complex structure of internal and external membranes contribute to its function. The same is true for the structure of the chloroplast.

The internal membrane of chloroplasts, called the **thylakoid membrane,** is a continuous phospholipid bilayer organized into flattened sacs that are found stacked on one another in columns called **grana** (singular, *granum*). The thylakoid membrane contains **chlorophyll** and other photosynthetic pigments for capturing light energy along with the machinery to make ATP. Connections between grana are termed *stroma lamella.*

Surrounding the thylakoid membrane system is a semiliquid substance called **stroma.** The stroma houses the enzymes needed to assemble organic molecules from CO_2 using energy from ATP coupled with reduction via NADPH. In the thylakoid membrane, photosynthetic pigments are clustered together to form **photosystems,** which show distinct organization within the thylakoid.

Each pigment molecule within the photosystem is capable of capturing photons, which are packets of energy. When light of a proper wavelength strikes a pigment molecule in the photosystem, the resulting excitation passes from one pigment molecule to another.

The excited electron is not transferred physically—rather, its *energy* passes from one molecule to another. The passage is similar to the transfer of kinetic energy along a row of upright dominoes. If you push the first one over, it falls against the next,

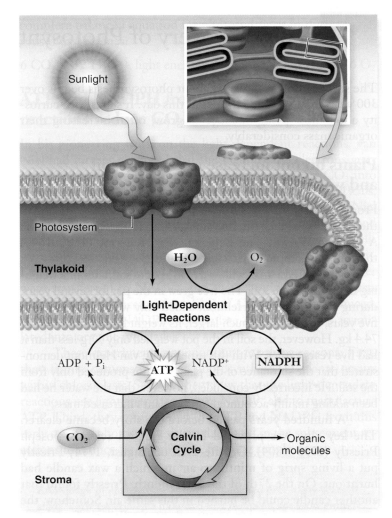

figure 8.2

The light-dependent reactions take place on the thylakoid membrane where photosystems absorb photons of light and use this energy to generate ATP and NADPH. Electrons lost from the photosystems are replaced by the oxidation of water, which produces O_2 as a by-product. The ATP and NADPH produced by the light reactions is used to fuel carbon fixation via the Calvin cycle. The stroma contains enzymes that carry out the Calvin cycle.

and that one against the next, and so on, until all of the dominoes have fallen down.

Eventually, the energy arrives at a key chlorophyll molecule that is in contact with a membrane-bound protein. The energy is transferred as an excited electron to that protein, which passes it on to a series of other membrane proteins that put the energy to work making ATP and NADPH. These compounds are then used to build organic molecules. The photosystem thus acts as a large antenna, gathering the light energy harvested by many individual pigment molecules.

Photosynthesis converts light energy into organic molecules. This process uses reactions, called light-dependent reactions, that require sunlight and others, that convert CO_2 into organic molecules in a process called carbon fixation. The overall reaction is essentially the reverse of respiration. The energy of sunlight is used to oxidize water with the electrons and protons used to reduce CO_2 to glucose. This also produces O_2 as a by-product.

uses part of the ancient glycolytic pathway, run in reverse. Also, the principal proteins involved in electron transport and ATP production in plants are evolutionarily related to those in mitochondria.

Photosynthesis is but one aspect of plant biology, although it is an important one. In chapters 36 through 42, we examine plants in more detail. We have discussed photosynthesis as a part of cell biology because photosynthesis arose long before plants did, and because most organisms depend directly or indirectly on photosynthesis for the energy that powers their lives.

Carbon fixation is the incorporation of inorganic CO_2 into an organic molecule. This is accomplished by the Calvin cycle reactions that takes place in the stroma of the chloroplast. The key intermediate is the five-carbon sugar RuBP that combines with CO_2 in the carbon fixation reaction. This reaction is catalyzed by the enzyme rubisco. The cycle can be broken down into three stages: carbon fixation, reduction, and regeneration of RuBP. The products of the light reactions, ATP and NADPH, provide energy and electrons for the reduction reactions.

8.7 Photorespiration

Evolution does not necessarily result in optimum solutions. Rather, it favors workable solutions that can be derived from features that already exist. Photosynthesis is no exception. Rubisco, the enzyme that catalyzes the key carbon-fixing reaction of photosynthesis, provides a decidedly suboptimal solution. This enzyme has a second enzymatic activity that interferes with carbon fixation, namely that of *oxidizing* RuBP. In this process, called **photorespiration,** O_2 is incorporated into RuBP, which undergoes additional reactions that actually release CO_2. Hence, photorespiration releases CO_2, essentially undoing carbon fixation.

Photorespiration reduces the yield of photosynthesis

The carboxylation and oxidation of RuBP are catalyzed at the same active site on rubisco, and CO_2 and O_2 compete with each other at this site. Under normal conditions at 25°C, the rate of the carboxylation reaction is four times that of the oxidation reaction, meaning that 20% of photosynthetically fixed carbon is lost to photorespiration.

This loss rises substantially as temperature increases, because under hot, arid conditions, specialized openings in the leaf called *stomata* (singular, *stoma*) (figure 8.18) close to conserve water. This closing also cuts off the supply of CO_2 entering the leaf and does not allow O_2 to exit (figure 8.19). As a result, the low-CO_2 and high-O_2 conditions within the leaf favor photorespiration.

Plants that fix carbon using only C_3 photosynthesis (the Calvin cycle) are called **C_3 plants** (figure 8.20*a*). Other plants add CO_2 to phosphoenolpyruvate (PEP) to form a four-carbon molecule. This reaction is catalyzed by the enzyme PEP *carboxylase*. This enzyme has two advantages over rubisco: it has a much greater affinity for CO_2 than rubisco, and it does not have oxidase activity.

The four-carbon compound produced by PEP carboxylase undergoes further modification, only to be eventually decarboxylated. The CO_2 released by this decarboxylation is then used by rubisco in the Calvin cycle. Because the source of CO_2 was

an organic compound and not atmospheric CO_2, the concentration of CO_2 relative to O_2 is increased and photorespiration is minimized. The four-carbon compound produced by PEP carboxylase allows CO_2 to be stored in an organic form, then later released to keep the level of CO_2 high relative to O_2.

The reduction in the yield of carbohydrate as a result of photorespiration is not trivial. C_3 plants lose between 25% and 50% of their photosynthetically fixed carbon in this way. The rate depends largely on temperature. In tropical climates, especially those in which the temperature is often above 28°C, the problem is severe, and it has a major effect on tropical agriculture.

The two main groups of plants that initially capture CO_2 using PEP carboxylase differ in how they maintain high levels of CO_2 relative to O_2. In **C_4 plants** (figure 8.20*b*), the capture of CO_2 occurs in one cell and the decarboxylation occurs in an

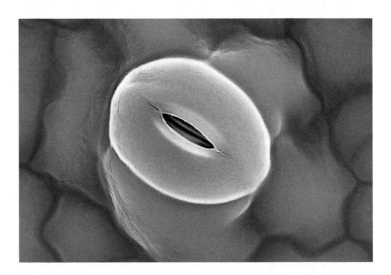

figure 8.18

STOMA. A closed stoma in the leaf of a tobacco plant. Each stoma is formed from two guard cells whose shape changes with turgor pressure to open and close. Under dry conditions plants close their stomata to conserve water.

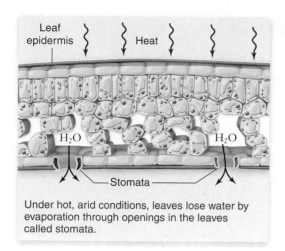

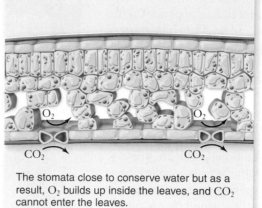

Under hot, arid conditions, leaves lose water by evaporation through openings in the leaves called stomata.

The stomata close to conserve water but as a result, O_2 builds up inside the leaves, and CO_2 cannot enter the leaves.

figure 8.19

CONDITIONS FAVORING PHOTORESPIRATION. In hot, arid environments, stomata close to conserve water, which also prevents CO_2 from entering and O_2 from exiting the leaf. The high-O_2/low-CO_2 conditions favor photorespiration.

adjacent cell. This represents a spatial solution to the problem of photorespiration. The second group, **CAM plants,** perform both reactions in the same cell, but capture CO_2 using PEP carboxylase at night, then decarboxylate during the day. CAM stands for **crassulacean acid metabolism,** after the plant family Crassulaceae (the stonecrops, or hens-and-chicks), in which it was first discovered. This mechanism represents a temporal solution to the photorespiration problem.

C₄ plants have evolved to minimize photorespiration

The C₄ plants include corn, sugarcane, sorghum and a number of other grasses. These plants initially fix carbon using PEP carboxylase in mesophyll cells. This reaction produces the organic acid oxaloacetate, which is converted to malate and transported to bundle-sheath cells that surround the leaf veins. Within the bundle-

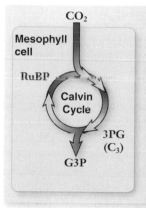

a. C₃ pathway

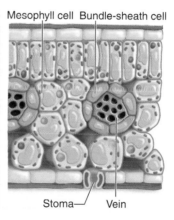

b. C₄ pathway

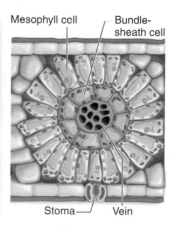

figure 8.20

COMPARISON OF C₃ AND C₄ PATHWAYS OF CARBON FIXATION. *a.* Carbon fixation by the C₃ pathway using the Calvin cycle as described in the text. All reactions occur in mesophyll cells using CO_2 that diffuses in through stomata. *b.* The C₄ pathway of carbon fixation in which one cell fixes CO_2 to produce a four-carbon molecule of malate in mesophyll cells. This is transported to the bundle sheath cells where it is converted back into CO_2 and pyruvate, creating a high local level of CO_2. This allows efficient carbon fixation by the Calvin cycle and avoids photorespiration.

sheath cells, malate is decarboxylated to produce pyruvate and CO_2 (figure 8.21). Because the bundle-sheath cells are impermeable to CO_2, the local level of CO_2 is high and carbon fixation by rubisco and the Calvin cycle is efficient. The pyruvate produced by decarboxylation is transported back to the mesophyll cells, where it is converted back to PEP, thereby completing the cycle.

The C_4 pathway, although it overcomes the problems of photorespiration, does have a cost. The conversion of pyruvate back to PEP requires breaking two high-energy bonds in ATP. Thus each CO_2 transported into the bundle-sheath cells cost the equivalent of two ATP. To produce a single glucose, this requires 12 additional ATP compared with the Calvin cycle alone. Despite this additional cost, C_4 photosynthesis is advantageous in hot dry climates where photorespiration would remove more than half of the carbon fixed by the usual C_3 pathway alone.

figure 8.22

CARBON FIXATION IN CAM PLANTS. CAM plants also use both C_4 and C_3 pathways to fix carbon and minimize photorespiration. In CAM plants, the two pathways occur in the same cell but are separated in time: The C_4 pathway is utilized to fix carbon at night, then CO_2 is released from these accumulated stores during the day to drive the C_3 pathway. This achieves the same effect of minimizing photorespiration while also minimizing loss of water by opening stomata at night when temperatures are lower.

The Crassulacean acid pathway splits photosynthesis into night and day

A second strategy to decrease photorespiration in hot regions has been adopted by the CAM plants. These include many succulent (water-storing) plants, such as cacti, pineapples, and some members of about two dozen other plant groups.

In these plants, the stomata open during the night and close during the day (figure 8.22). This pattern of stomatal opening and closing is the reverse of that in most plants. CAM plants initially fix CO_2 using PEP carboxylase to produce oxaloacetate. The oxaloacetate is often converted into other organic acids, depending on the particular CAM plant. These organic compounds accumulate during the night and are stored in the vacuole. Then during the day, when the stomata are closed, the organic acids are decarboxylated to yield high levels of CO_2. These high levels of CO_2 drive the Calvin cycle and minimize photorespiration.

Like C_4 plants, CAM plants use both C_3 and C_4 pathways. They differ in that they use both of these pathways in the same cell: the C_4 pathway at night and the C_3 pathway during the day. In C_4 plants the two pathways occur in different cells.

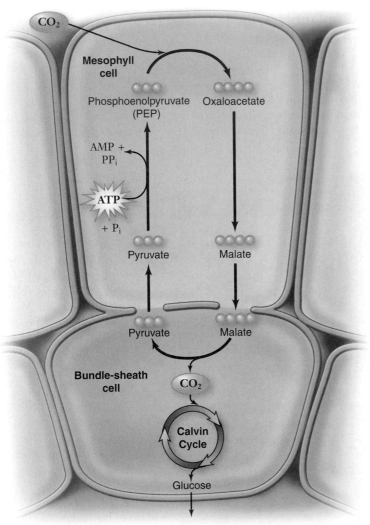

figure 8.21

CARBON FIXATION IN C_4 PLANTS. This process is called the C_4 pathway because the first molecule formed, oxaloacetate, contains four carbons. The oxaloacetate is converted to malate, which moves into bundle-sheath cells where it is decarboxylated back to CO_2 and pyruvate. This produces a high level of CO_2 in the bundle-sheath cells that can be fixed by the usual C_3 Calvin cycle with little photorespiration. The pyruvate diffuses back into the mesophyll cells, where it is converted back to PEP to be used in another C_4 fixation reaction.

In addition to catalyzing the carbon fixation reaction, the enzyme rubisco can oxidize RuBP. This photorespiration short-circuits photosynthesis under conditions of high O_2 and low CO_2. Plants in hot, dry environments have adapted to this by altering the usual dark reactions to store CO_2 in an organic molecule. This process is called C_4 metabolism to distinguish it from the usual C_3 metabolism. This involves CO_2 being stored in a four-carbon molecule in one cell, then transferred to another cell to release the CO_2 for use in the Calvin cycle. A further adaptation that splits photosynthesis into night and day reactions that occur in a single cell is found in CAM plants.

8.1 Overview of Photosynthesis

Photosynthesis is the conversion of light energy into chemical energy (figure 8.10).

- Photosynthesis has three stages: absorbing light energy, using the absorbed energy to synthesize ATP and NADPH, and using the ATP and NADPH to convert CO_2 to organic molecules.
- Photosynthesis has both light-dependent and light-independent reactions.
- In plant and algal cells, photosynthesis takes place in chloroplasts.
- Chloroplasts contain internal thylakoid membranes and a fluid matrix called stroma.

8.2 The Discovery of the Photosynthetic Processes

Knowledge about the photosynthetic process has accumulated over the last 300 years.

- Plants do not increase mass from soil and water alone as was originally thought.
- The rate of photosynthesis depends on the relative amounts of light, CO_2 concentration, and temperature.
- The O_2 produced by photosynthesis comes from H_2O.
- Light-dependent reactions produce O_2 from H_2O and generate ATP and NADPH that are used in carbon fixation reactions.

8.3 Pigments

For plants to utilize sunlight for photosynthesis pigments must be available that absorb light.

- Light is a form of energy that exists both as a wave and a particle called a photon.
- Light can remove electrons from some metals by the photoelectric effect.
- Photosynthetic pigments include chlorophyll *a*, chlorophyll *b*, and carotenoids; each has a characteristic absorption spectrum.
- Chlorophyll *a* is the primary pigment and the only pigment that can convert light energy into chemical energy.
- Chlorophyll *b* is an accessory pigment that increases the proportion of photons that can be harvested for photosynthesis.
- Carotenoids and other accessory pigments further increase a plants ability to harvest photons.

8.4 Photosystem Organization (figure 8.9)

Photosynthetic pigments are organized into photosystems that absorb photons of light and transfer electrons.

- A photosystem is a network of chlorophyll *a*, accessory pigments, and proteins embedded in the thylakoid membrane.
- Photosystems contain an antenna complex and a reaction center.
- *The antenna complex is composed of pigment molecules that harvest photons and feed light energy to the reaction center.*
- *The reaction center is composed of two chlorophyll a molecules in a protein matrix that pass an excited electron to an electron acceptor.*

8.5 The Light-dependent Reactions

Plants, algae, and photosynthetic bacteria use photosystems to trap and transfer energy.

- Anoxygenic photosynthetic bacteria use a single photosystem to generate ATP by cyclic photophosphorylation (figure 8.11).

- Cyclic photophosphorylation cycles excited electrons back to the pigment molecule to generate a proton gradient.
- Plants utilize two linked photosystems, photosystem I and photosystem II, acting to generate ATP and NADPH (figure 8.14).
- Photosystem I transfers electrons to $NADP^+$, reducing it to NADPH.
- Electrons lost by photosystem I are replaced by electrons from photosystem II.
- Electrons lost from photosystem II are replaced by electrons from the oxidation of water, which also produces O_2.
- Photosystem II and photosystem I are linked by an electron transport chain that pumps protons into the thylakoid space.
- The proton gradient is used by ATP synthase to phosphorylate ADP to ATP by a chemiosmotic mechanism similar to mitochondria.
- Plants can also make additional ATP by cyclic photophosphorylation.

8.6 Carbon Fixation: The Calvin Cycle (figure 8.16)

The Calvin cycle synthesizes organic molecules from inorganic carbon (CO_2).

- The Calvin cycle requires CO_2, ATP, and NADPH.
- The Calvin cycle occurs in three stages: carbon fixation, reduction, and regeneration.
- *Carbon fixation involves the enzyme rubisco, combining CO_2 and the five-carbon ribulose 1,5-bisphosphate (RuBP). The resulting compound splits into two 3-carbon 3-phosphoglycerates (PGA).*
- *Reduction converts the 3-carbon PGA to glyceraldehyde 3-phosphate (G3P) in a series of reactions that use ATP and NADPH.*
- *Regeneration involves using G3P to synthesize more RuBP.*
- Three turns of the cycle fix enough carbon for one new G3P.
- It takes six turns of the cycle to fix enough carbon to have two excess G3Ps that can be used to make one molecule of glucose.

8.7 Photorespiration

The enzyme rubisco, which catalyzes the carbon fixation reaction, can also catalyze the oxidation of RuBP, reversing carbon fixation.

- Dry, hot conditions cause the leaf stomata to close, resulting in lower CO_2 and higher O_2 concentrations within the leaf.
- As O_2 concentration increases, rubisco tends to bind oxygen and catalyze the oxidation of RuBP, eventually producing CO_2.
- Photorespiration can result in up to 50% reductions in glucose production.
- C_4 plants use an alternative mechanism of carbon fixation.
- C_4 plants fix carbon by adding CO_2 to a three-carbon phosphoenolpyruvate to form a four-carbon oxaloacetate. The CO_2 is released later for carbon fixation by the Calvin cycle.
- C_4 plants fix carbon in one cell by the C_4 pathway, then release CO_2 in another cell for the Calvin cycle (figure 8.21).
- CAM plants use the C_4 pathway during the day and the Calvin cycle at night in the same cell.

SELF TEST

1. The *light-dependent* reactions of photosynthesis are responsible for the production of—
   a. glucose
   b. CO_2
   c. ATP and NADPH
   d. H_2O

2. Which region of a chloroplast is associated with the capture of light energy?
   a. Thylakoid membrane
   b. Outer membrane
   c. Stroma
   d. Both a and c

3. The colors of light that are most effective for photosynthesis are—
   a. Red, blue, and violet
   b. Green, yellow, and orange
   c. Infrared and ultraviolet
   d. All colors of light are equally effective

4. The colors associated with pigments such as chlorophyll or carotenoids are a product of—
   a. The wavelengths of light absorbed by the pigment
   b. The wavelengths of light reflected by the pigment
   c. The energy transferred between pigments
   d. The wavelengths of light emitted by the pigment

5. Which of the following best describes a photosystem?
   a. A collection of pigment molecules
   b. A collection of pigments that transfer energy captured from light to a reaction center pigment
   c. A collection of thylakoid membranes assembled into a structure called a granum
   d. A collection of chlorophyll molecules that capture light energy and use it to make ATP

6. How is a reaction center pigment different from a pigment in the antenna complex?
   a. The reaction center pigment is a chlorophyll molecule.
   b. The antenna complex pigment can only reflect light.
   c. The reaction center pigment loses an electron when it absorbs light energy.
   d. The antenna complex pigments are not attached to proteins.

7. What happens to the energy from an excited reaction center electron in the cyclic photophosphorylation of sulfur bacteria?
   a. It is used to make ATP.
   b. It is used to phosphorylate proteins in an electron transport chain.
   c. It is used to generate a new pigment molecule.
   d. It is used to excite another pigment molecule within the photosystem.

8. During noncyclic photosynthesis, photosystem I functions to _____, and photosystem II functions to _____.
   a. synthesize ATP; produce O_2
   b. reduce $NADP^+$; oxidize H_2O
   c. reduce CO_2; oxidize NADPH
   d. restore an electron to its reaction center; gain an electron from water

9. Where in a chloroplast would you find the highest concentration of protons?
   a. In the stroma
   b. In the lumen of the thylakoid
   c. In the intermembrane space
   d. In the antenna complex

10. How does the reaction center of photosystem I regain an electron during noncyclic photosynthesis?
    a. The electron is recycled directly back to the reaction center pigment.
    b. The electron is donated from H_2O.
    c. The electron is donated from photosystem II.
    d. The electron is donated from NADPH.

11. Which of the following is NOT associated with the thylakoid membrane?
    a. Photosystem II
    b. ATP synthase
    c. Rubisco
    d. b_6-f complex

12. Carbon fixation occurs when a molecule of CO_2 reacts with a molecule of—
    a. ribulose 1,5-bisphosphate (RuBP)
    b. glyceraldehyde 3-phosphate (G3P)
    c. 3-phosphoglycerate (PGA)
    d. pyruvate

13. The function of the Calvin cycle is to—
    a. absorb light energy
    b. synthesize RuBP
    c. fix carbon
    d. convert glucose to CO_2, yielding energy

14. What is photorespiration?
    a. The production of chemical energy (ATP) using light energy (photons)
    b. Carbon fixation using the energy gained from the light reactions of photosynthesis
    c. The use of O_2 by plants as a final electron acceptor for photosynthesis
    d. The addition of O_2 to RuBP, leading to the loss of CO_2 and ATP

15. The adaptation of fixing CO_2 from the atmosphere at night is characteristic of—
    a. C_3 plants
    b. C_4 plants
    c. CAM plants
    d. all of the above

CHALLENGE QUESTIONS

1. Study the process of the Calvin cycle diagrammed in figure 8.16. Where do the ATP and NADPH used in this reaction come from? How can a chloroplast generate enough ATP to support the needs of the Calvin cycle?

2. Compare the process of photosynthesis in green plants and anoxygenic bacteria.

3. Do plant cells need mitochondria? Explain your answer.

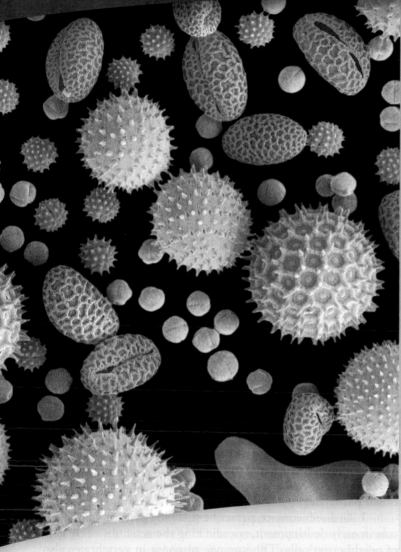

chapter 9

Cell Communication

introduction

SPRINGTIME IS A TIME OF REBIRTH and renewal. Trees that have appeared dead produce new leaves and buds, and flowers sprout from the ground. For sufferers of seasonal allergy, this is not quite such a pleasant time. The pollen in the micrograph and other allergens produced stimulate the immune system to produce the molecule histamine and other molecules that form cellular signals. These signals cause inflammation, mucus secretion, vasodilation, and other responses that together cause the runny nose, itching watery eyes, and other symptoms that make up the allergic reaction. We treat allergy symptoms by using drugs called antihistamines that interfere with this cellular signaling. The popular drug loratadine (better known as Claritin), for example, acts by blocking the receptor for histamine, thus preventing its action.

We will begin this chapter with a general overview of signaling, and the kinds of receptors cells use to respond to signals. Then we will look in more detail at how these different types of receptors can elicit a response from cells and finally how cells make connections with one another.

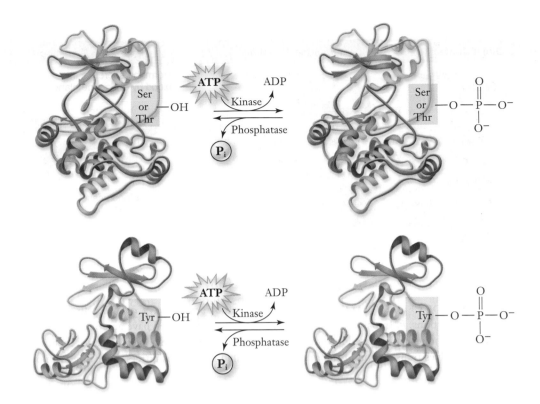

figure 9.3

PHOSPHORYLATION OF PROTEINS. Many proteins are controlled by their phosphorylation state: that is, they are activated by phosphorylation and deactivated by dephosphorylation or the reverse. The enzymes that add phosphate groups are called kinases. These form two classes depending on the amino acid the phosphate is added to, either serine/threonine kinases or tyrosine kinases. The action of kinases is reversed by another enzyme, a protein phosphatase.

As you learned in preceding chapters, the end result of the metabolic pathways of cellular respiration and photosynthesis was the phosphorylation of ADP to ATP. The ATP synthesized by these processes can donate phosphate groups to proteins. The phosphorylation of proteins alters their function, which allows them to transmit information from an extracellular signal through a signal transduction pathway.

Protein kinases

The class of enzyme that adds phosphate groups from ATP to proteins is called a *protein kinase*. These phosphate groups can be added to the three amino acids that have an OH as part of their R group, namely serine, threonine, and tyrosine. We categorize protein kinases based on which of these three substrates they alter (figure 9.3). Most cytoplasmic protein kinases fall into the serine/threonine kinase class.

Phosphatases

Part of the reason for the versatility of phosphorylation as a form of protein modification is that it is reversible. Another class of enzymes called **phosphatases** removes phosphate groups, reversing the action of kinases (see figure 9.3). Thus, a protein activated by a kinase can be deactivated by a phosphatase, or the reverse.

Cell communication involves chemical signals, or ligands, that bind to cellular receptors. Binding of ligand to receptor initiates signal transduction pathways that lead to a cellular response. Different cells may have the same response to one signal and the same signal can also elicit different responses in different cells. The phosphorylation-dephosphorylation of proteins is a common mechanism of controlling protein function found in signaling pathways.

9.2 Receptor Types

The first step in understanding cell signaling is to consider the receptors themselves. Cells must have a specific receptor to be able to respond to a particular signaling molecule. The interaction of a receptor and its ligand is an example of molecular recognition, a process in which one molecule fits specifically based on its complementary shape with another molecule. This interaction causes subtle changes in the structure of the receptor, thereby activating it. This is the beginning of any signal transduction pathway.

Receptors are defined by location

The nature of these receptor molecules depends on their location and on the kind of ligands they bind. The broadest classes of receptors are those that bind a ligand inside the cell (**intracellular receptors**), and

those that bind a ligand outside the cell (**cell surface receptors** or **membrane receptors**) (figure 9.1). Membrane receptors consist of transmembrane proteins that are in contact with both the cytoplasm and the extracellular environment. Table 9.1 summarizes the types of receptors and other communication mechanisms we will discuss in this chapter.

Membrane receptors include three subclasses

When a receptor is a transmembrane protein, the ligand can bind to the receptor outside of the cell and never actually cross the plasma membrane. In this case, the receptor itself, and not the signaling molecule is responsible for information crossing the membrane. Membrane receptors can be subdivided based on their structure and function.

TABLE 9.1	Receptors Involved in Cell Signaling		
Receptor Type	Structure	Function	Example
Intracellular Receptors	No extracellular signal-binding site	Receives signals from lipid-soluble or noncharged, nonpolar small molecules	Receptors for NO, steroid hormone, vitamin D, and thyroid hormone
Cell Surface Receptors			
Chemically gated ion channels	Multipass transmembrane protein forming a central pore	Molecular "gates" triggered chemically to open or close	Neurons
Enzymatic receptors	Single-pass transmembrane protein	Binds signal extracellularly; catalyzes response intracellularly	Phosphorylation of protein kinases
G protein-coupled receptors	Seven-pass transmembrane protein with cytoplasmic binding site for G protein	Binding of signal to receptor causes GTP to bind a G protein; G protein, with attached GTP, detaches to, deliver the signal inside the cell	Peptide hormones, rod cells in the eyes

Channel-linked receptors

Chemically gated ion channels are receptor proteins that allow the passage of ions (figure 9.4*a*). The receptor proteins that bind many neurotransmitters have the same basic structure. Each is a membrane protein with multiple transmembrane do-mains, meaning that the chain of amino acids threads back and forth across the plasma membrane several times. In the center of the protein is a pore that connects the extracellular fluid with the cytoplasm. The pore is big enough for ions to pass through, so the protein functions as an **ion channel.**

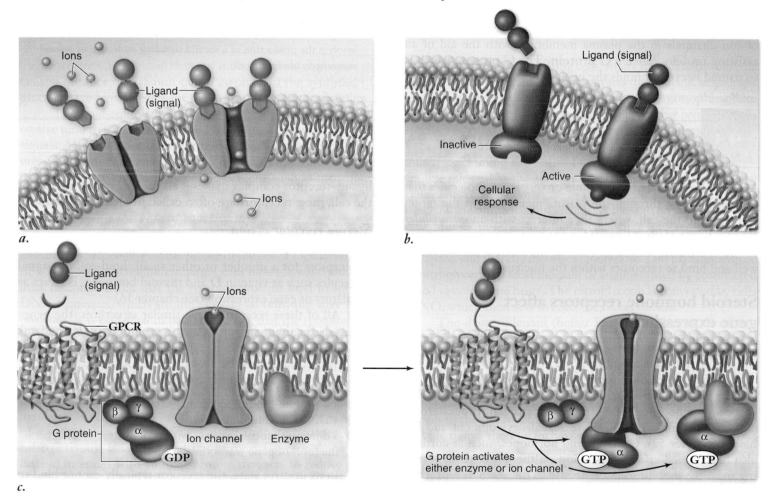

figure 9.4

CELL SURFACE RECEPTORS. *a.* Chemically gated ion channels are proteins that form a pore in the plasma membrane. This pore is opened or closed by chemical signals. They are usually selective, allowing the passage of only one type of ion. *b.* Enzymatic receptors are integral membrane proteins that bind the signal on the extracellular surface. A catalytic region on their cytoplasmic portion then transmits the signal by acting as an enzyme in the cytoplasm. *c.* G protein-coupled receptors (GPCR) bind to the signal outside the cell and to G proteins inside the cell. The G protein then activates an enzyme or ion channel, mediating the passage of a signal from the cell's surface to its interior.

Signal Transduction Through Receptor Kinases

Earlier you read that protein kinases phosphorylate proteins to alter protein function and that the most common kinases act on the amino acids serine, threonine, and tyrosine. The **receptor tyrosine kinases (RTK)** influence the cell cycle, cell migration, cell metabolism, and cell proliferation—virtually all aspects of the cell are affected by signaling through these receptors. Alterations to the function of these receptors and their signaling pathways can lead to cancers in humans and other animals.

Some of the earliest examples of cancer-causing genes, or oncogenes, involve RTK function (discussed in chapter 10). The cancer-causing simian sarcoma virus carries a gene for platelet-derived growth factor. When the virus infects a cell, the cell overproduces and secretes platelet-derived growth factor, causing overgrowth of the surrounding cells. Another virus, avian erythroblastosis virus, carries an altered form of the epidermal growth factor receptor that lacks most of its extracellular domain. When this virus infects a cell the altered receptors

produced are stuck in the "on" state. The continuous signaling from this receptor leads to cells that have lost the normal controls over growth.

Receptor tyrosine kinases recognize hydrophilic ligands and form a large class of membrane receptors in animal cells. Plants possess receptors with a similar overall structure and function, but they are serine–threonine kinases. These plant receptors have been named **plant receptor kinases.**

Because these receptors are performing similar functions in plant and animal cells but differ in their substrates, the duplication and divergence of each kind of receptor kinase probably occurred after the plant–animal divergence. The proliferation of these types of signaling molecules is thought to be coincident with the independent evolution of multicellularity in each group.

In this section, we will concentrate on the RTK family of receptors that has been extensively studied in a variety of animal cells.

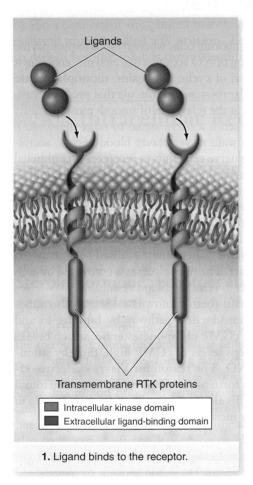

1. Ligand binds to the receptor.

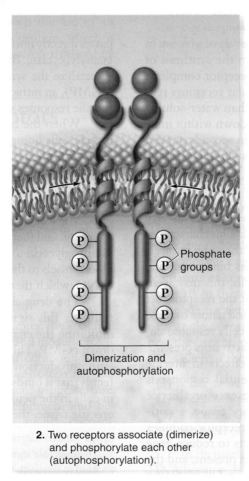

2. Two receptors associate (dimerize) and phosphorylate each other (autophosphorylation).

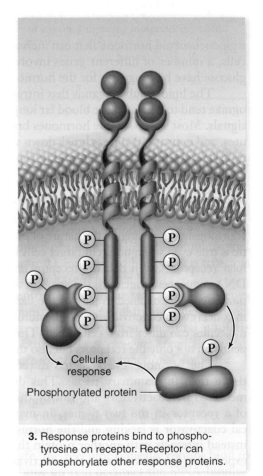

3. Response proteins bind to phosphotyrosine on receptor. Receptor can phosphorylate other response proteins.

figure 9.6

ACTIVATION OF A RECEPTOR TYROSINE KINASE (RTK). These membrane receptors bind hormones or growth factors that are hydrophilic and cannot cross the membrane. The receptor is a transmembrane protein with an extracellular ligand binding domain and an intracellular kinase domain. Signal transduction pathways begin with response proteins binding to phosphotyrosine on receptor, and by receptor phosphorylation of response proteins.

RTKs are activated by autophosphorylation

Receptor tyrosine kinases have a relatively simple structure consisting of a single transmembrane domain that anchors them in the membrane, an extracellular ligand-binding domain, and an intracellular kinase domain. This kinase domain contains the catalytic site of the receptor, which acts as a protein kinase that adds phosphate groups to tyrosines. On ligand binding to a specific receptor, two of these receptor–ligand complexes associate together (often referred to as dimerization) and phosphorylate each other, a process called *autophosphorylation* (figure 9.6).

The autophosphorylation event transmits across the membrane the signal that began with the binding of the ligand to the receptor. The next step, propagation of the signal in the cytoplasm, can take a variety of different forms. These forms include activation of the tyrosine kinase domain to phosphorylate other intracellular targets or interaction of other proteins with the phosphorylated receptor.

The cellular response after activation depends on the possible response proteins in the cell. Two different cells can have the same receptor yet a different response, depending on what response proteins are present in the cytoplasm. For example fibroblast growth factor stimulates cell division in fibroblasts but stimulates nerve cells to differentiate rather than to divide.

Phosphotyrosine domains mediate protein–protein interactions

One way that the signal from the receptor can be propagated in the cytoplasm is via proteins that bind specifically to phosphorylated tyrosines in the receptor. When the receptor is activated, regions of the protein outside of the catalytic site are phosphorylated. This creates "docking" sites for proteins that bind specifically to phosphotyrosine. The proteins that bind to these phosphorylated tyrosines can initiate intracellular events to convert the signal from the ligand into a response (see figure 9.6).

The insulin receptor

The use of docking proteins is illustrated by the insulin receptor. The hormone insulin is part of the body's control system to maintain a constant level of blood glucose. The role of insulin is to lower blood glucose, acting by binding to an RTK. Another protein called the *insulin response protein* binds to the phosphorylated receptor and is itself phosphorylated. The insulin response protein passes the signal on by binding to additional proteins that lead to the activation of the enzyme glycogen synthase, which converts glucose to glycogen (figure 9.7), thereby lowering blood glucose. Other proteins act to inhibit the synthesis of enzymes involved in making glucose.

Adapter proteins

Another class of proteins, **adapter proteins,** can also bind to phosphotyrosines. These proteins themselves do not participate in signal transduction but act as a link between the receptor and proteins that initiate downstream signaling events. For example, the Ras protein discussed later, is activated by adapter proteins binding to a receptor.

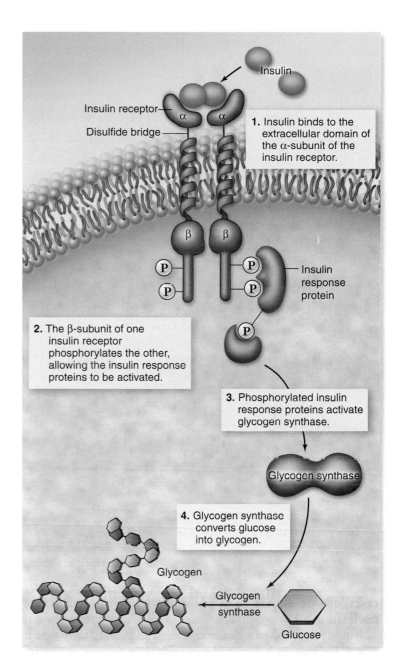

figure 9.7

THE INSULIN RECEPTOR. The insulin receptor is a receptor tyrosine kinase that initiates a variety of cellular responses related to glucose metabolism. One signal transduction pathway that this receptor mediates leads to the activation of the enzyme glycogen synthase. This enzyme converts glucose to glycogen.

Protein kinase cascades can amplify a signal

One important class of cytoplasmic kinases are **mitogen-activated protein (MAP) kinases.** A *mitogen* is a chemical that stimulates cell division by activating the normal pathways that control division. The MAP kinases are activated by a signaling module called a *phosphorylation cascade* or a **kinase cascade.** This module is a series of protein kinases that phosphorylate each other in succession. The final step in the

Signal Transduction Through G Protein-Coupled Receptors

The single largest category of receptor type in animal cells is **G protein-coupled receptors (GPCRs),** so named because the receptors act by coupling with a G protein. G proteins are proteins that bind guanosine nucleotides, such as Ras discussed previously.

It is estimated that 750 different genes encode GPCRs in the human genome. The sheer number and diversity of function of these receptors is overwhelming. In this section, we will concentrate on the basic mechanism of activation and some of the possible signal transduction pathways.

G proteins link receptors with effector proteins

The function of the G protein in signaling by GPCRs is to provide a link between a receptor that receives signals and effector proteins that produce cellular responses. The G protein functions as a switch that is turned on by the receptor. In its "on" state, the G protein activates effector proteins to cause a cellular response.

All G proteins are active when bound to GTP and inactive when bound to GDP. The main difference between the G proteins in GPCRs and the Ras protein described earlier is that these G proteins are composed of three subunits, called α, β, and γ. As a result, they are often called **heterotrimeric G proteins.** When a ligand binds to a GPCR and activates its associated G protein, the G protein exchanges GDP for GTP and dissociates into two parts consisting of the G_α subunit bound to GTP, and the G_β and G_γ subunits together ($G_{\beta\gamma}$). The signal can then be propagated by either the G_α or the $G_{\beta\gamma}$ components, thereby act-

ing to turn on effector proteins. The hydrolysis of bound GTP to GDP by G_α causes reassociation of the heterotrimer and restores the "off" state of the system (figure 9.11).

The effector proteins are usually enzymes. An effector protein might be a protein kinase that phosphorylates proteins to directly propagate the signal, or it may produce a second messenger to initiate a signal transduction pathway.

Effector proteins produce multiple second messengers

Often, the effector proteins activated by G proteins produce a second messenger. Two of the most common effectors are *adenylyl cyclase* and *phospholipase C*, which produce cAMP and IP$_3$ plus DAG, respectively.

Cyclic-AMP

All animal cells studied thus far use cAMP as a second messenger (chapter 46). When a signaling molecule binds to a GPCR that uses the enzyme **adenylyl cyclase** as an effector, a large amount of cAMP is produced within the cell (figure 9.12*a*). The cAMP then binds to and activates the enzyme protein kinase A (PKA), which adds phosphates to specific proteins in the cell (figure 9.13).

The effect of this phosphorylation on cell function depends on the identity of the cell and the proteins that are phosphorylated. In muscle cells, for example, PKA activates an enzyme necessary to break down glycogen and inhibits another enzyme

figure 9.11

THE ACTION OF G PROTEIN-COUPLED RECEPTORS.
G protein-coupled receptors act through a heterotrimeric G protein that links the receptor to an effector protein. When ligand binds to the receptor, it activates an associated G protein, exchanging GDP for GTP. The active G protein complex dissociates into G_α and $G_{\beta\gamma}$. The G_α subunit (bound to GTP) is shown activating an effector protein. The effector protein may act directly on cellular proteins or produce a second messenger to cause a cellular response. G_α can hydrolyze GTP inactivating the system, then reassociate with $G_{\beta\gamma}$.

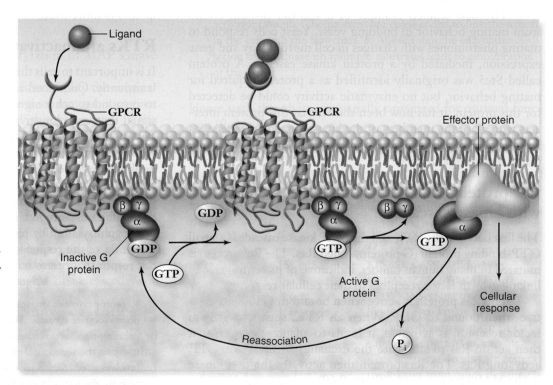

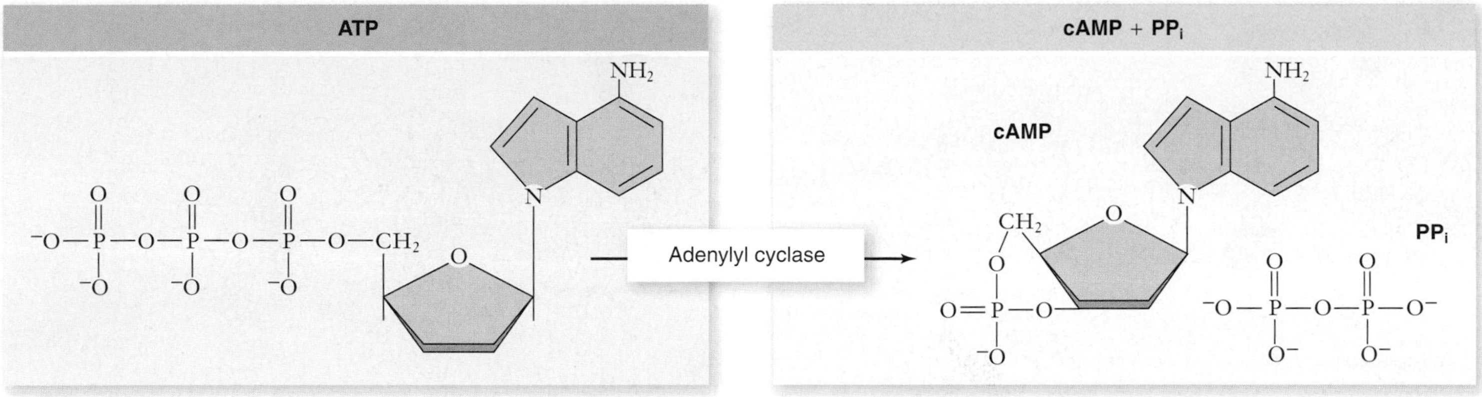

a.

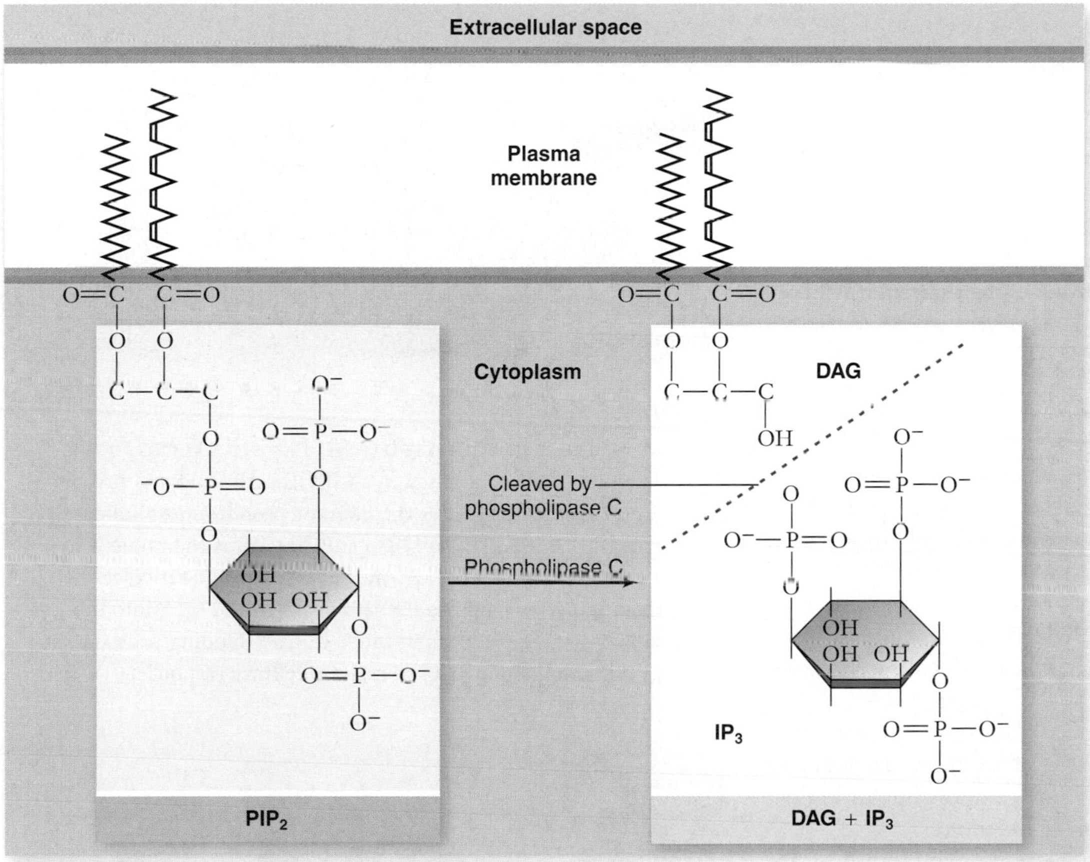

b.

figure 9.12

PRODUCTION OF SECOND MESSENGERS. Second messengers are signaling molecules produced within the cell. **a.** The nucleotide ATP is converted by the enzyme adenylyl cyclase into cyclic AMP, or cAMP. **b.** The inositol phospholipid is composed of two lipids and a phosphate attached to glycerol. The phosphate is also attached to the sugar inositol. This molecule can be cleaved by the enzyme phospholipase C to produce two different second messengers: DAG, made up of the glycerol with the two lipids, and IP$_3$, inositol triphosphate.

necessary to synthesize glycogen. This leads to an increase in glucose available to the muscle. By contrast, in the kidney the action of PKA leads to the production of water channels that can increase the permeability of tubule cells to water.

Disruption of cAMP signaling can have a variety of effects. The symptoms of the disease cholera are due to altered cAMP levels in cells in the gut. The bacterium *Vibrio cholerae* produces a toxin that binds to a GPCR in the epithelium of the gut, causing it to be locked into an "on" state. This causes a large increase in intracellular cAMP that, in these cells, causes Cl$^-$ ions to be transported out of the cell. Water follows the Cl$^-$, leading to the diarrhea and dehydration characteristic of the disease.

Inositol phosphates

A common second messenger is produced from the molecules called inositol phospholipids. These are inserted into the plasma membrane by their lipid ends and have the **inositol phosphate** portion protruding into the cytoplasm. The most common inositol phospholipid is phosphatidylinositol-4,5-bisphosphate (PIP$_2$). This molecule is a substrate of the effector protein phospholipase C, which cleaves PIP$_2$ to yield **diacylglycerol (DAG)** and **inositol-1,4,5-trisphosphate (IP$_3$)** (see figure 9.12*b*).

Both of these compounds then act as second messengers with a variety of cellular effects. DAG, like cAMP, can activate a protein kinase, in this case protein kinase C (PKC).

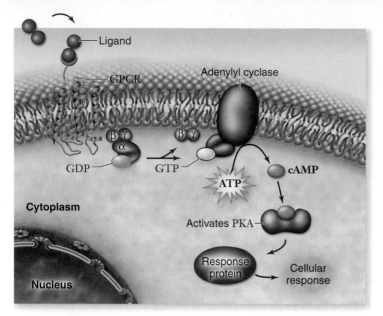

figure 9.13

cAMP SIGNALING PATHWAY. Extracellular signal binds to a GPCR, activating a G protein. The G protein then activates the effector protein adenylyl cyclase, which catalyzes the conversion of ATP to cAMP. The cAMP then activates protein kinase A (PKA), which phosphorylates target proteins to cause a cellular response.

Calcium

Calcium ions (Ca^{2+}) serve widely as second messengers. Ca^{2+} levels inside the cytoplasm are normally very low (less than 10^{-7} M), whereas outside the cell and in the endoplasmic reticulum, Ca^{2+} levels are quite high (about 10^{-3} M). The endoplasmic reticulum has receptor proteins that act as ion channels to release Ca^{2+}. One of the most common of these receptors can bind the second messenger IP_3 to release Ca^{2+}, linking signaling through inositol phosphates with signaling by Ca^{2+} (figure 9.14).

The result of the outflow of Ca^{2+} from the endoplasmic reticulum depends on the cell type. For example, in skeletal muscle cells Ca^{2+} stimulates muscle contraction but in endocrine cells it stimulates the secretion of hormones.

Ca^{2+} initiates some cellular responses by binding to *calmodulin*, a 148-amino-acid cytoplasmic protein that contains four binding sites for Ca^{2+} (figure 9.15). When four Ca^{2+} ions are bound to calmodulin, the calmodulin/Ca^{2+} complex is able to bind to other proteins to activate them. These proteins include protein kinases, ion channels, receptor proteins, and cyclic nucleotide phosphodiesterases. These many uses of Ca^{2+} make it one of the most versatile second messengers in cells.

Different receptors can produce the same second messengers

As mentioned previously, the two hormones glucagon and epinephrine can both stimulate liver cells to mobilize glucose. The reason that these different signals have the same effect is that they both act by the same signal transduction pathway to stimulate the breakdown and inhibit the synthesis of glycogen.

The binding of either hormone to its receptor activates a G protein that simulates adenylyl cyclase. The production of

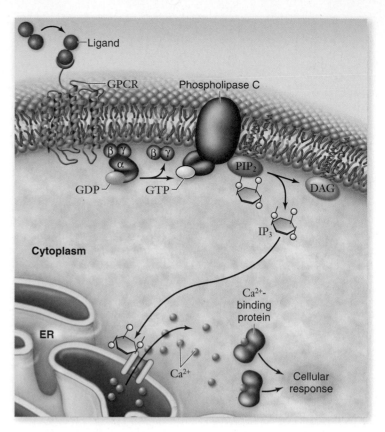

figure 9.14

INOSITOL PHOSPHOLIPID AND Ca^{2+} SIGNALING. Extracellular signal binds to a GPCR activating a G protein. The G protein activates the effector protein phospholipase C, which converts PIP_2 to DAG and IP_3. IP_3 is then bound to a channel-linked receptor on the endoplasmic reticular (ER) membrane, causing the ER to release stored Ca^{2+} into the cytoplasm. The Ca^{2+} then binds to Ca^{2+}-binding proteins such as calmodulin and PKC to cause a cellular response.

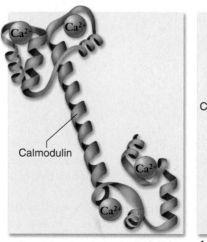 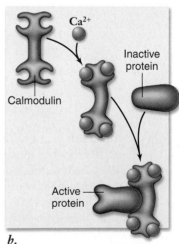

a.   b.

figure 9.15

CALMODULIN. *a.* Calmodulin is a protein containing 148 amino acid residues that mediates Ca^{2+} function. *b.* When four Ca^{2+} are bound to the calmodulin molecule, it undergoes a conformational change that allows it to bind to other cytoplasmic proteins and effect cellular responses.

cAMP leads to the activation of PKA, which in turn activates another protein kinase called phosphorylase kinase. Activated phosphorylase kinase then activates glycogen phosphorylase, which cleaves off units of glucose-6-phosphate from the glycogen polymer (figure 9.16). The action of multiple kinases again leads to amplification such that a few signaling molecules result in a large number of glucose molecules being released.

At the same time, PKA also phosphorylates the enzyme glycogen synthase, but in this case it inhibits the enzyme, thus preventing the synthesis of glycogen. In addition, PKA phosphorylates other proteins that activate the expression of genes encoding the enzymes needed to synthesize glucose. This convergence of signal transduction pathways from different receptors leads to the same result: Glucose is mobilized.

Receptor subtypes can lead to different effects in different cells

We also saw earlier how a single signaling molecule, epinephrine, can have different effects in different cells. One way this happens is through the existence of multiple forms of the same

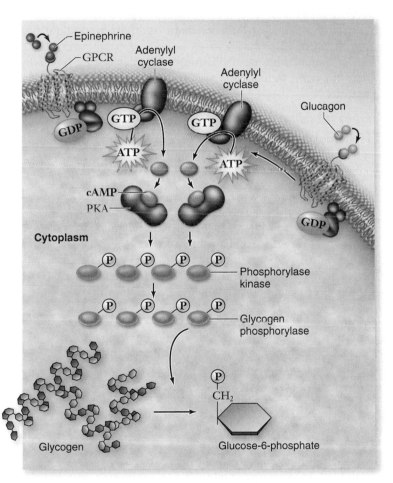

figure 9.16

DIFFERENT RECEPTORS CAN ACTIVATE THE SAME SIGNALING PATHWAY. The hormones glucagon and epinephrine both act through GPCRs. Each of these receptors acts via a G protein that activates adenylyl cyclase, producing cAMP. The activation of PKA begins a kinase cascade that leads to the breakdown of glycogen.

receptor. The receptor for epinephrine actually has nine different subtypes, or isoforms. These are encoded by different genes and are actually different receptor molecules. The sequences of these proteins are very similar, especially in the ligand-binding domain, which allows them to bind epinephrine. They differ mainly in their cytoplasmic domains, which interact with G proteins. This leads to different isoforms activating different G proteins, thereby leading to different signal transduction pathways.

Thus, in the heart, muscle cells have one isoform of the receptor that, when bound to epinephrine, activates a G protein that activates adenylyl cyclase, leading to increased cAMP. This increases the rate and force of contraction. In the intestine, smooth muscle cells have a different isoform of the receptor that, when bound to epinephrine, activates a different G protein that inhibits adenylyl cyclase, which decreases cAMP. This has the result of relaxing the muscle.

G protein-coupled receptors and receptor tyrosine kinases can activate the same pathways

Different receptor types can affect the same signaling module. For example, RTKs were shown to activate the MAP kinase cascade, but GPCRs can also activate this same cascade. Similarly, the activation of phospholipase C was mentioned previously in the context of GPCR signaling, but it can also be activated by RTKs.

This cross-reactivity may appear to introduce complications into cell function, but in fact it provides the cell with an incredible amount of flexibility. Cells have a large, but limited number of intracellular signaling modules, which can be turned on and off by different kinds of membrane receptors. This leads to signaling networks that interconnect possible cellular effectors with multiple incoming signals.

The Internet represents an example of a network in which many different kinds of computers are connected globally. This network can be broken down into subnetworks that are connected to the overall network. Because of the nature of the connections, when you send an e-mail message across the Internet, it can reach its destination through many different pathways. Likewise, the cell has interconnected networks of signaling pathways in which many different signals, receptors, and response proteins are interconnected. Specific pathways like the MAP kinase cascade, or signaling through second messengers like cAMP and Ca^{2+}, represent subnetworks within the global signaling network. A specific signal can activate different pathways in different cells, or different signals can activate the same pathway. We do not yet understand the cell at this level, but the field of systems biology is moving toward such global understanding of cell function.

Signaling through GPCRs uses a three-part system: a receptor, a G protein, and an effector protein. G proteins are active when bound to GTP and inactive when bound to GDP. A ligand binding to the receptor activates the G protein, which then activates the effector protein. Effector proteins include adenylyl cyclase, which produces the second messenger cAMP. Another effector protein, phospholipase C, cleaves the inositol phosphates and results in the release of Ca^{2+} from the ER.

Cell-to-Cell Interactions

In multicellular organisms, not only must cells be able to communicate with one another, but they must also be organized in specific ways. With the exception of a few primitive types of organisms, the hallmark of multicellular life is the organization of highly specialized groups of cells into *tissues*, such as blood and muscle. Remarkably, each cell within a tissue performs the functions of that tissue and no other, even though all cells of the body are derived from a single fertilized cell and contain the same genetic information—all of the genes found in the genome.

This kind of tissue organization requires that cells have both identity and specific kinds of cell-to-cell connections. As an organism develops, the cells acquire their identities by carefully controlling the *expression* of those genes, turning on the specific set of genes that encode the functions of each cell type. How do cells sense where they are, and how do they "know" which type of tissue they belong to? Table 9.2 provides a summary of the kinds of connections seen between cells that are explored in the following sections.

Surface proteins give cells identity

One key set of genes functions to mark the surfaces of cells, identifying them as being of a particular type. When cells make contact, they "read" each other's cell surface markers and react accordingly. Cells that are part of the same tissue type recognize each other, and they frequently respond by forming connections between their surfaces to better coordinate their functions.

Glycolipids

Most tissue-specific cell surface markers are glycolipids, that is, lipids with carbohydrate heads. The glycolipids on the surface of red blood cells are also responsible for the A, B, and O blood types.

MHC proteins

One example of the function of cell surface markers is the recognition of "self" and "nonself" cells by the immune system. This function is vital for multicellular organisms, which need to defend themselves against invading or malignant cells. The immune system of vertebrates uses a particular set of markers to distinguish self from nonself cells, encoded by genes of the *major histocompatibility complex* (*MHC*). Cell recognition and the immune system is covered in chapter 51.

Cell connections mediate cell-to-cell adhesion

Most cells in a multicellular organism are in physical contact with other cells at all times, usually as members of organized tissues such as those in a leaf or those in your lungs, heart, or gut. These cells and the mass of other cells clustered around them form long-lasting or permanent connections with one another called **cell junctions.**

The nature of the physical connections between the cells of a tissue in large measure determines what the tissue is like. Indeed, a tissue's proper functioning often depends critically on how the individual cells are arranged within it. Just as a house cannot maintain its structure without nails and cement, so a tissue cannot maintain its characteristic architecture without the appropriate cell junctions.

Cell junctions are divided into three categories, based on their functions: tight, anchoring, and communicating junctions (figure 9.17).

Tight junctions

Tight junctions connect the plasma membranes of adjacent cells in a sheet, preventing small molecules from leaking between the cells. This allows the sheet of cells to act as a wall within the organ, keeping molecules on one side or the other (figure 9.17a).

Creating sheets of cells. The cells that line an animal's digestive tract are organized in a sheet only one cell thick. One surface of the sheet faces the inside of the tract, and the other faces the extracellular space, where blood vessels are located. Tight junctions encircle each cell in the sheet, like a belt cinched around a person's waist. The junctions between neighboring cells are so

TABLE 9.2	Cell-to-Cell Connections and Cell Identity		
Type of Connection	**Structure**	**Function**	**Example**
Surface markers	Variable, integral proteins or glycolipids in plasma membrane	Identify the cell	MHC complexes, blood groups, antibodies
Tight junctions	Tightly bound, leakproof, fibrous protein seal that surrounds cell	Organizing junction; holds cells together such that materials pass *through* but not *between* the cells	Junctions between epithelial cells in the gut
Anchoring junction (Desmosome)	Intermediate filaments of cytoskeleton linked to adjoining cells through cadherins	Anchoring junction: binds cells together	Epithelium
Anchoring junction (Adherens junction)	Transmembrane fibrous proteins	Anchoring junction: connects extracellular matrix to cytoskeleton	Tissues with high mechanical stress, such as the skin
Communicating junction (Gap junction)	Six transmembrane connexon proteins creating a pore that connects cells	Communicating junction: allows passage of small molecules from cell to cell in a tissue	Excitable tissue such as heart muscle
Communicating junction (Plasmodesmata)	Cytoplasmic connections between gaps in adjoining plant cell walls	Communicating junction between plant cells	Plant tissues

securely attached that there is no space between them for leakage. Hence, nutrients absorbed from the food in the digestive tract must pass directly through the cells in the sheet to enter the blood because they cannot pass through spaces between cells.

Partitioning the sheet. The tight junctions between the cells lining the digestive tract also partition the plasma membranes of these cells into separate compartments. Transport proteins in the membrane facing the inside of the tract carry nutrients from that side to the cytoplasm of the cells. Other proteins, located in the membrane on the opposite side of the cells, transport those nutrients from the cytoplasm to the extracellular fluid, where they can enter the blood.

For the sheet to absorb nutrients properly, these proteins must remain in the correct locations within the fluid membrane.

Tight junctions effectively segregate the proteins on opposite sides of the sheet, preventing them from drifting within the membrane from one side of the sheet to the other. When tight junctions are experimentally disrupted, just this sort of migration occurs.

Anchoring junctions

Anchoring junctions mechanically attach the cytoskeleton of a cell to the cytoskeletons of other cells or to the extracellular matrix. These junctions are most common in tissues subject to mechanical stress, such as muscle and skin epithelium.

Cadherin and intermediate filaments *Desmosomes* connect the cytoskeletons of adjacent cells (figure 9.17b), and *hemidesmosomes* anchor epithelial cells to a basement membrane. Proteins called **cadherins,** most of which are single-pass transmembrane

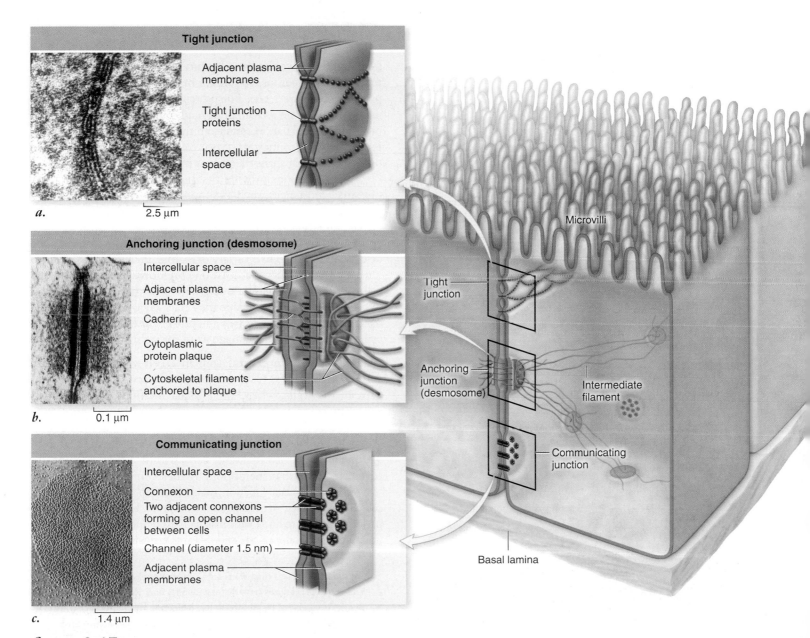

figure 9.17

AN OVERVIEW OF CELL JUNCTION TYPES. Here, the diagram of gut epithelial cells on the right illustrates the comparative structures and locations of common cell junctions. The detailed models on the left show the structures of the three major types of cell junctions: *(a)* tight junction; *(b)* anchoring junction, the example shown is a desmosome; *(c)* communicating junction, the example shown is a gap junction.

glycoproteins, create the critical link. Proteins link the short cytoplasmic end of a cadherin to the intermediate filaments in the cytoskeleton. The other end of the cadherin molecule projects outward from the plasma membrane, joining directly with a cadherin protruding from an adjacent cell similar to a firm handshake, binding the cells together. Connections between proteins tethered to the intermediate filaments are much more secure than connections between free-floating membrane proteins.

Cadherin and actin filaments. Cadherins can also connect the actin frameworks of cells in cadherin-mediated junctions (figure 9.18). When they do, they form less stable links between cells than when they connect intermediate filaments. Many kinds of actin-linking cadherins occur in different tissues. For example, during vertebrate development, the migration of neurons in the embryo is associated with changes in the type of cadherin expressed on their plasma membranes.

Integrin-mediated links. Anchoring junctions called **adherens junctions** connect the actin filaments of one cell with those of neighboring cells or with the extracellular matrix. The linking proteins in these junctions are members of a large superfamily of cell surface receptors called **integrins** that bind to a protein component of the extracellular matrix. At least 20 different integrins exist, having differently shaped binding domains.

Communicating junctions

Many cells communicate with adjacent cells through direct connections called **communicating junctions**. In these junctions, a chemical or electrical signal passes directly from one cell to an adjacent one. Communicating junctions permit small molecules or ions to pass from one cell to the other. In animals, these direct communication channels between cells are called *gap junctions*, and in plants, *plasmodesmata*.

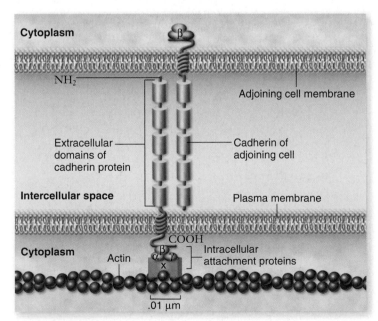

figure 9.18

A CADHERIN-MEDIATED JUNCTION. The cadherin molecule is anchored to actin in the cytoskeleton and passes through the membrane to interact with the cadherin of an adjoining cell.

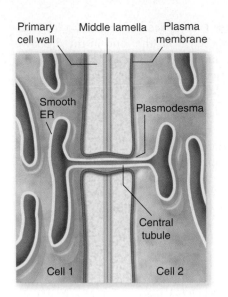

figure 9.19

PLASMODESMATA. Plant cells can communicate through specialized openings in their cell walls, called plasmodesmata, where the cytoplasm of adjoining cells are connected.

Gap junctions in animals. **Gap junctions** are composed of structures called connexons, complexes of six identical transmembrane proteins (see figure 9.17c). The proteins in a connexon are arranged in a circle to create a channel through the plasma membrane that protrudes several nanometers from the cell surface. A gap junction forms when the connexons of two cells align perfectly, creating an open channel that spans the plasma membranes of both cells.

Gap junctions provide passageways large enough to permit small substances, such as simple sugars and amino acids, to pass from one cell to the next, yet small enough to prevent the passage of larger molecules, such as proteins.

Gap junction channels are dynamic structures that can open or close in response to a variety of factors, including Ca^{2+} and H^+ ions. This gating serves at least one important function. When a cell is damaged, its plasma membrane often becomes leaky. Ions in high concentrations outside the cell, such as Ca^{2+}, flow into the damaged cell and close its gap junction channels. This isolates the cell and so prevents the damage from spreading to other cells.

Plasmodesmata in plants. In plants, cell walls separate every cell from all others. Cell–cell junctions occur only at holes or gaps in the walls, where the plasma membranes of adjacent cells can come into contact with one another. Cytoplasmic connections that form across the touching plasma membranes are called **plasmodesmata** (singular, plasmodesma) (figure 9.19). The majority of living cells within a higher plant are connected with their neighbors by these junctions.

Plasmodesmata function much like gap junctions in animal cells, although their structure is more complex. Unlike gap junctions, plasmodesmata are lined with plasma membrane and contain a central tubule that connects the endoplasmic reticulum of the two cells.

Cells in multicellular organisms are usually organized into tissues. This requires that cells have distinct identity and connections. Cell identity is conferred by glycoproteins, including MHC proteins, which are important in the immune system. Cell connections fall into three basic categories: tight, anchoring, and communicating junctions. Tight junctions help to make sheets of cells that form watertight seals, and anchoring junctions provide strength and flexibility. Communicating junctions include gap junctions in animals and plasmodesmata in plants.

9.1 Overview of Cell Communication (figure 9.1)

Cell communication requires signal molecules, called ligands, binding to specific receptor proteins producing a cellular response.

- Depending on the distance between signaling and responding cells there are four basic signaling mechanisms (figure 9.2)
  - *Direct contact—molecules on the plasma membrane of one cell contact the receptor molecules on an adjacent cell.*
  - *Paracrine signaling—short-lived signal molecules are released into the extracellular fluid and influence neighboring cells.*
  - *Endocrine signaling—long-lived hormones enter the circulatory system and are carried to target cells some distance away.*
  - *Synaptic signaling—short-lived neurotransmitters are released by neurons into the gap, called a synapse, between nerves and target cells.*
- Signal transduction refers to the intracellular events that result from a receptor binding a signal molecule.
- Proteins can be controlled by phosphate added by kinase and removed by phosphatase enzymes.

9.2 Receptor Types (figure 9.4)

Receptor proteins can be classified by both function and location.

- Receptors are broadly defined as intracellular or cell-surface receptors (membrane receptors).
- Membrane receptors are transmembrane proteins that transfer information across the membrane, but not the signal molecule.
- Membrane receptors are divided into three types:
  - *Channel-linked receptors are chemically gated ion channels that allow specific ions to pass through a central pore.*
  - *Enzymatic receptors are enzymes activated by binding a ligand; these enzymes are usually protein kinases.*
  - *G protein-coupled receptors interact with G proteins that control the function of effector proteins: enzymes or ion channels.*
- Some enzymatic and most G protein-coupled receptors produce second messengers, to relay messages in the cytoplasm.

9.3 Intracellular Receptors (figure 9.5)

Many cell signals are lipid-soluble and readily pass through the plasma membrane and bind to receptors in the cytoplasm or nucleus.

- Steroid hormones bind cytoplasmic receptors then are transported to the nucleus. Thus, they are called nuclear receptors.
- Nuclear receptors can directly affect gene expression, usually activating transcription of the genes they control.
- Nuclear receptors have three functional domains: hormone-binding, DNA-binding, and transcription-activating domains.
- Ligand binding changes receptor shape, releasing an inhibitor occupying the DNA-binding site.
- A cell's response to a lipid-soluble signal depends on the hormone–receptor complex and the other protein coactivators present.
- Some intracellular receptors activate cellular enzymes and do not affect gene expression.

9.4 Signal Transduction Through Receptor Kinases

Receptor kinases in plants and animals recognize hydrophilic ligands and influence the cell cycle, cell migration, cell metabolism, and cell proliferation.

- Because they are involved in growth control, alterations of receptor kinases and their signaling pathways can lead to cancer.

- Ligand binding causes the receptor to phosphorylate itself on specific tyrosines: this is called autophosphorylation.
- The activated receptor can also phosphorylate other intracellular proteins.
- Adapter proteins can bind to phosphotyrosine and act as links between the receptors and downstream signaling events.
- Protein kinase can amplify a signal because at each step of the cascade an enzyme acts on a number of substrate molecules.
- Scaffold proteins organize cascade proteins into a single complex so they can act sequentially and be optimally functional.
- Receptor tyrosine kinases are inactivated by dephosphorylation or internalization where they are degraded or recycled.

9.5 Signal Transduction Through G Protein-Coupled Receptors (figure 9.11)

G protein-coupled receptors function through activation of G proteins that link receptors to effector proteins.

- G proteins are active bound to GTP and inactive bound to GDP. Receptors promote exchange of GDP for GTP.
- The activated G protein dissociates into two parts, G_α and $G_{\beta\gamma}$, each of which can act on effector proteins.
- G_α also hydrolyzes GTP to GDP to inactivate the G protein.
- Effector proteins can produce second messengers.
- Two common effector proteins are adenylyl cyclase and phospholipase C, which produce second messengers known as cAMP, and DAG and IP_3, respectively.
- Ca^{2+} is also a second messenger. Ca^{2+} release is triggered by IP_3 binding to channel-linked receptors in the ER.
- Ca^{2+} can bind to a cytoplasmic protein calmodulin, which in turn activates other proteins, producing a variety of responses.
- Different receptors can activate the same effector, which will produce the same second messenger leading to the same response.
- Different receptor subtypes or isoforms lead to different effects in different cells.
- Different receptor types such as G protein-coupled receptors and receptor tyrosine kinases can activate the same signaling pathways.

9.6 Cell-to-Cell Interactions (figure 9.17)

In multicellular organisms, cells are organized in specific ways so they can function as tissues.

- Cells form long-lasting or permanent connections with one another called cell junctions.
- There are three categories of cell junctions:
  - *Tight junctions connect the plasma membranes of adjacent cells into sheets and prevent small molecules from leaking out of the cells.*
  - *Anchoring junctions connect adjacent cells.*
    - Desmosomes link the intermediate and actin filaments of neighboring cells by cadherin-mediated links.
    - Adherens junctions connect actin filaments of adjacent cells or attach to the extracellular matrix by integrin-mediated links.
  - *Communicating junctions allow cytoplasm to move between cells.*
    - In animals gap junctions composed of connexons allow the passage of small molecules between cells.
    - In plants junctions lined with plasma membrane called plasmodesmata penetrate the cell wall and connect cells.

SELF TEST

1. What is a ligand?
   a. An integral membrane protein associated with G proteins
   b. A DNA-binding protein that alters gene expression
   c. A cytoplasmic second messenger molecule
   d. A molecule or protein that can bind to a receptor

2. In the case of paracrine signaling the ligand is—
   a. produced by the cell itself
   b. secreted by neighboring cells
   c. present on the plasma membrane of neighboring cells
   d. secreted by distant cells

3. A neurotransmitter functions as a ligand in which type of signaling?
   a. Direct contact
   b. Endocrine
   c. Synaptic signaling
   d. Autocrine

4. The function of a _____ is to add phosphates to proteins, whereas a _____ functions to remove the phosphates.
   a. tyrosine; serine
   b. protein phosphatase; protein dephosphatase
   c. protein kinase; protein phosphatase
   d. receptor; ligand

5. Which of the following type(s) of membrane receptors functions by changing the phosphorylation state of proteins in the cell?
   a. Channel-linked receptor
   b. Enzymatic receptor
   c. G protein-coupled receptor
   d. Both b and c

6. How does the function of an intracellular receptor differ from that of a membrane receptor?
   a. The intracellular receptor binds a ligand.
   b. The intracellular receptor binds DNA.
   c. The intracellular receptor activates a kinase.
   d. The intracellular receptor functions as a second messenger.

7. During a protein kinase signal cascade—
   a. Sequential phosphorylation of different kinases leads to a change in gene expression.
   b. Multiple G proteins become activated.
   c. Phosphorylation of adapter proteins leads to the formation of second messengers.
   d. The number of MAP kinase proteins present in the cytoplasm is amplified.

8. What is the function of Ras during tyrosine kinase cell signaling?
   a. It activates the opening of channel-linked receptors.
   b. It synthesizes the formation of second messengers.
   c. It phosphorylates other enzymes as part of a pathway.
   d. It links the receptor protein to the MAP kinase pathway.

9. Which of the following best describes the immediate effect of ligand binding to a G protein-coupled receptor?
   a. The G protein trimer releases a GDP and binds a GTP.
   b. The G protein trimer dissociates from the receptor.
   c. The G protein trimer interacts with an effector protein.
   d. The α subunit of the G protein becomes phosphorylated.

10. The amplification of a cellular signal requires all of the following except—
    a. A ligand
    b. DNA
    c. A second messenger
    d. A protein kinase

11. *Adenylyl cyclase* is responsible for the production of which second-messenger molecule?
    a. Cyclic-AMP
    b. Calcium
    c. IP_3
    d. Calmodulin

12. The response to signaling through G protein-coupled receptors can vary in different cells because
    a. All receptors act through the same G protein.
    b. Different isoforms of a receptor bind the same ligand but activate different effectors.
    c. The amount of receptor in the membrane differs in different cell types.
    d. Different receptors can activate the same effector.

13. What is the function of the tight junctions in the formation of a tissue?
    a. Tight junctions connect one cell to the next, creating a barrier between the cells.
    b. Tight junctions form a strong anchor between two cells.
    c. Tight junctions allow for the movement of small molecules between cells.
    d. Tight junctions connect the cell to the extracellular matrix.

14. Cadherins and intermediate filament proteins are associated with _____, whereas connexons are associated with _____
    a. tight junctions; anchoring junctions
    b. cell surface markers; tight junctions
    c. desmosomes; gap junctions
    d. adherens junctions; plasmodesmata

15. Cells are able to anchor themselves to the extracellular matrix through the activity of—
    a. connexon proteins
    b. MHC proteins
    c. cadherin proteins
    d. integrin proteins

CHALLENGE QUESTIONS

1. Describe the common features found in all examples of cellular signaling discussed in this chapter. Provide examples to illustrate your answer.

2. The sheet of cells that form the gut epithelium folds into peaks called villi and valleys called crypts. The cells within the crypt region secrete a protein, Netrin-1, that becomes concentrated within the crypts. Netrin-1 is the ligand for a receptor protein that is found on the surface of all gut epithelial cells. Netrin-1 binding triggers a signal pathway that promotes cell growth. Gut epithelial cells undergo apoptosis (cell death) in the absence of Netrin-1 ligand binding.
   a. How would you characterize the type of signaling (autocrine, paracrine, endocrine) found in this system?
   b. Predict where the greatest amount of cell growth and cell death would occur in the epithelium.
   c. The loss of the Netrin-1 receptor is associated with some types of colon cancer. Suggest an explanation for the link between this signaling pathway and tumor formation.

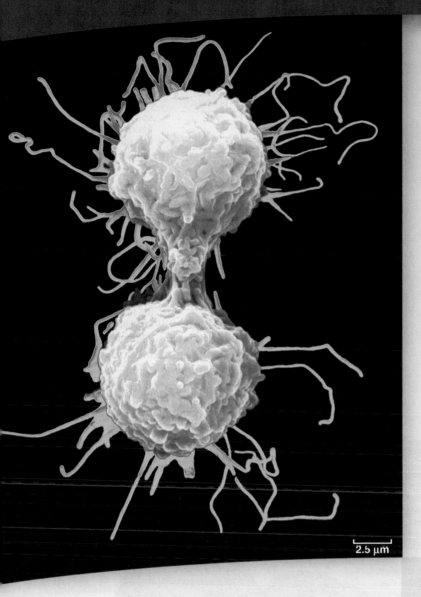

2.5 μm

introduction

ALL SPECIES OF ORGANISMS—bacteria, alligators, the weeds in a lawn—grow and reproduce. From the smallest creature to the largest, all species produce offspring like themselves and pass on the hereditary information that makes them what they are. In this chapter, we examine how cells, like the white blood cell shown in the figure, divide and reproduce. Cell division is necessary for the growth of organisms, for wound healing, and to replace cells lost regularly, like the cells in your skin and in the lining of your gut. The mechanism of cell reproduction and its biological consequences have changed significantly during the evolution of life on Earth. The process is complex in eukaryotes, involving both the replication of chromosomes and their separation into daughter cells. Much of what we are learning about the causes of cancer relates to how cells control this process, and in particular their propensity to divide, a mechanism that in broad outline remains the same in all eukaryotes.

concept outline

Eukaryotic Chromosomes

Chromosomes were first observed by the German embryologist Walther Flemming (1843–1905) in 1879, while he was examining the rapidly dividing cells of salamander larvae. When Flemming looked at the cells through what would now be a rather primitive light microscope, he saw minute threads within their nuclei that appeared to be dividing lengthwise. Flemming called their division **mitosis**, based on the Greek word *mitos*, meaning "thread."

Chromosome number varies among species

Since their initial discovery, chromosomes have been found in the cells of all eukaryotes examined. Their number may vary enormously from one species to another. A few kinds of organisms have only a single pair of chromosomes, whereas some ferns have more than 500 pairs (table 10.1). Most eukaryotes have between 10 and 50 chromosomes in their body cells.

Human cells each have 46 chromosomes, consisting of 23 nearly identical pairs (figure 10.4). Each of these 46 chromosomes contains hundreds or thousands of genes that play important roles in determining how a person's body develops and functions. For this reason, possession of all the chromosomes is essential to survival. Human embryos missing even one chromosome, a condition called *monosomy*, do not survive in most cases. Having an extra copy of any one chromosome, a condition called *trisomy*, is usually fatal except where the smallest chromosomes are involved. (You'll learn more about human chromosome abnormalities in chapter 13.)

Eukaryotic chromosomes exhibit complex structure

Researchers have learned a great deal about chromosome structure and composition in the more than 125 years since their discovery. But despite intense research, the exact structure of eukaryotic chromosomes during the cell cycle remains unclear. The structures described in this chapter represent the currently accepted model.

Composition of chromatin

Chromosomes are composed of **chromatin**, a complex of DNA and protein; most chromosomes are about 40% DNA and 60% protein. A significant amount of RNA is also associated with chromosomes because chromosomes are the sites of RNA synthesis.

The DNA of a single chromosome is one very long, double-stranded fiber that extends unbroken through the chromosome's entire length. A typical human chromosome contains

TABLE 10.1	Chromosome Number in Selected Eukaryotes
Group	**Total Number of Chromosomes**
FUNGI	
Neurospora (haploid)	7
Saccharomyces (a yeast)	16
INSECTS	
Mosquito	6
Drosophila	8
Honeybee	diploid females 32, haploid males 16
Silkworm	56
PLANTS	
Haplopappus gracilis	2
Garden pea	14
Corn	20
Bread wheat	42
Sugarcane	80
Horsetail	216
Adder's tongue fern	1262
VERTEBRATES	
Opossum	22
Frog	26
Mouse	40
Human	46
Chimpanzee	48
Horse	64
Chicken	78
Dog	78

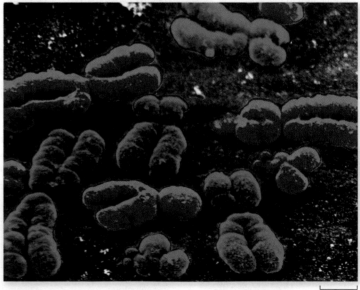

10.5 μm

figure 10.4

HUMAN CHROMOSOMES. This SEM micrograph shows human chromosomes as they appear immediately before nuclear division. Each DNA molecule has already replicated, forming identical copies held together at a visible constriction called the centromere. False color has been added to the chromosomes.

about 140 million (1.4×10^8) nucleotides in its DNA. If we think of each nucleotide as a "word," then the amount of information an average chromosome contains would fill about 280 printed books of 1000 pages each, with 500 "words" per page. If we could lay out the strand of DNA from a single chromosome in a straight line, it would be about 5 cm (2 in.) long. Fitting such a strand into a cell nucleus is like cramming a string the length of a football field into a baseball—and that's only 1 of 46 chromosomes! In the cell, however, the DNA is coiled, allowing it to fit into a much smaller space than would otherwise be possible.

The organization of chromatin in the nondividing nucleus is not well understood, but geneticists have recognized for years that some domains of chromatin, called **heterochromatin,** are not expressed, and other domains of chromatin, called **euchromatin,** are expressed. This genetically measurable state is also related to the physical state of chromatin, although researchers are just beginning to see the details.

Chromosome structure

If we gently disrupt a eukaryotic nucleus and examine the DNA with an electron microscope, we find that it resembles a string of beads (figure 10.5). Every 200 nucleotides (nt), the DNA duplex (double strand) is coiled around a core of eight histone proteins. Unlike most proteins, which have an overall negative charge, histones are positively charged because of an abundance of the basic amino acids arginine and lysine. Thus, they are strongly attracted to the negatively charged phosphate groups of the DNA, and the histone cores act as "magnetic forms" that promote and guide the coiling of the DNA. The complex of DNA and histone proteins is termed a **nucleosome.**

Further coiling occurs when the string of nucleosomes is wrapped into higher order coils called *solenoids.* The precise path of this higher order folding of chromatin is still a subject of some debate, but it leads to a fiber with a diameter of 30 nm and thus is often called the 30-nm fiber. This 30-nm fiber, or solenoid, is the usual state of interphase (nondividing) chromatin.

During mitosis the chromatin in the solenoid is arranged around a scaffold of protein assembled at this time to achieve maximum compaction of the chromosomes. This process prepares the chromosomes for the events of mitosis described later on. The exact nature of this compaction is unknown, but one long-standing model involves radial looping of the solenoid about the protein scaffold, aided by a complex of proteins called **condensin.** The protein scaffold itself is actually what gives mitotic chromosomes their distinctive shape.

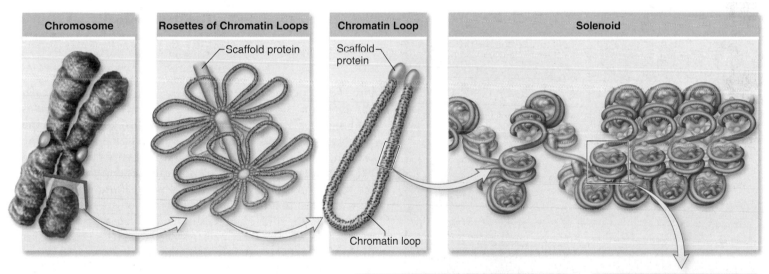

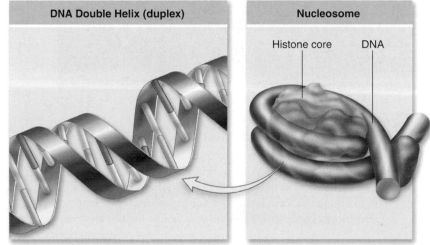

figure 10.5

LEVELS OF EUKARYOTIC CHROMOSOMAL ORGANIZATION. Each chromosome consists of a long double-stranded DNA molecule. These strands require further packaging to fit into the cell nucleus. The DNA duplex is tightly bound to and wound around proteins called histones. The DNA-wrapped histones are called nucleosomes. The nucleosomes are further coiled into a solenoid. This solenoid is then organized into looped domains. The final organization of the chromosome is unknown, but it appears to involve further radial looping into rosettes around a preexisting scaffolding of protein. The arrangement illustrated here is one of many possibilities.

Chromosome karyotypes

Chromosomes vary in size, staining properties, the location of the centromere (a constriction found on all chromosomes, described shortly), the relative length of the two arms on either side of the centromere, and the positions of constricted regions along the arms. The particular array of chromosomes an individual organism possesses is called its **karyotype.** The karyotype in figure 10.6 shows the set of chromosomes from a single human individual, exhibiting variations in size and structure.

When defining the number of chromosomes in a species, geneticists count the **haploid (*n*)** number of chromosomes. This refers to one complete set of chromosomes necessary to define an organism. For humans and many other species, the normal number of chromosomes in a cell is called the **diploid (2*n*)** number, which is twice the haploid number. For humans, the haploid number is 23 and the diploid number is 46. Diploid chromosomes reflect the equal genetic contribution that parents make to offspring. We refer to the maternal and paternal chromosomes as being **homologous,** and each one of the pair is termed a **homologue.**

Chromosome replication

Chromosomes as seen in a karyotype are only present for a brief period during cell division. Prior to replicating, each chromosome is composed of a single DNA molecule that is arranged

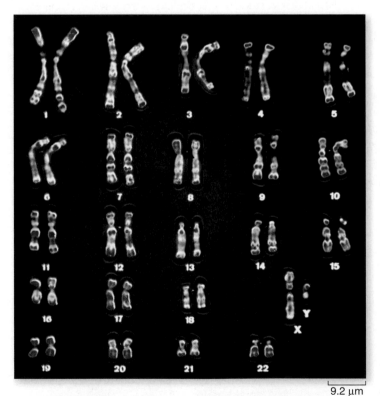

9.2 μm

figure 10.6

A HUMAN KARYOTYPE. The individual chromosomes that make up the 23 pairs differ widely in size and in centromere position. In this preparation, the chromosomes have been specifically stained to indicate differences in their composition and to distinguish them clearly from one another. Notice that members of a chromosome pair are very similar but not identical.

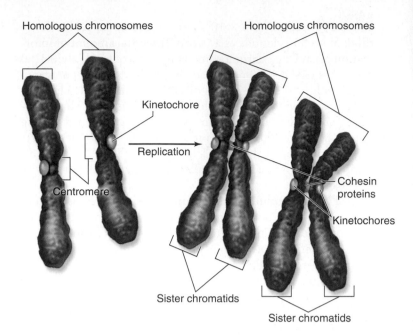

figure 10.7

THE DIFFERENCE BETWEEN HOMOLOGOUS CHROMOSOMES AND SISTER CHROMATIDS. Homologous chromosomes are the maternal and paternal copies of the same chromosome—say, chromosome number 16. Sister chromatids are the two replicas of a single chromosome held together at their centromeres by cohesin proteins after DNA replication. The kinetochore (described later in the chapter) is composed of proteins found at the centromere that attach to microtubules during mitosis.

into the 30-nm fiber described earlier. After replication, each chromosome is composed of two identical DNA molecules held together by a complex of proteins called **cohesins.** As the chromosomes become more condensed and arranged about the protein scaffold, they become visible as two strands that are held together. At this point, we still call this one chromosome, but it is composed of two sister **chromatids** (figure 10.7).

The fact that the products of replication are held together is critical to the division process. One problem that a cell must solve is how to ensure that each new cell receives a complete set of chromosomes. If we were designing a system, we might use some kind of label to identify each chromosome, much like most of us use when we duplicate files on a computer. The cell has no mechanism to label chromosomes; instead, it keeps the products of replication together until the moment of chromosome segregation, ensuring that one copy of each chromosome goes to each daughter cell. This separation of sister chromatids is the key event in the mitotic process described in detail shortly.

Eukaryotic chromosomes are complex structures that can be compacted for cell division. During interphase, DNA is coiled around proteins into a structure called a nucleosome. The string of nucleosomes is further coiled into a solenoid (30-nm fiber). After chromosome replication, the resulting chromatids are held together for the division process by proteins called cohesins.

Overview of the Eukaryotic Cell Cycle

Compared with prokaryotes, the increased size and more complex organization of eukaryotic genomes required radical changes in the partitioning of the two replicas of the genome into daughter cells. The overall process involves the duplication of the genome, its accurate segregation, and the division of cellular contents. These events make up the **cell cycle.**

The cell cycle is divided into five phases

The cell cycle is divided into phases based on the key events of genome duplication and segregation. The cell cycle is usually diagrammed using the metaphor of a clock face (figure 10.8).

Biologists separate the cell cycle into five main phases:

- **G_1 (gap phase 1)** is the primary growth phase of the cell. The term *gap phase* refers to its filling the gap between cytokinesis and DNA synthesis. For most cells, this phase encompasses the major portion of the cell cycle.
- **S (synthesis)** is the phase in which the cell synthesizes a replica of the genome.
- **G_2 (gap phase 2)** is the second growth phase, in which preparations are made for separation of the newly replicated genome. This phase fills the gap between DNA synthesis and the beginning of mitosis. During this phase, mitochondria and other organelles replicate, chromosomes prepare to condense, and microtubules begin to assemble at a spindle.

  G_1, S, and G_2 together constitute **interphase,** the portion of the cell cycle between cell divisions.
- **M (mitosis)** is the phase of the cell cycle in which the spindle apparatus assembles, binds to the chromosomes, and moves the sister chromatids apart. Mitosis is the essential step in the separation of the two daughter genomes. Although mitosis is a continuous process, it is traditionally subdivided into five stages: prophase, prometaphase, metaphase, anaphase, and telophase.
- **C (cytokinesis)** is the phase of the cell cycle when the cytoplasm divides, creating two daughter cells. In animal cells, the microtubule spindle helps position a contracting ring of actin that constricts like a drawstring to pinch the cell in two. In cells with a cell wall, such as plant cells, a plate forms between the dividing cells.

The duration of the cell cycle varies depending on cell type

The time it takes to complete a cell cycle varies greatly. Cells in animal embryos can complete their cell cycle in under 20 min; the shortest known animal nuclear division cycles occur in fruit fly embryos (8 min). Cells such as these simply divide their nuclei as quickly as they can replicate their DNA, without cell growth. Half of their cycle is taken up by S, half by M, and essentially none by G_1 or G_2.

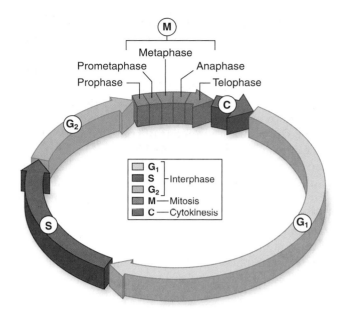

figure 10.8

THE CELL CYCLE. The cell cycle is depicted as a circle. The first gap phase, G_1, involves growth and preparation for DNA synthesis. During S phase, a copy of the genome is synthesized. The second gap phase, G_2, prepares the cell for mitosis. During mitosis, replicated chromosomes are partitioned. Cytokinesis divides the cell into two cells with identical genomes.

Because mature cells require time to grow, most of their cycles are much longer than those of embryonic tissue. Typically, a dividing mammalian cell completes its cell cycle in about 24 h, but some cells, such as certain cells in the human liver, have cell cycles lasting more than a year. During the cycle, growth occurs throughout the G_1 and G_2 phases (referred to as gap phases because they separate S from M), as well as during the S phase. The M phase takes only about an hour, a small fraction of the entire cycle.

Most of the variation in the length of the cell cycle between one organism or cell type and another occurs in the G_1 phase. Cells often pause in G_1 before DNA replication and enter a resting state called the **G_0 phase;** cells may remain in this phase for days to years before resuming cell division. At any given time, most of the cells in an animal's body are in G_0 phase. Some, such as muscle and nerve cells, remain there permanently; others, such as liver cells, can resume G_1 phase in response to factors released during injury.

Cell division in eukaryotes is a complex process that involves five phases: a first gap phase (G_1); DNA synthesis phase (S); a second gap phase (G_2); mitosis (M), in which chromosomes are separated; and cytokinesis (C) in which a cell becomes two separate cells.

Interphase: Preparation for Mitosis

The events that occur during interphase—the G_1, S, and G_2 phases—are very important for the successful completion of mitosis. During G_1, cells undergo the major portion of their growth. During the S phase, each chromosome replicates to produce two sister chromatids, which remain attached to each other at the centromere. In the G_2 phase, the chromosomes coil even more tightly.

The **centromere** is a point of constriction on the chromosome containing certain repeated DNA sequences that bind specific proteins. These proteins make up a disklike structure called the **kinetochore.** This disk functions as an attachment site for microtubules necessary to separate the chromosomes during cell division (figure 10.9). As seen in figure 10.6, each chromosome's centromere is located at a characteristic site along the length of the chromosome.

After the S phase, the sister chromatids appear to share a common centromere, but at the molecular level the DNA of the centromere has actually already replicated, so there are two complete DNA molecules. Functionally, however, the two chromatids have a single centromere due to their being attached by cohesin proteins at the centromere site (figure 10.10). In metazoan animals, the cohesins that hold sister chromatids together after replication appear to be replaced by condensin during the process of chromosome compaction. This leaves

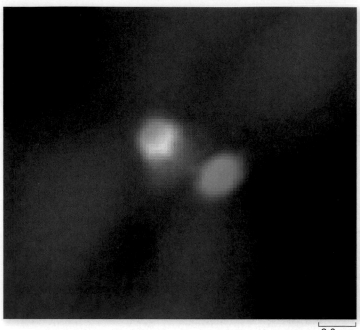

2.0 µm

figure 10.10

PROTEINS FOUND AT THE CENTROMERE. In this image DNA, a cohesin protein and a kinetochore protein have all been labeled with a different colored fluorescent dye. Cohesin (*red*), which holds centromeres together, lies between the sister chromatids (*blue*). Each sister chromatid has its own separate kinetochore (*green*).

the chromosomes still attached tightly at the centromere, but loosely attached elsewhere.

The cell grows throughout interphase. The G_1 and G_2 segments of interphase are periods of active growth, during which proteins are synthesized and cell organelles are produced. The cell's DNA replicates only during the S phase of the cell cycle.

After the chromosomes have replicated in S phase, they remain fully extended and uncoiled, which makes them invisible when viewed with the light microscope. In G_2 phase, they begin the process of **condensation,** coiling ever more tightly. Special *motor proteins* are involved in the rapid final condensation of the chromosomes that occurs early in mitosis. Also during G_2 phase, the cells begin to assemble the machinery they will later use to move the chromosomes to opposite poles of the cell. In animal cells, a pair of microtubule-organizing centers called **centrioles** replicate, producing one for each pole. All eukaryotic cells undertake an extensive synthesis of **tubulin,** the protein that forms microtubules.

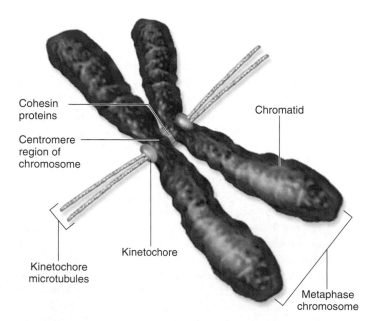

Cohesin proteins

Centromere region of chromosome

Kinetochore

Kinetochore microtubules

Chromatid

Kinetochore

Metaphase chromosome

figure 10.9

KINETOCHORES. Separation of sister chromatids during mitosis depends on microtubules attaching to proteins found in the kinetochore. These kinetochore proteins are assembled on the centromere of chromosomes. The centromeres of the two sister chromatids are held together by cohesin proteins.

Interphase includes the G_1, S, and G_2 phases of the cell cycle. During interphase, the cell grows; replicates chromosomes, organelles, and centrioles; and synthesizes components needed for mitosis, including tubulin.

Mitosis: Chromosome Segregation

The process of mitosis is one of the most dramatic and beautiful biological processes that can be readily observed. In our attempts to understand this process, we have divided it into discrete phases but it should always be remembered that this is a dynamic, continuous process, not a set of discrete steps. This process is shown both schematically and in micrographs in figure 10.11.

During prophase, the mitotic apparatus forms

When the chromosome condensation initiated in G_2 phase reaches the point at which individual condensed chromosomes first become visible with the light microscope, the first stage of mitosis, **prophase,** has begun. The condensation process continues throughout prophase; consequently, chromosomes that start prophase as minute threads appear quite bulky before its conclusion. Ribosomal RNA synthesis ceases when the portion of the chromosome bearing the rRNA genes is condensed.

The spindle and centrioles

The assembly of the **spindle** apparatus that will later separate the sister chromatids occurs during prophase. The normal microtubule structure in the cell disassembled in the G_2 phase is replaced by the spindle. In animal cells, the two centriole pairs formed during G_2 phase begin to move apart early in prophase, forming between them an axis of microtubules referred to as spindle fibers. By the time the centrioles reach the opposite poles of the cell, they have established a bridge of microtubules, called the spindle apparatus, between them. In plant cells, a similar bridge of microtubular fibers forms between opposite poles of the cell, although centrioles are absent in plant cells.

In animal cell mitosis, the centrioles extend a radial array of microtubules toward the nearby plasma membrane when they reach the poles of the cell. This arrangement of microtubules is called an **aster.** Although the aster's function is not fully understood, it probably braces the centrioles against the membrane and stiffens the point of microtubular attachment during the retraction of the spindle. Plant cells, which have rigid cell walls, do not form asters.

Breakdown of the nuclear envelope

During the formation of the spindle apparatus, the nuclear envelope breaks down, and the endoplasmic reticulum reabsorbs its components. At this point, the microtubular spindle fibers extend completely across the cell, from one pole to the other. Their orientation determines the plane in which the cell will subsequently divide, through the center of the cell at right angles to the spindle apparatus.

During prometaphase, chromosomes attach to the spindle

The transition from prophase to **prometaphase** occurs following the disassembly of the nuclear envelope. During prometaphase the condensed chromosomes become attached to the spindle by their kinetochores. Each chromosome possesses two kinetochores, one attached to the centromere region of each sister chromatid (see figure 10.9).

Microtubule attachment

As prometaphase continues, a second group of microtubules grow from the poles of the cell toward the centromeres. These microtubules are captured by the kinetochores on each pair of sister chromatids. This results in the kinetochores of each sister chromatid being connected to opposite poles of the spindle.

This bipolar attachment is critical to the process of mitosis; any mistakes in microtubule positioning can be disastrous. For example, the attachment of the kinetochores of both sister chromatids to the same pole leads to a failure of sister chromatid separation, and they will be pulled to the same pole ending up in the same daughter cell, with the other daughter cell missing that chromosome.

Movement of chromosomes to the cell center

With each chromosome attached to the spindle by microtubules from opposite poles to the kinetochores of sister chromatids, the chromosomes begin to move to the center of the cell. This movement is jerky, as if a chromosome is being pulled toward both poles at the same time. This process is called *congression,* and it eventually leads to all of the chromosomes being arranged at the equator of the cell with the sister chromatids of each chromosome oriented to opposite poles by their kinetochore microtubules.

The force that moves chromosomes has been of great interest since the process of mitosis was first observed. Two basic mechanisms have been proposed to explain this: (1) assembly and disassembly of microtubules provides the force to move chromosomes, and (2) motor proteins located at the kinetochore and poles of the cell pull on microtubules to provide force. Data have been obtained that support both mechanisms.

In support of the microtubule-shortening proposal isolated chromosomes can be pulled by microtubule disassembly. The spindle is a very dynamic structure, with microtubules being added to at the kinetochore and shortened at the poles, even during metaphase. In support of the motor protein proposal, multiple motor proteins have been identified as kinetochore proteins, and inhibition of the motor protein dynein slows chromosome separation at anaphase. Like many phenomena that we analyze in living systems, the answer is not a simple either–or choice; both mechanisms are probably at work.

In metaphase, the centromeres align

The alignment of the chromosomes in the center of the cell signals the third stage of mitosis, **metaphase.** When viewed with a light microscope, the chromosomes appear to array themselves in a circle along the inner circumference of the cell, just as the equator girdles the Earth (figure 10.12). An imaginary plane perpendicular to the axis of the spindle that passes through this circle is called the *metaphase plate.* The metaphase plate is not an actual structure, but rather an indication of the future axis of cell division.

Positioned by the microtubules attached to the kinetochores of their centromeres, all of the chromosomes line up on the metaphase plate. At this point their centromeres are neatly arrayed in

together. The destruction of Rec8 protein on the chromosome arms appears to be what allows homologues to be pulled apart at anaphase I.

This leaves the key distinction between meiosis and mitosis being the maintenance of sister chromatid cohesion at the centromere during all of meiosis I, while cohesion is lost from the chromosome arms during anaphase I (see figure 11.9). It is unclear how the cohesin complex at the centromere is protected from destruction, but it apparently depends on the replacement of Scc1 by the Rec8 protein. This meiosis-specific cohesin is critical to maintaining sister chromatid cohesion at the centromere until anaphase II.

Sister kinetochores are attached to the same pole during meiosis I

The cosegregation of sister centromeres requires that the kinetochores of sister chromatids are attached to the same pole during meiosis I. This attachment is in contrast to both mitosis (see figure 11.9) and meiosis II, in which sister kinetochores must become attached to opposite poles.

The underlying basis of this monopolar attachment of sister kinetochores is unclear, but it seems to be based on structural differences between centromere/kinetochore complexes in meiosis I and in mitosis. Mitotic kinetochores visualized with the electron microscope appear to be recessed, making bipolar attachment more likely. Meiosis I kinetochores protrude more, making monopolar attachment easier.

It is clear that both the maintenance of sister chromatid cohesion at the centromere and monopolar attachment are required for the segregation of homologues that distinguishes meiosis I from mitosis.

Replication is suppressed between meiotic divisions

After a mitotic division, a new round of DNA replication must occur before the next division. For meiosis to succeed in halving the number of chromosomes, this replication must be suppressed between the two divisions. The detailed mechanism of suppression of replication between meiotic division is un-

known. One clue is the observation that the level of one of the cyclins, cyclin B, is reduced between meiotic divisions, but is not lost completely, as it is between mitotic divisions.

During mitosis, the destruction of mitotic cyclin is necessary for a cell to enter another division cycle. The result of this maintenance of cyclin B between meiotic divisions in germ-line cells is the failure to form initiation complexes necessary for DNA replication to proceed. This failure to form initiation complexes appears to be critical to suppressing DNA replication.

Meiosis produces cells that are not identical

The daughter cells produced by mitosis are identical to the parental cell, at least in terms of their chromosomal constitution. This exact copying is critical to producing new cells for growth, for development, and for wound healing. Meiosis, because of the random orientation of different chromosomes at the first meiotic division and because of crossing over, rarely produces cells that are identical (figure 11.10). The gametes from meiosis all carry an entire haploid set of chromosomes, but these chromosomes are a mixture of maternal and paternal homologues; furthermore, the homologues themselves have exchanged material by crossing over. The resulting variation is essential for evolution and is the reason that sexually reproducing populations have much greater variation than asexually reproducing ones.

Meiosis is not only critical for the process of sexual reproduction, but is also the foundation for understanding the basis of heredity. The different cells produced by meiosis form the basis for understanding the behavior of observable traits in genetic crosses. In the next two chapters we will follow the behavior of traits in genetic crosses and see how this correlates with the behavior of chromosomes in meiosis.

Meiosis is characterized by homologue pairing and exchange; by loss of sister chromatid cohesion in the arms during the first division, but not at the centromere until the second division; and by the suppression of DNA replication between the two meiotic divisions. The haploid cells that result from meiosis are not identical and therefore allow variation in offspring.

figure 11.10

INDEPENDENT ASSORTMENT INCREASES GENETIC VARIABILITY. Independent assortment contributes new gene combinations to the next generation because the orientation of chromosomes on the metaphase plate is random. For example, in cells with three chromosome pairs, eight different gametes can result, each with different combinations of parental chromosomes. This is also increased by crossing over, or genetic recombination as this further shuffles the arrangements of genes on chromosomes.

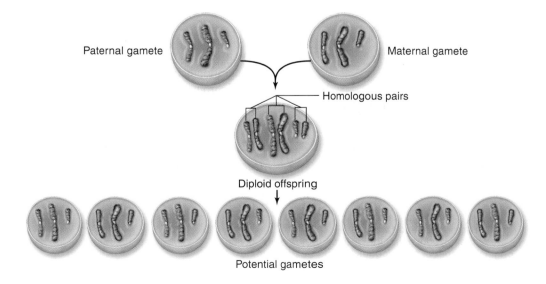

Paternal gamete   Maternal gamete

Homologous pairs

Diploid offspring

Potential gametes

11.1 Sexual Reproduction Requires Meiosis

(figure 11.2*b*)

Meiosis is a reductive division that converts a diploid cell into four haploid cells, each with one complete set of chromosomes.

- Egg and sperm are haploid and contain one set of all chromosomes.
- During fertilization, or syngamy, the fusion of two haploid gametes results in a diploid zygote, which contains two sets of chromosomes.
- Meiosis and fertilization constitute a cycle of reproduction in sexual organisms as they alternate between diploid and haploid chromosome numbers.
- Somatic cells divide by mitosis and form the body of an organism.
- Cells that form haploid gametes by meiosis are called germ-line cells.

11.2 Features of Meiosis

In diploid organisms meiosis consists of two rounds of division called meiosis I and meiosis II with no replication between divisions.

- The pairing of homologous chromosomes, called synapsis, occurs during early prophase I; synapsis does not occur in mitosis.
- During synapsis homologous chromosomes pair along their entire length, often joined by a structure called the synatonemal complex (figure 11.4).
- During synapsis, crossing over occurs between homologous chromosomes exchanging chromosomal material (figure 11.5).
- Because the homologues are paired, they move as a unit to the metaphase plate during metaphase I.
- During anaphase I, homologues of each pair are pulled to opposite poles, producing two cells each with one complete set of chromosomes.
- Meiosis II is like mitosis with no replication of DNA.

11.3 The Process of Meiosis (figures 11.8 and 11.9)

Meiosis depends on homologues exchanging chromosomal material by crossing over, which assists in holding homologues together during nuclear division.

- Meiotic cells have an interphase period similar to mitosis with G_1, S, and G_2 phases.
- During prophase I homologous chromosomes pair along their entire length in a process called synapsis.
- *The sister chromatids of each homologue are held together by cohesin proteins in a process called sister chromatid cohesion.*
- *While joined by the synaptonemal complex, recombinant nodules form, allowing homologues to exchange genetic material; crossing over between sister chromatids (of the same homologue) is suppressed.*
- *Sites of crossing over are called chiasmata.*
- *When crossing over is complete, the synaptonemal complex breaks down, leaving paired homologues joined together only at chiasmata.*
- *Sister chromatids of each homologue remain joined at their centromeres.*
- *Prior to metaphase the chiasmata between homologous pairs move to the ends of the chromosome arms where they become terminal chiasmata, holding the homologues together.*
- The nuclear envelope disperses and the spindle apparatus forms.
- During metaphase I homologous pairs align at the cell equator, or metaphase plate.
- *Spindle fibers attach to the kinetochores of the homologues and not those of the sister chromatids.*
- *Homologues of each pair become attached by kinetochore microtubules to opposite poles.*
- *The orientation of each homologous pair on the equator is random; either the maternal or paternal homologue may be oriented toward a given pole.*
- During anaphase I the homologues of each pair are pulled to opposite poles.
- *During anaphase I cohesin proteins joining sister chromatids are lost on the arms, but not between the centromeres that connect sister chromatids.*
- *Loss of sister chromatid cohesion on the arms but not the centromeres allows homologues to separate.*
- *Kinetochore microtubules shorten, pulling each homologue with its two sister chromatids to opposite poles.*
- *At the end of anaphase I each pole has a complete set of haploid chromosomes, consisting of one member of each homologous pair.*
- *As a result of the random orientation of homologous pairs at metaphase I, meiosis I results in the independent assortment of maternal and paternal chromosomes in gametes.*
- Telophase I is characterized by the reforming of the nuclear envelope around each daughter nucleus. This does not occur in all species.
- Cytokinesis may or may not occur after telophase I.
- A brief interphase II then occurs during which no DNA replication takes place.
- Meiosis II is similar to mitosis.
- The cohesin proteins at the centromeres holding sister chromatids together are destroyed, allowing them to migrate to opposite poles of the cell.
- The result of meiosis I and II is four cells, each containing haploid sets of chromosomes that are not identical.
- Once completed, the haploid cells may produce gametes or divide mitotically to produce even more gametes or haploid adults.
- Errors occur during meiosis because of nondisjunction, the failure of chromosomes to move to opposite poles.
- Nondisjunction results in one gamete with no chromosome and another gamete with two copies of a chromosome.
- Gametes with an improper number of chromosomes are called aneuploid gametes.

11.4 Summing Up: Meiosis Versus Mitosis

The basic machinery of meiosis and mitosis is the same, but the behavior of chromosomes is distinctly different during the first meiotic division.

- Four distinct features of meiosis I are not found in mitosis:
- *Maternal and paternal homologues pair, and exchange genetic information by crossing over.*
- *The kinetochores of sister chromatids function as a unit during meiosis I, allowing sister chromatids to cosegregate during anaphase I.*
- *Kinetochores of sister chromatids are connected to a single pole in meiosis I and to opposite poles in mitosis.*
- *DNA replication is suppressed between meiosis I and meiosis II.*
- Daughter cells produced by meiosis are not genetically identical because of independent assortment of homologues and crossing over.
- Meiosis is the foundation for understanding the basis of heredity.

SELF TEST

1. Gametes contain _____ the number of chromosomes found in somatic cells.
   a. the same
   b. twice
   c. half
   d. one fourth
2. Somatic cells are _____, whereas gametes are _____.
   a. haploid; diploid
   b. diploid; polyploid
   c. polyploidy; haploid
   d. diploid; haploid
3. An organism is said to be diploid if—
   a. It contains genetic information from two parents
   b. It is multicellular
   c. It reproduces
   d. Undergoes mitotic cell division
4. What are *homologous* chromosomes?
   a. The two halves of a replicated chromosome
   b. Two identical chromosomes from one parent
   c. Two genetically identical chromosomes, one from each parent
   d. Two genetically similar chromosomes, one from each parent
5. When homologous chromosomes form chiasmata, they are—
   a. exchanging genetic information
   b. reproducing their DNA
   c. separating their sister chromatids
   d. replicating their chromatids
6. Crossing over involves each of the following with the exception of—
   a. the transfer of DNA between two nonsister chromatids
   b. the transfer of DNA between two sister chromatids
   c. the formation of a synaptonemal complex
   d. the alignment of homologous chromosomes
7. Terminal chiasmata are seen during which phase of meiosis?
   a. Anaphase I
   b. Prophase I
   c. Metaphase I
   d. Metaphase II
8. Which of the following occurs during anaphase I?
   a. Sister chromatids are separated and move to the poles.
   b. Homologous chromosomes move to opposite poles.
   c. Homologous chromosomes align at the middle of the cell.
   d. All the chromosomes align independently at the middle of the cell.
9. Telophase I results in the production of—
   a. four cells containing one homologue of each homologous pair
   b. two cells containing both homologues of each homologous pair
   c. four cells containing both homologues of each homologous pair
   d. two cells containing one homologue of each homologous pair
10. Which of the following does *not* contribute to genetic diversity?
    a. Independent assortment
    b. Recombination
    c. Metaphase of meiosis II
    d. Metaphase of meiosis I
11. How does S phase following meiosis I differ from S phase in mitosis?
    a. DNA replication takes less time because the cell is haploid.
    b. DNA does not replicate during S phase following meiosis I.
    c. DNA replication takes more time due to cohesin proteins.
    d. There is no difference.
12. What occurs during anaphase of meiosis II?
    a. The homologous chromosomes align.
    b. Sister chromatids are pulled to opposite poles.
    c. Homologous chromosomes are pulled to opposite poles.
    d. The haploid chromosomes line up.
13. What is an aneuploid gamete?
    a. A diploid gamete cell
    b. A haploid gamete cell
    c. A gamete cell with the wrong number of chromosomes
    d. A haploid somatic cell
14. Which of the following is *not* a distinct feature of meiosis?
    a. Pairing and exchange of genetic material between homologous chromosomes
    b. Attachment of sister kinetochores to spindle microtubules
    c. Movement of sister chromatids to the same pole
    d. Suppression of DNA replication
15. Which phase of meiosis I is most similar to the comparable phase in mitosis?
    a. Prophase I
    b. Metaphase I
    c. Anaphase I
    d. Telophase I

CHALLENGE QUESTIONS

1. Diagram the process of meiosis for an imaginary cell with six chromosomes in a diploid cell.
   a. How many homologous pairs are present in this cell? Create a drawing that distinguishes between homologous pairs.
   b. Label each homologue to indicate whether it is maternal (M) or paternal (P).
   c. Draw a new cell showing how these chromosomes would arrange themselves during metaphase of meiosis I. Do all the maternal homologues have to line up on the same side of the cell?
   d. How would this picture differ if you were diagramming anaphase of meiosis II?
2. Mules are the offspring of the mating of a horse and a donkey. Mules are unable to reproduce. A horse has a total of 64 chromosomes, whereas donkeys have 62 chromosomes. Use your knowledge of meiosis to predict the diploid chromosome number of a mule. Propose a possible explanation for the inability of mules to reproduce.
3. Compare the processes of *independent assortment* and *crossing over*. Which process has the greatest influence on genetic diversity?
4. Aneuploid gametes are cells that contain the wrong number of chromosomes. Aneuploidy occurs as a result of *nondisjunction*, or lack of separation of the chromosomes during either phase of meiosis.
   a. At what point in meiotic cell division would nondisjunction occur?
   b. Imagine a cell had a diploid chromosome number of 4. Create a diagram to illustrate the effects of nondisjunction of one pair of homologous chromosomes in meiosis I versus meiosis II.

chapter 12

Patterns of Inheritance

introduction

EVERY LIVING CREATURE IS A PRODUCT of the long evolutionary history of life on Earth. All organisms share this history, but as far as we know, only humans wonder about the processes that led to their origin and investigate the possibilities. We are far from understanding everything about our origins, but we have learned a great deal. Like a partially completed jigsaw puzzle, the boundaries of this elaborate question have fallen into place, and much of the internal structure is becoming apparent. In this chapter, we discuss one piece of the puzzle—the enigma of heredity. Why do individuals, like the children in this picture, differ so much in appearance despite the fact that we are all members of the same species? And, why do members of a single family tend to resemble one another more than they resemble members of other families?

The Mystery of Heredity

As far back as written records go, patterns of resemblance among the members of particular families have been noted and commented on (figure 12.1), but there was no coherent model to explain these patterns. Before the 20th century, two concepts provided the basis for most thinking about heredity. The first was that heredity occurs within species. The second was that traits are transmitted directly from parents to offspring. Taken together, these ideas led to a view of inheritance as resulting from a blending of traits within fixed, unchanging species.

Inheritance itself was viewed as traits being borne through fluid, usually identified as blood, that led to their blending in offspring. This older idea persists today in the use of the term "bloodlines" when referring to the breeding of domestic animals such as horses.

Taken together, however, these two classical assumptions led to a paradox. If no variation enters a species from outside, and if the variation within each species blends in every generation, then all members of a species should soon have the same appearance. It is clear that this does not happen—individuals within most species differ from one another, and they differ in characteristics that are transmitted from generation to generation.

Early plant biologists produced hybrids and saw puzzling results

The first investigator to achieve and document successful experimental **hybridizations** was Josef Kölreuter, who in 1760 cross-fertilized (or crossed, for short) different strains of tobacco and obtained fertile offspring. The hybrids differed in appearance from both parent strains. When individuals within the hybrid generation were crossed, their offspring were highly variable. Some of these offspring resembled plants of the hybrid generation (their parents), but a few resembled the original strains (their grandparents).

Kölreuter's work represents the beginning of modern genetics. The patterns of inheritance observed in his hybrids contradicted the theory of direct transmission because of the variation observed in second-generation offspring.

Over the next hundred years, other investigators elaborated on Kölreuter's work. In one such series of experiments, carried out in 1823, T. A. Knight, an English landholder, crossed two varieties of the garden pea, *Pisum sativum* (figure 12.2). One of these varieties had green seeds, and the other had yellow seeds. Both varieties were **true-breeding,** meaning that the offspring produced from self-fertilization would remain uniform from one generation to the next.

All of the progeny (offspring) of the cross between the two varieties had yellow seeds. Among the offspring of these hybrids, however, some plants produced yellow seeds and others, less common, produced green seeds.

Other investigators made observations similar to Knight's, namely that alternative forms of observed characters were being distributed among the offspring. Referring to a heritable feature as a *character,* a modern geneticist would say the alternative forms of each character were **segregating** among the progeny of a mating, meaning that some offspring exhibited one form of a character (yellow seeds), and other offspring from the same mating exhibited a different form (green seeds). This segregation of alternative forms of a character, or **trait,** provided the clue that led Gregor Mendel to his understanding of the nature of heredity.

Within these deceptively simple results were the makings of a scientific revolution. Nevertheless, another century passed before the process of segregation was fully appreciated.

Mendel used mathematics to analyze his crosses

Born in 1822 to peasant parents, Gregor Mendel (figure 12.3) was educated in a monastery and went on to study science and mathematics at the University of Vienna, where he failed his examina-

figure 12.1

HEREDITY AND FAMILY RESEMBLANCE. Family resemblances are often strong—a visual manifestation of the mechanism of heredity.

figure 12.2

THE GARDEN PEA, *Pisum sativum.* Easy to cultivate and able to produce many distinctive varieties, the garden pea was a popular experimental subject in investigations of heredity as long as a century before Gregor Mendel's experiments.

tions for a teaching certificate. He returned to the monastery and spent the rest of his life there, eventually becoming abbot. In the garden of the monastery, Mendel initiated his own series of experiments on plant hybridization. The results of these experiments would ultimately change our views of heredity irrevocably.

Practical considerations for use of the garden pea

For his experiments, Mendel chose the garden pea, the same plant Knight and others had studied. The choice was a good one for several reasons. First, many earlier investigators had produced hybrid peas by crossing different varieties, so Mendel knew that he could expect to observe segregation of traits among the offspring.

Second, a large number of pure varieties of peas were available. Mendel initially examined 34 varieties. Then, for further study, he selected lines that differed with respect to seven easily distinguishable traits, such as round versus wrinkled seeds and yellow versus green seeds, the latter a trait that Knight had studied.

Third, pea plants are small and easy to grow, and they have a relatively short generation time. A researcher can therefore conduct experiments involving numerous plants, grow several generations in a single year, and obtain results relatively quickly.

A fourth advantage of studying peas is that both the male and female sexual organs are enclosed within each pea flower (figure 12.3), and gametes produced by the male and female parts of the same flower can fuse to form viable offspring, a process termed **self-fertilization.** This self-fertilization takes place automatically within an individual flower if it is not disturbed, resulting in offspring that are the progeny from a single individual. It is also possible to prevent self-fertilization by removing a flower's male parts before fertilization occurs, then introduce pollen from a different strain, thus performing *cross-pollination* that results in **cross-fertilization** (see figure 12.3).

Mendel's experimental design

Mendel was careful to focus on only a few specific differences between the plants he was using and to ignore the countless other differences he must have seen. He also had the insight to realize that the differences he selected must be comparable. For example, he appreciated that trying to study the inheritance of round seeds versus tall height would be useless.

Mendel usually conducted his experiments in three stages:

1. By allowing plants of a given variety to self-cross for multiple generations, Mendel was able to assure himself that the traits he was studying were indeed true-breeding, that is, transmitted unchanged from generation to generation.

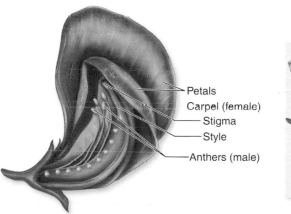

- Petals
- Carpel (female)
- Stigma
- Style
- Anthers (male)

1. The anthers are cut away on the purple flower.

2. Pollen is obtained from the white flower.

3. Pollen is transferred to the purple flower.

4. All progeny result in purple flowers.

figure 12.3

HOW MENDEL CONDUCTED HIS EXPERIMENTS. In a pea plant flower, petals enclose the male anther (containing pollen grains, which give rise to haploid sperm) and the female carpel (containing ovules, which give rise to haploid eggs). This ensures self-fertilization will take place unless the flower is disturbed. Mendel collected pollen from the anthers of a white flower, then placed that pollen onto the stigma of a purple flower with anthers removed. This cross fertilization yields all hybrid seeds that give rise to purple flowers. Using pollen from a white flower to fertilize a purple flower gives the same result.

inquiry

What confounding problems could have been seen if Mendel had chosen another plant with exposed male and female structures?

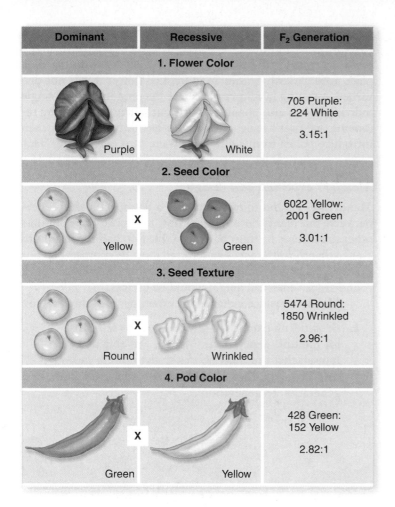

Dominant	Recessive	F₂ Generation
1. Flower Color		
Purple	White	705 Purple: 224 White 3.15:1
2. Seed Color		
Yellow	Green	6022 Yellow: 2001 Green 3.01:1
3. Seed Texture		
Round	Wrinkled	5474 Round: 1850 Wrinkled 2.96:1
4. Pod Color		
Green	Yellow	428 Green: 152 Yellow 2.82:1

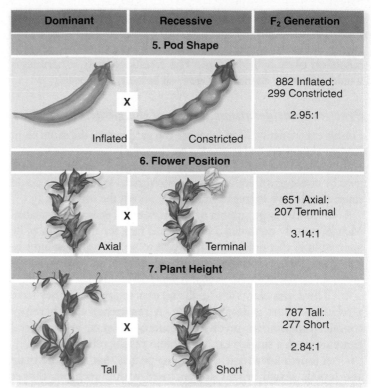

Dominant	Recessive	F₂ Generation
5. Pod Shape		
Inflated	Constricted	882 Inflated: 299 Constricted 2.95:1
6. Flower Position		
Axial	Terminal	651 Axial: 207 Terminal 3.14:1
7. Plant Height		
Tall	Short	787 Tall: 277 Short 2.84:1

figure 12.4

MENDEL'S SEVEN TRAITS. Mendel studied how differences among varieties of peas were inherited when the varieties were crossed. Similar experiments had been done before, but Mendel was the first to quantify the results and appreciate their significance. Results are shown for seven different monohybrid crosses. The F₁ generation is not shown in the table.

2. Mendel then performed crosses between true-breeding varieties exhibiting alternative forms of traits. He also performed **reciprocal crosses:** using pollen from a white-flowered plant to fertilize a purple-flowered plant, then using pollen from a purple-flowered plant to fertilize a white-flowered plant.

3. Finally, Mendel permitted the hybrid offspring produced by these crosses to self-fertilize for several generations, allowing him to observe the inheritance of alternative forms of a trait. Most important, he counted the numbers of offspring exhibiting each trait in each succeeding generation.

This quantification of results is what distinguished Mendel's research from that of earlier investigators, who only noted differences

in a qualitative way. Mendel's mathematical analysis of experimental results led to the inheritance model that we still use today.

> Ideas about inheritance before Mendel did not form a consistent model. The dominant view was of blending inheritance, but plant hybridizers before Mendel had already cast doubt on this model. Mendel followed up on the work of early plant hybridizers by systematizing and quantifying his observations.

 12.2 Monohybrid Crosses: The Principle of Segregation

A **monohybrid cross** is a cross that follows only two variations on a single trait, such as white- and purple-colored flowers. This deceptively simple kind of cross can lead to important conclusions about the nature of inheritance.

The seven characters Mendel studied in his experiments possessed two variants that differed from one another in ways that were easy to recognize and score (see figure 12.4). We examine in detail Mendel's crosses with flower color. His experiments with other characters were similar, and they produced similar results.

The F₁ generation exhibits only one of two traits, without blending

When Mendel crossed white-flowered and purple-flowered plants, the hybrid offspring he obtained did not have flowers of intermediate color, as the hypothesis of blending inheritance would predict. Instead, in every case the flower color of the offspring resembled that of one of their parents. These offspring are customarily referred to as the **first filial generation,**

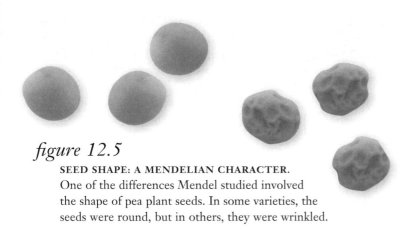

figure 12.5

SEED SHAPE: A MENDELIAN CHARACTER.
One of the differences Mendel studied involved
the shape of pea plant seeds. In some varieties, the
seeds were round, but in others, they were wrinkled.

or **F₁**. In a cross of white-flowered and purple-flowered
plants, the F_1 offspring all had purple flowers, as others had
reported earlier.

Mendel referred to the form of each trait expressed in the
F_1 plants as **dominant,** and to the alternative form that was not
expressed in the F_1 plants as **recessive.** For each of the seven
pairs of contrasting traits that Mendel examined, one of the
pair proved to be dominant and the other recessive.

The F₂ generation exhibits both traits in a 3:1 ratio

After allowing individual F_1 plants to mature and self-fertilize,
Mendel collected and planted the seeds from each plant to see
what the offspring in the **second filial generation,** or **F₂,** would
look like. He found that although most F_2 plants had purple flowers, some exhibited white flowers, the recessive trait. Although
hidden in the F_1 generation, the recessive trait had reappeared
among some F_2 individuals.

Believing the proportions of the F_2 types would provide
some clue about the mechanism of heredity, Mendel counted
the numbers of each type among the F_2 progeny. In the cross
between the purple-flowered F_1 plants, he obtained a total of
929 F_2 individuals. Of these, 705 (75.9%) had purple flowers,
and 224 (24.1%) had white flowers (see figure 12.4). Approximately 1/4 of the F_2 individuals, therefore, exhibited the recessive form of the character.

Mendel obtained the same numerical result with the other
six characters he examined: Of the F_2 individuals, 3/4 exhibited
the dominant trait, and 1/4 displayed the recessive trait. In other
words, the dominant-to-recessive ratio among the F_2 plants was
always close to 3:1. Mendel carried out similar experiments with
other traits, such as wrinkled versus round seeds (figure 12.5),
and obtained the same result.

The 3:1 ratio is actually 1:2:1

Mendel went on to examine how the F_2 plants passed traits
to subsequent generations. He found that plants exhibiting
the recessive trait were always true-breeding. For example,
the white-flowered F_2 individuals reliably produced white-
flowered offspring when they were allowed to self-fertilize.
By contrast, only 1/3 of the dominant, purple-flowered F_2
individuals (1/4 of all F_2 offspring) proved pure-breeding,

but 2/3 were not. This last class of plants produced dominant and recessive individuals in the third filial generation
(F_3) in a 3:1 ratio.

This result suggested that, for the entire sample, the 3:1
ratio that Mendel observed in the F_2 generation was really a
disguised 1:2:1 ratio: 1/4 true-breeding dominant individuals, 1/2 not-true-breeding dominant individuals, and 1/4 true-
breeding recessive individuals (figure 12.6).

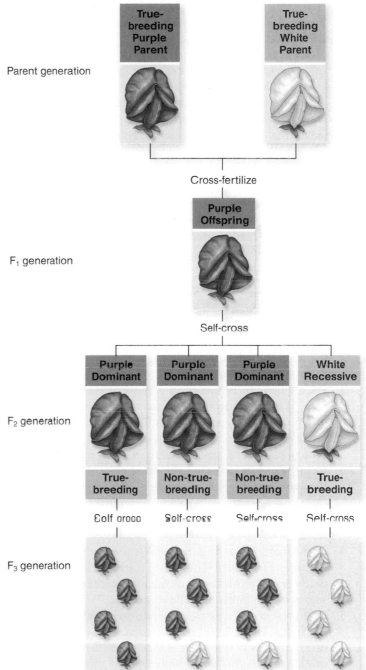

figure 12.6

THE F₂ GENERATION IS A DISGUISED 1:2:1 RATIO.
By allowing the F_2 generation to self-fertilize, Mendel found
from the offspring (F_3) that the ratio of F_2 plants was
1 pure-breeding dominant: 2 not-pure-breeding dominant:
and 1 pure-breeding recessive.

Mendel's Principle of Segregation explains monohybrid observations

From his experiments, Mendel was able to understand four things about the nature of heredity:

- The plants he crossed did not produce progeny of intermediate appearance, as a hypothesis of blending inheritance would have predicted. Instead, different plants inherited each trait intact, as a discrete characteristic.
- For each pair of alternative forms of a trait, one alternative was not expressed in the F_1 hybrids, although it reappeared in some F_2 individuals. *The trait that "disappeared" must therefore be latent (present but not expressed) in the F_1 individuals.*
- The pairs of alternative traits examined were segregated among the progeny of a particular cross, some individuals exhibiting one trait and some the other.
- These alternative traits were expressed in the F_2 generation in the ratio of 3/4 dominant to 1/4 recessive. This characteristic 3:1 segregation is referred to as the **Mendelian ratio** for a monohybrid cross.

Mendel's five-element model

To explain these results, Mendel proposed a simple model that has become one of the most famous in the history of science, containing simple assumptions and making clear predictions. The model has five elements:

1. Parents do not transmit physiological traits directly to their offspring. Rather, they transmit discrete information for the traits, what Mendel called "factors." We now call these factors **genes.**
2. Each individual receives two genes that encode each trait. We now know that the two factors are carried on chromosomes, and each adult individual is diploid. Gametes, produced by meiosis, are haploid.
3. Not all copies of a gene are identical. The alternative forms of a gene are called **alleles.** When two haploid gametes containing the same allele fuse during fertilization, the resulting offspring is said to be **homozygous.** When the two haploid gametes contain different alleles, the resulting offspring is said to be **heterozygous.**
4. The two alleles remain discrete—they neither blend with nor alter each other. Therefore, when the individual matures and produces its own gametes, the alleles segregate randomly into these gametes.
5. The presence of a particular allele does not ensure that the trait it encodes will be expressed. In heterozygous individuals, only one allele is expressed (the dominant one), and the other allele is present but unexpressed (the recessive one).

Geneticists now refer to the total set of alleles that an individual contains as the individual's **genotype.** The physical appearance or other observable characteristics of that individual, which result from an allele's expression, is termed the individual's **phenotype.** In other words, the genotype is the blueprint, and the phenotype is the visible outcome.

This also allows us to present Mendel's ratios in more modern terms. The 3:1 ratio of dominant to recessive is the monohybrid phenotypic ratio. The 1:2:1 ratio of homozygous dominant to heterozygous to homozygous recessive is the monohybrid genotypic ratio. The genotypic ratio "collapses" into the phenotypic ratio due to the action of the dominant allele making the heterozygote appear the same as homozygous dominant.

The principle of segregation

Mendel's model accounts for the ratios he observed in a neat and satisfying way. His main conclusion—that alternative alleles for a character segregate from each other during gamete formation and remain distinct—has since been verified in many other organisms. It is commonly referred to as Mendel's first law of heredity, or the **Principle of Segregation.** It can be simply stated as: *The two alleles for a gene segregate during gamete formation and are rejoined at random, one from each parent, during fertilization.*

The physical basis for allele segregation is the behavior of chromosomes during meiosis. As you saw in chapter 11, homologues for each chromosome disjoin during anaphase I of meiosis. The second meiotic division then produces gametes that contain only one homologue for each chromosome.

It is a tribute to Mendel's intellect that his analysis arrived at the correct scheme, even though he had no knowledge of the cellular mechanisms of inheritance; neither chromosomes nor meiosis had yet been described.

The Punnett square allows symbolic analysis

To test his model, Mendel first expressed it in terms of a simple set of symbols. He then used the symbols to interpret his results.

Consider again Mendel's cross of purple-flowered with white-flowered plants. By convention, we assign the symbol P (uppercase) to the dominant allele, associated with the production of purple flowers, and the symbol p (lowercase) to the recessive allele, associated with the production of white flowers.

In this system, the genotype of an individual that is true-breeding for the recessive white-flowered trait would be designated pp. Similarly, the genotype of a true-breeding purple-flowered individual would be designated PP. In contrast, a heterozygote would be designated Pp (dominant allele first). Using these conventions and denoting a cross between two strains with ×, we can symbolize Mendel's original purple × white cross as $PP \times pp$.

Because a white-flowered parent (pp) can produce only p gametes, and a true-breeding purple-flowered parent (PP, *homozygous dominant*) can produce only P gametes, the union of these gametes can produce only heterozygous Pp offspring in the F_1 generation. Because the P allele is dominant, all of these F_1 individuals are expected to have purple flowers.

When F_1 individuals are allowed to self-fertilize, the P and p alleles segregate during gamete formation to produce both P gametes and p gametes. Their subsequent union at fertilization to form F_2 individuals is random.

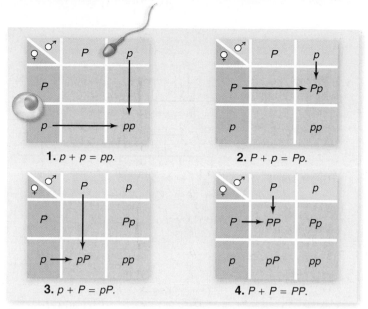

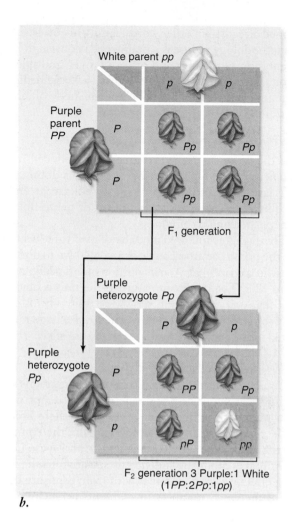

figure 12.7

USING A PUNNETT SQUARE TO ANALYZE MENDEL'S CROSS. *a.* To make a Punnett square, place the different possible types of female gametes along the side of a square and the different possible types of male gametes along the top. Each potential zygote is represented as the intersection of a vertical line and a horizontal line. *b.* In Mendel's cross of purple by white flowers, the original parents each only make one type of gamete. The resulting F$_1$ generation are all *Pp* heterozygotes with purple flowers. These F$_1$ then each make two types of gametes that can be combined to produce three kinds of F$_2$ offspring: *PP* homozygotes (purple flowers); *Pp* heterozygotes (also purple flowers); and *pp* homozygotes (white flowers). The ratio of dominant to recessive phenotypes is 3:1. The ratio of genotypes is 1:2:1 (1 *PP*: 2 *Pp*: 1 *pp*).

The F$_2$ possibilities may be visualized in a simple diagram called a **Punnett square,** named after its originator, the English geneticist R. C. Punnett (figure 12.7*a*). Mendel's model, analyzed in terms of a Punnett square, clearly predicts that the F$_2$ generation should consist of 3/4 purple-flowered plants and 1/4 white-flowered plants, a phenotypic ratio of 3:1 (figure 12.7*b*).

Some human traits exhibit dominant/recessive inheritance

A number of human traits have been shown to display both dominant and recessive inheritance (table 12.1 provides a sample of these). Researchers cannot perform controlled crosses in humans the way Mendel did with pea plants, so to analyze human inheritance, geneticists study crosses that have been

TABLE 12.1	Some Dominant and Recessive Traits in Humans		
Recessive Traits	**Phenotypes**	**Dominant Traits**	**Phenotypes**
Albinism	Lack of melanin pigmentation	Middigital hair	Presence of hair on middle segment of fingers
Alkaptonuria	Inability to metabolize homogentisic acid	Brachydactyly	Short fingers
Red-green color blindness	Inability to distinguish red or green wavelengths of light	Huntington disease	Degeneration of nervous system, starting in middle age
Cystic fibrosis	Abnormal gland secretion, leading to liver degeneration and lung failure	Phenylthiocarbamide (PTC) sensitivity	Ability to taste PTC as bitter
Duchenne muscular dystrophy	Wasting away of muscles during childhood	Camptodactyly	Inability to straighten the little finger
Hemophilia	Inability of blood to clot properly, some clots form but the process is delayed	Hypercholesterolemia (the most common human Mendelian disorder)	Elevated levels of blood cholesterol and risk of heart attack
Sickle cell anemia	Defective hemoglobin that causes red blood cells to curve and stick together	Polydactyly	Extra fingers and toes

performed already—in other words, family histories. The organized methodology we use is a **pedigree**, a consistent graphical representation of matings and offspring over multiple generations for a particular trait. The information in the pedigree may allow geneticists to deduce a model for the mode of inheritance of the trait.

A dominant pedigree: Juvenile glaucoma

One of the most extensive pedigrees yet produced traced the inheritance of a form of blindness caused by a dominant allele. The disease allele causes a form of hereditary juvenile glaucoma. The disease causes degeneration of nerve fibers in the optic nerve, leading to blindness.

This pedigree followed inheritance over three centuries following the origin back to a couple in a small town in northwestern France who died in 1495. A small portion of this pedigree is shown in figure 12.8. The dominant nature of the trait is obvious from the fact that every generation shows the trait. This is extremely unlikely for a recessive trait as it would require large numbers of unrelated individuals to be carrying the disease allele.

A recessive pedigree: Albinism

An example of inheritance of a recessive human trait is albinism, a condition in which the pigment melanin is not produced. Long thought to be due to a single gene, there are actually multiple genes that can all lead to albinism, the common feature is the loss of pigment from hair, skin, and eyes. The loss of pigment makes albinistic individuals sensitive to the sun. The tanning effect we are all familiar with from exposure to the sun is due to increased numbers of pigment-producing cells, and

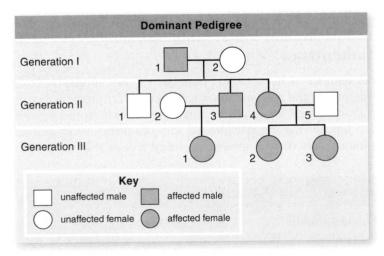

figure 12.8

DOMINANT PEDIGREE FOR HEREDITARY JUVENILE GLAUCOMA. Males are shown as squares and females are shown as circles. Affected individuals are shown shaded. The dominant nature of this trait can be seen in the trait appearing in every generation, a feature of dominant traits.

inquiry

? *If one of the affected females in the third generation married an unaffected male, could she produce unaffected offspring? If so, what are the chances of having unaffected offspring?*

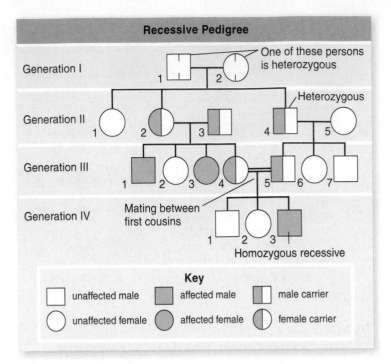

figure 12.9

RECESSIVE PEDIGREE FOR ALBINISM. One of the two individuals in the first generation must be heterozygous and individuals II-2 and II-4 must be heterozygous. Notice that for each affected individual, neither parent is affected, but both must be heterozygous (carriers). The double line indicates a consanguineous mating (between relatives) that, in this case. produced affected offspring.

inquiry

? *From a genetic disease standpoint, why is it never advisable for close relatives to mate and have children?*

increased production of pigment. This is lacking in albinistic individuals due to the lack of any pigment to begin with.

The pedigree in figure 12.9 is for a form of albinism due to a nonfunctional allele of the enzyme tyrosinase, which is required for the formation of melanin pigment. The genetic characteristics of this form of albinism are: females and males are affected equally, most affected individuals have unaffected parents, a single affected parent usually does not have affected offspring, and affected offspring are more frequent when parents are related. Each of these features can be see in figure 12.9, and all of this fits a recessive mode of inheritance quite well.

> Monohybrid crosses show that traits are due to factors inherited intact with no blending. Traits that appear in the F_1 generation are called dominant; traits that are not observed are called recessive. In the F_2 generation, both traits are observed in a predictable ratio of 3 dominant to 1 recessive. The Principle of Segregation states that during gamete formation, alleles segregate into different gametes and are then randomly combined during fertilization. Dominant/recessive inheritance is analyzed in humans using pedigrees.

Dihybrid Crosses: The Principle of Independent Assortment

The Principle of Segregation explains the behavior of alternative forms of a single trait in a monohybrid cross. The next step is to extend this to follow the behavior of two different traits in a single cross: a **dihybrid cross.**

With an understanding of the behavior of single traits, Mendel went on to ask if different traits behaved independently in hybrids. He first established a series of true-breeding lines of peas that differed in two of the seven characters he had studied. He then crossed contrasting pairs of the true-breeding lines to create heterozygotes. These heterozygotes are now doubly heterozygous, or dihybrid. Finally, he self-crossed the dihybrid F₁ plants to produce an F₂ generation, and counted all progeny types.

The F₁ generation displays two of four traits, without blending

Consider a cross involving different seed shape alleles (round, *R*, and wrinkled, *r*) and different seed color alleles (yellow, *Y*, and green, *y*). Crossing round yellow (*RR YY*) with wrinkled green (*rr yy*), produces heterozygous F₁ individuals having the same phenotype (namely round and yellow) and the same genotype (*Rr Yy*). Allowing these dihybrid F₁ individuals to self-fertilize produces an F₂ generation.

The F₂ generation exhibits four types of progeny in a 9:3:3:1 ratio

In analyzing these results, we first consider the number of possible phenotypes. We expect to see the two parental phenotypes: round yellow and wrinkled green. If the traits behave independently, then we can also expect one trait from each parent to produce plants with round green seeds and others with wrinkled yellow seeds.

Next consider what types of gametes the F₁ individuals can produce. Again, we expect the two types of gametes found in the parents: *R Y* and *r y*. If the traits behave independently, then we can also expect the gametes *R y* and *r Y*. Using modern language, two genes each with two alleles can be combined four ways to produce these gametes: *R Y*, *r y*, *R y*, and *r Y*.

A dihybrid Punnett square

We can then construct a Punnett square with these gametes to generate all possible progeny. This is a 4 × 4 square with 16 possible outcomes. Filling in the Punnett square produces all possible offspring (figure 12.10). From this we can see that there are 9 round yellow, 3 wrinkled yellow, 3 round green, and 1 wrinkled green. This predicts a phenotypic ratio of 9:3:3:1 for traits that behave independently.

Mendel's data

What did Mendel actually observe? From a total of 556 seeds from self-fertilized dihybrid plants, he observed the following results:

- 315 round yellow (signified *R__ Y__*, where the underscore indicates the presence of either allele),
- 108 round green (*R__ yy*),
- 101 wrinkled yellow (*rr Y__*), and
- 32 wrinkled green (*rr yy*).

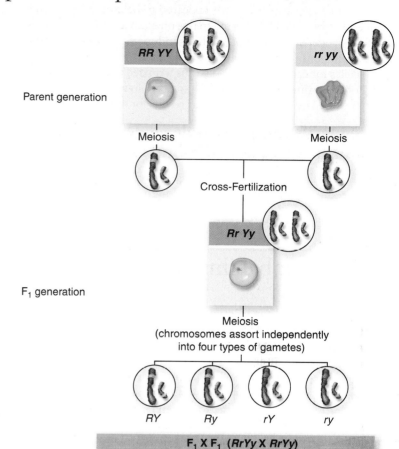

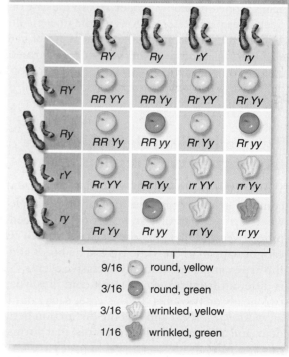

9/16	round, yellow
3/16	round, green
3/16	wrinkled, yellow
1/16	wrinkled, green

figure 12.10

ANALYZING A DIHYBRID CROSS. This Punnett square shows the results of Mendel's dihybrid cross between plants with round yellow seeds and plants with wrinkled green seeds. The ratio of the four possible combinations of phenotypes is predicted to be 9:3:3:1, the ratio that Mendel found.

These results are very close to a 9:3:3:1 ratio. (The expected 9:3:3:1 ratio from this many offspring would be 313:104:104:35.)

The alleles of two genes appeared to behave independently of each other. Mendel referred to this phenomenon as the traits assorting independently. Note that this **independent assortment** of different genes in no way alters the segregation of individual pairs of alleles for each gene. Round versus wrinkled seeds occur in a ratio of approximately 3:1 (423:133); so do yellow versus green seeds (416:140). Mendel obtained similar results for other pairs of traits.

Mendel's Principle of Independent Assortment explains dihybrid results

Mendel's discovery is often referred to as Mendel's second law of heredity, or the **Principle of Independent Assortment.**

This can also be stated simply: *In a dihybrid cross, the alleles of each gene assort independently.* Like segregation, independent assortment arises from the behavior of chromosomes during meiosis to produce haploid gametes (chapter 11)—in this case, the independent alignment of different homologous pairs during metaphase I.

Mendel's analysis of dihybrid crosses revealed that the segregation of allele pairs for different genes is independent, known as Mendel's Principle of Independent Assortment. When individuals that differ in two traits are crossed, and their progeny are intercrossed, the result is four different types that occur in a ratio of 9:3:3:1, or Mendel's dihybrid ratio.

Probability: Predicting the Results of Crosses

Probability allows us to predict the likelihood of the outcome of random events. Because the behavior of different chromosomes during meiosis is independent, we can use probability to predict the outcome of crosses. The probability of an event that is certain to happen is equal to 1. In contrast, an event that can never happen has a probability of 0. Therefore, probabilities for all other events have fractional values, between 0 and 1. For instance, when you flip a coin, two outcomes are possible; there is only one way to get the event "heads" so the probability of heads is one divided by two, or 1/2. In the case of genetics, consider a pea plant heterozygous for the flower color alleles P and p. This individual can produce two types of gametes in equal numbers, again due to the behavior of chromosomes during meiosis. There is one way to get a P gamete, so the probability of any particular gamete carrying a P allele is 1 divided by 2 or 1/2, just like the coin toss.

Two probability rules help predict monohybrid cross results

We can use probability to make predictions about the outcome of genetic crosses using only two simple rules. Before we describe these rules and their uses, we need another definition. We say that two events are *mutually exclusive* if both cannot happen at the same time. The heads and tails of a coin flip are examples of mutually exclusive events. Notice that this is different from two consecutive coin flips where you can get two heads or two tails. In this case, each coin flip represents an *independent event* and it is the distinction between independent and mutually exclusive events that forms the basis for our two rules.

The rule of addition

If we consider a six-sided die instead of a coin, for any roll of the die, only one outcome is possible; each of the possible outcomes are mutually exclusive. The probability of any particular number coming up is 1/6. The probability of either of two different numbers is the sum of the individual probabilities, or to restate as the **rule of addition:**

For two mutually exclusive events, the probability of either event occurring is the sum of the individual probabilities.

Probability of rolling either a 2 or a 6
is = 1/6 + 1/6 = 2/6 = 1/3

To apply this to our cross of heterozygous purple F_1, four mutually exclusive outcomes are possible: PP, Pp, pP, and pp. The probability of being heterozygous is the same as the probability of being either Pp or pP, or 1/4 plus 1/4, or 1/2.

Probability of F_2 heterozygote = $1/4Pp + 1/4pP = 1/2$

In the previous example, of 379 total offspring, we would expect about 190 to be heterozygotes. (The actual number is 189.5.)

The rule of multiplication

The second rule, and by far the most useful for genetics, deals with the outcome of independent events. This is called the **product rule,** or **rule of multiplication,** and it states that the probability of two independent events both occurring is the **product** of their individual probabilities.

We can apply this to a monohybrid cross where offspring are formed by gametes from each of two parents. For any particular outcome then, this is due to two independent events: the formation of two different gametes. Consider the purple F_1 parents from earlier. They are all Pp (heterozygotes), so the probability that a particular F_2 individual will be pp (homozygous recessive) is the probability of receiving a p gamete from the male (1/2) times the probability of receiving a p gamete from the female (1/2), or 1/4:

Probability of pp homozygote = $1/2p$ (male parent) $\times 1/2p$ (female parent) = $1/4pp$

This is actually the basis for the Punnett square that we used before. Each cell in the square was the product of the probabilities of the gametes that contribute to the cell. We then use the addition rule to sum the probabilities of the mutually exclusive events that make up each cell.

We can use the result of a probability calculation to predict the number of homozygous recessive offspring in a cross between heterozygotes. For example, out of 379 total offspring, we would expect about 95 to exhibit the homozygous recessive phenotype. (The actual calculated number is 94.75.)

Dihybrid cross probabilities are based on monohybrid cross probabilities

Probability analysis can be extended to the dihybrid case. For our purple F_1 by F_1 cross, there are four possible outcomes, three of which show the dominant phenotype. Thus the probability of any offspring showing the dominant phenotype is 3/4, and the probability of any offspring showing the recessive phenotype is 1/4. Now we can use this and the product rule to predict the outcome of a dihybrid cross. We will use our example of seed shape and color from earlier, but now examine it using probability.

If the alleles affecting seed shape and seed color segregate independently, then the probability that a particular pair of seed shape alleles would occur together with a particular pair of seed color alleles is the product of the individual probabilities for each pair. For example, the probability that an individual with wrinkled green seeds (*rryy*) would appear in the F_2 generation would be equal to the probability of obtaining wrinkled seeds (1/4) times the probability of obtaining green seeds (1/4), or 1/16.

Probability of *rryy* = 1/4 *rr* × 1/4 *yy* = 1/16 *rryy*

Because of independent assortment, we can think of the dihybrid cross of consisting of two independent monohybrid crosses; since these are independent events, the product rule applies. So, we can calculate the probabilities for each dihybrid phenotype:

Probability of round yellow (*R__ Y__*) =
3/4 *R__* × 3/4 *Y__* = 9/16

Probability of round green (*R__ yy*) =
3/4 *R__* × 1/4 *yy* = 3/16

Probability of wrinkled yellow (*rr Y__*) =
1/4 *rr* × 3/4 *Y_* = 3/16

Probability of wrinkled green (*rryy*) =
1/4 *rr* × 1/4 *yy* = 1/16

The hypothesis that color and shape genes are independently assorted thus predicts that the F_2 generation will display a 9:3:3:1 phenotypic ratio. These ratios can be applied to an observed total offspring to predict the expected number in each phenotypic group. The underlying logic and the results are the same as obtained using the Punnett square.

> The probability of either of two events occurring is the sum of the individual probabilities. The probability of two independent events both occurring is the product of the individual probabilities. These can be applied to genetic crosses to determine the probability of particular genotypes and phenotypes.

12.5 The Testcross: Revealing Unknown Genotypes

To test his model further, Mendel devised a simple and powerful procedure called the **testcross**. In a testcross, an individual with unknown genotype is crossed with the homozygous recessive genotype—that is, the recessive parental variety. The contribution of the homozygous recessive parent can be ignored, because this parent can contribute only recessive alleles.

Consider a purple-flowered pea plant. It is impossible to tell whether such a plant is homozygous or heterozygous simply by looking at it. To learn its genotype, you can per-

form a testcross to a white-flowered plant. In this cross, the two possible test plant genotypes will give different results (figure 12.11):

Alternative 1: Unknown individual is homozygous dominant (*PP*)
PP × *pp*: All offspring have purple flowers (*Pp*).

Alternative 2: Unknown individual is heterozygous (*Pp*)
Pp × *pp*: 1/2 of offspring have white flowers (*pp*), and 1/2 have purple flowers (*Pp*).

figure 12.11

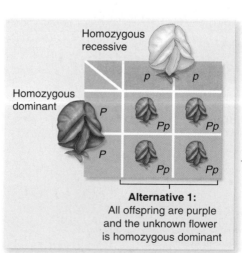

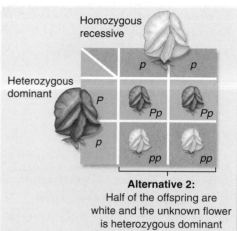

Dominant Phenotype (unknown genotype)

PP or *Pp*?

If *PP* then — If *Pp* then

Alternative 1:
All offspring are purple and the unknown flower is homozygous dominant

Alternative 2:
Half of the offspring are white and the unknown flower is heterozygous dominant

A TESTCROSS. To determine whether an individual exhibiting a dominant phenotype, such as purple flowers, is homozygous or heterozygous for the dominant allele, Mendel crossed the individual in question with a plant that he knew to be homozygous recessive— in this case, a plant with white flowers.

Put simply, the appearance of the recessive phenotype in the offspring of a testcross indicates that the test individual's genotype is heterozygous.

For each pair of alleles Mendel investigated, he observed phenotypic F_2 ratios of 3:1 (see figure 12.4) and testcross ratios of 1:1, just as his model had predicted. Testcrosses can also be used to determine the genotype of an individual when two genes are involved. Mendel often performed testcrosses to verify the genotypes of dominant-appearing F_2 individuals.

An F_2 individual exhibiting both dominant traits ($A\_\_ B\_\_$) might have any of the following genotypes: *AABB, AaBB, AABb,* or *AaBb*. By crossing dominant-appearing F_2 individuals with homozygous recessive individuals (that is, $A\_\_ B\_\_ \times aabb$), Mendel was able to determine whether either or both of the traits bred true among the progeny, and so to determine the genotype of the F_2 parent (table 12.2).

Testcrossing is a powerful tool that simplifies genetic analysis. We will use this method of analysis in the next chapter, when we explore genetic mapping.

TABLE 12.2 Dihybrid Testcross

Actual Genotype	Results of Testcross	
	Trait A	Trait B
AABB	Trait A breeds true	Trait B breeds true
AaBB	———	Trait B breeds true
AABb	Trait A breeds true	———
AaBb	———	———

Individuals showing the dominant phenotype can be either homozygous dominant or heterozygous. The genotype can be determined using a testcross, which involves crossing the individual of unknown genotype to a homozygous recessive individual. Heterozygous individuals produce both dominant and recessive phenotypes in equal numbers as a result of the testcross.

12.6 Extensions to Mendel

Although Mendel's results did not receive much notice during his lifetime, three different investigators independently rediscovered his pioneering paper in 1900, 16 years after his death. They came across it while searching the literature in preparation for publishing their own findings, which closely resembled those Mendel had presented more than 30 years earlier.

In the decades following the rediscovery of Mendel's ideas, many investigators set out to test them. However, scientists attempting to confirm Mendel's theory often had trouble obtaining the same simple ratios he had reported.

The reason that Mendel's simple ratios were not obtained had to do with the traits that others examined. A number of assumptions are built into Mendel's model that are oversimplifications. These assumptions include that each trait is specified by a single gene with two alternative alleles; that there are no environmental effects; and that gene products act independently. The idea of dominance also hides a wealth of biochemical complexity. In the following sections, you'll see

how Mendel's simple ideas can be extended to provide a more complete view of genetics (table 12.3).

In polygenic inheritance, more than one gene can affect a single trait

Often, the relationship between genotype and phenotype is more complicated than a single allele producing a single trait. Most phenotypes also do not reflect simple two-state cases like purple or white flowers.

Consider Mendel's crosses between tall and short pea plants. In reality, the "tall" plants actually have normal height, and the "short" plants are dwarfed by an allele at a single gene. But in most species, including humans, height varies over a continuous range, rather than having discrete values. This continuous distribution of a phenotype has a simple genetic explanation: the action of more than one gene. The mode of inheritance that takes place in this case is often called **polygenic inheritance.**

TABLE 12.3 When Mendel's Laws/Results May Not Be Observed

Genetic Occurrence	Definition	Examples
Polygenic inheritance	More than one gene can affect a single trait.	• Four genes are involved in determining eye color. • Human height
Pleiotropy	A single gene can affect more than one trait.	• A pleiotropic allele dominant for yellow fur in mice is recessive for a lethal developmental defect. • Cystic fibrosis • Sickle cell anemia
Multiple alleles for one gene	Genes may have more than two alleles.	ABO blood types in humans
Dominance is not always complete	• In incomplete dominance the heterozygote is intermediate. • In codominance no single allele is dominant, and the heterozygote shows some aspect of both homozygotes.	• Japanese four o'clocks • Human blood groups
Environmental factors	Genes may be affected by the environment.	Siamese cats
Gene interaction	Products of genes can interact to alter genetic ratios.	• The production of a purple pigment in corn • Coat color in mammals

In reality, few phenotypes result from the action of only one gene. Instead, most characters reflect multiple additive contributions to the phenotype by several genes. When multiple genes act jointly to influence a character, such as height or weight, the character often shows a range of small differences. When these genes segregate independently, a gradation in the degree of difference can be observed when a group consisting of many individuals is examined (figure 12.12). We call this gradation **continuous variation,** and we call such traits **quantitative traits.** The greater the number of genes that influence a character, the more continuous the expected distribution of the versions of that character.

This continuous variation in traits is similar to blending different colors of paint: Combining one part red with seven parts white, for example, produces a much lighter pink shade than does combining five parts red with three parts white. Different ratios of red to white result in a continuum of shades, ranging from pure red to pure white.

Often, variations can be grouped into categories, such as different height ranges. Plotting the numbers in each height category produces a curve called a *histogram,* such as that shown in figure 12.12. The bell-shaped histogram approximates an idealized *normal distribution,* in which the central tendency is characterized by the mean, and the spread of the curve indicates the amount of variation.

Even simple-appearing traits can have this kind of polygenic basis. For example, human eye colors are often described in simple terms with brown dominant to blue, but

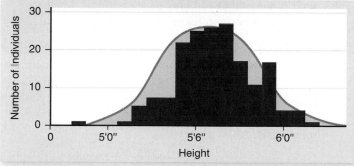

figure 12.12

HEIGHT IS A CONTINUOUSLY VARYING TRAIT. The photo and accompanying graph show variation in height among students of the 1914 class at the Connecticut Agricultural College. Because many genes contribute to height and tend to segregate independently of one another, the cumulative contribution of different combinations of alleles to height forms a *continuous* distribution of possible heights, in which the extremes are much rarer than the intermediate values. Variation can also arise due to environmental factors such as nutrition.

this is actually incorrect. Extensive analysis indicates that at least four genes are involved in determining eye color. This leads to more complex inheritance patterns than initially reported. For example, blue-eyed parents can have brown-eyed offspring, although it is rare.

In pleiotropy, a single gene can affect more than one trait

Not only can more than one gene affect a single trait, but a single gene can affect more than one trait. Considering the complexity of biochemical pathways and the interdependent nature of organ systems in multicellular organisms, this should be no surprise.

An allele that has more than one effect on phenotype is said to be **pleiotropic.** The pioneering French geneticist Lucien Cuenot studied yellow fur in mice, a dominant trait, and found he was unable to obtain a pure-breeding yellow strain by crossing individual yellow mice with each other. Individuals homozygous for the yellow allele died, because the yellow allele was pleiotropic: One effect was yellow coat color, but another was a lethal developmental defect.

A pleiotropic allele may be dominant with respect to one phenotypic consequence (yellow fur) and recessive with respect to another (lethal developmental defect). Pleiotropic effects are difficult to predict, because a gene that affects one trait often performs other, unknown functions.

Pleiotropic effects are characteristic of many inherited disorders in humans, including cystic fibrosis and sickle cell anemia (discussed in the following chapter). In these disorders, multiple symptoms (phenotypes) can be traced back to a single gene defect. Cystic fibrosis patients exhibit clogged blood vessels, overly sticky mucus, salty sweat, liver and pancreas failure, and a battery of other symptoms. It is often difficult to deduce the nature of the primary defect from the range of a gene's pleiotropic effects. As it turns out, all these symptoms of cystic fibrosis are pleiotropic effects of a single defect, a mutation in a gene that encodes a chloride ion transmembrane channel.

Genes may have more than two alleles

Mendel always looked at genes with two alternative alleles. Although any diploid individual can carry only two alleles for a gene, there may be more than two alleles in a population. The example of ABO blood types in humans, described later on, involves an allelic series with three alleles.

If you think of a gene as a sequence of nucleotides in a DNA molecule, then the number of possible alleles is huge because even a single nucleotide change could produce a new allele. In reality, the number of alleles possible for any gene is constrained, but usually more than two alleles exist for any gene in an outbreeding population. The dominance relationships of these alleles cannot be predicted, but can be determined by observing the phenotypes for the various heterozygous combinations.

Dominance is not always complete

Mendel's idea of dominant and recessive traits can seem hard to explain in terms of modern biochemistry. For example, if a recessive trait is caused by the loss of function of an enzyme encoded by the recessive allele, then why should a heterozygote, with only half the activity of this enzyme, have the same appearance as a homozygous dominant individual?

The answer is that enzymes usually act in pathways and not alone. These pathways, as you have seen in earlier chapters, can be highly complex in terms of inputs and outputs, and they can sometimes tolerate large reductions in activity of single enzymes in the pathway without reductions in the level of the end-product. When this is the case, complete dominance will be observed; however, not all genes act in this way.

Incomplete dominance

In **incomplete dominance,** the heterozygote is intermediate in appearance between the two homozygotes. For example, in a cross between red- and white-flowering Japanese four o'clocks, described in figure 12.13, all the F_1 offspring have pink flowers—indicating that neither red nor white flower color was dominant. Looking only at the F_1, we might conclude that this is a case of blending inheritance. But when two of the F_1 pink flowers are crossed, they produce red-, pink-, and white-flowered plants in a

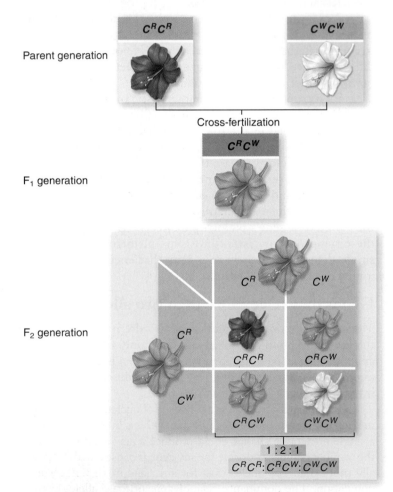

figure 12.13

INCOMPLETE DOMINANCE. In a cross between a red-flowered (genotype $C^R C^R$) Japanese four o'clock and a white-flowered one ($C^W C^W$), neither allele is dominant. The heterozygous progeny have pink flowers and the genotype $C^R C^W$. If two of these heterozygotes are crossed, the phenotypes of their progeny occur in a ratio of 1:2:1 (red:pink:white).

1:2:1 ratio. In this case the phenotypic ratio is the same as the genotypic ratio because all three genotypes can be distinguished.

Codominance

Most genes in a population possess several different alleles, and often no single allele is dominant; instead, each allele has its own effect, and the heterozygote shows some aspect of the phenotype of both homozygotes. The alleles are said to be **codominant.**

Codominance can be distinguished from incomplete dominance by the appearance of the heterozygote. In incomplete dominance, the heterozygote is intermediate between the two homozygotes, whereas in codominance, some aspect of both alleles is seen in the heterozygote. One of the clearest human examples is found in the human blood groups.

The different phenotypes of human blood groups are based on the response of the immune system to proteins on the surface of red blood cells. In homozygotes a single type of protein is found on the surface of cells, and in heterozygotes, two kinds of protein are found, leading to codominance.

The human ABO blood group system

The gene that determines ABO blood types encodes an enzyme that adds sugar molecules to proteins on the surface of red blood cells. These sugars act as recognition markers for the immune system (chapter 51). The gene that encodes the enzyme, designated I, has three common alleles: I^A, whose product adds galactosamine; I^B, whose product adds galactose; and i, which codes for a protein that does not add a sugar.

The three alleles of the I gene can be combined to produce six different genotypes. An individual heterozygous for the I^A and I^B alleles produces both forms of the enzyme and exhibits both galactose and galactosamine on red blood cells. Because both alleles are expressed simultaneously in heterozygotes, the I^A and I^B alleles are codominant. Both I^A and I^B are dominant over the i allele, because both I^A and I^B alleles lead to sugar addition, whereas the i allele does not. The different combinations of the three alleles produce four different phenotypes (figure 12.14):

1. Type A individuals add only galactosamine. They are either $I^A I^A$ homozygotes or $I^A i$ heterozygotes (two genotypes).
2. Type B individuals add only galactose. They are either $I^B I^B$ homozygotes or $I^B i$ heterozygotes (two genotypes).
3. Type AB individuals add both sugars and are $I^A I^B$ heterozygotes (one genotype).
4. Type O individuals add neither sugar and are ii homozygotes (one genotype).

These four different cell surface phenotypes are called the **ABO blood groups.**

A person's immune system can distinguish among these four phenotypes. If a type A individual receives a transfusion of type B blood, the recipient's immune system recognizes the "foreign" antigen (galactose) and attacks the donated blood cells, causing them to clump, or agglutinate. The same thing would happen if the donated blood is type AB. However, if the donated blood is type O, no immune attack occurs, because there are no galactose antigens.

In general, any individual's immune system will tolerate a transfusion of type O blood, and so type O is termed the "universal donor." Because neither galactose nor galactosamine

	Alleles	Blood Type	Sugars Exhibited	Donates and Receives
■	$I^A I^A$, $I^A i$ (I^A dominant to i)	A	Galactosamine	Receives A and O Donates to A and AB
■	$I^B I^B$, $I^B i$ (I^B dominant to i)	B	Galactose	Receives B and O Donates to B and AB
■	$I^A I^B$ (codominant)	AB	Both galactose and galactosamine	Universal receiver Donates to AB
☐	ii (i is recessive)	O	None	Receives O Universal donor

figure 12.14

ABO BLOOD GROUPS ILLUSTRATE BOTH CODOMINANCE AND MULTIPLE ALLELES. There are three alleles of the I gene: I^A, I^B and i. I^A and I^B are both dominant to i (see types A and B), but codominant to each other (see type AB). The genotypes that give rise to each blood type are shown with the associated phenotypes in terms of sugars added to surface proteins and the behavior in blood transfusions.

is foreign to type AB individuals (whose red blood cells have both sugars), those individuals may receive any type of blood, and type AB is termed the "universal recipient." Nevertheless, matching blood is preferable for any transfusion.

Genes may be affected by the environment

Another assumption, implicit in Mendel's work, is that the environment does not affect the relationship between genotype and phenotype. For example, the soil in the abbey yard where Mendel performed his experiments was probably not uniform, and yet its possible effect on the expression of traits was ignored. But in reality, although the expression of genotype produces phenotype, the environment can affect this relationship.

Environmental effects are not limited to the external environment. For example, the alleles of some genes encode heat-sensitive products, that are affected by differences in internal body temperature. The *ch* allele in Himalayan rabbits and Siamese cats encodes a heat-sensitive version of the enzyme tyrosinase, which as you may recall is involved in albinism (figure 12.15). The Ch version of the enzyme is inactivated at temperatures above about 33°C. At the surface of the torso and head of these animals, the temperature is above 33°C and tyrosinase is inactive, producing a whitish coat. At the extremities, such as the tips of the ears and tail, the temperature is usually below 33°C and the enzyme is active allowing production of melanin that turns the coat in these areas a dark color.

inquiry

Many studies of identical twins separated at birth have revealed phenotypic differences in their development (height, weight, etc.). If these are identical twins, can you propose an explanation for these differences?

In epistasis, interactions of genes alter genetic ratios

The last simplifying assumption in Mendel's model is that the products of genes do not interact. But the products of genes may not act independently of one another, and the interconnected behavior of gene products can change the ratio expected

by independent assortment, even if the genes are on different chromosomes that do exhibit independent assortment.

Given the interconnected nature of metabolism, it should not come as a surprise that many gene products are not independent. Genes that act in the same metabolic pathway, for example, should show some form of dependence at the level of function. In such cases, the ratio Mendel would predict is not readily observed, but it is still there in an altered form.

Epistasis in corn

In the tests of Mendel's ideas that followed the rediscovery of his work, scientists had trouble obtaining Mendel's simple ratios, particularly with dihybrid crosses. Sometimes, it was not possible to identify successfully each of the four phenotypic classes expected, because two or more of the classes looked alike.

An example of this comes from the analysis of particular varieties of corn, *Zea mays*. Some commercial varieties exhibit a purple pigment called anthocyanin in their seed coats, whereas others do not. In 1918, geneticist R. A. Emerson crossed two true-breeding corn varieties, each lacking anthocyanin pigment. Surprisingly, all of the F_1 plants produced purple seeds.

When two of these pigment-producing F_1 plants were crossed to produce an F_2 generation, 56% were pigment producers and 44% were not. This is clearly not the Mendelian expectation. Emerson correctly deduced that two genes were involved in producing pigment, and that the second cross had thus been a dihybrid cross. According to Mendel's theory, gametes in a dihybrid cross could combine in 16 equally possible ways—so the puzzle was to figure out how these 16 combinations could occur in the two phenotypic groups of progeny. Emerson multiplied the fraction that were pigment producers (0.56) by 16 to obtain 9, and multiplied the fraction that lacked pigment (0.44) by 16 to obtain 7. Emerson therefore had a **modified ratio** of 9:7 instead of the usual 9:3:3:1 ratio (figure 12.16).

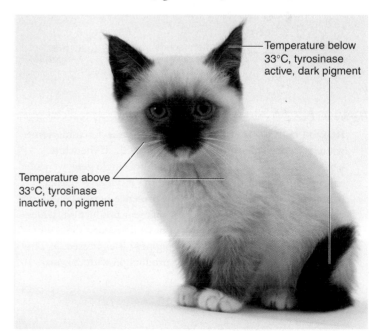

Temperature below 33°C, tyrosinase active, dark pigment

Temperature above 33°C, tyrosinase inactive, no pigment

figure 12.15

SIAMESE CAT. The pattern of coat color is due to an allele that encodes a temperature-sensitive form of the enzyme tyrosinase.

This modified ratio is easily rationalized by considering the function of the products encoded by these genes. When gene products act sequentially, as in a biochemical pathway, an allele expressed as a defective enzyme early in the pathway blocks the flow of material through the rest of the pathway. In this case, it is impossible to judge whether the later steps of the pathway are functioning properly. This type of gene interaction, in which one gene can interfere with the expression of another, is the basis of the phenomenon called **epistasis.**

The pigment anthocyanin is the product of a two-step biochemical pathway:

$$\text{starting molecule} \xrightarrow{\text{enzyme 1}} \text{intermediate} \xrightarrow{\text{enzyme 2}} \text{anthocyanin}$$
$$\text{(colorless)} \qquad\qquad \text{(colorless)} \qquad\qquad \text{(purple)}$$

To produce pigment, a plant must possess at least one functional copy of each enzyme's gene. The dominant alleles encode functional enzymes, and the recessive alleles encode nonfunctional enzymes. Of the 16 genotypes predicted by random assortment, 9 contain at least one dominant allele of both genes; they therefore produce purple progeny. The remaining 7 genotypes lack dominant alleles at *either or both* loci (3 + 3 + 1 = 7) and so produce colorless progeny, giving the phenotypic ratio of 9:7 that Emerson observed (see figure 12.16).

You can see that although this ratio is not the expected dihybrid ratio, it is a modification of the expected ratio.

Epistasis in Labrador retrievers

In many animals, coat color is the result of epistatic interactions among genes. Coat color in Labrador retrievers, a breed of dog, is due primarily to the interaction of two genes. The *E* gene determines whether a dark pigment, eumelanin, will be deposited in the fur. A dog having the genotype *ee* has no dark pigment deposited, and its fur is yellow. A dog having the genotype *EE* or *Ee* (*E__*) does have dark pigment deposited in the fur.

A second gene, the *B* gene, determines how dark the pigment will be. This gene controls the distribution of melanosomes in a hair. Dogs with the genotype *E__bb* have brown fur and are called chocolate labs. Dogs with the genotype *E__B__* have black fur.

Even in yellow dogs, however, the *B* gene does have some effect. Yellow dogs with the genotype *eebb* exhibit brown pigment on their nose, lips, and eye rims, but yellow dogs with the genotype *eeB__* have black pigment in these areas.

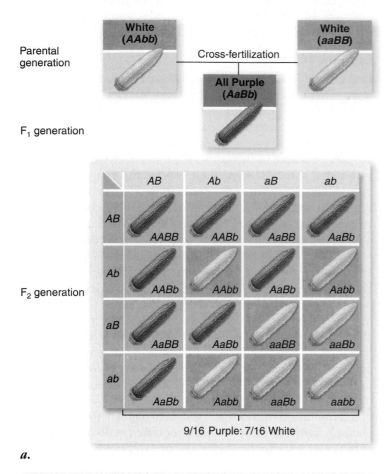

a.

b.

figure 12.16

HOW EPISTASIS AFFECTS GRAIN COLOR. *a.* Crossing some white varieties of corn yields an all purple F₁. If the white kernels were due to a recessive allele for a single gene we would expect white offspring. Self-crossing the F₁ yields 9 purple:7 white. This can be explained by the presence of two genes, each encoding an enzyme necessary for pigment production. Unless both enzymes are active (the plant has a dominant allele for each of the two genes, *A_B_*), no pigment is expressed. *b.* The biochemical pathway for pigment production with enzymes encoded by A and B genes.

Mendel's model is correct, but incomplete. Some traits are produced by the action of multiple genes (polygenic inheritance), and one gene can affect more than one trait (pleiotropy). Genes may have more than two alleles that may not show simple dominance. In incomplete dominance, the heterozygote is intermediate between the two homozygotes, and in codominance the heterozygote shows aspects of both homozygotes. The action of genes is also not always independent. This can lead to modified dihybrid ratios although the alleles of each gene are segregating independently. In epistasis, the action of one gene obscures the action of other genes.

12.1 The Mystery of Heredity

Our understanding of inheritance is the result of the scientific observations and Mendel's pea hybridization research.

- Traits, or characters, are transmitted directly to offspring, but they do not necessarily blend.

- Inherited characters can disappear in one generation only to reappear later, that is, the traits segregate among the offspring of a cross.

- Some traits are observed more often in the offspring of crosses.

- Mendel's experiments involved reciprocal crosses between pure-breeding pea varieties followed by one or more generations of self-fertilization.

- Mendel's mathematic analysis of experimental results lead to the present model of inheritance.

12.2 Monohybrid Crosses: The Principle of Segregation (figure 12.6)

A monohybrid cross follows only two forms of a single trait.

- Traits are determined by discrete factors we now call genes.

- Alleles are alternative forms of a gene that produce alternative forms of a trait.

- A genotype refers to the total set of alleles possessed by an individual.

- A phenotype refers to the physical appearance or other observable characteristic of the individual that is the result of the genotype's expression.

- The offspring of a parental cross (P) are the first filial generation (F_1).

- In crosses between pure-breeding parents the dominant trait is expressed and the alternative or recessive trait is not expressed until the F_2 generation.

- In the F_2 generation, the Mendelian ratio is expressed as 75% dominant to 25% recessive; also expressed as a 3:1 ratio.

- The F_2 generation disguises a 1:2:1 ratio in which 1/4 are true-breeding dominants, 2/4 (1/2) are not true-breeding and 1/4 are true-breeding recessives.

- The Principle of Segregation states that alternative alleles for a gene segregate during gamete formation and are randomly rejoined during fertilization.

- A homozygous individual carries two alleles of a gene that are the same.

- A heterozygous individual carries two alleles of a gene that are different.

- A trait determined by a dominant allele will be seen in both the homozygous dominant and the heterozygote.

- A trait determined by a recessive allele will only be seen in the homozygous recessive.

- The results of Mendelian crosses can be predicted with a Punnett square or by probability theory (figure 12.7).

- Human inheritance is studied using family pedigrees.

12.3 Dihybrid Crosses: The Principle of Independent Assortment (figure 12.10)

During meiosis, the segregation of different pairs of alleles is independent of each other.

- A dihybrid cross follows the behavior of two different traits during a single cross.

- During a dihybrid cross the F_1 generation displays only two of four possible combination of traits with no blending.

- The F_2 generation of a dihybrid cross exhibits all four possible combinations of traits in a 9:3:3:1 ratio.

- The Principle of Independent Assortment states that the alleles of each gene assort independently.

12.4 Probability: Predicting the Behavior of Crosses

Because the behavior of different chromosomes during meiosis is independent we can use probability to predict the outcome of crosses.

- Two rules of probability help predict genotypes and phenotypes from monohybrid cross results.

- The rule of addition states that the probability of two mutually exclusive events occuring is the **sum** of the individual probabilities.

- The rule of multiplication states that the probability of two independent events both occurring is the **product** of individual probabilities.

- Dihybrid cross probabilities are based on monohybrid cross probabilities using the product rule.

12.5 The Testcross: Revealing Unknown Genotypes (figure 12.11)

In a testcross, an unknown genotype is crossed with a homozygous recessive genotype.

- If the unknown genotype is homozygous dominant, the F_1 offspring will be the same.

- If the unknown genotype is heterozygous, the F_1 offspring will exhibit a 1:1 ratio.

- The results of a testcross support the Principle of Segregation.

12.6 Extensions to Mendel

In subsequent research scientists concluded that Mendel's basic model is correct, but it is incomplete and makes assumptions that are not valid.

- In polygenic inheritance more than one gene contributes to a phenotype.

- Many complex traits are due to multiple additive contributions by several genes, resulting in a continuous variation of quantitative traits.

- A pleiotropic effect occurs when an allele affects more than one trait and their effects are difficult to predict.

- Genes may have more than two (multiple), alleles.

- Incomplete dominance occurs when the heterozygous condition exhibits an intermediate phenotype resulting in a 1:2:1 ratio (figure 12.13).

- Codominant alleles each exhibit their own effect on the phenotype because one allele is not dominant over the other.

- The environment may affect the expression of a genotype, resulting in different phenotypes.

- In epistasis genes interact, and one gene interferes with the expression of a second.

SELF TEST

1. A true-breeding plant is one that—
   a. produces offspring that are different from the parent
   b. forms hybrid offspring through cross pollination
   c. produces offspring that are always the same as the parent
   d. can only reproduce with itself

2. What property distinguished Mendel's investigation from previous studies?
   a. Mendel used true-breeding pea plants.
   b. Mendel quantified his results.
   c. Mendel examined many different traits.
   d. Mendel examined the segregation of traits.

3. A monohybrid cross—
   a. is the same as self-fertilization
   b. examines a single variant of a trait
   c. produces a single offspring
   d. examines two variants of a single trait

4. What was the appearance of the F_1 generation of a monohybrid cross of purple (*PP*) and white (*pp*) flower pea plants?
   a. All the F_1 plants had white flowers.
   b. The F_1 plants had a light purple or blended appearance.
   c. All the F_1 plants had purple flowers.
   d. The most of the F_1 (3/4) had purple flowers, but 1/4 of the plants had white.

5. The F_1 plants from the previous question are allowed to self-fertilize. What will the phenotypic ratio be for this F_2?
   a. All purple
   b. 1 purple:1 white
   c. 3 purple:1 white
   d. 3 white:1 purple

6. Which of the following is *not* a part of Mendel's five-element model?
   a. Traits have alternative forms (what we now call alleles).
   b. Parents transmit discrete traits to their offspring.
   c. If an allele is present it will be expressed.
   d. Traits do not blend.

7. A *heterozygous* individual is one that carries—
   a. two completely different sets of genes
   b. two identical alleles for a particular gene
   c. only one functional allele
   d. two different alleles for a given gene

8. An organism's _____ is determined by its _____
   a. genotype; phenotype
   b. phenotype; genotype
   c. alleles; phenotype
   d. F_1 generation; alleles

9. Which of the following represent the phenotype for the recessive human trait, *albinism*?
   a. Absence of the pigment melanin
   b. Presence of a nonfunctional allele for the enzyme tyrosinase
   c. Absence of the enzyme tyrosinase from the individual's cells
   d. Both a and c

10. A dihybrid cross between a plant with long smooth leaves and a plant with short hairy leaves produces a long smooth F_1. If this F_1 is allowed to self-cross to produce an F_2, what would you predict for the ratio of F_2 phenotypes?
    a. 9 long smooth:3 long hairy:3 short hairy:1 short smooth
    b. 9 long smooth:3 long hairy:3 short smooth:1 short hairy
    c. 9 short hairy:3 long hairy:3 short smooth:1 long smooth
    d. 1 long smooth:1 long hairy:1 short smooth:1 short hairy

11. A testcross is used to determine if an individual is—
    a. homozygous dominant or heterozygous
    b. homozygous recessive or homozygous dominant
    c. heterozygous or homozygous recessive
    d. true-breeding

12. What is a polygenic trait?
    a. A set of multiple phenotypes determined by a single gene
    b. A single phenotypic trait determined by two alleles
    c. A single phenotypic trait determined by more than one gene
    d. The collection of traits possessed by an individual

13. When a single gene influences multiple phenotypic traits the effect is called—
    a. Codominance
    b. Epistasis
    c. Incomplete dominance
    d. Pleiotropy

14. What is the probability of obtaining an individual with the genotype *bb* from a cross between two individuals with the genotype *Bb*?
    a. 1/2
    b. 1/4
    c. 1/8
    d. 0

15. What is the probability of obtaining an individual with the genotype *CC* from a cross between two individuals with the genotypes *CC* and *Cc*?
    a. 1/2
    b. 1/4
    c. 1/8
    d. 1/16

CHALLENGE QUESTIONS

1. Create a Punnett square for the following crosses and use this to predict phenotypic ratio for dominant and recessive traits. Dominant alleles are indicated by uppercase letters and recessive are indicated by lowercase letters.
   a. A monohybrid cross between individuals with the genotype *Aa* and *Aa*
   b. A dihybrid cross between two individuals with the genotype *AaBb*
   c. A dihybrid cross between individuals with the genotype *AaBb* and *aabb*

2. Use probability to predict the following.
   a. What is the probability of obtaining an individual with the genotype *rr* from the self-fertilization of a plant with the genotype *Rr*?
   b. What is the probability that a testcross with a heterozygous individual will produce homozygous recessive offspring?
   c. A plant with the genotype *Gg* is self-fertilized. Use probability to determine the proportion of the offspring that will have the dominant phenotype.
   d. Use probability to determine the proportion of offspring from a dihybrid cross (*GgRr* × *GgRr*) that will have the phenotype *ggR_*.

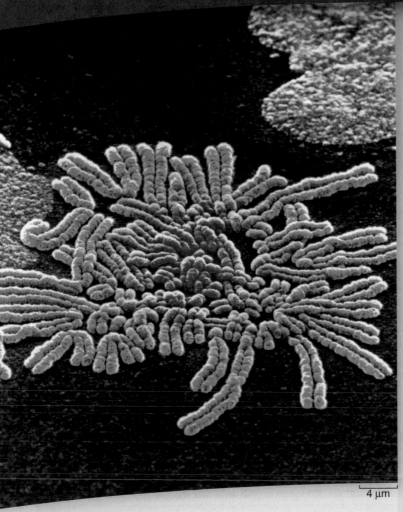

4 μm

chapter 13

Chromosomes, Mapping, and the Meiosis–Inheritance Connection

introduction

MENDEL'S EXPERIMENTS OPENED the door to understanding inheritance, but many questions remained. In the early part of the 20th century, we did not know the nature of the factors whose behavior Mendel had described. The next step, which involved many researchers in the early part of the century, was uniting information about the behavior of chromosomes, seen in the picture, and the inheritance of traits. The basis for Mendel's principles of segregation and independent assortment lie in events that occur during meiosis.

The behavior of chromosomes during meiosis not only explains Mendel's principles, but leads to new and different approaches to the study of heredity. The ability to construct genetic maps is one of the most powerful tools of classical genetic analysis. The tools of genetic mapping developed in flies and other organisms in combination with information from the human genome project now allow us to determine the location and isolate genes that are involved in genetic diseases.

concept outline

The exceptions to this rule actually provide support for this mechanism of sex determination. For example, movement of part of the Y chromosome to the X chromosome can cause otherwise XX individuals to develop as male. There is also a genetic disorder that causes a failure to respond to the androgen hormones (androgen insensitivity syndrome) that causes XY individuals to develop as female. Lastly, mutations in *SRY* itself can cause XY individuals to develop as females.

This form of sex determination seen in humans is shared among mammals, but is not universal in vertebrates. Among fishes and some species of reptiles, environmental factors can cause changes in the expression of this sex-determining gene, and thus in the sex of the adult individual.

Some human genetic disorders display sex linkage

From ancient times, people have noted conditions that seem to affect males to a greater degree than females. Red-green color blindness is one well-known condition that is more common in males because the gene affected is carried on the X chromosome.

Another example is **hemophilia,** a disease that affects a single protein in a cascade of proteins involved in the formation of blood clots. Thus, in an untreated hemophiliac, even minor cuts will not stop bleeding. This form of hemophilia is caused by an X-linked recessive allele; women who are heterozygous for the allele are asymptomatic carriers, and men who receive an X chromosome with the recessive allele exhibit the disease.

The allele for hemophilia was introduced into a number of different European royal families by Queen Victoria of England. Because these families kept careful genealogical records, we have an extensive pedigree for this condition. In the five generations after Victoria, ten of her male descendants have had hemophilia as shown in the pedigree in figure 13.3.

The Russian house of Romanov inherited this condition through Alexandra Feodorovna, a granddaughter of Queen Victoria. She married Czar Nicholas II, and their only son, Alexis,

figure 13.3

THE ROYAL HEMOPHILIA PEDIGREE. Queen Victoria, shown at the bottom center of the photo, was a carrier for hemophilia. Two of Victoria's four daughters, Alice and Beatrice, inherited the hemophilia allele from Victoria. Two of Alice's daughters are standing behind Victoria (wearing feathered boas): Princess Irene of Prussia (*right*) and Alexandra (*left*), who would soon become czarina of Russia. Both Irene and Alexandra were also carriers of hemophilia. From the pedigree, it is clear that Alice introduced hemophilia into the Russian and Prussian royal houses, and Victoria's daughter Beatrice introduced it into the Spanish royal house. Victoria's son Leopold, himself a victim, also transmitted the disorder in a third line of descent. Half-shaded symbols represent carriers with one normal allele and one defective allele; fully shaded symbols represent affected individuals.

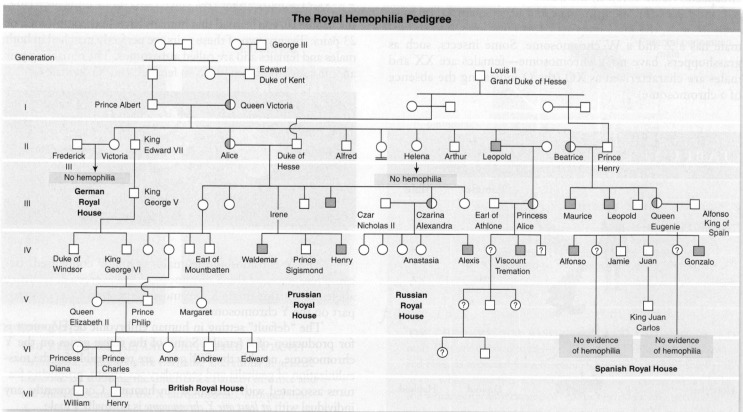

was afflicted with the disease. The entire family was executed during the Russian revolution. (Recently, a woman who had long claimed to be Anastasia, a surviving daughter, was shown not to be a Romanov using modern genetic techniques to test her remains.)

Ironically, this condition has not affected the current British royal family, because Victoria's son Edward, who became King Edward VII, did not receive the hemophilia allele. All of the subsequent rulers of England are his descendants.

Dosage compensation prevents doubling of sex-linked gene products

Although males have only one copy of the X chromosome and females have two, female cells do not produce twice as much of the proteins encoded by genes on the X chromosome. Instead, one of the X chromosomes in females is inactivated early in embryonic development, shortly after the embryo's sex is determined. This inactivation is an example of **dosage compensation,** which ensures an equal level of expression from the sex chromosomes despite a differing number of sex chromosomes in males and females. (In *Drosophila*, by contrast, dosage compensation is achieved by increasing the level of expression on the male X chromosome.)

Which X chromosome is inactivated in females varies randomly from cell to cell. If a woman is heterozygous for a sex-linked trait, some of her cells will express one allele and some the other. The inactivated and highly condensed X chromosome is visible as a darkly staining **Barr body** attached to the nuclear membrane.

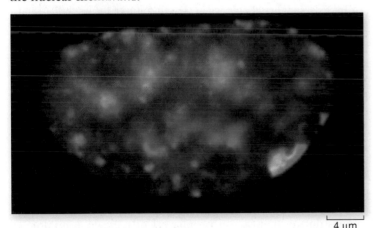

4 μm

X chromosome inactivation can lead to genetic mosaics

X chromosome inactivation to produce dosage compensation is not unique to humans but is true of all mammals. Females that are heterozygous for X chromosome alleles are **genetic mosaics:** Their individual cells may express different alleles, depending on which chromosome is inactivated.

One example is the calico cat, a female that has a patchy distribution of dark fur, orange fur, and white fur (figure 13.4). The dark fur and orange fur are due to heterozygosity for a gene on the X chromosome that determines pigment type. One allele results in dark fur, and another allele results in orange fur. Which of these colors is observed in any particular patch is

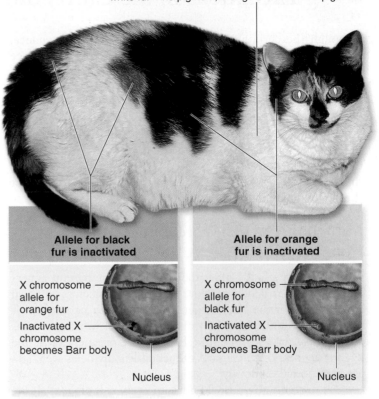

Second gene causes patchy distribution of pigment: white fur = no pigment, orange or black fur = pigment

Allele for black fur is inactivated

X chromosome allele for orange fur

Inactivated X chromosome becomes Barr body

Nucleus

Allele for orange fur is inactivated

X chromosome allele for black fur

Inactivated X chromosome becomes Barr body

Nucleus

figure 13.4

A CALICO CAT. The cat is heterozygous for alleles of a coat color gene that produce either black fur or orange fur. This gene is on the X chromosome, so the different-colored fur is due to inactivation of one X chromosome. The patchy distribution and white color is due to a second gene that is epistatic to the coat color gene and thus masks its effects.

due to inactivation of one X chromosome: If the chromosome containing the orange allele is inactivated, then the fur will be dark, and vice versa.

The patchy distribution of color, and the presence of white fur, is due to a second gene that is epistatic to the fur color gene (chapter 12). That is, the presence of this second gene produces a patchy distribution of pigment, with some areas totally lacking pigment. In the areas that lack pigment, the effect of either fur color allele is masked. Thus, in this one animal we can see an excellent example of both epistasis and X inactivation.

Not all organisms have the same sex chromosomes, but all have some difference in chromosomes between the sexes. In humans, male development depends on the presence of the Y chromosome. XY males will show recessive traits for alleles on the X chromosome, leading to sex-linked inheritance. Mammalian females inactivate one X chromosome to balance the levels of gene expression in males and females. This random inactivation can lead to genetic mosaics in females heterozygous for X chromosome genes.

Exceptions to the Chromosomal Theory of Inheritance

Although the chromosomal theory explains most inheritance, there are exceptions. Primarily, these are due to the presence of DNA in organelle genomes, specifically in mitochondria and chloroplasts. Non-Mendelian inheritance via organelles was studied in depth by Ruth Sager, who in the face of universal skepticism constructed the first map of chloroplast genes in *Chlamydomonas*, a unicellular green alga, in the 1960s and 1970s.

Mitochondria and chloroplasts are not partitioned with the nuclear genome by the process of meiosis. Thus any trait that is due to the action of genes in these organelles will not show Mendelian inheritance.

Mitochondrial genes are inherited from the female parent

Organelles are usually inherited from only one parent, generally the mother. When a zygote is formed, it receives an equal contribution of the nuclear genome from each parent, but it gets all of its mitochondria from the egg cell, which contains a great deal more cytoplasm (and thus the organelles). As the zygote divides, these original mitochondria divide as well, and are partitioned randomly.

As a result, the mitochondria in every cell of an adult organism can be traced back to the original maternal mitochondria present in the egg. This mode of uniparental (one-parent) inheritance from the mother is called **maternal inheritance.**

In humans, the disease Leber's hereditary optic neuropathy (LHON) shows maternal inheritance. The genetic basis of this disease is a mutant allele for a subunit of NADH dehydrogenase. The mutant allele reduces the efficiency of electron flow in the electron transport chain in mitochondria (see chapter 7), in turn reducing overall ATP production. Some nerve cells in the optic system are particularly sensitive to reduction in ATP production, resulting in neural degeneration.

A mother with this disease will pass it on to all of her progeny, whereas a father with the disease will not pass it on to any of his progeny. Note that this condition differs from sex-linked inheritance because males and females are equally affected.

Chloroplast genes may also be passed on uniparentally

The inheritance pattern of chloroplasts is also usually maternal, although both paternal and biparental inheritance of chloroplasts is also observed depending on the species. Carl Correns first hypothesized in 1909 that chloroplasts were responsible for inheritance of variegation (mixed green and white leaves) in the plant commonly known as the four o'clock (*Mirabilis jalapa*). The offspring exhibited the phenotype of the female parent, regardless of the male's phenotype.

In Sager's work on *Chlamydomonas*, resistance to the antibiotic streptomycin was shown to be transmitted via the DNA of chloroplasts from only one of the organism's two mating types (termed plus and minus).

> Organelles such as mitochondria and chloroplasts contain their own genomes. These organelles divide independently of the nucleus, and they are carried in the cytoplasm of the egg cell. Inheritance of traits in these genomes are therefore said to be maternally inherited. In some species, however, chloroplasts may also be passed on paternally or biparentally.

Genetic Mapping

We have seen that Mendelian characters are determined by genes located on chromosomes and that the independent assortment of Mendelian traits reflects the independent assortment of chromosomes in meiosis. This is fine as far as it goes, but it is still incomplete. Of Mendel's seven traits in figure 12.4, six are on different chromosomes and two are on the same chromosome, yet all show independent assortment with one another. The two on the same chromosome should not behave the same as those that are on different chromosomes. In fact, organisms will generally have many more genes that assort independently than the number of chromosomes. This means that independent assortment cannot be due only to the random alignment of chromosomes during meiosis.

inquiry

? *Mendel did not examine plant height and pod shape in his dihybrid crosses. The genes for these traits are very close together on the same chromosome. How would this have changed Mendel's results?*

The solution to this problem is found in an observation that was introduced in chapter 11: the crossing over of homologues during meiosis. In prophase I of meiosis, homologues appear to physically exchange material by crossing over (figure 13.5). In chapter 11, you saw how this was part of the mechanism that allows homologues, and not sister chromatids, to disjoin at anaphase I.

Genetic recombination exchanges alleles on homologues

Consider a dihybrid cross performed using the Mendelian framework. Two true-breeding parents that each differ with respect to two traits are crossed, producing doubly heterozygous F_1 progeny. If the genes for the two traits are on a single chromosome, then during meiosis we would expect alleles for both loci to segregate together and produce only gametes that resemble the two parental types. But if a crossover occurs between the two loci, then each homologue would carry one allele from each parent and produce gametes that combine these parental traits (see figure 13.5). We call gametes with this new combination of alleles *recombinant* gametes as they are formed by recombining the parental alleles.

The first investigator to provide evidence for this was Morgan, who studied three genes on the X chromosome of *Drosophila*. He found an excess of parental types, which he explained as due to the genes all being on the X chromosome and therefore coinherited (inherited together). He went further suggesting that the recombinant genotypes were due to crossing over between homologues during meiosis.

Experiments performed independently by Barbara McClintock and Harriet Creighton in maize and by Curt Stern in *Drosophila*, provided evidence for this physical exchange of genetic material. The experiment done by Creighton and McClintock is detailed in figure 13.6. In this experiment, they used a chromosome with two alterations visible under a microscope: a knob on one end of the chromosome and a part of a different chromosome attached to the other end. In addition to these cytological markers, this chromosome also carried two genetic markers: a gene that determines kernel color and one that determines kernel texture.

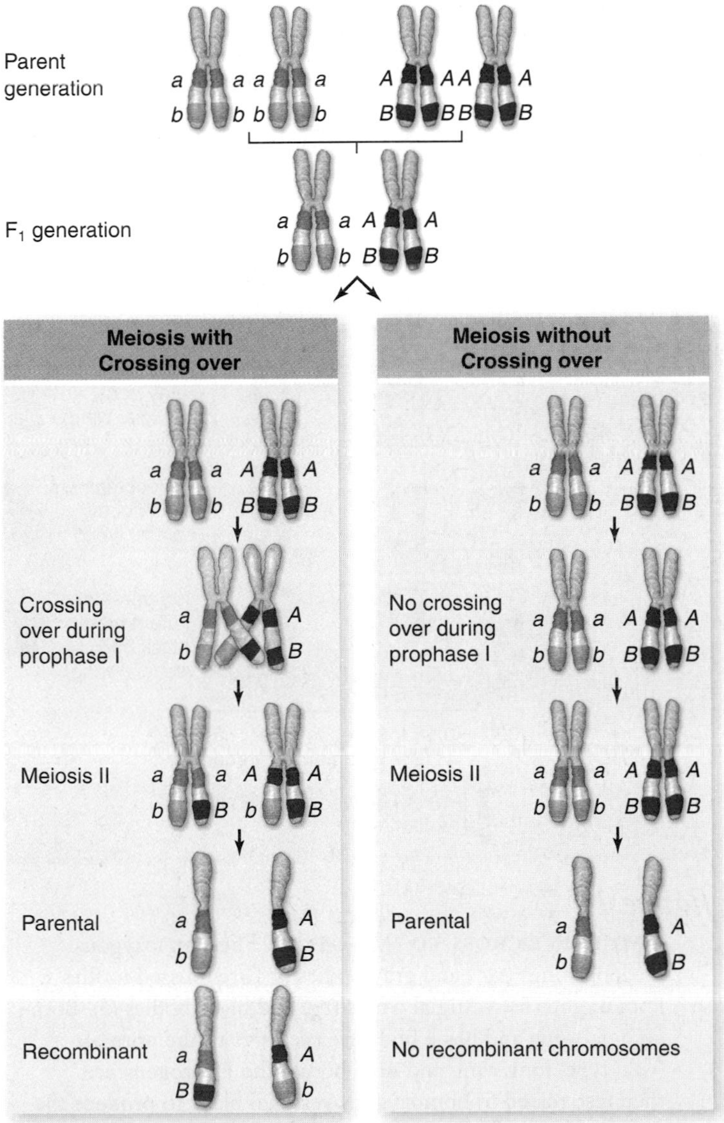

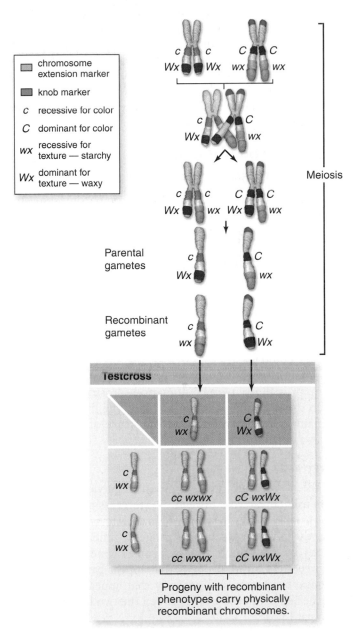

figure 13.5

CROSSING OVER EXCHANGES ALLELES ON HOMOLOGUES.
When a crossover occurs between two loci, it leads to the production of recombinant chromosomes. If no crossover occurs, then the chromosomes will carry the parental combination of alleles.

figure 13.6

THE CREIGHTON AND MCCLINTOCK EXPERIMENT.
This experiment first demonstrated that chromosomes physically exchange genetic material during recombination. The experimental design was to use chromosomal differences visible in the microscope and two unrelated genes on the same chromosome. When plants heterozygous for visible and genetic markers were testcrossed, progeny that are genetically recombinant have also exchanged visible markers. This shows that the chromosomes have physically exchanged genetic material.

The longer chromosome, which had the knob, carried the dominant colored allele for kernel color (*C*) and the recessive waxy allele for kernel texture (*wx*). Heterozygotes were produced with the altered chromosome paired with a normal chromosome carrying the recessive colorless allele for kernel color (*c*) and the dominant starchy allele for kernel texture (*Wx*) (see figure 13.6). These plants appeared colored and starchy because they were heterozygous for both loci, and they were also heterozygous for the two visibly distinct chromosomes.

A testcross was performed with these F₁ plants and colorless waxy plants. The progeny were analyzed for both physical recombination (using a microscope to observe chromosomes) and genetic recombination (by examining the phenotype of progeny). The results were striking: All of the progeny that exhibited the recombinant phenotype also now had only one of the chromosomal markers. That is, physical exchange was accompanied by the recombinant phenotype.

Recombination is the basis for genetic maps

The ability to map the location of genes on chromosomes using data from genetic crosses is one of the most powerful tools of genetics. The insight that allowed this technique, like many great insights, is so simple as to seem obvious in retrospect.

Morgan had already suggested that the frequency with which a particular group of recombinant progeny appeared was a reflection of the relative location of genes on the chromosome. An undergraduate in Morgan's laboratory, Alfred Sturtevant put this observation on a quantitative basis. Sturtevant reasoned that the frequency of recombination observed in crosses could be used as a measure of genetic distance. That is, as physical distance on a chromosome increases, so does the probability of recombination (crossover) occurring between the gene loci. Using this logic, the frequency of recombinant gametes produced is a measure of their distance apart on a chromosome.

Linkage data

To be able to measure recombination frequency easily, investigators used a testcross instead of intercrossing the F₁ progeny to produce an F₂ generation. In a testcross, as described earlier, the phenotypes of the progeny reflect the gametes produced by the doubly heterozygous F₁ individual. In the case of recombination, progeny that appear parental have not undergone crossover, and progeny that appear recombinant have experienced a crossover between the two loci in question (see figure 13.5).

When genes are close together, the number of recombinant progeny is much lower than the number of parental progeny, and the genes are defined on this basis as being **linked.** The number of recombinant progeny divided by total progeny gives a value defined as the **recombination frequency.** This value is converted to a percentage, and each 1% of recombination is termed a **map unit.** This unit has been named the centimorgan (cM) for T. H. Morgan, although it is also called simply a map unit (m.u.) as well.

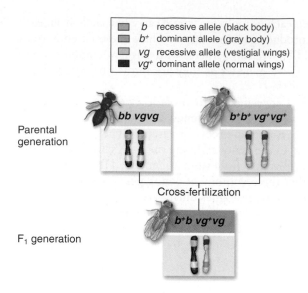

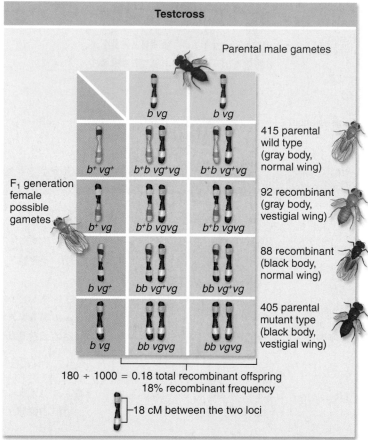

figure 13.7

TWO-POINT CROSS TO MAP GENES. Flies homozygous for long wings (*vg⁺*) and gray bodies (*b⁺*) are crossed to flies homozygous for vestigial wings (*vg*) and black bodies (*b*). Both vestigial wing and black body are recessive to the normal (wild type) long wing and grey body. The F₁ progeny are then testcrossed to homozygous vestigial black to produce the progeny for mapping. Data are analyzed in the text.

Constructing maps

Constructing genetic maps then becomes a simple process of performing testcrosses with doubly heterozygous individuals and counting progeny to determine percent recombination. This is best shown with an example using a two-point cross.

Drosophila homozygous for two mutations, vestigial wings (*vg*) and black body (*b*), are crossed to flies homozygous for the wild type, or normal alleles, of these genes (*vg*$^+$ *b*$^+$). The doubly heterozygous F$_1$ progeny are then testcrossed to homozygous recessive individuals (*vg b/vg b*), and progeny are counted (figure 13.7). The data are shown below:

vestigial wings, black body (*vg b*)	405 (parental)
long wings, gray body (*vg*$^+$ *b*$^+$)	415 (parental)
vestigial wings, gray body (*vg b*$^+$)	92 (recombinant)
long wings, black body (*vg*$^+$ *b*)	88 (recombinant)
Total Progeny	1000

The recombination frequency is 92 + 88 divided by 1000, or 0.18. Converting this number to a percentage yields 18 cM as the map distance between these two loci.

Multiple crossovers can yield independent assortment results

As the distance separating loci increases, the probability of recombination occurring between them during meiosis also increases. What happens when more than one recombination event occurs?

If homologues undergo two crossovers between loci, then the parental combination is restored. This leads to an underestimate of the true genetic distance because not all events can be noted. As a result, the relationship between true distance on a chromosome and the recombination frequency is not linear. It begins as a straight line, but the slope decreases; the curve levels off at a recombination frequency of 0.5 (figure 13.8).

At long distances, multiple events between loci become frequent. In this case, odd numbers of crossovers (1, 3, 5) produce recombinant gametes, and no crossover or even numbers of crossovers (0, 2, 4) produce parental gametes. At large enough distances, these frequencies are about equal, leading to the number of recombinant gametes being equal to the number of parental gametes, and the loci exhibit independent assortment! This is how Mendel could use two loci on the same chromosome and have them assort independently.

inquiry

What would Mendel have observed in a dihybrid cross if the two loci were 10 cM apart on the same chromosome? Is this likely to have led him to the idea of independent assortment?

Three-point crosses can be used to put genes in order

Because multiple crossovers reduce the number of observed recombinant progeny, longer map distances are not accurate. As a result, when geneticists try to construct maps from a series of two-point crosses, determining the order of genes is problematic. Using three loci instead of two, or a three-point cross, can help solve the problem.

In a three-point cross, the gene in the middle allows us to see recombination events on either side. For example, a double crossover for the two outside loci is actually a single crossover between each outside locus and the middle locus (figure 13.9).

The probability of two crossovers is equal to the product of the probability of each individual crossover, each of which is relatively low. Therefore, in any three-point cross, the class of offspring with two crossovers is the least frequent class. Analyzing these individuals to see which locus is recombinant identifies the locus that lies in the middle of the three loci in the cross (see figure 13.9).

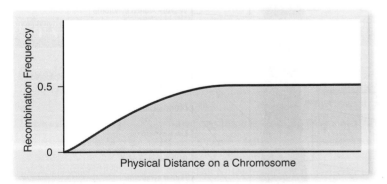

figure 13.8

RELATIONSHIP BETWEEN TRUE DISTANCE AND RECOMBINATION FREQUENCY. As distance on a chromosome increases, the recombinants are not all detected due to double crossovers. This leads to a curve that levels off at 0.5.

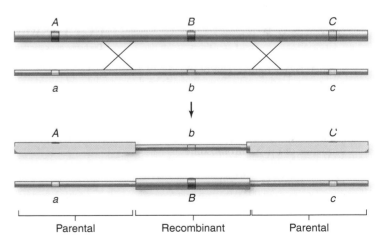

figure 13.9

USE OF A THREE-POINT CROSS TO ORDER GENES. In a two-point cross, the outside loci will appear parental for double crossovers. With the addition of a third locus, the two crossovers can still be detected because the middle locus will be recombinant. This double crossover class should be the least frequent, so whatever locus has recombinant alleles in this class must be in the middle.

Individuals with sickle-shaped red blood cells exhibit intermittent illness and reduced life span. At the molecular level, this condition is caused by a single amino acid, glutamic acid, in the 146-amino-acid β-globin chain being changed to another amino acid, valine. The affected amino acid site is not in the oxygen-binding region of the protein, but the change still has a catastrophic effect on hemoglobin's function. The replacement of glutamic acid, a charged amino acid, with nonpolar valine on the surface of the protein makes the protein sticky. This stickiness is due to the tendency of nonpolar amino acids to aggregate in water-based solutions such as blood plasma, leading to the stiff, rodlike structures seen in sickled red blood cells (figure 13.12).

Individuals heterozygous for the sickle cell allele are indistinguishable from normal individuals in a normal oxygen environment, although their red cells do exhibit reduced ability to carry oxygen.

The sickle cell allele is particularly prevalent in people of African descent. In some regions of Africa, up to 45% of the population is heterozygous for the trait, and 6% are homozygous. This proportion of heterozygotes is higher than would be expected on the basis of chance alone. It turns out that heterozygosity confers a greater resistance to the blood-borne parasite that causes the disease malaria. In regions of central Africa where malaria is endemic, the sickle cell allele also occurs at a high frequency.

The sickle cell allele is not the end of the story for the β-globin gene; a large number of other alterations of this gene have been observed that lead to anemias. In fact, for hemoglobin, which is composed of two α-globins and two β-globins, over 700 structural variants have been cataloged. It is estimated that 7% of the human population worldwide are carriers for different inherited hemoglobin disorders.

The Human Gene Mutation Database has cataloged the nature of many disease alleles, including the sickle cell allele. The majority of alleles seem to be simple changes. Almost 60% of the close to 28,000 alleles in the Human Gene

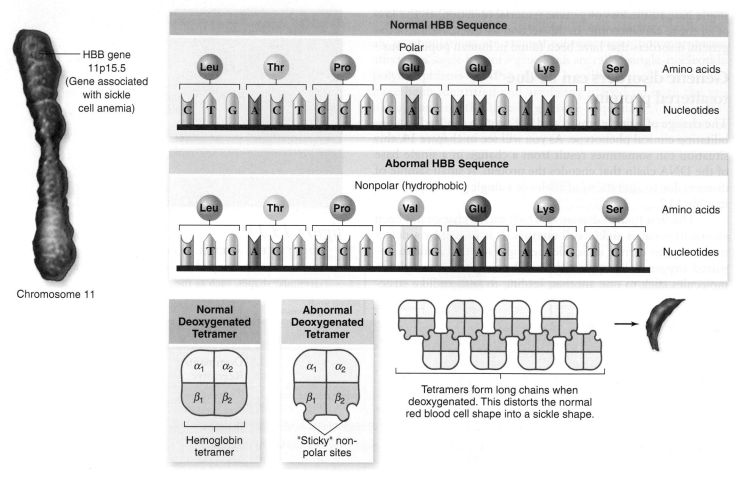

figure 13.12

SICKLE CELL ANEMIA IS CAUSED BY AN ALTERED PROTEIN. Hemoglobin is composed of a tetramer of two α-globin chains and two β-globin chains. Protein sequences are encoded in DNA in groups of three nucleotides (chapter 15 details the genetic code). The sickle cell allele of the β-globin gene has a single change in the DNA sequence that results in an amino acid substitution of valine for glutamic acid. The valine is hydrophobic creating regions on the surface of the protein that are "sticky." In an individual with sickle cell, mutant β-globin chains associate to form long chains that distort the red blood cell.

Mutation Database are single-base substitutions. Another 23% are due to small insertions or deletions of less than 20 bases. The rest of the alleles are made of more complex alterations. It is clear that simple changes in genes can have profound effects.

Nondisjunction of chromosomes changes chromosome number

The failure of homologues or sister chromatids to separate properly during meiosis is called **nondisjunction.** This failure leads to the gain or loss of a chromosome, a condition called **aneuploidy.** The frequency of aneuploidy in humans is surprisingly high, being estimated to occur in 5% of conceptions.

Nondisjunction of autosomes

Humans who have lost even one copy of an autosome are called **monosomics,** and generally they do not survive embryonic development. In all but a few cases, humans who have gained an extra autosome (called **trisomics**) also do not survive. Data from clinically recognized spontaneous abortions indicate levels of aneuploidy as high as 35%.

Five of the smallest human autosomes, however—those numbered 13, 15, 18, 21, and 22—can be present as three copies and still allow the individual to survive, at least for a time. The presence of an extra chromosome 13, 15, or 18 causes severe developmental defects, and infants with such a genetic makeup die within a few months. In contrast, individuals who have an extra copy of chromosome 21 or, more rarely, chromosome 22, usually survive to adulthood. In these people, the maturation of the skeletal system is delayed, so they generally are short and have poor muscle tone. Their mental development is also affected, and children with trisomy 21 are always mentally retarded to some degree.

The developmental defect produced by trisomy 21 (figure 13.13) was first described in 1866 by J. Langdon Down; for this reason, it is called **Down syndrome.** About 1 in every 750 children exhibits Down syndrome, and the frequency is comparable in all racial groups. Similar conditions also occur in chimpanzees and other related primates.

In humans, the defect occurs when a particular small portion of chromosome 21 is present in three copies instead of two. In 97% of the cases examined, all of chromosome 21 is present in three copies. In the other 3%, a small portion of chromosome 21 containing the critical segment has been added to another chromosome by a process called *translocation* (see chapter 15); it exists along with the normal two copies of chromosome 21. This latter condition is known as *translocation Down syndrome.*

In mothers younger than 20 years of age, the risk of giving birth to a child with Down syndrome is about 1 in 1700; in mothers 20 to 30 years old, the risk is only about 1 in 1400. However, in mothers 30 to 35 years old, the risk rises to 1 in 750, and by age 45, the risk is as high as 1 in 16 (figure 13.14).

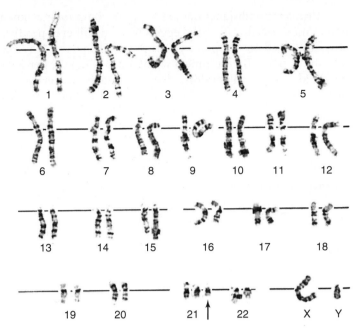

figure 13.13

DOWN SYNDROME. As shown in this male karyotype, Down syndrome is associated with trisomy of chromosome 21 (arrow shows third copy of chromosome 21).

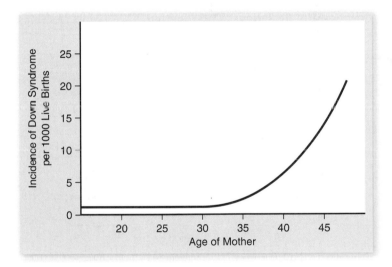

figure 13.14

CORRELATION BETWEEN MATERNAL AGE AND THE INCIDENCE OF DOWN SYNDROME. As women age, the chances they will bear a child with Down syndrome increase. After a woman reaches 35, the frequency of Down syndrome rises rapidly.

inquiry

? *Over a five-year period between ages 20 and 25, the incidence of Down syndrome increases 0.1 per thousand; over a five-year period between ages 35 and 40, the incidence increases to 8.0 per thousand, 80 times as great. The period of time is the same in both instances. What has changed?*

Primary nondisjunctions are far more common in women than in men because all of the eggs a woman will ever produce have developed to the point of prophase in meiosis I by the time she is born. By the time a woman has children, her eggs are as old as she is. Therefore, there is a much greater chance for cell-division problems of various kinds, including those that cause primary nondisjunction, to accumulate over time in female gametes. In contrast, men produce new sperm daily. For this reason, the age of the mother is more critical than that of the father for couples contemplating childbearing.

Nondisjunction of sex chromosomes

Individuals who gain or lose a sex chromosome do not generally experience the severe developmental abnormalities caused by similar changes in autosomes. Although such individuals have somewhat abnormal features, they often reach maturity and in some cases may be fertile.

X chromosome nondisjunction. When X chromosomes fail to separate during meiosis, some of the gametes produced possess both X chromosomes, and so are XX gametes; the other gametes have no sex chromosome and are designated "O" (figure 13.15).

If an XX gamete combines with an X gamete, the resulting XXX zygote develops into a female with one functional X chromosome and two Barr bodies. She may be taller in stature but is otherwise normal in appearance.

If an XX gamete instead combines with a Y gamete, the effects are more serious. The resulting XXY zygote develops

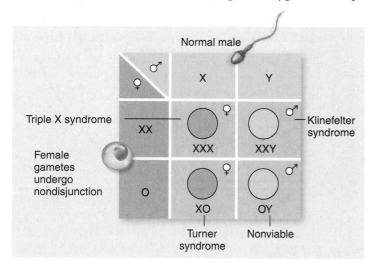

figure 13.15

HOW NONDISJUNCTION CAN PRODUCE ABNORMALITIES IN THE NUMBER OF SEX CHROMOSOMES. When nondisjunction occurs in the production of female gametes, the gamete with two X chromosomes (XX) produces Klinefelter males (XXY) and triple-X females (XXX). The gamete with no X chromosome (O) produces Turner females (XO) and nonviable OY males lacking any X chromosome.

inquiry

? *Can you think of two nondisjunction scenarios that would produce an XXY male?*

into a male who has many female body characteristics and, in some cases but not all, diminished mental capacity. This condition, called *Klinefelter syndrome*, occurs in about 1 out of every 500 male births.

If an O gamete fuses with a Y gamete, the resulting OY zygote is nonviable and fails to develop further; humans cannot survive when they lack the genes on the X chromosome. But if an O gamete fuses with an X gamete, the XO zygote develops into a sterile female of short stature, with a webbed neck and sex organs that never fully mature during puberty. The mental abilities of an XO individual are in the low-normal range. This condition, called *Turner syndrome*, occurs roughly once in every 5000 female births.

Y chromosome nondisjunction. The Y chromosome can also fail to separate in meiosis, leading to the formation of YY gametes. When these gametes combine with X gametes, the XYY zygotes develop into fertile males of normal appearance. The frequency of the XYY genotype (*Jacob syndrome*) is about 1 per 1000 newborn males.

Genomic imprinting depends on the parental origin of alleles

By the late 20th century, geneticists were confident that they understood the basic mechanisms governing inheritance. It came as quite a surprise when mouse geneticists found an important exception to classical Mendelian genetics that appears to be unique to mammals. In **genomic imprinting,** the phenotype caused by a specific allele is exhibited when the allele comes from one parent, but not from the other.

The basis for genomic imprinting is the expression of a gene depending on passage through maternal or paternal germ lines. Some genes are inactivated in the paternal germ line and therefore are not expressed in the zygote. Other genes are inactivated in the maternal germ line, with the same result. This condition makes the zygote effectively haploid for an imprinted gene. The expression of variant alleles of imprinted genes depends on the parent of origin. Furthermore, imprinted genes seem to be concentrated in particular regions of the genome. These regions include genes that are both maternally and paternally imprinted.

Prader–Willi and Angelman syndromes

An example of genomic imprinting in humans involves the two diseases Prader–Willi syndrome (PWS) and Angelman syndrome (AS). The effects of PWS include respiratory distress, obesity, short stature, mild mental retardation, and obsessive–compulsive behavior. The effects of AS include developmental delay, severe mental retardation, hyperactivity, aggressive behavior, and inappropriate laughter.

Genetic studies have implicated genes on chromosome 15 for both disorders, but the pattern of inheritance is complementary. The most common cause of both syndromes is a deletion of material on chromosome 15 and, in fact, the same deletion can cause either syndrome. The determining factor is the parental origin of the normal and deleted chromosomes. If the chromosome with the deletion is paternally inherited it causes PWS, if the chromosome with the deletion is maternally inherited it causes AS.

The region of chromosome 15 that is lost is subject to imprinting, with some genes being inactivated in the maternal germ line, and others in the paternal germ line. In PWS, genes are inactivated in the maternal germ line, such that deletion or other functional loss of paternally derived alleles produces the syndrome. The opposite is true for AS syndrome: Genes are inactivated in the paternal germ line, such that loss of maternally derived alleles leads to the syndrome.

Molecular basis of genomic imprinting

Although genomic imprinting is not well understood, at least one aspect seems clear: The basis for inactivating genes appears to be linked to modifications of the DNA itself. DNA can be modified by the addition of methyl groups, termed *methylation*. This modification is correlated with inactivity of genes. The proteins that are associated with chromosomes can also be modified, leading to effects on gene expression. The control of gene expression is discussed in more detail in the following chapters.

Some genetic defects can be detected early in pregnancy

Although most genetic disorders cannot yet be cured, we are learning a great deal about them, and progress toward successful therapy is being made in many cases. In the absence of a cure, however, the only recourse is to try to avoid producing children with these conditions. The process of identifying parents at risk for having children with genetic defects and of assessing the genetic state of early embryos is called **genetic counseling.**

Pedigree analysis

One way of assessing risks is through pedigree analysis, often employed as an aid in genetic counseling. By analyzing a person's pedigree, it is sometimes possible to estimate the likelihood that the person is a carrier for certain disorders. For example, if a counseling client's family history reveals that a relative has been afflicted with a recessive genetic disorder, such as cystic fibrosis, it is possible that the client is a heterozygous carrier of the recessive allele for that disorder.

When a couple is expecting a child, and pedigree analysis indicates that both of them have a significant chance of being heterozygous carriers of a deleterious recessive allele, the pregnancy is said to be high-risk. In such cases, a significant probability exists that their child will exhibit the clinical disorder.

Another class of high-risk pregnancy is that in which the mothers are more than 35 years old. As discussed earlier, the frequency of infants with Down syndrome increases dramatically in the pregnancies of older women (see figure 13.14).

Amniocentesis

When a pregnancy is diagnosed as high-risk, many women elect to undergo **amniocentesis,** a procedure that permits the prenatal diagnosis of many genetic disorders. In the fourth month of pregnancy, a sterile hypodermic needle is inserted into the expanded uterus of the mother, removing a small sample of the amniotic fluid that bathes the fetus (figure 13.16). Within the fluid are free-floating cells derived from the fetus, once removed, these cells can be grown in cultures in the laboratory.

During amniocentesis, the position of the needle and that of the fetus are usually observed by means of *ultrasound*.

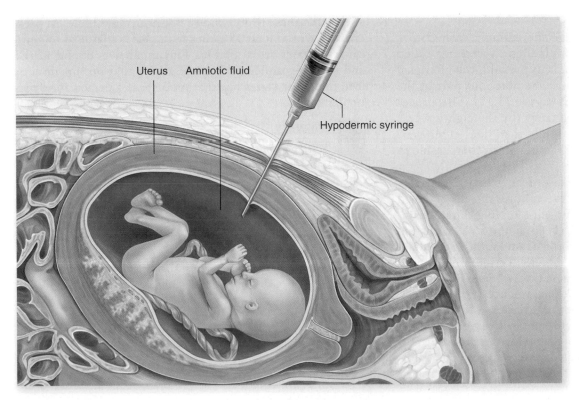

Uterus   Amniotic fluid

Hypodermic syringe

figure 13.16

AMNIOCENTESIS.
A needle is inserted into the amniotic cavity, and a sample of amniotic fluid, containing some free cells derived from the fetus, is withdrawn into a syringe. The fetal cells are then grown in culture, and their karyotype and many of their metabolic functions are examined.

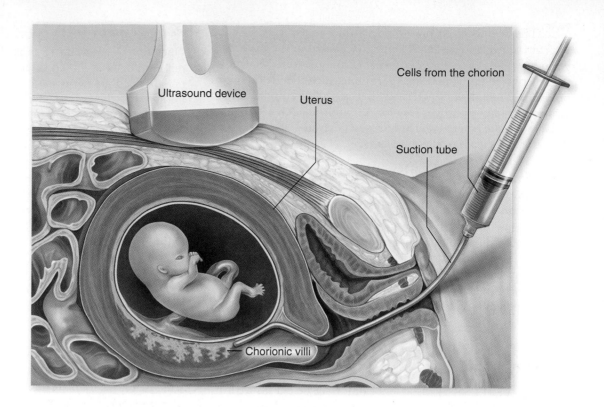

figure 13.17

CHORIONIC VILLI SAMPLING. Cells can be taken from the chorionic villi as early as the eighth to tenth week of pregnancy. Cells are removed by suction with a tube inserted through the cervix. These cells can then be grown in culture and examined for karyotypes and tested biochemically for defects.

The sound waves used in ultrasound are not harmful to mother or fetus, and they permit the person withdrawing the amniotic fluid to do so without damaging the fetus. In addition, ultrasound can be used to examine the fetus for signs of major abnormalities. However, about 1 out of 200 amniocentesis procedures may result in fetal death and miscarriage.

Chorionic villi sampling

In recent years, physicians have increasingly turned to a new, less invasive procedure for genetic screening called **chorionic villi sampling.** Using this method, the physician removes cells from the chorion, a membranous part of the placenta that nourishes the fetus (figure 13.17). This procedure can be used earlier in pregnancy (by the eighth week) and yields results much more rapidly than does amniocentesis. Risks from chorionic villi sampling are comparable to that for amniocentesis.

To test for certain genetic disorders, genetic counselors look for three characteristics in the cultures of cells obtained from amniocentesis or chorionic villi sampling. First, analysis of the karyotype can reveal aneuploidy (extra or missing chromosomes) and gross chromosomal alterations. Second, in many cases it is possible to test directly for the proper functioning of enzymes involved in genetic disorders. The lack of normal enzymatic activity signals the presence of the disorder. As examples, the lack of the enzyme responsible for breaking down phenylalanine indicates phenylketonuria (PKU); the absence of the enzyme responsible for the breakdown of gangliosides indicates Tay–Sachs disease; and so forth. Additionally, with information from the Human Genome Project,

more disease alleles for genetic disorders are known. If there are a small number of alleles for a specific disease in the population, these can be identified as well.

With the changes in human genetics brought about by the Human Genome Project (chapter 18), it is possible to design tests for many more diseases. Difficulties still exist in discerning the number and frequency of disease-causing alleles, but these problems are not insurmountable. At present, tests for at least 13 genes with alleles that lead to clinical syndromes are available. This number is bound to rise and to be expanded to include alleles that do not directly lead to disease states but that predispose a person for a particular disease.

 inquiry

Based on what you read in this chapter, what reasons could a mother have to undergo CVS, considering its small but potential risks?

Human genetic disorders may be caused by single-base mutations or by multiple changes, additions, or deletions in a gene's DNA. On the chromosome level, nondisjunction during meiosis can result in gametes with too few or too many chromosomes, most of which produce inviable offspring. Imprinting refers to inactivation of alleles depending on which parent the alleles come from. In parents with a high risk of bearing children with a genetic defect, testing can help provide information about the genetic health of a fetus.

13.1 Sex Linkage and the Chromosomal Theory of Inheritance

The chromosomal theory of inheritance proposed by Sutton states that hereditary traits are carried on chromosomes.

■ Morgan showed the trait for white eyes in *Drosophila* cosegregated with sex chromosomes, indicating that traits are associated with chromosomes (figure 13.2).

■ Traits associated with sex chromosomes are referred to as sex-linked.

13.2 Sex Chromosomes and Sex Determination

The structure and number of sex chromosomes vary in different organisms.

■ Sex determination in animals is usually associated with a chromosomal difference.

■ In some animals, for example mammals and flies, females have two similar sex chromosomes and males have sex chromosomes that differ.

■ In other species, for example birds and some reptiles, males have two similar sex chromosomes and females have sex chromosomes that differ (table 13.1).

■ The "default setting" in human embryonic development is for production of a female.

■ In humans the Y chromosome determines maleness.

■ The Y chromosome is highly condensed and does not have an active counterpart to most of the genes on the X chromosome.

■ The *SRY* gene on the Y chromosome is responsible for the masculinization of genitalia and secondary sex organs.

■ If an XX individual has part of the Y chromosome translocated to one of the X chromosomes, the embryo will develop as a male.

■ Mutations in the *SRY* gene or failure of the embryo to respond to androgen hormones can cause an XY individual to develop into sterile females.

■ Genetic disorders such as color blindness and hemophilia are sex linked (figure 13.3).

■ In mammalian females one of the X chromosomes is randomly inactivated during development.

■ This inactivated condensed chromosome, or Barr body, is an example of dosage compensation, which balances levels of gene expression in males and females.

■ X-chromosome inactivation can lead to genetic mosaics if the female is heterozygous for X-chromosome alleles. An example is calico cats (figure 13.4).

13.3 Exceptions to the Chromosomal Theory of Inheritance

Not all inheritance is explained by chromosomes.

■ Mitochondrial genes are usually inherited from the female parent.

■ Chloroplast genes are usually maternally inherited although paternal and biparental inheritance has also been observed.

13.4 Genetic Mapping

If two genes are linked, they both occur on a chromosome, and their inheritance behavior differ if the genes were on separate chromosomes.

■ Homologous chromosomes may exchange alleles during crossing over (figure 13.5).

■ The recombination of alleles during crossing over is the basis for constructing genetic maps.

■ The further apart two linked genes are, the greater the frequency of recombination due to crossing over between gene loci.

■ A map unit is expressed as the percentage of recombinant progeny out of total progeny.

■ The probability of multiple crossovers increases with distance between two linked genes and results in an underestimate of recombination frequency.

■ Maps constructed using crosses between three linked genes can be used to determine the order of genes (figure 13.9).

■ Longer map distances are calculated by adding shorter, more accurate, distances.

■ Human genetic mapping was difficult and usually involved disease-causing alleles until anonymous markers were developed.

■ Single-nucleotide polymorphisms (SNPs) can be used to detect differences among individuals.

13.5 Selected Human Genetic Disorders

Causes of human diseases range from an altered single base to the deletion of genetic material to the loss of an entire chromosome.

■ A change in a single amino acid can result in a debilitating clinical phenotype.

■ Nondisjunction, the failure of homologues or sister chromatids to separate during meiosis, results in a condition called aneuploidy.

■ Monosomics lose at least one copy of an autosome and generally do not survive embryonic development.

■ Trisomics gain an extra autosome and most often do not survive development.

■ Translocation occurs when part of one chromosome is joined to another chromosome, resulting in three copies of a chromosomal segment.

■ X-chromosome nondisjunction occurs when X chromosomes fail to separate during meiosis. The resulting gamete has either XX or O (zero sex chromosomes) (figure 13.15).

■ Y-chromosome nondisjunction results in YY gametes.

■ In genomic imprinting the expression of a gene depends on whether it passes through the maternal or paternal germ-line.

■ Imprinted genes are inactivated by methylation.

■ Genetic defects in populations can be determined by pedigree analysis, amniocentesis, or chorionic villi sampling.

SELF TEST

1. Why is the white-eye phenotype always observed in males carrying the white-eye allele?
   a. Because the trait is dominant
   b. Because the trait is recessive
   c. Because the allele is located on the X chromosome and males only have one X
   d. Because the allele is located on the Y chromosome and only males have Y chromosomes

2. An *autosome* is a chromosome that—
   a. contains genetic information to determine the sex of an organism
   b. determines all other traits of an organism other than sex
   c. is only inherited from the mother (maternal inheritance)
   d. has no matching chromosome within an organism's genome

3. Sex linkage in humans occurs when—
   a. an allele is located on both the X and Y chromosomes
   b. all allele is located on the X chromosome
   c. an allele is located on an autosome
   d. a phenotype is only observed in females

4. What are Barr bodies?
   a. X chromosomes inactivated to prevent overexpression of the alleles found on the X chromosome in females
   b. Highly condensed Y chromosomes in males
   c. X chromosomes inactivated to allow for expression of the males-specific phenotype
   d. Inactive autosomal chromosomes specific to females

5. How does maternal inheritance of mitochondrial genes differ from sex linkage?
   a. Mitochondrial genes do not contribute to the phenotype of an individual.
   b. Because mitochondria are inherited from the mother, only females are affected.
   c. Since mitochondria are inherited from the mother, females and males are equally affected.
   d. Mitochondrial genes must be dominant. Sex-linked traits are typically recessive.

6. What cellular process in responsible for genetic recombination?
   a. Independent assortment
   b. Separation of the homologues in meiosis 1
   c. Separation of the chromatids during meiosis II
   d. Crossing over

7. The number of map units between two genes is determined by—
   a. the recombination frequency
   b. the frequency of parental types
   c. the total number of genes within a given piece of DNA
   d. the number of linked genes within a chromosome

8. How many map units separate two alleles if the recombination frequency is 0.07?
   a. 700 cM
   b. 70 cM
   c. 7 cM
   d. 0.7 cM

9. Multiple crossovers lead to—
   a. restoration of the paternal combination of genes
   b. increased genetic diversity
   c. increased numbers of recombinant progeny
   d. anueploidy

10. The disease sickle cell anemia is caused by—
    a. altered expression of the HBB gene
    b. a change of a single amino acid in the protein hemoglobin
    c. a change in the HBB gene
    d. Both (b) and (c)

11. What determines whether an individual is a genetic mosaic?
    a. The presence of different alleles on the autosomal chromosomes
    b. The inactivation of an allele on an autosomal chromosome
    c. The inactivation of an allele on the X chromosome of a heterozygous female
    d. The inactivation of an allele on the X chromosome of a homozygous male

12. Down syndrome is the result of—
    a. a single-base substitution on human chromosome 21
    b. nondisjunction of chromosome 21 during meiosis
    c. inactivation of chromosome 21
    d. nondisjunction of chromosome 21 during mitosis in early development

13. Which of the following examples of nondisjunction of sex chromosomes is lethal?
    a. XXX
    b. XXY
    c. OY
    d. XO

14. What is genomic imprinting?
    a. The blending of a phenotype due to the genetic contribution of both parents
    b. The expression of a dominant allele
    c. The development of a phenotype in response to interactions between distinct alleles
    d. The expression of different phenotypes dependent upon the parental origin of alleles

15. Which of the following is *not* a method used in genetic counseling?
    a. ultrasound
    b. chorionic villi sampling
    c. amniocentesis
    d. pedigree analysis

CHALLENGE QUESTIONS

1. Color blindness is caused by a sex-linked, recessive gene. If a woman, heterozygous for the color blind allele, marries a man with normal color vision, what percentage of their children will be color blind? What sex will the color blind children be?

2. What conditions would have to exist to produce a color blind female?

3. Imagine that the genes for seed color and seed shape are located on the same chromosome. A cross was made between two, true-breeding plants. One plant produces green wrinkled seed *(rryy)* and the second parent produced round yellow seeds *(RRYY)*. A testcross is made between the F$_1$ generation with the following results:

green, wrinkled	645
green, round	36
yellow, wrinkled	29
yellow, round	590

   Calculate the distance between the two loci.

4. Is it possible to have a calico cat that is male? Why or why not?

chapter 14

DNA: The Genetic Material

introduction

THE REALIZATION THAT PATTERNS OF heredity can be explained by the segregation of chromosomes in meiosis raised a question that occupied biologists for over 50 years: What is the exact nature of the connection between hereditary traits and chromosomes? This chapter describes the chain of experiments that led to our current understanding of DNA, modeled in the picture, and of the molecular mechanisms of heredity. These experiments are among the most elegant in science. And, just as in a good detective story, each discovery has led to new questions. But however erratic and lurching the course of the experimental journey may appear, our picture of heredity has become progressively clearer, the image more sharply defined.

The Nature of the Genetic Material

In the previous two chapters, you learned about the nature of inheritance and how genes, which contain the information to specify traits, are located on chromosomes. This finding led to the question of what part of the chromosome actually contains the genetic information. Specifically, biologists wondered about the chemical identity of the genetic information. They knew that chromosomes are composed primarily of both protein and DNA. Which of these organic molecules actually comprises the genes?

Starting in the late 1920s and continuing for about 30 years, a series of investigations addressed this question. DNA consists of four chemically similar nucleotides. In contrast, protein contains 20 different amino acids that are much more chemically diverse than nucleotides. These characteristics seemed initially to indicate greater informational capacity in protein than in DNA.

Experiments began to reveal evidence in favor of DNA; we describe three of these major findings in this section.

Griffith finds that bacterial cells can be transformed

The first clue came in 1928 with the work of the British microbiologist Frederick Griffith. Griffith studied a pathogenic bacteria, *Streptococcus pneumoniae*, that causes pneumonia in mice. There are two forms of this bacteria: The normal virulent form that causes pneumonia, and a mutant, nonvirulent form that does not. The normal virulent form of this bacterium is referred to as the S form because it forms smooth colonies on a culture dish. The mutant, nonvirulent form, which lacks an enzyme needed to manufacture the polysaccharide coat, is called the R form because it forms rough colonies.

Griffith performed a series of simple experiments in which mice were infected with these bacteria, then monitored for disease symptoms (figure 14.1). Mice infected with the virulent S form died from pneumonia, whereas infection with the nonvirulent R form had no effect. This result shows that the polysaccharide coat is necessary for virulence. If the virulent S form is first heat-killed, infection does not harm the mice, showing that the coat itself is not sufficient to cause disease. Lastly, infecting mice with a mixture of heat-killed S form with live R form caused pneumonia and death in the mice. This was unexpected as neither treatment alone caused disease. Furthermore, high levels of live S form bacteria were found in the lungs of the dead mice.

Somehow, the information specifying the polysaccharide coat had passed from the dead, virulent S bacteria to the live, coatless R bacteria in the mixture, permanently altering the coatless R bacteria into the virulent S variety. Griffith called this transfer of virulence from one cell to another, **transformation.** Our modern interpretation is that genetic material was actually transferred between the cells.

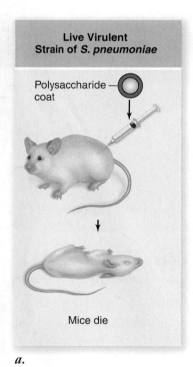

Live Virulent Strain of *S. pneumoniae*

Polysaccharide coat

Mice die

a.

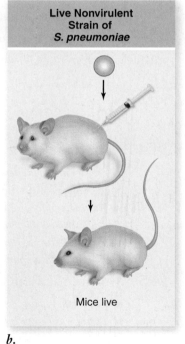

Live Nonvirulent Strain of *S. pneumoniae*

Mice live

b.

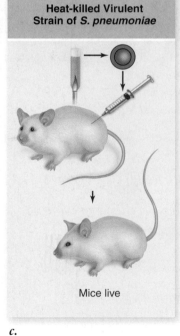

Heat-killed Virulent Strain of *S. pneumoniae*

Mice live

c.

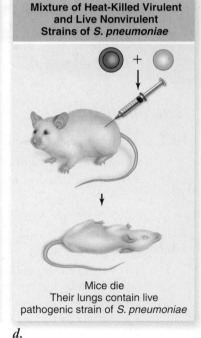

Mixture of Heat-Killed Virulent and Live Nonvirulent Strains of *S. pneumoniae*

Mice die
Their lungs contain live pathogenic strain of *S. pneumoniae*

d.

figure 14.1

GRIFFITH'S EXPERIMENT. Griffith was trying to make a vaccine against pneumonia and instead discovered transformation. *a.* Injecting live virulent bacteria into mice produces pneumonia. Injection of nonvirulent bacteria (*b*) or heat killed virulent bacteria (*c*) had no effect. *d.* However, a mixture of heat killed virulent and live nonvirulent bacteria produced pneumonia in the mice. This indicates the genetic information for virulence was transferred from dead, virulent cells to live, nonvirulent cells transforming them from nonvirulent to virulent.

Avery, MacLeod, and McCarty identify the transforming principle

The agent responsible for transforming *Streptococcus* went undiscovered until 1944. In a classic series of experiments, Oswald Avery and his coworkers Colin MacLeod and Maclyn McCarty identified the substance responsible for transformation in Griffith's experiment.

They first prepared the mixture of dead S *Streptococcus* and live R *Streptococcus* that Griffith had used. Then they removed as much of the protein as they could from their preparation, eventually achieving 99.98% purity. They found that despite the removal of nearly all protein, the transforming activity was not reduced.

Moreover, the properties of this substance resembled those of DNA in several ways:

1. The elemental composition agreed closely with that of DNA.
2. When spun at high speeds in an ultracentrifuge, it migrated to the same level (density) as DNA.
3. Extracting lipids and proteins did not reduce transforming activity.
4. Protein-digesting enzymes did not affect transforming activity, nor did RNA-digesting enzymes.
5. DNA-digesting enzymes destroyed all transforming activity.

These experiments supported the identity of DNA as the substance transferred between cells by transformation and indicated that the genetic material, at least in this bacterial species, is DNA.

Hershey and Chase demonstrate that phage genetic material is DNA

Avery's results were not widely accepted at first because many biologists continued to believe that proteins were the repository of hereditary information. But additional evidence supporting Avery's conclusion was provided in 1952 by Alfred Hershey and Martha Chase, who experimented with viruses that infect bacteria. These viruses are called **bacteriophages**, or more simply, **phages.**

Viruses, described in more detail in chapter 27, are much simpler than cells; they generally consist of genetic material (DNA or RNA) surrounded by a protein coat. The phage used in these experiments is called a *lytic* phage because infection causes the cell to burst, or lyse. When such a phage infects a bacterial cell, it first binds to the cell's outer surface and then injects its genetic information into the cell. There, the viral genetic information is expressed by the bacterial cell's machinery, leading to production of thousands of new viruses. The buildup of viruses eventually causes the cell to lyse, releasing progeny phage.

The phage used by Hershey and Chase, contains only DNA and protein, and therefore it provides the simplest possible system to differentiate the roles of DNA and protein. Hershey and Chase set out to identify the molecule that the phage injects into the bacterial cells. To do this, they needed a method to label both DNA and protein in unique ways that would allow them to be distinguished. Nucleotides contain phosphorus, but proteins do not, and some amino acids contain sulfur, but DNA does not. Thus, the radioactive ^{32}P isotope can be used to label DNA specifically, and the isotope ^{35}S can be used to label proteins specifically. The two isotopes are easily distinguished based on the particles they emit when they decay.

Two experiments were performed (figure 14.2). In one, viruses were grown on a medium containing ^{32}P, which was incorporated into DNA; in the other, viruses were grown on medium containing ^{35}S, which was incorporated into coat proteins. Each group of labeled viruses was then allowed to infect separate bacterial cultures.

After infection, the bacterial cell suspension was agitated in a blender to remove the infecting viral particles from the surfaces of the bacteria. This step ensured that only the part of the virus that had been injected into the bacterial cells—that is, the genetic material—would be detected.

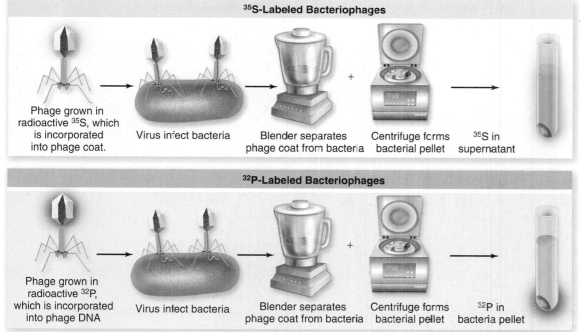

35S-Labeled Bacteriophages

Phage grown in radioactive ^{35}S, which is incorporated into phage coat. → Virus infect bacteria → Blender separates phage coat from bacteria → + Centrifuge forms bacterial pellet → ^{35}S in supernatant

32P-Labeled Bacteriophages

Phage grown in radioactive ^{32}P, which is incorporated into phage DNA → Virus infect bacteria → Blender separates phage coat from bacteria → + Centrifuge forms bacterial pellet → ^{32}P in bacteria pellet

figure 14.2

HERSHEY–CHASE EXPERIMENT. ^{35}S and ^{32}P are used as specific labels for protein and DNA respectively. The virus injects genetic material into the bacterial cell reprogramming it to produce progeny virus. The blender is used to separate phage coats from cells with injected genetic material. The presence of ^{32}P and lack of ^{35}S in cell pellet indicates that injected genetic material used to reprogram cell is DNA and not protein.

Each bacterial suspension was then centrifuged to produce a pellet of cells for analysis. In the ^{32}P experiment, a large amount of radioactive phosphorus was found in the cell pellet, but in the ^{25}S experiment, very little radioactive sulfur was found in the pellet (see figure 14.2). Hershey and Chase deduced that DNA, and not protein, constituted the genetic information that viruses inject into bacteria.

DNA and protein were both considered as possible candidates for genetic material. Experiments with pneumonia-causing bacteria showed that virulence could be passed from one cell to another, termed transformation. When the factor responsible for transformation was purified, it was shown to be DNA. Labeling experiments with phage also showed the genetic material to be DNA.

14.2 DNA Structure

A Swiss chemist, Friedrich Miescher, discovered DNA in 1869, only four years after Mendel's work was published—although it is unlikely that Miescher knew of Mendel's experiments.

Miescher extracted a white substance from the nuclei of human cells and fish sperm. The proportion of nitrogen and phosphorus in the substance was different from that found in any other known constituent of cells, which convinced Miescher that he had discovered a new biological substance. He called this substance "nuclein" because it seemed to be specifically associated with the nucleus. Because Miescher's nuclein was slightly acidic, it came to be called **nucleic acid.**

DNA's components were known, but its three-dimensional structure was a mystery

Although the three-dimensional structure of the DNA molecule was not elucidated until Watson and Crick, it was known that it contained three main components (figure 14.3):

1. a five-carbon sugar
2. a phosphate (PO_4) group
3. a nitrogen-containing (nitrogenous) base. The base may be a **purine** (adenine, A, or guanine, G), a two-ringed structure, or a **pyrimidine** (thymine, T, or cytosine, C),

a single-ringed structure. RNA contains the pyrimidine uracil (U) in place of thymine.

The convention in organic chemistry is to number the carbon atoms of a molecule and then to use these numbers to refer to any functional group attached to a carbon atom (chapter 3). In the ribose sugars found in nucleic acids, four of the carbon atoms together with an oxygen atom form a five-membered ring. As illustrated in figure 14.4, the carbon atoms are numbered 1' to 5',

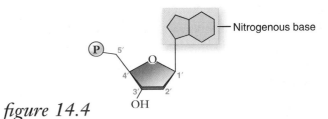

figure 14.4

NUMBERING THE CARBON ATOMS IN A NUCLEOTIDE. The carbon atoms in the sugar of the nucleotide are numbered 1' to 5', clockwise from the oxygen atom. The "prime" symbol (') indicates that the carbon belongs to the sugar rather than to the base.

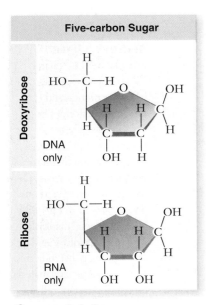
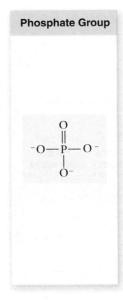
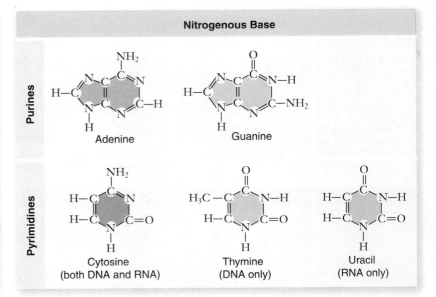

figure 14.3

NUCLEOTIDE SUBUNITS OF DNA AND RNA. The nucleotide subunits of DNA and RNA are composed of three components: (*left*) a five-carbon sugar (deoxyribose in DNA and ribose in RNA); (*middle*) a phosphate group; and (*right*) a nitrogenous base (either a purine or a pyrimidine).

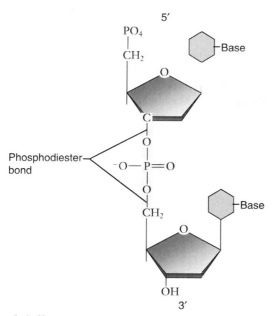

figure 14.5

A PHOSPHODIESTER BOND.

These findings are commonly referred to as **Chargaff's rules:**

1. The proportion of A always equals that of T, and the proportion of G always equals that of C, or: A = T, and G = C.
2. It follows that there is always an equal proportion of purines (A and G) and pyrimidines (C and T).

As mounting evidence indicated that DNA stored the hereditary information, investigators began to puzzle over how such a seemingly simple molecule could carry out such a complex coding function.

Franklin: X-ray diffraction patterns of DNA

Another line of evidence provided more direct information about the possible structure of DNA. The British chemist Rosalind Franklin (figure 14.6a) used the technique of X-ray diffraction to analyze DNA. In X-ray diffraction, a molecule is bombarded with a beam of X-rays. The rays are bent, or diffracted, by the molecules they encounter, and the diffraction pattern is recorded on photographic film. The patterns resemble the ripples created by tossing a rock into a smooth lake (figure 14.6b). When analyzed mathematically, the diffraction pattern can yield information about the three-dimensional structure of a molecule.

X-ray diffraction works best on substances that can be prepared as perfectly regular crystalline arrays. At the time Franklin conducted her analysis, it was impossible to obtain true crystals of natural DNA, so she had to use DNA in the form of fibers. Maurice Wilkins, another researcher working in the same laboratory, had been able to prepare more uniformly oriented DNA fibers than anyone else at the time. Using these fibers, Franklin succeeded in obtaining crude diffraction information on natural DNA. The diffraction patterns she obtained

proceeding clockwise from the oxygen atom; the prime symbol (') indicates that the number refers to a carbon in a sugar rather than to the atoms in the bases attached to the sugars.

Under this numbering scheme, the phosphate group is attached to the 5' carbon atom of the sugar, and the base is attached to the 1' carbon atom. In addition, a free hydroxyl (—OH) group is attached to the 3' carbon atom.

The 5' phosphate and 3' hydroxyl groups allow DNA and RNA to form long chains of nucleotides, by dehydration synthesis (chapter 3). The linkage is called a **phosphodiester bond** because the phosphate group is now linked to the two sugars by means of a pair of ester bonds (figure 14.5). Many thousands of nucleotides can join together via these linkages to form long nucleic acid polymers.

Linear strands of DNA or RNA, no matter how long, will almost always have a free 5' phosphate group at one end and a free 3' hydroxyl group at the other. Therefore, every DNA and RNA molecule has an intrinsic polarity, and we can refer unambiguously to each end of the molecule. By convention, the sequence of bases is usually written in the 5' to 3' direction.

Chargaff, Franklin, and Wilkins obtained some structural evidence

To understand the model that Watson and Crick proposed, we need to review the evidence that they had available to construct their model.

Chargaff's rules

A careful study carried out by Erwin Chargaff showed that the nucleotide composition of DNA molecules varied in complex ways, depending on the source of the DNA. This strongly suggested that DNA was not a simple repeating polymer and that it might have the information-encoding properties genetic material requires. Despite DNA's complexity, however, Chargaff observed an important underlying regularity in the ratios of the bases found in native DNA: *The amount of adenine present*

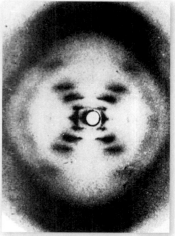

a.   b.

figure 14.6

ROSALIND FRANKLIN'S X-RAY DIFFRACTION PATTERNS.
a. Rosalind Franklin. *b.* This X-ray diffraction photograph of DNA fibers, made in 1953 by Rosalind Franklin was interpreted to show helical structure of DNA.

suggested that the DNA molecule had the shape of a helix, or corkscrew, with a consistent diameter of about 2 nm and a complete helical turn every 3.4 nm.

Tautomeric forms of bases

One piece of evidence important to Watson and Crick was the form of the bases themselves. Because of the alternating double and single bonds in the bases, they actually exist in equilibrium between two different forms when in solution. The different forms have to do with keto (C=O) versus enol (C—OH) groups and amino (—NH₂) versus imino (=NH) groups that are attached to the bases. These structural forms are called *tautomers*.

The importance of this distinction is that the two forms exhibit very different hydrogen-bonding possibilities. The predominant forms of the bases contain the keto and amino groups (see figure 14.3), but a prominent biochemistry text of the time actually contained the opposite, and incorrect, information. Legend has it that Watson learned the correct forms while having lunch with a biochemist friend.

The Watson–Crick model fit the evidence available

Learning informally of Franklin's results before they were published in 1953, James Watson and Francis Crick, two young investigators at Cambridge University, quickly worked out a likely structure for the DNA molecule (figure 14.7), which we now know was sub-

stantially correct. Watson and Crick did not perform a single experiment themselves related to DNA structure; rather, they built detailed molecular models based on the information available.

The key to their model was Watson and Crick's understanding that each DNA molecule is actually made up of *two* chains of nucleotides that are intertwined—the double helix.

The phosphodiester backbone

The two strands of the double helix are made up of long polymers of nucleotides, and as described earlier, each strand is made up of repeating sugar and phosphate units joined by phosphodiester bonds (figure 14.8). We call this the *phosphodiester backbone* of the molecule. The two strands of the backbone are then wrapped about a common axis forming a double helix (figure 14.9). The helix is often compared to a spiral staircase, in which the two strands of the double helix are the handrails on the staircase.

Complementarity of bases

Watson and Crick proposed that the two strands were held together by formation of hydrogen bonds between bases on opposite strands. These bonds would result in specific **base-pairs:** Adenine (A) can form two hydrogen bonds with thymine (T) to form an A–T base-pair, and guanine (G) can form three hydrogen bonds with cytosine (C) to form a G–C base-pair (figure 14.10).

Note that this configuration also pairs a two-ringed purine with a single-ringed pyrimidine in each case, so that the di-

figure 14.7

THE DNA DOUBLE HELIX. James Watson (*left*) and Francis Crick (*right*) deduced the structure of DNA in 1953 from Chargaff's rules, knowing the proper tautomeric forms of the bases and Franklin's diffraction studies.

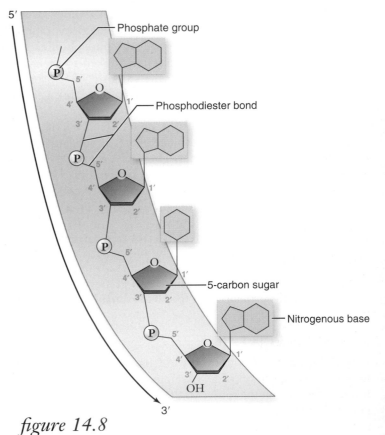

figure 14.8

STRUCTURE OF A SINGLE STRAND OF DNA. The phosphodiester backbone is composed of alternating sugar and phosphate groups. The bases are attached to each sugar.

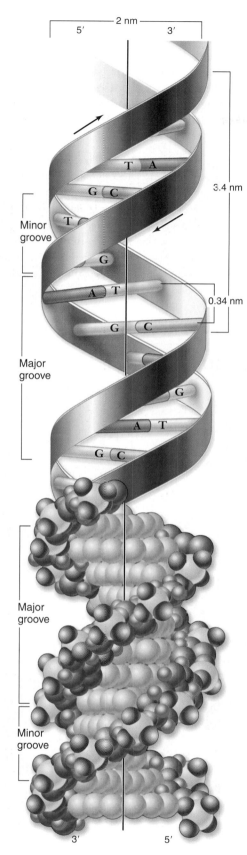

figure 14.9

THE DOUBLE HELIX. Shown with the phosphodiester backbone as a ribbon on top and a space-filling model on the bottom. The bases protrude into the interior of the helix where they hold it together by base-pairing. The backbone forms two grooves, the larger major groove and the smaller minor groove.

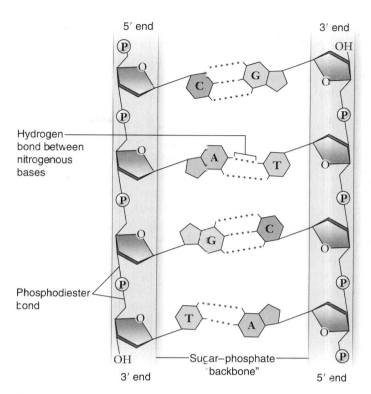

figure 14.10

BASE-PAIRING HOLDS STRANDS TOGETHER. The H-bonds that form between A and T and between G and C are shown with dashed lines. These produce AT and GC base-pairs that hold the two strands together. This always pairs a purine with a pyrimidine, keeping the diameter of the double helix constant.

inquiry

? *Does the Watson–Crick model account for all of the data discussed in the text?*

ameter of each base-pair is the same. This consistent diameter is indicated by the X-ray diffraction data.

We refer to this pattern of base-pairing as **complementary,** which means that although the strands are not identical, they each can be used to specify the other by base-pairing. If the sequence of one strand is ATGC, then the complementary strand sequence must be TACG. This characteristic becomes critical for DNA replication and expression, as you will see later in this chapter.

The Watson–Crick model also explained Chargaff's results: In a double helix, adenine forms two hydrogen bonds with thymine, but it will not form hydrogen bonds properly with cytosine. Similarly, guanine forms three hydrogen bonds with cytosine, but it will not form hydrogen bonds properly with thymine. Because of this base-pairing, adenine and thymine will always occur in the same proportions in any DNA molecule, as will guanine and cytosine.

Antiparallel configuration

As stated earlier, a single phosphodiester strand has an inherent polarity, meaning that one end terminates in a 3′ OH and the other end terminates in a 5′ PC₄. Strands are thus referred to as having either a 5′ to 3′ or a 3′ to 5′ polarity. Two strands could be put together in two ways: with the polarity the same in each (parallel) or with the polarity opposite (antiparallel).

Native double-stranded DNA always has the antiparallel configuration, with one strand running 5′ to 3′ and the other running 3′ to 5′ (see figure 14.10). In addition to its complementarity, this antiparallel nature also has important implications for DNA replication.

The Watson–Crick DNA molecule

In the Watson and Crick model, each DNA molecule is composed of two complementary phosphodiester strands that each form a helix with a common axis. These strands are antiparallel, with the bases extending into the interior of the helix. The bases from opposite strands form base-pairs with each other to join the two complementary strands (see figures 14.9 and 14.10).

Although the hydrogen bonds between each individual base-pair are low-energy bonds, the sum of bonds between the many base-pairs of the polymer has enough energy that the entire molecule is stable. To return to our spiral staircase analogy—the backbone is the handrails, the base-pairs are the stairs.

Although the Watson–Crick model provided a rational structural for DNA, researchers had to answer further questions about how DNA could be replicated, a crucial step in cell division, and also about how cells could repair damaged or otherwise altered DNA. We explore these questions in the rest of this chapter. (In the following chapter, we continue with the genetic code and the connection between the code and protein synthesis.)

> Chargaff's experiments on DNA structure revealed that the amount of adenine was equivalent to the amount of thymine, and similarly, the amount of guanosine was equivalent to the amount of cytosine. X-ray diffraction studies by Franklin and Wilkins indicated that DNA appeared to be helical. Watson and Crick built a model to elucidate the structure, consisting of two helical strands wrapped about a common axis, held together by hydrogen bonds. Adenosine pairs with thymine and guanosine pairs with cytosine, such that the strands are complementary.

14.3 Basic Characteristics of DNA Replication

The accurate replication of DNA prior to cell division is a basic and crucial function. Research has revealed that this complex process requires the participation of a large number of cellular proteins. Before geneticists could look for these details, however, they needed to perform some groundwork on the general mechanisms.

Meselson and Stahl demonstrate the semiconservative mechanism

The Watson–Crick model immediately suggested that the basis for copying the genetic information is complementarity. One chain of the DNA molecule may have any conceivable base sequence, but this sequence completely determines the sequence of its partner in the duplex.

In replication, the sequence of parental strands must be duplicated in daughter strands. That is, one parental helix with two strands must yield two daughter helices with four strands. The two daughter molecules are then separated during the course of cell division.

Three models of DNA replication are possible (figure 14.11):

1. In a **conservative model,** both strands of the parental duplex would remain intact (conserved), and new DNA copies would consist of all-new molecules. Both daughter strands would contain all-new molecules.

2. In a **semiconservative model,** one strand of the parental duplex remains intact in daughter strands (semiconserved); a new complementary strand is built for each parental strand consisting of new molecules. Daughter strands would consist of one parental strand and one newly synthesized strand.

3. In a **dispersive model,** copies of DNA would consist of mixtures of parental and newly synthesized strands; that is, the new DNA would be dispersed throughout each strand of both daughter molecules after replication.

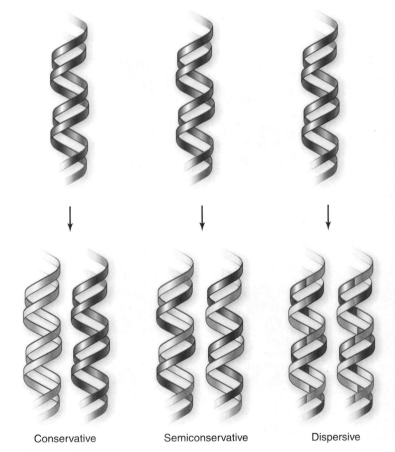

Conservative   Semiconservative   Dispersive

figure 14.11

THREE POSSIBLE MODELS FOR DNA REPLICATION. The conservative model produces one entirely new molecule and conserves the old. The semiconservative model produces two hybrid molecules of old and new strands. The dispersive model produces hybrid molecules with each strand a mixture of old and new.

Notice that these three models suggest general mechanisms of replication, without specifying any molecular details of the process.

The Meselson–Stahl experiment

The three models for DNA replication were evaluated in 1958 by Matthew Meselson and Franklin Stahl. To distinguish between these models, they labeled DNA and then followed the labeled DNA through two rounds of replication (figure 14.12).

The label Meselson and Stahl used though was a heavy isotope of nitrogen (^{15}N), not a radioactive label. Molecules containing ^{15}N have a greater density than those containing the common ^{14}N isotope. The technique of ultracentrifugation can be used to separate molecules that have different densities.

Bacteria were grown in a medium containing ^{15}N, which became incorporated into the bases of the bacterial DNA. After several generations, the DNA of these bacteria was denser than that of bacteria grown in a medium containing the normally available ^{14}N. Meselson and Stahl then transferred the bacteria from the ^{15}N medium to ^{14}N medium and collected the DNA at various time intervals.

The DNA for each interval was dissolved in a solution containing a heavy salt, cesium chloride. This solution was spun at very high speeds in an ultracentrifuge. The enormous centrifugal forces caused cesium ions to migrate toward the bottom of the centrifuge tube, creating a gradient of cesium concentration, and thus of density. Each DNA strand floated or sank in the gradient until it reached the point at which its density exactly matched the density of the cesium at that location. Because ^{15}N strands are denser than ^{14}N strands, they migrated farther down the tube.

The DNA collected immediately after the transfer of bacteria to new ^{14}N medium was all of one density equal to that of ^{15}N DNA alone. However, after the bacteria completed a first round of DNA replication, the density of their DNA had decreased to a value intermediate between ^{14}N-DNA alone and ^{15}N-DNA. After the second round of replication, two density classes of DNA were observed: one intermediate and one equal to that of ^{14}N-DNA (see figure 14.12).

Interpretation of the Meselson–Stahl findings

Meselson and Stahl compared their experimental data with the results that would be predicted on the basis of the three models.

1. The conservative model was not consistent with the data because after one round of replication, two densities should have been observed: DNA strands would either be all-heavy (parental) or all-light (daughter). This model is rejected.

2. The semiconservative model is consistent with all observations: After one round of replication, a single density would be predicted because all DNA molecules would have a light strand and a heavy strand. After two rounds of replication, half of the molecules would have two light strands, and half would have a light strand and a heavy strand—and so two densities would be observed. Therefore, the results support the semiconservative model.

3. The dispersive model was consistent with the data from the first round of replication, because in this model, every DNA

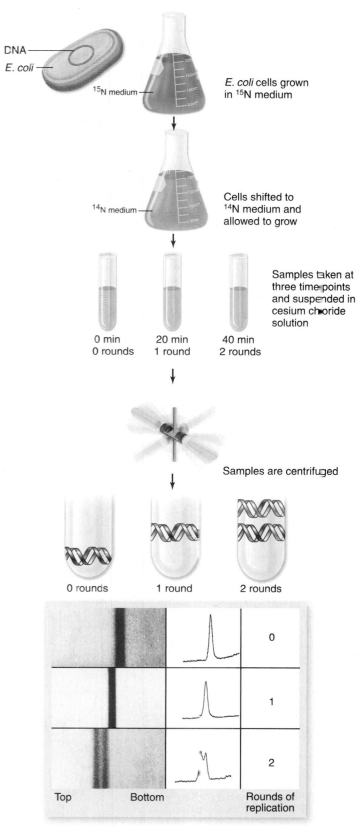

E. coli cells grown in ^{15}N medium

DNA
E. coli
^{15}N medium

^{14}N medium

Cells shifted to ^{14}N medium and allowed to grow

Samples taken at three time-points and suspended in cesium chloride solution

| 0 min | 20 min | 40 min |
| 0 rounds | 1 round | 2 rounds |

Samples are centrifuged

| 0 rounds | 1 round | 2 rounds |

Top   Bottom   Rounds of replication

figure 14.12

THE MESELSON–STAHL EXPERIMENT. Bacteria grown in heavy ^{15}N medium are shifted to light ^{14}N medium and grown for two rounds of replication. Samples are taken at time points corresponding to zero, one, and two rounds of replication and centrifuged in cesium chloride to form a gradient. The actual data are shown at the bottom with the interpretation of semiconservative replication shown schematically.

helix would consist of strands that are mixtures of 1/2 light (new) and 1/2 heavy (old) molecules. But after two rounds of replication, the dispersive model would still yield only a single density; DNA strands would be composed of 3/4 light and 1/4 heavy molecules. Instead, two densities were observed. Therefore, this model is also rejected.

The basic mechanism of DNA replication is semiconservative. At the simplest level, then, DNA is replicated by opening up a DNA helix and making copies of both strands to produce two daughter helices, each consisting of one old strand and one new strand.

The replication process: An overview

Replication requires three things: something to copy, something to do the copying, and the building blocks to make the copy. The parental DNA molecules serve as a template, enzymes perform the actions of copying the template, and the building blocks are nucleotide triphosphates.

The process of replication can be thought of as having a beginning where the process starts; a middle where the majority of building blocks are added; and an end where the process is terminated. We use the terms *initiation*, *elongation*, and *termination* to describe a biochemical process. Although this may seem overly simplistic, in fact, discrete functions are usually required for initiation and termination that are not necessary for elongation.

Initiation: The origin

Replication does not start at a random point on a DNA double strand, but rather has been found to begin at one or more sites called origins of replication. Initiator proteins recognize and bind to the origin, forming a complex that opens the helix to expose single-stranded templates used for the process of building a new strand.

Elongation: DNA polymerase

A number of enzymes work together to accomplish the task of assembling a new strand, but the enzyme that actually matches the existing DNA bases with complementary nucleotides and

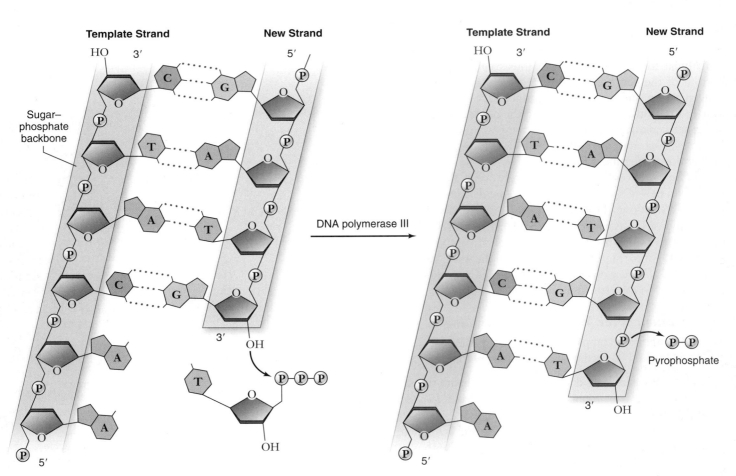

figure 14.13

ACTION OF DNA POLYMERASE. DNA polymerases add nucleotides to the 3′ end of a growing chain. The nucleotide added depends on the base that is in the template strand. Each new base must be complementary to the base in the template strand. With the addition of each new nucleotide triphosphate, two of its phosphates are cleaved off as pyrophosphate.

inquiry

Why do you think it is important that the sugar–phosphate backbone of DNA is held together by covalent bonds, and the cross-bridges between the two strands are held together by hydrogen bonds?

then links the nucleotides together to make the new strand is **DNA polymerase** (figure 14.13). As described shortly, many different types of DNA polymerases have been discovered.

All DNA polymerases that have been examined have several common features. They all add new bases to the 3' end of existing strands. That is, they synthesize in a 5' to 3' direction by extending a strand base-paired to the template. All DNA polymerases also require a **primer** to begin synthesis; they cannot begin without a strand of RNA or DNA base-paired to the template. RNA polymerases do not have this requirement, so they usually synthesize the primers.

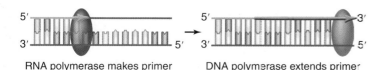

RNA polymerase makes primer   DNA polymerase extends primer

Termination

An end point for replication is as important as a starting point. In prokaryotes, which have circular DNA, the replication ends when the process comes around to the origin again. In eukaryotes, end points for each chromosome are indicated by *telomeres*, specific regions of repeated bases.

The details of this process are described in the sections that follow—first for prokaryotes, used for much of the initial research on DNA replication, and then for eukaryotes.

> Meselson and Stahl showed that the basic mechanism of replication is semiconservative: Each new DNA helix is composed of one old strand and one new strand. The process of replication involves three phases: initiation, elongation, and termination. Initiation and termination both occur at specific sites and require functions not found in elongation. Elongation is accomplished by DNA polymerase enzymes that synthesize in a 5' to 3' direction from a primer, usually RNA.

14.4 Prokaryotic Replication

To build up a more detailed picture of replication, we first concentrate on prokaryotic replication using *E. coli* as a model. We can then look at eukaryotic replication primarily in how it differs from the prokaryotic system.

Prokaryotic replication starts at a single origin

Replication in *E. coli* initiates at a specific site, the origin (called *oriC*), and ends at a specific site, the terminus. The sequence of *oriC* consists of repeated nucleotides that bind an initiator protein and an AT-rich sequence that can be opened easily during initiation of replication. (A–T base-pairs have only two hydrogen bonds, compared with the three hydrogen bonds in G–C base-pairs.)

After initiation, replication proceeds bidirectionally from this unique origin to the unique terminus (figure 14.14). The complete chromosome plus the origin can be thought of as a single functional unit called a **replicon.**

E. coli has at least three different DNA polymerases

DNA polymerase, as mentioned earlier, refers to a group of enzymes responsible for the building of a new DNA strand from the template. The first DNA polymerase isolated in *E. coli* was given the name **DNA polymerase I (pol I).** At first, investigators assumed this polymerase was responsible for the bulk synthesis of DNA during replication. A mutant was isolated, however, that had no pol I activity, but that still showed replication activity. Two additional polymerases were isolated from this strain of *E. coli* and were named **DNA polymerase II (pol II)** and **DNA polymerase III (pol III).** As with all other known polymerases, all three of these enzymes synthesize polynucleotide strands only in the 5' to 3' direction, and require a primer.

Many DNA polymerases have additional enzymatic activity that aids their function. This activity is a nuclease activity, or

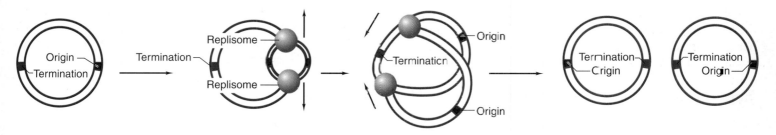

figure 14.14

REPLICATION IS BIDIRECTIONAL FROM A UNIQUE ORIGIN. Replication initiates from a unique origin. Two separate replisomes are loaded onto the origin and initiate synthesis in the opposite directions on the chromosome. These two replisomes continue in opposite directions until they come to a unique termination site.

the ability to break phosphodiester bonds between nucleotides. Nucleases are classified as either **endonucleases** (cut DNA internally) or **exonucleases** (chew away at an end of DNA). DNA pol I, pol II, and pol III have 3′ to 5′ exonuclease activity, which serves as a proofreading function because it allows the enzyme to remove a mispaired base. In addition, the DNA pol I enzyme also has a 5′ to 3′ exonuclease activity, the importance of which will become clear shortly.

The three different polymerases have different roles in the replication process. DNA pol III is the main replication enzyme, it is responsible for the bulk of DNA synthesis. DNA pol I acts on the lagging strand to remove primers and replace them with DNA. The pol II enzyme does not appear to play a role in replication but is involved in DNA repair processes.

For many years, these three polymerases were thought to be the only DNA polymerases in *E. coli*, but recently several new polymerases have been identified. There are now five known polymerases, although all are not active in DNA replication.

Unwinding DNA requires energy and causes torsional strain

Although some DNA polymerases can unwind DNA as they synthesize new DNA, another class of enzymes has the single function of unwinding DNA strands to make this process more efficient. Enzymes that use energy from ATP to unwind the DNA template are called **helicases.**

The single strands of DNA produced by helicase action are unstable because the process exposes the hydrophobic bases to water. Cells solve this problem by using a protein to coat exposed single strands called single-strand binding protein (SSB).

The unwinding of the two strands introduces torsional strain in the DNA molecule. Imagine two rubber bands twisted together. If you now unwind the rubber bands, what happens? The rubber bands, already twisted about each other, will further coil in space. When this happens with a DNA molecule it is called **supercoiling** (figure 14.15). The branch of mathematics that studies how forms twist and coil in space is called *topology*, and therefore we describe this coiling of the double helix as the *topological state* of DNA. This state describes how the double helix itself coils in space. You have already seen an example of this coiling with DNA wrapped about histone proteins in the nucleosomes of eukaryotic chromosomes (chapter 10).

Enzymes that can alter the topological state of DNA are called **topoisomerases.** Topoisomerase enzymes act to relieve the torsional strain caused by unwinding and to prevent this supercoiling from happening. **DNA gyrase** is the topoisomerase involved in DNA replication (see figure 14.15).

Replication is semidiscontinuous

Earlier, DNA was described as being antiparallel—meaning that one strand runs in the 3′ to 5′ direction, and its complementary strand runs in the 5′ to 3′ direction. The antiparallel nature of DNA combined with the nature of the polymerase enzymes puts constraints on the replication process. Because polymerases can synthesize DNA in only one direction, and the

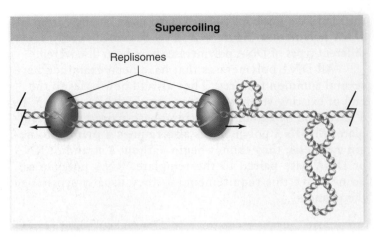

Supercoiling

Replisomes

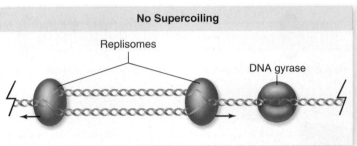

No Supercoiling

Replisomes

DNA gyrase

figure 14.15

UNWINDING THE HELIX CAUSES TORSIONAL STRAIN. If the ends of a linear DNA molecule are constrained, as they are in the cell, unwinding the helix produces torsional strain. This can cause the double helix to further coil in space (supercoiling). The enzyme DNA gyrase can relieve supercoiling.

two DNA strands run in opposite directions, polymerases on the two strands must be synthesizing DNA in opposite directions (figure 14.16).

The requirement of DNA polymerases for a primer means that on one strand primers will need to be added as the helix is opened up (see figure 14.16). This means that one strand can be synthesized in a continuous fashion from an initial primer, but the other strand must be synthesized in a discontinuous fashion with multiple priming events and short sections of DNA being assembled. The strand that is continuous is called the **leading strand,** and the strand that is discontinuous is the **lagging strand.** DNA fragments synthesized on the lagging strand are named **Okazaki fragments** in honor of the man who first experimentally demonstrated discontinuous synthesis. They introduce a need for even more enzymatic activity on the lagging strand, as is described next.

Synthesis occurs at the replication fork

The partial opening of a DNA helix to form two single strands has a forked appearance, and is thus called the **replication fork.** All of the enzymatic activities that we have discussed plus a few more are found at the replication fork (table 14.1). Synthesis on the leading strand and on the lagging strand proceed in different ways, however.

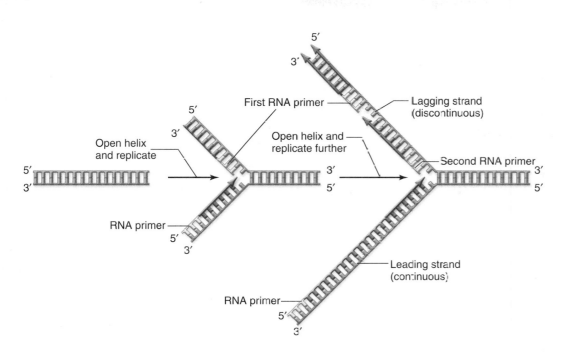

figure 14.16

REPLICATION IS SEMIDISCONTINUOUS. The 5′ to 3′ synthesis of the polymerase and the antiparallel nature of DNA mean that only one strand, the leading strand, can be synthesized continuously. The other strand, the lagging strand, must be made in pieces, each with its own primer.

Priming

The primers required by DNA polymerases during replication are synthesized by the enzyme DNA **primase.** This enzyme is an RNA polymerase that synthesizes short stretches of RNA 10–20 bp (base-pairs) long that function as primers for DNA polymerase. Later on, the RNA primer is removed and replaced with DNA.

Leading strand synthesis

Synthesis on the leading strand is relatively simple. A single priming event is required, and then the strand can be extended indefinitely by the action of DNA pol III. If the enzyme remains attached to the template, it can synthesize around the entire circular *E. coli* chromosome.

The ability of a polymerase to remain attached to the template is called **processivity.** The pol III enzyme is a large multisubunit enzyme that has high processivity due to the action of one subunit of the enzyme, called the β *subunit* (figure 14.17*a*).

The β subunit is made up of two identical protein chains that come together to form a circle. This circle can be loaded onto the template like a clamp to hold the pol III enzyme to the DNA (figure 14.17*b*). This structure is therefore referred to as the "sliding clamp," and a similar structure is found in eukaryotic polymerases as well.

a.

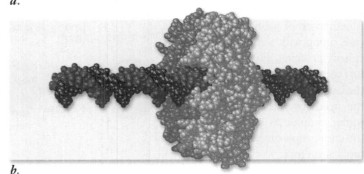

b.

figure 14.17

THE DNA POLYMERASE SLIDING CLAMP. *a.* The β subunit forms a ring that can encircle DNA. *b.* The β subunit is shown attached to the DNA. This forms the "sliding clamp" that keeps the polymerase attached to the template.

TABLE 14.1	DNA Replication Enzymes of *E. coli*		
Protein	**Role**	**Size (kDa)**	**Molecules per Cell**
Helicase	Unwinds the double helix	300	20
Primase	Synthesizes RNA primers	60	50
Single-strand binding protein	Stabilizes single-stranded regions	74	300
DNA gyrase	Relieves torque	400	250
DNA polymerase III	Synthesizes DNA	≈900	20
DNA polymerase I	Erases primer and fills gaps	103	300
DNA ligase	Joins the ends of DNA segments; DNA repair	74	300

Lagging strand synthesis

The discontinuous nature of synthesis on the lagging strand requires the cell to do much more work than on the leading strand (see figure 14.16). Primase is needed to synthesize primers for each Okazaki fragment, and then all these RNA primers need to be removed and replaced with DNA. Finally, the fragments need to be stitched together.

DNA pol III accomplishes the synthesis of Okazaki fragments. The removal and replacement of primer segments, however, is accomplished by DNA pol I. Using its 5′ to 3′ exonuclease activity, it can remove primers in front and then replace them by using its usual 5′ to 3′ polymerase activity. The synthesis is primed by the previous Okazaki fragment, which is composed of DNA and has a free 3′ OH that can be extended.

This leaves only the last phosphodiester bond to be formed where synthesis by pol I ends. This is done by **DNA ligase,** which seals this "nick," eventually joining the Okazaki fragments into complete strands. All of this activity on the lagging strand is summarized in figure 14.18.

inquiry

What is the role of DNA ligase? What would happen to DNA replication in a cell where this enzyme is not functional?

Termination

Termination occurs at a specific site located roughly opposite *oriC* on the circular chromosome. The last stages of replication produce two daughter molecules that are intertwined like two rings in a chain. These intertwined molecules are unlinked by the same enzyme that relieves torsional strain at the replication fork: DNA gyrase.

The replisome contains all the necessary enzymes for replication

The enzymes involved in DNA replication form a macromolecular assembly called the **replisome.** This assembly can be thought of as the "replication organelle," just as the ribosome is the protein synthesis organelle. The replisome is a protein machine capable of fast and accurate replication of DNA during cell division.

The replisome has two main subcomponents: the *primosome*, and a complex of two DNA pol III enzymes, one for each strand. The primosome is composed of primase and helicase, along with a number of accessory proteins. The need for constant priming on the lagging strand explains the need for the primosome complex as part of the entire replisome at the replication fork.

The two pol III complexes include two synthetic core subunits, each with its own β subunit, and a number of other proteins that hold the entire complex together.

Even given the difficulties with lagging-strand synthesis, the two pol III enzymes in the replisome are active on both leading and lagging

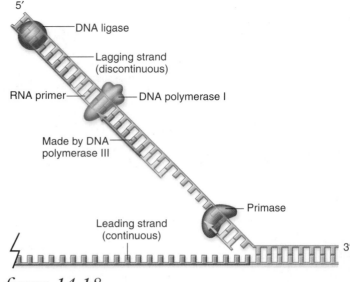

figure 14.18

LAGGING STRAND SYNTHESIS. The action of primase synthesizes the primers needed by DNA polymerase III (not shown). These primers are removed by DNA polymerase I using its 5′ to 3′ exonuclease activity and extending the previous Okazaki fragment to replace the RNA. The nick between Okazaki fragments after primer removal is sealed by DNA ligase.

figure 14.19

THE REPLICATION FORK. A model for the structure of the replication fork with two polymerase III enzymes held together by a large complex of accessory proteins. These include the "clamp loader," which loads the β-subunit sliding clamp periodically on the lagging strand. The polymerase III on the lagging strand periodically releases its template and reassociate along with the β-clamp. The loop in the lagging strand template allows both polymerases to move in the same direction despite DNA being antiparallel. Primase, which makes primers for the lagging strand fragments, and helicase are also associated with the central complex. Polymerase I removes primers and ligase joins the fragments together.

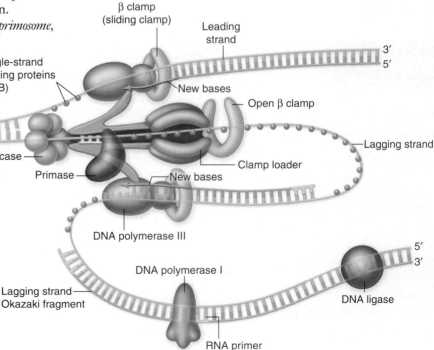

strands simultaneously. How can the two strands be synthesized in the same direction when the strands are antiparallel? The model first proposed, still with us in some form, involves a loop formed in the lagging strand, so that the polymerases can move in the same direction (figure 14.19). Current evidence also indicates that this replication complex is probably stationary, with the DNA strand moving through it like thread in a sewing machine, rather than the complex moving along the DNA strands. This stationary complex also pushes the newly synthesized DNA outward, which may aid in chromosome segregation. This process is summarized in figure 14.20.

DNA replication occurs at the replication fork, where the two strands are separated. Assembled here is a massive complex, the replisome, containing DNA polymerase III, primase, helicase, and other proteins. The lagging strand requires DNA polymerase I to remove the primers and replace them with DNA and ligase to join Okazaki fragments. The polymerases for each strand are actually part of a single complex made possible by a loop in the lagging strand. Replication starts at a unique site, the origin (oriC) and continues bidirectionally to a unique termination site.

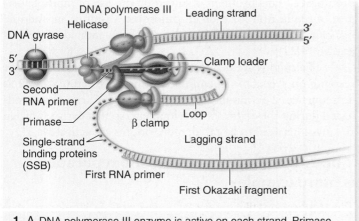

1. A DNA polymerase III enzyme is active on each strand. Primase synthesizes new primers for the lagging strand.

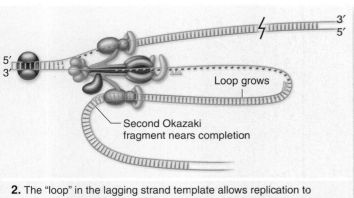

2. The "loop" in the lagging strand template allows replication to occur 5′ to 3′ on both strands, with the complex moving to the left.

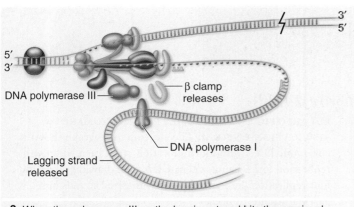

3. When the polymerase III on the lagging strand hits the previously synthesized fragment, it releases the β clamp and the template strand. DNA polymerase I attaches to remove the primer.

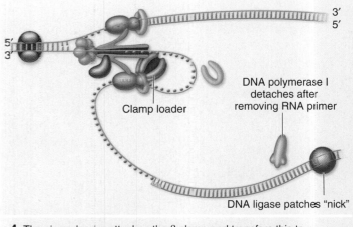

4. The clamp loader attaches the β clamp and transfers this to polymerase III, creating a new loop in the lagging strand template. DNA ligase joins the fragments after DNA polymerase I removes the primers.

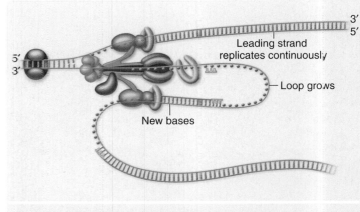

5. After the β clamp is loaded, the DNA polymerase III on the lagging strand adds bases to the next Okazaki fragment.

figure 14.20

DNA SYNTHESIS BY THE REPLISOME. The semidiscontinuous synthesis of DNA is illustrated in stages using the model from figure 14.19.

Eukaryotic Replication

Eukaryotic replication is complicated by two main factors: the larger amount of DNA organized into multiple chromosomes, and the linear structure of the chromosomes. This process requires new enzymatic activities only for dealing with the ends of chromosomes; otherwise the basic enzymology is the same.

Eukaryotic replication requires multiple origins

The sheer amount of DNA and how it is packaged constitute a problem for eukaryotes (figure 14.21). Eukaryotes usually have multiple chromosomes that are each larger than the *E. coli* chromosome. The machinery could in principle be the same; however, if only a single unique origin existed for each chromosome, the length of the time necessary for replication would be prohibitive. This problem is solved by the use of multiple origins of replication for each chromosome, resulting in multiple *replicons*, sections of DNA replicated from individual origins (figure 14.22).

The origins are not as sequence-specific as *oriC*, and their recognition seems to depend on chromatin structure as well as on sequence. The number of origins that "fire" can also be adjusted during the course of development, so that early on, when cell divisions need to be rapid, more origins are activated.

The enzymology of eukaryotic replication is more complex

The replication machinery of eukaryotes is similar to that found in *E. coli*, but it is larger and more complex. The initiation phase of replication requires more factors to assemble both

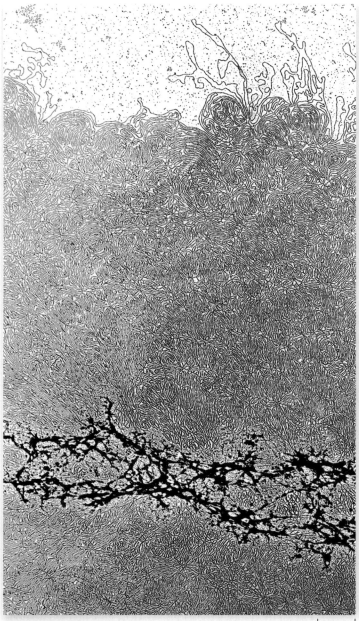

figure 14.21

9.09 μm

DNA OF A SINGLE HUMAN CHROMOSOME. This chromosome has been relieved of most of its packaging proteins leaving the DNA in its native form. The residual protein scaffolding appears as the dark material in the lower part of the micrograph.

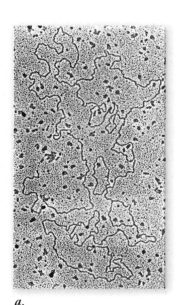

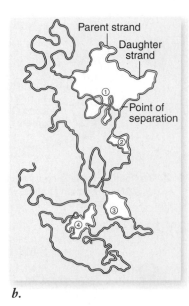

a.   *b.*

figure 14.22

EUKARYOTIC CHROMOSOMES POSSESS NUMEROUS REPLICATION UNITS. *a.* The electron micrograph shows eukaryotic DNA with four replication units, each with two replication forks, being replicated. *b.* The drawing shows the four replication units with newly synthesized strands in red and parental strands in black.

helicase and primase complexes onto the template, then load the polymerase with its sliding clamp unit.

The eukaryotic primase is interesting in that it is a complex of both an RNA polymerase and a DNA polymerase. It first makes short RNA primers, then extends these with DNA to produce the final primer. The reason for this added complexity is unclear.

The main replication polymerase itself is also a complex of two different enzymes that work together. One is called *DNA polymerase epsilon* (pol ε) and the other *DNA polymerase delta* (pol δ). The sliding clamp subunit that allows the enzyme complex to stay attached to the template also exists in eukaryotes, but it is called PCNA (for proliferating cell nuclear antigen). This unusual name reflects the fact that PCNA was first identified as an antibody-inducing protein in proliferating (dividing) cells. Despite the additional complexity, the action of the replisome seems to be similar to that described earlier in *E. coli*, and the replication fork has essentially the same components as well.

Linear chromosomes require different termination

The specialized structures found on the ends of eukaryotic chromosomes are called **telomeres.** These structures protect the ends of chromosomes from nucleases and maintain the integrity of linear chromosomes. These telomeres are composed of specific DNA sequences, but they are not made by the replication complex.

Replicating ends

The very structure of a linear chromosome causes a cell problems in replicating the ends. The directionality of polymerases, combined with their requirement for a primer, create this problem.

Consider a simple linear molecule like the one in figure 14.23. Replication of one end of each template strand is simple, namely the 5′ end of the leading-strand template. When the polymerase reaches this end, synthesizing in the 5′ to 3′ direction, it eventually runs out of template and is finished.

But on the other strand's end, the 3′ end of the lagging strand, removal of the last primer on this end leaves a gap. This gap cannot be primed, meaning that the polymerase complex cannot finish this end properly. The result would be a gradual shortening of chromosomes with each round of cell division (see figure 14.23).

The action of telomerase

When the sequence of telomeres was determined, they were found to be composed of short repeated sequences of DNA. This repeating nature is easily explained by their synthesis. They are made by an enzyme called **telomerase,**

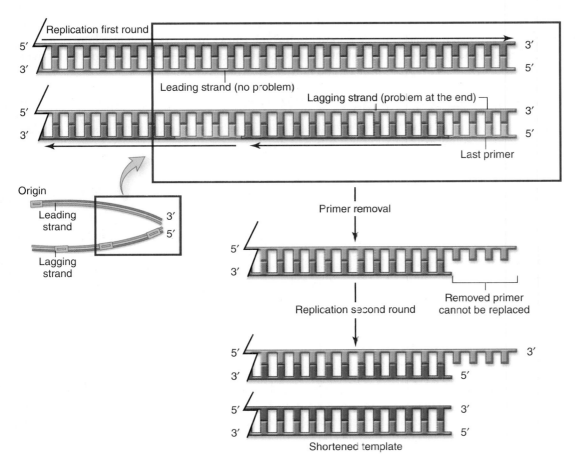

figure 14.23

REPLICATION OF THE END OF LINEAR DNA. Only one end is shown for simplicity, the problem exists at both ends. The leading strand can be completely replicated, but the lagging strand cannot be finished. When the last primer is removed, it cannot be replaced. During the next round of replication, when this shortened template is replicated, it will produce a shorter chromosome.

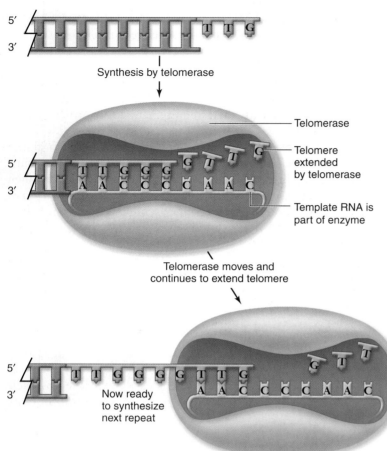

figure 14.24

ACTION OF TELOMERASE. Telomerase contains an internal RNA the enzyme uses as a template to extend the DNA of the chromosome end. Multiple rounds of synthesis by telomerase produce repeated sequences. This single strand is completed by normal synthesis using it as a template (not shown).

Labels in figure:
- Synthesis by telomerase
- Telomerase
- Telomere extended by telomerase
- Template RNA is part of enzyme
- Telomerase moves and continues to extend telomere
- Now ready to synthesize next repeat

which uses an internal RNA as a template and not the DNA itself (figure 14.24).

The use of the internal RNA template allows short stretches of DNA to be synthesized, composed of repeated nucleotide sequences complementary to the RNA of the enzyme. The other strand of these repeated units is synthesized by the usual action of the replication machinery copying the strand made by telomerase.

Telomerase, aging, and cancer

A gradual shortening of the ends of chromosomes occurs in the absence of telomerase activity. During embryonic and childhood development in humans, telomerase activity is high, but it is low in most somatic cells of the adult. The exceptions are cells that must divide as part of their function, such as lymphocytes. The activity of telomerase in somatic cells is kept low by preventing the expression of the gene encoding this enzyme.

Evidence for the shortening of chromosomes in the absence of telomerase was obtained by producing mice with no telomerase activity. These mice appear to be normal for up to six generations, but they show steadily decreasing telomere length that eventually leads to inviable offspring.

This finding indicates a relationship between cell senescence (aging) and telomere length. Normal cells undergo only a specified number of divisions when grown in culture. This limit is at least partially based on telomere length.

Support for the relationship between senescence and telomere length comes from experiments in which telomerase was introduced into fibroblasts in culture. These cells have their lifespan increased relative to controls with no added telomerase. Interestingly, these cells do not show the hallmarks of malignant cells, indicating that activation of telomerase alone does not make cells malignant.

A relationship has been found, however, between telomerase and cancer. Cancer cells do continue to divide indefinitely, and this would not be possible if their chromosomes were being continually shortened. Cancer cells generally show activation of telomerase, which allows them to maintain telomere length; but this is clearly only one aspect of conditions that allow them to escape normal growth controls.

inquiry

How does the structure of eukaryotic genomes affect replication? Does this introduce problems that are not faced by prokaryotes?

Eukaryotic replication uses the same basic enzymology as that in prokaryotes. Eukaryotes can replicate a large amount of DNA in a short time by using multiple origins of replication. Linear chromosomes end in telomeres, which cannot be constructed by the replication machinery. Another enzyme, telomerase, synthesizes the ends of chromosomes. Cancer cells show activation of telomerase.

14.6 DNA Repair

As you learned earlier, many DNA polymerases have 3′ to 5′ exonuclease activity that allows "proofreading" of added bases. This action increases the accuracy of replication, but errors still occur. Without error correction mechanisms, cells would accumulate errors at an unacceptable rate, leading to high levels of deleterious or lethal mutations. A balance must exist between the introduction of new variation by mutation, and the effects of deleterious mutations on the individual.

Cells are constantly exposed to DNA-damaging agents

In addition to errors in DNA replication, cells are constantly exposed to agents that can damage DNA. These agents include radiation, such as UV light and X-rays, and chemicals in the environment. Agents that damage DNA can lead to mutations, and any agent that increases the number of mutations above background levels is called a **mutagen.**

The number of potentially mutagenic agents that organisms encounter is huge. Sunlight itself includes radiation in the UV range and is thus mutagenic. Ozone normally screens out much of the harmful UV radiation in sunlight, but some remains. The relationship between sunlight and mutations is shown clearly by the increase in skin cancer in regions of the southern hemisphere that are underneath a seasonal "ozone hole."

Organisms also may encounter mutagens in their diet in the form of either contaminants in food or natural plant products that can damage DNA. When a simple test was designed to detect mutagens, screening of possible sources indicated an amazing diversity of mutagens in the environment and in natural sources. As a result, consumer products are now screened to reduce the load of mutagens we are exposed to, but we cannot escape natural sources.

DNA repair restores damaged DNA

Cells cannot escape exposure to mutagens, but systems have evolved that enable cells to repair some damage. These DNA repair systems are vital to continued existence, whether a cell is a free-living, single-celled organism or part of a complex multicellular organism.

The importance of DNA repair is indicated by the multiplicity of repair systems that have been discovered and characterized. All cells that have been examined show multiple pathways for repairing damaged DNA and for reversing errors that occur during replication. These systems are not perfect, but they do reduce the mutational load on organisms to an acceptable level. In the rest of this section, we illustrate the action of DNA repair by concentrating on two examples drawn from these multiple repair pathways.

Repair can be either specific or nonspecific

DNA repair falls into two general categories: specific and nonspecific. Specific repair systems target a single kind of lesion in DNA and repair only that damage. Nonspecific forms of repair use a single mechanism to repair multiple kinds of lesions in DNA.

Photorepair: A specific repair mechanism

Photorepair is specific for one particular form of damage caused by UV light, namely the **thymine dimer.** Thymine dimers are formed by a photochemical reaction of UV light and adjacent thymine bases in DNA. The UV radiation causes the thymines to react, covalently linking them together: a thymine dimer (figure 14.25).

Repair of these thymine dimers can be accomplished by multiple pathways, including photorepair. In photorepair, an enzyme called a *photolyase* absorbs light in the visible range and uses this energy to cleave the thymine dimer. This action restores the two thymines to their original state (see figure 14.25). It is interesting that sunlight in the UV range can

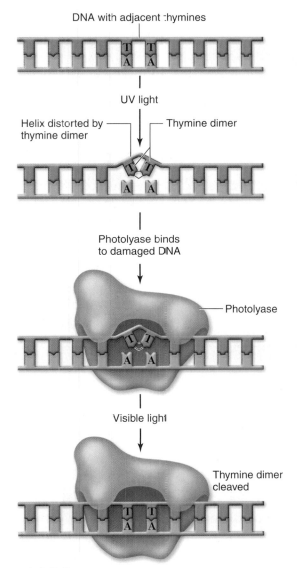

figure 14.25

REPAIR OF THYMINE DIMER BY PHOTOREPAIR. UV light can catalyze a photochemical reaction to form a covalent bond between two adjacent thymines, thereby creating a thymine dimer. A photolyase enzyme recognizes the damage and binds to the thymine dimer. The enzyme absorbs visible light and uses the energy to cleave the thymine dimer.

cause this damage, and sunlight in the visible range can be used to repair the damage. Photorepair does not occur in cells deprived of visible light.

The photolyase enzyme has been found in many different species, ranging from bacteria, to single-celled eukaryotes, to humans. The ubiquitous nature of this enzyme illustrates the importance of this form of repair. For as long as cells have existed on Earth, they have been exposed to UV light and its potential to damage DNA.

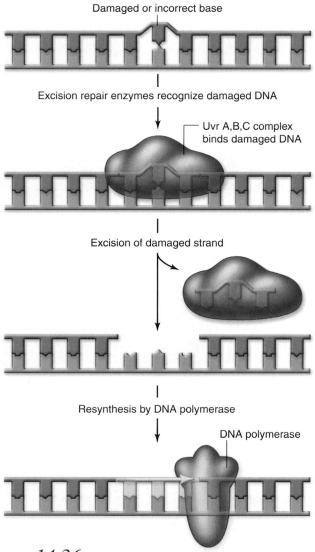

figure 14.26

REPAIR OF DAMAGED DNA BY EXCISION REPAIR. Damaged DNA is recognized by the uvr complex, which binds to the damaged region and removes it. Synthesis by DNA polymerase replaces the damaged region. DNA ligase finishes the process (not shown).

Excision repair: A nonspecific repair mechanism

A common form of nonspecific repair is **excision repair.** In this pathway, a damaged region is removed, or excised, and is then replaced by DNA synthesis (figure 14.26). In *E. coli*, this action is accomplished by proteins encoded by the *uvr A, B,* and *C* genes. Although these genes were identified based on mutations that increased sensitivity of the cell to UV light (hence the "uvr" in their names), their proteins can act on damage due to other mutagens.

Excision repair follows three steps: (1) recognition of damage, (2) removal of the damaged region, and (3) resynthesis using the information on the undamaged strand as a template (see figure 14.26). Recognition and excision are accomplished by the UvrABC complex. The UvrABC complex binds to damaged DNA and then cleaves a single strand on either side of the damage, removing it. In the synthesis stage, DNA pol I or pol II replaces the damaged DNA. This restores the original information in the damaged strand by using the information in the complementary strand.

Other repair pathways

Cells have other forms of nonspecific repair, and these fall into two categories: error-free and error-prone. It may seem strange to have an error-prone pathway, but it can be thought of as a last-ditch effort to save a cell that has been exposed to such massive damage that it has overwhelmed the error-free systems. In fact, this system in *E. coli* is part of what is called the "SOS response."

Cells can also repair damage that produces breaks in DNA. These systems use enzymes related to those that are involved in recombination during meiosis (chapter 11). It is thought that recombination uses enzymes that originally evolved for DNA repair.

The number of different systems and the wide spectrum of damage that can be repaired illustrate the importance of maintaining the integrity of the genome. Accurate replication of the genome is useless if a cell cannot reverse errors that can occur during this process or repair damage due to environmental causes.

inquiry

Cells are constantly exposed to DNA-damaging agents, ranging from UV light to by-products of oxidative metabolism. How does the cell deal with this, and what would happen if the cell had no way of dealing with this?

Cells have multiple repair pathways to reverse damage to DNA. Some of these systems are specific for one type of damage, such as photorepair that reverses thymine dimers caused by UV light. Other systems are nonspecific, such as excision repair that removes and replaces damaged regions.

14.1 The Nature of Genetic Material

Our knowledge of the molecular basis of the genetic material was built over a long history of experimentation.

■ Griffith's experiments showed genetic material can be transferred between cells, a process call transformation.

■ Avery, MacLeod, and McCarty demonstrated that DNA is the substance transferred between bacteria.

■ Hershey and Chase's research showed that DNA was the genetic material of bacteriophages.

14.2 DNA Structure

Miescher discovered nucleic acids, which contain three components: a five-carbon sugar, a phosphate group, and a nitrogenous base.

■ The sugar found in DNA is deoxyribose.

■ The nitrogenous bases in DNA are two-ringed purines—adenine (A) and guanine (G), and single-ringed pyrimidines—cytosine (C) and thymine (T).

■ Phosphodiester bonds are formed by linking the 5′ phosphate of one nucleotide to the 3′ hydroxyl of another nucleotide (figure 14.5).

■ Chargaff found that the proportion of adenine equals that of thymine and the proportion of cytosine equals that of guanine.

■ The bases can exist in two different tautomeric forms. The keto and enol forms predominate and this affects hydrogen bonding.

■ X-ray diffraction studies by Franklin and Wilkins indicated that DNA molecules had a helical structure.

■ Watson and Crick provided a rational structural model for DNA using available data and building models.

■ The Watson–Crick model consists of the following features (see figures 14.9 and 14.10).

• *DNA consists of two polynucleotide strands that form a double helix.*

• *The two strands are held together by hydrogen bonds, forming specific base pairs between adenine and thymine, and guanosine and cytosine.*

• *We say that the two strands are complementary because each strand determines the other by base pairing.*

• *Complementary phosphodiester strands are antiparallel to each other.*

14.3 Basic Characteristics of DNA Replication

Meselson and Stahl showed that DNA replication is semiconservative, resulting in two identical DNA molecules, each of which is composed of one original strand and one new strand (figure 14.11).

■ DNA replication can be divided into three phases:

• *Initiation of replication begins at a specific site called the origin.*

• *Elongation uses DNA polymerases that synthesize a new strand complementary to the template. These enzymes require a short base-paired primer and only synthesize in the 5′-to-3′ direction.*

• *Termination ends replication at a specific site, the terminus.*

14.4 Prokaryotic Replication

Prokaryotic replication involves a circular DNA template.

■ Prokaryotic replication begins at a unique site, the origin, and proceeds bidirectionally with two replication forks.

■ The complete prokaryotic chromosome with a single origin forms a single functional unit called a replicon.

■ There are three prokaryotic DNA polymerases, polymerase I, II and III, all of which synthesize DNA in a 5′-to-3′ direction.

■ DNA polymerases may also have the ability to degrade DNA from one end, a process called exonuclease activity.

■ Unwinding DNA uses the enzyme DNA helicase and energy.

■ Unwinding DNA introduces torsional strain that is removed by the enzyme DNA gyrase.

■ The antiparallel nature of DNA and the fact that the DNA polymerases only synthesize DNA in a 5′-to-3′ direction mean that replication is discontinuous on one strand (figure 14.16).

• *Only one strand, called the leading strand, is synthesized continuously.*

• *The other strand, called the lagging strand, is synthesized discontinuously.*

■ Synthesis occurs at the replication fork, where the two strands are being opened.

■ DNA polymerases require primers that are synthesized by DNA primase.

■ DNA polymerase III can remain attached to the template because of a subunit that acts as a sliding clamp.

■ Synthesis on the lagging strand is much more complex.

• *Polymerase III is still the main polymerase.*

• *DNA primase periodically synthesizes short primers.*

• *Each primer is extended by DNA polymerase III until it hits the previous fragment.*

• *The RNA primers are removed by DNA polymerase I and replaced with DNA.*

• *The fragments are joined by DNA ligase.*

■ All of the activities are coordinated in a complex called the replisome containing two copies of polymerase III, DNA primase, DNA helicase, and a number of accessory proteins.

■ The replisome moves in one direction by creating a loop in the lagging strand, allowing the antiparallel template strands to be copied in the same direction (figure 14.19).

14.5 Eukaryotic Replication

Eukaryotic replication is complicated by the large amount of DNA organized into multiple linear chromosomes.

■ Eukaryote chromosomes have multiple origins of replication.

■ The enzymology of eukaryotic replication is more complex, involving more enzymes.

■ The main replication polymerase is a complex of two enzymes.

■ The ends of linear chromosomes are called telomeres, and they protect the ends of chromosomes.

■ Linear chromosomes introduce problems finishing replication.

■ Telomeres are specialized structures that are made by the enzyme telomerase and not by the replication complex.

■ Telomerase contains an internal RNA that acts as a template to extend the DNA of the chromosome end.

■ Adult cells lack telomerase activity, and telomere shortening correlates with senescence.

14.6 DNA Repair

Detection and correction of DNA errors to reduce rates of mutations.

■ The error rate during replication is reduced by the proofreading ability of DNA polymerases.

■ Environmental mutagens damage DNA and increase the rate of mutations above background levels.

■ Cells have multiple specific and nonspecific pathways for repairing DNA damage.

■ In photorepair the enzyme photolyase absorbs visible light and uses this energy to specifically cleave thymine dimers caused by UV light.

■ Excision repair is nonspecific, and in prokaryotes a damaged region of DNA is removed by enzymes of the *uvr* system.

SELF TEST

1. What was the key finding from Griffith's experiments using live and heat-killed pathogenic bacteria?
   a. Bacteria with a smooth coat could kill mice
   b. Bacteria with a rough coat are not lethal
   c. Heat-killed smooth-coat bacteria would not cause death in mice.
   d. Heat-killed smooth-coat bacteria could transform the nonlethal live bacteria.

2. When Hershey and Chase differentially tagged the DNA and proteins of bacteriophages and allowed them to infect bacteria, what did the viruses transfer to the bacteria?
   a. Radioactive phosphorous and sulfur
   b. Radioactive sulfur
   c. DNA
   d. Both (b) and (c) are correct.

3. Which of the following is *not* a component of DNA?
   a. The pyrimidine uracil
   b. Five-carbon sugars
   c. The purine adenine
   d. Phosphate groups

4. What type of chemical bond allows DNA or RNA to form a long polymer?
   a. Hydrogen bonds
   b. Peptide bonds
   c. Ionic bonds
   d. Phosphodiester bonds

5. What is *Chargaff's rule*?
   a. The number of phosphate groups always equals the number of five-carbon sugars.
   b. The proportions of A equal that of C and G equals T.
   c. The proportions of A equal that of T and G equals C.
   d. Purines bind to pyrimidines.

6. The bonds that hold two complementary strands of DNA together are—
   a. Hydrogen bonds
   b. Peptide bonds
   c. Ionic bonds
   d. Phosphodiester bonds

7. If one strand of a DNA molecule has the sequence ATTGCAT, then the complementary strand will have the sequence—
   a. ATTGCAT
   b. TACGTTA
   c. TAACGTA
   d. GCCTAGC

8. Which of the following is *not* part of the Watson–Crick model of the structure of DNA?
   a. DNA is composed of two strands.
   b. The two DNA strands are oriented in parallel (5′ to 3′).
   c. Purines bind to pyrimidines.
   d. DNA forms a double helix.

9. Meselson and Stahl demonstrated that—
   a. DNA replication occurs in bacteria.
   b. DNA replication is dispersive.
   c. DNA replication is conservative.
   d. DNA replication is semiconservative.

10. Which of the following steps in DNA replication involves the formation of new phosphodiester bonds?
    a. Initiation at an origin of replication
    b. Elongation by a DNA polymerase
    c. Unwinding of the double helix
    d. Termination

11. The difference in leading- versus lagging-strand synthesis is a consequence of—
    a. the antiparallel configuration of DNA
    b. DNA polymerase III synthesizing only in the 5′-to-3′ direction
    c. the activity of DNA gyrase
    d. both (a) and (b)

12. Okazaki fragments are—
    a. synthesized in the 3′-to-5′ direction
    b. found on the lagging strand
    c. found on the leading strand
    d. made of RNA

13. Successful DNA synthesis requires all of the following *except*—
    a. helicase
    b. endonuclease
    c. DNA primase
    d. DNA ligase

14. What is a telomere?
    a. An A–T-rich region of DNA
    b. The point of DNA termination on a bacterial chromosome
    c. Regions of repeated sequences of DNA on the ends of eukaryotic chromosomes
    d. The sequence of RNA found on a replicating molecule of DNA

15. Which type of enzyme is involved in excision repair?
    a. photolyase
    b. DNA polymerase III
    c. endonuclease
    d. telomerase

CHALLENGE QUESTIONS

1. The work by Griffith provided the first indication that DNA was the genetic material. Review the four experiments outlined in figure 14.1. Predict the likely outcome for the following variations on this classic research.
   a. Heat-killed pathogenic and heat-killed nonpathogenic
   b. Heat-killed pathogenic and live nonpathogenic in the presence of an enzyme that digests proteins (proteases)
   c. Heat-killed pathogenic and live nonpathogenic in the presence of an enzyme that digests DNA (endonuclease)

2. Imagine that you have identified the sequence 5′-TTATAAAGCAATAGT-3′ in a eukaryotic chromosome. Could this region of the chromosome function as an origin of replication? Predict the sequence of an RNA primer that would be formed in association with this sequence.

3. Enzyme function is critically important for the proper replication of DNA. Predict the consequence of a loss of function for each of the following enzymes.
   a. DNA gyrase
   b. DNA polymerase III
   c. DNA ligase
   d. DNA polymerase I

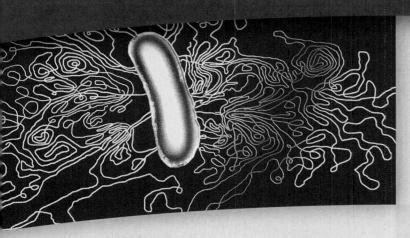

Genes and How They Work

introduction

YOU'VE SEEN HOW GENES SPECIFY TRAITS and how these traits can be followed in genetic crosses. You've also seen that the information in genes resides in the DNA molecule; the picture on the left shows the DNA that comprises the entire *E. coli* chromosome. Information in DNA is replicated by the cell and then partitioned equally during the process of cell division. The information in DNA is much like a blueprint for a building. The construction of the building uses the information in the blueprint, but requires building materials and, carpenters and other skilled laborers using a variety of tools and working together to actually construct the building. Similarly, the information in DNA requires nucleotide and amino acid building blocks, multiple forms of RNA, and many proteins acting in a coordinated fashion to make up the structure of a cell.

We now turn to the nature of the genes themselves and how cells extract the information in DNA in the process of **gene expression.** Gene expression can be thought of as the conversion of genotype into the phenotype.

15.1 The Nature of Genes

We know that DNA encodes proteins, but this knowledge alone tells us little about how the information in DNA can control cellular functions. Researchers had evidence that genetic mutations affected proteins, and in particular enzymes, long before the structure and code of DNA was known. In this section we review the evidence of the link between genes and enzymes.

Garrod concluded that inherited disorders can involve specific enzymes

In 1902, the British physician Archibald Garrod noted that certain diseases among his patients seemed to be more prevalent in particular families. By examining several generations of these families, he found that some of the diseases behaved as though they were the product of simple recessive alleles. Garrod concluded that these disorders were Mendelian traits, and that they had resulted from changes in the hereditary information in an ancestor of the affected families.

Garrod investigated several of these disorders in detail. In alkaptonuria, patients produced urine that contained homogentisic acid (alkapton). This substance oxidized rapidly when exposed to air, turning the urine black. In normal individuals, homogentisic acid is broken down into simpler substances. With considerable insight, Garrod concluded that patients suffering from alkaptonuria lack the enzyme necessary to catalyze this breakdown. He speculated that many other inherited diseases might also reflect enzyme deficiencies.

Beadle and Tatum showed that genes specify enzymes

From Garrod's finding, it took but a short leap of intuition to surmise that the information encoded within the DNA of chromosomes acts to specify particular enzymes. This point was not actually established, however, until 1941, when a series of experiments by George Beadle and Edward Tatum at Stanford University provided definitive evidence. Beadle and Tatum deliberately set out to create mutations in chromosomes and verified that they behaved in a Mendelian fashion in crosses. These alterations to single genes were analyzed for their effects on the organism (figure 15.1).

Neurospora crassa, *the bread mold*

One of the reasons Beadle and Tatum's experiments produced clear-cut results was their choice of experimental organism, the bread mold *Neurospora crassa*. This fungus can be grown readily in the laboratory on a defined medium consisting of only a carbon source (glucose), a vitamin (biotin), and inorganic salts. This type of medium is called "minimal" because it represent the minimal requirements to support growth. Any cells that can grow on minimal medium must be able to synthesize all necessary biological molecules.

Beadle and Tatum exposed *Neurospora* spores to X-rays, expecting that the DNA in some of the spores would experience damage in regions encoding the ability to make compounds needed for normal growth (see figure 15.1). Such a mutation would cause cells to be unable to grow on minimal medium.

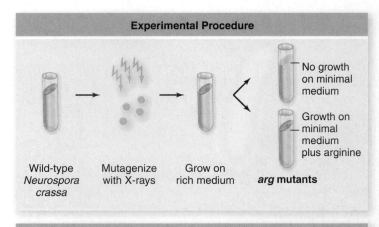

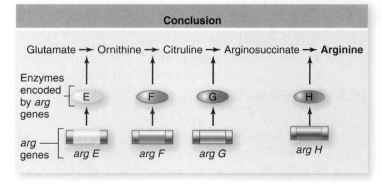

figure 15.1

THE BEADLE AND TATUM EXPERIMENT. Wild-type *Neurospora* were mutagenized with X-rays to produce mutants deficient in the synthesis of arginine (top panel). The specific defect in each mutant was identified by growing on medium supplemented with intermediates in the biosynthetic pathway for arginine (middle panel). A mutant will grow only on media supplemented with an intermediate produced after the defective enzyme in the pathway for each mutant. The enzymes in the pathway can then be correlated with genes on chromosomes (bottom panel).

Such mutations are called **nutritional mutations** because cells carrying them grow only if the medium is supplemented with additional nutrients.

Nutritional mutants

To identify mutations causing metabolic deficiencies, Beadle and Tatum placed subcultures of individual fungal cells grown on a rich medium onto minimal medium. Any cells that had lost the ability to make compounds necessary for growth would not grow on minimal medium. Using this approach, Beadle and Tatum succeeded in isolating and identifying many nutritional mutants.

Next, the researchers supplemented the minimal medium with different compounds to identify the deficiency in each mutant. This step allowed them to pinpoint the nature of the strain's biochemical deficiency. They concentrated in particular on mutants that would grow only in the presence of the amino acid arginine, dubbed *arg* mutants. When their chromosomal positions were located, the *arg* mutations were found to cluster in three areas.

One gene/one polypeptide

The next step was to determine where each mutation was blocked in the biochemical pathway for arginine biosynthesis. To do this, they supplemented the medium with each intermediate in the pathway to see which intermediate would support a mutant's growth. If the mutation affects an enzyme in the pathway that acts prior to the intermediate used as a supplement, then growth should be supported—but not if the mutation affects a step after the intermediate used (see figure 15.1). For each enzyme in the arginine biosynthetic pathway, Beadle and Tatum were able to isolate a mutant strain with a defective form of that enzyme. The mutation was always located at one of a few specific chromosomal sites, and each mutation had a unique location. Thus, each of the mutants they examined had a defect in a single enzyme, caused by a mutation at a single site on a chromosome.

Beadle and Tatum concluded that genes specify the structure of enzymes, and that each gene encodes the structure of one enzyme (see figure 15.1). They called this relationship the *one-gene/one-enzyme hypothesis*. Today, because many enzymes contain multiple polypeptide subunits, each encoded by a separate gene, the relationship is more commonly referred to as the **one-gene/one-polypeptide hypothesis.** This hypothesis clearly states the molecular relationship between genotype and phenotype.

As you learn more about genomes and gene expression, you'll find that this clear relationship is overly simple. As described later in this chapter, eukaryotic genes are more complex. In addition, some enzymes are composed at least in part of RNA, itself an intermediate in the production of proteins. Nevertheless, this one-gene/one-polypeptide concept is a useful starting point to think about gene expression.

The central dogma describes information flow in cells as DNA to RNA to protein

The conversion of genotype to phenotype requires information stored in DNA to be converted to protein. The nature of information flow in cells was first described by Francis Crick as

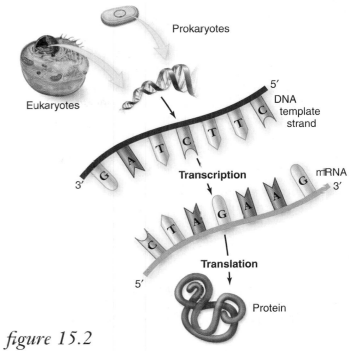

figure 15.2

THE CENTRAL DOGMA OF MOLECULAR BIOLOGY. DNA is transcribed to make mRNA, which is translated to make a protein.

the **central dogma of molecular biology.** Information passes in one direction from the gene (DNA) to an RNA copy of the gene, and the RNA copy directs the sequential assembly of a chain of amino acids into a protein (figure 15.2). Stated briefly,

$$DNA \longrightarrow RNA \longrightarrow protein$$

We can view this as a concise description of the process of gene expression, or the conversion of genotype to phenotype. We call the DNA to RNA step **transcription,** and the RNA to protein step **translation** (see figure 15.2). These processes will be explored in detail throughout the chapter.

This, again, is an oversimplification of how information flows in eukaryotic cells. A class of viruses called **retroviruses** has been discovered that can convert their RNA genome into a DNA copy, using the viral enzyme **reverse transcriptase.** This conversion violates the direction of information flow of the central dogma, and the discovery forced an updating of the dogma to include this "reverse" flow of information.

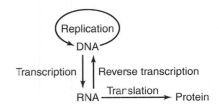

Metabolic disorders can be due to the presence of altered enzymes. Each gene encodes the information to make one polypeptide. The flow of information in cells, according to the central dogma, begins with information in a gene in DNA. DNA is transcribed into RNA, and this RNA copy is used to direct the synthesis of proteins.

15.2 The Genetic Code

How does the order of nucleotides in a DNA molecule encode the information that specifies the order of amino acids in a polypeptide? The answer to this essential question came in 1961, through an experiment led by Francis Crick and Sydney Brenner. That experiment was so elegant and the result so critical to understanding the genetic code that we describe it in detail.

The code is read in groups of three

Crick and Brenner reasoned that the genetic code most likely consisted of a series of blocks of information called **codons,** each corresponding to an amino acid in the encoded protein. They further hypothesized that the information within one codon was probably a sequence of three nucleotides. With four DNA nucleotides (G, C, T, and A), using two in each codon will produce only 4^2, or 16, different codons—not enough to code for 20 amino acids. However, three nucleotides results in 4^3, or 64, different combinations of three, more than enough.

Spaced or unspaced codons?

In theory, the sequence of codons in a gene could be punctuated with nucleotides between the codons that are not used, like the spaces that separate the words in this sentence. Alternatively, the codons could lie immediately adjacent to each other, forming a continuous sequence of nucleotides.

If the information in the genetic message is separated by spaces, then altering any single word would not affect the entire sentence. In contrast, if all of the words are run together but read in groups of three, then any alteration that is not in groups of three would alter the entire sentence. These two ways of using information in DNA imply different methods of translating the information into protein.

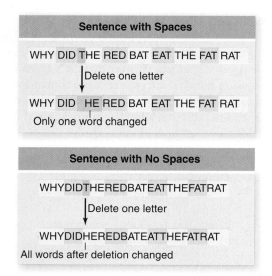

Determining that codons are unspaced

To choose between these alternative mechanisms, Crick and his colleagues used a chemical to create mutations that would delete one, two, or three nucleotides from a viral DNA molecule, which was then transcribed and translated into a polypeptide.

They then asked whether this action altered only a single amino acid or all amino acids after the deletions.

When they made a single deletion or two deletions near each other, the genetic message shifted, altering all of the amino acids after the deletion. When they made three deletions, however, the protein after the deletions was normal. They obtained the same results when they made additions to the DNA consisting of 1, 2, or 3 nt (nucleotides).

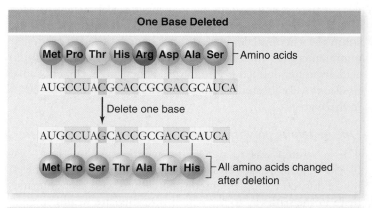

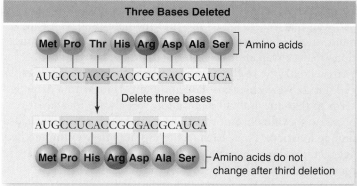

Thus, Crick and Brenner concluded that the genetic code is read in increments of three nucleotides (in other words, it is a triplet code), and that reading occurs continuously without punctuation between the 3-nt units.

These experiments indicate the importance of the **reading frame** for the genetic message. Because there is no punctuation, the reading frame established by the first codon in the sequence determines how all subsequent codons are read. We now call the kinds of mutations that Crick and Brenner used **frameshift mutations** because they alter the reading frame of the genetic message.

Nirenberg and others deciphered the code

The determination of which of the 64 possible codons encoded each particular amino acids was one of the greatest triumphs of 20th-century biochemistry. Accomplishing this decryption required two main developments to succeed. First, cell-free biochemical systems that would support protein synthesis from a defined RNA were needed. Second, the ability to produce synthetic, defined RNAs that could be used in the cell-free system was necessary.

During a five-year period from 1961 to 1966, work performed primarily in Marshall Nirenberg's laboratory led to the elucidation of the genetic code. Nirenberg's group first showed that adding the synthetic RNA molecule polyU (an RNA molecule consisting of a string of uracil nucleotides) to their cell-free systems produced the polypeptide polyphenylalanine (a string of phenylalanine amino acids). Therefore, UUU encodes phenylalanine.

Next they used enzymes to produce RNA polymers with more than one nucleotide. These polymers allowed them to infer the composition of many of the possible codons, but not the order of bases in each codon.

The researchers then were able to use enzymes to synthesize defined 3-base sequences that could be tested for binding to the protein synthetic machinery. This so-called *triplet-binding assay* allowed them to identify 54 of the 64 possible triplets.

The organic chemist H. Gobind Khorana provided the final piece of the puzzle by using organic synthesis to produce artificial RNA molecules of defined sequence, and then examining what polypeptides they directed in cell-free systems. The combination of all of these methods allowed the determination of all 64 possible three-nucleotide sequences, and the full genetic code was determined (table 15.1).

The code is degenerate but specific

Some obvious features of the code jump out of table 15.1. First, 61 of the 64 possible codons are used to specify amino acids. Three codons, UAA, UGA, and UAG, are reserved for another

function: they signal "stop" and are known as **stop codons.** The only other form of "punctuation" in the code is that AUG is used to signal "start" and is therefore the **start codon.** In this case the codon has a dual function as it also encodes the amino acid methionine (Met).

You can see that with 61 codons to encode only 20 amino acids, there are many more codons than amino acids. One way to deal with this abundance would be to use only 20 of the 61 codons, but this is not what cells do. In reality, all 61 codons are used, making the code **degenerate,** which means that some amino acids are specified by more than one codon. The reverse, however, in which a single codon would specify more than one amino acid, is never found.

This degeneracy is not uniform. Some amino acids have only one codon, and some have up to six codons. In addition, the degenerate base usually occurs in position 3 of a codon, such that the first two positions are the same, and two or four of the possible nucleic acids at position 3 encode the same amino acid. (The nature of protein synthesis on ribosomes explains how this codon usage works, and it is discussed later.)

The code is practically universal, but not quite

The genetic code is the same in almost all organisms. The universality of the genetic code is among the strongest evidence that all living things share a common evolutionary heritage. Because the code is universal, genes can be transferred from one organism to another and can be successfully expressed in

TABLE 15.1	The Genetic Code									

SECOND LETTER

First Letter	U			C			A			G			Third Letter
U	UUU	Phe	Phenylalanine	UCU	Ser	Serine	UAU	Tyr	Tyrosine	UGU	Cys	Cysteine	U
	UUC			UCC			UAC			UGC			C
	UUA	Leu	Leucine	UCA			UAA		"Stop"	UGA		"Stop"	A
	UUG			UCG			UAG		"Stop"	UGG	Trp	Tryptophan	G
C	CUU	Leu	Leucine	CCU	Pro	Proline	CAU	His	Histidine	CGU	Arg	Arginine	U
	CUC			CCC			CAC			CGC			C
	CUA			CCA			CAA	Gln	Glutamine	CGA			A
	CUG			CCG			CAG			CGG			G
A	AUU	Ile	Isoleucine	ACU	Thr	Threonine	AAU	Asn	Asparagine	AGU	Ser	Serine	U
	AUC			ACC			AAC			AGC			C
	AUA			ACA			AAA	Lys	Lysine	AGA	Arg	Arginine	A
	AUG	Met	Methionine; "Start"	ACG			AAG			AGG			G
G	GUU	Val	Valine	GCU	Ala	Alanine	GAU	Asp	Aspartate	GGU	Gly	Glycine	U
	GUC			GCC			GAC			GGC			C
	GUA			GCA			GAA	Glu	Glutamate	GGA			A
	GUG			GCG			GAG			GGG			G

A codon consists of three nucleotides read in the sequence shown. For example, ACU codes for threonine. The first letter, A, is in the First Letter column; the second letter, C, is in the Second Letter column; and the third letter, U, is in the Third Letter column. Each of the mRNA codons is recognized by a corresponding anticodon sequence on a tRNA molecule. Many amino acids are specified by more than one codon. For example, threonine is specified by four codons, which differ only in the third nucleotide (ACU, ACC, ACA, and ACG).

figure 15.3

TRANSGENIC PIG. The piglet on the right is a conventional piglet. The piglet on the left was engineered to express a gene from jellyfish that encodes green fluorescent protein. The color of this piglet's nose is due to expression of this introduced gene. Such transgenic animals indicate the universal nature of the genetic code.

their new host (figure 15.3). This universality of gene expression is central to many of the advances of genetic engineering discussed in chapter 17.

In 1979, investigators began to determine the complete nucleotide sequences of the mitochondrial genomes in humans, cattle, and mice. It came as something of a shock when these investigators learned that the genetic code used by these mammalian mitochondria was not quite the same as the "universal code" that has become so familiar to biologists.

In the mitochondrial genomes, what should have been a stop codon, UGA, was instead read as the amino acid tryptophan; AUA was read as methionine rather than as isoleucine; and AGA and AGG were read as stop codons rather than as arginine. Furthermore, minor differences from the universal code have also been found in the genomes of chloroplasts and in ciliates (certain types of protists).

Thus, it appears that the genetic code is not quite universal. Some time ago, presumably after they began their endosymbiotic existence, mitochondria and chloroplasts began to read the code differently, particularly the portion of the code associated with "stop" signals.

inquiry

? *The genetic code is almost universal. Why do you think it is nearly universal?*

The genetic codes was shown to be triplets with no punctuation: three bases determine one amino acid, and these groups of three are read in order with no "spaces." With 61 codons that specify amino acids (plus 3 that mean "stop," for 64 total), the code is degenerate: Some amino acids have more than one codon, but all codons encode only one amino acid. The code is practically universal, with some exceptions.

15.3 Overview of Gene Expression

The central dogma provides an intellectual framework that describes information flow in biological systems. We call the DNA-to-RNA step *transcription* because it produces an exact copy of the DNA, much as a legal transcription contains the exact words of a court proceeding. The RNA-to-protein step is termed *translation* because it requires translating from the nucleic acid to the protein "languages."

Transcription makes an RNA copy of DNA

The process of transcription produces an RNA copy of the information in DNA. That is, transcription is the DNA-directed synthesis of RNA. This process uses the principle of complementarity, described in the previous chapter, to use DNA as a template to make RNA (figure 15.4).

Because DNA is double-stranded and RNA is single-stranded, only one of the two DNA strands needs to be copied. We call the strand that is copied the **template strand.** The RNA transcript's sequence is complementary to the template strand. The strand of DNA not used as a template is called the **coding strand.** It has the same sequence as the RNA transcript, except that U in the RNA is T in the DNA-coding strand.

```
Coding   5′ –TCAGCCGTCAGCT– 3′ ⎤
                                ⎬ DNA
Template 3′ –AGTCGGCAGTCGA– 5′ ⎦
                  │
             Transcription
                  ↓
Coding   5′ –UCAGCCGUCAGCU– 3′  mRNA
```

The RNA transcript used to direct the synthesis of polypeptides is termed **messenger RNA (mRNA).** Its name reflects the recognition that some molecule must carry the DNA message to the ribosome for processing.

As with replication, DNA transcription can be thought of as having three parts: *initiation, elongation,* and *termination.*

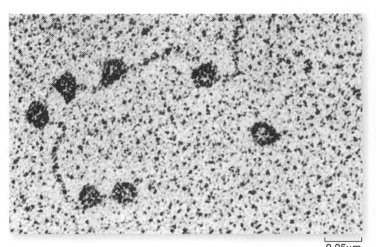

0.05µm

figure 15.4

RNA POLYMERASE. In this electron micrograph, the dark circles are RNA polymerase molecules synthesizing RNA from a DNA template.

Initiation of transcription

Initiation involves a number of components, which differ in prokaryotes and eukaryotes:

- DNA sequences called *promoters* provide attachment sites for the enzyme, *RNA polymerase*, that makes the RNA transcript.
- The *start site* on the DNA is the first base transcribed.
- In eukaryotes, one or more *transcription factors* are also involved in initiation.

Once RNA polymerase has become bound to the DNA at the promoter, transcription can begin at the start site.

Elongation of the transcript

During elongation, the RNA transcript is synthesized:

- New nucleotide triphosphates complementary to the template strand are joined by phosphodiester bonds in the 5′-to-3′ direction by the enzyme *RNA polymerase*, producing the new RNA chain.
- As transcription proceeds, DNA is unwound by RNA polymerase to allow transcription and rewound behind the enzyme. We call the region that is unwound by the enzyme a *transcription bubble*.

Termination of transcription

Elongation proceeds until a stop sequence is reached:
- DNA sequences called *terminators*, described later, cause the RNA polymerase to stop and release the DNA.
- The newly synthesized RNA transcript dissociates from the DNA, and the DNA rewinds.

Translation uses information in RNA to synthesize proteins

The process of translation is by necessity much more complex than transcription. In this case, RNA cannot be used as a direct template for a protein because there is no complementarity—that is, a sequence of amino acids cannot be aligned to an RNA template based on any kind of "chemical fit." Molecular geneticists suggested that some kind of adapter molecule must exist that can interact with both RNA and amino acids, and **transfer RNA (tRNA)** was found to fill this role. This need for an intermediary adds a level of complexity to the process that is not seen in either DNA replication or transcription of RNA.

Translation takes place on the ribosome, the cellular protein-synthetic machinery, and it requires the participation of multiple kinds of RNA and many proteins. Here we provide an outline of the processes; all are described in detail in the sections that follow.

Initiation of translation

Initiation depends on the presence of a start codon and the formation of an initiation complex:

- An *initiation complex* is formed containing the ribosome, mRNA, and the *initiator tRNA* bound to the amino acid methionine.
- The assembly of this complex requires the participation of a number of initiation factors.

Elongation of the polypeptide

The polypeptide grows as tRNA intermediaries bring individual amino acids to the ribosome complex. A tRNA molecule that is carrying an amino acid is called a *charged tRNA*. The ribosome must move along the mRNA strand and bind to charged tRNAs so their anticodon can hydrogen-bond to codons in the mRNA. The ribosome can bind to two tRNAs and form a peptide bond between the amino acids they are carrying.

- Charged tRNAs are brought to the ribosome. The charged tRNA's anticodon must be complementary to each new codon in the mRNA.
- The enzyme *peptidyl transferase* catalyzes formation of a peptide bond between each new amino acid and the growing chain.
- The ribosome complex moves along the mRNA, ejecting the empty tRNA and positioning the binding site for the new tRNA over the next codon in the mRNA.

Termination of translation

- Elongation proceeds until a *stop codon* is encountered.
- *Release factors* recognize the stop codon and cause dissociation of the peptide chain, releasing the last tRNA from the ribosome complex.

The sites at which transcription and translation occur are different in prokaryotes and eukaryotes because eukaryotes have a membrane-bounded nucleus—mRNA must exit the nucleus before translation can begin. In prokaryotes, by contrast, transcription and translation often occur in tandem. We will examine each of these processes in detail in the succeeding sections.

inquiry

? *It is widely accepted that RNA polymerase has no proofreading capacity. Would you expect high or low levels of error in transcription as compared with DNA replication? Why do you think it is more important for DNA polymerase than for RNA polymerase to proofread?*

RNA has multiple roles in gene expression

All RNAs are synthesized from a DNA template by transcription. Gene expression requires the participation of multiple kinds of RNA, each with different roles in the overall process. Here is a brief summary of these roles, which are described in detail later on.

Messenger RNA. Even before the details of gene expression were unraveled, geneticists recognized that there must be an intermediate form of the information in DNA that can be transported out of the eukaryotic nucleus to the cytoplasm for ribosomal processing. This hypothesis was called the "messenger hypothesis," and we retain this language in the name messenger RNA (mRNA).

Ribosomal RNA. The class of RNA found in ribosomes is called **ribosomal RNA (rRNA)**. There are multiple forms of rRNA, and rRNA is found in both ribosomal subunits. This rRNA is critical to the function of the ribosome.

Transfer RNA. The intermediary adapter molecule between mRNA and amino acids, as mentioned earlier, is transfer RNA **(tRNA).** Transfer RNA molecules have amino acids covalently attached to one end, and an anticodon that can base-pair with an mRNA codon at the other. The tRNAs act to interpret information in mRNA and to help position the amino acids on the ribosome.

Small nuclear RNA. Small nuclear RNAs **(snRNAs)** are part of the machinery that is involved in nuclear processing of eukaryotic "pre-mRNA." We discuss this splicing reaction later in the chapter.

SRP RNA. In eukaryotes where some proteins are synthesized by ribosomes on the RER, this process is mediated by the **signal recognition particle,** or **SRP,** described later in the chapter. The SRP is composed of both RNA and proteins.

Micro-RNA. A new class of RNA recently discovered is **micro-RNA (miRNA).** These very short RNAs escaped detection for many years because they were lost during the techniques typically used in nucleic acid purification. Their role is unclear, but one class, **small interfering RNAs (siRNAs)** appear to be involved in controlling gene expression and are part of a system to protect cells from viral attack.

> In transcription, the enzyme RNA polymerase synthesizes an RNA strand using DNA as a template. Only one strand of the DNA, the template strand, is copied; the other strand, which has the same sequence as the transcribed RNA, is called the coding strand. Translation takes place on ribosomes and utilizes tRNA as an adapter between mRNA codons and amino acids.

15.4 Prokaryotic Transcription

We begin a detailed examination of gene expression by describing the process of transcription in prokaryotes. The later description of eukaryotic transcription will concentrate on their differences from prokaryotes.

Prokaryotes have a single RNA polymerase

The single **RNA polymerase** of prokaryotes exists in two forms called *core polymerase* and *holoenzyme*. The core polymerase can synthesize RNA using a DNA template, but it cannot initiate synthesis accurately. The holoenzyme can initiate accurately.

The core polymerase is composed of four subunits: two identical α subunits, a β subunit, and a β′ subunit (figure 15.5a). The two α subunits help to hold the complex together and can bind to regulatory molecules. The active site of the enzyme is formed by the β and β′ subunits, which bind to the DNA template and the ribonucleotide triphosphate precursors.

The *holoenzyme* that can properly initiate synthesis is formed by the addition of a σ (sigma) subunit to the core polymerase (see figure 15.5a). Its ability to recognize specific signals in DNA allows RNA polymerase to locate the beginning of genes, which is critical to its function. Note that initiation of mRNA synthesis does not require a primer, in contrast to DNA replication.

Initiation occurs at promoters

Accurate initiation of transcription requires two sites in DNA: one called a **promoter** that forms a recognition and binding site for the RNA polymerase, and the actual **start site.** The polymerase also needs a signal to end transcription, which we call a **terminator.** We then refer to the region from promoter to terminator as a **transcription unit.**

The action of the polymerase moving along the DNA can be thought of as analogous to water flowing in a stream. We can speak of sites on the DNA as being "upstream" or "downstream" of the start site. We can also use this comparison to form a simple system for numbering bases in DNA to refer to positions in the transcription unit. The first base transcribed is called **+1,** and this numbering continues downstream until the last base transcribed. Any bases upstream of the start site receive negative numbers, starting at **–1.**

The promoter is a short sequence found upstream of the start site and is therefore not transcribed by the polymerase. Two 6-base sequences are common to bacterial promoters: One is located 35 nt upstream of the start site (–35), and the other is located 10 nucleotides upstream of the start site (–10) (figure 15.5b). These two sites provide the promoter with asymmetry; they indicate not only the site of initiation, but also the direction of transcription.

The binding of RNA polymerase to the promoter is the first step in transcription. Promoter binding is controlled by the σ subunit of the RNA polymerase holoenzyme, which recognizes the –35 sequence in the promoter and positions the RNA polymerase at the correct start site, oriented to transcribe in the correct direction.

figure 15.5

BACTERIAL RNA POLYMERASE AND TRANSCRIPTION INITIATION. *a.* RNA polymerase has two forms: core polymerase and holoenzyme. *b.* The σ subunit of holoenzyme recognizes promoter elements at –35 and –10 and binds to the DNA. The helix is opened at the –10 region, and transcription will begin at the start site at +1.

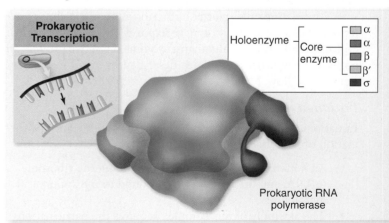

Prokaryotic Transcription

Holoenzyme — Core enzyme — α, α, β, β′, σ

Prokaryotic RNA polymerase

a.

Once bound to the promoter, the RNA polymerase begins to unwind the DNA helix at the –10 site (see figure 15.5*b*). The polymerase covers a region of about 75 bp but only unwinds about 12–14 bp.

inquiry

The prokaryotic promoter has two distinct elements that are not identical. How is this important to the initiation of transcription?

Elongation adds successive nucleotides

In prokaryotes, the transcription of the RNA chain usually starts with ATP or GTP. One of these forms the 5′ end of the chain, which grows in the 5′ to 3′ direction as ribonucleotides are added. As the RNA polymerase molecule leaves the promoter region, the σ factor is no longer required, although it may remain in association with the enzyme.

This process of leaving the promoter, called *clearance*, or *escape*, involves more than just synthesizing the first few nucleotides of the transcript and moving on, because the enzyme has made strong contacts to the DNA during initiation. It is necessary to break these contacts with the promoter region to be able to move progressively down the template. The enzyme goes through conformational changes during this clearance stage, and subsequently contacts less of the DNA than it does during the initial promoter binding.

The region containing the RNA polymerase, the DNA template, and the growing RNA transcript is called the **transcription bubble** because it contains a locally unwound "bubble" of DNA (figure 15.6). Within the bubble, the first 9 bases of the newly synthesized RNA strand temporarily form a helix with the template DNA strand. This stabilizes the positioning of the 3′ end of the RNA so it can interact with an incoming ribonucleotide triphosphate. The enzyme itself covers about 50 bp of DNA around this transcription bubble.

The transcription bubble created by RNA polymerase moves down the bacterial DNA at a constant rate, about 50 nt/sec, with the growing RNA strand protruding from the bubble. After the transcription bubble passes, the now-transcribed DNA is rewound as it leaves the bubble.

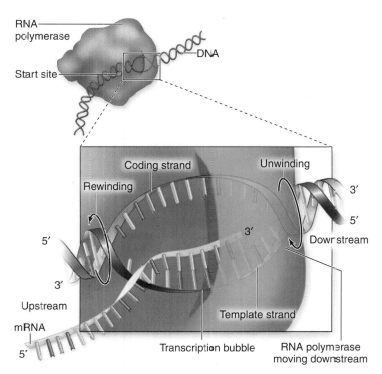

figure 15.6

MODEL OF A TRANSCRIPTION BUBBLE. The DNA duplex is unwound by the RNA polymerase complex, rewinding at the end of the bubble. One of the strands of DNA functions as a template, and nucleotide building blocks are assembled into RNA from this template. There is a short region of RNA–DNA hybrid within the bubble.

Termination occurs at specific sites

The end of a bacterial transcription unit is marked by terminator sequences that signal "stop" to the polymerase. Reaching these sequences causes the formation of phosphodiester bonds to cease, the RNA–DNA hybrid within the transcription bubble to dissociate, the RNA polymerase to release the DNA, and the DNA within the transcription bubble to rewind.

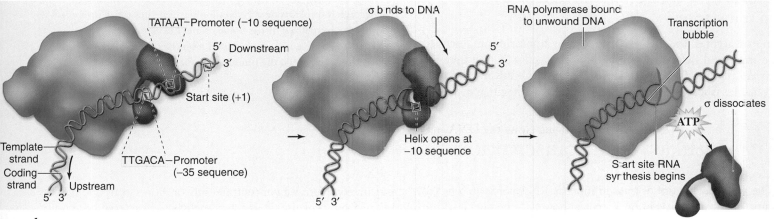

b.

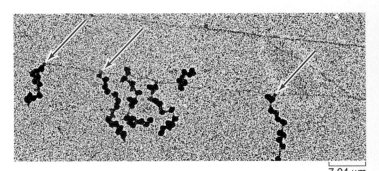

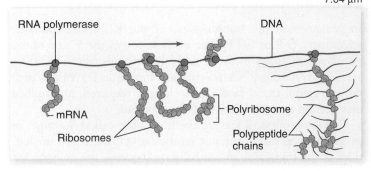

7.04 μm

figure 15.8

TRANSCRIPTION AND TRANSLATION ARE COUPLED IN BACTERIA. In this micrograph of gene expression in *E. coli*, translation is occurring during transcription. The arrows point to RNA polymerase enzymes and ribosomes are attached to the mRNAs extending from the polymerase. Polypeptides being synthesized by ribosomes, which are not visible in the micrograph, have been added to last mRNA in the drawing.

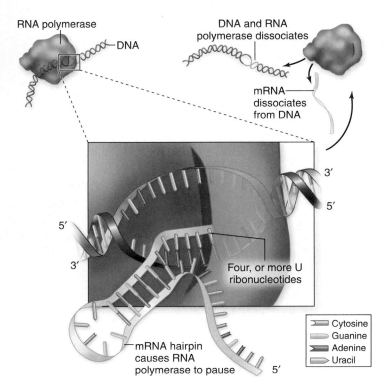

figure 15.7

BACTERIAL TRANSCRIPTION TERMINATOR. The self-complementary G–C region forms a double-stranded stem with a single stranded loop called a hairpin. The stretch of U's form a less stable RNA–DNA hybrid that falls off the enzyme.

The simplest terminators consist of a series of G–C base pairs followed by a series of A–T base pairs. The RNA transcript of this stop region can form a double-stranded structure in the GC region called a *hairpin*, which is followed by four or more uracil (U) ribonucleotides (figure 15.7). Formation of the hairpin causes the RNA polymerase to pause, placing it directly over the run of four uracils. The pairing of U with the DNA's A is the weakest of the four hybrid base-pairs, and it is not strong enough to hold the hybrid strands when the polymerase pauses. Instead, the RNA strand dissociates from the DNA within the transcription bubble, and transcription stops. A variety of protein factors also act at these terminators to aid in terminating transcription.

Prokaryotic transcription is coupled to translation

In prokaryotes, the mRNA produced by transcription begins to be translated before transcription is finished—that is, they are *coupled* (figure 15.8). As soon as a 5′ end of the mRNA becomes available, ribosomes are loaded onto this to begin translation.

(This coupling cannot occur in eukaryotes because transcription occurs in the nucleus, and translation occurs in the cytoplasm.)

Another difference between prokaryotic and eukaryotic gene expression is that the mRNA produced in prokaryotes may contain multiple genes. Prokaryotic genes are often organized such that genes encoding related functions are clustered together. This grouping of functionally related genes is referred to as an **operon.** An operon is a single transcripting unit that encodes multiple enzymes necessary for a biochemical pathway. By clustering genes by function, they can be regulated together, a topic that we return to in the next chapter.

> The bacterial RNA polymerase has two forms: a core polymerase with all synthetic activity, and a holoenzyme that can initiate accurately. Prokaryotic transcription starts at sites called promoters that are recognized by the holoenzyme. Elongation consists of synthesis by the core enzyme until it reaches a terminator where synthesis stops, and the transcript dissociates from the enzyme.

15.5 Eukaryotic Transcription

The basic mechanism of transcription by RNA polymerase is the same in eukaryotes as in prokaryotes; however, the details of the two processes differ enough that it is necessary to consider them

separately. Here we concentrate only on how eukaryotic systems differ from prokaryotic systems, such as the bacterial system just discussed. All other features may be assumed to be the same.

Eukaryotes have three RNA polymerases

Unlike prokaryotes, which have a single RNA polymerase enzyme, eukaryotes have three different RNA polymerases that are distinguished by both structure and function. The enzyme **RNA polymerase I** transcribes rRNA, **RNA polymerase II** transcribes mRNA and some small nuclear RNAs, and **RNA polymerase III** transcribes tRNA and some other small RNAs. Together, these three enzymes accomplish all transcription in the nucleus of eukaryotic cells.

Each polymerase has its own promoter

The existence of three different RNA polymerases requires different signals in the DNA to allow each polymerase to recognize where to begin transcription; each polymerase recognizes a different promoter structure.

RNA polymerase I promoters

RNA polymerase I promoters at first puzzled biologists, because comparisons of rRNA genes between species showed no similarities outside the coding region. The current view is that these promoters are also specific for each species, and for this reason, cross-species comparisons do not yield similarities.

RNA polymerase II promoters

The RNA polymerase II promoters are the most complex of the three types, probably a reflection of the huge diversity of genes that are transcribed by this polymerase. When the first eukaryotic genes were isolated, many had a sequence called the **TATA box** upstream of the start site. This sequence was similar to the prokaryotic −10 sequence, and it was assumed that the TATA

box was the primary promoter element. With the sequencing of entire genomes, many more genes have been analyzed, and this assumption has proved overly simple. It has been replaced by the idea of a "core promoter" that can be composed of a number of different elements, including the TATA box. Additional control elements allow for tissue-specific and developmental time–specific expression (chapter 16).

RNA polymerase III promoters

Promoters for RNA polymerase III also were a source of surprise for biologists in the early days of molecular biology who were examining the control of eukaryotic gene expression. A common technique for analyzing regulatory regions was to make successive deletions from the 5′ end of genes until enough was deleted to abolish specific transcription. The logic followed experiences with prokaryotes, in which the regulatory regions had been found at the 5′ end of genes. But in the case of tRNA genes, the 5′ deletions had no effect on expression! The promoters were found to actually be internal to the gene itself.

Initiation and termination differ from that in prokaryotes

The initiation of transcription at RNA polymerase II promoters is analogous to prokaryotic initiation but is more complex. Instead of a single factor allowing promoter recognition, eukaryotes use a host of **transcription factors.** These proteins are necessary to get the RNA polymerase II enzyme to a promoter and to initiate gene expression. A number of these transcription factors interact with RNA polymerase II to form an *initiation complex* at the promoter (figure 15.9). We explore this complex in detail in chapter 16 when we describe the control of gene expression.

The termination of transcription for RNA polymerase II also differs from that in prokaryotes. Although termination sites exist, they are not as well defined as are prokaryotic terminators. The end of the mRNA is also not formed by RNA polymerase II because the primary transcript is modified after transcription.

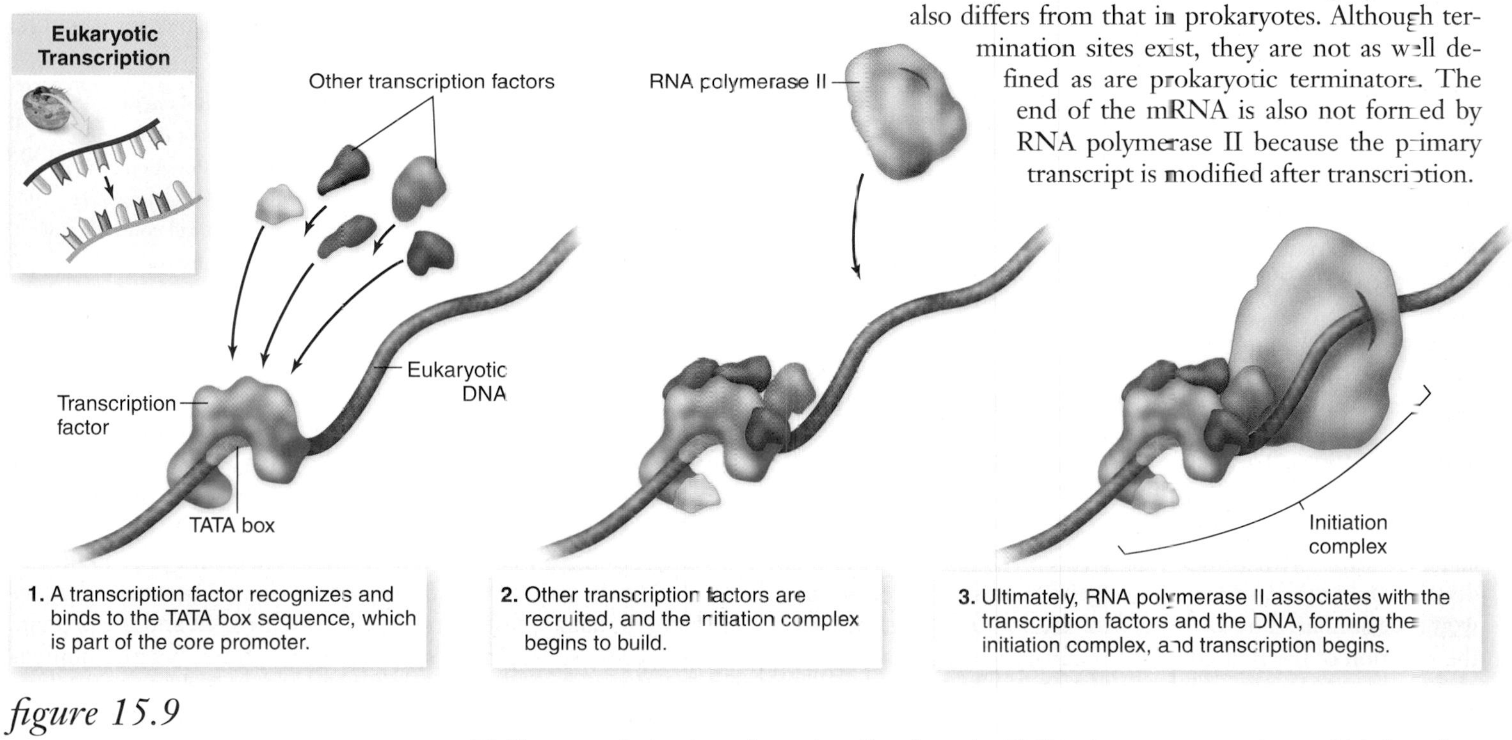

1. A transcription factor recognizes and binds to the TATA box sequence, which is part of the core promoter.

2. Other transcription factors are recruited, and the initiation complex begins to build.

3. Ultimately, RNA polymerase II associates with the transcription factors and the DNA, forming the initiation complex, and transcription begins.

figure 15.9

EUKARYOTIC INITIATION COMPLEX. Unlike transcription in prokaryotic cells, where the RNA polymerase recognizes and binds to the promoter, eukaryotic transcription requires the binding of transcription factors to the promoter before RNA polymerase II binds to the DNA. The association of transcription factors and RNA polymerase II at the promoter is called the initiation complex.

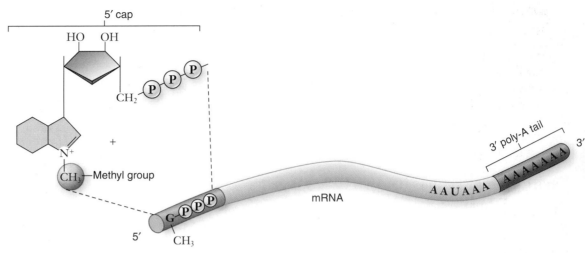

figure 15.10

POSTTRANSCRIPTIONAL MODIFICATIONS TO 5′ AND 3′ ENDS. Eukaryotic mRNA molecules are modified in the nucleus with the addition of a methylated GTP to the 5′ end of the transcript, called the 5′ cap, and a long chain of adenine residues to the 3′ end of the transcript, called the poly-A tail.

Eukaryotic transcripts are modified

A primary difference between prokaryotes and eukaryotes is the fate of the transcript itself. Between transcription in the nucleus and export of a mature mRNA to the cytoplasm, a number of modifications occur to the initial transcripts made by RNA polymerase II. We call the RNA synthesized by RNA polymerase II the **primary transcript** and the final processed form the **mature mRNA.**

The 5′ cap

Eukaryotic transcripts have an unusual structure that is added to the 5′ end of mRNAs. The first base in the transcript is usually an adenine (A) or a guanine (G), and this is further modified by the addition of GTP to the 5′ PO_4 group, forming what is known as a **5′ cap** (figure 15.10). The G nucleotide in the cap is joined to the transcript by its 5′ end; the only such 5′-to-5′ bond found in nucleic acids. The G in the GTP is also modified by the addition of a methyl group, so it is often called a *methyl-G cap.* The cap is added while transcription is still in progress. This cap protects the 5′ end of the mRNA from degradation and is also involved in translation initiation.

The 3′ poly-A tail

A major difference between transcription in prokaryotes and eukaryotes is that in eukaryotes, the end of the transcript is not the end of the mRNA. The eukaryotic transcript is cleaved downstream of a specific site (AAUAAA) prior to the termination site for transcription. A series of adenine (A) residues, called the **3′ poly-A tail,** are added after this cleavage by the enzyme poly-A polymerase. Thus the end of the mRNA is not created by RNA polymerase II and is not the end of the transcript (see figure 15.10).

The enzyme poly-A polymerase is part of a complex that recognizes the poly-A site, cleaves the transcript, then adds 1–200 A's to the end. The poly-A tail appears to play a role in the stability of mRNAs by protecting them from degradation (chapter 16).

Splicing of primary transcripts

Eukaryotic genes may contain noncoding sequences that have to be removed to produce the final mRNA. This process, called pre-mRNA splicing, is accomplished by an organelle called the *spliceosome.* This complex topic is discussed in the next section.

> Eukaryotes have three RNA polymerases, called polymerase I, II, and III. Each is responsible for the synthesis of different cellular RNAs and recognizes its own promoter. Polymerase II is responsible for mRNA synthesis. The primary mRNA transcript is modified by addition of a 5′ cap and a 3′ poly-A tail consisting of 1–200 adenines. Noncoding regions are removed by splicing.

15.6 Eukaryotic pre-mRNA Splicing

The first genes isolated were prokaryotic genes found in *E. coli* and its viruses. A clear picture of the nature and some of the control of gene expression emerged from these systems before any eukaryotic genes were isolated. It was assumed that although details would differ, the outline of gene expression in eukaryotes would be similar. The world of biology was in for a shock with the isolation of the first genes from eukaryotic organisms.

Eukaryotic genes may contain interruptions

Many eukaryotic genes appeared to contain sequences that were not represented in the mRNA. It is hard to exaggerate how unexpected this finding was. A basic tenet of molecular biology based on *E. coli* was that a gene was *colinear* with its protein product, that is, the sequence of bases in the gene corresponds to the sequence of bases in the mRNA, which in turn corresponds to the sequence of amino acids in the protein.

In the case of eukaryotes, genes can be interrupted by sequences that are not represented in the mRNA and the protein. The term "split genes" was used at the time, but the nomenclature that has stuck describes the unexpected nature of these sequences. We call the noncoding DNA that interrupts the sequence of the gene "intervening sequences," or **introns,** and we call the coding sequences **exons** because they are *ex*pressed (figure 15.11).

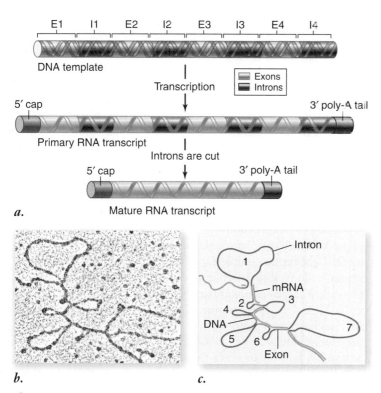

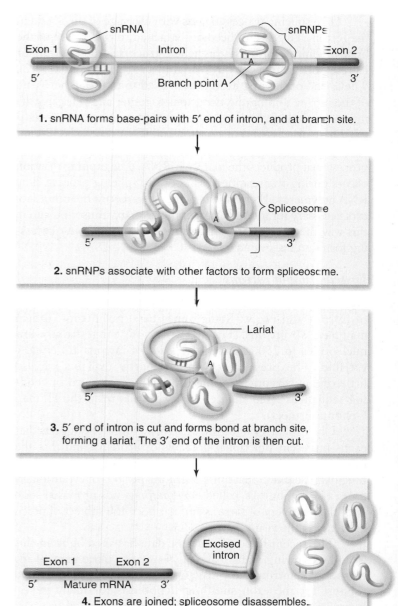

figure 15.11

EUKARYOTIC GENES CONTAIN INTRONS AND EXONS.
a. Eukaryotic genes contain sequences that will form the coding sequence called exons and intervening sequences called introns. *b.* An electron micrograph showing hybrids formed with the mRNA and the DNA of the ovalbumin gene, which has seven introns. Introns within the DNA sequence have no corresponding sequence in the mRNA and thus appear as seven loops. *c.* A schematic drawing of the micrograph.

inquiry

How can the same gene encode different transcripts?

The spliceosome is the splicing organelle

It is still true that the mature eukaryotic mRNA is colinear with its protein product, but a gene that contains introns is not. Imagine looking at an interstate highway from a satellite. Scattered randomly along the thread of concrete would be cars, some moving in clusters, others individually; most of the road would be bare. That is what a eukaryotic gene is like—scattered exons embedded within much longer sequences of introns.

In humans, only 1 to 1.5% of the genome is devoted to the exons that encode proteins; 24% is devoted to the noncoding introns within which these exons are embedded.

The splicing reaction

The obvious question is, how do eukaryotic cells deal with the noncoding introns? The answer is that the primary transcript is cut and put back together to produce the mature mRNA. The latter process is referred to as **pre-mRNA splicing,** and it occurs in the nucleus prior to the export of the mRNA to the cytoplasm.

The intron–exon junctions are recognized by **small nuclear ribonucleoprotein particles,** called **snRNPs** (pronounced "snurps"). The snRNPs are complexes composed of snRNA and

figure 15.12

PRE-mRNA SPLICING BY THE SPLICEOSOME. Particles called snRNPs contain snRNA that interacts with the 5' end of an intron and with a branch site internal to the intron. Several snRNPs come together with other proteins to form the spliceosome. As the intron forms a loop, the 5' end is cut and linked to a site near the 3' end of the intron. The intron forms a lariat that is excised, and the exons are spliced together. The spliceosome then disassembles and releases the spliced mRNA.

protein. These snRNPs then cluster together with other associated proteins to form a larger complex called the **spliceosome,** which is responsible for the splicing, or removal, of the introns.

For splicing to occur accurately, the spliceosome must be able to recognize intron–exon junctions. Introns all begin with the same 2-base sequence and end with another 2-base sequence that tags them for removal. In addition, within the intron there is a conserved A nucleotide, called the *branch point,* that is important for the splicing reaction (figure 15.12).

The splicing process begins with cleavage of the 5′ end of the intron. This 5′ end becomes attached to the 2′ OH of the branch point A, forming a branched structure called a *lariat* for its resemblance to a cowboy's lariat in a rope (see figure 15.12). The 3′ end of the first exon is then used to displace the 3′ end of the intron, joining the two exons together and releasing the intron as a lariat.

The process of transcription and RNA processing do not occur in a linear sequence, but are rather all part of a concerted process that produces the mature mRNA. The capping reaction occurs during transcription, as does the splicing process. The RNA polymerase II enzyme itself helps to recruit the other factors necessary for modification of the primary transcript, and in this way the process of transcription and pre-mRNA processing forms an integrated system.

Distribution of introns

No rules govern the number of introns per gene or the sizes of introns and exons. Some genes have no introns; others may have 50 introns. The sizes of exons range from a few nucleotides to 7500 nt, and the sizes of introns are equally variable. The presence of introns partly explains why so little of a eukaryotic genome is actually composed of "coding sequences" (see chapter 18 for results from the Human Genome Project).

One explanation for the existence of introns suggests that exons represent functional domains of proteins, and that the intron–exon arrangements found in genes represent the shuffling of these functional units over long periods of evolutionary time. This hypothesis, called *exon shuffling*, was proposed soon after the discovery of introns and has been the subject of much debate over the years.

The recent flood of genomic data has shed light on this issue by allowing statistical analysis of the placement of introns and on intron–exon structure. This analysis has provided support for the exon shuffling hypothesis for many genes; however, it is also clearly not universal, because all proteins do not show this kind of pattern. It is possible that introns do not have a single origin, and therefore cannot be explained by a single hypothesis.

Splicing can produce multiple transcripts from the same gene

One consequence of the splicing process is greater complexity in gene expression in eukaryotes. A single primary transcript can be spliced into different mRNAs by the inclusion of different sets of exons, a process called **alternative splicing.**

Evidence indicates that the normal pattern of splicing is important to an organism's function. It has been estimated that 15% of known human genetic disorders are due to altered splicing. Mutations in the signals for splicing can introduce new splice sites or can abolish normal patterns of splicing. (In chapter 16 we consider how alternative splicing can be used to regulate gene expression.)

Although many cases of alternative splicing have been documented, the recent completion of the draft sequence of the human genome, along with other large data sets of expressed sequences, now allow large-scale comparisons between sequences found in mRNAs and in the genome. Three different computer-based analyses have been performed, producing results that are in rough agreement. These initial genomic assessments indicate a range of 35 to 59% for human genes that exhibit some form of alternative splicing. If we pick the middle ground of around 40%, this result still vastly increases the number of potential proteins encoded by the 25,000 genes in the human genome.

It is important to note that these analyses are primarily computer-based, and the functions of the possible spliced products have been investigated for only a small part of the potentially spliced genes. These analyses, however, do explain how the 25,000 genes of the human genome can encode the more than 80,000 different mRNAs reported to exist in human cells. The emerging field of proteomics addresses the number and functioning of proteins encoded by the human genome.

> Eukaryotic genes contain exon regions that are expressed and intervening sequences, introns, that interrupt the exons. The introns are removed by the spliceosome, leaving the exons joined together. Alternative splicing can generate different mRNAs, and thus different proteins, from the same gene. Recent estimates are that as many as half of human genes may be alternatively spliced.

15.7 The Structure of tRNA and Ribosomes

The ribosome is the key organelle in translation, but it also requires the participation of mRNA, tRNA and a host of other factors. Critical to this process is the interaction of the ribosomes with tRNA and mRNA. To understand this, we first examine the structure of the tRNA adapter molecule and the ribosome itself.

Aminoacyl-tRNA synthetases attach amino acids to tRNA

Each amino acid must be attached to a tRNA with the correct anticodon for protein synthesis to proceed. This connection is accomplished by the action of activating enzymes called **aminoacyl-tRNA synthetases.** One of these enzymes is present for each of the 20 common amino acids.

tRNA structure

Transfer RNA is a bifunctional molecule that must be able to interact with mRNA and with amino acids. The structure of tRNAs is highly conserved in all living systems, and it can be formed into a cloverleaf type of structure based on intramolecular base-pairing that produces double-stranded regions. This primary structure is then folded in space to form an L-shaped molecule that has two functional ends: the **acceptor stem** and the **anticodon loop** (figure 15.13).

2D "Cloverleaf" Model	3D Ribbon-like Model	3D Space-filled Model	Icon

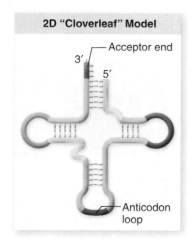 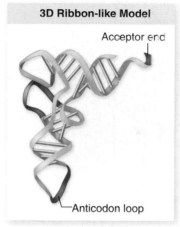  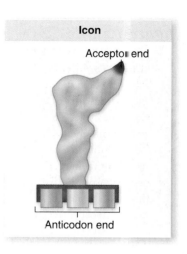

figure 15.13

THE STRUCTURE OF tRNA. Base-pairing within the molecule creates three stem and loop structures in a characteristic cloverleaf shape. The loop at the bottom of the cloverleaf contains the anticodon sequence, which can base-pair with codons in the mRNA. Amino acids are attached to the free, single-stranded —OH end of the acceptor stem. In its final three-dimensional structure, the loops of tRNA are folded into the final L-shaped structure.

The acceptor stem is the 3′ end of the molecule, which always ends in 5′ CCA 3′. The amino acid can be attached to this end of the molecule. The anticodon loop is the bottom loop of the cloverleaf, and it can base-pair with codons in mRNA.

The charging reaction

The aminoacyl-tRNA synthetases must be able to recognize specific tRNA molecules as well as their corresponding amino acids. Although 61 codons code for amino acids, there are actually not 61 tRNAs in cells, although the number varies from species to species. Therefore some aminoacyl-tRNA synthe-

tases must be able to recognize more than one tRNA—but each recognizes only a single amino acid.

The reaction catalyzed by the enzymes is called the tRNA **charging reaction,** and the product is an amino acid joined to a tRNA, now called a *charged tRNA*. An ATP molecule provides energy for this endergonic reaction. The charged tRNA produced by the reaction is an activated intermediate that can undergo the peptide bond-forming reaction without an additional input of energy.

The charging reaction joins the acceptor stem to the carboxyl terminus of an amino acid (figure 15.14). Keeping this

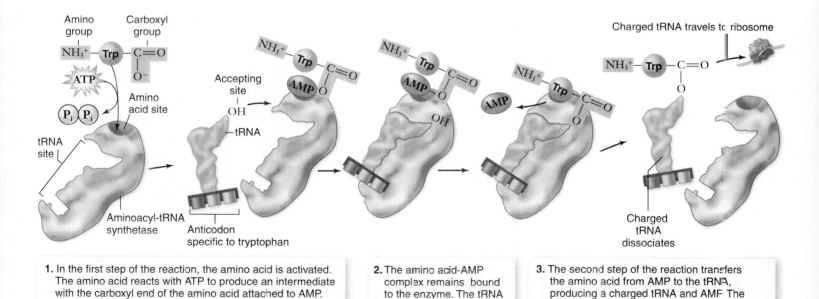

1. In the first step of the reaction, the amino acid is activated. The amino acid reacts with ATP to produce an intermediate with the carboxyl end of the amino acid attached to AMP. The two terminal phosphates (pyrophosphates) are cleaved from ATP in this reaction.

2. The amino acid-AMP complex remains bound to the enzyme. The tRNA next binds to the enzyme.

3. The second step of the reaction transfers the amino acid from AMP to the tRNA, producing a charged tRNA and AMP. The charged tRNA consists of a specific amino acid attached to the 3′ acceptor stem of its RNA.

figure 15.14

tRNA CHARGING REACTION. There are 20 different aminoacyl-tRNA synthetase enzymes each specific for one amino acid, such as tryptophan (Trp). The enzyme must also recognize and bind to the tRNA molecules with anticodons specifying that amino acid, ACC for tryptophan. The reaction uses ATP and produces an activated intermediate that will not require further energy for peptide bond formation.

figure 15.15

RIBOSOMES HAVE TWO SUBUNITS. Ribosome subunits come together and apart as part of a ribosome cycle. The smaller subunit fits into a depression on the surface of the larger one. Ribosomes have three tRNA-binding sites: aminoacyl site (A), peptidyl site (P), and empty site (E).

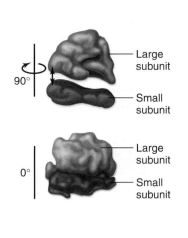

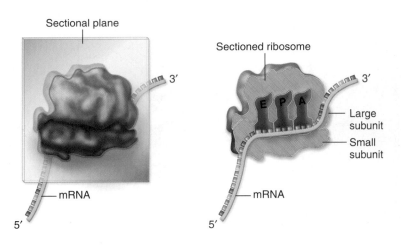

directionality in mind is critical to understanding the function of the ribosome, because each peptide bond will be formed between the amino group of one amino acid and the carboxyl group of another amino acid.

The correct attachment of amino acids to tRNAs is important because the ribosome does not verify this attachment. Ribosomes can only ensure that the codon–anticodon pairing is correct. In an elegant experiment, cysteine was converted chemically to alanine after the charging reaction, when the amino acid was already attached to tRNA. When this charged tRNA was used in an in vitro protein synthesis system, alanine was incorporated in the place of cysteine, showing that the ribosome cannot "proofread" the amino acids attached to tRNA.

In a very real sense, therefore, the charging reaction is the real translation step; amino acids are incorporated into a peptide based solely on the tRNA anticodon and its interaction with the mRNA.

The ribosome has multiple tRNA-binding sites

The synthesis of any biopolymer can be broken down into initiation, elongation, and termination—you have seen this division for DNA replication as well as for transcription. In the case of translation, or protein synthesis, all three of these steps take place on the ribosome, a large macromolecular assembly consisting of rRNA and proteins. Details of the process by which the two ribosome subunits are assembled during initiation are described shortly.

For the ribosome to function it must be able to bind to at least two charged tRNAs at once so that a peptide bond can be formed between their amino acids, as described in the previous overview. In reality, the bacterial ribosome contains three binding sites, summarized in figure 15.15:

- The **P site** (peptidyl) binds to the tRNA attached to the growing peptide chain.
- The **A site** (aminoacyl) binds to the tRNA carrying the next amino acid to be added.
- The **E site** (exit) binds the tRNA that carried the previous amino acid added (see figure 15.15).

Transfer RNAs move through these sites successively during the process of elongation. Relative to the mRNA, the sites are arranged 5′ to 3′ in the order E, P, and A. The incoming charged tRNAs enter the ribosome at the A site, transit through the P site, and then leave via the E site.

The ribosome has both decoding and enzymatic functions

The two functions of the ribosome involve decoding the transcribed message and forming peptide bonds. The decoding function resides primarily in the small subunit of the ribosome. The formation of peptide bonds requires the

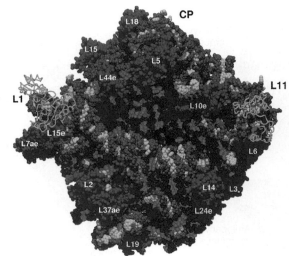

figure 15.16

3-D STRUCTURE OF PROKARYOTIC RIBOSOME. The complete atomic structure of a prokaryotic large ribosomal subunit has been determined at 2.4-Å resolution. Bases of RNA are white, the polynucleotide backbone is red and proteins are blue. The faces of each ribosomal subunit are lined with rRNA such that their interaction with tRNAs, amino acids, and mRNA all involve rRNA. Proteins are absent from the active site but abundant everywhere on the surface. The proteins stabilize the structure by interacting with adjacent RNA strands.

enzyme **peptidyl transferase**, which resides in the large subunit.

Our view of the ribosome has changed dramatically over time. Initially, molecular biologists assumed that the proteins in the ribosome carried out its function, and that the rRNA was a structural scaffold necessary to hold the proteins in the correct position. Now this view has mostly been reversed; the ribosome is seen instead as rRNAs that are held in place by proteins. The faces of the two subunits that interact with each other are lined with rRNA, and the parts of both subunits that interact with mRNA, tRNA, and amino acids are also primarily rRNA (figure 15.16). It is now thought that the peptidyl transferase activity resides in an rRNA in the large subunit.

> The tRNA is a bifunctional molecule with one end that can form a bond to an amino acid and another end that can base-pair with mRNA. The tRNA charging reaction joins the carboxyl end of an amino acid to the 3′ acceptor stem of its tRNA. This reaction is catalyzed by 20 different aminoacyl-tRNA synthetases, one for each amino acid. The ribosome has three different binding sites for tRNA, one for the growing chain (P site), one for the next charged tRNA (A site), and one for the last tRNA used (E site). The ribosome can be thought of as having a decoding function and an enzymatic function.

15.8 The Process of Translation

The process of translation is one of the most complex and energy-expensive tasks that cells perform. An overview of the process, as you saw earlier, is perhaps deceptively simple: The mRNA is threaded through the ribosome, while tRNAs carrying amino acids bind to the ribosome, where they interact with mRNA by base-pairing with the mRNA's codons. The ribosome and tRNAs position the amino acids such that peptide bonds can be formed between each new amino acid and the growing polypeptide.

Initiation requires accessory factors

As mentioned earlier, the start codon is AUG, which also encodes the amino acid methionine. The ribosome usually uses the first AUG it encounters in an mRNA strand to signal the start of translation.

Prokaryotic initiation

In prokaryotes, the **initiation complex** includes a special **initiator tRNA** molecule charged with a chemically modified methionine, *N-formylmethionine*. The initiator tRNA is shown as tRNA^{fMet}. The initiation complex also includes the small ribosomal subunit and the mRNA strand (figure 15.17). The small subunit is positioned correctly on the mRNA due to a conserved sequence in the 5′ end of the mRNA called the **ribosome-binding sequence (RBS)** that is complementary to the 3′ end of a small subunit rRNA.

A number of initiation factors mediate this interaction of the ribosome, mRNA and tRNA^{fMet} to form the initiation complex. These factors are involved in initiation only and are not part of the ribosome.

Once the complex of mRNA, initiator tRNA, and small ribosomal subunit is formed, the large subunit is added, and translation can begin. With the formation of the complete ribosome, the initiator tRNA is bound to the P site with the A site empty.

Eukaryotic initiation

Initiation in eukaryotes is similar, although it differs in two important ways. First, in eukaryotes, the initiating amino acid is methionine rather than *N*-formylmethionine. Second, the initiation complex is far more complicated than in prokaryotes, containing nine or more protein factors, many consisting of several subunits.

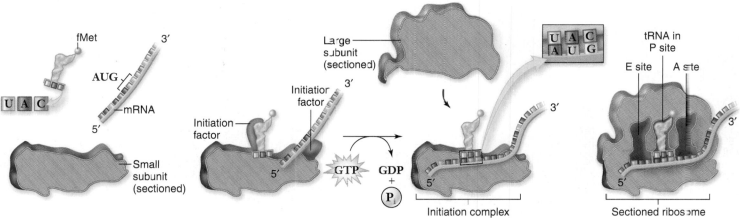

figure 15.17

INITIATION OF TRANSLATION. In prokaryotes, proteins called initiation factors play key roles in positioning the small ribosomal subunit and the initiator tRNA^{fMet}, molecule at the beginning of the mRNA. When the tRNA^{fMet} is positioned over the first AUG codon of the mRNA, the large ribosomal subunit binds, forming the E, P, and A sites where successive tRNA molecules bind to the ribosomes, and polypeptide synthesis begins.

Eukaryotic mRNAs also lack an RBS. The small subunit binds to the mRNA initially by binding to the 5′ cap of the mRNA.

Elongation adds successive amino acids

When the entire ribosome is assembled around the initiator tRNA and mRNA, the second charged tRNA can be brought to the ribosome and bind to the empty A site. This requires an **elongation factor** called **EF-Tu**, which binds to the charged tRNA and to GTP.

A peptide bond can then form between the amino acid of the initiator tRNA and the newly arrived charged tRNA in the A site. The geometry of this bond relative to the two charged tRNAs is critical to understanding the process. Remember that an amino acid is attached to a tRNA by its carboxyl terminus. The peptide bond is formed between the amino end of the incoming amino acid (in the A site) and the carboxyl end of the growing chain (in the P site) (figure 15.18).

The addition of successive amino acids is a series of events that occur in a cyclic fashion. Figure 15.19 shows the details of the elongation cycle.

1. **Matching tRNA anticodon with mRNA codon**
   Each new charged tRNA comes to the ribosome bound to EF-Tu and GTP. The charged tRNA binds to the A site if its anticodon is complementary to the mRNA codon in the A site.

   After binding, GTP is hydrolyzed, and EF-Tu–GDP dissociates from the ribosome where it is recycled by another factor. This two-step binding and hydrolysis of GTP is thought to increase the accuracy of translation since the codon–anticodon pairing can be checked twice.

2. **Peptide bond formation**
   Peptidyl transferase, located in the large subunit, forms a peptide bond between the amino group of the amino acid in the A site and the carboxyl group of the growing

chain. This reaction has the effect of transferring the growing chain to the tRNA in the A site, leaving the tRNA in the P site empty (no longer charged).

3. **Translocation of the ribosome**
   After the peptide bond has been formed, the ribosome moves relative to the mRNA and the tRNAs. The next codon in the mRNA shifts into the A site, and the tRNA with the growing chain moves to the P site. The uncharged tRNA formerly in the P site is now in the E site, and it will be ejected in the next cycle. This translocation step requires the accessory factor EF-G and the hydrolysis of another GTP.

   This elongation cycle continues with each new amino acid added. The ribosome moves down the mRNA in a 5′ to 3′ direction, reading successive codons. The tRNAs move through the ribosome in the opposite direction, from the A site to the P site and finally the E site, before they are ejected as empty tRNAs, which can be charged with another amino acid and then used again.

Wobble pairing

As mentioned, there are fewer tRNAs than codons. This situation is workable because the pairing between the 3′ base of the codon and the 5′ base of the anticodon is less stringent than normal. In some tRNAs, the presence of modified bases with less accurate pairing in the 5′ position of the anticodon enhances this flexibility. This effect is referred to as **wobble pairing** because these tRNAs can "wobble" a bit in the ribosome, so that a single tRNA can "read" more than one codon in the mRNA.

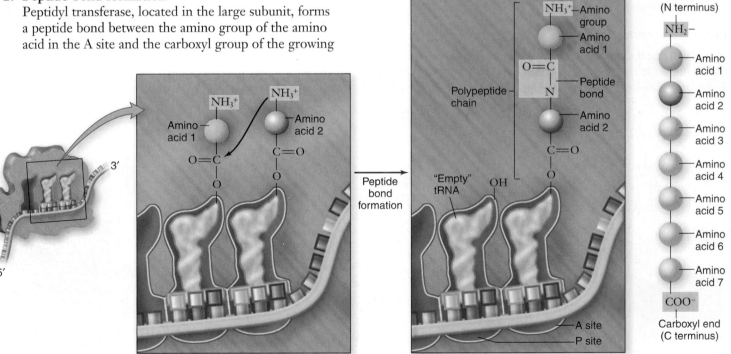

figure 15.18

PEPTIDE BOND FORMATION. Peptide bonds are formed on the ribosome between a "new" charged tRNA in the A site, and the growing chain attached to the tRNA in the P site. The bond forms between the amino group of the new amino acid and the carboxyl group of the growing chain. This transfers the growing chain to the A site as the new amino acid remains attached to its tRNA by its carboxyl terminus.

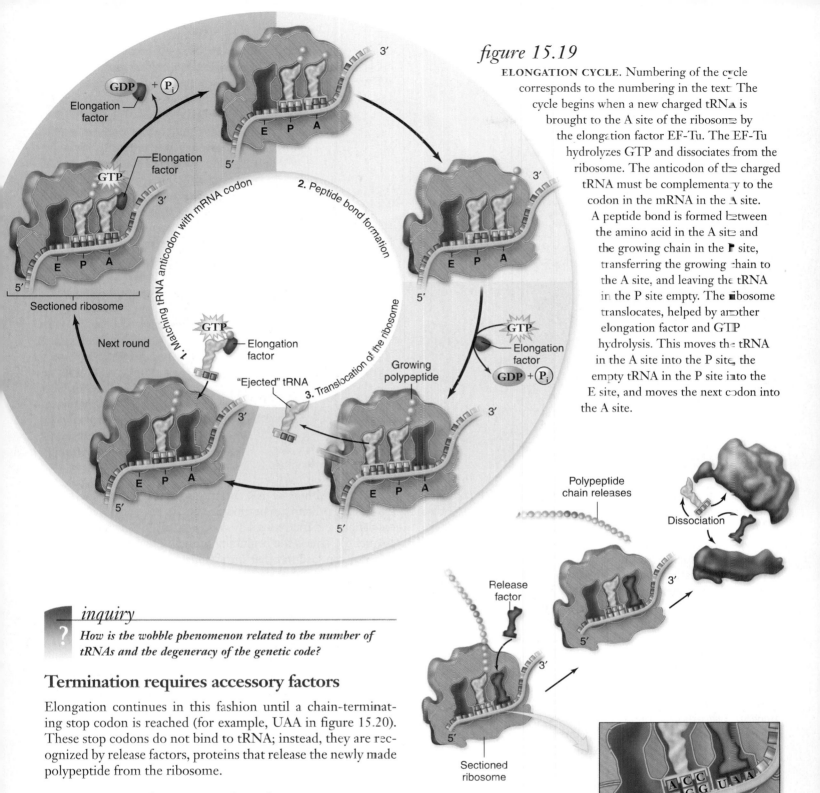

figure 15.19

ELONGATION CYCLE. Numbering of the cycle corresponds to the numbering in the text. The cycle begins when a new charged tRNA is brought to the A site of the ribosome by the elongation factor EF-Tu. The EF-Tu hydrolyzes GTP and dissociates from the ribosome. The anticodon of the charged tRNA must be complementary to the codon in the mRNA in the A site. A peptide bond is formed between the amino acid in the A site and the growing chain in the P site, transferring the growing chain to the A site, and leaving the tRNA in the P site empty. The ribosome translocates, helped by another elongation factor and GTP hydrolysis. This moves the tRNA in the A site into the P site, the empty tRNA in the P site into the E site, and moves the next codon into the A site.

inquiry

How is the wobble phenomenon related to the number of tRNAs and the degeneracy of the genetic code?

Termination requires accessory factors

Elongation continues in this fashion until a chain-terminating stop codon is reached (for example, UAA in figure 15.20). These stop codons do not bind to tRNA; instead, they are recognized by release factors, proteins that release the newly made polypeptide from the ribosome.

Proteins may be targeted to the ER

In eukaryotes, translation can occur either in the cytoplasm or on the rough endoplasmic reticulum (RER). The proteins being translated are targeted to the ER, based on their own initial amino acid sequence. The ribosomes found on the RER are not permanently bound to it.

A polypeptide that starts with a short series of amino acids called a **signal sequence** is specifically recognized and bound by a cytoplasmic complex of proteins called the **signal recognition particle (SRP).** The complex of signal sequence and SRP is in turn recognized by a receptor protein in the ER membrane. The

figure 15.20

TERMINATION OF PROTEIN SYNTHESIS. There is no tRNA with an anticodon complementary to any of the three termination signal codons, such as the UAA codon illustrated here. When a ribosome encounters a termination codon, it therefore stops translocating. A specific protein release factor facilitates the release of the polypeptide chain by breaking the covalent bond that links the polypeptide to the P site tRNA.

figure 15.21

SYNTHESIS OF PROTEINS ON RER. Proteins that are synthesized on RER arrive at the ER because of sequences in the peptide itself. A signal sequence in the amino terminus of the polypeptide is recognized by the signal recognition particle (SRP). This complex docks with a receptor associated with a channel in the ER. The peptide passes through the channel into the lumen of the ER as it is synthesized.

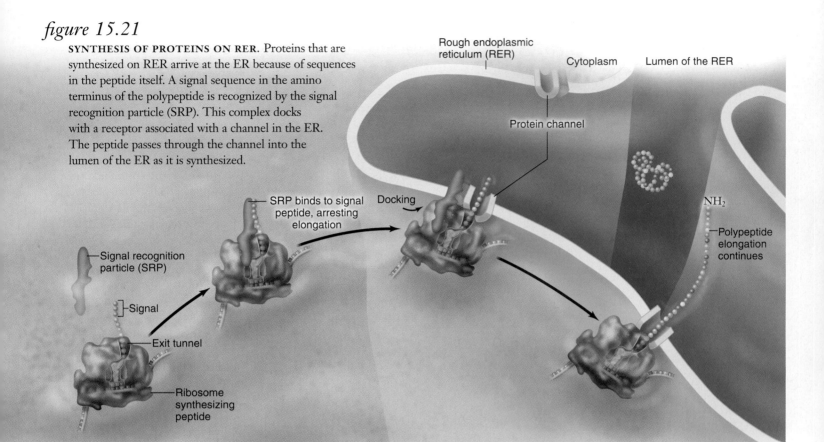

Rough endoplasmic reticulum (RER)
Cytoplasm
Lumen of the RER
Protein channel
SRP binds to signal peptide, arresting elongation
Docking
NH₂
Polypeptide elongation continues
Signal recognition particle (SRP)
Signal
Exit tunnel
Ribosome synthesizing peptide

binding of the ER receptor to the signal sequence/SRP complex holds the ribosome engaged in translation of the protein on the ER membrane, a process called *docking* (figure 15.21).

As the protein is assembled, it passes through a channel formed by the docking complex and into the interior ER compartment, the cisternal space. This is the basis for the docking metaphor—the ribosome is not actually bound to the ER itself, but with the newly synthesized protein entering the ER, the ribosome is like a boat tied to a dock with a rope.

The basic mechanism of protein translocation across membranes by the SRP and its receptor and channel complex has been conserved across all three cell types: eukaryotes, bacteria, and archaea. Given that only eukaryotic cells have an endomembrane system, this universality may seem curious; however, bacteria and archaea both export proteins through their plasma membrane, and the mechanism used is similar to the way in which eukaryotes move proteins into the cisternal space of the ER.

Once within the ER cisternal space, or lumen, the newly synthesized protein can be modified by the addition of sugars (glycosylation) and transported by vesicles to the Golgi apparatus (see chapter 5). This is the beginning of the protein-trafficking pathway that can lead to other intracellular targets, to incorporation into the plasma membrane, or to release outside of the cell itself.

> Translation initiation involves the interaction of the small ribosomal subunit with mRNA, and an initiator tRNA. The elongation cycle involves bringing in new charged tRNAs to the ribosomes A site, forming peptide bonds, and translocating the ribosome along the mRNA. The tRNAs transit through the ribosome from A to P to E sites during the process. In eukaryotes, signal sequences of newly formed polypeptides may target them to be moved to the endoplasmic reticulum, where they enter the cisternal space during synthesis.

15.9 Summarizing Gene Expression

Because of the complexity of the process of gene expression, it is worth stepping back to summarize some key points:

- The process of gene expression converts information in the genotype into the phenotype.
- A copy of the gene in the form of mRNA is produced by transcription, and the mRNA is used to direct the synthesis of a protein by translation.
- Both transcription and translation can be broken down into initiation, elongation, and termination cycles that produce their respective polymers. (The same is true for DNA replication.)
- Eukaryotic gene expression is much more complex than that of prokaryotes.

The nature of eukaryotic genes with their intron and exon components greatly complicates the process of gene expression by requiring additional steps between transcription and translation. The production and processing of eukaryotic mRNAs also takes place in the nucleus, whereas translation takes place in the cytoplasm. This necessitates the transport of the mRNA through the nuclear pores to the cytoplasm before translation can take place. The entire eukaryotic process is summarized in figure 15.22.

A number of differences can be highlighted between gene expression in prokaryotes and in eukaryotes. Table 15.2 (on p. 298) summarizes these main points.

The greater complexity of eukaryotic gene expression is related to the functional organization of the cell, with DNA in the nucleus and ribosomes in the cytoplasm. The differences in gene expression between prokaryotes and eukaryotes is mainly in detail, but some differences have functional significance.

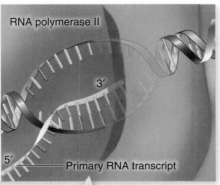

1. The primary transcript is produced by RNA polymerase II in the nucleus. The transcription reaction proceeds in the 5′ to 3′ direction by copying one strand of a DNA template.

RNA polymerase II

3′

5′

Primary RNA transcript

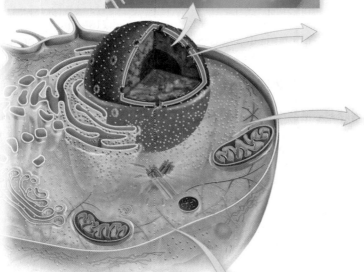

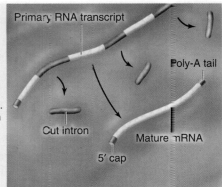

2. The primary transcript is processed to produce the mature mRNA. This involves addition of a 5′ methyl-G cap, cleavage and polyadenylation of the 3′ end and removal of introns by the spliceosome. All of these events occur in the nucleus. The mature mRNA is then exported through nuclear pores to the cytoplasm for translation.

Primary RNA transcript

Poly-A tail

Cut intron

Mature mRNA

5′ cap

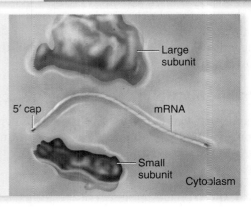

3. The mRNA associates with the ribosome in the cytoplasm. The 5′ cap binds to the small ribosomal subunit to begin the process of initiation. The initiator tRNA and the large subunit are added to complete initiation.

Large subunit

5′ cap

mRNA

Small subunit

Cytoplasm

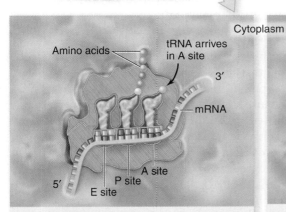

Cytoplasm

Amino acids

tRNA arrives in A site

3′

mRNA

A site

P site

E site

5′

4. Protein synthesis involves the ribosome cycle. The cycle begins with the growing peptide attached to the tRNA in the P site. The next tRNA binds to the A site with its anticodon complementary to the codon in the mRNA in the A site.

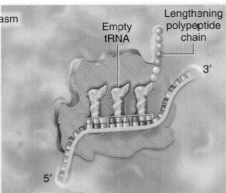

Empty tRNA

Lengthening polypeptide chain

3′

5′

5. Peptide bonds are formed between the amino terminus of the incoming tRNA and the carboxyl terminus of the growing peptide. This breaks the bond between the growing peptide and the tRNA in the P site leaving this tRNA "empty" and shifts the growing chain to the tRNA in the A site.

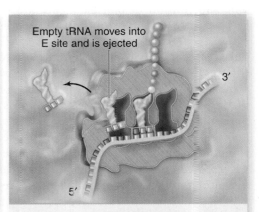

Empty tRNA moves into E site and is ejected

3′

5′

6. Ribosome translocation moves the ribosome relative to the mRNA and its bound tRNAs. This moves the empty tRNA into the E site, the tRNA with the growing peptide into the P site and leaves the A site unoccupied ready to accept the next charged tRNA.

figure 15.22

AN OVERVIEW OF GENE EXPRESSION IN EUKARYOTES.

TABLE 15.2 Differences Between Prokaryotic and Eukaryotic Gene Expression

Characteristic	Prokaryotes	Eukaryotes
Introns	No introns, although some archaeal genes possess them.	Most genes contain introns.
Number of genes in mRNA	Several genes may be transcribed into a single mRNA molecule. Often these have related functions and form an operon. This coordinates regulation of biochemical pathways.	Only one gene per mRNA molecule; regulation of pathways accomplished in other ways.
Site of transcription and translation	No membrane-bounded nucleus, transcription and translation are coupled.	Transcription in nucleus; mRNA moves out of nucleus for translation.
Initiation of translation	Begins at AUG codon preceded by special sequence that binds the ribosome.	Begins at AUG codon preceded by the 5′ cap (methylated GTP) that binds the ribosome.
Modification of mRNA after transcription	None; translation begins before transcription is completed.	A number of modifications while the mRNA is in the nucleus: Introns are removed and exons are spliced together; a 5′ cap is added; a poly-A tail is added.

15.10 Mutation: Altered Genes

One way to analyze the function of genes is to find or to induce mutations in a gene to see how this affects its function. In terms of the organism, however, inducing mutations is usually negative; most mutations have deleterious effects on the phenotype of the organism. In chapter 13, you saw how a number of genetic diseases, such as sickle cell anemia, are due to single base changes. We now consider mutations from the perspective of how the DNA itself is altered.

Point mutations affect a single site in the DNA

A mutation that alters a single base is termed a **point mutation.** The mutation can be either the substitution of one base for another, or the deletion or addition of a single base (or a small number of bases).

Base substitution

The substitution of one base pair for another in DNA is called a **base substitution mutation.** These are sometimes called **missense mutations** as the "sense" of the codon produced after transcription of the mutant gene will be altered (figure 15.23c). These fall into two classes, *transitions* and *transversions*. A transition does not change the type of bases in the base pair, that is, a pyrimidine is substituted for a pyrimidine, or purine for purine. In contrast, a transversion does change the type of bases in a base pair, that is, pyrimidine to purine or the reverse. Because of the degenerate nature of the genetic code, base substitution may or may not alter the amino acid encoded. If the new codon from the base substitution still encodes the same amino acid, we say the mutation is *silent* (figure 15.23b). A variety of human genetic diseases, including sickle cell anemia, are caused by base substitution.

Nonsense mutations

A special category of base substitution arises when a base is changed such that the transcribed codon is converted to a stop codon (figure 15.23d). We call these **nonsense mutations** because the mutation does not make "sense" to the translation apparatus. The stop codon results in premature termination of translation and leads to a truncated protein. How short the resulting protein is depends on where in the gene a stop codon has been introduced.

Frameshift mutations

The addition or deletion of a single base has much more profound consequences than does the substitution of one base for another. These mutations are called **frameshift mutations** because they alter the reading frame in the mRNA downstream of the mutation. This class of mutations was used by Crick and Brenner, as described earlier in the chapter, to infer the nature of the genetic code.

Changing the reading frame early in a gene, and thus in its mRNA transcript, means that the majority of the protein will be altered. Frameshifts also can cause premature termination of translation because 3 in 64 codons are stop codons, which represents a high probability in the sequence that has been randomized by the frameshift.

Triplet repeat expansion mutations

Given the long history of molecular genetics, and the relatively short time that molecular analysis has been possible on humans, it is surprising that a new kind of mutation was discovered in humans. However, one of the first genes isolated that was associated with human disease, the gene for *Huntington disease*, provided a new kind of mutation. The gene for Huntington contains a triplet sequence of DNA that is repeated, and this repeat unit is expanded in the disease allele relative to the normal allele. Since this initial discovery, at least 20 other human genetic diseases appear to be due to this mechanism. The prevalence of this kind of mutation is unknown, but at present humans and mice are the only organisms in which they have been observed, implying that they may be limited to vertebrates, or even mammals. No such mutation has ever been found in *Drosophila* for example.

The expansion of the triplet can occur in the coding region or in noncoding transcribed DNA. In the case of Huntington disease, the repeat unit is actually in the coding region of the gene where the triplet encodes glutamine and expansion results in a polyglutamine region in the protein. A num-

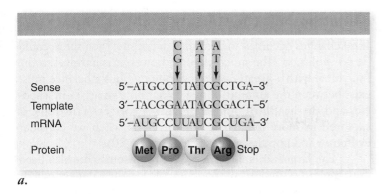

a.

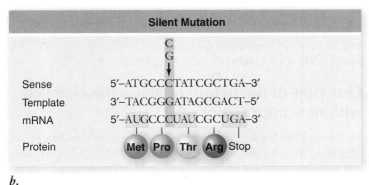

b.

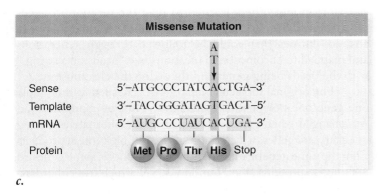

c.

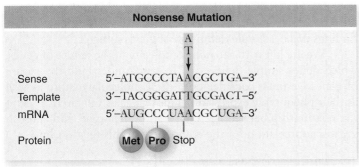

d.

figure 15.23

TYPES OF MUTATIONS. *a.* A hypothetical gene is shown with encoded mRNA and protein. Arrows above the gene indicate sites of mutations described in the rest of the figure. *b.* Silent mutation. A change in the third position of a codon is often silent due to degeneracy in the genetic code. In this case T/A to C/G mutation does not change the amino acid encoded (proline). *c.* Missesense mutation. The G/C to A/T mutation changes the amino acid encoded from arginine to histidine. *d.* Nonsense mutation. The T/A to A/T mutation produces a UAA stop codon in the mRNA.

ber of other neurodegenerative disorders also show this kind of mutation. In the case of fragile-X syndrome, an inherited form of mental retardation, the repeat is in noncoding DNA.

Chromosomal mutations change the structure of chromosomes

Point mutations affect a single site in a chromosome, but more extensive changes can alter the structure of the chromosome itself, resulting in **chromosomal mutations.** Many human cancers are associated with chromosomal abnormalities, so these are of great clinical relevance. We briefly consider possible alterations to chromosomal structure, all of which are summarized in figure 15.24.

Deletions

A **deletion** is the loss of a portion of a chromosome. Frameshifts can be caused by one or more small deletions, but much larger regions of a chromosome may also be lost. If too much information is lost, the deletion is usually fatal to the organism.

One human syndrome that is due to deletion is *cri-du-chat*, which is French for "cry of the cat" after the noise made by children with this syndrome. Cri-du-chat syndrome is caused by a large deletion from the short arm of chromosome 5. It usually results in early death, although many affected individuals show a normal lifespan. It has a variety of effects, including respiratory problems.

Duplications

The **duplication** of a region of a chromosome may or may not lead to phenotypic consequences. Effects depend upon the location of the "breakpoints" where the duplication occurred. If the duplicated region does not lie within a gene, there may be no effect. If the duplication occurs next to the original region, it is termed a *tandem duplication*. These tandem duplications are important in the evolution of families of related genes, such as the globin family that encode the protein hemoglobin.

Inversions

An **inversion** results when a segment of a chromosome is broken in two places, reversed, and put back together. An inversion may not have an effect on phenotype if the sites where the inversion occurs do not break within a gene. In fact, although humans all have the "same" genome, the order of genes in all individuals in a population is not precisely the same due to inversions that occur in different lineages.

Translocations

If a piece of one chromosome is broken off and joined to another chromosome, we call this a **translocation.** Translocations are complex because they can cause problems during meiosis, particularly when two different chromosomes try to pair with each other during meiosis I.

Translocations can also move genes from one chromosomal region to another in a way that changes the expression of genes in the region involved. Two forms of leukemia have been shown to be associated with translocations that move oncogenes into regions of a chromosome where they are expressed inappropriately in blood cells (see chapter 10).

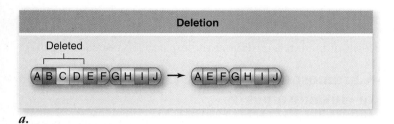

Deletion

a.

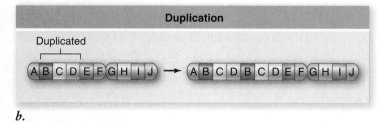

Duplication

b.

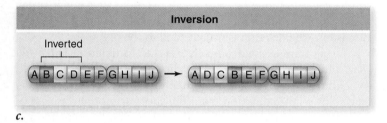

Inversion

c.

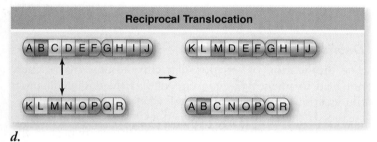

Reciprocal Translocation

d.

figure 15.24

CHROMOSOMAL MUTATIONS. Larger-scale changes in chromosomes are also possible. Material can be deleted *(a)*, duplicated *(b)*, and inverted *(c)*. Translocations occur when one chromosome is broken and becomes part of another chromosome. This often occurs where both chromosomes are broken and exchange material, an event called a reciprocal translocation *(d)*.

Mutations are the starting point of evolution

If no changes occurred in genomes over time, then there could be no evolution. Too much change, however, is harmful to the individual with a greatly altered genome. Thus a balance must exist between the amount of new variation that arises in a species and the health of individuals in the species. This topic is explored in more detail later in the book when we consider evolution and population genetics (chapter 20).

The larger scale alteration of chromosomes has also been important in evolution, although its role is poorly understood. It is clear that gene families arise by the duplication of an ancestral gene, followed by the functional divergence of the duplicated copies. It is also clear that even among closely related species, the number and arrangements of genes on chromosomes can differ. Large-scale rearrangements may have occurred.

Our view of the nature of genes has changed with new information

In this and the preceding chapters, we have seen multiple views of genes. Mendel followed traits determined by what we now call genes in his crosses. The behavior of these genes can be predicted based on the behavior of chromosomes during meiosis. Morgan and others learned to map the location of genes on chromosomes. These findings led to the view of genes as abstract entities that could be followed through generations and mapped to chromosomal locations like "beads on a string," with the beads being genes and the string the chromosome.

The original molecular analysis of genes led to the simple one-gene/one-polypeptide paradigm. This oversimplification was changed when geneticists observed the alternative splicing of eukaryotic genes, which can lead to multiple protein products from the same genetic information. Furthermore, some genes do not encode proteins at all, but only RNA, which can either be a part of the gene expression machinery (rRNA, tRNA, and other forms) or can itself act as an enzyme. Other stretches of DNA are important for regulating genes but are not expressed. All of these findings make a simple definition of genes difficult.

We are left with the rich complexity of the nature of genes, which defies simple definition. To truly understand the nature of genes we must consider both their molecular nature as well as their phenotypic expression. This brings us full circle, back to the relationship between genotype and phenotype, with a much greater appreciation for the complexity of this relationship.

15.1 The Nature of Genes (figure 15.2)

Evidence shows that genetic mutations affect proteins.

- Garrod showed alkaptonuria is due to an altered enzyme.
- Beadle and Tatum showed that genes specify enzymes.
- Information in cells flows from DNA through RNA to proteins.

15.2 The Genetic Code

The order of nucleotides in DNA encodes information to specify the order of amino acids in polypeptides.

- A codon consists of 3 nucleotides. There are $4^3 = 64$ possible codons.
- The code uses adjacent codons with no spaces.
- Three codons signal "stop," one codon signals "start" and also encodes methionine. Thus 61 codons encode the 20 amino acids.
- The code is degenerate (usually in the third position) but specific.
- The code is essentially universal.

15.3 Overview of Gene Expression

Transcription produces an RNA copy from a DNA template, and translation uses the RNA template to direct the synthesis of a protein.

- The strand copied during transcription is called the template strand.
- Transcription of RNA involves initiation at a promoter, elongation of the transcript, and termination at a terminator site.
- Translation involves formation of initiation complex, elongation by addition of amino acids, and termination at a stop codon.
- RNA plays multiple roles in gene expression (see pps. 283, 284).

15.4 Prokaryotic Transcription

Prokaryote gene expression is similar to that in eukaryotes, with some important differences.

- Prokaryotes have a single RNA polymerase that exists in two forms: core polymerase and holoenzyme.
- Core polymerase can synthesize RNA. Holoenzyme, core plus σ factor, can initiate RNA at a promoter (figure 15.5).
- A transcription unit begins with a promoter, contains one or more genes, and ends with a terminator.
- RNA polymerase unwinds a short region of DNA at promoters.
- Transcription of the mRNA chain grows in the 5' to 3' direction.
- A transcription bubble contains RNA polymerase, DNA template, and the growing mRNA transcript (figure 15.6).
- Terminators consist of complementary sequences that form a double-stranded hairpin loop where the polymerase pauses (figure 15.7).
- In prokaryotes the mRNA is translated into a polypeptide while it is being synthesized (transcription–translation coupling).

15.5 Eukaryotic Transcription

The transcription reaction in eukaryotes is the same as prokaryotes, but there are some distinct differences.

- Eukaryotes have three RNA polymerases: I transcribes rRNA; II transcribes mRNA and some snRNAs; III transcribes tRNA.
- Transcription by RNA polymerase II requires a host of transcription factors to form an initiation complex at the promoter.
- RNA polymerase III transcribes tRNA, and its promoters are found internal to the gene, not at the 5' end.
- In eukaryotes, the primary RNA transcript is modified (figure 15.10).
- *A methyl-GTP cap is added to the 5' end.*
- *A 3' poly-A tail is added by polyA polymerase at a specific site.*
- *Noncoding internal regions are also removed by splicing.*

15.6 Eukaryotic pre-mRNA Splicing

In eukaryotes introns are removed by splicing (figure 15.12).

- Coding DNA (exons) is interrupted by noncoding introns.
- Intron–exon junctions are recognized by snRNPs.
- The snRNPs recruit a larger complex called the spliceosome.
- During splicing the 5' end of the intron is cut and becomes bound to the branch site, forming a structure called a lariat.
- The 3' end of the first exon is joined to the 5' end of the next exon.
- One transcript can produce different mRNAs by alternative splicing.

15.7 The Structure of tRNA and Ribosomes

Although the ribosome is a key organelle in translation, it requires the participation of mRNA, tRNA, and other factors.

- The charging reaction attaches the carboxyl terminus of an amino acid to the 3' end of the correct tRNA (figure 15.14).
- This is catalyzed by enzymes called aminoacyl-tRNA synthetases.
- The anticodon loop of tRNAs can base-pair to codons in mRNA.
- The ribosome consists of two subunits: large and small.
- The small subunit binds to mRNA and is involved in decoding, while the large subunit contains the enzyme peptidyl transferase.
- The ribosome has three tRNA-binding sites (figure 15.15).
- *The P site binds to tRNA attached to the growing peptide chain.*
- *The A site binds to the tRNA carrying the next amino acid to be added.*
- *The E site binds to the tRNA that carried the previous amino acid.*

15.8 The Process of Translation

Protein synthesis is complex and energetically expensive.

- In prokaryotes the initiation complex forms with the small ribosomal subunit, mRNA, and a special initiator tRNA.
- The RBS in prokaryotic mRNA is complementary to rRNA in the small subunit. Eukaryotes use the 5' cap for the same function.
- Peptide bonds form between the amino end of the new amino acid and the carboxyl end of the growing chain (figure 15.18).
- Protein synthesis involves a cycle of events (figure 15.19):
- *New charged tRNAs are brought to the ribosomes by EF-Tu.*
- *A peptide bond forms between new amino acid and growing chain.*
- *The ribosome moves relative to the mRNA and bound tRNAs.*
- One tRNA can bind multiple codons by wobble pairing.
- Stop codons are recognized by termination factors.
- Proteins targeted to the ER have a signal sequence in their amino terminus that binds the SRP, and this complex docks the ribosome.
- The signal recognition particle (SRP) binds the signal sequence, and this complex is recognized by a receptor protein in the ER.

15.9 Summarizing Gene Expression

Gene expression converts information in the genome into proteins. This process differs between prokaryotes and eukaryotes (figure 15.22).

15.10 Mutation: Altered Genes

Mutations can be used to understand the function of genes.

- Point mutations involve the alteration of a single base.
- Nonsense mutations convert codons into stop codons.
- Frameshift mutations involve the addition or deletion of a base.
- Triplet-repeat expansion mutations can cause genetic diseases.
- Chromosomal mutations alter the structure of chromosomes.
- Mutations are the starting point of evolution.

SELF TEST

1. The experiments with nutritional mutants in *Neurospora* by Beadle and Tatum provided evidence that—
   a. Bread mold can be grown in a lab on minimal media.
   b. X-rays can damage DNA.
   c. Cells need enzymes.
   d. Genes specify enzymes.

2. What is the *central dogma* of molecular biology?
   a. DNA is the genetic material.
   b. Information passes from DNA to protein.
   c. Information passes from DNA to RNA to protein.
   d. One gene encodes only one polypeptide.

3. The manufacture of new proteins is termed _____, and the production of a messenger RNA corresponding to a specific gene is called _____.
   a. translation; transcription
   b. termination; translation
   c. transcription; translation
   d. transfer; translation

4. Each amino acid in a protein is specified by—
   a. multiple genes
   b. a promoter
   c. a codon
   d. a molecule of mRNA

5. The TATA box in eukaryotes is part of a—
   a. core promoter
   b. −35 sequence
   c. −10 sequence
   d. 5′ cap

6. What is the *coding strand*?
   a. The single DNA strand copied to produce a molecule of RNA
   b. The single-stranded RNA molecule that is transcribed from the DNA
   c. The DNA strand that is not copied to synthesize a molecule of RNA
   d. The region of a chromosome that contains a gene

7. An anticodon would be found on which of the following types of RNA?
   a. snRNA (small nuclear RNA)
   b. mRNA (messenger RNA)
   c. tRNA (transfer RNA)
   d. rRNA (ribosomal RNA)

8. RNA polymerase binds to a _____ to initiate _____.
   a. mRNA; translation
   b. promoter; transcription
   c. primer; transcription
   d. transcription factor; translation

9. Which of the following functions as a "stop" signal for a prokaryotic RNA polymerase?
   a. Formation of a transcription bubble
   b. Addition of a long chain of adenine nucleotides to the 3′ end
   c. Addition of a 5′ cap
   d. Formation of a GC hairpin

10. An *exon* is a sequence of RNA that—
    a. codes for protein
    b. is removed through the action of a spliceosome
    c. is part of a noncoding DNA sequence
    d. Both (b) and (c) are correct.

11. The job of a ribosome during translation can best be described as—
    a. targeting proteins to the rough endoplasmic reticulum
    b. determining the sequence of amino acids
    c. carrying amino acids to the mRNA
    d. catalyzing peptide bond formation between amino acids

12. What is the function of the *signal sequence*?
    a. It initiates transcription by triggering RNA polymerase binding.
    b. It initiates translation.
    c. It is the binding site of signal recognition particle.
    d. It signals the end of translation, resulting in the disassembly of the ribosome.

13. How can a point mutation lead to a nonsense mutation?
    a. Changing a single base has no effect on the protein.
    b. Changing a single base leads to a premature termination of translation.
    c. Changing a single base within a codon from an A to a C.
    d. The addition or deletion of a base alters the reading frame for the gene.

14. Which of the following is a consequence of a translocation?
    a. Genes move from one chromosome to another.
    b. RNA polymerase produces a molecule of mRNA.
    c. A molecule of mRNA interacts with a ribosome to produce a protein.
    d. A segment of a chromosome is broken, reversed, and reinserted.

15. What is the relationship between mutations and evolution?
    a. Mutations make genes better.
    b. Mutations can create new alleles.
    c. Mutations happened early in evolution, but not now.
    d. There is no relationship between evolution and genetic mutations.

CHALLENGE QUESTIONS

1. A template strand of DNA has the following sequence:
   3′ – CGTTACCCGAGCCGTACGATTAGG – 5′
   Use the sequence information to determine—
   a. the predicted sequence of the mRNA for this gene
   b. the predicted amino acid sequence of the protein

2. Describe how each of the following mutations will affect the final protein product. Name the type of mutation.
   Original template strand:
   3′ – CGTTACCCGAGCCGTACGATTAGG – 5′
   a. 3′ – CGTTACCCGAGCCGTAACGATTAGG – 5′
   b. 3′ – CGTTACCCGATCCGTACGATTAGG – 5′
   c. 3′ – CGTTACCCGAGCCGTTCGATTAGG – 5′

3. Predict whether gene expression (from initiation of transcription to final protein product) would be faster in a prokaryotic or eukaryotic cell. Explain your answer.

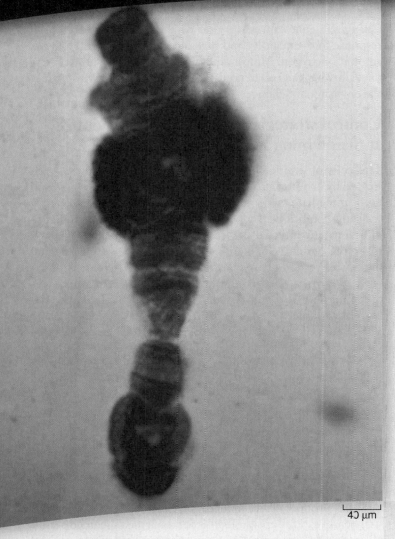

40 μm

chapter 16

Control of Gene Expression

introduction

IN MUSIC, DIFFERENT INSTRUMENTS PLAY their own parts at different times during a piece; a musical score determines which instruments play when. Similarly, in an organism different genes are expressed at different times, with a "genetic score," written in regulatory regions of the DNA, determining which genes are active when. The picture shows the expanded "puff" of this *Drosophila* chromosome, which represents genes that are being actively expressed. Gene expression and how it is controlled is our topic in this chapter.

16.1 Control of Gene Expression

Control of gene expression is essential to all organisms. In prokaryotes, it allows the cell to take advantage of changing environmental conditions. In multicellular eukaryotes, it is critical for directing development and maintaining homeostasis.

Control usually occurs at the level of transcription initiation

You learned in the previous chapter that gene expression is the conversion of genotype to phenotype—the flow of information from DNA to produce functional proteins that control cellular activities. We could envision controlling this process at any step along the way, and in fact, examples of control occur at most steps. The most logical place to control this process, however, is at the first step: production of mRNA from DNA by transcription.

Transcription itself could be controlled at any step, but again, the beginning is the most logical place. Although cells do not always behave in ways that conform to human logic, control of the initiation of transcription is common.

RNA polymerase is key to transcription, and it must have access to the DNA helix and must be capable of binding to the gene's promoter for transcription to begin. **Regulatory proteins** act by modulating the ability of RNA polymerase to bind to the promoter. This idea of controlling the access of RNA polymerase to a promoter is common to both prokaryotes and eukaryotes, but the details differ greatly, as you will see.

These regulatory proteins bind to specific nucleotide sequences on the DNA that are usually only 10–15 nt in length. (Even a large regulatory protein has a "footprint," or binding area, of only about 20 nt.) Hundreds of these regulatory sequences have been characterized, and each provides a binding site for a specific protein that is able to recognize the sequence. Binding of the protein either *blocks* transcription by getting in the way of RNA polymerase or *stimulates* transcription by facilitating the binding of RNA polymerase to the promoter.

Control strategies in prokaryotes are geared to adjust to environmental changes

Control of gene expression is accomplished very differently in prokaryotes than it is in eukaryotes. Prokaryotic cells have been shaped by evolution to grow and divide as rapidly as possible, enabling them to exploit transient resources. Proteins in prokaryotes turn over rapidly, allowing these organisms to respond quickly to changes in their external environment by changing patterns of gene expression.

In prokaryotes, the primary function of gene control is to adjust the cell's activities to its immediate environment.

Changes in gene expression alter which enzymes are present in response to the quantity and type of available nutrients and the amount of oxygen. Almost all of these changes are fully reversible, allowing the cell to adjust its enzyme levels up or down as the environment changes.

Control strategies in eukaryotes are aimed at maintaining homeostasis

The cells of multicellular organisms, in contrast, have been shaped by evolution to be protected from transient changes in their immediate environment. Most of them experience fairly constant conditions. Indeed, **homeostasis**—the maintenance of a constant internal environment—is considered by many to be the hallmark of multicellular organisms. Cells in such organisms respond to signals in their immediate environment (such as growth factors and hormones) by altering gene expression, and in doing so they participate in regulating the body as a whole.

Some of these changes in gene expression compensate for changes in the physiological condition of the body. Others mediate the decisions that actually produce the body, ensuring that the correct genes are expressed in the right cells at the right time during development. Later chapters deal with the details, but for now we can simplify by saying that the growth and development of multicellular organisms entail a long series of biochemical reactions, each catalyzed by a specific enzyme. Once a particular developmental change has occurred, these enzymes cease to be active, lest they disrupt the events that must follow.

To produce this sequence of enzymes, genes are transcribed in a carefully prescribed order, each for a specified period of time, following a fixed genetic program that may even lead to programmed cell death (**apoptosis**). The one-time expression of the genes that guide a developmental program is fundamentally different from the reversible metabolic adjustments prokaryotic cells make to the environment. In all multicellular organisms, changes in gene expression within particular cells serve the needs of the whole organism, rather than the survival of individual cells.

Unicellular eukaryotes also use different control mechanisms than prokaryotes. All eukaryotes have a membrane-bounded nucleus, use similar mechanisms to condense DNA into chromosomes, and have the same gene expression machinery, all of which differ from those of prokaryotes.

> Gene expression is usually controlled at the level of transcription initiation. Regulatory proteins that can bind to specific sites on DNA affect the binding of RNA polymerase to promoters. Prokaryotes and eukaryotes differ in the details of this process.

Regulatory Proteins

The ability of certain proteins to bind to *specific* DNA regulatory sequences provides the basic tool of gene regulation, the key ability that makes transcriptional control possible. To understand how cells control gene expression, it is first necessary to gain a clear picture of this molecular recognition process.

Proteins can interact with DNA through the major groove

Molecular biologists formerly thought that the DNA helix had to unwind before proteins could distinguish one DNA sequence from another; only in this way, they reasoned, could regulatory proteins gain access to the hydrogen bonds between base-pairs. We now know it is unnecessary for the helix to unwind because proteins can bind to its outside surface, where the edges of the base-pairs are exposed.

Careful inspection of a DNA molecule reveals two helical grooves winding around the molecule, one deeper than the other. Within the deeper groove, called the **major groove**, the nucleotides' hydrogen bond donors and acceptors are accessible. The pattern created by these chemical groups is unique for each of the four possible base-pair arrangements, providing a ready way for a protein nestled in the groove to read the sequence of bases (figure 16.1).

DNA-binding domains interact with specific DNA sequences

Protein–DNA recognition is an area of active research; so far, the structures of over 30 regulatory proteins have been analyzed. Although each protein is unique in its fine details, the part of the protein that actually binds to the DNA is much less variable. Almost all of these proteins employ one of a small set of **DNA-binding motifs.** A motif, as described in chapter 3, is a form of three-dimensional substructure that is found in many proteins. These DNA-binding motifs share the property of interacting with specific sequences of bases, usually through the major groove of the DNA helix.

DNA-binding motifs are the key structure within the DNA-binding domain of these proteins. This domain is a functionally distinct part of the protein necessary to bind to DNA in a sequence-specific manner. Regulatory proteins also need to be able to interact with the transcription apparatus, which is accomplished by a different regulatory domain.

Note that two proteins that share the same DNA-binding domain do not necessarily bind to the same DNA sequence. The similarities in the DNA-binding motifs appear in their 3-D structure, and not in the specific contacts that they make with DNA.

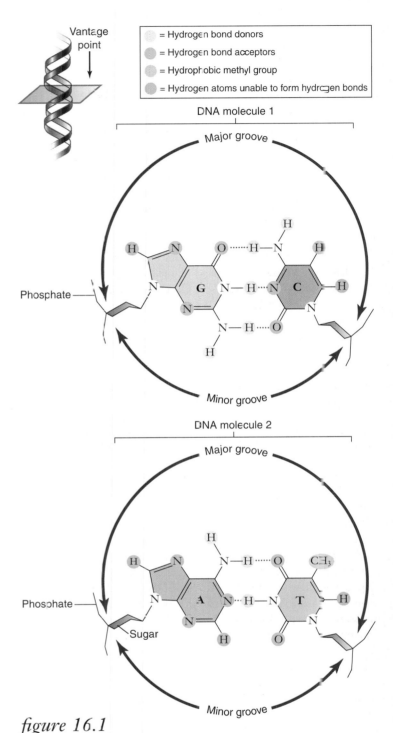

figure 16.1

READING THE MAJOR GROOVE OF DNA. Looking down into the major groove of a DNA helix, we can see the edges of the bases protruding into the groove. Each of the four possible base-pair arrangements (two are shown here) extends a unique set of chemical groups into the groove, indicated in this diagram by differently colored circles. A regulatory protein can identify the base-pair arrangement by this characteristic signature.

figure 16.2

MAJOR DNA-BINDING MOTIFS. A number of common DNA-binding motifs are pictured interacting with DNA. *a.* The helix-turn-helix motif illustrated here binds to DNA using one α helix, the recognition helix, to interact with the major groove and the other to position the recognition helix. Proteins with this motif are usually dimers, with two identical subunits, each containing the DNA-binding motif. The two copies of the motif (*red*) are separated by 3.4 nm, precisely the spacing of one turn of the DNA helix. This allows the regulatory proteins to slip into two adjacent portions of the major groove in DNA, providing a strong attachment. *b.* The homeodomain motif is common in proteins that regulate development and shares some structural similarity to the helix-turn-helix in (*a*). *c.* The zinc finger motif has two α helices that interact with the major groove. These DNA-binding motifs act like the fingers of a hand holding the DNA. *d.* The leucine zipper acts to hold two subunits in a multisubunit protein together allowing α-helical regions to interact with DNA.

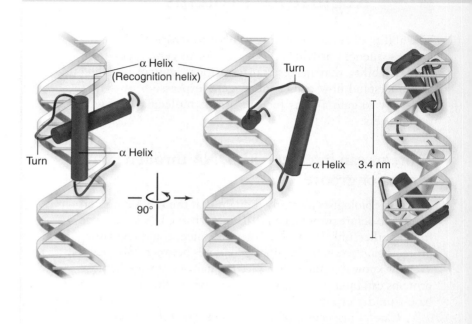

The Helix-Turn-Helix Motif

a.

Several common DNA-binding motifs are shared by many proteins

A limited number of common DNA-binding motifs have been described that are found in a wide variety of different proteins. Four of the best known are detailed in the following sections to give the flavor of how DNA-binding proteins interact with DNA.

The helix-turn-helix motif

The most common DNA-binding motif is the **helix-turn-helix,** constructed from two α-helical segments of the protein linked by a short, nonhelical segment, the "turn" (figure 16.2*a*). As the first motif recognized, the helix-turn-helix motif has since been identified in hundreds of DNA-binding proteins.

A close look at the structure of a helix-turn-helix motif reveals how proteins containing such motifs interact with the major groove of DNA. The helical segments of the motif interact with one another, so that they are held at roughly right angles. When this motif is pressed against DNA, one of the helical segments (called the *recognition helix*) fits snugly in the major groove of the DNA molecule, and the other butts up against the outside of the DNA molecule, helping to ensure the proper positioning of the recognition helix.

Most DNA regulatory sequences recognized by helix-turn-helix motifs occur in symmetrical pairs. Such sequences are bound by proteins containing two helix-turn-helix motifs separated by 3.4 nanometers (nm), the distance required for one turn of the DNA helix (figure 16.2*a*). Having *two* protein–DNA-binding sites doubles the zone of contact between protein and DNA and greatly strengthens the bond between them.

The homeodomain motif

A special class of helix-turn-helix motifs, the **homeodomain,** plays a critical role in development in a wide variety of eukaryotic organisms, including humans. These motifs were discovered when researchers began to characterize a set of homeotic mutations in *Drosophila* (mutations that cause one body part to be replaced by another). They found that the mutant genes encoded regulatory proteins. Normally these proteins would initiate key stages of development by binding to developmental switch-point genes. More than 50 of these regulatory proteins have been analyzed, and they all contain a nearly identical sequence of 60 amino acids, which was termed the *homeodomain* (figure 16.2*b*). The most conserved part of the homeodomain contains a recognition helix of a helix-turn-helix motif. The rest of the homeodomain forms the other two helices of this motif.

The zinc finger motif

A different kind of DNA-binding motif uses one or more zinc atoms to coordinate its binding to DNA. Called **zinc fingers** (figure 16.2*c*), these motifs exist in several forms. In one form, a zinc atom links an α-helical segment to a β-sheet segment (chapter 3) so that the helical segment fits into the major groove of DNA.

This sort of motif often occurs in clusters, the β sheets spacing the helical segments so that each helix contacts the major groove. The more zinc fingers in the cluster, the stronger the protein binds to the DNA. In other forms of the zinc finger motif, the β sheet's place is taken by another helical segment.

The leucine zipper motif

In yet another DNA-binding motif, two different protein subunits cooperate to create a single DNA-binding site. This motif is created where a region on one subunit containing several hydrophobic amino acids (usually leucines) interacts with a similar region on the other subunit. This interaction holds the two subunits together at those regions, while the rest of

The Homeodomain Motif	The Zinc Finger Motif	The Leucine Zipper Motif

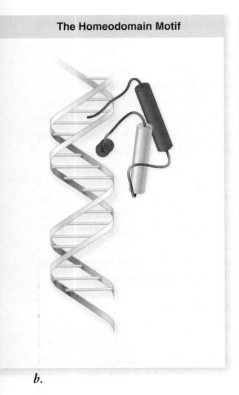

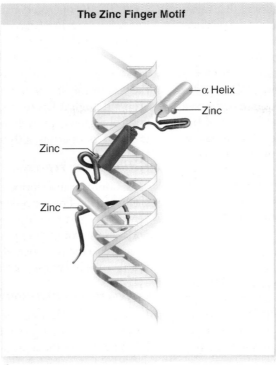

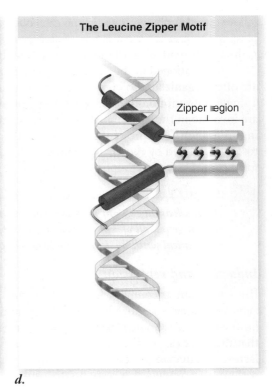

b. *c.* *d.*

the subunits remain separated. Called a **leucine zipper,** this structure has the shape of a Y, with the two arms of the Y being helical regions that fit into the major groove of DNA (figure 16.2*d*). Because the two subunits can contribute quite different helical regions to the motif, leucine zippers allow for great flexibility in controlling gene expression.

> Regulatory proteins must be able to bind to DNA to affect transcription. These proteins all contain one of a relatively small set of common DNA-binding motifs. These form the active part of the DNA-binding domain, and another domain of the protein interacts with the transcription apparatus.

16.3 Prokaryotic Regulation

The details of regulation can be revealed by examining mechanisms used by prokaryotes to control the initiation of transcription. Prokaryotes and eukaryotes share some common themes, but they have some profound differences as well. Later on we discuss eukaryotic systems and concentrate on how they differ from the simpler prokaryotic systems.

Control of transcription can be either positive or negative

Control at the level of transcription initiation can be either positive or negative. **Positive control** increases the frequency of initiation, and **negative control** decreases the frequency of initiation. Each of these forms of control are mediated by regulatory proteins, but the proteins have opposite effects.

Negative control by repressors

Negative control is mediated by proteins called **repressors.** Repressors are proteins that bind to regulatory sites on DNA called **operators** to prevent or decrease the initiation of transcription. They act as a kind of roadblock to prevent the polymerase from initiating effectively.

Repressors do not act alone; each responds to specific effector molecules. Effector binding can alter the conforma-

tion of the repressor to either enhance or abolish its binding to DNA. These repressor proteins are allosteric proteins with an active site that binds DNA and a regulatory site that binds effectors. Binding changes the conformation of allosteric proteins, as described in chapter 6.

Positive control by activators

Positive control is mediated by another class of regulatory, allosteric proteins called **activators** that can bind to DNA and stimulate the initiation of transcription. These activators enhance the binding of RNA polymerase to the promoter to increase the level of transcription initiation.

Activators are the logical and physical opposites to repressors. Effector molecules can either enhance or decrease activator binding.

Prokaryotes adjust gene expression in response to environmental conditions

Changes in the environments that bacteria and archaea encounter often result in changes in gene expression. In general, genes encoding proteins involved in catabolic pathways (breaking down molecules) respond oppositely from genes encoding proteins involved in anabolic pathways (building up molecules). In the discussion that follows, we describe enzymes in the catabolic pathway that transports and utilizes the sugar lactose. Later we describe the anabolic pathway that synthesizes the amino acid tryptophan.

As mentioned in the preceding chapter, prokaryotic genes are often organized into operons, multiple genes that are part of a single transcription unit having a single promoter. Genes that are involved in the same metabolic pathway are often organized in this fashion. The proteins necessary for the utilization of lactose are encoded by the **lac operon,** and the proteins necessary for the synthesis of tryptophan are encoded by the **trp operon.**

inquiry

What advantage might it be to a bacterium to link several genes, all of whose products contribute to a single biochemical pathway, into a single operon?

Induction and repression

If a bacterium encounters lactose, it begins to make the enzymes necessary to utilize lactose. When lactose is not present, however, there is no need to make these proteins. Thus, we say that the synthesis of the proteins is *induced* by the presence of lactose. **Induction** therefore occurs when enzymes for a certain pathway are produced in response to a substrate.

When tryptophan is available in the environment, a bacterium will not synthesize the enzymes necessary to make tryptophan. If tryptophan ceases to be available, then the bacterium begins to make these enzymes. **Repression** occurs when bacteria capable of making biosynthetic enzymes do not produce them. In the case of both induction and repression, the bacterium is adjusting to produce the enzymes that are optimal for its immediate environment.

Negative control

Knowing that control of gene expression is probably at the level of initiation of transcription does not tell us the nature of the control—it might be either positive or negative. On the surface, repression may appear to be negative and induction positive; but in the case of both the *lac* and *trp* operons, control is negative by a repressor protein. The key is that the effector proteins have opposite effects on the repressor in induction with those seen in repression.

For either mechanism to work, the molecule in the environment, such as lactose or tryptophan, must produce the proper effect on the gene being regulated. In the case of *lac* induction, the presence of lactose must *prevent* a repressor protein from binding to its regulatory sequence. In the case of *trp* repression, by contrast, the presence of tryptophan must *cause* a repressor protein to bind to its regulatory sequence.

These responses are opposite because the needs of the cell are opposite for anabolic versus catabolic pathways. Each pathway is examined in detail in the following sections to show how protein–DNA interactions allow the cell to respond to environmental conditions.

The *lac* operon is negatively regulated by the *lac* repressor

The control of gene expression in the *lac* operon was elucidated by the pioneering work of Jaques Monod and François Jacob.

The *lac* operon consists of the genes that encode functions necessary to utilize lactose: β-galactosidase (*lacZ*), lactose permease (*lacY*), and lactose transacetylase (*lacA*), plus the regulatory regions necessary to control the expression of these genes (figure 16.3). In addition, the gene for the *lac* repressor (*lacI*) is linked to the rest of the *lac* operon and is thus considered part of the operon although it has its own promoter. The arrangement of the control regions upstream of the coding region is typical of most prokaryotic operons, although the linked repressor is not.

Action of the repressor

Initiation of transcription of the *lac* operon is controlled by the *lac* repressor. The repressor binds to the operator, which is adjacent to the promoter (figure 16.4a). This binding prevents RNA polymerase from binding to the promoter. This DNA binding is sensitive to the presence of lactose: The repressor binds DNA in the absence of lactose, but not in the presence of lactose.

Interaction of repressor and effector

In the absence of lactose, the *lac* repressor binds to the operator, and the operon is repressed (figure 16.4a). The effector that controls the DNA binding of the repressor is a metabolite of lactose, allolactose, which is produced when lactose is available. Allolactose binds to the repressor, altering its conformation so that it no longer can bind to the operator (figure 16.4b). Induction of the operon begins.

As the level of lactose falls, allolactose will no longer be available to bind to the repressor, allowing the repressor to bind to DNA again. Thus this system of negative control by the *lac* repressor and its effector, allolactose, allow the cell to respond to changing levels of lactose in the environment.

Even in the absence of lactose, the *lac* operon is expressed at a very low level. When lactose becomes available, it is transported into the cell and enough allolactose is produced that induction of the operon can occur.

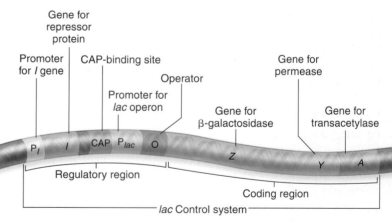

figure 16.3

THE *LAC* REGION OF THE *ESCHERICHIA COLI* CHROMOSOME. The *lac* operon consists of a promoter, an operator, and three genes (*lac Z, Y,* and *A*) that code for proteins required for the metabolism of lactose. In addition, there is a binding site for the catabolite activator protein (CAP), which affects RNA polymerase binding to the promoter. The *I* gene encodes the repressor protein, which will bind to the operator and block transcription of the *lac* genes.

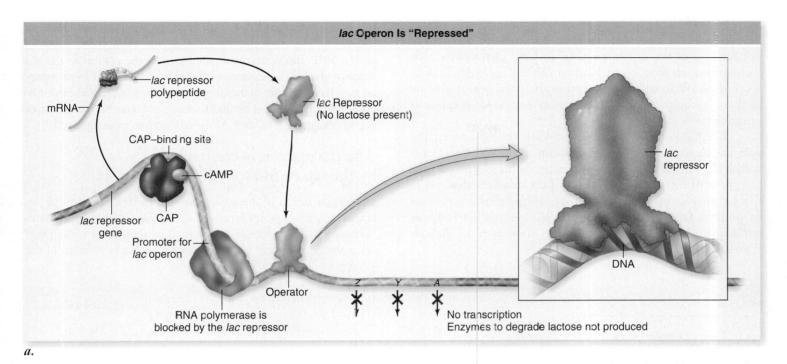

lac Operon Is "Repressed"

lac repressor
polypeptide

mRNA

CAP–binding site

cAMP

lac repressor
gene

CAP

Promoter for
lac operon

Operator

RNA polymerase is
blocked by the *lac* repressor

lac Repressor
(No lactose present)

lac
repressor

DNA

No transcription
Enzymes to degrade lactose not produced

a.

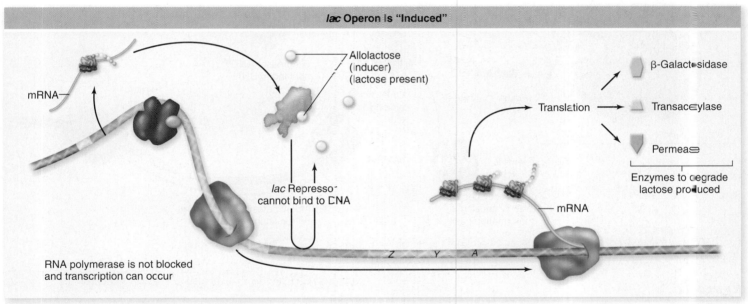

lac Operon Is "Induced"

mRNA

Allolactose
(inducer)
(lactose present)

lac Repressor
cannot bind to DNA

RNA polymerase is not blocked
and transcription can occur

Z   Y   A

mRNA

Translation

β-Galactosidase

Transacetylase

Permease

Enzymes to degrade
lactose produced

b.

figure 16.4

INDUCTION OF THE *LAC* OPERON. *a.* The *lac* repressor. Because the repressor fills the major groove of the DNA helix, RNA polymerase cannot fully attach to the promoter, and transcription is blocked. When the repressor protein is bound to the operator site the *lac* operon is shut down (repressed). Because promoter and operator sites overlap, RNA polymerase and the repressor cannot functionally bind at the same time, any more than two cars can occupy the same parking space. *b.* The *lac* operon is transcribed (induced) when CAP is bound and when the repressor is not bound. Allolactose binding to the repressor alters the repressor's shape such that it cannot bind to the operator site and block RNA polymerase activity.

The presence of glucose prevents induction of the *lac* operon

Glucose repression is the preferential use of glucose in the presence of other sugars such as lactose. If bacteria are grown in the presence of both glucose and lactose, the *lac* operon is not induced. When the glucose is used up, the *lac* operon is induced, allowing lactose to be used as an energy source.

Despite the name glucose repression, this mechanism involves an activator protein that can stimulate transcription from multiple catabolic operons, including the *lac* operon. This activator, **catabolite activator protein (CAP)** is an allosteric protein that has cAMP as an effector. This protein is also called **cAMP response protein (CRP)** because it binds cAMP, but we will use the name CAP to emphasize its role as a positive regulator. CAP alone does not bind to DNA, but binding of the effector cAMP to CAP changes its conformation such that it can bind to DNA (figure 16.5). The level of cAMP in cells is reduced in the presence of glucose so that no stimulation of transcription from CAP-responsive operons takes place.

The CAP–cAMP system was long thought to be the sole mechanism of glucose repression. But more recent research has indicated that the presence of glucose inhibits the transport of lactose into the cell. This deprives the cell of the *lac* operon inducer, allolactose, allowing the repressor to bind to the operator. This mechanism, called **inducer exclusion,** is now thought to be the main form of glucose repression of the *lac* operon.

Given that inducer exclusion occurs, the role of CAP in the absence of glucose seems superfluous. But in fact, the action of CAP–cAMP allows maximal expression of the operon in the absence of glucose. The positive control of CAP–cAMP is necessary because the promoter of the *lac* operon alone is not efficient in binding RNA polymerase. This inefficiency is overcome by the action of the positive control of the CAP–cAMP activator (see figure 16.5).

The *trp* operon is controlled by the *trp* repressor

The organization of the *trp* operon is similar to the *lac* operon in that a series of genes arranged in a sequence encode enzymes necessary to synthesize tryptophan. The regulatory region that controls transcription of these genes is located upstream of the

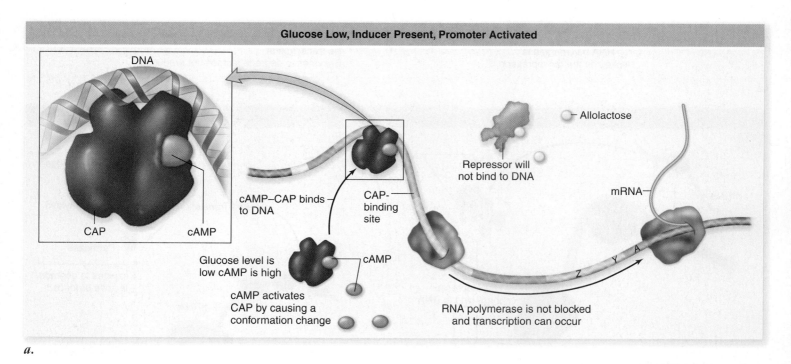

Glucose Low, Inducer Present, Promoter Activated

a.

figure 16.5

EFFECT OF GLUCOSE ON THE *LAC* OPERON. Expression of the *lac* operon is controlled by a negative regulator (repressor) and a positive regulator (CAP). The action of CAP is sensitive to glucose levels. *a.* For CAP to bind to DNA, it must bind to cAMP. When glucose levels are low, cAMP is abundant and binds to CAP. The CAP–cAMP complex causes the DNA to bend around it. This brings CAP into contact with RNA polymerase (not shown) making polymerase binding to the promoter more efficient. *b.* When glucose levels are high, there are two effects: cAMP is scarce so CAP is unable to activate the promoter, and the transport of lactose is blocked (inducer exclusion).

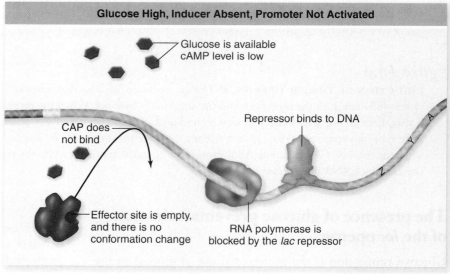

Glucose High, Inducer Absent, Promoter Not Activated

b.

genes. The *trp* repressor is encoded by a gene located outside the *trp* operon. The *trp* operon is expressed in the absence of tryptophan and is not expressed in the presence of tryptophan.

The *trp* repressor is a helix-turn-helix regulatory protein that binds to the operator site located adjacent to the *trp* promoter (figure 16.6). This repressor behaves in a manner opposite to the *lac* repressor. The *trp* repressor alone does not bind to its operator, but when it is bound to tryptophan (the *corepressor*) its conformation alters, allowing it to bind to its operator and prevent RNA polymerase from binding to the promoter. The binding of tryptophan to the repressor alters the orientation of a pair of helix-turn-helix motifs,

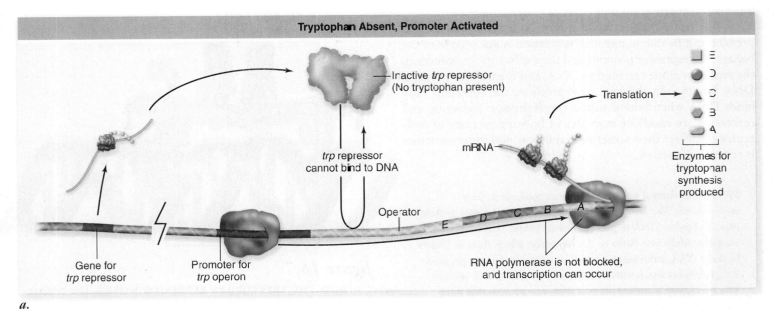

Tryptophan Absent, Promoter Activated

Inactive *trp* repressor (No tryptophan present)

trp repressor cannot bind to DNA

Translation →

mRNA

Enzymes for tryptophan synthesis produced

Operator   E   D   C   B   A

Gene for *trp* repressor

Promoter for *trp* operon

RNA polymerase is not blocked, and transcription can occur

a.

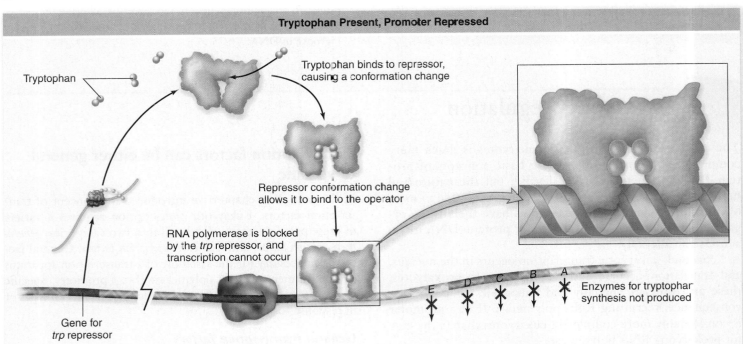

Tryptophan Present, Promoter Repressed

Tryptophan

Tryptophan binds to repressor, causing a conformation change

Repressor conformation change allows it to bind to the operator

RNA polymerase is blocked by the *trp* repressor, and transcription cannot occur

Gene for *trp* repressor

E   D   C   B   A

Enzymes for tryptophan synthesis not produced

b.

figure 16.6

HOW THE *TRP* OPERON IS CONTROLLED. The tryptophan operon encodes the enzymes necessary to synthesize tryptophan. *a.* The tryptophan repressor alone cannot bind to DNA. The promoter is free to function, and RNA polymerase transcribes the operon. *b.* When tryptophan is present, it binds to the repressor altering the repressor's conformation such that it now binds DNA. The tryptophan-repressor complex binds tightly to the operator, preventing RNA polymerase from initiating transcription.

causing their recognition helices to fit into adjacent major grooves of the DNA (figure 16.7).

When tryptophan is present and bound to the repressor, the repressor in turn is bound to the operator, the operon is said to be **repressed.** Transcription of the operon is shut off. As tryptophan levels fall, tryptophan is no longer bound to the repressor, so that the repressor can no longer bind to the operator. In this state, the operon is said to be **derepressed,** distinguishing this state from induction (see figure 16.6).

The key to understanding how both induction and repression can be due to negative regulation is knowledge of the behavior of repressor proteins and their effectors. In induction, the repressor alone can bind to DNA, and the inducer prevents DNA binding. In the case of repression, the repressor only binds DNA when bound to the corepressor. Induction and repression are excellent examples of how interactions of molecules can affect their structures, and how molecular structure is critical to function.

Prokaryotes control gene expression to conform to their environment. The *lac* operon is controlled by a repressor protein that can bind to DNA to prevent transcription. When lactose is available, allolactose binds to the repressor, which then no longer binds to DNA, inducing the synthesis of the *lac* operon proteins. This operon is also positively regulated by an activator protein. The *trp* operon is turned off by a repressor protein that must be bound to tryptophan to bind with DNA. In the absence of tryptophan, the repressor cannot bind with DNA, derepressing the operon.

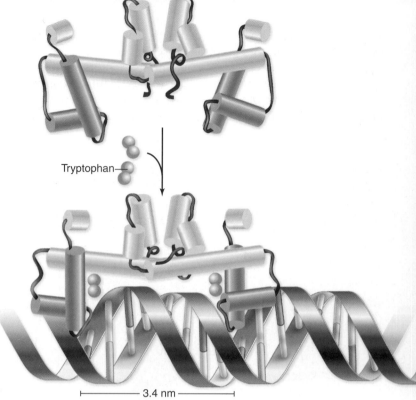

figure 16.7

HOW THE TRYPTOPHAN REPRESSOR WORKS. The binding of tryptophan to the repressor increases the distance between the two recognition helices in the repressor, allowing the repressor to fit snugly into two adjacent portions of the major groove in DNA.

16.4 Eukaryotic Regulation

The control of transcription in eukaryotes is much more complex than in prokaryotes. The basic concepts of protein–DNA interactions are still valid, but the nature and number of interacting proteins is much greater due to some obvious differences. First, eukaryotes have their DNA organized into chromatin, complicating protein–DNA interactions considerably.

Second, eukaryotic transcription occurs in the nucleus, and translation occurs in the cytoplasm; in prokaryotes, these processes are spatially and temporally coupled. As a consequence, recruiting RNA polymerase II to a promoter is considerably more complex in eukaryotes than is the case for prokaryotic RNA polymerase.

Because of these differences, the amount of DNA involved in regulating eukaryotic genes is much greater. The need for a fine degree of flexible control is especially important for multicellular eukaryotes, with their complex developmental programs and multiple tissue types. General themes, however, emerge from this complexity.

Transcription factors can be either general or specific

In the preceding chapter we introduced the concept of transcription factors. Eukaryotic transcription requires a variety of these protein factors, which fall into two categories: *general transcription factors* and *specific transcription factors.* General factors are necessary for the assembly of a transcription apparatus and recruitment of RNA polymerase II to a promoter. Specific factors increase the level of transcription in certain cell types or in response to signals.

General transcription factors

Transcription of RNA polymerase II templates (that is, genes that encode protein products) requires more than just RNA polymerase II to initiate transcription. A host of **general transcription factors** are also necessary to establish productive initiation. These factors are required for transcription to occur, but they do not increase the rate above this basal rate.

General transcription factors are named with letter designations that follow the abbreviation TFII, for "transcription factor RNA polymerase II." The most important of these factors, TFIID, contains the TATA-binding protein that recognizes the TATA box sequence found in many eukaryotic promoters (figure 16.8).

Binding of TFIID is followed by binding of TFIIE, TFIIF, TFIIA, TFIIB, and TFIIH and a host of accessory factors called *transcription-associated factors*, TAFs. The *initiation complex* that results (figure 16.9) is clearly much more complex than the bacterial RNA polymerase holoenzyme binding to a promoter. And there is yet another level of complexity: The initiation complex, although capable of initiating synthesis at a basal level, does not achieve transcription at a high level without the participation of other, specific factors.

Specific transcription factors

Specific transcription factors act in a tissue- or time-dependent manner to stimulate higher levels of transcription than the basal level. The number and diversity of these factors are overwhelming. Some sense can be made of this proliferation of factors by concentrating on the DNA-binding motif, as opposed to the specific factors.

A key common theme that emerges from the study of these factors is that specific transcription factors, called *activators,* have a domain organization, that is, each factor consists of a DNA-binding domain and a separate activating domain that interacts with the transcription apparatus. These domains are

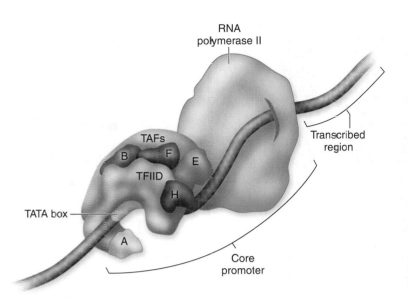

figure 16.9

FORMATION OF A EUKARYOTIC INITIATION COMPLEX. The general transcription factor, TFIID, binds to the TATA box and is joined by the other general factors, TFIIE, TFIIF, TFIIA, TFIIB, and TFIIH. This complex is added to by a number of transcription-associated factors (TAFs) that together recruit the RNA pol II molecule to the core promoter.

essentially independent in the protein, such that they can be "swapped" between different factors and still retain their function.

Promoters and enhancers are binding sites for transcription factors

Promoters, as mentioned in the preceding chapter, form the binding sites for general transcription factors. These factors then mediate the binding of RNA polymerase II to the promoter (and also the binding of RNA polymerases I and III to their specific promoters). In contrast, the holoenzyme portion of the RNA polymerase of prokaryotes can directly recognize a promoter and bind to it.

Enhancers were originally defined as DNA sequences necessary for high levels of transcription that can act independently of position or orientation. At first, this concept seemed counterintuitive, especially since molecular biologists had been conditioned by prokaryotic systems to expect control regions to be immediately upstream of the coding region. It turns out that enhancers are the binding site of the specific transcription factors. The ability of enhancers to act over large distances was at first puzzling, but investigators now think this action is accomplished by DNA bending to form a loop, positioning the enhancer closer to the promoter.

Although more important in eukaryotic systems, this looping was first demonstrated using prokaryotic DNA-binding

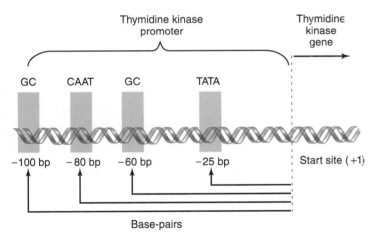

figure 16.8

A EUKARYOTIC PROMOTER. This promoter for the gene encoding the enzyme thymidine kinase. Formation of the transcription initiation complex begins with a general transcription factor binding to the TATA box. There are three other DNA sequences that direct the binding of other specific transcription factors.

proteins (figure 16.10). The important point is that the linear distance separating two sites on the chromosome does not have to translate to great physical distance, because the flexibility of DNA allows bending and looping. An activator bound to an enhancer can thus be brought into contact with the transcription factors bound to a distant promoter (figure 16.11).

Coactivators and mediators link transcription factors to RNA polymerase II

Other factors specifically mediate the action of transcription factors. These **coactivators** and **mediators** are also necessary for activation of transcription by the transcription factor. They act by binding the transcription factor and then binding to another part of the transcription apparatus. Mediators are essential to the function of some transcription factors, but not all transcription factors require them. The number of coactivators is much smaller than the number of transcription factors because the same coactivator can be used with multiple transcription factors.

The transcription complex brings things together

Although a few general principles apply to a broad range of situations, nearly every eukaryotic gene—or group of genes with coordinated regulation—represents a unique case. Virtu-

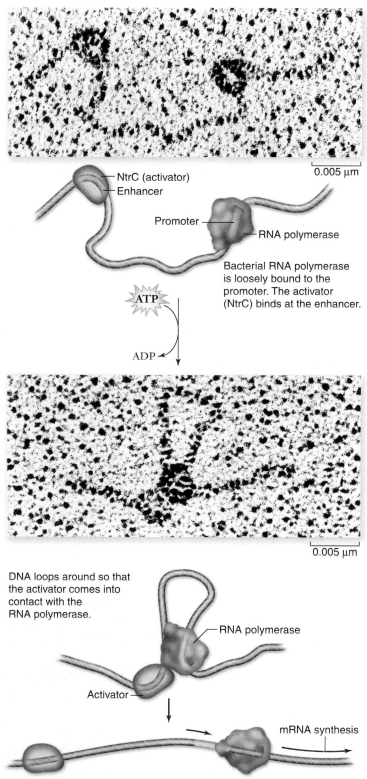

figure 16.10

DNA LOOPING CAUSED BY PROTEINS. When the bacterial activator NtrC binds to an enhancer, it causes the DNA to loop over to a distant site where RNA polymerase is bound, thereby activating transcription. Although such enhancers are rare in prokaryotes, they are common in eukaryotes.

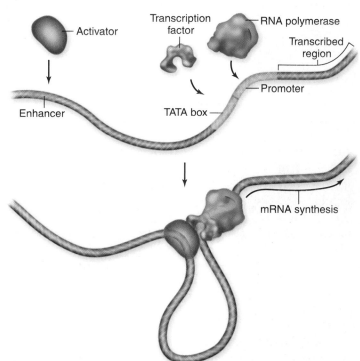

figure 16.11

HOW ENHANCERS WORK. The enhancer site is located far away from the gene being regulated. Binding of an activator (*gray*) to the enhancer allows the activator to interact with the transcription factors (*blue*) associated with RNA polymerase, stimulating transcription.

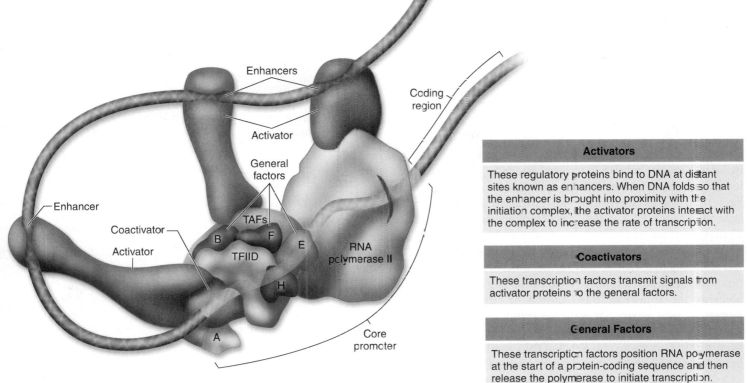

Activators

These regulatory proteins bind to DNA at distant sites known as enhancers. When DNA folds so that the enhancer is brought into proximity with the initiation complex, the activator proteins interact with the complex to increase the rate of transcription.

Coactivators

These transcription factors transmit signals from activator proteins to the general factors.

General Factors

These transcription factors position RNA polymerase at the start of a protein-coding sequence and then release the polymerase to initiate transcription.

figure 16.12

INTERACTIONS OF VARIOUS FACTORS WITHIN THE TRANSCRIPTION COMPLEX. All specific transcription factors bind to enhancer sequences that may be distant from the promoter. These proteins can then interact with the initiation complex by DNA looping to bring the factors into proximity with the initiation complex. As detailed in the text, some transcription factors, called activators, can directly interact with the RNA polymerase II or the initiation complex, whereas others require additional coactivators.

ally all genes that are transcribed by RNA polymerase II need the same suite of general factors to assemble an initiation complex, but the assembly of this complex and its ultimate level of transcription depend on specific transcription factors that in combination make up the **transcription complex** (figure 16.12).

The makeup of eukaryotic promoters, therefore, is either very simple, if we consider only what is needed for the initiation complex, or very complicated, if we consider all factors that may bind in a complex and affect transcription. This kind of combinatorial gene regulation leads to great flexibility because it can respond to the many signals a cell may receive affecting transcription, allowing integration of these signals.

inquiry

? *How do eukaryotes coordinate the activation of many genes whose transcription must occur at the same time?*

In eukaryotes, initiation of transcription requires general transcription factors that bind to the promoter and recruit RNA polymerase II to form an initiation complex. General factors result in the general level of transcription that can then be increased by the action of specific transcription factors that bind to enhancer sequences. Additional coactivators and mediators interact with both specific transcription factors and the rest of the transcription apparatus.

16.5 Eukaryotic Chromatin Structure

Eukaryotes have the additional gene expression hurdle of possessing DNA that is packaged into chromatin. The packaging of DNA first into nucleosomes and then into higher order chromatin structures is now thought to be directly related to the control of gene expression.

Chromatin structure at its lowest level is the organization of DNA and histone proteins into *nucleosomes* (see chapter 10).

These nucleosomes may block binding of transcription factors and RNA polymerase II at the promoter.

The higher order organization of chromatin, which is not completely understood, appears to depend on the state of the histones in nucleosomes. Histones can be modified to result in a greater condensation of chromatin, making promoters even less accessible for protein–DNA interactions. A chromatin remodeling complex exists that can make DNA more accessible.

Both DNA and histone proteins can be modified

Chemical **methylation** of the DNA was once thought to play a major role in gene regulation in vertebrate cells. The addition of a methyl group to cytosine creates 5-methylcytosine, but this change has no effect on its base-pairing with guanine (figure 16.13). Similarly, the addition of a methyl group to uracil produces thymine, which clearly does not affect base-pairing with adenine.

Many inactive mammalian genes are methylated, and it was tempting to conclude that methylation caused the inactivation. But methylation is now viewed as having a less direct role, blocking the accidental transcription of "turned-off" genes. Vertebrate cells apparently possess a protein that binds to clusters of 5-methylcytosine, preventing transcriptional activators from gaining access to the DNA. DNA methylation in vertebrates thus ensures that once a gene is turned off, it stays off.

The histone proteins that form the core of the nucleosome (chapter 10) can also be modified. This modification is correlated with active versus inactive regions of chromatin, similar to the methylation of DNA just described. Histones can also be methylated, and this alteration is generally found in inactive regions of chromatin. Finally, histones can be modified by the addition of an acetyl group, and this addition is correlated with active regions of chromatin.

Some transcription activators alter chromatin structure

The control of eukaryotic transcription requires the presence of many different factors to activate transcription. Some activators seem to interact directly with the initiation complex or with coactivators that themselves interact with the initiation complex, as described earlier. Other cases are not so clear. The emerging consensus is that some coactivators act by modifying the structure of chromatin by adding acetyl groups to amino acids, making DNA accessible to transcription factors.

Recently, some coactivators have been shown to be histone acetylases. In these cases, it appears that transcription is increased by removing higher order chromatin structure that

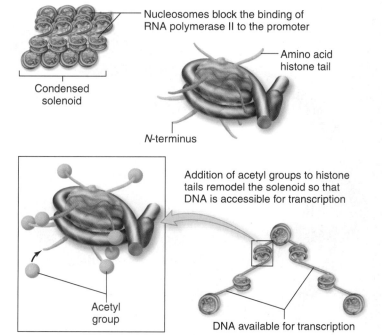

figure 16.14

HISTONE MODIFICATION AFFECTS CHROMATIN STRUCTURE. DNA in eukaryotes is organized first into nucleosomes and then into higher order chromatin structures. The histones that make up the nucleosome core have amino tails that protrude. These amino tails can be modified by the addition of acetyl groups. The acetylation alters the structure of chromatin, making it accessible to the transcription apparatus.

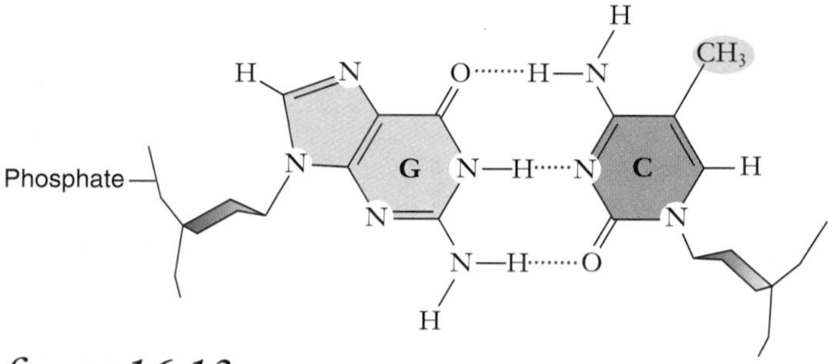

would prevent transcription (figure 16.14). Some corepressors have been shown to be histone deacetylases as well.

These observations have led to the suggestion that a "histone code" might exist, analogous to the genetic code. This histone code is postulated to underlie the control of chromatin structure and, thus, of access of the transcription machinery to DNA.

Chromatin-remodeling complexes also change chromatin structure

The outline of how alterations to chromatin structure can regulate gene expression are beginning to emerge. A key discovery is the existence of so-called **chromatin-remodeling complexes.** These large complexes of proteins include enzymes that modify histones and DNA and that also change chromatin structure itself. Chromatin remodeling complexes can move nucleosomes on DNA, repositioning them, and can even transfer nucleosomes from one part of DNA to another.

> Eukaryotic DNA is packaged into chromatin, adding another structural challenge to transcription. Changes in chromatin structure correlate with modification of DNA and histones, and access to DNA by transcriptional regulators requires changes in chromatin structure. Some transcriptional activators modify histones by acetylation. Large chromatin-remodeling complexes may alter the structure of chromatin and thus affect gene expression.

figure 16.13

DNA METHYLATION. Cytosine is methylated, creating 5-methylcytosine. Because the methyl group (*green*) is positioned to the side, it does not interfere with the hydrogen bonds of a G–C base-pair, but it can be recognized by proteins.

Eukaryotic Posttranscriptional Regulation

Up to this point we have discussed gene regulation entirely in terms of transcription initiation—that is, when and how often RNA polymerase starts "reading" a particular gene. Most gene regulation appears to occur at this point. However, in principle many other points after transcription exist where gene expression could be regulated, and all of them serve as control points for at least some eukaryotic genes. In general, these posttranscriptional control processes involve the recognition of specific sequences on the primary RNA transcript by regulatory proteins and small RNA molecules.

Small RNAs can affect gene expression

Recent experiments indicate that a class of RNA molecules loosely called *small RNAs* may play a major role in regulating gene expression by interacting directly with primary gene transcripts. Small RNAs are short segments of RNA, ranging in length from 21 to 28 nt—small interfering RNA and micro-RNA are two types that were described in the preceding chapter. Researchers focusing on far larger mRNA, tRNA, and rRNA had not noticed these far smaller bits, tossing them out during experiments.

The first hints of the existence of small RNAs emerged in 1993, when researchers reported in the nematode *Caenorhabditis elegans* the presence of tiny RNA molecules that don't encode any protein. These small RNAs appeared to regulate the activity of specific *C. elegans* genes.

Soon researchers found evidence of similar small RNAs in a wide range of other organisms. In the plant *Arabidopsis thaliana*, small RNAs seemed to be involved in the regulation of genes critical to early development, and in yeasts they were identified as the agents that silence genes in tightly packed regions of the genome. In the ciliated protozoan *Tetrahymena thermophila*, the loss of major blocks of DNA during development seems guided by small RNA molecules.

RNA interference

How can small fragments of RNA act to regulate gene expression? The first clue emerged in 1998, when researchers injected small stretches of double-stranded RNA into *C. elegans*. Double-stranded RNA forms when a single strand with ends having complementary nucleotide sequences folds back in a hairpin loop; base-pairing holds the strands together much as it does the strands of a DNA duplex (figure 16.15). The injected double-stranded RNA strongly inhibited the expression of the genes from which the double-stranded RNA had been generated. This kind of gene silencing, since seen in *Drosophila* and other organisms, is called **RNA interference.**

Mechanism of RNA interference

In 2001, researchers identified an enzyme, dubbed **Dicer,** that appears to generate the small RNAs in the cell. Dicer chops double-stranded RNA molecules into little pieces. Two types of small RNA result: **micro-RNAs (miRNAs)** and **small interfering RNAs (siRNAs).**

Micro-RNAs appear to act by binding directly to mRNAs and preventing their translation. Researchers have identified over 100 different miRNAs and are still trying to sort out how each functions and which miRNAs occur in which species.

siRNAs appear to be the main agents of RNA interference, acting to degrade particular messenger RNAs after they have been transcribed, but before they can be translated by the ribosomes. The exact way they achieve this degradation of selected gene transcripts is not yet known.

Current data suggest that dicer delivers siRNAs to an enzyme complex called the *RNA-induced silencing complex (RISC)*, which searches out and degrades any mRNA molecules with a complementary sequence (figure 16.16).

figure 16.15

SMALL RNAs FORM DOUBLE-STRANDED LOOPS. These three RNA molecules all contain self-complementary regions. The molecules fold back to form hairpin loops due to base-pairing of complementary regions.

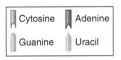

| | Cytosine | | Adenine |
| | Guanine | | Uracil |

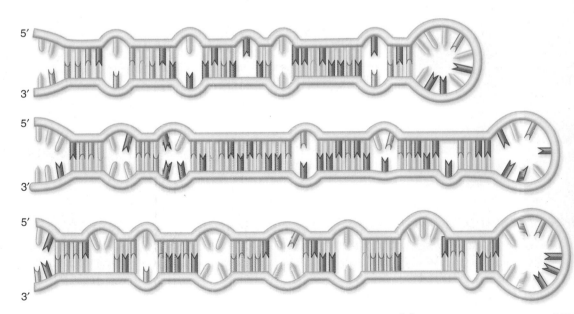

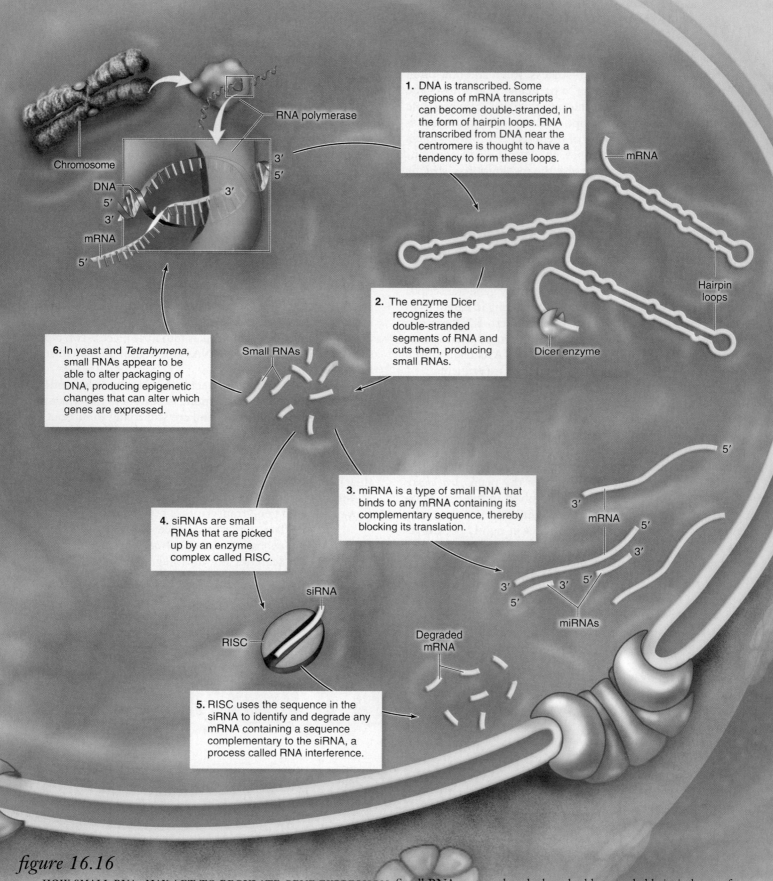

1. DNA is transcribed. Some regions of mRNA transcripts can become double-stranded, in the form of hairpin loops. RNA transcribed from DNA near the centromere is thought to have a tendency to form these loops.

2. The enzyme Dicer recognizes the double-stranded segments of RNA and cuts them, producing small RNAs.

3. miRNA is a type of small RNA that binds to any mRNA containing its complementary sequence, thereby blocking its translation.

4. siRNAs are small RNAs that are picked up by an enzyme complex called RISC.

5. RISC uses the sequence in the siRNA to identify and degrade any mRNA containing a sequence complementary to the siRNA, a process called RNA interference.

6. In yeast and *Tetrahymena*, small RNAs appear to be able to alter packaging of DNA, producing epigenetic changes that can alter which genes are expressed.

Chromosome

RNA polymerase

DNA

mRNA

5′ 3′

3′ 5′

3′

Small RNAs

mRNA

Hairpin loops

Dicer enzyme

mRNA

miRNAs

5′ 3′ 5′ 3′ 5′ 3′ 5′ 3′

siRNA

RISC

Degraded mRNA

figure 16.16

HOW SMALL RNAs MAY ACT TO REGULATE GENE EXPRESSION. Small RNAs are produced when double-stranded hairpin loops of an RNA transcript are cut. Although the details are not well understood, two types of small RNA, miRNA and siRNA, are thought to block gene expression within the nucleus at the level of the mRNA gene transcript, a process called RNA interference. As this diagram suggests, small RNAs are also thought to influence chromatin packaging in some organisms.

Alternative splicing can produce multiple proteins from one gene

As noted in the preceding chapter, splicing of pre-mRNA is one of the processes leading to mature mRNA. Many of these splicing events may produce different mRNAs from a single primary transcript by alternative splicing. This mechanism allows another level of control of gene expression.

Alternative splicing can change the splicing events that occur during different stages of development or in different tissues. An example of developmental differences is found in *Drosophila*, in which sex determination is the result of a complex series of alternative splicing events that differ in males and females.

An excellent example of tissue-specific alternative splicing in action is found in two different human organs: the thyroid gland and the hypothalamus. The thyroid gland is responsible for producing hormones that control processes such as metabolic rate. The hypothalamus, located in the brain, collects information from the body (for example, salt balance) and releases hormones that in turn regulate the release of hormones from other glands, such as the pituitary gland. (You'll learn more about these glands in chapter 46.)

These two organs produce two distinct hormones: *calcitonin* and *CGRP* (calcitonin gene-related peptide) as part of their function. Calcitonin controls calcium uptake and the balance of calcium in tissues such as bones and teeth. CGRP is involved in a number of neural and endocrine functions. Although these two hormones are used for very different physiological purposes, they are produced from the same transcript (figure 16.17).

The synthesis of one product versus another is determined by tissue-specific factors that regulate the processing of the primary transcript. In the case of calcitonin and CGRP, pre-mRNA splicing is controlled by different factors that are present in the thyroid and in the hypothalamus.

RNA editing alters mRNA after transcription

In some cases, the editing of mature mRNA transcripts can produce an altered mRNA that is not truly encoded in the genome—an unexpected possibility. RNA editing was first discovered as the insertion of uracil residues into some RNA transcripts in protozoa, and it was thought to be an anomaly.

RNA editing of a different sort has since been found in mammalian species, including humans. In this case, the editing involves chemical modification of a base to change its base-pairing properties, usually by deamination. For example, both deamination of cytosine to uracil and deamination of adenine to inosine have been observed (inosine pairs as G would during translation).

Apolipoprotein B

The human protein apolipoprotein B is involved in the transport of cholesterol and triglycerides. The gene that encodes this protein, *apoB*, is large and complex, consisting of 29 exons scattered across almost 50 kilobases (kb) of DNA.

The protein exists in two isoforms: a full-length APOB100 form and a truncated APOB48 form. The truncated form is due to an alteration of the mRNA that changes a codon for glutamine to one that is a stop codon. Furthermore, this editing occurs in a tissue-specific manner; the edited form appears only in the intestine, whereas the liver makes only the full-length form. The full-length APOB100 form is part of the low-density lipoprotein (LDL) particle that carries cholesterol. High levels of serum LDL are thought to be a major predictor of atherosclerosis in humans. It does not appear that editing has any effect on the levels of the intestine-specific transcript.

The 5-HT serotonin receptor

RNA editing has also been observed in some brain receptors for opiates in humans. One of these receptors, the serotonin (5-HT) receptor, is edited at multiple sites to produce a total of 12 different isoforms of the protein.

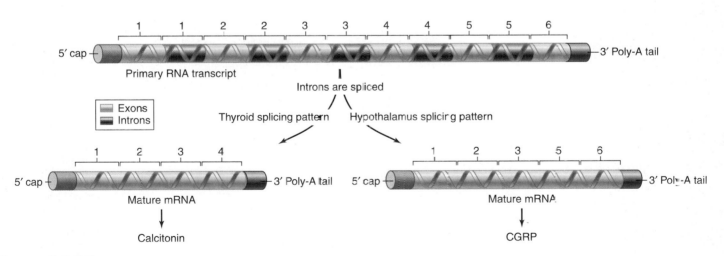

figure 16.17

ALTERNATIVE SPLICING. Many primary transcripts can be spliced in different ways to give rise to multiple mRNAs. In this example, in the thyroid the primary transcript is spliced to contain four exons encoding the protein calcitonin. In the hypothalamus the four h exon, which contains the polyA site used in the thyroid, is skipped and two additional exons are added to encode the protein calcitonin gene related product (CGRP).

It is unclear how widespread these forms of RNA editing are, but they are further evidence that the information encoded within genes is not the end of the story for protein production.

mRNA must be transported out of the nucleus for translation

Processed mRNA transcripts exit the nucleus through the nuclear pores (described in chapter 5). The passage of a transcript across the nuclear membrane is an active process that requires the transcript to be recognized by receptors lining the interior of the pores. Specific portions of the transcript, such as the poly-A tail, appear to play a role in this recognition.

The transcript cannot move through a pore as long as any of the splicing enzymes remain associated with the transcript, ensuring that partially processed transcripts are not exported into the cytoplasm.

There is little hard evidence that gene expression is regulated at this point, although it could be. On average, about 10% of primary transcripts consists of exons that will make up

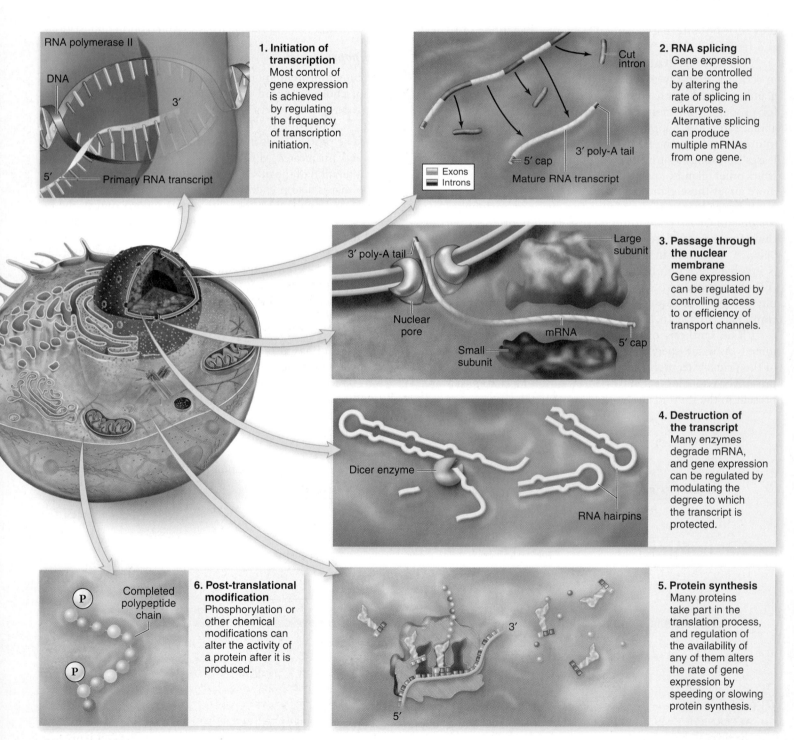

figure 16.18

MECHANISMS FOR CONTROL OF GENE EXPRESSION IN EUKARYOTES.

mRNA sequences, but only about 5% of the total mRNA produced as primary transcript ever reaches the cytoplasm. This observation suggests that about half of the exons in primary transcripts never leave the nucleus, but it is unclear whether the disappearance of this mRNA is selective.

Initiation of translation can be controlled

The translation of a processed mRNA transcript by ribosomes in the cytoplasm involves a complex of proteins called *translation factors*. In at least some cases, gene expression is regulated by modification of one or more of these factors. In other instances, **translation repressor proteins** shut down translation by binding to the beginning of the transcript, so that it cannot attach to the ribosome.

In humans, the production of ferritin (an iron-storing protein) is normally shut off by a translation repressor protein called aconitase. Aconitase binds to a 30-nt sequence at the beginning of the ferritin mRNA, forming a stable loop to which ribosomes cannot bind. When iron enters the cell, the binding of iron to aconitase causes the aconitase to dissociate from the ferritin mRNA, freeing the mRNA to be translated and increasing ferritin production 100-fold.

The degradation of mRNA is controlled

Another aspect that affects gene expression is the stability of mRNA transcripts in the cell cytoplasm. Unlike prokaryotic mRNA transcripts, which typically have a half-life of about 3 min, eukaryotic mRNA transcripts are very stable. For example, β-globin gene transcripts have a half-life of over 10 hr, an eternity in the fast-moving metabolic life of a cell.

The transcripts encoding regulatory proteins and growth factors, however, are usually much less stable, with half-lives of less than 1 hr. What makes these particular transcripts so unstable? In many cases, they contain specific sequences near their 3′ ends that make them targets for enzymes that degrade mRNA. A sequence of A and U nucleotides near the 3′ poly-A tail of a transcript promotes removal of the tail, which destabilizes the mRNA.

Histone transcripts, for example, have a half-life of about 1 hr in cells that are actively synthesizing DNA; at other times during the cell cycle, the poly-A tail is lost, and the transcripts are degraded within minutes.

Other mRNA transcripts contain sequences near their 3′ ends that are recognition sites for endonucleases, which causes these transcripts to be digested quickly. The short half-lives of the mRNA transcripts of many regulatory genes are critical to the function of those genes because they enable the levels of regulatory proteins in the cell to be altered rapidly.

A review of various methods of posttranscriptional control of gene expression is provided in figure 16.18.

> **Small RNAs may help to control gene expression by either selective degradation of mRNA, inhibition of translation, or alteration of chromatin structure. Multiple mRNAs can be formed from a single gene via alternative splicing, which can be tissue- and developmentally specific. The sequence of an mRNA can also be altered by RNA editing. All of these processes allow control of gene expression after transcription.**

16.7 Protein Degradation

If all of the proteins produced by a cell during its lifetime remained in the cell, serious problems would arise. Protein labeling studies in the 1970s indicated that eukaryotic cells turn over proteins in a controlled manner. That is, proteins are continually being synthesized and degraded. Although this protein turnover is not as rapid as in prokaryotes, it indicates that a system regulating protein turnover is important.

Proteins can become altered chemically, rendering them nonfunctional; in addition, the need for any particular protein may be transient. Proteins also do not always fold correctly, or they may become improperly folded over time. These changes can lead to loss of function or other chemical behaviors, such as aggregating into insoluble complexes. In fact, a number of neurodegenerative diseases, such as Alzheimer dementia, Parkinson disease, and mad cow disease, are related to proteins that aggregate, forming characteristic plaques in brain cells. Thus, in addition to normal turnover of proteins, cells need a mechanism to get rid of old, unused, and incorrectly folded proteins.

Enzymes called **proteases** can degrade proteins by breaking peptide bonds, converting a protein into its constituent amino acids. Although there is an obvious need for these enzymes, they clearly cannot be floating around in the cytoplasm active at all times.

One way that eukaryotic cells handle such problems is to confine destructive enzymes to a specific cellular compartment. You may recall from chapter 4 that lysosomes are vesicles that contain digestive enzymes, including proteases. Lysosomes are used to remove proteins and old or nonfunctional organelles, but this system is not specific for particular proteins. Cells need another regulated pathway to remove proteins that are old or unused, but leave the rest of cellular proteins intact.

Addition of ubiquitin marks proteins for destruction

Eukaryotic cells solve this problem by marking proteins for destruction, then selectively degrading them. The mark that cells use is the attachment of a **ubiquitin** molecule. Ubiquitin, so named because it is found in essentially all eukaryotic cells (that is, it is ubiquitous), is a 76–amino-acid protein that can exist as an isolated molecule or in longer chains that are attached to other proteins.

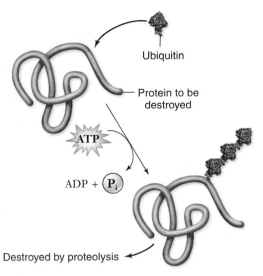

figure 16.19

UBIQUITINATION OF PROTEINS. Proteins that are to be degraded are marked with ubiquitin. The enzyme ubiquitin ligase, uses ATP to add ubiquitin to a protein. When a series of these have been added, the polyubiquitinated protein is destroyed.

The longer chains are added to proteins in a stepwise fashion by an enzyme called *ubiquitin ligase* (figure 16.19). This reaction requires ATP and other proteins, and it takes place in a multistep, regulated process. Proteins that have a ubiquitin chain attached are called *polyubiquitinated*, and this state is a signal to the cell to destroy this protein.

Two basic categories of proteins become ubiquitinated: those that need to be removed because they are improperly folded or nonfunctional, and those that are produced and degraded in a controlled fashion by the cell. An example of the latter are the cyclin proteins that help to drive the cell cycle (chapter 10). When these proteins have fulfilled their role in active division of the cell, they become polyubiquitinated and are removed. In this way, a cell can control entry into cell division or maintain a nondividing state.

The proteasome degrades polyubiquitinated proteins

The cellular organelle that degrades proteins marked with ubiquitin is the **proteasome,** a large cylindrical complex that proteins enter at one end and exit the other as amino acids or peptide fragments (figure 16.20).

The proteasome complex contains a central region that has protease activity and regulatory components at each end. Although not membrane-bounded, this organelle can be thought of as a form of compartmentalization on a very small scale. By using a two-step process, first to mark proteins for destruction, then to process them through a large complex, proteins to be degraded are isolated from the rest of the cytoplasm.

The process of ubiquitination followed by degradation by the proteasome is called the *ubiquitin–proteasome pathway*. It can be thought of as a cycle in that the ubiquitin added to proteins is not itself destroyed in the proteasome. As the proteins are degraded, the ubiquitin chain itself is simply cleaved back into ubiquitin units that can then be reused (figure 16.21).

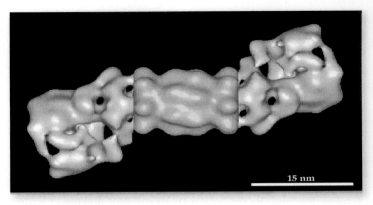

figure 16.20

THE *DROSOPHILA* PROTEASOME. The central complex contains the proteolytic activity, and the flanking regions act as regulators. Proteins enter one end of the cylinder and are cleaved to peptide fragments that exit the other end.

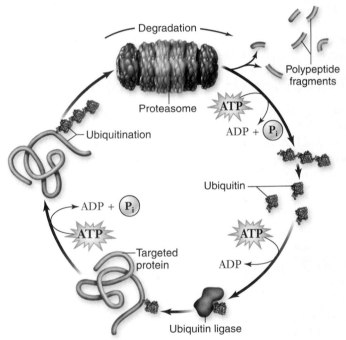

figure 16.21

DEGRADATION BY THE UBIQUITIN–PROTEASOME PATHWAY. Proteins are first ubiquitinated, then enter the proteasome to be degraded. In the proteasome, the polyubiquitin is removed and then is later "deubiquitinated" to produce single ubiquitin molecules that can be reused.

inquiry

What are two reasons a cell would polyubiquitinate a polypeptide?

Proteins are synthesized and degraded in a regulated fashion. Control of protein degradation in eukaryotes involves addition of the protein ubiquitin followed by proteolysis in the proteasome. The ubiquitin–proteasome pathway is a the major pathway to recycle old and improperly folded proteins.

16.1 Control of Gene Expression

Control of gene expression is essential to living organisms, allowing cells to respond to changing environmental conditions and enabling the development of complex multicellular organisms.

- Transcription is initiated by regulatory proteins that modulate the ability of RNA polymerase to bind to the promoter.
- Prokaryotes and eukaryotes differ in how they control gene expression.
- Control strategies in prokaryotes respond quickly to changing environmental conditions.
- In eukaryotes control strategies act to maintain homeostasis.

16.2 Regulatory Proteins

Regulatory proteins bind to specific DNA sequences and control whether transcription will occur.

- Regulatory proteins bind to the surface of the double helix and interact with base-pairs in the major groove.
- A DNA-binding motif refers to the three-dimensional structure of the region of the regulatory protein that binds to the DNA.
- Several different DNA-binding motifs are found in regulatory proteins (figure 16.2).

16.3 Prokaryotic Regulation

Although there are differences, prokaryotes and eukaryotes share many similarities in control of transcription.

- Negative control is mediated by allosteric proteins called repressors that prevent or decrease transcription.
- Positive control is mediated by another class of regulatory allosteric proteins called activators that stimulate transcription.
- Prokaryotes adjust gene expression in response to environmental conditions.
- The *lac* operon is induced in the presence of lactose; that is, the enzymes to utilize lactose are only produced in the presence of lactose.
- The *trp* operon is repressed; that is, the enzymes needed to produce trp are turned off in the presence of trp.
- In induction, the effector (allolactose) binds to the repressor, altering its conformation such that it no longer binds DNA (figure 16.4).
- In repression, the effector (called a corepressor) binds to the repressor, altering its conformation such that it can bind to DNA
- The presence of glucose prevents induction of the *lac* operon, a process callled glucose repression.
- Inducer exclusion occurs when an inducer molecule is prevented from entering the cell so operon activity remains suppressed. In the *lac* operon inducer exclusion is one mediator of glucose repression.
- The maximum expression of the *lac* operon requires positive control by the activitor: catabolite activator protein (CAP).
- CAP is active bound to cAMP, levels of which are high when glucose levels are low.
- The *trp* operon is also negatively controlled. The *trp* repressor does not bind to DNA, allowing expression in the absence of *trp*.
- The repressor binds to trp (the corepressor) and can then bind to DNA and shut off the operon when trp levels are high.

16.4 Eukaryotic Regulation

The control of transcription in eukaryotes is much more complex than in prokaryotes. Eukaryotic DNA is organized into chromatin and the nuclear membrane separates transcription and translation.

- Transcription factors can be either general or specific.
- General factors are necessary to assemble the transcription apparatus and recruitment of RNA polymerase II to the promoter.
- Specific factors act in a tissue- or time-dependent manner to stimulate higher rates of transcription.
- Promoters are binding sites for general transcription factors; enhancers are binding sites for specific transcription factors.
- Coactivators and mediators interact with specific transcription factors and the rest of the transcription apparatus (figure 16.12).
- Some transcription factors require a mediator but not all of them do.
- The number of coactivators is small because the same coactivator can be used with multiple transcription factors.
- Transcription by RNA polymerase II requires an initiation complex and specific transcription factors.

16.5 Eukaryotic Chromatin Structure

Expression of genes in eukaryotes is further complicated because their DNA is packaged into chromatin.

- In eukaryotes DNA is wrapped around proteins, called histones, forming nucleosomes and is not accessible for transcription.
- Methylation of DNA base-pairs, primarily cytosine, correlates with genes that have been "turned off."
- Methylation of histones is associated with inactive regions of chromatin.
- Acetylated histones are associated with active regions of chromatin.
- Transcription activators such as histone acetylases and deacetylases alter chromatin structure and accessibility for transcription
- Chromatin-remodeling complexes contain enzymes that move, reposition, and transfer nucleosomes.

16.6 Eukaryotic Posttranscriptional Regulation

Control of eukaryotic gene expression can occur after the initiation of transcription (figure 16.16).

- RNA interference is mediated by small RNAs that folds back on themselves to form a double-stranded RNA with a hairpin loop.
- An enzyme, called Dicer, chops double-stranded RNA into micro-RNA (miRNAs) and small interfering RNA (siRNAs).
- Micro-RNAs bind directly to mRNA and prevent translation.
- Small interfering RNAs degrade particular mRNAs after they have been formed by transcription.
- Currently it is thought that siRNA works with an enzyme called RISC, which degrades mRNA complementary to siRNA
- In response to tissue-specific factors alternative splicing of pre-mRNA from one gene can result in many different proteins.
- RNA editing involves modification of the mRNA, changing the base-pairing properties.
- An mRNA must be transported out of the nucleus for translation.
- The initiation of translation is controlled by translation factors and translation repressor proteins.
- Translation can be regulated by degradation of mRNA

16.7 Protein Degradation

Proteins are continuously synthesized and degraded.

- In eukaryotes, proteins targeted for destruction have ubiquitin added to them.
- Proteins are ubiquitinated when old, nonfunctional, or produced in a controlled fashion such as cyclins.
- A cell organelle—the proteasome—degrades ubiquitinated proteins.

SELF TEST

1. Control of gene expression can occur at which of the following steps?
   a. Splicing of pre-mRNA into mature mRNA
   b. Initiation of translation
   c. Initiation of transcription
   d. All of the above

2. Regulatory proteins interact with DNA by—
   a. unwinding the helix and changing the pattern of base-pairing
   b. the sugar–phosphate backbone of the double helix
   c. unwinding the helix and disrupting base-pairing
   d. binding to the major groove of the double helix and interacting with base-pairs

3. The two proteins subunits of a leucine zipper are held together—
   a. by β-sheet domains
   b. by the interactions of hydrophobic amino acids
   c. by two α-helical domains separated by a turn
   d. the interaction of atoms of zinc

4. Which domain of a helix-turn-helix protein is directly involved with binding a specific DNA sequence?
   a. The recognition helix
   b. The homeodomain
   c. The zinc finger
   d. The leucine zipper

5. Negative control of transcription in a prokaryotic cell involves _____ molecules that alter the conformation of _____ proteins that bind to DNA and prevent transcription.
   a. operator; repressor
   b. activator; RNA polymerase
   c. activator; operator
   d. effector; repressor

6. What is an operon?
   a. A region of DNA involved in regulation of transcription
   b. A cluster of genes that are expressed as a single unit
   c. A DNA-binding motif
   d. A regulator protein that enhances transcription

7. What effect would the presence of lactose have on a *lac* operon?
   a. The repressor would bind to the operator site of the operon.
   b. Lactose will bind to the operator site of the operon.
   c. The *lac* operon would be transcribed.
   d. It would have no effect.

8. How does the presence of glucose influence the regulation of the *lac* operon?
   a. Glucose reduces the amount of cAMP, which CAP requires for action.
   b. Glucose prevents the transport of lactose into the cell.
   c. Glucose binds to the repressor protein.
   d. Both (a) and (b)

9. What effect does the amino acid tryptophan have on the *trp* operon?
   a. Tryptophan binds to the repressor, resulting in transcription of the *trp* operon.
   b. Tryptophan binds to the repressor and prevents transcription.
   c. Tryptophan increases cAMP, activating catabolite activator protein (CAP).
   d. Tryptophan interacts with repressor, leading to derepression of the *trp* operon.

10. How do specific transcription factors differ from general transcription factors?
    a. Specific transcription factors increase the rate of transcription.
    b. Specific transcription factors bind to the TATA box sequence.
    c. Specific transcription factors form an initiation complex.
    d. Specific transcription factors bind to RNA polymerase.

11. DNA methylation—
    a. inhibits transcription by blocking base-pairing of cytosine and guanine
    b. inhibits transcription by blocking base-pairing of uracil and adenine
    c. prevents transcription by blocking the TATA box sequence
    d. is correlated with genes that are turned off

12. What is the function of small interfering RNAs?
    a. They bind to mRNA and block translation.
    b. They block transcription of complementary mRNAs.
    c. They trigger the destruction of complementary mRNAs.
    d. They compete with transfer RNAs during translation.

13. RNA editing is a consequence of—
    a. base-pair substitutions caused by a mutation in the DNA
    b. splicing of a pre-mRNA
    c. methylation of the mRNA
    d. modifications of a base within the mRNA

14. What is *ubiquitin*?
    a. A type of protease
    b. A posttranslational modification that targets proteins for destruction
    c. A protein involved in the transport of mRNAs out of the nucleus
    d. A posttranslational modification that degrades mRNAs

15. Which of the following is *not* a true statement about proteasomes?
    a. They are membrane-bounded organelles.
    b. They break proteins into amino acids.
    c. They do not degrade ubiquitin.
    d. They remove ubiquitin from proteins.

CHALLENGE QUESTIONS

1. Examples of positive and negative control of transcription can be found in the regulation of expression of the bacterial operons *lac* and *trp*. Use these two operon systems to describe the difference between positive and negative regulation.

2. How is an operator different from a promoter in the regulation of prokaryotic gene expression?

3. What forms of eukaryotic control of gene expression are unique to eukaryotes? Could prokaryotes use the mechanisms, or are they due to differences in these cell types?

4. The number and type of proteins found in a cell can be influenced by genetic mutation and regulation of gene expression. Discuss how these two processes differ.

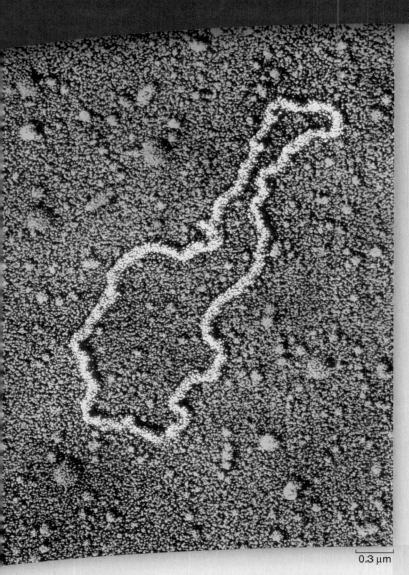

0.3 μm

Biotechnology

introduction

OVER THE PAST DECADES, the development of new and powerful techniques for studying and manipulating DNA has revolutionized biology. The knowledge gained in the last 25 years is greater than the rest of the history of biology. Biotechnology also affects more aspects of everyday life than any other area of biology. From the food on your table to the future of medicine, biotechnology touches your life.

The ability to isolate specific DNA sequences arose from the study and use of small DNA molecules found in bacteria, like the plasmid pictured here. In this chapter, we explore these technologies and consider how they apply to specific problems of practical importance.

DNA Manipulation

The ability to directly isolate and manipulate genetic material was one of the most profound changes in the field of biology in the late 20th century. The construction of **recombinant DNA** molecules, that is, a single DNA molecule made from two different sources, began in the mid-1970s. The development of this technology, which has led to the entire field of biotechnology, is based on enzymes that can be used to manipulate DNA.

Restriction enzymes cleave DNA at specific sites

The enzymes that catalyzed the molecular biology revolution were those able to cleave DNA at specific sites: these are called **restriction endonucleases.** As described in chapter 14, nucleases are enzymes that degrade DNA, and many were known prior to the isolation of the first restriction enzyme. But restriction endonucleases are different because they are able to fragment DNA at specific sites. If a DNA sequence were a rope, then restriction enzymes would be a knife that always cut that rope into specific lengths.

Discovery and significance of restriction endonucleases

This site-specific cleavage activity, long sought by molecular biologists, came out of basic research into why bacterial viruses can infect some cells but not others. This phenomenon was termed *host restriction*. The bacteria produce enzymes that can cleave the invading viral DNA at specific sequences. The host cells protect their own DNA from cleavage by modifying their DNA at the cleavage sites; the restriction enzymes do not cleave the modified DNA. Since the initial discovery of these restriction endonucleases, hundreds more have been isolated that recognize and cleave different **restriction sites.**

The ability to cut DNA at specific sites is significant in two ways: First, it allows a form of physical mapping that was previously impossible. Physical maps can be constructed based on the positioning of cleavage sites for restriction enzymes. These restriction maps provide crucial data for identifying and working with DNA molecules.

Second, restriction endonuclease cleavage allows the creation of recombinant molecules. The ability to construct recombinant molecules is critical to research, because many steps in the process of cloning and manipulating DNA require the ability to combine molecules from different sources.

How restriction enzymes work

There are two types of restriction enzymes: type I and type II. Type I enzymes make simple cuts across both DNA strands near, but not at, the recognition site. Because these do not cleave at precise locations, they are not often used in cloning and manipulating DNA.

Type II enzymes allow creation of recombinant molecules; these enzymes recognize a specific DNA sequence, ranging from 4 bases to 12 bases, and cleave the DNA at a specific base within this sequence (figure 17.1).

The recognition sites for type II enzymes are palindromes. A *palindrome* in language reads the same forward and in reverse, such as the sentence: "Madam I'm Adam." The palindromic

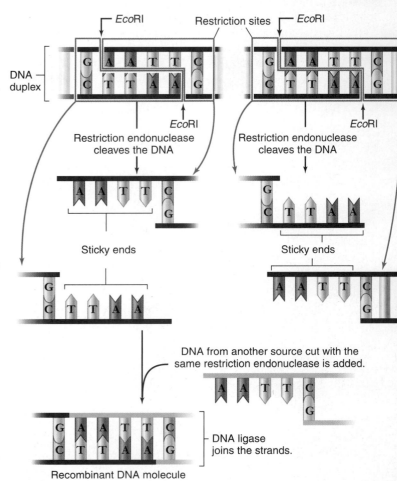

figure 17.1

MANY RESTRICTION ENDONUCLEASES PRODUCE DNA FRAGMENTS WITH "STICKY ENDS." The restriction endonuclease *Eco*RI always cleaves the sequence GAATTC between G and A. Because the same sequence occurs on both strands, both are cut. However, the two sequences run in opposite directions on the two strands. As a result, single-stranded tails called "sticky ends" are produced that are complementary to each other. These complementary ends can then be joined to a fragment from another DNA that is cut with the same enzyme. These two molecules can then be joined by DNA ligase to produce a recombinant molecule.

DNA sequence reads the same from 5' to 3' on one strand as it does on the complementary strand (see figure 17.1).

Given this kind of sequence, cutting the DNA at the same base on either strand can lead to staggered cuts that produce "sticky ends." These short, unpaired sequences will be the same for any DNA that is cut by this enzyme. Thus, these sticky ends allow DNAs from different sources to be easily joined together (see figure 17.1).

DNA ligase allows construction of recombinant molecules

As just described, the two ends of a DNA molecule cut by a type II restriction enzyme have complementary sequences, and so can pair to form a duplex. But to form a stable DNA molecule from the two fragments, an enzyme is needed to join the molecules. The enzyme **DNA ligase** catalyzes the formation of a phosphodiester bond between adjacent phosphate and hydroxyl groups of DNA nucleotides. The action of ligase is to seal nicks in one or both strands (see figure 17.1). This is the same enzyme that joins Okazaki fragments on the lagging strand during DNA replication (see chapter 14).

In the toolbox of the molecular biologist, the action of ligase is necessary to create stable recombinant molecules from the fragments that restriction enzymes make possible.

Gel electrophoresis separates DNA fragments

The fragments produced by restriction enzymes would not be much use if we could not also easily separate them for analysis. The most common separation technique used is gel electrophoresis. This technique takes advantage of the negative charge on DNA molecules by using an electrical field to provide the force necessary to separate DNA molecules based on size.

The gel, which is made of either agarose or polyacrylamide and spread thinly on supporting material, provides a three-dimensional matrix that separates molecules based on size. The gel is submerged in a buffer solution containing ions that can carry current and is subjected to an electrical field.

The strong negative charges from the phosphate groups in the backbone on DNA cause it to migrate toward the positive pole (figure 17.2b). The gel acts as a sieve to separate DNA molecules based on size: The larger the molecule, the slower it will move through the gel matrix. Over a given time period for electrophoresis, smaller molecules will migrate farther than larger molecules. The DNA in gels can be visualized using a fluorescent dye that binds to DNA (figure 17.2c, d).

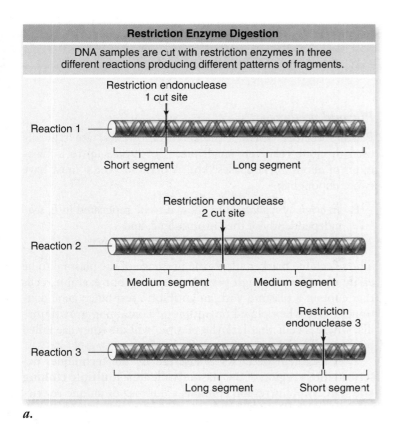

a.

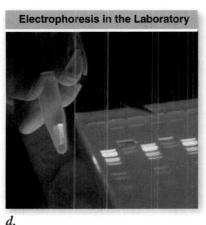

b.

figure 17.2

GEL ELECTROPHORESIS. *a.* Three restriction enzymes are used to fragment DNA into specific pieces depending on each enzyme's recognition sequence. *b.* The fragments are loaded into a gel (agarose or polyacrylamide), and an electrical current is applied. The DNA fragments migrate through the gel based on size, with larger ones moving more slowly. *c.* This results in a pattern of fragments separated based on size, with the smaller fragments migrating farther than larger ones. *d.* The fragments can be visualized by staining with the dye ethidium bromide. When the gel is exposed to UV light, the DNA with bound dye will fluoresce, appearing as pink bands in the gel. In the photograph, one band of DNA has been excised from the gel for further analysis and can be seen glowing in the tube the technician holds.

Electrophoresis is one of the most important methods in molecular biology, with uses ranging from DNA fingerprinting to DNA sequencing, both of which are described later on.

Transformation allows introduction of foreign DNA into *E. coli*

The construction of recombinant molecules is the first step toward genetic engineering. It is also necessary to be able to reintroduce these molecules into cells. In chapter 14 you learned that genetic material could be transferred between bacterial cells, as demonstrated by Frederick Griffith. This process, called *transformation*, is a natural process in the cells that Griffith was studying.

The bacterium *E. coli*, used routinely in molecular biology laboratories, does not undergo natural transformation; but artificial transformation techniques have been developed to allow introduction of foreign DNA into *E. coli*. In this way, re-combinant molecules can be propagated in a cell that will make many copies of the constructed molecules.

In general, the introduction of DNA from an outside source into a cell is referred to as transformation, and it is important in *E. coli* for molecular cloning and the propagation of cloned DNA. Researchers also want to be able to reintroduce DNA into the original cells from which it was isolated. In this case, if the transformed cell can also be used to form all or part of an organism, we call this a **transgenic** organism. Later in this chapter we explore the construction and uses of transgenic plants and animals.

Techniques to manipulate DNA include use of restriction endonucleases to cleave DNA and DNA ligase to construct recombinant molecules. Gel electrophoresis is used to separate DNA fragments. Foreign DNA can be introduced into *E. coli* through the process of artificial transformation.

17.2 Molecular Cloning

The term **clone** refers to a genetically identical copy. The technique of propagating plants by growing a new plant from a cutting of a donor plant is an early method of cloning widely used in agriculture and horticulture. The topic of cloning entire organisms is discussed in chapter 19. For now, we explore the idea of molecular cloning.

Molecular cloning involves the isolation of a specific sequence of DNA, usually one that encodes a particular protein product. This is sometimes called *gene cloning*, but the term *molecular cloning* is more accurate.

Host–vector systems allow propagation of foreign DNA in bacteria

Although short sequences of DNA can be synthesized in vitro, the cloning of large unknown sequences requires propagation of recombinant DNA molecules in vivo (in a cell). The enzymes and methods described earlier allow biologists to produce, separate, and then introduce foreign DNA into cells.

The ability to propagate DNA in a host cell requires a **vector** (something to carry the recombinant DNA molecule) that can replicate in the host when it has been introduced. Such host–vector systems are crucial to molecular biology.

The most flexible and common host used for molecular cloning is the bacterium *E. coli*, but many other hosts are now possible. Investigators routinely reintroduce cloned eukaryotic DNA, using mammalian tissue culture cells, yeast cells, and insect cells as host systems. Each kind of host–vector system allows particular uses of the cloned DNA.

The two most commonly used vectors are plasmids and phages. *Plasmids* are small, circular extrachromosomal DNAs that are dispensable to the bacterial cell. *Phages* are viruses that infect bacterial cells.

Plasmid vectors

Plasmid vectors (small, circular chromosomes) are typically used to clone relatively small pieces of DNA, up to a maximum of about 10 kilobases (kb). A plasmid vector must have two components:

1. an *origin of replication* to allow it to be replicated in *E. coli* independently of the chromosome, and
2. a *selectable marker*, usually antibiotic resistance.

The selectable marker allows the presence of the plasmid to be easily identified through genetic selection. For example, cells that contain a plasmid with an antibiotic resistance gene continue to live when plated on antibiotic-containing growth media, whereas cells that lack the plasmid will die (they are killed by the antibiotic).

A fragment of DNA is inserted by the techniques described into a region of the plasmid called the **multiple cloning site (MCS)**. This region contains a number of unique restriction sites such that when the plasmid is cut with restriction enzymes for these sites, the result is a linear plasmid. When DNA of interest is cut with the same restriction enzyme, it can then be ligated into this site. The plasmid is then introduced into cells by transformation (see figure 17.3*a*).

This region of the vector often has been engineered to contain another gene that becomes inactivated because it is now interrupted by the inserted DNA, so-called *insertional inactivation*. One of the first cloning vectors, pBR322, used another antibiotic resistance gene for insertional activation; resistance to one antibiotic and sensitivity to the other indicated the presence of inserted DNA.

More recent vectors use the gene for β-galactosidase, an enzyme that cleaves galactoside sugars such as lactose. When the enzyme cleaves the artificial substrate X-gal, a blue color is produced. In these plasmids, insertion of foreign DNA interrupts the β-galactosidase gene and a functional enzyme cannot be produced. When transformant cells are plated on medium containing both antibiotic (to select for plasmid-containing cells) and X-gal, they remain white, whereas transformants with no inserted DNA are blue (see figure 17.3a).

Phage vectors

Phage vectors are larger than plasmid vectors, and can take inserts up to 40 kb. Most phage vectors are based on the well-studied **phage lambda (λ).** Lambda-based vectors are used today primarily for *cDNA libraries*—collections of DNA fragments produced from mRNA (see below for details).

Although useful for cloning large fragments, the lambda vector has two requirements not shared with plasmid vectors:

- Both lambda and plasmids requires live cells for replication, but the virus will kill the cells such that you end up with a collection of viruses, not bacterial cells.
- The lambda genome is linear, so instead of "opening up" a circle, the middle part of the lambda genome is removed and replaced with the inserted DNA.

After the inserted DNA is ligated to the two lambda "arms," the lambda genome must be packaged into a phage head in vitro, and then used to infect *E. coli.* The two "arms" lacking any inserted DNA are not packaged efficiently into phage heads. This difference provides a kind of selection for recombinant phages, because the arms alone are not propagated (figure 17.3b).

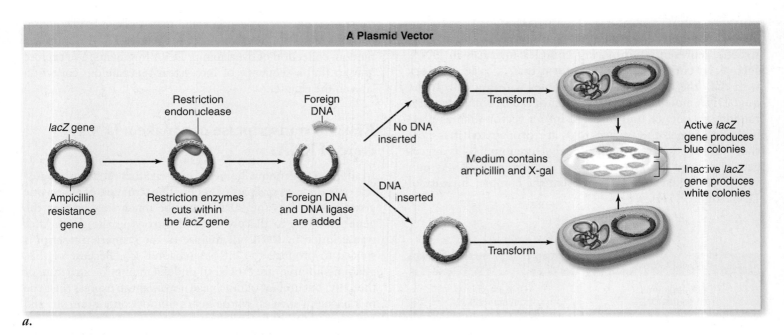

A Plasmid Vector

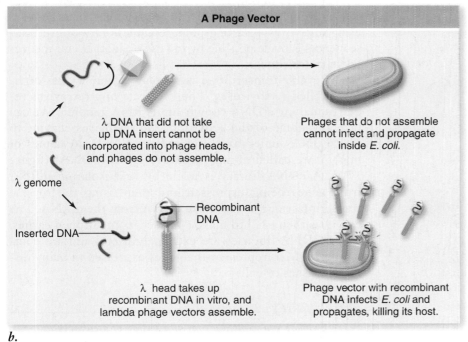

A Phage Vector

a.

b.

figure 17.3

USING PLASMID AND PHAGE VECTORS. *a.* Plasmids are cut within the β-galactosidase gene (*lacZ*), and foreign DNA and DNA ligase are added. Foreign DNA inserted into *lacZ* interrupts the coding sequence inactivating the gene. Plating cells on medium containing the antibiotic ampicillin selects for plasmid-containing cells. The medium also contains X-gal and when *lacZ* is intact (*top*), the expressed enzyme cleaves the X-gal producing blue colonies. When *lacZ* is inactivated (*bottom*), X-gal is not cleaved and colonies remain white. *b.* Phage vectors are selected for the presence of recombinant DNA by the ability of the phage to assemble in vitro, infect a host, and propagate inside its host. The phage has been engineered such that only phage genomes with inserted DNA are long enough for the packaging machinery to produce mature phage that can infect cells.

Artificial Chromosomes

The size of DNAs that can be cloned in either plasmid or phage vectors has limited the large-scale analysis of genomes. To deal with this, geneticists decided to follow the strategy of cells and construct chromosomes, leading to the development of yeast artificial chromosomes (YACs) and bacterial artificial chromosomes (BACs). Progress has also been made on mammalian artificial chromosomes. Use of artificial chromosomes is described in the next chapter.

inquiry

An investigator wishes to clone a 32-kb recombinant molecule. What do you think is the best vector to use?

DNA libraries contain the entire genome of an organism

The idea of molecular cloning depends on the ability to construct a representation of very complex mixtures in DNA, such as an entire genome, in a form that is easier to work with than the enormous chromosomes within a cell. If the huge DNA molecules in chromosomes can be converted into random fragments, and inserted into a vector such as plasmids or phages, then when they are propagated in a host they will together represent the whole genome. This aggregate is termed a **DNA library,** a collection of DNAs in a vector that taken together represent the complex mixture of DNA (figure 17.4).

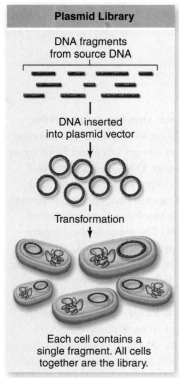

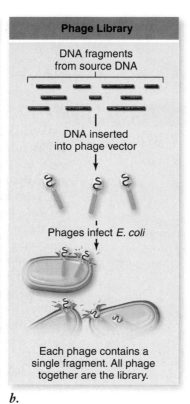

a.　　　　　　　　　　　*b.*

figure 17.4

CREATING DNA LIBRARIES. DNA libraries can be produced using *(a)* plasmid vectors or *(b)* phage vectors.

Conceptually the simplest kind of DNA library that can be made is a **genomic library,** a representation of the entire genome in a vector. This genome is randomly fragmented by partially digesting it with a restriction enzyme that cuts frequently. By not cutting the DNA to completion, not all sites are cleaved, and which sites are cleaved is random. The random fragments are then inserted into a vector and introduced into host cells.

Genomic libraries were originally made in phage λ because of the larger insertion sizes. Now with the advent of genomics and the analysis of entire genomes, these libraries are usually constructed in bacterial artificial chromosomes (BACs).

A variety of different kinds of libraries can be made depending on the source DNA used. Any particular clone in the library contains only a single DNA, and all of them together make up the library. Keep in mind that unlike a library full of books, which is organized and catalogued, a DNA library is a random collection of overlapping DNA fragments. We explore how to find a sequence of interest in this random collection later in the chapter.

Reverse transcriptase can make a DNA copy of RNA

In addition to genomic libraries, investigators often wish to isolate only the *expressed* part of genes. The structure of eukaryotic genes is such that the mRNA may be much smaller than the gene itself due to the presence of introns in the gene. After transcription by RNA polymerase II, the primary transcript is spliced to produce the mRNA (chapter 15). Because of this, genomic libraries are crucial to understanding the structure of the gene, but are not of much use if we want to express the gene in a bacterial species, whose genes do not contain introns and has no mechanism for splicing.

A library of only expressed sequences represents a much smaller amount of DNA than the entire genome, but it requires using mRNA as a starting point. Such a library of expressed sequences is made possible by the use of another enzyme: **reverse transcriptase.**

Reverse transcriptase was isolated from a class of viruses called retroviruses. The life cycle of a retrovirus requires making a DNA copy from its RNA genome. We can take advantage of the activity of the retrovirus enzyme to make DNA copies from isolated mRNA. DNA copies of mRNA are called **complementary DNA (cDNA)** (figure 17.5). A cDNA library is made by first isolating mRNA from genes being expressed and then using the reverse transcriptase enzyme to make cDNA from the mRNA. The cDNA is then used to make a library, as mentioned earlier. These cDNA libraries are extremely useful and are commonly made to represent the genes expressed in many different tissues or cells.

inquiry

Suppose you wanted a copy of a section of a eukaryotic genome that included the introns and exons. Would the creation of cDNA be a good way to go about this?

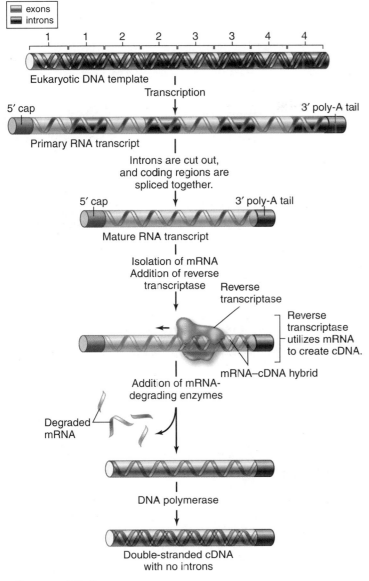

figure 17.5

THE FORMATION OF cDNA. A mature mRNA transcript is usually much smaller than the gene due to the loss of intron sequences by splicing. When mRNA is isolated from the cytoplasm of a cell the enzyme reverse transcriptase can use this as a template to make a DNA strand complementary to the mRNA. That newly made strand of DNA is the template for the enzyme DNA polymerase, which assembles a complementary DNA strand along it, producing cDNA, a double-stranded DNA version of the intron-free mRNA.

Hybridization allows identification of specific DNAs in complex mixtures

The technique of **molecular hybridization** is commonly used to identify specific DNAs in complex mixtures such as libraries. Hybridization, also called annealing, takes advantage of the specificity of base-pairing between the two strands of DNA. If a DNA molecule is denatured, that is, the two strands are separated, the strands can only reassociate with partners having the correct complementary sequence. Molecular

biologists can take advantage of this feature experimentally to use a known, specific DNA molecule to find its partner in a complex mixture.

Any single-stranded nucleic acid (DNA or RNA) can be labeled with radioactivity or another detectable label, such as a fluorescent dye. This can then be used as a probe to identify its complement in a complex mixture of DNA or RNA. This renaturing is termed *hybridization* because the combination of labeled probe and unlabeled DNA form a hybrid molecule through base-pairing.

Probes have been made historically by a variety of techniques. One technique involved isolating a protein of interest and then chemically sequencing the protein. With the protein sequence in hand, the DNA sequence could be predicted using the genetic code. This information could then be used to make a synthetic DNA for use as a probe.

Specific clones can be isolated from a library

The isolation of a specific clone from the random collection that is a DNA library is akin to finding the proverbial needle in a haystack. It requires some information about the gene of interest. For example, many of the first genes isolated were those that are highly expressed in a specific cell type, such as the globin genes that encode the proteins found in the oxygen carrier hemoglobin.

Hybridization is the most common way of identifying a clone within a DNA library. This procedure is outlined for a DNA library in a plasmid vector in figure 17.6.

In the early days of molecular biology, individual investigators made their own DNA libraries, as is shown earlier in figure 17.4. Now, genomic and cDNA libraries are commercially available for a large number of organisms. Screening such a library involves growing the library on agar plates, making a replica of the library, and screening for the cloned sequence of interest.

Stage 1: Plating the library

Physically, the library is either a collection of bacterial viruses that each contain an inserted DNA, or bacterial cells that each harbor a plasmid or artificial chromosome with inserted DNA. To find a specific clone, the library needs to be represented in an organized fashion. Figure 17.6 shows this representation for a plasmid vector. The library of bacteria containing plasmids is grown on agar plates at a high density, but not so high that individual colonies cannot be distinguished.

Stage 2: Replicating the library

Once the library has been grown on plates, a replica can be made by laying a piece of filter paper on the plate; some of the viruses or cells in each colony will stick to the filter, and some will be left on the plate. The result is a copy of the library on a piece of filter paper. The DNA can be affixed to the filter paper by baking or by cross-linking it to the filter using UV light.

Stage 3: Screening the library

Once a replica of the library has been formed on a filter, a specific clone can be identified by hybridization. The probe, which represents the specific sequence of interest, is labeled with a

figure 17.6

SCREENING A LIBRARY USING HYBRIDIZATION. This takes advantage of the ability of DNA to be denatured and renatured, with complementary strands finding each other. Cells containing the library are plated on agar plates. A replica of the plates is made using special filter paper, nitrocellulose or nylon, that binds to single-stranded DNA. The filter paper with replica colonies is treated to lyse the cells and denature the DNA such that there will now be a pattern of DNA bound to the filter corresponding to the pattern of colonies. When a radioactive probe is added, it will find complementary DNA and form hybrids at the site of colonies that contained the gene of interest.

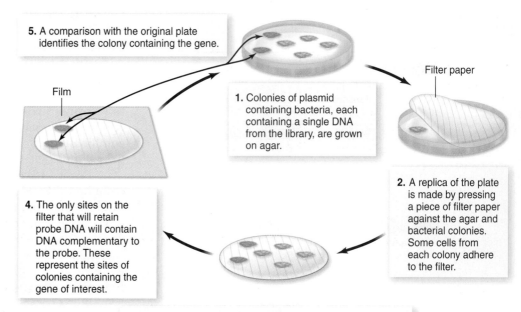

5. A comparison with the original plate identifies the colony containing the gene.

Film

1. Colonies of plasmid containing bacteria, each containing a single DNA from the library, are grown on agar.

Filter paper

2. A replica of the plate is made by pressing a piece of filter paper against the agar and bacterial colonies. Some cells from each colony adhere to the filter.

4. The only sites on the filter that will retain probe DNA will contain DNA complementary to the probe. These represent the sites of colonies containing the gene of interest.

3. The filter is washed with a solution to break the cells open and denature the DNA, which sticks to the filter at the site of each colony. The filter is incubated with a radioactively labeled probe that can form hybrids with complementary DNA in the gene of interest.

radioactive nucleotide. The probe is then added to the filters with the library replicated on them. Film sensitive to radioactive emissions is then placed in contact with the filters; where radioactivity is present, a dark spot appears on the film. When the film is aligned with the original plate, the clone of interest can be identified (see figure 17.6).

> Molecular cloning is the isolation of a specific DNA sequence. Host–vector systems allow us to propagate DNA in *E. coli* and other organisms. DNA libraries are representations of complex mixtures of DNA, such as an entire genome, in a host–vector system. DNA libraries are often screened for specific clones using molecular hybridization, which uses a labeled probe to find DNA complementary to the probe.

17.3 DNA Analysis

Molecular cloning provides specific DNA for further manipulation and analysis. The number of ways that DNA can be manipulated could fill the rest of this book, so for our purposes, we highlight a few important methods of analysis and uses of molecular clones.

Restriction maps provide molecular "landmarks"

If you are new to a city, the easiest way to find your way around is to obtain a map and compare that map with your surroundings. In a similar fashion, molecular biologists need maps to analyze and compare cloned DNAs.

The first kind of physical maps were restriction maps, composed of the location and order of sites cut by the battery of restriction enzymes available. Initially, these maps were created by cutting the DNA with different enzymes, separating the fragments by gel electrophoresis, and analyzing the resulting patterns. Although this method is still in use, many restriction maps are now generated by computer searching of known DNA sequences for the sites cut by restriction enzymes.

Southern blotting reveals DNA differences

Once a gene has been cloned, it may be used as a probe to identify the same or a similar gene in DNA isolated from a cell or tissue (figure 17.7). In this procedure, called a **Southern blot,** DNA from the sample is cleaved into fragments with a restriction endonuclease, and the fragments are separated by gel electrophoresis. The double-stranded helix of each DNA fragment is then denatured into single strands by making the pH of the gel basic, and the gel is "blotted" with a sheet of filter paper, transferring some of the DNA strands to the sheet.

Next, the filter is incubated with a labeled probe consisting of purified, single-stranded DNA corresponding to a specific gene (or mRNA transcribed from that gene). Any fragment that has a nucleotide sequence complementary to the probe's sequence hybridizes with the probe (see figure 17.7).

This kind of blotting technique has also been adapted for use with RNA and proteins. When mRNA is separated by electrophoresis, the technique is called a **Northern blot,** and the methodology is the same except for the starting material (mRNA instead of DNA) and that no denaturation step

1. Electrophoresis is performed, using radioactively labeled markers as a size guide in the first lane.

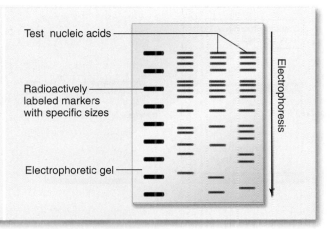

Test nucleic acids

Radioactively labeled markers with specific sizes

Electrophoretic gel

Electrophoresis

2. The gel is covered with a sheet of nitrocellulose and placed in a tray of buffer on top of a sponge. Alkaline chemicals in the buffer denature the DNA into single strands. The buffer wicks its way up through the gel and nitrocellulose into a stack of paper towels placed on top of the nitrocellulose.

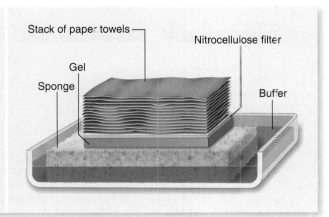

Stack of paper towels

Nitrocellulose filter

Gel

Sponge

Buffer

3. DNA in the gel is transferred, or "blotted," onto the nitrocellulose.

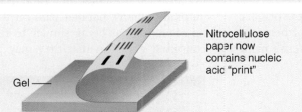

Nitrocellulose paper now contains nucleic acid "print"

Gel

4. Nitrocellulose with bound DNA is incubated with radioactively labeled nucleic acids and is then rinsed.

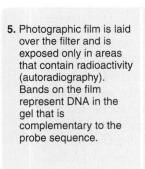

Sealed container

Radioactive probe (single-stranded DNA)

—AATGG—

—TTACC—

DNA fragments within bands

5. Photographic film is laid over the filter and is exposed only in areas that contain radioactivity (autoradiography). Bands on the film represent DNA in the gel that is complementary to the probe sequence.

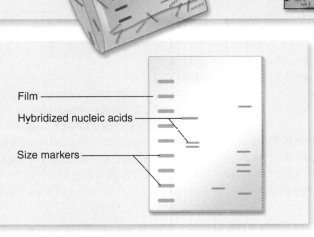

Film

Hybridized nucleic acids

Size markers

figure 17.7

THE SOUTHERN BLOT PROCEDURE. E. M. Southern developed this procedure in 1975 to enable DNA fragments of interest to be visualized in a complex sample containing many other fragments of similar size. In steps 1–3, the DNA is separated on a gel, and then transferred ("blotted") onto a solid support medium such as nitrocellulose paper or a nylon membrane. Sequences of interest can be detected by using a radioactively labeled probe. This probe (usually several hundred nucleotides in length) of single-stranded DNA (or an mRNA complementary to the gene of interest) is incubated with the filter containing the DNA fragments. All DNA fragments that contain nucleotide sequences complementary to the probe will form hybrids with the probe. Only a short segment of the probe and the complementary sequence are shown in panel 4. The fragments differ in size, with the smallest running the farthest in the gel. The fragments of interest are then detected using photographic film. A representative image is shown in panel 5. The use of film for detection is being replaced by phosphor imagers, computer-controlled devices that have electronic sensors for light or radioactive emissions.

figure 17.8

RESTRICTION FRAGMENT LENGTH POLYMORPHISM (RFLP) ANALYSIS. *a.* Three samples of DNA differ in their restriction sites due to a single base-pair substitution in one case and a sequence duplication in another case. *b.* When the samples are cut with a restriction endonuclease, different numbers and sizes of fragments are produced. *c.* Gel electrophoresis separates the fragments, and different banding patterns result.

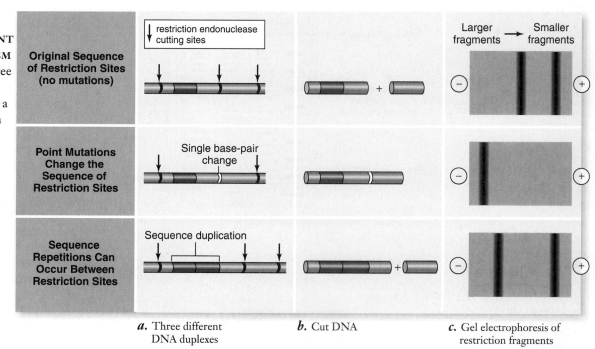

a. Three different DNA duplexes
b. Cut DNA
c. Gel electrophoresis of restriction fragments

is required. Proteins can also be separated by electrophoresis and blotted by a procedure called a **Western blot.** In this case both the electrophoresis and the detection step are different from Southern blotting. The detection, in this case, requires an antibody that can bind to one protein.

The names of these techniques all go back to the original investigator, whose last name was Southern; the Northern and Western blotting names were word play on Southern's name using the cardinal points of the compass.

RFLP analysis

In some cases, an investigator wants to do more than find a specific gene, but instead is looking for variation in the genes of different individuals. One powerful way to do this is by analyzing **restriction fragment length polymorphisms,** or **RFLPs,** using Southern blotting (figure 17.8).

Point mutations that change the sequence of DNA can eliminate sequences recognized by restriction enzymes or create new recognition sequences, changing the pattern of fragments seen in a Southern blot. Sequence repetitions may also occur between the restriction endonuclease sites, and differences in repeat number between individuals can also alter the length of the DNA fragments. These differences can all be detected with Southern blotting.

When a genetic disease has an associated RFLP, the RFLP can be used to diagnose the disease. Huntington disease, cystic fibrosis, and sickle cell anemia all have associated RFLPs that have been used as molecular markers for diagnosis.

DNA fingerprinting

This technique has been used in **DNA fingerprinting.** When a probe is made for DNA that is repetitive, it often detects a large number of fragments. These fragments are often not identical in different individuals. We say that the population is

polymorphic for these molecular markers. These markers can be used as DNA "fingerprints" in criminal investigations and other identification applications.

Figure 17.9 shows the DNA fingerprints a prosecuting attorney presented in a rape trial in 1987. They consist of autoradiographs, parallel bars on X-ray film. These bars can be thought of as being similar to the product price codes on consumer goods in that they may provide unique identifica-

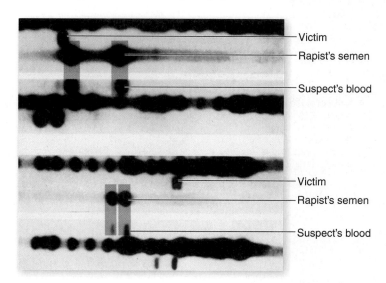

figure 17.9

TWO OF THE DNA PROFILES THAT LED TO THE CONVICTION OF TOMMIE LEE ANDREWS FOR RAPE IN 1987. The two DNA probes seen here were used to characterize DNA isolated from the victim, the semen left by the rapist, and the suspect. The dark channels are multiband controls. There is a clear match between the suspect's DNA and the DNA of the rapist's semen in these two profiles.

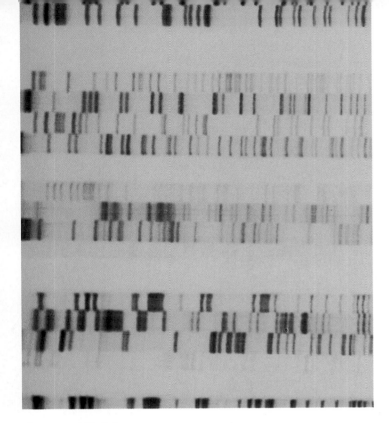

figure 17.10

LADDER OF FRAGMENTS USED IN DNA SEQUENCING. The photo shows the autoradiograph of the fragments generated by DNA-sequencing reactions. These fragments are generated by either organic reactions that cleave at specific bases or enzymatic reactions that terminate in specific bases. The gel can separate fragments that differ by a single base.

tion. Each bar represents the position of a DNA restriction endonuclease fragment produced by techniques similar to those described in figures 17.7 and 17.8. The lane with many bars represents a standardized control.

Two different probes were used to identify the restriction fragments. A vaginal swab had been taken from the victim within hours of her attack; from it, semen was collected and its DNA analyzed for restriction endonuclease patterns.

Compare the restriction endonuclease patterns of the semen to that of blood from the suspect. You can see that the suspect's two patterns match that of the rapist (and are not at all like those of the victim). The suspect was Tommie Lee Andrews, and on November 6, 1987, the jury returned a verdict of guilty. Andrews became the first person in the United States to be convicted of a crime based on DNA evidence.

Since the Andrews verdict, DNA fingerprinting has been admitted as evidence in more than 2000 court cases. Although some probes highlight profiles shared by many people, others are quite rare. Using several probes, the probability of identity can be calculated or identity can be ruled out.

Laboratory analyses of DNA samples, however, must be carried out properly—sloppy procedures could lead to a wrongful conviction. After widely publicized instances of questionable lab procedures, national standards are being developed.

DNA sequencing provides information about genes and genomes

The ultimate level of analysis is determination of the actual sequence of bases in a DNA molecule. The development of sequencing technology has paralleled the advancement of molecular biology. The field of genomics was born out of the ability to determine the sequence of an entire genome relatively rapidly.

The basic idea used in DNA sequencing is to generate a set of nested fragments that each begin with the same sequence and end in a specific base. When this set of fragments is separated by high-resolution gel electrophoresis, the result is a "ladder" of fragments (figure 17.10) in which each band consists of fragments that end in a specific base. By starting with the shortest fragment, one can then read the sequence by moving up the ladder.

The problem then became how to generate the sets of fragments that end in specific bases. In the early days of sequencing, both a chemical method and an enzymatic method were utilized. The chemical method involved organic reactions specific for the different bases that made breaks in the DNA chains at specific bases. The enzymatic method used DNA polymerase to synthesize chains, but it also included in the reaction modified nucleotides that could be incorporated but not extended: so-called *chain terminators*. The enzymatic method has proved more versatile, and it is easier to adapt to different uses.

Enzymatic sequencing

The enzymatic method of sequencing was developed by Fredrick Sanger, who also was the first to determine the complete sequence of a protein. This method uses dideoxynucleotides as chain terminators in DNA synthesis reactions. A **dideoxynucleotide** has H in place of OH at both the 2′ position and at the 3′ position.

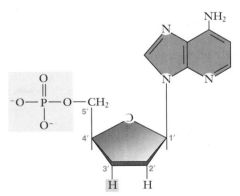

All DNA nucleotides lack —OH at the 2′ carbon of the sugar, but dideoxynucleotides have no 3′ —OH at which the enzyme can add new nucleotides. Thus the chain is terminated.

The experimenter must perform four separate reactions, each with a single dideoxynucleotide, to generate a set of fragments that terminate in specific bases. Thus all of the fragments produced in the A reaction incorporate dideoxyadenosine and must end in A, and the same for the other three reactions with different terminators. When these fragments are separated by high-resolution gel electrophoresis, each reaction is run in a different track, or lane, to generate a pattern of nested fragments that can be read from the smallest fragment to fragments that are each longer by one base (figure 17.11a).

Notice that since this is a DNA polymerase reaction, it requires a primer to begin synthesis. The vectors used for DNA sequencing have known regions next to the site where DNA is inserted. Short DNAs are then synthesized that are complementary to these regions that can be used as primers. This serves the dual purposes of providing a primer and ensuring that the first few bases sequenced are known because they are known in the vector itself. This allows the investigator to determine where the sequence of interest begins. As the se-

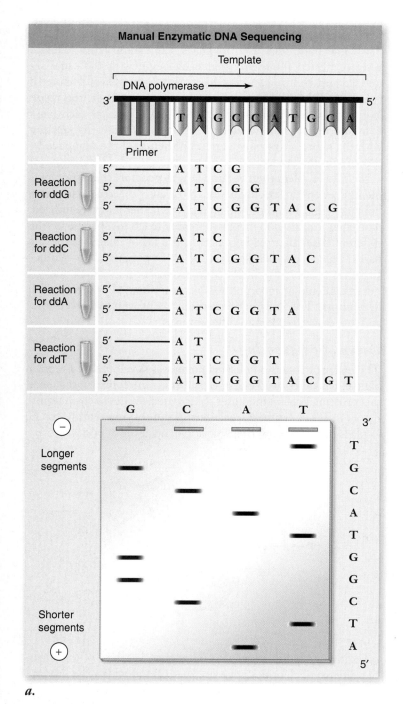

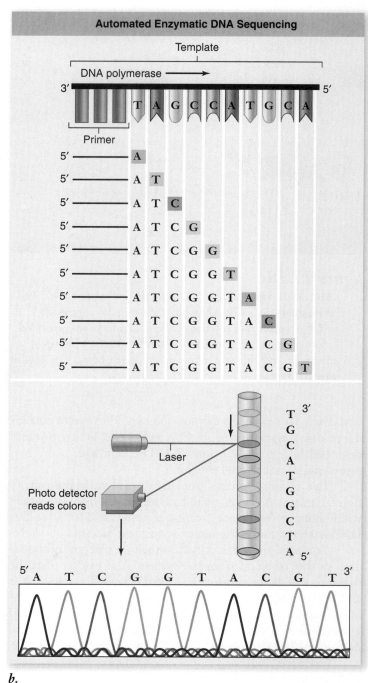

a.

b.

figure 17.11

MANUAL AND AUTOMATED ENZYMATIC DNA SEQUENCING. The sequence to be determined is shown at the top as a template strand for DNA polymerase with a primer attached. *a.* In the manual method, four reactions were done, one for each nucleotide. For example, the A tube would contain dATP, dGTP, dCTP, dTTP, and ddATP. This leads to fragments that end in A due to the dideoxy terminator. The fragments generated in each reaction are shown along with the results of gel electrophoresis. *b.* In automated sequencing, each ddNTP is labeled with a different color fluorescent dye, which allows the reaction to be done in a single tube. The fragments generated by the reactions are shown. When these are electrophoresed in a capillary tube, a laser at the bottom of the tube excites the dyes, and each will emit a different color that is detected by a photodetector.

quence is generated, new primers can be designed near the end of the known sequence and DNA synthesized to use as a primer to extend the region sequenced in the next set of reactions.

Automated sequencing

The technique of enzymatic sequencing is very powerful, but it is also labor-intensive and takes a significant amount of time. It requires a series of enzymatic manipulations, time for electrophoresis, then time to expose the gel to film. At the end of this, a skilled researcher can read around 300 bases of sequence reliably. The development of automated techniques made sequencing a much more practical and less human-intensive procedure.

Automated sequencing machines use fluorescent dyes instead of a radioactive label and separate the products of the sequencing reactions using gels in thin capillary tubes instead of the large slab gels. The tubes run in front of a laser that excites the dyes, causing them to fluoresce. With a different colored dye for each base, a photodetecter can determine the identity of each base by its color.

The data are assembled by a computer that generates a visual image consisting of different colored peaks; these are converted into the raw sequence data (figure 17.11b). The sequence data come directly from the electrophoresis, eliminating the time needed for exposing gels to film and for manual reading of the sequences. The use of different colored dyes also reduces handling and allows more sequence to be produced at one time.

With increases in the number of samples per run and the length of sequences able to be read, along with decreases in handling time, the amount of sequence information that can be generated is limited mainly by the number of machines that can be run at once.

The polymerase chain reaction accelerated the process of analysis

The next revolution in molecular biology was the development of the **polymerase chain reaction (PCR)**. Kary Mullis developed PCR in 1983 while he was a staff chemist at the Cetus Corporation; in 1993, he was awarded the Nobel Prize in chemistry for his discovery.

The idea of the polymerase chain reaction is simple: Two primers are used that are complementary to the opposite strands of a DNA sequence, oriented toward each other. When DNA polymerase acts on these primers and the sequence of interest, the primers produce complementary strands, each containing the other primer. If this procedure is done cyclically, the result is a large quantity of a sequence corresponding to the DNA that lies between the two primers (figure 17.12).

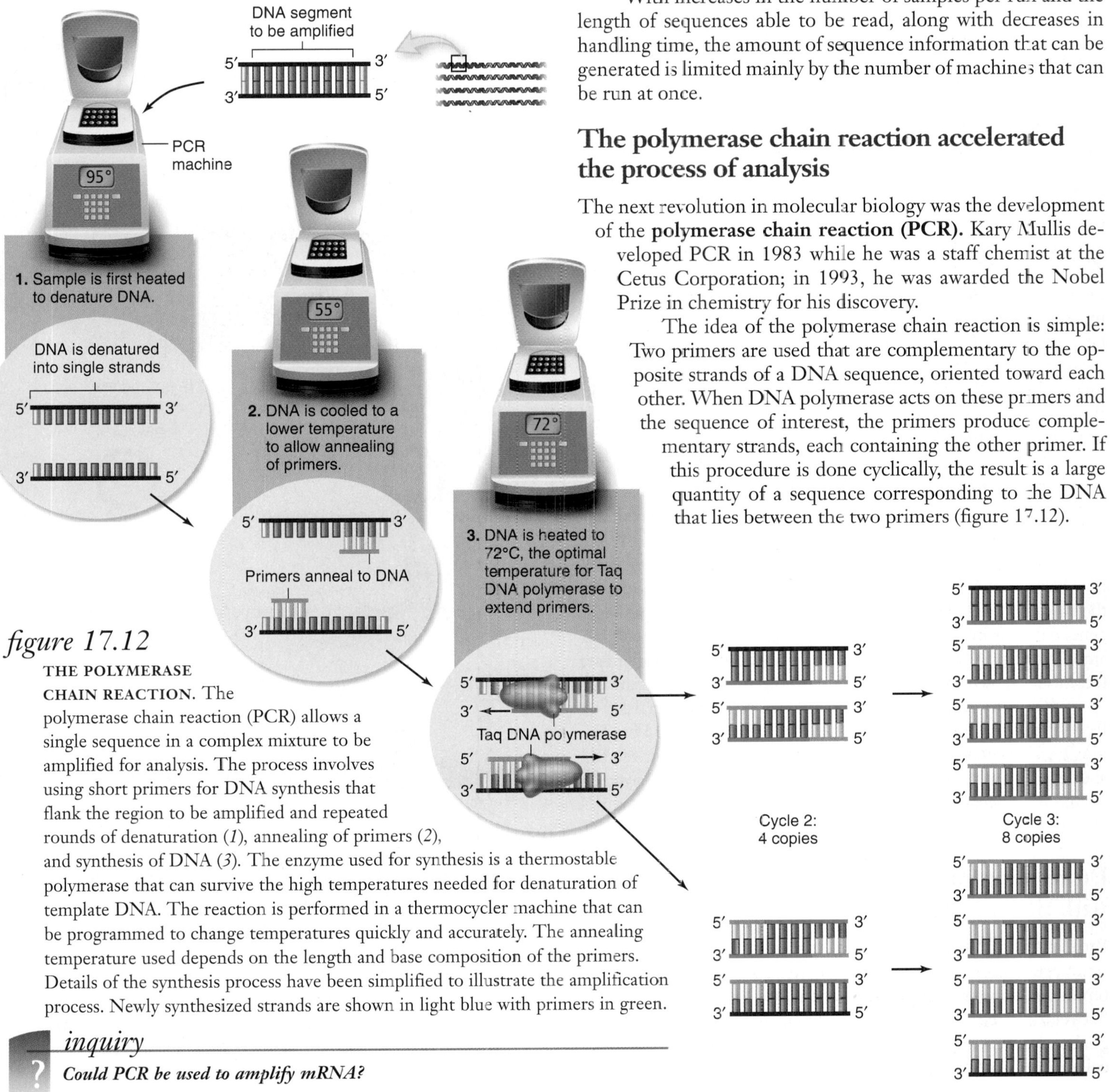

figure 17.12

THE POLYMERASE CHAIN REACTION. The polymerase chain reaction (PCR) allows a single sequence in a complex mixture to be amplified for analysis. The process involves using short primers for DNA synthesis that flank the region to be amplified and repeated rounds of denaturation (1), annealing of primers (2), and synthesis of DNA (3). The enzyme used for synthesis is a thermostable polymerase that can survive the high temperatures needed for denaturation of template DNA. The reaction is performed in a thermocycler machine that can be programmed to change temperatures quickly and accurately. The annealing temperature used depends on the length and base composition of the primers. Details of the synthesis process have been simplified to illustrate the amplification process. Newly synthesized strands are shown in light blue with primers in green.

DNA segment to be amplified

PCR machine — 95°

1. Sample is first heated to denature DNA.

DNA is denatured into single strands

2. DNA is cooled to a lower temperature to allow annealing of primers. — 55°

Primers anneal to DNA

3. DNA is heated to 72°C, the optimal temperature for Taq DNA polymerase to extend primers. — 72°

Taq DNA polymerase

Cycle 2: 4 copies

Cycle 3: 8 copies

inquiry

Could PCR be used to amplify mRNA?

The PCR procedure

Two developments turned this simple concept into a powerful technique. First, each cycle requires denaturing the DNA after each round of synthesis, which is easily done by raising the temperature; however, this destroys most polymerase enzymes. The solution was to isolate a DNA polymerase from a thermophilic, or heat-loving bacteria, *Thermus aquaticus*. This enzyme, called **Taq polymerase,** allows the reaction mixture to be repeatedly heated without destroying enzyme activity.

The second innovation was the development of machines with heating blocks that can be rapidly cycled over large temperature ranges with very accurate temperature control.

Thus each cycle of PCR involves three steps:

1. Denaturation (high temperature)
2. Annealing of primers (low temperature)
3. Synthesis (intermediate temperature)

Steps 1 to 3 are now repeated, and the two copies become four. It is not necessary to add any more polymerase, because the heating step does not harm Taq polymerase. Each complete cycle, which takes only 1–2 min, doubles the number of DNA molecules. After 20 cycles, a single fragment produces more than one million (2^{20}) copies!

In this way, the process of PCR allows the **amplification** of a single DNA fragment from a small amount of a complex mixture of DNA. This result is similar to what is isolated using molecular cloning, but in the case of PCR, the DNA cannot be reintroduced directly into a cell. The PCR product can be analyzed using electrophoresis, cloned into a vector for other manipulations, or directly sequenced. There are limitations on the size of the fragment that can be synthesized in this way, but it has been adapted for an amazing number of uses.

Applications of PCR

PCR, now fully automated, has revolutionized many aspects of science and medicine because it allows the investigation of minute samples of DNA. In criminal investigations, DNA fingerprints can now be prepared from the cells in a tiny speck of dried blood or at the base of a single human hair. In medicine, physicians can detect genetic defects in very early embryos by collecting a single cell and amplifying its DNA. Due to its sensitivity, speed, and ease of use, technicians now routinely use PCR methods for these applications.

PCR has even been used to analyze mitochondrial DNA from the early human species *Homo neanderthalensis*. This application provides the first glimpse of data from extinct related species. The amplification of ancient DNA has been a controversial field because contamination with modern DNA is difficult to avoid. But it remains an active area of genetic research.

Protein interactions can be detected with the two-hybrid system

Protein–protein interactions form the basis of many biological structures. Just as human society is ultimately dependent on interactions between people, cells are dependent on interactions between proteins. This observation has led to the large-scale goal of determining all interactions among proteins in different cells. This goal once would have been a dream, but it is now becoming a reality. The yeast two-hybrid system is one of the workhorses of this kind of analysis (figure 17.13).

The yeast two-hybrid system integrates much of the technology discussed in this chapter. It takes advantage of one feature of eukaryotic gene regulation, namely that the structure of proteins that turn on eukaryotic gene expression, transcription factors, have a modular structure.

The Gal4 gene of yeast encodes a transcriptional activator with modular structure consisting of a DNA-binding domain that binds sequences in Gal4-responsive promoters, and an activation domain that interacts with the transcription apparatus to turn on transcription. The system uses two vectors: one containing a fragment of the Gal4 gene that

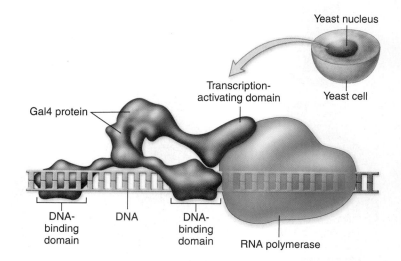

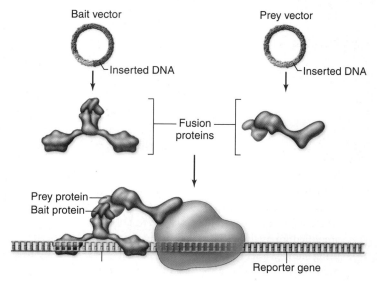

figure 17.13

THE YEAST TWO-HYBRID SYSTEM DETECTS INTERACTING PROTEINS. The Gal4 protein is a transcriptional activator (*top*). The Gal4 gene has been split and engineered into two different vectors such that one will encode only the DNA-binding domain (bait vector) and the other the transcription-activating domain (prey vector). When other genes are spliced into these vectors, they produce fusion proteins containing part of Gal4 and the proteins to be tested. If the proteins being tested interact, this will restore Gal4 function and activate expression of a reporter gene.

encodes the DNA binding domain, and another containing a fragment of the Gal4 gene that encodes the transcription activation domain. Neither of these alone can activate transcription.

When cDNAs are inserted into each of these two vectors in the proper reading frame, they are expressed as a single protein consisting of the protein of interest and part of the Gal4 activator protein (figure 17.13). These hybrid proteins are called *fusion proteins* since they are literally fused in the same polypeptide chain. The DNA-binding hybrid is called the *bait*, and the activating domain hybrid is called the *prey*.

These vectors are inserted into cells of different mating types that can be crossed. One of these vectors also contains a so-called *reporter gene* encoding a protein that can be assayed for enzymatic activity. The reporter gene is under control of a Gal4-responsive regulatory region, so that when active Gal4 is present, the reporter gene is expressed and can be detected by an enzymatic assay.

The DNA-binding hybrid binds to DNA adjacent to the reporter gene. When the two proteins in bait and prey interact, the prey hybrid brings the activating domain into position to turn on gene expression from the reporter gene (see figure 17.13).

The beauty of this system is that it is both simple and flexible. It can be used with two known proteins or with a known protein in the bait vector and entire cDNA libraries in the prey vector. In the latter case, all of the possible interactions in a cell type can be mapped.

It is already clear that there are even more protein interactions in cells than anticipated. In the future these data will form the basis for understanding the networks of protein interactions that make up the normal activities of a cell.

Restriction enzymes can be used to construct physical maps of DNA. The technique of Southern blotting allows the detection of DNA in complex mixtures, such as DNA isolated form cells or tissues, and can be used to anaylze differences between individuals. The ultimate level of analysis is to determine the actual DNA sequence. DNA sequencing uses a modified DNA polymerase reaction that contains chain terminators. The polymerase chain reaction (PCR) has changed molecular analysis allowing the production of a large amount of a specific DNA from a small amount of starting material. The yeast two-hybrid system is used to detect protein–protein interactions.

Genetic Engineering

The ability to clone individual genes for analysis ushered in an era of unprecedented research advancement. At the time, these advancements were not accompanied by grand announcements of potential medical breakthroughs and other applications. The ability to truly genetically engineer any kind of cell or organism was a long way off. But we are now approaching this ability, and it has generated much excitement and also controversy.

Expression vectors allow production of specific gene products

A variety of specialized vectors have been constructed since the development of cloning technology. One very important type of vector are the **expression vectors.** These vectors contain the sequences necessary to drive expression of inserted DNA in a specific cell type, namely the correct sequences to permit transcription and translation of the sequences. The production of recombinant proteins in bacteria, for example, uses expression vectors with bacterial promoters and other control regions. The bacteria transformed by such vectors synthesize large amounts of the protein encoded by the inserted DNA. A number of pharmaceuticals have been produced in this way, including the first, insulin, used to treat diabetes. (This type of application is discussed in more detail in the next section.)

Genes can be introduced across species barriers

The ability to reintroduce genes into an original host cell, or to introduce genes into another host, is true genetic engineering. An animal containing a gene that has been introduced without the use of conventional breeding is called a **transgenic animal.** We will explore a number of uses of transgenic animals in medicine and agriculture, but it is important to realize that their original use was for basic research.

The ability to engineer genes in context or out of context allows an experimenter to ask questions that could never be asked otherwise. A dramatic example was the use of the *eyeless* gene from mice in *Drosophila*. When this mouse gene was introduced into *Drosophila*, it was shown to be able to substitute for a *Drosophila* gene in organizing the formation of eyes. It could even cause the formation of eyes in incorrect locations when expressed in tissue that did not normally form eyes. This amazing result shows that the formation of the compound eye in an insect is not so different from the formation of the complex vertebrate eye.

Cloned genes can be used to construct "knockout" mice

One of the most important technologies for research purposes is **in vitro mutagenesis,** the ability to create mutations at any site in a cloned gene to examine the effect on function. Rather than depending on mutations induced by chemical agents or radiation in intact organisms, which is time- and labor-intensive, the DNA itself is directly manipulated. The ultimate use of this approach is to be able to replace the wild-type gene with a mutant copy to test the function of the mutated gene. Developed first in yeast, this technique has now been extended to the mouse.

In mice, this technique has produced **knockout mice** in which a known gene is inactivated ("knocked out"). The effect of loss of this function is then assessed in the adult mouse, or if it is lethal, the stage of development at which function fails can be determined. The idea is simple, but the technology is quite

figure 17.14

CONSTRUCTION OF A KNOCKOUT MOUSE. Steps in the construction of a knockout mouse. Some technical details have been omitted, but the basic concept is shown.

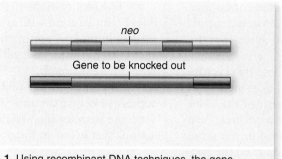

1. Using recombinant DNA techniques, the gene encoding resistance to *neomycin (neo)* is inserted into the gene of interest, disrupting it. The *neo* gene also confers resistance to the drug G418, which kills mouse cells. This construct is then introduced into ES cells.

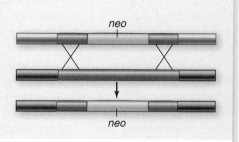

2. In some ES cells, the construct will recombine with the chromosomal copy of the gene to be knocked out. This replaces the chromosomal copy with the *neo* disrupted construct. This is the equivalent to a double crossover event in a genetic cross.

complex. A streamlined description of the steps in this type of experiment are outlined below and illustrated in figure 17.14:

1. The cloned gene is disrupted by replacing it with a marker gene using recombinant DNA techniques described earlier. The marker gene codes for resistance to the antibiotic neomycin in bacteria, and allows mouse cells to survive when grown in a medium containing the related drug G418. The construction is done such that the marker gene is flanked by the DNA normally flanking the gene of interest in the chromosome.

2. The interrupted gene is introduced into **embryonic stem cells (ES cells).** These cells are derived from early embryos and can develop into different adult tissues. In these cells, the gene can recombine with the chromosomal copy of the gene based on the flanking DNA. This is the same kind of recombination used to map genes (chapter 13). The knockout gene with the drug resistance gene does not have an origin of replication, and thus it will be lost if no recombination occurs. Cells are grown in medium containing G418 to select for recombination events. (Only those containing the marker gene can grow in the presence of G418.)

3. The embryonic stem cells containing the knocked-out gene is injected into a blastocyst stage embryo, which is then implanted into a pseudopregnant female (one that has been mated with a vasectomized male and as a result has a receptive uterus). The pups from this female have one copy of the gene of interest knocked out. Transgenic animals can then be crossed to generate homozygous lines. These homozygous lines can be analyzed for phenotypes.

In conventional genetics, genes are identified based on mutants that show a particular phenotype. Molecular genetic techniques are then used to find the gene and isolate a molecular clone for analysis. The use of knockout mice is an example of **reverse genetics:** One takes a cloned gene of unknown function, then uses it to make a mutant deficient in that gene. A geneticist can then assess the effect on the entire organism of eliminating a single gene.

Sometimes this approach leads to surprises, such as when the gene for the p53 tumor suppressor was knocked out. Because this protein is found mutated in many human cancers, and plays a key role in the regulation of the cell cycle (chapter 10), it was thought to be essential—the knockout was expected to be lethal. Instead, the mice were born normal; that is, development had proceeded normally. These mice do have a phenotype however, they exhibit an increased incidence of tumors in a variety of tissues as they age.

> Expression vectors that contain cloned genes allow the production of known proteins in different cells. This can be done for research purposes or to produce pharmaceuticals. Genes can also be introduced across species barriers. In mouse, mutations can be engineered in cloned genes, then reintroduced into the animal to create "knockout" mice deficient in specific genes.

17.5 Medical Applications

The early days of genetic engineering led to a rash of startup companies, many of which are no longer in business. At the same time, all of the major pharmaceutical companies either began research in this area or actively sought smaller companies with promising technology. The number of applications of this technology are far too numerous to mention here, so a few are highlighted; the section following discusses agricultural applications.

Human proteins can be produced in bacteria

The first and perhaps most obvious commercial application of genetic engineering was the introduction of genes that encode clinically important proteins into bacteria. Because bacterial cells can be grown cheaply in bulk, bacteria that incorporate recombinant genes can synthesize large amounts of the proteins those genes specify. This method has been used to produce several forms of

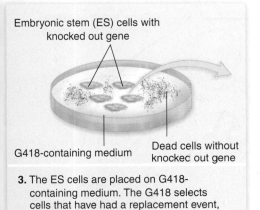

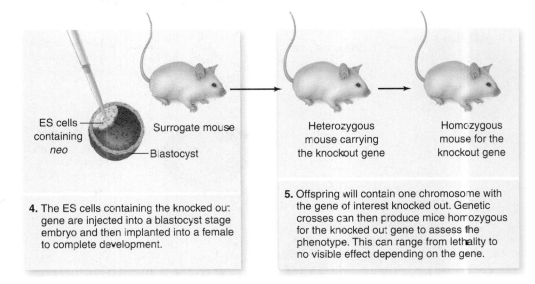

3. The ES cells are placed on G418-containing medium. The G418 selects cells that have had a replacement event, and now contain a copy of the knocked out gene.

4. The ES cells containing the knocked out gene are injected into a blastocyst stage embryo and then implanted into a female to complete development.

5. Offspring will contain one chromosome with the gene of interest knocked out. Genetic crosses can then produce mice homozygous for the knocked out gene to assess the phenotype. This can range from lethality to no visible effect depending on the gene.

human insulin and interferon, as well as other commercially valuable proteins, such as human growth hormone (figure 17.15) and erythropoietin, which stimulates red blood cell production.

Among the medically important proteins now manufactured by these approaches are **atrial peptides,** small proteins that may provide a new way to treat high blood pressure and kidney failure. Another is **tissue plasminogen activator (TPA),** a human protein synthesized in minute amounts that causes blood clots to dissolve and that if used within the first 3 hr after an ischemic stroke (i.e., one that blocks blood to the brain) can prevent catastrophic disability.

A problem with this approach has been the difficulty of separating the desired protein from the others the bacteria make. The purification of proteins from such complex mixtures is both time-consuming and expensive, but it is still easier than isolating the proteins from bulk processing of the tissues of animals, which is how such proteins were formerly obtained. For example, insulin previously was extracted from hog pancreases because hog insulin was similar to human insulin.

Recombinant DNA may simplify vaccine production

Another area of potential significance involves the use of genetic engineering to produce vaccines against communicable diseases. Two types of vaccines are under investigation: *subunit vaccines* and *DNA vaccines.*

Subunit vaccines

Subunit vaccines may be developed against viruses such as those that cause herpes and hepatitis. Genes encoding a part, or subunit, of the protein polysaccharide coat of the herpes simplex virus or hepatitis B virus are spliced into a fragment of the vaccinia (cowpox) virus genome (figure 17.16).

The vaccinia virus, which British physician Edward Jenner used more than 200 years ago in his pioneering vaccinations against smallpox, is now used as a vector to carry the herpes or hepatitis viral coat gene into cultured mammalian cells. These cells produce many copies of the recombinant vaccinia virus, which has the outside coat of a herpes or hepatitis virus. When this recombinant virus is injected into a mouse or rabbit, the immune system of the infected animal produces antibodies directed against the coat of the recombinant virus. It therefore develops an immunity to herpes or hepatitis virus.

Vaccines produced in this way are harmless because the vaccinia virus is benign, and only a small fragment of the DNA from the disease-causing virus is introduced via the recombinant virus.

The great attraction of this approach is that it does not depend on the nature of the viral disease. In the future, similar recombinant viruses may be used in humans to confer resistance to a wide variety of viral diseases.

DNA vaccines

In 1995, the first clinical trials began to test a novel new kind of **DNA vaccine,** one that depends not on antibodies but rather on the second arm of the body's immune defense, the so-called

figure 17.15

GENETICALLY ENGINEERED MOUSE WITH HUMAN GROWTH HORMONE. These two mice are from an inbred line and differ only in that the large one has one extra gene: the gene encoding human growth hormone. The gene was added to the mouse's genome and is now a stable part of the mouse's genetic endowment.

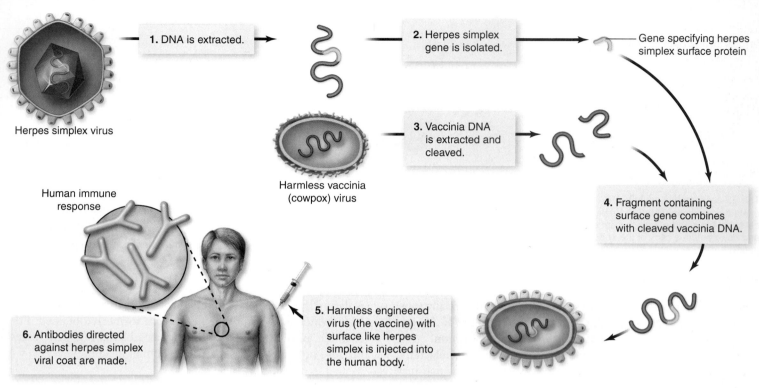

figure 17.16

STRATEGY FOR CONSTRUCTING A SUBUNIT VACCINE FOR HERPES SIMPLEX. Recombinant DNA techniques can be used to construct vaccines for a single protein from a virus or bacterium. In this example, the protein is a surface protein from the herpes simplex virus.

cellular immune response, in which blood cells known as killer T cells attack infected cells (chapter 51). The first DNA vaccines spliced an influenza virus gene encoding an internal nucleoprotein into a plasmid, which was then injected into mice. The mice developed a strong cellular immune response to influenza. Although new and controversial, the approach offers great promise.

Gene therapy can treat genetic diseases directly

In 1990, researchers first attempted to combat genetic defects by the transfer of human genes. When a hereditary disorder is the result of a single defective gene, an obvious way to cure the disorder would be to add a working copy of the gene. This approach is being used in an attempt to combat cystic fibrosis, and it offers potential for treating muscular dystrophy and a variety of other disorders (table 17.1).

One disease that illustrates both the potential and the problems with gene therapy is **severe combined immuno-deficiency disease (SCID).** There are multiple forms of this disease, including an X-linked form (X-SCID) and a form that lacks the enzyme adenosine deaminase (ADA-SCID).

Recent trials for both of these forms showed great initial promise, with patients exhibiting restoration of immune function. But then problems arose in the case of the X-SCID trial when a patient developed a rare leukemia. Since that time, two other patients have developed the same leukemia, and it appears to be due to the gene therapy itself. The vector used to introduce the X-SCID gene integrated into the genome next to

a proto-oncogene called *LMO2* in all three cases. Activation of this gene can cause childhood leukemias.

The insertion of a gene during gene therapy has always been a random event, and it has been a concern that the insertion could inactivate an essential gene, or turn on a gene

TABLE 17.1	Diseases Being Treated in Clinical Trials of Gene Therapy
Disease	
Cancer (melanoma, renal cell, ovarian, neuroblastoma, brain, head and neck, lung, liver, breast, colon, prostate, mesothelioma, leukemia, lymphoma, multiple myeloma)	
SCID (severe combined immunodeficiency)	
Cystic fibrosis	
Gaucher disease	
Familial hypercholesterolemia	
Hemophilia	
Purine nucleoside phosphorylase deficiency	
Alpha$_1$-antitrypsin deficiency	
Fanconi anemia	
Hunter syndrome	
Chronic granulomatous disease	
Rheumatoid arthritis	
Peripheral vascular disease	
Acquired immunodeficiency syndrome (AIDS)	

inappropriately. That effect had not been observed prior to the X-SCID trial, despite a large number of genes introduced into blood cells in particular. For leukemia to occur in 15% of the patients treated implies that some influence of the genetic background associated with X-SCID potentiates this development. This possibility is supported by the observation that the ADA-SCID patients treated have not been affected thus far.

On the positive side, 15 children treated successfully are still alive, 14 of them after more than four years, with functioning immune systems. On the negative side, three other children treated have developed leukemia.

When we understand the basis of the preferential integration in the case of X-SCID, it should be possible to overcome this unfortunate result. In the meantime, the investigators have halted the trial and are working on new vectors to reduce the possibility of this preferential integration.

> Medical applications of biotechnology include the production of proteins for pharmaceuticals and new ways to make vaccines. Genetic engineering can also be used to replace genes that cause genetic disease, a process called gene therapy. The technology for gene therapy has been controversial with some successes but some recent failures as well.

17.6 Agricultural Applications

Perhaps no area of genetic engineering touches all of us so directly as the applications that are being used in agriculture today. Crops are being modified to resist disease, to be tolerant to herbicides, and for changes in nutritional and other content in a variety of ways. Plant systems are also being used to produce pharmaceuticals by "biopharming," and domesticated animals are being genetically modified to produce biologically active compounds.

The Ti plasmid can transform broadleaf plants

In plants, the primary experimental difficulty has been identifying a suitable vector for introducing recombinant DNA. Plant cells do not possess the many plasmids that bacteria have, so the choice of potential vectors is limited.

The Ti plasmid

The most successful results thus far have been obtained with the **Ti (tumor-inducing) plasmid** of the plant bacterium *Agrobacterium tumefaciens*, which normally infects broadleaf plants such as tomato, tobacco, and soybean. Part of the Ti plasmid integrates into the plant DNA, and researchers have succeeded in attaching other genes to this portion of the plasmid (figure 17.17). The characteristics of a number of plants have been altered using this technique, which should be valuable in improving crops and forests.

Among the features scientists would like to affect are resistance to disease, frost, and other forms of stress; nutritional balance and protein content; and herbicide resistance. All of these traits have either been modified or are being modified.

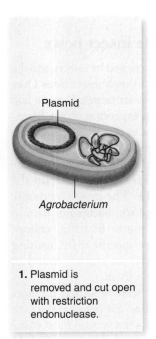
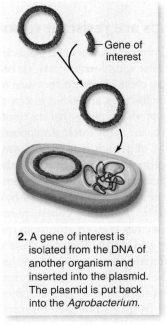
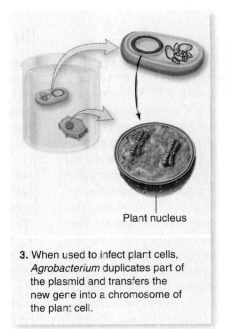

1. Plasmid is removed and cut open with restriction endonuclease.

2. A gene of interest is isolated from the DNA of another organism and inserted into the plasmid. The plasmid is put back into the *Agrobacterium*.

3. When used to infect plant cells, *Agrobacterium* duplicates part of the plasmid and transfers the new gene into a chromosome of the plant cell.

4. The plant cell divides, and each daughter cell receives the new gene. These cultured cells can be used to grow a new plant with the introduced gene.

figure 17.17

THE TI PLASMID. This *Agrobacterium tumefaciens* plasmid is used in plant genetic engineering.

Unfortunately, *Agrobacterium* normally does not infect cereals such as corn, rice, and wheat, but alternative methods can be used to introduce new genes into them.

Other methods of gene insertion

For cereal plants that are not normally infected by *Agrobacterium*, other methods have been used. One popular method, "the gene gun" uses bombardment with tiny gold or tungsten particles coated with DNA. This technique has the advantage of being possible for any species, but it does not allow as precise an engineering because the copy number of introduced genes is much harder to control.

Recently, modifications of the *Agrobacterium* system have allowed it to be used with cereal plants, so the gene gun technology may not be used as much in the future. A new bacterium has also been manipulated to function like *Agrobacterium*, offering another potential alternative method of engineering cereal crops.

It is clear that genetic modification of crop plants of all sorts has become a mature technology, which should accelerate the production of a variety of transgenic crops.

Case study: A better tomato?

One example of genetically manipulated fruit is Calgene's "Flavr Savr" tomato, which was genetically engineered to inhibit genes that cause cells to produce ethylene. In tomatoes and other plants, ethylene acts as a hormone to speed fruit ripening (chapter 41). In Flavr Savr tomatoes, inhibition of ethylene production delays ripening. The result is a tomato that can stay on the vine longer and resists overripening and rotting during transport to market.

The Flavr Savr tomato was a genetic engineering success, but was not a success in the marketplace. Its taste was not as good as other varieties, and it grew only in a limited area of the country and was not widely planted. The tomato was pulled from the market in 1997. Clearly there is more to producing a viable product than the genetic engineering of specific traits.

Herbicide-resistant crops allow no-till planting

Recently, broadleaf plants have been genetically engineered to be resistant to **glyphosate,** a powerful, biodegradable herbicide that kills most actively growing plants (figure 17.18). Glyphosate works by inhibiting an enzyme called EPSP synthetase, which plants require to produce aromatic amino acids.

Humans do not make aromatic amino acids; we get them from our diet, so we are unaffected by glyphosate. To make glyphosate-resistant plants, scientists used a Ti plasmid to insert extra copies of the EPSP synthetase gene into plants. These engineered plants produce 20 times the normal level of EPSP synthetase, enabling them to synthesize proteins and grow despite glyphosate's suppression of the enzyme. In later experiments, a bacterial form of the EPSP synthetase gene that differs from the plant form by a single nucleotide was introduced into plants via Ti plasmids; the bacterial enzyme is not inhibited by glyphosate.

These advances are of great interest to farmers because a crop resistant to glyphosate would not have to be weeded—the field could simply be treated with the herbicide. Because glyphosate is a broad-spectrum herbicide, farmers would no longer need to employ a variety of different herbicides, most of which kill only a few kinds of weeds. Furthermore, glyphosate

figure 17.18

GENETICALLY ENGINEERED HERBICIDE RESISTANCE. All four of these petunia plants were exposed to equal doses of the herbicide glyphosate. The two on the right were genetically engineered to be resistant to glyphosate, the active ingredient in glyphosate, but the two on the left were not.

breaks down readily in the environment, unlike many other herbicides commonly used in agriculture. A plasmid is actively being sought for the introduction of the EPSP synthetase gene into cereal plants, making them also glyphosate-resistant.

At this point four important crop plants have been modified to be glyphosate-resistant: maize (corn), cotton, soybeans, and canola. The use of glyphosate-resistant soy has been especially popular, accounting for 60% of the global area of GM (genetically modified) crops grown in nine countries worldwide. In the United States, 90% of soy currently grown is GM soy. Global variation in the use of GM crops has occurred, with the Americas, led by the United States, the largest adopter. The area currently with the largest growth in the use of GM crops is Asia, while Europe has been the slowest to move to their use.

Bt crops are resistant to some insect pests

Many commercially important plants are attacked by insects, and the usual defense against such attacks has been to apply insecticides. Over 40% of the chemical insecticides used today are targeted against boll weevils, bollworms, and other insects that eat cotton plants. Scientists have produced plants that are resistant to insect pests, removing the need to use many externally applied insecticides.

The approach is to insert into crop plants genes encoding proteins that are harmful to the insects that feed on the plants, but harmless to other organisms. The most commonly used protein is a toxin produced by the soil bacterium *Bacillus thuringiensis* (*Bt* toxin). When insects ingest Bt toxin, endogenous enzymes convert it into an insect-specific toxin, causing paralysis and death. Because these enzymes are not found in other animals, the protein is harmless to them.

The same four crops that have been modified for herbicide resistance have also been modified for insect resistance using the Bt toxin. The use of Bt maize is the second most common GM crop globally, representing 14% of global area of GM crops in nine countries. The global distribution of these crops is also similar to the herbicide resistant relatives.

Given the popularity of both of these types of crop modifications, it is not surprising that they have also been combined, so-called *stacked GM crops*, in both maize and cotton. Stacked crops now represent 9% of global area of GM crops.

Golden Rice shows potential of GM crops

One of the successes of GM crops is the development of Golden Rice. This rice has been genetically modified to produce β-carotene (provitamin A). The World Health Organization (WHO) estimates that vitamin A deficiency affects between 140 and 250 million preschool children worldwide. The deficiency is especially severe in developing countries where the major staple food is rice. Provitamin A in the diet can be converted by enzymes in the body to vitamin A, alleviating the deficiency.

The science that led to Golden Rice, so named for its distinctive color imparted by the presence of β-carotene in the endosperm (the outer layer of rice that has been milled). Rice does not normally make β-carotene in endosperm tissue, but does produce a precursor, geranyl geranyl diphosphate, that can be converted by three enzymes, phytoene synthase, phytoene desaturase, and lycopene β-cyclase, to β-carotene. These three genes were engineered to be expressed in endosperm and introduced into rice to complete the biosynthetic pathway producing β-carotene in endosperm (figure 17.19).

This is an interesting case of genetic engineering for two reasons. First, it introduces a new biochemical pathway in tissue of the transgenic plants. Second, it could not have been done by conventional breeding as no rice cultivar known produces these enzymes in endosperm. The original constructs used two genes from daffodil and one from a bacterium (see figure 17.19). There are many reasons to expect failure in the introduction of a biochemical pathway without disrupting normal metabolism. That the original form of Golden Rice makes significant amounts of β-carotene in an otherwise healthy plant is impressive. A second-generation version that makes much higher levels of β-carotene has also been produced by using the gene for phytoene synthase from maize in place of the original daffodil gene.

Golden Rice was originally constructed in a public facility in Switzerland and made available for free with no commercial entanglements. Since its inception, Golden Rice has been improved both by public groups and by industry scientists, and these improved versions are also being made available without commercial strings attached.

GM crops raise a number of social issues

The adoption of GM crops has been resisted in some areas for a variety of reasons. Questions are asked about the safety of these crops for human consumption, the movement of the introduced genes into wild relatives, and the possible loss of biodiversity associated with these crops.

Powerful forces form opposing sides in this debate. On the side in favor of the use of GM crops are the multinational companies that are utilizing this technology to produce seeds for the various GM crops. On the side questioning the use of GM crops are a variety of political organizations that are opposed to genetically modified foods. Scientists can be found on both sides of the controversy.

Issues originally centered on the safety of introduced genes for human consumption. In the United States, this issue has been "settled" for the crops already mentioned, and a large amount of GM soy and maize is consumed in this country. Although some still raise the issue of long-term use and allergic reactions, no negative effects have been documented. Existing crops will be monitored for adverse effects, and each new modification will require regulatory approval for human consumption.

Another issue has been fear about the spread of genes outside of the GM crops, but at this point there is no evidence for the introgression of introduced genes into wild relatives. A recent study indicated no evidence for the movement of genes from GM crops into native species in Mexico, despite earlier studies indicating significant movement of introduced genes.

This finding does not mean that such movement is impossible, but it does indicate that it seems not to have occurred at present. It is clear that this area requires more study. This issue will likely have to be considered on a case-by-case basis because the number of wild relatives, and the ease of hybridization, varies greatly among crop plants.

Pharmaceuticals can be produced by "biopharming"

The medicinal use of plants goes back as far as recorded history. In modern times, the pharmaceutical industry began by isolating biologically active compounds from plants. This approach began to change when in 1897, the Bayer company introduced acetyl salicylic acid, otherwise known as aspirin. This compound was

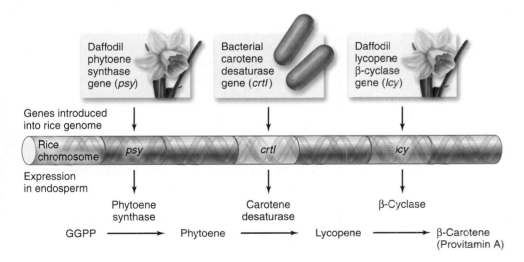

figure 17.19

CONSTRUCTION OF GOLDEN RICE. Rice does not normally express the enzymes needed to synthesize β-carotene in endosperm. Three genes were added to the rice genome to allow expression of the pathway for β-carotene in endosperm. The source of the genes and the pathway for synthesis of β-carotene is shown. The result is Golden Rice, which contains enriched levels of β-carotene in endosperm.

chapter 17 biotechnology **345**

Calvin and Hobbes

by Bill Watterson

a synthetic version of the compound salicylic acid, which was isolated from white willow bark. The production of pharmaceuticals has since been dominated more by organic synthesis and less by the isolation of plant products.

One exception to this trend is cancer chemotherapeutic agents such as taxol, vinblastine, and vincristine, all of which are isolated from plant sources. In an interesting closing of the historical loop, the industry is now looking at using transgenic plants for the production of useful compounds.

The first human protein to be produced in plants was human serum albumin, which was produced in 1990 by both genetically engineered tobacco and potato plants. Since that time more than 20 proteins have been produced in transgenic plants. This first crop of transgenic pharmaceuticals are now in the regulatory pipeline.

Recombinant subunit vaccines

One promising aspect of plant genetic engineering is the production of recombinant subunit vaccines discussed earlier. One of these, being produced in genetically modified potatoes, is a vaccine for Norwalk virus. Norwalk virus is not commonly known, but it reached the public consciousness when cruise ships were forced to cancel cruises due to outbreaks of the virus. The vaccine is now in clinical trials. A vaccine for rabies produced in transgenic spinach is also in clinical trials.

One obvious advantage of using plants for vaccine production is scalability. It has been estimated that 250 acres of greenhouse space could produce enough transgenic potato plants to supply Southeast Asia's need for hepatitis B vaccine.

Recombinant antibodies

Combining molecular cloning with immunology can be used to produce in transgenic plants the antibodies normally made by blood cells in vertebrates. The synthesis of monoclonal antibodies in plant systems is a promising use of transgenic plants.

A number of potentially therapeutic antibodies are being produced in plants, and some of these have reached clinical trial stage. One interesting example is an antibody against the bacterium responsible for dental caries, commonly known as tooth decay. It would make a visit to the dentist more pleasant to have a topical antibody applied instead of a drill.

Domesticated animals can also be genetically modified

Humans have been breeding and selecting domestic animals for hundreds of years. With the advent of genetic engineering, this process can be accelerated, and genes can be introduced from other species.

The production of transgenic livestock is in an early stage, and it is hard to predict where it will go. At this point, one of the uses of biotechnology is not to construct transgenics, but to use DNA markers to identify animals, and to map genes that are involved in such traits as palatability in food animals, texture of hair or fur, and other features of animal products. Molecular techniques combined with the ability to clone domestic animals (chapter 19) could produce improved animals for economically desirable traits.

Transgenic animal technology has not been as successful as initially predicted. Early on, pigs were engineered to overproduce growth hormone in the hope that this would lead to increased and faster growth. These animals proved to have only slightly increased growth, and they had lower fat levels, which reduces flavor, as well as showing other deleterious effects. The main use thus far has been engineering animals to produce pharmaceuticals in milk—another example of the biopharming concept.

One interesting idea for transgenics is the EnviroPig. This animal has been engineered with the introduction of the gene for phytase under the control of a salivary gland-specific promoter. The enzyme phytase breaks down phosphorus in the feed and can reduce phosphate excretion by up to 70%. Because phosphate is a major problem in pig waste, reducing its excretion could be a large environmental benefit.

As with GM crops, fears exist about the consumption of meat from transgenic animals. At this point, these fears do not seem to be based on sound science; nevertheless, every transgenic animal produced that is intended for consumption will need to be considered on a case-by-case basis.

> Genes can be introduced into plants with the Ti plasmid. Crops have been modified to resist herbicide, and to produce a bacterial toxin to kill insects. Golden Rice was modified to produce provitamin A in endosperm. Biopharming is the use of plants and animals to produce useful proteins. These technologies have raised ethical issues.

17.1 DNA Manipulation

The construction of recombinant DNA from molecules of two different sources led to the field of molecular biotechnology.

- Restriction endonucleases are used to fragment DNA molecules at specific restriction sites.
- Restriction endonucleases allowed for a form of physical mapping of DNA and the creation of recombinant molecules.
- Type II restriction enzymes recognize DNA sequences that are 4–12 bases long with a central axis of symmetry, and read the same way (5′ to 3′) on one strand that they do in the opposite direction (also 5′ to 3′).
- Cleaving such sequences at the same base on each strand results in fragments with "sticky ends" or overhanging complementary ends (figure 17.1).
- DNA ligase joins two fragments to form a stable DNA molecule.
- Gel electrophoresis separates fragments based on size, using an electric field to cause DNA to migrate through a gel matrix. Smaller fragments migrate farther than large fragments (figure 17.2).
- Foreign DNA is introduced into cells by a process called transformation.

17.2 Molecular Cloning

A clone is an identical copy. Molecular cloning involves the isolation of a specific sequence of DNA and the making of many identical copies.

- A vector is used to propagate recombinant DNA molecules in host cells.
- Plasmid vectors are small, extrachromosomal DNAs used to clone relatively small pieces of DNA.
- Phage λ vectors have a linear genome that can accept larger DNA molecules.
- Artificial chromosomes are used to clone very large DNA molecules.
- A DNA, or genomic, library is a collection of fragments from an entire genome that have been inserted into host cells.
- To obtain only the expressed parts of a genome, complementary DNA (cDNA) can be made from mRNA by using the enzyme reverse transcriptase (figure 17.5).
- DNA can be reversibly denatured and renatured. Renaturation of complementary strands from different sources is called hybridization.
- Hybridization is a powerful tool for finding DNAs in complex mixtures. Known DNA can be labeled, then used to find complementary strands by hybridization.
- Hybridization is the most common way of identifying a clone within a DNA library.

17.3 DNA Analysis

Molecular cloning provides specific DNA for further manipulation.

- The first physical maps of DNA molecules were the sites of restriction enzyme cleavage.
- These can be generated by enzyme digestion or by computers searching known DNA sequences for cleavage sites.
- Blotting is a procedure in which a complex mixture separated by electrophoresis is transferred to a piece of filter paper.
- The Southern blot uses DNA isolated from a cell or tissue with a filter that is hybridized with labeled, cloned DNA as a probe.
- Northern blots use mRNA instead of DNA, and Western blots use protein.

- Restriction fragment length polymorphisms (RFLPs) were the first method used to detect individual differences in DNA.
- DNA fingerprinting is a technique that uses probes to locate polymorphic repetitive DNA fragments.
- Determination of the actual sequence of bases in a DNA molecule is the ultimate level of analysis. This uses chain-terminating reagents and high-resolution electrophoresis (figure 17.11).
- The polymerase chain reaction (PCR) is used to amplify a single small DNA fragment using two short primers that flank the region to be amplified.
- The yeast two-hybrid system is used to study protein–protein interactions (figure 17.13).

17.4 Genetic Engineering

We can now genetically modify most plant and animal systems by introducing new DNA or modifying existing DNA in cells.

- Expression vectors contain the promoters and enhancers necessary to drive expression of inserted DNA.
- Transgenic organisms have DNA introduced across species "barriers."
- In vitro mutagenesis allows directed alteration of cloned genes, which can then be used to study gene function.
- Knockout mice are engineered to lack function for a specific gene. This allows a researcher to remove function for a gene and assess the phenotype (figure 17.14).

17.5 Medical Applications

There are many medical applications of genetic engineering.

- Human proteins, such as insulin, are produced by bacteria.
- Recombinant DNA may simplify vaccine production by producing harmless subunit vaccines and DNA vaccines that depend on the body's cellular immune response.
- Gene therapy, the adding of a copy of a functional gene, can be used to treat human genetic diseases.
- One problem with recent gene therapy trials was that the therapy caused leukemias in some patients.

17.6 Agricultural Applications

Crops are being modified to resist disease, tolerate herbicides, change nutritional value, and produce pharmaceuticals and biologically active compounds.

- The tumor-inducing plasmid from a plant bacterium is used to transfer genes into broad-leaf plants.
- Crops resistant to the herbicide glyphosate are a common genetic alteration. This allows for no-till planting.
- Bacterially derived insecticidal proteins have been transferred into crop plants to create pest-resistant crops.
- Golden Rice has been modified to have a higher concentration of provitamin A, important for diets in less-developed countries.
- The adoption of genetically modified (GM) crops has raised societal issues.
- "Biopharming" uses transgenic plants to produce pharmaceuticals such as serum albumin, subunit vaccines, and antibodies.
- Transgenic plant technologies are more successful than transgenic animal technologies.
- One recent successful GM pig produces an enzyme that reduces the excretion of harmful phosphates into the environment.

SELF TEST

1. A recombinant DNA molecule is one that is—
   a. produced through the process of crossing over that occurs in meiosis.
   b. constructed from DNA from different sources.
   c. constructed from novel combinations of DNA from the same source.
   d. produced through mitotic cell division.

2. Type II restriction endonucleases are useful because—
   a. they degrade DNA from the 5′ end.
   b. they cleave the DNA at random locations.
   c. they cleave the DNA at specific sequences.
   d. they only cleave modified DNA.

3. What is the basis of separation of DNA fragments by gel electrophoresis?
   a. The negative charge on DNA
   b. The size of the DNA fragments
   c. The sequence of the fragments
   d. The presence of a dye

4. How is the gene for β-galactosidase used in the construction of a plasmid?
   a. The gene is a promoter that is sensitive to the presence of the sugar, galactose.
   b. It is an origin of replication.
   c. It is a cloning site.
   d. It is a marker for insertion of DNA.

5. The basic logic of enzymatic DNA sequencing is to produce—
   a. a nested set of DNA fragments produced by restriction enzymes.
   b. a nested set of DNA fragments that each begin with different bases.
   c. primers to allow PCR amplification of the region between the primers.
   d. a nested set of DNA fragments that end with known bases.

6. A DNA library is—
   a. an orderly array of all the genes within an organism.
   b. a collection of vectors.
   c. the collection of plasmids found within a single *E. coli*.
   d. a collection of DNA fragments representing the entire genome of an organism.

7. Molecular hybridization is used to—
   a. generate cDNA from mRNA.
   b. introduce a vector into a bacterial cell.
   c. screen a DNA library.
   d. introduce mutations into genes.

8. The enzyme used in the polymerase chain reaction is—
   a. a restriction endonuclease.
   b. heat-resistant RNA polymerase.
   c. reverse transcriptase.
   d. a heat-resistant DNA polymerase.

9. How does the yeast two-hybrid system detect protein–protein interactions?
   a. Binding of fusion partners triggers a signal cascade that alters gene expression.
   b. Fusion partners are detected using radioactive probes of Western blots.
   c. Protein–protein binding of fusion partners triggers expression of a reporter gene.
   d. Protein–protein binding of fusion partners triggers expression of the *Gal4* gene.

10. In vitro mutagenesis is used to—
    a. produce large quantities of mutant proteins.
    b. create mutations at specific sites within a gene.
    c. create random mutations within multiple genes.
    d. create organisms that carry foreign genes.

11. Insertion of a gene for a surface protein from a medically important virus such as herpes into a harmless virus is an example of—
    a. a DNA vaccine.
    b. reverse genetics.
    c. gene therapy.
    d. a subunit vaccine.

12. What is a Ti plasmid?
    a. A vector that can transfer recombinant genes into plant genomes.
    b. A vector that can be used to produce recombinant proteins in yeast.
    c. A vector that is specific to cereal plants like rice and corn.
    d. A vector that is specific to embryonic stem cells.

13. Which of the following is *not* a possible benefit of genetically modified crops?
    a. Increased nutritional value for people.
    b. Enhanced resistance to insect pests.
    c. Enhanced resistance to broad-spectrum herbicides.
    d. Enhanced resistance to insecticides.

CHALLENGE QUESTIONS

1. Many human proteins, such as hemoglobin, are only functional as an assembly of multiple subunits. Assembly of these functional units occurs within the endoplasmic reticulum and Golgi apparatus of a eukaryotic cell. Discuss what limitations, if any exist to the large-scale production of genetically engineered hemoglobin.

2. Enzymatic sequencing of a short strand of DNA was completed using dideoxynucleotides. Use the gel shown to determine the sequence of that DNA?

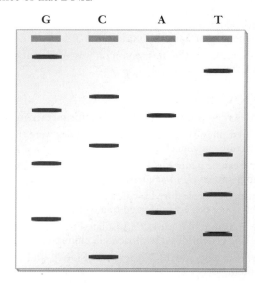

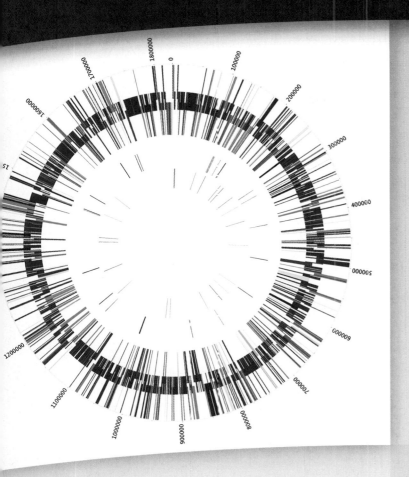

chapter 18

Genomics

introduction

THE PACE OF DISCOVERY IN BIOLOGY in the last 30 years has been like the exponential growth of a population. Starting with the isolation of the first genes in the mid-1970s, researchers had accomplished the first complete genome sequence by the mid-1990s—that of the bacteria species *Haemophilus influenzae*, shown in the picture (genes with similar functions are shown in the same color). By the turn of the twenty-first century, the molecular biology community had completed a draft sequence of the human genome.

Put another way, scientific accomplishments moved from cloning a single gene, to determining the sequence of a million base pairs in 20 years, then determining the sequence of a billion base pairs in another 5 years. The sequence of events is not linear since all are overlapping to some extent—but it is as though the first automobile had been invented on a Monday, then manufactured on an assembly line the following Wednesday, and by Friday, a formula 1 race car had been produced.

In the previous chapter you learned about the basic techniques of molecular biology. In this chapter you will see how those techniques have been applied to the analysis of whole genomes. This analysis integrates ideas from classical and molecular genetics with biotechnology.

concept outline

Mapping Genomes

We use maps to find out location, and depending on how accurately we wish to do this, we may use multiple maps with different resolutions. In genomics, we can locate a gene on a chromosome, in a subregion of a chromosome, and finally its precise location in the chromosome's DNA sequence. The DNA sequence level requires knowing the entire sequence of the genome, once out of our reach technologically. Knowing the entire sequence is useless, however, without other kinds of maps: Finding a single gene within the sequence of the human genome is like trying to find your house on a map of the world.

To overcome this difficulty, maps of genomes are constructed at different levels of resolution and using different kinds of information. We can make a distinction between *genetic maps* and *physical maps*. **Genetic maps** are abstract maps that place the relative location of genes on chromosomes based on recombination frequency (see chapter 13). **Physical maps** use landmarks within DNA sequences, ranging from restriction sites (described in the preceding chapter) to the ultimate level of detail: the actual DNA sequence.

Different kinds of physical maps can be generated

To make sense of genome mapping, it is important to have physical landmarks on the genome that are at a lower level of resolution than the entire sequence. In fact, long before the Human Genome Project was even conceived, physical maps of DNA were needed as landmarks on cloned DNA. Three types of physical maps are restriction maps, constructed using restriction enzymes; chromosome-banding patterns, generated by cytological dye methods; and radiation hybrid maps, created by using radiation to fragment chromosomes.

Restriction maps

Distances between "landmarks" on a physical map are measured in base-pairs (1000 base-pairs [bp] equal 1 kilobase, kb). It is not necessary to know the DNA sequence of a segment of DNA in order to create a physical map, or to know whether the DNA encompasses information for a specific gene.

The first physical maps were created by cutting genomic DNA with different restriction enzymes, both singly and with combinations of different enzymes (figure 18.1). The analysis of the patterns of fragments generated were used to generate a map.

In terms of larger pieces of DNA, this process is repeated and then used to put the pieces back together, based on size and overlap, into a contiguous segment of the genome called a **contig**. In an example of biological coincidence, the very first restriction enzymes to be isolated came from *Haemophilus*, which was also the first free-living genome to be completely sequenced.

Chromosome banding patterns

Cytologists studying chromosomes with light microscopes found that by using different stains, they could produce reproducible patterns of bands on the chromosomes. In this way,

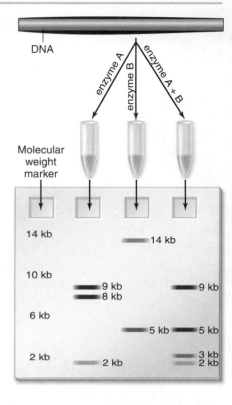

figure 18.1

RESTRICTION ENZYMES CAN BE USED TO CREATE A PHYSICAL MAP. DNA is digested with two different restriction enzymes singly and in combination, then electrophoresed to separate the fragments. The location of sites can be deduced by comparing the sizes of fragments from the individual reactions with the combined reaction.

they could identify all of the chromosomes and divide them into subregions based on banding pattern.

The use of different stains allow the construction of a cytological map of the entire genome. These large-scale physical maps are like a map of an entire country, in that they encompass the whole genome, but at low resolution.

Cytological maps have been used to characterize chromosomal abnormalities associated with human diseases, such as chronic myelogenous leukemia. In this disease, a reciprocal translocation has occurred between chromosome 9 and chromosome 22, resulting in an altered form of tyrosine kinase that is always turned on, causing white-cell proliferation.

The use of hybridization with cloned DNA has added to the utility of chromosome-banding analysis. In this case, because the hybridization involves whole chromosomes, it is called *in situ hybridization*. It is done using fluorescently labeled probes, and so its complete name is **fluorescent in situ hybridization (FISH)** (figure 18.2).

Radiation hybrid maps

Radiation hybrid maps use radiation to fragment chromosomes randomly, then recover the fragments by fusing the irradiated cell to another cell. To construct a radiation hybrid map of the human genome, a human cell in culture is lethally irradiated, then fused to a rodent cell. The chromosome fragments produced by the radiation become integrated into the rodent-cell chromosomes. These fragments can be identified based on their banding patterns and by using known genes for FISH.

For the purposes of mapping, a series of these hybrid cells have been constructed that have overlapping fragments of human chromosomes representing the entire genome. The use of radiation hybrids in mapping is discussed in more detail later on.

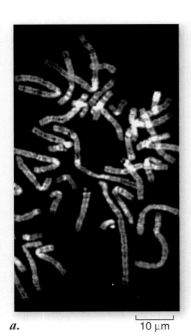 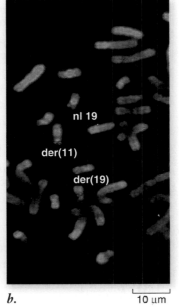

nl 19

der(11)

der(19)

a. 10 μm   *b.* 10 um

figure 18.2

USE OF FLUORESCENT IN SITU HYBRIDIZATION TO CORRELATE CLONED DNA WITH CYTOLOGICAL MAPS. *a.* Part of a karyotype of human chromosomes using G banding. The red bands indicate hybridization with cloned DNA. *b.* The probe used in panel (*a*) shows a translocation in this patient that led to multiple congenital malformations and mental retardation.

Sequence-tagged sites provide a common language for physical maps

The construction of a physical map for a large genome requires the efforts of many laboratories in different locations. A variety of difficulties arose in comparing data from different labs, as well as integrating different types of landmarks used on physical and genetic maps.

In the early days of the Human Genome Project, this problem was addressed by the creation of a common molecular language that could be used to describe the different types of landmarks.

Defining common markers

Since all genetic information is ultimately based on DNA sequence, it was important for this common language to be sequence-based, but not to require generating a large amount of sequence for any landmark. The solution was the **sequence-tagged site, or STS**. This site is a small stretch of DNA that is unique in the genome, that is, it only occurs once.

The boundary of the STS is defined by PCR primers, so the presence of the STS can be identified by PCR using any DNA as a template (see chapter 17). These sites need to be only 200–500 bp long, an amount of sequence that can be determined easily. The STS can contain any other kind of landmark—for example, part of a cloned gene that has been genetically mapped, or a restriction site that is polymorphic. Any marker that has been mapped can be converted to an STS by sequencing only 200–500 bp.

The use of STSs

As maps are generated, new STSs are identified, and added to a database. For each STS, the database indicates the sequence of the STS, its location in the genome, and the PCR primers needed to identify it. Any researcher is then able to identify the presence or absence of any STS in the DNA that he or she is analyzing.

Fragments of DNA can be pieced together using STSs by identifying overlapping regions in fragments. Because of the high density of STSs in the human genome and the relative ease of identifying an STS in a DNA clone, investigators were able to develop physical maps on the huge scale of the 3.2-gigabase genome in the mid-1990s (figure 18.3). STSs essentially provide a scaffold for assembling genome sequences.

Genetic maps provide a link to phenotypes

The first genetic (linkage) map was made in 1911 when Alfred Sturtevant mapped five genes in *Drosophila*. Distances on a genetic map are measured in centimorgans (cM) in honor of the geneticist T. H. Morgan. One centimorgan corresponds to 1% recombination between two loci. Today, 14,065 genes have been mapped on the *Drosophila* genome.

Linkage mapping can be done without knowing the DNA sequence of a gene. Computer programs make it possible to create a linkage map for a thousand genes at a time. But a few limitations to genetic maps exist. One is that distances between genes determined by recombination frequencies do not directly correspond with physical distance on a chromosome. The conformation of DNA between genes varies, and this conformation can affect the frequency

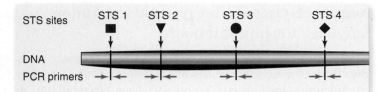

1. The location of 4 STSs in the genome is shown. PCR is used to amplify each STS from different clones in a library. Amplifying each STS by PCR generates a unique fragment that can be identified.

↓ PCR runs with four clones

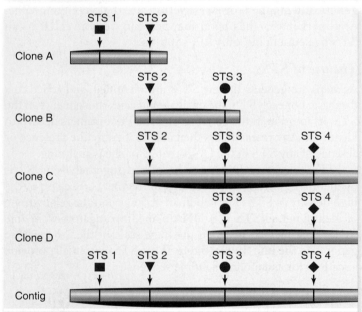

2. The products of the PCR reactions are separated by gel electrophoresis producing a different size fragment for each STS.

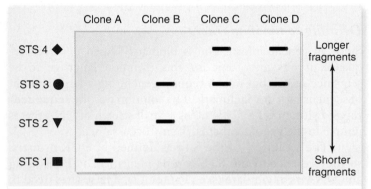

3. The presence or absence of each STS in the clones identifies regions of overlap. The final result is a contiguous sequence (contig) of overlapping clones.

figure 18.3

CREATING A PHYSICAL MAP WITH SEQUENCE-TAGGED SITES. The presence of landmarks called sequence-tagged sites, or STSs, in the human genome made it possible to begin creating a physical map large enough in scale to provide a foundation for sequencing the entire genome. (*1*) Primers (*green arrows*) that recognize unique STSs are added to a cloned segment of DNA, followed by DNA replication via PCR. (*2*) The PCR products from each reaction are separated based on size on a DNA gel, and the STSs contained in each clone are identified. (*3*) The cloned DNA segments are then aligned based on overlapping STSs to create a contig.

of recombination. Another limitation is that not all genes have obvious phenotypes that can be followed in segregating crosses.

As described in chapter 13, the human genetic map is quite dense, with a marker roughly every 1 cM. This level of detail would have been unheard of 20 years ago, and it was made possible by development of molecular markers that do not cause a phenotype change.

The most common type of markers are short repeated sequences, called short tandem repeats, or STR loci, that differ in repeat length between individuals. These repeats are identified by using PCR to amplify the region containing the repeat, then analyzing the products using electrophoresis. Once a map is constructed using these markers, genes with alleles that cause a disease state can be mapped relative to the molecular landmarks. Thirteen of these STR loci form the basis for modern DNA fingerprinting developed by the FBI. The alleles for these 13 loci are what is cataloged in the CODIS database used to identify criminal offenders.

Physical maps can be correlated with genetic maps

We need to be able to correlate genetic maps with physical maps, particularly genome sequences, to aid in finding physical sequences for genes that have been mapped genetically.

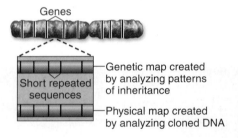

The problem in finding genes is that the resolution of genetic maps at present is not nearly as fine-grained as the genome sequence. Markers that are 1 cM apart may be as much as a million base pairs apart.

Since the markers used to construct genetic maps are now primarily molecular markers, they can be easily located within a genome sequence. Similarly, any gene that has been cloned can be placed within the genome sequence and can also be mapped genetically. This provides an automatic correlation between the two maps. The problem in terms of finding genes that have been mapped genetically but not isolated as molecular clones lies in the nature of genetic maps. Distances measured on genetic maps are not uniform due to variation in recombination frequency along the chromosome. So 1 cM of genetic distance will translate to different numbers of base-pairs in different regions.

Radiation hybrid maps provide an alternative to genetic maps and are easily correlated with physical maps. These radiation hybrid maps consist of simple binary data: the presence or absence of a particular molecular marker in each cell in a radiation hybrid panel (described earlier). The more similar the score for any two markers, the closer they are on a chromosome. This is due to the random nature of the fragmentation by radiation. If two markers are close together, they will often be found on the same fragment, the farther apart they are, the less likely they will be on the same fragment.

This technique allows any kind of molecular marker to be ordered in the genome, including markers that are not polymorphic and therefore not suitable for genetic mapping. This also allows the integration of genetic and physical maps as both types of markers can be placed on the same radiation hybrid map. This is most useful in the early stages of a large-scale sequencing project. Such maps have been constructed for most animal species of interest to basic researchers, as well as domestic animals that are of economic importance and even companion animals such as dogs and cats. Most animal genome-sequencing projects include this kind of analysis. Currently in humans, this technique is being used to identify the location in the genome of all known transcripts.

All of these different kinds of maps are then stored in databases so they can be aligned and viewed. The National Center for Biotechnology Information (NCBI) is a branch of the National Library of Medicine, and it serves as the U.S. repository for these data and more. Similar databases exist in Europe and Japan, and all are kept current. An enormous storehouse of information is available for use by biological researchers worldwide.

> Maps of genomes can by either physical maps or genetic maps. Physical maps include cytogenic maps of chromosome banding, restriction maps, or radiation hybrid maps. Genetic maps are correlated with physical maps by using DNA markers. Radiation hybrids can also be used to construct maps based on the probability of breakage by radiation occurring between two sites.

18.2 Whole-Genome Sequencing

The ultimate physical map is the base-pair sequence of the entire genome. In the early days of molecular biology, all sequencing was done manually, and was both time- and labor-intensive. As mentioned in chapter 17, the development of machines to automate this process increased the rate of sequence generation.

Large-scale genome sequencing requires the use of high-throughput automated sequencing and computer analysis (figure 18.4). Genome sequencing is one case in which technology drove the science, rather than the other way around. In a few hours, an automated sequencer can sequence the same number of base-pairs that a technician could manually sequence in a year—up to 50,000 bp. Without the automation of sequencing, it would have been impossible to sequence large, eukaryotic genomes like that of humans.

Genome sequencing requires larger molecular clones

Although it would be ideal to isolate DNA from an organism, add it to a sequencer, and then come back in a week or two to pick up a computer-generated printout of the genome sequence for that organism, scientific life is currently not quite that simple. Sequencers provide accurate sequences for DNA segments up to 800 bp long. Even then, errors are possible. So, 5–10 copies of a genome are sequenced to reduce errors.

Even with reliable sequence data in hand, each individual sequencing run produces a relatively small amount of sequence. Thus, the genome must be fragmented, and then individual molecular clones isolated for sequencing (see chapter 17).

Artificial chromosomes

As described in chapter 17, the development of artificial chromosomes has allowed scientists to clone larger pieces of DNA. The first generation of these new vectors were yeast artificial

chromosomes (YACs). These are constructed by using a yeast origin of replication and centromere sequence, then adding foreign DNA to this construct. The origin of replication allows the artificial chromosome to replicate independently of the rest of the genome, and the centromere sequences make the chromosome mitotically stable.

YACs were useful for cloning larger pieces of DNA but they had many drawbacks, including a tendency to rearrange, or to lose portions of DNA by deletion. Despite the difficulties, the

figure 18.4

AUTOMATED SEQUENCING. This sequence facility simultaneously runs multiple automated sequencers, each processing 96 samples at a time

YACs were used early on to construct physical maps by restriction enzyme digestion of the YAC DNA.

The artificial chromosomes most commonly used now, particularly for large-scale sequencing, are made in *E. coli*. These bacterial artificial chromosomes (BACs) are a logical extension of the use of bacterial plasmids. BAC vectors accept DNA inserts between 100 and 200 kb long. The downside of BAC vectors is that, like the bacterial chromosome, they are maintained as a single copy while plasmid vectors exist at high copy numbers.

Human artificial chromosomes

Human artificial chromosomes have been constructed to introduce large segments of human DNA into cultured cells. These artificial chromosomes are usually constructed by fragmentation of chromosomes and centromeric sequences. At present, they are circular, but some can still segregate correctly during mitosis up to 98% of the time. The construction of linear human artificial chromosomes is not yet possible.

Whole-genome sequencing is approached in two ways: clone-by-clone and shotgun

Sequencing an entire genome is an enormous task. Two ways of approaching this challenge have been developed, one a step-by-step approach, and one that attempts to take on the whole thing at once and depends on computers to sort out the data. The two techniques grew out of competing projects to sequence the human genome, as described in the next section.

Clone-by-clone sequencing

The cloning of large inserts in BACs facilitates the analysis of entire genomes. The strategy most commonly pursued is to construct a physical map first, and then use it to place the site of BAC clones for later sequencing.

Aligning large portions of a chromosome requires identifying regions that overlap between clones. This can be accomplished either by constructing restriction maps of each BAC clone, or by identifying STSs found in clones. If two BAC clones have the same STS, then they must overlap.

The alignment of a number of BAC clones results in a contiguous stretch of DNA called a *contig*. The individual BAC clones can then be sequenced 500 bp at a time to produce the sequence of the entire contig (figure 18.5*a*). This strategy of physical mapping followed by sequencing is called **clone-by-clone sequencing.**

Shotgun sequencing

The idea of **shotgun sequencing** is simply to randomly cut the DNA into small fragments, sequence all cloned fragments, and then use a computer to put together the overlaps (figure 18.5*b*). This terminology actually goes back to the early days of molecular cloning when the construction of a library of random cloned fragments was referred to as *shotgun cloning*. This approach is much less labor-intensive than the clone-by-clone method, but it requires much greater computer power to assemble the final sequence and very efficient algorithms to find overlaps.

Unlike the clone-by-clone approach, shotgun sequencing does not tie the sequence to any other information about the genome. Many investigators have used both clone-by-clone and shotgun-sequencing techniques, and such hybrid approaches are becoming the norm. This combination has the strength of tying the sequence to a physical map while greatly reducing the labor involved. The two methods are shown graphically in figure 18.5.

Assembler programs compare multiple copies of sequenced regions in order to assemble a **consensus sequence,** that is, a sequence that is consistent across all copies. Although computer assemblers are incredibly powerful, final human analysis is required after both clone-by-clone and shotgun sequencing to determine when a genome sequence is sufficiently accurate to be useful to researchers.

The Human Genome Project used both sequencing methods

The vast scale of genomics has ushered in a new way of doing biological research involving large teams. Although a single individual can isolate and manually sequence a molecular clone for a single gene, a huge genome like the human genome requires the collaborative efforts of hundreds of researchers.

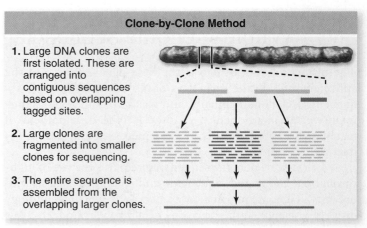

a.

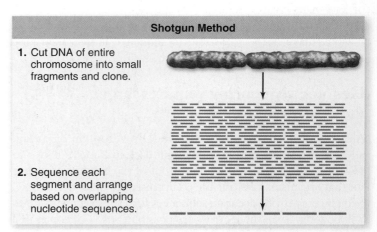

b.

figure 18.5

COMPARISON OF SEQUENCING METHODS. *a.* The clone-by-clone method uses large clones assembled into overlapping regions by STSs. Once assembled, these can be fragmented into smaller clones for sequencing. *b.* In the shotgun method the entire genome is fragmented into small clones and sequenced. Computer algorithms assemble the final DNA sequence based on overlapping nucleotide sequences.

The Human Genome Project originated in 1990 when a group of American scientists formed the International Human Genome Sequencing Consortium. The goal of this publicly funded effort was to use a clone-by-clone approach to sequence the human genome. Both genetic and physical maps were enhanced and published in the 1990s and used as scaffolding to sequence each chromosome.

Then, in May, 1998, Craig Venter, who had sequenced *Haemophilus influenzae*, announced that he had formed a private company to sequence the human genome. He proposed to shotgun-sequence the 3.2-gigabase genome in only two years. The consortium rose to the challenge, and the race to sequence the human genome began. The upshot was a tie of sorts. On June 26, 2000, the two groups jointly announced success, and each published its findings simultaneously in 2001. The consortium's draft alone included 248 names on its partial list of authors.

The draft sequence of the human genome was just the beginning. Gaps in the sequence are still being filled, and the map is constantly being refined. In 2004, the "finished" sequence was published and announced as the reference sequence (REF-SEQ) in the databases. This sequence is down to only 341 gaps, a 400-fold reduction in gaps, and it now includes 99% of the euchromatic sequence, up from 95%. The reference sequence has an error rate of 1 per 100,000 bases.

More significantly, research on the whole genome can move ahead. Now that the ultimate physical map is in place and is being integrated with the genetic map, diseases that result from defects in more than one gene, such as diabetes, can be addressed. Comparisons with other genomes are already changing our understanding of genome evolution (see chapter 24).

> Sequencing of entire genomes requires the use of automated sequencers running many samples in parallel. Larger fragments of DNA are needed for sequencing, and artificial chromosomes have provided a way of working with these large fragments. Two approaches were developed for whole genome sequencing, one that uses clones already aligned by physical mapping (clone-by-clone sequencing), and one that involves sequencing random clones and using a computer to assemble the final sequence (shotgun sequencing). In either case, significant computing power is necessary to assemble a final sequence.

18.3 Characterizing Genomes

Automated sequencing technology has produced huge amounts of sequence data, eventually allowing the sequencing of entire genomes. This has allowed researchers studying complex problems to move beyond approaches restricted to the analysis of individual genes. Sequencing projects in themselves are descriptive analysis that tells us nothing about the organization of genomes, let alone the function of gene products and how they may be interrelated. Additional research and evaluation has given us both answers and new puzzles.

The Human Genome Project found fewer genes than expected

For many years, geneticists had estimated the number of human genes to be around 100,000. This estimate, although based on some data, was really just a guess. Imagine researchers' surprise to find that the number now appears to be only around 25,000! This represents about twice as many genes as *Drosophila* and fewer genes than rice (figure 18.6). Clearly the complexity of an organism is not a simple function of the number of genes in its genome.

Finding genes in sequence data requires computer searches

Once a genome has been sequenced, the next step is to determine which regions of the genome contain which genes, and what those genes do. A lot of information can be mined from the sequence data. Using markers from physical maps and information from genetic maps, it is possible to find the sequence of the small percentage of genes that are identified by mutations with an observable (phenotypic) effect. Genes can

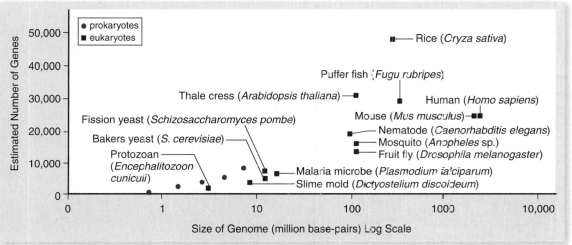

figure 18.6

SIZE AND COMPLEXITY OF GENOMES. In general, eukaryotic genomes are larger and have more genes than prokaryotic genomes, although the size of the organism is not the determining factor. The mouse genome is nearly as large as the human genome, and the rice genome contains more genes than the human genome.

also be found by comparing expressed sequences to genomic sequences. The analysis of expressed sequences is discussed later in this section.

Locating starts and stops

Information in the nucleotide sequence itself can also be used in the search for genes. A gene begins with a start codon, such as ATG, and it contains no stop codons (TAA, TGA, or TAG) for a distance long enough to encode a protein. This coding region is referred to as an **open reading frame (ORF).** Although these sequences between a start and a stop are likely to be genes, they may or may not actually be translated into a functional protein. Sequences for potential genes need to be tested experimentally to determine whether they have a function.

The addition of information to the basic sequence information, like identifying ORFs, is called sequence **annotation.** This process is what converts simple sequence data into something that we can recognize based on landmarks such as regions that are transcribed, and regions that are known or thought to encode proteins.

Inferring function across species: The BLAST algorithm

It is also possible to search genome databases for sequences that are homologous to known genes in other species. A researcher who has isolated a molecular clone for a gene of unknown function can search the database for similar sequences to infer function. The tool that makes this possible is a search algorithm called BLAST that is provided by the NCBI to search their databases. Using email and a networked computer, one can submit a sequence to the BLAST server and get back a reply with all possible similar sequences contained in the sequence database.

Using these techniques, sequences that are not part of ORFs have been identified that are conserved over millions of years of evolution. These sequences may be important for the regulation of the genes contained in the genome.

Using computer programs to search for genes, to compare genomes, and to assemble genomes are only a few of the new genomics approaches falling under the heading of **bioinformatics.**

Genomes contain both coding and noncoding DNA

When genome sequences are analyzed, regions that encode proteins and other regions that do not encode proteins are revealed. For many years investigators had known of the latter, but they did not know the extent and nature of the noncoding DNA. We first consider the types of coding DNA that have been found, then move on to look at types of noncoding DNA.

Protein-encoding DNA in eukaryotes

Four different classes of protein-encoding genes are found in eukaryotic genomes, differing largely in gene copy number.

Single-copy genes. Many genes exist as single copies on a particular chromosome. Most mutations in these genes result in recessive Mendelian inheritance.

Segmental duplications. Sometimes whole blocks of genes are copied from one chromosome to another, resulting in *segmental duplication*. Blocks of similar genes in the same order are found throughout the human genome. Chromosome 19 seems to have been the biggest borrower, sharing blocks of genes with 16 other chromosomes.

Multigene families. As more has been learned about eukaryotic genomes, many genes have been found to exist as parts of *multigene families*, groups of related but distinctly different genes that often occur together in clusters. These genes appear to have arisen from a single ancestral gene that duplicated during an uneven meiotic crossover in which genes were added to one chromosome and subtracted from the other. These multigene families may include silent copies called *pseudogenes*, inactivated by mutation.

Tandem clusters. Identical copies of genes can also be found in *tandem clusters*. These genes are all transcribed simultaneously, which increases the amount of mRNA available for protein production. Tandem clusters also include genes that do not encode proteins, such as rRNA genes that are typically present in clusters of several hundred copies.

Noncoding DNA in eukaryotes

Sequencing of several eukaryotic genomes has now been completed, and one of the most notable characteristics is the amount of noncoding DNA they possess. The Human Genome Project has revealed a particularly startling picture. Each of your cells has about 6 feet of DNA stuffed into it, but of that, less than 1 inch is devoted to genes! Nearly 99% of the DNA in your cells has little or nothing to do with the instructions that make you what you are.

True genes are scattered about the human genome in clumps among the much larger amount of noncoding DNA, like isolated hamlets in a desert. Six major sorts of noncoding human DNA have been described. (Table 18.1 shows the composition of the human genome, including noncoding DNA.)

Noncoding DNA within genes. As discussed in chapter 15, a human gene is not simply a stretch of DNA, like the letters of a word. Instead, a human gene is made up of numerous fragments of protein-encoding information (exons) embedded within a much larger matrix of noncoding DNA (introns). Together, introns make up about 24% of the human genome and exons less than 1.5%.

Structural DNA. Some regions of the chromosomes remain highly condensed, tightly coiled, and untranscribed throughout the cell cycle. Called *constitutive heterochromatin*, these portions tend to be localized around the centromere or located near the ends of the chromosome, at the telomeres.

TABLE 18.1 Classes of DNA Sequences Found in the Human Genome

Class	Frequency (%)	Description
Protein-encoding genes	1.5	Translated portions of the 25,000 genes scattered about the chromosomes
Introns	24	Noncoding DNA that constitutes the great majority of each human gene
Segmental duplications	5	Regions of the genome that have been duplicated
Pseudogenes (inactive genes)	2	Sequence that has characteristics of a gene but is not a functional gene
Structural DNA	20	Constitutive heterochromatin, localized near centromeres and telomeres
Simple sequence repeats	3	Stuttering repeats of a few nucleotides such as CGG, repeated thousands of times
Transposable elements	45	21%: Long interspersed elements (LINEs), which are active transposons 13%: Short interspersed elements (SINEs), which are active transposons 8%: Retrotransposons, which contain long terminal repeats (LTRs) at each end 3%: DNA transposon fossils

Simple sequence repeats. Scattered about chromosomes are **simple sequence repeats (SSRs).** An SSR is a one- to six-nucleotide sequence such as CA or CGG, repeated like a broken record thousands and thousands of times. SSRs can arise from DNA replication errors. SSRs make up about 3% of the human genome.

Segmental duplications. Blocks of genomic sequences composed of from 10,000 to 300,000 bp have duplicated and moved either within a chromosome or to a nonhomologous chromosome.

Pseudogenes. These are inactive genes that may have lost function because of mutation.

Transposable elements. Fully 45% of the human genome consists of mobile bits of DNA called *transposable elements.* Some of these elements code for proteins, but many do not. Because of the significance of these elements, we describe them more fully below.

Transposable elements: Mobile DNA

Discovered by Barbara McClintock in 1950, **transposable elements,** also termed *transposons* and *mobile genetic elements,* are bits of DNA that are able to move from one location on a chromosome to another. Barbara McClintock received the Nobel Prize in 1983 for discovery of these elements and their unexpected ability to change location.

Transposable elements move around in different ways. In some cases, the transposon is duplicated, and the duplicated DNA moves to a new place in the genome, so the number of copies of the transposon increases. Other types of transposons are excised without duplication and insert themselves elsewhere in the genome. The role of transposons in genome evolution is discussed in chapter 24.

Human chromosomes contain four sorts of transposable elements. Fully 21% of the genome consists of **long interspersed elements (LINEs).** These ancient and very successful elements are about 6000 bp long, and they contain all the equipment needed for transposition. LINEs encode a reverse transcriptase enzyme that can make a cDNA copy of the tran-scribed LINE RNA. The result is a double-stranded segment that can reinsert into the genome rather than undergo translation into a protein. Since these elements use an RNA intermediate, they are termed *retrotransposons.*

Short interspersed elements (SINEs) are similar to LINEs, but they cannot transpose without using the transposition machinery of LINEs. Nested within the genome's LINEs are over half a million copies of a SINE element called Alu (named for a restriction enzyme that cuts within the sequence). The Alu SINE represents 10% of the human genome. Like a flea on a dog, Alu moves with the LINE it resides within. Just as a flea sometimes jumps to a different dog, so Alu sometimes uses the enzymes of its LINE to move to a new chromosome location. Alu can jump right into genes, causing harmful mutations.

Two other sorts of transposable elements are also found in the human genome: 8% of the human genome is devoted to retrotransposons called **long terminal repeats (LTRs).** Although the transposition mechanism is a bit different from that of LINEs, LTRs also use reverse transcriptase to ensure that copies are double-stranded and can reintegrate into the genome.

Some 3% of the genome is devoted to dead transposons, elements that have lost the signals for replication and can no longer move.

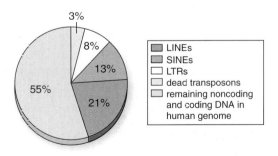

3%
8%
13%
55%
21%

- LINEs
- SINEs
- LTRs
- dead transposons
- remaining noncoding and coding DNA in human genome

inquiry

How do you think these repetitive elements would affect the determination of gene order?

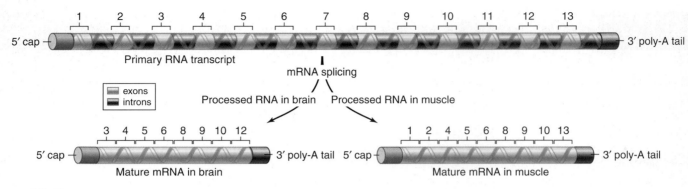

figure 18.7

ALTERNATIVE SPLICING CAN RESULT IN THE PRODUCTION OF DIFFERENT mRNAs FROM THE SAME CODING SEQUENCE. In some cells, exons can be excised along with neighboring introns, resulting in different proteins. Alternative splicing explains why 25,000 human genes can code for three to four times as many proteins.

Expressed sequence tags identify genes that are transcribed

Given the complexity of coding and noncoding DNA, it is important to be able to recognize regions of the genome that are actually expressed—that is, transcribed and then translated.

Because DNA is easier to work with than protein, one approach is to isolate mRNA, use this to make cDNA, then sequence one or both ends of as many cDNAs as possible. With automated sequencing, this task is not difficult, and these short sections of cDNA have been named **expressed sequence tags (ESTs).** An EST is another form of STS, and thus it can be included in physical maps. This technique does not tell us anything about the function of any particular EST, but it does provide one view, at the whole-genome level, of what genes are expressed, at least as mRNAs.

ESTs have been used to identify 87,000 cDNAs in different human tissues. About 80% of these cDNAs were previously unknown. You may wonder at this point how the estimated 25,000 genes of the human genome can result in 87,000 different cDNAs. The answer lies in the modularity of eukaryotic genes, which consist of exons interspersed with introns, as described in chapter 15.

Following transcription in eukaryotes, the introns are removed, and exons are spliced together. In some cells, some of the splice sites are skipped, and one or more exons is removed along with the introns. This process, called *alternative splicing* (figure 18.7), yields different proteins that can have different functions. Thus, the added complexity of proteins in the human genome comes not from additional genes, but from new ways to put existing parts of genes together.

SNPs are single-base differences between individuals

One fact becoming clear from analysis of the human genome is that a huge amount of genetic variation exists in our species. This information has practical use.

Single-nucleotide polymorphisms (SNPs) are sites where individuals differ by only a single nucleotide. To be classified as a polymorphism, an SNP must be present in at least 1% of the population. At present, the International SNP Map Working Group has identified 50,000 SNPs in coding regions of the genome and an additional 1.4 million in noncoding DNA. It is estimated that this represents about 10% of the variation available.

These SNPs are being used to look for associations between genes. We expect that the genetic recombination occurring during meiosis randomizes all but the most tightly linked genes. We call the tendency for genes *not* to be randomized **linkage disequilibrium.** This kind of association can be used to map genes.

The preliminary analysis of SNPs shows that many are in linkage disequilibrium. This unexpected result has led to the idea of genomic **haplotypes,** or regions of chromosomes that are not being exchanged by recombination. The existence of haplotypes allows the genetic characterization of genomic regions by describing a small number of SNPs (figure 18.8). If these haplotypes stand up to further analysis, they could greatly aid in mapping the genetic basis of disease. The Human Genome Project is now working on a haplotype map of the genome.

> The human genome contains far fewer genes than expected: about 25,000. A significant amount of noncoding DNA is found in all eukaryotic genomes. Coding sequences can be single copy, repeated in clusters, part of segmental duplications, or part of a gene family. Among the human genome are a variety of transposable elements that are repeated many times. These elements are capable of movement in the genome and are found in all eukaryotic genomes. The number and location of expressed genes can be estimated by sequencing the ends of randomly selected cDNAs to produce expressed sequence tags (ESTs).

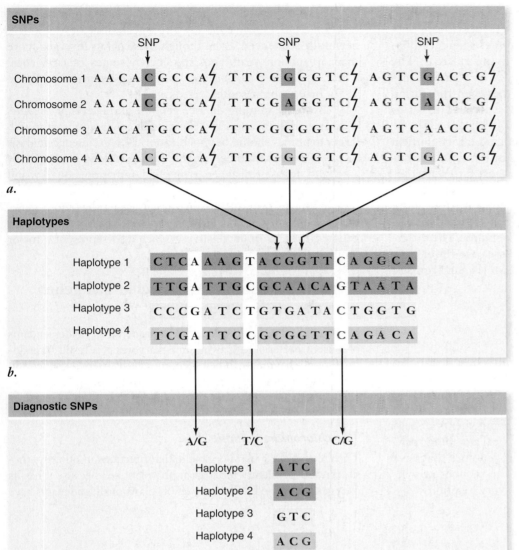

SNPs

Chromosome 1 A A C A **C** G C C A⚡ T T C G **G** G G T C⚡ A G T C **G** A C C G⚡
Chromosome 2 A A C A **C** G C C A⚡ T T C G **A** G G T C⚡ A G T C **A** A C C G⚡
Chromosome 3 A A C A **T** G C C A⚡ T T C G **G** G G T C⚡ A G T C **A** A C C G⚡
Chromosome 4 A A C A **C** G C C A⚡ T T C G **G** G G T C⚡ A G T C **G** A C C G⚡

a.

Haplotypes

Haplotype 1 **C T C A** A A G T **A C G G T T C A** G G **C A**
Haplotype 2 **T T G A** T T G C **G C A A C A G** T A A **T A**
Haplotype 3 **C C C G** A T C T **G T G A T A C** T G G **T G**
Haplotype 4 **T C G A** T T C C **G C G G T T C A** G A **C A**

b.

Diagnostic SNPs

A/G   T/C   C/G

Haplotype 1 **A T C**
Haplotype 2 **A C G**
Haplotype 3 G T C
Haplotype 4 **A C G**

c.

figure 18.8

CONSTRUCTION OF A HAPLOTYPE MAP. Single-nucleotide polymorphisms (SNPs) are single-base differences between individuals. Sections of DNA sequences from four individuals are shown in (*a*), with three SNPs indicated by arrows. *b.* These three SNPs are shown aligned along with 17 other SNPs from this chromosomal region. This represents a haplotype map for this region of the chromosome. Haplotypes are regions of the genome that are not exchanged by recombination during meiosis. *c.* Haplotypes can be identified using a small number of diagnostic SNPs that differ between the different haplotypes. In this case, 3 SNPs out of the 20 in this region are all that are needed to uniquely identify each haplotype. This greatly facilitates locating disease-causing genes, as haplotypes represent large regions of the genome that behave as a single site during meiosis.

18.4 Genomics and Proteomics

To fully understand how genes work, we need to characterize the proteins they produce. This information is essential in understanding cell biology, physiology, development, and evolution. In many ways, we continue to ask the same questions that Mendel asked, but at a much different level of organization.

Comparative genomics reveals conserved regions in genomes

With the large number of sequenced genomes, it is now possible to make comparisons at both the gene and genome level. One of the striking lessons learned from the sequence of the human genome is how very similar humans are to other organisms. More than half of the genes of *Drosophila* have human counterparts. Among mammals, the similarities are even greater. Humans have only 300 genes that have no counterpart in the mouse genome.

The flood of information from different genomes has given rise to a new field: **comparative genomics.** At this point, we have the complete sequences for close to 100 bacterial genomes. Among eukaryotes, we have the full genome sequences of both types of yeast used in genetics, *S. cerevisiae* and *S. pombe*, as well as the protist *Plasmodium*, the invertebrate animals *Drosophila* and *C. elegans*, and the vertebrates puffer fish (*Fugu* sp. and *Tetraodon* sp.), mouse, and human. In the plant kingdom, the genomes for *Arabidopsis* and rice have been completed. Most of these genomes, however, are draft sequences that include many gaps in regions of highly repetitive DNA.

The use of comparative genomics to ask evolutionary questions is also a field of great promise. The comparison of the many prokaryotic genomes already indicates a greater degree of lateral gene transfer than was previously suspected. The latest round of animal genomes sequenced has included the chimpanzee, our closest living relative. The draft sequence of the chimp (*Pan troglodytes*) genome has just been completed, and comparisons between the chimp and human genome may allow us to unravel what makes us uniquely human.

The early returns from this sequencing effort confirm that our genomes differ by only 1.23% in terms of nucleotide substitutions. At first glance, the largest difference between our genomes actually appears to be in transposable elements. In humans, the SINEs have been threefold more active than in the chimp, but the chimp has acquired two elements that are not found in the human genome. The differences due to insertion and deletion of bases are fewer than substitutions but account for about 1.5% of the euchromatic sequence being unique in each genome.

Synteny allows comparison of unsequenced genomes

Similarities and differences between highly conserved genes can be investigated on a gene-by-gene basis between species. Genome science allows for a much larger scale approach to comparing genomes by taking advantage of synteny.

Synteny refers to the conserved arrangements of segments of DNA in related genomes. Physical mapping techniques can be used to look for synteny in genomes that have not been sequenced. Comparisons with the sequenced, syntenous segment in another species can be very helpful.

To illustrate this, consider rice, already sequenced, and its grain relatives maize, barley, and wheat, none of which have been fully sequenced. Even though these plants diverged more than 50 million years ago, the chromosomes of rice, corn, wheat, and other grass crops show extensive synteny (figure 18.9). In a genomic sense, "rice is wheat."

By understanding the rice genome at the level of its DNA sequence, identification and isolation of genes from grains with larger genomes should be much easier. DNA sequence analysis of cereal grains could be valuable for identifying genes associated with disease resistance, crop yield, nutritional quality, and growth capacity.

The rice genome, as mentioned earlier, has more genes than the human genome. However, rice still has a much smaller genome than its grain relatives, which also represent a major food source for humans.

Organelle genomes have exchanged genes with the nuclear genome

Mitochondria and chloroplasts are considered to be descendants of ancient bacterial cells living in eukaryotes as a result of endosymbiosis (chapter 4). Their genomes have been sequenced in some species, and they are most like prokaryotic genomes. The chloroplast genome, having about 100 genes, is minute compared with the rice genome, with 32,000 to 55,000 genes.

The chloroplast genome

The chloroplast, a plant organelle that functions in photosynthesis, can independently replicate in the plant cell because it has its own genome. The DNA in the chloroplasts of all land plants have

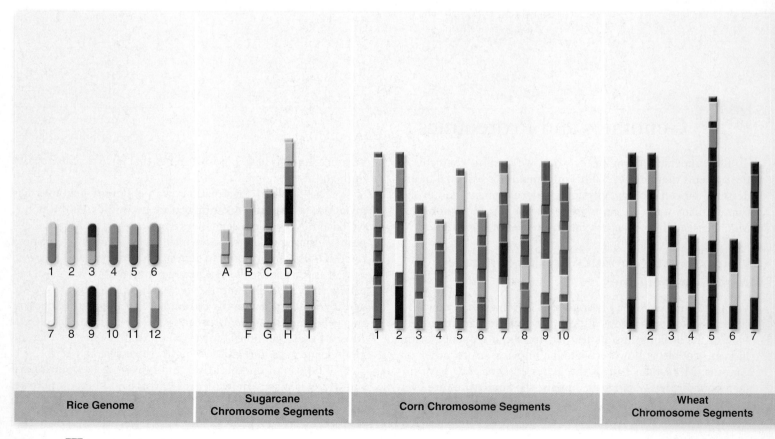

| Rice Genome | Sugarcane Chromosome Segments | Corn Chromosome Segments | Wheat Chromosome Segments |

about the same number of genes, and they are present in about the same order. In contrast to the evolution of the DNA in the plant cell nucleus, chloroplast DNA has evolved at a more conservative pace and therefore shows a more easily interpretable evolutionary pattern when scientists study DNA sequence similarities. Chloroplast DNA is also not subject to modification caused by transposable elements or to mutations due to recombination.

Over time, some genetic exchange appears to have occurred between the nuclear and chloroplast genomes. For example, Rubisco, the key enzyme in the Calvin cycle of photosynthesis (chapter 8), consists of large and small subunits. The small subunit is encoded in the nuclear genome. The protein it encodes has a targeting sequence that allows it to enter the chloroplast and combine with large subunits, which are coded for and produced by the chloroplast.

The mitochondrial genome

Mitochondria are also constructed of components encoded by both the nuclear genome and the mitochondrial genome. For example, the electron transport chain (chapter 7) is made up of proteins that are encoded by both nuclear and mitochondrial genomes—and the pattern varies with different species. This observation implies a movement of genes from the mitochondria to the nuclear genome with some lineage-specific variation.

The evolutionary history of the localization of these genes is a puzzle. Comparative genomics and their evolutionary implications are explored in detail in chapter 24, after we have established the fundamentals of evolutionary theory.

Functional genomics reveals gene function at the genome level

Bioinformatics takes advantage of high-end computer technology to analyze the growing gene databases, look for relationships among genomes, and then hypothesize functions of genes based on sequence. Genomics is now shifting gears and moving back to hypothesis-driven science, to **functional genomics**, the study of the function of genes and their products.

Like sequencing whole genomes, finding how these genomes work requires the efforts of a large team. For example, an international community of researchers has come together with a plan to assign function to all of the 20,000–25,000 *Arabidopsis* genes by 2010 (Project 2010). One of the first steps is to determine when and where these genes are expressed. Each step beyond that will require additional enabling technology.

DNA microarrays

The earlier description of ESTs indicated that we could locate sequences that are transcribed on our DNA maps—but this tells us nothing about when and where these genes are turned on. To be able to analyze gene expression at the whole-genome level requires a representation of the genome that can be manipulated experimentally. This has led to the creation of **DNA microarrays** or "gene chips" (figure 18.10).

Preparation of a microarray To prepare a particular microarray, fragments of DNA are deposited on a microscope slide by a robot at indexed locations (i.e., an array). Silicon chips instead of slides can also be arrayed. These chips can then

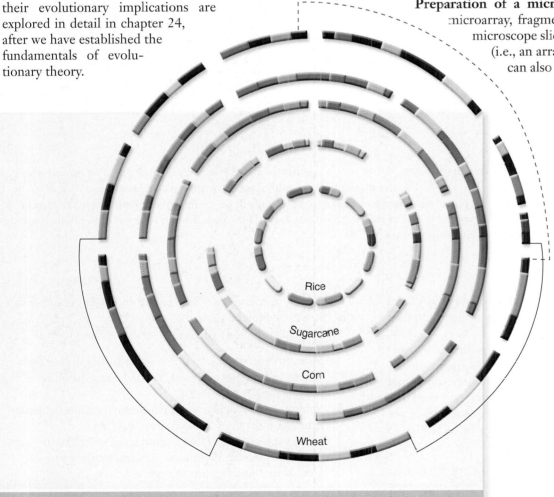

Genomic Alignment (Segment Rearrangement)

figure 18.9

GRAIN GENOMES ARE REARRANGEMENTS OF SIMILAR CHROMOSOME SEGMENTS. Shades of the same color represent pieces of DNA that are conserved among the different species but have been rearranged. By splitting the individual chromosomes of major grass species into segments and rearranging the segments, researchers have found that the genome components of rice, sugarcane, corn, and wheat are highly conserved. This implies that the order of the segments in the ancestral grass genome has been rearranged by recombination as the grasses have evolved.

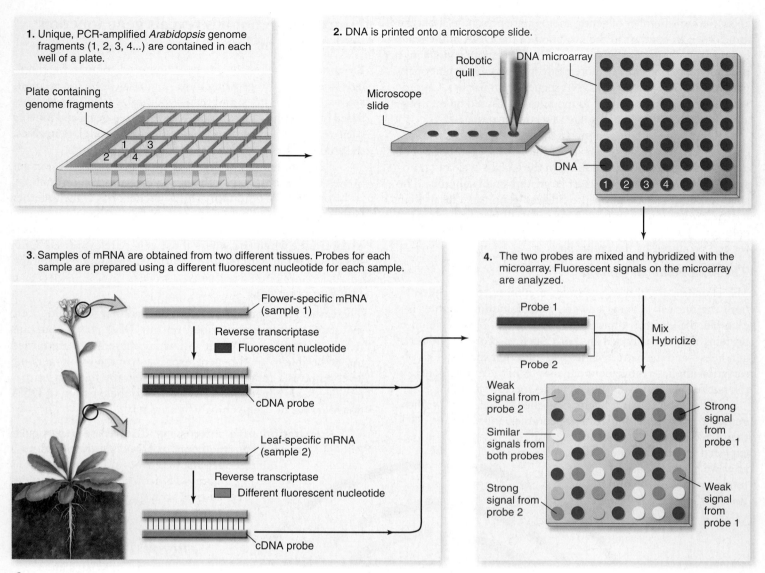

1. Unique, PCR-amplified *Arabidopsis* genome fragments (1, 2, 3, 4...) are contained in each well of a plate.

Plate containing genome fragments

2. DNA is printed onto a microscope slide.

Robotic quill
DNA microarray
Microscope slide
DNA

3. Samples of mRNA are obtained from two different tissues. Probes for each sample are prepared using a different fluorescent nucleotide for each sample.

Flower-specific mRNA (sample 1)
Reverse transcriptase
Fluorescent nucleotide
cDNA probe
Leaf-specific mRNA (sample 2)
Reverse transcriptase
Different fluorescent nucleotide
cDNA probe

4. The two probes are mixed and hybridized with the microarray. Fluorescent signals on the microarray are analyzed.

Probe 1
Mix Hybridize
Probe 2
Weak signal from probe 2
Strong signal from probe 1
Similar signals from both probes
Strong signal from probe 2
Weak signal from probe 1

figure 18.10

MICROARRAYS. Microarrays are created by robotically placing DNA onto a microscope slide. The microarray can then be probed with RNA from tissues of interest to identify expressed DNA. The microarray with hybridized probes is analyzed and often displayed as a false-color image. If a gene is frequently expressed in one of the samples, the fluorescent signal will be strong (*red* or *green*) where the gene is located on the microarray. If a gene is rarely expressed in one of the samples, the signal will be weak (*pink* or *light green*). A yellow color indicates genes that are expressed at similar levels in each sample.

be used in hybridization experiments with labeled mRNA from different sources. This gives a high-level view of genes that are active and inactive in specific tissues.

Researchers are currently using a chip with 24,000 *Arabidopsis* genes on it to identify genes that are expressed developmentally in certain tissues or in response to environmental factors. RNA from these tissues can be isolated and used as a probe for these microarrays. Only those sequences that are expressed in the tissues will be present and will hybridize to the microarray.

Microarray analysis and cancer One of the most exciting uses of microarrays has been the profiling of gene expression patterns in human cancers. Microarray analysis has revealed that different cancers have different gene expression patterns.

These findings are already being used to diagnose and design specific treatments for particular cancers.

From a large body of data, several patterns emerge:

1. Specific cancer types can be reliably distinguished from other cancer types and from normal tissue based on microarray data.
2. Subtypes of particular cancers often have different gene expression patterns in microarray data.
3. Gene expression patterns from microarray data can be used to predict disease recurrence, tendency to metastasize, and treatment response.

This represents an important step forward in both the diagnosis and treatment of human cancers.

Transgenics

How can we determine whether two genes from different species having similar sequences have the same function? And, how can we be sure that a gene identified by an annotation program actually functions as a gene in the organism? One way to address these questions is through **transgenics**—the creation of organisms containing genes from other species (transgenic organisms).

The technology for creating transgenic organisms was discussed in chapter 17; it is illustrated for plants in figure 18.11. For example, to test whether an *Arabidopsis* gene is homologous to a rice gene, the *Arabidopsis* gene can be inserted into cells of rice, which are then regenerated through cell culture into a rice plant (this type of cloning is discussed in chapter 42). Different markers can be incorporated into the gene so that its protein product can be visualized or isolated in the transgenic plant, demonstrating that the inserted gene is being transcribed. In some cases, the transgene (inserted foreign gene) may affect a visible phenotype. Of course, transgenics are but one of many ways to address questions about gene function.

Proteomics moves from genes to proteins

Proteins are much more difficult to study than DNA because of posttranslational modification and formation of protein complexes. And, as already mentioned, a single gene can code for multiple proteins using alternative splicing. Although all the DNA in a genome can be isolated from a single cell, only a portion of the proteome is expressed in a single cell or tissue.

Proteomics is the study of the **proteome**—all of the proteins encoded by the genome. Understanding the proteome for even a single cell will be a much more difficult task than determining the sequence of a genome. Because a single gene can produce more than one protein by alternative splicing, the first step is to characterize the **transcriptome**—all of the RNA that is present in a cell or tissue. Because of alternative splicing, both the transcriptome and the proteome are larger and more complex than the simple number of genes in the genome.

To make matters worse, a single protein can be modified posttranslationally to produce functionally different forms. The function of a protein can also depend on its association with other proteins. Nonetheless, since proteins perform most of the major functions of cells, understanding their diversity is essential.

inquiry

Why is the "proteome" likely to be different from simply the predicted protein products found in the complete genome sequence?

Predicting protein function

The use of new methods to quickly identify and characterize large numbers of proteins is the distinguishing feature between traditional protein biochemistry and proteomics. As with genomics, the challenge is one of scale.

Ideally, a researcher would like to be able to examine a nucleotide sequence and know what sort of functional

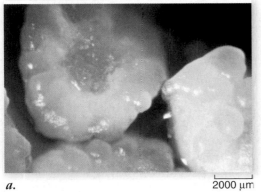

a.   2000 μm

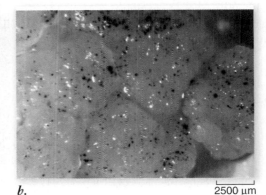

b.   2500 μm

c.   d.

figure 18.11

GROWTH OF A TRANSGENIC PLANT. DNA containing a gene for herbicide resistance was transferred into wheat (*Triticum aestivum*). The DNA also contains the *GUS* gene, which is used as a tag or label. The *GUS* gene produces an enzyme that catalyzes the conversion of a staining solution from clear to blue. *a.* Embryonic tissue just prior to insertion of foreign DNA. *b.* Following DNA transfer, callus cells containing the foreign DNA are indicated by color from the *GUS* gene (*blue spots*). *c.* Shoot formation in the transgenic plants growing on a selective medium. Here, the gene for herbicide resistance in the transgenic plants allows growth on the selective medium containing the herbicide. *d.* Comparison of growth on the selection medium for transgenic plants bearing the herbicide resistance gene (*left*) and a nontransgenic plant (*right*).

protein the sequence specifies. Databases of protein structures in different organisms can be searched to predict the structure and function of genes known only by sequence, as identified in genome projects. Analysis of these data provides a clearer picture of how gene sequence relates to protein structure and function. Having a greater number of DNA sequences available allows for more extensive comparisons as well as identification of common structural patterns as groups of proteins continue to emerge.

Fortunately, although there may be as many as a million different proteins, most are just variations on a handful of themes. The same shared structural motifs—barrels, helices, molecular zippers—are found in the proteins of plants, insects, and humans (figure 18.12; also see chapter 3 for more information on protein motifs). The maximum number of distinct motifs has been estimated to be fewer than 5000. About 1000 of these motifs have already been cataloged. Both publicly and privately financed efforts are now under way to detail the shapes of all the common motifs.

Protein microarrays

Protein microarrays, comparable to DNA microarrays, are being used to analyze large numbers of proteins simultaneously. Making a protein microarray starts with isolating the transcriptome of a cell or tissue. Then cDNAs are constructed and reproduced by cloning them into bacteria or viruses. Transcription and translation occur in the prokaryotic host, and micromolar quantities of protein are isolated and purified. These are then spotted onto glass slides.

Protein microarrays can be probed in at least three different ways. First, they can be screened with antibodies to specific proteins. Antibodies are labeled so that they can be detected, and the patterns on the protein array can be determined by computer analysis.

An array of proteins can also be screened with another protein to detect binding or other protein interactions. Thousands of interactions can be tested simultaneously. For example, calmodulin (which mediates Ca^{2+} function; see chapter 9) was labeled and used to probe a yeast proteome array with 5800 proteins. The screen revealed 39 proteins that bound calmodulin. Of those 39, 33 were previously unknown!

A third type of screen uses small molecules to assess whether they will bind to any of the proteins on the array. This approach shows promise for discovering new drugs that will inhibit proteins involved in disease.

Large-scale screens reveal protein–protein interactions

We often study proteins in isolation, compared with their normal cellular context. This approach is obviously artificial. One immediate goal of proteomics, therefore, is to map all the physical interaction between proteins in a cell. This is a daunting task that requires tools that can be automated, similarly to the way that genome sequencing was automated.

One approach is to use the yeast two-hybrid system discussed in the preceding chapter. This system can be automated once libraries of known cDNAs are available in each of the two vectors used.

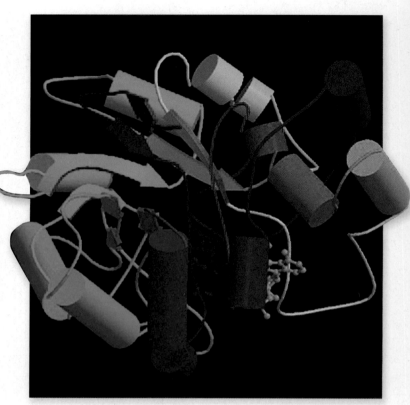

figure 18.12

COMPUTER-GENERATED MODEL OF AN ENZYME. Searchable databases contain known protein structures, including human aldose reductase shown here. Secondary structural motifs are shown in different colors.

inquiry

What is the relationship among genome, transcriptome, and proteome?

The use of two-hybrid screens has been applied to budding yeast to generate a map of all possible interacting proteins. This method is difficult to apply to more complex multicellular organisms, but in a technical tour-de-force, it has been applied to *Drosophila melanogaster* as well.

For humans, mice, and other vertebrates, the two-hybrid system is being applied more selectively at present. Useful data can still be collected by concentrating on a biologically significant process, such as signal transduction. The technique can then be used to map all of the interacting proteins in a specific signaling pathway.

Comparative genomics uses comparisons of different genomes to infer structural, functional, and evolutionary relationships between genes and proteins. Functional genomics attempts to infer functions using genomic level information. Microarrays can be used to look at gene expression for many genes at once. Proteomics is moving this to the level of proteins. Protein microarrays allow the analysis of many proteins at once, and interaction screens using the yeast two-hybrid system can be scaled up to analyze large-scale interaction of proteins in cells.

18.5 Applications of Genomics

Space allows us to highlight only a few of the myriad applications of genomics to show the possibilities. The tools being developed truly represent a revolution in biology that will likely have a lasting impact on the way that we think about living systems.

Genomics can help to identify infectious diseases

The genomics revolution has yielded millions of new genes to be investigated. The potential of genomics to improve human health is enormous. Mutations in a single gene can explain some, but not most, hereditary diseases. With entire genomes to search, the probability of unraveling human, animal, and plant diseases is greatly improved.

Although proteomics will likely lead to new pharmaceuticals, the immediate impact of genomics is being seen in diagnostics. Both improved technology and gene discovery are enhancing diagnosis of genetic abnormalities.

Diagnostics are also being used to identify individuals. For example, STRs were among the forensic diagnostic tools used to identify remains of victims of the September 11, 2001, terrorist attack on the World Trade Center in New York City.

The September 11 attacks were followed by an increased awareness and concern about biological weapons. When cases of anthrax began appearing in the fall of 2001, genome sequencing allowed exploration of possible sources of the deadly bacteria and determination of whether they had been genetically engineered to increase their lethality.

In addition, substantial effort has been turned toward the use of genomic tools to distinguish between naturally occurring infections and intentional outbreaks of disease. The Centers for Disease Control and Prevention (CDC) have ranked bacteria and viruses that are likely targets for bioterrorism (table 18.2).

figure 18.13

RICE FIELD. Most of the rice grown globally is directly consumed by humans and is the dietary mainstay of 2 billion people.

Genomics can help improve agricultural crops

Globally speaking, nutrition is the greatest impediment to human health. Much of the excitement about the rice genome project is based on its potential for improving the yield and nutritional quality of rice and other cereals worldwide. The development of Golden Rice (chapter 17) is an example of improved nutrition through genetic approaches. About one-third of the world population obtains half its calories from rice (figure 18.13). In some regions, individuals consume up to 1.5 kg of rice daily. More than 500 million tons of rice is produced each year, but this may not be enough in the future.

Due in large part to scientific advances in crop breeding and farming techniques, in the last 50 years world grain production has more than doubled, with an increase in cropland of only 1%. The world now farms a total area the size of South America, but without the scientific advances of the past 50 years, an area equal to the entire western hemisphere would need to be farmed to produce enough food for the world.

Unfortunately, water usage for crops has tripled in that time period, and quality farmland is being lost to soil erosion. Scientists are also concerned about the effects of global climate change on agriculture worldwide. Increasing the yield and quality of crops, especially on more marginal farmland, will depend on many factors—but genetic engineering, built on the findings of genomics projects, can contribute significantly to the solution.

TABLE 18.2	High-Priority Pathogens for Genomic Research	
Pathogen	**Disease**	**Genome***
Variola major	Smallpox	Complete
Bacillus anthracis	Anthrax	Complete
Yersinia pestis	Plague	Complete
Clostridium botulinum	Botulism	In progress
Francisella tularensis	Tularemia	Complete
Filoviruses	Ebola and Marburg hemorrhagic fever	Both are complete
Arenaviruses	Lassa fever and Argentine hemorrhagic fever	Both are complete

*There are multiple strains of these viruses and bacteria. "Complete" indicates that at least one has been sequenced. For example, the Florida strain of anthrax was the first to be sequenced.

figure 18.14

CORN CROP PRODUCTIVITY WELL BELOW ITS GENETIC POTENTIAL DUE TO DROUGHT STRESS. Corn production can be limited by water deficiencies due to the drought that occurs during the growing season in dry climates. Global climate change may increase drought stress in areas where corn is the major crop.

inquiry

The corn genome has not been sequenced. How could you use information from the rice genome sequence to try to improve drought tolerance in corn?

Most crops grown in the United States produce less than half of their genetic potential because of environmental stresses (salt, water, and temperature), herbivores, and pathogens (figure 18.14). Identifying genes that can provide resistance to stress and pests is the focus of many current genomics research projects. Having access to entire genomic sequences will enhance the probability of identifying critical genes.

Genomics raises ethical issues over ownership of genomic information

Genome science is also a source of ethical challenges and dilemmas. One example is the issue of gene patents. Actually, it is the use of a gene, not the gene itself, that is patentable. For a gene-related patent, the product and its function must be known.

The public genome consortia, supported by federal funding, have been driven by the belief that the sequence of genomes should be freely available to all and should not be patented. Private companies patent gene functions, but they often make sequence data available with certain restrictions. The physical sciences have negotiated the landscape of public and for-profit research for decades, but this is relatively new territory for biologists.

Another ethical issue involves privacy. How sequence data are used is the focus of thoughtful and ongoing discussions. The Universal Declaration on the Human Genome and Human Rights states, "The human genome underlies the fundamental unity of all members of the human family, as well as the recognition of their inherent dignity and diversity. In a symbolic sense, it is the heritage of humanity."

Although we talk about "the" human genome, each of us has subtly different genomes that can be used to identify us. Genetic disorders such as cystic fibrosis and Huntington disease can already be identified by screening, but genomics will greatly increase the number of identifiable traits. What if employers or insurance companies gain access to your personal SNP profile? Could you be discriminated against because your genome indicates you have a genetic tendency toward chemical addiction or heart disease? What sort of legal protections should be in place to prevent this sort of discrimination?

On a more positive note, the U.S. Armed Forces require DNA samples from members for possible casualty identification, and DNA-based identification brought peace of mind to some families of the World Trade Center victims. Forensic use of DNA-based identification has already been described.

Behavioral genomics is an area that is also rich with possibilities and dilemmas. Very few behavioral traits can be accounted for by single genes. Two genes have been associated with fragile-X mental retardation, and three with early-onset Alzheimer disease. Comparisons of multiple genomes will likely lead to the identification of multiple genes controlling a range of behaviors. Will this change the way we view acceptable behavior?

In Iceland, the parliament has voted to have a private company create a database from pooled medical, genetic, and genealogical information about all Icelanders, a particularly fascinating population from a genetic perspective. Because minimal migration or immigration has occurred there over the last 800 years, the information that can be mined from the Icelandic database is phenomenal. Ultimately, the value of that information has to be weighed against any possible discrimination or stigmatization of individuals or groups.

Genomics has been used in the identification of remains of victims of disasters and can be also be used to identify bioweapon agents. Genomics is being used to improve agricultural crops and domesticated animals. This field has also created controversy over who owns genetic/genomic information.

18.1 Mapping Genomes

Maps provide landmarks to find your way around a genome.

- Genetic maps provide relative locations of genes on a chromosome based on recombination frequency and provide a link to phenotypes.

- Physical maps are based on landmarks in the actual DNA

- There are several kinds of physical maps:

  - *Restriction maps are based on distances between restriction enzyme cleavage sites (figure 18.1).*

  - *Overlap between smaller fragments can be used to assemble them into contiguous segments called a contig.*

  - *Chromosomal banding using stains and hybridization results in low-resolution cytological maps.*

  - *Radiation hybrid maps are constructed by fusing an irradiated cell with another cell to identify overlapping fragments of chromosomes.*

  - *The ultimate physical map is the nucleotide sequence of a DNA.*

- Any physical site can be used as a sequence-tagged site (STS), based on a small stretch of a unique DNA sequence that allows unambiguous identification of a fragment.

- Physical and genetic maps can be correlated. Any gene that can be cloned can be placed within the genome sequence and mapped.

18.2 Whole-Genome Sequencing

Large-scale genome sequencing requires the use of high-throughput automated sequencing and computer analysis.

- Automated sequencing is accurate for DNA segments up to 800 bp long. Errors are possible, and 5–10 copies of the segments are sequenced and compared.

- Artificial chromosomes based on bacterial chromosomes allow the cloning of larger pieces of DNA.

- Clone-by-clone sequencing compares the overlapping of restriction maps using bacterial artificial chromosomes or identifying STS between clones (figure 18.3).

- Shotgun sequencing involves sequencing random clones, then using a computer to assemble the finished sequence.

- Whole-genome sequencing uses both clone-by-clone sequencing and shotgun sequencing (figure 18.5).

18.3 Characterizing Genomes

Genome sequences are descriptive and do not provide information on the organization of the genomes, their products, and how they are interrelated.

- Although eukaryotic genomes are larger and have more genes than prokaryotes, the size of the organism is not correlated with the size of the genome.

- Once a genome is sequenced, the genes are identified by looking for open-reading frames (ORF).

- An ORF begins with a start codon and contains no stop codon for a distance long enough to encode a protein.

- Bioinformatics uses computer programs to search for genes, compare genomes, and assemble genomes.

- Genomes contain both coding and noncoding DNA.

- In eukaryotes, protein-encoding DNA includes single-copy genes.

- Regions containing genes may be duplicated (segmental duplications).

- Protein-coding genes may be found as part of multigene families and tandem clusters.

- Noncoding DNA in eukaryotes makes up about 99% of DNA. Noncoding DNA can occur within genes (introns), be structural and untranscribable, or contain short repeated sequences.

- Protein-coding genes that are duplicated may accumulate mutations and become pseudogenes.

- Approximately 45% of the human genome is composed of mobile transposable elements that exist in many copies.

- The number and location of expressed genes can be estimated by sequencing the ends of randomly selected cDNAs to produce expressed sequence tags (ESTs).

- Alternative splicing of existing genes yields different proteins with different functions (figure 18.7).

- Genetic variation can exist as differences in a single nucleotide between individuals called single-nucleotide polymorphisms (SNPs).

- Genomic haplotypes are regions of chromosomes that are not exchanged by recombination. This is called linkage disequilibrium and can be used to map genes by association (figure 18.8).

18.4 Genomics and Proteomics

Proteomics characterizes the proteins produced by cells.

- The field of comparative genomics studies the conserved regions of genomes among organisms.

- The biggest difference between our genome and the chimpanzee genome is in transposable elements.

- Synteny refers to the conserved arrangements of segments of DNA in related genomes (figure 18.9).

- Chloroplasts and mitochondria contain components encoded by their own genomes and by the nuclear genome.

- Functional genomics, a hypothesis-driven science, studies gene function and gene products.

- DNA microarrays allow the expression of all of the genes in a cell to be monitored at once (figure 18.10).

- Proteomics characterizes all of the proteins produced by a cell. The transcriptome is all the mRNAs present in a cell at a specific time.

- Protein microarrays are used to identify and characterize large numbers of proteins.

- The yeast two-hybrid system is used to generate large-scale maps of interacting proteins.

18.5 Applications of Genomics

Genomics represents a revolution in biology that will have a lasting effect in the way we think about living systems.

- Genomics can help identify naturally occurring and intentional outbreaks of infectious diseases.

- Genomics can help improve domesticated animals, the nutritional value of crops, and their responses to environmental stresses.

- Genomics raises ethical issues over ownership of genomic information and personal privacy.

SELF TEST

1. A genetic map is based on—
   a. the sequence of the DNA.
   b. the relative position of genes on chromosomes.
   c. the location of sites of restriction enzyme cleavage.
   d. the banding pattern on a chromosome.

2. What is an STS?
   a. A unique sequence within the DNA that can be used for mapping
   b. A repeated sequence within the DNA that can be used for mapping
   c. An upstream element that allows for mapping of the 3′ region of a gene
   d. Both (b) and (c)

3. An artificial chromosome is useful because—
   a. it produces more consistent results than a natural chromosome.
   b. it allows for the isolation of larger DNA sequences.
   c. it provides a high copy number of a DNA sequence.
   d. it is linear.

4. Which of the following techniques relies on knowledge of overlapping sequences?
   a. Radiation hybrid mapping
   b. Shotgun method of genome sequencing
   c. FISH
   d. Clone-by-clone method of genome sequencing

5. Which number represents the total number of genes in the human genome?
   a. 2500
   b. 10,000
   c. 25,000
   d. 100,000

6. An open reading frame (ORF) is distinguished by the presence of—
   a. a stop codon.
   b. a start codon.
   c. a sequence of DNA long enough to encode a protein.
   d. All of the above.

7. What is a BLAST search?
   a. A mechanism for aligning consensus regions during whole-genome sequencing
   b. A search for similar gene sequences from other species
   c. A method of screening a DNA library
   d. A method for identifying ORFs

8. Which of the following is *not* an example of a protein-encoding gene?
   a. Single-copy gene
   b. Tandem clusters
   c. Pseudogene
   d. Multigene family

9. The duplication of a gene due to uneven meiotic crossing over is thought to lead to the production of a—
   a. segmental duplication.
   b. tandem duplication.
   c. simple sequence repeat.
   d. multigene family.

10. Which of the following is *not* an example of noncoding DNA?
    a. Promoter
    b. Intron
    c. Pseudogene
    d. Exon

11. Comparisons between genomes is made easier because of—
    a. synteny.
    b. haplotypes.
    c. transposons.
    d. expressed sequence tags.

12. What information can be obtained from a DNA microarray?
    a. The sequence of a particular gene
    b. The presence of genes within a specific tissue
    c. The pattern of gene expression
    d. Differences between genomes

13. Which of the following is true regarding microarray technology and cancer?
    a. A DNA microarray can determine the type of cancer.
    b. A DNA microarray can measure the response of a cancer to therapy.
    c. A DNA microarray can be used to predict whether the cancer will metastasize.
    d. All of the above.

14. What is a proteome?
    a. The collection of all genes encoding proteins
    b. The collection of all proteins encoded by the genome
    c. The collection of all proteins present in a cell
    d. The amino acid sequence of a protein

15. Which of the following techniques could be used to examine protein–protein interactions in a cell?
    a. Two-hybrid screens
    b. Protein structure databases
    c. Protein microarrays
    d. Both (a) and (c)

CHALLENGE QUESTIONS

1. You are in the early stages of a genome-sequencing project. You have isolated a number of clones from a BAC library and mapped the inserts in these clones using STSs. Use the STSs to align the clones into a contiguous sequence of the genome (a contig).

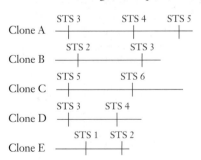

2. Genomic research can be used to determine if an outbreak of an infectious disease is natural or "intentional." Explain what a genomic researcher would be looking for in a suspected intentional outbreak of a disease like anthrax.

4000 µm

Cellular Mechanisms of Development

introduction

COUNTLESS GENERATIONS OF YOUNG children have reveled in the discovery of frog tadpoles during summer visits to freshwater ponds, have watched with fascination as a chick pecks its way out of its shell, or have delighted as the first flowers of spring push their way up through the soil. For thousands of years, the wonder of such events has inspired a desire to understand how organisms arise, grow, change, and mature.

We have explored gene expression from the perspective of individual cells, examining the diverse mechanisms cells employ to control the transcription of particular genes. Now we broaden our perspective and look at the unique challenge posed by the development of a single cell, the fertilized egg, into a multicellular organism such as is occurring in these fish embryos. In the course of this developmental journey, a pattern of decisions about gene expression takes place that causes particular lines of cells to proceed along different paths, spinning an incredibly complex web of cause and effect. Yet, for all its complexity, this developmental program works with impressive precision. In this chapter, we explore the mechanisms of development in multicellular organisms.

concept outline

19.1 Overview of Development

Development can be defined as the process of systematic, gene-directed changes through which an organism forms the successive stages of its life cycle. Development is a continuum, and explorations of development can be focused on any point along this continuum. The study of development plays a central role in unifying the understanding of both the similarities and diversity of life on Earth.

We can divide the overall process of development into four subprocesses:

- **Growth (cell division).** A developing plant or animal begins as a fertilized egg, or zygote, that must undergo cell division to form the body mass of the new individual.
- **Differentiation.** As cells divide, orchestrated changes in gene expression result in differences between cells that ultimately result in cell specialization. In differentiated cells, certain genes are expressed at particular times, but other genes may not be expressed at all.
- **Pattern Formation.** Cells in a developing embryo must become oriented to the body plan of the organism the embryo will become. Pattern formation involves cells' abilities to detect positional information that guides their ultimate fate.
- **Morphogenesis.** As development proceeds, the form of the body is generated, namely its organs and anatomical features. Morphogenesis may involve cell death as well as cell division and differentiation.

Despite the overt differences between groups of plants and animals, most multicellular organisms develop according to molecular mechanisms that are fundamentally very similar. This observation suggests that these mechanisms evolved very early in the history of multicellular life.

19.2 Cell Division

When a frog tadpole hatches out of its protective coats, it is roughly the same overall mass as the fertilized egg from which it came. Instead of being made up of just one cell, however, the tadpole comprises a million or so cells, which are organized into tissues and organs with different functions. Thus, the very first process that must occur during embryogenesis is cell division.

Immediately following fertilization, the diploid zygote undergoes a period of rapid mitotic divisions that ultimately result in an early embryo comprising dozens to thousands of diploid cells. In animal embryos, the timing and number of these divisions are species-specific and are controlled by a set of molecules that we examined in chapter 10: the **cyclins** and **cyclin-dependent kinases (Cdks)**. These molecules exert control over checkpoints in the cycle of mitosis.

Development begins with cell division

In animal embryos, the period of rapid cell division following fertilization is called **cleavage.** During cleavage, the enormous mass of the zygote is subdivided into a larger and larger number of smaller and smaller cells, called **blastomeres** (figure 19.1). Hence, cleavage is not accompanied by any increase in the overall size of the embryo. The G_1 and G_2 phases of the cell cycle, during which a cell increases its mass and size, are extremely shortened or eliminated during cleavage (figure 19.2).

Because of the absence of the two gap/growth phases, the rapid rate of mitotic divisions during cleavage is never again approached in the lifetime of any animal. For example, zebrafish blastomeres divide once every several minutes during cleavage, to create an embryo with a thousand cells in just under 3 hr! In contrast, cycling adult human intestinal epithelial cells divide on average only once every 19 hr.

When external sources of nutrients become available— for example, during larval feeding stages or after implantation of mammalian embryos—daughter cells can increase in size following cytokinesis, and an overall increase in the size of the organism occurs as more cells are produced.

Every cell division is known in the development of *C. elegans*

One of the most completely described models of development is the tiny nematode *Caenorhabditis elegans*. Only about 1 mm long, the adult worm consists of 959 somatic cells.

Because *C. elegans* is transparent, individual cells can be followed as they divide. By observing them, researchers have

figure 19.1

CLEAVAGE DIVISIONS IN A FROG EMBRYO. *a.* The first cleavage division divides the egg into two large blastomeres. *b.* After two more divisions, four small blastomeres sit on top of four large blastomeres, each of which continues to divide to produce (*c*) a compact mass of cells.

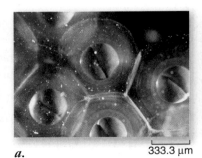
a.   333.3 µm

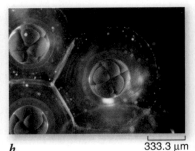
b.   333.3 µm

c.   333.3 µm

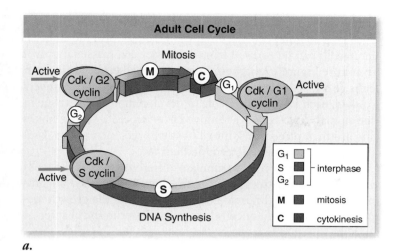

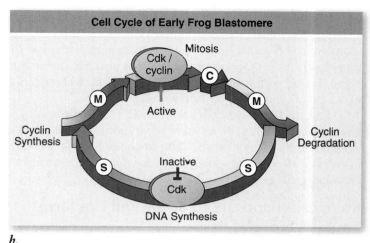

a.

b.

figure 19.2

CELL CYCLE OF ADULT CELL AND EMBRYONIC CELL. In contrast to the cell cycle of adult somatic cells (*a*), the dividing cells of early frog embryos lack G_1 and G_2 stages (*b*), enabling the cleavage stage nuclei to rapidly cycle between DNA synthesis and mitosis. Large stores of cyclin mRNA are present in the unfertilized egg. Periodic translation of this message produces cyclin proteins. Cyclin degradation and Cdk inactivation allows the cell to complete mitosis and initiate the next round of DNA synthesis.

learned how each of the cells that make up the adult worm is derived from the fertilized egg. As shown on the lineage map in figure 19.3*a*, the egg divides into two cells, and these daughter cells continue to divide. Each horizontal line on the map represents one round of cell division. The length of each vertical line represents the time between cell divisions, and the end of each vertical line represents one fully differentiated cell. In figure 19.3*b*, the major organs of the worm are color-coded to match the colors of the corresponding groups of cells on the lineage map.

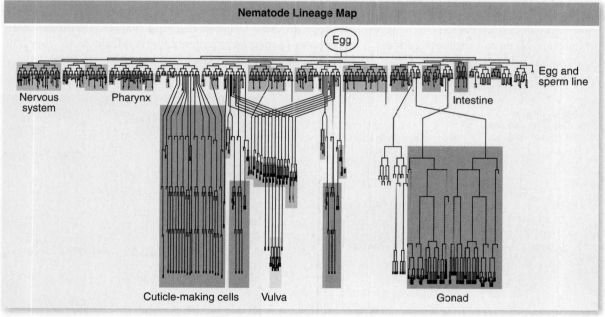

a.

b.

figure 19.3

STUDYING EMBRYONIC CELL DIVISION AND DEVELOPMENT IN THE NEMATODE. Development in *C. elegans* has been mapped out such that the fate of each cell from the single egg cell has been determined. *a.* The lineage map shows the number of cell divisions from the egg, and the color coding links their placement in (*b*) the adult organism.
M.E. Challinor illustration. From Howard Hughes Medical Institute © as published in *From Egg to Adult*, 1992. Reprinted by permission.

Some of these differentiated cells, such as some cells that generate the worm's external cuticle, are "born" after only 8 rounds of cell division; other cuticle cells require as many as 14 rounds. The cells that make up the worm's pharynx, or feeding organ, are born after 9 to 11 rounds of division, whereas cells in the gonads require up to 17 divisions.

Exactly 302 nerve cells are destined for the worm's nervous system. Exactly 131 cells are programmed to die, mostly within minutes of their "birth." The fate of each cell is the same in every *C. elegans* individual, except for the cells that will become eggs and sperm.

Stem cells continue to divide and can form multiple kinds of tissue

The blastomeres of cleavage stage mammalian embryos are nondifferentiated and can give rise to any tissue. As development proceeds, cells become limited in their fates, as discussed in the next section. Some cells, called **stem cells,** are set aside and will continue to divide while remaining undifferentiated. For example, one population of cells are set aside that will go on to form nerve cells, and another will go on to produce blood, and still others muscle. Each major tissue is represented by its own kind of **tissue-specific stem cell.** As development proceeds, these tissue-specific stem cells persist—even in adults.

Stem cells may give rise to a single cell type, such as muscle satellite cells that give rise to muscle cells, or give rise to multiple cell types, such as myeloid cells that give rise to different types of blood cells. Stem cells that form multiple cell types may be **totipotent,** meaning that they can become any cell type, or **pluripotent,** meaning that they can become multiple different cell types.

The cleavage stage in mammals continues for five or six days, producing a ball of cells called a **blastocyst.** This blastocyst consists of an outer layer that will become the placenta enclosing an inner cell mass that will go on to form the embryo. Stem cells can be isolated from this inner cell mass and grown in culture (figure 19.4); these cells are termed **embryonic stem cells (ES cells).** In mice, in whom these have been extensively studied, each ES cell is capable of developing into any tissue in the animal. If ES cells are removed from an early-stage embryo and are then placed back into a host embryo, interactions with the cells around them determine their fate.

Scientists can also stimulate differentiation of ES cells along different pathways in culture, by exposing the uncommitted ES cells to different chemical signals in the growth medium. The use of these cells is discussed later in the chapter.

Plant growth occurs in specific areas called meristems

A major difference between animals and plants is that most animals are mobile, at least in some phase of their life cycles, and therefore they can move away from unfavorable circumstances. Plants, in contrast, are anchored in position and must simply endure whatever environment they experience. Plants compensate for this restriction by allowing development to accommodate local circumstances.

Instead of creating a body in which every part is specified to have a fixed size and location, a plant assembles its body throughout its life span from a few types of modules, such as leaves, roots, branch nodes, and flowers. Each module has a rigidly controlled structure and organization, but how the modules are utilized is quite flexible—they can be adjusted to environmental conditions.

Plants develop by building their bodies outward, creating new parts from groups of stem cells that are contained in structures called **meristems.** As meristematic stem cells continually divide, they produce cells that can differentiate into the tissues of the plant.

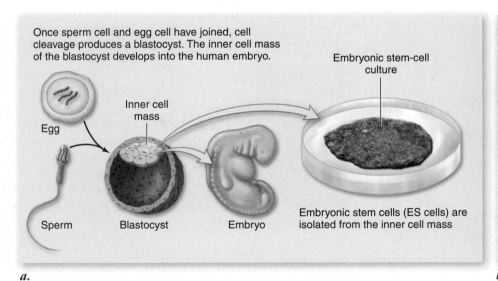

Once sperm cell and egg cell have joined, cell cleavage produces a blastocyst. The inner cell mass of the blastocyst develops into the human embryo.

Egg

Sperm

Inner cell mass

Blastocyst

Embryo

Embryonic stem-cell culture

Embryonic stem cells (ES cells) are isolated from the inner cell mass

a.

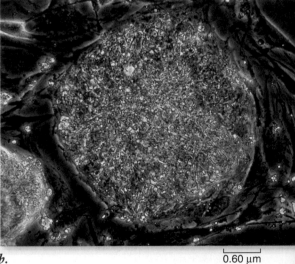

b.

0.60 μm

figure 19.4

ISOLATION OF EMBRYONIC STEM CELLS. *a.* Early cell divisions lead to the blastocyst stage that consists of an outer layer and an inner cell mass, which will go on to form the embryo. Embryonic stem cells (ES cells) can be isolated from this stage by disrupting the embryo and plating the cells. Stem cells removed from a six-day blastocyst can be established in culture and maintained indefinitely in an undifferentiated state. *b.* Human embryonic stem cells. This mass in the photograph is a colony of undifferentiated human embryonic stem cells surrounded by fibroblasts (elongated cells) that serve as a "feeder layer."

This simple scheme indicates a need to control the process of cell division. We know that cell-cycle control genes are present in both yeast (fungi) and animal cells, implying that these are a eukaryotic innovation—and in fact, the plant cell cycle is regulated by the same mechanisms, namely through cyclins and cyclin-dependent kinases. In one experiment, overexpression of a Cdk inhibitor in transgenic *Arabidopsis thaliana* plants resulted in strong inhibition of cell division in leaf meristems, leading to significant changes in leaf size and shape.

In animal embryos a series of rapid divisions convert the fertilized egg into many cells with no change in size. This is accomplished by eliminating G_1 and G_2 phases of mitosis. Every cell division that leads to the adult nematode *C. elegans* is known, and this pattern is invariant. Stem cells are nondifferentiated cells that have the potential to become a number of different tissues. In plants, growth is restricted to specific areas called meristems, where stem cells are retained.

19.3 Cell Differentiation

In chapter 16, we examined the mechanisms that control eukaryotic gene expression. These processes are critical for the development of multicellular organisms, in which life functions are carried out by different tissues and organs. In the course of development, cells become different from one another because of the differential expression of subsets of genes—not only at different times, but in different locations of the growing embryo. We now explore some of the mechanisms that lead to differential gene expression during development.

Cells become determined prior to differentiation

A human body contains more than 210 major types of differentiated cells. These differentiated cells are distinguishable from one another by the particular proteins that they synthesize, their morphologies, and their specific functions. A molecular decision to become a particular type of differentiated cell occurs prior to any overt changes in the cell. This molecular decision-making process is called **cell determination,** and it commits a cell to a particular developmental pathway.

Tracking determination

Determination is often not visible in the cell and can only be "seen" by experiment. The standard experiment to test whether a cell or group of cells is determined is to move the donor cell(s) to a different location in a host (recipient) embryo. If the cells of the transplant develop into the same type of cell as they would have if left undisturbed, then they are judged to be already determined (figure 19.5).

Determination has a time course; it depends on a series of intrinsic or extrinsic events, or both. For example, a cell in the prospective brain region of an amphibian embryo at the early gastrula stage (see chapter 53) has not yet been determined; if transplanted elsewhere in the embryo, it will develop according to the site of transplant. By the late gastrula stage, however, additional cell interactions have occurred, determination has taken place, and the cell will develop as neural tissue no matter where it is transplanted.

Determination often takes place in stages, with a cell first becoming partially committed, acquiring positional labels that reflect its location in the embryo. These labels can have a great influence on how the pattern of the body subsequently develops. In a chicken embryo, tissue at the base of the leg bud normally

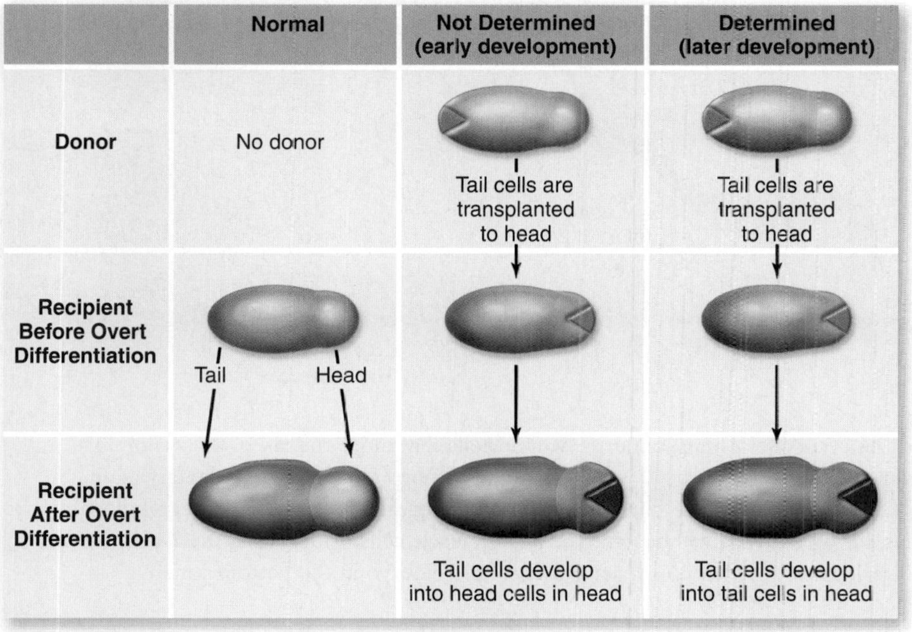

	Normal	Not Determined (early development)	Determined (later development)
Donor	No donor	Tail cells are transplanted to head	Tail cells are transplanted to head
Recipient Before Overt Differentiation	Tail Head		
Recipient After Overt Differentiation		Tail cells develop into head cells in head	Tail cells develop into tail cells in head

figure 19.5

THE STANDARD TEST FOR DETERMINATION. The gray ovals represent embryos at early stages of development. The cells to the right normally develop into head structures, whereas the cells to the left usually form tail structures. If prospective tail cells from an early embryo are transplanted to the opposite end of a host embryo, they develop according to their new position into head structures. These cells are not determined. At later stages of development, the tail cells are determined since they now develop into tail structures after transplantation into the opposite end of a host embryo!

gives rise to the thigh. If this tissue is transplanted to the tip of the identical-looking wing bud, which would normally give rise to the wing tip, the transplanted tissue will develop into a toe rather than a thigh. The tissue has already been determined as leg, but it is not yet committed to being a particular part of the leg. Therefore, it can be influenced by the positional signaling at the tip of the wing bud to form a tip (but in this case, a tip of leg).

The molecular basis of determination

Cells initiate developmental changes by using transcription factors to change patterns of gene expression. When genes encoding these transcription factors are activated, one of their effects is to reinforce their own activation. This reinforcement makes the developmental switch deterministic, initiating a chain of events that leads down a particular developmental pathway.

Cells in which a set of regulatory genes have been activated may not actually undergo differentiation until some time later, when other factors interact with the regulatory protein and cause it to activate still other genes. Nevertheless, once the initial "switch" is thrown, the cell is fully committed to its future developmental path.

Cells become committed to follow a particular developmental pathway in one of two ways:

(1) via the differential inheritance of cytoplasmic determinants, which are maternally produced and deposited into the egg during oogenesis; or
(2) via cell–cell interactions.

The first situation can be likened to a person's social status being determined by who his or her parents are and what he or she has inherited. In the second situation, the person's social standing is determined by interactions with his or her neighbors. Clearly both can be powerful factors in the development and maturation of that individual.

Determination can be due to cytoplasmic determinants

Many invertebrate embryos provide good visual examples of cell determination through the differential inheritance of cytoplasmic determinants. Tunicates are marine invertebrates (see chapter 35), and most adults have simple, saclike

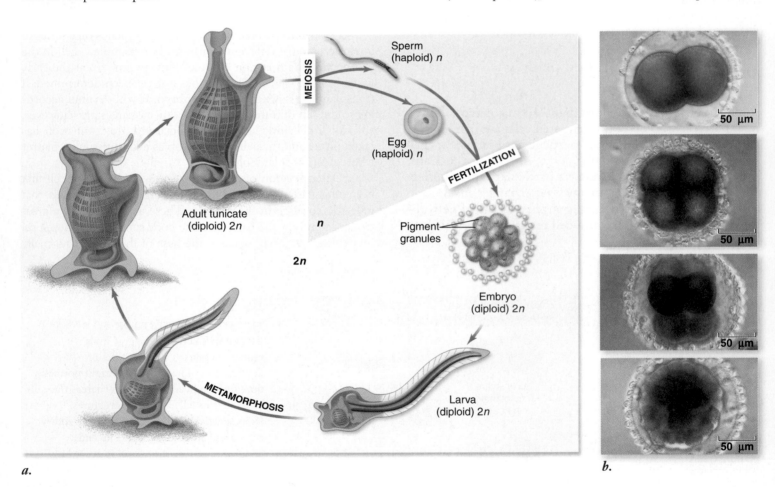

a.   b.

figure 19.6

MUSCLE DETERMINANTS IN TUNICATES. *a.* The life cycle of a solitary tunicate. Muscle cells that move the tail of the swimming tadpole are arranged on either side of the notochord and nerve cord. The tail is lost during metamorphosis into the sedentary adult. *b.* The egg of the tunicate *Styela* contains bright yellow pigment granules. These become asymmetrically localized in the egg following fertilization, and cells that inherit the yellow granules during cleavage will become the larval muscle cells. Embryos at the 2-cell, 4-cell, 8-cell, and 64-cell stages are shown. The tadpole tail will grow out from the lower region of the embryo in the bottom panel.

bodies that are attached to the underlying substratum. Tunicates are placed in the phylum Chordata, however, due to the characteristics of their swimming, tadpolelike larval stage, which has a dorsal nerve cord and notochord (figure 19.6a). The muscles that move the tail develop on either side of the notochord.

In many tunicate species, colored pigment granules become asymmetrically localized in the egg following fertilization and subsequently segregate to the tail muscle cell progenitors during cleavage (figure 19.6b). When these pigment granules are shifted experimentally into other cells that normally do not develop into muscle, their fate is changed and they become muscle cells. Thus, the molecules that flip the switch for muscle development appear to be associated with the pigment granules.

The next step is to determine the identity of the molecules involved. Experiments indicate that the female parent provides the egg with mRNA encoded by the *macho-1* gene. The elimination of *macho-1* function leads to a loss of tail muscle in the tadpole, and the misexpression of *macho-1* mRNA leads to the formation of additional (ectopic) muscle cells from nonmuscle lineage cells. The *macho-1* gene product has been shown to be a transcription factor that can activate the expression of several muscle-specific genes.

Induction can lead to cell differentiation

In chapter 9, we examined a variety of ways by which cells communicate with one another. We can demonstrate the importance of cell–cell interactions in development by separating the cells of an early frog embryo and allowing them to develop independently.

Under these conditions, blastomeres from one pole of the embryo (the "animal pole") develop features of ectoderm, and blastomeres from the opposite pole of the embryo (the "vegetal pole") develop features of endoderm. None of the two separated groups of cells ever develop features characteristic of mesoderm, the third main cell type. If animal-pole cells and vegetal-pole cells are placed next to each other, however, some of the animal-pole cells develop as mesoderm. The interaction between the two cell types triggers a switch in the developmental path of these cells. This change in cell fate due to interaction with an adjacent cell is called **induction.** Signaling molecules act to alter gene expression in the target cells, in this case, some of the animal-pole cells.

Another example of inductive cell interactions is the formation of the notochord and mesenchyme, a specific tissue, in tunicate embryos. Muscle, notochord, and mesenchyme all arise from mesodermal cells that form at the vegetal margin of the 32-cell stage embryo. These prospective mesodermal cells receive signals from the underlying endodermal precursor cells that lead to the formation of notochord and mesenchyme (figure 19.7).

The chemical signal is a member of the **fibroblast growth factor (FGF)** family of signaling molecules. It induces the overlying marginal zone cells to differentiate into either notochord (anterior) or mesenchyme (posterior). The FGF receptor on the marginal zone cells is a receptor tyrosine

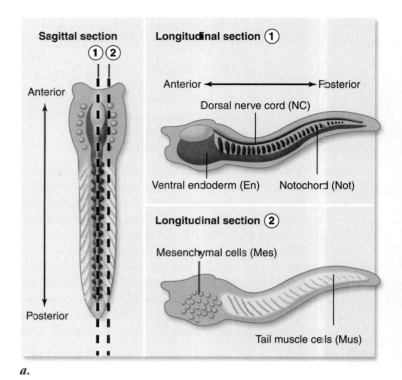

a.

b.

c.

figure 19.7

INDUCTIVE INTERACTIONS CONTRIBUTE TO CELL FATE SPECIFICATION IN TUNICATE EMBRYOS.

a. Internal structures of a tunicate larva. To the left is a sagittal section through the larva with dotted lines indicating two longitudinal sections. Section 1, through the midline of a tadpole, shows the dorsal nerve cord (NC), the underlying notochord (NC) and the ventral endoderm cells (En). Section 2, a more lateral section, shows the mesenchymal cells (Mes) and the tail muscle cells (Mus). *b.* View of the 32-cell stage looking up at the endoderm precursor cells. FGF secreted by these cells is indicated with light-green arrows. Only the surfaces of the marginal cells that directly border the endoderm precursor cells bind FGF signal molecules. Note that the posterior vegetal blastomeres also contain the *macho-1* determinants (red and white stripes). *c.* Cell fates have been fixed by the 64-cell stage. Colors are as in (*a*) Cells on the anterior margin of the endoderm precursor cells become notochord and nerve cord, respectively, whereas cells that border the posterior margin of the endoderm cells become mesenchyme and muscle cells, respectively.

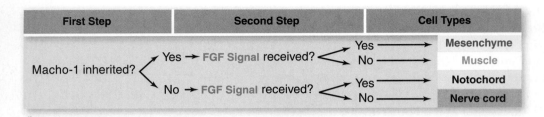

a.

figure 19.8

MODEL FOR CELL FATE SPECIFICATION BY Macho-1 MUSCLE DETERMINANT AND FGF SIGNALING. *a.* Two-step model of cell fate specification in vegetal marginal cells of the tunicate embryo. The first step is inheritance (or not) of muscle *macho-1* mRNA. The second step is receipt (or not) of the FGF signal from the underlying endoderm precursor cells. *b.* Posterior vegetal margin cells inherit *macho-1* mRNA. Signaling by FGF activates a Ras/MAP kinase pathway that produces the transcription factor T-Ets. Macho-1 protein and T-Ets together suppress muscle-specific genes and turn on mesenchyme specific genes (*green cells*). In cells with Macho-1 that do not receive the FGF signal, Macho-1 alone turns on muscle-specific cells (*yellow cells*). Anterior vegetal margin cells do not inherit *macho-1* mRNA. If these cells receive the FGF signal, T-Ets turns on notochord-specific genes (*purple cells*). In cells that lack Macho-1 and FGF signaling, notochord-specfic genes are suppressed and nerve cord-specific genes are activated (*gray cells*).

 inquiry

What dictates whether Macho-1 acts as a transcriptional repressor or a transcriptional activator?

b.

kinase that signals through a MAP kinase cascade to activate a transcription factor that turns on gene expression resulting in differentiation (figure 19.8).

This example is also a case of two cells responding differently to the same signal. The presence or absence of the *macho-1* muscle determinant discussed earlier controls this difference in cell fate. In the presence of *macho-1*, cells differentiate into mesenchyme; in its absence, cells differentiate into notochord. Thus, the combination of *macho-1* and FGF signaling leads to four different cell types (see figure 19.8)

Reversal of determination has allowed cloning

Experiments carried out in the 1950s showed that single cells from fully differentiated tissue of an adult plant could develop into entire, mature plants. The cells of an early cleavage stage mammalian embryo are also totipotent. When mammalian embryos naturally split in two, identical twins result. If individual blastomeres are separated from one another, any one of them can produce a completely normal individual. In fact, this type of procedure has been used to produce sets of four or eight identical offspring in the commercial breeding of particularly valuable lines of cattle.

Early research in amphibians

Until very recently, biologists thought determination and cell differentiation were irreversible processes in animals. Experiments carried out in the 1950s by Briggs and King and in the 1960s and 1970s by John Gurdon and colleagues made what seemed a convincing case.

Using very fine pipettes (hollow glass tubes), these researchers sucked the nucleus out of a frog or toad egg and replaced the egg nucleus with a nucleus sucked out of a body

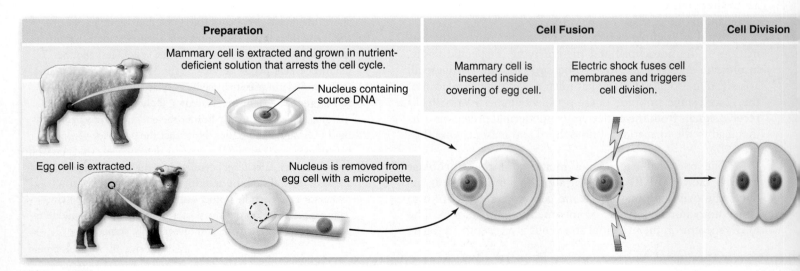

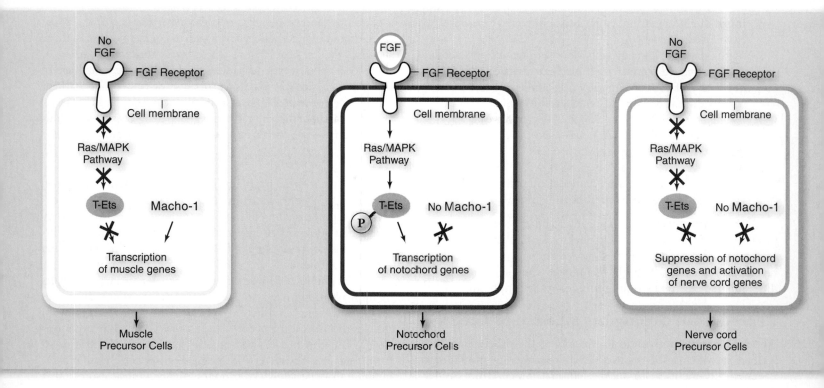

cell taken from another individual. If the transplanted nucleus was obtained from an advanced embryo, the egg went on to develop into a tadpole, but most died before becoming adult. Gurdon and colleagues also used donor nuclei from fully differentiated adult skin cells to direct tadpole development. However, all of these tadpoles died before the feeding stage.

Although these experiments showed that nuclei from adult cells have remarkable developmental potential, they did not provide evidence for the totipotency of these nuclei.

Successful nuclear transplant in mammals

Nuclear transplant experiments in mammals were attempted without success by many investigators, until finally, in 1984, a sheep was cloned using the nucleus from a cell of an early embryo. The key to this success was in picking a donor cell

very early in development. This exciting result was soon replicated by others in a host of other organisms, including pigs and monkeys. Only early embryo cells seemed to work, however.

Geneticists at the Roslin Institute in Scotland reasoned that the egg and donated nucleus would need to be at the same stage of the cell cycle for successful development. To test this idea, they performed the following procedure (figure 19.9):

1. They removed differentiated mammary cells from the udder of a six-year-old sheep. The cells were grown in tissue culture, and then the concentration of serum nutrients was substantially reduced for five days, causing them to pause at the beginning of the cell cycle.

Development	Implantation	Birth of Clone	Growth to Adulthood
Embryo begins to develop in vitro.	Embryo is implanted into surrogate mother.	After a five-month pregnancy, a lamb genetically identical to the sheep from which the mammary cell was extracted is born.	

Embryo

figure 19.9

PROOF THAT DETERMINATION IN ANIMALS IS REVERSIBLE. Scientists combined a nucleus from an adult mammary cell with an enucleated egg cell to successfully clone a sheep, named Dolly, who grew to be a normal adult and bore healthy offspring. This experiment, the first successful cloning of an adult animal, shows that a differentiated adult cell can be used to drive all of development.

2. In parallel preparation, eggs obtained from a ewe were enucleated.
3. Mammary cells and egg cells were surgically combined in a process called **somatic cell nuclear transfer (SCNT)** in January of 1996. Mammary cells and eggs were fused to introduce the mammary nucleus into egg.
4. Twenty-nine of 277 fused couplets developed into embryos, which were then placed into the reproductive tracts of surrogate mothers.
5. A little over five months later, on July 5, 1996, one sheep gave birth to a lamb named Dolly, the first clone generated from a fully differentiated animal cell.

Dolly matured into an adult ewe, and she was able to reproduce the old-fashioned way, producing six lambs. Thus, Dolly established beyond all dispute that determination in animals is reversible—that with the right techniques, the nucleus of a fully differentiated cell *can* be reprogrammed to be totipotent.

Reproductive cloning has inherent problems

The term **reproductive cloning** refers to the process just described, in which scientists use SCNT to create an animal that is genetically identical to another animal. Since Dolly's birth in 1997, scientists have successfully cloned one or more cats, rabbits, rats, mice, cattle, goats, pigs, and mules. All of these procedures used some form of adult cell.

Low success rate and age-associated diseases

The efficiency in all reproductive cloning is quite low—only 3–5% of adult nuclei transferred to donor eggs result in live births. In addition, many clones that are born usually die soon thereafter of liver failure or infections. Many become oversized,

a condition known as *large offspring syndrome (LOS)*. In 2003, three of four cloned piglets developed to adulthood, but all three suddenly died of heart failure at less than 6 months of age.

Dolly herself was euthanized at the relatively young age of six. Although she was put down because of virally induced lung cancer, she had been diagnosed with advanced-stage arthritis a year earlier. Thus, one difficulty in using genetic engineering and cloning to improve livestock is in producing enough healthy animals.

Lack of imprinting

The reason for these problems lies in a phenomenon discussed in chapter 13: **genomic imprinting.** Imprinted genes are expressed differently depending on parental origin—that is, they are turned off in either egg or sperm, and this "setting" continues through development into the adult. Normal mammalian development depends on precise genomic imprinting.

The chemical reprogramming of the DNA, which occurs in adult reproductive tissue, takes months for sperm and years for eggs. During cloning, by contrast, the reprogramming of the donor DNA must occur within a few hours. The organization of the chromatin in a somatic cell is also quite different from that in a newly fertilized egg. Significant chromatin remodeling of the transferred donor nucleus must also occur if the cloned embryo is to survive. Cloning fails because there is likely not enough time in these few hours to get the remodeling and reprogramming jobs done properly.

Therapeutic cloning is a promising possibility

One way to solve the problem of graft rejection, such as in skin grafts in severe burn cases, is to produce patient-specific lines of embryonic stem cells. Early in 2001, a

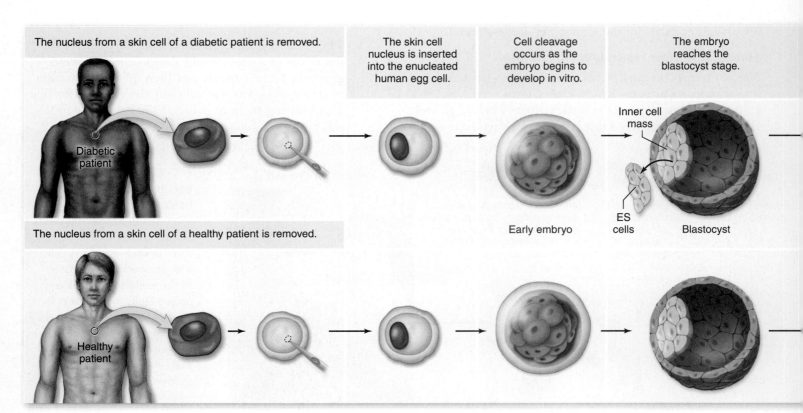

The nucleus from a skin cell of a diabetic patient is removed.

Diabetic patient

The skin cell nucleus is inserted into the enucleated human egg cell.

Cell cleavage occurs as the embryo begins to develop in vitro.

The embryo reaches the blastocyst stage.

Inner cell mass

Early embryo

ES cells

Blastocyst

The nucleus from a skin cell of a healthy patient is removed.

Healthy patient

research team at Rockefeller University devised a way to accomplish this feat.

First, skin cells are isolated; then, using the same SCNT procedure that created Dolly, an embryo is assembled. After removing the nucleus from the skin cell, they insert it into an egg whose nucleus has already been removed. The egg with its skin cell nucleus is allowed to form a blastocyst stage embryo. This artificial embryo is then destroyed, and its cells are used as embryonic stem cells for transfer to injured tissue (figure 19.10).

Using this procedure, termed **therapeutic cloning,** the researchers succeeded in converting cells from the tail of a mouse into the dopamine-producing cells of the brain that are lost in Parkinson disease. Therapeutic cloning successfully addresses the key problem that must be solved before stem cells can be used to repair human tissues damaged by heart attack, nerve injury, diabetes, or Parkinson disease—the problem of immune acceptance. Since stem cells are cloned from a person's own tissues in therapeutic cloning, they pass the immune system's "self" identity check, and the body readily accepts them.

Stem cell research has stimulated ethical debate

Human embryonic stem cells have enormous promise for treating a wide range of diseases. ES cells are derived from blastocyst stage embryos, and preimplantation stage human embryos can be obtained from fertility clinics, which routinely produce excess embryos when helping infertile couples to have children by in vitro fertilization.

In therapeutic cloning, an early embryo must be taken apart to create human embryonic stem cells. For this reason, stem cell research has raised profound ethical issues. The timeless question of when human life begins cannot be avoided. In addition, the question of whether reproductive cloning should be undertaken in humans, as it was in sheep to produce Dolly, is highly controversial.

In Britain, reproductive cloning is banned, but stem cell research and therapeutic cloning to obtain clinically useful stem cells are both permitted. Careful ethical supervision of all research is provided by a variety of governmental oversight committees. Britain's Human Fertilization and Embryology Authority (HFEA), for example, is a panel of scientists and ethicists accountable to Parliament, which oversees government-funded research. Similar arrangements are being established in Japan and France.

China, Thailand, and South Korea have permissive policies toward human ES cell research and cloning. Germany and most Latin American countries, by contrast, discourage such research.

In the United States, the first human ES cell lines were created in private research labs using private funds. After a long debate, federal funds were made available in the summer of 2001 for research on the small number of already existing human embryonic stem cell lines. But the George W. Bush administration specifically prohibited the use of federal funds to create *new* lines of human ES cells (which requires the destruction of human embryos).

Because patient-specific therapeutic cloning requires the creation of new human ES cell lines, federally funded U.S. researchers are currently prohibited from doing this research. Some states, notably California, have passed legislation to allow human embryonic stem cell research, while at the same time banning reproductive cloning of humans.

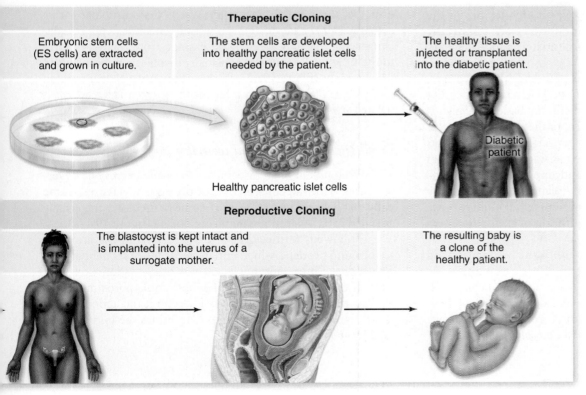

Therapeutic Cloning

Embryonic stem cells (ES cells) are extracted and grown in culture.

The stem cells are developed into healthy pancreatic islet cells needed by the patient.

The healthy tissue is injected or transplanted into the diabetic patient.

Healthy pancreatic islet cells

Diabetic patient

Reproductive Cloning

The blastocyst is kept intact and is implanted into the uterus of a surrogate mother.

The resulting baby is a clone of the healthy patient.

figure 19.10

HOW HUMAN EMBRYOS MIGHT BE USED FOR THERAPEUTIC CLONING. In therapeutic cloning after initial stages to reproductive cloning, the embryo is broken apart and its embryonic stem cells are extracted. These are grown in culture and used to replace the diseased tissue of the individual who provided the DNA. This is useful only if the disease in question is not genetic as the stem cells are genetically identical to the patient. In reproductive cloning the intact embryo is implanted in the uterus of a surrogate mother, where it might grow to term. Because of health issues for both the mother and the cloned fetus, most scientists agree that reproductive cloning of humans should be banned.

chapter **19** *cellular mechanisms of development*

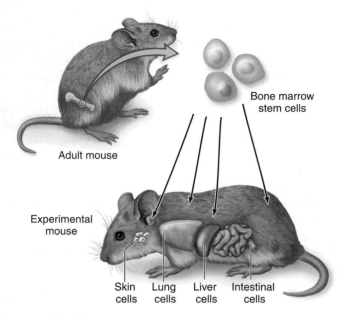

figure 19.11

MULTIPOTENT ADULT STEM CELLS. In May 2001, a single cell from the bone marrow of an adult mouse was claimed to have added functional cells to the lungs, liver, intestine, and skin of an experimental mouse. Cells isolated from adipose (fat) tissue may have similar capabilities. Since then these results have been challenged.

Adult stem cells may offer an alternative to ES cells

As discussed earlier, stem cells may be tissue specific, and they persist into adulthood in some tissues. Early reports on a variety of adult stem cells indicated that they may be reprogrammed to become other cell types than they normally would, that is, they are pluripotent (figure 19.11). These results have been challenged and the pluripotent status of adult stem cells is unclear. It also may be difficult to harvest these cells from the individual that you wish to treat. At this point the possible therapeutic use of both ES cells and adult stem cells is unclear.

Cell differentiation is preceded by determination when the cell is committed to a fate, but not yet differentiated. Differential inheritance of cytoplasmic factors can cause determination and differentiation, as can interactions between neighboring cells (induction). Inductive changes are mediated by signaling molecules that trigger signal transduction pathways. Determination may be reversible, as shown by reproductive cloning in some vertebrates. Cloned organisms, such as Dolly the sheep, have exhibited short life spans and early onset of disease, probably related to genomic imprinting. Human embryonic stem cells offer the possibility of replacing damaged or lost human tissues; however, the procedures are controversial and involve many ethical issues.

19.4 Pattern Formation

For cells in multicellular organisms to differentiate into appropriate cell types, they must gain information about their relative locations in the body. All multicellular organisms seem to use positional information to determine the basic pattern of body compartments and, thus, the overall architecture of the adult body. This positional information then leads to intrinsic changes in gene activity, so that cells ultimately adopt a fate appropriate for their location.

Pattern formation is an unfolding process. In the later stages, it may involve morphogenesis of organs (to be discussed later), but during the earliest events of development, the basic body plan is laid down, along with the establishment of the anterior–posterior (A/P, head-to-tail) axis and the dorsal–ventral (D/V, back-to-front) axis. Thus, pattern formation can be considered the process of taking a radially symmetrical cell and imposing two perpendicular axes to define the basic body plan, which in this way becomes bilaterally symmetrical. Developmental biologists use the term **polarity** to refer to the acquisition of axial differences in developing structures.

The fruit fly *Drosophila melanogaster* is the best understood animal in terms of the genetic control of early patterning. As described later, a hierarchy of gene expression that begins with maternally expressed genes controls the development of *Drosophila*. To understand the details of these gene interactions, we first need to briefly review the stages of *Drosophila* development.

Drosophila *embryogenesis produces a segmented larva*

Drosophila and many other insects produce two different kinds of bodies during their development: the first, a tubular eating machine called a **larva,** and the second, an adult flying sex machine with legs and wings. The passage from one body form to the other, called **metamorphosis,** involves a radical shift in development (figure 19.12). In this chapter, we concentrate on the process of going from a fertilized egg to a larva, which is termed *embryogenesis.*

Prefertilization maternal contribution

The development of an insect like *Drosophila* begins before fertilization, with the construction of the egg. Specialized **nurse cells** that help the egg grow move some of their own maternally encoded mRNAs into the maturing oocyte (figure 19.12*a*).

Following fertilization, the maternal mRNAs are transcribed into proteins, which initiate a cascade of sequential gene activations. Embryonic nuclei do not begin to function (that is, to direct new transcription of genes) until approximately 10 nuclear divisions have occurred. Therefore, the action of maternal, rather than zygotic, genes determines the initial course of *Drosophila* development.

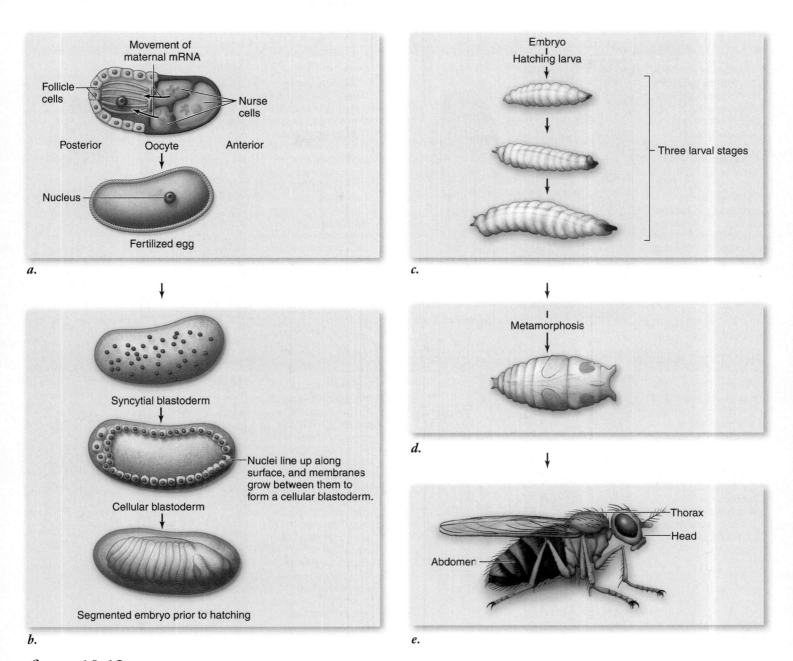

figure 19.12

THE PATH OF FRUIT FLY DEVELOPMENT. Major stages in the development of *Drosophila melanogaster* include formation of the (*a*) egg, (*b*) syncytial and cellular blastoderm, (*c*) larval instars, (*d*) pupa and metamorphosis into a (*e*) sexually mature adult.

Postfertilization events

After fertilization, 12 rounds of nuclear division without cytokinesis produce about 4000 nuclei, all within a single cytoplasm. All of the nuclei within this **syncytial blastoderm** (figure 19.12*b*) can freely communicate with one another, but nuclei located in different sectors of the egg encounter different maternal products.

Once the nuclei have spaced themselves evenly along the surface of the blastoderm, membranes grow between them to form the **cellular blastoderm.** Embryonic folding and primary tissue development soon follow, in a process fundamentally

similar to that seen in vertebrate development. Within a day of fertilization, embryogenesis creates a segmented, tubular body—which is destined to hatch out of the protective coats of the egg as a larva.

Morphogen gradients form the basic body axes in *Drosophila*

Pattern formation in the early *Drosophila* embryo requires positional information encoded in labels that can be read by cells. The unraveling of this puzzle, work that earned the 1995 Nobel Prize

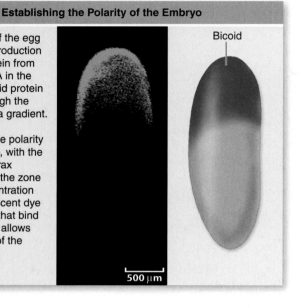

Establishing the Polarity of the Embryo

Fertilization of the egg triggers the production of bicoid protein from maternal RNA in the egg. The bicoid protein diffuses through the egg, forming a gradient. This gradient determines the polarity of the embryo, with the head and thorax developing in the zone of high concentration (*green* fluorescent dye in antibodies that bind bicoid protein allows visualization of the gradient).

Bicoid

500 µm

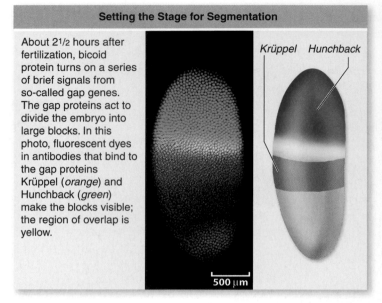

Setting the Stage for Segmentation

About 2½ hours after fertilization, bicoid protein turns on a series of brief signals from so-called gap genes. The gap proteins act to divide the embryo into large blocks. In this photo, fluorescent dyes in antibodies that bind to the gap proteins Krüppel (*orange*) and Hunchback (*green*) make the blocks visible; the region of overlap is yellow.

Krüppel   Hunchback

500 µm

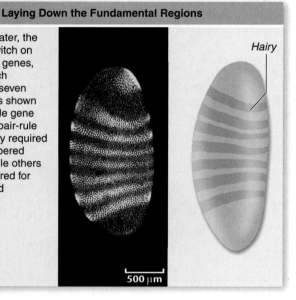

Laying Down the Fundamental Regions

About 0.5 hr later, the gap genes switch on the "pair-rule" genes, which are each expressed in seven stripes. This is shown for the pair-rule gene *hairy*. Some pair-rule genes are only required for even-numbered segments while others are only required for odd numbered segments.

Hairy

500 µm

Forming the Segments

The final stage of segmentation occurs when a "segment-polarity" gene called *engrailed* divides each of the seven regions into halves, producing 14 narrow compartments. Each compartment corresponds to one segment of the future body. There are three head segments (H, *bottom right*), three thoracic segments (T, *upper right*), and eight abdominal segments (A, *from top right to bottom left*).

A
T
H

Engrailed

500 µm

figure 19.13

BODY ORGANIZATION IN AN EARLY *DROSOPHILA* EMBRYO. In these fluorescent microscope images by 1995 Nobel laureate Christiane Nüsslein-Volhard and Sean Carroll, we watch a *Drosophila* egg pass through the early stages of development, in which the basic segmentation pattern of the embryo is established. The proteins in the photographs on the left were made visible by binding fluorescent antibodies to each specific protein. The drawings on the right help illustrate what is occurring in the photos.

for researchers Christiane Nüsslein-Volhard and Eric Wieschaus, is summarized in figure 19.13. We now know that two different genetic pathways control the establishment of A/P and D/V polarity in *Drosophila*.

Anterior–posterior axis

Formation of the A/P axis begins during maturation of the oocyte and is based on opposing gradients of two different proteins: **Bicoid** and **Nanos.** These protein gradients are established by an interesting mechanism.

Nurse cells in the ovary secrete maternally produced *bicoid* and *nanos* mRNAs into the maturing oocyte where they are dif-

ferentially transported along microtubules to opposite poles of the oocyte (figure 19.14*a*). This differential transport comes about due to the use of different motor proteins to move the two mRNAs. The *bicoid* mRNA then becomes anchored in the cytoplasm at the end of the oocyte closest to the nurse cells, and this end will develop into the anterior end of the embryo. *Nanos* mRNA becomes anchored to the opposite end of the oocyte, which will become the posterior end of the embryo. Thus, by the end of oogenesis, the *bicoid* and *nanos* mRNAs are already set to function as cytoplasmic determinants in the fertilized egg (figure 19.14*b*).

Following fertilization, translation of the anchored mRNA and diffusion of the proteins away from their respec-

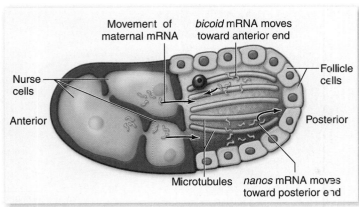

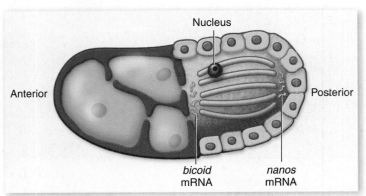

b.

figure 19.14

SPECIFYING THE A/P AXIS IN *DROSOPHILA* EMBRYOS I
a. In the ovary, nurse cells secrete maternal mRNAs into the cytoplasm of the oocyte. Clusters of microtubules direct oocyte growth and maturation. Motor proteins travel along the microtubules transporting molecules in two directions. *Bicoid* mRNAs are transported toward the anterior pole of the oocyte, *nanos* mRNA is transported toward the posterior pole of the oocyte. *b.* A mature oocyte, showing localization of *bicoid* mRNAs to the anterior pole and *nanos* mRNAs to the posterior pole.

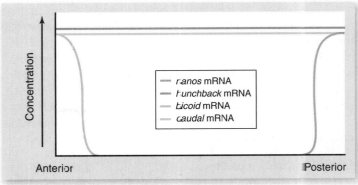

a. Oocyte mRNAs

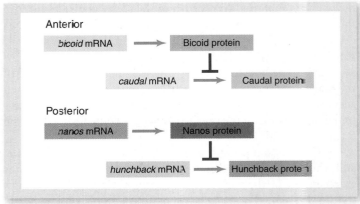

b. After fertilization

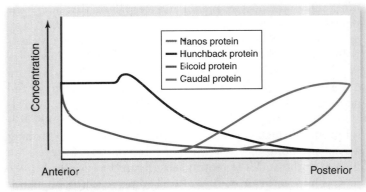

c. Early cleavage embryo proteins

figure 19.15

SPECIFYING THE A/P AXIS IN *DROSOPHILA* EMBRYOS II.
a. Unlike *bicoid* and *nanos*, *hunchback* and *caudal* mRNAs are evenly distributed throughout the cytoplasm of the oocyte. *b.* Following fertilization, *bicoid* and *nanos* mRNAs are translated into protein, making opposing gradients of each protein. Bicoid binds to and represses translation of *caudal* mRNAs (in anterior regions of the egg). Nanos binds to and represses translation of *hunchback* mRNAs (in posterior regions of the egg). *c.* Translation of *hunchback* mRNAs in anterior regions of the egg will create a Hunchback gradient that mirrors the Bicoid gradient. Translation of *caudal* mRNAs in posterior regions of the embryo will create a Caudal gradient that mirrors the Nanos gradient.

tive sites of synthesis create opposing gradients of each protein: Highest levels of bicoid protein are at the anterior pole of the embryo (figure 19.15*a*), and highest levels of the nanos protein are at the posterior pole. Concentration gradients of soluble molecules can specify different cell fates along an axis, and proteins that act in this way, like Bicoid and Nanos, are called **morphogens.**

The Bicoid and Nanos proteins control the translation of two other maternal messages, *hunchback* and *caudal*, that encode transcription factors. **Hunchback** activates genes required for the formation of anterior structures, and **Caudal** activates genes required for the development of posterior (abdominal) structures. The *hunchback* and *caudal* mRNAs are evenly distributed across the egg (figure 19.15*b*), so how is it that proteins translated from these mRNAs become localized?

The answer is that Bicoid protein binds to and inhibits translation of *caudal* mRNA. Therefore, *caudal* is only translated in the posterior regions of the egg where Bicoid is absent. Similarly, Nanos protein binds to and prevents translation of the *hunchback* mRNA. As a result, *hunchback* is only translated in the anterior regions of the egg (figure 19.15c). Thus, shortly after fertilization, four protein gradients exist in the embryo: anterior–posterior gradients of Bicoid and Hunchback proteins, and posterior–anterior gradients of Nanos and Caudal proteins (figure 19.15d).

Dorsal–ventral axis

The dorsal–ventral axis in *Drosophila* is established by actions of the *dorsal* gene product. Once again the process begins in the ovary, when maternal transcripts of the *dorsal* gene are put into the oocyte. However, unlike *bicoid* or *nanos*, the *dorsal* mRNA does not become asymmetrically localized. Instead, a series of steps are required for Dorsal to carry out its function.

First, the oocyte nucleus, which is located to one side of the oocyte, synthesizes *gurken* mRNA. The *gurken* mRNA then accumulates in a crescent between the nucleus and the membrane on that side of the oocyte (figure 19.16a). This will be the future dorsal side of the embryo.

The Gurken protein is a soluble cell-signaling molecule, and when it is translated and released from the oocyte, it binds to receptors in the membranes of the overlying follicle cells (figure 19.16b). These cells then differentiate into a dorsal morphology. Meanwhile, no Gurken signal is released from the other side of the oocyte, and the follicle cells on that side of the oocyte adopt a ventral fate.

Following fertilization, a signaling molecule is differentially activated on the ventral surface of the embryo in a complex sequence of steps. This signaling molecule then binds to a membrane receptor in the ventral cells of the embryo and activates a signal transduction pathway in those cells. Activation of this pathway results in the selected transport of the Dorsal protein (which is everywhere) into ventral nuclei, forming a gradient along the D/V axis. The Dorsal protein levels are highest in the nuclei of ventral cells (figure 19.16c).

The Dorsal protein is a transcription factor, and once it is transported into nuclei, it activates genes required for the proper development of ventral structures, simultaneously repressing genes that specify dorsal structures. Hence, the product of the *dorsal* gene ultimately directs the development of ventral structures.

(Note that many *Drosophila* genes are named for the mutant phenotype that results from a loss of function in that gene. A lack of *dorsal* function produces dorsalized embryos with no ventral structures.)

Although profoundly different mechanisms are involved, the unifying factor controlling the establishment of both A/P and D/V polarity in *Drosophila* is that *bicoid*, *nanos*, *gurken*, and *dorsal* are all maternally expressed genes. The polarity of the future embryo in both instances is therefore laid down in the oocyte using information coming from the maternal genome.

The preceding discussion simplifies events, but the outline is clear: Polarity is established by the creation of morpho-

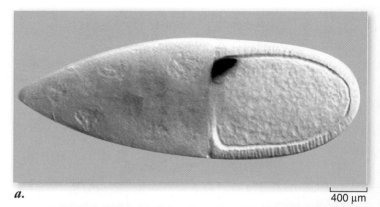

a.   400 µm

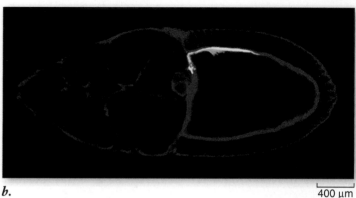

b.   400 µm

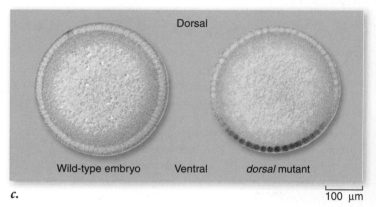

Dorsal

Wild-type embryo   Ventral   *dorsal* mutant

c.   100 µm

figure 19.16

SPECIFYING THE D/V AXIS IN *DROSOPHILA* EMBRYOS.
a. The *gurken* mRNA (*dark stain*) is concentrated between the oocyte nucleus (not visible) and the dorsal, anterior surface of the oocyte. **b.** In a more mature oocyte, Gurken protein (*yellow stain*) is secreted from the dorsal anterior surface of the oocyte, forming a gradient along the dorsal surface of the egg. Gurken then binds to membrane receptors in the overlying follicle cells. Double staining for actin (*red*) shows the cell boundaries of the oocyte, nurse cells, and follicle cells. **c.** For these images, cellular blastoderm stage embryos were cut in cross section, to visualize the nuclei of cells around the perimeter of the embryos. Dorsal protein (*dark stain*) is localized in nuclei on the ventral surface of the blastoderm in a wild-type embryo (*left*). The *dorsal* mutant on the right will not form ventral structures, and Dorsal is not present in ventral nuclei of this embryo.

gen gradients in the embryo based on maternal information in the egg. These gradients then drive the expression of the zygotic genes that will actually pattern the embryo. This reliance on a hierarchy of regulatory genes is a unifying theme for all of development.

The body plan is produced by sequential activation of genes

Let us now return to the process of pattern formation in *Drosophila* along the A/P axis. Determination of structures is accomplished by the sequential activation of three classes of **segmentation genes.** These genes create the hallmark segmented body plan of a fly, which consists of three fused head segments, three thoracic segments, and eight abdominal segments (see figure 19.12*e*).

To begin, Bicoid protein exerts its profound effect on the organization of the embryo by activating the translation and transcription of *hunchback* mRNA (which is the first mRNA to be transcribed after fertilization). *Hunchback* is a member of a group of nine genes called the **gap genes.** These genes map out the coarsest subdivision of the embryo along the A/P axis (see figure 19.13).

All of the gap genes encode transcription factors, which, in turn, activate the expression of eight or more **pair-rule genes.** Each of the pair-rule genes, such as *hairy*, produces seven distinct bands of protein, which appear as stripes when visualized with fluorescent reagents (see figure 19.13). These bands subdivide the broad gap regions and establish boundaries that divide the embryo into seven zones. When mutated, each of the pair-rule genes alters every other body segment.

All of the pair-rule genes also encode transcription factors, and they, in turn, regulate the expression of each other and of a group of nine or more **segment polarity genes.** The segment polarity genes are each expressed in 14 distinct bands of cells, which subdivide each of the seven zones specified by the pair-rule genes (see figure 19.13). The *engrailed* gene, for example, divides each of the seven zones established by *hairy* into anterior and posterior compartments. The segment polarity genes encode proteins that function in cell–cell signaling pathways. Thus, they function in inductive events—which occur *after* the syncytial blastoderm is divided into cells—to fix the anterior and posterior fates of cells within each segment.

In summary, within 3 hr after fertilization, a highly orchestrated cascade of segmentation gene activity transforms the broad gradients of the early embryo into a periodic, segmented structure with A/P and D/V polarity. The activation of the segmentation genes depends on the free diffusion of maternally encoded morphogens, which is only possible within the syncytial blastoderm of the early *Drosophila* embryo.

Segment identity arises from the action of homeotic genes

With the basic body plan laid down, the next step is to give identity to the segments of the embryo. A highly interesting class of *Drosophila* mutants has provided the starting point for understanding the creation of segment identity.

In these mutants, a particular segment seems to have changed its identity—that is, it has characteristics of a different segment. In wild-type flies, a pair of legs emerges from each of the three thoracic segments, but only the second thoracic segment has wings. Mutations in the *Ultrabithorax* gene cause a fly to grow an extra pair of wings, as though it has two second thoracic segments (figure 19.17). Even more bizarre are mutations in *Antennapedia*, which cause legs to grow out of the head in place of antennae!

Thus, mutations in these genes lead to the appearance of perfectly normal body parts in inappropriate places. Such mutants are termed *homeotic mutants* because the transformed body part looks similar (homeotic) to another. The genes in which such mutants occur are therefore called **homeotic genes.**

Homeotic gene complexes

In the early 1950s, geneticist and Nobel laureate Edward Lewis discovered that several homeotic genes, including *Ultrabithorax*, map together on the third chromosome of *Drosophila* in a tight cluster called the **bithorax complex.** Mutations in these genes all affect body parts of the thoracic and abdominal segments, and Lewis concluded that the genes of the bithorax complex control the development of body parts in the rear half of the thorax and all of the abdomen.

Interestingly, the order of the genes in the bithorax complex mirrors the order of the body parts they control, as though the genes are activated serially. Genes at the beginning of the cluster switch on development of the thorax; those in the middle control the anterior part of the abdomen; and those at the end affect the posterior tip of the abdomen.

A second cluster of homeotic genes, the **Antennapedia complex,** was discovered in 1980 by Thomas Kaufmann. The Antennapedia complex governs the anterior end of the fly, and

figure 19.17

MUTATIONS IN HOMEOTIC GENES. Three separate mutations in the Bithorax complex caused this fruit fly to develop an additional second thoracic segment, with accompanying wings.

the order of genes in this complex also corresponds to the order of segments they control (figure 19.18*a*).

The homeobox

An interesting relationship was discovered after the genes of the bithorax and Antennapedia complexes were cloned and sequenced. These genes all contain a conserved sequence of 180 nucleotides that codes for a 60-amino-acid, DNA-binding domain. Because this domain was found in all of the homeotic genes, it was named the *homeodomain*, and the DNA that encodes it is called the homeobox. Thus, the term **Hox gene** now refers to a homeobox-containing gene that specifies the identity of a body part. These genes function as transcription factors that bind DNA using their homeobox domain.

Clearly, the homeobox distinguishes portions of the genome that are devoted to pattern formation. How the *Hox* genes do this is the subject of much current research. Scientists believe that the ultimate targets of *Hox* gene function must be genes that control cell behaviors associated with organ morphogenesis.

Evolution of homeobox-containing genes

A large amount of research has been devoted to analyzing the clustered complexes of *Hox* genes in other organisms. These investigations have led to a fairly coherent view of homeotic gene evolution.

It is now clear that the *Drosophila* bithorax and Antennapedia complexes represent two parts of a single cluster of genes. In vertebrates, there are four copies of *Hox* gene clusters. As in *Drosophila*, the spatial domains of *Hox* gene expression correlate with the order of the genes on the chromosome (figure 19.18*b*). The existence of four *Hox* clusters in vertebrates is viewed by many as evidence that two duplication events of the entire genome have occurred in the vertebrate lineage.

This idea raises the issue of when the original cluster arose. To answer this question, researchers have turned to more primitive organisms, such as *Amphioxus* (now called *Branchiostoma*), a lancelet chordate (see chapter 35). The finding of only one cluster of *Hox* genes in *Amphioxus* implies that indeed there have been two duplications in the vertebrate lineage, at least of the *Hox* cluster. Given the single cluster in arthropods, this finding

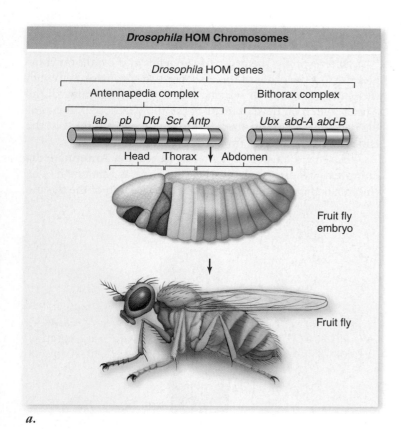

a.

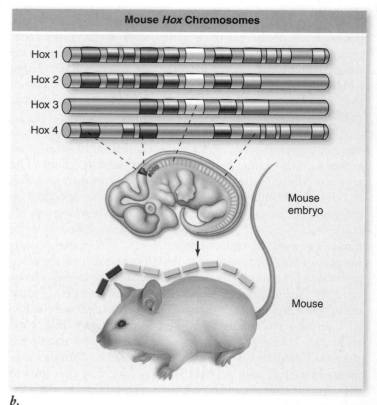

b.

figure 19.18

A COMPARISON OF HOMEOTIC GENE CLUSTERS IN THE FRUIT FLY *Drosophila melanogaster* AND THE MOUSE *Mus musculus*. *a.* *Drosophila* homeotic genes. Called the homeotic gene complex, or HOM complex, the genes are grouped into two clusters: the Antennapedia complex (anterior) and the bithorax complex (posterior). *b.* The *Drosophila* HOM genes and the mouse *Hox* genes are related genes that control the regional differentiation of body parts in both animals. These genes are located on a single chromosome in the fly and on four separate chromosomes in mammals. In this illustration, the genes are color-coded to match the parts of the body along the A/P axis in which they are expressed. Note that the order of the genes along the chromosome(s) is mirrored by their pattern of expression in the embryo and in structures in the adult fly.

implies that the common ancestor to all animals with bilateral symmetry had a single *Hox* cluster as well.

The next logical step is to look at even more-primitive animals: the radially symmetrical cnidarians such as *Hydra* (see chapter 33). Thus far, *Hox* genes have been found in a number of cnidarian species, and recent sequence analyses suggest that cnidarian *Hox* genes are also arranged into clusters. Thus, the appearance of the ancestral *Hox* cluster likely preceded the divergence between radial and bilateral symmetries in animal evolution.

Pattern formation in plants is also under genetic control

The evolutionary split between plant and animal cell lineages occurred about 1.6 BYA, before the appearance of multicellular organisms with defined body plans. The implication is that multicellularity evolved independently in plants and animals. Because of the activity of meristems, additional modules can be added to plant bodies throughout their lifetimes. In addition, plant flowers and roots have a radial organization, in contrast to the bilateral symmetry of most animals. We may therefore expect that the genetic control of pattern formation in plants is fundamentally different from that of animals.

Although plants have homeobox-containing genes, they do not possess complexes of *Hox* genes similar to the ones that determine regional identity of developing structures in animals. Instead, the predominant homeotic gene family in plants appears to be the **MADS-box** genes.

MADS-box genes are a family of transcriptional regulators found in most eukaryotic organisms, including plants, animals, and fungi. The MADS-box is a conserved DNA-binding and dimerization domain, named after the first five genes to be discovered with this domain. Only a small number of MADS-box genes are found in animals, where their functions include the control of cell proliferation and tissue-specific gene expression in postmitotic muscle cells. They do not appear to play a role in the patterning of animal embryos.

In contrast, the number and functional diversity of MADS-box genes increased considerably during the evolution of land plants, and there are more than 100 MADS-box genes in the *Arabidopsis* genome. In flowering plants, the MADS-box genes dominate the control of development, regulating such processes as the transition from vegetative to reproductive growth, root development, and floral organ identity.

Although distinct from genes in the *Hox* clusters of animals, homeodomain-containing transcription factors in plants do have important developmental functions. One such example is the family of *knottedlike homeobox* (*knox*) genes, which are important regulators of shoot apical meristem development in both seed-bearing and nonseed-bearing plants. Mutations that affect expression of *knox* genes produce changes in leaf and petal shape, suggesting that these genes play an important role in generating leaf form.

Pattern formation in animals involves the coordinated expression of a hierarchy of genes. Gradients of morphogens in *Drosophila* specify A/P and D/V axes, then lead to sequential activation of segmentation genes that subdivide the embryo in progressively more defined segments. The action of homeotic genes acts to provide segment identity. Genes with a DNA-binding homeodomain sequence are called *Hox* genes (for *homeobox* genes), and they are organized into clusters. Plants also change gene expression to control development, but they use a different set of control genes called MADS-box genes.

19.5 Morphogenesis

At the end of cleavage, the *Drosophila* embryo still has a relatively simple structure: It comprises several thousand identical-looking cells, which are present in a single layer surrounding a central yolky region. The next step in embryonic development is **morphogenesis**—the generation of ordered form and structure.

Morphogenesis is the product of changes in cell structure and cell behavior. Animals regulate the following processes to achieve morphogenesis:

- The number, timing, and orientation of cell divisions;
- Cell growth and expansion;
- Changes in cell shape;
- Cell migration; and
- Cell death.

Plant and animal cells are fundamentally different in that animal cells have flexible surfaces and can move, but plant cells are immotile and encased within stiff cellulose walls. Each cell in a plant is fixed into position when it is created. Thus, animal cells use cell migration extensively during development while plants use the other four mechanisms but lack cell migration. We consider the morphogenetic changes in animals first, and then those that occur in plants.

Cell division during development may result in unequal cytokinesis

The orientation of the mitotic spindle determines the plane of cell division in eukaryotic cells. The coordinated function of microtubules and their motor proteins determines the respective position of the mitotic spindle within a cell (see chapter 10). If the spindle is centrally located in the dividing cell, two equal-sized daughter cells will result. If the spindle is off to one side, one large daughter cell and one small daughter cell will result.

The great diversity of cleavage patterns in animal embryos is determined by differences in spindle placement. In many cases, the fate of a cell is determined by its relative placement in the embryo during cleavage. For example, in preimplantation mammalian embryos, cells on the outside of the embryo usually differentiate into trophectoderm cells, which form only extraembryonic structures later in development (for example, a part of the placenta). In contrast, the embryo proper is derived from the inner cell mass, cells which, as the name implies, are in the interior of the embryo.

Cells change shape and size as morphogenesis proceeds

In animals, cell differentiation is often accompanied by profound changes in cell size and shape. For example, the large nerve cells that connect your spinal cord to the muscles in your big toe develop long processes called *axons* that span this entire distance. The cytoplasm of an axon contains microtubules, which are used for motor-driven transport of materials along the length of the axon.

As another example, muscle cells begin as *myoblasts*, undifferentiated muscle precursor cells. They eventually undergo conversion into the large, multinucleated *muscle fibers* that make up mammalian skeletal muscles. These changes begin with the expression of the *MyoD1* gene, which encodes a transcription factor that binds to the promoters of muscle-determining genes to initiate these changes.

Programmed cell death is a necessary part of development

Not every cell produced during development is destined to survive. For example, human embryos have webbed fingers and toes at an early stage of development. The cells that make up the webbing die in the normal course of morphogenesis. As another example, vertebrate embryos produce a very large number of neurons, ensuring that enough neurons are available to make the necessary synaptic connections, but over half of these neurons never make connections and die in an orderly way as the nervous system develops.

Unlike accidental cell deaths due to injury, these cell deaths are planned—and indeed required—for proper development and morphogenesis. Cells that die due to injury typically swell and burst, releasing their contents into the extracellular fluid. This form of cell death is called **necrosis.** In contrast, cells programmed to die shrivel and shrink in a process called **apoptosis,** which means "falling away," and their remains are taken up by surrounding cells.

Genetic control of apoptosis

Apoptosis occurs when a "death program" is activated. All animal cells appear to possess such programs. In *C. elegans*, the same 131 cells always die during development in a predictable and reproducible pattern.

Work on *C. elegans* showed that three genes are central to this process. Two (*ced-3* and *ced-4*) activate the death program itself; if either is mutant, those 131 cells do not die, and go on instead to form nervous tissue and other tissue. The third gene (*ced-9*) represses the death program encoded by the other two: All 1090 cells of the *C. elegans* embryo die in *ced-9* mutants. In *ced-9/ced-3* double mutants, all 1090 cells live, which suggests that *ced-9* inhibits cell death by functioning prior to *ced-3* in the apoptotic pathway (figure 19.19a).

The mechanism of apoptosis appears to have been highly conserved during the course of animal evolution. In human nerve cells, the *Apaf1* gene is similar to *ced-4* of *C. elegans* and activates the cell death program, and the human *bcl-2* gene acts similarly to *ced-9* to repress apoptosis. If a copy of the human *bcl-2* gene is transferred into a nematode with a defective *ced-9* gene, *bcl-2* suppresses the cell death program of *ced-3* and *ced-4*.

The mechanism of apoptosis

The product of the *C. elegans ced-4* gene is a protease that activates the product of the *ced-3* gene, which is also a protease. The human *Apaf1* gene is actually named for its role: *A*poptotic *p*rotease *a*ctivating *f*actor. It activates two proteases called caspases that have a role similar to the Ced-3 protease in *C. elegans* (figure 19.19b). When the final proteases are activated, they chew up proteins in important cellular structures such as the cytoskeleton and the nuclear lamina, leading to cell fragmentation.

The role of Ced-9/Bcl-2 is to inhibit this program. Specifically, it inhibits the activating protease, preventing the activation of the destructive proteases. The entire process is thus controlled by an inhibitor of the death program.

Both internal and external signals control the state of the Ced-9/Bcl-2 inhibitor. For example, in the human nervous system, neurons have a cytoplasmic inhibitor of Bcl-2 that allows the death program to proceed (figure 19.19b). In the presence of nerve growth factor, a signal transduction pathway leads to the cytoplasmic inhibitor being inactivated, allowing Bcl-2 to inhibit apoptosis and the nerve cell to survive.

Cell migration gets the right cells to the right places

The migration of cells is important during many stages of animal development. The movement of cells involves both adhesion and the loss of adhesion. Adhesion is necessary for cells to get "traction," but cells that are initially attached to others must lose this adhesion to be able to leave a site.

Cell movement also involves cell-to-substrate interactions, and the extracellular matrix may control the extent or route of cell migration. The central paradigm of morphogenetic cell movements in animals is a change in cell adhesiveness, which is mediated by changes in the composition of macromolecules in the plasma membranes of cells or in the extracellular matrix. Cell-to-cell interactions are often mediated through cadherins, but cell-to-substrate interactions often involve integrin-to-extracellular-matrix (ECM) interactions.

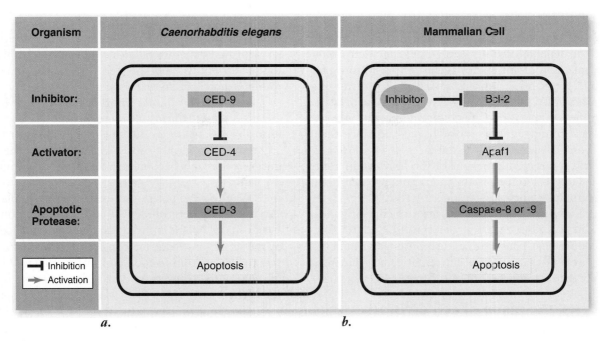

Organism	*Caenorhabditis elegans*	Mammalian Cell
Inhibitor:	CED-9	Inhibitor ⊣ Bcl-2
Activator:	CED-4	Apaf1
Apoptotic Protease:	CED-3	Caspase-8 or -9
	Apoptosis	Apoptosis

⊣ Inhibition
→ Activation

a.　　　　　*b.*

figure 19.19

PROGRAMMED CELL DEATH PATHWAY. Apoptosis, or programmed cell death, is necessary for the normal development of all animals. *a.* In the developing nematode, for example, two genes, *ced-3* and *ced-4*, code for proteins that cause the programmed cell death of 131 specific cells. In the other (surviving) cells of the developing nematode, the product of a third gene, *ced-9*, represses the death program encoded by *ced-3* and *ced-4*. *b.* The mammalian homologues of the apoptotic genes in *C. elegans* are *bcl-2* (*ced-9* homologue), *Apaf1* (*ced-4* homologue), and *caspase-8* or *-9* (*ced-3* homologues). In the absence of any cell survival factor, Bcl-2 is inhibited and apoptosis occurs. In the presence of nerve growth factor (NGF) and NGF receptor binding, Bcl-2 is activated, thereby inhibiting apoptosis.

Cadherins

Cadherins are a large gene family, with over 80 members identified in humans. In the genomes of *Drosophila*, *C. elegans*, and humans, the cadherins can be sorted into several subfamilies that exist in all three genomes.

The cadherin proteins are all transmembrane proteins that share a common motif, the *cadherin domain*, a 110-amino-acid domain in the extracellular portion of the protein that mediates Ca^{2+}-dependent binding between like cadherins (homophilic binding).

Experiments in which cells are allowed to sort in vitro illustrate the function of cadherins. Cells with the same cadherins adhere specifically to one another, while not adhering to other cells with different cadherins. If cell populations with different cadherins are dispersed and then allowed to reaggregate, they sort into two populations of cells based on the nature of the cadherins on their surface.

An example of the action of cadherins can be seen in the development of the vertebrate nervous system. All surface ectoderm cells of the embryo express E-cadherin. The formation of the nervous system begins when a central strip of cells on the dorsal surface of the embryo turns off E-cadherin expression and turns on N-cadherin expression. In the process of **neurulation,** the formation of the neural tube (see chapter 53), the central strip of N-cadherin-expressing cells folds up to form the tube. The neural tube pinches off from the overlying cells, which continue to express E-cadherin. The surface cells outside the tube differentiate into the epidermis of the skin, whereas the neural tube develops into the brain and spinal cord of the embryo.

Integrins

In some tissues, such as connective tissue, much of the volume of the tissue is taken up by the spaces *between* cells. These spaces are filled with a network of molecules secreted by surrounding cells, termed a *matrix*. In connective tissue such as cartilage, long polysaccharide chains are covalently linked to proteins (proteoglycans), within which are embedded strands of fibrous protein (collagen, elastin, and fibronectin). Migrating cells traverse this matrix by binding to it with cell surface proteins called **integrins.**

Integrins are attached to actin filaments of the cytoskeleton and protrude out from the cell surface in pairs, like two hands. The "hands" grasp a specific component of the matrix, such as collagen or fibronectin, thus linking the cytoskeleton to the fibers of the matrix. In addition to providing an anchor, this binding can initiate changes within the cell, alter the growth of the cytoskeleton, and activate gene expression and the production of new proteins.

The process of **gastrulation,** during which the hollow ball of animal embryonic cells folds in on itself to form a multilayered structure, depends on fibronectin–integrin interactions. For example, injection of antibodies against either fibronectin or integrins into salamander embryos blocks binding of cells to fibronectin in the ECM and inhibits gastrulation. The result is like a huge traffic jam following a major accident on a freeway: Cells (cars) keep coming, but they get backed up since they cannot get beyond the area of inhibition (accident site) (figure 19.20). Similarly, a targeted knockout of the fibronectin gene in mice resulted in gross defects in the migration, proliferation, and differentiation of embryonic mesoderm cells.

Thus, cell migration is largely a matter of changing patterns of cell adhesion. As a migrating cell travels, it continually extends projections that probe the nature of its environment. Tugged this way and that by different tentative attachments, the cell literally feels its way toward its ultimate target site.

In seed plants, the plane of cell division determines morphogenesis

The form of a plant body is largely determined by the plane in which cells divide. The first division of the fertilized egg in a flowering plant is off-center, so that one of the daughter cells is small, with dense cytoplasm (figure 19.21*a*). That cell, the future embryo, begins to divide repeatedly, forming a ball of cells. The other daughter cell also divides repeatedly, forming an elongated structure called a *suspensor,* which links the embryo to the nutrient tissue of the seed. The suspensor also provides a route for nutrients to reach the developing embryo.

Just as many animal embryos acquire their initial axis as a cell mass formed during cleavage divisions, so the plant embryo forms its root–shoot axis at this time. Cells near the suspensor are destined to form a root, whereas those at the other end of the axis ultimately become a shoot, the aboveground portion of the plant.

The relative position of cells within the plant embryo is also a primary determinant of cell differentiation. The outermost cells in a plant embryo become epidermal cells. The bulk of the embryonic interior consists of ground tissue cells that eventually function in food and water storage. Finally, cells at the core of the embryo are destined to form the future vascular tissue (figure 19.21*b*). (Plant tissues and development are described in detail in chapters 36 and 37.)

Soon after the three basic tissues form, a flowering plant embryo develops one or two seed leaves called *cotyledons*. At this point, development is arrested, and the embryo is either surrounded by nutritive tissue or has amassed stored food in its cotyledons (figure 19.21*c*). The resulting package, known as a *seed*, is resistant to drought and other unfavorable conditions.

A seed germinates in response to favorable changes in its environment. The embryo within the seed resumes development and grows rapidly, its roots extending downward and its leaf-bearing shoots extending upward (figure 19.21*d*). Plant development exhibits its great flexibility during the assembly of the modules that make up a plant body. Apical meristems at the root and shoot tips generate the large numbers of cells needed to form leaves, flowers, and all other components of the mature plant (figure 19.21*e*).

Growth within the developing flower is controlled by a cascade of transcription factors. A key member of this cascade is the *AINTEGUMENTA* (*ANT*) gene. Loss of ANT function reduces the number and size of floral organs, and inappropriate expression leads to larger floral organs.

Plant body form is also established by controlled changes in cell shape as cells expand osmotically after they form. Plant growth-regulating hormones and other factors influence the orientation of bundles of microtubules on the interior of the plasma membrane. These microtubules seem to guide cellulose deposition as the cell wall forms around the outside of a new cell. The orientation of the cellulose fibers, in turn, determines how the cell will elongate as it increases in volume due to osmosis, and so determines the cell's final shape.

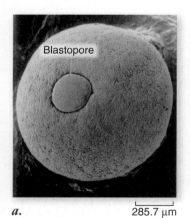
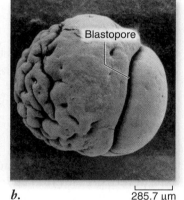

Blastopore

Blastopore

a.   285.7 μm   *b.*   285.7 μm

figure 19.20

REAGENTS THAT INTERFERE WITH CELL BINDING TO FIBRONECTIN INHIBIT GASTRULATION OF AMPHIBIAN EMBRYOS. *a.* Scanning electron micrograph of a normal salamander embryo during gastrulation. This embryo was injected with a control saline solution at the blastula stage. Cells have moved into the interior of the embryo around the circumference of the blastopore, allowing the outer cells to spread evenly over the surface of the embryo. *b.* Micrograph of a salamander embryo of the same age that was previously injected with fibronectin antibodies, which block binding of the migrating cells to the extracellular matrix. In this embryo, the cells lack traction to move into the interior of the embryo, and they pile up on the surface forming deep convolutions. Note also that the circumference of the blastopore has not decreased in these embryos.

> Morphogenesis is the generation of ordered form and structure. Morphogenesis occurs by cell growth, cell shape change, cell death (apoptosis), and cell migration. Because plant cells cannot move, cell division and cell expansion are the primary morphogenetic processes in plants.

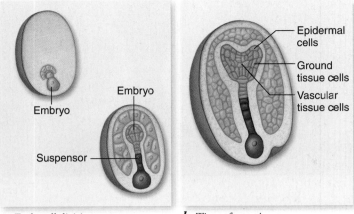

a. Early cell division

b. Tissue formation

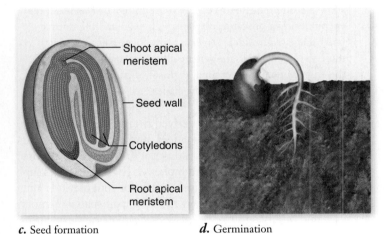

c. Seed formation

d. Germination

e. Meristematic development and morphogenesis

figure 19.21

THE PATH OF PLANT DEVELOPMENT. The developmental stages of *Arabidopsis thaliana* are (*a*) early embryonic cell division, (*b*) embryonic tissue formation, (*c*) seed formation, (*d*) germination, and (*e*) meristematic development and morphogenesis.

19.6 Environmental Effects on Development

In seed plants, embryogenesis is a brief stage in the life of the plant and results in the production of a seed. The environment affects essentially every subsequent step after embryogenesis, from seed dispersal to flower formation. For example, seed dispersal in jack pines can only occur following a fire because the heat of the fire causes the tightly closed cones to open up and release their seeds.

Germination of a dormant seed proceeds when soil conditions, temperature, and daylight hours are favorable. Similarly, a combination of environmental factors determines the timing of flower production in adult angiosperms.

Plant development is also influenced by interactions with other organisms. For example, the survival of a plant that is munched on by an animal depends on the quick regrowth afforded by meristems. Plant development is also guided by symbiotic relationships; the roots of legumes and a few other plant species develop nodules to house the nitrogen-fixing bacteria *Rhizobium*.

The effects of the environment on animal development are perhaps not so intuitive. Organisms such as *C. elegans* and *Drosophila* were selected as model systems for studying animal development because they develop so regularly under typical laboratory conditions. Organisms living in the wild, however, are subject to many different environmental changes, which may produce different phenotypes from a single genotype.

In mammals, embryonic and fetal development has a longer time course and is subject to the effects of blood-borne agents in the mother. The prescription of the sedative drug thalidomide to pregnant women in the 1950s and 1960s illustrated the profound effects that drugs can have on human development. Many of the women who took thalidomide gave birth to children with limb defects. Environmental agents such as lead compounds also affect the postnatal growth and maturation of children, leading to cognitive defects and mental retardation.

The environment affects normal development

The environment controls many aspects of normal animal development. The larvae of certain marine invertebrates, for example, do not metamorphose into adults until they encounter a specific surface on which to settle down. As with plants, animal development is often influenced by interactions with other organisms. For example, the small water flea *Daphnia* will change its morphology by doubling the size of the "helmet" on the top of its head after encountering a predatory fly larva (figure 19.22). Finally, when mice or zebrafish are reared under germ-free conditions, their guts are devoid of bacteria that normally colonize the gut. As a result, there are defects in intestinal differentiation and function in both species.

A particularly clear example of environmental effects on development is temperature-dependent sex determination (TSD). In many reptiles, the temperature of the soil in which the eggs incubate determines the sex of the hatchlings. In some species, one sex dominates at intermediate temperatures, whereas only the opposite sex develops at temperatures at either extreme of the natural thermal range. In other species, one end of the natural thermal range usually produces all males, the other end of the range produces all females, and intermediate temperatures are usually gender-neutral.

One possible hazard of TSD is that an increase in global temperatures may eventually skew sex ratios in wild populations of animals, leading to their demise. Some researchers have even speculated that the extinction of dinosaurs could be attributable to the effects of global temperature changes on sex determination and sex ratios in dinosaurs.

Endocrine disrupters can perturb development

The large family of **endocrine hormones** includes the androgens and estrogens that control sexual differentiation and function in animals. Endogenous endocrine activity is essential for the normal development and homeostasis of all complex animals. For example, hormones trigger metamorphosis in both frogs and flies, and if cells in the pituitary gland that produce human growth hormone malfunction during childhood, dwarfism or gigantism will result.

Although a number of endocrine disorders have a genetic basis, recent studies have shown that many environmental chemicals interfere with endocrine signaling. An **endocrine-disrupting chemical (EDC)** is any exogenous compound that interferes with the production, transport, or receptor binding of endogenous hormones.

Perhaps the best known endocrine disrupter is diethylstilbestrol (DES), which was prescribed to millions of women in the United States from 1938 to 1971 to prevent miscarriages and preterm births. The children of women who took DES while pregnant exhibited an increased incidence of abnormalities in differentiation of their reproductive organs, and DES daughters are at an increased risk of a rare kind of vaginal and cervical cancer.

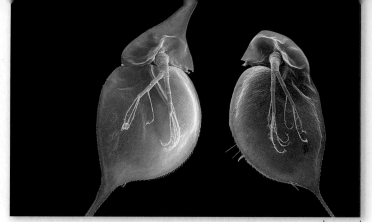

500 µm

figure 19.22

PREDATOR-INDUCED MORPHOLOGICAL CHANGES IN THE WATER FLEA *DAPHNIA*. These scanning electron micrographs show the difference between the morphology of *Daphnia* after encountering a predatory fly larva (left side) and the normal body shape (right side). *Daphnia* reproduces asexually, and these two individuals are genetic clones of one another. Thus, the environment can act on a *single* set of genes to induce the formation of *two different* body shapes.

Environmental EDCs are derived from three main sources: industrial wastes, agricultural practices, and the effluent of municipal sewage-treatment plants. Industrial EDCs include dioxins, heavy metals, and polychlorinated biphenyls (PCBs). Dioxins suppress immune system function in humans for long periods after exposure. PCBs, dioxins, and heavy metals impair spatial memory, learning, and other cognitive processes in primates and rodents.

Agricultural EDCs include the pesticides atrazine and DDT. The historic decline in bald eagle populations in North America was correlated with high levels of DDT in adults; females with high levels of DDT laid eggs with extremely weak shells, which broke easily. The subsequent banning of DDT use in the United States is thought to be the major reason for the rebounding of bald eagle populations from the verge of extinction.

A recent rise has been reported in the incidence of morphological defects in the urinary and reproductive structures of human males, which include abnormally placed urethral openings and undescended testes. A global decline in sperm count and quality and increased infertility has also been reported in men. All of these problems are exacerbated in areas with high EDC production.

A 2005 study in *Science* magazine showed that brief exposure of pregnant rats to two different EDCs—one a fungicide and the other a pesticide—caused decreased sperm number and viability and an increased incidence of infertility in adult male offspring of these mothers. Moreover, the reduced male fertility was passed on to nearly every male in future generations (examined F_1 to F_4). Thus, this startling set of experiments showed that the effects of EDCs can reach beyond affected individuals to affect multiple generations of offspring.

> Both plant and animal development is affected by environmental influences. Sex determination in reptiles is controlled by temperature. Human development may be affected by environmental pollutants that mimic the effects of steroid hormones.

19.1 Overview of Development

- Development is the successive process of systematic, gene-directed changes throughout a life cycle.
- Development occurs in four subprocesses: growth, cell differentiation, pattern formation, and morphogenesis.

19.2 Cell Division

Early growth occurs by mitotic cell division, resulting in many undifferentiated cells.

- In animals, cleavage stage divisions divide the fertilized egg into numerous smaller cells called blastomeres.
- During cleavage the G_1 and G_2 phases of the cell cycle are shortened or eliminated (figure 19.2).
- The lineage of 959 adult somatic *Caenorhabditis elegans* cells is invariant.
- Stem cells can divide indefinitely and give rise to multiple cell types.
- Totipotent cells can give rise to any cell type; pluripotent cells can give rise to multiple cell types.
- Embryonic stem cells are derived from the inner cell mass of the blastocyst and are pluripotent (figure 19.4).
- Plant growth continues throughout the life span from meristematic stem cells that can differentiate into any plant tissue.

19.3 Cell Differentiation

During development, cells assume different fates due to both temporal and spatial differences in gene expression in the growing embryo.

- Cells that are committed to a particular developmental pathway are determined.
- Cells can become committed to a particular developmental pathway by the inheritance of cytoplasmic determinants or cell–cell interactions.
- Cytoplasmic determinants such as encoded maternal mRNA are produced during oogenesis and deposited into the egg.
- Induction occurs when one cell type produces signal molecules that induce gene expression in neighboring target cells.
- The nucleus of a fully differentiated cell can be reprogrammed to be totipotent (figure 19.9).
- Reproductive cloning suffers from a low success rate, and age-associated diseases.
- Therapeutic cloning uses stem cells from the recipient, solving the problem of tissue rejection in tissue and organ transplants.

19.4 Pattern Formation

For cells in multicellular organisms to differentiate into appropriate cell types, they must gain information about their relative locations in the body before their fate is determined.

- Pattern formation produces two perpendicular axes, or anterior–posterior and dorsal–ventral polarity, to a bilaterally symmetrical organism.
- Positional information leads to changes in gene activity so cells adopt a fate appropriate for their location.
- Fruit fly embryogenesis illustrates the genetic control of early patterning.
- Maternally encoded mRNAs are deposited in a maturing oocyte by nurse cells and they initiate a cascade of sequential gene activations.

- Formation of the anterior–posterior axis is based on opposing gradients of morphogens, Bicoid and Nanos, synthesized from maternal mRNA (figures 19.14, 19.15).
- The dorsal–ventral axis is established by a gradient of the Dorsal transcription factor.
- Gap genes encode transcription factors, which, in turn, activate the expression of pair-rule genes that divide the embryo into seven zones.
- Pair-rule genes regulate the expression of each other and the segment polarity genes that finish defining the embryonic segments.
- Homeotic genes give identity to segments of the embryo. They contain a DNA sequence called the homeobox and are called *Hox* genes.
- *Hox* genes are found in four clusters in vertebrates.
- Instead of *Hox* genes, plants have MADS box genes that control the transition from vegetative to reproductive growth, root development, and floral organ identity.

19.5 Morphogenesis

Morphogenesis is the product of changes in cell structure and cell behavior.

- Depending on the orientation of the mitotic spindle, cells of equal or different sizes can arise.
- Morphogenesis can arise by changes in cell shape, cell size, and by cell migration.
- Apoptosis, the programmed death of cells, is important in development to remove structures (figure 19.19).
- The migration of cells requires both the adhesion and loss of adhesion between cells and their substrate.
- Cell-to-cell interactions are often mediated by cadherin proteins, whereas cell-to-substrate interactions may involve integrin-to-extracellular-matrix interactions.
- Integrins bind to fibers found in the extracellular matrix. This can alter the cytoskeleton and activate gene expression.
- In plants, the primary morphogenetic processes are cell division, relative position of cells within the embryo, and changes in cell shape.
- Plant development stages begin with cell division and end with meristematic development and morphogenesis (figure 19.21).
- Relative position of cells in the plant embryo is the main determinant of cell differentiation.

19.6 Environmental Effects on Development

Both plant and animal development are affected by environmental factors.

- Seed dispersal, germination, and plant development are influenced by abiotic and biotic factors.
- In plants and animals, different phenotypic expressions of a single genotype can be influenced by the environment.
- In animals, blood-borne agents and environmental contaminants can affect embryonic development.
- The environment controls normal animal development by influencing attributes such as morphology and sex determination.
- Human development may be affected by exogenous compounds, called endocrine-disrupting chemicals, such as dioxin and PCBs, which interfere with the production, transport, or receptor binding of endogenous hormones.

SELF TEST

1. Which of the following developmental stages is associated with the generation of organs?
   a. Growth
   b. Pattern formation
   c. Differentiation
   d. Morphogenesis

2. Growth of the developing embryo involves rapid _____ cell divisions.
   a. mitotic
   b. meiotic
   c. binary fission
   d. sexual

3. The reduced size of a blastomere is the consequence of a shortened—
   a. M phase.
   b. S phase.
   c. G_1 and G_2 phases.
   d. All of the above.

4. A pluripotent cell is one that can—
   a. become any cell type.
   b. produce an indefinite supply of a single cell type.
   c. produce a limited amount of a specific cell type.
   d. produce multiple cell types.

5. Which of the following statements is *not* true regarding embryonic stem (ES) cells?
   a. They retain the ability to develop into any cell type.
   b. They are isolated from the inner cell mass of a developing embryo.
   c. They are tissue-specific.
   d. They are totipotent.

6. Plant meristems—
   a. are only present during development.
   b. contain stem cells.
   c. undergo meiosis.
   d. All of the above.

7. What is the common theme in cell determination by induction or cytoplasmic determinants?
   a. The activation of transcription factors
   b. The activation of cell signaling pathways
   c. A change in gene expression
   d. Both (a) and (c)

8. Which of the following is *not* a limitation to reproductive cloning?
   a. Efficiency of the process
   b. Ethical considerations
   c. Sources for donor DNA
   d. Genetic imprinting of the DNA

9. How do the products of therapeutic cloning differ from those of reproductive cloning?
   a. Therapeutic cloning provides a source of embryonic stem cells.
   b. Therapeutic cloning produces an embryo that can be implanted into a uterus.
   c. Therapeutic cloning produces whole tissue or organs.
   d. Therapeutic cloning provides a source of proteins.

10. The anterior–posterior axis of the fruit fly *Drosophila* is determined by—
    a. growth factors.
    b. zygotic RNA.
    c. morphogens.
    d. cellular blastoderm formation.

11. Which of the following best describes a morphogen?
    a. A cell that secretes a diffusible signal that specifies cell fate
    b. A diffusible signal that functions to determine cell fate
    c. A protein that helps mediate cell–cell interactions, altering cell fate
    d. A protein that allows a cell to become totipotent

12. Suppose that during a mutagenesis screen to isolate mutations in *Drosophila*, you come across a fly with legs growing out of its head. What gene cluster is likely affected?
    a. *bicoid*
    b. *hunchback*
    c. Bithorax
    d. Antennapedia

13. What would be the likely result of a mutation of the *bcl-2* gene on the level of apoptosis?
    a. No change
    b. A decrease in apoptosis
    c. An increase in apoptosis
    d. An initial increase, followed by a decrease in apoptosis

14. How is the body plan of a plant first determined?
    a. The activity of MADS box genes
    b. The first cell division following fertilization
    c. Gastrulation
    d. Both (a) and (b)

15. How does an endocrine-disrupting chemical (EDC) affect development?
    a. It alters the normal pathway of endocrine hormone activity.
    b. It induces mutations.
    c. It alters the sex determination of the developing embryo.
    d. Both (a) and (b).

CHALLENGE QUESTIONS

1. The fate map for *C. elegans* (refer to figure 19.3) diagrams development of a multicellular organism from a single cell. Use this fate map to determine the number of cell divisions required to establish the population of cells that will become (a) the nervous system and (b) the gonads.

2. Carefully examine the *C. elegans* fate map in figure 19.3. Notice that some of the branchpoints (daughter cells) do *not* go on to produce more cells. What is the cellular mechanism underlying this pattern?

3. You have generated a set of mutant embryonic mouse cells. Predict the developmental consequences for each of the following mutations.
   a. Knockout mutation for N-cadherin
   b. Knockout mutation for integrin
   c. Deletion of the cytoplasmic domain of integrin

chapter 20

Genes Within Populations

introduction

NO OTHER HUMAN BEING is exactly like you (unless you have an identical twin). Often the particular characteristics of an individual have an important bearing on its survival, on its chances to reproduce, and on the success of its offspring. Evolution is driven by such consequences, as different alleles rise and fall in populations. These deceptively simple matters lie at the core of evolutionary biology, which is the topic of this chapter and chapters 21 through 25.

concept outline

20.1 Genetic Variation and Evolution

Genetic variation, that is, differences in alleles of genes found within individuals of a population, provides the raw material for natural selection, which will be described shortly. Natural populations contain a wealth of such variation. In plants, insects, and vertebrates, many genes exhibit some level of variation. In this chapter, we explore genetic variation in natural populations and consider the evolutionary forces that cause allele frequencies in natural populations to change.

The word **evolution** is widely used in the natural and social sciences. It refers to how an entity—be it a social system, a gas, or a planet—changes through time. Although development of the modern concept of evolution in biology can be traced to Darwin's landmark work, *On the Origin of Species*, the first five editions of his book never actually used the term. Rather, Darwin used the phrase "descent with modification."

Although over time many more complicated definitions have been proposed, Darwin's phrase probably best captures the essence of biological evolution: Through time, species accumulate differences; as a result, descendants differ from their ancestors. In this way, new species arise from existing ones.

Natural selection is an important mechanism of evolutionary change

You have already learned about the development of Darwin's ideas in chapter 1. Darwin was not the first to propose a theory of evolution. Rather, he followed a long line of earlier philosophers and naturalists who deduced that the many kinds of organisms around us were produced by a process of evolution.

Unlike his predecessors, however, Darwin proposed **natural selection** as the mechanism of evolution. Natural selection produces evolutionary change when some individuals in a population possess certain inherited characteristics and then produce more surviving offspring than individuals lacking these characteristics. As a result, the population gradually comes to include more and more individuals with the advantageous characteristics. In this way, the population evolves and becomes better adapted to its local circumstances.

A rival theory, championed by the prominent biologist Jean-Baptiste Lamarck, was that evolution occurred by the **inheritance of acquired characteristics.** According to Lamarck, individuals passed on to their offspring bodily and behavorial changes acquired during their lives. For example, Lamarck proposed that ancestral giraffes with short necks tended to stretch their necks to feed on tree leaves, and this extension of the neck was passed on to subsequent generations, leading to the long-necked giraffe (figure 20.1*a*). In Darwin's theory, by contrast, the variation is not created by experience, but is the result of preexisting genetic differences among individuals (figure 20.1*b*).

One way to monitor how populations change through time is to look at changes in the frequencies of alleles of a gene from one generation to the next. Natural selection, by favoring individuals with certain alleles, can lead to change in such *allele frequencies*, but it is not the only process that can do so. Allele frequencies can also change when mutations occur repeatedly, changing one allele to another, and when migrants bring alleles into a population. In addition, when populations are small, the frequencies of alleles can change randomly as the result of chance events. Often, natural selection overwhelms the effects of these other processes, but as you will see later in this chapter, this is not always the case.

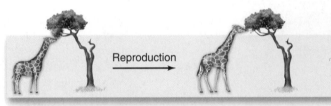

a. Lamarck's theory: acquired variation is passed on to descendants.

Some individuals born happen to have longer necks due to genetic differences.

Individuals pass on their traits to next generation.

Over many generations, longer-necked individuals are more successful, perhaps because they can feed on taller trees, and pass the long-neck trait on to their offspring.

b. Darwin's theory: natural selection or genetically-based variation leads to evolutionary change.

figure 20.1

TWO IDEAS OF HOW GIRAFFES MIGHT HAVE EVOLVED LONG NECKS.

Evolution can result from any process that causes a change in the genetic composition of a population. We cannot talk about evolution, therefore, without also considering **population genetics,** the study of the properties of genes in populations.

It is best to start by looking at the genetic variation present among individuals within a species. This is the raw material available for the selective process.

As you saw in chapter 12, a natural population can contain a great deal of genetic variation. How much variation usually occurs? Humans are representative of most—but not all—species in that human populations contain substantial amounts of genetic variation. For example:

1. **Genes that influence blood groups.** Chemical analysis has revealed the existence of more than 30 blood group genes in humans, in addition to the ABO locus. At least one-third of these genes are routinely found in several alternative allelic forms in human populations. In addition to these, more than 45 variable genes encode other proteins in human blood cells and plasma that are not considered blood groups. In short, many genetically variable genes are present in this one system alone.

2. **Genes that influence enzymes.** Alternative alleles of genes specifying particular enzymes are easy to distinguish by measuring how fast the alternative proteins migrate in an electrical field (a process called *electrophoresis*—see chapter 17). A great deal of variation exists at enzyme-specifying loci. About 5% of the enzyme loci of a typical human are heterozygous: If you picked an individual at random, and in turn selected one of the enzyme-encoding genes of that individual at random, the chances are 1 in 20 (5%) that the gene you selected would be heterozygous in that individual.

Considering the entire genome, it is fair to say that all humans are different from one another except for identical twins. This is also true of other organisms, except for those that reproduce asexually. In nature, genetic variation is the rule.

Enzyme polymorphism

Many loci in a particular population have more than one allele at frequencies significantly greater than would occur due to mutation alone. Researchers refer to such a locus as **polymorphic** (figure 20.2). The extent of such variation within natural populations was not even suspected a few decades ago, when modern techniques such as protein electrophoresis made it possible to examine enzymes and other proteins directly.

We now know that most populations of insects and plants are polymorphic at more than half of their enzyme-encoding loci, that is, the loci have more than one allele occurring at a frequency greater than 5%. Vertebrates are somewhat less polymorphic. **Heterozygosity,** the probability that a randomly selected gene will be heterozygous in a randomly selected individual, is about 15% in *Drosophila* and other invertebrates, between 5% and 8% in vertebrates, and around 8% in outcrossing plants (values of heterozygosity tend to be lower than the proportion of loci that are polymorphic because for loci

figure 20.2

POLYMORPHIC VARIATION. This natural population of loosestrife, *Lythrum salicaria*, exhibits considerable variation in flower color. Individual differences are inherited and passed on to offspring.

that are polymorphic, many individuals within the population will be homozygous). These high levels of genetic variability provide ample supplies of raw material for evolution.

DNA sequence polymorphism

The advent of gene technology has made it possible to assess genetic variation even more directly by sequencing the DNA itself. For example, when the *ADH* genes (which encode for alcohol dehydrogenase) of 11 *Drosophila melanogaster* individuals were sequenced, scientists found 43 variable sites, only one of which had been detected by protein electrophoresis!

Numerous other studies of variation at the DNA level have confirmed these findings: Abundant variation exists in both the coding regions of genes and in their nontranslated introns—considerably more variation than we can detect by examining enzymes with electrophoresis.

Darwin proposed that species undergo descent with modification, which is the concept underlying the theory of evolution.

Natural selection, by which some alleles are favored over others and leave more offspring, is one process by which species evolve.

Natural populations contain considerable amounts of genetic variation—more than can be accounted for by mutation alone.

Changes in Allele Frequency

Genetic variation within natural populations was a puzzle to Darwin and his contemporaries in the mid-1800s. The way in which meiosis produces genetic segregation among the progeny of a hybrid had not yet been discovered. And, although Mendel performed his experiments during this same time period, his work was largely unknown. Selection, scientists then thought, should always favor an optimal form, and so tend to eliminate variation. Moreover, the theory of *blending inheritance*—in which offspring were expected to be phenotypically intermediate relative to their parents—was widely accepted. If blending inheritance were correct, then the effect of any new genetic variant would quickly be diluted to the point of disappearance in subsequent generations.

The Hardy–Weinberg principle describes stable populations

Following the rediscovery of Mendel's research, two people in 1908 solved the puzzle of why genetic variation persists—Godfrey H. Hardy, an English mathematician, and Wilhelm Weinberg, a German physician. These workers were initially confused about why, after many generations, a population didn't come to be composed solely of individuals with the dominant phenotype. The conclusion they independently came to was that the original proportions of the genotypes in a population will remain constant from generation to generation, as long as the following assumptions are met:

1. No mutation takes place.
2. No genes are transferred to or from other sources (no immigration or emigration takes place)
3. Random mating is occurring.
4. The population size is very large.
5. No selection occurs.

Because the genotypes' proportions do not change, they are said to be in **Hardy–Weinberg equilibrium.**

The Hardy–Weinberg equation with two alleles: A binomial expansion

In algebraic terms, the Hardy–Weinberg principle is written as an equation. Consider a population of 100 cats in which 84 are black and 16 are white. The frequencies of the two phenotypes would be 0.84 (or 84%) black and 0.16 (or 16%) white. Based on these phenotypic frequencies, can we deduce the underlying frequency of genotypes?

If we assume that the white cats are homozygous recessive for an allele we designate as b, and the black cats are either homozygous dominant BB or heterozygous Bb, we can calculate the **allele frequencies** of the two alleles in the population from the proportion of black and white individuals, assuming that the population is in Hardy–Weinberg equilibrium.

Let the letter p designate the frequency of the B allele and the letter q the frequency of the alternative allele. Because there are only two alleles, p plus q must always equal 1 (that is, the total population). In addition, we know that the sum of the three genotypes must also equal 1. If the frequency of the B allele is p, then the probability that an individual will have two B alleles is simply the probability that each of its alleles is a B. The probability of two events happening independently is the product of the probability of each event; in this case, the probability that the individual received a B allele from its father is p, and the probability the individual received a B allele from its mother is also p, so the probability that both happened is $p \times p = p^2$ (figure 20.3). By the same reasoning, the probability that an individual will have two b alleles is q^2.

What about the probability that an individual will be a heterozygote? There are two ways this could happen: The indi-

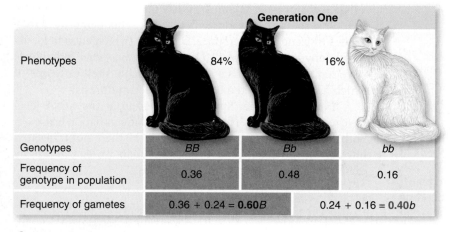

Generation One			
Phenotypes	84%	16%	
Genotypes	BB	Bb	bb
Frequency of genotype in population	0.36	0.48	0.16
Frequency of gametes	0.36 + 0.24 = **0.60**B	0.24 + 0.16 = **0.40**b	

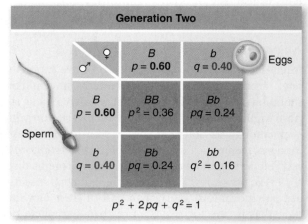

Generation Two

	♀	B $p = 0.60$	b $q = 0.40$	Eggs
♂				
B $p = 0.60$		BB $p^2 = 0.36$	Bb $pq = 0.24$	
Sperm				
b $q = 0.40$		Bb $pq = 0.24$	bb $q^2 = 0.16$	

$$p^2 + 2pq + q^2 = 1$$

figure 20.3

THE HARDY–WEINBERG EQUILIBRIUM. In the absence of factors that alter them, the frequencies of gametes, genotypes, and phenotypes remain constant generation after generation.

inquiry

If all white cats died, what proportion of the kittens in the next generation would be white?

vidual could receive a *B* from its father and a *b* from its mother, or vice versa. The probability of the first case is $p \times q$ and the probability of the second case is $q \times p$. Because the result in either case is that the individual is a heterozygote, the probability of that outcome is the sum of the two probabilities, or $2pq$.

So, to summarize, if a population is in Hardy–Weinberg equilibrium with allele frequencies of *p* and *q*, then the probability that an individual will have each of the three possible genotypes is $p^2 + 2pq + q^2$. You may recognize this as the *binomial expansion*:

$$(p + q)^2 = p^2 + 2pq + q^2$$

Finally, we may use these probabilities to predict the distribution of genotypes in the population. If the probability that any individual is a heterozygote is $2pq$, then we would expect the proportion of heterozygous individuals in the population to be $2pq$; similarly, the frequency of *BB* and *bb* homozygotes would be expected to be p^2 and q^2.

Let us return to our example. Remember that 16% of the cats are white. If white is a recessive trait, then this means that such individuals must have the genotype *qq*. If the frequency of this genotype is $q^2 = 0.16$ (the frequency of white cats), then the frequency of $q = 0.4$. Because $p + q = 1$, therefore, *p*, the frequency of allele *B*, would be $1.0 - 0.4 = 0.6$ (remember, the frequencies must add up to 1). We can now easily calculate the **genotype frequencies:** homozygous dominant *BB* cats would make up the p^2 group, and the value of $p^2 = (0.6)^2 = 0.36$, or 36 homozygous dominant *BB* individuals in a population of 100 cats. The heterozygous cats have the *Bb* genotype and would have the frequency corresponding to $2pq$, or $(2 \times 0.6 \times 0.4) = 0.48$, or 48 heterozygous *Bb* individuals.

Using the Hardy–Weinberg equation to predict frequencies in subsequent generations

The Hardy–Weinberg equation is a simple extension of the Punnett square described in chapter 12, with two alleles assigned frequencies, *p* and *q*. Figure 20.3 allows you to trace genetic reassortment during sexual reproduction and see how it affects the frequencies of the *B* and *b* alleles during the next generation.

In constructing this diagram, we have assumed that the union of sperm and egg in these cats is random, so that all combinations of *b* and *B* alleles occur. The alleles are therefore mixed randomly and are represented in the next generation in proportion to their original occurrence. Each individual egg or sperm in each generation has a 0.6 chance of receiving a *B* allele ($p = 0.6$) and a 0.4 chance of receiving a *b* allele ($q = 0.4$).

In the next generation, therefore, the chance of combining two *B* alleles is p^2, or 0.36 (that is, 0.6×0.6), and approximately 36% of the individuals in the population will continue to have the *BB* genotype. The frequency of *bb* individuals is q^2 (0.4×0.4) and so will continue to be about 16%, and the frequency of *Bb* individuals will be $2pq$ ($2 \times 0.6 \times 0.4$), or on average, 48%.

Phenotypically, if the population size remains at 100 cats, we would still see approximately 84 black individuals (with either *BB* or *Bb* genotypes) and 16 white individuals (with the *bb* genotype). Allele, genotype, and phenotype frequencies have remained unchanged from one generation to the next, despite the reshuffling of genes that occurs during meiosis and sexual reproduction. Dominance and recessiveness of alleles can therefore be

seen only to affect how an allele is expressed in an individual and not how allele frequencies will change through time.

Hardy–Weinberg predictions can be applied to data to find evidence of evolutionary processes

The lesson from the example of black and white cats is that if all five of the assumptions listed earlier hold true, the allele and genotype frequencies will not change from one generation to the next. But in reality, most populations in nature will not fit all five assumptions. The primary utility of this method is to determine whether some evolutionary process or processes are operating in a population and, if so, to suggest hypotheses about what they may be.

Suppose, for example, that the observed frequencies of the *BB* and *bb* genotypes in a different population of cats were each 0.45, and that of *Bb* was 0.10. We can calculate that allele frequencies for *B* and *b* are both 0.5, and thus the genotype frequencies should be *BB/Bb/bb* = 0.25/0.50/0.25. In this case, the observed population is not in Hardy–Weinberg equilibrium, and specifically, there are too many homozygotes and too few heterozygotes.

What could cause such an excess of homozygotes? A number of possibilities exist, including (1) natural selection favoring homozygotes over heterozygotes, (2) individuals choosing to mate with genetically similar individuals (because $BB \times BB$ and $bb \times bb$ matings always produce homozygous offspring, but only half of $Bb \times Bb$ produce heterozygous offspring, such mating patterns would lead to an excess of homozygotes), or (3) an influx of homozygous individuals from outside populations (or conversely, emigration of heterozygotes to other populations). By detecting a lack of Hardy–Weinberg equilibrium, we can generate potential hypotheses that we can then investigate directly.

The operation of evolutionary processes can be detected in a second way. As discussed previously, if all of the Hardy–Weinberg assumptions are met, then allele frequencies will stay the same from one generation to the next. Changes in allele frequencies between generations would indicate that one of the assumptions is not met.

Suppose, for example, that the frequency of *b* was 0.53 in one generation and 0.61 in the next. Again, there are a number of possible explanations: For example, (1) selection favoring individuals with *b* over *B*, (2) immigration of *b* into the population or emigration of *B* out of the population, or (3) high rates of mutation that more commonly occur from *B* to *b* than vice versa. Another possibility is that the population is a small one, and that the change represents the random fluctuations that result because, simply by chance, some individuals pass on more of their genes than others. We will discuss how each of these processes is studied in the rest of the chapter.

> The Hardy–Weinberg principle states that in a large population mating at random and in the absence of other forces that would change allele proportions, the process of sexual reproduction alone will not change these proportions.
>
> Finding that populations are not in Hardy–Weinberg equilibrium or that allele frequencies have changed from one generation to the next indicates that one or more evolutionary agents are operating in a population.

Five Agents of Evolutionary Change

The five assumptions of the Hardy–Weinberg principle also indicate the five agents that can lead to evolutionary change in populations. They are mutation, gene flow, nonrandom mating, genetic drift in small populations, and the pressures of natural selection. Any one of these may bring about changes in allele or genotype proportions.

Mutation changes alleles

Mutation from one allele to another can obviously change the proportions of particular alleles in a population. Mutation rates are generally so low that they have little effect on the Hardy–Weinberg proportions of common alleles. A typical gene mutates about once per 100,000 cell divisions. Because this rate is so low, other evolutionary processes are usually more important in determining how allele frequencies change.

Nonetheless, mutation is the ultimate source of genetic variation and thus makes evolution possible (figure 20.4a). It is important to remember, however, that the likelihood of a particular mutation occurring is not affected by natural selection; that is, mutations do not occur more frequently in situations in which they would be favored by natural selection.

Gene flow occurs when alleles move between populations

Gene flow is the movement of alleles from one population to another. It can be a powerful agent of change. Sometimes gene flow is obvious, as when an animal physically moves from one place to another. If the characteristics of the newly arrived individual differ from those of the animals already there, and if the newcomer is adapted well enough to the new area to survive

and mate successfully, the genetic composition of the receiving population may be altered.

Other important kinds of gene flow are not as obvious. These subtler movements include the drifting of gametes or the immature stages of plants or marine animals from one place to another (figure 20.4b). Pollen, the male gamete of flowering plants, is often carried great distances by insects and other animals that visit flowers. Seeds may also blow in the wind or be carried by animals to new populations far from their place of origin. In addition, gene flow may also result from the mating of individuals belonging to adjacent populations.

Consider two populations initially different in allele frequencies: In population 1, $p = 0.2$ and $q = 0.8$; in population 2, $p = 0.8$ and $q = 0.2$. Gene flow will tend to bring the rarer allele into each population. Thus, allele frequencies will change from generation to generation, and the populations will not be in Hardy–Weinberg equilibrium. Only when allele frequencies reach 0.5 for both alleles in both populations will equilibrium be attained. This example also indicates that gene flow tends to homogenize allele frequencies among populations.

Nonrandom mating shifts genotype frequencies

Individuals with certain genotypes sometimes mate with one another more commonly than would be expected on a random basis, a phenomenon known as *nonrandom mating* (figure 20.4c). **Assortative mating,** in which phenotypically similar individuals mate, is a type of nonrandom mating that causes the frequencies of particular genotypes to differ greatly from those predicted by the Hardy–Weinberg principle.

Assortative mating does not change the frequency of the individual alleles, but rather increases the proportion of homo-

Mutation	Gene Flow	Nonrandom Mating	Genetic Drift	Selection
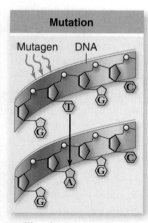		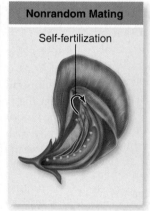		

a. The ultimate source of variation. Individual mutations occur so rarely that mutation alone usually does not change allele frequency much.

b. A very potent agent of change. Individuals or gametes move from one population to another.

c. Inbreeding is the most common form. It does not alter allele frequency but changes the proportion of heterozygotes.

d. Statistical accidents. The random fluctuation in allele frequencies increases as population size decreases.

e. The only agent that produces *adaptive* evolutionary changes.

figure 20.4

FIVE AGENTS OF EVOLUTIONARY CHANGE. *a.* Mutation, (*b*) gene flow, (*c*) nonrandom mating, (*d*) genetic drift, and (*e*) selection.

zygous individuals because phenotypically similar individuals are likely to be genetically similar and thus are also more likely to produce offspring with two copies of the same allele. This is why populations of self-fertilizing plants consist primarily of homozygous individuals.

By contrast, **disassortative mating,** in which phenotypically different individuals mate, produces an excess of heterozygotes.

Genetic drift may alter allele frequencies in small populations

In small populations, frequencies of particular alleles may change drastically by chance alone. Such changes in allele frequencies occur randomly, as if the frequencies were drifting from their values. These changes are thus known as **genetic drift** (figure 20.4d). For this reason, a population must be large to be in Hardy–Weinberg equilibrium.

If the gametes of only a few individuals form the next generation, the alleles they carry may by chance not be representative of the parent population from which they were drawn, as illustrated in figure 20.5. In this example, a small number of individuals are removed from a bottle containing many. By chance, most of the individuals removed are green, so the new population has a much higher population of green individuals than the parent generation had.

A set of small populations that are isolated from one another may come to differ strongly as a result of genetic drift, even if the forces of natural selection are the same for both. Because of genetic drift, sometimes harmful alleles may increase in frequency in small populations, despite selective disadvantage, and favorable alleles may be lost even though they are selectively advantageous. It is interesting to realize that humans have lived in small groups for much of the course of their evolution; consequently, genetic drift may have been a particularly important factor in the evolution of our species.

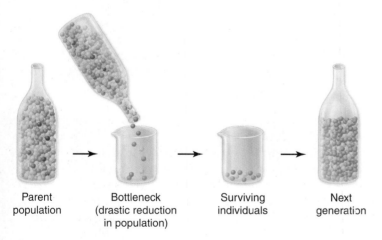

figure 20.5

GENETIC DRIFT: A BOTTLENECK EFFECT. The parent population contains roughly equal numbers of green and yellow individuals and a small number of red individuals. By chance, the few remaining individuals that contribute to the next generation are mostly green. The bottleneck occurs because so few individuals form the next generation, as might happen after an epidemic or a catastrophic storm.

Parent population — Bottleneck (drastic reduction in population) — Surviving individuals — Next generation

Even large populations may exhibit the effect of genetic drift. Large populations may have been much smaller in the past, and genetic drift may have greatly altered allele frequencies at that time. Imagine a population containing only two alleles of a gene, B and b, in equal frequency (that is, $p = q = 0.50$). In a large Hardy–Weinberg population, the genotype frequencies are expected to be 0.25 BB, 0.50 Bb, and 0.25 bb. If only a small sample of individuals produces the next generation, large deviations in these genotype frequencies can occur simply by chance.

Suppose, for example, that four individuals form the next generation, and that by chance they are two Bb heterozygotes and two BB homozygotes—that is, the allele frequencies in the next generation would be $p = 0.75$ and $q = 0.25$. In fact, if you were to replicate this experiment 1000 times, each time randomly drawing four individuals from the parental population, then in about 8 of the 1000 experiments, one of the two alleles would be missing entirely.

This result leads to an important conclusion: Genetic drift can lead to the loss of alleles in isolated populations. Alleles that initially are uncommon are particularly vulnerable (see figure 20.5a).

Although genetic drift occurs in any population, it is particularly likely in populations that were founded by a few individuals or in which the population was reduced to a very small number at some time in the past.

The Founder Effect

Sometimes one or a few individuals disperse and become the founders of a new, isolated population at some distance from their place of origin. These pioneers are not likely to carry all the alleles present in the source population. Thus, some alleles may be lost from the new population, and others may change drastically in frequency. In some cases, previously rare alleles in the source population may be a significant fraction of the new population's genetic endowment. This phenomenon is called the **founder effect.**

Founder effects are not rare in nature. Many self-pollinating plants start new populations from a single seed. Founder effects have been particularly important in the evolution of organisms on distant oceanic islands, such as the Hawaiian and Galápagos Islands. Most of the organisms in such areas probably derive from one or a few initial founders.

In a similar way, isolated human populations begun by relatively few individuals are often dominated by genetic features characteristic of their founders. Amish populations in the United States, for example, have unusually high frequencies of a number of conditions, such as polydactylism (the presence of a sixth finger).

The bottleneck effect

Even if organisms do not move from place to place, occasionally their populations may be drastically reduced in size. This may result from flooding, drought, epidemic disease, and other natural forces, or from changes in the environment. The few surviving individuals may constitute a random genetic sample of the original population (unless some individuals survive specifically because of their genetic makeup). The resulting alterations and loss of genetic variability have been termed the **bottleneck effect.**

The genetic variation of some living species appears to be severely depleted, probably as the result of a bottleneck effect in the past. For example, the northern elephant seal, which breeds on the western coast of North America and nearby islands, was nearly hunted to extinction in the nineteenth century and was reduced to a single population containing perhaps no more than 20 individuals on the island of Guadalupe off the coast of Baja, California (figure 20.6). As a result of this bottleneck, the species has lost almost all of its genetic variation, even though the seal populations have rebounded and now number in the tens of thousands and breed in locations as far north as near San Francisco.

Any time a population becomes drastically reduced in numbers, such as in endangered species, the bottleneck effect is a potential problem. Even if population size rebounds, the lack of variability may mean that the species remains vulnerable to extinction—a topic we will return to in chapter 59.

Selection favors some genotypes over others

As Darwin pointed out, some individuals leave behind more progeny than others, and the rate at which they do so is affected by phenotype and behavior. We describe the results of this process as **selection** (see figure 20.4e). In **artificial selection**, a breeder selects for the desired characteristics. In **natural selection**, environmental conditions determine which individuals in a population produce the most offspring.

For natural selection to occur and to result in evolutionary change, three conditions must be met:

1. **Variation must exist among individuals in a population.** Natural selection works by favoring individuals with some traits over individuals with alternative traits. If no variation exists, natural selection cannot operate.

2. **Variation among individuals must result in differences in the number of offspring surviving in the next generation.** This is the essence of natural selection. Because of their phenotype or behavior, some individuals are more successful than others in producing offspring. Although many traits are phenotypically variable, individuals exhibiting variation do not always differ in survival and reproductive success.

3. **Variation must be genetically inherited.** For natural selection to result in evolutionary change, the selected differences must have a genetic basis. Not all variation has a genetic basis—even genetically identical individuals may be phenotypically quite distinctive if they grow up in different environments. Such environmental effects are common in nature. In many turtles, for example, individuals that hatch from eggs laid in moist soil are heavier, with longer and wider shells, than individuals from nests in drier areas.

When phenotypically different individuals do not differ genetically, then differences in the number of their offspring will not alter the genetic composition of the population in the next generation, and thus, no evolutionary change will have occurred.

It is important to remember that natural selection and evolution are not the same—the two concepts often are incorrectly equated. Natural selection is a process, whereas evolution is the historical record, or outcome, of change through time. Natural selection (the process) can lead to evolution (the outcome), but natural selection is only one of several processes that can result in evolutionary change. Moreover, natural selection can occur without producing evolutionary change; only if variation is genetically based will natural selection lead to evolution.

Selection to avoid predators

The result of evolution driven by natural selection is that populations become better adapted to their environment. Many of the most dramatic documented instances of adaptation involve genetic changes that decrease the probability of capture by a predator. The caterpillar larvae of the common sulphur butterfly *Colias eurytheme*

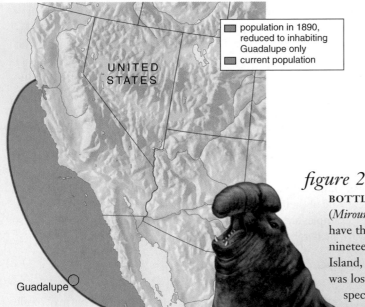

population in 1890, reduced to inhabiting Guadalupe only

current population

UNITED STATES

Guadalupe

MEXICO

figure 20.6

BOTTLENECK EFFECT: CASE STUDY. Because the Northern Elephant Seal (*Mirounga angustirostris*) lives in very cold waters, these, the world's largest seals, have thick layers of fat, for which they were hunted nearly to extinction late in the nineteenth century. At the low point, only one population remained on Guadalupe Island, with perhaps as few as 20 individuals; during this time, genetic variation was lost through the process of random genetic drift. Since being protected, the species has reclaimed most of its original range and now numbers in the tens of thousands, but genetic variation will only recover slowly over time as mutations accumulate.

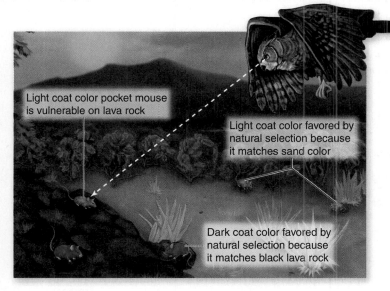

figure 20.7

POCKET MICE FROM THE TULAROSA BASIN OF NEW MEXICO WHOSE COLOR MATCHES THEIR BACKGROUND. Black lava formations are surrounded by desert, and selection favors coat color in pocket mice that matches their surroundings.

usually exhibit a pale green color, providing excellent camouflage against the alfalfa plants on which they feed. An alternative bright blue color morph is kept at very low frequency because this color renders the larvae highly visible on the food plant, making it easier for bird predators to see them (see figure 20.4e).

One of the most dramatic examples of background matching involves ancient lava flows in the deserts of the American Southwest. In these areas, the black rock formations produced when the lava cooled contrast starkly with the surrounding bright glare of the desert sand. Populations of many species of animals occurring on these rocks—including lizards, rodents, and a variety of insects—are dark in color, whereas sand-dwelling populations in surrounding areas are much lighter (figure 20.7).

Predation is the likely cause for these differences in color. Laboratory studies have confirmed that predatory birds such as owls are adept at picking out individuals occurring on backgrounds to which they are not adapted.

Selection to match climatic conditions

Many studies of selection have focused on genes encoding enzymes, because in such cases the investigator can directly assess the consequences to the organism of changes in the frequency of alternative enzyme alleles.

Often investigators find that enzyme allele frequencies vary with latitude, so that one allele is more common in northern populations, but is progressively less common at more southern locations. A superb example is seen in studies of a fish, the mummichog (*Fundulus heteroclitus*), which ranges along the eastern coast of North America. In this fish, geographic variation occurs in allele frequencies for the gene that produces the enzyme lactate dehydrogenase, which catalyzes the conversion of pyruvate to lactate (see section 9.4).

Biochemical studies show that the enzymes formed by these alleles function differently at different temperatures, thus explaining their geographic distributions. The form of the enzyme more frequent in the north is a better catalyst at low tem-

peratures than is the enzyme from the south. Moreover, studies indicate that at low temperatures, individuals with the northern allele swim faster, and presumably survive better, than individuals with the alternative allele.

Selection for pesticide resistance

A particularly clear example of selection in natural populations is provided by studies of pesticide resistance in insects. The widespread use of insecticides has led to the rapid evolution of resistance in more than 500 pest species.

In the housefly, the resistance allele at the *pen* gene decreases the uptake of insecticide, whereas alleles at the *kdr* and *dld-r* genes decrease the number of target sites, thus decreasing the binding ability of the insecticide (figure 20.8). Other alleles enhance the ability of the insects' enzymes to identify and detoxify insecticide molecules.

Single genes are also responsible for resistance in other organisms. For example, Norway rats are normally susceptible to the pesticide warfarin, which diminishes the clotting ability of the rat's blood and leads to fatal hemorrhaging. However, a resistance allele at a single gene reduces the ability of warfarin to bind to its target enzyme and thus renders it ineffective.

Five factors can bring about a deviation from the proportions of homozygotes and heterozygotes predicted by the Hardy–Weinberg principle. Only selection regularly produces adaptive evolutionary change, but the genetic constitution of populations, and thus the course of evolution, can also be affected by mutation, gene flow, nonrandom mating, and genetic drift.

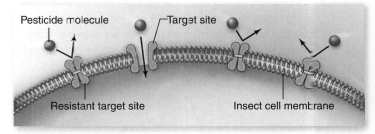

a. Insect cells with resistance allele at *pen* gene: decreased uptake of the pesticide.

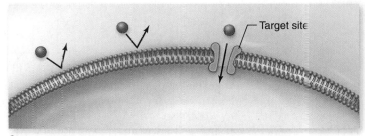

b. Insect cells with resistance allele at *kdr* gene: decreased number of target sites for the pesticide.

figure 20.8

SELECTION FOR PESTICIDE RESISTANCE. Resistance alleles at genes such as *pen* and *kdr* allow insects to be more resistant to pesticides. Insects that possess these resistance alleles have become more common through selection.

Selection occurs when individuals with one phenotype leave more surviving offspring in the next generation than individuals with an alternative phenotype. Evolutionary biologists quantify reproductive success as **fitness,** the number of surviving offspring left in the next generation.

Fitness is a relative concept; the most fit phenotype is simply the one that produces, on average, the greatest number of offspring.

A phenotype with greater fitness usually increases in frequency

Suppose, for example, that in a population of toads, two phenotypes exist: green and brown. Suppose, further, that green toads leave, on average, 4.0 offspring in the next generation, but brown toads leave only 2.5. By custom, the most fit phenotype is assigned a fitness value of 1.0, and other phenotypes are expressed as relative proportions. In this case, the fitness of the green phenotype would be 4.0/4.0 = 1.000, and the fitness of the brown phenotype would be 2.5/4.0 = 0.625. The difference in fitness would therefore be 1.000 − 0.625 = 0.375. A difference in fitness of 0.375 is quite large; natural selection in this case strongly favors the green phenotype.

If differences in color have a genetic basis, then we would expect evolutionary change to occur; the frequency of green toads should be substantially greater in the next generation. Further, if the fitness of two phenotypes remained unchanged, we would expect alleles for the brown phenotype eventually to disappear from the population.

inquiry

Why might the frequency of green toads not increase in the next generation, even if color differences have a genetic basis?

Fitness may consist of many components

Although selection is often characterized as "survival of the fittest," differences in survival are only one component of fitness.

Even if no differences in survival occur, selection may operate if some individuals are more successful than others in attracting mates. In many territorial animal species, for example, large males mate with many females, and small males rarely get to mate. Selection with respect to mating success is termed *sexual selection;* we describe this topic more fully in the discussion of behavioral biology in chapter 54.

In addition, the number of offspring produced per mating is also important. Large female frogs and fish lay more eggs than do smaller females, and thus they may leave more offspring in the next generation.

Fitness is therefore a combination of survival, mating success, and number of offspring per mating. Selection favors phenotypes with the greatest fitness, but predicting fitness from a single component can be tricky because traits favored for one component of fitness may be at a disadvantage for others. As an example, in water striders, larger females lay more eggs per day (figure 20.9). Thus, natural selection at this stage favors large size. However, larger females also die at a younger age and thus have fewer opportunities to reproduce; consequently, smaller females have a survival advantage. Overall, the two opposing directions of selection cancel each other out, and the intermediate-sized females, on average, leave the most offspring in the next generation.

> An organism's reproductive success is affected by how long it survives, how often it mates, and how many offspring it produces per mating.

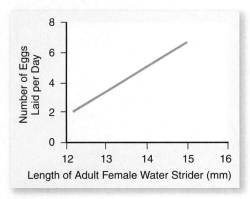 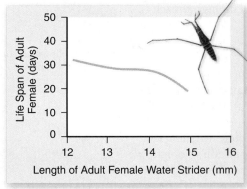 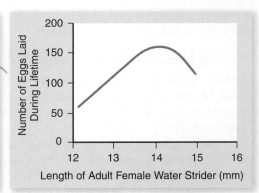

figure 20.9

BODY SIZE AND EGG-LAYING IN WATER STRIDERS. Larger female water striders lay more eggs per day (left panel), but also survive for a shorter period of time (center panel). As a result, intermediate-sized females produce the most offspring over the course of their entire lives and thus have the highest fitness (right panel).

inquiry

What evolutionary change in body size might you expect? If the number of eggs laid per day was not affected by body size, would your prediction change?

Interactions Among Evolutionary Forces

The amount of genetic variation in a population may be determined by the relative strength of different evolutionary processes. Sometimes these processes act together, and in other cases they work in opposition.

Mutation and genetic drift may counter selection

In theory, if allele *B* mutates to allele *b* at a high enough rate, allele *b* could be maintained in the population, even if natural selection strongly favored allele *B*. In nature, however, mutation rates are rarely high enough to counter the effects of natural selection.

The effect of natural selection also may be countered by genetic drift. Both of these processes may act to remove variation from a population. But selection is a nonrandom process that operates to increase the representation of alleles that enhance survival and reproductive success, whereas genetic drift is a random process in which any allele may increase. Thus, in some cases, drift may lead to a decrease in the frequency of an allele that is favored by selection. In some extreme cases, drift may even lead to the loss of a favored allele from a population.

Remember, however, that the magnitude of drift is inversely related to population size; consequently, natural selection is expected to overwhelm drift, except when populations are very small.

Gene flow may promote or constrain evolutionary change

Gene flow can be either a constructive or a constraining force. On one hand, gene flow can spread a beneficial mutation that arises in one population to other populations. On the other hand, gene flow can impede adaptation within a population by the continual flow of inferior alleles from other populations.

Consider two populations of a species that live in different environments. In this situation, natural selection might favor different alleles—*B* and *b*—in the two populations. In the absence of other evolutionary processes such as gene flow, the frequency of *B* would be expected to reach 100% in one population and 0% in the other. However, if gene flow occurred between the two populations, then the less favored allele would continually be reintroduced into each population. As a result, the frequency of the two alleles in each population would reflect a balance between the rate at which gene flow brings the inferior allele into a population, and the rate at which natural selection removes it.

A classic example of gene flow opposing natural selection occurs on abandoned mine sites in Great Britain. Although mining activities ceased hundreds of years ago, the concentration of metal ions in the soil is still much greater than in surrounding areas. Large concentrations of heavy metals are generally toxic to plants, but alleles at certain genes confer the ability to grow on soils high in heavy metals. The ability to tolerate heavy metals comes at a price, however; individuals with the resistance allele exhibit lower growth rates on nonpolluted soil. Consequently, we would expect the resistance allele to occur with a frequency of 100% on mine sites and 0% elsewhere.

Heavy-metal tolerance has been studied intensively in the slender bent grass *Agrostis tenuis*, in which the resistance allele occurs at intermediate levels in many areas (figure 20.10). The explanation relates to the reproductive system of this grass, in which pollen, the floral equivalent of sperm, is dispersed by the wind. As a result, pollen grains—and the alleles they carry—can move great distances, leading to levels of gene flow between mine sites and unpolluted areas high enough to counteract the effects of natural selection.

In general, the extent to which gene flow can hinder the effects of natural selection should depend on the relative strengths of the two processes. In species in which gene flow is generally strong, such as in birds

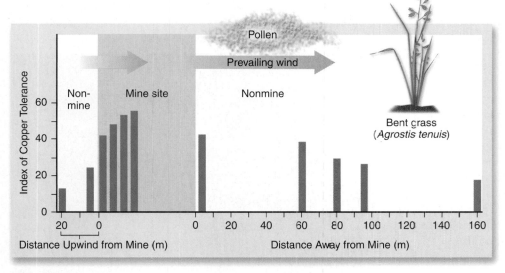

figure 20.10

DEGREE OF COPPER TOLERANCE IN GRASS PLANTS ON AND NEAR ANCIENT MINE SITES. Individuals with tolerant alleles have decreased growth rates on unpolluted soil. Thus, we would expect copper tolerance to be 100% on mine sites and 0% on nonmine sites. However, prevailing winds blow pollen containing nontolerant alleles onto the mine site and tolerant alleles beyond the site's borders. The amount of pollen received decreases with distance, which explains the changes in levels of tolerance. The index of copper tolerance is calculated as the growth rate of a plant on soil with high concentrations of copper relative to growth rate on soils with low levels of copper; the higher the index, the more tolerant the plant is of heavy metal pollution.

inquiry

? *Would you expect the frequency of copper tolerance to be affected by distance from the mine site?*

and wind-pollinated plants, the frequency of the less favored allele may be relatively high. In more sedentary species that exhibit low levels of gene flow, such as salamanders, the favored allele should occur at a frequency near 100%.

Allele frequencies sometimes reflect a balance between opposing processes, such as gene flow and natural selection. In such cases, observed frequencies will depend on the relative strength of the processes.

20.6 Maintenance of Variation

In the previous pages, natural selection has been discussed as a process that removes variation from a population by favoring one allele over others at a gene locus. However, in some circumstances, selection can do exactly the opposite and actually maintain population variation.

Frequency-dependent selection may favor either rare or common phenotypes

In some circumstances, the fitness of a phenotype depends on its frequency within the population, a phenomenon termed **frequency-dependent selection.** This type of selection favors certain phenotypes depending on how commonly or uncommonly they occur.

Negative frequency-dependent selection

In negative frequency-dependent selection, rare phenotypes are favored by selection. Assuming a genetic basis for phenotypic variation, such selection will have the effect of making rare alleles more common, thus maintaining variation.

Negative frequency-dependent selection can occur for many reasons. For example, it is well known that animals or people searching for something form a "search image." That is, they become particularly adept at picking out certain objects. Consequently, predators may form a search image for common prey phenotypes. Rare forms may thus be preyed upon less frequently.

An example is fish predation on an insect, the water boatman, which occurs in three different colors. Experiments indicate that each of the color types is preyed upon disproportionately when it is the most common one; fish eat more of the common-colored insects than would occur by chance alone (figure 20.11a).

Another cause of negative frequency dependence is resource competition. If genotypes differ in their resource requirements, as occurs in many plants, then the rarer genotype will have fewer competitors. When the different resource types are equally abundant, the rarer genotype will be at an advantage relative to the more common genotype.

Positive frequency-dependent selection

Positive frequency-dependent selection has the opposite effect; by favoring common forms, it tends to eliminate variation from a population. For example, predators don't always select common individuals. In some cases, "oddballs" stand out from the rest and attract attention (figure 20.11b).

The strength of selection should change through time as a result of frequency-dependent selection. In negative frequency-dependent selection, rare genotypes should become increasingly common, and their selective advantage will decrease correspondingly. Conversely, in positive frequency dependence, the rarer a genotype becomes, the greater the chance it will be selected against.

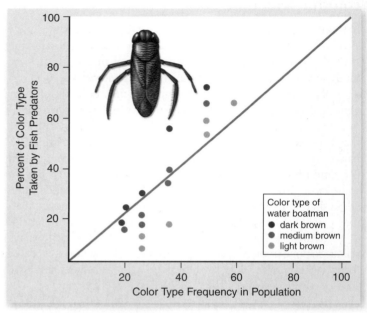

a. Negative frequency-dependent selection

b. Positive frequency-dependent selection

figure 20.11

FREQUENCY-DEPENDENT SELECTION. *a.* Predators often form search images for the most common prey. In one experiment, fish were placed in an enclosure with water boatmen (aquatic insects) of three different colors. When each color was common, the fish disproportionately captured boatmen of that color. By contrast, when each color type was uncommon, it was rarely taken (negative frequency-dependent selection). *b.* However, in some cases, rare individuals stand out from the rest and draw the attention of predators; thus, in these cases, common phenotypes have the advantage (positive frequency-dependent selection).

In oscillating selection, the favored phenotype changes as the environment changes

In some cases, selection favors one phenotype at one time and another phenotype at another time, a phenomenon called **oscillating selection.** If selection repeatedly oscillates in this fashion, the effect will be to maintain genetic variation in the population.

One example, discussed in chapter 21, concerns the medium ground finch of the Galápagos Islands. In times of drought, the supply of small, soft seeds is depleted, but there are still enough large seeds around. Consequently, birds with big bills are favored. However, when wet conditions return, the ensuing abundance of small seeds favors birds with smaller bills.

Oscillating selection and frequency-dependent selection are similar because in both cases the form of selection changes through time. But it is important to recognize that they are not the same: In oscillating selection, the fitness of a phenotype does not depend on its frequency; rather, environmental changes lead to the oscillation in selection. In contrast, in frequency-dependent selection, it is the change in frequencies themselves that leads to the changes in fitness of the different phenotypes

In some cases, heterozygotes may exhibit greater fitness than homozygotes

If heterozygotes are favored over homozygotes, then natural selection actually tends to maintain variation in the population. This **heterozygote advantage** favors individuals with copies of both alleles, and thus works to maintain both alleles in the population.

Some evolutionary biologists believe that heterozygote advantage is pervasive and can explain the high levels of polymorphism observed in natural populations. Others, however, believe that it is relatively rare.

The best documented example of heterozygote advantage is sickle cell anemia, a hereditary disease affecting hemoglobin in humans. Individuals with sickle cell anemia exhibit symptoms of severe anemia and abnormal red blood cells that are irregular in shape, with a great number of long, sickle-shaped cells (figure 20.12). Chapter 13 discusses why the sickle cell mutation (S) causes red blood cells to sickle.

The average incidence of the S allele in central African populations is about 0.12, far higher than that found among African Americans. From the Hardy–Weinberg principle, you can calculate that 1 in 5 central African individuals is heterozygous at the S allele, and 1 in 100 is homozygous and develops the fatal form of the disorder. People who are homozygous for the sickle cell allele almost never reproduce because they usually die before they reach reproductive age.

Why, then, is the S allele not eliminated from the central African population by selection rather than being maintained at such high levels? As it turns out, one of the leading causes of illness and death in central Africa, especially among young children, is malaria. People who are heterozygous for the sickle cell allele (and thus do not suffer from sickle cell anemia) are much less susceptible to malaria. The reason is that when the parasite that causes malaria, *Plasmodium falciparum*, enters a red blood cell, it causes extremely low oxygen tension in the cell, which leads to sickling in cells of individuals either homozygous or heterozygous for the sickle cell allele (but not in individuals that do

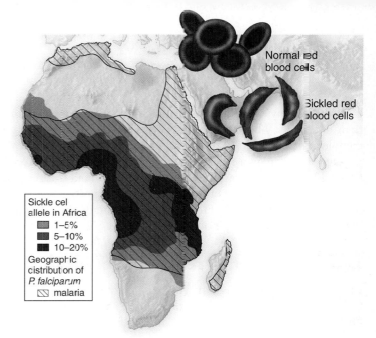

figure 20.12

FREQUENCY OF SICKLE CELL ALLELE AND DISTRIBUTION OF *PLASMODIUM FALCIPARUM* MALARIA. The red blood cells of people homozygous for the sickle cell allele collapse into sickled shapes when the oxygen level in the blood is low. The distribution of the sickle cell allele in Africa coincides closely with that of *P. falciparum* malaria.

not have the sickle cell allele). Such cells are quickly filtered out of the bloodstream by the spleen, thus eliminating the parasite. (The spleen's filtering effect is what leads to anemia in persons homozygous for the sickle cell allele because large numbers of red blood cells are removed; in the case of malaria, only those cells containing the *Plasmodium* parasite sickle, whereas the remaining cells are not affected, and thus anemia does not occur.)

Consequently, even though most homozygous recessive individuals die before they have children, the sickle cell allele is maintained at high levels in these populations because it is associated with resistance to malaria in heterozygotes and also, for reasons not yet fully understood, with increased fertility in female heterozygotes. Figure 20.12 shows the overlap between regions where sickle cell anemia is found and where malaria is prevalent.

For people living in areas where malaria is common, having the sickle cell allele in the heterozygous condition has adaptive value (see figure 20.12). Among African Americans, however, many of whose ancestors have lived for many generations in a country where malaria is now essentially absent, the environment does not place a premium on resistance to malaria. Consequently, no adaptive value counterbalances the ill effects of the disease; in this nonmalarial environment, selection is acting to eliminate the S allele. Only 1 in 375 African Americans develops sickle cell anemia, far less than in central Africa.

Selection can maintain variation within populations in a number of ways. Negative frequency-dependent selection tends to favor rare phenotypes. Oscillating selection favors different phenotypes at different times, depending on environmental conditions. And in some cases, heterozygotes have a selective advantage that may keep deleterious alleles in a population.

In nature, many traits—perhaps most—are affected by more than one gene. The interactions between genes are typically complex, as you saw in chapter 12. For example, alleles of many different genes play a role in determining human height (see figure 12.16). In such cases, selection operates on all the genes, influencing most strongly those that make the greatest contribution to the phenotype. How selection changes the population depends on which genotypes are favored.

Disruptive selection removes intermediates

In some situations, selection acts to eliminate intermediate types, a phenomenon called **disruptive selection** (figure 20.13*a*). A clear example is the different beak sizes of the African black-bellied seedcracker finch *Pyrenestes ostrinus* (figure 20.14). Populations of these birds contain individuals with large and small beaks, but very few individuals with intermediate-sized beaks.

As their name implies, these birds feed on seeds, and the available seeds fall into two size categories: large and small. Only large-beaked birds can open the tough shells of large seeds, whereas birds with the smaller beaks are more adept at handling small seeds. Birds with intermediate-sized beaks are at a disadvantage with both seed types—they are unable to open large seeds and too clumsy to efficiently process small seeds. Consequently, selection acts to eliminate the intermediate phenotypes, in effect partitioning (or "disrupting") the population into two phenotypically distinct groups.

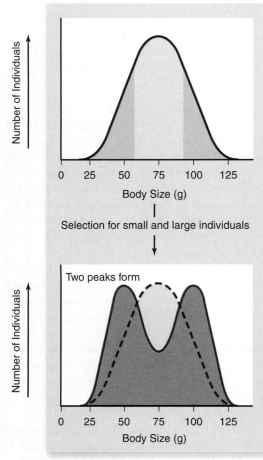

a. Disruptive selection

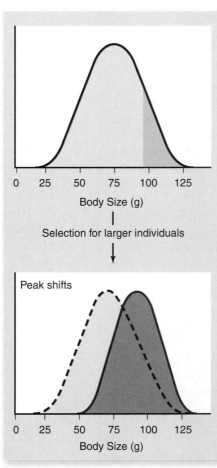

b. Directional selection

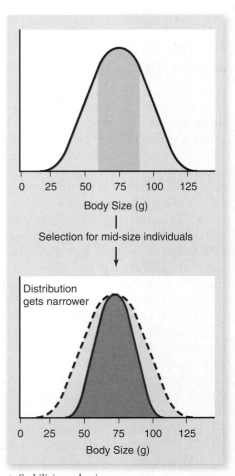

c. Stabilizing selection

figure 20.13

THREE KINDS OF SELECTION. The top panels show the populations before selection has occurred (under the solid red line). Within the population, those favored by selection are shown in light brown. The bottom panels indicate what the populations would look like in the next generation. The dashed red lines are the distribution of the original population and the solid, dark brown lines are the true distribution of the population in the next generation. *a.* In disruptive selection, individuals in the middle of the range of phenotypes of a certain trait are selected against, and the extreme forms of the trait are favored. *b.* In directional selection, individuals concentrated toward one extreme of the array of phenotypes are favored. *c.* In stabilizing selection, individuals with midrange phenotypes are favored, with selection acting against both ends of the range of phenotypes.

figure 20.14

DISRUPTIVE SELECTION FOR LARGE AND SMALL BEAKS. Differences in beak size in the black-bellied seedcracker finch of west Africa are the result of disruptive selection.

Directional selection eliminates phenotypes on one end of a range

When selection acts to eliminate one extreme from an array of phenotypes, the genes promoting this extreme become less frequent in the population. This form of selection is called **directional selection** (see figure 20.13*b*). Thus, in the *Drosophila* population illustrated in figure 20.15, eliminating flies that move toward light causes the population over time to contain fewer individuals with alleles promoting such behavior. If you were to pick an individual at random from a later generation of flies, there is a smaller chance that the fly would spontaneously move toward light than if you had selected a fly from the original population. Artificial selection has changed the population in the direction of being less attracted to light.

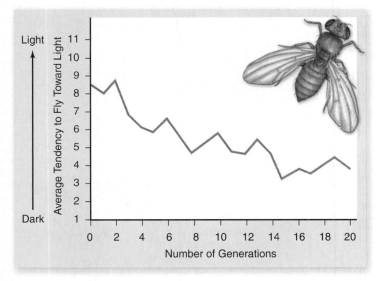

figure 20.15

DIRECTIONAL SELECTION FOR NEGATIVE PHOTOTROPISM IN *DROSOPHILA*. Flies that moved toward light were discarded, and only flies that moved away from light were used as parents for the next generation. This procedure was repeated for 20 generations, producing substantial evolutionary change.

 inquiry

What would happen if after 20 generations, experimenters started keeping flies that moved toward the light and discarded the others?

Stabilizing selection favors individuals with intermediate phenotypes

When selection acts to eliminate both extremes from an array of phenotypes, the result is to increase the frequency of the already common intermediate type. This form of selection is called **stabilizing selection** (see figure 20.13*c*). In effect, selection is operating to prevent change away from this middle range of values. Selection does not change the most common phenotype of the population, but rather makes it even more common by eliminating extremes. Many examples are known. In humans, infants with intermediate weight at birth have the highest survival rate (figure 20.16). In ducks and chickens, eggs of intermediate weight have the highest hatching success.

> Selection on traits affected by many genes can favor both extremes of the trait, only one extreme, or intermediate values.

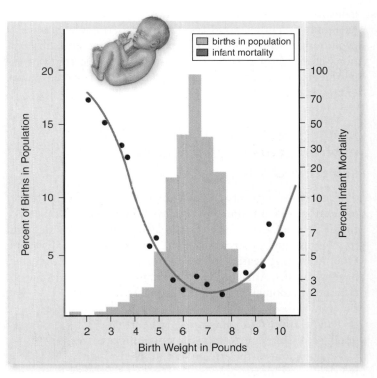

figure 20.16

STABILIZING SELECTION FOR BIRTH WEIGHT IN HUMANS. The death rate among babies (*red curve; right y-axis*) is lowest at an intermediate birth weight; both smaller and larger babies have a greater tendency to die than those around the most frequent weight (*tan area; left y-axis*) of between 7 and 8 pounds. Recent medical advances have reduced mortality rates for small and large babies.

 inquiry

As improved medical technology leads to decreased infant mortality rates, how would you expect the distribution of birth weights in the population to change?

Experimental Studies of Natural Selection

To study evolution, biologists have traditionally investigated what has happened in the past, sometimes many millions of years ago. To learn about dinosaurs, a paleontologist looks at dinosaur fossils. To study human evolution, an anthropologist looks at human fossils and, increasingly, examines the "family tree" of mutations that have accumulated in human DNA over millions of years. In this traditional approach, evolutionary biology is similar to astronomy and history, relying on observation rather than experimentation to examine ideas about past events.

Nonetheless, evolutionary biology is not entirely an observational science. Darwin was right about many things, but one area in which he was mistaken concerns the pace at which evolution occurs. Darwin thought that evolution occurred at a very slow, almost imperceptible pace. But in recent years many case studies have demonstrated that in some circumstances, evolutionary change can occur rapidly. Consequently, experimental studies can be devised to test evolutionary hypotheses.

Although laboratory studies on fruit flies and other organisms have been common for more than 50 years, scientists have only recently started conducting experimental studies of evolution in nature. One excellent example of how observations of the natural world can be combined with rigorous experiments in the lab and in the field concerns research on the guppy, *Poecilia reticulata*.

Guppy color variation in different environments suggests natural selection at work

The guppy is a popular aquarium fish because of its bright coloration and prolific reproduction. In nature, guppies are found in small streams in northeastern South America and in many mountain streams on the nearby island of Trinidad. One interesting feature of several of the streams is that they have waterfalls. Amazingly, guppies and some other fish are capable of colonizing portions of the stream above the waterfall.

The killifish, *Rivulus hartii*, is a particularly good colonizer; apparently on rainy nights, it will wriggle out of the stream and move through the damp leaf litter. Guppies are not so proficient, but they are good at swimming upstream. During flood seasons, rivers sometimes overflow their banks, creating secondary channels that move through the forest. On these occasions, guppies may be able to swim upstream in the secondary channels and invade the pools above waterfalls.

By contrast, some species are not capable of such dispersal and thus are only found in streams below the first waterfall. One species whose distribution is restricted by waterfalls is the pike cichlid, *Crenicichla alta*, a voracious predator that feeds on other fish, including guppies.

Because of these barriers to dispersal, guppies can be found in two very different environments. In pools just below the waterfalls, predation by the pike cichlid is a substantial risk, and rates of survival are relatively low. But in similar pools just above the waterfall, the only predator present is the killifish, which rarely preys on guppies.

Guppy populations above and below waterfalls exhibit many differences. In the high-predation pools, guppies exhibit drab coloration. Moreover, they tend to reproduce at a younger age and attain relatively smaller adult sizes. Male fish above the waterfall, in contrast, are colorful (figure 20.17), mature later, and grow to larger sizes.

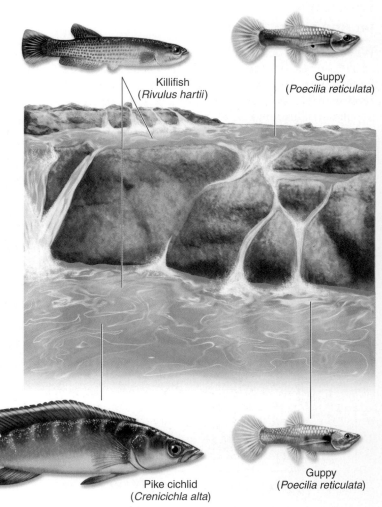

Killifish
(*Rivulus hartii*)

Guppy
(*Poecilia reticulata*)

Pike cichlid
(*Crenicichla alta*)

Guppy
(*Poecilia reticulata*)

figure 20.17

THE EVOLUTION OF PROTECTIVE COLORATION IN GUPPIES. In pools below waterfalls where predation is high, male guppies (*Poecilia reticulata*) are drab in color. In the absence of the highly predatory pike cichlid (*Crenicichla alta*) in pools above waterfalls, male guppies are much more colorful and attractive to females. The killifish (*Rivulus hartii*) is also a predator, but it only rarely eats guppies. The evolution of these differences in guppies can be experimentally tested.

These differences suggest the operation of natural selection. In the low-predation environment, males display gaudy colors and spots that they use to court females. Moreover, larger males are most successful at holding territories and mating with females, and larger females lay more eggs. Thus, in the absence of predators, larger and more colorful fish may have produced more offspring, leading to the evolution of those traits.

In pools below the waterfall, natural selection would favor different traits. Colorful males are likely to attract the attention of the pike cichlid, and high predation rates mean that most fish live short lives. Individuals that are more drab and shunt energy into early reproduction, rather than growth to a larger size, are therefore likely to be favored by natural selection.

Experimentation reveals the agent of selection

Although the differences between guppies living above and below the waterfalls suggest evolutionary responses to differences in the strength of predation, alternative explanations are possible. Perhaps, for example, only very large fish are capable of crawling past the waterfall to colonize pools. If this were the case, then a founder effect would occur in which the new population was established solely by individuals with genes for large size. The only way to rule out such alternative possibilities is to conduct a controlled experiment.

The laboratory experiment

The first experiments were conducted in large pools in laboratory greenhouses. At the start of the experiment, a group of 2000 guppies was divided equally among 10 large pools. Six months later, pike cichlids were added to four of the pools and killifish to another four, with the remaining two pools left to serve as "no-predation" controls.

Fourteen months later (which corresponds to 10 guppy generations), the scientists compared the populations. The guppies in the killifish and control pools were indistinguishable—brightly colored and large. In contrast, the guppies in the pike cichlid pools were smaller and drab in coloration.

These results established that predation can lead to rapid evolutionary change, but do these laboratory experiments reflect what occurs in nature?

The field experiment

To find out whether the laboratory results were an accurate reflection of natural processes, the scientists located two streams that had guppies in pools below a waterfall, but not above it. As in other Trinidadian streams, the pike cichlid was present in the lower pools, but only the killifish was found above the waterfalls.

The scientists then transplanted guppies to the upper pools and returned at several-year intervals to monitor the populations. Despite originating from populations in which predation levels were high, the transplanted populations

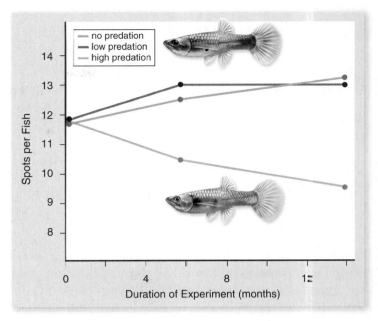

figure 20.18

EVOLUTIONARY CHANGE IN SPOT NUMBER. Guppy populations raised for 10 generations in low-predation or no-predation environments in laboratory greenhouses evolved a greater number of spots, whereas selection in more dangerous environments, such as the pools with the highly predatory pike cichlid, led to less conspicuous fish. The same results are seen in field experiments conducted in pools above and below waterfalls.

inquiry

How do these results depend on the manner by which the guppy predators locate their prey?

rapidly evolved the traits characteristic of low-predation guppies: They matured late, attained greater size, and had brighter colors. The control populations in the lower pools, by contrast, continued to be drab and to mature early and at a smaller size (figure 20.18). Laboratory analysis confirmed that the variations between the populations were the result of genetic differences.

These results demonstrate that substantial evolutionary change can occur in less than 12 years. More generally, these studies indicate how scientists can formulate hypotheses about how evolution occurs and then test these hypotheses in natural conditions. The results give strong support to the theory of evolution by natural selection.

Evolutionary biology is a historical science. Nonetheless, in some circumstances, experiments can be conducted in nature to test hypotheses about how evolution occurs. Such studies often reveal that natural selection can lead to rapid evolutionary change.

The Limits of Selection

Although selection is the most powerful of the principal agents of genetic change, there are limits to what it can accomplish. These limits result from multiple phenotypic effects of alleles, lack of genetic variation upon which selection can act, and interactions between genes.

Genes have multiple effects

Alleles often affect multiple aspects of a phenotype (the phenomenon of *pleiotropy*; see chapter 12). These multiple effects tend to set limits on how much a phenotype can be altered.

For example, selecting for large clutch size in chickens eventually leads to eggs with thinner shells that break more easily. For this reason, we could never produce chickens that lay eggs twice as large as the best layers do now. Likewise, we cannot produce gigantic cattle that yield twice as much meat as our leading breeds, or corn with an ear at the base of every leaf, instead of just at the bases of a few leaves.

Evolution requires genetic variation

Over 80% of the gene pool of the thoroughbred horses racing today goes back to 31 known ancestors from the late eighteenth century. Despite intense directional selection on thoroughbreds, their performance times have not improved for more than 50 years (figure 20.19). Years of intense selection presumably have removed variation from the population at a rate greater than it could be replenished by mutation, such that now little genetic variation remains, and evolutionary change is not possible.

In some cases, phenotypic variation for a trait may never have had a genetic basis. The compound eyes of insects are made up of hundreds of visual units, termed ommatidia (described in chapter 34). In some individuals, the left eye contains

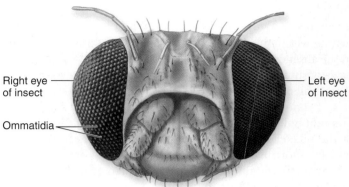

figure 20.20

PHENOTYPIC VARIATION IN INSECT OMMATIDIA. In some individuals, the number of ommatidia in the left eye is greater than the number in the right.

more ommatidia than the right. In other individuals, the right eye contains more than the left (figure 20.20). However, despite intense selection experiments in the laboratory, scientists have never been able to produce a line of fruit flies that consistently has more ommatidia in the left eye than in the right.

The reason is that separate genes do not exist for the left and right eyes. Rather, the same genes affect both eyes, and differences in the number of ommatidia result from differences that occur as the eyes are formed in the development process. Thus, despite the existence of phenotypic variation, no underlying genetic variation is available for selection to favor.

Gene interactions affect fitness of alleles

As discussed in chapter 12, *epistasis* is the phenomenon in which an allele for one gene may have different effects, depending on alleles present at other genes. Because of epistasis, the selective advantage of an allele at one gene may vary from one genotype to another. If a population is polymorphic for a second gene, then selection on the first gene may be constrained because different alleles are favored in different individuals of the same population.

Studies on bacteria illustrate how selection on alleles for one gene can depend on which alleles are present at other genes. In *E. coli*, two biochemical pathways exist to break down gluconate, each using enzymes produced by different genes. One gene produces the enzyme 6-PGD, for which there are several alleles. When the common allele for the second gene, which codes for the other biochemical pathway, is present, selection does not favor one allele over another at the 6-PGD gene. In some *E. coli*, however, an alternative allele at the second gene occurs that is not functional. The bacteria with this alternative allele are forced to rely only on the 6-PGD pathway, and in this case, selection favors one 6-PGD allele over another. Thus, epistatic interactions exist between the two genes, and the outcome of natural selection on the 6-PGD gene depends on which alleles are present at the second gene.

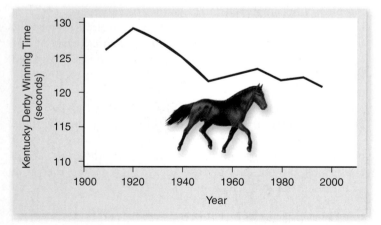

figure 20.19

SELECTION FOR INCREASED SPEED IN RACEHORSES IS NO LONGER EFFECTIVE. Kentucky Derby winning speeds have not improved significantly since 1950.

inquiry

What might explain the lack of change in winning speeds?

The ability of selection to produce evolutionary change is hindered by a variety of factors, including multiple effects of a single gene, lack of genetic variation, and gene interactions.

20.1 Genetic Variation and Evolution

Genetic variation, that is, differences in allele frequencies, provides the raw material for natural selection.

- Natural selection is the primary mechanism of evolution, which refers to change through time.
- Natural selection favors certain alleles and can lead to change in allele frequencies. Allelic frequencies can also change because of other processes.
- Loci are polymorphic if they contain more than one allele with frequencies greater than would occur due to mutation alone.

20.2 Changes in Allele Frequency (figure 20.3)

The Hardy–Weinberg principle predicts stable allele frequencies within a population as long as certain assumptions are met:

- No mutation occurs.
- There is no gene flow because migration between populations does not occur.
- Mating is random.
- Population size is very large.
- No natural selection occurs.

The Hardy–Weinberg principle provides evidence for evolution.

- If allele frequencies are not the same in subsequent generations, then one of the Hardy-Weinberg assumptions is not being met and evolution is occurring.
- If the frequency of two alleles in a population meeting the preceding assumptions is the same in subsequent generations then evolution is not occurring.
- The frequency of two alleles, p and q, in a Hardy–Weinberg population is described by $p^2 + 2pq + q^2 = 1$.

20.3 Five Agents of Evolutionary Change
(figure 20.4)

- Allele mutations are the ultimate raw material for evolution.
- Migration of new alleles into a population can be a powerful agent of change.
- Gene flow homogenizes allele frequencies among populations.
- Assortative mating occurs when phenotypically similar individuals mate and may increase the proportion of homozygous individuals.
- Disassortative mating occurs when phenotypically dissimilar individuals mate and may result in an excess of heterozygous individuals.
- A founder effect occurs when a small portion of a population with a different distribution of alleles leaves and inhabits a new area some distance from the parent population.
- A bottleneck effect occurs when a large population is catastrophically reduced in size and the remaining population has a distribution of alleles different from the parent population (figure 20.5).
- For natural selection to occur, genetic variation must exist in a population, variation must result in differential reproductive success—due to increased survival, mating success or fecundity— and variation must be genetically inherited.
- The outcome of natural selection is that future generations are better adapted to their environment.

20.4 Fitness and Its Measurement

Fitness is defined as the reproductive success of a phenotype and may consist of many components.

- Some phenotypes may survive better than others.
- Sexual selection refers to differences between phenotypes in mating success.
- Some phenotypes may have more offspring per mating than others.

20.5 Interactions Among Evolutionary Forces

The amount of genetic variation in a population may reflect a balance between opposing forces.

- Mutation rates are rarely high enough to counter the effects of natural selection.
- Genetic drift may lead to an increase in the frequency of alleles not favored by natural selection.
- Gene flow can spread a beneficial mutation to other populations.
- Gene flow can impede adaptations because of the influx of inferior genes.

20.6 Maintenance of Variation

- Negative frequency-dependent selection favors rare phenotypes.
- Positive frequency-dependent selection tends to eliminate variation.
- Oscillating selection favors different phenotypes at different times because of changing environmental conditions.
- Heterozygote advantage favors individuals with both alleles even if one allele is deleterious when homozygous.

20.7 Selection Acting on Traits Affected by Multiple Genes (figure 20.13)

Selection acts in different ways.

- Disruptive selection eliminates intermediate phenotypes, resulting in two phenotypically distinct extremes.
- Directional selection favors one extreme while progressively eliminating the other extreme.
- Stabilizing selection eliminates both extremes and increases the frequency of a common intermediate type.

20.8 Experimental Studies of Natural Selection

Laboratory and field studies test not only how evolution can occur, but also how rapidly it can occur.

20.9 The Limits of Selection

Although selection is powerful, there are limits to what it can accomplish.

- Alleles that have multiple effects set limits on how much a phenotype can be altered.
- Intense selection pressure may remove genetic variation, the basis for evolution, from a population.
- Gene interactions can affect the fitness of alleles when the phenotypic effect of one allele depends on which allele occurs at a second gene; this is called epistasis. Selection of the first allele may be constrained if the second allele is polymorphic.

SELF TEST

1. Assortative mating
   a. affects genotype frequencies expected under Hardy–Weinberg equilibrium.
   b. affects allele frequencies expected under Hardy–Weinberg equilibrium.
   c. has no effect on the genotypic frequencies expected under Hardy–Weinberg equilibrium because it does not affect the relative proportion of alleles in a population.
   d. increases the frequency of heterozygous individuals above Hardy–Weinberg expectations.

2. In a population of red (dominant allele) or white flowers in Hardy–Weinberg equilibrium, the frequency of red flowers is 91%. What is the frequency of the red allele?
   a. 9%
   b. 30%
   c. 91%
   d. 70%

3. Genetic drift and natural selection can both lead to rapid rates of evolution. However,
   a. genetic drift works fastest in large populations.
   b. only drift leads to adaptation.
   c. natural selection requires genetic drift to produce new variation in populations.
   d. both processes of evolution can be slowed by gene flow.

4. When the environment changes from year to year and different phenotypes have different fitness in different environments
   a. natural selection will operate in a frequency-dependent manner.
   b. the effect of natural selection may oscillate from year to year, favoring alternative phenotypes in different years.
   c. genetic variation is not required to get evolutionary change by natural selection.
   d. none of the above.

5. The frequency of the sickle cell allele (S) is about 0.12 in African locales exposed to malaria but has dropped over about 15 generations to about 0.003 in places where malaria has been eradicated (USA). Given the mechanism for heterozygote advantage how long must malaria be suppressed in Africa before we would expect S to be purged?
   a. It will never be lost.
   b. Less than 15 generations.
   c. More than 15 generations.
   d. Not enough information to predict.

6. What would happen to average birth weight if over the next several years advances in medical technology reduced infant mortality rates of large babies to equal that for intermediate sized babies (see the following figure, red line). Assume that differences in birth weight have a genetic basis.
   a. Over time, average birth weight would only increase.
   b. Over time, average birth weight would only decrease.
   c. Both (a) and (b).
   d. None of the above.

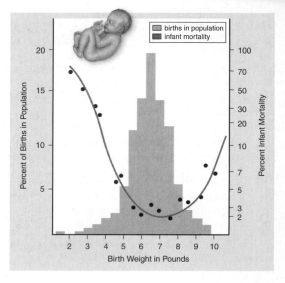

7. Many factors can limit the ability of natural selection to cause evolutionary change, including
   a. a conflict between reproduction and survival as seen in Trinidadian guppies.
   b. lack of genetic variation.
   c. pleiotropy.
   d. all of the above.

8. Stabilizing selection differs from directional selection because
   a. in the former, phenotypic variation is reduced but the average phenotype stays the same, whereas in the latter both the variation and the mean phenotype change.
   b. the former requires genetic variation, but the latter does not.
   c. intermediate phenotypes are favored in directional selection.
   d. none of the above.

9. Founder effects and bottlenecks
   a. are expected only in large populations.
   b. are mechanisms that increase genetic variation in a population.
   c. are two different modes of natural selection.
   d. are forms of genetic drift.

CHALLENGE QUESTIONS

1. In Trinidadian guppies a combination of elegant laboratory and field experiments builds a very compelling case for predator-induced evolutionary changes in color and life history traits. It is still possible, though not likely, that there are other differences between the sites above and below the falls aside from whether predators are present. What additional studies could strengthen the interpretation of the results?

2. Based on a consideration of how strong artificial selection has helped eliminate genetic variation for speed in thoroughbred horses, we are left with the question of why, for many traits like speed (continuous traits), there is usually abundant genetic variation. This is true even for traits we know are under strong selection. Where does genetic variation ultimately come from, and how does the rate of production compare with the strength of natural selection? What other mechanisms can maintain and increase genetic variation in natural populations?

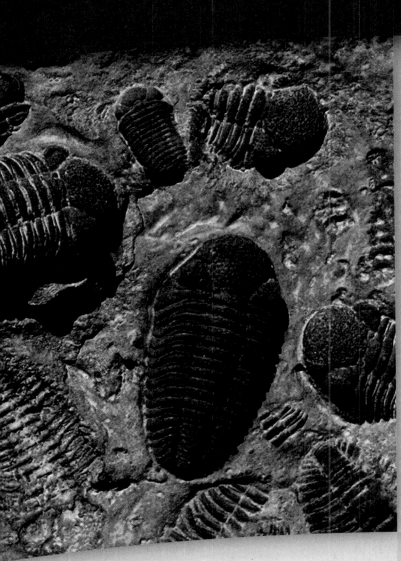

chapter # 21

The Evidence
for Evolution

introduction

AS WE DISCUSSED IN CHAPTER 1, when Darwin proposed his
revolutionary theory of evolution by natural selection, little actual
evidence existed to bolster his case. Instead, Darwin relied on
observations of the natural world, logic, and results obtained by
breeders working with domestic animals. Since his day, however,
the evidence for Darwin's theory has become overwhelming.

The case is built upon two pillars: first, evidence that
natural selection can produce evolutionary change, and
second, evidence from the fossil record that evolution has
occurred. In addition, information from many different areas of
biology—fields as different as anatomy, molecular biology, and
biogeography—is only interpretable scientifically as being the
outcome of evolution.

As you learned in the preceding chapter, a variety of processes can produce evolutionary change. Most evolutionary biologists, however, agree with Darwin's thinking that natural selection is the primary process responsible for evolution. Although we cannot travel back through time, modern-day evidence confirms the power of natural selection as an agent of evolutionary change. This evidence comes from both the field and the laboratory and from both natural and human-altered situations.

Darwin's finches are a classic example of evolution by natural selection. When he visited the Galápagos Islands off the coast of Ecuador in 1835, Darwin collected 31 specimens of finches from three islands. Darwin, not an expert on birds, had trouble identifying the specimens, believing by examining their bills that his collection contained wrens, "gross-beaks," and blackbirds.

Upon Darwin's return to England, ornithologist John Gould informed Darwin that his collection was in fact a closely related group of distinct species, all similar to one another except for their bills. In all, 14 species are now recognized.

Galápagos finches exhibit variation related to food gathering

Figure 21.1 illustrates the diversity of Darwin's finches. The ground finches feed on seeds that they crush in their powerful beaks, species with smaller and narrower bills such as the warbler finch eat insects. Other species include fruit and bud eaters, and species that feed on cactus fruits and the insects they attract; some populations of the sharp-beaked ground finch even include "vampires" that creep up on seabirds and use their sharp beaks to pierce the seabirds' skin and drink their blood. Perhaps most remarkable are the tool users, woodpecker finches that pick up a twig, cactus spine, or leaf stalk, trim it into shape with their bills, and then poke it into dead branches to pry out grubs.

The correspondence between the beaks of the finch species and their food source immediately suggested to Darwin that natural selection had shaped them. In *The Voyage of the Beagle*, Darwin wrote, "Seeing this gradation and diversity of structure in one small, intimately related group of birds, one might really fancy that from an original paucity of birds in this archipelago, one species has been taken and modified for different ends."

Modern research has verified Darwin's selection hypothesis

Darwin's observations suggest that differences among species in beak size and shape have evolved as the species adapted to use different food resources, but can this hypothesis be tested? In chapter 20, you read that the theory of evolution by natural selection requires that three conditions be met:

1. Variation must exist in the population;
2. This variation must lead to differences among individuals in lifetime reproductive success; and

Woodpecker finch (*Cactospiza pallida*)

Large ground finch (*Geospiza magnirostris*)

Cactus finch (*Geospiza scandens*)

Warbler finch (*Certhidea olivacea*)

Vegetarian tree finch (*Platyspiza crassirostris*)

figure 21.1

DARWIN'S FINCHES. These species show differences in bills and feeding habits among Darwin's finches. This diversity arose when an ancestral finch colonized the islands and diversified in habitats lacking other types of small birds. The bills of several species resemble those of different families of birds on the mainland. For example, the warbler finch has a beak very similar to warblers, to which it is not closely related.

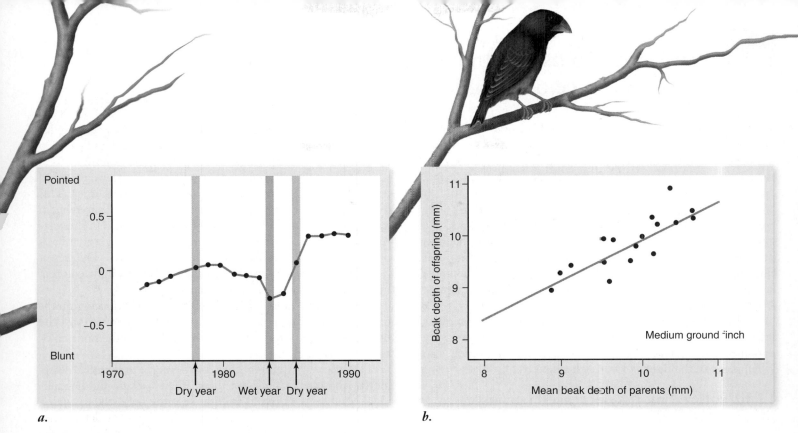

a. *b.*

figure 21.2

EVIDENCE THAT NATURAL SELECTION ALTERS BEAK SHAPE IN *GEOSPIZA FORTIS*. *a.* In dry years, when only large, tough seeds are available, the mean beak depth increases. In wet years, when many small seeds are available, mean beak depth decreases. *b.* Beak depth is inherited from parents to offspring.

inquiry

Suppose a bird with a large bill mates with a bird with a small bill. Would the bills of the pair's offspring tend to be larger or smaller than the bills of offspring from a pair of birds with medium-sized bills?

3. Variation among individuals must be genetically transmitted to the next generation.

The key to successfully testing Darwin's proposal proved to be patience. For more than 30 years, starting in 1973, Peter and Rosemary Grant of Princeton University and their students have studied the medium ground finch, *Geospiza fortis*, on a tiny island in the center of the Galápagos called Daphne Major. These finches feed preferentially on small, tender seeds, produced in abundance by plants in wet years. The birds resort to larger, drier seeds, which are harder to crush, only when small seeds become depleted during long periods of dry weather, when plants produce few seeds.

The Grants quantified beak shape among the medium ground finches of Daphne Major by carefully measuring beak depth (width of beak, from top to bottom, at its base) on individual birds. Measuring many birds every year, they were able to assemble for the first time a detailed portrait of evolution in action. The Grants found that not only did a great deal of variation in beak depth exist among members of the population, but the average beak depth changed from one year to the next in a predictable fashion.

During droughts, plants produced few seeds, and all available small seeds were quickly eaten, leaving large seeds as the major remaining source of food. As a result, birds with deeper, more powerful beaks survived better, because they were better able to break open these large seeds. Consequently, the aver-

age beak depth of birds in the population increased the next year. Then, when normal rains returned, average beak depth decreased to its original size (figure 21.2*a*).

Conversely, in particularly wet years, plants flourished, producing an abundance of small seeds; as a result, small-beaked birds were favored, and beak depth decreased greatly.

Could these changes in beak dimension reflect the action of natural selection? An alternative possibility might be that the changes in beak depth do not reflect changes in gene frequencies, but rather are simply a response to diet—for example, perhaps crushing large seeds causes a growing bird to develop a larger beak.

To rule out this possibility, the Grants measured the relation of parent beak size to offspring beak size, examining many broods over several years. The depth of the beak was passed down faithfully from one generation to the next, regardless of environmental conditions (figure 21.2*b*), suggesting that the differences among individuals in beak size reflect genetic differences, and therefore that the year-to-year changes in average beak depth represent evolutionary change resulting from natural selection.

Among Darwin's finches, natural selection adjusts the shape of the beak in response to the nature of the available food supply. Such adjustments can be seen to occur even today.

Peppered Moths and Industrial Melanism: More Evidence of Selection

When the environment changes, natural selection often may favor different traits in a species. One classic example concerns the peppered moth, *Biston betularia*. Adults come in a range of shades, from light gray with black speckling (hence the name "peppered" moth) to jet black (melanic).

Extensive genetic analysis has shown that the moth's body color is a genetic trait that reflects different alleles of a single gene. Black individuals have a dominant allele, one that was present but very rare in populations before 1850. From that time on, dark individuals increased in frequency in moth populations near industrialized centers until they made up almost 100% of these populations.

Biologists soon noticed that in industrialized regions where the dark moths were common, the tree trunks were darkened almost black by the soot of pollution, which also killed many of the light-colored lichens on tree trunks.

Light-colored moths decreased because of selection by predation

Why did dark moths gain a survival advantage around 1850? In 1896, an amateur moth collector named J. W. Tutt proposed what became the most commonly accepted hypothesis explaining the decline of the light-colored moths. He suggested that peppered forms were more visible to predators on sooty trees that have lost their lichens. Consequently, birds ate the peppered moths resting on the trunks of trees during the day. The black forms, in contrast, were at an advantage because they were camouflaged (figure 21.3).

Although Tutt initially had no evidence, British ecologist Bernard Kettlewell tested the hypothesis in the 1950s by releasing equal numbers of dark and light individuals into two sets of woods: one near heavily polluted Birmingham, and the other in unpolluted Dorset. Kettlewell then set up traps in the woods to see how many of both kinds of moths survived. To evaluate his results, he had marked the released moths with a dot of paint on the underside of their wings, where birds could not see it.

In the polluted area near Birmingham, Kettlewell trapped only 19% of the light moths, but 40% of the dark ones. This indicated that dark moths had a far better chance of surviving in these polluted woods, where tree trunks were dark. In the relatively unpolluted Dorset woods, Kettlewell recovered 12.5% of the light moths but only 6% of the dark ones. This result indicated that where the tree trunks were still light-colored, light moths had a much better chance of survival.

Kettlewell later solidified his argument by placing moths on trees and filming birds looking for food. Sometimes the birds actually passed right over a moth that was the same color as its background.

Kettlewell's finding that birds more frequently detect moths whose color does not match their background has subsequently been confirmed in eight separate field studies, with a variety of experimental designs and corrections for deficiencies in Kettlewell's initial design. These results, combined with the recapture studies, provide strong evidence for the action of natural selection and implicate birds as the agent of selection in the case of the peppered moth.

When environmental conditions reverse, so does selection pressure

In industrialized areas throughout Eurasia and North America, dozens of other species of moths have evolved in the same way as the peppered moth. The term **industrial melanism** refers to the phenomenon in which darker individuals come to predominate over lighter ones. In the second half of the twentieth century, with the widespread implementation of pollution

figure 21.3

TUTT'S HYPOTHESIS EXPLAINING INDUSTRIAL MELANISM. These photographs show preserved specimens of the peppered moth (*Biston betularia*) placed on trees. Tutt proposed that the dark melanic variant of the moth is more visible to predators on unpolluted trees (*left*), while the light "peppered" moth is more visible to predators on bark blackened by industrial pollution (*right*).

controls, the trend toward melanism began reversing for many species of moths throughout the northern continents.

In England, the air pollution promoting industrial melanism began to reverse following enactment of the Clean Air Act in 1956. Beginning in 1959, the *Biston* population at Caldy Common outside Liverpool has been sampled each year. The frequency of the melanic (dark) form has dropped from a high of 93% in 1959 to a low of 15% in 1995 (figure 21.4).

The drop correlates well with a significant drop in air pollution, particularly with a lowering of the levels of sulfur dioxide and suspended particulates, both of which act to darken trees. The drop is consistent with a 15% selective disadvantage acting against moths with the dominant melanic allele.

Interestingly, the same reversal of melanism occurred in the United States. Of 576 peppered moths collected at a field station near Detroit from 1959 to 1961, 515 were melanic, a frequency of 89%. The American Clean Air Act, passed in 1963, led to significant reductions in air pollution. Resampled in 1994, the Detroit field station peppered moth population had only 15% melanic moths (see figure 21.4). The moth populations in Liverpool and Detroit, both part of the same natural experiment, exhibit strong evidence for natural selection.

The agent of selection may be difficult to pin down

Although the evidence for natural selection in the case of the peppered moth is strong, Tutt's hypothesis about the agent of selection is currently being reevaluated. Researchers have noted that the recent selection against melanism does not appear to correlate with changes in tree lichens.

At Caldy Common, the light form of the peppered moth began to increase in frequency long before lichens began to reappear on the trees. At the Detroit field station, the lichens never changed significantly as the dark moths first became dominant and then declined over the last 30 years. In fact, investigators have not been able to find peppered moths on Detroit trees at all, whether covered with lichens or not. Some evidence suggests the moths rest on leaves in the treetops during the day, but no one is sure. Could poisoning by pollution rather than predation by birds be the agent of natural selection on the moths? Perhaps—but to date, only predation by birds is backed by experimental evidence.

Researchers supporting the bird predation hypothesis point out that a bird's ability to detect moths may depend less on the presence or absence of lichens, and more on other ways in which the environment is darkened by industrial pollution. Pollution tends to cover all objects in the environment with a fine layer of particulate dust, which tends to decrease how much

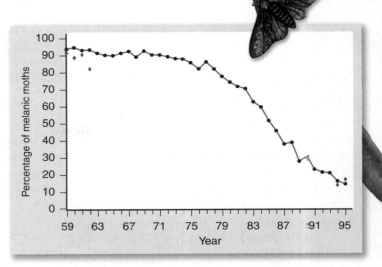

figure 21.4

SELECTION AGAINST MELANISM. The circles indicate the frequency of melanic *Biston betularia* moths at Caldy Common in England, sampled continuously from 1959 to 1995. Diamonds indicate frequencies of melanic *B. betularia* in Michigan from 1959 to 1962 and from 1994 to 1995.

inquiry

What can you conclude from the fact that the frequency of melanic moths decreased to the same degree in the two locations?

light surfaces reflect. In addition, pollution has a particularly severe effect on birch trees, which are light in color. Both effects would tend to make the environment darker, and thus would favor darker moths by protecting them from predation by birds.

Despite this uncertainty over the agent of selection, the overall pattern is clear. Kettlewell's experiments established indisputably that selection favors dark moths in polluted habitats and light moths in pristine areas. The increase and subsequent decrease in the frequency of melanic moths, correlated with levels of pollution independently on two continents, demonstrates clearly that this selection drives evolutionary change.

The current reconsideration of the agent of natural selection illustrates well the way in which scientific progress is achieved: Hypotheses, such as Tutt's, are put forth and then tested. If rejected, new hypotheses are formulated, and the process begins anew.

> **Natural selection has favored the dark form of the peppered moth in areas subject to severe air pollution, perhaps because on darkened trees they are less easily seen by moth-eating birds. Selection has in turn favored the light form as pollution has abated.**

21.3 Artificial Selection: Human-Initiated Change

Humans have imposed selection upon plants and animals since the dawn of civilization. Just as in natural selection, artificial selection operates by favoring individuals with certain phenotypic traits, allowing them to reproduce and pass their genes on to the next generation. Assuming that phenotypic differences are genetically determined, this directional selection should lead to evolutionary change, and indeed it has.

Artificial selection, imposed in laboratory experiments, agriculture, and the domestication process, has produced substantial change in almost every case in which it has been applied. This success is strong proof that selection is an effective evolutionary process.

Experimental selection produces changes in populations

With the rise of genetics as a field of science in the 1920s and 1930s, researchers began conducting experiments to test the hypothesis that selection can produce evolutionary change. A favorite subject was the laboratory fruit fly, *Drosophila melanogaster*. Geneticists have imposed selection on just about every conceivable aspect of the fruit fly—including body size, eye color, growth rate, life span, and exploratory behavior—with a consistent result: Selection for a trait leads to strong and predictable evolutionary response.

In one classic experiment, scientists selected for fruit flies with many bristles (stiff, hairlike structures) on their abdomens. At the start of the experiment, the average number of bristles was 9.5. Each generation, scientists picked out the 20% of the population with the greatest number of bristles and allowed them to reproduce, thus establishing the next generation. After 86 generations of this directional selection, the average number of bristles had quadrupled, to nearly 40! In another experiment, fruit flies in one population were selected for high numbers of bristles, while fruit flies in the other cage were selected for low numbers of bristles. Within 35 generations, the populations did not overlap at all in range of variation (figure 21.5).

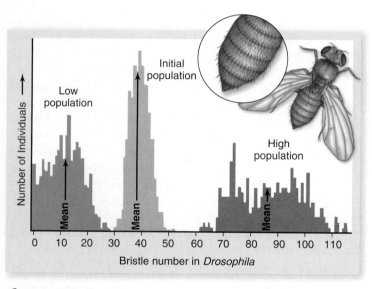

figure 21.5

ARTIFICIAL SELECTION IN THE LABORATORY. In this experiment, one population of *Drosophila* was selected for low numbers of bristles and the other for high numbers. Note that not only did the means of the populations change greatly in 35 generations, but also all individuals in both experimental populations lie outside the range of the initial population. Selection can move a population beyond its original range of variation because mutation and recombination alternately introduce new variation into populations.

 inquiry

What would happen if, within a population, both small and large individuals were allowed to breed, but middle-sized ones were not?

Similar experiments have been conducted on a wide variety of other laboratory organisms. For example, by selecting for rats that were resistant to tooth decay, in less than 20 generations scientists were able to increase the average time for onset of decay from barely over 100 days to greater than 500 days.

Agricultural selection has led to extensive modification of crops and livestock

Familiar livestock, such as cattle and pigs, and crops, such as corn and strawberries, are greatly different from their wild ancestors (figure 21.6). These differences have resulted from generations of human selection for desirable traits, such as greater milk production and larger corn ear size.

An experiment with corn demonstrates the ability of artificial selection to rapidly produce major change in crop plants. In 1896, agricultural scientists began selecting for the oil content of corn kernels, which initially was 4.5%. Just as in the fruit fly experiments, the top 20% of all individuals were allowed to reproduce. By 1986, at which time 90 generations had passed, average oil content of the corn kernels had increased approximately 450%.

Domesticated breeds have arisen from artificial selection

Human-imposed selection has produced a great variety of breeds of cats, dogs (figure 21.7), pigeons, and other domestic animals. In some cases, breeds have been developed for particular purposes. Greyhound dogs, for example, resulted from selection for maximal running ability, resulting in an animal with long legs, a long tail for balance, an arched back to increase stride length, and great muscle mass. By contrast, the odd proportions of the ungainly dachshund resulted from selection for dogs that could enter narrow holes in pursuit of badgers. In other cases, varieties have been selected primarily for their appearance, such as the many colorful breeds of pigeons or cats.

Domestication also has led to unintentional selection for some traits. In recent years, as part of an attempt to domesticate

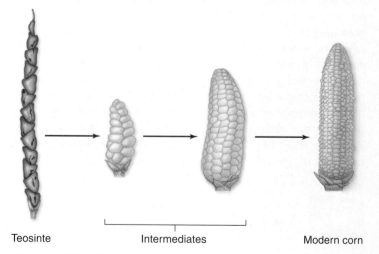

Teosinte   Intermediates   Modern corn

figure 21.6

CORN LOOKS VERY DIFFERENT FROM ITS ANCESTOR. Teosinte, which can be found today in a remote part of Mexico, is very similar to the ancestor of modern corn. Artificial selection has transformed it into the form we know today.

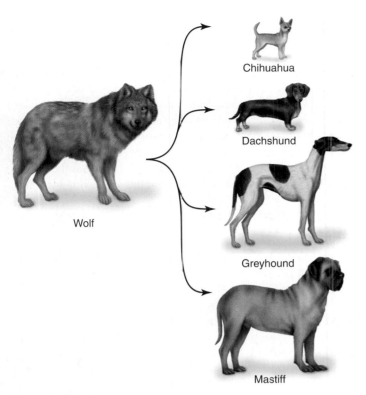

figure 21.7

BREEDS OF DOGS. The differences among dog breeds are greater than the differences displayed among wild species of canids.

Chihuahua

Dachshund

Greyhound

Mastiff

Wolf

figure 21.8

DOMESTICATED FOXES. After 40 years of selectively breeding the tamest individuals, artificial selection has produced silver foxes that are not only as friendly as domestic dogs, but also exhibit many physical traits seen in dog breeds.

the silver fox, Russian scientists have chosen the most docile animals in each generation and allowed them to reproduce. Within 40 years, most foxes were exceptionally tame, not only allowing themselves to be petted, but also whimpering to get attention and sniffing and licking their caretakers (figure 21.8). In many respects, they had become no different from domestic dogs.

It was not only their behavior that changed, however. These foxes also began to exhibit other traits seen in some dog breeds, such as different color patterns, floppy ears, curled tails, and shorter legs and tails. Presumably, the genes responsible for docile behavior either affect these traits as well or are closely linked to the genes for these other traits (the phenomena of pleiotropy and linkage, which are discussed in chapters 12 and 13).

Can selection produce major evolutionary changes?

Given that we can observe the results of selection operating over a relatively short time, most scientists think that natural selection is the process responsible for the evolutionary changes documented in the fossil record. Some critics of evolution accept that selection can lead to changes within a species, but contend that such changes are relatively minor in scope and not equivalent to the substantial changes documented in the fossil record. In other words, it is one thing to change the number of bristles on a fruit fly or the size of an ear of corn, and quite another to produce an entirely new species.

This argument does not fully appreciate the extent of change produced by artificial selection. Consider, for example, the existing breeds of dogs, all of which have been produced since wolves were first domesticated, perhaps 10,000 years ago. If the various dog breeds did not exist and a paleontologist found fossils of animals similar to dachshunds, greyhounds, mastiffs, and chihuahuas, there is no question that they would be considered different species. Indeed, the differences in size and shape exhibited by these breeds are greater than those between members of different genera in the family Canidae—such as coyotes, jackals, foxes, and wolves—which have been evolving separately for 5 to 10 million years. Consequently, the claim that artificial selection produces only minor changes is clearly incorrect. If selection operating over a period of only 10,000 years can produce such substantial differences, it should be powerful enough, over the course of many millions of years, to produce the diversity of life we see around us today.

Artificial selection often leads to rapid and substantial results over a short time, thus demonstrating the power of selection to produce major evolutionary change.

21.4 Fossil Evidence of Evolution

The most direct evidence that evolution has occurred is found in the fossil record. Today we have a far more complete understanding of this record than was available in Darwin's time.

Fossils are the preserved remains of once-living organisms. They include specimens preserved in amber, Siberian permafrost, and dry caves, as well as the more common fossils preserved as rocks.

Rock fossils are created when three events occur. First, the organism must become buried in sediment; then, the calcium in bone or other hard tissue must mineralize; and finally, the surrounding sediment must eventually harden to form rock.

The process of fossilization probably occurs rarely. Usually, animal or plant remains decay or are scavenged before the process can begin. In addition, many fossils occur in rocks that are inaccessible to scientists. When they do become available, they are often destroyed by erosion and other natural processes before they can be collected. As a result, only a very small fraction of the species that have ever existed (estimated by some to be as many as 500 million) are known from fossils. Nonetheless, the fossils that have been discovered are sufficient to provide detailed information on the course of evolution through time.

The age of fossils is estimated by rates of radioactive decay

By dating the rocks in which fossils occur, we can get an accurate idea of how old the fossils are. In Darwin's day, rocks were dated by their position with respect to one another (*relative dating*); rocks in deeper strata are generally older. Knowing the relative positions of sedimentary rocks and the rates of erosion of different kinds of sedimentary rocks in different environments, geologists of the nineteenth century derived a fairly accurate idea of the relative ages of rocks.

Today, geologists take advantage of radioactive decay to age rocks (*absolute dating*). Many types of rock, such as the igneous rocks formed when lava cools, contain radioactive elements such as uranium-238. These isotopes transform at a precisely known rate into nonradioactive forms. For example, for U^{238} the *half-life* (that is, the amount of time needed for one-half of the original amount to be transformed) is 4.5 billion years. Once a rock is formed, no additional radioactive isotopes are added. Therefore, by measuring the ratio of the radioactive isotope to its derivative, "daughter" isotope (figure 21.9), geologists can determine the age of the rock. If a fossil is found between two layers of rock, each of which can be dated, then the age at which the fossil formed can be determined.

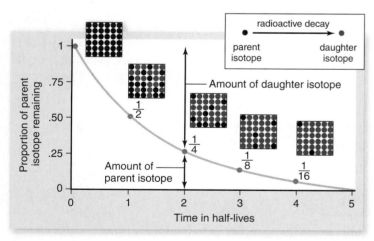

figure 21.9

RADIOACTIVE DECAY. Radioactive elements decay at a known rate, called their half-life. After one half-life, one-half of the original amount of parent isotope has transformed into a nonradioactive daughter isotope. After each successive half-life, one-half of the remaining amount of parent isotope is transformed.

Fossils present a history of evolutionary change

When fossils are arrayed according to their age (figure 21.10), from oldest to youngest, they often provide evidence of successive evolutionary change. At the largest scale, the fossil record documents the course of life through time, from the origin of first prokaryotic and then eukaryotic organisms, through the evolution of fishes, the rise of land-dwelling organisms, the reign of the dinosaurs, and on to the origin of humans. In addition, the fossil record shows the waxing and waning of biological diversity through time, such as the periodic mass extinctions that have reduced the number of living species.

Fossils document evolutionary transition

Given the low likelihood of fossil preservation and recovery, it is not surprising that there are gaps in the fossil record. Nonetheless, intermediate forms are often available to illustrate how the major transitions in life occurred.

Undoubtedly the most famous of these is the oldest known bird, *Archaeopteryx* (meaning "ancient feather") which lived around 165 million years ago (figure 21.11). This specimen is clearly intermediate between birds and dinosaurs. Its feathers, similar in many respects to those of birds today, clearly reveal

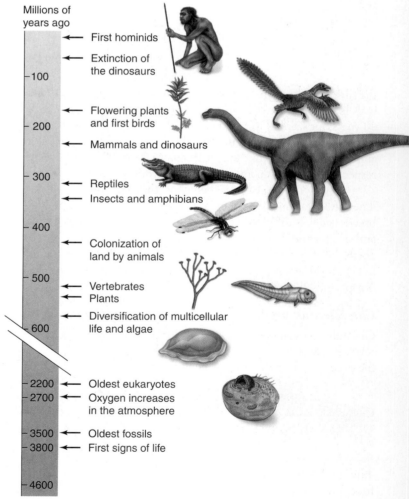

figure 21.10

HISTORY OF EVOLUTIONARY CHANGE AS REVEALED BY THE FOSSIL RECORD.

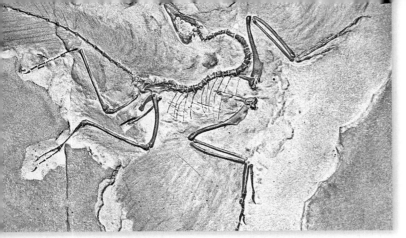

figure 21.11

FOSSIL OF *ARCHAEOPTERYX*, THE FIRST BIRD. The remarkable preservation of this specimen reveals soft parts usually not preserved in fossils; the presence of feathers makes clear that *Archaeopteryx* was a bird, despite the presence of many dinosaurian traits.

that it is a bird. Nonetheless, in many other respects—for example, possession of teeth, a bony tail, and other anatomical characteristics—it is indistinguishable from some carnivorous dinosaurs. Indeed, it is so similar to these dinosaurs that several specimens lacking preserved feathers were misidentified as dinosaurs and lay in the wrong natural history museum cabinet for several decades before the mistake was discovered!

Archaeopteryx reveals a pattern commonly seen in intermediate fossils—rather than being intermediate in every trait, such fossils usually exhibit some traits like their ancestors and others like their descendants. In other words, traits evolve at different rates and different times; expecting an intermediate form to be intermediate in every trait would not be correct.

The first *Archaeopteryx* fossil was discovered in 1859, the year Darwin published *On the Origin of Species*. Since then, paleontologists have continued to fill in the gaps in the fossil record. Today, the fossil record is far more complete, particularly among the vertebrates; fossils have been found linking all the major groups.

Recent years have seen spectacular discoveries, closing some of the major remaining gaps in our understanding of vertebrate evolution. For example, a four-legged aquatic mammal was discovered only recently that provides important insights concerning the evolution of whales and dolphins from land-dwelling, hoofed ancestors (figure 21.12). Similarly, a fossil snake with legs has shed light on the evolution of snakes, which are descended from lizards that gradually became more and more elongated with the simultaneous reduction and eventual disappearance of the limbs. In chapter 35, we discuss the most recent such discovery, *Tiktaalik*, a species that bridged the gap between fish and the first amphibians.

On a finer scale, evolutionary change within some types of animals is known in exceptional detail. For example, about 200 million years ago, oysters underwent a change from small, curved shells to larger, flatter ones, with progressively flatter fossils seen in the fossil record over a period of 12 million years. A host of other examples illustrate similar records of successive change. The demonstration of this successive change is one of the strongest lines of evidence that evolution has occurred.

The evolution of horses is a prime example of evidence from fossils

One of the most studied cases in the fossil record concerns the evolution of horses. Modern-day members of the family Equidae include horses, zebras, donkeys, and asses, all of which are large, long-legged, fast-running animals adapted to living on open grasslands. These species, all classified in the genus *Equus*, are the last living descendants of a long lineage that has produced 34 genera since its origin in the Eocene period, approximately 55 million years ago. Examination of these fossils has provided

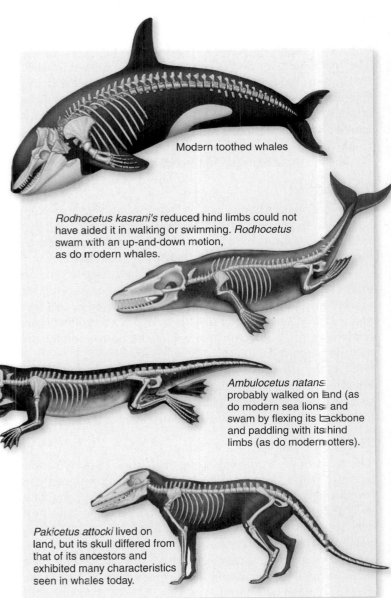

Modern toothed whales

Rodhocetus kasrani's reduced hind limbs could not have aided it in walking or swimming. *Rodhocetus* swam with an up-and-down motion, as do modern whales.

Ambulocetus natans probably walked on land (as do modern sea lions and swam by flexing its backbone and paddling with its hind limbs (as do modern otters).

Pakicetus attocki lived on land, but its skull differed from that of its ancestors and exhibited many characteristics seen in whales today.

figure 21.12

WHALE "MISSING LINKS." The recent discoveries of *Ambulocetus*, *Rodhocetus*, and *Pakicetus* have filled in the gaps between whales and their hoofed mammal ancestors. The features of *Pakicetus* illustrate that intermediate forms are not intermediate in all characteristics; rather, some traits evolve before others. In the case of the evolution of whales, changes occurred in the skull prior to evolutionary modification of the limbs. All three fossil forms occurred in the Eocene period, 45–55 million years ago.

a particularly well-documented case of how evolution has proceeded through adaptation to changing environments.

The first horse

The earliest known members of the horse family, species in the genus *Hyracotherium*, didn't look much like modern-day horses at all. Small, with short legs and broad feet, these species occurred in wooded habitats, where they probably browsed on leaves and herbs and escaped predators by dodging through openings in the forest vegetation. The evolutionary path from these diminutive creatures to the workhorses of today has involved changes in a variety of traits, including size, toe reduction, and tooth size and shape (figure 21.13).

Changes in size

The first species of horses were the size of dogs or smaller. By contrast, modern equids can weigh more than a half ton. Examination of the fossil record reveals that horses changed little in size for their first 30 million years, but since then, a number of different lineages have exhibited rapid and substantial increases. However, trends toward decreased size were also exhibited in some branches of the equid evolutionary tree.

Toe reduction

The feet of modern horses have a single toe enclosed in a tough, bony hoof. By contrast, *Hyracotherium* had four toes on its front feet and three on its hind feet. Rather than hooves, these toes were encased in fleshy pads like those of dogs and cats.

Examination of fossils clearly shows the transition through time: a general increase in length of the central toe, development of the bony hoof, and reduction and loss of the other toes (see figure 21.13). As with body size, these trends occurred concurrently on several different branches of the horse evolutionary tree and were not exhibited by all lineages.

At the same time as toe reduction was occurring, these horse lineages were evolving changes in the length and skeletal structure of their limbs, leading to animals capable of running long distances at high speeds.

Tooth size and shape

The teeth of *Hyracotherium* were small and relatively simple in shape. Through time, horse teeth have increased greatly in length and have developed a complex pattern of ridges on their molars and premolars. The effect of these changes is to produce teeth better capable of chewing tough and gritty vegetation, such as grass, which tends to wear teeth down.

Accompanying these changes have been alterations in the shape of the skull that strengthened the skull to withstand the stresses imposed by continual chewing. As with body size, evolutionary change has not been constant through time. Rather, much of the change in tooth shape has occurred within the past 20 million years and changes have not been constant among all horse lineages.

All of these changes may be understood as adaptations to changing global climates. In particular, during the late Miocene and early Oligocene epochs (approximately 20 to 25 million years ago), grasslands became widespread in North America, where much of horse evolution occurred. As horses adapted to these habitats, high-speed locomotion probably became more

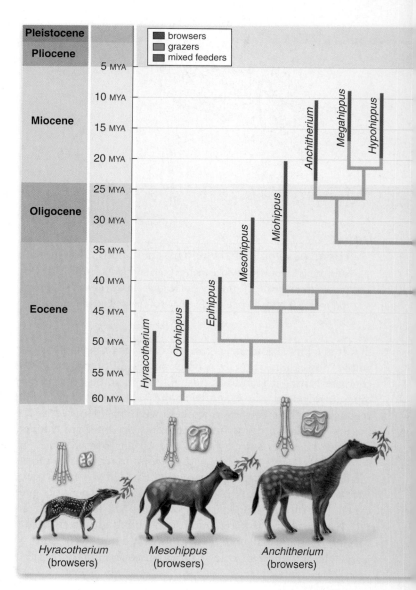

figure 21.13

EVOLUTIONARY CHANGE IN BODY SIZE OF HORSES. Lines indicate evolutionary relationships of the horse family. Horse evolution is more like a bush than a single-trunk tree; diversity was much greater in the past than it is today. In general, there has been a trend toward larger size, more complex molar teeth, and fewer toes, but this trend has exceptions. For example, a relatively recent form, *Nannippus*, evolved in the opposite direction, toward decreased size.

 inquiry

Why might the evolutionary line leading to Nannippus have experienced an evolutionary decrease in body size?

important to escape predators. By contrast, the greater flexibility provided by multiple toes and shorter limbs, which was advantageous for ducking through complex forest vegetation, was no longer beneficial. At the same time, horses were eating grasses and other vegetation that contained more grit and other hard substances, thus favoring teeth and skulls better suited for withstanding such materials.

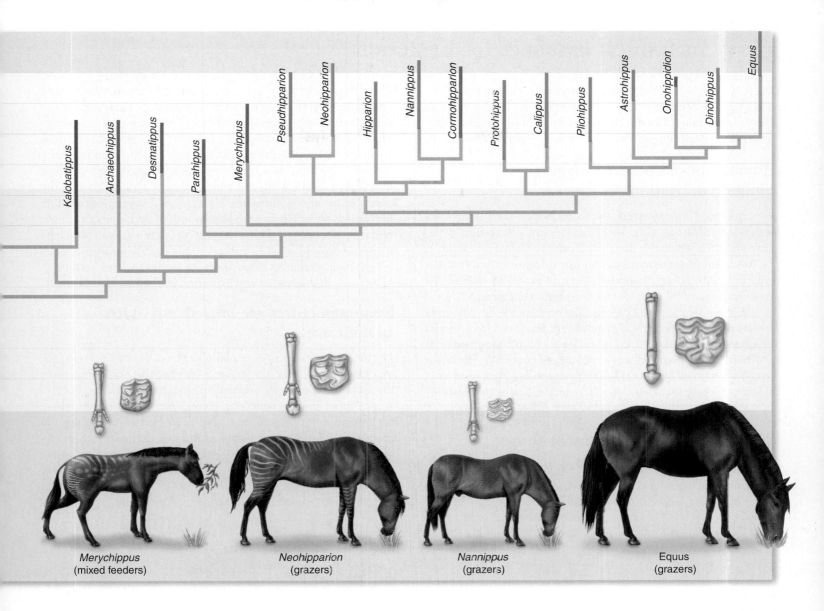

Merychippus
(mixed feeders)

Neohipparion
(grazers)

Nannippus
(grazers)

Equus
(grazers)

Evolutionary trends

For many years, horse evolution was held up as an example of constant evolutionary change through time. Some even saw in the record of horse evolution evidence for a progressive, guiding force, consistently pushing evolution in a single direction. We now know that such views are misguided, and that the course of evolutionary change over millions of years is rarely so simple.

Rather, the fossils demonstrate that even though overall trends have been evident in a variety of characteristics, evolutionary change has been far from constant and uniform through time. Instead, rates of evolution have varied widely, with long periods of little observable change and some periods of great change. Moreover, when changes happen, they often occur simultaneously in different lineages of the horse evolutionary tree.

Finally, even when a trend exists, exceptions, such as the evolutionary decrease in body size exhibited by some lineages, are not uncommon. These patterns are usually discovered for any group of plants and animals for which we have an extensive fossil record, as you will see when we discuss human evolution in chapter 35.

Horse diversity

One reason that horse evolution was originally conceived of as linear through time may be that modern horse diversity is relatively limited. For this reason it is easy to mentally picture a straight line from *Hyracotherium* to modern-day *Equus*. But today's limited horse diversity—only one surviving genus—is unusual. In fact, at the peak of horse diversity in the Miocene epoch, 13 genera of horses could be found in North America alone. These species differed in body size and in a wide variety of other characteristics. Presumably, they lived in different habitats and exhibited different dietary preferences. Had this diversity existed to modern times, early evolutionary biologists presumably would have had a different outlook on horse evolution.

The fossil record provides a clear record of the major evolutionary transitions that have occurred through time. The extensive fossil record for horses provides a detailed view of the evolutionary diversification of this group, from small forest dwellers to large and fast modern grassland species.

Anatomical Evidence for Evolution

Much of the power of the theory of evolution is its ability to provide a sensible framework for understanding the diversity of life. Many observations from throughout biology simply cannot be understood in any meaningful way except as a result of evolution.

Homologous structures suggest common derivation

As vertebrates have evolved, the same bones have sometimes been put to different uses. Yet the bones are still recognizable, their presence betraying their evolutionary past. For example, the forelimbs of vertebrates are all **homologous structures**—structures with different appearances and functions that all derived from the same body part in a common ancestor.

You can see in figure 21.14 how the bones of the forelimb have been modified in different ways for different mammals. Why should these very different structures be composed of the same bones? If evolution had not occurred, this would indeed be a riddle. But when we consider that all of these animals are descended from a common ancestor, it is easy to understand that natural selection has modified the same initial starting blocks to serve very different purposes.

Early embryonic development shows similarities in some groups

Some of the strongest anatomical evidence supporting evolution comes from comparisons of how organisms develop. Embryos of different types of vertebrates, for example, often are similar early on, but become more different as they develop. Early in their development vertebrate embryos possess pharyngeal pouches (figure 21.15), which develop into different structures. In humans, for example, they become various glands and ducts; in fish, they turn into gill slits. At a later stage, every human embryo has a long bony tail, the vestige of which we carry to adulthood as the coccyx at the end of our spine. Human fetuses even possess a fine fur (called *lanugo*) during the fifth month of development. These relict developmental forms suggest strongly that our development has evolved, with new instructions modifying ancestral developmental patterns. We will return to the topic of vertebrate embryonic development and evolution in chapter 53.

Some structures are imperfectly suited to their use

Because natural selection can only work on the variation present in a population, it should not be surprising that some organisms do not appear perfectly adapted to their environments. For example, most animals with long necks have many neck vertebrae for enhanced flexibility: Geese have up to 25, and plesiosaurs, the long-necked reptiles that patrolled the seas during the age of dinosaurs, had as many as 76. By contrast, almost all mammals have only 7 neck vertebrae, even the giraffe. In the absence of variation in vertebrae number, selection led to an evolutionary increase in vertebra size to produce the long neck of the giraffe.

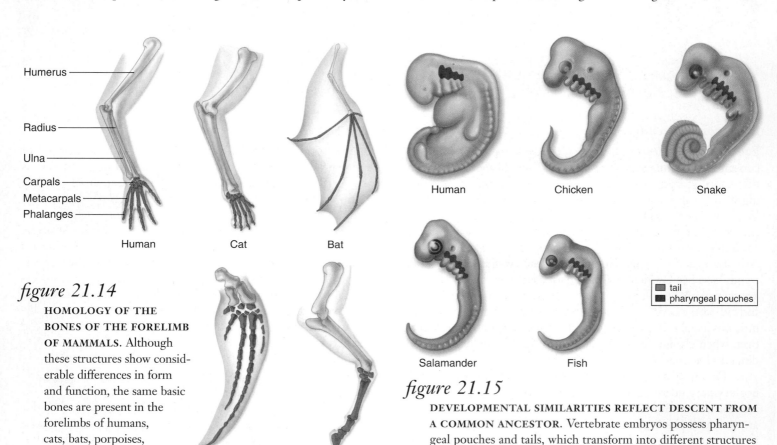

figure 21.14

HOMOLOGY OF THE BONES OF THE FORELIMB OF MAMMALS. Although these structures show considerable differences in form and function, the same basic bones are present in the forelimbs of humans, cats, bats, porpoises, and horses.

figure 21.15

DEVELOPMENTAL SIMILARITIES REFLECT DESCENT FROM A COMMON ANCESTOR. Vertebrate embryos possess pharyngeal pouches and tails, which transform into different structures during development or sometimes disappear entirely.

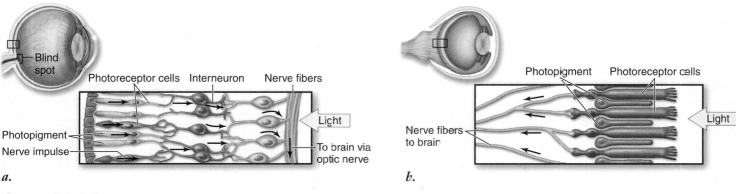

figure 21.16

THE EYES OF VERTEBRATES AND MOLLUSKS. ***a.*** Photoreceptors of vertebrates point backward, whereas (***b***) those of mollusks face forward. As a result, vertebrate nerve fibers pass in front of the photoreceptor; and where they bundle together and exit the eye, a blind spot is created. Mollusks' eyes have neither of these problems.

An excellent example of an imperfect design is the eye of vertebrate animals, in which the photoreceptors face backward, toward the wall of the eye (figure 21.16*a*). As a result, the nerve fibers extend not backward, toward the brain, but forward into the eye chamber, where they slightly obstruct light. Moreover, these fibers bundle together to form the optic nerve, which exits through a hole at the back of the eye, creating a blind spot.

By contrast, the eye of mollusks—such as squid and octopuses—are more optimally designed: The photoreceptors face forward, and the nerve fibers exit at the back, neither obstructing light nor creating a blind spot (figure 21.16*b*).

Such examples illustrate that natural selection is like a tinkerer, working with whatever material is available to craft a workable solution, rather than like an engineer, who can design and build the best possible structure for a given task. Workable, but imperfect, structures such as the vertebrate eye are an expected outcome of evolution by natural selection.

Vestigial structures can be explained as holdovers from the past

Many organisms possess **vestigial structures** that have no apparent function, but resemble structures their ancestors possessed. Humans, for example, possess a complete set of muscles for wiggling their ears, just like many other mammals do. Although these muscles allow other mammals to move their ears to pinpoint sounds such as the movements or growl of a predator, they have little purpose in humans other than amusement.

As other examples, boa constrictors have hip bones and rudimentary hind legs. Manatees (a type of aquatic mammal often referred to as "sea cows") have fingernails on their fins, which evolved from legs. Blind cave fish, which never see the light of day, have small, nonfunctional eyes. Figure 21.17 illustrates the skeleton of a baleen whale, which contains pelvic bones, as other mammal skeletons do, even though such bones serve no known function in the whale.

The human vermiform appendix is apparently vestigial; it represents the degenerate terminal part of the cecum, the blind pouch or sac in which the large intestine begins. In other mammals, such as mice, the cecum is the largest part of the large intestine and functions in storage—usually of bulk cellulose in herbivores. Although some functions have been suggested, it is difficult to assign any current function to the human vermiform

appendix. In many respects, it can be a dangerous organ: appendicitis, which results from infection of the appendix, can be fatal.

It is difficult to understand vestigial structures such as these as anything other than evolutionary relics, holdovers from the past. However, the existence of vestigial structures argues strongly for the common ancestry of the members of the groups that share them, regardless of how different those groups have subsequently become.

All of these anatomical lines of evidence—homology, development, and imperfect and vestigial structures—are readily understandable as a result of descent with modification, that is, evolution.

Comparisons of the anatomy of different living animals often reveal evidence of shared ancestry. In some instances, the same organ has evolved to carry out different functions; in others, an organ loses its function altogether.

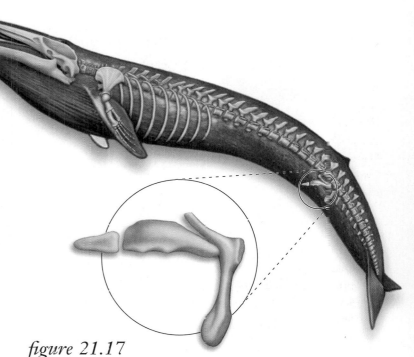

figure 21.17

VESTIGIAL STRUCTURES. The skeleton of a whale reveals the presence of pelvic bones. These bones resemble those of other mammals, but are only weakly developed in the whale and have no apparent function.

Convergent Evolution and the Biogeographical Record

Biogeography, the study of the geographic distribution of species, reveals that different geographical areas sometimes exhibit groups of plants and animals of strikingly similar appearance, even though the organisms may be only distantly related.

It is difficult to explain so many similarities as the result of coincidence. Instead, natural selection appears to have favored parallel evolutionary adaptations in similar environments. Because selection in these instances has tended to favor changes that made the two groups more alike, their phenotypes have converged. This form of evolutionary change is referred to as **convergent evolution.**

Marsupials and placentals demonstrate convergence

In the best known case of convergent evolution, two major groups of mammals—marsupials and placentals—have evolved in very similar ways in different parts of the world. Marsupials are a group in which the young are born in a very immature condition and held in a pouch until they are ready to emerge into the outside world. In placentals, by contrast, offspring are not born until they can safely survive in the external environment (with varying degrees of parental care).

Australia separated from the other continents more than 70 million years ago; at that time, both marsupials and placental mammals had evolved, but in different places. In particular, only marsupials occurred in Australia. As a result of this continental separation, the only placental mammals in Australia today are bats and a few colonizing rodents (which arrived relatively recently), and Australia is dominated by marsupials.

What are the Australian marsupials like? To an astonishing degree, they resemble the placental mammals living today on the other continents (figure 21.18). The similarity between some individual members of these two sets of mammals argues strongly that they are the result of convergent evolution, similar forms having evolved in different, isolated areas because of similar selective pressures in similar environments.

Convergent evolution is a widespread phenomenon

When species interact with the environment in similar ways, they often are exposed to similar selective pressures, and they therefore frequently develop the same evolutionary adaptations. Consider, for example, fast-moving marine predators (figure 21.19). The hydrodynamics of moving through water require a streamlined body shape to minimize friction. It is no coincidence that dolphins, sharks, and tuna—among the fastest of marine species—have all evolved to have the same basic shape. We can speculate as well that ichthyosaurs—marine reptiles that lived during the Age of the Dinosaurs—exhibited a similar lifestyle.

Island trees exhibit a similar phenomenon. Most islands are covered by trees (or were until the arrival of humans). Careful inspection of these trees, however, reveals that they are not closely related to the trees with which we are familiar. Although they have all the characteristics of trees, such as being tall and having a tough outer covering, in many cases island trees are members of plant families that elsewhere exist only as flowers,

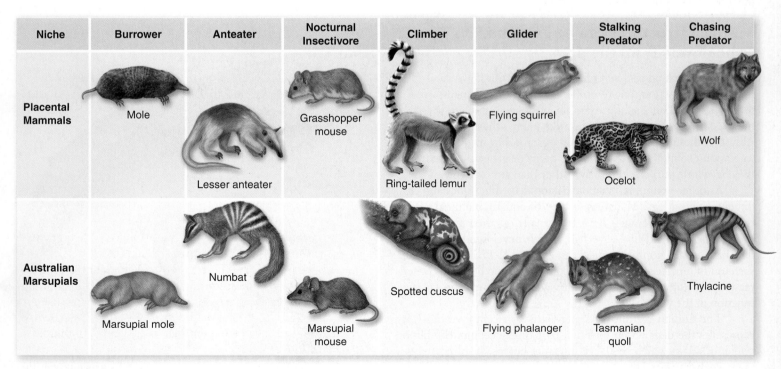

Niche	Burrower	Anteater	Nocturnal Insectivore	Climber	Glider	Stalking Predator	Chasing Predator
Placental Mammals	Mole	Lesser anteater	Grasshopper mouse	Ring-tailed lemur	Flying squirrel	Ocelot	Wolf
Australian Marsupials	Marsupial mole	Numbat	Marsupial mouse	Spotted cuscus	Flying phalanger	Tasmanian quoll	Thylacine

figure 21.18

CONVERGENT EVOLUTION. Many marsupial species in Australia resemble placental mammals occupying similar ecological niches elsewhere in the rest of the world. Marsupials evolved in isolation after Australia separated from other continents.

figure 21.19

CONVERGENCE AMONG FAST-SWIMMING PREDATORS. Fast movement through water requires a streamlined body form, which has evolved numerous times.

shrubs, or other small bushes. For example, on many islands, the native trees are members of the sunflower family.

Why do these plants evolve into trees on islands? Probably because seeds from trees rarely make it to isolated islands. As a result, those species that do manage to colonize distant islands face an empty ecological landscape upon arrival. In the absence of other treelike plants, natural selection often would favor individual plants that could capture the most sunlight for photosynthesis, and the result is the evolution of similar treelike forms on islands throughout the world.

Biogeographical studies provide further evidence of evolution

Darwin made several important observations during his voyage around the world. He noted that islands often are missing plants and animals common on continents, such as frogs and land mammals. Accidental human introductions have proven that these species can survive if they are released on islands, so lack of suitable habitat is not the cause. In addition, those species that are present on islands often have diverged from their continental relatives and sometimes—as with Darwin's finches and the island trees just discussed—occupy ecological niches used by other species on continents. Lastly, island species usually are more closely related to species on nearby continents, even though the environment there is often not very similar to that on island.

Darwin deduced the explanation for these phenomena. Many islands have never been connected to continental areas. The species that occur there arrived by dispersing across the water; dispersal from nearby areas is more likely than from more distant sources, though long-distance colonization does occur occasionally. Some species, those that can fly, float, or swim are more likely to get to the island than others. Some, like frogs, are particularly vulnerable to dehydration in saltwater and have almost no chance of island colonization.

The absence of some types of plants and animals provides opportunity to those that do arrive; as a result colonizers often evolve into many species exhibiting great ecological and morphological diversity.

Geographic proximity is not always a good predictor of evolutionary relationships, however. Earth's continents have not always been where they are today; rather, continents are constantly moving because of the process known as *continental drift*. Although the pace is slow, on the order of several centimeters per year, the configuration of the continents can, and has, changed considerably over geologic time. As a result, closely related species that at one time occurred near each other may now be separated by thousands of miles. Many examples occur on the southern continents, which were last united as the supercontinent Gondwana more than 100 million years ago. One such example is the southern beech tree, which is found in Chile, and also in Australia and New Zealand. In cases such as this, Earth's history and evolutionary history must be considered jointly to make sense of geographical distributions.

> Convergence is the evolution of similar forms in different lineages when exposed to the same selective pressures. The biogeographical distribution of species often reflects the outcome of evolutionary diversification.

21.7 Darwin's Critics

In the century since he proposed it, Darwin's theory of evolution by natural selection has become nearly universally accepted by biologists, but has been a source of controversy among some members of the general public. Here we discuss seven principle objections that critics raise to the teaching of evolution as biological fact, along with some answers that scientists present in response.

1. **Evolution is not solidly demonstrated.** "Evolution is just a theory," Darwin's critics point out, as though *theory* meant a lack of knowledge, or some kind of guess.

   Scientists, however, use the word *theory* in a very different sense than the general public does. Theories are the solid ground of science—that about which we are most certain. Few of us doubt the theory of gravity because it is "just a theory."

2. **There are no fossil intermediates.** "No one ever saw a fin on the way to becoming a leg," critics claim, pointing to the many gaps in the fossil record in Darwin's day.

   Since that time, however, many fossil intermediates in vertebrate evolution have indeed been found. A clear line of fossils now traces the transition between hoofed mammals and whales, between reptiles and mammals, between dinosaurs and birds, and between apes and humans. The fossil evidence of evolution between major forms is compelling.

3. **The intelligent design argument.** "The organs of living creatures are too complex for a random process to have produced—the existence of a clock is evidence of the existence of a clockmaker."

Evolution by natural selection is not a random process. Quite the contrary, by favoring those variations that lead to the highest reproductive fitness, natural selection is a nonrandom process that can construct highly complex organs by incrementally improving them from one generation to the next.

For example, the intermediates in the evolution of the mammalian ear can be seen in fossils, and many intermediate "eyes" are known in various invertebrates. These intermediate forms arose because they have value—being able to detect light slightly is better than not being able to detect it at all. Complex structures such as eyes evolved as a progression of slight improvements. Moreover, inefficiencies of certain designs, such as the vertebrate eye and the existence of vestigial structures, do not support the idea of an intelligent designer.

4. **Evolution violates the Second Law of Thermodynamics.** "A jumble of soda cans doesn't by itself jump neatly into a stack—things become more disorganized due to random events, not more organized."

Biologists point out that this argument ignores what the second law really says: Disorder increases in a closed system, which the Earth most certainly is not. Energy continually enters the biosphere from the Sun, fueling life and all the processes that organize it.

5. **Proteins are too improbable.** "Hemoglobin has 141 amino acids. The probability that the first one would be leucine is 1/20, and that all 141 would be the ones they are by chance is $(1/20)^{141}$, an impossibly rare event."

This argument illustrates a lack of understanding of probability and statistics—probability cannot be used to argue backwards. The probability that a student in a classroom has a particular birthdate is 1/365; arguing this way, the probability that everyone in a class of 50 would have the birthdates that they do is $(1/365)^{50}$, and yet there the class sits, all with their actual birthdates.

6. **Natural selection does not imply evolution.** "No scientist has come up with an experiment in which fish evolve into frogs and leap away from predators."

Can we extrapolate from our understanding that natural selection produces relatively small changes that are observable in populations *within* species to explain the major differences observed *between* species? Most biologists who have studied the problem think so. The differences between breeds produced by artificial selection—such as chihuahuas, mastiffs, and greyhounds—are more distinctive than the differences between some wild species, and laboratory selection experiments sometimes create forms that cannot interbreed and thus would in nature be considered different species. Thus, production of radically different forms has indeed been observed, repeatedly. To object that evolution still does not explain really major differences, such as those between fish and amphibians, simply takes us back to point number 2. These changes take millions of years, and they are seen clearly in the fossil record.

7. **The irreducible complexity argument.** Because each part of a complex cellular mechanism such as blood clotting is essential to the overall process, the intricate machinery of the cell cannot be explained by evolution from simpler stages.

What's wrong with this argument is that each part of a complex molecular machine evolves as part of the whole system. Natural selection can act on a complex system because at every stage of its evolution, the system functions. Parts that improve function are added. Subsequently, other parts may be modified or even lost, so that parts that were not essential when they first evolved become essential. In this way, an "irreducible complex" structure can evolve by natural selection. The same process works at the molecular level.

For example, snake venom initially evolved as enzymes to increase the ability of snakes to digest large prey items, which were captured by biting the prey and then constricting them with coils. Subsequently, the digestive enzymes evolved to become increasingly lethal. Rattlesnakes kill large prey by injecting them with venom, letting them go, and then tracking them down and eating them after they die. To do so, they have evolved extremely toxic venom, highly modified syringe-like front teeth, and many other characteristics. Take away the fangs or the venom and the rattlesnakes can't feed—what initially evolved as nonessential parts are now indispensable; irreducible complexity has evolved by natural selection.

The mammalian blood clotting system, for example, has evolved from much simpler systems. The core clotting system evolved at the dawn of the vertebrates more than 500 million years ago, and it is found today in primitive fishes such as lampreys. One hundred million years later, as vertebrates continued to evolve, proteins were added to the clotting system, making it sensitive to substances released from damaged tissues. Fifty million years later, a third component was added, triggering clotting by contact with the jagged surfaces produced by injury. At each stage, as the clotting system evolved to become more complex, its overall performance came to depend on the added elements. Thus, blood clotting has become "irreducibly complex" as the result of Darwinian evolution.

Statements that various structures could not have been built by natural selection have repeatedly been made over the past century. In each case, after detailed scientific study, the likely path by which such structures have evolved has been discovered.

Darwin's theory of evolution has proven controversial among some of the general public, although the commonly raised objections are without scientific merit.

21.1 The Beaks of Darwin's Finches: Evidence for Natural Selection

Natural selection is the primary process contributing to evolution.

- Galápagos finches vary in food procurement.
- Differences among finch beak size and shape have evolved as descendants of the ancestral species adapted to different food resources.
- Over multiple generations, the beak shapes of contemporary finches change in response to the available food supply and these changes reflect genetic differences (figure 21.2a).
- For evolution to occur, genetically determined differences must lead to differences in reproductive success.

21.2 Peppered Moths and Industrial Melanism: More Evidence of Selection

When the environment changes, natural selection may favor different traits in a species.

- Natural selection changes allele frequencies as the environment changes.
- Pollution may indirectly affect a change in phenotypes. In industrial melanism darker moths predominate. Initial studies suggested that birds selected moths that were lighter in color and stood out against their background.
- When environmental conditions reverse, so do selection pressures (figure 21.4).
- Agents of selection may be difficult to identify.

21.3 Artificial Selection: Human-Initiated Change
(figure 21.5)

Humans impose artificial selection on species by favoring certain phenotypic traits.

- Directional selection of favored traits leads to evolutionary change in populations.
- Artificial selection in laboratory experiments and agriculture produces rapid and substantial changes.
- If artificial selection can rapidly create change then it is reasonable to assume that natural selection created the diversity of life in the world over millions of years.

21.4 Fossil Evidence for Evolution

The most direct evidence that evolution has occurred is found in the fossil record.

- Specimens are fossilized in different ways: embedded in amber, frozen in permafrost, mummified in dry caves, and mineralized in rock.
- Rock fossils are created by burial of body parts in sediment, mineralization, and solidification as rock.
- Only a very small fraction of species are preserved because decay, scavengers, accessibility, and erosion prevent fossilization.
- Relative dating of fossils refers to the dating of remains by their position in sedimentary rock strata.
- Absolute dating of fossils is more precise and relies on radioactive decay rates.
- When fossils are arrayed from oldest to youngest they provide insight into the history of evolutionary changes.
- Gaps in the fossil record may be filled by the discovery of intermediate life forms that contain combined traits found in their ancestors and descendants.
- The fossil record has been used to document major evolutionary transitions over time such as those found in horses.

21.5 Anatomical Evidence for Evolution

The evidence for evolution supports the concept of descent with modification.

- Homologous structures may have different appearances and functions even though they are derived from the same common ancestral body part (figure 21.14).
- Embryonic development shows similarity in developmental patterns among species whose adult phenotype is very different.
- Natural selection can only influence the variation present in a population and because of this often results in workable, but imperfect structures.
- The existence of vestigial structures supports the concept of common ancestry among organisms that share them.

21.6 Convergent Evolution and the Biogeographical Record

Natural selection favors parallel evolutionary adaptations in similar environments.

- Biogeography, the study of geographic distributions of species, reveals that groups of organisms can have similar appearances even though they are distantly related.
- Convergent evolution may occur in distantly related species that are exposed to similar selective pressures.
- Early colonizers to new habitats usually come from nearby areas and often evolve into diverse species because there is no competition from other preexisting species.
- Island species usually are closely related to species on nearby continents even if the environments are different.
- Evolutionary relationships cannot always be predicted by current geographic proximity. Continental drift has separated landmasses that were originally close to each other.

21.7 Darwin's Critics

Darwin's theory of evolution by natural selection is almost universally accepted by all biologists because the data have scientific merit.

- Scientists accept evolution as a theory grounded in facts derived from using the scientific method; evolution is not based on guesswork.
- The geologic record contains intermediate forms between major life forms; evolution can no longer be dismissed by supposed missing data.
- The complexity of structures evolved over time as a progression of small improvements; evolution is not a random process.
- Because the Sun provides continuous energy to maintain life and all the processes that organize it, it is not a closed system; evolution does not violate the Second Law of Thermodynamics.
- Correctly used, statistics support evolution; evolution cannot be discounted by using statistics in reverse.
- The progression of small observable changes within a species can be used to explain major differences between species; change resulting from artificial selection cannot be dismissed.
- Evolution of whole systems occurs at all levels of biological complexity; natural selection is not restricted to a particular level of biological organization.
- Natural selection can build complex structures; moreover, the existence of non-optimally designed and vestigial structures argues against an "intelligent designer."

SELF TEST

1. In Darwin's finches,
   a. occurrence of wet and dry years preserves genetic variation for beak size.
   b. increasing beak size over time proves that beak size is inherited.
   c. large beak size is always favored.
   d. all of the above.

2. In the case of peppered moths, a parallel case in the United States
   a. is best considered as a coincidence.
   b. adds support to the industrial melanism hypothesis.
   c. demonstrates that birds don't eat moths.
   d. none of these.

3. Artificial selection is different from natural selection because
   a. artificial selection is not capable of producing large changes.
   b. artificial selection does not require genetic variation.
   c. natural selection cannot produce new species.
   d. breeders (people) choose which individuals reproduce based on desirability of traits.

4. Gaps in the fossil record
   a. demonstrate our inability to date geological sediments.
   b. are expected since the probability that any organism will fossilize is extremely low.
   c. have not been filled in as new fossils have been discovered.
   d. weaken the theory of evolution.

5. Intermediate fossils illustrating the evolution of whales from their hoofed ancestors include
   a. *Pakicetus.*
   b. *Archaeopteryx.*
   c. *Equus.*
   d. all of these.

6. The evolution of modern horses (*Equus*) is best described as
   a. the constant change and replacement of one species by another over time.
   b. a complex history of lineages that changed over time, with many going extinct.
   c. a simple history of lineages that have always resembled extant horses.
   d. none of these.

7. Homologous structures
   a. are structures in two or more species that originate as the same structure in a common ancestor.
   b. are structures that look the same in different species.
   c. cannot serve different functions in different species.
   d. must serve different functions in different species.

8. Convergent evolution
   a. involves species that are closely related evolutionarily.
   b. depends on natural selection to produce similar phenotypic responses in unrelated lineages.
   c. occurs only on islands.
   d. is expected when different lineages are exposed to vastly different selective environments.

9. Convergent evolution
   a. occurs when natural selection produces similar characteristics in unrelated species.
   b. is a weakness in evolutionary theory.
   c. is the best explanation for why there are gaps in the fossil record.
   d. always involves homologous characteristics.

10. Which of the following represents valid criticism that weakens the theory of evolution?
    a. Evolution violates the Second Law of Thermodynamics.
    b. There are no fossil intermediates.
    c. Irreducible complexity of biological structures.
    d. All of the above.
    e. None of the above.

11. Darwin's finches are a noteworthy case study of evolution by natural selection because evidence suggests
    a. they are descendants of many different species that colonized the Galápagos.
    b. they radiated from a single species that colonized the Galápagos.
    c. they are more closely related to mainland species than to one another.
    d. none of the above.

12. Vestigial structures
    a. are difficult to explain given evolutionary theory.
    b. always provide evidence of convergent evolution.
    c. can help establish patterns of common ancestry.
    d. must always severely affect survivorship or reproduction.

13. An example of convergent evolution is
    a. Australian marsupials and placental mammals.
    b. the flippers in fish, penguins, and dolphins.
    c. the wings in birds, bats, and insects.
    d. all of these.

14. *Archaeopteryx*
    a. is an intermediate form between extinct and extant horses.
    b. is an intermediate form between extinct and extant whales.
    c. is an intermediate form between dinosaurs and birds that is intermediate in all its characteristics.
    d. shows a common pattern for intermediate forms of mixtures of some traits resembling descendants and some traits resembling ancestors.

15. The shape of the beaks of Darwin's finches, industrial melanism, and the changes in horse teeth are all examples of
    a. artificial selection.
    b. natural selection.
    c. sympatric speciation.
    d. stabilizing selection.

CHALLENGE QUESTIONS

1. What conditions are necessary for evolution by natural selection?

Refer to figure 21.2 for the following two questions.

2. Explain how data shown in figure 21.2*a* and *b* relate to the conditions identified by you in question 1.

3. On figure 21.2*b* draw the relationship between offspring beak depth and parent beak depth, assuming that there is no genetic basis to beak depth in the medium ground finch.

4. Refer to figure 21.5, artificial selection in the laboratory. In this experiment, one population of *Drosophila* was selected for low numbers of bristles and the other for high numbers. Note that not only did the means of the populations change greatly in 35 generations, but also all individuals in both experimental populations lie outside the range of the initial population.

   What would happen if, within a population, both small and large individuals were allowed to breed, but middle-sized ones were not?

The Origin of Species

introduction

ALTHOUGH DARWIN TITLED HIS BOOK *On the Origin of Species*, he never actually discussed what he referred to as that "mystery of mysteries"—how one species gives rise to another. Rather, his argument concerned evolution by natural selection; that is, how one species evolves through time to adapt to its changing environment. Although an important mechanism of evolutionary change, the process of adaptation does not explain how one species becomes another, a process we call **speciation.** As we shall see, adaptation may be involved in the speciation process, but it does not have to be.

Before we can discuss how one species gives rise to another, we need to understand exactly what a species is. Even though the definition of a species is of fundamental importance to evolutionary biology, this issue has still not been completely settled and is currently the subject of considerable research and debate.

concept outline

22.1 The Nature of Species

Any concept of a species must account for two phenomena: the distinctiveness of species that occur together at a single locality, and the connection that exists among different populations belonging to the same species.

Sympatric species inhabit the same locale but remain distinct

Put out a birdfeeder on your balcony or back porch and you will attract a wide variety of birds (especially if you include different kinds of foods). In the midwestern United States, for example, you might routinely see cardinals, blue jays, downy woodpeckers, house finches—even hummingbirds in the summer.

Although it might take a few days of careful observation, you would soon be able to readily distinguish the many different species. The reason is that species that occur together (termed **sympatric**) are distinctive entities that are phenotypically different, utilize different parts of the habitat, and behave separately. This observation is generally true not only for birds, but also for most other types of organisms.

Occasionally, two species occur together that appear to be nearly identical. In such cases, we need to go beyond visual similarities. When other aspects of the phenotype are examined, such as the mating calls or the chemicals exuded by each species, they usually reveal great differences. In other words, even though we have trouble distinguishing them, the organisms themselves have no such difficulties.

Populations of a species exhibit geographic variation

Within a single species, individuals in populations that occur in different areas may be distinct from one another. Such groups of distinctive individuals may be classified as **subspecies** (the

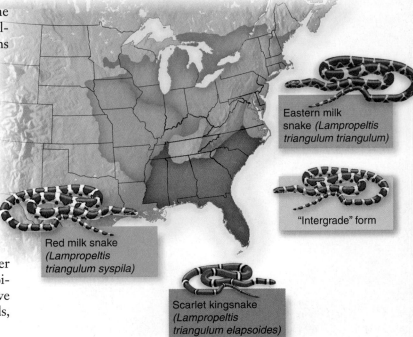

Eastern milk snake *(Lampropeltis triangulum triangulum)*

"Intergrade" form

Red milk snake *(Lampropeltis triangulum syspila)*

Scarlet kingsnake *(Lampropeltis triangulum elapsoides)*

figure 22.1

GEOGRAPHIC VARIATION IN THE MILK SNAKE, *Lampropeltis triangulum.* Although subspecies appear phenotypically quite distinctive from one another, they are connected by populations that are phenotypically intermediate.

vague term "race" has a similar connotation, but is no longer commonly used). In areas where these populations approach one another, individuals often exhibit combinations of features characteristic of both populations (figure 22.1). In other words, even though geographically distant populations may appear distinct, they are usually connected by intervening populations that are intermediate in their characteristics.

22.2 The Biological Species Concept

What can account for both the distinctiveness of sympatric species and the connectedness of geographically separate populations of the same species? One obvious possibility is that each species exchanges genetic material only with other members of its species. If sympatric species commonly exchanged genes, which they generally do not, we might expect such species to rapidly lose their distinctions, as the **gene pools** (that is, all of the alleles present in a species) of the different species became homogenized. Conversely, the ability of geographically distant populations of a single species to share genes through the process of gene flow may keep these populations integrated as members of the same species.

Based on these ideas, in 1942 the evolutionary biologist Ernst Mayr set forth the **biological species concept,** which defines *species* as ". . . groups of actually or potentially inter-

breeding natural populations which are reproductively isolated from other such groups."

In other words, the biological species concept says that a species is composed of populations whose members mate with each other and produce fertile offspring—or would do so if they came into contact. Conversely, populations whose members do not mate with each other or who cannot produce fertile offspring are said to be **reproductively isolated** and, therefore, are members of different species.

What causes reproductive isolation? If organisms cannot interbreed or cannot produce fertile offspring, they clearly belong to different species. However, some populations that are considered separate species can interbreed and produce fertile offspring, but they ordinarily do not do so under natural conditions. They are still considered reproductively isolated in that genes from one species generally will not enter the gene pool of the other species.

Table 22.1 summarizes the steps at which barriers to successful reproduction may occur. Such barriers are termed **reproductive isolating mechanisms** because they prevent genetic exchange between species. We will discuss examples of these next, beginning with those that prevent the formation of zygotes, which are called **prezygotic isolating mechanisms**.

Mechanisms that prevent the proper functioning of zygotes after they form are called **postzygotic isolating mechanisms**.

Prezygotic isolating mechanisms prevent the formation of a zygote

Mechanisms that prevent formation of a zygote include ecological or environmental isolation, behavioral isolation, temporal isolation, mechanical isolation, and prevention of gamete fusion.

Ecological isolation

Even if two species occur in the same area, they may utilize different portions of the environment and thus not hybridize because they do not encounter each other. For example, in India, the ranges of lions and tigers overlapped until about 150 years ago. Even so, there were no records of natural hybrids. Lions stayed mainly in the open grassland and hunted in groups called prides; tigers tended to be solitary creatures of the forest (figure 22.2). Because of their ecological and behavioral differences, lions and tigers rarely came into direct contact with each other, even though their ranges overlapped over thousands of square kilometers.

In another example, the ranges of two toads, *Bufo woodhousei* and *B. americanus*, overlap in some areas. Although these two species can produce viable hybrids, they usually do not interbreed because they utilize different portions of the habitat for breeding. *B. woodhousei* prefers to breed in streams, and *B. americanus* breeds in rainwater puddles.

Similar situations occur among plants. Two species of oaks occur widely in California: the valley oak, *Quercus lobata*, and the scrub oak, *Q. dumosa*. The valley oak, a graceful deciduous tree that can be as tall as 35 m, occurs in the fertile soils of open grassland on gentle slopes and valley floors. In contrast, the scrub oak is an evergreen shrub, usually only 1 to 3 m tall, which often forms the kind of dense scrub known as chaparral. The scrub oak is found on steep slopes in less fertile soils. Hybrids between these different oaks do occur and are fully fertile, but they are rare. The sharply distinct habitats of their parents limit their occurrence together, and there is little intermediate habitat where the hybrids might flourish.

TABLE 22.1	Reproductive Isolating Mechanisms	
Mechanism		**Description**
PREZYGOTIC ISOLATING MECHANISMS		
Geographic isolation	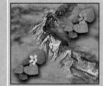	Species occur in different areas, which are often separated by a physical barrier such as a river or mountain range.
Ecological isolation		Species occur in the same area, but they occupy different habitats and rarely encounter each other.
Behavioral isolation		Species differ in their mating rituals.
Temporal isolation		Species reproduce in different seasons or at different times of the day.
Mechanical isolation		Structural differences between species prevent mating.
Prevention of gamete fusion		Gametes of one species function poorly with the gametes of another species or within the reproductive tract of another species.
POSTZYGOTIC ISOLATING MECHANISMS		
Hybrid inviability or infertility		Hybrid embryos do not develop properly, hybrid adults do not survive in nature, or hybrid adults are sterile or have reduced fertility.

figure 22.2

LIONS AND TIGERS ARE ECOLOGICALLY ISOLATED. The ranges of lions and tigers overlap in India. However, lions and tigers do not hybridize in the wild because they utilize different portions of the habitat. Lions live in open grassland, whereas tigers are solitary animals that live in the forest. Hybrids, such as this tiglon, have been successfully produced in captivity, but hybridization does not occur in the wild.

figure 22.3

DIFFERENCES IN COURTSHIP RITUALS CAN ISOLATE RELATED BIRD SPECIES. These Galápagos blue-footed boobies select their mates only after an elaborate courtship display. This male is lifting his feet in a ritualized high-step that shows off his bright blue feet. The display behavior of the two other species of boobies that occur in the Galápagos is very different, as is the color of their feet.

Behavioral isolation

Chapter 54 describes the often elaborate courtship and mating rituals of some groups of animals. Related species of organisms such as birds often differ in their courtship rituals, which tends to keep these species distinct in nature even if they inhabit the same places (figure 22.3). For example, mallard and pintail ducks are perhaps the two most common freshwater ducks in North America. In captivity, they produce completely fertile hybrid offspring, but in nature they nest side by side and only rarely hybridize.

Sympatric species avoid mating with members of the wrong species in a variety of ways; every mode of communication imaginable appears to be used by some species. Differences in visual signals, as just discussed, are common; however, other types of animals rely more on other sensory modes for communication. Many species, such as frogs, birds and a variety of insects, use sound production to attract mates. Predictably, sympatric species of these animals produce different calls. Similarly, the "songs" of lacewings are produced when they vibrate their abdomens against the surface on which they are sitting, and sympatric species produce different vibration patterns (figure 22.4).

Other species rely on the detection of chemical signals, called **pheromones.** The use of pheromones in moths has been particularly well studied. When female moths are ready to mate, they emit a pheromone that males can detect at great distances. Sympatric species differ in the pheromone they produce: Either they use different chemical compounds, or, if using the same compounds the proportions used are different. Laboratory studies indicate that males are remarkably adept at distinguishing the pheromones of their own species from those of other species or even from synthetic compounds similar, but not identical, to that of their own species.

Some species even use electroreception. African electric fish have specialized organs in their tails that produce electrical discharges and electroreceptors on their skins to detect them. These discharges are used to communicate in social interactions; field experiments indicate that males can distinguish between signals produced by their own and other species, probably on the basis of differences in the timing of the electrical pulses.

Temporal isolation

Lactuca graminifolia and *L. canadensis*, two species of wild lettuce, grow together along roadsides throughout the southeastern United States. Hybrids between these two species are easily made experimentally and are completely fertile. But these hybrids are rare in nature because *L. graminifolia* flowers in early spring and *L. canadensis* flowers in summer. When their blooming periods overlap, as happens occasionally, the two species do form hybrids, which may become locally abundant.

Many species of closely related amphibians have different breeding seasons that prevent hybridization. For example, five species of frogs of the genus *Rana* occur together in most of the eastern United States, but hybrids are rare because the peak breeding time is different for each of them.

Mechanical isolation

Structural differences prevent mating between some related species of animals. Aside from such obvious features as size, the structure of the male and female copulatory organs may be incompatible. In many insect and other arthropod groups, the sexual organs, particularly those of the male, are so diverse that they are used as a primary basis for distinguishing species.

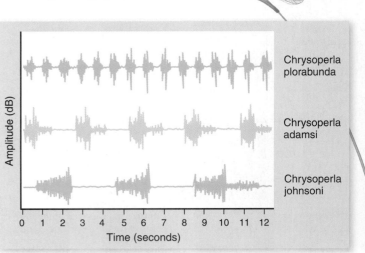

figure 22.4

DIFFERENCES IN COURTSHIP SONG OF SYMPATRIC SPECIES OF LACEWINGS. Lacewings are small insects that rely on auditory signals produced by moving their abdomens to vibrate the surface on which they are sitting to attract mates. As these recordings indicate, the vibration patterns produced by sympatric species differ greatly. Females, which detect the calls as they are transmitted through solid surfaces such as branches, are able to distinguish calls of different species and only respond to individuals producing their own species' call.

Similarly, flowers of related species of plants often differ significantly in their proportions and structures. Some of these differences limit the transfer of pollen from one plant species to another. For example, bees may carry the pollen of one species on a certain place on their bodies; if this area does not come into contact with the receptive structures of the flowers of another plant species, the pollen is not transferred.

Prevention of gamete fusion

In animals that shed gametes directly into water, the eggs and sperm derived from different species may not attract or fuse with one another. Many land animals may not hybridize successfully because the sperm of one species functions so poorly within the reproductive tract of another that fertilization never takes place. In plants, the growth of pollen tubes may be impeded in hybrids between different species. In both plants and animals, isolating mechanisms such as these prevent the union of gametes, even following successful mating.

Postzygotic isolating mechanisms prevent normal development into reproducing adults

All of the factors we have discussed so far tend to prevent hybridization. If hybrid matings do occur and zygotes are produced, many factors may still prevent those zygotes from developing into normally functioning, fertile individuals.

As you saw in chapter 19, development is a complex process. In hybrids, the genetic complements of two species may be so different that they cannot function together normally in embryonic development. For example, hybridization between sheep and goats usually produces embryos that die in the earliest developmental stages.

Leopard frogs (*Rana pipiens* complex) of the eastern United States are a group of similar species, assumed for a long time to constitute a single species (figure 22.5). Careful examination, however, revealed that although the frogs appear similar, successful mating between them is rare because of problems that occur as the fertilized eggs develop. Many of the hybrid combinations cannot be produced even in the laboratory.

Examples of this kind, in which similar species have been recognized only as a result of hybridization experiments, are common in plants. Sometimes the hybrid plant embryos can be removed at an early stage and grown in an artificial medium. When these hybrids are supplied with extra nutrients or other supplements that compensate for their weakness or inviability, they may complete their development normally.

Even when hybrids survive the embryo stage, they may still not develop normally. If the hybrids are less physically fit than their parents, they will almost certainly be eliminated in nature. Even if a hybrid is vigorous and strong, as in the case of the mule, which is a hybrid between a female horse and a male donkey, it may still be sterile and thus incapable of contributing to succeeding generations.

Hybrids may be sterile because the development of sex organs is abnormal, because the chromosomes derived from the respective parents cannot pair properly during meiosis, or due to a variety of other causes.

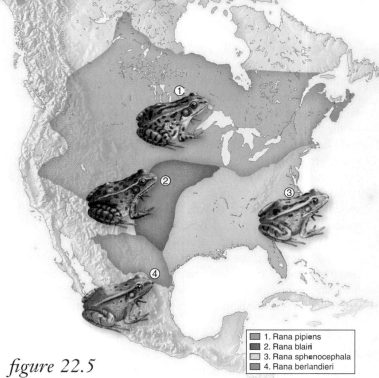

figure 22.5

POSTZYGOTIC ISOLATION IN LEOPARD FROGS. These four species resemble one another closely in their external features. Their status as separate species was first suspected when hybrids between some pairs of these species were found to produce defective embryos in the laboratory. Subsequent research revealed that the mating calls of the four species differ substantially, indicating that the species have both pre- and postzygotic isolating mechanisms.

1. *Rana pipiens*
2. *Rana blairi*
3. *Rana sphenocephala*
4. *Rana berlandieri*

The biological species concept does not explain all observations

The biological species concept has proven to be an effective way of understanding the existence of species in nature. Nonetheless, it fails to take into account all observations, leading some biologists to propose alternative species concepts.

One criticism of the biological species concept concerns the extent to which all species truly are reproductively isolated. By definition, under the biological species concept, species should not interbreed and produce fertile offspring. But in recent years, biologists have detected much greater amounts of interspecific hybridization than was previously thought to occur between populations that seem to coexist as distinct biological entities.

Botanists have always been aware that plant species often undergo substantial amounts of hybridization. More than 50% of California plant species included in one study, for example, were not well defined by genetic isolation. This coexistence without genetic isolation can be long-lasting: Fossil data show that balsam poplars and cottonwoods have been phenotypically distinct for 12 million years, but they also have routinely produced hybrids throughout this time. Consequently, many botanists have long felt that the biological species concept applies only to animals.

New evidence, however, increasingly indicates that hybridization is not all that uncommon in animals, either. In recent years, many cases of substantial hybridization between animal species have been documented. One recent survey indicated that almost 10% of the world's 9500 bird species are known to have hybridized in nature.

The Galápagos finches provide a particularly well-studied example. Three species on the island of Daphne Major—the medium ground finch, the cactus finch, and the small ground finch—are clearly distinct morphologically, and they occupy different ecological niches. Studies over the past 20 years by Peter and Rosemary Grant found that, on average, 2% of the medium ground finches and 1% of the cactus ground finches mated with other species every year. Furthermore, hybrid offspring appeared to be at no disadvantage in terms of survival or subsequent reproduction. This is not a trivial amount of genetic exchange, and one might expect to see the species coalesce into one genetically variable population—but the species are maintaining their distinctiveness.

Hybridization is not rampant throughout the animal world, however. Most bird species do not hybridize, and probably even fewer experience significant amounts of hybridization. Still, hybridization is common enough to cast doubt on whether reproductive isolation is the only force maintaining the integrity of species.

Natural selection and the ecological species concept

An alternative hypothesis proposes that the distinctions among species are maintained by natural selection. The idea is that each species has adapted to its own specific part of the environment. Stabilizing selection, described in chapter 20, then maintains the species' adaptations. Hybridization has little effect because alleles introduced into one species' gene pool from other species are quickly eliminated by natural selection.

You probably recall from chapter 20 that the interaction between gene flow and natural selection can have many outcomes. In some cases, strong selection can overwhelm any effects of gene flow—but in other situations, gene flow can prevent populations from eliminating less successful alleles from a population.

As a general explanation, then, natural selection is not likely to have any fewer exceptions than does the biological species concept, although the ecological species concept may prove to be a more successful description for certain types of organisms or habitats.

Other weaknesses of the biological species concept

The biological species concept has been criticized for other reasons as well. For example, it can be difficult to apply the concept to populations that are geographically separated in nature. Because individuals of these populations do not encounter each other, it is not possible to observe whether they would interbreed naturally.

Although experiments can determine whether fertile hybrids can be produced, this information is not enough. Many species that coexist without interbreeding in nature will readily hybridize in the artificial settings of the laboratory or zoo. Consequently, evaluating whether such populations constitute different species is ultimately a judgment call. In addition, the concept is more limited than its name would imply. Many organisms are asexual and reproduce without mating; reproductive isolation has no meaning for such organisms.

For these reasons, a variety of other ideas have been put forward to establish criteria for defining species. Many of these are specific to a particular type of organism, and none has universal applicability. In reality, there may be no single explanation for what maintains the identity of species. Given the incredible variation evident in plants, animals, and microorganisms in all aspects of their biology, it would not be surprising to find that different processes are operating in different organisms.

In addition, some scientists have turned from emphasizing the processes that maintain species distinctions to examining the evolutionary history of populations. These genealogical species concepts are currently a topic of great debate and are discussed further in chapter 23.

Species are populations of organisms that are (1) distinct from other, co-occurring species, and (2) interconnected geographically. The ability to exchange genes can explain these phenomena.

Prezygotic isolating mechanisms lead to reproductive isolation by preventing the formation of hybrid zygotes.

Postzygotic isolating mechanisms are those in which hybrid zygotes fail to develop or develop abnormally, cannot survive in nature, or are unable to reproduce.

The surprisingly high incidence of hybridization in plants and animals has caused some researchers to seek alternatives to the biological species concept. Because of the diversity of living organisms, no single definition of what constitutes a species may be universally applicable.

22.3 The Evolution of Reproductive Isolation

One of the oldest questions in the field of evolution is: How does one ancestral species become divided into two descendant species (a process termed **cladogenesis**)? If species are defined by the existence of reproductive isolation, then the process of speciation is identical to the evolution of reproductive isolating mechanisms.

Selection may reinforce isolating mechanisms

The formation of species is a continuous process, and as a result, two populations may only be partially reproductively isolated. For example, because of behavioral or ecological differences, individuals of two populations may be more likely to mate with members of their own population, and yet between-population matings may still occur. If mating occurs and fertilization produces a zygote, postzygotic barriers may also be incomplete: developmental problems may result in lower embryo survival or reduced fertility, but some individuals may survive and reproduce.

What happens when two populations come into contact thus depends on the extent to which isolating mechanisms have already evolved. If isolating mechanisms have not evolved at all, then the two populations will interbreed freely, and whatever other differences have evolved between them should disappear over the course of time, as genetic exchange homogenizes the populations. Conversely, if the populations are completely reproductively isolated, then no genetic exchange will occur, and the two populations will be different species.

How reinforcement can complete the speciation process

The intermediate state, in which reproductive isolation has partially evolved but is not complete, is perhaps the most interesting situation. If the hybrids are partly sterile, or not as well adapted to the existing habitats as their parents, they will be at a disadvantage. Selection would favor any alleles in the parental populations that prevented hybridization, because individuals that did not engage in hybridization would produce more successful offspring.

The result would be the continual improvement of prezygotic isolating mechanisms until the two populations were completely reproductively isolated. This process is termed **reinforcement** because initially incomplete isolating mechanisms are reinforced by natural selection until they are completely effective.

An example of reinforcement is provided by pied and collared flycatchers. Throughout much of eastern and central Europe, these two species are geographically separated **(allopatric)** and are very similar in color (figure 22.6). However, in the Czech Republic and Slovakia, the two species occur together and occasionally hybridize, producing offspring that usually have very low fertility. At those sites, the species have evolved to look very different from each other, and birds prefer to mate with individuals with their own species' coloration. In contrast, birds from the allopatric populations prefer the allopatric color pattern. As a consequence of the color differences where the species are sympatric, the rate of hybridization is extremely low. These results indicate that when populations of the two species came into contact, natural selection led to the evolution of differences in color patterns, resulting in the evolution of behavioral, prezygotic isolation.

How gene flow may counter speciation

Reinforcement is not inevitable, however. When incompletely isolated populations come together, gene flow immediately begins to occur between them. Although hybrids may be inferior, they are not completely inviable or infertile—if they were, the species would already be completely reproductively isolated. When these surviving hybrids reproduce with members of either population, they will serve as a conduit of genetic exchange from one population to the other, and the two populations will tend to lose their genetic distinctiveness. Thus, a race ensues: Can complete reproductive isolation evolve before gene flow

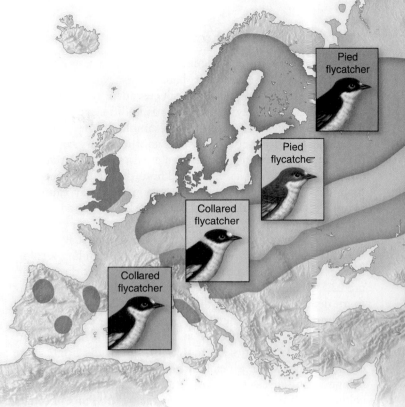

figure 22.6

REINFORCEMENT IN EUROPEAN FLYCATCHERS. The pied flycatcher (*Ficedula hypoleuca*) and the collared flycatcher (*F. albicollis*) appear very similar when they occur alone. However, in places where the two species occur sympatrically (indicated by the yellow color), they have evolved differences in color and pattern, which allow individuals to choose mates from their own species and thus avoid hybridizing.

erases the differences between the populations? Experts disagree on the likely outcome, but many consider reinforcement to be the much less common outcome.

 inquiry

How might the initial degree of reproductive isolation affect the probability that reinforcement will occur?

22.4 The Role of Genetic Drift and Natural Selection in Speciation

What role does natural selection play in the speciation process? Certainly, the process of reinforcement is driven by natural selection favoring the evolution of complete reproductive isolation. But reinforcement may not be common. In situations other than reinforcement, does natural selection play a role in the evolution of reproductive isolating mechanisms?

Random changes may cause reproductive isolation

As mentioned in chapter 20, populations may diverge for purely random reasons. Genetic drift in small populations, founder effects, and population bottlenecks all may lead to changes in traits that cause reproductive isolation.

For example, in the Hawaiian Islands, closely related species of *Drosophila* often differ greatly in their courtship behavior. Colonization of new islands by these fruit flies probably involved a founder effect, in which one or a few flies—perhaps only a single pregnant female—was blown by strong winds to the new island. Changes in courtship behavior between ancestor and descendant populations may be the result of such founder events.

Given enough time, any two isolated populations will diverge because of genetic drift (remember that even large populations experience drift, but at a lower rate than in small populations). In some cases, this random divergence may affect traits responsible for reproductive isolation, and speciation may occur.

Adaptation can lead to speciation

Although random processes may sometimes be responsible, in many cases natural selection probably plays a role in the speciation process. As populations of a species adapt to different circumstances, they likely accumulate many differences that may lead to reproductive isolation. For example, if one population of flies adapts to wet conditions and another to dry conditions, then natural selection will produce a variety of differences in physiological and sensory traits. These differences may promote ecological and behavioral isolation, and may cause any hybrids the two populations produce to be poorly adapted to either habitat.

Selection might also act directly on mating behavior. Male *Anolis* lizards, for example, court females by extending a colorful flap of skin, called a *dewlap*, located under their throats (figure 22.7). The ability of one lizard to see the dewlap of another lizard depends not only on the color of the dewlap, but on the environment in which the lizards occur. A light-colored dewlap, for example, is most effective in reflecting light in a dim forest, whereas dark colors are more apparent in the bright glare of open habitats. As a result, when these lizards occupy new habitats, natural selection favors evolutionary change in dewlap color because males whose dewlaps cannot be seen will attract few mates. But the lizards also distinguish members of their own species from other species by the color of the dewlap. Adaptive

change in mating signals in new environments could therefore have the incidental consequence of producing reproductive isolation from populations in the ancestral environment.

Laboratory scientists have conducted experiments on fruit flies and other fast-reproducing organisms in which they isolate populations in different laboratory chambers and measure how much reproductive isolation evolves. These experiments indicate that genetic drift by itself can lead to some degree of reproductive isolation, but in general, reproductive isolation evolves more rapidly when the populations are forced to adapt to different laboratory environments (such as temperature or food type). Although natural selection does not directly favor traits because they lead to reproductive isolation, the incidental effect of adaptive divergence is that populations in different environments become reproductively isolated. For this reason, some biologists believe that the term *isolating mechanisms* is misguided, because it implies that the traits evolved specifically for the purpose of genetically isolating a species, which in most cases is probably incorrect.

> Reproductive isolating mechanisms can evolve either through random changes or as an incidental by-product of adaptive evolution. Under some circumstances, however, natural selection can directly select for traits that increase the reproductive isolation of a species.

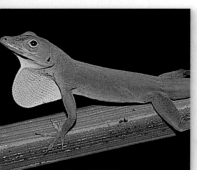

figure 22.7

DEWLAPS OF DIFFERENT SPECIES OF CARIBBEAN *ANOLIS* LIZARDS. Males use their dewlaps in both territorial and courtship displays. Coexisting species almost always differ in their dewlaps, which are used in species recognition. Some dewlaps are easier to see in open habitats, whereas others are more visible in shaded environments.

22.5 The Geography of Speciation

Speciation is a two-part process. First, initially identical populations must diverge, and second, reproductive isolation must evolve to maintain these differences. The difficulty with this process, as we have seen, is that the homogenizing effect of gene flow between populations will constantly be acting to erase any differences that may arise, either by genetic drift or natural selection. Gene flow only occurs between populations that are in contact, however, and populations can become geographically isolated for a variety of reasons (figure 22.8). Consequently, evolutionary biologists have long recognized that speciation is much more likely in geographically isolated populations.

Allopatric speciation takes place when populations are geographically isolated

Ernst Mayr was the first biologist to demonstrate that geographically separated, or *allopatric*, populations appear much more likely to have evolved substantial differences leading to speciation. Marshalling data from a wide variety of organisms and localities, Mayr made a strong case for allopatric speciation as the primary means of speciation.

For example, the Papuan kingfisher, *tanysiptera hydrocharis*, varies little throughout its wide range in New Guinea, despite

a.   *b.*   *c.*

figure 22.8

POPULATIONS CAN BECOME GEOGRAPHICALLY ISOLATED FOR A VARIETY OF REASONS. *a.* Colonization of remote areas by one or a few individuals can establish populations in a distant place. *b.* Barriers to movement can split an ancestral population into two isolated populations. *c.* Extinction of intermediate populations can leave the remaining populations isolated from one another.

the great variation in the island's topography and climate. By contrast, isolated populations on nearby islands are strikingly different from one another and from the mainland population (figure 22.9). Thus, geographic isolation seems to have been an important prerequisite for the evolution of differences between populations.

Many other examples indicate that speciation can occur under allopatric conditions. Because we would expect isolated populations to diverge over time, by either drift or selection, this result is not surprising. Rather, the more intriguing question becomes: Is geographic isolation *required* for speciation to occur?

Sympatric speciation occurs without geographic separation

For decades, biologists have debated whether one species can split into two at a single locality, without the two new species ever having been geographically separated. Investigators have suggested that this sympatric speciation could occur either instantaneously or over the course of multiple generations. Although most of the hypotheses suggested so far are highly controversial, one type of instantaneous sympatric speciation is known to occur commonly, as the result of polyploidy.

Instantaneous Speciation Through Polyploidy

Instantaneous sympatric speciation occurs when an individual is born that is reproductively isolated from all other members of its species. In most cases, a mutation that would cause an individual to be greatly different from others of its species would have many adverse pleiotropic side effects, and the individual would not survive. One exception often seen in plants, however, occurs through the process of **polyploidy,** which produces individuals that have more than two sets of chromosomes.

Polyploid individuals can arise in two ways. In **autopolyploidy,** all of the chromosomes may arise from a single species.

This might happen, for example, due to an error in meiosis that causes an individual to have four sets of chromosomes. Such individuals, termed *tetraploids,* could fertilize themselves or mate with other tetraploids, but could not mate and produce fertile offspring with normal diploids. The reason is that the

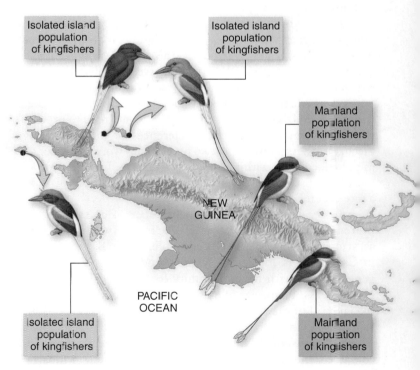

Isolated island population of kingfishers

Isolated island population of kingfishers

Mainland population of kingfishers

NEW GUINEA

PACIFIC OCEAN

Isolated island population of kingfishers

Mainland population of kingfishers

figure 22.9

PHENOTYPIC DIFFERENTIATION IN THE PAPUAN KING-FISHER *Tanysiptera hydrocharis* IN NEW GUINEA. Isolated island populations (*left*) are quite distinctive, showing variation in tail feather structure and length, plumage coloration, and bill size, whereas kingfishers on the mainland (*right*) show little variation.

offspring from such a mating would be triploid (having three sets of chromosomes) and would be sterile due to problems with chromosome pairing during meiosis.

A more common type of polyploid speciation is **allopolyploidy,** which may happen when two species hybridize (figure 22.10). The resulting offspring, having one copy of the chromosomes of each species, is usually infertile because the chromosomes do not pair correctly in meiosis. However, such individuals are often otherwise healthy, can reproduce asexually, and can even become fertile through a variety of events. For example, if the chromosomes of such an individual were to spontaneously double, as just described, the resulting tetraploid would have two copies of each set of chromosomes. Consequently, pairing would no longer be a problem in meiosis. As a result, such tetraploids would be able to interbreed, and a new species would have been created.

It is estimated that about half of the approximately 260,000 species of plants have a polyploid episode in their history, including many of great commercial importance, such as bread wheat, cotton, tobacco, sugarcane, bananas, and potatoes. Speciation by polyploidy is also known to occur in a variety of animals, including insects, fish, and salamanders, although much more rarely than in plants.

Sympatric speciation by disruptive selection

Some investigators believe that sympatric speciation can occur over the course of multiple generations through the process of disruptive selection. As noted in chapter 20, disruptive selection can cause a population to contain individuals exhibiting two different phenotypes.

One might think that if selection is strong enough, these two phenotypes would evolve over a number of generations into different species. But before the two phenotypes could become different species, they would have to evolve reproductive isolating mechanisms. Initially, the two phenotypes would not be reproductively isolated at all, and genetic exchange between individuals of the two phenotypes would tend to prevent genetic divergence in mating preferences or other isolating mechanisms. As a result, the two phenotypes would be retained as polymorphisms within a single population. For this reason, most biologists consider sympatric speciation of this type to be a rare event.

In recent years, however, a number of cases have appeared that are difficult to interpret in any way other than as sympatric speciation. For example, Lake Barombi Mbo in Cameroon is an extremely small and ecologically homogeneous lake, with no opportunity for within-lake isolation. Nonetheless, 11 species of closely related cichlid fish occur in the lake; all of the species are more closely related evolutionarily to one another than to any species outside of the lake. The most reasonable explanation is that an ancestral species colonized the lake and subsequently underwent sympatric speciation multiple times.

> Speciation occurs much more readily in the absence of gene flow between populations. Sympatric speciation can occur, however, by means of polyploidy and perhaps by disruptive selection.

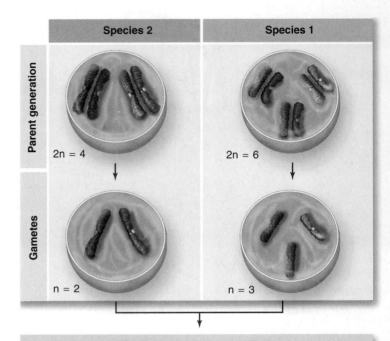

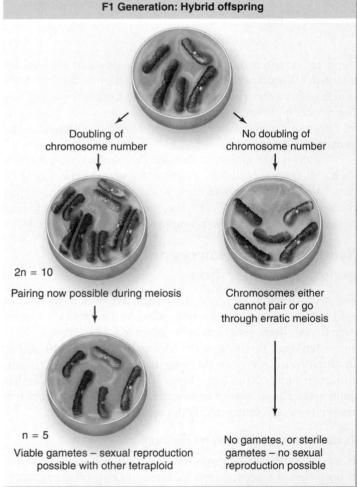

figure 22.10

ALLOPOLYPLOID SPECIATION. Hybrid offspring from parents with different numbers of chromosomes often cannot reproduce asexually. Sometimes, the number of chromosomes in such hybrids doubles to produce a tetraploid individual which can undergo meiosis and reproduce with similar tetraploid individuals.

Species Clusters: Evidence of Rapid Evolution

One of the most visible manifestations of evolution is the existence of groups of closely related species that have recently evolved from a common ancestor by adapting to different parts

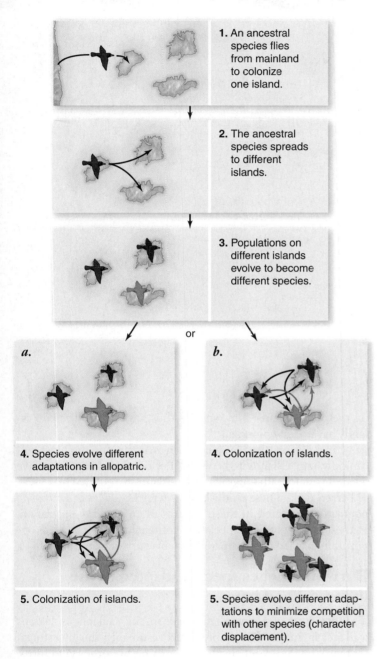

1. An ancestral species flies from mainland to colonize one island.

2. The ancestral species spreads to different islands.

3. Populations on different islands evolve to become different species.

or

a.

4. Species evolve different adaptations in allopatric.

5. Colonization of islands.

b.

4. Colonization of islands.

5. Species evolve different adaptations to minimize competition with other species (character displacement).

figure 22.11

CLASSIC MODEL OF ADAPTIVE RADIATION ON ISLAND ARCHIPELAGOES. (*1*) An ancestral species colonizes an island in an archipelago. Subsequently, the population colonizes other islands (*2*), after which the populations on the different island speciate in allopatry (*3*). Then some of these new species colonize other islands, leading to local communities of two or more species. Adaptive differences can either evolve when species are in allopatry in response to different environmental conditions (*a*) or as the result of ecological interactions between species after (*b*) by the process of character displacement.

of the environment. These **adaptive radiations** are particularly common in situations in which a species occurs in an environment with few other species and many available resources. One example is the creation of new islands through volcanic activity, such as the Hawaiian and Galápagos Islands. Another example is a catastrophic event leading to the extinction of most other species, a situation we will discuss soon in greater detail.

Adaptive radiation can also result when a new trait, called a **key innovation,** evolves within a species allowing it to use resources or other aspects of the environment that were previously inaccessible. Classic examples of key innovation leading to adaptive radiation are the evolution of lungs in fish and of wings in birds and insects, both of which allowed descendant species to diversify and adapt to many newly available parts of the environment.

Adaptive radiation requires both speciation and adaptation to different habitats. A classic model postulates that a species colonizes multiple islands in an archipelago. Speciation subsequently occurs allopatrically, and then the newly arisen species colonize other islands, producing multiple species per island (figure 22.11).

Adaptation to new habitats can occur either during the allopatric phase, as the species respond to different environments on the different islands, or after two species become sympatric. In the latter case, this adaptation may be driven by the need to minimize competition for available resources with other species. Populations on different islands evolve to become different species. In this process, termed **character displacement,** natural selection in each species favors those individuals that use resources not used by the other species. Because those individuals will have greater fitness, whatever traits cause the differences in resource use will increase in frequency (assuming that a genetic basis exists for these differences), and, over time, the species will diverge (figure 22.12).

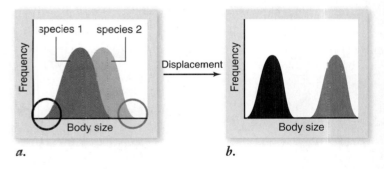

a. *b.*

figure 22.12

CHARACTER DISPLACEMENT. *a.* Two species are initially similar and thus overlap greatly in resource use, as might happen if the two species were similar in size (in many species, body size and food size are closely related). Individuals in each species that are most different from the other species (circled) will be favored by natural selection, because they will not have to compete with the other species. For example, the smallest individuals of one species and the largest of the other would not compete with the other species for food and thus would be favored. *b.* As a result, the species will diverge in resource use and minimize competition between the species.

chapter 22 the origin of species

An alternative possibility is that adaptive radiation occurs through repeated instances of sympatric speciation, producing a suite of species adapted to different habitats. As discussed earlier such scenarios are hotly debated.

In the following sections we discuss four exemplary cases of adaptive radiation.

inquiry

How would the scenario for adaptive radiation differ depending on whether speciation is allopatric? What is the relationship between character displacement and sympatric speciation?

Hawaiian *Drosophila* exploited a rich, diverse habitat

More than 1000 species in the fly genus *Drosophila* occur on the Hawaiian Islands. New species of *Drosophila* are still being discovered in Hawaii, although the rapid destruction of the native vegetation is making the search more difficult.

Aside from their sheer number, Hawaiian *Drosophila* species are unusual because of their incredible diversity of morphological and behavioral traits (figure 22.13). Evidently, when their ancestors first reached these islands, they encountered many "empty" habitats that other kinds of insects and other animals occupied elsewhere. As a result, the species have adapted to all manners of fruit fly life and include predators, parasites, and herbivores, as well as species specialized for eating the detritus in leaf litter and the nectar of flowers. The larvae of various species live in rotting stems, fruits, bark, leaves, or roots, or feed on sap. No comparable diversity of *Drosophila* species is found anywhere else in the world.

The great diversity of Hawaiian species is a result of the geological history of these islands. New islands have continually arisen from the sea in this region. As they have done so, they appear to have been invaded successively by the various *Drosophila* groups present

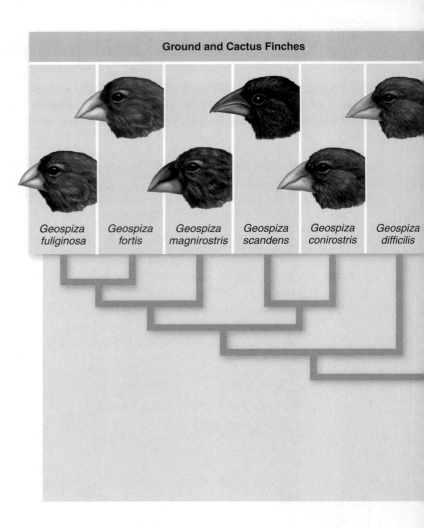

Ground and Cactus Finches

Geospiza fuliginosa   *Geospiza fortis*   *Geospiza magnirostris*   *Geospiza scandens*   *Geospiza conirostris*   *Geospiza difficilis*

on the older islands. New species thus have evolved as new islands have been colonized.

In addition, the Hawaiian Islands are among the most volcanically active islands in the world. Periodic lava flows often have created patches of habitat within an island surrounded by a "sea" of barren rock. These land islands are termed *kipukas. Drosophila* populations isolated in these kipukas often undergo speciation. In these ways, rampant speciation combined with ecological opportunity has led to an unparalleled diversity of insect life.

Darwin's finches species adapted to use different food types

The diversity of Darwin's finches on the Galápagos Islands was first mentioned in chapter 21. Presumably, the ancestor of Darwin's finches reached these islands before other land birds, and many of the types of habitats which other types of birds use on the mainland were unoccupied.

As the new arrivals moved into these vacant ecological niches and adopted new lifestyles, they were subjected to many different sets of selective pressures. Under these circumstances, and aided by the geographic isolation afforded by the many islands of the Galápagos archipelago, the ancestral finches rapidly split into a series of diverse populations, some of which evolved into separate species. These species now occupy many different habitats on the Galápagos Islands, which are comparable to the habitats several distinct groups of birds occupy on

a.   *b.*

figure 22.13

HAWAIIAN *Drosophila*. The hundreds of species that have evolved on the Hawaiian Islands are extremely variable in appearance, although genetically almost identical. *a. Drosophila heteroneura. b. Drosophila digressa.*

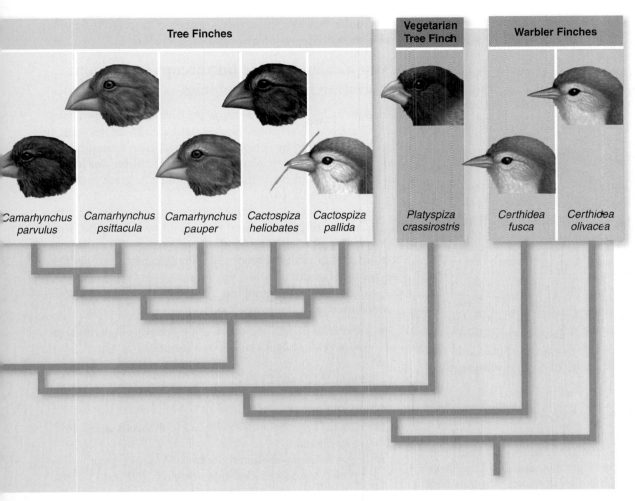

Tree Finches

Camarhynchus parvulus
Camarhynchus psittacula
Camarhynchus pauper
Cactospiza heliobates
Cactospiza pallida

Vegetarian Tree Finch

Platyspiza crassirostris

Warbler Finches

Certhidea fusca
Certhidea olivacea

figure 22.14

AN EVOLUTIONARY TREE OF DARWIN'S FINCHES. This evolutionary tree, derived from examination of DNA sequences, suggests that warbler finches are an early offshoot. Ground and tree finches subsequently diverged, and then species within each group specialized to use different resources. Recent studies have shown, surprisingly, that the two warbler finches are not each other's closest relatives. Rather, *Certhidea fusca* is more closely related to the remaining Darwin's finches than it is to *C. olivacea*.

the mainland. As illustrated in figure 22.14, the 14 species fall into four groups:

1. **Ground finches.** There are six species of *Geospiza* ground finches. Most of the ground finches feed on seeds. The size of their bills is related to the size of the seeds they eat. Some of the ground finches feed primarily on cactus flowers and fruits, and they have a longer, larger, and more pointed bill than the others.
2. **Tree finches.** There are five species of insect-eating tree finches. Four species have bills suitable for feeding on insects. The woodpecker finch has a chisel-like beak. This unusual bird carries around a twig or a cactus spine, which it uses to probe for insects in deep crevices.
3. **Vegetarian finch.** The very heavy bill of this species is used to wrench buds from branches.
4. **Warbler finches.** These unusual birds play the same ecological role in the Galápagos woods that warblers play on the mainland, searching continuously over the leaves and branches for insects. They have slender, warblerlike beaks.

Recently, scientists have examined the DNA of Darwin's finches to study their evolutionary history. These studies suggest that the deepest branches in the finch evolutionary tree lead to warbler finches, which implies that warbler finches were among the first types to evolve after colonization of the islands. All of the ground species are closely related to one another, as are all of

the tree finches. Nonetheless, within each group, species differ in beak size and other attributes, as well as in resource use.

Field studies, conducted in conjunction with those discussed in chapter 21, demonstrate that ground species compete for resources; the differences between species likely resulted from character displacement as initially similar species diverged to minimize competitive pressures.

Lake Victoria cichlid fishes diversified very rapidly

Lake Victoria is an immense, shallow, freshwater sea about the size of Switzerland in the heart of equatorial East Africa. Until recently, the lake was home to an incredibly diverse collection of over 300 species of cichlid fishes.

Geologically recent radiation

The cluster of cichlid species appears to have evolved recently and quite rapidly. By sequencing the cytochrome *b* gene in many of the lake's fish, scientists have been able to estimate that the first cichlids entered Lake Victoria only 200,000 years ago, colonizing from the Nile.

Dramatic changes in water level encouraged species formation. As the lake rose, it flooded new areas and opened up new habitats. Many of the species may have originated after the lake dried down 14,000 years ago, isolating local populations in small lakes until the water level rose again.

Cichlid diversity

Cichlids are small, perchlike fishes ranging from 5 to 25 centimeters (cm) in length, and the males come in endless varieties of colors. The ecological and morphological diversity of these fish is remarkable, particularly given the short span of time over which they have evolved.

We can gain some sense of the vast range of types by looking at how different species eat. There are mud biters, algae scrapers, leaf chewers, snail crushers, zooplankton eaters, insect eaters, prawn eaters, and fish eaters. Snail shellers pounce on slow-crawling snails and spear their soft parts with long, curved teeth before the snail can retreat into its shell. Scale scrapers rasp slices of scales off other fish. There are even cichlid species that are "pedophages," eating the young of other cichlids.

Cichlid fish have a remarkable key innovation that may have been instrumental in their evolutionary radiation: They carry a second set of functioning jaws (figure 22.15). This trait occurs in many other fish, but in cichlids it is greatly enlarged. The ability of these second jaws to manipulate and process food has freed the oral jaws to evolve for other purposes, and the result has been the incredible diversity of ecological roles filled by these fish.

Abrupt extinction in the last several decades

Recently, much of the cichlid diversity has disappeared. In the 1950s, the Nile perch, a large commercial fish with a voracious appetite, was introduced on the Ugandan shore of Lake Victoria. Since then, it has spread through the lake, eating its way through the cichlids.

By 1990, many of the open-water cichlid species had become extinct, as well as others living in rocky shallow regions. Over 70% of all the named Lake Victoria cichlid species had disappeared, as well as untold numbers of species that had yet to be described. We will revisit the story of Lake Victoria when we discuss conservation biology in chapter 59.

New Zealand alpine buttercups underwent speciation in glacial habitats

Adaptive radiations such as those we have described in Hawaiian *Drosophila*, Galápagos finches, and cichlid fishes seem to have been favored by periodic isolation. A clear example of the role periodic isolation plays in species formation can be seen in the alpine buttercups that grow among the glaciers of New Zealand (figure 22.16).

More species of alpine buttercups grow on the two main islands of New Zealand than in all of North and South America combined. The evolutionary mechanism responsible for this diversity is recurrent isolation associated with the recession of glaciers.

The 14 species of alpine buttercups occupy five distinctive habitats within glacial areas:

- *snowfields*—rocky crevices among outcrops in permanent snowfields at 2130- to 2740-m elevation;
- *snowline fringe*—rocks at lower margin of snowfields between 1220 and 2130 m;
- *stony debris*—slopes of exposed loose rocks at 610 to 1830 m;
- *sheltered situations*—shaded by rock or shrubs at 305 to 1830 m; and
- *boggy habitats*—sheltered slopes and hollows, poorly drained tussocks at elevations between 760 and 1525 m.

Buttercup speciation and diversification have been promoted by repeated cycles of glacial advance and retreat. As the glaciers retreat up the mountains, populations become isolated on mountain peaks, permitting speciation (figure 22.16). In the next glacial advances, these new species can expand throughout the mountain

figure 22.15

CICHLID FISHES OF LAKE VICTORIA. These fishes have evolved adaptations to use a variety of different habitats. The enlarged second set of jaws located in the throat of these fish has provided evolutionary flexibility, allowing oral jaws to be modified in many ways.

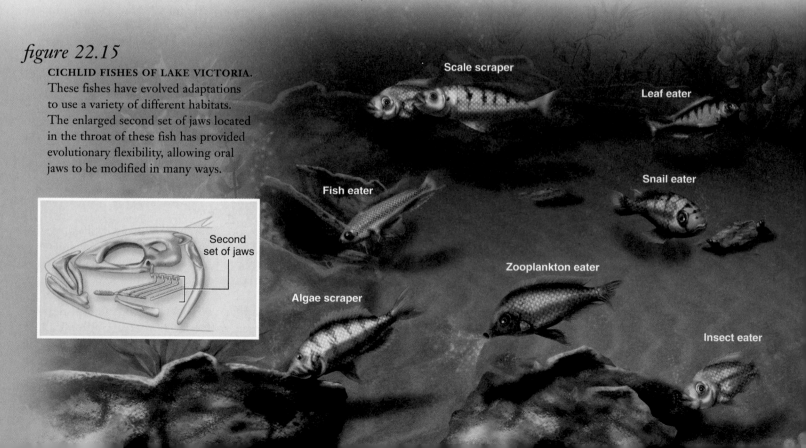

Second set of jaws

Scale scraper

Leaf eater

Fish eater

Snail eater

Zooplankton eater

Algae scraper

Insect eater

snowfield   snowline fringe   stony debris   sheltered   boggy

a.

Glaciers link alpine zones into one
continuous range.

→ Glaciers recede

Mountain populations become isolated,
permitting divergence and speciation.

→ Glaciation

Alpine zones are reconnected. Separately
evolved species come back into contact.

b.

figure 22.16

NEW ZEALAND ALPINE BUTTERCUPS (genus *Ranunculus*). Periodic glaciation encouraged species formation among alpine buttercups in New Zealand. *a.* Fourteen species of alpine *Ranunculus* grow among the glaciers and mountains of New Zealand. *b.* The formation of extensive glaciers during the Pleistocene epoch linked the alpine zones (*white*) of many mountains together. When the glaciers receded, these alpine zones were isolated from one another, only to become reconnected with the advent of the next glacial period. During periods of isolation, populations of alpine buttercups diverged in the isolated habitats.

range, coming into contact with their close relatives. In this way, one initial species could give rise to many descendants. Moreover, on isolated mountaintops during glacial retreats, species have convergently evolved to occupy similar habitats; these distantly related but ecologically similar species have then been brought back into contact in subsequent glacial advances.

> Adaptive radiation occurs when a species diversifies, producing descendant species that are adapted to use many different parts of the environment. Adaptive radiation is facilitated by recurrent isolation, which increases the rate at which speciation occurs, and by occupation of areas, such as volcanic islands, where there are few competitors and many types of available resources.

22.7 The Pace of Evolution

We have discussed the manner in which speciation may occur, but we haven't yet considered the relationship between speciation and the evolutionary change that occurs within a species. Two hypotheses, *gradualism* and *punctuated equilibrium*, have been advanced to explain the relationship.

Gradualism is the accumulation of small changes

For more than a century after the publication of *On the Origin of Species*, the standard view was that evolutionary change occurred extremely slowly. Such change would be nearly imperceptible

from generation to generation, but would accumulate such that, over the course of thousands and millions of years, major changes could occur. This view is termed **gradualism** (figure 22.17*a*).

Punctuated equilibrium is long periods of stasis followed by relatively rapid change

Gradualism was challenged in 1972 by paleontologists Niles Eldredge of the American Museum of Natural History in New York and Stephen Jay Gould of Harvard University, who argued that species experience long periods of little or no evolutionary change (termed **stasis**), punctuated by bursts of evolutionary

a. Gradualism   **b.** Punctuated equilibrium

figure 22.17

TWO VIEWS OF THE PACE OF MACROEVOLUTION.
a. Gradualism suggests that evolutionary change occurs slowly through time and is not linked to speciation, whereas **(b)** punctuated equilibrium surmises that phenotypic change occurs in bursts associated with speciation, separated by long periods of little or no change.

change occurring over geologically short time intervals. They called this phenomenon **punctuated equilibrium** (figure 22.17*b*) and argued that these periods of rapid change occurred only during the speciation process.

Initial criticism of the punctuated equilibrium hypothesis focused on whether rapid change could occur over short periods of time. As we have seen in the last two chapters, however,

when natural selection is strong, rapid and substantial evolutionary change can occur. A more difficult question involves the long periods of lack of change: Why would species exist for thousands, even millions, of years without changing?

Although a number of possible reasons have been suggested, most researchers now believe that a combination of stabilizing and oscillating selection is responsible for stasis. If the environment does not change over long periods of time, or if environmental changes oscillate back and forth, then selection may favor stasis, even for long periods. One factor that may enhance this stasis is the ability of species to shift their ranges; for example, during the ice ages, when the global climate cooled, the geographic ranges of many species shifted southward, so that the species continued to experience similar environmental conditions.

inquiry

? *Why would changes in geographic ranges promote evolutionary stasis?*

Evolution may include both types of change

Eldredge and Gould's proposal prompted a great deal of research. Some well-documented groups, such as African mammals, clearly have evolved gradually, not in spurts. Other groups, such as marine bryozoa, seem to show the irregular pattern of evolutionary change predicted by the punctuated equilibrium model. It appears, in fact, that gradualism and punctuated equilibrium are two ends of a continuum. Although some groups appear to have evolved solely in a gradual manner and others only in a punctuated mode, many other groups show evidence of both gradual and punctuated episodes at different times in their evolutionary history.

The idea that speciation is necessarily linked to phenotypic change has not been supported, however. It is now clear that speciation can occur without substantial phenotypic change, and that phenotypic change can occur within species in the absence of speciation.

Evolutionary change can be slow and gradual (gradualism). Or, it may be rapid and discontinuous, separated by long periods of stasis (punctuated equilibrium). The latter may result from a combination of stabilizing and oscillating selection. The link between speciation and phenotypic change, however, has not been supported.

22.8 ## Speciation and Extinction Through Time

Biological diversity has increased vastly since the Cambrian period, but the trend has been far from consistent. After a rapid rise, diversity reached a plateau for about 200 million years, but since then has risen steadily. Because changes in the number of species reflect the rate of origin of new species relative to the rate at which existing species disappear, this long-term trend reveals that speciation has, in general, surpassed extinction.

Nonetheless, speciation has not always outpaced extinction. In particular, interspersed in the long-term increase in species diversity have been a number of sharp declines, termed **mass extinctions.**

Five mass extinctions have occurred in the distant past

Five major mass extinctions have been identified, the most severe one occurring at the end of the Permian period, approximately 250 million years ago (figure 22.18). At that time, more than half of all plant and animal families and as much as 96% of all species may have perished.

The most famous and well-studied extinction, although not as drastic, occurred at the end of the Cretaceous period (65 million years ago), at which time the dinosaurs and a variety of other organisms went extinct. Recent findings have supported the hypothesis that this extinction event was triggered when a large asteroid slammed into Earth, perhaps causing global forest fires and obscuring the Sun for months by throwing particles into the air. The cause of other mass extinction events is less certain. Some scientists suggest that asteroids may have played a role in at least some of the other mass extinction events; other theories implicate global climate change and other causes.

One important result of mass extinctions is that not all groups of organisms are affected equally. For example, in the extinction at the end of the Cretaceous, not only dinosaurs, but also marine and flying reptiles, and ammonites (a type of mollusk) went extinct. Marsupials, flowering plants, birds, and some forms of plankton were greatly reduced in diversity. In contrast, turtles, crocodilians, and amphibians seemed to have been unscathed. Why some groups were harder hit than others is not clear, but one theory suggests that survivors were those animals that could shelter underground or in water, and that could either scavenge or required little food in the cool temperatures that resulted from the blockage of sunlight.

A consequence of mass extinctions is that previously dominant groups may perish, thus changing the course of evolution. This is certainly true of the Cretaceous extinction. During the Cretaceous period, placental mammals were a minor group composed of species that were mostly no larger than a house cat. When the dinosaurs, the dominant animal of the world for more than 100 million years, disappeared at the end of this period, the placental mammals underwent a significant adaptive radiation. It is humbling to think that humans might never have arisen had that asteroid not struck Earth 65 million years ago.

As the world around us illustrates today, species diversity does rebound after mass extinctions, but this recovery is not rapid. Examination of the fossil record indicates that rates of speciation do not immediately increase after an extinction pulse, but rather take about 10 million years to reach their maximum. The cause of this delay is not clear, but it may result because it takes time for ecosystems to recover and for the processes of speciation and adaptive diversification to begin. Consequently, species diversity may require 10 million years, or even much longer, to attain its previous level.

A sixth extinction is underway

The number of species in the world in recent times is greater than it has ever been. Unfortunately, that number is decreasing at an alarming rate due to human activities (see chapter 59).

Some estimate that as much as one-fourth of all species will become extinct in the near future, a rate of extinction not seen on Earth since the Cretaceous mass extinction. Moreover, the rebound in species diversity may be even slower than following previous mass extinction events because, instead of the ecologically impoverished, but energy-rich environment that existed after previous mass extinction events, a large proportion of the world's resources will be already taken up by human activities, leaving few resources available for adaptive radiation.

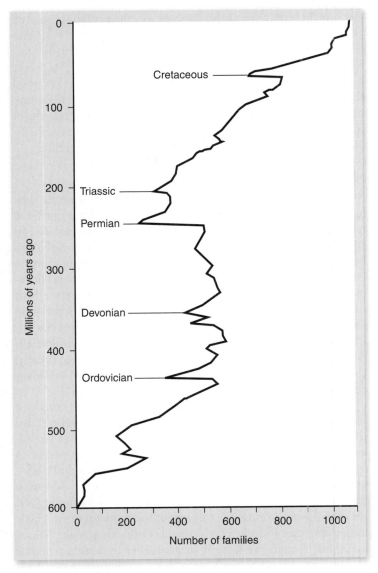

figure 22.18

BIODIVERSITY THROUGH TIME. The taxonomic diversity of families of marine animals has increased since the Cambrian period, although occasional dips have occurred. The fossil record is most complete for marine organisms because they are more readily fossilized than terrestrial species. Families are shown, rather than species, because many species are known from only one specimen, thus introducing error into estimates of the time of extinction. Arrows indicate the five major mass extinction events.

The number of species has increased through time, although not at a constant rate. Several major extinction events have substantially, though briefly, reduced the number of species. Diversity rebounds, but the recovery is not rapid, and the groups making up that diversity are not the same as those that existed before the extinction event.

The Future of Evolution

In this chapter and in chapters 20 and 21, we have discussed the results of evolution through time. What does the future hold? Global biodiversity seems headed for a major extinction event from which recovery will be slow. Does this mean the end of evolution? We can use what we know about evolutionary processes to predict how evolution will occur in the future, both for diversity in general and for the human species in particular.

The future operation of evolutionary processes

Human influences on the environment affect the evolutionary process in many ways. Most obviously, by changing the environment, humans are changing the patterns of natural selection. In many cases, these changes will be so drastic that populations will be unable to adapt. But for those species that can survive, natural selection will act on genetic variation to produce evolutionary change. Global climate change, in particular, will be a major challenge, leading either to evolutionary change or extinction for many species.

Other factors will also lead to evolutionary change. Decreased population sizes will increase the likelihood of genetic drift, and geographic isolation of formerly connected populations will remove the homogenizing effect of gene flow, allowing these populations to evolve differences as adaptations to local environments. Chemicals and radiation released into the environment could even increase the mutation rate.

Consequently, for those species able to survive, evolutionary processes will continue and in some cases will even be accelerated. But what about species diversity?

Extinction rates are increasing vastly, but it is also possible that speciation rates may increase, at least in some circumstances. The reason is that many formerly widespread species now exist only as geographically isolated populations (figure 22.19). Moreover, humans have introduced species to localities in which they formerly didn't occur, thus creating isolated populations. Given the importance of allopatry in the speciation process, these actions are likely to increase speciation rates for some species.

This is not to say that geographic fragmentation is a good thing: Many, perhaps most, small populations will go extinct long before they can undergo speciation, and any increased rate of speciation that does occur is not likely to make up for the vastly increased rate of extinction, at least not for a very long time.

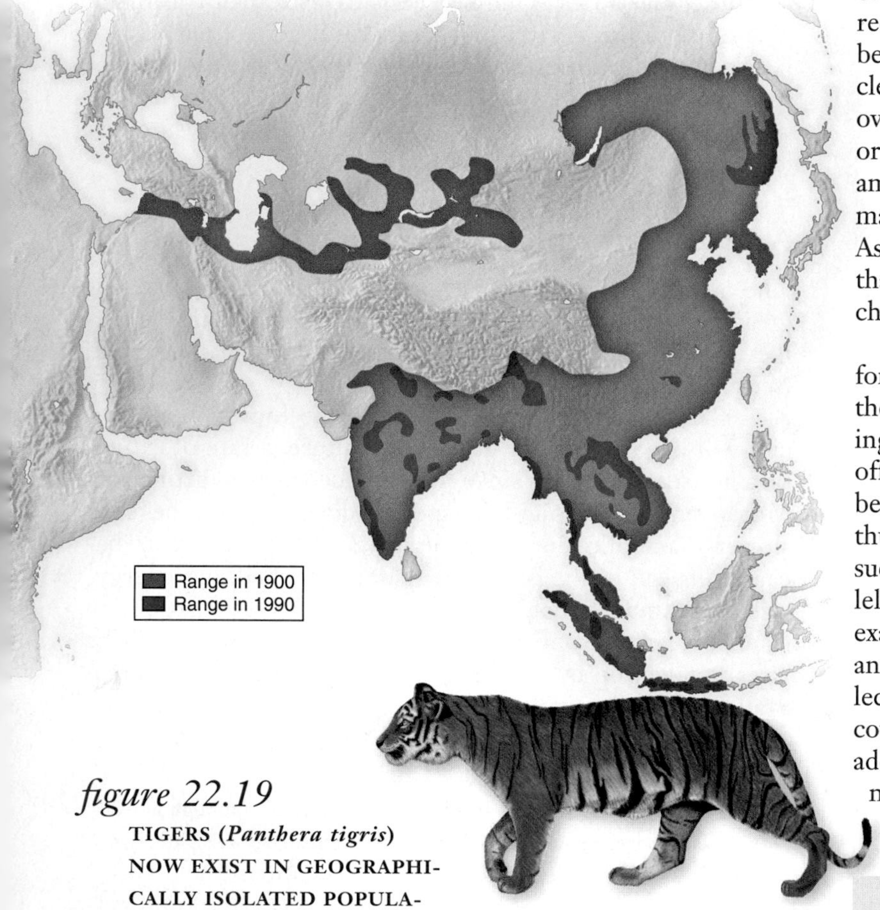

Human evolutionary future

Many science fiction writers have speculated about where evolution will take the human species, but consideration of the evolutionary process suggests that these ideas are fanciful. In recent times, the movement of people around the globe has begun to erase regional differences in human populations, a clear example of the homogenizing effect of gene flow. Moreover, to an ever-increasing extent, people with different ethnic origins are reproducing, further diminishing differentiation among human populations. Because of the huge size of the human population, genetic drift is unlikely to be very important. Assuming that the rate of mutations doesn't increase greatly, that leaves only natural selection as an engine of evolutionary change in humans.

The question then becomes: Do the conditions necessary for evolution by natural selection exist in humans? That is, are there phenotypic traits that both affect the number of surviving offspring and that are genetically inherited from parent to offspring? Certainly, in one way, human populations will evolve because many genetic diseases that used to be fatal, and would thus remove alleles from the population, can now be treated successfully. As a result, we can expect the frequency of such alleles to increase in future generations. Other than such obvious examples, though, we leave it to the reader to ponder whether and in what cases the conditions for evolution by natural selection are likely to be met in future human populations. Of course, the advent of the genomics revolution (see chapter 19) adds another dimension to this discussion. Will future technological advances allow us to alter the human gene pool directly? And, if so, will that be a good idea?

figure 22.19

TIGERS (*Panthera tigris*) NOW EXIST IN GEOGRAPHICALLY ISOLATED POPULATIONS. Human activities, such as hunting and habitat destruction, have greatly reduced populations of the tiger and have fragmented the species' range into many small and isolated populations.

- ■ Range in 1900
- ■ Range in 1990

For those species able to avoid extinction, human-caused changes may lead to evolutionary adaptation and, in some cases, to the formation of new species.

22.1 The Nature of Species

Any species concept must account for the distinctiveness of co-occurring species and the cohesion among populations of the same species.

- Sympatric species that co-occur differ phenotypically in their behavior and utilize different parts of the habitat.

- Subspecies are geographically distant populations of the same species that are distinct from one another.

- Populations separated geographically may have intermediate populations between them.

22.2 The Biological Species Concept

Biological species are generally defined as groups of populations that have either the potential to or do interbreed with each other and produce fertile offspring.

- Populations that either do not mate with each other or if they do produce infertile offspring are reproductively isolated.

- Reproductive-isolating mechanisms prevent genetic exchange between species.

- Prezygotic isolation prevents mating and formation of zygotes between two species.

- Postzygotic isolation occurs after zygote formation and prevents development of reproducing adults (figure 22.5).

- The high incidence of natural and artificial hybridization among species has generated alternative definitions of species and has led many scientists to look at the evolutionary history of populations.

22.3 The Evolution of Reproductive Isolation

If species are defined by the existence of reproductive isolation, then the process of speciation is the same as the evolution of reproductive-isolating mechanisms.

- Populations may evolve complete reproductive isolation in allopatry.

- If populations that have evolved only partial reproductive isolation come into contact, either natural selection can lead to the evolution of increased reproductive isolation or gene flow can lead to homogenization of the populations.

22.4 The Role of Genetic Drift and Natural Selection in Speciation

In addition to reinforcement, natural selection can play other roles in the evolution of reproductive-isolating mechanisms.

- In small populations, random genetic drift, whether by founder effect or bottleneck effect, may result in reproductive isolation.

- Adaptation to different situations or environments may incidentally lead to reproductive isolation through the accumulation of differences.

- Natural selection can directly select for traits that increase reproductive isolation.

22.5 The Geography of Speciation (figure 22.9)

Speciation occurs in two parts: population divergence and reproductive isolation.

- Allopatric, or geographically isolated, populations are much more likely to evolve into separate species because there is no gene flow.

- Sympatric speciation occurs without geographical separation by polyploidy and disruptive selection.

- Autopolyploidy can instantaneously create a new species because of errors in meiosis, resulting in tetraploidy.

- Allopolyploidy can create a new species with a mixture of genes, and the hybrids reproduce asexually. If there is subsequent autopolyploidy, a new sexually reproducing species can be created.

- Disruptive selection may result in sympatric speciation.

22.6 Species Clusters: Evidence of Rapid Evolution

Groups of closely related species evolve from a common ancestor by adapting to different parts of a new accessible environment (figure 22.12).

- Adaptive radiation is common when a species inhabits a new environment with many resources and few competing species or if there is abrupt extinction of many species resulting in increased accessibility to new environments.

- Character displacement involves the evolution of divergent adaptations that minimizes competition between two species for available resources.

22.7 The Pace of Evolution

The relationship between speciation and evolutionary change is on a continuum between gradualism and punctuated equilibrium.

- Gradualism is defined as the slow and steady accumulation of changes over long periods of time.

- Punctuated equilibrium is the relatively quick change in speciation followed by long periods with little or no evolutionary change because of stabilizing or oscillating selection.

22.8 Speciation and Extinction Through Time

In general, speciation rates have surpassed extinction rates (figure 22.18).

- Mass extinctions have occurred five times in the distant past due to asteroids hitting the Earth and global climate change, among other events.

- Human activities are creating a sixth mass extinction.

- During mass extinctions not all species are equally affected, allowing adaptive radiation to occur among the remaining species.

- Species diversity slowly rebounds after mass extinctions.

22.9 The Future of Evolution

Human influences on the environment affect evolutionary processes.

- Decreased population size due to habitat fragmentation and pollution will increase genetic drift and isolate previously connected populations. Small populations may evolve or become extinct.

- The introduction of alien species into new environments has increased the isolation of populations and may increase speciation.

- The human gene pool is becoming homogenized as people from different ethnic origins reproduce.

SELF TEST

1. Prezygotic isolating mechanisms include all of the following except
   a. hybrid sterility.
   b. courtship rituals.
   c. habitat separation.
   d. seasonal reproduction.
2. Reproductive isolation
   a. is a result of individuals not mating with each other.
   b. is a specific type of postzygotic isolating mechanism.
   c. is required by the biological species concept.
   d. none of the above.
3. Leopard frogs from different geographic populations of the *Rana pipiens* complex
   a. are members of a single species because they look very similar to one another.
   b. are different species shown to have pre- and postzygotic isolating mechanisms.
   c. frequently interbreed to produce viable hybrids.
   d. are genetically identical due to effective reproductive isolation.
4. _____ isolating mechanisms include improper development of hybrids and failure of hybrids to become established in nature.
   a. Prezygotic
   b. Postzygotic
   c. Temporal
   d. Mechanical
5. Problems with the biological species concept include the fact that
   a. many species reproduce asexually.
   b. postzygotic isolating mechanisms decrease hybrid viability.
   c. prezygotic isolating mechanisms are extremely rare.
   d. all of these.
6. Cladogenesis
   a. is a type of prezygotic isolating mechanism.
   b. is a type of postzygotic isolating mechanism.
   c. only occurs in plants.
   d. none of the above.
7. If reinforcement is weak and hybrids are not completely infertile,
   a. genetic divergence between populations may be overcome by gene flow.
   b. speciation will occur 100% of the time.
   c. gene flow between populations will be impossible.
   d. the speciation will be more likely than if hybrids were completely infertile.
8. Natural selection
   a. can enhance the probability of speciation.
   b. can enhance reproductive isolation.
   c. can act against hybrid survival and reproduction.
   d. all of these.
9. Allopatric speciation
   a. is less common than sympatric speciation.
   b. involves geographic isolation of some kind.
   c. is the only kind of speciation that occurs in plants.
   d. requires polyploidy.
10. Cichlid diversity can be attributed to
    a. adaptive radiation.
    b. new habitats and geographic isolation.
    c. a second set of jaws in the throat of the fish.
    d. All of the above contributed to cichlid diversity.
11. The hypothesis that evolution occurs in spurts, with great amounts of evolutionary change followed by periods of stasis, is
    a. punctuated equilibrium.
    b. allopatric speciation.
    c. gradualism.
    d. Hardy–Weinberg equilibrium.
12. Gradualism and punctuated equilibrium are
    a. two ends of continuum of the rate of evolutionary change over time.
    b. mutually exclusive views about how all evolutionary change takes place.
    c. mechanisms of reproductive isolation.
    d. none of the above.
13. During the history of life on Earth
    a. there have been major extinction events.
    b. species diversity has steadily increased.
    c. species diversity has stayed relatively constant.
    d. extinction rates have been completely offset by speciation rates.
14. Character displacement
    a. arises through competition and natural selection, favoring divergence in resource use.
    b. arises through competition and natural selection, favoring convergence in resource use.
    c. does not promote speciation.
    d. reduced speciation rates in Galápagos finches.
15. Hybridization between incompletely isolated populations
    a. always leads to reinforcement due to the inferiority of hybrids.
    b. can serve as a mechanism for preserving gene flow between populations.
    c. only occurs in plants.
    d. never affects rates of speciation.

CHALLENGE QUESTIONS

1. Natural selection can lead to the evolution of prezygotic isolating mechanisms, but not postzygotic isolating mechanisms? Explain.
2. If there is no universally accepted definition of a species, what good is the term? Will the idea of and need for a "species concept" be eliminated in the future?
3. Refer to figure 22.6. In Europe, pied and collared flycatchers are dissimilar in sympatry, but very similar in allopatry, consistent with character divergence in coloration. In this case, there is no competition for ecological resources as in other cases of character divergence discussed. How might this example work?
4. Refer to figure 22.14. *Geospiza fuliginosa* and *Geospiza fortis* are found in sympatry on at least one island in the Galápagos and in allopatry on several islands in the same archipelago. Compare your expectations about degree of morphological similarity of the two species in these two contexts, given the hypothesis that competition for food played a large role in the adaptive radiation of this group. Would your expectations be the same for a pair of finch species that are not as closely related? Explain.

Systematics and the Phylogenetic Revolution

introduction

ALL ORGANISMS SHARE MANY biological characteristics. They are composed of one or more cells, carry out metabolism and transfer energy with ATP, and encode hereditary information in DNA. Yet, there is also a tremendous diversity of life, ranging from bacteria and amoebas to blue whales and sequoia trees. For generations, biologists have tried to group organisms based on shared characteristics. The most meaningful groupings are based on the study of evolutionary relationships among organisms. New methods for constructing evolutionary trees and a sea of molecular sequence data are leading to improved evolutionary hypotheses to explain life's diversification.

23.1 Systematics

One of the great challenges of modern science is to understand the history of ancestor–descendant relationships that unites all forms of life on Earth, from the earliest single-celled organisms to the complex organisms we see around us today. If the fossil record were perfect, we could trace the evolutionary history of species and examine how each arose and proliferated; however, as discussed in chapter 21, the fossil record is far from complete. Although it answers many questions about life's diversification, it leaves many others unsettled.

Consequently, scientists must rely on other types of evidence to establish the best hypothesis of evolutionary relationships. Bear in mind that the outcomes of such studies *are* hypotheses, and as such, they require further testing. All hypotheses may be disproven by new data, leading to the formation of better, more accurate scientific ideas.

The reconstruction and study of evolutionary relationships is called **systematics.** By looking at the similarities and differences between species, systematists can construct an evolutionary tree, or **phylogeny,** which represents a hypothesis about patterns of relationship among species.

Branching diagrams depict evolutionary relationships

Darwin envisioned that all species were descended from a single common ancestor, and that the history of life could be depicted as a branching tree (figure 23.1). In Darwin's view, the twigs of the tree represent existing species. As one works down the tree, the joining of twigs and branches reflects the pattern of common ancestry back in time to the single common ancestor of all life. The process of descent with modification from common ancestry results in all species being related in this branching, hierarchical fashion, and their evolutionary history can be depicted using branching diagrams or phylogenetic trees. Figure 23.1*b* shows how evolutionary relationships are depicted with a branching diagram. Humans and chimpanzees are descended from a common ancestor and are each other's closest living relative (the position of this common ancestor is indicated by the node labeled 1). Humans, chimps, and gorillas share an older common ancestor (node 2), and all great apes share a more distant common ancestor (node 3).

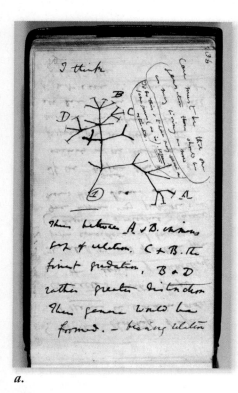

a.

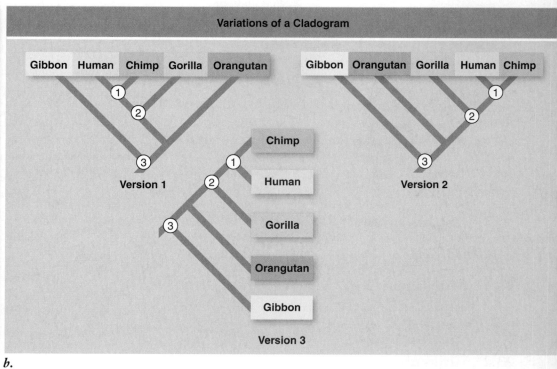

b.

figure 23.1

PHYLOGENIES DEPICT EVOLUTIONARY RELATIONSHIPS. *a.* A drawing from one of Darwin's notebooks, written in 1837 as he developed his ideas that led to *On the Origin of Species.* Darwin viewed life as a branching process akin to a tree, with species on the twigs, and evolutionary change represented by the branching pattern displayed by a tree as it grows. *b.* An example of a phylogeny. Humans and chimpanzees are more closely related to each other than they are to any other living species. This is apparent because they share a common ancestor (the node labeled 1) that was not an ancestor of other species. Similarly, humans, chimpanzees, and gorillas are more closely related to each other than any of them is to orangutans because they share a common ancestor (node 2) that was not ancestral to orangutans. Node 3 represents the common ancestor of all apes. Note that these three figure versions convey the same information despite the differences in arrangement of species and orientation.

One key to interpreting a phylogeny is to look at how recently species share a common ancestor, rather than looking at the arrangement of species across the top of the tree. If you compare the three versions of the phylogeny of figure 23.1*b*, you can see that the relationships are the same: Regardless of where they are positioned, chimpanzees and humans are still more closely related to each other than to any other species.

Moreover, even though humans are placed next to gibbons in version 1 of figure 23.1*b*, the pattern of relationships still indicates that humans are more closely related (that is, share a more recent common ancestor) with gorillas and orangutans than with gibbons. Phylogenies are also sometimes displayed on their side, rather than upright figure 23.1*b* (version 3), but this arrangement also does not affect its interpretation.

Similarity may not accurately predict evolutionary relationships

We might expect that the greater the time since two species diverged from a common ancestor, the more different they would be. Early systematists relied on this reasoning and constructed phylogenies based on overall similarity. If, in fact, species evolved at a constant rate, then the amount of divergence between two species would be a function of how long they had been diverging, and thus phylogenies based on degree of similarity would be accurate. As a result, we might think that chimps and gorillas are more closely related to each other than either is to humans.

But as chapter 22 revealed, evolution can occur very rapidly at some times and very slowly at others. In addition, evolution is not unidirectional—sometimes species' traits evolve in one direction, and then back the other way (a result of oscillating selection; see chapter 20). Species invading new habitats are likely to experience new selective pressures and may change greatly; those staying in the same habitats as their ancestors may change only a little. For this reason, similarity is not necessarily a good predictor of how long it has been since two species shared a common ancestor.

A second fundamental problem exists as well: Evolution is not always divergent. In chapter 21, we discussed **convergent evolution,** in which two species independently evolve the same features. Often, species evolve convergently because they use similar habitats, in which similar adaptations are favored. As a result, two species that are not closely related may end up more similar to each other than they are to their close relatives. Evolutionary reversal, the process in which a species re-evolves the characteristics of an ancestral species, also has this effect.

Systematics is the study of evolutionary relationships. Phylogenies, or phylogenetic trees, are graphic representations of relationships among species. Similarity of organisms alone does not necessarily correlate with their relatedness because evolutionary change is not constant in rate and direction.

Cladistics

For these reasons, most systematists no longer construct their phylogenetic hypotheses solely on the basis of similarity. Rather, they distinguish similarity that is inherited from the most recent common ancestor of an entire group, which is called **derived,** from similarity that arose prior to the common ancestor of the group, which is termed **ancestral.** In this approach, termed **cladistics,** only **shared derived characters** are considered informative in determining evolutionary relationships.

The cladistic method requires that character variation be identified as ancestral or derived

To employ the method of cladistics, systematists first gather data on a number of characters for all the species in the analysis. Characters can be any aspect of the phenotype, including morphology, physiology, behavior, and DNA. As chapters 18 and 24 show, the revolution in genomics should soon provide a vast body of data that may revolutionize our ability to identify and study character variation.

To be useful, the characters should exist in recognizable **character states.** For example, consider the character "teeth" in amniote vertebrates (namely birds, reptiles, and mammals; see chapter 35). This character has two states: presence in most mammals and reptiles and absence in birds and a few other groups such as turtles.

Examples of ancestral versus derived characters

The presence of hair is a shared derived feature of mammals (figure 23.2); in contrast, the presence of lungs in mammals is an ancestral feature because it is also present in amphibians and reptiles (represented by a salamander and a lizard) and therefore presumably evolved prior to the common ancestor of mammals (see figure 23.2). The presence of lungs, therefore, does not tell us that mammal species are all more closely related to each other than to reptiles or amphibians, but the shared, derived feature of hair suggests that all mammal species share a common ancestor that existed more recently than the common ancestor of mammals, amphibians, and reptiles.

To return to the question concerning the relationships of humans, chimps, and gorillas, a number of morphological and DNA characters exist that are derived and shared by chimps and humans, but not by gorillas or other great apes. These characters suggest that chimps and humans diverged from a common ancestor (see figure 23.1*b*, node 1) that existed more recently than the common ancestor of gorillas, chimps, and humans (node 2).

Determination of ancestral versus derived

Once the data are assembled, the first step in a manual cladistic analysis is to **polarize** the characters—that is, to determine whether particular character states are ancestral or derived. To polarize the character "teeth," for example, systematists must determine which state—presence or absence—was exhibited by the most recent common ancestor of this group.

Usually, the fossils available do not represent the most recent common ancestor—or we cannot be confident that they do. As a result, the method of **outgroup comparison** is used to assign character polarity. To use this method, a species or group of species that is closely related to, but not a member of, the group under study is designated as the **outgroup.** When the group under study exhibits multiple character states, and one of those states is exhibited by the outgroup, then that state is considered to be ancestral and other states are considered to be derived. However, outgroup species also evolve from their ancestors, so the outgroup species will not always exhibit the ancestral condition.

Polarity assignments are most reliable when the same character state is exhibited by several different outgroups. In the preceding example, teeth are generally present in the nearest outgroups of amniotes—amphibians and fish—as well as in many species of amniotes themselves. Consequently, the presence of teeth in mammals and reptiles is considered ancestral, and their absence in birds and turtles is considered derived.

Construction of a cladogram

Once all characters have been polarized, systematists use this information to construct a **cladogram,** which depicts a hypothesis of evolutionary relationships. Species that share a common ancestor, as indicated by the possession of shared derived characters, are said to belong to a **clade.** Clades are thus evolutionary units and refer to a common ancestor and all of its descendants. A derived character shared by clade members is called a **synapomorphy** of that clade. Figure 23.2 illustrates that a simple cladogram is a nested set of clades, each characterized by its own synapomorphies. For example, amniotes are a clade for which the evolution of an amniotic membrane is a synapomorphy. Within that clade, mammals are a clade, with hair as a synapomorphy, and so on.

Ancestral states are also called **plesiomorphies,** and shared ancestral states are called **symplesiomorphies.** In contrast to synapomorphies, symplesiomorphies are not informative about phylogenetic relationships.

Consider, for example, the character state "presence of a tail," which is exhibited by lampreys, sharks, salamanders, lizards, and tigers. Does this mean that tigers are more closely related to—and shared a more recent common ancestor with—lizards and sharks than to apes and humans, their fellow mammals? The answer, of course, is no: Because symplesiomorphies reflect character states inherited from a distant ancestor, they do not imply that species exhibiting that state are closely related.

Homoplasy complicates cladistic analysis

In real-world cases, phylogenetic studies are rarely as simple as the examples we have shown so far. The reason is that in some cases, the same character has evolved independently in several species. These characters would be categorized as shared derived characters, but they would be false signals of a close evolutionary relationship. In addition, derived characters may sometimes be lost as species within a clade re-evolve to the ancestral state.

Homoplasy refers to a shared character state that has not been inherited from a common ancestor exhibiting that character state. Homoplasy can result from convergent evolution or from evolutionary reversal. For example, adult frogs do not have a tail. Thus, absence of a tail is a synapomorphy that unites not only gorillas and humans, but also frogs. However, frogs

Traits: Organism	Jaws	Lungs	Amniotic Membrane	Hair	No Tail	Bipedal
Lamprey	0	0	0	0	0	0
Shark	1	0	0	0	0	0
Salamander	1	1	0	0	0	0
Lizard	1	1	1	0	0	0
Tiger	1	1	1	1	0	0
Gorilla	1	1	1	1	1	0
Human	1	1	1	1	1	1

a.

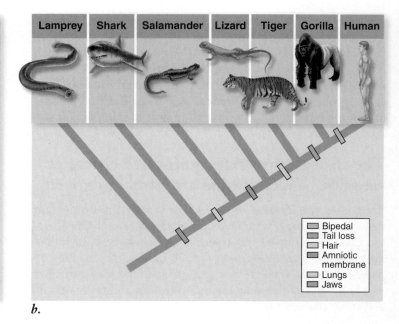

b.

figure 23.2

A CLADOGRAM. *a.* Morphological data for a group of seven vertebrates are tabulated. A "1" indicates possession of the derived character state, and a "0" indicates possession of the ancestral character state (note that the derived state for character "no tail" is the absence of a tail; for all other traits, absence of the trait is the ancestral character state). *b.* A tree, or cladogram, diagrams the relationships among the organisms based on the presence of derived characters. The derived characters between the cladogram branch points are shared by all organisms above the branch points and are not present in any below them. The outgroup (in this case, the lamprey) does not possess any of the derived characters.

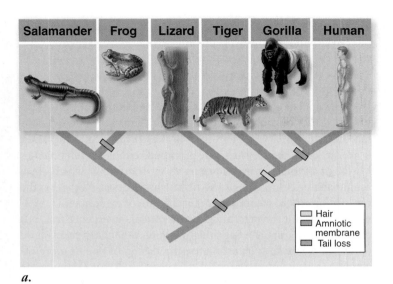
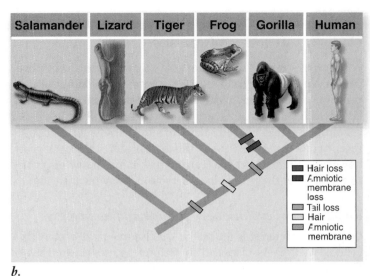

a.

b.

figure 23.3

PARSIMONY AND HOMOPLASY. a. The placement of frogs as closely related to salamanders requires that tail loss evolved twice, an example of homoplasy. **b.** If frogs are closely related to gorillas and humans, then tail loss only had to evolve once. However, this arrangement would require two additional evolutionary changes: Frogs would have had to have lost the amniotic membrane and hair (alternatively, hair could have evolved independently in tigers and the clade of humans and gorillas; this interpretation would require two evolutionary changes in the hair character, just like the interpretation shown in the figure, in which hair evolved only once, but then was lost in frogs). Based on the principle of parsimony, the cladogram that requires the fewest number of evolutionary changes is favored; in this case the cladogram in (**a**) requires four changes, whereas that in (**b**) requires five, so (**a**) is considered the preferred hypothesis of evolutionary relationships.

have neither an amniotic membrane nor hair, both of which are synapomorphies for clades that contain gorillas and humans.

In cases such as this, when there are conflicts among the characters, systematists rely on the **principle of parsimony,** which favors the hypothesis that requires the fewest assumptions. As a result, the phylogeny that requires the fewest evolutionary events is considered the best hypothesis of phylogenetic relationships (figure 23.3). In the example just stated, therefore, grouping frogs with salamanders is favored because it requires only one instance of homoplasy (the multiple origins of taillessness), whereas

a phylogeny in which frogs were most closely related to humans and gorillas would require two homoplastic evolutionary events (the loss of both amniotic membranes and hair in frogs).

The examples presented so far have all involved morphological characters, but systematists increasingly use DNA sequence data to construct phylogenies. Cladistics analyzes sequence data in the same manner as any other type of data: Character states are polarized by reference to the sequence of an outgroup, and a cladogram is constructed that minimizes the amount of character evolution required (figure 23.4).

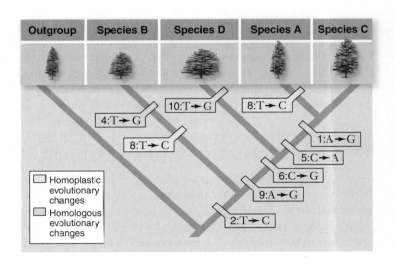

figure 23.4

CLADISTIC ANALYSIS OF DNA SEQUENCE DATA. Sequence data are analyzed just like any other data. The most parsimonious interpretation of the DNA sequence data requires eight evolutionary changes. Each of these changes is indicated on the phylogeny. Change in site 8 is homoplastic: Species A and B independently evolved from thymine to cytosine at that site.

chapter 23 systematics and the phylogenetic revolution **457**

Other phylogenetic methods work better than cladistics in some situations

If characters evolve from one state to another at a slow rate compared with the frequency of speciation events, then the principle of parsimony works well in reconstructing evolutionary relationships. In this situation, the principle's underlying assumption—that shared derived similarity is indicative of recent common ancestry—is usually correct. In recent years, however, systematists have realized that some characters evolve so rapidly that the principle of parsimony may be misleading.

Rapid rates of evolutionary change and homoplasy

Of particular interest is the rate at which some parts of the DNA genome evolve. As discussed in chapter 18, some stretches of DNA do not appear to have any function. As a result, mutations that occur in these parts of the DNA are not eliminated by natural selection, and thus the rate of evolution of new character states can be quite high in these regions as a result of genetic drift.

Moreover, because only four character states are possible for any nucleotide base (A, C, G, or T), there is a high probability that two species will independently evolve the same derived character state at any particular base position. If such homoplasy dominates the character data set, then the assumptions of the principle of parsimony are violated, and as a result, phylogenies inferred using this method are likely to be inaccurate.

inquiry

Why do high rates of evolutionary change and a limited number of character states cause problems for parsimony analyses?

Statistical approaches

For this reason, systematists in recent years have been exploring other methods based on statistical approaches, such as maximum likelihood, to infer phylogenies. These methods start with an assumption of the rate at which characters evolve and then fit the data to these models to derive the phylogeny that best accords with these assumptions.

One advantage of these methods is that different assumptions of rate of evolution can be used for different characters. If some DNA characters evolve more slowly than other parts of the DNA—for example, because they are constrained by natural selection—then the methods can employ different models

of evolution for the different characters. This approach is more effective than parsimony in dealing with homoplasy when rates of evolutionary changes are high.

The molecular clock

In general, cladograms such as the one in figure 23.2 only indicate the order of evolutionary branching events; they do not contain information about the timing of these events. In some cases, however, branching events can be timed, either by reference to fossils, or by making assumptions about the rate at which characters change. One widely used but controversial method is the **molecular clock,** which states that the rate of evolution of a molecule is constant through time. In this model, divergence in DNA can be used to calculate the times at which branching events have occurred. To make such estimates, the timing of one or more divergence events must be confidently estimated. For example, the fossil record may indicate that two clades diverged from a common ancestor at a particular time. Alternatively, the timing of separation of two clades may be estimated from geological events that likely led to their divergence, such as the rise of a mountain that now separates the two clades. With this information, the amount of DNA divergence separating two clades can be divided by the length of time separating the two clades, which produces an estimate of the rate of DNA divergence per unit of time (usually, per million years). Assuming a molecular clock, this rate can then be used to date other divergence events in a cladogram.

Although the molecular clock appears to hold true in some cases, in many others the data indicate that rates of evolution have not been constant through time across all branches in an evolutionary tree. For this reason, evolutionary dates derived from molecular data must be treated cautiously. Recently, methods have been developed to date evolutionary events without assuming that molecular evolution has been clocklike. These methods hold great promise for providing more reliable estimates of evolutionary timing.

> In cladistics, derived character states are distinguished from ancestral character states, and species are grouped based on shared derived character states. All descendants of a common ancestor are said to belong to the same clade. Homoplasies, shared character states that are the result of convergent evolution or evolutionary reversal, may give a false picture of relationships. Other methods sometimes have advantages over cladistics when assumptions about rates of evolution do not hold.

23.3 Systematics and Classification

Whereas systematics is the reconstruction and study of evolutionary relationships, **classification** refers to how we place species and higher groups—genus, family, class, and so forth—into the taxonomic hierarchy (a topic we discuss in greater detail in chapter 26).

Systematics and traditional classification are not always congruent; to understand why, we need to consider how species may be grouped based on their phylogenetic relationships. A **monophyletic** group includes the most recent common ancestor of the group and all of its descendants. By definition, a clade is a mono-

phyletic group. A **paraphyletic** group includes the most recent common ancestor of the group, but not all its descendants, and a **polyphyletic** group does not include the most recent common ancestor of all members of the group (figure 23.5).

Taxonomic hierarchies are based on shared traits, and ideally they should reflect evolutionary relationships. Traditional taxonomic groups, however, do not always fit well with new understanding of phylogenetic relationships. For example, birds have historically been placed in the class Aves, and dinosaurs

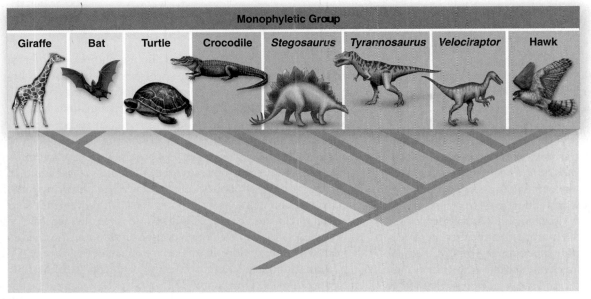

Monophyletic Group

| Giraffe | Bat | Turtle | Crocodile | *Stegosaurus* | *Tyrannosaurus* | *Velociraptor* | Hawk |

a.

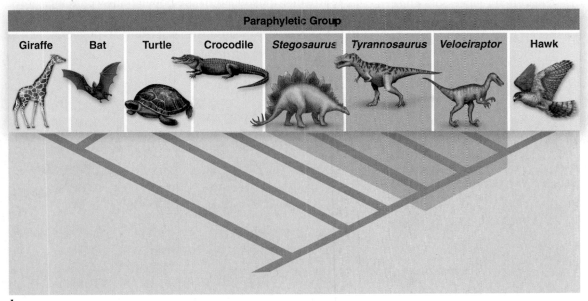

Paraphyletic Group

| Giraffe | Bat | Turtle | Crocodile | *Stegosaurus* | *Tyrannosaurus* | *Velociraptor* | Hawk |

b.

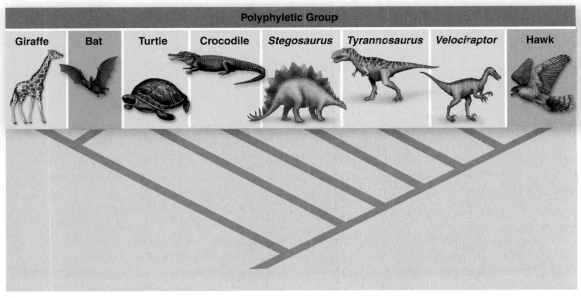

Polyphyletic Group

| Giraffe | Bat | Turtle | Crocodile | *Stegosaurus* | *Tyrannosaurus* | *Velociraptor* | Hawk |

c.

figure 23.5

MONOPHYLETIC, PARAPHYLETIC, AND POLYPHYLETIC GROUPS. *a.* A monophyletic group consists of the most recent common ancestor and all of its descendants. For example, the name "Archosaurs" is given to the monophyletic group that includes a crocodile, *Stegosaurus*, *Tyrannosaurus*, *Velociraptor*, and a hawk. *b.* A paraphyletic group consists of the most recent common ancestor and some of its descendants. For example, some, but not all, taxonomists traditionally give the name "dinosaurs" to the paraphyletic group that includes *Stegosaurus*, *Tyrannosaurus*, and *Velociraptor*. This group is paraphyletic because one descendant of the most recent ancestor of these species, the bird, is not included in the group. Other taxonomists include birds within the Dinosauria because *Tyrannosaurus* and *Velociraptor* are more closely related to birds than to other dinosaurs. *c.* A polyphyletic group does not contain the most recent common ancestor of the group. For example, bats and birds could be classified in the same group because they have similar shapes, anatomical features, and habitats. However, their similarities reflect convergent evolution, not common ancestry.

have been considered part of the class Reptilia. But recent phylogenetic advances make clear that birds evolved from dinosaurs. The last common ancestor of all birds and a dinosaur was a meat-eating dinosaur (see figure 23.5).

Therefore, having two separate monophyletic groups, one for birds and one for reptiles (including dinosaurs and crocodiles, as well as lizards, snakes, and turtles), is not possible based on phylogeny. And yet the terms Aves and Reptilia are so familiar and well established that suddenly referring to birds as a type of dinosaur, and thus a type of reptile, is difficult for some. Nonetheless, biologists increasingly refer to birds as a type of dinosaur and hence a type of reptile.

Situations like this are not uncommon. Another example concerns the classification of plants. Traditionally, three major groups were recognized: green algae, bryophytes, and vascular plants (figure 23.6). However, recent research reveals that neither the green algae nor the bryophytes constitute monophyletic groups. Rather, some bryophyte groups are more closely related to vascular plants than they are to other bryophytes, and some green algae are more closely related to bryophytes and vascular plants than they are to other green algae. As a result, systematists no longer recognize green algae or bryophytes as evolutionary groups, and the classification system has been changed to reflect evolutionary relationships.

The phylogenetic species concept (PSC) focuses on shared derived characters

In the preceding chapter, you read about a number of different ideas concerning what determines whether two populations belong to the same species. The **biological species concept (BSC)** defines species as groups of interbreeding populations that are reproductively isolated from other groups. In recent years, a phylogenetic perspective has emerged and has been applied to the question of species concepts. Advocates of the **phylogenetic species concept (PSC)** propose that the term *species* should be applied to groups of populations that have been evolving independently of other groups of populations. Moreover, they suggest that phylogenetic analysis is the way to identify such species. In this view, a species is a population or set of populations characterized by one or more shared derived characters.

This approach solves two of the problems with the BSC that were discussed in chapter 22. First, the BSC cannot be applied to allopatric populations because scientists cannot determine whether individuals of the populations would interbreed and produce fertile offspring if they ever came together. The PSC solves this problem: Instead of trying to predict what will

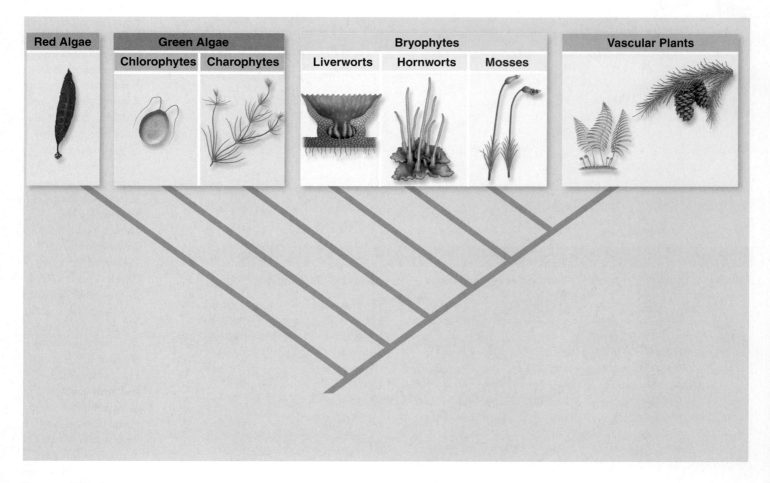

figure 23.6

PHYLOGENETIC INFORMATION TRANSFORMS PLANT CLASSIFICATION. The traditional classification included two groups that we now realize are not monophyletic: the green algae and bryophytes. For this reason, plant systematists have developed a new classification of plants that does not include these groups (discussed in chapter 30).

happen in the future if allopatric populations ever come into contact, the PSC looks to the past to determine whether a population (or groups of populations) has evolved independently for a long enough time to develop its own derived characters.

Second, the PSC can be applied equally well to both sexual and asexual species, in contrast to the BSC, which deals only with sexual forms.

The PSC also has drawbacks

The PSC is controversial, however, for several reasons. First, some critics contend that it will lead to the recognition of every slightly differentiated population as a distinct species. In Missouri, for example, open, desertlike habitat patches called glades are distributed throughout much of the state. These glades contain a variety of warmth-loving species of plants and animals that do not occur in the forests that separate the glades. Glades have been isolated from one another for a few thousand years, allowing enough time for populations on each glade to evolve differences in some rapidly evolving parts of the genome. Does that mean that each of the hundreds, if not thousands, of Missouri glades contains its own species of lizards, grasshoppers, and scorpions? Some scientists argue that if one takes the PSC to its logical extreme, that is exactly what would result.

A second problem is that species may not always be monophyletic, contrary to the definition of some versions of the phylogenetic species concept. Consider, for example, a species composed of five populations, with evolutionary relationships like those indicated in figure 23.7. Suppose that population C becomes isolated and evolves differences that make it qualify as a species by any concept (for example, reproductively isolated, ecologically differentiated). But this distinction would mean that the remaining populations, which might still be perfectly capable of exchanging genes, would be paraphyletic, rather than monophyletic. Such situations probably occur often in the natural world.

Phylogenetic species concepts, of which there are many different permutations, are increasingly used, but are also contentious for the reasons just discussed. Evolutionary biologists are trying to find ways to reconcile the historical perspective of the PSC with the process-oriented perspective of the BSC and other species concepts.

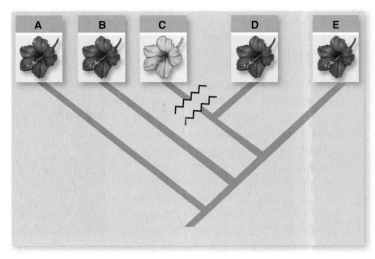

figure 23.7

PARAPHYLY AND THE PHYLOGENETIC SPECIES CONCEPT. The five populations initially were all members of the same species, with their historical relationships indicated by the cladogram. Then, population C evolved in some ways to become greatly differentiated ecologically and reproductively from the other populations. By all species concepts, this population would qualify as a different species. However, the remaining four species do not form a clade; they are paraphyletic because population C has been removed and placed in a different species. Although the phylogenetic species concept does not recognize paraphyletic species, this scenario may occur commonly in nature.

Systematics and traditional taxonomic classification sometimes conflict when new information about evolutionary relationships becomes available. The phylogenetic species concept emphasizes the possession of shared derived characters, in contrast to the biological species concept, which focuses on reproductive isolation. The PSC solves some problems of the BSC but has difficulties of its own.

Phylogenetics and Comparative Biology

Phylogenies not only provide information about evolutionary relationships among species, but they are also indispensable for understanding how evolution has occurred. By examining the distribution of traits among species in the context of their phylogenetic relationships, much can be learned about how and why evolution may have proceeded. In this way, phylogenetics is the basis of all comparative biology.

Homologous features are derived from the same ancestral source; homoplastic features are not

In chapter 21, we pointed out that **homologous structures** are those that are derived from the same body part in a common ancestor. Thus, the flipper of a dolphin and the leg of a horse are homologous because they are derived from the same bones in an ancestral vertebrate. By contrast, the wings of birds and those of dragonflies are homoplastic structures because they are derived from different ancestral structures. Phylogenetic analysis can help determine whether structures are homologous or homoplastic.

Homologous parental care in dinosaurs, crocodiles, and birds

Recent fossil discoveries have revealed that many species of dinosaurs exhibited parental care. They incubated eggs laid in nests and took care of growing baby dinosaurs, many of which could not have fended for themselves. Some recent fossils show dinosaurs sitting on a nest in exactly the same posture used

figure 23.8

PARENTAL CARE IN DINOSAURS AND CROCODILES _a._ Fossil dinosaur incubating its eggs. This remarkable fossil of _Oviraptor_ shows the dinosaur sitting on its nest of eggs just as chickens do today. Not only is the dinosaur squatting on the nest, but its forelimbs are outstretched, perhaps to shade the eggs. _b._ Crocodile exhibiting parental care. Female crocodilians build nests and then remain nearby, guarding them, while the eggs incubate. When they are ready to hatch, the baby crocodiles vocalize; females respond by digging up the eggs and carrying the babies to the water.

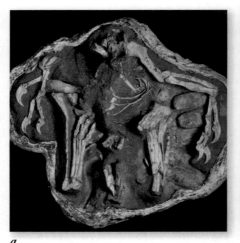

_a._

_b._

by birds today (figure 23.8_a_)! Initially, these discoveries were treated as remarkable and unexpected—dinosaurs apparently had independently evolved behaviors similar to those of modern-day organisms. But examination of the phylogenetic position of dinosaurs (see figure 23.5) indicates that they are most closely related to two living groups of animals—crocodiles and birds—both of which exhibit parental care (figure 23.8_b_).

It appears likely, therefore, that the parental care exhibited by crocodiles, dinosaurs, and birds did not evolve convergently from different ancestors that did not exhibit parental care; rather, the behaviors are homologous, inherited by each of these groups from their common ancestor that cared for its young.

Homoplastic convergence: Saber teeth and plant conducting tubes

In other cases, by contrast, phylogenetic analysis can indicate that similar traits have evolved independently in different clades. This convergent evolution from different ancestral sources indicates that such traits represent homoplasies. As one example, the fossil record reveals that extremely elongated canine teeth (saber teeth) occurred in a number of different groups of extinct carnivorous mammals. Although how these teeth were actually used is still debated, all saber-toothed carnivores had body proportions similar to those of cats, which suggests that these different types of carnivores all evolved into a similar predatory lifestyle. Examination of the saber-toothed character state in a phylogenetic context reveals that it most likely evolved independently at least three times (figure 23.9).

Conducting tubes in plants provide a similar example. The tracheophytes, a large group of land plants discussed in chapter 30, transport photosynthetic products, hormones, and other molecules over long distances through elongated, tubular cells that have perforated walls at the end. These structures are stacked upon each other to create a conduit called a sieve tube. Sieve tubes facilitate long-distance transport that is essential for the survival of tall plants on land.

Most members of the brown algae, which includes kelp, also have sieve elements (see figure 23.10 for a comparison of the sieve plates in brown algae and angiosperms) that aid in the rapid transport of materials. The land plants and brown algae are distantly related (see figure 23.10), and their last common ancestor was a single-celled organism that could not have had a multicellular transport system. This indicates that the strong structural and functional similarity of sieve elements in these plant groups is an example of convergent evolution.

Complex characters evolve through a sequence of evolutionary changes

Most complex characters do not evolve, fully formed, in one step. Rather, they are often built up, step-by-step, in a series of evolutionary transitions. Phylogenetic analysis can help discover these evolutionary sequences.

Modern-day birds—with their wings, feathers, light bones, and breastbone—are exquisitely adapted flying machines. Fossil discoveries in recent years now allow us to reconstruct the evolution of these features. When the fossils are arranged phylogenetically, it becomes clear that the features characterizing living birds did not evolve simultaneously. Figure 23.11 shows how the features important to flight evolved sequentially, probably over a long period of time, in the ancestors of modern birds.

One important finding often revealed by studies of the evolution of complex characters is that the initial stages of a character evolved as an adaptation to some environmental selective pressure different from that for which the character is currently adapted. Examination of figure 23.11 reveals that the first feathery structures evolved deep in the theropod phylogeny, in animals with forearms clearly not modified for flight. Therefore, the initial featherlike structures must have evolved for some other reason, perhaps serving as insulation or decoration. Through time, these structures have become modified to the extent that modern feathers produce excellent aerodynamic performance.

Phylogenetic methods can be used to distinguish between competing hypotheses

Understanding the causes of patterns of biological diversity observed today can be difficult because a single pattern often could have resulted from several different processes. In many cases, scientists can use phylogenies to distinguish between competing hypotheses.

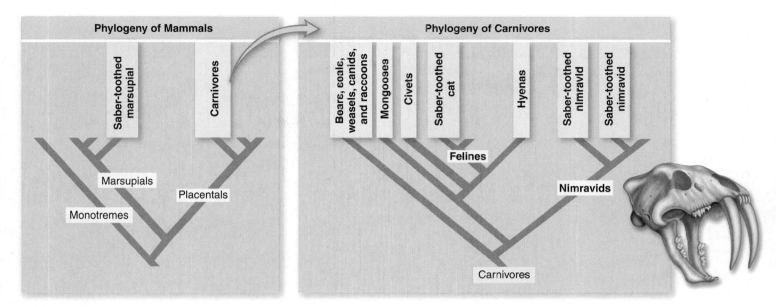

figure 23.9

DISTRIBUTION OF SABER-TOOTHED MAMMALS. Saber teeth have evolved at least three times in mammals—once within marsupials, once in felines, and at least once in a now-extinct group of cat-like carnivores called nimravids. It is possible that the condition evolved twice in nimravids, but another possibility that requires the same number of evolutionary changes (and thus is equally parsimonious) is that saber teeth evolved only once in the ancestor of nimravids and then were subsequently lost in one group of nimravids (not all of the branches within marsupials and placentals are shown).

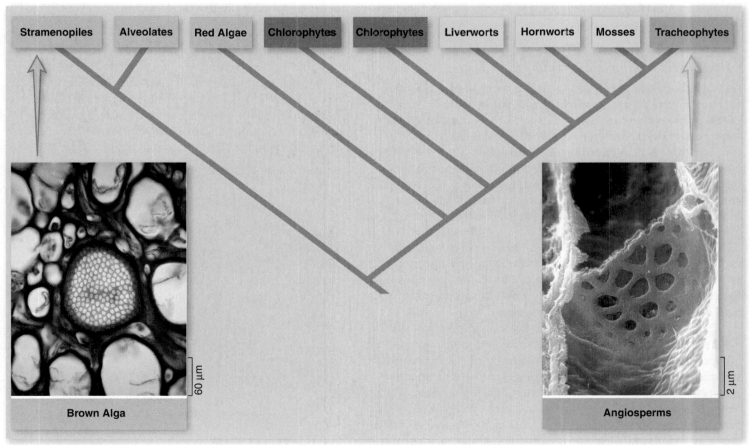

figure 23.10

CONVERGENT EVOLUTION OF CONDUCTING TUBES. Sieve tubes, which transport hormones and other substances throughout the plant, have evolved in two distantly related plant groups (brown algae are stramenopiles and angiosperms are tracheophytes).

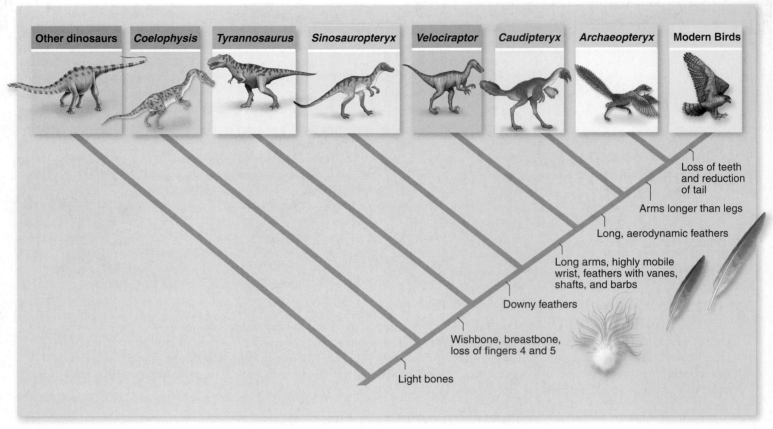

figure 23.11

THE EVOLUTION OF BIRDS. The traits we think of as characteristic of modern birds have evolved in stages over many millions of years.

From Richard O. Prum and Alan H. Brush, "The Evolutionary Origin and Diversification of Feathers," *Quarterly Review of Biology*, September 2002. Reprinted with permission of the University of Chicago Press.

Larval dispersal in marine snails

An example of this use of phylogenetic analysis concerns the evolution of larval forms in marine snails. Most species of snails produce microscopic larvae that drift in the ocean currents, sometimes traveling hundreds or thousands of miles before becoming established and transforming into adults. Some species, however, have evolved larvae that settle to the ocean bottom very quickly and thus don't disperse far from their place of origin. Studies of fossil snails indicate that the proportion of species that produce nondispersing larvae has increased through geological time (figure 23.12).

Two processes could produce an increase in nondispersing larvae through time. First, if evolutionary change from dispersing to nondispersing occurs more often than change in the opposite direction, then the proportion of species that are nondispersing would increase through time.

Alternatively, if species that are nondispersing speciate more frequently, or become extinct less frequently, than dispersing species, then through time the proportion of nondispersers would also increase (assuming that the descendants of nondispersing species also were nondispersing). This latter case is a reasonable hypothesis because nondispersing species probably have lower amounts of gene flow than dispersing species, and thus might more easily become geographically isolated, increasing the likelihood of allopatric speciation (chapter 22).

These two processes would result in different phylogenetic patterns. If evolution from a dispersing ancestor to a nondispers- ing descendant occurred more often than the reverse, then an excess of such changes should be evident in the phylogeny, as shown by more D ⟶ N branchpoints in figure 23.13a. In contrast, if nondispersing species underwent greater speciation, then clades of nondispersing species would contain more species than clades of dispersing species, as shown in figure 23.13b.

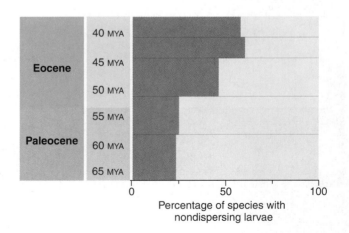

figure 23.12

Increase through time in the proportion of species whose larvae do not disperse far from their place of birth.

Evidence for both processes was revealed in an examination of the phylogeny of marine snails in the genus *Conus*, in which 30% of species are nondispersing (figure 23.13c). The phylogeny indicates that possession of dispersing larvae was the ancestral state; nondispersing larvae are inferred to have evolved eight times, with no evidence for evolutionary reversal from nondispersing to dispersing larvae.

At the same time, clades of nondispersing larvae tend to have on average 3.5 times as many species as dispersing larvae, which suggests that in nondispersing species, rates of speciation are higher, rates of extinction are lower, or both.

This analysis therefore indicates that the evolutionary increase in nondispersing larvae through time may be a result both of a bias in the direction in which evolution proceeds plus an increase in rate of diversification (that is, speciation rate minus extinction rate) in nondispersing clades.

The lack of evolutionary reversal is not surprising because when larvae evolve to become nondispersing, they often lose a variety of structures used for feeding while drifting in the ocean current. In most cases, once a structure is lost, it rarely re-evolves, and thus the standard view is that the evolution of nondispersing larvae is a one-way street, with few examples of re-evolution of dispersing larvae.

Loss of the larval stage in marine invertebrates

A related phenomenon in many marine invertebrates is the loss of the larval stage entirely. Most marine invertebrates—in groups as diverse as snails, sea stars, and anemones—pass through a larval stage as they develop. But in a number of different types of organisms, the larval stage is omitted, and the eggs develop directly into adults.

The evolutionary loss of the larval stage has been suggested as another example of a nonreversible evolutionary change because once the larval stages are lost, they are difficult to re-evolve—or so the argument goes. A recent study on one group of marine limpets, shelled marine organisms related to snails, shows that this is not necessarily the case. Among these limpets, direct development has evolved many times; however,

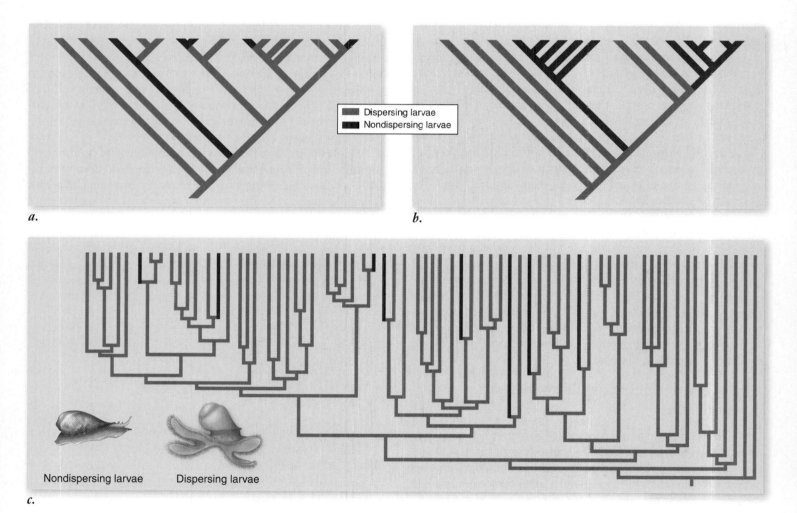

a. *b.*

Dispersing larvae
Nondispersing larvae

Nondispersing larvae   Dispersing larvae

c.

figure 23.13

PHYLOGENETIC INVESTIGATION OF THE EVOLUTION OF NONDISPERSING LARVAE. *a.* In this phylogeny the evolutionary transition from dispersing to nondispersing larvae occurs more frequently (four times) than the converse (once). By contrast, in (**b**) clades that have nondispersing larvae diversify to a greater extent due to higher rates of speciation or lower rates of extinction (assuming that extinct forms are not shown). *c.* Phylogeny for *Conus*, a genus of marine snails. Nondispersing larvae have evolved eight separate times from dispersing larvae, with no instances of evolution in the reverse direction. This phylogeny does not show all species, however; nondispersing clades contain on average 3.5 times as many species as dispersing clades.

figure 23.14

EVOLUTION OF DIRECT DEVELOPMENT IN A FAMILY OF LIMPETS. *a.* Direct development evolved many times (indicated by beige lines stemming from a red ancestor), and three instances of reversed evolution from direct development to larval development are indicated (red lines from a beige ancestor). *b.* A less parsimonious interpretation of evolution in the clade in the light blue box is that, rather than two evolutionary reversals, six instances of the evolution of development occurred without any evolutionary reversal.

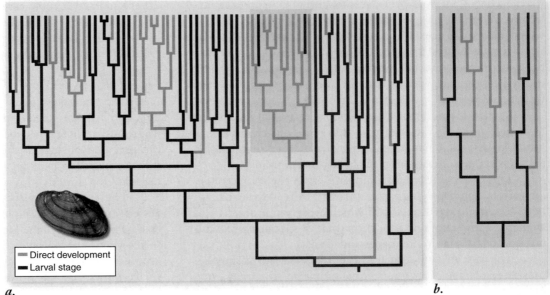

Direct development
Larval stage

a.

b.

in three cases, the phylogeny strongly suggests that evolution reversed and a larval stage re-evolved (figure 23.14*a*).

It is important to remember that patterns of evolution suggested by phylogenetic analysis are not always correct—evolution does not necessarily occur parsimoniously. In the limpet study, for example, it is always possible that within the clade in the light blue box, presence of a larva was retained throughout as the ancestral state, and direct development evolved independently six times (figure 23.14*b*). Phylogenetic analysis cannot rule out this possibility, even if it is less phylogenetically parsimonious.

If the re-evolution of lost traits seems unlikely, then the alternative hypothesis that direct development evolved six times independently—rather than only evolving once at the base of the clade, with two instances of evolutionary reversal—should be considered. For example, detailed studies of the morphology or embryology of direct-developing species might shed insight on whether such structures are homologous or convergent. In some cases, artificial selection experiments in the laboratory or genetic manipulations can test the hypothesis that lost structures are difficult to re-evolve. Conclusions from phylogenetic analyses are always bolstered when they are supported by the results of other types of studies.

Phylogenetics helps explain species diversification

One of the central goals of evolutionary biology is to explain patterns of species diversity: Why do some types of plants and animals exhibit more **species richness**—a greater number of species per clade—than others? Phylogenetic analysis can be used both to suggest and to test hypotheses about such differences.

Species richness in beetles

Beetles (order Coleoptera) are the most diverse group of animals. Approximately 60% of all animal species are insects, and approximately 80% of all insect species are beetles. Among beetles, families that are herbivorous are particularly species rich.

Examination of the phylogeny provides insight into beetle evolutionary diversification (figure 23.15). Among the

Phytophaga, the clade which contains most herbivorous beetle species, the deepest branches belong to beetle families that specialize on conifers. This finding agrees with the fossil record because conifers were among the earliest seed plant groups to evolve. By contrast, the flowering plants (angiosperms) evolved more recently, in the Cretaceous, and beetle families specializing on them have shorter evolutionary branches, indicating their more recent evolutionary appearance.

This correspondence between phylogenetic position and timing of plant origins suggests that beetles have been remarkably conservative in their diet. The family Nemonychidae, for example, appears to have remained specialized on conifers since the beginning of the Jurassic, approximately 210 MYA.

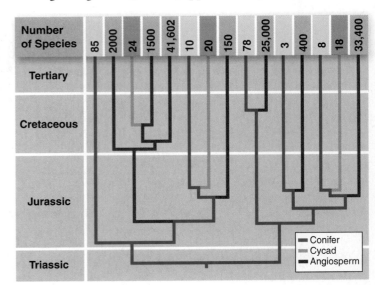

figure 23.15

EVOLUTIONARY DIVERSIFICATION OF THE PHYTOPHAGA, THE LARGEST CLADE OF HERBIVOROUS BEETLES. Clades that originated deep in the phylogenetic tree feed on conifers; clades that feed on angiosperms, which evolved more recently, originated more recently. Age of clades is established by examination of fossil beetles.

Phylogenetic explanations for beetle diversification

The phylogenetic perspective suggests factors that may be responsible for the incredible diversity of beetles. The phylogeny for the Phytophaga indicates that it is not the evolution of herbivory itself that is linked to great species richness. Rather, specialization on angiosperms seems to have been a prerequisite for great species diversification. Specialization on angiosperms appears to have arisen five times independently within herbivorous beetles; in each case, the angiosperm-specializing clade is substantially more species-rich than the clade to which it is most closely related (termed a *sister clade*) and which specializes on some other type of plant.

Why specialization on angiosperms has led to great species diversity is not yet clear and is the focus of much current research. One possibility is that this diversity is linked to the great species-richness of angiosperms themselves. With more than 250,000 species of angiosperms, beetle clades specializing on them may have had a multitude of opportunities to adapt to feed on individual species, thus promoting divergence and speciation.

> Homologous traits are derived from the same ancestral character states, whereas homoplastic traits are not, although they may have similar function. Phylogenetic analysis can help determine whether homology or homoplasy has occurred in trait evolution. Examination of a phylogeny can also be used to test hypotheses about trait evolution and species diversification.

23.5 Phylogenetics and Disease Evolution

The examples so far have illustrated the use of phylogenetic analysis to examine relationships among species. Such analyses can also be conducted on virtually any group of biological entities, as long as evolutionary divergence in these groups occurs by a branching process, with little or no genetic exchange between different groups. No example illustrates this better than recent attempts to understand the evolution of the virus that causes autoimmune deficiency syndrome (AIDS).

HIV has evolved from a simian viral counterpart

AIDS was first recognized in the early 1980s, and it rapidly became epidemic in the human population. Current estimates are that more than 39 million people are infected with the human immunodeficiency virus (HIV) and more than 3 million die each year.

At first, scientists were perplexed about where HIV had come from and how it had infected humans. In the mid-1980s, however, scientists discovered a related virus in laboratory monkeys, termed simian immunodeficiency virus (SIV). In biochemical terms, the viruses are very similar, although genetic differences exist. At last count, SIV has been detected in 36 species of primates, but only in species found in sub-Saharan Africa. Interestingly, SIV—which appears to be transmitted sexually—does not appear to cause any illness in these primates.

Based on the degree of genetic differentiation among strains of SIV, scientists estimate that SIV may have been around for more than a million years in these primates, perhaps providing enough time for these species to adapt to the virus and thus prevent it from having adverse effects.

Phylogenetic analysis identifies the path of transmission

Phylogenetic analysis of strains of HIV and SIV reveals three clear findings. First, HIV obviously descended from SIV. All strains of HIV are phylogenetically nested within clades of SIV strains, indicating that HIV is derived from SIV (figure 23.16).

Second, a number of different strains of HIV exist, and they appear to represent independent transfers from different primate species. Each of the human strains is more closely related to a strain of SIV than it is to other HIV strains, indicating separate origins of the HIV strains.

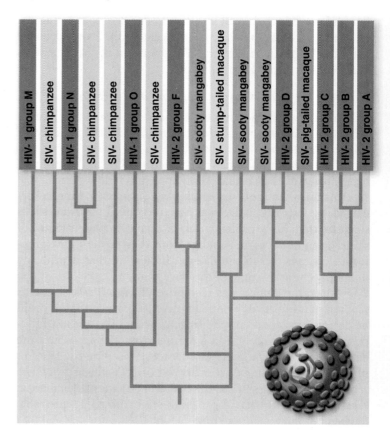

figure 23.16

EVOLUTION OF HIV AND SIV. HIV has evolved multiple times and from strains of SIV in different primate species (each primate species indicated by a different color). The three-way branching event on the right side of the phylogeny results because the data do not clearly indicate the relationships among the three clades.

Finally, humans have acquired HIV from different host species. HIV-1, which is the virus responsible for the global epidemic, has three subtypes. Each of these subtypes is most closely related to a different strain of chimpanzee SIV, indicating that the transfer occurred from chimps to humans. By contrast, subtypes of HIV-2, which is much less widespread (in some cases known from only one individual), are related to SIV found in West African monkeys, primarily the sooty mangabey (*Cercocebus atys*). Moreover, the subtypes of HIV-2 also appear to represent several independent cross-species transmissions to humans.

Transmission from other primates to humans

Several hypotheses have been proposed to explain how SIV jumped from chimps and monkeys to humans. The most likely idea is that transmission occurred as the result of blood-to-blood contact that may occur when humans kill and butcher monkeys and apes. Recent years have seen a huge increase in the rate at which primates are hunted for the "bushmeat" market, particularly in central and western Africa. This increase has resulted from a combination of increased human populations desiring ever greater amounts of protein, combined with increased access to the habitats in which these primates live as the result of road building and economic development. The unfortunate result is that population sizes of many primate species, including our closest relatives, are plummeting toward extinction. A second consequence of this hunting is that humans are increasingly brought into contact with bodily fluids of these animals, and it is easy to imagine how during the butchering process, blood from a recently killed animal might enter the human bloodstream through cuts in the skin, perhaps obtained during the hunting process.

Establishing the crossover timeline and location

Where and when did this cross-species transmission occur? HIV strains are most diverse in Africa, and the incidence of HIV is higher there than elsewhere in the world. Combined with the evidence that HIV is related to SIV in African primates, it seems certain that AIDS appeared first in Africa.

As for when the jump from other primates to humans occurred, the fact that AIDS was not recognized until the 1980s suggests that HIV probably arose recently. Descendants of slaves brought to North America from West Africa in the nineteenth century lacked the disease, indicating that it probably did not occur at the time of the slave trade.

Once the disease was recognized in the 1980s, scientists scoured repositories of blood samples to see whether HIV could be detected in blood samples from the past. The earliest HIV-positive result was found in a sample from 1959, pushing the date of origin back at least two decades. Based on the amount of genetic difference between strains of HIV-1, including the 1959 sample, and assuming the operation of a molecular clock, scientists estimate that the deadly strain of AIDS probably crossed into humans some time before 1940.

Phylogenies can be used to track the evolution of AIDS among individuals

The AIDS virus evolves extremely rapidly, so much so that different strains can exist within different individuals in a single population. As a result, phylogenetic analysis can be applied to

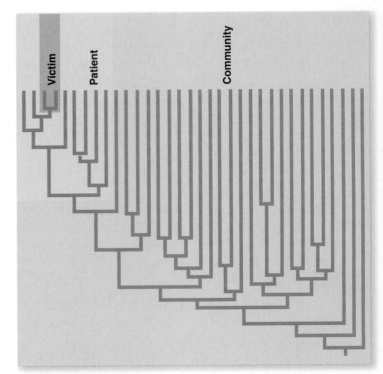

figure 23.17

EVOLUTION OF HIV STRAINS REVEALS THE SOURCE OF INFECTION. HIV mutates so rapidly that HIV-infected individuals often contain multiple genotypes in their body. As a result, it is possible to create a phylogeny of HIV strains and to identify the source of infection of a particular individual. In this case, the HIV strains of the victim (V) clearly are derived from strains in the body of another individual, the patient. Other HIV strains are from HIV-infected individuals in the local community.

answer very specific questions; just as phylogeny proved useful in determining the source of HIV, it can also pinpoint the source of infection for particular individuals.

This ability became apparent in a court case in Louisiana in 1998, in which a dentist was accused of injecting his former girlfriend with blood drawn from an HIV-infected patient. The dentist's records revealed that he had drawn blood from the patient and had done so in a suspicious manner. Scientists sequenced the viral strains from the victim, the patient, and from a large number of HIV-infected people in the local community. The phylogenetic analysis clearly demonstrated that the victim's viral strain was most closely related to the patient's (figure 23.17). This analysis, which for the first time established phylogenetics as a legally admissible form of evidence in courts in the United States, helped convict the dentist, who is now serving a 50-year sentence for attempted murder.

Modern phylogenetic techniques and analysis can be used to track the evolution of disease strains, uncovering sources and progression. The HIV virus provides a prime example of application of phylogenetic analysis to human disease.

23.1 Systematics

One of the greatest challenges to biologists is to understand the history of ancestor–descendant relationships that unites all forms of life on Earth.

- Systematics is the study of evolutionary relationships.

- A phylogeny is a hypothesis about the evolutionary relationships among species.

- Similarities may not accurately predict evolutionary relationships because the rate of evolution varies, evolution is not unidirectional and evolution is not always divergent.

- Evolution may be convergent, in which case similarities do not reflect ancestry but similar morphological changes.

23.2 Cladistics

Cladistics is an approach to studying evolutionary relationships that emphasizes the shared possession of derived characters.

- Cladistics examines the distribution of character states among species.

- Derived character states are those that are different from the character state possessed by ancestors.

- Ancestral character states are those that are the same as the character state possessed by ancestor.

- Character polarity uses an outgroup comparison in which the character states of the species in the group being studied are compared with that of a closely related species or group of species.

- Character states exhibited by the outgroup are assumed to be ancestral, and other character states are considered derived.

- Cladograms depict a hypothesis of evolutionary relationships (figure 23.2).

- The derived characters shared by members of a clade and not by their nearest relative are called synapomorphies.

- Homoplasy refers to a shared character state that has not been inherited from a common ancestor exhibiting that character state.

- The principle of parsimony works best when evolutionary change occurs relatively slowly. The principle states that the phylogeny with the fewest assumptions is favored when conflicts arise among characters.

23.3 Systematics and Classification

Systematics is the reconstruction and study of evolutionary relationships, and classification is how species are organized into taxonomic hierarchies.

- A monophyletic group consists of the most recent common ancestor and all of its descendants.

- A paraphyletic group consists of the most recent common ancestor and some of its descendants.

- A polyphyletic group does not contain the most recent ancestor of the group.

- The phylogenetic species concept emphasizes the possession of shared derived characters, whereas the biological species concept focuses on reproductive isolation.

23.4 Phylogenetics and Comparative Biology

Phylogenies not only provide information about evolutionary relationships among species, but they are also indispensable for understanding how evolution occurred.

- Homologous features are derived from the same ancestor, but homoplastic features are not.

- Phylogenies illustrate how complex characters can evolve through a series of intermediate stages (figure 23.11).

- Phylogenies can be used to test hypotheses about trait evolution and species diversification.

23.5 Phylogenetics and Disease Evolution

The evolution of disease strains can be tracked using modern phylogenetic techniques because evolutionary divergence occurs with little or no genetic exchange between different groups.

- Based on phylogenetic studies, the HIV virus descended from the SIV virus, the different strains of the HIV virus have different origins and humans acquired HIV from different primate hosts (figure 23.16).

- It is hypothesized that humans contracted HIV by handling meat from butchered primates.

- Phylogenetic techniques can be used to pinpoint the particular source of an HIV infection (figure 23.17).

SELF TEST

1. Overall similarity of phenotypes may not always reflect evolutionary relationships
   a. due to convergent evolution.
   b. because of variation in rates of evolutionary change of different kinds of characters.
   c. due to homoplasy.
   d. all of the above

2. Cladistics
   a. is based on overall similarity of phenotypes.
   b. requires distinguishing similarity due to inheritance from a common ancestor from other reasons for similarity.
   c. is not affected by homoplasy.
   d. none of the above

3. The principle of parsimony
   a. helps evolutionary biologists distinguish among competing phylogenetic hypotheses.
   b. does not require that the polarity of traits be determined.
   c. is a way to avoid having to use outgroups in a phylogenetic analysis.
   d. cannot be applied to molecular traits.

4. The phylogenetic species concept (PSC)
   a. depends on whether individuals from different populations can successfully breed.
   b. is indistinguishable from the biological species concept.
   c. does not apply to allopatric populations.
   d. is based on evolutionary independence among populations.

5. Parsimony suggests that parental care in birds, crocodiles, and some dinosaurs
   a. evolved independently multiple times by convergent evolution.
   b. evolved once in an ancestor common to all three groups.
   c. is a homoplastic trait.
   d. is not a homologous trait.

6. The re-evolution of lost traits, especially if they are complex
   a. can be identified with phylogenetic analyses.
   b. never happens.
   c. is not an example of a reversal.
   d. does not affect interpretation of evolutionary relationships.

7. The term *molecular clock* in the context of evolutionary biology and phylogenetics
   a. refers to a group of proteins that induce endogenous circadian rhythms in animals.
   b. is an undisputed assumption that all biological molecules evolve at a constant rate.
   c. may help provide a way to estimate the absolute timing of historical events in evolution.
   d. applies only to organisms that reproduce sexually.

8. A taxonomic group that contains a common ancestor, but leaves out a descendant group
   a. is paraphyletic.
   b. is monophyletic.
   c. is polyphyletic.
   d. is a good cladistic group.

9. The forelimb of a bird and the forelimb of a rhinoceros
   a. are homologous and symplesiomorphic.
   b. are not homologous but are symplesiomorphic.
   c. are homologous and synapomorphic.
   d. are not homologous but are synapomorphic.

10. In order to determine polarity for different states of a character
    a. there must be a fossil record of the groups in question.
    b. genetic sequence data must be available.
    c. an appropriate name for the taxonomic group must be selected.
    d. an outgroup must be identified.

11. A paraphyletic group
    a. includes an ancestor and all of its descendants.
    b. an ancestor and some of its descendants.
    c. descendants of more than one common ancestor.
    d. all of the above

12. Sieve tubes and sieve elements
    a. are homoplastic because they have different function.
    b. are homologous because they have similar function.
    c. are homoplastic because their common ancestor was single-celled.
    d. are structures involved in transport within animals.

13. The phylogeny of dinosaurs leading to birds
    a. demonstrates that the first function of feathers was flight.
    b. demonstrates that feathers and wings evolved simultaneously.
    c. suggests that complex characters evolve rapidly, in one step.
    d. reveals many transitional forms between modern birds and their ancestors.

14 . A phylogenetic analysis of HIV suggests.
    a. a single origin of HIV from primates.
    b. multiple origins of HIV from different primate species.
    c. multiple origins of HIV from a single primate species.
    d. that SIV originated from HIV.

CHALLENGE QUESTIONS

1. List the synapomorphy and the taxa defined by that synapomorphy for the groups pictured in figure 23.2. Name each group defined by a set of synapomorphies in a way that might be construed as informative about what kind of characters define the group

2. Identifying "outgroups" is a central component of cladistic analysis. As described on page 456, a group is chosen that is closely related to, but not a part of the group under study. If one does not know the relationships of members of the group under study, how can one be certain that an appropriate outgroup is chosen? Can you think of any approaches that would minimize the effect of a poor choice of outgroup?

3. As noted in your reading, cladistics is a widely utilized method of systematics, and our classification system (taxonomy) is increasingly becoming reflective of our knowledge of evolutionary relationships. Using birds as an example, discuss the advantages and disadvantages of recognizing them as reptiles versus as a group separate and equal to reptiles.

4. Across many species of limpets, loss of larval development and reversal from direct development appears to have occurred multiple times. Under the simple principle of parsimony are changes in either direction merely counted equally in evaluating the most parsimonious hypothesis? If it is much more likely to lose a larval mode than to re-evolve it from direct development, should that be taken into account? How?

Genome Evolution

introduction

GENOMES CONTAIN THE RAW MATERIAL for evolution, and many clues to evolution are hidden in the ever-changing nature of genomes. As more genomes have been sequenced, the new and exciting field of comparative genomics has emerged and has yielded some surprising results and many, many questions. Comparing whole genomes, not just individual genes, enhances our ability to understand the workings of evolution, to improve crops and to identify the genetic basis of disease so that we may develop more effective treatments with minimal side effects. The focus of this chapter is on how comparative genomics is enhancing our understanding of genome evolution and how this new knowledge can be applied to improve our lives.

Comparative Genomics

A key challenge of modern evolutionary biology is to find a way to link changes in DNA sequences, which we are now able to study in great detail, with the evolution of the complex morphological characters used to construct a traditional phylogeny. Many different genes contribute to complex characters—such as feathers, described in the preceding chapter. Making the connection between a specific change in a gene and a modification in a morphological character is particularly difficult.

Comparing genomes (entire DNA sequences) of different species provides a powerful new tool for exploring the evolutionary divergence among organisms in an effort to connect changes at the DNA level to morphological differences. Genomes are more than instruction books for building and maintaining an organism; they also contain vast amounts of information on the history of life. As you saw in chapter 18, the growing number of fully sequenced genomes in all kingdoms is leading to a revolution in comparative evolutionary biology (table 24.1). It is now possible to explore the genetic differences between species in a very direct way, examining the footprints on the evolutionary path between different species.

Evolutionary differences accumulate over long periods

Genomes of viruses and bacteria can evolve in a matter of days, whereas complex eukaryotic species evolve over millions of years. To illustrate this point, we will compare three vertebrate genomes: human, tiger pufferfish (*Fugu rubripes*), and mouse (*Mus musculus*).

Comparison between human and pufferfish genomes

The draft (preliminary) sequence of the tiger pufferfish was completed in 2002; it was only the second vertebrate genome to be sequenced. For the first time, we were able to compare the genomes of two vertebrates: humans and pufferfish. These two animals last shared a common ancestor 450 million years ago (MYA).

TABLE 24.1	Milestones for Comparative Eukaryotic Genomics			
Organism		**Estimated Genome Size (Mb)**	**Estimated Number of Genes**	**Year Sequenced**
Vertebrates				
Homo sapiens (human)		2,900	20,000–25,000	2001
Mus musculus (mouse)		2,600	30,000	2002
Fugu rubripes (pufferfish)		365	33,609	2002
Rattus norvegicus (rat)		2,750	20,973	2004
Pan troglodytes (chimpanzee)		3,100	20,000–25,000	2005

TABLE 24.1	Milestones for Comparative Eukaryotic Genomics, continued		
Organism	Estimated Genome Size (Mb)	Estimated Number of Genes	Year Sequenced
Vertebrates, continued			
Gallus gallus (red jungle fowl)	 1,000	20,000–23,000	2004
Invertebrates			
Drosophila melanogaster (fruit fly)	 137	13,600	2000
Anopheles gambiae (mosquito)	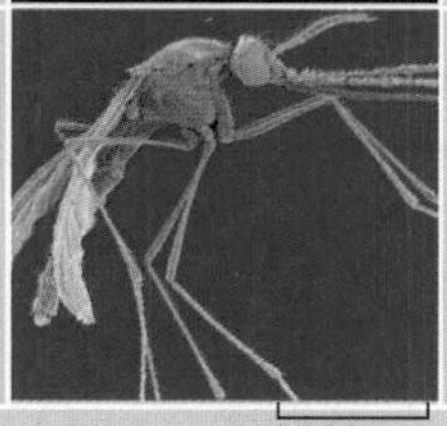 278	46,000–56,000	2002
Fungi			
Schizosaccharomyces pombe (fission yeast)	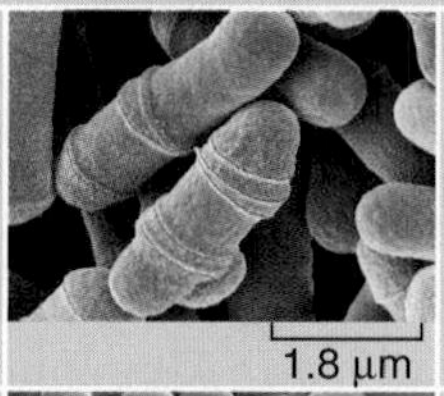 13.8	4,824	2002
Saccharomyces cerevisiae (brewer's yeast)	 12.7	5,805	1997
Plants			
Arabidopsis thaliana (wall cress)	 125	25,498	2000
Oryza sativa (rice)	 430	41,000	2002
Protists			
Plasmodium falciparum (malaria parasite)	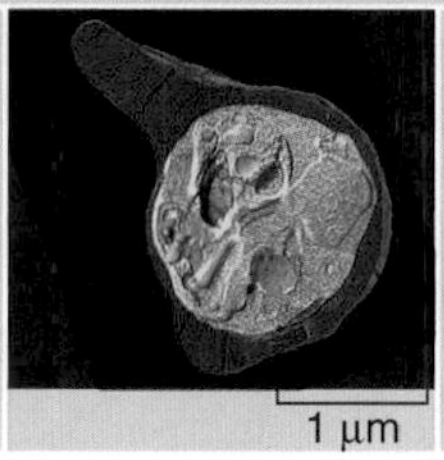 23	5,300	2002

2500 μm

1.8 μm

23.25 μm

1 μm

Some human and pufferfish genes have been conserved during evolution, but others are unique to each species. About 25% of human genes have no counterparts in *Fugu*. Also, extensive genome rearrangements have occurred during the 450 million years since the mammal lineage and the teleost fish diverged, indicating a considerable scrambling of gene order. Finally, the human genome is 97% repetitive DNA (chapter 18), but repetitive DNA accounts for less than one-sixth of the *Fugu* sequence.

Comparison between human and mouse genomes

Later in 2002, a draft sequence of the mouse genome was completed by an international consortium of investigators, allowing for the first time a comparison of two mammalian genomes. In contrast to the human–pufferfish genome comparison, the differences between these two mamalian genomes are miniscule.

The human genome has about 400 million more nucleotides than that of the mouse. A comparison of the genomes reveals that both have about 25,000 genes, and that they share the bulk of them; in fact, the human genome shares 99% of its genes with mice. Humans and mice diverged about 75 million years ago, approximately one-sixth of the amount of time that separates humans from pufferfish. There are only 300 genes unique to either organism, constituting about 1% of the genome.

From a human perspective, 75 million years is a vast amount of time, yet there is striking similarity between the mouse and human genomes. Even 450 million years after last sharing a common ancestor, 75% of the genes found in humans have counterparts in pufferfish. Although conservation of genes is high over evolutionary time, rearrangements of chromosomal regions large and small are not unusual.

Comparison between human and chimpanzee genomes

Humans and chimpanzees, *Pan troglodytes*, diverged only about 35 MYA, leaving even less time for their genomes to accumulate mutational differences. The chimp genome was sequenced in 2005, providing a comparative window between us and our sibling species. Comparisons of single nucleotide substitutions reveal that only 1.06% of the two genomes have fixed (apparently not changing) differences in single nucleotides. A 1.5% difference in insertions and deletions (indels) is found between chimps and humans. Fifty-three of the potentially human-specific indels lead to loss-of-function changes that might correlate with some of the traits that distinguish us from chimps, including a larger cranium and lack of body hair. As will be discussed later in this chapter, mutations leading to differences in the patterns of gene expression are particularly important in understanding why chimps are chimps and humans are humans.

Mutations in coding DNA are classified into two groups: those that alter the amino acids coded for in the sequence (non-synonymous changes) and those that do not alter the coded amino acids (synonymous changes, refer to table 15.1). A comparison of the mouse and rat genomes reveals a smaller ratio of nonsynonymous to synonymous changes than a comparison of humans and chimps. The higher ratio in the primates indicates that fewer nonsynonymous mutations have been removed by natural selection than has occurred in mice and rats. The removal of nonsynonymous genes during evolution is called purifying selection because the mutations are more likely to be deleterious, and purifying selection removes these mutations. A likely answer to the puzzle of less purifying selection in the primates is that their population sizes are much smaller. Purifying selection is less effective in small populations.

Genomes evolve at different rates

Comparison of the mouse and human genomes reveals that since mice and humans last shared a common ancestor, about 75 MYA, mouse DNA has mutated about twice as fast as human DNA. The fruit fly *Drosophila* and the mosquito *Anopheles* are separated by approximately 250 million years of evolution, and they appear to have evolved more rapidly over that interval than have the vertebrates. The extent of similarity between these two insects is comparable to that between humans and pufferfish, which diverged 450 MYA.

These fascinating observations require an explanation. One hypothesis, for which there now is a great deal of support, is that differences in generation time accounts for the different rates of genome evolution. For example, mice which are capable of reproducing every six weeks, have more germ line divisions and opportunities for recombination over any period of time than would humans. The rate of mutation in the germ line of mice and humans would be the same in each generation, but there would be more generations of mice.

Plant, fungal, and animal genomes have unique and shared genes

We now step back further and consider genomic differences among the eukaryotic kingdoms that diverged long before the examples just discussed. You have already seen that many genes are highly conserved in animals. Are plant genes also highly conserved, and if so, are they similar to animal and fungal genes?

Comparison between two plant genomes

The first plant genome to be sequenced was *Arabidopsis thaliana*, the wall cress, a tiny member of the mustard family often used as a model organism for studying flowering plant molecular genetics and development. Its genome sequence, largely completed in 2000, revealed 25,948 genes, about as many as humans have, in a genome with a size of only 125 million base-pairs, a 30-fold smaller genome than that of humans.

Rice, *Oryza sativa*, belongs to the grass family, which includes maize (corn), wheat, barley, sorghum, and sugarcane. Unlike most grasses, rice has a relatively small genome of 430 million base-pairs. Even in this small genome there are 41,000 genes.

Although rice and *Arabidopsis* are distant relatives, they share many genes. More than 80% of the genes found in rice, including duplicates, are also found in *Arabidopsis*. Among the other 20% are genes that may be responsible for some of the physiological and morphological differences between rice (a monocot) and *Arabidopsis* (a eudicot), two very different kinds of flowering plants. It is probable that many of the other differences between the two species reflect differences in gene expression, as discussed later in this chapter. (The morphological and physiological distinctions are described in chapter 30.)

Comparison of plants with animals and fungi

About one-third of the genes in *Arabidopsis* and rice appear to be in some sense "plant" genes—that is, genes not found in any animal or fungal genome sequenced so far. These include the many thousands of genes involved in photosynthesis and photosynthetic anatomy. Few plant genomes have been sequenced to date, however.

Among the remaining genes found in plants are many that are very similar to those found in animal and fungal genomes, particularly the genes involved in basic intermediary metabolism, in genome replication and repair, and in RNA transcription and protein synthesis. Prior to the availability of whole-genome sequences, assessment of the extent of genetic similarity and difference among diverse organisms had been difficult at best.

> Genome evolution may take millions of years, or, in some cases, a few days, and does not occur at a uniform rate in all species. Although many genes are highly conserved across kingdoms, many genes, including one-third of the plant genome, distinguish one kingdom from another.

24.2 Evolution of Whole Genomes

Polyploidy (three or more chromosome sets) can give rise to new species, as you learned in chapter 22. Polyploidy can result from either genome duplication in one species or from hybridization of two different species. In **autopolyploids,** the genome of one species is duplicated through a meiotic error, resulting in four copies of each chromosome. **Allopolyploids** result from the hybridization and subsequent duplication of the genomes of two different species (figure 24.1). The origins of

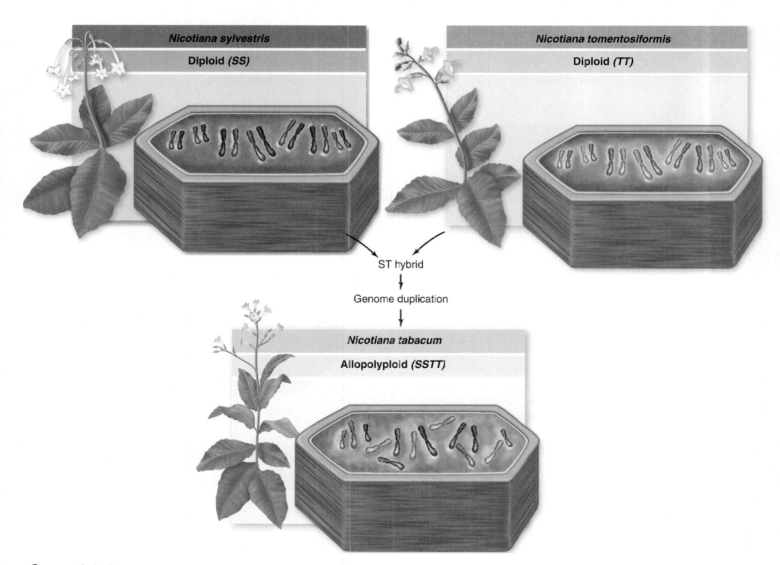

figure 24.1

ALLOPOLYPLOIDY occurred in tobacco 5 MYA, but can be approximated by crossing the progenitor species and initiating a doubling of chromosomes, often through tissue culture followed by plant regeneration, which can lead to chromosome doubling. Tobacco species have many chromosomes, not all of which are visible in a single plane of a cell.

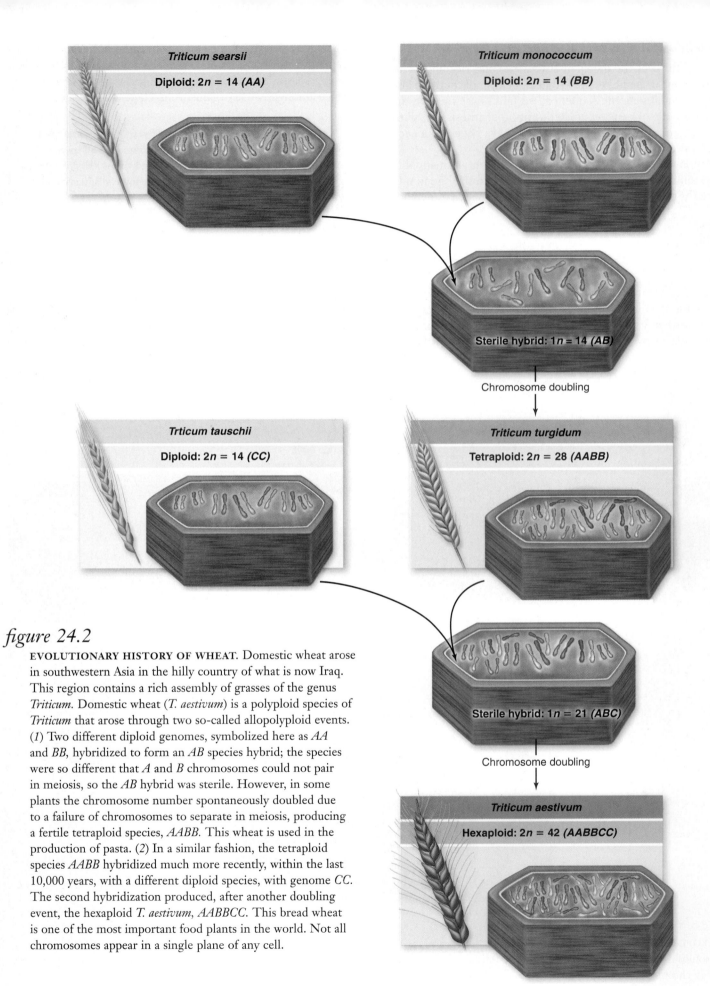

figure 24.2

EVOLUTIONARY HISTORY OF WHEAT. Domestic wheat arose in southwestern Asia in the hilly country of what is now Iraq. This region contains a rich assembly of grasses of the genus *Triticum*. Domestic wheat (*T. aestivum*) is a polyploid species of *Triticum* that arose through two so-called allopolyploid events. (1) Two different diploid genomes, symbolized here as *AA* and *BB*, hybridized to form an *AB* species hybrid; the species were so different that *A* and *B* chromosomes could not pair in meiosis, so the *AB* hybrid was sterile. However, in some plants the chromosome number spontaneously doubled due to a failure of chromosomes to separate in meiosis, producing a fertile tetraploid species, *AABB*. This wheat is used in the production of pasta. (2) In a similar fashion, the tetraploid species *AABB* hybridized much more recently, within the last 10,000 years, with a different diploid species, with genome *CC*. The second hybridization produced, after another doubling event, the hexaploid *T. aestivum*, *AABBCC*. This bread wheat is one of the most important food plants in the world. Not all chromosomes appear in a single plane of any cell.

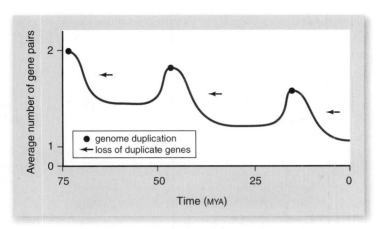

figure 24.3

SEQUENCE COMPARISONS OF NUMEROUS GENES IN A POLYPLOID GENOME TELL US HOW LONG AGO ALLOPOLYPLOIDY OR AUTOPOLYPLOIDY EVENTS OCCURRED. Complex analyses of sequence divergence among duplicate gene pairs and presence or absence of duplicate gene pairs provide information about when both genome duplication and gene loss occurred. The graph reveals multiple polyploidy events over evolutionary time.

 inquiry

Why is there a decrease in the number of duplicate genes after multiple rounds of polyploidization?

wheat, illustrated in figure 24.2, involve two successive allopolyploid events.

Ancient and newly created polyploids guide studies of genome evolution

Two avenues of research have lead to intriguing insights into genome alterations following polyploidization. The first method studies ancient polyploids, called **paleopolyploids.** Sequence comparisons and phylogenetic tools establish the time and patterns of polyploidy events (figure 24.3). Sequence divergence

between homologues, as well as the presence or absence of duplicated gene pairs from the hybridization, can be used for historical reconstructions of genome evolution. Specific examples are provided in chapter 25. All copies of duplicate gene pairs arising through polyploidy are not necessarily found thousands or millions of years after polyploidization. The loss of duplicate genes is considered later in this section.

The second approach is to create **synthetic polyploids** by crossing plants most closely related to the ancestral species and then chemically inducing chromosome doubling. Unless the hybrid genome becomes doubles, the plant will be sterile because it will lack homologous chromosomes that need to pair during metaphase I of meiosis.

Because meiosis requires an even number of chromosome sets, species with ploidy levels that are multiples of two can reproduce sexually. However, meiosis would be a disaster in a 3n organism such as the banana since three sets of chromosomes can't be evenly divided between two cells. Breeders have taken advantage of this in commercial bananas (not wild ones), which are seedless. The aborted ovules appear as the little brown dots in the center of any cross section of a banana.

 inquiry

Sketch out what would happen in meiosis in a 3n banana cell, referring back to chapter 11 if necessary. Commercial banana plants rely on asexual means of propagation.

In the following sections we examine further the effect of polyploidization on genomes. Plant examples have been selected to illustrate key points in this section because polyploidization occurs more frequently in plants. The somewhat surprising findings, however, are not limited to the plant kingdom.

Plant polyploidy is ubiquitous, with multiple common origins

Polyploidy has occurred numerous times in the evolution of the flowering plants (figure 24.4). The legume clade that includes the soybean (*Glycine max*), the plant *Medicago truncatula* (a forage

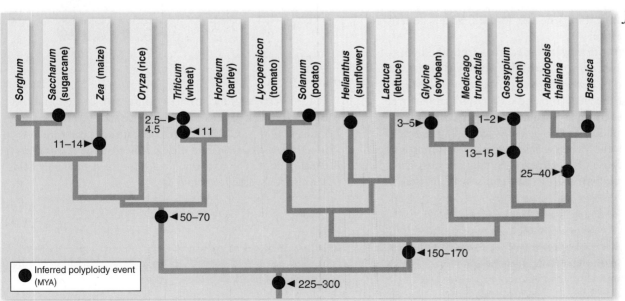

figure 24.4

POLYPLOIDY HAS OCCURRED NUMEROUS TIMES IN THE EVOLUTION OF THE FLOWERING PLANTS.

chapter 24 *genome evolution* **477**

figure 24.5

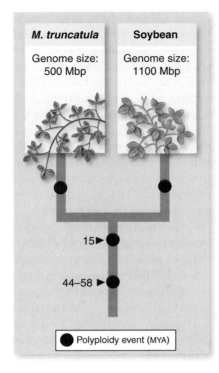

GENOME DOWNSIZING.
Genome downsizing must have occurred in *M. truncatula.*

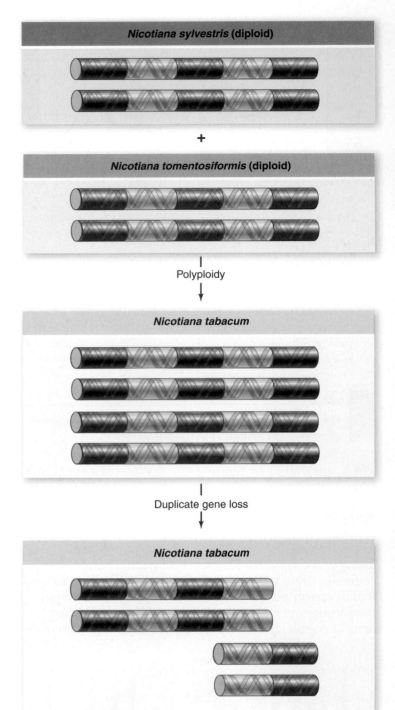

legume often used in research), and the garden pea (*Pisum sativum*) underwent a major polyploidization event 44–58 MYA and again 15 MYA (figure 24.5).

A quick comparison of the genomes of soybean and *M. truncatula* reveal a huge difference in genome size. In addition to increasing genome size through polyploidization, genomes like *M. trunculata* definitely downsized over evolutionary time as well. The total size of a genome cannot be explained solely on the basis of polyploidization.

Polyploidy induces elimination of duplicated genes

Formation of an allopolyploid from two different species is often followed by a rapid loss of genes (figure 24.6) or even whole chromosomes, although in some polyploids loss of one copy of many duplicated genes arises over a much longer period. In some species there is a great deal of gene loss in the first few generations after polyploidization.

Modern tobacco, *Nicotiana tabacum* arose from the hybridization and genome duplication of a cross between *Nicotiana sylvestris* (female parent) and *N. tomentosiformis* (male parent) (see figure 24.1). To complement the analysis based on a cross that occurred over 5 MYA, researchers constructed synthetic *N. tabacum* and observed the chromosome loss that followed. Curiously, the loss of chromosomes is not even. More *N. tomentosiformis* chromosomes were jettisoned than those of *N. sylvestris*. Similar unequal chromosome loss has been observed in synthetic wheat hybrids, in

figure 24.6

POLYPLOIDY MAY BE FOLLOWED BY THE UNEQUAL LOSS OF DUPLICATE GENES FROM THE COMBINED GENOMES. In the case of *N. tabacum*, more duplicate gene pairs are lost from the *N. tomentosiformis* male parent than the *N. sylvestris* female parent. The conclusion holds for the natural polyploids and also for synthetic polyploids.

which 13% of the genome of one parent is lost in contrast to 0.5% of the other parental genome. It is possible that different rates of genome replication could explain the differential loss, as is true for synthetic human–mouse hybrid cultured cells.

Polyploidy can alter gene expression

A striking discovery is the change in gene expression that occurs in the early generations after polyploidization. Some of this may be connected to an increase in the methylation of cytosines in the DNA. Methylated genes cannot be transcribed, as described in chapter 16. Simply put, polyploidization can lead to a short-term silencing of some genes. In subsequent generations, there is a decrease in methylation.

Transposons jump around following polyploidization

Barbara McClintock, in her Nobel Prize-winning work on transposable (mobile) genetic elements, referred to these jump-ing DNA regions as *controlling elements*. She hypothesized that transposons could respond to genome shock and jump into a new position in the genome. Depending on where the transposon moved, new phenotypes could emerge.

Recent work on transposon activity following hybridization supports McClintock's hypothesis. Again, during the early generations following a polyploidization event, new transposon insertions occur because of unusually active transposition. These new insertions may cause gene mutations, changes in gene expression, and chromosomal rearrangements, all of which provide additional genetic variation on which evolution might act.

> Polyploidization can lead to major changes in genome structure. These changes range from discarding genes to alteration of gene expression through methylation of DNA to increased transposon hopping and chromosomal rearrangements. Because polyploidy is so common, especially in plants, it is considered important in the generation of biodiversity and adaptation.

24.3 Evolution Within Genomes

From individual genes to whole chromosomes, duplications of portions of genomes contribute to evolution. Duplications provide opportunities for genes with the same function to diverge because a "backup pair" of genes is in place. As with all mutations, most are deleterious or neutral—only a small percentage increase the fitness of the individual at the time of the mutation, perpetuating the modified genome.

Individual chromosomes may be duplicated

Aneuploidy, you may recall, refers to the duplication or loss of an individual chromosome rather than of an entire genome (see chapter 13). Failure of a pair of homologous chromosomes or sister chromatids to separate during meiosis is the most common way that aneuploids occur.

In general, plants are better able to tolerate aneuploidy than animals, but the explanation for this difference is elusive.

DNA segments may be duplicated

One of the greatest sources of novel traits in genomes is duplication of segments of DNA. When a gene duplicates, the most likely fates of the duplicate gene are: (1) losing function through subsequent mutation, (2) gaining a novel function through subsequent mutation, and (3) having the total function of the ancestral gene partitioned into the two duplicates. In reality, however, most duplicate genes lose function, some quite quickly following genome doubling and others more slowly over evolutionary time.

How then can researchers claim that gene duplication is a major evolutionary force for gene innovation, that is, genes gaining new function? One piece of the answer can be found by noting where in the genome gene duplication is most likely to occur. In humans, the highest rates of duplication have occurred in the three most gene-rich chromosomes of the genome. The seven chromosomes with the fewest genes also show the least amount of duplication. (Remember, having fewer genes does not mean that there is less total DNA.)

Even more compelling, certain types of human genes appear to be more likely to be duplicated: growth and development genes, immune system genes, and cell-surface receptors. About 5% of the human genome consists of segmental duplications (figure 24.7). Finally, and importantly, gene duplication is thought to be a major evolutionary force for gene innovation because the duplicated genes are found to have different patterns of gene expression (see chapter 25 for examples). For example,

figure 24.7

SEGMENTAL DUPLICATION ON THE HUMAN Y CHROMOSOME. Each red region has 98% sequence similarity with a sequence on a different human chromosome. Each dark blue region has 98% sequence similarity with a sequence elsewhere on the Y chromosome.

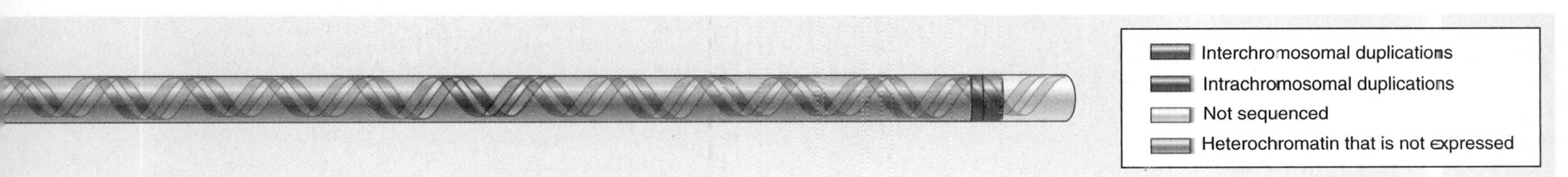

■ Interchromosomal duplications
■ Intrachromosomal duplications
□ Not sequenced
■ Heterochromatin that is not expressed

the two duplicated copies may be expressed in different or over-lapping sets of tissues or organs during development.

As more species are compared, it can be seen that gene duplication rates appear to vary for different groups of organisms. *Drosophila* has about 31 new duplicates per genome per million years, which is equivalent to 0.0023 duplications per gene per million years. The rate is about 10 times faster for the nematode *Caenorhabditis elegans*. Two genes within an organism that have arisen from the duplication of a single gene in an ancestor are called **paralogues**. In contrast, **orthologues** reflect the conservation of a single gene from a common ancestor.

Genomes may become rearranged

Humans have one fewer chromosome than chimpanzees, gorillas, and orangutans (figure 24.8). It's not that we have lost a chromosome. Rather, at some point in time, two midsized ape chromosomes fused to make what is now human chromosome 2, the second largest chromosome in our genome.

The fusion leading to human chromosome 2 is an example of the sort of genome reorganization that has occurred in many species. Rearrangements like this can provide evolutionary clues, but they are not always definitive proof of how closely related two species are.

Consider the organization of known orthologues shared by humans, chickens, and mice. One study estimated that 72 chromosome rearrangements had occurred since the chicken and human last shared a common ancestor. This number is substantially less than the estimated 128 rearrangements between chicken and mouse, or 171 between mouse and human.

This does not mean that chickens and humans are more closely related than mice and humans or mice and chickens.

What these data actually show is that chromosome rearrangements have occurred at a much lower frequency in the lineages that led to humans and to chickens than in the lineages leading to mice. Chromosomal rearrangements in mouse ancestors seem to have occurred at twice the rate seen in the human line. These different rates of changes help counter the notion that humans existed hundreds of millions of years ago.

Genomes that have undergone relatively slow chromosome change are the most helpful in reconstructing the hypothetical genomes of ancestral vertebrates. If regions of chromosomes have changed little in distantly related vertebrates over the last 300 million years, then we can reasonably infer that the common ancestor of these vertebrates had genomic similarities.

Variation in the organization of genomes is as intriguing as gene sequence differences. Chromosome rearrangements are common, yet over long segments of chromosomes, the linear order of mouse and human genes is the same—the common ancestral sequence has been preserved in both species. This **conservation of synteny** (see chapter 18) was anticipated from earlier gene mapping studies, and it provides strong evidence that evolution actively shapes the organization of the eukaryotic genome. As seen in figure 24.9, the conservation of synteny allows researchers to more readily locate a gene in a different species using information about synteny, thus underscoring the power of a comparative genomic approach.

Gene inactivation results in pseudogenes

The loss of gene function is another important way genomes evolve. Consider the olfactory receptor (OR) genes that are responsible for our sense of smell. These genes code for receptors that bind odorants, initiating a cascade of signaling events that eventually lead to our perception of scents.

figure 24.8

LIVING GREAT APES. All living great apes, with the exception of humans, have a haploid chromosome number of 24. Humans have not lost a chromosome; rather, two smaller chromosomes fused to make a single chromosome.

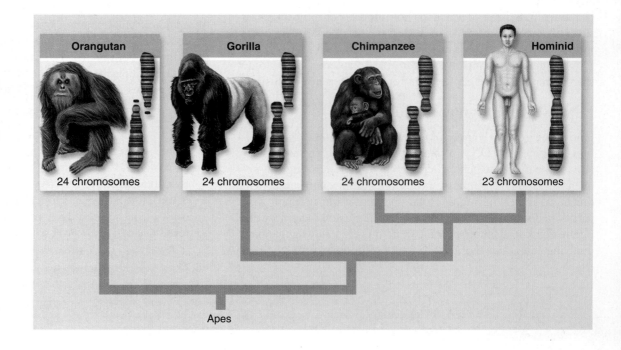

Orangutan — 24 chromosomes

Gorilla — 24 chromosomes

Chimpanzee — 24 chromosomes

Hominid — 23 chromosomes

Apes

Soybean

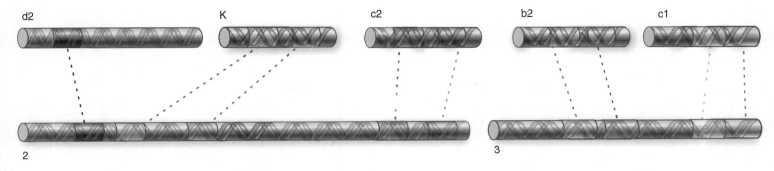

M. truncatula

figure 24.9

SYNTENY AND GENE IDENTIFICATION. Genes sequenced in the model legume, *Medicago truncatula*, can be used to identify homologous genes in soybean, *Glycine max*, because large regions of the genomes are syntenic as illustrated for some of the linkage groups (chromosomes) of the two species. Regions of the same color represent homologous genes.

Gene inactivation seems to be the best explanation for our reduced sense of smell relative to that of the great apes and other mammals. Primate genomes have over 1000 copies of OR genes (figure 24.10). An estimated 70% of human OR genes are inactive **pseudogenes** (sequences of DNA that are very similar to functional genes but do not produce a functional product because they have premature stop codons, missense mutations, or deletions that prevent the production of an active protein). In contrast, half the chimpanzee and gorilla OR genes function effectively, and over 95% of New World monkey OR genes and probably all mouse OR genes are working quite well. The most likely explanation for these differences is that humans came to rely on other senses, reducing the selection pressure against loss of OR gene function by random mutation.

An older question about the possibility of positive selection for OR genes in chimps was resolved with the completion of the chimp genome. A careful analysis indicated that both humans and chimps are gradually losing OR genes to pseudogenes and that there is no evidence to suppport positive selection for any of the OR genes in the chimp.

Horizontal gene transfer complicates matters

Evolutionary biologists build phylogenies on the assumption that genes are passed from generation to generation, a process called **vertical gene transfer (VGT)**. Hitchhiking genes from other species, referred to as **horizontal gene transfer (HGT)**, sometimes called *lateral gene transfer*, can lead to phylogenetic complexity. Horizontal gene transfer was likely most prevalent very early in the history of life, when the boundaries between individual cells and species seem to have been less firm than they are now and DNA more readily moved among different organisms. Although earlier in the history of life, gene swapping between species was rampant, HGT continues today in prokaryotes and eukaryotes. An intriguing example of more recent HGT between moss and a flowering plant is described in chapter 26.

Gene swapping in early lineages

The extensive gene swapping among early organisms has caused many researchers to reexamine the base of the tree of life. Early phylogenies based on ribosomal RNA (rRNA) sequences indicate that an early prokaryote gave rise to two major domains: the Bacteria

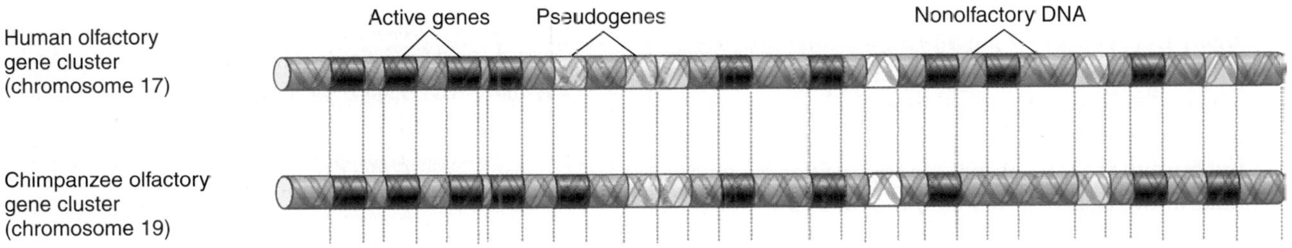

Human olfactory gene cluster (chromosome 17)

Active genes Pseudogenes Nonolfactory DNA

Chimpanzee olfactory gene cluster (chromosome 19)

figure 24.10

GENE INACTIVATION. Although almost all mouse olfactory receptor genes are functional, an evolutionary loss of olfactory receptors has occurred in primates, which rely less on their sense of smell. Comparisons of the olfactory receptor genes in humans and chimpanzees reveal that humans have more pseudogenes (inactive genes) than do chimpanzees.

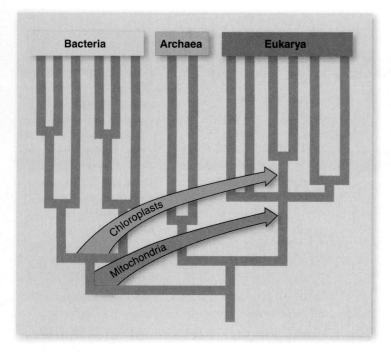

figure 24.11

PHYLOGENY BASED ON A UNIVERSAL COMMON ANCESTOR.
The three domains share a common ancestor, and the tree of
life is firmly rooted. Exchange of genetic information between
domains occurred through multiple endosymbiotic events.

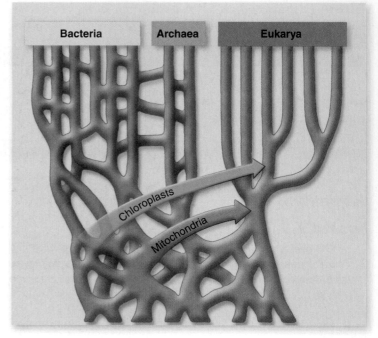

figure 24.12

HORIZONTAL GENE TRANSFER. Early in the history of life,
organisms may have freely exchanged genes beyond multiple
endosymbiotic events. To a lesser extent, this transfer continues
today. The tree of life may be more like a web or a net.

and the Archaea. From one of these lineages, the domain Eukarya
emerged, its organelles originated as unicellular organisms engulfed specialized prokaryotes (figure 24.11).

This somewhat straightforward rRNA phylogeny is being
revised as more microbial genomes are sequenced; by 2005, 225
microbial genomes had been sequenced. Phylogenies built with
rRNA sequences suggest that the domain Archaea is more closely
related to the Eukarya than to the Bacteria. But as more microbial
genomes are sequenced, investigators find bacterial and archaeal
genes showing up in the same organism! The most likely conclusion is that organisms swapped genes, possibly even absorbing
DNA obtained from a food source. Perhaps the base of the tree of
life is better viewed as a web than a branch (figure 24.12).

Gene swapping evidence in the human genome

Let's move closer to home and look at the human genome,
which is riddled with foreign DNA, often in the form of transposons. The many transposons of the human genome provide a
paleontological record over several hundred million years.

Comparisons of versions of a transposon that has duplicated
many times allow researchers to construct a "family tree" to identify the ancestral form of the transposon. The percent of sequence
divergence found in duplicates allows an estimate of the time at

which that particular transposon originally invaded the human genome. In humans, most of the DNA hitchhiking seems to have
occurred millions of years ago in very distant ancestor genomes.

Our genome carries many more ancient transposons,
making it quite different from other genomes that have been
studied, such as those of *Drosophila*, *C. elegans*, and *Arabidopsis*. One explanation for the observed low level of transposons
in *Drosophila* is that fruit flies somehow eliminate unnecessary
DNA from their genome 75 times faster than humans do. Our
genome has simply hung on to hitchhiking DNA more often.

The human genome has had minimal transposon activity
in the past 50 million years; mice, by contrast, are continuing
to acquire new transposable elements. This difference may explain in part the more rapid change in chromosome organization in mice than in humans.

Segmental duplication, genome rearrangement, and loss of
gene function have all contributed to the evolution of genomes.
Horizontal gene transfer has led to an unexpected mixing of
genes among organisms. This gene swapping creates many
phylogenetic questions, particularly with regard to the origins of
the three major domains.

24.4 Gene Function and Expression Patterns

Gene function can be inferred by comparing genes in different species. You saw earlier that the function of 1000 human
genes was understood once the mouse genome was sequenced.
One of the major puzzles arising from comparative genomics

is that organisms with very different forms can share so many
conserved genes in their genomic toolkit.

As an example, let's consider once again mice versus humans. Most of the nearly 150 genes found in mice but not in

humans are associated with either the sense of smell, which is highly developed in rodents, or with reproduction. The genomes of humans and mice are so similar that one wonders why mice and humans are so different.

The best explanation for why a mouse develops into a mouse and not a human is that genes are expressed at different times, in different tissues, and in different amounts and combinations. The cystic fibrosis gene, which has been identified in both species and affects a chloride ion channel illustrates this point. Defects in the human cystic fibrosis gene cause especially devastating effects in the lungs, but mice with the cystic fibrosis mutant gene do not have lung symptoms. Mutations in another mouse gene can cause lung symptoms. Possibly variations in expression of gene combinations between mouse and human explain the difference in lung symptoms when the cystic fibrosis gene is defective.

Chimp and human gene transcription patterns differ

Humans and chimps diverged from a common ancestor only about 5 MYA—too little time for much genetic differentiation to evolve, but enough for significant morphological and behavioral differences to have developed. Sequence comparisons indicate that chimp DNA is 98.7% identical to human DNA. If just the gene sequences encoding proteins are considered, the similarity increases to 99.2%. How could two species differ so much in body and behavior, and yet have almost equivalent sets of genes?

One potential answer to this question is based on the observation that chimp and human genomes show very different patterns of gene transcription activity, at least in brain cells. Investigators used microarrays containing up to 18,000 human genes to analyze RNA isolated from cells in the fluid extracted from several regions of living brains of chimps and humans. (See figure 18.10 for a summary of this technique.) The RNA was linked with a fluorescent tag and then incubated with the microarray under conditions that allow the formation of DNA–RNA hybrids if the sequences are complementary. If the transcript of a particular gene is present in the cells, then the microarray spot corresponding to that gene lights up under UV light. The more copies of the RNA, the more intense the signal.

Because the chimp genome is so similar to that of humans, the microarray could detect the activity of chimp genes reasonably well. Although the same genes were transcribed in chimp and human brain cells, the patterns and levels of transcription varied widely. It would seem that much of the difference between human and chimp brains lies in which genes are transcribed, and when and where that transcription occurs.

inquiry

? **You are given a microarray of ape genes and RNA from both human and ape brain cells. Using the experimental technique described for comparison of humans and chimps, what would you expect to find in terms of genes being transcribed? What about levels of transcription?**

Posttranscriptional differences may also play a role in building distinct organisms from similar genomes. As research continues to push the frontiers of proteomics and functional genomics, a more detailed picture of the subtle differences in the developmental and physiological processes of closely related species will be revealed. The integration of development and genome evolution is explored in depth in the following chapter.

Speech is uniquely human: An example of complex expression

Development of human culture is closely tied to the capacity to control the larynx and mouth to produce speech. Humans with a single point mutation in the transcription factor gene *FOXP2* have impaired speech and grammar but not impaired language comprehension.

The *FOXP2* gene is also found in chimpanzees, gorillas, orangutans, rhesus macaques, and even the mouse, yet none of these mammals speak. The gene is expressed in areas of the brain that affect motor function, including the complex coordination needed to create words.

FOXP2 protein in mice and humans differs by only three amino acids. There is only a single amino acid difference between mouse and chimp, gorilla, and rhesus macaque, which all have identical amino acid sequences for FOXP2. Two more amino acid differences exist between humans and the sequence shared by chimp, gorilla, and rhesus macaque. The difference of only two amino acids between human and other primate FOXP2 appears to have made it possible for language to arise. Evidence points to strong selective pressure for the two *FOXP2* mutations that allow brain, larynx, and mouth to coordinate to produce speech.

It is possible that two amino acid changes lead to speech, language, and ultimately human culture? This box of mysteries will take a long time to unpack, but hints indicate that the changes are linked to signaling and gene expression. The two amino acids that are altered may change the ability of FOXP2 transcription factor to be phosphorylated. One way signaling pathways operate is through activation or inactivation of an existing transcription factor by phosphorylation.

Comparative genomics efforts are now extending beyond primates. A role for *FOXP2* in songbird singing and vocal learning has been proposed. Mice communicate via squeaks, with lost young mice emitting high-pitched squeaks, and *FOXP2* mutations leave mice squeakless. For both mice and songbirds, it is a stretch to claim that *FOXP2* is a language gene—but it likely is needed in the neuromuscular pathway to make sounds.

Diverse life forms emerge from highly similar toolkits of genes. To understand functional differences, one must look beyond sequence similarity and ask questions about the time and place of gene expression. Even small changes in a protein can affect gene function, as seen with the FOXP2 factor and speech in humans.

24.5 Nonprotein-Coding DNA and Regulatory Function

So far, we have primarily compared genes that code for proteins. As more genomes are sequenced, we learn that much of the genome is composed of noncoding DNA. The repetitive DNA is often retrotransposon DNA, contributing to as much as 30% of animal genomes and 40–80% of plant genomes. (Refer to chapter 18 for more information on repetitive DNA in genomes.)

Perhaps the most unexpected finding in comparing the mouse and human genomes lies in the similarities between the repetitive DNA, mostly retrotransposons, in the two species. This DNA does not code for proteins. A survey of the location of retrotransposon DNA in both species shows that it has independently ended up in comparable regions of the genome.

At first glance it appeared that all this extra DNA was "junk" DNA, DNA just along for the ride. But it is beginning to look like this nonprotein-coding DNA may have more of a function than was previously assumed. The possibility that this DNA is rich in regulatory RNA sequences, such as those described in chapter 18, is being actively investigated. RNAs that are not translated can play several roles, including silencing other genes. Small RNAs can form double-stranded RNA with complementary mRNA sequences, blocking translation. They can also participate in the targeted degradation of RNAs.

In one study, researchers collected almost all of the RNA transcripts made by mouse cells taken from every tissue. Although most of the transcripts coded for mouse proteins, as many as 4280 could not be matched to any known mouse protein. This finding suggests that a large part of the transcribed genome consists of genes that do not code for proteins—that is, transcripts that function as RNA. Perhaps this function can explain why a single retrotransposon can cause heritable differences in coat color in mice.

DNA that does not code for protein may regulate gene expression, often through its RNA transcript. Nonprotein-coding sequences can be found in retrotransposon-rich regions of the genome.

24.6 Genome Size and Gene Number

Genome size was a major factor in selecting which genomes would be sequenced first. Practical considerations led to the choice of organisms with relatively small genomes. Considering genome size, the original gene count for the human genome was estimated at 100,000 genes.

As sequence data were analyzed, the predicted number of genes started to decrease. A very different picture emerged. Our genome has only 25% of the 100,000 anticipated genes, approximately the same number of genes as the tiny *Arabidopsis* plant. Humans have nine times the amount of DNA found in the 365-million-base-pairs pufferfish genome, but about the same number of genes. Keep in mind that the number of genes may not correspond to the number of proteins. For example, alternative splicing (see chapters 15, 16, and 18) can produce multiple, distinct transcripts from a single gene.

Noncoding DNA inflates genome size

Why do humans have so much extra DNA? Much of it appears to be in the form of introns, noncoding segments within a gene's sequence, that are substantially bigger than those in pufferfish. The *Fugu* genome has only a handful of "giant" genes containing long introns; studying them should provide insight into the evolutionary forces that have driven the change in genome size during vertebrate evolution. (Introns are described in chapter 15.)

As described earlier, large expanses of retrotransposon DNA contribute to the differences in genome size from one species to another. Although part of the genome, nonprotein-coding DNA does not contain genes in the usual sense. As another example, *Drosophila* exhibits less noncoding DNA than *Anopheles*, although the evolutionary force driving this reduction in noncoding regions is unclear. A correlation does not exist between the number of genes and genome size.

Plants have widely varying genome size

Plants have an even greater range of genome sizes. As much as a 200-fold difference has been found, yet all these plants weigh in with about 30,000 to 40,000 genes. Tulips for example, have 170 times more DNA than *Arabidopsis*.

Both rice and *Arabidopsis* have higher *copy numbers* for gene families (multiple slightly divergent copies of a gene) than are seen in animals or fungi, suggesting that these plants have undergone numerous episodes of polyploidy, segmental duplication, or both during the 150–200 million years since rice and *Arabidopsis* diverged from a common ancestor.

Whole-genome duplication is insufficient to explain the size of some genomes. Wheat and rice are very closely related and have similar gene content, and yet the wheat genome is 40 times larger than the rice genome. This difference cannot be explained solely by the fact that bread wheat is a hexaploid (6*n*) and rice is a diploid (2*n*).

Now that the rice genome is fully sequenced, attention has shifted to sequencing the other cereal grains, especially wheat. The wheat genome apparently contains lots of repetitive DNA, which has increased its DNA content, but not necessarily its gene content. Comparisons between the rice and wheat genomes should provide clues about the genome of their common ancestor and the dynamic evolutionary balance between opposing forces that increase genome size (polyploidy, transposable element proliferation, and gene duplication) and those that decrease genome size (mutational loss).

Genome size has varied over evolutionary time, but increases or decreases in size do not correlate as expected with number of genes. Polyploidy in plants does not by itself explain differences in genome size. A greater amount of DNA is explained by the presence of introns and nonprotein-coding sequences than gene duplicates.

Comparisons among individual human genomes continue to provide information on genetic disease detection and the best course of treatment. An even broader array of possibilities arise when comparisons are made among species. There are advantages to comparing both closely and distantly related pairs of species, as well as comparing the genomes of a pathogen and its host. Examples of the benefits of each type of genome comparison follow.

Distantly related genomes offer clues for causes of disease

Sequences that are conserved between humans and pufferfish provide valuable clues for understanding the genetic basis of many human diseases. Amino acids critical to protein function tend to be preserved over the course of evolution, and changes at such sites within genes are more likely to cause disease.

It is difficult to distinguish functionally conserved sites when comparing human proteins with those of other mammals because not enough time has elapsed for sufficient changes to accumulate at nonconserved sites. Because the pufferfish genome is only distantly related to humans, conserved sequences are far more easily distinguished.

Closely related organisms enhance medical research

It is much easier to design experiments to identify gene function in an experimental system like the mouse than it is in humans. Comparing mouse and human genomes quickly revealed the function of 1000 previously unidentified human genes. The effects of these genes can be studied in mice, and the results can be used in potential treatments for human diseases.

A draft of the rat genome has been completed, and even more exciting news about the evolution of mammalian genomes may emerge from comparisons of these species. One of the most exciting aspects of comparing rat and mice genomes is the potential to capitalize on the extensive research on rat physiology, especially heart disease, and the long history of genetics in mice. Linking genes to disease has become much easier.

Humans have been found to contain segmental duplications that are absent in the chimp. Some of these duplications correspond to human disease. These differences can aid medical researchers in developing treatments for genetic disease, but ethical issues surrounding chimpanzee research and protection should be of paramount importance.

Pathogen–host genome differences reveal drug targets

With genome sequences in hand, pharmaceutical researchers are more likely to find suitable drug targets to eliminate pathogens without harming the host. Diseases in many developing countries—including malaria and Chagas disease—have both human and insect hosts. Both these infections are caused by protists (chapter 29). The value of comparative genomics in drug discovery is illustrated as follows for both malaria and Chagas disease.

Malaria

Anopheles gambiae, the malaria-carrying mosquito, along with *Plasmodium falciparum*, the protistan parasite it transmits, together have an enormous effect on human health, resulting in 1.7 to 2.5 million deaths each year from malaria. The genomes of both *Anopheles* and *Plasmodium* were sequenced in 2002.

The parasitic protist *Plasmodium falciparum*, which causes malaria, has a relatively small genome of 24.6 million base-pairs that proved very difficult to sequence. It has an unusually high proportion of adenine and thymine, making it hard to distinguish one portion of the genome from the next. The project took five years to complete. *P. falciparum* appears to have about 5300 genes, with those of related function clustered together, suggesting that they might share the same regulatory DNA.

P. falciparum is a particularly crafty organism that hides from our immune system inside red blood cells, regularly changing the proteins it presents on the surface of the red blood cell. This "cloaking" has made it particularly difficult to develop a vaccine or other treatment for malaria.

Recently, a link to chloroplast-like structures in *P. falciparum* has raised other possibilities for treatment. An odd subcellular component called the *apicoplast*, found only in *Plasmodium* and its relatives, appears to be derived from a chloroplast appropriated from algae engulfed by the parasite's ancestor (figure 24.13).

Analysis of the *Plasmodium* genome reveals that about 12% of all the parasite's proteins, encoded by the nuclear genome, head for the apicoplast. These proteins act there to produce fatty acids. The apicoplast is the only location in which the parasite makes the fatty acids, suggesting that drugs targeted at this biochemical pathway might be very effective against malaria.

Another disease-prevention possibility is to look at chloroplast-specific herbicides, which might kill *Plasmodium* by targeting the chloroplast-derived apicoplast.

Chagas disease

Trypanosoma cruzi, an insect-borne protozoan, kills about 21,000 people in Central and South America each year, with up to 18 million suffering from the infection with symptoms that include damage to the heart and other internal organs. Genome sequencing of *T. cruzi* was completed in 2005.

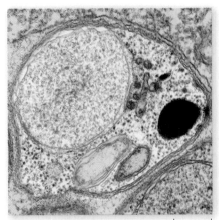

figure 24.13

PLASMODIUM APICOPLAST. Drugs targeting enzymes used for fatty acid biosynthesis within *Plasmodium* apicoplasts (colored dark green) offer hope for treating malaria.

2 μm

figure 24.14

COMPARATIVE GENOMICS MAY AID IN DRUG DEVELOPMENT. Chagas disease, African sleeping sickness, and Leishmania, which claim millions of lives in developing nations each year, share 6200 core genes. Drug development targeted at proteins encoded by the shared core genes could yield a single treatment for all three diseases.

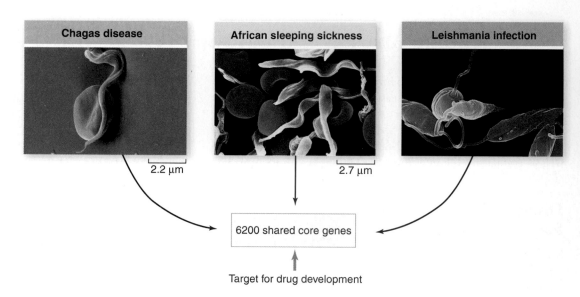

A surprising and hopeful finding is that a common core of 6200 genes is shared among *T. cruzi* and two other insect-borne pathogens: *T. brucei* and *Leishmania major*. *T. brucei* causes African sleeping sickness, and *L. major* infections result in lesions on the limbs as well as the face. These core genes are being considered as possible targets for drug treatments.

Currently, no effective vaccines and only a few drugs with limited effectiveness are available to treat any of these diseases.

The genomic similarities may not only aid in targeting drug development, but perhaps also result in a treatment or vaccine that is effective against all three devastating illnesses (figure 24.14).

> Genome comparisons enable drug development targeted at genetically based disease and at organisms causing infections and diseases that are very difficult to prevent or treat.

24.8 Crop Improvement Through Genome Analysis

Farmers and researchers have long relied on genetics for crop improvement. Whole-genome sequences offer even more information for research on artificial selection for crop improvement. Highly conserved genes can be characterized in a model system and then used to identify orthologues in crop species.

Model plant genomes provide links to genetics of crop plants

Arabidopsis is mainly an experimental flowering plant, with no commercial significance. The second plant genome for which a draft sequence has been prepared—rice—is of enormous economic significance. Rice, as mentioned earlier, belongs to the grass family, which includes a number of important cereal crop plants. Together, these crops provide most of the world's food and animal feed.

Unlike most grasses, rice has a relatively small genome of 430 million base-pairs, in contrast to the maize (corn) genome (2500 million base-pairs) and that of barley (an enormous 4900 million base-pairs). Two different subspecies of rice have been sequenced, yielding similar results. The proportion of the rice nuclear genome devoted to repetitive DNA, for example, was 42% in one variety, and 45% in the other.

Genome sequencing is underway for maize and another model, *Medicago truncatula*. *M. truncatula* has a much smaller genome than its close relative, soybean, making it much easier to sequence. Large regions of *M. truncatula* DNA are syntenous with soybean DNA, increasing the odds of finding agriculturally important soybean genes using the *M. truncatula* genome (see figure 24.9).

Beneficial bacterial genes can be located and utilized

Genome sequences of beneficial microbes may also improve crop yield. *Pseudomonas fluorescens* naturally protects plant roots from disease by excreting protective compounds. In 2005, *P. fluorescens* became the first biological control agent to have its genome sequenced. Work on identifying chemical pathways that produce protective compounds should proceed rapidly, given the bacterium's small genome size. Understanding these pathways can lead to more effective methods of protecting crops from disease—for example, isolation of a protective gene (or genes) could lead to being able to insert this beneficial gene into a crop plant's genome, so that the plant could protect itself directly.

> Comparative genomics extends the benefits of sequenced genomes in model plant species to other species important as crop plants. A growing understanding of microbial genomes may also be used to improve crop yield.

24.1 Comparative Genomics

Comparing entire genomes (DNA sequences) of different species allows scientists to explore the evolutionary divergence among organisms and attempt to connect changes at the DNA level to morphological differences.

- By comparing genomes, scientists conclude that evolutionary differences accumulate over very long time periods.
- Humans and mice, and humans and chimpanzees, share approximately 99% of the same genes.
- DNA mutations are divided into nonsynonymous, mutations that alter amino acid codes, and synonymous, mutations that do not alter the coded amino acids.
- The shorter the time period for evolution and the smaller a population size, the slower the rate of nonsynonymous gene loss.
- Genomes evolve at different rates, and evolution may occur more quickly in organisms with short generation times.
- Plants, animals, and fungi have approximately 70% of their genes in common. These common genes are involved in intermediary metabolism, genome replication and repair, and protein synthesis.
- Kingdoms are distinguished by differences in approximately one-third of the respective genomes.

24.2 Evolution of Whole Genomes

Major changes in genome structure occur by polyploidization or changes in chromosome sets.

- Polyploidy can occur by genome duplication within a species or by hybridization of two different species.
- Autopolyploidy results in doubling of the genome because of an error in meiosis.
- Allopolyploidy is the result of hybridization and subsequent duplication of the new genome (figure 24.1).
- Historical reconstruction of genome evolution involves paleopolyploids, in which sequence divergence and the presence and absence of gene pairs are studied, and synthetic polyploids, in which plants are crossed with closely related ancestral species.
- Polyploidy has occurred numerous times in the evolution of flowering plants, and downsizing of genomes is common.
- Allopolyploidy induces elimination of duplicated genes, and the loss is unequal. More genes are lost from one contributor to the hybrid than the other (figure 24.6).
- Polyploidization can lead to short-term silencing of genes by methylation of cytosines in the DNA.
- Polyploidization causes transposons to jump into new positions in the genome, possibly leading to new phenotypes.

24.3 Evolution Within Genomes

Duplication of portions of genomes provides opportunities for genes with the same function to diverge.

- Aneuploidy, the duplication of an individual chromosome, creates problems in gamete formation and is tolerated better in plants than in animals.
- Segments of DNA can be duplicated (figure 24.7).
- Duplicated DNA is common for genes associated with growth and development, immunity, and cell-surface receptors.
- Duplicated genes may have different levels of expression, depending on their location and when they are active during development.
- Paralogues refer to duplicated ancestral genes; orthologues are conserved ancestral genes.

- Genomes may be rearranged by moving gene locations within a chromosome or by the fusion of two chromosomes.
- Conservation of synteny refers to the preservation of long segments of ancestral chromosome sequences (figure 24.9).
- Genome evolution includes the formation of pseudogenes, genes that have been inactivated, perhaps because of a reduction in selection pressures (figure 24.10).
- The horizontal swapping of genes creates many phylogenetic questions, especially the origins of the three major domains (figure 24.12).

24.4 Gene Function and Expression Patterns

Although gene function can be inferred by comparing genes in different species, the comparison does not tell the whole story.

- Different species may have the same genes, but they are expressed at different times, in different tissues, and in different amounts and combinations.
- Posttranscriptional differences may also play a role in building distinct organisms from similar genomes.
- Small changes in a protein can alter its function.

24.5 Nonprotein-Coding DNA and Regulatory Function

Much of a genome is repetitive nonprotein-coding DNA.

- Noncoding DNA may be rich in regulatory RNA sequences that can silence genes and block translation.
- Nonprotein-coding sequences can be found in retrotransposon-rich regions of the genome.

24.6 Genome Size and Gene Number

Small genome size was a major factor in deciding which genomes would be sequenced first.

- Genome size does not correlate with the expected number of genes.
- Polyploidy does not by itself explain differences in genome size.
- Larger genomes result most often from the presence of introns and nonprotein-coding sequences rather than duplicated genes.

24.7 Genome Analysis and Disease Prevention and Treatment

Genome comparisons enable targeted drug development and disease treatment.

- Amino acids critical to protein function tend to be preserved over the course of evolution, and changes in amino acid sequences are more likely to cause diseases.
- By comparing closely related organisms, researchers can focus on genes that cause diseases in humans.
- Genomic comparisons enable drug development targeted at organisms that cause infections and diseases that are very difficult to prevent or treat.

24.8 Crop Improvement Through Genomic Analysis

Whole genome sequences offer more information for research on artificial selection for crop improvement.

- Model plant genomes provide links to crop plant genetics.
- Genomic sequences of beneficial and pathogenic microbes may provide new information that can be used to improve crop yield.

SELF TEST

1. Humans and pufferfish diverged from a common ancestor about 450 MYA, and these two genomes have
   a. very few of the same genes in common.
   b. all the same genes.
   c. share about 75% of the genes in the genomes.
   d. no nucleotide divergence.
2. Genome comparisons have suggested that mouse DNA has mutated about twice as fast as human DNA. What is a possible explanation for this discrepancy?
   a. Mice are much smaller than humans.
   b. Mice live in much less sanitary conditions than humans and are therefore exposed to a wider range of mutation-causing substances.
   c. Mice have a smaller genome size.
   d. Mice have a much shorter generation time.
3. Polyploidy in plants
   a. has only arisen once and therefore is very rare.
   b. only occurs naturally when there is a hybridization event between two species.
   c. is common, but never occurs in animals.
   d. is common, and does occur in some animals.
4. Homologous genes in distantly related organisms can often be easily located on chromosomes due to
   a. horizontal gene transfer.
   b. conservation of synteny.
   c. gene inactivation.
   d. pseudogenes.
5. All of the following are believed to contribute to genomic diversity among various species, *except*
   a. gene duplication.
   b. gene transcription.
   c. lateral gene transfer.
   d. chromosomal rearrangements.
6. What is the fate of *most* duplicated genes?
   a. Gene inactivation
   b. Gain of a novel function through subsequent mutation
   c. They are transferred to a new organism using lateral gene transfer.
   d. They become orthologues.
7. Chimp and human DNA is close to 99% similar; morphological differences
   a. must be due largely to gene expression.
   b. must be due exclusively to environmental differences.
   c. cannot be explained with current genetic theory.
   d. are caused by random effects during development.
8. The noncoding repetitive DNA common in both plants and animals
   a. is probably "junk" DNA.
   b. directly produces protein through mechanisms other than transcription.
   c. is still usually translated.
   d. may often produce RNA transcripts that themselves have regulatory function.
9. In general, as genome size increases there is
   a. a proportional increase in the number of genes.
   b. a proportional decrease in the number of genes.
   c. an increase in the amount of DNA.
   d. a decrease in the amount of DNA.
10. An herbicide that targets the chloroplast might be effective against malaria because—
    a. *Plasmodium* needs a functional apicoplast.
    b. the main vector for malaria is a plant.
    c. mosquitoes require plant leaves for food.
    d. none of the above.

CHALLENGE QUESTIONS

1. The loss of functional olfactory receptor (OR) genes in humans compared with chimps may be explained by reliance on other sensory systems and therefore weakened selection against individuals that lose olfactory capabilities. Could natural selection work in any other way to decrease the number of human OR genes without selecting for OR function in chimps?
2. One of the common misconceptions about sequencing projects (especially the high-profile Human Genome Project) is that creating a complete roadmap of the DNA will lead directly to cures for genetically based diseases. Given the percentage similarity in DNA between humans and chimps, is this simplistic view justified? Explain.
3. How does horizontal gene transfer (HGT) complicate phylogenetic analysis?
4. What are the consequences of polyploidy for rates of speciation? Explain.

Evolution of Development

introduction

HOW IS IT THAT CLOSELY related species of frogs can have completely different patterns of development? One frog goes from fertilized egg to adult frog with no intermediate tadpole stage. The sister species has an extra developmental stage neatly slipped in between early development and the formation of limbs—the tadpole stage. The answer to this and other such evolutionary differences in development that yield novel phenotypes are now being investigated with modern genetic and genomic tools. Research findings are accentuating the biological paradox that many developmental genes are highly conserved, and a tremendous diversity of life shares this basic toolkit of developmental genes. In this chapter, we explore the emerging field of developmental evolution, a field that brings together previously distinct fields of biology.

25.1 The Evolutionary Paradox of Development

Ultimately, to explain the differences that occur among species, we need to look at changes in developmental processes. That is, changes in genes have their effects by altering development, thus producing a different phenotype.

Phenotypic diversity could either result from many different genes or the differences could be explained by how a smaller set of genes are deployed and regulated. Based on our current understanding of the evolution of new phenotypes, the latter explanation is most probable.

Closely related sea urchins have been discovered that have very distinctive developmental patterns (figure 25.1). The direct-developing urchin never makes a pluteus larva—it just jumps ahead to its adult form. We could speculate that the two forms have different developmental genes, but it turns out that this is not the case. Instead, the two forms have undergone dramatic changes in patterns of developmental gene expression, even though their adult form is nearly the same. Here lies the evolutionary paradox: Genes have not changed, but patterns of expression have.

Highly conserved genes produce diverse morphologies

A relatively small number of gene families, about two dozen, regulate animal and plant development. The developmental roles of several of these families, including *Hox* gene transcription factors, are described in chapter 19.

Hox (homeobox) genes establish the body plan in animals by specifying when and where genes are expressed. These genes code for proteins with a highly conserved homeodomain that binds to the regulatory region of other genes to activate or repress these genes' expression. *Hox* genes appeared before the divergence of plants and animals; in plants, they have a role in shoot growth and leaf development, and in animals they establish body plans.

Another family of transcription factors, *MADS* box genes, are found throughout the eukaryotes. The *MADS* box codes for a DNA-binding motif. Large numbers of *MADS* box genes establish the body plan of plants, especially the flowers. Although the *MADS* box region is highly conserved, variation exists in other regions of the coding sequence. Later in this chapter, we consider how there came to be so many *MADS* box genes in plants and how such similar genes can have very different functions.

Transcription factors and genes involved in signaling pathways are responsible for coordination of development. As you saw in chapter 9, key elements of kinase and G protein-signaling pathways are also highly conserved among organisms. Even subtle changes in a signaling pathway can alter the enzyme that is activated or repressed, the transcription factor that is activated or repressed, or the activation or repression of gene expression. Any of these changes can have dramatic effects on the development of an organism.

Developmental mechanisms exhibit evolutionary change

Understanding how development evolves requires integration of knowledge about genes, gene expression, development, and evolution. As just mentioned, either transcription factors or signal-

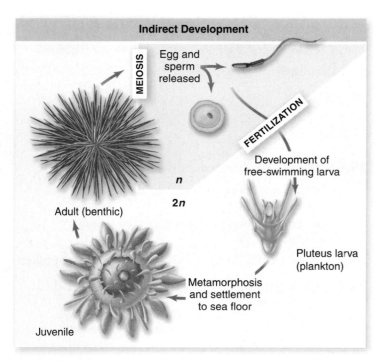
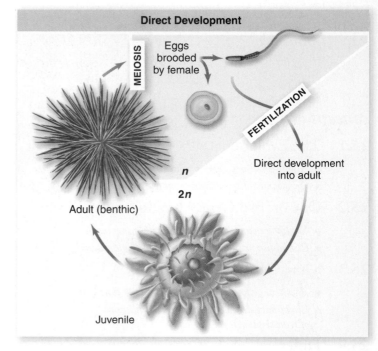

figure 25.1

DIRECT AND INDIRECT SEA URCHIN DEVELOPMENT. Phylogenetic analysis shows that indirect development was the ancestral state. Direct-developing sea urchins have lost an intermediate stage of development.

ing molecules can be modified during evolution. In both cases, it is possible to alter the timing or position of gene expression and, as a result, gene function within the developing organism.

Heterochrony

Alterations in timing of developmental events due to a genetic change are called **heterochrony.** A heterochronic mutation could affect a gene that controls when a plant transitions from the juvenile to the adult stage, at which point it can produce reproductive organs. A mutation in a gene that delays flowering in plants can result in a small plant that flowers quickly rather than requiring months or years of growth.

Most mutations that affect developmental regulatory genes are lethal, but every so often a novel phenotype emerges that persists because of increased fitness. If a mutation leading to early flowering increased the fitness of a plant, the new phenotype will persist. For example, a tundra plant that flowers earlier could have increased fitness over an individual of the same species that flowers later, just as the short summer comes to a close.

Homeosis

Alternations in the spatial pattern of gene expression can result in **homeosis.** The *Bithorax Drosophila* fly that you were introduced to in chapter 19 is an example of a homeotic mutation in which gene expression patterns shift, producing two, rather than one, pairs of wings.

The *Drosophila Antennapedia* mutant, which has a leg where an antenna should be, is another example of a homeotic mutation. Mutations in genes such as *Antennapedia* and *Bithorax* can arise spontaneously in the natural world or by mutagenesis in the laboratory, but their bizarre phenotypes would have little survival value in nature.

Changes in translated regions of transcription factors

The coding sequence of a gene can contain multiple regions with different functions (figure 25.2). The DNA-binding motifs, exemplified by *MADS* box and *Hox* genes, could be altered so that they no longer bind to their target genes; as a result, that developmental pathway would cease to function. Alternatively, the modified transcription factor might bind to a different target and initiate a new sequence of developmental events.

The regulatory region of a transcription factor must also be considered in the evolution of developmental mechanisms. A sequence change could alter the transcription complex that forms at a regulatory region, resulting in novel expression patterns. Either the time or place of gene expression could be affected, giving rise to heterochrony or homeosis. In this case, the downstream targets might be the same, but the cells that express the target genes or the time at which these target genes are expressed could change.

Changes in signaling pathways

Coordinating information about neighboring cells and the external environment is essential for successful development. Signaling pathways are essential for cell-to-cell communication. If the structure of a ligand changes, it may no longer bind to its target receptor, or it could bind to a different receptor or no receptor at all. If, as a result of a genetic change, a receptor is produced in a different cell type, a homeotic phenotype may

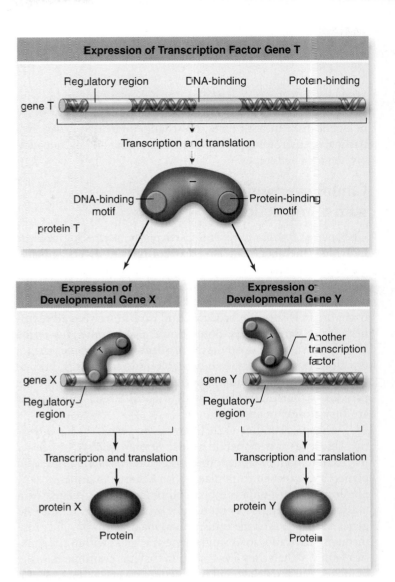

figure 25.2

TRANSCRIPTION FACTORS HAVE A KEY ROLE IN THE EVOLUTION OF DEVELOPMENT. Altering different regions of the coding and regulatory sequences of a transcription factor can alter development and the phenotype of an organism. A single gene can code for a protein with multiple binding sites

appear. And, as mentioned earlier, small changes in signaling molecules can alter their targets.

The sections that follow use specific examples of the evolution of diverse morphologies. For each example, consider how the mechanism of development has been altered and what the outcome is. Keep in mind that these are the successful examples; most morphological novelties that arise quickly, but do not improve fitness, go extinct.

> Highly conserved genes can undergo small changes in their coding or regulatory regions that alter the place or time of gene expression and function, resulting in new body plans. Changes in transcription factors and signaling pathways are the most common source of new morphologies.

One or Two Gene Mutations, New Form

Here we consider two examples of single gene mutations with altered morphology: that of wild cabbage, and that of jaw shape in cichlid fish. In both cases, the change increased fitness in a particular environment at a specific time, enhancing selection of the new phenotypes.

Cauliflower and broccoli began with a stop codon

The species *Brassica oleracea* is particularly fascinating because individual members can have extraordinarily diverse phenotypes. The diversity in form is so great that *B. oleracea* members are divided into subspecies (figure 25.3).

Wild cabbage, kale, tree kale, red cabbage, green cabbage, brussels sprouts, broccoli, and cauliflower are all members of the same species. Some flower early, some late. Some have long stems and others have short stems. Some form a few flowers and others, like broccoli and cauliflower, initiate many flowers but development of the flowers is arrested. The paradox is that these plants with such different appearances are very closely related.

One piece of the puzzle lies with the gene *CAL* (*Cauliflower*), which was first cloned in a close *Brassica* relative, *Arabidopsis*. In combination with another mutation, *Apetala1*, *Arabidopsis* plants can be turned from plants with a limited number of simple flowers into miniature broccoli or cauliflower plants with masses of arrested flower meristems or flower buds. These two genes are needed for the transition to making flowers and arose through duplication of a single ancestral gene within the brassica family. When they are absent, meristems continue to make branches, but are delayed in producing flowers.

The *CAL* gene was cloned from large numbers of *B. oleracea* subspecies, and a stop codon, TAG, was found in the middle of the *CAL* coding sequences of broccoli and cauliflower. A phylogenetic analysis of *B. oleracea* coupled with the *CAL* sequence analysis leads to the conclusion that this stop codon appeared after the ancestors of broccoli and cauliflower diverged from other subspecies members, but before broccoli and cauliflower diverged from each other (see figure 25.3).

inquiry

Knowing that cauliflower and broccoli have a stop codon in the middle of the CAL gene-coding sequence, predict the wild-type function of CAL. What additional evolutionary events may have occurred since broccoli and cauliflower diverged?

This study points out the importance of having well-documented phylogenies in place for the analysis of developmental pattern evolution. A second, somewhat unusual, feature of this example is that the driving selective force for these subspecies was artificial. Wild relatives are still found scattered along the rocky coasts of Spain and the Mediteranean region. The most likely scenario is that humans found

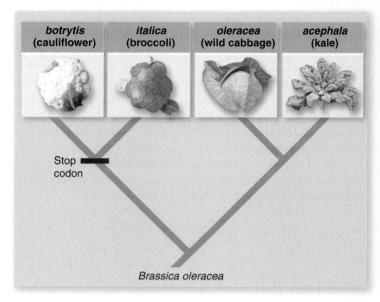

figure 25.3

EVOLUTION OF CAULIFLOWER AND BROCCOLI. A point mutation that converted an amino acid-coding region into a stop codon resulted in the extensive reproductive branching pattern that was artificially selected for in two crop plants that are subspecies of *Brassica oleracea*.

a *cal* mutant and selected for that phenotype through cultivation. The large heads of broccoli and cauliflower offer a larger amount of a vegetable material than the wild kale plants and a tasty alternative to *Brassica* leaves.

Cichlid fish jaws demonstrate morphological diversity

Our second example of how a single gene can change form and function comes from natural selection of cichlid fish in Lake Malawi in East Africa. In less than a few million years, hundreds of species have evolved in the lake from a common ancestor. The rapid speciation of cichlids is discussed in chapter 22.

One explanation for the successful speciation is that different species have acquired different niches based on feeding habits. There are bottom eaters, biters, and rammers. The rammers have particularly long snouts with which to ram their prey; biters have an intermediate snout; and the bottom feeders have short snouts adapted to scrounging for food at the base of the lake (figure 25.4).

How did these fish acquire such different snout forms? An extensive genetic analysis revealed that two genes, of yet unknown function, are likely responsible for the shape and size of the jaw. The results of crossing long- and short-snouted

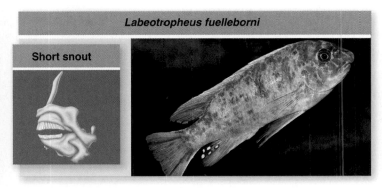

figure 25.4

DIVERSITY OF CICHLID FISH JAWS. A difference in one gene is responsible for a short-snout in *Labeotropheus fuelleborni* and a long-snout in *Metriaclima zebra*. Genes that affect jaw length can affect body shape as well because of the constraints the size of the jaw places on muscle development.

cichlids indicate the importance of a single gene in determining jaw length and height.

Regulating the length versus the height of the jaw may well be an early and important developmental event. The overall size of the fish and the extent of muscle development both hinge on the form of the jaw. The range of jaw forms appears to have persisted because the cichlids establish unique niches for feeding within the lake.

Our examples have focused on a few genes that underlie a development change. It is more common, however, that many genes with additive effects control large differences in phenotype.

A single or a few mutations can dramatically alter body plans. Although most mutations are lethal, some confer a fitness advantage under natural or artificial selection.

25.3 Same Gene, New Function

In the preceding chapter, we discussed the similarity between human and mouse genomes. If all but 300 of humans' 20,000 to 25,000 genes are shared with mice, why are mice and humans so different? Part of the answer is that genes with similar sequences in two different species may work in slightly or even dramatically different ways.

Ancestral genes may be co-opted for new functions

The evolution of chordates can partially be explained by the co-option of an existing gene for a new function. Ascidians are basal chordates that have a notochord but no vertebrae (chapter 35). The *Brachyury* gene of ascidians encodes a transcription factor, and it is expressed in the developing notochord (figure 25.5).

Brachyury is not a novel gene that appeared in animals as vertebrates evolved. It is found in invertebrates as well. A mollusk homologue of *Brachyury* is associated with anterior–posterior axis specification. Most likely, an ancestral *Brachyury* gene was co-opted for a new role in notochord development. *Brachyury* is a member of a gene family with a specific domain, that is, a conserved sequence of base-pairs within the gene. A region of *Brachyury* encodes a protein domain called the **T box,** which is a transcription factor. So, *Brachyury*-encoded protein turns on a gene or genes The details of which genes are regulated by *Brachyury* are only now being discovered.

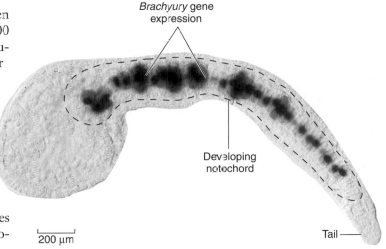

200 µm

figure 25.5

CO-OPTING A GENE FOR A NEW FUNCTION. *Brachyury* is a gene found in invertebrates that has been used for notochord development in this ascidian, a basal chordate. By attaching the *Brachyury* promoter to a gene with a protein product that stains blue, it is possible to see that *Brachyury* gene expression in ascidians is associated with the development of the notochord, a novel function compared with its function in organisms lacking a notochord. The orthologue in the nematode *Caenorhabditis elegans* is important for hind gut and male tail development, but there is no evidence of a notochord precursor.

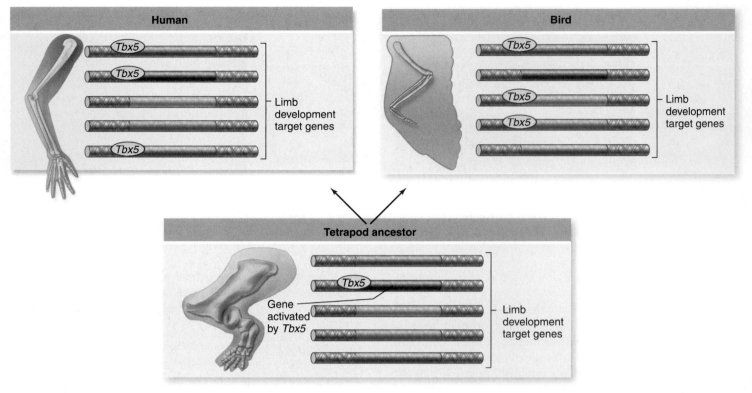

figure 25.6

TBX5 REGULATES WING AND ARM DEVELOPMENT. Wings and arms are very different, but the development of each depends on *Tbx5*. Why the difference? *Tbx5* turns on different genes in birds and humans.

In mice and dogs, a mutation in *Brachyury* causes a short tail to develop. Humans lack tails, but have wild-type copies of *Brachyury*. Genes in addition to *Brachyury* must be needed to make a tail.

The paradox that puzzles evolutionary developmental biologists, as already mentioned, is how a single genetic toolkit can be used to build an insect, a bird, a bat, a whale, or a human. This paradox is illustrated with the *Brachyury* gene. One explanation is that *Brachyury* turns on different genes or combinations of genes in different animals. Although there is not yet enough data to sort out the details of *Brachyury*, we can look at limb formation for an explanation of how such a change could have evolved.

Limbs have developed through modification of transcriptional regulation

All tetrapods have four limbs—two hindlimbs and two forelimbs. The forelimb in a bird is actually the wing. Our forelimb is the arm. Clearly, these are two very different structures, but they have a common evolutionary origin. As you learned in chapter 23, these are termed *homologous structures*.

At the genetic level, humans and birds both express the *Tbx5* gene in developing forelimb buds. Like *Brachyury*, *Tbx5* is a member of a transcription factor gene family with T box domain, that is, a conserved sequence of base-pairs within the gene. So, *Tbx5*-encoded protein turns on a gene or genes that are needed to make a limb. Mutations in the human *Tbx5*

gene cause Holt–Oram syndrome, resulting in forelimb and heart abnormalities.

What seems to have changed as birds and humans evolved are the genes that are transcribed because of the Tbx5 protein (figure 25.6). In the ancestral tetrapod, perhaps Tbx5 protein bound to only one gene and triggered transcription. In humans and birds, genes are expressed in response to Tbx5 protein, but they are different genes.

The story of the evolution of limb development is far more complex than *Tbx5*, which gets limb development started. The genes that *Tbx5* regulates in turn may affect the expression of other genes. Gleaning information about genome sequences for different organisms will be essential in identifying all the genes involved.

Note also that development occurs in four dimensions—namely, in three-dimensional space and over time. Changing the timing of gene expression, as well as the genes that are expressed, can result in dramatic changes in shape.

During evolution, genes have been co-opted for new functions, one example being the use of the *Brachyury* gene for notochord development. In limb development, the Tbx5 protein binds to different genes in birds and in humans, resulting in the development of very different limbs. Changes in genes expressed and in the timing of expression can result in dramatic morphological differences.

Different Genes, Convergent Function

Homoplastic structures, also known as analogous structures, have the same or similar functions, but arose independently—unlike homologous structures that arose once from a common ancestor. Phylogenies reveal convergent events, but the origin of the convergence may not be easily understood. In many cases different developmental pathways have been modified, as is the case with the spots on butterfly wings. In other cases, such has flower shape, it is not always as clear whether the same or different genes are responsible for convergent evolution.

Insect wing patterns demonstrate homoplastic convergence

Insect wings, especially those of moths and butterflies, have beautiful patterns that can protect them from predation and allow them to thermoregulate (figure 25.7). The origins of these patterns are best explained by the recruitment of existing regulatory programs for new functions.

Sensory bristles in *Drosophila* and butterfly wing scales that produce amazing colors are regulated by the same underlying mechanism. Scales are believed to be evolutionarily derived from bristles. Both bristle and scale development is initiated by the transcription factor *achaete-scute*. In the development of scales, structures that have their origins in bristles are inverted, and the daughter cell that previously gave rise to neuronal connections dies because of programmed cell death. Later in development, pigment production is triggered.

Not all insects have co-opted the same sets of genes for these new functions, but all the evolutionary pathways have converged around production of these novel, highly patterned wings.

Flower shapes have also altered in a convergent way

Flowers exhibit two types of symmetry. Looking down on a **radially symmetrical** flower, you see a circle. No matter how you cut that flower, as long as you have a straight line that intersects with the center, you end up with two identical parts. Examples are daisies, roses, tulips, and many other flowers.

Bilaterally symmetrical flowers have mirror-image halves on each side of a single central axis. If they are cut in any other direction, two nonsimilar shapes result. Plants with bilaterally symmetrical flowers include snapdragons, mints, and peas.

Bilaterally symmetrical flowers are attractive to their pollinators, and the shape may have been an important factor in their evolutionary success. At the crossroads of evolution and development, the questions that arise are first, what genes are involved in bilateral symmetry? And second, are the same genes involved in the numerous, independent origins of asymmetrical flowers?

Cycloidia (*CYC*) is a snapdragon gene responsible for the bilateral symmetry of the flower. Snapdragons with mutations in *CYC* have radially symmetrical flowers (see figure 42.17b). Beginning with robust phylogenies, researchers have selected flowers that evolved bilateral symmetry independently from snapdragons and cloned the *CYC* gene. The *CYC* gene in closely related symmetrical flowers has also been sequenced.

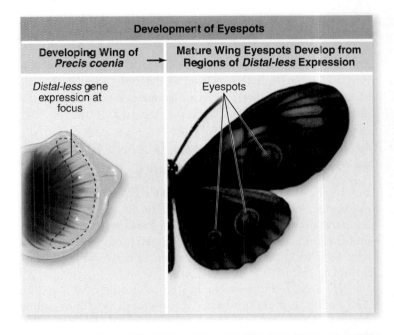

Development of Eyespots

| Developing Wing of *Precis coenia* | Mature Wing Eyespots Develop from Regions of *Distal-less* Expression |

Distal-less gene expression at focus

Eyespots

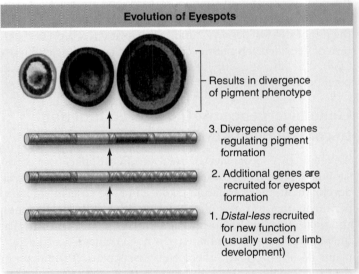

Evolution of Eyespots

Results in divergence of pigment phenotype

3. Divergence of genes regulating pigment formation

2. Additional genes are recruited for eyespot formation

1. *Distal-less* recruited for new function (usually used for limb development)

figure 25.7

BUTTERFLY EYESPOT EVOLUTION. The *Distal-less* gene, usually used for limb development, was recruited for eyespot development on butterfly wings. Distal-less initiates the development of different colored spots in different butterfly species by regulating different pigment genes in different species. Eyespots can protect butterflies by startling predators.

Comparisons of the *CYC* gene sequence among phylogenetically diverse flowers indicate that both radial symmetry and bilateral symmetry evolved in multiple ways in flowers. Although radial symmetry is the ancestral condition, some radially symmetrical flowers have a bilaterally symmetrical ancestor. Loss of *CYC* function accounts for the loss of bilateral symmetry in some of these plants.

Gain of bilateral symmetry arose independently among some species because of the *CYC* gene. This change is an example of convergent evolution through mutations of the same gene. In other cases, *CYC* is not clearly responsible for the bilateral symmetry. Other genes also played a role in the convergent evolution of bilaterally symmetrical flowers.

> Similar morphologies can be produced using different developmental pathways through convergent evolution.

25.5 Gene Duplication and Divergence

You encountered the role of gene duplication in genome evolution in chapter 24. In this section, we explore a specific example of the evolution of development through gene duplication and divergence in flower form.

Gene duplications of *paleoAP3* led to flowering-plant morphology

Before the flowering plants originated, a *MADS* box gene duplicated, giving rise to genes called *PI* and *paleoAP3*. In ancestral flowering plants, these genes affected stamen development, and this function has been retained. (Stamens, you may already know, are the male reproductive structures of flowering plants.)

The *paleoAP3* gene duplicated to produce *AP3* and an *AP3* duplicate some time after members of the poppy family last shared a common ancestor with the clade of plants called the eudicots (plants like apple, tomato, and *Arabidopsis*). This clade of eudicots is distinguished on the genome level by both the duplication of *paleoAP3* and by the origins of a precise pattern of petal development in their last common ancestor (figure 25.8). The phylogenetic inference is that *AP3* gained a role in petal development.

Gene divergence of *AP3* altered function to control petal development

Although the occurrence of *AP3* through duplication corresponds with a uniform development process for specifying petal development, the correlation could simply be coincidental. Experiments that mix and match parts of the *AP3* and *PI* genes, and then introduce them into *ap3* mutant plants, confirm that the phylogenetic correspondence is not a coincidence. The *ap3* plants do not produce either petals or stamens. A summary of the experiments is shown in figure 25.9.

Earlier in this chapter, the *MADS* box transcription factor gene family was introduced. One region of a *MADS* gene codes for a DNA-binding motif; other regions of the resulting protein have different functions, including protein–protein binding. The PI and AP3 proteins can bind to each other, and as a result, can regulate the transcription of genes needed for stamen and petal formation.

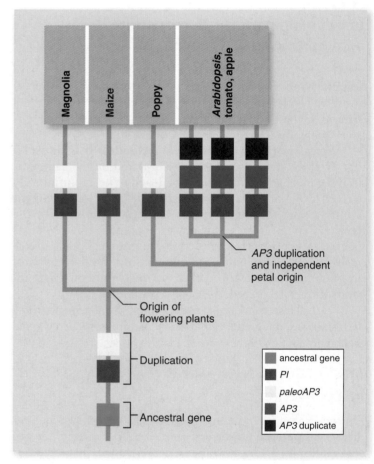

figure 25.8

PETAL EVOLUTION THROUGH GENE DUPLICATION. Two gene duplications resulted in the *AP3* gene in the eudicots that has acquired a role in petal development.

The C terminus of AP3

Both AP3 and PI have distinct sequences at the C (carboxy) protein terminus (coded for by the 3′ end of the genes). The C-terminus sequence of the AP3 protein is essential for specifying petal function, and it contains a conserved sequence shared

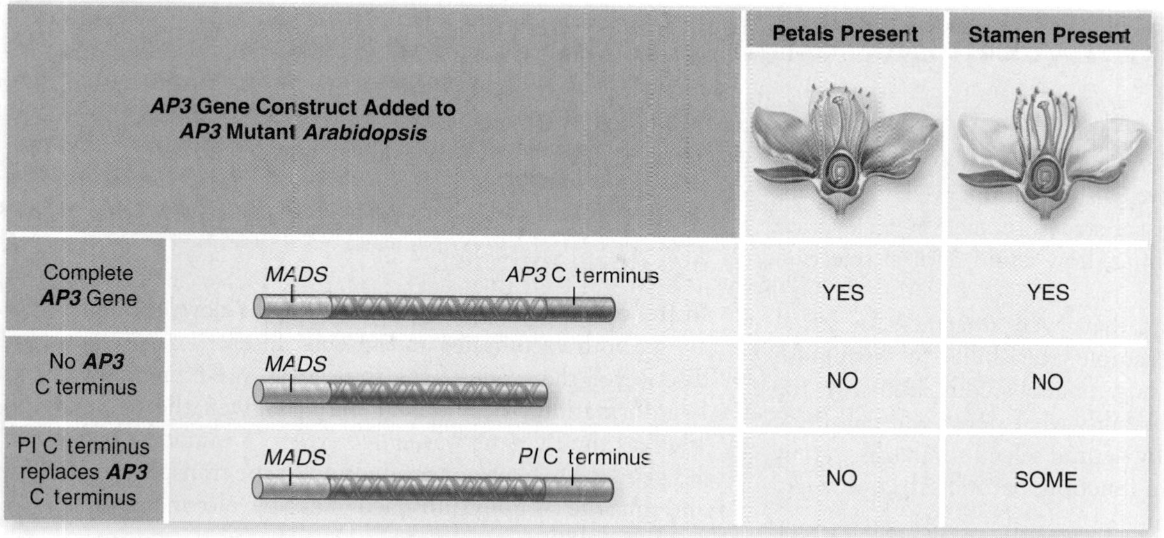

AP3 Gene Construct Added to AP3 Mutant *Arabidopsis*			Petals Present	Stamen Present
Complete **AP3** Gene	MADS .... AP3 C terminus		YES	YES
No **AP3** C terminus	MADS		NO	NO
PI C terminus replaces **AP3** C terminus	MADS .... PI C terminus		NO	SOME

figure 25.9

AP3 HAS ACQUIRED A DOMAIN NECESSARY FOR PETAL DEVELOPMENT. The *AP3* gene includes MADS box encoding a DNA-binding domain and a highly specific sequence near the C terminus. Without the 3′ region of the *AP3* gene, the *Arabidopsis* plant will not make petals.

among the eudicots. The *AP3* C-terminus DNA sequence was deleted from the wild-type gene and the new construct was inserted into *ap3* plants to create a transgenic plant. Other transgenic plants were created by inserting the complete *AP3* sequence into *ap3* plants. The complete *AP3* sequence rescued the mutant, and petals were produced. No petals formed when the C-terminus motif was absent.

AP3 is also needed for stamen (male floral organ) development, an ancestral trait found in *paleoAP3*. Plants lacking *AP3* fail to produce either stamens or petals. Transgenic plants with the C-terminus deletion construct also failed to produce stamens.

The C terminus of PI

The *pi* mutant phenotype also lacks stamens and petals. To test whether the *PI* C terminus could substitute for the *AP3* C terminus in specifying petal formation, the *PI* C terminus was added to the truncated *AP3* gene. No petals formed, but stamen development was partially rescued. These experiments demonstrate that *AP3* has acquired an essential role in petal development, encoded in a sequence at the 3′ end of the gene.

Through gene duplication and divergence, the *AP3* gene gained a role in petal development in eudicot flowering plants.

25.6 Functional Analysis of Genes Across Species

Comparative genomics is amazingly useful in understanding morphological diversity. Limitations exist, however, in the inferences about evolution of development that we can draw from sequence comparison alone. Functional analysis includes a range of experiments designed to test the actual function of a gene in different species.

Sequence comparisons among organisms are essential for both phylogenetic and comparative developmental studies. Careful analysis is needed to distinguish paralogues from orthologues. Rapidly evolving research using bioinformatics, which utilizes computer programming to analyze DNA and protein data, leads to hypotheses that can be tested experimentally.

You have already seen how this can work with highly conserved genes such as *Tbx5*. However, a single base mutation can change an active gene into an inactive pseudogene. Therefore, although function can be inferred from sequence data, experiments are necessary to demonstrate the actual function of the gene. This type of work is called *functional genomics* and is explained in chapter 17.

inquiry

? *Explain how functional analysis was used to support the claim that petal development evolved through the acquisition of petal function in the AP3 gene of Arabidopsis.*

Tools for functional analysis exist in model systems but need to be developed in other organisms on the tree of life if we are going to piece together evolutionary history. Model systems such as yeast, the flowering plant *Arabidopsis*, the nematode *Caenorhabditis elegans*, *Drosophila*, and the mouse have been selected because they are easy to manipulate in the laboratory, have short life cycles, and have well-delineated genomes. Also, it is possible to visualize gene expression within parts of the organism using labeled markers and to create transgenic organisms that contain and express foreign genes.

Functional analysis is necessary to determine the actual function of similar genes in different species.

Diversity of Eyes in the Natural World: A Case Study

The eye is one of the most complex organs. Biologists have studied it for centuries. Indeed, explaining how such a complicated structure could evolve was one of the great challenges facing Darwin. If all parts of a structure such as an eye are required for proper functioning, how could natural selection build such a structure?

Darwin's response was that even intermediate structures—which provide, for example, the ability to distinguish light from dark—would be advantageous compared with the ancestral state of no visual capability whatsoever, and thus these structures would be favored by natural selection. In this way, by incremental improvements in function, natural selection could build a complicated structure.

Morphological evidence indicates eyes evolved at least twenty times

Comparative anatomists long have noted that the structures of the eyes of different types of animals are quite different. Consider, for example, the difference in the eyes of a vertebrate, an insect, a mollusk (octopus), and a planarian (figure 25.10). The eyes of these organisms are extremely different in many ways, ranging from compound eyes, to simple eyes, to mere eyespots.

Consequently, these eyes are examples of convergent evolution and are homoplastic (analogous) rather than homologous. For this reason, evolutionary biologists traditionally viewed the eyes of different organisms as having independently evolved, perhaps as many as 20 times. Moreover, this view holds that the most recent common ancestor

of all these forms was a primitive animal with no ability to detect light.

The same gene, *Pax6*, initiates fly and mouse eye development

In the early 1990s, biologists studied the development of the eye in both vertebrates and insects. In each case, a gene was discovered that codes for a transcription factor important in lens formation; the mouse gene was given the name *Pax6*, whereas the fly gene was called *eyeless*. A mutation in the *eyeless* gene led to a lack of production of the transcription factor, and thus the complete absence of eye development, giving the gene its name.

When these genes were sequenced, it became apparent that they were highly similar; in essence, the homologous *Pax6* gene was responsible for triggering lens formation in both insects and vertebrates. A stunning demonstration of this homology was conducted by the Swiss biologist Walter Gehring, who inserted the mouse version of the *Pax6* into the genome of a fruit fly, creating a transgenic fly. In this fly, the *Pax6* gene was turned on by regulatory factors in the fly's leg. *Pax6* was expressed, and an eye formed on the leg of the fly (figure 25.11)!

These results were truly shocking to the evolutionary biology community. Insects and vertebrates diverged from a common ancestor more than 500 MYA. Moreover, given the large differences in structure of the vertebrate eye and insect eye, the standard assumption was that the eyes evolved independently, and thus that their develop-

figure 25.10

A DIVERSITY OF EYES. Morphological and anatomical comparisons of eyes are consistent with the hypothesis of independent, convergent evolution of eyes in diverse species such as flies and humans.

figure 25.11

MOUSE *Pax6* MAKES AN EYE ON THE LEG OF A FLY. *Pax6* and *eyeless* are functional homologues. The *Pax6* master regulator gene can initiate compound eye development in a fruit fly or simple eye development in a mouse.

ment would be controlled by completely different genes. That eye development was affected by the same homologous gene, and that these genes were so similar that the vertebrate gene seemed to function normally in the insect genome, was completely unexpected.

The *Pax6* story extends to eyeless fish found in caves (figure 25.12). Fish that live in dark caves need to rely on senses other than sight. In cavefish, *Pax6* gene expression is greatly reduced. Eyes start to develop, but then degenerate.

Ribbon worms, but not planaria, use *Pax6* for eye development

Recent discoveries have yielded further surprises about the *Pax6* gene. Even the very simple ribbon worm, *Lineus sanguineus*, relies on *Pax6* for development of its eyespots. A *Pax6* homologue has been cloned and has shown to be expressed at the sites where eyespots develop. In contrast, planarian worms do not rely on *Pax6* for eyespot development.

Ribbon worm eyespot regeneration

The simple marine ribbon worm evolved later than the flatworm planaria. Just like planaria, ribbon worms can regenerate their head region if it is removed. In an elegant experiment, the head of a ribbon worm was removed, and biologists followed the regeneration of eyespots. At the same time, the expression of the *Pax6* homologue was observed using in situ hybridization.

To observe *Pax6* gene expression, an antisense RNA sequence of the *Pax6* was made and labeled with a color marker. When the regenerating ribbon worms were exposed to the antisense *Pax6* probe, the antisense RNA paired with expressed

Surface Dweller — Has *Pax6*

a.

Cave Dweller — Loss of *Pax6*

b.

figure 25.12

CAVEFISH HAVE LOST THEIR SIGHT. Mexican tetras (*Astyanax mexicanus*) have **(a)** surface-dwelling members and **(b)** cave-dwelling members of the same species. The cavefish have very tiny eyes, partly because of reduced expression of *Pax6*.

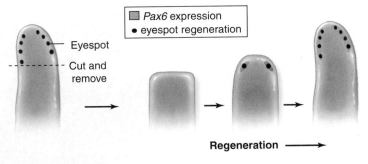

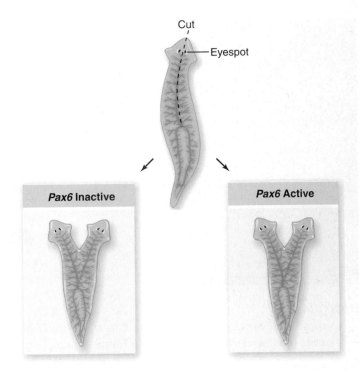

figure 25.13

Pax6 EXPRESSION CORRELATES WITH RIBBON WORM EYESPOT REGENERATION. *Pax6* is expressed at the same time and place as eyespots on regenerating ribbon worms. The presence of *Pax6* transcripts was visualized through in situ hybridization with a colored antisense probe.

Pax6 RNA transcripts and could be seen as colored spots under the microscope (figure 25.13).

Planarian eyespot regeneration

Similar experiments were tried with planaria species that are phylogenetically related to ribbon worms, but the conclusion was quite different from in ribbon worms. If a planaria is cut in half lengthwise, it can regenerate its missing half, including the second eyespot, but no *Pax6* gene expression is associated with regenerating the eyespots.

Planaria do have *Pax6*-related genes, but inactivating those genes does not stop eye regeneration (figure 25.14). These *Pax6*-related genes are, however, expressed in the central nervous system. A *Pax6*-responsive element, P3-enhancer, has also been identified and shown to be active in planaria. Perhaps some clues as to the origin of *Pax6*'s role in eye development will be uncovered as comparisons between ribbon worm and planarian eyespot regeneration continue.

The initiation of eye development may have evolved just once

Several explanations are possible for these findings. One is that eyes in different types of animals evolved truly independently, as originally believed. But if this is the case, why is *Pax6* so structurally similar and able to play a similar role in so many different groups? Proponents of this opinion point out that *Pax6* is involved not only in development of the eye, but also in development of the entire forehead region of many organisms. Consequently, it is possible that if *Pax6* had a regulatory role in the forehead of early animals, perhaps it has been independently co-opted time and time again to serve a role in eye development. This role would be consistent with the data on planarians (see figure 25.14).

Many other biologists find this interpretation unlikely. The consistent use of *Pax6* in eye development in so many organisms, the fact that it functions in the same role in each case, and the great similarity in DNA sequence and even functional replaceability suggest to many that *Pax6* acquired its evolutionary role in eye development only a single time, in

figure 25.14

Pax6 IS NOT REQUIRED FOR PLANARIA EYESPOT REGENERATION. Planaria can regenerate their heads and eyespots when cut in half longitudinally. Unlike ribbon worms, *Pax6* does not appear to play a role in planaria eyespot regeneration. When both *Pax6*-related genes in planaria were prevented from producing a protein product, eyespots still formed.

the common ancestor of all extant organisms that use *Pax6* in eye development.

Given the great dissimilarity among eyes of different groups, how can this be? One hypothesis is that the common ancestor of these groups was not completely blind, as traditionally assumed. Rather, that organism may have had some sort of rudimentary visual system—maybe no more than a pigmented photoreceptor cell, maybe a slightly more elaborate organ that could distinguish light from dark.

Whatever the exact phenotype, the important point is that some sort of basic visual system existed that used *Pax6* in its development. Subsequently, the descendants of this ancestor diversified independently, evolving the sophisticated and complex image-forming eyes exhibited by different animal groups today.

Most evolutionary and developmental biologists today support some form of this hypothesis. Nonetheless, no independent evidence exists that the common ancestor of most of today's animal groups, a primitive form that lived probably more than 500 MYA, had any ability to detect light. The reason for this belief comes not from the fossil record, but from a synthesis of phylogenetic and molecular developmental data.

Understanding the evolution of eyes illustrates the power of multidisciplinary approaches to clarify the evolutionary history of the world's biological diversity. The *Pax6* gene and its many homologues indicate that eye development, although highly diverse in outcome, may have a single evolutionary origin.

25.1 The Evolutionary Paradox of Development

Highly conserved genes can undergo small changes that affect the time and place of gene expression and function.

- Understanding development requires an understanding of genes, gene expression, development, and evolution or natural selection.

- Changes in *Hox* gene transcription factors and genes involved in signaling pathways are responsible for new morphologies.

- Heterochrony refers to alteration of timing of developmental events due to genetic changes.

- Homeosis refers to the alterations in the spatial pattern of gene expression.

- Modifications of different parts of the coding and regulatory sequences of a transcription factor can alter development and phenotypic expression (figure 25.2).

25.2 One or Two Gene Mutations, New Form

A single or a few gene mutations may increase fitness and result in the persistence of a new phenotype.

- The wide diversity of cabbage subspecies is due to a simple mutation of one gene (figure 25.3).

- Although most mutations are lethal, some confer a fitness advantage under natural or artificial selection.

25.3 Same Gene, New Function (figure 25.6)

Genes with similar sequences in different species may work in different ways.

- Ancestral genes may be co-opted for a new function; the same gene may turn on different genes or combinations of genes in different species.

- Changes in the location and the timing of gene expression can result in dramatic morphological differences.

25.4 Different Genes, Convergent Function

Homoplastic or analogous structures have the same or similar functions but they did not arise from a common ancestor.

- Different morphologies can be produced using different developmental pathways through convergent evolution.

25.5 Gene Duplication and Divergence

Through gene duplication and divergence, ancestral genes gain new functions.

- Gene duplication of *paleoAF3* was important in the evolution of flowers (figure 25.8).

- Gene divergence of *AP3* altered function to control petal development (figure 25.9).

25.6 Functional Analysis of Genes Across Species

Function analysis is necessary to determine the actual function of similar genes in different species.

- Sequence comparisons are essential for both phylogenetic and comparative development studies, but we can only infer function from this information.

- Some tools for studying function include labeled markers for genes and proteins and for transgenic organisms.

- Careful phylogenetic analysis is needed to distinguish paralogues from orthologues.

25.7 Diversity of Eyes in the Natural World: A Case Study

The eye is one of the most complex organs whose function has been incrementally improved over time.

- The diversity of eyes is an example of homeoplasy and convergent evolution.

- The same *Pax6* gene initiates lens development in insects and vertebrates even though eye morphology is different.

- In ribbon worms, the *Pax6* gene regulates eyespot regeneration, but in planarians this gene is expressed in the central nervous system.

- The *Pax6* gene and its homologues indicate that eye development may have had a single evolutionary origin.

SELF TEST

1. Heterochrony is
   a. the alteration of the spatial pattern of gene expression.
   b. a change in the relative position of a body part.
   c. a change in the relative timing of developmental events.
   d. a change in a signaling pathway.

2. Vast differences in the phenotypes of organisms as different as fruit flies and humans
   a. must result from differences among many thousands of genes controlling development.
   b. have apparently arisen largely through manipulation of the timing and regulation of expression of probably less than 100 highly conserved genes.
   c. can be entirely explained by heterochrony.
   d. can be entirely explained by homeotic factors.

3. Homoplastic structures
   a. can involve convergence of completely unrelated developmental pathways.
   b. always are morphologically distinct.
   c. are produced by divergent evolution of homologous structures.
   d. are derived from the same structure in a shared common ancestor.

4. Why was it important to create transgenic plants to determine the role of *AP3* in petal formation?
   a. It provided a functional test of the role of *AP3* in petal development.
   b. Duplication of *AP3* could not be resolved on the phylogeny.
   c. To check if the phylogenetic position of *AP3* is really derived.
   d. Because tests already established the role of *paleoAP3* in stamen development.

5. Eyes of vertebrates and invertebrates
   a. are two examples of a structure that has probably evolved completely independently more than 10 times.
   b. are similar in structure due to convergence.
   c. share no similarities in their pattern of development.
   d. may be homologous at the level of the initiation of development of a visual receptor.

6. *Hox* genes
   a. are found in both plants and animals.
   b. are found only in animals.
   c. are found only in plants.
   d. are only associated with genes in the *MADS* complex.

7. The *Brachyury* gene in vertebrates and the *Ap3* gene in flowering plants
   a. are examples of *Hox* genes.
   b. are examples of co-opting a gene for a new function.
   c. are homologues for determining the body plan of eukaryotes.
   d. help regulate the formation of a photoreceptor organ.

8. Which of the following organisms is not considered to be a model genetic system?
   a. Mice
   b. Fruit flies
   c. Humans
   d. Yeast

9. Which of the following statements about *Pax6* is false?
   a. *Pax6* has a similar function in mice and flies.
   b. *Pax6* is involved in eyespot formation in ribbon worms.
   c. *Pax6* is required for eye formation in *Drosophila*.
   d. *Pax6* is required for eyespot formation in planaria.

10. Which of the following statements about *Tbx5* is true?
    a. *Tbx5* is found only in tetrapods.
    b. *Tbx5* is involved in limb development in vertebrates.
    c. *Tbx5* is only found in the ancestors of tetrapods.
    d. *Tbx5* interacts with the same set of genes across different tetrapod species.

11. Homeosis
    a. refers to a maintained and unchanging genetic environment.
    b. is a temporal change in gene expression.
    c. is a spatial change in gene expression.
    d. is not an important genetic mechanism in development.

12. Transcription factors
    a. are genes.
    b. are sequences of RNA.
    c. are proteins that affect the expression of genes.
    d. none of the above.

13. Independently derived mutations of the *CYC* gene in plants
    a. suggests bilateral floral symmetry among all plants is homologous.
    b. establishes that radial floral symmetry is preferred by pollinators.
    c. establishes that radial floral symmetry is derived for all plants.
    d. none of the above

CHALLENGE QUESTIONS

1. Several examples of genes being co-opted for new functions were given in the chapter. Cite two and explain what is meant by *co-opted*.

2. From the chapter on evolution of development it would seem that the generation of new developmental patterns would be fairly easy and fast, leading to the ability of organisms to adapt quickly to environmental changes. Why then does it take millions of years, typically, for many of the traits examined to evolve?

3. There are several ways in which phenotypic diversity among major groups of organisms can be explained. On one end of the spectrum, such differences could arise out of differences in many genes that control development. On the other end, small sets of genes might differ in how they regulate the expression of various parts of the genome. Which view represents our current understanding? Explain.

chapter 26

The Tree of Life

introduction

IN PRECEDING CHAPTERS, you've seen that many common features are found in living things. To name a few, they are composed of one or more cells, they carry out metabolism and transfer energy with ATP, and they encode hereditary information in DNA. Yet there is also a tremendous diversity of life, ranging from bacteria and amoebas to blue whales and sequoia trees. Coral reefs, such as the one pictured here, are microcosms of diversity, comprising many life forms and sheltering an enormous array of life. For generations, biologists have tried to group organisms based on shared characteristics. The most meaningful groupings are based on the study of evolutionary relationships among organisms. These phylogenetic approaches and a sea of molecular sequence data are leading to new evolutionary hypotheses to explain life's diversification. In this chapter and those that follow in this unit, we explore the diversity of the living world.

concept outline

26.1 Origins of Life

The cell is the basic unit of life and today all cells come from preexisting cells. But, how can we explain the origins of the tremendous diversity of life on Earth today? The Earth formed as a hot mass of molten rock about 4.5 billion years ago (BYA). As the Earth cooled, much of the water vapor present in its atmosphere condensed into liquid water that accumulated on the surface in chemically rich oceans. One scenario for the origin of life is that it originated in this dilute, hot, smelly soup of ammonia, formaldehyde, formic acid, cyanide, methane, hydrogen sulfide, and organic hydrocarbons. Whether at the oceans' edges, in hydrothermal deep-sea vents, or elsewhere, the consensus among researchers is that life arose spontaneously from these early waters. Although the way in which this happened remains a puzzle, we cannot escape a certain curiosity about the earliest steps that eventually led to the origin of all living things on Earth, including ourselves. How did organisms evolve from the complex molecules that swirled in the early oceans?

All organisms share fundamental properties of life

Before we can address the origins of life, we must consider what qualifies something as "living." Biologists have found that the following set of properties are common to all organisms on Earth, with heredity playing a particularly key role.

Cellular organization. All organisms consist of one or more *cells*—complex, organized assemblages of molecules enclosed within membranes (figure 26.1).

Sensitivity. All organisms respond to stimuli—though not always to the same stimuli in the same ways.

Growth. All living things assimilate energy and use it to maintain order and grow, a process called **metabolism.** Plants, algae, and some bacteria use sunlight to create covalent carbon–carbon bonds from CO_2 and H_2O through photosynthesis. This transfer of the energy in covalent bonds is essential to all life on Earth.

Development. Both unicellular and multicellular organisms undergo systematic, gene-directed changes as they grow and mature.

Reproduction. Organisms reproduce, passing on genes from one generation to the next.

Regulation. All organisms have regulatory mechanisms that coordinate internal processes.

Homeostasis. All living things maintain relatively constant internal conditions, different from their environment.

Heredity. All organisms on Earth possess a *genetic system* that is based on the replication of a long, complex molecule called DNA. This mechanism allows for adaptation and evolution over time and is a distinguishing characteristic of living organisms.

Long before there were cells with the properties of life, organic (carbon-based) molecules formed from inorganic molecules. The formation of proteins, nucleic acids, carbohydrates, and lipids were essential, but not sufficient for life. The evolu-

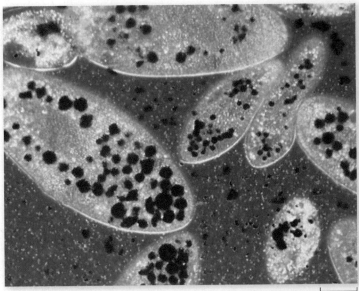

figure 26.1

CELLULAR COMPARTMENTALIZATION. These complex, single-celled organisms called *Paramecia* are classified as protists. The yeasts, stained red in this photograph, have been consumed and are enclosed within membrane-bounded sacs called digestive vacuoles.

tion of cells required early organic molecules to assemble into a functional, interdependent unit.

Life may have had extraterrestrial origins

Life may not have originated on Earth at all; instead, life may have infected Earth from some other planet. This hypothesis, called **panspermia,** proposes that meteors or cosmic dust may have carried significant amounts of complex organic molecules to Earth, kicking off the evolution of life. Hundreds of thousands of meteorites and comets are known to have slammed into the early Earth, and recent findings suggest that at least some may have carried organic materials. Nor is life on other planets ruled out. For example, the discovery of liquid water under the surface of Jupiter's ice-shrouded moon Europa and suggestions of fossils in rocks from Mars lend some credence to this idea. The Endurance Crater on Mars appears to have once contained salty water. As of June 2006 the Mars rover Spirit was experiencing difficulty with a wheel, but the other rover, Opportunity, continues to gather data about the watery environment that might have harbored life (figure 26.2).

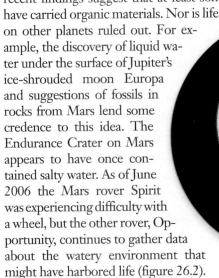
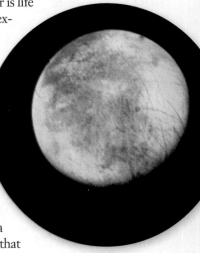

figure 26.2

The Mars rovers sent gigabytes of information and tens of thousands of images of Mars' surface back to Earth, providing clues about the possibility of ancient life on Mars.

Life may have originated on early Earth

Conditions on early Earth

The more we learn about the Earth's early history, the more likely it seems that Earth's first organisms emerged and lived at very high temperatures. Rubble from the forming solar system slammed into the early Earth beginning about 4.6 BYA, keeping the surface molten hot. As the bombardment slowed down, temperatures dropped. By about 3.8 BYA, ocean temperatures are thought to have dropped to a hot 49°–88°C (120°–190°F). Between 3.8 and 2.5 BYA, life first appeared, promptly after the Earth was inhabitable. Thus, as intolerable as early Earth's infernal temperatures seem to us today, they gave birth to life.

Very few geochemists agree on the exact composition of the early atmosphere. One popular view is that it contained principally carbon dioxide (CO_2) and nitrogen gas (N_2), along with significant amounts of water vapor (H_2O). It is possible that the early atmosphere also contained hydrogen gas (H_2) and compounds in which hydrogen atoms were bonded to the other light elements (sulfur, nitrogen, and carbon), producing hydrogen sulfide (H_2S), ammonia (NH_3), and methane (CH_4).

We refer to such an atmosphere as a *reducing atmosphere* because of the ample availability of hydrogen atoms and their electrons. In a reducing atmosphere, it would not take as much energy as it would today to form the carbon-rich molecules from which life evolved.

Exactly where on Earth life originated is also an open question. Possible locations include the ocean's edge, under frozen oceans, deep in the Earth's crust, within clay, or at deep-sea vents.

Organic molecules on early Earth

An early attempt to see what kinds of organic molecules might have been produced on the early Earth was carried out in 1953 by Stanley L. Miller and Harold C. Urey. In what has become a classic experiment, they attempted to reproduce the conditions in the Earth's primitive oceans under a reducing atmosphere. Even if this assumption proves incorrect—the jury is still out on this—their experiment is critically important because it ushered in the whole new field of prebiotic chemistry.

To carry out their experiment, Miller and Urey (1) assembled a reducing atmosphere rich in hydrogen and excluding gaseous oxygen; (2) placed this atmosphere over liquid water; (3) maintained this mixture at a temperature somewhat below 100°C; and (4) simulated lightning by bombarding it with energy in the form of sparks (figure 26.3).

They found that within a week, 15% of the carbon originally present as methane gas (CH_4) had converted into other simple carbon compounds. Among these compounds were formaldehyde (CH_2O) and hydrogen cyanide (HCN). These compounds then combined to form simple molecules, such as

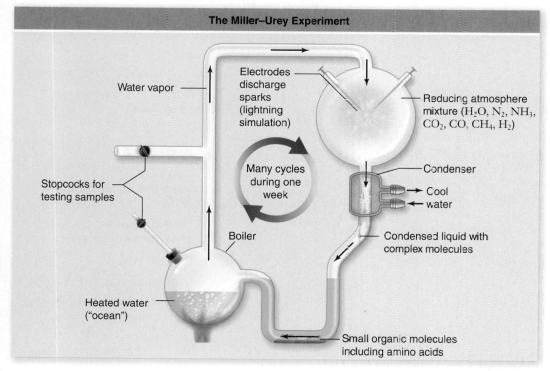

figure 26.3

THE MILLER–UREY EXPERIMENT. The apparatus consisted of a closed tube connecting two chambers. The upper chamber contained a mixture of gases thought to resemble the primitive Earth's atmosphere. Electrodes discharged sparks through this mixture, simulating lightning. Condensers then cooled the gases, causing water droplets to form, which passed into the second heated chamber, the "ocean." Any complex molecules formed in the atmosphere chamber would be dissolved in these droplets and carried to the ocean chamber, from which samples were withdrawn for analysis.

formic acid (HCOOH) and urea (NH_2CONH_2), and more complex molecules containing carbon–carbon bonds, including the amino acids glycine and alanine.

In similar experiments performed later by other scientists, more than 30 different carbon compounds were identified, including the amino acids glycine, alanine, glutamic acid, valine, proline, and aspartic acid. As we saw in chapter 3, amino acids are the basic building blocks of proteins, and proteins are one of the major kinds of molecules of which organisms are composed. Other biologically important molecules were also formed in these experiments.

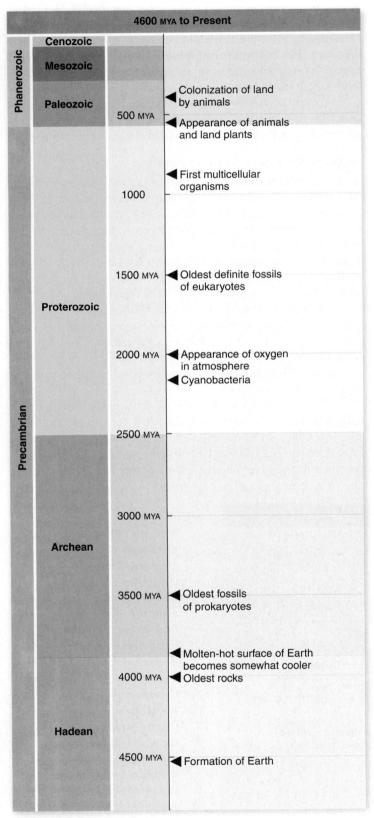

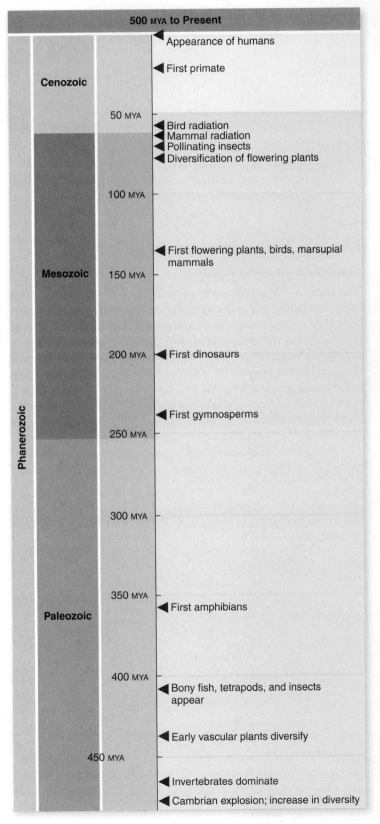

figure 26.4

GEOLOGICAL TIMESCALE AND THE EVOLUTION OF LIFE ON EARTH.

For example, hydrogen cyanide contributed to the production of a complex ring-shaped molecule called adenine—one of the bases found in DNA and RNA. Thus, the key molecules of life could have formed in the atmosphere of the early Earth.

Cells evolved from the functional assembly of organic molecules

Organic molecules can convey information or provide energy to maintain life through metabolism. Although DNA is the hereditary information molecule today, RNA that both could act as an enzyme used in self-replication (a ribozyme) and may have been the first genetic material. Amino acids polymerized to form proteins. Metabolic pathways emerged.

Constraining organic molecules to a physical space within a lipid or protein bubble could lead to an increased concentration of specific molecules. This in turn could increase the probability of metabolic reactions occurring. At some point, these bubbles became living cells with cell membranes and all the properties of life described earlier. As detailed in chapter 27, 3.5-billion-year-old prokaryotic fossils have been identified. For most of the history of life on Earth, these single-celled organisms were the only life forms. Several major evolutionary innovations—eukaryotic cells, sexual reproduction, and multicellularity—contributed to the diverse forms of life on Earth today (figure 26.4). We will continue with an overview of the amazing diversity of life on Earth and the evolutionary relationships among organisms.

Life began when organic molecules assembled in a coordinated manner within a cell membrane and began reproducing. Whether the organic molecules formed on Earth or elsewhere and were transported to Earth within meteors is an open question. Although it is impossible to fully reconstruct the conditions on early Earth, it is likely that the temperatures were extreme and the atmosphere had a very different gaseous composition than today's atmosphere.

26.2 Classification of Organisms

People have known from the earliest times that differences exist between organisms. Early humans learned that some plants could be eaten, but others were poisonous. Some animals could be hunted or domesticated; others were clearly dangerous hunters themselves. In this section, we review the history of formal scientific classification and the system in use today.

Taxonomy is a quest for identity and relationships

More than 2000 years ago, the Greek philosopher Aristotle formally categorized living things as either plants or animals. The Greeks and Romans expanded this simple system and grouped animals and plants into basic units such as cats, horses, and oaks. Eventually, these units began to be called genera (singular, *genus*), the Latin word for "groups." Starting in the Middle Ages, these names began to be systematically written down in Latin, the language used by scholars at that time. Thus, cats were assigned to the genus *Felis*, horses to *Equus*, and oaks to *Quercus*.

Linnaeus instituted the use of binomial names

Until the mid-1700s, whenever biologists wanted to refer to a particular kind of organism, which they called a species, they added a series of descriptive terms to the name of the genus; this was a polynomial, or "many names" system.

A much simpler system of naming organisms stemmed from the work of the Swedish biologist Carolus Linnaeus (1707–1778). In the 1750s, Linnaeus used the polynomial names *Apis pubescens, thorace subgriseo, abdomine fusco, pedibus posticis glabris utrinque margine ciliates* to denote the European honeybee. But as a kind of shorthand, he also included a two-part name the honeybee; he designated it *Apis mellifera*. These two-part names, or **binomials**, have become our standard way of designating species. You have already encountered many binomial names in earlier chapters.

Taxonomy is the science of classifying living things. A group of organisms at a particular level in a classification system is called a *taxon* (plural, *taxa*). By agreement among taxonomists throughout the world, no two organisms can have the same scientific name. The scientific name of an organism is the same anywhere in the world and avoids the confusion caused by common names (figure 26.5).

Also by agreement, the first word of the binomial name is the genus to which the organism belongs. This word is always capitalized. The second word refers to the particular species and is not capitalized. The two words together are called the species name (or scientific name) and are written in italics—for example, *Homo sapiens*. Once a genus has been used in the body of a text, it is often abbreviated in later uses. For example, the dinosaur *Tyrannosaurus rex* becomes *T. rex*.

figure 26.5

COMMON NAMES MAKE POOR LABELS. In North America, the common names "bear" and "corn" bring clear images to our minds, but the images are very different for someone living in Europe or Australia.

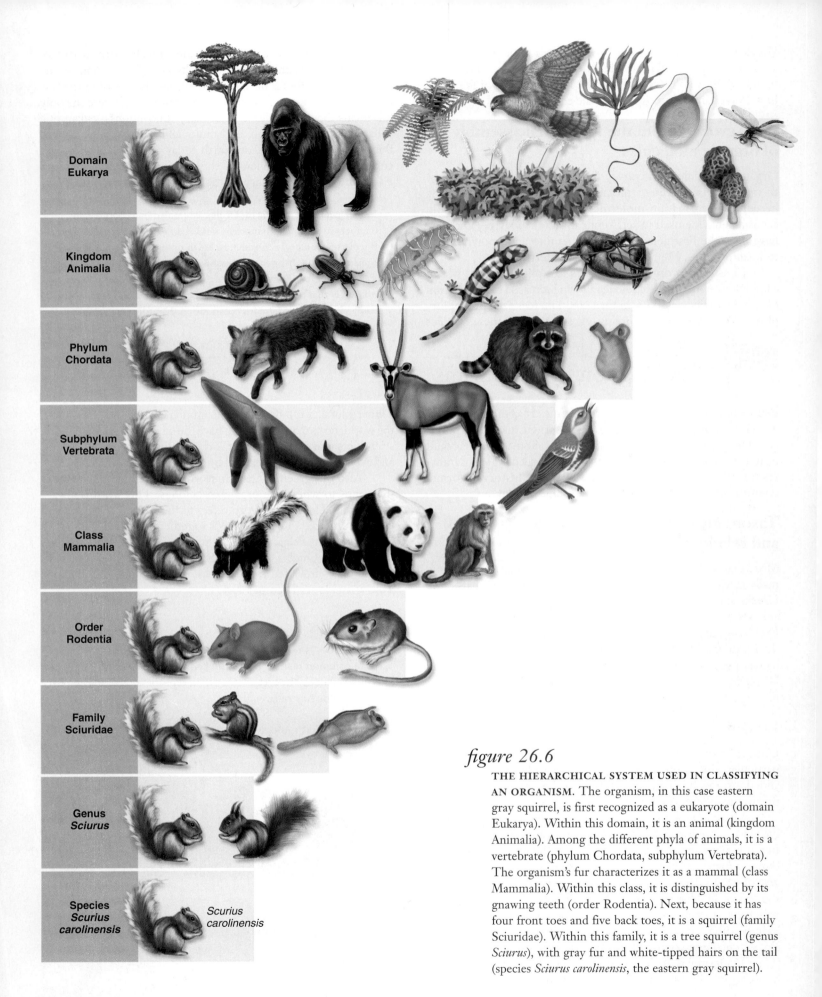

Domain
Eukarya

Kingdom
Animalia

Phylum
Chordata

Subphylum
Vertebrata

Class
Mammalia

Order
Rodentia

Family
Sciuridae

Genus
Sciurus

Species
Scurius carolinensis

Scurius carolinensis

figure 26.6

THE HIERARCHICAL SYSTEM USED IN CLASSIFYING AN ORGANISM. The organism, in this case eastern gray squirrel, is first recognized as a eukaryote (domain Eukarya). Within this domain, it is an animal (kingdom Animalia). Among the different phyla of animals, it is a vertebrate (phylum Chordata, subphylum Vertebrata). The organism's fur characterizes it as a mammal (class Mammalia). Within this class, it is distinguished by its gnawing teeth (order Rodentia). Next, because it has four front toes and five back toes, it is a squirrel (family Sciuridae). Within this family, it is a tree squirrel (genus *Sciurus*), with gray fur and white-tipped hairs on the tail (species *Sciurus carolinensis*, the eastern gray squirrel).

Taxonomic hierarchies have limitations

Named species are organized into larger groups based on shared characteristics. As discussed in chapter 23, sound evolutionary hypotheses can be constructed when organisms are grouped based on derived characters, not ancestral characters. Early taxonomists were not aware that the distinction between derived and ancestral characters could make a difference; as a result, many hierarchies are now being re-examined. As the phylogenetic and systematic revolution continues, other limitations of the original levels of taxonomic organization, called the *Linnaean taxonomy*, are being revealed.

The Linnaean hierarchy

In the decades following Linnaeus, taxonomists began to group organisms into larger, more inclusive categories. Genera with similar characters were grouped into a cluster called a **family**, and similar families were placed into the same **order** (figure 26.6). Orders with common properties were placed into the same **class**, and classes with similar characteristics into the same **phylum** (plural, *phyla*). Finally, the phyla were assigned to one of several great groups, the **kingdoms.** These kingdoms include two kinds of prokaryotes (Archaea and Bacteria), a largely unicellular group of eukaryotes (Protista), and three multicellular groups (Fungi, Plantae, and Animalia).

In addition, an eighth level of classification, called a **domain,** is frequently used. Biologists recognize three domains, which will be discussed in section 26.3. The names of the taxonomic units higher than the genus level and higher are capitalized.

The categories at the different levels may include many, a few, or only one taxon. For example, there is only one living genus of the family Hominidae (namely *Homo*), but several living genera of Fagaceae (the birch family). To someone familiar with classification or having access to the appropriate reference books, each taxon implies both a set of characteristics and a group of organisms belonging to the taxon.

To return to the example of the European honeybee, we can analyze the bee's taxonomic classification as follows:

1. **Species level:** *Apis mellifera*, which means the honey-bearing bee.
2. **Genus level:** *Apis*, a genus of bees.
3. **Family level:** Apidae, a bee family. All members of this family are bees—some solitary, some living in colonies as *A. mellifera* does.
4. **Order level:** Hymenoptera, a grouping that includes bees, wasps, ants, and sawflies.
5. **Class level:** Insecta, a very large class that comprises animals with three major body segments, three pairs of legs attached to the middle segment, and wings.
6. **Phylum level:** Arthropoda. Animals in this phylum have a hard exoskeleton made of chitin and jointed appendages.
7. **Kingdom level:** Animalia. The animals are multicellular heterotrophs with cells that lack cell walls.

Limitations of the hierarchy

In chapter 23, we discussed the modern phylogenetic approach, which distinguishes relationships between different species based on evolutionary history. Emerging phylogenies, frequently based on molecular data, reveal that the Linnaean hierarchy is inadequate for recognizing the hierarchical relationships among taxa that result naturally from a history of common ancestory and descent. New evolutionary hypotheses are developing.

One problem with the Linnaean system is that many higher taxonomic ranks are not monophyletic (for example, Reptilia) and therefore do not represent natural groups. A common ancestor and all of its descendants is a natural group that results from descent from a common ancestor, but any other type of group (paraphyletic or polyphyletic) is an artificial group created by taxonomists.

In addition, Linnaean ranks, as currently recognized, are not equivalent in any meaningful way. For example, two families may not represent clades that originated at the same time. One family may have diverged 70 million years before another family, and therefore these families have had vastly different amounts of time to diverge and develop evolutionary adaptations. Two groups that diverged from a common ancestor at the same time may be given different ranks. Thus, comparisons using Linnaean categories may be misleading. It is much better to use hypotheses of phylogenetic relationships in such instances.

One result of all these differences is that families demonstrate different degrees of biological diversity. As just one example, it is difficult to say that the legume family with 16,000 species represents the same level of taxonomic organization as the cat family with only 36 species. The differences across a single rank, whether it is class, order, or family, limit the usefulness of taxonomic hierarchies in making evolutionary predictions.

By convention, a species is given a binomial name. The first part of the name identifies the genus, and the second part the individual species. The Linnaean taxonomic hierarchy groups species into genera, then families, orders, classes, phyla, and kingdoms. Traditional classification systems are based on similar traits, but because they include a mix of derived and ancestral traits, they do not take into account evolutionary relationships.

26.3 Grouping Organisms

In this section, we examine the largest groupings of organisms: kingdoms and domains. The earliest classification systems recognized only two kingdoms of living things, namely animals and plants. But as biologists discovered microorganisms and learned more about other multicellular organisms, they added kingdoms in recognition of certain fundamental differences. The six-kingdom system was first proposed by Carl Woese of the University of Illinois (figure 26.7b).

The six kingdoms are not necessarily monophyletic

In the six-kingdom system, four of the kingdoms consist of eukaryotic organisms. The two most familiar kingdoms, **Animalia** and **Plantae**, contain only organisms that are multicellular during most of their life cycle. The kingdom **Fungi** contains multicellular forms and single-celled yeasts, which are thought to have had multicellular ancestors.

Fundamental differences divide these three kingdoms. Plants are mainly stationary, but some have motile sperm; most fungi lack motile cells; animals are mainly motile or mobile organisms. Animals ingest their food, plants manufacture it, and fungi digest and absorb it by means of secreted extracellular enzymes. Each of these kingdoms probably evolved from a different single-celled ancestor.

The large number of eukaryotes that do not fit in any of the three eukaryotic kingdoms are arbitrarily grouped into a single kingdom called **Protista** (see chapter 29). Most protists are unicellular or, in the case of some algae, have a unicellular phase in their life cycle. This kingdom reflects the current controversy between taxonomic and phylogenetic approaches. The protists

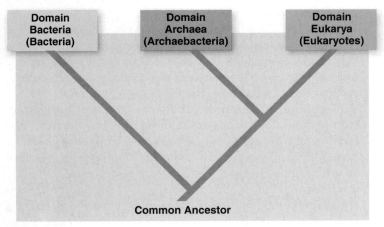

a.

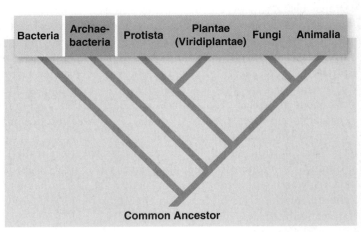

b.

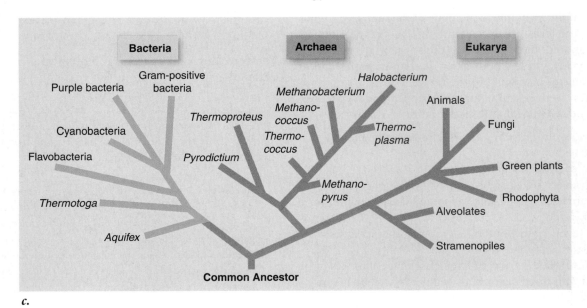

c.

figure 26.7

DIFFERENT APPROACHES TO CLASSIFYING LIVING ORGANISMS. *a.* Bacteria and Archaea are so distinct that they have been assigned to separate domains distinct from the Eukarya. Members of the domain Bacteria are thought to have diverged early from the evolutionary line that gave rise to the archaea and eukaryotes. *b.* Eukarya are grouped into four kingdoms, but these, especially the protists, are not necessarily monophyletic groups. *c.* This phylogeny is prepared from rRNA analyses. The base of the tree was determined by examining genes that are duplicated in all three domains, the duplication presumably having occurred in the common ancestor. Archaea and eukaryotes diverged later than bacteria and are more closely related to each other than either is to bacteria. Bases of trees constructed with other traits are often less clear because of lateral gene transfer (see chapter 24).

are a paraphyletic group, containing several nonmonophyletic adaptive lineages with distinct evolutionary origins.

The remaining two kingdoms, **Archaea** and **Bacteria,** consist of prokaryotic organisms, which are vastly different from all other living things (see chapter 28). Archaea are a diverse group that includes the methanogens and extreme thermophiles, and its members differ from the other prokaryotes, Bacteria.

The three domains probably are monophyletic

As biologists have learned more about the Archaea, it has become increasingly clear that this group is very different from all other organisms. When the full genomic DNA sequences of an archaean and a bacterium were first compared in 1996, the differences proved striking. Archaea are as different from bacteria as bacteria are from eukaryotes.

Recognizing this, biologists are increasingly adopting a classification of living organisms that recognizes three **domains,** a taxonomic level higher than kingdom (figure 26.7a). Archaea are in one domain **(Domain Archaea),** bacteria in a second **(Domain Bacteria),** and eukaryotes in the third **(Domain Eukarya).** Phylogenetically each of these domains form a clade.

inquiry

? *Why would the Archaea be considered a clade?*

In the remainder of this section, we preview the major characteristics of the three domains, along with a brief discussion about viruses. One view of our current understanding of the "tree of life" is presented in figure 26.7c. The oldest divergences represent the deepest-rooted branches in the tree. The archaea and eukaryotes are more closely related to each other than to bacteria and are on a separate evolutionary branch of the tree, even though archaea and bacteria are both prokaryotes.

Bacteria are more numerous than any other organism

The bacteria are the most abundant organisms on Earth. There are more living bacteria in your mouth than there are mammals living on Earth.

Although too tiny to see with the unaided eye, bacteria play critical roles throughout the biosphere. They extract from the air all the nitrogen used by organisms, and they play key roles in cycling carbon and sulfur. Much of the world's photosynthesis is carried out by bacteria. In contrast, however, certain groups of bacteria are also responsible for many forms of disease. Understanding bacterial metabolism and genetics is a critical part of modern medicine.

Bacteria are highly diverse, and the evolutionary links among species are not well understood. Although taxonomists disagree about the details of bacterial classification, most recognize 12–15 major groups of bacteria. Comparisons of the nucleotide sequences of ribosomal RNA (rRNA) molecules are beginning to reveal how these groups are related to one another and to the other two domains.

Archaea may live in extreme environments

The archaea seem to have diverged very early from the bacteria and are more closely related to eukaryotes than to bacteria (figure 26.7c.). This conclusion comes largely from comparisons of genes that encode ribosomal RNAs.

Horizontal gene transfer in microorganisms

Comparing whole-genome sequences from microorganisms has led evolutionary biologists to a variety of phylogenetic trees, some of which contradict each other. It appears that during their early evolution, microorganisms swapped genetic information via horizontal gene transfer (HGT), as you learned in chapter 24. The potential for gene swapping makes constructing phylogenetic trees for microorganisms very difficult.

Consider the archaean *Thermotoga,* a thermophile found on Volcano Island off the coast of Italy. The sequence of one of its RNAs places it squarely within the bacteria near an ancient microbe called *Aquifex.* Recent DNA sequencing, however, fails to support any consistent relationship between the two microbes.

Over the next few years, we can expect to see considerable change in accepted viewpoints as more and more data are brought to bear.

Archaean characteristics

Although they are a diverse group, all archae share certain key characteristics (table 26.1). Their cell walls lack peptidoglycan (an important component of the cell walls of bacteria); the lipids in the cell membranes of archae have a different structure from those in all other organisms; and archae have distinctive ribosomal RNA sequences. Some of their genes possess introns, unlike those of bacteria. Both archae and eukaryotes lack the peptidoglycan cell wall found in bacteria.

The archae are grouped into three general categories—methanogens, extremophiles, and nonextreme archae—based

TABLE 26.1	Features of the Domains of Life		
	DOMAIN		
Feature	**Archaea**	**Bacteria**	**Eukarya**
Amino acid that initiates protein synthesis	Methionine	Formyl-methionine	Methionine
Introns	Present in some genes	Absent	Present
Membrane-bounded organelles	Absent	Absent	Present
Membrane lipid structure	Branched	Unbranched	Unbranched
Nuclear envelope	Absent	Absent	Present
Number of different RNA polymerases	Several	One	Several
Peptidoglycan in cell wall	Absent	Present	Absent
Response to the antibiotics streptomycin and chloramphenicol	Growth not inhibited	Growth inhibited	Growth not inhibited

primarily on the environments in which they live or on their specialized metabolic pathways. The word *extreme* refers to our current environment. When archae first appeared on the scene their now extreme habitats may have been typical.

Methanogens obtain their energy by using hydrogen gas (H_2) to reduce carbon dioxide (CO_2) to methane gas (CH_4). They are strict anaerobes, poisoned by even traces of oxygen. They live in swamps, marshes, and the intestines of mammals. Methanogens release about 2 billion tons of methane gas into the atmosphere each year.

Extremophiles are able to grow under conditions that seem extreme to us. There are several types of extremophiles:

- Thermophiles, which live in temperatures ranging from 60° to 80°C. Many of these are autotrophs with a sulfur-based metabolism.
- Cold-adapted, which live in glacier ice and alpine lakes.
- Halophiles, which live in very salty environments including the Great Salt Lake and the Dead Sea. These organisms require water with a salinity of 15% to 20%.
- pH-tolerant archaea, growing in highly acidic (pH = 0.7) or highly basic (pH = 11) environments.
- Pressure-tolerant archaea found in the ocean depths. These archaeans require at least 300 atmospheres (atm) of pressure to survive and tolerate up to 800 atm. To experience a pressure of 300 atm (300 times the pressure of our atmosphere) you would need to dive 3000 m below the surface of the ocean (not a good idea unless you were in a deep-sea submersible). The deepest recorded skin dive is 127 m and a 145-m record is reported for SCUBA diving.

Nonextreme archaea grow in the same environments bacteria do. As the genomes of archaea have become better known, microbiologists have been able to identify signature sequences of DNA present only in archaea. The newly discovered microbe *Nanoarchaeum equitens* was identified as an archaean based on a signature sequence. This odd Icelandic microbe may have the smallest known genome, only 500 bp.

Eukaryotes have compartmentalized cells

For at least 1 billion years, prokaryotes ruled the Earth. No other types of organisms existed to eat them or compete with them, and their tiny cells formed the world's oldest fossils. Members of the third great domain of life, the eukaryotes, appear in the fossil record much later, only about 2.5 BYA. But despite the metabolic similarity of eukaryotic cells to prokaryotic cells, their structure and function enabled these cells to be larger, and eventually, allowed multicellular life to evolve.

Endosymbiosis and the origin of eukaryotes

The hallmark of eukaryotes is complex cellular organization, highlighted by an extensive endomembrane system that subdivides the eukaryotic cell into functional compartments (chapter 4). Not all cellular compartments, however, are derived from the endomembrane system.

With few exceptions, modern eukaryotic cells possess the energy-producing organelles termed *mitochondria*, and some eukaryotic cells possess *chloroplasts*, the energy-harvesting organelles. Mitochondria and chloroplasts are both believed to have entered early eukaryotic cells by a process called **endosymbiosis**, which is described in chapter 4 and discussed in more detail in chapter 29 (figure 26.8).

Mitochondria are the descendants of purple nonsulfur bacteria that were incorporated into eukaryotic cells early in the history of the group, and chloroplasts are derived from cyanobacteria (figure 26.9). As shown in figures 26.8 and 26.9, the red and green algae acquired their chloroplasts by directly engulfing a cyanobacterium. The brown algae most likely engulfed red algae to obtain chloroplasts.

The four kingdoms of eukaryotes

The first eukaryotes were unicellular organisms. A wide variety of unicellular eukaryotes exist today, grouped together in the kingdom Protista (along with some multicellular descendants) on the basis that they do not fit into any of the other three kingdoms of

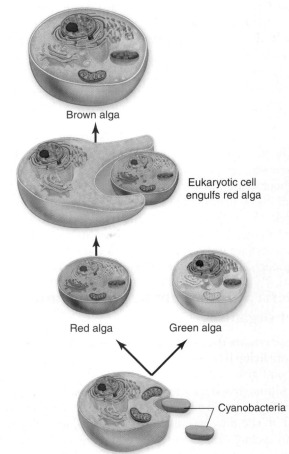

Brown alga

Eukaryotic cell engulfs red alga

Red alga   Green alga

Cyanobacteria

Eukaryotic cell with mitochondria engulfs cyanobacteria

figure 26.8

ALL CHLOROPLASTS ARE MONOPHYLETIC. The same cyanobacteria were engulfed by multiple hosts that were ancestral to the red and green algae. Brown algae share the same ancestral chloroplast DNA, but most likely gained it by engulfing red algae.

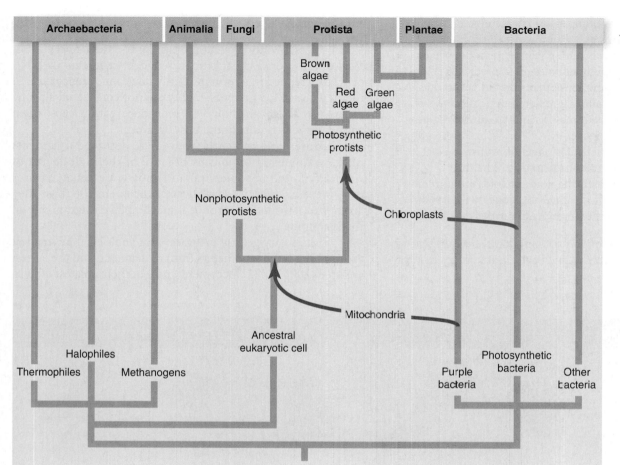

figure 26.9

HYPOTHESIS FOR EVOLUTIONARY RELATIONSHIPS AMONG THE SIX KINGDOMS OF ORGANISMS. The colored lines indicate symbiotic events.

eukaryotes. Fungi, plants, and animals are largely multicellular kingdoms, each a distinct evolutionary line from a single-celled ancestor that would be classified in the kingdom Protista.

Because of the size and ecological dominance of plants, animals, and fungi, and because they are predominantly multicellular, we recognize them as kingdoms distinct from Protista, even though the amount of diversity among the protists is much greater than that within or between the fungi, plants, and animals.

Key characteristics of the eukaryotes

Although eukaryotic organisms are extraordinarily diverse, they share three characteristics that distinguish them from prokaryotes: compartmentalization; multicellularity in many, but not all, eukaryotes; and sexual reproduction.

Compartmentalization. Discrete compartments provide evolutionary opportunities for increased specialization within the cell, as we see with chloroplasts and mitochondria. The evolution of a nuclear membrane, not found in prokaryotes, also accounts for increased complexity in eukaryotes. In eukaryotes, RNA transcripts from nuclear DNA are processed and transported across the nuclear membrane into the cytosol, where translation occurs. The physical separation of transcription and translation in eukaryotes adds additional levels of control to the process of gene expression.

Multicellularity. The unicellular body plan has been tremendously successful, with unicellular prokaryotes and eukaryotes constituting about half of the biomass on Earth. But a single cell has limits. The evolution of multicellularity allowed organisms to deal with their environments in novel ways through differentiation of cell types into tissues and organs.

True multicellularity, in which the activities of individual cells are coordinated and the cells themselves are in contact, occurs only in eukaryotes and is one of their major characteristics. Bacteria and many protists form colonial aggregates of many cells, but the cells in the aggregates have little differentiation or integration of function.

Other protists—the red, brown, and green algae, for example—have independently attained multicellularity. One lineage of multicellular green algae was the ancestor of the plants (see chapters 29 and 30), and most taxonomists now place its members in the green plant kingdom, the **Viridiplantae.**

The multiple origins of multicellularity are also seen in the fungi and the animals, which arose from unicellular protist ancestors with different characteristics. As you will see in subsequent chapters, the groups that seem to have given rise to each of these kingdoms are still in existence.

Sexual Reproduction. Another major characteristic of eukaryotic organisms as a group is sexual reproduction. Although some interchange of genetic material occurs in bacteria, it is certainly not a regular, predictable mechanism in the same sense that sex is in eukaryotes. Sexual reproduction allows greater genetic diversity through the processes of meiosis and crossing over, as you learned in chapter 13.

In many of the unicellular phyla of protists, sexual reproduction occurs only occasionally. The first eukaryotes were probably haploid; diploids seem to have arisen on a number of separate occasions by the fusion of haploid cells, which then eventually divided by mitosis.

The characteristics of the six kingdoms are outlined in table 26.2; note that the archaea and bacteria are grouped in the same column.

Viruses are a special case

Viruses possess only a portion of the properties of organisms. Viruses are literally "parasitic" chemicals, segments of DNA or RNA wrapped in a protein coat. They cannot reproduce on their own, and for this reason they are not considered alive by biologists. They can, however, reproduce within cells, often with disastrous results to the host organism.

Viruses are currently viewed as detached fragments of the genomes of organisms because of the high degree of similarity found in some viral and eukaryotic genes. Viruses thus present a special classification problem. Because they are not organisms, we cannot logically place them in any of the kingdoms.

Viruses vary greatly in appearance and size. The smallest are only about 17 nanometers (nm) in diameter, and the largest are up to 1000 nm (1 micrometer; μm) in their greatest dimen-

TABLE 26.2	Characteristics of the Six Kingdoms and Three Domains				
	Archaea and Bacteria	Protista	Plantae	Fungi	Animalia
Cell Type	Prokaryotic	Eukaryotic	Eukaryotic	Eukaryotic	Eukaryotic
Nuclear Envelope	Absent	Present	Present	Present	Present
Transcription and Translation	Occur in same compartment	Occur in different compartments	Occur in different compartments	Occur in different compartments	Occur in different compartments
Histone Proteins Associated with DNA	Absent	Present	Present	Present	Present
Cytoskeleton	Absent	Present	Present	Present	Present
Mitochondria	Absent	Present (or absent)	Present	Present	Present
Chloroplasts	None (photosynthetic membranes in some types)	Present (some forms)	Present	Absent	Absent
Cell Wall	Noncellulose (polysaccharide plus amino acids)	Present in some forms, various types	Cellulose and other polysaccharides	Chitin and other noncellulose polysaccharides	Absent
Means of Genetic Recombination, If Present	Conjugation, transduction, transformation	Fertilization and meiosis	Fertilization and meiosis	Fertilization and meiosis	Fertilization and meiosis
Mode of Nutrition	Autotrophic (chemosynthetic, photosynthetic) or heterotrophic	Photosynthetic or heterotrophic, or combination of both	Photosynthetic, chlorophylls *a* and *b*	Absorption	Ingestion
Motility	Bacterial flagella, gliding or nonmotile	9 + 2 cilia and flagella; amoeboid, contractile fibrils	None in most forms; 9 + 2 cilia and flagella in gametes of some forms	Both motile and nonmotile	9 + 2 cilia and flagella, contractile fibrils
Multicellularity	Absent	Absent in most forms	Present in all forms	Present in most forms	Present in all forms
Nervous System	None	Primitive mechanisms for conducting stimuli in some forms	A few have primitive mechanisms for conducting stimuli	None	Present (except sponges), often complex

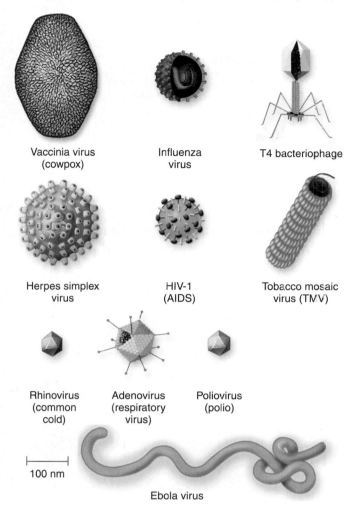

Vaccinia virus (cowpox)

Influenza virus

T4 bacteriophage

Herpes simplex virus

HIV-1 (AIDS)

Tobacco mosaic virus (TMV)

Rhinovirus (common cold)

Adenovirus (respiratory virus)

Poliovirus (polio)

100 nm

Ebola virus

figure 26.10

VIRAL DIVERSITY. Viruses exhibit extensive diversity in shape and size. At the scale these sample viruses are shown, a human hair would be nearly 8 m thick.

sion (figure 26.10). The largest viruses are barely visible with a light microscope, but viral morphology is best revealed using the electron microscope.

Biologists first began to suspect the existence of viruses near the end of the nineteenth century. European scientists were attempting to isolate the infectious agent responsible for hoof-and-mouth disease in cattle, and they concluded that it was smaller than a bacterium. The true nature of viruses was discovered in 1933, when biologist Wendell Stanley prepared an extract of the tobacco mosaic virus (TMV) and attempted to purify it. To his great surprise, the purified TMV preparation precipitated in the form of crystals—the virus was acting like a chemical off the shelf rather than like an organism. Stanley concluded that TMV is best regarded as just that—chemical matter rather than a living organism.

Within a few years, scientists disassembled the TMV virus and found that Stanley was right. TMV was not cellular but chemical. Each particle of TMV virus is in fact a mixture of two chemicals: RNA and protein. The TMV virus consists of a tube made of protein with an RNA core. If these two components are separated and then reassembled, the reconstructed TMV particles are fully able to infect healthy tobacco plants.

Because eukaryote diversity is so vast, we next take a brief look at the three kingdoms of the eukaryote domain.

Bacteria and archaea are tiny but numerous unicellular organisms that lack compartmentalization. Eukaryotic cells are highly compartmentalized, and they acquired mitochondria and chloroplasts by endosymbiosis. The complex differentiation we associate with many life-forms depends on multicellularity and sexual reproduction. Viruses are not organisms classified in the kingdoms of life, but instead chemical assemblies that can infect cells and replicate within them.

Making Sense of the Protists

In reading this chapter and chapter 23, you may have sensed some tension between traditional classification systems and systems based on evolutionary relationships, such as cladistics and phylogenetic analysis. The kingdom Protista illustrates well the source of this tension. This kingdom is the weakest area of the six-kingdom classification system shown in figure 26.7.

Eukaryotes diverged rapidly in a world that was shifting from anaerobic to aerobic conditions. We may never be able to completely sort out the relationships among different lineages during this major evolutionary transition. Molecular systematics, however, clearly shows that the protists are a paraphyletic group (figure 26.11). Although biologists continue to use the term *protist* as a catchall for any eukaryote that is not a plant, fungus, or animal, this grouping is not based on evolutionary relationships.

The six main branchings of protists, shown at the base of figure 26.11, represent a current working hypothesis, although at least 60 protists do not seem to fit in any of the six groups. Choanoflagellates are most closely related to sponges, and indeed, to all animals. The green algae can be

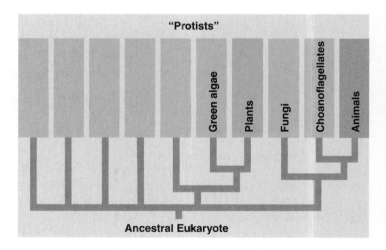

figure 26.11

THE FALL OF KINGDOM PROTISTA. Systematists have shown that protists as a group are not monophyletic. Note how some lineages are actually more closely related to plants or animals than they are to other protists.

split into two monophyletic groups, one of which gave rise to land plants. Many sytematists are calling for a new kingdom called Viridiplantae, or the green plant kingdom, which would include all the green algae (not red or brown algae) and the land plants. Thus, the definition of a plant has been expanded beyond those species that made it onto land. Although the kingdom Protista is in ruins, our understanding of the evolutionary relationships among these early eukaryotes is growing exponentially.

> Molecular systematics and cladistics have led to a new understanding of the relationships among organisms formerly classified as members of kingdom Protista.

26.5 Origin of Plants

The origin of land plants from a green algal ancestor has long been recognized as a major evolutionary event. Molecular phylogenetics reveals that land plants arose from an ancestral green alga, and that the evolution of land plants occurred only once, an indication of the incredible challenges involved in the move onto land.

Molecular phylogenetics has identified the closest living relatives of land plants

The phylogenetic relationships among the algae and the first land plants have been fuzzy and subject to long debate. Cell biology, biochemistry, and molecular systematics have provided surprising new evolutionary hypotheses.

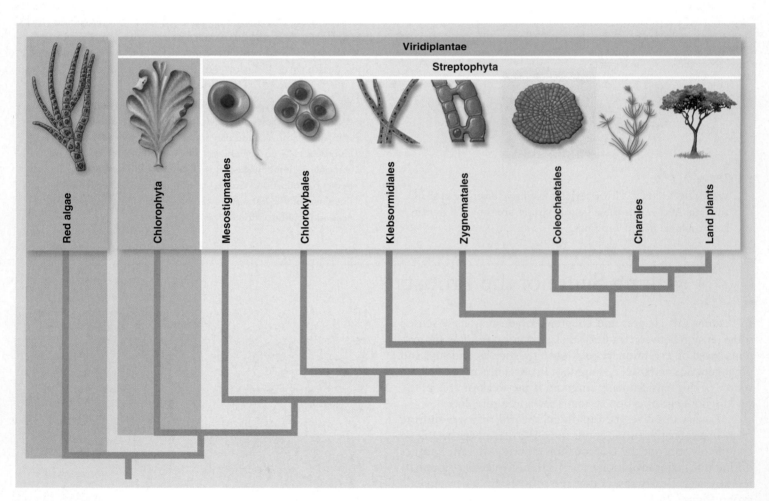

figure 26.12

A NEW HYPOTHESIS FOR LAND PLANT EVOLUTION. Kingdom Plantae (land plants) has been reduced to a clade within the green algal branch Streptophyta, and a new kingdom, Viridiplantae, which includes the green algal branches Chlorophyta and Streptophyta, has been proposed. Within the Streptophyta, the relatively complex Charales are believed to be the sister clade to the land plants. Contrast this phylogeny with the one predicted by the six-kingdom system in figure 26.7.

The green algae consist of two monophyletic groups, the **Chlorophyta** and the **Streptophyta** (chapter 29). Land plants are actually members of the Streptophyta, rather than being a separate kingdom. In terms of the Linnaean hierarchy, this new phylogenetic information demoted the land plants from constituting a kingdom to being a branch within the algal group Streptophyta. The Streptophyta along with the sister green algal clade, Chlorophyta, are now considered by most to make up the kingdom Viridiplantae, mentioned previously.

In the past decade, new hypotheses about relationships within the Streptophyta have emerged, with the current phylogeny composed of seven clades as shown in figure 26.12. What was the earliest streptophyte? Conflicting answers have been obtained with different phylogenetic analyses, but growing evidence supports the hypothesis that the scaly, unicellular flagellate Mesostigma (order Mesostigmatales) represents the earliest streptophyte branch.

Which of the Streptophyta clades contains the closest living relative of land plants? The two contenders have been the Charales, with about 300 species, and the Coleochaetales, with about 30 species. Both lineages are freshwater algae, but the Charales are huge compared with the microscopic Coleochae-

tales. At the moment, the Charales appear to be the sister clade to land plants, with the Coleochaetales the next closest relatives. Charales fossils dating back 420 million years indicate that the common ancestor of land plants was a relatively complex freshwater alga.

Horizontal gene transfer occurred in land plants

The shrub *Amborella trichopoda* is the closest living relative to the earliest flowering plants (angiosperms). Its clade is a sister clade to all other flowering plants, yet at least one copy of 20 out of 31 of its known mitochondrial protein genes hopped into the mitochondrial genome from other land plants through horizontal gene transfer (HGT). In addition, three different moss species contributed to the mix (figure 26.13).

Amborella is not typical of most extant flowering plants. It is the only existing member of its genus and is native only to the tropical rain forests of New Caledonia, an island group east of Australia that has been isolated for some 70 million years and contains many ancient endemic species. Here parasitic plants called *epiphytes* (plants that derive

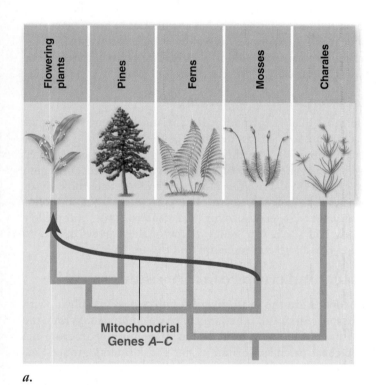

a.

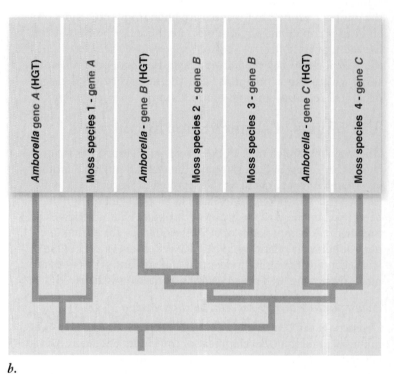

b.

figure 26.13

THE FLOWERING PLANT *AMBORELLA* ACQUIRED THREE MOSS GENES THROUGH HORIZONTAL GENE TRANSFER.
a. Phylogenetic relationship of *Amborella* to other land plants. As shown by the arrow connecting moss and the flowering plants, HGT is the only plausible explanation for the presence of moss mitochondrial genes in *Amborella*. *b.* Phylogenetic relationships among the horizontally transferred gene.

inquiry

? *Explain why a phylogenetic tree based on comparisons of a single gene could result in an inaccurate evolutionary hypothesis.*

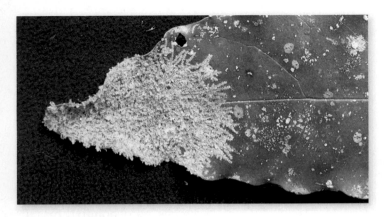

figure 26.14

CLOSE CONTACT BETWEEN SPECIES CAN LEAD TO HGT.
Here moss are growing on the base of an *Amborella* leaf with lichens scattered on the rest of the leaf.

nutrients from other plants) are common. Close contact with parasitic plants could increase the probability of HGT (figure 26.14).

An open question is whether the foreign genes in *Amborella* have functions. About half the genes are intact and could be transcribed and translated into a protein. The protein would be similar to an existing protein in the plant, but its function, if any, remains to be determined.

inquiry

? *How would you determine if a moss gene in Amborella had a function? (Hint: Refer to chapter 25.)*

Molecular phylogenetics has changed the way algae and plants are classified. Evidence for HGT in land plants needs to be considered when using gene sequences to construct phylogenies.

26.6 Sorting Out the Animals

Molecular systematics is leading to a revision of our understanding of evolutionary history in all kingdoms, including the animals. Some phylogenies are changing, and others, including mammalian phylogenies, are actually being written for the first time. In this section, we explore three examples: the relationship between annelids and arthropods, relationships within the arthropods, and the discovery of phylogenetic relationships among mammals.

The origins of segmentation are puzzling

The arthropod phylum is a huge group of invertebrates that includes the insects and crustaceans; the annelid phylum, another invertebrate group, contains the segmented worms such as the earthworm. Morphological traits such as segmentation have been used in the past to group arthropods and annelids close together, but comparisons of rRNA sequences are raising questions about their relationship. As rRNA sequences are obtained, it is becoming increasingly clear that annelids and arthropods are more distantly related than taxonomists previously believed.

Evolutionary occurrences of segmentation

Distinctions can be made among eukaryote animals on the basis of timing of embryonic development of the mouth and anus. Annelids and arthropods belong to the **protostome** group, in which the mouth develops before the anus. Chordates, including humans, fall into the **deuterostome** group, in which the anus forms first. (You'll learn more about these divisions in chapter 32.)

With the addition of newly available molecular traits, annelids and arthropods fell into two distinct protostome branches (figure 26.15): **lophotrochozoans** and **ecdysozoans.** These two branches have been evolving independently since ancient times. Lophotrochozoans include flatworms, mollusks, and annelids. Two ecdysozoan phyla have been particularly successful: roundworms (nematodes) and arthropods.

In the new protostome phylogeny, annelids and arthropods do not constitute a monophyletic group, as they had in the past. The implication is that segmentation arose twice, not once, in the protostomes, as had been believed originally. Segmentation then arose independently once again in the deuterostomes, specifically in the chordates.

Molecular details of segmentation

The most likely explanation for the independent appearance of segmentation is that members of the same family of genes were co-opted at least three times. Segmentation is regulated by the *Hox* gene family that contains a homeodomain region (see chapter 19). The *Hox* ancestral genes predate the ecdysozoans and lophotrochozoans. The ancient ancestor of the lophotrochozoans, ecdysozoans, and deuterostomes most likely already had seven *Hox* genes. Some of these genes appear to have evolved a role in segmentation (figure 26.15).

Insects and crustaceans are sister groups

Arthropods are the most diverse of all the animal phyla, with more species than all other animal phyla combined. Within the arthropods, insects have traditionally been set apart from the crustaceans (such as, shrimp, crabs, and lobsters), grouped instead with the myriapods (centipedes and millipedes).

This phylogeny, still widely employed, dates back to benchmark work by Robert Snodgrass in the 1930s. He pointed out that insects, centipedes, and millipedes are united by several seemingly powerful attributes, including **uniramous** (single-branched) appendages. All crustacean appendages, by contrast, are basically **biramous,** or "two-branched" (figure 26.16), although some of these appendages have become single-branched by reduction in the course of their evolution.

Taxonomists have traditionally assumed a character such as two-branched appendages to be a fundamental one, conserved

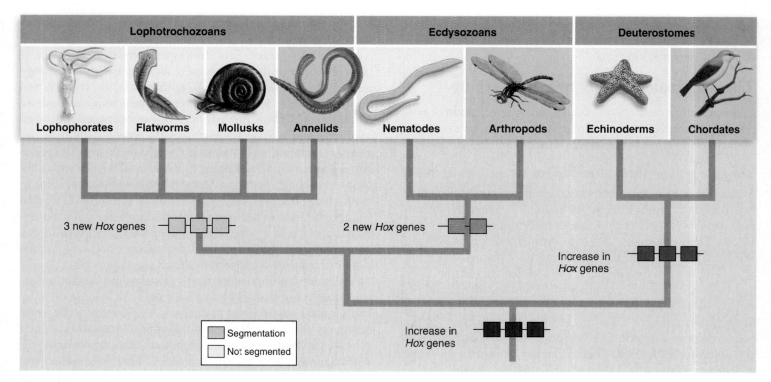

figure 26.15

MULTIPLE ORIGINS OF SEGMENTATION. New phylogenies based on ribosomal RNA show that segmentation in arthropods and annelids arose independently. Segmentation in both appears to be regulated by some of the *Hox* genes.

Misleading Morphology	
Crayfish maxilliped (biramous)	**Insect appendage (uniramous)**

figure 26.16

BRANCHED AND SINGLE APPENDAGES. Development of a biramous leg in a crustacean (crayfish) and a uniramous leg in an insect are both initiated by the *Distal-less* gene even though their adult morphologies are distinct.

over the course of evolution, and thus suitable for making taxonomic distinctions. As molecular methods have been developed, however, this assumption has become questionable.

Hox genes and appendages

The patterning of appendages among arthropods is orchestrated by *Hox* genes. A single one of these *Hox* genes, called *Distal-less*, has been shown to initiate development of unbranched limbs in insects and branched limbs in crustaceans. The same *Distal-less* gene is found in many animal phyla, including the vertebrates.

Distal-less appears to be necessary to initiate limb development, and it turns on genes that are more directly involved in the development of the limb itself. Evolutionary changes in the genes that *Distal-less* acts upon most likely account for differences in limb morphology.

A change in taxonomic relationship?

In recent years, a mass of accumulating morphological and molecular data has led many taxonomists to suggest new arthropod phylogenies. The most revolutionary of these, championed by Richard Brusca of Columbia University, considers crustaceans to be the basal arthropod group, and insects a close sister group. Molecular phylogenies all place insects as a sister group to crustaceans, and not to myriapods. This relationship suggests that insects are "flying crustaceans."

These conclusions are certain to engender lively discussion since they are in conflict with 150 years of morphology-based phylogenetic inference.

The mammalian family tree is emerging

In the preceding arthropod examples, our interpretation of evolutionary history has been rewritten. In mammals, however, parts of the phylogeny are just emerging, based on molecular data.

The four groups of placental mammals

Among the vertebrate classes, mammals are unique because they have mammary glands to feed their young. The majority of mammals—over 90%—are **eutherians,** or **placental mammals** (chapter 35). There are at least 18 extant (living) orders of eutherians, which are now divided into four major groups (figure 26.17).

The first major split occurred between the African clade and the other placental mammals when South America and Africa separated about 100 MYA. Aardvarks and elephants are part of this African lineage, called the Afrotheria, a clade we did not even recognize a decade ago. In South America, anteaters and armadillos soon appeared. Then two other branches arose—one includes ungulates with an even number of toes (camels, llamas, and other artiodactyls), odd-toed ungulates (perissodactyls such as horses and rhinoceros), and carnivores, and the other, primates and rodents. Sorting out the relationships within these branches is an ongoing challenge.

Whales and hippos

The origin and relationships of whales provides a good example. Whales were initially thought to be relatives of carnivores based on morphological information from fossils and extant animals, primarily the bones of the skull and shape of the teeth.

DNA sequence data, however, revealed a particularly close relationship between whales and hippopotamuses, suggesting that whales were derived from within the group Artiodactyla. Whales and hippopotamuses appear to be much more closely related than, for example, hippopotamuses and cows. With this new phylogenetic information, the possibility arises that some adaptations to aquatic environments in both species had a common origin. Recent finds of fossil whales with hind limbs have confirmed the artiodactyl origin of whales. Prior to these recent discoveries, no fossil whales with hind limbs had been found, so the key character uniting whales and artiodactyls, the shape of a bone in the ankle, was not known. In this case, molecular data provided insight into whale origins that was later confirmed by fossil evidence.

Understanding evolutionary relationships among organisms does more than provide biologists with a sense of order and a logical way to name organisms. A phylogenetically based taxonomy allows researchers to ask important questions about physiology, behavior, and development using information already known about a related species. This information not only enriches our understanding of how biological complexity evolved, but also provides novel insights that lead to progress in our understanding of the history and origins of important features and functions.

Molecular systematics and cladistic approaches are providing new information about the evolutionary relationships among animals, including members of our own class, the mammals.

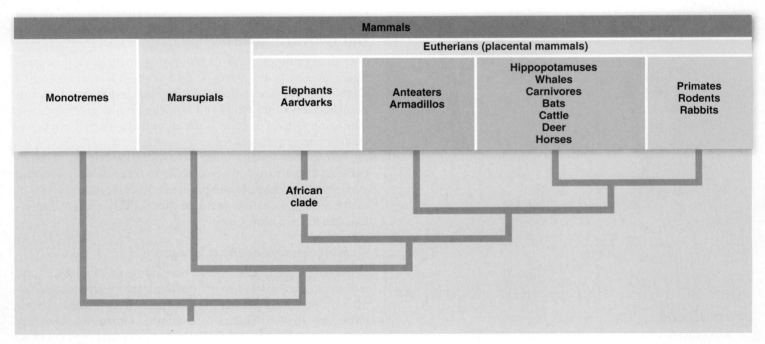

figure 26.17

MAJOR GROUPS OF MAMMALS.

26.1 Origins of Life

The consensus is that life arose spontaneously from pools of water rich in molecules.

- Living organisms share common properties: cells, responsiveness, growth, development, reproduction, regulation, homeostasis, and heredity.

- Panspermia proposes that complex organic molecules came from extraterrestrial sources and initiated the evolution of life on Earth.

- Many scientists agree that the early atmosphere of the Earth was reducing and contained the basic atoms for the creation of life.

- The Miller–Urey experiment exposed an approximate early reducing atmosphere of the Earth to heat and simulated lightning and produced basic organic molecules (figure 26.3).

- Life began when organic molecules assembled in a coordinated manner within a cell membrane and began reproducing.

26.2 Classification of Organisms

Since early times humans have classified organisms to better understand, study, and use them.

- Taxonomy is the science of assigning organisms to a particular level of classification called a taxon.

- Linnaeus proposed the binomial method of naming species.

- A binomial name begins with the capitalized genus, followed by a species name. The genus and species names are italicized.

- Taxonomic hierarchies are based on shared characteristics.

- The taxonomic hierarchies, from most to least shared characteristics, are: domain, kingdom, phylum, class, order, family, genus, and species (figure 26.6).

- Traditional classifications are limited because they are based on similar traits and do not take into account evolutionary relationships.

26.3 Grouping Organisms

The grouping of organisms is changing based on new technologies, including molecular techniques.

- The six kingdoms proposed by Woese are not necessarily monophyletic, but the three domains are probably monophyletic (figures 26.7 and 26.9).

- Four of the six kingdoms contain eukaryotes and are found within one domain, whereas the other two domains each contain prokaryotes.

- Bacteria are the most abundant and diverse organisms on Earth; they can be beneficial or cause disease.

- Archaea are prokaryotes that are closely related to eukaryotes and often are extremophiles.

- Archaean characteristics include cell walls that lack peptidoglycan, distinct rRNA sequences, and cell membrane lipids that differ from those found in other organisms.

- Eukaryotes are highly compartmentalized, but prokaryotes are not. Eukaryotes acquired mitochondria and chloroplasts by endosymbiosis (figure 26.8).

- Eukaryotes can be multicellular, and most undergo sexual reproduction.

- Viruses are diverse chemical assemblies that cannot reproduce on their own (figure 26.10).

26.4 Making Sense of the Protists

Molecular systematics and cladistics have lead to a new understanding of the relationships among organisms formally classified as Protista (figure 26.11).

- Molecular systematics shows that protists are a paraphyletic group.

- Protists are divided into six groups, but at least 60 protists do not fit into any of these groups.

- A new kingdom, Viridiplantae, has been suggested to include green algae and all aquatic and land plants.

26.5 Origin of Plants

Land plants originated from a green algal ancestor (figure 26.12).

- Green algae consist of two monophyletic groups: the Chlorophyta and the Streptophyta. The latter group gave rise to land plants.

- Some land plants show evidence of horizontal gene (figure 26.13).

26.6 Sorting out the Animals

Molecular systematics are leading to a revision of evolutionary relationships among animals.

- Phylogeny based on rRNA shows that segmentation in arthropods, annelids, and chordates arose independently at least three times (figure 26.15).

- Segmentation in animals is regulated by the *Hox* gene family.

- Based on molecular data, mammalian phylogeny continues to emerge. Whales are more closely related to hippopotamuses than to presumptive carnivorous ancestors.

SELF TEST

1. The Miller–Urey experiment demonstrated that
   a. life originated on Earth.
   b. that organic molecules could have originated in the early atmosphere.
   c. that the early genetic material on the planet was DNA.
   d. that the early atmosphere contained large amounts of oxygen.

2. Which of the following properties of life would have to be significantly different for an organism that evolved on a planet located far from its sun?
   a. Homeostasis
   b. Reproduction
   c. Growth
   d. Sensitivity

3. Which of the following represents a limitation of the Linnaean system of classification?
   a. The degree of biological diversity may differ significantly between two evolutionarily different families.
   b. Some taxonomic groups are not monophyletic.
   c. The ranks used in the Linnaean system are not evolutionarily equivalent.
   d. All of the above are limitations.

4. Which of the following would not belong to the domain Eukarya?
   a. Photosynthetic plants
   b. Multicellular fungi
   c. Thermophile archae
   d. Multicellular animals

5. Which of the following kingdoms presented the greatest challenge to the acceptance of a six-kingdom system?
   a. Plantae
   b. Animalia
   c. Protista
   d. Archaea

6. Which of the following statements is false?
   a. Brown and red algae are not closely related phylogenetically.
   b. Chloroplasts in brown and red algae are monophyletic.
   c. Brown algae gained chloroplasts by engulfing green algae (endosymbiosis).
   d. None of the above statements are false.

7. Which of the following events occurred first in eukaryotic evolution?
   a. Endosymbiosis and mitochondria evolution
   b. Endosymbiosis and chloroplast evolution
   c. Compartmentalization and formation of the nucleus
   d. Formation of multicellular organisms

8. Biologists include viruses in which of the following kingdoms?
   a. Archaea
   b. Fungi
   c. Bacteria
   d. None of the above

9. As a researcher you discover a new species that is eukaryotic, motile, possesses a cell wall made of chitin, but lacks any evidence of a nervous system. To what kingdom of life would you assign it?
   a. Protista
   b. Animalia
   c. Fungi
   d. Plantae

10. Kingdom Plantae is being replaced by a new kingdom named Viridiplantae based on evidence obtained from which of the following?
    a. Molecular phylogenetics
    b. Newly discovered fossils
    c. Biochemical differences
    d. All of the above

11. Research in which of the following areas has the greatest potential to increase our understanding of land plant evolution.
    a. Photosynthetic pigments
    b. Chloroplast endosymbiosis
    c. Horizontal gene transfer
    d. Changes in cell wall composition

12. In animals, the process of _____ is under control of the *Hox* genes.
    a. multicellularity
    b. sexual reproduction
    c. cellular compartmentalization
    d. segmentation

13. Based on ribosomal RNA information, which of the following groups is most closely related to the Arthropods?
    a. Echinoderms
    b. Nematodes
    c. Mollusks
    d. Annelids

14. The study of the *Distal-less Hox* gene refuted what morphological evidence of arthropod evolution?
    a. Protostome and deuterostome classification
    b. Limb morphology and development
    c. Metamorphosis
    d. Development of eyes

15. The reclassification of whales within the Eutherians is primarily based on which of the following?
    a. Morphological information from fossils
    b. Morphological information from hippopotamuses
    c. Morphological evidence from carnivores
    d. DNA sequence data

CHALLENGE QUESTIONS

1. The conditions on Mars, Jupiter's moon Europa, and Saturn's moon Titan mimic those that are believed to have occurred on the early Earth. Yet, these places are also different from our early planet. For example, both Europa and Titan are located far from the Sun. Suppose that someday in the future scientists discover bacteria on these moons that are very similar biochemically to the early bacteria on Earth. Explain how this information would support the theory of panspermia. What if the life was biochemically different?

2. You are part of a research team that has recently discovered evidence of a single-celled prokaryotic organism on Mars. As you begin your study of the organism, you wish to use a species from Earth as a comparison. What domain of life should you obtain your reference species from, and why?

3. In the past, classification has relied primarily on the evolution of morphological characteristics. Modern approaches are relying more heavily on molecular analysis. Why is the molecular approach so important in developing evolutionary hypotheses?

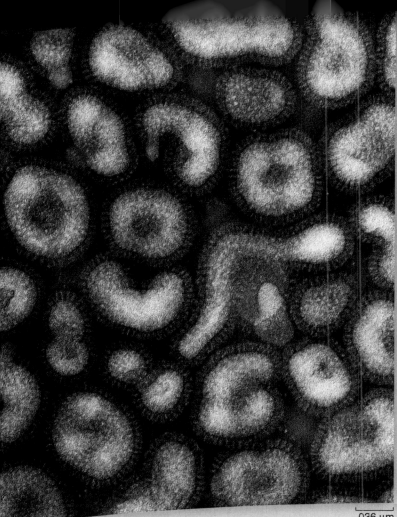

.036 μm

chapter 27

Viruses

introduction

WE BEGIN OUR EXPLORATION of the diversity of life with viruses. Viruses are genetic elements enclosed in protein; they are not considered organisms since they lack many of the features associated with life, including cellular structure, and *independent* metabolism or replication. For this reason viral particles are not called viral cells, but **virions,** and they are generally not described as living or dead but as active or inactive. Because of their disease-producing potential, however, viruses are important biological entities. The virus particles pictured here produce influenza—flu for short. In the flu season of 1918–1919, an influenza pandemic killed approximately 20–50 million people worldwide, twice as many as were killed in combat during World War I. Other viruses cause such diseases as AIDS, SARS, and hemorrhagic fever, and some cause certain forms of cancer.

For more than four decades, viral studies have been thoroughly intertwined with those of genetics and molecular biology. Classic studies using viruses that infect bacteria (known as **bacteriophages**) have led to the discovery of restriction enzymes and the identification of nucleic acid, not protein as the hereditary material. Currently, viruses are one of the principal tools used to experimentally carry genes from one organism to another. Applications of this technology could include treating genetic illnesses and fighting cancer.

27.1 The Nature of Viruses

All viruses have the same basic structure: a core of nucleic acid surrounded by protein. This structure lacks cytoplasm, and it is not a cell. Individual viruses contain only a single type of nucleic acid, either DNA or RNA. The DNA or RNA genome may be linear or circular, and single-stranded or double-stranded.

RNA viruses may be segmented, with multiple RNA molecules within a virion, or nonsegmented, with a single RNA molecule. Viruses are classified, in part, by the nature of their genomes: RNA viruses, DNA viruses, or retroviruses.

Viruses are strands of nucleic acids encased in a protein coat

Nearly all viruses form a protein sheath, or **capsid,** around their nucleic acid core (figure 27.1). The capsid is composed of one to a few different protein molecules repeated many times.

In several viruses, specialized enzymes are stored with the nucleic acid, inside the capsid. One example would be reverse transcriptase, which is required for retroviruses to complete their cycle and is not found in the host. This enzyme is needed early in the infection process and is carried within each virion.

Many animal viruses have an **envelope** around the capsid that is rich in proteins, lipids, and glycoprotein molecules. The lipids found in the envelope are derived from the host cell; however, the proteins found in a viral envelope are generally virally encoded.

Viral hosts include virtually every kind of organism

Viruses occur as obligate intracellular parasites in every kind of organism that has been investigated for their presence. Viruses infect fungal cells, bacterial cells, and protists as well as cells of plants and animals; however, each type of virus can replicate in only a very limited number of cell types. A virus that infects bacteria would be ill-equipped to infect a human or plant cell.

The suitable cells for a particular virus are collectively referred to as its **host range.** Once inside a multicellular host, many viruses also exhibit **tissue tropism,** targeting only a specific set of cells. For example, rabies grows within neurons, and hepatitis virus replicates within liver cells. Once inside a host cell, some viruses, such as the highly dangerous Ebola virus, wreak havoc on the cells they infect; others produce little or no damage. Still other viruses remain dormant until a specific signal or event triggers their expression.

As one example, a person can get chicken pox as a child, recover, and develop the disease shingles decades later. Both chicken pox and shingles are caused by the same virus, varicella zoster. This virus can remain dormant or **latent** for years. Stresses to the immune system may trigger an outbreak of shingles in people who have had chicken pox in the past. This is caused by the same virus, but the infection may be called herpes zoster because the virus is actually a herpes virus.

Any given organism may often be susceptible to more than one kind of virus. This observation suggests that many more kinds of viruses may exist than there are kinds of organisms—perhaps trillions of different viruses. Only a few thousand viruses have been described at this point.

figure 27.1

STRUCTURE OF VIRIONS. Viruses are characterized as helical, icosahedral, binal or polymorphic depending on the symmetry of the virion. *a.* The capsid may have helical symmetry such as the tobacco mosaic virus (TMV) shown here. TMV infects plants and consists of 2130 identical protein molecules (*green*) that form a cylindrical coat around the single strand of RNA (*red*). *b.* The capsid of icosahedral viruses has 20 facets made of equilateral triangles. These viruses can come in many different sizes all based on the same basic shape. *c.* Bacteriophage come in a variety of shapes but binal symmetry is exclusively seen in phages such as the T4 phage of *E. coli*. This form of symmetry is characterized by an icosahedral head, which contains the viral genome, and a helical tail. *d.* Viruses can also have an envelope surrounding the capsid such as the influenza virus. This virus has eight RNA segments, each within a helical capsid. This gives the virus a polymorphic shape.

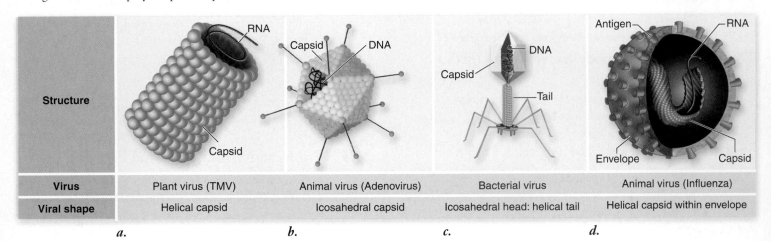

Structure				
Virus	Plant virus (TMV)	Animal virus (Adenovirus)	Bacterial virus	Animal virus (Influenza)
Viral shape	Helical capsid	Icosahedral capsid	Icosahedral head: helical tail	Helical capsid within envelope
	a.	*b.*	*c.*	*d.*

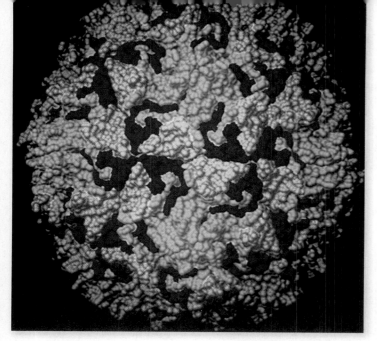

figure 27.2

ICOSAHEDRAL VIRION. The poliovirus shown has icosahedral symmetry. The capsid is formed from multiple copies of four different proteins shown in different colors. (One protein is internal and cannot be seen.)

Viruses replicate by taking over host machinery

An infecting virus can be thought of as a set of instructions, not unlike a computer program. A cell is normally directed by chromosomal DNA-encoded instructions, just as a computer's operation is controlled by the instructions in its operating system. A virus is simply a set of instructions, the viral genome, that can trick the cell's replication and metabolic enzymes into making copies of the virus. Computer viruses get their name because they perform similar actions, taking over a computer

and directing its activities. Like a computer with a virus, a cell with a virus is often damaged by infection.

Viruses can reproduce only when they enter cells. When they are outside of a cell, viral particles are called *virions* and they are metabolically inert. Viruses lack ribosomes and the enzymes necessary for protein synthesis and most, if not all of the enzymes for nucleic acid replication. Inside cells, the virus hijacks the transcription and translation systems to produce viral proteins from *early genes*, which are the genes in the viral genome expressed first. This is followed by the expression of *middle genes* and eventually *late genes*. This cascade of gene expression leads to replication of viral nucleic acid and production of viral capsid proteins. The late genes generally code for proteins important in assembly and release of viral particles from a host cell.

Most viruses come in two simple shapes

Most viruses have an overall structure that is either **helical** or **icosahedral.** Helical viruses, such as the tobacco mosaic virus in figure 27.1*a*, have a rodlike or threadlike appearance. Icosahedral viruses have a soccer ball shape, the geometry of which is revealed only under the highest magnification with an electron microscope.

The **icosahedron** is a structure with 20 equilateral triangular facets. Most animal viruses are icosahedral in basic structure (figure 27.1*b*). The icosahedron is the basic design of the geodesic dome, and it is the most efficient symmetrical arrangement that subunits can take to form a shell with maximum internal capacity (figure 27.2).

Some viruses are complex, such as the T-even bacteriophage shown in figure 27.3. Complex viruses have a *binal*, or two-fold, symmetry that is not either purely icosahedral or helical. In the case of the T-even phage shown, there is a head structure that is an elongated icosahedron. A collar connects the head to a hollow tube with helical symmetry that ends in a complex baseplate with tail fibers. Although animal viruses do not have this binal symmetry, some, such as the poxviruses, do have a complex multilayered capsid

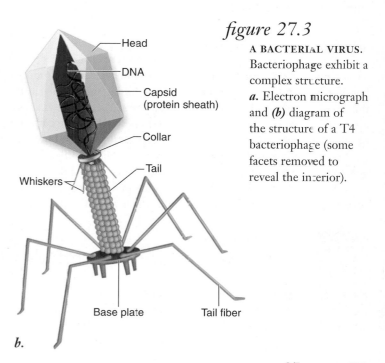

a.

b.

figure 27.3

A BACTERIAL VIRUS. Bacteriophage exhibit a complex structure. *a.* Electron micrograph and (*b*) diagram of the structure of a T4 bacteriophage (some facets removed to reveal the interior).

Head
DNA
Capsid (protein sheath)
Collar
Whiskers
Tail
Base plate
Tail fiber

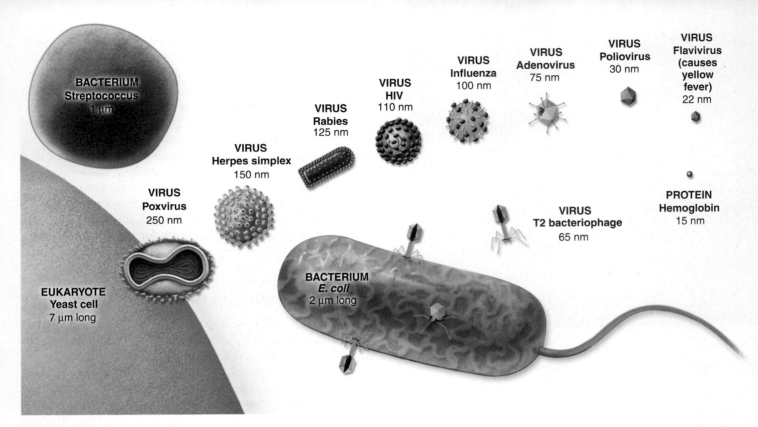

figure 27.4

VIRUSES VARY IN SIZE AND SHAPE. Note the dramatic differences in the size of a eukaryotic yeast cell, prokaryotic bacterial cells, and the many different viruses.

structure. Some enveloped viruses, such as influenza, are *polymorphic*, having no distinctive symmetry.

Viruses also vary greatly in size. As shown in figure 27.4, the very smallest viruses, such as the polio virus, have actually been synthesized in a lab using nothing more than sequence data and a machine capable of synthesizing nucleic acids from nucleotides. The larger viruses, such as the poxviruses, generally carry more genes, have more complex structures and tend to have a very short cycle time between entry of viral particles and release of newly formed virions.

Viral genomes exhibit great variation

Viral genomes vary greatly in both type of nucleic acid and number of strands (table 27.1). Some viruses, including those that cause flu, measles, and AIDS, possess RNA genomes. Most RNA viruses are single-stranded and are replicated and assembled in the cytosol of infected eukaryotic cells. RNA virus replication is error-prone, leading to high rates of mutation. This makes them difficult targets for the host immune system, vaccines, and antiviral drugs.

In single-stranded RNA viruses, if the genome has the same base sequence as the mRNA used to produce viral proteins, then the genomic RNA can serve as the mRNA. Such viruses are called *positive-strand viruses*. In contrast, if the genome is complementary to the viral mRNA, then the virus is called a *negative-strand virus*.

A special class of RNA viruses, called **retroviruses**, have an RNA genome that is reverse-transcribed into DNA by the enzyme **reverse transcriptase**. The DNA fragments produced by reverse transcription are often integrated into a host's chromosomal DNA. **Human immunodeficiency virus (HIV)**, the agent that causes **acquired immune deficiency syndrome (AIDS)**, is a retrovirus. (We describe HIV in detail later on.)

Other viruses, such as the viruses causing smallpox and herpes, have DNA genomes. Most DNA viruses are double-stranded, and their DNA is replicated in the nucleus of eukaryotic host cells.

> Viruses have a very simple structure that includes a nucleic acid genome encased in a protein coat. Viruses replicate by taking over a host's cell systems and are thus obligate intracellular parasites. Viruses show diverse genomes that are composed of DNA or RNA, which may be single- or double-stranded.

27.2 Bacteriophage: Bacterial Viruses

Bacteriophage (both singular and plural) are viruses that infect bacteria. They are diverse, both structurally and functionally, and are united solely by their occurrence in bacterial hosts. Many of these types of bacteriophage, called *phage* for short, are large and complex, with relatively large amounts of DNA and proteins.

Viruses have been found in archaea that resemble the binal symmetry phage with icosahedral heads and helical tails.

TABLE 27.1

Important Human Viral Diseases

Disease	Pathogen	Genome	Vector/Epidemiology
Chicken pox	Varicella zoster	Double-stranded DNA	Spread through contact with infected individuals. No cure. Rarely fatal. Vaccine approved in U.S. in early 1995.
Hepatitis B (viral)	Hepadnavirus	Double-stranded DNA	Highly infectious through contact with infected body fluids. Approximately 1% of U.S. population infected. Vaccine available. No cure. Can be fatal.
Herpes	Herpes simplex virus	Double-stranded DNA	Blisters; spread primarily through skin-to-skin contact with cold sores/blisters. Very prevalent worldwide. No cure. Exhibits latency—the disease can be dormant for several years.
Mononucleosis	Epstein–Barr virus	Double-stranded DNA	Spread through contact with infected saliva. May last several weeks; common in young adults. No cure. Rarely fatal.
Smallpox	Variola virus	Double-stranded DNA	Historically a major killer; the last recorded case of smallpox was in 1977. A worldwide vaccination campaign wiped out the disease completely.
AIDS	HIV	(+) Single-stranded RNA (two copies)	Destroys immune defenses, resulting in death by infection or cancer. As of 2005, WHO estimated that 40 million people are living with AIDS; 4.1 million new HIV infections were predicted and 2.8 million deaths were expected. More than 25 million have died from AIDS since 1981.
Polio	Enterovirus	(+) Single-stranded RNA	Acute viral infection of the CNS that can lead to paralysis and is often fatal. Prior to the development of Salk's vaccine in 1954, 60,000 people a year contracted the disease in the U.S. alone.
Yellow fever	Flavivirus	(+) Single-stranded RNA	Spread from individual to individual by mosquito bites; a notable cause of death during the construction of the Panama Canal. If untreated, this disease has a peak mortality rate of 60%.
Ebola	Filoviruses	(−) Single-stranded RNA	Acute hemorrhagic fever; virus attacks connective tissue, leading to massive hemorrhaging and death. Peak mortality is 50–90% if untreated. Outbreaks confined to local regions of central Africa.
Influenza	Influenza viruses	(−) Single-stranded RNA (eight segments)	Historically a major killer (20–50 million died during 18 months in 1918–1919); wild Asian ducks, chickens, and pigs are major reservoirs. The ducks are not affected by the flu virus, which shuffles its antigen genes while multiplying within them, leading to new flu strains.
Measles	Paramyxoviruses	(−) Single-stranded RNA	Extremely contagious through contact with infected individuals. Vaccine available. Usually contracted in childhood, when it is not serious; more dangerous to adults.
SARS	Coronavirus	(−) Single-stranded RNA	Acute respiratory infection; an emerging disease, can be fatal, especially in the elderly. Commonly infected animals include bats, foxes, skunks, and racoons. Domestic animals can be infected.
Rabies	Rhabdovirus	(−) Single-stranded RNA	An acute viral encephalomyelitis transmitted by the bite of an infected animal. Fatal if untreated. Commonly infected animals include bats, foxes, skunks, and racoons. Domestic animals can be infected.

Other archaeal viruses have complex symmetry and do not resemble any known viruses. Not much is known about these viruses and they will not be discussed further.

E. coli-infecting viruses were among the first bacteriophage to be discovered and are still some of the best studied. Some of these viruses that infect *E. coli* have been named as members of a "T" series (T1, T2, and so forth); others have been given different types of names. To illustrate the diversity of these viruses, T3 and T7 phage are icosahedral and have short tails. In contrast, the so-called T-even phage (T2, T4, and T6) have an icosahedral head, a capsid that consists primarily of three proteins, a connecting neck with a collar and long "whiskers," a long tail, and a complex base plate (see figure 27.3).

Bacterial viruses exhibit two reproductive cycles

During the process of bacterial infection by phage T4, at least one of the tail fibers of the phage—they are normally held near the phage head by the "whiskers"—contacts proteins of the host bacterial cell wall. The other tail fibers set the phage perpendicular to the surface of the bacterium and bring the base plate into contact with the cell surface.

Contact with the host

Different phages may target different parts of the outer surface of a bacterial cell. This first step is called **attachment, or adsorption.** The next step, release of the phage genome into the host, is best understood in the binal phage, such as T4. Once contact is established, the tail contracts, and the tail tube passes through an opening that appears in the base plate, piercing the bacterial cell wall. The contents of the head, the DNA genome, are then injected into the host cytoplasm. This step is called **penetration, or injection.**

Once inside the bacterial cell a phage may immediately take over the cell's replication and protein synthesis enzymes to synthesize viral components. This is the **synthesis** phase. Once the components are made, they are assembled (**assembly**) and **release** of mature virus particles occur, either through the action of enzymes that lyse the host cell or by budding through the host cell wall.

The time between adsorption and the formation of new viral particles is called an *eclipse period* because if a cell is lysed at this point, few if any active virions can be released.

The lytic cycle

When a virus lyses the infected host cell in which it is replicating, the reproductive cycle is referred to as a **lytic cycle** (figure 27.5, *left*). The basic steps of a lytic bacteriophage cycle are similar to those of a nonenveloped animal virus. The T-series bacteriophages are all **virulent** or **lytic phage,** multiplying within infected cells and eventually lysing (rupturing) them.

The lysogenic cycle

In contrast to the rather simple lytic cycles, some bacteriophages do not immediately kill the cells they infect, instead they integrate their nucleic acid into the genome of the infected host cell. This integration gives them a distinct advantage; because all viruses require a living host cell to replicate within, integration allows a virus to remain within a host cell and be replicated along with the host cell's DNA as the host divides. These viruses are called **temperate,** or **lysogenic, phage.** The DNA segment that is integrated into a host cell's genome is called a **prophage** and the resulting cell is called a **lysogen.**

Among the bacteriophages that do this is the binal phage lambda (λ) of *E. coli.* We know as much about this bacteriophage as we do about virtually any other biological particle; the complete sequence of its 48,502 bases has been determined. At least 23 proteins are associated with the development and maturation of phage λ, and other enzymes are involved in integrating this virus into the host genome.

When phage λ infects a cell, the early events constitute a genetic switch that will determine whether the virus will replicate and destroy the cell or become a lysogen and be passively replicated with the cell's genome. This lysis/lysogeny "decision" depends on the expression of early genes. Two regulatory proteins are produced early that will compete for binding to sites on the phage's DNA. Depending on which protein "wins" either the genes necessary for replication of the genome will be expressed beginning the lytic cycle, or the enzymes necessary for integrating the viral genome into the chromosome will be expressed and the **lysogenic cycle** initiated (figure 27.5, *right*).

The integration of a bacteriophage into a cellular genome is called **lysogeny.** A lysogenic phage has the expression of its genome repressed (see chapter 16) by one of the two viral regulatory proteins mentioned earlier. This is not a permanent state, however; in times of cell stress, the prophage can be derepressed and the enzymes necessary for excision of the genome expressed. The viral genome then is in the same state as the initial stage of infection and the lytic cycle can commence with early gene expression, viral DNA replication, and then expression of late genes, leading to formation of viral particles and lysis of the cell.

The switch from a lysogenic prophage to a lytic cycle is called **induction.** It can be stimulated in the laboratory by stressors such as starvation or ultraviolet radiation. The molecular events of induction take advantage of host proteins that respond to stress to produce a protease that can destroy the repressor protein that is keeping the viral genome silent. The normal function of this protease is to degrade a host repressor that controls DNA repair genes. The two repressor proteins are similar enough that both are degraded by the protease.

Bacteriophages can contribute genes to the host genome

During the integrated portion of a lysogenic reproductive cycle, a few viral genes may be expressed at the same time as host cell genes. Sometimes the expression of these genes has an important effect on the host cell, altering it in novel ways. When the phenotype or characteristics of the lysogenic bacterium is altered by the prophage, the alteration is called **phage conversion.**

Phage conversion of the cholera-causing bacterium

The bacterium *Vibrio cholerae* usually exists in a harmless form, but a second, disease-causing form also occurs. In this latter form, the bacterium is responsible for the deadly disease cholera, but how the bacteria changed from harmless to deadly was not known until recently.

Research now shows that a lysogenic bacteriophage that infects *V. cholerae* introduces into the host bacterial cell a gene that codes for the cholera toxin. This gene, along with the rest of the phage genome, becomes incorporated into the bacterial chromosome. The toxin gene is expressed along with the other host genes, thereby converting the benign bacterium to a disease-causing agent.

The receptors used by this toxin-encoding phage are pili (hairlike projections) found on the outer surface of *V. cholerae* (chapter 28); in recent experiments, it was determined that mutant bacteria that did not have pili were resistant to infection by the bacteriophage. This discovery has important implications in efforts to develop vaccines against cholera, which have been unsuccessful up to this

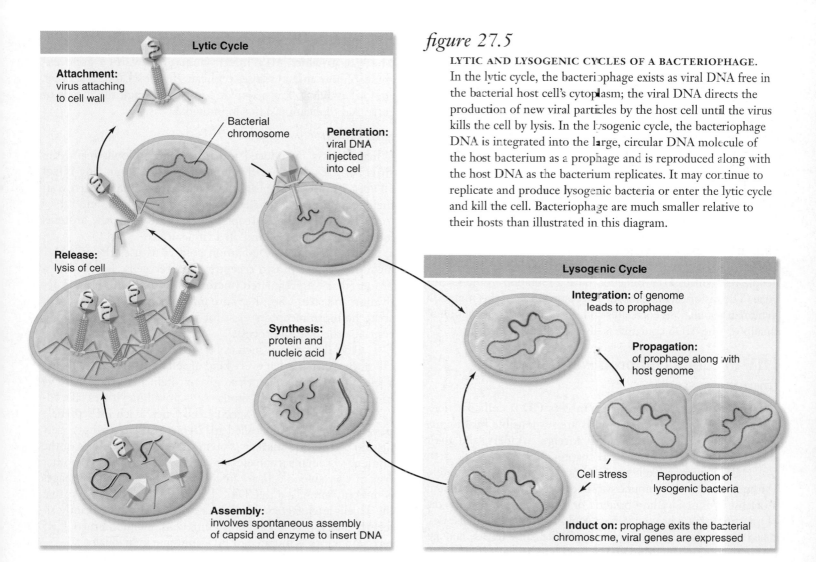

figure 27.5

LYTIC AND LYSOGENIC CYCLES OF A BACTERIOPHAGE. In the lytic cycle, the bacteriophage exists as viral DNA free in the bacterial host cell's cytoplasm; the viral DNA directs the production of new viral particles by the host cell until the virus kills the cell by lysis. In the lysogenic cycle, the bacteriophage DNA is integrated into the large, circular DNA molecule of the host bacterium as a prophage and is reproduced along with the host DNA as the bacterium replicates. It may continue to replicate and produce lysogenic bacteria or enter the lytic cycle and kill the cell. Bacteriophage are much smaller relative to their hosts than illustrated in this diagram.

Lytic Cycle

Attachment: virus attaching to cell wall

Bacterial chromosome

Penetration: viral DNA injected into cell

Release: lysis of cell

Synthesis: protein and nucleic acid

Assembly: involves spontaneous assembly of capsid and enzyme to insert DNA

Lysogenic Cycle

Integration: of genome leads to prophage

Propagation: of prophage along with host genome

Cell stress

Reproduction of lysogenic bacteria

Induction: prophage exits the bacterial chromosome, viral genes are expressed

---

point. Phage conversion could change any pili-expressing, nontoxigenic *V. cholerae* into a toxin-producing, potentially deadly form.

Although the phage conversion of *V. cholerae* is a classic example, it is far from the only example of phage conversion in human disease. The toxin found in *Corynebacterium diphtheriae*, also named for the disease it causes, is the product of phage conversion, as are the changes to the outer surface of certain infectious *Salmonella* species.

Bacteriophages are viruses that infect bacteria. They have two major types of life cycles: the lytic cycle that results in immediate death of the host and the lysogenic cycle in which the virus becomes part of the host genome and is transmitted vertically by cell division. Conditions can cause the lysogenic phage to switch to the lytic cycle. Phages that are lysogenic can contribute genes to the host, as in the case with *V. cholera* in which the toxin involved in the disease cholera came from a phage.

Human Immunodeficiency Virus (HIV)

A diverse array of viruses occurs among animals. A good way to gain a general idea of the characteristics of these viruses is to look at one animal virus in detail. Here we examine the virus responsible for a comparatively new and fatal viral disease, *acquired immune deficiency syndrome (AIDS)*.

AIDS is caused by HIV

The disease now known as AIDS was first reported in the United States in 1981, although a few dozen people in the United States had likely died of AIDS prior to that time and had not been diagnosed. Frozen plasma samples and estimates based on evolutionary speed and current diversity of HIV strains in the human population place the origins of HIV in the human population to Africa in the 1950s. It was not long before the infectious agent, a retrovirus, was identified by laboratories in France. Study of HIV revealed it to be closely related to a chimpanzee virus, suggesting a recent host expansion to humans from chimpanzees in central Africa.

Infected humans have varying degrees of resistance to HIV. Some have little resistance to infection and rapidly progress from having HIV-positive status to developing AIDS and eventually die. Others, even after repeated exposure, fail to become HIV-positive or may become HIV-positive without developing AIDS.

A relatively recent hypothesis to explain this great variability in susceptibility is genetic variation among these groups due to the selective pressure put on the human population by the smallpox virus, variola major, over the centuries. Because of successful vaccination and immunization, smallpox has been eradicated from the human population; however, before its eradication, it caused billions of deaths worldwide.

In order for smallpox to infect a cell, the cell must have a receptor protein in its plasma membrane that the virus can bind to. Individuals with mutated receptors would have been more resistant to smallpox and would have passed their genes on to their offspring. It has been suggested that one of the receptors used by HIV, CCR5, is also a receptor for smallpox. It is known that people resistant to HIV infection have a mutation in the CCR5 gene. The historical appearance and distribution of this mutation in human populations correlates with the historical distribution of smallpox. The AIDS epidemic is discussed further in chapter 51.

HIV infection compromises the host immune system

In AIDS patients, HIV primarily targets **CD4$^+$ cells,** particularly helper T cell. **Helper T cells** are responsible for mounting the immune response against foreign invaders, and their action is described more fully in chapter 51.

HIV infects and kills the CD4$^+$ cells until very few are left. Without these crucial immune system cells, the body cannot mount a defense against invading bacteria or viruses. AIDS patients die of infections that a healthy person could fight off. These diseases, called *opportunistic infections*, normally do not cause disease and are part of the progression from HIV infection to having AIDS.

Clinical symptoms typically do not begin to develop until after a long latency period, generally 8–10 years after the initial infection with HIV. Some individuals, however, may develop symptoms in as few as two years. During latency, HIV particles are not in circulation, but the virus can be found integrated within the genome of macrophages and CD4$^+$ T cells as a provirus (equivalent to a prophage in bacteria).

HIV testing

HIV tests do not test for the presence of circulating virus but rather for the presence of antibody against HIV. Because only those people exposed to HIV in their bloodstream at one time or another would have anti-HIV antibodies, this screening provides an effective way to determine whether further testing is needed to confirm HIV-positive status.

The spread of AIDS

Although carriers of HIV have no clinical symptoms during the long latency period, they are apparently fully infectious, which makes the spread of HIV very difficult to control. The reason HIV remains hidden for so long seems to be that its infection cycle continues throughout the 8–10 year latency period without doing serious harm to the infected person because of an effective immune response. Eventually, however, a random mutational event in the virus or a failure of the immune response allows the virus to quickly overcome the immune defense, beginning the course of AIDS.

HIV infects key immune-system cells

The way in which HIV infects humans provides a good example of how animal viruses replicate (figure 27.6). Most other viral infections follow a similar course, although the details of entry and replication differ in individual cases.

Attachment

When HIV is introduced into the human bloodstream, the virus particles circulate throughout the body but only infect CD4$^+$ cells. Most other animal viruses are similarly narrow in their requirements; hepatitis goes only to the liver, and rabies to the brain. This tissue tropism is determined by the proteins found on a cell surface and on a viral surface.

For example, the common cold virus uses the ICAM-1 membrane protein as a receptor to enter cells. ICAM-1 is a protein that is upregulated (increased) in times of immune activation and stress. So, the more inflammation and stress in an area, the more receptors exist for the virus to enter a cell and continue the disease process.

How does a virus such as HIV recognize a target cell? Recall from chapter 9 that every kind of cell in the human body has a specific array of cell-surface glycoprotein markers that serve to identify them to other, similar cells. Invading viruses take advantage of this to bind to specific cell types. Each HIV particle possesses a glycoprotein called gp120 on its surface that precisely fits the cell-surface marker protein CD4 on the surfaces of the immune system macrophages and T cells. Macrophages, another type of white blood cell, are infected first. Because macrophages commonly interact with CD4$^+$ T cells, this may be one way that the T cells are infected. Several coreceptors also significantly affect the likelihood of viral entry into cells, including the CCR5 receptor, which is mutated in HIV-immune individuals.

Entry of virus

After docking onto the CD4 receptor of a cell, HIV requires a coreceptor such as CCR5, to pull itself across the cell membrane. After gp120 binds to CD4, it goes through a conformational change that allows it to then bind the coreceptor. Receptor binding is thought to ultimately result in fusion of the viral and target cell membranes and entry of the virus through a fusion pore. The coreceptor, CCR5, is hypothesized to have been used by the smallpox virus as was mentioned earlier.

Replication

Once inside the host cell, the HIV particle sheds its protective coat. This leaves viral RNA floating in the cytoplasm, along with the reverse transcriptase enzyme that was also within the virion. Reverse transcriptase synthesizes a double strand of DNA complementary to the virus RNA, often making mistakes and introducing new mutations. This double-stranded DNA then enters the nucleus along with a viral enzyme that incorporates the viral DNA into the host cell's DNA. After a variable period of dormancy the HIV provirus directs the host cell's machinery to produce many copies of the virus.

As is the case with most enveloped viruses, HIV does not directly rupture and kill the cells it infects. Instead, the new viruses are released from the cell by *budding,* a process much like exocytosis. HIV synthesizes large numbers of viruses in this way, challenging the immune system over a period of years. In contrast, naked viruses,

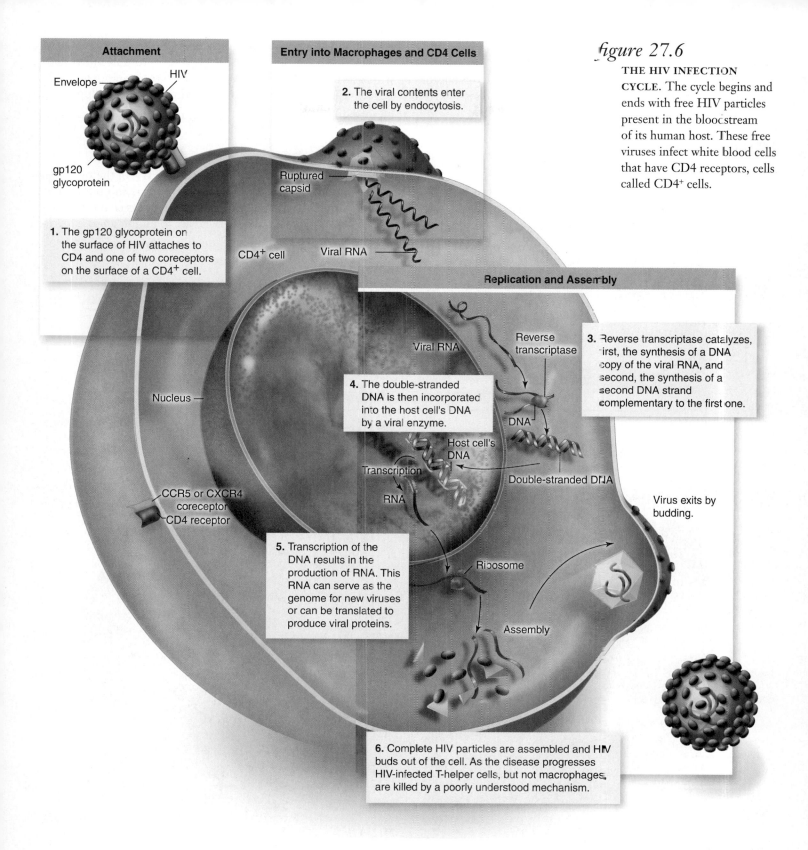

figure 27.6

THE HIV INFECTION CYCLE. The cycle begins and ends with free HIV particles present in the bloodstream of its human host. These free viruses infect white blood cells that have CD4 receptors, cells called CD4+ cells.

Attachment

Envelope — HIV

gp120 glycoprotein

1. The gp120 glycoprotein on the surface of HIV attaches to CD4 and one of two coreceptors on the surface of a CD4+ cell.

Entry into Macrophages and CD4 Cells

2. The viral contents enter the cell by endocytosis.

Ruptured capsid

Viral RNA

CD4+ cell

Nucleus —

CCR5 or CXCR4 coreceptor
CD4 receptor

4. The double-stranded DNA is then incorporated into the host cell's DNA by a viral enzyme.

Transcription

RNA

Host cell's DNA

5. Transcription of the DNA results in the production of RNA. This RNA can serve as the genome for new viruses or can be translated to produce viral proteins.

Replication and Assembly

Viral RNA

Reverse transcriptase

DNA

Double-stranded DNA

3. Reverse transcriptase catalyzes, first, the synthesis of a DNA copy of the viral RNA, and second, the synthesis of a second DNA strand complementary to the first one.

Virus exits by budding.

Ribosome

Assembly

6. Complete HIV particles are assembled and HIV buds out of the cell. As the disease progresses HIV-infected T-helper cells, but not macrophages, are killed by a poorly understood mechanism.

those lacking an envelope, generally lyse the host cell in order to exit. Some enveloped viruses may produce enzymes that damage the host cell enough to kill it or may produce lytic enzymes as well.

Evolution of HIV during infection

During an infection, HIV is constantly replicating and mutating. The reverse transcriptase enzyme is less accurate than DNA polymerases, leading to a high mutation rate. Eventually, by chance, variants in the gene for gp120 arise that causes the gp120 protein to alter its second-receptor partner. This new form of gp120 protein will bind to a different second receptor, for example CXCR4, instead of CCR5. During the early phase of an infection, HIV primarily targets immune cells with the CCR5 receptor. Eventually the virus mutates to infect a broader range of cells. Ultimately, infection results in the destruction and loss of critical T helper cells.

This destruction of T cells blocks the body's immune response and leads directly to the onset of AIDS, with cancers and opportunistic infections free to invade the defenseless victim. Most deaths due to AIDS are not a direct result of HIV, but are from other diseases that normally do not harm a host with a normal immune system.

HIV treatment is a highly active field of research

New discoveries about how HIV works continue to fuel research on devising ways to counter HIV. For example, scientists are testing drugs and vaccines that act on HIV receptors,

researching the possibility of blocking CCR5, and looking for defects in the structures of HIV receptors in individuals who are infected with HIV but have not developed AIDS. Figure 27.7 summarizes some of the recent developments and discoveries.

Combination drug therapy

Two main categories of drugs can inhibit HIV in the test tube: nucleoside analogues such as AZT and protease inhibitors. The first are similar to normal nucleotides but act as chain terminators to halt replication. The second act to inhibit a protease re-

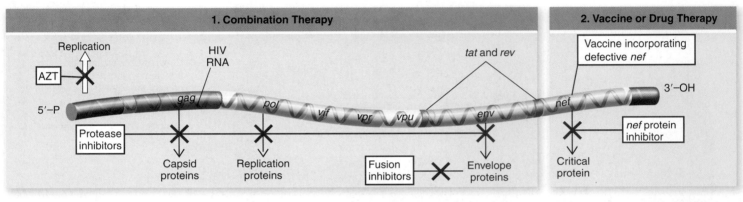

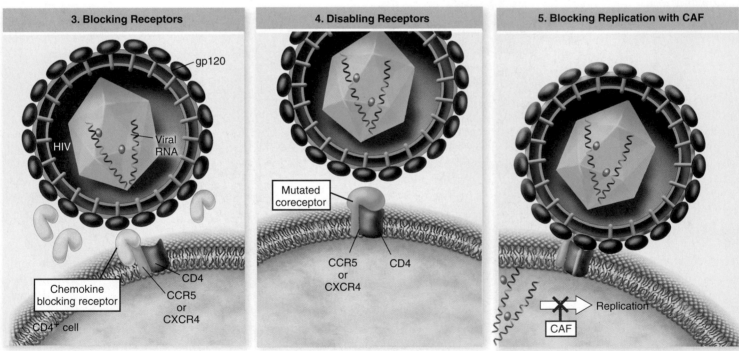

figure 27.7

POTENTIAL NEW TREATMENTS FOR HIV. The HIV genetic map is shown above and attachment to a host cell below. Research is currently under way in the following five areas: (*1*) The currently used combination therapy involves using two types of drugs: AZT to block replication of the virus and protease inhibitors to block the production of critical viral proteins. Recently, synthetic proteins have been developed that block fusion of the HIV membrane and the host cell membrane. (*2*) Using a defective form of the viral gene *nef*, scientists may be able to construct an HIV vaccine. Also, drug therapy that inhibits *nef*'s protein product is being tested. (*3*) Other research focuses on the use of chemokine agents to block coreceptors (CXCR4 and CCR5), thereby disabling the mechanism HIV uses to enter CD4+ cells. (*4*) Producing mutations that will disable receptors may also be possible. (*5*) Finally, CAF, an antiviral factor that acts inside the CD4+ cell, may be able to block replication of HIV.

quired for cleavage of large polyproteins into functional capsid, enzyme and envelope proteins during the normal life cycle.

When combinations of these drugs are administered to people with HIV in controlled studies, their condition can be improved for variable periods of time. A combination of a protease inhibitor and two AZT-type drugs have entirely eliminated the HIV virus from many patients' bloodstreams. All of these patients began to receive the combination drug therapy within three months of contracting the virus, before their bodies had an opportunity to develop tolerance to any single drug. Widespread use of this **combination therapy** or **highly active antiretroviral therapy (HAART)** has cut the U.S. AIDS death rate by three-fourths since its introduction in the mid-1990s.

Unfortunately, this sort of combination therapy does not appear to actually eliminate HIV from the body. Although the virus disappears from the bloodstream, traces of it can still be detected in lymph tissue of the patients. When combination therapy is discontinued, virus levels in the bloodstream once again rise.

Because of expense, demanding therapy schedules, and many side effects, long-term combination therapy does not seem a promising approach. In addition, although HIV is a significant disease problem in the United States, in Africa it has reached critical proportions. Over 95% of the people currently infected with HIV worldwide are living in a developing country. The HAART therapies with the cocktail of specialized, expensive drugs are not a viable treatment plan for people living in these countries.

Vaccine therapy: Using a defective HIV gene to combat AIDS

Recently, five people in Australia who are HIV-positive but have not developed AIDS in 14 years have been found to have all received a blood transfusion from a single HIV-positive person, who also has not developed AIDS. This finding led scientists to believe that the strain of virus transmitted to these people has some sort of genetic defect that prevents it from effectively disabling the human immune system. Thus, they all have low viral loads and do not develop AIDS.

In subsequent research, a defect was found in one of the nine genes present in this strain of HIV. This gene is called *nef*, named for "negative factor," and the defective version of *nef* in the HIV strain that infected the six Australians seems to be missing some pieces. Viruses with the defective gene may have reduced reproductive capability, allowing the immune system to keep the virus in check.

This finding has exciting implications for developing a vaccine against AIDS. Before this, scientists had been unsuccessful in trying to produce a harmless strain of AIDS that can elicit an effective immune response. The Australian strain with the defective *nef* gene has the potential for use in such a vaccine.

Another potential application of this discovery is in the possibility of developing drugs to inhibit HIV proteins that speed virus replication. It seems that the protein produced from the *nef* gene is one of these critical HIV proteins, because viruses with defective forms of *nef* do not reproduce efficiently,

as seen in the cases of the six Australians. Research is currently under way to develop a drug that targets the *nef* protein.

Blocking replication: Chemokines and CAF

In the laboratory, chemicals called **chemokines,** natural immune cell regulators, appear to inhibit HIV infection by binding to and blocking the CCR5 and CXCR4 coreceptors. The CCR in CCR5 and CXCR5 stands for chemokine coreceptor. These are natural receptors for chemokines. One might expect, people long infected with the HIV virus who have not developed AIDS would have high levels of chemokines in their blood, or to have low levels of CCR5 and CXR4 coreceptors. The search for HIV-inhibiting chemokines has therefore been intense. Not all results are promising. Researchers report that in their tests, the levels of chemokines were *not* different between patients in which the disease was not progressing and those in which it was rapidly progressing. More promising, levels of another factor called CAF (CD8[+] cell antiviral factor) *are* different between these two groups. Researchers have not yet succeeded in isolating CAF, which seems not to block receptors that HIV uses to gain entry to cells, but instead, to prevent replication of the virus once it has infected the cells.

One problem with using chemokines as drugs is that they are also involved in the inflammatory response of the immune system. Chemokines attract white blood cells to areas of infection; these chemicals work well in small amounts and in local areas, but in mass numbers they can cause an inflammatory response that is worse than the original infection. In addition, recruiting more macrophages and T helper cells would provide more local targets for HIV infection! Therefore scientists warn that injection of chemokines could make patients *more* susceptible to infections.

Blocking or disabling receptors

Variants in the gene that encodes the CCR5 receptor have been identified. One allele with a 32-bp deletion gives rise to cells that are less readily infected. Individuals at high risk of HIV infection who are homozygous for this mutation rarely develop AIDS. In one study involving 1955 people, scientists found no infection in individuals who were homozygous for the mutated allele. Heterozygotes for this allele also have some protection in that progression of the disease is slowed.

The allele seems to be more common in Caucasian populations (10–11%) than in African-American populations (2%), and it is absent in African and Asian populations. Treatment for AIDS involving disruption of CCR5 looks promising because research indicates that people live perfectly well without CCR5. Attempts to block or disable CCR5 are being sought in numerous laboratories.

> The disease AIDS is caused by the virus HIV. The virus primarily infects host CD4[+] T cells and thereby compromises the immune system. HIV is a retrovirus that enters cells via membrane fusion. HIV infection ultimately results in the massive death of CD4[+] T cells. Combination drug therapy is a treatment modality in developed countries, and much research is being done to develop vaccines or to find agents that can prevent infection.

Other Viral Diseases

Humans have known and feared diseases caused by viruses for thousands of years. Among the diseases that viruses cause (see table 27.1) are influenza, smallpox, hepatitis, yellow fever, polio, AIDS, and SARS. In addition, viruses have been implicated in some cancers including leukemias. Viruses not only cause many human diseases, but also cause major losses in agriculture, forestry, and the productivity of natural ecosystems.

The flu is caused by influenza virus

Perhaps the most lethal virus in human history has been the influenza virus. As mentioned, some 20–50 million people worldwide died of flu within 18 months in 1918 and 1919.

Types and subtypes

Flu viruses are enveloped segmented RNA viruses that infect animals. An individual flu virus resembles a rod studded with spikes composed of two kinds of protein. The three general "types" of flu virus are distinguished by their capsid protein, which surrounds the viral RNA segments and is different for each type: **Type A flu virus** causes most of the serious flu epidemics in humans and also occurs in mammals and birds. Type B and type C viruses are restricted to humans and rarely cause serious health problems.

Different strains of flu virus, called subtypes, differ in their protein spikes. One of these proteins, hemagglutinin (H), aids the virus in gaining access to the cell interior. The other, neuraminidase (N), helps the daughter viruses break free of the host cell once virus replication has been completed.

Parts of the H molecule contain "hotspots" that display an unusual tendency to change as a result of mutation of the viral RNA during imprecise replication. Point mutations cause changes in these spike proteins in 1 of 100,000 viruses during the course of each generation. These highly variable segments of the H molecule are targets against which the body's antibodies are directed. These constantly changing H-molecule regions improve the reproductive capacity of the virus and hinder our ability to make effective vaccines.

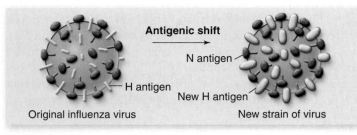

Because of accumulating changes in the H and N molecules, different flu vaccines are required to protect against different subtypes. Type A flu viruses are currently classified into 13 distinct **H subtypes** and 9 distinct **N subtypes,** each of which requires a different vaccine to protect against infection. Thus, the type A virus that caused the 1918 influenza pandemic (that is, worldwide epidemic) has type 1H and type 1N and is designated A(H1N1).

The importance of recombination

The greatest problem in combating flu viruses arises not through mutation, but through recombination. Viral RNA segments are readily reassorted by genetic recombination when two different subtypes simultaneously infect the same cell. This may put together novel combinations of H and N spikes unrecognizable by human antibodies specific for the old configuration.

Viral recombination of this kind seems to have been responsible for the three major flu pandemics that occurred in the twentieth century, by producing drastic shifts in H–N combinations. The "Spanish flu" of 1918, A(H1N1), killed 20–50 million people worldwide. The Asian flu of 1957, A(H2N2), killed over 100,000 Americans. The Hong Kong flu of 1968, A(H3N2), infected 50 million people in the United States alone, of which 70,000 died.

Origin of new strains

It is no accident that new strains of flu usually originate in the Far East. The most common hosts of influenza virus are ducks, chickens, and pigs, which in Asia often live in close proximity to each other and to humans. Pigs are subject to infection by both bird and human strains of the virus, and individual animals are often simultaneously infected with multiple strains. This creates conditions favoring genetic recombination between strains, producing new combinations of H and N subtypes.

The Hong Kong flu, for example, arose from recombination between A(H3N8) from ducks and A(H2N2) from humans. The new strain of influenza, in this case A(H3N2), then passed back to humans, creating an epidemic because the human population has never experienced that H–N combination before.

In 1997, a form of avian influenza, A(H5N1), was discovered that could infect humans. Avian influenza or "bird flu" is highly contagious and deadly among domestic bird populations, and it is now clear that this H5N1 strain is transmitted between domestic birds, which live in close contact with humans, and wild birds that have worldwide migratory patterns. Currently the "bird flu" has high mortality in humans, with over 100 deaths reported, but does not appear to spread between humans. Because of the extreme variability in the influenza genome, the spread of "bird flu" from human to human is a possibility. Given the ease of travel for both people and livestock, a global pandemic is possible. As a result, the spread of the H5N1 strain is being carefully monitored by epidemiologists. In addition, scientists are working on the development of a vaccine and governments are stockpiling antiviral drugs in preparation for this possibility.

New viruses emerge by infecting new hosts

Sometimes viruses that originate in one organism pass to another, thus expanding their host range. Often, this expansion is deadly to the new host. HIV, for example, is thought to have arisen in chimpanzees and relatively recently passed to humans. Influenza is fundamentally a bird virus. Viruses that originate in one organism and then pass to another and cause disease are called **emerging viruses.** They represent a considerable threat in an age when airplane travel potentially allows infected individuals to move about the world quickly, spreading an infection.

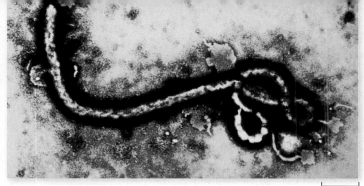

figure 27.8

THE EBOLA VIRUS. This virus, with a fatality rate that can exceed 90%, appears sporadically in West Africa. The natural host of the virus currently is unknown.

0.3 μm

Hantavirus

An emerging virus caused a sudden outbreak of a deadly pneumonia in the southwestern United States in 1993. This disease was traced to a species of **hantavirus** and was called the *sin nombre* or *no-name virus*. Hantavirus is a single-stranded RNA virus associated with rodents. This virus was eventually traced to deer mice. The deer mouse hantavirus is transmitted to humans through fecal and urine contamination in areas of human habitation. Controlling the deer mouse population has limited the disease.

Hemorrhagic fever: Ebola

Sometimes the origin of an emerging virus is unknown, making an outbreak more difficult to control. Among the most lethal of emerging viruses are a collection of filamentous viruses arising in central Africa that cause severe hemorrhagic fever. With lethality rates in excess of 50%, these so-called **filoviruses** are among the most lethal infectious diseases known. One, **Ebola virus** (figure 27.8), has exhibited lethality rates in excess of 90% in isolated outbreaks in central Africa. The outbreak of Ebola virus in the summer of 1995 in Zaire killed 245 people out of 316 infected—a mortality rate of 78%. A recent (2004) outbreak of Ebola in Yambio, southern Sudan, caused 17 infections and 7 deaths. This outbreak was rapidly contained by isolating patients from family members as soon as symptoms appeared. The natural host of Ebola is unknown.

SARS

A recently emerged species of coronavirus (figure 27.9) was responsible for a worldwide outbreak in 2003 of **severe acute respiratory syndrome (SARS)**, a respiratory infection with pneumonia-like symptoms that is fatal in over 8% of cases. When the 29,751-nucleotide RNA genome of the SARS coronavirus was sequenced, it proved to be a completely new form of coronavirus, not closely related to any of the three previously described forms.

Virologists suspect that the SARS coronavirus most likely came from civets (a weasel-like mammal) and possibly other wild animals that live in China and are eaten as delicacies. If the SARS virus indeed exists in natural populations, future outbreaks will be difficult to prevent without an effective vaccine. Recent data have implicated bats as the natural reservoir for SARS virus. The significance of this finding for control of the virus is unclear at present.

Genome sequences have been analyzed from SARS patients at various stages of the outbreak, and these analyses indicate that the virus's mutation rate is low, in marked contrast to HIV, another RNA virus. The stable genome of the SARS virus should make development of a SARS vaccine practical. The

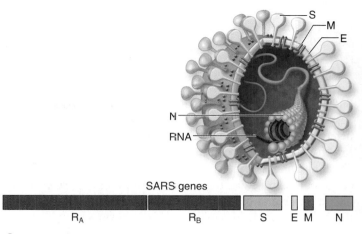

figure 27.9

SARS CORONAVIRUS. The 29,751-nucleotide SARS genome is composed of RNA and contains six principle genes: R_A and R_B–replicases; S–spike proteins; E–envelope glycoproteins; M–membrane glycoprotein; N–nucleocapsid protein.

lessons learned from developing antiviral agents against other RNA viruses, such as HIV and influenza have helped develop drugs to treat SARS. Several anti-SARS agents and vaccines are currently being tested in laboratories across the world.

Viruses can cause cancer

Through epidemiological studies and research, scientists have established a link between some viral infections and the subsequent development of cancer. Examples include the association between chronic hepatitis B infections and the development of liver cancer, and the development of cervical carcinoma following infections with certain strains of papillomaviruses.

Viruses may contribute to about 15% of all human cancer cases worldwide. They are capable of altering the growth properties of human cells they infect by triggering the expression of cancer-causing genes called oncogenes (see chapter 10). Changes in the normal function of these genes leads to cancer.

These changes can occur because viral proteins intefere with the regulation of oncogene expression. Alternatively, the integration of a viral genome into a host chromosome may disrupt a gene required to control the cell cycle. Viruses themselves may encode these oncogenes as well. Virus-induced cancer involves complex interactions with cellular genes and requires a series of events in order to develop. The association of viruses with some forms of cancer has led to research on vaccine development for the prevention of these cancers. In June 2006, the FDA approved the use of a new HPV vaccine in women and young girls from the age of 11 to prevent cervical cancer.

Many types of viruses have caused disease in humans for as long as we have recorded history. Some of these, such as influenza, have involved pandemics that have been responsible for millions of deaths worldwide. Recombination is common in the influenza virus, making natural immunity and development of vaccines problematic. Emerging diseases can be due to viruses that switch hosts, that is, jump from another species to humans. Hantavirus, Ebola, and SARS all fall into this category. Virus infection has also been linked to the development of certain cancers.

Prions and Viroids: Subviral Particles

For decades, scientists have been fascinated by a peculiar group of fatal brain diseases. These diseases have an unusual property: Years and often decades pass after infection before the disease is detected in infected individuals. In fact, when these diseases were first discovered, they were thought to be caused by *slow viruses*.

The brains of infected individuals develop numerous small cavities as neurons die, producing a marked spongy appearance. Called **transmissible spongiform encephalopathies (TSEs)**, these diseases include scrapie in sheep, bovine spongiform encephalopathy (BSE) or mad cow disease in cattle, chronic wasting disease in deer and elk, and kuru, Creutzfeldt–Jakob disease (CJD), and variant Creutzfeldt–Jakob disease (vCJD) in humans.

TSEs can be transmitted experimentally by injecting infected brain tissue into a recipient animal's brain. TSEs can also spread via tissue transplants and, apparently, food. The disease kuru was common in the Fore people of Papua, New Guinea, because they practiced ritual cannibalism, eating the brains of infected individuals. Mad cow disease spread widely among the cattle herds of England in the 1990s because cows were fed bonemeal prepared from sheep and cattle carcasses to increase the protein content of their diet. Like the Fore, the British cattle were eating the tissue of cattle that had died of the disease.

In the years following the outbreak of BSE, there has been a significant increase in CJD incidence in England. Some cases of CJD appear to be genetic. Mysteriously, patients with no family history of CJD were being diagnosed with the disease. This led to the discovery of a new form of CJD called variant CJD or vCJD that is acquired from eating meat of BSE-infected animals. Concern exists that vCJD may be transmitted from person to person through blood products, similar to the transmission of HIV through blood and blood products.

Prion replication was a heretical suggestion

In the 1960s, British researchers T. Alper and J. Griffith noted that infectious TSE preparations remained infectious even after exposure to radiation that would destroy DNA or RNA. They suggested that the infectious agent was a protein. They speculated that the protein usually preferred one folding pattern, but could sometimes misfold, and then catalyze other proteins to do the same, the misfolding spreading like a chain reaction.

This "heretical" suggestion was not accepted by the scientific community, because it violated a key tenet of molecular biology: Only DNA or RNA act as hereditary material, transmitting information from one generation to the next.

Evidence has accumulated that prions cause TSEs

In the early 1970s, physician Stanley Prusiner began to study TSEs. Try as he might, Prusiner could find no evidence of nucleic acids or viruses in the infectious TSE preparations. He concluded, as Alper and Griffith had, that the infectious agent was a *protein*, which in a 1982 paper he named a **prion,** for "proteinaceous infectious particle."

Prusiner went on to isolate a distinctive prion protein, and to amass evidence that prions play a key role in triggering TSEs. Every

host tested to date expresses a normal prion protein (PrP^c) in their cells. The disease-causing prions are the same protein, but folded differently (PrP^{sc}). These misfolded proteins have been shown in vitro to serve as a template for normal PrP to misfold (figure 27.10). The misfolded PrP proteins are very resistant to degradation, making it possible for them to pass through the acidic digestive tract intact and therefore to be transmitted orally, by ingestion.

Experimental evidence has accumulated to support this idea. Injection of prions with different abnormal conformation into hosts leads to the same abnormal conformations as the parent prions. Mice genetically engineered to lack PrP^c are immune to TSE infection. If brain tissue with the prion protein is grafted into the mice, the grafted tissue—but not the rest of the brain—can then be infected with TSE. However, the nucleation or misfolding of PrP^c by PrP^{sc} has not been conclusively demonstrated in vivo. The mechanism of pathogenesis also remains controversial.

Viroids are infectious RNA with no protein coat

Viroids are tiny, naked molecules of circular RNA, only a few hundred nucleotides long, that are important infectious disease agents in plants. A recent viroid outbreak killed over 10 million coconut palms in the Philippines.

It is unclear how viroids cause disease. One clue is that viroid nucleotide sequences resemble the sequences of introns within ribosomal RNA genes. These sequences are capable of catalyzing excision from DNA—perhaps the viroids are catalyzing the destruction of chromosomal integrity, which leads to massive cell death. Another theory is that viroids interfere with gene expression through interactions with cellular mRNAs, thus targeting the mRNA for degradation before it can be translated.

> Prions and viroids are smaller and simpler than viruses. Prions are infectious particles that do not seem to contain any nucleic acid. Prions are misfolded proteins that are thought to cause related cellular proteins to also become misfolded. Prions are the causative agent of TSEs. Viroids are infectious RNAs that are implicated in some plant diseases.

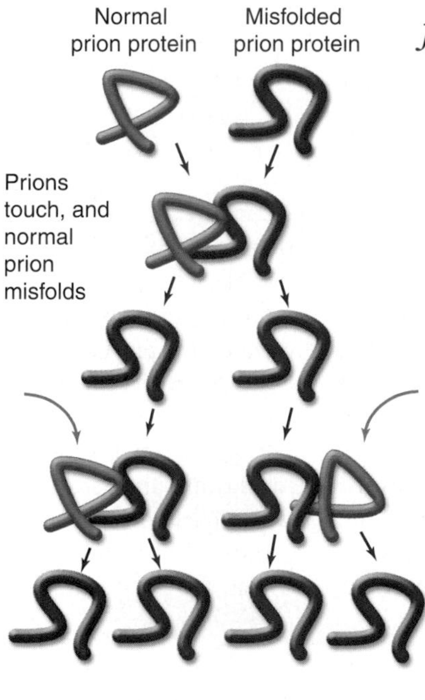

Normal prion protein   Misfolded prion protein

Prions touch, and normal prion misfolds

figure 27.10

HOW PRIONS CAUSE DISEASE. Misfolded prions seem to cause normal prion protein to misfold simply by contacting them. When prions misfolded in different ways (*red*) contact normal prion protein (*blue*), the normal prion protein misfolds in the same way.

27.1 The Nature of Viruses

Viruses all have the same basic structure: a core of nucleic acid surrounded by protein.

- Viral genomes can consist of either DNA or RNA and can be classified as DNA viruses, RNA viruses, or retroviruses.

- Most viruses have a protein sheath or capsid around their nucleic acid core.

- Some viruses also have enzymes inside their capsid that are important in early infection.

- Many animal viruses have an envelope around the capsid. The envelope is composed of proteins that are virally encoded, lipids from the host cell, and glycoproteins.

- Each virus has a limited host range and many also exhibit tissue tropism.

- Viruses are obligate intracellular parasites because they lack ribosomes and proteins needed for replication.

- Viruses replicate by taking over host machinery and directing their own nucleic acid replication and protein synthesis.

- Viruses vary in size and come in two simple shapes: helical (rodlike) or icosahedral (spherical) (figures 27.1 and 27.4).

- Viral genomes exhibit great variation. The DNA or RNA viral genome may be linear or circular, single- or double-stranded.

- RNA viruses may be segmented, with multiple RNA molecules, or nonsegmented, with one RNA molecule.

- Retroviruses contain RNA that is transcribed into DNA by reverse transcriptase.

27.2 Bacteriophages: Bacterial Viruses
(figure 27.3)

Bacteriophages are highly variable viruses that only infect bacteria.

- Bacterial viruses exhibit two reproductive cycles: the lytic cycle that kills the host cell and the lysogenic cycle in which the virus is incorporated into the host genome as a prophage (figure 27.5).

- A cell containing a prophage is called a lysogen.

- The prophage can be induced by DNA damage and other environmental cues to reenter the lytic cycle.

- For most phage, steps in infection include attachment, injection of DNA, macromolecular synthesis, assembly of new phage, and release of progeny phage.

- Phage conversion occurs when foreign DNA is contributed to the host by a bacterial virus.

27.3 Human Immunodeficiency Virus (HIV)

Human immunodeficiency virus (HIV) causes acquired immunodeficiency syndrome (AIDS) and it is a good example of how an animal virus works (figure 27.6).

- HIV specifically targets macrophages and CD4 cells, a type of helper T lymphocyte cell. With the loss of these cells the human body cannot fight off opportunistic infections, which ultimately kill the host.

- The viral glycoprotein, gp120, precisely fits on the cell-surface marker protein CD4 on macrophages and T cells.

- HIV infection involves two receptors, CD4 and CCR5, when HIV attaches to these two receptors, receptor-mediated endocytosis is activated, bringing the HIV particle into the cell.

- Once inside the cell, the protective coat is shed, releasing viral RNA and reverse transcriptase into the cytoplasm.

- Reverse transcriptase makes double-stranded DNA that is complementary to the viral RNA. This DNA may be incorporated into the host DNA.

- Replicated viruses are budded off the host cell by exocytosis.

- HIV has a high mutation rate because the reverse transcriptase enzyme is much less accurate than DNA polymerases.

- Mutations lead to an altered glycoprotein gp120, which now binds instead to the CXCR4 receptor found only on the surface of CD4 cells. Incorporation of the altered HIV particle leads to a rapid decline in T cells and immune response.

- Combination drug therapy using nucleoside analogues and protease inhibitors eliminates HIV from the bloodstream but not totally from the body.

- Potential infection-fighting therapies include vaccines made from a defective HIV gene and chemicals that block or disable the cell-surface receptors that bind to the HIV particle.

27.4 Other Viral Diseases

Humans have long known and feared diseases caused by viruses.

- One of the most lethal viruses in human history is type A influenza virus. These can also infect other mammals and birds.

- Genes in influenza viruses recombine readily so they are not recognized by antibodies against past infections. Thus we have yearly flu shots and not a single vaccine.

- Viruses can extend their host range by passing to another species. Examples include hantavirus, hemorrhagic fever, and SARS (figure 27.9).

- Viruses have been linked to cancers, including liver cancer and cervical papillomas.

27.5 Prions and Viroids: Subviral Particles

Infectious diseases can also be caused by proteins and naked molecules of RNA.

- Prions are infectious particles that are misfolded proteins that serve as templates for normal proteins to misfold.

- Prions are responsible for transmissible spongiform encephalopathies (figure 27.10).

- Viroids are circular, naked molecules of RNA that infect plants. They resemble intron sequences of ribosomal RNA genes that can catalyze excision from DNA.

SELF TEST

1. The reverse transcriptase enzyme is active in which class of viruses?
   a. Positive-strand RNA viruses
   b. Double-stranded DNA viruses
   c. Retroviruses
   d. Negative-strand RNA viruses

2. Which of the following is not part of a virus?
   a. Capsid
   b. Ribosomes
   c. Genetic material
   d. All of the above are found in viruses

3. Which of the following is common in animal viruses but not in bacteriophage?
   a. DNA
   b. Capsid
   c. Envelope
   d. Icosahedral shape

4. Which of the following would not be part of the life cycle of a lytic virus?
   a. Macromolecular synthesis
   b. Attachment to host cell
   c. Assembly of progeny virus
   d. Integration into the host genome

5. A process by which a virus may change a benign bacteria into a virulent strain is called _____.
   a. induction.
   b. phage conversion.
   c. lysogeny.
   d. replication.

6. Prior to entry, the _____ glycoprotein of the HIV virus recognizes the _____ receptor on the surface of the macrophage.
   a. CCR5; gp120
   b. CXCR4; CCR5
   c. CD4; CCR5
   d. gp120; CD4

7. The varying degrees of resistence to HIV in populations has been suggested to be related to the patterns of smallpox outbreaks over human history. This explanation hinges on
   a. the similarity in the genomes of the two viruses.
   b. the fact that both viruses use reverse transcriptase.
   c. both viruses using the same receptor to bind to host cells.
   d. both viruses compromise the immune system.

8. Current research has focused on blocking what receptor on the T cell lymphocyte in order to protect it from infection by the HIV virus?
   a. gp120
   b. CD4
   c. CCR5
   d. CD5

9. Which of the following HIV treatments prevents the formation of a functional capsid?
   a. Vaccine therapy
   b. Chemokines
   c. Nucleoside analogues
   d. Protease inhibitors

10. Which of the following viruses routinely exhibit antigenic shifts, making vaccination programs difficult?
    a. HIV
    b. Influenza
    c. Hantavirus
    d. Filoviruses

11. Activation of oncogenes to cause cancer may result from the action of a
    a. virus.
    b. viroid.
    c. prion.
    d. bacteria.

12. Prions are responsible for which of the following?
    a. Mad cow disease
    b. Scrapie
    c. Variant Creutzfeldt–Jakob disease
    d. All of the above

13. An infectious RNA without a capsid or envelope is called a
    a. prion.
    b. bacteriophage.
    c. viroid.
    d. virus.

14. SARS is associated with what form of virus?
    a. Hantavirus
    b. Coronavirus
    c. Filovirus
    d. Influenza virus

15. A nonliving, infectious particle that lacks any nucleic acids or an envelope, and contains only amino acids would be classified with the
    a. prions.
    b. viruses.
    c. viroids.
    d. bacteria.

CHALLENGE QUESTIONS

1. *E. coli* lysogens derived from infection by phage λ can be induced to form progeny viruses by exposure to radiation. The inductive event is the destruction of a repressor protein that keeps the prophage genome unexpressed. What might be the normal role of the protein that recognizes and destroys the λ repressor?

2. Most biologists believe that viruses evolved following the origin of the first cells? Why do you think that this is the case?

3. In 1971 President Nixon declared a "war on cancer." Although significant advances have been made in the past three decades, the war is far from over. The discovery that viruses may be responsible for some forms of cancer have both helped and hindered the battle. Why?

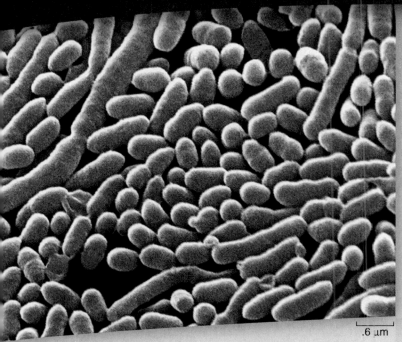

.6 µm

chapter 28

Prokaryotes

introduction

ONE OF THE HALLMARKS OF LIVING organisms is their cellular organization. You learned earlier that living things come in two basic cell types: **prokaryotes** and **eukaryotes.** To review, prokaryotes lack the membrane-bounded nucleus found in all eukaryotes, and they also have a much less complex cellular structure, lacking many of the organelles seen in eukaryotes (chapter 4). Prokaryotes are considerably smaller and more numerous than their eukaryotic counterparts. If we examined a human being closely, we would discover that there are approximately 10 prokaryotic cells living in or on the human body for every single human cell present—and there are trillions of human cells.

Prokaryotic microbes play an important role in global ecology as well. Most biologists think prokaryotes were the first organisms to evolve. The diversity of eukaryotic organisms that currently live on Earth could not exist without prokaryotes because they make possible many of the essential functions of ecosystems. Prokaryotic photosynthesis, for example, is thought to have been the source for the oxygen in the ancient Earth's atmosphere, and it still contributes significantly to oxygen production today. An understanding of prokaryotes is essential to understanding all life on Earth, past and present.

The First Cells

As no human was present at the formation of life, we are left with indirect evidence for the earliest life-forms. The most direct evidence we have are fossils, but these can be difficult to interpret, especially since we are looking for microscopic evidence of life. We also can analyze the composition of carbon-containing rocks to look for signs of life acting on organic material, as indicated by a change in isotopic ratios. Finally, we can look for the presence of organic chemicals, which are of biological origin.

Microfossils indicate that the first cells were probably prokaryotic

Evidence of life in the form of microfossils is difficult both to find and to interpret. Rocks of age greater than 3 billion years old are rarely unchanged by geologic action over time. Two main formations of 3.5- to 3.8-billion-year-old rocks have been found that are mostly intact: the Kaapvaal craton in South Africa and the Pilbara craton in western Australia. (A *craton* is a rock layer of undisturbed continental crust.) Structures have been found in each of these formations and others that are interpreted to be biological in origin. Although this interpretation has been controversial, the accumulation of evidence over time favors these structures as being true fossil cells.

Microfossils are fossilized forms of microscopic life. Many microfossils are small (1–2 μm in diameter) and appear to be single-celled, lack external appendages, and have little evidence of internal structure (figure 28.1). Thus, microfossils seem to resemble present-day prokaryotes.

The currently oldest microfossils are 3.5 billion years old. The claim that these microfossils are the remains of living organisms is supported by isotopic data (described shortly) and by spectroscopic analysis that indicates they do contain complex carbon molecules. Whether these microscopic structures are true fossil cells is still controversial, and the identity of the prokaryotic groups represented by the various microfossils is still unclear. Arguments have been made for various bacteria, including cyanobacteria (described later on), but definitive interpretation is difficult.

figure 28.2

STROMATOLITES. Mats of bacterial cells that trap mineral deposits and form the characteristic dome shapes seen here.

In addition to these microfossils, indirect evidence for ancient life can be found in the form of sedimentary deposits called **stromatolites.** These structures are commonly interpreted as a combination of sedimentary deposits and precipitated material that are held in place by mats of microorganisms. The microorganisms that make up the mats are thought to be cyanobacteria. Formations of stromatolites are as old as 2.7 billion years. Because relatively modern stromatolites are also known, the formation and biological nature of these structures is less contentious (figure 28.2).

Isotopic data indicate that carbon fixation is an ancient process

Another way to ask when life began is to look for the signature of living systems in the geological record. Living systems alter their environments, and sometimes this change can be detected. The most obvious change is that living systems are selective in the isotopes of carbon in compounds they use. Living organisms incorporate carbon-12 into their cells before any other carbon isotope, and thus they can alter the ratios of these isotopes in the atmosphere. They also have a higher level of carbon-12 in their fossilized bodies than does the nonorganic rock around them.

Much work has been done on dating and analyzing carbon compounds in the oldest rocks, looking for signatures of life. Although this work is controversial, it has been argued that carbon signatures indicate carbon fixation, the incorporation of inorganic carbon into organic form, was active as much as 3.8 BYA.

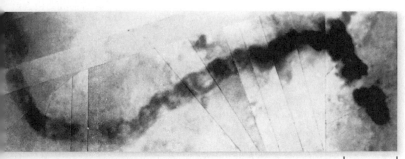

20 μm

figure 28.1

EVIDENCE OF BACTERIAL FOSSILS. Rocks approximately 3.5 billion years old to 1 billion years old have tiny fossils resembling bacterial cells embedded within them.

There are four possible pathways for the ancient fixation of carbon. The most common pathway for carbon fixation is the Calvin cycle (chapter 8). This is the pathway used by cyanobacteria, algae, and modern land plants that perform oxygenic photosynthesis using two photosystems. The Calvin cycle is also active in green and purple sulfur bacteria that perform anoxygenic photosynthesis using a single photosystem. This anoxygenic form of photosynthesis could account for ancient carbon fixation

To date, the entire Calvin cycle has not been demonstrated in the domain Archaea, although the key enzyme for this pathway has been identified in a few archaeal isolates. Instead, some archaea use a reductive version of the Krebs cycle (chapter 7). This pathway of carbon fixation is also used by some lithotrophic bacteria, which derive energy from the oxidation of inorganic compounds, and by the green sulfur bacteria. Two other pathways may also occur in the lithotrophs, archaea, and the green nonsulfur bacteria. Evidence suggests that the ability to fix carbon has evolved multiple times over the course of evolution.

Some hydrocarbons found in ancient rocks may have biological origins

Another way to look for evidence of ancient life is to look for organic molecules, which are clearly of biological origin; such molecules are called *biomarkers*. Although the process sounds simple, it has proved difficult to find such markers. One type of molecule that has been used is hydrocarbons, which can be analyzed for their carbon isotope ratios to indicate biological origin. The analysis of extractable hydrocarbons from the Pilbara formation in Australia found lipids that are indicative of cyanobacteria as long ago as 2.7 billion years. The search for definitive chemical markers for living systems in the oldest rocks and in meteorites is an area of intense interest.

Arguments for the oldest microfossils have been supported by analysis of the carbon isotope ratios in carbonaceous material from the same formations. If these fossils indeed represent living cells, it implies that life was much more abundant 3.5 BYA than previously thought. Although much of this work is still being debated, it pushes the possible origin of life back well beyond 3.5 BYA.

Evidence for the earliest cells exists in microfossils. There is controversy over the earliest microfossils, but they are at least 3.5 billion years old. Other evidence for early life includes isotopic ratios that are skewed by biological activity, which would indicate carbon fixation is an ancient process. Some hydrocarbons appear to be biomarkers and may therefore also indicate an ancient origin of life.

28.2 Prokaryotic Diversity

Although thousands of different kinds of prokaryotes are currently recognized, many thousands more await proper identification. New molecular techniques have allowed scientists to identify and study microorganisms without culturing them. As a result, microbiologists have discovered thousands of new species that were never discovered or characterized because they could not be maintained in culture.

It is estimated that only between 1 and 10% of all prokaryotic species are known and characterized, leaving between 90 and 99% unknown and undescribed. Every place microbiologists look, new species are being discovered, often altering the way we think about prokaryotes. In the 1970s and 1980s, a new type of prokaryote was discovered and analyzed that eventually led to the division of prokaryotes into two groups: the **Archaea** (formerly called Archaebacteria) and the **Bacteria** (sometimes also called Eubacteria).

Archaea and bacteria are the oldest, structurally simplest, and most abundant forms of life. They are also the only organisms with prokaryotic cellular organization. Prokaryotes were abundant for over a billion years before eukaryotes appeared in the world. Early photosynthetic bacteria (cyanobacteria) altered the Earth's atmosphere by producing oxygen, which stimulated extreme bacterial and eukaryotic diversity.

Prokaryotes are ubiquitous and live everywhere eukaryotes do; they are also able to thrive in places no eukaryote could live. Bacteria and archaea have been found in deep-sea caves, volcanic rims, and deep within glaciers. Some of the extreme environments in which prokaryotes can be found would be lethal to any other life-form.

Many archaeans are **extremophiles**. They live in hot springs that would cook other organisms, in hypersaline environments that would dehydrate other cells, and in atmospheres rich in otherwise-toxic gases such as methane or hydrogen sulfide—and they have even been recovered living beneath 435 m of ice in Antarctica!

These harsh environments may be similar to the conditions present on the early Earth when life first began. It is likely that prokaryotes evolved to dwell in these harsh conditions early on and have retained the ability to exploit these areas as the rest of the atmosphere has changed.

Prokaryotes are fundamentally different from eukaryotes

Prokaryotes differ from eukaryotes in numerous important features. These differences represent some of the most fundamental distinctions that separate any groups of organisms.

Unicellularity. With a few exceptions prokaryotes are fundamentally single-celled (figure 28.3). In some types, individual cells adhere to one another within a matrix and form filaments; however, the cells retain their individuality. Cyanobacteria, in particular, are likely to form such associations, but their cytoplasm is not directly interconnected, as is often the case in multicellular eukaryotes.

Recent evidence suggests that in their natural environments, most bacteria are capable of forming a complex community of different species called a **biofilm.** Although not a multicellular organism, a biofilm is more resistant to antibiotics, dessication, and other environmental stressors than is a simple colony of a single type of microbe, such as a laboratory culture.

Cell size. As new species of prokaryotes are discovered, investigators are finding that the size of prokaryotic cells varies tremendously, by as much as five orders of magnitude. The largest bacterial cells currently characterized are from *Thiomargarita namibia*. A single cell from this species is up to 750 micrometers (μm) across, which is visible to the naked eye and is roughly the size of the eye of a bumblebee. Most prokaryotic cells, however, are only 1 μm or less in diameter, whereas most eukaryotic cells are well over 10 times that size. This generality is misleading, however, because there are very small eukaryotes as well as very large prokaryotes.

Chromosomes. Eukaryotic cells have a membrane-bounded nucleus containing linear chromosomes made up of both nucleic acids and histone proteins. Prokaryotes do not have membrane-bounded nuclei; instead they usually have a single circular chromosome made up of DNA and histonelike proteins in a **nucleoid** region of the cell. An exception to this single chromosome includes *Vibrio cholerae*, which has two circular chromosomes. Prokaryotic cells often have accessory DNA molecules called **plasmids** as well. Plasmids are genetic elements that can sometimes be transferred between prokaryotic cells.

Cell division and genetic recombination. Cell division in eukaryotes takes place by mitosis and involves spindles made up of microtubules. Cell division in prokaryotes takes place mainly by binary fission (chapter 10), which is also a form of asexual reproduction. True sexual reproduction occurs only in eukaryotes and involves the production of haploid gametes that fuse to form a diploid zygote that grows to adulthood, producing more gametes and starting the cycle over again (chapter 11).

Despite their asexual mode of reproduction, prokaryotes do have mechanisms that lead to the transfer of genetic material and generation of genetic diversity. These mechanisms are collectively called **horizontal gene transfer** and are not a form of reproduction.

Internal compartmentalization. In eukaryotes, the enzymes for cellular respiration are packaged in mitochondria. In prokaryotes, the corresponding enzymes are not packaged separately, but instead are bound to the cell membranes or are in the cytosol. The cytoplasm of prokaryotes, unlike that of eukaryotes, contains no

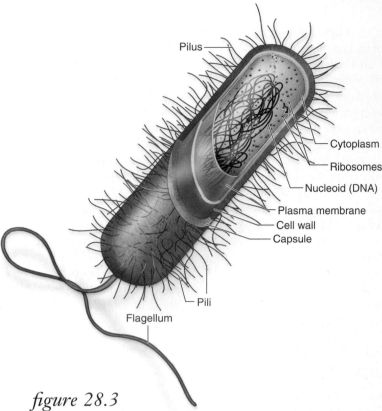

figure 28.3

THE STRUCTURE OF A PROKARYOTIC CELL. Prokaryotic cells, such as this bacterium, are small and lack membrane-bounded organelles. The plasma membrane is encased within a rigid cell wall, and DNA is not contained in a nuclear membrane. In addition to a flagellum, prokaryotes may also possess projections called pili that aid in attachment to surfaces or other cells. They may also possess a capsule, composed primarily of carbohydrate, that surrounds the cell. A capsule aids in adhesion as well as making the bacteria difficult for immune cells to recognize.

internal compartments and no membrane-bounded organelles. Ribosomes are found in both prokaryotes and eukaryotes, but differ significantly in structure. (See chapter 4 for a review of cell structure.)

Flagella. Prokaryotic flagella are simple in structure, composed of a single fiber of the protein flagellin. Eukaryotic flagella and cilia are complex, having a 9 + 2 structure of microtubules (see figure 4.23). Bacterial flagella also function differently, being rigid and spinning like propellers, whereas eukaryotic flagella have a whiplike motion (described in more detail later and in figure 28.9).

Metabolic diversity. Only one kind of photosynthesis occurs in eukaryotes, and it involves the release of oxygen. Photosynthetic bacteria have two basic patterns of photosynthesis: **oxygenic,** producing oxygen, and **anoxygenic,** nonoxygen producing. Anoxygenic photosynthesis involves the formation of products such as sulfur and sulfate instead of oxygen.

Prokaryotic cells can also be **chemolithotrophic,** meaning that they use the energy stored in chemical bonds of inorganic molecules to synthesize carbohydrates; eukaryotes are not capable of this metabolic process.

Despite similarities, Bacteria and Archaea differ fundamentally

Archaea and Bacteria are similar in that both have a prokaryotic cellular structure, but they vary considerably at the biochemical and molecular levels. They differ in four key areas: plasma membranes, cell walls, DNA replication, and gene expression.

Plasma membranes. All prokaryotes have plasma membranes with a fluid mosaic architecture (chapter 5). The plasma membranes of archaea differ from both bacteria and eukaryotes. Archaean membrane lipids are composed of glycerol linked to hydrocarbon chains by ether linkages, not the ester linkages seen in bacteria and eukaryotes (figure 28.4a). These hydrocarbons may also be branched, and they may be organized as tetraethers that form a monolayer instead of a bilayer (figure 28.4b).

   In the case of some hyperthermophiles, the majority of the membrane may be this tetraether monolayer. This structural feature is part of what allows these archaeans to withstand high temperatures.

Cell wall. Both kinds of prokaryotes typically have cell walls covering the plasma membrane that strengthen the cell. The cell walls of bacteria are constructed, minimally, of **peptidoglycan,** which is formed from carbohydrate polymers linked together by peptide cross-bridges. The peptide cross-bridges also contain D-amino acids, which are never found in cellular protein. The cell walls of archaea lack peptidoglycan, although some have **pseudomurein,** which is similar to peptidoglycan in structure and function. This wall layer is also a carbohydrate polymer with peptide cross-bridges, but the carbohydrates are different and the peptide cross-bridge structure also differs. Other archaeal cell walls have been found to be composed of a variety of proteins and carbohydrates, making generalizations difficult.

DNA replication. Although both archaea and bacteria have a single replication origin, the nature of this origin and the proteins that act there are quite different. Archaeal initiation of DNA replication is more similar to that of eukaryotes. (See chapter 14.)

Gene expression. The machinery used for gene expression is also different. The archaea may have more than one RNA polymerase, and these enzymes more closely resemble the eukaryotic RNA polymerases than they do the single bacterial RNA polymerase. Some of the translation machinery is also more similar to that of eukaryotes. (See chapter 16.)

Most prokaryotes have not been characterized

Prokaryotes are not easily classified according to their forms, and only recently has enough been learned about their biochemical and metabolic characteristics to develop a satisfactory overall classification scheme comparable to that used for other organisms.

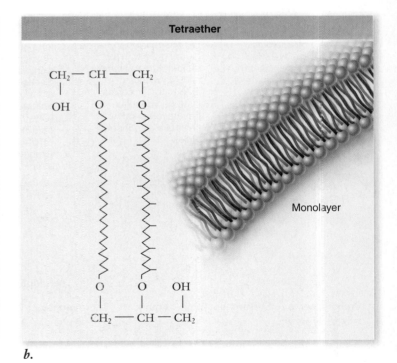

a.   *b.*

figure 28.4

ARCHAEA MEMBRANE LIPIDS. *a.* Archaea membrane lipids are formed on a glycerol skeleton similar to bacterial and eukaryotic lipids, but the hydrocarbon chains are connected to the glycerol by ether linkages not ester linkages. The hydrocarbons can also be branched and even contain rings. *b.* These lipids can also form as tetraethers instead of diethers. The tetraether forms a monolayer as it includes two polar regions connected by hydrophobic hydrocarbons.

Early classification characteristics

Early systems for classifying prokaryotes relied on differential stains such as the Gram stain and differences in the observable phenotype of the organism. Key characteristics once used in classifying prokaryotes were as follows:

1. photosynthetic or nonphotosynthetic;
2. motile or nonmotile;
3. unicellular or colony-forming or filamentous;
4. formation of spores or division by transverse binary fission; and
5. importance as human pathogens or not.

Molecular approaches to classification

With the development of genetic and molecular approaches, prokaryotic classifications can at last reflect true evolutionary relatedness. Molecular approaches include:

1. the analysis of the amino acid sequences of key proteins;
2. the analysis of nucleic acid–base sequences by establishing the percent of guanine (G) and cytosine (C);
3. nucleic acid hybridization, which is essentially the mixing of single-stranded DNA from two species and determining the amount of base-pairing (closely related species will have more bases pairing);
4. gene and RNA sequencing, especially looking at ribosomal RNA; and
5. whole-genome sequencing.

The three-domain, or Woese, system of phylogeny (figure 28.5) relies on all of these molecular methods, but emphasizes the comparison of rRNA sequences to establish the evolutionary relatedness of all organisms.

Based on these sorts of molecular data, several groupings of prokaryotes have been proposed. The most widely accepted is that presented in *Bergey's Manual of Systematic*

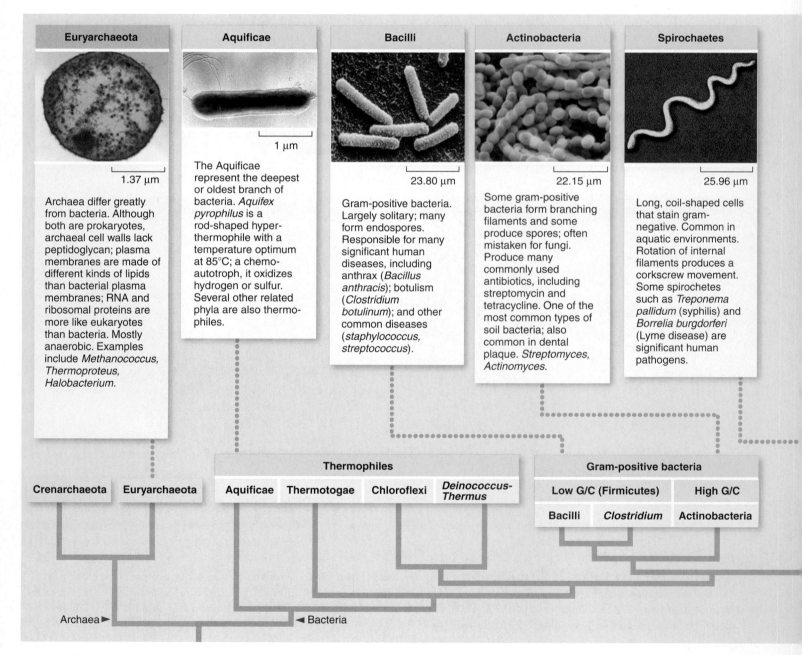

Euryarchaeota	Aquificae	Bacilli	Actinobacteria	Spirochaetes
1.37 μm	1 μm	23.80 μm	22.15 μm	25.96 μm
Archaea differ greatly from bacteria. Although both are prokaryotes, archaeal cell walls lack peptidoglycan; plasma membranes are made of different kinds of lipids than bacterial plasma membranes; RNA and ribosomal proteins are more like eukaryotes than bacteria. Mostly anaerobic. Examples include *Methanococcus*, *Thermoproteus*, *Halobacterium*.	The Aquificae represent the deepest or oldest branch of bacteria. *Aquifex pyrophilus* is a rod-shaped hyper-thermophile with a temperature optimum at 85°C; a chemo-autotroph, it oxidizes hydrogen or sulfur. Several other related phyla are also thermo-philes.	Gram-positive bacteria. Largely solitary; many form endospores. Responsible for many significant human diseases, including anthrax (*Bacillus anthracis*); botulism (*Clostridium botulinum*); and other common diseases (*staphylococcus*, *streptococcus*).	Some gram-positive bacteria form branching filaments and some produce spores; often mistaken for fungi. Produce many commonly used antibiotics, including streptomycin and tetracycline. One of the most common types of soil bacteria; also common in dental plaque. *Streptomyces*, *Actinomyces*.	Long, coil-shaped cells that stain gram-negative. Common in aquatic environments. Rotation of internal filaments produces a corkscrew movement. Some spirochetes such as *Treponema pallidum* (syphilis) and *Borrelia burgdorferi* (Lyme disease) are significant human pathogens.

Crenarchaeota	Euryarchaeota		Thermophiles				Gram-positive bacteria		
		Aquificae	Thermotogae	Chloroflexi	*Deinococcus-Thermus*		Low G/C (Firmicutes)		High G/C
							Bacilli	*Clostridium*	Actinobacteria

Archaea ▶   ◀ Bacteria

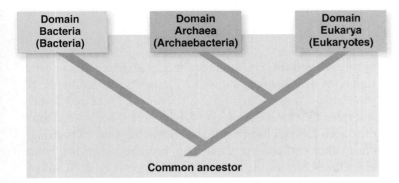

figure 28.5

THE THREE DOMAINS OF LIFE. The two prokaryotic domains, Archaea and Bacteria, are not closely related, though both are prokaryotes. In many ways (see text), archaea more closely resemble eukaryotes than bacteria. This tree is based on rRNA sequences.

Bacteriology, second edition, vol. 1, 2001 (figure 28.6). As of 2005, the combined number of bacterial and archaeal species was approximately 7000. Because such a large percentage of bacteria cannot be cultured, however, the number of actual species may be much higher, in the range of 100,000 different species.

Prokaryotes are distinctly different from eukaryotes, lacking a membrane-bounded nucleus and diverse organelles. They also reproduce by a fundamentally different mechanism, binary fission. Bacteria and archaea are also clearly different from each other based on both structure and metabolism. Classification of prokaryotes has been aided by the use of DNA analysis, but a vast number of prokaryotes remain unidentified because they cannot be cultured.

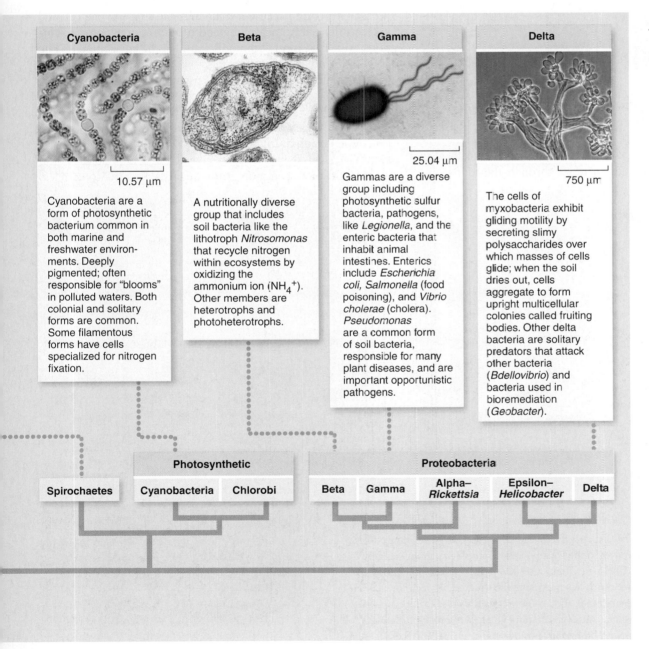

figure 28.6

SOME MAJOR CLADES OF PROKARYOTES. The classification adopted here is that of *Bergey's Manual of Systematic Bacteriology,* second edition, 2001.

Cyanobacteria

10.57 μm

Cyanobacteria are a form of photosynthetic bacterium common in both marine and freshwater environments. Deeply pigmented; often responsible for "blooms" in polluted waters. Both colonial and solitary forms are common. Some filamentous forms have cells specialized for nitrogen fixation.

Beta

A nutritionally diverse group that includes soil bacteria like the lithotroph *Nitrosomonas* that recycle nitrogen within ecosystems by oxidizing the ammonium ion (NH_4^+). Other members are heterotrophs and photoheterotrophs.

Gamma

25.04 μm

Gammas are a diverse group including photosynthetic sulfur bacteria, pathogens, like *Legionella*, and the enteric bacteria that inhabit animal intestines. Enterics include *Escherichia coli*, *Salmonella* (food poisoning), and *Vibrio cholerae* (cholera). *Pseudomonas* are a common form of soil bacteria, responsible for many plant diseases, and are important opportunistic pathogens.

Delta

750 μm

The cells of myxobacteria exhibit gliding motility by secreting slimy polysaccharides over which masses of cells glide; when the soil dries out, cells aggregate to form upright multicellular colonies called fruiting bodies. Other delta bacteria are solitary predators that attack other bacteria (*Bdellovibrio*) and bacteria used in bioremediation (*Geobacter*).

28.3 Prokaryotic Cell Structure

Prokaryotic cells are relatively simple, but they can be categorized based on cell shape. They also have some variation in structure that give them different staining properties for certain dyes. Other features are found in some types of cells but not in others.

Prokaryotes have three basic forms: Rods, cocci, and spiral

Most prokaryotes exhibit one of three basic shapes: **rod**-shaped, often called a **bacillus** (plural, *bacilli*); **coccus** (plural, *cocci*), spherical- or ovoid-shaped; and **spirillum** (plural, *spirilla*), long and helical-shaped; these bacteria are also called *spirochetes*.

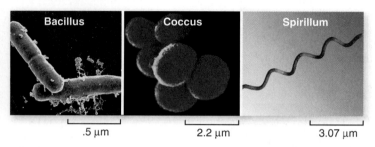

Bacillus	Coccus	Spirillum
.5 µm	2.2 µm	3.07 µm

The bacterial cell wall is the single most important contributor to cell shape. Bacteria that normally lack cell walls, such as the mycoplasmas, do not have a set shape.

As diverse as their shapes may be, prokaryotic cells also have many different methods to move through their environ-ment. A **flagellum** or several flagella may be found on the outer surface of many prokaryotic cells. These structures are used to propel the organisms in a fluid environment. Some rod-shaped and spherical bacteria form colonies, adhering end-to-end after they have divided, forming chains. Some bacterial cells change into stalked structures or grow long, branched filaments. Some filamentous bacteria are capable of a gliding motion on solid surfaces, often combined with rotation around a longitudinal axis.

Prokaryotes have a tough cell wall and other external structures

The prokaryotic **cell wall** is often complex, consisting of many layers. Minimally it consists of peptidoglycan, a polymer unique to bacteria. This polymer forms a rigid network of polysaccharide strands cross-linked by peptide side chains. It is an important structure because it maintains the shape of the cell and protects the cell from swelling and rupturing in hypotonic solutions, which are most commonly found in the environment. The archaea do not possess peptidoglycan, but some have a similar structure called pseudomurein, or pseudopeptidoglycan.

Gram-positive and gram-negative bacteria

Two types of bacteria can be identified using a staining process called the **Gram stain,** hence their names. **Gram-positive** bacteria have a thicker peptidoglycan wall and stain a purple color, whereas the more common **gram-negative** bacteria contain less

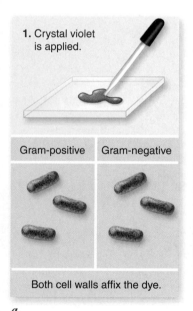
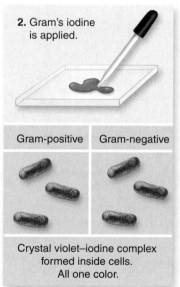
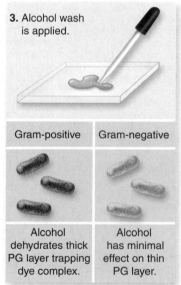
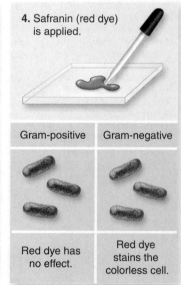

a.

figure 28.7

THE GRAM STAIN. *a.* The thick peptidoglycan (PG) layer encasing gram-positive bacteria traps crystal violet dye, so the bacteria appear purple in a gram-stained smear (named after Hans Christian Gram, who developed the technique). Because gram-negative bacteria have much less peptidoglycan (located between the plasma membrane and an outer membrane), they do not retain the crystal violet dye and so exhibit the red counterstain (usually a safranin dye). *b.* A micrograph showing the results of a Gram stain with both gram-positive and gram-negative cells.

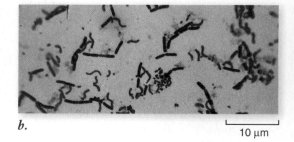

b.

10 µm

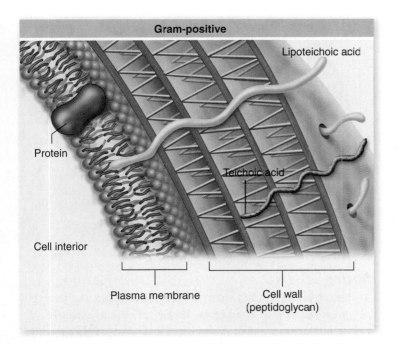

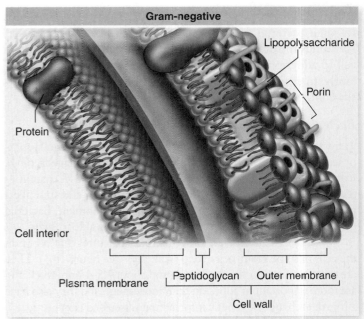

figure 28.8

THE STRUCTURE OF GRAM-POSITIVE AND GRAM-NEGATIVE CELL WALLS. The gram-positive cell wall is much simpler, composed of a thick layer of cross-linked peptidoglycan chains. Molecules of lipoteichoic acid and teichoic acid are also embedded in the wall and exposed on the surface of the cell. The gram-negative cell wall is composed of multiple layers. The peptidoglycan layer is thinner than in gram-positive bacteria and is surrounded by an additional membrane composed of lipopolysaccharide. Porin proteins form aqueous pores in the outer membrane. The space between the outer membrane and peptidoglycan is called the periplasmic space.

peptidoglycan and do not retain the purple-colored dye. These gram-negative bacteria can be stained with a red counterstain and then appear dark pink (figure 28.7).

In the gram-positive bacteria, the peptidoglycan forms a thick, complex network around the outer surface of the cell. This network also contains lipoteichoic and teichoic acid, which protrudes from the cell wall. In the gram-negative bacteria, a thin layer of peptidoglycan is sandwiched between the plasma membranes and a second outer membrane (figure 28.8). The outer membrane contains large molecules of **lipopolysaccharide,** lipids with polysaccharide chains attached. The outer membrane layer makes gram-negative bacteria resistant to many antibiotics that interfere with cell-wall synthesis in gram-positive bacteria. For example, penicillin acts to inhibit the cross-linking of peptidoglycan in a gram-positive cell wall, killing growing bacterial populations as they attempt to reproduce.

S-layer

In some bacteria and archaea, an additional protein or glycoprotein layer forms a rigid paracrystalline surface called an **S-layer** outside of the peptidoglycan or outer membrane layers of gram-positive and gram-negative bacteria, respectively. Among the archaea, the S-layer is almost universal and can be found outside of a pseudopeptidoglycan layer or, in contrast to the bacteria, may be the only rigid layer surrounding the cell. The functions of S-layers are diverse and variable but often involve adhesion to surfaces or protection.

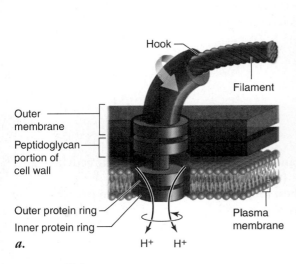

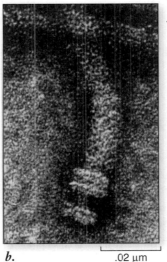

figure 28.9

THE FLAGELLAR MOTOR OF A GRAM-NEGATIVE BACTERIUM. *a.* A protein filament, composed of the protein flagellin, is attached to a protein rod that passes through a sleeve in the outer membrane and through a hole in the peptidoglycan layer to rings of protein anchored in the cell wall and plasma membrane, like rings of ballbearings. The rod rotates when the inner protein ring attached to the rod turns with respect to the outer ring fixed to the cell wall. The inner ring is an H^+ ion channel, a proton pump that uses the flow of protons into the cell to power the movement of the inner ring past the outer one. The membrane wall anchor of the flagellum is called the basal body. *b.* Electron micrograph of bacterial flagellum.

The capsule

In some bacteria, an additional gelatinous layer, the **capsule**, surrounds the other wall layers. A capsule enables a prokaryotic cell to adhere to surfaces and to other cells, and, most importantly, to evade an immune response. Therefore, a capsule often contributes to the ability of bacteria to cause disease.

Bacterial flagella and pili

Many kinds of prokaryotes have slender, rigid, helical flagella composed of the protein **flagellin** (figure 28.9). These flagella range from 3 to 12 μm in length and are very thin—only 10–20 nanometers (nm) thick. They are anchored in the cell wall and spin like a propeller, moving the cell through a liquid environment. Bacterial cells that have lost the genes for flagellin are not able to swim.

Pili (singular, *pilus*) are other hairlike structures that occur on the cells of some gram-negative prokaryotes (see figure 28.3). They are shorter than prokaryotic flagella and about 7.5–10 nm thick. Pili are more important in adhesion than movement, and they also have a role in exchange of genetic information (discussed later).

Endospore formation

Some prokaryotes are able to form **endospores**, developing a thick wall around their genome and a small portion of the cytoplasm when they are exposed to environmental stress. These endospores are highly resistant to environmental stress, especially heat, and when environmental conditions improve, they can germinate and return to normal cell division to form new individuals after decades or even centuries.

The bacteria that cause tetanus, botulism, and anthrax are all capable of forming spores. With a puncture wound, tetanus endospores may be driven deep into the skin where conditions are favorable for the endospores to germinate and cause disease, or even death.

The interior of prokaryotic cells is organized

The most fundamental characteristic of prokaryotic cells is their simple interior organization. Prokaryotic cells lack the extensive functional compartmentalization seen within eukaryotic cells, but they do have the following structures:

Internal membranes. Many prokaryotes possess invaginated regions of the plasma membrane that function in respiration or photosynthesis (figure 28.10).

Nucleoid region. Prokaryotes lack nuclei and generally do not possess linear chromosomes. Instead, their genes are

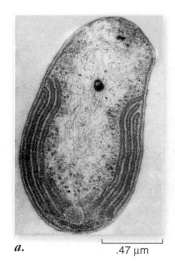 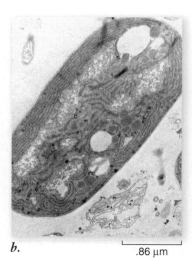

a.   .47 μm   *b.*   .86 μm

figure 28.10

PROKARYOTIC CELLS OFTEN HAVE COMPLEX INTERNAL MEMBRANES. *a.* This aerobic bacterium exhibits extensive respiratory membranes within its cytoplasm not unlike those seen in mitochondria. *b.* This cyanobacterium has thylakoid-like membranes that provide a site for photosynthesis.

encoded within a single double-stranded ring of DNA that is highly condensed to form a visible region of the cell known as the **nucleoid region.** Many prokaryotic cells also possess plasmids, which as described earlier are small, independently replicating circles of DNA. Plasmids contain only a few genes, and although these genes may confer a selective advantage, they are not essential for the cell's survival.

Ribosomes. Prokaryotic ribosomes are smaller than those of eukaryotes and differ in protein and RNA content. Antibiotics such as tetracycline and chloramphenicol can tell the difference, however—they bind to prokaryotic ribosomes and block protein synthesis, but they do not bind to eukaryotic ribosomes.

> The three basic shapes of prokaryotes are rod-shaped, spherical cocci, and spiral-shaped spirilla. Although prokaryotes do not have membrane-bounded organelles, the interior of the cell is organized and may include extensive infolding of the plasma membrane. Prokaryotic DNA is also organized into a nucleoid region despite the lack of a surrounding membrane.

28.4 Prokaryotic Genetics

In sexually reproducing populations, traits are transferred vertically from parent to child. Prokaryotes do not reproduce sexually, but they can exchange DNA between different cells. This horizontal gene transfer occurs when genes move from one cell to another by **conjugation**, requiring cell-to-cell contact, or by means of viruses **(transduction)**. Some species of bacteria can also pick up genetic material directly from the environment **(transformation)**.

All of these processes have been observed in archaea, but the study of archaeal genetics is still in its infancy because of the difficulty in culturing most species. We concentrate here on bacterial systems, and primarily *E. coli*, which has been studied extensively.

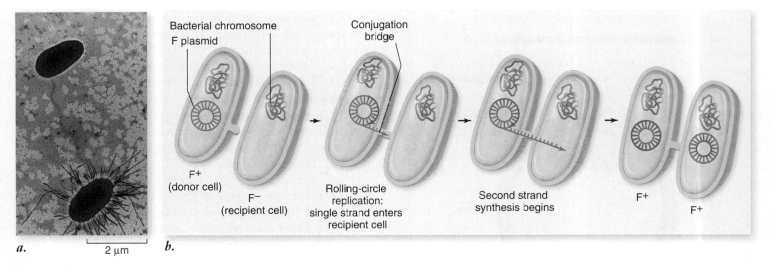

a. 　2 µm 　*b.*

Bacterial chromosome
F plasmid
Conjugation bridge

F+ (donor cell)
F− (recipient cell)
Rolling-circle replication: single strand enters recipient cell
Second strand synthesis begins
F+
F+

figure 28.11

CONJUGATION BRIDGE AND TRANSFER OF F PLASMID BETWEEN F⁺ AND F⁻ CELL. *a.* The electron micrograph shows two *E. coli* cells caught in the act of conjugation. The connection between the cells is the extended F pilus. *b.* F⁻ cells are converted to F⁺ cells by the transfer of the F plasmid. The cells are joined by a conjugation bridge and the plasmid is replicated in the donor cell, displacing one parental strand. The displaced strand is transferred to the recipient cell then replicated. After successful transfer, the recipient cell becomes an F⁺ cell able to express genes for the F pilus and act as a donor.

Conjugation depends on the presence of a conjugative plasmid

Plasmids may encode functions that can confer an advantage to the cell, such as antibiotic resistance, on which natural selection can operate—but they are not required for normal function. In some cases, plasmids can be transferred from one cell to another via conjugation. The best known plasmid capable of transfer is called the **F plasmid,** for fertility factor; cells containing F plasmids are termed F⁺ cells, and cells that lack the F plasmid are F⁻ cells. The F plasmid occurs in *E. coli* and, like all plasmids, acts as an independent genetic entity that nevertheless depends on the cell for replication. Studies involving the F plasmid were critical to our current understanding of bacterial genetics and the organization of the *E. coli* chromosome.

F plasmid transfer

The F plasmid contains a DNA replication origin and several genes that promote its transfer to other cells. These genes encode protein subunits that assemble on the surface of the bacterial cell, forming a hollow pilus that is necessary for the transfer process (figure 28.11*a*).

First, the F plasmid binds to a site on the interior of the F⁺ cell just beneath the pilus, now called a *conjugation bridge.* Then, by a process called *rolling-circle replication,* the F plasmid begins to copy its DNA at the binding point. As it is replicated, the displaced single strand of the plasmid passes into the other cell. There, a complementary strand is added, creating a new, stable F plasmid (figure 28.11*b*).

Recombination between the F plasmid and host chromosome

The F plasmid can integrate into the host chromosome by recombining with it (chapter 13). The molecular events in this process are similar to events during meiosis in eukaryotes when crossing over (recombination) exchanges material between chromosomes. This

process is also called homologous recombination. In the case of the F plasmid and the *E. coli* chromosome, a single recombination event between two circles produces a larger circle, consisting of the chromosome and the integrated plasmid. This integration is actually mediated by host-encoded proteins, but it takes advantage of regions in

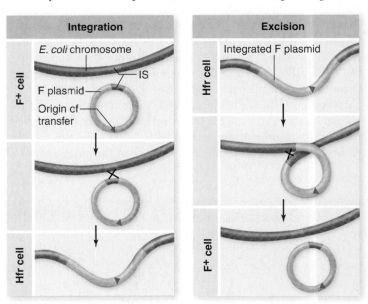

Integration

E. coli chromosome
IS
F plasmid
Origin of transfer

F⁺ cell

Hfr cell

Excision

Integrated F plasmid

Hfr cell

F⁺ cell

figure 28.12

INTEGRATION AND EXCISION OF F PLASMID. The F plasmid contains short IS sequences that also exist in the chromosome. This allows the plasmid to pair with the chromosome, and a single recombination event between two circles leads to a larger circle. This integrates the plasmid into the chromosome, creating an Hfr cell, as shown on the left. The process is reversible because the IS sequences in the integrated plasmid can pair and now a recombination event will return the two circles and convert the Hfr back to an F⁺ cell as shown on the right.

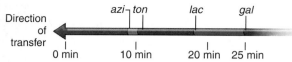

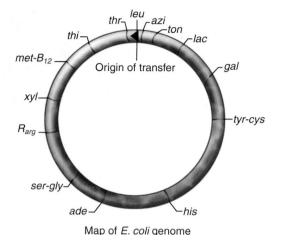

a. Time elapsed from beginning of conjugation until interruption

b. Map of *E. coli* genome

figure 28.13

GENETIC MAP OF *E. coli*. The genetic map of *E. coli* was originally developed by interrupted mating experiments in which an Hfr cell transfers chromosomal genes to the recipient cell. *a.* A region of the chromosome is shown beginning with the origin of transfer and showing the time in minutes to transfer a series of markers. *b.* A simplified version of the *E. coli* genetic map.

the F plasmid called insertion sequences (IS) that also exist in the *E. coli* chromosome. These IS elements are actually transposable elements that probably moved from the chromosome to the F plasmid.

When the F plasmid is integrated into the chromosome, the cell is called an **Hfr cell** for high frequency of recombination (figure 28.12), because now transfer by the F plasmid will include chromosomal DNA. The site on the F plasmid where transfer initiates is located in the middle of the integrated plasmid, so that the entire chromosome would have to be transferred to also transfer all of the integrated plasmid. The transfer of the entire chromosome takes around 100 min, and the conjugation bridge is usually broken before that time. This leads to transfer of portions of donor chromosome that can then replace regions of the recipient chromosome by homologous recombination. This occurs by *two* recombination events between the linear piece and the circular chromosome, similar to a double crossover in eukaryotic meiosis.

Geneticists have taken advantage of this process to map the order of genes in the *E. coli* chromosome. Genes close to the origin of transfer are transferred early in the process, and those far from the origin are transferred later. If the process of mating is experimentally interrupted at different times, then gene order can be mapped based on time of entry of each gene (figure 28.13). The entry of genes can be detected by using a donor with wild-type alleles that can replace mutant alleles in the recipient by homologous recombination as described. These experiments have shown that the *E. coli* chromosome is indeed circular, and the genetic map is therefore circular. The units of the map are minutes, and the entire map is 100 min long.

The F plasmid can also excise itself by reversing the integration process. In this case, the IS elements bounding the integrated plasmid pair and now a single recombination event will restore the two circles (see figure 28.12). If excision is inaccurate, the F plasmid can pick up some chromosomal DNA in the process. This creates what is called an **F′ plasmid** that can then be transferred rapidly and in its entirety to another cell. In this case, the cell will already have the same genetic material in its chromosome as that carried by the F′. This makes the cell a **partial diploid**, sometimes called a **merodiploid**. Merodiploids can be used to determine if new isolated mutations are alleles of known genes. This is done by using wild types of alleles of known genes of the F′ plasmid to provide normal function heterozygous to unknown mutant alleles in the chromosome.

Viruses transfer DNA by transduction

Horizontal transfer of DNA can also be mediated by bacteriophage. In **generalized transduction**, virtually any gene can be transferred between cells; in **specialized transduction**, only a few genes are transferred.

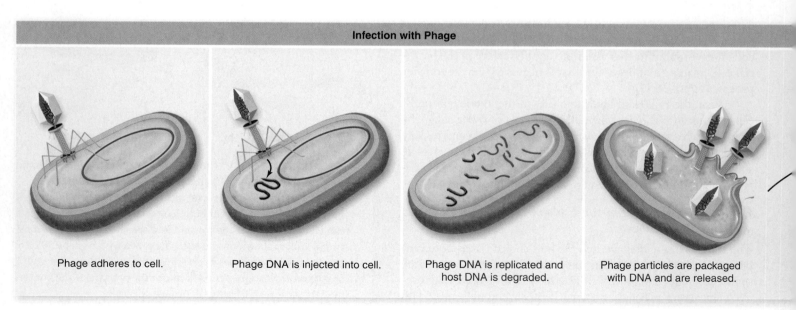

Infection with Phage

Phage adheres to cell.

Phage DNA is injected into cell.

Phage DNA is replicated and host DNA is degraded.

Phage particles are packaged with DNA and are released.

Generalized transduction

Generalized transduction can be thought of as an accident of the biology of some types of lytic phage (chapter 27). In these viruses, after the viral genome is replicated and the phage head is constructed, the phage packaging machinery stuffs DNA into the phage head until no more fits, so-called headfull packaging. Sometimes the phage begins with bacterial DNA instead of phage DNA and packages this DNA into a phage head (figure 28.14). When this viral particle goes on to infect another cell, it injects the bacterial DNA into the infected cell instead of viral DNA. This DNA can then be incorporated into the recipient chromosome by homologous recombination. Similar to transfer by Hfr cells described earlier two recombination events are necessary to integrate the linear piece of DNA into the circular chromosome (see figure 28.14).

Generalized transduction has also been used for mapping purposes in *E. coli*, although the logic is different from that in conjugation. In transduction, the closer together two genes are, the more likely it is that they will be transferred in a single transduction event. This can be expressed mathematically as the *cotransduction frequency*. Correlation of maps from the two methods allows an empirical conversion between cotransduction frequency and minutes in the genetic map.

Specialized transduction

Specialized transduction is limited to phage that exhibit a lysogenic life cycle (chapter 27). The prototype for this is phage lambda (λ) from *E. coli*. When λ infects a cell and its genome integrates into the host chromosome, it does not destroy the cell but is passed on by cell division. This integration event is similar to the integration of the F plasmid, except that in the case of λ the recombination is a site-specific event mediated by phage-encoded proteins.

In this lysogenic state, the phage is called a prophage and it is dormant. The prophage encodes the functions necessary to eventually excise itself and undergo lytic growth, leading to the death of the cell. If this excision event is imprecise, it may take some chromosomal DNA with it, in the process making a specialized transducing phage. These phage carry both phage genes and chromosomal genes, unlike generalized transducing phage that carry only chromosomal DNA.

Because the phage head can carry only as much DNA as is found in the phage genome, imprecise excision results in deletion of phage genes. Thus specialized transducing phage may be defective if genes necessary for phage growth are lost in the process.

Specialized transducing phage particles can then integrate into the chromosome, just like wild-type phage, also making the cell diploid for the genes carried by the phage. Phage particles that can integrate as prophages may become trapped in the host genome if the genes necessary for excision become defective by mutation or are lost. The *E. coli* genome contains a number of such cryptic prophage, some of which encode functions important to the cell and must now be considered part of the host genome.

Transformation is the uptake of DNA directly from the environment

Transformation is a naturally occurring process in some species, such as the bacteria that were studied by Frederick Griffith (see chapter 14). Griffith discovered the process despite not knowing what chemical component was transferred. Transformation occurs when one bacterial cell has died and ruptured, spilling its fragmented DNA into the surrounding environment. This

figure 28.14

TRANSDUCTION BY GENERALIZED TRANSDUCING PHAGE. When some phage infect cells, they degrade the host DNA into pieces. When the phage package their DNA, they can package host DNA in place of phage DNA to produce a transducing phage as shown on the left. When a transducing phage infects a cell, it injects host DNA that can then be integrated into the host genome by homologous recombination. With a linear piece of DNA, it requires two recombination events, which will replace the chromosomal DNA with the transducing DNA as shown on the right. If the new allele is different from the old, the cell's phenotype will change.

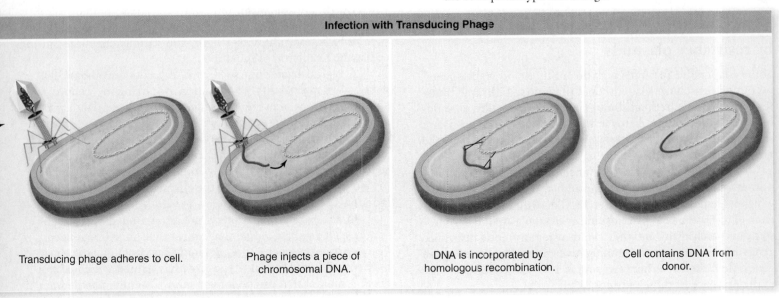

Infection with Transducing Phage

Transducing phage adheres to cell.

Phage injects a piece of chromosomal DNA.

DNA is incorporated by homologous recombination.

Cell contains DNA from donor.

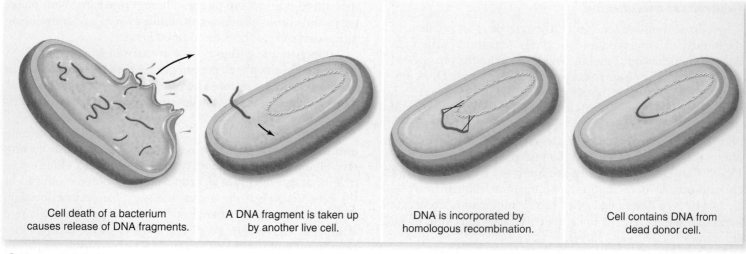

| Cell death of a bacterium causes release of DNA fragments. | A DNA fragment is taken up by another live cell. | DNA is incorporated by homologous recombination. | Cell contains DNA from dead donor cell. |

figure 28.15

NATURAL TRANSFORMATION. Natural transformation occurs when one cell dies and releases its contents to the surrounding environment. The DNA is usually fragmented, and small pieces can be taken up by other, living cells. The DNA taken up can replace chromosomal DNA by homologous recombination as in conjugation and transduction. If the new DNA contains different alleles from the chromosome, the phenotype of the cell changes, possibly providing a selective advantage.

DNA can be taken up by another cell and incorporated into its genome, thereby transforming it (figure 28.15). When the uptake occurs under natural conditions, it is termed natural transformation. Some species of both gram-positive and gram-negative bacteria exhibit natural transformation, although the mechanisms seem to differ between the groups.

The proteins involved in the process of natural transformation are all encoded by the bacterial chromosome. The implication is that natural transformation may be the only one of the mechanisms of DNA exchange that evolved as part of normal cellular machinery. The transfer of chromosomal DNA by either conjugation or transduction can be thought of as accidents of plasmid or phage biology, respectively.

Transformation is also important in molecular cloning, but *E. coli* does not exhibit natural transformation. When transformation is accomplished in the laboratory it is called artificial transformation. Artificial transformation is useful for cloning and DNA manipulation (see chapter 17).

Antibiotic resistance can be transferred by resistance plasmids

Some conjugative plasmids pick up antibiotic resistance genes, becoming resistance plasmids, or **R plasmids.** The rapid transfer of newly acquired, antibiotic resistance genes by plasmids has been an important factor in the appearance of the resistant strains of the pathogen *Staphylococcus aureus* discussed in the next section.

The means by which resistance plasmids acquire antibiotic resistance genes is often through transposable elements, which were described in chapter 18. These elements can move from chromosome to chromosome or from plasmid to chromosome and back again, and they can transfer antibiotic resistance genes in the process. If a conjugative plasmid picks up these genes, then the bacterium carrying it has a selective advantage in the presence of those antibiotics.

An important example in terms of human health involves the Enterobacteriaceae, the family of bacteria to which the common intestinal bacterium *E. coli* belongs. This family contains many pathogenic bacteria, including the organisms that cause dysentery, typhoid, and other major diseases. At times, some of the genetic material from these pathogenic species is exchanged with or transferred to *E. coli* by transmissible plasmids or bacteriophage. Because of its abundance in the human digestive tract, *E. coli* poses a special threat if it acquires harmful traits, as seen by the outbreaks of the food-borne O157:H7 strain of *E. coli*. Infection with this strain of *E. coli* can lead to serious illness. This is a new strain of *E. coli* that evolved by acquiring genes for pathogenic traits. Evidence suggests this occurred by both transduction and the acquisition of a large virulence plasmid by conjugation.

Variation can also arise by mutation

Just as with any organism, mutations can arise spontaneously in bacteria. Certain factors, especially those that damage DNA, such as radiation, ultraviolet light, and various chemicals, increase the likelihood of mutation.

A typical bacterium such as *E. coli* contains about 5000 genes. The probability of mutation occurring by chance is about in one out of every million copies of a gene. With 5000 genes in a bacterium, we can predict that approximately 1 out of every 200 bacteria will have a mutation. With adequate food and nutrients, a population of *E. coli* can double in 20 min. Because bacteria multiply so rapidly, mutations can spread rapidly in a population and can change the characteristics of that population in a relatively short time.

In the laboratory, bacteria are grown on different substrates, called *growth media*, that reflect their nutritional needs. For a particular species, the medium that contains only those nutrients required for wild-type growth is termed a *minimal medium*. A mutant that can no longer survive on minimal medium and needs particular nutritional supplements, such as an amino acid, is called an **auxotroph. Replica plating** allows identification of bacterial auxotrophs from a master plate of rich growth media by isolating individual colonies and observing their growth (or failure to grow) on different supplemented media. The technique is somewhat like using a rubber stamp—an impression of colonies growing in a petri plate is made on a velvet surface, and then this surface is pressed onto different media in other plates. The impression contains many thousands if not millions of cells from each colony—and each colony has grown from a single cell. In this way, a bacterium with a highly specific mutation can be isolated, identified, and grown.

The ability of prokaryotes to change rapidly in response to new challenges often has adverse effects on humans. Recently, a number of antibiotic-resistant strains of the bacterium *Staphylococcus aureus* (termed methicilin-resistant *staphylococcus aureus*, or MRSA) have appeared, some of them with alarming frequency. These bacteria are associated with serious infections and even deaths in hospitalized patients, and they are often resistant to more than one antibiotic. Of most concern among these strains is **vancomycin-resistant** *staphylococcus aureus* (VRSA). This appears to have arisen rapidly by mutation and is alarming because vancomycin is the drug of last resort, making these strains and the infections they cause very difficult to stop. *Staphylococcus* infections, or "staph" infections for short, provide an excellent example of the way in which mutation and intensive selection can bring about rapid change in bacterial populations.

Despite the lack of sexual reproduction, prokaryotes can still exchange DNA. This exchange is horizontal, from donor cell to recipient cell. DNA can be exchanged by conjugation via plasmids, by transduction via viruses, and by transformation through the direct uptake of DNA from the environment. Variation in prokaryotes also arises by mutation. Transfer of genes that confer resistance to antibiotics can have negative consequences for humans and other organisms.

28.5 Prokaryotic Metabolism

The variation seen in prokaryotes manifests itself in biochemical rather than morphological diversity. Wide variation has been found in the types of metabolism prokaryotes exhibit, especially in the means of acquiring energy and carbon.

Prokaryotes acquire carbon and energy in four basic ways

Prokaryotes have evolved many mechanisms to acquire the energy and carbon they need for growth and reproduction. Many are **autotrophs** that obtain their carbon from inorganic CO_2. Other prokaryotes are **heterotrophs** that obtain at least some of their carbon from organic molecules, such as glucose. Depending on the method by which they acquire energy, autotrophs and heterotrophs are categorized as follows.

Photoautotrophs. Many bacteria carry out photosynthesis, using the energy of sunlight to build organic molecules from carbon dioxide. The **cyanobacteria** use chlorophyll *a* as the key light-capturing pigment and H_2O as an electron donor, releasing oxygen gas as a by-product. They are therefore oxygenic, and their method of photosynthesis is very similar to that found in algae and plants. The chlorophyll *a* gives them a blue-green (cyan) color.

Other bacteria use bacteriochlorophyll as their light-capturing pigment and H_2S as an electron donor, leaving elemental sulfur as the by-product. These bacteria do not produce oxygen (anoxygenic) and have a simpler method of photosynthesis. These are the purple and green sulfur bacteria.

Chemolithoautotrophs. Some prokaryotes obtain energy by oxidizing inorganic substances. Nitrifiers, for example, oxidize ammonia or nitrite to obtain energy, producing the nitrate that is taken up by plants. This process is called **nitrification,** and it is essential in terrestrial ecosystems because plants can only absorb nitrogen in the form of nitrate.

Other chemolithoautotrophs oxidize sulfur, hydrogen gas, and other inorganic molecules. On the dark ocean floor at depths of 2500 m, entire ecosystems subsist on prokaryotes that oxidize hydrogen sulfide as it escapes from thermal vents.

Photoheterotrophs. The so-called purple and green nonsulfur bacteria use light as their source of energy but obtain carbon from organic molecules, such as carbohydrates or alcohols that have been produced by other organisms.

Chemoheterotrophs. The majority of prokaryotes obtain both carbon atoms and energy from organic molecules. These include decomposers and most pathogens. Human beings and all nonphotosynthetic eukaryotes are chemoheterotrophs as well.

Some bacteria can attack other cells directly

In the 1980s, researchers found that cells of a species of *Yersinia* produced and secreted large amounts of proteins. (*Yersinia pestis* is the bacterial species responsible for bubonic plague.) Most proteins secreted by gram-negative bacteria have special signal sequences that allow them to pass through the bacterium's double membrane. The proteins secreted by *Yersinia* lacked a key signal sequence that two known secretion mechanisms require for transport. The proteins must therefore have been secreted by means of a third type of system, which researchers called the *type III system*.

As more bacterial species are studied, the genes coding for the type III system are turning up in other gram-negative animal pathogens, and even in more distantly related plant pathogens. The genes seem more closely related to one another than are the bacteria. Furthermore, the genes are similar to those that code for bacterial flagella.

These proteins are used to transfer other virulence proteins, such as toxins, into nearby eukaryotic cells. Given the similarity of the type III genes to the genes that code for flagella, the transfer proteins may form a flagellum-like structure that shoots virulence proteins into the host cells. Once in the eukaryotic cells, the virulence proteins affect the host's response to the pathogen.

In *Yersinia*, proteins secreted by the type III system are injected into macrophages; the proteins disrupt signals that tell the macrophages to engulf bacteria. *Salmonella* and *Shigella* use their type III proteins to enter the cytoplasm of eukaryotic cells, and thus they are protected from the immune system of their host. The proteins secreted by certain strains of *E. coli* alter the cytoskeleton of nearby intestinal eukaryotic cells, resulting in a bulge onto which the bacterial cells can tightly bind.

Bacteria are costly plant pathogens

Although the majority of commercially relevant plant pathogens are fungi, many diseases of plants are associated with particular heterotrophic bacteria. Almost every kind of plant is susceptible to one or more kinds of bacterial disease, including blights, soft rots, and wilts. Fire blight, which destroys pear and apple trees and related plants, is a well-known example of bacterial disease.

The symptoms of these plant diseases vary, but they are commonly manifested as spots of various sizes on the stems, leaves, flowers, or fruits. Most bacteria that cause plant diseases are members of the group of rod-shaped gram-negative bacteria known as pseudomonads.

> Prokaryotes exhibit amazing metabolic diversity with both autotrophic and heterotrophic species. Photoautotrophs use light as an energy source; chemolithoautotrophs oxidize inorganic compounds. Photoheterotrophs use light as an energy source and organic compounds as carbon sources. Chemoheterotrophs use organic compounds for both. Some bacteria make a living as pathogens, infecting other organisms.

28.6 Human Bacterial Disease

In the early twentieth century, before the discovery and widespread use of antibiotics, infectious diseases killed nearly 20% of all U.S. children before they reached the age of five. Sanitation and antibiotics considerably improved the situation. In recent years, however, we have seen the appearance or reappearance of many bacterial diseases, including cholera, leprosy, tetanus, bacterial pneumonia, whooping cough, diphtheria, and Lyme disease (table 28.1). Members of the genus *Streptococcus* are associated with scarlet fever, rheumatic fever, pneumonia, "flesh-eating disease," and other infections. Tuberculosis (TB), another bacterial disease, is still a leading cause of death in humans worldwide.

Bacteria have many different methods to spread through a susceptible population. Tuberculosis and many other bacterial diseases of the respiratory tract are mostly spread through the air in droplets of mucus or saliva. Diseases such as typhoid fever, paratyphoid fever, and bacillary dysentery are spread by fecal contamination of food or water. Lyme disease and Rocky Mountain spotted fever are spread to humans by tick vectors.

Tuberculosis has infected humans for all of recorded history

Tuberculosis (TB) has been a scourge to humanity for thousands of years. There is evidence that peoples from ancient Egypt and pre-Columbian South America died from tuberculosis; the TB bacillus has been identified in prehistoric mummies. The bacteria causing tuberculosis afflict the respiratory system, thwart the immune system, and are easily transmitted from person to person through the air.

The spread of tuberculosis

Currently, about one-third of all people worldwide are regularly exposed to *Mycobacterium tuberculosis*, the tuberculosis bacterium (figure 28.16). An estimated 9 million new cases are diagnosed each year, and 1.7 million deaths occurred in 2004. In 2006, the World Health Organization reported the incidence of TB falling in five of six WHO regions, but the numbers continue to rise in Africa driven by the spread of HIV.

Since the mid-1980s, the United States has been experiencing a dramatic resurgence of tuberculosis. The causes of this resurgence include social factors such as poverty, crowding, homelessness, and incarceration (the same factors that have always promoted the spread of TB). Tuberculosis may be passed from one individual to another with astonishing ease—as few as 10 tubercule bacilli are required to start an infection in a healthy adult. The increasing prevalence of HIV

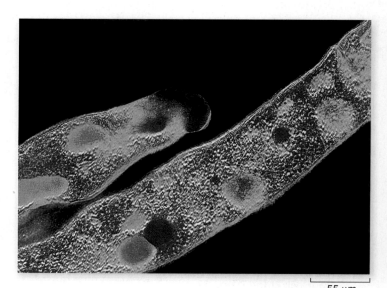

.55 μm

figure 28.16

MYCOBACTERIUM TUBERCULOSIS. This color-enhanced image shows the rod-shaped bacterium responsible for tuberculosis in humans.

TABLE 28.1   Important Human Bacterial Diseases

Disease	Pathogen	Vector/Reservoir	Epidemiology
Anthrax	*Bacillus anthracis*	Animals, including processed skins	Bacterial infection that can be transmitted through contact or ingestion. Rare except in sporadic outbreaks. May be fatal.
Botulism	*Clostridium botulinum*	Improperly prepared food	Contracted through ingestion or contact with wound. Produces acute toxic poison; can be fatal.
Chlamydia	*Chlamydia trachomatis*	Humans, STD	Urogenital infections with possible spread to eyes and respiratory tract. Increasingly common over past 20 years.
Cholera	*Vibrio cholerae*	Human feces, plankton	Causes severe diarrhea that can lead to death by dehydration; 50% peak mortality if untreated. A major killer in times of crowding and poor sanitation; over 100,000 died in Rwanda in 1994 outbreak.
Dental caries	*Streptococcus mutans, Streptococcus sobrinus*	Humans	A dense collection of these bacteria on the surface of teeth leads to secretion of acids that destroy minerals in tooth enamel; sugar alone will not cause caries.
Diphtheria	*Corynebacterium diphtheriae*	Humans	Acute inflammation and lesions of respiratory mucous membranes. Spread through respiratory droplets. Vaccine available.
Gonorrhea	*Neisseria gonorrhoeae*	Humans only	STD, on the increase worldwide. Usually not fatal.
Hansen disease (leprosy)	*Mycobacterium leprae*	Humans, feral armadillos	Chronic infection of the skin; worldwide incidence about 10–12 million, especially in southeast Asia. Spread through contact with infected individuals.
Lyme disease	*Borrelia burgdorferi*	Ticks, deer, small rodents	Spread through bite of infected tick. Lesion followed by malaise, fever, fatigue, pain, stiff neck, and headache.
Peptic ulcers	*Helicobacter pylori*	Humans	Originally thought to be caused by stress or diet, most peptic ulcers now appear to be caused by this bacterium; good news for ulcer sufferers because it can be treated with antibiotics.
Plague	*Yersinia pestis*	Fleas of wild rodents: rats and squirrels	Killed one-fourth of the population of Europe in the fourteenth century; endemic in wild rodent populations of the western United States today.
Pneumonia	*Streptococcus, Mycoplasma, Chlamydia, Haemophilus*	Humans	Acute infection of the lungs; often fatal without treatment. Vaccine for streptococcal pneumonia available.
Tuberculosis	*Mycobacterium tuberculosis*	Humans	An acute bacterial infection of the lungs, lymph, and meninges. Its incidence is on the rise, complicated by the development of new strains of the bacterium that are resistant to antibiotics.
Typhoid fever	*Salmonella typhi*	Humans	A systemic bacterial disease of worldwide incidence. Fewer than 500 cases a year are reported in the United States. Spread through contaminated water or foods (such as improperly washed fruits and vegetables). Vaccines are available for travelers.
Typhus	*Rickettsia typhi*	Lice, rat fleas, humans	Historically a major killer in times of crowding and poor sanitation; transmitted from human to human through the bite of infected lice and fleas. Peak untreated mortality rate of 70%.

infection is also a significant contributing factor. People with AIDS are much more likely to develop TB than people with healthy immune systems.

Tuberculosis treatment

Most TB patients are placed on multiple, expensive antibiotics for up to six months. Alarming outbreaks of **multidrug-resistant (MDR) strains** of tuberculosis have occurred, however, in the United States and worldwide. These MDR strains are resistant to most of the best available anti-TB medications. MDR TB is of particular concern because it requires much more time to treat, is more expensive to treat, and may prove fatal.

This spread of MDR tuberculosis is likely due to the extremely long course of antibiotics required to treat TB. Patients often quit taking the antibiotics before completing the course, setting up conditions in their bodies to allow drug-resistant bacteria to thrive.

The basic principles of TB treatment and control are to make sure all patients complete a full course of medication, so that all of the bacteria causing the infection are killed and drug-resistant strains do not develop. Great efforts are being made to ensure that high-risk individuals who are infected but not yet sick receive preventive therapy under observation. Such programs are approximately 90% effective in reducing the likelihood of developing active TB and spreading it to others. These efforts are having an effect, since tuberculosis is on the decline in the United States, decreasing by 3.3% from 2003 to 2004.

Bacterial biofilms are involved in tooth decay

Bacteria and other organisms may form mixed cultures on certain surfaces that are extremely difficult to treat. On teeth, this biofilm, or plaque, consists largely of bacterial cells surrounded by a polysaccharide matrix. Most of the bacteria in plaque are filaments of rod-shaped cells classified as various species of *Actinomyces*, which extend out perpendicular to the surface of the tooth. Many other bacterial species are also present in plaque.

Tooth decay, or **dental caries,** is caused by the bacteria present in the plaque, which persist especially in places that are difficult to reach with a toothbrush. Diets that are high in simple sugars are especially harmful to teeth because certain bacteria, notably *Streptococcus sobrinus* and *S. mutans*, ferment the sugars to lactic acid. This acid production reduces the pH in the area around the plaque, breaking down the structure of the hydroxyapatite that makes tooth enamel hard. As the enamel degenerates, the remaining soft matrix of the tooth becomes vulnerable to bacterial attack.

Bacteria can cause ulcers

Bacteria can also be the cause of disease states that on the surface appear to have no infectious basis. Peptic ulcer disease is due to craterlike lesions in the gastrointestinal tract that are exposed to peptic acid. Ulcers can be caused by drugs, such as nonsteroidal anti-inflammatory drugs, and also by some tumors of the pancreas that cause an oversecretion of peptic acid. In 1982, a bacterium named *Campylobacter pylori* (now named *Helicobacter pylori*) was isolated from gastric juices. Over the years evidence has accumulated that this bacterium is actually the causative agent in the majority of cases of peptic ulcer disease.

Antibiotic therapy can now eliminate *H. pylori*, treating the cause of the disease not just the symptoms. The discovery of the action of this bacterial species illustrates how even disease states that appear to be unrelated to infectious disease may actually be caused by cryptic (unknown) infection.

Many sexually transmitted diseases are bacterial

A number of bacteria cause sexually transmitted diseases (STDs). Three that are particularly important are gonorrhea, syphilis, and chlamydia (figure 28.17).

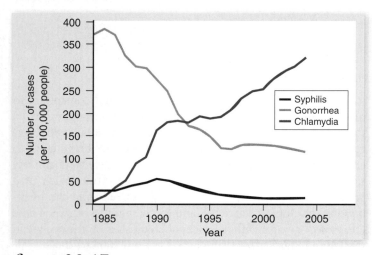

figure 28.17

TRENDS IN SEXUALLY TRANSMITTED DISEASES IN THE UNITED STATES.

inquiry

? *How is it possible for the incidence of one STD (chlamydia) to rise as another (gonorrhea) falls?*

Gonorrhea

Gonorrhea is one of the most prevalent communicable diseases in North America. Caused by the bacterium *Neisseria gonorrhoeae*, gonorrhea can be transmitted through sexual intercourse or any other sexual contacts in which body fluids are exchanged, such as oral or anal intercourse. It can also pass from mother to baby during delivery through the birth canal. Gonorrhea can spread to the eyes and internal organs, causing conjunctivitis (a severe infection of the eyes) and arthritic meningitis (an infection of the joints). Conjunctivitis in newborns is now routinely prevented with antibacterial eye drops.

Left untreated in women, gonorrhea can cause pelvic inflammatory disease (PID), a condition in which the fallopian tubes become scarred and blocked. Over time, PID can eventually lead to sterility.

The incidence of gonorrhea has been on the decline in the United States, but it remains a serious threat worldwide. Of particular concern is the appearance of antibiotic-resistant strains of *N. gonorrhoeae*.

Syphilis

Syphilis, a very destructive STD, was once prevalent and deadly but is now less common due to the advent of blood-screening procedures and antibiotics. Syphilis is caused by a spirochete bacterium, *Treponema pallidum*, transmitted during sexual intercourse or through direct contact with an open syphilis chancre sore. The bacterium can also be transmitted from a mother to her fetus, often causing damage to the heart, eyes, and nervous system of the baby.

Once inside the body, the disease progresses in four distinct stages. The first, or primary stage, is characterized by the appearance of a small, painless, often unnoticed sore called a *chancre*. The chancre resembles a blister and occurs at the location where the bacterium entered the body about three weeks following exposure. This stage of the disease is highly infectious, and an infected person may unwittingly transmit the disease to others. This sore heals without treatment in approximately four weeks, deceptively indicating a "cure" of the disease, although the bacterium remains in the body.

The second stage of syphilis, or secondary syphilis, is marked by a rash, a sore throat, and sores in the mouth. The bacteria can be transmitted at this stage through kissing or contact with an open sore. Commonly at this point, the disease enters the third stage, a latent period. This latent stage of syphilis is symptomless and may last for several years. At this point, the person is no longer infectious, but the bacteria are still present in the body, attacking the internal organs.

The final stage of syphilis is the most debilitating, as the damage done by the bacteria in the third stage becomes evident. Sufferers at this stage of syphilis experience heart disease, mental deficiency, and nerve damage, which may include loss of motor functions or blindness.

Chlamydia

Chlamydia is caused by an unusual bacterium. *Chlamydia trachomatis* is genetically a bacterium but is an obligate intracellular parasite, much like a virus in this respect. It is susceptible to antibiotics but it depends on its host to replicate its genetic material. The bacterium is transmitted through vaginal, anal, or oral intercourse with an infected person.

Chlamydia is called the "silent STD" because women usually experience no symptoms until after the infection has become established. In part because of this symptomless nature, the incidence of chlamydia has skyrocketed, increasing by more than sevenfold in the United States since 1984.

The effects of an established chlamydia infection on the female body are extremely serious. Chlamydia can cause pelvic inflammatory disease (PID), which can lead to sterility.

It has recently been established that infection of the male or female reproductive tract by chlamydia can cause heart disease. Chlamydiae produce a peptide similar to one produced by cardiac muscle. As the body's immune system tries to fight off the infection, it recognizes and reacts to this peptide. The similarity between the bacterial and cardiac peptides confuses the immune system, and T cells attack cardiac muscle fibers, inadvertently causing inflammation of the heart and other problems.

Within the last few years, two types of tests for chlamydia have been developed. The treatment for chlamydia is antibiotics, usually tetracycline, which can penetrate the eukaryotic plasma membrane to attack the bacterium. Any woman who experiences the symptoms associated with this STD or is at risk of developing an STD should be tested for the presence of the chlamydia bacterium; otherwise, her fertility may be at risk.

This discussion of bacterial STDs and the description of HIV in the preceding chapter may give the impression that sexual activity is fraught with danger—and in terms of disease transmission, it is. Using precautions against the spread of STDs is a responsible course of action for every individual.

Bacteria have been developed into bioweapons

With the success of antibiotics in treating high-mortality bacterial diseases such as typhus and cholera, many have been lulled into thinking that the battle against infectious bacteria has been won. To defeat an infectious disease, however, you must control its transmission, and unfortunately the new century has seen the introduction of a more deadly way for disease to spread—by the deliberate action of people using disease as a weapon.

In 2001, bioterrorists struck at the United States with anthrax endospores, adding them to letters sent through the mail. Twenty-two people developed anthrax infections and five of them died of the disease. Although this number is small, it is almost 20% of those infected and 50% of those who had the inhalation variety of the infection.

The attack points out the dangerous potential of biological weapons. Anthrax and the smallpox virus are thought to pose the greatest immediate threats. Bioweapons are discussed at length in the enhancement chapter *Infectious Disease and Bioterrorism*, which you can read at www.ravenbiology.com.

> Bacteria cause a variety of diseases in humans, many of which, such as tuberculosis, seem to have been with us for all of recorded history. Bacteria are also involved in tooth decay and are a major cause of sexually transmitted disease. Bacteria have even been turned into potential weapons.

28.7 Beneficial Prokaryotes

Prokaryotes were largely responsible for creating the properties of the atmosphere and the soil over billions of years. Today, they affect the Earth and human life in many important ways.

Prokaryotes are involved in cycling important elements

Life on Earth is critically dependent on the cycling of chemical elements between organisms and the physical environments in which they live—that is, between the living and nonliving elements of ecosystems. Prokaryotes, algae, and fungi play many key roles in this chemical cycling, a process discussed in detail in chapter 57.

Decomposition

The carbon, nitrogen, phosphorus, sulfur, and other atoms of biological systems all have come from the physical environment, and when organisms die and decay, these elements all return to it. The prokaryotes and fungi that carry out the decomposition portion of chemical cycles, releasing a dead organism's atoms to the environment, are called **decomposers.**

Fixation

Other prokaryotes play important roles in fixation, the other half of chemical cycles, helping to return elements from inorganic forms to organic forms that heterotrophic organisms can use.

Carbon The role of photosynthetic prokaryotes in fixing carbon is obvious. The organic compounds that plants, algae, and photosynthetic prokaryotes produce from CO_2 pass up through food chains to form the bodies of all the ecosystem's heterotrophs. Ancient cyanobacteria are thought to have added oxygen to the Earth's atmosphere as a by-product of their photosynthesis. Modern photosynthetic prokaryotes continue to contribute to the production of oxygen.

Nitrogen Less obvious, but no less critical to life, is the role of prokaryotes in recycling nitrogen. The nitrogen in the Earth's atmosphere is in the form of N_2 gas. A triple covalent bond links the two nitrogen atoms and is not easy to break. Among the Earth's organisms, only a very few species of prokaryotes are able to accomplish this feat, reducing N_2 to ammonia (NH_3), which is used to build amino acids and other nitrogen-containing biological molecules. When the organisms that contain these molecules die, other prokaryotes, called *denitrifiers*, return the nitrogen to the atmosphere, completing the cycle.

To fix atmospheric nitrogen, prokaryotes employ an enzyme complex called nitrogenase, encoded by a set of genes called *nif* ("nitrogen fixation") genes. The nitrogenase complex is extremely sensitive to oxygen and is found in a wide range of free-living prokaryotes.

In aquatic environments, nitrogen fixation is carried out largely by cyanobacteria such as *Anabaena*, which forms long chains of cells. Because the nitrogen fixation process is strictly anaerobic, individual cyanobacteria cells may develop into **heterocysts,** specialized nitrogen-fixing cells impermeable to oxygen.

In soil, nitrogen fixation occurs in the roots of plants that harbor symbiotic colonies of nitrogen-fixing bacteria. These associations include *Rhizobium* (a genus of proteobacteria; see figure 28.6) with legumes, *Frankia* (an actinomycete) with many woody shrubs, and *Anabaena* with water ferns.

Prokaryotes may live in symbiotic associations with eukaryotes

Many prokaryotes live in symbiotic association with eukaryotes. **Symbiosis** refers to the ecological relationship between different species that live in direct contact with each other. The symbiotic association of nitrogen-fixing bacteria with plant roots is an example of *mutualism,* a form of symbiosis in which both parties benefit. The bacteria supply the plant with useful nitrogen, and the plant supplies the bacteria with sugars and other organic nutrients (chapter 39).

Many bacteria live symbiotically within the digestive tracts of animals, providing nutrients to their hosts. Cattle and other grazing mammals are unable to digest cellulose in the grass and plants they eat because they lack the required cellulase enzyme. Colonies of cellulase-producing bacteria inhabiting the gut allow cattle to digest their food (see chapter 48 for a fuller account). Similarly, humans maintain large colonies of bacteria in the large intestine that produce vitamins—particularly B_{12} and K—that the body cannot make.

Many bacteria inhabit the outer surfaces of animals and plants without doing damage. These associations are examples of *commensalism,* in which one organism (the bacterium) receives benefits while the animal or plant is neither benefited nor harmed.

Parasitism is a form of symbiosis in which one member (in this case, the bacterium) is benefited, and the other (the infected animal or plant) is harmed. Infection might be considered a form of parasitism.

Bacteria are used in genetic engineering

Because the genetic code is universal, a gene from a human can be inserted into a bacterial cell, and the bacterium will produce a human protein. The use of bacteria in genetic engineering was discussed in chapter 17, and it is a large part of modern molecular biology.

In addition to the production of pharmaceutical agents such as insulin, discussed in chapter 17, applying genetic engineering methods to produce improved strains of bacteria for commercial use holds promise for the future. Bacteria are now widely used as "biofactories" in the commercial production of a variety of enzymes, vitamins, and antibiotics. Immense cultures of bacteria, often genetically modified to enhance performance, are used to produce commercial acetone and other industrially important compounds.

Bacteria can be used for bioremediation

The use of organisms to remove pollutants from water, air, and soil is called *bioremediation.* The normal functioning of sewage

figure 28.18

USING BACTERIA TO CLEAN UP OIL SPILLS. Bacteria can often be used to remove environmental pollutants, such as petroleum hydrocarbons and chlorinated compounds. In rocky areas contaminated by the *Exxon Valdez* oil spill (*left*), oil-degrading bacteria produced dramatic results (*right*).

treatment plants depends on the activity of microorganisms. In sewage treatment plants, the solid matter from raw sewage is broken down by bacteria and archaea naturally present in the sewage. The end product, methane gas (CH_4) is often used as an energy source to heat the treatment plant.

Biostimulation, that is, the addition of nutrients such as nitrogen and phosphorus sources, has been used to encourage the growth of naturally occurring microbes that can degrade crude oil spills. This approach was used successfully to clean up the Alaskan shoreline after the crude oil spill of the Exxon Valdez in 1998 (figure 28.18). Similarly, biostimulation has been used to encourage the growth of naturally occurring microbial flora in contaminated groundwater. Current efforts include those concentrated on the use of endogenous microbes such as *Geobacter* (see figure 28.6) to eliminate radioactive uranium from groundwater contaminated during the cold war.

Chlorinated compounds released into the environment by a variety of sources are another serious pollutant. Some bacteria can actually use these compounds for energy by performing reductive halogenation that is linked to electron transport, a process termed *halorespiration.* Although still at the development stage, the use of such bacteria to remove halogenated compounds from toxic waste holds great promise.

Prokaryotes are vital to ecosystems for both recycling elements and fixation, or making elements available in organic form. Bacteria are involved in fixation of both carbon and nitrogen and are the only organisms that can fix nitrogen. These nitrogen-fixing bacteria may live in symbiotic association with plants. Bacteria are a key component of waste treatment, and they are also being used in bioremediation to remove toxic compounds introduced into the environment.

28.1 The First Cells

Evidence of the first cells includes microfossils and their deposits, changes in isotopic ratios, and the presence of organic chemicals.

- The oldest microfossils are 3.5 billion years old.

- Stromatolites are a combination of sedimentary deposits and precipitated materials. Stromatolite formations are as old as 2.7 billion years.

- Relatively higher levels of carbon-12 in fossils compared with neighboring rocks indicate the action of ancient carbon fixation.

- Biomarkers, such as lipids, indicate that cyanobacteria are at least 2.7 billion years old.

28.2 Prokaryotic Diversity

Prokaryotes are a highly diverse group that is distinctly different from eukaryotes.

- Prokaryotic features include unicellularity, small circular DNA, cell division by binary fission, lack of internal compartmentalization, singular flagellum, and metabolic diversity.

- Despite similarities, Bacteria and Archaea differ in four key areas: plasma membranes, cell walls, DNA replication, and gene expression.

- Archaeal lipids have ether instead of ester linkages and can form tetraether monolayers.

- The cell walls of Bacteria contain peptidoglycans, but Archaea do not. Some archaeal cell walls contain pseudomurien.

- Both Bacteria and Archaea DNA have a single replication origin, but the origin and the replication proteins are different.

- Archaeal initiation of DNA replication is similar to eukaryotes.

- Archaeal RNA polymerases more closely resemble eukaryotic RNA polymerases than does the single bacterial RNA polymerase.

- Nine clades of prokaryotes have been found so far, but many bacteria have not been studied (figure 28.4).

28.3 Prokaryotic Cell Structure

Prokaryotic cells are relatively simple, but they can be categorized by cell shape and structural variations.

- Prokaryotes have three basic forms: rod-shaped (bacillus), cocci (spherical or ovoid), and spirillum (long helical shape).

- Bacteria are classified as either gram-positive or gram-negative based on the Gram stain (figure 28.8).

- Gram-positive bacteria have a thick peptidoglycan layer in the cell wall that contains techoic acid (figure 28.8).

- Gram-negative bacteria have a thin peptidoglycan layer and an outer membrane containing lipopolysaccharides in their cell wall (figure 28.8).

- Some bacteria have a gelatinous layer, the capsule, enabling the bacterium to adhere to surfaces and evade an immune response.

- Many bacteria have a slender, rigid, helical flagellum composed of flagellin. This flagellum can rotate to drive movement (figure 28.9).

- Some bacteria have hairlike pili that are important in adhesion and exchange of genetic information.

- Some bacteria form highly resistant endospores in response to environmental stress.

- In prokaryotes, invaginated regions of the plasma membrane function in respiration and photosynthesis.

- The nucleoid region contains a compacted circular DNA with no bounding membrane.

- Prokaryotic ribosomes that are smaller than eukaryotic.

- Some antibiotics work by binding to prokaryotic ribosomes and blocking protein synthesis.

28.4 Prokaryotic Genetics

Prokaryotes do not reproduce sexually but they can exchange DNA between different cells.

- DNA can be exchanged by conjugation, cell-to-cell contact, and transduction, by viruses, or be picked up from the environment by transformation (figures 28.11, 28.12, and 28.14).

- Conjugation depends on the presence of conjugative plasmids like the F plasmid in *E coli*.

- A replicated F$^+$ plasmid from the donor cell enters the F$^-$ recipient cell through a conjugation bridge by a process called rolling-circle replication.

- The F plasmid can integrate into the bacterial genome. Excision may be imprecise so that the F plasmid carries genetic information from the host.

- Generalized transduction occurs when viruses package host DNA and transfer it upon subsequent infection.

- Specialized transduction is limited to phage that exhibit lysogenic life cycles.

- Antibiotic resistance can be transferred by resistance, or R, plasmids.

- Mutations can occur spontaneously in bacteria due to radiation, UV, and various chemicals.

28.5 Prokaryotic Metabolism

Prokaryotes acquire carbon and energy using light or chemical reactions via autotrophic or heterotrophic pathways.

- Photoautotrophs carry out photosynthesis and obtain carbon from carbon dioxide.

- Chemolithoautotrophs obtain energy by oxidizing inorganic substances.

- Photoheterotrophs use light for energy but obtain carbon from organic molecules.

- Most bacteria are chemoheterotrophs and obtain carbon and energy from organic molecules.

- Some bacteria release proteins through their cell walls, and these proteins may transfer other virulent proteins into eukaryotic cells.

- Gram-negative bacteria known as pseudomonads are responsible for most plant diseases.

28.6 Human Bacterial Disease

In recent years many bacterial diseases have appeared or reappeared in human populations (table 28.1).

- Bacterial diseases are spread through mucus or saliva droplets, contaminated food and water, and insect vectors.

- The tuberculosis-causing bacterium has infected humans for all of recorded history and is currently a leading cause of death.

- Bacterial biofilms are involved in tooth decay.

- Most ulcers are caused by infection with *Helicobacter pylori*.

- The sexually transmitted diseases gonorrhea, syphilis, and chlamydia are caused by bacteria.

28.7 Beneficial Prokaryotes

Prokaryotes are largely responsible for creating the upper atmosphere and soils on Earth.

- Prokaryotes are involved in the recycling of important elements such as carbon and nitrogen.

- Prokaryotes form important symbiotic associations with eukaryotes.

- Genetically engineered prokaryotes can be used to produce human pharmaceutical agents and other useful products

- Bacteria can be used for bioremediation of biological wastes and removal of toxic compounds introduced into the environment.

SELF TEST

1. Which of the following would be an example of a biomarker?
   a. A microfossil found in a meteorite
   b. A hydrocarbon found in an ancient rock layer
   c. An area that is high in carbon-12 concentration in a rock layer
   d. A newly discovered formation of stromatolites

2. Isotopic dating is a technique used in
   a. carbon fixation studies.
   b. dating of microfossils.
   c. study of biomarkers.
   d. all of the above.

3. In a volcanic vent, rich in hydrogen sulfide, you discover a new single-celled nonphotosynthetic organism that lacks a nucleus. Based on these characteristics, you initially decide to classify it with the
   a. cyanobacteria.
   b. bacteria.
   c. eukaryotes.
   d. archaea.

4. Which of the following is typically not associated with a prokaryote?
   a. Horizontal transfer of genetic information
   b. A lack of internal compartmentalization
   c. Multiple, linear chromosomes
   d. A cell size of 1 μm

5. Which of the following characteristics is unique to the Archaea?
   a. A fluid mosaic model of plasma membrane structure
   b. The use of an RNA polymerase during gene expression
   c. Ether-linked phospholipids
   d. A single origin of DNA replication

6. Which of the following is present only in a gram-positive cell?
   a. Peptidoglycan
   b. Techoic acid
   c. Lipopolysaccharides
   d. Plasma membrane

7. The _____ contains the genetic information of a prokaryotic cell.
   a. nucleoid region
   b. pili
   c. capsule
   d. nucleus

8. Generalized transduction arises from
   a. DNA released from dead cells.
   b. infection by a lysogenic phage.
   c. phage packaging host DNA instead of phage DNA.
   d. transfer of host DNA by the F plasmid.

9. The horizontal transfer of DNA using a plasmid is an example of
   a. generalized transduction.
   b. binary fission.
   c. transformation.
   d. conjugation.

10. A chemolithoautotroph bacterium gets its carbon from _____ and its energy from _____.
   a. carbon dioxide; sunlight
   b. organic molecules; organic molecules
   c. carbon dioxide; inorganic molecules
   d. organic molecules; sunlight

11. Which of the following bacterial species is transmitted through sexual intercourse?
   a. *Yersinia pestis*
   b. *Salmonella typhi*
   c. *Clostridium botulinum*
   d. *Chlamydia trachomatis*

12. Which of the following diseases is *not* caused by bacteria?
   a. Peptic ulcers
   b. The flu
   c. Tuberculosis
   d. Dental carries

13. Bacteria lack independent internal membrane systems, but are able to perform photosynthesis and respiration, both of which use membranes. They are able to perform these functions because
   a. they actually have internal membranes, but only for these functions.
   b. invaginations of the plasma membrane can provide an internal membrane surface.
   c. they take place outside of the cell between the membrane and the cell wall.
   d. they use protein-based structures to take the place of internal membranes.

14. Plants cannot fix nitrogen, yet some plants do not need nitrogen from the soil. This is because
   a. of a symbiotic association with a bacterium that can fix nitrogen.
   b. these plants are the exceptions that can fix nitrogen.
   c. they have been infected by a parasitic virus that can fix nitrogen.
   d. they are able to obtain nitrogen from the air.

15. Which of the following processes involves the removal of toxic compounds from the environment using a bacterial species?
   a. Commensalism
   b. Decomposition
   c. Nitrogen fixation
   d. Bioremediation

CHALLENGE QUESTIONS

1. If a new form of carbon fixation was discovered that was not biased toward carbon-12, would this affect our analysis of the earliest evidence for life?

2. Frederick Griffith's experiments (see Chapter 14) played an important role in the recognition that DNA is the genetic material. Griffith showed that heat-killed smooth (virulent) bacteria mixed with live rough (nonvirulent) bacteria would cause pneumonia when injected into mice. Furthermore, live rough bacteria could be cultured from the infected mice. The difference between the two strains is a polysaccharide capsule found in the smooth strain. Given what you have learned in this chapter, how would you explain these observations.

3. In the 1960s, it was common practice to prescribe multiple antibiotics to fight bacterial infections. It is also often the case that patients do not always take the entire "course" of their antibiotics. Antibiotic resistance genes are often found on conjugative plasmids. How do these factors affect the evolution of antibiotic resistance and of resistance to multiple antibiotics in particular?

4. Soil-based nitrogen-fixing bacteria appear to be highly vulnerable to exposure to UV radiation. Suppose that the ozone level continues to be depleted, what are the long-term effects on the planet?

500 µm

Protists

introduction

FOR MORE THAN HALF OF the long history of life on Earth, all life was microscopic. The biggest organisms that existed for over 2 billion years were single-celled bacteria fewer than 6 µm thick. These prokaryotes lacked internal membranes, except for invaginations of surface membranes in photosynthetic bacteria.

The first evidence of a different kind of organism is found in tiny fossils in rock 1.5 billion years old. These fossil cells are much larger than bacteria (up to 10 times larger) and contain internal membranes and what appear to be small, membrane-bounded structures. The complexity and diversity of form among these single cells is astonishing. The step from relatively simple to quite complex cells marks one of the most important events in the evolution of life, the appearance of a new kind of organism, the **eukaryote.** Eukaryotes that are clearly not animals, plants, or fungi have been lumped together and called protists.

29.1 Defining Protists

Protists are the most diverse of the four kingdoms in the domain Eukarya. Compartmentalization is the key feature that distinguishes protists and other eukaryotes from archaea and bacteria. The kingdom Protista contains many unicellular, colonial, and multicellular groups. The origin of eukaryotes, which began with ancestral protists, is among the most significant events in the evolution of life.

Protista is not monophyletic

One of the most important statements we can make about the kingdom Protista is that it is paraphyletic and not a kingdom at all; as a matter of convenience, single-celled eukaryotic organisms have typically been grouped together and called protists. This lumps 200,000 different and only distantly related forms together. The "single-kingdom" classification of the Protista is artificial and not representative of any evolutionary relationships.

Monophyletic clades have been identified among the protists

New applications of a wide variety of molecular methods are providing important insights into the relationships among the protists. Many questions about how to classify the protists are being addressed with these techniques. Are protists best considered as several different kingdoms, each of equal rank with animals, plants, and fungi? Are some of the protists actually members of other kingdoms? While these questions continue to be debated, ever-increasing information is becoming available concerning which organisms among the protists are most likely to be monophyletic.

In this chapter, we group the 15 major protist phyla into seven major monophyletic groups, based on our current understanding of phylogeny (figure 29.1). Although these lineages may change, this approach allows us to examine groups with many shared traits. Keep in mind that about 60 of the protist lineages cannot yet be placed on the tree of life with any confidence! Protists exemplify the challenges and excitement of the revolutionary changes in taxonomy and phylogeny we explored in chapter 26. Understanding the evolution of protists is key to understanding the origins of plants, fungi, and animals.

Because green algae and land plants form a monophyletic clade, the green algae are explored in more detail in the following chapter on plant diversity. The characteristics of green algae and land plants are best understood when considered in concert because of their shared evolutionary history. The remaining six monophyletic clades that are loosely called protists are examined in this chapter.

The taxonomy of the protists is in a state of flux as new information shapes our understanding of this kingdom.

figure 29.1

THE CHALLENGE OF PROTISTAN CLASSIFICATION. Our understanding of the evolutionary relationships among protists is currently in flux. The most recent data support seven major, monophyletic groups within the protists. Consider this a working model, not fact. The green algae (Chlorophyta) are not truly monophyletic in that another branch, Streptophyta, gave rise to the land plants. Protist lineages are shaded in blue.

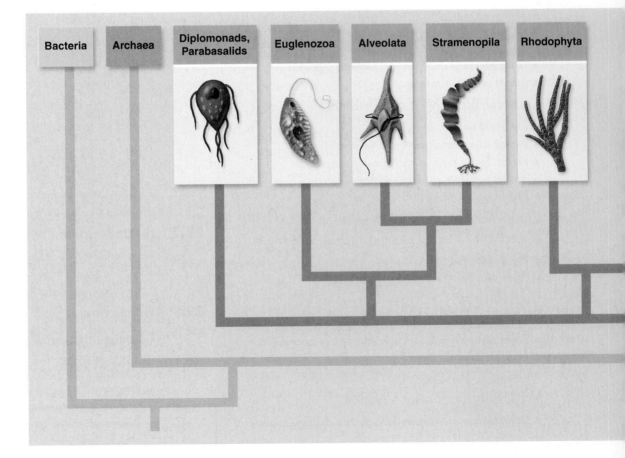

Eukaryotic Origins and Endosymbiosis

Eukaryotic cells are distinguished from prokaryotes by the presence of a cytoskeleton and compartmentalization that includes a nuclear envelope and organelles. The exact sequence of events that led to large, complex eukaryotic cells is unknown. Loss of a rigid cell wall allowed membranes to fold inward, increasing surface area. Membrane flexibility also made it possible for one cell to engulf another.

Fossil evidence dates the origins of eukaryotes

Indirect chemical traces hint that eukaryotes may go as far back as 2.7 billion years, but no fossils as yet support such an early appearance. In rocks about 1.5 billion years old, we begin to see the first microfossils that are noticeably different in appearance from the earlier, simpler forms, none of which were more than 6 μm in diameter (figure 29.2). These cells are much larger than those of prokaryotes and have internal membranes and thicker walls.

These early fossils mark a major event in the evolution of life: A new kind of organism had appeared. These new cells are called **eukaryotes,** from the Greek words meaning "true nucleus," because they possess an internal structure called a nucleus. All organisms other than prokaryotes are eukaryotes.

In the sections that follow, the origins of eukaryotic internal structure are considered. Keep in mind that, as discussed in chapter 24, horizontal gene transfer occurred frequently while eukaryotic cells were evolving. Eukaryotic cells evolved not only through horizontal gene transfer, but through infolding of membranes and

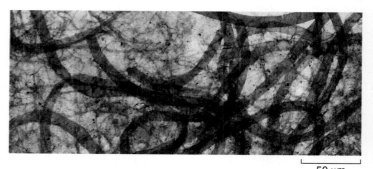

figure 29.2

EARLY EUKARYOTIC FOSSIL. Fossil algae that lived in Siberia 1000 MYA.

engulfing other cells. Today's eukaryotic cell is the result of cutting and pasting of DNA and organelles from different species.

The nucleus and ER arose from membrane infoldings

Many prokaryotes have infoldings of their outer membranes extending into the cytoplasm and serving as passageways to the surface. The network of internal membranes in eukaryotes called endoplasmic reticulum (ER), and the nuclear envelope, an extension of the ER network that isolates and protects the nucleus, is thought to have evolved from such infoldings (figure 29.3).

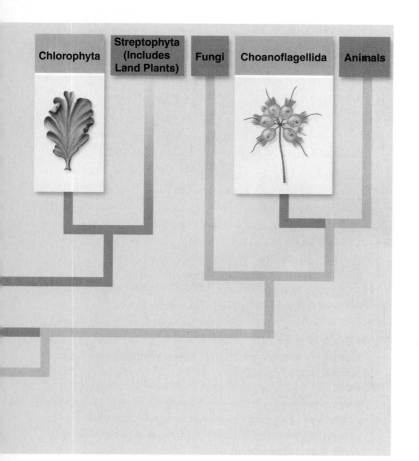

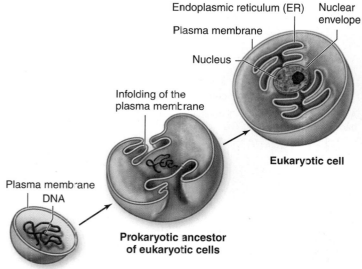

figure 29.3

ORIGIN OF THE NUCLEUS AND ENDOPLASMIC RETICULUM. Many prokaryotes today have infoldings of the plasma membrane (see also figure 27.6). The eukaryotic internal membrane system, called the endoplasmic reticulum (ER), and the nuclear envelope may have evolved from such infoldings of the plasma membrane, encasing the DNA of prokaryotic cells that gave rise to eukaryotic cells.

Mitochondria evolved from engulfed aerobic bacteria

Bacteria that live within other cells and perform specific functions for their host cells are called *endosymbiotic bacteria*. Their widespread presence in nature led biologist Lynn Margulis in the early 1970s to champion the theory of endosymbiosis which was first proposed by Konstantin Mereschkowsky in 1905. **Endosymbiosis** means living together in close association. You may recall this theory from discussion of the biology of the cell in chapter 4.

Endosymbiosis, now widely accepted, suggests that a critical stage in the evolution of eukaryotic cells involved endosymbiotic relationships with prokaryotic organisms. According to this theory, energy-producing bacteria may have come to reside within larger bacteria, eventually evolving into what we now know as mitochondria (figure 29.4).

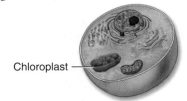

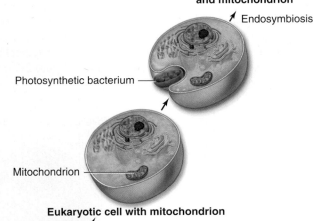

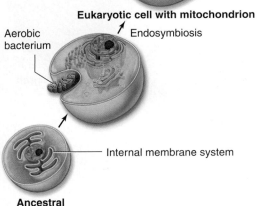

figure 29.4

THE THEORY OF ENDOSYMBIOSIS. Scientists propose that ancestral eukaryotic cells, which already had an internal system of membranes, engulfed aerobic bacteria, which then became mitochondria in the eukaryotic cell. Chloroplasts also originated this way, with eukaryotic cells engulfing photosynthetic bacteria.

Chloroplasts evolved from engulfed photosynthetic bacteria

Photosynthetic bacteria may have come to live within other larger bacteria, leading to the evolution of chloroplasts, the photosynthetic organelles of plants and algae (see figure 29.4). The history of chloroplast evolution is an example of the care that must be taken in phylogenetic studies. All chloroplasts are likely derived from a single line of cyanobacteria, but the organisms that host these chloroplasts are not monophyletic. This apparent paradox is resolved by considering the possibility of secondary, and even tertiary endosymbiosis. Figure 26.8 explains how red and green algae both obtained their chloroplasts by engulfing photosynthetic cyanobacteria. The brown algae most likely obtained their chloroplasts by engulfing one or more red algae, a process called **secondary endosymbiosis** (figure 29.5). (As mentioned, green algae are considered in the following chapter even though they are protists.)

A phylogenetic tree based only on chloroplast gene sequences from red and green algae reveals an incredibly close evolutionary relationship. This tree is misleading, however, because it is not possible to tell just from these data how much

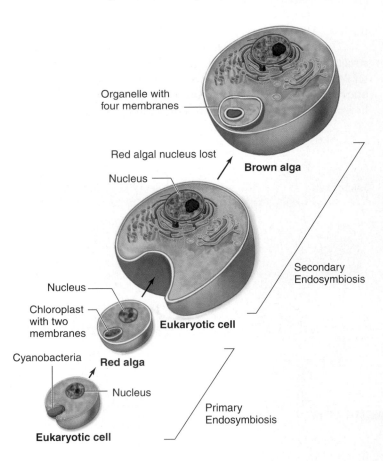

figure 29.5

ENDOSYMBIOTIC ORIGINS OF CHLOROPLASTS IN RED AND BROWN ALGAE.

inquiry

? *How could you distinguish between primary and secondary endosymbiosis by looking at micrographs of cells with chloroplasts?*

the two algal lines had diverged at the time they engulfed the same line of cyanobacteria. Morphological and chemical traits are more helpful than chloroplast gene sequences in sorting out red and green algal relations. More data and analyses are still needed to confirm the position of red algae in figure 29.1.

Endosymbiosis was not a rare occurrence

Many eukaryotic cells contain other endosymbiotic bacteria in addition to mitochondria. Bacteria with flagella, long whiplike cellular appendages used for propulsion, may have become symbiotically involved with nonflagellated bacteria to produce larger, motile cells. Centrioles, organelles associated with the assembly of microtubules, resemble in many respects spirochete bacteria, and they contain bacteria-like DNA involved in the production of their structural proteins.

The fact that we now witness so many symbiotic relationships lends general support to the endosymbiotic theory. Even stronger support comes from the observation that present-day organelles such as mitochondria, chloroplasts, and centrioles contain their own DNA, which is remarkably similar to the DNA of bacteria in size and character.

Genes have migrated from endosymbiotic organelles

During the billion and a half years in which mitochondria have existed as endosymbionts within eukaryotic cells, most of their genes have been transferred to the chromosomes of the host cells—but not all. Each mitochondrion still has its own genome, a circular, closed molecule of DNA similar to that found in bacteria, on which are located genes encoding the essential proteins of oxidative metabolism. These genes are transcribed within the mitochondrion, using mitochondrial ribosomes that are smaller than those of eukaryotic cells, very much like bacterial ribosomes in size and structure.

Mitochondria divide by simple fission, just as bacteria do, and they also replicate and sort their DNA much as bacteria do. However, nuclear genes direct the process, and mitochondria cannot be grown outside of the eukaryotic cell, in cell-free culture.

Mitosis evolved in eukaryotes

The mechanisms of mitosis and cytokinesis, now so common among eukaryotes, did not evolve all at once. Traces of very different, and possibly intermediate, mechanisms survive today in some of the eukaryotes. In fungi and in some groups of protists, for example, the nuclear membrane does not dissolve, as it does in plants, animals, and most other protists, and mitosis is confined to the nucleus. When mitosis is complete in these organisms, the nucleus divides into two daughter nuclei, and only then does the rest of the cell divide. We do not know whether mitosis without nuclear membrane dissolution represents an intermediate step on the evolutionary journey, or simply a different way of solving the same problem. We cannot see the interiors of dividing cells well enough in fossils to be able to trace the history of mitosis.

The theory of endosymbiosis proposes that mitochondria, chloroplasts, and possibly other organelles originated as symbiotic bacteria. Over time, genes from chloroplasts and mitochondria have moved into the nucleus, and the organelle genomes are no longer as comprehensive as free-living bacteria.

29.3 General Biology of the Protists

Protists are united on the basis of a single negative characteristic: They are eukaryotes that are not fungi, plants, or animals. In all other respects, they are highly variable, with no uniting features. Many are unicellular, but numerous colonial and multicellular groups exist. Most are microscopic, but some are as large as trees. They represent all symmetries and exhibit all types of nutrition.

Protist cell surfaces vary widely

Protists possess a varied array of cell surfaces. Some protists, such as amoebas, are surrounded only by their plasma membrane. All other protists have a plasma membrane with extracellular material (ECM) deposited on the outside of the membrane. Some ECM forms strong cell walls. Diatoms and forams secrete glassy shells of silica.

Many protists with delicate surfaces are successful in quite harsh habitats. How do they manage to survive so well? They form cysts, dormant forms of a cell with resistant outer coverings in which cell metabolism is more or less completely shut down. Not all cysts are so sturdy. Vertebrate parasitic amoebas, for example, form cysts that are quite resistant to gastric acidity, but will not tolerate desiccation or high temperature.

Protists have several means of locomotion

Movement in protists is also accomplished by diverse mechanisms. Protists move chiefly by either flagellar rotation or pseudopodial movement. Many protists wave one or more flagella to propel themselves through the water, and others use banks of short, flagella-like structures called cilia to create water currents for their feeding or propulsion. Pseudopods (Greek, "false feet") are the chief means of locomotion among amoebas, whose pseudopods are large, blunt extensions of the cell body called lobopodia. Other related protists extend thin, branching protrusions called filopodia. Still other protists extend long, thin pseudopods called axopodia supported by axial rods of microtubules. Axopodia can be extended or retracted. Because the tips can adhere to adjacent surfaces, the cell can move by a rolling motion, shortening the axopodia in front and extending those in the rear.

Protists have a range of nutritional strategies

Protists employ every form of nutritional acquisition except the chemoautotrophic form, which has so far been observed only in prokaryotes. Some protists are photosynthetic and are called

phototrophs. Others are heterotrophs that obtain energy from organic molecules synthesized by other organisms.

Among the heterotrophic protists, those that ingest visible particles of food are called **phagotrophs.** Phagotrophs ingest food particles into intracellular vesicles called food vacuoles or phagosomes. Lysosomes fuse with the food vacuoles, introducing enzymes that digest the food particles within. Digested molecules are absorbed across the vacuolar membrane.

Protists that ingest food in soluble form are called **osmotrophs.** Protists can exhibit tremendous flexibility as seen in **mixotrophs,** protists that are both phototrophic and heterotrophic.

Protists reproduce asexually and sexually

Protists typically reproduce asexually, although some protists have an obligate sexual reproductive phase and others undergo sexual reproduction at times of stress, including food shortages.

Asexual reproduction

Asexual reproduction involves mitosis, but the process is often somewhat different from the mitosis that occurs in multicellular animals. For example, the nuclear membrane often persists throughout mitosis, with the microtubular spindle forming within it.

In some species, a cell simply splits into nearly equal halves after mitosis. Sometimes the daughter cell is considerably smaller than its parent and then grows to adult size, this type of cell division is called **budding.** In **schizogony,** common among some protists, cell division is preceded by several nuclear divisions, so that cytokinesis produces several individuals almost simultaneously.

Sexual reproduction

Most eukaryotic cells also possess the ability to reproduce sexually, something prokaryotes cannot do effectively. Meiosis (see chapter 11) is a major evolutionary innovation that arose in ancestral protists and allows for the production of haploid cells from diploid cells. **Sexual reproduction** is the process of producing offspring by fertilization, the union of two haploid cells. The great advantage of sexual reproduction is that it allows for frequent genetic recombination, which generates the variation that is the starting point of evolution. Not all eukaryotes reproduce sexually, but most have the capacity to do so. The evolution of meiosis and sexual reproduction led to the tremendous explosion of diversity among the eukaryotes.

Protists are the bridge to multicellularity

Diversity was also promoted by the development of **multicellularity.** Some single eukaryotic cells began living in association with others, in colonies. Eventually, individual members of the colony began to assume different duties, and the colony began to take on the characteristics of a single individual. Multicellularity has arisen many times among the eukaryotes. Practically every organism big enough to be seen with the unaided eye, including all animals and plants, is multicellular. The great advantage of multicellularity is that it fosters specialization; some cells devote all of their energies to one task, other cells to another. Few innovations have had as great an influence on the history of life as the specialization made possible by multicellularity.

Protists exhibit a wide range of forms, locomotion, nutrition, and reproduction.

29.4 Diplomonads and Parabasalids: Flagellated Protists Lacking Mitochondria

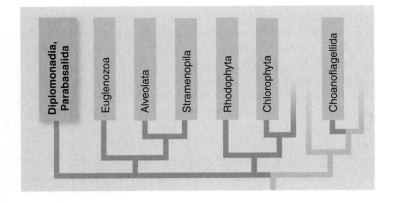

What was the first eukaryote like? We cannot be sure, but the **diplomonads** and the **parabasalids** likely had early eukaryotic ancestors. Although these groups have similar features, their differences put them into separate clades.

Diplomonads have two nuclei

Diplomonads are unicellular and move with flagella. This group lacks mitochondria, but has two nuclei. *Giardia intestinalis* is an example of a diplomonad (figure 29.6). *Giardia* is a parasite that can pass from human to human via contaminated water and cause diarrhea. Mitochondrial genes are found in their nuclei, leading to the conclusion that *Giardia* evolved from aerobes. Thus, *Giardia* is unlikely to represent an early protist.

Parabasalids have undulating membranes

Parabasalids contain an intriguing array of species. Some live in the gut of termites and digest cellulose, the main component of the termite's wood-based diet. The symbiotic relationship is one layer more complex because these parabasalids have a symbiotic relationship with bacteria that also aid in the digestion of cellulose. The persistent activity of

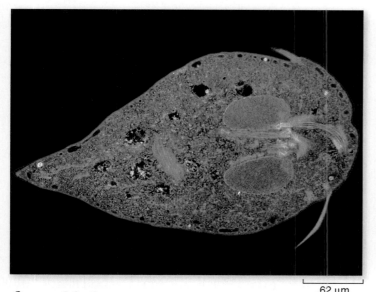

figure 29.6

    Giardia intestinalis. This parasitic diplomonad lacks a mitochondrion.

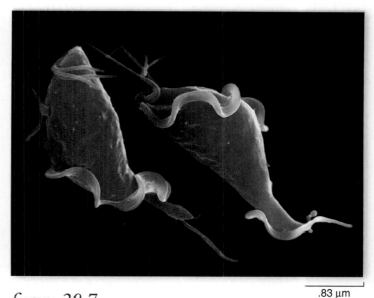

figure 29.7

    UNDULATING MEMBRANE CHARACTERISTIC OF PARABASALIDS. Vaginitis can be caused by this parasite species, *Trichomonas vaginalis.*

these three symbiotic organisms from three different kingdoms can lead to the collapse of a home built of wood or recycle tons of fallen trees in a forest. Another parabasalid, *Trichomonas vaginalis*, causes a sexually transmitted disease in humans.

    Parabasalids have undulating membranes that assist in locomotion (figure 29.7). Like diplomonads, parabasalids also use fla-gella to move and lack mitochondria. The lack of mitochondria is now believed to be a derived rather than an ancestral trait.

> Diplomonads and parabasalids are closely related to the now extinct, early eukaryotes. They lack mitochondria, but may have lost them rather than never acquired them.

29.5 Euglenozoa: A Diverse Group in Which Some Members Have Chloroplasts

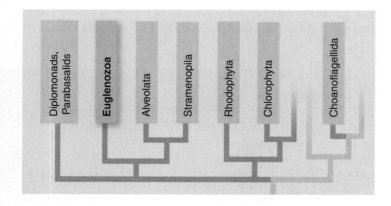

Among their distinguishing features, a number of the **Euglenozoa** have acquired chloroplasts through endosymbiosis. None of the algae are closely related to Euglenozoa, a reminder that endosymbiosis is widespread. At one time, these organisms were considered animals, which is why their name contains *-zoa.*

Euglenoids are free-living eukaryotes with anterior flagella

Euglenoids diverged early and were among the earliest free-living eukaryotes to possess mitochondria. Euglenoids clearly illustrate the impossibility of distinguishing "plants" from "animals" among the protists. About one-third of the approximately 40 genera of euglenoids have chloroplasts and are fully autotrophic; the others lack chloroplasts, ingest their food, and are heterotrophic.

    Some euglenoids with chloroplasts may become heterotrophic in the dark; the chloroplasts become small and nonfunctional. If they are put back in the light, they may become green within a few hours. Photosynthetic euglenoids may sometimes feed on dissolved or particulate food.

    Individual euglenoids range from 10 to 500 μm long and are highly variable in form. Interlocking proteinaceous strips arranged in a helical pattern form a flexible structure called the *pellicle,* which lies within the plasma membrane of

figure 29.8

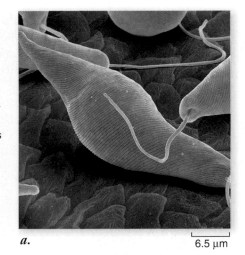

EUGLENOIDS.
a. Micrograph of *Euglena gracilis.*
b. Diagram of *Euglena.* Paramylon granules are areas where food reserves are stored.

a.   6.5 μm

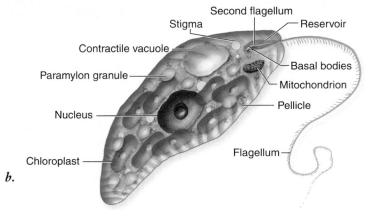

Second flagellum
Stigma — Reservoir
Contractile vacuole —
Paramylon granule — Basal bodies
— Mitochondrion
Nucleus — Pellicle
Chloroplast — Flagellum —
b.

the euglenoids. Because its pellicle is flexible, a euglenoid is able to change its shape.

Reproduction in this phylum occurs by mitotic cell division. The nuclear envelope remains intact throughout the process of mitosis. No sexual reproduction is known to occur in this group.

Euglena, the best known euglenoid

In *Euglena* (figure 29.8), the genus for which the phylum is named, two flagella are attached at the base of a flask-shaped opening called the *reservoir*, which is located at the anterior end of the cell. One of the flagella is long and has a row of very fine, short, hairlike projections along one side. A second, shorter flagellum is located within the reservoir but does not emerge from it. Contractile vacuoles collect excess water from all parts of the organism and empty it into the reservoir, which apparently helps regulate the osmotic pressure within the organism. The stigma, an organ that also occurs in the green algae (phylum Chlorophyta), is light-sensitive and helps these photosynthetic organisms move toward light.

Cells of *Euglena* contain numerous small chloroplasts. These chloroplasts, like those of the green algae and plants, contain chlorophylls *a* and *b*, together with carotenoids. Although the chloroplasts of euglenoids differ somewhat in structure from those of green algae, they probably had a common origin. It seems likely that euglenoid chloroplasts ultimately evolved from a symbiotic relationship through ingestion of green algae. Recent phylogenetic evidence indicates that *Euglena* had multiple origins within the Euglenoids, and the concept of a single *Euglena* genus is now being debated.

Kinetoplastids are parasitic

A second major group within the Euglenozoa is the **kinetoplastids.** The name kinetoplastid refers to a unique, single mitochondrion in each cell. The mitochondria have two types of DNA: minicircles and maxicircles. (Remember that prokaryotes have circular DNA, and mitochondria had prokaryotic origins.) This mitochondrial DNA is responsible for very rapid glycolysis and also for an unusual kind of editing of the DNA by guide RNAs encoded in the minicircles.

Trypanosomes: Disease-causing kinetoplastids

Parasitism has evolved multiple times within the kinetoplastids. Trypanosomes are a group of kinetoplastids that cause many serious human diseases, the most familiar being trypanosomiasis, also known as African sleeping sickness, which causes extreme lethargy and fatigue (figure 29.9).

Leishmaniasis, which is transmitted by sand flies, is a trypanosomic disease that causes skin sores and in some cases can affect internal organs, leading to death. About 1.5 million new cases are reported each year. The rise in leishmaniasis in South America correlates with the move of infected individuals from rural to urban environments, where there is a greater chance of spreading the parasite.

Chagas disease is caused by *Trypanosoma cruzi.* At least 90 million people, from the southern United States to Argentina, are at risk of contracting *T. cruzi* from small wild mammals that carry the parasite and can spread it to other mammals and humans through skin contact with urine and feces. Blood transfusions have also increased the spread of the infection. Chagas disease can lead to severe cardiac and digestive problems in humans and domestic animals, but it appears to be tolerated in the wild mammals.

Control is especially difficult because of the unique attributes of these organisms. For example, tsetse fly–transmitted trypanosomes have evolved an elaborate genetic mechanism for repeatedly changing the antigenic nature of their protective glycoprotein coat, thus dodging the antibodies their hosts produce against them (see chapter 51). Only a single one out of

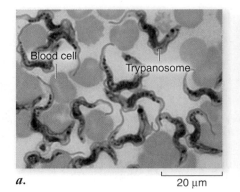
Blood cell
Trypanosome

a.   20 μm   *b.*

figure 29.9

A KINETOPLASTID. *a. Trypanosoma* among red blood cells. The nuclei (dark-staining bodies), anterior flagella, and undulating, changeable shape of the trypanosomes are visible in this photograph. *b.* The tsetse fly, shown here sucking blood from a human arm, can carry trypanosomes.

some 1000–2000 variable antigen genes is expressed at a time. Rearrangements of these genes during the asexual cycle of the organism allow for the expression of a seemingly endless variety of different antigen genes.

In the guts of the flies that spread them, trypanosomes are noninfective. When they are ready to transfer to the skin or bloodstream of their host, trypanosomes migrate to the salivary glands and acquire the thick coat of glycoprotein antigens that protect them from the host's antibodies. Later, when they are taken up by a tsetse fly, the trypanosomes again shed their coats.

The production of vaccines against such a system is complex, but tests are under way. Releasing sterilized flies to impede the reproduction of populations is another technique attempted to control the fly population. Traps made of dark cloth and scented like cows, but poisoned with insecticides, have likewise proved effective.

The recent sequencing of the genomes of the three kinetoplastids described earlier revealed a core of common genes in all three, as described in chapter 24. The devastating toll of all three on human life could be alleviated by the development of a single drug targeted at one or more of the core proteins shared by the three parasites.

> Euglenozoa include free-living and parasitic protists that move with flagella. Euglenoids have chloroplasts obtained via endosymbiosis. Kinetoplastids, such as trypanosomes, have unusual mitochondria that use RNA editing.

Alveolata: Protists With Submembrane Vesicles

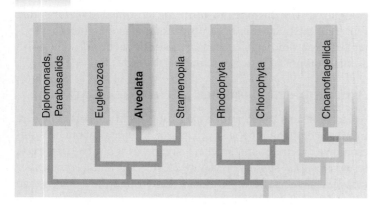

Members of the **Alveolata** include the **dinoflagellates,** **apicomplexans,** and **ciliates,** all of which have a common lineage despite their diverse modes of locomotion. One common trait is the presence of flattened vesicles called alveoli (hence the name alveolata) that are stacked in a continuous layer below their plasma membranes (figure 29.10). The alveoli may function in membrane transport, similarly to Golgi bodies.

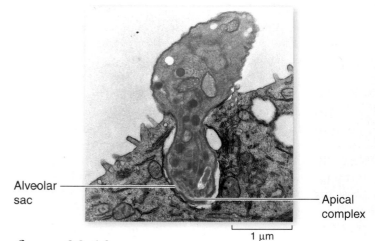

Alveolar sac — — Apical complex

1 µm

figure 29.10

ALVEOLI ARE A CONTINUUM OF VESICLES JUST BELOW THE PLASMA MEMBRANE OF DINOFLAGELLATES, APICOMPLEXANS, AND CILIATES. The apical complex of apicomplexans forces the parasite into host cells.

Dinoflagellates are photosynthesizers with distinctive features

Most **dinoflagellates** are photosynthetic unicells with two flagella. Dinoflagellates live in both marine and freshwater environments. Some dinoflagellates are luminous and contribute to the twinkling or flashing effects seen in the sea at night, especially in the tropics.

The flagella, protective coats, and biochemistry of dinoflagellates are distinctive, and the dinoflagellates do not appear to be directly related to any other phylum. Plates made of a cellulose-like material, often encrusted with silica, encase the dinoflagellate cells (figure 29.11). Grooves at the junctures of these plates usually house the flagella, one encircling the cell like a belt, and the other perpendicular to it. By beating in their grooves, these flagella cause the dinoflagellate to spin as it moves.

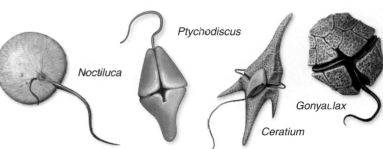

Noctiluca   Ptychodiscus   Ceratium   Gonyaulax

figure 29.11

SOME DINOFLAGELLATES. *Noctiluca,* which lacks the heavy cellulose armor characteristic of most dinoflagellates, is one of the bioluminescent organisms that cause the waves to sparkle in warm seas. In the other three genera, the shorter, encircling flagellum is seen in its groove, with the longer one projecting away from the body of the dinoflagellate. (Not drawn to scale.)

chapter 29 protists   **569**

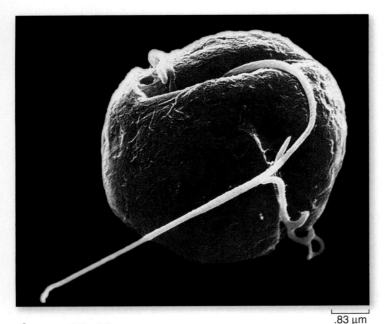

figure 29.12

RED TIDE. Although small in size, huge populations of dinoflagellates, including this *Gymnopodium* species, can color the sea red and release toxins into the water.

.83 µm

Most dinoflagellates have chlorophylls *a* and *c*, in addition to carotenoids, so that in the biochemistry of their chloroplasts, they resemble the diatoms and the brown algae. Possibly this lineage acquired such chloroplasts by forming endosymbiotic relationships with members of those groups.

Red tide: Overgrowth of dinoflagellates

The poisonous and destructive "red tides" that occur frequently in coastal areas are often associated with great population explosions, or "blooms," of dinoflagellates, whose pigments color the water (figure 29.12). Red tides have a profound, detrimental effect on the fishing industry in the United States. Some 20 species of dinoflagellates produce powerful toxins that inhibit the diaphragm and cause respiratory failure in many vertebrates. When the toxic dinoflagellates are abundant, many fishes, birds, and marine mammals may die.

Although sexual reproduction does occur under starvation conditions, dinoflagellates reproduce primarily by asexual cell division. Asexual cell division relies on a unique form of mitosis in which the permanently condensed chromosomes divide longitudinally within a permanent nuclear envelope. After the numerous chromosomes duplicate, the nucleus divides into two daughter nuclei.

Also, the dinoflagellate chromosome is unique among eukaryotes in that the DNA is not generally complexed with histone proteins. In all other eukaryotes, the chromosomal DNA is complexed with histones to form nucleosomes, structures that represent the first order of DNA packaging in the nucleus (chapter 10). How dinoflagellates maintain distinct chromosomes with a small amount of histones remains a mystery.

Apicomplexans include the malaria parasite

Apicomplexans are spore-forming parasites of animals. They are called apicomplexans because of a unique arrangement of fibrils, microtubules, vacuoles, and other cell organelles at one end of the cell, termed an *apical complex* (see figure 29.10). The apical complex is a cytoskeletal and secretory complex that enables the apicomplexan to invade its host. The best known apicomplexan is the malarial

figure 29.13

THE LIFE CYCLE OF *PLASMODIUM*. *Plasmodium*, the apicomplexan that causes malaria, has a complex life cycle that alternates between mosquitoes and mammals.

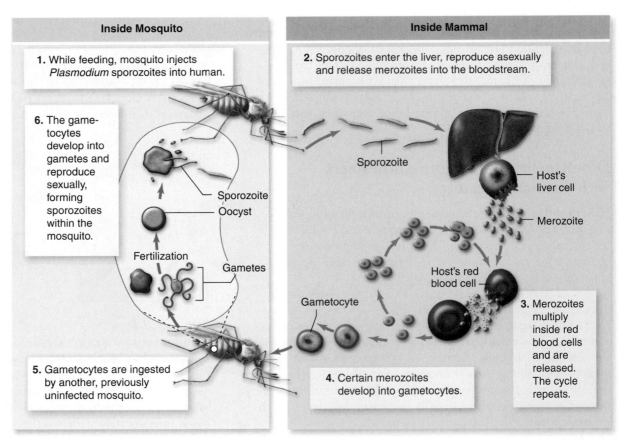

Inside Mosquito

1. While feeding, mosquito injects *Plasmodium* sporozoites into human.

6. The gametocytes develop into gametes and reproduce sexually, forming sporozoites within the mosquito.

Sporozoite
Oocyst

Fertilization

Gametes

5. Gametocytes are ingested by another, previously uninfected mosquito.

Inside Mammal

2. Sporozoites enter the liver, reproduce asexually and release merozoites into the bloodstream.

Sporozoite

Host's liver cell

Merozoite

Gametocyte

Host's red blood cell

3. Merozoites multiply inside red blood cells and are released. The cycle repeats.

4. Certain merozoites develop into gametocytes.

parasite *Plasmodium*. (The use of the genome sequence of the parasite and the mosquito that carries it is discussed in chapter 24.)

Plasmodium and malaria

Plasmodium glides inside the red blood cells of its host with amoeboid-like contractility. Like other apicomplexans, *Plasmodium* has a complex life cycle involving sexual and asexual phases and alternation between different hosts, in this case mosquitoes and humans (figure 29.13). Even though *Plasmodium* has mitochondria, it grows best in a low-O_2, high-CO_2 environment.

Efforts to eradicate malaria have focused on (1) eliminating the mosquito vectors; (2) developing drugs to poison the parasites that have entered the human body; and (3) developing vaccines. From the 1940s to the 1960s, wide-scale applications of dichlorodiphenyltrichloroethane (DDT) killed mosquitoes in the United States, Italy, Greece, and certain areas of Latin America. For a time, the worldwide elimination of malaria appeared possible. But this hope was soon crushed by the development of DDT-resistant mosquitoes in many regions. Furthermore, the use of DDT has had serious environmental consequences. In addition to the problems with resistant strains of mosquitoes, strains of *Plasmodium* have appeared that are resistant to the drugs historically used to kill them, including quinine.

An experimental vaccine containing a surface protein of one malaria-causing parasite, *P. falciparum*, seems to induce the immune system to defend against future infections. In tests, six out of seven vaccinated people did not get malaria after being fed upon by mosquitoes that carried *P. falciparum*. Many are hopeful that this new vaccine may be able to fight malaria. (Chapter 24 contains a discussion of the genome sequences of both *Plasmodium* and its mosquito vector.)

Gregarines

Gregarines are another group of apicomplexans that use their distinctive apical complex to attach themselves in the intestinal epithelium of arthropods, annelids, and mollusks. Most of the gregarine body, aside from the apical complex, is in the intestinal cavity, and nutrients appear to be obtained through the apicomplex attachment to the cell (figure 29.14).

Toxoplasma

Using its apical complex, *Toxoplasma gondii* invades the epithelial cells of the human gut. Most individuals infected with the parasite mount an immune response and no permanent damage occurs. In the absence of a fully functional immune system, *Toxoplasma* can damage brain (figure 29.15), heart, and skeletal tissues, in addition to gut and lymph tissue, during extended infections. Individuals with AIDS are particularly susceptible to *Toxoplasma* infection. *Toxoplasma* parasites can find their way from a cat litterbox into pregnant women where they cross the placental barrier and harm the developing fetus with an immature immune system.

Ciliates are characterized by their mode of locomotion

As the name indicates, most **ciliates** feature large numbers of cilia (tiny beating hairs). These heterotrophic, unicellular protists are 10–3000 μm long. Their cilia are usually arranged either in longitudinal rows or in spirals around the cell. Cilia

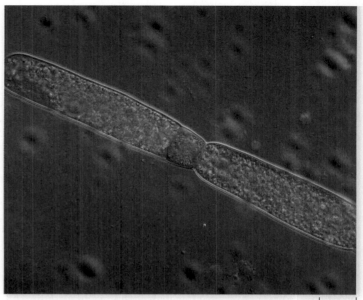

figure 29.14

GREGARINE ENTERING A CELL.

100 μm

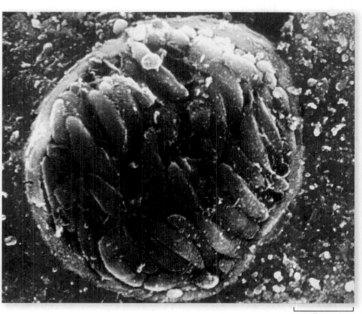

figure 29.15

TOXOPLASMA CAN ENTER THE BRAIN AND FORM CYSTS FILLED WITH SLOWLY REPLICATING PARASITES. Micrograph of a cyst filled with *Toxoplasma*.

5 μm

are anchored to microtubules beneath the plasma membrane (see chapter 5), and they beat in a coordinated fashion. In some groups, the cilia have specialized functions, becoming fused into sheets, spikes, and rods that may then function as mouths, paddles, teeth, or feet.

The ciliates have a pellicle, a tough but flexible outer covering, that enables them to squeeze through or move around obstacles.

Micronucleus and macronucleus

All known ciliates have two different types of nuclei within their cells: a small **micronucleus** and a larger **macronucleus** (figure 29.16). Macronuclei divide by mitosis and are essential for the physiological function of the well-known ciliate *Paramecium*. The micronucleus of some individuals of *Tetrahymena pyriformis*, a common laboratory species, was ex-perimentally removed in the 1930s, and their descendants continue to reproduce asexually to this day! *Paramecium*, however, is not immortal. The cells divide asexually for about 700 generations and then die if sexual reproduction has not occurred. The micronucleus in ciliates is evidently needed only for sexual reproduction.

Vacuoles

Ciliates form vacuoles for ingesting food and regulating water balance. Food first enters the gullet, which in *Paramecium* is lined with cilia fused into a membrane (see figure 29.16). From the gullet, the food passes into food vacuoles, where enzymes and hydrochloric acid aid in its digestion. Afterward, the vacuole empties its waste contents through a special pore in the pellicle called the **cytoproct,** which is essentially an exocytotic vesicle that appears periodically when solid particles are ready to be expelled.

The contractile vacuoles, which regulate water balance, periodically expand and contract as they empty their contents to the outside of the organism.

Conjugation: Exchange of micronuclei

Like most ciliates, *Paramecium* undergoes a sexual process called **conjugation,** in which two individual cells remain attached to each other for up to several hours (figure 29.17).

Paramecia have multiple mating types. Only cells of two different genetically determined mating types can conjugate.

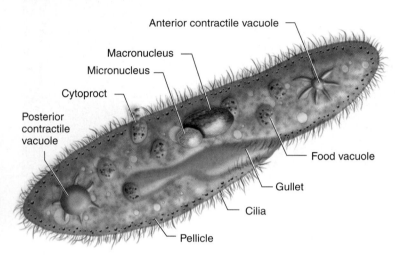

figure 29.16

PARAMECIUM. The main features of this familiar ciliate include cilia, two nuclei, and numerous specialized organelles.

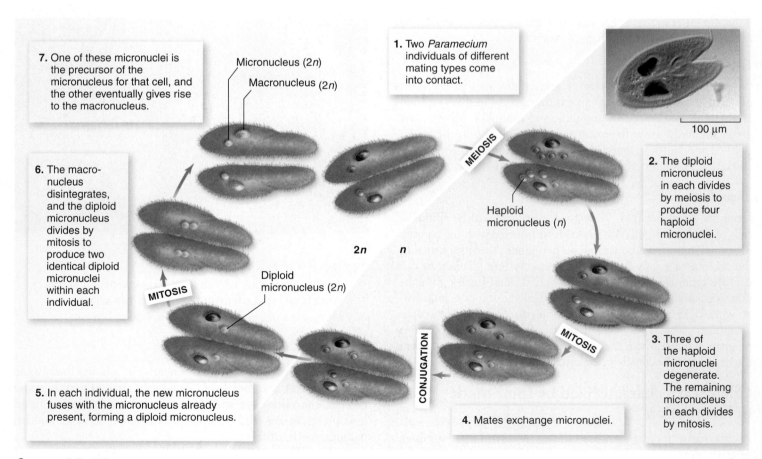

figure 29.17

LIFE CYCLE OF *PARAMECIUM*. In sexual reproduction, two mature cells fuse in a process called conjugation.

Meiosis in the micronuclei of each individual produces several haploid micronuclei, and the two partners exchange a pair of these micronuclei through a cytoplasmic bridge between them.

In each conjugating individual, the new micronucleus fuses with one of the micronuclei already present in that individual, resulting in the production of a new diploid micronucleus. After conjugation, the macronucleus in each cell disintegrates, and the new diploid micronucleus undergoes mitosis, thus giving rise to two new identical diploid micronuclei in each individual.

One of these micronuclei becomes the precursor of the future micronuclei of that cell, while the other micronucleus undergoes multiple rounds of DNA replication, becoming the new macronucleus. This complete segregation of the genetic material is unique to the ciliates and makes them ideal organisms for the study of certain aspects of genetics.

"Killer" strains

Paramecium strains that kill other, sensitive strains of *Paramecium* long puzzled researchers. Initially, killer strains were believed to have genes coding for a substance toxic to sensitive strains. The true source of the toxin turned out to be an endosymbiotic bacterium in the "killer" strains. If this bacterium is engulfed by a "nonkiller" strain, the toxin is released, and the sensitive *Paramecium* dies.

Alveolata comprise what is believed to be a monophyletic group of organisms with varied forms of locomotion and reproduction and characteristic submembrane vesicles.

29.7 Stramenopila: Protists With Fine Hairs

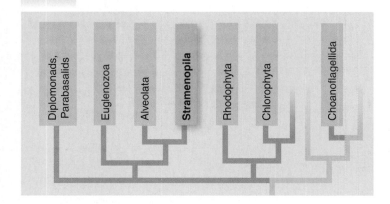

Stramenopiles include **brown algae, diatoms,** and the **oomycetes** (water molds). The name *stramenopila* refers to unique, fine hairs (figure 29.18) found on the flagella of members of this group, although a few species have lost their hairs during evolution.

Brown algae include large seaweeds

Brown algae are the most conspicuous seaweeds in many northern regions (figure 29.19). The life cycle of the brown algae is marked by an alternation of generations between a multicellular

figure 29.19

BROWN ALGA. The giant kelp, *Macrocystis pyrifera*, grows in relatively shallow water along the coasts throughout the world and provides food and shelter for many different kinds of organisms.

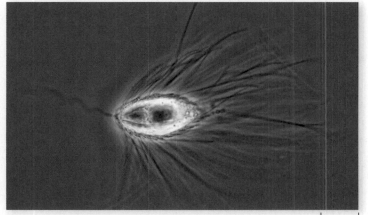

20 µm

figure 29.18

STRAMENOPILES HAVE VERY FINE HAIRS ON THEIR FLAGELLA.

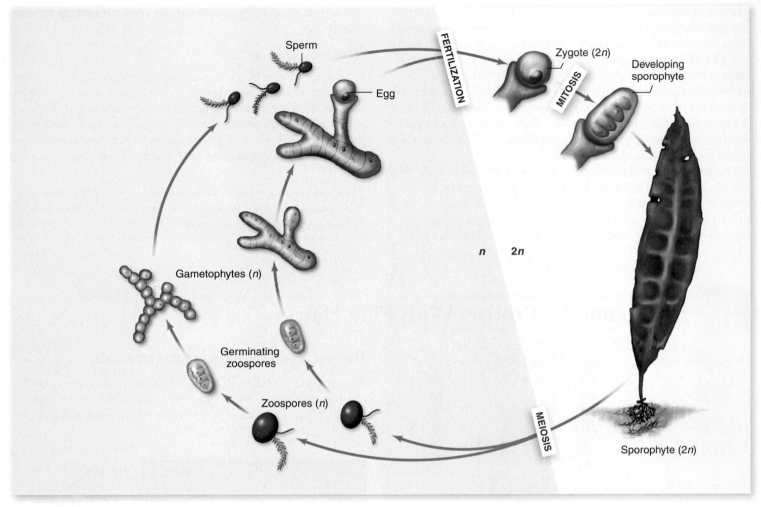

figure 29.20

LIFE CYCLE OF *LAMINARIA*, A BROWN ALGA. Multicellular haploid and diploid stages are found in this life cycle, although the male and female gametophytes are quite small.

sporophyte (diploid) and a multicellular gametophyte (haploid) (figure 29.20). Some sporophyte cells go through meiosis and produce spores. These spores germinate and undergo mitosis to produce the large individuals we recognize, such as the kelps. The gametophytes are often much smaller, filamentous individuals, perhaps a few centimeters in width.

Even in an aquatic environment, tranport can be a challenge for the very large brown algal species. Distinctive transport cells that stack one upon the other enhance transport within some species (see figure 23.10).

Diatoms are unicellular organisms with double shells

Diatoms, members of the phylum Chrysophyta, are photosynthetic, unicellular organisms with unique double shells made of opaline silica, which are often strikingly and characteristically marked (figure 29.21). The shells of diatoms are like small boxes with lids, one half of the shell fitting inside the other. Their chloroplasts, containing chlorophylls *a* and *c*, as well as carotenoids, resemble those of the brown algae and dinoflagellates. Diatoms produce a unique carbohydrate called chrysolaminarin.

60 µm

figure 29.21

DIATOMS. These different radially symmetrical diatoms have unique silica, two-part shells.

Some diatoms move by using two long grooves, called *raphes*, which are lined with vibrating fibrils (figure 29.22). The exact mechanism is still being unraveled and may involve the ejection of mucopolysaccharide streams from the raphe that propel the diatom. Pencil-shaped diatoms can slide back and forth on each other, creating an ever-changing shape.

Oomycetes, the "water molds," have some pathogenic members

All **oomycetes** are either parasites or saprobes (organisms that live by feeding on dead organic matter). At one time, these organisms were considered fungi, which is the origin of the term *water mold* and why their name contains -*mycetes*.

They are distinguished from other protists by the structure of their motile spores, or zoospores, which bear two unequal flagella, one pointed forward and the other backward. Zoospores are produced asexually in a sporangium. Sexual reproduction involves the formation of male and female reproductive organs that produce gametes. Most oomycetes are found in water, but their terrestrial relatives are plant pathogens.

Phytophthora infestans, which causes late blight of potatoes, was responsible for the Irish potato famine of 1845 and 1847. During the famine, about 400,000 people starved to death or died of diseases complicated by starvation. About 2

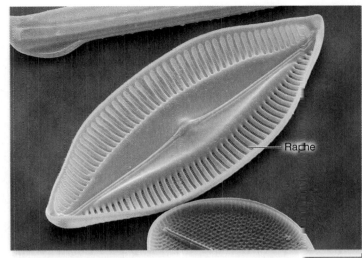

figure 29.22
4.1 μm

DIATOM RAPHE ARE LINED WITH FIBRILS THAT AID IN LOCOMOTION.

million Irish immigrated to the United States and elsewhere as a result of this disaster.

> The diverse stramenopiles are distinguished by fine hairs a derived trait that was later lost in some species.

29.8 Rhodophyta: Red Algae

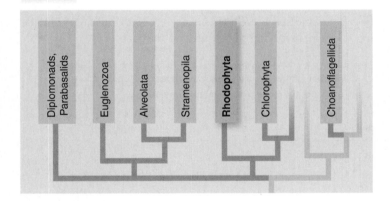

Rhodophyta, the red algae, range in size from microscopic organisms to *Schizymenia borealis* with blades as long as 2 m (figure 29.23). Sushi rolls are wrapped in nori, a red alga. Red algal polysaccharides are used commercially to thicken ice cream and cosmetics.

This lineage lacks flagella and centrioles, and has the accessory photosynthetic pigments phycoerythrin, phycocyanin, and allophycocyanin, which are arranged within structures called *phycobilisomes*. They reproduce using alternation of generations.

The origin of the over 7000 species of Rhodophyta has been a source of controversy. Evidence supporting very early eukaryotic origins and a common ancestry with green algae has been considered. Molecular comparisons of the chloroplasts in red and green algae support a single endosymbiotic origin for both.

Comparisons of the nuclear DNA coding for the large subunit of RNA polymerase II from two red algae, a green alga,

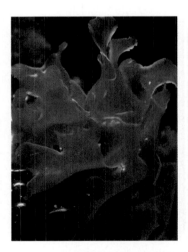

figure 29.23
RED ALGAE COME IN MANY FORMS AND SIZES.

and another protist support the conclusion that the Rhodophyta emerged before the evolutionary lineage that led to plants, animals, and fungi.

How can we reconcile the data from plastid and nuclear DNA? The host cells and the cyanobacterial symbionts probably did not follow congruent evolutionary pathways. The host cell that gave rise to red algae may have been distinct from the one that gave rise to plants. One possibility is that different host cells engulfed the same bacterial symbiont. Tentatively, we will treat Rhodophyta and Chlorophyta (the green algae; see chapter 30) as sister clades based on the substantial amount of chloroplast data.

> Red algae are marine or freshwater organisms capable of photosynthesis. Some members have commercial uses in foods and other products.

29.9 Choanoflagellida: Possible Animal Ancestors

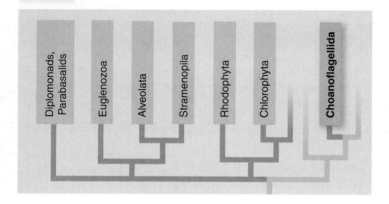

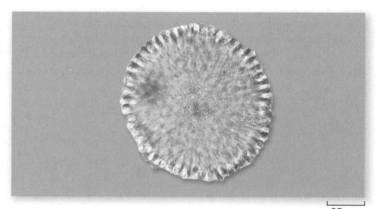

figure 29.24

COLONIAL CHOANOFLAGELLATES RESEMBLE THEIR CLOSE ANIMAL RELATIVES, THE SPONGES.

Choanoflagellates are most like the common ancestor of the sponges and, indeed, all animals. Choanoflagellates have a single emergent flagellum surrounded by a funnel-shaped, contractile collar composed of closely placed filaments, a structure that is exactly matched in the sponges, which are animals. These protists feed on bacteria strained out of the water by their collar. Colonial forms resemble freshwater sponges (figure 29.24).

The close relationship of choanoflagellates to animals was further demonstrated by the strong homology between a surface receptor (a tyrosine kinase receptor) found in choano- flagellates and sponges. This surface receptor initiates a signaling pathway involving phosphorylation (chapter 9).

> Choanoflagellates are the closest relatives of animals. They have features homologous to some found in sponges.

29.10 Protists Without a Clade

Many protists remain to be placed on the tree of life. The following examples are of particular importance to human health and the environment.

Amoebas are paraphyletic

So far, we have organized the protists based on their closest relatives. Some lineages vary tremendously if you consider just a single trait. For example, the stramenopiles include autotrophic, marine algae and terrestrial plant pathogens. As seen in chapter 25, it is also possible for unrelated organisms to acquire similar traits. That is the case with amoebas, which have similar cell morphology, but are not monophyletic.

Rhizopoda: True amoebas

Amoebas move from place to place by means of their pseudopods. Pseudopods are flowing projections of cytoplasm that extend and pull the amoeba forward or engulf food particles, a process called cytoplasmic streaming. An amoeba puts a pseudopod forward and then flows into it (figure 29.25). Microfilaments of actin and myosin similar to those found in muscles are associated with these movements. The

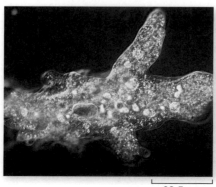

figure 29.25

Amoeba proteus. The projections are pseudopods; an amoeba moves by flowing into them.

62.5 μm

pseudopods can form at any point on the cell body so that it can move in any direction.

Actinopoda: Radiolarians

The pseudopods of amoeboid cells give them truly amorphous bodies. One group, however, has more distinct structures. Members of the phylum Actinopoda, often called **radiolarians,** secrete glassy exoskeletons made of silica. These skeletons give the unicellular organisms a distinct shape, exhibiting either bilateral or radial symmetry. The shells of different species form many elaborate and beautiful shapes, with pseudopods extruding outward along spiky projections of the skeleton (figure 29.26). Microtubules support these cytoplasmic projections.

Foraminifera fossils created huge limestone deposits

Members of the phylum **Foraminifera** are heterotrophic marine protists. They range in diameter from about 20 μm to several centimeters. They resemble tiny snails and can form 3-m-deep layers in marine sediments. Characteristic of the group are pore-studded shells (called *tests*) composed of organic materials usually reinforced with grains of inorganic matter. These grains may be calcium carbonate, sand, or even plates from the shells of echinoderms or spicules (minute needles of calcium carbonate) from sponge skeletons.

Depending on the building materials they use, foraminifera—often informally called "forams"—may have shells of very different appearance. Some of them are brilliantly colored red, salmon, or yellow-brown.

Most foraminifera live in sand or are attached to other organisms, but two families consist of free-floating planktonic organisms. Their tests may be single-chambered, but are more often multichambered, and they sometimes have a spiral shape resembling that of a tiny snail. Thin cytoplasmic projections called *podia* emerge through openings in the tests (figure 29.27). Podia are used for swimming, gathering materials for the tests, and feeding. Forams eat a wide variety of small organisms.

The life cycles of foraminifera are extremely complex, involving alternation between haploid and diploid generations.

figure 29.27

A REPRESENTATIVE OF THE FORAMINIFERA. Podia, thin cytoplasmic projections, extend through pores in the calcareous test, or shell, of this living foram.

8.3 μm

Forams have contributed massive accumulations of their tests to the fossil record for more than 200 million years. Because of the excellent preservation of their tests and the striking differences among them, forams are very important as geological markers. The pattern of occurrence of different forams is often used as a guide in searching for oil-bearing strata. Limestones all over the world, including the famous White Cliffs of Dover in southern England, are often rich in forams (figure 29.28).

Slime molds exhibit "group behavior"

Slime molds originated at least three distinct times, and the three lineages are very distantly related. Like water molds, these organisms were once considered fungi. We will explore two lineages: the plasmodial slime molds, which are huge, single-celled, multinucleate, oozing masses, and the cellular slime molds, in which single cells combine and differentiate, creating an early model of multicellularity.

figure 29.28

WHITE CLIFFS OF DOVER. The limestone that forms these cliffs is composed almost entirely of fossil shells of protists, including foraminifera.

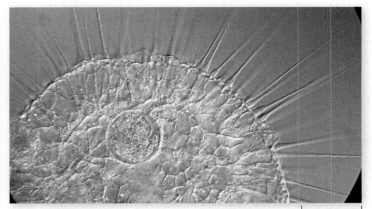

33.3 μm

figure 29.26

ACTINOSPHAERIUM WITH NEEDLELIKE PSEUDOPODS.

figure 29.29

A PLASMODIAL PROTIST. This multinucleate pretzel slime mold, *Hemitrichia serpula*, moves about in search of the bacteria and other organic particles that it ingests.

figure 29.30

SPORANGIA OF A PLASMODIAL SLIME MOLD. These *Arcyria* sporangia are found in the phylum Myxomycota.

Plasmodial slime molds

Plasmodial slime molds stream along as a **plasmodium,** a non-walled, multinucleate mass of cytoplasm that resembles a moving mass of slime (figure 29.29). This form is called the *feeding phase,* and the plasmodia may be orange, yellow, or another color.

Plasmodia show a back-and-forth streaming of cytoplasm that is very conspicuous, especially under a microscope. They are able to pass through the mesh in cloth or simply to flow around or through other obstacles. As they move, they engulf and digest bacteria, yeasts, and other small particles of organic matter.

A multinucleated *Plasmodium* cell undergoes mitosis synchronously, with the nuclear envelope breaking down, but only at late anaphase or telophase. Centrioles are absent in cellular slime molds.

When either food or moisture is in short supply, the plasmodium migrates relatively rapidly to a new area. Here it stops moving and either forms a mass in which spores differentiate or divides into a large number of small mounds, each of which produces a single, mature **sporangium,** the structure in which spores are produced. These sporangia are often beautiful and extremely complex in form

(figure 29.30). The spores are highly resistant to unfavorable environmental influences and may last for years if kept dry.

Cellular slime molds

The cellular slime molds have become an important group for the study of cell differentiation because of their relatively simple developmental systems (figure 29.31). The individual organisms behave as separate amoebas, moving through the soil and ingesting bacteria. When food becomes scarce, the individuals aggregate to form a moving "slug." Cyclic adenosine monophosphate (cAMP) is sent out in pulses by some of the cells, and other cells move in the direction of the cAMP to form the slug. In the cellular slime mold *Dictyostelium discoideum,* this slug goes through morphogenesis to make stalk and spore cells. The spores then go on to form a new amoeba if they land in a moist habitat.

> The evolutionary origins of some protists, including amoebas and slime molds, are less well understood, and these organisms may have arisen independently more than once.

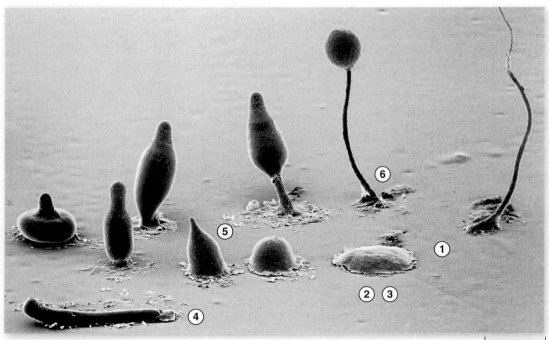

figure 29.31

DEVELOPMENT IN *Dictyostelium discoideum,* **A CELLULAR SLIME MOLD.** *1.* First, a spore germinates, forming an amoeba. The amoebas feed and reproduce until the food runs out. *2.* The amoebas aggregate and move toward a fixed center. *3.* Next, they form a multicellular "slug," 2–3 mm long, that migrates toward light. *4.* The slug stops moving and begins to differentiate into (*5*) a spore-forming body called a sorocarp. *6.* Within the heads of the sorocarps, amoebas become encysted as spores.

.01 μm

29.1 Defining Protists

Protists are the most diverse of the four kingdoms in the domain Eukarya, containing unicellular, colonial, and multicellular groups.

- Protista are paraphyletic and grouped for convenience; they are not a monophyletic kingdom.

- Based on molecular methods, the 15 major protist phyla are grouped into seven major monophyletic groups. About 60 protist lineages cannot accurately be placed in the tree of life (figure 29.1).

29.2 Eukaryotic Origins and Endosymbiosis

Eukaryotes are distinguished from prokaryotes by the presence of a cytoskeleton and compartmentalization that includes a nuclear envelope and organelles.

- Differences in the appearance of microfossils occurred about 1.5 BYA.

- Eukaryotic cells evolved through horizontal gene transfer, infolding of membranes to form the nuclear envelope and endoplasmic reticulum, and endosymbiosis (figures 29.3, 29.4, and 29.5).

- Endosymbiosis involved the incorporation of prokaryotes to form mitochondria, chloroplasts, and probably centrioles.

- Over time, genes from endosymbiotic organisms moved into the eukaryotic nucleus, preventing these organelles from living in a cell-free culture.

29.3 General Biology of the Protists

Protists are united on the basis of a single negative characteristic—they do not belong to the other kingdoms.

- Protist cell surfaces include a plasma membrane or a membrane covered with extracellular material.

- Fragile protists can survive in harsh environments by forming protective cysts and shutting down metabolism.

- Protists can move by flagella, cilia, or pseudopods such as lobopods, filopods, or axopods.

- Protists employ all nutritional strategies except chemoautotrophy. They can be phototrophs, heterotrophs that engulf either particulate matter (phagotrophs) or soluble foods (osmotrophs), or mixotrophs that can utilize both photosynthesis and heterotrophy.

- Protists can reproduce asexually by mitosis, budding, or schizogony or sexually by forming gametes by meiosis.

- Protists may live in colonies, which promotes the development of multicellularity and division of labor.

29.4 Diplomonads and Parabasalids: Flagellated Protists Lacking Mitochondria

The lack of mitochondria in diplomonads and parabasalids is believed to be a derived, not an ancestral, trait.

- Diplomonads move with flagella and have two nuclei.

- In addition to a flagellum, parabasalids have an undulating membrane to aid in locomotion.

29.5 Euglenozoa: A Diverse Group in Which Some Members Have Chloroplasts

The Euglenozoa acquired chloroplasts by endosymbiosis.

- Euglenoids are free-living eukaryotes that have a pellicle, anterior flagella, and may have chloroplasts (figure 29.8).

- Kinetoplastids are parasitic and have a single, unique mitochondrion with two types of circular DNA, one of which is involved in RNA editing.

29.6 Alveolata: Protists With Submembrane Vesicles

The diverse members of the Alveolata share a common trait—flattened vesicles called alveoli that function like Golgi bodies below their plasma membrane (figure 29.10).

- Most dinoflagellates are photosynthetic unicells with two flagella. Most contain chlorophylls *a* and *c*, in addition to carotenoids. They may produce red tides that are toxic to vertebrates.

- Apicomplexans have a unique arrangement of organelles at one end of the cell, the apical complex used to invade a host, and are spore-forming parasites.

- Ciliates move by cilia and have a macronucleus and a micronucleus, which are exchanged during conjugation.

29.7 Strameopila: Protists With Fine Hairs

Strameopilas contain fine hairs on their flagella although a few members of this group have lost their flagella.

- Brown algae are conspicuous seaweeds with multicellular alternation of generation, and meiosis produces spores, not gametes (figure 29.20).

- Diatoms are unique unicellular organisms with double shells made of silica. They move using two long grooves called raphes that are lined with vibrating fibrils (figure 29.22).

- Oomycetes resemble fungi. They are either parasites or saprobes and are distinguished from other protists by their motile zoospores, which bear unequal flagella.

29.8 Rhodophyta: Red Algae

Red algae vary greatly in size, and their origin is the subject of controversy.

- Red algae lack flagella and centrioles.

- Red algae have accessory photosynthetic pigments: phycoerythrin, phycocyanin, and allophycocyanin located in phycobilisomes.

- Comparison of the DNA code for RNA polymerase II supports the conclusion that red algae emerged before the lineage that led to plants, animals, and fungi.

29.9 Choanoflagellida: Possible Animal Ancestors

Choanoflagellates are the closest relatives of animals.

- Choanoflagellates have a single emergent flagellum surrounded by a funnel-shaped, contractile collar that is the same as found in sponges.

- Choanoflagellates also have a surface tyrosine kinase receptor that is found in sponges.

29.10 Protists Without a Clade

Not all protists can be placed on the tree of life at this time.

- Amoebas have similar morphologies, but they are paraphyletic.

- Foraminifera are heterotrophic marine protists with pore-studded shells or tests (figure 29.27).

- Slime molds originated at three times with distinct lineages. Two of these lineages include plasmodial slime molds that stream along as a plasmodium, a nonwalled, multinucleate mass and cellular slime molds that behave like amoeba until food becomes scarce, at which time they aggregate to form the sorocarp (figures 29.29 and 29.31).

SELF TEST

1. Which of the following is correct regarding the phylogenetic nature of kingdom Protista?
   a. Scientists have placed all of the known protist species into at least one of the major monophyletic clades.
   b. All members of the kingdom share characteristics that are unique to Protists.
   c. The kingdom as a whole is paraphyletic.
   d. All of the above are correct.

2. Which of the following events is believed to have occurred first in eukaryotic evolution?
   a. Secondary endosymbiosis of chloroplasts
   b. Endosymbiosis of a cyanobacteria
   c. Endosymbiosis of an energy-producing bacteria
   d. Infolding of the plasma membrane

3. Which of the following would be a key identifying characteristic of an organelle that has undergone secondary endosymbiosis?
   a. A single plasma membrane
   b. A set of two plasma membranes
   c. Four plasma membranes
   d. None of the above

4. The evolution of which of the following provides for higher levels of genetic recombination and variation?
   a. Schizogony
   b. Sexual reproduction
   c. Asexual reproduction
   d. Budding

5. A protist that lacks mitochondria would most likely be classified with the
   a. Rhodoyphyta.
   b. Parabasalida.
   c. Euglenozoa.
   d. Stramenophila.

6. Which of the following statements best describes the term *kinetoplastid*?
   a. An organism with a single mitochondrion in each cell
   b. A form of locomotion in the Alveolata
   c. A form of asexual reproduction in the euglenoids
   d. A Golgi-like organelle found in the Alveolata

7. A species that is photosynthetic, with fine hairs on the flagella, would be classified with which of the following groups?
   a. Chlorophyta
   b. Euglenozoa
   c. Alveolata
   d. Stramenophila

8. The organism that causes red tide is a(n) _____ and belongs to the clade _____.
   a. apicomplexan; Alveolata
   b. kinetoplastid; Euglenozoa
   c. dinoflagellate; Alveolata
   d. brown algae; Stramenopila

9. What characteristic separates the Rhodophyta and the Chlorophyta?
   a. Method of locomotion
   b. Lack of mitochondria
   c. Method of reproduction
   d. Form of photosynthetic pigments

10. Which of the following represents a monophyletic group?
    a. Euglenoids
    b. Amoebas
    c. Forams
    d. Slime molds

11. The disease malaria is caused by a species of the genus _____, which belongs to the _____.
    a. *Giardia;* diplomonads
    b. *Plasmodium;* apicomplexans
    c. *Paramecium;* ciliates
    d. *Trypanosoma;* kinetoplastids

12. In the following list, which multicellular eukaryotic kingdom is *not* linked to its correct protistan ancestor?
    a. Viridiplantae—Chlorophyta
    b. Fungi—Oomycetes
    c. Animals—Choanoflagellates
    d. All of the above are linked correctly.

13. One of the earliest examples of multicellularity occurs in the
    a. euglenoids.
    b. amoebas.
    c. red algae.
    d. cellular slime molds.

14. The origin of which of the following does *not* represent an endosymbiotic event?
    a. Mitochondria
    b. Chloroplasts
    c. Nucleus
    d. All of the above are endosymbiotic events.

15. In *Paramecium*, removal of which of the following organelles would inhibit sexual reproduction?
    a. The alveoli
    b. The micronucleus
    c. The macronucleus
    d. The cytoproct

CHALLENGE QUESTIONS

1. Many species of diplomonads and parabasalids lack mitochondria, yet remain aerobic. In chapter 9 you were introduced to the mitochondria as the site of aerobic respiration. If these species retained their aerobic capabilities, what must have happened to the metabolic pathways?

2. Review the diversity of photosynthetic pigments in the Protista. Can photosynthesis be used as a basis of classification? Explain your answer.

3. Most scientists will agree that the protist kingdom is a mess. Some scientists believe that these organisms should be placed within the multicellular eukaryotic kingdoms (plants, fungi, and animals). Do you think that this approach is practical? Support your answer.

4. You are viewing a previously unidentified alga under a high-powered microscope. The chloroplast appears to have four membranes surrounding it. Which group of algae would you predict is most closely related to this new species? Why?

chapter 30

Overview of Green Plants

introduction

PLANT EVOLUTION IS THE STORY of adaptation to terrestrial life by green algal ancestors. All green algae and land plants share a common ancestor, composing a monophyletic group called the green plants. For about 500 million years, algae were confined to a watery domain, limited by the need for water, which was necessary for reproduction, structural support, prevention of water loss, and some protection from the sun's ultraviolet irradiation. Numerous evolutionary solutions to terrestrial challenges have resulted in over 300,000 species of plants dominating all terrestrial communities today, from forests to alpine tundra and from agricultural fields to deserts. Most plants are photosynthetic, and we rely on plants for food, clothing, wood for shelter and fuel, chemicals, and many medicines. This chapter explores the evolutionary history and strategies of the green plants.

30.1 Defining Plants

As you saw in chapter 26, the phylogenetic revolution has completely altered our definition of a plant. We now know that all green algae and the land plants shared a common ancestor a little over a billion years ago, and the two groups are now recognized as a kingdom or crown group referred to as the **Viridiplantae,** or simply the **green plants.** DNA sequence data are consistent with the claim that a single "Eve" gave rise to all plants. Thus, plants are all members of the Viridiplantae.

The definition of a plant is broad, but it excludes the red and brown algae. All algae—red, brown, and green—shared a primary endosymbiotic event 1500 MYA. But sharing an ancestral chloroplast lineage is not the same as being monophyletic. Red and green algae last shared a common ancestor about 1400 MYA. Brown algae became photosynthetic through endosymbiosis with a eukaryotic red alga that had itself already acquired a photosynthetic cyanobacterium, as described in the preceding chapter.

Plants are also not fungi, which are more closely related to metazoan animals (see chapter 31). Fungi, however, were essential to the colonization of land by the green plants.

Land plants evolved from freshwater algae

Some saltwater algae evolved to thrive in a freshwater environment. Just a single species of freshwater green algae gave rise to the entire terrestrial plant lineage, from mosses through the flowering plants (angiosperms). Given the incredibly harsh conditions of life on land, it is not surprising that all land plants share a single common ancestor. Exactly what this ancestral alga was is still a mystery, but close relatives, the Charales (members of the charophytes), exist in freshwater lakes today.

The green algae split into two major clades: the chlorophytes, which never made it to land, and the charophytes, which are sister to all the land plants (figure 30.1). Land plants, although diverse, have certain characteristics in common. For example, all of them afford some protection to their embryos, and all have multicellular haploid and diploid phases. Over time, the trend has been toward more embryo protection and a smaller haploid stage in the life cycle. Recessive, deleterious genes affect the phenotype and survival of the haploid phase, but are masked in the diploid phase.

Land plants have adapted to terrestrial life

Unlike their freshwater ancestors, most land plants have only limited amounts of water available. As an adaptation to living on land, most plants are protected from **desiccation**—the tendency of organisms to lose water to the air—by a waxy **cuticle** that is secreted onto their exposed surfaces. The cuticle is relatively impermeable, preventing water loss. This solution, however, limits the gas exchange essential for respiration and photosynthesis. Gas diffusion into and out of a plant occurs through tiny mouth-shaped openings called **stomata** (singular, *stoma*), which allows water to diffuse out at the same time. Stomata can be closed at times to limit water loss.

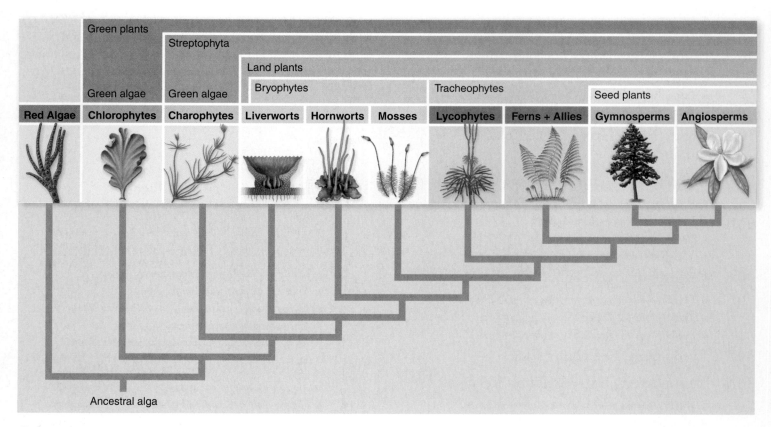

figure 30.1

GREEN PLANT PHYLOGENY.

Members of the land plants can be distinguished based on the presence or absence of **tracheids,** specialized cells that facilitate the transport of water and minerals (chapter 36). **Tracheophytes** have tracheids and have evolved highly efficient transport systems: water-conducting xylem and food-conducting phloem strands of tissues in their stems, roots, and leaves.

Two additional adaptations also allowed larger land plants to flourish. The evolution of leaves, which may have occurred multiple times, resulted in an increased photosynthetic surface area. The shift to a dominant diploid generation, accompanied by the structural support of conducting tissue, allowed plants to take advantage of the vertical dimension of the terrestrial environment, making the evolution of trees possible. For the purposes of discussion in this chapter, we subdivide the green plants into phylogenetic groups emphasizing the acquisition of traits that led to the remarkable colonization of land and to the vast diversity of green plants today.

> Green plants include all green algae and the land plants. A single freshwater green alga successfully invaded land, and its descendants eventually developed reproductive strategies, conducting systems, stomata, and cuticles that adapt them well for life on land.

30.2 Plant Life Cycles

Multicellular plants undergo mitosis after both gamete fusion and meiosis. The result is a multicellular haploid individual and a multicellular diploid individual—unlike in the human life cycle, in which gamete fusion directly follows meiosis. Humans have a **diplontic** life cycle, meaning that only the diploid stage is multicellular; by contrast, the plant life cycle is **haplodiplontic,** having multicellular haploid and diploid stages.

The haplodiplontic cycle produces alternation of generations

The basic haplodiplontic cycle is summarized in figure 30.2. Brown, red, and green algae are also haplodiplontic. Humans produce gametes via meiosis, but land plants actually produce gametes by *mitosis* in a multicellular, haploid individual. The diploid generation, or **sporophyte,** alternates with the haploid generation, or **gametophyte.** Sporophyte means "spore plant," and gametophyte means "gamete plant." These terms indicate the kinds of reproductive cells the respective generations produce.

The diploid sporophyte produces haploid spores (not gametes) by meiosis. Meiosis takes place in structures called **sporangia,** where diploid **spore mother cells (sporocytes)** undergo meiosis, each producing four haploid **spores.** Spores are the first cells of the gametophyte generation. Spores divide by mitosis, producing a multicellular, haploid gametophyte.

The haploid gametophyte is the source of gametes. When the gametes fuse, the zygote they form is diploid and is the first cell of the next sporophyte generation. The zygote grows into a diploid sporophyte by mitosis and produces sporangia in which meiosis ultimately occurs.

The relative sizes of haploid and diploid generations vary

All plants are haplodiplontic; however, the haploid generation consumes a much larger portion of the life cycle in mosses and ferns than it does in gymnosperms and angiosperms. In mosses, liverworts, and ferns, the gametophyte is photosynthetic and free-living. When you look at mosses, what you see is largely gametophyte tissue; the sporophytes are usually

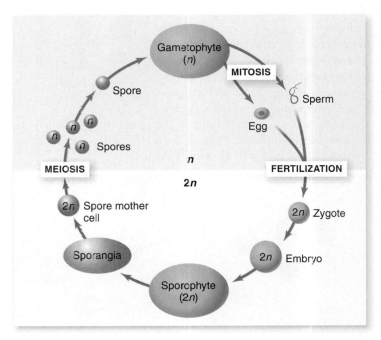

figure 30.2

A GENERALIZED MULTICELLULAR PLANT LIFE CYCLE. Note that both haploid and diploid individuals can be multicellular. Also, spores are produced by meiosis, while gametes are produced by mitosis.

smaller, brownish or yellowish structures attached to the tissues of the gametophyte. In other plants, the gametophyte is usually nutritionally dependent on the sporophyte. When you look at a gymnosperm or angiosperm, such as most trees, the largest, most visible portion is a sporophyte.

Although the sporophyte generation can get very large, the size of the gametophyte is limited in all plants. The gametophyte generation of mosses produces gametes at its tips. The egg is stationary, and sperm lands near the egg in a droplet of water. If the moss were the height of a sequoia, not only would vascular tissue be needed for conduction and support, but the sperm would have to swim up the tree! In contrast, the small fern gametophyte develops on the forest floor where gametes can meet. Tree ferns are especially abundant in Australia; the haploid spores the sporophyte trees produce fall to the ground and develop into gametophytes.

Having completed an overview of plant life cycles, we next consider the major plant groups. As we proceed, you will see a reduction of the gametophyte from group to group, a loss of multicellular **gametangia** (structures in which gametes are produced), and increasing specialization for life on land, including the remarkable structural adaptations of the flowering plants, the dominant plants today.

> Multicellular plants have haplodiplontic life cycles. Diploid multicellular sporophytes produce haploid spores by meiosis. Spores develop into haploid, multicellular gametophytes by mitosis and produce haploid gametes by mitosis.

30.3 Chlorophytes: Aquatic Green Algae

Green algae have two distinct lineages: the **chlorophytes,** discussed here, and another lineage, the **streptophytes,** that gave rise to the land plants (see figure 30.1). The chlorophytes are of special interest here because of their unusual diversity and lines of specialization. The chlorophytes have an extensive fossil record dating back 900 million years. Modern chlorophytes closely resemble land plants, especially in their chloroplasts, which are biochemically similar to those of the plants. They contain chlorophylls *a* and *b*, as well as carotenoids.

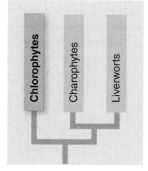

Chlorophytes can be unicellular

The green alga *Chlamydomonas* probably represents a primitive state for green algae (figure 30.3). Individuals are microscopic (usually less than 25 μm long), green, and rounded, and they have two flagella at the anterior end. They move rapidly in water by beating their flagella in opposite directions. Each individual has an eyespot, which contains about 100,000 molecules of rhodopsin, the same pigment employed in vertebrate eyes. Light received by this eyespot is used by the alga to help direct its swimming. Most individuals of *Chlamydomonas* are haploid. *Chlamydomonas* reproduces asexually as well as sexually (see figure 30.3).

Several lines of evolutionary specialization have been derived from organisms such as *Chlamydomonas*. The first is the evolution of nonmotile, unicellular green algae. *Chlamydomonas* is capable of retracting its flagella and settling down as an immobile unicellular organism if the pond in which it lives dries out. Some common algae found in soil and bark, such as *Chlorella*, are essentially like *Chlamydomonas* in this trait, but they do not have the ability to form flagella. *Chlorella* is widespread in

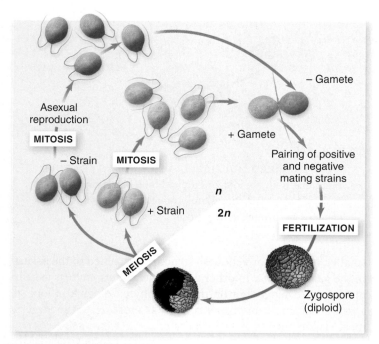

figure 30.3

***Chlamydomonas* LIFE CYCLE.** This single-celled chlorophyte has both asexual and sexual reproduction. Unlike multicellular green plants, gamete fusion is not followed by mitosis.

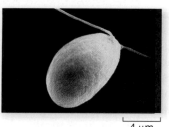

4 μm

figure 30.4

Volvox. This chlorophyte forms a colony where some cells specialize for reproduction. *Volvox* represents an intermediate stage on the way to multicellularity.

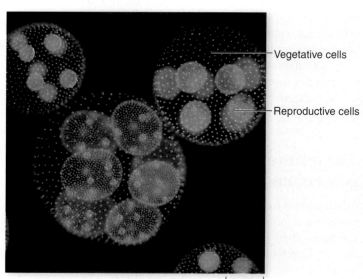

Vegetative cells

Reproductive cells

1666 μm

both fresh and salt water as well as in soil, and is only known to reproduce asexually.

Colonial chlorophytes have some cell specialization

Another major line of specialization from cells like those of *Chlamydomonas* concerns the formation of motile, colonial organisms. In these genera of green algae, the *Chlamydomonas*-like cells retain some of their individuality.

The most elaborate of these organisms is *Volvox* (figure 30.4), a hollow sphere made up of a single layer of 500 to 60,000 individual cells, each cell having two flagella. Only a small number of the cells are reproductive. Some reproductive cells may divide asexually, bulge inward, and give rise to new colonies that initially remain within the parent colony. Others produce gametes.

Multicellular chlorophytes can have haplodiplontic life cycles

Haplodiplontic life cycles are found in some chlorophytes and the streptophytes, which include both charophytes and land plants. *Ulva*, a multicellular chlorophyte, has identical gametophyte and sporophyte generations that consist of flattened sheets two cells thick (figure 30.5). Unlike the charophytes, none of the ancestral chlorophytes gave rise to land plants.

> Nonmotile, unicellular algae and multicellular, flagellated colonies have been derived from green algae such as *Chlamydomonas*—a biflagellated, unicellular organism. Multicellular chlorophytes, including *Ulva*, have also been derived from simpler, unicellular ancestors. Chlorophytes did not give rise to land plants.

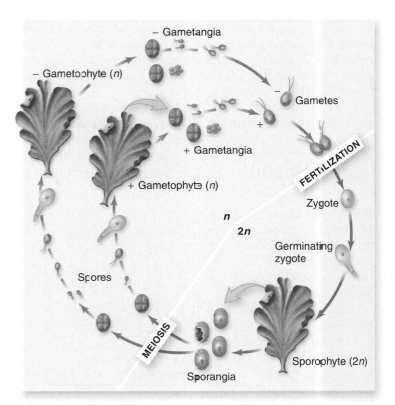

figure 30.5

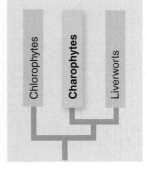

LIFE CYCLE OF *Ulva*. This chlorophyte alga has a haplodiplontic life cycle. The gametophyte and sporophyte are multicellular and identical in appearance.

inquiry

? *Are Ulva gametes formed by meiosis? Explain your response.*

30.4 Charophytes: Green Algae Related to Land Plants

Charophytes, a clade of streptophytes, are also green algae, distinguished from chlorophytes by their close phylogenetic relationship to the land plants. Identifying which of the charophyte clades is sister (most closely related) to the land plants puzzled biologists for a long time. Currently, the molecular evidence from rRNA and DNA sequences favors the charophytes as the green algal clade in the streptophytes.

The two candidate Charophyta clades have been the Charales, with about 300 species, and the Coleochaetales, with about 30 species (figure 30.6). Both lineages are primarily freshwater algae, but the Charales are huge relative to the microscopic Coleochaetales. Both clades have similarities to land plants. *Coleochaete* and its relatives have cytoplasmic linkages

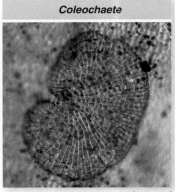

figure 30.6

***Chara* AND *Coleochaete* REPRESENT THE TWO CLADES MOST CLOSELY RELATED TO LAND PLANTS.** *Chara* and its relatives are now believed to share a common ancestor with the alga that gave rise to all terrestrial plant life.

100 μm

between cells called *plasmodesmata*, which are found in land plants. *Chara* undergoes mitosis and cytokinesis like land plant cells. Both charophyte clades formed green mats around the edges of freshwater ponds and marshes, and one species must have successfully inched its way onto land.

At the moment, the Charales appear to be the sister clade to land plants, with the Coleochaetales the next closest

relatives. Charales fossils dating back 420 million years indicate that the common ancestor of land plants was a relatively complex freshwater alga.

Charales are green algae that are most likely sister to the land plants.

30.5 Bryophytes: Nontracheophyte Green Plants

Bryophytes are the closest living descendants of the first land plants. Plants in this group are also called nontracheophytes because they lack the derived transport cell called a *tracheid*.

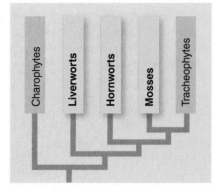

Fossil evidence and molecular systematics can be used to reconstruct early terrestrial plant life. Water and gas availability were limiting factors. These plants likely had little ability to regulate internal water levels and likely tolerated dessication, traits found in most extant mosses, although some are aquatic.

Algae, including the Charales, lack roots. Fungi and early land plants cohabitated, and the fungi formed close associations with the plants that enhanced water uptake. The tight symbiotic relationship between fungi and plants, called **mycorrhizal associations,** are also found in many existing bryophytes.

Bryophytes are unspecialized but successful in many environments

The approximately 24,700 species of bryophytes are simple but highly adapted to a diversity of terrestrial environments, even deserts. Most bryophytes are small; few exceed 7 cm in height. Bryophytes have conducting cells other than tracheids for water and nutrients. The tracheid is a derived trait that characterizes the tracheophytes, all land plants but the bryophytes. Bryophytes are sometimes called nonvascular plants, but *nontracheophyte* is a more accurate term because they do have conducting cells of different types.

Scientists now agree that bryophytes consist of three quite distinct clades of relatively unspecialized plants: **liverworts, hornworts,** and **mosses.** Their gametophytes are photosynthetic and are more conspicuous than the sporophytes. Sporophytes are attached to the gametophytes and depend on them nutritionally in varying degrees. Some of the sporophytes are completely enclosed within gametophyte tissue; others are not and usually turn brownish or straw-colored at maturity. Like ferns and certain other vascular (tracheophyte) plants, bryophytes require water (such as rainwater) to reproduce sexually, tracing back to their aquatic regions. It is not surprising that they are especially common in moist places, both in the tropics and temperate regions.

Liverworts are an ancient phylum

The Old English word *wyrt* means "plant" or "herb." Some common **liverworts** (phylum Hepaticophyta) have flattened gametophytes with lobes resembling those of liver—hence the name "liverwort." Although the lobed liverworts are the best known representatives of this phylum, they constitute only about 20% of the species (figure 30.7). The other 80% are leafy and superficially resemble mosses. The gametophytes are prostrate instead of erect, and the rhizoids are one-celled.

Some liverworts have air chambers containing upright, branching rows of photosynthetic cells, each chamber having a pore at the top to facilitate gas exchange. Unlike stomata, the pores are fixed open and cannot close.

Sexual reproduction in liverworts is similar to that in mosses. Lobed liverworts may form gametangia in umbrella-like structures. Asexual reproduction occurs when lens-shaped pieces of tissue that are released from the gametophyte grow to form new gametophytes.

Hornworts developed stomata

The origin of **hornworts** (phylum Anthocerotophyta) is a puzzle. They are most likely among the earliest land plants, yet the earliest hornwort fossil spores date from the Cretaceous period (65–145 MYA), when angiosperms were emerging.

Female gametophyte

figure 30.7

A COMMON LIVERWORT, *Marchantia* (PHYLUM HEPATICOPHYTA). The microscopic sporophytes are formed by fertilization within the tissues of the umbrella-shaped structures that arise from the surface of the flat, green, creeping gametophyte.

figure 30.8

HORNWORTS (PHYLUM ANTHOCEROTOPHYTA). Hornwort sporophytes are seen in this photo. Unlike the sporophytes of other bryophytes, most hornwort sporophytes are photosynthetic.

The small hornwort sporophytes resemble tiny green broom handles or horns, rising from filmy gametophytes usually less than 2 cm in diameter (figure 30.8). The sporophyte base is embedded in gametophyte tissue, from which it derives some of its nutrition. However, the sporophyte has stomata to regulate gas exchange, is photosynthetic, and provides much of the energy needed for growth and reproduction. Hornwort cells usually have a single large chloroplast.

Mosses have rhizoids and water-conducting tissue

Unlike other bryophytes, the gametophytes of **mosses** typically consist of small, leaflike structures (not true leaves, which contain vascular tissue) arranged spirally or alternately around a stemlike axis (figure 30.9); the axis is anchored to its substrate

figure 30.9

A HAIR-CUP MOSS, *Polytrichum* **(PHYLUM BRYOPHYTA).** The leaflike structures belong to the gametophyte. Each of the yellowish-brown stalks with a capsule, or sporangium, at its summit is a sporophyte.

by means of **rhizoids.** Each rhizoid consists of several cells that absorb water, but not nearly the volume of water that is absorbed by a vascular plant root.

Moss leaflike structures have little in common with leaves of vascular plants, except for the superficial appearance of the green, flattened blade and slightly thickened midrib that runs lengthwise down the middle. Only one cell layer thick (except at the midrib), they lack vascular strands and stomata, and all the cells are haploid.

Water may rise up a strand of specialized cells in the center of a moss gametophyte axis. Some mosses also have specialized food-conducting cells surrounding those that conduct water.

Moss reproduction

Multicellular gametangia are formed at the tips of the leafy gametophytes (figure 30.10). Female gametangia (**archegonia**) may develop either on the same gametophyte as the male gametangia (**antheridia**) or on separate plants. A single egg is produced in the swollen lower part of an archegonium, whereas numerous sperm are produced in an antheridium.

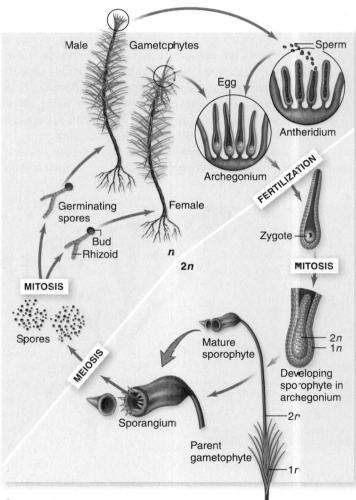

figure 30.10

LIFE CYCLE OF A TYPICAL MOSS. The majority of the life cycle of a moss is in the haploid state. The leafy gametophyte is photosynthetic, but the smaller sporophyte is not, and is nutritionally dependent on the gametophyte. Water is required to carry sperm to the egg.

When sperm are released from an antheridium, they swim with the aid of flagella through a film of dew or rainwater to the archegonia. One sperm (which is haploid) unites with an egg (also haploid), forming a diploid zygote. The zygote divides by mitosis and develops into the sporophyte, a slender, basal stalk with a swollen capsule, the *sporangium*, at its tip. As the sporophyte develops, its base is embedded in gametophyte tissue, its nutritional source.

The sporangium is often cylindrical or club-shaped. Spore mother cells within the sporangium undergo meiosis, each producing four haploid spores. In many mosses at maturity, the top of the sporangium pops off, and the spores are released. A spore that lands in a suitable damp location may germinate and grow, using mitosis, into a threadlike structure, which branches to form rhizoids and "buds" that grow upright. Each bud develops into a new gametophyte plant consisting of a leafy axis.

Moss distribution

In the Arctic and the Antarctic, mosses are the most abundant plants, boasting the largest number of individuals in these harsh regions. The greatest diversity of moss species, however, is found in the tropics. Many mosses are able to withstand prolonged periods of drought, although mosses are not common in deserts.

Most mosses are highly sensitive to air pollution and are rarely found in abundance in or near cities or other areas with high levels of air pollution. Some mosses, such as the peat mosses (*Sphagnum*), can absorb up to 25 times their weight in water and are valuable commercially as a soil conditioner or as a fuel when dry.

> The three major clades of nontracheophyte plants, liverworts, hornworts, and mosses, are all relatively unspecialized, but well suited for diverse terrestrial environments.

30.6 Features of Tracheophyte Plants

The first tracheophytes for which we have a relatively complete record belonged to the phylum Rhyniophyta. They flourished some 410 MYA, but are now extinct. We are not certain what the very earliest of these vascular plants looked like, but fossils of *Cooksonia* provide some insight into their characteristics (figure 30.11).

Cooksonia, the first known vascular land plant, appeared in the late Silurian period about 420 MYA. It was successful partly because it encountered little competition as it spread out over vast tracts of land. The plants were only a few centimeters tall and had

no roots or leaves. They consisted of little more than a branching axis, the branches forking evenly and expanding slightly toward the tips. They were **homosporous** (producing only one type of spore). Sporangia formed at branch tips. Other ancient vascular plants that followed evolved more complex arrangements of sporangia. Leaves began to appear as protuberances from stems.

Vascular tissue allows for distribution of nutrients

Cooksonia and the other early plants that followed it became successful colonizers of the land by developing efficient water- and food-conducting systems called *vascular tissues*. These tissues consist of strands of specialized cylindrical or elongated cells that form a network throughout a plant, extending from near the tips of the roots, through the stems, and into true leaves, defined by the presence of vascular tissue in the blade. One type of vascular tissue, **xylem**, conducts water and dissolved minerals upward from the roots; another type of tissue, **phloem**, conducts sucrose and hormonal signals throughout the plant. Vascular tissue enables enhanced height and size in the tracheophytes. It is important to note that vascular tissue develops in the sporophyte, but (with a few exceptions) not in the gametophyte. (Vascular tissue structure is discussed more fully in chapter 38.) The presence of a cuticle and stomata are also characteristic of vascular plants.

 inquiry

Explain why tracheophytes may have had a selective advantage during the evolution of land plants.

Tracheophytes include seven extant phyla grouped in three clades

Three clades of vascular plants exist today: (1) **lycophytes** (club mosses), (2) **pterophytes** (ferns and their relatives), and (3) **seed plants.** Advances in molecular systematics have changed

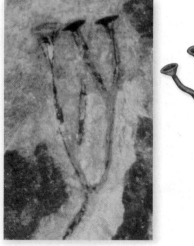
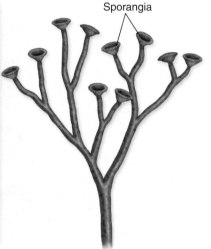

Sporangia

figure 30.11

***Cooksonia*, THE FIRST KNOWN VASCULAR LAND PLANT.**
This fossil represents a plant that lived some 410 MYA. *Cooksonia* belongs to phylum Rhyniophyta, consisting entirely of extinct plants. Its upright, branched stems, which were no more than a few centimeters tall, terminated in sporangia, as seen here. It probably lived in moist environments such as mudflats, had a resistant cuticle, and produced spores typical of vascular plants.

the way we view the evolutionary history of vascular plants. Whisk ferns and horsetails were long believed to be distinct phyla that were transitional between bryophytes and vascular plants. Phylogenetic evidence now shows they are the closest living relatives to ferns, and they are grouped as pterophytes.

The seven living phyla of vascular plants (table 30.1) dominate terrestrial habitats everywhere, except for the highest mountains and the tundra. The haplodiplontic life cycle persists, but the gametophyte has been reduced in size relative to the sporophyte during the evolution of tracheophytes. A similar reduction in multicellular gametangia has occurred as well.

Seeds are another innovation in some phyla

Accompanying this reduction in size and complexity of the gametophytes has been the appearance of the seed. **Seeds** are highly resistant structures well suited to protect a plant embryo from drought and to some extent from predators. In addition, almost all seeds contain a supply of food for the young plant. Seeds occur only in **heterosporous** plants (plants that produce two types of spores—that develop into haploid male and female gametophytes). Heterospory is believed to have arisen multiple times in the evolution of plants from homospory. Zygotes of

TABLE 30.1	The Seven Phyla of Extant Vascular Plants		
Phylum	**Examples**	**Key Characteristics**	**Approximate Number of Living Species**
SEED PLANTS			
Anthophyta	Flowering plants (angiosperms)	Heterosporous. Sperm not motile; conducted to egg by a pollen tube. Seeds enclosed within a fruit. Leaves greatly varied in size and form. Herbs, vines, shrubs, trees. About 14,000 genera.	250,000
Coniferophyta	Conifers (including pines, spruces, firs, yews, redwoods, and others)	Heterosporous seed plants. Sperm not motile; conducted to egg by a pollen tube. Leaves mostly needlelike or scalelike. Trees, shrubs. About 50 genera. Many produce seeds in cones.	601
Cycadophyta	Cycads	Heterosporous. Sperm flagellated and motile but confined within a pollen tube that grows to the vicinity of the egg. Palmlike plants with pinnate leaves. Secondary growth slow compared with that of the conifers. Ten genera. Seeds in cones.	206
Gnetophyta	Gnetophytes	Heterosporous. Sperm not motile; conducted to egg by a pollen tube. The only gymnosperms with vessels. Trees, shrubs, vines. Three very diverse genera (*Ephedra, Gnetum, Welwitschia*).	65
Ginkgophyta	*Ginkgo*	Heterosporous. Sperm flagellated and motile but conducted to the vicinity of the egg by a pollen tube. Deciduous tree with fan-shaped leaves that have evenly forking veins. Seeds resemble a small plum with fleshy, foul-smelling outer covering. One genus.	1
SEEDLESS VASCULAR PLANTS			
Pterophyta	Ferns	Primarily homosporous (a few heterosporous). Sperm motile. External water necessary for fertilization. Leaves uncoil as they mature. Sporophytes and virtually all gametophytes are photosynthetic. About 365 genera.	11,000
	Horsetails	Homosporous. Sperm motile. External water necessary for fertilization. Stems ribbed, jointed, either photosynthetic or nonphotosynthetic. Leaves scalelike, in whorls; nonphotosynthetic at maturity. One genus.	15
	Whisk ferns	Homosporous. Sperm motile. External water necessary for fertilization. No differentiation between root and shoot. No leaves; one of the two genera has scalelike extensions and the other leaflike appendages.	6
Lycophyta	Club mosses	Homosporous or heterosporous. Sperm motile. External water necessary for fertilization. About 12–13 genera.	1,150

homosporous plants may be produced by gametes from the same gametophyte or two different gametophytes. Zygotes of heterosporous plants are produced by two different gametophytes. Heterospory increases the chance of new genetic combinations during sexual reproduction.

Fruits in the flowering plants add a layer of protection to seeds and attract animals that assist in seed dispersal, expanding the potential range of the species. Flowers, which evolved among the angiosperms, attract pollinators. Flowers allow plants to secure the benefits of wide outcrossing in promoting genetic diversity.

> Most tracheophytes have well-developed conducting tissues and specialized stems, leaves, roots, cuticles, and stomata. Many have seeds, which protect embryos until conditions are suitable for further development.

30.7 Lycophytes: The Club Mosses

The earliest vascular plants lacked seeds. Members of four phyla of living vascular plants also lack seeds, as do at least three other phyla known only from fossils. As we explore the adaptations of the vascular plants, we focus on both reproductive strategies and the advantages of increasingly complex transport systems.

The **club mosses** are relict species of an ancient past when vascular plants first evolved (figure 30.12). They are the sister group to all vascular plants. Several genera of club mosses, some of them treelike, became extinct about 270 MYA. Today, club mosses are worldwide in distribution but are most abundant in the tropics and moist temperate regions.

Members of the 12–13 genera and about 1150 living species of club mosses superficially resemble true mosses, but once their internal structure and reproductive processes became known, it was clear that these vascular plants are quite unrelated to mosses. Modern club mosses are either homosporous

figure 30.12
A CLUB MOSS. *Lycopodium clavatum* grows on moist forest floors.

or heterosporous. The sporophytes have leafy stems that are seldom more than 30 cm long.

> Lycophytes are basal to all other vascular plants. They superficially resemble bryophytes, but are not related to them.

30.8 Pterophytes: Ferns and Their Relatives

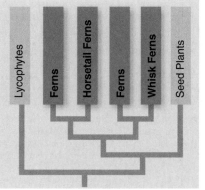

The phylogenetic relationships among ferns and their near relations are still being sorted out. A common ancestor gave rise to two clades: One clade diverged to produce a line of ferns and horsetails; the other diverged to yield another line of ferns and whisk ferns, ancient-looking plants.

Whisk ferns and horsetails are close relatives of ferns. Like lycophytes and bryophytes, they all form antheridia and archegonia. Free water is required for the process of fertilization, during which the sperm, which have flagella, swim to and unite with the eggs. In contrast, most seed plants have nonflagellated sperm.

Whisk ferns lost their roots and leaves secondarily

In **whisk ferns,** which occur in the tropics and subtropics, the sporophytic generation consist merely of evenly forking green stems without roots (figure 30.13). The two or three species of the genus *Psilotum* do, however, have tiny, green, spirally arranged flaps of tissue lacking veins and stomata. Another genus, *Tmespiteris*, has more leaflike appendages. Currently, systematists believe that whisk ferns lost leaves and roots when they diverged from others in the fern lineage.

Given the simple structure of whisk ferns, it was particularly surprising to discover that they are monophyletic with

figure 30.13

A WHISK FERN. Whisk ferns have no roots or leaves. The green, photosynthetic stems have yellow sporangia attached.

ferns. The gametophytes of whisk ferns are essentially colorless and are less than 2 mm in diameter, but they can be up to 18 mm long. They form symbiotic associations with fungi, which furnish their nutrients. Some develop elements of vascular tissue and have the distinction of being the only gametophytes known to do so.

Horsetails have jointed stems with brushlike leaves

The 15 living species of **horsetails** are all homosporous. They constitute a single genus, *Equisetum*. Fossil forms of *Equisetum* extend back 300 million years to an era when some of their relatives were treelike. Today, they are widely scattered around the world, mostly in damp places. Some that grow among the coastal redwoods of California may reach a height of 3 m, but most are less than a meter tall (figure 30.14).

Horsetail sporophytes consist of ribbed, jointed, photosynthetic stems that arise from branching underground *rhi-*

figure 30.14

A HORSETAIL, *Equisetum telmateia*. This species forms two kinds of erect stems; one is green and photosynthetic, and the other, which terminates in a spore-producing "cone," is mostly light brown.

zomes with roots at their nodes. A whorl of nonphotosynthetic, scalelike leaves emerges at each node. The stems, which are hollow in the center, have silica deposits in the epidermal cells of the ribs, and the interior parts of the stems have two sets of vertical, tubular canals. The larger outer canals, which alternate with the ribs, contain air, while the smaller inner canals opposite the ribs contain water. Horsetails are also called scouring rushes because pioneers of the American West used them to scrub pans.

Ferns have fronds that bear sori

Ferns are the most abundant group of seedless vascular plants, with about 11,000 living species. Recent research indicates that they may be the closest relatives to the seed plants.

The fossil record indicates that ferns originated during the Devonian period about 350 MYA and became abundant and varied in form during the next 50 million years. Their apparent ancestors were established on land as much as 375 MYA. Today, ferns flourish in a wide range of habitats throughout the world; however, about 75% of the species occur in the tropics.

The conspicuous sporophytes may be less than a centimeter in diameter—as seen in small aquatic ferns such as *Azolla*—or more than 24 m tall, with leaves up to 5 m or longer in the tree ferns (figure 30.15). The sporophytes and the much smaller gametophytes, which rarely reach 6 mm in diameter, are both photosynthetic.

figure 30.15

A TREE FERN (PHYLUM PTEROPHYTA) IN THE FORESTS OF MALAYSIA. The ferns are by far the largest group of seedless vascular plants.

The fern life cycle (figure 30.16) differs from that of a moss primarily in the much greater development, independence, and dominance of the fern's sporophyte. The fern sporophyte is structurally more complex than the moss sporophyte, having vascular tissue and well-differentiated roots, stems, and leaves. The gametophyte, however, lacks the vascular tissue found in the sporophyte.

Fern morphology

Fern sporophytes, like horsetails, have rhizomes. The leaves, referred to as *fronds*, usually develop at the tip of the rhizome as tightly rolled-up coils ("fiddleheads") that unroll and expand (figure 30.17). Fiddleheads are considered a delicacy, but some species contain secondary compounds linked to stomach cancer.

Many fronds are highly dissected and feathery, making the ferns that produce them prized as ornamentals. Some ferns, such as *Marsilea*, have fronds that resemble a four-leaf clover, but *Marsilea* fronds still begin as coiled fiddleheads. Other ferns produce a mixture of photosynthetic fronds and nonphotosynthetic reproductive fronds that tend to be brownish in color.

Tightly Coiled Fern | **Uncoiling Fern**

figure 30.17

FERN "FIDDLEHEAD." Fronds develop in a coil and slowly unfold on ferns, including the tree fern fronds in these photos.

Fern reproduction

Most ferns are homosporous, producing distinctive sporangia, usually in clusters called **sori** (singular, *sorus*), typically on the underside of the fronds. Sori are often protected during their

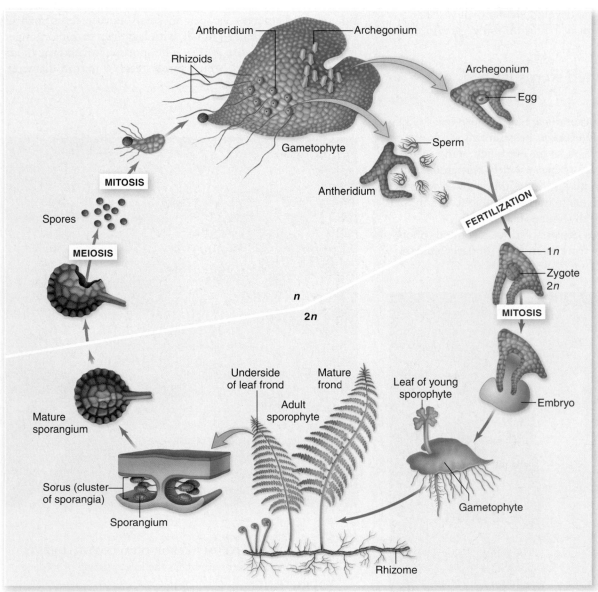

figure 30.16

LIFE CYCLE OF A TYPICAL FERN. Both the gametophyte and sporophyte are photosynthetic and can live independently. Water is necessary for fertilization. Sperm are released on the underside of the gametophyte and swim in moist soil to neighboring gametophytes. Spores are dispersed by wind.

development by a transparent, umbrella-like covering. (At first glance, one might mistake the sori for an infection on the plant.) Diploid spore mother cells in each sporangium undergo meiosis, producing haploid spores.

At maturity, the spores are catapulted from the sporangium by a snapping action, and those that land in suitable damp locations may germinate, producing gametophytes that are often heart-shaped, are only one cell layer thick (except in the center), and have rhizoids that anchor them to their substrate. These rhizoids are not true roots because they lack vascular tissue, but they do aid in transporting water and nutrients from the soil. Flask-shaped archegonia and globular antheridia are produced on either the same or a different gametophyte.

The sperm formed in the antheridia have flagella, with which they swim toward the archegonia when water is present,

often in response to a chemical signal secreted by the archegonia. One sperm unites with the single egg toward the base of an archegonium, forming a zygote. The zygote then develops into a new sporophyte, completing the life cycle (figure 30.16).

Multicellular gametangia still develop in ferns. As discussed earlier, the shift to a dominant sporophyte generation in tracheophytes allows ferns to achieve significant height without interfering with sperm swimming to the egg. The multicellular archegonia provide some protection for the developing embryo.

Ferns and their relatives have a large and conspicuous sporophyte with vascular tissue. Many have well-differentiated roots, stems, and leaves. The shift to a dominant sporophyte led to the evolution of trees.

The Evolution of Seed Plants

Seed plants, which have additional embryo protection, first appeared about 305–465 MYA and were the ancestors of gymnosperms and angiosperms. Seed plants appear to have evolved from spore-bearing plants known as **progymnosperms.** Progymnosperms shared several features with modern gymnosperms, including secondary vascular tissues (which allow for an increase in girth later in development). Some progymnosperms had leaves. Their reproduction was very simple, and it is not certain which particular group of progymnosperms gave rise to seed plants.

The seed protects the embryo

From an evolutionary and ecological perspective, the seed represents an important advance. The embryo is protected by an extra layer of sporophyte tissue, creating the ovule. During development, this tissue hardens to produce the seed coat. In addition to protecting the embryo from drought, the seed can be easily dispersed. Perhaps even more significantly, the presence of seeds introduces into the life cycle a dormant phase that allows the embryo to survive until environmental conditions are favorable for further growth.

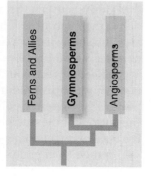

625 μm

A pollen grain is the male gametophyte

Seed plants produce two kinds of gametophytes—male and female—each of which consists of just a few cells. **Pollen grains,** multicellular male gametophytes, are conveyed to the egg in the female gametophyte by wind or by a pollinator. In some seed plants, the sperm moves toward the egg through a growing **pollen tube.** This eliminates the need for external water. In contrast to the seedless plants, the whole male gametophyte, rather than just the sperm, moves to the female gametophyte.

A female gametophyte develops within an ovule. In angiosperms, the ovules are completely enclosed within diploid sporophyte tissue (ovaries that develop into the fruit). In gymnosperms (mostly cone-bearing seed plants), the ovules are not completely enclosed by sporophyte tissue at the time of pollination.

A common ancestor that had seeds gave rise to the gymnosperms and the angiosperms. Seeds protect the embryo and can allow for an extended pause in the life cycle until environmental conditions are more optimal. Seed plants produce male and female gametophytes; the male gametophyte is a pollen grain, which is carried to the female gametophyte by wind or other means.

Gymnosperms: Plants with "Naked Seeds"

There are four groups of living **gymnosperms,** namely coniferophytes, cycadophytes, gnetophytes, and ginkgophytes, all of which lack the flowers and fruits of angiosperms. In all of them, the **ovule,** which becomes a seed, rests exposed on a scale (a modified shoot or leaf) and is not completely enclosed by sporophyte tissues at the time of pollination. The name *gymnosperm*

Ferns and Allies
Gymnosperms
Angiosperms

literally means "naked seed." Although the ovules are naked at the time of pollination, the seeds of gymnosperms are sometimes enclosed by other sporophyte tissues by the time they are mature.

Details of reproduction vary somewhat in gymnosperms, and their forms vary greatly. For example, cycads and *Ginkgo* have motile sperm, whereas conifers and gnetophytes have sperm with no flagella. All sperm are carried within a pollen tube. The female cones range from tiny, woody structures weighing less than 25 g and having a diameter of a few millimeters, to massive structures produced in some cycads,

figure 30.18

CONIFERS.
Slash pines,
Pinus palustris,
in Florida are
representative
of the
Coniferophyta,
the largest
phylum of
gymnosperms.

weighing more than 45 kg and growing to lengths of more than a meter.

Conifers are the largest gymnosperm phylum

The most familiar gymnosperms are **conifers** (phylum Coniferophyta), which include pines (figure 30.18), spruces, firs, cedars, hemlocks, yews, larches, cypresses, and others. The coastal redwood (*Sequoia sempervirens*), a conifer native to northwestern California and southwestern Oregon, is the tallest living vascular plant; it may attain a height of nearly 100 m (300 ft). Another conifer, the bristlecone pine (*Pinus longaeva*) of the White Mountains of California, is the oldest living tree; one specimen is 4900 years of age.

Conifers are found in the colder temperate and sometimes drier regions of the world. Various species are sources of timber, paper, resin, taxol (used to treat cancer), and other economically important products.

Pines are an exemplary conifer genus

More than 100 species of pines exist today, all native to the northern hemisphere, although the range of one species does extend a little south of the equator. Pines and spruces, which belong to the same family, are members of the vast coniferous forests that lie between the arctic tundra and the temperate deciduous forests and prairies to their south. During the past century, pines have been extensively planted in the southern hemisphere.

Pine morphology

Pines have tough, needlelike leaves produced mostly in clusters of two to five. The leaves, which have a thick cuticle and recessed stomata, represent an evolutionary adaptation for retarding water loss. This strategy is important because many of the trees grow in areas where the topsoil is frozen for part of the year, making it difficult for the roots to obtain water.

The leaves and other parts of the sporophyte have canals into which surrounding cells secrete resin. The resin deters insect and fungal attacks. The resin of certain pines is harvested commercially for its volatile liquid portion, called *turpentine*, and for the solid *rosin*, which is used on bowed stringed instruments. The wood of pines lacks some of the more rigid cell types found in other trees, and it is considered a "soft" rather than a "hard" wood. The thick bark of pines is an adaptation for surviving fires and subzero temperatures. Some cones actually depend on fire to open them, releasing seed to reforest burned areas.

Reproductive structures

As mentioned earlier, all seed plants are heterosporous, so the spores give rise to two types of gametophytes (figure 30.19). The male gametophytes (pollen grains) of pines develop from microspores, which are produced in male cones that develop in clusters of 30 to 70, typically at the tips of the lower branches; there may be hundreds of such clusters on any single tree.

The male pine cones generally are 1–4 cm long and consist of small, papery scales arranged in a spiral or in whorls. A pair of microsporangia form as sacs within each scale. Numerous **microspore mother cells** in the microsporangia undergo meiosis, each becoming four microspores. The microspores develop into four-celled pollen grains with a pair of air sacs that give them added buoyancy when released into the air. A single cluster of male pine cones may produce more than a million pollen grains.

Female pine cones typically are produced on the upper branches of the same tree that produces male cones. Female cones are larger than male cones, and their scales become woody.

Two ovules develop toward the base of each scale. Each ovule contains a megasporangium called the **nucellus.** The nucellus itself is completely surrounded by a thick layer of cells called the **integument** that has a small opening (the **micropyle**) toward one end. One of the layers of the integument later becomes the seed coat. A single **megaspore mother cell** within each megasporangium undergoes meiosis, becoming a row of four megaspores. Three of the megaspores break down, but the remaining one, over the better part of a year, slowly develops into a female gametophyte. The female gametophyte at maturity may consist of thousands of cells, with two to six archegonia formed at the micropylar end. Each archegonium contains an egg so large it can be seen without a microscope.

Fertilization and seed formation

Female cones usually take two or more seasons to mature. At first they may be reddish or purplish in color, but they soon turn green, and during the first spring, the scales spread apart. While the scales are open, pollen grains carried by the wind drift down between them, some catching in sticky fluid oozing out of the micropyle. The pollen grains within the sticky fluid are slowly drawn down through the micropyle to the top of the nucellus, and the scales close shortly thereafter.

The archegonia and the rest of the female gametophyte are not mature until about a year later. While the female gametophyte is developing, a pollen tube emerges from a pollen grain at the bottom of the micropyle and slowly digests its way through the nucellus to the archegonia. During growth of the pollen tube, one of the pollen grain's four cells, the *generative cell*, divides by mitosis, with one of the resulting two cells dividing once more.

figure 30.19

LIFE CYCLE OF A TYPICAL PINE. The male and female gametophytes are dramatically reduced in size in these plants. Wind generally disperses the male gametophyte (pollen), which produces sperm. Pollen tube growth delivers the sperm to the egg on the female cone. Additional protection for the embryo is provided by the integument, which develops into the seed coat.

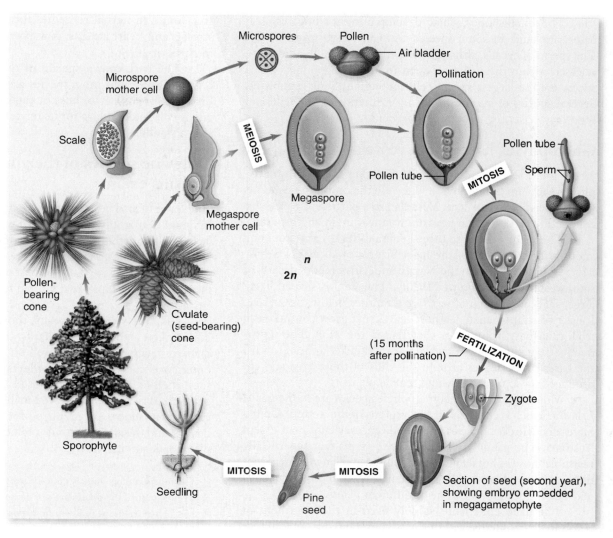

These last two cells function as sperm. The germinated pollen grain with its two sperm is the mature male gametophyte, a very limited haploid phase compared with fern gametophytes.

About 15 months after pollination, the pollen tube reaches an archegonium and discharges its contents into it. One sperm unites with the egg, forming a zygote. The other sperm and cells of the pollen grain degenerate. The zygote develops into an embryo within the seed. After dispersal and germination of the seed, the young sporophyte of the next generation develops into a tree.

Cycads resemble palms, but are not flowering plants

Cycads (phylum Cycadophyta) are slow-growing gymnosperms of tropical and subtropical regions. The sporophytes of most of the 100 known species resemble palm trees (figure 30.20a) with trunks that can attain heights of 15 m or more in height. Unlike palm trees, which are flowering plants, cycads produce cones and have a life cycle similar to that of pines.

a.

b.

c.

figure 30.20

THREE PHYLA OF GYMNOSPERMS.
a. A cycad, *Cycas circinalis*. *b. Welwitschia mirabilis* represents one of the three genera of gnetophytes.
c. Maidenhair tree, *Ginkgo biloba*, the only living representative of the phylum Ginkgophyta.

The female cones, which develop upright among the leaf bases, are huge in some species and can weigh up to 45 kg. The sperm of cycads, although formed within a pollen tube, are released within the ovule to swim to an archegonium. These sperm are the largest sperm cells among all living organisms. Several species of cycads are facing extinction in the wild and soon may exist only in botanical gardens.

Gnetophytes have xylem vessels

There are three genera and about 65 living species of **gnetophytes** (phylum Gnetophyta). They are the only gymnosperms with vessels in their xylem. **Vessels** are a particularly efficient conducting cell type that is a common feature in angiosperms.

The members of the three genera differ greatly from one another in form. One of the most bizarre of all plants is *Welwitschia*, which occurs in the Namib and Mossamedes deserts of southwestern Africa (figure 30.20*b*). The stem is shaped like a large, shallow cup that tapers into a taproot below the surface. It has two strap-shaped, leathery leaves that grow continuously from their base, splitting as they flap in the wind. The reproductive structures of *Welwitschia* are conelike, appear toward the bases of the leaves around the rims of the stems, and are produced on separate male and female plants.

More than half of the gnetophyte species are in the genus *Ephedra*, which is common in arid regions of the western United States and Mexico. Species are found on every continent except Australia. The plants are shrubby, with stems that superficially resemble those of horsetails, being jointed and having tiny, scale-like leaves at each node. Male and female reproductive structures may be produced on the same or different plants.

The drug ephedrine, widely used in the treatment of respiratory problems, was in the past extracted from Chinese species of *Ephedra*, but it has now been largely replaced with synthetic preparations (pseudoephedrine). Because ephedrine found in herbal remedies for weight loss was linked to strokes and heart attacks, it was withdrawn from the market in April of 2004.

The best known species of the third genus, *Gnetum*, is a tropical tree, but most species are vinelike. All species have broad leaves similar to those of angiosperms. One *Gnetum* species is cultivated in Java for its tender shoots, which are cooked as a vegetable.

Only one species of the ginkgophytes remains extant

The fossil record indicates that members of the **ginkgophytes** (phylum Ginkgophyta) were once widely distributed, particularly in the northern hemisphere; today, only one living species, *Ginkgo biloba*, remains (figure 30.20*c*). This tree, which sheds its leaves in the fall, was first encountered by Europeans in cultivation in Japan and China; it apparently no longer exists in the wild.

Like the sperm of cycads, those of *Ginkgo* have flagella. The ginkgo is **dioecious**—that is, the male and female reproductive structures are produced on separate trees. The fleshy outer coverings of the seeds of female ginkgo plants exude the foul smell of rancid butter, caused by the presence of butyric and isobutyric acids. As a result, male plants vegetatively propagated from shoots are preferred for cultivation. Because of its beauty and resistance to air pollution, *Ginkgo* is commonly planted along city streets.

Gymnosperms are mostly cone-bearing seed plants. In gymnosperms, the ovules are not completely enclosed by sporophyte tissue at pollination, hence the name that means "naked seed." The four groups of gymnosperms are conifers, cycads, gnetophytes, and ginkgophytes.

Angiosperms: The Flowering Plants

The 255,000 known species of flowering plants are called **angiosperms** because their ovules, unlike those of gymnosperms, are enclosed within diploid tissues at the time of pollination. The *carpel*, a modified leaf that encapsulates seeds, develops into the fruit, a unique angiosperm feature. Although some gymnosperms, including the yew (*Taxus* spp.), have fleshlike tissue around their seeds, it is of a different origin and not a true fruit.

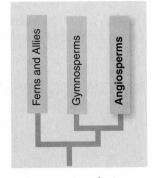

Angiosperm origins are a mystery

The origins of the angiosperms puzzled even Darwin (he referred to their origin as an "abominable mystery"). Recent fossil pollen and plants accompanied by molecular sequence data have provided exciting clues about basal angiosperms, indicating origins as early as 145–208 MYA.

In the remote Liaoning province of China, a complete angiosperm fossil that is at least 125 million years old has been found (figure 30.21). The fossil may represent a new, basal, and extinct angiosperm family, Archaefructaceae, with two species: *Archaefructus liaoningensis* and *A. sinensis*. *Archaefructus* was a herbaceous, aquatic plant. This family is proposed to be the sister clade to all other angiosperms, but there is a lively debate about the validity of this claim.

Archaefructus fossils have both male and female reproductive structures; however, they lack the sepals and petals that evolved in later angiosperms to attract pollinators. The fossils were so well preserved that fossil pollen could be examined using scanning electron microscopy. Although *Archaefructus* is ancient, it is unlikely to be the very first angiosperm. Still, the incredibly well-preserved fossils provide valuable detail on

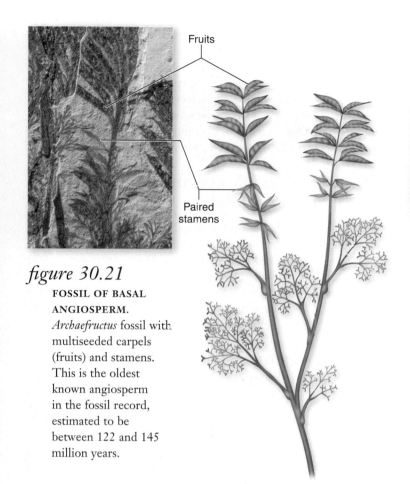

figure 30.21

FOSSIL OF BASAL ANGIOSPERM. *Archaefructus* fossil with multiseeded carpels (fruits) and stamens. This is the oldest known angiosperm in the fossil record, estimated to be between 122 and 145 million years.

Fruits

Paired stamens

figure 30.22

AN ANCIENT LIVING ANGIOSPERM, *Amborella trichopoda*. This plant is believed to be the closest living relative to the original angiosperm.

angiosperms in the Upper Jurassic/Lower Cretaceous period, when dinosaurs roamed the Earth.

Consensus has also been growing on the most basal living angiosperm—*Amborella trichopoda* (figure 30.22). *Amborella*, with small, cream-colored flowers, is even more primitive than water lilies. This small shrub, found only on the island of New Caledonia in the South Pacific, is the last remaining species of the earliest extant lineage of the angiosperms, arising about 135 MYA.

Although *Amborella* is not the original angiosperm, it is sufficiently close that studying its reproductive biology may help us understand the early radiation of the angiosperms. The angiosperm phylogeny reflects an evolutionary hypothesis that is driving new research on angiosperm origins (figure 30.23).

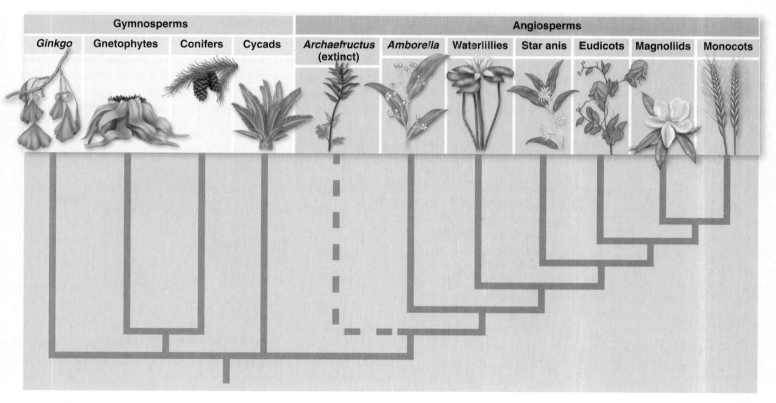

Gymnosperms				Angiosperms						
Ginkgo	Gnetophytes	Conifers	Cycads	*Archaefructus* (extinct)	*Amborella*	Waterlillies	Star anis	Eudicots	Magnoliids	Monocots

figure 30.23

Archaefructus **MAY BE THE SISTER CLADE TO ALL OTHER ANGIOSPERMS.** All members of the *Archaefructus* lineage are extinct, leaving *Amborella* as the most basal, living angiosperm. Gymnosperms species labels are shaded green.

Flowers house the gametophyte generation of angiosperms

Flowers are considered to be modified stems bearing modified leaves. Regardless of their size and shape, they all share certain features (figure 30.24). Each flower originates as a **primordium** that develops into a bud at the end of a stalk called a **pedicel.** The pedicel expands slightly at the tip to form the **receptacle,** to which the remaining flower parts are attached.

Flower morphology

The other flower parts typically are attached in circles called **whorls.** The outermost whorl is composed of **sepals.** Most flowers have three to five sepals, which are green and somewhat leaflike. The next whorl consists of **petals** that are often colored, attracting pollinators such as insects and birds. The petals, which also commonly number three to five, may be separate, fused together, or missing altogether in wind-pollinated flowers.

The third whorl consists of **stamens** and is collectively called the **androecium.** This whorl is where the male gametophytes, pollen, are produced. Each stamen consists of a pollen-bearing **anther** and a stalk called a **filament,** which may be missing in some flowers.

At the center of the flower is the fourth whorl, the **gynoecium,** where the small female gametophytes are housed; the gynoecium consists of one or more **carpels.** The first carpel is believed to have been formed from a leaflike structure with ovules along its margins. Primitive flowers can have several to many separate carpels, but in most flowers, two to several carpels are fused together. Such fusion can be seen in an orange sliced in half; each segment represents one carpel.

Structure of the carpel

A carpel has three major regions (figure 30.24*a*). The **ovary** is the swollen base, which contains from one to hundreds of ovules; the ovary later develops into a **fruit.** The tip of the carpel is called a **stigma.** Most stigmas are sticky or feathery, causing pollen grains that land on them to adhere. Typically, a neck or stalk called a **style** connects the stigma and the ovary; in some flowers, the style may be very short or even missing.

Many flowers have nectar-secreting glands called *nectaries*, often located toward the base of the ovary. Nectar is a fluid containing sugars, amino acids, and other molecules that attracts insects, birds, and other animals to flowers.

Most species use flowers to attract pollinators and reproduce

Eudicots (about 175,000 species) include the great majority of familiar angiosperms—almost all kinds of trees and shrubs, snapdragons, mints, peas, sunflowers, and other plants. Monocots (about 65,000 species) include the lilies, grasses, cattails, palms, agaves, yuccas, orchids, and irises and share a common ancestor with the eudicots (see figure 30.23). Some of the monocots, including maize, rely on wind rather than pollinators to reproduce.

The angiosperm life cycle includes double fertilization

During development of a flower bud, a single megaspore mother cell in the ovule undergoes meiosis, producing four megaspores (see figure 30.24*b*). In most flowering plants, three of the megaspores soon disappear; the nucleus of the remaining megaspore divides mitotically, and the cell slowly expands until it becomes many times its original size.

The female gametophyte

While the expansion of the megaspore is occurring, each of the daughter nuclei divides twice, resulting in eight haploid nuclei arranged in two groups of four. At the same time, two layers of the ovule, the integuments, differentiate and become the *seed coat* of a seed. The integuments, as they develop, form the

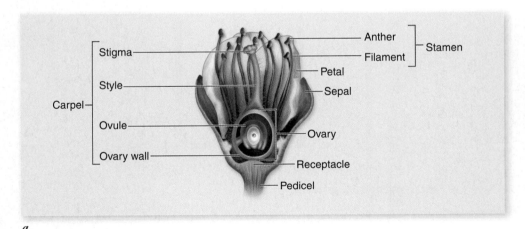

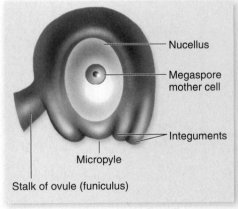

a.   *b.*

figure 30.24

DIAGRAM OF AN ANGIOSPERM FLOWER. *a.* The main structures of the flower are labeled. *b.* Details of an ovule. The ovary as it matures will become a fruit; as the ovule's outer layers (integuments) mature, they will become a seed coat.

micropyle, a small gap or pore at one end that was described earlier (see figure 30.24*b*).

One nucleus from each group of four migrates toward the center, where they function as **polar nuclei.** Polar nuclei may fuse together, forming a single diploid nucleus, or they may form a single cell with two haploid nuclei. Cell walls also form around the remaining nuclei. In the group closest to the micropyle, one cell functions as the **egg;** the other two nuclei are called **synergids.** At the other end, the three cells are now called **antipodals;** they have no apparent function and eventually break down and disappear.

The large sac with eight nuclei in seven cells is called an **embryo sac;** it constitutes the female gametophyte. Although it is completely dependent on the sporophyte for nutrition, it is a multicellular, haploid individual.

Pollen production

While the female gametophyte is developing, a similar but less complex process takes place in the anthers (figure 30.25). Most anthers have patches of tissue (usually four) that eventually become chambers lined with nutritive cells. The tissue in each patch is composed of many diploid microspore mother cells that undergo meiosis more or less simultaneously, each producing four microspores.

The four microspores at first remain together as a quartet or tetrad, and the nucleus of each microspore divides once; in

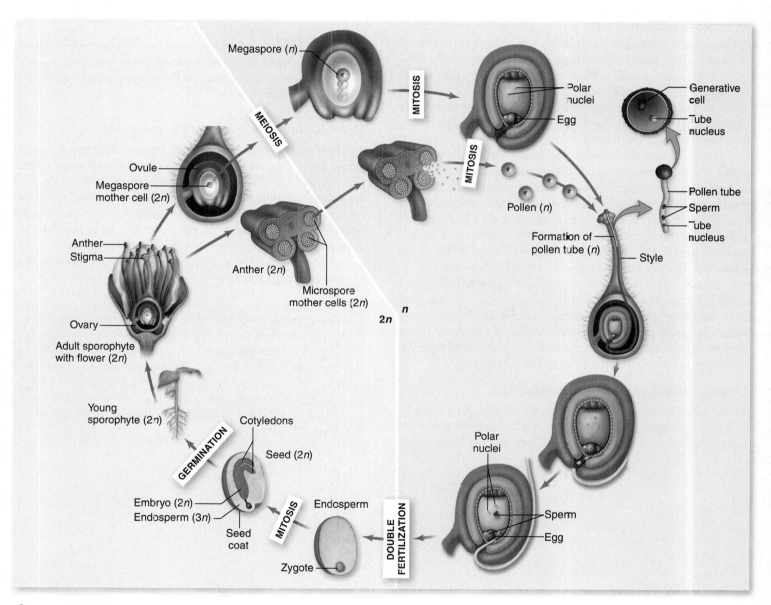

figure 30.25

LIFE CYCLE OF A TYPICAL ANGIOSPERM. As in pines, external water is no longer required for fertilization. In most species of angiosperms, animals carry pollen to the carpel. The outer wall of the carpel forms the fruit, which often entices animals to disperse the seed.

most species, the microspores of each quartet then separate. At the same time, a two-layered wall develops around each microspore. As the microspore-containing anther continues to mature, the wall between adjacent pairs of chambers breaks down, leaving two larger sacs. At this point, the binucleate microspores have become pollen grains.

The outer pollen grain wall layer often becomes beautifully sculptured, and it contains chemicals that may react with others in a stigma to signal whether development of the male gametophyte should proceed to completion. The pollen grain has areas called *apertures*, through which a pollen tube may later emerge.

Pollination and the male gametophyte

Pollination is simply the mechanical transfer of pollen from its source (an anther) to a receptive area (the stigma of a flowering plant). Most pollination takes place between flowers of different plants and is brought about by insects, wind, water, gravity, bats, and other animals. In as many as one-quarter of all angiosperms, however, a pollen grain may be deposited directly on the stigma of its own flower, and self-pollination occurs. Pollination may or may not be followed by *fertilization*, depending on the genetic compatibility of the pollen grain and the flower on whose stigma it has landed.

If the stigma is receptive, the pollen grain's dense cytoplasm absorbs substances from the stigma and bulges through an aperture. The bulge develops into a pollen tube that responds to chemical and mechanical stimuli that guide it to the embryo sac. It follows a diffusion gradient of the chemicals and grows down through the style and into the micropyle. The pollen tube usually takes several hours to two days to reach the micropyle, but in a few instances, it may take up to a year.

One of the pollen grain's two cells, the *generative cell*, lags behind. Its nucleus divides in the pollen grain or in the pollen tube, producing two sperm cells. Unlike sperm in mosses, ferns, and some gymnosperms, the sperm of flowering plants have no flagella. At this point, the pollen grain with its tube and sperm has become a mature male gametophyte.

Double fertilization and seed production

As the pollen tube enters the embryo sac, it destroys a synergid in the process and then discharges its contents. Both sperm are functional, and an event called **double fertilization** follows. One sperm unites with the egg and forms a zygote, which develops into an embryo sporophyte plant. The other sperm and the two polar nuclei unite, forming a triploid primary endosperm nucleus.

The primary endosperm nucleus begins dividing rapidly and repeatedly, becoming triploid **endosperm tissue** that may soon consist of thousands of cells. Endosperm tissue can become an extensive part of the seed in grasses such as corn, and it provides nutrients for the embryo in most flowering plants (see figure 37.12).

Until recently, the nutritional, triploid endosperm was believed to be the ancestral state in angiosperms. A recent analysis of extant, basal angiosperms revealed that diploid endosperms were also common. The female gametophyte in these species has four, not eight nuclei. At the moment, it is unclear whether diploid or triploid endosperms are the most primitive.

inquiry

? *If the endosperm failed to develop in a seed, how do you think the fitness of that seed's embryo would be affected? Explain your answer.*

Germination and growth of the sporophyte

As mentioned earlier, a seed may remain dormant for many years, depending on the species. When environmental conditions become favorable, the seed undergoes germination, and the young sporophyte plant emerges. Again depending on the species, the sporophyte may grow and develop for many years before becoming capable of reproduction, or it may quickly grow and produce flowers in a single growing season.

We present a more detailed description of reproduction in plants in chapter 42.

Angiosperms are characterized by ovules that at pollination are enclosed within an ovary at the base of a carpel, a structure unique to the phylum; a fruit develops from the ovary. Evolutionary innovations—including flowers to attract pollinators, fruits to protect and aid in embryo dispersal, and double fertilization providing additional nutrients for the embryo in the form of endosperm—have all contributed to the widespread success of this group.

30.1 Defining Plants

All plants, except red and brown algae, are placed in the kingdom Viridiplantae.

- All green plants arose from a single freshwater green algal species (figure 30.1).
- Green algae are split into two clades: the chlorophytes, which never made it to land, and the charophytes, which did.
- Land plants have two major common characteristics: protected embryos and multicellular haploid and diploid phases.
- As an adaptation to land, most plants are protected from desiccation by a waxy cuticle and stomata that can be opened and closed.
- Land plants can be distinguished based on the presence or absence of tracheids, specialized cells that facilitate the transport of water and minerals.
- Two additional adaptations that allowed larger land plants to flourish are leaves and a shift to a dominant, vertical diploid generation.

30.2 Plant Life Cycles

Plants have a haplodiplontic life cycle, in which both phases are multicellular (figure 30.2).

- The diploid sporophyte is formed by the fusion of gametes. In the sporangium the diploid spore mother cells each produce four haploid spores by meiosis.
- The haploid gametophyte grows and produces gametes by mitosis.
- As some plants evolved to have greater complexity, the dominant portion of the life cycle shifted from the haploid to the diploid stage, the gametophyte became more limited in size, and the sporophyte shifted from a nutritionally dependent structure to one that is independent.

30.3 Chlorophytes: Aquatic Green Algae (figure 30.5)

The ancestors of the plant kingdom were multicellular green algae.

- There are two distinct lineages of green algae: the chlorophytes, which gave rise to aquatic algae, and the streptophytes, which now include land plants.

30.4 Charophytes: Green Algae Related to Land Plants

Charophytes are also green algae and are closely related to land plants.

- Both candidate Streptophyta clades—Charales and Coleochaetales—have plasmodesmata, cytoplasmic links between cells, and they undergo mitosis and cytokinesis like terrestrial plants. Charales are most closely related to land plants.

30.5 Bryophytes: Nontracheophyte Green Plants

Bryophytes are the closest living descendants of the first land plants. They are called nontracheophytes because they lack tracheids.

- Even though bryophytes do not have roots or tracheids, they do have conducting cells for movement of water and minerals, as well as sugars.
- The nonphotosynthetic sporophyte is nutritionally dependent on the gametophyte.
- Bryophytes consist of three distinct clades: liverworts, hornworts, and mosses.

30.6 Features of Tracheophyte Plants (table 30.1)

The vascular plants found today exist in three clades: lycophytes, pterophytes, and seed plants.

- Vascular plants have a much reduced gametophyte, small multicellular gametangia, and seeds.
- Seed arose only in heterosporous plants, and they are highly resistant structures that protect the plant embryo.
- Fruits in flowering plants add a layer of protection to seeds and attract animals that assist in seed dispersal.

30.7 Lycophytes: The Club Mosses

Lycophytes are the earliest vascular plants.

- Lycophytes lack seeds.

30.8 Pterophytes: Ferns and Their Relatives (figure 30.17)

Phylogenetic relationships among ferns and their relatives are in the process of being sorted out, but the ancestors did give rise to two clades: one line of ferns and horsetails, and a second line of ferns and whisk ferns.

- Pterophytes form antheridia and archegonia and require water for fertilization.

30.9 The Evolution of Seed Plants

Seed plants appeared to evolve from spore-bearing plants known as progymnosperms.

- All seed plants are heterosporous.
- Progymnosperms and modern gymnosperms share secondary vascular tissues that facilitate an increase in girth.
- The seed protects the embryo and provides a dormant stage that pauses the life cycle until environmental conditions are favorable.
- Pollen grains, the male gametophyte, are dispersed by wind or other means.
- The female gametophyte develops within an ovule that is enclosed in sporophyte tissue.

30.10 Gymnosperms: Plants with "Naked Seeds" (figure 30.23)

Gymnosperms have naked ovules at the time of pollination.

- Coniferophytes, cycadophytes, gnetophytes, and ginkgophytes are gymnosperms, all of which lack flowers and true fruits.

30.11 Angiosperms: The Flowering Plants (figure 30.23)

Angiosperms are distinct from gymnosperms and other plants because their ovules are enclosed with diploid tissue called the ovary at the time of pollination and they form fruits.

- Flower parts are organized into four whorls: sepals, petals, androecium, and gynoecium (figure 30.24a).
- The male androecium consists of the stamens where haploid pollen, the male gametophyte, is produced.
- The female gynoecium consists of one or more carpels that contain the gametophyte.
- Most angiosperm species use flowers to attract pollinators, but some are pollinated by wind

SELF TEST

1. Which of the following plant structures is not matched to its correct function?
   a. Stomata—allow gas transfer
   b. Tracheids—allow the movement of water and mineral
   c. Cuticle—prevents desiccation
   d. All of the above are matched correctly.

2. In the following diagram, which box represents reduction division from a *2n* state to a *1n* state?
   a. Box a
   b. Box b
   c. Box c
   d. None of the boxes represent reduction division.

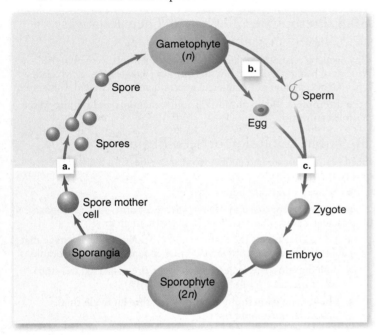

3. Which of the following species most likely directly gave rise to the land plants?
   a. *Volvox*
   b. *Chlamydomonas*
   c. *Ulva*
   d. *Chara*

4. Which of the following would not be found in a member of the bryophytes?
   a. Mycorrhizal associations
   b. Rhizoids
   c. Tracheid cells
   d. Photosynthetic gametophytes

5. Which of the following statements is correct regarding the bryophytes?
   a. The bryophytes represent a monoplyletic clade.
   b. The sporophyte stage of all bryophytes is photosynthetic.
   c. Archegonium and antheridium represent haploid structures that produce reproductive cells.
   d. Stomata are common to all bryophytes.

6. The lack of seeds is a characteristic of all
   a. lycophytes.
   b. gymnosperms.
   c. tracheophytes.
   d. gnetophytes.

7. Which of the following adaptations allowed plants to pause their life cycle until environmental conditions are optimal?
   a. Stomata
   b. Phloem and xylem
   c. Seeds
   d. Flowers

8. Which of the following gymnosperms possesses a form of vascular tissue that is similar to that found in the angiosperms.
   a. Cycads
   b. Gnetophytes
   c. Ginkgophytes
   d. Conifers

9. In a pine tree, the microspores and megaspores are produced by the process of
   a. fertilization.
   b. mitosis.
   c. fusion.
   d. meiosis.

10. Which of the following terms is *not* associated with a male portion of a plant?
    a. Megaspore
    b. Antheridium
    c. Pollen grains
    d. Microspore

11. Which of the following whorls contains the carpels?
    a. The sepal
    b. The androecium
    c. The gynoecium
    d. The petals

12. In double fertilization, one sperm produces a diploid _____, and the other produces a triploid _____.
    a. zygote; primary endosperm
    b. primary endosperm; microspore
    c. antipodal; zygote
    d. polar nuclei; zygote

13. Which of the following potentially represents the oldest known living species of angiosperm?
    a. *Cooksonia*
    b. *Chlamydomonas*
    c. *Archaefructus*
    d. *Ambroella*

CHALLENGE QUESTIONS

1. You have been hired as a research assistant to investigate the origins of the angiosperms, specifically the boundary between a gymnosperm and an angiosperm. What characteristics would you use to clearly define a new fossil as a gymnosperm? An angiosperm?

2. What are the benefits and drawbacks of self-pollination for a flowering plant? Explain your answer.

3. The relationship between flowering plants and pollinators is often used as an example of coevolution. Many flowering plant species have flower structures that are adaptive to a single species of pollinator. What are the benefits and drawbacks of using such a specialized relationship?

chapter 31

Fungi

introduction

THE FUNGI, AN OFTEN-OVERLOOKED group of unicellular and multicellular organisms, have a profound influence on ecology and human health. Along with bacteria, they are important decomposers and disease-causing organisms. Fungi are found everywhere—from the tropics to the tundra and in both terrestrial and aquatic environments. Fungi made it possible for plants to colonize land by associating with rootless stems and aiding in the uptake of nutrients and water. Mushrooms and toadstools are the multicellular spore-producing part of fungi that grow rapidly under proper conditions. A single *Armillaria* fungus can cover 15 hectares underground and weigh 100 tons. Some puffball fungi are almost a meter in diameter and may contain 7 trillion spores—enough to circle the Earth's equator! Yeasts are used to make bread and beer, but other fungi cause disease in plants and animals. These fungal killers are particularly problematic because fungi are animals' closest relatives. Drugs that can kill fungi often have toxic effects on animals, including humans. In this chapter we present the major groups of this intriguing life form.

31.1 Defining Fungi

Mycologists, scientists who study fungi as well as fungi-like protists, believe there may be as many as 1.5 million fungal species. Currently, taxonomists classify the **fungi** into six main groups: the **chytrids,** the **zygomycetes,** the **glomeromycetes,** the **ascomycetes,** the **basidiomycetes,** and the **deuteromycetes** (figure 31.1). The deuteromycetes, formerly called the *imperfect fungi,* lack a phylogenetic position because sexual reproduction has not been observed and insufficient data are available to determine which fungi are most closely related to specific deuteromycetes.

Recent phylogenetic analysis of DNA sequences and protein sequences indicates that fungi are more closely related to animals than to plants. The last common ancestor with animals was a single cell, and unique solutions to multicellularity evolved both in fungi and in animals. Although the fungi are incredibly diverse, they share some characteristics:

1. **Fungi are heterotrophs that absorb nutrients.** Fungi obtain their food by secreting digestive enzymes into substrates, which may include anything from fallen logs to the skin of frogs. Then they absorb the organic molecules released by the action of the enzymes. Thus, fungi actually live in their food.
2. **Fungi have a number of different cell types.** Multicellular fungi are primarily filamentous in their growth form; their bodies consist of long, slender filaments called **hyphae.** These hyphae, however, may be packed together to form complex structures, such as a mushroom. Hyphal cells have a range of morphologies. Some unicellular fungi have a flagellum.
3. **Fungi have cell walls that include chitin.** The cell walls of fungi are built of polysaccharides and **chitin,** the same tough, insoluble material a crab shell is made of.
4. **Some fungi have a dikaryon stage.** Many sexually reproducing fungi undergo a stage in which two haploid nuclei coexist in a single cell **(dikaryon)** for some period of time before fusing to make a diploid nucleus.
5. **Fungi undergo nuclear mitosis.** Mitosis in fungi is different from that in plants and animals in one key respect: The nuclear envelope does not break down and re-form. Instead, mitosis takes place within the nucleus. A spindle apparatus forms there, dragging chromosomes to opposite poles of the nucleus (not the cell, as in most other eukaryotes). This type of mitosis is found in some protists as well (see chapter 29).

> Fungi are characterized by their mode of nutrition, their cell types and body form, the chitin in their cell walls, and unique nuclear mitosis. Fungi are more closely related to animals than to plants.

Chytrid	Zygomycete	Glomeromycete	Ascomycete	Basidiomycete

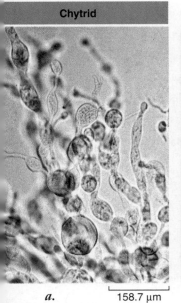 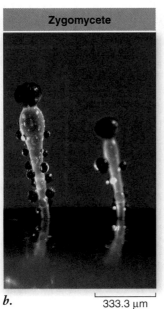   

a. 158.7 μm *b.* 333.3 μm *c.* 300 μm *d.* *e.*

figure 31.1

REPRESENTATIVES OF FUNGAL PHYLA. *a.* Some chytrids, including members of the genus *Hypochytrium,* are plant parasites while others are free living. *b.* *Pilobolus,* a zygomycete, grows on animal dung and also on culture medium. Stalks about 10 mm long contain dark, spore-bearing sacs. *c.* Spores of *Glomus intraradices,* a glomeromycete associated with roots. *d.* The cup fungus *Cookeina tricholoma* is an ascomycete from the rain forest of Costa Rica. In the cup fungi, the spore-producing structures line the cup; in basidiomycetes that form mushrooms such as *Amanita,* they line the gills beneath the cap of the mushroom. *e.* *Amanita muscaria,* the fly agaric, is a toxic basidiomycete. All visible structures of fungi, including the ones shown here, arise from an extensive network of filamentous hyphae that penetrates and is interwoven with the substrate on which they grow.

General Biology of the Fungi

Fungi exist either as single-celled yeasts or in multicellular form with several different cell types. Their reproduction may be either sexual or asexual, and they exhibit an unusual form of mitosis. They are specialized to extract and absorb nutrients from their surroundings through external secretion of enzymes. We begin by looking at fungal morphology.

The body of a fungus is a mass of connected hyphae

Some hyphae are continuous or branching tubes filled with cytoplasm and multiple nuclei. Other hyphae are typically made up of long chains of cells joined end-to-end and divided by cross-walls called **septa** (singular, *septum*). The septa rarely form a complete barrier, except when they separate the reproductive cells. Even fungi with septa can be considered one long cell.

Cytoplasm characteristically flows or streams freely throughout the hyphae, passing through major pores in the septa (figure 31.2). Because of this streaming, proteins synthesized throughout the hyphae may be carried to their actively growing tips. As a result, fungal hyphae may grow very rapidly when food and water are abundant and the temperature is optimum. For example, you may have seen mushrooms suddenly appear in a lawn overnight after a rain in summer.

The mycelium

A mass of connected hyphae is called a **mycelium** (plural, *mycelia*). (This word and the term **mycology**, the study of fungus, are both derived from the Greek word *mykes*.) The mycelium of a fungus (figure 31.3) constitutes a system that may, in the aggregate, be many meters long. This mycelium grows into the soil, wood, or other material in which the fungus is growing, and digestion of the material begins quickly. All parts of such a fungus are metabolically active.

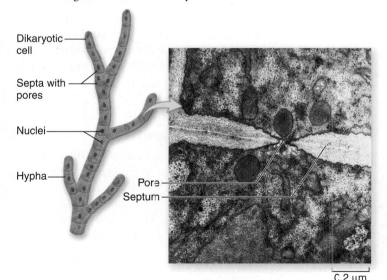

Dikaryotic cell
Septa with pores
Nuclei
Hypha
Pore
Septum
C.2 μm

figure 31.2

A SEPTUM. This transmission electron micrograph of a section through a hypha of the basidiomycete *Inonotus tomentosus* shows a pore through which the cytoplasm streams.

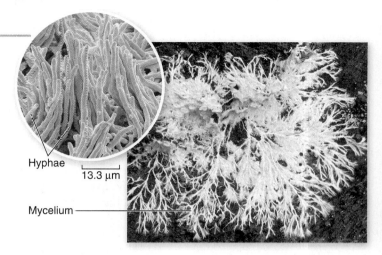

Hyphae
13.3 μm
Mycelium

figure 31.3

FUNGAL MYCELIUM. This mycelium, composed of hyphae, is growing through leaves on the forest floor in Maryland.

In two of the four major groups of fungi, reproductive structures formed of interwoven hyphae, such as mushrooms, puffballs, and morels, are produced at certain stages of the life cycle. These structures expand rapidly because of rapid inflation of the hyphae.

Cell walls with chitin

The cell walls of fungi are formed of polysaccharides, including chitin, not cellulose, as are those of plants and many groups of protists. As you learned in chapter 3, chitin is a modified cellulose consisting of linked glucose units to which nitrogen groups have been added; this polymer is then cross-linked with proteins. Chitin is the same material that makes up the major portion of the hard shells, or exoskeletons, of arthropods, a group of animals that includes insects and crustaceans (chapter 34). Chitin is one of the shared traits that has led scientists to believe that fungi and animals are more closely related than fungi and plants.

Fungal cells may have more than one nucleus

Fungi are different from most animals and plants in that each cell (or hypha) can house one, two, or more nuclei. A hypha that has only one nucleus is called **monokaryotic;** a hypha with two nuclei is **dikaryotic.** In a dikaryotic cell, the two haploid nuclei exist independently. Dikaryotic hyphae have some of the genetic properties of diploids, because both genomes are transcribed.

Sometimes, many nuclei intermingle in the common cytoplasm of a fungal mycelium, which can lack distinct cells. If a dikaryotic or multinucleate hypha has nuclei that are derived from two genetically distinct individuals, the hypha is called **heterokaryotic.** Hyphae whose nuclei are genetically similar to one another are called **homokaryotic.**

Mitosis divides the nucleus, not the hypha

Mitosis in multicellular fungi differs from that in most other organisms. Because of the linked nature of the cells, the cell itself is not the relevant unit of reproduction; instead, the nucleus is. The nuclear envelope does not break down and re-form; instead, the spindle apparatus is formed *within* it.

Centrioles are absent in all fungi; instead, fungi regulate the formation of microtubules during mitosis with small, relatively amorphous structures called *spindle plaques*. This unique combination of features strongly suggests that fungi originated from some unknown group of single-celled eukaryotes with these characteristics.

Fungi can reproduce both sexually and asexually

Many fungi are capable of producing both sexual and asexual spores. When a fungus reproduces sexually, two haploid hyphae of compatible mating types may come together and fuse.

The dikaryon stage

In animals, plants, and some fungi, the fusion of two haploid cells immediately results in a diploid cell (2*n*). But in other fungi, namely basidiomycetes and ascomycetes, an intervening dikaryotic stage (1*n* + 1*n*) occurs before the parental nuclei fuse and form a diploid nucleus. In ascomycetes, this **dikaryon** stage is brief, occurring in only a few cells of the sexual reproductive structure. In basidiomycetes, however, it can last for most of the life of the fungus, including both the feeding and sexual spore-producing structures.

Reproductive structures

Some fungus species produce specialized mycelial structures to house the production of spores. Examples are the mushrooms we see above ground, the "shelf" mycelium that appears on the trunks of dead trees, and puffballs, which can house billions of spores.

As noted, the cytoplasm in fungal hyphae normally flows through perforated septa or moves freely in their absence. Reproductive structures are an important exception to this general pattern. When reproductive structures form, they are cut off by complete septa that lack perforations or that have perforations that soon become blocked.

Spores

Spores are the most common means of reproduction among fungi. They may form as a result of either asexual or sexual processes, and they are often dispersed by the wind. When spores land in a suitable place, they germinate, giving rise to a new fungal mycelium.

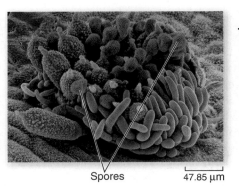

figure 31.4

FUNGAL SPORES. Scanning electron micrograph of fungal spores that infect rose plants.

Spores   47.85 μm

Because the spores are very small, between 2 and 75 μm in diameter (figure 31.4), they can remain suspended in the air for a long time. Unfortunately, many of the fungi that cause diseases in plants and animals are spread rapidly by such means. The spores of other fungi are routinely dispersed by insects or other small animals. Only one group of fungi, the chytrids, retain the ancestral flagella and have motile zoospores.

Biologists had believed for a long time that the worldwide presence of fungal species could be accounted for, on an evolutionary timescale, by the almost limitless, long-distance dispersal of fungal spores. Recent biogeographic studies, however, have examined the phylogenetic relationships among fungi in distant parts of the world and disproved this long-held assumption.

Fungi are heterotrophs that absorb nutrients

All fungi obtain their food by secreting digestive enzymes into their surroundings and then absorbing the organic molecules produced by this **external digestion.** The fungal body plan reflects this approach. Unicellular fungi have the greatest surface area-to-volume ratio of any fungus, maximizing the surface area for absorption. Extensive networks of hyphae also provide an enormous surface area for absorptive nutrition in a fungal mycelium.

Many fungi are able to break down the cellulose in wood, cleaving the linkages between glucose subunits and then absorbing the glucose molecules as food. Fungi also digest lignin, an insoluble organic compound that strengthens plant cell walls. The specialized metabolic pathways of fungi allow them to obtain nutrients from dead trees and from an extraordinary range of organic compounds, including tiny roundworms called nematodes (figure 31.5*a*).

a.   277.7 μm   *b.*

figure 31.5

CARNIVOROUS FUNGI. *a.* Fungus obtaining nutrients from a nematode. *b.* The oyster mushroom *Pleurotus ostreatus* not only decomposes wood but also immobilizes nematodes, which the fungus uses as a source of nitrogen.

The mycelium of the edible oyster mushroom *Pleurotus ostreatus* (figure 31.5*b*) excretes a substance that paralyzes nematodes that feed on the fungus. When the worms become sluggish and inactive, the fungal hyphae envelop and penetrate their bodies. Then the fungus secretes digestive juices and absorbs the nematode's nutritious contents, just like it would from a plant source.

This fungus usually grows within living trees or on old stumps, obtaining the bulk of its glucose through the enzymatic digestion of cellulose and lignin from plant cell walls. The nematodes it consumes apparently serve mainly as a source of nitrogen—a substance almost always in short supply in biological systems. Other fungi are even more active predators than *Pleurotus*, snaring, trapping, or firing projectiles into nematodes, rotifers, and other small animals on which they prey.

Because of their ability to break down almost any carbon-containing compound—even jet fuel—fungi are of interest for use in bioremediation, using organisms to clean up soil or water that is environmentally contaminated. As one example, some fungal species can remove selenium, an element that is toxic in high accumulations, from soils by combining it with other harmless volatile compounds.

> A fungus consists of a mass of hyphae (cells) termed a mycelium; cell walls contain chitin, a substance also found in arthropod animals. Mitosis divides the nucleus but not the hypha itself. Sexual reproduction may occur when hyphae of two different mating types fuse. Haploid nuclei from each type may persist separately in some groups, which is termed a dikaryon stage. Spores are produced sexually or asexually and are spread by wind or animals. Fungi secrete digestive enzymes onto organic matter and then absorb the products of the digestion.

31.3 Phylogenetic Relationships

Fungi are clustered in this chapter based on phylogenetic relationships. In some cases, it is not yet possible to accurately assign fungi to specific lineages, especially when sexual reproduction has not been observed. Mode of sexual reproduction is one of the defining characters of members within a phylum.

New fungal phylogenies are emerging

The understanding of fungal phylogeny is going through rapid and exciting changes, aided by increasing molecular sequence data (figure 31.6 and table 31.1). We now believe that fungi

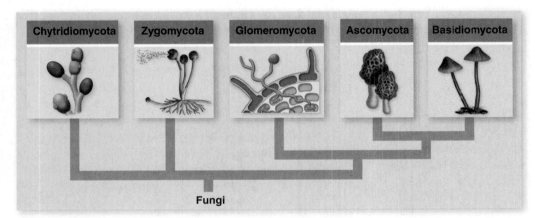

figure 31.6

FIVE MAJOR PHYLA OF FUNGI. Chytridiomycota and Zygomycota are not monophyletic, but Glomeromycota, Ascomycota and Basidiomycota are. The Deuteromycetes are not shown because they are not monophyletic.

TABLE 31.1	**Fungi**		
Group	**Typical Examples**	**Key Characteristics**	**Approximate Number of Living Species**
Chytridiomycota	*Allomyces*	Aquatic, flagellated fungi that produce haploid gametes in sexual reproduction or diploid zoospores in asexual reproduction	1,000
Zygomycota	*Rhizopus, Pilobolus*	Multinucleate hyphae lack septa, except for reproductive structures; fusion of hyphae leads directly to formation of a zygote in zygosporangium, in which meiosis occurs just before it germinates; asexual reproduction is most common	1,050
Glomeromycota	*Glomus*	Form arbuscular mycorrhizae. Multinucleate hyphae lack septa. Reproduce asexually.	150
Ascomycota	Truffles, morels	In sexual reproduction, ascospores are formed inside a sac called an ascus; asexual reproduction is also common	45,000
Basidiomycota	Mushrooms, toadstools, rusts	In sexual reproduction, basidiospores are borne on club-shaped structures called basidia; asexual reproduction occurs occasionally	22,000

are more closely related to animals than to plants. The relationships between the fungal groups are less clear. To illustrate the diversity within this kingdom, we will look at five phyla of fungi—Chytridiomycota, Zygomycota, Glomeromycota, Basidiomycota, and Ascomycota.

Not all groups are monophyletic

The Glomeromycota, Ascomycota, and Basidiomycota are monophyletic, but the other two groups are not. Microsporidia, also monophyletic, are obligate animal parasites that may belong to the kingdom Fungi. Deuteromycetes, formerly called imperfect fungi, are indeed fungi, but their relationships to one another and other fungi remains to be determined. Several other groups that historically were considered fungi, such as the slime molds and water molds, now are classed as protists, not fungi (see chapter 29). As additional molecular evidence is acquired, analyzed, and integrated with other traits, a new fungal phylogeny is likely to appear.

> Fungal phylogenies are changing rapidly. The fungi are provisionally placed in five groups—Chytridiomycota, Zygomycota, Glomeromycota, Basidiomycota, and Ascomycota. Of these, Ascomycota, Glomeromycota, and Basidiomycota are monophyletic.

31.4 Chytridiomycetes: Aquatic Fungi with Flagellated Zoospores

Members of phylum Chytridiomycota, the **chitridiomycetes** or **chytrids,** are aquatic, flagellated fungi that are most closely related to ancestral fungi. Motile zoospores are a distinguishing character of this fungal group (figure 31.7).

Because only chytrids have flagella, this trait must have been lost among ancestors of modern groups. Also, since most chytrids are aquatic, it is likely that fungi first appeared in the water, as did plants and animals. The presence of chitin in the cell walls of chytrids is a unifying feature of all fungi.

Fossils and molecular data indicate that animals and fungi last shared a common ancestor at least 460 MYA, but inconsistencies remain. The oldest fossil resembles extant members of the genus *Glomus* that arose within the Glomeromycota. One DNA analysis placed the animal/fungal divergence at 1500 MYA. This latter estimate is not generally accepted, but many researchers believe that the last common ancestor existed around close to 670 MYA based on an analysis of multiple genes.

Regardless of the time of origin, chytrids play a major role in ecosystems. Details of their symbiotic relationships are considered later on. Chytrids have been implicated in die-off of frogs and other amphibians in many countries.

> Chytrids are the closest living relatives of the first fungal ancestors.

figure 31.7

***Allomyces,* A CHYTRID THAT GROWS IN THE SOIL. *a.* The** spherical sporangia can produce either diploid zoospores via mitosis or haploid zoospores via meiosis. *b.* Life cycle of *Allomyces,* which has both haploid and diploid multicellular stages (alternation of generations).

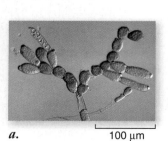

a.   100 μm   *b.*

Zygomycetes: Fungi That Produce Zygotes

Zygomycetes (phylum Zygomycota) include only about 1050 named species, but they are incredibly diverse. Among them are some of the more common bread molds (figure 31.8), as well as a variety of others found on decaying organic material, including strawberries and other fruits. A few human pathogens are in this group.

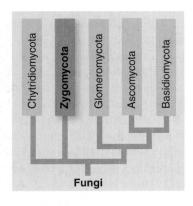

In sexual reproduction, zygotes form inside a zygosporangium

The zygomycetes lack septa in their hyphae except when they form sporangia or gametangia (structures in which spores or gametes are produced). The group is named after a characteristic feature of the sexual phase of the life cycle, the formation of diploid zygote nuclei.

Sexual reproduction begins with the fusion of gametangia, which contain numerous nuclei. The gametangia are cut off from the hyphae by complete septa. These gametangia may be formed on hyphae of different mating types or on a single hypha.

The haploid nuclei fuse to form a diploid zygote nuclei, a process called karyogamy. The area where the fusion has taken place develops into a **zygosporangium** (figure 31.9b), within which a **zygospore** develops. The zygospore, which may contain one or more diploid nuclei, acquires a thick coat that helps the fungus survive conditions not favorable for growth.

Meiosis, followed by mitosis, occurs during the germination of the zygospore, which releases haploid spores. Haploid hyphae grow when these haploid spores germinate. Except for the zygote nuclei, all nuclei of the zygomycetes are haploid.

Asexual reproduction is more common

Asexual reproduction occurs much more frequently than sexual reproduction in the zygomycetes. During asexual reproduction, hyphae produce clumps of erect stalks, called **sporangiophores.** The tips of the sporangiophores form sporangia, which are separated by septa. Thin-walled haploid spores are produced within the sporangia. These spores are shed above the food substrate, in a position where they may be picked up by the wind and dispersed to a new food source.

> Many zygomycetes form characteristic resting structures, called zygosporangia, which contain zygospores with one or more diploid nuclei. The hyphae of zygomycetes are multinucleate, with septa only where gametangia or sporangia are separated.

a.

667 μm

figure 31.8

Rhizopus, **A ZYGOMYCETE THAT GROWS ON SIMPLE SUGARS.** This fungus is often found on moist bread or fruit. *a.* The dark, spherical, spore-producing sporangia are on hyphae about 1 cm tall. The rootlike hyphae anchor the sporangia. *b.* Life cycle of *Rhizopus.* The Zygomycota group is named for the zygosporangia characteristic of *Rhizopus.*

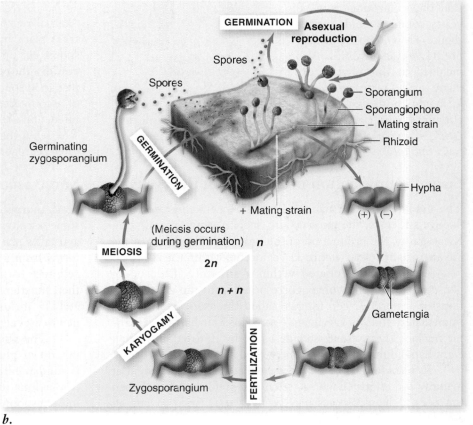

b.

31.6 Glomeromycetes: Asexual Plant Symbionts

The Glomeromycetes, a tiny group of fungi with approximately 150 described species, likely made the evolution of terrestrial plants possible. Tips of hyphae grow within the root cells of most trees and herbaceous plants, forming a branching structure that allows nutrient exchange. The intracellular associations with plant roots are called **arbuscular mycorrhizae.** The specifics of arbuscular mycorrhizal associations and other forms of mycorrhizal interactions are detailed in section 31.10.

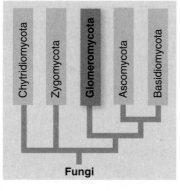

Glomeromycetes cannot survive in the absence of a host plant. The symbiotic relationship is mutualistic, with the glomeromycetes providing essential minerals, especially phosphorous, and the plants providing carbohydrates.

The Glomeromycota are challenging to characterize, in part, because their is no evidence of sexual reproduction. These fungi exemplify our emerging understanding of fungal phylogeny. Like zygomycetes, glomeromycetes lack septae in their hyphae and were once grouped with the zygomycetes. However, comparisons of DNA sequences of small subunit rRNAs reveal that glomeromycetes are a monophyletic clade that is phylogenetically distinct from zygomycetes. Unlike zygomycetes, glomeromycetes lack zygospores. Glomeromycota originated at least 600–620 MYA, well before the split of the Ascomycota and Basidiomycota, which we will consider next.

> Glomeromycetes are a monophyletic fungal lineage. Their obligatory symbiotic relationship with the roots of many plants appears to be ancient and may have made it possible for terrestrial plants to evolve.

31.7 Ascomycetes: The Sac (Ascus) Fungi

The **ascomycetes** (phylum Ascomycota) contain about 75% of the known fungi. Among the ascomycetes are such familiar and economically important fungi as bread yeasts, common molds, morels (figure 31.9a), cup fungi (figure 31.9b), and truffles. Also included in this phylum are many serious plant pathogens, including those that produce

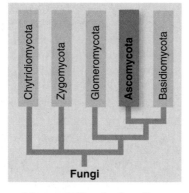

chestnut blight, *Cryphonectria parasitica*, and Dutch elm disease, *Ophiostoma ulmi*. Penicillin-producing ascomycetes are in the genus *Penicillium*.

Sexual reproduction occurs within the ascus

The ascomycetes are named for their characteristic reproductive structure, the microscopic, saclike **ascus** (plural, *asci*). **Karyogamy,** the production of the only diploid nucleus of the ascomycete life cycle (figure 31.9c), occurs within the ascus.

Asci are differentiated within a structure made up of densely interwoven hyphae, corresponding to the visible portions of a morel or cup fungus, called the **ascocarp.** Meiosis immediately follows karyogamy, forming four haploid daughter nuclei. These usually divide again by mitosis, producing eight haploid nuclei that become walled **ascospores.**

In many ascomycetes, the ascus becomes highly turgid at maturity and ultimately bursts, often at a preformed area. When this occurs, the ascospores may be thrown as far as 31 cm, an amazing distance considering that most ascospores are only about

10 μm long. This would be equivalent to throwing a baseball (diameter 7.5 cm) 1.25 km—about 10 times the length of a home run!

Asexual reproduction occurs within conidiophores

Asexual reproduction is very common in the ascomycetes. It takes place by means of **conidia** (singular, *conidium*), asexual spores cut off by septa at the ends of modified hyphae called **conidiophores.** Conidia allow for the rapid colonization of a new food source. Many conidia are multinucleate. The hyphae of ascomycetes are divided by septa, but the septa are perforated, and the cytoplasm flows along the length of each hypha. The septa that cut off the asci and conidia are initially perforated, but later become blocked.

Some ascomycetes have yeast morphology

Yeast morphology (a single-celled lifestyle) is found among some ascomycetes. Most of yeasts' reproduction is asexual and takes place by cell fission or budding, when a smaller cell forms from a larger one (figure 31.10). Sometimes two yeast cells fuse, forming one cell containing two nuclei. This cell may then function as an ascus, with karyogamy followed immediately by meiosis. The resulting ascospores function directly as new yeast cells.

The ability of yeasts to ferment carbohydrates, breaking down glucose to produce ethanol and carbon dioxide, is fundamental in the production of bread, beer, and wine. About four billion of these tiny, powerful organisms can fit in a teaspoon. Many different strains of yeast have been domesticated and selected for these processes, using the sugars

a.

b.

figure 31.9

ASCOMYCETES. *a.* This morel, *Morchella esculenta*, is a delicious edible ascomycete that appears in early spring. *b.* A cup fungus. *c.* Life cycle of an ascomycete. Haploid ascospores form within the ascus.

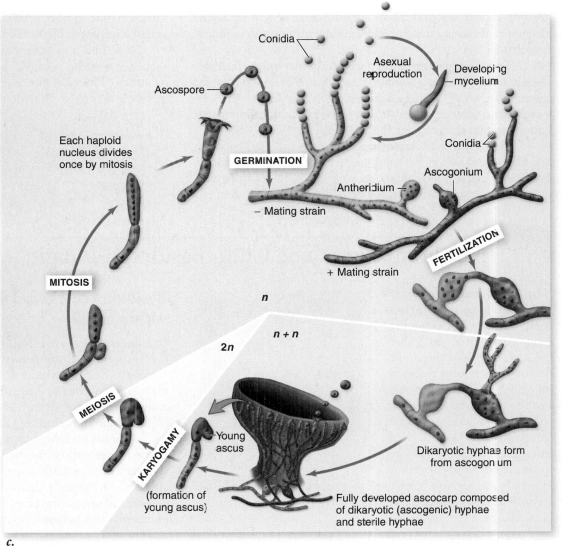

Conidia

Asexual reproduction

Developing mycelium

Ascospore

Conidia

GERMINATION

Each haploid nucleus divides once by mitosis

Antheridium

Ascogonium

MITOSIS

– Mating strain

FERTILIZATION

+ Mating strain

n

n + n

2n

MEIOSIS

KARYOGAMY

Young ascus

Dikaryotic hyphae form from ascogonium

(formation of young ascus)

Fully developed ascocarp composed of dikaryotic (ascogenic) hyphae and sterile hyphae

c.

---

in rice, barley, wheat, and corn. Wild yeasts—ones that occur naturally in the areas where wine is made—were important in wine making historically, but domesticated cultured yeasts are normally used now.

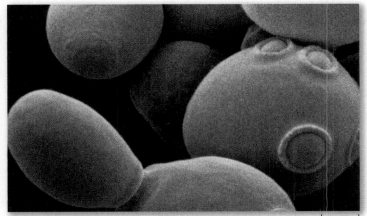

5.18 µm

figure 31.10

BUDDING IN *Saccharomyces.* As shown in this scanning electron micrograph, the cells tend to hang together in chains, a feature that calls to mind the derivation of single-celled yeasts from multicellular ancestors.

Wild yeast, often *Candida milleri*, is still important in making sourdough bread. Unlike most breads that are made with pure cultures of yeast, sourdough uses an active culture of wild yeast and bacteria that generate acid. This culture is maintained, and small amounts—"starter" cultures—are used for each batch of bread. The combination of yeast and acid-producing bacteria is needed for fermentation and gives sourdough bread its unique flavor.

The most important yeast in baking, brewing, and wine making is *Saccharomyces cerevisiae*. This yeast has been used by humans throughout recorded history. Yeast is used as a nutritional supplement because it contains high levels of B vitamins, and because about 50% of yeast is protein.

Ascomycete genetics and genomics have practical applications

Over the past few decades, yeasts have become increasingly important in genetic research. They were the first eukaryotes to be manipulated extensively by the techniques of genetic engineering, and they still play the leading role as models for research in eukaryotic cells. In 1996, the genome sequence of *S. cerevisiae*, the first eukaryote to be sequenced entirely, was completed. Yeast artificial chromosomes (YACs) have been used for genome sequencing, and the yeast two-hybrid system has been an important component of research on protein interactions (see chapter 17).

The fungal genome initiative is now under way to provide sequence information on other fungi. Initially 15 fungi were selected for whole-genome sequencing. These fungi were selected based on their effects on human health, including plant pathogens that threaten our food supply.

The ascomycete *Coccidiodes posadasii* was included because it is endemic in soil in the southwestern portion of the United States, can cause a fatal infection, and has been considered a possible bioterrorism threat. In the United States the annual infection rate is about 100,000 individuals, although only a small percentage of infected individuals die.

A second important criterion in selecting which fungi to sequence was the potential to provide information on fungal evolution. This new information will complement and expand our understanding of the diverse fungal kingdom.

> Ascomycetes undergo karyogamy within a characteristic saclike structure, the ascus. Meiosis follows, resulting in the production of ascospores. Yeasts within this group generally reproduce asexually by budding.

31.8 Basidiomycetes: The Club (Basidium) Fungi

The **basidiomycetes** (phylum Basidiomycota) include some of the most familiar fungi. Among the basidiomycetes are not only the mushrooms, toadstools, puffballs, jelly fungi, and shelf fungi, but also many important plant pathogens, including rusts and smuts (figure 31.11*a*). Rust infections resemble rusting metals, and smut infections appear black and powdery due to the spores. Many mushrooms are used as food, but others are hallucinogenic or deadly poisonous.

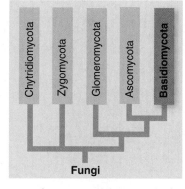

Basidiomycetes sexually reproduce within basidia

Basidiomycetes are named for their characteristic sexual reproductive structure, the club-shaped **basidium** (plural, *basidia*). Karyogamy occurs within the basidium, giving rise to the only diploid cell of the life cycle (figure 31.11*b*). Meiosis occurs immediately after karyogamy. In the basidiomycetes, the four haploid products of meiosis are incorporated into **basidiospores.**

In most members of this phylum, the basidiospores are borne at the end of the basidia on slender projections. The structure of a basidium therefore differs from that of an ascus, although functionally the two are identical. Recall that the ascospores of the ascomycetes are borne internally in asci instead of externally as in basidiospores.

figure 31.11

BASIDIOMYCETES. *a*. The death cap mushroom, *Amanita phalloides*. When eaten, these mushrooms are usually fatal. *b*. Life cycle of a basidiomycete. The basidium is the reproductive structure.

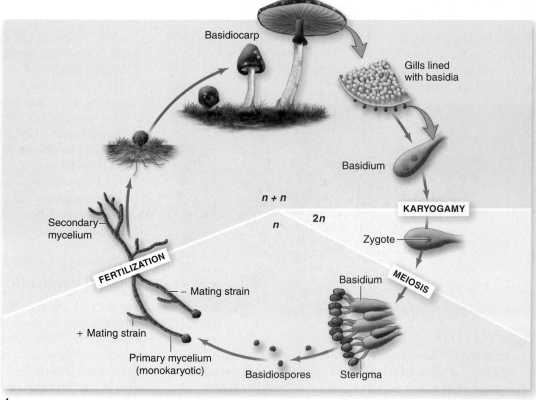

The secondary mycelium of basidiomycetes is heterokaryotic

The life cycle of a basidiomycete continues with the production of monokaryotic hyphae after spore germination. These hyphae lack septa early in development. Eventually, septa form between the nuclei of the monokaryotic hyphae. A basidiomycete mycelium made up of monokaryotic hyphae is called a *primary mycelium*.

Different mating types of monokaryotic hyphae may fuse, forming a dikaryotic mycelium, or *secondary mycelium*. Such a mycelium is heterokaryotic, with two nuclei representing the two different mating types, between each pair of septa. This stage is the dikaryon stage described earlier as being a distinguishing feature of fungi. The maintenance of two genomes in the heterokaryon allows for more genetic plasticity than in a diploid cell with one nucleus. One genome may compensate for mutations in the other.

The **basidiocarps**, or mushrooms, are formed entirely of secondary (dikaryotic) mycelium. Gills, sheets of tissue on the undersurface of the cap of a mushroom, form vast numbers of minute spores. It has been estimated that a mushroom with a cap measuring 7.5 cm in diameter produces as many as 40 million spores per hour!

Basidiomycetes undergo karyogamy, after which meiosis occurs within club-shaped basidia, the part of the fungus that we can identify. The secondary mycelium of basidiomycetes is the dikaryon stage, in which two nuclei exist within a single hyphal segment.

31.9 Deuteromycetes: A Polyphyletic Group That Includes Most Molds

Most of the **deuteromycetes** are fungi in which the sexual reproductive stages have not been observed. Many appear to be related to the ascomycetes, although some have clear affinities to the other phyla. The group of fungi from which a particular nonsexual strain has been derived usually can be determined by features of its hyphae and its asexual reproduction, but most of the standards rely on sexual reproduction. Some species of both *Penicillium* and *Aspergillus* form ascocarps, but the genera are still classified primarily as deuteromycetes because the ascocarps are found rarely in only a few species.

Deuteromycetes have limited genetic recombination

There are some 15,000 described species of deuteromycetes (figure 31.12). Even though sexual reproduction is absent, a certain amount of genetic recombination occurs. This becomes possible when hyphae of different genetic types fuse, as sometimes happens spontaneously.

Within the heterokaryotic hyphae that arise from such fusion, a special kind of genetic recombination called *parasexuality* may occur. In parasexuality, genetically distinct nuclei within a common hypha exchange portions of chromosomes. Recombination of this sort also occurs in other groups of fungi and seems to be responsible for some of the new pathogenic strains of wheat rust.

Deuteromycetes include economically important genera

Among the economically important genera of the deuteromycetes are *Penicillium* and *Aspergillus*. Some species of *Penicillium* are the source of the well-known antibiotic penicillin, and other species of the genus give the characteristic flavors and aromas to cheeses such as Roquefort and Camembert. Species of *Aspergillus* are used to ferment soy sauce and soy paste, processes in which certain bacteria and yeasts also play important roles. Not all deuteromycete interactions with food are to human advantage, however. *Fusarium* species growing on spoiled food produce highly toxic substances, as described later on.

Deuteromycetes are fungi in which no sexual reproduction has been observed. The majority are likely related to ascomycetes. They include commercially important genera such as *Penicillium* and *Aspergillus*.

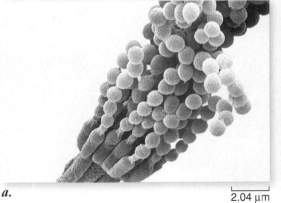

a.

2.04 μm

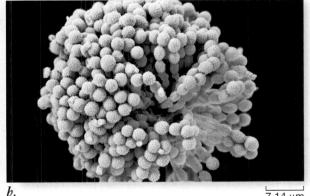

b.

7.14 μm

figure 31.12

DEUTEROMYCETES. *a. Penicillium* and (*b*) *Aspergillus* producing spores asexually.

Ecology of Fungi

Fungi, together with bacteria, are the principal decomposers in the biosphere. They break down organic materials and return the substances locked in those molecules to circulation in the ecosystem. Fungi are virtually the only organisms capable of breaking down cellulose and lignin, an insoluble organic compound that is one of the major constituents of wood. By breaking down such substances, fungi release carbon, nitrogen, and phosphorus from the bodies of living or dead organisms and make them available to other organisms.

In addition to their role as decomposers, fungi have entered into fascinating relationships with a variety of life forms. Interactions between different species are described in chapter 56 where we discuss community ecology, but we cover fungal relationships briefly here because of their unique character.

Fungi have a range of symbioses

The interactions, or **symbioses,** between fungi and other living organisms fall into a broad range of categories. In some cases, the symbiosis is an **obligate symbiosis** (essential for survival), and in other cases it is a **facultative symbiosis** (the fungus can survive without the host). Within a group of closely related fungi, several different types of symbiosis can be found.

First, here is a summary of ways in which living things can interact. **Pathogens** and **parasites** gain resources from their host, but they have a negative effect on the host that can even lead to death. The difference between pathogens and parasites is that pathogens cause disease, but parasites do not, except in extreme cases.

Commensal relationships benefit one partner but do not harm the other. Fungi that are in a **mutualistic** relationship benefit both themselves and their hosts. Many of these relationships are described in the discussion that follows.

Endophytes live inside plants and may protect plants from parasites

Endophytic fungi live inside plants, actually in the intercellular spaces. Found throughout the plant kingdom, many of these relationships may be examples of parasitism or commensalism.

There is growing evidence that some of these fungi protect their hosts from herbivores by producing chemical toxins or deterrents. Most often, the fungus synthesizes alkaloids that protect the plant. As you will learn in chapter 40, plants also synthesize a wide range of alkaloids, many of which serve to defend the plant.

One way to assess whether an endophyte is enhancing the health of its host plant is to grow plots of plants with and without the same endophyte. An experiment with Italian ryegrass, *Lolium multiflorum*, demonstrated that it is more resistant to aphid feeding when an endophytic fungus, *Neotyphodium*, is present (figure 31.13).

Lichens are an example of symbiosis between different kingdoms

Lichens (figure 31.14) are symbiotic associations between a fungus and a photosynthetic partner. The term *symbiosis* was coined to describe this relationship. Although many lichens are excellent examples of mutualism, some fungi are parasitic on their photosynthetic host.

Composition of a lichen

Ascomycetes are the fungal partners in all but about 20 of the approximately 15,000 species of lichens estimated to exist. Most of the visible body of a lichen consists of its fungus, but between the filaments of that fungus are cyanobacteria, green algae, or sometimes both (figure 31.15).

Specialized fungal hyphae penetrate or envelop the photosynthetic cell walls within them and transfer nutrients

figure 31.13

EFFECT OF THE FUNGAL ENDOPHYTE *Neotyphodium* **ON THE APHID POPULATION LIVING ON ITALIAN RYE GRASS (***Lolium multiflorum***).**

inquiry

? *Explain why or why not you are convinced that the endophyte decreases aphid predation of rye grass.*

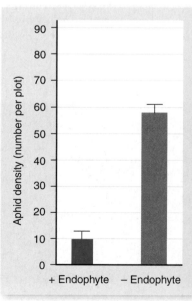

Plant With Endophyte

Plant Without Endophyte

Fruticose Lichen	Foliose Lichen	Crustose Lichen

  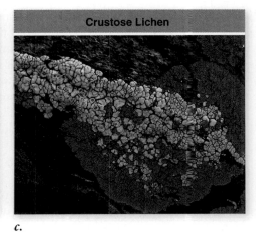

a.   *b.*   *c.*

figure 31.14

LICHENS ARE FOUND IN A VARIETY OF HABITATS. *a.* A fruticose lichen, growing in the soil. *b.* A foliose ("leafy") lichen, growing on the bark of a tree in Oregon. *c.* A crustose lichen, growing on rocks leading to the breakdown of rock into soil.

directly to the fungal partner. Note that although fungi penetrate the cell wall, they do not penetrate the plasma membrane. Biochemical signals sent out by the fungus apparently direct its cyanobacterial or green algal component to produce metabolic substances that it does not produce when growing independently of the fungus.

The fungi in lichens are unable to grow normally without their photosynthetic partners, and the fungi protect their partners from strong light and desiccation. When fungal components of lichens have been experimentally isolated from their photosynthetic partner, they survive, but grow very slowly.

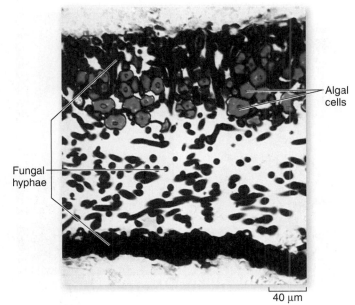

Algal cells

Fungal hyphae

40 μm

figure 31.15

STAINED SECTION OF A LICHEN. This section shows fungal hyphae (*purple*) more densely packed into a protective layer on the top and, especially, the bottom layer of the lichen. The blue cells near the upper surface of the lichen are those of a green alga. These cells supply carbohydrate to the fungus.

Ecology of lichens

The durable construction of the fungus combined with the photosynthetic properties of its partner have enabled lichens to invade the harshest habitats—the tops of mountains, the farthest northern and southern latitudes, and dry, bare rock faces in the desert. In harsh, exposed areas, lichens are often the first colonists, breaking down the rocks and setting the stage for the invasion of other organisms.

Lichens are often strikingly colored because of the presence of pigments that probably play a role in protecting the photosynthetic partner from the destructive action of the sun's rays. These same pigments may be extracted from the lichens and used as natural dyes. The traditional method of manufacturing Scotland's famous Harris tweed used fungal dyes.

Lichens vary in sensitivity to pollutants in the atmosphere, and some species are used as bioindicators of air quality. Their sensitivity results from their ability to absorb substances dissolved in rain and dew. Lichens are generally absent in and around cities because of automobile traffic and industrial activity, but some are adapted to these conditions. As pollution decreases, lichen populations tend to increase.

Mycorrhizae are fungi associated with roots of plants

The roots of about 90% of all known species of vascular plants normally are involved in mutualistic symbiotic relationships with certain kinds of fungi. It has been estimated that these fungi probably amount to 15% of the total weight of the world's plant roots. Associations of this kind are termed **mycorrhizae**, from the Greek words for fungus and root.

The fungi in mycorrhizal associations function as extensions of the root system. The fungal hyphae dramatically increase the amount of soil contact and total surface area for absorption. When mycorrhizae are present, they aid in the direct transfer of phosphorus, zinc, copper, and other mineral nutrients from the soil into the roots. The plant, on the other hand, supplies organic carbon to the fungus, so the system is an example of mutualism.

There are two principal types of mycorrhizae (figure 31.16). In **arbuscular mycorrhizae,** the fungal hyphae penetrate the outer cells of the plant root, forming coils, swellings, and minute branches, and also extend out into the surrounding soil. In **ectomycorrhizae,** the hyphae surround but do not penetrate the cell walls of the roots. In both kinds of mycorrhizae, the mycelium extends far out into the soil. A single root may associate with many fungal species, dividing the root at a millimeter-by-millimeter level.

Arbuscular mycorrhizae

Arbuscular mycorrhizae are by far the more common of the two types, involving roughly 70% of all plant species (figure 31.16a). The fungal component in them are Glomeromycetes, a monophyletic group that arose within one of the zygomycete lineages. The Glomeromycetes are associated with more than 200,000 species of plants.

Unlike mushrooms, none of the Glomeromycetes produce aboveground fruiting structures, and as a result, it is difficult to arrive at an accurate count of the number of extant species. Arbuscular mycorrhizal fungi are being studied intensively because they are potentially capable of increasing crop yields with lower phosphate and energy inputs.

The earliest fossil plants often show arbuscular mycorrhizal roots. Such associations may have played an important role in allowing plants to colonize land. The soils available at such times would have been sterile and lacking in organic matter. Plants that form mycorrhizal associations are particularly successful in infertile soils; considering the fossil evidence, it seems reasonable that mycorrhizal associations helped the earliest plants succeed on such soils. In addition, the closest living relatives of early vascular plants surviving today continue to depend strongly on mycorrhizae.

Some nonphotosynthetic plants also have mycorrhizal associations, but the symbiosis is one-way because the plant has no photosynthetic resources to offer. Instead of a two-partner symbiosis, a tripartite symbiosis is established. The fungal mycelium extends between a photosynthetic plant and a nonphotosynthetic, parasitic plant. This third, nonphotosynthetic member of the symbiosis is called an *epiparasite*. Not only does it obtain phosphate from the fungus, but it uses the fungus to channel carbohydrates from the photosynthetic plant to itself. Epiparasitism also occurs in ectomycorrhizal symbiosis.

Ectomycorrhizae

Ectomycorrhizae (see figure 31.16b) involve far fewer kinds of plants than do arbuscular mycorrhizae—perhaps a few thousand. Most ectomycorrhizal hosts are forest trees, such as pines, oaks, birches, willows, eucalyptus, and many others. The fungal components in most ectomycorrhizae are basidiomycetes, but some are ascomycetes.

Most ectomycorrhizal fungi are not restricted to a single species of plant, and most ectomycorrhizal plants form associations with many ectomycorrhizal fungi. Different combinations have different effects on the physiological characteristics of the plant and its ability to survive under different environmental conditions. At least 5000 species of fungi are involved in ectomycorrhizal relationships.

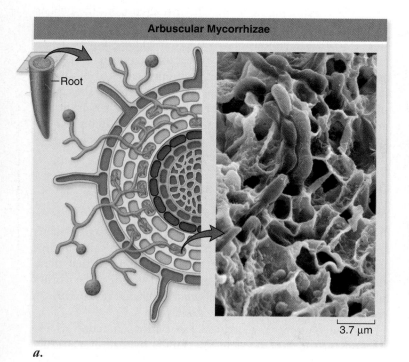

a.

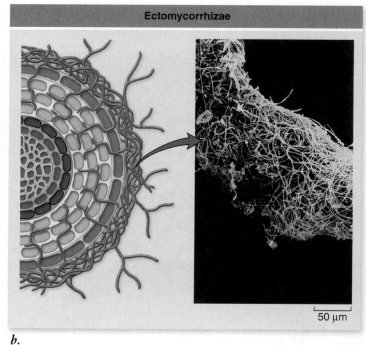

b.

figure 31.16

ARBUSCULAR MYCORRHIZAE AND ECTOMYCORRHIZAE. *a.* In arbuscular mycorrhizae, fungal hyphae penetrate the root cell wall of plants but not the plant membranes. ***b.*** Ectomycorrhizae on the roots of a *Eucalyptus* tree do not penetrate root cells, but grow around and extend between the cells.

Fungi also form mutual symbioses with animals

A range of mutualistic fungal–animal symbioses has been identified. For example, ruminant animals host fungi in their gut. Grassy diets have a high content of cellulose and lignin that cannot be digested by the grazing animal. Fungal enzymes, however, release nutrients that would otherwise be unavailable to the animal. The fungus gains a nutrient-rich environment in exchange.

One tripartite symbiosis involves ants, plants, and fungi. Leaf-cutter ants are the dominant herbivore in the New World tropics. These ants, members of the phylogenetic tribe Attini, have an obligate symbiosis with specific fungi that they have domesticated and maintain in an underground garden. The ants provide fungi with leaves to eat and protection from pathogens and other predators (figure 31.17). The fungi are the ants' food source.

Depending on the species of ant, the ant nest can be as small as a golf ball or as large as 50 cm in diameter and many feet deep. Some nests are inhabited by millions of leaf-cutter ants that maintain fungal gardens. These social insects have a caste system, and different ants have specific roles. Traveling on trails as long as 200 m, leaf-cutter ants search for foliage for their fungi. A colony of ants can defoliate an entire tree in a day. This ant farmer–fungi symbiosis has evolved multiple times and may have occurred as early as 50 MYA.

figure 31.17

ANT–FUNGAL SYMBIOSIS. Ants farming their fungal garden.

Fungi are the primary decomposers in ecosystems. A range of symbiotic relationships have evolved between fungi and other plants. Endophytes live inside tissues of a plant and may protect it from parasites. Lichens are a complex symbiosis between fungus species and cyanobacteria or green algae. Mycorrhizal associations between fungi and plant roots are mutually beneficial and in some cases are obligate symbioses. Fungi have also coevolved with animals in mutualistic relationships.

31.11 Fungal Parasites and Pathogens

Fungi can destroy a crop of plants and create significant problems for human health. A major problem in treatment and prevention is that fungi are eukaryotes, as are plants and animals. Understanding how fungi are distinct from these other two eukaryotic kingdoms may lead to safer and more efficient means of treating diseases caused by fungal parasites and pathogens.

Fungal infestation can harm plants and those who eat them

Fungal species cause many diseases in plants (figure 31.18), and they are responsible for billions of dollars in agricultural losses every year. Not only are fungi among the most harmful pests

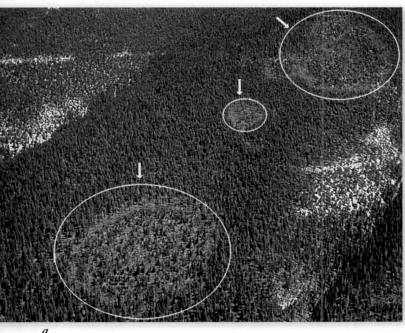

a.

b.

c.

figure 31.18

WORLD'S LARGEST ORGANISM?
a. Armillaria, a pathogenic fungus shown here afflicting three discrete regions of coniferous forest in Montana, grows out from a central focus as a single, circular clone. The large patch at the bottom of the picture is almost 8 hectares. *b.* Closeup of tree destroyed by *Armillaria. c. Armillaria* growing on a tree.

of living plants, but they also spoil food products that have been harvested and stored. In addition, fungi often secrete substances into the foods they are infesting that make these foods unpalatable, carcinogenic, or poisonous.

Pathogenic fungal–plant symbioses are numerous, and fungal pathogens of plants can also harm the animals that consume the plants. *Fusarium* species growing on spoiled food produce highly toxic substances, including vomitoxin, which has been implicated in brain damage in humans and animals in the southwestern United States.

Aflatoxins, which are among the most carcinogenic compounds known, are produced by some *Aspergillus flavus* strains growing on corn, peanuts, and cotton seed (figure 31.19). Aflatoxins can also damage the kidneys and the nervous system of animals, including humans. Most developed countries have legal limits on the concentration of aflatoxin permitted in different foods. More recently, aflatoxins have been considered as possible bioterrorism agents.

Hot, humid summers are especially conducive to the growth of this fungus. Monoculture has enhanced its spread; crop rotation with resistant crops can help control the spread of *Aspergillus*.

Fungal infections are difficult to treat in humans and other animals

Human and animal diseases can also be fungal in origin. Some common diseases, such as ringworm (which is not a worm but a fungus), athlete's foot, and nail fungus, can be treated with topical antifungal ointments and in some cases with oral medication.

Fungi can create devastating human diseases that are often difficult to treat because of the close phylogenetic relationship between fungi and animals. Yeast ascomycetes are important pathogens that cause diseases such as thrush, an infection of the mouth; the yeast *Candida* causes common oral or vaginal infections. *Pneumocystis jiroveci* (formerly *P. carinii*) invades the lungs, disrupting breathing, and can spread to other organs. In immune-suppressed AIDS patients, this infection can lead to death.

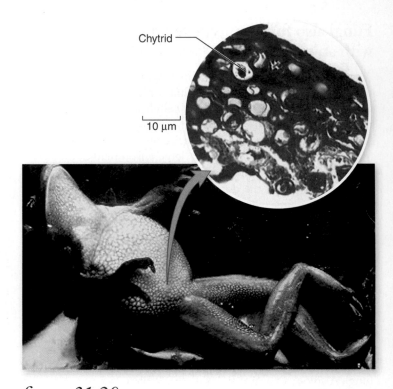

figure 31.20

FROG KILLED BY CHYTRIDIOMYCOSIS. Lesions formed by the chytrid can be seen on the abdomen of this frog.

Mold allergies are common, and mold-infested "sick" buildings pose concerns for inhabitants. Individuals with suppressed immune systems and people undergoing steroid treatments for inflammation disorders are particularly at risk for fungal disease.

An example of a parasitic fungal–animal symbiosis is **chytridiomycosis**, first identified in 1998 as an emerging infectious disease of amphibians. Amphibian populations have been declining worldwide for over three decades. The decline correlates with the presence of the chytrid *Batrachochytrium dendrobatidis* (see figure 31.1*a*), identified after extensive studies of frog carcasses. Sick and dead frogs were more likely than healthy frogs to have flasklike structures encased in their skin, which proved to be associated with chytrid spore production (figure 31.20).

The connection with *B. dendrobatidis* has been supported by DNA sequence data, by isolating and culturing the chytrid, and by infecting healthy frogs with the organism and replicating disease symptoms. It is unclear how the fungus kills the amphibians. Its primary symptoms are observed in the skin, which implies that the fungus may affect gas and water exchange—or it may be that the fungus produces a toxin. Bathing frogs in antifungal drugs can halt the disease or even eliminate the chytrids.

How the disease emerged simultaneously on different continents is a yet unsolved mystery. Both environmental change and carriers are being considered.

a.   *b.*   4.50 μm

figure 31.19

***Aspergillus flavus* INFECTS MAIZE AND CAN PRODUCE AFLATOXINS THAT ARE HARMFUL TO ANIMALS. *a.* Maize** (corn) infected with the fungus. *b.* A photomicrograph of *Aspergillus flavus* conidia.

Fungi can severely harm or kill both plants and animals. Treatment of fungal disease and parasitism in animals is made difficult by the close relationship between fungi and animals.

31.1 Defining Fungi

Fungi are classified into six main groups.

- Fungi are more closely related to animals than plants.

- Fungi are heterotrophic and have hyphal cells; their cell walls contain chitin, may have a dikaryon stage, and undergo nuclear mitosis.

31.2 General Biology of the Fungi

Fungi exist either as single cells or as a multicellular form with several different cell types.

- Hyphae can be continuous and multinucleate, or they may be divided into long chains of cells separated by cross-walls called septa.

- A mycelium, a mass of connected hyphae, penetrates a food source, and digestion of the material begins very quickly.

- The chitin found in fungi is the same material found in the exoskeletons of arthropods. Chitin is not cellulose; it is composed of linked glucose molecules with nitrogen groups and cross-linked with proteins.

- In multinucleated hyphae, if the nuclei are from two genetically distinct individuals, the hyphae are called heterokaryotic; if the nuclei are similar to each other, the hyphae are called homokaryotic.

- Fungi can reproduce sexually by fusion of hyphae from two compatible mating types.

- Spores can form either by asexual or sexual reproduction and are usually dispersed by the wind.

- Fungi obtain their nutrients through external digestion.

31.3 Phylogenetic Relationships

Fungi are clustered into five major phyla based primarily on their evolutionary relationships and a sixth group that is difficult to classify (figure 31.6).

- The Glomeromycota, Ascomycota, and Basidiomycota are monophyletic, but the Chytridiomycota and Zygomycota are not.

- The relationship of the non-monophyletic Deuteromycota to other fungi has not been determined.

31.4 Chytridiomycetes: Aquatic Fungi with Flagellated Zoospores

The Chytridiomycetes, or chytrids, are aquatic, flagellated fungi that are closely related to ancestral fungi (figure 31.7).

- Chytrids are the only fungi with flagellated sperm.

- Chytrids form symbiotic relationships and have been implicated in the decline of amphibian species.

31.5 Zygomycetes: Fungi That Produce Zygotes

Zygomycetes are incredibly diverse, but they have a characteristic zygosporangium.

- Zygomycetes lack septa in their hyphae except when they form sporangia or gametangia.

- Before two hyphae fuse, each forms gametangia within which gametes are formed.

- Sexual reproduction begins with the fusion of gametangia.

- Karyogamy, a process in which haploid nuclei fuse to form a diploid zygote nuclei, occurs after the gametangia fuse.

- The zygosporangium develops after fusion of the gametangia.

- Diploid zygospores develop within the zygosporangium.

- During asexual reproduction hyphae produce clumps of erect sporangiophores. The tips of the sporangiophores form sporangia within which haploid spores are produced by mitosis.

31.6 Glomeromycetes: Asexual Plant Symbionts

Glomeromycetes are a monophyletic lineage that forms obligatory relationships with plant roots.

- Glomeromycete hyphae form intracellular associations with plant roots are called arbuscular mycorrhizae.

- The Glomeromycetes show no evidence of sexual reproduction.

31.7 Ascomycetes: The Sac (Ascus) Fungi

The Ascomycetes are named for their characteristic reproductive structure, the saclike ascus that forms within the ascocarp (figure 31.9).

- Karyogamy occurs only in the ascus and results in the only diploid nucleus of the ascomycete life cycle.

- Meiosis and mitosis occur immediately after karyogamy, resulting in eight haploid nuclei that become walled ascospores.

- Asexual reproduction is very common and occurs by means of conidia formed at the end of modified hyphae called conidiophores.

- Yeasts are ascomycetes that usually reproduce by cell fission or budding.

31.8 Basidiomycetes: The Club (Basidium) Fungi

Basidiomycetes are identified by their club-shaped reproductive structure (figure 31.11).

- Karyogamy occurs within the basidia, giving rise to the only true diploid cell in the basidiomycete life cycle.

- Immediately after karyogamy meiosis results in four haploid cells being incorporated into the basidiospores, which form mycelium.

31.9 Deuteromycetes: A Polyphyletic Group That Includes Most Molds

The sexual reproductive stages of the deuteromycetes have not been observed, but they do have a limited genetic recombination.

- Heterokaryotic hyphae exhibit parasexuality because genetically distinct nuclei exchange portions of chromosomes.

31.10 Ecology of Fungi

Fungi, along with bacteria, are the principal decomposers in the biosphere that recycle nutrients such as carbon, nitrogen, and phosphorus.

- Fungi are capable of breaking down cellulose and lignin, an insoluble organic compound found in wood.

- Fungi form obligate or facultative symbiotic relationships with other living organisms.

- Fungi exhibit a range of symbioses; they can be pathogenic or parasitic, commensal or mutualistic.

31.11 Fungal Parasites and Pathogens

Fungi can create substantial agricultural human health issues.

- Fungi can secrete chemicals that make food unpalatable, carcinogenic, or poisonous.

- Treatment of fungal diseases is difficult because of their similarity to animals.

SELF TEST

1. Which of the following is not a characteristic of a fungus?
   a. Cell walls made of chitin
   b. A form of mitosis different from plants and animals
   c. Ability to conduct photosynthesis
   d. Filamentous structure

2. A fungal cell that contains two genetically different nuclei would be classified as
   a. monokaryotic.
   b. bikaryotic.
   c. homokaryotic.
   d. heterokaryotic.

3. Which of the following groups of fungi is *not* monophyletic?
   a. Zygomycota
   b. Basidiomycota
   c. Glomeromycota
   d. Ascomycota

4. Based on physical characteristics, the _____ represent the most ancient phylum of fungi.
   a. Basidiomycota
   b. Zygomycota
   c. Ascomycota
   d. Chytridiomycota

5. Which of the following groups of fungi does not represent a true phylogenic relationship, but rather a classification due to a lack of scientific information?
   a. Basidiomycota
   b. Ascomycota
   c. Deuteromycetes
   d. Glomeromycota

6. The early evolution of terrestrial plants was made possible by mycorrhizal relationships with the
   a. Zygomycetes.
   b. Glomeromycota.
   c. Ascomycota.
   d. Basidiomycota.

7. In a culture of hyphae of unknown origin you notice that the hyphae lack septa and that the fungi reproduce asexually by using clumps of erect stalks. However, at times sexual reproduction can be observed. To what group of fungi would you assign it?
   a. Deuteromycetes
   b. Basidiomycota
   c. Ascomycota
   d. Zygomycota

8. In a life cycle of a typical Basidiomycota, where would you expect to find a dikaryotic cell?
   a. Primary mycelium
   b. Secondary mycelium
   c. In the basidiospores
   d. In the zygote

9. Which of the following is correct regarding the yeast *Saccharomyces cerevisiae*?
   a. It reproduces asexually by a process called budding.
   b. It produces an ascocarp during reproduction.
   c. It belongs in the group Deuteromycetes.
   d. All of the above are correct.

10. If biologists were to abandon the use of Deuteromycetes as a method of classification, to which phylum would the majority of the Deuteromycetes be assigned?
    a. Ascomycota
    b. Basidiomycota
    c. Zygomycota
    d. None of the above

11. *Penicillium* and *Aspergillus* are both classified as
    a. Basidiomycota.
    b. Deuteromycetes.
    c. Ascomycota.
    d. Zygomycota.

12. Symbiotic relationships occur between the fungi and
    a. plants.
    b. bacteria.
    c. animals.
    d. all of the above.

13. A fungal relationship between a forest tree and a basidiomycetes would most likely be classified as an example of
    a. parasitism only.
    b. an arbuscular mycorrhizae.
    c. ectomycorrhizae.
    d. a lichen.

14. Symbiotic relationships between animals and fungi are focused on which of the following?
    a. Protection from bacteria
    b. Colonization of land
    c. Protection from desiccation
    d. Exchange of nutrients

15. Which of the following species of fungi is not associated with diseases in humans?
    a. *Pneumocystis jiroveci*
    b. *Aspergillus flavus*
    c. *Candida albicans*
    d. *Batrachochytrium dendrobatidis*

CHALLENGE QUESTIONS

1. Historically fungi have been classified as being more plantlike despite their lack of photosynthetic ability. Although we now know that fungi are more closely related to the animals than the plants, what characteristics would have initially led scientists to place them closer to the plants?

2. The importance of fungi in the evolution of terrestrial life is typically understated. Explain the importance of fungi in the colonization of land.

3. Based on your understanding of fungi, why won't antibiotics work in the treatment of a fungal infection?

Do you need additional review? *Visit* www.ravenbiology.com *for practice quizzes, animations, videos, and activities designed to help you master the material in this chapter.*

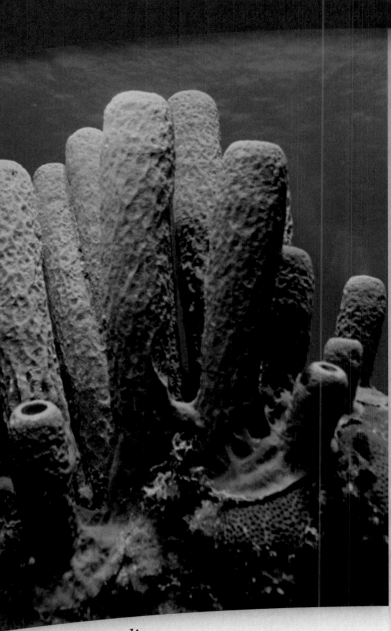

chapter 32

Overview of Animal Diversity

introduction

WE NOW EXPLORE THE GREAT diversity of modern animals, the result of a long evolutionary history. Animals are among the most abundant living organisms. Found in almost every conceivable habitat, they bewilder us with their diversity. More than a million species have been described, and several million more are thought to await discovery. Despite their great diversity, all animals have much in common. For example, locomotion is a distinctive characteristic of higher animals, although not all animals move about. Early naturalists thought that sponges and corals were plants because the adults seemed rooted to one place and did not move about.

Biologists have long debated the evolutionary relationships among animal groups and have traditionally used information derived from their anatomy and especially from their early embryonic development. For the last two decades, molecular tools have provided new insights; together with the recent synthesis of evolutionary and developmental biology (chapter 25) and the discovery of new and mysterious forms of life, these new data are leading biologists to reconsider the animal tree of life.

32.1 Some General Features of Animals

Animals are the eaters, or consumers, of the Earth. Animals are a very diverse group—no one criterion fits all—but several characteristics are of major importance:

- Animals are heterotrophs and must ingest other organisms for nourishment.
- All animals are multicellular. Unlike plants, fungi, and certain protists, animal cells lack cell walls; in addition, unlike protists, animals have specialized cells that form tissues or organs.
- Most animals are able to move from place to place, which requires the development of musculature and a nervous system.

- Animals are very diverse in form and habitat.
- Most animals reproduce sexually and have a special type of **haploid gamete.**
- Animals have a characteristic pattern of embryonic development and possess unique tissues.

Table 32.1 describes the general features of animals.

> Animals are complex multicellular organisms typically characterized by high mobility and heterotrophy. Most animals also possess internal tissues and reproduce sexually.

TABLE 32.1	**General Features of Animals**
Heterotrophs. Unlike autotrophic plants and algae, animals cannot construct organic molecules from inorganic chemicals. All animals are heterotrophs—that is, they obtain energy and organic molecules by ingesting other organisms. Some animals (herbivores) consume autotrophs; other animals (carnivores) consume heterotrophs; and still others (detritivores) consume decomposing organisms.	
Multicellular. All animals are multicellular, often with complex bodies like that of this brittlestar (phylum Echinodermata). The unicellular heterotrophic organisms called Protozoa, which were at one time regarded as simple animals, are now considered members of the large and diverse kingdom Protista, discussed in chapter 29.	
No Cell Walls. Animal cells are distinct among those of other multicellular organisms because they lack rigid cell walls and are usually quite flexible, as are these cancer cells. The many cells of animal bodies are held together by extracellular frames of structural proteins such as collagen. Other proteins form a collection of unique intercellular junctions between animal cells.	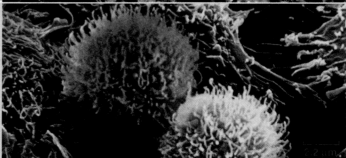
Active Movement. The ability of animals to move more rapidly and in more complex ways than members of other kingdoms is perhaps their most striking characteristic, one that is directly related to the flexibility of their cells and the evolution of nerve and muscle tissues. A remarkable form of movement unique to animals is flying, an ability that is well developed among vertebrates and insects such as this butterfly (phylum Arthropoda). Many animals are sessile although they have muscles or muscle fibers that allow sudden movements. Sponges have little capacity for movement.	

| TABLE 32.1 | General Features of Animals, *continued* |

Diverse in Form. Almost all animals (99%) are **invertebrates,** which, like this millipede (phylum Arthropoda), lack a backbone. Of the estimated 10 million living animal species, only 42,500 have a backbone and are referred to as **vertebrates.** Animals are very diverse in form, ranging in size from organisms too small to see with the unaided eye to enormous whales and giant squids.

Diverse in Habitat. The animal kingdom includes about 36 phyla, most of which, like these jellyfish (phylum Cnidaria), occur in the sea. Far fewer phyla occur in fresh water, and fewer still occur on land. Members of three successful marine phyla, Arthropoda, Mollusca, and Chordata, also dominate animal life on land. Only one animal phylum, Onychophora (velvet worms) is entirely terrestrial; at least some members of all other phyla are aquatic.

Sexual Reproduction. Most animals reproduce sexually, as these tortoises (phylum Chordata) are doing. Animal eggs, which are nonmotile, are much larger than the small, usually flagellated sperm. In animals, cells formed in meiosis function directly as gametes. The haploid cells do not divide by mitosis first, as they do in plants and fungi, but rather fuse directly with each other to form the zygote. Consequently, there is no counterpart among animals to the alternation of haploid (gametophyte) and diploid (sporophyte) generations characteristic of plants.

Embryonic Development. Most animals have a similar pattern of embryonic development. The zygote first undergoes a series of mitotic divisions, called *cleavage,* and like this dividing frog's egg, becomes a solid ball of cells, the **morula,** and then a hollow ball of cells, the **blastula.** In most animals, the blastula folds inward at one point to form a hollow sac with an opening at one end called the **blastopore.** An embryo at this stage is called a **gastrula.** The subsequent growth and movement of the cells of the gastrula differ widely from one phylum of animals to another.

Unique Tissues. The cells of all animals except sponges are organized into structural and functional units called **tissues,** collections of cells that have joined together and are specialized to perform a specific function. Animals are unique in having two tissues associated with movement: (1) muscle tissue, which powers animal movement, and (2) nervous tissue, which conducts signals among cells. Neuromuscular junctions, where nerves connect with muscle tissue, are shown here.

Evolution of the Animal Body Plan

The features described in the preceding section evolved over the course of millions of years. We can trace some of these developments and make hypotheses about others by examining the different types of animal bodies and body plans present in fossils and in existence today. Five key transitions can be noted in animal evolution:

1. The evolution of **tissues,** allowing specialization
2. The evolution of different forms of **symmetry**
3. The evolution of a **body cavity**
4. The evolution of different patterns of **development**
5. The evolution of **segmentation,** or repeated units in the body

Each of these transitions is explained in the sections that follow.

Tissues evolved, allowing specialized functions

The simplest animals, the Parazoa (chapter 33), lack both clearly defined tissues and organs, although they have a sophisticated degree of cell specialization not found in protists. Characterized by the sponges, these animals exist as aggregates of cells that can differentiate and dedifferentiate. Sponges have the ability to disaggregate and aggregate their cells, which means that the cells can be separated from one another, and can then gather back together again. All other animals, the Eumetazoa, have distinct and well-defined tissues, and the differentiation of cells is irreversible for most cell types.

Most animals exhibit radial or bilateral symmetry

Sponges also lack any definite symmetry, in most cases growing asymmetrically as irregular masses. Virtually all other animals have a definite shape and symmetry that can be defined along an imaginary axis drawn through the animal's body. The two main types of symmetry are *radial* and *bilateral.*

Radial symmetry

Symmetrical bodies first evolved in marine animals belonging to the phylum Cnidaria (jellyfish, sea anemones, and corals; chapter 33). The bodies of members of this phylum exhibit **radial symmetry,** a body design in which the parts of the body are arranged around a central axis in such a way that any plane passing through the central axis divides the organism into halves that are approximate mirror images (figure 32.1*a*).

Bilateral symmetry

The bodies of most other animals, the Bilateria, are marked by a fundamental **bilateral symmetry,** a body design in which the body has a right and a left half that are mirror images of each other (figure 32.1*b*). A bilaterally symmetrical body plan has a top and a bottom, better known respectively as the *dorsal* and *ventral* portions of the body. It also has a

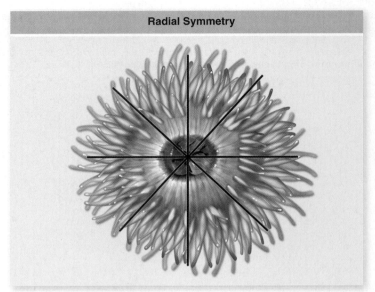

Radial Symmetry

a.

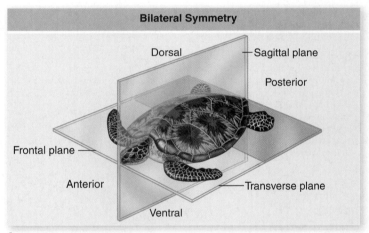

Bilateral Symmetry

Dorsal — Sagittal plane

Posterior

Frontal plane —

Anterior

— Transverse plane

Ventral

b.

figure 32.1

A COMPARISON OF RADIAL AND BILATERAL SYMMETRY.
a. Radially symmetrical animals, such as this sea anemone (phylum Cnidaria), can be bisected into equal halves in any two-dimensional plane. *b.* Bilaterally symmetrical animals, such as this turtle (phylum Chordata), can only be bisected into equal halves in one plane (the sagittal plane).

front, or *anterior* end, and a back, or *posterior* end. In some higher animals, such as echinoderms (sea stars), the adults are radially symmetrical, but even in this group the larvae are bilaterally symmetrical.

Bilateral symmetry constitutes a major evolutionary advance in the animal body plan. This unique form of organization allows different organs to concentrate in different parts of the body, with sensory organs generally concentrated in the anterior end. Also, bilaterally symmetrical animals have a greater ability to move through their environments than radially symmetrical ones.

Much of the nervous system in bilaterally symmetrical animals is in the form of major longitudinal nerve cords. In a very early evolutionary advance, nerve cells became grouped in the anterior end of the body. These nerve cells probably first functioned mainly to transmit impulses from the anterior sense organs to the rest of the nervous system. This trend ultimately led to the evolution of a definite brain area and a head, a process called **cephalization,** as well as to the increasing dominance and specialization of these organs. Cephalization is often seen as a consequence of the development of bilateral symmetry.

A body cavity made possible the development of advanced organ systems

Most animals produce three germ layers: an outer **ectoderm,** an inner **endoderm,** and a third layer, the **mesoderm,** between the ectoderm and endoderm. In general, the outer coverings of the body and the nervous system develop from the ectoderm; the digestive organs and intestines develop from the endoderm; and the skeleton and muscles develop from the mesoderm. Cnidarians have only two layers, the endoderm and the ectoderm, and sponges lack any germ layers.

A third key transition in the animal body plan was the evolution of the body cavity, which is a space surrounded by mesodermal tissue that is formed during development. The evolution of efficient organ systems within the animal body was not possible until a body cavity evolved for supporting organs, distributing materials, and fostering complex developmental interactions.

Kinds of body cavities

Three basic kinds of body plans evolved multiple times in the Bilateria (figure 32.2). **Acoelomates** have no body cavity because the space between the mesoderm and the endoderm is filled with cells and organic materials. **Pseudocoelomates** have a body cavity called the **pseudocoel** located between the mesoderm and endoderm. In animals having the third type of body plan, called **coelomates,** a fluid-filled body cavity develops not between the endoderm and mesoderm, but rather entirely within the mesoderm. Such a body cavity is called a **coelom.**

In coelomates, the heart, along with other organ systems, is suspended within the coelom; the coelom, in turn, is surrounded by a layer of epithelial cells entirely derived from the mesoderm (see figure 32.2).

The circulatory system

The development of the coelom poses a problem—circulation of nutrients and removal of wastes. In pseudocoelomates, the problem is solved by churning the fluid within the body cavity. Coelomates, in contrast, have developed a **circulatory system,** a network of vessels that carry fluids to and from the

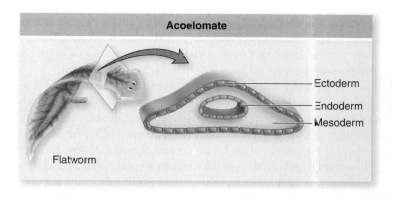

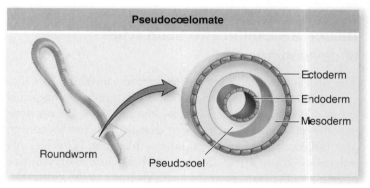

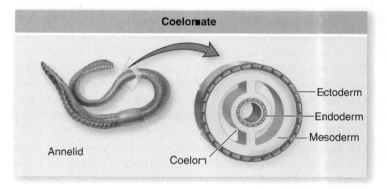

figure 32.2

THREE BODY PLANS FOR BILATERALLY SYMMETRICAL ANIMALS. Acoelomates, such as flatworms, have no body cavity between the digestive tract (endoderm) and the musculature layer (mesoderm). Pseudocoelomates have a body cavity, the pseudocoel, between the endoderm and the mesoderm. Coelomates have a body cavity, the coelom, that develops entirely within the mesoderm, and so is lined on both sides by mesoderm tissue.

parts of the body. The circulating fluid, or blood, carries nutrients and oxygen to the tissues and removes wastes and carbon dioxide.

Blood is usually pushed through the circulatory system by contraction of one or more muscular hearts. In an **open circulatory system,** the blood passes from vessels into sinuses, mixes with body fluid, and then reenters the vessels later in another location. In a **closed circulatory system,** the blood is physically separated from other body fluids and can be separately controlled. Also, blood moves through a closed circulatory system faster and more efficiently than it does through an open system.

For many years zoologists assumed that animal evolution proceeded from the simpler acoelomates toward the more complex coelomate body plan, passing through an intermediate pseudocoelomate stage. However, as you saw in chapter 21, evolution rarely occurs in such a linear and directional way. Rather, it appears that both the pseudocoelomate and coelomate stages evolved several times, and that many other advanced animals have become acoelomate secondarily.

Bilaterians exhibit two main types of development

The processes of embryonic development in animals is discussed fully in chapter 53. Briefly, bilaterally symmetrical animals exhibit a pattern of development that begins with mitotic cell divisions of the egg that lead to the formation of a hollow ball of cells, called the **blastula.** The blastula indents to form a two-layer-thick ball with a **blastopore** opening to the outside and a primitive gut cavity called the **archenteron.**

Bilaterians can be divided into two groups based on differences in the basic pattern of development. One group is called the **protostomes.** In this group, the mouth develops before the anus; *protostome* means "first mouth." Protostomes include most bilaterians, such as the flatworms, nematodes, mollusks, annelids, and arthropods. Two outwardly dissimilar groups, the echinoderms and the chordates, together with a few other smaller related phyla, constitute the second group, the **deuterostomes.** In this group, the mouth develops after the anus; *deuterostome* means "second mouth." Protostomes and deuterostomes differ in several aspects of their embryology, including cleavage patterns, determinate versus indeterminate development, the fate of the blastopore as development proceeds, and how the coelom is formed.

Cleavage patterns

The progressive division of cells during embryonic growth is called **cleavage.** The cleavage pattern relative to the embryo's polar axis determines how the cells will array. In some protostomes, each new cell buds off at an angle oblique to the polar axis. As a result, a new cell nestles into the space between the older ones in a closely packed array. This pattern is called **spiral cleavage** because a line drawn through a sequence of dividing cells spirals outward from the polar axis (figure 32.3 *top*). Spiral cleavage is characteristic of annelids, mollusks, nemerteans, and related phyla, known as Spiralia.

In deuterostomes, by contrast, the cells divide parallel to and at right angles to the polar axis. As a result, the pairs of cells from each division are positioned directly above and below one another; this process gives rise to a loosely packed array of cells. This pattern is called **radial cleavage** because a line drawn through a sequence of dividing cells describes a radius outward from the polar axis (figure 32.3 *bottom*).

Determinate versus indeterminate development

Many protostomes exhibit **determinate development.** In this type of development, the type of tissue each embryonic cell will form in the adult is predetermined. Before cleavage begins, the molecules that act as developmental signals are localized in different regions of the egg. Consequently, the cell divisions that occur after fertilization separate different molecular signals into different daughter cells. This process specifies the fate of even the very earliest embryonic cells.

Deuterostomes, on the other hand, display **indeterminate development.** The first few cell divisions of the egg produce identical daughter cells. Any one of these cells, if separated from the others, can develop into a complete organism, that is, their fate is indeterminate. The molecules that signal the embryonic cells to develop differently are not localized until later in the embryo's development.

The fate of the blastopore

In protostomes, the mouth (stoma) of the animal develops from or near the blastopore. If such an animal has a distinct anus or anal pore, it develops either from the blastopore or it forms later in another region of the embryo. So, in many protostomes such as annelids, nematodes, or onychophorans (velvet worms), both the mouth and the anus form from the embryonic blastopore.

In deuterostomes, the blastopore gives rise to the organism's anus, and the mouth always develops from a second pore that arises in the blastula later in development.

Formation of the coelom

In all coelomate animals the coelom originates from mesoderm. In protostomes, this development occurs simply and directly: The cells simply move away from one another as the coelomic cavity expands within the mesoderm.

In deuterostomes, whole groups of cells usually move around to form new tissue associations. The coelom is normally produced by an invagination of the archenteron, which is the central tube within the gastrula and is also called the primitive gut. This tube, lined with endoderm, opens to the outside via the blastopore and eventually becomes the gut cavity.

Deuterostomes evolved from protostomes more than 500 MYA, and the consistency of deuterostome development, along with its distinctiveness from that of the protostomes, suggests that it evolved once, in an ancestor common to all of the deuterostome phyla. The mode of development in protostomes is more diverse, but the previously discussed spiral development also evolved once, in the common ancestor to all spiralian phyla.

Segmentation allowed redundant systems and more efficient locomotion

Another key transition in the evolution of animal body plans involved the subdivision of the body into segments. Segmented animals are "assembled" from a succession of segments that look alike but have the possibility of specialization. During the animal's early development, these segments become most obvious in the mesoderm but later are reflected in the ectoderm and

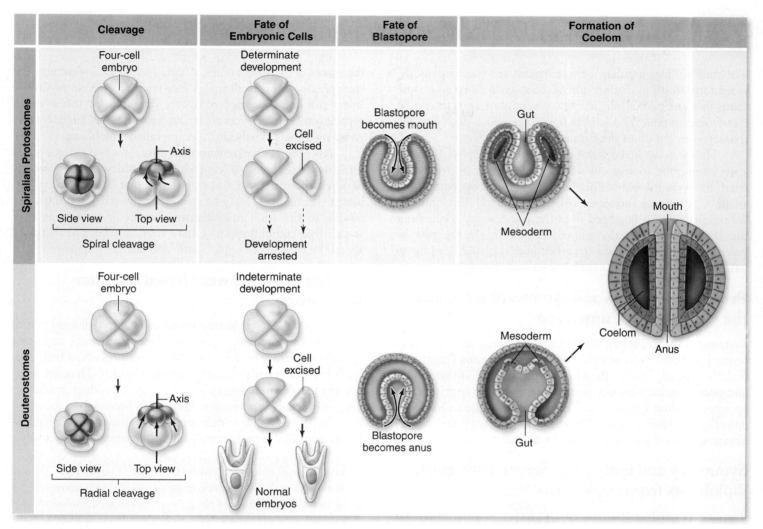

| Cleavage | Fate of Embryonic Cells | Fate of Blastopore | Formation of Coelom |

Spiralian Protostomes

Four-cell embryo

Side view — Top view — Axis

Spiral cleavage

Determinate development

Cell excised

Development arrested

Blastopore becomes mouth

Gut

Mesoderm

Mouth

Deuterostomes

Four-cell embryo

Side view — Top view — Axis

Radial cleavage

Indeterminate development

Cell excised

Normal embryos

Blastopore becomes anus

Mesoderm

Gut

Coelom

Anus

figure 32.3

EMBRYONIC DEVELOPMENT IN PROTOSTOMES AND DEUTEROSTOMES. In spiralian protostomes, embryonic cells cleave in a spiral pattern and exhibit determinate development; the blastopore becomes the animal's mouth, and the coelom originates from a mesodermal split. In deuterostomes, embryonic cells cleave radially and exhibit indeterminate development; the blastopore becomes the animal's anus, and the coelom originates from an invagination of the archenteron.

endoderm as well. Two advantages result from early embryonic segmentation:

1. In annelids, such as earthworms, and in other highly segmented animals, each segment may go on to develop a more or less complete set of adult organ systems. Such systems are termed *redundant systems.* Damage to any one segment need not be fatal because the other segments duplicate that segment's functions.

2. Locomotion is far more efficient when individual segments can move independently because the animal as a whole has more flexibility of movement. Because the separations isolate each segment into an individual skeletal unit, each is able to contract or expand autonomously. Therefore, a long body can move in ways that are often quite complex.

Segmentation underlies the organization of all advanced animal body plans. In some adult arthropods, the segments are fused, but segmentation is usually apparent in their embryological development. In vertebrates, the backbone and muscular areas are segmented, although segmentation is often disguised in the adult form.

Although in the past zoologists considered that true segmentation was found in only three phyla, namely the annelids, the arthropods, and the chordates, it is now widely recognized that segmentation is more widespread than previously thought. Some other animals, such as onychophorans (velvet worms), tardigrades (water bears), and kinorhynchs (mud dragons), are also segmented.

Key transitions in body design are responsible for most of the differences we see among the major animal phyla: the evolution of (1) tissues, (2) bilateral symmetry, (3) a body cavity, (4) spiralian and deuterostome development, and (5) segmentation.

32.3 The Traditional Classification of Animals

The multicellular animals, or metazoans, are traditionally divided into 36 or so distinct phyla (zoologists disagree on the status of some animal phyla). The diversity of animals can be clearly seen in table 32.2 (on pp. 630–631), which describes key characteristics of 20 of these animal phyla.

How can we make sense of this great diversity? Systematists attempting to sort out which phyla are more closely related have traditionally created phylogenies by comparing anatomical features and aspects of embryological development. A broad consensus emerged over the last century concerning the main branches of the animal tree of life. In the past 30 years, however, more data have become available, leading to new classification schemes (to be described shortly).

Presence of tissues and symmetry separated the Parazoa and Eumetazoa

Systematists traditionally divided the kingdom Animalia (also termed Metazoa) into two main branches: **Parazoa** ("near animals")—animals that for the most part lack a definite symmetry and possess neither tissues nor organs, mostly comprising the sponges, phylum Porifera; and **Eumetazoa** ("true animals")—animals that have a definite shape and symmetry and tissues organized into organs and organ systems.

Symmetry and embryonic layers distinguished diploblasts from triploblasts

The eumetazoan branches of animal phylogeny were distinguished by the type of symmetry and by the nature of the embryonic layers that form during development and go on to differentiate into the tissues of the adult animal. Animals with radial symmetry, the ctenophores and the cnidarians (see table 32.2), have two tissue layers that form during embryonic development (often called germ layers). These layers are an outer *ectoderm* and an inner *endoderm*, and these animals are therefore called diploblastic.

All other eumetazoans that have bilateral symmetry at some developmental stage, produce a third layer, the *mesoderm*, between the ectoderm and endoderm; that is, they are triploblastic (see figure 32.2). The mesoderm gives rise, among other tissues, to the adult musculature. Recent investigations have shown that ctenophores also have true muscles and therefore should be considered triploblastic animals.

Further divisions were based on other key features

Systematists identified further branches of the traditional animal phylogeny by comparing traits that seemed profoundly important to the evolutionary history of phyla, key features of the body plan shared by most animals belonging to that branch. Thus, the bilateral animals were split into two main groups depending on whether the embryonic blastopore (see figure 32.3) becomes the mouth or the anus (or both) in the adult animal. This major split divided bilateral animals into protostomes and deuterostomes, respectively.

> Animals are traditionally classified into some 36 phyla. The evolutionary relationships among the animal phyla are inferred by assuming relatedness of phyla that share certain fundamental morphological and molecular characters, which are assumed to have arisen only once.

32.4 A New Look at the Metazoan Tree of Life

The traditional animal phylogeny, although accepted by a broad consensus of biologists for almost a century, is now being reevaluated. Its simple organization based on information from single or few organ systems has always presented certain problems—puzzling minor groups, for example, do not fit well into the standard scheme.

The myzostomids have defied classification: A case study

The myzostomids, an enigmatic and anatomically bizarre group of marine animals, are parasites of echinoderms (figure 32.4). Myzostomid fossils are found associated with echinoderms since the Ordovician period, so the myzostomid–echinoderm relationship is a very ancient one. Their long history of obligate association has led to the loss or simplification of many myzostomid body elements, leaving them, for example, with no body cavity (they are acoelomates) and only incomplete segmentation.

This character loss has led to considerable disagreement among systematists. But although taxonomists have disagreed about the details, all have generally allied myzostomids in some fashion with the annelids—sometimes within the polychaetes and sometimes as a separate phylum closely allied to the annelids.

Molecular data

Recently, this view has been challenged. New comparisons using molecular data have come to very different conclusions. Researchers have examined two components of the protein synthesis machinery—the small ribosomal subunit rRNA gene and an elongation factor gene, elongation factor-1α. The phylogeny they obtain based on sequence data does not place the myzostomids with the annelids. Indeed, investigators find that the myzostomids have no close links to the annelids at all. Instead, they are more closely allied with the flatworms—organisms such as planaria and tapeworms.

Implications of molecular reclassification

The result of the molecular analysis on myzostomids hints strongly that the key morphological characters that biologists have traditionally used to construct animal phylogenies—segmentation,

figure 32.4

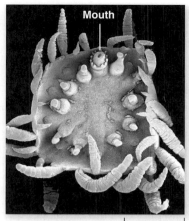

A SYSTEMATIC PUZZLE. Ventral view of *Myzostoma mortenensi*, a myzostomid. Myzostomids have no body cavity and segmentation is incomplete. Such animals present a classification challenge, causing systematists to reconsider traditional animal phylogenies.

Mouth

4 μm

coeloms, jointed appendages, and the like—are not the conservative characters we had supposed. Among the myzostomids, these features appear to have been gained and lost again during the course of their evolution. If this nonconservative evolutionary pattern should prove general, our view of the evolution of the animal body plan, and how the various animal phyla relate to one another, will soon be in need of major revision.

Molecular systematics is changing our understanding of animal phylogeny

The last decade has seen a wealth of new molecular DNA sequence data on the various animal groups. The field of **molecular systematics** uses unique sequences within certain genes to identify clusters of related groups. Expressed in the terminology of cladistics, the shared derived sequence characters unique to a group and its ancestors defines clusters of monophyletic taxa that make up clades (see chapter 23). The animal phylogenetic tree viewed in these terms is a hierarchy of clades nested within larger clades.

A variety of phylogenies have been produced in the last decades. Although they differ from one another in some respects, the new phylogenies share some deep branch structure with the traditional animal tree of life. Most new phylogenies agree, however, on one revolutionary difference: the separation of annelids and arthropods into different clades. These two groups were previously considered related due to the presence of segmentation. Instead, arthropods are now grouped with a series of mostly pseudocoelomate protostomes that molt their cuticles at least once during the life cycle. These animals are termed *ecdysozoans*, which means "molting animals" (chapter 34).

At present, molecular phylogenetic analysis of the animal kingdom is still inconclusive, but work done so far provides strong support for certain clades. Some zoologists had been concerned because phylogenies developed separately from different molecules sometimes suggested quite different evolutionary relationships. It is now clear, however, that the best picture emerges when the information from different molecules is combined. Over the next few years, a mountain of additional molecular data can be anticipated. As more data are brought to bear, the confusion can be expected to lessen.

Figure 32.5 presents a summary animal tree of life developed from DNA sequence data, as of the year 2005, including ribosomal RNAs and protein-coding genes. In it, the traditional protostome group is broken into Ecdysozoa and Spiralia, and the latter is broken into Lophotrochozoa and Platyzoa. This new view of the metazoan tree of life is only a rough outline; it is already clear, however, that major groups are related in very different ways in molecular phylogenies than in the more traditional morphological one.

The use of molecular data to construct phylogenies has significantly altered our understanding of relationships among the animal phyla.

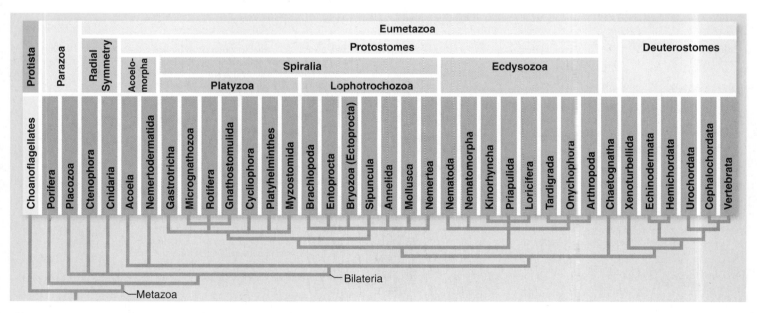

figure 32.5

PROPOSED REVISION OF THE ANIMAL TREE OF LIFE. This phylogeny reflects the current consensus based on interpretation of new anatomical and developmental data as well as results derived from molecular phylogenetic studies. Whether Chaetognatha is a protostome or a deuterostome is unclear.

TABLE 32.2 **The Major Animal Phyla Including the 3 Most Recently Discovered Phyla**

Phylum	Typical Examples		Key Characteristics	Approximate Number of Named Species
Arthropoda (arthropods)	Beetles, other insects, crabs, spiders, scorpions, centipedes, millipedes		Most successful of all animal phyla; chitinous exoskeleton covering segmented bodies with paired, jointed appendages; many insect groups have wings.	1,000,000
Mollusca (mollusks)	Snails, oysters, octopuses, sea slugs		Soft-bodied animals whose bodies are divided into three parts: head-foot, visceral mass, and mantle; many have shells; almost all possess a unique rasping tongue, called a radula; 35,000 species are terrestrial.	110,000
Chordata (chordates)	Mammals, fish, reptiles, birds, amphibians		Segmented coelomates with a notochord; possess a dorsal nerve cord, pharyngeal slits, and a postanal tail at some stage of life; in vertebrates, the notochord is replaced during development by the spinal column; 20,000 species are terrestrial.	56,000
Platyhelminthes (flatworms)	Planarians, tapeworms, liver flukes		Compact, unsegmented, bilaterally symmetrical worms; no body cavity; digestive cavity has only one opening. Many species are parasites and can lose the digestive cavity.	20,000
Nematoda (roundworms)	*Ascaris*, pinworms, hookworms, *Filaria*		Pseudocoelomate or acoelomate, unsegmented, bilaterally symmetrical worms; complete tubular digestive tract with mouth and anus; live in great numbers in soil and aquatic sediments; some are important animal parasites.	25,000
Annelida (segmented worms)	Earthworms, polychaetes, tube worms, leeches		Coelomate, serially segmented, bilaterally symmetrical worms; complete digestive tract; most have bristles called setae on each segment that anchor them during crawling.	16,000
Cnidaria (cnidarians)	Jellyfish, *Hydra*, corals, sea anemones		Soft, gelatinous, radially symmetrical bodies whose digestive cavity has a single opening; possess tentacles armed with stinging cells called cnidocytes that shoot sharp harpoons called nematocysts; most species are marine.	10,000
Echinodermata (echinoderms)	Sea stars, sea urchins, sand dollars, sea cucumbers		Deuterostomes with pentaradial symmetry in the adults; five-part body plan and unique water-vascular system with tube feet; able to regenerate lost body parts; marine. Endoskeleton of calcium plates.	7,000
Porifera (sponges)	Barrel sponges, boring sponges, basket sponges, vase sponges		Asymmetrical bodies without distinct tissues or organs; saclike body consists of two layers breached by many pores; internal cavity lined with food-filtering cells called choanocytes; most are marine (150 species live in fresh water).	5,150
Bryozoa (moss animals)	*Bowerbankia*, *Plumatella*, sea mats, sea moss		Microscopic, aquatic coelomates that form branching colonies; possess U-shaped row of ciliated tentacles for feeding called a lophophore that usually protrudes through pores in a hard exoskeleton; also called Ectoprocta because the anus, or proct, is external to the lophophore; marine or freshwater.	4,500

TABLE 32.2 The Major Animal Phyla, *continued*

Phylum	Typical Examples		Key Characteristics	Approximate Number of Named Species
Rotifera (wheel animals)	Rotifers		Small, aquatic pseudocoelomates with a crown of cilia around the mouth resembling a wheel and a set of complex jaws; many freshwater species.	2,000
Chaetognatha (arrow worms)	*Sagitta*		Coelomates of uncertain affinity, unclear whether they are protostomes or deuterostomes; bilaterally symmetrical; large eyes (some) and powerful jaws. Marine worms with dorsal *and* ventral nerve cords.	100
Hemichordata (acorn worms)	*Ptychodera*		Marine coelomates, deuterostomes with gill slits but without a notochord. Free living or colonial.	85
Onychophora (velvet worms)	*Peripatus*		Segmented protostomes with a chitinous soft exoskeleton and protrusible appendages. All species are terrestrial, although their Cambrian ancestors were marine.	110
Nemertea (ribbon worms)	*Lineus*		Coelomate protostomes, bilaterally symmetrical worms with long, extendable proboscis. Most species are marine, but a few live in fresh water and even fewer are terrestrial.	900
Brachiopoda (lamp shells)	*Lingula*		Like bryozoans, possess a lophophore, but within two clamlike shells; more than 30,000 species known as fossils.	300
Ctenophora (sea walnuts)	Comb jellies, sea walnuts		Gelatinous, almost transparent, often bioluminescent marine animals; eight bands of cilia; largest animals that use cilia for locomotion; complete digestive tract with anal pore.	100
Loricifera (loriciferans)	*Nanaloricus mysticus*		Tiny, bilaterally symmetrical, marine pseudocoelomates that live in spaces between grains of sand; mouth on tip of a unique flexible tube. Loriciferans were discovered in 1983.	10
Cycliophora (cycliophorans)	*Symbion*		Microscopic metazoans that live in the mouthparts of claw lobsters. Cycliophorans were discovered in 1995.	3
Micrognathozoa (micrognathozoans)	*Limnognathia*		Microscopic animals with complicated jaws. They were discovered in 2000 in Greenland.	1

Evolutionary Developmental Biology and the Roots of the Animal Tree of Life

Some of the most exciting contributions of molecular systematics are being made to our understanding of the base of the animal family tree—the origins of the major clades of animals.

Metazoans appear to have evolved from colonial protists

Most taxonomists agree that the animal kingdom is monophyletic—that is, that parazoans and eumetazoans have a common ancestor. This ancestor was presumably a protist (see chapter 29), but it is not clear from which line of protists animals evolved. Three prominent hypotheses currently exist for the origin of metazoans from single-celled protists:

- The **multinucleate hypothesis** suggests that metazoans arose from a multinuclear protist similar to today's ciliates. The cells later became compartmentalized into the multicellular condition.
- The **colonial flagellate hypothesis,** first proposed by Haeckel in 1874, states that metazoans descended from colonial protists, which are composed of hollow spherical colonies of flagellated cells. Some of the cells of sponges are strikingly like those of choanoflagellate protists.
- The **polyphyletic origin hypothesis** proposes that sponges evolved independently from the eumetazoans.

Molecular systematics based on ribosomal RNA sequences settles this argument clearly in favor of the colonial flagellate hypothesis. The molecular evidence excludes the multinucleate ciliate hypothesis because metazoans are molecularly closer to eukaryotic algae than to ciliates. The polyphyletic origin hypothesis is excluded because metazoans are found to represent a monophyletic group.

Molecular analysis may explain the Cambrian explosion

A second contentious issue in animal phylogeny is being addressed successfully with molecular systematics. Study of the fossil record reveals that the great diversity of animals evolved quite rapidly in geological terms around the beginning of the Cambrian period—an event known as the **Cambrian explosion.** Nearly all the major animal body plans can be seen in Cambrian rocks dating from 543 to 525 MYA.

In rocks from the earlier Ediacaran period as old as 565 million years, fossil cnidarians are found, along with what appear to be fossil mollusks and the burrows of worms. This implies that the earliest branches of the animal family tree arose before the Cambrian period.

In the half-billion years since the early Cambrian, no significant new innovations in animal body plan have occurred. Biologists have long debated what caused this enormous expansion of animal diversity (figure 32.6). Many have argued that the emergence of new body plans was the consequence of the emergence of predatory lifestyles, which encouraged an arms race be-

figure 32.6

DIVERSITY OF ANIMALS THAT EVOLVED DURING THE CAMBRIAN EXPLOSION. The Cambrian saw an astonishing variety of body plans, many of which gave origin to the extant animals we find today. Some of these strange creatures are: (1) *Amiskwia,* (2) *Odontogriphus,* (3) *Eldonia,* (4) *Halichondrites,* (5) *Anomalocaris,* (6) *Pikaia,* (7) *Canadia,* (8) *Marrella,* (9) *Opabinia,* (10) *Ottoia,* (11) *Wiwaxia,* (12) *Yohoia,* (13) *Xianguangia,* (14) *Aysheaia,* (15) *Sidneyia,* (16) *Dinomischus,* and (17) *Hallucigenia.* The natural history of these species is open to speculation.

tween defenses, such as armor, and innovations that improved mobility and hunting success. Others have attributed the rapid diversification in body plans to geological factors, such as the buildup of dissolved oxygen and minerals in the oceans.

A third possibility arises from molecular studies being carried out by biologists in the new field of evolutionary developmental biology. Much of the variation in animal body plans is associated with changes in the location or time of expression of *Hox* genes within developing animal embryos (see chapters 19 and 25). Briefly, *Hox* genes, also known as *homeobox* genes, specify the identity of developing body parts, such as the legs, thorax, antenna, and so forth. Perhaps the Cambrian explosion reflects the evolution of the *Hox* developmental gene complex, which provides a tool that can produce rapid changes in body plan.

> The animal kingdom is monophyletic and probably arose from a colonial flagellated protist. The great diversity in the animal body plan arose quickly and became prominent during the Cambrian period, possibly as the result of the evolution of *Hox* genes.

32.1 Some General Features of Animals

Animals are a very diverse group that share some important major characteristics.

- Animals are heterotrophic and depend on other organisms for organic molecules.

- All animals are multicellular, and their cells do not have cell walls.

- Except for the sponges, animals have structural and functional units called tissues.

- Most animals are capable of rapid movement because of their flexibility and the evolution of nerve and muscle tissues.

- Most animals reproduce sexually, and the organism is a multicellular diploid life stage. The only haploid cells are the gametes.

- Animals are divided into invertebrates, which do not have a back bone, and vertebrates, which have a backbone.

- Animals are found in very diverse habitats.

32.2 Evolution of the Animal Body Plan

Except for the Parazoa, eumetazoans have tissues that carry out specialized functions. There are five key transitions in the evolution of animal body plans.

- Most sponges are asymmetrical but other members of the Eumetazoa are bilaterally or radially symmetrical at some time in their life course (figure 32.1).

  • *Radially symmetrical animals have their body parts arranged around a central axis, and any plane passing through the central axis will divide the animal into halves that are approximately mirror images.*

  • *Bilaterally symmetrical animals have a body design in which the left and right sides are mirror images. The single plane of division can only pass through the sagittal plane.*

  • *Bilaterally symmetrical organisms exhibit cephalization and greater ability to move.*

- A body cavity makes possible the development of advanced organ systems.

  • *Three basic body plans have evolved several times in the Bilateria. They include acoelomate, pseudocoelomate, and coelomate animals (figure 32.2).*

  • *Development of a coelom created problems with circulation of nutrients and removal of wastes. Coelomate organisms have evolved a circulatory system to assist with these problems.*

    ■ In open circulatory systems, the blood passes from vessels into sinuses where it mixes with body fluids before it returns to the vessels.

    ■ Animals with closed circulatory systems have the blood moving continuously through vessels that are separated from the body fluids.

  • *Most animals, except sponges, produce three germ layers.*

    ■ The outer ectoderm develops into body coverings and the nervous system.

    ■ The inner endoderm develops into digestive organs and the intestines.

    ■ The middle mesoderm develops into skeleton and muscles.

  • *Body cavities originate from mesoderm and their development also varies in the Bilateria.*

    ■ In protostomes the coelom is formed by spaces created by migration of mesoderm cells to opposite positions.

    ■ In deuterostomes the coelom is usually formed by an evagination of the archenteron or primitive gut.

- Bilaterians exhibit two types of embryonic development depending on whether the mouth or anus is formed first (figure 32.3).

  • *Protostomes develop the mouth first, and it develops from or near the blastophore.*

  • *Deuterostomes develop the anus first from the blastophore*

- Cleavage, the progressive division of cells during embryonic growth, also varies within the Bilateria.

  • *Protostomes exhibit spiral cleavage and determinate development.*

  • *Deuterostomes exhibit radial cleavage and indeterminate development.*

- Segmentation is highly convergent and it appeared at least three times in the evolution of animals.

  • *Segmentation allows redundant organ systems in adults such as occurs in the annelids.*

  • *Segmentation allows for more efficient and flexible movement because each segment can move independently.*

32.3 The Traditional Classification of Animals

Animals have been traditionally classified into 36 phyla.

- This traditional classification was based on shared morphological and embryological characteristics.

- Symmetry and embryonic tissues distinguish diploblasts and triploblasts.

32.4 A New Look at the Metazoan Tree of Life (figure 32.5)

The traditional animal phylogeny, although accepted by many scientists, is being re-evaluated using molecular data.

- Molecular data have resulted in the separation of annelids and arthropods into different clades. Arthropods are now grouped with some pseudocoelomate protostomes.

- The traditional protostomes are divided into Ecdysozoa and Lophotrochozoa. Lophotrochozoa are further divided into Spiralia and Platyzoa.

32.5 Evolutionary Developmental Biology and the Roots of the Animal Tree of Life

Molecular biology has helped scientists understand the origins of major animal clades.

- Systematics based on ribosomal RNA supports the hypothesis that the Eumetazoa are monophyletic and arose from colonial flagellates.

- Molecular analysis supports the rapid diversification during the Cambrian explosion, possibly due to the evolution of the *Hox* genes.

SELF TEST

1. Which of the following characteristics is unique to all members of the animal kingdom?
   a. Sexual reproduction
   b. Multicellularity
   c. Lack of cell walls
   d. Heterotrophy

2. Animals are unique in the fact that they possess _____ for movement and _____ for conducting signals between cells.
   a. brains; muscles
   b. muscle tissue; nervous tissue
   c. limbs; spinal cords
   d. flagella; nerves

3. In animal sexual reproduction the gametes are formed by the process of
   a. meiosis.
   b. mitosis.
   c. fusion.
   d. binary fission.

4. The evolution of bilateral symmetry was a necessary precursor for the evolution of
   a. tissues.
   b. segmentation.
   c. a body cavity.
   d. cephalization.

5. A fluid-filled cavity that develops completely within the mesoderm is a characteristic of a
   a. coelomate.
   b. pseudocoelomate.
   c. acoelomate.
   d. all of the above

6. The following diagram is of the blastopore stage of embryonic development. Based on the information in the diagram, which of the following statements is correct?
   a. It is a diagram of a protostome.
   b. It would have formed by radial cleavage.
   c. It would exhibit determinate development.
   d. All of the above are correct.

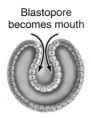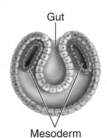

Blastopore becomes mouth

Gut

Mesoderm

7. Which of the following statements is not true regarding segmentation?
   a. Segmentation allows the evolution of redundant systems.
   b. Segmentation is a requirement for a closed circulatory system.
   c. Segmentation enhances locomotion.
   d. Segmentation represents an example of convergent evolution.

8. Which of the following characteristics is used to distinguish between a parazoan and an eumetazoan.
   a. Presence of a true coelom
   b. Segmentation
   c. Cephalization
   d. Tissues

9. With regard to classification in the animals, the study of which of the following is changing the organization of the kingdom?
   a. Molecular systematics
   b. Origin of tissues
   c. Patterns of segmentation
   d. Evolution of morphological characteristics.

10. Which of the following characteristics would not apply to a species in the Ecdysozoa?
    a. Bilateral
    b. Deuterostome
    c. Molt cuticles at least once in their life cycle
    d. Metazoan

11. The most recent phylum to be added to the Animalia is
    a. Bryozoa.
    b. Annelida.
    c. Micrognathoza.
    d. Loricifera.

12. The _____ contain the greatest number of known species.
    a. Chordata
    b. Arthropoda
    c. Porifera
    d. Mollusca

13. Which hypothesis for the evolution of the Metazoa suggests that they evolved from organisms similar to modern ciliates?
    a. Polyphyletic origin hypothesis
    b. Multinucleate hypothesis
    c. Colonial flagellate hypothesis
    d. None of the above

14. The evolution of which of the following occurs after the Cambrian explosion?
    a. Cephalization
    b. True coelom
    c. Segmentation
    d. None of the above

15. A coelomate organism may have which of the following characteristics?
    a. A circulatory system
    b. Internal skeleton
    c. Larger size than a pseudocoelomate
    d. All of the above

CHALLENGE QUESTIONS

1. Worm evolution represents an excellent means of understanding the evolution of a body cavity. Using only the following phyla of worms (nematodes, annelids, platyhelminthes, nemetera, hemichordates), construct a phylogenetic tree that is only based on the form of body cavity (refer to figure 32.2 and table 32.2 for assistance). How does this relate to the material in figure 32.5? Should body cavity be used as the sole characteristic for classifying a worm?

2. Most students find it hard to believe that the echinoderms and chordates are closely related phyla. If it were not for their method of forming a body cavity, where would you place the Echinodermata in the animal kingdom? Defend your answer.

3. In the rain forest you discover a new species that is terrestrial, has determinate development, molts during its lifetime, and possesses jointed appendages. To which phyla of animals should it be assigned?

 **Do you need additional review?** *Visit www.ravenbiology.com for practice quizzes, animations, videos, and activities designed to help you master the material in this chapter.*

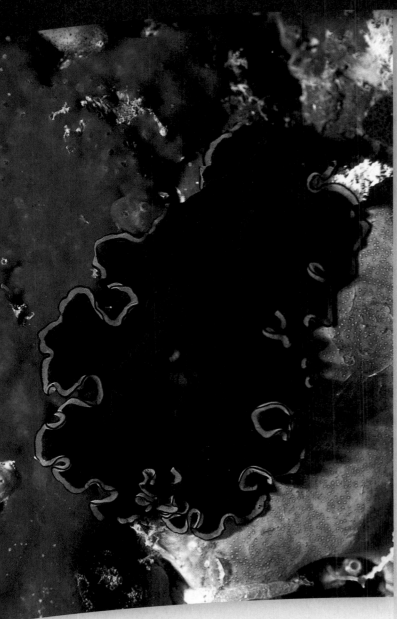

chapter 33

Noncoelomate Invertebrates

introduction

WE START OUR EXPLORATION of the great diversity of animals with the simplest members of the animal kingdom—sponges, jellyfish, and simple worms. These animals lack a body cavity (coelom), and they are thus called noncoelomates. The major organization of the animal body first evolved in these animals, a basic body plan on which all the rest of animal evolution has depended. Although traditionally systematists considered noncoelomate animals to be closely related, we now realize that this is not the case, as we discussed in the preceding chapter. In chapter 34, we consider the invertebrate animals that have a coelom, and in chapter 35 the vertebrates. You will see that all animals, despite their great diversity, have much in common.

concept outline

A Revolution in Invertebrate Phylogeny

There is little disagreement among biologists about the taxonomic classification of animals. For example, an annelid worm would be classified in the phylum Annelida by any competent taxonomist. Great disagreement exists, however, about how the animal phyla are related to one another. Depending on what aspects of the phyla are compared, different biologists draw quite different family trees.

Both traditional and novel phylogenies agree on major groupings

For many years, biologists have based their reconstructions of the animal tree of life on key aspects of body architecture, lumping together those phyla that share fundamental aspects of body plan. For the better part of a century, biologists agreed on the principal aspects of this tree, basing the phylogeny largely on anatomical and embryological comparisons.

However, as you learned in chapter 32, a new synthesis of animal phylogeny has been proposed in the last decade by researchers employing molecular comparisons in addition to anatomical ones, focusing particularly on ribosomal RNA sequences and a series of nuclear protein-encoding genes.

Like the more traditional trees based solely on morphological and anatomical data, these new trees incorporating molecular information place the sponges (phylum **Porifera**), one of the few animals without tissues, as the sister group to all animals with tissues, or Eumetazoa. Among the eumetazoans, both approaches find that cnidarians (hydras, sea jellies, and corals) and cteno-

phores (comb jellies) branch out early, before the origin of animals with bilateral symmetry. The bilaterally symmetrical eumetazoans are assigned to one of two groups of animals that differ in their embryological development: *protostomes* or *deuterostomes*. The new molecular phylogenies differ radically from the traditional phylogeny in how they construct the protostome branch of the animal family tree.

The traditional phylogeny focuses on the state of the coelom

As noted in chapter 32, the traditional animal family tree divides the bilaterally symmetrical animals into three great branches, based on the nature of the body cavity: (1) acoelomates (such as the phylum Platyhelminthes), which have no body cavity; (2) pseudocoelomates (such as the phylum Nematoda), which have a pseudocoel—a body cavity separating mesoderm from endoderm; and (3) coelomates (such as the phylum Annelida), which have a coelomic body cavity encased within mesoderm (figure 33.1). Coelomates can be either protostomes or deuterostomes; acoelomates and pseudocoelomates are always protostomes.

The novel protostome phylogeny distinguishes spiralians from ecdysozoans

Phylogenies based on the combined analysis of morphological features and sequences from rRNAs and other genes suggest a very different lineage of the protostome phyla. Two major

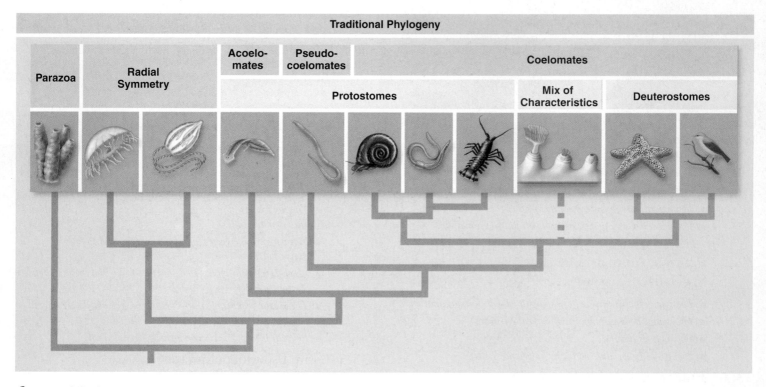

figure 33.1

THE TRADITIONAL PROTOSTOME PHYLOGENY. Some biologists have traditionally separated the bilaterally symmetrical animals into three groups that differ with respect to their body cavity: acoelomates, pseudocoelomates, and coelomates.

clades are recognized as having evolved independently since ancient times: the spiralians and the ecdysozoans (figure 33.2).

Spiralians

Spiralian animals grow the same way you do, by adding additional mass to an existing body. Most live in water, and propel themselves through it using cilia or using contractions of the body musculature. Many spiralians undergo spiral cleavage (see figure 32.3).

The spiralians are divided into two main groups: *Lophotrochozoa* and *Platyzoa*. Lophotrochozoans include most coelomate protostome phyla; they move by muscular contractions, and have a special type of free-living larva known as a **trochophore.** Platyzoans are mostly acoelomates; they are flat, and they move by ciliary action. Some platyzoans have a set of complicated jaws, such as the ones found in rotifers, gnathostomulids, or most prominently in the recently discovered phylum Micrognathozoa.

Four major kinds of protostomes are assigned to the Spiralia; the first three are lophotrochozoans, and the fourth is a platyzoan.

The lophophorate phyla, which include **Bryozoa** and **Brachiopoda,** are all coelomates with a horseshoe-shaped crown of ciliated tentacles around their mouths called a **lophophore.** Lophophorates are **sessile** (anchored in place), and therefore their lophophore is used for filter feeding.

Phylum Mollusca. Mollusks are unsegmented lophotrochozoans, often considered coelomates, although their coelom is reduced to a hemocoel (open circulatory system) and some other small body spaces. They display a wide variety of body forms; included in this group are octopuses, snails, and clams and a diverse set of other body forms that constitute the eight molluscan classes. (Mollusks are described in the following chapter with other coelomates.)

Phylum Annelida. Annelids are segmented coelomate worms. This group includes marine polychaetes, terrestrial earthworms, and leeches.

Phylum Platyhelminthes. Platyhelminths are acoelomate spiralians belonging to Platyzoa. They have a simple body plan with no enclosed body cavity and no circulatory organs. This group includes marine and freshwater planarians as well as a large number of parasitic forms such as flukes and tapeworms.

Ecdysozoans

Ecdysozoans are the molting animals. They increase in size by molting their external skeletons, an ability that seems to have evolved only once in the animal kingdom. Molting animals have been successful in nearly all environments, and all move about by means other than ciliary action. Although the developmental pattern of some ecdysozoans is extremely well known, as in the model organisms *Drosophila melanogaster* (phylum Arthropoda) and *Caenorhabditis elegans* (phylum Nematoda), their mode of embryonic cleavage is not spiral, as described in chapter 32.

Of the numerous phyla of protostomes assigned to the Ecdysozoa, two have been particularly successful:

Phylum Nematoda. Roundworms are pseudocoelomate worms that lack special circulatory or gas exchange

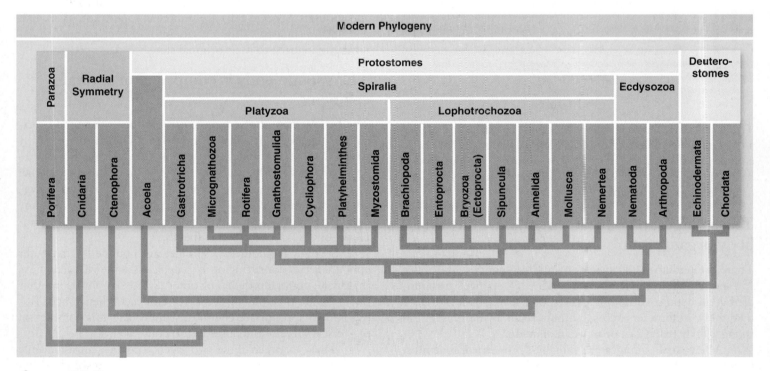

figure 33.2

A MODERN PROTOSTOME PHYLOGENY. New phylogenies based on the combination of anatomical, developmental, and molecular evidence suggest that segmented protostomes (annelids and arthropods) are not closely related and that arthropods might better be grouped with other animals that grow by molting (ecdysozoans).

structures, and their body wall has only longitudinal muscles. They grow by shedding their hard cuticle as they grow to adult size. Nematodes inhabit marine, freshwater, and terrestrial environments, and many species are parasitic in plants or animals. It is one of the most abundant and widely distributed of all animal phyla.

Phylum Arthropoda. Arthropods are coelomate animals with jointed appendages and segmented external skeletons composed of chitin. The most successful of all animal phyla, arthropods include insects, spiders, crustaceans, and centipedes, among many others. Arthropods have colonized almost all habitats as they are found from the ocean floor to the air and all terrestrial environments.

As more complete anatomical and developmental information as well as genomic comparisons become available, our picture of the evolutionary lineage of the protostomes will undoubtedly become clearer. In this text, we adopt a recent synthesis based on molecular, anatomical, and developmental information with the understanding that, like any phylogeny, it is a hypothesis of phylogenetic relationships that may change as new data are acquired (see figure 32.5).

> The protostomes were traditionally grouped according to the nature of their body cavity and the presence of segmentation, but recent molecular and anatomical evidence suggests that these characters may have evolved convergently in several groups and that, instead, protostomes should be grouped based on whether they molt.

33.2 Parazoa: Animals That Lack Specialized Tissues

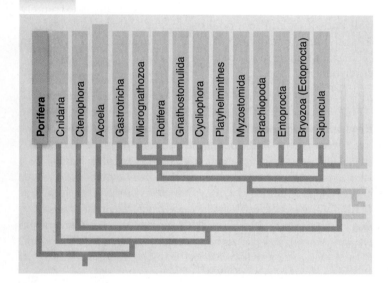

The sponges are parazoans, animals that lack tissues and organs and a definite symmetry. But like all animals, sponges have true, complex *multicellularity*. The body of a sponge contains several distinctly different types of cells whose activities are loosely coordinated with one another.

The sponges, phylum Porifera, have a loose body organization

There are perhaps 5000 species of marine sponges, and about 150 species that live in fresh water. In the sea, sponges are abundant at all depths. Although some sponges are small (no more than a few millimeters across), others, such as the loggerhead sponges, may reach 2 m or more in diameter.

A few small sponges are radially symmetrical, but most members of this phylum completely lack symmetry. Many sponges are colonial. Some have a low and encrusting form and can grow covering all sorts of surfaces; others may be erect and lobed, sometimes in complex patterns (figure 33.3*a*). Although

larval sponges are free-swimming, adults are anchored onto rocks or other submerged objects.

Sponges, like all animals, are composed of multiple cell types (figure 33.3*b*). At first sight, a sponge seems to be little more than a mass of cells embedded in a gelatinous matrix, but these cells are specialized for different functions and recognize one another with a high degree of fidelity. In some cases, they can coordinate their actions to rapidly contract their oscula (the holes through water is discharged), synchronize their reproduction, or to build complex, organized meshworks of spicules and spongin. Sponge cells are unique among those of animals in that they can easily differentiate into other cell types and dedifferentiate back to their original state.

The sponge body is composed of several cell types

The basic structure of a sponge can best be understood by examining the form of a young individual. A small, anatomically simple sponge first attaches to a substrate and then grows into a vaselike shape. The walls of the "vase" have three functional layers. First, facing into the internal cavity are specialized flagellated cells called **choanocytes,** or collar cells. These cells line either the entire body interior or, in many large and more complex sponges, specialized chambers.

Second, the bodies of sponges are bounded by an outer epithelial layer consisting of flattened cells somewhat like those that make up the epithelia, or outer layers, of animals in other phyla. Some portions of this layer contract when touched or exposed to appropriate chemical stimuli, and this contraction may cause some of the pores in the body to close.

Third, between these two layers, sponges consist mainly of a gelatinous, protein-rich matrix called the **mesohyl,** within which various types of amoeboid cells occur. In addition, many kinds of sponges have minute needles of calcium carbonate or silica known as **spicules,** fibers of a tough protein called **spongin,**

figure 33.3

PHYLUM PORIFERA: SPONGES. *a. Aplysina longissima.* This beautiful, bright orange and purple elongated sponge is found on deep coral reefs. *b.* Sponges are composed of several distinctly different cell types, whose activities are coordinated with one another. The sponge body is not symmetrical and has no organized tissues.

or both, within this matrix. Spicules and spongin strengthen the bodies of the sponges. The spongin skeleton of a real sponge was the original bathtub sponge. "Sponges" now sold for cleaning are made of cellulose or plastic.

Sponges feed in a unique way. The beating of flagella from the choanocytes that line the inside of the sponge draws water in through numerous small pores; the name of the phylum, Porifera, refers to this system of pores. Plankton and other small organisms are filtered from the water, which flows through passageways and eventually is forced out through an **osculum**, a specialized, larger pore.

Because as adults they become firmly attached to the substrate, sponges use chemical cues for settlement and to avoid being covered by other benthic (bottom-dwelling) organisms. Sponges are often investigated by pharmaceutical companies interested in the numerous metabolites produced by these primitive animals.

Choanocytes circulate water through the sponge

Each choanocyte closely resembles a protist with a single flagellum (see figure 33.3b), a similarity that reflects its evolutionary derivation. The beating of the flagella of the many choanocytes that line the body interior is a major force that draws water in through the pores and through the sponge, thus bringing in food and oxygen and expelling wastes. Each choanocyte flagellum beats independently, and the pressure they create collectively in the cavity forces water out of the osculum. In some sponges, the inner wall of the body interior is highly convoluted, increasing the surface area and, therefore, the number of flagella that can drive the water. In such a sponge, 1 cm³ of sponge can propel more than 20 L of water per day.

Sponges can reproduce both asexually and sexually

Some sponges will re-form themselves once they have passed through a silk mesh. Thus, as you might suspect, sponges frequently reproduce by simply breaking into fragments. If a sponge breaks up, the resulting fragments usually are able to reconstitute whole new individuals.

Sexual reproduction is also exhibited by sponges, with some mature individuals producing eggs and sperm. Larval sponges may undergo their initial stages of development within the parent. They are externally ciliated and use the cilia for swimming. After a short planktonic stage, they settle down on a suitable substrate, where they begin their transformation into adults.

Sponges probably represent the most primitive animals, possessing multicellularity but neither tissue-level development nor body symmetry. Their cellular organization hints at the evolutionary ties between the unicellular protists and the multicellular animals. Sponges possess choanocytes, special flagellated cells whose beating drives water through the body cavities.

Eumetazoa: Animals with True Tissues

The Eumetazoa contains animals that evolved the first key transition in the animal body plan: distinct *tissues*. Two distinct cell layers form in the embryos of these animals: an outer ectoderm and an inner endoderm. These embryonic tissues give rise to the basic body plan, differentiating into the many tissues of the adult body.

Typically, the outer covering of the body (called the epidermis) and the nervous system develop from the ectoderm, and the layer of digestive tissue (called the **gastrodermis**) develops from the endoderm. A layer of gelatinous material, called the **mesoglea,** lies between the epidermis and gastrodermis in cnidarians and ctenophores. In bilateral animals, a third body layer, the mesoderm, forms between endoderm and ectoderm, and forms the muscles of most eumetazoans.

Eumetazoans also evolved true body symmetry. Initially, metazoans living sessile on the ocean floor or free-living in the water were radially symmetrical organisms. Two radially symmetrical groups exist today: the phylum Cnidaria (pronounced ni-DAH-ree-ah), or the cnidarians, is composed of hydroids, jellyfish, sea anemones, and corals, and the phylum Ctenophora (pronounced ten-of'-o-rah) is composed of the comb jellies. All other eumetazoans are in the Bilateria and exhibit a fundamental bilateral symmetry.

The cnidarians, phylum Cnidaria, exhibit internal extracellular digestion

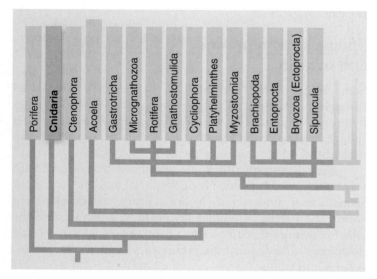

Cnidarians are nearly all marine, although a few live in fresh water. These fascinating and simply constructed animals are basically gelatinous in composition. They differ markedly from the sponges in organization; their bodies are made up of distinct tissues, although they have not evolved true organs. These animals are carnivores and capture their prey (which includes fishes, crustaceans, and many other kinds of animals) with the tentacles that ring their mouths.

Basic body plans

Cnidarians may have two basic body forms: *polyps* and *medusae* (figure 33.4). **Polyps** are cylindrical and are usually found attached to a firm substrate. They may be solitary or colonial. In a polyp, the body opening, which acts both as a mouth and anus, faces away from the substrate on which the animal is growing, and therefore often faces upward. Many polyps build a chitinous or calcareous (calcium carbonate) external or internal skeleton, or both. Only a few polyps are free-living.

In contrast, most **medusae** are free-living and many are umbrella-shaped, with tentacles surrounding their mouth. Medusae, particularly those of the class Scyphozoa, are commonly known as jellyfish or sea jellies because their mesoglea is thick and jellylike.

The cnidarian life cycle

Many cnidarians occur only as polyps, but others exist only as medusae; still others alternate between these two phases during their life cycles, both phases consisting of diploid individuals. Polyps may reproduce sexually or asexually; asexual reproduction may produce either new polyps or medusae. Medusae reproduce sexually.

In most cnidarians, fertilized eggs give rise to free-swimming ciliated larvae known as **planulae.** Planulae are common in the plankton at times and may be dispersed widely in the currents.

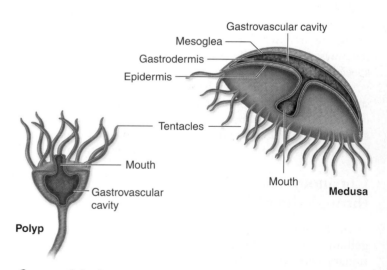

figure 33.4

TWO BODY FORMS OF CNIDARIANS, THE MEDUSA AND THE POLYP.

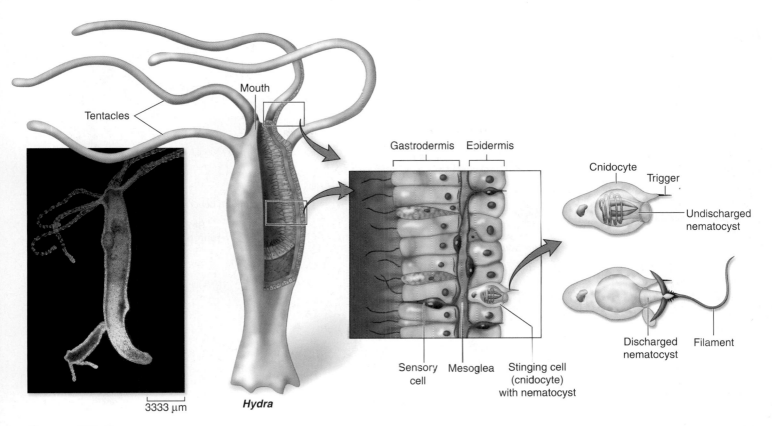

Tentacles

Mouth

3333 μm

Hydra

Gastrodermis **Epidermis**

Sensory cell **Mesoglea** **Stinging cell (cnidocyte) with nematocyst**

Cnidocyte **Trigger**

Undischarged nematocyst

Discharged nematocyst **Filament**

figure 33.5

PHYLUM CNIDARIA: CNIDARIANS. The cells of a cnidarian such as this *Hydra* are organized into specialized tissues. The interior gut cavity is specialized for extracellular digestion—that is, digestion begins within the gut cavity rather than within individual cells. The epidermis contains cnidocytes for defense and for capturing prey.

Cnidarian body structure

A major evolutionary innovation in cnidarians, compared with sponges, is the internal extracellular digestion of food (figure 33.5). Digestion takes place within a gut cavity, rather than only within individual cells. Digestive enzymes, released from cells lining the walls of the cavity, partially break down food. Cells lining the gut subsequently engulf food fragments by phagocytosis.

The extracellular fragmentation that precedes phagocytosis and intracellular digestion allows cnidarians to digest animals larger than individual cells, an important improvement over the strictly intracellular digestion that occurs in sponges. Cnidarians have no blood vessels, respiratory system, or excretory organs.

Cnidocytes and nematocysts

On their tentacles and sometimes on their body surface, cnidarians bear specialized cells called **cnidocytes.** The name of the phylum, Cnidaria, refers to these cells, which are highly distinctive and occur in no other group of organisms. A special type of cnidocyte, the nematocyst, contains a small but powerful "harpoon." Each nematocyst features a coiled, threadlike tubule that may be barbed. The tubule of a nematocyst entwines or spears prey and may deliver a toxin; the

cnidarian then draws back the prey with a tentacle containing the cnidocyte.

To propel the harpoon, the nematocyst uses water pressure. Before firing, the nematocyst builds up a very high internal osmotic pressure by active transport to build a high concentration of ions inside, while keeping the nematocyst's cell wall impermeable to water.

When the nematocyst is stimulated to discharge, it opens and the tubule everts (turns inside out). Nematocyst discharge is one of the fastest cellular processes in nature. The tubule everts so explosively that the barb of some species can penetrate even the hard shell of a crab. The tubule of some types delivers a toxic protein that can produce a stinging sensation, causing some cnidarians to be called "stinging nettles" because their effect is similar to that of plants of this name. In some cases, the results can be fatal to humans.

Cnidarians are grouped into four classes

There are four classes of cnidarians: Hydrozoa (hydroids), Scyphozoa (jellyfish), Cubozoa (sea wasps and box jellyfish), and Anthozoa (anemones and corals).

figure 33.6

THE LIFE CYCLE OF A SPECIES OF THE GENUS *OBELIA*, A MARINE COLONIAL HYDROID. Polyps reproduce by asexual budding, forming colonies. They may also give rise to medusae, which reproduce sexually via gametes. These gametes fuse, producing zygotes that develop into planula larvae, which in turn settle down to produce polyps.

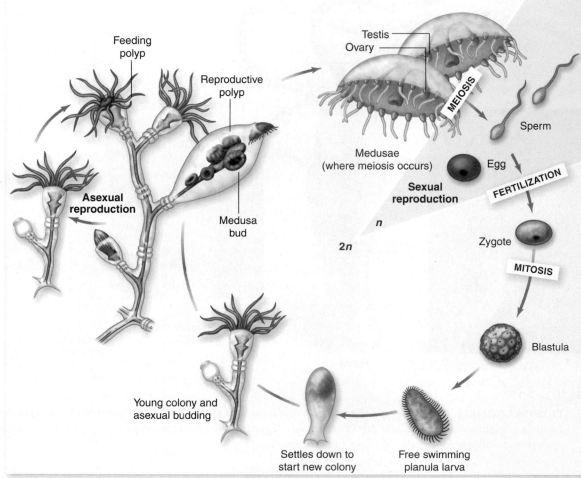

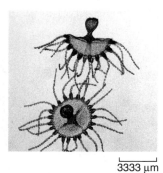

3333 μm

Class Hydrozoa: The hydroids

Most of the approximately 2700 species of hydroids (class **Hydrozoa**) have both polyp and medusa stages in their life cycle (figure 33.6). Most of these animals are marine and colonial, such as the members of the genus *Obelia* and the very unusual Portuguese man-of-war. Some marine hydroids are bioluminescent.

A well-known hydroid is the abundant freshwater genus *Hydra*, which is exceptional in having no medusa stage and existing as a solitary polyp. Each polyp sits on a basal disk, which the hydra can use to glide around, aided by mucous secretions. It can also move by somersaulting—bending over and attaching itself to the substrate by its tentacles, and then looping over to a new location. If the polyp detaches itself from the substrate, it can float to the surface.

Class Scyphozoa: The jellyfish (sea jellies)

The approximately 200 species of jellyfish (class **Scyphozoa**) are transparent or translucent marine organisms, some of a striking orange, blue, or pink color (figure 33.7). In all of them, the medusa stage is dominant—much larger and more complex than the polyp stage. The medusae are bell-shaped, with hanging tentacles around their margins. The polyp stage is small, inconspicuous, and simple in structure.

The outer layer, or epithelium, of a jellyfish contains a number of specialized cells, each of which can contract individually. Together, the cells form a muscular ring around the margin of the bell that pulses rhythmically and propels the animal through the water.

Jellyfish have separate male and female individuals. After fertilization, planulae form, which then attach and develop into polyps. The polyps can reproduce asexually as well as budding off medusae. In some jellyfish that live in the open ocean, the polyp stage is absent, and planulae develop directly into medusae.

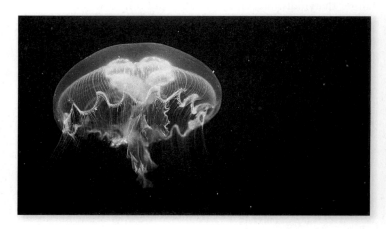

figure 33.7

CLASS SCYPHOZOA. *Aurelia aurita*, a jellyfish.

figure 33.8

CLASS CUBOZOA.
Chironex fleckeri,
a box jelly.

Class Cubozoa: The box jellyfish (box jellies)

As their name implies, the cnidarians in the class **Cubozoa** are box-shaped medusae; the polyp stage is inconspicuous and in many cases not known. Most are only a few centimeters in height, although some are 25 cm tall. A tentacle or group of tentacles is found at each corner of the box (figure 33.8). Box jellies are strong swimmers and voracious predators of fish in tropical and subtropical waters. The stings of some species can be fatal to humans.

Class Anthozoa: The sea anemones and corals

By far the largest class of cnidarians is **Anthozoa,** the "flower animals." The approximately 6200 species of this group are solitary or colonial marine animals. They include stonelike corals, soft-bodied sea anemones, and other groups known by such fanciful names as sea pens, sea pansies, sea fans, and sea whips (figure 33.9). All of these names reflect a plantlike body, each polyp topped by a tuft or crown of hollow tentacles. Like other cnidarians, anthozoans use these tentacles in feeding.

Fertilized eggs of most anthozoans develop into planulae that settle as polyps; no medusae are formed. The internal body space of an anthozoan polyp is divided into compartments by sheets of tissue, unlike that of polyps of the other cnidarian classes. When touched, polyps of many anthozoans retract their hollow tentacles into their bodies.

Sea anemones, a large group of highly muscular and relatively complex soft-bodied anthozoans, live in waters of all depths throughout the world. They range from a few millimeters to over a meter in diameter, and many are equally long.

Most corals are anthozoans. "True corals" secrete outer skeletons, or exoskeletons, of calcium carbonate, which give them their stony texture. "Soft corals" may have small calcium carbonate needles embedded in their tissues, and a horny exoskeleton around which the members of a colony grow that provides support while allowing flexibility (an example of such a colony is a sea fan). Some hard corals are important builders of coral reefs, which are shallow-water limestone ridges or mounds that occur in warm seas. Most waters in which coral reefs develop are nutrient-poor, but the corals are able to grow well because they contain within their cells symbiotic dinoflagellates (zooxanthellae) that photosynthesize to provide energy for the animals (see chapter 29).

The comb jellies, phylum Ctenophora, use cilia for movement

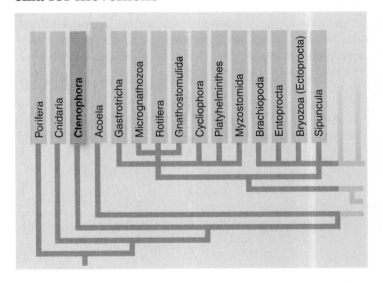

The members of the small phylum **Ctenophora** range from spherical to ribbonlike and are known as comb jellies, sea walnuts, or sea gooseberries. Traditionally, the roughly 100 marine species of ctenophores were considered closely related to the cnidarians. However, ctenophores are structurally more complex than cnidarians. They have anal pores, so that water and other substances pass completely through the animal.

Comb jellies, abundant in the open ocean, are transparent and usually only a few centimeters long, but some species reach 1 m in length. Most have two long, retractable tentacles that they use to capture prey using a special type of cell, a **colloblast,** that on contact with zooplankton prey bursts and discharges a strong adhesive material.

Ctenophores propel themselves through water with eight rows of comblike plates of fused cilia that beat in a coordinated fashion (figure 33.10). They are the largest animals that use cilia for locomotion. Many ctenophores are bioluminescent, giving off bright flashes of light particularly evident in the open ocean at night.

Traditionally the members of the phylum Ctenophora have been considered diploblasts with radial symmetry, as are cnidarians.

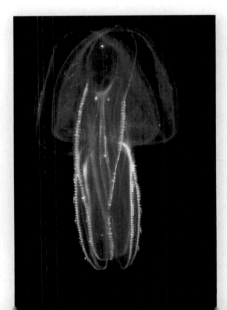

figure 33.10

A COMB JELLY (PHYLUM CTENOPHORA). Note the iridescent comb plates.

figure 33.9

CLASS ANTHOZOA. Crimson anemone, *Cribnnopsia fernaldi.*

Recent developmental studies, however, have shown that ctenophores have true muscle cells derived from the mesoderm, and therefore must be considered triploblasts, like the bilaterians. It has also been shown that ctenophores have three main axes of symmetry, none of which produces identical halves, and therefore their mode of symmetry is not strictly radial as in cnidarians.

> Cnidarians have a specialized kind of cell called a cnidocyte and are abundant in the oceans where they can contribute to building reefs. Ctenophores propel themselves through the water by means of eight rows of comblike plates of fused cilia and have special cells called colloblasts.

33.4 The Bilaterian Acoelomates

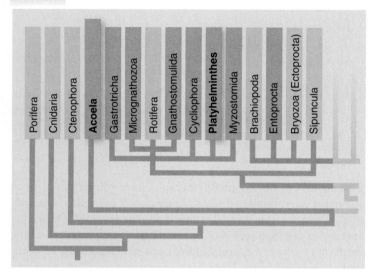

The Bilateria are characterized by the second key transition in the animal body plan, *bilateral symmetry*, which allowed animals to achieve high levels of specialization within parts of their bodies—such as the concentration of sensory structures in the anterior portion of the body.

As we have discussed, bilaterians are traditionally classified by the condition of their coelom: acoelomates, pseudocoelomates, and coelomates. Although recent studies indicate that the acoelomate and pseudocoelomate condition have each evolved multiple times, we retain the traditional groupings to discuss their biology. In this chapter, we cover the acoelomates and pseudocoelomates, and in the next chapter the coelomates.

Structurally, the simplest bilaterians are the acoelomates; they lack any internal cavity other than the digestive tract. As discussed earlier, all bilaterians have three embryonic germ layers formed during development: ectoderm, endoderm, and mesoderm. We focus our discussion of the acoelomate body architecture on the largest phylum of the group, the Platyhelminthes.

The flatworms, phylum Platyhelminthes, have an incomplete gut or none at all

The phylum Platyhelminthes consists of some 20,000 species. These ciliated soft-bodied animals are flattened dorsoventrally, from top to bottom. Flatworms were considered among the simplest of bilaterally symmetrical animals, but they do have a definite head at the anterior end and certain complex structures, like their reproductive apparatus. Their bodies are compact; the only internal space consists of an incomplete (blind) digestive cavity (figure 33.11).

Flatworms range in length from 1 mm or less to many meters, as in some tapeworms. Many species of flatworms are parasitic, occurring within the bodies of many other kinds of animals. Other flatworms are free-living, occurring in a wide variety of marine and freshwater habitats, as well as in moist places on land.

Free-living flatworms are carnivores and scavengers; they eat various small animals and bits of organic debris. They move from place to place by means of ciliated epithelial cells, which are particularly concentrated on their ventral surfaces, but they also have well-developed musculature.

Digestion in flatworms

Flatworms have a digestive cavity with an incomplete gut, one with only one opening. As a result, they cannot feed, digest, and eliminate undigested particles of food simultaneously, and thus, flatworms cannot feed continuously. Their mouth opening is located ventrally toward the mid part of the animal, and not in the anterior end as in many other bilaterians. Muscular contractions in the upper end of the gut, the pharynx, cause a strong sucking force, allowing flatworms to ingest their food and tear it into small bits.

In many species the gut is branched and extends throughout the body, functioning in both digestion and transport of food particles. Cells that line the gut engulf most of the food particles by phagocytosis and digest them; but, as in the cnidarians and in most bilaterians, some of these particles are partly digested extracellularly. Tapeworms, which are parasitic flatworms, lack digestive systems. They absorb their food directly through their body walls.

Excretion and osmoregulation

Unlike cnidarians, flatworms have an excretory and osmoregulatory system, which consists of a network of fine tubules that runs throughout the body. Cilia line the hollow centers of bulblike **flame cells** located on the side branches of the tubules. Flame cells were named because of the flickering movements of the tuft of cilia within them.

Cilia in the flame cells move water and excretory substances into the tubules and then to exit pores located between the epidermal cells. Flame cells primarily regulate the water balance of the organism; their excretory function appears to be secondary. A large proportion of the metabolic wastes excreted by flatworms diffuses directly into the gut and is eliminated through the mouth.

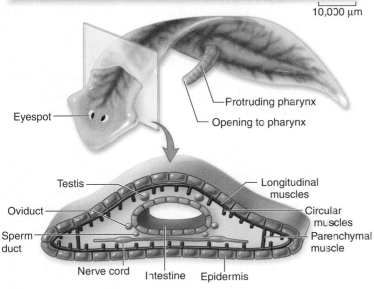

Protruding pharynx

Opening to pharynx

Eyespot

Testis

Oviduct

Sperm duct

Longitudinal muscles

Circular muscles

Parenchymal muscle

Nerve cord   Intestine   Epidermis

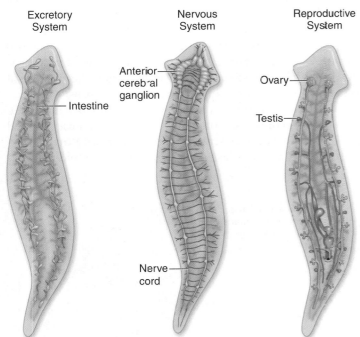

Excretory System

Intestine

Nervous System

Anterior cerebral ganglion

Nerve cord

Reproductive System

Ovary

Testis

10,000 μm

figure 33.11

ARCHITECTURE OF A FLATWORM. An idealized species of the genus *Dugesia*, the familiar freshwater planaria of many ponds and rivers. Schematic on top shows an animal and a transverse section through the anterior part of the body. Schematics below show the digestive, central nervous, and reproductive systems.

Nervous system and sensory organs

Like sponges, cnidarians, and ctenophores, flatworms lack circulatory systems for the transport of oxygen and food molecules. Consequently, all flatworm cells must be within diffusion distance of oxygen and food. Flatworms have thin bodies, and many have highly branched digestive cavities that make such a system possible.

The nervous system of flatworms is composed of a central nervous system comprising an anterior cerebral ganglion and nerve cords that run down the body but in a ladderlike shape.

Free-living members of this phylum have eyespots on their heads. These are inverted, pigmented cups containing light-sensitive cells connected to the nervous system. These eyespots enable the worms to distinguish light from dark; flatworms tend to move away from strong light.

Flatworm reproduction

The reproductive systems of flatworms are complex. Most flatworms are **hermaphroditic,** with each individual containing both male and female sexual structures. In many of them, fertilization is internal. When they mate, each partner deposits sperm in the copulatory sac of the other. The sperm travel along special tubes to reach the eggs.

In most freshwater flatworms, fertilized eggs are laid in cocoons strung in ribbons and hatch into miniature adults. In contrast, some marine species undergo indirect development; the egg divides following a typical spiral cleavage pattern and can give origin to a larva that will swim about until settling in the proper substrate.

Flatworms are also well known for their extraordinary capacity for regeneration. In some species, when a single individual is divided into two or more parts, each part can regenerate an entirely new flatworm.

Flatworms have been traditionally grouped into four major classes

All free-living flatworms are grouped into the class Turbellaria, although recent studies show that this is not a monophyletic group. The parasitic flatworms have traditionally been placed into three classes: Monogenea and Trematoda (fluke classes), and Cestoda (tapeworms). New studies indicate that parasitic lifestyle only evolved once in platyhelminths, from within the Turbellaria. These studies group all three parasitic classes into a single group, Neodermata. This name refers to the "new skin" that replaces the primitively ciliated epidermis of the free-living forms.

Class Turbellaria: Free-living flatworms

One of the most familiar members of the class **Turbellaria** are the freshwater members of the genus *Dugesia*, the common planaria used in biology laboratory exercises. Other turbellarians are widespread and often abundant in lakes, ponds, and the sea. Some close relatives of the freshwater planarians also occur in moist places on land. In the case of being introduced accidentally, terrestrial planarians can pose an important ecological threat to other soil organisms as happened with the decline of earthworm populations in Great Britain, caused by the introduction of a land planaria from New Zealand.

Classes Monogenea and Trematoda: The flukes

Three groups of parasitic flatworms live as ectoparasites or endoparasites in the bodies of other animals: flukes (**Monogenea** and **Trematoda**) and tapeworms (**Cestoda**). The endoparasitic groups of these worms have epithelial layers resistant to the digestive enzymes and immune defenses produced by their hosts—an important feature in their parasitic way of life. They lack certain features of the free-living flatworms, however, such as cilia in the adult stage, eyespots, and other sensory organs. These features have no adaptive significance for an organism that lives within the body of another animal.

Flukes take in food through their mouth, just like their free-living relatives. There are more than 10,000 named species of flukes, ranging in length from less than 1 mm to more than 8 cm. Flukes attach themselves within the bodies of their hosts by means of suckers, anchors, or hooks. Some have a life cycle that involves only one host, usually a fish, but most have life cycles involving two or more hosts. Their larvae almost always occur in snails, and there may be other intermediate hosts. The final host of these flukes is almost always a vertebrate.

Flukes important to human health

To human beings, one of the most important flatworms is the oriental liver fluke, *Clonorchis sinensis.* It lives in the bile passages of the liver of humans, cats, dogs, and pigs. It is especially common in Asia.

The worms are 1–2 cm long and have a complex life cycle. Although they are hermaphroditic, cross-fertilization usually occurs between different individuals. Eggs, each containing a complete, ciliated first-stage larva, the **miracidium,** are passed in the feces (figure 33.12). If they reach water, they may be ingested by a snail. Within the snail, an egg transforms into a **sporocyst,** a baglike structure with embryonic germ cells. Within the sporocysts, **rediae** (singular, *redia*), elongated, non-ciliated larvae, are produced. These larvae continue growing within the snail, giving rise to several individuals of the tadpolelike next larval stage, the **cercaria.**

Cercariae escape into the water, where they swim about freely. When they encounter a fish of the family Cyprinidae—the family that includes carp and goldfish—they bore into the muscles or under the scales, lose their tails, and transform into **metacercariae** within cysts in the muscle tissue. If a human or other mammal eats raw infected fish, the cysts dissolve in the intestine, and the young flukes migrate to the bile duct, where they mature. An individual fluke may live for 15 to 30 years in the liver. In humans, a heavy infestation of liver flukes may cause cirrhosis of the liver and death.

Other very important flukes are the blood flukes of the genus *Schistosoma.* They afflict about 1 in 20 of the world's population, more than 200 million people throughout tropical Asia, Africa, Latin America, and the Middle East. Three species of *Schistosoma* cause the disease called **schistosomiasis,** or bilharzia. Some 800,000 people die each year from this disease.

Recently, a great deal of effort has gone into controlling schistosomiasis. The worms protect themselves from the body's immune system in part by coating themselves with a variety of the host's own antigens that effectively render the worm immunologically invisible (see chapter 51). Despite this difficulty, the search is on for a vaccine that would cause the host to develop antibodies to one of the antigens of the young worms before they protect themselves with host antigens. This vaccine would protect humans from infection.

Class Cestodes: The tapeworms

Members of Cestodes, like many of the flukes, live as parasites within the bodies of other animals. In contrast to flukes, adult tapeworms simply hang on to the inner walls of their hosts by means of specialized terminal attachment organs and absorb food through their epithelium. Tapeworms lack digestive cavities as well as digestive enzymes. They are extremely specialized in relation to their parasitic way of life. Most species of tapeworms occur in the intestines of vertebrates, about a dozen of them regularly in humans.

The long, flat bodies of tapeworms are divided into three zones: the **scolex,** or attachment organ; the unsegmented neck; and a series of repetitive sections, the **proglottids** (figure 33.13). The scolex usually bears four suckers and may also

figure 33.12

LIFE CYCLE OF THE ORIENTAL LIVER FLUKE, *Clonorchis sinensis*.

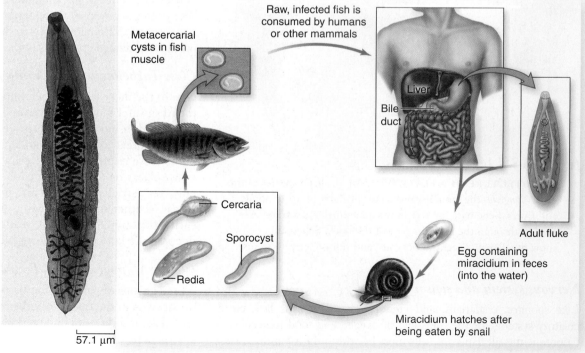

Metacercarial cysts in fish muscle

Raw, infected fish is consumed by humans or other mammals

Liver

Bile duct

Adult fluke

Cercaria

Sporocyst

Redia

Egg containing miracidium in feces (into the water)

Miracidium hatches after being eaten by snail

57.1 μm

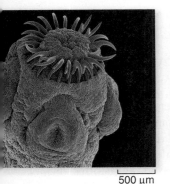

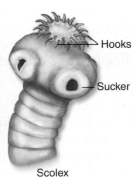

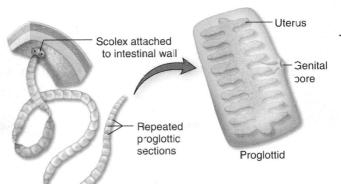

Hooks

Sucker

Scolex

Scolex attached to intestinal wall

Repeated proglottic sections

Uterus

Genital pore

Proglottid

500 μm

figure 33.13

TAPEWORMS. Acoelomate flatworms such as this beef tapeworm, *Taenia saginata*, live parasitically in the intestine of mammals.

have hooks. Each proglottid is a complete hermaphroditic unit, containing both male and female reproductive organs. Proglottids are formed continuously in an actively growing zone at the base of the neck, with maturing ones moving farther back as new ones are formed in front of them. Ultimately, the proglottids near the end of the body form mature eggs. As these eggs are fertilized, the zygotes in the very last segments begin to differentiate, and these segments fill with embryos, break off, and leave their host with the host's feces. Embryos, each surrounded by a shell, emerge from the proglottid through a pore or the ruptured body wall. They are deposited on leaves, in water, or in other places where they may be picked up by another animal.

The beef tapeworm, *Taenia saginata*, occurs as a juvenile in the intermuscular tissue of cattle but as an adult in the intestines of human beings. A mature adult beef tapeworm may reach a length of 10 m or more. These worms attach themselves to the intestinal wall of their host by a scolex with four suckers. Segments shed from the end of the worm pass from the human in the feces and may crawl onto vegetation. These segments ultimately rupture and scatter the embryos. Embryos may remain viable for up to five months. If ingested by cattle, they burrow through the wall of the intestine and ultimately reach muscle tissues through the blood or lymph vessels. About 1% of the cattle in the United States are infected, and some 20% of the beef consumed is not federally inspected. Thus, when humans eat infected beef that is cooked "rare," infection by these tapeworms is possible. As a result, the beef tapeworm is a frequent parasite of humans.

Acoel flatworms appear to be distinct from Platyhelminthes: A case study

Acoel flatworms (Acoela) (figure 33.14) were once considered basal members of the phylum Platyhelminthes. They have a primitive nervous system that consists of a simple network of nerves with a minor concentration of neurons in the anterior body end. Acoels lack a permanent digestive cavity, and instead the pharynx leads to a solid mass of digestive cells.

These primitive characteristics were used to place the whole of the Platyhelminthes at the base of the Bilateria (see figure 33.1). However, recent molecular studies show that acoels evolved early before the split between protostomes and deuterostomes, and that they are not related to the members of the phylum Platyhelminthes, which are legitimate members of the spiralian protostome clade.

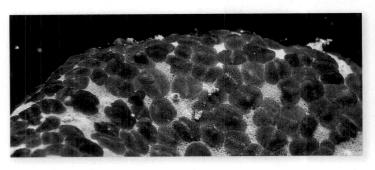

figure 33.14

PHYLUM ACOELA. An acoel flatworm of the genus *Waminoa*. These flatworms, long thought to be relatives of the Platyhelminthes, have a primitive nervous sytem and lack a permanent digestive cavity.

Acoels are now classified as members of their own phylum, Acoela, or as members of the phylum Acoelomorpha, which includes the nemertodermatids, another group of primitive bilaterians once also considered part of the flatworms.

The ribbon worms, phylum Nemertea, are not closely related to other acoelomates

The phylogenetic relationships of the phylum **Nemertea** (figure 33.15) have been reconsidered recently. Nemerteans are often called ribbon worms or proboscis worms. The ribbon worm body plan resembles that of a flatworm, without internal body cavities and with networks of fine tubules that constitute the excretory system. But they also possess a fluid-filled sac, called a **rhynchocoel**,

figure 33.15

PHYLUM NEMERTEA: A RIBBON WORM OF THE GENUS *LINEUS.* Nemerteans are long animals that can stretch to several meters in length.

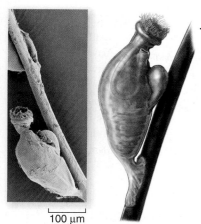

100 μm

figure 33.16

PHYLUM CYCLIOPHORA. About the size of the period at the end of this sentence, these acoelomates live on the mouthparts of claw lobsters. One feeding stage (and part of a second one) of the species *Symbion pandora* are shown attached to the mouthpart of a lobster.

merteans are now classified closely with other coelomate protostomes in the Lophotrochozoa.

Phylum Cycliophora is a relatively new phylum

In December 1995, Danish biologists Peter Funch and Reinhardt Kristensen reported the discovery of a strange new kind of acoelomate creature, about the size of a period on a printed page. The tiny organism had a striking circular mouth surrounded by a ring of cilia, and its anatomy and life cycle are so unusual that their discoverers assigned it to an entirely new phylum, **Cycliophora** (figure 33.16).

The newest phylum to be named before Cycliophora, the Loricifera, was discovered also by Reinhard Kristensen in 1983, and he and Peter Funch discovered yet another new phylum in 2000, the Micrognathozoa.

Cycliophorans live on the mouthparts of claw lobsters on both sides of the North Atlantic. When the lobster to which they are attached starts to molt, the tiny symbiont begins a bizarre form of sexual reproduction. Dwarf males emerge, composed of nothing but brains and reproductive organs. Each dwarf male seeks out another female symbiont on the molting lobster and fertilizes its eggs, generating free-swimming individuals that can seek out another lobster and renew their life cycle.

which constitutes a true coelomic cavity. This sac serves as a hydraulic power source for their proboscis, a long muscular tube that can be thrust out quickly from a sheath to capture prey.

The phylum Nemertea consists of about 900 species. Shaped like a thread or a ribbon, nemerteans are mostly marine, with a few species living in fresh water and humid terrestrial habitats. Ribbon worms are long, often 10–20 cm, although the animals are difficult to measure because they can stretch. One species, *Lineus longissimus*, has been reported to measure 60 m in length—the longest animal!

Due to their compact body resembling that of the acoelomates, nemerteans were traditionally considered the simplest animals that possess a **complete digestive system,** one that has two separate openings, a mouth and an anus—a major difference with the flatworms. Ribbon worms also exhibit a closed circulatory system in which blood flows in vessels. The closed circulatory system together with the rhynchocoel now make it clear that nemerteans are not related to flatworms. Instead, ne-

> The acoelomates, typified by flatworms, are compact, bilaterally symmetrical animals, but they are not necessarily primitive and they are placed in many branches of the animal tree of life. New phyla continue to emerge, one of the most recent being Cycliophora. Representatives of the phylum live on the mouthparts of lobsters and have a unique form of sexual reproduction.

33.5 The Pseudocoelomates

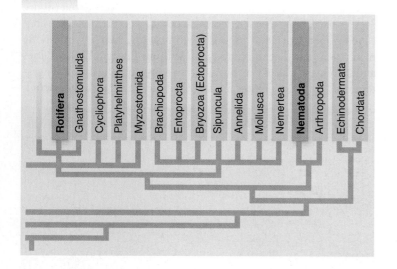

All bilaterians that do not have an acoelomate body plan possess an internal body cavity. A number of bilaterian phyla are characterized by their possession of a pseudocoel, which is a cavity between the mesoderm and endoderm (chapter 32; see figure 32.2).

Although their evolutionary relationships were once considered unclear, most pseudocoelomates are now considered members of the Ecdysozoa (in the case of nematodes and related taxa) or the Platyzoa (in the case of rotifers). The pseudocoel serves as a hydrostatic skeleton—one that gains its rigidity from being filled with fluid under pressure. The animals' muscles can work against this "skeleton," thus making the movement of the pseudocoelomates far more efficient than that of the acoelomates. Pseudocoelomates lack a defined circulatory system; this role is performed by fluids that move within the pseudocoel.

The pseudocoelomate condition seems to be an adaptive solution to the demands of a larger body size; the smallest members of phyla traditionally considered pseudocoelomates are often acoelomate, as is the case of the smallest nematodes. In this section, we focus on two significant pseudocoelomate phyla.

The roundworms, phylum Nematoda, are ecdysozoans comprising many species

Vinegar eels, eelworms, and other roundworms constitute a large phylum, **Nematoda,** with some 20,000 recognized species. Scientists estimate that the actual number might approach

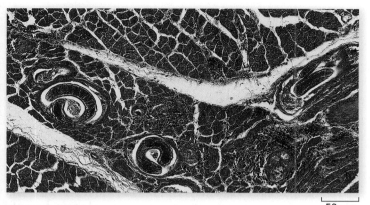

figure 33.17

TRICHINELLA NEMATODE ENCYSTED IN PORK. The serious disease trichinosis can result from eating undercooked pork or bear meat containing such cysts.

100 times that many. Members of this phylum are found everywhere. Nematodes are abundant and diverse in marine and freshwater habitats, and many members of this phylum are parasites of animals (figure 33.17) and plants. Many nematodes are microscopic and live in soil. A spadeful of fertile soil may contain, on the average, a million nematodes.

Nematode structure

Nematodes are bilaterally symmetrical, unsegmented worms. They are covered by a flexible, thick cuticle, which is molted four times as they grow. Their muscles constitute a layer beneath the epidermis and extend along the length of the worm (longitudinal musculature), rather than encircling its body (they lack circular musculature). These longitudinal muscles pull against both the cuticle and the pseudocoel, which forms a hydrostatic skeleton. Nematodes lack specialized respiratory organs and exchange oxygen through their cuticles. They possess a well-developed digestive system and feed on a diversity of food sources.

Near the mouth of a nematode, at its anterior end, are usually 16 raised, hairlike, sensory organs. The mouth is often equipped with piercing organs called **stylets.** Food passes through the mouth as a result of the sucking action produced by the rhythmic contraction of a muscular chamber called the **pharynx** at the worm's anterior end. After passing through a short corridor into the pharynx, food continues through the intestine; undigested material is eliminated through the anus (figure 33.18).

Reproduction and development

Reproduction in nematodes is usually sexual, with the sexes separate (as opposed to being hermaphroditic) and exhibiting many differences. Their development is direct, meaning that a small juvenile, not a larva, emerges from the egg. The egg cleavage pattern is unique among animals, but it is well studied.

The adults of some species consist of a fixed number of cells, a phenomenon known as *eutely.* For this reason, nematodes have become extremely important subjects for genetic and developmental studies (see chapter 19). The 1-mm-long *Caenorhabditis elegans* matures in three days; its body is transparent, and it has only 959 cells. It is the only animal whose complete developmental cellular anatomy is known.

Nematode lifestyles

Many nematodes are active hunters, preying on protists and other small animals. Many species of nematodes are parasites of plants or live within the bodies of larger animals. Almost every species of plant and animal that has been studied has been

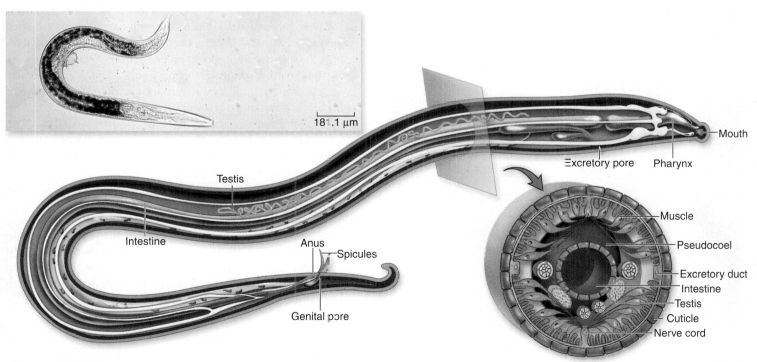

figure 33.18

PHYLUM NEMATODA: ROUNDWORMS. Roundworms such as this male nematode possess a body cavity between the gut and the body wall called the pseudocoel. It allows nutrients to circulate throughout the body and prevents organs from being deformed by muscle movements.

found to have at least one parasitic species of nematode living in it. The largest known nematode, reaching a length of 9 m, is a parasite in the placenta of sperm whales.

About 50 species of nematodes, including several that are rather common in the United States, regularly parasitize human beings. For example, hookworms, mostly of the genus *Necator*, can be common in southern states. By sucking blood through the intestinal wall, they can produce anemia if untreated.

Nematodes are related to other minor pseudocoelomate phyla (phyla Nematomorpha, Priapulida, Kinorhyncha and Loricifera), and they form the large group Ecdysozoa together with the arthropods and their relatives (phyla Onychophora and Tardigrada). They are not related to other pseudocoelomates such as the rotifers; thus, the pseudocoelomate condition has evolved several times independently.

Nematode-caused human diseases

The most serious and common nematode-caused disease in temperate regions is trichinosis, caused by worms of the genus *Trichinella*. These worms live in the small intestine of pigs, where fertilized female worms burrow into the intestinal wall. Once a female has penetrated these tissues, it produces about 1500 live young. The young enter the lymph channels and travel to muscle tissue throughout the body, where they mature and form highly resistant, calcified cysts.

Infection in humans or other animals arises from eating undercooked or raw pork in which the cysts of *Trichinella* are present. If the worms are abundant, a fatal infection can result, but such infections are rare. In the United States, only about 20 deaths have been attributed to trichinosis during the past decade.

Pinworms, *Enterobius vermicularis*, are abundant throughout the United States, where it is estimated they infect about 30% of all children and about 16% of adults. Adult pinworms live in the human rectum. Fortunately, the symptoms they cause are not severe, and the worms can easily be controlled by drugs.

The intestinal roundworm *Ascaris lumbricoides* infects approximately one of six people worldwide, but it is rare in areas with modern plumbing. Like pinworms, these worms live in the intestine. Their fertilized eggs are spread in feces, and can remain viable for years in the soil. Adult females, which are up to 30 cm long, contain up to 30 million eggs, and can release up to 20,000 of them each day.

Other nematode-caused diseases are extremely serious in the tropics. Filariasis is caused by several species of nematodes that infect at least 250 million people worldwide. Up to 10 cm long, filarids live in the lymphatic system, which they may seriously obstruct, causing severe inflammation and swelling. Extreme infection by the filarid species *Wuchereria bancrofti* produces the condition known as elephantiasis in which extremities may swell to disfiguring proportions. These parasites require an intermediate host, typically a blood-sucking insect.

The rotifers, phylum Rotifera, move by rapidly beating cilia

Rotifers (phylum **Rotifera**) are bilaterally symmetrical, unsegmented pseudocoelomates. Although they are pseudocoelomates, rotifers are very unlike nematodes and belong to a different branch of the animal tree of life. Several features suggest their ancestors may have resembled flatworms and therefore are classified with them in the spiralian Platyzoa.

Rotifers are very small; at 50–500 μm long, they are smaller than some ciliate protists. But rotifers have complex bodies with three cell layers and highly developed internal organs. A complete gut passes from mouth to anus. An extensive pseudocoel acts as a hydroskeleton to provide body rigidity.

Rotifer locomotion and distribution

Rotifers are aquatic animals that propel themselves through the water by rapidly beating cilia, like a boat with oars. Rotifers are often called "wheel animals" because the cilia, when beating together, resemble the movement of spokes radiating from a wheel. There are about 1800 named species of this phylum. Although a few rotifers live in soil or in the capillary water in cushions of mosses, most occur in fresh water, and they are common everywhere. Most live no longer than 1 or 2 weeks, but some species can survive in a desiccated, inactive state on the leaves of plants; when rain falls, the rotifers regain activity and actively feed in the film of water that temporarily covers the leaf. A few rotifer species are marine.

Food gathering

Rotifers have a well-developed food-processing apparatus. A conspicuous organ on the tip of the head, called the *corona*, gathers food (figure 33.19). It is composed of a circle of cilia that sweeps food into the rotifer's mouth, with a complex jaw. The corona is also used for locomotion, although many species attach to the substrate with the aid of toes and adhesive glands.

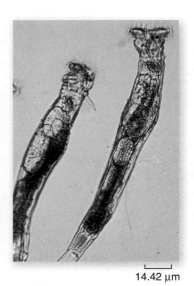
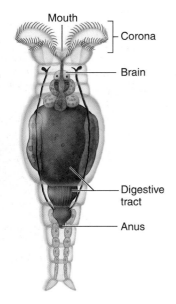

14.42 μm

figure 33.19

PHYLUM ROTIFERA. Microscopic in size, rotifers are smaller than some ciliate protists, and yet have complex internal organs.

Mouth — Corona — Brain — Digestive tract — Anus

Pseudocoelomates have fluid-filled body cavities lined by endoderm and mesoderm. Nematodes are bilaterally symmetrical, unsegmented worms. Other pseudocoelomates have very different body plans. Rotifers, while tiny, are very complex.

33.1 A Revolution in Invertebrate Phylogeny

There is little disagreement among scientists about the taxonomic classification of 36 phyla of animals; however, there is great disagreement about how these 36 phyla are related to one another.

- The Porifera first separate from Eumetazoa and the Cnidaria and Ctenophora branch out early before the bilaterally symmetrical animals.

- The bilaterally symmetrical animals are assigned to two groups of animals that differ in their embryology: protostomes and deuterostomes.

- The ancestry of the protostome lineage is constructed differently using molecular data instead of morphological data.

- Traditionally the branches of animals were based on the presence of a coelom.

- Coelomates can be either protostomes or deuterostomes; acoelomates and pseudocoelomates are always protostomes (figure 33.2).

- Two major clades of protostomes, the Spiralia and Ecdysozoa, evolved independently.

- Spiralians have a free-living larval stage called the trochophore and grow by adding mass to an existing body. They are divided into Lophotrochozoa and Platyzoa that include many phyla such as the coelomate Lophophorata, Mollusca, Annelida, and acoelomate Platyhelminthes.

- Ecdysozoans grow by molting their external skeleton and they include numerous phyla, including the very successful pseudocoelomate Nematoda and coelomate and segmented Arthropoda.

33.2 Parazoa: Animals That Lack Specialized Tissues

The multicellular sponges are parazoans, animals that lack tissues and organs and a definite symmetry.

- Larval sponges are free-swimming, and the adults are anchored onto submerged objects.

- Sponges are composed of three layers, an external protective epithelial layer, a central protein-rich matrix called mesohyl, and an inner layer of choanocytes that are used in food acquisition and water circulation (figure 33.3b).

- The mesohyl may contain spicules or fibers (or both) of a tough protein called spongin that strengthen the body of the sponge.

- Sponge reproduction can be asexual by fragmentation or sexual.

33.3 Eumetazoa: Animals with True Tissues

The subkingdom Eumetazoa contains animals that have distinct tissues and true body symmetry.

- The carnivorous and radially symmetrical Cnidaria have distinct tissues but no true organs.

- The ectoderm gives rise to the epidermis and nervous system, and the endoderm gives rise to the digestive tissue, gastrodermis. The mesoglea lies between these two basic tissues and contains the muscles (figure 33.5).

- Cnidarians have two basic body forms: a sessile, cylindrical polyp and free-floating medusae (figure 33.4).

- The polyp can reproduce asexually or sexually, but the medusae reproduce sexually, forming the planulae after fertilization (figure 33.6).

- Digestion begins with extracellular fragmentation followed by phagocytosis and internal digestion.

- A nerve net coordinates contraction of cnidarian myoepithelial cells, but there is no central control.

- Cnidarians have no circulatory, excretory, or respiratory systems.

- Cnidarians have specialized cells called cnidocytes that contain nematocysts, which are used in food acquisition and defense.

- Cnidarians exist as polyps or medusae and are grouped into four classes: Hydrozoa or hydroids, Scyphozoa or jellyfish, the Cubozoa or box jellyfish, and the Anthozoa or sea anemones and corals.

- Ctenophora is a small phylum called comb jellies that are triploblastic and have true muscle cells derived from mesoderm. They capture prey with a colloblast that contains an adhesive, are not strictly radially symmetrical, and use fused cilia for locomotion.

33.4 The Bilaterian Acoelomates

The Bilateria are characterized by bilateral symmetry, which allows for a higher degree of specialization.

- The flatworms, phylum Platyhelminthes have a definite head and move by ciliated epithelial cells (figure 33.11).

- The incomplete gut has only one opening, and flatworms cannot feed continuously or simultaneously feed, digest, and eliminate undigested foods at the same time.

- Flatworms have an excretory and osmoregulatory system containing a fine network of tubules with flame cells. The primary function of this system is water balance, and most metabolic wastes are excreted into the gut.

- The flattened body shape and highly branched gut utilize diffusion; there is no circulatory system.

- Flatworms reproduce sexually and are hermaphroditic. They also have the capacity for asexual regeneration.

- Flatworms are divided into four major classes: the free-living Turbellaria, the parasitic flukes, Monogenea and Trematoda, and the Cestoda (or parasitic tapeworms). The flukes and tapeworms are important repercussions on human health.

- Acoel flatworms, Acoela, were once considered basal to the phylum Platyhelminthes, but molecular studies indicate that they evolved before the split between protostomes and deuterostomes.

- Nemertea, or ribbon worms, resemble flatworms except that they have a true coelom called a rhynchocoel, a complete digestive tract, and a closed circulatory system.

33.5 The Pseudocoelomates

The pseudocoelomates possess a pseudocoelom, a cavity between the mesoderm and the endoderm.

- The pseudocoelom is fluid-filled and acts as a hydrostatic skeleton against which the muscles work.

- The pseudocoelomates lack a circulatory system and this role is performed by the movement of fluids in the pseudocoelom.

- Oxygen is exchanged through the cuticle.

- Nematodes are ecdysozoans that reproduce sexually and exhibit sexual dimorphism.

- The pseudocoelomate Rotifera are not related to the Nematoda and are classified as a spiralian Platyzoa.

- Rotifers propel themselves and gather food with cilia, break down food with a complex jaw, and the pseudocoelom acts as a hydrostatic skeleton. They are either free-swimming or attach themselves to substrate by their toes and adhesive glands.

- The phylum Cycliophora is one of three new phyla with a unique form of sexual reproduction.

SELF TEST

1. In modern phylogenetic analysis of the animals, the protostomes are divided into what two major groups based on what characteristic?
   a. Their symmetry
   b. Formation of a head
   c. Their ability to molt
   d. The presence or absence of vertebrae

2. Which of the following classifications best describes a species that does not molt, possesses a coelom, and whose larval stage utilizes a trochophore?
   a. Ecdysozoan
   b. Parazoans
   c. Platyzoa
   d. Lophotrochozoa

3. A chitin exoskeleton and jointed appendages are key characteristics of which phylum?
   a. Arthropoda
   b. Nematoda
   c. Chordata
   d. Mollusca

4. All animals have which of the following characteristics?
   a. Body symmetry
   b. Tissues
   c. Multicellularity
   d. Body cavity

5. Which of the following cell types of a sponge possesses a flagella?
   a. Choanocyte
   b. Amoebocyte
   c. Epithelial
   d. Spicules

6. Spicules and sponging are found _____ of a sponge.
   a. within the osculum
   b. within the mesohyl
   c. between the choanocytes
   d. outside of the epithelial cells

7. The larval stage of a cnidarian is known as a
   a. medusa.
   b. planulae.
   c. polyp.
   d. cnidocyte.

8. Which of the following cell layers is not necessary to be considered a eumetazoan?
   a. Ectoderm
   b. Endoderm
   c. Mesoderm
   d. All of the above are found in all eumetazoans.

9. Which of the following classes has an endosymbiotic relationship with a dinoflagellate?
   a. Anthozoa
   b. Cubozoa
   c. Hydrozoa
   d. Scyphozoa

10. Which of the following phyla possesses true bilateral symmetry?
    a. Cnidaria
    b. Porifera
    c. Platyhelminthes
    d. Ctenophora

11. In the flatworm, the flame cells are involved in what metabolic process?
    a. Reproduction
    b. Digestion
    c. Locomotion
    d. Osmoregulation

12. Which of the following would not be considered a Neodermata?
    a. Class Cestodes
    b. Class Turbellaria
    c. Class Trematoda
    d. Class Monogenea

13. The "simplest" animal with a complete digestive system belongs to
    a. Class Turbellaria.
    b. Phylum Nemertea.
    c. Phylum Acoela.
    d. Class Cestodes.

14. Nematodes used to be classified close to the Rotifera due to the presence of a pseudocoel, but are not located closer to the Arthropods due to the presence of
    a. molting.
    b. jointed appendages.
    c. wings.
    d. a complete coelom.

15. Which of the following is a disease caused by a nematode?
    a. Filariasis
    b. Pinworm
    c. Trichinosis
    d. All of the above

CHALLENGE QUESTIONS

1. What do the phyla Acoela and Cycliophora tell you about our understanding of noncoelomate invertebrates? Do you think that the phylogeny presented in figure 33.2 is complete? Explain your answer.

2. Why would being a hermaphrodite be a benefit for a parasitic species?

3. Does the lack of a digestive system in tapeworms indicate that it is a primitive, ancestral form of platyhelminthes? Explain your answer.

chapter 34

Coelomate Invertebrates

introduction

ALTHOUGH ACOELOMATES AND PSEUDOCOELOMATES have proven very successful, a third way of organizing the animal body has also evolved, one that occurs in many protostomes and in all deuterostomes. We begin our discussion of the coelomate invertebrate animals with mollusks, which include animals such as clams, snails, slugs, and octopuses. Annelids, represented by earthworms, leeches, and seaworms, also have a coelomic body cavity, as well as segmented bodies. Arthropods, such as the paper wasp pictured here, evolved segmentation and jointed appendages, and they have become the most successful of all animal groups. Echinoderms are exclusively marine animals that exhibit deuterostome development and endoskeletons, two evolutionary innovations that they share with the chordates, which are the subject of chapter 35.

concept outline

Phylum Mollusca: The Mollusks

The origin of the coelom, as was mentioned in the preceding chapter, was a significant advance in the structure of the animal body because it allowed for increased body size. Coelomates have a body design that repositions the body's fluid and allows complex tissues and organs to develop. This new body plan also made it possible for a wide variety of different body architectures to evolve and the body to grow to much larger sizes than most acoelomate or pseudocoelomate animals. Currently, there is little doubt that coeloms evolved multiple times during animal evolution.

Mollusks are extremely diverse

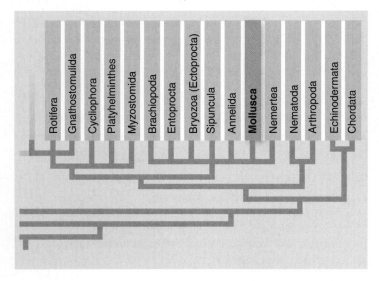

Mollusks (phylum **Mollusca**) are an extremely diverse animal phylum, second only to the arthropods, with over 110,000 described species. Mollusks exhibit a variety of body forms and live in many different environments. They include snails, slugs, clams, scallops, oysters, cuttlefish, octopuses, and many other familiar animals (figures 34.1 and 34.2). The durable shells of some mollusks are often beautiful and elegant; they have long been favorite objects for professional scientists and amateurs alike to collect, preserve, and study. Chitons, aplacophorans, and nudibranchs are less familiar marine mollusks because

figure 34.2

GIANT CLAM. Second only to the arthropods in number of described species, members of the phylum Mollusca occupy almost every habitat on Earth. This giant clam, *Tridacna maxima*, has a green color caused by the presence of symbiotic dinoflagellates (zooxanthellae). Through photosynthesis, the dinoflagellates probably contribute most of the food supply of the clam, although it remains a filter feeder like most bivalves. Some individual giant clams may be nearly 1.5 m long and weigh up to 270 kg.

a.

b.

c.

d.

figure 34.1

MOLLUSK DIVERSITY. Mollusks exhibit a broad range of variation. *a.* The flame scallop, *Lima Scabra*, is a filter feeder. *b.* The blue-ringed octopus, *Hapalochlaena maculosa*, is one of the few mollusks dangerous to humans. Strikingly beautiful, it is equipped with a sharp beak that can deliver a strong poisonous bite—divers give it a wide berth! *c.* Nautiluses, such as this chambered nautilus, *Nautilus pompilius*, have been around since before the age of the dinosaurs. *d.* The banana slug, *Ariolimax columbianus*, native to the Pacific Northwest, is the second largest slug in the world, attaining a length of 25 cm.

they lack a true shell. Mollusks are characterized by a coelom surrounding the heart, and although there is extraordinary diversity in this phylum, many of the basic components of the mollusk body plan can be seen in figure 34.3.

Mollusks range in size from almost microscopic to huge. Although most measure a few millimeters to centimeters in their largest dimension, the giant squid may grow up to 10 m long! Weighing up to 250 kg, the giant squid is therefore one of the heaviest invertebrates (although nemerteans can be much longer, as discussed in chapter 33). Other large mollusks are the bivalve (two-shelled) mollusks of the genus *Tridacna*, the giant clam, which may be as long as 1.5 m and may weigh as much as 270 kg (see figure 34.2).

Like most animal phyla, mollusks evolved in the oceans, and most groups have remained there. Marine mollusks are widespread and often abundant. Some groups of mollusks that have invaded freshwater and terrestrial habitats include the snails and slugs, and the freshwater pearl mussels. Terrestrial mollusks are often abundant in places that are at least seasonally moist. Some of these places, such as the crevices of desert rocks, may appear very dry, but even these habitats have at least a temporary supply of water at certain times.

As a group, mollusks are an important source of food for humans. Oysters, clams, scallops, mussels, octopuses, and squids are among the culinary delicacies that belong to this large phylum. Mollusks are also economically significant in many other ways. For example, pearls are produced in oysters, and the material called mother-of-pearl, often used in jewelry and other decorative objects, is produced in the shells of a number of different mollusks, but most notably in the abalone.

Mollusks can also be pests. Bivalve mollusks called shipworms burrow through wood submerged in the sea, damaging boats, docks, and pilings. The zebra mussel (*Dreissena polymorpha*) has recently invaded many freshwater ecosystems via the ballast water of cargo ships, wreaking havoc in many aquatic ecosystems (see figure 59.16). Slugs and snails often cause extensive damage to flowers, vegetable gardens, and crops. Other mollusks serve as hosts to the intermediate stages of many serious parasites, including several nematodes and flatworms, discussed in chapter 33.

The mollusk body plan is complex and varied

In their basic body plan (see figure 34.3), mollusks are bilaterally symmetrical, although this symmetry disappeared during the evolution of gastropods (snails and their relatives). Although traditionally considered as coelomates, the coelom of mollusks is highly reduced and is limited to small spaces around the excretory organs, heart, and part of the intestine.

Internal organs

Mollusks' digestive, excretory, and reproductive organs are concentrated in a **visceral mass,** and a muscular foot is their primary mechanism of locomotion, although this has changed radically in the cephalopods (octopuses, squid, and chambered *Nautilus*). They may also have a differentiated head at the anterior end of the body. The **mantle,** a thick

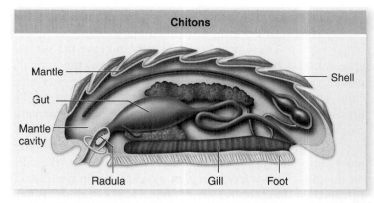

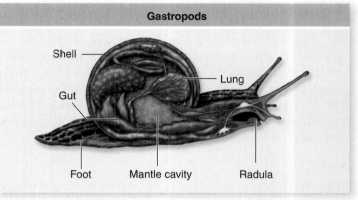

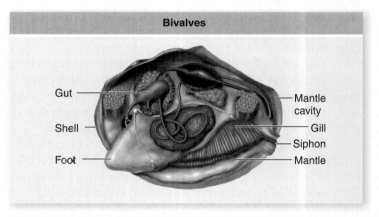

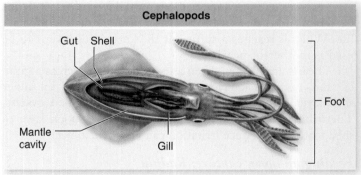

figure 34.3
BODY PLANS OF THE MOLLUSKS.

epidermal sheet of skin, covers the dorsal side of the body and forms a cavity in which are housed the respiratory organs (*ctenidia*, or gills), and the openings of the excretory, reproductive, and digestive organs.

The **ctenidia** are specialized portions of the mantle that usually consist of a system of filamentous projections rich in blood vessels. These projections greatly increase the surface area available for gas exchange and, therefore, the animal's overall respiratory potential. Mollusk ctenidia are very efficient, and many gilled mollusks extract 50% or more of the dissolved oxygen from the water that passes through the mantle cavity.

In aquatic mollusks, a continuous stream of water passes into and out of the mantle cavity, drawn by the cilia on the gills. This water brings in oxygen and, in the case of the bivalves, food; it also carries out waste materials. When gametes are produced, they are frequently carried out in the same stream.

The muscular foot of a mollusk may be adapted for locomotion, attachment, food capture (in squids and octopuses), or various combinations of these functions. Slugs and snails secrete mucus, forming a path that they glide along on their foot. In cephalopods the foot is divided into tentacles. In some mollusks that occur in the open ocean the foot is modified into winglike projections to increase flotation surface.

Shells

In most members of this phylum, the outer surface of the mantle is responsible for secreting the shell, one of the best known characters of the phylum. Not all mollusks have shells, but the so-called higher mollusks have a dorsal shell that grows with them thanks to the secretory action of the mantle.

The shell of mollusks serves primarily for protection; most species can withdraw into their shells. But shells come in many fashions, and many lineages of mollusks have internalized or reduced their shells, as is the case of many cephalopods (cuttlefish, squids, and octopuses) and gastropods (slugs).

A typical mollusk shell consists of calcium carbonate produced extracellularly, laid down in layers, and often covered by a thin organic coating rich in the protein conchin. This outer layer protects the two underlying calcium-rich layers from erosion. The middle layer consists of densely packed crystals of calcium carbonate. The inner layer is pearly in appearance and increases in thickness throughout the animal's life. When it reaches a sufficient thickness, this layer can be used as mother-of-pearl.

Pearls are formed when a foreign object, such as a grain of sand, becomes lodged between the mantle and the inner shell layer of bivalve mollusks, including clams and oysters. The mantle coats the foreign object with layer upon layer of shell material to reduce irritation caused by the object.

Feeding and prey capture: Radula

One of the most characteristic features of most mollusks is the **radula,** a rasping, tonguelike organ used for feeding. The radula consists primarily of dozens to hundreds of microscopic, chitinous teeth arranged in rows (figure 34.4). Benthic (bottom dwelling) mollusks use their radula to scrape algae and other food materials off their substrates and then to convey this food to the digestive tract.

Certain gastropods are active predators, some using a modified radula to drill through the shells of prey and extract the food. The small holes often seen in clam shells are produced by gastropods, primarily moon snails, that have bored holes with their radulae to kill the bivalve and extract its body for food. *Conus* gastropods have transformed the radula into a harpoonlike structure with an associated venom gland that can be used to capture large prey such as fish; in some cases the venom can be harmful to humans.

The radula is absent in all bivalves, which instead have adapted to use the gills to filter food particles from water currents. Primitive bivalves are deposit feeders, but they do not have a radula either.

Removal of wastes

Nitrogenous wastes are removed from the mollusk body by the **nephridia,** a special type of excretory structure. A typical nephridium has an open funnel, the **nephrostome,** which is lined with cilia. A coiled tubule runs from the nephrostome into a bladder, which in turn connects to an excretory pore.

Wastes are gathered by the nephridia from the coelomic cavity and discharged into the mantle cavity. The wastes are then expelled from the mantle cavity by the continuous pumping of the gills. Sugars, salts, water, and other materials are reabsorbed by the walls of the nephridia and returned to

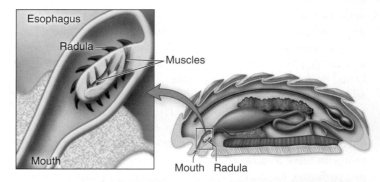

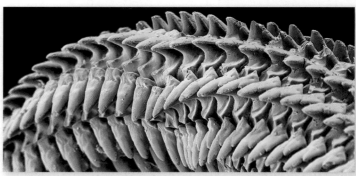

25 µm

figure 34.4

STRUCTURE OF THE RADULAR APPARATUS IN A CHITON. The radula consists of chitin and is covered with rows of teeth that extend backward. As the animal feeds, its mouth opens, and the radula brings food in by scraping backward.

the animal's body as needed to achieve an appropriate osmotic balance.

The open circulatory system

Although mollusks are coelomate protostomes, the coelom is greatly reduced. The main coelomic cavity is an open circulatory space or hemocoel, which comprises several sinuses and a network of vessels in the gills, where gas exchange takes place. This is the so-called **open circulatory system** of all mollusks except cephalopods, which consists of a heart and an open system in which blood circulates freely. The mollusk heart usually has three chambers, two that collect aerated blood from the gills, and a third that pumps it to the other body tissues, but other models exist. Cephalopods have a **closed circulatory system**, where the blood does not enter in contact with tissues.

Mollusk reproduction

Most mollusks have distinct male and female individuals, although a few bivalves and many freshwater and terrestrial gastropods are hermaphroditic. Even in hermaphroditic mollusks, cross-fertilization is most common. Remarkably, some sea slugs and oysters are able to change from one sex to the other several times during a single season.

    Most aquatic mollusks engage in external fertilization. The males and females release their gametes into the water, where they mix and fertilization occurs. Gastropods, however, more often have internal fertilization, with the male inserting sperm directly into the female's body. Internal fertilization and an efficient excretory system that prevents dessication are some of the key adaptations that allowed gastropods to colonize the land.

The trochophore larva

Many marine mollusks develop through a spiral cleavage pattern, grouping them with spiralians. The embryos give rise to free-swimming larvae called **trochophores** (figure 34.5a) that closely resemble the larval stage of many marine annelids and other lophotrochozoans. Trochophores swim by means of a row of cilia that encircles the middle of their body.

    In most marine snails and in bivalves, a second free-swimming stage, the **veliger,** follows the trochophore stage. This veliger stage has the beginnings of a foot, shell, and mantle (figure 34.5b). Trochophores and veligers drift widely in the ocean currents, allowing mollusks to disperse to new areas.

Four classes of mollusks show the diversity of the phylum

Of the eight recognized classes of mollusks, we examine four as representatives of the phylum: (1) Polyplacophora—chitons; (2) Gastropoda—limpets, snails, slugs, and their relatives; (3) Bivalvia—clams, oysters, scallops, and their relatives; and (4) Cephalopoda—squids, octopuses, cuttlefishes, and the chambered *Nautilus.*

Mollusk Trochophore Larva

a.

Gastropod Veliger Stage

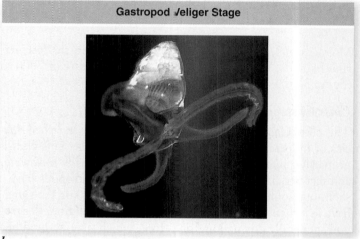

b.

figure 34.5

STAGES IN THE MOLLUSCAN LIFE CYCLE. *a.* The trochophore larva of a mollusk. Similar larvae are characteristic of some annelid worms as well as a few other phyla. *b.* The veliger stage of a gastropod.

By studying living mollusks and the fossil record, some scientists have deduced that the ancestral mollusk was probably a bilateral, dorsoventrally flattened, unsegmented, wormlike animal that glided on its ventral surface. This animal may also have had a chitinous cuticle that secreted calcareous spicules. Other scientists believe that mollusks arose from segmented ancestors and became unsegmented secondarily.

Class Polyplacophora: Chitons

Chitons (class **Polyplacophora**) are marine mollusks that have oval bodies with eight overlapping dorsal calcareous plates. Underneath the plates, the body is not segmented, but chitons do have eight sets of dorsoventral pedal retractor muscles and serially repeated gills.

    Chitons creep along using a broad, flat foot surrounded by a groove or mantle cavity in which the gills are arranged.

figure 34.6

THE NOBLE CHITON, *Eudoxochiton nobilis*, FROM NEW ZEALAND. The foot of the chiton can grip the substrate very strongly, making chitons hard to dislodge by waves or predators.

Most chitons are grazing herbivores that live in shallow marine habitats, but some live at depths of more than 7000 m (figure 34.6).

Class Gastropoda: Snails and slugs

The class **Gastropoda** contains about 40,000 described species of limpets, snails, slugs, and similar animals. This class is primarily a marine group, but it also contains many freshwater and terrestrial mollusks (figure 34.7). Most gastropods have a shell, but some, such as slugs and nudibranchs (or sea slugs), have lost their shells through the course of evolution. Gastropods generally creep along on a foot, which may be modified for swimming.

The heads of most gastropods have a pair of tentacles with eyes at the base. In the typical garden snails, there may be two sets of tentacles, with one of them bearing the eyes at the ends. Other gastropods have additional tentacles in other parts of the body, including in the mantle or along the foot. These tentacles are sensory structures that serve as chemo- or mechanosensors.

During embryological development, gastropods undergo torsion. **Torsion,** known only in the gastropods, is a process by which the mantle cavity and anus are moved from a posterior location to the front of the body, closer to where the mouth is located. Torsion is brought about by a disproportionate growth of the lateral muscles; that is, one side of the larva grows much more rapidly than the other. It often leads to the reduction or the disappearance of the left posttorsional nephridia, gonad, or other internal organs. Gastropods are therefore not bilaterally symmetrical.

Another process superimposed on torsion is the so-called **coiling,** or spiral winding of the shell. This process has occurred in other groups of mollusks, such as in the cephalopods, and the fossil record suggests that the first gastropods were coiled but not torsioned.

Sea slugs, or nudibranchs, are active predators; a few species of nudibranchs have the extraordinary ability to extract the nematocysts from the cnidarian polyps they eat, transfer them through their digestive tract to the surface of their own bodies intact, and use them for their own protection. Nudibranchs get their name from their gills, which, instead of being enclosed within the mantle cavity, are exposed along the dorsal surface (figure 34.8).

In terrestrial gastropods, the empty mantle cavity, which was occupied by gills in their aquatic ancestors, is extremely rich in blood vessels and in effect serves as a lung. This structure has evolved in environments with plentiful oxygen; a lung absorbs oxygen from the air much more effectively than a gill could, but is not as effective under water.

Class Bivalvia: Bivalves

Members of the class **Bivalvia** include the clams, scallops, mussels, and oysters, among others. About 10,000 species of bivalves are extant. Most species are marine, although many also live in fresh water. Over 500 species of pearly freshwater mussels, or naiads, occur in the rivers and lakes of North America.

Bivalves have two lateral (left and right) shells (valves) hinged together dorsally (figure 34.9). A ligament hinges

figure 34.7

A GASTROPOD MOLLUSK. The terrestrial Oregon forest snail *Allogona townsendiana*.

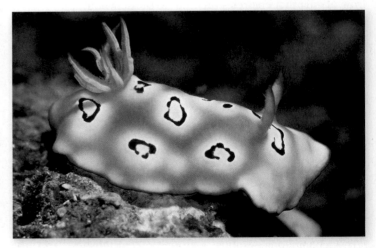

figure 34.8

A NUDIBRANCH (OR SEA SLUG). The bright colors of many nudibranchs alert predators to their potent sting.

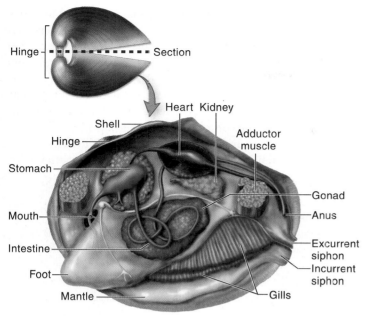

figure 34.9

DIAGRAM OF A CLAM. Internal organs and the foot are shown. The left shell and mantle are removed. Bivalves such as this clam circulate water through their gills and filter out food particles.

the shells together and causes them to gape open. Pulling against this ligament are one or two large adductor muscles that can draw the shells together. The mantle secretes the shells and ligament and envelops the internal organs within the pair of shells.

figure 34.10

PROBLEM-SOLVING BY AN OCTOPUS. This two-month-old common octopus (*Octopus vulgaris*) was presented with a crab in a jar. It tried to unscrew the lid to get to the crab. Although it failed in this attempt, in some cases it was successful.

The mantle is frequently drawn out to form two siphons, one for an incoming stream of water and one for an outgoing stream; these are termed **inhalant** and **exhalant siphons.** The siphons often function as snorkels to allow bivalves to filter water through their body while remaining almost completely buried in sediments. A complex folded gill lies on each side of the visceral mass. These gills consist of pairs of filaments that contain many blood vessels. Rhythmic beating of cilia on the gills creates a pattern of water circulation.

Most bivalves are filter feeders that live cemented to the substrate or live buried in the sediment. They extract small organisms from the water that passes through their mantle cavity. Bivalves do not have distinct heads or radulas, differing from all other mollusks in this respect (see figure 34.3). Most have a wedge-shaped foot, however, that may be adapted, in species not attached to the substrate, for creeping, burrowing, cleansing the animal, or anchoring it in its burrow. Some species of clams can dig into sand or mud very rapidly by means of muscular contractions of their foot.

Bivalves disperse from place to place largely as larvae. Although most adults are adapted to a burrowing way of life, some species of scallops and file clams can move swiftly through the water by using their large adductor muscles to clap their shells together. These muscles are what we usually eat as "scallops." The edge of a scallop's body is lined with tentacle-like projections with complex eyes between them.

Class Cephalopoda: Octopuses, squids, and nautiluses

The more than 600 species of the class **Cephalopoda**—octopuses, squids, and chambered nautiluses—live entirely in the oceans. They are active marine predators that swim, often swiftly, and compete successfully with fish. The foot has evolved into a series of arms equipped with suction cups, adhesive structures, or hooks that seize prey efficiently. Squids have 10 arms; octopuses, as indicated by their name, have 8; and the chambered nautilus has about 30 to 90. After snaring prey with its arms, a cephalopod bites the prey with its strong, beaklike paired jaws and then pulls it into its mouth by the tonguelike action of the radula.

Cephalopods have highly developed nervous systems, and their brains are unique among invertebrates. Their eyes are very elaborate, and have a structure much like that of vertebrate eyes, although they evolved separately (see chapter 45). Many cephalopods exhibit complex patterns of behavior and a high level of intelligence; octopuses can be easily trained to distinguish among classes of objects, and they are capable of escaping from one tank to seize prey in another tank, and then return to their original one (see figure 34.10). Cephalopods are the only mollusks that have closed circulatory systems.

Although they evolved from shelled ancestors as seen in the chambered nautiluses and in the many fossil cephalopods such as ammonites and belemnites, most living cephalopods lack an external shell. Cephalopods became extraordinarily successful by being able to spend most of their time in the open water instead of on the bottom like most other mollusks. This led to the evolutionary reduction and eventual loss of their heavy shells. The cuttlebone of the cuttlefish and the lighter plume of the squid are internalized shells that give these animals some rigidity

figure 34.11

INK DEFENSE BY A GIANT PACIFIC OCTOPUS (*Octopus doflein*) WHEN BOTHERED BY A SCUBA DIVER. Octopuses and squids expel a dark cloudy liquid when threatened by predators.

without adding much weight. Finally, the internal shell disappeared completely in the lineage that gave origin to octopuses.

Like other mollusks, cephalopods take water into the mantle cavity and expel it, in this case through a siphon. Cephalopods have modified this system into a means of jet propul-

sion. When threatened, they eject water violently and shoot themselves through the water. Many cephalopods have an ink sac that contains a brown-black fluid that can be expelled through the anus to create an ink cloud to confuse predators (figure 34.11).

Most octopuses and squids are capable of changing color and texture to suit their background or display messages to one another. They accomplish this feat using their *chromatophores*, pouches of pigments embedded in the epithelium. Some cephalopods can host luminescent bacteria that they can use for counterillumination to avoid predation.

> Mollusks have an efficient excretory system, ctenidia for respiration, and a special feeding structure, the radula. The mantle of mollusks not only secretes their protective shell but also forms a cavity that is essential to respiration, excretion, and release of gametes. Most mollusk groups reproduce via external fertilization, with some exceptions. The trochophore larva resembles that of other lophotrochozoans.
>
> Four representative classes of mollusks exemplify their variations in morphology. Chitons have overlapping dorsal plates and creep along the bottom. Gastropods include the snails and slugs, many of which live in terrestrial habitats. Bivalves are the two-shelled mollusks, including clams, oysters, mussels, scallops, and many others, some of which live in fresh water. Cephalopods are the squids, octopuses, and chambered nautiluses.

34.2 Phylum Annelida: The Annelids

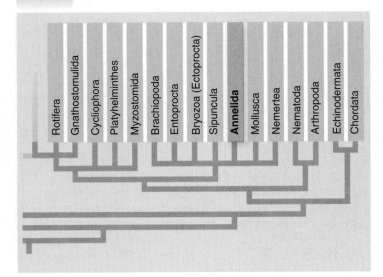

An important innovation in the animal body plan was the origin of *segmentation*, the building of a body from a series of similar segments, or repeated units (see chapter 32). Segmentation evolved multiple times during animal evolution, and therefore different modes of segmentation are exhibited in different phyla. Annelid worms, phylum **Annelida** (figure 34.12) are one of these examples.

One advantage of having a body built from segments is that the development and function of these units can be more

figure 34.12

A POLYCHAETE ANNELID. *Nereis virens* is a wide-ranging, predatory, marine polychaete worm equipped with feathery parapodia for movement and respiration, as well as jaws for hunting. You may have purchased *Nereis* as fishing bait!

precisely controlled, namely at the level of individual segments or groups of segments—something we know as specialization. For example, different segments may possess different combinations of organs or perform different functions relating to reproduction, feeding, locomotion, respiration, or excretion.

Two-thirds of all annelids live in the sea (about 8000 species), and most of the rest, some 3100 species, are earthworms. Although they have been traditionally considered a natural group, the monophyly of annelids is not well established. For example, myzostomes were once considered annelids, but new evidence suggests that they may be related to certain acoelomates (see chapter 32).

Annelids have distinct ringlike segments

In general, the annelid body is broken into distinct and sometimes ringlike segments.

Repeated segments

The body of an annelid worm is composed of a series of ringlike segments running the length of the body, looking like a stack of doughnuts or a roll of coins (figure 34.13). Internally, the segments are divided from one another by partitions called **septa,** just as bulkheads separate the segments of a submarine. In each of the cylindrical segments, the excretory and locomotor organs are repeated.

The fluid within the coelom of each segment creates a hydrostatic (liquid-supported) skeleton that gives the segment rigidity, like an inflated balloon (see chapter 47). Muscles within each segment push against the fluid in the coelom. Because each segment is separate, each can expand or contract independently. This lets the annelid move in complex ways. A number of unsegmented forms can be found, however, such as the echiurans that live buried in mudflats or some microscopic annelids that live in the spaces between sand grains.

Specialized segments

The anterior (front) segments of annelids have become modified to contain specialized sensory organs. Some are sensitive to light, and elaborate eyes with lenses and retinas have evolved in some annelids. A well-developed cerebral ganglion, or brain, is contained in one anterior body region.

Segment connections

Although partitions separate the segments, materials and information do pass between segments. Annelids have a dorsal closed circulatory system that carries blood from one segment to another. A ventral nerve cord connects the nerve centers or ganglia in each segment with one another and with the brain. These neural connections are critical features that allow the worm to function and behave as a unified and coordinated organism.

Annelids move by contracting their segments

The basic annelid body plan is a tube within a tube, with the internal digestive tract—a tube running from mouth to anus—suspended within the coelom, which is lined by mesodermal tissue. The tube that makes up the digestive tract has several portions—the pharynx, esophagus, crop, gizzard, and intestine—that are specialized for different functions.

Annelids make use of their hydrostatic skeleton for locomotion (see chapter 47). To move, annelids contract circular muscles running around each segment. Doing so squeezes the segment, causing the coelomic fluid to squirt outward,

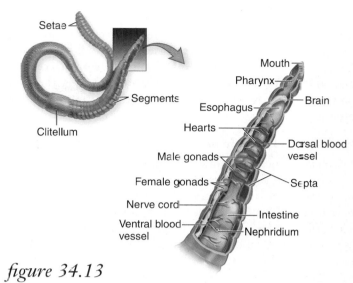

figure 34.13

PHYLUM ANNELIDA: AN OLIGOCHAETE. Earthworms show a body plan based on repeated body segments. Segments are separated internally from each other by septa.

like a tube of toothpaste. The segment elongates and gets much thinner. Next, contraction of longitudinal muscles that run along the length of the worm return the segment to its original shape. In most annelid groups, each segment typically possesses **chaetae,** bristles of chitin that help anchor the worms during locomotion. By extending the chaetae in some segments so that they anchor in the substrate and retracting them in other segments, the worm can squirt its body, section by section, in either direction.

Unlike the arthropods and the noncephalopod mollusks, most annelids have a closed circulatory system. Annelids exchange oxygen and carbon dioxide with the environment through their body surfaces; most lack gills or lungs. Much of their oxygen supply reaches the different parts of their bodies through the blood vessels. Some of these vessels at the anterior end of the worm body are enlarged and heavily muscular, serving as hearts that pump the blood. Earthworms have five pulsating blood vessels on each side that serve as hearts, helping to move blood from the main dorsal vessel, the major pumping structure, to the main ventral vessel.

The excretory system of annelids consists of ciliated, funnel-shaped nephridia generally similar to those of mollusks. These nephridia—each segment has a pair—collect waste products and transport them out of the body through the coelom by way of specialized excretory tubes. Some polychaetes have protonephridia, similar to those of planarians or larval mollusks.

Annelids have traditionally been grouped into three main classes

The roughly 12,000 described species of annelids occur in many different habitats. They range in length from as little as 0.5 mm to more than 3 m in some polychaetes and giant Australian earthworms. Although traditionally annelids have been classified into three classes, namely Polychaeta (mostly marine worms), Oligochaeta (mostly terrestrial, including earthworms), and Hirudinea (leeches), the current status of the group is in a state of flux because the monophyly of polychaetes and Annelida is not well established. For this reason, the classification of annelids is likely to change in the near future; for now, however, we retain the traditional division into three classes.

Traditional class Polychaeta: Polychaetes

Polychaetes include clamworms, plume worms, scaleworms, lugworms, twin-fan worms, sea mice, peacock worms, deep sea tubeworms, and many others. These worms are often surprisingly beautiful, with unusual forms and sometimes iridescent colors (figures 34.12, 34.14). Polychaetes are often a crucial part of marine food chains, because they are extremely abundant in certain habitats.

Some polychaetes live in tubes or permanent burrows of hardened mud, sand, mucuslike secretions, or calcium carbonate. These sedentary polychaetes project a set of feathery tentacles that sweep the water for food, making them primarily filter feeders. Other polychaetes are active swimmers, crawlers, or burrowers. Many are active predators and have powerful jaws.

Polychaetes have paired, fleshy, paddlelike lateral projections, called **parapodia,** on most of their segments. These parapodia, which bear bristle-like chaetae, are used in swimming, burrowing, or crawling. They can also play an important role in gas exchange because they greatly increase the surface area of the body. Some polychaetes that live in burrows or tubes may have parapodia featuring hooks to help anchor the worm; other parapodia may be transformed in gill-like structures.

Several species of deep-sea tubeworms are gutless as adults and are able to culture sulfur-oxidizing bacteria that allow the worm to grow enormously. These form the tubeworm fields in hydrothermal vents, as is the case in species of the genus *Riftia* (figure 34.15).

The sexes of polychaetes are usually separate, and fertilization is often external, occurring in the water and away from both parents. Unlike other annelids, polychaetes usually lack

figure 34.15

GIANT TUBEWORMS LIVING IN DEEP-SEA HYDROTHERMAL VENTS NEAR THE GALÁPAGOS. These tubeworms and associated animals (notice the small white crab) are an example of a community dependent on hydrogen sulfide, rather than the sun, as an energy source.

permanent gonads. They produce their gametes directly from germ cells in the lining of the coelom or in their septa.

Fertilization results in spiral cleavage followed by the production of ciliated, mobile trochophore larvae similar to the larvae of mollusks. The trochophores develop for long periods in the plankton before beginning to add segments from a posterior growth zone.

Traditional classes Oligochaeta and Hirudinea: Earthworms and leeches

Oligochaetes include earthworms and other forms of segmented worms that have evolved from within this group, such as the leeches. Leeches were formerly considered a separate class, the Hirudinea, but they are now considered a subcategory of Oligochaeta. We will discuss first the body plan of earthworms followed by that of the leeches.

Earthworms. The body of a typical earthworm consists of 100 to 175 similar segments, with a mouth on the first and an anus on the last. Earthworms seem to eat their way through the soil because they ingest soil by expanding their strong pharynx. Everything they ingest passes through their long, straight digestive tract. One region of this tract, the gizzard, grinds up the organic material with the help of soil particles. The material that passes through an earthworm is deposited outside the opening of its burrow in the form of *castings* that consist of irregular mounds. In this way, earthworms contribute to loosening, aeration, and enrichment of the soil.

In view of their underground lifestyle, it is not surprising that earthworms have no eyes. But earthworms do have numerous light-, chemo-, and touch-sensitive cells, mostly concentrated in segments near each end of the body—those regions most likely to encounter light or other stimuli. Earthworms have fewer setae than polychaetes, and they lack parapodia and a differentiated head region.

figure 34.14

A POLYCHAETE. The shiny bristleworm *Oenone fulgida.*

Earthworms are hermaphroditic, another way in which they differ from most polychaetes. When they mate (figure 34.16), their anterior ends point in opposite directions, and their ventral surfaces touch. The *clitellum* is a thickened band on an earthworm's body; the mucus it secretes holds the worms together during copulation. Sperm cells are released from pores in specialized segments of one partner into the sperm receptacles of the other, the process going in both directions simultaneously.

Two or three days after the worms separate, the clitellum of each worm secretes a mucus cocoon, surrounded by a protective layer of chitin. As this sheath passes over the female pores of the body—a process that takes place as the worm moves—it receives eggs and incorporates the deposited sperm. Fertilization of the eggs takes place within the cocoon. When the cocoon finally passes over the end of the worm, its ends pinch together. Within the cocoon, the fertilized eggs develop directly into young worms similar to adults.

Leeches. Leeches occur mostly in fresh water, although a few are marine and some tropical leeches occupy terrestrial habitats. Most leeches are 2–6 cm long, but one tropical species reaches up to 30 cm. Leeches are usually flattened dorsoventrally, like flatworms. They are hermaphroditic, and develop a clitellum during the breeding season; cross-fertilization is obligatory because they are unable to self-fertilize.

A leech's coelomic cavity is reduced and continuous throughout the body, not divided into individual segments as in the polychaetes and other oligochaetes. Leeches have evolved suckers at one or both ends of the body that they use for both locomotion and to attach to their prey. Those that have suckers at both ends move by attaching first one and then the other end to the substrate, looping along. Many species are also capable of swimming. Except for one species, leeches have no chaetae.

Leeches have evolved the ability to suck blood or other fluids from their hosts, an ability found in more than half of the known species of leeches. Many freshwater parasitic leeches remain on their hosts for long periods and suck the hosts' blood from time to time.

One of the best-known leeches is the medicinal leech, *Hirudo medicinalis* (figure 34.17). Individuals of this species are 10–12 cm long and have bladelike, chitinous jaws that rasp

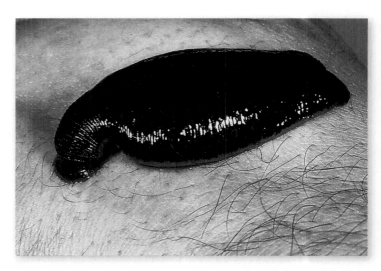

figure 34.17

A LEECH. *Hirudo medicinalis*, the medicinal leech, is seen here feeding on a human arm. Leeches use chitinous, bladelike jaws to make an incision to access blood and secrete an anticoagulant to keep the blood from clotting. Both the anticoagulant and the leech itself have made important contributions to modern medicine.

through the skin of the victim. The leech secretes an anticoagulant into the wound to prevent the blood from clotting, and its powerful sucking muscles pump the blood out quickly once the hole has been opened. Leeches were used in medicine for hundreds of years to suck blood out of patients whose diseases were mistakenly believed to be caused by an excess of blood. Today, European pharmaceutical companies still raise and sell leeches, but they are used to remove excess blood after certain surgeries or to keep blood from coagulating in severed appendages that have been reattached. Following the surgery, blood may accumulate because veins may function improperly and fail to circulate the blood. The accumulating blood "turns off" the arterial supply of fresh blood, and the tissue often dies. When leeches remove the excess blood, new capillaries form in about a week, and the tissues remain healthy.

To feed, leeches use vasodilators and powerful anesthetics to avoid being detected while piercing the skin. The anticoagulant and anesthetic properties of the leech bite are investigated by pharmaceutical companies. Those of you that have encoutered leeches in the wild will have noticed that they can go undetected while they take their blood meal, and most often they are only detected once they detach and the blood starts to flow from the leftover Y-shaped wound. Leeches detect their hosts by sensing gradients of carbon dioxide in the environment. In some tropical forests after a few seconds of stopping in a place one can see dozens of leeches approaching from all directions toward the center of a circle where the prey rests.

figure 34.16

EARTHWORMS MATING. The anterior ends are pointing in opposite directions.

> Annelids are a diverse group of coelomate animals characterized by serial segmentation. Each segment in the annelid body has its own excretory and locomotor elements and is connected to other segments by a common circulatory and nervous system.

The Lophophorates: Bryozoa and Brachiopoda

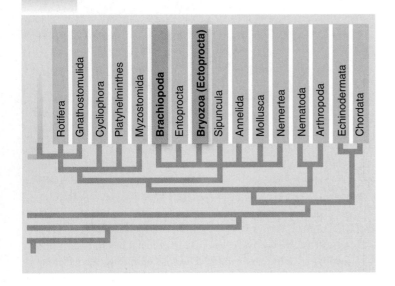

Two phyla of mostly marine animals—**Bryozoa** and **Brachiopoda**—are characterized by a **lophophore,** a circular or U-shaped ridge around the mouth bearing one or two rows of ciliated tentacles. Because of this unusual feature, they were thought to be related to one another, but recent data indicate that they belong to different branches of the animal tree of life.

Although the coelomic cavity of both bryozoans and brachiopods extends into the lophophore and its tentacles, the structures probably evolved convergently. The lophophore functions as a surface for gas exchange and as a food-collection organ. The cilia of the lophophore serve to capture the organic detritus and plankton on which the animals feed.

Brachiopods share some features with mollusks and annelids (protostomes), and share others with deuterostomes. Cleavage in both brachiopods and bryozoans is mostly radial, as in deuterostomes. The formation of the coelom varies. In the phoronids (once considered a phylum on their own but now considered part of Brachiopoda), the mouth forms from the blastopore, whereas in the rest of brachiopods and in the bryozoans, it forms from the end of the embryo opposite to the blastopore.

Molecular evidence shows that the groups of lophophorates are protostome-related, lending strength to the argument for placing them within the protostome phyla. But because of the discrepancies between anatomical and developmental characters with molecular-based phylogenies, their relationships continue to present a fascinating puzzle.

The bryozoans, phylum Bryozoa, are colonial and produce a chitinous chamber

Bryozoans (phylum **Bryozoa** or **Ectoprocta**) are small—usually less than 0.5 mm long—and live in colonies that look like patches of moss on the surfaces of rocks, seaweed, or other submerged objects. Their common name, bryozoans, means "moss-animals" (figure 34.18). The digestive system is U-shaped with the anus opening near the mouth, as in many sessile animals. The alternative name Ectoprocta refers to this location of the anus (proct), which is external to the lophophore.

The 4000 species of bryozoans include both marine and freshwater forms. Individual bryozoans secrete a tiny chitinous chamber called a **zoecium** that attaches to rocks or other substrates such as the leaves of marine plants and algae, and to other members of the colony since bryozoans are colonial organisms. Many can deposit calcium carbonate in the zoecium.

Colonies often have polyps specialized for different functions such as feeding, reproduction, or defense. Individuals in the colony communicate chemically through pores between zoecia. Bryozoans develop as deuterostomes, with the mouth developing secondarily and the anus originating from the blastopore; cleavage is radial. Asexual reproduction occurs frequently by budding.

figure 34.18

BRYOZOANS (PHYLUM BRYOZOA). *a.* This drawing depicts a small portion of a colony of the freshwater bryozoan genus *Plumatella*, which grows on the underside of rocks. The individual at the left has a fully extended lophophore. The tiny individuals disappear into their shells when disturbed. ***b.*** *Plumatella repens*, a freshwater bryozoan.

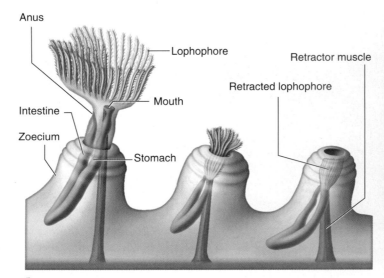

a.

b.

The brachiopods and phoronids, phylum Brachiopoda, are solitary lophophorates

Brachiopods, or lamp shells, superficially resemble clams because they have two calcified shells (figure 34.19). Instead of being lateral shells, as in bivalves, the valves are dorsal and ventral. Many species attach to rocks or sand by a stalk that protrudes through an opening in one shell. Others cement to the substrate and lack stalks. The gut is generally U-shaped as in bryozoans, and the lophophore lies within the shell and functions when the brachiopod's shells are opened slightly.

Although a little more than 300 species of brachiopods exist today, more than 30,000 species of this phylum are known as fossils. Because brachiopods were common in the Earth's oceans for millions of years and because their shells fossilize readily, they are often used as index fossils to define a particular time period or sediment type.

Once classified as a separate phylum, **phoronids** (figure 34.20) are now considered to be phylogenetically nested within the Brachiopoda. Each phoronid secretes a chitinous tube and lives out its life within it. They can extend their lophophore tentacles to feed and quickly withdraw them into the tube when disturbed, as some polychaete worms do.

Only about 10 phoronid species are known, ranging in length from a few millimeters to 30 cm. Some species lie buried in sand; others are attached to rocks, either singly or in groups, forming loose colonies. Phoronids, unlike brachiopods, develop as protostomes, with radial cleavage and the anus developing secondarily.

> The two phyla of lophophorates probably share a common ancestor, and they show a mixture of protostome and deuterostome characteristics.

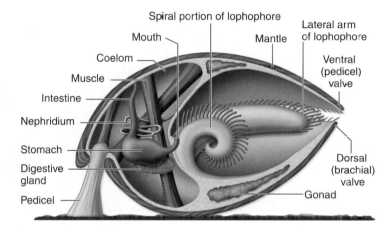

figure 34.19

BRACHIOPODS (PHYLUM BRACHIOPODA). *a.* All the body structures except the pedicel lay within two calcified shells, or valves. *b.* The brachiopod *Terebratulina septentrionalis* is shown here slightly opened so that the lophophore is visible.

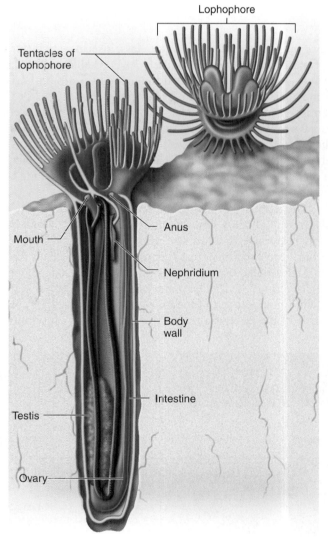

figure 34.20

PHORONIDS. A phoronid lives in a chitinous tube that the animal secretes. The lophophore consists of two parallel, horseshoe-shaped ridges of tentacles and can be withdrawn into the tube when the animal is disturbed.

Phylum Arthropoda: The Arthropods

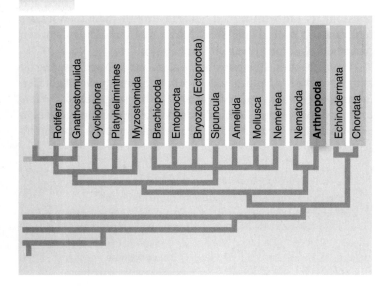

Arthropods—especially the largest class, insects—are by far the most successful of all animals (table 34.1). Well over 1,000,000 species—about two-thirds of all the named species on Earth—are members of the phylum **Arthropoda** (figure 34.21). One scientist recently estimated that the insect class alone may comprise as many as 30 million species. About 200 million individual insects are alive at any one time for each human! Insects and other arthropods (figure 34.22) abound in every habitat on the planet, but they especially dominate the land, along with flowering plants and vertebrates.

The majority of arthropod species consist of small animals, mostly about a few millimeters in length. Members of the phylum range in adult size from about 80 μm long (some parasitic mites) to 3 m across (the great Japanese spider crabs).

Arthropods, especially insects and mites, are of enormous economic importance and affect all aspects of human life. They compete with humans for food of every kind, play a key role in the pollination of certain crops, and cause billions of dollars of damage to crops, before and after harvest. They are by far the most important herbivores in all terrestrial ecosystems and are a valuable food source as well. Virtually every kind of plant is eaten by one or more species of insect. Diseases spread by insects cause enormous financial damage each year and strike every kind of domesticated animal and plant, as well as human beings.

Although the groups of arthropods and their relationships to other groups may shift with new findings from molecular data, taxonomists recognize four major classes of arthropods. In some systems, these classes may be differently grouped and considered to be separate phyla in the Ecdysozoan lineage. They include arachnids, myriapods, crustaceans, and insects.

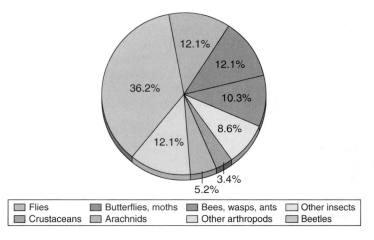

figure 34.21

ARTHROPODS ARE A SUCCESSFUL GROUP. About two-thirds of all named species are arthropods. From those, about 80% of all arthropods are insects, and about half of the named species of insects are beetles.

TABLE 34.1	Major Groups Within the Traditional Classification of the Phylum Arthropoda	
Group	**Characteristics**	**Members**
Arachnids	Mouthparts are chelicerae (pincers or fangs)	Spiders, mites, ticks, scorpions, daddy long-legs
Myriapods	Mouthparts are mandibles; bodies consist of a head and numerous body segments bearing paired uniramous ("single-branched") appendages; one pair of antennae	Centipedes, millipedes
Crustaceans	Mouthparts are mandibles (biting jaws); appendages are biramous ("two-branched"); two pairs of antennae	Lobsters, crabs, shrimps, isopods, barnacles
Insects	Mouthparts are mandibles; bodies consist of three regions: a head, a thorax, and an abdomen; appendages are uniramous; one pair of antennae	Beetles, bees, flies, fleas, true bugs, grasshoppers, butterflies, termites

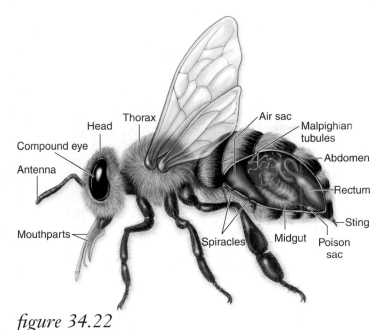

figure 34.22

PHYLUM ARTHROPODA. This bee, like all insects and other arthropods (phylum Arthropoda), has a segmented body and jointed appendages. An insect body is composed of three body regions: head, thorax, and abdomen. All arthropods have an exoskeleton made of chitin. Insects evolved wings that permit them to fly.

We consider each group separately later on; first, we review general features of arthropod morphology.

The arthropod body plan features jointed appendages and an exoskeleton

Part of the arthropods' success is explained by the modularity of their segmented body plan and the presence of *jointed appendages* and an *exoskeleton*. The presence of jointed appendages has allowed arthropods to develop many efficient modes of locomotion, both in the oceans where they originated and on land, which they conquered early in the Devonian period. Exoskeletons conferred special protection against the different predators that appeared during the Cambrian explosion.

Jointed appendages

The name *arthropod* means "jointed feet," and all arthropods have jointed appendages. The numbers of these appendages are reduced to only specific body regions in some members of the phylum. Individual appendages may be modified into antennae, mouthparts of various kinds, or legs. Some appendages, such as the wings of certain insects, are not homologous to the other appendages; rather, insect wings evolved separately.

Jointed appendages have advantages, one of which is that they can be extended and retracted because they are able to bend. Imagine how difficult life would be if your arms and legs could not bend. In addition, joints serve as a fulcrum, or stable point, for appendage movement, so *leverage* is possible. A small muscle force on a lever can produce a large movement; for example, extending your lower arm takes advantage of the fulcrum of the elbow. A small contraction distance in your muscles moves your

hand through a large arc. The same principle is at work in the jointed appendages of insects.

An exterior skeleton

The arthropod body plan has a second major innovation: a rigid external skeleton, or **exoskeleton,** made of secreted chitin and protein. In any animal, the skeleton provides places for muscle attachment. In arthropods, the muscles attach to the interior surface of the hard exoskeleton, which also protects the animal from predators and impedes water loss.

As you learned in chapter 3, chitin is chemically similar to cellulose, the dominant structural component of plants, and shares similar properties of toughness and flexibility. Together, the chitin and protein provide an external covering that is very strong but also capable of flexing in response to the contraction of muscles attached to it. In most crustaceans and millipedes, the exoskeleton is made even tougher, although less flexible, with deposits of calcium salts.

Exoskeletons have inherent limitations, however. An exoskeleton must be much thicker to bear the pull of the muscles in large insects than in small ones. That is why we don't see beetles as big as eagles, or crabs the size of a cow—the exoskeleton would be so thick that the animal couldn't move its great weight. As a result, few terrestrial arthropods weigh more than a few grams.

Arthropod bodies are segmented like those of annelids, although these phyla are not directly related. Members of some classes of arthropods have many body segments that look alike. In others, the segments have become specialized into functional groups, or **tagmata** (singular, *tagma*), such as the head, thorax, and abdomen of an insect. This process, known as *tagmatization,* is of central importance in the evolution of arthropods.

In some arthropods, different segments may fuse in the adult individuals, although the original segments can be always distinguished during larval development. All arthropods have a distinct head, sometimes fused with the thorax to form a tagma called the **cephalothorax** or **prosoma,** as is the case of many crustaceans (crabs, lobsters, shrimp) and chelicerates (a group that includes spiders).

Molting: Growth with an exoskeleton

Because the arthropod body is encased in a rigid skeleton, arthropods periodically undergo **ecdysis,** or molting, the shedding of the outer cuticular layer. This process is controlled by ecdysteroid hormones (chapter 46).

As arthropods grow, they form a new exoskeleton underneath the existing one. When the new exoskeleton is complete, it becomes separated from the old one by fluid. This fluid dissolves the chitin and protein (and calcium carbonate, if present) from the old exoskeleton. The fluid increases in volume until the original exoskeleton cracks open, usually along the back, and is shed. The arthropod emerges, clothed in a new, pale, and still somewhat soft exoskeleton. The arthropod then "puffs itself up," ultimately expanding to full size. The blood circulation to all parts of the body aids in this expansion, and many insects and spiders take in air to assist as well. The expanded exoskeleton subsequently hardens. While the exoskeleton is soft, the animal is especially vulnerable and it may hide from predators until the new exoskeleton hardens.

The compound eye

Another important structure in many arthropods is the **compound eye** (figure 34.23). Compound eyes are present in many insects, crustaceans, house centipedes, and in the extinct trilobites. They are composed of many independent visual units, often thousands of them, called **ommatidia** (singular, *ommatidium*). Each ommatidium is covered with a lens and includes a complex of eight retinular cells and a light-sensitive central core, or **rhabdom.**

Simple eyes, or **ocelli** (singular, *ocellus*) with single lenses are found in the other arthropod groups, and they sometimes occur together with compound eyes, as is often the case in insects. Ocelli distinguish light from darkness. The ocelli of some flying insects, namely locusts and dragonflies, function as horizon detectors and help the insect visually stabilize its course in flight.

Circulatory system

The circulatory system of arthropods is open; their blood flows through cavities between the internal organs and not through closed vessels. The principal component of an insect's circulatory system is a longitudinal, muscular vessel called the heart. This vessel runs near the dorsal surface of the thorax and abdomen (figure 34.24).

When the heart contracts, blood flows into the head region of the insect. When an insect's heart relaxes, blood returns to it through a series of valves located in the posterior region of the heart. These valves allow the blood to flow inward only. Thus, blood from the head and other anterior portions of the insect gradually flows through the spaces between the tissues toward the posterior end and then back through the one-way valves into the heart.

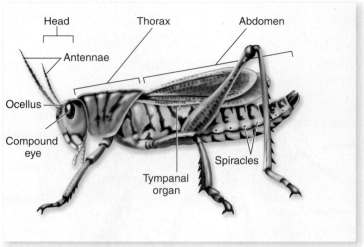

a.

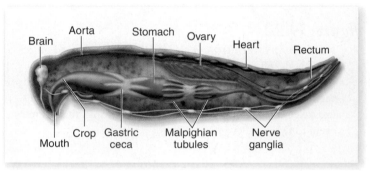

b.

figure 34.24

A GRASSHOPPER (ORDER ORTHOPTERA). This grasshopper illustrates the major structural features of the insects, the most numerous group of arthropods. *a.* External anatomy. *b.* Internal anatomy.

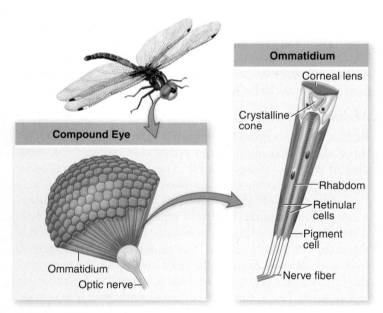

figure 34.23

THE COMPOUND EYE. The compound eyes in insects are complex structures composed of many independent visual units called ommatidia.

Nervous system

The central feature of the arthropod nervous system is a double chain of segmented ganglia running along the animal's ventral surface. At the anterior end of the animal are three fused pairs of dorsal ganglia, which constitute the brain; however, much of the control of an arthropod's activities is relegated to ventral ganglia (generally a pair per segment). Therefore, the animal can carry out many functions, including eating, moving, and copulating, even if the brain has been removed. The brain of arthropods seems to be a control point, or inhibitor, for various actions, rather than a stimulator, as it is in vertebrates.

Respiratory system

Insects depend on their respiratory system rather than their circulatory system to carry oxygen to their tissues, although this is not necessarily the case in other arthropods. Therefore, all parts of the body need to be near a respiratory passage to obtain oxygen. Along with the thickness of their chitin exoskeletons, this feature places severe limitations on arthropod size compared to that of vertebrates.

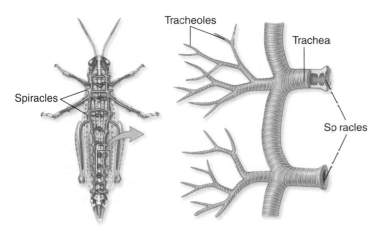

figure 34.25

TRACHEAE AND TRACHEOLES. Tracheae and tracheoles are connected to the exterior by specialized openings called spiracles and carry oxygen to all parts of a terrestrial insect's body.

Unlike most animals, the arthropods have no single major respiratory organ. The respiratory system of most terrestrial arthropods consists of small, branched, cuticle-lined air ducts called **tracheae** (singular, *trachea*) (figure 34.25). These tracheae, which ultimately branch into very small **tracheoles,** are a series of tubes that transmit oxygen throughout the body. Tracheoles are in direct contact with individual cells, and oxygen diffuses directly across the plasma membranes.

Air passes into the tracheae by way of specialized openings in the exoskeleton called **spiracles,** which, in most insects, can be opened and closed by valves. The ability to prevent water loss by closing the spiracles was a key adaptation that facilitated the arthropod invasion of land. Marine arthropods such as crustaceans have respiratory gills, and many chelicerates (horseshoe crabs, scorpions, and spiders) have so-called *book lungs* and lunglike structures. Yet some tiny arthropods lack any structures for exchanging oxygen, and their skin or gut have a respiratory function.

Excretory system

Although various kinds of excretory systems are found in different groups of arthropods, we focus here on the unique excretory system consisting of **Malpighian tubules,** which evolved in terrestrial insects, myriapods, and arachnids. Malpighian tubules are slender projections from the digestive tract that are attached at the junction of the midgut and hindgut (see figure 50.9). Fluid passes through the walls of the Malpighian tubules to and from the blood in which the tubules are bathed. As this fluid passes through the tubules toward the hindgut, nitrogenous wastes are precipitated as concentrated uric acid or guanine. These substances are then emptied into the hindgut and eliminated.

Most of the water and salts in the fluid are reabsorbed by the hindgut and rectum and returned to the arthropod's body. Malpighian tubules are an efficient mechanism for water conservation and were another key adaptation facilitating invasion of the land by arthropods.

Arachnids have specialized anterior appendages, the chelicerae

The largely terrestrial arthropod class, **Arachnida,** with some 57,000 named species, occupies a distinct evolutionary line of arthropods in which the most anterior appendages, called **chelicerae** (singular, *chelicera*), often function as fangs or pincers. Arachnids include familiar arthropods such as spiders, ticks, mites, scorpions, and daddy long-legs, and they have the body divided into two main body regions, or tagmata. The anterior body region is called the prosoma and bears all the appendages. Arachnids have a pair of chelicerae, a pair of pedipalps, and four pairs of walking legs in their prosoma. The posterior body region, or **opisthosoma,** contains the reproductive organs.

The **pedipalps** (often simply called palps), found posteriorly to the chelicerae, resemble legs but have one less segment, and they are not used for locomotion. In male spiders, the pedipalps are specialized copulatory organs. In scorpions, they are large pincers. In most other arachnids the pedipalps have a sensorial function, acting like the antennae of other arthropods.

Most arachnids are carnivorous. The main exception is mites, which are largely herbivorous. Most arachnids can ingest only preliquified food, which they often digest externally by secreting enzymes into their prey. They can then suck up the digested material with their muscular, pumping pharynx. Daddy long-legs, however, can ingest food as little particles. Arachnids are primarily, but not exclusively, terrestrial. Some 4000 known species of mites and one species of spider live in freshwater habitats, and a few mites live in the sea.

Many spiders have a unique respiratory system that involves **book lungs,** a series of leaflike plates within a chamber. Air is drawn into and expelled out of this chamber by muscular contraction. Book lungs may exist alongside tracheae, or they may function instead of tracheae.

Horseshoe crabs are close relatives to arachnids, and both groups are classified as Chelicerata due to the common presence of chelicerae. Horseshoe crabs can be extremely abundant in some areas and comprise four species that live near the North American Atlantic coast and in Southeast Asia.

Order Araneae: Spiders

There are about 35,000 named species of spiders (order Araneae). These animals play a major role in virtually all terrestrial ecosystems. They are particularly important as predators of insects and other small animals. Spiders hunt their prey or catch it in silk webs of remarkable diversity. The silk is formed from a fluid protein that is forced out of **spinnerets** on the posterior portion of the spider's abdomen. The webs and habits of spiders are often distinctive.

Many kinds of spiders, including the familiar wolf spiders and tarantulas, do not spin webs but instead hunt their prey actively. Others, called trap-door spiders, construct silk-lined burrows with lids, seizing their prey as it passes by.

All spiders have poison glands leading through their chelicerae, which are pointed and used to bite and paralyze prey. The bites of some members of this order, such as the western black widow (*Latrodectus hesperus*) and brown recluse (*Loxosceles*

Black Widow

Brown Recluse

a.   *b.*

figure 34.26

TWO COMMON POISONOUS SPIDERS. *a.* The southern black widow, *Latrodectus mactans.* *b.* The brown recluse spider, *Loxosceles reclusa.* Both species are common throughout temperate and subtropical North America, but bites are rare in humans.

reclusa) spiders (figure 34.26), can be fatal to humans and other large mammals.

Order Acari: Mites and ticks

The order Acari, the mites and ticks, is the largest in terms of number of species and the most diverse of the arachnids. Although only about 30,000 species of mites and ticks have been named, scientists that study the group estimate that more than a million members of this order may exist. Mites and ticks are found in virtually every habitat and feed on a variety of organisms as predators and parasites.

Most mites are small, less than 1 mm long, but adults of different species range from 100 nm to 2 cm. In most mites, the cephalothorax and abdomen are fused into an unsegmented, ovoid body. Respiration occurs either by means of tracheae or directly through the body surface. Many mites pass through

several distinct stages during their life cycle. In most, an inactive eight-legged prelarva gives rise to an active six-legged larva, which in turn produces a succession of three eight-legged stages and, finally, the adult males and females.

Ticks are blood-eating parasites that attach to the surface of their host. They are larger than most other Acari and cause discomfort by sucking the blood of humans and other animals. Ticks can carry many diseases, including some caused by viruses, bacteria, and protozoa. The spotted fevers (for example, Rocky Mountain spotted fever) are caused by bacteria carried by ticks, and Lyme disease is caused by spirochaetes transmitted by ticks. Red-water fever, or Texas fever, is an important tick-borne protozoan disease of cattle, horses, sheep, and dogs.

Centipedes and millipedes are segmented and have a large number of legs

The centipedes (class **Chilopoda**) and millipedes (class **Diplopoda**) both have bodies that consist of a head region followed by numerous segments, all more or less similar and nearly all bearing one or two sets of paired appendages.

Although the name *centipede* would imply an animal with 100 legs and the name *millipede* one with 1000, adult centipedes usually have fewer than 100 legs (most centipedes have 15, 21, or 23 pairs of legs); adult millipedes never reach 1000 legs, and most have 100 or fewer. Centipedes have one pair of legs on each body segment (figure 34.27*a*), and millipedes have two on some or all segments (figure 34.27*b*). Each segment of a millipede is a tagma that originated during the group's evolution when two ancestral segments fused. This explains why millipedes have twice as many legs per segment as centipedes.

In both centipedes and millipedes, fertilization is internal and takes place by direct transfer of sperm. The sexes are separate, and all species lay eggs. Young millipedes usually hatch with three pairs of legs; they experience a number of growth stages, adding segments and legs as they mature, but do not change in general appearance. Centipedes have different types of development, and some species hatch with their final number of legs whereas others keep adding legs after hatching. Centipedes that

Centipede

Millipede

a.   *b.*

figure 34.27

MYRIAPODS. *a.* Centipedes, such as this member of the genus *Scolopendra*, are active predators. *b.* Millipedes, such as this member of the genus *Sigmoria*, are important herbivores and detritivores. Centipedes have one pair of legs per body segment, millipedes have two pairs per segment.

don't add legs as they grow tend to take care of their young, a behavior rather uncommon among invertebrates.

Centipedes, of which some 2500 species are known, are all carnivorous and feed mainly on insects. The appendages of the first trunk segment are modified into a pair of poison fangs. The poison is often quite toxic to humans, and many centipede bites are extremely painful, although never fatal.

In contrast, most millipedes are herbivores, feeding mainly on decaying vegetation such as leaf litter, rotting logs, or in the soil, although a few species are carnivorous. Many millipedes can roll their bodies into a flat coil or sphere as a defense mechanism. More than 10,000 species of millipedes have been named, but this is estimated to be no more than one-sixth of the actual number of species that exists.

In each segment of their body, many millipedes have a pair of complex glands that produce a bad-smelling fluid. This fluid is exuded for defensive purposes out through openings along the sides of the body. The chemistry of the external secretions of different millipedes has become a subject of considerable interest because of the chemical diversity of the compounds involved and their effectiveness in protecting millipedes from attack. Some species produce cyanide gas from segments near their head end. Millipedes live primarily in damp, protected places, such as leaf litter, in rotting logs, under bark or stones, or in the soil.

Crustaceans are mostly aquatic and have biramous (branched) appendages

The crustaceans (class **Crustacea**, which in some systems is considered a subphylum) are a large group of primarily aquatic organisms, consisting of some 35,000 species of crabs, shrimps, lobsters, crayfish, barnacles, water fleas, pillbugs, and related groups.

Crustacean body plans

Typical crustaceans have three tagmata, and the two most anterior ones can fuse to form a cephalothorax. Most crustaceans have two pairs of antennae, three pairs of appendages for chewing and manipulating food, and various pairs of legs. Most crustacean appendages, with the possible exception of the first pair of antennae, are basically biramous ("two-branched"). In some crustaceans, appendages appear to have only a single branch; in those cases, one of the branches has been lost during the course of evolutionary specialization.

Crustaceans differ from insects but resemble centipedes and millipedes in that they have appendages on their abdomen as well as on their thorax. They are the only arthropods with two pairs of antennae. Their **mandibles** (biting jaws) likely originated from a pair of limbs that took on a chewing function during the course of evolution, a process that apparently occurred once in the common ancestor of myriapods, crustaceans, and insects.

Large crustaceans have feathery gills for respiration near the bases of their legs. In smaller members of this class, gas exchange takes place directly through the thinner areas of the cuticle or the entire body. The oxygen extracted from the gills is distributed through the circulatory system.

Crustacean reproduction

Most crustaceans have separate sexes. Many different kinds of specialized copulation occur among the crustaceans, and the members of some orders carry their eggs with them, either singly or in egg pouches, until they hatch. The majority of crustaceans develop through a **nauplius** stage (figure 34.28), which provides evidence that all members of this diverse group descend from a common ancestor.

The nauplius hatches with three pairs of appendages, and it undergoes metamorphosis through several stages before reaching maturity. In many groups, this nauplius stage is passed in the egg, and development of the hatchling to the adult form is direct.

Habitats

Although most crustaceans are marine, many occur in freshwater habitats, and a few have become terrestrial. These last include pillbugs and sowbugs, the terrestrial members of a large order of crustaceans known as the isopods (order Isopoda). About half of the estimated 4500 species of this order are terrestrial and live primarily in places that are moist, at least seasonally. Sand fleas or beach fleas (order Amphipoda) are other familiar crustaceans, many of which are semiterrestrial (intertidal) species.

Along with the larvae of larger species, minute crustaceans are abundant in the plankton as well as in the interstices of sand grains. Especially significant are the tiny copepods (order Copepoda; figure 34.29), which are among the most abundant multicellular organisms on Earth.

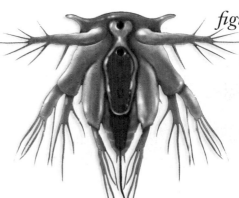

figure 34.28

THE NAUPLIUS LARVA. Although crustaceans are diverse, they have fundamentally similar developmental stages. The nauplius of a crustacean is an important unifying feature found in most members of this group.

figure 34.29

FRESHWATER CRUSTACEAN. A copepod with attached eggs. Copepods are members of a group of marine and freshwater crustaceans (order Copepoda), important components of plankton, most of which are a few millimeters long.

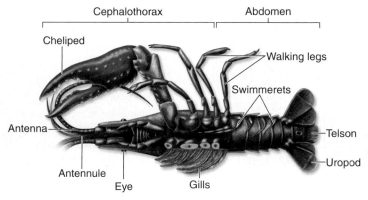

figure 34.30

DECAPOD CRUSTACEAN. Ventral view of a lobster, *Homarus americanus*, with its principal features labeled.

figure 34.31

GOOSENECK BARNACLES, *Lepas anatifera*, FEEDING. These are stalked barnacles; others lack a stalk.

Decapod crustaceans: Shrimps, lobsters, crabs, and crayfish

Large, primarily marine crustaceans such as shrimps, lobsters, and crabs, along with their freshwater relatives, the crayfish, are *decapod crustaceans* (figure 34.30). The term *decapod* means "ten-footed." In these animals, the exoskeleton is usually reinforced with calcium carbonate. Most of their body segments are fused into a cephalothorax covered by a dorsal shield, or carapace, which arises from the head. The crushing pincers common in many decapod crustaceans are used in obtaining food—for example, by crushing mollusk shells.

In lobsters and crayfish, appendages called **swimmerets** occur in lines along the ventral surface of the abdomen and are used in reproduction and swimming. In addition, flattened appendages known as **uropods** form a kind of compound "paddle" at the end of the abdomen. These animals may also have a **telson,** or tail spine. By snapping its abdomen, the animal can propel itself through the water rapidly and forcefully.

Crabs differ from lobsters in several ways, one of which is that their carapace is much larger and broader, and the abdomen is tucked under it.

Sessile crustaceans: barnacles

Barnacles (order Cirripedia; figure 34.31) are a group of crustaceans that are sessile as adults. Barnacles have free-swimming larvae that ultimately attach their heads to a piling, rock, or other submerged or floating object and then stir food into their mouths with their feathery legs. Calcareous plates protect the barnacle's body, and these plates are usually attached directly and solidly to the substrate. Although most crustaceans have separate sexes, barnacles are hermaphroditic, but they generally cross-fertilize.

Insects exhibit vast diversity and exist in huge numbers

The insects are members of the class **Hexapoda** and are by far the largest group of animals on Earth, whether measured in terms of numbers of species or numbers of individuals. Insects live in every conceivable habitat on land and in fresh water, and a few have even invaded the sea. More than half of all the named animal species are insects, and the actual proportion is doubtless much higher because millions of additional forms await detection, classification, and naming.

Approximately 90,000 described species occur in the United States and Canada, and the actual number of species in this region probably approaches 125,000. A hectare of lowland tropical forest is estimated to be inhabited by as many as 41,000 species of insects, and many suburban gardens may have 1500 or more species. It has been estimated that approximately a billion billion (10^{18}) individual insects are alive at any one time. A glimpse at the enormous diversity of insects is presented in figure 34.32 and also later in table 34.2.

Order: Lepidoptera

a.

Order: Homoptera

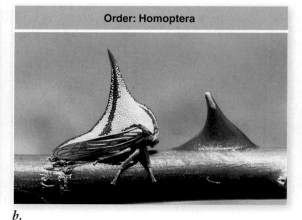

b.

Order: Coleoptera

c.

		TABLE 34.2	**Major Orders of Insects**	

Order	Typical Examples		Key Characteristics	Approximate Number of Named Species
Coleoptera	Beetles		The most diverse animal order; two pairs of wings; front pair of wings is hard and protects the rear pair of flying wings; heavily armored exoskeleton; biting and chewing mouthparts; complete metamorphosis	350,000
Diptera	Flies		Some that bite people and other mammals are considered pests; front flying wings are transparent; hindwings are reduced to knobby balancing organs; sucking, piercing, or lapping mouthparts; complete metamorphosis	120,000
Lepidoptera	Butterflies, moths		Two pairs of broad, scaly, flying wings, often brightly colored; hairy body; tubelike, sucking mouthparts; complete metamorphosis	120,000
Hymenoptera	Bees, wasps, ants		Often social; two pairs of transparent flying wings; mobile head and well-developed compound eyes; often possess stingers; chewing and sucking mouthparts; complete metamorphosis	100,000
Hemiptera and Homoptera	True bugs, bedbugs, leafhoppers, aphids, cicadas		Often live on blood; some are plant-eaters; two pairs of wings or wingless; piercing, sucking mouthparts; simple metamorphosis	60,000
Orthoptera	Grasshoppers, crickets		Third pair of legs modified for jumping; two pairs of wings or wingless; among the largest insects; biting and chewing mouthparts in adults. simple metamorphosis.	20,000
Odonata	Dragonflies		Primitive insect order; two pairs of transparent flying wings that do not fold backwards; large, long, and slender body; chewing mouthparts; simple metamorphosis	5,000
Isoptera	Termites		One of the few types of animals able to eat wood; two pairs of wings, but some stages are wingless; social insects; there are several body types with division of labor; chewing mouthparts; simple metamorphosis	2,000
Siphonaptera	Fleas		Small, known for their irritating bites; wingless; small, flattened body with jumping legs; piercing and sucking mouthparts; complete metamorphosis	1,200

figure 34.32

INSECT DIVERSITY. *a.* Luna moth, *Actias luna.* Luna moths and their relatives are among the most spectacular insects (order Lepidoptera). *b.* A thorn-shaped treehopper, *Umbonia crassicornis* (order Homoptera). *c.* Boll weevil, *Anthonomus grandis.* Weevils are one of the largest groups of beetles (order Coleoptera). *d.* Soldier fly, *Ptecticus trivittatus* (order Diptera). *e.* Lubber grasshopper, *Romalea guttata* (order Orthoptera). *f.* Termites, like ants, have several castes with individuals specialized for different tasks. The individual on the left is a soldier; its large jaws aid in defending the colony. These Peruvian termites belong to a species in the *Macrotermes.*

Order: Diptera

Order: Orthoptera

e.

Order: Isoptera

f.

figure 34.33

MODIFIED MOUTHPARTS IN THREE KINDS OF INSECTS. Mouthparts are modified for **(a)** piercing in this mosquito of the genus *Culex;* **(b)** sucking nectar from flowers in the alfalfa butterfly of the genus *Colias;* and **(c)** sopping up liquids in the housefly, *Musca domestica.*

a.

b.

c.

External features

Insects are primarily a terrestrial group, and most, if not all, of the aquatic insects probably had terrestrial ancestors. Most insects are relatively small, ranging from 0.1 mm to about 30 cm in length or wingspan. Insects have three body regions, the head, thorax, and abdomen; three pairs of legs, all attached to the thorax; and one pair of antennae. In addition, they may have one or two pairs of wings. Insect mouthparts all have the same basic structure but are modified in different groups in relation to their feeding habits (figure 34.33). Most insects have compound eyes, and many have ocelli as well. The insect thorax consists of three segments, each with a pair of legs. Legs are completely absent in the larvae of certain groups—for example, in most flies (order

figure 34.34

LARVAE OF A MOSQUITO, *Culex pipiens.* The aquatic larvae of mosquitoes are quite active. They breathe through tubes at the surface of the water, as shown here. Covering the water with a thin film of oil causes them to drown.

Diptera) and mosquitoes (figure 34.34). The two pairs of wings attach to the middle and posterior segments of the thorax. The thorax is almost entirely filled with muscles that operate the legs and wings.

The wings of insects arise as saclike outgrowths of the body wall. In adult insects, the wings are solid except for the veins. Insect wings are not homologous to the other appendages. Basically, insects have two pairs of wings, but in some groups, such as flies, the second set has been reduced to a pair of balancing knobs, called halteres, during the course of evolution. Most insects can fold their wings over their abdomen when they are at rest; but a few, such as the dragonflies and damselflies (order Odonata), keep their wings erect or outstretched at all times.

Insect forewings may be tough and hard, as in beetles. If they are, they form a cover for the hindwings and usually open during flight. The tough forewings also serve a protective function in the order Orthoptera, which includes grasshoppers and crickets. The wings of insects are made of sheets of chitin and protein; their strengthening veins are tubules of chitin and protein. Moths and butterflies have wings covered with detachable scales that provide most of their bright colors (figure 34.35). In some wingless hexapods, such as the springtails or silverfish, wings never evolved. Other wingless groups, such as fleas and lice, are derived from ancestral groups of insects that had wings and therefore have lost the wings secondarily.

Internal organization

The internal features of insects resemble those of the other arthropods in many ways. The digestive tract is a tube, usually somewhat coiled. It is often about the same length as the body. However, in leafhoppers, cicadas, and related groups, and in many flies, the digestive tube may be greatly coiled and several times longer than the body. Such long digestive

figure 34.35

SCALES ON THE WING OF *Parnassius imperator*, A BUTTERFLY FROM CHINA. Scales of this sort account for most of the colored patterns on the wings of butterflies and moths.

tracts are generally found in insects that have sucking mouthparts and feed on juices rather than on protein-rich solid foods.

The anterior and posterior regions of an insect's digestive tract are lined with cuticle. Digestion takes place primarily in the stomach, or midgut, and excretion of circulating wastes takes place through Malpighian tubules. Digestive enzymes are mainly secreted from the cells that line the midgut, although some are contributed by the salivary glands near the mouth.

The tracheae of insects extend throughout the body and permeate the different tissues. In many winged insects, the tracheae are dilated in various parts of the body, forming air sacs. These air sacs are surrounded by muscles and form a kind of bellows system to force air deep into the tracheal system. The spiracles through which air enters the tracheal system, a maximum of 10 on each side of the insect, are paired and located on or between the segments along the sides of the thorax and abdomen. In most insects, the spiracles can be opened by muscular action. In some parasitic and aquatic groups of insects, the spiracles are permanently closed. In these groups, the tracheae run just below the surface of the insect, and gas exchange takes place by diffusion.

Sensory receptors

In addition to the eyes, insects have several characteristic kinds of sense receptors. These include **sensory setae,** hairlike structures which are usually widely distributed over their bodies. The sensory setae are linked to nerve cells and are sensitive to mechanical and chemical stimulation. They are particularly abundant on the antennae and legs—the parts of the insect most likely to come into contact with other objects.

Sound, which is of vital importance to insects, is detected by tympanal organs in groups such as grasshoppers and crickets, cicadas, and some moths. These organs are paired structures composed of a thin membrane, the **tympanum,** associated with the tracheal air sacs. In many other groups of insects, sound waves are detected by sensory hairs. Male mosquitoes use thousands of sensory hairs on their antennae to detect the sounds made by the vibrating wings of female mosquitoes.

In addition to using sound, nearly all insects communicate by means of chemicals or mixtures of chemicals known as **pheromones.** These extremely diverse compounds are sent forth into the environment, where they convey a variety of messages, including mating signals and trail markers.

Insect life histories

During the course of their development, many insects undergo **metamorphosis.** In *simple metamorphosis,* seen in grasshoppers, immature stages are quite similar to adults, but gradually get bigger and more developed over a series of molts. In *complete metamorphosis,* seen in moths and butterflies, immature larva are often wormlike and active in feeding. A resting stage, during which the insect is called a **pupa** or **chrysalis,** immediately precedes the final molt into the adult form.

Arthropods are segmented animals with jointed appendages. All arthropods have a rigid exoskeleton of chitin and protein, which requires molting (ecdysis) for an individual to grow in size. Arthropods also have an open circulatory system. Insects have many adaptations for terrestrial living.

Scorpions, spiders, and mites are all arachnids. Their body is divided into prosoma and opisthosoma. The prosoma carries a pair of chelicerae, a pair of pedipalps, and four pairs of walking legs. Centipedes are segmented hunters with one pair of legs on each segment. Millipedes are segmented herbivores with two pairs of legs on most segments.

Crustaceans include marine, freshwater, and terrestrial forms and exhibit great diversity of forms. Most possess a nauplius larval stage and branched appendages. Decapod crustaceans include shrimps, lobsters, crabs, and crayfish. Barnacles are sessile crustaceans.

All insects possess three body segments (tagmata): the head, the thorax, and the abdomen. The three pairs of legs and two pairs of wings are attached to the thorax; the abdomen lacks appendages. Most insects have compound eyes, ocelli, and other sophisticated means of sensing their environment. Many insects undergo simple or complete metamorphosis.

Phylum Echinodermata: The Echinoderms

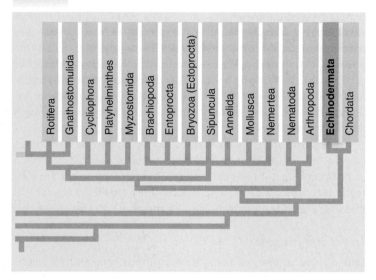

The mollusks, annelids, and arthropods discussed so far are protostomes. Members of phylum **Echinodermata,** described in this section, are characterized by *deuterostome development* and an **endoskeleton.** The term *echinoderm* means "spiny skin," referring to the endoskeleton of hard, calcium-rich plates just beneath the delicate skin (figure 34.36a).

Echinoderms undergo deuterostome development and have an endoskeleton

Echinoderms are an ancient group of marine animals that appeared nearly 600 MYA and that comprise about 6000 living species. As with other deuterostomes, their mouth forms secondarily during embryonic development.

When the endoskeletal plates of echinoderms first form, they are enclosed in living tissue and so truly constitute an endoskeleton—although in the adults of some of the best known groups of echinoderms the plates frequently fuse, forming a hard shell.

Another innovation in echinoderms is the development of a hydraulic system to aid in movement or feeding. Called a **water-vascular system,** this fluid-filled system is composed of a central ring canal from which five radial canals extend out into the body and arms.

Some of the most familiar animals seen along the seashore, sea stars (starfish), brittle stars, sea urchins, sand dollars, and sea cucumbers, are echinoderms. All are radially symmetrical as adults, although their radial symmetry is an unusual one that is based on the presence of five axes of symmetry, and therefore they are said to have **pentaradial symmetry.** Although some other kinds of animals are radially symmetrical, none have the complex organ systems of adult echinoderms.

Even though an excellent fossil record is available extending back into the Cambrian period, the origin of echinoderms remains unclear. They are thought to have evolved from bilaterally symmetrical ancestors because echinoderm larvae are bilaterally symmetrical. The radial symmetry that is the hallmark of echinoderms develops later, in the adult body.

The echinoderm body plan is bilateral in the larvae but pentaradial in adults

The body plan of echinoderms undergoes a fundamental shift during development, from bilateral symmetry to radial symmetry. Because of echinoderms' radially symmetrical bodies, the terms used to describe a bilaterally symmetrical animal's body are not applicable: Dorsal, ventral, anterior, and posterior have no meaning without a head or tail. Instead, the body structure of echinoderms is discussed in reference to their mouths, which define their **oral surface.**

Most echinoderms crawl along on their oral surfaces, although in sea cucumbers, the animal's axis lies horizontally, and they crawl with the oral surface in front. In sea lilies and feather stars, the oral surface is located opposed to the substrate, contrary to most other echinoderms.

The nervous systems consist of a central **nerve ring** from which branches arise. The animals are capable of complex response patterns, but there is no centralization of function.

The endoskeleton

Echinoderms have a delicate epidermis containing thousands of neurosensory cells that is stretched over an endoskeleton composed of either movable or fixed calcium-rich (calcite) plates called **ossicles.** In some echinoderms, such as sea stars and sea cucumbers, the ossicles are widely scattered, and the body wall is flexible. In others, especially the echinoids (sea urchins and sand dollars), the ossicles become fused and form a rigid shell. In many cases, these plates bear spines.

Another important feature of this phylum is the presence of mutable collagenous tissue, which can range in texture from tough and rubbery to weak and fluid. This amazing tissue accounts for many of the special attributes of echinoderms, such as the ability to rapidly autotomize (cut off) body parts. A sea star can lose an arm if necessary for survival and grow a new one. This tissue is also responsible for the different textures observed in sea cucumbers that can change from almost rigid to flaccid in a matter of seconds.

The plates in certain portions of the body of some echinoderms are perforated by pores. Through these pores extend **tube feet,** part of the water-vascular system that is a unique feature of this phylum.

The water-vascular system

The water-vascular system of an echinoderm radiates from a ring canal that encircles the animal's esophagus. Five **radial canals,** their positions determined early in the development of the embryo, extend into each of the five parts of the body and determine its basic symmetry (figure 34.36).

Water enters the water-vascular system through a **madreporite,** a sievelike plate on the animal's surface, and flows to the ring canal through a tube, or *stone canal,* so named because of the surrounding rings of calcium carbonate. The five radial canals in turn extend out through short side branches into the hollow tube feet (figure 34.36b). In some echinoderms, each tube foot has a sucker at its end; in others, suckers are absent.

At the base of each tube foot is a muscular sac, the **ampulla,** which contains fluid. When the ampulla contracts, the fluid is prevented from entering the radial canal by a one-way valve and is forced into the tube foot, thus extending it. When extended, the foot can attach itself to the substrate or to other objects. Longitudinal muscles in the tube foot wall then contract, causing the tube feet to bend. Relaxation of the muscles in the ampulla allows the fluid to flow back into the ampulla, which pulls the body toward the foot. By the concerted action of a very large number of small, individually weak tube feet, the animal can move across the seafloor.

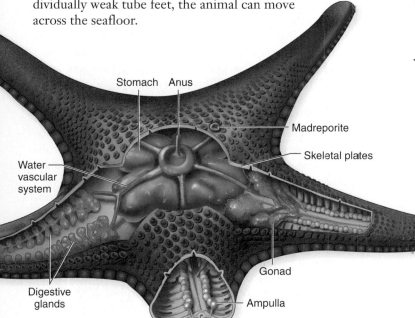

a.

b.   Papulae

figure 34.36

PHYLUM ECHINODERMATA. *a.* Echinoderms, such as sea stars (class Asteroidea), are coelomates with a deuterostomate pattern of development and an endoskeleton made of calcium-rich plates. The water-vascular system of an echinoderm is shown in detail. Radial canals allow water to flow into the tube feet. As the ampulla in each tube foot contracts, the tube extends and can attach to the substrate. Subsequently, muscles in the tube feet contract, bending the tube foot and pulling the animal forward. *b.* Extended nonsuckered tube feet of the sea star *Ludia magnifica.*

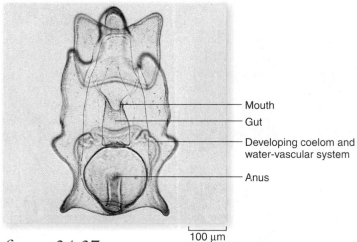

100 μm

figure 34.37

THE FREE-SWIMMING LARVA OF THE COMMON STARFISH, *Asterias rubens.* Such bilaterally symmetrical larvae suggest that the ancestors of the echinoderms were not radially symmetrical, like the living members of the phylum.

Sea cucumbers usually have five rows of tube feet on the body surface that are used in locomotion. They also have modified tube feet around their mouth cavity that are used in feeding. In sea lilies, the tube feet arise from the branches of the arms, which extend from the margins of an upward-directed cup. With these tube feet, the animals take food from the surrounding water. In brittle stars, the tube feet are pointed and specialized for feeding.

Body cavity

In echinoderms, the coelom, which is proportionately large, connects with a complicated system of tubes and helps provide circulation and respiration. In sea stars, respiration and waste removal occur across the skin through small, fingerlike extensions of the coelom called *papulae* (see figure 34.36). They are covered with a thin layer of skin and protrude through the body wall to function as gills.

Regeneration and reproduction

Many echinoderms are able to regenerate lost parts, and some, especially sea stars and brittle stars, drop various parts when under attack (autotomy). In a few echinoderms, asexual reproduction takes place by splitting, and the broken parts of sea stars can sometimes regenerate whole animals. Some of the smaller brittle stars, especially tropical species, regularly reproduce by breaking into two equal parts; each half then regenerates a whole animal.

Despite the ability of many echinoderms to break into parts and regenerate, most reproduction in the phylum is sexual and external. The sexes in most echinoderms are separate, although there are few external differences between both sexes. The gametes are generally released into the water column, where fertilization occurs. Fertilized eggs of echinoderms usually develop into free-swimming, bilaterally symmetrical larvae (see figure 34.37), which differ considerably from the trochophore larvae of spiralian lophotrochozoans such as mollusks and annelids. These

Class: Asteroidea

a.

Class: Holothuroidea

b.

Class: Echinoidea

c.

Class: Crinoidea

d.

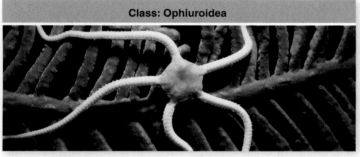

Class: Ophiuroidea

e.

figure 34.38

DIVERSITY IN ECHINODERMS. *a.* Sea star, *Oreaster occidentalis* (class Asteroidea), in the Gulf of California, Mexico. *b.* Warty sea cucumber, *Parastichopus parvimensis* (class Holothuroidea), Philippines. *c.* Tropical sea urchin of the genus *Echinometra* (class Echinoidea). *d.* Feather star (class Crinoidea), *Comatheria*, from Indonesia. *e.* Gaudy brittle star, *Ophioderma ensiferum* (class Ophiuroidea).

larvae develop in the plankton until they metamorphose through a series of stages into the more sedentary adults.

Echinoderms have five extant classes

There are more than 20 extinct classes of echinoderms and an additional five with living members: (1) Crinoidea, sea lilies and feather stars; (2) Asteroidea, sea stars, or starfish and sea daisies; (3) Ophiuroidea, brittle stars; (4) Echinoidea, sea urchins and sand dollars; and (5) Holothuroidea, sea cucumbers (figure 34.38). All the classes of echinoderms exhibit a basic five-part body plan, although the basic symmetry of the phylum is not obvious in some. Even sea cucumbers, which are shaped like their plant namesake, have five radial grooves running along their bodies. We describe three of the living classes here. Two of

the remaining classes, **Crinoidea** (sea lilies and feather stars) and **Holothuroidea** (sea cucumbers) have been briefly discussed in the preceding sections.

Class Asteroidea: Sea stars

Sea stars, or starfish, class **Asteroidea,** are perhaps the most familiar echinoderms. Among the most important predators in many marine ecosystems, they range in size from a centimeter to a meter across. They are abundant in the intertidal zone, but they also occur at depths as great as 10,000 m. Around 1500 species of sea stars occur throughout the world.

The body of a sea star is composed of a central disc that merges gradually with the arms. Although most sea stars have five arms, members of some families have many more, typically in multiples of five. The body is somewhat flattened, flexible, and covered with a pigmented epidermis. The Asteroidea now includes sea daisies, which were discovered in 1986 and were once considered their own echinoderm class.

Class Ophiuroidea: Brittle stars

Brittle stars (class **Ophiuroidea**) are the largest class of echinoderms in numbers of species (about 2000), and they are probably the most abundant also. Secretive, they avoid light and are more active at night.

Brittle stars have slender, branched arms. The most mobile of echinoderms, brittle stars move by pulling themselves along, "rowing" over the substrate by moving their arms, often in pairs or groups, from side to side. Some brittle stars use their arms to swim, a very unusual habit among echinoderms.

Like sea stars, brittle stars have five arms. They are closely related to the sea stars, but have several morphological differences, one of which is that their tube feet lack ampullae, have no suckers, and are used for feeding, not locomotion.

Class Echinoidea: Sea urchins and sand dollars

The members of the class **Echinoidea,** sand dollars and sea urchins, lack distinct arms but have the same five-part body plan as all other echinoderms. Five rows of tube feet protrude through the plates of the calcareous skeleton. Sea urchins have rounded spines protruding from their skeleton, which help to protect them from predation.

About 950 living species constitute the class Echinoidea. Because of their calcareous plates, sea urchins and sand dollars are well preserved in the fossil record, with more than 5000 additional ancient species described.

Echinoderms are deuterostomes characterized by a pentaradial adult symmetry. They have characteristic calcium-rich plates called ossicles that form the endoskeleton and a unique water-vascular system that includes hollow tube feet. Many echinoderms can regenerate lost parts, and whole animals may sometimes develop from lost parts. Sexual reproduction is external, producing bilaterally symmetrical planktonic larvae.

The five classes of Echinodermata are Asteroidea, Ophiuroidea, Echinoidea, Crinoidea, and Holothuroidea. All exhibit pentaradial symmetry.

34.1 Phylum Mollusca: The Mollusks

Mollusks are extraordinarily diverse and characterized by a reduced coelom surrounding the heart.

■ The mollusk body plan is complex. Mollusks are bilaterally symmetrical although this symmetry disappeared during the evolution of gastropods (figure 34.3).

■ The digestive, excretory, and reproductive organs are concentrated in a visceral mass (figure 34.3).

■ The mantle, a dorsal thick epidermal sheet of skin, covers the body and forms a mantle cavity, which houses the respiratory organs or ctenidia, and the openings to the digestive, excretory, and reproductive organs.

■ The foot is used in locomotion, attachment, food capture, or a combination of these functions.

■ The calcium-rich shell is secreted by the outer covering of the mantle and is primarily protective. Some mollusks have internalized or reduced shells.

■ Most mollusks except bivalves have a radula, a grasping, tonguelike organ used for feeding. Bivalves use gills to filter food from the water (figure 34.4).

■ Most mollusks have an open circulatory system.

■ Most mollusks are distinct male and female, and, except for gastropods, fertilization is external. Fertilized eggs undergo spiralian cleavage. The embryos give rise to free-swimming trochopores and may develop into a free-swimming veliger before they settle out of the water.

■ The phylum Mollusca has four classes: Polyplacophora, Gastropoda, Bivalvia, and Cephalopoda.

34.2 Phylum Annelida: The Annelids

The Annelida body plan consists of repeated ringlike segments; however, members of this phylum are not monophyletic (figure 34.13).

■ The segments run the length of the body and they are separated by septa. The fluid-filled segments provide a hydrostatic skeleton, and each contains paired nephridia.

■ A ventral nerve cord connects the ganglia in each segment to one another and to the brain. Anterior and posterior segments contain specialized light-, chemo- and touch receptors.

■ Annelids have a dorsal closed circulatory system connected to ventral vessels by five pumping hearts.

■ The annelids have specialized digestive organs and nephridia and exchange oxygen and carbon dioxide through the skin.

■ Each segment typically possesses chaetae, chitin bristles that help anchor the worm.

■ Traditionally annelids are grouped into three main classes, but this classification will likely change soon. The classical groups are Polychaeta, Oligochaeta, and Hirudinea.

34.3 The Lophophorates: Bryozoa and Brachiopoda

The Bryozoa and Brachiopoda have convergently evolved a lophophore, a circular or U-shaped ridge around the mouth bearing one or two rows of ciliated tentacles. The lophophore functions in gas exchange and as a food collection organ.

■ Most members of both phyla exhibit radial cleavage and develop as deuterostomes.

■ In the Bryozoa or Ectoprocta, the anus opens external to the lophophore (figure 34.18).

■ Bryozoa are colonial and produce a chitinous chamber called a zoecium.

■ The Brachiopoda and Phoronida are solitary lophophorates. The anus opens internal to the lophophore. The body is enclosed between two calcified shells that are dorsal and ventral, not lateral as in bivalves (figures 34.19 and 34.20).

34.4 Phylum Arthropoda: The Arthropods

The Arthropoda are characterized by jointed appendages and an exoskeleton with the muscles attached inside the exoskeleton (figures 34.22 and 34.24).

■ Jointed appendages may be modified into mouth parts, antennae, or legs. Joints serve as a fulcrum to provide leverage.

■ The exoskeleton is composed of chitin and protein, and it must be molted during ecdysis if the arthropod is to grow. Some arthropod exoskeletons may contain calcium crystals.

■ Arthropods are segmented, and some segments are specialized into functional groups called tagmata, such as the head, thorax, and abdomen.

■ Many arthropods have compound eyes composed of independent ommatidia; others have simple eyes or ocelli (figure 34.23).

■ The arthropods have an open circulatory system and a ventral nervous system controlled by an inhibitory brain.

■ Terrestrial insects, myriapods, and arachnids have a unique excretory system consisting of Malpighian tubules that eliminate uric acid or guanine.

■ The arachnids have specialized anterior appendages, the chelicerae, that function as fangs or pincers. Posterior to the chelicerae are the pedipalps, which can be specialized as copulatory organs, pincers, or sensory organs.

■ Crustacea have three tagmata, and the two anterior ones fuse to form a cephalothorax. They have two pairs of antennae, three pairs of appendages for eating, various pairs of legs, and appendages on their abdomen. Most crustaceans develop through a nauplius stage (figures 34.28, 34.30).

■ Crustaceans include free-swimming and terrestrial members, decapods, and sessile barnacles.

■ Insects, the Hexapoda, are extraordinarily diverse. They have three pairs of legs and may have wings. The internal features are similar to those of other arthropods, except the fore- and hindgut are lined with cuticle, and chemical digestion occurs in the midgut. In addition to compound eyes, insects have sensory setae and a tympanum for sound, and they use pheromones to communicate.

34.5 Phylum Echinodermata: The Echinoderms

The Echinodermata are characterized by deuterostome development and an endoskeleton.

■ The larvae of echinoderms are bilateral, but the adult is pentaradial. The adult body structure is discussed in reference to the ventral mouth (figure 34.36).

■ Echinoderms have a water-vascular system to aid in movement and feeding.

■ Echinoderms have a delicate epidermis containing neurosensory cells stretched over an endoskeleton composed of movable or fixed calcium-rich ossicles.

■ The large coelom helps provide circulation and respiration.

■ Echinoderm reproduction is sexual and external, although many can regenerate lost parts or pieces, which can develop into entire adult animals.

■ There are five classes of echinoderms: Crinoidea, Asteroidea, Ophiuroidea, Echinoidea, and Holothuroidea.

SELF TEST

1. The _____ of a mollusk is a highly efficient respiratory structure.
   a. nephridium
   b. radula
   c. ctenidia
   d. veliger

2. Torsion is a unique characteristic of the
   a. bivalves.
   b. gastropods.
   c. chitons.
   d. cephalopods.

3. Elevated intelligence and complex behaviors are characteristics of the
   a. cephalopods.
   b. polychaetes.
   c. brachiopods.
   d. echinoderms.

4. Serial segmentation is a key characteristic of which of the following phyla?
   a. Mollusks
   b. Brachiopods
   c. Bryozoa
   d. Annelids

5. Which of the following forms of annelids is not hermaphroditic?
   a. Earthworms
   b. Leeches
   c. Polychaetes
   d. All annelids are hermaphroditic.

6. In annelids, which of the following chitin-based structures is used for the process of locomotion?
   a. Clitellum
   b. Chaetae
   c. Septa
   d. Parapodia

7. The distinguishing feature of the Bryozoa and Brachiopoda is the presence of _____
   a. a coelum.
   b. segmentation.
   c. chaetae.
   d. a lophophore.

8. In terms of numbers, the most successful phylum on the planet is the
   a. Mullusca
   b. Arthropoda
   c. Echinodermata
   d. Annelida

9. Which of the following characteristics is not found in the arthropods?
   a. Jointed appendages
   b. Segmentation
   c. Closed circulatory system
   d. Segmented ganglia

10. The fact that Arthropods molt means that they are considered
    a. spiralia.
    b. ecdysozoans.
    c. deuterostomes.
    d. parazoans.

11. Which of the following classes of arthropod possess chelicerae?
    a. Chilopoda
    b. Crustacea
    c. Hexapoda
    d. Arachnida

12. Decapods are
    a. centipedes and millipedes.
    b. barnacles.
    c. ticks and mites.
    d. lobsters and crayfish.

13. What characteristic separates the Diptera, Lepidoptera, and Hymenoptera?
    a. Type of wings
    b. Type of mouthparts
    c. Type of legs
    d. Method of reproduction

14. Based on protostomes–deuterostome characteristics, which of the following phyla is the closest relative to the chordates?
    a. Annelida
    b. Arthropoda
    c. Echinodermata
    d. Mollusca

15. Which of the following structures is not a component of the water-vascular system of an echinoderm?
    a. Ossicles
    b. Ampulla
    c. Radial canals
    d. Madreporite

CHALLENGE QUESTIONS

1. Scientists along the Chesapeake Bay have discovered that the decline in bivalve populations in the bay (specifically clams and scallops) has led to a drastic increase in the levels of pollution in the bay. What characteristic of this group could be attributed to this observation?

2. Chitin is present in a number of invertebrate coelomates, as well as the fungi. What does this tell you about the origin and importance of this substance?

chapter 35

Vertebrates

introduction

MEMBERS OF THE PHYLUM CHORDATA exhibit great changes in the endoskeleton compared with what is seen in echinoderms. As you saw in chapter 34, the endoskeleton of echinoderms is functionally similar to the exoskeleton of arthropods; it is a hard shell that encases the body, with muscles attached to its inner surface. Chordates employ a very different kind of endoskeleton, one that is truly internal. Members of the phylum Chordata are characterized by a flexible rod that develops along the back of the embryo. Muscles attached to this rod allowed early chordates to swing their bodies from side to side, swimming through the water. This key evolutionary advance, attaching muscles to an internal element, started chordates along an evolutionary path that led to the vertebrates—and, for the first time, to truly large animals.

35.1 The Chordates

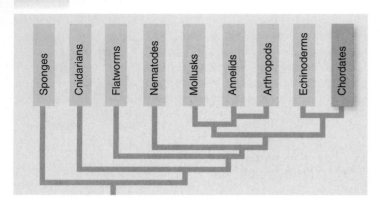

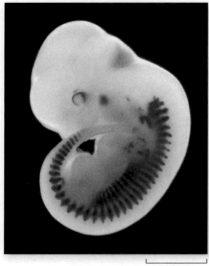

figure 35.2

A MOUSE EMBRYO. At 11.5 days of development, the mesoderm is already divided into segments called somites (stained dark in this photo), reflecting the fundamentally segmented nature of all chordates.

500 μm

Chordates (phylum **Chordata**) are deuterostome coelomates; their nearest relatives in the animal kingdom are the echinoderms, the only other deuterostomes. There are some 56,000 species of chordates, a phylum that includes fishes, amphibians, reptiles, birds, and mammals.

Four features characterize the chordates and have played an important role in the evolution of the phylum (figure 35.1):

1. A single, hollow **nerve cord** runs just beneath the dorsal surface of the animal. In vertebrates, the dorsal nerve cord differentiates into the brain and spinal cord.
2. A flexible rod, the **notochord,** forms on the dorsal side of the primitive gut in the early embryo and is present at some developmental stage in all chordates. The notochord is located just below the nerve cord. The notochord may persist in some chordates; in others it is replaced during embryonic development by the vertebral column that forms around the nerve cord.
3. **Pharyngeal slits** connect the **pharynx,** a muscular tube that links the mouth cavity and the esophagus, with the external environment. In terrestrial vertebrates, the slits do not actually connect to the outside and are better termed **pharyngeal pouches.** Pharyngeal pouches are

present in the embryos of all vertebrates. They become slits, open to the outside in animals with gills, but disappear in those lacking gills. The presence of these structures in all vertebrate embryos provides evidence of their aquatic ancestry.

4. Chordates have a **postanal tail** that extends beyond the anus, at least during their embryonic development. Nearly all other animals have a terminal anus.

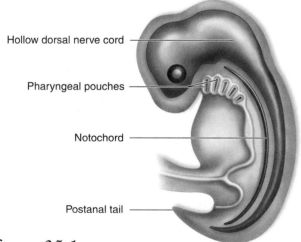

figure 35.1

THE FOUR PRINCIPAL FEATURES OF THE CHORDATES, AS SHOWN IN A GENERALIZED EMBRYO.

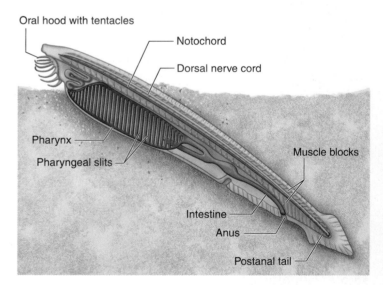

figure 35.3

PHYLUM CHORDATA: CHORDATES. Vertebrates, tunicates, and lancelets are chordates (phylum Chordata), coelomate animals with a flexible rod, the notochord, that provides resistance to muscle contraction and permits rapid lateral body movements. Chordates also possess pharyngeal pouches or slits (reflecting their aquatic ancestry and present habitat in some) and a hollow dorsal nerve cord. In vertebrates, the notochord is replaced during embryonic development by the vertebral column.

All chordates have all four of these characteristics at some time in their lives. For example, humans as embryos have pharyngeal pouches, a dorsal nerve cord, a postanal tail, and a notochord. As adults, the nerve cord remains, and the notochord is replaced by the vertebral column. All but one pair of pharyngeal pouches are lost; this remaining pair forms the Eustachian tubes that connect the throat to the middle ear. The postanal tail regresses, forming the tail bone (coccyx).

A number of other characteristics also fundamentally distinguish the chordates from other animals. Chordate muscles are arranged in segmented blocks that affect the basic organization of the chordate body and can often be clearly seen in embryos of this phylum (figure 35.2). Most chordates have an internal skeleton against which the muscles work. Either this internal skeleton or the notochord (figure 35.3) makes possible the extraordinary powers of locomotion characteristic of this group.

> Chordates are characterized by a hollow dorsal nerve cord, a notochord, pharyngeal pouches, and a postanal tail at some point in their development. The flexible notochord anchors internal muscles and allows rapid, versatile movement.

35.2 The Nonvertebrate Chordates

Phylum Chordata can be divided into three subphyla. Two of these, Urochordata and Cephalochordata, are nonvertebrate; the third subphylum is Vertebrata. The nonvertebrate chordates do not form vertebrae or other bones, and in the case of the urochordates, their adult form is greatly different from what we expect chordates to look like.

Tunicates have clearly chordate larval forms

The tunicates and salps (subphylum **Urochordata**) are a group of about 1250 species of marine animals. Most of them are immobile as adults, with only the larvae having a notochord and nerve cord. As adults, they exhibit neither a major body cavity nor visible signs of segmentation (figure 35.4a, b). Most species occur in shallow waters, but some are found at great depths. In some tunicates, adults are colonial, living in masses on the ocean floor. The pharynx is lined with numerous cilia; beating of the cilia draws a stream of water into the pharynx, where microscopic food particles become trapped in a mucous sheet secreted from a structure called an *endostyle*.

The tadpolelike larvae of tunicates plainly exhibit all of the basic characteristics of chordates and mark the tunicates as having the most primitive combination of features found in any chordate (figure 35.4c). The larvae do not feed and have a poorly developed gut. They remain free-swimming for only a few days before settling to the bottom and attaching themselves to a suitable substrate by means of a sucker.

Tunicates change so much as they mature and adjust developmentally to an immobile, filter-feeding existence that it would be difficult to discern their evolutionary relationships solely by examining an adult. Many adult tunicates secrete a **tunic**, a tough sac composed mainly of cellulose, a substance frequently found in the cell walls of plants and algae but rarely found in animals. The tunic surrounds the animal and gives the

a.

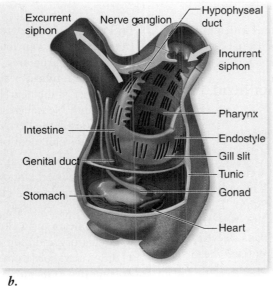

b.

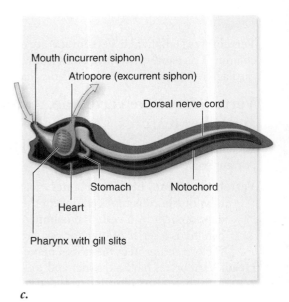

c.

figure 35.4

TUNICATES (PHYLUM CHORDATA, SUBPHYLUM UROCHORDATA). *a.* The sea peach, *Halocynthia auranthium*, like other tunicates, does not move as an adult, but rather is firmly attached to the seafloor. *b.* Diagram of the structure of an adult tunicate. *c.* Diagram of the structure of a larval tunicate, showing the characteristic tadpolelike form. Larval tunicates resemble the postulated common ancestor of the chordates.

subphylum its name. Colonial tunicates may have a common sac and a common opening to the outside.

One group of Urochordates, the Larvacea, retains the tail and notochord into adulthood. One theory of vertebrate origins involves a larval form, perhaps that of a tunicate, which acquired the ability to reproduce.

Lancelets are small marine chordates

Lancelets (subphylum **Cephalochordata**) were given their English name because they resemble a lancet—a small, two-edged surgical knife. These scaleless chordates, a few centimeters long, occur widely in shallow water throughout the oceans of the world. There are about 23 species of this subphylum. Most of them belong to the genus *Branchiostoma*, formerly called *Amphioxus*, a name still used widely. In lancelets, the notochord runs the entire length of the dorsal nerve cord and persists throughout the animal's life.

Lancelets spend most of their time partly buried in sandy or muddy substrates, with only their anterior ends protruding (figure 35.5). They can swim, although they rarely do so. Their muscles can easily be seen through their thin, transparent skin as a series of discrete blocks, called myomeres. Lancelets have many more pharyngeal gill slits than fishes do. They lack pigment in their skin, which has only a single layer of cells, unlike the multilayered skin of vertebrates. The lancelet body is pointed at both ends. There is no distinguishable head or sensory structures other than pigmented light receptors.

Lancelets feed on microscopic plankton, using a current created by beating cilia that line the oral hood, pharynx, and gill slits. The gill slits provide an exit for the water and are an

figure 35.5

LANCELETS. Two lancelets, *Branchiostoma lanceolatum* (phylum Chordata, subphylum Cephalochordata), partly buried in shell gravel, with their anterior ends protruding. The muscle segments are clearly visible.

adaptation for filter feeding. The oral hood projects beyond the mouth and bears sensory tentacles, which also ring the mouth.

The recent discovery of fossil forms similar to living lancelets in rocks 550 million years old argues for the antiquity of this group. Recent studies by molecular systematists further support the hypothesis that lancelets are the closest relatives of vertebrates.

> Nonvertebrate chordates, including tunicates and lancelets, have notochords but no vertebrae or bones. They are the closest relatives of vertebrates.

35.3 The Vertebrate Chordates

Vertebrates (subphylum **Vertebrata**) are chordates with a spinal column. The name *vertebrate* comes from the individual bony or cartilaginous segments called vertebrae that make up the spine.

Vertebrates have vertebrae, a distinct head, and other features

Vertebrates differ from the tunicates and lancelets in two important respects:

Vertebral column. In all vertebrates except the earliest fishes, the notochord is replaced during the course of embryonic development by a vertebral column (figure 35.6). The column is a series of bony or cartilaginous vertebrae that enclose and protect the dorsal nerve cord like a sleeve.

Head. Vertebrates have a distinct and well-differentiated head with three pairs of well developed sensory organs; the brain is fully encased within a protective box, the skull, or cranium, made of bone or cartilage.

In addition to these two key characteristics, vertebrates differ from other chordates in other important respects (figure 35.7):

Neural crest. A unique group of embryonic cells called the **neural crest** contributes to the development of many vertebrate structures. These cells develop on the crest of the neural tube as it forms by invagination and pinching together of the neural plate (see chapter 53 for a detailed account). Neural crest cells then migrate to various locations in the developing embryo, where they participate in the development of many different structures.

Internal organs. Internal organs characteristic of vertebrates include a liver, kidneys, and endocrine glands. The ductless endocrine glands secrete hormones that help regulate many of the body's functions. All vertebrates have a heart and a closed circulatory system. In both their circulatory and their excretory functions, vertebrates differ markedly from other animals.

Endoskeleton. The endoskeleton of most vertebrates is made of cartilage or bone. Cartilage and bone are specialized tissue containing fibers of the protein collagen compacted together (see chapter 47). Bone also contains crystals of a calcium phosphate salt. The great advantage of bone over chitin as a structural material

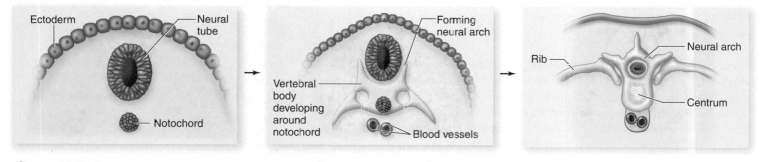

figure 35.6

EMBRYONIC DEVELOPMENT OF A VERTEBRA. During the evolution of animal development, the flexible notochord is surrounded and eventually replaced by a cartilaginous or bony covering, the centrum. The neural tube is protected by an arch above the centrum. The vertebral column functions as a strong, flexible rod that the muscles pull against when the animal swims or moves.

is that bone is a dynamic, living tissue that is strong without being brittle. The vertebrate endoskeleton makes possible the great size and extraordinary powers of movement that characterize this group.

Vertebrates evolved half a billion years ago: An overview

The first vertebrates evolved in the oceans about 545 MYA, during the Cambrian period. Many of them looked like a flattened hot dog, with a mouth at one end and a fin at the other. The appearance of a hinged jaw was a major advance, opening up new food-gathering options, and jawed fishes became the dominant creatures in the sea. Their descendants, the amphibians, invaded the land. Amphibians, in turn, gave rise to the first reptiles about 300 MYA. Within 50 million years, reptiles, better suited to living out of water, replaced amphibians as the dominant land vertebrates.

With the success of reptiles, vertebrates truly came to dominate the surface of the Earth. Many kinds of reptiles evolved, ranging in size from smaller than a chicken to bigger than a truck, and including some that flew and others that swam. Among them evolved reptiles that gave rise to the two remaining great lines of terrestrial vertebrates: birds and mammals.

Dinosaurs and mammals appear at about the same time in the fossil record, 220 MYA. For over 150 million years, dinosaurs dominated the face of the Earth. Over all these million-and-a-half centuries, the largest mammal was no bigger than a medium-sized dog. Then, in the Cretaceous mass extinction, about 65 MYA, the dinosaurs abruptly disappeared. In their absence, mammals and birds quickly took their place, becoming abundant and diverse.

The history of vertebrates has been a series of evolutionary advances that have allowed vertebrates to first invade the land and then the air. In this chapter, we examine the key evolutionary advances that permitted vertebrates to invade the land successfully. As you will see, this invasion was a staggering evolutionary achievement, involving fundamental changes in many body systems.

> Vertebrates, the principal chordate group, are characterized by a vertebral column and a distinct head, along with other distinctions such as a closed circulatory system and certain organs. Vertebrates appeared about 470 MYA.

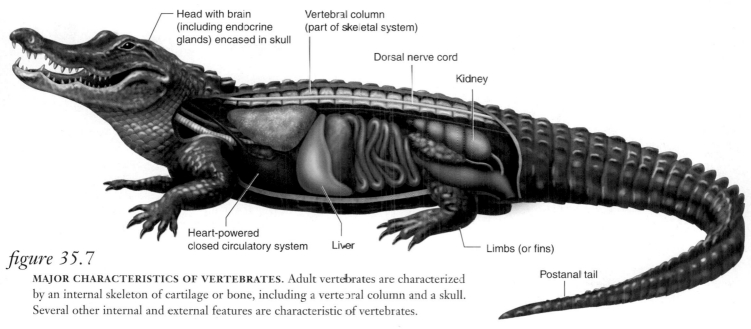

figure 35.7

MAJOR CHARACTERISTICS OF VERTEBRATES. Adult vertebrates are characterized by an internal skeleton of cartilage or bone, including a vertebral column and a skull. Several other internal and external features are characteristic of vertebrates.

35.4 Fishes

Over half of all vertebrates are fishes. The most diverse vertebrate group, fishes provided the evolutionary base for invasion of land by amphibians. In many ways, amphibians can be viewed as transitional—"fish out of water."

The story of vertebrate evolution started in the ancient seas of the Cambrian period (545–490 MYA), when the first backboned animals appeared. Figure 35.8 shows the key vertebrate characteristics that evolved subsequently. Wriggling through the water, jawless and toothless, the first fishes sucked up small food particles from the ocean floor like miniature vacuum cleaners. Most were less than a foot long, respired with gills, and had no paired fins or vertebrae (although some had rudimentary vertebrae); they did have a head and a primitive tail to push them through the water.

For 50 million years, during the Ordovician period (490–438 MYA), these simple fishes were the only vertebrates. By the end of this period, fish had developed primitive fins to help them swim and massive shields of bone for protection. Jawed fishes first appeared during the Silurian period (438–408 MYA), and along with them came a new mode of feeding.

Fishes exhibit five key characteristics

From whale sharks 18 m long to tiny gobies no larger than your fingernail, fishes vary considerably in size, shape, color, and appearance (figure 35.9). Some live in freezing arctic seas, others in warm, freshwater lakes, and still others spend a lot of time entirely out of water. However varied, all fishes have important characteristics in common:

1. **Vertebral column.** Fish have an internal skeleton with a bony or cartilaginous spine surrounding the dorsal nerve cord, and a bony or cartilaginous skull encasing the

figure 35.9

FISH. Fish are the most diverse vertebrates and include more species than all other kinds of vertebrates combined. Top, ribbon eel, *Rhinomuraena quaesita*; bottom left, leafy sea-dragon, *Phycodurus eques*; bottom right, yellowfin tuna, *Thunnus albacares*.

brain. Exceptions are the jawless hagfish and lampreys. In hagfish, a cartilaginous skull is present, but vertebrae are not; the notochord persists and provides support. In lampreys, a cartilaginous skeleton and notochord are present, but rudimentary cartilaginous vertebrae also surround the notochord in places.

2. **Jaws and paired appendages.** Fishes other than lampreys and hagfish all have jaws and paired appendages, features that are also seen in tetrapods (see figure 35.8). Jaws allowed these fish to capture larger and more active

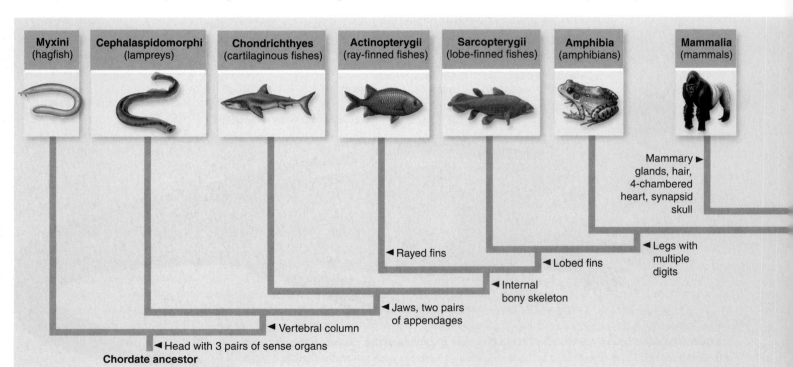

TABLE 35.1   Major Classes of Fishes

Class	Typical Examples		Key Characteristics	Approximate Number of Living Species
Actinopterygii	Ray-finned fishes		Most diverse group of vertebrates; swim bladders and bony skeletons; paired fins supported by bony rays	30,000
Sarcopterygii	Lobe-finned fishes		Largely extinct group of bony fishes; ancestral to amphibians; paired lobed fins	8
Chondrichthyes	Sharks, skates, rays		Cartilaginous skeletons; no swim bladders; internal fertilization	750
Myxini	Hagfishes		Jawless fishes with no paired appendages; scavengers; mostly blind, but having a well-developed sense of smell	30
Cephalaspidomorphi	Lampreys		Largely extinct group of jawless fishes with no paired appendages; parasitic and nonparasitic types; all breed in fresh water	35
Placodermi	Armored fishes		Jawed fishes with heavily armored heads; many were quite large	Extinct
Acanthodii Ostracodermi	Spiny fishes		Fishes with jaws; all now extinct; paired fins supported by sharp spines; head shields made of bone; rest of skeleton cartilaginous	Extinct

prey. Most fishes have two pairs of fins: a pair of pectoral fins at the shoulder, and a pair of pelvic fins at the hip. In the lobe-finned fish, these pairs of fins became jointed.

3. **Internal gills.** Fishes are water-dwelling creatures and must extract oxygen dissolved in the water around them. They do this by directing a flow of water through their mouths and across their gills (see chapter 49). The gills are composed of fine filaments of tissue that are rich in blood vessels.

4. **Single-loop blood circulation.** Blood is pumped from the heart to the gills. From the gills, the oxygenated blood passes to the rest of the body, and then returns to the heart. The heart is a muscular tube-pump made of four chambers that contract in sequence.

5. **Nutritional deficiencies.** Fishes are unable to synthesize the aromatic amino acids (phenylalanine, tryptophan, and tyrosine; see chapter 3), and they must consume them in their foods. This inability has been inherited by all their vertebrate descendants.

The first fishes

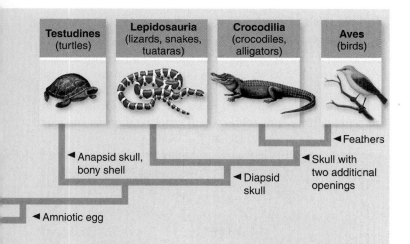

figure 35.8

PHYLOGENY OF THE LIVING VERTEBRATES. Some of the key characteristics that evolved among the vertebrate groups are shown in this phylogeny.

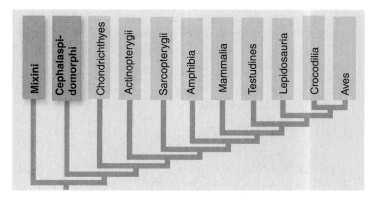

The first fishes did not have jaws, and instead had only a mouth at the front end of the body that could be opened to take in food. One group, the jawless Agnatha, survives today as hagfish (class Myxini; table 35.1) and lampreys (class Cephalaspidomorphi).

Another group were the ostracoderms (a word meaning "shell-skinned"). Only their head-shields were made of bone;

their elaborate internal skeletons were constructed of cartilage. Many ostracoderms were bottom-dwellers, with a jawless mouth underneath a flat head, and eyes on the upper surface. Ostracoderms thrived in the Ordovician and Silurian periods (490–408 MYA), only to become almost completely extinct at the close of the Devonian period (408–360 MYA).

Evolution of the jaw

A fundamentally important evolutionary advance that occurred in the late Silurian period was the development of jaws. Jaws evolved from the most anterior of a series of arch-supports made of cartilage, which were used to reinforce the tissue between gill slits to hold the slits open (figure 35.10). This transformation was not as radical as it might at first appear.

Each gill arch was formed by a series of several cartilages (which later evolved to become bones) arranged somewhat in the shape of a V turned on its side, with the point directed outward. Imagine the fusion of the front pair of arches at top and bottom, with hinges at the points, and you have the primitive vertebrate jaw. The top half of the jaw is not attached to the skull directly except at the rear. Teeth developed on the jaws from modified scales on the skin that lined the mouth.

Armored fishes called placoderms and spiny fishes called acanthodians both had jaws. Spiny fishes were very common during the early Devonian period, largely replacing ostracoderms, but they became extinct themselves at the close of the Permian. Like ostracoderms, they had internal skeletons made of cartilage, but their scales contained small plates of bone, foreshadowing the much larger role bone would play in the future of vertebrates. Spiny fishes were jawed predators and far better swimmers than ostracoderms, with as many as seven fins to aid their swimming. All of these fins were reinforced with strong spines, giving these fishes their name. No spiny fishes survive today.

By the mid-Devonian period, the heavily armored placoderms became common. A very diverse and successful group, seven orders of placoderms dominated the seas of the late Devonian, only to become extinct at the end of that period. The placoderm jaw was much improved over the primitive jaw of spiny fishes, with the upper jaw fused to the skull and the skull hinged on the shoulder. Many of the placoderms grew to enormous sizes, some over 30 feet long, with 2-foot skulls that had an enormous bite.

Sharks, with cartilaginous skeletons, became top predators

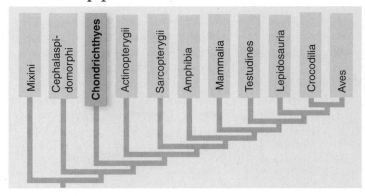

At the end of the Devonian period, essentially all of these pioneer vertebrates disappeared, replaced by sharks and bony fishes in one of several mass extinctions that occurred during Earth's history (see chapter 22). Sharks and bony fishes first evolved in the early Devonian, 400 MYA. In these fishes, the jaw was improved even further, with the upper part of the first gill arch behind the jaws being transformed into a supporting strut or prop, joining the rear of the lower jaw to the rear of the skull. This allowed the mouth to open much wider than was previously possible.

During the Carboniferous period (360–280 MYA), sharks became the dominant predators in the sea. Sharks (class **Chondrichthyes**) have a skeleton made of cartilage, like primitive fishes, but it is "calcified," strengthened by granules of calcium carbonate deposited in the outer layers of cartilage. The result is a very light and strong skeleton.

Streamlined, with paired fins and a light, flexible skeleton, sharks are superior swimmers (figure 35.11). Their pectoral fins are particularly large, jutting out stiffly like airplane wings—and that is how they function, adding lift to compensate for the downward thrust of the tail fin. Sharks are very aggressive predators, and some early sharks reached enormous size.

The evolution of teeth

Sharks were among the first vertebrates to develop teeth. These teeth evolved from rough scales on the skin and are not set into the jaw as human teeth are, but rather sit atop it. They are not firmly anchored and are easily lost. In a shark's mouth, the teeth are arrayed in up to 20 rows; the teeth in front do the biting and cutting, and behind them other teeth grow and wait

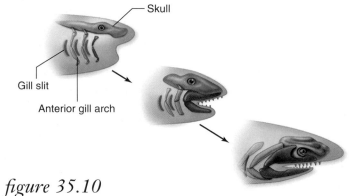

figure 35.10

EVOLUTION OF THE JAW. Jaws evolved from the anterior gill arches of ancient, jawless fishes.

figure 35.11

CHONDRICHTHYES. Members of the class Chondrichthyes, such as this blue shark, *Prionace glauca*, are mainly predators or scavengers.

their turn. When a tooth breaks or is worn down, a replacement from the next row moves forward. A single shark may eventually use more than 20,000 teeth.

A shark's skin is covered with tiny, toothlike scales, giving it a rough "sandpaper" texture. Like the teeth, these scales are constantly replaced throughout the shark's life.

The lateral line system

Sharks, as well as bony fishes, possess a fully developed lateral line system. The lateral line system consists of a series of sensory organs that project into a canal beneath the surface of the skin. The canal runs the length of the fish's body and is open to the exterior through a series of sunken pits. Movement of water past the fish forces water through the canal. The pits are oriented so that some are stimulated no matter what direction the water moves. Details of the lateral line system's function are described in chapter 45. In a very real sense, the lateral line system is a fish's equivalent of hearing.

Reproduction in cartilaginous fishes

Reproduction among the class Chondrichthyes differs from that of other fishes. Shark eggs are fertilized internally. During mating, the male grasps the female with modified fins called claspers, and sperm run from the male into the female through grooves in the claspers. Although a few species lay fertilized eggs, the eggs of most species develop within the female's body, and the pups are born alive.

In recent times, this reproductive system has worked to the detriment of sharks. Because of their long gestation periods and relatively few offspring, shark populations are not able to recover quickly from population declines. Unfortunately, in recent times sharks have been fished very heavily because shark fin soup has become popular in Asia and elsewhere. As a result, shark populations have declined greatly, and there is concern that many species may soon face extinction.

Shark evolution

Many of the early evolutionary lines of sharks died out during the great extinction at the end of the Permian period (248 MYA). The survivors thrived and underwent a burst of diversification during the Mesozoic era (248–65 MYA), when most of the modern groups of sharks appeared. Skates and rays, which are dorsoventrally flattened relatives of sharks, evolved at this time, some 200 million years after the sharks first appeared.

Bony fishes dominate the waters

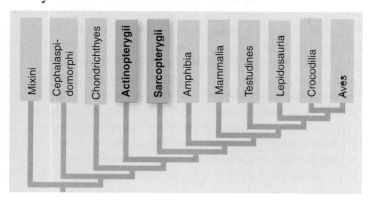

Bony fishes evolved at the same time as sharks, some 400 MYA, but took quite a different evolutionary road. Instead of gaining speed through lightness, as sharks did, bony fishes adopted a heavy internal skeleton made completely of bone.

Bone is very strong, providing a base against which very strong muscles can pull. Not only is the internal skeleton ossified, but so is the outer covering of plates and scales. Most bony fishes have highly mobile fins, very thin scales, and completely symmetrical tails (which keep the fish on a straight course as it swims through the water). Bony fishes are the most species-rich group of fishes, indeed of all vertebrates. There are several dozen orders containing more than 30,000 living species.

The remarkable success of the bony fishes has resulted from a series of significant adaptations that have enabled them to dominate life in the water. These include the swim bladder and the gill cover (figure 35.12).

Swim bladder

Although bones are heavier than cartilaginous skeletons, bony fishes are still buoyant because they possess a **swim bladder**, a gas-filled sac that allows them to regulate their buoyant density and so remain suspended at any depth in the water effortlessly. Sharks, by contrast, must move through the water or sink because, lacking a swim bladder, their bodies are denser than water.

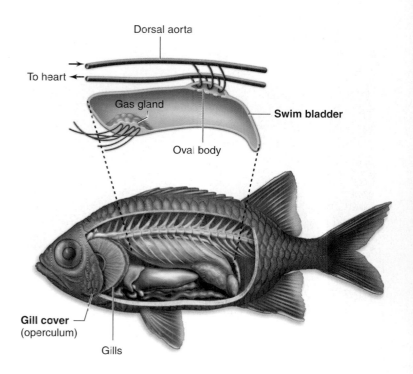

figure 35.12

DIAGRAM OF A SWIM BLADDER. The bony fishes use this structure, which evolved as a dorsal outpocketing of the pharynx, to control their buoyancy in water. The swim bladder can be filled with or drained of gas to allow the fish to control buoyancy. Gases are taken from the blood, and the gas gland secretes the gases into the swim bladder; gas is released from the bladder by a muscular valve, the oval body.

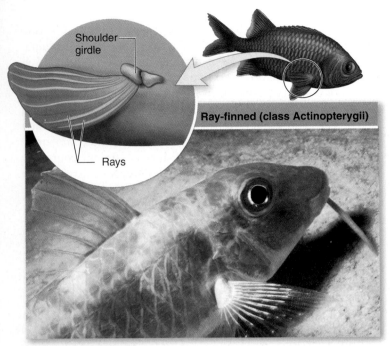

Ray-finned (class Actinopterygii)

Shoulder girdle

Rays

a.

Shoulder girdle

Rays

Central core of bones in fleshy lobe

Lobe-finned (class Sarcopterygii)

b.

figure 35.13

RAY-FINNED AND LOBE-FINNED FISHES. *a.* The ray-finned fishes, such as this Korean angelfish, are characterized by fins of only parallel bony rays. *b.* By contrast, the fins of lobe-finned fish have a central core of bones as well as rays. The coelacanth, *Latimeria chalumnae*, a lobe-finned fish (class Sarcopterygii), was discovered in the western Indian Ocean in 1938. This coelacanth represents a group of fishes thought to have been extinct for about 70 million years. Scientists who studied living individuals in their natural habitat at depths of 100 to 200 m observed them drifting in the current and hunting other fishes at night. Some individuals are nearly 3 m long; they have a slender, fat-filled swim bladder.

In primitive bony fishes, the swim bladder is a dorsal out-pocketing of the pharynx behind the throat, and these species fill the swim bladder by simply gulping air at the surface of the water. In most of today's bony fishes, the swim bladder is an independent organ that is filled and drained of gases, mostly nitrogen and oxygen, internally.

How do bony fishes manage this remarkable trick? It turns out that the gases are harvested from the blood by a unique gland that discharges the gases into the bladder when more buoyancy is required. To reduce buoyancy, gas is reabsorbed into the bloodstream through a structure called the oval body. A variety of physiological factors control the exchange of gases between the bloodstream and the swim bladder.

Gill cover

Most bony fishes have a hard plate called the **operculum** that covers the gills on each side of the head. Flexing the operculum permits bony fishes to pump water over their gills. The gills are suspended in the pharyngeal slits that form a passageway between the pharynx and the outside of the fish's body. When the operculum is closed, it seals off the exit.

When the mouth is open, closing the operculum increases the volume of the mouth cavity, so that water is drawn into the mouth. When the mouth is closed, opening the operculum decreases the volume of the mouth cavity, forcing water past the gills to the outside. Using this very efficient bellows, bony fishes can pass water over the gills while remaining stationary in the water. That is what a goldfish is doing when it seems to be gulping in a fish tank.

The evolutionary path to land ran through the lobe-finned fishes

Two major groups of bony fish are the ray-finned fishes (class **Actinopterygii**; figure 35.13*a*) and lobe-finned fishes (class **Sarcopterygii**). The groups differ in the structure of their fins (figure 35.13*b*). In ray-finned fishes, the internal skeleton of the fin is composed of parallel bony rays that support and stiffen each fin. There are no muscles within the fins; rather, the fins are moved by muscles within the body.

By contrast, lobe-finned fishes have paired fins that consist of a long fleshy muscular lobe (hence their name), supported by a central core of bones that form fully articulated joints with one another. There are bony rays only at the tips of each lobed fin. Muscles within each lobe can move the fin rays independently of one another, a feat no ray-finned fish could match.

Lobe-finned fishes evolved 390 MYA, shortly after the first bony fishes appeared. Only eight species survive today, two species of coelacanth (figure 35.13*b*) and six species of lungfish. Although rare today, lobe-finned fishes played an important part in the evolutionary story of vertebrates. Amphibians almost certainly evolved from the lobe-finned fishes.

Fishes were the first vertebrates. Fishes are characterized by gills and a simple, single-loop circulatory system. Cartilaginous fishes, such as sharks, are fast swimmers having evolved as top predators. The very successful bony fishes have unique characteristics such as swim bladders and gill covers, as well as ossified skeletons.

Amphibians

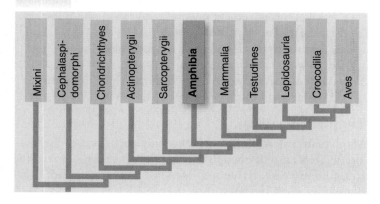

Frogs, salamanders, and caecilians, the damp-skinned vertebrates, are direct descendants of fishes. They are the sole survivors of a very successful group, the amphibians (class **Amphibia**), the first vertebrates to walk on land. Most present-day amphibians are small and live largely unnoticed by humans, but they are among the most numerous of terrestrial vertebrates. Throughout the world, amphibians play key roles in terrestrial food chains.

Living amphibians have five distinguishing features

Biologists have classified living species of amphibians into three orders (table 35.2): 5000 species of frogs and toads in 22 families make up the order Anura ("without a tail"); 500 species of salamanders and newts in 9 families make up the order Caudata ("visible tail"); and 170 species (6 families) of wormlike, nearly blind organisms called caecilians that live in the tropics make up the order Apoda ("without legs"). These amphibians have several key characteristics in common:

1. **Legs.** Frogs and most salamanders have four legs and can move about on land quite well. Legs were one of the key adaptations to life on land. Caecilians have lost their legs during the course of adapting to a burrowing existence.
2. **Lungs.** Most amphibians possess a pair of lungs, although the internal surfaces have much less surface area than do reptilian or mammalian lungs. Amphibians breathe by lowering the floor of the mouth to suck in air, and then raising it back to force the air down into the lungs (see chapter 49).
3. **Cutaneous respiration.** Frogs, salamanders, and caecilians all supplement the use of lungs by respiring through their skin, which is kept moist and provides an extensive surface area.
4. **Pulmonary veins.** After blood is pumped through the lungs, two large veins called pulmonary veins return the aerated blood to the heart for repumping. In this way aerated blood is pumped to the tissues at a much higher pressure.
5. **Partially divided heart.** A dividing wall helps prevent aerated blood from the lungs from mixing with nonaerated blood being returned to the heart from the rest of the body. The blood circulation is thus divided into two separate paths, pulmonary and systemic. The separation is imperfect, however, because no dividing wall exists in one chamber of the heart, the ventricle (see chapter 49).

Several other specialized characteristics are shared by all present-day amphibians. In all three orders, there is a zone of weakness between the base and the crown of the teeth. They also have a peculiar type of sensory rod cell in the retina of the eye called a "green rod." The function of this rod is unknown.

Amphibians overcame terrestrial challenges

The word *amphibia* means "double life," and it nicely describes the essential quality of modern-day amphibians, reflecting their ability to live in two worlds—the aquatic world of their fish ancestors and the terrestrial world they first invaded. Here, we review the checkered history of this group, almost all of whose members have been extinct for the last 200 million years. Then we examine in more detail what the few kinds of surviving amphibians are like.

The successful invasion of land by vertebrates posed a number of major challenges:

- Because amphibian ancestors had relatively large bodies, supporting the body's weight on land as well as enabling

TABLE 35.2	Orders of Amphibians			
Order	Typical Examples		Key Characteristics	Approximate Number of Living Species
Anura	Frogs, toads		Compact, tailless body; large head fused to the trunk; rear limbs specialized for jumping	5000
Caudata	Salamanders, newts		Slender body; long tail and limbs set out at right angles to the body	500
Apoda	Caecilians		Tropical group with a snakelike body; no limbs; little or no tail	170

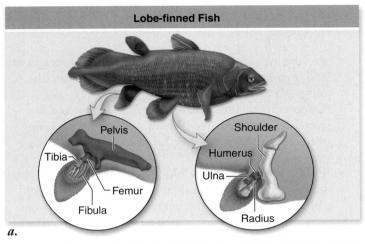

Lobe-finned Fish

Pelvis

Tibia

Femur

Fibula

Shoulder

Humerus

Ulna

Radius

a.

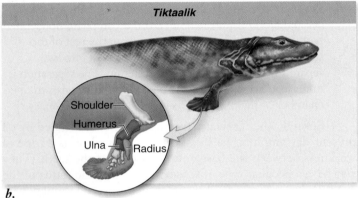

Tiktaalik

Shoulder

Humerus

Ulna

Radius

b.

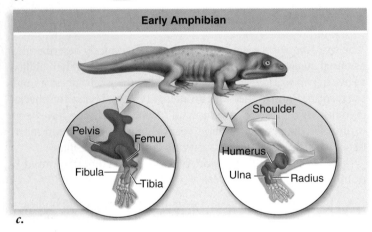

Early Amphibian

Pelvis

Femur

Fibula

Tibia

Shoulder

Humerus

Ulna

Radius

c.

figure 35.14

A COMPARISON BETWEEN THE LIMBS OF A LOBE-FINNED FISH, *TIKTAALIK*, AND A PRIMITIVE AMPHIBIAN. *a.* A lobe-finned fish. Some of these animals could probably move on land. *b. Tiktaalik.* The shoulder and limb bones are like those of an amphibian, but the fins are like those of a lobe-finned fish. The fossil of *Tiktaalik* did not contain the hindlimbs. *c.* A primitive amphibian. As illustrated by their skeletal structure, the legs of such an animal clearly could function better than those of its ancestors for movement on land.

movement from place to place was a challenge (figure 35.14). Legs evolved to meet this need.

- Even though far more oxygen is available to gills in air than in water, the delicate structure of fish gills requires the buoyancy of water to support them, and they will not

function in air. Therefore, other methods of obtaining oxygen were required.

- Delivering great amounts of oxygen to the larger muscles needed for movement on land required modifications to the heart and circulatory system.
- Reproduction still had to be carried out in water so that eggs would not dry out.
- Most importantly, the body itself had to be prevented from drying out.

The first amphibian

Amphibians solved these problems only partially, but their solutions worked well enough that amphibians have survived for 350 million years. Evolution does not insist on perfect solutions, only workable ones.

Paleontologists agree that amphibians evolved from lobe-finned fish. *Ichthyostega*, one of the earliest amphibian fossils (figure 35.15), was found in a 370-million-year-old rock in Greenland. At that time, Greenland was part of what is now the North American continent and lay near the equator. All amphibian fossils from the next 100 million years are found in North America. Only when Asia and the southern continents merged with North America to form the supercontinent Pangaea did amphibians spread throughout the world.

Ichthyostega was a strongly built animal, with sturdy forelegs well supported by shoulder bones. Unlike the bone structure of fish, the shoulder bones were no longer attached to the skull, so the limbs could support the animal's weight. Because the hindlimbs were flipper-shaped, *Ichthyostega* probably moved like a seal, with the forelimbs providing the propulsive force for locomotion and the hindlimbs being dragged along with the rest of the body. To strengthen the backbone further, long, broad ribs that overlap each other formed a solid cage for the lungs and heart. The rib cage was so solid that it probably couldn't expand and contract for breathing. Instead, *Ichthyostega* probably obtained oxygen as many amphibians do today, by lowering the floor of the mouth to draw air in, and then raising it to push air down the windpipe into the lungs.

In 2006, an important transitional fossil between fish and *Ichthyostega* was discovered in northern Canada. *Tiktaalik*, which lived 375 million years ago, had gills and scales like a fish, but a neck like an amphibian. Particularly significant, however, was the form of its forelimbs (figure 35.14): its shoulder, forearm, and wrist bones were like those of amphibians, but at the end of the limb was a lobed fin, rather than the toes of an amphibian. Ecologically, the 3-m long *Tiktaalik* was probably also intermediate between fish and amphibians, spending most of its time in the water, but capable of hauling itself out onto land to capture food or escape predators.

The rise and fall of amphibians

By moving onto land, amphibians were able to utilize many resources and to access many habitats. Amphibians first became common during the Carboniferous period (360–280 MYA). Fourteen families of amphibians are known from the early Carboniferous, nearly all of them aquatic or semiaquatic, like *Ichthyostega* (figure 35.15). By the late Carboniferous, much of North America was covered by low-lying tropical swamplands, and 34 families of amphibians thrived in this wet terrestrial environment, sharing it with pelycosaurs and other early reptiles.

figure 35.15

AMPHIBIANS WERE THE FIRST VERTEBRATES TO WALK ON LAND. *Ichthyostega*, one of the first amphibians, had efficient limbs for crawling on land, an improved olfactory sense associated with a lengthened snout and a relatively advanced ear structure for picking up airborne sounds. Despite these features, *Ichthyostega*, which lived about 350 MYA, was still quite fishlike in overall appearance and may have spent much of its life in water.

In the early Permian period that followed (280–248 MYA), a remarkable change occurred among amphibians—they began to leave the marshes for dry uplands. Many of these terrestrial amphibians had bony plates and armor covering their bodies and grew to be very large, some as big as a pony. Both their large size and the complete covering of their bodies indicate that these amphibians did not use the skin respiratory system of present-day amphibians, but rather had an impermeable leathery skin to prevent water loss. Consequently, they must have relied entirely on their lungs for respiration. By the mid-Permian period, there were 40 families of amphibians. Only 25% of them were still semiaquatic like *Ichthyostega*; 60% of the amphibians were fully terrestrial, and 15% were semiterrestrial. This was the peak of amphibian success, sometimes called the Age of Amphibians.

By the end of the Permian period, reptiles had evolved from amphibians. One group, therapsids, had become common and ousted the amphibians from their newly acquired niche on land. Following the mass extinction event at the end of the Permian, therapsids were the dominant land vertebrate, and most amphibians were aquatic. This trend continued in the following Triassic period (248–213 MYA), which saw the virtual extinction of amphibians from land.

Modern amphibians belong to three groups

All of today's amphibians descended from the three families of amphibians that survived the Age of the Dinosaurs. During the Tertiary period (65–2 MYA), these moist-skinned amphibians accomplished a highly successful invasion of wet habitats all over the world, and today there are over 5600 species of amphibians in 37 different families, comprising the orders Anura, Caudata, and Apoda.

Order Anura: Frogs and toads

Frogs and toads, amphibians without tails, live in a variety of environments, from deserts and mountains to ponds and puddles (figure 35.16*a*). Frogs have smooth, moist skin, a broad body, and long hind legs that make them excellent jumpers. Most frogs live in or near water, although some tropical species live in trees.

Unlike frogs, toads have a dry, bumpy skin and short legs, and are well adapted to dry environments. Toads do not form

Order Anura	Order Caudata	Order Apoda

a.	*b.*	*c.*

figure 35.16

CLASS AMPHIBIA. *a.* Red-eyed tree frog, *Agalychnis callidryas* (order Anura). *b.* An adult tiger salamander, *Ambystoma tigrinum* (order Caudata). *c.* A caecilian, *Caecilia tentaculata* (order Apoda).

a monophyletic group; that is, all toads are not more closely related to each other than they are to some other frogs. Rather, the term *toad* is applied to those anurans that have adapted to dry environments by evolving a suite of adaptive characteristics; this convergent evolution has occurred many times among distantly related anurans.

Most frogs and toads return to water to reproduce, laying their eggs directly in water. Their eggs lack watertight external membranes and would dry out quickly on land. Eggs are fertilized externally and hatch into swimming larval forms called tadpoles. Tadpoles live in the water, where they generally feed on minute algae. After considerable growth, the body of the tadpole gradually undergoes metamorphosis into that of an adult frog.

Order Caudata: Salamanders

Salamanders have elongated bodies, long tails, and smooth, moist skin (figure 35.16*b*). They typically range in length from a few inches to a foot, although giant Asiatic salamanders of the genus *Andrias* are as much as 1.5 m long and weigh up to 33 kg. Most salamanders live in moist places, such as under stones or logs, or among the leaves of tropical plants. Some salamanders live entirely in water.

Salamanders lay their eggs in water or in moist places. Most species practice a type of internal fertilization in which the female picks up sperm packets deposited by the male. Like anurans, many salamanders go through a larval stage before metamorphosing into adults. However, unlike anurans, in which the tadpole is strikingly different from the adult frog, larval salamanders are quite similar to adults, although most live in water and have external gills which disappear at metamorphosis.

Order Apoda: Caecilians

Caecilians, members of the order Apoda (also called Gymnophiona), are a highly specialized group of tropical burrowing amphibians (figure 35.16*c*). These legless, wormlike creatures average about 30 cm long, but can be up to 1.3 m long. They have very small eyes and are often blind. They resemble worms but have jaws with teeth. They eat worms and other soil invertebrates. Fertilization is internal.

> Amphibians ventured onto land some 370 MYA. They are characterized by moist skin, legs (secondarily lost in some species), lungs, and a more complex and divided circulatory system. Most species rely on a water habitat for reproduction.

Reptiles

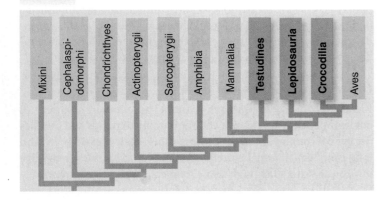

If we think of amphibians as a first draft of a manuscript about survival on land, then reptiles are the finished book. For each of the five key challenges of living on land, reptiles improved on the innovations of amphibians. The arrangement of legs evolved to support the body's weight more effectively, allowing reptile bodies to be bigger and to run. Lungs and heart became more efficient. The skin was covered with dry plates or scales to minimize water loss, and watertight coverings evolved for eggs.

Over 7000 species of reptiles (class **Reptilia**) now live on Earth (table 35.3). They are a highly successful group in today's world; there are more living species of snakes and lizards than there are of mammals.

Reptiles exhibit three key characteristics

All living reptiles share certain fundamental characteristics, features they retain from the time when they replaced amphibians as the dominant terrestrial vertebrates. Among the most important are:

1. **Amniotic egg.** Amphibians eggs must be laid in water or a moist setting to avoid drying out. Most reptiles lay watertight eggs that contain a food source (the yolk) and a series of four membranes: the yolk sac, the amnion, the allantois, and the chorion (figure 35.17). Each membrane plays a role in making the egg an independent life-support system. All modern reptiles, as well as birds and mammals, show exactly this same

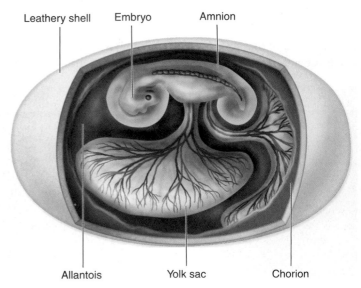

Leathery shell   Embryo   Amnion

Allantois   Yolk sac   Chorion

figure 35.17

THE WATERTIGHT EGG. The amniotic egg is perhaps the most important feature that allows reptiles to live in a wide variety of terrestrial habitats.

TABLE 35.3

Major Orders of Reptiles

Order	Typical Examples	Key Characteristics	Approximate Number of Living Species
Squamata, suborder Sauria	Lizards	Lizards; limbs set at right angles to body; anus is in transverse (sideways) slit; most are terrestrial	3800
Squamata, suborder Serpentes	Snakes	Snakes; no legs; move by slithering; scaly skin is shed periodically; most are terrestrial	3000
Rhynchocephalia	Tuataras	Sole survivors of a once successful group that largely disappeared before dinosaurs; fused, wedgelike, socketless teeth; primitive third eye under skin of forehead	2
Chelonia	Turtles, tortoises, sea turtles	Armored reptiles with shell of bony plates to which vertebrae and ribs are fused; sharp, horny beak without teeth	250
Crocodylia	Crocodiles, alligators	Advanced reptiles with four-chambered heart and socketed teeth; anus is a longitudinal (lengthwise) slit; closest living relatives to birds	25
Ornithischia	Stegosaur	Dinosaurs with two pelvic bones facing backward, like a bird's pelvis; herbivores; legs under body	Extinct
Saurischia	Tyrannosaur	Dinosaurs with one pelvic bone facing forward, the other back, like a lizard's pelvis; both plant- and flesh-eaters; legs under body; birds evolved from Saurischian dinosaurs.	Extinct
Pterosauria	Pterosaur	Flying reptiles; wings were made of skin stretched between fourth fingers and body; wingspans of early forms typically 60 cm, later forms nearly 8 m	Extinct
Plesiosaura	Plesiosaur	Barrel-shaped marine reptiles with sharp teeth and large, paddle-shaped fins; some had snakelike necks twice as long as their bodies	Extinct
Ichthyosauria	Ichthyosaur	Streamlined marine reptiles with many body similarities to sharks and modern fishes	Extinct

pattern of membranes within the egg. These three classes are called **amniotes.**

The outermost membrane of the egg is the **chorion,** which lies just beneath the porous shell. It allows exchange of respiratory gases but retains water. The **amnion** encases the developing embryo within a fluid-filled cavity. The **yolk sac** provides food from the yolk for the embryo via blood vessels connecting to the embryo's gut. The **allantois** surrounds a cavity into which waste products from the embryo are excreted.

2. **Dry skin.** Most living amphibians have moist skin and must remain in moist places to avoid drying out. Reptiles

figure 35.18

SKULLS OF REPTILE
GROUPS. Reptile groups
are distinguished by the
number of holes on the
side of the skull behind the
eye orbit: 0 (anapsids), 1
(synapsids), or 2 (diapsids).
Turtles are the only living
anapsids, although several
extinct groups also had this
condition.

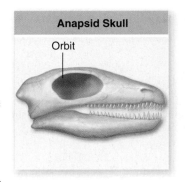

Anapsid Skull
Orbit

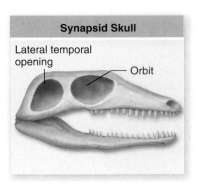

Synapsid Skull
Lateral temporal
opening
Orbit

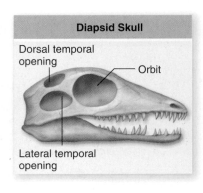

Diapsid Skull
Dorsal temporal
opening
Orbit
Lateral temporal
opening

have dry, watertight skin. A layer of scales covers their
bodies, preventing water loss. These scales develop as
surface cells fill with keratin, the same protein that forms
claws, fingernails, hair, and bird feathers.

3. **Thoracic breathing.** Amphibians breathe by squeezing
their throat to pump air into their lungs; this limits
their breathing capacity to the volume of their mouths.
Reptiles developed pulmonary breathing, expanding and
contracting the rib cage to suck air into the lungs and
then force it out. The capacity of this system is limited
only by the volume of the lungs.

Reptiles dominated the Earth for 250 million years

During the 250 million years that reptiles were the dominant
large terrestrial vertebrates, a series of different reptile groups
appeared and then disappeared.

Synapsids

An important feature of reptile classification is the presence and
number of openings behind the eyes (figure 35.18). Reptiles'
jaw muscles were anchored to these holes, which allowed them
to bite more powerfully. The first group to rise to dominance
were the **synapsids,** whose skulls had a pair of temporal holes
behind the openings for the eyes.

Pelycosaurs, an important group of early synapsids, were
dominant for 50 million years and made up 70% of all land vertebrates; some species weighed as much as 200 kg. With long,
sharp, "steak knife" teeth, these pelycosaurs were the first land
vertebrates to kill beasts their own size (figure 35.19).

About 250 MYA, pelycosaurs were replaced by another
type of synapsid, the therapsids (figure 35.20). Some evidence
indicates that they may have been endotherms, able to produce
heat internally, and perhaps even possessed hair. This would
have permitted therapsids to be far more active than other vertebrates of that time, when winters were cold and long.

For 20 million years, therapsids (also called "mammal-like
reptiles") were the dominant land vertebrate, until they were
largely replaced 230 MYA by another group of reptiles, the diapsids. Most therapsids became extinct 170 MYA, but one group
survived and has living descendants today—the mammals.

figure 35.19

A PELYCOSAUR. *Dimetrodon*, a carnivorous pelycosaur, had a
dorsal sail that is thought to have been used to regulate body
temperature by dissipating body heat or gaining it by basking.

figure 35.20

A THERAPSID. This small, weasel-like cynodont therapsid,
Megazostrodon, may have had fur. Living in the late Triassic
period, this therapsid is so similar to modern mammals that
some paleontologists consider it the first mammal.

Archosaurs

Diapsids have skulls with two pairs of temporal holes, and like
amphibians and early reptiles, they were ectotherms. A variety of
different diapsids occurred in the Triassic period (213–248 MYA),
but one group, the archosaurs, were of particular evolutionary
significance because they gave rise to crocodiles, pterosaurs, dinosaurs, and birds (figure 35.21).

Among the early archosaurs were the largest animals the
world had seen, up to that point, and the first land vertebrates

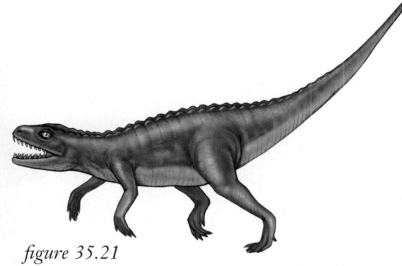

figure 35.21

AN EARLY ARCHOSAUR. *Euparkeria* had rows of bony plates along the sides of the backbone, as seen in modern crocodiles and alligators.

to be bipedal—to stand and walk on two feet. By the end of the Triassic period, however, one archosaur group rose to prominence: the dinosaurs.

Dinosaurs evolved about 220 MYA. Unlike previous bipedal diapsids, their legs were positioned directly underneath their bodies (figure 35.22). This design placed the weight of the body directly over the legs, which allowed dinosaurs to run with great speed and agility. Subsequently, a number of types of dinosaur evolved enormous size and reverted to a four-legged posture to support their massive weight. Dinosaurs went on to become the most successful of all land vertebrates, dominating for more than 150 million years. All dinosaurs became extinct rather abruptly 65 MYA, apparently as a result of an asteroid's impact.

Important characteristics of modern reptiles

As you might imagine from the structure of the amniotic egg, reptiles and other amniotes do not practice external fertilization as most amphibians do. Sperm would be unable to pen-

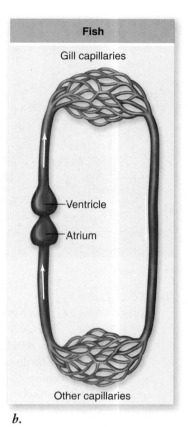
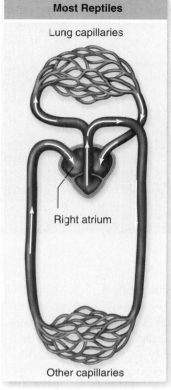

a.   b.

figure 35.23

A COMPARISON OF REPTILE AND FISH CIRCULATION.
a. In most reptiles, oxygenated blood (red) is repumped after leaving the lungs, and circulation to the rest of the body remains vigorous. *b.* The blood in fishes flows from the gills directly to the rest of the body, resulting in slower circulation.

figure 35.22

MOUNTED SKELETON OF *AFROVENATOR*. This bipedal carnivore was about 30 feet long and lived in Africa about 130 MYA.

etrate the membrane barriers protecting the egg. Instead, the male places sperm inside the female, where sperm fertilizes the egg before the protective membranes are formed. This is called internal fertilization.

The circulatory system of reptiles is improved over that of fish and amphibians, providing oxygen to the body more efficiently (figure 35.23). The improvement is achieved by extending the septum within the heart from the atrium partway across the ventricle. This septum creates a partial wall that tends to lessen mixing of oxygen-poor blood with oxygen-rich blood within the ventricle. In crocodiles, the septum completely divides the ventricle, creating a four-chambered heart, just as it does in birds and mammals (and probably did in dinosaurs).

All living reptiles are **ectothermic,** obtaining their heat from external sources. In contrast, **endothermic** animals are able to generate their heat internally (see chapter 50). Although they are ectothermic, that does not mean that reptiles cannot control their body temperature. Many species are able to precisely regulate their temperature by moving in and out of the Sun. In this way, some desert lizards can keep their bodies at a constant temperature throughout the course of an entire day. Of course, on cloudy days, or for species that live in shaded habitats, such thermoregulation is not possible, and in these cases, body temperature is the same as the temperature of the surrounding environment.

a.

b.

figure 35.24

LIVING ORDERS OF REPTILES. *a.* **Chelonia.** The red-bellied turtle, *Pseudemys rubriventris* (*left*) is shown basking, an effective means by which many ectotherms can precisely regulate their body temperature. The domed shells of tortoises, such as this Sri Lankan Star tortoise, *Geochelone elegans*, provide protection against predators for these entirely terrestrial chelonians. *b.* **Tuatara (*Sphenodon punctatus*).** The sole living members of the ancient group Rhynchocephalia. Although they look like lizards, the common ancestor of rhynchocephalians and lizards diverged more than 250 MYA. *c.* **Squamata.** A collared lizard, *Crotaphytus collaris*, is shown left, and a smooth green snake, *Liochlorophis vernalis*, on the right. *d.* **Crocodylia.** Most crocodilans, such as the crocodile, *Crocodylus acutus*, and the gharial, *Gavialis gangeticus* (*right*), resemble birds and mammals in having four-chambered hearts; all other living reptiles have three-chambered hearts. Like birds, crocodiles are more closely related to dinosaurs than to any of the other living reptiles.

Modern reptiles belong to four groups

The four surviving orders of reptiles contain about 7000 species. Reptiles occur worldwide except in the coldest regions, where it is impossible for ectotherms to survive. Reptiles are among the most numerous and diverse of terrestrial vertebrates.

Order Chelonia: Turtles and tortoises

The order Chelonia (figure 35.24*a*) consists of about 250 species of turtles (most of which are aquatic) and tortoises (which are terrestrial). Turtles and tortoises lack teeth but have sharp beaks. They differ from all other reptiles because their bodies are encased within a protective shell. Many of them can pull their head and legs into the shell as well, for total protection from predators.

The shell consists of two basic parts. The carapace is the dorsal covering, and the plastron is the ventral portion. In a fundamental commitment to this shell architecture, the verte-

brae and ribs of most turtle and tortoise species are fused to the inside of the carapace. All of the support for muscle attachment comes from the shell.

Whereas most tortoises have a dome-shaped shell into which they can retract their head and limbs, water-dwelling turtles have a streamlined, disc-shaped shell that permits rapid turning in water. Freshwater turtles have webbed toes; in marine turtles, the forelimbs have evolved into flippers.

Although marine turtles spend their lives at sea, they must return to land to lay their eggs. Many species migrate long distances to do this. Atlantic green turtles (*Chelonia mydas*) migrate from their feeding grounds off the coast of Brazil to Ascension Island in the middle of the South Atlantic—a distance of more than 2000 km—to lay their eggs on the same beaches where they themselves hatched.

Order Rhynchocephalia: Tuataras

Today, the order Rhynchocephalia contains only two species of tuataras, large, lizardlike animals about half a meter long (figure 35.24*b*). The only place in the world where these endangered species are found is a cluster of small islands off the coast of New Zealand. The limited diversity of modern rhynchocephalians belies their rich evolutionary past: In the Triassic period, rhynchocephalians experienced a great adaptive radiation, producing many species that differed greatly in size and habitat.

An unusual feature of the tuatara (and some lizards) is the inconspicuous "third eye" on the top of its head, called a parietal eye. Concealed under a thin layer of scales, the eye has a lens and a retina and is connected by nerves to the brain. Why have an eye, if it is covered up? The parietal eye may function to alert the tuatara when it has been exposed to too much Sun, protecting it against overheating. Unlike most reptiles, tuataras are most active at low temperatures. They burrow during the day and feed at night on insects, worms, and other small animals.

Rhynchocephalians are the closest relatives of snakes and lizards and with them form the group Lepidosauria.

Order Squamata	Order Crocodylia

c.   d.

Order Squamata: Lizards and snakes

The order Squamata (figure 35.24c) includes 3800 species of lizards and about 3000 species of snakes. A distinguishing characteristic of this order is the presence of paired copulatory organs in the male. In addition, changes to the morphology of the head and jaws allow greater strength and mobility. Most lizards and snakes are carnivores, preying on insects and small animals, and these improvements in jaw design have made a major contribution to their evolutionary success.

Snakes, which evolved from a lizard ancestor, are characterized by the lack of limbs, movable eyelids, and external ears, as well as a great number of vertebrae (sometimes more than 300). Limblessness has actually evolved more than a dozen times in lizards; snakes are simply the most extreme case of this evolutionary trend.

Common lizards include iguanas, chameleons, geckos, and anoles. Most are small, measuring less than a foot in length. The largest lizards belong to the monitor family. The largest of all monitor lizards is the Komodo dragon of Indonesia, which reaches 3 m in length and can weigh more than 100 kg. Snakes also vary in length from only a few inches to more than 10 m.

Many lizards and snakes rely on agility and speed to catch prey and elude predators. Only two species of lizard are venomous, the Gila monster of the southwestern United States and the beaded lizard of western Mexico. Similarly, most species of snakes are nonvenomous. Of the 13 families of snakes, only 4 contain venomous species: the elapids (cobras, kraits, and coral snakes); the sea snakes; the vipers (adders, bushmasters, rattlesnakes, water moccasins, and copperheads); and some colubrids (African boomslang and twig snake).

Many lizards, including anoles, skinks, and geckos, have the ability to lose their tails and then regenerate a new one. This ability allows these lizards to escape from predators.

Order Crocodylia: Crocodiles and alligators

The order Crocodylia is composed of 25 species of large, primarily aquatic reptiles (figure 35.24d). In addition to crocodiles and alligators, the order includes two less familiar animals: the caimans and the gavials. Although all crocodilians are fairly

similar in appearance today, much greater diversity existed in the past, including species that were entirely terrestrial and others that achieved a total length in excess of 50 feet.

Crocodiles are largely nocturnal animals that live in or near water in tropical or subtropical regions of Africa, Asia, and the Americas. The American crocodile (*Crocodylus acutus*) is found in southern Florida, Cuba, and throughout tropical Central America. Nile crocodiles (*Crocodylus niloticus*) and estuarine crocodiles (*Crocodylus porosus*) can grow to enormous size and are responsible for many human fatalities each year.

There are only two species of alligators: one living in the southern United States (*Alligator mississippiensis*) and the other a rare endangered species living in China (*Alligator sinensis*). Caimans, which resemble alligators, are native to Central America. Gharials, or Gavials, are a group of fish-eating crocodilians with long, slender snouts that live only in India and Burma.

All crocodilians are carnivores. They generally hunt by stealth, waiting in ambush for prey and then attacking ferociously. Their bodies are well adapted for this form of hunting with eyes on top of their heads and their nostrils on top of their snouts, so they can see and breathe while lying quietly submerged in water. They have enormous mouths, studded with sharp teeth, and very strong necks. A valve in the back of the mouth prevents water from entering the air passage when a crocodilian feeds underwater.

In many ways, crocodiles resemble birds far more than they do other living reptiles. For example, crocodiles build nests and care for their young (traits they share with at least some dinosaurs), and they have a four-chambered heart, as birds do. Why are crocodiles more similar to birds than to other living reptiles? Most biologists agree that birds are in fact the direct descendants of dinosaurs. Both crocodiles and birds are more closely related to dinosaurs, and to each other, than they are to lizards and snakes.

Many major reptile groups that dominated life on land for 250 million years are now extinct. The four living orders of reptiles include the turtles, tuataras, lizards and snakes, and crocodiles.

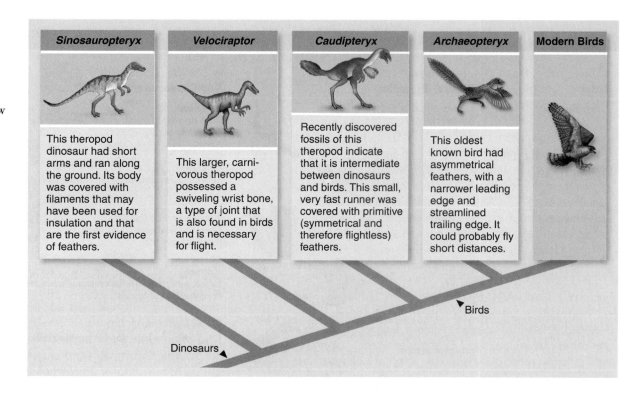

figure 35.27

THE EVOLUTIONARY PATH TO THE BIRDS. Almost all paleontologists now accept the theory that birds are the direct descendants of theropod dinosaurs.

Sinosauropteryx

This theropod dinosaur had short arms and ran along the ground. Its body was covered with filaments that may have been used for insulation and that are the first evidence of feathers.

Velociraptor

This larger, carnivorous theropod possessed a swiveling wrist bone, a type of joint that is also found in birds and is necessary for flight.

Caudipteryx

Recently discovered fossils of this theropod indicate that it is intermediate between dinosaurs and birds. This small, very fast runner was covered with primitive (symmetrical and therefore flightless) feathers.

Archaeopteryx

This oldest known bird had asymmetrical feathers, with a narrower leading edge and streamlined trailing edge. It could probably fly short distances.

Modern Birds

Dinosaurs

Birds

solid, not hollow like a bird's. Also, it had a long, reptilian tail and no enlarged breastbone such as modern birds use to anchor flight muscles. Finally, the skeletal structure of the forelimbs were nearly identical to those of theropods.

Because of its many dinosaur features, several *Archaeopteryx* fossils were originally classified as the coelurosaur *Compsognathus*, a small theropod dinosaur of similar size—until feathers were discovered on the fossils. What makes *Archaeopteryx* distinctly avian is the presence of feathers on its wings and tail.

The remarkable similarity of *Archaeopteryx* to *Compsognathus* has led almost all paleontologists to conclude that *Archaeopteryx* is the direct descendant of dinosaurs—indeed, that today's birds are "feathered dinosaurs." Some even speak flippantly of "carving the dinosaur" at Thanksgiving dinner. The recent discovery of fossils of a feathered dinosaur in China lends strong support to this inference. The dinosaur *Caudipteryx*, for example, is clearly intermediate between *Archaeopteryx* and dinosaurs, having large feathers on its tail and arms but also many features of dinosaurs like *Velociraptor* (figure 35.27). Because the arms of *Caudipteryx* were too short to use as wings, feathers probably didn't evolve for flight, but instead served as insulation, much as fur does for mammals.

Flight is an ability certain kinds of dinosaurs achieved as they evolved longer arms. We call these dinosaurs birds. Despite their close affinity to dinosaurs, birds exhibit three evolutionary novelties: feathers, hollow bones, and physiological mechanisms such as superefficient lungs that permit sustained, powered flight.

By the early Cretaceous period, only a few million years after *Archaeopteryx* lived, a diverse array of birds had evolved, with many of the features of modern birds. Fossils in Mongolia, Spain, and China discovered within the last few years reveal a diverse collection of toothed birds with the hollow bones and breastbones necessary for sustained flight (figure 35.28). Other fossils reveal highly specialized,

figure 35.28

A FOSSIL BIRD FROM THE EARLY CRETACEOUS. *Confuciornis* had long tail feathers. Some fossil specimens of this species lack the long tail feathers, suggesting that this trait was present in only one sex, as in some modern birds.

flightless diving birds. The diverse birds of the Cretaceous shared the skies with pterosaurs for 70 million years.

Because the impression of feathers is rarely fossilized and modern birds have hollow, delicate bones, the fossil record of birds is incomplete. Relationships among the 166 families of modern birds are mostly inferred from studies of the anatomy and degree of DNA similarity among living birds.

Modern birds are diverse but share several characteristics

The most ancient living birds appear to be the flightless birds, such as the ostrich. Ducks, geese, and other waterfowl evolved next, in the early Cretaceous, followed by a diverse group of woodpeckers, parrots, swifts, and owls. The largest of the bird orders, Passeriformes, evolved in the mid-Cretaceous and comprise 60% of species alive today. Overall, there are 28 orders of birds, the largest consisting of over 5000 species (figure 35.29).

You can tell a great deal about the habits and food of a bird by examining its beak and feet. For instance, carnivorous birds such as owls have curved talons for seizing prey and sharp beaks for tearing apart their meal. The beaks of ducks are flat for shoveling through mud, and the beaks of finches are short, thick seed-crushers.

Many adaptations enabled birds to cope with the heavy energy demands of flight, including respiratory and circulatory adaptations and endothermy.

Efficient respiration

Flight muscles consume an enormous amount of oxygen during active flight. The reptilian lung has a limited internal surface area, not nearly enough to absorb all the oxygen needed. Mammalian lungs have a greater surface area, but bird lungs satisfy this challenge with a radical redesign.

When a bird inhales, the air goes past the lungs to a series of air sacs located near and within the hollow bones of the back; from there, the air travels to the lungs and then to a set of anterior air sacs before being exhaled. Because air passes all the way through the lungs in a single direction, gas exchange is highly efficient. Respiration in birds is described in more detail in chapter 49.

Efficient circulation

The revved-up metabolism needed to power active flight also requires very efficient blood circulation, so that the oxygen captured by the lungs can be delivered to the flight muscles quickly. In the heart of most living reptiles, oxygen-rich blood coming from the lungs mixes with oxygen-poor blood returning from the body because the wall dividing the ventricle into two chambers is not complete. In birds, the wall dividing the ventricle is complete, and the two blood circulations do not mix—so flight muscles receive fully oxygenated blood (see chapter 49).

In comparison with reptiles and most other vertebrates, birds have a rapid heartbeat. A hummingbird's heart beats about 600 times a minute, and an active chickadee's heart beats 1000 times a minute. In contrast, the heart of the large, flightless ostrich averages 70 beats per minute—the same rate as the human heart.

Endothermy

Birds, like mammals, are endothermic. Many paleontologists believe the dinosaurs from which birds evolved were endothermic as well. Birds maintain body temperatures significantly higher than those of most mammals, ranging from 40° to 42°C (human body temperature is 37°C). Feathers provide excellent insulation, helping to conserve body heat.

The high temperatures maintained by endothermy permit metabolism in the bird's flight muscles to proceed at a rapid pace, to provide the ATP necessary to drive rapid muscle contraction.

Birds have the greatest diversity of species of all terrestrial vertebrates. *Archaeopteryx*, the oldest fossil bird, exhibited many traits shared with theropod dinosaurs. Modern birds are characterized by feathers, scales on the legs and feet, a thin, hollow skeleton, auxiliary air sacs, and a four-chambered heart. Birds lay amniotic eggs and are endothermic.

Order: Passeriformes

a. *b.* *c.* *d.*

figure 35.29

DIVERSITY OF *Passeriformes***, THE LARGEST ORDER OF BIRDS.** *a.* Summer tanager, *Piranga rubra*; (*b*) Indigo bunting, *Passerina cyanea*; (*c*) Stellar's jay, *Cyanositta stelleri*; (*d*) Bobolink, *Dolichonyx oryzivorus*.

35.8 Mammals

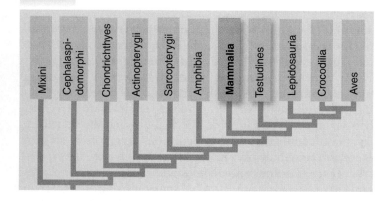

There are about 4500 living species of mammals (class **Mammalia**), the smallest number of species in any of the five classes of vertebrates. Most large, land-dwelling vertebrates are mammals. When we look out over an African plain, we see the big mammals—the lions, zebras, gazelles, and antelope. But the typical mammal is not that large. Of the 4500 species of mammals, 3200 are rodents, bats, shrews, or moles.

Mammals have hair, mammary glands, and other characteristics

Mammals are distinguished from all other classes of vertebrates by two fundamental characteristics—hair and mammary glands—and are marked by several other notable features:

1. **Hair.** All mammals have hair. Even apparently hairless whales and dolphins grow sensitive bristles on their snouts. The evolution of fur and the ability to regulate body temperature enabled mammals to invade colder climates that ectothermic reptiles do not inhabit. Mammals are endothermic animals, and typically maintain body temperatures higher than the temperature of their surroundings. The dense undercoat of many mammals reduces the amount of body heat that escapes.

   Another function of hair is camouflage. The coloration and pattern of a mammal's coat usually matches its background. A little brown mouse is practically invisible against the brown leaf litter of a forest floor, and the orange and black stripes of a Bengal tiger disappear against the orange-brown color of the tall grass in which it hunts. Hairs also function as sensory structures. The whiskers of cats and dogs are stiff hairs that are very sensitive to touch. Mammals that are active at night or live underground often rely on their whiskers to locate prey or to avoid colliding with objects. Finally, hair can serve as a defensive weapon. Porcupines and hedgehogs protect themselves with long, sharp, stiff hairs called quills.

   Unlike feathers, which evolved from modified reptilian scales, mammalian hair is a completely different form of skin structure. An individual mammalian hair is a long, protein-rich filament that extends like a stiff thread from a bulblike foundation beneath the skin known as a hair follicle. The filament is composed mainly of dead cells filled with the fibrous protein keratin.

2. **Mammary glands.** All female mammals possess mammary glands that secrete milk. Newborn mammals, born without teeth, suckle this milk as their primary food. Even baby whales are nursed by their mother's milk. Milk is a very high-calorie food (human milk has 750 kcal per liter), important because of the high energy needs of a rapidly growing newborn mammal. About 50% of the energy in the milk comes from fat.

3. **Endothermy.** As stated previously, mammals are endothermic, a crucial adaptation that has allowed them to be active at any time of the day or night and to colonize severe environments, from deserts to ice fields. Also, more efficient blood circulation provided by the four-chambered heart and more efficient respiration provided by the *diaphragm* (a special sheet of muscles below the rib cage that aids breathing) make possible the higher metabolic rate on which endothermy depends.

4. **Placenta.** In most mammal species, females carry their developing young internally in a uterus, nourishing them through the placenta, and give birth to live young. The **placenta** is a specialized organ that brings the bloodstream of the fetus into close contact with the bloodstream of the mother (figure 35.30). Food, water, and oxygen can pass across from mother to child, and wastes can pass over to the mother's blood and be carried away.

In addition to these main characteristics, the mammalian lineage gave rise to several other adaptations in certain groups.

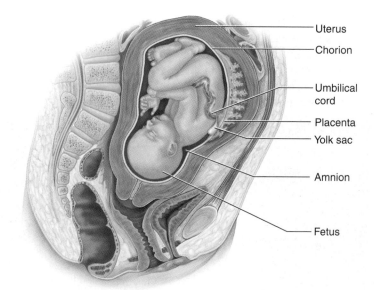

figure 35.30

THE PLACENTA. The placenta is characteristic of the largest group of mammals, the placental mammals. It evolved from membranes in the amniotic egg. The umbilical cord evolved from the allantois. The chorion, or outermost part of the amniotic egg, forms most of the placenta itself. The placenta serves as the provisional lungs, intestine, and kidneys of the fetus, without ever mixing maternal and fetal blood.

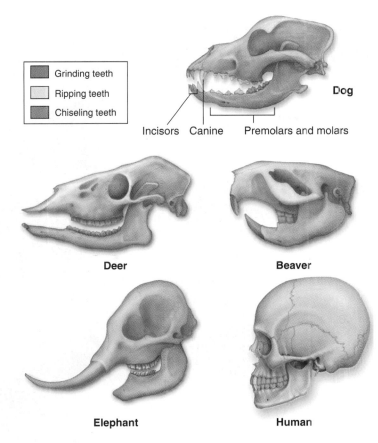

Incisors Canine Premolars and molars

Dog

Deer

Beaver

Elephant

Human

figure 35.31

MAMMALS HAVE DIFFERENT TYPES OF SPECIALIZED TEETH. Carnivores, such as dogs, have canine teeth that are able to rip food; some of the premolars and molars in dogs are also ripping teeth. Herbivores, such as deer, have incisors to chisel off vegetation and molars designed to grind up the plant material. In the beaver, the chiseling incisors dominate. In the elephant, the incisors have become specialized weapons, and molars grind up vegetation. Humans are omnivores; we have all three types: grinding, ripping, and chiseling teeth.

These include specialized teeth, the ability of grazing animals to digest plants, hooves and horns made of keratin, and adaptations for flight in the bats.

Specialized teeth

Mammals have different types of teeth that are highly specialized to match particular eating habits (figure 35.31). It is usually possible to determine a mammal's diet simply by examining its teeth. A dog's long canine teeth, for example, are well suited for biting and holding prey, and some of its premolar and molar teeth are triangular and sharp for ripping off chunks of flesh.

In contrast, large herbivores such as deer lack canine teeth; instead, deer clip off mouthfuls of plants with flat, chisel-like incisors on its lower jaw. The deer's molars are large and covered with ridges to effectively grind and break up tough plant tissues.

Digestion of plants

Most mammals are herbivores, eating mostly or only plants. Cellulose forms the bulk of a plant's body and is a major source of food for mammalian herbivores. Mammals do not have en-

zymes, however, that can break the links between glucose molecules in cellulose. Herbivorous mammals rely on a mutualistic partnership with bacteria in their digestive tracts that have the necessary cellulose-splitting enzymes.

Mammals such as cows, buffalo, antelopes, goats, deer, and giraffes have huge, four-chambered fermentation vats derived from the esophagus and stomach. The first chamber is the largest and holds a dense population of cellulose-digesting bacteria. Chewed plant material passes into this chamber, where the bacteria attack the cellulose. The material is then digested further in the other three chambers.

Rodents, horses, rabbits, and elephants, by contrast, have relatively small stomachs, and instead digest plant material in their large intestine, like a termite. The bacteria that actually carry out the digestion of the cellulose live in a pouch called the cecum that branches from the end of the small intestine.

Even with these complex adaptations for digesting cellulose, a mouthful of plant is less nutritious than a mouthful of meat. Herbivores must consume large amounts of plant material to gain sufficient nutrition. An elephant eats 135–150 kg (300–330 pounds) of plant foods each day.

Development of hooves and horns

Keratin, the protein of hair, is also the structural building material in claws, fingernails, and hooves. Hooves are specialized keratin pads on the toes of horses, cows, sheep, antelopes, and other running mammals. The pads are hard and horny, protecting the toe and cushioning it from impact.

The horns of cattle, sheep, and antelope are composed of a core of bone surrounded by a sheath of keratin. The bony core is attached to the skull, and the horn is not shed.

Deer antlers are made not of keratin, but of bone. Male deer grow and shed a set of antlers each year. While growing during the summer, antlers are covered by a thin layer of skin known as velvet.

Flying mammals: Bats

Bats are the only mammals capable of powered flight (figure 35.32). Like the wings of birds and pterosaurs, bat wings are modified forelimbs. The bat wing is a leathery membrane of

figure 35.32

GREATER HORSESHOE BAT, *Rhinolophus ferrumequinum*. Bats are the only mammal capable of true flight.

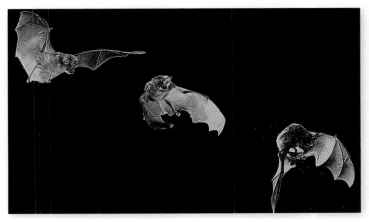

skin and muscle stretched over the bones of four fingers. The edges of the membrane attach to the side of the body and to the hind leg. When resting, most bats prefer to hang upside down by their toe claws.

After rodents, bats are the second largest order of mammals. They have been a particularly successful group because many species have been able to utilize a food resource that most birds do not use—night-flying insects.

How do bats navigate in the dark? Late in the eighteenth century, the Italian biologist Lazzaro Spallanzani showed that a blinded bat could fly without crashing into things and still capture insects. Clearly another sense other than vision was being used by bats to navigate in the dark. When Spallanzani plugged the ears of a bat, it was unable to navigate and collided with objects. Spallanzani concluded that bats "hear" their way through the night world. (Chapter 45 describes bats' use of echolocation for night navigation.)

Mammals diverged about 220 MYA

Mammals have been around since the time of the dinosaurs, about 220 MYA. Tiny, shrewlike creatures that lived in trees eating insects, the earliest mammals were only a minor element in a land that quickly came to be dominated by dinosaurs. Fossils reveal that these early mammals had large eye sockets, evidence that they may have been active at night. Early mammals also had a single lower jawbone. The fossil record shows a change in therapsids (the ancestors of mammals) from the reptile lower jaw, having several bones, to a jaw closer to the mammalian type. Two of the bones forming the therapsid jaw joint moved into the middle ear of mammals, linking with a bone already there to produce a three-bone structure that amplifies sound better than the reptilian ear.

The age of mammals

At the end of the Cretaceous period, 65 MYA, the dinosaurs and numerous other land and marine animals became extinct, but mammals survived, possibly because of the insulation their fur provided. In the Tertiary period (lasting from 65 MYA to 2 MYA), mammals rapidly diversified, taking over many of the ecological roles once dominated by dinosaurs.

Mammals reached their maximum diversity late in the Tertiary period, about 15 MYA. At that time, tropical conditions existed over much of the world. During the last 15 million years, world climates have changed, and the area covered by tropical habitats has decreased, causing a decline in the total number of mammalian species (table 35.5).

Modern mammals are placed into three groups, of which placental mammals are the largest

For 155 million years, while the dinosaurs flourished, mammals were a minor group of small insectivores and herbivores. The most primitive mammals were members of the subclass **Prototheria.** Most prototherians were small and resembled modern shrews. All prototherians laid eggs, as did their synapsid ancestors. The only prototherians surviving today are the monotremes.

TABLE 35.5	Some Groups of Extinct Mammals
Group	**Description**
Cave bears	Numerous in the ice ages; this enormous bear had a largely vegetarian diet and slept through the winter in large groups.
Irish elk	Neither Irish nor an elk (it is a kind of deer), *Megaloceros* was the largest deer that ever lived, with horns spanning 12 feet. Seen in French cave paintings, they became extinct about 2500 years ago.
Mammoths	Although only two species of elephants survive today, the elephant family was far more diverse during the late Tertiary. Many were cold-adapted mammoths with long, shaggy fur.
Giant ground sloths	*Megatherium* was a giant, 20-foot ground sloth that weighed three tons and was as large as a modern elephant.
Sabertooth cats	The jaws of these large, lionlike cats opened an incredible 120° to allow the animal to drive its huge upper pair of saber teeth into prey.

The other major mammalian group is the subclass **Theria.** Therians are viviparous (that is, their young are born alive). The two living therian groups are marsupials, or pouched mammals (including kangaroos, opossums, and koalas), and the placental mammals (dogs, cats, humans, horses, and most other mammals).

Monotremes: Egg-laying mammals

The duck-billed platypus (*Ornithorhynchus anatinus*) and two species of echidna are the only living **monotremes** (figure 35.33*a*). Among living mammals, only monotremes lay shelled eggs. The structure of their shoulder and pelvis is more similar to that of the early reptiles than to any other living mammal. Also like reptiles, monotremes have a cloaca, a single opening through which feces, urine, and reproductive products leave the body.

Monotremes

a.

Marsupials

b.

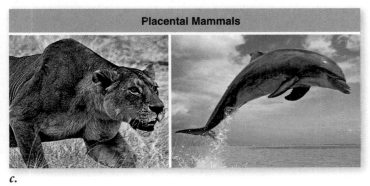

Placental Mammals

c.

figure 35.33

TODAY'S MAMMALS. *a.* Monotremes, the short-nosed echidna, *Tachyglossus aculeatus* (*left*), and the duck-billed platypus, *Ornithorhynchus anatinus* (*right*); (*b*) Marsupials, the red kangaroo, *Macropus rufus* (*left*) and the opossum, *Didelphis virginiana*, (*right*); (*c*) Placental mammals, the lion, *Panthera leo* (*left*) and the bottle-nosed dolphin, *Tursiops truncatus* (*right*).

Despite the retention of some reptilian features, monotremes have the diagnostic mammalian characters: a single bone on each side of the lower jaw, fur, and mammary glands. Young monotremes drink their mother's milk after they hatch from eggs. Females lack well-developed nipples; instead, the milk oozes onto the mother's fur, and the babies lap it off with their tongues.

The platypus, found only in Australia, lives much of its life in the water and is a good swimmer. It uses its bill much as a duck does, rooting in the mud for worms and other soft-bodied animals. Echidnas of Australia (*Tachyglossus aculeatus*, the short-nosed echidna) and New Guinea (*Zaglossus bruijni*, the long-nosed echidna) have very strong, sharp claws, which they use for burrowing and digging. The echidna probes with its snout for insects, especially ants and termites.

Marsupials: Pouched mammals

The major difference between **marsupials** (figure 35.33*b*) and other mammals is their pattern of embryonic development. In marsupials, a fertilized egg is surrounded by chorion and amniotic membranes, but no shell forms around the egg as it does in monotremes. During most of its early development, the marsupial embryo is nourished by an abundant yolk within the egg. Shortly before birth, a short-lived placenta forms from the chorion membrane. Soon after, sometimes within eight days of fertilization, the embryonic marsupial is born. It emerges tiny and hairless, and crawls into the marsupial pouch, where it latches onto a mammary-gland nipple and continues its development.

Marsupials evolved shortly before placental mammals, about 125 MYA. Today, most species of marsupials live in Australia and South America, areas that have undergone long periods of geographic isolation. Marsupials in Australia and New Guinea have diversified to fill ecological positions occupied by placental mammals elsewhere in the world (see figure. 22.20). The placental mammals in Australia and New Guinea today arrived relatively recently and include some introduced by humans. The only marsupial found in North America is the Virginia opossum (*Didelphis virginiana*), which has migrated north from Central America within the last three million years.

Placental mammals

A placenta that nourishes the embryo throughout its entire development forms in the uterus of placental mammals (figure 35.33*c*). Most species of mammals living today, including humans, are in this group. Of the 19 orders of living mammals, 17 are placental mammals (although some scientists recognize four orders of marsupials, rather than one). Table 35.6 (p. 708) shows some of these orders. They are a very diverse group, ranging in size from 1.5-g pygmy shrews to 100,000-kg whales.

Early in the course of embryonic development, the placenta forms. Both fetal and maternal blood vessels are abundant in the placenta, and substances can be exchanged efficiently between the bloodstreams of mother and offspring (see figure 35.30). The fetal placenta is formed from the membranes of the chorion and allantois. In placental mammals, unlike in marsupials, the young undergo a considerable period of development before they are born.

Mammals were not a major group until the dinosaurs disappeared. Mammals are the only animals with hair and mammary glands. Other mammalian specializations include the placenta, a tooth design suited to diet, and specialized sensory systems. Today three subgroups of mammals are recognized: monotremes, marsupials, and placental mammals.

| | | | TABLE 35.6 | Major Orders of Placental Mammals |

Order	Typical Examples		Key Characteristics	Approximate Number of Living Species
Rodentia	Beavers, mice, porcupines, rats		*Small plant-eaters* Chisel-like incisor teeth	1814
Chiroptera	Bats		*Flying mammals* Primarily fruit- or insect-eaters; elongated fingers; thin wing membrane; mostly nocturnal; navigate by sonar	986
Insectivora	Moles, shrews		*Small, burrowing mammals* Insect-eaters; the most primitive placental mammals; spend most of their time underground	390
Carnivora	Bears, cats, raccoons, weasels, dogs		*Carnivorous predators* Teeth adapted for shearing flesh; no native families in Australia	274
Primates	Apes, humans, lemurs, monkeys		*Tree-dwellers* Large brain size; binocular vision; opposable thumb; group that evolved from a line that branched off early from other mammals	233
Artiodactyla	Cattle, deer, giraffes, pigs		*Hoofed mammals with two or four toes* Most species are herbivorous ruminants	211
Cetacea	Dolphins, porpoises, whales		*Fully marine mammals* Streamlined bodies; front limbs modified into flippers; no hindlimbs; blowholes on top of head; no hair except on muzzle	79
Lagomorpha	Rabbits, hares, pika		*Rodentlike jumpers* Four upper incisors (rather than the two seen in rodents); hind legs often longer than forelegs, an adaptation for jumping	69
Edentata	Anteaters, armadillos, sloths		*Toothless insect-eaters* Many are toothless, but some have degenerate, peglike teeth	30
Perissodactyla	Horses, rhinoceroses, tapirs		*Hoofed mammals with odd number of toes* Herbivorous teeth adapted for chewing	17
Proboscidea	Elephants		*Long-trunked herbivores* Two upper incisors elongated as tusks; largest living land animal	2

35.9 Evolution of the Primates

Primates are the mammalian group that gave rise to our own species. Primates evolved two distinct features that allowed them to succeed in as arboreal (tree-dwelling) insectivores.

1. **Grasping fingers and toes.** Unlike the clawed feet of tree shrews and squirrels, primates have grasping hands and feet that enable them to grip limbs, hang from branches, seize food, and in some primates, use tools. The first digit in many primates, namely the thumb, is opposable, and at least some, if not all, of the digits have nails.

2. **Binocular vision.** Unlike the eyes of shrews and squirrels, which sit on each side of the head, the eyes of primates are shifted forward to the front of the face. This produces overlapping binocular vision that lets the brain judge distance precisely—important to an animal moving through the trees and trying to grab or pick up food items.

Other mammals have binocular vision—for example, carnivorous predators—but only primates have both binocular vision and grasping hands, making them particularly well adapted to their arboreal environment.

The anthropoid lineage led to the earliest humans

About 40 MYA, the earliest primates split into two groups: the prosimians and the anthropoids. The **prosimians** ("before monkeys") looked something like a cross between a squirrel and a cat and were common in North America, Europe, Asia, and Africa. Only a few prosimians survive today—lemurs, lorises, and tarsiers (figure 35.34). In addition to grasping digits and binocular vision, prosimians have large eyes with increased visual acuity. Most prosimians are nocturnal, feeding on fruits, leaves, and flowers, and many lemurs have long tails for balancing.

Anthropoids

Anthropoids include monkeys, apes, and humans. Anthropoids are almost all diurnal—that is, active during the day—feeding mainly on fruits and leaves. Natural selection favored

figure 35.34

A PROSIMIAN. This tarsier, *Tarsius*, a prosimian native to tropical Asia, shows the characteristic features of primates: grasping fingers and toes and binocular vision.

many changes in eye design, including color vision, that were adaptations to daytime foraging. An expanded brain governs the improved senses, with the braincase forming a larger portion of the head.

Anthropoids, like the relatively few diurnal prosimians, live in groups with complex social interactions. They tend to care for their young for prolonged periods, allowing for a long childhood of learning and brain development.

About 30 MYA, some anthropoids migrated to South America. Their descendants, known as the New World monkeys (figure 35.35a), are easy to identify: All are arboreal; they have flat, spreading noses; and many of them grasp objects with long, prehensile tails.

Anthropoids that remained in Africa gave rise to two lineages: the Old World monkeys (figure 35.35b) and the hominoids (apes and humans, figure 35.35c). Old World monkeys include ground-dwelling as well as arboreal species. None of them have prehensile tails, their nostrils are close together, their noses point downward, and some have toughened pads of skin on their rumps for prolonged sitting.

Hominoids

The **hominoids** include the apes and the **hominids** (humans and their direct ancestors). The living apes consist of the gibbon (genus *Hylobates*), orangutan (*Pongo*), gorilla (*Gorilla*), and chimpanzee (*Pan*). Apes have larger brains than monkeys, and they lack tails. With the exception of the gibbon, which is small, all living apes are larger than any monkey. Apes exhibit

New World Monkeys	Old World Monkeys	Hominoids

a.	*b.*	*c.*

figure 35.35

ANTHROPOIDS. *a.* New World Monkey, the squirrel monkey, *Saimiri oerstedii;* *(b)* Old World Monkey, the mandrill, *Mandrillus sphinx;* *(c)* hominoids, gorilla, *Gorilla gorilla* (*left*) and human, *Homo sapiens* (*right*).

the most adaptable behavior of any mammal except human beings. Once widespread in Africa and Asia, apes are rare today, living in relatively small areas. No apes ever occurred in North or South America.

Studies of ape DNA have explained a great deal about how the living apes evolved. The Asian apes evolved first. The line of apes leading to gibbons diverged from other apes about 15 MYA, whereas orangutans split off about 10 MYA (figure 35.36). Neither group is closely related to humans.

The African apes evolved more recently, between 6 and 10 MYA. These apes are the closest living relatives to humans. The taxonomic group "apes" is a paraphyletic group; some apes are more closely related to hominids than they are to other apes. For this reason, some taxonomists have advocated placing humans and the African apes in the same zoological family, the Hominidae.

Fossils of the earliest hominids (humans and their direct ancestors), described later in this section, suggest that the common ancestor of the hominids was more like a chimpanzee than a gorilla. Based on genetic differences, scientists estimate that gorillas diverged from the line leading to chimpanzees and humans some 8 MYA.

Soon after the gorilla lineage diverged, the common ancestor of all hominids split off from the chimpanzee line to begin the evolutionary journey leading to humans. Because this split was so recent, few genetic differences between humans and chimpanzees have had time to evolve. For example, a human hemoglobin molecule differs from its chimpanzee counterpart in only a single amino acid. In general, humans and chimpanzees exhibit a level of genetic similarity normally found between closely related species of the same genus!

Comparing apes with hominids

The common ancestor of apes and hominids is thought to have been an arboreal climber. Much of the subsequent evolution of the hominoids reflected different approaches to locomotion. Hominids became bipedal, walking upright; in contrast, the apes evolved knuckle-walking, supporting their weight on the dorsal sides of their fingers. (Monkeys, by contrast, walk using the palms of their hands.)

Humans depart from apes in several areas of anatomy related to bipedal locomotion. Because humans walk on two legs, their vertebral column is more curved than an ape's, and the human spinal cord exits from the bottom rather than the back of the skull. The human pelvis has become broader and more bowl-shaped, with the bones curving forward to center the weight of the body over the legs. The hip, knee, and foot have all changed proportions.

Being bipedal, humans carry much of the body's weight on the lower limbs, which comprise 32–38% of the body's weight and are longer than the upper limbs; human upper limbs do not bear the body's weight and make up only 7–9% of human body weight. African apes walk on all fours, with the upper and lower limbs both bearing the body's weight; in gorillas, the longer upper limbs account for 14–16% of body weight, the somewhat shorter lower limbs for about 18%.

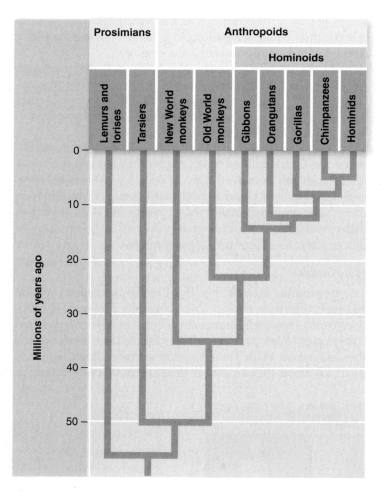

figure 35.36

A PRIMATE EVOLUTIONARY TREE. Prosimians diverged early in primate evolution, whereas hominids diverged much more recently. Apes constitute a paraphyletic group because some apes are more closely related to nonape species (hominids) than they are to other apes.

Australopithecines were early hominids

Five to 10 MYA, the world's climate began to get cooler, and the great forests of Africa were largely replaced with savannas and open woodland. In response to these changes, a new kind of hominoid was evolving, one that was bipedal. These new hominoids are classified as hominids—that is, of the human line.

The major groups of hominids include three to seven species of the genus *Homo* (depending how you count them), seven species of the older, smaller-brained genus *Australopithecus*, and several even older lineages. In every case where the fossils allow a determination to be made, the hominids are bipedal, the hallmark of hominid evolution.

In recent years, anthropologists have found a remarkable series of early hominid fossils extending as far back as 6–7 million years. Often displaying a mixture of primitive and modern traits, these fossils have thrown the study of early hominids into turmoil. Although the inclusion of these fossils among the hominids seems warranted, only a few specimens have been discovered, and they do not provide enough information to determine with certainty their relationships to australopithecines and humans. The search for additional early hominid fossils continues.

Early australopithecines

Our knowledge of australopithecines is based on hundreds of fossils, all found in South and East Africa (except for one specimen from Chad in West Africa). Australopithecines may have lived over a much broader area of Africa, but rocks of the proper age that might contain fossils are not exposed elsewhere. The evolution of hominids seems to have begun with an initial radiation of numerous species. The seven species identified so far provide ample evidence that australopithecines were a diverse group.

These early hominids weighed about 18 kg and were about 1 m tall. Their dentition was distinctly hominid, but their brains were no larger than those of apes, generally 500 cubic centimeters (cc) or less. *Homo* brains, by comparison, are usually larger than 600 cc; modern *H. sapiens* brains average 1350 cc.

The structure of australopithecine fossils clearly indicates that they walked upright. Evidence of bipedalism includes a set of some 69 hominid footprints found at Laetoli, East Africa. Two individuals, one larger than the other, walked upright side-by-side for 27 m, their footprints preserved in a layer of 3.7-million-year-old volcanic ash. Importantly, the big toe is not splayed out to the side as in a monkey or ape, indicating that these footprints were clearly made by hominids.

Bipedalism

The evolution of **bipedalism** marks the beginning of hominids. Bipedalism seems to have evolved as australopithecines left dense forests for grasslands and open woodland.

Whether larger brains or bipedalism evolved first was a matter of debate for some time. One school of thought proposed that hominid brains enlarged first, and then hominids became bipedal. Another school of thought saw bipedalism as a precursor to larger brains, arguing that bipedalism freed the forelimbs to manufacture and use tools, leading to the evolution of bigger brains. Recently, fossils unearthed in Africa have settled the debate. These fossils demonstrate that bipedalism extended back 4 million years; knee joint, pelvis, and leg bones all exhibit the hallmarks of an upright stance. Substantial brain expansion, on the other hand, did not appear until roughly 2 MYA. In hominid evolution, upright walking clearly preceded large brains.

The reason bipedalism evolved in hominids remains a matter of controversy. No tools appeared until 2.5 MYA, so tool-making seems an unlikely cause. Alternative ideas suggest that walking upright is faster and uses less energy than walking on four legs; that an upright posture permits hominids to pick fruit from trees and see over tall grass; that being upright reduces the body surface exposed to the Sun's rays; that an upright stance aided the wading of semiaquatic hominids; and that bipedalism frees the forelimbs of males to carry food back to females, encouraging pair-bonding. All of these suggestions have their proponents, and none is universally accepted. The origin of bipedalism, the key event in the evolution of hominids, remains a mystery.

> Primates include the prosimians and anthropoids. The apes, monkeys, and hominids make up the anthropoids; apes and hominids are the hominoids. The evolution of bipedalism—walking upright—marks the beginning of hominid evolution, although no one is quite sure why bipedalism evolved. The root of the hominid evolutionary tree is only imperfectly known, but it appears to have led through the australopithecines.

The genus *Homo* arose roughly 2 MYA

The first humans (genus *Homo*) evolved from australopithecine ancestors about 2 MYA. The exact ancestor has not been clearly identified, but is commonly thought to be *Australopithecus afarensis*. Only within the last 30 years have a significant number of fossils of early *Homo* been uncovered. An explosion of interest has fueled intensive field exploration, and new finds are announced regularly; every year, our picture of the base of the human evolutionary tree grows clearer. The following historical account will undoubtedly be supplanted by future discoveries, but it provides a good example of science at work.

The first human: Homo habilis

In the early 1960s, stone tools were found scattered among hominid bones close to the site where *A. boisei* had been unearthed. Although the fossils were badly crushed, painstaking reconstruction of the many pieces suggested a skull with a brain volume of about 680 cc, larger than the australopithecine range of 400–550 cc. Because of its association with tools, this early human was called *Homo habilis*, meaning "handy man." Partial skeletons discovered in 1986 indicate that *H. habilis* was small in stature, with arms longer than its legs and a skeleton much like that of *Australopithecus*. Because of its general similarity to australopithecines, many researchers at first questioned whether this fossil was human.

How diverse was early Homo?

Because so few fossils of early *Homo* have been found, lively debate has ensued concerning whether they should all be lumped into *H. habilis* or split into three species: *H. rudolfensis*, *H. habilis*, and *H. ergaster*. If the three species designations are accepted, a view held by increasing numbers of researchers, it would appear that *Homo* underwent an adaptive radiation, with *H. rudolfensis* the most ancient species, followed by *H. habilis* and then *H. ergaster*. Because of its

Genetic Similarity	Skin Pigmentation
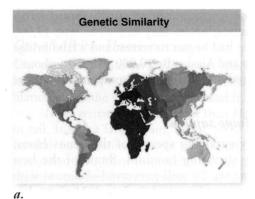	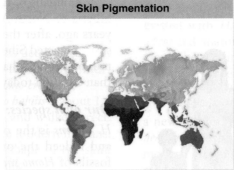
a.	*b.*

figure 35.41

PATTERNS OF GENETIC VARIATION IN HUMAN POPULATIONS DIFFER FROM PATTERNS OF SKIN COLOR VARIATION.
a. Genetic variation among *Homo sapiens.* The more similar areas are in color, the more similar they are genetically based on many enzyme and blood group genetic loci. *b.* Similarity among *Homo sapiens* based on skin color. The color of an area represents the skin pigmentation of the people native to that region.

Human races

Human beings, like all other species, have differentiated in their characteristics as they have spread throughout the world. Local populations in one area often appear significantly different from those that live elsewhere. For example, northern Europeans often have blond hair, fair skin, and blue eyes, whereas Africans often have black hair, dark skin, and brown eyes. These traits may play a role in adapting the particular populations to their environments. Blood groups may be associated with immunity to diseases more common in certain geographical areas, and dark skin shields the body from the damaging effects of ultraviolet radiation, which is much stronger in the tropics than in temperate regions.

All human beings are capable of mating with one another and producing fertile offspring. The reasons that they do or do not choose to associate with one another are purely psychological and behavioral (cultural).

The number of groups into which the human species might logically be divided has long been a point of contention. Some contemporary anthropologists divide people into as many as 30 "races," others as few as three: Caucasoid, Negroid, and Oriental. American Indians, Bushmen, and Aborigines are examples of particularly distinctive subunits that are sometimes regarded as distinct groups.

The problem with classifying people or other organisms into races in this fashion is that the characteristics used to define the races are usually not well correlated with one another, and so the determination of race is always somewhat arbitrary. Humans are visually oriented; consequently, we have relied on visual cues—primarily skin color—to define races. However, when other types of characteristics, such as blood groups, are examined, patterns of variation correspond very poorly with visually determined racial classes. Indeed, if one were to break the human species into subunits based on overall genetic similarity, the groupings would be very different from those based on skin color or other visual features (figure 35.41).

In human beings, it is simply not possible to delimit clearly defined races that reflect biologically differentiated and well-defined groupings. The reason is simple: Different groups of people have constantly intermingled and interbred with one another during the entire course of history. This constant gene flow has prevented the human species from fragmenting into highly differentiated subspecies. Those characteristics that are differentiated among populations, such as skin color, represent classic examples of the antagonism between gene flow and natural selection. As you saw in chapter 20, when selection is strong enough, as it is for dark coloration in tropical regions, populations can differentiate even in the presence of gene flow. However, even in cases such as this, gene flow will still ensure that populations are relatively homogeneous for genetic variation at other loci.

For this reason, relatively little of the variation in the human species represents differences between the described races. Indeed, one study calculated that only 8% of all genetic variation among humans could be accounted for as differences that exist among racial groups; in other words, the human racial categories do a very poor job in describing the vast majority of genetic variation that exists in humans. For this reason, most modern biologists reject human racial classifications as reflecting patterns of biological differentiation in the human species. This is a sound biological basis for dealing with each human being on his or her own merits and not as a member of a particular "race."

Several species of *Homo* evolved in Africa, and some migrated from there to Europe and Asia. *Homo sapiens*, our own species, seems to have evolved in Africa and then, like *H. erectus* before it, migrated to Europe and Asia. *Homo sapiens* is proficient at conceptual thought and tool use and is the only animal that uses symbolic language.

Human races do not reflect significant patterns of underlying biological differentiation.

35.1 The Chordates

Chordates are deuterostome coelomates most closely related to echinoderms.

- Chordates share four features at some time during their development: a single, hollow nerve tube, a flexible notochord, pharyngeal slits, and a postanal tail (figure 35.1).

35.2 The Nonvertebrate Chordates

The phylum Chordata can be divided into three subphyla: the Vertebrata and two, the Urochordata and Cephalochordata, which are not vertebrates.

- The mobile Urochordata larvae have a notochord and a nerve cord, but the adults are immobile, have no major body cavity, nor any visible signs of segmentation. Many have a tunic composed mainly of cellulose (figure 35.4).

- The Cephalchordata have a persistent notochord that runs the length of the dorsal nerve cord, segmented myomeres, and a single cell layer of skin, but they do not have bones or a distinct head as adults (figure 35.3).

35.3 The Vertebrate Chordates

The Vertebrata are chordates with a spinal column composed of bony or cartilaginous vertebrae.

- Vertebrates are separate from the other chordate subphyla because they have a vertebral column that encloses and protects the dorsal nerve cord and a distinct and well-differentiated head with sensory organs.

- Vertebrates also have a neural crest during embryonic development, internal organs and an endoskeleton composed of calcium phosphate (figures 35.6 and 35.7).

35.4 Fishes

Over half of all vertebrates are fishes (figure 35.8)

- Fishes exhibit several key characteristics: vertebral column of bone or cartilage, jaws and paired appendages, internal gills, and a closed circulatory system.

- The jaw evolved from the anterior gill arches of ancient jawless fishes (figure 35.10).

- Fishes have a lateral line system that senses changes in pressure waves.

- Most bony fishes have an operculum that covers the gills, but cartilaginous fishes do not.

- Bony fishes belong either to the ray-finned fishes, the Actinopterygii, or the lobe-finned fishes, the Sarcopterygii.

- Ray-finned fishes have fins stiffened with bony parallel rays; the lobe-finned fishes contain muscle lobes and bones that form fully articulated joints with one another (figure 35.13).

35.5 Amphibians

The amphibians are damp-skinned vertebrates directly descended from fishes.

- Living amphibians have five distinguishing features: legs, lungs, cutaneous respiration, pulmonary veins, and a partially divided heart.

- Amphibian invasion of land posed many problems to organisms whose ancestors lived in water, such as supporting large bodies, and breathing out of water, and preventing dessication.

- Modern amphibians belong to three groups: the Anura or frogs and toads, that do not have tails as adults; the elongated Caudata, or salamanders; and the legless Apoda, or caecilians.

35.6 Reptiles

Reptiles are primarily terrestrial species with a dry, scaly skin.

- Reptiles exhibit three key characteristics: an amniotic egg that is watertight; dry, watertight skin; and thoracic breathing (figure 35.17).

- Reptiles use negative pressure to ventilate their lungs created by expanding the rib cage and sucking in air.

- Modern reptiles practice internal fertilization and are ectothermic, obtaining their heat from external sources.

- Modern reptiles belong to four groups: Chelonia, or turtles; Rhynchocephalia, or tuataras; Squamata, or lizards and snakes; and Crocodylia, or crocodiles and alligators.

35.7 Birds

Birds are the most diverse of the terrestrial vertebrates and possess a unique adaptation, the feather (figure 35.25 and table 35.4).

- The two key characteristics of birds are a modified reptilian scale, or feather, which provides lift and conserves heat, and a lightweight flight skeleton.

- Birds evolved from theropod dinosaurs (figure 35.27).

- Modern birds share several characteristics: efficient respiration and circulation, and endothermy.

35.8 Mammals

Mammals evolved from therapsid reptiles and are easily distinguished from all other classes of vertebrates.

- Mammals are distinguished by fur and mammary glands.

- Mammals are also endothermic, and for most, embryonic development occurs in a uterus and the fetus is connected to the mother by the placenta.

- Mammals have teeth matched to their food source, and most are herbivores.

- Modern mammals are placed either in the Prototheria, or monotremes that lay shelled eggs, or the viviparous Theria.

- The Theria consist of marsupials, in which the embryo develops primarily externally in a pouch, and the placental mammals.

35.9 Evolution of the Primates (figure 35.36)

Primates are the mammalian group that gave rise to our own species.

- Primates share two innovations grasping fingers and toes, and binocular vision.

- The earliest primates gave rise to the prosimians, including lemurs, lorises and tarsiers, and the anthropoids including monkeys, apes, and humans.

- Hominoids include apes and hominids, or humans.

- The hallmark of hominid evolution is upright posture and bipedal locomotion. Apes use knuckle walking.

- The genus *Homo* arose roughly 2 MYA from australopithecine ancestors.

- Common features of early *Homo* species include a larger body and brain size.

- *Homo sapiens* is the only surviving species of the genus *Homo* and is proficient at conceptual thought, tool use, and symbolic language.

SELF TEST

1. Which of the following statements regarding all species of chordates is false?
   a. Chordates are deuterostomes.
   b. A notochord is present in the embryo.
   c. The notochord is surrounded by bone or cartilage.
   d. All possess a postanal tail during embryonic development.

2. In the following figure, item A is the ___ and item B is the ___.
   a. complete digestive system; notochord
   b. spinal cord; nerve cord
   c. notochord; nerve cord
   d. pharyngeal slits; notochord

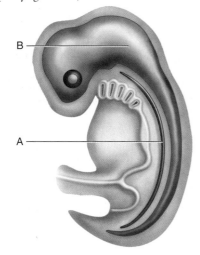

3. During embryonic development, a neural crest would be found in all of the following chordates, except
   a. cephalochordates.
   b. reptiles.
   c. birds.
   d. mammals.

4. A single-loop closed circulatory system is a characteristic of all
   a. amphibians.
   b. birds.
   c. reptiles.
   d. fish.

5. The _____ of the bony fish evolved to counter the effects of increased bone density.
   a. gills
   b. jaws
   c. swim bladder
   d. teeth

6. From which group of fish did the amphibians originate?
   a. Ray-finned
   b. Lobe-finned
   c. Cartilaginous
   d. Acanthodians

7. *Icthyostega* represents
   a. the first chordate.
   b. a feathered reptile.
   c. the first amphibian.
   d. an early cartilaginous fish.

8. Why was the evolution of the pulmonary veins important for amphibians?
   a. To move oxygen to and from the lungs
   b. To increase the metabolic rate
   c. For increased blood circulation to the brain
   d. None of the above

9. The first group of animals to utilize an amniotic egg are the
   a. birds.
   b. mammals.
   c. amphibians.
   d. reptiles.

10. Which of the following groups lacks a four-chambered heart?
    a. Birds
    b. Reptiles
    c. Mammals
    d. Amphibians

11. All of the following are characteristics of reptiles, except
    a. cutaneous respiration.
    b. amniotic egg.
    c. thoracic breathing.
    d. dry, watertight skin.

12. Which of the following evolutionary adaptations allows the birds to become efficient at flying?
    a. Structure of the feather
    b. High metabolic temperatures
    c. Increased respiratory efficiency
    d. All of the above

13. The group of mammals that is most closely related to the reptiles is the
    a. therapsids.
    b. marsupials.
    c. monotremes.
    d. placentals.

14. Which of the following groups only includes the apes and humans and their direct ancestors?
    a. Hominoids
    b. Primates
    c. Arthropoids
    d. Placentals

15. Which of the following species was the first hominid to move from Africa in social groups?
    a. *Homo habilis*
    b. *Homo erectus*
    c. *Homo sapiens*
    d. *Homo floresiensis*

CHALLENGE QUESTIONS

1. Some scientists believe the feathers did not evolve initially for flight, but rather for insulation. What benefits would this have had for early flightless birds?

2. Some people state that the dinosaurs have not "gone extinct," they are with us today. What evidence can be used to support this statement?

3. Some people may state that humans are evolved from apes, yet this statement is not truly correct. Why?

chapter 36

Plant Form

introduction

ALTHOUGH THE SIMILARITIES AMONG a cactus, an orchid, and a hardwood tree might not be obvious at first sight, most plants have a basic unity of structure. This unity is reflected in how the plants are constructed; in how they grow, manufacture, and transport their food; and in how their development is regulated. This chapter addresses the question of how a vascular plant is "built." We will focus on the cells, tissues, and organs that compose the adult plant body. The roots and shoots that give the adult plant its distinct above- and below-ground architecture are the final product of a basic body plan first established during embryogenesis, a process we will explore in detail in this chapter

Organization of the Plant Body: An Overview

As you learned in chapter 30, the plant kingdom has great diversity, not only among its many phyla but even within species. The earliest vascular plants, many of which are extinct, did not have clear differentiation of the plant body into specialized organs such as roots and leaves.

Among modern vascular plants, the presence of these organs reflects increasing specialization, particularly in relation to the demands of a terrestrial existence. Obtaining water, for example, is a major challenge on land and roots are adapted for water absorption from the soil. Leaves, roots, branches, and flowers all exhibit variations in size and number from plant to plant. The development of the form and structure of these parts may be precisely controlled, but some aspects of leaf, stem, and root development are quite flexible. This chapter emphasizes the unifying aspects of plant form, using the flowering plants as a model.

Vascular plants have roots and shoots

A vascular plant consists of a root system and a shoot system (figure 36.1). Roots and shoots grow at their tips, which are called apices (singular, **apex**).

The **root system** anchors the plant and penetrates the soil, from which it absorbs water and ions crucial for the plant's nutrition. Root systems are often extensive, and growing roots can exert great force to move matter as they elongate and expand. Roots developed later than the shoot system as an adaptation to living on land.

The **shoot system** consists of the stems and their leaves. Stems serve as a scaffold for positioning the leaves, the principal sites of photosynthesis. The arrangement, size, and other features of the leaves are critically important in the plant's production of food. Flowers, other reproductive organs, and ultimately, fruits and seeds are also formed on the shoot (flower morphology and plant reproduction is covered in chapter 42).

The iterative (repeating) unit of the vegetative shoot consists of the internode, node, leaf, and axillary bud, but not reproductive structures. An axillary bud is a lateral shoot apex that allows the plant to branch or replace the main shoot if it is munched by an herbivore. A vegetative axillary bud has the capacity to reiterate the development of the primary shoot. When the plant has shifted to the reproductive phase of development, these axillaries may produce flowers or floral shoots.

Roots and shoots are composed of three types of tissues

Roots, shoots, and leaves all contain three basic types of tissues: **dermal, ground,** and **vascular tissue.** Like organs in animal bodies, these tissues consist of one or more cell types. Because each of these tissues extend through the root and shoot systems, they are called **tissue systems.**

Plant cell types can be distinguished by the size of their vacuoles, whether they are living or not at maturity, and by the thickness of secretions found in their cellulose cell walls, a distinguishing feature of plant cells (see chapter 4 to review cell structure). Some cells have only a primary cell wall of cellulose, synthesized at the protoplast (cell membrane). Microtubules align

within the cell and determine the orientation of the cellulose fibers (figure 36.2a). Cells that support the plant body have more heavily reinforced cell walls with multiple layers of cellulose. Cellulose layers are laid down at angles to adjacent layers like plywood; this enhances the strength of the cell wall (figure 36.2b).

Dermal tissue, primarily *epidermis,* is one cell layer thick in most plants, and it forms an outer protective covering for the plant. In most plants, a layer of wax is added to the outer layer of the epidermis to limit water loss and ultraviolet light damage, a key adaptation to land. Desert succulents have additional

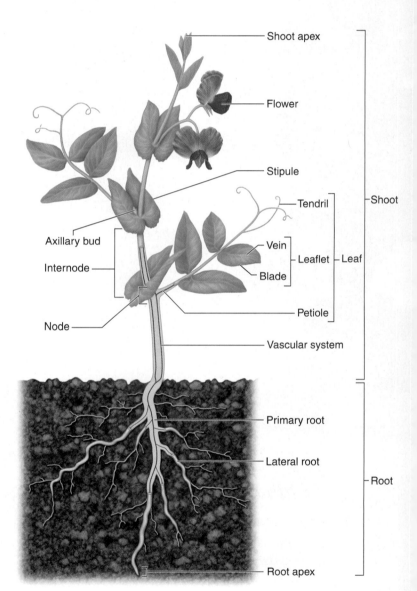

figure 36.1

DIAGRAM OF A PLANT BODY. Branching root and shoot systems create the plant's architecture. Each root and shoot has an apex that extends growth. Leaves are initiated at the nodes of the shoot, which also contain axillary buds that can remain dormant, grow to form lateral branches, or make flowers. A leaf can be a simple blade or consist of multiple parts as shown here. Roots, shoots, and leaves are all connected with vascular (conducting) tissue.

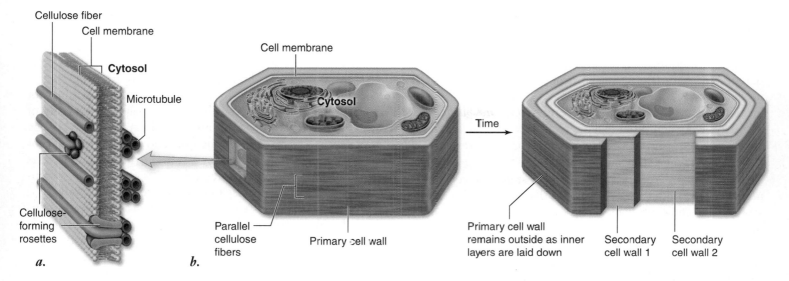

figure 36.2

SYNTHESIS OF A PLANT CELL WALL. *a.* Cellulose is a glucose polymer that is produced at the cellulose-forming rosettes in the cell membrane to form the cell wall. Cellulose fibers are laid down parallel to microtubules inside the cell membrane. Additional substances that strengthen and waterproof the cell wall are added to the cell wall in some cell types. *b.* Some cells extrude additional layers of cellulose, increasing the mechanical strength of the wall. Because new cellulose is produced at the cell, the oldest layers of cellulose are on the outside of the cell wall. All cells have a primary cell wall. Additional layers of cellulose and lignin contribute to the secondary cell wall.

layers of wax. In some cases, the dermal tissue is more extensive and forms the bark of trees.

Some *ground tissue* cells function in storage, photosynthesis, and secretion. Other ground tissue includes fibers that serve to support and protect the plant. In cotton clothing, the fabric is composed primarily of fibers from cotton ground tissue.

Vascular tissue conducts fluids and dissolved substances throughout the plant body. One type of vascular tissue, the *xylem*, carries water and dissolved minerals; the other type, *phloem*, conducts a solution containing nutrients, including sucrose.

Each of these tissues and their many functions are described in more detail in later sections.

Meristems elaborate the body plan throughout the plant's life

When a seed sprouts, only a tiny portion of the adult plant exists. Although embryo cells can undergo division and differentiation to form many cell types, the fate of most adult cells is more restricted. Further development of the plant body depends on the activities of *meristems*, specialized cells found in shoot and root apices, as well as other parts of the plant.

Overview of meristems

Meristems are clumps of small cells with dense cytoplasm and proportionately large nuclei that act as stem cells do in animals. That is, one cell divides to give rise to two cells, of which one remains meristematic, while the other undergoes differentiation and contributes to the plant body (figure 36.3). In this way, the population of meristem cells is continually renewed.

Molecular genetic evidence supports the hypothesis that animal stem cells and plant meristem cells may also share some

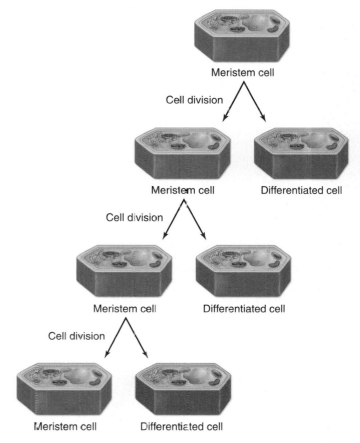

figure 36.3

MERISTEM CELL DIVISION. Plant meristems consist of cells that divide to give rise to a differentiating daughter cell and a cell that persists as a meristem cell.

common pathways of gene expression. Plant biologists use the term meristem cells, rather than stem cells, to avoid confusion with the stems of the shoot system.

Extension of both root and shoot takes place as a result of repeated cell divisions and subsequent elongation of the cells produced by the **apical meristems.** In some vascular plants, including shrubs and most trees, **lateral meristems** produce an increase in root and shoot diameter.

Apical meristems

Apical meristems are located at the tips of stems and roots (figure 36.4). During periods of growth, the cells of apical meristems divide and continually add more cells at the tips. Tissues derived from apical meristems are called **primary tissues,** and

Young leaf
primordium

Shoot apical
meristem

Older leaf
primordium

Lateral bud
primordium

100 μm

☐ dermal tissue
☐ ground tissue
☐ vascular tissue

Root apical
meristem

Root cap

400 μm

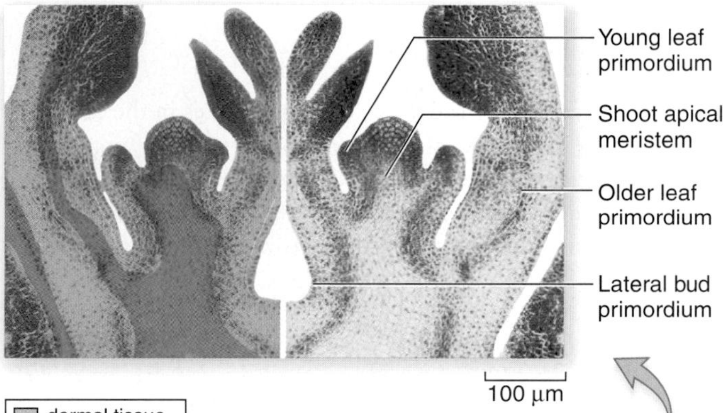

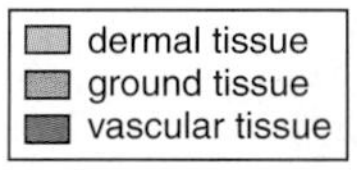

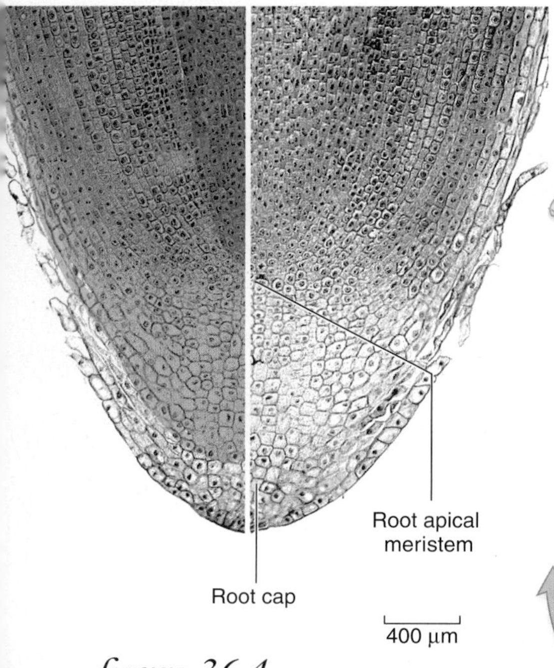

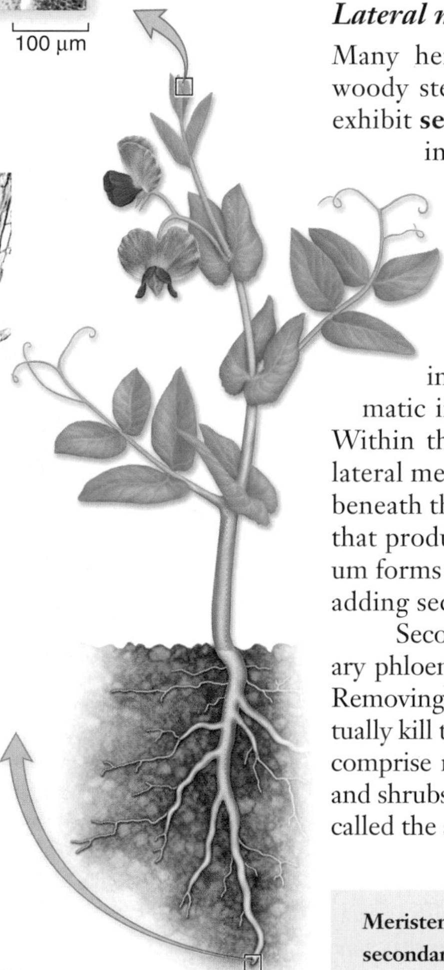

figure 36.4

APICAL MERISTEMS. Shoot and root apical meristems extend the plant body above and below ground. Leaf primordia protect the fragile shoot meristem, while the root meristem produces a protective root cap in addition to new root tissue.

the extension of the root and stem forms what is known as the **primary plant body.** The primary plant body comprises the young, soft shoots and roots of a tree or shrub, or the entire plant body in some plants.

Both root and shoot apical meristems are composed of delicate cells that need protection (see figure 36.4). The root apical meristem is protected by the root cap, the anatomy of which is described later on. Root cap cells are produced by the root meristem and are sloughed off and replaced as the root moves through the soil. In contrast, leaf primordia shelter the growing shoot apical meristem, which is particularly susceptible to desiccation because of its exposure to air and sun.

The apical meristem gives rise to the three tissue systems by first initiating **primary meristems.** The three primary meristems are the **protoderm,** which forms the epidermis; the **procambium,** which produces primary vascular tissues (primary xylem and primary phloem); and the **ground meristem,** which differentiates further into ground tissue. In some plants, such as horsetails and corn, **intercalary meristems** arise in stem internodes (spaces between leaf attachments), adding to the internode lengths. If you walk through a cornfield on a quiet summer night when the corn is about knee high, you may hear a soft popping sound. This sound is caused by the rapid growth of the intercalary meristems. The amount of stem elongation that occurs in a very short time is quite surprising.

Lateral meristems

Many herbaceous plants (that is, plants with fleshy, not woody stems) exhibit only primary growth, but others also exhibit **secondary growth,** which may result in a substantial increase of diameter. Secondary growth is accomplished by the lateral meristems, peripheral cylinders of meristematic tissue within the stems and roots that increase the girth (diameter) of gymnosperms and most angiosperms, with monocots being the major exception (figure 36.5).

Although secondary growth increases girth in many nonwoody plants, its effects are most dramatic in woody plants, which have two lateral meristems. Within the bark of a woody stem is the **cork cambium,** a lateral meristem that produces the outer bark of the tree. Just beneath the bark is the **vascular cambium,** a lateral meristem that produces secondary vascular tissue. The vascular cambium forms between the xylem and phloem in vascular bundles, adding secondary vascular tissue to both of its sides.

Secondary xylem is the main component of wood. Secondary phloem is very close to the outer surface of a woody stem. Removing the bark of a tree damages the phloem and may eventually kill the tree. Tissues formed from lateral meristems, which comprise most of the trunk, branches, and older roots of trees and shrubs, are known as **secondary tissues** and are collectively called the **secondary plant body.**

Meristems elaborate the shoot and root systems of the primary and secondary plant body. Cells derived from the meristems in both roots and shoots differentiate into one of three tissue systems: the dermal, ground, and vascular tissue systems.

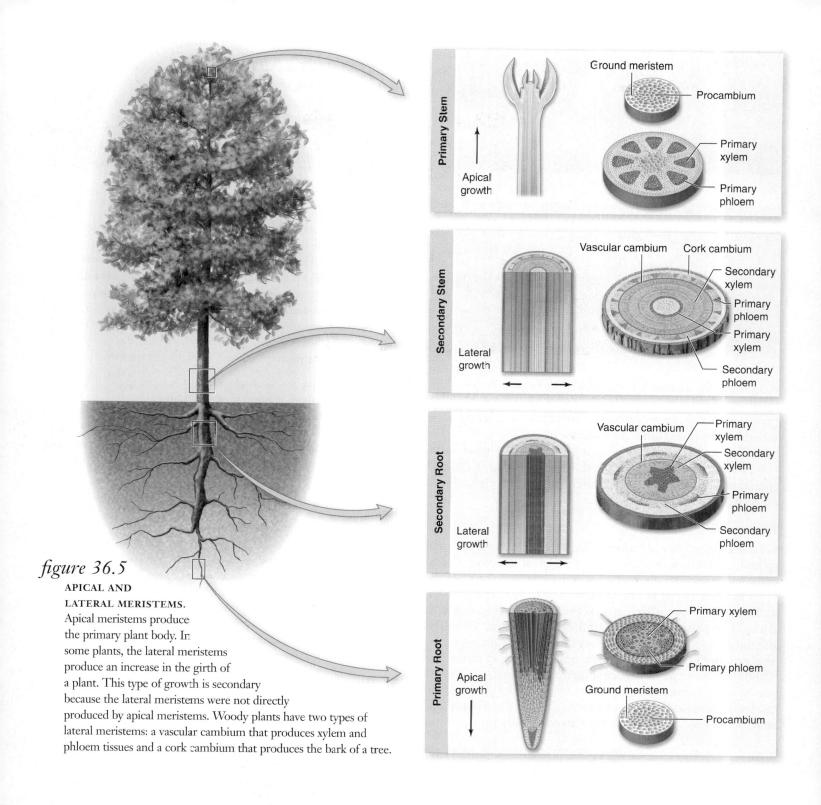

figure 36.5

APICAL AND LATERAL MERISTEMS. Apical meristems produce the primary plant body. In some plants, the lateral meristems produce an increase in the girth of a plant. This type of growth is secondary because the lateral meristems were not directly produced by apical meristems. Woody plants have two types of lateral meristems: a vascular cambium that produces xylem and phloem tissues and a cork cambium that produces the bark of a tree.

Plant Tissues

Three main categories of tissue can be distinguished in the plant body. These are (1) *dermal tissue* on external surfaces that serves a protective function, (2) *ground tissue* that forms several different internal tissue types, and that can participate in photosynthesis, serve a storage function, or provide structural support, and (3) *vascular tissue* that conducts water and nutrients.

Dermal tissue forms a protective interface with the environment

Dermal tissue derived from an embryo or apical meristem forms the **epidermis.** This tissue is one cell layer thick in most plants and forms the outer protective covering of the plant. In young, exposed parts of the plant, the epidermis is covered with a fatty **cutin** layer

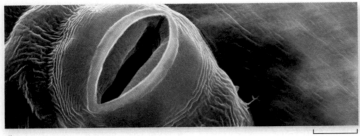

a.

4 μm

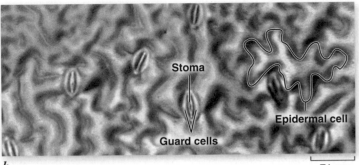

Stoma

Epidermal cell

Guard cells

b.

71 μm

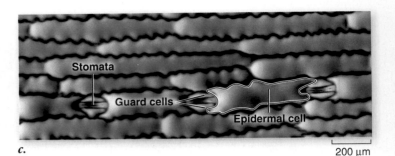

Stomata

Guard cells

Epidermal cell

c.

200 μm

figure 36.6

STOMATA. *a.* A stoma is the space between two guard cells that regulate the size of the opening. Stomata are evenly distributed within the epidermis of monocots and eudicots, but the patterning is quite different. *b.* A pea (eudicot) leaf with a random arrangement of stomata. *c.* A maize (corn, a monocot) leaf with stomata evenly spaced in rows. These photomicrographs also show the variety of cell shapes in plants. Some plant cells are boxlike, as seen in maize *(c)*, while others are irregularly shaped, as seen in the jigsaw puzzle shapes of the pea epidermal cells, *(b)*.

constituting the **cuticle;** in plants such as the desert succulents, several layers of wax may be added to the cuticle to limit water loss and protect against ultraviolet damage. In some cases, the dermal tissue is more extensive and forms the bark of trees.

Epidermal cells, which originate from the protoderm, cover all parts of the primary plant body. A number of types of specialized cells occur in the epidermis, including *guard cells, trichomes,* and *root hairs.*

Guard cells

Guard cells are paired, sausage-shaped cells flanking a **stoma** (plural, *stomata*), a mouth-shaped epidermal opening. Guard cells, unlike other epidermal cells, contain chloroplasts.

Stomata occur in the epidermis of leaves (figure 36.6*a*) and sometimes on other parts of the plant, such as stems or fruits. The passage of oxygen and carbon dioxide, as well as the diffusion of water in vapor form, takes place almost exclusively through the stomata. There are from 1000 to more than 1 million stomata per square centimeter of leaf surface. In many plants, stomata are more numerous on the lower epidermis of the leaf than on the upper epidermis, a factor that helps minimize water loss. Some plants have stomata only on the lower

epidermis, and a few, such as water lilies, have them only on the upper epidermis to maximize gas exchange.

Guard cell formation is the result of an asymmetrical cell division producing a guard cell and a subsidiary cell that aids in the opening and closing of the stoma. The patterning of these asymmetrical divisions that results in stomatal distribution has intrigued developmental biologists (figure 36.6*b,c*).

Research on mutants that get "confused" about where to position stomata is providing information on the timing of stomatal initiation and the kind of intercellular communication that triggers guard cell formation. For example, the *too many mouths (tmm)* mutation that occurs in *Arabidopsis* disrupts the normal pattern of cell division that spatially separates stomata (figure 36.7). Investigations of this and other stomatal patterning genes revealed a coordinated network of cell–cell communication (chapter 9) that informs cells of their position relative to other cells and determines cell fate. The *TMM* gene encodes a membrane-bound receptor that is part of a signaling pathway controlling asymmetrical cell division.

Trichomes

Trichomes are cellular or multicellular hairlike outgrowths of the epidermis (figure 36.8). They occur frequently on stems,

figure 36.7

THE *TOO MANY MOUTHS* STOMATAL MUTANT. This *Arabidopsis* mutant plant lacks an essential signal for spacing stomata. Usually a differentiating guard cell pair will inhibit differentiation of a nearby cell into a guard cell.

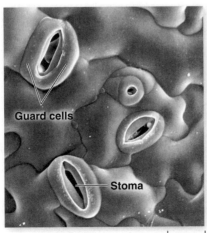

Guard cells

Stoma

272 μm

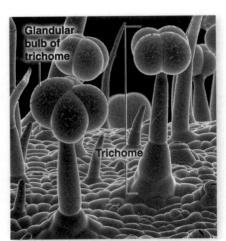

Glandular bulb of trichome

Trichome

34.62 μm

figure 36.8

TRICHOMES. The trichomes with tan, bulbous tips on this tomato plant are glandular trichomes. These trichomes secrete substances that can literally glue insects to the trichome.

figure 36.9

TRICHOME PATTERNING.
Mutants have revealed genes involved in regulating the spacing and development of trichomes in *Arabidopsis*. *a.* Wild type. *b. glaborous3* mutant which fails to initiate trichome development. *c.* When there is sufficient GL3 in a cell and the levels of trichome inhibiting proteins are sufficiently low, that cell will develop a trichome. Once a cell begins trichome initiation, it signals neighboring cells and inhibits their ability to develop trichomes.

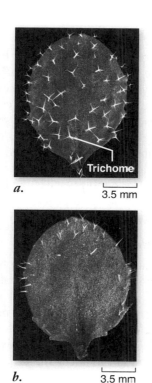

a. 3.5 mm

b. 3.5 mm

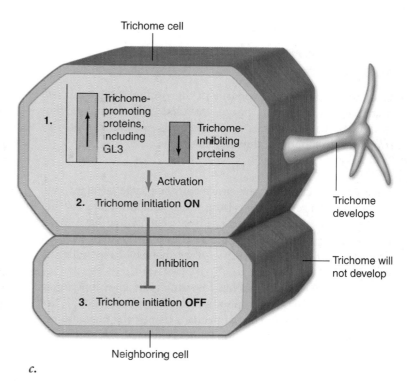

c.

leaves, and reproductive organs. A "fuzzy" or "woolly" leaf is covered with trichomes that can be seen clearly with a microscope under low magnification. Trichomes keep leaf surfaces cool and reduce evaporation by covering stomatal openings. Trichomes can vary greatly in form; some consist of a single cell, while others are multicellular. Some are glandular, often secreting sticky or toxic substances to deter herbivory.

Genes that regulate trichome development have been identified, including *GLABROUS3* (*GL3*) (figure 36.9). When trichome-initiating proteins, like GL3, reach a threshold level compared with trichome-inhibiting proteins, an epidermal cell becomes a trichome. Signals from this trichome cell now prevent neighbor cells from expressing trichome-promoting genes (see figure 36.9).

Root hairs

Root hairs, which are tubular extensions of individual epidermal cells, occur in a zone just behind the tips of young, growing roots (figure 36.10). Because a root hair is simply an extension of an epidermal cell and not a separate cell, no cross-wall isolates the hair from the rest of the cell. Root hairs keep the root in intimate contact with the surrounding soil particles and greatly increase the root's surface area and efficiency of absorption.

As a root grows, the extent of the root hair zone remains roughly constant as root hairs at the older end slough off while new ones are produced at the apex. Most of the absorption of water and minerals occurs through root hairs, especially in herbaceous plants. Root hairs should not be confused with lateral roots, which are multicellular structures and originate deep within the root.

The first land plants lacked roots, which later evolved from shoots. Given this common ancestry, it is not surprising that some of the genes needed for trichome and stomatal differentiation in shoot epidermal cells also play a role in root hair development.

Root hairs are not found when the dermal tissue system is extended by the cork cambium, which contributes to the periderm (outer bark) of a tree trunk or root. Periderm replaces the

epidermis as it gets stretched and broken with the radial expansion of the axis by the vascular cambium. The periderm consists of cork cells, cork cambium, and parenchyma cells called phelloderm produced from the cork cambium.

inquiry

Identify three dermal tissue traits that are adaptive for a terrestrial lifestyle and explain why these traits are advantageous.

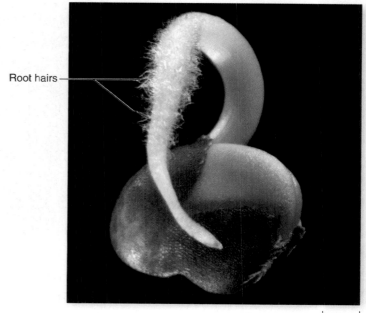

Root hairs

1250 µm

figure 36.10

ROOT HAIRS. Root hair cells are a type of epidermal cell that increase surface area of the root to enhance water and mineral uptake.

Ground tissue cells perform many functions, including storage, photosynthesis, and support

Ground tissue consists primarily of thin-walled *parenchyma cells* that function in storage, photosynthesis, and secretion. Other ground tissue, composed of *collenchyma cells* and *sclerenchyma cells*, provide support and protection.

Parenchyma

Parenchyma cells are the most common type of plant cell. They have large vacuoles, thin walls, and are initially (but briefly) more or less spherical. These cells, which have living protoplasts, push up against each other shortly after they are produced, however, and assume other shapes, often ending up with 11 to 17 sides.

Parenchyma cells may live for many years; they function in storage of food and water, photosynthesis, and secretion. They are the most abundant cells of primary tissues and may also occur, to a much lesser extent, in secondary tissues (figure 36.11a). Most parenchyma cells have only primary walls, which are walls laid down while the cells are still maturing. Parenchyma are less specialized than other plant cells, although there are many variations that do have special functions, such as nectar and resin secretion or storage of latex, proteins, and metabolic wastes.

Parenchyma cells have functional nuclei and are capable of dividing, and they usually remain alive after they mature; in some plants (for example, cacti), they may live to be over 100 years old. The majority of cells in fruits such as apples are parenchyma. Some parenchyma contain chloroplasts, especially in leaves and in the outer parts of herbaceous stems. Such photosynthetic parenchyma tissue is called *chlorenchyma*.

Collenchyma

If celery "strings" have ever been caught between your teeth, you are familiar with tough, flexible **collenchyma cells.** Like parenchyma cells, collenchyma cells have living protoplasts and may live for many years. These cells, which are usually a little longer than wide, have walls that vary in thickness (figure 36.11b).

Flexible collenchyma cells provide support for plant organs, allowing them to bend without breaking. They often form strands or continuous cylinders beneath the epidermis of stems or leaf petioles (stalks) and along the veins in leaves. Strands of collenchyma provide much of the support for stems in the primary plant body.

Sclerenchyma

Sclerenchyma cells have tough, thick walls. Unlike collenchyma and parenchyma, they usually lack living protoplasts at maturity. Their secondary cell walls are often impregnated with **lignin,** a highly branched polymer that makes cell walls more rigid; for example, lignin is an important component in wood. Cell walls containing lignin are said to be *lignified*. Lignin is common in the walls of plant cells that have a structural or mechanical function. Some kinds of cells have lignin deposited in primary as well as secondary cell walls.

Sclerenchyma is present in two general types: fibers and sclereids. *Fibers* are long, slender cells that are usually grouped together in strands. Linen, for example, is woven from strands of sclerenchyma fibers that occur in the phloem of flax (*Linum* spp.) plants. *Sclereids* are variable in shape but often branched. They may occur singly or in groups; they are not elongated, but may have many different forms, including that of a star. The gritty texture of a pear is caused by groups of sclereids that occur throughout the soft flesh of the fruit (figure 36.11c). Both

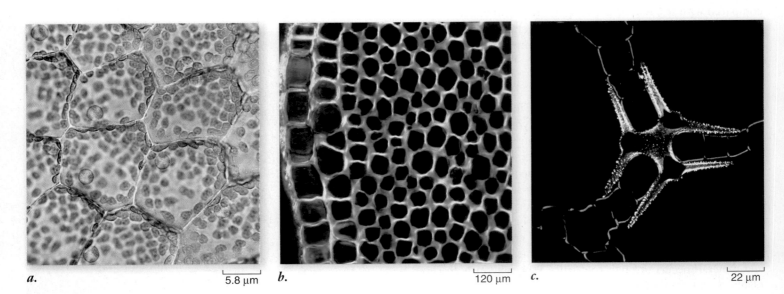

a.   5.8 µm   *b.*   120 µm   *c.*   22 µm

figure 36.11

THE THREE TYPES OF GROUND TISSUE. *a.* Parenchyma cells. Only primary cell walls are seen in this cross section of parenchyma cells from grass. *b.* Collenchyma cells. Thickened side walls are seen in this cross section of collenchyma cells from a young branch of elderberry (*Sambucus*). In other kinds of collenchyma cells, the thickened areas may occur at the corners of the cells or in other kinds of strips. *c.* Sclereids. Clusters of sclereids ("stone cells"), stained red in this preparation, in the pulp of a pear. The surrounding thin-walled cells, stained green, are parenchyma. These sclereid clusters give pears their gritty texture. Sclereids are one type of sclerenchyma tissue, which also contains fibers.

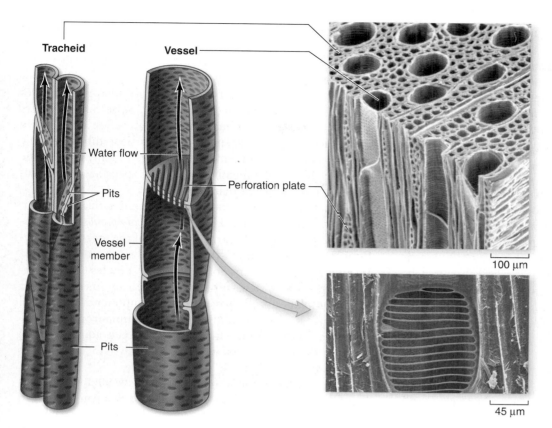

figure 36.12

COMPARISON BETWEEN TRACHEIDS AND VESSEL MEMBERS. In tracheids, the water passes from cell to cell by means of pits. In vessel members, water moves by way of perforation plates (as seen in the photomicrograph in this figure) bars of wall material. In gymnosperm wood, tracheids both conduct water and provide support; in most kinds of angiosperms, vessels are present in addition to tracheids. These two types of cells conduct water, and fibers provide additional support. The wood of red maple, *Acer rubrum*, contains both tracheids and vessels as seen in the electron micrographs in this figure.

Tracheid Vessel Water flow Pits Vessel member Pits Perforation plate 100 μm 45 μm

of these tough, thick-walled cell types serve to strengthen the tissues in which they occur.

Vascular tissue conducts water and nutrients throughout the plant

Vascular tissue, as mentioned earlier, includes two kinds of conducting tissues: *xylem*, which conducts water and dissolved minerals, and *phloem*, which conducts a solution of carbohydrates—mainly sucrose—used by plants for food. The phloem also transports hormones, amino acids, and other substances that are necessary for plant growth. Xylem and phloem differ in structure as well as in function.

Xylem

Xylem, the principal water-conducting tissue of plants, usually contains a combination of *vessels*, which are continuous tubes formed from dead, hollow, cylindrical cells arranged end-to-end, and *tracheids*, which are dead cells that taper at the ends and overlap one another (figure 36.12). Primary xylem is derived from the procambium produced by the apical meristem. Secondary xylem is formed by the vascular cambium, a lateral meristem. Wood consists of accumulated secondary xylem.

In some plants (but not angiosperms), tracheids are the only water-conducting cells present; water passes in an unbroken stream through the xylem from the roots up through the shoot and into the leaves. When the water reaches the leaves, much of it diffuses in the form of water vapor into the intercellular spaces and out of the leaves into the surrounding air, mainly through the stomata. This diffusion of water vapor from

a plant is known as **transpiration** (chapter 38). In addition to conducting water, dissolved minerals, and inorganic ions such as nitrates and phosphates throughout the plant, xylem supplies support for the plant body.

Vessel members tend to be shorter and wider than tracheids. When viewed with a microscope, they resemble beverage cans with both ends removed. Both vessel members and tracheids have thick, lignified secondary walls and no living protoplasts at maturity. Lignin is produced by the cell and secreted to strengthen the cellulose cell walls before the protoplast dies, leaving only the cell wall.

Tracheids contain *pits*, which are small, mostly rounded-to-elliptical areas where no secondary wall material has been deposited. The pits of adjacent cells occur opposite one another; the continuous stream of water flows through these pits from tracheid to tracheid. In contrast, vessel members, which are joined end-to-end, may be almost completely open or may have bars or strips of wall material across the open ends (see figure 36.12). Vessels appear to conduct water more efficiently than do the overlapping strands of tracheids. We know this partly because vessel members have evolved from tracheids independently in several groups of plants, suggesting that they are favored by natural selection.

In addition to conducting cells, xylem typically includes fibers and parenchyma cells (ground tissue cells). It is probable that some types of fibers have evolved from tracheids, becoming specialized for strengthening rather than conducting. The parenchyma cells, which are usually produced in horizontal rows called *rays* by special *ray initials* of the vascular cambium, function in lateral conduction and food storage. (An *initial* is another term for a meristematic cell. It divides to produce another initial and a cell that differentiates.)

In cross sections of woody stems and roots, the rays can be seen radiating out from the center of the xylem like the spokes of a wheel. Fibers are abundant in some kinds of wood, such as oak (*Quercus* spp.), and the wood is correspondingly dense and heavy. The arrangements of these and other kinds of cells in the xylem make it possible to identify most plant genera and many species from their wood alone.

Over 2000 years ago paper as we recognize it today was made in China by mashing herbaceous plants in water and separating out a thin layer of phloem fibers on a screen. Not until the third century of the common era did the secret of making paper make its way out of China. Today the ever-growing demand for paper is met by extracting xylem fibers from wood, including spruce, that is relatively soft, having fewer ray fibers than oak. The lignin-rich cell walls yield brown paper which is often bleached. In addition, many other plants have been developed as sources of paper, including kenaf and hemp. United States paper currency is 75% cotton and 25% flax.

Phloem

Phloem, which is located toward the outer part of roots and stems, is the principal food-conducting tissue in vascular plants. If a plant is *girdled* (by removing a substantial strip of bark down to the vascular cambium), the plant eventually dies from starvation of the roots.

Food conduction in phloem is carried out through two kinds of elongated cells: sieve cells and sieve-tube members. Seedless vascular plants and gymnosperms have only sieve cells; most angiosperms have sieve-tube members. Both types of cells have clusters of pores known as sieve areas because the cell walls resemble sieves. Sieve areas are more abundant on the overlapping ends of the cells and connect the protoplasts of adjoining sieve cells and sieve-tube members. Both of these types of cells are living, but most sieve cells and all sieve-tube members lack a nucleus at maturity.

In sieve-tube members, some sieve areas have larger pores and are called sieve plates (figure 36.13). Sieve-tube members occur end-to-end, forming longitudinal series called sieve tubes. Sieve cells are less specialized than sieve-tube members, and the pores in all of their sieve areas are roughly of the same diameter. Sieve-tube members are more specialized, and presumably, more efficient than sieve cells.

Each sieve-tube member is associated with an adjacent, specialized parenchyma cell known as a *companion cell.* Companion cells apparently carry out some of the metabolic functions needed to maintain the associated sieve-tube member. In angiosperms, a common initial cell divides asymmetrically to produce a sieve-tube member cell and its companion cell. Companion cells have all the components of normal parenchyma cells, including nuclei, and numerous **plasmodesmata** (cytoplasmic connections between adjacent cells) connect their cytoplasm with that of the associated sieve-tube members.

Sieve cells in nonflowering plants have albuminous cells that function as companion cells. Unlike a companion cell, an albuminous cell is not necessarily derived from the same mother cell as its associated sieve cell. Fibers and parenchyma cells are often abundant in phloem.

> Dermal, ground, and vascular tissue systems are composed of diverse cell types. Dermal tissue provides protection. Vascular tissue enhances transport throughout the plant, and ground tissues have metabolic, structural, and storage functions.

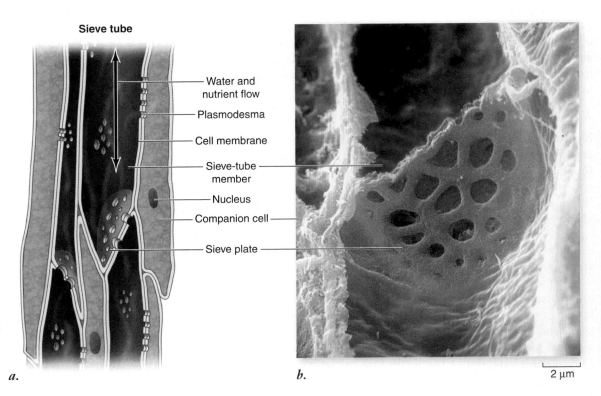

figure 36.13

A SIEVE-TUBE MEMBER. *a.* Sieve-tube member cells are stacked, with sieve plates forming the connection. The narrow cell with the nucleus at the right of the sieve-tube member is a companion cell. This cell nourishes the sieve-tube members, which have plasma membranes, but no nuclei. *b.* Looking down into sieve plates in squash phloem reveals the perforations through which sucrose and hormones move.
Photo © Dr. Richard Kessel and Dr. Gene Shih/Visuals Unlimited

Sieve tube

- Water and nutrient flow
- Plasmodesma
- Cell membrane
- Sieve-tube member
- Nucleus
- Companion cell
- Sieve plate

a.

b.

2 µm

Roots: Anchoring and Absorption Structures

Roots have a simpler pattern of organization and development than stems, and we will consider them first. Keep in mind, however, that roots evolved after shoots and are a major innovation for terrestrial living.

Roots are adapted for growing underground and absorbing water and solutes

Four regions are commonly recognized in developing roots: the *root cap*, the *zone of cell division*, the *zone of elongation*, and the *zone of maturation* (figure 36.14). In these last three zones, the boundaries are not clearly defined.

When apical initials divide, daughter cells that end up on the tip end of the root become root cap cells. Cells that divide in the opposite direction pass through the three other zones before they finish differentiating. As you consider the different zones, visualize the tip of the root moving deeper into the soil, actively growing. This will counter the static image of a root that diagrams and photos convey.

The root cap

The **root cap** has no equivalent in stems. It is composed of two types of cells: the inner *columella cells* (they look like columns), and the outer, lateral *root cap cells*, which are continuously replenished by the root apical meristem. In some plants with larger roots, the root cap is quite obvious. Its main function is to protect the delicate tissues behind it as growth extends the root through mostly abrasive soil particles.

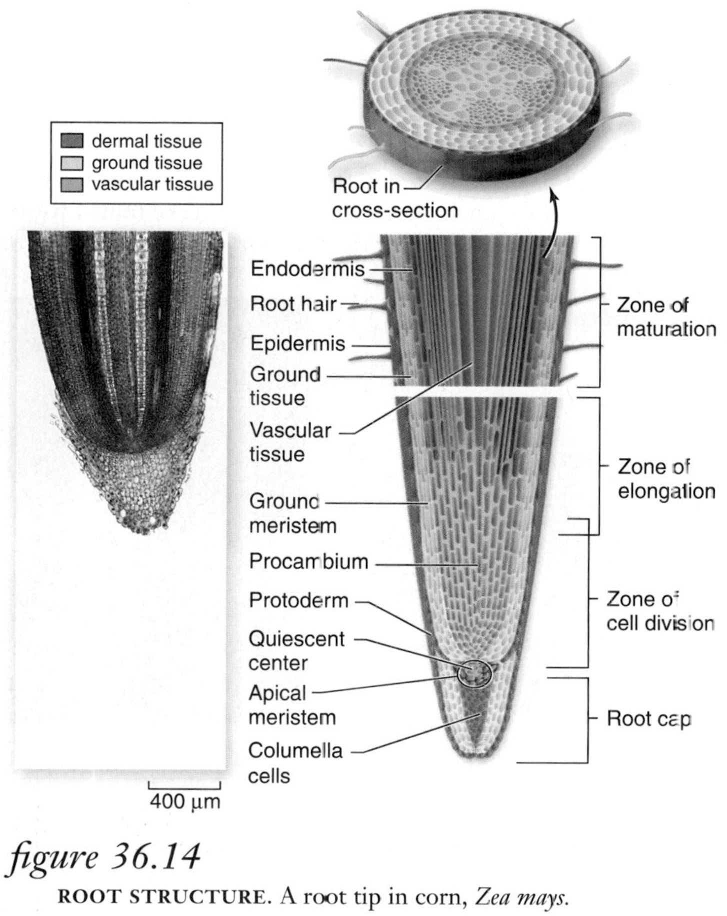

Golgi bodies in the outer root cap cells secrete and release a slimy substance that passes through the cell walls to the outside. The root cap cells, which have an average life of less than a week, are constantly being replaced from the inside, forming a mucilaginous lubricant that eases the root through the soil. The slimy mass also provides a medium for the growth of beneficial nitrogen-fixing bacteria in the roots of plants such as legumes. A new root cap is produced when an existing one is artificially or accidentally removed from a root.

The root cap also functions in the perception of gravity. The columella cells are highly specialized, with the endoplasmic reticulum in the periphery and the nucleus located at either the middle or the top of the cell. They contain no large vacuoles. Columella cells contain *amyloplasts* (plastids with starch grains) that collect on the sides of cells facing the pull of gravity. When a potted plant is placed on its side, the amyloplasts drift or tumble down to the side nearest the source of gravity, and the root bends in that direction.

Lasers have been used to ablate (kill) individual columella cells in *Arabidopsis*. It turns out that only two columella cells are sufficient for gravity sensing! The precise nature of the gravitational response is unknown, but some evidence indicates that calcium ions in the amyloplasts influence the distribution of growth hormones (auxin in this case) in the cells. Multiple signaling mechanisms may exist, because bending has been observed in the absence of auxin. A current hypothesis is that an electrical signal moves from the columella cell to cells in the elongation zone (the region closest to the zone of cell division).

The zone of cell division

The apical meristem is located in the center of the root tip in the area protected by the root cap. Most of the activity in this **zone of cell division** takes place toward the edges of the meristem, where the cells divide every 12 to 36 hours, often rhythmically, reaching a peak of division once or twice a day.

Most of the cells are essentially cuboidal, with small vacuoles and proportionately large, centrally located nuclei. These rapidly dividing cells are daughter cells of the apical meristem. A group of cells in the center of the root apical meristem, termed the *quiescent center*, divide only very infrequently. The presence of the quiescent center makes sense if you think about a solid ball expanding—the outer surface would have to increase far more rapidly than the very center.

The apical meristem daughter cells soon subdivide into the three primary tissues previously discussed: protoderm, procambium, and ground meristem. Genes have been identified in the relatively simple root of *Arabidopsis* that regulate the patterning of these tissue systems. The patterning of these cells begins in this zone, but the anatomical and morphological expression of this patterning is not fully revealed until the cells reach the zone of maturation.

Legend:
- dermal tissue
- ground tissue
- vascular tissue

Root in cross-section

Endodermis
Root hair
Epidermis
Ground tissue
Vascular tissue
Ground meristem
Procambium
Protoderm
Quiescent center
Apical meristem
Columella cells

Zone of maturation
Zone of elongation
Zone of cell division
Root cap

400 µm

figure 36.14

ROOT STRUCTURE. A root tip in corn, *Zea mays*.

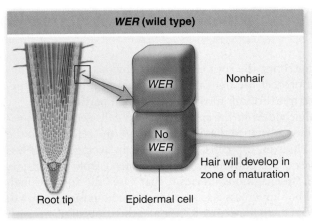
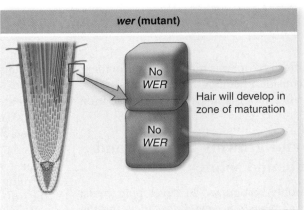
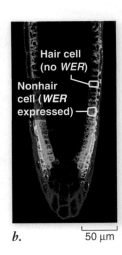

WER (wild type)

WER

No WER

Nonhair

Hair will develop in zone of maturation

Root tip   Epidermal cell

a.

wer (mutant)

No WER

No WER

Hair will develop in zone of maturation

Hair cell (no *WER*)

Nonhair cell (*WER* expressed)

b.   50 μm

figure 36.15

TISSUE-SPECIFIC GENE EXPRESSION. *a.* The *WEREWOLF* gene of *Arabidopsis* is expressed in some, but not all, epidermal cells and suppresses root hair development. The *wer* mutant is covered with root hairs. *b.* The *WER* promoter was attached to a gene coding for a green fluorescent protein and used to make a transgenic plant. The green fluorescence shows the nonhair epidermal cells where the gene is expressed. The red visually indicates cell boundaries.

For example, the *WEREWOLF* (*WER*) gene is required for the patterning of the two root epidermal cell types, those with and those without root hairs (figure 36.15). Plants with the *wer* mutation have an excess of root hairs because *WER* is needed to prevent root hair development in nonhair epidermal cells. Similarly, the *SCARECROW* (*SCR*) gene is necessary in ground cell differentiation (figure 36.16). A ground meristem cell undergoes an asymmetrical cell division that gives rise to two nested cylinders of cells from one if *SCR* is present. The outer cell layer becomes ground tissue and serves a storage function. The inner cell layer forms the endodermis, which regulates the intercellular flow of water and solutes into the vascular core of the root (see figure 36.5). The *scr* mutant, in contrast, forms a single layer of cells that have both endodermal and ground cell traits.

SCR illustrates the importance of the orientation of cell division. If a cell's position changes because of a mistake in cell division or the ablation of another cell, the cell develops ac-

cording to its new position. The fate of most plant cells is determined by their position relative to other cells.

The zone of elongation

In the **zone of elongation,** roots lengthen because the cells produced by the primary meristems become several times longer than wide, and their width also increases slightly. The small vacuoles present merge and grow until they occupy 90% or more of the volume of each cell. No further increase in cell size occurs above the zone of elongation. The mature parts of the root, except for increasing in girth, remain stationary for the life of the plant.

The zone of maturation

The cells that have elongated in the zone of elongation become differentiated into specific cell types in the **zone of maturation** (see figure 36.14). The cells of the root surface cylinder mature

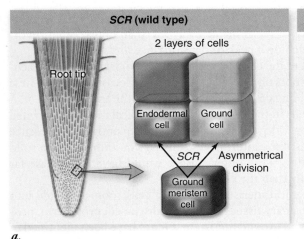
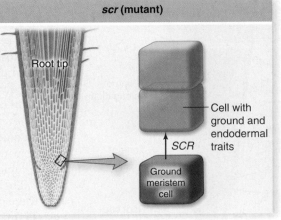

SCR (wild type)

Root tip

2 layers of cells

Endodermal cell   Ground cell

SCR   Asymmetrical division

Ground meristem cell

a.

scr (mutant)

Root tip

Cell with ground and endodermal traits

SCR

Ground meristem cell

SCR is expressed only in endodermal cells

b.   400 μm

figure 36.16

SCARECROW REGULATES ASYMMETRICAL CELL DIVISION. *a.* *SCR* is needed for an asymmetrical cell division leading to the differentiation of daughter cells into endodermal and ground cells. *b.* The *SCR* promoter was attached to a gene coding for a green fluorescent protein. *SCR* is only expressed in the endodermal cells, not the ground cells.

into *epidermal cells*, which have a very thin cuticle, and include both root hair and nonhair cells. Although the root hairs are not visible until this stage of development, their fate was established much earlier, as you saw with the expression patterns of *WER* (see figure 36.15).

Root hairs can number over 37,000 cm² of root surface and many billions per plant; they greatly increase the surface area and therefore the absorptive capacity of the root. Symbiotic bacteria that fix atmospheric nitrogen into a form usable by legumes enter the plant via root hairs and "instruct" the plant to create a nitrogen-fixing nodule around it (see chapter 39).

Parenchyma cells are produced by the ground meristem immediately to the interior of the epidermis. This tissue, called the **cortex,** may be many cell layers wide and functions in food storage. As just described, the inner boundary of the cortex differentiates into a single-layered cylinder of **endodermis,** after an asymmetrical cell division regulated by *SCR* (figures 36.16 and 36.17). Endodermal primary walls are impregnated with *suberin*, a fatty substance that is impervious to water. The suberin is produced in bands, called **Casparian strips,** that surround each adjacent endodermal cell wall perpendicular to the root's surface (see figure 36.17). These strips block transport between cells. The two surfaces that are parallel to the root surface are the only way into the vascular tissue of the root, and the plasma membranes control what passes through.

All the tissues interior to the endodermis are collectively referred to as the **stele.** Immediately adjacent and interior to the endodermis is a cylinder of parenchyma cells

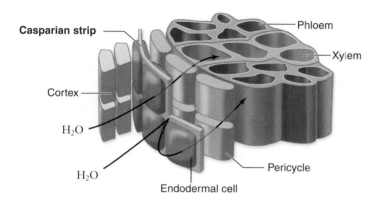

figure 36.17

CROSS SECTIONS OF THE ZONE OF MATURATION OF ROOTS. Both monocot and eudicot roots have a Casparian strip as seen in the cross-section of greenbriar (*Smilax*), a monocot, and buttercup (*Ranunculus*), a eudicot. The Casparian strip is a water-proofing band that forces water and minerals to pass through the plasma membranes, rather than through the spaces in the cell walls.

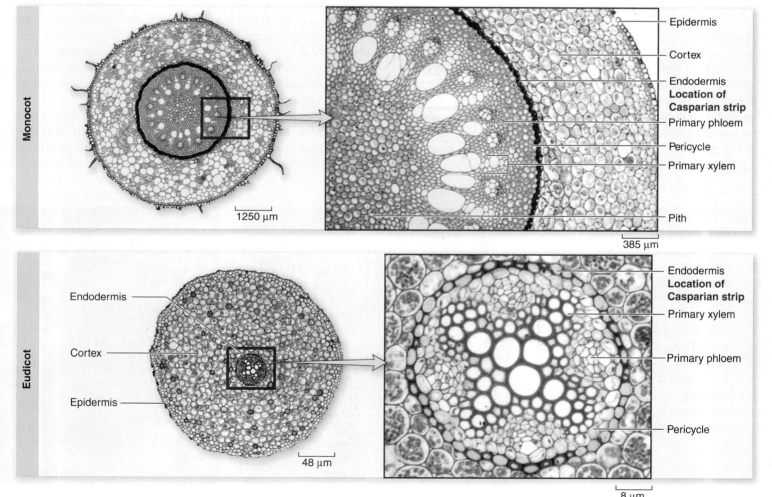

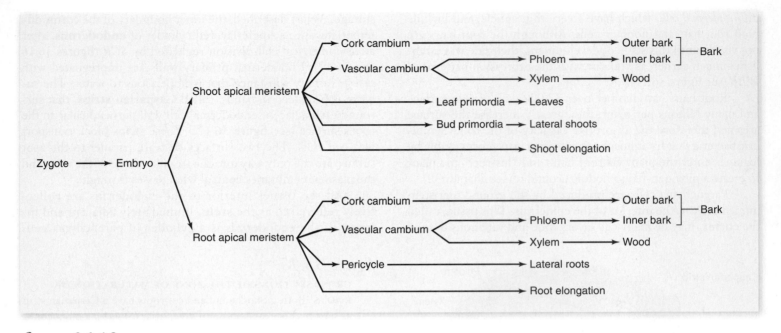

figure 36.18

STAGES IN THE DIFFERENTIATION OF PLANT TISSUES.

known as the **pericycle.** Pericycle cells divide, even after they mature. They can give rise to lateral (branch) roots or, in eudicots, to the two lateral meristems, the vascular cambium and the cork cambium.

The water-conducting cells of the primary xylem are differentiated as a solid core in the center of young eudicot roots. In a cross section of a eudicot root, the central core of primary xylem often is somewhat star-shaped, having from two to several radiating arms that point toward the pericycle (see figure 36.17). In monocot (and a few eudicot) roots, the primary xylem is in discrete vascular bundles arranged in a ring, which surrounds parenchyma cells, called *pith*, at the very center of the root (see figure 36.17). Primary phloem, composed of cells involved in food conduction, is differentiated in dis-

crete groups of cells adjacent to the xylem in both eudicot and monocot roots.

In eudicots and other plants with secondary growth, part of the pericycle and the parenchyma cells between the phloem patches and the xylem become the root vascular cambium, which starts producing secondary xylem to the inside and secondary phloem to the outside. Eventually, the secondary tissues acquire the form of concentric cylinders. The primary phloem, cortex, and epidermis become crushed and are sloughed off as more secondary tissues are added.

In the pericycle of woody plants, the cork cambium contributes to the outer bark, which will be discussed in more detail when we look at stems. In the case of secondary growth in eudicot roots, everything outside the stele is lost and replaced

figure 36.19

FIVE TYPES OF MODIFIED ROOTS.
a. Maize (corn) prop roots originate from the stem and keep the plant upright. *b.* Epiphytic orchids attach to trees far above the tropical soil. Their roots are adapted to obtain water from the air rather than the soil. *c.* Pneumatophores (*foreground*) are spongy outgrowths from the roots below. *d.* A water storage root weighing over 25 kilograms (60 pounds). *e.* Buttress roots of a tropical fig tree.

a.

b.

with bark. Figure 36.18 summarizes the process of differentiation that occurs in plant tissue.

Modified roots accomplish specialized functions

Most plants produce either a taproot system, characterized by a single large root with smaller branch roots, or a fibrous root system, composed of many smaller roots of similar diameter. Some plants, however, have intriguing root modifications with specific functions in addition to those of anchorage and absorption.

Not all roots are produced by preexisting roots. Any root that arises along a stem or in some place other than the root of the plant is called an **adventitious root.** For example, climbing plants such as ivy produce roots from their stems; these can anchor the stems to tree trunks or to a brick wall. Adventitious root formation in ivy depends on the developmental stage of the shoot. When the shoot enters the adult phase of development, it is no longer capable of initiating these roots. Below we investigate functions of modified roots.

Prop roots. Some monocots, such as corn, produce thick adventitious roots from the lower parts of the stem. These so-called prop roots grow down to the ground and brace the plants against wind (figure 36.19a).

Aerial roots. Plants such as epiphytic orchids, which are attached to tree branches and grow unconnected to the ground (but are not parasites), have roots that extend into the air (figure 36.19b). Some aerial roots have an epidermis that is several cell layers thick, an adaptation to reduce water loss. These aerial roots may also be green and photosynthetic, as in the vanilla orchid (*Vanilla planifolia*).

Pneumatophores. Some plants that grow in swamps and other wet places may produce spongy outgrowths called *pneumatophores* from their underwater roots (figure 36.19c). The pneumatophores commonly extend several centimeters above water, facilitating oxygen uptake in the roots beneath (figure 36.19c).

Contractile roots. The roots from the bulbs of lilies and from several other plants, such as dandelions, contract by spiraling to pull the plant a little deeper into the soil each year, until they reach an area of relatively stable temperature. The roots may contract to one-third their original length as they spiral like a corkscrew due to cellular thickening and constricting.

Parasitic roots. The stems of certain plants that lack chlorophyll, such as dodder (*Cuscuta* spp.), produce peglike roots called *haustoria* that penetrate the host plants around which they are twined. The haustoria establish contact with the conducting tissues of the host and effectively parasitize their host.

Food storage roots. The xylem of branch roots of sweet potatoes and similar plants produce at intervals many extra parenchyma cells that store large quantities of carbohydrates. Carrots, beets, parsnips, radishes, and turnips have combinations of stem and root that also function in food storage. Cross sections of these roots reveal multiple rings of secondary growth.

Water storage roots. Some members of the pumpkin family (Cucurbitaceae), especially those that grow in arid regions, may produce water storage roots weighing 50 or more kilograms (figure 36.19d).

Buttress roots. Certain species of fig and other tropical trees produce huge buttress roots toward the base of the trunk, which provide considerable stability (figure 36.19e).

The root system develops from the apical meristem, protected by the root cap. Both lateral roots and root hairs increase the amount of water and minerals that can be transported through the vascular tissue to the rest of the plant. Modified roots enhance one or more of the key characteristics of the root system.

c.

d.

e.

The supporting structure of a vascular plant's shoot system is the mass of stems that extend from the root system below ground into the air, often reaching great height. Stiff stems capable of rising upward against gravity are an ancient adaptation that allowed plants to move into terrestrial ecosystems.

Stems carry leaves and flowers and support the plant's weight

Like roots, stems contain the three types of plant tissue. Stems also undergo growth from cell division in apical and lateral meristems. The stem may be thought of as an axis from which other stems or organs grow. The shoot apical meristems are capable of producing these new stems and organs.

External stem structure

The shoot apical meristem initiates stem tissue and intermittently produces bulges **(primordia)** that are capable of developing into leaves, other shoots, or even flowers (figure 36.20). Leaves may be arranged in a spiral around the stem, or they may be in pairs opposite or alternate to one another; they also may occur in whorls (circles) of three or more (figure 36.21). The spiral arrangement is the most common, and for reasons still not understood, sequential leaves tend to be placed 137.5° apart. This angle relates to the golden mean, a mathematical ratio found in nature. The angle of coiling in shells of some gastropods is the same. The golden mean has been used in classical architecture (the Greek Parthenon wall dimensions), and even in modern art (for example, in paintings by Mondrian). In plants, this pattern of leaf arrangement, called **phyllotaxy,** may optimize the exposure of leaves to the sun.

The region or area of leaf attachment to the stem is called a **node;** the area of stem between two nodes is called an **internode.** A leaf usually has a flattened blade and sometimes a petiole (stalk). The angle between a leaf's petiole (or blade) and the stem is called an **axil.** An **axillary bud** is produced

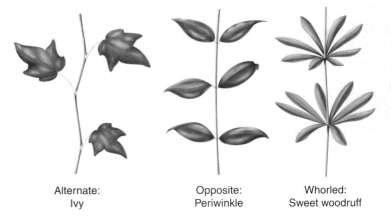

figure 36.21

TYPES OF LEAF ARRANGEMENTS. The three common types of leaf arrangements are alternate, opposite, and whorled.

in each axil. This bud is a product of the primary shoot apical meristem, and it is itself a shoot apical meristem. Axillary buds frequently develop into branches with leaves or may form flowers.

Neither monocots nor herbaceous eudicot stems produce a cork cambium. The stems in these plants are usually green and photosynthetic, with at least the outer cells of the cortex containing chloroplasts. Herbaceous stems commonly have stomata, and may have various types of trichomes (hairs).

Woody stems can persist over a number of years and develop distinctive markings in addition to the original organs that form (figure 36.22). Terminal buds usually extend the length of the shoot system during the growing season. Some buds, such as those of geraniums, are unprotected, but most

figure 36.20

A SHOOT APEX. Scanning electron micrograph of the apical meristem of wheat (*Triticum*).

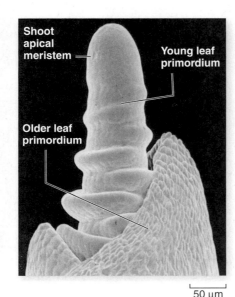

50 μm

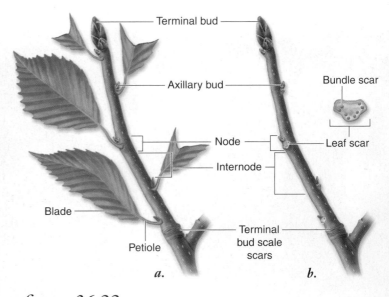

figure 36.22

A WOODY TWIG. *a.* In summer. *b.* In winter.

buds of woody plants have protective winter bud scales that drop off, leaving tiny bud scale scars as the buds expand.

Some twigs have tiny scars of a different origin. A pair of butterfly-like appendages called *stipules* (part of the leaf) develop at the base of some leaves. The stipules can fall off and leave stipule *scars*. When the leaves of deciduous trees drop in the fall, they leave leaf scars with tiny bundle scars, marking where vascular connections were. The shapes, sizes, and other features of leaf scars can be distinctive enough to identify deciduous plants in winter, when they lack leaves (see figure 36.22).

Internal stem structure

A major distinguishing feature between monocot and eudicot stems is the organization of the vascular tissue system (figure 36.23). Most monocot vascular bundles are scattered throughout the ground tissue system, whereas eudicot vascular tissue is arranged in a ring with internal ground tissue (*pith*) and external ground tissues (*cortex*). The arrangement of vascular tissue is directly related to the ability of the stem to undergo secondary growth. In eudicots, a vascular cambium develops between the primary xylem and primary phloem (figure 36.24). In many ways, this is a connect-the-dots game whereby the vascular

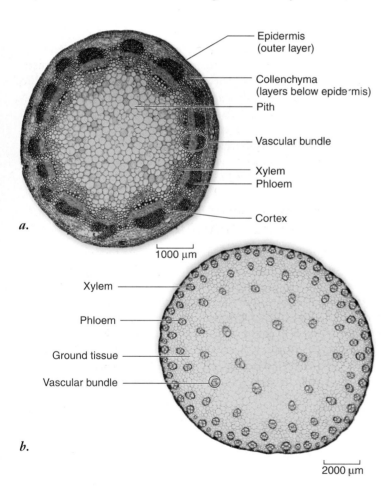

Epidermis (outer layer)

Collenchyma (layers below epidermis)

Pith

Vascular bundle

Xylem
Phloem

Cortex

a.

1000 µm

Xylem

Phloem

Ground tissue

Vascular bundle

b.

2000 µm

figure 36.23

STEMS. Transverse sections of a young stem in (*a*) a eudicot, the common sunflower (*Helianthus annuus*), in which the vascular bundles are arranged around the outside of the stem; and (*b*) a monocot, corn (*Zea mays*), with scattered vascular bundles characteristic.

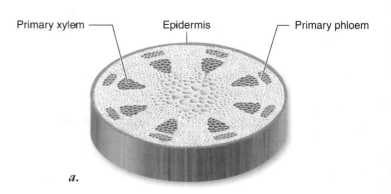

Primary xylem — Epidermis — Primary phloem

a.

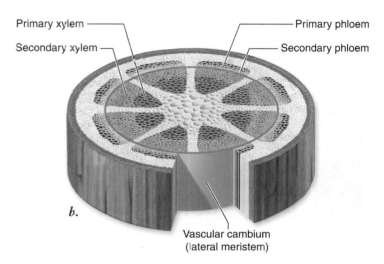

Primary xylem — Primary phloem
Secondary xylem — Secondary phloem

b.

Vascular cambium (lateral meristem)

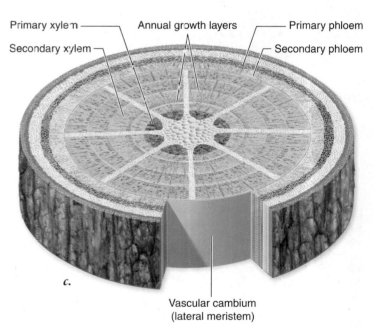

Primary xylem — Annual growth layers — Primary phloem
Secondary xylem — Secondary phloem

c.

Vascular cambium (lateral meristem)

figure 36.24

SECONDARY GROWTH. *a.* Before secondary growth begins in eudicot stems, primary tissues continue to elongate as the apical meristems produce primary growth. *b.* As secondary growth begins, the vascular cambium produces secondary tissues, and the stem's diameter increases. *c.* In this four-year-old stem, the secondary tissues continue to widen, and the trunk has become thick and woody. Note that the vascular cambium forms a cylinder that runs axially (up and down) in the roots and shoots that have them.

cambium connects the ring of primary vascular bundles. There is no logical way to connect primary monocot vascular tissue that would allow a uniform increase in girth. Lacking a vascular cambium, therefore, monocots do not have secondary growth.

Rings in the stump of a tree reveal annual patterns of vascular cambium growth; cell size varies, depending on growth conditions (figure 36.25). Large cells form under favorable conditions such as the summer months in temperate climates. Rings of smaller cells mark the seasons where growth is limited. In woody eudicots and gymnosperms, a second cambium, the cork cambium, arises in the outer cortex (occasionally in the epidermis or phloem); it produces boxlike cork cells to the outside and also may produce parenchyma-like phelloderm cells to the inside (figure 36.26).

The cork cambium, cork, and phelloderm are collectively referred to as the *periderm* (see figure 36.26). Cork tissues, the

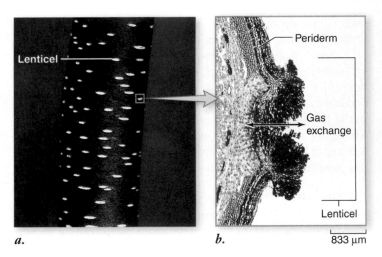

a. *b.* 833 µm

figure 36.27

LENTICELS. *a.* Lenticels, the numerous small, pale, raised areas shown here on cherry tree bark (*Prunus cerasifera*), allow gas exchange between the external atmosphere and the living tissues immediately beneath the bark of woody plants.
b. Transverse section through a lenticel in a stem of elderberry, *Sambucus canadensis*.

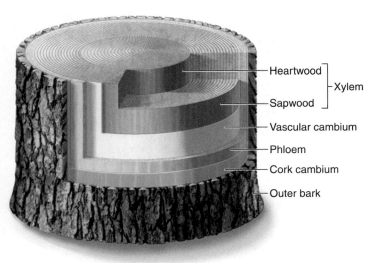

figure 36.25

TREE STUMP. The vascular cambium produces rings of xylem (sapwood and nonconducting heartwood) and phloem, while the cork cambium produces the cork.

cells of which become impregnated with water-repellent suberin shortly after they are formed and which then die, constitute the *outer bark*. The cork tissue cuts off water and food to the epidermis, which dies and sloughs off. In young stems, gas exchange between stem tissues and the air takes place through stomata, but as the cork cambium produces cork, it also produces patches of unsuberized cells beneath the stomata. These unsuberized cells, which permit gas exchange to continue, are called *lenticels* (figure 36.27).

Modified stems carry out vegetative propagation and store nutrients

Although most stems grow erect, some have modifications that serve special purposes, including natural vegetative propagation. In fact, the widespread artificial vegetative propagation of plants, both commercial and private, frequently involves cutting modified stems into segments, which are then planted, producing new plants. As you become acquainted with the following modified stems, keep in mind that stems have leaves at nodes, with internodes between the nodes, and buds in the axils of the leaves, whereas roots have no leaves, nodes, or axillary buds.

Bulbs. Onions, lilies, and tulips have swollen underground stems that are really large buds with adventitious roots at the base (figure 36.28a). Most of a bulb consists of fleshy leaves attached to a small, knoblike stem. In onions, the fleshy leaves are surrounded by papery, scalelike leaf bases of the long, green, above-ground leaves.

Corms. Crocuses, gladioluses, and other popular garden plants produce corms that superficially resemble bulbs. Cutting a corm in half, however, reveals no fleshy leaves. Instead, almost all of a corm consists of stem, with a few papery, brown nonfunctional leaves on the outside, and adventitious roots below.

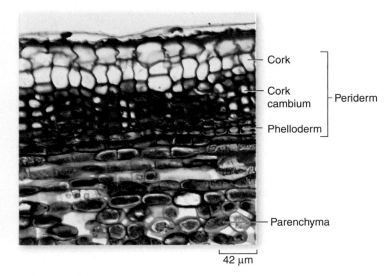

42 µm

figure 36.26

SECTION OF PERIDERM. An early stage in the development of periderm in cottonwood, *Populus* sp.

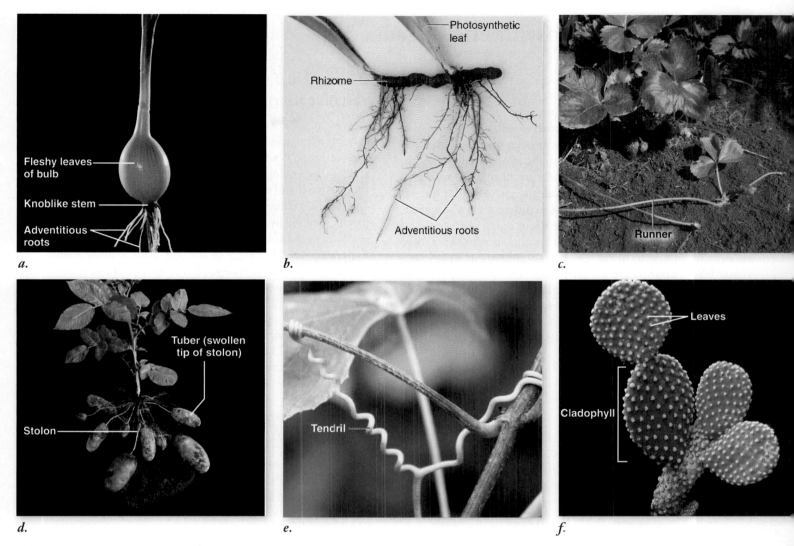

a.
Fleshy leaves of bulb
Knoblike stem
Adventitious roots

b.
Photosynthetic leaf
Rhizome
Adventitious roots

c.
Runner

d.
Tuber (swollen tip of stolon)
Stolon

e.
Tendril

f.
Leaves
Cladophyll

figure 36.28 TYPES OF MODIFIED STEMS.

Rhizomes. Perennial grasses, ferns, bearded iris, and many other plants produce rhizomes, which typically are horizontal stems that grow underground, often close to the surface (figure 36.28b). Each node has an inconspicuous scalelike leaf with an axillary bud; much larger photosynthetic leaves may be produced at the rhizome tip. Adventitious roots are produced throughout the length of the rhizome, mainly on the lower surface.

Runners and stolons. Strawberry plants produce horizontal stems with long internodes that unlike rhizomes, usually grow along the surface of the ground. Several runners may radiate out from a single plant (figure 36.28c). Some biologists use the term stolon synonymously with runner; others reserve the term stolon for a stem with long internodes (but no roots) that grows underground, as seen in potato plants (*Solanum* sp.). A potato itself, however, is another type of modified stem—a tuber.

Tubers. In potato plants, carbohydrates may accumulate at the tips of rhizomes, which swell, becoming tubers; the rhizomes die after the tubers mature (figure 36.28d). The "eyes" of a potato are axillary buds formed in the axils of scalelike leaves. The scalelike leaves, which are present when the potato is starting to form, soon drop off; the tiny ridge adjacent to each "eye" of a mature potato is a leaf scar.

Crop potatoes are not grown from seeds produced by potato flowers, but propagated vegetatively from "seed potatoes." A tuber is cut up into pieces that contain at least one eye, and these pieces are planted. The eye then grows into a new potato plant.

Tendrils. Many climbing plants, such as grapes and English ivy, produce modified stems known as tendrils that twine around supports and aid in climbing (figure 36.28e). Some other tendrils, such as those of peas and pumpkins, are actually modified leaves or leaflets.

Cladophylls. Cacti and several other plants produce flattened, photosynthetic stems called cladophylls that resemble leaves (figure 36.28f). In cacti, the real leaves are modified as spines (see the following section).

The stem system expands the primary and secondary plant body by producing axillary shoots and organs including leaves, tendrils, and flowers. Modified stems can increase storage and photosynthetic capacity.

Leaves: Photosynthetic Organs

Leaves, which are initiated as primordia by the apical meristems (see figure 36.20), are vital to life as we know it because they are the principal sites of photosynthesis on land, providing the base of the food chain. Leaves expand by cell enlargement and cell division. Like arms and legs in humans, they are determinate structures, which means their growth stops at maturity. Because leaves are crucial to a plant, features such as their arrangement, form, size, and internal structure are highly significant and can differ greatly. Different patterns have adaptive value in different environments.

Leaves are an extension of the shoot apical meristem and stem development. When they first emerge as primordia, they are not committed to being leaves. Experiments in which very young leaf primordia are isolated from fern and coleus plants and grown in culture have demonstrated this feature: If the primordia are young enough, they will form an entire shoot rather than a leaf. The positioning of leaf primordia and the initial cell divisions occur before those cells are committed to the leaf developmental pathway.

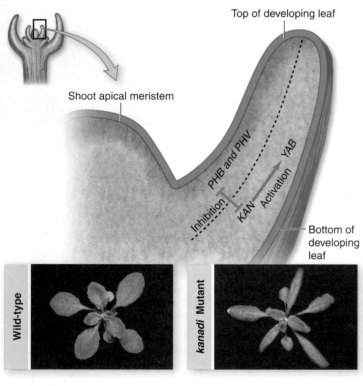

figure 36.29

ESTABLISHING TOP AND BOTTOM IN LEAVES. Several genes, including *PHABULOSA (PHB), PHAVOLUTA (PHV), KANADI (KAN),* and *YABBY (YAB)* make a flattened *Arabidopsis* leaf with a distinct upper and lower surface. *PHB* and *PHV* RNAs are restricted to the top, while *KAN* and *YAB* are expressed in the bottom cells of a leaf. *PHB* and *KAN* have an antagonistic relationship, restricting expression of each to separate leaf regions. *KAN* leads to *YABBY* expression and lower leaf development. Without *KAN,* both sides of the leaf develop like the top portion.

External leaf structure reflects vascular morphology

Leaves fall into two different morphological groups, which may reflect differences in evolutionary origin. A **microphyll** is a leaf with one vein branching from the vascular cylinder of the stem and not extending the full length of the leaf; microphylls are mostly small and are associated primarily with the phylum Lycophyta (see chapter 30). Most plants have leaves called **megaphylls,** which have several to many veins.

Most eudicot leaves have a flattened *blade* and a slender stalk, the *petiole.* The flattening of the leaf blade reflects a shift from radial symmetry to dorsal–ventral (top–bottom) symmetry. Leaf flattening increases the photosynthetic surface. Plant biologists are just beginning to understand how this shift occurs by analyzing mutants lacking distinct tops and bottoms (figure 36.29).

In addition, a pair of stipules may be present at the base of the petiole. The stipules, which may be leaflike or modified as spines (as in the black locust, *Robinia pseudo-acacia*) or glands (as in the purple-leaf plum tree *Prunus cerasifera*), vary considerably in size from the microscopic to almost half the size of the leaf blade. Grasses and other monocot leaves usually lack a petiole; these leaves tend to sheathe the stem toward the base.

Veins (a term used for the vascular bundles in leaves) consist of both xylem and phloem and are distributed throughout the leaf blades. The main veins are parallel in most monocot leaves; the veins of eudicots, on the other hand, form an often intricate network (figure 36.30).

Leaf blades come in a variety of forms, from oval to deeply lobed to having separate leaflets. In **simple leaves** (figure 36.31*a*), such as those of lilacs or birch trees, the blades are undivided, but simple leaves may have teeth, indentations, or lobes of various sizes, as in the leaves of maples and oaks.

In **compound leaves** (figure 36.31*b,c*), such as those of ashes, box elders, and walnuts, the blade is divided into *leaflets.*

figure 36.30

EUDICOT AND MONOCOT LEAVES. *a.* The leaves of eudicots, such as this African violet relative from Sri Lanka, have netted, or reticulate, veins. *b.* Those of monocots, such as this cabbage palmetto, have parallel veins. The eudicot leaf has been cleared with chemicals and stained with a red dye to make the veins show more clearly.

Axillary bud

a.   *b.*   *c.*

figure 36.31

SIMPLE VERSUS COMPOUND LEAVES. *a.* A simple leaf, its margin deeply lobed, from the oak tree (*Quercus robur*). *b.* A pinnately compound leaf, from a mountain ash (*Sorbus* sp.). A compound leaf is associated with a single lateral bud, located where the petiole is attached to the stem. *c.* Palmately compound leaves of a Virginia creeper (*Parthenocissus quinquefolia*).

The relationship between the development of compound and simple leaves is an open question. Two explanations are being debated: (1) A compound leaf is a highly lobed simple leaf, or (2) a compound leaf utilizes a shoot development program, and each leaflet was once a leaf. To address this question, researchers are using single mutations that are known to convert compound leaves to simple leaves.

If the leaflets are arranged in pairs along a common central axis, the leaf is pinnately compound (figure 36.31*b*). If, however, the leaflets radiate out from a common point at the blade end of the petiole, the leaf is palmately compound (figure 36.31*c*). Palmately compound leaves occur in buckeyes (*Aesculus* spp.) and Virginia creeper (*Parthenocissus quinquefolia*). The leaf blades themselves may have similar arrangements of their veins and are said to be pinnately or palmately veined.

Internal leaf structure regulates gas exchange and evaporation

The entire surface of a leaf is covered by a transparent epidermis, and most of these epidermal cells have no chloroplasts. As described earlier, the epidermis has a waxy cuticle and may have different types of glands and trichomes present. Also, the lower epidermis (and occasionally the upper epidermis) of most leaves contains numerous slitlike or mouth-shaped stomata flanked by guard cells (figure 36.32).

The tissue between the upper and lower epidermis is called **mesophyll.** Mesophyll is interspersed with veins of various sizes.

Most eudicot leaves have two distinct types of mesophyll. Closest to the upper epidermis are one to several (usually two) rows of tightly packed, barrel-shaped to cylindrical chlorenchyma cells (parenchyma with chloroplasts) that constitute

the palisade mesophyll (figure 36.33). Some plants, including species of *Eucalyptus*, have leaves that hang down, rather than extending horizontally. They have palisade mesophyll on both sides of the leaf, and there is, in effect, no upper side.

Nearly all eudicot leaves have loosely arranged spongy mesophyll cells between the palisade mesophyll and the lower epidermis, with many air spaces throughout the tissue. The interconnected intercellular spaces, along with the stomata, function in gas exchange and the passage of water vapor from the leaves.

The mesophyll of monocot leaves often is not differentiated into palisade and spongy layers, and there is often little distinction between the upper and lower epidermis. Instead, cells surrounding the vascular tissue are distinctive and are the site of carbon fixation. This anatomical difference often correlates with a modified photosynthetic pathway, C_4 *photosynthesis*, that maximizes the amount of CO_2 relative to O_2 to reduce energy loss through photorespiration (see chapter 9). The anatomy of a leaf directly relates to its juggling act of balancing water loss, gas exchange, and transport of photosynthetic products to the rest of the plant.

Modified leaves are highly versatile organs

As plants colonized a wide variety of environments, from deserts to lakes to tropical rain forests, plant organ modifications arose that would adapt the plants to their specific habitats. Leaves, in particular, have evolved some remarkable adaptations. A brief discussion of a few of these modifications follows:

Floral leaves (bracts). Poinsettias and dogwoods have relatively inconspicuous, small, greenish-yellow flowers. However, both plants produce large modified leaves called *bracts* (mostly colored red in poinsettias and white or pink in dogwoods). These bracts surround the true flowers and perform the same function as showy petals. In other plants, however, bracts can be quite small and inconspicuous.

Spines. The leaves of many cacti, barberries, and other plants are modified as *spines* (see figure 36.28*f*). In cacti, having less leaf surface reduces water loss, and the sharp spines also may deter predators. Spines should not be confused with *thorns*, such as those on the honey locust (*Gleditsia triacanthos*), which are modified stems, or with the prickles on raspberries, which are simply outgrowths from the epidermis or the cortex just beneath it.

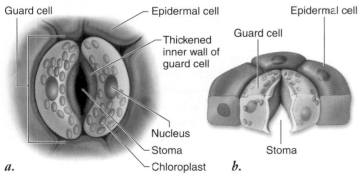

Guard cell   Epidermal cell   Epidermal cell

Thickened inner wall of guard cell

Guard cell

Nucleus

Stoma   Stoma

a.   Chloroplast   *b.*

figure 36.32

A STOMA. *a.* Surface view. *b.* View in cross section.

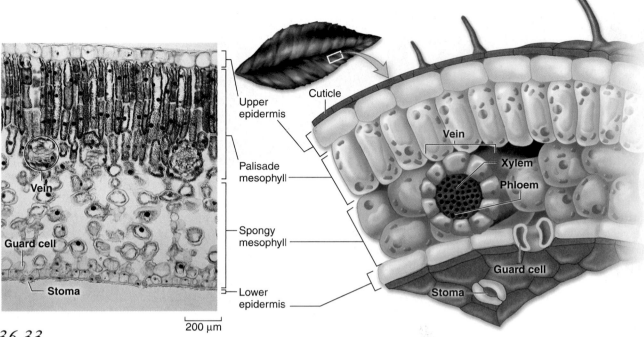

figure 36.33

A LEAF IN CROSS SECTION. Transection of a leaf showing the arrangement of palisade and spongy mesophyll, a vascular bundle or vein, and the epidermis with paired guard cells flanking the stoma.

Reproductive leaves. Several plants, notably *Kalanchoë*, produce tiny but complete plantlets along their margins. Each plantlet, when separated from the leaf, is capable of growing independently into a full-sized plant. The walking fern (*Asplenium rhizophyllum*) produces new plantlets at the tips of its fronds. Although many species can regenerate a whole plant from isolated leaf tissue, this in vivo regeneration is found among just a few species.

Window leaves. Several genera of plants growing in arid regions produce succulent, cone-shaped leaves with transparent tips. The leaves often become mostly buried in sand blown by the wind, but the transparent tips, which have a thick epidermis and cuticle, admit light to the hollow interiors. This strategy allows photosynthesis to take place beneath the surface of the ground.

Shade leaves. Leaves produced in the shade, where they receive little sunlight, tend to be larger in surface area, but thinner and with less mesophyll than leaves on the same tree receiving more direct light. This plasticity in development is remarkable. Environmental signals can have a major effect on development.

Insectivorous leaves. Almost 200 species of flowering plants are known to have leaves that trap insects; some plants digest the insects' soft parts. Plants with insectivorous leaves often grow in acid swamps that are deficient in needed elements or contain elements in forms not readily available to the plants; this inhibits the plants' capacities to maintain metabolic processes needed for their growth and reproduction. Their needs are met, however, by the supplementary absorption of nutrients from the animal kingdom.

Pitcher plants (for example, *Sarracenia*, *Darlingtonia*, or *Nepenthes* spp.) have cone-shaped leaves in which rainwater can accumulate. The insides of the leaves are very smooth, but stiff, downward-pointing hairs line the rim. An insect falling into such a leaf finds it very difficult to escape and eventually drowns. The leaf absorbs the nutrients released when bacteria, and in most species the plant's own digestive enzymes, decompose the insect bodies. Other plants, such as sundews (*Drosera*), have glands that secrete sticky mucilage that traps insects, which are then digested by enzymes.

The Venus flytrap (*Dionaea muscipula*) produces leaves that look hinged at the midrib. When tiny trigger hairs on the leaf blade are stimulated by a moving insect, the two halves of the leaf snap shut, and digestive enzymes break down the soft parts of the trapped insect into nutrients that can be absorbed through the leaf surface. Nitrogen is the most common nutrient needed. Curiously, the Venus flytrap cannot survive in a nitrogen-rich environment, perhaps as a result of a biochemical trade-off during the intricate evolutionary process that developed its ability to capture and digest insects.

Leaves come in a range of forms to maximize photosynthetic capacity and, in some cases, to provide protection or obtain other nutrients. A delicate balance must be struck in leaves between gas exchange for photosynthesis and water loss through the stomata. This chapter has focused on structure of the plant body, including gross morphology and tissue structure. In the chapters that follow, we present details of plant function, beginning with vegetative plant development in the next chapter.

36.1 Organization of the Plant Body: An Overview
(figure 36.1)

Modern vascular plants have many specialized structures that are adaptations to a terrestrial environment.

- Vascular plants consist of a root system and a shoot system.
  - *The root system is below ground and both anchors the plant and is used to acquire water and minerals.*
  - *The shoot system is above ground and consists of supporting stems, photosynthetic leaves, and reproductive flowers.*
- Roots, shoots, and leaves contain three basic tissue systems: dermal, ground, and vascular tissues (discussed in the following section).
- Meristems contain stem cells that act like animal stem cells. When they divide, one cell remains meristematic and the other differentiates into a specific tissue.
  - *Apical meristems are located on the tips of stems and near the tips of roots, and they give rise to three primary meristems: protoderm, procambium, and ground meristem.*
  - *Lateral meristems are found in plants that exhibit secondary growth and give rise to the vascular cambium and the cork cambium, resulting in an increase in stem diameter.*

36.2 Plant Tissues

Plants contain three basic tissue systems: dermal, ground, and vascular tissues.

- Dermal tissue is usually one cell thick and forms the protective epidermis that is covered with a fatty cutin to retard water loss. Specialized cells in the epidermis include:
  - *Guard cells flank the stomata and regulate the passage of water and gases.*
  - *Trichomes are hair-like growths of the shoot epidermis that keep leaves cool and reduce evaporation. Some are glandular and secrete a substance to deter herbivory.*
  - *Root hairs are extensions of the root epidermis that increase the surface area and efficiency of absorption.*
- Ground tissues have multiple functions including support, storage, and photosynthesis.
- Vascular tissue consists of xylem and phloem that transports materials through a plant.
  - *Xylem cells are dead and are the principal water- and mineral-conducting tissue of the plant (figure 36.12).*
  - *Phloem cells are alive and are the principal food-conducting tissue in vascular plants (figure 36.13).*

36.3 Roots: Anchoring and Absorption Structures

Roots are a major adaptation to terrestrial environments, and they anchor the plant and absorb plant nutrients and water (figure 36.14).

- Developing roots have four regions: root cap, zone of cell division, zone of elongation, and zone of maturation.
- Interior to the epidermis, the ground meristem produces parenchyma cells for food storage, and the endodermis is banded with suberin or the Casparian strip.
- All tissues interior to the endodermis are collectively referred to as the stele.
- Between the endodermis and the phloem is a cylinder of cells called the pericycle, which gives rise to lateral roots in monocots or the vascular cambium and cork cambium in eudicots.

- Eudicots often have a central tap root that extends deep into the soil, and monocots have a shallow fibrous root system.
- Adventitious roots arise primarily from the stem, not the root system, and they help anchor the plant.
- Roots may be highly modified for support and stability, acquisition of oxygen, storage of water and food, or to parasitize a host plant.

36.4 Stems: Support for Above-Ground Organs

Stems extend into the air above the root system and support plant mass against gravity.

- The apical shoot meristem produces stem tissues, leaf primordia, and bud primordia that develop into leaves, shoots, and flowers.
- Leaves are attached to stems at the nodes, and the space between nodes is called an internode.
- The axil is the area between the leaf and stem, and an axillary bud develops in axils of eudicots.
- Monocots and herbaceous eudicots contain chloroplasts and stomata, may have trichomes, and do not produce cork.
- Woody stems contain terminal buds along their length, and deciduous trees have stipules when leaves are lost in autumn.
- The vascular bundles of monocots are randomly scattered throughout the ground tissue, whereas in eudicots the bundles are arranged in a ring with vascular cambium between the inner xylem and the outer phloem that allows for secondary growth.
- Pith refers to the parenchyma cells located in the middle of the stem.
- The periderm contains cork cambium and dead, water-repellent suberin impregnated cork.
- Cork cambium also produces lenticels, unsuberized cells that allow for gas exchange.
- Modified stems (including bulbs, corms, rhizomes, runners and stolons, tubers, tendrils, and cladophylls) reproduce vegetatively and store nutrients.

36.5 Leaves: Photosynthetic Organs

Leaves are the principal sites of photosynthesis, and they exhibit many different morphological adaptations (figure 36.33).

- There are two basic types of leaves: microphylls and megaphylls.
- The veins or vascular bundles are parallel in monocots and form a network in eudicots.
- The leaves of most eudicots have a flattened blade and a slender petiole, whereas monocots usually do not have a petiole and sheathe the stem.
- Leaf blades may be simple and undivided or divided into leaflets that are arranged in a variety of patterns.
- The tissues of the leaf include the epidermis with chlorophyll-containing guard cells, vascular tissue, and two layers of chlorenchyma cells (palisade mesophyll and spongy mesophyll), which are located between the upper and lower epidermis.
- Leaves are remarkably adapted to serve many different functions: floral bracts, spines, reproductive units, protection, and carnivory.

SELF TEST

1. Fifteen years ago, your parents hung a swing from the lower branch of a large tree growing in your yard. When you go and sit in it today, you realize it is exactly the same height off the ground as it was when you first sat in it 15 years ago. The reason the swing is not higher off the ground as the tree has grown is that
   a. the tree trunk is only showing secondary growth.
   b. the tree trunk is part of the primary growth system of the plant, but elongation is no longer occurring in that part of the tree.
   c. trees lack apical meristems and so do not get taller.
   d. you are hallucinating, because it is impossible for the swing not to have been raised off the ground as the tree grew.

2. A unique feature of plants is indeterminate growth. Indeterminate growth is possible because:
   a. meristematic regions for primary growth occur throughout the entire plant body.
   b. all cell types in a plant often give rise to meristematic tissue.
   c. meristematic cells continually replace themselves.
   d. all cells in a plant continue to divide indefinitely.

3. If you were to relocate the pericycle of a plant root to the epidermal layer, how would it affect root growth?
   a. Secondary growth in the mature region of the root would not occur.
   b. The root apical meristem would produce vascular tissue in place of dermal tissue.
   c. Nothing would change because the pericycle is normally located near the epidermal layer of the root.
   d. Lateral roots would grow from the outer region of the root and fail to connect with the vascular tissue.

4. In vascular plants, one difference between root and shoot systems is that:
   a. root systems cannot undergo secondary growth.
   b. root systems undergo secondary growth, but do not form bark.
   c. root systems contain pronounced zones of cell elongation, whereas shoot systems do not.
   d. root systems can store food reserves, whereas stem structures do not

5. When you peel your potatoes for dinner, you are removing the majority of their
   a. dermal tissue.
   b. vascular tissue.
   c. ground tissue.
   d. Only (a) and (b) are removed with the peel.
   e. All of these are removed with the peel.

6. You can determine the age of an oak tree by counting the annual rings of _____ formed by the _____.
   a. primary xylem/apical meristem
   b. secondary phloem/vascular cambium
   c. dermal tissue/cork cambium
   d. secondary xylem/vascular cambium

7. Many vegetables are grown today through hydroponics, where the plant roots exist primarily in an aqueous solution. Which of the following root structures is no longer beneficial in hydroponics?
   a. epidermis.
   b. xylem.
   c. root cap.
   d. root hairs.

8. Root hairs and lateral roots are similar in each respect except:
   a. both increase the absorptive surface area of the root system.
   b. both are generally long-lived.
   c. both are multicellular.
   d. (b) and (c).

9. Which of the following statements is not true of the stems of vascular plants?
   a. Stems are composed of repeating segments, including nodes and internodes.
   b. Primary growth only occurs at the shoot apical meristem.
   c. Vascular tissues may be arranged on the outside of the stem or scattered throughout the stem.
   d. Stems can contain stomata.

10. Plant organs form by
    a. cell division in gamete tissue.
    b. cell division in meristematic tissue.
    c. cell migration into the appropriate position in the tissue.
    d. rearranging the genetic material in the precursor cells so that the organ-specific genes are activated.

11. Which of the following plant cell type is mismatched to its function?
    a. xylem; conducts mineral nutrients.
    b. phloem; serves as part of the bark.
    c. trichomes; reduces evaporation.
    d. collenchyma; performs photosynthesis.

12. Which is the correct sequence of cell types encountered in an oak tree, moving from the center of the tree out?
    a. pith, secondary xylem, primary xylem, vascular cambium, primary phloem, secondary phloem, cork cambium, cork.
    b. pith, primary xylem, secondary xylem, vascular cambium, secondary phloem, primary phloem, cork cambium, cork
    c. pith, primary xylem, secondary xylem, vascular cambium, secondary phloem, primary phloem, cork, cork cambium
    d. pith, primary phloem secondary phloem, vascular cambium, secondary xylem, primary xylem, cork cambium, cork.

13. You've just bought a house with a great view of the mountains, but you have a neighbor who planted a bunch of trees which are now blocking your view. In an attempt to ultimately remove the trees and remain unlinked to the deed, you begin training several porcupines to enter the yard under the cover of night and perform a stealth operation. In order to most effectively kill the trees, you should train the porcupines to completely remove:
    a. the vascular cambium
    b. the cork
    c. the cork cambium
    d. the primary phloem

CHALLENGE QUESTIONS

1. Plant organs undergo many modifications to deal with environmental challenges. Define your favorite modified root, shoot, and leaf, and make a case for why it is the best example of a modified plant organ.

2. You have identified a mutant maize plant that cannot differentiate vessel cells. How would this affect the functioning of the plant?

3. Increasing human population on the planet is stretching our ability to produce sufficient food to support the world's population. If you could engineer the perfect crop plant, what features might it possess?

Vegetative Plant Development

introduction

HOW DOES A FERTILIZED EGG DEVELOP into a complex adult plant body? Because plant cells cannot move, the timing and directionality of each cell division must be carefully orchestrated. Cells need information about their location relative to other cells so that cell specialization is coordinated. The developing embryo is quite fragile, and numerous protective structures have evolved since plants first colonized land.

Only a portion of the plant has actually formed when it first emerges from the soil. New plant organs develop throughout the plant's life.

37.1 Embryo Development

Embryo development begins once the egg cell is fertilized. As described briefly in chapter 30, the growing pollen tube from a pollen grain enters the angiosperm embryo sac through one of the synergids, releasing two sperm cells (figure 37.1). One sperm cell fertilizes the central cell with its polar nuclei, and the resulting cell division produces a nutrient source, the **endosperm,** for the embryo. The other sperm cell fertilizes the egg to produce a **zygote,** and cell division soon follows, creating the **embryo.**

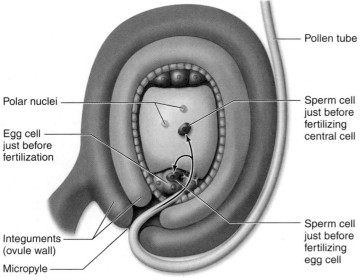

figure 37.1

FERTILIZATION TRIGGERS EMBRYOGENESIS. The egg cell, within the embryo sac, is fertilized by one sperm cell released from the pollen tube. The second sperm cell fertilizes the central cell and initiates endosperm development. This diagram shows sperm just before fertilization.

A single cell divides to produce a three-dimensional body plan

The first division of the zygote (fertilized egg) in a flowering plant is asymmetrical and generates cells with two different fates (figure 37.2). One daughter cell is small, with dense cytoplasm. That cell, which is destined to become the embryo, begins to divide repeatedly in different planes, forming a ball of cells. The other, larger daughter cell divides repeatedly, forming an elongated structure called a **suspensor,** which links the embryo to the nutrient tissue of the seed. The suspensor also provides a route for nutrients to reach the developing embryo. The root–shoot axis also forms at this time; cells near the suspensor are destined to form a root, while those at the other end of the axis ultimately become a shoot.

Investigating mechanisms for establishing asymmetry in plant embryo development is difficult because the zygote is embedded within the female gametophyte, which is surrounded by sporophyte tissue (ovule and carpel tissue) (chapter 30). To understand the cell biology of the first asymmetrical division of zygotes, biologists have studied the brown alga *Fucus*. We must be cautious about inferring too much about angiosperm asymmetrical divisions from the brown algae because the last common ancestor of brown algae and the angiosperm line was a single-celled organism. Nevertheless, asymmetrical division extends far back in the tree of life, even to bacteria.

Zygote development in Fucus

In the brown alga *Fucus*, the egg is released prior to fertilization, so no extra tissues surround the zygote, making its development easier to observe. A bulge that develops on one side of the zygote establishes the vertical axis. Cell division occurs, and the original bulge becomes the smaller of the two daughter

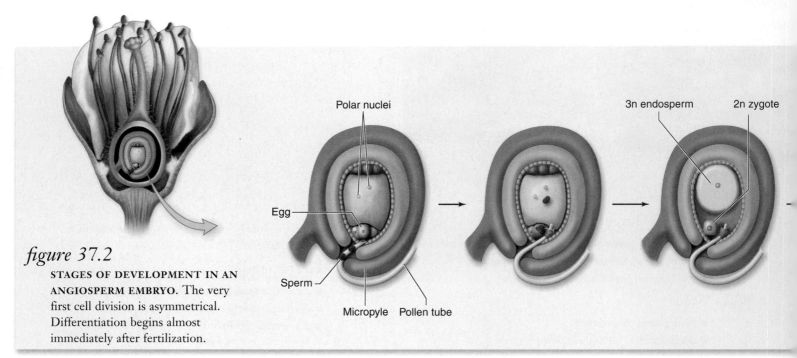

figure 37.2

STAGES OF DEVELOPMENT IN AN ANGIOSPERM EMBRYO. The very first cell division is asymmetrical. Differentiation begins almost immediately after fertilization.

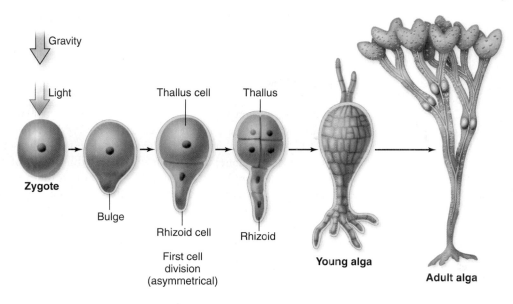

Gravity

Light

Zygote

Bulge

Thallus cell

Rhizoid cell

First cell
division
(asymmetrical)

Thallus

Rhizoid

Young alga

Adult alga

figure 37.3

ASYMMETRICAL CELL DIVISION IN A *FUCUS* ZYGOTE. An unequal distribution of material in the zygote leads to a bulge where the first cell division will occur. This division results in a smaller cell that will go on to divide and produce the rhizoid that anchors the alga; the larger cell divides to form the thallus, or main algal body. The point of sperm entry determines where the smaller rhizoid cell will form, but light and gravity can modify this to ensure that the rhizoid will point downward where it can anchor this brown alga. Calcium-mediated currents set up an internal gradient of charged molecules, which leads to a weakening of the cell wall where the rhizoid will form. The fate of the two resulting cells is held "in memory" by cell wall components.

cells. The smaller cell develops into a rhizoid that anchors the alga, and the larger cell develops into the main body, or thallus, of the sporophyte (figure 37.3).

This axis is first established by the point of sperm entry, but it can be changed by environmental signals, especially light and gravity, which ensure that the rhizoid is down and the thallus is up. Internal gradients are established that specify where the rhizoid forms in response to environmental signals.

The ability to "remember" where the rhizoid will form depends on the presence of the cell wall. In experiments, enzymatic removal of the cell wall in *Fucus* cells that were destined to form either rhizoids or plant body resulted in cells that could give rise to both. Cell walls contain a wide variety of carbohydrates and proteins attached to the wall's structural fibers. Attempting to pin down the identities of suspected developmental signals in the cell wall is an area of active research.

Studies of developmental mutants

Genetic approaches make it possible to explore asymmetrical development in angiosperms. Studies of mutants reveal what can go wrong in development, which often makes it possible to infer normal developmental mechanisms. For example, the suspensor mutant in *Arabidopsis* undergoes aberrant development in the embryo followed by embryo-like

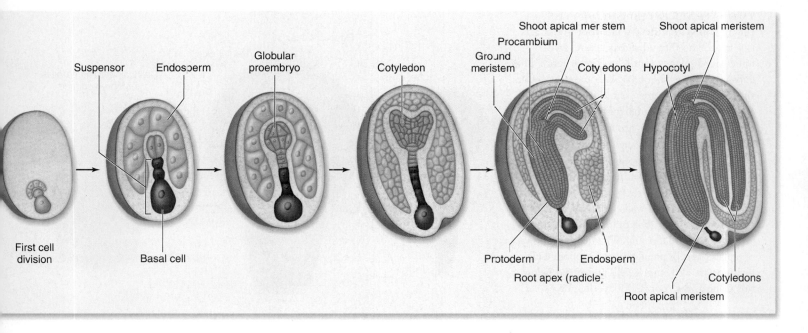

First cell division

Suspensor

Endosperm

Basal cell

Globular proembryo

Cotyledon

Shoot apical meristem

Procambium

Ground meristem

Protoderm

Root apex (radicle)

Cotyledons

Shoot apical meristem

Hypocotyl

Endosperm

Root apical meristem

Cotyledons

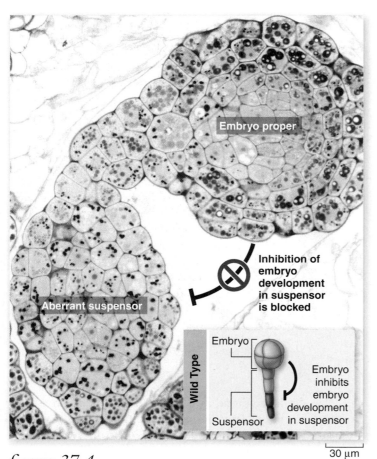

figure 37.4

THE EMBRYO SUPPRESSES DEVELOPMENT OF THE SUSPENSOR AS A SECOND EMBRYO. This *suspensor* (*sus*) mutant of *Arabidopsis* has a defect in embryo development. Aborted embryo development is followed by embryo-like development of the suspensor. *SUS* is required to suppress embryo development in suspensor cells.

development of the suspensor (figure 37.4). Analysis of this mutant led to the conclusion that the presence of the embryo normally prevents the suspensor from developing into a second embryo.

A simple body plan emerges during embryogenesis

In plants, three-dimensional shape and form arise by regulating the amount and the pattern of cell division. We have just described how a vertical axis (root–shoot axis) becomes established at a very early stage; the same is true for establishment of a radial axis (inner–outer axis) (figure 37.5). Although the first cell division gives rise to a single row of cells, cells soon begin dividing in different directions, producing a three-dimensional solid ball of cells. The root–shoot axis lengthens as cells divide, with new cell walls forming parallel to the ground.

To create the radial axis, cells alternate between synchronous cell divisions, producing cell walls parallel to and perpendicular to the surface of the embryo (see figure 37.5). The cells must divide in two directions in the radial plane in order to maintain the circumference in early development. The body plan that emerges is shown in figure 37.6. Apical meristems, the actively dividing cell regions at the tips of roots and shoots, establish the root–shoot axis in the globular stage, and the three basic tissue systems arise: *dermal, ground,* and *vascular* tissue (see chapter 36). These tissues are organized radially around the root–shoot axis.

Root and shoot formation

Both the shoot and root meristems are apical meristems, but their formation is controlled independently. This conclusion is supported by mutant analysis in *Arabidopsis*, in which the *shootmeristemless* (*stm*) mutant fails to produce a viable shoot, but does produce a root (figure 37.7). Thus, *STM* is necessary for shoot meristem development, but has no role in root meristem development.

figure 37.5

TWO AXES ARE ESTABLISHED IN THE DEVELOPING EMBRYO. The root–shoot axis is vertical, and the radial axis creates two-dimensional planes parallel to the ground. The ends of the root–shoot axis become the root and shoot apical meristems. Three tissue systems develop around the radial axis. Embryos form concentric rings of cells around the root–shoot axis by regulating the planes of cell division. Early in embryogenesis cells alternate between coordinated divisions that produce new cell walls parallel to the embryo surface and cell divisions that produce new cell walls perpendicular to the embryo surface. The orange lines show new cell walls that are forming. The diagram shows one plane of cells parallel to the ground. Cell divisions are also adding cells above and below this plane as the root–shoot axis lengthens.

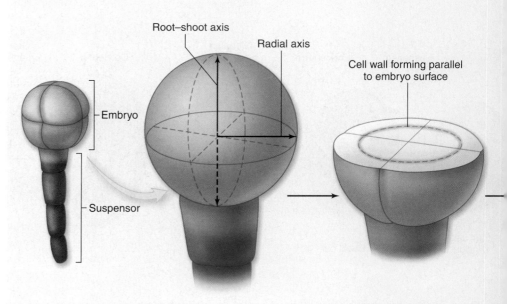

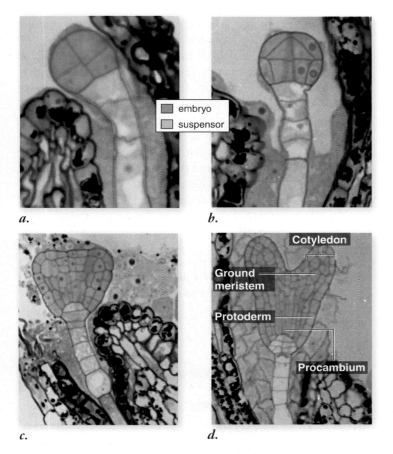

embryo
suspensor

a.　　　　　　　*b.*

Cotyledon

Ground
meristem

Protoderm

Procambium

c.　　　　　　　*d.*

figure 37.6

EARLY DEVELOPMENTAL STAGES OF *Arabidopsis thaliana.*
a. Early cell division has produced the embryo and suspensor.
b. Globular stage results from cell divisions along both the root–
shoot and radial axes. Cell differentiation, including establishment
of the root and shoot apical meristems occurs at this stage. *c, d.*
Heart-shaped stage. The cotyledons (seed leaves) are now visible,
and the three tissue systems continue to differentiate.

figure 37.7

**SHOOTMERISTEMLESS IS NEEDED FOR SHOOT
FORMATION.** Shoot-specific genes specify formation of
the shoot apical meristem, but are not necessary for root
development. The *stm* mutant of *Arabidopsis* (shown on top)
has a normal root meristem but fails to produce a shoot
meristem between its two cotyledons. The *STM* wild type is
shown below the *stm* mutant for comparison.

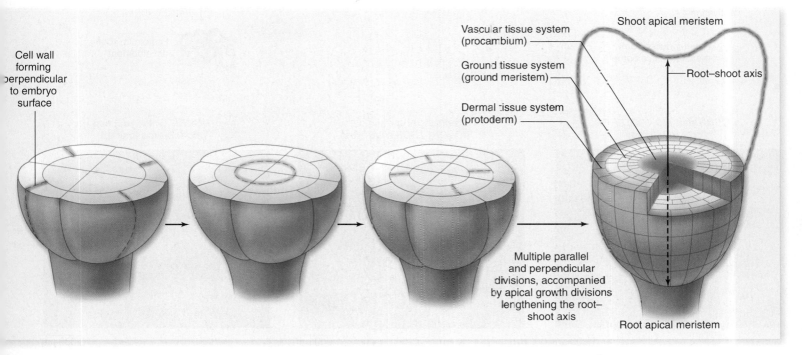

Cell wall
forming
perpendicular
to embryo
surface

Vascular tissue system
(procambium)

Ground tissue system
(ground meristem)

Dermal tissue system
(protoderm)

Shoot apical meristem

Root–shoot axis

Multiple parallel
and perpendicular
divisions, accompanied
by apical growth divisions
lengthening the root–
shoot axis

Root apical meristem

The *STM* gene codes for a transcription factor with a homeobox region, sharing a common evolutionary origin with the *Hox* genes that are important in establishing animal body plans (see chapters 19 and 25). Compared with animals, however, *Hox*-like genes have a more limited role in regulating plant body plans. Other gene families, encoding different transcription factors, have been identified as also playing key roles in patterning in plants.

Unlike *STM*, *HOBBIT* is required for root meristem formation, but not shoot meristem formation in *Arabidopsis* (figure 37.8). Cell divisions in *hobbit* embryos occur in the wrong directions, and no root meristem forms. Plants with a *hobbit* mutation accummulate a biochemical repressor of genes that are induced by auxin (a plant hormone). Based on the mutant phenotype, the function of *HOBBIT* appears to be to repress the production of the repressor of auxin-induced genes. Or, more simply stated, HOBBIT protein allows auxin to induce the expression of a gene or genes needed for correct cell division to make a root meristem. (Auxin is one of seven classes of hormones that regulate plant development and function; we will explore each of these classes of hormones in chapter 41.)

One way that auxin induces gene expression is by activating a transcription factor. *MONOPTEROS* (MP) is a gene that codes for an auxin-induced transcription factor (see figure 37.8), and like *HOBBIT*, it is necessary for root formation, but not shoot formation, in *Arabidopsis*. Once activated, MP protein binds to the promoter of another gene, leading to transcription of a gene or genes needed for root meristem formation.

figure 37.8

GENETIC CONTROL OF EMBRYONIC ROOT DEVELOPMENT IN *Arabidopsis*. *a*. HOBBIT both activates auxin-induced root development and blocks repression of the auxin response. ***b*.** MONOPTEROS cannot act as a transcription factor when it is bound by a repressor. Auxin releases the repressor from Monopteros which then activates transcription of a root development gene. The mechanism that involves auxin-mediated release of the monoptuous repressor is complex. For details see Figure 41.23. ***c*.** A wild-type seedling depends on auxin-induced genes for normal root initiation during embryogenesis. ***d*.** The *hobbit* seedling has a stub rather than a root because abnormal cell divisions prevent root meristem formation. ***e*.** The *monopteros* seedling has a basal peg, not a root.

inquiry

? *Referring to (e), **explain why this mutant fails to develop an embryonic root.***

inquiry

? *Predict the phenotype of a plant with a mutation in the **MP** gene that results in an **MP** protein that can no longer bind its repressor.*

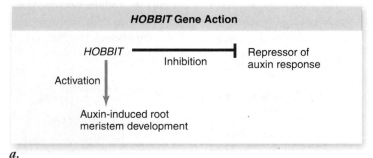

HOBBIT Gene Action

a.

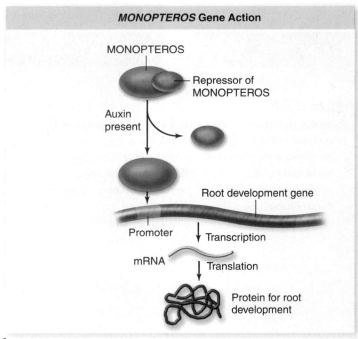

MONOPTEROS Gene Action

b.

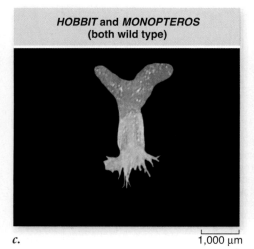

HOBBIT and MONOPTEROS (both wild type)

c.   1,000 µm

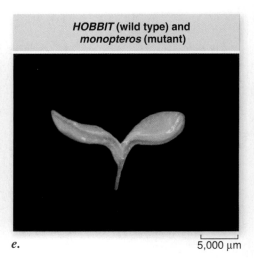

hobbit (mutant) and MONOPTEROS (wild type)

d.   1,000 µm

HOBBIT (wild type) and monopteros (mutant)

e.   5,000 µm

Formation of the three tissue systems

Three basic tissues, called *primary meristems*, differentiate while the plant embryo is still a ball of cells (called the globular stage; see figure 37.6), but no cell movements are involved. The **protoderm** consists of the outermost cells in a plant embryo and will become *dermal tissue* (see chapter 36). These cells almost always divide with their cell plate perpendicular to the body surface, thus perpetuating a single outer layer of cells. Dermal tissue protects the plant from desiccation, and the stomata that open and close to facilitate gas exchange and minimize water loss are derived from cells of this tissue.

A **ground meristem** gives rise to the bulk of the embryonic interior, consisting of *ground tissue* cells that eventually function in food and water storage.

Finally, **procambium** at the core of the embryo is destined to form the future *vascular tissue*, which is responsible for water and nutrient transport.

As you study embryo development, you will notice that many of the same patterns of tissue differentiation are seen in the apical meristems of germinated plants described in chapter 36. Cell fates are generally more limited after embryogenesis, however, when embryo-specific genes are not expressed. For example, the *LEAFY COTYLEDON* gene in *Arabidopsis* is active in early and late embryo development, and it may be respon-

sible for maintaining an embryonic environment. It is possible to turn this gene on later in development using recombinant DNA techniques described in chapter 16. When it is turned on, embryos can form on leaves!

Morphogenesis

The globular stage gives rise to a heart-shaped embryo with two bulges in one group of angiosperms (the eudicots, such as *A. thaliana* in figure 37.6c,d), and a ball with a bulge on a single side in another group (the monocots). These bulges are **cotyledons** ("first leaves") and are produced by the embryonic cells, and not by the shoot apical meristem that begins forming during the globular stage. This process, called **morphogenesis** (generation of form), results from changes in planes and rates of cell division (see figure 37.5).

Because plant cells cannot move, the form of a plant body is largely determined by the plane in which its cells divide. It is also controlled by changes in cell shape as cells expand osmotically after they form (figure 37.9). The position of the cell plate determines the direction of division, and both microtubules and actin play a role in establishing the cell plate's position. Plant hormones and other factors influence the orientation of bundles of microtubules on the interior of the plasma membrane. These microtubules also guide cellulose deposition as the cell wall forms around the outside of a new cell (see figure 36.2) where four of the six sides are reinforced more heavily with cellulose; the cell tends to expand and grow in the direction of the two sides having less reinforcement (figure 37.9b).

Much is being learned about morphogenesis at the cellular level from mutants that are able to divide, but cannot control their plane of cell division or the direction of cell expansion. The lack of root meristem development in *hobbit* mutants is just one such example. As the procambium begins differentiating in the root, a critical division parallel to the root's surface is regulated by the gene *WOODEN LEG* (*WOL*, figure 37.10). Without that division, the cylinder of cells that would form phloem is missing. Only xylem forms in the vascular tissue system, giving the root a "wooden leg."

Early in embryonic development, most cells can give rise to a wide range of cell and organ types, including leaves. As

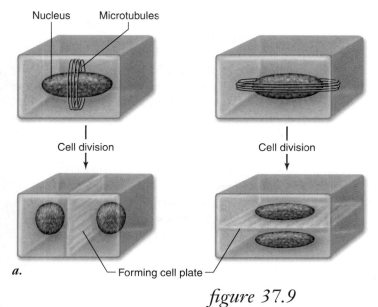

a.

Nucleus   Microtubules

Cell division

Forming cell plate

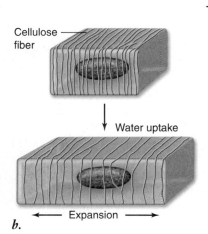

Cellulose fiber

Water uptake

Expansion

b.

figure 37.9

CELL DIVISION AND EXPANSION. *a.* Orientation of microtubules determine the orientation of cell plate formation and thus the new cell wall. *b.* Not all sides of a plant cell have the same amount of cellulose reinforcement. With water uptake, cells expand in directions that have the least amount of cell wall reinforcement.

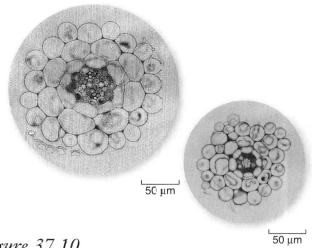

50 μm

50 μm

figure 37.10

***WOODEN LEG* IS NEEDED FOR PHLOEM DEVELOPMENT.** The *wol* mutant (*right*) has less vascular tissue than wild-type *Arabidopsis* (*left*), but all of it is xylem.

development proceeds, the cells with multiple potentials are mainly restricted to the meristem regions. Many meristems have been established by the time embryogenesis ends and the seed becomes dormant. After germination, apical meristems continue adding cells to the growing root and shoot tips. Apical meristem cells of corn, for example, divide every 12 hours, producing half a million cells per day in an actively growing corn plant. Lateral meristems can cause an increase in the girth of some plants, while intercalary meristems in the stems of grasses allow for elongation.

Food reserves form during embryogenesis

While the embryo is developing, three other critical events are occurring in angiosperms. The first event is the establishment of a food supply that will support the embryo during germination, while it gains photosynthetic capacity. In angiosperms, double fertilization produces endosperm for nutrition; in gymnosperms, the megagametophyte is the food source (see chapter 30). A second critical event is the differentiation of ovule tissue (from the parental sporophyte) to form a hard, protective covering around the embryo. The seed then enters a dormant phase, signaling the end of embryogenesis. In angiosperms, the third critical event is the development of the carpel wall into the fruit also accompanies embryo development. Seed development and germination, as well as fruit development, are addressed later in this chapter. In this section, we focus on nutrient reserves.

Throughout embryogenesis, starch, lipids, and proteins are synthesized. The seed storage proteins are so abundant that the genes coding for them were the first cloning targets for plant molecular biologists. Providing nutritional resources is part of the evolutionary trend toward enhancing embryo survival.

The sporophyte transfers nutrients via the suspensor in angiosperms. (In gymnosperms, the suspensor serves only to push the embryo closer to the megagametophytic nutrient source.) This happens concurrently with the development of the endosperm, which is present only in angiosperms (although double fertilization has been observed in the gymnosperm *Ephedra*). Endosperm formation may be extensive or minimal.

Endosperm in coconut includes the "milk," a liquid. In corn, the endosperm is solid, and in popcorn, it expands with heat

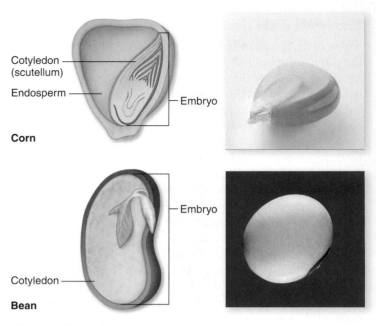

figure 37.11

ENDOSPERM IN MAIZE AND BEAN. The maize kernel has endosperm that is still present at maturity, while the endosperm in the bean has disappeared. The bean embryo's cotyledons take over food storage functions.

to form the white edible part of popped corn. In peas and beans, the endosperm is used up during embryo development, and nutrients are stored in thick, fleshy cotyledons (figure 37.11).

Because the photosynthetic machinery is built in response to light, it is critical that seeds have stored nutrients to aid in germination until the growing sporophyte can photosynthesize. A seed buried too deeply in the soil will use up all its reserves in cellular respiration before reaching the surface and sunlight.

> The root–shoot and radial axes form during embryogenesis. The mature embryo has a simple body plan and the potential to build root and shoot systems from apical meristems after germination. In addition, nutrients are stored in the endosperm or cotyledons.

37.2 Seeds

Early in the development of an angiosperm embryo, a profoundly significant event occurs: The embryo stops developing. In many plants, development of the embryo is arrested soon after the meristems and cotyledons differentiate. The integuments—the outer cell layers of the ovule—develop into a relatively impermeable **seed coat,** which encloses the seed with its dormant embryo and stored food (figure 37.12).

Seeds protect the embryo

The seed is a vehicle for dispersing the embryo to distant sites. Being encased in the protective layers of a seed allows a plant embryo to survive in environments that might kill a mature plant.

Seeds are an important adaptation in at least four ways:

1. Seeds maintain dormancy under unfavorable conditions and postpone development until better conditions arise. If conditions are marginal, a plant can "afford" to have some seeds germinate, because some of those that germinate may survive, while others remain dormant.

2. Seeds afford maximum protection to the young plant at its most vulnerable stage of development.

3. Seeds contain stored food that permits a young plant to grow and develop before photosynthetic activity begins.

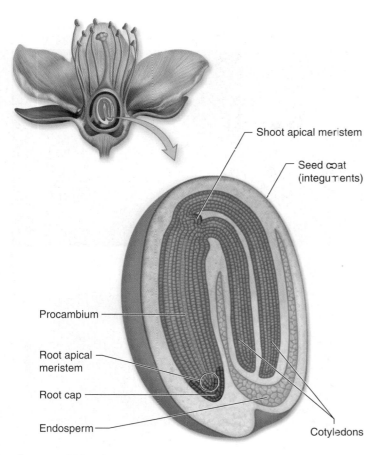

figure 37.12

SEED DEVELOPMENT. The integuments of this mature angiosperm ovule are forming the seed coat. Note that the two cotyledons have grown in a bent shape to accommodate the tight confines of the seed. In some embryos, the shoot apical meristem will have already initiated a few leaf primordia as well.

inquiry

? *Is this embryo a monocot or a eudicot?*

figure 37.13

FIRE INDUCES SEED RELEASE IN SOME PINES. Fire can destroy adult jack pines, but stimulate growth of the next generation. *a.* The cones of a jack pine are tightly sealed and cannot release the seeds protected by the scales. *b.* High temperatures lead to the release of the seeds.

4. Perhaps most important, seeds are adapted for dispersal, facilitating the migration of plant genotypes into new habitats.

Once a seed coat forms, most of the embryo's metabolic activities cease. A mature seed contains only about 5 to 20% water. Under these conditions, the seed and the young plant within it are very stable; its arrested growth is primarily due to the progressive and severe desiccation of the embryo and the associated reduction in metabolic activity. Germination cannot take place until water and oxygen reach the embryo; meanwhile, the seed coat may crack by abrasion or alternate freezing and thawing. Seeds of some plants have been known to remain viable for hundreds and, in rare instances, thousands of years.

Specialized seed adaptations improve survival

Specific adaptations often help ensure that seeds will germinate only under appropriate conditions. Sometimes, seeds lie within tough cones that do not open until they are exposed to

the heat of a fire (figure 37.13). This strategy causes the seed to germinate in an open, fire-cleared habitat where nutrients are relatively abundant, having been released from plants burned in the fire.

Seeds of other plants germinate only when inhibitory chemicals leach from their seed coats, thus guaranteeing their germination when sufficient water is available. Still other seeds germinate only after they pass through the intestines of birds or mammals or are regurgitated by them, which both weakens the seed coats and ensures dispersal. Sometimes seeds of plants thought to be extinct in a particular area may germinate under unique or improved environmental circumstances, and the plants may then become reestablished.

Seeds allow embryos to stay dormant for an extended period of time, protected from harsh environmental conditions. Embryo dispersal is enhanced because of seeds.

37.3 Fruits

Survival of angiosperm embryos depends on fruit development as well as seed development. **Fruits** are most simply defined as mature ovaries (carpels). During seed formation, the flower ovary begins to develop into fruit (figure 37.14). In some cases, the event of pollen landing on the stigma can initiate fruit development, but more frequently the coordination of fruit, seed coat, embryo, and endosperm development follow fertilization.

It is possible for fruits to develop without seed development. Bananas for example have aborted seed development, but do produce mature, edible ovaries. Bananas are propagated asexually since no embryo develops.

Fruit morphology exhibits environmental adaptations

Fruits form in many ways and exhibit a wide array of adaptations for dispersal. Three layers of ovary wall, also called the *pericarp*, can have distinct fates, which account for the diversity of fruit types from fleshy to dry and hard. The differences among some of the fruit types are shown in figure 37.15.

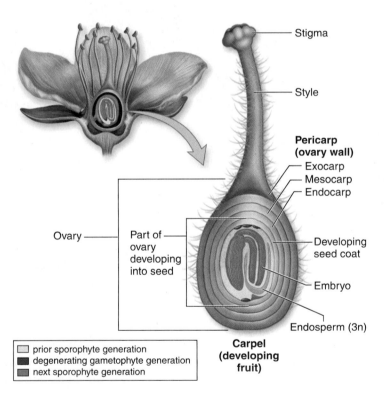

- ☐ prior sporophyte generation
- ■ degenerating gametophyte generation
- ▩ next sporophyte generation

figure 37.14

FRUIT DEVELOPMENT. The carpel (specifically the ovary) wall is composed of three layers, the exocarp, mesocarp, and endocarp. One, some, or all of these layers develops to contribute to the recognized fruit in different species. The seed matures within this developing fruit.

 inquiry

Three generations are represented in this diagram. Label the ploidy levels of the tissues of different generations shown here.

True Berries

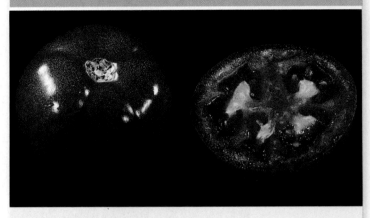

The entire pericarp is fleshy, although there may be a thin skin. Berries have multiple seeds in either one or more ovaries. The tomato flower had four carpels that fused. Each carpel contains multiple ovules that develop into seeds.

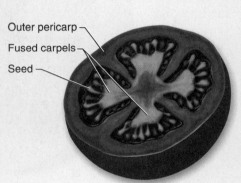

- Outer pericarp
- Fused carpels
- Seed

Legumes

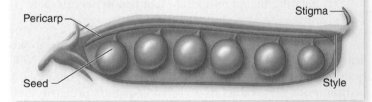

Split along two carpel edges (sutures) with seeds attached to edges; peas, beans. Unlike fleshy fruits, the three tissue layers of the ovary do not thicken extensively. The entire pericarp is dry at maturity.

- Pericarp
- Seed
- Stigma
- Style

figure 37.15

EXAMPLES OF SOME KINDS OF FRUITS. Distinguishing features of each of these fruit types are diagrammed and described below each photo. Legumes and samaras are examples of dry fruits. Legumes open to release their seeds, while samara do not. Drupes, true berries, and hesperidia are simple fleshy fruits; they develop from a flower with a single pistil composed of one or more carpels. Aggregate and multiple fruits are compound fleshy fruits; they develop from flowers with more than one pistil or from more than one flower.

Drupes

Single seed enclosed in a hard pit; peaches, plums, cherries. Each layer of the pericarp has a different structure and function, with the endocarp forming the pit.

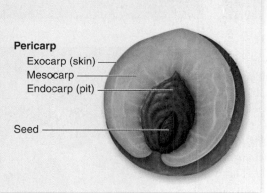

Pericarp
Exocarp (skin)
Mesocarp
Endocarp (pit)

Seed

Aggregate Fruits

Derived from many ovaries of a single flower; strawberries, blackberries. Unlike tomato, these ovaries are not fused and covered by a continuous pericarp.

Sepals of a single flower

Ovary

Seed

Samaras

Not split and with a wing formed from the outer tissues; maples, elms, ashes.

Pericarp

Seed

Multiple Fruits

Individual flowers form fruits around a single stem. The fruits fuse as seen with pineapple.

Main stem

Pericarp of individual flower

a.

b.

c.

figure 37.16

ANIMAL-DISPERSED FRUITS. *a.* The bright red berries of this honeysuckle, *Lonicera hispidula*, are highly attractive to birds. After eating the fruits, birds may carry the seeds they contain for great distances either internally or, because of their sticky pulp, stuck to their feet or other body parts. *b.* You will know if you have ever stepped on the fruits of *Cenchrus incertus;* their spines adhere readily to any passing animal. *c.* False dandelion, *Pyrrhopappus carolinianus*, has "parachutes" that widely disperse the fruits in the wind, much to the gardener's despair. *d.* This fruit of the coconut palm, *Cocos nucifera*, is sprouting on a sandy beach. Coconuts, one of the most useful fruits for humans in the tropics, have become established on other islands by drifting there on the waves.

d.

Developmentally, fruits are fascinating organs that contain three genotypes in one package. The fruit and seed coat are from the prior sporophyte generation. Remnants of the gametophyte generation that produced the egg are found in the developing seed, and the embryo represents the next sporophyte generation.

Fruits allow angiosperms to colonize large areas

Aside from the many ways fruits can form, they also exhibit a wide array of specialized dispersal methods. Fruits with fleshy coverings, often shiny black or bright blue or red, normally are dispersed by birds or other vertebrates (figure 37.16*a*). Like red flowers, red fruits signal an abundant food supply. By feeding on these fruits, birds and other animals may carry seeds from place to place and thus transfer plants from one suitable habitat to another. Such seeds require a hard seed coat to resist stomach acids and digestive enzymes.

Fruits with hooked spines, such as those of burrs (figure 37.16*b*), are typical of several genera of plants that occur in the northern deciduous forests. Such fruits are often disseminated by mammals, including humans, when they hitch a ride on fur or clothing. Squirrels and similar mammals disperse and bury fruits such as acorns and other nuts. Some of these sprout when conditions become favorable, such as after the spring thaw.

Other fruits, such as those of maples, elms, and ashes, have wings that aid in their distribution by the wind. Orchids have minute, dustlike seeds, which are likewise blown away by the wind. The dandelion provides another familiar example of a fruit type that is wind-dispersed (figure 37.16*c*), and the dispersal of seeds from plants such as milkweeds, willows, and cottonwoods is similar. Water dispersal adaptations include air-filled chambers surrounded by impermeable membranes to prevent the entrance of H_2O.

Coconuts and other plants that characteristically occur on or near beaches are regularly spread throughout a region by floating in water (figure 37.16*d*). This sort of dispersal is especially important in the colonization of distant island groups, such as the Hawaiian Islands.

It has been calculated that the seeds of about 175 angiosperms, nearly one-third from North America, must have reached Hawaii to have evolved into the roughly 970 species found there today. Some of these seeds blew through the air, others were transported on the feathers or in the guts of birds, and still others floated across the Pacific. Although the distances are rarely as great as the distance between Hawaii and the mainland, dispersal is just as important for mainland plant species that have discontinuous habitats, such as mountaintops, marshes, or north-facing cliffs.

Fruits are an angiosperm innovation that form from the carpel wall protecting seeds and greatly enhancing dispersal of the embryo.

When conditions are satisfactory, the embryo emerges from its previously desiccated state, utilizes food reserves, and resumes growth. Although **germination** is a process characterized by several stages, it is often defined as the emergence of the **radicle** (first root) through the seed coat.

External signals and conditions trigger germination

Germination begins when a seed absorbs water and its metabolism resumes. The amount of water a seed can absorb is phenomenal, and osmotic pressure creates a force strong enough to break the seed coat. At this point, it is important that oxygen be available to the developing embryo because plants, like animals, require oxygen for cellular respiration. Few plants produce seeds that germinate successfully under water, although some, such as rice, have evolved a tolerance to anaerobic conditions.

Even though a dormant seed may have imbibed a full supply of water and may be respiring, synthesizing proteins and RNA, and apparently carrying on normal metabolism, it may fail to germinate without an additional signal from the environment. This signal may be light of the correct wavelength and intensity, a series of cold days, or simply the passage of time at temperatures appropriate for germination. The seeds of many plants will not germinate unless they have been **stratified**—held for periods of time at low temperatures. This phenomenon prevents the seeds of plants that grow in seasonally cold areas from germinating until they have passed the winter, thus protecting their tender seedlings from harsh, cold conditions.

Germination can occur over a wide temperature range (5°–30°C), although certain species and habitats may have relatively narrow optimum ranges. Some seeds will not germinate even under the best conditions. In some species, a significant fraction of a season's seeds remain dormant for an indeterminate length of time, providing a gene pool of great evolutionary significance to the future plant population. The presence of ungerminated seeds in the soil of an area is referred to as the **seed bank.**

Nutrient reserves sustain the growing seedling

Germination occurs when all internal and external requirements are met. Germination and early seedling growth require the utilization of metabolic reserves stored in the starch grains of **amyloplasts** (colorless plastids that store starch) and protein bodies. Fats and oils also are important food reserves in some kinds of seeds. These food sources can readily be digested during germination, producing glycerol and fatty acids, which yield energy through cellular respiration; they can also be converted to glucose. Depending on the kind of plant, any of these reserves may be stored in the embryo or in the endosperm.

In the kernels of cereal grains, the single cotyledon is modified into a relatively massive structure called the **scutellum** (figure 37.17). The abundant food stored in the scutellum

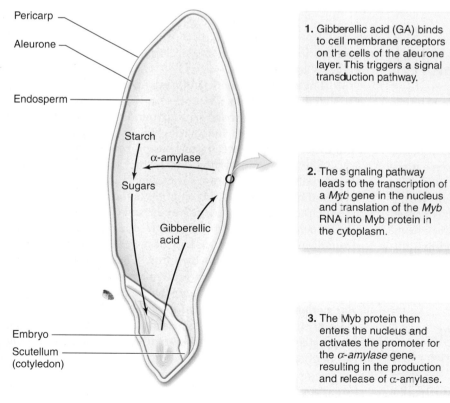

1. Gibberellic acid (GA) binds to cell membrane receptors on the cells of the aleurone layer. This triggers a signal transduction pathway.

2. The signaling pathway leads to the transcription of a *Myb* gene in the nucleus and translation of the *Myb* RNA into Myb protein in the cytoplasm.

3. The Myb protein then enters the nucleus and activates the promoter for the *α-amylase* gene, resulting in the production and release of α-amylase.

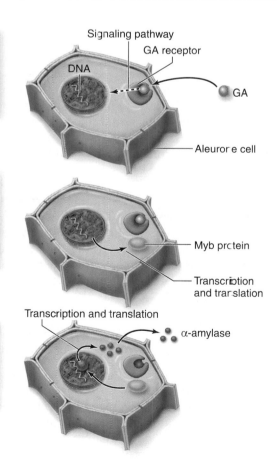

figure 37.17

HORMONAL REGULATION OF SEEDLING GROWTH.

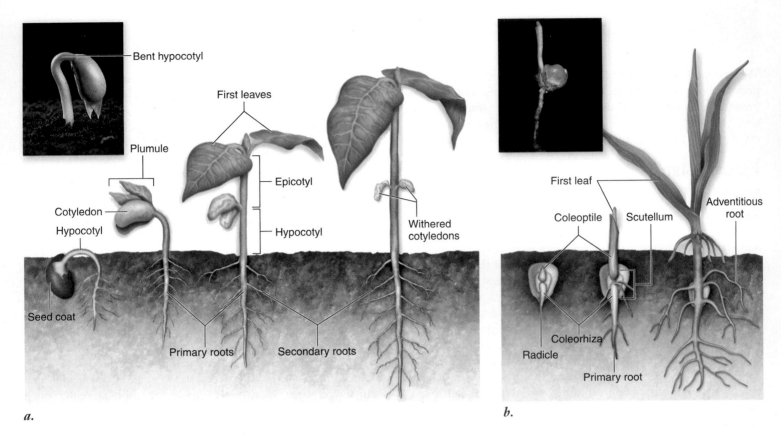

a. *b.*

figure 37.18

GERMINATION. The stages shown are for (*a*) a eudicot, the common bean (*Phaseolus vulgaris*), and (*b*) a monocot, maize (*Zea mays*). Note that the bending of the hypocotyl (region below the cotyledons) protect the delicate bean shoot apex as it emerges through the soil. Maize radicles are protected by a protective layer of tissue called the coleorhiza, in addition to the root cap found in both bean and maize. A sheath of cells called the coleoptile, rather than a hypocotyl tissue, protects the emerging maize shoot tip.

is used up first during germination. Later, while the seedling is becoming established, the scutellum serves as a nutrient conduit from the endosperm to the rest of the embryo.

The utilization of stored starch by germinating plants is one of the best examples of how hormones modulate plant development (see figure 37.17). The embryo produces gibberellic acid, a hormone, that signals the outer layer of the endosperm, called the **aleurone,** to produce α-amylase. This enzyme is responsible for breaking down the endosperm's starch, primarily amylose, into sugars that are passed by the scutellum to the embryo. Abscisic acid, another plant hormone, which is important in establishing dormancy, can inhibit starch breakdown. Abscisic acid levels may be reduced when a seed beginning to germinate absorbs water. (The action of plant hormones is covered in chapter 41.)

The seedling becomes oriented in the environment, and photosynthesis begins

As the sporophyte pushes through the seed coat, it orients with the environment so that the root grows down and the shoot grows up. New growth comes from delicate meristems that are protected from environmental rigors. The shoot becomes photosynthetic, and the postembryonic phase of growth and development is under way. Figure 37.18 shows the process of

germination and subsequent development of the plant body in eudicots and monocots.

The emerging shoot and root tips are protected by additional tissue layers in the monocots—the *coleoptile* surrounding the shoot, and the *coleorhiza* surrounding the radicle. Other protective strategies include having a bent shoot emerge so tissues with more rugged cell walls push through the soil.

The emergence of the embryonic root and shoot from the seed during germination varies widely from species to species. In most plants, the root emerges before the shoot appears and anchors the young seedling in the soil (see figure 37.18). In plants such as peas and corn, the cotyledons may be held below ground; in other plants, such as beans, radishes, and onions, the cotyledons are held above ground. The cotyledons may become green and contribute to the nutrition of the seedling as it becomes established, or they may shrivel relatively quickly. The period from the germination of the seed to the establishment of the young plant is critical for the plant's survival; the seedling is unusually susceptible to disease and drought during this period.

Internal and external signals trigger germination followed by the release of nutrients to nourish the embryo until the shoot becomes photosynthetic.

37.1 Embryo Development

In angiosperms, embryonic development begins once the egg cell is fertilized and the zygote begins cell division. A second fertilization produces a nutrient source, the endosperm (figure 37.1).

- The first zygotic cell division is asymmetrical, resulting in two daughter cells: one large and one small, both of which will continue to divide.

- Division of the large daughter cell will develop into an elongated suspensor that will transport nutrients to the developing embryo (figure 37.2).

- Early division of the small daughter cell will result in a globular embryo that will establish two axes of the embryo (figure 37.5).

- The root–shoot axis depends on the location of cells relative to the suspensor. Cells located near the suspensor will become the root apical meristem, whereas cells at the other end will become the shoot apical meristem. Development of each meristem is independently controlled.

- The radial axis is formed as cells subsequently divide parallel and perpendicular to the surface of the embryo.

- Three basic tissue systems are radially organized around the root–shoot axis: procambium, ground meristem, and protoderm.

- The inner procambium forms the vascular tissue to transport water and nutrients.

- The ground meristem gives rise to the bulk of embryonic tissue with functions including food and water storage.

- The protoderm will develop into dermal tissue that will protect the plant from desiccation and stomata that regulate gas exchange and minimize water loss.

- Plant embryos undergo morphogenesis because of changes in the planes and rates of cell division and cell shape.

- The form of the plant is determined by the plane in which its cells divide.

- The globular stage will develop bulges called cotyledons, one for monocots and two for eudicots.

- Apical meristems become established at the end of embryogenesis, and the seed becomes dormant.

- During embryo development angiosperms will undergo three other critical events: storage of food in the cotyledons or endosperm, differentiation of ovule tissue to form a seed coat and differentiation of carpel (ovary) tissue to form the fruit.

37.2 Seeds (figure 37.12)

The impermeable seed coat protects the dormant embryo and prevents germination until environmental conditions are favorable.

- Seeds are adaptive in four ways. They (1) maintain dormancy during unfavorable conditions, (2) protect the young plant when it is most vulnerable, (3) provide food for the embryo until it can produce its own food, (4) and facilitate dispersal.

- Germination requires that the seed coat becomes permeable to allow water and oxygen to reach the embryo.

- Specific adaptations ensure germination under appropriate conditions. Seed coats can be weakened by fire, passage through a digestive tract, or alternate freezing and thawing.

- In some plants germination will only occur when sufficient water is available to leach inhibitory chemicals from the seed coat.

37.3 Fruits (figure 37.14)

Survival of angiosperm embryos depends on seeds maturing within a developing fruit.

- Fruits are mature ovaries and fruit development is synchronized with embryo, endosperm, and seed coat development.

- The type of the fruit depends on the fate of the pericarp (carpel wall). Fruits can be dry or fleshy, and they can be simple (single carpel), aggregate (multiple carpels), or multiple (multiple flowers).

- Fruits are unique in that they contain tissues including the seed coat from the prior sporophyte, remnants of the gametophyte generation that produced the egg, and the embryo that is the next sporophyte generation.

- Fruits exhibit a wide array of dispersal mechanisms. They are ingested and transported by animals, are buried in caches by herbivores, hitch a ride with hooked spines on birds and mammals, are blown by the wind, or float and drift in water.

37.4 Germination

Germination is triggered by external and internal cues and is defined by the emergence of the radical or first root through the seed coat, which anchors most young seedlings to the soil.

- Germination begins when the seed absorbs water and oxygen is available for metabolism.

- Germination often requires an additional environmental signal, such as a specific wavelength of light, an appropriate germination temperature, or stratification (periods of low temperature exposure).

- Germination requires an energy source, such as starch stored in amyloplasts, proteins, or fats and oils.

- Starch metabolism is initiated by gibberellic acid, which signals the outer layer of the endosperm, the aleurone, to produce α-amylase. Starch metabolism can also be inhibited by abscisic acid.

- Emerging shoots and roots are protected by additional tissues in monocots. In many eudicots the tough bent shoot emerges before the growing tip (figure 37.18).

- Cotyledons may be held above or below ground and can either become photosynthetic or simply shrivel as food reserves are utilized.

- When the emerging shoot becomes photosynthetic the seedling enters the postembryonic phase of growth and development.

SELF TEST

1. Loss-of-function mutations in the *suspensor* gene in *Arabidopsis* lead to the development of two embryos in a seed. After analyzing the expression of this gene in early wild-type embryos, you find high levels of mRNA transcribed from the *suspensor* gene in the developing suspensor cells. What is the likely function of the suspensor protein?
   a. Suspensor protein likely stimulates development of the embryonic tissue.
   b. Suspensor protein likely stimulates development of the suspensor tissue.
   c. Suspensor protein likely inhibits embryonic development in the suspensor.
   d. Suspensor protein likely inhibits suspensor development in the embryo.

2. How would plant development change if the functions of the genes SHOOTMERISTEMLESS (STM) and MONOPTEROUS (MP) were reversed?
   a. The embryo–suspensor axis would be reversed.
   b. The embryo–suspensor axis would be duplicated.
   c. The root–shoot axis would be reversed.
   d. The root–shoot axis would be duplicated.

3. The most obvious difference between plant embryonic development and animal embryonic development is that
   a. plants develop from unfertilized eggs, and animals develop from fertilized eggs.
   b. whereas plant cells retain their relative positions after cell division, animal morphogenesis involves movement of cells within the embryo.
   c. plant embryos have an available source of nutrients, but animal embryos must begin feeding to obtain nutrients.
   d. plant embryos produce their own nutrients through photosynthesis.

4. Which of the following is *not* evident from looking at a plant embryo?
   a. You can tell if the plant is a monocot or dicot.
   b. You can tell where the shoot will form.
   c. You can tell where the root will form.
   d. You can tell when the seed will germinate.

5. Which of the following is not true of mature angiosperm seeds?
   a. Food reserves may be stored in the endosperm or cotyledons.
   b. Both shoot and root meristems exist.
   c. Mature tissues have formed in the embryo.
   d. The embryo remains attached to the suspensor.

6. The longest period of time that a seed can remain dormant is
   a. days.
   b. weeks.
   c. months.
   d. years.

7. Fruits are complex organs that are specialized for dispersal of seeds. Which of the following plant tissues does *not* contribute to mature fruit?
   a. sporophytic tissue from the previous generation
   b. gametophytic tissue from the previous generation
   c. sporophytic tissue from the next generation
   d. gametophytic tissue from the next generation

8. If you want to ensure that a seed fails to germinate, which of the following strategies would be most effective?
   a. prevent imbibition
   b. prevent desiccation
   c. prevent fertilization
   d. prevent dispersal

9. How would a loss-of-function mutation in the α-amylase gene affect seed germination?
   a. The seed could not imbibe water.
   b. The embryo would starve.
   c. The seed coat would not rupture.
   d. The seed would germinate prematurely.

10. Which of the following statements is not true of embryogenesis?
    a. Plant hormones play a minor role in proper embryo formation.
    b. Many aspects of embryogenesis are genetically determined.
    c. The precursors to the root and shoot develop during embryogenesis.
    d. None of the above

11. Which of the following are imperative for seed germination?
    a. Oxygen.
    b. Carbon dioxide
    c. Light
    d. a and b

12. Cotyledons:
    a. arise from the shoot apical meristem
    b. remain active for much of a plant's life
    c. exist in pairs in eudicots and singly in monocots
    d. none of the above.

13. A plant lacking the WOODEN LEG gene will likely:
    a. be incapable of transporting water to its leaves
    b. lack xylem and phloem
    c. be incapable of transporting photosynthate
    d. all of the above

14. A germinating seed that does not contain a suspensor will likely:
    a. not develop because nutrients will not be transferred from the endosperm to the embryo
    b. begin photosynthesizing immediately
    c. grow quickly in order to emerge from the soil
    d. develop properly

CHALLENGE QUESTIONS

1. You are writing a science fiction screenplay in which the best of animal and plant lifestyles are used to make a "superspecies." Discuss the aspects of plant development you would include in this new species.

2. The oldest known seeds to successfully germinate were found in the Yukon Territory in the Canadian Arctic in the 1950s. Radiocarbon dating of lemming droppings accompanying these seeds indicated these seeds were approximately 10,000 years old. Discuss the mechanisms that seeds use to remain dormant for long periods of time.

3. How might the reproductive success of angiosperms have changed if seeds had developed without fruit?

chapter 38

Transport in Plants

introduction

TERRESTRIAL PLANTS FACE TWO major challenges: maintaining water and nutrient balance, and providing sufficient structural support for upright growth. The vascular system transports water, minerals, and organic molecules over great distances. Whereas the secondary growth of vascular tissue allows trees to achieve great heights, water balance alone keeps herbaceous plants upright. Think of a plant cell as a water balloon pressing against the insides of a soft-sided box, with many other balloon/box cells stacked on top. If the balloon springs a leak, the support is gone, and the box can collapse. How water, minerals, and organic molecules move between the roots and shoots of small and tall plants is the topic of this chapter.

Transport Mechanisms

How does water get from the roots to the top of a 10-story-high tree? Throughout human existence, curious people have wondered about this question. Plants lack muscle tissue or a circulatory system like animals have to pump fluid throughout a plant's body. Nevertheless, water moves through the cell wall spaces between the protoplasts of cells, through plasmodesmata (connections between cells), through plasma membranes, and through the interconnected, conducting elements extending throughout a plant (figure 38.1). Water first enters the roots and then moves to the xylem, the innermost vascular tissue of plants. Water rises through the xylem because of a combination of factors, and some of that water exits through the stomata in the leaves (figure 38.2).

Local changes result in long-distance movement of materials

The greatest distances traveled by water molecules and dissolved minerals are in the xylem. Once water enters the xylem of a redwood, for example, it can move upward as much as 100 m. Some "pushing" from the pressure of water entering the roots is involved. However, most of the force is "pulling" caused by **transpiration,** evaporation from thin films of water in the stomata. This pulling occurs because water molecules stick to each other (cohesion) and to the walls of the tracheid or xylem vessel (adhesion). The result is an unusually stable column of liquid reaching great heights.

The movement of water at the cellular level plays a significant role in bulk water transport in the plant as well, although over much shorter distances. Although water can diffuse through plasma membranes, charged ions and organic compounds, including sucrose, depend on protein transporters to cross membranes (see figure 38.1). Some membrane proteins form channels that allow molecules to diffuse. Other protein transporters require energy to move minerals and other nutrients against a concentration gradient. ATP-dependent hydrogen ion pumps often fuel active transport. They create a hydrogen ion gradient across a membrane. Unequal concentrations of solutes (for example, ions and organic molecules), in turn, affect the movement of water across membranes. Using information about the concentration of solutes inside and outside a cell, you can predict which way water will move.

Water potential regulates movement of water through the plant

Plant biologists explain the forces that act on water within a plant in terms of potentials. *Potentials* are a way of representing free energy (the potential to do work; see chapter 6). **Water potential,** abbreviated by the Greek letter psi (Ψ_w), is used to predict which way water will move. The key is to remember that water will move from a cell or solution with higher water potential to a cell or solution with lower water potential. Water potential is measured in units of pressure called **megapascals (MPa).** If you turn on your kitchen or bathroom faucet full blast, the water pressure should be between 0.2 and 0.3 MPa.

Movement of water by osmosis

If a single plant cell is placed into water, then the concentration of solutes inside the cell is greater than that of the external solution, and water moves into the cell by the process of **osmosis,** which you may recall from the discussion of membranes in chapter 5. The cell expands and presses against the cell wall, a condition known as *turgid* because of the cell's increased turgor pressure. By contrast, if the cell is placed into a solution with a very high concentration of sucrose, water leaves the cell and turgor pressure drops. The cell membrane pulls away from the cell wall as

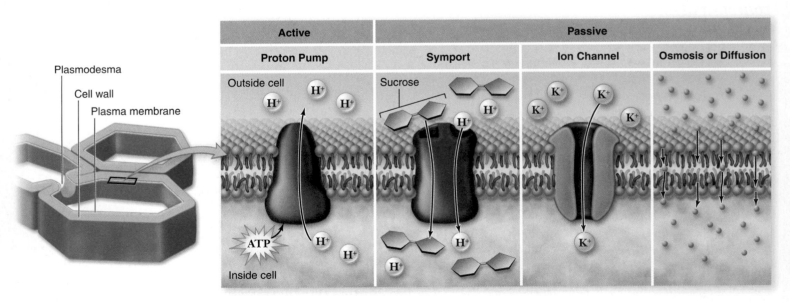

figure 38.1

TRANSPORT BETWEEN CELLS. Water, minerals, and organic molecules can diffuse across membranes, be actively or passively transported by membrane-bound transporters, or move through plasmodesmata. Details of membrane transport are found in chapter 6.

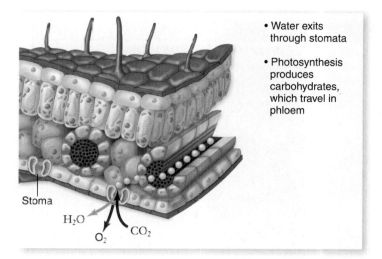

- Water exits through stomata
- Photosynthesis produces carbohydrates, which travel in phloem

Stoma

H_2O

O_2 CO_2

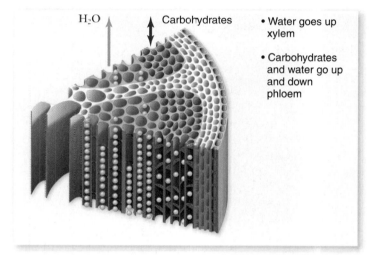

H_2O Carbohydrates

- Water goes up xylem
- Carbohydrates and water go up and down phloem

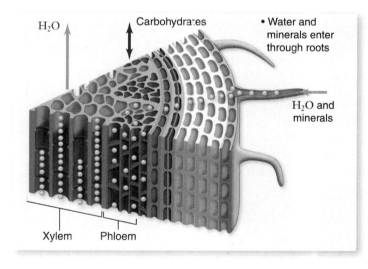

H_2O Carbohydrates

- Water and minerals enter through roots

H_2O and minerals

Xylem   Phloem

CO_2 and light

H_2O

O_2

H_2O and minerals

figure 38.2

WATER AND MINERAL MOVEMENT THROUGH A PLANT. This diagram illustrates the path of water and inorganic materials as they move into, through, and out of the plant body.

the volume of the cell shrinks. This process is called **plasmolysis,** and if the cell loses too much water it will die. Even a tiny change in cell volume causes large changes in turgor pressure. When the turgor pressure falls to zero, most plants will wilt.

Calculation of water potential

A change in turgor pressure can be predicted more accurately by calculating the water potential of the cell and the surrounding solution. There are two components to water potential: (1)

physical forces, such as a plant cell wall or gravity, and (2) the concentration of solute in each solution.

In terms of physical forces, the contribution of gravity to water potential is so small that it is generally not included in calculations until you are considering a very tall tree. The turgor pressure, resulting from pressure against the cell wall, is referred to as **pressure potential (Ψ_p).** As turgor pressure increases, Ψ_p increases. A beaker of water containing dissolved sucrose, however, is not bounded by a cell membrane or a cell wall. Solutions

chapter 38 transport in plants **759**

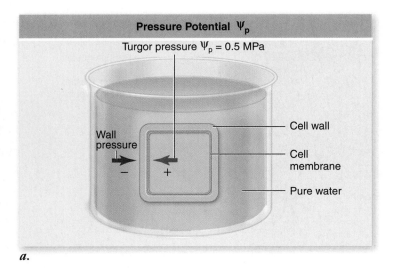

Pressure Potential Ψₚ

Turgor pressure Ψ_p = 0.5 MPa

Wall pressure

Cell wall

Cell membrane

Pure water

a.

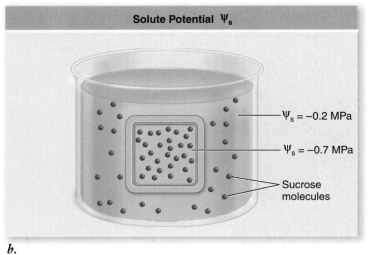

Solute Potential Ψₛ

Ψ_s = −0.2 MPa

Ψ_s = −0.7 MPa

Sucrose molecules

b.

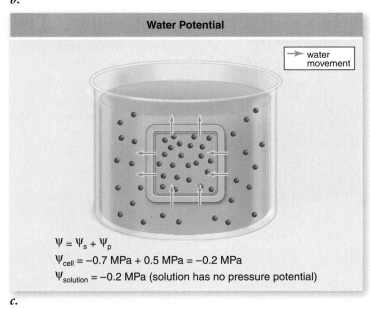

Water Potential

→ water movement

$\Psi = \Psi_s + \Psi_p$

Ψ_{cell} = −0.7 MPa + 0.5 MPa = −0.2 MPa

$\Psi_{solution}$ = −0.2 MPa (solution has no pressure potential)

c.

figure 38.3

DETERMINING WATER POTENTIAL. *a.* Cell walls exert pressure in the opposite direction of cell turgor pressure. *b.* Using the given solute potentials, predict the direction of water movement based only on solute potential. *c.* Total water potential is the sum of Ψ_s and Ψ_p. Water will move into the cell because the Ψ_w of the solution is less than the Ψ_w of the cell.

alone, therefore, cannot have turgor pressure, and they always have a Ψ_p of 0 MPa (figure 38.3*a*).

Water potential also arises from an uneven distribution of a solute on either side of a membrane, which results in osmosis. Applying pressure on the side of the membrane that has the greater concentration of solute prevents osmosis. The smallest amount of pressure needed to stop osmosis is proportional to the osmotic or **solute potential (Ψₛ)** of the solution (figure 38.3*b*). Pure water has a solute potential of zero. As a solution increases in solute concentration, it decreases in Ψ_s (< 0 MPa). A solution with a higher solute concentration has a more negative Ψ_s. Water flows from the solution with the less negative Ψ_s to the more negative Ψ_s. In figure 38.3*c*, the cell has a more negative Ψ_s than the solution, and water moves into the cell.

The total water potential (Ψ_w) of a plant cell is the sum of its pressure potential and solute potential; it represents the total potential energy of the water in the cell:

$$\Psi_w = \Psi_p + \Psi_s$$

When a cell is placed into pure water, which has a water potential of zero unless under pressure, water moves into the cell because the water potential of the cell is relatively negative (figure 38.3*c*). When a cell is placed into a solution with a different Ψ_w, the tendency is for water to move in the direction that eventually results in equilibrium—both the cell and the solution have the same Ψ_w (figure 38.4). The Ψ_p and Ψ_s values may differ for cell and solution, but the sum (=Ψ_w) should be the same.

Aquaporins enhance osmosis

For a long time, scientists believed that water moved across plasma membranes only by osmosis through the lipid bilayer. Water, however, was found to move more rapidly than predicted by osmosis alone. We now know that osmosis is enhanced by membrane water channels called **aquaporins,** which you first encountered in chapter 5 (figure 38.5). These transport channels occur in both plants and animals; in plants, they exist in vacuole and plasma membranes and also allow for bulk flow across the membrane.

At least 30 different genes code for aquaporin-like proteins in *Arabidopsis*. Aquaporins speed up osmosis, but they do not change the direction of water movement. They are important in maintaining water balance within a cell and in moving water into the xylem.

A water potential gradient from roots to shoots enables transport

Water potential regulates movement of water through a whole plant, as well as across cell membranes. Roots are the entry point. Water moves from the soil into the plant only if water potential of the soil is greater than in the root. Too much fertilizer or drought conditions lower the Ψ_w of the soil and limit water flow into the plant. Water in a plant moves along a water potential gradient from the soil (where the Ψ_w may be close to zero under wet conditions) to successively more negative water potentials in the roots, stems, leaves, and atmosphere (figure 38.6 on p. 762).

Evaporation of water in a leaf creates negative pressure or tension (which is the same as a negative water potential) in the xylem, which literally pulls water up the stem from the roots. The extremely negative water potential in leaves cannot be explained

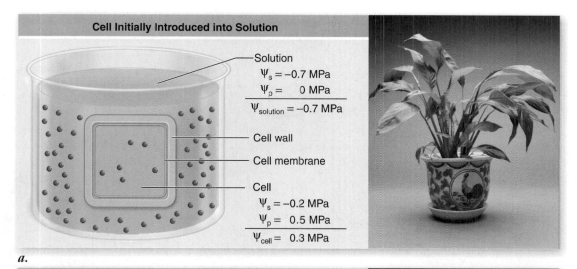

Cell Initially Introduced into Solution

Solution
$\Psi_s = -0.7$ MPa
$\Psi_p = 0$ MPa
$\overline{\Psi_{solution} = -0.7}$ MPa

Cell wall

Cell membrane

Cell
$\Psi_s = -0.2$ MPa
$\Psi_p = 0.5$ MPa
$\overline{\Psi_{cell} = 0.3}$ MPa

a.

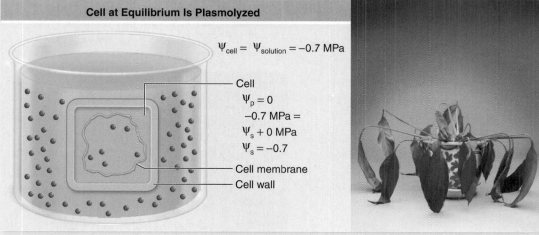

Cell at Equilibrium Is Plasmolyzed

$\Psi_{cell} = \Psi_{solution} = -0.7$ MPa

Cell
$\Psi_p = 0$
-0.7 MPa $=$
$\Psi_s + 0$ MPa
$\Psi_s = -0.7$

Cell membrane
Cell wall

b.

figure 38.4

WATER POTENTIAL AT EQUILIBRIUM. *a.* This cell initially had a larger Ψ_w than the solution surrounding it. *b.* At osmotic equilibrium, the Ψ_w of the cell and the solution should be the same. We assume that the cell is in a very large volume of solution of constant concentration. The final Ψ_w of the cell should therefore equal the initial Ψ_w of the solution. When a cell is plasmolyzed, $\Psi_p = 0$. As the cell loses water, the cell's solution becomes concentrated.

? *inquiry*

What would Ψ_w, Ψ_s, and Ψ_p of the cell in (a) be at equilibrium if it had been placed in a solution with a Ψ_s of −0.5?

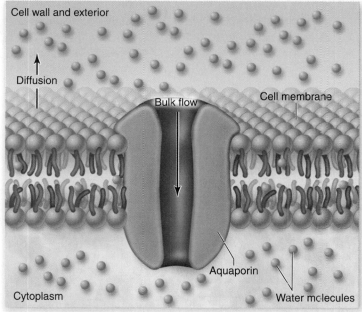

figure 38.5

AQUAPORINS. Aquaporins are water-selective pores in the plasma membrane that increase the rate of osmosis because they allow bulk flow across the membrane. They do not alter the direction of water movement, however.

by evaporation alone. As water diffuses from the xylem of tiny, branching veins in a leaf, it forms a thin film along mesophyll cell walls. If the surface of the air–water interface is fairly smooth (flat), the water potential is higher than if the surface becomes rippled.

The driving force for transpiration is the gradient in vapor pressure from 100% relative humidity inside the leaf to much less than 100% relative humidity outside the stomata. Molecules diffusing from the xylem replace evaporating water molecules. As the rate of evaporation increases, diffusion cannot replace all the water molecules. The film is pulled back into the cell walls and becomes rippled rather than smooth. The change decreases the water potential, increasing the pull on the column of water in the xylem, and concurrently increasing the rate of transpiration. A 50-fold change in water potential is not unusual.

Water potential forms a foundation for understanding local and long-distance transport in plants. The remaining sections of this chapter explore transport within and among different tissues and organs of the plant in more depth.

Water moves in the direction of lower water potential that is determined by solute concentration and physical pressure. Aquaporins can enhance the rate at which water moves across a membrane without changing the direction. Water potential gradients, fueled by transpiration, drive the movement of water up the xylem.

figure 38.6

WATER POTENTIAL IS HIGHER IN SOIL AND ROOTS THAN AT THE SHOOT TIP. Water evaporating from the leaves through the stomata causes additional water to move upward in the xylem and also to enter the plant through the roots. Water potential drops substantially in the leaves due to transpiration.

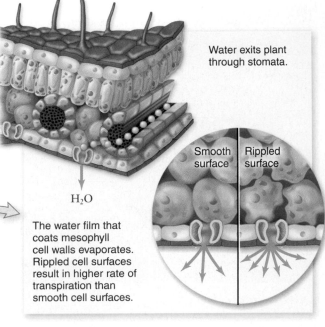

Water exits plant through stomata.

H_2O

The water film that coats mesophyll cell walls evaporates. Rippled cell surfaces result in higher rate of transpiration than smooth cell surfaces.

Smooth surface | Rippled surface

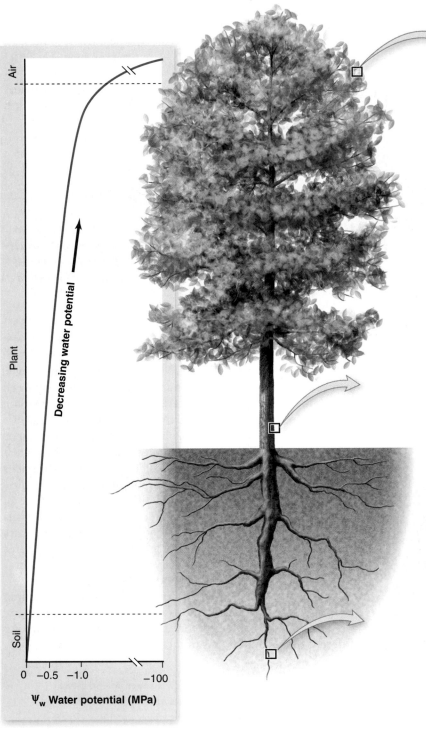

Air

Plant

Soil

Decreasing water potential

0   −0.5   −1.0   −100

Ψ_w **Water potential (MPa)**

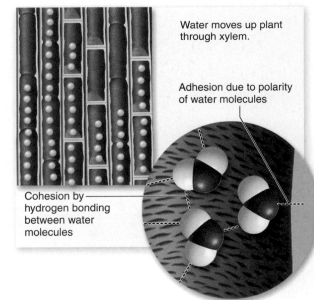

Water moves up plant through xylem.

Adhesion due to polarity of water molecules

Cohesion by hydrogen bonding between water molecules

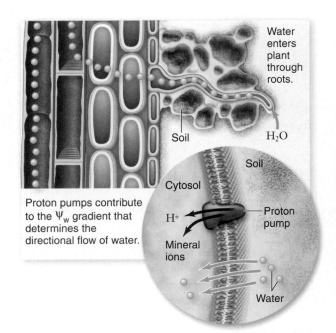

Water enters plant through roots.

Soil

H_2O

Soil

Cytosol

H^+

Mineral ions

Proton pump

Water

Proton pumps contribute to the Ψ_w gradient that determines the directional flow of water.

Water and Mineral Absorption

Most of the water absorbed by the plant comes in through **root hairs,** which collectively have an enormous surface area (figure 38.7). As you learned in chapter 36, root hairs are extensions of root epidermal cells located just behind the tips of growing roots. Root hairs are almost always turgid because their water potential is greater than that of the surrounding soil.

Because the mineral ion concentration in the soil water is usually much lower than it is in the plant, an expenditure of energy (supplied by ATP) is required for these ions to accumulate in root cells. The plasma membranes of root hair cells contain a variety of protein transport channels, through which proton pumps transport specific ions against even larger concentration gradients (refer to figure 38.1). Once in the roots, the ions, which are plant nutrients, are transported via the xylem throughout the plant. These actively transported minerals are responsible for the larger solute potential of root hairs and thus their turgidity.

Surface area for the absorption of water and minerals is further increased in many species of plants by interacting with mycorrhizal fungi. These fungi extend the absorptive net far beyond that of root hairs and are particularly helpful in the uptake of phosphorous in the soil. Mycorrhizae are discussed in detail in chapter 31.

Once absorbed through root hairs, water and minerals must move across cell layers until they reach the vascular tissues; water and dissolved ions then enter the xylem and move throughout the plant.

Three transport routes exist through cells

Water and minerals can follow three pathways to the vascular tissue of the root (figure 38.8). The **apoplast route** includes movement through the cell walls and the space between cells. Transport through the apoplast avoids membrane transport. The **symplast route** is the continuum of cytoplasm between cells connected by plasmodesmata. Once molecules are inside a cell, they can move between cells through plasmodesmata without crossing a plasma membrane. The **transmembrane route** involves membrane transport between cells and also across the membranes of vacuoles within cells. This route permits each cell the greatest amount of control over what substances enter and leave. These three routes are not exclusive, and molecules can change pathways at any time, until reaching the endodermis of the root.

Transport through the endodermis is selective

Eventually, on their journey inward, molecules reach the endodermis. Any further passage through the cell walls is blocked by the Casparian strips. As described in chapter 36, all cells in cylinder of endodermis have connecting walls embedded

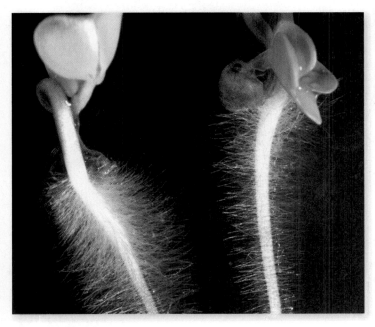

figure 38.7

ROOT HAIRS INCREASE SURFACE AREA FOR ABSORPTION OF MINERALS AND WATER.

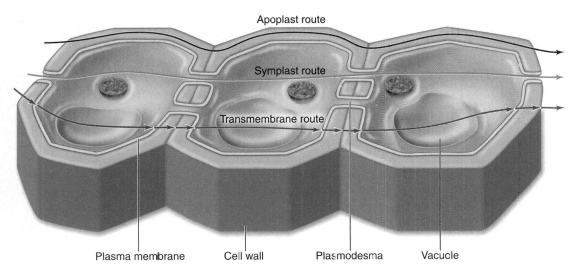

Apoplast route

Symplast route

Transmembrane route

Plasma membrane   Cell wall   Plasmodesma   Vacuole

figure 38.8

TRANSPORT ROUTES BETWEEN CELLS.

inquiry

Which route would be the fastest for water movement? Would this always be the best way to move nutrients into the plant?

figure 38.9

THE PATHWAYS OF MINERAL TRANSPORT IN ROOTS. Minerals are absorbed at the surface of the root, mainly by the root hairs. In passing through the cortex, they must either follow the cell walls and the spaces between them or go directly through the plasma membranes and the protoplasts of the cells, passing from one cell to the next by way of the plasmodesmata. When they reach the endodermis, however, their further passage through the cell walls is blocked by the Casparian strips, and they must pass through the membrane and protoplast of an endodermal cell before they can reach the xylem.

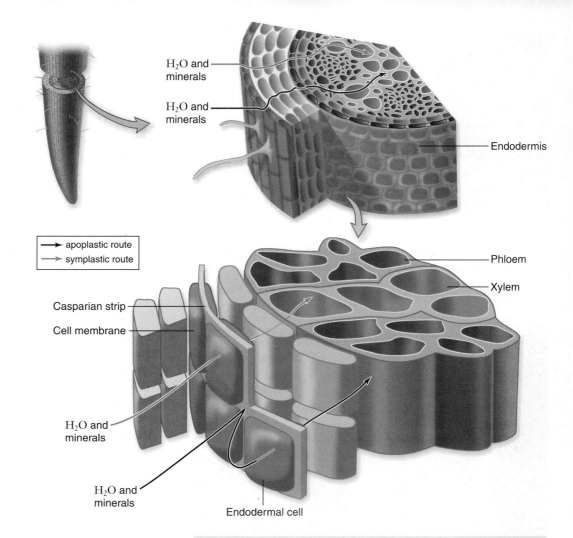

with the waterproof material suberin (figure 38.9). Molecules must pass through the plasma membranes and protoplasts of the endodermal cells to reach the xylem. The endodermis, with its unique structure, along with the cortex and epidermis, controls water and nutrient flow to the xylem to regulate water potential and helps limit leakage of water out of the root.

> Minerals, along with water, enter the root with its surface area greatly enlarged by root hairs and mycorrhizal fungi. Water and minerals can move between cell walls, through plasmodesmata, or across cell membranes. The Casparian strips force water to move across endodermal cell membranes for more precise control of water flow.

38.3 Xylem Transport

The aqueous solution that passes through the membranes of endodermal cells enters the plant's vascular tissues and into the tracheids and vessel members of the xylem. As ions are actively pumped into the root or move via facilitated diffusion, their presence increases the water potential and increases turgor pressure in the roots due to osmosis.

Root pressure is present even when transpiration is not occurring

Root pressure, which often occurs at night, is caused by the continued accumulation of ions in the roots at times when transpiration from the leaves is very low or absent. This accumulation results in an increasingly high ion concentration within the cells, which in turn causes more water to enter the root hair cells by osmosis. Ion transport further decreases the Ψ_s of the roots. The result is movement of water into the plant and up the xylem columns despite the absence of transpiration.

Under certain circumstances, root pressure is so strong that water will ooze out of a cut plant stem for hours or even days. When root pressure is very high, it can force water up to the leaves, where it may be lost in a liquid form through a process known as **guttation.** Guttation cannot move water up great heights or at rapid speeds. It does not take place through the stomata, but instead occurs through special groups of cells located near the ends of small veins that function only in this process. Guttation produces what is more commonly called dew on leaves.

Root pressure alone, however, is insufficient to explain xylem transport. Transpiration provides the main force for moving water and ionic solutes from roots to leaves.

Vessels and tracheids accommodate bulk flow

Water has an inherent **tensile strength** that arises from the cohesion of its molecules, their tendency to form hydrogen bonds with one another (see chapter 2). These two factors are the basis of the cohesion–tension theory of the bulk flow of water in the xylem. The tensile strength of a column of water varies inversely with the diameter of the column; that is, the smaller the diameter of the column, the greater the tensile strength. Because plant tracheids and vessels are tiny in diameter, the cohesive force of water in them is strong. The water molecules also adhere to the sides of the tracheid or xylem vessels, further stabilizing the long column of water.

Given that a narrower column of water has greater tensile strength, it is intriguing that vessels, having diameters that are larger than tracheids, are found in so many plants. The difference in diameter between vessels and tracheids has a larger effect on the mass of water that can pass through the column than on the tensile strength of the column. The volume of liquid that can pass a particular point in a column per second is proportional to r^4, where r is equal to the radius of the column. A two-fold increase in radius would result in a 16-fold increase in the volume of liquid moving through the column. Given equal cross-sectional areas of xylem, a plant with larger diameter vessels can move much more water up its stems than a plant with narrower tracheids.

inquiry

If a mutation increased the radius of a xylem vessel three-fold, how would the movement of water through the plant be affected?

The effect of cavitation

Tensile strength depends on the continuity of the water column; air bubbles introduced into the column when a vessel is broken or cut would cause the continuity and the cohesion to fail. A gas-filled bubble can expand and block the tracheid or vessel, a phenomenon called **cavitation** or embolism. Cavitations stop water transport and can lead to dehydration and death of part or all of a plant (figure 38.10).

Anatomical adaptations can compensate for the problem of cavitation, including the presence of alternative pathways that can be used if one path is blocked. Individual tracheids and vessel members are connected to a number of other tracheids or vessels by one or more pores in their walls, and air bubbles are generally larger than these openings. In this way, bubbles cannot pass through the pores to further block transport. Freezing or deformation of cells can

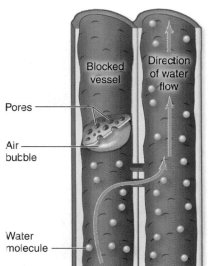

figure 38.10

CAVITATION. An air bubble can break the tensile strength of the water column. Bubbles are larger than pores and can block transport to the next tracheid or vessel. Multiple connections among tracheids or vessels provide alternative pathways, minimizing the damage from cavitation.

also cause small bubbles of air to form within xylem cells. This discontinuity is more likely to occur with seasonal temperature changes. Cavitation is one of the reasons that older xylem no longer conducts water.

Mineral transport

Tracheids and vessels are essential for the bulk transport of minerals. Ultimately, the minerals that are actively transported into the roots are removed and relocated through the xylem to other metabolically active parts of the plant. Phosphorus, potassium, nitrogen, and sometimes iron may be abundant in the xylem during certain seasons. In many plants, this pattern of ionic concentration helps conserve these essential nutrients, which may move from mature deciduous parts such as leaves and twigs to areas of active growth, namely meristem regions.

Keep in mind that minerals that are relocated via the xylem must move with the generally upward flow through the xylem. Not all minerals can reenter the xylem conduit once they leave. Calcium, an essential nutrient, cannot be transported elsewhere once it has been deposited in a particular plant part. But some other nutrients can be transported in the phloem.

Vessels or tracheids connect to form hollow tubes from root to shoot. Water molecules connect in a continuous column within the xylem and move upward more rapidly in larger diameter vessels. Breaking the water column with an air bubble, cavitation, prevents further transport within the connected vessels or tracheids.

38.4 The Rate of Transpiration

More than 90% of the water taken in by the roots of a plant is ultimately lost to the atmosphere. Water moves from the tips of veins into mesophyll cells, and from the surface of these cells it evaporates into pockets of air in the leaf. As discussed in chapter 36, these intercellular spaces are in contact with the air outside the leaf by way of the stomata.

Stomata open and close to balance H_2O and CO_2 needs

Water is essential for plant metabolism, but it is continuously being lost to the atmosphere. At the same time, photosynthesis requires a supply of CO_2 entering the chlorenchyma cells from the atmosphere. Plants therefore face two somewhat conflicting

chapter 38 transport in plants **765**

figure 38.11

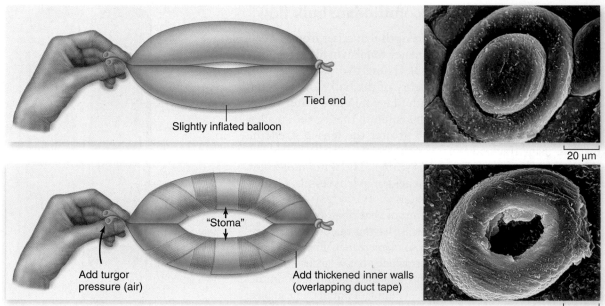

UNEQUAL
CELL WALL
THICKENINGS
ON GUARD
CELLS RESULT
IN THE OPENING
OF STOMATA
WHEN THE
GUARD CELLS
EXPAND.

Tied end

Slightly inflated balloon

20 µm

"Stoma"

Add turgor
pressure (air)

Add thickened inner walls
(overlapping duct tape)

20 µm

requirements: the need to minimize the loss of water to the atmosphere and the need to admit carbon dioxide. Structural features such as stomata and the cuticle have evolved in response to one or both of these requirements.

The rate of transpiration depends on weather conditions, including humidity and the time of day. As stated earlier, transpiration from the leaves decreases at night, when the vapor pressure gradient between the leaf and the atmosphere is less. During the day, sunlight increases the temperature of the leaf, while transpiration cools the leaf through evaporative cooling.

On a short-term basis, closing the stomata can control water loss. This occurs in many plants when they are subjected to water stress. But the stomata must be open at least part of the time so that CO_2 can enter. As CO_2 enters the intercellular spaces, it dissolves in water before entering the plant's cells. The gas dissolves mainly in water on the walls of the intercellular spaces below the stomata. The continuous stream of water that reaches the leaves from the roots keeps these walls moist.

Turgor pressure in guard cells causes stomata to open and close

The two sausage-shaped guard cells on each side of a stoma stand out from other epidermal cells not only because of their shape, but also because they are the only epidermal cells containing chloroplasts. Their distinctive wall construction, which

is thicker on the inside and thinner elsewhere, results in a bulging out and bowing when they become turgid.

You can make a model of this for yourself by taking two elongated balloons, tying the closed ends together, and inflating both balloons slightly. When you hold the two open ends together, there should be very little space between the two balloons. Now wrap duct tape around both balloons as shown in figure 38.11 (without releasing any air) and inflate each one a bit more. Hold the open ends together again. You should now be holding a roughly doughnut-shaped pair of "guard cells" with a "stoma" in the middle. Real guard cells rely on the influx and efflux of water, rather than air, to open and shut.

Turgor in guard cells results from the active uptake of potassium (K^+), chloride (Cl^-), and malate. As solute concentration increases, water potential decreases in the guard cells, and water enters osmotically. As a result, these cells accumulate water and become turgid, opening the stomata (figure 38.12). The energy required to move the ions across the guard cell membranes comes from the ATP-driven H^+ pump shown in figure 38.1.

The guard cells of many plant species regularly become turgid in the morning, when photosynthesis occurs, and lose turgor in the evening, regardless of the availability of water. During the course of a day, sucrose accumulates in the photosynthetic guard cells. The active pumping of sucrose out of guard cells in the evening leads to loss of turgor and guard cell closing.

figure 38.12

HOW A STOMA OPENS. When ions from surrounding cells are pumped into guard cells, the guard cell turgor pressure increases as water enters by osmosis. The increased turgor pressure causes the guard cells to bulge, with the thick walls on the inner side causing each guard cell to bow outward, thereby opening the stoma.

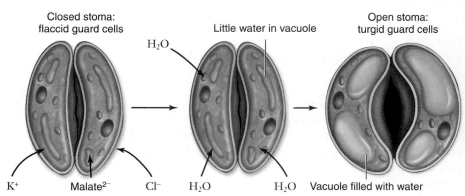

Closed stoma:
flaccid guard cells

H_2O

Little water in vacuole

Open stoma:
turgid guard cells

K^+ Malate^{2-} Cl^- H_2O H_2O Vacuole filled with water

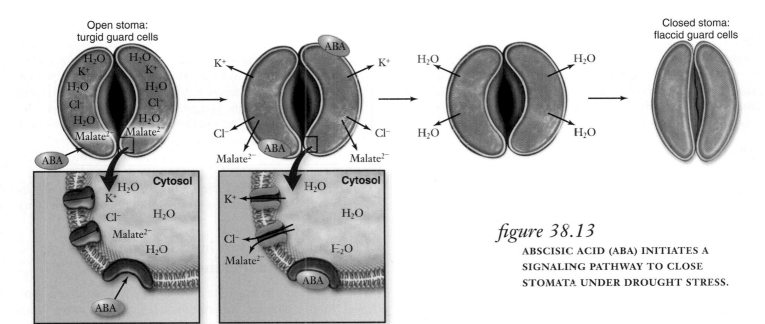

figure 38.13

ABSCISIC ACID (ABA) INITIATES A SIGNALING PATHWAY TO CLOSE STOMATA UNDER DROUGHT STRESS.

Environmental factors affect transpiration rates

Transpiration rates increase with temperature and wind velocity because water molecules evaporate more quickly. As humidity increases, the water potential between the leaf and the atmosphere decreases, but even at 95% relative humidity in the atmosphere, the vapor pressure gradient can sustain full transpiration. On a catastrophic level, when a whole plant wilts because insufficient water is available, the guard cells may lose turgor, and as a result, the stomata may close. Fluctuations in transpiration rate are tempered by opening or closing stomata.

Experimental evidence has indicated that several pathways regulate stomatal opening and closing. **Abscisic acid (ABA),** a plant hormone discussed in chapter 42, plays a primary role in allowing K^+ to pass rapidly out of guard cells, causing the stomata to close in response to drought. ABA binds to receptor sites in the plasma membranes of guard cells, triggering a signaling pathway that opens K^+, Cl^-, and malate ion channels. Turgor pressure decreases as water loss follows, and the guard cells close (figure 38.13).

CO_2 concentration, light, and temperature also affect stomatal opening. When CO_2 concentrations are high, the guard cells of many plant species are triggered to decrease the stomatal opening. Additional CO_2 is not needed at such times, and water is conserved when the guard cells are closed.

Blue light regulates stomatal opening. This helps increase turgor to open the stomata when sunlight increases the evaporative cooling demands. K^+ transport against a concentration gradient is promoted by light. Blue light in particular triggers proton (H^+) transport, creating a proton gradient that drives the opening of K^+ channels.

The stomata may close when the temperature exceeds 30° to 34°C and water relations are unfavorable. To ensure sufficient gas exchange, these stomata open when it is dark and the temperature has dropped. Some plants are able to collect CO_2 at night in a modified form to be utilized in photosynthesis during daylight hours. In chapter 9, you learned about **Crassulacean acid metabolism (CAM),** which occurs in succulent plants such as cacti. In this process, CO_2 is taken in at night and stored in organic compounds. These compounds are decarboxylated during the day, providing a source of CO_2 for fixation when stomata are closed. CAM plants are able to conserve water in dry environments.

> Transpiration rates vary to balance competing needs for water conservation and gas exchange at the stomata. Alternative photosynthetic pathways reduce water loss through transpiration.

38.5 Water-Stress Responses

Because plants cannot simply move on when water availability or salt concentrations change, adaptations have evolved to allow plants to cope with environmental fluctuations, including drought, flooding, and changing salinity.

Plant adaptations to drought include strategies to limit water loss

Many mechanisms for controlling the rate of water loss have evolved in plants. Regulating the opening and closing of stomata provides an immediate response. Morphological adaptations provide longer term solutions to drought periods. For example, for some plants dormancy occurs during dry times of the year; another mechanism involves loss of leaves, limiting transpiration. Deciduous plants are common in areas that periodically experience severe drought. In a broad sense, annual plants conserve water when conditions are unfavorable simply by going into "dormancy" as seeds.

Thick, hard leaves often with relatively few stomata—and frequently with stomata only on the lower side of the leaf—lose water far more slowly than large, pliable leaves with abundant stomata. Leaves covered with masses of wooly-looking trichomes (hairs) reflect more sunlight and thereby reduce the heat load on the leaf and the demand for transpiration for evaporative cooling.

Plants in arid or semiarid habitats often have their stomata in crypts or pits in the leaf surface (figure 38.14). Within these depressions, the water surface tensions are altered, reducing the rate of water loss.

Plant responses to flooding include short-term hormonal changes and long-term adaptations

Plants can also receive too much water, in which case they ultimately "drown." Flooding rapidly depletes available oxygen in the soil and interferes with the transport of minerals and carbohydrates in the roots. Abnormal growth often results. Hormone levels change in flooded plants; ethylene, a hormone associated with suppression of root elongation, increases, while gibberellins and cytokinins, which enhance growth of new roots, usually decrease (see chapter 42). Hormonal changes contribute to the abnormal growth patterns.

Oxygen deprivation is among the most significant problems because it leads to decreased cellular respiration. Standing water has much less oxygen than moving water. Generally, standing-water flooding is more harmful to a plant (riptides excluded). Flooding that occurs when a plant is dormant is much less harmful than flooding when it is growing actively.

Physical changes that occur in the roots as a result of oxygen deprivation may halt the flow of water through the plant. Paradoxically, even though the roots of a plant may be standing in water, its leaves may be drying out. Plants can respond to flooded conditions by forming larger lenticels (which facilitate gas exchange) and adventitious roots that reach above flood level for gas exchange.

Whereas some plants survive occasional flooding, others have adapted to living in fresh water. One of the most frequent adaptations among plants to growing in water is the formation of **aerenchyma,** loose parenchymal tissue with large air spaces in it (figure 38.15). Aerenchyma is very prominent in water lilies and many other aquatic plants. Oxygen may be transported from the parts of the plant above water to those below by way of passages in the aerenchyma. This supply of oxygen allows oxidative respiration to take place even in the submerged portions of the plant.

Some plants normally form aerenchyma, whereas others, subject to periodic flooding, can form it when necessary. In corn, increased ethylene due to flooding induces aerenchyma formation.

Plant adaptations to high salt concentration include elimination methods

The algal ancestors of plants adapted to a freshwater environment from a saltwater environment before the "move" onto land. This adaptation involved a major change in controlling salt balance.

Growth in salt water

Plants such as mangroves that grow in areas normally flooded with salt water must not only provide a supply of oxygen to their submerged parts, but also control their salt balance. The salt must be excluded, actively secreted, or diluted as it enters. The black mangrove (*Avicennia germinans*) has long, spongy, air-filled roots that emerge above the mud. These roots, called *pneumatophores* (see chapter 36), have large lenticels on their above-water portions through which oxygen enters; it is then transported to the submerged roots (figure 38.16).

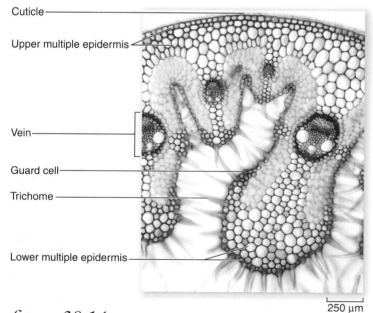

Cuticle

Upper multiple epidermis

Vein

Guard cell

Trichome

Lower multiple epidermis

250 μm

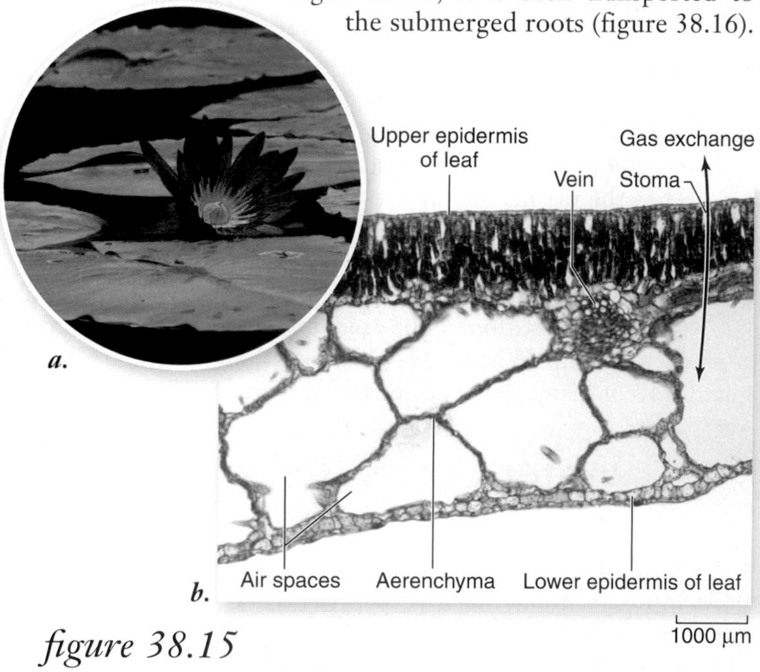

a.

Upper epidermis of leaf

Gas exchange

Vein   Stoma

b.   Air spaces   Aerenchyma   Lower epidermis of leaf

1000 μm

figure 38.14

ANATOMICAL PROTECTION FROM DROUGHT IN LEAVES. Deeply embedded stomata, extensive trichomes, and multiple layers of epidermis minimize water loss in this leaf, shown in cross section.

figure 38.15

AERENCHYMA. This tissue facilitates gas exchange in aquatic plants. *a.* Water lilies float on the surface of ponds, collecting oxygen and then transporting it to submerged portions of the plant. *b.* Large air spaces in the leaves of the water lily add buoyancy. The specialized parenchyma tissue that forms these open spaces is called aerenchyma. Gas exchange occurs through stomata found only on the upper surface of the leaf.

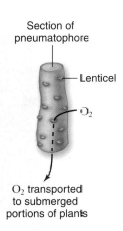

Section of
pneumatophore

— Lenticel

— O₂

O₂ transported
to submerged
portions of plants

figure 38.16

HOW MANGROVES GET OXYGEN TO THEIR SUBMERGED PARTS. The black mangrove (*Avicennia germinans*) grows in areas that are commonly flooded, and much of each plant is usually submerged. However, modified roots called pneumatophores supply the submerged portions of the plant with oxygen because these roots emerge above the water and have large lenticels. Oxygen diffuses into the roots through the lenticels, pass into the abundant aerenchyma, and move to the rest of the plant.

In addition, the succulent leaves of mangroves contain large quantities of water, which dilute the salt that reaches them. Many plants that grow in such conditions also either secrete large quantities of salt or block salt uptake at the root level.

Growth in saline soil

Soil salinity is increasing, often caused by salt accumulation from irrigation. Currently 23% of the world's cultivated land has high levels of saline that reduce crop yield. The low water potential of saline soils results in water-stressed crops. Some plants, called **halophytes** (salt lovers), can tolerate soils with high salt concentrations. Mechanisms for salt tolerance are being studied with the goal of breeding more salt-tolerant plants. Some halophytes produce high concentrations of organic molecules within their roots to alter the water potential gradient between the soil and the root so that water flows into the root.

> Standing water can limit oxygen availability to the roots. Pneumatophores are one innovation to increase root exposure to oxygen. High salt concentration changes the water potential gradient in the plant, preventing transpiration and even leading to plant death. Some plants have strategies to enhance water uptake by roots in salt water.

38.6 Phloem Transport

Most carbohydrates manufactured in leaves and other green parts are distributed through the phloem to the rest of the plant. This process, known as **translocation,** provides suitable carbohydrate building blocks for the roots and other actively growing regions of the plant. Carbohydrates concentrated in storage organs such as tubers, often in the form of starch, are also converted into transportable molecules, such as sucrose, and moved through the phloem. In this section we discuss the ways by which carbohydrate- and nutrient-rich fluid, termed **sap,** is moved through the plant body.

Organic molecules are transported up and down the plant

The pathway that sugars and other substances travel within the plant has been demonstrated precisely by using radioactive tracers, despite the fact that living phloem is delicate and that the process of transport within it is easily disturbed. Radioactive carbon dioxide ($^{14}CO_2$) can be incorporated into glucose as a result of photosynthesis. Glucose molecules are used to make the disaccharide sucrose, which is transported in the phloem. Such studies have shown that sucrose moves both up and down in the phloem.

Aphids, a group of insects that extract plant sap for food, have been valuable tools in understanding translocation. Aphids thrust their stylets (piercing mouthparts) into phloem cells of leaves and stems to obtain the abundant sugars there. When a feeding aphid is removed by cutting its stylet, the liquid from the phloem continues to flow through the detached mouthpart and is thus available in pure form for analysis (figure 38.17). The

figure 38.17

FEEDING ON PHLOEM. *a.* Aphids, including this individual shown on the edge of a leaf, feed on the food-rich contents of the phloem, which they extract through (*b*) their piercing mouthparts, called stylets. When an aphid is separated from its stylet and the cut stylet is left in the plant, the phloem fluid oozes out of it and can then be collected and analyzed.

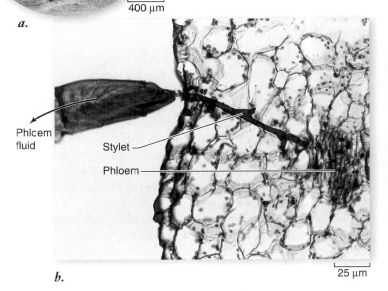

400 μm

a.

Phloem
fluid

Stylet

Phloem

25 μm

b.

figure 38.18

DIAGRAM OF MASS FLOW. In this diagram, *red* dots represent sucrose molecules and *blue* dots symbolize water molecules. After moving from the mesophyll cells of a leaf or another part of the plant into the conducting cells of the phloem, the sucrose molecules are transported to other parts of the plant by mass flow and unloaded where they are required.

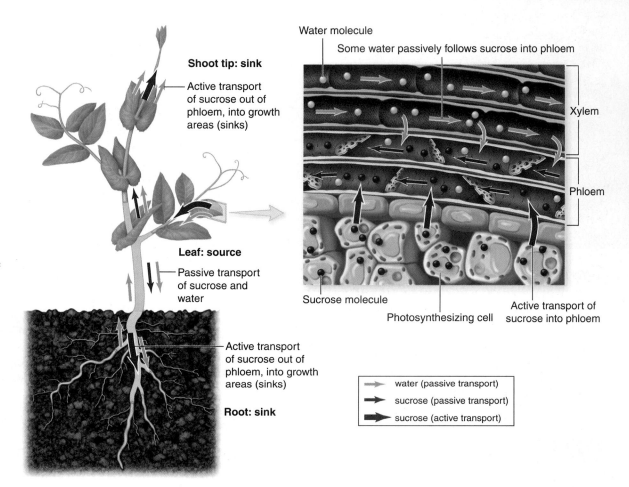

Shoot tip: sink
Active transport of sucrose out of phloem, into growth areas (sinks)

Leaf: source
Passive transport of sucrose and water

Active transport of sucrose out of phloem, into growth areas (sinks)

Root: sink

Water molecule
Some water passively follows sucrose into phloem
Xylem
Phloem
Sucrose molecule
Photosynthesizing cell
Active transport of sucrose into phloem

→ water (passive transport)
➡ sucrose (passive transport)
➡ sucrose (active transport)

liquid in the phloem, when evaporated, contains 10% to 25% of dry-weight matter, almost all of which is sucrose. Using aphids to obtain the critical samples and radioactive tracers to mark them, plant biologists have demonstrated that substances in phloem can move remarkably fast, as much as 50 to 100 cm/h.

Phloem also transports plant hormones, and as will be explored in chapter 42, environmental signals can result in the rapid translocation of hormones in the plant. Recent evidence also indicates that mRNA can move through the phloem, providing a previously unknown mechanism for long-distance communication among cells.

Turgor pressure differences drive phloem transport

The most widely accepted model of how carbohydrates in solution move through the phloem has been called the **pressure–flow theory.** Dissolved carbohydrates flow from a *source* and are released at a *sink*, where they are utilized. Carbohydrate sources include photosynthetic tissues, such as the mesophyll of leaves. Food-storage tissues, such as the cortex of roots, can be either sources or sinks. Sinks also occur at the growing tips of roots and stems and in developing fruits. Also, because sources and sinks can change through time as needs change, the direction of phloem flow can change

In a process known as **phloem loading,** carbohydrates (mostly sucrose) enter the sieve tubes in the smallest veins at the source. Some sucrose travels from mesophyll cells to the companion and sieve cells via the symplast (see figure 38.8). Much of the sucrose arrives at the sieve cell through apoplas-

tic transport and is moved across the membrane via a sucrose and H^+ symporter (see chapter 5). This energy-requiring step is driven by a proton pump (see figure 38.1). Companion cells and parenchyma cells adjacent to the sieve tubes provide the ATP energy to drive this transport. Unlike vessels and tracheids, sieve cells must be alive to participate in active transport.

Bulk flow occurs in the sieve tubes without additional energy requirements. Because of the difference between the water potential in the sieve tubes and in the nearby xylem cells, water flows into the sieve tubes by osmosis. Turgor pressure in the sieve tubes thus increases, and this pressure drives the fluid throughout the plant's system of sieve tubes. At the sink, sucrose and hormones are actively removed from the sieve tubes, and water follows by osmosis. The turgor pressure at the sink drops, causing a mass flow from the stronger pressure at the source to the weaker pressure at the sink (figure 38.18). Most of the water at the sink then diffuses back into the xylem, where it may either be recirculated or lost through transpiration.

Transport of sucrose and other carbohydrates within sieve tubes does not require energy. But the pressure needed to drive the movement is created through energy-dependent loading and unloading of these substances from the sieve tubes.

Carbohydrates and hormones are transported through the phloem from a source to a sink via a pressure–flow transport mechanism. Active transport is required to load the materials at the source. Phloem loading alters the water potential, leading to an influx of water that moves the materials to the sink. Unlike xylem transport, phloem transport is bidirectional.

38.1 Transport Mechanisms (figure 38.2)

The movement of water through a plant involves several factors that do not require a pumping system.

- To understand how water moves through a plant one needs to consider the properties of water, osmosis, and events at the cellular level.

- Although water can be "pushed" by root pressure, the major force is the "pulling" of water by transpiration. The movement of water involves cohesion, adhesion, and osmosis.

- Water potential predicts which way water will move. Water moves to a cell or solution that has a lower water potential, which corresponds to a higher osmotic concentration (figures 38.3 and 38.4).

- In plants, water potential is the sum of pressure potential or turgor pressure against the cell wall and solute or osmotic potential of the cell and surrounding environment.

- Plant cells placed in a solution with high water potential become turgid, and cells placed in a solution with low water potential exhibit shrinkage or plasmolysis.

- Aquaporins speed up osmosis but do not change the direction of water movement (figure 38.5).

- In summary, water moves into plants if the soil water potential is greater than that of the root and evaporation of water from the leaf creates a negative water potential pulling water up the xylem (figure 38.6).

38.2 Water and Mineral Absorption

Root hairs and mycorrhizal fungi increase the root absorption surface area for water and minerals.

- Minerals are actively transported from the soil into the roots, causing a greater internal solute potential and attraction for water.

- Once inside the root, water and minerals follow three paths to enter the vascular tissue of the root (figure 38.8).

- The apoplast pathway includes movement through cell walls and spaces between cells.

- The symplast pathway is a continuum of cytoplasm between cells connected by plasmodesmata.

- The transmembrane route permits control over what substances enter and leave a cell.

- As water and minerals move toward the transport tissues, Casparian strips force water to move across endodermal cell membranes for precise control of nutrients and water flow.

38.3 Xylem Transport

The aqueous solution that passes through the membranes of the endodermal cells enters the cells of the xylem.

- Root pressure results from the active transport of ions into the root cells and resultant osmosis. It is a minor contributor to water transport.

- Guttation, or the production of dew, is the loss of water from leaves when root pressure is very high.

- Tensile strength of water is dependent on cohesion and adhesion properties of water.

- The amount of water that can be transported by the xylem vessels is a function of the radius of the transport vessels. The smaller the diameter of the xylem column the greater the tensile strength, resulting in the movement of more water. However, the greater the radius of the column, the more water that can be transported over a given distance.

- If the water column is disrupted by an air bubble, cavitation occurs, and the water column is no longer continuous and the plant will dehydrate and die.

- Minerals are distributed to actively growing tissues as water moves up the xylem.

38.4 The Rate of Transpiration

Even though transpiration rates can be moderated, more than 90% of the water absorbed by the roots is lost by evaporation.

- The primary loss of water occurs through the stomata when they are open to obtain CO_2 and release O_2 (figure 38.12).

- Turgor pressure of the guard cells causes the stomata to open and close. As the guard cells actively take up solutes their water potential decreases and water enters by osmosis. When the guard cells become turgid the stomata open. At night sucrose is pumped out of the guard cells, and they become flaccid and close.

- Transpiration rates increase as temperature and wind velocity increase. Rates decrease as humidity increases.

- The stomata close at high temperatures or when CO_2 concentrations increase; they open when blue wavelengths of light promote uptake of K^+ by the guard cells.

- Alternative photosynthetic pathways, such as CAM, reduce transpiration.

38.5 Water-Stress Responses

Adaptations have evolved for plants to cope with changes in water availability.

- Plants adaptations to minimize water loss include closing stomata, reducing the number of stomata, having the stomata located in pits on the leaf surface, loss of leaves, dormancy, and covering the leaf with cuticles and wooly trichomes.

- Plants have adapted to flooding conditions that deprive the roots of oxygen. These adaptations include forming lenticels and adventitious roots.

- Aquatic plants form aerenchyma to ensure oxygen transport to parts of the plant below water (figure 38.15).

- Plants found in saline waters have pneumatophores for gas exchange and secrete salts.

- Halophytes enhance water uptake from saline soils by decreasing the water potential of their roots with high concentrations of organic molecules.

38.6 Phloem Transport

Most carbohydrates formed during photosynthesis as well as plant hormones are translocated through the phloem.

- Organic molecules are transported up *and down* the phloem.

- Mass flow of carbohydrates from a source to a sink occurs because of changes in water potential of the phloem contents. At the source active transport of sugars into the phloem causes a reduction in water potential. As water moves into the phloem, turgor pressure drives the contents to the sink where the sugar is actively transported into the cells. Water diffuses back into the xylem to be reused (figure 38.18).

SELF TEST

1. Which of the following statements is inaccurate?
   a. Water moves to areas of low water potential.
   b. Xylem transports materials up the plant while phloem transports materials down the plant.
   c. Water movement in the xylem is largely due to the cohesive and adhesive properties of water.
   d. Water movement across membranes is often due to differences in solute concentrations.

2. Water movement from the soil, through the plant and into the atmosphere requires:
   a. decreasing water potential from soil to plant to atmosphere
   b. increasing water potential from soil to plant to atmosphere
   c. equal water potentials in soil, plant and atmosphere
   d. none of the above

3. If you could override the control mechanisms that open stomata and force them to remain closed, what would you expect to happen to the plant?
   a. Sugar synthesis would likely slow down.
   b. Water transport would likely slow down.
   c. Both of these could be the result of keeping stomata closed.
   d. Neither of these would be the result of keeping stomata closed.

4. What will happen if a cell with a solute potential of −0.4 MPa and a pressure potential of 0.2 MPa is placed in a chamber filled with pure water that is pressurized with 0.5 MPa?
   a. Water will flow out of the cell.
   b. Water will flow into the cell.
   c. The cell will be crushed.
   d. The cell will explode.

5. If you were able to remove the aquaporins from cell membranes, which of the following would be the likely consequence?
   a. Water would no longer move across membranes.
   b. Plants would no longer be able to control the direction of water movement across membranes.
   c. The rate of movement of water across the membrane would greatly decrease.
   d. Cells would no longer stay turgid.

6. A molecule of water is traveling through the plant. Which of the following processes would not provide a driving force for it to move at either a cellular level or over longer distances through the plant?
   a. Osmosis
   b. Diffusion
   c. Transpiration
   d. All the above are driving forces for water movement.

7. Which of the following statements about water and mineral transport between cells in roots is incorrect?
   a. The apoplast pathway moves water and minerals through cell walls and in the space between cells.
   b. The symplast pathway moves materials water and minerals through plasmodesmata.
   c. The transmembrane route utilizes membrane transport across cells.
   d. All of the above statements are correct.

8. What would be the consequence of removing the Casparian strip?
   a. Water and mineral nutrients would not be able to reach the xylem.
   b. There would be less selectivity as to what passed into the xylem.
   c. Water and mineral nutrients would be lost from the xylem back into the soil.
   d. Water and mineral nutrients would no longer be able to pass through the cell walls of the endodermis.

9. The movement of water in the xylem relies on
   a. the ability of water molecules to hydrogen-bond with one another
   b. active transport
   c. evaporation of water from the leaf surface
   d. Both (a) and (c) are correct

10. When stomata close at night:
    a. water potential in the xylem no longer exists
    b. water cannot exit the leaves
    c. the roots cease to draw water from the soil
    d. none of the above.

11. If you wanted to force stomata to open, which of the following would work?
    a. Treat the plant with abscisic acid.
    b. Stimulate water movement into the guard cells.
    c. Stimulate water movement out of the guard cells.
    d. Force the dermal cells around the stomata to dehydrate, thereby pulling the guard cells apart.

12. Blowing water up through a drinking straw is most like
    a. guttation
    b. diffusion
    c. bulk flow in xylem
    d. osmosis

13. Sucrose enters a phloem sieve-tube cell because of
    a. osmosis
    b. water potential
    c. active transport
    d. a process regulated by auxin

14. Which of the following statements accurately describes the mass flow hypothesis?
    a. Carbohydrates move from a source to a sink.
    b. The movement of carbohydrates into the phloem occurs passively, whereas removal from the phloem is done actively.
    c. Carbohydrates in the phloem near the leaves increases the density of the phloem contents, thereby causing it to flow downward to the roots.
    d. Answers (a) and (b) are both correct.

CHALLENGE QUESTIONS

1. Roots are highly specialized to acquire water from the environment, yet plants that grow in wet, boggy environments have roots that are specialized to acquire oxygen! Discuss these structural adaptations and why they are important for survival of the plant.

2. Cavitation presents considerable challenges to the function of xylem in plants. Plants can cope with this because water can be rerouted around the blocked passage. Some plants are capable of removing air bubbles by producing enough positive pressure to cause the air bubble to dissolve back into the water. How might plants do this?

Plant Nutrition and Soils

introduction

VAST ENERGY INPUTS ARE REQUIRED for the ongoing construction of a plant. In this chapter, you'll learn what inputs, besides energy from the Sun, a plant needs to survive. Plants, like animals, need various nutrients to remain healthy. The lack of an important nutrient may slow a plant's growth or make the plant more susceptible to disease or even death. Plants acquire these nutrients mainly through photosynthesis and from the soil. In addition to contributing nutrients, the soil hosts bacteria and fungi that aid plants in obtaining nutrients in a usable form. Getting sufficient nitrogen is particularly problematic because plants cannot directly convert atmospheric nitrogen into amino acids. A few plants are able to capture animals and secrete digestive juices to make nitrogen available for absorption.

Soils: The Substrates on Which Plants Depend

Much of the activity that supports plant life is hidden within the soil. **Soil** is the highly weathered outer layer of the Earth's crust. It is composed of a mixture of ingredients, which may include sand, rocks of various sizes, clay, silt, humus, and various other forms of mineral and organic matter. Pore spaces containing water and air occur between the particles.

Soil is composed of minerals, organic matter, water, air, and organisms

The mineral fraction of soils varies according to the composition of the rocks. The Earth's crust includes about 92 naturally occurring elements (chapter 2). Most elements are found in the form of inorganic compounds called *minerals;* most rocks consist of several different minerals.

The soil is also full of microorganisms that break down and recycle organic debris. For example, about 5 metric tons of carbon is tied up in the organisms present in the soil under a hectare of wheat land in England—an amount that approximately equals the weight of 100 sheep!

Most roots are found in **topsoil** (figure 39.1), which is a mixture of mineral particles of varying size (most less than

2 mm in diameter), living organisms, and **humus,** which consists of partly decayed organic material. Topsoils are characterized by relative amounts of sand, silt, and clay. Soil composition determines the degree of water and nutrient binding to soil particles. Sand binds molecules minimally, but clay adsorbs (binds) water and nutrients quite tightly.

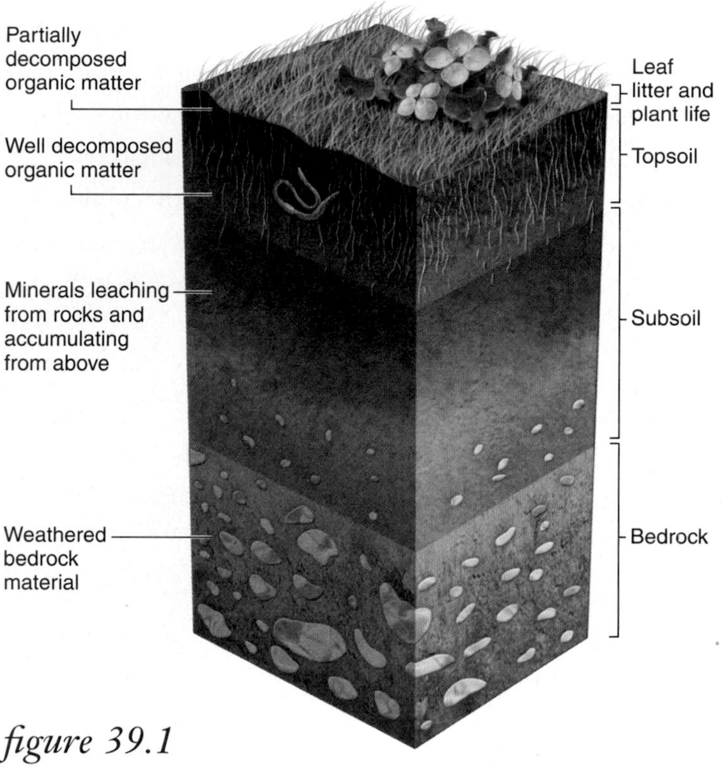

Water and mineral availability is determined by soil characteristics

Only minerals that are dissolved in water in the spaces or pores among soil particles are available for uptake by roots. Both mineral and organic soil particles tend to have negative charges, so they attract positively charged molecules and ions. The negatively charged anions stay in solution, creating a charge gradient between the soil solution and the root cells, so that positive ions would normally tend to move out of the cells. Proton pumps move H$^+$ out of the root to form a strong membrane potential ($\approx$ −160 mV). The strong electrochemical gradient then causes K$^+$ and other ions to enter via ion channels. Some ions, especially anions, use cotransporters (figure 39.2). The membrane potential maintained by the root, as well as the water potential difference inside and outside the root, affects root transport. (Water potential is described in chapter 38.)

About half of the total soil volume is occupied by pores, which may be filled with air or water, depending on moisture conditions (figure 39.3). Some of the soil water is unavailable to plants. Some of the water that reaches a soil drains through it immediately due to gravity. In sandy soil, the amount of water drainage is substantial. Another fraction of the water is held in small soil pores, which are generally less than about 50 μm in diameter. This water is readily available

Partially decomposed organic matter

Well decomposed organic matter

Minerals leaching from rocks and accumulating from above

Weathered bedrock material

Leaf litter and plant life

Topsoil

Subsoil

Bedrock

figure 39.1

MOST ROOTS GROW IN THE TOPSOIL. Leaf litter and animal remains cover the uppermost layer in soil called topsoil. Topsoil contains organic matter, such as roots, small animals, humus, and mineral particles of various sizes. Subsoil lies underneath the topsoil and contains larger mineral particles and relatively little organic matter. Beneath the subsoil are layers of bedrock, the raw material from which soil is formed over time and through weathering.

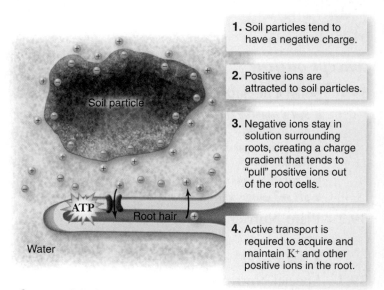

1. Soil particles tend to have a negative charge.

2. Positive ions are attracted to soil particles.

3. Negative ions stay in solution surrounding roots, creating a charge gradient that tends to "pull" positive ions out of the root cells.

4. Active transport is required to acquire and maintain K$^+$ and other positive ions in the root.

Soil particle

ATP

Root hair

Water

figure 39.2

ROLE OF SOIL CHARGE IN TRANSPORT. Active transport is required to move positively charged ions into a root hair.

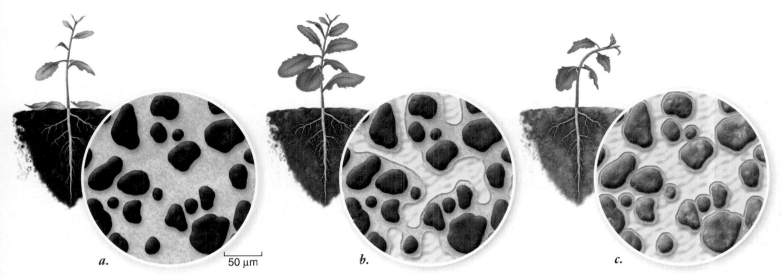

figure 39.3

WATER AND AIR FILL PORES AMONG SOIL PARTICLES. *a.* Without some space for air circulation in the soil, roots cannot respire. *b.* A balance of air and water in the soil is essential for root growth. *c.* Too little water decreases the soil water potential and prevents transpiration in plants.

to plants. When this water is depleted through evaporation or root uptake, the plant will wilt and eventually die unless more water is added to the soil. However, as plants deplete water near the roots, the soil water potential decreases. This helps to move more water toward the roots since the soil water further away has a higher water potential.

Soils have widely varying composition, and any particular soil may provide more or fewer plant nutrients. In addition, acidity and salinity, described shortly, can affect the availability of nutrients and water.

Cultivation can result in soil loss and nutrient depletion

When topsoil is lost because of erosion or poor landscaping, both the water-holding capacity and the nutrient relationships of the soil are adversely affected. Up to 50 billion tons of topsoil have been lost from fields in the United States in a single year.

Whenever the vegetative cover of soil is disrupted, such as by plowing and harvesting, erosion by water and wind increases—sometimes dramatically, as was the case in the 1930s in the southwestern Great Plains of the United States. This region became known as the "Dust Bowl" when a combination of poor farming practices and several years of drought caused susceptibility to wind erosion (figure 39.4*a*).

New approaches to cultivation are aimed at reducing soil loss. Intercropping, mixing of crops in a field, and conservation tillage, and not tilling fall crops are all erosion-prevention measures. Conservation tillage includes minimal till and even no-till approaches to farming.

Overuse of fertilizers in agriculture, lawns, and gardens can cause significant water pollution and its associated negative effects, such as overgrowth of algae in lakes (chapters 56 and 57). Maintaining nutrient levels in the soil and preventing nutrient runoff into lakes, streams, and rivers improves crop growth and minimizes ecosystem damage.

a.

b.

figure 39.4

SOIL DEGRADATION. *a.* Drought and poor farming practices lead to wind erosion of farmland in the southwestern Great Plains of the United States in the 1930s. *b.* Draining marshland in Iraq resulted in a salty desert.

One approach, site-specific farming, uses variable-rate fertilizer applicators guided by a computer and the global positioning system (GPS). Variable-rate application relies on information about local soil nutrient levels, based on analysis of soil samples. Another approach, integrated nutrient management, maximizes nutritional inputs using "green manure" (such as, alfalfa tilled back into the soil), animal manure, and inorganic fertilizers. Green manures and animal manure have the advantage of releasing nutrients slowly as they are broken down by decomposer organisms, so that nutrients may be utilized before leaching away. All these conservation approaches may be used together.

pH and salinity affect water and mineral availability

Anything that alters water pressure differences or ionic gradient balance between soil and roots can affect the ability of plants to absorb water and nutrients. Acid soils (having low pH) and saline soils (high in salts) are challenging habitats.

Acid soils

The pH of a soil affects the release of minerals from weathering rock. For example, at low pH, aluminum is released from rocks; this element is toxic to many plants. Furthermore, aluminum can also combine with other nutrients and make them inaccessible to plants.

Most plants grow best at a neutral pH, but about 26% of the world's arable land is acidic. In the tropical Americas, 68% of the soil is acidic. Aluminum toxicity in acid soils in Colombian fields can reduce maize (corn) yield fourfold.

Breeding efforts in Columbia are producing aluminum-tolerant plants, and crop yields have increased 33%. In a few test fields, the yield increases have been as high as 70% compared with that for nontolerant plants. The ability of plants to take up toxic metals can also be used to clean up polluted soil, a topic explored later in this chapter.

Salinity

The accumulation of salt ions, usually Na^+ and Cl^-, in soil alters water potential, leading to the loss of turgor in plants. Approximately 23% of all arable land has salinity levels that limit plant growth. Saline soil is most common in dry areas where salts are introduced through irrigation. In such areas, precipitation is insufficient to remove the salts, which gradually accumulate in the soil.

One of the more dramatic examples of soil salinity is in the cradle of civilization, Mesopotamia. The region once called the Fertile Crescent for its abundant agriculture is now largely a desert. Desertification has been witnessed over a shorter time frame in southern Iraq, where Saddam Hussein drained most of 20,000 km² of marshlands by redirecting water flow, turning the marshes into a salty desert (figure 39.4b). With the fall of Hussein's regime, dams were destroyed, allowing water to enter the marshlands once again. Recovery of the marshlands is not guaranteed, but in areas where the entering water has lowered the salinity there is hope.

> Plants depend on nutrients in soils for survival. Soil composition, pH, water, and salinity determine the availability of these nutrients to the plant.

39.2 Plant Nutrients

The major source of plant nutrition is the fixation of atmospheric carbon dioxide (CO_2) into simple sugars using the energy of the Sun. CO_2 enters through the stomata; oxygen (O_2) is a waste product of photosynthesis and an atmospheric component that also moves through the stomata. Oxygen is used in cellular respiration to support growth and maintenance in the plant.

CO_2 and light energy are not sufficient, however, for the synthesis of all the molecules a plant needs. Plants require a number of inorganic nutrients as well. Some of these are **macronutrients,** which plants need in relatively large amounts, and others are **micronutrients,** required in trace amounts (table 39.1).

Plants require nine macronutrients and seven micronutrients

The nine macronutrients are carbon, oxygen, and hydrogen—the three elements found in all organic compounds—plus nitrogen (essential for amino acids), potassium, calcium, magnesium (the center of the chlorophyll molecule), phosphorus, and sulfur. Each of these nutrients approaches or, as in the case of carbon, may greatly exceed 1% of the dry weight of a healthy plant.

The seven micronutrient elements—chlorine, iron, manganese, zinc, boron, copper, and molybdenum—constitute from less than one to several hundred parts per million in most plants. A deficiency of any one can have severe effects on plant growth (figure 39.5). The macronutrients were generally discovered in the last century, but the micronutrients have been detected much more recently as technology developed to identify and work with such small quantities.

Nutritional requirements are assessed by growing plants in hydroponic cultures in which the plant roots are suspended in aerated water containing nutrients. For the purposes of testing, the solutions contain all the necessary nutrients in the right proportions, but with certain known or suspected nutrients left out. The plants are then allowed to grow and are studied for the presence of abnormal symptoms that might indicate a need for the missing element

TABLE 39.1 — Essential Nutrients in Plants

Element	Principal Form in Which Element Is Absorbed	Approximate Percent of Dry Weight	Examples of Important Functions
MACRONUTRIENTS			
Carbon	CO_2	44	Major component of organic molecules
Oxygen	O_2, H_2O	44	Major component of organic molecules
Hydrogen	H_2O	6	Major component of organic molecules
Nitrogen	NO_3^-, NH_4^+	1–4	Component of amino acids, proteins, nucleotides, nucleic acids, chlorophyll, coenzymes, enzymes
Potassium	K^+	0.5–6	Protein synthesis, operation of stomata
Calcium	Ca^{2+}	0.2–3.5	Component of cell walls, maintenance of membrane structure and permeability; activates some enzymes
Magnesium	Mg^{2+}	0.1–0.8	Component of chlorophyll molecule, activates many enzymes
Phosphorus	$H_2PO_4^-$, HPO_4^-	0.1–0.8	Component of ADP and ATP, nucleic acids, phospholipids, several coenzymes
Sulfur	SO_4^{2-}	0.05–1	Components of some amino acids and proteins, coenzyme A
MICRONUTRIENTS (CONCENTRATIONS in ppm)			
Chlorine	Cl^-	100–10,000	Osmosis and ionic balance
Iron	Fe^{2+}, Fe^{3+}	25–300	Chlorophyll synthesis, cytochromes, nitrogenase
Manganese	Mn^{2+}	15–800	Activator of certain enzymes
Zinc	Zn^{2+}	15–100	Activator of many enzymes; active in fomation of chlorophyll
Boron	BO_3^-, $B_4O_7^-$, or $H_2BO_3^-$	5–75	Possibly involved in carbohydrate transport, nucleic acid synthesis
Copper	Cu^2 or Cu^+	4–30	Activator or component of certain enzymes
Molybdenum	MoO_4^-	0.1–5	Nitrogen fixation, nitrate reduction

a. *b.* *c.* *d.*

figure 39.5

MINERAL DEFICIENCIES IN PLANTS. *a.* Leaves of a healthy wheat plant. *b.* Chlorine-deficient plants with necrotic leaves (leaves with patches of dead tissue). *c.* Copper-deficient plant with dry, bent leaf tips. *d.* Zinc-deficient plant with stunted growth and chlorosis (loss of chlorophyll) in patches on leaves. The agricultural implications of deficiencies such as these are obvious, a trained observer can determine the nutrient deficiencies affecting a plant simply by inspecting it.

figure 39.6

IDENTIFYING NUTRITIONAL REQUIREMENTS OF PLANTS. A seedling is first grown in a complete nutrient solution. The seedling is then transplanted to a solution that lacks one suspected essential nutrient. The growth of the seedling is studied for the presence of abnormal symptoms, such as discolored leaves or stunted growth. If the seedling's growth is normal, the nutrient that was left out may not be essential; if the seedling's growth is abnormal, the lacking nutrient is essential for growth.

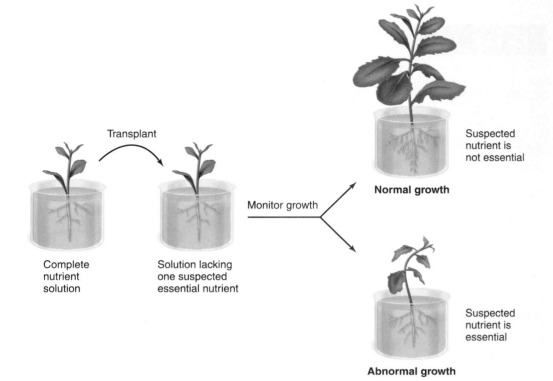

(figure 39.6). To give an idea of how small the needed quantities of micronutrients may be, the standard dose of molybdenum added to seriously deficient soils in Australia amounts to about 34 g (about one handful) per hectare (a square 100 meters on a side—about 2.5 acres), once every 10 years!

Most plants grow satisfactorily in hydroponic culture, if the roots are properly aerated. The method, although expensive, is occasionally practical for commercial purposes (figure 39.7). Analytical chemistry has made it much easier to test plant material for levels of different molecules.

Food security is related to crop productivity and nutrient levels

Nutrient levels and crop productivity are a significant human concern. **Food security,** avoiding starvation, is a global issue. Increasing the nutritional value of crop species, especially in developing countries, could have tremendous human health benefits.

Food fortification is an active area of research focused on ways to increase plants' uptake of minerals and the storage of minerals in roots and shoots for later human consumption. Phosphate uptake can be increased, for example, if it is more soluble in the soil. Some plants have been genetically modified to secrete citrate, an organic acid that solubilizes phosphate. As an added benefit, the citrate binds to aluminum, which can be toxic to plants and animals, and thus limits the uptake of aluminum into plants.

For other nutrients, such as iron, manganese, and zinc, plasma membrane transport is a limiting factor. Genes coding for these plasma membrane transporters have been cloned in other species and are being incorporated into crop plants. Eventually, breakfast cereals may be fortified with additional

nutrients while the grains are growing in the field, as opposed to when they are processed in the factory.

> Plants require both macronutrients and micronutrients, which are often accumulated in the roots through active transport. Genetic modification of plants to increase nutrient uptake may enhance the nutritional value of plants for human consumption.

figure 39.7

HYDROPONICS. Soil provides nutrients and support, but both of these functions can be replaced in hydroponic systems. Here, tomato plants are suspended in the air and the roots rotate through a nutrient bath.

39.3 Special Nutritional Strategies

Bacteria living in close association with roots can provide nitrogen

Plants need ammonia (NH_3) or nitrate (NO_3^-) to build amino acids, but most of the nitrogen in the atmosphere is in the form of gaseous nitrogen (N_2). Plants lack the biochemical pathways (including the enzyme nitrogenase) necessary to convert to ammonia, but some bacteria have this capacity. Symbiotic relationships between some plant groups and these bacteria have evolved.

Some of these bacteria live in close association with the roots of plants. Others end up being housed in tissues the plant grows especially for this purpose, called **nodules** (figure 39.8). Legumes and a few other plants can form root nodules. Hosting these bacteria costs the plant energy, but is well worth it when the soil lacks nitrogen compounds. To conserve energy, legume root hairs will not respond to bacterial signals when nitrogen levels are high.

Nitrogen fixation is the most energetically expensive reaction known to occur in any cell. Why should it be so difficult to add H_2 to N_2? The answer lies in the strength of the triple bond in N_2. Nitrogenase requires 16 ATPs to make two molecules of NH_3. Making NH_3 without nitrogenase requires a contained system maintained at 450°C and 500 atm pressure—far beyond the maximums under which plants can survive.

Rhizobium bacteria require oxygen and carbohydrates to support their energetically expensive lifestyle as nitrogen fixers. Carbohydrates are supplied through the vascular tissue of the plant, and leghemoglobin, which is structurally similar to animal hemoglobin, is produced by the plant to regulate oxygen availability to the bacteria. Without oxygen, the bacteria will die; within the bacteria, however, nitrogenase has to be isolated from oxygen, which inhibits its activity. Leghemoglobin binds oxygen and controls its availability within the nodule to optimize both nitrogenase activity and cellular respiration.

Just how do legumes and nitrogen-fixing *Rhizobium* bacteria get together (figure 39.9)? Extensive signaling between the bacteria and the legume not only lets each organism know the

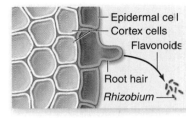

1. Pea roots produce flavonoids (a group of molecules used for plant defense and for making reddish pigment, among numerous other functions). The flavonoids are transported into the rhizobial cells.

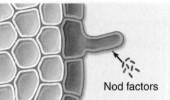

2. Flavonoids signal rhizobia to produce sugar-containing compounds called Nod (nodulation) factors.

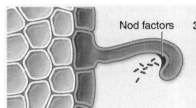

3. Nod factors bind to the surface of root hairs and signal the root hair to grow so that it curls around the rhizobia.

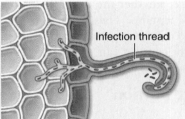

4. Rhizobia make an infection thread that grows in the root hair and moves into the cortex of the root. The rhizobia take control of cell division in the cortex and pericycle cells of the root (see chapter 35).

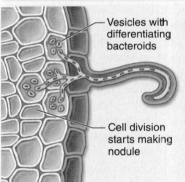

5. Rhizobia change shape and are now called bacteroids. Bacteroids produce an O_2-binding heme group that combines with a globin group from the pea to make leghemoglobin. Leghemoglobin gives a pinkish tinge to the nodule, and its function is much like that of hemoglobin, bringing O_2 to the rapidly respiring bacteroids, but isolating O_2 from nitrogenase.

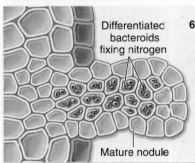

6. Bacteroids produce nitrogenase and begin fixing atmospheric nitrogen for the plant's use. In return, the plant provides organic compounds.

figure 39.8

NITROGEN-FIXING NODULE.
A root hair of alfalfa hosts *Rhizobium*, a bacterium that fixes nitrogen in exchange for carbohydrates.

500 μm

figure 39.9

NODULE FORMATION IN *Rhizobium*.

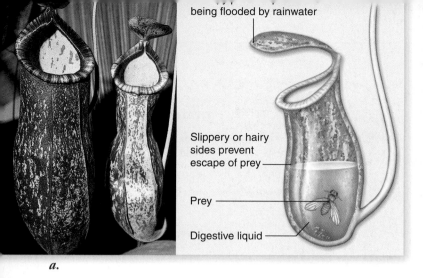

a.

being flooded by rainwater

Slippery or hairy
sides prevent
escape of prey

Prey

Digestive liquid

b.

Prey

After two
trichomes are
touched, leaf
trap closes

Digestive juices
are secreted

figure 39.10

NUTRITIONAL ADAPTATIONS. *a*. Asian pitcher plant, *Nepenthes*. Insects enter this carnivorous plant and are trapped and digested. Complex communities of invertebrate animals and protists inhabit the pitchers. ***b*.** Venus flytrap, *Dionaea*. If this fly touches two of the trichomes (hairs) on this modified leaf in a short time span, the trap will close. The plant will secrete digestive enzymes that release nitrogen compounds from the fly, which will then be absorbed by the flytrap. ***c*.** Sundew, *Drosera*, traps insects with sticky secretions and then excretes digestive enzymes to obtain nutrients from the insect's body. ***d*.** Aquatic waterwheel, *Aldrovanda*. This close relative of the Venus flytrap snaps shut to capture and digest small aquatic animals. This aquatic plant's ancestor was a land dweller.

other is present, but also checks whether the bacterium is the correct species for the specific legume. These highly evolved symbiotic relationships depend on exact species matches. Soybean and garden peas are both legumes, but each requires its own species of symbiotic *Rhizobium*.

Mycorrhizae aid a large portion of terrestrial plants

Nitrogen is not the only nutrient that is difficult for plants to obtain without assistance. Whereas symbiotic relationships with nitrogen-fixing bacteria are rare, symbiotic associations with mycorrhizal fungi are found in about 90% of vascular plants. These fungi have been described in detail in chapter 31. In terms of plant nutrition, mycorrhizae play a significant role in enhancing phosphorus transfer to the plant, and the uptake of some of the micronutrients is also facilitated. Functionally, the mycorrhizae extend the surface area available for nutrient uptake substantially.

Fungi most likely aided early rootless plants in colonizing land. Evidence now indicates that the signaling pathways that lead to plant symbiosis with some mycorrhizae may have been exploited to bring about the *Rhizobium*-legume symbiosis that evolved later on.

Carnivorous plants trap and digest animals to extract additional nutrients

Some plants are able to obtain nitrogen directly from other organisms, just as animals do. These carnivorous plants often grow in acidic soils, such as bogs, that lack organic nitrogen. By capturing and digesting small animals, primarily insects, directly, such plants obtain adequate nitrogen supplies and are able to grow in these seemingly unfavorable environments. Carnivorous plants have modified leaves adapted for luring and

trapping prey. The plants often digest their prey with enzymes secreted from specialized types of glands.

Pitcher plants (*Nepenthes* spp.) attract insects by the bright, flowerlike colors within their pitcher-shaped leaves by scents, and perhaps also by sugar-rich secretions (figure 39.10*a*). Once inside the pitcher, insects slide down into the cavity of the leaf, which is filled with water and digestive enzymes. This passive mechanism provides pitcher plants with a steady supply of nitrogen.

The Venus flytrap (*Dionaea muscipula*), which grows in the bogs of coastal North and South Carolina, has three sensitive hairs on each side of each leaf that, when touched, trigger the two halves of the leaf to snap together in about 100 ms (figure 39.10*b*). The speed of trap closing has puzzled biologists as far back as Darwin. Turgor pressure changes can account for the movement; the speed, however, depends on the curved geometry of the leaf, which can snap between convex and concave shapes.

Once the Venus flytrap enfolds prey within a leaf, enzymes secreted from the leaf surfaces digest the prey. These flytraps use a growth mechanism to close, not just a decrease in turgor pressure. As a result, they can only open and close a limited number of times.

In the sundews (*Drosera* spp.), another carnivorous group, glandular trichomes secrete both sticky mucilage, which traps small animals, and digestive enzymes; they do not close rapidly (figure 39.10*c*). Venus flytraps and the sundews share a common ancestor that lacked the snap-trap mechanism characteristic of the flytrap lineage (figure 39.11).

Aldrovanda vesicular, the aquatic waterwheel, is a closer relative of the flytraps. The waterwheel is a rootless plant that uses trigger hairs and a snap-trap mechanism like that of the Venus flytrap to capture and digest small animals (figure 39.10*d*). Molecular phylogenetic studies indicate that Venus flytraps are sister species with sundews, forming a sister clade. It appears that the snap-trap mechanism evolved only once in

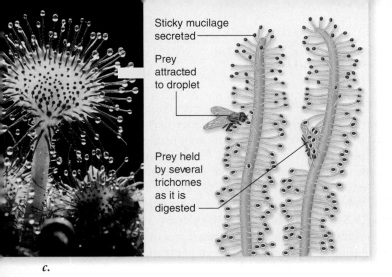

Sticky mucilage secreted

Prey attracted to droplet

Prey held by several trichomes as it is digested

c.

Prey

d.

descendants of a sundew ancestor. Therefore, the waterwheel's common ancestor must have been a terrestrial plant that made its way back into the water.

Bladderworts (*Utricularia*) are aquatic, but appear to have different origins from the waterwheel, as well as a different mechanism for trapping organisms. Small animals are swept into their bladderlike leaves by the rapid action of a springlike trapdoor; then the leaves digest these animals.

Parasitic plants exploit resources of other plants

Parasitic plants come in photosynthetic and nonphotosynthetic varieties. In total, at least 3000 types of plants are known to tap into the nutrient resources of other plants. Adaptations include structures that tap into the vascular tissue of the host plant so

that nutrients can be siphoned into the parasite. One example is dodder (*Cuscuta* spp.), which looks like brown twine wrapped around its host. Dodder lacks chlorophyll and relies totally on its host for all its nutritional needs.

Indian pipe, *Hypopitys uniflora*, also lacks chlorophyll. This parasitic plant hooks into host trees through the fungal hyphae of the host's mycorrhizae (figure 39.12). The above-ground portion of the plant consists of flowering stems.

Innovations in nutritional strategies allow some plants to harvest nitrogen from bacteria and phophorous from fungi. Unusual strategies include capturing and digesting animals, while other plants tap into the nutrient resources of host plants.

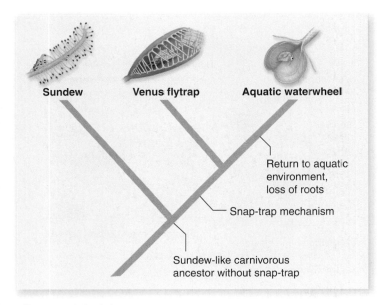

Sundew Venus flytrap Aquatic waterwheel

Return to aquatic environment, loss of roots

Snap-trap mechanism

Sundew-like carnivorous ancestor without snap-trap

figure 39.11

PHYLOGENETIC RELATIONSHIPS AMONG CARNIVOROUS PLANTS. The snap-trap mechanism was acquired by a common ancestor of the Venus flytrap and the aquatic waterwheel. Pitcher plants are not related to this clade.

figure 39.12

INDIAN PIPE, *Hypopitys uniflora*. This plant lacks chlorophyll and depends completely on nutrient transfer through the invasion of mycorrhizae and associated roots of other plants. Indian pipes are frequently found in the forests of the northeastern United States.

Carbon–Nitrogen Balance and Global Change

The Intergovernmental Panel on Climate Change (IPCC), established by the United Nations and the World Meteorological Organization, has concluded that CO_2 is probably at its highest concentration in the atmosphere in at least 20 million years. In only the last 250 years atmospheric CO_2 has increased 31%, which correlates with increases in many human activities, including the burning of fossil fuels.

The long-term effects of elevated CO_2 are complex and are not fully understood, but are associated with increased temperatures. The IPCC predicts the average global surface temperatures will continue to increase to between 1.4°C and 5.8°C above 1990 levels, by 2100. Chapter 57 explores the causal link between elevated CO_2 and global warming. Here, we consider how increased CO_2 may alter nutrient balance within plants, specifically the carbon and nitrogen balance.

The ratio of carbon to nitrogen in a plant is important for both plant health and the health of herbivores. Altering this ratio could alter plant–pest interactions as well as have an impact on human nutrition.

Elevated CO_2 levels can alter photosynthesis and carbon levels in plants

First, we investigate the relationship between photosynthesis and the relative concentration of atmospheric CO_2. The two questions to be addressed in this section are (1) Does elevated CO_2 increase the rate of photosynthesis? and (2) Will elevated levels of CO_2 change the ratio of carbohydrates and proteins in plants?

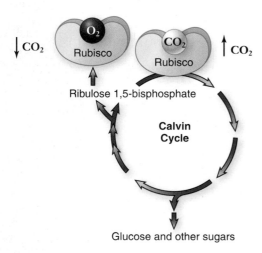

The rate of photosynthesis

The **Calvin cycle** of photosynthesis fixes atmospheric CO_2 into sugar (chapter 8). The first step of the Calvin cycle stars the most abundant protein on Earth, **ribulose 1,5-bisphosphate carboxylase/oxygenase (rubisco).** The active site of this enzyme can bind either CO_2 or O_2, and it catalyzes the addition of either molecule to a five-carbon molecule, **ribulose 1,5-bisphosphate (RuBP)** (figure 39.13). CO_2 is used to produce a three-carbon sugar that can in turn be used to synthesize glucose and sucrose; in contrast, O_2 is used in photorespiration, which results in neither nutrient nor energy storage. Photorespiration is a wasteful process.

You may recall that C_4 plants have evolved a novel anatomical and biochemical strategy to reduce photorespiration (figure 39.14). CO_2 does not enter the Calvin cycle until it has been transported via another pathway to cells surrounding the vascular tissue. Here the level of CO_2 is increased relative to O_2 levels, and thus CO_2 has less competition for rubisco's binding site.

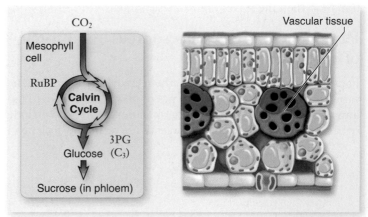

a. C_3 leaf

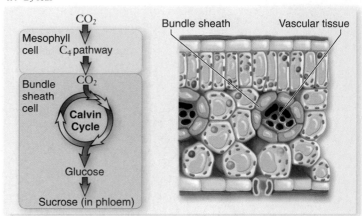

b. C_4 leaf (Kranz anatomy)

figure 39.13

PHOTORESPIRATION. CO_2 and O_2 compete for the same site on the enzyme that catalyzes the first step in the Calvin cycle. If CO_2 binds, a three-carbon sugar is produced that can make glucose and sucrose. If O_2 binds, photorespiration occurs, and energy is used to break down a five-carbon molecule without yielding any useful product. As the ratio of CO_2 to O_2 increases, the Calvin cycle can produce more sugar.

figure 39.14

C_4 plants reduce photorespiration by limiting the Calvin cycle to cells surrounding the vascular tissue where O_2 levels are reduced. ***a.*** C_3 photosynthesis occurs in the mesophyll cells. ***b.*** C_4 photosynthesis uses an extra biochemical pathway to shuttle carbon deep within the leaf.

In C_3 plants, as the relative amount of CO_2 increases, the Calvin cycle becomes more efficient. Thus, it is reasonable to hypothesize that the global increase in CO_2 should lead to increased photosynthesis and increased plant growth. Assuming that nutrient availability in the soil remains the same, the more rapidly growing plants should have lower levels of nitrogen-containing compounds, such as proteins, and also lower levels of minerals obtained from the soil. The ratio of carbon to nitrogen should increase.

The optimal way to determine how CO_2 concentrations affect plant nutrition is to grow plants in an environment where CO_2 levels can be precisely controlled. Experiments with potted plants in growth chambers are one approach, but far more information can be obtained in natural areas enriched with CO_2. For example, the Duke Experimental Forest has rings of towers that release CO_2 toward the center of the ring (figure 39.15). These rings are 30 m in diameter and allow studies to be conducted at the ecosystem level. Such facilities allow for long-term studies of the effects of altered atmospheric conditions on ecosystems.

Such studies have complex results. Potatoes grown in a European facility had a 40% higher photosynthetic rate when the concentration of CO_2 was approximately doubled. Potted plants often show an initial increase in photosynthesis, followed by a decrease over time that is associated with lower levels of rubisco production. Different species of plants in a Florida oak-shrub system showed different responses to elevated CO_2 levels, while over three years in the Duke Experimental Forest, plants achieved more biomass in the CO_2 enclosures than outside the enclosures, if the soil contained sufficient nitrogen availability to support enhanced growth. In general, increased CO_2 corresponds to increased biomass, but also to an increase in the carbon-to-nitrogen ratio.

The ratio of proteins and carbohydrates

You learned earlier in this chapter that nitrogen availability limits plant growth. As CO_2 levels increase, relatively less nitrogen and other macronutrients may be found in leaves. In that event, herbivores would need to eat more biomass to obtain adequate nutrients, particularly protein. This situation would be of significant concern in agriculture, and it could affect human health. Insect infestations could be more devastating if each herbivore consumed more biomass. Protein deficiencies in human diets could result from decreased nitrogen in crops. Again, it is difficult to generalize for all plants.

The relative decreases in nitrogen in some plants is greater than would be predicted by an increase in CO_2 fixation alone. The additional decrease in nitrogen incorporation into proteins has been accounted for by a decrease in photorespiration in plants using NO_3^- as their primary nitrogen source, but not in plants using ammonia. It is possible that energy-wasting photorespiration may actually be necessary for nitrogen to be incorporated into proteins in some plants.

This example illustrates how interdependent the biochemical pathways are that regulate carbon and nitrogen levels. Although global change is an ecosystem-level problem, predictions about long-term effects hinge on understanding the physiological complexities of plant nutrition.

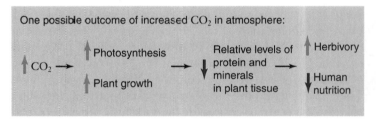

figure 39.15

EXPERIMENTALLY ELEVATING CO_2. CO_2 rings in the Duke Experimental Forest provide ecosystem-level comparisons of plants grown in ambient and elevated CO_2 environments. **a.** Each ring is 30 m in diameter. **b.** Towers surrounding the rings blow CO_2 inward under closely monitored conditions.

a.

b.

Elevated temperature can affect respiration and carbon levels in plants

As much as half of all the carbohydrates produced from photosynthesis each day can be used in plant respiration that same day. The amount of carbohydrate available for respiration may be affected by atmospheric CO_2 and photosynthesis, as just discussed. Furthermore, the anticipated rise in temperature over the next century may affect the rate of respiration in other ways. Altered respiration rates can affect overall nutrient balance and plant growth.

inquiry

Why is plant respiration affected by both short-term and long-term temperature changes.

Biologists have known for a long time that respiration rates are temperature-sensitive in a broad range of plant species. Why does respiration rate change with temperature? One important factor is the effect of temperature on enzyme activity (chapter 3). This effect is particularly important at very low temperatures and also very high temperatures that lead to protein denaturation.

Many responses of respiration rate to temperature change may be short-term, rather than long-term. There is growing evidence that respiration rate acclimates to a temperature increase over time, especially in leaves and roots that develop after the temperature shift. Over a long period at an elevated temperature, a plant could end up respiring at the same rate at which it had previously respired at a lower temperature.

> **Carbon–nitrogen balance affects plant growth and herbivory.**
> Global climate change is predicted to alter the ratio of carbon and nitrogen via increased CO_2 levels and increased temperature, which will affect both respiration and photsynthetic rates.

39.5 Phytoremediation

Some root cell membrane channels and transporters lack absolute specificity and can take up heavy metals like aluminum and other toxins. Although in most cases uptake of toxins is lethal or growth limiting, some plants have evolved the ability to sequester or release these compounds into the atmosphere. These plants have potential for **phytoremediation,** the use of plants to concentrate or breakdown pollutants (figure 39.16).

Phytoremediation can work in a number of ways with both aquatic and soil pollutants. Plants may secrete a substance from their roots that breaks down the contaminant. More often, the harmful chemical enters the roots and preferably is transported to the shoot system, which is easier to remove from the site. Some substances are simply stored by the plant; later, the plant material is harvested, dried, and removed to a storage site.

For example, after the nuclear reactor disaster at Chernobyl in northern Ukraine, sunflowers effectively removed radioactive cesium from nearby lakes. The plants were floated in foam supports on the surface of the lakes and later collected. Because up to 85% of the weight of herbaceous plants can be water, drying down phytoremediators can restrict toxins like radioactive cesium to a small area.

In this section, we will explore several examples of soil phytoremediation.

Trichloroethylene may be removed by poplar trees

Trichloroethylene (TCE) is a volatile solvent that has been widely used as spot remover in the dry-cleaning industry, for degreasing engine turbines, as an ingredient in paints and cosmetics, and even as an anesthetic in human and veterinary medicine. Unfortunately, TCE is also a confirmed carcinogen, and chronic exposure can damage the liver.

In 1980, the Environmental Protection Agency (EPA) established a Superfund to clean up contamination in the United States. Forty percent of all sites funded by the Superfund include TCE contamination. How can we clean up 1900 hectares of soil in a Marine Corps Air Station in Orange County, California, that contain TCE once used to clean fighter jets? Landfills can isolate, but not eliminate, this volatile substance. Burning eliminates it from the site, but may release harmful substances into the atmosphere. A promising approach is to use plants to remove TCE from the soil.

Plants may take up a toxin from soil, allowing the toxin to be removed and concentrated elsewhere; but an even more successful strategy is for the plant to break down the contaminant into nontoxic by-products. Poplar trees (genus *Populus*) may provide just such a solution for TCE-contaminated sites (figure 39.17). Poplars naturally take up TCE from the soil and metabolize it into CO_2 and chlorine.

Other plant species can break down TCE as well, but poplars have the advantage of size and rapid transpiration. A five-year-old poplar can move between 100 and 200 L of water from its roots out through its leaves in a day. A plant that transpires less would not be able to remove as much TCE in a day.

Although removing TCE with poplar trees sounds like the perfect solution, this method has some limitations. Not all the TCE is metabolized, and given the rapid rate of transpiration in the poplar, some of the TCE enters the atmosphere via

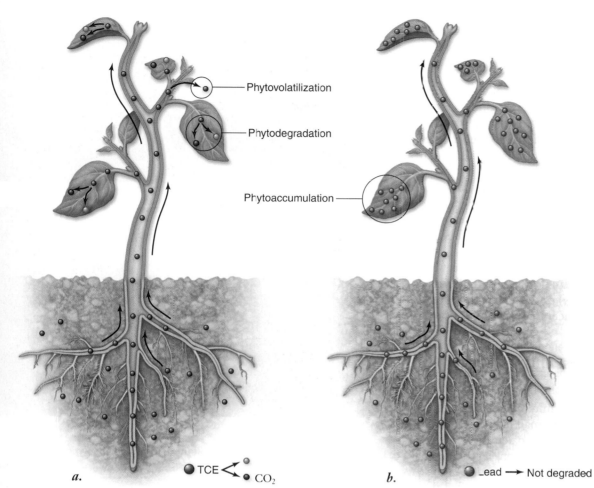

Phytovolatilization

Phytodegradation

Phytoaccumulation

a. ● TCE ⟨ ○ / ● CO_2

b. ● Lead → Not degraded

figure 39.16

PHYTOREMEDIATION. Plants can use the same mechanisms to remove both nutrients and toxins from the soil. *a.* TCE (trichloroethylene) can be taken up by plants and degraded into CO_2 and chlorine before being released into the atmosphere. This process is called *phytodegradation*. Some of the TCE moves so rapidly through the xylem that it is not degraded before it is released through the stomata as a gas in a process called *phytovolatilization*. *b.* Other toxins, including heavy metals such as lead, can be taken up by plants, but not degraded. Such *phytoaccumulation* is particularly effective in removing toxins if they are stored in the shoot, where they can more easily be harvested.

the leaves. Once in the air, TCE has a half-life of 9 hr (half of it will break down into smaller molecules every 9 hr). Clearly, more risk assessment is needed before poplars are planted on every TCE-containing Superfund site.

The TCE that remains in the plant is metabolized quickly, and it is possible that the wood could be used after remediation is complete. It has been suggested that any remaining TCE would be eliminated if the wood were processed to make paper.

Genetically modified poplars have been shown to metabolize about four times as much TCE as nonmodified poplars, so perhaps greater metabolic rates can be obtained.

As with any phytoremediation plan, it is critically necessary to estimate how much of a contaminant can be removed from a site by plants, and arriving at this estimate can be a challenge. Possible risks, particularly when genetic modification is involved, must be weighed against the dangers posed by the contaminant.

figure 39.17

PHYTOREMEDIATION FOR TCE. The U.S. Air Force is testing phytoremediation technology to clean up TCE at a former Air Force base in Fort Worth, Texas.

SELF TEST

1. Which of the following affect the availability of nutrients for plants?
   a. Soil pH
   b. Soil salinity
   c. Soil composition
   d. All of the above

2. If you wanted to conduct an experiment to determine the effects of varying levels of macronutrients on plant growth and did so in your small greenhouse at home, which of the following macronutrients would be the most difficult to regulate?
   a. Carbon
   b. Nitrogen
   c. Potassium
   d. Phosphorus

3. You are performing an experiment to determine the nutrient requirements for a newly discovered plant and find that for some reason your plants die if you leave boron out of the growth medium but do fine with as low as 5 parts per million in solution. This suggests that boron is
   a. an essential macronutrient
   b. a nonessential micronutrient
   c. an essential micronutrient
   d. a nonessential macronutrient

4. Which of the following might you do to increase nutrient uptake by crop plants?
   a. Decrease the solubility of nutrients.
   b. Create nutrients as positive ions.
   c. Frequently plow the soil.
   d. Genetically modify plants to increase the density of plasma membrane transporters in root cells.

5. Which of the following would decrease nitrogen availability for a pea plant?
   a. Inability of the plant to produce flavonoids
   b. Formation of Nod factors
   c. Presence of oxygen in the soil
   d. Production of leghemoglobin

6. Different soils contain varying amounts of space between the soil particles. Which of the following statements is correct?
   a. Some of the space in soils must be filled by air in order for plants to survive.
   b. The amount of water a soil can hold is the same amount of water a plant can extract.
   c. Even though sandy soils have a lot of space between particles, alot of water is lost because of drainage due to gravity.
   d. All of these statements are correct.

7. Some plants, such as the Venus flytrap, have evolved the ability to digest insects. This benefits the plants because
   a. they gain energy from the digested insect and can therefore photosynthesize less.
   b. they live in nutrient-poor environments and can gain valuable macronutrients such as nitrogen.
   c. they are prone to herbivory from insects and thus can defend themselves from insects.
   d. they gain carbon from the insects, enabling them to increase photosynthetic rates.

8. There is concern about increased atmospheric carbon dioxide levels as it may relate to global warming. However, increasing atmospheric carbon dioxide should theoretically help plant growth. Which of the following could be drawbacks for plant from increased carbon dioxide?
   a. Increased protein to carbohydrate ratios in plants
   b. Increased herbivory
   c. Increased photorespiration
   d. All of the above

9. If you were asked how to clean up a trichloroethylene (TCE) spill without having to resort to burning or other chemical methods, how would you do it?
   a. Plant poplar trees to phytoremediate the soil.
   b. Plant bean plants to replace the TCE with fixed nitrogen.
   c. Plant *Brassica* plants to phytoaccumulate the TCE.
   d. Plant Indian pipe because it is not adversely affected by TCE in the soil.

10. Which of the following is most common among plants as nutritional adaptations?
    a. Forming symbiotic relationships with nitrogen-fixing bacteria
    b. Becoming carnivorous
    c. Becoming parasitic
    d. Forming symbiotic relationships with mycorrhizae

11. Why must plants actively transport positive ions into their roots?
    a. It is the positive ions in the roots that allow water to be drawn from the soil.
    b. The soil solution contains primarily negatively-charged ions, so positive ions are continually drawn out of the root.
    c. Negative ions in the plant root must be balanced by positive ions.
    d. None of the above.

12. Saline soils are detrimental to plant growth for which of the following reasons?
    a. Salt in soils does not allow active transport of mineral nutrients into the plant.
    b. Salt in soils prevents water from moving into the plant via osmosis.
    c. Salt is toxic to plants.
    d. None of the above.

13. Most plants are limited in their growth because nitrogen is often limiting in the environment, yet plants are bathed in an atmosphere that is full of nitrogen. Why can't plants utilize this nitrogen source?
    a. Nitrogen gas (N_2) is held together by very strong bonds, which plants cannot break.
    b. N_2 is not soluble.
    c. N_2 cannot be broken down by any organisms.
    d. All of the above.

CHALLENGE QUESTIONS

1. If you were to eat one ton (1000 kg) of potatoes, approximately how much of the following minerals would you eat?
   a. Copper between 0.4 and 3 g
   b. Zinc between 1.5 and 10 g
   c. Potassium between 0.5 and 6%
   d. Iron between 2.5 and 30 g

2. Using what you know about phytoremediation, design a strategy to speculate for gold that would not require any digging or disruption of the soil.

3. Given increased carbon dioxide levels in the atmosphere, predict how a grassland community, composed of C_3 and C_4 plants, might change over time.

Do you need additional review? *Visit www.ravenbiology.com for practice quizzes, animations, videos, and activities designed to help you master the material in this chapter.*

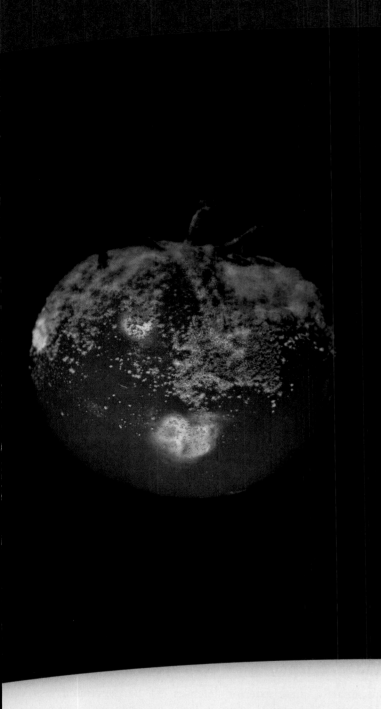

chapter 40

Plant Defense Responses

introduction

PLANTS ARE CONSTANTLY UNDER ATTACK by viruses, bacteria, fungi, animals, and even other plants. An amazing array of defense mechanisms has evolved to block or temper an invasion. Many plant–pest relationships undergo coevolution, with the plant winning some times and the pest winning with a new offensive adaptation at other times. The first line of plant defense is thick cell walls covered with a strong cuticle. Bark, thorns, and even trichomes can deter a hungry insect. When that first line of defense fails, a chemical arsenal of toxins is waiting. Many of these molecules have no effect on the plant. Some are modified by microbes in the intestine of an herbivore into a poisonous compound. Maintaining a toxin arsenal is energy-intensive; thus, an alternative means of defense uses induced responses to protect and prevent future attacks.

Physical Defenses

There are no tornado shelters for trees. Storms and changing environmental conditions present life-threatening challenges to plants. Structurally, trees can often withstand high winds and the weight of ice and snow, but there are limits. Winds can uproot a tree, or snap the main shoot off a small plant. Axillary buds give many plants a second chance as they grow out and replace the lost shoot (figure 40.1).

Although abiotic factors such as weather constitute genuine threats to a plant, even greater daily threats exist in the form of viruses, bacteria, fungi, animals, and other plants. These enemies can tap into the nutrient resources of plants or use their DNA-replicating mechanisms to self-replicate. Some invaders kill the plant cells immediately, leading to necrosis (brown, dead tissue). Certain insects may tap into the phloem of a plant seeking carbohydrates, but leave behind a hitchhiking virus or bacterium.

The threat of these attackers is reduced when they have natural predators themselves. One of the greatest problems with

figure 40.2

ALFALFA PLANT BUG. This invasive species is an agricultural problem because it arrived without any natural predators and feeds on alfalfa.

figure 40.1

SHOOTS IN RESERVE. Axillary shoots give plants a second chance when the terminal shoot breaks off, as is the case with this storm-felled tree.

nonnative invasive species, such as the alfalfa plant bug (figure 40.2), is the lack of natural predators in the new environment.

Dermal tissue provides first-line defense

The first defense all plants have is the dermal tissue system (see chapter 36). Epidermal cells throughout the plant secrete wax, which is a mixture of hydrophobic lipids, and layers of lipid material protect exposed plant surfaces from water loss and attack. Above-ground plant parts are also covered with **cutin,** a macromolecule consisting of long-chain fatty acids linked together. **Suberin,** another version of linked fatty acid chains, is found in cell walls of subterranean plant organs; suberin forms the water-impermeable Casparian strips of roots. Silica inclusions, trichomes, bark, and even thorns can also protect the nutrient-rich plant interior.

Invaders can penetrate dermal defenses

Unfortunately, these exterior defenses can be penetrated in many ways. Mechanical wounds leave an open passageway through which microbial organisms can penetrate. Parasitic nematodes use their sharp mouthparts to get through the plant cell walls. Their actions either trigger the plant cells to divide, forming a tumorous growth, or, in species that attach to a single plant cell, cause the cell to enlarge and transfer carbohydrates from the plant to the hungry nematode (figure 40.3).

In some cases simply having bacteria on the leaf surface can increase the risk of frost damage. The bacteria function as sites for ice nucleation; the resulting ice crystals severely damage the leaves.

Fungi strategically seek out the weak spot in the dermal system, the stomatal openings, to enter the plant. Some fungi have coevolved with a monocot that has evenly spaced stomata. These fungi appear to be able to measure distance to locate these evenly spaced stomatal openings and invade the plant.

figure 40.3

NEMATODES ATTACK THE ROOTS OF CROP PLANTS. *a.* A nematode breaks through the epidermis of the root. *b.* Root-knot nematodes form tumors on roots.

Knots

a.   *b.*

Figure 40.4 shows the phases of fungal invasion, which can include the following:

1. Windblown spores land on leaves. A germ tube emerges from the spore. Host recognition is necessary for the infection to proceed.

2. The spore germinates and forms an adhesion pad, allowing it to stick to the leaf.
3. Hyphae grow through cell walls and press against the cell membrane.
4. Hyphae differentiate into specialized structures called haustoria. They expand, surrounded by cell membrane, and nutrient transfer begins.

Bacteria and fungi can also be beneficial to plants

Mutualistic and parasitic relationships are often just opposite sides of the evolutionary coin. In chapters 31 and 39, you saw how mycorrhizal fungi use a mechanism similar to the one just described to the mutual benefit of both the plant and the fungus. In the case of the relationship between legumes and nitrogen-fixing bacteria, the *Rhizobium* bacteria seeks out a root hair, infects the root hair and other tissues, and forms a root nodule. Other soil bacteria can also enhance plant growth, and are called plant growth-promoting rhizobacteria (PGPR). The term *rhizobacteria* refers to bacteria that live around the root system and often benefit from root exudates. In return they provide substances that support plant growth. *Azospirillum* species, for example, provide gibberellins, growth hormones, for rice plants when living in close proximity to the root system. PGPR can also limit the growth of pathogenic soil bacteria.

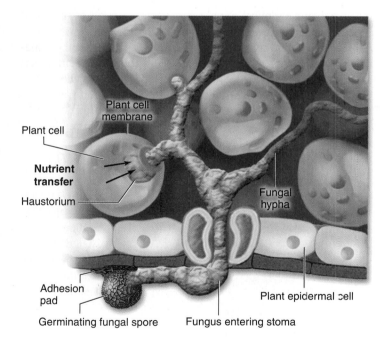

Plant cell membrane

Plant cell

Nutrient transfer

Haustorium

Fungal hypha

Adhesion pad

Germinating fungal spore

Fungus entering stoma

Plant epidermal cell

figure 40.4

FUNGI SNEAK IN THROUGH STOMATA. Fungal hyphae penetrate cell walls, but not plasma membranes. The close contact between the fungal hyphae and the plant plasma membrane allows for ready transfer of plant nutrients to the fungus.

The dermal tissue system is the first line of defense. Suberin and cutin secreted by the dermal tissues form barriers that prevent invasion and lessen water loss, but invaders have evolved numerous strategies to overcome these barriers. Not all bacteria and fungi are harmful; some provide plants with compounds that enhance growth.

Toxin Defenses

Many plants are filled with toxins that kill herbivores or, at the very least, make them quite ill. One example is the production of cyanide, (HCN). Over 3000 species of plants produce cyanide-containing compounds called *cyanogenic glycosides* that break down into cyanide when ingested. Cyanide stops electron transport, blocking cellular respiration.

Cassava (genus *Manihot*), a major food staple for many Africans, is filled with cyanogenic glycosides (specifically, manihotoxins) in the outer layers of the edible root. Unless these outer layers are scrubbed off, the cumulative effect of eating primarily cassava can be deadly.

In addition to toxins that kill, plants can produce other toxins that make potential herbivores ill or that repel them with strong flavors or odors.

Plants maintain chemical arsenals

How did the biosynthetic pathways that produce these toxins evolve? Growing evidence indicates that the metabolic pathways needed to sustain life in plants have taken some evolutionary side trips, leading to the production of a stockpile of chemicals known as **secondary metabolites.** Many of these secondary metabolites affect herbivores as well as humans (table 40.1).

Alkaloids, including caffeine, nicotine, cocaine, and morphine, can affect multiple cellular processes; if a plant cannot kill its attackers, it can overstimulate them with caffeine or sedate them with morphine. For example, the tobacco hornworm (*Manduca sexta*) can level a field of tobacco (figure 40.5); however, wild species of tobacco appear to have elevated levels of nicotine that are lethal to tobacco hornworms.

Tannins bind to proteins and inactivate them. For example, some act by blocking enzymes that digest proteins, which reduces the nutritional value of the plant tissue. An insect that gets sick from a strong dose of tannins is likely to associate the flavor with illness and to avoid having that type of plant for lunch another time. Small doses, equivalent to the amount a

human gets from eating apples or blackberries, are unlikely to cause any major digestive difficulties. Animals, including humans, can avoid many of the cumulative toxic effects of secondary metabolites by eating a varied diet.

Plant oils, particularly those found in plants of the mint family, which includes peppermint, sage, pennyroyal and many others, repel insects with their strong odors. At high concentrations, some of these oils can also be toxic if ingested.

Why don't the toxins kill the plant? One strategy is for a plant to sequester a toxin in a membrane-bound structure, so that it does not come into contact with the cell's metabolic processes. The second solution is for the plant to produce a compound that is not toxic unless it is metabolized, often by microorganisms, in the intestine of an animal. Cyanogenic glycosides are a good example of the latter solution. The plant produces a sugar-bound cyanide compound that does not affect electron transport chains. Once an animal ingests cyanogenic glycoside, the compound is enzymatically broken down, releasing the toxic hydrogen cyanide.

Coevolution has led to defenses against some plant toxins. A tropical butterfly, *Helioconius sara*, can sequester the cyanogenic glycosides it ingests from its sole food source, the passion vine. Even more intriguing is a biochemical pathway that allows the butterfly to safely break down cyanogenic glycosides and use the released nitrogen in its own protein metabolism.

Plants can poison other plants

Some chemical toxins protect plants from other plants. **Allelopathy** occurs when a chemical signal secreted by the roots of one plant blocks the germination of nearby seeds or inhibits the growth of a neighboring plant. This strategy minimizes shading and competition for nutrients, while it maximizes the ability of a plant to use radiant sunlight for photosynthesis. Allelopathy works with both a plant's own species and different species. Black walnut trees (*Juglans nigra*) are a good example. Very little vegetation will grow under a black walnut tree because of allelopathy (figure 40.6).

figure 40.5

HERBIVORES CAN KILL PLANTS. Tobacco hornworms, *Manduca sexta*, consume huge amounts of tobacco leaf tissue, as well as tomato leaves.

figure 40.6

BLACK WALNUTS ARE ALLELOPATHS. Seedlings die when their roots come in contact with the root secretions of a black walnut tree.

TABLE 40.1 Secondary Metabolites

Compound	Source	Structure	Effect on Humans
Manihotoxin (cyanogenic glycoside)	Cassava, *Manihot esculenta*		Metabolized to release lethal cyanide
Genistein (phytoestrogen)	Soybean, *Glycine max*		Estrogen mimic
Taxol (terpenoid)	Pacific yew, *Taxus brevifolia*		Anticancer drug
Quinine (alkaloid)	Quinine bark, *Cinchona officinalis*		Antimalarial drug
Morphine (alkaloid)	Opium poppy, *Papaver somniferum*		Narcotic pain killer

Humans are susceptible to plant toxins

Not only have humans been inadvertently poisoned by plants, but throughout much of human history, they have also been intentionally poisoned by other humans using plant products. Socrates, an important Greek philosopher who lived 2400 years ago, was sentenced to death in Athens, and he died after he drank a hemlock extract containing an alkaloid that paralyzes motor nerve endings.

Ricin, an alkaloid found in castor beans (*Ricinus communis*), is six times more lethal than cyanide and twice as lethal as cobra venom. A single seed from the plant, which is still grown in flower gardens, can kill a young child if ingested. Death occurs because ricin functions as a ribosome-binding protein that inhibits translation (figure 40.7).

Ricin is found in the endosperm of the seed as a heterodimer composed of ricin A and ricin B, joined by a single disulfide bond. This heterodimer (proricin) is nontoxic, but when the disulfide bond is broken in humans or other animals, ricin A targets the GAGA sequence of the 28s rRNA of the ribosome (remember that ribosomes are composed of rRNA and protein). A single ricin molecule can inactivate 1500 ribosomes per minute, blocking translation of proteins.

In 1978, Bulgarian expatriate and dissident Georgi Markov was about to board a bus in London on his way to work at the BBC when he felt a sharp stabbing pain in his thigh. A man near him picked up an umbrella from the ground and hurriedly left. Markov had been injected via a mechanism in the umbrella tip with a pinhead-sized metal sphere containing 0.2 mg of ricin. He died four days later. After the collapse of the Soviet Union, former KGB officers revealed that the KGB had set up the assassination at the behest of the Bulgarian Communist Party leadership.

inquiry

? *Explain how ricin led to Markov's death.*

Secondary metabolites may have medicinal value

Major research efforts on plant secondary metabolites are in progress because of their potential benefits, as well as dangers, to human health (see table 40.1).

Soy and phytoestrogens

One example is the presence of **phytoestrogens,** compounds very similar to the human hormone estrogen, in soybean products. In soybean plants, genistein is one of the major phytoestrogens.

Comparative studies between Asian populations that consume large amounts of soy foods and populations with lower dietary intake of soy products are raising intriguing questions and some conflicting results. For example, the lower rate of prostate cancer in Asian males might be accounted for by the down-regulation of androgen and estrogen receptors by a phytoestrogen. Soy is being marketed as a means for minimizing menopausal symptoms caused by declining estrogen levels in older women.

In humans, dietary phytoestrogens cross the placenta and can be found in the amniotic fluid during the second trimester of pregnancy. Questions have been raised about the effect of phytoestrogens on developing fetuses and even on babies who consume soy-based formula because of allergies to cow's milk formula. Because hormonal signaling is so complex, much more research is needed to fully understand how or even if phytoestrogens affect human physiology and development.

Taxol and breast cancer

Taxol, a secondary metabolite found in the Pacific yew (*Taxus brevifolia*), is effective in fighting cancer, especially breast cancer. The discovery of taxol's pharmaceutical value raised an environmental challenge. The very existence of the Pacific yew was being threatened as the shrubs were destroyed so that taxol could be extracted. Fortunately, it became possible to synthesize taxol in the laboratory.

figure 40.7

RICIN BLOCKS TRANSLATION. When the ricin A subunit is released from proricin, it binds to rRNA in ribosomes and prevents mRNA from being translated into protein.

Taxol is not an isolated case of drug discovery in plants. The hidden pharmaceutical value of many plants may lead to increased conservation efforts to protect plants that have the potential to make contributions toward human health. Although the plant pharmaceutical industry is growing, it is certainly not a new field. Until recent times, almost all medicines used by humans came from plants.

Quinine and malaria

In the 1600s, the Incas of Peru were treating malaria with a drink made from the bark of *Cinchona* trees. Malaria is caused by four types of human malaria parasites in the genus *Plasmodium*, which are carried by female *Anopheles* mosquitoes. *Plasmodium falciparum* is the most lethal of the four types. Symptoms include severe fevers and vomiting. The parasite feeds on red blood cells, and death can result from anemia or blocking of blood flow to the brain.

By 1820, the active ingredient in the bark of *Cinchona* trees, **quinine,** had been identified (see table 40.1). In the nineteenth century, British soldiers in India used quinine-containing "tonic water" to fight malaria. They masked the bitter taste of quinine with gin, creating the first gin and tonic drinks. In 1944, Robert Woodward and William Doering synthesized quinine. Now several other synthetic drugs are available to treat malaria.

Exactly how quinine and synthetic versions of this drug family work has puzzled researchers for a long time. Quinine can affect DNA replication, and also, when *P. falciparum* breaks down hemoglobin from red blood cells in its digestive vacuole, an intermediary toxic form of heme is released. Quinine may interfere with the subsequent polymerization of these hemes, leading to a build up of toxic hemes that poison the parasite.

Unfortunately, even today malaria is a major threat to human health, causing over a million deaths per year. Ninety percent of these deaths are in sub-Saharan Africa. An estimated 300,000,000 individuals are infected. *P. falciparum* strains have acquired resistance to synthetic drugs, and quinine is once again the drug of choice in some cases.

Herbal remedies have been used for centuries in most cultures. A resurgence of interest in plant-based remedies is resulting in a growing and unregulated industry. Although herbal remedies have great promise, we need to be aware that each plant contains many secondary metabolites, many of which have evolved to cause harm to herbivores including humans.

> **Modified metabolic pathways lead to the production of toxins that protect plants from herbivores. Secondary metabolites also have tremendous pharmaceutical potential.**

40.3 Animals that Protect Plants

Not only do individual species and their traits evolve over time, but so do relationships between species. For example, evolution of chemicals to deter herbivores may often be accompanied over time by adaptation on the part of herbivores to withstand these chemicals. This evolutionary pattern is called coevolution. In chapter 56 we cover details of interactions between species in a community. Here we consider two cases of relationships between animal species and plant species that benefit both, an interaction termed **mutualism.**

Acacia trees and ants. Several species of ants provide small armies to protect some species of *Acacia* trees from other herbivores. These stinging ants may inhabit a home in an enlarged thorn of the tree; they attack other insects (figure 40.8) and sometimes small mammals and epiphytic plants. Some of the *Acacia* species provide their ants with sugar in nectaries located away from the flowers, and even with lipid food bodies at the tips of leaves.

The only problem with ants chasing away other insects is that acacia trees depend on bees to pollinate their flowers. What keeps the ants from swarming and stinging a bee that stops by to pollinate? Evidence indicates that when a flower opens on an acacia tree, it produces some type of chemical ant deterrent that does not deter the bees. This chemical has not yet been identified.

Parasitoid wasps, caterpillars, and leaves. Caterpillars fill up on leaf tissue before they metamorphose into a moth or a butterfly. In some cases, proteinase inhibitors in leaves are sufficient to deter very hungry caterpillars. But some

figure 40.8

ANTS ATTACKING A KATYDID TO PROTECT "THEIR" *Acacia*. Through coevolution, ants are sheltered by acacia trees and attack otherwise harmful herbivores.

1. A volatile signal is released as the caterpillar eats a leaf.

Volatile signal

2. Female wasp is attracted by the volatile signal, finds caterpillar, and lays eggs.

3. Wasp larvae feed on the caterpillar and then emerge.

4. Larvae continue to feed on the caterpillar after it dies, but not the plant. The larvae then spin cocoons to pupate.

Larvae

figure 40.9

PARASITOID WASPS PROVIDE PROTECTION FROM HERBIVORY.

plants have developed another strategy: As the caterpillar chews away, a wound response in the plant leads to the release of a volatile compound. This compound wafts through the air, and if a female parasitoid wasp happens to be in the neighborhood, it is immediately attracted to the source. Parasitoid wasps are so named because they are parasitic on caterpillars. The wasp lays her fertilized eggs in the body of the caterpillar that is feeding on the leaf of the plant. These eggs hatch, and the emerging larvae kill and eat the caterpillar (figure 40.9).

Complex coevolution between plants and animals has resulted in associations that protect the plant from other animals.

40.4 Systemic Responses to Invaders

So far, we have focused mainly on static plant responses to threats. Most of the deterrent chemicals such as toxins are maintained at steady-state levels. In addition, the morphological structures such as thorns or trichomes that help defend plants are part of the normal developmental program. Because these defenses are maintained whether an herbivore or other invader is present or not, they have an energetic downside. By contrast, resources could be conserved if the response to being under seige was inducible—that is, if the defense response could be launched only when a threat had been recognized. In this section, we explore these inducible defense mechanisms.

Wound responses protect plants from herbivores

As you just learned from the example of the parasitoid wasp, a **wound response** may occur when a leaf is chewed or injured. One induced outcome is the rapid production of proteinase inhibitors. These chemical toxins do not exist in the stockpile of defenses, but instead are produced in response to wounding.

Proteinase inhibitors bind to digestive enzymes in the gut of the herbivore. The proteinase inhibitors are produced throughout the plant, and not just at the wound site. How are cells in distant parts of the plant signaled to produce proteinase inhibitors? In tomato plants, the following sequence of events is responsible for this systemic response (figure 40.10):

1. Wounded leaves produce an 18-amino-acid peptide called **systemin** from a larger precursor protein.
2. Systemin moves through the apoplast (the space between cell walls) of the wounded tissue and into the nearby phloem. This small peptide-signaling molecule then moves throughout the plant in the phloem.
3. Cells with a systemin receptor bind the systemin, which leads to the production of **jasmonic acid.**
4. Jasmonic acid signals gene expression, which leads to the production of a proteinase inhibitor.

Although we know the most about the signaling pathway involving jasmonic acid, other molecules are involved in wound response as well. **Salicylic acid,** which is found in the bark of plants such as the white willow (*Salix alba*) is one example. Cell

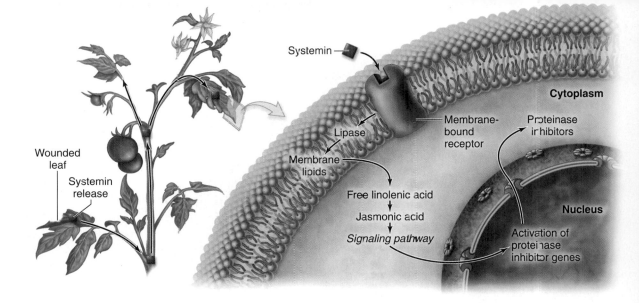

figure 40.10

WOUND RESPONSE IN TOMATO.
Wounding a tomato leaf leads to the production of jasmonic acid in other parts of the plants. Jasmonic acid initiates a signaling pathway that turns on genes needed to synthesize a proteinase inhibitor.

Wounded leaf

Systemin release

Systemin

Cytoplasm

Lipase

Membrane-bound receptor

Proteinase inhibitors

Membrane lipids

Free linolenic acid

Jasmonic acid

Signaling pathway

Nucleus

Activation of proteinase inhibitor genes

wall fragments also appear to be important signals for triggering an induced response, as is discussed shortly.

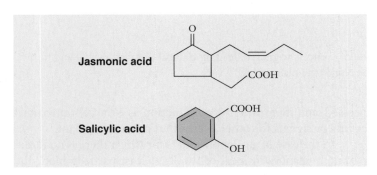

Jasmonic acid

Salicylic acid

Mechanical damage separate from herbivore attack also elicits wound responses, which presents a challenge in designing plant experiments that involve cutting or otherwise mechanically damaging the tissue. Experimental controls, which should be cut or manipulated in the same way but without the test treatment, are especially important to ensure that any changes observed are not due only to the wound response.

Defense responses can be pathogen-specific

Wound responses are independent of the type of herbivore or other agent causing the damage, but other responses are triggered by a specific pathogen that carries a specific allele in its genome.

Pathogen recognition

Half a century ago, the geneticist H. H. Flor proposed the existence of a plant resistance gene (*R*), the product of which interacts with the product of an avirulence gene (*avr*) carried by a pathogen. *Avirulent* means not virulent (disease causing). An **avirulent pathogen** is one that can utilize host resources for its own use and reproduction without causing severe damage or death. The product of this pathogen's avr protein interacts with the plant's R protein to signal the pathogen's presence. In this way, the plant under attack can mount defenses, thus ensuring that the pathogen remains avirulent. If the pathogen's avr protein is not recognized by the plant, disease symptoms appear.

Flor's proposal is called the **gene-for-gene hypothesis** (figure 40.11), and several pairs of *avr* and *R* genes have been cloned in different species pathogenized by microbes, fungi, and even

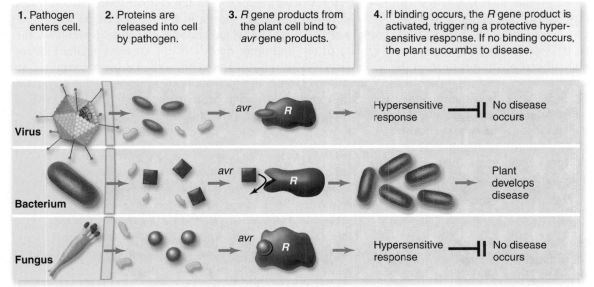

1. Pathogen enters cell.

2. Proteins are released into cell by pathogen.

3. *R* gene products from the plant cell bind to *avr* gene products.

4. If binding occurs, the *R* gene product is activated, triggering a protective hypersensitive response. If no binding occurs, the plant succumbs to disease.

Virus — avr — R — Hypersensitive response — No disease occurs

Bacterium — avr — R — Plant develops disease

Fungus — avr — R — Hypersensitive response — No disease occurs

figure 40.11

GENE-FOR-GENE HYPOTHESIS.
Flor proposed that pathogens have an avirulence (*avr*) gene that recognizes the product of a plant resistance gene (*R*). If the virus, bacterium, fungus, or insect has an *avr* gene product that matches the *R* gene product, a defense response will occur.

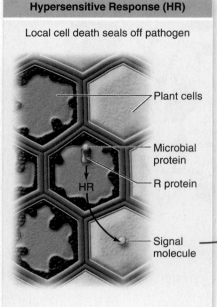

Hypersensitive Response (HR)

Local cell death seals off pathogen

Plant cells

Microbial protein

R protein

HR

Signal molecule

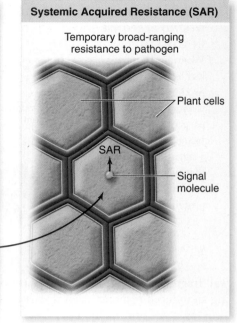

Systemic Acquired Resistance (SAR)

Temporary broad-ranging resistance to pathogen

Plant cells

SAR

Signal molecule

figure 40.12

PLANT DEFENSE RESPONSES. In the gene-for-gene response, a cascade of events is triggered, leading to local cell death (HR) and to the production of a mobile signal that provides longer term resistance in the rest of the plant (SAR).

insects in one case. This research has been motivated partially by the agronomic benefit of identifying genes that can be added via gene technology to crop plants to protect them from invaders.

Evidently plants and many of their pathogens have "worked out" this compromise, by which the avirulent invader can be detected and recognized. By contrast, virulent pathogens overcome a plant's defenses and kill it—often leading to the pathogen's demise as well.

Specific defenses and the hypersensitive response

Much is now known about the signal transduction pathways that follow the recognition of the pathogen by the *R* gene product. These pathways lead to the triggering of the **hypersensitive response (HR),** which leads to rapid cell death around the source of the invasion and also to a longer term, whole-plant resistance (figures 40.11 and 40.12). A gene-for-gene response does not always occur, but plants still have defense responses to pathogens in general as well as to mechanical wounding. Some of the response pathways may be similar. Also, fragments of cell wall carbohydrates may serve as recognition and signaling molecules.

When a plant is attacked and a gene-for-gene recognition occurs, the HR leads to very rapid cell death around the site of attack. This seals off the wounded tissue to prevent the pathogen or pest from moving into the rest of the plant. Hydrogen peroxide and nitric oxide are produced and may signal a cascade of biochemical events resulting in the localized death of host cells. These chemicals may also have negative effects on the pathogen, although protective mechanisms have coevolved in some pathogens.

Other antimicrobial agents produced include the **phytoalexins,** which are antimicrobial chemical defense agents. A variety of pathogenesis-related genes (*PR* genes) are also ex-

pressed, and their proteins can function as either antimicrobial agents or signals for other events that protect the plant.

In the case of virulent invaders for which there is no *R* recognition, changes in local cell walls at least partially block the pathogen or pest from moving further into the plant. In this case, an HR does not occur, and the local plant cells do not die.

Long-term protection

In addition to the HR or other local responses, plants are capable of a systemic response to a pathogen or pest attack, called a **systemic acquired resistance (SAR)** (see figure 40.12). Several pathways lead to broad-ranging resistance that lasts for a period of days.

The long-distance signal that induces SAR is likely salicylic acid, rather than systemin, which is the long-distance signal in wound responses. At the cellular level, jasmonic acid (which was mentioned earlier in the context of the wound response pathways) is involved in SAR signaling. SAR allows the plant to respond more quickly if it is attacked again. This response, however, is not the same as the human or mammalian immune response, in which antibodies (proteins) that recognize specific antigens (foreign proteins) persist in the body. SAR is neither as specific nor as long-lasting.

Wounding triggers the release of a small peptide-signaling molecule called systemin that travels throughout the plant to initiate the production of proteinase inhibitors. Plants can also defend themselves from specific invaders in ways reminiscent of the animal immune system. When an invader is recognized, via a gene-for-gene mechanism, the hypersensitive response is triggered. Systemic acquired resistance follows and provides plant-wide protection from future invasions.

40.1 Physical Defenses

Biotic threats can be more detrimental to plants than abiotic factors because they tap into the nutrient resources of the plant, replicate by taking over plant cell DNA, or kill the plant outright.

- Dermal tissues are covered with lipids, such as cutin and suberin, that protect the plant from water loss and attack.
- Silica inclusions and morphological protrusions such as trichomes, bark, and thorns protect some plants from damage.
- In spite of defense mechanisms damage can occur by piercing, chewing, or entering through the stomata. Bacteria can also cause damage because they provide sites for ice nucleation.
- Mycorrhizal fungi and rhizobacteria form beneficial relationships with plants and provide them with nutrients.

40.2 Toxin Defenses

Plants produce toxins that repel predators, make them ill, or even kill them (table 40.1, figure 40.7).

- Plants may produce secondary metabolites such as alkaloids, tannins, and oils to protect themselves from predators.
- Plants protect themselves by sequestering toxins in vesicles or produce metabolites that are not toxic until they are ingested by a predator.
- Allelopathic plants secrete chemicals to block seed germination or inhibit growth of nearby plants to minimize competition for resources.
- Plant secondary metabolites such as phytoestrogens, taxol, and quinine have pharmaceutical value for humans.

40.3 Animals that Protect Plants

Complex coevolution between plants and animals has resulted in mutualistic associations.

- Ants and acacia trees coevolved an association in which ants physically protect the trees from other invaders and the acacia provides the ants with food and shelter.
- In some plants volatile chemicals are released during caterpillar herbivory that attracts parasitoid wasps, which lay their eggs in the caterpillar and kill it (figure 40.9).

40.4 Systemic Responses to Invaders

Energy resources are conserved if plants produce defense mechanisms only when they are needed.

- Wound responses are generalized reactions independent of the agent that causes the damage. During these responses a signal spreads throughout the plant, inducing production of proteinase inhibitors that bind to predator digestive enzymes (figure 40.10).
- The production of proteinase inhibitors utilizes signaling pathways involving systemin, jasmonic acid, and salicylic acid.
- In many plants a compound produced by an avirulent pathogen gene may be recognized by a plant *R* gene protein; this is called the gene-for-gene hypothesis even though it is the recognition of the proteins encoded by the genes that is critical (figure 40.11).
- Recognition of a pathogen by the *R* gene product triggers the hypersensitive response (HR) that leads to rapid cell death around the site of attack, thereby preventing further invasion of the pathogen and protecting the plant.
- Plants can also produce antimicrobial agents such as phytoalexins.
- Plants are capable of a long-term protection to pathogens; this is called a systemic acquired resistance (SAR).

SELF TEST

1. If you were a plant pathogen, what would be the first obstacle to invading a host plant that you would have to overcome?
   a. Chemical toxins on the surface of a plant
   b. Physical barriers on the exterior of a plant
   c. Animal guardians of the host plant
   d. Immune proteins in the plant tissue

2. Plant dermal cells can secrete which of the following?
   a. Suberin
   b. Wax
   c. Cutin
   d. All of the above

3. Some plants are recognized by fungal pathogens on the basis of their stomatal pores. Which of the following would provide these plants immunity from fungal infection?
   a. Removing all of the stomata from the plant
   b. Changing the spacing of stomatal pores in these plants
   c. Reinforcing the cell wall in the guard cells of stomatal pores
   d. Increasing the number of trichomes on the surfaces of these plants

4. Your friend informs you that it is highly likely all of the plants in your yard are "infected" with some kind of fungi or bacteria. The plants look perfectly healthy to you at this time. The most prudent thing for you to do would be:
   a. Remove all your plants because they are likely to die.
   b. Spray your plants with chemicals to remove all bacteria and fungi.
   c. Remove all your plants and replace the soil.
   d. Do nothing because many of these bacteria and fungi may be beneficial.

5. Eating unscrubbed cassava root would likely
   a. lead to indigestion because the skin of the cassava plant is very difficult to digest
   b. make you sick because the soil on the surface may contain harmful microbes
   c. make you sick because the skin contains cyanogenic glycosides that would produce cyanide in the digestive track
   d. harm your teeth because of the small stones that would be on the surface of the root

6. You decide to plant a garden with a beautiful black walnut at one end and a majestic white oak at the other end. You are quite disappointed, however, when none of the seeds you plant around the walnut tree grow. What might explain this observation?
   a. The walnut tree filters out too much light, so the seeds fail to germinate.
   b. The roots of the walnut tree deplete all of the nutrients from the soil, so the new seedlings starve.
   c. The walnut tree produces chemical toxins that prevent seed germination.
   d. The roots of the walnut tree deplete all of the water from the soil and thereby prevent seed germination.

7. Vegetarians whose diet consists largely of soy are less likely to develop prostate cancer because
   a. Soy contains the anticancer drug taxol.
   b. Eating meat increases the probability of developing prostate cancer, so eliminating it from your diet reduces the chances of developing the disease.
   c. Soy protein prevents accumulation of the prostate-specific antigen (PSA) associated with prostate cancer.
   d. Soy contains a phytoestrogen that may down-regulate estrogen and androgen receptors in males who consume diets high in soy.

8. Tonic water, containing quinine, was developed primarily
   a. as an enjoyable beverage
   b. to help fight malaria
   c. to reduce the risk of cancer
   d. as a weak sedative

9. Some plants have developed a mutualistic relationship with parasitoid wasps. This mutualistic relationship would not occur if
   a. the plant quit producing nectar for the wasp
   b. the wasp ceased to live on the plant
   c. the plant quit producing volatile compounds that attract the wasp
   d. the plant attracted too many caterpillars

10. Tomato leaves are not good to eat because
    a. when their tissue is damaged, it generates a foul odor that makes people sick
    b. they do not contain any useful nutrients for animals
    c. they contain chemical toxins that will make animals sick
    d. when their tissue is damaged, they produce proteinase inhibitors that will prevent digestive enzyme function in animals

11. A plant lacking *R* genes would likely
    a. be unable to carry out photosynthesis
    b. be susceptible to infection by pathogens
    c. be susceptible to predation by herbivores
    d. paralyze animals who ingested it

12. If a pathogen contains an *avr* gene not recognized by a plant, the plant will most likely
    a. develop a disease
    b. eliminate the pathogen because it is unrecognized
    c. develop proteinase inhibitors
    d. develop a different *R* gene

13. Both plant and animal immune systems can
    a. develop memory of past pathogens to more effectively deal with subsequent infections
    b. initiate expression of proteins to help fight the infection
    c. kill their own cells to prevent spread of the infection
    d. all of the above

14. In the hypersensitive response, plants might
    a. quickly kill their own cells at the site of infection
    b. release hydrogen peroxide and nitric oxide
    c. release antimicrobial chemicals
    d. all of the above

CHALLENGE QUESTIONS

1. Herbal medicine has long relied on plant extracts to cure human disease. Find three examples of herbal remedies that have been exploited by the pharmaceutical industry in modern drug production.

2. Diagram the events surrounding infection of a plant by a fungal pathogen. Include strategies used both by the pathogen and by the host plant.

3. If you wanted to find new phytopharmaceuticals that could fight cancer, how would you go about identifying new compounds?

4. Herbivory and infestation by pests and pathogens are detrimental to plants. Many plants have developed toxins, or secondary compounds, to harm or deter these activities. Why don't all plants do this, and why don't plants develop a whole arsenal of defenses?

chapter 41

Sensory Systems in Plants

introduction

ALL ORGANISMS SENSE AND INTERACT with their environments. This is particularly true of plants. Plant survival and growth are critically influenced by abiotic factors, including water, wind, and light. The effect of the local environment on plant growth also accounts for much of the variation in adult form within a species. In this chapter, we explore how a plant senses such factors and transduces these signals to elicit an optimal physiological, growth, or developmental response. Although responses can be observed on a macroscopic scale, the mechanism of response occurs at the level of the cell. Signals are perceived when they interact with a receptor molecule, causing a shape change and altering the receptor's ability to interact with signaling molecules. Hormones play an important role in the internal signaling that brings about environmental responses and are keyed in many ways to the environment.

Responses to Light

In chapter 8 we covered the details of photosynthesis, the process by which plants convert light energy into chemical bond energy. We described pigments, molecules that are capable of absorbing light energy; you learned that chlorophylls are the primary pigment molecules of photosynthesis. Plants contain other pigments as well, and one of the functions of these other pigments is to detect light and to mediate plants' response to light by passing on information.

Several environmental factors, including light, can initiate seed germination, flowering, and other critical developmental events in the life of a plant. **Photomorphogenesis** is the term used for nondirectional, light-triggered development. It can result in complex changes in form, including flowering.

Unlike photomorphogenesis, **phototropisms** are directional growth responses to light. Both photomorphogenesis and phototropisms compensate for the plant's inability to walk away from unfavorable environmental conditions.

Phytochromes initiate signal transduction

The pigment-containing protein **phytochrome (P)** consists of two parts: a smaller part that is sensitive to light, called the *chromophore*, and a larger portion called the *apoprotein* (figure 41.1). The apoprotein initiates a signal transduction pathway leading to a particular biological response.

Phytochrome is present in all groups of plants and in a few genera of green algae, but not in other protists, bacteria, or fungi. Phytochrome systems probably evolved among the green algae and were present in the common ancestor of the plants.

The phytochrome molecule exists in two interconvertible forms: The first form, P_r, absorbs red light at 660 nm wavelength; the second, P_{fr}, absorbs far-red light at 730 nm. P_r is biologically inactive; it is converted into P_{fr}, the active form, when red photons are present. P_{fr} is converted back into P_r when far-red photons are available. In other words, biological reactions that are affected by phytochrome occur when P_{fr} is present. When most of the P_{fr} has been replaced by P_r, the reaction will not occur (figure 41.2).

Chlorophyll also absorbs red light, but it is not a receptor like phytochrome. Unlike receptors that transduce information, chlorophyll transduces energy.

The amount of P_{fr} is also regulated by degradation. Ubiquitin is a protein that tags P_{fr} for transport to the **proteasome**, a protein shredder composed of 28 proteins. The proteasome

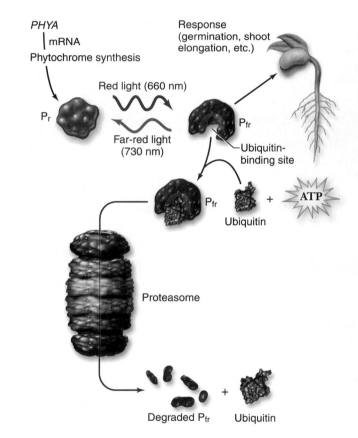

figure 41.2

HOW PHYTOCHROME WORKS. *PHYA* is one of the five *Arabidopsis* phytochrome genes. When exposed to red light, P_r changes to P_{fr}, the active form that elicits a response in plants. P_{fr} is converted to P_r when exposed to far-red light. The amount of P_{fr} is regulated by protein degradation. The protein ubiquitin tags P_{fr} for degradation in the proteasome.

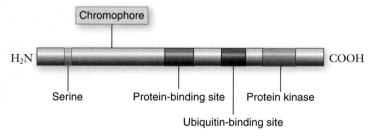

figure 41.1

PHYTOCHROME. Different parts of the phytochrome molecule have distinct roles in light regulation of growth and development. Phytochrome changes conformation when the chromophore responds to relative amounts of red and far-red light. The shape change affects the ability of phytochrome to bind to other proteins that participate in the signaling process. The ubiquitin-binding sites allow for degradation, and the protein kinase domain allows for further signaling via phosphorylation.

has a channel in the center, and as proteins pass through, they are clipped into amino acids that can be used to build other proteins. The process of tagging and recycling P_{fr} is precisely regulated to maintain needed amounts of phytochrome in the cell.

Although we often refer to phytochrome as a single molecule here, several different phytochromes have been identified that appear to have specific functions. In *Arabidopsis* five forms of phytochrome, PHYA to PHYE, have been characterized, each playing overlapping but distinct roles in the light regulation of growth and development.

Many growth responses are linked to phytochrome action

Phytochrome is involved in a number of plant growth responses, including seed germination, shoot elongation, and detection of plant spacing.

Seed germination

Seed germination is inhibited by far-red light and stimulated by red light in many plants. Because chlorophyll absorbs red light strongly but does not absorb far-red light, light filtered through the green leaves of canopy trees above a seed contains a reduced amount of red light. The far-red light inhibits seed germination by converting P_{fr} into the biologically inactive P_r form.

Consequently, seeds on the ground under deciduous plants, which lose their leaves in winter, are more apt to germinate in the spring after the leaves have decomposed and the seeds are exposed to direct sunlight and a greater amount of red light. This adaptation greatly improves the chances that seedlings will become established before leaves on taller plants shade the seedlings and reduce sunlight available for photosynthesis.

Shoot elongation

Elongation of the shoot in an etiolated seedling (one that is pale and slender from having been kept in the dark) is caused by a lack of red light. The morphology of such plants becomes normal when they are exposed to red light, increasing the amount of P_{fr}.

Etiolation is an energy conservation strategy to help plants growing in the dark reach the light before they die. They don't green up until light becomes available, and they divert energy to internode elongation. This strategy is useful for seedlings when they have sprouted underground or under leaf cover.

The de-etiolated (*det2*) *Arabidopsis* mutant has a poor etiolation response; seedlings fail to elongate in the dark (figure 41.3). The *det2* mutants are defective in an enzyme necessary for biosynthesis of a brassinosteroid hormone, leading researchers to suspect that brassinosteroids play a role in plant responses to light through phytochrome. (Brassinosteroids and other hormones are discussed later in this chapter.)

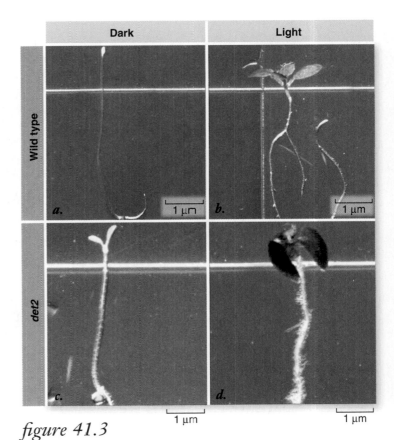

figure 41.3

ETIOLATION IS REGULATED BY LIGHT AND THE *DET2* GENE IN *Arabidopsis*. *DET2* is needed for etiolation in dark grown plants.

Detection of plant spacing

Red and far-red light also signal plant spacing. Again, leaf shading increases the amount of far-red light relative to red light. Plants somehow measure the amount of far-red light bounced back to them from neighboring plants. The closer together plants are, the more far-red relative to red light they perceive and the more likely they are to grow tall, a strategy for outcompeting others for sunshine. If their perception is distorted by putting a light-blocking collar around the stem, the elongation response no longer occurs.

P_{fr} facilitates expression of light-response genes

Over 2500 genes, 10% of the *Arabidopsis* genome, are involved in biological responses that begin with a conformational change in one of the phytochromes in response to red light. Phytochromes are involved in numerous signaling pathways that lead to gene expression. Some pathways also involve protein kinases or G proteins (described in chapter 8).

Phytochrome is found in the cytoplasm, but enters the nucleus to facilitate transcription of light response genes. When P_r is converted to P_{fr}, it can move into the nucleus. Once

figure 41.4

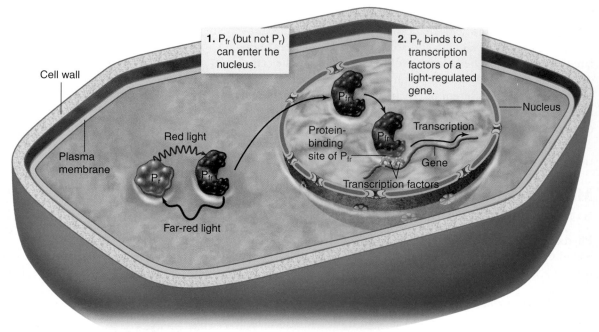

1. P_{fr} (but not P_r) can enter the nucleus.

2. P_{fr} binds to transcription factors of a light-regulated gene.

in the nucleus, P_{fr} binds with other proteins that form a transcription complex, leading to the expression of light-regulated genes (figure 41.4). Phytochrome's protein-binding site (see figure 41.1) is essential for interactions with proteins that serve as transcription factors.

Phytochrome also works through protein kinase-signaling pathways. When phytochrome converts to the P_{fr} form, its protein kinase domain may phosphorylate a serine and connect it to the amino (N) terminus of the phytochrome itself (autophosphorylation), or it may phosphorylate the serine of another protein involved in light signaling (figure 41.5). Phosphorylation initiates a signaling cascade that can activate transcription factors and lead to the transcription of light-regulated genes.

Although phytochrome is involved in multiple signaling pathways, it does not directly initiate the expression of that 10% of the *Arabidopsis* plant genome. Rather, phytochrome initiates expression of master regulatory genes that manage the complex interactions leading to photomorphogenesis and phototropisms. Gene expression is just the first step, with hormones playing important roles as well.

inquiry

? *You are given seed of a plant with a mutation in the protein kinase domain of phytochrome. Would you expect to see any red-light–mediated responses when you germinate the seed? Explain your answer.*

Light affects directional growth

Phototropisms, directional growth responses, contribute to the variety of overall plant shape we see within a species as shoots grow toward light. Tropisms are particularly intriguing because they challenge us to connect environmental signals with cellular perception of the signal, transduction into biochemical pathways, and ultimately an altered growth response.

figure 41.5

In this example,
signaling leads to the
release of a transcription
factor from a protein
complex.

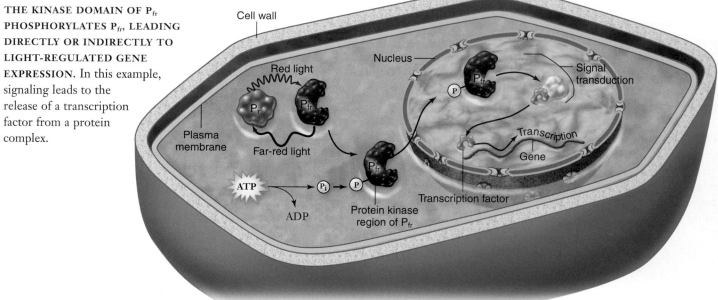

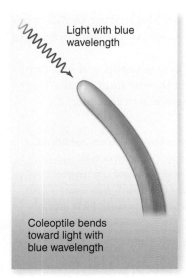
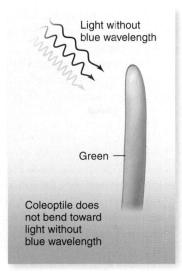

figure 41.6

PHOTOTROPISM. Oat coleoptiles growing toward light with blue wavelengths. Colors indicate the color of light shining on coleoptiles. Arrows indicate the direction of light.

Positive phototropism in stems

Phototropic responses include the bending of growing stems and other plant parts toward sources of light with blue wavelengths (460-nm range) (figure 41.6). In general, stems are positively phototropic, growing toward a light source, but most roots do not respond to light, or in exceptional cases, exhibit only a weak negative phototropic response.

The phototropic reactions of stems are clearly of adaptive value, giving plants greater exposure to available light. They are also important in determining the development of plant organs and, therefore, the appearance of the plant. Individual leaves may also display phototropic responses; the position of leaves is important to the photosynthetic efficiency of the plant. A plant

hormone called *auxin*, discussed in a later section, is probably involved in most, if not all, of the phototropic growth responses of plants.

Blue-light receptors

The recent identification of blue-light receptors in plants is leading to exciting discoveries of how the light signal can ultimately be connected with a phototropic response. A blue-light receptor **phototropin 1 (PHOT1)** was identified through the characterization of a nonphototropic mutant.

The phot1 protein has two light-sensing regions, and they change conformation in response to blue light. This change activates another region of the protein that is a kinase. Both PHOT1 and a similar receptor, PHOT2, are receptor kinases unique to plants. A portion of PHOT1 is a kinase that autophosphorylates (figure 41.7). Currently, only the early steps in this signal transduction are understood. It will be intriguing to watch the story of the phot1 signal transduction pathway unfold, leading to an explanation of how plants grow toward the light.

Circadian clocks are independent of light but are entrained by light

Although shorter and much longer naturally occurring rhythms also exist, **circadian rhythms** ("around the day") are particularly common and widespread among eukaryotic organisms. They relate the day–night cycle on Earth, although they are not exactly 24 hr in duration.

Jean de Mairan, a French astronomer, first identified circadian rhythms in 1729. He studied the sensitive plant (*Mimosa pudica*), which closes its leaflets and leaves at night. When de Mairan put the plants in total darkness, they continued "sleeping" and "waking" just as they had when exposed to night and day. This is one of four characteristics of a circadian rhythm—it must continue to run in the absence of external inputs. Plants with a circadian rhythm do not actually have to be experiencing a pattern of daylight and darkness for their cycle to occur.

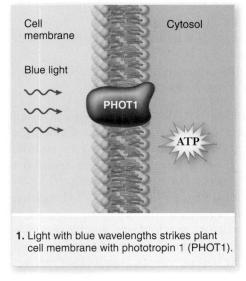
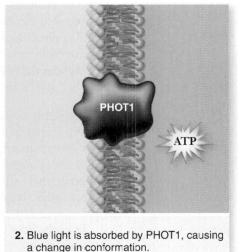
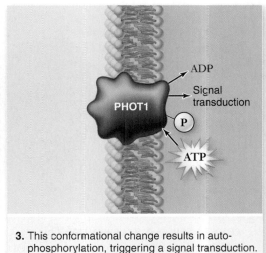

1. Light with blue wavelengths strikes plant cell membrane with phototropin 1 (PHOT1).

2. Blue light is absorbed by PHOT1, causing a change in conformation.

3. This conformational change results in autophosphorylation, triggering a signal transduction.

figure 41.7

BLUE-LIGHT RECEPTOR. Blue light activates the light-sensing region of PHOT1, which in turn stimulates the kinase region of PHOT1 to autophosphorylate. This is just the first step in a signal transduction pathway that leads to phototropic growth.

In addition, a circadian rhythm must be about 24 hr in duration, and the cycle can be reset or entrained. Although plants kept in darkness will continue the circadian cycle, the cycle's period may gradually move away from the actual day–night cycle, becoming desynchronized. In the natural environment, the cycle is entrained to a daily cycle through the action of phytochrome and blue-light photoreceptors.

Other eukaryotes, including humans, have circadian rhythms, and perhaps you have experienced jet lag when you traveled by airplane across a few time zones. Recovery from jet lag involves entrainment to the new time zone.

Another characteristic of a circadian cycle is that the clock can compensate for differences in temperature, so that the duration remains unchanged. This characteristic is unique, considering what we know about biochemical reactions, because most rates of reactions vary significantly based on temperature. Circadian clocks exist in many organisms, and they appear to have evolved independently multiple times.

The reversible circadian rhythm changes in leaf movements are typically brought about by alteration of cells' turgor pressure; we describe these changes in a later section.

> Although plants cannot move away or toward optimal conditions, they can grow and develop in response to environmental signals. Red light changes the shape of phytochrome and can trigger photomorphogenesis. Phototropisms are growth responses toward a unidirectional, often blue, light source. Circadian clocks are endogenous timekeepers that keep plant movements and other responses synchronized with the environment.

41.2 Responses to Gravity

When a potted plant is tipped over and left in place, the shoot bends and grows upward. The same thing happens when a storm pushes plants over in a field. These are examples of **gravitropism,** the response of a plant to the gravitational field of the Earth (figure 41.8; see also chapter opener). Because plants also grow in response to light, separating out phototropic effects is important in the study of gravitropism.

figure 41.8

PLANT RESPONSE TO GRAVITY. This plant was placed horizontally and allowed to grow for seven days. Note the negative gravitational response of the shoot.

inquiry

Where would you expect to find the highest concentration of auxin?

Plants align with the gravitational field: An overview

Gravitropic responses are present at germination, when the root grows down and the shoot grows up. Why does a shoot have a negative gravitropic response (growth away from gravity), while a root has a positive gravitropic response? Auxins play a primary role in gravitropic responses, but they may not be the only way gravitational information is sent through the plant.

The opportunity to experiment on the Space Shuttle in a gravity-free environment has accelerated research in this area. Analysis of gravitropic mutants is also adding to our understanding of gravitropism. Investigators propose that four general steps lead to a gravitropic response:

1. Gravity is perceived by the cell.
2. A mechanical signal is transduced into a physiological signal in the cell that perceives gravity.
3. The physiological signal is transduced inside the cell and externally to other cells.
4. Differential cell elongation occurs, affecting cells in the "up" and "down" sides of the root or shoot.

Currently researchers are debating the steps involved in perception of gravity. In shoots, gravity is sensed along the length of the stem in the endodermal cells that surround the vascular tissue (figure 41.9*a*), and signaling occurs toward the outer epidermal cells. In roots, the cap is the site of gravity perception, and a signal must trigger differential cell elongation and division in the elongation zone (figure 41.9*b*).

In both shoots and roots, **amyloplasts,** plastids that contain starch, sink toward the center of the gravitational field and thus may be involved in sensing gravity. Amyloplasts interact with the cytoskeleton. Auxin evidently plays a role in transmitting a signal from the gravity-sensing cells that contain amyloplasts and the site where growth occurs. The link between amyloplasts and auxin is not fully understood.

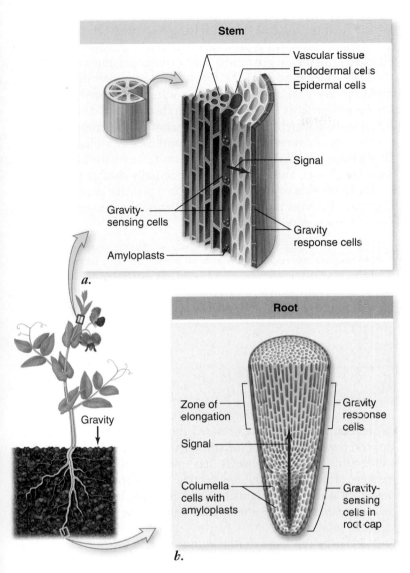

figure 41.9

SITES OF GRAVITY SENSING AND RESPONSE IN ROOTS AND SHOOTS.

Stems bend away from a center of gravity

Increased auxin concentration on the lower side in stems causes the cells in that area to grow more than the cells on the upper side. The result is a bending upward of the stem against the force of gravity—in other words, a *negative gravitropic response*. Such differences in hormone concentration have not been as well documented in roots. Nevertheless, the upper sides of roots oriented horizontally grow more rapidly than the lower sides, causing the root ultimately to grow downward; this phenomenon is known as *positive gravitropic response*.

Two *Arabidopsis* mutants, *scarecrow* (*scr*) and *short root* (*shr*), were initially identified by aberrant root phenotypes, but they also affect shoot gravitropism (figure 41.10). Both genes are needed for normal endodermal development (see figure 36.16). Without a fully functional endodermis, stems lack a normal gravitropic response. These endodermal cells carry amyloplasts in the stems, and in the mutants, stem endodermis fails to differentiate and produce gravity-sensing amyloplasts.

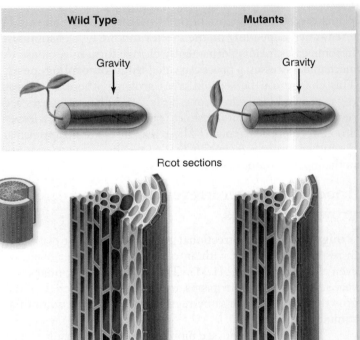

figure 41.10

AMYLOPLASTS IN STEM ENDODERM IS NEEDED FOR GRAVITROPISMS. *a.* The *scr* and *shr* mutants of *Arabidopsis* have abnormal root development because they lack a fully differentiated endodermal layer. *b.* The endodermal defect extends into the stem, eliminating the positive gravitropic response of wild-type stems.

Roots bend toward a center of gravity

In roots, the gravity-sensing cells are located in the root cap, and the cells that actually undergo asymmetrical growth are in the distal elongation zone, which is closest to the root cap. How the information is transferred over this distance is an intriguing question. Auxin may be involved, but when auxin transport is suppressed, a gravitropic response still occurs in the distal elongation zone. Some type of electrical signaling involving membrane polarization has been hypothesized, and this idea was tested aboard the Space Shuttle. So far, the verdict is still not in on the exact mechanism.

The growing number of auxin mutants in roots do confirm that auxin has an essential role in root gravitropism, even if it may not be the long-distance signal between the root cap and the elongation zone. Mutations that affect both auxin influx and efflux can eliminate the gravitropic response by altering the directional transport of this hormone.

It may surprise you to learn that in tropical rain forests, the roots of some plants may grow up the stems of neighboring plants, instead of exhibiting the normal positive gravitropic responses typical of other roots. It appears that rainwater dissolves nutrients, both while passing through the lush upper canopy of the forest, and subsequently while trickling down the tree trunks. This water is a more reliable source of nutrients for the roots than the nutrient-poor rain forest soils in which the plants are anchored. Explaining this observation in terms of current hypotheses is a challenge. It has been proposed that roots are more sensitive to auxin than are shoots, and that auxin may actually inhibit growth on the lower side of a root, resulting in a positive gravitropic response. Perhaps in these tropical plants, the sensitivity to auxin in roots is reduced.

> Gravitropism, the response of a plant to gravity, generally causes shoots to grow up (negative gravitropism) and roots to grow down (positive gravitropism).

41.3 Responses to Mechanical Stimuli

Plants respond to touch and other mechanical stimuli in different ways, depending on the species and the type of stimulus. In some cases, plants permanently change form in response to mechanical stresses, a process termed **thigmomorphogenesis.** This change can be seen in trees growing where an almost constant wind blows from one direction. Other responses are reversible and occur in the short term, as when mimosa leaves droop in response to touch. These responses are not tropisms, but rather turgor movements that come about due to changes in the internal water pressure of cells.

Touch can trigger irreversible growth responses

A **thigmotropism** is directional growth of a plant or plant part in response to contact with an object, animal, other plant, or even the wind (figure 41.11). **Thigmonastic responses** are very similar to thigmotropisms, except that the direction of the growth response is the same regardless of the direction of the stimulus.

Tall, slender plants are more likely to snap during a wind or rain storm than are plants with short, wide internodes. Environmental signals such as regularly occurring winds or the rubbing of one plant against another are sufficient to induce morphogenetic change leading to thicker, shorter internodes. In some cases, even repeated touching of a plant with a finger is enough to cause a change in plant growth.

Tendrils are modified stems that some species use to anchor themselves in the environment. When a tendril makes contact with an object, specialized epidermal cells perceive the contact and promote uneven growth, causing the tendril to curl around the object, sometimes within only 3 to 10 min. Two hormones, auxin and ethylene, appear to be involved in tendril movements, and they can induce coiling even in the absence of any contact stimulus. Curiously, the tendrils of some species coil toward the site of the stimulus (thigmotropic growth), while tendrils of other species may always coil clockwise, regardless of the side of the tendril that makes contact with an object. In some other plants, such as clematis, bindweed, and dodder, leaf petioles or unmodified stems twine around other stems or solid objects.

Perhaps the most dramatic touch response is the snapping of a Venus flytrap. As discussed in chapter 39, the modi-

figure 41.11

THIGMOTROPISM. The thigmotropic response of these twining stems causes them to coil around the object with which they have come in contact.

fied leaves of the flytrap close in response to a touch stimulus, trapping insects or other potential sources of protein. A flytrap can shut in a mere 0.5 sec. The enlarged epidermal or mesophyll cells of the flytrap cause the trap to close. The speed of trap closure is enhanced by the shape of the leaf, which flips between a concave and convex form.

What is particularly amazing about this response is that the outer cells actually grow. The cell walls may soften in response to an electrical signal that moves through the leaf when the trigger hairs are touched, and the high pressure (turgor) of the water inside the cells pushes against the softened walls to enlarge the cell. This growth mechanism is distinct from other turgor movements (to be discussed shortly) because the water is already within the cell, not transferred into it in response to the electrical signal.

If digestible prey is caught, the trap will open about 24 hr later through the growth of inner cells of the flytrap. This growth response can only be triggered about four times before the leaf dies, presumably because so much energy is required and the individual flytrap leaf runs out of energy.

Arabidopsis is proving valuable as a model system to explore plant responses to touch. A gene has been identified that is expressed in 100-fold higher levels 10 to 30 min after touch. The gene codes for a calmodulin-like protein that binds Ca^{2+}, which is involved in a number of plant physiological processes. Given the value of a molecular genetics approach in dissecting the pathways leading from an environmental signal to a growth response, the touch gene provides a promising first step in understanding how plants respond to touch.

Reversible responses to touch and other stimuli involve turgor pressure

Unlike tropisms, some touch-induced plant movements are not based on growth responses, but instead result from reversible changes in the turgor pressure of specific cells. Turgor, as described in chapter 38, is pressure within a living cell resulting from diffusion of water into it. If water leaves turgid cells, the cells may collapse, causing plant movement; conversely, water entering a limp cell may also cause movement as the cell once more becomes turgid.

Many plants, including those of the legume family (Fabaceae), exhibit leaf movements in response to touch or other stimuli. After exposure to a stimulus, the changes in leaf orientation are mostly associated with rapid turgor pressure changes in **pulvini** (singular, pulvinus), two-sided multicellular swellings located at the base of each leaf or leaflet. When leaves with pulvini, such as those of the sensitive plant (*Mimosa pudica*), are stimulated by wind, heat, touch, or in some instances, intense light, an electrical signal is generated. The electrical signal is translated into a chemical signal, with potassium ions followed by water migrating from the cells in one-half of a pulvinus to the intercellular spaces in the other half.

The loss of turgor in half of the pulvinus causes the leaf to "fold." The movements of the leaves and leaflets of a sensitive plant are especially rapid; the folding occurs within a second or two after the leaves are touched (figure 41.12). Over a span of about 15 to 30 min after the leaves and leaflets have folded,

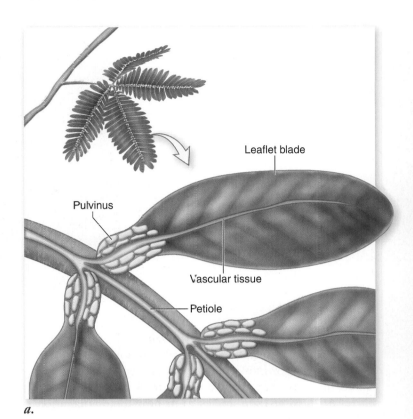

a.

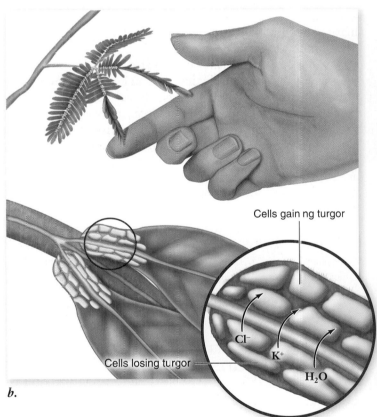

b.

figure 41.12

SENSITIVE PLANT (*Mimosa pudica*). *a.* The blades of *Mimosa* leaves are divided into numerous leaflets; at the base of each leaflet is a swollen structure called a pulvinus. *b.* Changes in turgor cause leaflets to fold in response to a stimulus. When leaves are touched (center two leaves), ions move to the outer side of the pulvinus, water follows by osmosis, and the decreased interior turgor pressure leads to folding.

figure 41.13

HELIOTROPISM. These sunflowers track the movement of the Sun every day.

water usually diffuses back into the same cells from which it left, and the leaf returns to its original position.

Some turgor movements are triggered by light. For example, the leaves of some plants may track the Sun, with their blades oriented at right angles to it; how their orientation is directed is, however, poorly understood. Such leaves can move quite rapidly (as much as 15 degrees an hour). This movement maximizes photosynthesis and is analogous to solar panels designed to track the Sun (figure 41.13).

Some of the most familiar reversible changes due to turgor pressure are the circadian rhythms seen in leaves and flowers that open during the day and close at night, or vice versa. For example, the flowers of four o'clocks open at 4 P.M., and evening primrose petals open at night. As described earlier, sensitive plant leaves also close at night. Bean leaves are horizontal during the day when their pulvini are turgid, but become more or less vertical at night as the pulvini lose turgor (figure 41.14). These sleep movements reduce water loss from transpiration during the night, but maximize photosynthetic surface area during the day.

Thigmotropic and thigmonastic movements are growth responses of plants to contact. Turgor movements of plants are reversible and involve changes in the turgor pressure of specific cells.

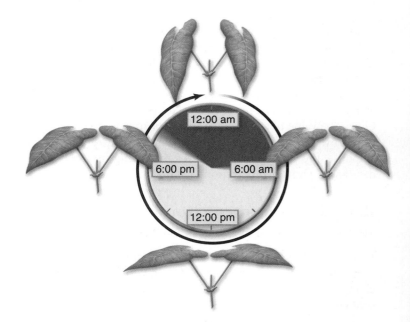

figure 41.14

SLEEP MOVEMENTS IN BEAN LEAVES. In the bean plant, leaf blades are oriented horizontally during the day and vertically at night.

41.4 Responses to Water and Temperature

Sometimes, modifying the direction of growth is not enough to protect a plant from harsh conditions. The ability to cease growth and go into a dormant stage when conditions become unfavorable, such as during seasonal changes in temperate climates, provides a survival advantage. The extreme example is seed dormancy, but there are intermediate approaches to waiting out the bad times as well.

Plants also have developed adaptations to more short-term fluctuations in temperature, such as might occur during a heat wave or cold snap. These strategies include changes in membrane composition and the production of heat shock proteins.

Dormancy is a response to water, temperature, and light

In temperate regions, we generally associate dormancy with winter, when freezing temperatures and the accompanying unavailability of water make it impossible for most plants to grow.

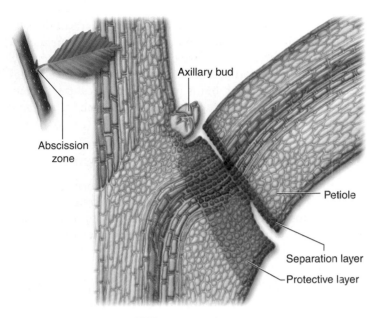

figure 41.15

LEAF ABSCISSION. Hormonal changes in the leaf's abscission zone cause abscission. Two layers of cells in the abscission zone differentiate into a protective layer and a separation layer. As pectins in the separation layer break down, wind and rain can easily separate the leaf from the stem.

During this season, buds of deciduous trees and shrubs remain dormant, and apical meristems remain well protected inside enfolding scales. Perennial herbs spend the winter underground, existing as stout stems or roots packed with stored food. Many other kinds of plants, including most annuals, pass the winter as seeds. Often dormancy begins with the dropping of leaves, which you have probably seen occur in deciduous trees in the autumn.

Organ abscission

Deciduous leaves are often shed as the plant enters dormancy. The process by which leaves or petals are shed is called **abscission.**

Abscission can be useful even before dormancy is established. For example, shaded leaves that are no longer photosynthetically productive can be shed. Petals, which are modified leaves, may senesce once pollination occurs. Orchid flowers remain fresh for long periods of time, even in a florist shop; however, once pollination occurs, a hormonal change is triggered that leads to petal senescence. This strategy makes sense in terms of allocation of energy resources because the petals are no longer necessary to attract a pollinator. One advantage of organ abscission, therefore, is that nutrient sinks can be discarded, conserving resources.

On a larger scale, deciduous plants in temperate areas produce new leaves in the spring and then lose them in the fall. In the tropics, however, the production and subsequent loss of leaves in some species is correlated with wet and dry seasons. Evergreen plants, such as most conifers, usually have a complete change of leaves every two to seven years, periodically losing some but not all of their leaves.

Abscission involves changes that take place in an *abscission zone* at the base of the petiole (figure 41.15). Young leaves produce hormones (especially cytokinins) that inhibit the de-

velopment of specialized layers of cells in this zone. Hormonal changes take place as the leaf ages, however, and two layers of cells become differentiated. A *protective layer,* which may be several cells wide, develops on the stem side of the petiole base. These cells become impregnated with suberin, which you may recall is a fatty substance impervious to moisture. A *separation layer* develops on the leaf-blade side; the cells of the separation layer sometimes divide, swell, and become gelatinous.

When temperatures drop, when the duration and intensity of light diminishes, or when other environmental changes occur, enzymes break down the pectins in the middle lamellae of the separation cells. Wind and rain can then easily separate the leaf from the stem. Left behind is a sealed leaf scar that is protected from invasion by bacteria and other disease organisms.

As the abscission zone develops, the green chlorophyll pigments present in the leaf break down, revealing the yellows and oranges of other pigments, such as carotenoids, that previously had been masked by the intense green colors. At the same time, water-soluble red or blue pigments called *anthocyanins* and *betacyanins* may also accumulate in the vacuoles of the leaf cells—all contributing to an array of fall colors in leaves (figure 41.16).

Seed dormancy

The extraordinary evolutionary innovation of the seed plants is the dormant seed that allows plant offspring to wait until conditions for germination are optimal. Sometimes the seeds can

figure 41.16

LEAF COLOR CHANGES DURING ABSCISSION.

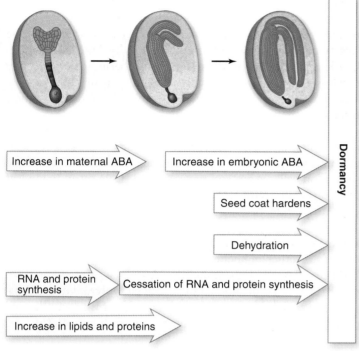

figure 41.17

SEED DORMANCY. Accummulating food reserves, forming a protective seed coat, and dehydration are essential steps leading to dormancy. Abscisic acid (ABA) from both maternal and embryonic tissue is necessary for dormancy.

endure a wait of hundreds of years (figure 41.17). In seasonally dry climates, seed dormancy occurs primarily during the dry season, often the summer. Rainfalls trigger germination when conditions for survival are more favorable.

Annual plants occur frequently in areas of seasonal drought. Seeds are ideal for allowing annual plants to bypass the dry season, when there is insufficient water for growth. When it rains, these seeds can germinate, and the plants can grow rapidly, having adapted to the relatively short periods when water is available.

Chapter 37 covered some of the mechanisms involved in breaking seed dormancy and allowing germination under favorable circumstances. These include water leaching away the chemicals that inhibit germination or mechanically cracking the seed coats due to osmotic swelling, a procedure particularly suitable for promoting growth in seasonally dry areas.

Seeds may remain dormant for a surprisingly long time. Many legumes have tough seeds that are virtually impermeable to water and oxygen. These seeds often last decades and even longer without special care; they will eventually germinate when their seed coats have been cracked and water is available. Seeds that are thousands of years old have been successfully germinated!

Favorable temperatures, day length, and amounts of water can release buds, underground stems and roots, and seeds from a dormant state. Requirements vary among species. For example, some weed seeds germinate in cooler parts of the year and are inhibited from germinating by warmer temperatures. Day length differences can have dramatic effects on dormancy. For example, tree dormancy is common in temperate climates when the days are short, but is unusual in tropical trees growing near the equator, where day length remains about the same regardless of season.

Plants can survive temperature extremes

Sometimes temperatures change rapidly, and dormancy is not possible. How do plants survive temperature extremes? A number of adaptations, including some rapid response strategies, help plants overcome sudden chilling or extreme heat.

Chilling

Knowing the lipid composition of a plant's membranes can help predict whether the plant will be sensitive or resistant to chilling. Saturated lipids solidify at a higher temperature because they pack together more closely (chapter 5), so the more unsaturated the membrane lipids are, the more resistant the plant is to chilling. *Arabidopsis* plants genetically modified to contain a higher percentage of saturated fatty acids have proved to be more sensitive to chilling.

When chilling occurs, the enzyme desaturase converts the single bonds in the saturated lipids to double bonds. This process lowers the temperature at which the membrane becomes rigid and cannot function properly.

Even highly unsaturated membranes are not enough to protect plants from freezing temperatures. At freezing, ice crystals form and the cells die from dehydration—not enough liquid water is available for metabolism. Some plants, however, have the ability to undergo deep **supercooling** and survive temperatures as low as −40°C. Supercooling occurs when ice crystal formation is limited, and the crystals occur in extracellular spaces where they cannot damage cell organelles. Furthermore, the cells of these plants must be able to withstand gradual dehydration.

Acquiring tolerance to chilling or freezing as the temperature drops can be explained by increased solute concentration. In addition, antifreeze proteins prevent ice crystals from forming. Ice crystals can also form (nucleate) around bacteria naturally found on the leaf surface. Some bacteria have been genetically engineered so that they do not nucleate ice crystals. Spraying leaves with these modified bacteria can provide frost tolerance in some crops.

High temperatures

High temperatures can be harmful because proteins denature and lose their function when heated. If temperatures suddenly rise 5° to 10°C, **heat shock proteins (HSPs)** are produced. These proteins can stabilize other proteins so that they don't unfold or misfold at higher temperatures. In some cases, HSPs induced by temperature increases can also protect plants from other stresses, including chilling.

Plants can survive otherwise lethal temperatures if they are gradually exposed to increasing temperature. These plants have *acquired thermotolerance*. More is being learned about temperature acclimation by isolating mutants that fail to acquire thermotolerance, including the aptly named *hot* mutants in *Arabidopsis*. One of the *HOT* genes codes for an HSP. Characterization of other *HOT* genes indicates that thermotolerance requires more than the synthesis of HSPs; some *HOT* genes stabilize membranes and are necessary for protein activity.

Mature plants may become dormant in dry or cold seasons that are unfavorable for growth. Dormant plants usually lose their leaves and produce drought-resistant winter buds. Long unfavorable periods may be bypassed through the production of dormant seeds. Chilling and freezing acclimation rely on high levels of unsaturated fatty acids, supercooling, and synthesis of antifreeze proteins. HSPs stabilize proteins at high temperatures.

41.5 Hormones and Sensory Systems

Sensory responses that alter morphology rely on complex physiological networks. Many internal signaling pathways involve plant hormones, which are the focus of this section. Hormones are involved in responses to the environment, as well as in internally regulated development (see chapter 36).

The hormones that guide growth are keyed to the environment

Hormones are chemical substances produced in small, often minute quantities in one part of an organism and then transported to another part where they bring about physiological or developmental responses. How hormones act in a particular instance is influenced both by the hormone and the tissue that receives the message.

In animals, hormones are usually produced at definite sites, most commonly in organs such as glands. In plants, hormones are not produced in specialized tissues but, instead, in tissues that also carry out other, usually more obvious functions. Seven major kinds of plant hormones have been identified: auxin, cytokinins, gibberellins, brassinosteroids, oligosaccharins, ethylene, and abscisic acid (table 41.1, page 814). Current research is focused on the biosynthesis of hormones and on characterizing the hormone receptors involved in signal transduction pathways. Much of the molecular basis of hormone function remains enigmatic.

Because hormones are involved in so many aspects of plant function and development, we have chosen to integrate examples of hormone activity with specific aspects of plant biology throughout the text. In this section, our goal is to give a brief overview of these hormones.

Auxin allows elongation and organizes the body plan

More than a century ago, an organic substance known as **auxin** became the first plant hormone to be discovered. Auxin increases the plasticity of plant cell walls and is involved in elongation of stems. Cells can enlarge in response to changes in turgor pressure, but cell walls must be fairly plastic for this expansion to occur. Auxin plays a role in softening cell walls. The discovery of auxin and its role in plant growth is an elegant example of thoughtful experimental design and is recounted here for that reason.

Discovery of auxin

Later in life, the great evolutionist Charles Darwin became increasingly devoted to the study of plants. In 1881, he and his son Francis published a book called *The Power of Movement*

of Plants. In this book, the Darwins reported their systematic experiments on the response of growing plants to light—the responses that came to be known as phototropisms. They used germinating oat and canary grass seedlings in their experiments and made many observations in this field.

Charles and Francis Darwin knew that if light came primarily from one direction, seedlings would bend strongly toward it. If they covered the tip of a shoot with a thin glass tube, the shoot would bend as if it were not covered. However, if they used a metal foil cap to exclude light from the plant tip, the shoot would not bend (figure 41.18). They also found that

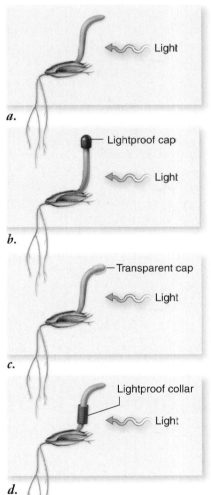

figure 41.18

THE DARWINS' EXPERIMENT. *a.* Young grass seedlings normally bend toward the light. *b.* The bending did not occur when the tip of a seedling was covered with a lightproof cap. *c.* Bending did occur when it was covered with a transparent one. *d.* When a collar was placed below the tip, the characteristic light response took place. From these experiments, the Darwins concluded that, in response to light, an "influence" that caused bending was transmitted from the tip of the seedling to the area below, where bending normally occurs.

		TABLE 41.1	Functions of the Major Plant Hormones		

TABLE 41.1 Functions of the Major Plant Hormones

Hormone		Major Functions	Where Produced or Found in Plant
Auxins		Promotion of stem elongation and growth; formation of adventitious roots; inhibition of leaf abscission; promotion of cell division (with cytokinins); inducement of ethylene production; promotion of lateral bud dormancy	Apical meristems; other immature parts of plants
Cytokinins		Stimulation of cell division, but only in the presence of auxin; promotion of chloroplast development; delay of leaf aging; promotion of bud formation	Root apical meristems; immature fruits
Gibberellins		Promotion of stem elongation; stimulation of enzyme production in germinating seeds	Roots and shoot tips; young leaves; seeds
Brassinosteroids		Overlapping functions with auxins and gibberellins	Pollen, immature seeds, shoots, leaves
Oligosaccharins		Pathogen defense, possibly reproductive development	Cell walls
Ethylene		Control of leaf, flower, and fruit abscission; promotion of fruit ripening	Roots, shoot apical meristems; leaf nodes; aging flowers; ripening fruits
Abscisic acid		Inhibition of bud growth; control of stomatal closure; some control of seed dormancy; inhibition of effects of other hormones	Leaves, fruits, root caps, seeds

figure 41.19

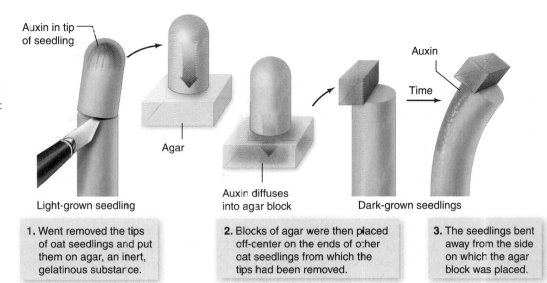

FRITS WENT'S EXPERIMENT.
Went concluded that a substance he named *auxin* promoted the elongation of the cells and that it accumulated on the side of an oat seedling away from the light.

Auxin in tip of seedling

Agar

Auxin diffuses into agar block

Auxin

Time

Light-grown seedling

Dark-grown seedlings

1. Went removed the tips of oat seedlings and put them on agar, an inert, gelatinous substance.

2. Blocks of agar were then placed off-center on the ends of other oat seedlings from which the tips had been removed.

3. The seedlings bent away from the side on which the agar block was placed.

using an opaque collar to exclude light from the stem below the tip did not keep the area above the collar from bending.

In explaining these unexpected findings, the Darwins hypothesized that when the shoots were illuminated from one side, they bent toward the light in response to an "influence" that was transmitted downward from its source at the tip of the shoot.

For some 30 years, the Darwins' perceptive experiments remained the sole source of information about this interesting phenomenon. Then the Danish plant physiologist Peter Boysen-Jensen and the Hungarian plant physiologist Arpad Paal independently demonstrated that the substance causing the shoots to bend was a chemical. They showed that if the tip of a germinating grass seedling was cut off and then replaced, with a small block of agar separating it from the rest of the seedling, the seedling would still grow as if there had been no change. Something evidently was passing from the tip of the seedling through the agar into the region where the bending occurred.

On the basis of these observations under conditions of either uniform illumination or darkness, Paal suggested that an unknown substance continually moves down from the tips of grass seedlings and promotes growth on all sides. Such a light pattern would not, of course, cause the shoot to bend.

inquiry

Propose a mechanism to explain how seedlings could bend in the light using what Paal discovered.

Then, in 1926, the Dutch plant physiologist Frits Went carried Paal's experiments a step further. Went cut off the tips of oat seedlings that had been illuminated normally and set these tips on agar. He then took oat seedlings that had been grown in the dark and cut off their tips in a similar way. Finally, Went cut tiny blocks from the agar on which the tips of the light-grown seedlings had been placed and placed them off-center on the tops of the decapitated dark-grown seedlings (figure 41.19). Even though these seedlings had not been exposed to the light themselves, they bent away from the side on which the agar blocks were placed.

As an experimental control, Went put blocks of pure agar on the decapitated stem tips and noted either no effect or a slight bending toward the side where the agar blocks were placed.

Finally, Went cut sections out of the lower portions of the light-grown seedlings. He placed these sections on the tips of decapitated, dark-green oat seedlings and again observed no effect.

As a result of his experiments, Went was able to show that the substance that had diffused into the agar from the tips of light-grown oat seedlings could make seedlings bend when they otherwise would have remained straight. He also showed that this chemical messenger caused the cells on the side of the seedling into which it flowed to grow more than those on the opposite side (figure 41.20). In other words, the chemical enhanced rather than retarded cell elongation. He named the substance that he had discovered *auxin*.

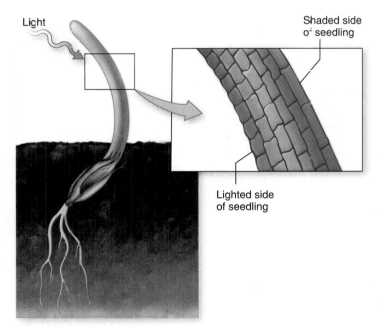

Light

Shaded side of seedling

Lighted side of seedling

figure 41.20

AUXIN CAUSES CELLS ON THE DARK SIDE TO ELONGATE.
Plant cells that are in the shade have more auxin and grow faster than cells on the lighted side, causing the plant to bend toward light. Further experiments showed exactly why there is more auxin on the shaded side of a plant.

Went's experiments provided a basis for understanding the responses that the Darwins had obtained some 45 years earlier. The oat seedlings bent toward the light because of differences in the auxin concentrations on the two sides of the shoot. The side of the shoot that was in the shade had more auxin, and its cells therefore elongated more than those on the lighted side, bending the plant toward the light.

The effects of auxin

Auxin acts to adapt the plant to its environment in a highly advantageous way by promoting growth and elongation. Environmental signals directly influence the distribution of auxin in the plant. How does the environment—specifically, light—exert this influence? Theoretically, light might destroy the auxin, might decrease the cells' sensitivity to auxin, or might cause the auxin molecules to migrate away from the light into the shaded portion of the shoot. This last possibility has proved to be the case.

In a simple but effective experiment, Winslow Briggs inserted a thin sheet of transparent mica vertically between the half of the shoot oriented toward the light and the half of the shoot oriented away from it (figure 41.21). He found that light from one side does not cause a shoot with such a barrier to bend. When Briggs examined the illuminated plant, he found equal auxin levels on both the light and dark sides of the barrier. He concluded that a normal plant's response to light from one direction involves auxin migrating from the light side to the dark side, and that the mica barrier prevented a response by blocking the migration of auxin.

The effects of auxin are numerous and varied. Auxin promotes the activity of the vascular cambium and the vascular tissues. Also, auxin is present in pollen in large quantities and plays a key role in the development of fruits. Synthetic auxins are used commercially for the same purpose. Fruits will normally not develop if fertilization has not occurred and seeds are not present, but frequently they will develop if auxin is applied. Pollination may trigger auxin release in some species, leading to fruit development even before fertilization has taken place.

How auxin works

In spite of this long history of research, auxin's molecular basis of action has been an enigma. The chemical structure of the most common auxin, **indoleacetic acid (IAA),** resembles that of the amino acid tryptophan, from which it is probably synthesized by plants (figure 41.22). Although other forms of auxin exist, IAA is the most common natural auxin.

An auxin binding protein (ABP1) was identified two decades ago. ABP1 is found in the cytoplasm and its role in auxin response is still unclear. Mutants that lack ABP1 do not make it past embryogenesis because cell elongation is inhibited and the basic body plan described in chapter 37 is not organized. But, the *abp1* mutant cells divide, which indicates that part of the auxin pathway is still functioning.

More recently, two families of proteins that mediate rapid, auxin-induced changes in gene expression have been identified: the auxin response factors (ARFs) and the Aux/IAA proteins. Transcription can be either enhanced or suppressed by ARFs which are

figure 41.21

PHOTOTROPISM AND AUXIN: THE WINSLOW BRIGGS EXPERIMENTS. Directional light causes the accumulation of auxin in the dark side of the shoot tip, which can move down the stem. Barriers inserted in the tip revealed that light affects auxin displacement at rather different rates of auxin production on the light and dark sides.

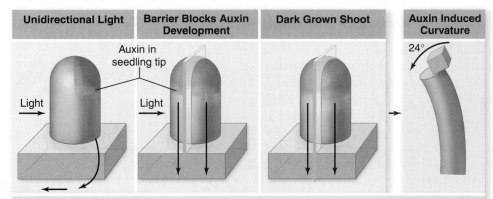

The same amount of total auxin is produced by a shoot tip grown with directional light, even when a barrier divides the shoot tip, and a shoot tip grown in the dark. All three blocks of agar cause the same amount of curvature in a tipless shoot.

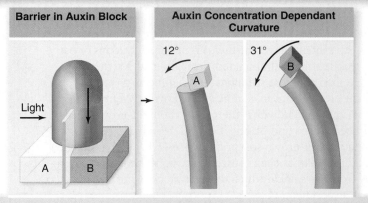

Separating the base of the shoot tip and the agar block results in two agar blocks with different concentrations of auxin that produce different degrees of curvature in tipless shoots.

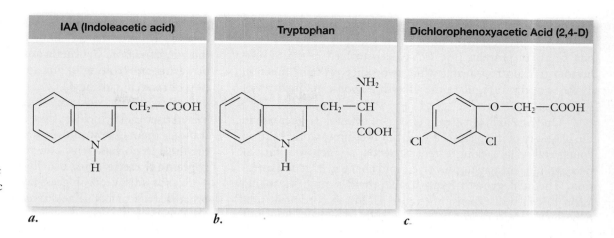

figure 41.22

AUXINS.
a. Indoleacetic acid (IAA), the principal naturally occurring auxin. *b.* Tryptophan, the amino acid from which plants probably synthesize IAA. *c.* Dichlorophenoxyacetic acid (2,4-D), a synthetic auxin, is a widely used herbicide.

known to bind DNA. The Aux/IAA proteins function a bit earlier in the auxin response pathway and have been shown to bind to and repress proteins that activate the expression of *ARF* genes.

ARF genes are activated when Aux/IAA proteins are degraded by ubiquitin tagging and protein degradation in the proteasome. Auxin binding to ARF protein is not sufficient to initiate gene expression in response to auxin signaling because of Aux/IAA repression of ARF activity. How then does a plant sense auxin and degrade Aux/IAA proteins?

The identification of the elusive auxin receptor in 2005 hints at how plants sense and respond to auxin. Auxin binds directly to a protein called the transport inhibitor response protein 1 (TIR1). TIR1 is the enigmatic auxin receptor. It is part of a protein complex known as SCF which is found throughout eukaryotes. SCF is short-hand for the three of the polypeptide subunits found in the complex. Auxin binds to TIR1 in the SCF complex if Aux/IAA

proteins are present. Once auxin binds, the SCF complex degrades the Aux/IAA proteins through the ubiquitin pathway.

Five steps lead from auxin perception to auxin-induced gene expression (figure 41.23)

1. Auxin binds TIR1 in the SCF complex if Aux/IAA is present.
2. The SCF complex tags Aux/IAA proteins with ubiquitin.
3. Aux/IAA proteins are degraded in the proteasome.
4. Aux/IAA proteins no longer bind and repress transcriptional activators of *ARF* genes.
5. Transcription of *ARF* genes leads to an auxin response.

Unlike with animal hormones, a specific signal is not sent to specific cells, eliciting a predictable response. Most likely, multiple auxin perception sites are present. Auxin is also unique among the plant hormones in that it is transported toward the

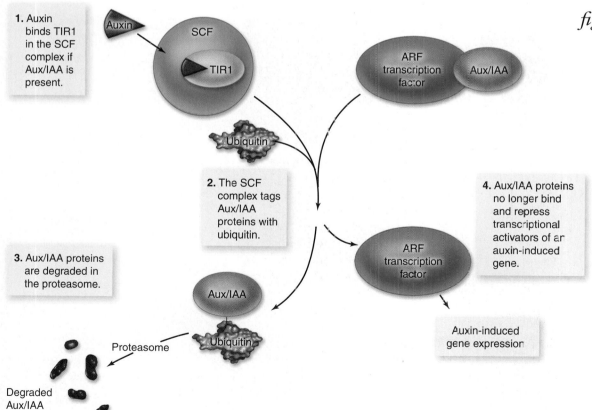

1. Auxin binds TIR1 in the SCF complex if Aux/IAA is present.

2. The SCF complex tags Aux/IAA proteins with ubiquitin.

3. Aux/IAA proteins are degraded in the proteasome.

4. Aux/IAA proteins no longer bind and repress transcriptional activators of an auxin-induced gene.

Auxin-induced gene expression

Degraded Aux/IAA proteins

Proteasome

figure 41.23

AUXIN REGULATION OF GENE EXPRESSION. Auxin activates a ubiquitination pathway that releases ARF transcription factors from repression by Aux/IAA proteins. The result is auxin-induced gene expression.

base of the plant. Two families of genes have been identified in *Arabidopsis* that are involved in auxin transport. For example, one family of proteins (the *PINs*) are involved in the top-to-bottom transport of auxin, while two other proteins function in the root tip to regulate the growth response to gravity, described earlier.

One of the downstream effects of auxin is an increase in the plasticity of the plant cell wall, but this effect works only on young cell walls lacking extensive secondary cell wall formation and may or may not involve rapid changes in gene expression. The **acid growth hypothesis** provides a model linking auxin to cell wall expansion (figure 41.24). According to this hypothesis, auxin causes responsive cells to actively transport hydrogen ions from the cytoplasm into the cell wall space. This decreases the pH, which activates enzymes that can break the bonds between cell wall fibers.

This hypothesis has been experimentally supported in several ways. Buffers that prevent cell wall acidification block cell expansion. And, other compounds that release hydrogen ions from the cell can also cause cell expansion. Finally, the movement of hydrogen ions has been observed in response to auxin treatment. The snapping of the Venus flytrap is postulated to involve an acid growth response that allows cells to grow in just 0.5 sec and close the trap.

Synthetic auxins

Synthetic auxins, such as **naphthalene acetic acid (NAA)** and **indolebutyric acid (IBA),** have many uses in agriculture and horticulture. One of their most important uses is based on their prevention of abscission. Synthetic auxins are used to prevent fruit drop in apples before they are ripe and to hold berries on holly that is being prepared for shipping during the winter season. Synthetic auxins are also used to promote flowering and fruiting in pineapples and to induce the formation of roots in cuttings.

Synthetic auxins are routinely used to control weeds. When used as herbicides, they are applied in higher concentrations than IAA would normally occur in plants. One of the most important synthetic auxin herbicides is **2,4-dichlorophenoxyacetic acid,** usually known as **2,4-D** (see figure 41.22*c*). It kills weeds in grass lawns by selectively eliminating broad-leaved dicots. The stems of the dicot weeds cease all axial growth.

The herbicide 2,4,5-trichlorophenoxyacetic acid, better known as 2,4,5-T, is closely related to 2,4-D. 2,4,5-T was widely used as a broad-spectrum herbicide to kill weeds and the seedlings of woody plants. It became notorious during the Vietnam War as a component of a jungle defoliant known as Agent Orange. When 2,4,5-T is manufactured, it is unavoidably contaminated with minute amounts of dioxin. Dioxin, in doses as low as a few parts per billion, has produced liver and lung diseases, leukemia, miscarriages, birth defects, and even death in laboratory animals. This chemical was banned in 1979 for most uses in the United States.

Cytokinins stimulate cell division and differentiation

Cytokinins comprise another group of naturally occurring growth hormones in plants. Studies by Gottlieb Haberlandt of Austria around 1913 demonstrated the existence of an un-

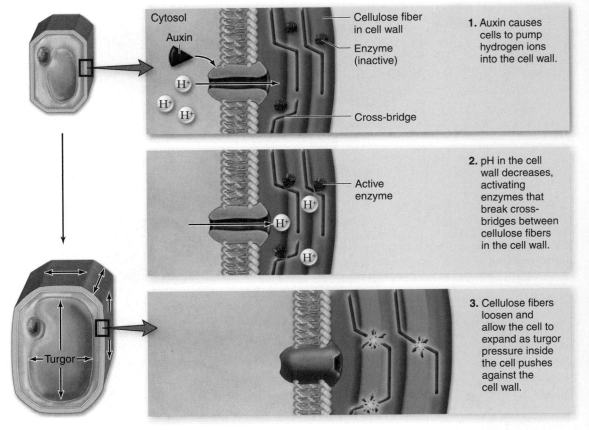

figure 41.24

ACID GROWTH HYPOTHESIS. Auxin stimulates the release of hydrogen ions from the target cells, which alters the pH of the cell wall. This optimizes the activity of enzymes that break bonds in the cell wall, allowing the wall to expand.

Cytosol
Auxin
Cellulose fiber in cell wall
Enzyme (inactive)
Cross-bridge

1. Auxin causes cells to pump hydrogen ions into the cell wall.

Active enzyme

2. pH in the cell wall decreases, activating enzymes that break cross-bridges between cellulose fibers in the cell wall.

Turgor

3. Cellulose fibers loosen and allow the cell to expand as turgor pressure inside the cell pushes against the cell wall.

figure 41.25

Kinetin	6-Benzylamino Purine (BAP)	Adenine

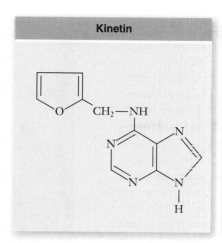
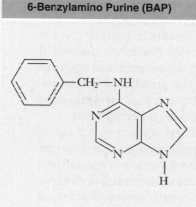
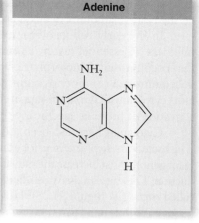

SOME CYTOKININS. Two commonly used synthetic cytokinins: kinetin and 6-benzylamino purine. Note their resemblance to the purine base adenine.

known chemical in various tissues of vascular plants that, when applied to cut potato tubers, would cause parenchyma cells to become meristematic, and would induce the differentiation of a cork cambium. In other research, coconut milk, subsequently found to contain cytokinins was used to promote the differentiation of organs in masses of plant tissue growing in culture. Subsequent studies have focused on the role cytokinins play in the differentiation of tissues from callus.

A cytokinin is a plant hormone that, in combination with auxin, stimulates cell division and differentiation. Most cytokinins are produced in the root apical meristems and transported throughout the plant. Developing fruits are also important sites of cytokinin synthesis. In mosses, cytokinins cause the formation of vegetative buds on the gametophyte. In all plants, cytokinins, working with other hormones, seem to regulate growth patterns.

Cytokinins are purines that appear to be derivatives of adenine, or at least have molecule side chains similar to adenine (figure 41.25). Other chemically diverse molecules, not known to occur naturally, have effects similar to those of cytokinins. Cytokinins promote the growth of lateral buds into branches (figure 41.26). Conversely, cytokinins inhibit the formation of lateral roots, while auxins promote their formation.

As a consequence of these relationships, the balance between cytokinins and auxin, along with many other factors, determines the form of a plant. In addition, the application of cytokinins to leaves detached from a plant retards their yellowing. Therefore, they function as antiaging hormones.

The action of cytokinins, like that of other hormones, has been studied in terms of its effects on the growth and differentiation of masses of tissue growing in defined media. Plant tissue

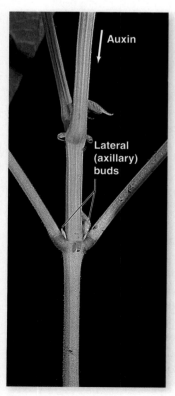

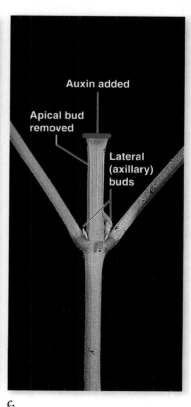

figure 41.26

CYTOKININS STIMULATE LATERAL BUD GROWTH.
a. When the apical meristem of a plant is intact, auxin from the apical bud will inhibit the growth of lateral buds. *b.* When the apical bud is removed, cytokinins are able to induce the growth of lateral buds into branches. *c.* When the apical bud is removed and auxin is added to the cut surface, lateral bud outgrowth is suppressed.

a.   *b.*   *c.*

can form shoots, roots, or an undifferentiated mass, depending on the relative amounts of auxin and cytokinin (figure 41.27).

In the early cell-growth experiments in culture, coconut milk was an essential factor. Eventually, researchers discovered that coconut milk is not only rich in amino acids and other reduced nitrogen compounds required for growth, but it also contains cytokinins. Cytokinins apparently promote the synthesis or activation of proteins specifically required for **cytokinesis.**

Cytokinins have also been used against plants by pathogens. The bacterium *Agrobacterium*, for example, introduces genes into the plant genome that increase the rate of cytokinin, as well as auxin, production. This causes massive cell division and the formation of a tumor called *crown gall* (figure 41.28). How these hormone–biosynthesis genes ended up in a bacterium is an intriguing evolutionary question. Coevolution does not always work to a plant's advantage.

Gibberellins enhance plant growth and nutrient utilization

Gibberellins are named after the fungus *Gibberella fujikuroi*, which causes rice plants, on which it is parasitic, to grow abnormally tall. The Japanese plant pathologist Eiichi Kurosawa in-

figure 41.28

CROWN GALL TUMOR. Sometimes cytokinins can be used against the plant by a pathogen. In this case, *Agrobacterium tumefaciens* (a bacterium) has incorporated a piece of its DNA into the plant genome. This DNA contains genes coding for enzymes necessary for cytokinin and auxin biosynthesis. The increased levels of these hormones in the plant cause massive cell division and the formation of a tumor.

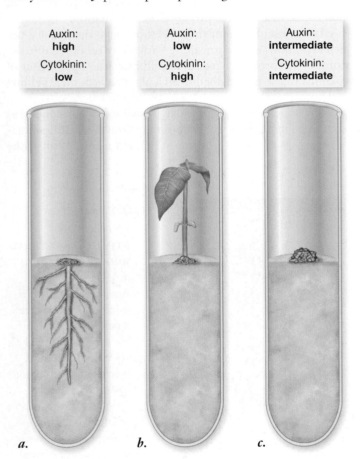

Auxin: **high**	Auxin: **low**	Auxin: **intermediate**
Cytokinin: **low**	Cytokinin: **high**	Cytokinin: **intermediate**

a.   *b.*   *c.*

figure 41.27

RELATIVE AMOUNTS OF CYTOKININS AND AUXIN AFFECT ORGAN REGENERATION IN CULTURE. In tobacco, *a.* high auxin-to-cytokinin ratios favor root development; *b.* high cytokinin-to-auxin ratios favor shoot development; and *c.* intermediate concentrations result in the formation of undifferentiated cells. These developmental responses to cytokinin–auxin ratios in culture are species-specific.

vestigated bakanae ("foolish seedling") disease in the 1920s. He grew *Gibberella* in culture and obtained a substance that, when applied to rice plants, produced bakanae. This substance was isolated and its structural formula identified by Japanese chemists in 1939. British chemists reconfirmed the formula in 1954.

Although such chemicals were first thought to be only a curiosity, they have since turned out to belong to a large class of more than 100 naturally occurring plant hormones. All are acidic and are usually abbreviated GA (for gibberellic acid), with a different subscript (GA_1, GA_2, and so forth) to distinguish each one.

Gibberellins, which are synthesized in the apical portions of stems and roots, have important effects on stem elongation. The elongation effect is enhanced if auxin is also present. The application of gibberellins to certain dwarf mutants is known to restore normal growth and development in many plants (figure 41.29). Some dwarf mutants produce insufficient amounts of gibberellin and respond to GA applications; others lack the ability to respond to gibberellin.

The large number of gibberellins are all part of a complex biosynthetic pathway that has been unraveled using gibberellin-deficient mutants in maize (corn). While many of these gibberellins are intermediate forms in the production of GA_1, recent work shows that some forms may have specific biological roles.

In chapter 37, we noted the role of gibberellins in stimulating the production of α-amylase and other hydrolytic enzymes needed for utilization of food resources during germination and establishment of cereal seedlings. How is transcription of the genes encoding these enzymes regulated? GA is used as a signal from the embryo that turns on transcription of one or more genes encoding hydrolytic enzymes in the aleurone

figure 41.29

EFFECTS OF GIBBERELLINS. This rapid-cycling member of the mustard family (*Brassica rapa*) will "bolt" and flower because of increased gibberellin levels. Mutants such as the rosette mutant (*left*) are defective in producing gibberellins. They can be rescued by applying gibberellins to the shoot tip (*right*). Other mutants have been identified that are defective in perceiving gibberellins, and they will not respond to gibberellin applications.

layer. The GA receptor has been identified. When GA binds to its receptor, it frees GA-dependent transcription factors from a repressor. These transcription factors can now directly affect gene expression (figure 41.30). Synthesis of DNA does not seem to occur during the early stages of seed germination, but it becomes important when the radicle has grown through the seed coats.

Gibberellins also affect a number of other aspects of plant growth and development. In some cases, GAs hasten seed germination, apparently by substituting for the effects of cold or light requirements. Gibberellins are used commercially to increase space between grape flowers by extending internode length, so that the fruits have more room to grow. The result is a larger bunch of grapes containing larger individual fruits (figure 41.31).

Although gibberellins function endogenously as hormones, they also function as pheromones in ferns. In ferns, gibberellin-like compounds released from one gametophyte can trigger the development of male reproductive structures on a neighboring gametophyte.

Brassinosteroids are structurally similar to animal hormones

Although plant biologists have known about **brassinosteroids** for 30 years, it is only recently that they have claimed their place as a class of plant hormones. They were first discovered in *Brassica* spp. pollen, hence the name. Their historical absence in discussions of hormones may be partially due to their functional overlap with other plant hormones, especially auxins and gibberellins. Additive effects among these three classes have been reported.

The application of molecular genetics to the study of brassinosteroids has led to tremendous advances in our understanding of how they are made and, to some extent, how they function in signal transduction pathways. What is particularly intriguing about brassinosteroids is their similarity to animal

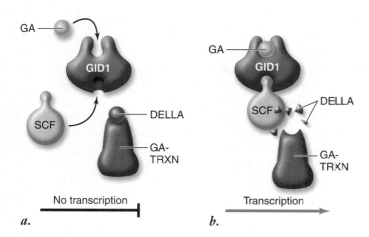

figure 41.30

GIBBERELLINS ACTIVATE GIBBERELLIN-DEPENDENT TRANSCRIPTION FACTORS (GA-TRXN). *a.* GA-TRXN cannot bind to a promotor when they are bound to DELLA proteins. ***b.*** GA activates a protein complex that degrades DELLA proteins, freeing GA-TRXN to bind to a promoter, inducing gene transcription.

figure 41.31

APPLICATIONS OF GIBBERELLINS INCREASE THE SPACE BETWEEN GRAPES. Larger grapes (*right*) develop because there is more room between individual grapes.

figure 41.32

BRASSINOSTEROIDS.
Brassinolide and other
brassinosteroids have
structural similarities
to animal steroid
hormones. Cortisol,
testosterone, and
estradiol (not shown)
are animal steroid
hormones.

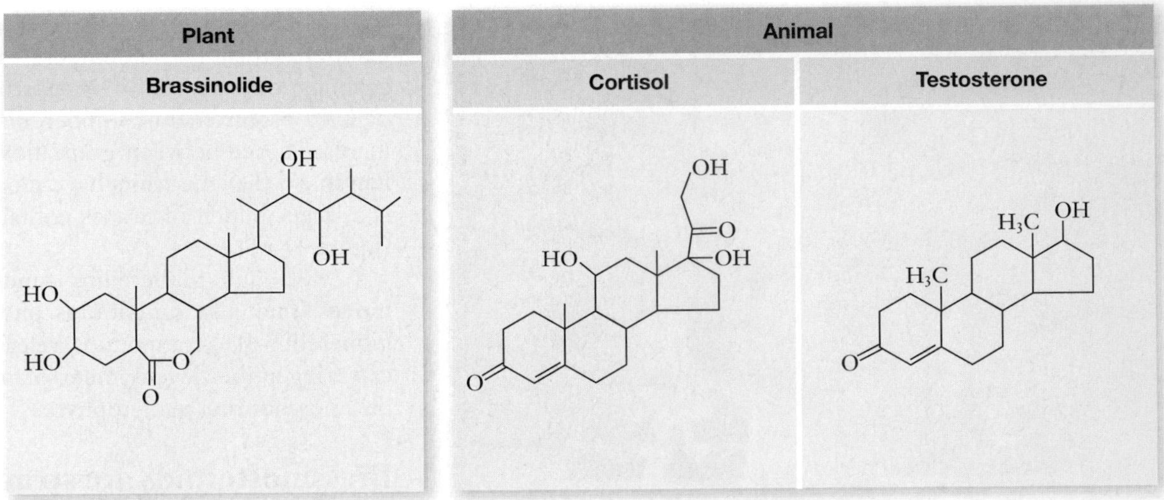

steroid hormones (figure 41.32). One of the genes coding for an enzyme in the brassinosteroid biosynthetic pathway has significant similarity to an enzyme used in the synthesis of testosterone and related steroids. Brassinosteroids have also been identified in algae, and they appear to be ubiquitous among the plants. It is plausible that their evolutionary origin predated the plant–animal split.

Brassinosteroids have a broad spectrum of physiological effects—elongation, cell division, bending of stems, vascular tissue development, delayed senescence, membrane polarization, and reproductive development. Environmental signals can trigger brassinosteroid actions. Mutants have been identified that alter the response to a brassinosteroid, but signal transduction pathways remain to be uncovered. From an evolutionary perspective, it will be quite interesting to see how these pathways compare with animal steroid signal transduction pathways.

Oligosaccharins act as defense-signaling molecules

Plant cell walls are composed not only of cellulose but also of numerous complex carbohydrates called *oligosaccharides.* Some evidence indicates that these cell wall components (when degraded by pathogens) function as signaling molecules as well as structural wall components. Oligosaccharides that are proposed to have a hormonelike function are called **oligosaccharins.**

Oligosaccharins can be released from the cell wall by enzymes secreted by pathogens. These carbohydrates are believed to signal defense responses, such as the hypersensitive response (HR) discussed in chapter 40.

Another oligosaccharin has been shown to inhibit auxin-stimulated elongation of pea stems. These molecules are active at concentrations one to two orders of magnitude less than those of the traditional plant hormones; you have seen how auxin and cytokinin ratios can affect organogenesis in culture (see figure 41.27).

Oligosaccharins also affect the phenotype of regenerated tobacco tissue, inhibiting root formation and stimulating flower production in tissues that are competent to regenerate flowers. How the culture results translate to in vivo systems remains an open question.

Ethylene induces fruit ripening and aids plant defenses

Long before its role as a plant hormone was appreciated, the simple, gaseous hydrocarbon **ethylene** ($H_2C—CH_2$) was known to defoliate plants when it leaked from gaslights in streetlamps. Ethylene is, however, a natural product of plant metabolism that, in minute amounts, interacts with other plant hormones.

When auxin is transported down from the apical meristem of the stem, it stimulates the production of ethylene in the tissues around the lateral buds and thus retards their growth. Ethylene also suppresses stem and root elongation, probably in a similar way. An ethylene receptor has been identified and characterized, and it appears to have evolved early in the evolution of photosynthetic organisms, sharing features with environmental-sensing proteins identified in bacteria.

Ethylene plays a major role in fruit development. At first, auxin, which is produced in significant amounts in pollinated flowers and developing fruits, stimulates ethylene production; this, in turn, hastens fruit ripening. Complex carbohydrates are broken down into simple sugars, chlorophylls are broken down, cell walls become soft, and the volatile compounds associated with flavor and scent in ripe fruits are produced.

One of the first observations that led to the recognition of ethylene as a plant hormone was the premature ripening in bananas produced by gases coming from oranges. Such relationships have led to major commercial uses of ethylene. For example, tomatoes are often picked green and artificially ripened later by the application of ethylene.

Ethylene is widely used to speed the ripening of lemons and oranges as well. Carbon dioxide has the opposite effect of arresting ripening; fruits are often shipped in an atmosphere of carbon dioxide.

Also, a biotechnology solution has been developed in which one of the genes necessary for ethylene biosynthesis has been cloned, and its antisense copy inserted into the tomato genome (figure 41.33). The antisense copy of the gene is a nucleotide sequence that is complementary to the sense copy of the gene. In this transgenic plant, both the sense and antisense sequences for the ethylene biosynthesis gene are transcribed. The sense and antisense mRNA sequences then pair with each other. This pairing blocks translation, which requires single-stranded RNA; as a result, ethylene is not synthesized, and the transgenic tomatoes do not ripen. In this way, the sturdy green tomatoes can be shipped without ripening and rotting. Exposing these tomatoes to ethylene later induces them to ripen.

Studies have shown that ethylene plays an important ecological role. Ethylene production increases rapidly when a plant is exposed to ozone and other toxic chemicals, temperature extremes, drought, attack by pathogens or herbivores, and other stresses. The increased production of ethylene that occurs can accelerate the loss of leaves or fruits that have been damaged by these stresses. Some of the damage associated with exposure to ozone is due to the ethylene produced by the plants.

The production of ethylene by plants attacked by herbivores or infected with pathogens may be a signal to activate the defense mechanisms of the plants and may include the production of molecules toxic to the pests.

Abscisic acid suppresses growth and induces dormancy

Abscisic acid appears to be synthesized mainly in mature green leaves, fruits, and root caps. The hormone earned its name because applications of it appear to stimulate fruit abscission in cotton, but there is little evidence that it plays an important role in this process. Ethylene is actually the chemical that promotes senescence and abscission.

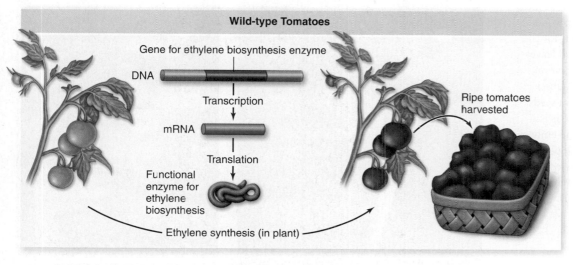

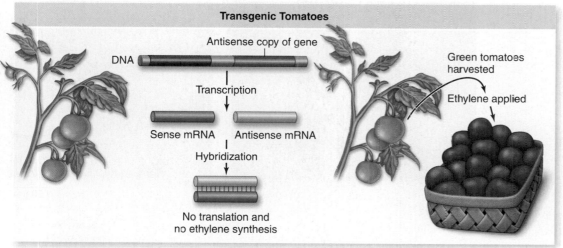

figure 41.33

GENETIC REGULATION OF FRUIT RIPENING. An antisense copy of the gene for ethylene biosynthesis prevents the formation of ethylene and subsequent ripening of transgenic fruit. The antisense strand is complementary to the sequence for the ethylene biosynthesis gene. After transcription, the antisense mRNA pairs with the sense mRNA, and the double-stranded mRNA cannot be translated into a functional protein. Ethylene is not produced, and the fruit does not ripen. The fruit is sturdier for shipping in its unripened form and can be ripened later with exposure to ethylene. Thus, while wild-type tomatoes may already be rotten and damaged by the time they reach stores, transgenic tomatoes stay fresh longer.

Abscisic acid probably induces the formation of winter buds—dormant buds that remain through the winter. The conversion of leaf primordia into bud scales follows (figure 41.34*a*). Like ethylene, abscisic acid may also suppress growth of dormant lateral buds. It appears that abscisic acid, by suppressing growth and elongation of buds, can counteract some of the effects of gibberellins; it also promotes senescence by counteracting auxin.

Abscisic acid plays a role in seed dormancy and is antagonistic to gibberellins during germination. Abscisic acid levels in seeds rise during embryogenesis (see figure 41.17). As maize embryos develop in the kernels on the cob, abscisic acid is necessary to induce dormancy and prevent precocious germination, called vivipary (figure 41.34*b*). It is also important in controlling the opening and closing of stomata (figure 41.34*c*).

Found to occur in all groups of plants, abscisic acid apparently has been functioning as a growth-regulating substance since early in the evolution of the plant kingdom. Relatively little is known about the exact nature of its physiological and biochemical effects, but these effects are very rapid—often taking place within a minute or two—and therefore they must be at least partly independent of gene expression.

All of the genes have been sequenced in *Arabidopsis*, making it easier to identify which genes are transcribed in response to abscisic acid. Abscisic acid levels become greatly elevated when the plant is subject to stress, especially drought. Like other plant hormones, abscisic acid will probably prove to have valuable commercial applications when its mode of action is better understood.

> The seven major kinds of plant hormones are auxin, cytokinins, gibberellins, brassinosteroids, oligosaccharins, ethylene, and abscisic acid. These hormones interact with sensory systems and each other to regulate growth and development in response to the environment.

figure 41.34

EFFECTS OF ABSCISIC ACID. *a.* Abscisic acid plays a role in the formation of these winter buds of an American basswood. These buds will remain dormant for the winter, and bud scales—modified leaves—will protect the buds from desiccation. *b.* In addition to bud dormancy, abscisic acid is necessary for dormancy in seeds. This viviparous mutant in maize is deficient in abscisic acid, and the embryos begin germinating on the developing cob. *c.* Abscisic acid also affects the closing of stomata by influencing the movement of potassium ions out of guard cells.

Dormant bud

a.

Seedling shoot

b.

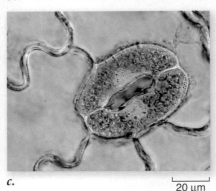

c.

20 μm

41.1 Responses to Light

Pigments, other than those used in photosynthesis, detect light and mediate nondirectional photomorphogenic and directional phototropic responses (figure 41.2).

- Phytochrome, a pigment containing protein, consists of a chromophore that is light-receptive and an apoprotein that initiates a signal transduction pathway.

- Phytochrome consists of two interconvertible forms: the inactive form P_r responds to red light and the active form P_{fr} responds to far-red light.

- P_{fr} is involved in seed germination, shoot elongation, and detection of plant spacing.

  - *Far-red inhibits and red light stimulates seed germination.*
  - *Etiolation occurs when shoot internodes elongate because red light is not available.*
  - *Crowded plants receive far-red light bounced from neighboring plants, which increases plant height in competition for sunlight.*

- P_{fr} can enter the nucleus and bind with other proteins to form a transcription complex, leading to expression of light regulated genes (figure 41.4).

- P_{fr} can also work through a protein kinase signaling pathway by phosphorylation of an amino acid that in turn causes a cascade of transcription factors (figure 41.5).

- The amount of P_{fr} is regulated by protein degradation in the proteosome.

- Phototropisms are unidirectional growth responses of stems toward light with blue wavelengths.

- Circadian rhythms are independent of light but entrain to the daily cycle through the action of phytochrome and blue-light photoreceptors.

41.2 Responses to Gravity

Gravitropism is the response of a plant to a gravitational field.

- Gravitropism is believed to be due to the combined action of amyloplasts that sink toward the center of gravity in plant cells and the hormone auxin (figures 41.9 and 41.10).

- Shoots exhibit negative gravitropism because auxin accumulates on the lower side of the stem, resulting in asymmetrical cell elongation and curvature of the stem upward.

- Roots exhibit positive gravitropism because the lower cells in horizontally oriented root cap are less elongated than the cells on the upper side of the root.

41.3 Responses to Mechanical Stimuli

Plant responses to touch and mechanical stimuli can be permanent or reversible.

- Thigmotropism is a permanent directional growth of a plant in response to physical stimuli, resulting in thigmomorphogenesis.

- Thigmonastic responses are independent of the direction of the stimuli and are usually the result of changes in turgor pressure. In some plants, thigmonastic responses will be followed by changes in growth.

- Touch-induced responses result from changes in turgor pressure resulting from an electrical signal followed by a loss of K^+ from cells of the pulvini.

- Light can induce changes in turgor pressure resulting in leaf tracking of sunlight, flower opening, and sleep movements of leaves.

41.4 Responses to Water and Temperature

When water availability and temperature affect plants, the responses can be short-term or long-term.

- Dormancy results in the cessation of growth when environmental conditions become challenging.

- Leaf abscission occurs in deciduous trees as they enter dormancy during periods of cold temperatures and limited water availability.

- Seed dormancy allows plant offspring to survive long periods of time until environmental conditions are optimal (figure 41.17).

- Plants respond to cold temperatures by increasing the number of unsaturated lipids in the plasma membrane, limiting ice crystal formation to extracellular spaces, or producing antifreeze proteins.

- Heat shock proteins are produced by plants exposed to very rapid increases in temperature.

- Plants can survive otherwise lethal temperatures by developing an acquired thermal tolerance as temperatures are slowly increased.

41.5 Hormones and Sensory Systems

Hormones that guide plant growth are keyed to changes in the environment.

- Hormones are chemicals produced in one part of a plant and transported to another where they bring about physiological or developmental responses.

- Auxins are produced in apical meristems and immature parts of a plant and affect DNA transcription by binding to proteins. They promote stem elongation, formation of adventitious roots, inhibit leaf abscission, promote cell division, induce ethylene production and promote lateral bud dormancy. Synthetic auxins have been used in agriculture and horticulture to control plant and fruit development and as herbicides.

- Cytokinins are purines that are produced in root apical meristems and immature fruits. They promote the synthesis or activation of proteins necessary for mitosis when auxins are present, promote chloroplast development, delay leaf aging, and promote bud formation (figure 41.27).

- Gibberellins are produced by root and shoot tips, young leaves, and seeds. They promote stem elongation and stimulate enzyme production in germinating seeds. In ferns gibberellins function as pheromones.

- Brassinosteroids are steroids produced in pollen, immature seeds, shoots, and leaves. They have overlapping functions with auxins and gibberellins and affect vascular tissue development and membrane polarization.

- Oligosaccharins are released from cell walls by enzymes secreted from pathogens and they signal pathogen defense responses. They also can inhibit auxin-stimulated elongation and affect the phenotype of regenerated tobacco tissue, inhibit root formation, and stimulate flower production.

- Ethylene is produced by roots, shoot apical meristems, aging flowers, and ripening fruits. It controls leaf, flower, and fruit abscission; promotes fruit ripening; suppresses stem and root elongation; and may activate response to attacks by pathogens and herbivores.

- Abscisic acid is produced by mature green leaves, fruits, root caps, and seeds. It inhibits bud growth, induces seed dormancy, controls closure of the stomata, and inhibits the effects of other hormones.

SELF TEST

1. If you exposed seeds to a series of red light versus far-red light treatments, which of the following exposure treatments would result in seed germination?
   a. Red; far-red
   b. Far-red; red
   c. Red; far-red; red; far-red; red; far-red; red; far-red
   d. None of the above

2. Which of the following statements provides a true example of both photomorphogenesis and phototropism?
   a. Phototropism is growth toward blue light, and photomorphogenesis is growth toward red light.
   b. Phototropism is growth toward blue light, and photomorphogenesis is germination triggered by red light.
   c. Phototropism is growth toward red light, and photomorphogenesis is germination triggered by blue light.
   d. Phototropism is movement toward blue light that does not involve growth; photomorphogenesis is movement toward red light that does involve growth.

3. If you were to plant a de-etiolated (*det2*) mutant *Arabidopsis* seed and keep it in a dark box, what would you expect to happen?
   a. The seed would germinate normally, but the plant would not become tall and spindly while it sought a light source.
   b. The seed would fail to germinate because it would not have light.
   c. The seed would germinate, and the plant would become tall and spindly while it sought a light source.
   d. The seed would germinate, and the plant would immediately die because it could not make sugar in the dark.

4. Which of the following statements is not true of phytochrome?
   a. P_r is converted to P_{fr} when exposed to red light.
   b. P_{fr} is the biologically active form of phytochrome.
   c. Phytochrome triggers many light responses by regulating gene expression.
   d. Phytochrome controls most seed germination in plants.

5. We often have the misconception that plants are unable to move in their environment. Many plants, however, (such as, bean leaves) display daily movements to maximize their capacity to absorb light energy. These daily changes in shape are caused by
   a. changes in turgor in specific cells
   b. growth of specific cells
   c. muscle contraction in the leaves
   d. temperature changes in the environment

6. When Charles and Francis Darwin investigated phototropisms in plants, they discovered that
   a. auxin was responsible for light-dependent growth
   b. light was detected at the shoot tip of a plant
   c. light was detected below the shoot tip of a plant
   d. only red light stimulated phototropism

7. Auxin promotes a plant to grow toward a light source by
   a. increasing the rate of cell division on the shaded side of the stem
   b. shortening the cells on the light side of the stem
   c. causing cells on the shaded side of the stem to elongate
   d. decreasing the rate of cell division on the light side of the stem

8. Which of the following statements is not true of auxin?
   a. It usually travels downward in a plant.
   b. We do not know how auxin is involved in cell signaling.
   c. It likely initiates cell wall acidification.
   d. Indoleacetic acid is its most common form.

9. You have come up with a brilliant idea to stretch your grocery budget by buying green fruit in bulk and then storing it in a bag that you have blown up like a balloon. As you need fruit, you would take it out of the bag, and it would miraculously ripen. How would this work?
   a. The bag would block light from reaching the fruit, so it would not ripen.
   b. The bag would keep the fruit cool, so it would not ripen.
   c. The high CO_2 levels in the bag would prevent ripening.
   d. The high O_2 levels in the bag would prevent ripening.

10. If you were to accidentally plant a mutant strain of barley that could not synthesize the plant hormone abscisic acid (ABA), what would you expect to happen?
    a. The shoots would elongate too much and fall over because they could not support themselves.
    b. The shoots would not elongate normally, and you would get short plants.
    c. The seeds would germinate prematurely.
    d. The leaves would fall off the plant.

11. Just as auxin and cytokinin counterbalance each other in many respects, gibberellins are countered by which plant hormone?
    a. Ethylene
    b. Brassinosteroids
    c. Abscisic acid
    d. Oligosaccharins

12. Gibberellins are used to increase productivity in grapes because they
    a. cause fruits to be larger by promoting cell division within the fruit
    b. increase the internode length so the fruits have more room to grow
    c. increase the number of flowers produced, thus increasing the number of fruits
    d. all of the above

13. Which of the following might not be observed in a plant that is grown on the Space Shuttle in space?
    a. Phototropism
    b. Photomorphogenesis
    c. Circadian rhythms
    d. Gravitropism

CHALLENGE QUESTIONS

1. Discuss the similarities and differences between thigmotropism and turgor movement.

2. Compare the mechanisms that animals and plants use to survive harsh environments by thinking of an equivalent animal response for each of the following:
   a. dormancy
   b. thigmomorphogenesis
   c. abscission
   d. phototropism

3. Tree burls, often caused by tumor growths on the trunk and branches of trees, are highly prized and very valuable to woodworkers because of their beautiful grain pattern. If you wanted to start a "tree burl farm," how would you go about doing so?

Plant Reproduction

introduction

THE REMARKABLE EVOLUTIONARY SUCCESS of flowering plants can be linked to their novel reproductive strategies. In this chapter, we explore the reproductive strategies of the angiosperms and how their unique features—flowers and fruits—have contributed to their success. This is, in part, a story of coevolution between plants and animals that ensures greater genetic diversity by dispersing plant gametes widely. In a stable environment, however, there are advantages to maintaining the status quo genetically; asexual reproduction, for example, is a strategy that produces cloned individuals. An unusual twist to sexual reproduction in some flowering plants is that senescence and death of the parent plant follow.

concept outline

42.1 Reproductive Development

In chapter 30, we noted that angiosperms represent an evolutionary innovation with their production of flowers and fruits. In chapter 36, we outlined the development of form, or **morphogenesis,** that a germinating seed undergoes to become a vegetative plant. In this section, we describe the additional changes that occur in a vegetative plant to produce the elaborate structures associated with flowering (figure 42.1).

Plants go through developmental changes leading to reproductive maturity just as many animals do. This shift from juvenile to adult development is seen in the metamorphosis of a tadpole to an adult frog or a caterpillar to a butterfly that can then reproduce. Plants undergo a similar metamorphosis that leads to the production of a flower. Unlike the juvenile frog, which loses its tail, plants just keep adding structures to existing structures with their meristems.

Carefully regulated processes determine when and where flowers will form. Moreover, plants must often gain competence to respond to internal or external signals regulating flowering. Once plants are competent to reproduce, a combination of factors—including light, temperature, and both promotive and inhibitory internal signals—determines when a flower is produced (figure 42.2). These signals turn on genes that specify formation of the floral organs—sepals, petals, stamens, and car-

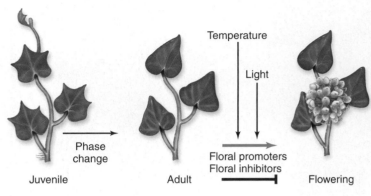

figure 42.2

FACTORS INVOLVED IN INITIATING FLOWERING. This model depicts the environmentally cued and internally processed events that result in a shoot meristem initiating flowers. During phase change, the plant acquires competence to respond to flowering signals.

pels. Once cells have instructions to become a specific floral organ, yet another developmental cascade leads to the three-dimensional construction of flower parts. We describe details of this process in the following sections.

The transition to flowering competence is termed phase change

At germination, most plants are incapable of producing a flower, even if all the environmental cues are optimal. Internal developmental changes allow plants to obtain competence to respond to external or internal signals (or both) that trigger flower formation. This transition is referred to as **phase change.**

Phase change can be morphologically obvious or very subtle. Take a look at a red or white oak tree in the winter: Leaves will still be clinging to the lower branches until spring when the new buds push them off, while leaves on the upper branches will have fallen earlier (figure 42.3*a*). Those lower branches were initiated by a juvenile meristem. The fact that they did not respond to environmental cues and drop their leaves indicates that they are juvenile branches and have not made a phase change. Although the lower branches are older, their juvenile state was established when they were initiated and will not change.

Ivy also has distinct juvenile and adult phases of growth (figure 42.3*b*). Stem tissue produced by a juvenile meristem initiates adventitious roots that can cling to walls. If you look at very old brick buildings covered with ivy, you will notice that the uppermost branches are falling off because they have transitioned to the adult phase of growth and have lost the ability to produce adventitious roots.

It is important to note that even though a plant has reached the adult stage of development, it may or may not produce reproductive structures. Other factors may be necessary to trigger flowering.

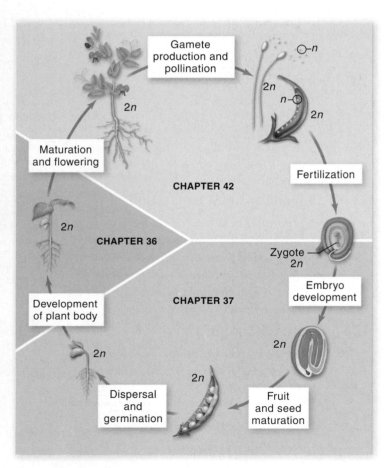

figure 42.1

LIFE CYCLE OF A FLOWERING PLANT (*ANGIOSPERM*).

figure 42.3

PHASE CHANGE. a. The lower branches of this oak tree represent the juvenile phase of development; they cling to their leaves in the winter. The lower leaves are not able to form an abscission layer and break off the tree in the fall. Such visible changes are marks of phase change, but the real test is whether or not the plant is able to flower. **b.** Juvenile ivy (*right*) makes adventitious roots and has an alternating leaf phyllotaxy. Mature ivy (*left*) lacks adventitious roots, has spiral phyllotaxy, and can make flowers.

a.

b.

Mutations have clarified how phase change is controlled

Generally it is easier to get a plant to revert from an adult to juvenile state than to induce phase change experimentally. Applications of the plant hormone gibberellin and severe pruning can cause reversion. In the latter case, new vegetative growth occurs, as when certain shrubs are cut back and put out lush new growth.

The *embryonic flower* (*emf*) mutant of *Arabidopsis* flowers almost immediately (figure 42.4), which is consistent with the hypothesis that the wild-type allele suppresses flowering. As the wild-type plant matures, *EMF* expression decreases. This finding suggests that flowering is the default state, and that mechanisms have evolved to delay flowering. This delay presumably allows the plant to store more energy to be allocated for reproduction.

An example of inducing the juvenile-to-adult transition comes from overexpressing a gene necessary for flowering that is found in many species. This gene, *LEAFY* (*LFY*), was cloned in *Arabidopsis*, and its promoter was replaced with a viral promoter that results in constant, high levels of *LFY* transcription. *LFY* with its viral promoter was then introduced into cultured aspen cells that were used to regenerate plants. When *LFY* is overexpressed in aspen, flowering occurs in weeks instead of years (figure 42.5).

Phase change requires both sufficient promotive signal and the ability to perceive the signal. Some plants acquire com-

petence in the shoot to perceive a signal of a certain intensity. Others acquire competence to produce sufficient promotive signal(s) or to decrease inhibitory signal(s).

Phase change, as we said earlier, results in an adult plant, but not necessarily a flowering plant. The ability to reproduce is distinct from actual reproductive development. Flower production depends on a number of factors, which we explore next.

> Reproductive development in plants involves a phase change from juvenile to adult form. This change results in a plant that is competent to produce flowers.

Normal Flowering

Accelerated Flowering

a.

b.

figure 42.4

EMBRYONIC FLOWER (*EMF*) PREVENTS EARLY FLOWERING. Mutant plants that lack EMF protein flower as soon as they germinate. The flowers have malformed carpels and other defective floral structures close to the roots.

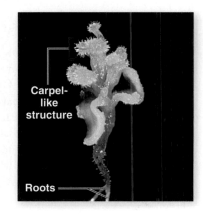

Carpel-like structure

Roots

figure 42.5

OVEREXPRESSION OF A FLOWERING GENE CAN ACCELERATE PHASE CHANGE. a. Normally, an aspen tree grows for several years before producing flowers (see inset). **b.** Overexpression of the *Arabidopsis* flowering gene, *LEAFY*, causes rapid flowering in a transgenic aspen (see inset).

Flower Production

Four genetically regulated pathways to flowering have been identified: (1) the light-dependent pathway, (2) the temperature-dependent pathway, (3) the gibberellin-dependent pathway, and (4) the autonomous pathway.

Plants can rely primarily on one pathway, but all four pathways can be present.

The environment can promote or repress flowering, and in some cases, it can be relatively neutral. For example, increasing light duration can be a signal that long summer days have arrived in a temperate climate and that conditions are favorable for reproduction. In other cases, plants depend on light to accumulate sufficient amounts of sucrose to fuel reproduction, but flower independently of day length.

Temperature can also be used as a signal. **Vernalization,** the requirement for a period of chilling of seeds or shoots for flowering, affects the temperature-dependent pathway. Clearly, reproductive success would be unlikely in the middle of a blizzard.

Assuming that regulation of reproduction first arose in more constant tropical environments, many of the day-length and temperature controls would have evolved as plants colonized more temperate climates.

The complexity of the flowering pathways has been explored physiologically. Just as with phase change, analysis of flowering mutants is providing insight into the molecular mechanisms of the flowering pathways. The redundancy of pathways to flowering ensures that there will be another generation.

The light-dependent pathway is geared to the photoperiod

Flowering requires much energy accumulated via photosynthesis. Thus, all plants require light for flowering, but this is distinct from the **photoperiodic,** or light-dependent, flowering pathway. Aspects of growth and development in most plants are keyed to changes in the proportion of light to dark in the daily 24-hr cycle (day length).

In the preceding chapter, you learned that many plants have circadian rhythms that affect movements of leaves and other structures. The 24-hr circadian cycle, however, is separate from the photoperiodic changes affecting flowering. This sensitivity to the photoperiod provides a mechanism for organisms to respond to seasonal changes in the relative length of day and night. Day length changes with the seasons; the farther a region is from the equator, the greater the variation in day length.

Short-day and long-day plants

The flowering responses of plants to day length fall into several basic categories. In **short-day plants,** flowering is initiated when daylight becomes shorter than a critical length (figure 42.6). In **long-day plants,** flowering begins when daylight becomes longer. Other plants, such as snapdragons, roses, and many native to the tropics, flower when mature regardless of day length, as long as they have received enough light for normal growth. These are referred to as **day-neutral plants.** Still other plants, including ivy, have two critical photoperiods; they

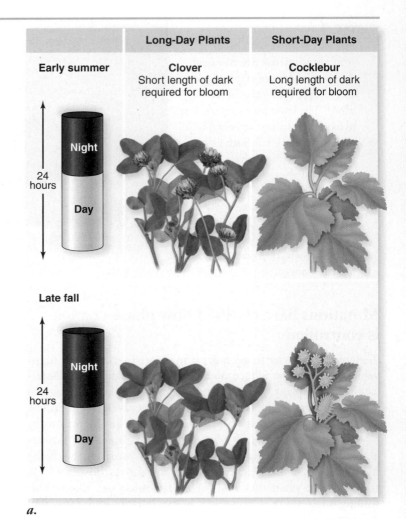

a.

b.

figure 42.6

HOW FLOWERING RESPONDS TO DAY LENGTH. *a.* Clover (*center panels*) is a long-day plant that is stimulated by short nights to flower in the spring. Cocklebur (*right-hand panels*) is a short-day plant that, throughout its natural distribution in the northern hemisphere, is stimulated by long nights to flower in the fall. *b.* If the long night of late fall is artificially interrupted by a flash of light, the cocklebur will not flower, and the clover will. Although the terms refer to the length of day, in each case, it is the duration of uninterrupted darkness that determines when flowering will occur.

will not flower if the days are too long, and they also will not flower if the days are too short.

Although plants are referred to as long-day or short-day plants, it is actually the amount of darkness that determines whether a plant will flower. In **obligate long-** or **short-day species,** there is a sharp distinction between short and long nights, respectively. Flowering occurs in obligate long-day plants when the night length is less than the maximal amount of required darkness (critical night length) for that species. For obligate short-day plants, the amount of darkness must exceed the critical night length for the species.

In other long- or short-day plants, flowering occurs more rapidly or slowly depending on the length of day. These plants, which rely on other flowering pathways as well, are called **facultative long-** or **short-day** plants because the photoperiodic requirement is not absolute. The garden pea is an example of a facultative long-day plant.

Advantages of photoperiodic control of flowering

Using light as a cue permits plants to flower when abiotic environmental conditions are optimal, pollinators are available, and competition for resources with other plants may be less. For example, the spring herbaceous plants termed *ephemerals* flower in the woods before the tree canopy leafs out and blocks the sunlight necessary for photosynthesis. An example is the trailing arbutus (*Epigaea repens*) of the Northeast woods, which is also known as mayflower because of the time of year in which it blooms.

At middle latitudes, most long-day plants flower in the spring and early summer; examples of such plants include clover, irises, lettuce, spinach, and hollyhocks. Short-day plants usually flower in late summer and fall; these include chrysanthemums, goldenrods, poinsettias, soybeans, and many weeds, such as ragweed. Commercial plant growers use these responses to day length to bring plants into flower at specific times. For example, photoperiod is manipulated in greenhouses so that poinsettias flower just in time for the winter holidays (figure 42.7). The geographic distribution of certain plants may be determined by their flowering responses to day length.

The mechanics of light signaling

Photoperiod is perceived by several different forms of phytochrome and also by a blue-light–sensitive molecule (cryptochrome). Another type of blue-light–sensitive molecule (phototropin) was discussed in chapter 41. Phototropin affects photomorphogenesis, and cryptochrome affects photoperiodic responses.

The conformational change in a phytochrome or cryptochrome light-receptor molecule triggers a cascade of events that leads to the production of a flower. There is a link between light and the circadian rhythm regulated by an internal clock that facilitates or inhibits flowering. At a molecular level, the gaps between light signaling and production of flowers are rapidly filling in, and the control mechanisms have been found to be quite complex.

Photoperiodic Regulation of Transcription of the CO Gene

Arabidopsis, which as you know is commonly used in plant studies, is a facultative long-day plant that flowers in response to both far-red and blue light. Phytochrome and cryptochrome, the red-

Bract

Flowers

figure 42.7

FLOWERING TIME CAN BE ALTERED. Manipulation of photoperiod in greenhouses ensures that short-day poinsettias flower in time for the winter holidays. Even after flowering is induced, many developmental events must occur in order to produce species-specific flowers.

and blue-light receptors, respectively, regulate flowering via the gene *CONSTANS (CO)*. Precise levels of CO protein are maintained in accordance with the circadian clock, and phytochrome regulates the transcription of *CO*. Levels of *CO* mRNA are low at night and increase at daybreak. In addition, CO protein levels are modulated through the action of cryptochrome.

inquiry

If levels of CO mRNA follow a circadian pattern, how could you determine whether protein levels are modulated by a mechanism other than transcription? Why would an additional level of control even be necessary?

The importance of posttranslational regulation of CO activity became apparent through studies of transgenic *Arabidopsis* plants. These plants contain a *CO* gene fused to a viral promoter that is always on and produces high levels of *CO* mRNA regardless of whether it is day or night. The regulation of *CO* gene expression by phytochrome A is therefore eliminated when this viral promoter is fused to the gene. Curiously, CO protein levels still follow a circadian pattern.

Although CO protein is produced day and night, levels of CO are lower at night because of targeted protein degradation. Ubiquitin tags the CO protein, and it is degraded by the proteasome as was described in chapter 41 for phytochrome degradation. Blue light acting via cryptochrome stabilizes CO during the day and protects it from ubiquitination and subsequent degradation.

CO *and* LFY *Expression*

CO is a transcription factor that turns on other genes, which results in the expression of *LFY*. As discussed in connection with phase change earlier in this section, *LFY* is one of the key genes that "tells" a meristem to switch over to flowering. We will see that other pathways also converge on this important gene. Genes that are regulated by *LFY* are discussed later in this chapter.

Florigen—The elusive flowering hormone

A considerable amount of evidence demonstrates the existence of substances that promote flowering and substances that inhibit it. Grafting experiments have shown that these substances can move from leaves to shoots. The complexity of their interactions, as well as the fact that multiple chemical messengers are evidently involved, has made this scientifically and commercially interesting search very difficult. Much like the historical search for the "Holy Grail," the existence of a flowering hormone remains strictly hypothetical even after a scientific quest of 50 years.

One intriguing possibility is that CO protein is a graft-transmissible flowering signal, or that it affects such a signal. CO has been found in the phloem that moves throughout the plant body. When *co* mutant shoots are grafted to stocks that produce CO, flowering occurs. Because CO is found in the phloem, it is possible that this is the protein that moves in the grafted plant to cause flowering. Equally likely, however, is the possibility that CO directly or indirectly affects a separate graft-transmissible factor that is essential for flowering.

Clearly, information about day length gathered by pigments and photosystems in leaves is transmitted to shoot apices.

Given that there are multiple pathways to flowering, several signals may be facilitating communication between leaves and shoots. We also know that roots can be a source of floral inhibitors, affecting shoot development.

The temperature-dependent pathway is linked to cold

Cold temperatures can accelerate or permit flowering in many species. As with light, this environmental connection ensures that plants flower at more optimal times.

Some plants require a period of chilling before flowering, called *vernalization*. This phenomenon was discovered in the 1930s by the Ukrainian scientist T. D. Lysenko while trying to solve the problem of winter wheat rotting in the fields. Because winter wheat would not flower without a period of chilling, Lysenko chilled the seeds and then planted them in the spring. The seeds successfully sprouted, grew, and produced grain.

Although this discovery was scientifically significant, Lysenko erroneously concluded that he had converted one species, winter wheat, to another, spring wheat, by simply altering the environment. This point of view was supported by the communist philosophy of the time, which held that people could easily manipulate nature to increase production. Unfortunately, a great many problems resulted, including mistreatment of legitimate geneticists in the former Soviet Union. In addition, genetics and Darwinian evolution were suspect in the Soviet Union until the mid-1960s.

Vernalization is necessary for some seeds or plants in later stages of development. Analysis of mutants in *Arabidopsis* and pea plants indicate that vernalization is a separate flowering pathway.

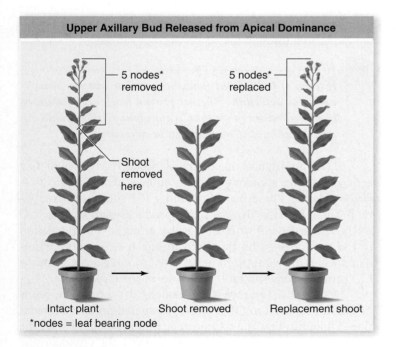

Upper Axillary Bud Released from Apical Dominance

5 nodes* removed — 5 nodes* replaced — Shoot removed here

Intact plant — Shoot removed — Replacement shoot
*nodes = leaf bearing node

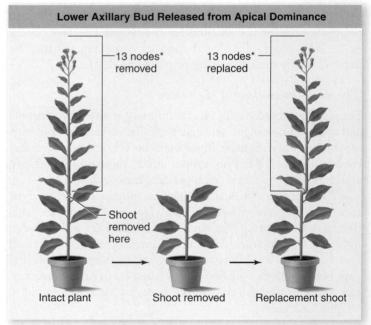

Lower Axillary Bud Released from Apical Dominance

13 nodes* removed — 13 nodes* replaced — Shoot removed here

Intact plant — Shoot removed — Replacement shoot

figure 42.8

PLANTS CAN "COUNT." When axillary buds of flowering, day-neutral tobacco plants are released from apical dominance by removing the main shoot, they replace the number of nodes that were initiated by the main shoot.

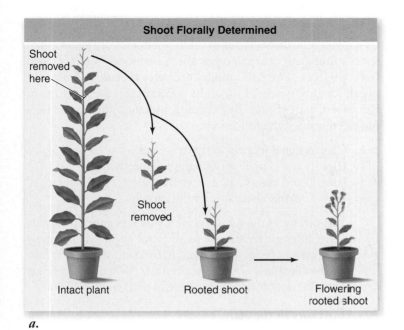

Shoot Florally Determined

Shoot removed here

Shoot removed

Intact plant   Rooted shoot   Flowering rooted shoot

a.

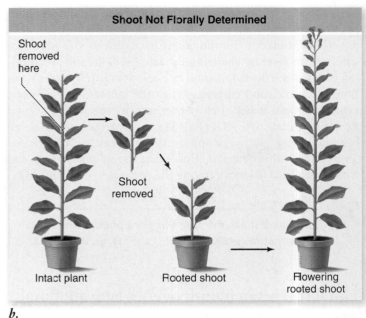

Shoot Not Florally Determined

Shoot removed here

Shoot removed

Intact plant   Rooted shoot   Flowering rooted shoot

b.

figure 42.9

PLANTS CAN "REMEMBER." At a certain point in the flowering process, shoots become committed to making a flower. This is called floral determination. *a.* Florally determined shoots "remember" their position when rooted in a pot. That is, they produce the same number of nodes that they would have if they had grown out on the plant, and then they flower. *b.* Shoots that are not yet florally determined cannot remember how many nodes they have left, so they start counting again. That is, they develop like a seedling and then flower.

The gibberellin-dependent pathway requires an increased hormone level

In *Arabidopsis* and some other species, decreased levels of gibberellins delay flowering. Thus the gibberellin pathway is proposed to promote flowering, but to date the only components that have been identified are enzymes required for the biosynthesis of gibberellins.

It is known that gibberellins enhance the expression of *LFY*. Gibberellin actually binds the promoter of the *LFY* gene, so its effect on flowering is direct.

The autonomous pathway is independent of environmental cues

The autonomous pathway to flowering does not depend on external cues except for basic nutrition. Presumably, this was the first pathway to evolve. Day-neutral plants often depend primarily on the autonomous pathway, which allows plants to "count" and "remember."

As an example, a field of day-neutral tobacco plants will produce a uniform number of nodes before flowering. If the shoots of these plants are removed at different positions, axillary buds will grow out and produce the same number of nodes as the removed portion of the shoot (figure 42.8). The upper axillary buds of flowering tobacco will remember their position when rooted or grafted. The terminal shoot tip becomes committed or determined to flower about four nodes before it actually initiates a flower (figure 42.9). In some other species, this commitment is less stable or it occurs later.

How do shoots "know" where they are and at some point "remember" that information? It has become clear that inhibi-

tory signals are sent from the roots. When bottomless pots are continuously placed over a growing tobacco plant and filled with soil, flowering is delayed by the formation of adventitious roots (figure 42.10). Control experiments with leaf removal show that the addition of roots, and not the loss of leaves, delays flowering. A balance between floral promoting and inhibiting

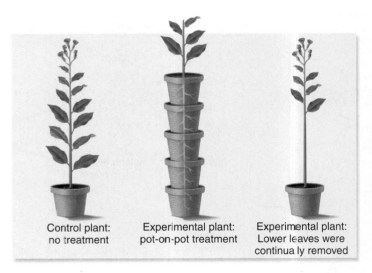

Control plant: no treatment   Experimental plant: pot-on-pot treatment   Experimental plant: Lower leaves were continually removed

figure 42.10

ROOTS CAN INHIBIT FLOWERING. Adventitious roots formed as bottomless pots were continuously placed over growing tobacco plants, delaying flowering. The delay in flowering is caused by the roots, not by the loss of the leaves. This was shown by removing leaves on plants at the same time and in the same position as leaves on experimental plants that became buried as pots were added.

signals may regulate when flowering occurs in the autonomous pathway and the other pathways as well.

Determination for flowering is tested at the organ or whole-plant level by changing the environment and ascertaining whether the developmental fate has changed. In *Arabidopsis*, floral determination correlates with the increase of *LFY* gene expression, and it has already occurred by the time a second flowering gene, *APETALA1* (*AP1*), is expressed. Because all four flowering pathways appear to converge with increased levels of *LFY*, this determination event should occur in species with a variety of balances among the pathways (figure 42.11).

inquiry

? *Why would it be advantageous for a plant to have four distinct pathways that all affect the expression of LFY?*

Floral meristem identity genes activate floral organ identity genes

Arabidopsis and snapdragons are valuable model systems for identifying flowering genes and understanding their interactions. The four flowering pathways discussed earlier in this section lead to an adult meristem becoming a floral meristem by either activating or repressing the inhibition of **floral meristem identity genes** (see figure 42.11). Two of the key floral meristem identity genes are *LFY* and *AP1*. These genes establish the meristem as a flower meristem. They then turn on **floral organ identity genes.** The floral organ identity genes define four concentric whorls, moving inward in the floral meristem, as sepal, petal, stamen, and carpel.

The ABC model

To explain how three classes of floral organ identity genes could specify four distinct organ types, the ABC model was developed (figure 42.12). The ABC model proposes that three classes of organ identity genes (*A, B,* and *C*) specify the floral organs in the four floral whorls. By studying mutants, the researchers have determined the following:

1. Class A genes alone specify the sepals.
2. Class A and class B genes together specify the petals.
3. Class B and class C genes together specify the stamens.
4. Class C genes alone specify the carpels.

The beauty of the ABC model is that it is entirely testable by making different combinations of floral organ identity mutants. Each class of genes is expressed in two whorls, yielding four different combinations of the gene products. When any one class is missing, aberrant floral organs occur in predictable positions.

Modifications to the ABC model

As compelling as the ABC model is, it cannot fully explain specification of floral meristem identity. Class D genes that are essential for carpel formation have been identified, but even this discovery did not explain why a plant lacking A, B, and C gene function produced four whorls of sepals rather than four whorls of leaves. Floral parts are thought to have evolved from leaves; therefore, if the floral organ identity genes are removed, whorls of leaves, rather than sepals, would be predicted.

The answer to this puzzle is found in the more recently discovered class E genes, *SEPALATA1* (*SEP1*) through *SEPALATA4* (*SEP4*). The triple mutant *sep1 sep2 sep 3* and the *sep4* mutant both produce four whorls of leaves. The *SEP* genes proteins can

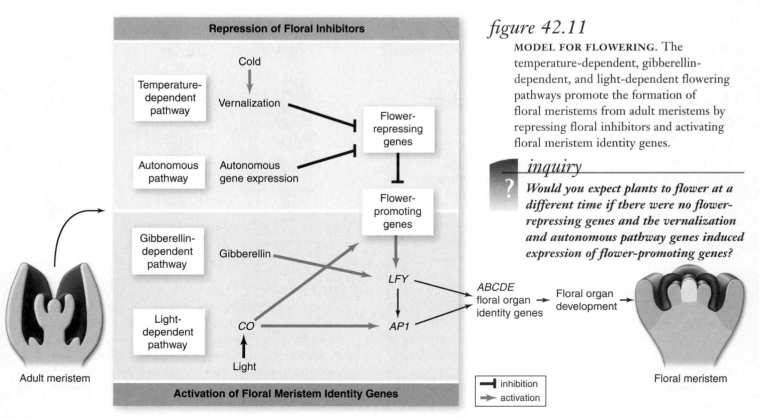

figure 42.11

MODEL FOR FLOWERING. The temperature-dependent, gibberellin-dependent, and light-dependent flowering pathways promote the formation of floral meristems from adult meristems by repressing floral inhibitors and activating floral meristem identity genes.

inquiry

? *Would you expect plants to flower at a different time if there were no flower-repressing genes and the vernalization and autonomous pathway genes induced expression of flower-promoting genes?*

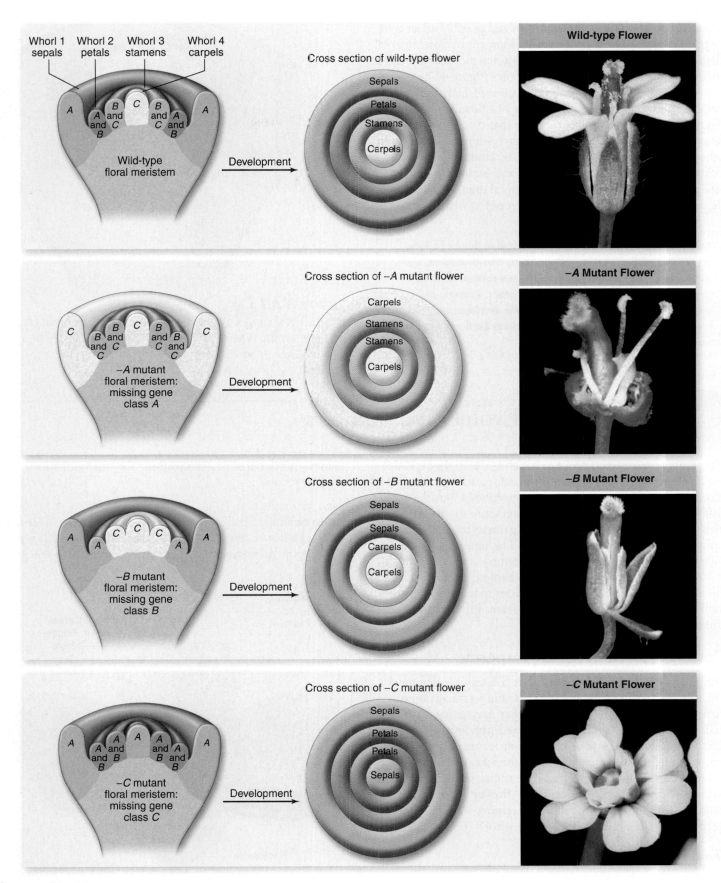

figure 42.12

ABC MODEL FOR FLORAL ORGAN SPECIFICATION. Letters labeling whorls indicate which gene classes are active. When *A* function is lost (−*A*), *C* expands to the first and second whorls. When *B* function is lost (−*B*), both outer two whorls have just *A* function, and both inner two whorls have just *C* function; none of the whorls have dual gene function. When *C* function is lost (−*C*), *A* expands into the inner two whorls. These new combinations of gene expression patterns alter which floral structures form in each whorl.

interact with class A, B, and C proteins and possibly affect transcription of genes needed for the development of floral organs. Identification of the *SEP* genes lead to a new floral organ identity model that includes these class E genes (figure 42.13).

It is important to recognize that the *ABCDE* genes are actually only the beginning of the making of a flower. These organ identity genes are transcription factors that turn on many more genes that actually give rise to the three-dimensional flower. Other genes "paint" the petals—that is, complex biochemical pathways lead to the accumulation of anthocyanin pigments in petal cell vacuoles. These pigments can be orange, red, or purple, and the actual color is influenced by pH as well.

> Four pathways have been identified that lead to flowering: light-dependent, temperature-dependent, gibberellin-dependent, and autonomous. Floral meristem identity genes turn on floral organ identity genes that specify where sepals, petals, stamens, and carpels will form. This is followed by organ development, which involves many complex pathways that account for floral diversity among species.

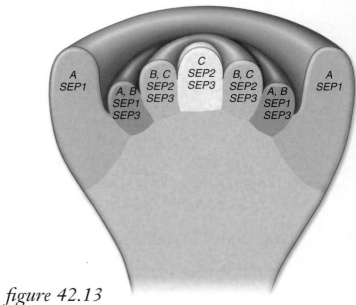

figure 42.13

CLASS E GENES ARE NEEDED TO SPECIFY FLORAL ORGAN IDENTITY. When all three *SEP* genes are mutated, four whorls of leaves are produced.

Structure and Evolution of Flowers

The complex and elegant process that gives rise to the reproductive structure called the flower is often compared with metamorphosis in animals. It is indeed a metamorphosis, but the subtle shift from mitosis to meiosis in the megaspore mother cell that leads to the development of a haploid, gamete-producing gametophyte is perhaps even more critical. The same can be said for pollen formation in the anther of the stamen.

The flower not only houses the haploid generations that will produce gametes, but it also functions to increase the probability that male and female gametes from different (or sometimes the same) plants will unite.

Flowers evolved in the angiosperms

The evolution of the angiosperms is a focus of chapter 30. The diversity of angiosperms is partly due to the evolution of a great variety of floral phenotypes that may enhance the effectiveness of pollination. As mentioned previously, floral organs are thought to have evolved from leaves. In some early angiosperms, these organs maintain the spiral developmental pattern often found in leaves. The trend has been toward four distinct whorls of parts. A **complete flower** has four whorls (calyx, corolla, androecium, and gynoecium), while an **incomplete flower** lacks one or more of the whorls (figure 42.14).

Flower morphology

In both complete and incomplete flowers, the **calyx** usually constitutes the outermost whorl; it consists of flattened appendages, called **sepals,** which protect the flower in the bud. The petals collectively make up the **corolla** and may be fused. Many petals function to attract pollinators. Although these two outer whorls of floral organs are not involved directly in

gamete production or fertilization, they can enhance reproductive success.

Male structures

Androecium is a collective term for all the **stamens** (male structures) of a flower. Stamens are specialized structures that bear the angiosperm microsporangia. Similar structures bear the micro-

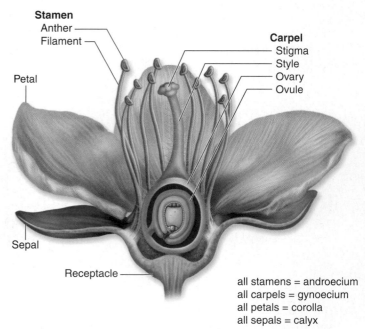

all stamens = androecium
all carpels = gynoecium
all petals = corolla
all sepals = calyx

figure 42.14

STRUCTURE OF A COMPLETE ANGIOSPERM FLOWER. Many flowers have multiple carpels that may fuse.

sporangia in the pollen cones of gymnosperms. Most living angiosperms have stamens with **filaments** ("stalks") that are slender and often threadlike; four microsporangia are evident at the apex in a swollen portion, the **anther.** Some of the more primitive angiosperms have stamens that are flattened and leaflike, with the sporangia produced from the upper or lower surface.

Female structures

The **gynoecium** is a collective term for all the female parts of a flower. In most flowers, the gynoecium, which is unique to angiosperms, consists of a single **carpel** or two or more fused carpels. Single or fused carpels are often referred to as simple or compound pistils, respectively. Most flowers with which we are familiar—for example, those of tomatoes and oranges—have a compound pistil. Other, less specialized flowers—for example, buttercups and stonecups—may have several to many separate, simple pistils, each formed from a single carpel.

Ovules (which develop into seeds) are produced in the pistil's swollen lower portion, the **ovary,** which usually narrows at the top into a slender, necklike **style** with a pollen-receptive **stigma** at its apex. Sometimes the stigma is divided, with the number of stigma branches indicating how many carpels compose the particular pistil.

Carpels are essentially rolled floral leaves with ovules along the margins. It is possible that the first carpels were leaf blades that folded longitudinally; the leaf margins, which had hairs, did not actually fuse until the fruit developed, but the hairs interlocked and were receptive to pollen. In the course of evolution, evidence indicates that the hairs became localized into a stigma; a style was formed; and the fusing of the carpel margins ultimately resulted in a pistil. In many modern flowering plants, the carpels have become highly modified and are not visually distinguishable from one another unless the pistil is cut open.

Trends of floral specialization

Two major evolutionary trends led to the wide diversity of modern flowering plants: (1) Separate floral parts have grouped together, or fused, and (2) floral parts have been lost or reduced (figure 42.15).

In the more advanced angiosperms, the number of parts in each whorl has often been reduced from many to few. The spiral patterns of attachment of all floral parts in primitive angiosperms have, in the course of evolution, given way to a single whorl at each level. The central axis of many flowers has short-

figure 42.16

BILATERAL SYMMETRY IN AN ORCHID. While more basal flowers are usually radially symmetrical, flowers of many derived groups, such as the orchid family (Orchidaceae), are bilaterally symmetrical.

ened, and the whorls are close to one another. In some evolutionary lines, the members of one or more whorls have fused with one another, sometimes joining into a tube. In other kinds of flowering plants, different whorls may be fused together.

Whole whorls may even be lost from the flower, which may lack sepals, petals, stamens, carpels, or various combinations of these structures. Modifications often relate to pollination mechanisms, and in plants such as the grasses, wind has replaced animals for pollen dispersal.

Trends in floral symmetry

Other trends in floral evolution have affected the symmetry of the flower. Primitive flowers such as those of buttercups are **radially symmetrical;** that is, one could draw a line anywhere through the center and have two roughly equal halves. Flowers of many advanced groups are **bilaterally symmetrical;** they are divisible into two equal parts along only a single plane. Examples of such flowers are snapdragons, mints, and orchids (figure 42.16). Bilaterally symmetrical flowers are also common among violets and peas. In these groups, they are often associated with advanced and highly precise pollination systems.

Bilateral symmetry has arisen independently many times. In snapdragons, the *CYCLOIDIA* gene regulates floral symmetry, and in its absence flowers are more radial (figure 42.17).

figure 42.15

TRENDS IN FLORAL SPECIALIZATION. Wild geranium, *Geranium maculatum,* a typical eudicot. The petals are reduced to five each, the stamens to ten, compared to early angiosperms.

a.

b.

figure 42.17

GENETIC REGULATION OF ASYMMETRY IN FLOWERS.
a. Snapdragon flowers normally have bilateral symmetry.
b. The *CYCLOIDIA* gene regulates floral symmetry, and *cycloidia* mutant snapdragons have radially symmetrical flowers.

Here the experimental alteration of a single gene is sufficient to cause a dramatic change in morphology. Whether the same gene or functionally similar genes arose naturally in parallel in other species is an open question.

The human impact on flower morphology

Although much floral diversity is the result of natural selection related to pollination, it is important to recognize the impact of breeding (artificial selection) on flower morphology. Humans have selected for practical or aesthetic traits that may have little adaptive value to species in the wild. For example, maize (corn) has been bred to satisfy the human palate. Human intervention ensures the reproductive success of each generation; however, in a natural setting, modern corn would not have the same protection from herbivores as its ancestors, and the fruit dispersal mechanism would be quite different).

Gametes are produced in the gametophytes of flowers

Reproductive success depends on uniting the gametes (egg and sperm) found in the embryo sacs and pollen grains of flowers. As you learned in chapter 30, plant sexual life cycles are characterized by an *alternation of generations*, in which a diploid sporophyte generation gives rise to a haploid gametophyte generation. In angiosperms, the gametophyte generation is very small and is completely enclosed within the tissues of the parent sporophyte. The male gametophytes, or microgametophytes, are **pollen grains.** The female gametophyte, or megagametophyte, is the **embryo sac.** Pollen grains and the embryo sac both are produced in separate, specialized structures of the angiosperm flower.

Like animals, angiosperms have separate structures for producing male and female gametes (figure 42.18), but the reproductive organs of angiosperms are different from those of animals in two ways. First, both male and female structures usually occur together in the same individual flower. Second, angiosperm reproductive structures are not permanent parts of the adult individual. Angiosperm flowers and reproductive organs develop seasonally, at times of the year most favorable for pollination. In some cases, reproductive structures are produced only once, and the parent plant dies. And, as you learned earlier in this chapter, the germ line in angiosperms is not set aside early on, but forms quite late during phase change.

Pollen formation

Anthers contain four microsporangia which produce microspore mother cells (*2n*). Microspore mother cells produce microspores (*n*) through meiotic cell division. As the microspores, through mitosis and wall differentiation, become pollen, the

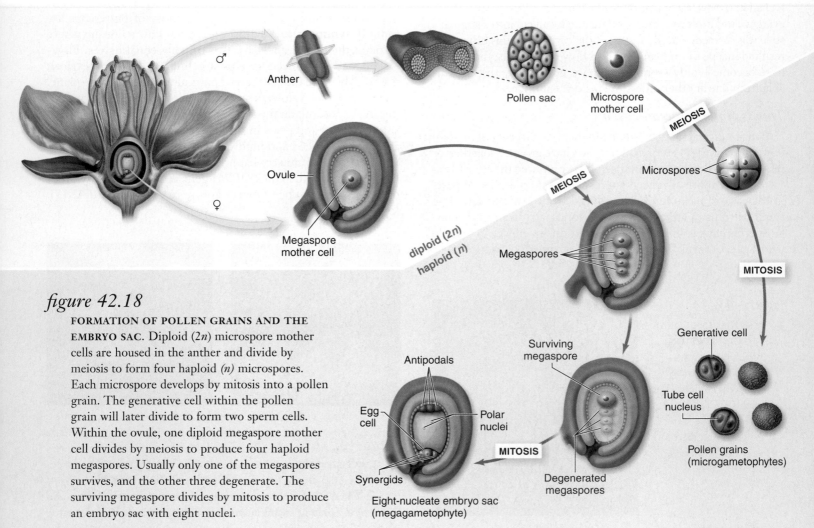

figure 42.18

FORMATION OF POLLEN GRAINS AND THE EMBRYO SAC. Diploid (*2n*) microspore mother cells are housed in the anther and divide by meiosis to form four haploid (*n*) microspores. Each microspore develops by mitosis into a pollen grain. The generative cell within the pollen grain will later divide to form two sperm cells. Within the ovule, one diploid megaspore mother cell divides by meiosis to produce four haploid megaspores. Usually only one of the megaspores survives, and the other three degenerate. The surviving megaspore divides by mitosis to produce an embryo sac with eight nuclei.

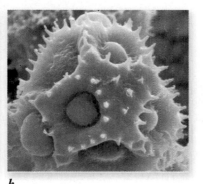

a.   *b.*

figure 42.19

POLLEN GRAINS. *a.* In the Easter lily, *Lilium candidum*, the pollen tube emerges from the pollen grain through the groove or furrow that occurs on one side of the grain. *b.* In a plant of the sunflower family, *Hyoseris longiloba*, three pores are hidden among the ornamentation of the pollen grain. The pollen tube may grow out through any one of them.

two microsporangia on each side become pollen sacs. Inside each pollen grain is a generative cell; this cell will later divide to produce two sperm cells.

Pollen grain shapes are specialized for specific flower species. As discussed in more detail later in this section, fertilization requires that the pollen grain grow a tube that penetrates the style until it encounters the ovary. Most pollen grains have a furrow or pore from which this pollen tube emerges; some grains have three furrows (figure 42.19).

Embryo sac formation

Eggs develop in the ovules of the angiosperm flower. Within each ovule is a megaspore mother cell. Just as in pollen production, the megaspore mother cell undergoes meiosis to produce four haploid megaspores. In most plants, however, only one of these megaspores survives; the rest are absorbed by the ovule. The lone remaining megaspore enlarges and undergoes repeated mitotic divisions to produce eight haploid nuclei that are enclosed within a seven-celled embryo sac.

Within the embryo sac, the eight nuclei are arranged in precise positions. One nucleus is located near the opening of the embryo sac in the egg cell. Two others are located together in a single cell in the middle of the embryo sac; these are called

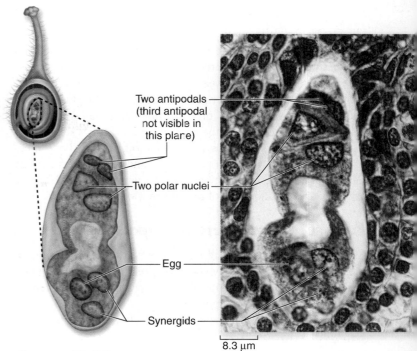

Two antipodals (third antipodal not visible in this plane)

Two polar nuclei

Egg

Synergids

8.3 μm

figure 42.20

A MATURE EMBRYO SAC OF A LILY. Eight nuclei are produced by mitotic divisions of the haploid megaspore. One is in the egg, two are polar nuclei, two occur in synergid cells, and three are in antipodal cells. The micrograph is false-colored.

polar nuclei. Two more nuclei are contained in individual cells called synergids that flank the egg cell; the other three nuclei reside in cells called the antipodals, located at the end of the sac, opposite the egg cell (figure 42.20).

The first step in uniting the two sperm cells in the pollen grain with the egg and polar nuclei is the germination of pollen on the stigma of the carpel and its growth toward the embryo sac.

Flowers have contributed to the evolutionary success of the angiosperms. Floral morphology comprises four whorls: the calyx, the corolla, the androecium (male reproductive organs), and the gynoecium (female reproductive organs). The male gametophyte consists of pollen, which contains two sperm cells; the female gametophyte is the embryo sac, which contains eight haploid nuclei, of which one is the nucleus of the egg cell.

42.4 Pollination and Fertilization

Pollination is the process by which pollen is placed on the stigma. Pollen may be carried to the flower by wind or by animals, or it may originate within the individual flower itself. When pollen from a flower's anther pollinates the same flower's stigma, the process is called *self-pollination*. When pollen from the anther of one flower pollinates the stigma of a different flower, the process is termed *cross-pollination*, or *outcrossing*.

As you just learned, pollination in angiosperms does not involve direct contact between the pollen grain and the ovule.

When pollen reaches the stigma, it germinates, and a pollen tube grows down, carrying the sperm nuclei to the embryo sac. After double fertilization takes place, development of the embryo and endosperm begins. The seed matures within the ripening fruit; eventually, the germination of the seed initiates another life cycle.

Successful pollination in many angiosperms depends on the regular attraction of **pollinators,** such as insects, birds, and other animals, which transfer pollen between plants of the same

chapter 42 *plant reproduction* **839**

species. When animals disperse pollen, they perform the same function for flowering plants that they do for themselves when they actively search out mates.

The relationship between plant and pollinator can be quite intricate. Mutations in either partner can block reproduction. If a plant flowers at the "wrong" time, the pollinator may not be available. If the morphology of the flower or pollinator is altered, the result may be physical barriers to pollination. Clearly, floral morphology has coevolved with pollinators, and the result is a much more complex and diverse morphology, going beyond the simple initiation and development of four distinct whorls of organs.

Early seed plants were wind-pollinated

Early seed plants were pollinated passively, by the action of the wind. As in present-day conifers, great quantities of pollen were shed and blown about, occasionally reaching the vicinity of the ovules of the same species.

Individual plants of any wind-pollinated species must grow relatively close to one another for such a system to operate efficiently. Otherwise, the chance that any pollen will arrive at an appropriate destination is very small. The vast majority of windblown pollen travels less than 100 m. This short distance is significant compared with the long distances pollen is routinely carried by certain insects, birds, and other animals.

Flowers and animal pollinators have coevolved

The spreading of pollen from plant to plant by pollinators visiting flowers of an angiosperm species has played an important role in the evolutionary success of the group. It now seems clear that the earliest angiosperms, and perhaps their ancestors also, were insect-pollinated, and the coevolution of insects and plants has been important for both groups for over 100 million years. Such interactions have also been important in bringing about increased floral specialization. As flowers become increasingly specialized, so do their relationships with particular groups of insects and other animals.

Bees

Among insect-pollinated angiosperms, the most numerous groups are those pollinated by bees (figure 42.21). Like most insects, bees initially locate sources of food by odor and then orient themselves on the flower or group of flowers by its shape, color, and texture.

Flowers that bees characteristically visit are often blue or yellow. Many have stripes or lines of dots that indicate the location of the nectaries, which often occur within the throats of specialized flowers. Some bees collect nectar, which is used as a source of food for adult bees and occasionally for larvae. Most of the approximately 20,000 species of bees visit flowers to obtain pollen, which is used to provide food in cells where bee larvae complete their development.

Except for a few hundred species of social and semisocial bees and about 1000 species that are parasitic in the nests of other bees, the great majority of bees—at least 18,000 species—are solitary. Solitary bees in temperate regions characteristically produce only a single generation in the course of a year. Often, they are active as adults for as little as a few weeks a year.

figure 42.21

POLLINATION BY A BUMBLEBEE. As this bumblebee, *Bombus* sp., squeezes into the bilaterally symmetrical, advanced flower of a member of the mint family, the stigma contacts its back and picks up any pollen that the bee may have acquired during a visit to a previous flower.

Solitary bees often use the flowers of a particular group of plants almost exclusively as sources of their larval food. The highly constant relationships of such bees with those flowers may lead to modifications, over time, in both the flowers and the bees. For example, the time of day when the flowers open may correlate with the time when the bees appear; the mouthparts of the bees may become elongated in relation to tubular flowers; or the bees' pollen-collecting apparatuses may be adapted to the anthers of the plants that they normally visit. When such relationships are established, they provide both an efficient mechanism of pollination for the flowers and a constant source of food for the bees that "specialize" on them.

Insects other than bees

Among flower-visiting insects other than bees, a few groups are especially prominent. Flowers such as phlox, which are visited regularly by butterflies, often have flat "landing platforms" on which butterflies perch. They also tend to have long, slender floral tubes filled with nectar that is accessible to the long, coiled proboscis characteristic of Lepidoptera, the order of insects that includes butterflies and moths.

Flowers such as jimsonweed (*Datura stramonium*), evening primrose (*Oenothera biennis*), and others visited regularly by moths are often white, yellow, or some other pale color; they also tend to be heavily scented, making the flowers easy to locate at night (figure 42.22).

Birds

Several interesting groups of plants are regularly visited and pollinated by birds, especially the hummingbirds of North and South America and the sunbirds of Africa (figure 42.23). Such

figure 42.22

MOTHS AS POLLINATORS.

figure 42.23

HUMMINGBIRDS AND FLOWERS. A long-tailed hermit hummingbird extracts nectar from the flowers of *Heliconia imbricata* in the forests of Costa Rica. Note the pollen on the bird's beak. Hummingbirds of this group obtain nectar primarily from long, curved flowers that more or less match the length and shape of their beaks.

plants must produce large amounts of nectar because birds will not continue to visit flowers if they do not find enough food to maintain themselves. But flowers producing large amounts of nectar have no advantage in being visited by insects, because an insect could obtain its energy requirements at a single flower and would not cross-pollinate the flower. How are these different selective forces balanced in flowers that are "specialized" for hummingbirds and sunbirds?

The answer involves the evolution of flower color. Ultraviolet light is highly visible to insects. Carotenoids, the yellow or orange pigments we described in chapter 8 in the context of photosynthesis, are responsible for the colors of many flowers, including sunflowers and mustard. Carotenoids reflect both in the yellow range and in the ultraviolet range, the mixture resulting in a distinctive color called "bee's purple." Such yellow flowers may also be marked in distinctive ways normally invisible to us, but highly visible to bees and other insects (figure 42.24). These markings can be in the form of a bull's-eye or a landing strip.

In contrast, red does not stand out as a distinct color to most insects, but it is a very conspicuous color to birds. To most insects, the red upper leaves of poinsettias look just like the other leaves of the plant. Consequently, even though the flowers produce abundant supplies of nectar and attract hummingbirds, insects tend to bypass them. Thus, the red color both signals to birds the presence of abundant nectar and makes that nectar as inconspicuous as possible to insects. Red is also seen again in fruits that are dispersed by birds (see chapter 37).

Other animal pollinators

Other animals, including bats and small rodents, may aid in pollination. The signals here are also species-specific. As an example, the saguaro cactus (*Carnegeia gigantea*) of the Sonoran desert is pollinated by bats that feed on nectar at night, as well as by birds and insects.

a.

b.

figure 42.24

HOW A BEE SEES A FLOWER. *a.* The yellow flower of *Ludwigia peruviana* (Peruvian primrose) photographed in normal light and *(b)* with a filter that selectively transmits ultraviolet light. The outer sections of the petals reflect both yellow and ultraviolet, a mixture of colors called "bee's purple"; the inner portions of the petals reflect yellow only and therefore appear dark in the photograph that emphasizes ultraviolet reflection. To a bee, this flower appears as if it has a conspicuous central bull's-eye.

figure 42.25

STAMINATE AND PISTILLATE FLOWERS OF A BIRCH, *Betula*
sp. Birches are monoecious; their staminate flowers hang down
in long, yellowish tassels, while their pistillate flowers mature
into clusters of small, brownish, conelike structures.

These animals may also assist in dispersing the seeds and
fruits that result from pollination. Monkeys are attracted to orange and yellow, and thus can be effective in dispersing fruits of
this color in their habitats.

Some flowering plants continue to use wind pollination

A number of groups of angiosperms, are wind-pollinated—a
characteristic of early seed plants. Among these groups are
oaks, birches, cottonwoods, grasses, sedges, and nettles. The
flowers of these plants are small, greenish, and odorless; their
corollas are reduced or absent (figures 42.25 and 42.26). Such
flowers often are grouped together in fairly large numbers and
may hang down in tassels that wave about in the wind and shed
pollen freely.

Many wind-pollinated plants have stamen- and carpel-containing flowers separated between individuals or
physically separated on a single individual. Maize is a good
example, with pollen-producing tassels at the top of the
plant and axillary shoots with female flowers lower down.
Separation of pollen-producing and ovule-bearing flowers
is a strategy that greatly promotes outcrossing, since pollen
from one flower must land on a different flower for fertilization to have any chance of occurring. Some wind-pollinated
plants, especially trees and shrubs, flower in the spring, before the development of their leaves can interfere with the
wind-borne pollen. Wind-pollinated species do not depend

on the presence of a pollinator for species survival, which
may be another survival advantage.

Self-pollination is favored in stable environments

Thus far we have considered examples of pollination that tend
to lead to outcrossing, which is as highly advantageous for
plants and for eukaryotic organisms generally. Nevertheless,
self-pollination also occurs among angiosperms, particularly
in temperate regions. Most self-pollinating plants have small,
relatively inconspicuous flowers that shed pollen directly onto
the stigma, sometimes even before the bud opens.

You might logically ask why many self-pollinated plant
species have survived if outcrossing is as important genetically
for plants as it is for animals. Biologists propose two basic reasons for the frequent occurrence of self-pollinated angiosperms:

1. Self-pollination is ecologically advantageous under certain
   circumstances because self-pollinators do not need to
   be visited by animals to produce seed. As a result, self-
   pollinated plants expend less energy in producing pollinator
   attractants and can grow in areas where the kinds of insects
   or other animals that might visit them are absent or very
   scarce—as in the Arctic or at high elevations.
2. In genetic terms, self-pollination produces progenies that
   are more uniform than those that result from outcrossing.
   Remember that because meiosis is involved, recombination
   still takes place, as described in chapter 11—and therefore

figure 42.26

**WIND-POLLINATED
FLOWERS.** The
large yellow anthers,
dangling on very
slender filaments, are
hanging out, about to
shed their pollen to
the wind. Later, these
flowers will become
pistillate, with long,
feathery stigmas—
well suited for
trapping windblown
pollen—sticking far
out of them. Many
grasses, like this
one, are therefore
dichogamous.

the offspring will not be identical to the parent. However, such progenies may contain high proportions of individuals well-adapted to particular habitats.

Self-pollination in normally outcrossing species tends to produce large numbers of ill-adapted individuals because it brings together deleterious recessive alleles—but some of these combinations may be highly advantageous in particular habitats. In these habitats, it may be advantageous for the plant to continue self-pollinating indefinitely.

Several evolutionary strategies promote outcrossing

Outcrossing, as we have stressed, is critically important for the adaptation and evolution of all eukaryotic organisms, with a few exceptions. Often, flowers contain both stamens and pistils, which increases the likelihood of self-pollination. One general strategy to promote outcrossing, therefore, is to separate stamens and pistils. Another strategy involves self-incompatibility that prevents self-fertilization.

Separation of male and female structures in space or in time

In a number of species—for example, willows and some mulberries—staminate and pistillate flowers may occur on separate plants. Such plants, which produce only ovules or only pollen, are called **dioecious,** meaning "two houses." These plants clearly cannot self-pollinate and must rely exclusively on outcrossing. In other kinds of plants, such as oaks, birches, corn (maize), and pumpkins, separate male and female flowers may both be produced on the same plant. Such plants are called **monoecious,** meaning "one house" (see figure 42.25). In monoecious plants, the separation of pistillate and staminate flowers, which may mature at different times, greatly enhances the probability of outcrossing.

Even if, as usually is the case, functional stamens and pistils are both present in each flower of a particular plant species, these organs may reach maturity at different times. Plants in which this occurs are called **dichogamous.** If the stamens mature first, shedding their pollen before the stigmas are receptive, the flower is effectively staminate at that time. Once the stamens have finished shedding pollen, the stigma or stigmas may become receptive, and the flower may become essentially pistillate (figures 42.26 and 42.27). This separation in time has the same effect as if individuals were dioecious; the outcrossing rate is thereby significantly increased.

Many flowers are constructed such that the stamens and stigmas do not come in contact with each other. With this arrangement, the natural tendency is for the pollen to be transferred to the stigma of another flower, rather than to the stigma of its own flower, thereby promoting outcrossing.

Self-incompatibility

Even when a flower's stamens and stigma mature at the same time, genetic **self-incompatibility,** which is widespread in flowering plants, increases outcrossing. Self-incompatibility results

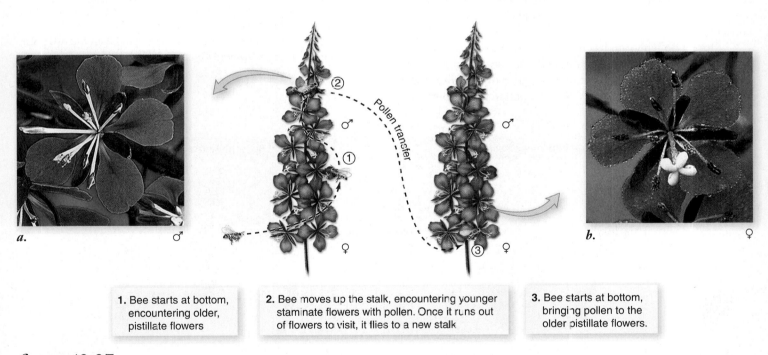

1. Bee starts at bottom, encountering older, pistillate flowers

2. Bee moves up the stalk, encountering younger staminate flowers with pollen. Once it runs out of flowers to visit, it flies to a new stalk

3. Bee starts at bottom, bringing pollen to the older pistillate flowers.

figure 42.27

DICHOGAMY, AS ILLUSTRATED BY THE FLOWERS OF FIREWEED, *Epilobium angustifolium.* More than 200 years ago (in the 1790s) fireweed, which is outcrossing, was one of the first plant species to have its process of pollination described. First, the anthers shed pollen, and then the style elongates above the stamens while the four lobes of the stigma curl back and become receptive. Consequently, the flowers are functionally staminate at first, becoming pistillate about two days later. The flowers open progressively up the stem, so that the lowest are visited first, promoting outcrossing. Working up the stem, the bees encounter pollen-shedding, staminate-phase flowers and become covered with pollen, which they then carry to the lower, functionally pistillate flowers of another plant. Shown here are flowers in (*a*) the staminate phase and (*b*) the pistillate phase.

when the pollen and stigma recognize each other as being genetically related, and pollen tube growth is blocked (figure 42.28).

Self-incompatibility is controlled by the *S* (self-incompatibility) locus. Many alleles at the *S* locus regulate recognition responses between pollen and stigma. Researchers have identified two types of self-incompatibility. *Gametophytic self-incompatibility* depends on the haploid *S* locus of the pollen and the diploid *S* locus of the stigma. If either of the *S* alleles in the stigma matches the pollen's *S* allele, pollen tube growth stops before it reaches the embryo sac. Petunias exhibit gametophytic self-incompatibility.

In *sporophytic self-incompatibility*, as occurs in broccoli, both *S* alleles of the pollen parent, not just the *S* allele of the pollen itself, are important. If the alleles in the stigma match either of the pollen parent's *S* alleles, the haploid pollen will not germinate.

Much is being learned about the molecular and biochemical basis of recognition mechanisms and the signal transduction pathways that block the successful growth of the pollen tube. Pollen-recognition mechanisms may have originated in a common ancestor of the gymnosperms. Fossils with pollen tubes from the Carboniferous period are consistent with the hypothesis that they had highly evolved pollen-recognition systems.

Angiosperms undergo double fertilization

Fertilization in angiosperms is a complex, somewhat unusual process in which two sperm cells are utilized in a unique process called **double fertilization.** Double fertilization results in two key developments: (1) the fertilization of the egg, and (2) the formation of a nutrient substance called **endosperm** that nourishes the embryo.

Once a pollen grain has been spread by wind, by animals, or through self-pollination, it adheres to the sticky, sugary substance that covers the stigma and begins to grow a **pollen tube** that pierces the style (figure 42.29). The pollen tube, nourished

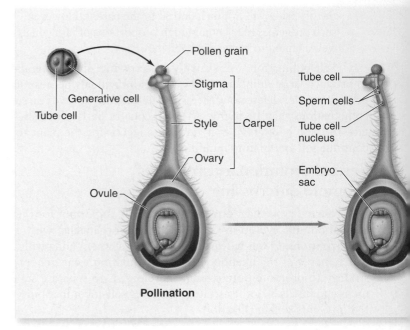

figure 42.29

THE FORMATION OF THE POLLEN TUBE AND DOUBLE FERTILIZATION. When pollen lands on the stigma of a flower, the pollen tube cell grows toward the embryo sac, forming a pollen tube. While the pollen tube is growing, the generative cell divides to form two sperm cells. When the pollen tube reaches the embryo sac, it enters one of the synergids and releases the sperm cells. In a process called double fertilization, one sperm cell nucleus fuses with the egg cell to form the diploid ($2n$) zygote, and the other sperm cell nucleus fuses with the two polar nuclei to form the triploid ($3n$) endosperm nucleus.

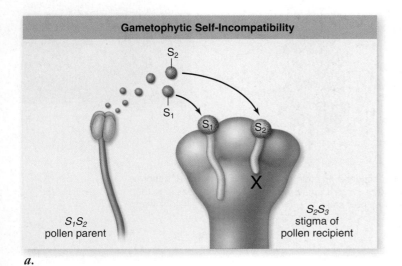

a.

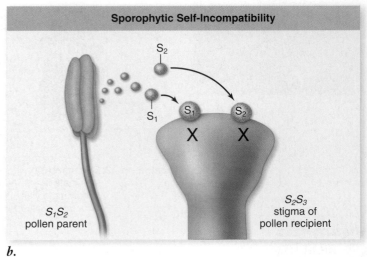

b.

figure 42.28

SELF-POLLINATION CAN BE GENETICALLY CONTROLLED SO SELF-POLLINATION IS BLOCKED. *a.* Gametophytic self-incompatibility is determined by the haploid pollen genotype. *b.* Sporophytic self-incompatibility recognizes the genotype of the diploid pollen parent, not just the haploid pollen genotype. The pollen contains proteins produced by the S_1S_2 parent. In both cases, the recognition is based on the *S* locus, which has many different alleles. The subscript numbers indicate the *S* allele genotype. In gametophytic self-incompatibility, the block comes after pollen tube germination. In sporophytic self-incompatibility, the pollen tube fails to germinate.

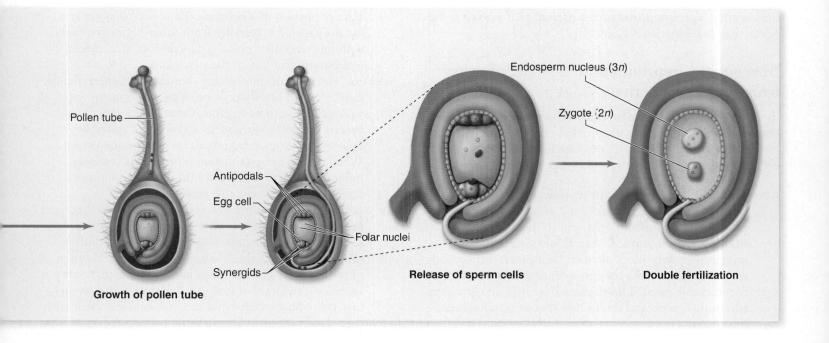

Pollen tube

Antipodals
Egg cell
Polar nuclei
Synergids

Growth of pollen tube

Release of sperm cells

Endosperm nucleus (3n)
Zygote (2n)

Double fertilization

by the sugary substance, grows until it reaches the ovule in the ovary. Meanwhile, the generative cell within the pollen grain tube cell divides to form two sperm cells.

The pollen tube eventually reaches the embryo sac in the ovule. At the entry to the embryo sac, one of the nuclei flanking the egg cell degenerates, and the pollen tube enters that cell. The tip of the pollen tube bursts and releases the two sperm cells. One of the sperm cells fertilizes the egg cell, forming a zygote. The other sperm cell fuses with the two polar nuclei located at the center of the embryo sac, forming the triploid ($3n$) primary endosperm nucleus. The primary endosperm nucleus eventually develops into the endosperm.

Once fertilization is complete, the embryo develops as its cells divide numerous times. Meanwhile, protective tissues en-

close the embryo, resulting in the formation of the seed. The seed, in turn, is enclosed in another structure, called the fruit. These typical angiosperm structures evolved in response to the need for seeds to be dispersed over long distances to ensure genetic variability.

Flowers, with their ability to attract pollinators, contribute to the evolutionary success of the angiosperms. While wind pollination and self-crossing are advantageous in some situations, outcrossing promotes genetic diversity, the raw material of evolution. Double fertilization produces a diploid zygote and a triploid nutrient tissue, the endosperm.

Asexual Reproduction

Self-pollination reduces genetic variability, but asexual reproduction results in genetically identical individuals because only mitotic cell divisions occur. In the absence of meiosis, individuals that are highly adapted to a relatively unchanging environment persist for the same reasons that self-pollination is favored. Should conditions change dramatically, there will be less variation in the population for natural selection to act upon, and the species may be less likely to survive.

Asexual reproduction is also used in agriculture and horticulture to propagate a particularly desirable plant with traits that would be altered by sexual reproduction or even by self-pollination. Most roses and potatoes, for example, are vegetatively (asexually) propagated.

Apomixis involves development of diploid embryos

In certain plants, including some citruses, certain grasses (such as Kentucky bluegrass), and dandelions, the embryos in the seeds may be produced asexually from the parent plant. This kind of asexual reproduction is known as **apomixis.** Seeds produced in this way give rise to individuals that are genetically identical to their parents.

Although these plants reproduce by cloning diploid cells in the ovule, they also gain the advantage of seed dispersal, an adaptation usually associated with sexual reproduction. Asexual reproduction in plants is far more common in harsh or marginal environments, where there is little leeway for variation.

For example, a greater proportion of asexual plants occur in the Arctic than in temperate regions.

In vegetative reproduction, new plants arise from nonreproductive tissues

In a very common form of asexual reproduction called **vegetative reproduction,** new plant individuals are simply cloned from parts of adults (figure 42.30). The forms of vegetative reproduction in plants are many and varied.

Runners or stolons. Some plants reproduce by means of *runners* (also called stolons)—long, slender stems that grow along the surface of the soil. In the cultivated strawberry, for example, leaves, flowers, and roots are produced at every other node on the runner. Just beyond each second node, the tip of the runner turns up and becomes thickened. This thickened portion first produces adventitious roots and then a new shoot that continues the runner.

Rhizomes. Underground horizontal stems, or *rhizomes,* are also important reproductive structures, particularly in grasses and sedges. Rhizomes invade areas near the parent plant, and each node can give rise to a new flowering shoot. The noxious character of many weeds results from this type of growth pattern, and many garden plants, such as irises, are propagated almost entirely from rhizomes. Corms and bulbs are vertical underground stems. Tubers are also stems specialized for storage and reproduction. Tubers are the terminal storage portion of a rhizome. Potatoes (*Solanum* spp.) are propagated artificially from tuber segments, each with one or more "eyes." The eyes, or "seed pieces," of a potato give rise to the new plant.

Suckers. The roots of some plants—for example, cherry, apple, raspberry, and blackberry—produce *suckers,* or sprouts, which give rise to new plants. Commercial varieties of banana do not produce seeds and are propagated by suckers that develop from buds on underground stems. When the root of a dandelion is broken, as it may be if one attempts to pull it from the ground, each root fragment may give rise to a new plant.

Adventitious plantlets. In a few plant species, even the leaves are reproductive. One example is the houseplant *Kalanchoë daigremontiana* (see figure 42.30), familiar to many people as the "maternity plant," or "mother of thousands." The common names of this plant are based on the fact that numerous plantlets arise from meristematic tissue located in notches along the leaves. The maternity plant is ordinarily propagated by means of these small plants, which, when they mature, drop to the soil and take root.

Plants can be cloned from isolated cells in the laboratory

Whole plants can be cloned by regenerating plant cells or tissues on nutrient medium with growth hormones. This is another form of asexual reproduction. Cultured leaf, stem, and root tissues can undergo organogenesis in culture and form

figure 42.30

VEGETATIVE REPRODUCTION. Small plants arise from notches along the leaves of the house plant *Kalanchoë daigremontiana.* The plantlets can fall off and grow into new plants, an unusual form of vegetative reproduction.

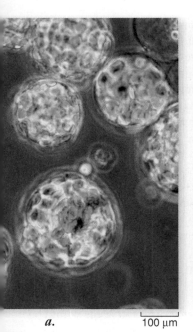   

a. 100 μm *b.* 1 μm *c.* 1 μm *d.* 1 μm

figure 42.31

PROTOPLAST REGENERATION. Different stages in the recovery of intact plants from single plant protoplasts of evening primrose. *a.* Individual plant protoplasts. *b.* Regeneration of the cell wall and the beginning of cell division. *c.* Production of somatic cell embryos from the callus. *d.* Recovery of a plantlet from the somatic cell embryo in culture. The plant can later be rooted in soil.

roots and shoots. In some cases, individual cells can also give rise to whole plants in culture.

Individual cells can be isolated from tissues with enzymes that break down cell walls, leaving behind the *protoplast*, a plant cell enclosed only by a plasma membrane. Plant cells have greater developmental plasticity than most vertebrate animal cells, and many, but not all, cell types in plants maintain the ability to generate organs or an entire organism in culture. Consider the limited number of adult stem cells in vertebrates and the challenges associated with cloning discussed in chapter 19.

When single plant cells are cultured, wall regeneration takes place. Cell division follows to form a *callus*, an undifferentiated mass of cells (figure 42.31). Once a callus is formed,

whole plants can be produced in culture. Whole-plant development can go through an embryonic stage or can start with the formation of a shoot or root.

Tissue culture has many agricultural and horticultural applications. Virus-free raspberries and sugarcane can be propagated by culturing meristems, which are generally free of viruses, even in an infected plant. As with other forms of asexual reproduction, genetically identical individuals can be propagated.

Plants that reproduce asexually clone new individuals from portions of the root, stem, leaves, or ovules of adult individuals. Asexually produced progeny are genetically identical to the parent.

42.6 Plant Life Spans

Once established, plants live for highly variable periods of time, depending on the species. Life span may or may not correlate with reproductive strategy. Woody plants, which have extensive secondary growth, nearly always live longer than herbaceous plants, which have limited or no secondary growth. Bristlecone pine, for example, can live upward of 4000 years.

Some herbaceous plants send new stems above the ground every year, producing them from woody underground structures. Others germinate and grow, flowering just once before they die. Shorter-lived plants rarely become very woody because there is not enough time for secondary tissues to accumulate. Depending on the length of their life cycles, herbaceous plants may be annual, biennial, or perennial, whereas woody plants are generally perennial (figure 42.32).

Determining life span is even more complicated for clonally reproducing organisms. Aspen trees (*Populus tremuloides*) form huge clones from asexual reproduction of their roots. Collectively, an aspen clone may form the largest "organism" on Earth. Other asexually reproducing plants may cover less territory but live for thousands of years. Creosote bushes (*Larrea tridentata*) in the Mojave Desert have been identified that are up to 12,000 years old!

Perennial plants live for many years

Perennial plants continue to grow year after year and may be herbaceous (as are many woodland, wetland, and prairie wildflowers), or woody (as are trees and shrubs). The majority of vascular plant species are perennials. Perennial plants in general are able to flower and produce seeds and fruit for an indefinite number of growing seasons.

Herbaceous perennials rarely experience any secondary growth in their stems; the stems die each year after a period of relatively rapid growth and food accumulation. Food is often stored in the plants' roots or underground stems, which can become quite large in comparison with their less substantial aboveground counterparts.

Trees and shrubs generally flower repeatedly, but there are exceptions. Bamboo lives for many seasons as a nonreproducing

a.

figure 42.32

ANNUAL AND PERENNIAL PLANTS. Plants live for very different lengths of time. *a.* Desert annuals complete their entire life span in a few weeks, flowering just once. *b.* Some trees, such as the giant redwood (*Sequoiadendron giganteum*), which occurs in scattered groves along the western slopes of the Sierra Nevada in California, live 2000 years or more, and flower year after year.

b.

plant, but senesces and dies after flowering. The same is true for at least one tropical tree (*Tachigali versicolor*), which achieves great heights before flowering and senescing. Considering the tremendous amount of energy that goes into the growth of a tree, this particular reproductive strategy is quite curious.

Trees and shrubs are either *deciduous*, with all the leaves falling at one particular time of year and the plants remaining bare for a period, or *evergreen*, with the leaves dropping throughout the year and the plants never appearing completely bare. In northern temperate regions, conifers are the most familiar evergreens, but in tropical and subtropical regions, most angiosperms are evergreen, except where there is severe seasonal drought. In these areas, many angiosperms are deciduous, losing their leaves during the drought and thus conserving water.

Annual plants grow, reproduce, and die in a single year

Annual plants grow, flower, and form fruits and seeds within one growing season and die when the process is complete. Many crop plants are annuals, including corn, wheat, and soybeans. Annuals generally grow rapidly under favorable conditions and in proportion to the availability of water or nutrients. The lateral meristems of some annuals, such as sunflowers or giant ragweed, do produce some secondary tissues for support, but most annuals are entirely herbaceous.

Annuals typically die after flowering once; the developing flowers or embryos use hormonal signaling to reallocate nutrients, so the parent plant literally starves to death. This can be demonstrated by comparing a population of bean plants in which the beans are continually picked with a population in which the beans are left on the plant. The frequently picked population will continue to grow and yield beans much longer than the untouched population. The process that leads to the death of a plant is called **senescence.**

Biennial plants follow a two-year life cycle

Biennial plants, which are much less common than annuals, have life cycles that take two years to complete. During the first year, biennials store the products of photosynthesis in underground storage organs. During the second year of growth, flowering stems are produced using energy stored in the underground parts of the plant. Certain crop plants, including carrots, cabbage, and beets, are biennials, but these plants generally are harvested for food during their first season, before they flower. They are grown for their leaves or roots, not for their fruits or seeds.

Wild biennials include evening primroses, Queen Anne's lace (*Daucus carota*), and mullein (*Verbascum thapsis*). Many plants that are considered biennials actually do not flower until they are three or more years of age, but all biennial plants flower only once before they die.

Plants have different strategies for dealing with the intense energy demands of reproduction. Perennials flower repeatedly and live for many years. Annual plants complete their whole growth cycle within a single year. Biennial plants flower only once, normally after two seasons of growth.

42.1 Reproductive Development (figure 42.1)

Plants go through developmental stages leading to reproductive maturity by adding structures to existing structures with their meristems.

- Plant life cycles are characterized by an alternation of generations in which the diploid sporophyte gives rise to a haploid gametophyte.
- Before flower formation can occur, plants must undergo a phase change to prepare a plant to respond to external or internal signals.
- Once plants are mature, flowers are produced by a combination of factors: light, temperature, and both promotive and inhibitory signals.

42.2 Flower Production

Flower production can require up to four genetically regulated pathways: the light-dependent pathway, the temperature-dependent pathway, the gibberellin pathway, and the autonomous pathway.

- The photoperiodic, or light-dependent, pathway is sensitive to the amount of darkness a plant receives each 24-hr period.
- Flowering may be regulated by a balance of floral-promoting and -inhibiting signals regardless of pathway.
- The temperature-dependent pathway requires vernalization, a period of cold before flowering.
- The gibberellin-dependent pathway requires an increase in this hormone for flowering to occur.
- The autonomous pathway typical of day-neutral plants depends on nutrition and is independent of environmental cues. These plants "count" nodes and "remember" node location by a balance between floral-promoting and -inhibiting signals.
- The four flowering pathways lead an adult meristem to become a floral meristem by activating floral meristem identity genes.
- Floral meristem identity genes activate floral organ identity genes that can in turn produce four floral organs using the ABCDE gene class model (figure 42.11).

42.3 Structure and Evolution of Flowers

(figures 42.14, 42.18)

Flowers house the haploid gametophyte generation and function to increase the probability that male and female gametes, usually from different plants, will unite.

- Floral organs are believed to have evolved from leaves.
- Complete flowers have four whorls: calyx, corolla, androecium, and gynoecium, whereas incomplete flowers lack one or more of the whorls.
- Modification and loss of floral parts is often related to pollination mechanisms.
- Angiosperms may have either radially or bilaterally symmetrical flowers.
- Pollen grains are the male gametophytes, or microgametophytes.
- *Pollen grains form in the anther by meiosis.*
- *Subsequently the haploid microspores divide by mitosis to form four pollen grains, which undergo further mitosis.*
- *Each pollen grain consists of the generative cell that will divide to produce two sperm cells and a cell with the pollen tube nucleus.*
- *Pollen grains contain furrows or pores from which the pollen tube will emerge.*
- The female gametophyte or megagametophyte is the embryo sac.
- *Eggs develop in the ovules from megaspore mother cells.*

- *Megaspore mother cells undergo meiosis to form four haploid megaspores, and three of these megaspores usually disintegrate.*
- *The remaining megaspore undergoes three mitotic divisions to form eight nuclei enclosed in a seven-celled embryo sac.*
- *One cell becomes the egg, and it is flanked by two haploid cells called synergids.*
- *Three haploid antipodal cells are located opposite the egg.*
- *Two polar nuclei are located in one of the seven cells and ultimately become the endosperm after fertilization.*

42.4 Pollination and Fertilization (figure 42.29)

Pollination is the process by which pollen comes in contact with the stigma of the flower.

- Flowers are pollinated by the wind, animals, or within the flower itself.
- Self-pollination occurs when the pollen from an anther lands on the stigma of the same flower.
- Cross-pollination, or outcrossing, occurs when the pollen from one flower lands on the stigma of a flower of a different plant.
- Flowers and animal pollinators have coevolved, resulting in specialized relationships.
- Many wind-pollinated plants have stamen- and carpel-containing flowers on separate individuals or are physically separated on a single individual.
- Self-pollination is advantageous in stable environments, especially where pollinators are scarce, because these plants expend no energy in attracting pollinators and their progeny are more uniform and probably better adapted to the environment.
- Outcrossing is promoted in plants in which male and female structures are separated in space and time.
- Self-incompatibility prevents self-fertilization and promotes outcrossing by blocking pollen tube growth from genetically related plants.
- Angiosperms undergo double fertilization: fertilization of the egg to produce a diploid zygote and formation of the triploid endosperm that nourishes the embryo.

42.5 Asexual Reproduction

Asexual reproduction results in genetically identical individuals because progeny are produced by mitosis.

- Apomixis occurs in plants that asexually produce diploid embryos in seeds that can be dispersed.
- Vegetative reproduction occurs when new individuals are cloned from parts of an adult plant. Examples include runners, rhizomes, suckers, and adventitious plantlets.
- Plants can be cloned by regenerating plant cells or tissues grown on a medium with nutrients and hormones.

42.6 Plant Life Spans

Plants live for highly variable periods of time. Woody plants usually live longer than herbaceous species, and asexual clones may live for thousands of years.

- Perennial plants are able to flower and produce seeds and fruit for a varying number of growing seasons. They can be woody or herbaceous plants.
- Annual plants grow, flower, produce seeds and fruits, and typically die within one growing season. Annuals are typically herbaceous plants.
- Biennial plants complete their life cycle in two years. They store energy the first year and flower the second year.

SELF TEST

1. Based on the finding from the embryonic flower mutant of *Arabadopsis* and use of *Arabadopsis* for expression of the *LEAFY* gene, it has been show that:
   a. flowering is entirely controlled by external cues.
   b. genes expressed later in plant development signal flowering.
   c. gene expression early in development suppresses flowering.
   d. (b) and (c).

2. Your roommate is taking biology with you this semester and thinks he understands short- and long-day plants. He purchases one plant of each type and decides to see the difference himself by first trying to cause the short-day plant to flower. He places both plants under the same conditions and exposes each to a regimen of 10-hr days, expecting that the short-day plant will flower, and the long-day plant will not. You play a trick on your roommate and reverse the outcome. Specifically, what did you have to do?
   a. Lengthen the time each is exposed to light
   b. Shorten the time each is exposed to light
   c. Quickly expose the plants to light during the middle of the night
   d. None of the above

3. Which of the following does not serve as a cue for initiation of flowering?
   a. Circadian cycle
   b. Photoperiod
   c. Gibberellin levels
   d. Temperature

4. Which of the following would likely prevent flowering in a plant like tobacco (*Nicotiana tobacum*) that follows the autonomous pathway?
   a. Removal of the lower leaves
   b. Removal of the apical meristem
   c. Continual upward-moving formation of adventitious roots on the stem
   d. None of the above because the autonomous pathway sets a predetermined time when flowering occurs

5. Cryptochrome is responsible for:
   a. blue-light–regulated responses.
   b. phototropisms.
   c. photoperiodic responses.
   d. (a) and (c).

6. In Iowa, a company called Team Corn works to ensure that fields of seed corn outcross so that hybrid vigor can be maintained. They do this by removing the staminate (that is, pollen-producing) flowers from the corn plants. In an attempt to put Team Corn out of business, you would like to develop genetically engineered corn plants that
   a. contain Z genes to prevent germination of pollen on the stigmatic surface.
   b. contain S genes to stop pollen tube growth during self-fertilization.
   c. express B-type homeotic genes throughout developing flowers.
   d. express A-type homeotic genes throughout developing flowers.

7. Monoecious plants such as corn have either staminate or carpelate flowers. Knowing what you do about the molecular mechanisms of floral development, which of the following might explain the development of single-sex flowers?
   a. Expression of B-type genes in the presumptive carpel whorl will generate staminate flowers.
   b. Loss of A-type genes in the presumptive petal whorl will allow C-type and B-type genes to produce stamens instead of petals in that whorl.
   c. Restricting B-type gene expression to the presumptive petal whorl will generate carpelate flowers.
   d. All of these are correct.

8. You have been asked to collect plant sperm for a new plant-breeding program involving in vitro fertilization. Which of the following tissues would be a good source of sperm?
   a. Anthers
   b. Ovaries
   c. Stigma
   d. Spores

9. If you wanted to create a super tobacco plant to increase the number of leaves per acre on a tobacco farm, which of the following strategies would potentially work?
   a. Suppress more root growth in the plants.
   b. Decrease expression of the *LEAFY* gene in the shoot apical meristem.
   c. Harvest lower leaves as the plant is growing so that flowering is delayed.
   d. Remove flowers so that the plant will produce more vegetative internodes than normal.

10. One of the most notable differences between gamete formation in most animals and gamete formation in plants is that
    a. plants produce gametes in somatic tissue, whereas animals produce gametes in germ tissue.
    b. plants produce gametes by mitosis, whereas animals produce gametes by meiosis.
    c. plants produce only one of each gamete, but animals produce many gametes.
    d. plants produce gametes that are diploid, but animals produce gametes that are haploid.

11. If you were to discover a flower that was small, white, and heavily scented, its most likely pollinator would be
    a. bees.
    b. birds.
    c. humans.
    d. moths.

12. Under which of the following conditions would pollen from an S_2S_5 plant successfully pollinate an S_1S_5 flower?
    a. Using pollen from a carpelate flower to fertilize a staminate flower would be successful.
    b. If the plants used gametophytic self-incompatibility, half of the pollen would be successful.
    c. If the plants used sporophytic self-incompatibility, half of the pollen would be successful.
    d. Pollen from an S_2S_5 plant can never pollinate an S_1S_5 flower.

CHALLENGE QUESTIONS

1. We often have the impression that plants lack the ability to move around in the environment. This, however, is far from the truth. Discuss the variety of ways that plants successfully move throughout the environment.

2. Identify the pros and cons of plants that are wind-pollinated versus those that are animal-pollinated.

3. Compare the ecological advantages and disadvantages of plants that reproduce sexually, those that reproduce via apomixis, and those that have vegetative reproduction.

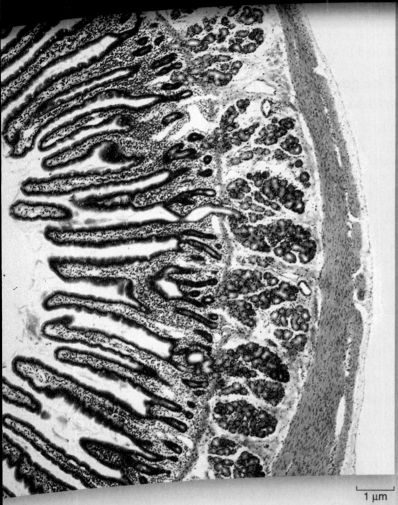

1 μm

chapter **43**

The Animal Body and Principles of Regulation

introduction

WHEN PEOPLE THINK OF ANIMALS, they may think of pet dogs and cats, the animals in a zoo, on a farm, in an aquarium, or wild animals living outdoors. When thinking about the diversity of animals, people may picture the differences between the predatory lions and tigers and the herbivorous deer and antelope, or between a ferocious-looking shark and a playful dolphin. Despite the differences among these animals, they are all vertebrates. All vertebrates share the same basic body plan, with similar tissues and organs that operate in much the same way. The micrograph shows a portion of the duodenum, part of the digestive system, which is made up of multiple types of tissues. In this chapter, we begin a detailed consideration of the biology of the vertebrates and the fascinating structure and function of their bodies. We conclude this chapter by exploring the principles involved in regulation and control of complex functional systems.

Organization of the Vertebrate Body

There are four levels of organization in the vertebrate body: (1) cells, (2) tissues, (3) organs, and (4) organ systems. Like those of all animals, the bodies of vertebrates are composed of different cell types. Depending on the group, between 50 and several hundred different kinds of cells contribute to the adult vertebrate body. Humans have 210 different cell types.

Tissues are groups of cells of a single type and function

Groups of cells that are similar in structure and function are organized into **tissues.** Early in development, the cells of the growing embryo differentiate into the three fundamental embryonic tissues, called **germ layers.** From the innermost to the outermost layers, these are the **endoderm, mesoderm,** and **ectoderm.** Each germ layer, in turn, differentiates into the scores of different cell types and tissues that are characteristic of the vertebrate body.

In adult vertebrates, there are four principal kinds of tissues, or **primary tissues:** They are **epithelial, connective, muscle,** and **nerve tissue,** and each type is discussed in separate sections of this chapter.

Organs and organ systems provide specialized functions

Organs are body structures composed of several different types of tissues that form a structural and functional unit (figure 43.1). One example is the heart, which contains cardiac muscle, connective tissue, and epithelial tissue. Nerve tissue connects the brain and spinal cord to the heart and helps regulate the heartbeat.

An **organ system** is a group of organs that cooperate to perform the major activities of the body. For example, the circulatory system is composed of the heart and blood vessels (arteries, capillaries, and veins) (see chapter 49). These organs cooperate in the transport of blood and help distribute substances about the body. The vertebrate body contains 11 principal organ systems.

The general body plan of vertebrates is a tube within a tube, with internal support

The bodies of all vertebrates have the same general architecture. The body plan is essentially a tube suspended within a tube. The inner tube is the digestive tract, a long tube that travels from the mouth to the anus. An internal skeleton made of jointed bones or cartilage that grows as the body grows supports the outer tube, which forms the main vertebrate body. The outermost layer of the vertebrate body is the integument, or the skin, and its many accessory organs and parts such as hair, feathers, scales, and sweat glands.

Vertebrates have both dorsal and ventral body cavities

Inside the main vertebrate body are two identifiable cavities. The **dorsal body cavity** forms within a bony skull and a column of bones, the vertebrae. The skull surrounds the brain, and within the stacked vertebrae is a channel that contains the spinal cord.

The **ventral body cavity** is much larger and extends anteriorly from the area bounded by the rib cage and vertebral column posteriorly to the area contained within the ventral body muscles (the abdominals) and the pelvic girdle. In mammals, a sheet of muscle, the diaphragm, breaks the ventral body cavity anteriorly into the **thoracic cavity,** which contains the heart and lungs, and posteriorly into the **abdominopelvic cavity,** which contains many organs, including the stomach, intestines, liver, kidneys, and urinary bladder (figure 43.2a).

Recall from the discussion of the animal body plan in chapter 32 that a coelom is a fluid-filled body cavity completely

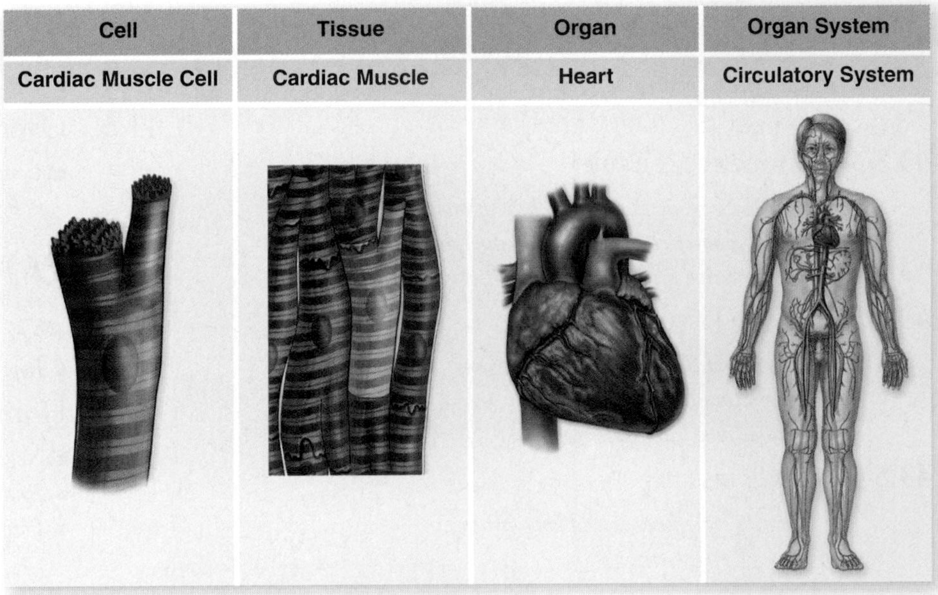

figure 43.1

LEVELS OF ORGANIZATION WITHIN THE BODY. Similar cell types operate together and form tissues. Tissues functioning together form organs. An organ system consists of several organs working together to carry out a function for the body. An example of an organ system is the circulatory system, which consists of the heart, blood vessels, and blood. The heart is composed primarily of cardiac muscle with a lining of epithelial tissue. Cardiac muscle is made of cardiac muscle cells.

Cell	Tissue	Organ	Organ System
Cardiac Muscle Cell	Cardiac Muscle	Heart	Circulatory System

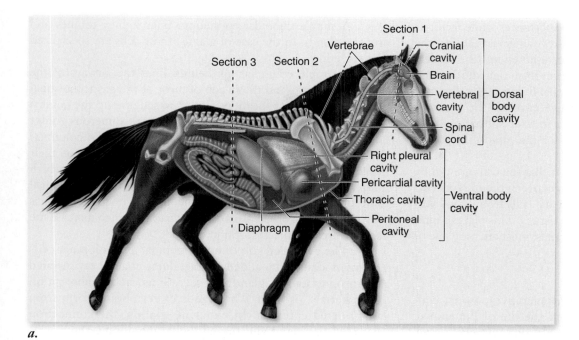

Section 1
Vertebrae
Section 3   Section 2

Cranial cavity
Brain
Vertebral cavity — Dorsal body cavity
Spinal cord

Right pleural cavity
Pericardial cavity
Thoracic cavity — Ventral body cavity
Diaphragm
Peritoneal cavity

a.

figure 43.2

ARCHITECTURE OF THE VERTEBRATE BODY. *a.* All vertebrates have dorsal and ventral body cavities. The dorsal cavity divides into the cranial (contains the brain) and vertebral (contains the spinal cord) cavities. In mammals, a muscular diaphragm divides the ventral cavity into the thoracic and abdominopelvic cavities. *b.* Cross-sections through three body regions show the relationships between body cavities, major organs, and coeloms (pericardial, pleural, and peritoneal cavities).

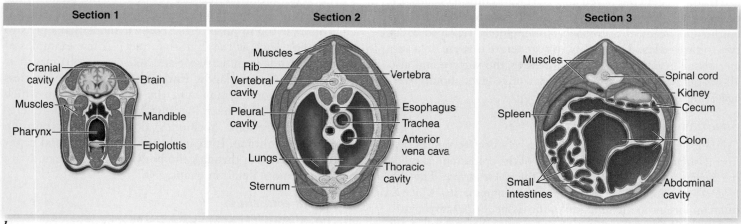

Section 1
Cranial cavity — Brain
Muscles
Pharynx — Mandible
— Epiglottis

Section 2
Muscles
Rib
Vertebral cavity
Pleural cavity — Vertebra
Lungs — Esophagus
Sternum — Trachea
Anterior vena cava
Thoracic cavity

Section 3
Muscles
Spleen — Spinal cord
Kidney
Cecum
Colon
Small intestines — Abdominal cavity

b.

formed within the embryonic mesoderm layer of some animals (vertebrates included). The coelom is still present in adults, only it is constricted, folded, and subdivided. The mesodermal layer that lines the coelom extends from the body wall to envelop and suspend several organs within the ventral body cavity (figure 43.2*b*). In the abdominopelvic cavity, the coelomic space is the **peritoneal cavity.**

In the thoracic cavity, the heart and lungs invade and greatly constrict the coelomic space. The thin space within mesodermal layers around the heart is the **pericardial cavity,** and

the two thin spaces around the lungs are the **pleural cavities** (figure 43.2*b*).

The body's cells are organized into tissues, which in turn are organized into organs and organ systems. The bodies of humans and other mammals contain dorsal and ventral cavities. The ventral body cavity is divided by the diaphragm into thoracic and abdominopelvic cavities. The adult coelom subdivides into the peritoneal, pericardial, and pleural cavities.

43.2   Epithelial Tissue

An epithelial membrane, or **epithelium** (plural, *epithelia*), covers every surface of the vertebrate body. Epithelial membranes can come from any of the three germ layers. For example, the epidermis, derived from ectoderm, constitutes the outer portion of the skin. An epithelium derived from endoderm lines the inner surface of the digestive tract, and the inner surfaces of blood vessels derive from mesoderm. Some epithelia change

in the course of embryonic development into glands, which are specialized for secretion.

Epithelium forms a barrier

Because epithelial membranes cover all body surfaces, a substance must pass through an epithelium in order to enter or leave the body. Epithelial membranes thus provide a barrier that can impede the passage of some substances while facilitating the passage of others. For land-dwelling vertebrates, the relative impermeability of the surface epithelium (the epidermis) to water offers essential protection from dehydration and from airborne pathogens. The epithelial lining of the digestive tract, in contrast, must allow selective entry of the products of digestion while providing a barrier to toxic substances. The epithelium of the lungs must allow for the rapid diffusion of gases into and out of the blood.

A characteristic of all epithelia is that the cells are tightly bound together, with very little space between them. Nutrients and oxygen must diffuse to the epithelial cells from blood vessels supplying underlying connective tissues. This places a limit on the thickness of epithelial membranes; most are only one or a few cell layers thick.

Epithelial regeneration

Epithelium possesses remarkable regenerative powers, constantly replacing its cells throughout the life of the animal. For example, the liver, a gland formed from epithelial tissue, can readily regenerate, even after surgical removal of substantial portions. The epidermis renews every two weeks, and the epithelium inside the stomach is completely replaced every two to three days. This ability to regenerate is useful in a surface tissue because it constantly renews the surface and also allows quick replacement of the protective layer should damage or injury occur.

Structure of epithelial tissues

Epithelial tissues attach to underlying connective tissues by a fibrous membrane. The secured side of the epithelium is called the *basal surface*, and the free side is the *apical surface*. This difference gives epithelial tissues an inherent polarity, which is often important in the function of the tissue. For example, proteins stud the basal surfaces of some epithelial tissues in the kidney tubules; these proteins actively transport Na$^+$ into the intercellular spaces, creating an osmotic gradient that helps return water to the blood (see chapter 50).

Epithelial types reflect their function

The two general classes of epithelial membranes are termed *simple* (single layer of cells) and *stratified* (multiple layers of cells). These classes are further subdivided into squamous, cuboidal, and columnar, based on the shape of the cells (table 43.1). **Squamous cells** are flat, **cuboidal cells** are about as wide as they are tall, and **columnar cells** are taller than they are wide.

Simple epithelium

As mentioned, **simple epithelial membranes** are one cell thick. A simple squamous epithelium is composed of squamous epithelial cells that have a flattened shape when viewed in cross-section. Examples of such membranes are those that line the lungs and blood capillaries, where the thin, delicate nature of these membranes permits the rapid movement of molecules (such as the diffusion of gases).

A simple cuboidal epithelium lines kidney tubules and several glands. In the case of glands, these cells are specialized for secretion.

A simple columnar epithelium lines the airways of the respiratory tract and the inside of most of the gastrointestinal tract, among other locations. Interspersed among the columnar epithelial cells of mucous membranes are numerous **goblet cells,** which are specialized to secrete mucus. The columnar epithelial cells of the respiratory airways contain cilia on their apical surface (the surface facing the lumen, or cavity), which move mucus and dust particles toward the throat. In the small intestine, the apical surface of the columnar epithelial cells forms fingerlike projections called *microvilli*, which increase the surface area for the absorption of food.

The expanded size of both cuboidal and columnar cells accommodates the added intracellular machinery needed for production of glandular secretions, active absorption of materials, or both. The glands of vertebrates form from invaginated epithelia. In **exocrine glands,** the connection between the gland and the epithelial membrane remains as a duct. The duct channels the product of the gland to the surface of the epithelial membrane, and thus to the external environment (or to an interior compartment that opens to the exterior, such as the digestive tract). A few examples of exocrine glands include sweat and sebaceous (oil) glands as well as the salivary glands. **Endocrine glands** are ductless glands; their connections with the epithelium from which they are derived has been lost during development. Therefore, their secretions (hormones) do not channel onto an epithelial membrane. Instead, hormones enter blood capillaries and circulate through the body. Endocrine glands are covered in more detail in chapter 46.

Stratified epithelium

Stratified epithelial membranes are two to several cell layers thick and are named according to the features of their apical cell layers. For example, the epidermis is a *stratified squamous epithelium*; its properties are discussed in chapter 51. In terrestrial vertebrates, the epidermis is further characterized as a *keratinized epithelium* because its upper layer consists of dead squamous cells and is filled with a water-resistant protein called *keratin*.

The deposition of keratin in the skin increases in response to repeated abrasion, producing calluses. The water-resistant property of keratin is evident when comparing the skin of the face to the red portion of the lips, which can easily become dried and chapped. Lips are covered by a nonkeratinized, stratified squamous epithelium.

> Epithelial tissues include membranes that cover all body surfaces and glands. The epidermis of the skin is an epithelial membrane specialized for protection, whereas membranes that cover the surfaces of hollow organs are often specialized for transport and secretion. Epithelial tissues are classified as simple (one cell layer) or stratified (two or more cell layers).

TABLE 43.1 | **Epithelial Tissue**

SIMPLE EPITHELIUM

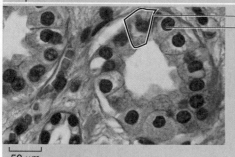

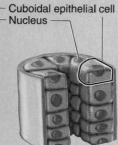

Simple squamous
epithelial cell

Nucleus

40 μm

Squamous
Typical Location
Lining of lungs, capillary walls, and blood vessels
Function
Cells very thin; provides thin layer across which diffusion can readily occur
Characteristic Cell Types
Epithelial cells

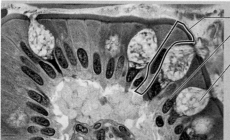

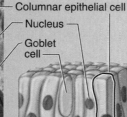

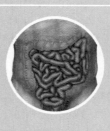

Cuboidal epithelial cell
Nucleus

50 μm

Cuboidal
Typical Location
Lining of some glands and kidney tubules; covering of ovaries
Function
Cells rich in specific transport channels; functions in secretion and absorption
Characteristic Cell Types
Gland cells

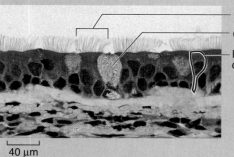

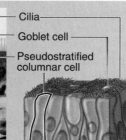

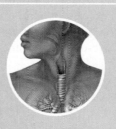

Columnar epithelial cell
Nucleus
Goblet cell

40 μm

Columnar
Typical Location
Surface lining of stomach, intestines, and parts of respiratory tract
Function
Thicker cell layer; provides protection and functions in secretion and absorption
Characteristic Cell Types
Epithelial cells

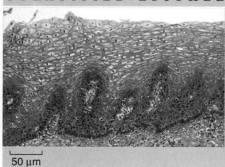

Cilia
Goblet cell
Pseudostratified columnar cell

40 μm

Pseudostratified Columnar
Typical Location
Lining of parts of the respiratory tract
Function
Secretes mucus; dense with cilia that aid in movement of mucus; provides protection
Characteristic Cell Types
Gland cells; ciliated epithelial cells

STRATIFIED EPITHELIUM

50 μm

Squamous
Typical Location
Outer layer of skin; lining of mouth
Function
Tough layer of cells; provides protection
Characteristic Cell Types
Epithelial cells

Connective Tissues

Connective tissues derive from embryonic mesoderm and occur in many different forms (table 43.2). We divide these various forms into two major classes: **connective tissue proper,** which further divides into loose and dense connective tissues, and **special connective tissues,** which include cartilage, bone, and blood.

At first glance, it may seem odd that such diverse tissues are in the same category. Yet all connective tissues share a common structural feature: They all have abundant extracellular material because their cells are spaced widely apart. This extracellular material is called the **matrix** of the tissue. In bone, the matrix contains crystals that make the bones hard; in blood, the matrix is plasma, the fluid portion of the blood. The matrix itself consists of protein fibers and **ground substance,** the fluid material between cells and fibers containing a diverse array of proteins and polysaccharrides.

Connective tissue proper may be either loose or dense

During the development of both loose and dense connective tissues, cells called **fibroblasts** produce and secrete the extracellular matrix. Loose connective tissue contains other cells as well, including mast cells and macrophages, cells of the immune system.

Loose connective tissue

Loose connective tissue consists of cells scattered within a matrix that contains a large amount of ground substance. This gelatinous material is strengthened by a loose scattering of protein fibers such as collagen, which supports the tissue by forming a collagenous meshwork (figure 43.3), elastin, which makes the tissue elastic, and reticulin, which helps support the network of collagen. The flavored gelatin of certain desserts consists primarily of extracellular material extracted from the loose connective tissues of animals.

Adipose cells, more commonly termed fat cells, are important for nutrient storage, and they also occur in loose con-

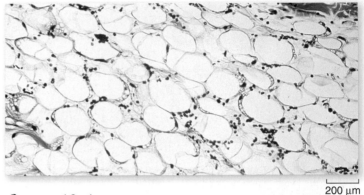

figure 43.4

ADIPOSE TISSUE. Fat is stored in globules of adipose tissue, a type of loose connective tissue. As a person gains or loses weight, the size of the fat globules increases or decreases. A person cannot decrease the number of fat cells by losing weight.

nective tissue. In certain areas of the body, including under the skin, in bone marrow, around the kidneys these cells can develop in large groups, forming **adipose tissue** (figure 43.4).

Each adipose cell contains a droplet of triglycerides within a storage vesicle. When needed for energy, the adipose cell hydrolyzes its stored triglyceride and secretes fatty acids into the blood for oxidation by the cells of the muscles, liver, and other organs. Adipose cells cannot divide; the number of adipose cells in an adult is generally fixed. When a person gains weight, the cells become larger, and when weight is lost, the cells shrink.

Dense connective tissue

Dense connective tissue with less ground substance, contains tightly packed collagen fibers, making it stronger than loose connective tissue. It consists of two types: regular and irregular. The collagen fibers of *dense regular connective tissue* line up in parallel, like the strands of a rope. This is the structure of tendons, which bind muscle to bone, and ligaments, which bind bone to bone.

In contrast, the collagen fibers of *dense irregular connective tissue* have many different orientations. This type of connective tissue produces the tough coverings that package organs, such as the capsules of the kidneys and adrenal glands. It also covers muscle, nerves, and bones.

Special connective tissues have unique characteristics

The special connective tissues—cartilage, bone, and blood—each have unique cells and matrices that allow them to perform their specialized functions.

Cartilage

Cartilage (see table 43.2) is a specialized connective tissue in which the ground substance forms from a characteristic type of glycoprotein, called *chondroitin*, and collagen fibers laid down along lines of stress in long, parallel arrays. The result is a firm

figure 43.3

COLLAGEN FIBERS. These fibers, shown under an electron microscope, are composed of many individual collagen strands and can be very strong under tension.

TABLE 43.2	Connective Tissue

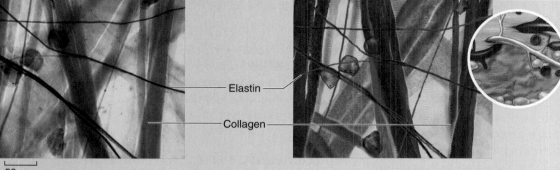

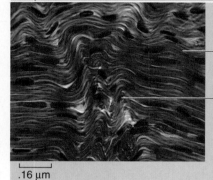
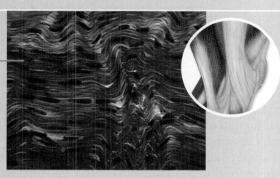

Elastin

Collagen

58 μm

Loose Connective Tissue
Typical Location
Beneath skin; between organs
Function
Provides support, insulation, food storage, and nourishment for epithelium
Characteristic Cell Types
Fibroblasts, macrophages, mast cells, fat cells

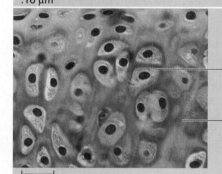
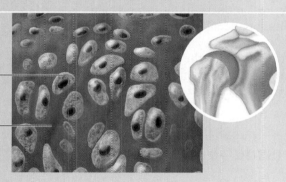

Collagen fibers

Nuclei of fibroblasts

.16 μm

Dense Connective Tissue
Typical Location
Tendons; sheath around muscles; kidney; liver; dermis of skin
Function
Provides flexible, strong connections
Characteristic Cell Types
Fibroblasts

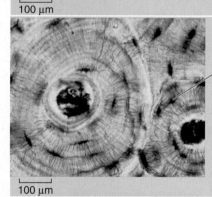

Chondrocyte

Ground substance

100 μm

Cartilage
Typical Location
Spinal disks; knees and other joints; ear; nose; tracheal rings
Function
Provides flexible support, shock absorption, and reduction of friction on load-bearing surfaces
Characteristic Cell Types
Chondrocytes

Osteocyte

100 μm

Bone
Typical Location
Most of skeleton
Function
Protects internal organs; provides rigid support for muscle attachment
Characteristic Cell Types
Osteocytes

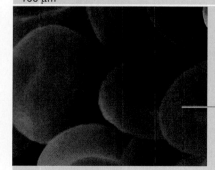
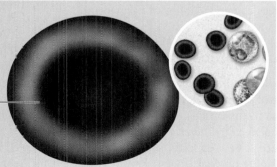

Red blood cell

5.8 μm

Blood
Typical Location
Circulatory system
Function
Functions as highway of immune system and primary means of communication between organs
Characteristic Cell Types
Erythrocytes, leukocytes

and flexible tissue that does not stretch, is far tougher than loose or dense connective tissue, and has great tensile strength.

Cartilage makes up the entire skeletal system of the modern agnathans and cartilaginous fishes (see chapter 34). In most adult vertebrates, however, cartilage is restricted to the joint surfaces of bones that form freely movable joints and certain other locations. In humans, for example, the tip of the nose, the outer ear, the intervertebral disks of the backbone, the larynx, and a few other structures are composed of cartilage.

Chondrocytes, the cells of cartilage, live within spaces called **lacunae** within the cartilage ground substance. These cells remain alive even though there are no blood vessels within the cartilage matrix; they receive oxygen and nutrients by diffusion through the cartilage ground substance from surrounding blood vessels. This diffusion can only occur because the cartilage matrix is well hydrated and not calcified, as is bone.

Bone

Bone cells, or **osteocytes,** remain alive even though the extracellular matrix becomes hardened with crystals of calcium phosphate. Blood vessels travel through central canals into the bone, providing nutrients and removing wastes. Osteocytes extend cytoplasmic processes toward neighboring osteocytes through tiny canals, or *canaliculi.* Osteocytes communicate with the blood vessels in the central canal through this cytoplasmic network. Bone is described in more detail in chapter 47 along with muscle.

In the course of fetal development, the bones of vertebrate fins, arms, and legs, among other appendages, are first "modeled" in cartilage. The cartilage matrix then calcifies at particular locations, so that the chondrocytes are no longer able to obtain oxygen and nutrients by diffusion through the matrix. Living bone replaces the dying and degenerating cartilage.

Blood

We classify **blood** as a connective tissue because it contains abundant extracellular material, the fluid plasma. The cells of blood are *erythrocytes,* or red blood cells, and *leukocytes,* or white blood cells. Blood also contains platelets, or *thrombocytes,* which are fragments of a type of bone marrow cell. We discuss blood more fully in chapter 49.

All connective tissues have similarities

Although the descriptions of the types of connective tissue suggest numerous different functions for these tissues, they have some similarities. As mentioned, connective tissues originate as embryonic mesoderm, and they all contain abundant extracellular material called matrix; however, the extracellular matrix material is different in different types of connective tissue. Embedded within the extracellular matrix of each tissue type are varieties of cells, each with specialized functions.

Abundant extracellular materials forming a matrix between loosely organized cells characterize connective tissues. Connective tissue proper is classified as either loose or dense. Special connective tissues each have a unique extracellular matrix between cells. The matrix of cartilage is composed of organic materials, whereas calcium crystals impregnate bone. The matrix of blood is a fluid, the plasma.

43.4 Muscle Tissue

Muscles are the motors of the vertebrate body. The characteristic that makes muscle cells unique is the relative abundance and organization of actin and myosin filaments within them. Although these filaments form a fine network in all eukaryotic cells, where they contribute to movement of materials within the cell, they are far more abundant and organized in muscle cells, which are specialized for contraction.

Vertebrates possess three kinds of muscle: *smooth, skeletal,* and *cardiac* (table 43.3). Skeletal and cardiac muscles are also known as *striated muscles* because their cells appear to have transverse stripes when viewed in longitudinal section under the microscope. The contraction of each skeletal muscle is under voluntary control, whereas the contraction of cardiac and smooth muscles is generally involuntary.

Smooth muscle is found in most organs

Smooth muscle was the earliest form of muscle to evolve, and it is found throughout most of the animal kingdom. In vertebrates, smooth muscle occurs in the organs of the internal environment, or *viscera,* and is also called *visceral muscle.* Smooth muscle tissue is arranged into sheets of long, spindle-shaped cells, each cell containing a single nucleus. In some tissues, the cells contract only when a nerve stimulates them—and then all of the cells in the sheet contract as a unit.

In vertebrates, muscles of this type line the walls of many blood vessels and make up the iris of the eye, which contracts in bright light. In other smooth muscle tissues, such as those in the wall of the digestive tract, the muscle cells themselves may spontaneously initiate electrical impulses, leading to a slow, steady contraction of the tissue. Here nerves regulate, rather than cause, the activity.

Skeletal muscle moves the body

Skeletal muscles are usually attached to bones by tendons, so that their contraction causes the bones to move at their joints. A skeletal muscle is made up of numerous, very long muscle cells called **muscle fibers,** which have multiple nuclei. The fibers lie parallel to each other within the muscle and are connected to the tendons on the ends of the muscle. Each skeletal muscle fiber is stimulated to contract by a motor neuron.

The nervous system controls the overall strength of a skeletal muscle contraction by controlling the number of motor neurons that fire, and therefore the number of muscle fibers

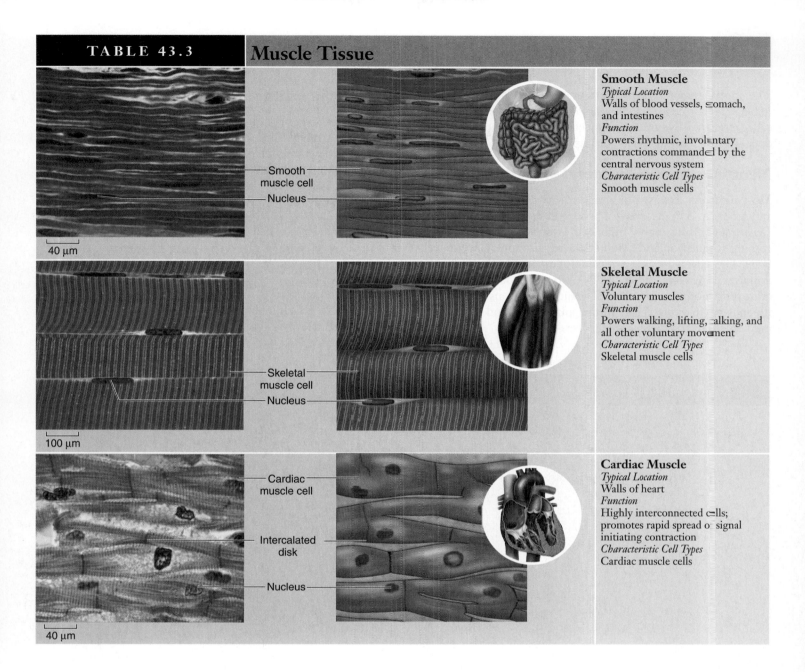

TABLE 43.3 Muscle Tissue

Smooth Muscle
Typical Location
Walls of blood vessels, stomach, and intestines
Function
Powers rhythmic, involuntary contractions commanded by the central nervous system
Characteristic Cell Types
Smooth muscle cells

Smooth muscle cell
Nucleus
40 µm

Skeletal Muscle
Typical Location
Voluntary muscles
Function
Powers walking, lifting, talking, and all other voluntary movement
Characteristic Cell Types
Skeletal muscle cells

Skeletal muscle cell
Nucleus
100 µm

Cardiac Muscle
Typical Location
Walls of heart
Function
Highly interconnected cells; promotes rapid spread of signal initiating contraction
Characteristic Cell Types
Cardiac muscle cells

Cardiac muscle cell
Intercalated disk
Nucleus
40 µm

---

stimulated to contract. Each muscle fiber contracts by means of substructures called **myofibrils** containing highly ordered arrays of actin and myosin myofilaments. These filaments give the muscle fiber its striated appearance.

Skeletal muscle fibers are produced during development by the fusion of several cells, end to end. This embryological development explains why a mature muscle fiber contains many nuclei. The structure and function of skeletal muscle is explained in more detail in chapter 47.

Cardiac muscle composes the heart

The hearts of vertebrates are made up of striated muscle cells arranged very differently from the fibers of skeletal muscle. Instead of having very long, multinucleate cells running the length of the muscle, **cardiac muscle** consists of smaller, interconnected cells, each with a single nucleus. The interconnections between adjacent cells appear under the microscope

as dark lines called **intercalated disks.** In reality, these lines are regions where gap junctions link adjacent cells. As noted in chapter 7, gap junctions have openings that permit the movement of small substances and ions from one cell to another. These interconnections enable the cardiac muscle cells to form a single functioning unit.

Certain specialized cardiac muscle cells can generate electrical impulses spontaneously, but the nervous system usually regulates the rate of impulse activity. The impulses generated by the specialized cell groups spread across the gap junctions from cell to cell, synchronizing the heart's contraction. Chapter 49 describes this process more fully.

Muscle tissue is of three types: smooth, skeletal, and cardiac. Smooth muscles provide a variety of visceral functions. Skeletal muscles enable the vertebrate body to move. Cardiac muscle forms a muscular pump, the heart.

Nerve Tissue

The fourth major class of vertebrate tissue is **nerve tissue** (table 43.4). Its cells include *neurons* and their supporting cells, called *neuroglia*. Neurons are specialized to produce and conduct electrochemical events, or impulses.

Neurons sometimes extend long distances

Most **neurons** consist of three parts: a cell body, dendrites, and an axon. The *cell body* of a neuron contains the nucleus. *Dendrites* are thin, highly branched extensions that receive incoming stimulation and conduct electrical impulses to the cell body. The *axon* is a single extension of cytoplasm that conducts impulses away from the cell body. Axons and dendrites can be quite long. For example, the cell bodies of neurons that control the muscles in your feet lie in the spinal cord, and their axons may extend over a meter to your feet.

Neuroglia provide support for neurons

Neuroglia do not conduct electrical impulses, but instead support and insulate neurons and eliminate foreign materials in and around neurons. In many neurons, neuroglia cells associate with the axons and form an insulating covering, a *myelin sheath*, produced by successive wrapping of the membrane around the axon. Gaps in the myelin sheath, known as *nodes of Ranvier*, serve as sites for accelerating an impulse (see chapter 44).

Two divisions of the nervous system coordinate activity

The nervous system is divided into the **central nervous system (CNS),** which includes the brain and spinal cord, and the **peripheral nervous system (PNS),** which includes *nerves* and

---

TABLE 43.4 **Nerve Tissue**

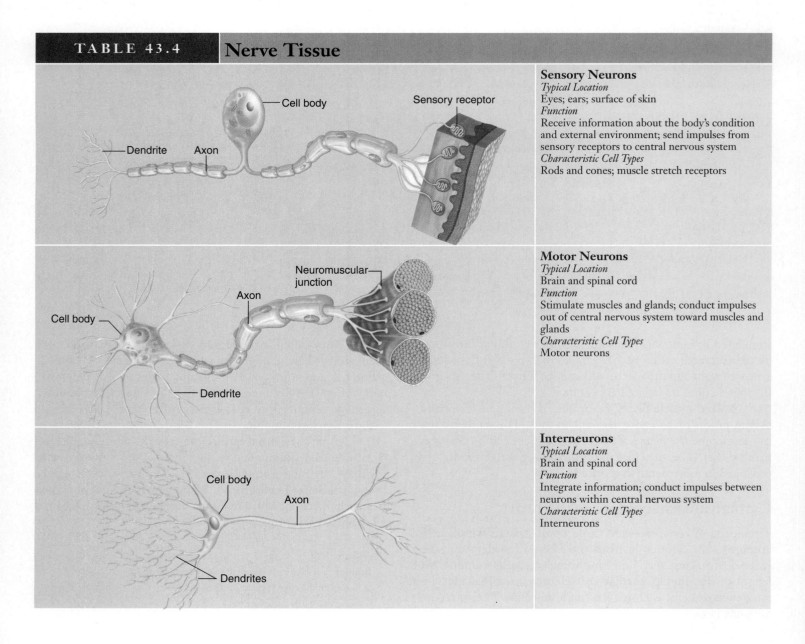

Sensory Neurons
Typical Location
Eyes; ears; surface of skin
Function
Receive information about the body's condition and external environment; send impulses from sensory receptors to central nervous system
Characteristic Cell Types
Rods and cones; muscle stretch receptors

Motor Neurons
Typical Location
Brain and spinal cord
Function
Stimulate muscles and glands; conduct impulses out of central nervous system toward muscles and glands
Characteristic Cell Types
Motor neurons

Interneurons
Typical Location
Brain and spinal cord
Function
Integrate information; conduct impulses between neurons within central nervous system
Characteristic Cell Types
Interneurons

ganglia. Nerves consist of axons in the PNS that are bundled together in much the same way as wires are bundled together in a cable. Ganglia are collections of neuron cell bodies. The CNS generally has the role of integration and interpretation of input, such as input from the senses; the PNS communicates signals to and from the CNS to the rest of the body, such as to muscle cells or endocrine glands.

Nerve tissue is composed of neurons and neuroglia. Neurons are specialized to receive and conduct electrical signals. Neuroglia do not conduct electrical impulses but have various support functions, including insulating axons to accelerate an electrical impulse. Both neurons and neuroglia are present in the CNS and the PNS.

43.6 Overview of Vertebrate Organ Systems

In the chapters that follow, we look closely at the major organ systems of vertebrates (figure 43.5). In each chapter, you will be able to see the intimate relationship of structure and function. We approach the organ systems by placing them in the following functional groupings:

- Communication and integration
- Support and movement
- Regulation and maintenance
- Defense
- Reproduction and development

Communication and integration sense and respond to the environment

Three organ systems detect external and internal stimuli and coordinate the body's responses. The **nervous system,** which consists of the brain, spinal cord, nerves, and sensory organs, detects external stimuli such as light, sound, and touch. This information is collected and integrated, and then the appropriate response is made. The **sensory systems,** a subset of the nervous system, include the organs and tissues that sense stimuli, such as vision, hearing, smell, and so on.

Working in parallel with nervous and sensory systems, the **endocrine system** issues chemical signals that regulate and fine-tune the myriad chemical processes taking place in all other organ systems.

Skeletal support and movement are vital to all animals

The **musculoskeletal system** consists of two interrelated organ systems. Muscles are most obviously responsible for movement, but without something to pull on, a muscle is useless. The skeletal system is the rigid framework against which most muscles pull. Vertebrates have internal skeletons, but many other animals exhibit external skeletons (such as insects) or hydrostatic skeletons (earthworms). Together, these two organ systems enable animals to exhibit a wide array of finely controlled movements.

Regulation and maintenance of the body's chemistry ensures continued life

The organ systems grouped under regulation and maintenance participate in nutrient acquisition, waste disposal, material distribution, and maintenance of the internal environment. The chapter on the **digestive system** describes how we eat, absorb nutrients, and eliminates solid wastes. The heart and vessels of the **circulatory system** pumps and distributes blood, carrying nutrients and other substances throughout the body. The body acquires oxygen and expels carbon dioxide via the **respiratory system.**

Finally, vertebrates tightly regulate internal body temperature and the concentration of their body fluids. We explore these processes in the chapter on temperature and osmoregulation, the latter of which is largely carried out by the **urinary system.**

The body can defend itself from attackers and invaders

Every animal faces assault by bacteria, viruses, fungi, protists, and even other animals. The body's first line of defense is the **integumentary system**—intact skin. Disease-causing agents that penetrate the first defense encounter a host of other protective **immune system** responses, including the production of antibodies and specialized cells that attack invading organisms.

Reproduction and development ensure the continuity of species

The biological continuity of vertebrates is the province of the **reproductive system.** Male and female reproductive systems consist of organs where male and female gametes develop, as well as glands and tubes that nurture gametes and allow gametes of complementary sexes to come into contact with one another. The female reproductive system in many vertebrates also has systems for nurturing the developing embryo and fetus.

After gametes have fused to form a *zygote*, an elaborate process of cell division and development must take place to change this beginning diploid cell into a multicellular adult, capable of reproducing on its own. We explore this process in the animal development chapter, which concludes this unit.

The organ systems of vertebrate bodies are grouped functionally based on their roles in communication and integration, support and movement, regulation and maintenance, defense, and reproduction and development.

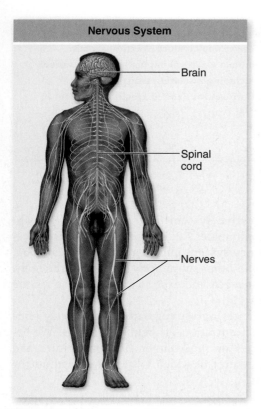

Nervous System

- Brain
- Spinal cord
- Nerves

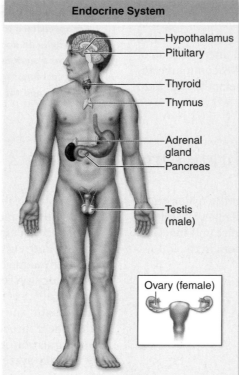

Endocrine System

- Hypothalamus
- Pituitary
- Thyroid
- Thymus
- Adrenal gland
- Pancreas
- Testis (male)
- Ovary (female)

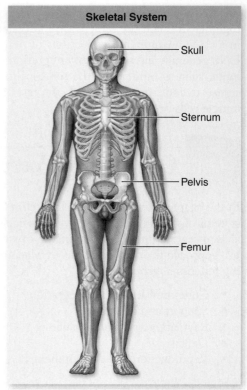

Skeletal System

- Skull
- Sternum
- Pelvis
- Femur

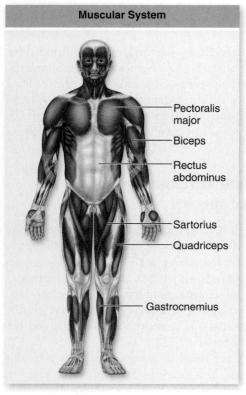

Muscular System

- Pectoralis major
- Biceps
- Rectus abdominus
- Sartorius
- Quadriceps
- Gastrocnemius

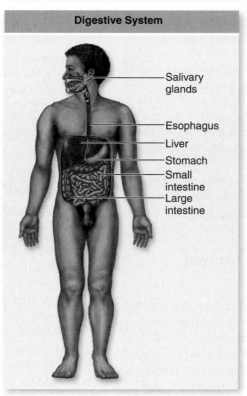

Digestive System

- Salivary glands
- Esophagus
- Liver
- Stomach
- Small intestine
- Large intestine

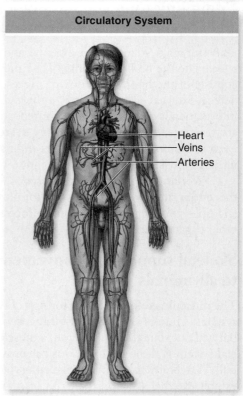

Circulatory System

- Heart
- Veins
- Arteries

figure 43.5

VERTEBRATE ORGAN SYSTEMS. Shown are the 11 principal organ systems of the human body, including both male and female reproductive systems.

Respiratory System

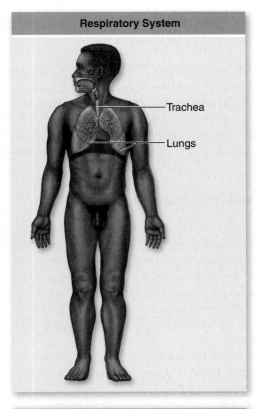

- Trachea
- Lungs

Urinary System

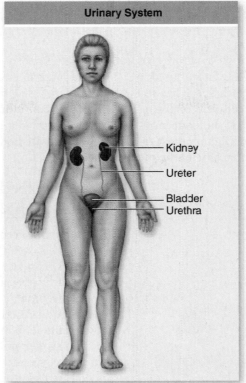

- Kidney
- Ureter
- Bladder
- Urethra

Integumentary System

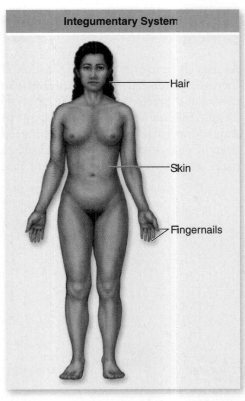

- Hair
- Skin
- Fingernails

Lymphatic/Immune System

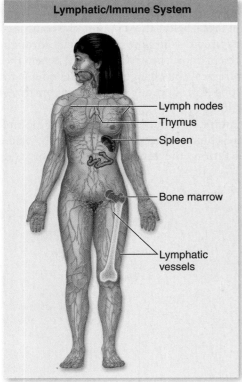

- Lymph nodes
- Thymus
- Spleen
- Bone marrow
- Lymphatic vessels

Reproductive System (male)

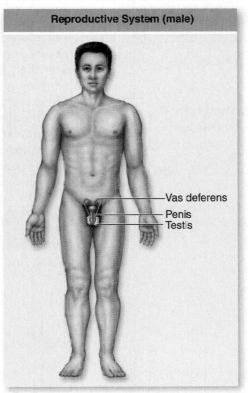

- Vas deferens
- Penis
- Testis

Reproductive System (female)

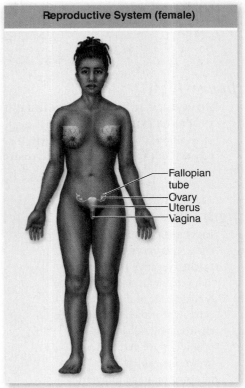

- Fallopian tube
- Ovary
- Uterus
- Vagina

Homeostasis

As animals have evolved, specialization of body structures has increased. Each cell is a sophisticated machine, finely tuned to carry out a precise role within the body. Such specialization of cell function is possible only when extracellular conditions stay within narrow limits. Temperature, pH, the concentrations of glucose and oxygen, and many other factors must remain relatively constant for cells to function efficiently and interact properly with one another. The dynamic constancy of the internal environment is called **homeostasis.** The term *dynamic* is used because conditions are never constant, but fluctuate continuously within narrow limits. Homeostasis is essential for life, and most of the regulatory mechanisms of the vertebrate body are involved with maintaining homeostasis (figure 43.6).

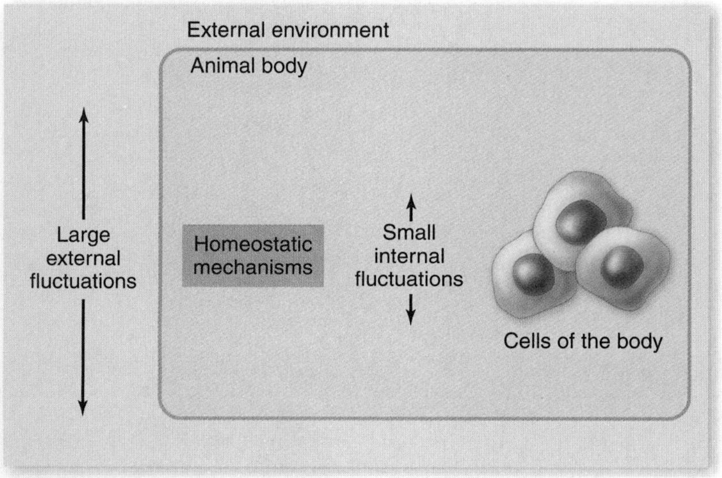

Negative feedback mechanisms keep values within a range

To maintain internal constancy, the vertebrate body uses a type of control system known as a **negative feedback.** In negative feedback, conditions within the body as well as outside it are detected by specialized sensors, which may be cells or membrane receptors. If conditions deviate too far from a set point, biochemical reactions are initiated to change conditions back toward the set point.

This **set point** is analogous to the temperature setting on a space heater. When room temperature drops, the change is detected by a temperature-sensing device inside the heater controls—the **sensor.** The temperature dial on which you have indicated the set point for the heater is the **comparator;** when the sensor information drops below the set point, the comparator closes an electrical circuit. The flow of electricity through the heater then produces more heat. Conversely, when the room temperature increases, the change causes the circuit to open, and heat is no longer produced. Figure 43.7 summarizes the negative feedback loop.

In a similar manner, the human body has set points for body temperature, blood glucose concentration, electrolyte (ion) concentration, the tension on a tendon, and

figure 43.6

HOMEOSTATIC MECHANISMS HELP MAINTAIN STABLE INTERNAL CONDITIONS. Even though conditions outside of an animal's body may vary widely, the inside stays relatively constant due to many finely tuned control systems.

figure 43.7

GENERALIZED DIAGRAM OF A NEGATIVE FEEDBACK LOOP. Negative feedback loops maintain a state of homeostasis, or dynamic constancy of the internal environment. Changing conditions are detected by sensors, which feed information to an integrating center that compares conditions to a set point. Deviations from the set point lead to a response to bring internal conditions back to the set point. Negative feedback to the sensor terminates the response.

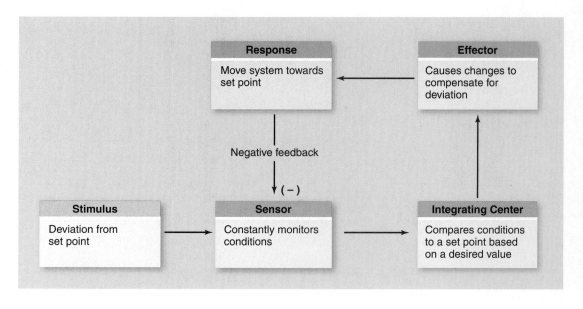

so on. The integrating center is often a particular region of the brain or spinal cord, but in some cases, it can also be cells of endocrine glands. When a deviation in a condition occurs, a message is sent to increase or decrease the activity of particular target organs, termed **effectors.** Effectors are generally muscles or glands, and their actions can change the value of the condition in question back toward the set point value.

Mammals and birds are *endothermic;* they can maintain relatively constant body temperatures independent of the environmental temperature. In humans, when the blood temperature exceeds 37°C (98.6°F), neurons in a part of the brain called the **hypothalamus** detect the temperature change. Acting through the control of motor neurons, the hypothalamus responds by promoting the dissipation of heat through sweating, dilation of blood vessels in the skin, and other mechanisms. These responses tend to counteract the rise in body temperature.

Antagonistic effectors act in opposite directions

The negative feedback mechanisms that maintain homeostasis often oppose each other to produce a finer degree of control. Most factors in the internal environment are controlled by several effectors, which often have antagonistic (opposing) actions. Control by antagonistic effectors is sometimes described as "push–pull," in which the increasing activity of one effector is accompanied by decreasing activity of an antagonistic effector. This affords a finer degree of control than could be achieved by simply switching one effector on and off.

Room temperature, to go back to an earlier example, can be maintained by just turning a heater on and off, or turning an air conditioner on and off. A much more stable temperature is possible, however, if a thermostat controls both the air conditioner and heater. Then the heater turns on when the air conditioner shuts off, and vice versa (figure 43.8*a*).

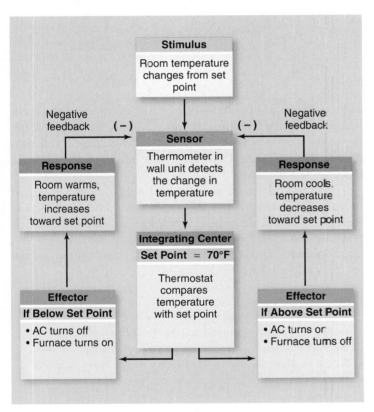

a.

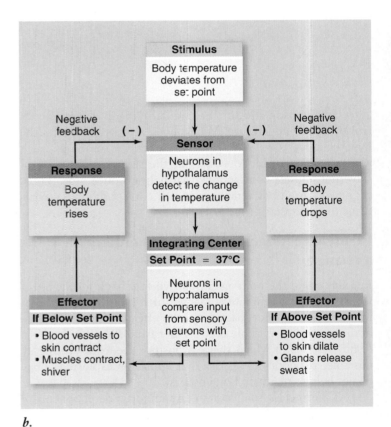

b.

figure 43.8

ROOM AND BODY TEMPERATURE ARE MAINTAINED BY NEGATIVE FEEDBACK AND ANTAGONISTIC EFFECTORS. *a.* If a thermostat senses a low temperature (as compared with a set point), the furnace turns on and the air conditioner turns off. If the temperature is too high, the air conditioner turns on and the furnace turns off. *b.* The hypothalamus of the brain detects an increase or decrease in body temperature. The comparator (also in the hypothalamus) then processes the information and activates effectors, such as surface blood vessels, sweat glands, and skeletal muscles. Negative feedback results in a reduction in the difference of the body's temperature compared with the set point. Consequently, the stimulation of the effectors by the comparator is also reduced.

figure 43.9

POSITIVE FEEDBACK DURING CHILDBIRTH. This is one of the few examples of positive feedback in the vertebrate body.

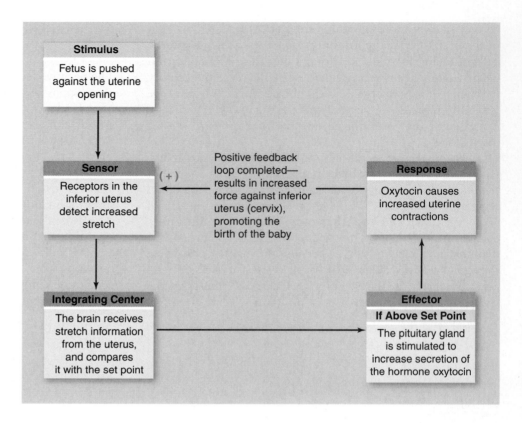

Antagonistic effectors are similarly involved in the control of body temperature. When body temperature falls, the hypothalamus coordinates a different set of responses, such as constriction of blood vessels in the skin and initiation of shivering, muscle contractions that help produce heat. These responses raise body temperature and correct the initial challenge to homeostasis (figure 43.8*b*).

Positive feedback mechanisms enhance a change

In a few cases, the body uses **positive feedback** mechanisms, which push or accentuate a change further in the same direction. In a positive feedback loop, the effector drives the value of the controlled variable even farther from the set point. As a result, systems in which there is positive feedback are highly unstable, analogous to a spark that ignites an explosion. They do not help to maintain homeostasis.

Nevertheless, such systems are important components of some physiological mechanisms. For example, positive feedback occurs in blood clotting, in which one clotting factor activates another in a cascade that leads quickly to the formation of a clot. Positive feedback also plays a role in the contractions of the uterus during childbirth (figure 43.9). In this case, stretching of the uterus by the fetus stimulates contraction, and contraction causes further stretching; the cycle continues until the uterus expels the fetus.

In the body, most positive feedback systems act as part of some larger mechanism that maintains homeostasis. In the examples we have described, formation of a blood clot stops bleeding and therefore tends to keep blood volume constant, and expulsion of the fetus reduces the contractions of the uterus, stopping the cycle.

Homeostasis can be thought of as the dynamic constancy of an organism's internal environment. Negative feedback mechanisms correct deviations from a set point for different internal variables. In this way, body conditions stay within a normal range. Effectors that act antagonistically to each other are more effective than effectors that act alone. Positive feedback mechanisms that accentuate changes are less common and have specialized functions in the body.

43.1 Organization of the Vertebrate Body

There are four levels of organization of the vertebrate body: cells, tissues, organs, and organ systems (figure 43.1).

- Tissues are groups of cells with a similar structure and function.

- Adult tissues are derived from three embryonic tissues: endoderm, mesoderm, and ectoderm.

- Adult vertebrate primary tissues are epithelial, connective, muscle, and nerve tissues.

- Organs are body structures composed of several different types of tissues that form a structural and functional unit.

- Organ systems are a group of organs that cooperate to perform the major activities of the body.

- The general body plan of vertebrates is a digestive tract or tube surrounded by skeleton, integument, and accessory organs (figure 43.2).

- Vertebrates have both a dorsal body cavity formed within the skull and vertebrae and a ventral body cavity bounded by the rib cage and lower abdominal muscles.

- The ventral body cavity is divided by the diaphragm into two parts: the thoracic cavity, which contains the heart and lungs, and the abdominopelvic cavity, which contains most of the organs.

- The coelomic space of the abdominopelvic cavity is called the peritoneal cavity, the space around the heart the pericardial cavity, and the spaces around the lungs the pleural cavities.

43.2 Epithelial Tissue

An epithelial membrane or epithelium covers every surface of the vertebrate body.

- The epithelial cells are tightly bound together, forming a protective barrier that selectively impedes or facilitates the passage of substances.

- Epithelium, which is attached to underlying connective tissues, has an inherent polarity that affects tissue function and it rapidly regenerates.

- Epithelium is divided into two general classes: simple and stratified (multiple cells thick). These classes are further divided into squamous, cuboidal, and columnar cells based on their shape (table 43.1).

- Simple epithelium is one cell thick and regulates the passage of materials. It may be secretory or invaginated to form exocrine glands with ducts or ductless endocrine glands.

- Stratified epithelium is two or more cells thick and is named by its apical cell layers.

43.3 Connective Tissue

Connective tissues are derived from embryonic mesoderm and contain an extracellular matrix composed of protein fibers and a ground substance.

- Connective tissue is divided into connective tissue proper and specialized connective tissue.

- Connective tissue proper is divided into loose and dense connective tissue, and the matrix is produced and secreted by fibroblasts.

- Loose connective tissue, such as adipose, is composed of a large amount of ground substance and a loose scattering of collagen, elastin, and reticulin fibers.

- Dense connective tissue has less ground substance and tightly packed collagen fibers arranged in regular parallel strands or in irregular randomly oriented fibers.

- Specialized connective tissues have unique cells and matrices that allow them to perform specialized functions. They include flexible cartilage, rigid bone, and blood (table 43.2).

43.4 Muscle Tissue

Muscles perform work, and they contain a unique organization of actin and myosin filaments.

- Vertebrates possess three types of muscle: smooth, skeletal, and cardiac (table 43.3).

- Involuntary smooth muscle occurs in the viscera, is composed of long, spindle-shaped cells, each with a single nucleus, and its activity is regulated by nerves.

- Voluntary skeletal or striated muscle is usually attached by tendons to bones, the cells have multiple nuclei and contain contractile myofibrils.

- The nervous system controls the overall strength of contraction by the number of muscle fibers stimulated to contract.

- Cardiac muscle consists of striated muscle cells connected by gap junctions that allow the cardiac muscle cells to form a single functioning unit.

- Specialized cardiac muscle cells can generate spontaneous electrical impulses, but the nervous system regulates the rate of impulse activity.

43.5 Nerve Tissue

Nerve tissue is specialized to produce and conduct electrochemical impulses (table 43.4).

- Nervous tissue is composed of neurons and supporting cells called neuroglia.

- Neurons consist of three parts: a cell body with the nucleus; dendrites, which receive incoming information to the cell body; and an axon, which conducts impulses away from the cell body.

- Neuroglia support and insulate neurons and regulate the neuronal environment. In some cells they form the myelin sheaths that accelerate impulses.

- The two main divisions of the nervous system are the central nervous system, which includes the brain and spinal cord, and the peripheral nervous system, which includes neurons and ganglia.

43.6 Overview of Vertebrate Organ Systems

Organ systems are composed groups of organs that perform unique functions (figure 43.5).

- Three organ systems are involved in communication and integration of information: the nervous, sensory, and endocrine systems.

- The musculoskeletal system is composed of two interrelated organ systems involved in support and movement.

- Four organ systems are involved in the regulation and maintenance of body chemistry: the digestive, circulatory, respiratory, and urinary systems.

- The body defends itself with two organ systems: the integumentary and immune systems.

- The reproductive system ensures the continuation of a species.

43.7 Homeostasis

Homeostasis refers to the dynamic constancy of the internal environment and is essential for life.

- Regulation of a relatively constant parameter requires negative feedback to return conditions to a set point.

- Homeostatic negative feedback loops include a sensor that monitors internal and external conditions, an integration center to compare conditions to a set point, and effectors that cause changes to compensate for deviation from the set point.

- Negative feedback mechanisms often occur in antagonistic pairs that oppose each other to dampen deviations from set points.

- Positive feedback increases a deviation from a set point and is usually detrimental.

SELF TEST

1. Connective tissues, although quite diverse in structure and location, do share a common theme; the connection between other types of tissues. Although all of the following seem to fit that criterion, one of the tissues listed is not a type of connective tissue. Which one?
   a. Blood
   b. Muscle
   c. Adipose tissue
   d. Cartilage

2. What do all the organs of the body have in common?
   a. Each contains the same kinds of cells.
   b. Each is composed of several different kinds of tissue.
   c. Each is derived from ectoderm.
   d. Each can be considered part of the circulatory system.

3. Which of the following cavities would contain your stomach?
   a. Peritoneal
   b. Pericardial
   c. Pleural
   d. Thoracic

4. Epithelial tissues do all of the following *except*—
   a. form barriers or boundaries.
   b. absorb nutrients in the digestive tract.
   c. transmit information in the central nervous system.
   d. allow exchange of gases in the lung.

5. Endocrine and exocrine glands form from—
   a. epithelial tissue.
   b. connective tissue.
   c. nervous tissue.
   d. muscle tissue.

6. Connective tissues include a diverse group of cells, yet they all share—
   a. cuboidal shape.
   b. the ability to produce hormones.
   c. the ability to contract.
   d. the presence of an extracellular matrix.

7. Rheumatoid arthritis is an autoimmune disease that attacks the linings of joints within the body. The cells that line these joints, and whose destruction causes the symptoms of arthritis, are known as—
   a. osteocytes.
   b. erythrocytes.
   c. chondrocytes.
   d. thrombocytes.

8. Skeletal muscle cells differ from the "typical" mammalian cell in that they—
   a. contain multiple nuclei.
   b. have mitochondria.
   c. have no plasma membrane.
   d. are not derived from embryonic tissue.

9. Examples of smooth muscle sites include—
   a. lining of blood vessels.
   b. iris of the eye.
   c. wall of the digestive tract.
   d. all of these.

10. Suppose that a strange alien virus arrives on Earth. This virus causes damage to the nervous system by attacking the structures of neurons. Which of the following structures would be immune from attack?
    a. Axon
    b. Dendrite
    c. Neuroglia
    d. All of these would be attacked by the virus.

11. The function of neuroglia is to—
    a. carry messages from the PNS to the CNS.
    b. support and protect neurons.
    c. stimulate muscle contraction.
    d. store memories.

12. Systems involved in communication and integration of information include all of the following *except*—
    a. the nervous system.
    b. the immune system.
    c. the endocrine system.
    d. the sensory system.

13. Another way to describe the functions of the digestive, respiratory, and circulatory systems is as the _____ systems
    a. defense
    b. communication and integration
    c. support and movement
    d. regulation and maintenance

14. Homeostasis—
    a. is a dynamic process.
    b. describes the maintenance of the internal environment of the body.
    c. is essential to life.
    d. all of these.

15. Which of the following scenarios correctly describes positive feedback?
    a. If the temperature increases in your room, your furnace increases its output of warm air.
    b. If you drink too much water, you produce more urine.
    c. If the price of gasoline increases, drivers decrease the length of their trips.
    d. If you feel cold, you start to shiver.

CHALLENGE QUESTIONS

1. Suppose that you discover a new disease that affects nutrient absorption in the gut as well as causing problems with the skin. Is it possible that one disease could involve the same tissues? How could this occur?

2. Which organ systems are involved in regulation and maintenance? Why do you think they are linked together in this way?

3. We have all experienced hunger pangs. Is hunger a positive or negative feedback stimulus? Describe the steps involved in the response to this stimulus.

4. Why is homeostasis described as a dynamic process?

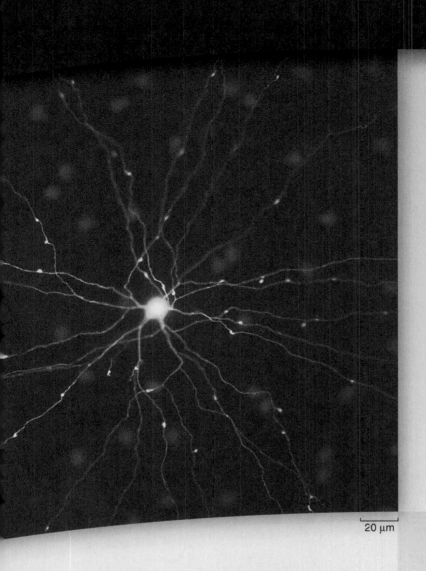

20 μm

chapter 44

The Nervous System

introduction

ALL ANIMALS EXCEPT SPONGES use a network of nerve cells to gather information about the body's condition and the external environment, to process and integrate that information, and to issue commands to the body's muscles and glands. As we saw in chapter 43, homeostasis of the animals body is accomplished by negative feedback loops that maintain conditions within narrow limits. Negative feedback implies not only detection of appropriate stimuli but also communication of information to begin a response. The nervous system, composed of neurons, such as the one pictured here, is a fast communication system and a part of many feedback systems in the body.

concept outline

44.1 Nervous System Organization

An animal must be able to respond to environmental stimuli. A fly escapes a flyswatter; the antennae of a crayfish detect food and the crayfish moves toward it. To accomplish these actions, animals must have **sensory receptors** that can detect the stimulus and **motor effectors** that can respond to it. In most invertebrate phyla and in all vertebrate classes, sensory receptors and motor effectors are linked by way of the nervous system.

The central nervous system is the "command center"

As described in chapter 43, the nervous system consists of neurons and supporting cells. Figure 44.1 shows the three types of neurons. In vertebrates, **sensory neurons** (or afferent neurons) carry impulses from sensory receptors to the **central nervous system (CNS),** which is composed of the brain and spinal cord. **Motor neurons** (or efferent neurons) carry impulses from the CNS to effectors—muscles and glands. A third type of neuron is present in the nervous systems of most invertebrates and all vertebrates: **interneurons** (or association neurons). Interneurons are located in the brain and spinal cord of vertebrates, where they help provide more complex reflexes and higher associative functions, including learning and memory.

The peripheral nervous system collects information and carries out responses

Together, sensory and motor neurons constitute the **peripheral nervous system (PNS)** in vertebrates. Motor neurons that stimulate skeletal muscles to contract make up the **somatic nervous system;** those that regulate the activity of the smooth muscles, cardiac muscle, and glands compose the **autonomic nervous system.**

The autonomic nervous system is further broken down into the **sympathetic** and **parasympathetic** divisions. These divisions counterbalance each other in the regulation of many organ systems. Figure 44.2 illustrates the relationships among the different parts of the vertebrate nervous system.

The structure of neurons supports their function

Despite their varied appearances, most neurons have the same functional architecture (figure 44.3). The **cell body** is an enlarged region containing the nucleus. Extending from the cell body are one or more cytoplasmic extensions called **dendrites.** Motor and association neurons possess a profu-

figure 44.1

THREE TYPES OF NEURONS. The brain and spinal cord form the central nervous system of vertebrates while sensory and motor neurons form the peripheral nervous system. *Sensory neurons* of the peripheral nervous system carry information about the environment to the central nervous system. *Interneurons* in the central nervous system provide links between sensory and motor neurons. *Motor neurons* of the peripheral nervous system carry impulses or "commands" to muscles and glands (effectors).

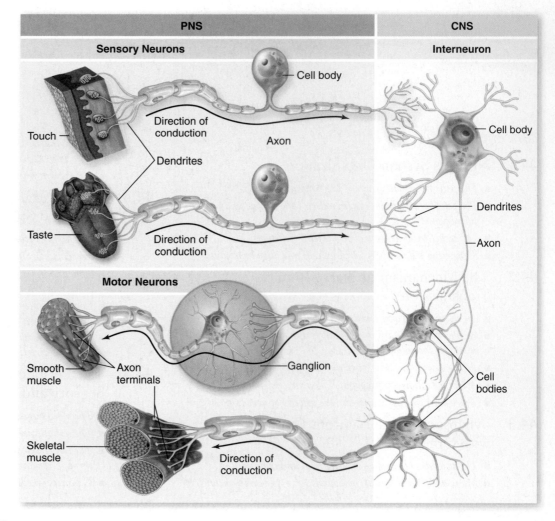

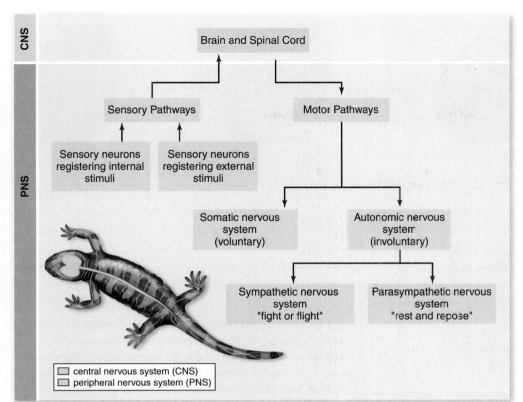

figure 44.2

DIVISIONS OF THE VERTEBRATE NERVOUS SYSTEM. The major divisions are the central and peripheral nervous systems. The brain and spinal cord make up the central nervous system. The peripheral nervous system is made of everything outside the central nervous system and is broken into sensory and motor pathways. Sensory pathways can detect either external or internal stimuli. Motor pathways are divided into the somatic nervous system that activates voluntary muscles (such as those that control movement of the skeleton) and the autonomic nervous system that activate involuntary muscles (such as the smooth muscle that controls movement of food along the digestive tract). The sympathetic and parasympathetic nervous systems are subsets of the autonomic nervous system that trigger opposing actions.

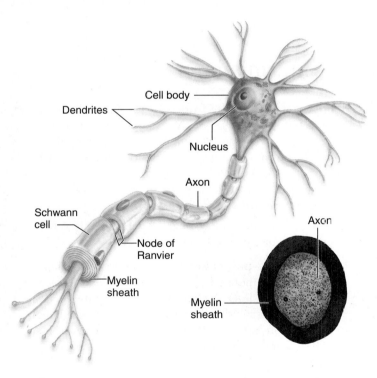

figure 44.3

STRUCTURE OF A TYPICAL VERTEBRATE NEURON. Extending from the cell body are many dendrites, which receive information and carry it to the cell body. A single axon transmits impulses away from the cell body. Many axons are encased by a myelin sheath, with multiple membrane layers that insulate the axon. Small gaps, called nodes of Ranvier, interrupt the sheath at regular intervals. Schwann cells form myelin sheaths in the PNS (as shown for this motor neuron); extensions of oligodendrocytes form myelin sheaths in the CNS.

sion of highly branched dendrites, enabling those cells to receive information from many different sources simultaneously. Some neurons have extensions from the dendrites called *dendritic spines* that increase the surface area available to receive stimuli.

The surface of the cell body integrates the information arriving at its dendrites. If the resulting membrane excitation is sufficient, it triggers the conduction of impulses away from the cell body along an **axon.** Each neuron has a single axon leaving its cell body, although an axon may also branch to stimulate a number of cells. An axon can be quite long: The axons controlling the muscles in a person's feet can be more than a meter long, and the axons that extend from the skull to the pelvis in a giraffe are about 3 m long.

Supporting cells include Schwann cells and oligodendrocytes

Neurons are supported both structurally and functionally by supporting cells, which are collectively called **neuroglia.** These cells are one-tenth as big and 10 times more numerous than neurons, and they serve a variety of functions, including supplying the neurons with nutrients, removing wastes from neurons, guiding axon migration, and providing immune functions.

Two of the most important kinds of neuroglia in vertebrates are **Schwann cells** and **oligodendrocytes,** which produce **myelin sheaths** that surround the axons of many neurons. Schwann cells produce myelin in the PNS, and oligodendrocytes produce myelin in the CNS. During development, these cells wrap themselves around each axon several times to form the myelin sheath,

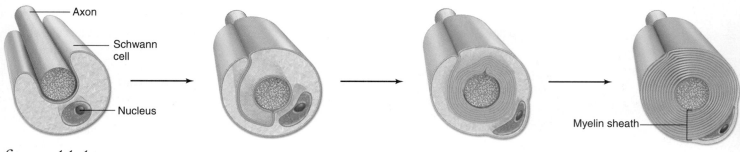

figure 44.4

THE FORMATION OF THE MYELIN SHEATH AROUND A PERIPHERAL AXON. The myelin sheath forms by successive wrappings of Schwann cell membranes around the axon.

an insulating covering consisting of multiple layers of compacted membrane (figure 44.4).

Axons that have myelin sheaths are said to be myelinated, and those that don't are unmyelinated. In the CNS, myelinated axons form the **white matter,** and the unmyelinated dendrites and cell bodies form the **gray matter.** In the PNS, myelinated axons are bundled together, much like wires in a cable, to form **nerves.**

Small gaps, known as **nodes of Ranvier** (see figure 44.3), interrupt the myelin sheath at intervals of 1 to 2 μm. We dis-

cuss the role of the myelin sheath in impulse conduction in the next section.

> Neurons and neuroglia make up the central and peripheral nervous systems in vertebrates. Sensory neurons, motor neurons, and interneurons play different roles in the nervous system. Most neurons have similar functional components of dendrites, cell body, and axon. The neuroglia aid neuron function, in part by producing myelin sheaths.

44.2 The Mechanism of Nerve Impulse Transmission

Neuron function depends on a changeable permeability to ions. Upon stimulation, electrical changes in the plasma membrane spread or propagate from one part of the cell to another. The architecture of the neuron provides the mechanisms for the generation and spread of these membrane electrical potentials.

The unique mechanisms of neurons primarily depend on the presence of specialized membrane transport proteins, where they are located, and how they are activated. First, we examine some of the basic electrical properties common to the plasma membranes of most animal cells, and then we look at how these properties operate in neurons.

An electrical difference exists across the plasma membrane

You first learned about membrane potential in chapter 5, where transport of ions across the cell membrane was discussed. Membrane potential is similar to the electrical potential difference that exists between the two poles of a flashlight or automobile battery. One pole is positive, and the other is negative. Similarly, a potential difference exists across every cell's plasma membrane. The side of the membrane exposed to the cytoplasm is the negative pole, and the side exposed to the extracellular fluid is the positive pole.

When a neuron is not being stimulated, it maintains a **resting potential.** A cell is very small, and so its membrane potential is very small. The resting membrane potential of many

vertebrate neurons ranges from −40 to −90 millivolts (mV), or 0.04 to 0.09 volts (V). For the examples and figures in this chapter, we use an average resting membrane potential value of −70 mV. The minus sign indicates that the inside of the cell is negative with respect to the outside.

Contributors to membrane potential

The inside of the cell is more negatively charged in relation to the outside because of two factors:

(1) The **sodium–potassium pump,** described in chapter 5, brings two potassium ions (K^+) into the cell for every three sodium ions (Na^+) it pumps out (figure 44.5). This helps establish and maintain concentration gradients resulting in high K^+ and low Na^+ concentrations inside the cell, and high Na^+ and low K^+ concentrations outside the cell.

(2) **Ion leakage channels** in the cell membrane are more numerous for K^+ than for Na^+. Ion leakage channels are membrane proteins that form pores through the membrane, allowing the flow of specific ions (such as K^+ or Na^+) in and out of the cell. Because there are more ion channels for K^+, it is easier for this ion to diffuse out of the cell than for Na^+ to diffuse into the cell.

Two major forces act on ions in establishing the resting membrane potential: (1) The electrical potential produced by unequal distribution of charges, and (2) the concentration

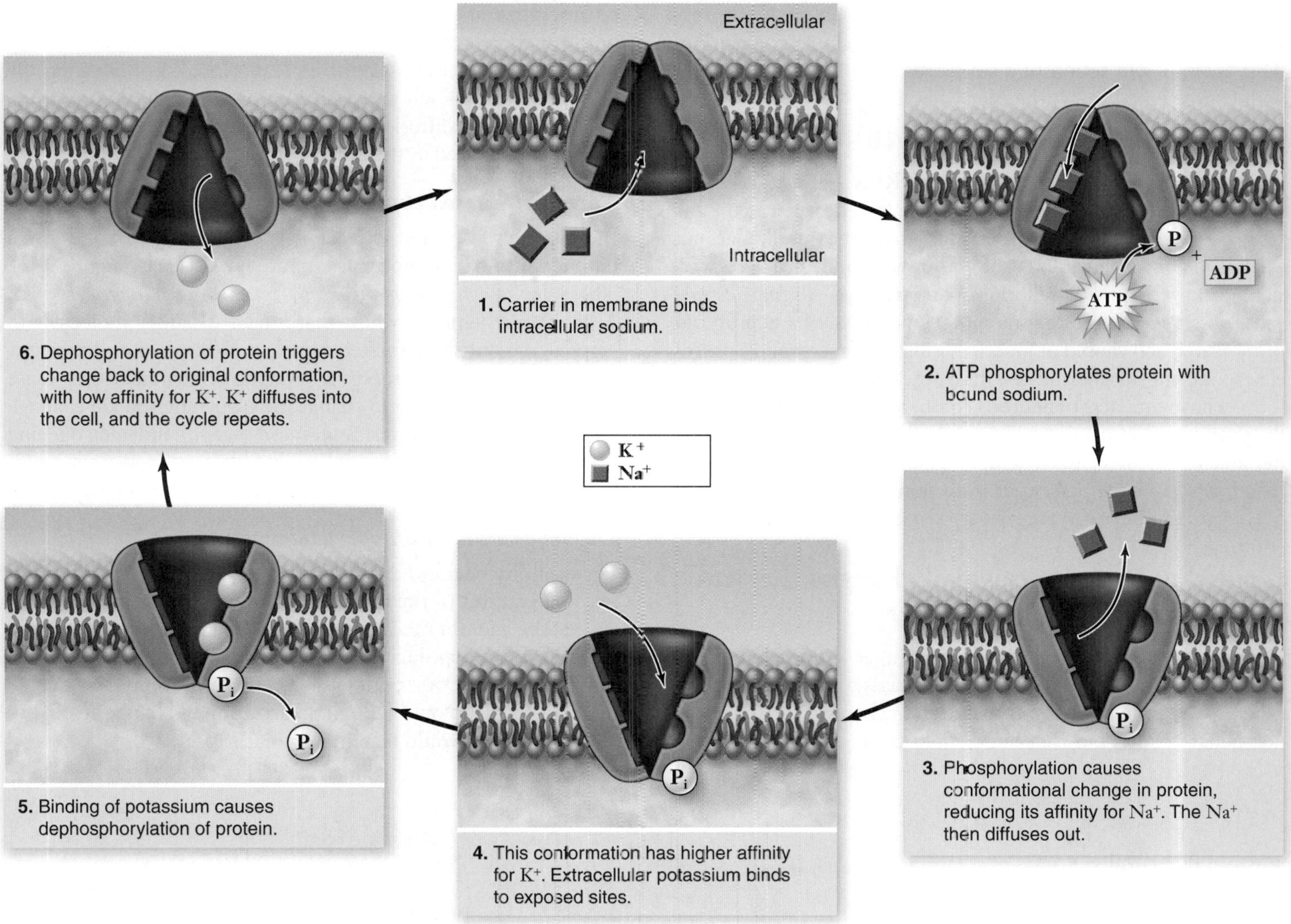
Extracellular
Intracellular
1. Carrier in membrane binds intracellular sodium.
2. ATP phosphorylates protein with bound sodium.
ATP
P
ADP
3. Phosphorylation causes conformational change in protein, reducing its affinity for Na+. The Na+ then diffuses out.
4. This conformation has higher affinity for K+. Extracellular potassium binds to exposed sites.
5. Binding of potassium causes dephosphorylation of protein.
6. Dephosphorylation of protein triggers change back to original conformation, with low affinity for K+. K+ diffuses into the cell, and the cycle repeats.
K+
Na+

figure 44.5

THE SODIUM–POTASSIUM PUMP. This pump transports three Na^+ to the outside of the cell and simultaneously transports two K^+ to the inside of the cell. This is an active transport carrier requiring the (phosphorylating) energy of ATP.

gradient produced by unequal concentrations of molecules from one side of the membrane to the other.

The resting potential: Balance between two gradients

The resting potential arises due to the action of the sodium–potassium pump and the differential permeability of the membrane to Na^+ and K^+ due to leakage channels. The pump moves three Na^+ outside for every two K^+ inside, which creates a small imbalance in cations outside the cell. This has only a minor effect, however, but the concentration gradients created by the pump are significant. The concentration of K^+ is much higher inside the cell than outside, leading to leakage of K^+ through open K^+ channels. Since the membrane is not permeable to the negative ions that could counterbalance this (mainly organic phosphates, amino acids, and proteins) it leads to a buildup of positive charge outside the membrane and negative charge inside the membrane. This electrical potential then is an attractive force to bring the K^+ ions back inside the cell. The balance between the diffusional force and the electrical force leads to the **equilibrium potential** (table 44.1). By relating the work

TABLE 44.1 The Ionic Composition of Cytoplasm and Extracellular Fluid (ECF)

Ion	Concentration in ECF (mM)	Concentration in Cytoplasm (mM)	Ratio (ECF:cytoplasm)	Equilibrium Potential (mV)
Na^+	150	15	10:1	+ 60
K^+	5	150	1:30	– 90
Cl^-	110	7	15:1	– 70

done by each type of force, we can derive a quantitative expression for this equilibrium potential called the Nernst equation. This assumes the action of a single ion, and for a positive ion with charge equal to +1, the Nernst equation is:

$$E_K = 58 \text{ mV } \log([K^+]_{out}/[K^+]m_{in})$$

The calculated equilibrium potential for K^+ is –90 mV (see table 44.1), close to the measured value of –70 mV. The calculated value for Na^+ is +60 mV, clearly not at all close to the measured value, but the leakage of a small amount of Na^+ back into the cell is responsible for lowering the equilibrium potential of K^+ to the –70 mV value observed. The resting membrane potential of a neuron can be measured and viewed or graphed using a voltmeter and a pair of electrodes, one outside and one inside the cell (figure 44.6).

The uniqueness of neurons compared with other cells is not the production and maintenance of the resting membrane potential, but rather the sudden temporary disruptions to the resting membrane potential that occur in response to stimuli. Two types of changes can be observed: *graded potentials* and *action potentials*.

Graded potentials are small changes that can reinforce or negate each other

Graded potentials, small transient changes in membrane potential, are caused by the activation of a class of channel proteins called **gated ion channels.** Introduced in chapter 9, gated channels behave like a door that can open or close, unlike ion leakage channels that are always open. The structure of gated ion channels is such that they have alternative conformations that can be open, allowing the passage of ions, or closed, not allowing the passage of ions. Each gated channel is selective, that is, when open they allow diffusion of only one type of ion. Most gated channels are closed in the normal resting cell.

Chemically gated channels

In most neurons, gated ion channels in dendrites respond to the binding of signaling molecules (figure 44.7, see also figure 9.4*a*). These are referred to as *chemically gated* or *ligand-gated channels*. *Ligands* are chemical groups that attach to larger molecules to regulate or contribute to their function. When ligands temporarily bind to membrane receptor proteins or channels, they cause the shape of the protein to change, thus opening the ion channel. Hormones and neurotransmitters act as ligands, inducing opening of ligand-gated channels, and causing changes in plasma membrane permeability that leads to changes in membrane voltage.

Depolarization and hyperpolarization

Permeability changes are measurable as depolarizations or hyperpolarizations of the membrane potential. A **depolarization** makes the membrane potential less negative (more positive), whereas a **hyperpolarization** makes the membrane potential more negative. For example, a change in potential from –70 mV to –65 mV would be a depolarization; a change from –70 mV to –75 mV would be a hyperpolarization.

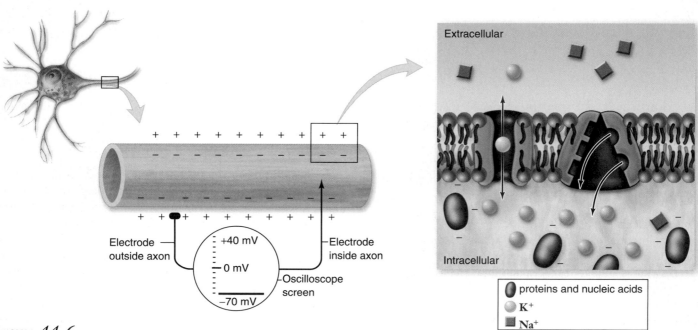

figure 44.6

ESTABLISHMENT OF THE RESTING MEMBRANE POTENTIAL. A voltmeter placed with one electrode inside an axon and the other outside the membrane. The electric potential inside is –70 mV relative to the outside of the membrane. K^+ leaves the cell through leakage channels due to diffusion down its concentration gradient. Negatively charged proteins and nucleic acids inside the cell cannot leave the cell and attract cations from outside the cell, such as K^+. This balance of electrical and diffusional forces produces the resting potential. The sodium-potassium pump maintains cell equilibrium by counteracting the effects of Na^+ leakage into the cell and contributes to the resting potential by moving 3 Na^+ outside for every 2 K^+ moved inside.

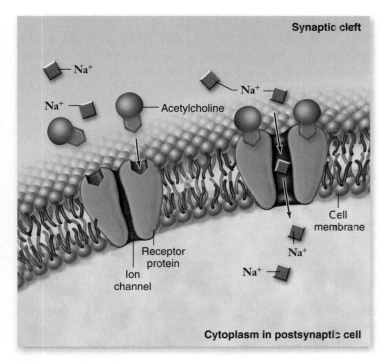

figure 44.7

A CHEMICALLY GATED ION CHANNEL. The acetylcholine (ACh) receptor is a chemically gated channel that can bind the neurotransmitter ACh. Binding of ACh causes the channel to open allowing Na^+ ions to flow into the cell by diffusion.

These small changes in membrane potential result in *graded potentials* because their size depends on either the strength of the stimulus or the amount of ligand available to bind with their receptors. These potentials diminish in amplitude as they spread from their point of origin. Depolarizing or hyperpolarizing potentials can add together to amplify or reduce their effects, just as two waves can combine to make a bigger one when they meet in synchronization or can cancel each other out when a trough meets with a crest. The ability of graded potentials to combine is called **summation** (figure 44.8). We will return to this topic in the next section after we discuss the nature of action potentials.

Action potentials result when depolarization reaches a threshold

When a particular level of depolarization is reached (about −55 mV in some mammalian axons), a nerve impulse, or **action potential,** is produced in the region where the axon arises from the cell body. The level of depolarization needed to produce an action potential is called the **threshold potential.** Depolarizations bring a neuron closer to the threshold, and hyperpolarizations move the neuron further from the threshold.

The action potential is caused by another class of ion channels: **voltage-gated ion channels.** These channels open and close in response to changes in membrane potential; the flow of ions controlled by these channels creates the action potential. Voltage-gated channels are found in neurons and in muscle cells. Two different channels are used to create an action potential in neurons: **voltage-gated Na^+ channels** and **voltage-gated K^+ channels.**

Sodium and potassium voltage-gated channels

The behavior of the voltage-gated Na^+ channel is more complex than that of the K^+ channel, so we will consider it first. The channel has two gates: an activation gate and an inactivation gate. In its resting state the activation gate is closed and the inactivation gate is open. When the threshold voltage is reached, the activation gate opens rapidly, leading to an influx of Na^+ ions due to both concentration and voltage gradients. After a short period the inactivation gate closes, stopping the influx of Na^+ ions and leaving the channel in a temporarily inactivated state. The channel is returned to its resting state by the activation gate closing and the inactivation gate opening. The result of this is a transient influx of Na^+ that depolarizes the membrane in response to a threshold voltage.

The K^+ channel has a single activation gate that is closed in the resting state. In response to a threshold voltage, it opens slowly. With the high concentration of K^+ inside the cell, and the membrane now far from the equilibrium potential, this leads to an efflux of K^+. The positive charge now leaving the cell counteracts the effect of the Na^+ channel and repolarizes the membrane.

Tracing an action potential's changes

Let us now put all of this together and see how the changing flux of ions leads to an action potential. The action potential has three phases: a *rising phase,* a *falling phase,* and an *undershoot*

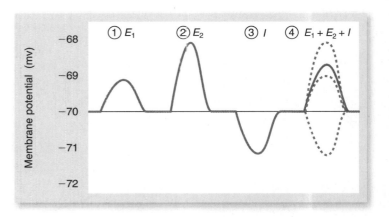

figure 44.8

GRADED POTENTIALS. Graded potentials are the summation of subthreshold potentials produced by the opening of different chemically gated channels. (*1*) A weak excitatory stimulus, E_1, elicits a smaller depolarization than (*2*) a stronger stimulus, E_2. (*3*) An inhibitory stimulus, I, produces a hyperpolarization. (*4*) If all three stimuli occur very close together, the resulting polarity change will be the sum of the three individual changes.

phase (figure 44.9). When a threshold potential is reached, the rapid opening of the Na$^+$ channel causes an influx of Na$^+$ that shifts the membrane potential toward the equilibrium potential for Na (+60 mV). This appears as the rising phase on an oscilloscope. This never quite reaches +60 mV because the inactivation gate of the Na$^+$ channel closes, terminating the rising phase. At the same time, the opening of the K$^+$ channel leads to K$^+$ diffusing out of the cell, repolarizing the membrane in the falling phase. The K$^+$ channels remain open longer than necessary to restore the resting potential, resulting in a slight undershoot. This entire sequence of events for a single action potential takes about a millisecond.

The nature of action potentials

Action potentials are always separate, all-or-none events with the same amplitude. An action potential occurs if the threshold voltage is reached, but not while the membrane remains below threshold. Action potentials do not add together or interfere with one another, as graded potentials can. After Na$^+$ channels "fire" they remain in an inactivated state for an additional millisecond until the inactivation gate reopens, preventing any summing of effects. It is the frequency not amplitude of an action potential that the nervous system uses to code the intensity of a stimulus.

The production of an action potential results entirely from the passive diffusion of ions. However, at the end of each

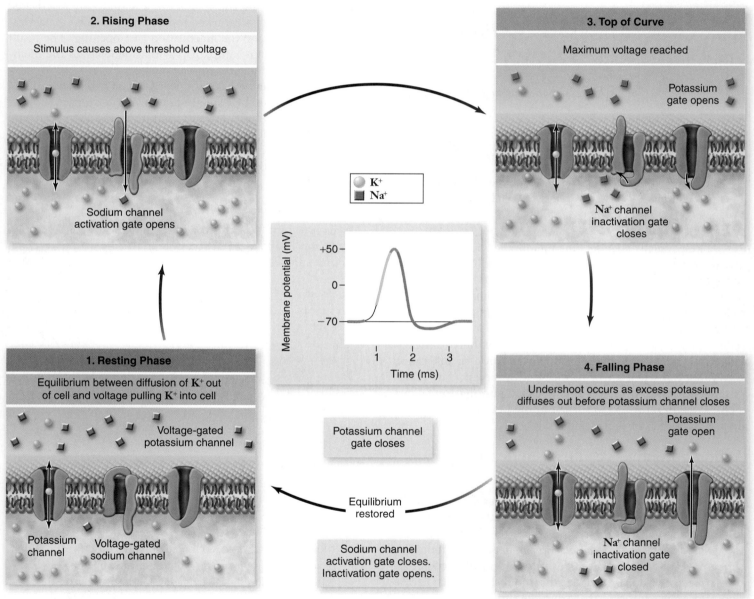

figure 44.9

THE ACTION POTENTIAL. (*1*) At resting membrane potential, voltage-gated ion channels are closed, but there is some leakage of K$^+$. In response to a stimulus, the cell begins to depolarize, and once the threshold level is reached, an action potential is produced. (*2*) Rapid depolarization occurs (the rising portion of the spike) because voltage-gated sodium channel activation gates open, allowing Na$^+$ to diffuse into the axon. (*3*) At the top of the spike, Na$^+$ channel inactivation gates close, and voltage-gated potassium channels that were previously closed begin to open. (*4*) With the K$^+$ channels open, repolarization occurs because of the diffusion of K$^+$ out of the axon. An undershoot occurs before the membrane returns to its original resting potential.

action potential, the cytoplasm contains a little more Na$^+$ and a little less K$^+$ than it did at rest. Although the number of ions moved by a single action potential is tiny relative to the concentration gradients of Na$^+$ and K$^+$, eventually this would have an effect. The constant activity of the sodium-potassium pump compensates for these changes. Thus, although active transport is not required to produce action potentials, it is needed to maintain the ion gradients.

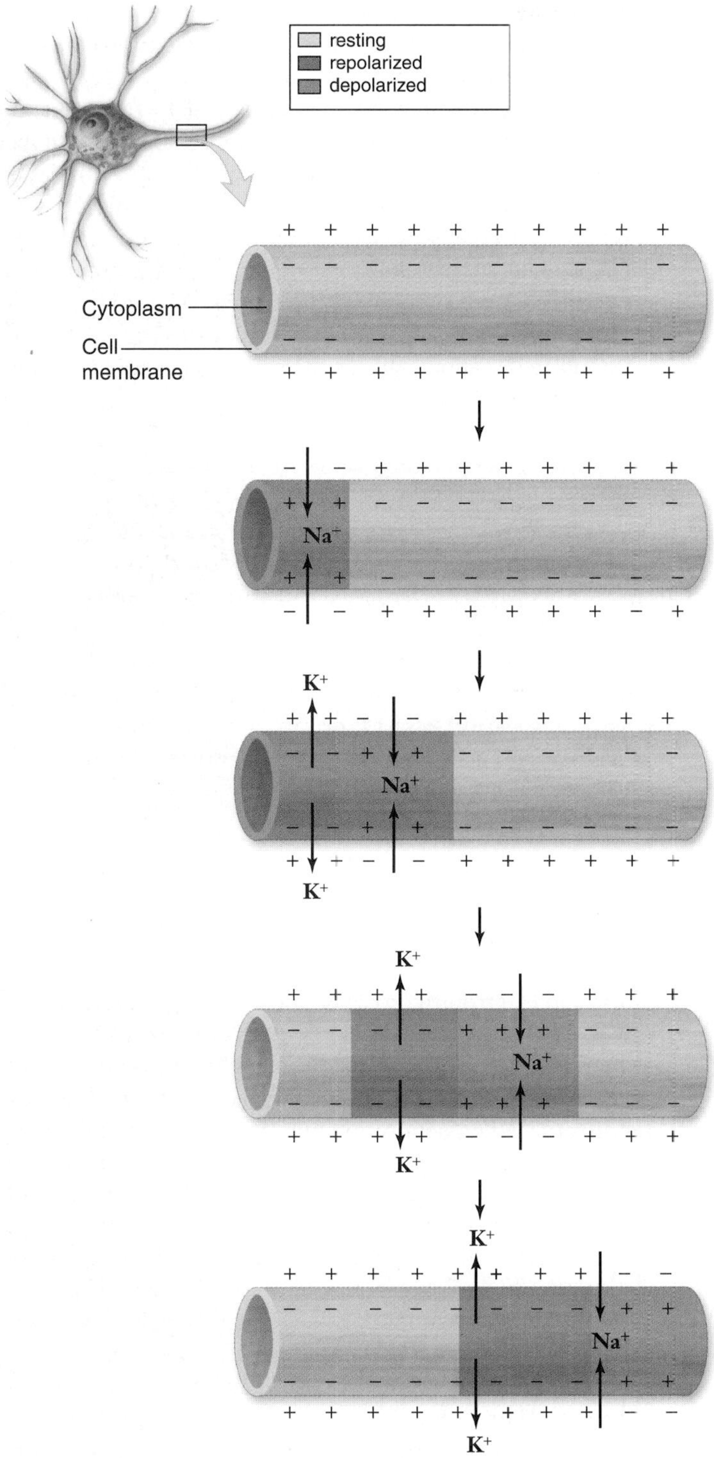

Action potentials are propagated along axons

The movement of an action potential through an axon is not generated by ions flowing from the base of the axon to the end. Instead an action potential originates at the base of the axon, and is then recreated in adjacent stretches of membrane along the axon.

Each action potential, during its rising phase, reflects a reversal in membrane polarity. The positive charges due to influx of Na$^+$ can depolarize the adjacent region of membrane to threshold, so that the next region produces its own action potential (figure 44.10). Meanwhile, the previous region of membrane repolarizes back to the resting membrane potential. The signal does not back up because the Na$^+$ channels that have just "fired" are still in an inactivated state and are refractory (resistant) to stimulation.

The propagation of an action potential is similar to people in a stadium performing the "wave": Individuals stay in place as they stand up (depolarize), raise their hands (peak of the action potential), and sit down again (repolarize). The wave travels around the stadium, but the people stay in place.

There are two ways to increase the velocity of nerve impulses

Action potentials are conducted without decreasing in amplitude, so the last action potential at the end of an axon is just as large as the first action potential. Animals have evolved two ways to increase the velocity of nerve impulses. The velocity of conduction is greater if the diameter of the axon is large or if the axon is myelinated (table 44.2).

Increasing the diameter of an axon increases the velocity of nerve impulses due to the electrical property of resistance. Electrical resistance is inversely proportional to cross-sectional area, which is a function of diameter, so larger diameter axons have less resistance to current flow. The positive charges carried by Na$^+$ will flow farther in a larger diameter axon, leading to a higher than threshold voltage farther from the origin of Na$^+$ influx.

figure 44.10

PROPAGATION OF AN ACTION POTENTIAL IN AN UNMYELINATED AXON. When one region produces an action potential and undergoes a reversal of polarity, it serves as a depolarization stimulus for the next region of the axon. In this way, action potentials regenerate along each small region of the unmyelinated axon membrane.

TABLE 44.2	Conduction Velocities of Some Axons		
	Axon Diameter (μm)	Myelin	Conduction Velocity (m/s)
Squid giant axon	500	No	25
Large motor axon to human leg muscle	20	Yes	120
Axon from human skin pressure receptor	10	Yes	50
Axon from human skin temperature receptor	5	Yes	20
Motor axon to human internal organ	1	No	2

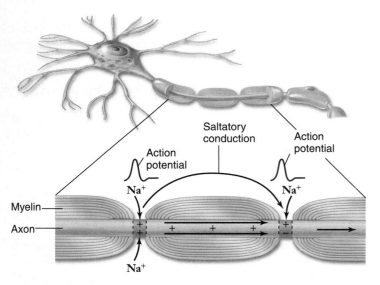

figure 44.11

SALTATORY CONDUCTION IN A MYELINATED AXON.
Action potentials are only produced at the nodes of Ranvier
in a myelinated axon. One node depolarizes the next node so
that the action potentials can skip between nodes. As a result,
saltatory ("jumping") conduction in a myelinated axon is more
rapid than conduction in an unmyelinated axon.

Larger diameter axons are found primarily in invertebrates.
For example, in the squid, the escape response is controlled by
a so-called, giant axon. This large axon conducts nerve impulses
faster than other smaller squid axons, allowing a rapid escape response. The squid giant axon was used by Hodgkin and Huxley
in their pioneering studies of nerve transmission.

Myelinated axons conduct impulses more rapidly than
unmyelinated axons because the action potentials in myelinated
axons are only produced at the nodes of Ranvier. One action
potential still serves as the depolarization stimulus for the next,
but the depolarization at one node spreads quickly beneath the
insulating myelin to trigger opening of voltage-gated channels
at the next node. The impulses therefore seem to jump from
node to node (figure 44.11) in a process called **saltatory conduction** (Latin *saltare*, "to jump").

To see how saltatory conduction speeds impulse transmission,
let's return for a moment to the stadium wave analogy to describe
propagation of an action potential. The wave moves across the seats
of a crowded stadium as fans seeing the people in the adjacent section stand up are triggered to stand up in turn. Because the wave
skips sections of empty bleachers, it actually progresses around the
stadium even faster with more empty sections. The wave doesn't
have to "wait" for the missing people to stand, so it simply moves to
the next populated section—just as the action potential jumps the
nonconducting regions of myelin between exposed nodes.

An unequal distribution of charge across the plasma membrane
produces a resting potential. Inside the membrane is negatively
charged relative to the outside (resting potential is about -70 mV).
Graded potentials, due to opening ligand-gated channels, can
depolarize or hyperpolarize the membrane. These can be combined
(summation). Action potentials occur when membrane depolarization
exceeds a threshold value. Action potentials are all-or-none events,
not subject to summation. Action potentials propagate along an axon
with one action potential serving as the depolarization stimulus for
the next. Velocity of nerve impulses can be increased by increasing
axon diameter and by myelination of axons.

44.3 Synapses: Where Neurons Communicate with Other Cells

An action potential passing down an axon eventually reaches the
end of the axon and all of its branches. These branches may form
junctions with the dendrites of other neurons, with muscle cells,
or with gland cells. Such intercellular junctions are called **synapses.** The neuron whose axon transmits action potentials to the
synapse is termed the *presynaptic cell*, and the cell receiving the
signal on the other side of the synapse is the *postsynaptic cell*.

The two types of synapses are electrical and chemical

Two basic types of synapses exist in the nervous systems of animals: electrical and chemical. **Electrical synapses** involve direct cytoplasmic connections formed by gap junctions between
the pre- and postsynaptic neurons (chapter 9; see figure 9.17).
Membrane potential changes, including action potentials, pass
directly and rapidly from one cell to the other through the gap
junctions. Electrical synapses are common in invertebrate nervous systems, but are somewhat rare in vertebrates.

The vast majority of vertebrate synapses are **chemical
synapses.** When synapses are viewed under a light microscope,
the presynaptic and postsynaptic cells appear to touch, but when

viewed with an electron microscope most have a **synaptic cleft,**
a narrow space that separates these two cells (figure 44.12).

The end of the presynaptic axon is swollen and contains numerous **synaptic vesicles,** each packed with chemicals called **neurotransmitters.** When action potentials arrive at the end of the
axon, they stimulate the opening of voltage-gated calcium (Ca^{2+})
channels, causing a rapid inward diffusion of Ca^{2+}. This influx of
Ca^{2+} triggers a complex series of events that leads to the fusion
of synaptic vesicles with the plasma membrane and the release of
neurotransmitter by exocytosis (chapter 5; see figure 44.13).

The higher the frequency of action potentials in the presynaptic axon, the greater the number of vesicles that release
their contents of neurotransmitters. The neurotransmitters
diffuse to the other side of the cleft and bind to chemical- or
ligand-gated **receptor proteins** in the membrane of the postsynaptic cell. The action of these receptors produces graded
potentials in the postsynaptic membrane.

Neurotransmitters are chemical signals in an otherwise
electrical system, requiring tight control over the duration of
their action. Neurotransmitters must be rapidly removed from
the synaptic cleft to allow new signals to be transmitted. This is
accomplished by a variety of mechanisms, including enzymatic

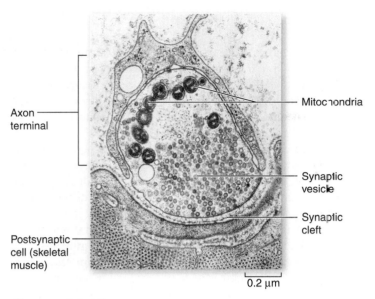

Axon terminal

Mitochondria

Postsynaptic cell (skeletal muscle)

Synaptic vesicle

Synaptic cleft

0.2 μm

figure 44.12

A SYNAPTIC CLEFT. An electron micrograph showing a neuromuscular synapse. Synaptic vesicles have been colored green.

10 μm

figure 44.14

NEUROMUSCULAR JUNCTIONS. A light micrograph shows axons branching to make contact with several individual muscle fibers.

digestion in the synaptic cleft, reuptake of neurotransmitter molecules by the neuron, and uptake by glial cells.

Several different types of neurotransmitters have been identified, and they act in different ways. We next consider the action of a few of the important neurotransmitter chemicals.

Many different chemical compounds serve as neurotransmitters

No single chemical characteristic defines a neurotransmitter, although we can group certain types according to chemical similarities. Some, such as acetylcholine, have wide use in the nervous system, particularly where nerves connect with mus-

cles. Other neurotransmitters are found only in very specific types of junctions, such as in the CNS.

Acetylcholine

Acetylcholine (ACh) is the neurotransmitter that crosses the synapse between a motor neuron and a muscle fiber. This synapse is called a **neuromuscular junction** (figures 44.13, 44.14). Acetylcholine binds to its receptor proteins

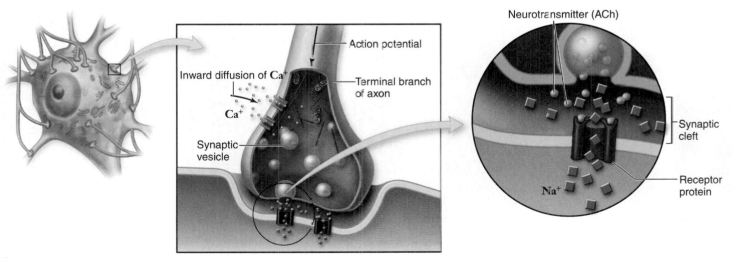

Action potential

Inward diffusion of Ca⁺

Ca⁺

Synaptic vesicle

Terminal branch of axon

Neurotransmitter (ACh)

Synaptic cleft

Na⁺

Receptor protein

figure 44.13

THE RELEASE OF NEUROTRANSMITTER. Action potentials arriving at the end of an axon trigger inward diffusion of Ca^{2+}, which causes synaptic vesicles to fuse with the plasma membrane and release their neurotransmitters (acetylcholine [ACh] in this case). Neurotransmitter molecules diffuse across the synaptic gap and bind to ligand-gated receptors in the postsynaptic membrane.

in the postsynaptic membrane and causes ligand-gated ion channels within these proteins to open (see figure 44.7). As a result, that site on the postsynaptic membrane produces a depolarization (figure 44.15a) called an **excitatory post-synaptic potential (EPSP).** The EPSP, if large enough, can open the voltage-gated channels for Na+ and K+ that are responsible for action potentials. Because the postsynaptic cell in this case is a skeletal muscle fiber, the action potentials it produces stimulate muscle contraction through mechanisms discussed in chapter 47.

For the muscle to relax, ACh must be eliminated from the synaptic cleft. **Acetylcholinesterase (AChE),** an enzyme in the postsynaptic membrane, eliminates ACh. This enzyme, one of the fastest known, cleaves ACh into inactive fragments. Nerve gas and the agricultural insecticide parathion are potent inhibitors of AChE; in humans, they can produce severe spastic paralysis and even death if paralysis affects the respiratory muscles. Although ACh acts as a neurotransmitter between motor neurons and skeletal muscle cells, many neurons also use ACh as a neurotransmitter at their synapses with the dendrites or cell bodies of other neurons.

Amino acids

Glutamate is the major excitatory neurotransmitter in the vertebrate CNS. Excitatory neurotransmitters act to stimulate action potentials by producing EPSPs. Some neurons in the brains of people suffering from Huntington disease undergo changes that render them hypersensitive to glutamate, leading to neurodegeneration.

Glycine and **GABA** (γ-aminobutyric acid) are inhibitory neurotransmitters. These neurotransmitters cause the opening of ligand-gated channels for chloride ion (Cl−), which has a concentration gradient favoring its diffusion into the neuron. Because Cl− is negatively charged, it makes the inside of the membrane even more negative than it is at rest—for example, from −70 mV to −85 mV (see figure 44.15b). This hyperpolarization is called an **inhibitory postsynaptic potential (IPSP),** and it is very important for neural control of body movements and other brain functions. The drug diazepam (Valium) causes its sedative and other effects by enhancing the binding of GABA to its receptors and thereby increasing the effectiveness of GABA at the synapse.

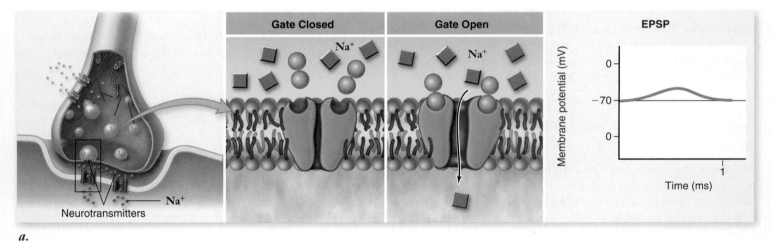

a.

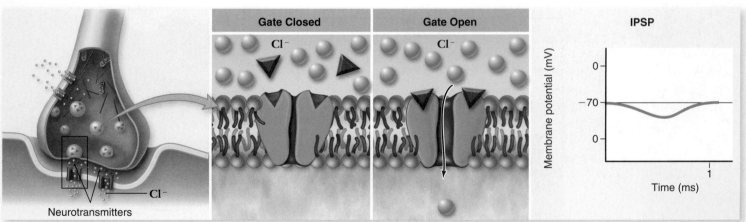

b.

figure 44.15

DIFFERENT NEUROTRANSMITTERS CAN HAVE DIFFERENT EFFECTS. *a.* An excitatory neurotransmitter promotes a depolarization, or excitatory postsynaptic potential (EPSP). *b.* An inhibitory neurotransmitter promotes a hyperpolarization, or inhibitory postsynaptic potential (IPSP).

Biogenic amines

The **biogenic amines** include the hormone epinephrine (adrenaline), together with the neurotransmitters dopamine, norepinephrine, and serotonin. Epinephrine, norepinephrine, and dopamine are derived from the amino acid tyrosine and are included in the subcategory of *catecholamines*. Serotonin is a biogenic amine derived from a different amino acid, tryptophan.

Epinephrine is released into the blood as a hormonal secretion, while **norepinephrine** is released at synapses of neurons in the sympathetic nervous system (discussed in detail later on). The effects of these neurotransmitters on target receptors are responsible for the "fight or flight" response, including faster and stronger heartbeat, increased blood glucose concentration, and diversion of blood flow into the muscles and heart.

Dopamine is a very important neurotransmitter used in some areas of the brain controlling body movements and other functions. Degeneration of particular dopamine-releasing neurons produces the resting muscle tremors of Parkinson disease, and people with this condition are treated with L-dopa (an acronym for L-dihydroxyphenylalanine), a precursor from which dopamine can be produced. Additionally, studies suggest that excessive activity of dopamine-releasing neurons in other areas of the brain is associated with schizophrenia. As a result, drugs that block the production of dopamine, such as the dopamine antagonist chlorpromazine (Thorazine), sometimes help patients with schizophrenia.

Serotonin is a neurotransmitter involved in the regulation of sleep, and it is also implicated in various emotional states. Insufficient activity of neurons that release serotonin may be one cause of clinical depression. Antidepressant drugs, such as fluoxetine (Prozac), block the elimination of serotonin from the synaptic cleft; these drugs are termed *selective serotonin reuptake inhibitors*, or SSRIs.

Other neurotransmitters

Axons also release various polypeptides, called **neuropeptides,** at synapses. These neuropeptides may have a typical neurotransmitter function, or they may have more subtle, long-term action on the postsynaptic neurons. In the latter case, they are often called **neuromodulators.** A given axon generally releases only one kind of neurotransmitter, but many can release both a neurotransmitter and a neuromodulator.

Substance P is an important neuropeptide released at synapses in the CNS by sensory neurons activated by painful stimuli. The perception of pain, however, can vary depending on circumstances. An injured football player may not feel the full extent of his trauma, for example, until he is out of the game.

The intensity with which pain is perceived partly depends on the effects of neuropeptides called *enkephalins* and *endorphins.* **Enkephalins,** released by axons descending from the brain into the spinal cord, inhibit the passage of pain information back up to the brain. **Endorphins,** released by neurons in the brain stem, also block the perception of pain. Opium and its derivatives, morphine and heroin, have an analgesic (pain-reducing) effect because they are similar enough in chemical structure to bind to the receptors normally used by enkephalins and endorphins. For this reason, the enkephalins and the endorphins are referred to as *endogenous opiates.*

Nitric oxide (NO) is the first gas known to act as a regulatory molecule in the body. Because NO is a gas, it diffuses through membranes, so it cannot be stored in vesicles. It is produced as needed from the amino acid arginine. Nitric oxide diffuses out of the presynaptic axon and into neighboring cells by simply passing through the lipid portions of the plasma membranes.

In the PNS, nitric oxide is released by some neurons that innervate the gastrointestinal tract, penis, respiratory passages, and cerebral blood vessels. These autonomic neurons cause smooth-muscle relaxation in their target organs. This relaxation can produce the engorgement of the spongy tissue of the penis with blood, causing an erection. The drug sildenafil (Viagra) increases the release of nitric oxide in the penis, prolonging an erection. The brain releases nitric oxide as a neurotransmitter, where it appears to participate in the processes of learning and memory.

A postsynaptic neuron must integrate input from many synapses

Different types of input from a number of presynaptic neurons influence the activity of a postsynaptic neuron in the brain and spinal cord of vertebrates. For example, a single motor neuron in the spinal cord can have in excess of 50,000 synapses from presynaptic axons.

Each postsynaptic neuron may receive both excitatory and inhibitory synapses (figure 44.16). The EPSPs (depolarizations) and IPSPs (hyperpolarizations) from these synapses interact with each other when they reach the cell body of the neuron. Small EPSPs add together to bring the membrane potential closer to

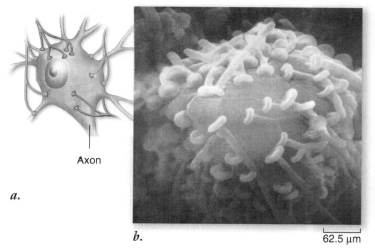

a.

b.   62.5 µm

figure 44.16

INTEGRATION OF EPSPS AND IPSPS TAKES PLACE ON THE NEURONAL CELL BODY. *a.* The synapses made by some axons are excitatory (*green*); the synapses made by other axons are inhibitory (*red*). The summed influence of all of these inputs determines whether the axonal membrane of the postsynaptic cell will be sufficiently depolarized to produce an action potential. *b.* Micrograph of a neuronal cell body with numerous synapses.

the threshold, and IPSPs subtract from the depolarizing effect of the EPSPs, deterring the membrane potential from reaching threshold. This process is called **synaptic integration.**

Because of the all-or-none character of an action potential, a postsynaptic neuron is like a switch that is either turned on or remains off. Information may be encoded in the pattern of firing over time, but each neuron can only fire or not fire when it receives a signal.

The events that determine whether a neuron fires may be extremely complex, and involve many presynaptic neurons. There are two ways the membrane can reach the threshold voltage: by many different dendrites producing EPSPs that sum to the threshold voltage, or by one dendrite producing repeated EPSPs that sum to the threshold voltage. We call the first **spatial summation** and the second **temporal summation.**

In spatial summation, graded potentials due to dendrites from different presynaptic neurons produced at the same time add together to produce an above threshold voltage. All of this input does not need to be in the form of EPSPs, just that the potential produced by summing all of the EPSPs and IPSPs is greater than the threshold voltage. When the membrane at the base of the axon is depolarized above the threshold, it produces an action potential and a nerve impulse is sent down the axon.

In temporal summation, a single dendrite can produce sufficient depolarization to produce an action potential if it produces EPSPs that are close enough in time to sum to a greater than threshold depolarization. A typical EPSP can last for 15 ms, so for temporal summation to occur, the next impulse must arrive in less time. If enough EPSPs are produced to raise the membrane at the base of the axon above threshold, then an impulse will be sent.

The distinction between these two methods of summation is like filling a hole in the ground with soil: you can have many shovels that add soil to the hole until it is filled, or a single shovel that adds soil at a faster rate to fill the hole. When the hole is filled, the axon will fire.

Neurotransmitters play a role in drug addiction

When certain cells of the nervous system are exposed to a constant stimulus that produces a chemically mediated signal for a prolonged period, the cells may lose their ability to respond to that stimulus, a process called **habituation.** You are familiar with this loss of sensitivity—when you sit in a chair, for example, your awareness of the chair diminishes after a certain length of time.

Some nerve cells are particularly prone to this loss of sensitivity. If receptor proteins within synapses are exposed to high levels of neurotransmitter molecules for prolonged periods, the postsynaptic cell often responds by decreasing the number of receptor proteins in its membrane. This feedback is a normal function in all neurons, one of several mechanisms that have evolved to make the cell more efficient. In this case, the cell adjusts the number of receptors downward because plenty of stimulating neurotransmitter is available.

In the case of artificial neurotransmitter effects produced by drugs, long-term drug use means that more of the drug is needed to obtain the same effect.

Cocaine

The drug cocaine causes abnormally large amounts of neurotransmitter to remain in the synapses for long periods. Cocaine affects neurons in the brain's "pleasure pathways" (the *limbic system*, described later). These cells use the neurotransmitter dopamine. Cocaine binds tightly to the transporter proteins on presynaptic membranes that normally remove dopamine from the synaptic cleft. Eventually the dopamine stays in the cleft, firing the receptors repeatedly. New signals add more and more dopamine, firing the pleasure pathway more and more often (figure 44.17).

With prolonged exposure to dopamine, limbic system neurons simply reduce the number of receptors (figure 44.18). The cocaine user is now addicted. With so few dopamine re-

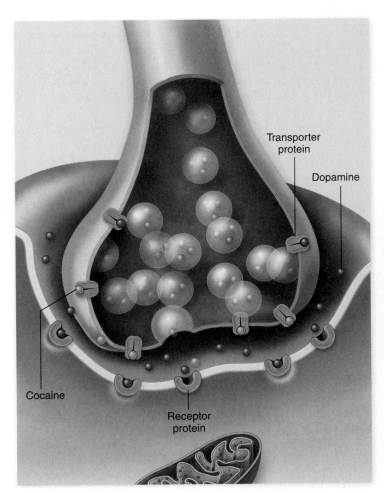

figure 44.17

HOW COCAINE ALTERS EVENTS AT THE SYNAPSE. When cocaine binds to the dopamine transporters, it prevents reuptake of dopamine so the neurotransmitter survives longer in the synapse and continues to stimulate the postsynaptic cell. Cocaine thus acts to intensify pleasurable sensations.

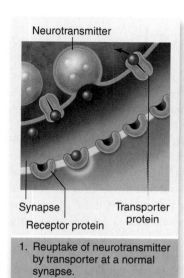
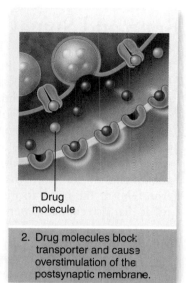
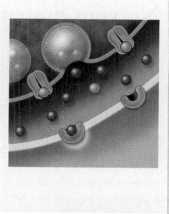
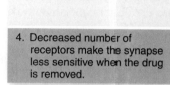

Neurotransmitter

Synapse
Receptor protein
Transporter protein

Drug molecule

1. Reuptake of neurotransmitter by transporter at a normal synapse.

2. Drug molecules block transporter and cause overstimulation of the postsynaptic membrane.

3. Neuron adjusts to overstimulation by decreasing the number of receptors.

4. Decreased number of receptors make the synapse less sensitive when the drug is removed.

figure 44.18

DRUG ADDICTION. (*1*) Some drugs affect neurons by altering the normal reuptake of neurotransmitter. (*2*) If a drug molecule blocks the reuptake of a neurotransmitter, the resulting excess leads to overstimulation of the neuron. (*3*) The central nervous system adjusts to the increased firing by producing fewer receptors in the postsynaptic membrane. The result is addiction. (*4*) When the drug is removed, normal reuptake of the neurotransmitter resumes, and the decreased number of receptors creates a less-sensitive nerve pathway. Physiologically, the only way a person can then maintain normal functioning is to continue to take the drug. If the drug is removed permanently, the nervous system will eventually adjust and restore the original amount of receptors.

ceptors available, the cocaine user now needs the drug to maintain even normal levels of limbic activity.

When an addict stops using cocaine, he or she often experiences a crushing depressed state in which feelings of pleasure are essentially impossible. The cocaine addict is driven by neurochemistry to seek relief with more of the drug. Breaking this cycle is extremely difficult, with only a 50% success rate.

Nicotine

Nicotine has been found to have no affinity for proteins on the presynaptic membrane, as cocaine does; instead, it binds directly to a specific receptor on postsynaptic neurons of the brain. Because nicotine does not normally occur in the brain, why should it have a receptor there?

Researchers have found that "nicotine receptors" are a class of receptors that normally bind the neurotransmitter acetylcholine. Nicotine evolved in tobacco plants as a secondary compound—it affects the CNS of herbivorous insects, and therefore helps to protect the plant. It is an "accident of nature" that nicotine is also able to bind to some human ACh receptors. When neurobiologists compare the nerve cells in the brains of smokers with those of nonsmokers, they find changes in both the number of nicotine receptors and the levels of RNA used to make the receptors. The brain adjusts to prolonged, chronic exposure to nicotine by "turning down the volume" in two ways: (1) by making fewer receptor proteins to which nico-

tine can bind; and (2) by altering the pattern of activation of the nicotine receptors—that is, their sensitivity to stimulation by neurotransmitters.

After prolonged use, nicotine alters the release of many neurotransmitters, including acetylcholine, dopamine, and serotonin. Adjustments are also made in the sensitivity and number of receptors in postsynaptic neurons. Removal of nicotine from this altered system can cause severe symptoms of withdrawal, both physical and psychological, including intense craving for the drug. The symptoms of withdrawal do diminish over time as long as nicotine is avoided completely. The nervous system readjusts to the absence of the drug and resumes normal function.

Having summarized the physiology and chemistry of neurons and synapses, we turn now to the structure of the vertebrate nervous system, beginning with the CNS and then the PNS.

Electrical synapses are common in invertebrates, and chemical synapses predominate in vertebrates. Electrical synapses transmit signals through gap junctions. Chemical synapses release neurotransmitters that diffuse across a narrow synaptic cleft. Many neurotransmitters have been discovered, including acetylcholine, epinephrine, glycine, GABA, and the biogenic amines. The effects of many synapses and different neurotransmitters are integrated through summation of depolarizations and hyperpolarizations.

The Central Nervous System: Brain and Spinal Cord

The complex nervous system of vertebrate animals has a long evolutionary history. In this section we describe the structures making up the CNS, namely the brain and the spinal cord. First, it is helpful to review the origin and development of the vertebrate nervous system.

As animals became more complex, so did their nervous systems

Among the noncoelomate invertebrates (chapter 33), sponges are the only major phylum that lack nerves. The simplest nervous systems occur among cnidarians (figure 44.19), in which all neurons are similar and linked to one another in a web, or **nerve net.** There is no associative activity, no control of complex actions, and little coordination.

The simplest animals with associative activity in the nervous system are the free-living flatworms, phylum Platyhelminthes. Running down the bodies of these flatworms are two nerve cords, from which peripheral nerves extend outward to the muscles of the body. The two nerve cords converge at the front end of the body, forming an enlarged mass of nervous tissue that also contains interneurons with synapses connecting neurons to one another. This primitive "brain" is a rudimen-

Cnidarian 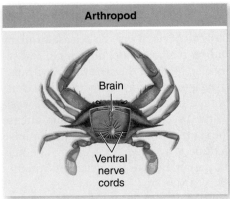	**Earthworm** 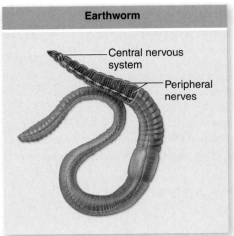	**Human** 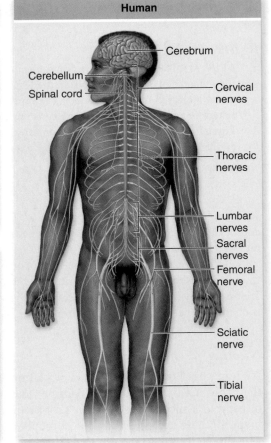
Arthropod 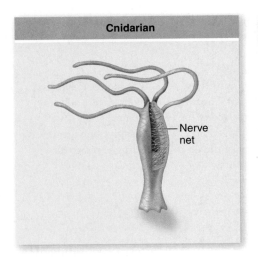	**Echinoderm** 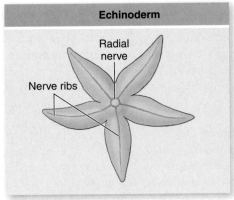	
Mollusk 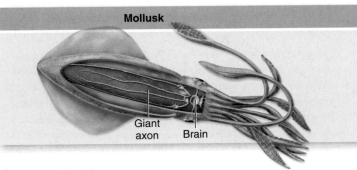	**Flatworm** 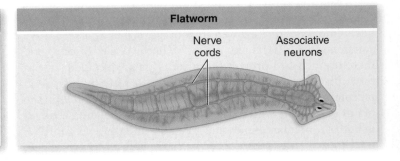	

figure 44.19

DIVERSITY OF NERVOUS SYSTEMS. Nervous systems in animals range from simple nerve nets to paired nerve cords with primitive brains to elaborate brains and sensory systems. Bilateral symmetry is correlated with the concentration of nervous tissue and sensory structures in the front end of the nerve cord. This evolutionary process is referred to as cephalization.

tary central nervous system and permits a far more complex control of muscular responses than is possible in cnidarians.

All of the subsequent evolutionary changes in nervous systems can be viewed as a series of elaborations on the characteristics already present in flatworms. For example, among coelomate invertebrates (chapter 34), earthworms exhibit a central nervous system that is connected to all other parts of the body by peripheral nerves. And, in arthropods, the central coordination of complex responses is increasingly localized in the front end of the nerve cord. As this region evolved, it came to contain a progressively larger number of interneurons, and to develop tracts, which are major information highways within the brain.

Vertebrate brains have three basic divisions

Casts of the interior braincases of fossil agnathans, fishes that swam 500 MYA (chapter 35), have revealed much about the early evolutionary stages of the vertebrate brain. Although small, these brains already had the three divisions that characterize the brains of all contemporary vertebrates:

(1) the **hindbrain,** or rhombencephalon;
(2) the **midbrain,** or mesencephalon; and
(3) the **forebrain,** or prosencephalon (figure 44.20 and table 44.3).

The hindbrain in fishes

The hindbrain was the major component of these early brains, as it still is in fishes today. Composed of the **cerebellum, pons,** and **medulla oblongata,** the hindbrain may be considered an extension of the spinal cord devoted primarily to coordinating motor reflexes. Tracts containing large numbers of axons run like cables up and down the spinal cord to the hindbrain. The hindbrain, in turn, integrates the many sensory signals coming from the muscles and coordinates the pattern of motor responses.

Major Subdivision	Function
TABLE 44.3	**Subdivisions of the Central Nervous System**
SPINAL CORD	Spinal reflexes; relays sensory and motor information
BRAIN	
Hindbrain (Rhombencephalon)	
Medulla oblongata	Sensory nuclei; reticular-activating system; autonomic functions
Pons	Reticular-activating system; autonomic functions
Cerebellum	Coordination of movements; balance
Midbrain (Mesencephalon)	Reflexes involving eyes and ears
Forebrain (Prosencephalon)	
Diencephalon	
Thalamus	Relay station for ascending sensory and descending motor tracts; autonomic functions
Hypothalamus	Autonomic functions; neuroendocrine control
Telencephalon (cerebrum)	
Basal ganglia	Motor control
Corpus callosum	Connects and relays information between the two hemispheres
Hippocampus (limbic system)	Memory; emotion
Cerebral cortex	Higher cognitive functions; integrates and interprets sensory information; organizes motor output

Much of this coordination is carried on within a small extension of the hindbrain called the cerebellum ("little cerebrum"). In more advanced vertebrates, the cerebellum plays an increasingly important role as a coordinating center for movement and it is correspondingly larger than it is in the fishes. In all vertebrates, the cerebellum processes data on the current position and movement of each limb, the state of relaxation or contraction of the muscles involved, and the general position of the body and its relation to the outside world.

The midbrain and forebrain of fishes

In fishes, the remainder of the brain is devoted to the reception and processing of sensory information. The midbrain is composed primarily of the **optic tectum,** which receives and processes visual information, whereas the forebrain is devoted to the processing of olfactory (smell) information.

The brains of fishes continue growing throughout their lives. This continued growth is in marked contrast to the brains of other classes of vertebrates, which generally complete their development by infancy. The human brain continues to develop through early childhood, but few new neurons are produced once development has ceased. One exception is the hippocampus, which has control over which experiences are filed away into long-term memory and which are forgotten. The extent

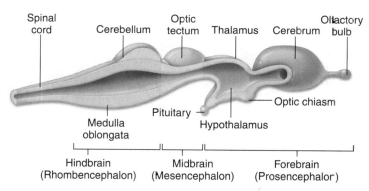

Spinal cord Cerebellum Optic tectum Thalamus Cerebrum Olfactory bulb

Optic chiasm

Medulla oblongata Pituitary Hypothalamus

Hindbrain (Rhombencephalon) Midbrain (Mesencephalon) Forebrain (Prosencephalon)

figure 44.20

THE BASIC ORGANIZATION OF THE VERTEBRATE BRAIN CAN BE SEEN IN THE BRAINS OF PRIMITIVE FISHES. The brain is divided into three regions that are found in differing proportions in all vertebrates: the hindbrain, which is the largest portion of the brain in fishes; the midbrain, which in fishes is devoted primarily to processing visual information; and the forebrain, which is concerned mainly with olfaction (the sense of smell) in fishes. In terrestrial vertebrates, the forebrain plays a far more dominant role in neural processing than it does in fishes.

of neurogenesis (production of new neurons) in adult brains is controversial, and one area of active current research.

The dominant forebrain in more recent vertebrates

Starting with the amphibians and continuing more prominently in the reptiles, processing of sensory information is increasingly centered in the forebrain. This pattern was the dominant evolutionary trend in the further development of the vertebrate brain (figure 44.21).

The forebrain in reptiles, amphibians, birds, and mammals is composed of two elements that have distinct functions. The **diencephalon** consists of the thalamus and hypothalamus. The **thalamus** is an integration and relay center between incoming sensory information and the cerebrum. The **hypothalamus** participates in basic drives and emotions and controls the secretions of the pituitary gland. The **telencephalon** or "end brain," is located at the front of the forebrain and is devoted largely to associative activity. In mammals, the telencephalon is called the **cerebrum.** The telencephalon also includes structures we discuss later on when describing the human brain.

The expansion of the cerebrum

In examining the relationship between brain mass and body mass among the vertebrates, a remarkable difference is observed between fishes and reptiles on the one hand, and birds and mammals on the other. Mammals have brains that are particularly large relative to their body mass. This is especially true of porpoises and humans.

The increase in brain size in mammals largely reflects the great enlargement of the cerebrum, the dominant part of the mammalian brain. The **cerebrum** is the center for correlation, association, and learning in the mammalian brain. It receives sensory data from the thalamus and issues motor commands to the spinal cord via descending tracts of axons.

In vertebrates, the central nervous system is composed of the brain and the spinal cord (see table 44.3). These two structures are responsible for most of the information processing within the nervous system and they consist primarily of interneurons and neuroglia. Ascending tracts carry sensory information to the brain. Descending tracts carry impulses from the brain to the motor neurons and interneurons in the spinal cord that control the muscles of the body.

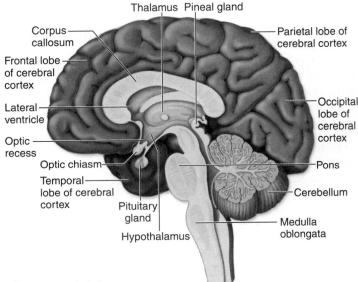

figure 44.22

A SECTION THROUGH THE HUMAN BRAIN. In this sagittal section showing one cerebral hemisphere, the corpus callosum, a fiber tract connecting the two cerebral hemispheres, can be clearly seen.

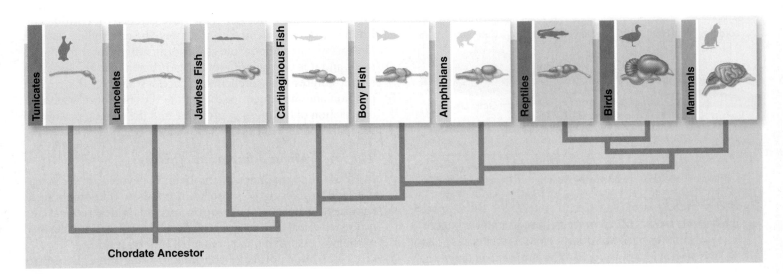

figure 44.21

EVOLUTION OF THE VERTEBRATE BRAIN. The relative sizes of different brain regions have changed as vertebrates have evolved. In sharks and other fishes, the hindbrain is predominant, and the rest of the brain serves primarily to process sensory information. In amphibians and reptiles, the forebrain is far larger, and it contains a larger cerebrum devoted to associative activity. In birds, which evolved from reptiles, the cerebrum is even more pronounced. In mammals, the cerebrum covers the optic tectum and is the largest portion of the brain. The dominance of the cerebrum is greatest in humans, in whom it envelops much of the rest of the brain.

The human forebrain exhibits exceptional information-processing ability

The human cerebrum is so large that it appears to envelop the rest of the brain. It is split into right and left **cerebral hemispheres,** which are connected by a tract called the **corpus callosum** (figure 44.22). The hemispheres are further divided into the *frontal, parietal, temporal,* and *occipital lobes.*

Each hemisphere primarily receives sensory input from the opposite, or contralateral, side of the body and exerts motor control primarily over that side. Therefore, a touch on the right hand is relayed primarily to the left hemisphere, which may then initiate movement of the right hand in response to the touch. Damage to one hemisphere due to a stroke often results in a loss of sensation and paralysis on the contralateral side of the body.

The cerebral cortex

Much of the neural activity of the cerebrum occurs within a layer of gray matter only a few millimeters thick on its outer surface. This layer, called the **cerebral cortex,** is densely packed with nerve cells. In humans, it contains over 10 billion nerve cells, amounting to roughly 10% of all the neurons in the brain. The surface of the cerebral cortex is highly convoluted; this is particularly true in the human brain, where the convolutions increase the surface area of the cortex threefold.

The activities of the cerebral cortex fall into one of three general categories: motor, sensory, and associative. Each of its regions correlates with a specific function (figure 44.23). The **primary motor cortex** lies along the *gyrus* (convolution) on the posterior border of the frontal lobe, just in front of the central *sulcus* (crease). Each point on the surface of the motor cortex is associated with the movement of a different part of the body (figure 44.24, *right*).

Just behind the central sulcus, on the anterior edge of the parietal lobe, lies the **primary somatosensory cortex.** Each point in this area receives input from sensory neurons serving skin and muscle senses in a particular part of the body (figure 44.24, *left*). Large areas of the primary motor cortex and primary somatosensory cortex are devoted to the fingers, lips, and tongue because of the need for manual dexterity and speech. The auditory cortex lies within the temporal lobe, and

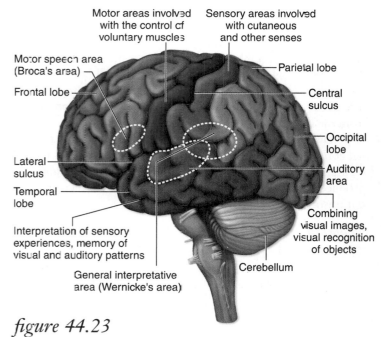

figure 44.23

THE CEREBRUM. This diagram shows the lobes of the cerebrum and indicates some of the known regions of specialization.

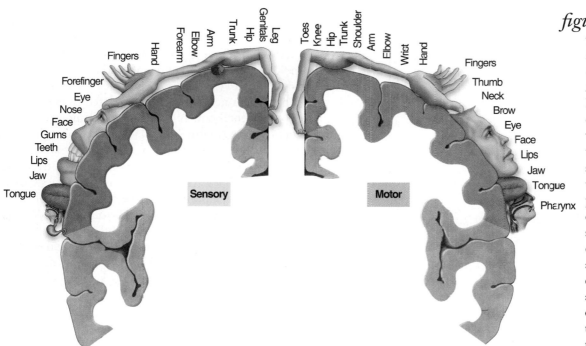

figure 44.24

THE PRIMARY SOMATOSENSORY CORTEX (*LEFT*) AND THE PRIMARY MOTOR CORTEX (*RIGHT*). Each of these regions of the cerebral cortex is associated with a different region of the body, as indicated in this stylized map. The areas of the body are drawn in relative proportion to the amount of cortex dedicated to their sensation or control. For example, the hands have large areas of sensory and motor control, and the pharynx has a considerable area of motor control but little area devoted to the sensations of the pharynx.

different regions of this cortex deal with different sound frequencies. The visual cortex lies on the occipital lobe, with different sites processing information from different positions on the retina, equivalent to particular points in the visual fields of the eyes.

The portion of the cerebral cortex that is not occupied by these motor and sensory cortices is referred to as the **association cortex.** The site of higher mental activities, the association cortex reaches its greatest extent in primates, especially humans, where it makes up 95% of the surface of the cerebral cortex.

Basal ganglia

Buried deep within the white matter of the cerebrum are several collections of cell bodies and dendrites that produce islands of gray matter. These aggregates of neuron cell bodies, which are collectively termed the **basal ganglia,** receive sensory information from ascending nerve tracts and motor commands from the cerebral cortex and cerebellum.

Outputs from the basal ganglia are sent down the spinal cord, where they participate in the control of body movements. Damage to specific regions of the basal ganglia can produce the resting tremor of muscles that is characteristic of Parkinson disease.

Thalamus and hypothalamus

The thalamus is a primary site of sensory integration in the brain. Visual, auditory, and somatosensory information is sent to the thalamus, where the sensory tracts synapse with association neurons. The sensory information is then relayed via the thalamus to the occipital, temporal, and parietal lobes of the cerebral cortex, respectively. The transfer of each of these types of sensory information is handled by specific aggregations of neuron cell bodies within the thalamus.

The hypothalamus integrates the visceral activities. It helps regulate body temperature, hunger and satiety, thirst, and—along with the limbic system—various emotional states. The hypothalamus also controls the pituitary gland, which in turn regulates many of the other endocrine glands of the body. By means of its interconnections with the cerebral cortex and with control centers in the brainstem (a term used to refer collectively to the midbrain, pons, and medulla oblongata), the hypothalamus helps coordinate the neural and hormonal responses to many internal stimuli and emotions.

The **hippocampus** and **amygdala,** along with the hypothalamus, are the major components of the **limbic system**—an evolutionarily ancient group of linked structures deep within the cerebrum that are responsible for emotional responses, as described earlier. The hippocampus is also believed to be important in the formation and recall of memories.

Complex functions of the human brain may be controlled in specific areas

Although studying brain function is difficult, it has long fascinated researchers. The distinction between sleep and waking, the use and acquisition of language, spatial recognition, and memory are all areas of active research. Although far from understood, one generalization that emerged was the regionalization of function.

Sleep and arousal

The brainstem contains a diffuse collection of neurons referred to as the *reticular formation.* One part of this formation, the **reticular-activating system,** controls consciousness and alertness. All of the sensory pathways feed into this system, which monitors the information coming into the brain and identifies important stimuli. When the reticular-activating system has been stimulated to arousal, it increases the level of activity in many parts of the brain. Neural pathways from the reticular formation to the cortex and other brain regions are depressed by anesthetics and barbiturates.

The reticular-activating system controls both sleep and the waking state. It is easier to sleep in a dark room than in a lighted one because there are fewer visual stimuli to stimulate the reticular-activating system. In addition, activity in this system is reduced by serotonin, a neurotransmitter discussed earlier. Serotonin causes the level of brain activity to fall, bringing on sleep.

Brain state can be monitored by means of an electroencephalogram (EEG), a recording of electrical activity. Awake but relaxed individuals with eyes closed exhibit a brain pattern of large, slow waves termed *alpha waves.* In an alert individual with eyes open, the waves are more rapid (*beta waves*) and more desynchronized as sensory input is being received. *Theta waves* and *delta waves* are very slow waves seen during sleep. When an individual is in REM sleep—characterized by rapid eye movements with the eyes closed—the EEG is more like that of an awake, relaxed individual.

Language

Although the two cerebral hemispheres seem structurally similar, they are responsible for different activities. The most thoroughly investigated example of this lateralization of function is language.

The left hemisphere is the "dominant" hemisphere for language in 90% of right-handed people and nearly two-thirds of left-handed people. (By *dominant,* we mean it is the hemisphere in which most neural processing related to language is performed.) Different brain regions control language in the dominant hemisphere (figure 44.25). Wernicke's area, located in the parietal lobe between the primary auditory and visual areas, is important for language comprehension and the formulation of thoughts into speech (see figure 44.23). Broca's area, found near the part of the motor cortex controlling the face, is responsible for the generation of motor output needed for language communication.

Damage to these brain areas can cause language disorders known as *aphasias.* For example, if Wernicke's area is damaged, the person's speech is rapid and fluid but lacks meaning; words are tossed together as in a "word salad."

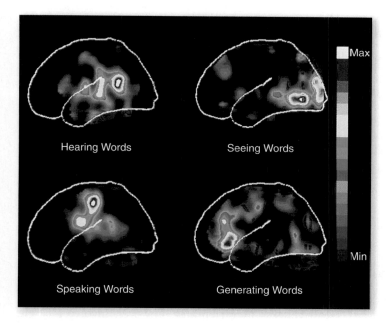

figure 44.25

DIFFERENT BRAIN REGIONS CONTROL VARIOUS LANGUAGE ACTIVITIES. This illustration shows how the brain reacts in human subjects asked to listen to a spoken word, to read that same word silently, to repeat the word out loud, and then to speak a word related to the first. Regions of white, red, and yellow show the greatest activity. Compare this with figure 44.24 to see how regions of the brain are mapped.

Spatial recognition

Whereas the dominant hemisphere for language is adept at sequential reasoning, like that needed to formulate a sentence, the nondominant hemisphere (the right hemisphere in most people) is adept at spatial reasoning, the type of reasoning needed to assemble a puzzle or draw a picture. It is also the hemisphere primarily involved in musical ability—a person with damage to Broca's speech area in the left hemisphere may not be able to speak but may retain the ability to sing.

Damage to the nondominant hemisphere may lead to an inability to appreciate spatial relationships and may impair musical activities such as singing. Even more specifically, damage to the inferior temporal cortex in that hemisphere eliminates the capacity to recall faces, known as prosopagnosia. Reading, writing, and oral comprehension remain normal, and patients with this disability can still recognize acquaintances by their voices. The nondominant hemisphere is also important for the consolidation of memories of nonverbal experiences.

Memory and learning

One of the great mysteries of the brain is the basis of memory and learning. Memory appears dispersed across the brain. Specific cortical sites cannot be identified for particular memories because relatively extensive cortical damage does not selectively remove memories. Although memory is im-paired if portions of the brain, particularly the temporal lobes, are removed, it is not lost entirely. Many memories persist in spite of the damage, and the ability to access them is gradually recovered with time.

Fundamental differences appear to exist between short-term and long-term memory. Short-term memory is transient, lasting only a few moments. Such memories can readily be erased by the application of an electrical shock, leaving previously stored long-term memories intact. This result suggests that short-term memories are stored in the form of a transient neural excitation. Long-term memory, in contrast, appears to involve structural changes in certain neural connections within the brain.

Two parts of the temporal lobes, the hippocampus and the amygdala, are involved in both short-term memory and its consolidation into long-term memory. Damage to these structures impairs the ability to process recent events into long-term memories. Synapses that are used intensively for a short time display more effective synaptic transmission on subsequent use. This phenomenon is called **long-term potentiation (LTP)**. During LTP, the presynaptic neuron may release increased amounts of neurotransmitter with each action potential, and the postsynaptic neuron may become increasingly sensitive to the neurotransmitter. It is believed that these changes in synaptic transmission may be responsible for some aspects of memory storage.

Alzheimer disease: Degeneration of brain neurons

In the past, little was known about **Alzheimer disease,** a condition in which the memory and thought processes of the brain become dysfunctional. Scientists disagree about the biological nature of the disease and its cause. Two hypotheses have been proposed: One suggests that nerve cells in the brain are killed from the outside in, and the other that the cells are killed from the inside out.

In the first hypothesis, external proteins called β-amyloid peptides kill nerve cells. A mistake in protein processing produces an abnormal form of the peptide, which then forms aggregates, or plaques. The plaques begin to fill in the brain and then damage and kill nerve cells. However, these amyloid plaques have been found in autopsies of people who did not exhibit Alzheimer disease.

The second hypothesis maintains that the nerve cells are killed by an abnormal form of an internal protein. This protein, called tau (τ), normally functions to maintain protein transport microtubules. Abnormal forms of τ-protein assemble into helical segments that form tangles, which interfere with the normal functioning of the nerve cells. Researchers continue to study whether tangles and plaques are causes or effects of Alzheimer disease. Progress has been made in identifying genes that increase the likelihood of developing Alzheimer disease and genes that, when mutated, can cause the disorder. Most Alzheimer patients do not have these mutated genes, but for those that do, the symptoms of Alzheimer disease are expressed much earlier in life.

The spinal cord conveys messages and controls some responses directly

The spinal cord is a cable of neurons extending from the brain down through the backbone (figure 44.26). It is enclosed and protected by the vertebral column and layers of membranes called *meninges*, which also cover the brain. Inside the spinal cord are two zones.

The inner zone is gray matter and primarily consists of the cell bodies of interneurons, motor neurons, and neuroglia. The outer zone is white matter and contains cables of sensory axons in the dorsal columns and motor axons in the ventral columns. These nerve tracts may also contain the dendrites of other nerve cells. Messages from the body and the brain run up and down the spinal cord, the body's "information highway."

In addition to relaying messages, the spinal cord also functions in **reflexes,** the sudden, involuntary movement of muscles. A reflex produces a rapid motor response to a stimulus because the sensory neuron passes its information to a motor neuron in the spinal cord, without higher level processing. One of the most frequently used reflexes in your body is blinking, a reflex that protects your eyes. If an object such as an insect or a cloud of dust approaches your eye, the eyelid blinks before you realize what has happened. The reflex occurs before the cerebrum is aware the eye is in danger.

Because they pass information along only a few neurons, reflexes are very fast. A few reflexes, such as the knee-jerk reflex (figure 44.27), are monosynaptic reflex arcs. In these, the sensory nerve cell makes synaptic contact directly with a motor neuron in the spinal cord whose axon travels directly back to the muscle.

Most reflexes in vertebrates, however, involve a single connecting interneuron between the sensory neuron and the motor neuron (figure 44.28). The withdrawal of a hand from a hot stove or the blinking of an eye in response to a puff of air involves a relay of information from a sensory neuron through one or more interneurons to a motor neuron. The motor neuron then stimulates the appropriate muscle to contract. Notice that the sensory neuron may also connect to other interneurons to send signals to the brain. Although you jerked your hand away from the stove, you will still feel pain.

Spinal cord regeneration

In the past, scientists have tried to repair severed spinal cords by installing nerves from another part of the body to bridge the gap and act as guides for the spinal cord to regenerate. But most of these experiments have failed. Although axons may

figure 44.26

A VIEW DOWN THE HUMAN SPINAL CORD. Pairs of spinal nerves can be seen extending from the spinal cord. Along these nerves, as well as the cranial nerves that arise from the brain, the central nervous system communicates with the rest of the body.

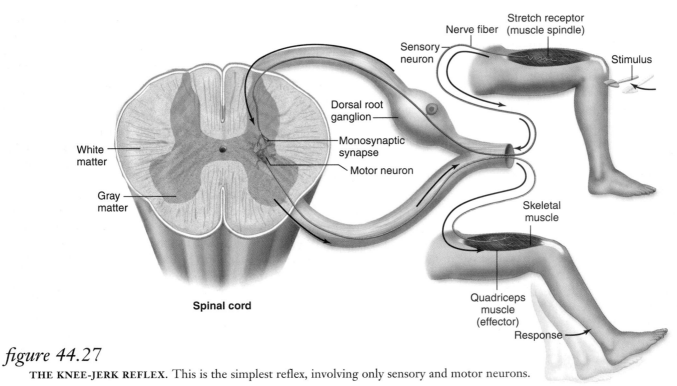

figure 44.27

THE KNEE-JERK REFLEX. This is the simplest reflex, involving only sensory and motor neurons.

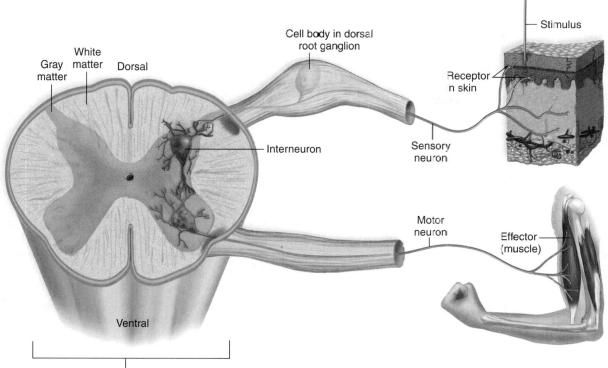

figure 44.28

A CUTANEOUS SPINAL REFLEX. This reflex is more complex than a knee-jerk reflex because it requires interneurons as well as sensory and motor neurons. Interneurons connect a sensory neuron with a motor neuron to cause muscle contraction as shown. Other interneurons inhibit motor neurons, allowing antagonistic muscles to relax.

regenerate through the implanted nerves, they cannot penetrate the spinal cord tissue once they leave the implant. Also, a factor that inhibits nerve growth is present in the spinal cord.

After discovering that fibroblast growth factor stimulates nerve growth, neurobiologists working with rats tried "gluing" the nerves on, from the implant to the spinal cord, with fibrin that had been mixed with the fibroblast growth factor. Three months later, rats with the nerve bridges began to show movement in their lower bodies. Dye tests indicated that the spinal cord nerves had regrown from both sides of the gap.

Many scientists are encouraged by the potential to use a similar treatment in human medicine. But most spinal cord injuries in humans do not involve a completely severed spinal cord; often, nerves are crushed, which results in different tissue damage. Also, even though the rats with nerve bridges did re-

gain some ability to move, tests indicated that they were barely able to walk or stand.

The vertebrate brain consists of three primary regions: the forebrain, midbrain, and hindbrain. In vertebrates more advanced than fishes, the processing of information is increasingly centered in the forebrain. The cerebrum is composed of two cerebral hemispheres. Each hemisphere consists of the gray matter of the cerebral cortex overlying white matter and islands of gray matter (nuclei) called the basal ganglia. These areas are involved in the integration of sensory information, control of body movements, and associative functions such as learning and memory. The spinal cord relays messages to and from the brain and processes some sensory information directly.

The Peripheral Nervous System: Sensory and Motor Neurons

The PNS consists of nerves, the cablelike collections of axons (figure 44.29), and **ganglia** (singular, *ganglion*), aggregations of neuron cell bodies located outside the CNS. To review, the function of the PNS is to receive information from the environment, convey it to the CNS, and to carry responses to effectors such as muscle cells.

The PNS has somatic and autonomic systems

At the spinal cord, a spinal nerve separates into sensory and motor components. The axons of sensory neurons enter the dorsal surface of the spinal cord and form the **dorsal root** of the spinal nerve, whereas motor axons leave from the ventral surface of the spinal cord and form the **ventral root** of the

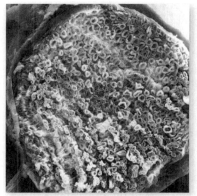

6.25 μm

figure 44.29

NERVES IN THE PERIPHERAL NERVOUS SYSTEM. Photomicrograph showing a cross section of a bullfrog nerve. The nerve is a bundle of axons bound together by connective tissue. Many myelinated axons are visible, each looking somewhat like a doughnut.

TABLE 44.4	Comparison of the Somatic and Autonomic Nervous Systems	
Characteristic	**Somatic**	**Autonomic**
Effectors	Skeletal muscle	Cardiac muscle
		Smooth muscle Gastrointestinal tract Blood vessels Airways
		Exocrine glands
Effect on motor nerves	Excitation	Excitation or inhibition
Innervation of effector cells	Always single	Typically dual
Number of sequential neurons in path to effector	One	Two
Neurotransmitter	Acetylcholine	Acetylcholine, norepinephrine

spinal nerve. The cell bodies of sensory neurons are grouped together outside each level of the spinal cord in the **dorsal root ganglia.** The cell bodies of somatic motor neurons, on the other hand, are located within the spinal cord and so are not located in ganglia.

As mentioned earlier, somatic motor neurons stimulate skeletal muscles to contract, and autonomic motor neurons innervate involuntary effectors—smooth muscles, cardiac muscle, and glands. A comparison of the somatic and autonomic nervous systems is provided in table 44.4; we discuss each system in turn.

The somatic nervous system controls movements

Somatic motor neurons stimulate the skeletal muscles of the body to contract in response to conscious commands and as part of reflexes that do not require conscious control. Voluntary control of skeletal muscles is achieved by activation of tracts of axons that descend from the cerebrum to the appropriate level of the spinal cord. Some of these descending axons stimulate spinal cord motor neurons directly, and others activate interneurons that in turn stimulate the spinal motor neurons.

When a particular muscle is stimulated to contract, however, its antagonist must be inhibited. In order to flex the arm, for example, the flexor muscles must be stimulated while the antagonistic extensor muscle is inhibited (chapter 47). Descending motor axons produce this necessary inhibition by causing hyperpolarizations (IPSPs) of the spinal motor neurons that innervate the antagonistic muscles.

The autonomic nervous system controls involuntary functions through two divisions

The autonomic nervous system is composed of the *sympathetic* and *parasympathetic* divisions plus the medulla oblongata of the hindbrain, which coordinates this system. Although they dif-

fer, the sympathetic and parasympathetic divisions share several features. In both, the efferent motor pathway involves two neurons: The first has its cell body in the CNS and sends an axon to an autonomic ganglion; it is called *preganglionic neuron.* These neurons release acetylcholine at their synapses.

The second neuron has its cell body in the autonomic ganglion and sends its axon to synapse with a smooth muscle, cardiac muscle, or gland cell (figure 44.30). This second neuron is termed the *postganglionic neuron.* Those in the parasympathetic division release ACh, and those in the sympathetic division release norepinephrine.

The sympathetic division

In the sympathetic division, the preganglionic neurons originate in the thoracic and lumbar regions of the spinal cord (figure 44.31, *left*). Most of the axons from these neurons synapse in two parallel chains of ganglia immediately outside the spinal cord. These structures are usually called the *sympathetic chain* of ganglia. The sympathetic chain contains the cell bodies of postganglionic neurons, and it is the axons from these neurons that innervate the different visceral organs.

There are some exceptions to this general pattern, however. The axons of some preganglionic sympathetic neurons pass

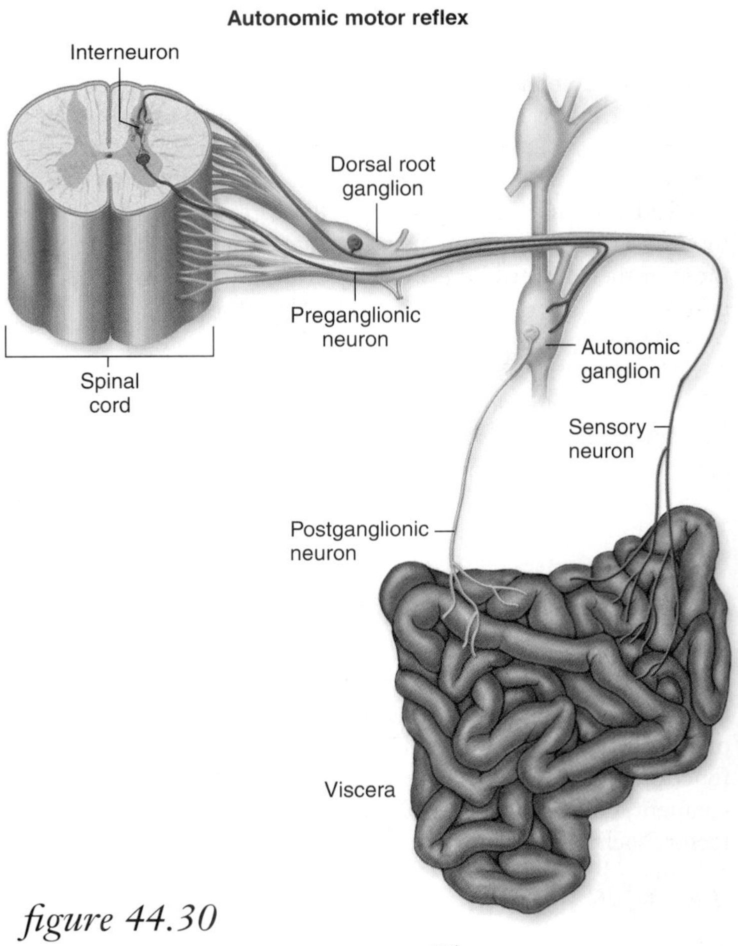

Autonomic motor reflex

figure 44.30

AN AUTONOMIC NEURAL PATH. There are two motor neurons in the efferent pathway. The first, or preganglionic neuron, exits the CNS and synapses at an autonomic ganglion. The second, or postganglionic neuron, exits the ganglion and regulates the visceral effectors (smooth muscle, cardiac muscle, or glands).

through the sympathetic chain without synapsing and, instead, terminate within the medulla of the adrenal gland (chapter 46). In response to action potentials, the adrenal medulla cells secrete the hormone epinephrine (adrenaline). At the same time, norepinephrine is released at the synapses of the postganglionic neurons. As described earlier, both of these neurotransmitters prepare the body for action by heightening metabolism and blood flow.

The parasympathetic division

The actions of the sympathetic division are antagonized by the parasympathetic division. Preganglionic parasympathetic neurons originate in the brain and sacral regions of the spinal cord (see figure 44.31, *right*). Because of this origin, there cannot be a chain of parasympathetic ganglia analogous to the sympathetic chain. Instead, the preganglionic axons, many of which travel in the vagus (tenth cranial) nerve, terminate in ganglia located near or even within the internal organs. The postganglionic neurons then regulate the internal organs by releasing ACh at their synapses. Para-

sympathetic nerve effects include a slowing of the heart, increased secretions and activities of digestive organs, and so on. Table 44.5 compares the actions of the sympathetic and parasympathetic divisions.

G proteins mediate cell responses to autonomic signals

You might wonder how release of ACh can slow the heart rate—an inhibitory effect—when it has excitatory effects elsewhere. The answer is simple, the cells involved in each case have different receptors for ACh that produce different effects. In the neuromuscular junction, the receptor for ACh was a ligand-gated Na^+ channel that when open allowed an influx of Na^+ that depolarized the membrane. In the case of the heart, the inhibitory effect on the pacemaker cells is produced because the ACh receptor causes K^+ channels to open, leading to the outward diffusion of K^-, hyperpolarizing the membrane. This receptor is a member of the class of receptors called G protein-coupled receptors.

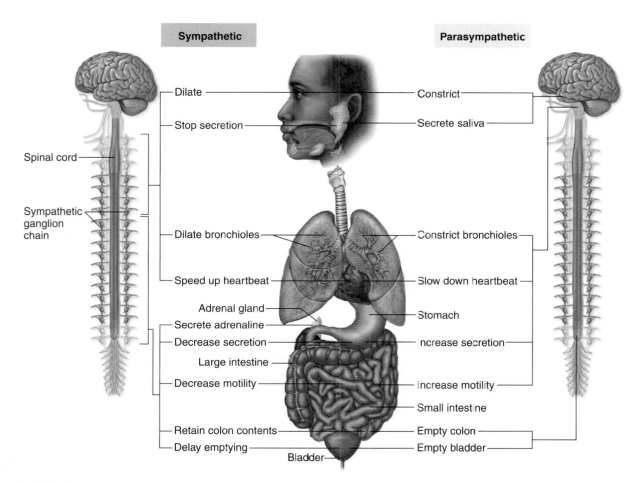

figure 44.31

THE SYMPATHETIC AND PARASYMPATHETIC DIVISIONS OF THE AUTONOMIC NERVOUS SYSTEM. The preganglionic neurons of the sympathetic division exit the thoracic and lumbar regions of the spinal cord, and those of the parasympathetic division exit the brain and sacral region of the spinal cord. The ganglia of the sympathetic division are located near the spinal cord; and those of the parasympathetic division are located near the organs they innervate. Most of the internal organs are innervated by both divisions.

TABLE 44.5	Autonomic Innervation of Target Tissues	
Target Tissue	**Sympathetic Stimulation**	**Parasympathetic Stimulation**
Pupil of eye	Dilation	Constriction
Glands		
Salivary	Vasoconstriction; slight secretion	Vasodilation; copious secretion
Gastric	Inhibition of secretion	Stimulation of gastric activity
Liver	Stimulation of glucose secretion	Inhibition of glucose secretion
Sweat	Sweating	None
Gastrointestinal tract		
Sphincters	Increased tone	Decreased tone
Wall	Decreased tone	Increased motility
Gallbladder	Relaxation	Contraction
Urinary bladder		
Muscle	Relaxation	Contraction
Sphincter	Contraction	Relaxation
Heart muscle	Increased rate and strength	Decreased rate
Lungs	Dilation of bronchioles	Constriction of bronchioles
Blood vessels		
In muscles	Dilation	None
In skin	Constriction	None
In viscera	Constriction	Dilation

In chapter 9, you learned that G protein-coupled receptors consist of a membrane receptor and effector protein that are coupled by the action of a G protein. The receptor is activated by binding to its ligand, in this case ACh, and the receptor activates a G protein that in turn activates an effector protein, in this case a K^+ channel (figure 44.32).

This kind of system can also lead to excitation in other organs if the G protein acts on different effector proteins. For example, the parasympathetic nerves that innervate the stomach can cause increased gastric secretions and contractions.

The sympathetic nerve effects also are mediated by the action of G protein-coupled receptors. Stimulation by norepinephrine from sympathetic nerve endings and epinephrine from the adrenal medulla requires G proteins to activate the

target cells. We describe these interactions in more detail, together with hormone action, in chapter 46.

A spinal nerve contains sensory neurons and motor neurons. The somatic nervous system controls reflexes and voluntary movements. The autonomic nervous system controls involuntary functions. The sympathetic division of the autonomic nervous system, together with the adrenal medulla, activates the body for fight-or-flight responses; the parasympathetic division generally promotes relaxation and digestion. When both systems innervate the same effector, they generally have antagonistic effects. G proteins play a role as intermediaries in PNS responses.

figure 44.32

THE PARASYMPATHETIC EFFECTS OF ACh REQUIRE THE ACTION OF G PROTEINS. The binding of ACh to its receptor causes dissociation of a G protein complex, releasing some components of this complex to move within the membrane and bind to other proteins that form ion channels. Shown here are the effects of ACh on the heart, where the G protein components cause the opening of K^+ channels. This leads to outward diffusion of potassium and hyperpolarization, slowing the heart rate.

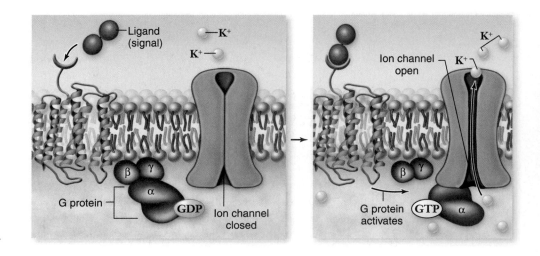

44.1 Nervous System Organization

For an organism to respond to stimuli the nervous system must link sensory receptors to motor effectors (figure 44.1).

- The central nervous system (CNS) consists of the brain and spinal cord.
- The peripheral nervous system (PNS) is made up of sensory neurons that carry impulses to the CNS and motor neurons that carry impulses from the CNS (figure 44.2).
- The autonomic nervous system is composed of the antagonistic sympathetic and parasympathetic divisions.
- Neurons have a cell body, dendrites that receive information, and a long axon that conducts impulses from the cell.
- Schwann cells and oligodendrocytes produce myelin sheaths that surround and insulate axons (figure 44.3).

44.2 The Mechanism of Nerve Impulse Transmission

Nerve impulse transmission depends on the permeability of nerves to ions through gated ion channels.

- Nerve function depends on membrane potential, which is due to differences in ion concentration across the cell membrane.
- The sodium–potassium pump moves Na^+ outside the cell and K^+ inside the cell.
- Leakage of K^+ moves positive charge outside the cell, making the inside negative relative to the outside.
- A balance between K^+ diffusion out and negative charge attracting K^+ into the cell produces the resting potential (figure 44.6).
- Neurons exhibit graded potentials due to the activation of ligand-gated ion channels (figure 44.7).
- Summation is the ability of graded potentials to combine.
- Action potentials result when depolarization reaches a threshold, and they are all-or-nothing events.
- An action potential consists of a rising phase, a falling phase, and an undershoot phase (figure 44.9).
- The intensity of a stimulus is coded by the frequency of action potentials.
- The influx of Na^+ from an action potential causes the adjacent region to depolarize, producing its own action potential (figure 44.10).
- The speed of propagation of nerve impulses increases as the diameter of the axon increases and by insulation from myelin sheaths, which results in saltatory conduction (figure 44.11).

44.3 Synapses: Where Neurons Communicate with Other Cells

An action potential terminates at the end of the axon, and information must bridge the synapse—a gap between neurons and other cells.

- Electrical synapses have gap junctions between pre- and postsynaptic cells, chemical synapses use neurotransmitters.
- The arrival of a nerve impulse opens Ca^{2+} voltage-gated channels causing release of neurotransmitter (figure 44.13).
- Neurotransmitters diffuse across the synaptic cleft and bind to ligand-gated receptors, altering membrane potential.
- Neurotransmitter types include amino acids, biogenic amines, neuropeptides, and a gas—nitric oxide.
- Excitatory postsynaptic potentials (EPSP) depolarize the membrane, and inhibitory postsynaptic potentials (IPSP) hyperpolarize the membrane (figure 44.15).
- The action of neurotransmitters is terminated by enzymatic degradation and reuptake into the presynaptic cell.
- Graded potentials can be spatially or temporally summed to produce an action potential.

44.4 The Central Nervous System: Brain and Spinal Cord

The evolution of the nervous system had led to the capacity for associative activity—the control of complex actions.

- The nervous system has evolved from a nerve net composed of linked nerves, to nerve cords with association nerves, and the development of coordination centers (figure 44.19).
- Vertebrate brains have three main divisions: the hindbrain, midbrain, and forebrain (figure 44.20).
- The hindbrain coordinates motor reflexes.
- The midbrain receives and processes visual information.
- The forebrain is responsive to olfaction and is divided further into the diencephalon and telencephalon.
- The diencephalon consists of the thalamus, which integrates and relays sensory information, and the hypothalamus, which controls secretions of the pituitary gland.
- The telencephalon, called the cerebrum in mammals, is the center for correlation, association, and learning (table 44.3).
- In humans the cerebrum is divided into right and left hemispheres connected by the corpus callosum (figure 44.22).
- Each hemisphere is divided into frontal, parietal, temporal, and occipital lobes (figure 44.23).
- Each hemisphere receives sensory input from the opposite side of the body and controls activities on that side.
- The cerebrum contains the primary motor cortex and the primary somatosensory cortex (figure 44.24).
- Basal ganglia receive sensory input from ascending nerve tracts and output motor commands down the spinal cord.
- The limbic system consists of the hypothalamus, hippocampus, and amygdala and it is responsible for emotional states.
- Complex actions of the brain may be controlled in specific areas.
- Short-term memory may be stored as transient neural excitement, whereas long-term memory involves structural changes in neural connections.
- Reflexes are the sudden, involuntary movement of muscles (figure 44.27 and figure 44.28).

44.5 The Peripheral Nervous System: Sensory and Motor Neurons

The peripheral nervous system (PNS) is a collection of axons and aggregations of cell bodies called ganglia located outside the CNS.

- Sensory axons enter the dorsal surface of the spinal cord and form the dorsal root of the spinal nerve.
- Motor axons leave the ventral surface of the spinal cord and form the ventral root of the spinal nerve.
- Cell bodies of sensory neurons are grouped outside the spinal cord and form the dorsal root ganglia.
- Cell bodies of motor neurons are located in the spinal cord.
- Somatic motor neurons stimulate skeletal muscles in response to conscious commands and involuntary reflexes.
- Involuntary functions are controlled by the autonomic nervous system.
- Sympathetic preganglionic neurons originate in the thoracic and lumbar regions of the spinal cord and synapse at an autonomic ganglion outside the spinal cord (figure 44.30).
- Parasympathetic preganglionic neurons originate in the brain and sacral regions of the spinal cord and terminate in ganglia located near or within internal organs (figure 44.31 and table 44.5).

SELF TEST

1. Which of the following best describes the electrical state of a neuron at rest?
   a. The inside of a neuron is more negatively charged than the outside.
   b. The outside of a neuron is more negatively charged than the inside.
   c. The inside and the outside of a neuron have the same electrical charge.
   d. Potassium ions leak into a neuron at rest.

2. The _____ cannot be controlled by conscious thought.
   a. motor neurons
   b. somatic nervous system
   c. autonomic nervous system
   d. skeletal muscles

3. A fight-or-flight response in the body is controlled by the—
   a. sympathetic division of the nervous system.
   b. parasympathetic division of the nervous system.
   c. release of acetylcholine from postganglionic neurons.
   d. somatic nervous system.

4. Imagine that you are doing an experiment on the movement of ions across neural membranes. Which of the following plays a role in determining the equilibrium concentration of ions across these membranes?
   a. ion concentration gradients
   b. ion pH gradients
   c. ion electrical gradients
   d. Both (a) and (c)

5. The Na^+/K^+ ATPase pump—
   a. is not required for action potential firing.
   b. is important for long-term maintenance of resting potential.
   c. is important only at the synapse.
   d. is used to stimulate graded potentials.

6. Botox, a derivative of the botulinum toxin that causes food poisoning, inhibits the release of acetylcholine at the neuromuscular junction. How could this strange-sounding treatment produce desired cosmetic effects?
   a. By inhibiting the parasympathetic branch of autonomic nervous system
   b. By inhibiting the sympathetic branch of autonomic nervous system
   c. By causing paralysis of facial muscles, which decreases wrinkles in the face
   d. By causing facial muscles to contract, whereby the skin is stretched tighter reducing wrinkles

7. The following is a list of the components of a chemical synapse. A mutation in the structure of which of these would affect only the reception of the message, not its release or the response?
   a. membrane proteins in the postsynaptic cell
   b. proteins in the presynaptic cell
   c. cytoplasmic proteins in the postsynaptic cell
   d. Both (a) and (b)

8. Inhibitory neurotransmitters—
   a. hyperpolarize postsynaptic membranes.
   b. hyperpolarize presynaptic membranes.
   c. depolarize postsynaptic membranes.
   d. depolarize presynaptic membranes.

9. Suppose that you stick your finger with a sharp pin. The area affected is very small and only one pain receptor fires. However, it fires repeatedly at a rapid rate (it hurts!). This is an example of—
   a. temporal summation.
   b. spatial summation.
   c. habituation.
   d. repolarization.

10. You are dissecting a rat that died peacefully in its sleep at an advanced age. Which structure would you not expect to find in its brain?
    a. Forebrain
    b. Midbrain
    c. Hindbrain
    d. Ventral nerve cords

11. White matter is _____ , and gray matter is _____ .
    a. comprised of axons; comprised of cell bodies and dendrites
    b. myelinated; unmyelinated
    c. found in the CNS; also found in the CNS
    d. All of these are correct

12. A functional reflex requires—
    a. only a sensory neuron and a motor neuron.
    b. a sensory neuron, the thalamus, and a motor neuron.
    c. the cerebral cortex and a motor neuron.
    d. only the cerebral cortex and the thalamus.

13. As you sit quietly reading this sentence, the part of the nervous system that is most active is the—
    a. somatic nervous system.
    b. sympathetic nervous system.
    c. parasympathetic nervous system.
    d. None of these choices is correct.

14. G protein-coupled receptors are involved in the nervous system by—
    a. controlling the release of neurotransmitters.
    b. controlling the opening and closing of Na^+ channels during an action potential.
    c. controlling the opening and closing of K^+ channels during an action potential.
    d. acting as receptors for neurotransmitters on postsynaptic cells.

15. During an action potential—
    a. the rising phase is due to an influx of Na^+.
    b. the falling phase is due to an influx of K^+.
    c. the falling phase is due to an efflux of K^+.
    d. both (a) and (c).

CHALLENGE QUESTIONS

1. Tetraethylammonium (TEA) is a drug that blocks voltage-gated K^+ channels. What effect would TEA have on the action potentials produced by a neuron? If TEA could be applied selectively to a presynaptic neuron that releases an excitatory neurotransmitter, how would it alter the synaptic effect of that neurotransmitter on the postsynaptic cell?

2. Describe the status of the Na^+ and K^+ channels at each of the following stages: rising, falling and undershoot.

3. Describe the steps required to produce an excitatory post-synaptic potential (EPSP). How would these differ at an inhibitory synapse?

4. Your friend Karen loves caffeine. However, lately she has been complaining that she needs to drink more caffeinated beverages in order to get the same effect that she used to. Excellent student of biology that you are, you tell her that this is to be expected. Why?

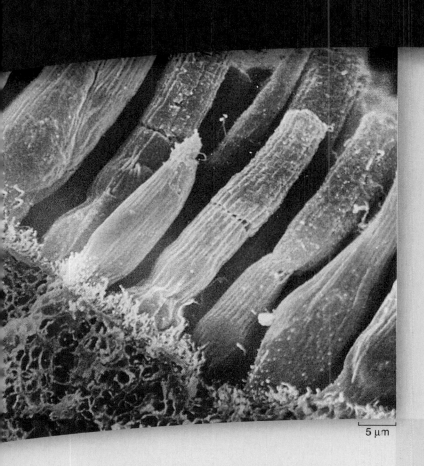

5 μm

chapter 45

Sensory Systems

introduction

ALL INPUT FROM SENSORY NEURONS to the central nervous system arrives in the same form, as action potentials. Sensory neurons receive input from a variety of different kinds of sense receptor cells, such as the rod and cone cells found in the vertebrate eye shown in the micrograph. Different sensory neurons lead to different brain regions, and so are associated with the different senses. The intensity of the sensation depends on the frequency of action potentials conducted by the sensory neuron. The brain distinguishes a sunset, a symphony, and searing pain only in terms of the identity of the sensory neuron carrying the action potentials and the frequency of these impulses. Thus, if the auditory nerve is artificially stimulated, the brain perceives the stimulation as sound. But if the optic nerve is artificially stimulated in exactly the same manner and degree, the brain perceives a flash of light.

In this chapter, we examine sensory systems, primarily in vertebrates. We also compare some of these systems with their counterparts in invertebrates.

45.1 Overview of Sensory Receptors

When we think of sensory receptors, the senses of vision, hearing, taste, smell, and touch come to mind—those senses that provide information about our environment. Certainly this external information is crucial to the survival and success of animals—but sensory receptors also provide information about internal states, such as stretching of muscles, position of the body, and blood pressure. In this section, we take a general look at types of receptors and how they work.

TABLE 45.1	Sensory Transduction Among the Vertebrates			
Stimulus	**Receptor**	**Location**	**Structure**	**Transduction Process**
INTEROCEPTION				
Temperature	Heat receptors and cold receptors	Skin, hypothalamus	Free nerve ending	Temperature change opens or closes ion channels in membrane
Touch	Meissner corpuscles, hair follicle receptors, Merkel cells, Ruffini corpuscles	Skin epithelium	Nerve ending within elastic capsule	Rapid or extended change in pressure deforms membrane
Vibration	Pacinian corpuscles	Deep within skin	Nerve ending within elastic capsule	Severe change in pressure deforms membrane
Pain	Nociceptors	Throughout body	Free nerve ending	Chemicals or changes in pressure or temperature open or close ion channels in membrane
Muscle stretch	Stretch receptors	Within muscles	Spiral nerve endings wrapped around muscle spindle	Stretch of spindle deforms membrane
Blood pressure	Baroreceptors	Arterial branches	Nerve endings over thin part of arterial wall	Stretch of arterial wall deforms membrane
EXTEROCEPTION				
Gravity	Statocysts	Outer chambers of inner ear	Otoliths and hair cells	Otoliths deform hair cells
Motion	Cupula	Semicircular canals of inner ear	Collection of hair cells	Fluid movement deforms hair cells
	Lateral line organ	Within grooves on body surface of fish	Collection of hair cells	Fluid movement deforms hair cells
Taste	Taste bud cells	Mouth; skin of fish	Chemoreceptors: epithelial cells with microvilli	Chemicals bind to membrane receptors
Smell	Olfactory neurons	Nasal passages	Chemoreceptors: ciliated neurons	Chemicals bind to membrane receptors
Hearing	Organ of Corti	Cochlea of inner ear	Hair cells between basilar and tectorial membranes	Sound waves in fluid move hairs, deform membranes
Vision	Rod and cone cells	Retina of eye	Array of photosensitive pigments	Light initiates process that closes ion channels
Heat	Pit organ	Face of snake	Temperature receptors in two chambers	Receptors compare temperatures of surface and interior chambers
Electricity	Ampullae of Lorenzini	Within skin of fishes	Closed vesicles with asymmetrical ion channel distribution	Electrical field alters ion distribution on membranes
Magnetism	Unknown	Unknown	Unknown	Deflection at magnetic field initiates nerve impulses?

Sensory receptors detect both external and internal stimuli

Exteroceptors are receptors that sense stimuli that arise in the external environment. Almost all of a vertebrate's exterior senses evolved in water before vertebrates invaded the land. Consequently, many senses of terrestrial vertebrates emphasize stimuli that travel well in water, using receptors that have been retained in the transition from sea to land. Mammalian hearing, for example, converts an airborne stimulus into a waterborne one, using receptors similar to those that originally evolved in the water.

A few vertebrate sensory systems that function well in the water, such as the electrical organs of fish, cannot function in the air and are not found among terrestrial vertebrates. In contrast, some land-dwellers have sensory systems that could not function in an aquatic environment, for example infrared detectors.

Interoceptors sense stimuli that arise from within the body. These internal receptors detect stimuli related to muscle length and tension, limb position, pain, blood chemistry, blood volume and pressure, and body temperature. Many of these receptors are simpler than those that monitor the external environment and are believed to bear a closer resemblance to primitive sensory receptors. In the rest of this chapter, we consider the different types of exteroceptors and interoreceptors according to the kind of stimulus each is specialized to detect (table 45.1).

Receptors can be grouped into three categories

Sensory receptors differ with respect to the nature of the environmental stimulus that best activates their sensory dendrites. Broadly speaking, we can recognize three classes of receptors:

1. **Mechanoreceptors** are stimulated by mechanical forces such as pressure. These include receptors for touch, hearing, and balance.
2. **Chemoreceptors** detect chemicals or chemical changes. The senses of smell and taste rely on chemoreceptors.
3. **Energy-detecting receptors** react to electromagnetic and thermal energy. The photoreceptors of the eyes that detect light energy are an example, and so are the thermal receptors found in certain reptiles.

The simplest sensory receptors are free nerve endings that respond to bending or stretching of the sensory neuron's membrane to changes in temperature or to chemicals such as oxygen in the extracellular fluid. Other sensory receptors are more complex, involving the association of the sensory neurons with specialized epithelial cells.

Sensory information is conveyed in a four-step process

Sensory information picked up by sensory neurons is conveyed to the CNS, where the impulses are perceived in a four-step process (figure 45.1):

1. *Stimulation.* A physical stimulus impinges on a sensory neuron or an associated, but separate, sensory receptor.
2. *Transduction.* The stimulus energy is transformed into graded potentials in the dendrites of the sensory neuron.
3. *Transmission.* Action potentials develop in the axon of the sensory neuron and are conducted to the CNS along an afferent nerve pathway.
4. *Interpretation.* The brain creates a sensory perception from the electrochemical events produced by afferent stimulation. We actually perceive the five senses with our brains, not with our sense organs.

Sensory transduction involves gated ion channels

Sensory cells respond to stimuli because they possess **stimulus-gated ion channels** in their membranes. The sensory stimulus causes these ion channels to open or close, depending on the sensory system involved. In most cases, the sensory stimulus produces a depolarization of the receptor cell, analogous

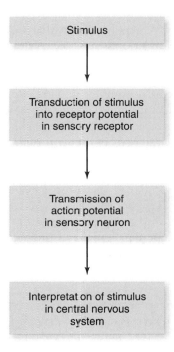

figure 45.1

THE PATH OF SENSORY INFORMATION. Sensory stimuli are transduced into receptor potentials, which can trigger sensory neuron action potentials that are conducted to the brain for interpretation.

figure 45.2

EVENTS IN SENSORY TRANSDUCTION.
a. Depolarization of a free nerve ending leads to a receptor potential that spreads by local current flow to the axon. **b.** Action potentials are produced in the axon in response to a sufficiently large receptor potential.

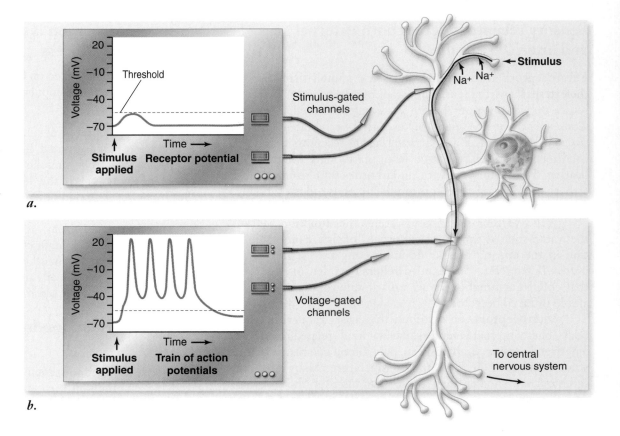

to the excitatory postsynaptic potential (EPSP, described in chapter 44) produced in a postsynaptic cell in response to a neurotransmitter. A depolarization that occurs in a sensory receptor upon stimulation is referred to as a **receptor potential** (figure 45.2a).

Like an EPSP, a receptor potential is a graded potential: The larger the sensory stimulus, the greater the degree of depolarization. Receptor potentials also decrease in size with distance from their source. This prevents small, irrelevant stimuli from reaching the cell body of the sensory neuron. If the receptor potential or the summation of receptor potentials is great enough to generate a threshold level of depolarization, an action potential is produced that propagates along the sensory axon into the CNS (figure 45.2b).

The greater the sensory stimulus, the greater the depolarization of the receptor potential and the higher the frequency of action potentials. (Remember that frequency of action po-

tentials, not their summation, is responsible for conveying the intensity of the stimulus.)

Generally, a logarithmic relationship exists between stimulus intensity and action potential frequency—for example, a particular sensory stimulus that is 10 times greater than another sensory stimulus produces action potentials at twice the frequency of the other stimulus. This relationship allows the CNS to interpret the strength of a sensory stimulus based on the frequency of incoming signals.

> Sensory receptors convey information about the environment (exteroreceptors) and the body's internal state (interoreceptors). Both types of receptors may be grouped into mechanoreceptors, chemoreceptors, or energy-detecting receptors. Frequency of action potentials from sensory receptors informs the CNS of the intensity of the stimulus.

45.2 Mechanoreceptors: Touch and Pressure

Although the receptors of the skin, called the **cutaneous receptors,** are classified as interoceptors, they in fact respond to stimuli at the border between the external and internal environments. These receptors serve as good examples of the specialization of receptor structure and function, responding to heat, cold, pain, touch, and pressure. Here we discuss pain, touch, and pressure, which are sensed by cutaneous mechanoreceptors. Detection of heat and cold are discussed in a later section.

Pain receptors alert the body to damage or potential damage

A stimulus that causes or is about to cause tissue damage is perceived as pain. The receptors that transmit impulses perceived as pain are called **nociceptors,** so named because they can be sensitive to noxious substances as well as tissue damage. Although specific nociceptors exist, many hyperstimulated sensory receptors can also produce the perception of pain in the brain.

Most nociceptors consist of free nerve endings located throughout the body, especially near surfaces where damage is most likely to occur. Different nociceptors may respond to extremes in temperature, very intense mechanical stimulation such as a hard impact, or specific chemicals in the extracellular fluid, including some that are released by injured cells. The thresholds of these sensory cells vary; some nociceptors are sensitive only to actual tissue damage, but others respond before damage has occurred.

Transient receptor potential ion channels

One kind of tissue damage can be due to extremes of temperature, and in this case the molecular details of how a noxious stimulus can result in sensation of pain are becoming clear. A class of ion channel protein found in nociceptors, the transient receptor potential (TRP) ion channel, can be stimulated by temperature to produce an inward flow of cations, primarily Na^+ and Ca^{2+}. This depolarizing current causes the sensory neuron to fire, leading to the release of glutamate and an EPSP in neurons in the spinal cord, ultimately producing the pain response.

TRP channels that respond to both hot and cold have been found. Differences have also been found in the sensitivity of TRP channels to degree of temperature change, with some responding only to temperature changes that damage tissues and others that respond to milder changes. Thus, we can respond to the feelings of hot and cold as well as feel pain associated with extremes of hot and cold.

The first such TRP channel identified responds to the chemical capsaicin, found in chili peppers, as well as to heat. This explains the sensation of heat we feel when we eat chili peppers, as well as the associated pain! A cold-responsive TRP receptor also responds to the chemical menthol, explaining how this substance is perceived as "cold." Chemical stimulation of TRP channels can reduce the body's pain response by desensitizing the sensory neuron. This analgesic response is why menthol is found in cough drops.

Thermoreceptors detect changes in heat

The skin contains two populations of **thermoreceptors,** which are naked dendritic endings of sensory neurons that are sensitive to changes in temperature. (Nociceptors are similar in that they consist of free nerve endings.) These thermoreceptors contain TRP ion channels that are responsive to hot and cold.

Cold receptors are stimulated by a fall in temperature and are inhibited by warming, whereas *warm receptors* are stimulated by a rise in temperature and inhibited by cooling. Cold receptors are located immediately below the epidermis; they are three to four times more numerous than are warm receptors. Warm receptors are typically located slightly deeper, in the dermis.

Thermoreceptors are also found within the hypothalamus of the brain, where they monitor the temperature of the circulating blood and thus provide the CNS with information on the body's internal (core) temperature. Information from the hypothalamic thermoreceptors alter metabolism and stimulate other responses to increase or decrease core temperature as needed.

Different receptors detect touch, depending on intensity

Several types of mechanoreceptors are present in the skin, some in the dermis and others in the underlying subcutaneous tissue (figure 45.3). These receptors contain sensory cells with ion channels that open in response to mechanical distortion of the membrane. They detect various forms of physical contact, known as the sense of **touch.**

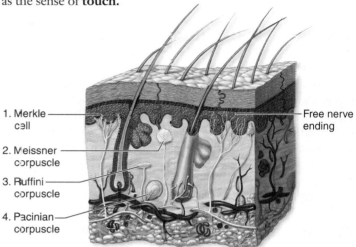

1. Merkle cell
2. Meissner corpuscle
3. Ruffini corpuscle
4. Pacinian corpuscle
Free nerve ending

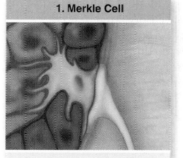

1. Merkle Cell

Tonic receptors located near the surface of the skin that are sensitive to touch pressure and duration.

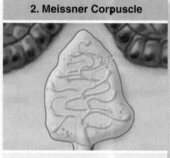

2. Meissner Corpuscle

Receptors sensitive to fine touch, concentrated in hairless skin.

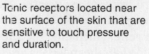

3. Ruffini Corpuscle

Tonic receptors located near the surface of the skin that are sensitive to touch pressure and duration.

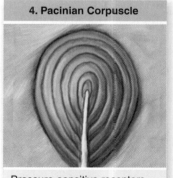

4. Pacinian Corpuscle

Pressure-sensitive receptors deep below the skin in the subcutaneous tissue.

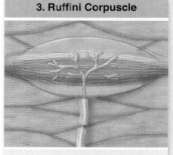

figure 45.3

SENSORY RECEPTORS IN HUMAN SKIN. Cutaneous receptors may be free nerve endings or sensory dendrites in association with other supporting structures.

Morphologically specialized receptors that respond to fine touch are most concentrated on areas such as the fingertips and face. They are used to localize cutaneous stimuli very precisely. These receptors can be either phasic (intermittently activated) or tonic (continuously activated). The phasic receptors include *hair follicle receptors* and *Meissner corpuscles*, which are present on surfaces that do not contain hair, such as the fingers, palms, and nipples.

The tonic receptors consist of *Ruffini corpuscles* in the dermis and *touch dome endings* (*Merkel disks*) located near the surface of the skin. These receptors monitor the duration of a touch and the extent to which it is applied.

Deep below the skin in the subcutaneous tissue lie phasic, pressure-sensitive receptors called *Pacinian corpuscles*. Each of these receptors consists of the end of an afferent axon surrounded by a capsule of alternating layers of connective tissue cells and extracellular fluid. When sustained pressure is applied to the corpuscle, the elastic capsule absorbs much of the pressure, and the axon ceases to produce impulses. Pacinian corpuscles thus monitor only the onset and removal of pressure, as may occur repeatedly when something that vibrates is placed against the skin.

Muscle length and tension are monitored by proprioceptors

Buried within the skeletal muscles of all vertebrates except the bony fishes are **muscle spindles,** sensory stretch receptors that lie in parallel with the rest of the fibers in the muscle (figure 45.4). Each spindle consists of several thin muscle fibers wrapped together and innervated by a sensory neuron, which becomes activated when the muscle, and therefore the spindle, is stretched.

Muscle spindles, together with other receptors in tendons and joints, are known as **proprioceptors.** These sensory receptors provide information about the relative position or movement of the animal's body parts. The sensory neurons conduct action potentials into the spinal cord, where they synapse with somatic motor neurons that innervate the muscle. This pathway constitutes the **muscle stretch reflex,** including the knee-jerk reflex mentioned in chapter 44. When the muscle is briefly stretched by tapping the patellar ligament with a rubber mallet, the *muscle spindle apparatus* is also stretched. The spindle apparatus is embedded within the muscle, and, like the muscle fibers outside the spindle, is stretched along with the muscle. The result is the action potential that activates the somatic motor neurons and causes the leg to jerk.

When a muscle contracts, it exerts tension on the tendons attached to it. The **Golgi tendon organs,** another type of proprioceptor, monitor this tension. If it becomes too high, they elicit a reflex that inhibits the motor neurons innervating the muscle. This reflex helps ensure that muscles do not contract so strongly that they damage the tendons to which they are attached.

Baroreceptors detect blood pressure

Blood pressure is monitored at two main sites in the body. One is the *carotid sinus*, an enlargement of the left and right internal carotid arteries that supply blood to the brain. The other is the *aortic arch*, the portion of the aorta very close to its emergence from the heart. The walls of the blood vessels at both sites

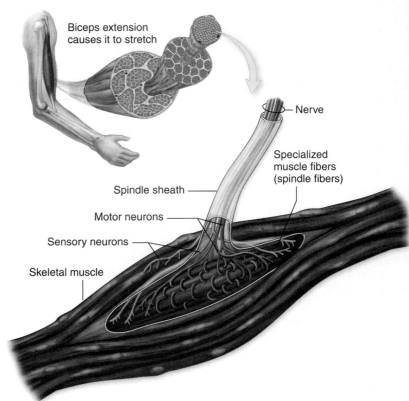

figure 45.4

HOW A MUSCLE SPINDLE WORKS. A muscle spindle is a stretch receptor embedded within skeletal muscle. Stretching of the muscle elongates the spindle fibers and stimulates the sensory dendritic endings wrapped around them. This causes the sensory neurons to send impulses to the CNS, where they synapse with interneurons and, in some cases, motor neurons.

contain a highly branched network of afferent neurons called **baroreceptors,** which detect tension or stretch in the walls.

When the blood pressure decreases, the frequency of impulses produced by the baroreceptors decreases. The CNS responds to this reduced input by stimulating the sympathetic division of the autonomic nervous system, causing an increase in heart rate and vasoconstriction. Both effects help raise the blood pressure, thus maintaining homeostasis. A rise in blood pressure increases baroreceptor impulses, which conversely reduces sympathetic activity and stimulates the parasympathetic division, slowing the heart and lowering the blood pressure.

Mechanical distortion of the plasma membrane of mechanoreceptors produces nerve impulses. Different cutaneous receptors respond to touch, pressure, and pain. Some of these receptors are naked nerve endings; others have supporting structures. Transient receptor potential ion channels are stimulated by heat and cold to depolarize sensory neurons. These channels are also found in thermoreceptors in the skin and other locations. These thermoreceptors sense changes in heat energy and produce responses to reestablish homeostasis. Proprioceptors monitor muscle length, preventing overstretching, and baroreceptors monitor blood pressure within arteries.

Hearing, Vibration, and Detection of Body Position

Hearing, the detection of sound waves, actually works better in water than in air because water transmits pressure waves more efficiently. Despite this limitation, hearing is widely used by terrestrial vertebrates to monitor their environments, communicate with other members of their species, and detect possible sources of danger. Auditory stimuli travel farther and more quickly than chemical ones, and auditory receptors provide better directional information than do chemoreceptors. Auditory stimuli alone, however, provide little information about distance.

Sound is a result of vibration, or waves, traveling through a medium, such as water or air. Detection of sound waves is possible through the action of specialized mechanoreceptors that first evolved in aquatic organisms. The cells that are involved in the detection of sound are also evolutionarily related to the gravity-sensing systems discussed in the end of this section.

The lateral line system in fish detects low-frequency vibrations

In addition to hearing, the **lateral line system** in fish provides a sense of "distant touch," enabling them to sense objects that reflect pressure waves and low-frequency vibrations. This enables a fish to detect prey, for example, and to swim in synchrony with the rest of its school. It also enables a blind cave fish to sense its environment by monitoring changes in the patterns of water flow past the lateral line receptors.

The lateral line system is found in amphibian larvae, but is lost at metamorphosis and is not present in any terrestrial vertebrate. The sense provided by the lateral line system supplements the fish's sense of hearing, which is performed by a different sensory structure.

The lateral line system consists of hair cells within a longitudinal canal in the fish's skin that extends along each side of the body and within several canals in the head (figure 45.5*a*). The hair cells' surface processes project into a gelatinous membrane called a *cupula*. The hair cells are innervated by sensory neurons that transmit impulses to the brain.

Hair cells have several hairlike processes of approximately the same length, called **stereocilia,** and one longer process called a **kinocilium** (figure 45.5*b*). The stereocilia are actually microvilli containing actin fibers, and the kinocilium is a true cilium that contains microtubules. Vibrations carried through the fish's environment produce movements of the cupula, which cause the processes to bend. When the stereocilia bend in the direction of the kinocilium, the associated sensory neurons are stimulated and generate a receptor potential. As a result, the frequency of action potentials produced by the sensory neuron is increased. In contrast, if the stereocilia are bent in the opposite direction, then the activity of the sensory neuron is inhibited.

Ear structure is specialized to detect vibration

The structure of the ear allows pressure waves to be transduced into nerve impulses based on mechanosensory cells like those in the lateral line system. We will first consider the structure of the ear in fish, which is related to the lateral line system that

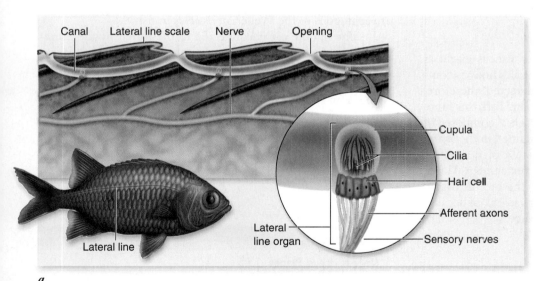

a.

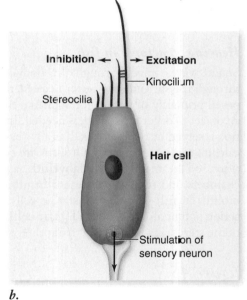

b.

figure 45.5

THE LATERAL LINE SYSTEM. *a.* This system consists of canals running the length of the fish's body beneath the surface of the skin. Within these canals are sensory structures containing hair cells with cilia that project into a gelatinous cupula. Pressure waves traveling through the water in the canals deflect the cilia and depolarize the sensory neurons associated with the hair cells. ***b.*** Hair cells are mechanoreceptors with hairlike cilia that project into a gelatinous membrane. The hair cells of the lateral line system (and the membranous labyrinth of the vertebrate inner ear) have a number of smaller cilia called stereocilia and one larger kinocilium. When the cilia bend in the direction of the kinocilium, the hair cell releases a chemical transmitter that depolarizes the associated sensory neuron. Bending of the cilia in the opposite direction has an inhibitory effect.

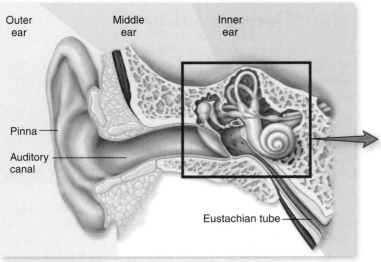

a.

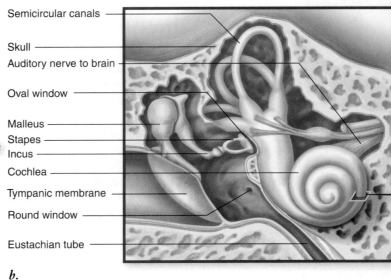

b.

figure 45.6

STRUCTURE AND FUNCTION OF THE HUMAN EAR.
The structure of the human ear is shown in successive enlargements illustrating functional parts (*a* to *d*). Sound waves passing through the ear canal produce vibrations of the tympanic membrane, which causes movement of the middle ear ossicles (the malleus, incus, and stapes) against an inner membrane (the oval window). This vibration creates pressure waves in the fluid in the vestibular and tympanic canals of the cochlea. These pressure waves cause cilia in hair cells to bend, producing signals from sensory neurons.

senses pressure waves in water. Then we will consider how the structure of the ear of terrestrial vertebrates allows the sensing of pressure waves in air.

Hearing structures in fish

Sound waves travel through the body of a fish as easily as through the surrounding water because the fish's body is composed primarily of water. For sound to be detected, therefore, an object of different density is needed. In many fish, this function is served by the **otoliths**, literally "ear rocks," composed of calcium carbonate crystals. Otoliths are contained in the *otolith organs* of the membranous **labyrinth,** a system of fluid-filled chambers and tubes also present in other vertebrates. When otoliths in fish vibrate against hair cells in the otolith organ, action potentials are produced. **Hair cells** are so-called because of the stereocilia that project from their surface.

figure 45.7

FREQUENCY LOCALIZATION IN THE COCHLEA.
The cochlea is shown unwound, so that the length of the basilar membrane can be seen. The fibers within the basilar membrane vibrate in response to different frequencies of sound, related to the pitch of the sound. Thus, regions of the basilar membrane show maximum vibrations in response to different sound frequencies. *a.* Notice that high-frequency (pitch) sounds vibrate the basilar membrane more toward the base whereas medium frequencies (*b*) and low frequencies (*c*) cause vibrations more toward the apex.

In catfish, minnows, and suckers, however, this function is served by an air-filled swim bladder that vibrates with the sound. A chain of small bones, Weberian ossicles, then transmits the vibrations to the labyrinth hair cells in some of these fish.

Hearing structures of terrestrial vertebrates

In the ears of terrestrial vertebrates, vibrations in air may be channeled through an ear canal to the eardrum, or *tympanic membrane*. These structures are part of the **outer ear.** Vibrations of the tympanic membrane cause movement of three small bones (ossicles)—the **malleus** (hammer), **incus** (anvil), and **stapes** (stirrup)—that are located in a bony cavity known as the **middle ear** (figure 45.6*a*, *b*). These middle ear ossicles are analogous to the Weberian ossicles in fish.

The middle ear is connected to the throat by the *Eustachian tube*, also known as the auditory tube, which equalizes the air pressure between the middle ear and the external environment. The "ear popping" you may have experienced when flying in an airplane or driving on a mountain is caused by pressure equalization between the two sides of the eardrum.

The stapes vibrates against a flexible membrane, the *oval window*, which leads into the **inner ear.** Because the oval window is smaller in diameter than the tympanic membrane, vibrations against it produce more force per unit area, transmitted into the

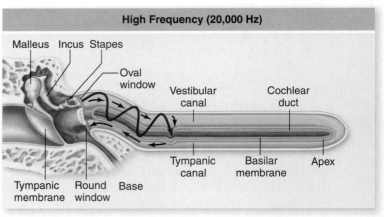

a.

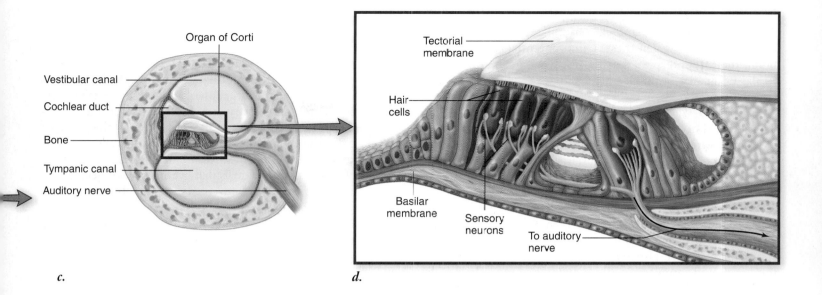

Organ of Corti

Vestibular canal

Cochlear duct

Bone

Tympanic canal

Auditory nerve

Tectorial membrane

Hair cells

Basilar membrane

Sensory neurons

To auditory nerve

c.

d.

inner ear. The inner ear consists of the **cochlea,** a bony structure containing part of the membranous labyrinth called the *cochlear duct.* The cochlear duct is located in the center of the cochlea; the area above the cochlear duct is the *vestibular canal,* and the area below is the *tympanic canal* (figure 45.6*c*). All three chambers are filled with fluid. The oval window opens to the upper vestibular canal, so that when the stapes causes it to vibrate, it produces pressure waves of fluid. These pressure waves travel down to the tympanic canal, pushing another flexible membrane, the *round window,* that transmits the pressure back into the middle ear cavity.

Transduction occurs in the cochlea

As pressure waves are transmitted through the cochlea to the round window, they cause the cochlear duct to vibrate. The bottom of the cochlear duct, called the *basilar membrane,* is quite flexible and vibrates in response to these pressure waves. The surface of the basilar membrane contains sensory hair cells. The stereocilia from the hair cells project into an overhanging gelatinous membrane, the *tectorial membrane.* This sensory apparatus, consisting of the basilar membrane, hair cells with associated sensory neurons, and tectorial membrane, is known as the **organ of Corti** (45.6*d*).

As the basilar membrane vibrates, the cilia of the hair cells bend in response to the movement of the basilar mem-

brane relative to the tectorial membrane. The bending of these stereocilia in one direction depolarizes the hair cells. Bending in the opposite direction repolarizes or even hyperpolarizes the membrane. The hair cells, in turn, stimulate the production of action potentials in sensory neurons that project to the brain, where they are interpreted as sound.

Frequency localization in the cochlea

The basilar membrane of the cochlea consists of elastic fibers of varying length and stiffness, like the strings of a musical instrument, embedded in a gelatinous material. At the base of the cochlea (near the oval window), the fibers of the basilar membrane are short and stiff. At the far end of the cochlea (the apex), the fibers are 5 times longer and 100 times more flexible. Therefore, the resonant frequency of the basilar membrane is higher at the base than at the apex; the base responds to higher pitches, the apex to lower pitches.

When a wave of sound energy enters the cochlea from the oval window, it initiates an up-and-down motion that travels the length of the basilar membrane. However, this wave imparts most of its energy to that part of the basilar membrane with a resonant frequency near the frequency of the sound wave, resulting in a maximum deflection of the basilar membrane at that point (figure 45.7). As a result, the

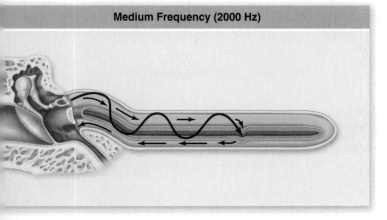

Medium Frequency (2000 Hz)

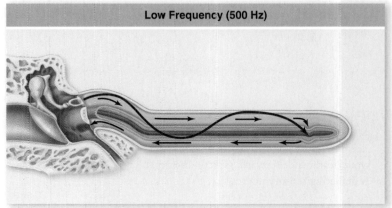

Low Frequency (500 Hz)

c.

chapter 45 *sensory systems*

hair cell depolarization is greatest in that region, and the afferent axons from that region are stimulated more than those of other regions. When these action potentials arrive in the brain, they are interpreted as representing a sound of a particular frequency, or pitch.

The range of terrestrial vertebrate hearing

The flexibility of the basilar membrane limits the frequency range of human hearing to between approximately 20 and 20,000 cycles per second (hertz, Hz) in children. Our ability to hear high-pitched sounds decays progressively throughout middle age. Other vertebrates can detect sounds at frequencies lower than 20 Hz and much higher than 20,000 Hz. Dogs, for example, can detect sounds at 40,000 Hz, enabling them to hear high-pitched dog whistles that seem silent to a human listener.

Hair cells are also innervated by efferent axons from the brain, and impulses in those axons can make hair cells less sensitive. This central control of receptor sensitivity can increase an individual's ability to concentrate on a particular auditory signal (for example, a single voice) in the midst of background noise, which is effectively "tuned out" by the efferent axons.

Some vertebrates have the ability to navigate by sound

Because terrestrial vertebrates have two ears located on opposite sides of the head, the information provided by hearing can be used to determine the direction of a sound source with some precision. Sound sources vary in strength, however, and sounds are weakened and reflected to varying degrees by the presence of objects in the environment. For these reasons, auditory sensors do not provide a reliable measure of distance.

A few groups of mammals that live and obtain their food in dark environments have circumvented the limitations of darkness. A bat flying in a completely dark room easily avoids objects placed in its path—even a wire less than a millimeter in diameter. Shrews use a similar form of "lightless vision" beneath the ground, as do whales and dolphins beneath the sea. All of these mammals are able to perceive presence and distance of objects by sound.

These mammals emit sounds and then determine the time it takes these sounds to reach an object and return to the animal. This process is called **echolocation**. A bat, for example, produces clicks that last 2–3 ms and are repeated several hundred times per second. The three-dimensional imaging achieved with such an auditory sonar system is quite sophisticated.

Being able to "see in the dark" has opened a new ecological niche to bats, one largely closed to birds because birds must rely on vision. There are no truly nocturnal birds; even owls rely on vision to hunt, and do not fly on very dark nights.

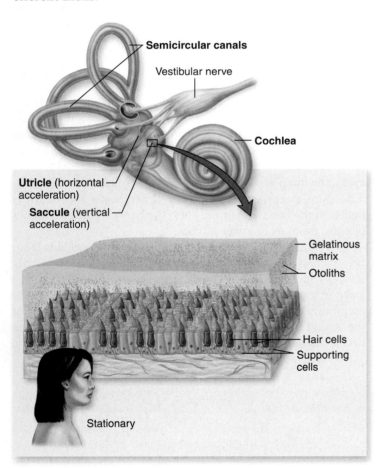

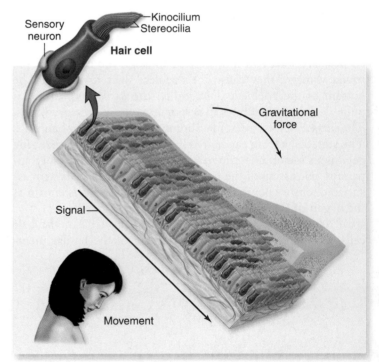

a.

b.

figure 45.8

STRUCTURE AND FUNCTION OF THE UTRICLE AND SACCULE. *a.* The relative positions of the utricle and saccule within the membranous labyrinth of the human inner ear. Enlargement shows the gelatinous matrix containing otoliths covering hair cells. *b.* When your head bends forward, gravity distorts the matrix in the direction of movement. This causes the stereocilia in hair cells to bend, stimulating sensory neurons.

Because bats are able to be active and efficient in total darkness, they are one of the most numerous and widespread of all orders of mammals.

The human invention of *sonar* (*sound navigation and ranging*), used on submarines, relies on the principles of echolocation for underwater navigation and location of objects.

Body position and movement are detected by systems associated with hearing systems

The evolutionary strategy of using internal calcium carbonate crystals as a way to detect vibration has also allowed the development of sensory organs that detect body position in space and movements such as acceleration.

Most invertebrates can orient themselves with respect to gravity due to a sensory structure called a **statocyst.** Statocysts generally consist of ciliated hair cells with the cilia embedded in a gelatinous membrane containing crystals of calcium carbonate. These stones, or **statoliths,** increase the mass of the gelatinous membrane so that it can bend the cilia when the animal's position changes. If the animal tilts to the right, for example, the statolith membrane will bend the cilia on the right side and activate associated sensory neurons.

A similar structure is found in the membranous labyrinth of the inner ear of vertebrates. This labyrinth is surrounded by bone and perilymph, which is similar in ionic content to interstitial fluid. Inside, the chambers and tubes are filled with endolymph fluid, which is similar in ionic content to intracellular fluid. Though intricate, the entire structure is very small; in a human, it is about the size of a pea.

Structure of the labyrinth and semicircular canals

The receptors for gravity in vertebrates consist of two chambers of the membranous labyrinth called the **utricle** and **saccule** (figure 45.8). Within these structures are hair cells with stereocilia and a kinocilium, similar to those in the lateral line system of fish. The hairlike processes are embedded within a gelatinous membrane, the otolith membrane, containing calcium carbonate crystals. Because the otolith organ is oriented differently in the utricle and saccule, the utricle is more sensitive to horizontal acceleration (as in a moving car) and the saccule to vertical acceleration (as in an elevator). In both cases, the acceleration causes the stereocilia to bend, and consequently produces action potentials in an associated sensory neuron.

The membranous labyrinth of the utricle and saccule is continuous with three **semicircular canals,** oriented in different planes so that angular acceleration in any direction can be detected (figure 45.9). At the ends of the canals are swollen chambers called *ampullae*, into which protrude the cilia of another group of hair cells. The tips of the cilia are embedded within a sail-like wedge of gelatinous material called a *cupula* (similar to the cupula of the fish lateral

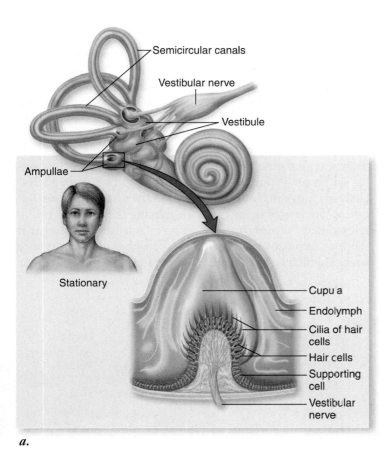

Semicircular canals

Vestibular nerve

Vestibule

Ampullae

Stationary

Cupula

Endolymph

Cilia of hair cells

Hair cells

Supporting cell

Vestibular nerve

a.

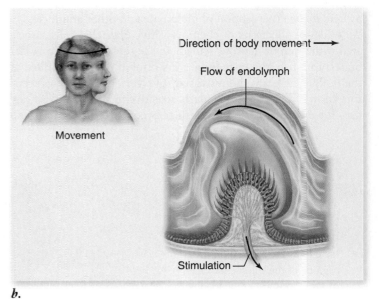

Direction of body movement ⟶

Flow of endolymph

Movement

Stimulation

b.

figure 45.9

THE STRUCTURE OF THE SEMICIRCULAR CANALS. The position of the semicircular canals in relation to the rest of the inner ear. *a.* Enlargement of a section of one ampulla, showing how hair cell cilia insert into the cupula. *b.* Angular acceleration in the plane of the semicircular canal causes bending of the cupula, thereby stimulating the hair cells.

line system) that protrudes into the endolymph fluid of each semicircular canal.

Action of the vestibular apparatus

When the head rotates, the fluid inside the semicircular canals pushes against the cupula and causes the cilia to bend. This bending either depolarizes or hyperpolarizes the hair cells, depending on the direction in which the cilia are bent. This is similar to the way the lateral line system works in a fish: If the stereocilia are bent in the direction of the kinocilium, a receptor potential is produced, which stimulates the production of action potentials in associated sensory neurons.

The saccule, utricle, and semicircular canals are collectively referred to as the **vestibular apparatus.** The saccule and utricle provide a sense of linear acceleration, and the semicircu-

lar canals provide a sense of angular acceleration. The brain uses information that comes from the vestibular apparatus about the body's position in space to maintain balance and equilibrium.

The middle ear ossicles vibrate in response to sound waves, creating fluid vibrations within the inner ear. This causes the hair cells to bend, transducing the sound into action potentials. The basilar membrane is more responsive to high frequency at the base and low frequency at the apex, allowing different frequencies of sound to be distinguished. Some mammals, such as bats, use sound to locate themselves and other objects by echolocation. Hair cells in the lateral line organ of fishes detect water movements, and hair cells in the vestibular apparatus of terrestrial vertebrates provide a sense of acceleration and balance.

45.4 Chemoreceptors: Taste, Smell, and pH

Some sensory cells, called *chemoreceptors*, contain membrane proteins that can bind to particular chemicals or ligands in the extracellular fluid. In response to this chemical interaction, the membrane of the sensory neuron becomes depolarized and produce action potentials. Chemoreceptors are used in the senses of taste and smell and are also important in monitoring the chemical composition of the blood and cerebrospinal fluid.

Taste detects and analyzes potential food

The perception of **taste (gustation),** like the perception of color, is a combination of physical and psychological factors. This is commonly broken down into five categories: *sweet, sour, salty, bitter,* and *umami* (perception of glutamate and other amino acids). *Taste buds*—collections of chemosensitive epithelial cells associated with afferent neurons—mediate the sense of taste in vertebrates. In a fish, the taste buds are scattered over the surface of the body. These are the most sensitive vertebrate chemoreceptors known. They are particularly sensitive to amino acids; a catfish, for example, can distinguish between two different amino

acids at a concentration of less than 100 parts per billion (1 g in 10,000 L of water)! The ability to taste the surrounding water is very important to bottom-feeding fish, enabling them to sense the presence of food in an often murky environment.

The taste buds of all terrestrial vertebrates are located in the epithelium of the tongue and oral cavity, within raised areas called *papillae* (figure 45.10). Taste buds are onion-shaped structures of between 50 and 100 taste cells; each cell has fingerlike projections called microvilli that poke up through an opening at the top of the taste bud, called the *taste pore* (figure 45.10c). Chemicals from food dissolve in saliva and contact the taste cells through the taste pore.

Within a taste bud, the chemicals that produce salty and sour tastes act directly through ion channels. The prototypical salty taste is due to Na^+ ions, which diffuse through Na^+ channels into cells in receptor cells in the taste bud. This Na^+ influx depolarizes the membrane, causing the receptor cell to release neurotransmitter and activate a sensory neuron that sends an impulse to the brain. The cells that detect sour taste act in a similar fashion except that the ion detected is H^+. Sour

figure 45.10

TASTE. *a.* Human tongues have projections called papillae that bear taste buds. Different sorts of taste buds are located on different regions of the tongue. *b.* Groups of taste buds are embedded within a papilla. *c.* Individual taste buds are bulb-shaped collections of chemosensitive receptors that open out into the mouth through a pore. *d.* Photomicrograph of taste buds in papillae.

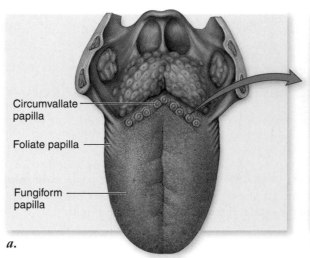

Circumvallate papilla

Foliate papilla

Fungiform papilla

a.

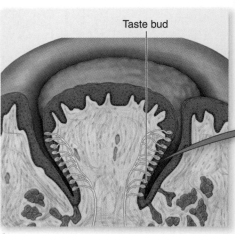

Taste bud

b.

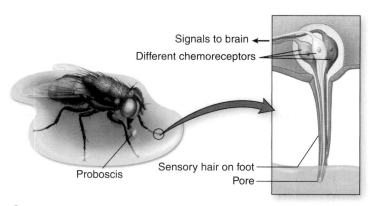

figure 45.11

MANY INSECTS TASTE WITH THEIR FEET. In the blowfly shown here, chemoreceptors extend into the sensory hairs on the foot. Different chemoreceptors detect different types of food molecules. When the fly steps in a food substance, it can taste the different food molecules and extend its proboscis for feeding.

tastes are associated with increased concentration of protons that can also depolarize the membrane when they diffuse through ion channels.

The mechanism of detection of sweet, bitter, and umami are indirect. In this case substances that fall into these categories can bind to G protein-coupled receptors (chapter 9) specific for each category. The nature and distribution of these receptors is an area of active investigation, but recent data indicate that individual receptor cells in the taste bud express only one type of receptor. This leads to cells that have receptors for sweet, for bitter or for umami tastes. Activation of any of these G protein-coupled receptors then converges on a single signaling pathway that leads the release of neurotransmitter from receptor cells to activate a sensory neuron and send an impulse to the brain. There they interact with other sensory neurons carrying information related to smell, described next. In this model, the different tastes are encoded to the brain based on which receptor cells are activated.

Like vertebrates, many arthropods also have taste chemoreceptors. For example, flies, because of their mode of searching for food, have taste receptors in sensory hairs located on

their feet. The sensory hairs contain a variety of chemoreceptors that are able to detect sugars, salts, and other tastes by the integration of stimuli from these chemoreceptors (figure 45.11). If they step on potential food, their proboscis (the tubular feeding apparatus) extends to feed.

Smell can identify a vast number of complex molecules

In terrestrial vertebrates, the sense of **smell (olfaction)** involves chemoreceptors located in the upper portion of the nasal passages (figure 45.12). These receptors, whose dendrites end in tassels of cilia, project into the nasal mucosa, and their axons project directly into the cerebral cortex. A terrestrial vertebrate

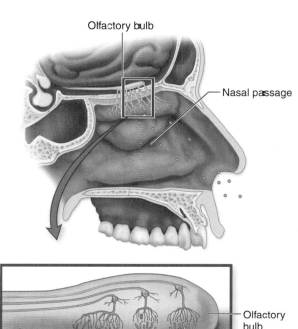

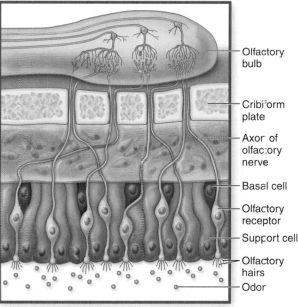

figure 45.12

SMELL. Humans detect smells by means of olfactory neurons (receptor cells) located in the lining of the nasal passages. The axons of these neurons transmit impulses directly to the brain via the olfactory nerve. Basal cells regenerate new olfactory neurons to replace dead or damaged cells. Olfactory neurons typically live about one month.

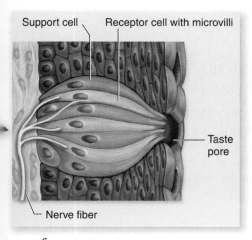

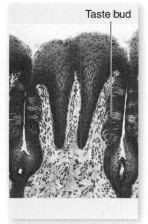

c.   *d.*

uses its sense of smell in much the same way that a fish uses its sense of taste—to sample the chemical environment around it.

Because terrestrial vertebrates are surrounded by air, their sense of smell has become specialized to detect airborne particles—but these particles must first dissolve in extracellular fluid before they can activate the olfactory receptors. The sense of smell can be extremely acute in many mammals, so much so that a single odorant molecule may be all that is needed to excite a given receptor.

Although humans can detect only four modalities of taste, they can discern thousands of different smells. New research suggests that as many as a thousand different genes may code for different receptor proteins for smell. The particular set of olfactory neurons that respond to a given odor might serve as a "fingerprint" the brain can use to identify the odor.

Internal chemoreceptors detect pH and other characteristics

Sensory receptors within the body detect a variety of chemical characteristics of the blood or fluids derived from the blood, including cerebrospinal fluid. Included among these

receptors are the **peripheral chemoreceptors** of the aortic and carotid bodies, which are sensitive primarily to plasma pH, and the **central chemoreceptors** in the medulla oblongata of the brain, which are sensitive to the pH of cerebrospinal fluid. When the breathing rate is too low, the concentration of plasma CO_2 increases, producing more carbonic acid and causing a fall in the blood pH. The carbon dioxide can also enter the cerebrospinal fluid and lower the pH, thereby stimulating the central chemoreceptors. This stimulation indirectly affects the respiratory control center of the brainstem, which increases the breathing rate. The aortic bodies can also respond to a lowering of blood oxygen concentrations, but this effect is normally not significant unless a person goes to a high altitude where the partial pressure of oxygen is lower.

Receptors that sense chemicals originating outside the body are responsible for the senses of taste and smell. Internal chemoreceptors help to monitor chemicals, particularly acid balance, within the body and are needed for the regulation of breathing.

45.5 Vision

The ability to perceive objects at a distance is important to most animals. Predators locate their prey, and prey avoid their predators, based on the three long-distance senses of hearing, smell, and vision. Of these, vision can act most distantly; with the naked eye, humans can see stars thousands of light years away—and a single photon is sufficient to stimulate a cell of the retina to send an action potential.

Vision senses light and light changes at a distance

Vision begins with the capture of light energy by **photoreceptors.** Because light travels in a straight line and arrives virtually instantaneously at distances on the Earth, visual information can be used to determine both the direction and the distance of an object. No other stimulus provides as much detailed information.

Invertebrate eyes

Many invertebrates have simple visual systems with photoreceptors clustered in an eyespot. Simple eyespots can be made sensitive to the direction of a light source by the addition of a pigment layer that shades one side of the eye. Flatworms have a screening pigmented layer on the inner and back sides of both eyespots, allowing stimulation of the photoreceptor cells only by light from the front of the animal (figure 45.13). The flatworm will turn and swim in the direction in which the photoreceptor cells are the least stimulated. Although an eyespot can perceive the direction of light, it cannot be used to construct a visual image.

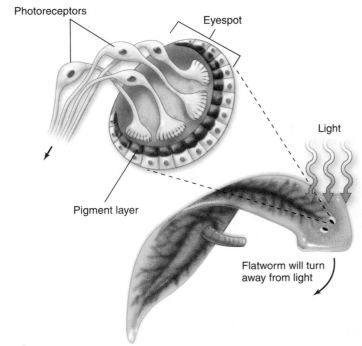

figure 45.13

SIMPLE EYESPOTS IN THE FLATWORM. Eyespots can detect the direction of light because a pigmented layer on one side of the eyespot screens out light coming from the back of the animal. Light is thus detected more readily coming from the front of the animal; flatworms respond by turning away from the light.

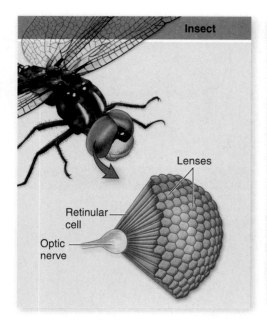
Insect

Lenses

Retinular cell

Optic nerve

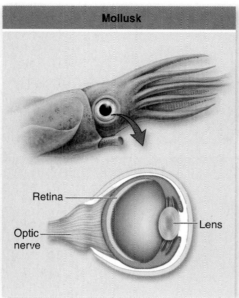
Mollusk

Retina

Optic nerve

Lens

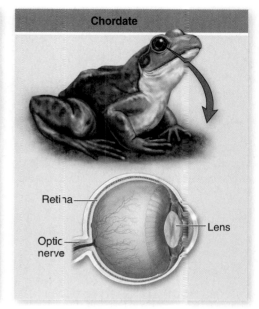
Chordate

Retina

Optic nerve

Lens

figure 45.14

EYES IN THREE PHYLA OF ANIMALS. Although they are superficially similar, these eyes differ greatly in structure and are not homologous. Each has evolved separately and, despite the apparent structural complexity, has done so from simpler structures.

The members of four phyla—annelids, mollusks, arthropods, and chordates—have evolved well-developed, image-forming eyes. True image-forming eyes in these phyla, although strikingly similar in structure, are believed to have evolved independently, an example of convergent evolution (figure 45.14). Interestingly, the photoreceptors in all of these image-forming eyes use the same light-capturing molecule, suggesting that not many alternative molecules are able to play this role.

Structure of the vertebrate eye

The human eye is typical of the vertebrate eye (figure 45.15). The "white of the eye" is the **sclera,** formed of tough connective tissue. Light enters the eye through a transparent **cornea,** which begins to focus the light. Focusing occurs because light is refracted (bent) when it travels into a medium of different density. The colored portion of the eye is the **iris;** contraction of the iris muscles in bright light decreases the size of its opening, the pupil. Light passes through the pupil to the **lens,** a transparent structure that completes the focusing of the light onto the retina at the back of the eye. The lens is attached by the *suspensory ligament* to the ciliary muscles.

The shape of the lens is influenced by the amount of tension in the suspensory ligament, which surrounds the lens and attaches it to the circular ciliary muscle. When the ciliary muscle contracts, it puts slack in the suspensory ligament, and the lens becomes more rounded and bends light more strongly. This rounding is required for close vision. In distance vision, the ciliary muscles relax, moving away from the lens and tightening the suspensory ligament. The lens thus becomes more flattened and bends light less, keeping the image focused on the

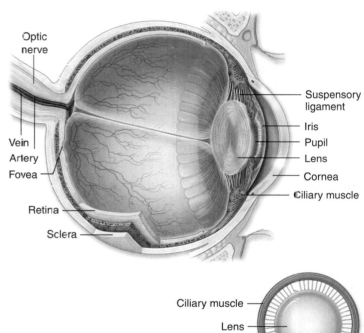

Optic nerve

Vein

Artery

Fovea

Retina

Sclera

Suspensory ligament

Iris

Pupil

Lens

Cornea

Ciliary muscle

Ciliary muscle

Lens

Suspensory ligament

figure 45.15

STRUCTURE OF THE HUMAN EYE. The transparent cornea and lens focus light onto the retina at the back of the eye, which contains the photoreceptors (rods and cones). The center of each eye's visual field is focused on the fovea. Focusing is accomplished by contraction and relaxation of the ciliary muscle, which adjusts the curvature of the lens.

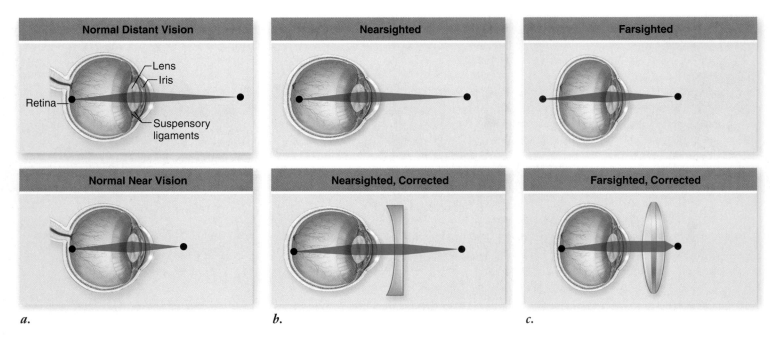

| Normal Distant Vision | Nearsighted | Farsighted |

a.   *b.*   *c.*

| Normal Near Vision | Nearsighted, Corrected | Farsighted, Corrected |

figure 45.16

FOCUSING THE HUMAN EYE. *a.* In people with normal vision, the image remains focused on the retina in both near and far vision because of changes produced in the curvature of the lens. When a person with normal vision stands 20 feet or more from an object, the lens is in its least convex form, and the image is focused on the retina. *b.* In nearsighted people, the image comes to a focus in front of the retina, and the image thus appears blurred. *c.* In farsighted people, the focus of the image would be behind the retina because the distance from the lens to the retina is too short. Corrective lenses adjust the angle of the light as it enters the eye, focusing it on the retina.

retina. People who are nearsighted or farsighted do not properly focus the image on the retina (figure 45.16). Interestingly, the lens of an amphibian or a fish does not change shape; these animals instead focus images by moving their lens in and out, just as you would do to focus a camera.

Vertebrate photoreceptors are rod cells and cone cells

The vertebrate retina contains two kinds of photoreceptor cells, called *rods* and *cones* (figure 45.17). **Rods,** which get their name from the shape of their outer segment, are responsible for black-and-white vision when the illumination is dim. In contrast, **cones** are responsible for high visual acuity (sharpness) and color vision; cones have a cone-shaped outer segment. Humans have about 100 million rods and 3 million cones in each retina. Most of the cones are located in the central region of the retina known as the **fovea,** where the eye forms its sharpest image. Rods are almost completely absent from the fovea.

Structure of rods and cones

Rods and cones have the same basic cellular structure. An inner segment rich in mitochondria contains numerous vesicles filled with neurotransmitter molecules. It is connected by a narrow stalk to the outer segment, which is packed with hundreds of flattened disks stacked on top of one another. The light-capturing molecules, or photopigments, are located on the membranes of these disks (see figure 45.17).

In rods, the photopigment is called **rhodopsin.** It consists of the protein opsin bound to a molecule of *cis*-retinal, which is

produced from vitamin A. Vitamin A is derived from carotene, a photosynthetic pigment in plants.

The photopigments of cones, called **photopsins,** are structurally very similar to rhodopsin. Humans have three kinds of cones, each of which possesses a photopsin consisting of *cis-*

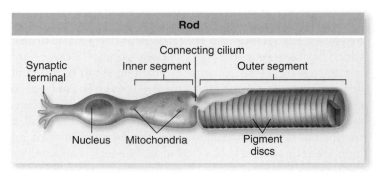

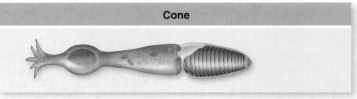

figure 45.17

RODS AND CONES. The pigment-containing outer segment in each of these cells is separated from the rest of the cell by a partition through which there is only a narrow passage, the connecting cilium.

retinal bound to a protein with a slightly different amino acid sequence. These differences shift the *absorption maximum*, the region of the electromagnetic spectrum that is best absorbed by the pigment (figure 45.18). The absorption maximum of the *cis*-retinal in rhodopsin is 500 nanometers (nm); in contrast, the absorption maxima of the three kinds of cone photopsins are 420 nm (blue-absorbing), 530 nm (green-absorbing), and 560 nm (red-absorbing). These differences in the light-absorbing properties of the photopsins are responsible for the different color sensitivities of the three kinds of cones, which are often referred to as simply blue, green, and red cones.

The **retina**, the inside surface of the eye, is made up of three layers of cells (figure 45.19): The layer closest to the external surface of the eyeball consists of the rods and cones; the next layer contains **bipolar cells;** and the layer closest to the cavity of the eye is composed of **ganglion cells.** Thus, light must first pass through the ganglion cells and bipolar cells in order to reach the photoreceptors. The rods and cones synapse with the bipolar cells, and the bipolar cells synapse with the ganglion cells, which transmit impulses to the brain via the optic nerve. Ganglion cells are the only neurons of the retina capable of sending action potentials to the brain. The flow of sensory information in the retina is therefore opposite to the path of light through the retina.

The retina contains two additional types of neurons called *horizontal cells* and *amacrine cells.* Stimulation of horizontal cells by photoreceptors at the center of a spot of light on the retina can inhibit the response of photoreceptors peripheral to the center. This lateral inhibition enhances contrast and sharpens the image.

Most vertebrates, particularly those that are diurnal (active during the day), have color vision, as do many insects.

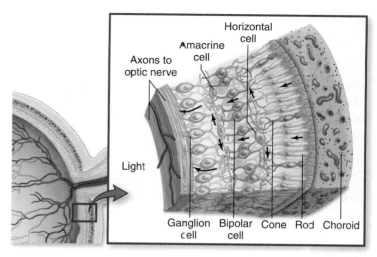

figure 45.19

STRUCTURE OF THE RETINA. Note that the rods and cones are at the rear of the retina, not the front. Light passes through four other types of cells (ganglion, amacrine, bipolar, and horizontal) in the retina before it reaches the rods and cones. Once the photoreceptors are activated, they stimulate bipolar cells, which in turn stimulate ganglion cells. The flow of sensory information in the retina is thus opposite to the direction of light.

Indeed, honeybees can see light in the near-ultraviolet range, which is invisible to the human eye. Color vision requires the presence of more than one photopigment in different receptor cells, but not all animals with color vision have the three-cone system characteristic of humans and other primates. Fish, turtles, and birds, for example, have four or five kinds of cones; the "extra" cones enable these animals to see near-ultraviolet light. Many mammals, for example, squirrels, have only two types of cones.

Sensory transduction in photoreceptors

The transduction of light energy into nerve impulses follows a sequence that is the opposite of the usual way that sensory stimuli are detected. In the dark, the photoreceptor cells release an inhibitory neurotransmitter that hyperpolarizes the bipolar neurons. This prevents the bipolar neurons from releasing excitatory neurotransmitter to the ganglion cells that signal to the brain. In the presence of light, the photoreceptor cells stop releasing their inhibitory neurotransmitter, in effect, stimulating bipolar cells. The bipolar cells in turn stimulate the ganglion cells, which transmit action potentials to the brain.

The production of inhibitory neurotransmitter by photoreceptor cells is due to the presence of ligand-gated Na^+ channels. In the dark, many of these channels are open, allowing an influx of Na^+. This flow of Na^+ in the absence of light, called the *dark current*, depolarizes the membrane of photoreceptor cells. In this state, the cells produce inhibitory neurotransmitter that hyperpolarizes the membrane of bipolar cells. In the light, the Na^+ channels in the photoreceptor cell rapidly close, reducing the dark current and causing the photoreceptor to hyperpolarize. In this state, they no longer produce inhibitory neurotransmitter. In the absence of inhibition, the membrane

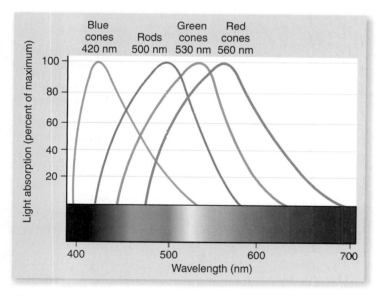

figure 45.18

COLOR VISION. The absorption maximum of *cis*-retinal in the rhodopsin of rods is 500 nm. However, the "blue cones" have their maximum light absorption at 420 nm; the "green cones" at 530 nm; and the "red cones" at 560 nm. The brain perceives all other colors from the combined activities of these three cones' systems.

of the bipolar cells is depolarized, causing them to release excitatory neurotransmitter to the ganglion cells.

The control of the dark current depends on the ligand for the Na$^+$ channels in the photoreceptor cells: the nucleotide **cyclic guanosine monophosphate (cGMP)**. In the dark, the level of cGMP is high, and the channels are open. The system is made sensitive to light by the nature and structure of the photopigments. Photopigments in the eye are actually G protein-coupled receptor proteins that are activated by absorbing light. When a photopigment absorbs light, *cis*-retinal isomerizes and dissociates from the receptor protein, opsin, in what is known as the *bleaching reaction*. As a result of this dissociation, the opsin receptor protein changes shape, activating its associated G protein. The activated G protein then activates its effector protein, the enzyme phosphodiesterase, which cleaves cGMP to GMP. The loss of cGMP causes the cGMP-gated Na$^+$ channels to close, reducing the dark current (figure 45.20). Each opsin is associated with over 100 regulatory G proteins, which, when activated, will release subunits that activate hundreds of molecules of the phosphodiesterase enzyme. Each enzyme molecule can convert thousands of cGMP to GMP, closing the Na$^+$ channels at a rate of about 1000 per second and inhibiting the dark current.

The absorption of a single photon of light can block the entry of more than a million Na$^+$, without changing K$^+$ per-

figure 45.20

SIGNAL TRANSDUCTION IN THE VERTEBRATE EYE. In the absence of light, cGMP keeps Na$^+$ channels open causing a Na$^+$ influx that leads to the release of inhibitory neurotransmitter. Light is absorbed by the retinal in Rhodopsin, changing its structure. This causes Rhodopsin to associate with a G protein. The activated G protein stimulates phosphodiesterase, which converts cGMP to GMP. Loss of cGMP closes Na$^+$ channels and prevents release of inhibitory neurotransmitter, which causes bipolar cells to stimulate ganglion cells.

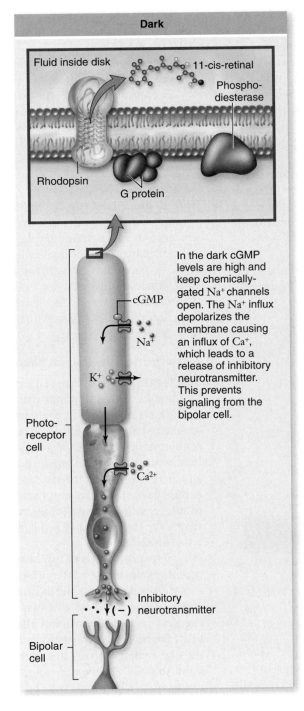

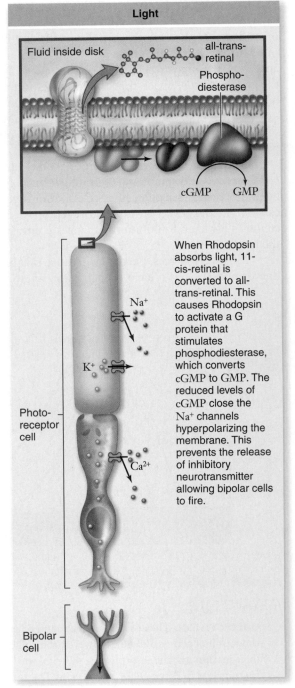

meability—the photoreceptor becomes hyperpolarized and releases less inhibitory neurotransmitter. Freed from inhibition, the bipolar cells activate the ganglion cells, which send impulses to the brain (figure 45.21).

Visual processing takes place in the cerebral cortex

Action potentials propagated along the axons of ganglion cells are relayed through structures called the **lateral geniculate nuclei** of the thalamus and projected to the occipital lobe of the cerebral cortex (see figure 45.21). There the brain interprets this information as light in a specific region of the eye's receptive field. The pattern of activity among the ganglion cells across the retina encodes a point-to-point map of the receptive field, allowing the retina and brain to image objects in visual space.

The frequency of impulses in each ganglion cell provides information about the light intensity at each point. At the same time, the relative activity of ganglion cells connected (through bipolar cells) with the three types of cones provides color information.

Visual acuity

The relationship between receptors, bipolar cells, and ganglion cells varies in different parts of the retina. In the fovea, each cone makes a one-to-one connection with a bipolar cell, and

each bipolar cell synapses, in turn, with one ganglion cell. This point-to-point relationship is responsible for the high acuity of foveal vision.

Outside the fovea, many rods can converge on a single bipolar cell, and many bipolar cells can converge on a single ganglion cell. This convergence permits the summation of neural activity, making the area of the retina outside the fovea more sensitive to dim light than the fovea, but at the expense of acuity and color vision. This is why dim objects, such as faint stars at night, are best seen when you don't look directly at them. It has been said that we use the periphery of the eye as a detector, and the fovea as an inspector.

Color blindness is due to an inherited lack of one or more types of cones. People with normal color vision are *trichromats*, having all three cones; those with only two types of cones are dichromats. For example, people with red/green color blindness may lack red cones and have difficulty distinguishing red from green. Color blindness is a sex-linked recessive trait (see chapter 13), and therefore it is most often exhibited in males.

Binocular vision

Primates (including humans) and most predators have two eyes, one located on each side of the face. When both eyes are trained on the same object, the image that each eye sees is slightly different because the views have a slightly different angle. This slight displacement of the images (an effect called *parallax*) permits **binocular vision,** the ability to perceive three-dimensional images and to sense depth. Having eyes facing forward maximizes the field of overlap in which this stereoscopic vision occurs.

In contrast, prey animals generally have eyes located to the sides of the head, preventing binocular vision but enlarging the overall receptive field. It seems that natural selection has favored the detection of potential predators over depth perception in many prey species. The eyes of the American woodcock (*Scolopax minor*), for example, are located at exactly opposite sides of its skull so that it has a 360° field of view without turning its head.

Most birds have laterally placed eyes and, as an adaptation, have two foveas in each retina. One fovea provides sharp frontal vision, like the single fovea in the retina of mammals, and the other fovea provides sharper lateral vision.

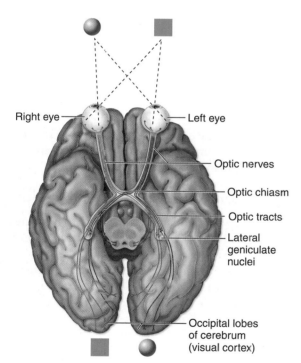

figure 45.21

THE PATHWAY OF VISUAL INFORMATION. Action potentials in the optic nerves are relayed from the retina to the lateral geniculate nuclei, and from there to the visual cortex of the occipital lobes. Note that half the optic nerves (the medial fibers arising from the inner portion of the retinas) cross to the other side at the optic chiasm, so that each hemisphere of the cerebrum receives input from both eyes.

Annelids, mollusks, arthropods, and chordates have independently evolved image-forming eyes. The vertebrate eye admits light through a pupil and then focuses this light by means of an adjustable lens onto the retina at the back of the eye. Photoreceptor rods and cones contain the photopigment *cis*-retinal, which dissociates in response to light and indirectly activates bipolar neurons and then ganglion cells. Ganglion cells transmit action potentials to the thalamus, which in turn relays visual information to the occipital lobe of the brain. The fovea of the retina provides high visual acuity, whereas the area outside the fovea provides high sensitivity to dim light. Binocular vision with overlapping visual fields provides depth perception.

Vision is the primary sense used by all vertebrates that live in a light-filled environment, but visible light is by no means the only part of the electromagnetic spectrum that vertebrates use to sense their environment.

Some snakes have receptors capable of sensing infrared radiation

Electromagnetic radiation with wavelengths longer than those of visible light is too low in energy to be detected by photoreceptors. Radiation from this infrared portion of the spectrum is what we normally think of as radiant heat.

Heat is an extremely poor environmental stimulus in water because water readily absorbs heat. Air, in contrast, has a low thermal capacity, so heat in air is a potentially useful stimulus. The only vertebrates known to have the ability to sense infrared radiation, however, are the snakes known as pit vipers.

The pit vipers possess a pair of heat-detecting **pit organs** located on either side of the head between the eye and the nostril (figure 45.22). These exteroceptors locate heat sources in the environment, and they enable a rattlesnake to locate and strike at prey in darkness—such as in an underground den or cave, or at night.

Each pit organ is composed of two chambers separated by a membrane. The infrared radiation falls on the membrane and warms it. Thermal receptors on the membrane are stimulated. The nature of these receptors is not known; they probably consist of temperature-sensitive neurons innervating the two chambers. The paired pit organs appear to provide stereoscopic information, in much the same way that two eyes do. In fact, in snakes the information transmitted from the pit organs is processed in the brain by a structure homologous to the visual center in other vertebrates.

Some vertebrates can sense electrical currents

Although air does not readily conduct an electrical current, water is a good conductor. All aquatic animals generate electrical currents from contractions of their muscles. A number of different groups of fishes can detect these electrical currents. The so-called electrical fish even have the ability to produce electrical discharges from specialized electrical organs. Electrical fish use these weak discharges to locate their prey and mates and to construct a three-dimensional image of their environment, even in murky water.

The elasmobranchs (sharks, rays, and skates) have electroreceptors called the **ampullae of Lorenzini.** The receptor cells are located in sacs that open through jelly-filled canals to pores on the body surface. The jelly is a very good conductor, so a negative charge in the opening of the canal can depolarize the receptor at the base, causing the release of neurotransmitter and increased activity of sensory neurons. This allows sharks, for example, to detect the electrical fields generated by the muscle contractions of their prey. Although the ampullae of Lorenzini were lost in the evolution of teleost fish (most of

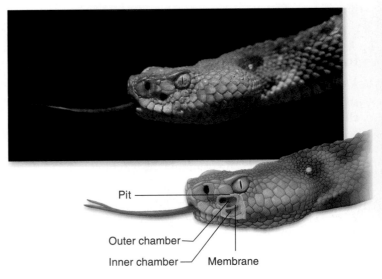

figure 45.22

"SEEING" HEAT. The depression between the nostril and the eye of this rattlesnake opens into the pit organ. In the cutaway portion of the diagram, you can see that the organ is composed of two chambers separated by a membrane. Snakes known as pit vipers have this unique ability to sense infrared radiation (heat).

the bony fish), electroreception reappeared in some groups of teleost fish that developed analogous sensory structures. Electroreceptors evolved yet another time, independently, in the duck-billed platypus, an egg-laying mammal. The receptors in its bill can detect the electrical currents created by the contracting muscles of shrimp and fish, enabling the mammal to detect its prey at night and in muddy water.

Some organisms detect magnetic fields

Eels, sharks, bees, and many birds appear to navigate along the magnetic field lines of the Earth. Even some bacteria use such forces to orient themselves.

Birds kept in blind cages, with no visual cues to guide them, peck and attempt to move in the direction in which they would normally migrate at the appropriate time of the year. They do not do so, however, if the cage is shielded from magnetic fields by steel. In addition, if the magnetic field of a blind cage is deflected 120° clockwise by an artificial magnet, a bird that normally orients to the north will orient toward the east-southeast.

The nature of magnetic receptors in these vertebrates is the subject of much speculation, but the mechanism remains very poorly understood.

Pit vipers can locate warm prey by infrared radiation (heat), and many aquatic vertebrates can locate prey and ascertain the contours of their environment by means of electroreceptors. Magnetic receptors may aid in bird migration.

45.1 Overview of Sensory Receptors

Sensory receptors provide information from our internal and external environments that is crucial for our survival and success (table 45.1).

- Exteroreceptors sense stimuli from the external environment. Interoreceptors sense stimuli from the internal environment.

- There are three categories of receptors: mechanoreceptors, chemoreceptors, and energy-detecting receptors.

- Sensory information is conveyed in four steps: stimulation, transduction, transmission, and interpretation (figure 45.1).

- Sensory transduction involves stimulus-gated ion channels that produce a graded receptor potential.

- If the receptor potential or the sum of receptor potentials is above threshold, an action potential is produced (figure 45.2).

- There is a logarithmic relationship between stimulus intensity and action potential frequency.

45.2 Mechanoreceptors: Touch and Pressure

Mechanoreceptors are stimulated by mechanical forces such as pressure.

- Nociceptors are free nerve endings located in the skin that respond to damaging stimuli, which is perceived as pain.

- Extreme temperatures can affect transient receptor potential (TRP) ion channels and cause depolarization by inflow of Na^+ and Ca^{2-}.

- Thermoreceptors are naked dendritic endings of sensory neurons that also contain TRP ion channels and respond to cold or heat.

- The sense of touch is produced by various receptors in the skin that respond to mechanical distortion of the membrane (figure 45.3).

- Muscle length is monitored by proprioreceptors.

- Baroreceptors, located in the carotid sinus and aortic arch, monitor blood pressure.

45.3 Hearing, Vibration, and Detection of Body Position

Hearing, the detection of sound or pressure waves, works best in water and provides directional information.

- Pressure waves are transduced into nerve impulses by deflection of hair cells that result in action potentials.

- The lateral line system of fishes senses pressure waves and low-frequency vibrations (figure 45.5).

- The ear of a fish uses an otolith, a calcium carbonate crystal, that vibrates against a hair cell to transduce sound.

- The outer ear of terrestrial vertebrates channels vibrations in the air to the eardrum, or tympanic membrane (figure 45.6).

- Sound waves are transmitted from the eardrum via the malleus, incus, and stapes to the oval window. Sound waves are transduced in the cochlea by the organ of Corti.

- The basilar membrane of the cochlea consists of fibers that respond to different frequencies of sound (figure 45.7).

- Echolocation allows some species to navigate by sound.

- Body position is detected by statocysts, ciliated hair cells embedded in a gelatinous matrix containing statoliths (figure 45.8).

- Body movement is detected by hair cells located in the saccule, utricle, so angular acceleration can be detected (figure 45.9).

45.4 Chemoreceptors: Taste, Smell, and pH

Chemoreceptors contain membrane proteins that bind to particular ligands, leading to depolarization and formation of action potentials.

- The perception of taste, or gustation, is a combination of physical and psychological factors.

- Taste buds are collections of chemosensitive epithelial cells located on papillae (figure 45.10).

- Tastes are broken down into five categories: sweet, sour, salty, bitter, and umami.

- Salty and sour tastes act directly through ion channels, and the other tastes act indirectly by binding to G protein-coupled receptors.

- Smell, or olfaction, involves chemoreceptors located in the upper portion of the nasal passages (figure 45.12).

- The dendrites of chemoreceptors end in tassels of cilia that project directly into the nasal mucosa and their axons go directly into the cerebral cortex.

- Internal peripheral chemoreceptors of the aorta detect changes in blood pH, and central chemoreceptors in the medulla oblongata are sensitive to the pH of the cerebrospinal fluid.

45.5 Vision

Energy-detecting receptors allow animals to perceive objects at long distances.

- Vision begins with the capture of light energy.

- Four phyla—annelids, mollusks, arthropods, and chordates—have independently evolved image-forming eyes (figure 45.14).

- In the vertebrate eye, light enters through a transparent cornea, with intensity controlled by the iris. The lens focuses the light on the retina at the back of the eye (figure 45.15).

- The ciliary muscle controls the shape of the lens for near and far vision.

- The two types of vertebrate photoreceptors are rods, detect black and white; and cones, necessary for visual acuity and color vision (figure 45.17).

- In the retina, the photoreceptors synapse with the bipolar cells, which in turn synapse with the ganglion cells, which send action potentials to the brain (figure 45.19).

- In the absence of light, cGMP keeps Na^+ channels open leading to Na^+ influx called the dark current, which causes release of inhibitory neurotransmitters.

- Light absorbed by photoreceptors activates a G protein that stimulates phosphodiesterase, which cleaves cGMP.

- Loss of cGMP prevents the influx of Na^+ and consequent release of inhibitory neurotransmitters. This causes bipolar cells to stimulate the ganglion cells (figure 45.20).

- Visual processing takes place in the occipital lobe of the cerebral cortex (figure 45.21).

- High visual acuity is due to the fovea, a region of the retina, where each cone cell is connected to one bipolar cell/ganglion cell.

- In dim light, acuity decreases because of the summation of many rods converge on each bipolar cell, and many bipolar cells converge on each ganglion cell.

- Primates and most predators have binocular vision—images from each eye overlap to produce a three-dimensional image.

45.6 The Diversity of Sensory Experiences

Visible light is not the only part of the electromagnetic spectrum that vertebrates use to sense their environment.

- Pit vipers detect infrared radiation with heat-detecting pit organs.

- Electroreceptors in elasmobranches and the duck-billed platypuses can detect electrical currents.

- Many organisms appear to navigate along magnetic field lines, but the mechanisms remains poorly understood.

SELF TEST

1. Which of these is *not* a method by which sensory receptors receive information about the internal or external environment?
   a. Changes in pressure
   b. Light or heat changes
   c. Changes in molecular concentration
   d. All of these are used by sensory receptors.

2. Which of the following correctly lists the steps of perception?
   a. Interpretation, stimulation, transduction, transmission
   b. Stimulation, transduction, transmission, interpretation
   c. Interpretation, transduction, stimulation, transmission
   d. Transduction, interpretation, stimulation, transmission

3. Your Aunt Hattie stood up suddenly at Sunday dinner and then fainted. The doctor said that it might be a problem involving her baroreceptors. What is the function of baroreceptors?
   a. They detect changes in blood pressure.
   b. They detect muscle contractions and the movement of limbs.
   c. They are exteroceptors.
   d. They detect changes in blood chemistry.

4. All sensory receptors are able to initiate nerve impulses by opening or closing—
   a. voltage-gated ion channels.
   b. exteroceptors.
   c. interoceptors.
   d. stimulus-gated ion channels.

5. In the fairytale, Sleeping Beauty fell asleep after pricking her finger. What kind of receptor responds to that kind of painful stimulus?
   a. Mechanoreceptor
   b. Nociceptor
   c. Thermoreceptor
   d. Touch receptor

6. The ear detects sound by the movement of—
   a. the basilar membrane.
   b. the tectorial membrane.
   c. the Eustachian tube.
   d. fluid in the semicircular canals.

7. A friend who knows that you are afraid of snakes told you that you should wrap yourself in insulation when you go into the woods so that the snakes won't be able to find you. You now realize, after reading the chapter, that she was right (sort-of) because snakes use their _____ for detecting infrared radiation as heat.
   a. photoreceptors
   b. cochlea
   c. semicircular canals
   d. pit organs

8. Hair cells in the vestibular apparatus of terrestrial vertebrates—
   a. measure temperature changes within the body.
   b. sense sound in very low range of hearing.
   c. provide a sense of acceleration and balance.
   d. measure changes in blood pressure.

9. Think back to the meal you ate for lunch. The ability to taste food relies on—
   a. external chemoreceptors.
   b. internal chemoreceptors.
   c. olfactory receptors.
   d. nociceptors.

10. What do the sensory systems of annelids, mollusks, arthropods, and chordates have in common?
    a. They all use the same stimuli for taste.
    b. They all use neurons to detect vibration.
    c. They all have image-forming eyes that evolved independently.
    d. They all use chemoreceptors in their skin to detect food.

11. While honeymooning in Borneo you find a previous unidentified species of mammal. You proceed to explain to your very patient spouse that the placement of the eyes on the same side of the head (face) such that the visual fields overlap—
    a. is found in primates and predators.
    b. is called binocular vision.
    c. allows for depth perception.
    d. all of these.

12. You arise to a beautiful day. As you raise your eyelids and catch your first glimpse of the brand new day, the structure that actually admits light into your eye is the—
    a. retina.
    b. fovea.
    c. pupil.
    d. lens.

13. _____ is the photopigment contained within both rods and cones of the eye.
    a. Carotene
    b. *Cis*-Retinal
    c. Photochrome
    d. Chlorophyll

14. Which of the following is *not* a method used by vertebrates to gather information about their environment?
    a. Infrared radiation
    b. Magnetic fields
    c. Electrical currents
    d. All of these are methods used for sensory reception.

15. The lobe of the brain that recognizes and interprets visual information is the—
    a. occipital lobe.
    b. frontal lobe.
    c. parietal lobe.
    d. temporal lobe.

CHALLENGE QUESTIONS

1. When blood pH falls too low, a potentially fatal condition known as acidosis results. Among the variety of responses to this condition, the body will change the breathing rate. How does the body sense this change? How does the breathing rate change? How does this increase pH?

2. The function of the vertebrate eye is unusual compared with other processes found within the body. For example, the direction in which sensory information flows is actually opposite to path that light takes through the retina. Explain the sequence of events involved in the movement of light and information through the structures of the eye, and explain why they move in opposite directions.

3. How would the otolith organs of an astronaut respond to zero gravity? Would the astronaut still have a subjective impression of motion? Would the semicircular canals detect angular acceleration equally well at zero gravity?

chapter 46

The Endocrine System

introduction

DIABETES IS A DISEASE IN WHICH well-fed patients appear to starve to death. The disease was known to Roman and Greek physicians, who described a "melting away of flesh" coupled with excessive urine production "like the opening of aqueducts." Until 1922, the diagnosis of diabetes in children was effectively a death sentence. In that year, Frederick Banting and Charles Best extracted the molecule insulin from the pancreas. Injections of insulin into the bloodstream dramatically reversed the symptoms of the disease. This served as an impressive confirmation of a new concept: that certain internal organs produced powerful regulatory chemicals that were distributed via the blood.

We now know that the tissues and organs of the vertebrate body cooperate to maintain homeostasis through the actions of many regulatory mechanisms. Two systems, however, are devoted exclusively to the regulation of the body organs: the nervous system and the endocrine system. Both release regulatory molecules that control the body organs by binding to receptor proteins on or in the cells of those organs. In this chapter, we examine the regulatory molecules of the endocrine system, the cells and glands that produce them, and how they function to regulate the body's activities.

Regulation of Body Processes by Chemical Messengers

In chapter 9, when discussing cell-to-cell interaction, we described the four mechanisms of cell communication, namely direct contact, synaptic signaling, endocrine signaling, and paracrine signaling. Here we are concerned with the signaling methods of communication; we begin by reviewing the three signaling mechanisms.

As discussed in chapter 44, the axons of neurons secrete chemical messengers called neurotransmitters into the synaptic cleft. These chemicals diffuse only a short distance to the postsynaptic membrane, where they bind to their receptor proteins and stimulate the postsynaptic cell. Synaptic transmission generally affects only the postsynaptic cell that receives the neurotransmitter.

A **hormone**, in contrast, is a regulatory chemical that is secreted into extracellular fluid and carried by the blood, and can therefore act at a distance from its source. Organs that are specialized to secrete hormones are called **endocrine glands,** but some organs, such as the liver and the kidney, can produce hormones in addition to performing other functions. The organs and tissues that produce hormones are collectively called the **endocrine system.**

The blood carries hormones to every cell in the body, but only target cells with the appropriate receptor for a given hormone can respond to it. Hormone receptor proteins function in a similar manner to neurotransmitter receptors. The receptor proteins specifically bind the hormone and activate signal transduction pathways that produce a response to the hormone. The highly specific interaction between hormones and their receptors enable hormones to be active at remarkably small concentrations. It is not unusual to find hormones circulating in the blood at concentrations of 10^{-8} to 10^{-10} molar. In addition to the chemical messengers released as neurotransmitters and as hormones, other chemical molecules are released and act within an organ as local regulators. These chemicals are termed **paracrine regulators.** They act in a way similar to endocrine hormones, but they do not travel through the blood. In this way, the cells of an organ regulate one another.

Chemical communication is not limited to cells within an organism. **Pheromones** are chemicals released into the environment to communicate among individuals of a single species. These messengers aid in the communication between animals and may alter the behavior or physiology of the receiver, but are not involved in the normal metabolic regulation within an animal.

A comparison of the different types of chemical messengers used for internal regulation is given in figure 46.1.

Some neurotransmitters also act as circulating hormones

Blood delivery of hormones enables endocrine glands to coordinate the activity of large numbers of target cells distributed throughout the body. A chemical regulator called norepineph-rine, for example, which is released as a neurotransmitter by sympathetic nerve endings, is also secreted into the blood by the adrenal glands. Norepinephrine acts as a hormone to coordinate the activity of the heart, liver, and blood vessels during response to stress. Neurons can secrete chemicals called **neurohormones** that are carried by blood. Some specialized regions of the brain contain not only neurotransmitting neurons, but also clusters of neurons producing neurohormones. In this way, neurons can deliver chemical messages beyond the nervous system itself.

The secretory activity of many endocrine glands is controlled by the nervous system. The major site that carries out neural regulation of the endocrine system is the brain. As you will see, the hypothalamus controls the hormonal secretions of the anterior pituitary gland, which in turn regulates other endocrine glands.

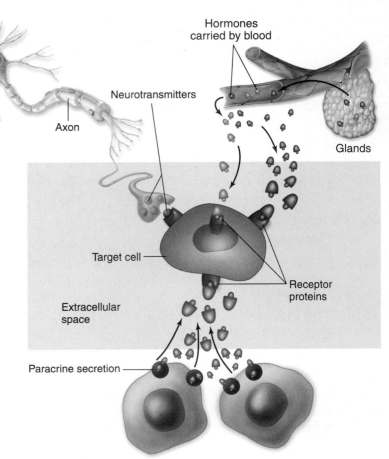

figure 46.1

DIFFERENT TYPES OF CHEMICAL MESSENGERS. The functions of organs are influenced by neural, paracrine, and endocrine regulators. Each type of chemical regulator binds to specific receptor proteins on the surface or within the cells of target organs.

The secretion of a number of hormones, however, can be independent of neural control. For example, the release of insulin by the pancreas and aldosterone by the adrenal cortex is stimulated by increases in the blood concentrations of glucose and potassium (K^+), respectively.

Endocrine glands produce three chemical classes of hormones

The endocrine system (figure 46.2) includes all of the organs that secrete hormones—the thyroid gland, pituitary gland, adrenal glands, and so on (table 46.1). Cells in these organs secrete hormones into extracellular fluid, where it diffuses into surrounding blood capillaries. For this reason, hormones are referred to as endocrine secretions. In contrast, cells of some glands secrete their products, such as saliva or milk, into a duct for transport to the outside. These glands are termed **exocrine glands,** and their secretions are exocrine secretions.

Molecules that function as hormones must exhibit two basic characteristics. First, they must be sufficiently complex to convey regulatory information to their targets. Simple molecules such as carbon dioxide, or ions such as calcium ion, do not function as hormones. Second, hormones must be adequately stable to resist destruction prior to reaching their target cells. Three primary chemical categories of molecules meet these requirements.

Peptides and proteins are composed of chains of amino acids. Some important examples of peptide hormones include antidiuretic hormone (ADH, 9 amino acids), insulin (51 amino acids), and growth hormone (191 amino acids). These hormones are encoded in DNA and produced by the same cellular machinery responsible for transcription and translation of other peptide molecules. The most complex are **glycoproteins,** composed of two peptide chains with attached carbohydrates. Examples include thyroid stimulating hormone (TSH) and luteinizing hormone (LH).

Amino acid derivatives are hormones manufactured by enzymatic modification of specific amino acids; this group comprises the biogenic amines discussed in chapter 44. They include hormones secreted by the adrenal medulla (the inner portion of the adrenal gland), thyroid, and pineal glands. Those secreted by the adrenal medulla are derived from tyrosine. Known as **catecholamines,** they include epinephrine (adrenaline) and norepinephrine (noradrenaline). Other hormones derived from tyrosine are the **thyroid hormones,** secreted by the thyroid gland. The pineal gland secretes a different amine hormone, **melatonin,** derived from tryptophan.

Steroids are lipids manufactured by enzymatic modifications of cholesterol. They include the hormones testosterone, estradiol, progesterone, aldosterone, and cortisol. Steroid hormones can be subdivided into **sex steroids,** secreted by the testes, ovaries, placenta, and adrenal cortex, and **corticosteroids,** secreted only by the adrenal cortex (the outer portion of the adrenal gland).

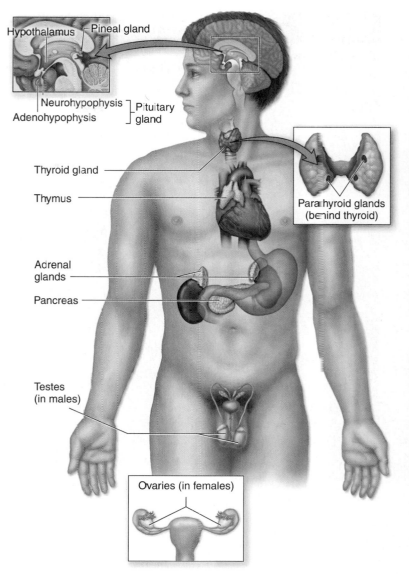

figure 46.2

THE HUMAN ENDOCRINE SYSTEM. The major endocrine glands are shown, but many other organs secrete hormones in addition to their primary functions.

Hormones can be categorized as lipophilic or hydrophilic

The manner in which hormones are transported and interact with their targets differs depending on their chemical nature. Hormones may be categorized as **lipophilic** (nonpolar), which are fat-soluble, or **hydrophilic** (polar), which are water-soluble. The lipophilic hormones include the steroid hormones and thyroid hormones. All other hormones are hydrophilic.

This distinction is important in understanding how these hormones regulate their target cells. Hydrophilic hormones are freely soluble in blood, but cannot enter target cells. They must therefore activate their receptors from outside the cell membrane. In contrast, lipophilic hormones travel in the

TABLE 46.1

Principal Mammalian Endocrine Glands and Their Hormones*

Endocrine Gland and Hormone	Target Tissue	Principal Actions	Chemical Nature
Hypothalamus			
Releasing hormones	Adenohypophysis	Activate release of adenohypophysial hormones	Peptides
Inhibiting hormones	Adenohypophysis	Inhibit release of adenohypophysial hormones	Peptides (except prolactin-inhibiting factor, which is dopamine)
Neurohypophysis			
Antidiuretic hormone (ADH)	Kidneys	Conserves water by stimulating its reabsorption from urine	Peptide (9 amino acids)
Oxytocin	Uterus	Stimulates contraction	Peptide (9 amino acids)
	Mammary glands	Stimulates milk ejection	
Adenohypophysis			
Adrenocorticotropic hormone (ACTH)	Adrenal cortex	Stimulates secretion of adrenal cortical hormones such as cortisol	Peptide (39 amino acids)
Melanocyte-stimulating hormone (MSH)	Skin	Stimulates color change in reptiles and amphibians; various functions in mammals	Peptide (two forms; 13 and 22 amino acids)
Growth hormone (GH)	Many organs	Stimulates growth by promoting bone growth, protein synthesis, and fat breakdown	Protein
Prolactin (PRL)	Mammary glands	Stimulates milk production	Protein
Thyroid-stimulating hormone (TSH)	Thyroid gland	Stimulates thyroxine secretion	Glycoprotein
Luteinizing hormone (LH)	Gonads	Stimulates ovulation and corpus luteum formation in females; stimulates secretion of testosterone in males	Glycoprotein
Follicle-stimulating hormone (FSH)	Gonads	Stimulates spermatogenesis in males; stimulates development of ovarian follicles in females	Glycoprotein
Thyroid Gland			
Thyroid hormones (thyroxine and triiodothyronine)	Most cells	Stimulates metabolic rate; essential to normal growth and development	Amino acid derivative (iodinated)
Calcitonin	Bone	Inhibits loss of calcium from bone	Peptide (32 amino acids)

*These are hormones released from endocrine glands. Hormones are released from organs that have additional, nonendocrine functions, such as the liver, kidney, and intestine.

TABLE 46.1	continued			

Endocrine Gland and Hormone	Target Tissue		Principal Actions	Chemical Nature
Parathyroid Glands				
Parathyroid hormone	Bone, kidneys, digestive tract		Raises blood calcium level by stimulating bone breakdown; stimulates calcium reabsorption in kidneys; activates vitamin D	Peptide (34 amino acids)
Adrenal Medulla				
Epinephrine (adrenaline) and norepinephrine (noradrenaline)	Smooth muscle, cardiac muscle, blood vessels		Initiates stress responses; raises heart rate, blood pressure, metabolic rate; dilates blood vessels; mobilizes fat; raises blood glucose level	Amino acid derivatives
Adrenal Cortex				
Glucocorticoids (e.g., cortisol)	Many organs		Adaptation to long-term stress; raises blood glucose level; mobilizes fat	Steroid
Mineralocorticoids (e.g., aldosterone)	Kidney tubules		Maintains proper balance of Na^+ and K^+ excretion	Steroid
Pancreas				
Insulin	Liver, skeletal muscles, adipose tissue		Lowers blood glucose level; stimulates glycogen, fat, protein synthesis	Peptide (51 amino acids)
Glucagon	Liver, adipose tissue		Raises blood glucose level; stimulates breakdown of glycogen in liver	Peptide (29 amino acids)
Ovary				
Estradiol	General		Stimulates development of female secondary sex characteristics	Steroid
	Female reproductive structures		Stimulates growth of sex organs at puberty and monthly preparation of uterus for pregnancy	
Progesterone	Uterus		Completes preparation for pregnancy	Steroid
	Mammary glands		Stimulates development	
Testis				
Testosterone	Many organs		Stimulates development of secondary sex characteristics in males and growth spurt at puberty	Steroid
	Male reproductive structures		Stimulates development of sex organs; stimulates spermatogenesis	
Pineal Gland				
Melatonin	Gonads, brain, pigment cells		Regulates biological rhythms	Amino acid derivative

blood attached to transport proteins (figure 46.3). Their lipid solubility enables them to cross cell membranes and bind to intracellular receptors.

Both types of hormonal molecules are eventually destroyed or otherwise deactivated after their use, eventually being excreted in bile or urine. However, hydrophilic hormones are deactivated more rapidly than lipophilic hormones. As a result of these chemical differences, hydrophilic hormones tend to act over relatively brief periods of time (minutes to hours), whereas lipophilic hormones generally are active over prolonged periods, such as days to weeks.

Paracrine regulators exert powerful effects within tissues

Paracrine regulation occurs in most organs and among the cells of the immune system. **Growth factors,** proteins that promote growth and cell division in specific organs, are among the most important paracrine regulators. Growth factors play a critical role in regulating mitosis (see chapter 10), and they are produced in embryos to regulate the proper differentiation of cells. For example, *epidermal growth factor* activates mitosis

of skin and development of connective tissue cells, whereas *nerve growth factor* stimulates the growth and survival of neurons. *Insulin-like growth factor* stimulates cell division in developing bone as well as protein synthesis in many other tissues. **Cytokines** (described in chapter 51) are growth factors specialized to control cell division and differentiation in the immune system, whereas **neurotropins** are growth factors that regulate the nervous system.

The importance of growth factor function is underscored by the observation that damage to the genes coding for growth factors or their receptors can lead to the unregulated cell division and development of tumors.

Paracrine regulation of blood vessels

The gas nitric oxide (NO), which can function as a neurotransmitter (chapter 44), is also produced by the endothelium of blood vessels. In this context, it is a paracrine regulator because it diffuses to the smooth muscle layer of the blood vessel and promotes vasodilation. One of its major roles involves the control of blood pressure by dilating arteries. The endothelium of blood vessels is a rich source of paracrine regulators, including *endothelin*, which stimulates vasoconstriction, and *bradykinin*, which promotes vasodilation.

figure 46.3

THE LIFE OF HORMONES. Endocrine glands produce both hydrophilic and lipophilic hormones, which are transported to targets through the blood. Lipophilic hormones bind to transport proteins that make them soluble in blood. Target cells have membrane receptors for hydrophilic hormones, and intracellular receptors for lipophilic hormones. Hormones are eventually destroyed by their target cells or cleared from the blood by the liver or the kidney.

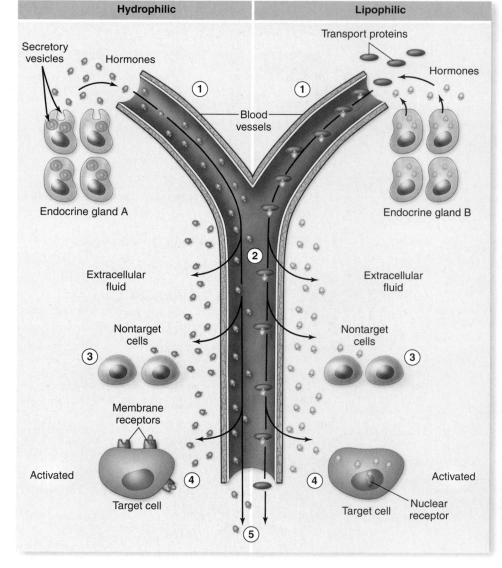

1. Hormones secreted into extracellular fluid and diffuse into bloodstream.

2. Hormones distributed by blood to all cells. Diffuse from blood to extracellular fluid.

3. Nontarget cells lack receptors, and cell stimulation does not occur.

4. Target cells possess receptors, and are activated by hormones.

5. Unused, deactivated hormones are removed by the liver and kidney.

Paracrine regulation supplements the regulation of blood vessels by autonomic nerves, enabling vessels to respond to local conditions, such as increased pressure or reduced oxygen.

Prostaglandins

A particularly diverse group of paracrine regulators are the **prostaglandins.** A prostaglandin is a 20-carbon-long fatty acid that contains a five-membered carbon ring. This molecule is derived from the precursor molecule *arachidonic acid,* released from phospholipids in the cell membrane under hormonal or other stimulation. Prostaglandins are produced in almost every organ and participate in a variety of regulatory functions. Some prostaglandins are active in promoting smooth muscle contraction. Through this action, they regulate reproductive functions such as gamete transport, labor, and possibly ovulation. Excessive prostaglandin production may be involved in premature labor, endometriosis, or dysmenorrhea (painful menstrual cramps). They also participate in lung and kidney regulation through effects on smooth muscle.

In fish, prostaglandins have been found to function as both a hormone and a paracrine regulator. Prostaglandins produced in the ovary during ovulation can travel to the brain to synchronize associated spawning behavior.

Prostaglandins are produced at locations of tissue damage, where they promote many aspects of inflammation, including swelling, pain, and fever. This effect of prostaglandins has been well studied. Drugs that inhibit prostaglandin synthesis, such as aspirin, help alleviate these symptoms.

Aspirin is the most widely used of the **nonsteroidal antiinflammatory drugs (NSAIDs),** a class of drugs that also includes indomethacin and ibuprofen. These drugs act to inhibit two related enzymes, cyclooxygenase-1 and 2 (cox-1 and cox-2). The antiinflammatory effects are due to the inhibition of cox-2, which is necessary for the production of prostaglandins from arachidonic acid. This reduces inflammation and associated pain from the action of prostaglandins. Unfortunately, the inhibition of cox-1 produces unwanted side effects, including gastric bleeding and prolonged clotting time.

More recently developed pain relievers, called *cox-2 inhibitors,* selectively inhibit cox-2 but not cox-1. Cox-2 inhibitors may be of potentially great benefit to arthritis sufferers and others who must use pain relievers regularly, but concerns have been raised that they may also affect other aspects of prostaglandin function in the cardiovascular system. Some cox-2 inhibitors were removed from the market when a greater risk of heart attack and stroke was detected. Some have remained in use, however, and others may be reintroduced upon FDA approval. Aside from the possibly lessened gastrointestinal side effects, cox-2 inhibitors are not more effective for pain than the older NSAIDs.

> Regulatory molecules released by axons at a synapse are neurotransmitters, those released by endocrine glands and carried by blood are hormones, and those that act within the organ in which they are produced are paracrine regulators. Any organ that secretes a hormone carried by blood is part of the endocrine system. Hormonal molecules, which belong to three chemical classes, work to coordinate the activity of specific target cells throughout the body.

46.2 Actions of Lipophilic Versus Hydrophilic Hormones

As mentioned previously, hormones can be divided into those that are lipophilic (lipid-soluble) and those that are hydrophilic (water-soluble). The receptors and actions of these two broad categories have notable differences, which we explore in this section.

Lipophilic hormones activate intracellular receptors

The lipophilic hormones include all of the steroid hormones and thyroid hormones (figure 46.4) as well as other lipophilic regulatory molecules including the retinoids, or vitamin A.

Lipophilic hormones can enter cells because the lipid portion of the plasma membrane does not present a barrier. Once

figure 46.4

CHEMICAL STRUCTURES OF LIPOPHILIC HORMONES. Steroid hormones are derived from cholesterol. The two steroid hormones shown, cortisol and testosterone, differ slightly in chemical structure yet have widely different effects on the body. The thyroid hormone, thyroxine, is formed by coupling iodine to the amino acid tyrosine.

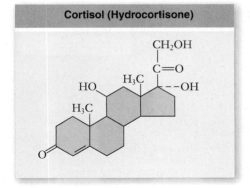

Cortisol (Hydrocortisone)

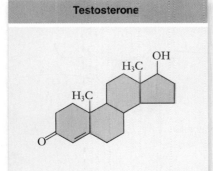

Testosterone

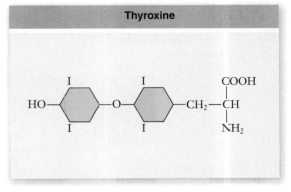

Thyroxine

inside the cell the lipophilic regulatory molecules all have a similar mechanism of action.

Transport and receptor binding

These hormones circulate bound to transport proteins (see figure 46.3), which prolong their survival in the blood. When the hormones arrive at their target cells, they dissociate from their transport proteins and pass through the plasma membrane of the cell (figure 46.5). The hormone then binds to an intracellular receptor protein.

Some steroid hormones bind to their receptors in the cytoplasm, and then move as a hormone-receptor complex into the nucleus. Other steroids and the thyroid hormones travel directly into the nucleus before encountering their receptor proteins. Whether the hormone finds its receptor in the nucleus or translocates with its receptor into the nucleus from the cytoplasm, the rest of the story is the same.

Activation of transcription in the nucleus

The hormone receptor, activated by binding to the hormone, is now also able to bind to specific regions of the DNA. These DNA regions, located in the promoters of specific genes, are known as **hormone response elements.** The binding of the hormone-receptor complex has a direct effect on the level of transcription at that site by activating, or in some cases deactivating, gene transcription. Receptors therefore function as *hormone-activated transcription factors* (see chapters 9 and 16).

The proteins that result from activation of these transcription factors often have activity that changes the metabolism of the target cell in a specific fashion; this change constitutes the cell's response to hormone stimulation. When estrogen binds to its receptor in liver cells of chickens, for example, it activates the cell to produce the protein vitellogenin, which is then transported to the ovary to form the yolk of eggs. In contrast, when thyroid hormone binds to its receptor in the anterior pituitary of humans, it inhibits the expression of the gene for thyrotropin, a mechanism of negative feedback (described later).

figure 46.5

THE MECHANISM OF LIPOPHILIC HORMONE ACTION. Lipophilic hormones diffuse through the plasma membrane of cells and bind to intracellular receptor proteins. The hormone-receptor complex then binds to specific regions of the DNA (hormone response elements), regulating the production of messenger RNA (mRNA). Most receptors for these hormones reside in the nucleus; if the hormone is one that binds to a receptor in the cytoplasm, the hormone-receptor complex moves together into the nucleus.

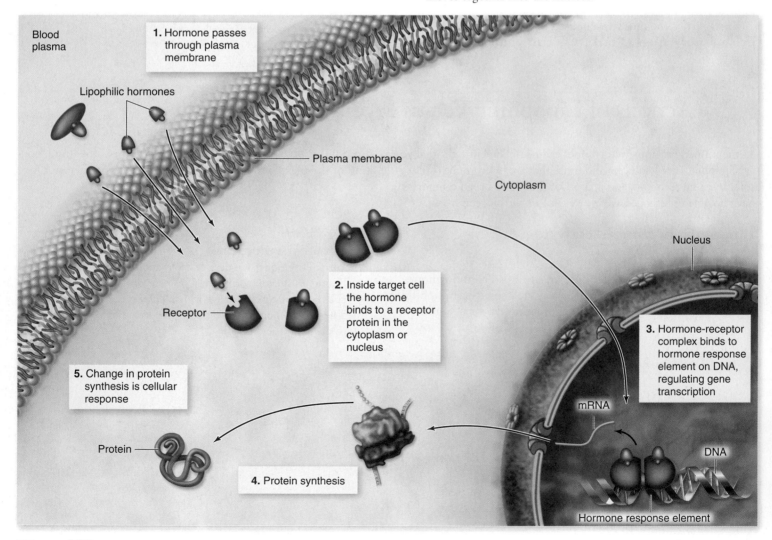

Blood plasma

1. Hormone passes through plasma membrane

Lipophilic hormones

Plasma membrane

Cytoplasm

Nucleus

Receptor

2. Inside target cell the hormone binds to a receptor protein in the cytoplasm or nucleus

3. Hormone-receptor complex binds to hormone response element on DNA, regulating gene transcription

5. Change in protein synthesis is cellular response

mRNA

DNA

Protein

4. Protein synthesis

Hormone response element

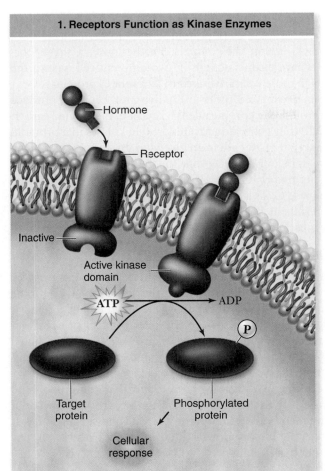

1. Receptors Function as Kinase Enzymes

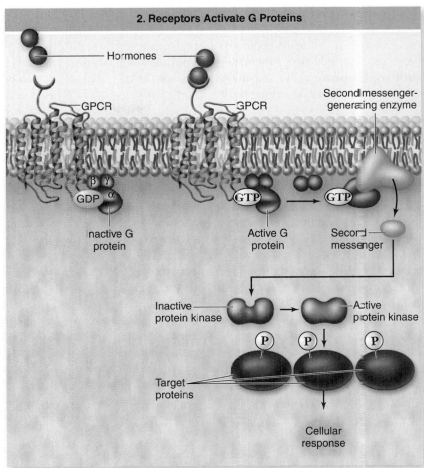

2. Receptors Activate G Proteins

figure 46.6

THE ACTION OF HYDROPHILIC HORMONES. Hydrophilic hormones cannot enter cells and must therefore work extracellularly via activation of transmembrane receptor proteins. (*1*) These receptors can function as kinase enzymes, activating phosphorylation of other proteins inside cells. (*2*) Alternatively, acting through intermediary G proteins, the hormone-bound receptor activates production of a second messenger. The second messenger activates protein kinases that phosphorylate and thereby activate other proteins. GPCR = G protein-coupled receptor.

Because this activation and transcription process requires alterations in gene expression, it often takes several hours before the response to lipophilic hormone stimulation is apparent in target cells.

Hydrophilic hormones activate receptors on target cell membranes

Hormones that are too large or too polar to cross the plasma membranes of their target cells include all of the peptide, protein, and glycoprotein hormones, as well as the catecholamine hormones. These hormones bind to receptor proteins located on the outer surface of the plasma membrane. This binding must then activate the hormone response inside the cell, initiating the process of signal transduction. The cellular response is most often achieved through receptor-dependent activation of the powerful intracellular enzymes called *protein kinases*. As described in chapter 9, protein kinases are critical regulatory enzymes that activate or deactivate intracellular proteins by phosphorylation. By regulating protein kinases, hydrophilic hormone receptors exert a powerful influence over the broad range of intracellular functions.

Receptor kinases

For some hormones, such as insulin, the receptor itself is a kinase (figure 46.6), and it can directly phosphorylate intracellular proteins that alter cellular activity. In the case of insulin, this action results in the placement in the plasma membrane of glucose transport proteins that enable glucose to enter cells. Other peptide hormones, such as growth hormone, work through similar mechanisms, although the receptor itself is not a kinase. Instead, the hormone-bound receptor recruits and activates intracellular kinases, which then initiate the cellular response.

Second-messenger systems

Many hydrophilic hormones, such as epinephrine, work through second-messenger systems. A number of different molecules in the cell can serve as second messengers, as you saw in chapter 9. The interaction between the hormone and its receptor activates mechanisms in the plasma membrane that increase the concentration of the second messengers within the target cell cytoplasm.

In the early 1960s, Earl Sutherland showed that cyclic adenosine monophosphate, or **cyclic AMP (cAMP),** serves as a second messenger when epinephrine binds to receptors on the plasma membranes of liver cells. The cAMP second-messenger system was the first such system to be described. Since that time, another hormonally regulated second-messenger system has been described that generates two lipid messengers: **inositol triphosphate (IP₃)** and **diacyl glycerol (DAG).** These systems were described in chapter 9.

The action of G proteins

Receptors that activate second messengers do not manufacture the second messenger themselves. Rather, they are linked to a second-messenger-generating enzyme via membrane proteins called *G proteins* [that is, they are G-protein-coupled receptors (GPCR); see chapter 9]. The binding of the hormone to its receptor causes the G protein to shuttle within the plasma membrane from the receptor to the second-messenger-generating enzyme (see figure 46.6). When the G protein activates the enzyme, the result is an increase in second-messenger molecules inside the cell.

In the case of epinephrine, the G protein activates an enzyme called *adenylyl cyclase*, which catalyzes the formation of the second messenger cAMP from ATP. The second messenger formed at the inner surface of the plasma membrane then diffuses within the cytoplasm, where it binds to and activates protein kinases.

The identities of the proteins that are subsequently phosphorylated by the protein kinases vary from one cell type to the next and include enzymes, membrane transport proteins, and transcription factors. This diversity provides hormones with distinct actions in different tissues. In liver cells, for example, cAMP-dependent protein kinases activate enzymes that convert glycogen into glucose. In contrast, cardiac muscle cells express a different set of cellular proteins such that a cAMP increase activates an increase in the rate and force of cardiac muscle contraction.

Activation versus inhibition

The cellular response to a hormone depends on the type of G protein activated by the hormone's receptor. Some receptors are linked to G proteins that activate second-messenger-producing enzymes, whereas other receptors are linked to G proteins that inhibit their second-messenger-generating enzyme. As a result, some hormones stimulate protein kinases in their target cells, while others inhibit their targets. Furthermore, a single hormone can have distinct actions in two different cell types if the receptors in those cells are linked to different G proteins.

Epinephrine receptors in the liver, for example, produce cAMP through the enzyme adenylyl cyclase, mentioned earlier. The cAMP they generate activates protein kinases that promote the production of glucose from glycogen. In smooth muscle, by contrast, epinephrine receptors can be linked through a different stimulatory G protein to the IP₃-generating enzyme phopholipase C. As a result, epinephrine stimulation of smooth muscle results in IP₃-regulated release of intracellular calcium, causing muscle contraction.

Duration of hydrophilic hormone effects

The binding of a hydrophilic hormone to its receptor is reversible and usually very brief; hormones soon dissociate from receptors or are rapidly deactivated by their target cells after binding. Additionally, target cells contain specific enzymes that rapidly deactivate second messengers and protein kinases. As a result, hydrophilic hormones are capable of stimulating immediate responses within cells, but often have a brief duration of action (minutes to hours).

Recently, investigators have found that some lipophilic hormones, such as the steroid progesterone, are also capable of activating membrane receptors and second-messenger systems, illustrating the importance of these signaling pathways in endocrine control systems.

> Lipophilic hormones pass through the target cell's plasma membrane and bind to intracellular receptor proteins. The hormone-receptor complex then binds to specific regions of DNA, regulating gene expression in the target cell. Hydrophilic hormones cannot pass through the plasma membrane; they bind to membrane receptors that activate protein kinases directly or through second-messenger systems.

46.3 The Pituitary and Hypothalamus: The Body's Control Centers

The **pituitary gland,** also known as the **hypophysis,** hangs by a stalk from the hypothalamus at the base of the brain posterior to the optic chiasm. The **hypothalamus** is a part of the CNS that has a major role in regulation of body processes. Both these structures were described in chapter 44; here we discuss in detail how they work together to bring about homeostasis and changes in body processes.

The pituitary is a compound endocrine gland

A microscopic view reveals that the gland consists of two parts, one of which appears glandular and is called the **anterior pituitary,** or **adenohypophysis.** The other portion appears fibrous and is called the **posterior pituitary,** or **neurohypophysis.** These two portions of the pituitary gland have different embryonic origins, secrete different hormones, and are regulated by different control systems. These two regions are conserved in all vertebrate animals, suggesting an ancient and important function of each.

The posterior pituitary stores and releases two neurohormones

The posterior pituitary appears fibrous because it contains axons that originate in cell bodies within the hypothalamus and that extend along the stalk of the pituitary as a tract

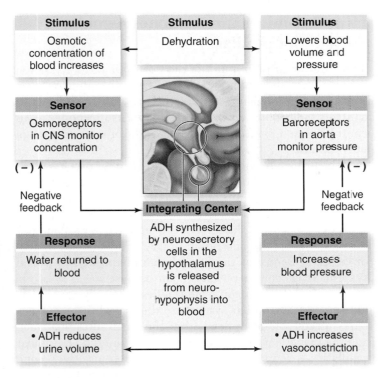

figure 46.7

THE EFFECTS OF ANTIDIURETIC HORMONE (ADH).
Dehydration increases the osmotic concentration of the blood and lowers blood pressure, stimulating the neurohypophysis to secrete ADH. ADH increases reabsorption of water by the kidneys and causes vasoconstriction, increasing blood pressure. Decreased blood osmolarity and increased blood pressure complete negative feedback loops to maintain homeostasis.

of fibers. This anatomical relationship results from the way the posterior pituitary is formed in embryonic development. As the floor of the third ventricle of the brain forms the hypothalamus, part of this neural tissue grows downward to produce the posterior pituitary. The hypothalamus and posterior pituitary thus remain directly interconnected by a tract of axons.

Antidiuretic hormone

The endocrine role of the posterior pituitary first became evident in 1912, when a remarkable medical case was reported: A man who had been shot in the head developed the need to urinate every 30 minutes or so, 24 hours a day. The bullet had lodged in his posterior pituitary. Subsequent research demonstrated that removal of this portion of the pituitary produces the same symptoms.

In the early 1950s investigators isolated a peptide from the posterior pituitary, **antidiuretic hormone (ADH).** ADH stimulates water reabsorption by the kidneys (figure 46.7), and in doing so inhibits diuresis (urine production). When ADH is missing, as it was in the shooting victim, the kidneys do not reabsorb as much water, and excessive quantities of urine are produced. This is why the consumption of alcohol, which inhibits ADH secretion, leads to frequent urination.

Oxytocin

The posterior pituitary also secretes **oxytocin,** a second peptide neurohormone that, like ADH, is composed of nine amino acids. In mammals, oxytocin stimulates the milk ejection reflex. During suckling, sensory receptors in the nipples send impulses to the hypothalamus, which triggers the release of oxytocin. Oxytocin is also needed to stimulate uterine contractions in women during childbirth.

Oxytocin secretion continues after childbirth in a woman who is breast-feeding; as a result, the uterus of a nursing mother contracts and returns to its normal size after pregnancy more quickly than the uterus of a mother who does not breast-feed.

A related posterior pituitary neurohormone, *arginine vasotocin,* exerts similar effects in nonmammalian species. For example, in chickens and sea turtles, arginine vasotocin activates oviduct contraction during egg laying.

More recently, oxytocin has been identified as an important regulator of reproductive behavior. In both men and women, it is thought to be involved in promoting pair bonding (leading to its being called the "cuddle hormone") as well as regulating sexual responses, including arousal and orgasm. For these effects, it most likely functions in a paracrine fashion inside the central nervous system, much like a neurotransmitter.

Hypothalamic production of the neurohormones

ADH and oxytocin are actually produced by neuron cell bodies located in the hypothalamus. These two neurohormones are transported along the axon tract that runs from the hypothalamus to the posterior pituitary, where they are stored. In response to the appropriate stimulation—increased blood plasma osmolality in the case of ADH, the suckling of a baby in the case of oxytocin—the neurohormones are released by the posterior pituitary into the blood.

Because this reflex control involves both the nervous and the endocrine systems, ADH and oxytocin are said to be secreted by a **neuroendocrine reflex.**

The anterior pituitary produces seven hormones

The anterior pituitary, unlike the posterior pituitary, does not develop from growth of the brain; instead, it develops from a pouch of epithelial tissue that pinches off from the roof of the embryo's mouth. In spite of its proximity to the brain, it is not part of the nervous system.

Because it forms from epithelial tissue, the anterior pituitary is an independent endocrine gland. It produces at least seven essential hormones, many of which stimulate growth of their target organs, as well as production and secretion of other hormones from additional endocrine glands. Therefore, several hormones of the anterior pituitary are collectively termed *tropic hormones* or *tropins.* When the target organ of a tropic hormone is another endocrine gland, that gland is stimulated by the tropic hormone to secrete its own hormones.

The hormones produced and secreted by different cell types in the anterior pituitary can be categorized into three structurally similar families: the *peptide hormones,* the *protein hormones,* and the *glycoprotein hormones.*

Peptide hormones

The **peptide hormones** of the anterior pituitary are cleaved from a single precursor protein, and therefore they share some common sequence. They are fewer than 40 amino acids in size.

Adrenocorticotropic hormone (ACTH, or *corticotropin*) stimulates the adrenal cortex to produce corticosteroid hormones, including cortisol (in humans) and corticosterone (in many other vertebrates). These hormones regulate glucose homeostasis and are important in the response to stress.

Melanocyte-stimulating hormone (MSH) stimulates the synthesis and dispersion of melanin pigment, which darkens the epidermis of some fish, amphibians, and reptiles, and can control hair pigment color in mammals.

Protein hormones

The **protein hormones** each comprise a single chain of approximately 200 amino acids, and they share significant structural similarities.

Growth hormone (GH, or *somatotropin*) stimulates the growth of muscle, bone (indirectly), and other tissues, and it is also essential for proper metabolic regulation.

Prolactin (PRL) is best known for stimulating the mammary glands to produce milk in mammals; however, it has diverse effects on many other targets, including regulation of ion and water transport across epithelia, stimulation of a variety of organs that nourish young, and activation of parental behaviors.

Glycoprotein hormones

The largest and most complex hormonal molecules known, the **glycoprotein hormones** are dimers, containing alpha (α) and beta (β) subunits, each around 100 amino acids in size, with covalently linked sugar residues. The α subunit is common to all three hormones. The β subunit differs, endowing each hormone with a different target specificity.

Thyroid-stimulating hormone (TSH, or *thyrotropin*) stimulates the thyroid gland to produce the hormone thyroxine, which in turn regulates development and metabolism.

Luteinizing hormone (LH) stimulates the production of estrogen and progesterone by the ovaries and is needed for ovulation in female reproductive cycles (see chapter 52). In males, it stimulates the testes to produce testosterone, which is needed for sperm production and for the development of male secondary sexual characteristics.

Follicle-stimulating hormone (FSH) is required for the development of ovarian follicles in females. In males, it is required for the development of sperm. FSH and LH are collectively referred to as *gonadotropins* because they stimulate function of the gonads.

Hypothalamic neurohormones regulate the anterior pituitary

The anterior pituitary, unlike the posterior pituitary, is not derived from the brain and does not receive an axon tract from the hypothalamus. Nevertheless, the hypothalamus controls the production and secretion of its hormones. This control is itself exerted hormonally rather than by means of nerve axons.

Neurons in the hypothalamus secrete two types of neurohormones, **releasing hormones** and **inhibiting hormones,** that diffuse into blood capillaries at the base of the hypothalamus (figure 46.8). These capillaries drain into small veins that run within the stalk of the pituitary to a second bed of capillaries in the anterior pituitary. This unusual system of vessels is known as the **hypothalamohypophysial portal system.** It is called a portal system because it has a second capillary bed downstream from the first; the only other body locations with a similar system are the liver, in which capillaries receive blood drained from the gastrointestinal tract, and the kidneys, in which the glomerulus and the peritubular capillaries are in series (see chapter 50). Because the second bed of capillaries receive little oxygen from such vessels, the vessels are likely delivering something else of importance.

Releasers

Each neurohormone released by the hypothalamus into the portal system regulates the secretion of a specific hormone in the anterior pituitary. Releasing hormones are peptide neurohormones that stimulate release of other hormones; specifically, *thyrotropin-releasing hormone* (TRH) stimulates the release of TSH; *corticotropin-releasing hormone* (CRH) stimulates the release of ACTH; and *gonadotropin-releasing hormone* (GnRH) stimulates the release of FSH and LH. A releasing hormone for growth hormone, called *growth hormone-releasing hormone* (GHRH), has also been discovered, and a releasing hormone for prolactin has been postulated but has thus far not been identified.

Inhibitors

The hypothalamus also secretes neurohormones that inhibit the release of certain anterior pituitary hormones. To date, three such neurohormones have been discovered: *Somatostatin*, or *growth hormone-inhibiting hormone* (GHIH), which inhibits the secretion of GH; *prolactin-inhibiting factor* (PIF), which inhibits the secretion of prolactin and has been found to be the neurotransmitter dopamine; and *MSH-inhibiting hormone* (MIH), which inhibits the secretion of MSH.

Feedback from peripheral endocrine glands regulates anterior pituitary hormones

Because hypothalamic hormones control the secretions of the anterior pituitary, and because the hormones of the anterior pituitary in turn control the secretions of other endocrine

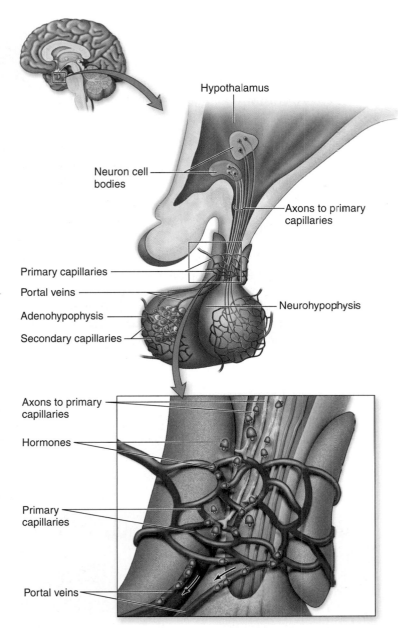

Hypothalamus

Neuron cell bodies

Axons to primary capillaries

Primary capillaries

Portal veins

Adenohypophysis

Secondary capillaries

Neurohypophysis

Axons to primary capillaries

Hormones

Primary capillaries

Portal veins

figure 46.8

HORMONAL CONTROL OF THE ADENOHYPOPHYSIS BY THE HYPOTHALAMUS. Neurons in the hypothalamus secrete hormones that are carried by portal blood vessels directly to the adenohypophysis, where they either stimulate or inhibit the secretion of hormones from the adenohypophysis.

glands, it may seem that the hypothalamus is in charge of hormonal secretion for the whole body. This does not, however, provide a complete picture of endocrine regulation, for two reasons.

First, a number of endocrine organs, such as the adrenal medulla and the pancreas, are not directly regulated by this hypothalamic control system. Second, the hypothalamus and the anterior pituitary are themselves partially controlled by the very hormones whose secretion they stimulate. In most cases,

this control is inhibitory (figure 46.9). This type of control system is called *negative feedback*, and it acts to maintain relatively constant levels of the target cell hormone.

An example of negative feedback: Thyroid gland control

To illustrate how important the negative feedback mechanism is, let's consider the hormonal control of the thyroid gland. The hypothalamus secretes TRH into the hypothalamohypophysial portal system, which stimulates the anterior pituitary to secrete TSH. TSH in turn causes the thyroid gland to release **thyroxine.** Thyroxine and other thyroid hormones affect metabolic rate, as described in the following section.

Among thyroxine's many target organs are the hypothalamus and the anterior pituitary themselves. Thyroxine acts on these organs to inhibit their secretion of TRH and TSH, respectively. This negative feedback inhibition is essential for homeostasis because it keeps the thyroxine levels fairly constant.

The hormone thyroxine contains the element iodine; without iodine, the thyroid gland cannot produce thyroxine. Individuals living in iodine-poor areas (such as central prairies distant from seacoasts) lack sufficient iodine to manufacture thyroxine, so the hypothalamus and anterior pituitary receive far less negative feedback inhibition than is normal. This reduced inhibition results in elevated secretion of TRH and TSH.

High levels of TSH stimulate the thyroid gland, whose cells enlarge in a futile attempt to manufacture more thyroxine, but they cannot without iodine. The consequence is an

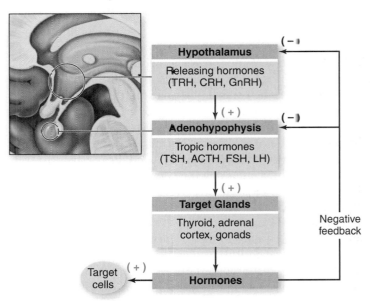

figure 46.9

NEGATIVE FEEDBACK INHIBITION. The hormones secreted by some endocrine glands feed back to inhibit the secretion of hypothalamic releasing hormones and adenohypophysis tropic hormones.

figure 46.10

A WOMAN WITH A GOITER. This condition is caused by a lack of iodine in the diet. As a result, thyroxine secretion is low, so there is less negative feedback inhibition of TSH. The elevated TSH secretion, in turn, stimulates the thyroid to enlarge in an effort to produce additional thyroxine.

enlarged thyroid gland, a condition known as a goiter (figure 46.10). Goiter size can be reduced by providing iodine in the diet. In most countries, goiter is prevented through the addition of iodine to table salt.

An example of positive feedback: Ovulation

Positive feedback in the control of the hypothalamus and anterior pituitary by the target glands is uncommon because positive feedback causes deviations from homeostasis. Positive feedback accentuates change, driving the change in the same direction. One example is the control of **ovulation,** the explosive release of a mature egg (an oocyte) from the ovary.

As the oocyte grows, follicle cells surrounding it produce increasing levels of the steroid hormone estrogen, resulting in a progressive rise in estrogen in the blood. Peak estrogen levels signal the hypothalamus that the oocyte is ready to be ovulated. Estrogen then exerts positive feedback on the hypothalamus and pituitary, resulting in a surge of LH release from the anterior pituitary. This LH surge causes the follicle cells to rupture and release the oocyte to the oviduct, where it can potentially be fertilized. The positive feedback cycle is then terminated because the tissue that produced the estrogen, the ovarian follicle, is itself destroyed by the LH surge. This process is discussed in more detail in chapter 52.

Hormones of the anterior pituitary work directly and indirectly

Early in the twentieth century, experimental techniques were developed for surgical removal of the pituitary gland (a procedure called *hypophysectomy*). Hypophysectomized animals exhibited a number of deficits, including reduced growth and development, diminished metabolism, and failure of reproduction. These powerful and diverse effects earned the pituitary a reputation as the "master gland." Indeed, many of these are *direct effects*, resulting from anterior-pituitary hormones activating receptors in nonendocrine targets, such as liver, muscle, and bone. These hormones also have *indirect effects*, however, through their ability to activate other endocrine glands, such as the thyroid, adrenal glands, and gonads. Of the seven anterior-pituitary hormones, growth hormone, prolactin, and MSH work primarily through direct effects, whereas ACTH, TSH, LH, and FSH have endocrine glands as their exclusive targets.

Effects of growth hormone

The importance of the anterior pituitary is illustrated by a condition known as **gigantism,** characterized by excessive growth of the entire body or any of its parts. The tallest human being ever recorded, Robert Wadlow, had gigantism (figure 46.11). Born in 1928, he stood 8 feet 11 inches tall, weighed 485

figure 46.11

THE ALTON GIANT. This photograph of Robert Wadlow of Alton, Illinois, taken on his 21st birthday, shows him at home with his father and mother and four siblings. Born normal size, he developed a growth-hormone-secreting pituitary tumor as a young child and never stopped growing during his 22 years of life, reaching a height of 8 ft 11 in.

pounds, and was still growing before he died from an infection at the age of 22.

We now know that gigantism is caused by the excessive secretion of GH in a growing child. By contrast, a deficiency in GH secretion during childhood results in **pituitary dwarfism,** failure to achieve normal stature.

GH stimulates protein synthesis and growth of muscles and connective tissues; it also indirectly promotes the elongation of bones by stimulating cell division in the cartilaginous epiphyseal growth plates of bones (chapter 47). Researchers found that this stimulation does not occur in the absence of blood plasma, suggesting that GH must work in concert with another hormonal molecule to exert its effects on bone. We now know that GH stimulates the production of **insulin-like growth factors,** which liver and bone produce in response to stimulation by GH. The insulin-like growth factors then stimulate cell division in the epiphyseal growth plates, and thus the elongation of the bones.

Although GH exhibits its most dramatic effects on juvenile growth, it also functions in adults to regulate protein, lipid, and carbohydrate metabolism. Recently a peptide hormone named **ghrelin,** produced by the stomach between meals, was identified as a potent stimulator of GH release, establishing an important linkage between nutrient intake and GH production.

Because human skeletal growth plates transform from cartilage into bone at puberty, GH can no longer cause an increase in height in adults. Excessive GH secretion in an adult results in a form of gigantism called **acromegaly,** characterized by bone and soft tissue deformities such as a protruding jaw, elongated fingers, and thickening of skin and facial features. Our knowledge of the regulation of GH has led to the development of drugs that can control its secretion, for example through activation of somatostatin, or by mimicking ghrelin. As a result, gigantism is much less common today.

Animals that have been genetically engineered to express additional copies of the GH gene grow to larger than normal size (see figure 17.15), making agricultural applications of GH manipulation an active area of investigation. Among other actions, GH has been found to increase milk yield in cows, promote weight gain in pigs, and increase the length of fish. The growth-promoting actions of GH thus appear to have been conserved throughout the vertebrates.

Other hormones of the anterior pituitary

Like growth hormone, prolactin acts on organs that are not endocrine glands. In contrast to GH, however, the actions of prolactin appear to be very diverse. In addition to stimulating production of milk in mammals, prolactin has been implicated in the regulation of tissues important in birds for the nourishment and incubation of young, such as the crop (which produces "crop milk," a nutritional fluid fed to chicks by regurgitation) and the brood patch (a vascular area on the abdomen of birds used to warm eggs).

In amphibians, prolactin promotes transformation of salamanders from terrestrial forms to aquatic breeding adults. Associated with these reproductive actions is an ability of prolactin to activate associated behaviors, such as parental care in mammals, broodiness in birds, and "water drive" in amphibians.

Prolactin also has varied effects on electrolyte balance through actions on the kidneys of mammals, the gills of fish, and the salt glands of marine birds (discussed in chapter 50). This variation suggests that although prolactin may have an ancient function in the regulation of salt and water movement across membranes, its actions have diversified with the appearance of new vertebrate species. The field of **comparative endocrinology** studies questions about the actions of hormones across diverse species, with the objective of understanding the mechanisms of hormone evolution.

Unlike growth hormone and prolactin, the other adenohypophysial hormones act on relatively few targets. Thyroid-stimulating hormone stimulates only the thyroid gland, and ACTH stimulates only the adrenal cortex (outer portion of the adrenal glands). The gonadotropins, FSH and LH, act only on the gonads (testes and ovaries). Although both FSH and LH act on the gonads, they each target different cells in the gonads of both females and males (see chapter 52). These hormones all share the common characteristic of activating target endocrine glands.

The final pituitary hormone, MSH regulates the activity of pigment-containing cells called **melanophores,** which contain the black pigment **melanin.** In response to MSH, melanin is dispersed throughout these cells, darkening the skin of reptiles, amphibians, or fish. In mammals, which lack melanophores but have similar cells called **melanocytes,** MSH (and the structurally related hormone ACTH) can darken hair by increasing melanin deposition in the developing hair shaft.

> The posterior pituitary contains axons originating from neurons in the hypothalamus. These neurons produce neurohormones that are released to the general circulation at the neural lobe. Hypothalamic neurons also control the anterior pituitary by producing releasing and inhibiting hormones. In response, the anterior pituitary secretes seven hormones, four of which stimulate target endocrine glands. Both the hypothalamus and the anterior pituitary are controlled by hormones from the endocrine glands that they stimulate through negative feedback inhibition.

46.4 The Major Peripheral Endocrine Glands

Although the pituitary produces an impressive array of hormones, many endocrine glands are found in other locations. Some of these may be controlled by tropic hormones of the pituitary, but others are independent of pituitary control. Several endocrine glands develop from derivatives of the primitive pharynx, which is the most anterior segment of the digestive tract (chapter 48). These glands, which include the *thyroid* and *parathyroid* glands, produce hormones that regulate processes associated with nutrient uptake, such as carbohydrate, lipid, protein, and mineral metabolism.

The thyroid gland regulates basal metabolism and development

The **thyroid gland** varies in shape in different vertebrate species, but is always found in the neck area, anterior to the heart. In humans it is shaped like a bow tie and lies just below the Adam's apple in the front of the neck.

The thyroid gland secretes three hormones, primarily thyroxine, smaller amounts of **triiodothyronine** (collectively referred to as thyroid hormones), and calcitonin. As described earlier, thyroid hormones are unique in being the only molecules in the body containing iodine (thyroxine contains four iodine atoms, triiodothyronine contains three).

Thyroid-related disorders

Thyroid hormones work by binding to nuclear receptors located in most cells in the body, influencing the production and activity of a large number of cellular proteins. The importance of thyroid hormones first became apparent from studies of human thyroid disorders. Adults with **hypothyroidism** have low metabolism due to underproduction of thyroxine, including a re-

duced ability to utilize carbohydrates and fats. As a result, they are often fatigued, overweight, and feel cold. Hypothyroidism is particularly concerning in infants and children, where it impairs growth, brain development, and reproductive maturity. Fortunately, because thyroid hormones are small, simple molecules, people with hypothyroidism can take thyroxine orally as a pill.

People with **hyperthyroidism,** by contrast, often exhibit opposite symptoms: weight loss, nervousness, high metabolism, and overheating because of overproduction of thyroxine. Drugs are available which block thyroid hormone synthesis in the thyroid gland, but in some cases portions of the thyroid gland must be removed surgically.

Actions of thyroid hormones

Thyroid hormones regulate enzymes controlling carbohydrate and lipid metabolism in most cells, promoting the appropriate use of these fuels for maintaining the body's basal metabolic rate. Thyroid hormones often function cooperatively, or *synergistically*, with other hormones, promoting the activity of growth hormone, epinephrine, and reproductive steroids. Through these actions, thyroid hormones function to ensure that adequate cellular energy is available to support metabolically demanding activities.

In humans, which exhibit a relatively high metabolic rate at all times, thyroid hormones are maintained in the blood at con-

figure 46.12

THYROXINE TRIGGERS METAMORPHOSIS IN AMPHIBIANS. In tadpoles at the premetamorphic stage, the hypothalamus stimulates the adenohypophysis to secrete TSH (thyroid-stimulating hormone). TSH then stimulates the thyroid gland to secrete thyroxine. Thyroxine binds to its receptor and initiates the changes in gene expression necessary for metamorphosis. As metamorphosis proceeds, thyroxine reaches its maximal level, after which the forelimbs begin to form and the tail is reabsorbed.

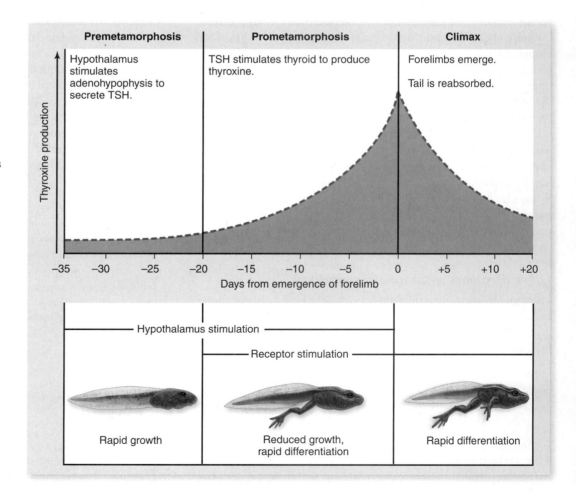

stantly elevated levels. In contrast, in reptiles, amphibians, and fish, which undergo seasonal cycles of activity, thyroid hormone levels in the blood increase during periods of metabolic activation (such as growth, reproductive development, migration, or breeding) and diminish during periods of inactivity in cold months.

Some of the most dramatic effects of thyroid hormones are observed in their regulation of growth and development. In developing humans, for example, thyroid hormones promote growth of neurons and stimulate maturation of the central nervous system. Children born with hypothyroidism are stunted in their growth and suffer severe mental retardation, a condition called *cretinism*. Early detection through measurement of thyroid hormone levels allows this condition to be treated with thyroid hormone administration.

The most impressive demonstration of the importance of thyroid hormones in development is displayed in amphibians. Thyroid hormones direct the metamorphosis of tadpoles into frogs, a process that requires the transformation of an aquatic, herbivorous larva into a terrestrial, carnivorous juvenile (figure 46.12). If the thyroid gland is removed from a tadpole, it will not change into a frog. Conversely, if an immature tadpole is fed pieces of a thyroid gland, it will undergo premature metamorphosis and become a miniature frog. This illustrates the powerful actions thyroid hormones can elicit by regulating the expression of multiple genes.

Calcium homeostasis is regulated by several hormones

Calcium is a vital component of the vertebrate body both because of its being a structural component of bones and because of its role in ion-mediated processes such as muscle contraction. Glands that regulate calcium include the thyroid and parathyroid facilitated by the effects of vitamin D.

Calcitonin secretion by the thyroid

In addition to the thyroid hormones, the thyroid gland also secretes **calcitonin,** a peptide hormone that plays a role in maintaining proper levels of calcium (Ca^{2+}) in the blood. When the blood Ca^{2+} concentration rises too high, calcitonin stimulates the uptake of calcium into bones, thus lowering its level in the blood. Although calcitonin may be important in the physiology of some vertebrates, it appears less important in the day-to-day regulation of Ca^{2+} levels in adult humans. It may, however, play an important role in bone remodeling in rapidly growing children.

Parathyroid hormone (PTH)

The **parathyroid glands** are four small glands attached to the thyroid. Because of their size, researchers ignored them until well into the twentieth century. The first suggestion that these organs have an endocrine function came from experiments on dogs: If their parathyroid glands were removed, the Ca^{2+} concentration in the dogs' blood plummeted to less than half the normal value. The Ca^{2+} concentration returned to normal when an extract of parathyroid gland was administered. However, if too much of the extract was administered, the dogs' Ca^{2+} levels rose far above normal as the calcium phosphate crystals in their bones were dissolved.

It was clear that the parathyroid glands produce a hormone that stimulates the release of calcium from bone.

The hormone produced by the parathyroid glands is a peptide called **parathyroid hormone (PTH).** PTH is synthesized and released in response to falling levels of Ca^{2+} in the blood. This decline cannot be allowed to continue uncorrected because a significant fall in the blood Ca^{2+} level can cause severe muscle spasms. A normal blood Ca^{2+} level is important for the functioning of muscles, including the heart, and for proper functioning of the nervous and endocrine systems.

PTH stimulates the osteoclasts (bone cells) in bone to dissolve the calcium phosphate crystals of the bone matrix and release Ca^{2+} into the blood (figure 46.13). PTH also stimulates

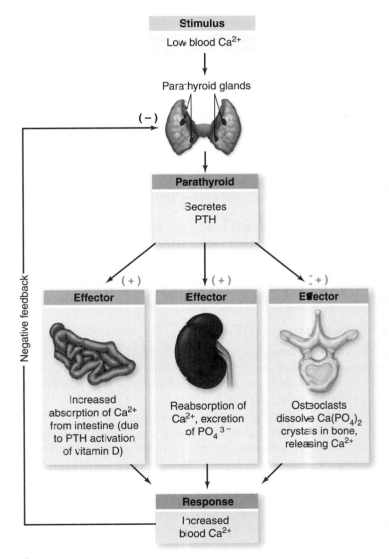

figure 46.13

REGULATION OF BLOOD Ca^{2+} LEVELS BY PARATHYROID HORMONE (PTH). When blood Ca^{2+} levels are low, PTH is released by the parathyroid glands. PTH directly stimulates the dissolution of bone and the reabsorption of Ca^{2+} by the kidneys. PTH indirectly promotes the intestinal absorption of Ca^{2+} by stimulating the production of the active form of vitamin D.

the kidneys to reabsorb Ca^{2+} from the urine and leads to the activation of vitamin D, needed for the absorption of calcium from food in the intestine.

Vitamin D

Vitamin D is produced in the skin from a cholesterol derivative in response to ultraviolet light. It is called an essential vitamin because in temperate regions of the world a dietary source is needed to supplement the amount produced by the skin. (In the tropics, people generally receive enough exposure to sunlight to produce adequate vitamin D.) Diffusing into the blood from the skin, vitamin D is actually an inactive form of a hormone. In order to become activated, the molecule must gain two hydroxyl groups (—OH); one of these is added by an enzyme in the liver, the other by an enzyme in the kidneys.

The enzyme needed for this final step is stimulated by PTH, thereby producing the active form of vitamin D known as 1,25-dihydroxyvitamin D. This hormone stimulates the intestinal absorption of Ca^{2+} and thereby helps raise blood Ca^{2+} levels so that bone can become properly mineralized. A diet deficient in vitamin D thus leads to poor bone formation, a condition called rickets.

To ensure adequate amounts of this essential hormone, vitamin D is now added to commercially produced milk in the United States and some other countries. This is certainly a preferable alternative to the prior method of vitamin D administration, the dreaded dose of cod liver oil.

The adrenal gland releases both catecholamine and steroid hormones

The **adrenal glands** are located just above each kidney (figure 46.14). Each gland is composed of an inner portion, the *adrenal medulla*, and an outer layer, the *adrenal cortex*.

The adrenal medulla

The **adrenal medulla** receives neural input from axons of the sympathetic division of the autonomic nervous system, and it secretes the catecholamines **epinephrine** and **norepinephrine** in response to stimulation by these axons. The actions of these hormones trigger "alarm" responses similar to those elicited by the sympathetic division, helping to prepare the body for extreme efforts. Among the effects of these hormones are an increased heart rate, increased blood pressure, dilation of the bronchioles, elevation in blood glucose, reduced blood flow to the skin and digestive organs, and increased blood flow to the heart and muscles. The actions of epinephrine, released as a hormone, supplement those of neurotransmitters released by the sympathetic nervous system.

The adrenal cortex

The hormones from the adrenal cortex are all steroids and are referred to collectively as **corticosteroids**. *Cortisol* (also called hydrocortisone) and related steroids secreted by the adrenal cortex act on various cells in the body to maintain glucose

homeostasis. In mammals, these hormones are referred to as glucocorticoids, and their secretion is primarily regulated by ACTH from the anterior pituitary.

The glucocorticoids stimulate the breakdown of muscle protein into amino acids, which are carried by the blood to the liver. They also stimulate the liver to produce the enzymes needed for gluconeogenesis, which can convert amino acids into glucose. Glucose synthesis from protein is particularly important during very long periods of fasting or exercise, when blood glucose levels might otherwise become dangerously low.

Whereas glucocorticoids are important in the daily regulation of glucose and protein, they, like the adrenal medulla hormones, are also secreted in large amounts in response to stress. It has been suggested that during stress they activate the production of glucose at the expense of protein and fat synthesis.

In addition to regulating glucose metabolism, the glucocorticoids modulate some aspects of the immune response. The physiological significance of this action is still unclear, and it may be apparent only under circumstances where glucocorticoids are maintained at elevated levels for long periods of time (such as long-term stress). Glucocorticoids are given medically to suppress the immune system in persons with immune disorders, such as rheumatoid arthritis, and to prevent the immune system from rejecting transplants. Derivatives of cortisol, such as prednisone, have widespread medical use as antiinflammatory agents.

Aldosterone, the other major corticosteroid, is classified as a mineralocorticoid because it helps regulate mineral balance. The secretion of aldosterone from the adrenal cortex is activated by angiotensin II, a product of the renin–angiotensin system described in chapter 50, as well as high blood K^+. Angiotensin II activates aldosterone secretion when blood pressure falls.

A primary action of aldosterone is to stimulate the kidneys to reabsorb Na^+ from the urine. (Blood levels of Na^+ decrease if Na^+ is not reabsorbed from the urine.) Sodium is the major extracellular solute; it is needed for the maintenance of normal blood volume and pressure, as well as for the generation of action potentials in neurons and muscles. Without aldosterone, the kidneys would lose excessive amounts of blood Na^+ in the urine.

Aldosterone-stimulated reabsorption of Na^+ also results in kidney excretion of K^+ in the urine. Aldosterone thus prevents K^+ from accumulating in the blood, which would lead to malfunctions in electrical signaling in nerves and muscles. Because of these essential functions performed by aldosterone, removal of the adrenal glands, or diseases that prevent aldosterone secretion, are invariably fatal without hormone therapy.

Pancreatic hormones are primary regulators of carbohydrate metabolism

The pancreas is located adjacent to the stomach and is connected to the duodenum of the small intestine by the pancreatic duct. It secretes bicarbonate ions and a variety of digestive enzymes into the small intestine through this duct (chapter 48), and for a long time the pancreas was thought to be solely an exocrine gland.

Insulin

In 1869, however, a German medical student named Paul Langerhans described some unusual clusters of cells scattered throughout the pancreas; these clusters came to be called *islets of Langerhans.* Laboratory workers later observed that the surgical removal of the pancreas caused glucose to appear in the urine, the hallmark of the disease diabetes mellitus. This led to the discovery that the pancreas, specifically the islets of Langerhans, produced a hormone that prevents this disease.

That hormone is **insulin,** secreted by the beta (β) cells of the islets. Insulin was not isolated until 1922 when Banting and Best succeeded where many others had not. On January 11, 1922, they injected an extract purified from beef pancreas into a 13-year-old diabetic boy, whose weight had fallen to 65 pounds and who was not expected to survive. With that single injection, the glucose level in the boy's blood fell 25%. A more potent extract soon brought the level down to near normal. The doctors had achieved the first instance of successful insulin therapy.

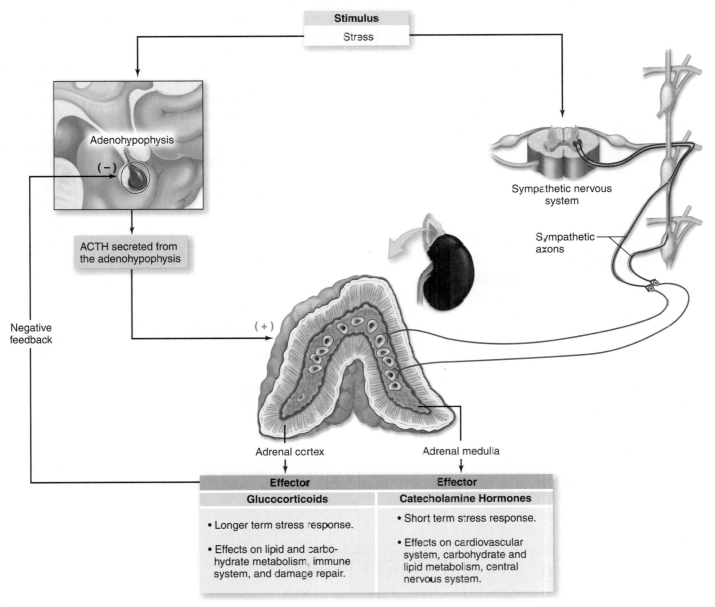

figure 46.14

THE ADRENAL GLANDS. The adrenal medulla produces the catecholamines epinephrine and norepinephrine, which initiate a response to acute stress. The adrenal cortex produces steroid hormones, including the glucocorticoid cortisol. In response to stress, cortisol secretion increases glucose production and stimulates the immune response.

Glucagon

The islets of Langerhans produce another hormone; the alpha (α) cells of the islets secrete **glucagon,** which acts antagonistically to insulin (figure 46.15). When a person eats carbohydrates, the blood glucose concentration rises. Blood glucose directly activates the secretion of insulin by the β cells and inhibits the secretion of glucagon by the α cells. Insulin promotes the cellular uptake of glucose into liver, muscle, and fat cells. It also activates the storage of glucose as glycogen in liver and muscle or as fat in fat cells. Between meals, when the concentration of blood glucose falls, insulin secretion decreases, and glucagon secretion increases. Glucagon promotes the hydrolysis of stored glycogen in the liver and fat in adipose tissue. As a result, glucose and fatty acids are released into the blood and can be taken up by cells and used for energy.

Treatment of diabetes

Although many hormones (for example, growth hormone, glucocorticoids, glucagon) favor the movement of glucose out of cells, insulin is the only hormone that promotes movement of glucose from blood into cells. For this reason, disruptions in insulin signaling can have serious consequences. People with *type I*, or insulin-dependent, diabetes mellitus, lack the insulin-secreting β cells and consequently produce no insulin. Treatment for these patients consists of insulin injections. (Because insulin is a peptide hormone, it would be digested if taken orally and must instead be injected into the blood.)

In the past, only insulin extracted from the pancreas of pigs or cattle was available, but today people with insulin-dependent diabetes can inject themselves with human insulin produced by genetically engineered bacteria. Active research on the possibility of transplanting islets of Langerhans holds much promise of a lasting treatment for these patients.

Most diabetic patients, however, have *type II*, or noninsulin-dependent, diabetes mellitus. They generally have normal or even above-normal levels of insulin in their blood, but their cells have a reduced sensitivity to insulin. These people may not require insulin injections and can often control their diabetes through diet and exercise. It is estimated that over 90% of the cases of diabetes in North America are type II. Worldwide at least 171 million suffer from diabetes, and it is expected that this number will grow. Type II diabetes is especially common in developed countries, and it has been suggested that there is a linkage between type II diabetes and obesity.

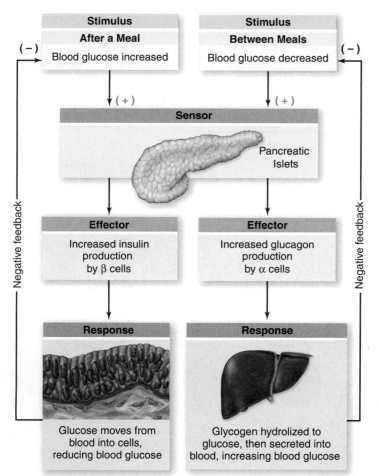

figure 46.15

THE ANTAGONISTIC ACTIONS OF INSULIN AND GLUCAGON ON BLOOD GLUCOSE. Insulin stimulates the cellular uptake of blood glucose into skeletal muscles, adipose cells, and the liver after a meal. Glucagon stimulates the hydrolysis of liver glycogen between meals, so that the liver can secrete glucose into the blood. These antagonistic effects help to maintain homeostasis of the blood glucose concentration.

Hormones from peripheral endocrine glands regulate diverse processes. Thyroid hormones, adrenal glucocorticoids, and pancreatic hormone control carbohydrate, lipid, and protein metabolism. Mineral metabolism, including blood levels of Ca^{2+}, Na^+, and K^+, are regulated by calcitonin, parathyroid hormone, vitamin D, and mineralocorticoids. Epinephrine and norepinephrine from the adrenal medulla, as well as glucocorticoids, are released during stress to mobilize nutrients for survival.

46.5 Other Hormones and Their Effects

A variety of vertebrate and invertebrate processes are regulated by hormones and other chemical messengers, and in this section we review the most important ones.

Sex steroids regulate reproductive development

The ovaries and testes in vertebrates are important endocrine glands, producing the sex steroid hormones, including estrogens, progesterone, and testosterone (to be described in detail in chapter 52). Estrogen and progesterone are the primary "female" sex steroids, and testosterone and its immediate derivatives are the primary "male" sex steroids, or androgens. Both types of hormone can be found in both genders, however.

During embryonic development, testosterone production in the male embryo is critical for the development of male sex organs. In mammals, sex steroids are responsible for the development of secondary sexual characteristics at puberty. These characteristics include breasts in females, body hair, and increased muscle mass in males. Because of this latter effect, some athletes have misused androgens to increase muscle mass. Use of steroids for this purpose has been condemned by virtually all major sports organizations, and it can cause liver disorders as well as a number of other serious side effects.

In females, sex steroids are especially important in maintaining the sexual cycle. Estrogen and progesterone produced in the ovaries are critical regulators of the menstrual and ovarian cycles. During pregnancy, estrogen production in the placenta maintains the uterine lining, which protects and nourishes the developing embryo.

Melatonin is crucial to circadian cycles

Another major endocrine gland is the **pineal gland,** located in the roof of the third ventricle of the brain in most vertebrates (see figure 44.22). It is about the size of a pea and is shaped like a pinecone, which give it its name.

The pineal gland evolved from a medial light-sensitive eye (sometimes called a "third eye," although it could not form images) at the top of the skull in primitive vertebrates. This pineal eye is still present in primitive fish (cyclostomes) and some modern reptiles. In other vertebrates, however, the pineal gland is buried deep in the brain, and it functions as an endocrine gland by secreting the hormone melatonin.

Melatonin was named for its ability to cause blanching of the skin of lower vertebrates by reducing the dispersal of melanin granules. We now know, however, that it serves as an important timing signal delivered through the blood. Melatonin levels in the blood increase in darkness and fall during the daytime.

The secretion of melatonin is regulated by activity of the *suprachiasmatic nucleus* (*SCN*) of the hypothalamus. The SCN is known to function as the major biological clock in vertebrates, entraining (synchronizing) various body processes to a circadian rhythm—one that repeats every 24 hr. Through regulation by the SCN, the secretion of melatonin by the pineal gland is activated in the dark.

This daily cycling of melatonin release regulates sleep/wake and temperature cycles. Disruptions of these cycles, as occurs with jet lag or night shift work, can sometimes be minimized by melatonin administration. Melatonin also helps regulate reproductive cycles in some vertebrate species that have distinct breeding seasons.

Some hormones are not produced by endocrine glands

A variety of hormones are secreted by organs that are not exclusively endocrine glands. The thymus is the site of T cell production in many vertebrates and T cell maturation in mammals. It also secretes a number of hormones that function in the regulation of the immune system.

The right atrium of the heart secretes *atrial natriuretic hormone,* which stimulates the kidneys to excrete salt and water in the urine. This hormone acts antagonistically to aldosterone, which promotes salt and water retention.

The kidneys secrete *erythropoietin,* a hormone that stimulates the bone marrow to produce red blood cells. Other organs, such as the liver, stomach, and small intestine, also secrete hormones, and as mentioned earlier, the skin secretes vitamin D.

Insect hormones control molting and metamorphosis

Most invertebrate groups produce hormones as well; these control reproduction, growth, and color change. A dramatic action of hormones in insects is similar to the role of thyroid hormones in amphibian metamorphosis.

As insects grow during postembryonic development, their hardened exoskeletons do not expand. To overcome this problem, insects undergo a series of **molts** wherein they shed their old exoskeleton (figure 46.16) and secrete a new, larger one. In some insects, a juvenile insect, or larva, undergoes a radical transformation to the adult form during a single molt. This process is called **metamorphosis.**

figure 46.16

A MOLTING CICADA. This adult insect is emerging from its old cuticle. Molting is under hormonal control.

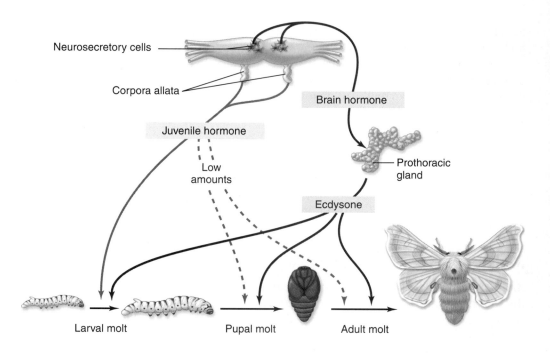

figure 46.17

HORMONAL CONTROL OF METAMORPHOSIS IN THE SILKWORM MOTH, *Bombyx mori.* Molting hormone, ecdysone, controls when molting occurs. Brain hormone stimulates the prothoracic gland to produce ecdysone. Juvenile hormone determines the result of a particular molt. Juvenile hormone is produced by bodies near the brain called the corpora allata. High levels of juvenile hormone inhibit the formation of the pupa. Low levels of juvenile hormone are necessary for the pupal molt and metamorphosis.

Neurosecretory cells

Corpora allata

Juvenile hormone

Low amounts

Brain hormone

Prothoracic gland

Ecdysone

Larval molt   Pupal molt   Adult molt

Hormonal secretions influence both molting and metamorphosis in insects. Prior to molting, neurosecretory cells on the surface of the brain secrete a small peptide, **prothoracicotropic hormone (PTTH),** which in turn stimulates a gland in the thorax called the prothoracic gland to produce **molting hormone,** or **ecdysone** (figure 46.17). High levels of ecdysone bring about the biochemical and behavioral changes that cause molting to occur.

Another pair of endocrine glands near the brain, called the *corpora allata,* produce a hormone called **juvenile hormone.** High levels of juvenile hormone prevent the transformation to the adult and result in a larval-to-larval molt. If the level of juvenile hormone is low, however, the molt will result in metamorphosis.

Cancer cells may alter hormone production or have altered hormonal responses

Hormones and paracrine secretions actively regulate growth and cell division. Normally, hormone production is kept under precise control, but malfunctions in signaling systems can sometimes occur. Unregulated hormone stimulation can then lead to serious physical consequences.

Tumors that develop in endocrine glands, such as the anterior pituitary or the thyroid, can produce excessive amounts of hormones, causing conditions such as gigantism or hyper-

thyroidism. Spontaneous mutations can damage receptors or intracellular signaling proteins, with the result that target cell responses are activated even in the absence of hormone stimulation. Mutations in growth factor receptors, for example, can activate excessive cell division, resulting in tumor formation. Some tumors that develop in steroid-responsive tissues, such as the breast and prostate, remain sensitive to hormone stimulation. Blocking steroid hormone production can therefore diminish tumor growth.

The important effects of hormones on development and differentiation are illustrated by the case of diethystilbestrol (DES). DES is a synthetic estrogen that was given to pregnant women from 1940 to 1970 to prevent miscarriage. It was subsequently discovered that daughters who had been exposed to DES as fetuses had an elevated probability of developing a rare form of cervical cancer later in life. Developmental alterations elicited by hormone treatment may thus take many years to become apparent.

Sex steroid hormones from the gonads regulate reproduction, melatonin secreted by the pineal gland helps regulate circadian rhythms, and thymus hormones help regulate the vertebrate immune system. Metamorphosis is regulated by thyroid hormone in amphibians and by ecdysone and juvenile hormone in insects. In cancer, hormone production or hormone sensitivity may be altered.

46.1 Regulation of Body Processes by Chemical Messengers

Chemicals are used to communicate between cells, within an organism, and between organisms.

- Hormones are signaling molecules carried by the blood.
- Pheromones are chemicals released into the environment to communicate between individuals of the same species.
- Some neurotransmitters are distributed by the blood and act as neurohormones.
- Hormone production and release is often controlled directly or indirectly by the nervous system.
- The endocrine system is composed of endocrine glands that produce three classes of hormones: peptides and proteins, amino acid derivatives, and steroids (table 46.1).
- Exocrine glands secrete substances directly into a duct that transports them out of the body.
- Hormones may be either lipophilic (nonpolar, fat-soluble) or hydrophilic (polar, water-soluble).
- Hormonal molecules are destroyed or deactivated after use and excreted through bile or urine.
- Paracrine regulation occurs in most organs and among cells of the immune system.
- Prostaglandins are a diverse group of fatty acids involved in the inflammatory response.
- Prostaglandin synthesis is a target of many pain relievers and nonsteroidal antiinflammatory drugs.

46.2 Actions of Lipophilic Versus Hydrophilic Hormones

The receptors for, and the action of, lipophilic and hydrophilic hormones are notably different.

- Lipophilic hormones, such as steroids and the thyroid hormones, pass through the plasma membrane and activate intracellular receptors.
- Circulating lipophilic hormones are transported bound to transport proteins (figure 46.3).
- Depending on the steroid hormone it can bind with an intracellular receptor protein and then move into the nucleus or diffuse directly into the nucleus to bind a nuclear receptor (figure 46.5).
- Steroid hormone receptors act by binding to promoter regions of specific genes known as hormone response elements to activate transcription.
- Hydrophilic hormones activate receptors located on the outer surface of target cell membranes. This binding initiates signal transduction pathways that may produce a second messenger (figure 46.6).
- Receptor tyrosine kinase can phosphorylate other proteins on ligand binding.
- Receptor tyrosine kinases can activate signaling through the MAP kinase cascade, which involves sequential activation of kinase enzymes.
- When a hormone binds to a GPCR, the G protein activates an enzyme that generates a second messenger such as cAMP.
- The binding of a hydrophilic hormone to a receptor is short term, and the hormone is deactivated.

46.3 The Pituitary and Hypothalamus: The Body's Control Centers

The main body control areas are the compound endocrine pituitary gland and the hypothalamus of the central nervous system.

- The pituitary gland, also known as the hypophysis, hangs from the base of the hypothalamus.
- The anterior pituitary (adenohypophysis) is composed of glandular tissue. The posterior pituitary (neurohypophysis) releases neurohormones produced in the hypothalamus.
- The anterior pituitary produces seven hormones that stimulate target tissue growth or other endocrine glands (table 46.1).
- Hypothalamic neurohormones pass to the anterior pituitary through a portal system and regulate the anterior pituitary by specific releasing and inhibiting hormones (figure 46.8).
- The activity of the anterior pituitary hormones is usually regulated by negative feedback (figure 46.9).

46.4 The Major Peripheral Endocrine Glands

Major peripheral endocrine glands do not include the pituitary gland, and they are found in other locations.

- Some endocrine glands are controlled by tropic hormones of the pituitary, others are independent of pituitary control.
- The thyroid gland secretes thyroxine and triiodothyronine to regulate basal metabolism in vertebrates and trigger metamorphosis in amphibians (figure 46.12).
- Blood calcium is regulated by calcitonin, which lowers blood calcium levels, and parathyroid hormone, which raises blood calcium levels (figure 46.13).
- Catecholamines, epinephrine and norepinephrine, from the adrenal medulla trigger "alarm" responses (figure 46.14).
- Glucocorticosteroids maintain glucose homeostasis and modulate some aspects of the immune response.
- Mineralocorticoids such as aldosterone regulate mineral balance by stimulating the kidneys to reabsorb sodium and excrete potassium.
- The pancreas secretes insulin, which reduces blood glucose and glucagon, which raises blood glucose (figure 46.15).

46.5 Other Hormones and Their Effects

A variety of vertebrate and invertebrate processes are regulated by hormones and other chemical messengers.

- The gonads produce sex steroids that regulate, among other things, secondary sex characteristics.
- The ovaries produce estrogen and progesterone, which in mammals are responsible for female secondary sex characteristics, follicle and uterine development, and libido.
- The testes produce testosterone, which is responsible for male secondary sex characteristics, sperm production, and libido.
- The pineal gland produces melatonin, which can control the dispersion of pigment granules and daily wake–sleep cycles.
- The thymus secretes a number of hormones that function in the regulation of the immune system.
- The right atrium of the heart secretes atrial natriuretic hormone, which acts antagonistically to aldosterone.
- The kidneys also secrete erythropoietin, a hormone that stimulates the bone marrow to make red blood cells.
- Other organs, such as the liver, stomach, and small intestine, also secrete hormones, and the skin secretes vitamin D.
- In insects the hormone ecdysone stimulates molting, and juvenile hormone levels control the nature of the molt.
- Tumors in endocrine glands can alter hormone production.

SELF TEST

1. Which of the following best describes hormones?
   a. Hormones are relatively unstable and work only in the area adjacent to the gland that produced them.
   b. Hormones are long-lasting chemicals released from glands.
   c. All hormones are lipid-soluble.
   d. Hormones are chemical messengers that are released into the environment.

2. You suspect that one of your teammates is using anabolic steroids to build muscle. You know that continued use of steroids can cause profound changes in cell function. This is due in part to the fact that the receptors for steroid hormones lie—
   a. in the cytoplasm or nucleus.
   b. within the plasma membrane.
   c. within the mitochondria.
   d. in the blood plasma.

3. The adrenal cortex releases _____ , which stimulates Na$^+$ reabsorption by the kidneys.
   a. epinephrine
   b. aldosterone
   c. glucose
   d. cortisol

4. What is a neurohormone?
   a. A hormone that affects only the central nervous system
   b. A hormone that is produced by the nervous system
   c. A hormone that mimics the effect of a specific neurotransmitter
   d. A neurotransmitter that acts as a hormone

5. Your Uncle Sal likes to party. When he goes out drinking, he complains that he needs to urinate more often. You explain to him that this is because alcohol suppresses the release of the hormone _____ , which regulates the kidney's retention of water.
   a. prolactin
   b. oxytocin
   c. thyroxine
   d. vasopressin (ADH)

6. Second messengers are activated in response to—
   a. steroid hormones.
   b. thyroxine.
   c. peptide hormones
   d. all of these.

7. Your new research project is to design a pesticide that will disrupt the endocrine systems of arthropods without harming humans and other mammals. Which of the following substances should be the target of your investigations?
   a. Insulin
   b. Vasopressin
   c. Juvenile hormone
   d. Cortisol

8. At dinner last night you ate a well-balanced meal that included a portion of steak, a food known to be rich in cholesterol and protein. Having just read this chapter, you remark to your friends that you should be able to manufacture plenty of which of the following types of hormones?
   a. Peptide hormones
   b. Steroid hormones
   c. Both peptide and steroid hormones
   d. Neither peptide nor steroid hormones

9. Which of the following is true about lipophilic hormones?
   a. They are freely soluble in the blood.
   b. They require a transport protein in the bloodstream.
   c. They cannot enter their target cells.
   d. They are rapidly deactivated after binding to their receptors.

10. What do growth factors, cytokines, and prostaglandins have in common?
   a. They are all steroid hormones.
   b. They are all amino acid- (phenylalanine) based hormones.
   c. They are all peptide/protein hormones.
   d. They are all paracrines or local regulators.

11. An organ is classified as part of the endocrine system if—
   a. it produces cholesterol.
   b. it is capable of converting amino acids into hormones.
   c. it has intracellular receptors for hormones.
   d. it secretes hormones into the circulatory system.

12. Which of these can easily pass through the plasma membrane?
   a. Thyroid hormones
   b. Estrogen
   c. Epinephrine
   d. Both (a) and (b)

13. Hormones released from the pituitary gland have two different sources. Those that are produced by the neurons of the hypothalamus are released through the _____ , and those produced within the pituitary are released through the _____ .
   a. thalamus, hippocampus
   b. neurohypophysis, adenohypophysis
   c. right pituitary, left pituitary
   d. cortex, medulla

14. Which of the following conditions is unrelated to the production of growth hormone?
   a. Control of blood calcium
   b. Pituitary dwarfism
   c. Increased milk production in cows
   d. Acromegaly

15. Epinephrine, norepinephrine, and glucocorticoids—
   a. are all hydrophilic hormones.
   b. are all released by the anterior pituitary.
   c. are all used to mobilize nutrients during stress reactions.
   d. are all precursors of the sex steroids.

CHALLENGE QUESTIONS

1. How can blocking hormone production decrease cancerous tumor growth?

2. Suppose that two different organs, such as the liver and heart, are sensitive to a particular hormone (such as epinephrine). The cells in both organs have identical receptors for the hormone, and hormone-receptor binding produces the same intracellular second messenger in both organs. However, the hormone produces different effects in the two organs. Explain how this can happen.

3. Many physiological parameters, such as blood Ca^{2+} concentration and blood glucose levels, are controlled by two hormones that have opposite effects. What is the advantage of achieving regulation in this manner instead of by using a single hormone that changes the parameters in one direction only?

The Musculoskeletal System

introduction

THE ABILITY TO MOVE is so much a part of our daily lives that we tend to take it for granted. It is made possible by the combination of a semirigid skeletal system, joints that act as hinges, and a muscular system that can pull on this skeleton. Animal locomotion can be thought of as muscular action that produces a change in body shape, which places a force on the outside environment. When a race horse runs down the track, its legs move forward and backward. As its feet contact the ground, the force they exert move its body forward at a considerable speed. In a similar way, when a bird takes off into flight, its wings exert force on the air; a swimming fish's movements push against the water. In this chapter, we will examine the nature of the muscular and skeletal systems that allow animal movement.

Types of Skeletal Systems

Muscles have to pull against something to produce the changes that cause movement. This necessary form of supporting structure is called a skeletal system. Zoologists commonly recognize three types of skeletal systems in animals: *hydrostatic skeletons*, *exoskeletons*, and *endoskeletons*.

Hydrostatic skeletons use water pressure inside a body wall

Hydrostatic skeletons are found primarily in soft-bodied terrestrial invertebrates, such as earthworms and slugs, and soft-bodied aquatic invertebrates, such as jellyfish, squids, octopuses, and so on.

Musculoskeletal action in earthworms

In these animals a fluid-filled central cavity is encompassed by two sets of muscles in the body wall: circular muscles that are repeated in segments and run the length of the body and longitudinal muscles that oppose the action of the circular muscles.

Muscles act on the fluid in the body's central space, which represents the hydrostatic skeleton. As locomotion begins (figure 47.1) the anterior circular muscles contract, pressing on the inner fluid, and forcing the front of the body to become thin as the body wall in this region extends forward.

On the underside of a worm's body are short, bristle-like structures called **chaetae.** When circular muscles act, the chaetae of that region are pulled up close to the body and lose contact with the ground. Circular-muscle activity is passed backward, segment by segment, to create a backward wave of contraction.

As this wave continues, the anterior circular muscles now relax, and the longitudinal muscles take over, thickening the front end of the worm and allowing the chaetae to protrude and regain contact with the ground. The chaetae now prevent that body section from slipping backward. This locomotion process proceeds as waves of circular muscle contraction are followed by waves of longitudinal muscle effects.

Musculoskeletal action in aquatic invertebrates

Some marine animals use a fluid-based movement that does not fit perfectly under the definition of locomotion through means of a hydrostatic skeleton. Ejection of fluid taken in from the environment is used in these cases, using a method that some call "jetting." For example, the locomotion system of jellyfishes (chapter 33) includes the bell of the animal, the contractile fibers around the edge of the bell, and a jellylike mass, the **mesoglea,** that occupies much of the animal's interior (figure 47.2*a*). The contractile fibers produce regular pulsations that squeeze the bell. Each such contraction presses out some of the water contained beneath it.

This jetting action of the water is not very forceful, but it keeps the jellyfish afloat and can also carry it horizontally through the water depending on the tilt of the bell. Each contraction also squeezes the mesoglea, which rebounds elastically between pulses, and "resets" the bell to its fully open shape.

The gentle movements of a jellyfish are nothing compared with what a squid can do. Squid use undulating movements of their fins for routine swimming. But they can also engage in strong jet-swimming that, as in the case of jellyfish, does not use internal body fluids but seawater from the external environment. Inside the mantle is a large, central cavity that can be filled with water (figure 47.2*b*). The entrance into the cavity is not through the siphon, but through a relatively small intake opening just next to it, between the head and the mantle.

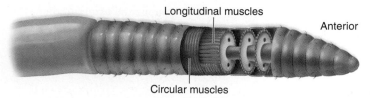

Longitudinal muscles

Anterior

Circular muscles

Longitudinal muscles contracted

Longitudinal muscles contract, and segments catch up

Circular muscles contracted

Circular muscles contract, and anterior end moves forward

Circular muscles contract, and anterior end moves forward

figure 47.1

LOCOMOTION IN EARTHWORMS. The hydrostatic skeleton of the earthworm uses muscles to move fluid within the segmented body cavity, changing the shape of the animal. When circular muscles contract the pressure in the fluid rises. At the same time the longitudinal muscles relax, and the body becomes longer and thinner. When the longitudinal muscles contract and the circular muscles relax, the chaetae of the worm's lower surface extend to prevent backsliding. A wave of circular followed by longitudinal muscle contractions down the body produces forward movement.

a.

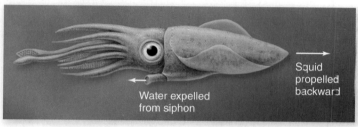

b.

figure 47.2

JET SWIMMERS, JELLYFISH, AND SQUID. *a.* Jellyfish can swim through pulsations in the bell, squeezing against the seawater beneath it. *b.* Squid can fill mantle cavity with seawater and expel it forcibly through the siphon, to jet-swim backward.

Muscles running within the mantle have the effect of thinning the mantle walls, and thus expanding the inner cavity. Water is taken in, but is prevented from leaking back out again by a valve in the intake entrance. Circular muscles within the mantle walls then contract, expelling water forcefully through the siphon, and the animal shoots backward. Jetting is typically the method of making a rapid escape, and the circular muscles contract with so much power that the acceleration of some species of squid may be nearly as great as some of the fastest marine fishes.

Exoskeletons consist of a rigid outer covering

Exoskeletons surround the body as a rigid, hard case. Arthropods, such as crustaceans and insects, have exoskeletons made of the polysaccharide *chitin* (figure 47.3*a*). As you learned in earlier chapters, chitin is found in the cell walls of fungi and some protists as well as in the exoskeletons of arthropods.

A chitinous exoskeleton resists bending and thus acts as the skeletal framework of the body; it also provides protection of the internal organs and attachment sites for the muscles,

which lie inside the exoskeletal casing. But in order to grow, the animal must periodically molt, shedding the exoskeleton (see chapter 34). The animal is vulnerable to predation until the new (slightly larger) exoskeleton forms. Molting crabs and lobsters often hide until the process is completed. If these animals were not buoyed up by water, the larger, heavier ones would collapse under their own weight during this period.

Exoskeletons have other disadvantages. The chitinous framework is not as strong as a bony, internal one. This fact by itself would set a limit for insect size, but there is a more important limitation: Insects breathe through openings in their body that lead into tiny tubes, and as insect size increases beyond a certain limit, the ratio between the inside surface area of the tubes and the volume of the body overwhelms this sort of respiratory system. Finally, when muscles are confined within an exoskeleton, they cannot enlarge in size and power with increased use, as they can in animals with endoskeletons.

Endoskeletons are composed of hard, internal structures

Endoskeletons, found in vertebrates and echinoderms, are rigid internal skeletons that form the body's framework, and offer surfaces for muscle attachment. Echinoderms, such as sea urchins

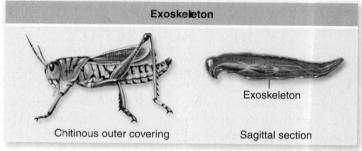

a.

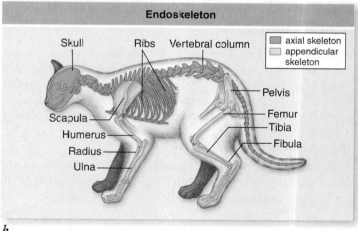

b.

figure 47.3

EXOSKELETON AND ENDOSKELETON. *a.* The hard, tough outer covering of an arthropod, such as this grasshopper, is its exoskeleton and is composed of chitin. *b.* Vertebrates, such as this cat, have endoskeletons formed of bone and cartilage. Some of the major bony features are labeled.

and sand dollars, have skeletons made of calcite, a crystalline form of calcium carbonate. This calcium compound is different from that in bone, which is based on calcium phosphate.

Vertebrate skeletal tissues

The vertebrate endoskeleton includes fibrous dense connective tissue along with the more rigid special connective tissues, cartilage or bone (see chapter 43). Cartilage is strong and slightly flexible, a characteristic important in such functions as padding the ends of bones where they come together in a joint. Although some large, active animals such as sharks have totally cartilaginous skeletons, bone is the main component in vertebrate skeletons. Bone is much stronger than cartilage and much less flexible.

Unlike chitin, both cartilage and bone are living tissues. Bone particularly can have high metabolic activity, especially if bone cells are present throughout the matrix, a common but not universal condition. Bone, and to some extent cartilage, can change and remodel itself in response to injury or to physical stresses.

Vertebrate skeletal structure

Vertebrate endoskeletons are divided into axial and appendicular portions (figure 47.3b). As the name implies, the bones of the **axial skeleton** form the axis of the body beginning with

the skull in front and ending with the tail behind. Besides the skull, the axial portion includes vertebrae, ribs, and the sternum. It supports the body as a basic framework, allows for jaw movements and body bending, and also affords protection of the organs of the head, neck, and trunk.

The **appendicular skeleton** is the set of limb bones (or in fishes, fin bones) and their associated *pectoral girdle* (forelimbs) or *pelvic girdle* (hindlimbs). In bony fishes, the pectoral girdle is connected to the skull. This attachment is lost in terrestrial vertebrates, whose pectoral girdles are tied to the ribs through fibrous connective tissue and muscle. In those same terrestrial vertebrates, the pelvic girdle is connected to one or more specialized sacral vertebrae. This is never the case in fishes, which lack any sacral specialization of the vertebral column.

> The three types of animal skeletons are the hydrostatic skeleton, exoskeleton, and endoskeleton. Hydrostatic skeletons are found in worms and marine invertebrates. Invertebrate exoskeletons are composed of chitin and as the animal grows must be periodically renewed by molting. Endoskeletons in vertebrates are composed of fibrous dense connective tissue, bone, and/or cartilage, and they are typically organized into axial and appendicular portions.

47.2 A Closer Look at Bone

Bone is a hard but resilient tissue that is unique to vertebrate animals. This connective tissue first appeared over 520 MYA in strange little eel-like creatures called conodonts. Since that time, bone has been found in many diverse fossil vertebrates, including the "true eels," a subgroup of bony fishes that arose millions of years after the conodonts.

Bones can be classified by two modes of development

Bone tissue itself can be of several types classified in a few different ways. The most common system is based on the way in which bone develops. The two fundamental methods of bone development are *intramembranous development*, the simplest method, and *endochondral development*, which is more complex.

Intramembranous development

In **intramembranous development,** bones form within a layer of connective tissue. Many of the bones that make up the skull, particularly its external portions, are intramembranous.

Typically, the site of the intramembranous bone-to-be begins in a designated region in the dermis of the skin. During embryonic development, the dermis is formed largely of **mesenchyme,** a loose tissue consisting of undifferentiated mesenchyme cells and other cells that have arisen from them, along with collagen fibers. Some of the undifferentiated mesenchyme cells differentiate to become specialized cells called **osteoblasts** (figure 47.4). These osteoblasts arrange themselves along the collagenous fibers and begin to

secrete the enzyme alkaline phosphatase, which causes calcium phosphate salts to form in a crystalline configuration called **hydroxyapatite.** The crystals merge along the fibers to encase them.

The crystals give the bone its hardness, but without the resilience afforded by the collagen, bone would be rigid but dangerously brittle. Typical bones have roughly equal volumes of collagen and hydroxyapatite, but hydroxyapatite contributes about 65% to the bone's weight.

As the osteoblasts continue to make bone crystals, some become trapped in the bone matrix and undergo dramatic changes in shape and function, now becoming cells called **osteocytes** (figure 47.4). They lie in tight spaces within the bone matrix called **lacunae.** Little canals extending from the lacunae, called **canaliculi,** permit contact of the starburstlike extensions of each osteocyte with those of its neighbors (figure 47.4). In this way, many cells within bones can participate in intercellular communication.

As an intramembranous bone grows, it requires alterations of shape. Imagine that you were modeling with clay, and you wanted to take a tiny clay bowl and make it larger. Simply putting more clay on the outside would not work; you would need to remove clay from the inside to increase the bowl's capacity as well. As bone grows, it must also undergo a remodeling process, with matrix being added in some regions and removed in others. This is where **osteoclasts** come in. These unusual cells are formed from the fusion of monocytes, a type of white blood cell, to form large multinucleate cells. Their function is to break down the bone matrix.

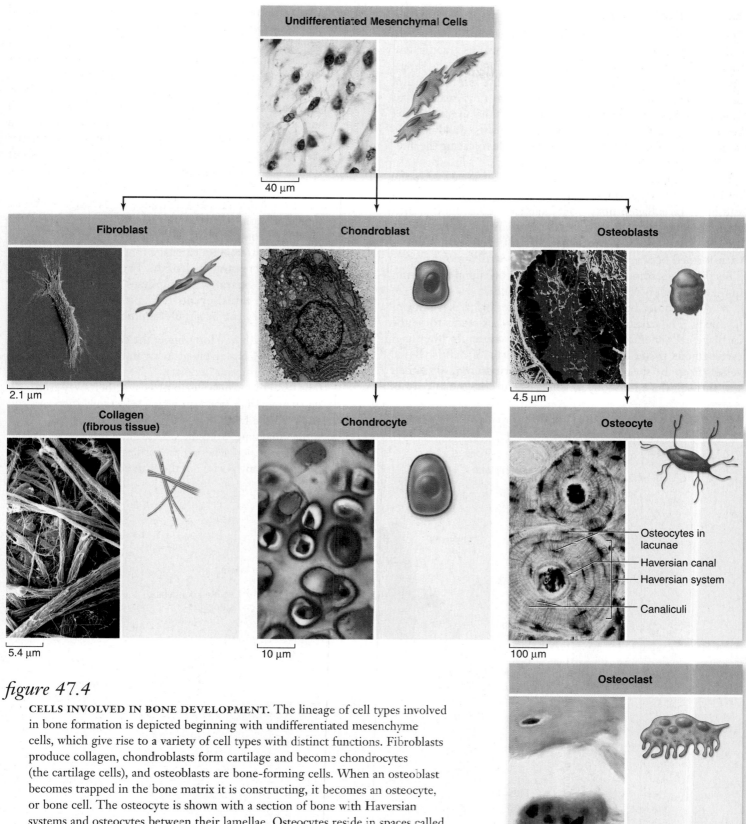

figure 47.4

CELLS INVOLVED IN BONE DEVELOPMENT. The lineage of cell types involved in bone formation is depicted beginning with undifferentiated mesenchyme cells, which give rise to a variety of cell types with distinct functions. Fibroblasts produce collagen, chondroblasts form cartilage and become chondrocytes (the cartilage cells), and osteoblasts are bone-forming cells. When an osteoblast becomes trapped in the bone matrix it is constructing, it becomes an osteocyte, or bone cell. The osteocyte is shown with a section of bone with Haversian systems and osteocytes between their lamellae. Osteocytes reside in spaces called lacunae. Small canals (canaliculi) radiate out from the central lacunar space, which contains the arms of the osteocyte. Osteoclasts, bone-removing cells, are not derived from mesenchyme cells but are formed by fusion of monocytes, a type of white blood cell.

Endochondral development

Bones that form through **endochondral development** are typically those that are deeper in the body and form its architectural framework. Examples include vertebrae, ribs, bones of the shoulder and pelvis, long bones of the limbs, and the most internal of the skull bones. Endochondral bones begin as tiny, cartilaginous models that have the rough shape of the bones that eventually will be formed. Bone development of this kind consists of adding bone to the outside of the cartilaginous model, while replacing the interior cartilage with bone.

Bone added to the outside of the model is produced in the fibrous sheath that envelopes the cartilage. This sheath is tough and made of collagen fibers, but it also contains undifferentiated mesenchyme cells. Osteoblasts arise and sort themselves out along the fibers in the deepest part of the sheath. Bone is then formed between the sheath and the cartilaginous matrix. This process is somewhat similar to what occurs in the dermis in the production of intramembranous bone.

As the outer bone is formed, the interior cartilage begins to calcify. The calcium source for this process seems to be the cartilage cells themselves. As calcification continues, the inner cartilaginous tissue breaks down into pieces of debris. Blood vessels from the sheath, now called the **periosteum,** force their way through the outer bony jacket, thus entering the interior of the cartilaginous model, and cart off the debris. Again, trapped osteoblasts transform into osteocytes, and osteoclasts for bone remodeling arise from cell fusions in the same manner as occurs in intramembranous bone. Growth in bone thickness occurs by adding additional bone layers just beneath the periosteum whereas the medullary cavity is hollowed out by osteoclasts. The increasing length of endochondral bones, however, is an intriguing and complicated process. As an example, consider a long bone such as a mammalian humerus (in humans, the upper arm bone). Like many limb bones, it is formed of a slender **shaft** with widened ends, called **epiphyses** (figure 47.5). The cartilage that remains after replacement by bone is crucial for the bone's continued growth and proper function. The cartilage left at the surfaces of the epiphyses provides a padding to the ends of the bone, which come into contact with other bones in joints.

Other regions contain cartilage that remains only as long as the bone is increasing in length. These are the *epiphyseal growth plates* that separate the epiphyses from the shaft itself (figure 47.5). The actual events taking place in the plates are not simple, but they can be simply summarized.

1. During growth of a long bone, the cartilage of the growth plates is actively growing in the lengthwise direction to thicken the plate.
2. This growth pushes the epiphysis farther away from the slender shaft portion, which effectively increases the length of the bone.
3. At the same time, from the shaft's side, a process of cartilage calcification encroaches on the cartilaginous growth plate, so that the bony portion of the shaft elongates.

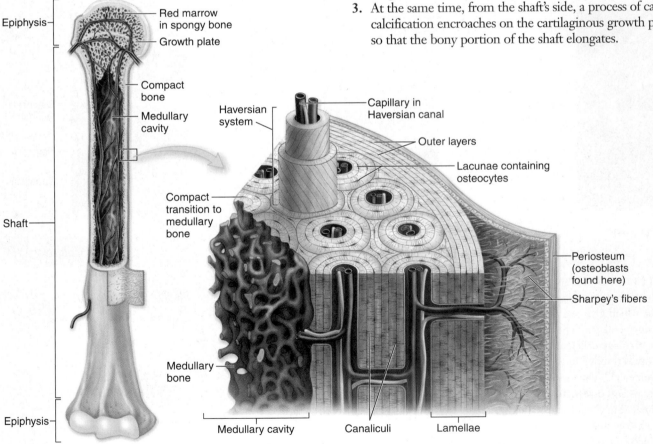

figure 47.5

THE STRUCTURE OF BONE. A mammalian humerus is partly opened to show its interior on the left. A section has been removed and magnified on the right to show the difference in structure between the outer compact bone and the inner spongy bone that lines the medullary cavity. Details of basic layers, Haversian canals, and osteocytes in lacunae can be seen here.

As long as the rate of new cartilage thickening stays ahead of the creeping calcification, the bone continues to grow in length. Eventually the cartilaginous expansion slows, and is overtaken by the calcification, which obliterates this region of growth.

Growth in length usually ceases in humans by late adolescence. Although growth of the bone length is curtailed at this time, growth in width is not. The diameter of the shaft can be enhanced by bone addition just beneath the periosteum.

Bone structure may include blood vessels and nerves

Developing bone often has an internal blood supply, which is especially evident in endochondral bones. The internal blood routes, however, do not necessarily remain after the bones have completed development. In most mammals the endochondral bones retain internal blood vessels and are called **vascular bones.** Vascular bone is also found in many reptiles and a few amphibians. *Cellular bones* contain osteocytes, and many such bones are also vascular. This bone remains metabolically active (see figure 47.5).

In fishes and birds, bones are **avascular.** Typically avascular bone does not contain osteocytes and is termed *acellular bone.* This type of bone is fairly inert except for its surface, where the periosteum with its mesenchyme cells is capable of repairing the bone.

Many bones, particularly the endochondral long bones, contain a central cavity termed the *medullary cavity.* In many vertebrates, the medullary cavity houses the bone marrow, important in the manufacture of red blood cells. In such cases this cavity is termed the **marrow cavity.** Not all medullary cavities contain marrow, however. Light-boned birds, for example, have huge interior cavities, but they are empty of marrow. Birds depend on stem cells in other body locations to produce red blood cells.

Bone lining the medullary cavities differs from the smooth, dense bone found closer to the outer surface. Based on density and texture, bone falls into three categories: the outer dense **compact bone,** the **medullary bone** that lines the internal cavity, and **spongy bone** that has a honeycomb structure and typically forms the epiphyses inside a thick shell of compact bone. Both compact and spongy bone contribute to a bone's strength. Medullary cavities are lined with thin tissues called the **endosteum,** which contains no collagenous fibers but does possess other constituents including mesenchyme cells.

Vascular bone usually has a special internal organization called the **Haversian system.** Beneath the outer basic layers, endochondral bone is constructed of concentric layers called *Haversian lamellae.* These concentric tubes are laid down around narrow channels called *Haversian canals* that run parallel to the length of the bone. Haversian canals may contain nerve fibers but always contain blood vessels that keep the osteocytes alive even though they are entombed in the bony matrix.

The small vessels within the canals include both arterioles and venules, or more often, capillaries, and they connect to larger vessels that extend internally from both the periosteum and endosteum and that run in canals perpendicular to the Haversian canals.

Bone remodeling allows bone to respond to use or disuse

It is easy to think of bones as being inert, especially since we rarely encounter them except as the skeletons of dead animals.

But just as muscles, skin, and other body tissues may change depending on the stresses of the environment, bone also is a dynamic tissue that can change with demands made on it.

Mechanical stresses such as compression at joints, the forces of muscles on certain portions and features of a bone, and similar effects may all be remodeling factors that not only shape the bone during its embryonic development, but after birth as well. Depending on the directions and magnitudes of forces impinging on a bone, it may thicken; the size and shape of surface features to which muscles, tendons, or ligaments attach may change in size and shape; even the direction of the tiny bony struts that make up spongy bone may be altered. Exercise and frequent use of muscles for a particular task change more than just the muscles; blood vessels and fibrous connective tissue increases, and the skeletal frame becomes more robust through bone thickening and enhancement.

The phenomenon of **remodeling** is known for all bones, but it is easiest to demonstrate in a long bone. Small forces may not have much of an effect on the bone, but larger ones—if frequent enough—can initiate remodeling (figure 47.6). In the example shown, larger compressive forces may tend to bend a bone, even if the bend is imperceptible to the eye. This bending stress promotes bone formation that thickens the compact bone. As the bone becomes thicker the amount of bending is reduced (figure 47.6c). Further bone addition produces sufficient bone thickness to entirely prevent significant bending (figure 47.6d). Once this point has been attained, the bone addition stops. This is another example of a negative-feedback system.

The mechanism of remodeling is not well understood. One possible explanation stems from the structure of hydroxyapatite itself. Many types of crystals can have their structure slightly pushed out of alignment by an outside force, with the

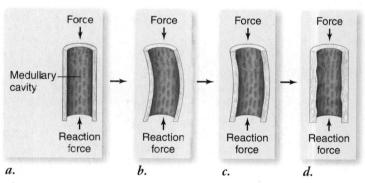

figure 47.6

MODEL OF STRESS AND REMODELING IN A LONG BONE. This figure shows a diagrammatic section of a long bone, such as a leg bone. The section is placed under a load or force, which causes a reaction force from the ground the leg is standing upon. *a.* Under a mild compressive load the bone does not bend. *b.* If the load is large enough, and the compact bone is not sufficiently thick, the bone will bend (the bending shown is exaggerated for clarity). *c.* Osteoblasts are signaled by the stresses in the bending section to produce additional bone. As additional compact bone builds up, the degree of bending is reduced. *d.* When sufficient bone is added to prevent significant bending, the production of new osteoblasts stops and no more bone is added.

result that they lose electrical neutrality and become slightly charged. This phenomenon is termed the **piezoelectric effect.** It is possible that in bone, mechanical stress deforms the hydroxyapatite crystal and produces a piezoelectric effect.

One view is that the charges in some way stimulate the production of osteoblasts, and thus the production of new bone at the site where force is applied. In remodeling, bone may need to be removed from some other region, which would depend on osteoclasts. What exactly would trigger their production is still uncertain.

Bone may be classified on the basis of its mode of development as either intramembranous or endochondral. Bone is formed by osteoblasts that arrange themselves on connective tissue strands and secrete crystals of calcium phosphate called hydroxyapatite. Mature bone consists of a combination of collagen and hydroxyapatite crystals. Many bones have an internal blood circulation and living cells within their substance. Bones undergo reshaping in response to physical stresses.

47.3 Joints and Skeletal Movement

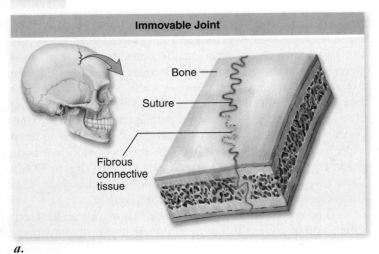
a.

Movements of the endoskeleton are powered by the skeletal musculature. These skeletal muscles are attached to bones by several means. Muscle fibers may connect directly to the periosteum of bones, or the connective tissue sheets within skeletal muscles may join to form a strong, fibrous cord, a **tendon,** that connects the aggregation of muscle fibers to a bone's periosteum. The skeletal movements that respond to muscle action occur at **joints,** or articulations, where one bone meets another.

There are three main classes of joints:

1. **Immovable joints,** which include the *sutures* that join the bones of the human skull (figure 47.7*a*).
2. **Slightly movable joints** can include two general types, based on the binding material between the connected bones. One type involves fibrous connective tissue, such

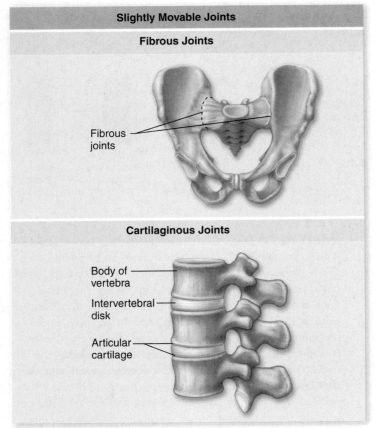
b.

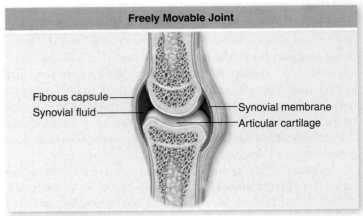
c.

figure 47.7

THREE TYPES OF JOINTS. *a.* The sutures of the skull are an immovable joint. *b.* Slightly movable joints include the sacroiliac joint of the mammalian pelvis, which is fibrous (*top*), and the cartilaginous joints between the vertebrae (*bottom*). *c.* Freely movable joints are synovial joints, such as a finger joint.

as that in the sacroiliac joint, where the pelvis meets the vertebral column. The other employs cartilage; an example is the intervertebral disks between vertebrae in the spine, which provide cushioning (figure 47.7*b*).

3. **Freely movable joints** are common in vertebrates. They are known as *synovial joints* because the articulating ends of different bones are joined to one another with an intervening *synovial capsule* filled with a lubricating fluid. The ends of the bones are capped with cartilage, and the outer wall of the synovial capsule is composed of thick, tough, fibrous tissue that holds the articulating bones in place. Synovial joints include those of limbs, the jaw joint, and fingers and toe joints (figure 47.7*c*).

Moveable joints have different ranges of motion, depending on type

Each movable joint within the skeleton has a characteristic range of motion. Four basic joint movement patterns can be distinguished: *ball-and-socket*, *hinge*, *gliding*, and *combination*.

Ball-and-socket joints are like those of the hip, where the upper leg bone forms a ball fitting into a socket in the pelvis. This type of joint can perform universal movement in all directions, plus twisting of the ball (figure 47.8*a*).

The simplest type of joint is the **hinge joint,** such as the knee, where movement of the lower leg is restricted to rotate forward or backward, but not side to side. The suturelike fibrous joint through the top of the lizard skull is also a hinge joint (figure 47.8*b*).

Gliding joints can be found in the skulls of a number of nonmammalian vertebrates, but are also present between the lateral vertebral projections in many of them and in mammals as well (figure 47.8*c*). The vertebral projections are paired and extend from the front and back of each vertebra. The projections in front are a little lower, and each can slip along the undersurface of the posterior projection from the vertebra just ahead of it. This sliding synovial joint gives stability to the vertebral column while allowing some flexibility of movement between vertebrae.

Combination joints are, as you might suppose, those that have movement characteristics of two or more joint types. The typical mammalian jaw joint is a good example. Jaw joints in all vertebrates are synovial, but they differ significantly in the kinds of movement that they permit. Reptile jaw joints, for example, function basically as a hinge. Most mammals chew food into small pieces, which is a rare thing for most vertebrates, which can swallow prey whole or in large hunks. To chew food well, the lower jaw needs to move from side to side to get the best contact between upper and lower teeth. The lower jaw can also slip forward and backward to some extent. At the same time, the jaw joint must be shaped to allow the hingelike opening and closing of the mouth. The mammalian joint conformation thus combines features from hinge and gliding joints (figure 47.8*d*).

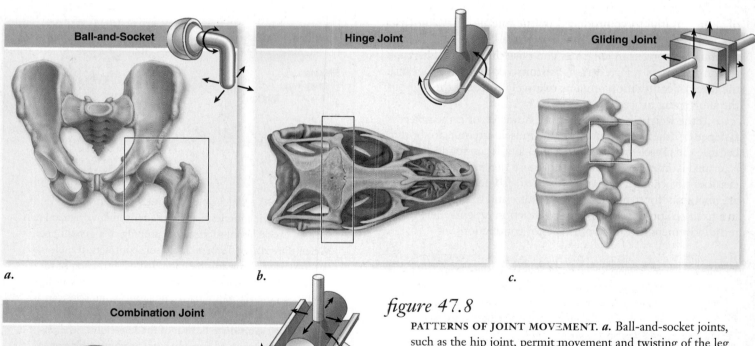

a.   *b.*   *c.*   *d.*

figure 47.8

PATTERNS OF JOINT MOVEMENT. *a.* Ball-and-socket joints, such as the hip joint, permit movement and twisting of the leg within the hip socket. ***b.*** A hinge joint, as the term implies, allows movement in only one plane. Examples include the fibrous joint through the top of a lizard skull (shown here) or a knee joint. ***c.*** Gliding joints are well represented by the lateral vertebral joints (not the central ones) that permit sliding of one surface on another. ***d.*** Combination joints can be represented by several sorts of joints; the one here is the mammalian jaw joint that allows rotation but also side-to-side sliding.

Skeletal muscles pull on bones to produce movement at joints

Skeletal muscles produce movement of the skeleton when they contract. Usually, the two ends of a skeletal muscle are attached to different bones, although some may be attached to other structures, such as skin. There are two means of bone attachment: Muscle fibers may connect directly to the periosteum, the bone's fibrous covering, or the muscle might continue as a dense connective tissue strap or cord, called a *tendon* that attaches to the periosteum (figure 47.9*b*).

One attachment of the muscle, the **origin,** remains relatively stationary during a contraction. The other end, the **insertion,** is attached to a bone that moves when the muscle contracts. For example, contraction of the quadriceps muscles of the leg causes the lower leg to rotate forward relative to the upper leg section.

Typically, muscles are arranged so that any movement produced by one muscle can be reversed by another. The leg flexor muscles (figure 47.9*b*) draw the lower leg back and upward, bending the knee. This muscle group in action would be termed the **agonist.** Its movement is countered by the quadriceps muscles, which is termed the **antagonist** in this case. Notice that the quadriceps could also be termed the agonist, in which case the flexors would be the antagonist. The important concept is that two muscles or muscle groups can be mutually antagonistic, and that generally each agonist has an antagonist.

When you lift a weight by contracting the biceps muscles of your arm, the force produced by the muscle is greater than the force of gravity on the object you are lifting. In this case, the muscle shortens in length as you raise the object. This type of contraction is referred to as **isotonic contraction** because the force of contraction remains relatively constant throughout the shortening process.

If the weight is too great to lift, however, or if the object is fastened or otherwise immovable, the muscle attempts to shorten because the muscle fibers are activated, but it cannot do so. This is termed **isometric contraction** because the length of the muscle does not change as force is exerted. Isometric contractions are important for maintaining our posture and to support objects in a fixed position. Both types of contraction expend energy, and most movement involves both types of contractions.

> Joints confer movability to a rigid skeleton, allowing a range of motion determined by the type of joint, such as ball-and-socket, hinge, gliding, or combination joint. Muscles, positioned across joints, contract and cause the movement of the body. Mutually antagonistic muscles have opposite actions.

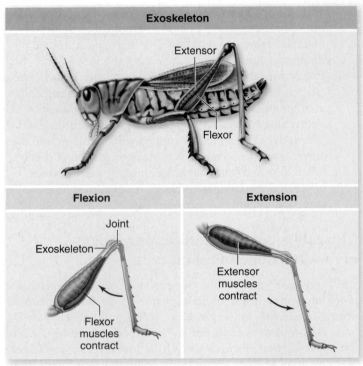

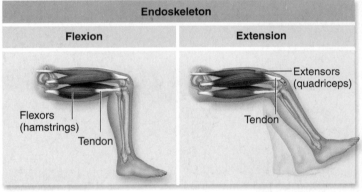

figure 47.9

FLEXOR AND EXTENSOR MUSCLES OF THE LEG.
a. Antagonistic muscles control the movement of an animal with an exoskeleton, such as the jumping of a grasshopper. When the smaller flexor tibia muscle contracts, it pulls the lower leg in toward the upper leg. Contraction of the extensor tibia muscles straightens out the leg and sends the insect into the air. *b.* Similarly, antagonistic muscles can act on an endoskeleton. In humans, the hamstrings, a group of three muscles, produce flexion of the knee joint, whereas the quadriceps, a group of four muscles, produce extension.

47.4 Muscle Contraction

This section concentrates on the skeletal muscle of vertebrates. Vertebrate muscle has enjoyed the most attention and is thus the best understood of animal muscular function. Each skeletal muscle contains numerous **muscle fibers,** as described in chapter 43. Each muscle fiber encloses a bundle of 4 to 20 elongated structures called **myofibrils.** Each myofibril, in turn, is composed of thick and thin **myofilaments** (figure 47.10).

Under a microscope, the myofibrils have alternating dark and light bands, which give a skeletal muscle fiber its striated, or striped, appearance. The thick myofilaments are stacked together to produce the dark bands, called *A bands*; the thin filaments alone are found in the light bands, or *I bands*.

Each I band in a myofibril is divided in half by a disc of protein called a *Z line* because of its appearance in electron micrographs. The thin filaments are anchored to these disks. In an electron micrograph of a myofibril (figure 47.11), the structure of the myofibril can be seen to repeat from Z line to Z line. This repeating structure, called a **sarcomere,** is the smallest subunit of muscle contraction.

Muscle fibers contract as overlapping filaments slide together

The thin filaments stick partway into, and overlap with, thick filaments on each side of an A band, but in a resting muscle, they do not project all the way to the center of the A band. As a result, the center of an A band (called an *H band*) is lighter than the areas on each side, which have interdigitating thick and thin filaments. This appearance of the sarcomeres changes when the muscle contracts.

A muscle contracts and shortens because its myofibrils contract and shorten. When this occurs, the myofilaments do *not* shorten; instead, the thick and thin myofilaments slide relative to each other (see figure 47.11). The thin filaments slide deeper into the A bands, making the H bands narrower until, at maximal shortening, they disappear entirely. This also makes the I bands narrower, as the dark A bands are brought closer together. This is the **sliding filament mechanism** of contraction.

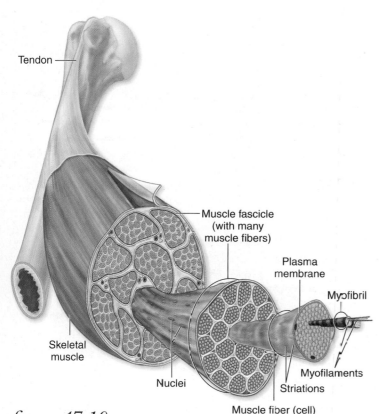

figure 47.10

THE ORGANIZATION OF VERTEBRATE SKELETAL MUSCLE. Each muscle is composed of many fascicles, which are bundles of muscle cells, or fibers. Each fiber is composed of many myofibrils, which are each, in turn, composed of myofilaments.

The sliding filament mechanism

Electron micrographs reveal **cross-bridges** that extend from the thick to the thin filaments, suggesting a mechanism that might cause the filaments to slide. To understand how this is accomplished requires examining the thick and thin filaments at a molecular level. Biochemical studies show that

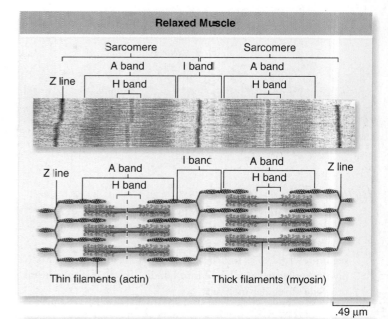

figure 47.11

THE STRUCTURE OF SARCOMERES IN RELAXED AND CONTRACTED MUSCLES. Two sarcomeres are shown in micrographs and as drawings of thick and thin filaments. The Z lines form the borders of each sarcomere and the A bands represent thick filaments. The thin filaments are within the I bands and extend into the A bands interdigitated with thick filaments. The H band is the lighter-appearing central region of the A band containing only thick filaments. The muscle on the top is shown relaxed. In the contracted muscle in the bottom, the Z lines have moved closer together, with the I bands and H bands becoming shorter. The A band does not change in size as it contains the thick filaments, which do not change in length.

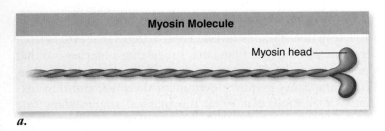

Myosin Molecule

Myosin head

a.

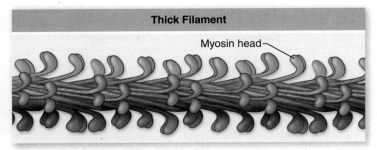

Thick Filament

Myosin head

b.

figure 47.12

THICK FILAMENTS ARE COMPOSED OF MYOSIN. *a.* Each myosin molecule consists of two polypeptide chains wrapped around each other; at the end of each chain is a globular region referred to as the "head." *b.* Thick filaments consist of myosin molecules combined into bundles from which the heads protrude at regular intervals.

each thick filament is composed of many subunits of the protein **myosin** packed together. The myosin protein consists of two subunits, each with a head region that protrudes from a long filament, with the filaments twisted together. Thick filaments are composed of many copies of myosin arranged with heads protruding from along the length of the fiber (figure 47.12). The myosin heads form the cross-bridges seen in electron micrographs.

Each thin filament consists primarily of many globular **actin** proteins arranged into two fibers twisted into a double helix (figure 47.13). If we were able to see a sarcomere at the molecular level, it would have the structure depicted in figure 47.14.

Myosin is a member of the class of protein called *motor proteins* that are able to convert the chemical energy in ATP into mechanical energy (see chapter 4). This occurs by a series of events called the **cross-bridge cycle** (figure 47.15). When the myosin heads hydrolyze ATP into ADP and P_i, the conforma-

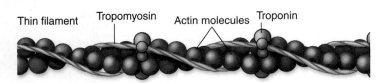

Thin filament   Tropomyosin   Actin molecules   Troponin

figure 47.13

THIN FILAMENTS ARE COMPOSED OF GLOBULAR ACTIN PROTEINS. Two rows of actin proteins are twisted together in a helix to produce the thin filaments. Other proteins, tropomyosin and troponin, associate with the strands of actin and are involved in muscle contraction. These other proteins are discussed later in the chapter.

tion of myosin is changed, activating it for the later power stroke. The ADP and P_i both remain attached to the myosin head, keeping it in this activated conformation. The analogy to a hammer on a revolver being cocked is often used to describe this action. In this cocked position, the myosin heads bind to actin, forming cross-bridges. When a myosin head binds to actin, it releases the ADP and P_i and undergoes another conformational change, pulling the thin filament toward the center of the sarcomere in the *power stroke* (see figures 47.14*b*, 47.15). At the end of the power stroke, the myosin head binds to a new molecule of ATP, which displaces it from actin. This cross-bridge cycle repeats as long as the muscle is stimulated to contract. This sequence of events can be thought of like pulling a rope hand-over-hand. The myosin heads are the hands and the actin fibers the rope.

In death, the cell can no longer produce ATP, and therefore the cross-bridges cannot be broken—causing the muscle stiffness of death called *rigor mortis*. A living cell, however, always has enough ATP to allow the myosin heads to detach from actin. How, then, is the cross-bridge cycle arrested so that the muscle can relax? We discuss the regulation of contraction and relaxation next.

Contraction depends on calcium ion release following a nerve impulse

When a muscle is relaxed, its myosin heads are in the activated conformation bound to ADP and P_i, but they are unable to bind to actin. In the relaxed state, the attachment sites for the myosin heads on the actin are physically blocked

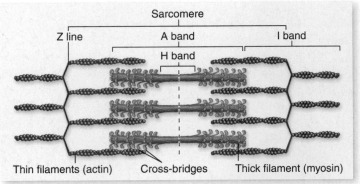

Sarcomere

Z line   A band   I band

H band

Thin filaments (actin)   Cross-bridges   Thick filament (myosin)

a.

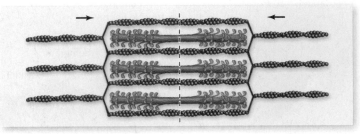

b.

figure 47.14

THE INTERACTION OF THICK AND THIN FILAMENTS IN STRIATED MUSCLE SARCOMERES. *a.* The heads on the two ends of the thick filaments are oriented in opposite directions so that the cross-bridges pull the thin filaments and the Z lines on each side of the sarcomere toward the center. *b.* This sliding of the filaments produces muscle contraction.

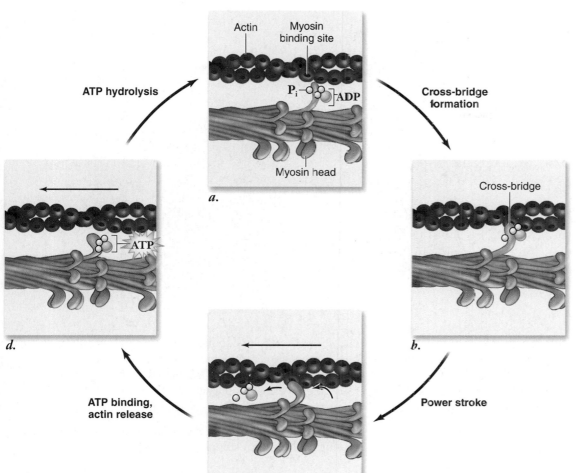

figure 47.15

THE CROSS-BRIDGE CYCLE IN MUSCLE CONTRACTION. *a.* Hydrolysis of ATP by myosin causes a conformational change that moves the head into an energized state. The ADP and P_i remain bound to the myosin head, which can bind to actin. *b.* Myosin binds to actin forming a cross-bridge. *c.* During the power stroke, myosin returns to its original conformation, releasing ADP and P_i. *d.* ATP binds to the myosin head breaking the cross-bridge. ATP hydrolysis returns the myosin head to its energized conformation, allowing the cycle to begin again.

by another protein, known as **tropomyosin,** in the thin filaments. Cross-bridges therefore cannot form and the filaments cannot slide.

For contraction to occur, the tropomyosin must be moved out of the way so that the myosin heads can bind to the uncovered actin-binding sites. This requires the action of **troponin,** a regulatory protein complex that holds tropomyosin and actin

together. The regulatory interactions between troponin and tropomyosin are controlled by the calcium ion (Ca^{2+}) concentration of the muscle fiber cytoplasm.

When the Ca^{2+} concentration of the cytoplasm is low, tropomyosin inhibits cross-bridge formation (figure 47.16a). When the Ca^{2+} concentration is raised, Ca^{2+} binds to troponin, altering its conformation and shifting the troponin–tropomyosin

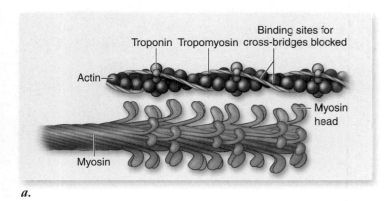

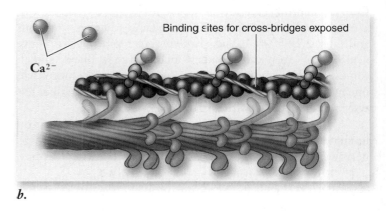

figure 47.16

HOW CALCIUM CONTROLS STRIATED MUSCLE CONTRACTION. *a.* When the muscle is at rest, a long filament of the protein tropomyosin blocks the myosin-binding sites on the actin molecule. Because myosin is unable to form cross-bridges with actin at these sites, muscle contraction cannot occur. *b.* When Ca^{2+} binds to another protein, troponin, the Ca^{2+}–troponin complex displaces tropomyosin and exposes the myosin-binding sites on actin, permitting cross-bridges to form and contraction to occur.

figure 47.17

RELATIONSHIP BETWEEN THE MYOFIBRILS, TRANSVERSE TUBULES, AND SARCOPLASMIC RETICULUM. Neurotransmitter released at a neuromuscular junction binds chemically-gated Na⁺channels, causing the muscle cell membrane to depolarize. This depolarization is conducted along the muscle cell membrane and down the transverse tubules to stimulate the release of Ca^{2+} from the sarcoplasmic reticulum. Ca^{2+} diffuses through the cytoplasm to myofibrils, causing contraction.

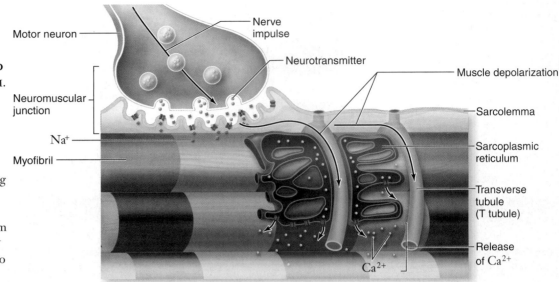

complex. This shift in conformation exposes the myosin-binding sites on the actin. Cross-bridges can thus form, undergo power strokes, and produce muscle contraction (figure 47.16*b*).

Muscles need a reliable supply of Ca^{2+}. Muscle fibers store Ca^{2+} in a modified endoplasmic reticulum called a **sarcoplasmic reticulum (SR)** (figure 47.17). When a muscle fiber is stimulated to contract, the membrane of the muscle fiber becomes depolarized. This is transmitted deep into the muscle fiber by invaginations of the cell membrane called the **transverse tubules (T tubules).** This depolarization of the T tubules causes Ca^{2+} channels in the SR to open, releasing Ca^{2+} into the cytosol. Ca^{2+} then diffuses into the myofibrils, where it binds to troponin, altering its conformation and allowing contraction. The involvement of Ca^{2+} in muscle contraction is called **excitation–contraction coupling** because it is the release of Ca^{2+} that links the excitation of the muscle fiber by the motor neuron to the contraction of the muscle.

Nerve impulses from motor neurons

Muscles are stimulated to contract by motor neurons. The motor neurons that stimulate skeletal muscles are called *somatic motor neurons*. The axon of a somatic motor neuron extends from the neuron cell body and branches to make synapses with a number of muscle fibers. These synapses between neurons and muscle cells are called *neuromuscular junctions* (see figure 47.17). One axon can stimulate many muscle fibers, and in some animals, a muscle fiber may be innervated by more than one motor neuron. However, in humans, each muscle fiber has only a single synapse with a branch of one axon.

When a somatic motor neuron delivers electrochemical impulses, it stimulates contraction of the muscle fibers it innervates (makes synapses with) through the following events:

1. The motor neuron, at the neuromuscular junction, releases the neurotransmitter acetylcholine (ACh). ACh binds to receptors in the muscle cell membrane to open Na⁺ channels. The influx of Na⁺ ions depolarizes the muscle cell membrane.

2. The impulses spread along the membrane of the muscle fiber and are carried into the muscle fibers through the T tubules.

3. The T tubules conduct the impulses toward the sarcoplasmic reticulum, opening Ca^{2+} channels and releasing Ca^{2+}. The Ca^{2+} binds to troponin, exposing the myosin-binding sites on the actin myofilaments and stimulating muscle contraction.

When impulses from the motor neuron cease, it stops releasing ACh, in turn stopping the production of impulses in the

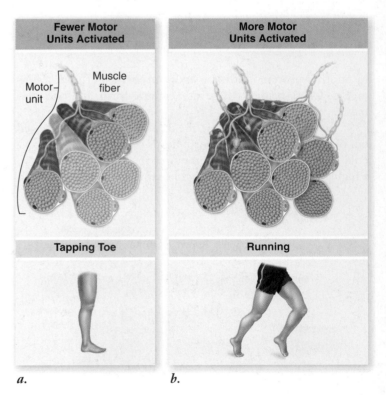

a.   *b.*

figure 47.18

THE NUMBER AND SIZE OF MOTOR UNITS. A motor unit consists of a motor neuron and all of the muscle fibers it innervates. *a.* Precise muscle contractions require smaller motor units. *b.* Large muscle movements require larger motor units. The more motor units activated, the stronger the contraction.

muscle fiber. Another membrane protein in the SR then uses energy from ATP hydrolysis to pump Ca^{2+} back into the SR by active transport. Troponin is no longer bound to Ca^{2+}, so tropomyosin returns to its inhibitory position, allowing the muscle to relax.

Motor units and recruitment

A single muscle fiber can produce variable tension depending on the frequency of stimulation. The response of an entire muscle depends on the number of individual fibers involved and their degree of tension. The set of muscle fibers innervated by all the axonal branches of a motor neuron, plus the motor neuron itself, is defined as a **motor unit** (figure 47.18).

Every time the motor neuron produces impulses, all muscle fibers in that motor unit contract together. The division of the muscle into motor units allows the muscle's strength of contraction to be finely graded, a requirement for coordinated movements. Muscles that require a finer degree of control, such as those that move the eyes, have smaller motor units (fewer muscle fibers per neuron). Muscles that require less precise control but must exert more force, such as the large muscles of the legs, have more fibers per motor neuron.

Most muscles contain motor units in a variety of sizes, and these can be selectively activated by the nervous system. The weakest contractions of a muscle involve activation of a few small motor units. If a slightly stronger contraction is necessary, additional small motor units are also activated. The initial increments of increased force are therefore relatively small. As ever greater forces are required, more units and larger units are brought into action, and the force increments become larger. This cumulative increase of numbers and sizes of motor units to produce a stronger contraction is termed **recruitment**.

The two main types of muscle fibers are slow-twitch and fast-twitch

An isolated skeletal muscle can be studied by stimulating it artificially with electric shocks. A muscle stimulated with a single electric shock quickly contracts and relaxes in a response called a **twitch.** Increasing the stimulus voltage increases the strength of the twitch up to a maximum. If a second electric shock is delivered immediately after the first, it produces a second twitch that may partially "ride piggyback" on the first. This cumulative response is called **summation** (figure 47.19).

An increasing frequency of electric shocks shortens the relaxation time between successive twitches as the strength of contraction increases. Finally, at a particular frequency of stimulation, no visible relaxation occurs between successive twitches. Contraction is smooth and sustained, as it is during normal muscle contraction in the body. This sustained contraction is called **tetanus.** (The disease known as tetanus gets its name because the muscles of its victims go into an agonizing state of contraction.)

Skeletal muscle fibers can be divided on the basis of their contraction speed into *slow-twitch*, or *type I*, *fibers* and *fast-twitch*, or *type II*, *fibers*. The muscles that move the eyes, for example, have a high proportion of fast-twitch fibers and reach maximum tension in about 7.3 milliseconds (msec); the soleus muscle in the leg, by contrast, has a high proportion of slow-twitch fibers and requires about 100 msec to reach maximum tension (figure 47.20).

Slow-twitch fibers

Slow-twitch fibers have a rich capillary supply, numerous mitochondria and aerobic respiratory enzymes, and a high concentration of **myoglobin** pigment. Myoglobin is a red pigment similar to the hemoglobin in red blood cells, but its higher affinity for oxygen improves the delivery of oxygen to the slow-twitch fibers. Because of their high myoglobin content, slow-twitch

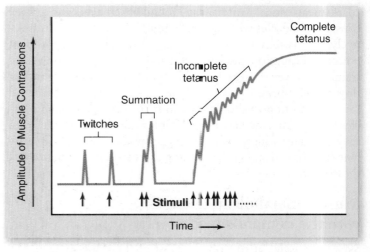

figure 47.19

SUMMATION. Muscle twitches summate to produce a sustained, tetanic contraction. This pattern is produced when the muscle is stimulated electrically or naturally by neurons. Tetanus, a smooth, sustained contraction, is the normal type of muscle contraction in the body.

 inquiry

What determines the maximum amplitude of a summated muscle contraction?

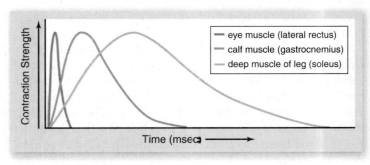

figure 47.20

SKELETAL MUSCLES HAVE DIFFERENT PROPORTIONS OF FAST-TWITCH AND SLOW-TWITCH FIBERS. The muscles that move the eye contain mostly fast-twitch fibers, whereas the deep muscle of the leg (the soleus) contains mostly slow-twitch fibers. The calf muscle (gastrocnemius) is intermediate in its composition.

 inquiry

How would you determine if the calf muscle contains a mix of fast-twitch and slow-twitch fibers, or instead is composed of an intermediate form of fiber?

fibers are also called *red fibers*. These fibers can sustain action for a long period of time without fatigue.

Fast-twitch fibers

The thicker **fast-twitch fibers** have fewer capillaries and mitochondria than slow-twitch fibers and not as much myoglobin; hence, these fibers are also called *white fibers*. Fast-twitch fibers are adapted to respire anaerobically by using a large store of glycogen and high concentrations of glycolytic enzymes. The "dark meat" and "white meat" found in chicken and turkey consists of muscles with primarily red and white fibers, respectively. Fast-twitch fibers are adapted for the rapid generation of power and can grow thicker and stronger in response to weight training; however, they lack the endurance characteristics of slow-twitch fibers.

In addition to the type I and type II fibers, human muscles have an intermediate form of fibers that are fast-twitch, but they also have a high oxidative capacity, and so are more resistant to fatigue. Endurance training increases the proportion of these fibers in muscles.

Muscle metabolism changes with the demands made on it

Skeletal muscles at rest obtain most of their energy from the aerobic respiration of fatty acids (see chapter 7). During use of the muscle, such as during exercise, muscle stores of glycogen and glucose delivered by the blood are also used as energy sources. The energy obtained by cellular respiration is used to make ATP, which is needed for (1) the movement of the cross-bridges during muscle contraction and (2) the pumping of Ca^{2+} back into the sarcoplasmic reticulum during muscle relaxation.

Skeletal muscles respire anaerobically for the first 45 to 90 sec of moderate-to-heavy exercise because the cardiopulmonary system requires this amount of time to increase the oxygen supply to the muscles. If exercise is moderate, aerobic respiration contributes the major portion of the skeletal muscle energy requirements following the first 2 min of exercise.

Whether exercise is light, moderate, or intense for a particular individual depends on that person's maximal capacity for aerobic exercise. The maximum rate of oxygen consumption in the body is called the *aerobic capacity*. The intensity of exercise can also be defined by the lactate threshold. This threshold is the percentage of the aerobic capacity at which a significant rise in blood lactate levels occurs because of anaerobic respiration. For example, in an average healthy person, a significant amount of blood lactate appears when exercise is performed at about 50 to 70% of the aerobic capacity.

Physical training increases aerobic capacity and muscle strength

Muscle fatigue refers to the use-dependent decrease in the ability of a muscle to generate force. Fatigue is highly variable and can arise from a number of causes. The intensity of contraction as well as duration of contraction are involved. In addition, fatigue is affected by cellular metabolism: aerobic or anaerobic. In the case of short-duration maximal exertion, fatigue was long thought to be caused by a buildup of lactic acid (from anaerobic metabolism). More recent data also implicate a buildup in inorganic phosphate from the breakdown of creatine phosphate, which also occurs during anaerobic metabolism. In longer term, lower intensity exertion, fatigue appears to result from depletion of glycogen.

Because the depletion of muscle glycogen places a limit on exercise, any adaptation that spares muscle glycogen will improve physical endurance. Trained athletes have an increased proportion of energy derived from the aerobic respiration of fatty acids, resulting in a slower depletion of their muscle glycogen reserve. Athletes also have greater muscle vascularization, which facilitates both oxygen delivery and lactic acid removal. Because the aerobic capacity of endurance-trained athletes is higher than that of untrained people, athletes can perform more effort before muscle fatigue occurs.

Endurance training does not increase muscle size. Muscle enlargement is produced only by frequent periods of high-intensity exercise in which muscles work against high resistance, as in weight lifting. Resistance training increases the thickness of type II (fast-twitch) muscle fibers, causing skeletal muscles to grow by **hypertrophy** (increased cell size) rather than by cell division and an increased number of cells.

> Muscle cells contain myofibrils, which shorten during contraction. This is due to sliding of interdigitated myofilaments. Sliding of myofilaments involves the motor protein myosin, which forms cross-bridges on actin fibers. The process is controlled by Ca^{2+} ions that bind to another protein, troponin. Troponin controls the position of tropomyosin, which can block myosin-binding sites in actin. This is initiated by an action potential arriving from a neuron that leads to release of stored Ca^{2+} from the sarcoplasmic reticulum. A single stimulation of muscle causes a twitch. These can sum to produce sustained contraction.

47.5 Modes of Animal Locomotion

Animals are unique among multicellular organisms in their ability to move actively from one place to another. Locomotion requires both a propulsive mechanism and a control mechanism. There are a wide variety of propulsive mechanisms, most involving contracting muscles to generate the necessary force. Ultimately, it is the nervous system that activates and coordinates the muscles used in locomotion. In large animals, active locomotion is almost always produced by appendages that oscillate—*appendicular locomotion*—or by bodies that undulate, pulse, or undergo peristaltic waves—*axial locomotion*.

Although animal locomotion occurs in many different forms, the general principles remain much the same in all groups. The physical constraints to movement—gravity and frictional drag—are the same in every environment, differing only in degree.

Aquatic animals demonstrate a number of locomotion adaptations

Many aquatic invertebrates move along the bottom of a body of water using the same form of locomotion employed by terrestrial animals moving over the land surface. Flatworms employ ciliary activity to creep forward, roundworms perform a peristaltic slither, and leeches utilize a contract-anchor-extend creeping not unlike terrestrial earthworms. Crabs walk using limbs to pull themselves along; mollusks use a muscular foot, and sea stars use unique tube feet to do the same thing.

Moving directly through the water, swimming, presents quite a different challenge. Water's buoyancy reduces the influence of gravity. The primary force retarding forward movement is frictional drag, so body shape is important in reducing the friction and turbulence produced by swimming through the water.

Some marine invertebrates move about using hydraulic propulsion. For example, scallops clap the two sides of their shells together forcefully; squids and octopuses squirt water like a marine jet, as described earlier.

All aquatic vertebrates swim. Swimming involves using the body or its appendages to push against the water. An eel swims by sinuous undulations of its whole body (figure 47.21*a*). The undulating body waves of eel-like swimming are created by waves of muscle contraction alternating between the left and right axial musculature. As each body segment in turn pushes against the water, the moving wave forces the eel forward.

Fish use the similar mechanics as the eel but generate most of their propulsive from the posterior part of the body using the caudal (rear) fin (figure 47.21*b*). This also allows considerable specialization in the front end of the body without sacrificing propulsive force. Reptiles, such as alligators, depend almost entirely on tail undulations.

Whales and other marine mammals also swim using undulating body waves, but unlike any of the fishes, the waves pass from top to bottom and not from side to side. The body musculature of eels and fishes is highly segmental so that each set of two adjacent vertebrae is spanned by a muscle segment. As the muscle segments overlap through the length of the body, this arrangement permits the smooth passage of undulatory waves along the body. Whales are unable to produce lateral undulations because the mammalian vertebral column is structured to emphasize dorsal–ventral rather than side-to-side bending.

Many terrestrial tetrapod vertebrates are able to swim, usually through movement of their limbs. Most birds that swim, such as ducks and geese, propel themselves through the water by pushing against it with their hind legs, which typically have webbed feet. Frogs, turtles, and most marine mammals also swim with their hind legs and have webbed feet. Tetrapod vertebrates that swim with their forelegs usually have these limbs modified as flippers and pull themselves through the water; examples include sea turtles, penguins, and fur seals. A few principally terrestrial tetrapod vertebrates, such as polar bears, platypuses, and humans, swim with walking forelimbs not modified for swimming.

Terrestrial locomotion must deal primarily with gravity

The three great groups of terrestrial animals—mollusks, arthropods, and vertebrates—each move over land in different ways.

Mollusk locomotion is far less efficient than that of the other groups. Snails, slugs, and other terrestrial mollusks secrete a path of mucus that they glide along, pushing with a muscular foot.

Only vertebrates and arthropods (insects, spiders, and crustaceans) have developed a means of rapid surface locomotion. In both groups, the body is raised above the ground and moved forward by pushing against the ground with a series of jointed appendages, the legs.

Because legs must provide support as well as propulsion, it is important that the sequence of their movements not shove the body's center of gravity outside the legs' zone of support, unless the duration of such imbalance is short. Otherwise, the animal will fall. The need to maintain stability determines the sequence of leg movements, which are similar in vertebrates and arthropods.

The apparent differences in the walking gaits of these two groups reflect the differences in leg number. Vertebrates are tetrapods; all arthropods have six or more limbs. Although having many legs increases stability during locomotion, it also appears to reduce the maximum speed that can be attained.

The basic walking pattern of quadrupeds, from salamanders to most mammals, is left hind leg, right foreleg, right hind leg, left foreleg. This produces a diagonal pattern of foot falls. The highest running speeds of quadruped mammals, such as the gallop of a horse, may involve the animal being supported by only one leg, or even none at all. This is because mammals have evolved changes in the structure of

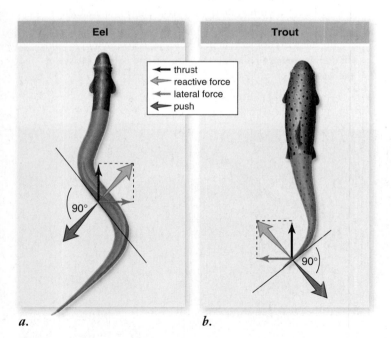

a.   **b.**

figure 47.21

MOVEMENTS OF SWIMMING FISHES. *a*. An eel pushes against the water with its whole body, whereas (*b*) a trout pushes only with its posterior half.

both their axial and appendicular skeleton that permit running by a series of leaps.

Many insects, such as grasshoppers, greatly emphasize leaping from their strong rear legs. Vertebrates such as kangaroos, rabbits, and frogs are also effective leapers (figure 47.22). Antelope and other grazing animals of open savannas may use running and leaping as effective strategies to avoid predators.

Flying uses air for support

Flight has evolved among the animals four times: in insects, pterosaurs (extinct flying reptiles), birds, and bats. In all four groups, active flying takes place in much the same way. Propulsion is achieved by pushing down against the air with wings. This alone provides enough lift to keep insects in the air. Vertebrates, being larger, need greater lift, obtaining it with wings whose upper surface is more convex (in cross section) than the lower. Because air travels farther over the top surface, it moves faster. A fluid, like air, decreases its internal pressure the faster it moves. Thus, there is a lower pressure on top of the wing and higher pressure on the bottom of the wing. This is the same principle used by airplane wings.

In birds and most insects, the raising and lowering of the wings is achieved by the alternate contraction of extensor muscles (elevators) and flexor muscles (depressors). Four insect orders (including those containing flies, mosquitoes, wasps, bees, and beetles) beat their wings at frequencies ranging from 100 to more than 1000 times per second, faster than nerves can carry successive impulses!

In these insects, the flight muscles are not attached to the wings at all, but rather to the stiff wall of the thorax, which is distorted in and out by their contraction. The reason these muscles can beat so fast is that the contraction of one muscle set stretches the other set, triggering its contraction in turn without waiting for the arrival of a nerve impulse.

Among vertebrates, flight first evolved some 200 MYA in flying reptiles called pterosaurs (figure 47.23). A very successful and diverse group, pterosaurs ranged in size from individuals no bigger than sparrows to some the size of a fighter plane. For much of this time, they shared the skies with birds,

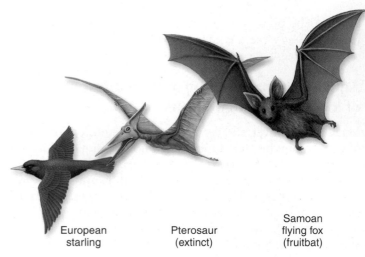

European starling   Pterosaur (extinct)   Samoan flying fox (fruitbat)

figure 47.23

WINGS HAVE EVOLVED THREE TIMES AMONG THE VERTEBRATES. These three very different vertebrates all have lightened bones and forelimbs transformed into wings.

which most paleontologists believe evolved from feathered dinosaurs about 150 MYA.

Bats, flying mammals that evolved after pterosaurs went extinct, are night fliers. By flying at night, bats are able to shop in a store with few other customers and a wealth of food in the form of night-flying insects such as mosquitoes and moths. Bats have adapted to nocturnal activity by developing unique sensory organs that use echolocation (see chapter 45), rather than relying on vision. It has proven to be a very successful approach. One-quarter of all mammal species are bats.

Locomotion in larger animals is almost always produced by appendages that push against the surroundings in some fashion. In water, this involves undulations of the entire body in eels and of mainly the posterior section in fish. Flying is achieved by creating lift due to pressure differences on the top and bottom surface of the wing.

figure 47.22

ANIMALS THAT HOP OR LEAP USE THEIR REAR LEGS TO PROPEL THEMSELVES THROUGH THE AIR. The powerful leg muscles of this frog allow it to explode from a crouched position to a takeoff in about 100 msec.

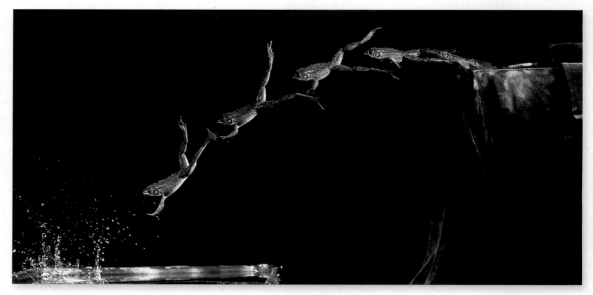

47.1 Types of Skeletal Systems

Changes in movement occur because muscles pull against a support structure.

- Hydrostatic skeletons use muscular contraction to exert water pressure inside a body wall (figure 47.1).

- Jetting is used in locomotion by the expulsion of water in jellyfish and squid (figure 47.2).

- Exoskeletons are rigid, hard cases that protect the body. The exoskeleton must be shed for the organism to grow, and the use of trachea for respiration limit body size (figure 47.3*a*).

- Endoskeletons of vertebrates are living tissues that contain calcium phosphate, are hard, rigid, and internal (figure 47.3*b*).

- Vertebrate skeletons are divided into two portions: axial, which makes up the axis of the body, and appendicular, which is composed of limbs and associated pectoral and pelvic girdles (figure 47.3*b*).

47.2 A Closer Look at Bone

Bones is a hard but resilient living tissue that is unique to vertebrate animals.

- Bones develop in two different ways: intramembranous development and the more complex endochondral development.

- In intramembranous development the bone forms on an underlying connective tissue by osteoblasts and osteocytes. This is common in skull bone formation (figure 47.4).

- In endochondral development bone calcification occurs deep in the body beginning in the fibrous sheath that envelops cartilage, later called the periosteum, and progresses inward.

- Bones grow by lengthening and widening. Cartilage remaining after development of the epiphyses serves as a pad between bone surfaces (figure 47.5).

- In birds and fishes, bone is avascular and basically acellular. In other vertebrates, bone contains osteoblasts, osteocytes, blood capillaries, and nerves traversing through the Haversian canals.

- Three cells are involved in bone development: osteoblasts initiate bone development; osteocytes form from osteoblasts embedded in bone; and osteoclasts, which resorb bone.

- Bone structure can be remodeled depending on use and forces impinging on the bone (figure 47.6).

47.3 Joints and Skeletal Movement

Skeletal movements that respond to muscle action occur at joints where one bone meets another bone.

- Muscles attach directly to the periosteum, a dense connective tissue that surrounds bones, or by tendons, which attach muscles to the periosteum.

- There are three classes of joints: immovable joints, which join bones; slightly movable joints with fibrous connective tissue or cartilage; and freely movable or synovial joints filled with an articulating fluid (figure 47.7).

- Movable joints have a characteristic range of motion. Ball-and-socket joints can perform universal movement in all directions, hinge joints have restricted movement, gliding joints slide and provide stability and flexibility, and combination joints allow rotation and sliding (figure 47.8).

- Skeletal muscles occur in antagonistic pairs in which the action of one muscle counters the action of the other muscle (figure 47.9).

47.4 Muscle Contraction

Skeletal muscle contains numerous muscle fibers composed of myofibrils containing myofilaments (figure 47.10).

- Myofibrils of skeletal or striated muscles have a distinct morphology: A bands composed of actin and myosin, I bands composed of actin, and Z lines located at each end of the sarcomere (figure 47.11).

- Muscle contraction occurs when actin and myosin filaments slide relative to each other and form cross-bridges (figure 47.11).

- Thick filaments are composed of many myosin filaments. Each myosin molecule has two subunits: a head region and long filament that is bundled with other long filaments (figure 47.12).

- Myosin is a motor protein that converts chemical energy, ATP, into mechanical energy by changing the myosin head conformation.

- A thin filament contains two twisted chains of actin, a globular proteins and tropomyosin and troponin (figure 47.13).

- The globular head of myosin forms a cross-bridge with actin when ATP is hydrolyzed to ADP and P_i (figure 47.15).

- Tropomyosin physically blocks the formation of a cross-bridge, and this protein is attached to actin by another protein troponin.

- Troponin is controlled by calcium, and a troponin–calcium complex displaces tropomyosin, allowing cross-bridges to form (figure 47.16).

- Calcium is stored in the sarcoplasmic reticulum, and the release of acetylcholine from the nerves depolarizes the muscle cell membrane.

- Membrane depolarization is carried into the muscle fiber by transverse tubules, and calcium is released into the cytosol. This process is called excitation-contraction coupling (figure 47.18).

- A single muscle fiber can produce variable muscle tension depending on the frequency of nerve stimulation, but the response of an entire muscle depends on the number of muscle fibers involved and their degree of tension.

- Motor units of various sizes are composed of all the muscle fibers innervated by branches of a single motor neuron and the motor neuron itself.

- Recruitment produces stronger muscle contractions as the number and size of motor units that are stimulated also increase.

- A twitch refers to the interval between contraction and relaxation of single muscle stimulation. Summation is a cumulative response when a second twitch "piggy-backs" on the first twitch and tetanus occurs when there is no relaxation between twitches (figure 47.19).

- Skeletal muscles have two major types of muscle fibers: slow-twitch, or type I fibers, which can sustain activity for long periods of time, and fast-twitch, or type II fibers, which are adapted for power but lack endurance.

- At rest skeletal muscles obtain energy by metabolism of fatty acids, whereas under exercise energy comes from glucose and glycogen.

- Muscle fatigue comes with a use-dependent decrease in the ability to generate force.

- Endurance training does not increase muscle size, but high-intensity exercise with resistance increases the size of the muscle.

47.5 Modes of Animal Locomotion

Animals exhibit a wide variety of propulsive mechanisms activated and controlled by the nervous system.

- In large animals, locomotion is appendicular by oscillations of appendages or axial by undulation or peristaltic waves (figure 47.21).

- Modes of terrestrial locomotion mollusks push themselves along a mucus path with a muscular foot other animals walk or leap by pushing the foot against the ground, and others fly by creating lift due to pressure differences on the surfaces of the wing.

SELF TEST

1. Exoskeletons and endoskeletons differ in that
   a. an exoskeleton is rigid, and an endoskeleton is flexible.
   b. endoskeletons are found only in vertebrates.
   c. exoskeletons are composed of calcium, and endoskeletons are built from chitin.
   d. exoskeletons are external to the soft tissues, and endoskeletons are internal.

2. Worms and marine invertebrates use a hydrostatic skeleton to generate movement. How do they do this?
   a. Their bones are filled with water, which provides the weight of the skeleton.
   b. The change in body structure is caused by contraction of muscles compressing the watery body fluid.
   c. The muscles contain water vacuoles, which, when filled, provide a rigid internal structure.
   d. The term *hydrostatic* simply refers to moist environment. They generate movement just as arthropods do.

3. You take X-rays of two individuals. Ray has been a weightlifter and body builder for 30 years, Ben has led a mostly sedentary life. What differences would you expect in their X-rays?
   a. No difference, they would both have thicker bones than a younger person due to natural thickening with age.
   b. No difference, lifestyle does not affect bone density.
   c. Ray would have thicker bones due to reshaping as a result of physical stress.
   d. Ben would have thicker bones because bone accumulates like fat tissue from a sedentary lifestyle.

4. Bone develops by one of two mechanisms depending on the underlying scaffold. Which pairing correctly describes these mechanisms?
   a. Intramembranous and extramembranous
   b. Endochondral and exochondral
   c. Extramembranous and exochondral
   d. Endochondral and intramembranous

5. Which of the following statements best describes the sliding filament mechanism of muscle contraction?
   a. Actin and myosin filaments do not shorten, but rather, slide past each other.
   b. Actin and myosin filaments shorten and slide past each other.
   c. As they slide past each other, actin filaments shorten, but myosin filaments do not shorten.
   d. As they slide past each other, myosin filaments shorten, but actin filaments do not shorten.

6. You have identified a calcium storage disease in rats. How would this inability to store Ca^{2+} affect muscle contraction?
   a. Ca^{2+} would be unable to bind to tropomyosin, which enables troponin to move and reveal binding sites for cross-bridges.
   b. Ca^{2+} would be unable to bind to troponin, which enables tropomyosin to move and reveal binding sites for cross-bridges.
   c. Ca^{2+} would be unable to bind to tropomyosin, which enables troponin to release ATP.
   d. Ca^{2+} would be unable to bind to troponin, which enables tropomyosin to release ATP.

7. Motor neurons stimulate muscle contraction via the release of—
   a. Ca^{2+}.
   b. ATP.
   c. acetylcholine.
   d. hormones.

8. Which of the following statements about muscle metabolism is false?
   a. Skeletal muscles at rest obtain most of their energy from muscle glycogen and blood glucose.
   b. ATP can be quickly obtained by combining ADP with phosphate derived from creatine phosphate.
   c. Exercise intensity is related to the maximum rate of oxygen consumption.
   d. ATP is required for the pumping of the Ca^{2+} back into the sarcoplasmic reticulum.

9. If you wanted to study the use of ATP during a single contraction cycle within a muscle cell, which of the following processes would you use?
   a. Summation
   b. Twitch
   c. Treppe
   d. Tetanus

10. Place the following events in the correct order:
    1. Sarcoplasmic reticulum releases Ca^{2+}.
    2. Myosin binds to actin.
    3. Action potential arrives from neuron.
    4. Ca^{2+} binds to troponin.
    a. 1, 2, 3, 4
    b. 3, 1, 2, 4
    c. 2, 4, 3, 1
    d. 3, 1, 4, 2

11. How do the muscles move your hand through space?
    a. By contraction
    b. By attaching to two bones across a joint
    c. By lengthening
    d. Both (a) and (b) are correct

12. Differences in which of the following permit animal flight?
    a. Gravity
    b. Humidity
    c. Pressure
    d. Temperature

13. How can osteocytes remain alive within bone?
    a. Bones are composed of only dead or dormant cells.
    b. Haversian canals are bone structures that contain blood vessels that provide materials for the osteocytes.
    c. Osteocytes have membrane extensions that protrude from bone and allow them to exchange materials with the surrounding fluids
    d. Bones are hollow in the middle and the low pressure there draws fluid from the blood that nourishes the osteocytes.

CHALLENGE QUESTIONS

1. You are designing a space-exploration vehicle to use on a planet with a gravity greater than Earth. Given a choice between a hydrostatic or an exoskeleton, which would you choose? Why?

2. You start running as fast as you can. Then, you settle into a jog that you can easily maintain. How do energy sources utilized by your skeletal muscles change during the switch? Why?

3. The nerve gas methylphosphonofluoridic acid (sarin) inhibits the enzyme acetylcholinesterase, required to break down acetylcholine. Based on this information, what are the likely effects of this nerve gas on muscle function?

chapter 48

The Digestive System

introduction

PLANTS AND OTHER PHOTOSYNTHETIC ORGANISMS
can produce the organic molecules they need from inorganic
components. Therefore, they are autotrophs, or self-sustaining.
Animals, such as the chipmunk shown, are heterotrophs: They
must consume organic molecules present in other organisms.
The molecules heterotrophs eat must be digested into smaller
molecules in order to be absorbed into the animal's body. Once
these products of digestion enter the body, the animal can use
them for energy in cellular respiration or for the construction
of the larger molecules that make up its tissues. The process of
animal digestion is the focus of this chapter.

concept outline

Types of Digestive Systems

Heterotrophs are divided into three groups on the basis of their food sources. Animals that eat plants exclusively are classified as **herbivores;** common examples include algae-eating snails, sap-sucking insects, and vertebrates such as cattle, horses, rabbits, and sparrows. Animals that eat other animals, such as crabs, squid, many insects, cats, eagles, trout, and frogs, are **carnivores.** Animals that eat both plants and other animals are **omnivores.** Humans are omnivores, as are pigs, bears, and crows.

Invertebrate digestive systems are bags or tubes

Single-celled organisms as well as sponges digest their food intracellularly. Other multicellular animals digest their food extracellularly, within a digestive cavity. In this case, the digestive

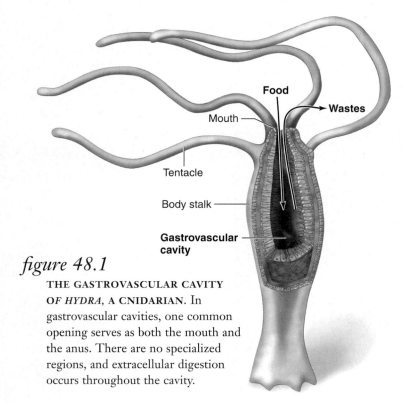

figure 48.1

THE GASTROVASCULAR CAVITY OF *HYDRA*, A CNIDARIAN. In gastrovascular cavities, one common opening serves as both the mouth and the anus. There are no specialized regions, and extracellular digestion occurs throughout the cavity.

enzymes are released into a cavity that is continuous with the animal's external environment. In cnidarians and in flatworms such as planarians, the digestive cavity has only one opening that serves as both mouth and anus (see chapter 33). There is no specialization within this type of digestive system, called a *gastrovascular cavity*, because every cell is exposed to all stages of food digestion (figure 48.1).

Specialization occurs when the digestive tract, or alimentary canal, has a separate mouth and anus, so that transport of food is one-way. The most primitive digestive tract is seen in nematodes (phylum Nematoda), where it is simply a tubular *gut* lined by an epithelial membrane. Earthworms (phylum Annelida) have a digestive tract specialized in different regions for the ingestion, storage, fragmentation, digestion, and absorption of food. All more complex animal groups, including all vertebrates, show similar specializations (figure 48.2).

The ingested food may be stored in a specialized region of the digestive tract or it may first be subjected to physical fragmentation. This fragmentation may occur through the chewing action of teeth (in the mouth of many vertebrates) or the grinding action of pebbles (in the gizzard of earthworms and birds). Chemical digestion then occurs, breaking down the larger food molecules of polysaccharides and disaccharides, fats, and proteins into their smallest subunits.

Chemical digestion involves hydrolysis reactions that liberate the subunit molecules—primarily monosaccharides, amino acids, and fatty acids—from the food. These products of chemical digestion pass through the epithelial lining of the gut into the blood, in a process known as *absorption*. Any molecules in the food that are not absorbed cannot be used by the animal. These waste products are excreted, or defecated, from the anus.

Vertebrate digestive systems include highly specialized structures molded by diet

In humans and other vertebrates, the digestive system consists of a tubular gastrointestinal tract and accessory digestive organs (figure 48.3).

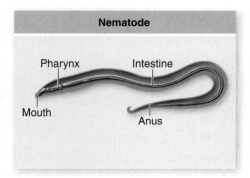

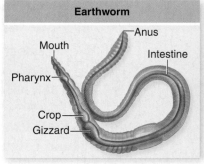

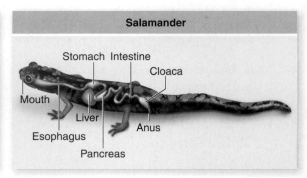

figure 48.2

THE ONE-WAY DIGESTIVE TRACT OF NEMATODES, EARTHWORMS, AND VERTEBRATES. One-way movement through the digestive tract allows different regions of the digestive system to become specialized for different functions.

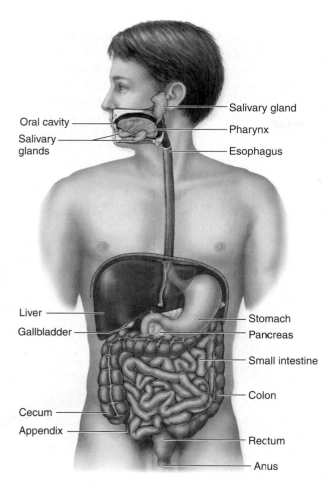

figure 48.3

THE HUMAN DIGESTIVE SYSTEM. The human digestive system consists of the oral cavity, esophagus, stomach, small intestine, large intestine, rectum, and anus; and is aided by accessory organs.

Overview of the digestive tract

The initial components of the gastrointestinal tract are the mouth and the pharynx, which is the common passage of the oral and nasal cavities. The pharynx leads to the esophagus, a muscular tube that delivers food to the stomach, where some preliminary digestion occurs.

From the stomach, food passes to the small intestine, where a battery of digestive enzymes continues the digestive process. The products of digestion, together with minerals and most imbibed water, are absorbed across the wall of the small intestine into the bloodstream. What remains is emptied into the large intestine, where some of the remaining water and minerals are absorbed.

In most vertebrates other than mammals, the waste products emerge from the large intestine into a cavity called the cloaca (see figure 48.2), which also receives the products of the urinary and reproductive systems. In mammals, the urogenital products are separated from the fecal material in the large intestine; the fecal material enters the rectum and is expelled through the anus.

The accessory digestive organs include the liver, which produces *bile* (a green solution that emulsifies fat), the gallblad-

der, which stores and concentrates the bile, and the pancreas. The pancreas produces *pancreatic juice*, which contains digestive enzymes and bicarbonate buffer. Both bile and pancreatic juice are secreted into the first region of the small intestine, the duodenum, where they aid digestion.

Tissues of the digestive tract

The tubular gastrointestinal tract of a vertebrate has a characteristic layered structure (figure 48.4). The innermost layer is the **mucosa,** an epithelium that lines the interior, or **lumen,** of the tract. The next major tissue layer, made of connective tissue, is called the **submucosa.**

Just outside the submucosa is the **muscularis,** which consists of a double layer of smooth muscles. The muscles in the inner layer have a circular orientation, and those in the outer layer are arranged longitudinally. Another epithelial tissue layer, the **serosa,** covers the external surface of the tract. Nerve networks, intertwined in *plexuses* between muscle layers, are located in the submucosa and help regulate the gastrointestinal activities.

Variations in digestive tract structure

In general, carnivores have shorter intestines for their size than do herbivores. A short intestine is adequate for a carnivore, but herbivores ingest a large amount of plant cellulose, which resists digestion. These animals have a long, convoluted small intestine.

In addition, mammals called **ruminants** (such as cattle) that consume grass and other vegetation have several prestomach chambers, where bacteria digest cellulose. Other

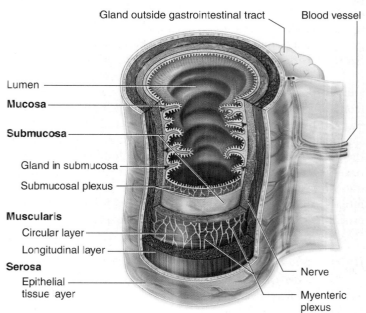

figure 48.4

THE LAYERS OF THE GASTROINTESTINAL TRACT. The mucosa contains an epithelial lining; the submucosa is composed of connective tissue; and the muscularis consists of smooth muscles. Glands secrete substances via ducts into specific regions of the tract.

herbivores, including rabbits and horses, have cellulose-digesting bacteria in a pouch called the **cecum** located at the juncture between the small and large intestine. Variations in digestive systems of vertebrates are described in more detail in section 48.6.

In the rest of this chapter, we focus on the details of the vertebrate digestive system's structure and function. We close the chapter with discussion of nutrients that are essential to vertebrates.

Most animals digest their food extracellularly. The digestive tract, with a one-way transport of food and specialization of regions for different functions, allows food to be ingested, physically fragmented, chemically digested, and absorbed. The vertebrate digestive system consists of a tubular gastrointestinal tract composed of a series of tissue layers and including several accessory organs. Vertebrates exhibit digestive tract adaptations geared to their diets.

48.2 # The Mouth and Teeth: Food Capture and Bulk Processing

Specializations of the digestive systems in different kinds of vertebrates reflect the way these animals live. Fishes have a large **pharynx** (throat) with gill slits, whereas air-breathing vertebrates have a greatly reduced pharynx. Many vertebrates have teeth (figure 48.5), used for chewing, or *mastication*, that break up food into small particles and mix it with fluid secretions.

Birds, which lack teeth, break up food in their two-chambered stomachs (figure 48.6). In one of these chambers, called the *gizzard*, small pebbles ingested by the bird are churned together with the food by muscular action. This churning grinds up the seeds and other hard plant material into smaller chunks that can be digested more easily.

Vertebrate teeth are adapted to types of food items

Carnivorous mammals have pointed teeth that lack flat grinding surfaces. Such teeth are adapted for cutting and shearing. Carnivores often tear off pieces of their prey but have little need to chew them, because digestive enzymes can act directly on animal cells. By contrast, grass-eating herbivores must pulverize the cellulose cell walls of plant tissue before the bacteria in their rumens or ceca can digest it. These animals have large, flat teeth with complex ridges well suited to grinding.

Human teeth are specialized for eating both plant and animal food. Viewed simply, humans are carnivores in the front of the mouth and herbivores in the back (see figure 48.5). The four front teeth in the upper and lower jaws are sharp, chisel-shaped incisors used for biting. On each side of the incisors are sharp, pointed teeth called cuspids (sometimes referred to as "canine" teeth), which are used for tearing food. Behind the canines are two premolars and three molars, all with flattened, ridged surfaces for grinding and crushing food (figure 48.7). Children have only 20 teeth, but

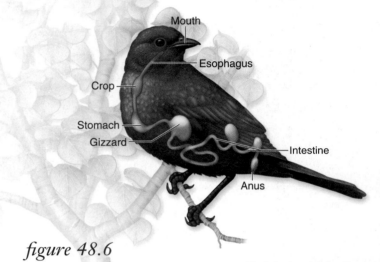

figure 48.6

THE DIGESTIVE TRACT OF BIRDS. Birds lack teeth but have a muscular chamber called the gizzard that works to break down food. Birds swallow gritty objects or pebbles that lodge in the gizzard and pulverize food before it passes into the intestine. Food is stored in the crop.

these deciduous teeth are lost during childhood and are replaced by 32 adult teeth.

The mouth is a chamber for ingestion and initial processing

Inside the mouth, the tongue mixes food with a mucous solution, **saliva**. In humans, three pairs of salivary glands secrete saliva into the mouth through ducts in the mouth's mucosal lining. Saliva moistens and lubricates the food so that it is easier to swallow and does not abrade the tissue it passes on its way through the esophagus.

figure 48.5

PATTERNS OF DENTITION DEPEND ON DIET. Different vertebrates (herbivore, carnivore, or omnivore) have evolved specific variations from a generalized pattern of dentition depending on their nutritional source.

◼ Incisors ◻ Premolars
◻ Canines ◼ Molars

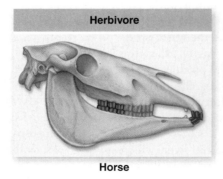
Horse

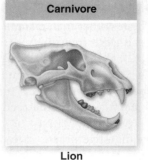
Lion

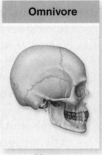
Human

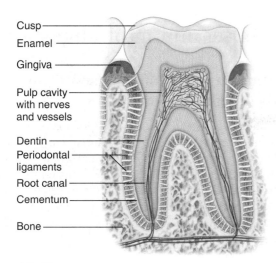

figure 48.7

ANATOMY OF A HUMAN TOOTH. Humans have carnivore-like teeth in the front of their mouths and herbivore-like teeth in the back. Each tooth is alive, with a central pulp containing nerves and blood vessels. The actual chewing surface is a hard enamel layered over the softer dentin, which forms the body of the tooth.

Saliva also contains the hydrolytic enzyme **salivary amylase,** which initiates the breakdown of the polysaccharide starch into the disaccharide maltose. This digestion is usually minimal in humans, however, because most people don't chew their food very long.

Stimulation of salivation

The secretions of the salivary glands are controlled by the nervous system, which in humans maintains a constant flow of about half a milliliter per minute when the mouth is empty of food. This continuous secretion keeps the mouth moist.

The presence of food in the mouth triggers an increased rate of secretion. Taste buds as well as olfactory (smell) neurons send impulses to the brain, which responds by stimulating the salivary glands (see chapter 45). The most potent stimuli are acidic solutions; lemon juice, for example, can increase the rate of salivation eightfold. The sight, sound, or smell of food can stimulate salivation markedly in dogs, but in humans, thinking or talking about food can also stimulate salivation.

Swallowing

Swallowing is initiated by voluntary action, then is continued under involuntary control. When food is ready to be swallowed, the tongue moves it to the back of the mouth. In mammals, the process of swallowing begins when the soft palate elevates, pushing against the back wall of the pharynx (figure 48.8). Elevation of the soft palate seals off the nasal cavity and prevents food from entering it. Pressure against the pharynx triggers an automatic, involuntary response, the **swallowing reflex.** Because it is a reflex, swallowing cannot be stopped once it is initiated.

Neurons within the walls of the pharynx send impulses to the swallowing center in the brain. In response, electrical impulses in motor neurons stimulate muscles to contract and raise the **larynx** (voice box). This pushes the **glottis,** the opening from the larynx into the trachea (windpipe), against a flap of tissue called the **epiglottis.** These actions keep food out of the respiratory tract, directing it instead into the esophagus.

> In many vertebrates, ingested food is fragmented through the tearing or grinding action of specialized teeth. In birds, this is accomplished through the grinding action of pebbles in the gizzard. Taste and scent of food stimulate production of saliva, which lubricates food. This also stimulates the swallowing response as food mixed with saliva is swallowed entering the esophagus.

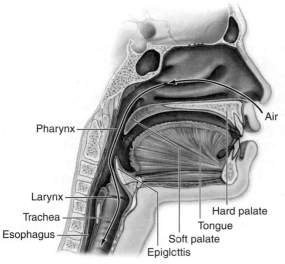

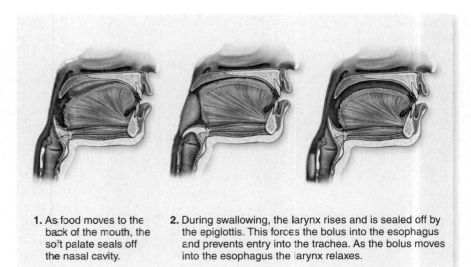

1. As food moves to the back of the mouth, the soft palate seals off the nasal cavity.

2. During swallowing, the larynx rises and is sealed off by the epiglottis. This forces the bolus into the esophagus and prevents entry into the trachea. As the bolus moves into the esophagus the larynx relaxes.

figure 48.8

THE MECHANICS OF SWALLOWING. Cross section through head and throat showing relevant structures (*left*). During swallowing (*right*), the tongue pushes the palate upwards and the soft palate seals off the nasal cavity. Elevation of the larynx causes the epiglottis to seal off the trachea preventing food from entering the airway.

The Esophagus and the Stomach: The Beginning of Digestion

Swallowed food enters a muscular tube called the esophagus, which connects the pharynx to the stomach. The esophagus actively moves a processed lump of food, called a **bolus,** through the action of muscles. Food from a meal is stored in the stomach and undergoes initial digestion.

Muscular contractions of the esophagus move food to the stomach

In adult humans, the **esophagus** is about 25 cm long; the upper third is enveloped in skeletal muscle for voluntary control of swallowing, whereas the lower two-thirds is surrounded by involuntary smooth muscle. The swallowing center stimulates successive one-directional waves of contraction in these muscles that move food along the esophagus to the stomach. These rhythmic waves of muscular contraction are called **peristalsis** (figure 48.9); they enable humans and other vertebrates to swallow even if they are upside down.

In many vertebrates, the movement of food from the esophagus into the stomach is controlled by a ring of circular smooth muscle, or a *sphincter,* that opens in response to the pressure exerted by the food. Contraction of this sphincter prevents food in the stomach from moving back into the esophagus. Rodents and horses have a true sphincter at this site, and as a result they cannot regurgitate; humans lack a true sphincter. Normally, the esophagus is closed off except during swallowing.

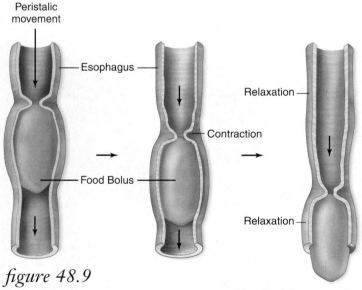

figure 48.9

THE ESOPHAGUS AND PERISTALSIS. After food has entered the esophagus, rhythmic waves of muscular contraction, called peristalsis, move the food down to the stomach.

The stomach is a "holding station" involved in acidic breakdown of food

The **stomach** (figure 48.10) is a saclike portion of the digestive tract. Its inner surface is highly convoluted, enabling it to fold up when empty and open out like an expanding balloon as it fills with food. For example the human stomach has a volume of only about 50 mL when empty, but, it may expand to contain 2 to 4 L of food when full. Carnivores that engage in sporadic gorging as an important survival strategy possess stomachs that are able to distend even more.

Secretory systems

The stomach contains a third layer of smooth muscle for churning food and mixing it with **gastric juice,** an acidic secretion of the tubular gastric glands of the mucosa (see figure 48.10). These exocrine glands contain two kinds of secretory cells: *parietal cells,* which secrete hydrochloric acid (HCl), and *chief cells,* which secrete **pepsinogen,** the inactive form (zymogen) of the protease (protein-digesting enzyme) **pepsin.**

Pepsinogen has 44 additional amino acids that block the active site. HCl causes pepsinogen to unfold, exposing the active site, which then acts to remove the 44 amino acids. This yields the active protease, pepsin. This process of secreting an inactive form that is then converted into an active enzyme outside the cell prevents the chief cells from digesting themselves.

In adult humans, only proteins are partially digested in the stomach—no significant digestion of carbohydrates or fats occurs.

In addition to producing HCl, the parietal cells of the stomach also secrete **intrinsic factor,** a polypeptide needed for the intestinal absorption of vitamin B_{12}. Because this vitamin is required for the production of red blood cells, persons who lack sufficient intrinsic factor develop a type of anemia (low red blood cell count) called *pernicious anemia.*

Action of acid

The human stomach produces about 2 L of HCl and other gastric secretions every day, creating a very acidic solution inside the stomach. The concentration of HCl in this solution is about 10 millimolar (mM), corresponding to a pH of 2. Thus, gastric juice is about 250,000 times more acidic than blood, whose normal pH is 7.4.

The low pH in the stomach helps denature food proteins, making them easier to digest, and keeps pepsin maximally active. Active pepsin hydrolyzes food proteins into shorter chains of polypeptides that are not fully digested until the mixture enters the small intestine. The mixture of partially digested food and gastric juice is called **chyme.**

The acidic solution within the stomach also kills most of the bacteria that are ingested with the food. The few bacteria that survive the stomach and enter the intestine intact are able to grow and multiply there, particularly in the large intestine. In fact, vertebrates harbor thriving colonies of bacteria within their intes-

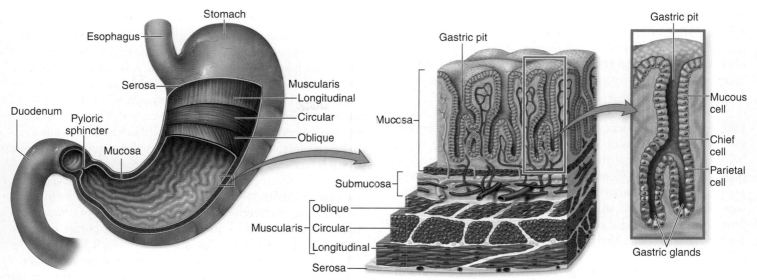

figure 48.10

THE STOMACH AND DUODENUM. Food enters the stomach from the esophagus. A ring of smooth muscle called the pyloric sphincter controls the entrance to the duodenum, the upper part of the small intestine. The epithelial walls of the stomach are dotted with deep infoldings called gastric pits that contain gastric glands. The gastric glands consist of mucous cells, chief cells that secrete pepsinogen, and parietal cells that secrete HCl. Gastric pits are the openings of the gastric glands.

tines, and bacteria are a major component of feces. As we discuss later, bacteria that live within the digestive tracts of ruminants play a key role in the ability of these mammals to digest cellulose.

Ulcers

Overproduction of gastric acid can occasionally eat a hole through the wall of the stomach or the duodenum, causing a peptic ulcer. Although we once blamed consumption of spicy food, the most common cause of peptic ulcers is now thought to be infection with the bacterium *Heliobacter pylori*.

H. pylori can grow on the lining of the human stomach, surviving the acid pH by secreting substances that buffer the pH of its immediate surroundings. While infection with *H. pylori* is common in the United States (about 20% of people younger than 40 and 50% older than 60), most people are asymptomatic. However, in some cases, infection by *H. pylori* can cause a reduction or weakening of the mucosal layer in the stomach or duodenum, allowing acidic secretions to attack the underly-

ing epithelium. Antibiotic treatment of the infection can reduce symptoms and often even cure the ulcer.

Leaving the stomach

Chyme leaves the stomach through the *pyloric sphincter* (see figure 48.10) to enter the small intestine. This is where all terminal digestion of carbohydrates, lipids, and proteins occurs and where the products of digestion—amino acids, glucose, and so on—are absorbed into the blood. Only some of the water in chyme and a few substances, such as aspirin and alcohol, are absorbed through the wall of the stomach.

> Peristaltic waves of contraction propel food along the esophagus to the stomach. Gastric juice contains strong hydrochloric acid and the protease enzyme pepsin, which begins the digestion of proteins into shorter polypeptides. The acidic chyme is then transferred through the pyloric sphincter into the small intestine.

48.4 The Small Intestine: Breakdown and Absorption

The capacity of the small intestine is limited, and its digestive processes take time. Consequently, efficient digestion requires that only relatively small amounts of chyme be introduced from the stomach into the small intestine at any one time. Coordination between gastric and intestinal activities is regulated by neural and hormonal signals, which we will describe in section 48.7.

The structure of the small intestine is specialized for digestion and nutrient uptake

The **small intestine** is approximately 4.5 m long in a living person, but 6 m long at autopsy when all the muscles have relaxed. The first 25 cm is the **duodenum;** the remainder of the small intestine is divided into the **jejunum** and the **ileum.**

The duodenum receives acidic chyme from the stomach, digestive enzymes and bicarbonate from the pancreas, and bile from the liver and gallbladder. Enzymes in the pancreatic juice digest larger food molecules into smaller fragments. This digestion occurs primarily in the duodenum and jejunum.

The epithelial wall of the small intestine is covered with tiny, fingerlike projections called **villi** (singular, *villus;* figure 48.11). In turn, each of the epithelial cells lining the villi is covered on its apical surface (the side facing the lumen) by many foldings of the plasma membrane that form cytoplasmic extensions called **microvilli.** These are quite tiny and can be seen clearly only with an electron microscope. In a light micrograph, the microvilli resemble the bristles of a brush, and for that reason the epithelial wall of the small intestine is also called a *brush border.*

The villi and microvilli greatly increase the surface area of the small intestine; in humans, this surface area is 300 m²—about 3200 square feet! It is over this vast surface that the products of digestion are absorbed.

The microvilli also participate in digestion because a number of digestive enzymes are embedded within the epithelial cells' plasma membranes, with their active sites exposed to the chyme. These brush border enzymes include those that hydrolyze the disaccharides lactose and sucrose, among others. Many adult humans lose the ability to produce the brush border enzyme **lactase** and therefore cannot digest lactose (milk sugar), a rather common condition called *lactose intolerance.* The brush border enzymes complete the digestive process that started with the action of salivary amylase in the mouth.

Accessory organs secrete enzymes into the small intestine

The main organs that aid digestion are the pancreas, liver, and gallbladder. They empty their secretions, primarily enzymes, through ducts directly into the small intestine.

Secretions of the pancreas

The **pancreas** (figure 48.12), a large gland situated near the junction of the stomach and the small intestine, secretes pancreatic fluid into the duodenum through the *pancreatic duct;* thus, the pancreas functions as an exocrine gland. This fluid contains a host of enzymes, including **trypsin** and **chymotrypsin,** which digest proteins; **pancreatic amylase,** which digests starch; and **lipase,** which digests fat. These enzymes are released into the duodenum primarily as inactive enzymes, called zymogens, and are then activated by trypsin, which is first activated by a brush border enzyme of the intestine.

Pancreatic enzymes digest proteins into smaller polypeptides, polysaccharides into shorter chains of sugars, and fats into free fatty acids and monoglycerides. The further digestion of proteins and carbohydrates is then completed by the brush border enzymes. Pancreatic fluid also contains bicarbonate, which neutralizes the HCl from the stomach and gives the chyme in the duodenum a slightly alkaline pH. The digestive enzymes and bicarbonate are produced by clusters of secretory cells known as **acini.**

In addition to its exocrine role in digestion, the pancreas also functions as an endocrine gland, secreting several hormones into the blood that control the blood levels of glucose and other nutrients. These hormones are produced in the **islets of Langerhans,** clusters of endocrine cells scattered throughout the pancreas. The two most important pancreatic hormones, insulin and glucagon, were described in chapter 46; their actions are also discussed later on.

Liver and gallbladder

The **liver** is the largest internal organ of the body (see figure 48.3). In an adult human, the liver weighs about 1.5 kg and is the size of a football. The main exocrine secretion of the liver is bile, a fluid mixture consisting of *bile pigments* and *bile salts* that is delivered into the duodenum during the digestion of a meal.

The bile pigments do not participate in digestion; they are waste products resulting from the liver's destruction of old

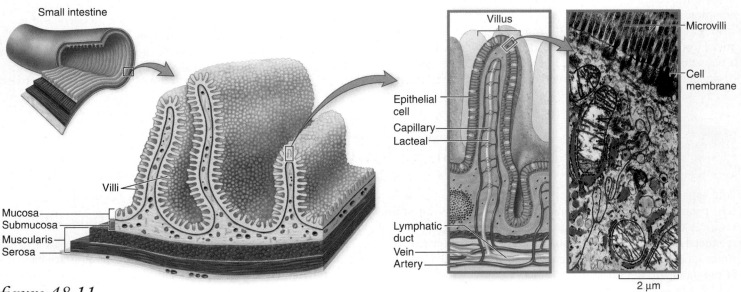

figure 48.11

THE SMALL INTESTINE. Successive enlargements show folded epithelium studded with villi that increase surface area. The micrograph shows an epithelial cell with numerous microvilli.

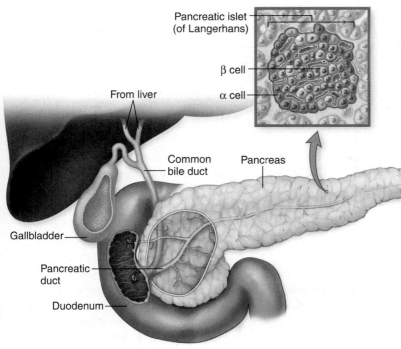

figure 48.12

THE PANCREAS. The pancreatic duct and bile duct empty into the duodenum. The pancreas secretes pancreatic juice into the pancreatic duct. The pancreatic islets of Langerhans secrete hormones into the blood; α cells secrete glucagon, and β cells secrete insulin. The liver secretes bile, which consists of bile pigments (waste products from the liver) and bile salts. Bile salts play a role in the digestion of fats. Bile is concentrated and stored in the gallbladder until it is needed in the duodenum on the arrival of fatty food.

red blood cells and are ultimately eliminated with the feces. If the excretion of bile pigments by the liver is blocked, the pigments can accumulate in the blood and cause a yellow staining of the tissues known as *jaundice*.

In contrast, the bile salts play a very important role in preparing fats for subsequent enzymatic digestion. Because fats are insoluble in water, they enter the intestine as drops within the watery chyme. The bile salts, which are partly lipid-soluble and partly water-soluble, work like detergents, dispersing the large drops of fat into a fine suspension of smaller droplets. This emulsification action produces a greater surface area of fat for the action of lipase enzymes, and thus allows the digestion of fat to proceed more rapidly.

After bile is produced in the liver, it is stored and concentrated in the gallbladder. The arrival of fatty food in the duodenum triggers a neural and endocrine reflex that stimulates the gallbladder to contract, causing bile to be transported through the common bile duct and injected into the duodenum. (These reflexes are the topic of a later section.) Gallstones are hardened precipitates of cholesterol that form in some individuals. If these stones block the bile duct, contraction of the gallbladder causes intense pain, often felt in the back. In severe cases of blockage, surgical removal of the gallbladder may be performed.

Absorbed nutrients move into blood or lymph capillaries

After their enzymatic breakdown, proteins and carbohydrates are absorbed as amino acids and monosaccharides, respectively. They are transported across the brush border into the epithelial cells that line the intestine by a combination of active transport and facilitated diffusion (figure 48.13*a*).

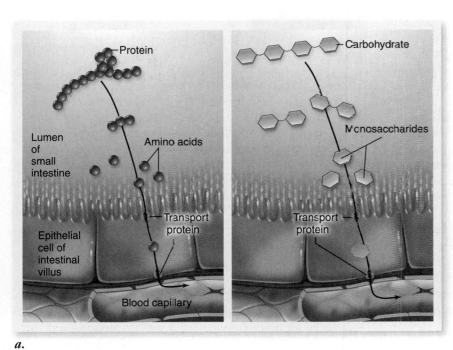

a.

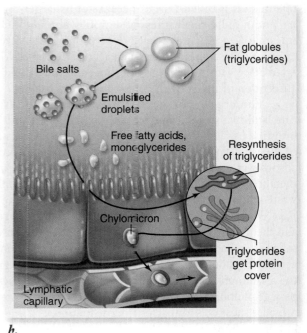

b.

figure 48.13

ABSORPTION OF THE PRODUCTS OF DIGESTION. *a.* Monosaccharides and amino acids are transported into blood capillaries. *b.* Fatty acids and monoglycerides within the intestinal lumen are absorbed and converted within the intestinal epithelial cells into triglycerides. These are then coated with proteins to form structures called chylomicrons, which enter lymphatic capillaries.

Glucose is transported by coupled transport with Na⁺ ions (also called secondary active transport). Fructose, found in most fruit, is transported by facilitated diffusion. Most amino acids are transported by active transport using a variety of different transporters. Some of these carrier proteins use cotransport with Na⁺ ions; others transport only amino acids. Once they have entered epithelial cells across the apical membrane, these monosaccharides and amino acids move through the cytoplasm and are transported across the basolateral membrane and into the blood capillaries within the villi.

The blood carries these products of digestion from the intestine to the liver via the **hepatic portal vein.** A portal vein connects two beds of capillaries instead of returning to the heart. In this case, the intestine is connected to the liver by the hepatic portal vein, thus the liver receives blood-borne molecules from the intestine. Because of the hepatic portal vein, the liver is the first organ to receive most of the products of digestion, except for fat.

The products of fat digestion are absorbed by a different mechanism (figure 48.13*b*). Fats (triglycerides) are hydrolyzed into fatty acids and monoglycerides by digestion. These fatty acids and monoglycerides are nonpolar and can thus enter epithelial cells by simple diffusion. Once inside the intestinal epithelial cells they are reassembled into triglycerides. The triglycerides then combine with proteins to form small particles called **chylomicrons,** which are too bulky to enter blood capillaries in the intestine. Instead of entering the hepatic portal circulation, the chylomicrons are absorbed into lymphatic capillaries (see the following chapter), which empty their contents into the blood in veins near the neck. Chylomicrons can make the blood plasma appear cloudy if a sample of blood is drawn after a fatty meal.

The amount of fluid passing through the small intestine in a day is startlingly large: approximately 9 L. However, almost all of this fluid is absorbed into the body rather than eliminated in the feces: About 8.5 L is absorbed in the small intestine and an additional 350 mL in the large intestine. Only about 50 g of solid and 100 mL of liquid leaves the body as feces. The normal fluid absorption efficiency of the human digestive tract approaches 99%, which is very high indeed.

> Digestion occurs primarily in the duodenum with the aid of pancreatic and liver secretions. The small intestine provides a large surface area for absorption. Glucose and amino acids from food are absorbed by both active transport and facilitated diffusion. They enter the bloodstream via the hepatic portal vein, going directly to the liver. Fat from food is absorbed directly through the intestinal cells by simple diffusion, then enter the lymphatic system as chylomicrons.

48.5 The Large Intestine: Elimination of Waste Material

The **large intestine,** or **colon,** is much shorter than the small intestine, occupying approximately the last meter of the digestive tract; it is called "large" because of its larger diameter, not its length. The small intestine empties directly into the large intestine at a junction where two vestigial structures, the **cecum** and the **appendix,** remain (figure 48.14). No digestion takes place within the large intestine, and only about 4% of the absorption of fluids by the intestine occurs there.

The large intestine is not as convoluted as the small intestine, and its inner surface has no villi. Consequently, the large intestine has less than one-thirtieth the absorptive surface area of the small intestine. The function of the large intestine is to absorb water, remaining electrolytes, and products of bacterial metabolism (including vitamin K). The large intestine prepares waste material to be expelled from the body.

Many bacteria live and reproduce within the large intestine, and the excess bacteria are incorporated into the refuse material, called *feces*. Bacterial fermentation produces gas within the colon at a rate of about 500 mL per day. This rate increases greatly after the consumption of beans or other types of vegetables because the passage of undigested plant material (fiber) into the large intestine provides substrates for bacterial fermentation.

The human colon has evolved to process food with a relatively high fiber content. Diets that are low in fiber, which are

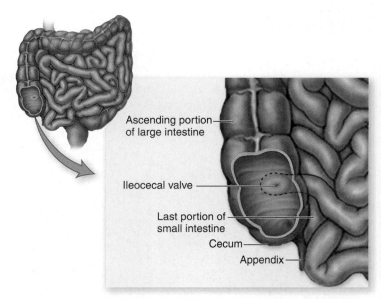

figure 48.14

THE JUNCTION OF THE SMALL AND LARGE INTESTINES IN HUMANS. The large intestine, or colon, starts with the cecum, which is relatively small in humans compared with that in other mammals. A vestigial structure called the appendix extends from the cecum.

common in the United States and other developed countries, result in a slower passage of food through the colon. Low dietary fiber content is thought to be associated with the level of colon cancer in the United States, which is among the highest in the world.

Compacted feces, driven by peristaltic contractions of the large intestine, pass from the large intestine into a short tube called the rectum. From the rectum, the feces exit the body through the anus. Two sphincters control passage through the anus. The first is composed of smooth muscle and opens involuntarily in response to pressure inside the rectum. The second, composed of striated muscle, can be controlled voluntarily by the brain, thus permitting a conscious decision to delay defecation.

In most vertebrates, the reproductive and urinary tracts empty together with the digestive tract into a common cavity, the cloaca. In some reptiles and birds, additional water from either the feces or urine may be absorbed in the cloaca before the products are expelled from the body. The exception is most mammals, which do not have a cloaca. The reproductive and urinary tracts empty separately from the digestive tract.

> The large intestine concentrates wastes for excretion by absorbing water. Some ions and vitamin K are also absorbed by the large intestine.

48.6 Variations in Vertebrate Digestive Systems

Animals lack the enzymes necessary to digest cellulose, but the digestive tracts of some animals contain bacteria and protists that convert cellulose into substances the host can absorb. Although digestion by gastrointestinal microorganisms plays a relatively small role in human nutrition, it is an essential element in the nutrition of many other kinds of animals, including insects such as termites and cockroaches, and a few groups of herbivorous mammals. The relationships between these microorganisms and their animal hosts are mutually beneficial and provide an excellent example of symbiosis (see chapter 56).

Ruminants rechew regurgitated food

Ruminants have a four-chambered stomach (figure 48.15). The first three portions include the reticulum, the rumen, and the omasum. These are followed by the true stomach, the abomasum.

The rumen, which may hold up to 50 gallons, serves as a fermentation vat where bacteria and protists convert cellulose and other molecules into a variety of simpler compounds. The location of the rumen at the front of the four chambers is important because it allows the animal to regurgitate and rechew the contents of the rumen, an activity called *rumination*, or "chewing the cud." This breaks tougher fiber in the diet into smaller particles, increasing surface area for microbial attachment

After chewing, the cud is swallowed for further microbial digestion in the rumen, then passes to the omasum and then to the abomasum, where it is finally mixed with gastric juice. This process leads to far more efficient digestion of cellulose in ruminants than in mammalian herbivores such as horses, that lack a rumen.

Other herbivores have alternative strategies for digestion

In some animals, such as rodents, horses, deer, and lagomorphs (rabbits and hares), the digestion of cellulose by microorganisms takes place in the cecum, which is greatly

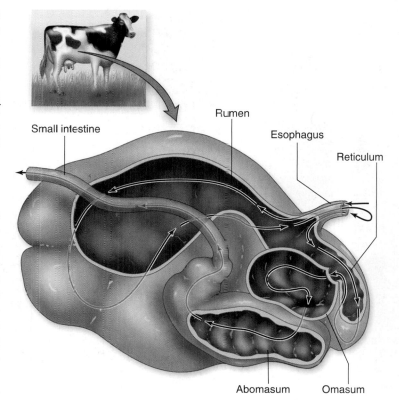

figure 48.15

FOUR-CHAMBERED STOMACH OF A RUMINANT. Grass and other plants eaten by ruminants enter the rumen, where they are partially digested. The rumen contains bacteria that break down cellulose from the plant cell walls. Before moving into a second chamber, the reticulum, the food may be regurgitated and rechewed. The food is then transferred to the rear two chambers: the omasum and abomasum. Only the abomasum secretes gastric juice as in the human stomach.

enlarged (figure 48.16). Because the cecum is located beyond the stomach, regurgitation of its contents is impossible.

Rodents and lagomorphs have evolved another way to capture nutrients from cellulose that achieves a degree of efficiency similar to ruminant digestion. They do this by eating their feces, a practice known as **coprophagy**—thus passing the food through their digestive tract a second time. The second passage allows the animal to absorb the nutrients produced by the microorganisms in its cecum. Coprophagic animals cannot remain healthy if they are prevented from eating their feces.

Animals with diets that don't include cellulose, such as insectivores or carnivores, don't have a cecum, or if they do, it is greatly reduced.

Wax digestion

Cellulose is not the only plant product that vertebrates can use as food thanks to intestinal microorganisms. Wax, a substance indigestible by most terrestrial animals, is digested by symbiotic bacteria living in the gut of African honey guides, *Prodotiscus insignis*, birds that eat the wax in bee nests.

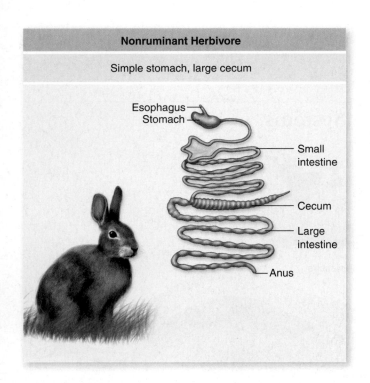

Nonruminant Herbivore

Simple stomach, large cecum

Esophagus
Stomach
Small intestine
Cecum
Large intestine
Anus

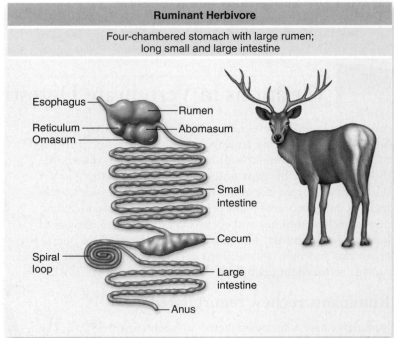

Ruminant Herbivore

Four-chambered stomach with large rumen; long small and large intestine

Esophagus
Reticulum
Omasum
Rumen
Abomasum
Small intestine
Spiral loop
Cecum
Large intestine
Anus

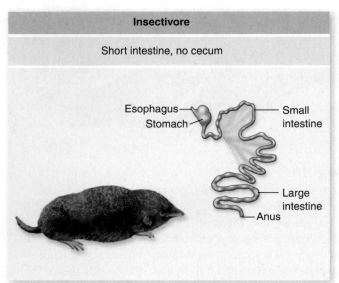

Insectivore

Short intestine, no cecum

Esophagus
Stomach
Small intestine
Large intestine
Anus

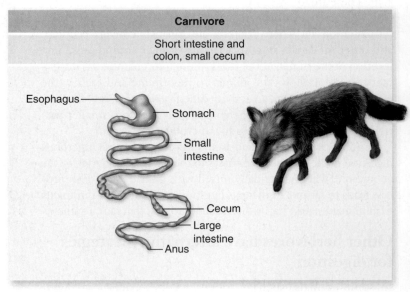

Carnivore

Short intestine and colon, small cecum

Esophagus
Stomach
Small intestine
Cecum
Large intestine
Anus

figure 48.16

THE DIGESTIVE SYSTEMS OF DIFFERENT MAMMALS REFLECT THEIR DIETS. Herbivores, such as rabbits and deer, require long digestive tracts with specialized compartments for the breakdown of plant matter. Protein diets are more easily digested; thus, insectivorous and carnivorous mammals, such as voles and foxes respectively, have short digestive tracts with few specialized pouches.

In the marine food chain, wax is a major constituent of copepods (crustaceans found in plankton), and many marine fish and birds appear able to digest wax with the aid of symbiotic microorganisms.

Vitamin K

Another example of the way intestinal microorganisms function in the metabolism of their animal hosts is provided by the synthesis of vitamin K. All mammals rely on intestinal bacteria to synthesize this vitamin, which is necessary for the clotting of blood. Birds, which lack these bacteria, must consume the required quantities of vitamin K in their food.

In humans, prolonged treatment with antibiotics greatly reduces the populations of bacteria in the intestine; under such circumstances, it may be necessary to provide supplementary vitamin K. Restoring the normal flora of the digestive tract with beneficial bacteria may also help replace vitamin K.

Much of the food value of plants is tied up in cellulose, and the digestive tract of many animals harbors colonies of cellulose-digesting microorganisms. Intestinal microorganisms also produce molecules such as vitamin K that are important to the well-being of their vertebrate hosts.

48.7 Neural and Hormonal Regulation of the Digestive Tract

The activities of the gastrointestinal tract are coordinated by the nervous system and the endocrine system. The nervous system, for example, stimulates salivary and gastric secretions in response to the sight, smell, and consumption of food. When food arrives in the stomach, proteins in the food stimulate the secretion of a stomach hormone called **gastrin,** which in turn stimulates the secretion of pepsinogen and HCl from the gastric glands (figure 48.17). The secreted HCl then lowers the pH of the gastric juice, which acts to inhibit further secretion of gastrin in a negative feedback loop. In this way, the secretion of gastric acid is kept under tight control.

The passage of chyme from the stomach into the duodenum of the small intestine inhibits the contractions of the stomach, so that no additional chyme can enter the duodenum until the previous amount can be processed. This stomach or gastric inhibition is mediated by a neural reflex and by duodenal hormones secreted into the blood. These hormones are collectively known as the **enterogastrones.**

The major enterogastrones include **cholecystokinin (CCK), secretin,** and **gastric inhibitory peptide (GIP).** Chyme with high fat content is the strongest stimulus for CCK and GIP secretions, whereas increasing chyme acidity primarily influences the release of secretin. All three of these enterogastrones inhibit gastric motility (churning action) and gastric juice secretions; the result is that fatty meals remain in the stomach longer than nonfatty meals.

In addition to gastric inhibition, CCK and secretin have other important regulatory functions in digestion. CCK also stimulates increased pancreatic secretions of digestive enzymes and gallbladder contractions. Gallbladder contractions inject more bile into the duodenum, which enhances the emulsification and

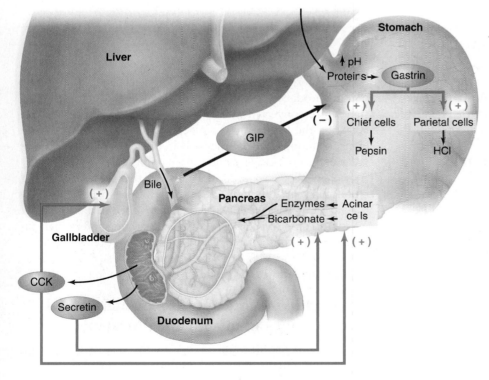

figure 48.17

HORMONAL CONTROL OF THE GASTROINTESTINAL TRACT. Gastrin, secreted by the mucosa of the stomach, stimulates the secretion of HCl and pepsinogen (which is converted into pepsin). The duodenum secretes three hormones: cholecystokinin (CCK), which stimulates contraction of the gallbladder and secretion of pancreatic enzymes; secretin, which stimulates secretion of pancreatic bicarbonate; and gastric inhibitory peptide (GIP), which inhibits stomach emptying.

efficient digestion of fats. The other major function of secretin is to stimulate the pancreas to release more bicarbonate, which neutralizes the acidity of the chyme. Secretin has the distinction of being the first hormone ever discovered. Table 48.1 summarizes the actions of the digestive hormones and enzymes.

Neural and hormonal reflexes regulate the activity of the digestive system. The stomach's secretions are regulated by food and by gastrin. Enterogastrones secreted by the duodenum inhibit gastric functions and promote digestion in the duodenum.

TABLE 48.1 Hormones and Enzymes of Digestion

HORMONES

Hormone	Class	Source	Stimulus	Action	Note
Gastrin	Polypeptide	Pyloric portion of stomach	Entry of food into stomach	Stimulates secretion of HCl and pepsinogen by stomach	Acts on same organ that secretes it
Cholecystokinin (CCK)	Polypeptide	Duodenum	Fatty chyme in duodenum	Stimulates gallbladder contraction and secretion of digestive enzymes by pancreas	Structurally similar to gastrin
Gastric inhibitory peptide (GIP)	Polypeptide	Duodenum	Fatty chyme in duodenum	Inhibits stomach emptying	Also stimulates insulin secretion
Secretin	Polypeptide	Duodenum	Acidic chyme in duodenum	Stimulates secretion of bicarbonate by pancreas	The first hormone to be discovered (1902)

ENZYMES

Location	Enzymes	Substrates	Digestion Products
Salivary glands	Amylase	Starch, glycogen	Disaccharides
Stomach	Pepsin	Proteins	Short peptides
Pancreas	Lipase	Triglycerides	Fatty acids, monoglycerides
	Trypsin, chymotrypsin	Proteins	Peptides
	DNase	DNA	Nucleotides
	RNase	RNA	Nucleotides
Small intestine (brush border)	Peptidases	Short peptides	Amino acids
	Nucleases	DNA, RNA	Sugars, nucleic acid bases
	Lactase, maltase, sucrase	Disaccharides	Monosaccharides

48.8 Accessory Organ Function

The liver and pancreas both have critical roles beyond the production of digestive enzymes. The liver is a key organ in breakdown of toxins, and the pancreas secretes hormones that regulate the blood glucose level, in part through actions on liver cells.

The liver modifies chemicals to maintain homeostasis

Because the hepatic portal vein carries blood from the stomach and intestine directly to the liver, the liver is in a position to chemically modify the substances absorbed in the gastrointestinal tract before they reach the rest of the body. For example, ingested alcohol and other drugs are taken into liver cells and metabolized; this is one reason that the liver is often damaged as a result of alcohol and drug abuse.

The liver also removes toxins, pesticides, carcinogens, and other poisons, converting them into less toxic forms. An important example of this is the liver's conversion of the toxic ammonia produced by intestinal bacteria into urea, a compound that can be contained safely and carried by the blood at higher concentrations.

Similarly, the liver regulates the levels of many compounds produced within the body. Steroid hormones, for instance, are converted into less active and more water-soluble forms by the liver. These molecules are then included in the bile and eliminated from the body in the feces, or are carried by the blood to the kidneys and excreted in the urine.

The liver also produces most of the proteins found in blood plasma. The total concentration of plasma proteins is significant because it must be kept within certain limits to maintain osmotic balance between blood and interstitial (tissue) fluid. If the concentration of plasma proteins drops too low, as can happen as a result of liver disease such as cirrhosis, fluid accumulates in the tissues, a condition called *edema*.

Blood glucose concentration is maintained by the actions of insulin and glucagon

The neurons in the brain obtain energy primarily from the aerobic respiration of glucose obtained from the blood plasma. It is therefore vitally important that the blood glucose concentration not fall too low, as might happen during fasting cr prolonged exercise. It is also important that the blood glucose concentration not stay at too high a level, as it does in people with untreated diabetes mellitus, because too high a level can lead to tissue damage.

After a carbohydrate-rich meal, the liver and skeletal muscles remove excess glucose from the blood and store it as the polysaccharide **glycogen.** This process is stimulated by the hormone insulin, secreted by the β (beta) cells in the pancreatic islets of Langerhans (figure 48.18).

When blood glucose levels decrease, as they do between meals, during periods of fasting, and during exercise, the liver secretes glucose into the blood. This glucose is obtained in part from the breakdown of liver glycogen to glucose-6-phosphate, a process called **glycogenolysis.** The phosphate group is then removed, and free glucose is secreted into the blood. Skeletal muscles lack the enzyme needed to remove the phosphate group, and so, even though they have glycogen stores, they cannot secrete glucose into the blood. However, muscle cells can use this glucose directly for energy metabolism because glucose-6-phosphate is actually the product of the first reaction in glycolysis. The breakdown of liver glycogen is stimulated by another hormone, glucagon, which is secreted by the α (alpha) cells of the islets of Langerhans in the pancreas (see figure 48.18).

If fasting or exercise continues, the liver begins to convert other molecules, such as amino acids and lactic acid, into glucose. This process is called **gluconeogenesis** ("new formation of glucose"). The amino acids used for gluconeogenesis are obtained from muscle protein, which explains the severe muscle wasting that occurs during prolonged fasting.

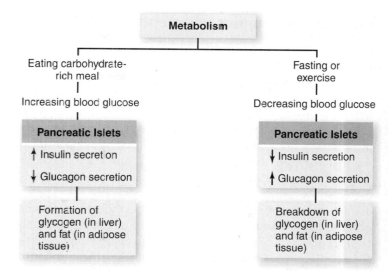

figure 48.18

THE ACTIONS OF INSULIN ANE GLUCAGON. After a meal, an increased secretion of insulin by the β cells of the pancreatic islets promotes the deposition of glycogen and fat. During fasting or exercising, increased glucagon secretion by the α cells of the pancreatic islets and decreased insulin secretion promote the breakdown (through hydrolysis reactions) of glycogen and fat.

The liver is responsible for chemical modification of many molecules, including toxins and steroid hormones. Pancreatic hormones and the liver regulate blood glucose concentrations. Insulin stimulates the formation of glycogen and fat in the liver. Glucagon stimulates the breakdown of glycogen in the liver, which releases glucose into the blood.

48.9 Food Energy, Energy Expenditure, and Essential Nutrients

The ingestion of food serves two primary functions: It provides a source of energy and it provides raw materials the animal is unable to manufacture for itself.

Even an animal that is completely at rest requires energy to support its metabolism; the minimum rate of energy consumption under defined resting conditions is called the **basal metabolic rate (BMR).** The BMR is relatively constant for a given individual, depending primarily on the person's age, sex, and body size.

Exertion increases metabolic rate

Physical exertion raises the metabolic rate above the basal levels, so the amount of energy the body consumes per day is determined not only by the BMR but also by the level of physical activity. If food energy taken in is greater than the energy consumed per day, the excess energy will be stored in glycogen and fat. Because glycogen reserves are limited, however, continued ingestion of excess food energy results primarily in the accumulation of fat.

The intake of food energy is measured in **kilocalories** (1 kilocalorie = 1000 calories; nutritionists use Calorie with a capital C instead of kilocalorie). The measurement of kilo-

calories in food is determined by the amount of heat generated when the food is "burned," either literally, in a testing device called a calorimeter, or in the body, when the food is digested and later oxidized during cellular respiration. Caloric intake can be altered by the choice of foods, and the amount of energy expended can be changed by the choice of lifestyle.

The daily energy expenditures (metabolic rates) of humans vary between 1300 and 5000 kilocalories per day, depending on the person's BMR and level of physical activity. When the total kilocalories ingested exceeds the metabolic rate for a sustained period, a person accumulates an amount of fat that is deleterious to health, a condition called **obesity.** Obesity is currently classified using the metric **body mass index (BMI).** This is a ratio of height and weight that is intended to estimate body fat without directly measuring it. BMI is calculated as weight in kilograms divided by the square of height in meters. A BMI of over 30 is considered obese. In the United States, about 34% of all adults between 40 and 59 are classified as obese. Among men ages 40 to 59, about 31% are classified as obese and among women ages 40 to 59 about 37% are classified as obese. If 20–40-year-olds are added to this group, the percentage of obese individuals drops some but is still fully 30%.

Food intake is under neuroendocrine control

For many years the neuronal and hormonal basis of appetite was a mystery. Experiments with fasting and overfeeding in rats showed an increase in food intake when fasting ends. This increase restores lost bodyweight to baseline values and food intake then drops. These experiments indicated the existence of control mechanisms to link food intake to energy balance. The presence of a hormonal *satiety factor* produced by adipose tissue was hypothesized to explain these observations. It has also been shown that regions of the hypothalamus are involved in feeding behavior. Other studies in rodent models had identified a number of genes that can lead to obesity. As modern molecular genetics has allowed the cloning of many of these genes, the outlines of a model to link dietary intake to energy balance have emerged. This model involves afferent signaling tied to adipose tissue and feeding behavior, and efferent signaling tied to energy expenditure, storage, reproduction and feeding behavior. We will discuss the relevant hormones first, then show how they fit into an overall control circuit.

Leptin

One of the rodent models for obesity, the obese mouse, is caused by a mutation in a single gene named *ob* (for obese). Animals homozygous for the recessive mutant allele become obese compared with wild-type mice (figure 48.19). When the gene responsible for this dramatic phenotype was isolated, it proved to encode a peptide hormone named **leptin.** When *ob/ob* animals are injected with leptin, they stop overeating and are not obese (see figure 48.19). These experiments identified leptin as the main satiety factor, and the key to the control of appetite. The gene for the leptin receptor (*db*) has also been isolated and it is expressed in brain neurons in the hypothalamus involved in energy intake.

Leptin is now thought to be the main signaling molecule in the afferent portion of the control circuit for energy sensing, food intake, and energy expenditure. Leptin is produced by

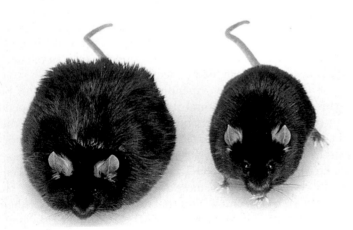

figure 48.19

EFFECTS OF THE HORMONE LEPTIN. Both of these mice are homozygous for the recessive *ob* mutation, which causes obesity. The *ob* gene encodes the peptide hormone leptin. The mouse on the right has been injective with leptin. It lost 30% of its body weight in just two weeks, with no apparent side effects.

adipose tissue in response to feeding, and leptin levels correlate with feeding behavior and amount of body fat. Dietary restriction reduces leptin levels, signaling the brain that food intake is necessary, refeeding after fasting leads to rapid increase in leptin levels. The efferent part of this control circuit is complex and includes control of energy expenditure, energy storage, and feeding behavior. Reproduction is even affected by this system as reproduction is inhibited under starvation conditions.

The leptin gene has also been isolated in humans and leptin appears to function in humans much as it does in mice. More recent studies in humans show that the activity of the *ob* gene and the blood concentrations of leptin are actually higher in obese than in lean people, and that the leptin produced by obese people appears to be normal. It has been suggested that most cases of human obesity may result from a reduced sensitivity to the actions of leptin in the brain, rather than from reduced leptin production by adipose cells. Research on leptin in humans is ongoing and is of great interest to both academic scientists and to the pharmaceutical industry.

Insulin

Although the extreme obesity associated with the loss-of-function mutations in the *ob* gene indicate that other hormonal signals cannot substitute for leptin signaling, other hormones are also involved. Insulin has been implicated in signaling satiety as well, and insulin levels also fall with fasting and rise with obesity. As insulin's primary role is homeostasis of blood glucose, as described earlier, its role in the control circuit of energy balance is complex.

Gut hormones

The gut produces a number of hormones that control the physiology of digestion described earlier. Several of these have also been implicated in the regulation of food intake. They are produced directly in response to feeding, necessary for their role in digestion.

The hormones GIP and CCK have receptors in the hypothalamus and seem to send the same kind of inhibitory signals to the brain as leptin and insulin. The levels of these gut hormones also vary with feeding behavior in a pattern similar to leptin and insulin.

The gut hormone ghrelin has the opposite effect of these appetite-suppressing hormones. Ghrelin also has receptors in the hypothalamus, but ghrelin appears to stimulate food intake. This role is supported by studies in rats showing that chronic administration of ghrelin leads to obesity. Ghrelin levels appear to rise before feeding and may be involved in initiating feeding behavior. One of the treatments for severe obesity, gastric bypass surgery, leads to reduced levels of ghrelin. It has been suggested that this is one of the reasons for the suppression of appetite seen after this surgery.

Neuropeptides

The efferent control over feeding and energy balance is less clear than the afferent control detailed earlier. The central regulator is the hypothalamus and two brain neuropeptides have been implicated: **neuropeptide Y (NPY)** and melanocyte-stimulating hormone (α-MSH). These two peptides are antagonistic, with NPY inducing feeding activity and α-MSH suppressing it.

Evidence for this comes from experiments that show that production and release of α-MSH is stimulated by leptin and that administration of α-MSH suppresses feeding. Loss of function for the α-MSH receptor also leads to obesity. In contrast, the expression of NPY is negatively regulated by leptin and administration of NPY stimulates feeding behavior

Model for energy balance

The current model for energy balance and feeding behavior is summarized in figure 48.20. The role of both leptin and insulin is long-term regulation of the afferent portion of this signaling network. Leptin and insulin are produced by adipose tissue and the pancreas respectively in response to the effects of feeding behavior, not as a direct response to feeding itself. This leads to circulating levels of leptin that correlate with the amount of adipose tissue. The extreme example of this is the very high level of leptin seen in obesity. High levels of leptin and insulin then act on the hypothalamus to increase levels of α-MSH and reduce levels of NPY. This causes a reduction in appetite and increased energy expenditure and allows reproduction and growth. Low levels of these hormones act on the hypothalamus to reduce α-MSH levels and increase NPY levels. This leads to increased appetite and decreased energy expenditure. If very low levels of leptin persist, this can inhibit reproduction and growth.

The gut hormones CCK and GIP are produced in response to feeding and represent short-term regulators of the afferent portion of the energy balance control circuit. Their ac-tion is the same as that of leptin and insulin. The gut hormone ghrelin is also a short-term regulator that stimulates feeding.

Eating disorders

In the United States, serious eating disorders have become much more common since the mid-1970s. The most common of these disorders are anorexia nervosa, a condition in which the afflicted individuals literally starve themselves, and bulimia, in which individuals gorge themselves and then induce vomiting, so that their weight stays constant. Ninety to 95% of those suffering from these disorders are female; researchers estimate that 2 to 5% of the adolescent girls and young women in the United States have eating disorders.

Essential nutrients are those that the body cannot manufacture

Over the course of evolution, many animals have lost the ability to synthesize certain substances that nevertheless continue to play critical roles in their metabolism. Substances that an animal cannot manufacture for itself but that are necessary for its health and must be obtained in the diet are referred to as **essential nutrients.**

Included among the essential nutrients are **vitamins,** certain organic substances required in trace amounts. For example, humans, apes, monkeys, and guinea pigs have lost the ability to synthesize ascorbic acid (vitamin C). If vitamin C is not supplied in sufficient quantities in their diets, these

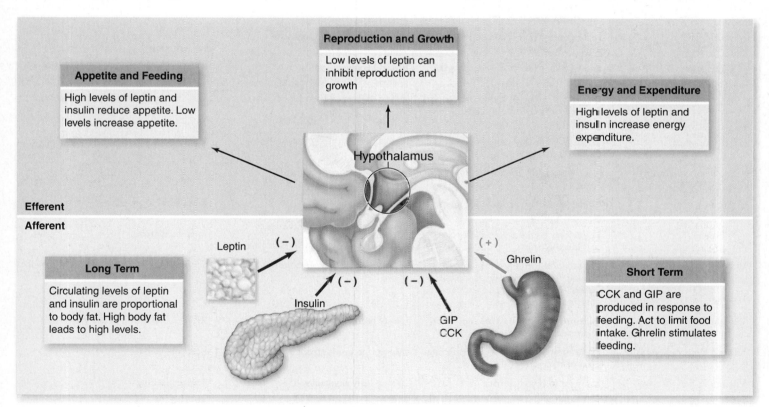

figure 48.20

HORMONAL CONTROL OF FEEDING BEHAVIOR. The control of feeding behavior is under both long-term control related to the amount of adipose tissue, and short-term control related to the act of feeding. This control is mediated by the CNS. The major brain region involved is the hypothalamus.

mammals develop scurvy, a potentially fatal disease caused by degeneration of connective tissues. Humans require at least 13 different vitamins (table 48.2).

Some essential nutrients are required in more than trace amounts. Many vertebrates, for example, are unable to synthesize one or more of the 20 amino acids. These *essential amino acids* must be obtained from food they eat. Humans require nine amino acids. People who are strict vegetarians must choose their foods so that the essential amino acids in one food complement those in another. Vegetarians may also need supplements to provide certain vitamins not found in large amounts in plants, such as some B vitamins.

In addition, all vertebrates have lost the ability to synthesize certain long-chain unsaturated fatty acids and therefore must obtain them in food. In contrast, some essential nutrients that vertebrates can synthesize cannot be manufactured by the members of other animal groups. For example, vertebrates can synthesize cholesterol, a key component of steroid hormones, but some carnivorous insects cannot.

Food also supplies **essential minerals** such as calcium, magnesium, phosphorus, and other inorganic substances, including a wide variety of *trace elements* such as zinc and molybdenum, which are required in very small amounts. Animals obtain trace elements either directly from plants or from animals that have eaten plants.

The amount of caloric energy expended by the body depends on the basal metabolic rate and the additional calories consumed by exercise. Obesity results if the ingested food energy exceeds the energy expenditure by the body over a prolonged period. Control of feeding behavior is linked to energy expenditure by the hormone leptin. Leptin levels are linked to body fat. The hypothalamus is involved in control of feeding behavior and energy storage and expenditure. The body requires vitamins and minerals obtained in food. Also, food must provide particular essential amino acids and fatty acids that the body cannot manufacture by itself.

TABLE 48.2   Major Vitamins Required by Humans

Vitamin	Function	Source	Deficiency Symptoms
Vitamin A (retinol)	Used in making visual pigments, maintaining epithelial tissues	Green vegetables, milk products, liver	Night blindness, flaky skin
B-complex vitamins			
B_1	Coenzyme in CO_2 removal during cellular respiration	Meat, grains, legumes	Beriberi, weakening of heart, edema
B_2 (riboflavin)	Part of coenzymes FAD and FMN, which play metabolic roles	Many different kinds of foods	Inflammation and breakdown of skin, eye irritation
B_3 (niacin)	Part of coenzymes NAD^+ and $NADP^+$	Liver, lean meats, grains	Pellagra, inflammation of nerves, mental disorders
B_5 (pantothenic acid)	Part of coenzyme-A, a key connection between carbohydrate and fat metabolism	Many different kinds of foods	Rare: fatigue, loss of coordination
B_6 (pyridoxine)	Coenzyme in many phases of amino acid metabolism	Cereals, vegetables, meats	Anemia, convulsions, irritability
B_{12} (cyanocobalamin)	Coenzyme in the production of nucleic acids	Red meats, dairy products	Pernicious anemia
Biotin	Coenzyme in fat synthesis and amino acid metabolism	Meat, vegetables	Rare: depression, nausea
Folic acid	Coenzyme in amino acid and nucleic acid metabolism	Green vegetables	Anemia, diarrhea
Vitamin C	Important in forming collagen, cement of bone, teeth, connective tissue of blood vessels; may help maintain resistance to infection	Fruit, green leafy vegetables	Scurvy, breakdown of skin, blood vessels
Vitamin D (calciferol)	Increases absorption of calcium and promotes bone formation	Dairy products, cod liver oil	Rickets, bone deformities
Vitamin E (tocopherol)	Protects fatty acids and cell membranes from oxidation	Margarine, seeds, green leafy vegetables	Rare
Vitamin K	Essential to blood clotting	Green leafy vegetables	Severe bleeding

48.1 Types of Digestive Systems

Animals are grouped based on food sources: herbivores eat plants, carnivores eat other animals, and omnivores eat both plants and animals.

- In cnidarians and flatworms the digestive or gastrovascular cavity has only one opening (figure 48.1).
- A one-way digestive system allows specialization in regions for ingestion, storage, fragmentation, chemical digestion, and absorption.
- Chemical digestion breaks food molecules into smaller subunits.
- The gastrointestinal tract is the mouth, pharynx, esophagus, stomach, small and large intestines, and cloaca or rectum (figure 48.3).
- The gastrointestinal tract has four layers (figure 48.4).

48.2 The Mouth and Teeth: Food Capture and Bulk Processing

Specializations in digestive systems reflect the way animals live.

- Many vertebrates have teeth to grind food. Birds have a gizzard where small pebbles grind food into smaller particles.
- The teeth of mammals reflect their feeding habits (figure 48.5).
- When food enters the mouth, salivary glands secrete saliva that moistens and lubricates food and contains amylase.
- The process of swallowing begins as a voluntary action and ends with involuntary control (figure 48.8).

48.3 The Esophagus and the Stomach: The Beginning of Digestion

Swallowed food leaves the pharynx, and enters the esophagus and then the stomach where food is stored and initial digestion occurs.

- Rhythmic muscular contractions, called peristalsis, propel a bolus of food to the stomach.
- In the stomach, food is mixed with hydrochloric acid and inactive pepsinogen, which is activated by low pH (figure 48.10).
- Pepsinogen is converted into pepsin, an active protease.
- Parietal cells produce intrinsic factor that promotes the absorption of vitamin B_{12}.
- Low pH in the stomach helps denature food proteins.
- The mixture of partially digested food and gastric juice, chyme, is transferred through the pyloric sphincter to the small intestine.
- Infection with *Helicobacter pylori* can weaken the lining of the stomach and the duodenum allowing stomach acid to cause ulcers.

48.4 The Small Intestine: Breakdown and Absorption

The terminal digestion of carbohydrates, lipids, and proteins and most of the absorption occurs in the small intestine.

- The duodenum receives digestive enzymes and bicarbonate from the pancreas and bile from the liver and gallbladder.
- The surface area of the small intestine is increased by fingerlike projections called villi (figure 48.11).
- Accessory organs for digestion include the pancreas, liver, and gallbladder (figure 48.12).
- The pancreas secretes digestive enzymes and bicarbonate.
- Bile salts disperse fats into small droplets.
- The liver secretes bile and the gallbladder stores it.
- Amino acids and monosaccharides move into epithelial cells by active transport and facilitated diffusion (figure 48.13).
- Fats are hydrolyzed to fatty acids and monoglycerides, which diffuse into epithelial cells. They are reassembled into chylomicrons.

- Chylomicrons enter the lymphatic system and later join the circulatory system (figure 48.13).
- Absorbed nutrients are transported to the liver through the hepatic portal vein.

48.5 The Large Intestine: Elimination of Waste Material

The small intestine joins the large intestine at a junction with two vestigial structures, the cecum and appendix (figure 48.14).

- Some intestinal water, vitamin K, and ions are absorbed here.
- The primary function is the concentration of waste material.
- Compacted feces is stored in the rectum until it can be eliminated.
- Most mammals have a rectum but most vertebrates have a common cavity, the cloaca, where the urinary, reproductive, and gastrointestinal tracts join.

48.6 Variations in Vertebrate Digestive Systems

Most animals cannot digest cellulose, but some have bacteria and protists in their digestive tracts to break down cellulose (figure 48.16).

- Ruminants have a four-chambered stomach sequentially divided into the rumen, reticulum, omasum, and abomasum.
- Food is initially processed in the rumen, a fermentation vat containing bacteria and protists (figure 48.15).
- In some herbivores the digestion of cellulose by microorganisms takes place in the cecum, which is located beyond the stomach.
- To take advantage of nutrients released by microbial activity in the cecum, rodents and lagomorphs ingest their feces (coprophagy).
- Some animals digest waxes with the assistance of microorganisms.

48.7 Neural and Hormonal Regulation of the Digestive Tract

The activities of the gastrointestinal tract are coordinated by the nervous and endocrine systems (figure 48.17 and table 48.1).

- In the stomach, proteins stimulate the release of gastrin, which stimulates the release of HCl and pepsinogen.
- Duodenal hormones inhibit stomach contractions and prevent additional chyme from entering the duodenum.
- High fat content in the chyme stimulates the release of CCK and GIP, low chyme pH stimulates the release of secretin.
- CCK stimulates release of pancreatic enzymes and gallbladder contractions. Secretin stimulates release of bicarbonate.

48.8 Accessory Organ Function

The liver and pancreas contribute more than the digestion of enzymes.

- The liver is involved in detoxification, regulation of steroid hormone levels, and production of proteins found in the blood plasma.
- The pancreatic hormones, insulin and glucagon, regulate blood glucose levels and synthesis of glycogen, which is stored in the liver.

48.9 Food Energy, Energy Expenditure, and Essential Nutrients

The ingestion of food provides a source of energy and raw materials.

- The basal metabolic rate is the minimum rate of energy consumption under resting conditions.
- Food intake is regulated by the hormones, leptin and insulin, gut hormones, and neuropeptides (figure 48.20).
- Essential nutrients cannot be synthesized by animals (table 48.2).

SELF TEST

1. The intestines of herbivores are _____ compared with those of carnivores.
   a. longer
   b. shorter
   c. about the same size
   d. less convoluted

2. Which of the following is *not* a function of the digestive system?
   a. Ingestion
   b. Digestion (physical and chemical)
   c. Absorption
   d. All of these are functions of the digestive system

3. For birds, what replaces the action of teeth in grinding up food?
   a. Hard, bony ridges in the beak
   b. Pebbles in the gizzard
   c. An extremely muscular esophagus
   d. The normal bird diet does not require grinding for digestion

4. How is the digestion of fats different from that of proteins and carbohydrates?
   a. Fat digestion occurs in the small intestine, and the digestion of proteins and carbohydrates occurs in the stomach.
   b. Fats are absorbed into the cells as fatty acids and monoglycerides but are then modified for absorption; amino acids and glucose are not modified further.
   c. Fats enter the hepatic portal circulation, but digested proteins and carbohydrates enter the lymphatic system.
   d. Digested fats are absorbed in the large intestine, and digested proteins and carbohydrates are absorbed in the small intestine.

5. Although the stomach is normally thought of as the major player in the digestive process, the bulk of chemical digestion actually occurs in the—
   a. mouth.
   b. appendix.
   c. duodenum.
   d. large intestine.

6. After being absorbed through the intestinal mucosa, glucose and amino acids—
   a. are absorbed directly into the systemic circulation.
   b. are used to build glycogen and peptides before being released to the body cells.
   c. are transported directly to the liver by the hepatic portal vein.
   d. are further digested by bile before release into the circulation.

7. The small intestine is specialized for absorption because it—
   a. is the last section of the digestive tract and retains food the longest.
   b. has saclike extensions along its length that collect food.
   c. has no outlet so food remains within it for longer periods of time.
   d. has an extremely large surface area that allows extended exposure to food.

8. Which of these pairings is incorrect?
   a. Fat transport/lymphatic system
   b. Glucose transport/lymphatic system
   c. Amino acid transport/circulatory system
   d. All of these pairings are correct.

9. The primary function of the large intestine is to concentrate wastes into solid form (feces) for release from the body. How does it accomplish this?
   a. By adding additional cells from the mucosal layer
   b. By absorbing water
   c. By releasing salt
   d. All of these are methods used by the large intestine

10. Which of these is not a method used by vertebrates to increase the nutritive value of the food they eat?
    a. Regurgitation and rechewing
    b. Eating feces
    c. Adding minerals from the blood
    d. Increasing length and surface area of the digestive tract

11. Intestinal microorganisms aid digestion and absorption by—
    a. digesting cellulose.
    b. producing glucose.
    c. synthesizing vitamin K.
    d. Both (a) and (c).

12. Suppose that you join the Homeland Security Administration after medical school. Your job is to study the potential effects of toxin exposure on the digestive system. Which of the following organs is the site for toxin modification?
    a. Duodenum
    b. Liver
    c. Pancreas
    d. Gallbladder

13. The _____ and _____ play important roles in the digestive process by producing chemicals that are required to digest proteins, lipids, and carbohydrates.
    a. liver; pancreas
    b. liver; gallbladder
    c. kidneys; appendix
    d. pancreas; gallbladder

14. Which of the following represents the action of insulin?
    a. Increases blood glucose levels by the hydrolysis of glycogen
    b. Increases blood glucose levels by stimulating glucagon production
    c. Decreases blood glucose levels by forming glycogen
    d. Increases blood glucose levels by promoting cellular uptake of glucose

15. On a medical TV show, you see a story about a man who could not stop eating. The program said that this behavior was the result of a brain tumor. Which section of the brain is involved in controlling feeding behavior and therefore likely to be the site of the tumor?
    a. Cerebral cortex
    b. Cerebellum
    c. Occipital lobe
    d. Hypothalamus

CHALLENGE QUESTIONS

1. Many birds possess crops, although few mammals do. Suggest a reason for this difference between birds and mammals.

2. Suppose that you wanted to develop a new treatment for obesity based on the hormone leptin. What structures in the body produce leptin? What does it do? Should your treatment cause an increase in blood levels of leptin or a decrease? Could this treatment affect any other systems in the body?

3. How could a drop in plasma proteins and a decrease in bile production be related to alcohol and drug abuse?

chapter **49**

The Circulatory and Respiratory Systems

introduction

EVERY CELL IN THE ANIMAL BODY must exchange materials with its surrounding environment. In single-celled organisms, this exchange occurs directly across the cell membrane to and from the external environment. In multicellular organisms, however, most cells are not in contact with the external environment and must rely on specialized systems for transport and exchange. Although these systems aid in bulk transport, the properties of transport across the plasma membrane do not change. Many structural adaptations throughout the animal kingdom increase surface areas where transport occurs, so that the needs of every cell are met. The interface between air from the environment and blood in the mammalian lungs provides an excellent example of the efficiency associated with increased surface area. In the time it takes you to breathe in, trillions of oxygen molecules have been transported across 80 m² of alveolar membrane into blood capillaries. In this chapter, we describe circulation and respiration, the two systems that directly support the other organ systems and tissues of the body.

Invertebrate Circulatory Systems

The nature of the circulatory system in multicellular invertebrates is directly related to the size, complexity, and lifestyle of the organism in question. Sponges and most cnidarians utilize water from the environment as a circulatory fluid. Sponges pass water through a series of channels in their bodies, and *Hydra* and other cnidarians circulate water through a **gastrovascular cavity** (figure 49.1*a*). Because the body wall in *Hydra* species is only two cell layers thick, each cell layer is in direct contact with either the external environment or the gastrovascular cavity.

Pseudocoelomate invertebrates (roundworms, rotifers) use the fluids of the body cavity for circulation. Most of these invertebrates are quite small, or are long and thin, and therefore adequate circulation is accomplished by movements of the body against the body fluids, which are in direct contact with the internal tissues and organs. Larger animals, however, have tissues that are several cell layers thick, so that many cells are too far away from the body surface or digestive cavity to directly exchange materials with the environment. Instead, oxygen and nutrients are transported from the environment and digestive cavity to the body cells by an internal fluid within a circulatory system.

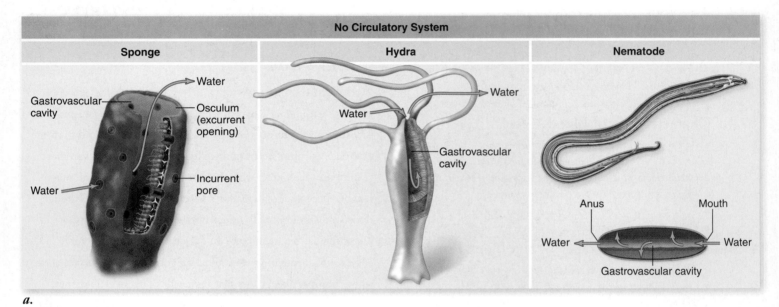

a.

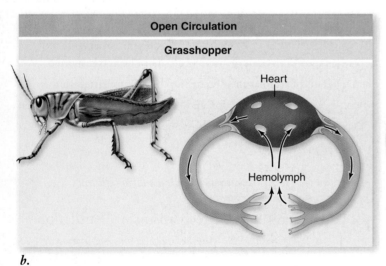

b.

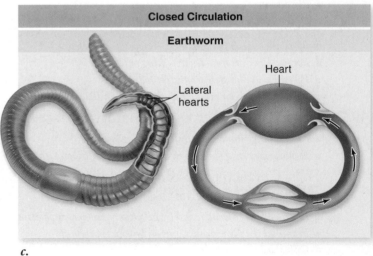

c.

figure 49.1

CIRCULATORY SYSTEMS OF THE ANIMAL KINGDOM. *a.* Sponges (*left panel*) do not have a separate circulatory system. They circulate water using many incurrent pores and one excurrent pore. The gastrovascular cavity of a hydra (*middle panel*) serves as both a digestive and a circulatory system, delivering nutrients directly to the tissue cells by diffusion from the digestive cavity. The nematode (*right panel*) is thin enough that the digestive tract can also be used as a circulatory system. Larger animals require a separate circulatory system to carry nutrients to and wastes away from tissues. *b.* In the open circulation of an insect, hemolymph is pumped from a tubular heart into cavities in the insect's body; the hemolymph then returns to the blood vessels so that it can be recirculated. *c.* In the closed circulation of the earthworm, blood pumped from the hearts remains within a system of vessels that returns it to the hearts. All vertebrates also have closed circulatory systems.

Open circulatory systems move fluids in a one-way path

The two main types of circulatory systems are *open* and *closed*. In an **open circulatory system,** such as that found in most mollusks and in arthropods (figure 49.1*b*), there is no distinction between the circulating fluid and the extracellular fluid of the body tissues. This fluid is thus called **hemolymph.**

In insects, a muscular tube, or **heart,** pumps hemolymph through a network of channels and cavities in the body. The fluid then drains back into the central cavity.

Closed circulatory systems move fluids in a loop

In a **closed circulatory system,** the circulating fluid, or blood, is always enclosed within blood vessels that transport it away from and back to the heart (figure 49.1*c*). Some invertebrates, such as cephalopod mollusks and annelids (see chapter 34), and all vertebrates have a closed circulatory system.

In annelids such as earthworms, a dorsal vessel contracts rhythmically to function as a pump. Blood is pushed through five small connecting arteries, which also function as pumps, to a ventral vessel, which transports the blood posteriorly until it eventually reenters the dorsal vessel. Smaller vessels branch from each artery to supply the tissues of the earthworm with oxygen and nutrients and to remove waste products.

> The need to move nutrients and waste materials to and from all cells in an increasingly large and complex body led to the development of circulatory systems. In invertebrates, open circulatory systems pump hemolymph into tissues, which then drains into a central cavity. Closed circulatory systems move fluid in a loop to and from a pumping region such as a heart.

49.2 Vertebrate Circulatory Systems

As the body size and physiological complexity of animals increased, so did the need for more surface area both to deliver nutrients and oxygen and to remove wastes and carbon dioxide from the growing mass of tissues. Vertebrates have evolved a remarkable set of adaptations inextricably linking circulation and respiration, which has allowed the development of large body size and locomotion on land.

In fishes, more efficient circulation developed concurrently with gills

Chordates ancestral to the vertebrates are thought to have had simple tubular hearts, similar to those now seen in lancelets (see chapter 35). The heart was little more than a specialized zone of the ventral artery that was more heavily muscled than the rest of the arteries; it contracted in simple peristaltic waves.

The development of gills by fishes required a more efficient pump, and in fishes we see the evolution of a true chamber-pump heart. The fish heart is, in essence, a tube with four structures arrayed one after the other to form two pumping chambers (figure 49.2). The first two structures—the **sinus venosus** and **atrium**—form the first chamber; the second two, the **ventricle** and **conus arteriosus,** form the second chamber. The sinus venosus is the first to contract, followed by the atrium, the ventricle, and finally the conus arteriosus.

Despite shifts in the relative positions of these structures, this heartbeat sequence is maintained in all vertebrates. In fish, the electrical impulse that produces the contraction is initiated in the sinus venosus; in other vertebrates, the electrical impulse is initiated by a structure homologous to the sinus venosus—the **sinoatrial (SA) node.**

After blood leaves the conus arteriosus, it moves through the gills, becoming oxygenated. Blood leaving the gills then flows through a network of arteries to the rest of the body, finally returning to the sinus venosus. This simple loop has one serious limitation: in passing through the capillaries in the gills, blood

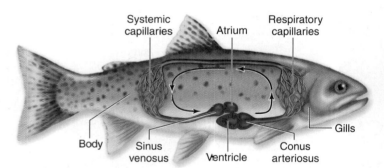

figure 49.2

THE HEART AND CIRCULATION OF A FISH. Diagram of a fish heart, showing the structures in series with each other (sinus venosus; atrium; ventricles; conus arteriosus) that form two pumping chambers. Blood is pumped by the ventricle through the gills and then to the body. Blood rich in oxygen (oxygenated) is shown in red; blood low in oxygen (deoxygenated) is shown in blue.

pressure drops significantly. This slows circulation from the gills to the rest of the body, and can limit oxygen delivery to tissues.

In amphibians and most reptiles, lungs required a separate circulation

The advent of lungs in amphibians involved a major change in the pattern of circulation, a second pumping circuit. After blood is pumped by the heart through the *pulmonary arteries* to the lungs, it does not go directly to the tissues of the body. Instead, it is returned via the *pulmonary veins* to the heart. Blood leaves the heart a second time to be circulated through other tissues. This system is termed **double circulation:** One system, the **pulmonary circulation,** moves blood between heart and lungs, and another, the **systemic circulation,** moves blood between the heart and the rest of the body.

Amphibian circulation

Optimally, oxygenated blood from lungs would go directly to tissues, rather than being mixed in the heart with deoxygenated blood returning from the body. The amphibian heart has two structural features that significantly reduce this mixing (figure 49.3). First, the atrium is divided into two chambers: The right atrium receives deoxygenated blood from the systemic circulation, and the left atrium receives oxygenated blood from the lungs. These two types of blood, therefore, do not mix in the atria.

Because an amphibian heart has a single ventricle, the separation of the pulmonary and systemic circulations is incomplete. The extent of mixing when the contents of each atrium enter the ventricle is reduced by internal channels created by recesses in the ventricular wall. The conus arteriosus is partially separated by a dividing wall, which directs deoxygenated blood into the pulmonary arteries and oxygenated blood into the *aorta*, the major artery of the systemic circulation.

Amphibians living in water can obtain additional oxygen by diffusion through their skin. Thus, amphibians have a *pul-mocutaneous circuit* that sends blood to both the lungs and the skin. Cutaneous respiration is also seen in many aquatic reptiles such as turtles.

Reptilian circulation

Among reptiles, additional modifications have further reduced the mixing of blood in the heart. In addition to having two separate atria, reptiles have a septum that partially subdivides the ventricle. This separation is complete in one order of reptiles, the crocodilians, which have two separate ventricles divided by a complete septum (see the following section). Another change in the circulation of reptiles is that the conus arteriosus has become incorporated into the trunks of the large arteries leaving the heart.

Mammals, birds, and crocodilians have two completely separated circulatory systems

Mammals, birds, and crocodilians have a four-chambered heart with two separate atria and two separate ventricles (figure 49.4). The right atrium receives deoxygenated blood from the body

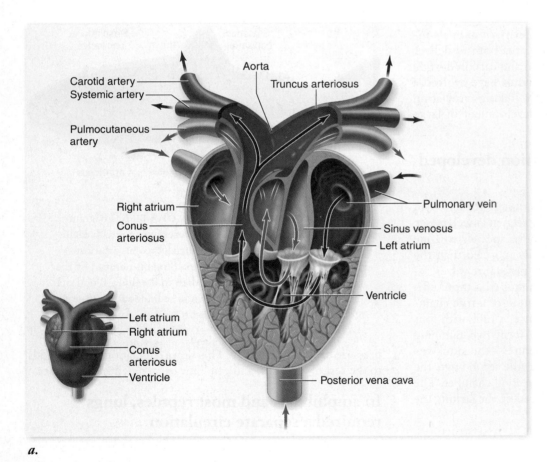
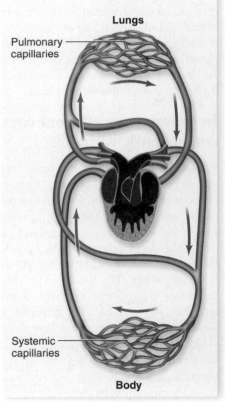

a.

b.

figure 49.3

THE HEART AND CIRCULATION OF AN AMPHIBIAN. *a.* The frog has a three-chambered heart with two atria but only one ventricle, which pumps blood both to the lungs and to the body. *b.* Despite the potential for mixing, the oxygenated and deoxygenated bloods (*red* and *blue* lines, respectively) mix very little as they are pumped to the body and lungs. Oxygenation of blood also occurs by gas exchange through the skin.

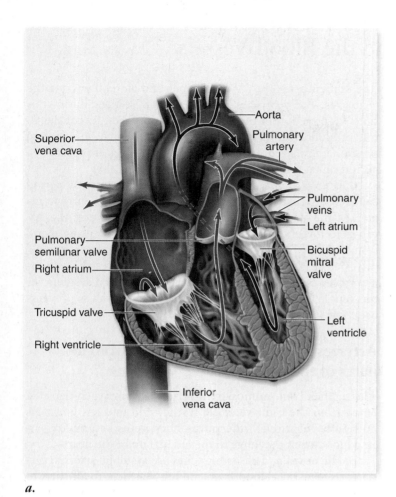

a.

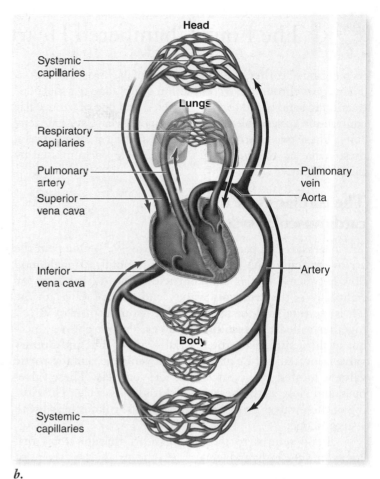

b.

figure 49.4

THE HEART AND CIRCULATION OF MAMMALS AND BIRDS. *a.* The path of blood through the four-chambered heart. *b.* The right side of the heart receives deoxygenated blood and pumps it to the lungs; the left side of the heart receives oxygenated blood and pumps it to the body. In this way, the pulmonary and systemic circulations are kept completely separate.

and delivers it to the right ventricle, which pumps the blood to the lungs. The left atrium receives oxygenated blood from the lungs and delivers it to the left ventricle, which pumps the oxygenated blood to the rest of the body.

The heart in these vertebrates is a two-cycle pump. Both atria fill with blood and simultaneously contract, emptying their blood into the ventricles. Both ventricles also contract at the same time, pushing blood simultaneously into the pulmonary and systemic circulations.

The increased efficiency of the double circulatory system in mammals and birds is thought to have been important in the evolution of endothermy. More efficient circulation is necessary to support the high metabolic rate required for maintenance of internal body temperature about a set point.

Throughout the evolutionary history of the vertebrate heart, the sinus venosus has served as a pacemaker, the site where the impulses that initiate the heartbeat originate. Although the sinus

venosus constitutes a major chamber in the fish heart, it is reduced in size in amphibians and is further reduced in reptiles. In mammals and birds, the sinus venosus is no longer present as a separate chamber, although some of its tissue remains in the wall of the right atrium. This tissue, the sinoatrial (SA) node, is still the site where each heartbeat originates as detailed later in the chapter.

All vertebrates have a closed circulatory system. The fish heart is a modified tube, which consists of two chambers. Blood is pumped from the heart to the gills, and then to the rest of the body. Amphibians and reptiles have two circuits—pulmonary and systemic—that deliver blood to the lungs and to the rest of the body, respectively. Because these vertebrates have a single ventricle, some mixing occurs between oxygenated and deoxygenated blood. In birds, mammals, and crocodilians, the division of ventricles is complete, and mixing does not occur.

The Four-Chambered Heart and the Blood Vessels

As mentioned earlier, the heart of mammals, birds, and crocodilians goes through two contraction cycles, one of atrial contraction to send blood to the ventricles, and one of ventricular contraction to send blood to the pulmonary and systemic circuits. These two contractions plus the resting period between these make up the complete **cardiac cycle** encompassed by the heartbeat.

The cardiac cycle drives the cardiovascular system

The heart has two pairs of valves (figure 49.5). One pair, the **atrioventricular (AV) valves,** maintains unidirectional blood flow between the atria and ventricles. The AV valve on the right side is the **tricuspid valve,** and the AV valve on the left is the **bicuspid,** or **mitral, valve.** Another pair of valves, together called the **semilunar valves,** ensure one-way flow out of the ventricles to the arterial systems. The **pulmonary valve** is located at the exit of the right ventricle, and the **aortic valve** is located at the exit of the left ventricle. These valves open and close as the heart goes through its cycle. The closing of these valves produces the "lub-dub" sounds heard with a stethoscope.

Blood returns to the resting heart through veins that empty into the right and left atria. As the atria fill and the pres-

sure in them rises, the AV valves open and blood flows into the ventricles. The ventricles become about 80% filled during this time. Contraction of the atria tops up the final 20% of the 80 mL of blood the ventricles receives, on average, in a resting person. These events occur while the ventricles are relaxing, a period called ventricular **diastole.**

After a slight delay, the ventricles contract, a period called ventricular **systole.** Contraction of each ventricle increases the pressure within each chamber, causing the AV valves to forcefully close (the "lub" sound), preventing blood from backing up into the atria. Immediately after the AV valves close, the pressure in the ventricles forces the semilunar valves open and blood flows into the arterial systems. As the ventricles relax, closing of the semilunar valves prevents backflow (the "dub" sound).

Arteries and veins branch to and from all parts of the body

The right and left **pulmonary arteries** deliver oxygen-depleted blood from the right ventricle to the right and left lungs. As previously mentioned, the **pulmonary veins** return oxygenated blood from the lungs to the left atrium of the heart.

The **aorta** and all its branches are systemic arteries, carrying oxygen-rich blood from the left ventricle to all parts of the body. The **coronary arteries** are the first branches off the aorta; these supply oxygenated blood to the heart muscle itself (figure 49.5*b*). Other systemic arteries branch from the aorta as it makes an arch above the heart and as it descends and traverses the thoracic and abdominal cavities.

The blood from the body's organs, now lower in oxygen, returns to the heart in the systemic veins. These eventually empty into two major veins: the **superior vena cava,** which

figure 49.5

HEART VALVES AND CORONARY CIRCULATION. *a.* The four heart valves are shown with atrioventricular valves open (*left*), and aortic and pulmonary valves open (*right*). These valves prevent backflow during heart function. *b.* The first branches out of the aorta are the coronary arteries that provide blood to the heart. Blood returns directly to the heart via the coronary veins.

Posterior

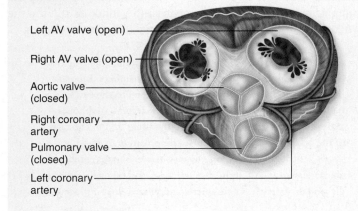
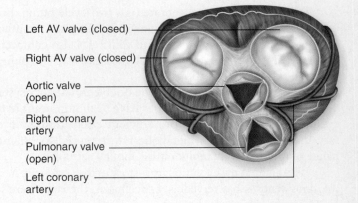

Left AV valve (open)
Right AV valve (open)
Aortic valve (closed)
Right coronary artery
Pulmonary valve (closed)
Left coronary artery

Left AV valve (closed)
Right AV valve (closed)
Aortic valve (open)
Right coronary artery
Pulmonary valve (open)
Left coronary artery

Anterior

a. The valves of the heart

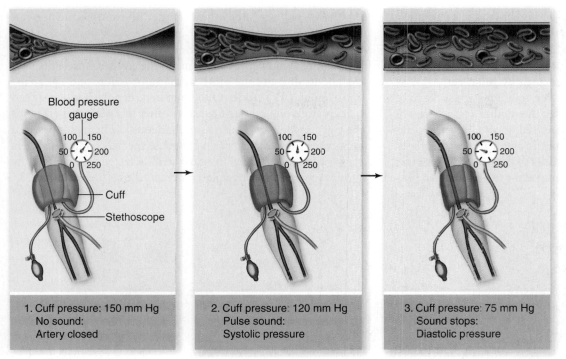

figure 49.6

Blood pressure gauge

100 150
50 — 200
0 250

Cuff

Stethoscope

1. Cuff pressure: 150 mm Hg
No sound:
Artery closed

100 150
50 — 200
0 250

2. Cuff pressure: 120 mm Hg
Pulse sound:
Systolic pressure

100 150
50 — 200
0 250

3. Cuff pressure: 75 mm Hg
Sound stops:
Diastolic pressure

MEASUREMENT OF BLOOD PRESSURE. The blood pressure cuff is tightened to stop the blood flow through the brachial artery. As the cuff is loosened, the systolic pressure eventually becomes greater than the cuff pressure and thus a pulse is heard at the stethoscope. The pressure at this point is recorded as the systolic pressure. As the cuff pressure continues to drop, the blood vessel eventually is no longer distorted and silent laminar flow returns. The diastolic pressure is recorded as the pressure at which a sound is no longer heard.

drains the upper body, and the **inferior vena cava,** which drains the lower body. These veins empty into the right atrium, completing the systemic circulation.

The flow of blood through the arteries, capillaries, and veins is driven by the pressure generated by ventricular contraction. The ventricles must contract forcefully enough to move the blood through the entire circulatory system.

Arterial blood pressure can be measured

As the ventricles contract, great pressure is generated within them and transferred through the arteries once the aortic valve opens. The pulse that you can detect in your wrist or neck results

from changes in pressure as elastic arteries expand and contract with the periodic blood flow. Doctors use blood pressure as an general indicator of cardiovascular health because a variety of conditions can cause increases or decreases in pressure.

The measuring device used, called a *sphygmomanometer,* measures the blood pressure in the brachial artery found on the inside part of the arm, above the elbow (figure 49.6). A cuff wrapped around the upper part of the arm is tightened enough to stop the flow of blood to the lower part of the arm. As the cuff is slowly loosened, blood begins pulsating through the artery, and this sound can be detected using a stethoscope. The point at which this pulsing sound begins marks the peak pressure, or **systolic pressure,** at which ventricles are contracting. As the cuff is loosened further, the vessel is no longer distorted and the pulsing sound stops. This point marks the minimum pressure between heartbeats or **diastolic pressure,** at which the ventricles are relaxed.

The blood pressure is written as a ratio of systolic over diastolic pressure, and for a healthy person in his or her twenties, a typical blood pressure is 120/75 (measured in millimeters of mercury, or mm Hg). The medical condition called **hypertension** (high blood pressure) is defined as either a systolic pressure greater than 150 mm Hg or a diastolic pressure greater than 90 mm Hg.

Contraction of heart muscle is initiated by autorhythmic cells

As in other types of muscle, contraction of heart muscle is stimulated by membrane depolarization (see chapters 44 and 47). In skeletal muscles, only nerve impulses from motor neurons can normally initiate depolarization. The heart, by contrast, contains specialized "self-excitable" muscle cells called **autorhythmic fibers,** which can initiate periodic action potentials without neural activation.

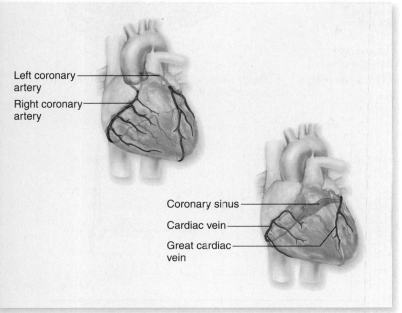

Left coronary artery

Right coronary artery

Coronary sinus

Cardiac vein

Great cardiac vein

b. The coronary circulation of the heart

The most important group of autorhythmic cells is the sinoatrial (SA) node, described earlier (figure 49.7). Located in the wall of the right atrium, the SA node acts as a pacemaker for the rest of the heart by producing spontaneous action potentials at a faster rate than other autorhythmic cells. These spontaneous action potentials are due to a constant leakage of Na^+ ions into the cell that depolarize the membrane. When the threshold is reached, an action potential occurs. At the end of the action potential, the membrane is again below threshold and the process begins again. The cells of the SA node generate an action potential every 0.6 sec, equivalent to about 100 a min-

ute. As we will see later in the chapter, the autonomic nervous system can modulate this rate.

Each depolarization initiated by this pacemaker is transmitted through two pathways: one to the cardiac muscle fibers of the left atrium, and the other to the right atrium and the **atrioventricular (AV) node.** Once initiated, depolarizations spread quickly from one muscle fiber to another in a wave that envelops the right and left atria nearly simultaneously. The rapid spread of depolarization is made possible because special conducting fibers are present and because the cardiac muscle cells are coupled by groups of gap junctions called *intercalated disks* (see chapter 43).

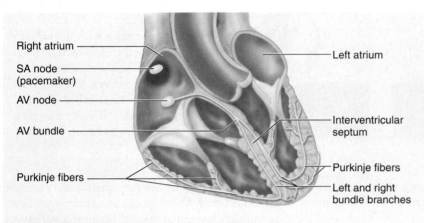

1. The impulse begins at the SA node and travels to the AV node.

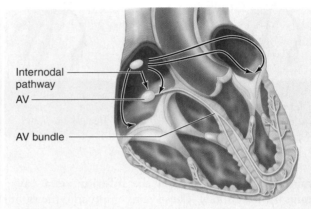

2. The impulse is delayed at the AV node. It then travels to the AV bundle.

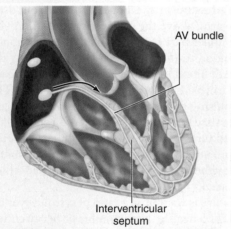

3. From the AV bundle, the impulse travels down the interventricular septum.

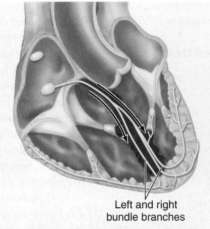

4. The impulse spreads to branches from the interventricular septum.

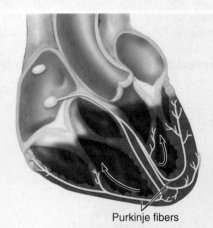

5. Finally reaching the Purkinje fibers, the impulse is distributed throughout the ventricles.

figure 49.7

THE PATH OF ELECTRICAL EXCITATION IN THE HEART. The events occurring during contraction of the heart are correlated with the measurement of electrical activity by an electrocardiogram (ECG also called EKG). The depolarization/contraction of the atrium is shown in green above and corresponds to the P wave of the ECG (also in green). The depolarization/contraction of the ventricle is shown in red above and corresponds to the QRS wave of the ECG (also in red). The T wave on the ECG corresponds to the repolarization of the ventricles. The atrial repolarization is masked by the QRS wave.

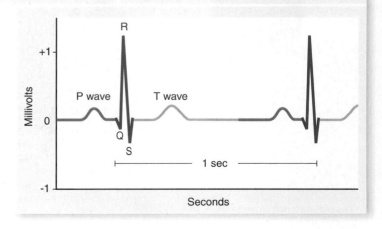

A sheet of connective tissue separating the atria from the ventricles blocks the spread of excitation through muscle fibers from one chamber to the other. The AV node provides the only pathway for conduction of the depolarization from the atria to the ventricles. The fibers of the AV node slow down the conduction of the depolarizing signals, delaying the contraction of the ventricle by about 0.1 sec. This delay permits the atria to finish contracting and emptying their blood into the ventricles before the ventricles contract.

From the AV node, the wave of depolarization is conducted rapidly over both ventricles by a network of fibers called the **atrioventricular bundle**, or **bundle of His**. These fibers relay the depolarization to **Purkinje fibers**, which directly stimulate the myocardial cells of the left and right ventricles, causing their almost simultaneous contraction.

The stimulation of myocardial cells produces an action potential that leads to contraction. Contraction is controlled by Ca^{2+} and the troponin/tropomyosin system similar to skeletal muscle (see chapter 47), but the shape of the action potential is different. The initial rising phase due to an influx of Na^+ from voltage-gated Na^+ channels is followed by a plateau phase that leads to more sustained contraction. The plateau phase is due to the opening of voltage-gated Ca^{2+} channels. The resulting influx of Ca^{2+} keeps the membrane depolarized when the Na^+ channels inactivate. This, in turn, leads to more voltage-gated Ca^{2+} channels in the sarcoplasmic reticulum opening and the additional Ca^{2+} in the cytoplasm produces a more sustained

contraction. The Ca^{2+} is removed from the cytoplasm by a pump in the sarcoplasmic reticulum similar to skeletal muscle, and an additional carrier in the plasma membrane pumps Ca^{2+} into the interstitial space.

The electrical activity of the heart can be recorded from the surface of the body with electrodes placed on the limbs and chest. The recording, called an **electrocardiogram (ECG or EKG)**, shows how the cells of the heart depolarize and repolarize during the cardiac cycle (see figure 49.7). Depolarization causes contraction of the heart, and repolarization causes relaxation.

The first peak in the recording, P, is produced by the depolarization of the atria, and is associated with atrial systole. The second, larger peak, QRS, is produced by ventricular depolarization; during this time, the ventricles contract (ventricular systole). The last peak, T is produced by ventricular repolarization; at this time, the ventricles begin diastole.

The cardiac cycle consists of systole and diastole; the ventricles contract at systole and relax at diastole. The SA node in the right atrium initiates waves of depolarization that stimulate first the atria and then the ventricles to contract. The contractions are more sustained than skeletal muscle due to voltage-gated Ca^{2+}; channels that release additional Ca^{2+}; prolonging contraction. The electrocardiogram (ECG) traces the depolarization occurring during the cardiac cycle.

49.4 Characteristics of Blood Vessels

You already know that blood leaves the heart through vessels known as **arteries**. These continually branch, forming a hollow "tree" that enters each organ of the body. The finest, microscopic branches of the arterial tree are the **arterioles**. Blood from the arterioles enters the **capillaries**, an elaborate latticework of very narrow, thin-walled tubes. After traversing the capillaries, the blood is collected into microscopic **venules**, which lead to larger vessels called **veins**, and these carry blood back to the heart.

Larger vessels are composed of four tissue layers

Arteries, arterioles, veins, and venules all have the same basic structure (figure 49.8). The innermost layer is an epithelial sheet called the *endothelium*. Covering the endothelium is a thin layer of elastic fibers, a smooth muscle layer, and a connective tissue layer. The walls of these vessels, therefore, are thick

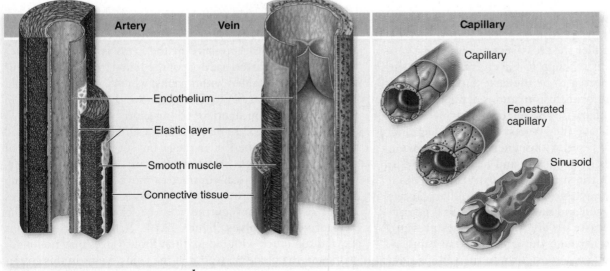

a.   *b.*   *c.*

figure 49.8

THE STRUCTURE OF BLOOD VESSELS. Arteries (*a*) and veins (*b*) have the same tissue layers but the smooth muscle layer in arteries is much thicker and there are two elastic layers. *c.* Capillaries are composed of only a single layer of endothelial cells. (Not to scale.)

enough to significantly reduce exchange of materials between the blood and the tissues outside the vessels.

The walls of capillaries, in contrast, are made up only of endothelium, so molecules and ions can leave the blood plasma by diffusion, by filtration through pores between the cells of the capillary walls, and by transport through the endothelial cells. Therefore, exchange of gases and metabolites between the blood and the interstitial fluids and cells of the body takes place through the capillaries.

Arteries and arterioles have evolved to withstand pressure

The larger arteries contain more elastic fibers in their walls than other blood vessels, allowing them to recoil each time they receive a volume of blood pumped by the heart. Smaller arteries and arterioles are less elastic, but their relatively thick smooth muscle layer enables them to resist bursting.

The narrower the vessel, the greater the frictional resistance to flow. In fact, a vessel that is half the diameter of another has *16 times* the frictional resistance. Resistance to blood flow is inversely proportional to the fourth power of the radius of the vessel. Therefore, within the arterial tree, the small arteries and arterioles provide the greatest resistance to blood flow.

Contraction of the smooth muscle layer of the arterioles results in **vasoconstriction,** which greatly increases resistance and decreases flow. Relaxation of the smooth muscle layer results in **vasodilation,** decreasing resistance and increasing blood flow to an organ (figure 49.9). Chronic vasoconstriction of the arterioles can result in hypertension, or high blood pressure.

Capillaries form a vast network for exchange of materials

The huge number and extensive branching of the capillaries ensure that every cell in the body is within 100 micrometers (μm) of a capillary. On the average, capillaries are about 1 mm long and 8 μm in diameter, this diameter is only slightly larger than a red blood cell (5–7 μm in diameter). Despite the close fit, normal red blood cells are flexible enough to squeeze through capillaries without difficulty.

The rate of blood flow through vessels is governed by hydrodynamics. The smaller the cross-sectional area of a vessel, the faster fluid moves through it. Given this, flow in the capillaries would be expected to be the fastest in the system. This would not be ideal for diffusion, and is actually not the case. Although each capillary is very narrow, so many of them exist that the capillaries have the greatest *total* cross-sectional area of any other type of vessel. Consequently, blood moving through capillaries goes more slowly and has more time to exchange materials with the surrounding extracellular fluid. By the time the blood reaches the end of a capillary, it has released some of its oxygen and nutrients and picked up carbon dioxide and other waste products. Blood loses pressure and velocity as it moves through the arterioles and capillaries, but as cross-sectional area goes down in the venous side, velocity increases.

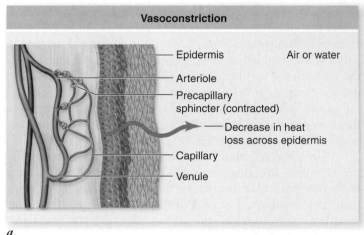

a.

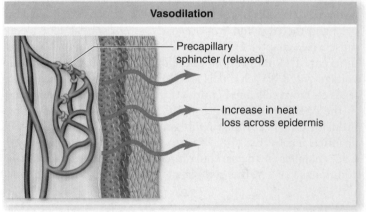

b.

figure 49.9

REGULATION OF HEAT LOSS. The amount of heat lost at the body's surface can be regulated by controlling the flow of blood to the surface. *a.* Constriction of surface blood vessels limits flow and heat loss; (*b*) dilation of these vessels increases flow and heat loss.

Venules and veins have less muscle in their walls

Venules and veins have the same tissue layers as arteries, but they have a thinner layer of smooth muscle. Less muscle is needed because the pressure in the veins is only about one-tenth that in the arteries. Most of the blood in the cardiovascular system is contained within veins, which can expand to hold additional amounts of blood. You can see the expanded veins in your feet when you stand for a long time.

The venous pressure alone is not sufficient to return blood to the heart from the feet and legs, but several other sources of pressure provide help. Most significantly, skeletal muscles surrounding the veins can contract to move blood by squeezing the veins, a mechanism called the **venous pump.** Blood moves in one direction through the veins back to the heart with the help of **venous valves** (figure 49.10). When a person's veins expand too much with blood, the venous valves may no longer work and the blood may pool in the veins. Veins in this condition are known as varicose veins.

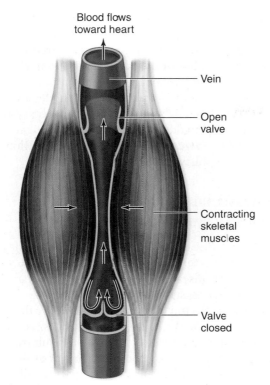

figure 49.10

ONE-WAY FLOW OF BLOOD THROUGH VEINS. Venous valves ensure that blood moves through the veins in only one direction, back to the heart.

The lymphatic system is an open, one-way system

The cardiovascular system is considered a closed system because all its vessels are connected with one another—none are simply open-ended. But a significant amount of water and solutes in the blood plasma filter through the walls of the capillaries to form the interstitial (tissue) fluid. Most of the fluid leaves the capillaries near their arteriolar ends, where the blood pressure is higher; it is returned to the capillaries near their venular ends (figure 49.11).

Return of fluid occurs by osmosis (see chapter 5). Most of the plasma proteins cannot escape through the capillary pores because of their large size, and so the concentration of proteins in the plasma is greater than the protein concentration in the interstitial fluid. The difference in protein concentration produces an osmotic pressure gradient that causes water to move into the capillaries from the interstitial space.

High capillary blood pressure can cause too much interstitial fluid to accumulate. In pregnant women, for example, the enlarged uterus, carrying the fetus, compresses veins in the abdominal cavity and thereby increases the capillary blood pressure in the woman's lower limbs. The increased interstitial fluid can cause swelling of the tissues, or **edema,** of the feet.

Edema may also result if the plasma protein concentration is too low. Fluids will not return to the capillaries, but will remain as interstitial fluid. Low protein concentration in the plasma may be caused either by liver disease, because the liver

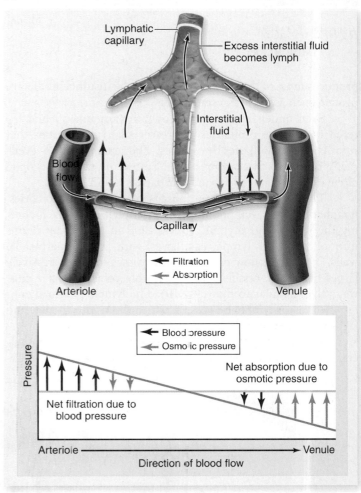

figure 49.11

RELATIONSHIP BETWEEN BLOOD, LYMPH, AND INTERSTITIAL FLUID. *a.* Vessels of the circulatory and lymphatic systems with arrows indicating the direction of flow of fluid in the vessels. *b.* Plasma fluid, minus proteins, is filtered out of capillaries, forming interstitial fluid that bathes tissues. Much of this fluid is returned to the capillaries by osmosis due to the higher protein concentration in plasma. Excess interstitial fluid drains into open-ended lymphatic capillaries, which ultimately return the fluid to the cardiovascular system.

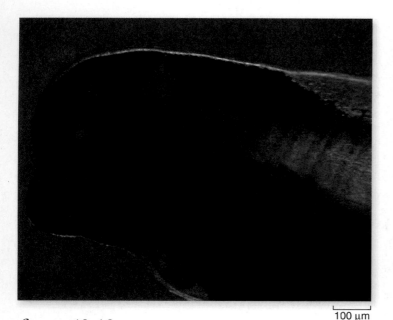

figure 49.12

A LYMPH HEART.

100 μm

produces most of the plasma proteins, or by insufficient dietary protein such as occurs in starvation.

Even under normal conditions, the amount of fluid filtered out of the capillaries is greater than the amount that returns to the capillaries by osmosis. The remainder does eventually return to the cardiovascular system by way of an open circulatory system called the **lymphatic system.**

The lymphatic system consists of lymphatic capillaries, lymphatic vessels, lymph nodes, and lymphatic organs, including the spleen and thymus. Excess fluid in the tissues drains into blind-ended lymph capillaries with highly permeable walls. This fluid, now called **lymph,** passes into progressively larger lymphatic vessels, which resemble veins and have one-way valves (similar to figure 49.10). The lymph eventually enters two major lymphatic vessels, which drain into the left and right subclavian veins located under the collarbones.

Movement of lymph in mammals is accomplished by skeletal muscles squeezing against the lymphatic vessels, a mecha-

nism similar to the venous pump that moves blood through veins. In some cases, the lymphatic vessels also contract rhythmically. In many fishes, all amphibians and reptiles, bird embryos, and some adult birds, movement of lymph is propelled by **lymph hearts** (figure 49.12).

As the lymph moves through lymph nodes and lymphatic organs, it is modified by phagocytic cells that line the channels of those organs. In addition, the lymph nodes and lymphatic organs contain *germinal centers*, where the activation and proliferation of lymphocytes occurs.

Cardiovascular diseases affect the delivery system

Cardiovascular diseases are the leading cause of death in the United States; more than 42 million people have some form of cardiovascular disease. Many disease conditions result from problems in arteries, such as blockage or rupture.

Heart attacks (myocardial infarctions) are the main cause of cardiovascular deaths in the United States, accounting for about one-fifth of all deaths. Heart attacks result from an insufficient supply of blood to one or more parts of the heart muscle, which causes myocardial cells in those parts to die. Heart attacks may be caused by a blood clot forming somewhere in the coronary arteries and may also result if an artery is blocked by atherosclerosis. Recovery from a heart attack is possible if the portion of the heart that was damaged is small enough that the heart can still contract as a functional unit.

Angina pectoris, which literally means "chest pain," occurs for reasons similar to those that cause heart attacks, but it is not as severe. The pain may occur in the heart and often also in the left arm and shoulder. Angina pectoris is a warning sign that the blood supply to the heart is inadequate but is still sufficient to avoid myocardial cell death.

Strokes are caused by an interference with the blood supply to the brain. They may occur when a blood vessel bursts in the brain (hemorrhagic stroke), when blood flow in a cerebral artery is blocked by a blood clot, or by atherosclerosis (ischemic stroke). The effects of a stroke depend on the severity of the damage and where in the brain the stroke occurs.

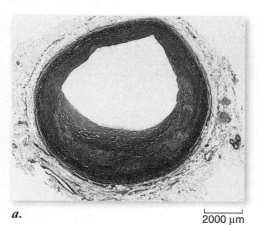
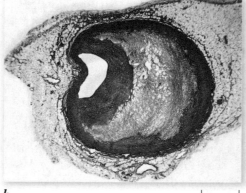
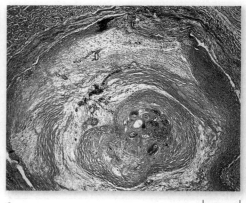

a.   2000 μm   *b.*   2500 μm   *c.*   1000 μm

figure 49.13

ATHEROSCLEROSIS. *a.* The coronary artery shows only minor blockage. *b.* The artery exhibits severe atherosclerosis—much of the passage is blocked by buildup on the interior walls of the artery. *c.* The coronary artery is essentially completely blocked.

Atherosclerosis is an accumulation within the arteries of fatty materials, abnormal amounts of smooth muscle, deposits of cholesterol or fibrin, or various kinds of cellular debris. These accumulations cause an increase in vascular resistance, which leads to a decrease in blood flow (figure 49.13). The lumen (interior) of the artery may be further narrowed by a clot that forms as a result of the atherosclerosis. In the severest cases, the artery becomes completely blocked.

The accumulation of cholesterol in vessels is affected by a number of factors including total serum cholesterol and the levels of different cholesterol carrier proteins. Because cholesterol is not very water-soluble, it is carried in blood in the form of lipoprotein complexes. Two main forms are observed that differ in density: low-density lipoproteins (**LDL**) and high-density lipoproteins (**HDL**). These are often called "good cholesterol" (HDL) and "bad cholesterol" (LDL). The reason for this is that HDLs tend to take cholesterol out of circulation, transporting it to the liver for elimination, and LDL is the carrier that brings cholesterol to all cells in the body. The problem arises when cells have enough cholesterol. This causes a reduction in the amount of LDL receptors, leading to high levels of circulating LDLs, which can end up being deposited in blood vessels.

Atherosclerosis is promoted by genetic factors, smoking, hypertension (high blood pressure), and the effects of cholesterol just discussed. Stopping smoking is the single most effective action a smoker can take to reduce the risk of atherosclerosis.

Arteriosclerosis, or hardening of the arteries, occurs when calcium is deposited in arterial walls. It tends to occur when atherosclerosis is severe. Not only do such arteries have restricted blood flow, but they also lack the ability to expand as normal arteries do. This decrease in flexibility forces the heart to work harder because blood pressure increases to maintain flow.

Blood is pumped from the heart into the arterial system, which branches into fine arterioles. Arterioles deliver blood into the thin-walled capillaries, where exchanges with the tissues occur. Blood returns to the heart through venules and veins. Blood pressure moves fluid out of the arteries, and much of this fluid returns by osmosis into veins. Excess interstitial fluid, called lymph, is returned to the cardiovascular system via the lymphatic system. Cardiovascular diseases result from failure of blood supply and include heart attack, stroke, atherosclerosis, and arteriosclerosis.

49.5 Regulation of Blood Flow and Blood Pressure

Although the autonomic nervous system does not initiate the heartbeat, it does modulate its rhythm and force of contraction. In addition, several mechanisms regulate characteristics of the cardiovascular system, including cardiac output, blood pressure, and blood volume.

The nervous system may speed up or slow down heart rate

Heart rate is under the control of the autonomic nervous system. The cardiac center of the medulla oblongata (a part of the hindbrain; chapter 44) consists of two neuronal centers that modulate heart rate. The **cardioacceleratory center** sends signals by way of the sympathetic cardiac accelerator nerves to the SA node, AV node, and myocardium. These nerves secrete norepinephrine, which increases the heart rate. Sympathetic nervous system stimulation can also increase contractility of the heart muscle itself, thus ejecting more blood per contraction (stroke volume).

The **cardioinhibitory center** sends signals via the parasympathetic fibers in the vagus nerve to the SA and AV nodes. The vagus nerve secretes acetylcholine, which inhibits the development of action potentials so the heart slows down.

Cardiac output increases with exertion

Cardiac output is the volume of blood pumped by each ventricle per minute. It is calculated by multiplying the heart rate by the *stroke volume*, which is the volume of blood ejected by each ventricle per beat. For example, if the heart rate is 72 beats per minute and the stroke volume is 70 mL, the cardiac output is 5 L/min, which is about average in a resting human.

Cardiac output increases during exertion because of an increase in both heart rate and stroke volume. When exertion begins, such as running, the heart rate increases up to about 100 beats per minute. As movement becomes more intense, skeletal muscles squeeze on veins more vigorously, returning blood to the heart more rapidly. In addition, the ventricles contract more strongly, so they empty more completely with each beat.

During exercise, the cardiac output increases to a maximum of about 25 L/min in an average young adult. Although the cardiac output has increased fivefold, not all organs receive five times the blood flow; some receive more, others less. Arterioles in some organs, such as in the digestive system, constrict, while the arterioles in the working muscles and heart dilate.

The baroreceptor reflex maintains homeostasis in blood pressure

The arterial blood pressure depends on two factors: the cardiac output (CO) and the resistance (R) to blood flow in the vascular system. This relationship can be expressed as:

$$BP = CO \times R$$

An increased blood pressure, therefore, could be produced by an increase in either heart rate or blood volume (because both increase the cardiac output), or by vasoconstriction, which increases the resistance to blood flow. Conversely, blood pressure falls if the heart rate slows or if the blood volume is reduced—for example, by dehydration or excessive bleeding (hemorrhage).

Changes in arterial blood pressure are detected by **baroreceptors** located in the arch of the aorta and in the carotid arteries (see chapter 45). These sensors are stretch receptors sensitive to expansion and contraction of arteries. When the baroreceptors detect a fall in blood pressure, the number of impulses to the cardiac center is decreased, resulting in increased sympathetic stimulation and decreased parasympathetic stimulation of the heart and other targets. This increases heart rate and stroke volume to increase cardiac output. This also causes vasoconstriction of blood vessels in the skin and viscera, increasing resistance. These combine to increase blood pressure, closing the feedback loop in this direction (figure 49.14, top).

When baroreceptors detect an increase in blood pressure, the number of impulses to the cardiac center is increased. This has the opposite effect decreasing sympathetic stimulation and increasing parasympathetic simulation of the heart. This decreases heart rate and stroke volume to reduce cardiac output. The cardiac center also sends signals causing vasodilation of blood vessels in the skin and viscera, lowering resistance. These combine to decrease blood pressure closing the feedback loop in this direction. Thus, the baroreceptor reflex forms a negative feedback loop responding to changes in blood pressure (figure 49.14, bottom).

Blood volume is regulated by hormones

Blood pressure depends in part on the total blood volume because this can affect the cardiac output. A decrease in blood volume decreases blood pressure, if all else remains equal. Blood volume regulation involves the effects of four hormones:

(1) antidiuretic hormone, (2) aldosterone, (3) atrial natriuretic hormone, and (4) nitric oxide.

Antidiuretic hormone (ADH), also called *vasopressin*, is secreted by the posterior pituitary gland in response to an increase in the osmolarity of the blood plasma (see chapter 46). Dehydration, for example, causes the blood volume to decrease. Osmoreceptors in the hypothalamus promote thirst and stimulate ADH secretion from the posterior pituitary gland. ADH, in turn, stimulates the kidneys to retain more water in the blood, excreting less in the urine. A dehydrated person thus drinks more and urinates less, helping to raise the blood volume and restore homeostasis.

Whenever the kidneys experience a decreased blood flow, a group of kidney cells initiate the release of an enzyme known as renin into the blood. Renin activates a blood protein, angiotensin, which stimulates vasoconstriction throughout the body while it also stimulates the adrenal cortex to secrete **aldosterone.** This steroid hormone acts on the kidneys to promote the retention of Na^+ and water in the blood (see chapter 46).

When excess Na^+ is present, less aldosterone is secreted by the adrenals, so that less Na^+ is retained by the kidneys. In recent years, scientists have learned that Na^+ excretion in the urine is promoted by another hormone, **atrial natriuretic hormone.** This hormone is secreted by the right atrium of the heart in response to stretching caused by an increased blood volume. The action of atrial natriuretic hormone completes a negative feedback loop, lowering the blood volume and pressure.

Nitric oxide (NO) is a gas produced by endothelial cells of blood vessels. As described in chapter 46, it is one of a

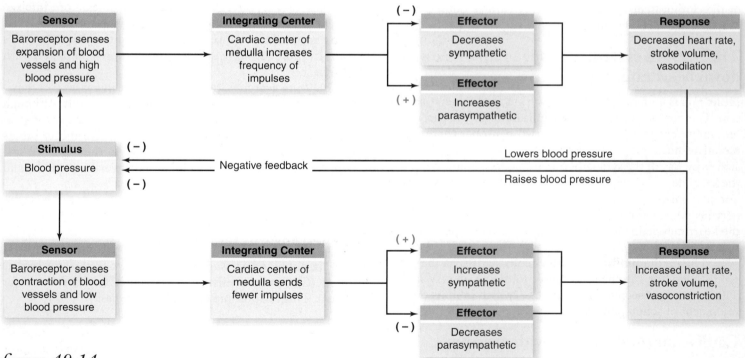

figure 49.14

BARORECEPTOR NEGATIVE FEEDBACK LOOPS CONTROL BLOOD PRESSURE. Baroreceptors form the afferent portion of a feedback loop controlling blood pressure. The frequency of nerve impulses from these stretch receptors correlates with blood pressure. This information is processed in the cardiac center of the medulla. The efferent portion of the loop involves sympathetic and parasympathetic nerves that innervate the heart. This control can raise or lower heart rate and stroke volume to raise and lower blood pressure in response to baroreceptors signaling.

number of paracrine regulators of blood vessels. In solution, NO passes outward through the cell layers of the vessel, causing the smooth muscles that encase it to relax and the blood vessels to dilate (become wider). For over a century, heart patients have been prescribed nitroglycerin to relieve chest pain, but only now has it become clear that nitroglycerin acts by releasing nitric oxide.

Cardiac output depends on the rate of the heartbeat and how much blood is ejected per beat. Blood flow is regulated by the degree of constriction of the arteries, which affects the resistance to flow. Blood pressure is influenced by blood volume; the volume of water retained in the vascular system is regulated by hormones that act on the kidneys and blood vessels.

49.6 The Components of Blood

Blood is a connective tissue composed of a fluid matrix, called **plasma,** and several different kinds of cells and other **formed elements** that circulate within that fluid (figure 49.15). Blood **platelets,** although included in figure 49.15, are not complete cells; rather, they are fragments of cells that are produced in the bone marrow. (We describe the action of platelets in blood clotting in a later section.)

The functions of circulating blood can be summarized as follows:

1. **Transportation.** All of the substances essential for cellular metabolism are transported by blood. Red blood cells transport oxygen attached to hemoglobin, nutrient molecules are carried in the plasma, sometimes bound to carriers, and metabolic wastes are eliminated as blood passes through the liver and kidneys.
2. **Regulation.** The cardiovascular system transports regulatory hormones from the endocrine glands and also participates in temperature regulation. Contraction and dilation of blood vessels near the surface of the body, beneath the epidermis, helps to conserve or to dissipate heat as needed (see figure 49.9).
3. **Protection.** The circulatory system protects against injury and foreign microbes or toxins introduced into the body. Blood clotting helps to prevent blood loss when vessels are damaged. White blood cells, or leukocytes, help to disarm or disable invaders such as viruses and bacteria (see chapter 51).

Blood plasma is a fluid matrix

Blood plasma is the matrix in which blood cells and platelets are suspended. Interstitial (extracellular) fluids originate from the fluid present in plasma.

Although plasma is 92% water, it also contains the following solutes:

1. **Nutrients, wastes, and hormones.** Dissolved within the plasma are all of the nutrients resulting from digestive breakdown that can be used by cells, including glucose, amino acids, and vitamins. Also dissolved in the plasma are wastes such as nitrogen compounds and CO_2 produced by metabolizing cells. Endocrine hormones released from glands are also carried through the blood to their target cells.
2. **Ions.** Blood plasma is a dilute salt solution. The predominant plasma ions are Na^+, Cl^-, and

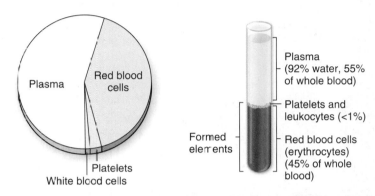

figure 49.15
COMPOSITION OF BLOOD.

Blood Plasma	Red Blood Cells	Platelets
Plasma proteins (7%) Albumin (54%) Globulins (38%) Fibrinogen (7%) All others (1%) **Water** (91.5%)	4 million–6 million/ mm³ blood	150,000–300,000/ mm³ blood
Other solutes (1.5%)	**Neutrophils**	**Eosinophils**
Electrolytes Nutrients Gases Regulatory substances Waste products	60–70%	2–4%
Monocytes	**Basophils**	**Lymphocytes**
3–8%	0.5–1%	20–25%

bicarbonate ions (HCO_3^-). In addition, plasma contains trace amounts of other ions such as Ca^{2+}, Mg^{2+}, Cu^{2+}, K^+, and Zn^{2+}.
3. **Proteins.** As mentioned earlier, the liver produces most of the plasma proteins, including **albumin,** which constitutes most of the plasma protein; the alpha (α) and beta (β) **globulins,** which serve as carriers of lipids and steroid hormones; and **fibrinogen,** which is required for blood clotting. Blood plasma with the fibrinogen removed is called **serum.**

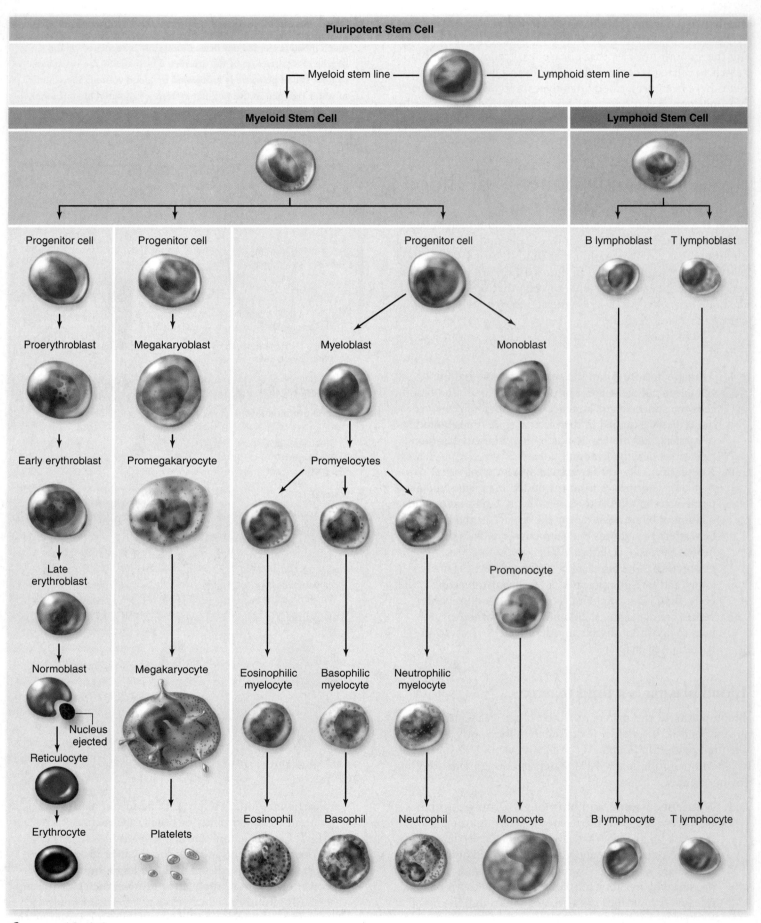

figure 49.16

STEM CELLS AND THE PRODUCTION OF FORMED ELEMENTS.

Formed elements include circulating cells and platelets

The **formed elements** of blood cells and cell fragments include red blood cells, white blood cells, and platelets. Each element has a specific function in maintaining the body's health and homeostasis.

Erythrocytes

Each microliter of blood contains about 5 million **red blood cells,** or **erythrocytes.** The fraction of the total blood volume that is occupied by erythrocytes is called the blood's *hematocrit;* in humans, the hematocrit is typically around 45%.

Each erythrocyte resembles a doughnut-shaped disk with a central depression that does not go all the way through. Mature mammalian erythrocytes lack nuclei. The erythrocytes of vertebrates contain hemoglobin, a pigment that binds and transports oxygen. (Hemoglobin is described more fully later in this chapter when we discuss respiration.) In vertebrates, hemoglobin is found only in erythrocytes. In invertebrates, the oxygen-binding pigment (not always hemoglobin) is also present in plasma.

Leukocytes

Less than 1% of the cells in human blood are **white blood cells,** or **leukocytes;** there are only 1 or 2 leukocytes for every 1000 erythrocytes. Leukocytes are larger than erythrocytes and have nuclei. Furthermore, leukocytes are not confined to the blood as erythrocytes are, but can migrate out of capillaries through the intercellular spaces into the surrounding interstitial (tissue) fluid.

Leukocytes come in several varieties, each of which plays a specific role in defending against invading microorganisms and other foreign substances, as described in chapter 51. **Granular leukocytes** include neutrophils, eosinophils, and basophils, which are named according to the staining properties of granules in their cytoplasm. **Nongranular leukocytes** include monocytes and lymphocytes. In humans, neutrophils are the most numerous of the leukocytes, followed in order by lymphocytes, monocytes, eosinophils, and basophils.

Platelets

Platelets are cell fragments that pinch off from larger cells in the bone marrow. They are approximately 3 μm in diameter, and following an injury to a blood vessel, platelets release clotting factors (proteins) into the blood. In the presence of these clotting factors, fibrinogen is converted into insoluble threads of **fibrin.** Fibrin then aggregates to form the clot.

Formed elements arise from stem cells

The formed elements of blood each have a finite life span and therefore must be constantly replaced. Many of the old cell fragments are digested by phagocytic cells of the spleen; however, many products from the old cells, such as iron and amino acids, are incorporated into new formed elements. The creation of new formed elements begins in the bone marrow, as mentioned in chapter 47.

All of the formed elements develop from **pluripotent stem cells** (see chapter 19). The production of blood cells occurs in the bone marrow and is called **hematopoiesis.** This process generates two types of stem cells with more restricted fate: a lymphoid stem cell that gives rise to lymphocytes and a myeloid stem cell that gives rise to the rest of the blood cells (figure 49.16).

When the oxygen available in the blood decreases, the kidney converts a plasma protein into the hormone **erythropoietin.** Erythropoietin then stimulates the production of erythrocytes from the myeloid stem cells through a process called *erythropoiesis.*

In mammals, maturing erythrocytes lose their nuclei prior to release into circulation. In contrast, the mature erythrocytes of all other vertebrates remain nucleated. *Megakaryocytes* are examples of committed cells formed in bone marrow from stem cells. Pieces of cytoplasm are pinched off the megakaryocytes to form the platelets.

inquiry

Why do you think the use of erythropoietin as a drug is banned in the Olympics and in some other sports?

Blood clotting is an example of an enzyme cascade

When a blood vessel is broken or cut, smooth muscle in the vessel walls contracts, causing the vessel to constrict. Platelets then accumulate at the injured site and form a plug by sticking to one another and to the surrounding tissues (figure 49.17). A cascade of enzymatic reactions is triggered by

figure 49.17

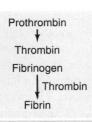

1. Vessel is damaged, exposing surrounding tissue to blood.

2. Platelets adhere and become sticky, forming a plug.

Prothrombin
↓
Thrombin

Fibrinogen
↓ Thrombin
Fibrin

3. Cascade of enzymatic reactions is triggered by platelets, plasma factors, and damaged tissue.

4. Threads of fibrin trap erythrocytes and form a clot.

5. Once tissue damage is healed, the clot is dissolved.

BLOOD CLOTTING. Fibrin is formed from a soluble protein, fibrinogen, in the plasma. This reaction is catalyzed by the enzyme thrombin, which is formed from an inactive enzyme called prothrombin. The activation of thrombin is the last step in a cascade of enzymatic reactions that produces a blood clot when a blood vessel is damaged.

the platelets, plasma factors, and molecules released from the damaged tissue.

One of the results of this cascade is that fibrinogen, normally dissolved in the plasma, comes out of solution in a reaction that forms fibrin. The platelet plug is then reinforced by fibrin threads, which contract to form a tighter mass. The tightening plug of platelets, fibrin, and often trapped erythrocytes constitutes a blood clot.

Once the tissue damage is healed, the careful process of dissolving the blot clot begins. This process is significant because if a clot breaks loose and travels in the circulatory system, it may end up blocking a blood vessel in the brain, causing a stroke, or in the heart, causing a myocardial infarction—heart attack.

Plasma, the liquid portion of the blood, contains different types of nutrients, wastes, and hormones, as well as proteins and ions. Formed elements in blood include cells and cell fragments. Erythrocytes contain hemoglobin and serve in oxygen transport. Leukocytes have specialized functions that protect the body from invading pathogens, and platelets participate in blood clotting. This involves a cascade of enzymatic reactions to produce fibrin from fibrinogen.

49.7 Gas Exchange Across Respiratory Surfaces

One of the primary functions of the circulatory system is acquisition, distribution, and removal of gases to support the tissues of the body. Indeed, one of the major physiological challenges facing all multicellular animals is obtaining sufficient oxygen and disposing of excess carbon dioxide (figure 49.18). Oxygen is used in mitochondria for cellular respiration, and this process also produces CO_2 as waste (see chapter 7). Respiration at the body system level involves a host of processes not found at the cellular level, ranging from the mechanics of breathing to the exchange of oxygen and carbon dioxide in respiratory organs.

Invertebrates display a wide variety of respiratory organs, including the epithelium, tracheae, and gills. Some vertebrates, such as fish and larval amphibians, also use gills; adult amphibians use their skin or other epithelia either as a supplemental or primary external respiratory organ.

Many adult amphibians, reptiles, birds, and mammals have lungs to perform external respiration. In both aquatic and terrestrial animals, these highly vascularized respiratory organs are the site at which oxygen diffuses into the blood, and carbon dioxide diffuses out. In the body tissues, the direction of gas diffusion is the reverse of that in the respiratory organs.

The mechanics, structure, and evolution of respiratory systems, along with the principles of gas diffusion between the blood and tissues, are the subjects of the following pages.

Gas exchange involves diffusion across membranes

Because plasma membranes must be surrounded by water to be stable, the external environment in gas exchange is always aqueous. This is true even in terrestrial vertebrates; in these cases, oxygen from air dissolves in a thin layer of fluid that covers the respiratory surfaces.

In vertebrates, the gases diffuse into the aqueous layer covering the epithelial cells that line the respiratory organs. The diffusion process is passive, driven only by the difference in O_2 and CO_2 concentrations on the two sides of the membranes and their relative solubilities in the plasma membrane. For dissolved gases, concentration is usually expressed as pressure; we explain this more fully a little later.

In general, the rate of diffusion between two regions is governed by a relationship known as **Fick's law of diffusion.** Fick's law states that for a dissolved gas, the rate of diffusion (R) is directly proportional to the pressure difference (Δp) between the two sides of the membrane and the area (A) over which the diffusion occurs. Furthermore, R is inversely proportional to the distance (d) across which the diffusion must occur. A molecule-specific diffusion constant, D, accounts for the size of molecule, membrane permeability, and temperature. Shown as a formula, Fick's law is stated as:

$$R = \frac{DA\Delta p}{d}$$

Major evolutionary changes in the mechanism of respiration have occurred to optimize the rate of diffusion (see figure 49.18).

figure 49.18

ELEPHANT SEALS ARE RESPIRATORY CHAMPIONS. Diving to depths greater than those reached by all other marine animals, including sperm whales and sea turtles, elephant seals can hold their breath for over 2 hr, descend and ascend rapidly in the water, and endure repeated dives without suffering any apparent respiratory distress.

R can be optimized by changes that (1) increase the surface area, A; (2) decrease the distance, d; or (3) increase the concentration difference, as indicated by Δp. The evolution of respiratory systems has involved changes in all of these factors.

inquiry

? *What part of the vertebrate cardiovascular system maximizes surface area?*

Evolutionary strategies have maximized gas diffusion

The levels of oxygen needed for cellular respiration cannot be obtained by diffusion alone over distances greater than about 0.5 mm. This restriction severely limits the size and structure of organisms that obtain oxygen entirely by diffusion from the environment. Bacteria, archaea, and protists are small enough that such diffusion can be adequate, even in some colonial forms (figure 49.19a), but most multicellular animals require structural adaptations to enhance gas exchange.

Increasing pressure difference

Most phyla of invertebrates lack specialized respiratory organs, but they have developed means of improving diffusion. Many organisms create a water current that continuously replaces the water over the respiratory surfaces; often, beating cilia produce this current. Because of this continuous replenishment of water, the external oxygen concentration does not decrease along the diffusion pathway. Although some of the oxygen molecules that pass into the organism have been removed from the surrounding water, new water continuously replaces the oxygen-depleted water. This maximizes the concentration difference—the Δp of the Fick equation.

Increasing area and decreasing distance

More complex invertebrates (mollusks, arthropods, echinoderms) and vertebrates, possess respiratory organs that increase the surface area available for diffusion—such as gills, tracheae, and lungs. These adaptations also bring the external environment (either water or air) close to the internal fluid, which

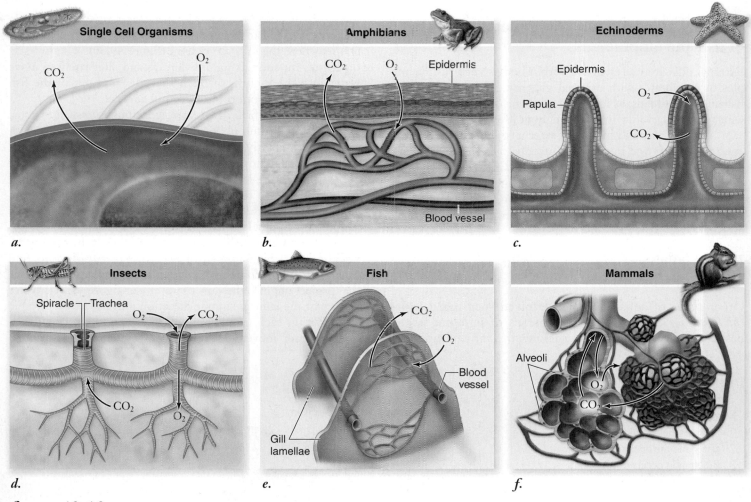

figure 49.19

DIFFERENT GAS EXCHANGE SYSTEMS IN ANIMALS. *a.* Gases diffuse directly into single-celled organisms. *b.* Amphibians and many other animals respire across their skin. Amphibians also exchange gases via lungs. *c.* Echinoderms have protruding papulae, which provide an increased respiratory surface area. *d.* Insects respire through an extensive tracheal system. *e.* The gills of fishes provide a very large respiratory surface area and countercurrent exchange. *f.* The alveoli in mammalian lungs provide a large respiratory surface area but do not permit countercurrent exchange.

is usually circulated throughout the body—such as blood or hemolymph. The respiratory organs thus increase the rate of diffusion by maximizing surface area (*A*) and decreasing the distance (*d*) the diffusing gases must travel.

> The exchange of oxygen and carbon dioxide between an organism and its environment is maximized by increasing the concentration gradient and the surface area and by decreasing the distance that the diffusing gases must travel.

49.8 Gills, Cutaneous Respiration, and Tracheal Systems

Gills are specialized extensions of tissue that project into water. Gills can be simple, as in the papulae of echinoderms (see figure 49.19*c*), or complex, as in the highly convoluted gills of fish (figure 49.19*e*). The great increase in diffusion surface area that gills provide enables aquatic organisms to extract far more oxygen from water than would be possible from their body surface alone. In this section we concentrate on gills found in vertebrate animals.

Moist surfaces are also involved in gas exchange in some vertebrates and invertebrates. Gas exchange across the skin is a common strategy in many amphibian groups, and terrestrial arthropods such as insects have developed tracheal systems that allow gas exchange through their hard exoskeletons.

External gills are found in immature forms of fish and amphibians

External gills are not enclosed within body structures. Examples of vertebrates with external gills are the larvae of many fish and amphibians, as well as amphibians such as the axolotl, which retains larval features throughout life.

One of the disadvantages of external gills is that they must constantly be moved to ensure contact with fresh water having high oxygen content. The highly branched gills, however, offer significant resistance to movement, making this form of respiration ineffective except in smaller animals. Another disadvantage is that external gills, with their thin epithelium for gas exchange, are easily damaged.

Branchial chambers protect gills of some invertebrates

Other types of aquatic animals evolved specialized *branchial chambers*, which provide a means of pumping water past stationary gills. The internal *mantle cavity* of mollusks

opens to the outside and contains the gills. Contraction of the muscular walls of the mantle cavity draws water in and then expels it.

In crustaceans, the branchial chamber lies between the bulk of the body and the hard exoskeleton of the animal. This chamber contains gills and opens to the surface beneath a limb. Movement of the limb draws water through the branchial chamber, thus creating currents over the gills.

Gills of bony fishes are covered by the operculum

The gills of bony fishes are located between the oral cavity, sometimes called the buccal (mouth) cavity, and the *opercular cavities* where the gills are housed (figure 49.20). The two sets of cavities function as pumps that expand alternately to move water into the mouth, through the gills, and out of the fish through the open **operculum,** or gill cover.

Some bony fishes that swim continuously, such as tuna, have practically immobile opercula. These fishes swim with their mouths partly open, constantly forcing water over the gills in what is known as *ram ventilation* (figure 49.20). Most bony fishes, however, have flexible gill covers. For example, the remora, a fish that rides "piggyback" on sharks, uses ram ventilation while the shark is swimming, but uses the pumping action of its opercula when the shark stops swimming.

There are four **gill arches** on each side of the fish's head. Each gill arch is composed of two rows of *gill filaments*, and each gill filament contains thin membranous plates, or *lamellae*, that project out into the flow of water (see figure 49.21). Water flows past the lamellae in one direction only.

Within each lamella, blood flows opposite to the direction of water movement. This arrangement is called **coun-**

figure 49.20

HOW MOST BONY FISHES RESPIRE. The gills are suspended between the buccal (mouth) cavity and the opercular cavity. Respiration occurs in two stages. The oral valve in the mouth is opened and the jaw is depressed, drawing water into the buccal cavity while the opercular cavity is closed. The oral valve is closed and the operculum is opened, drawing water through the gills to the outside.

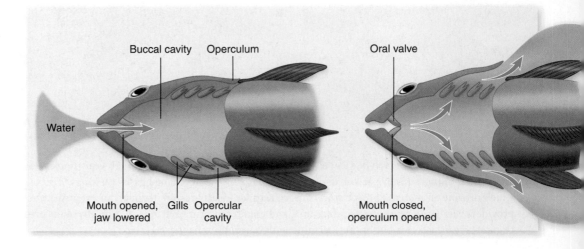

Buccal cavity · Operculum · Water · Mouth opened, jaw lowered · Gills · Opercular cavity · Oral valve · Mouth closed, operculum opened

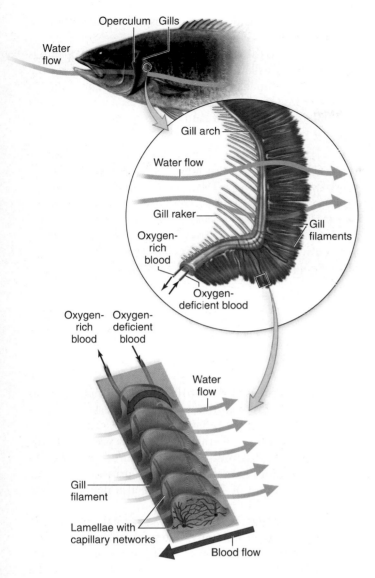

figure 49.21

STRUCTURE OF A FISH GILL. Water passes from the gill arch over the filaments (from left to right in the diagram). Water always passes the lamellae in a direction opposite to the direction of blood flow through the lamellae. The success of the gill's operation critically depends on this countercurrent flow of water and blood.

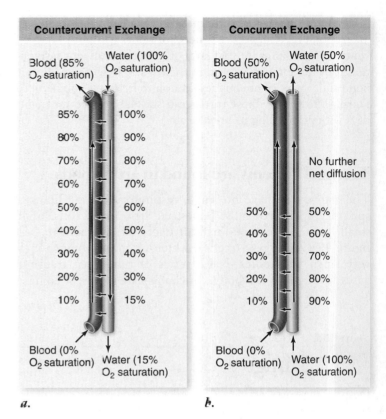

a.   *b.*

figure 49.22

COUNTERCURRENT EXCHANGE. This process allows for the most efficient blood oxygenation. When blood and water flow in opposite directions (*a*), the initial oxygen concentration difference between water and blood is small, but is sufficient for oxygen to diffuse from water to blood. As more oxygen diffuses into the blood, raising the blood's oxygen concentration, the blood encounters water with ever higher oxygen concentrations. At every point, the oxygen concentration is higher in the water, so that diffusion continues. In this example, blood attains an oxygen concentration of 85%. When blood and water flow in the same direction (*b*), oxygen can diffuse from the water into the blood rapidly at first, but the diffusion rate slows as more oxygen diffuses from the water into the blood, until finally the concentrations of oxygen in water and blood are equal. In this example, blood's oxygen concentration cannot exceed 50%.

tercurrent flow, and it acts to maximize the oxygenation of the blood by increasing the concentration gradient of oxygen along the pathway for diffusion, increasing Δp in Fick's law of diffusion. The advantages of a countercurrent flow system are illustrated in figure 49.22*a*. Countercurrent flow ensures that an oxygen concentration gradient remains between blood and water throughout the length of the gill lamellae. This permits oxygen to continue to diffuse all along the lamellae, so that the blood leaving the gills has nearly as high an oxygen concentration as the water entering the gills.

If blood and water flowed in the same direction, the flow would be *concurrent* (figure 49.22*b*). In this case, the concentration difference across the gill lamellae would fall rapidly as the water lost oxygen to the blood, and net diffusion of oxygen would cease when equilibrium was reached.

Because of the countercurrent exchange of gases, fish gills are the most efficient of all respiratory organs.

Cutaneous respiration requires constant moisture

Oxygen and carbon dioxide are able to diffuse across cutaneous (skin) surfaces in some vertebrates (see figure 49.19*b*). Most commonly, these vertebrates are aquatic, such as amphibians and some turtles, and they have highly vascularized areas of thin epidermis. The process of exchanging oxygen and carbon dioxide across the skin is called **cutaneous respiration.** In amphibians, cutaneous respiration supplements the action of lungs. Although not common, some terrestrial amphibians, such as plethodontid salamanders, rely on cutaneous respiration exclusively.

Within the aquatic reptiles, soft-shelled turtles can remain submerged in river sediment for hours without having to ventilate their lungs. At that level of activity, cutaneous respiration provides enough oxygen to the tissues. Even the common pond slider uses cutaneous respiration to help stay submerged. During the winter, these turtles can stay submerged for many days without needing to breathe air.

Tracheal systems are found in arthropods

The arthropods have no single respiratory organ. The respiratory system of most terrestrial arthropods consists of small, branched cuticle-lined air ducts called **tracheae** (see figure 49.19*d*). These trachea, which ultimately branch into very small **tracheoles,** are a series of tubes that transmit gases throughout the body. Tracheoles are in direct contact with individual cells, and oxygen diffuses directly across the plasma membranes.

Air passes into the trachea by way of specialized openings in the exoskeleton called **spiracles,** which, in most terrestrial arthropods, can be opened and closed by valves. The ability to prevent water loss by closing the spiracles was a key adaptation that facilitated the invasion of land by arthropods.

Gills are highly subdivided structures allowing a large surface area for exchange. In the gills of bony fishes, blood flows in a direction opposite to the flow of water. This countercurrent flow maximizes gas exchange, making the fish's gill an efficient respiratory organ. Some amphibians rely on cutaneous respiration, especially in larval forms. In arthropods, the evolution of a tracheal system with spicules allowed gas exchange with an exoskeleton and the move onto land.

49.9 Lungs

Despite the high efficiency of gills as respiratory organs in aquatic environments, gills were replaced in terrestrial animals for two principal reasons:

1. **Air is less supportive than water.** The fine membranous lamellae of gills lack inherent structural strength and rely on water for their support. A fish out of water, although awash in oxygen, soon suffocates because its gills collapse into a mass of tissue. Unlike gills, internal air passages such as trachaea and lungs can remain open because the body itself provides the necessary structural support.
2. **Water evaporates.** Air is rarely saturated with water vapor, except immediately after a rainstorm. Consequently, terrestrial organisms constantly lose water to the atmosphere. Gills would provide an enormous surface area for water loss.

The **lung** minimizes evaporation by moving air through a branched tubular passage. The tracheal system of arthropods also uses internal tubes to minimize evaporation.

The air drawn into the respiratory passages becomes saturated with water vapor before reaching the inner regions of the lung. In these areas, a thin, wet membrane permits gas exchange. The lungs of all terrestrial vertebrates except birds use a uniform pool of gases that is in contact with the gas exchange surface. Unlike the one-way flow of water that is so effective in the respiratory function of gills, gases moves in and out of lungs by way of the same airway passages, a two-way flow system. Birds have an exceptional respiratory system, as described later on.

Breathing of air takes advantage of partial pressures of gases

Dry air contains 78.09% nitrogen, 20.95% oxygen, 0.93% argon and other inert gases, and 0.03% carbon dioxide. Convection currents cause the atmosphere to maintain a constant composition to altitudes of at least 100 km, although the *amount* (number of molecules) of air that is present decreases with altitude (figure 49.23).

Because of the force of gravity, air exerts a pressure downward. An apparatus that measures air pressure is called a *barometer,* and 760 mm Hg is the barometric pressure of the air at sea level. A pressure of 760 mm Hg is also defined as one **atmosphere** (1.0 atm) of pressure.

Each type of gas contributes to the total atmospheric pressure according to its fraction of the total molecules present. The pressure contributed by a gas is called its **partial pressure,** and it is indicated by P_{N_2}, P_{O_2}, P_{CO_2}, and so on. At sea level, the partial pressures of N_2, O_2, and CO_2 are as follows:

$$P_{N_2} = 760 \times 79.02\% = 600.6 \text{ mm Hg}$$

$$P_{O_2} = 760 \times 20.95\% = 159.2 \text{ mm Hg}$$

$$P_{CO_2} = 760 \times 0.03\% = 0.2 \text{ mm Hg}$$

Humans do not survive for long at altitudes above 6000 m. Although the air at these altitudes still contains 20.95% oxygen, the atmospheric pressure is only about 380 mm Hg, so the P_{O_2} is only 80 mm Hg (380 × 20.95%), half the amount of oxygen available at sea level.

In the following sections, we describe respiration in vertebrates with lungs, beginning with reptiles and amphibians. We then summarize mammalian lungs and the highly adapted and specialized lungs of birds.

Lungs of amphibians and reptiles are specialized outgrowths of the gut

The lungs of amphibians are formed as saclike outpouchings of the gut (figure 49.24). Although the internal surface area of these sacs is increased by folds, much less surface area is available for gas exchange in amphibian lungs than in the lungs of other terrestrial vertebrates. Each amphibian lung is connected

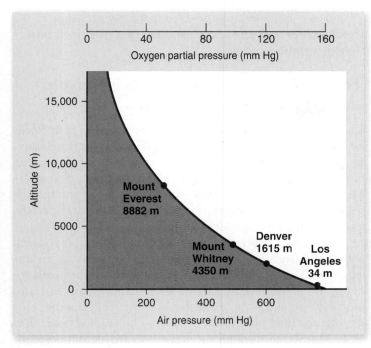

figure 49.23

THE RELATIONSHIP BETWEEN AIR PRESSURE AND
ALTITUDE ABOVE SEA LEVEL. Air pressure at high altitudes
(characteristic of mountaintops) is much less than at sea level.
At the top of Mount Everest, the world's highest mountain, the
air pressure is only one-third that at sea level.

inquiry

*What is the difference in the percentage of atmospheric
oxygen between Mount Everest and Mount Whitney?*

to the rear of the oral cavity, or pharynx, and the opening to
each lung is controlled by a valve, the glottis.

Amphibians do not breathe the same way other terrestrial vertebrates do. Amphibians force air into their lungs; they
fill their oral cavity with air (figure 49.24*a*), close their mouth
and nostrils, and then elevate the floor of their oral cavity. This
pushes air into their lungs in the same way that a pressurized
tank of air is used to fill balloons (figure 49.24*b*). This is called
positive pressure breathing; in humans, it would be analogous to forcing air into a person's lungs by performing mouth-
to-mouth resuscitation.

Terrestrial reptiles have dry, tough, scaly skins that not
only prevent desiccation, but also prohibit cutaneous respiration, which is utilized by many amphibians. Reptiles expand
their rib cages by muscular contraction. This action creates a
lower pressure inside the lungs compared with the atmosphere,
and the greater atmospheric pressure moves air into the lungs.
This type of ventilation is termed **negative pressure breathing** because of the air being "pulled in" by the animal rather
than "pushed in"

Reptiles' lungs have somewhat more surface area than the
lungs of amphibians and so are more efficient at gas exchange.
Cutaneous respiration, however, can occur in some reptiles,
such as marine sea snakes.

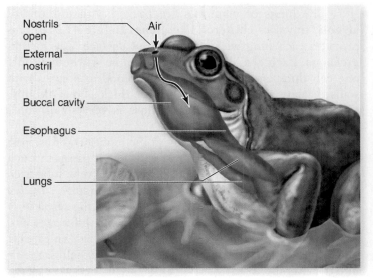

a.

b.

figure 49.24

AMPHIBIAN LUNGS. Each lung of this frog is an outpouching
of the gut and is filled with air by the creation of a positive
pressure in the buccal cavity. *a.* The buccal cavity is
expanded and air flows through the open nostrils. *b.* The
nostrils are closed and the buccal cavity is compressed,
thus creating the positive pressure that fills the lungs. The
amphibian lung lacks the structures present in the lungs of
other terrestrial vertebrates that provide an enormous surface
area for gas exchange, and so are not as efficient as the lungs
of other vertebrates.

Mammalian lungs have greatly increased surface area

Endothermic animals, such as birds and mammals, have consistently higher metabolic rates. Both these vertebrate groups
exhibit more complex and efficient respiratory systems than
ectothermic animals. The evolution of more efficient respiratory systems accommodate the increased demands on cellular
respiration of endothermy.

The lungs of mammals are packed with millions of **alveoli,** tiny sacs clustered like grapes (figure 49.25). This provides each lung with an enormous surface area for gas exchange. Each alveolus is composed of an epithelium only one cell thick, and is surrounded by blood capillaries with walls that are also only one cell layer thick. Thus, the distance d across which gas must diffuse is very small—only 0.5–1.5 μm.

Inhaled air is taken in through the mouth and nose past the pharynx to the **larynx** (voice box), where it passes through an opening in the vocal cords, the *glottis,* into a tube supported by C-shaped rings of cartilage, the **trachea** (windpipe). *Trachea* is used both for the vertebrate windpipe and the respiratory tubes of arthropods. The mammalian trachea bifurcates into right and left **bronchi** (singular, *bronchus*), which enter each lung and further subdivide into **bronchioles** that deliver the air into the alveoli.

The alveoli are surrounded by an extensive capillary network. All gas exchange between the air and blood takes place across the walls of the alveoli. The branching of bronchioles and the vast number of alveoli combine to increase the respiratory surface area far above that of amphibians or reptiles. In humans, there are about 300 million alveoli in each of the two lungs, and the total surface area available for diffusion can be as much as 80 m², or about 42 times the surface area of the body. Details of gas exchange at the alveolar interface with blood capillaries is described in sections that follow.

The respiratory system of birds is a highly efficient flow-through system

The avian respiratory system is a unique structure that affords birds the most efficient respiration of all terrestrial vertebrates. Unlike the mammalian lung, which ends in blind alveoli, the bird lung channels air through tiny air vessels called **parabronchi,** where gas exchange occurs. Air flows through the parabronchi in one direction only. This flow is similar to the unidirectional flow of water through a fish gill.

In other terrestrial vertebrates, inhaled fresh air is mixed with "old," oxygen-depleted air left from the previous breathing cycle. The lungs of amphibians, reptiles, and mammals are never completely empty of the gases within them. In birds, only fresh air enters the parabronchi of the lung, and the "old" air exits the lung by a different route. The unidirectional flow of air is achieved through the action of anterior and posterior air sacs that are unique to birds (figure 49.26a). When these sacs are expanded during inhalation, they take in air, and when they are compressed during exhalation, they push air into and through the lungs.

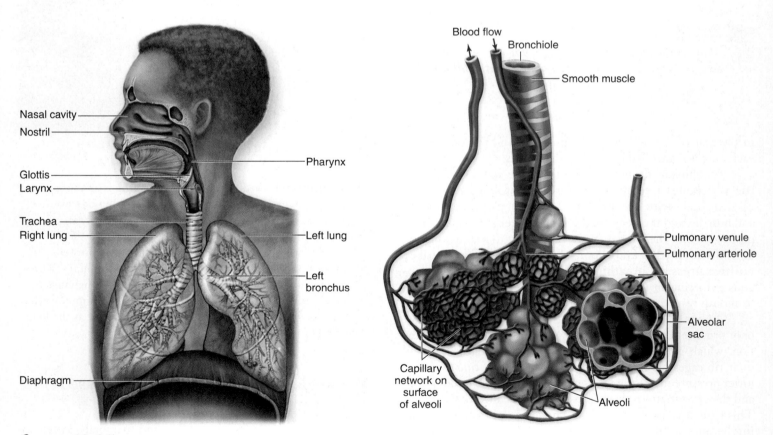

figure 49.25

THE HUMAN RESPIRATORY SYSTEM AND THE STRUCTURE OF THE MAMMALIAN LUNG. The lungs of mammals have an enormous surface area because of the millions of alveoli that cluster at the ends of the bronchioles. This provides for efficient gas exchange with the blood.

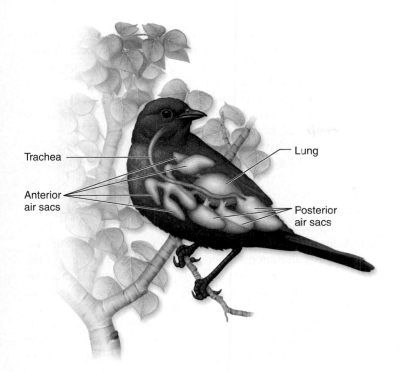

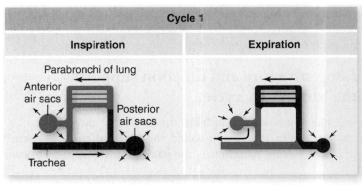

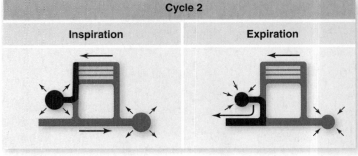

a.

b.

figure 49.26

HOW A BIRD BREATHES. *a.* Birds have a system of air sacs, divided into an anterior group and posterior group, that extend between the internal organs and into the bones. *b.* Breathing occurs in two cycles. *Cycle 1:* Inhaled air (shown in *red*) is drawn from the trachea into the posterior air sacs and then is exhaled into the lungs. *Cycle 2:* Air is drawn from the lungs into the anterior air sacs and then is exhaled through the trachea. Passage of air through the lungs is always in the same direction, from posterior to anterior (right to left in this diagram).

Respiration in birds occurs in two cycles (figure 49.26*b*). Each cycle has an inhalation and exhalation phase—but the air inhaled in one cycle is not exhaled until the second cycle.

Upon inhalation, both the anterior and posterior air sacs expand. The inhaled air, however, only enters the posterior air sacs; the anterior air sacs fill with air pulled from the lungs. Upon exhalation, the air forced out of the anterior air sacs is released outside the body, but the air forced out of the posterior air sacs now enters the lungs. This process is repeated in the second cycle.

The unidirectional flow of air also permits further respiratory efficiency: The flow of blood through the avian lung runs at a 90° angle to the air flow. This crosscurrent flow is not as efficient as the 180° countercurrent flow in fishes' gills, but it has a greater capacity to extract oxygen from the air than does a mammalian lung.

Because of these respiratory adaptations, a sparrow can be active at an altitude of 6000 m, whereas a mouse, which has a similar body mass and metabolic rate, would die from lack of oxygen in a fairly short time.

Terrestrial vertebrates take air into saclike lungs that provide a huge surface area for gas exchange. Some amphibians and aquatic reptiles supplement their respiratory gas exchange by using cutaneous respiration. The avian respiratory system is the most efficient among terrestrial vertebrates because it has unidirectional air flow and cross-current blood flow through the lungs.

49.10 Structures and Mechanisms of Ventilation

About 30 billion capillaries can be found in each lung, roughly 100 capillaries per alveolus. Thus, an alveolus can be visualized as a microscopic air bubble whose entire surface is bathed by blood. Gas exchange occurs very rapidly at this interface.

Blood returning from the systemic circulation, depleted in oxygen, has a partial oxygen pressure (P_{O_2}) of about 40 mm Hg. By contrast, the gas mixture in the alveoli is about 105 mm Hg. The difference in pressures, namely the Δp of Fick's law, is 65 mm Hg, leading to oxygen moving into the blood. The blood leaving the lungs, as a result of this gas exchange, normally contains a partial oxygen pressure (P_{O_2}) of about 100 mm Hg. As you can see, the lungs do a very effective, but not perfect, job of oxygenating the blood. These changes in the P_{O_2} of the blood, as well as the changes in plasma carbon dioxide (indicated as the P_{CO_2}), are shown in figure 49.27.

Lung structure and function supports the respiratory cycle

In humans and other mammals, the outside of each lung is covered by a thin membrane called the **visceral pleural membrane.** A second membrane, the **parietal pleural membrane,** lines the inner wall of the thoracic cavity. The space between these two membrane sheets, the **pleural cavity,** is normally very small and filled with fluid. This fluid causes the two membranes to adhere, effectively coupling the lungs to the thoracic cavity. The pleural membranes package each lung separately—if one lung collapses due to a perforation of the membranes, the other lung can still function.

During inhalation, the thoracic volume is increased through contraction of two sets of muscles: the *external intercostal muscles* and the *diaphragm.* Contraction of the external intercostal muscles between the ribs raises the ribs and expands the rib cage. Contraction of the **diaphragm,** a convex sheet of striated muscle separating the thoracic cavity from the abdominal cavity, causes the diaphragm to lower and assume a more flattened shape. This expands the volume of the thorax and lungs, bringing about negative pressure ventilation, while it increases the pressure on the abdominal organs (figure 49.28*a*).

The thorax and lungs have a degree of elasticity; expansion during inhalation places these structures under elastic tension. The relaxation of the external intercostal muscles and diaphragm produces unforced exhalation because the elastic tension is released, allowing the thorax and lungs to recoil. You can force a greater exhalation by contracting your abdominal muscles—such as when blowing up a balloon (figure 49.28*b*).

Ventilation efficiency depends on lung capacity and breathing rate

A variety of terms are used to describe the volume changes of the lung during breathing. In a person at rest, each breath moves a **tidal volume** of about 500 mL of air into and out of the lungs. About 150 mL of the tidal volume is contained in the tubular passages (trachea, bronchi, and bronchioles), where no gas exchange occurs—termed the *anatomical dead space.* The gases in this space mix with fresh air during inspiration. This mixing is one reason that respiration in mammals is not as efficient as in birds, where air flow through the lungs is one-way.

The maximum amount of air that can be expired after a forceful, maximum inspiration is called the **vital capacity.** This measurement, which averages 4.6 L in young men and 3.1 L in young women, can be clinically important because an abnormally low vital capacity may indicate damage to the alveoli in various pulmonary disorders.

The rate and depth of breathing normally keeps the blood P_{O_2} and P_{CO_2} within a normal range. If breathing is insufficient to maintain normal blood gas measurements (a rise in the blood P_{CO_2} is the best indicator), the person is **hypoventilating.** If breathing is excessive, so that the blood P_{CO_2} is abnormally lowered, the person is said to be **hyperventilating.**

The increased breathing that occurs during moderate exertion is not necessarily hyperventilation because the faster and more forceful breathing is matched to the higher metabolic rate and blood gas measurements remain normal.

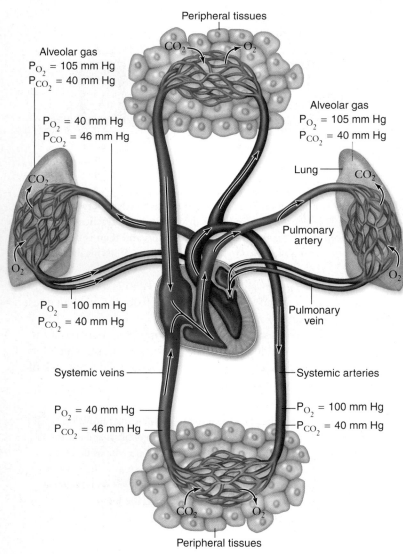

Peripheral tissues

Alveolar gas
P_{O_2} = 105 mm Hg
P_{CO_2} = 40 mm Hg

P_{O_2} = 40 mm Hg
P_{CO_2} = 46 mm Hg

Alveolar gas
P_{O_2} = 105 mm Hg
P_{CO_2} = 40 mm Hg

Lung

Pulmonary artery

Pulmonary vein

P_{O_2} = 100 mm Hg
P_{CO_2} = 40 mm Hg

Systemic veins

Systemic arteries

P_{O_2} = 40 mm Hg
P_{CO_2} = 46 mm Hg

P_{O_2} = 100 mm Hg
P_{CO_2} = 40 mm Hg

Peripheral tissues

figure 49.27

GAS EXCHANGE IN THE BLOOD CAPILLARIES OF THE LUNGS AND SYSTEMIC CIRCULATION. As a result of gas exchange in the lungs, the systemic arteries carry oxygenated blood with a relatively low carbon dioxide concentration. After the oxygen is unloaded to the tissues, the blood in the systemic veins has a lowered oxygen content and an increased carbon dioxide concentration.

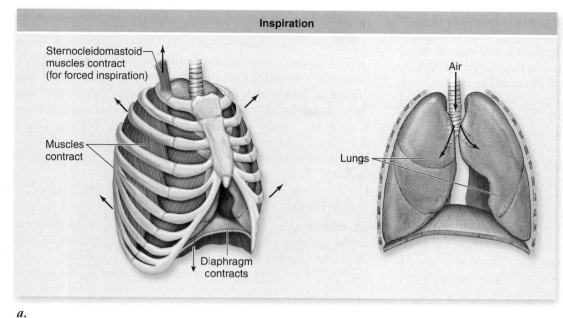

Inspiration

Sternocleidomastoid muscles contract (for forced inspiration)

Muscles contract

Diaphragm contracts

Air

Lungs

a.

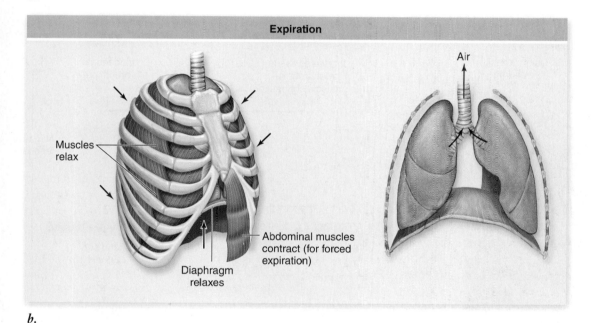

Expiration

Muscles relax

Air

Abdominal muscles contract (for forced expiration)

Diaphragm relaxes

b.

figure 49.28

HOW A HUMAN BREATHES. *a.* Inspiration. The diaphragm contracts and the walls of the chest cavity expand, increasing the volume of the chest cavity and lungs. As a result of the larger volume, air is drawn into the lungs. *b.* Expiration. The diaphragm and chest walls return to their normal positions as a result of elastic recoil, reducing the volume of the chest cavity and forcing air out of the lungs through the trachea. Note that inspiration can be forced by contracting accessory respiratory muscles (such as the sternocleidomastoid), and expiration can be forced by contracting abdominal muscles.

Next, we describe how breathing is regulated to keep pace with metabolism.

Ventilation is under nervous system control

Each breath is initiated by neurons in a *respiratory control center* located in the medulla oblongata. These neurons stimulate the diaphragm and external intercostal muscles to contract, causing inhalation. When these neurons stop producing impulses, the inspiratory muscles relax and exhalation occurs. Although the muscles of breathing are skeletal muscles, they are usually controlled automatically. This control can be voluntarily overridden, however, as in hypoventilation (breath holding) or hyperventilation.

Neurons of the medulla oblongata must be responsive to changes in blood P_{O_2} and P_{CO_2} in order to maintain homeostasis. You can demonstrate this mechanism by simply holding your breath. Your blood carbon dioxide level immediately rises, and your blood oxygen level falls. After a short time, the urge to breathe induced by the changes in blood gases becomes overpowering. The rise in blood carbon dioxide, as indicated by a rise in P_{CO_2}, is the primary initiator, rather than the fall in oxygen levels.

A rise in P_{CO_2} causes an increased production of carbonic acid (H_2CO_3), which lowers the blood pH. A fall in blood pH stimulates chemosensitive neurons in the **aortic** and **carotid bodies,** in the aorta and the carotid artery (figure 49.29*a*). These peripheral receptors send impulses to the respiratory control center, which then stimulates increased breathing. The brain also contains central chemoreceptors that are stimulated by a drop in the pH of cerebrospinal fluid (figure 49.29*b*).

A person cannot voluntarily hyperventilate for too long. The decrease in plasma P_{CO_2} and increase in pH of plasma and CSF caused by hyperventilation suppress the reflex drive to breathe. Deliberate hyperventilation allows people to hold their breath longer—not because it increases oxygen in the blood, but because the carbon dioxide level is lowered and takes longer to build back up, postponing the need to breathe.

In people with normal lungs, P_{O_2} becomes a significant stimulus for breathing only at high altitudes, where the P_{O_2} of the atmosphere is low. The symptoms of low oxygen at high altitude are known as mountain sickness, which may include feelings of weakness, headache, nausea, vomiting, and reduced mental function. All of these symptoms are related to the low partial pressure of oxygen, and breathing supplemental oxygen may remove all symptoms.

Respiratory diseases restrict gas exchange

Chronic obstructive pulmonary disease (COPD) refers to any disorder that obstructs airflow on a long-term basis. The major COPDs are asthma, chronic bronchitis, and emphysema. In **asthma,** an allergen triggers the release of histamine and other inflammatory chemicals that cause intense constriction of the bronchi and sometimes suffocation. Other COPDs are commonly caused by cigarette smoking but can also result from air pollution or occupational exposure to airborne irritants.

Emphysema

In **emphysema,** alveolar walls break down and the lung exhibits larger but fewer alveoli. The lungs also become fibrotic and less elastic. The air passages open adequately during inspiration but they tend to collapse and obstruct the outflow of air. People with emphysema become exhausted because they

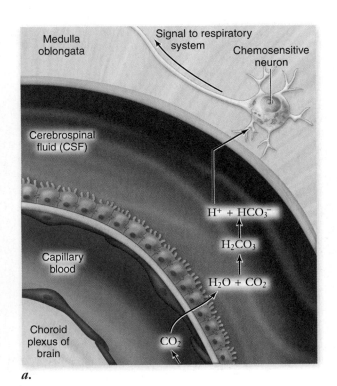

a.

figure 49.29

REGULATION OF BREATHING BY pH SENSITIVE CHEMORECEPTORS. *a.* Changes in the pH of the cerebrospinal fluid (CSF) are detected by H$^+$ ion chemoreceptors in the brain that help regulate breathing. *b.* Peripheral and central chemoreceptors sense a fall in the pH of blood and cerebrospinal fluid, respectively, when the blood CO$_2$ levels rise as a result of inadequate breathing. In response, they stimulate the respiratory control center in the medulla oblongata, which directs an increase in breathing. As a result, the blood CO$_2$ concentration is returned to normal, completing the negative feedback loop.

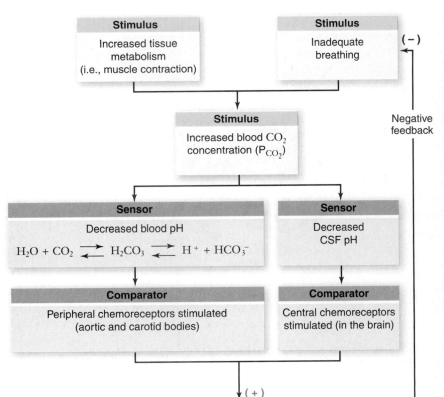

b.

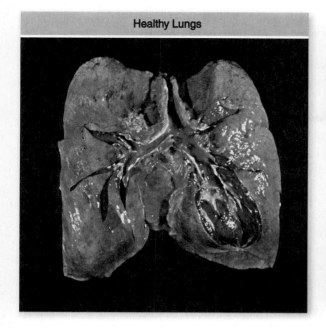

Healthy Lungs

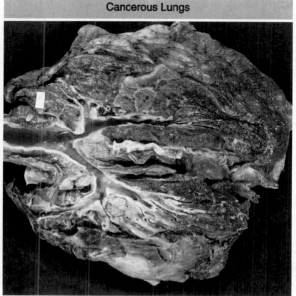

Cancerous Lungs

figure 49.30
COMPARISON OF HEALTHY LUNG (*a*) AND A LUNG WITH CANCER (*b*).

expend three to four times the normal amount of energy just to breathe. Eighty to ninety percent of emphysema deaths are caused by cigarette smoking.

inquiry

 How does emphysema affect the diffusion of gases in and out of the lung, based on Fick's law?

Lung cancer

Lung cancer accounts for more deaths than any other form of cancer. The most important cause of lung cancer is cigarette smoking, distantly followed by air pollution (figure 49.30). Lung cancer follows or accompanies COPD.

Over 90% of lung tumors originate in the mucous membranes of the large bronchi. As a tumor invades the bronchial wall and grows around it, it compresses the airway and may cause collapse of more distal parts of the lung.

Growth of a tumor often produces coughing, but coughing is such an everyday occurrence for smokers, it seldom causes much alarm. Often, the first sign of serious trouble is the coughing up of blood.

Lung cancer metastasizes (spreads) so rapidly that it has usually invaded other organs by the time it is diagnosed. The chance of recovery is poor, with only 7% of patients surviving for 5 years after diagnosis.

Humans, like other mammals, inhale through contraction of the diaphragm and intercostal muscles, producing negative pressure ventilation. Exhalation is produced primarily by muscle relaxation and elastic recoil. Ventilation keeps the blood gases and pH in the normal range and is under the reflex control of chemoreceptors. Respiratory diseases such as COPD limit gas exchange. Lung cancer is associated with cigarette smoking and has a low survival rate.

49.11 Transport of Gases in Body Fluids

The amount of oxygen that can be dissolved in the blood plasma depends directly on the P_{O_2} of the air in the alveoli, as explained earlier. When mammalian lungs are functioning normally, the blood plasma leaving the lungs has almost as much dissolved oxygen as is theoretically possible, given the P_{O_2} of the air. Because of oxygen's low solubility, however, blood plasma can contain a maximum of only about 3 mL of O_2 per liter. But whole blood normally carries almost 200 mL of O_2 per liter. Most of the oxygen in the blood is bound to molecules of hemoglobin inside red blood cells.

Respiratory pigments bind oxygen for transport

Hemoglobin is a protein composed of four polypeptide chains and four organic compounds called *heme groups*. At the center of each heme group is an atom of iron, which can bind to

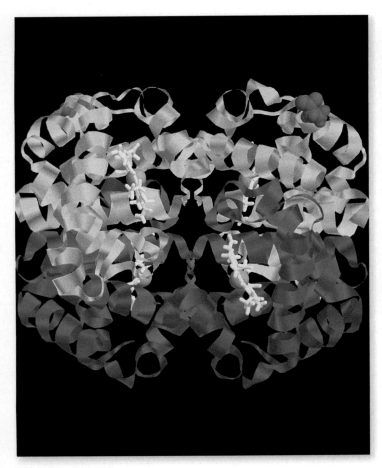

figure 49.31

THE STRUCTURE OF THE HEMOGLOBIN PROTEIN.
Hemoglobin consists of four polypeptide chains: two α chains and two β chains. Each chain is associated with a heme group, and each heme group has a central iron atom, which can bind to a molecule of O_2.

a molecule of oxygen (figure 49.31). Thus, each hemoglobin molecule can carry up to four molecules of oxygen.

Hemoglobin loads up with oxygen in the alveolar capillaries of the pulmonary circulation, forming **oxyhemoglobin.** This molecule has a bright red color. As blood passes through capillaries in the systemic circulation, some of the oxyhemoglobin releases oxygen, becoming **deoxyhemoglobin.** Deoxyhemoglobin has a darker red color; but it imparts a bluish tinge to tissues. Illustrations of the cardiovascular system show vessels carrying oxygenated blood with a red color and vessels that carry oxygen-depleted blood with a blue color.

Hemoglobin is an ancient protein; it is not only the oxygen-carrying molecule in all vertebrates, but is also used as an oxygen carrier by many invertebrates, including annelids, mollusks, echinoderms, flatworms, and even some protists. Many other invertebrates, however, employ different oxygen carriers, such as **hemocyanin.** In hemocyanin, the oxygen-binding atom is copper instead of iron. Hemocyanin is not found associated with blood cells, but is instead one of the free proteins in the circulating fluid (hemolymph) of arthropods and some mollusks.

inquiry

? *If oxygen-depleted vessels have a bluish color, does this mean that all veins in the body have a bluish color? Why or why not?*

Hemoglobin provides an oxygen reserve

At a blood P_{O_2} of 100 mm Hg, the level found in blood leaving the alveoli, approximately 97% of the hemoglobin within red blood cells is in the form of oxyhemoglobin—indicated as a percent oxyhemoglobin saturation of 97%.

In a person at rest, blood that returns to the heart in the systemic veins has a P_{O_2} that is decreased to about 40 mm Hg. At this lower P_{O_2}, the percent saturation of hemoglobin is only 75%. In a person at rest, therefore, 22% (97% minus 75%) of the oxyhemoglobin has released its oxygen to the tissues. Put another way, roughly one-fifth of the oxygen is unloaded in the tissues, leaving four-fifths of the oxygen in the blood as a reserve. A graphic representation of these changes is called an oxyhemoglobin dissociation curve (figure 49.32).

This large reserve of oxygen serves an important function. It enables the blood to supply the body's oxygen needs during exertion as well as at rest. During exercise, for example, the muscles' accelerated metabolism uses more oxygen and decreases the venous blood P_{O_2}. The P_{O_2} of the venous blood could drop to 20 mm Hg; in this case, the percent saturation of hemoglobin would be only 35% (see figure 49.32). Because arterial blood would still contain 97% oxyhemoglobin, the amount of oxygen unloaded would now be 62% (97% minus 35%), instead of the 22% at rest.

In addition to this function, the oxygen reserve also ensures that the blood contains enough oxygen to maintain life for 4–5 min if breathing is interrupted or if the heart stops pumping.

inquiry

? *Based on the preceding information, would an otherwise healthy person benefit significantly from breathing 100% oxygen following a bout of intense exercise such as a 400-m sprint?*

Hemoglobin's affinity for oxygen is affected by pH and temperature

Oxygen transport in the blood is affected by other conditions including temperature and pH. The CO_2 produced by metabolizing tissues combines with H_2O to form carbonic acid (H_2CO_3). H_2CO_3 dissociates into bicarbonate (HCO_3^-) and H^+, thereby lowering blood pH. This reaction occurs primarily inside red blood cells, where the lowered pH reduces hemoglobin's affinity for oxygen, causing it to release oxygen more readily.

The effect of pH on hemoglobin's affinity for oxygen, known as the **Bohr shift,** is the result of H^+ binding to hemoglobin. It is shown graphically by a shift of the oxyhemoglobin dissociation curve to the right (figure 49.33*a*).

Increasing temperature has a similar effect on hemoglobin's affinity for oxygen (figure 49.33*b*). Because skeletal muscles produce carbon dioxide more rapidly during exercise,

figure 49.32

THE OXYHEMOGLOBIN DISSOCIATION CURVE. Hemoglobin combines with O_2 in the lungs, and this oxygenated blood is carried by arteries to the body cells. After oxygen is removed from the blood to support cellular respiration, the blood entering the veins contains less oxygen.

inquiry

How would you determine how much oxygen was unloaded to the tissues?

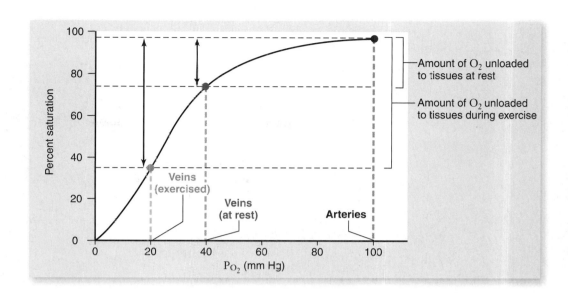

and because active muscles produce heat, the blood unloads a higher percentage of the oxygen it carries during exercise.

Carbon dioxide is primarily transported as bicarbonate ion

About 8% of the CO_2 in blood is simply dissolved in plasma; another 20% is bound to hemoglobin. Because CO_2 binds to the protein portion of hemoglobin, and not to the iron atoms of the heme groups, it does not compete with oxygen; however, it does cause hemoglobin's shape to change, lowering its affinity for oxygen.

The remaining 72% of the CO_2 diffuses into the red blood cells, where the enzyme **carbonic anhydrase** catalyzes the combining of CO_2 with water to form H_2CO_3. H_2CO_3 dissociates into HCO_3^- and H^+ ions. The H^+ binds to deoxyhemoglobin, and the HCO_3^- moves out of the erythrocyte into the plasma via a transporter that exchanges one Cl^- for a HCO_3^- (this is called the "chloride shift").

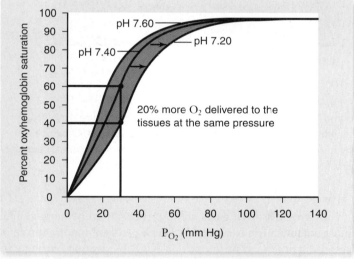

a. pH shift

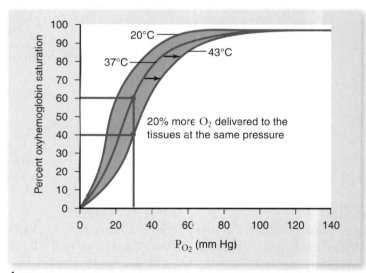

b. Temperature shift

figure 49.33

THE EFFECT OF pH AND TEMPERATURE ON THE OXYHEMOGLOBIN DISSOCIATION CURVE. *a.* Lower blood pH and *(b)* higher blood temperatures shift the oxyhemoglobin dissociation curve to the right, facilitating oxygen unloading. In this example, this can be seen as a lowering of the oxyhemoglobin percent saturation from 60% to 40%, indicating that the difference of 20% more oxygen is unloaded to the tissues.

inquiry

What effect does high blood pressure have on oxygen unloading to the tissues during exercise?

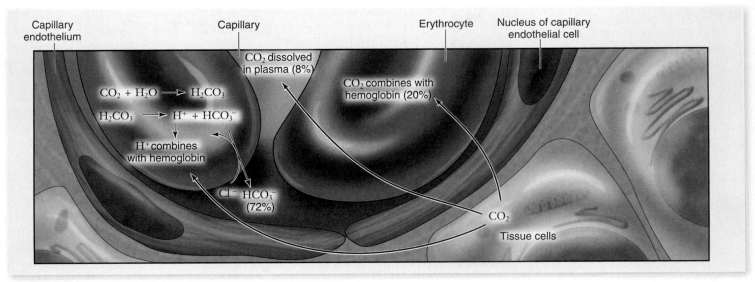

a.

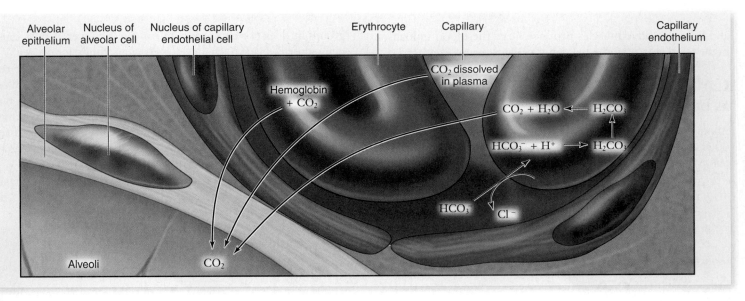

b.

figure 49.34

THE TRANSPORT OF CARBON DIOXIDE BY THE BLOOD. *a.* Passage into bloodstream. CO_2 is transported in three ways: dissolved in plasma, bound to the protein portion of hemoglobin, and as bicarbonate, which forms in red blood cells. The reaction of CO_2 with H_2O to form H_2CO_3 (carbonic acid) is catalyzed by the enzyme carbonic anhydrase in red blood cells. *b.* Removal from bloodstream. When the blood passes through the pulmonary capillaries, these reactions are reversed so that CO_2 gas is formed, which is exhaled.

This reaction removes large amounts of CO_2 from the plasma, maintaining a diffusion gradient that allows additional CO_2 to move into the plasma from the surrounding tissues (figure 49.34*a*). The formation of H_2CO_3 is also important in maintaining the acid–base balance of the blood; HCO_3^- serves as the major buffer of the blood plasma.

In the lungs, the lower P_{CO_2} of the gas mixture inside the alveoli causes the carbonic anhydrase reaction to proceed in the reverse direction, converting H_2CO_3 into H_2O and CO_2 (figure 49.34*b*). The CO_2 diffuses out of the red blood cells and into the alveoli, so that it can leave the body in the next exhalation.

Other dissolved gases are also transported by hemoglobin, most notably nitric oxide (NO), which plays an important role in vessel dilation. Carbon monoxide (CO) binds more strongly to hemoglobin than does oxygen, which is why carbon monoxide poisoning can be deadly. Victims of carbon monoxide poisoning often have bright red skin due to hemoglobin's binding with CO.

Deoxyhemoglobin combines with oxygen in the lungs to form oxyhemoglobin, which dissociates in the tissue capillaries to release its oxygen. Carbon dioxide is transported in the blood in three ways: dissolved in the plasma, bound to hemoglobin, and primarily as bicarbonate in the plasma following an enzyme-catalyzed reaction in the red blood cells.

49.1 Invertebrate Circulatory Systems

The circulation system of multicellular organisms is directly related to size, complexity, and organism lifestyle (figure 49.1).

- Sponges pass water through channels, and cnidarians circulate water through a gastrovascular cavity.
- Small animals can use body cavity fluids for circulation.
- Closed systems have a distinct circulatory fluid enclosed in blood vessels and transported in a loop.

49.2 Vertebrate Circulatory Systems

Increases in size and complexity require more surface area to deliver nutrients and oxygen, and to remove wastes and carbon dioxide.

- Fishes have a linear heart with two pumping chambers to increase efficiency of blood flow through the gills (figure 49.2).
- Pulmonary circulation pumps blood to the lungs and systemic circulation pumps blood to the body.
- Amphibian hearts have two atria separating blood flow to the lungs and body, and one ventricle (figure 49.3).
- Reptiles have a heart with a septum that partially divides the ventricle, reducing the mixing of oxygenated and deoxygenated blood.
- Mammals, birds, and crocodilians have two ventricles (figure 49 4).

49.3 The Four-Chambered Heart and the Blood Vessels

The four-chambered heart uses a complete cardiac cycle with a rest period and two contraction cycles.

- The unidirectional flow of blood through the heart is maintained by two atrioventricular valves (figure 49.5).
- During diastole ventricles relax and atria contract; during systole ventricles contract.
- Arteries and arterioles carry oxygenated blood to the body, veins and venules return deoxygenated blood to the heart (figure 49.4).
- Contraction is initiated by the SA node (figure 49.7).

49.4 Characteristics of Blood Vessels

Blood vessels, except for capillaries, have the same basic structure.

- Arteries and veins consist of endothelium, elastic fibers, smooth muscle, and connective tissues (figure 49.8).
- Capillaries have only one layer of endothelium.
- Arteries and arterioles can withstand changes in blood pressure and control blood flow due to elastic fibers in their walls.
- Exchange of materials in capillaries is rapid (figure 49.9).
- The return of blood to the heart through veins is facilitated by skeletal muscle contractions and one-way valves (figure 49.10).
- Fluid from plasma is filtered out of capillaries, then returns by osmosis and via separate lymphatic system (figure 49.11).
- Lymph moves through lymphatic vessels to lymph nodes and lymphatic organs and returns to the heart via the subclavian veins.

49.5 Regulation of Blood Flow and Blood Pressure

Blood flow and pressure are modulated by the autonomic nervous system (figure 49.14).

- Norepinephrine, from sympathetic neurons, increases heart rate; acetylcholine, from parasympathetic neurons, decreases rate.
- Cardiac output, the product of beats per unit time and stroke volume, increases with exertion.
- Arterial blood pressure is monitored by baroreceptors in the aortic arch and carotid arteries.
- Blood volume is regulated by hormones.

49.6 The Components of Blood

Blood is a connective tissue composed of a fluid matrix, plasma, and formed elements including cells and cell fragments (figure 49.16).

- Plasma is 92% water plus nutrients, wastes, hormones, ions, and protein (figure 49.15).
- Blood cells include erythrocytes, leukocytes, and platelets.
- Blood cells are derived from pluripotent stem cells in bone marrow by hematopoiesis (figure 49.16).
- Erythrocytes contain hemoglobin for oxygen transport.
- Leukocytes are part of the immune system.
- Platelets produce blood clots (figure 49.17).

49.7 Gas Exchange Across Respiratory Surfaces

One of the primary functions of the circulatory system is the acquisition, distribution and removal of gases to support metabolic activity.

- Gas exchange involves diffusion of gases across moist membranes.
- Diffusion is a passive process and the rate of diffusion (R) is measured by Fick's law of diffusion (page 1000).

49.8 Gills, Cutaneous Respiration, and Tracheal Systems

Evolution has maximized gas diffusion in gills and lungs (figure 49.19).

- Gills increase the respiratory surface area for gas exchange.
- In bony fishes diffusion is maximized by countercurrent exchange (figures 49.20, and 49.21).
- Many amphibians use cutaneous respiration for gas exchange.
- Insects have tracheoles that carry oxygen directly to the cells.

49.9 Lungs

Lungs replaced gills in terrestrial organisms because supportive structures are needed, and water evaporates rapidly.

- Lungs move air through branched tubular passages to minimize evaporation (figure 49.25).
- Lungs are ventilated either by positive pressure or negative pressure (figure 49.28).
- The surface area of lungs is enormous due to numerous alveoli, encased by an extensive capillary network (figure 49.25).
- The respiratory system of birds is very efficient (figure 49.26).

49.10 Structures and Mechanisms of Ventilation

Gas exchange depends on pressure differences and ventilation of the lungs.

- Gas exchange is driven by differences in partial pressures (figure 49.27).
- Lungs are filled by contraction of the diaphragm and external intercostal muscles, which creates negative pressure (figure 49.28).
- Lung ventilation is under nervous system control (figure 49.29).

49.11 Transport of Gases in Body Fluids

The amount of oxygen in the blood depends on the partial pressure of oxygen. The low solubility of oxygen requires oxygen carriers.

- Hemoglobin increases the ability of the blood to transport oxygen and also provide an oxygen reserve (figure 49.32).
- The affinity of hemoglobin for oxygen decreases as pH decreases and temperature increases (figure 49.33).
- Carbon dioxide is transported primarily as HCO_3^- (figure 49.34).

SELF TEST

1. You have access to a set of pigments that can be added to blood and a color-sensitive machine to measure these within the body. If you injected red pigment into the systemic circulation and yellow into the pulmonary circulation, in which organisms could the pigments mix to create an orange signal?
   a. Birds
   b. Mammals
   c. Amphibians
   d. Crocodilians

2. Advantages of a closed circulatory system include all of the following except—
   a. separation of circulating and extracellular fluids.
   b. transport of oxygen.
   c. efficient delivery to specific areas of the body.
   d. increased body size and complexity.

3. An ECG measures—
   a. changes in electrical potential during the cardiac cycle.
   b. Ca^{2+} concentration of the ventricles in diastole.
   c. the force of contraction of the atria during systole.
   d. the volume of blood being pumped during the contraction cycle.

4. Systole is vitally important to heart function and begins in the heart with the—
   a. activation of the AV node.
   b. activation of the SA node.
   c. opening of the voltage-gated potassium gates.
   d. opening of the semilunar valves.

5. Which of the following is the correct sequence of events in the circulation of blood?
   a. Heart → arteries → arterioles → capillaries → venules → lymph → heart
   b. Heart → arteries → arterioles → capillaries → veins → venules → heart
   c. Heart → arteries → arterioles → capillaries → venules → veins → heart
   d. Heart → arterioles → arteries → capillaries → venules → veins → heart

6. Which of the following statements is not true?
   a. Only arteries carry oxygenated blood.
   b. Both arteries and veins have a layer of smooth muscle.
   c. Both arteries and veins branch out into capillary beds.
   d. Precapillary sphincters regulate blood flow through capillaries.

7. The lymphatic system is like the circulatory system in that they both—
   a. have nodes that filter out pathogens.
   b. have a network of arteries.
   c. have capillaries.
   d. are closed systems.

8. A molecule of CO_2 that is generated in the cardiac muscle of the left ventricle would *not* pass through which of the following structures before leaving the body?
   a. Right atrium
   b. Left atrium
   c. Right ventricle
   d. Left ventricle

9. In vertebrate hearts, atria contract from the top, and ventricles contract from the bottom. How is this accomplished?
   a. Depolarization from the SA node proceeds across the atria from the top; depolarization from the AV node is carried to the bottom of the ventricles before it emanates over ventricular tissue.
   b. The depolarization from the SA node is initiated from motor neurons coming down from our brain; depolarization from the AV node is initiated from motor neurons coming up from our spinal cord.
   c. Gravity carries the depolarization from the SA node down from the top of the heart; contraction of the diaphragm forces depolarization from the AV node from the bottom up.
   d. This statement is false; both contract from the bottom.

10. When you take a deep breath, your stomach moves out because—
    a. swallowing air increases the volume of the thoracic cavity.
    b. your stomach shouldn't move out when you take a deep breath because you want the volume of your chest cavity to increase, not your abdominal cavity.
    c. contracting your abdominal muscles pushes your stomach out, generating negative pressure in your lungs.
    d. when your diaphragm contracts, it moves down, pressing your abdominal cavity out.

11. If you hold your breath for a long time, body CO_2 levels are likely to _____ , and the pH of body fluids is likely to _____ .
    a. increase; increase
    b. decrease; increase
    c. increase; decrease
    d. decrease; decrease

12. Which pairing of structure and function is incorrect?
    a. Erythrocytes: oxygen transport
    b. Platelets: blood clotting
    c. Plasma: waste transport
    d. All of these are correct.

13. Increased efficiency of gas exchange in vertebrates has been brought about by all of the following mechanisms except—
    a. cutaneous respiration.
    b. unidirectional air flow.
    c. cross-current blood flow.
    d. cartilaginous rings in the trachea.

14. Which of the following is the primary method by which carbon dioxide is transported to the lungs?
    a. Dissolved in plasma
    b. Bound to hemoglobin
    c. As carbon monoxide
    d. As bicarbonate

CHALLENGE QUESTIONS

1. Humans have a number of mechanisms that help to maintain blood pressure, particularly when it falls too low. Explain how the kidney and the endocrine systems help to maintain blood pressure.

2. Explain why a sparrow can fly above a mountain peak at a height of 6000 m, but a mouse of similar size would expire from lack of oxygen.

3. Your roommate has just returned from a 5-km run. She is breathing rapidly and sweating profusely. At the beginning of her run her body began using more glucose and releasing more carbon dioxide than when she was at rest. How does the body compensate for this increase in CO_2 during exercise?

introduction

ON A COLD WINTER DAY, the temperature in many regions will be near freezing. Yet when you leave your house, your core body temperature will not immediately drop. The reason is that you are producing heat internally, and you have a thermostat in your brain with a temperature set point. In addition, the majority of your body weight is actually water, and you exist in a very dehydrating environment by comparison. You are able to do this because of elaborate mechanisms that allow you to retain water and to control the osmotic strength of your blood and extracellular fluids. Both the regulation of internal temperature and the regulation of internal fluid and its composition are examples of *homeostasis*—the ability of living organisms to maintain internal conditions within an optimal range. In this chapter, we discuss these two kinds of regulation. Animals exhibit a number of adaptations to help regulate temperature, including behavior such as that of the elephant pictured. We also describe the osmoregulatory systems of a number of animals, including the mammalian *urinary system*. These organ systems maintain the water and ionic balance of fluids in the body.

50.1 Regulating Body Temperature

Temperature is one of the most important aspects of the environment that all organisms must contend with. As we will see, some organisms have a body temperature that conforms to the environment and others regulate their body temperature. First, let's consider why temperature is so important.

Q_{10} is a measure of temperature sensitivity

The rate of any chemical reaction is affected by temperature: The rate increases with increasing temperature, and it decreases with decreasing temperature. Reactions catalyzed by enzymes show the same kinetic effects, but the enzyme itself is also affected by temperature.

We can make this temperature dependence quantitative by examining the rate of a reaction at two different temperatures. The ratio between the rates of a reaction at two temperatures that differ by 10°C is called the Q_{10} for the enzyme:

$$Q_{10} = R_{T+10}/R_T$$

For most enzymes the Q_{10} value is around 2, which means for every 10°C increase in temperature, the rate of the reaction doubles. Obviously, this cannot continue forever since at high temperatures the enzyme's structure is affected and it can no longer be active.

The Q_{10} concept can also be applied to overall metabolism. The equation remains the same, but instead of the rate of a single reaction, the overall metabolic rate is used. When this has been measured, most organisms have a Q_{10} for metabolic rate around 2 to 3. This observation implies that the effect of temperature is mainly on the enzymes that make up metabolism.

In rare cases—for example, in some intertidal invertebrates—the Q_{10} is close to 1. Notice that this value means no change in metabolic rate with temperature. In the case of these intertidal invertebrates, they are exposed to large temperature fluctuations as they are alternately flooded with relatively cold water and exposed to direct sunlight and much higher air temperatures. These organisms have adapted to deal with these large temperature fluctuations, probably through the evolution of different enzymes in a single metabolic pathway that have large differences in optimal temperature. This allows one enzyme to "make up" for other enzymes with decreased activity at a particular temperature.

Temperature is determined by internal and external factors

Body temperature appears simple, yet there are a large number of variables that influence it. These variables include both internal and external factors, as well as behavior. As you may recall from chapter 7, the second law of thermodynamics indicates that no energy transaction is 100% efficient. Thus the reactions that make up metabolism are constantly producing heat as a result of this inefficiency. This heat must either be dissipated or can be used to raise body temperature.

Overall metabolic rate and body temperature are interrelated. Lower body temperatures do not allow high metabolic rates because of the temperature dependence of enzymes discussed earlier. Conversely, high metabolic rates may cause unacceptable heating of the body, requiring cooling.

At the same time, external temperature plays a part as well. Consider a very cold environment that results in great heat loss. As body temperature gets lower, it is more difficult to generate metabolic heat to raise body temperature.

Organisms therefore must deal with external and internal factors that relate body heat, metabolism, and the environment. The simplest model for temperature, more accurately body heat, is:

body heat = heat produced + (heat gained − heat lost)

we can simplify this further to:

body heat = heat produced + heat transferred

Notice that the heat transferred can be either positive or negative, that is, it can be used for both heating and cooling.

We recognize four mechanisms of heat transfer that are relevant to biological systems: radiation, conduction, convection, and evaporation (figure 50.1).

- **Radiation.** The transfer of heat by electromagnetic radiation, such as from the Sun, does not require direct contact. Heat is transferred from hotter bodies to colder bodies by radiation.
- **Conduction.** The direct transfer of heat between two objects is called conduction. It is literally a direct transfer of kinetic energy between the molecules of the two objects in contact. Energy is transferred from hotter objects to colder ones.
- **Convection.** Convection is the transfer of heat brought about by the movement of a gas or liquid. This movement may be externally caused (wind) or may be due to density differences related to heating and cooling—for example, heated air is less dense and rises; the same is true for water.
- **Evaporation.** All substances have a heat of vaporization, that is, the amount of energy needed to change them from a liquid to a gas phase. Water, as you saw in chapter 2, has a high heat of vaporization, and many animals use this attribute of water as a source of cooling.

Other factors

The overall rate of heat transfer by the methods just listed depends on a number of factors that influence these physical processes. These factors include surface area, temperature difference, and specific heat conduction. Taking these in order, the larger the surface area relative to overall mass, the greater the conduction of heat. Thus, small organisms have a relatively larger surface area for their mass, and they gain or lose heat more readily to the surroundings. This can be affected to a small extent by changing posture, and by extending or pulling in the limbs.

Temperature difference is also important; the greater the difference between ambient temperature and body temperature, the greater the heat transfer. The closer an animal's temperature is to the ambient temperature, the less heat is gained or lost.

Finally, an animal with high heat conductance tends to have a body temperature close to the ambient temperature. For animals that regulate temperature, surrounding the body with a substance with lower heat conductance has an advantage: It acts as insulation. Insulating substances includes such features as feathers, fur, and blubber. For animals that regulate body temperature through behavior, a high heat conductance can maximize heat transfer.

Organisms are classified based on heat source

For many years, physiologists classified animals according to whether they maintained a constant body temperature, or their body temperature fluctuated with environment. Animals that regulated their body temperature about a set point were called *homeotherms*, and those that allow their body temperature to conform to the environment were called *poikilotherms*.

Because homeotherms tended to maintain their body temperature above the ambient temperature, they were also colloquially called "warm-blooded"; poikilotherms were termed "cold-blooded." The problem with this terminology is that a poikilotherm in an environment with a stable temperature (for example, many deep-sea fish species) has a more constant body temperature than some homeotherms.

These limitations to the dichotomy based on temperature regulation led to another view based on how body heat is generated. Animals that use metabolism to generate body heat and maintain their temperatures above the ambient temperature are called **endotherms.** Animals with a relatively low metabolic rate that do not use metabolism to produce heat and have a body temperature that conforms to the ambient temperature are called **ectotherms.** Endotherms tend to have a lower thermal conductivity due to insulating mechanisms, and ectotherms tend to have high thermal conductivity and lack insulation.

These two terms represent ideal end points of a spectrum of physiology and adaptations. Many animals fall in between these extremes and can be considered **heterotherms.** It is a matter of judgment how a particular animal is classified if they exhibit characteristics of each group.

Ectotherms regulate temperature using behavior

Despite having low metabolic rates, ectotherms can regulate their temperature using behavior. Most invertebrates use behavior to adjust their temperature. Many butterflies, for example, must reach a certain body temperature before they can

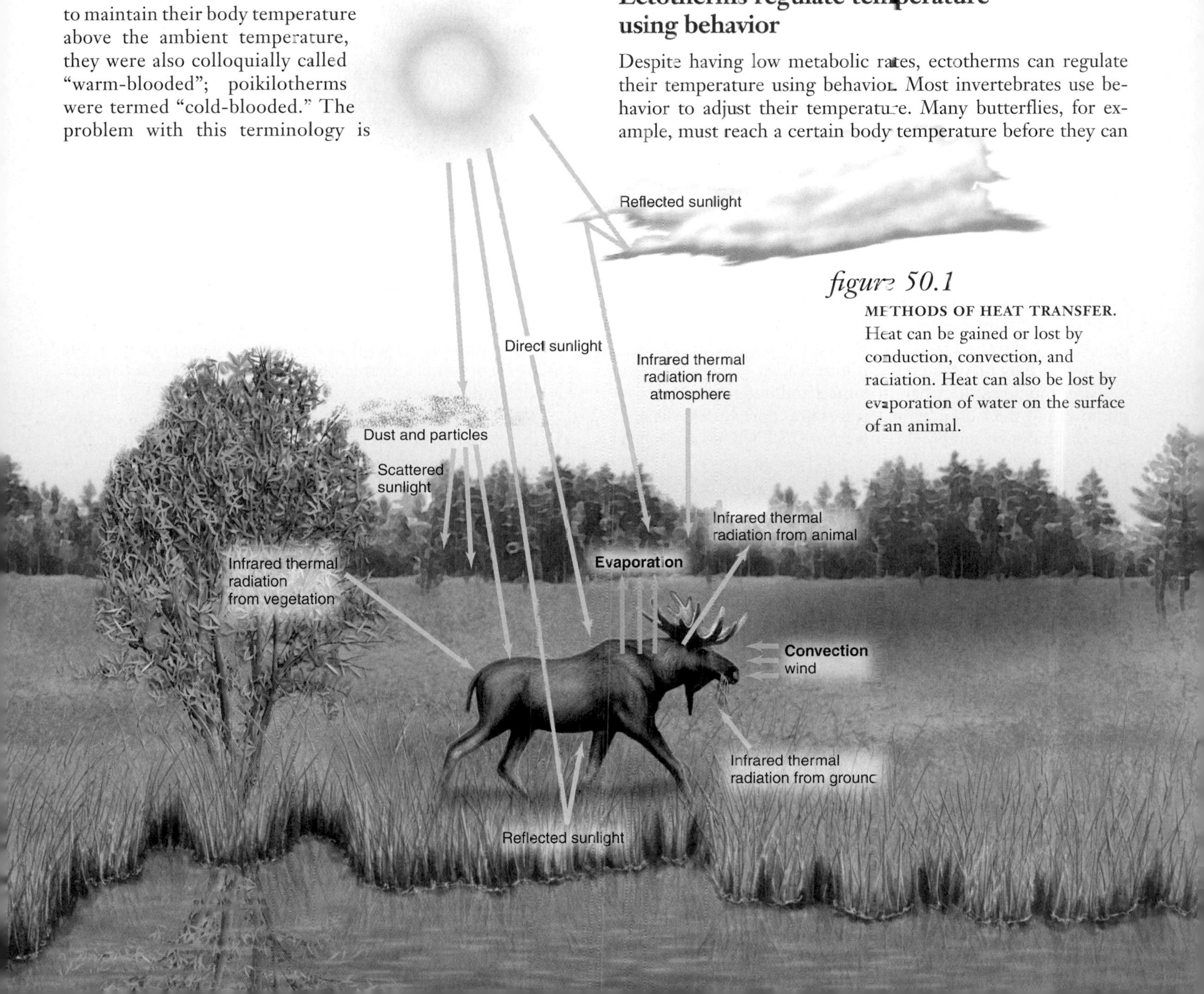

figure 50.1

METHODS OF HEAT TRANSFER.
Heat can be gained or lost by conduction, convection, and radiation. Heat can also be lost by evaporation of water on the surface of an animal.

Reflected sunlight

Direct sunlight

Infrared thermal radiation from atmosphere

Dust and particles

Scattered sunlight

Infrared thermal radiation from animal

Infrared thermal radiation from vegetation

Evaporation

Convection wind

Infrared thermal radiation from ground

Reflected sunlight

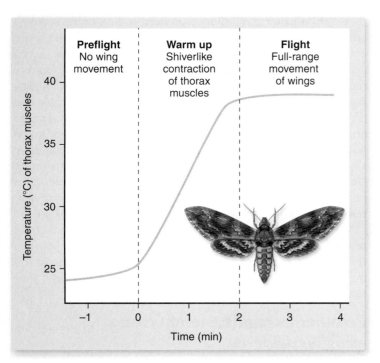

Preflight
No wing movement

Warm up
Shiverlike contraction of thorax muscles

Flight
Full-range movement of wings

figure 50.2

THERMOREGULATION IN INSECTS. Some insects, such as the sphinx moth, contract their thoracic muscles to warm up for flight.

inquiry

? Why does muscle temperature stop warming after 2 min?

fly. In the cool of the morning, they orient their bodies so as to maximize their absorption of sunlight. Moths and many other insects use a shivering reflex to warm their thoracic flight muscles so that they may take flight (figure 50.2).

Vertebrates other than mammals and birds are also ectothermic, and their body temperatures are more or less dependent on the environmental temperature. This does not mean that these animals cannot maintain high and relatively constant body temperatures, but they must use behavior to do this. Many ectothermic vertebrates are able to maintain temperature homeostasis, that is, are homeothermic ectotherms.

For example, certain large fish, including tuna, swordfish, and some sharks, can maintain parts of their body at a significantly higher temperature than that of the water. They do so using **countercurrent heat exchange.** This circulatory adaptation allows the cooler blood in the veins to be warmed through radiation of heat from the warmer blood in the arteries located close to the veins. The arteries carry warmer blood from the center of the body (figure 50.3).

Reptiles attempt to maintain a constant body temperature through behavioral means—by placing themselves in varying locations of sunlight and shade. That's why you frequently see lizards basking in the Sun. Some reptiles can maximize the effect of behavioral regulation by also controlling blood flow. The marine iguana can increase and decrease its heart rate and control the extent of dilation or contraction of blood vessels to regulate the amount of blood available for heat transfer by conduction. Increased heart rate and vasodilation allows them to maximize heating when on land, whereas decreased heart rate and vasoconstriction minimize cooling when diving for food.

In general, ectotherms have low metabolic rates, which has the advantage of correspondingly low intake of energy (food). It is estimated that a lizard (ectotherm) needs only 10% of the energy intake of a mouse (endotherm) of comparable size. The tradeoff is that ectotherms are not capable of sustained high-energy activity.

Endotherms create internal metabolic heat for conservation or dissipation

For endotherms, the generation of internal heat via high metabolic rate can be used to warm the organism if it is cold, but also represents a source of heat that must be dissipated at higher temperatures.

The simplest response that affects heat transfer is to control the amount of blood flow to the surface of the animal. Di-

figure 50.3

COUNTERCURRENT HEAT EXCHANGE. Many marine animals, such as this killer whale, limit heat loss in cold water using countercurrent heat exchange. The warm blood pumped from within the body in arteries loses heat to the cooler blood returning from the skin in veins. This warms the venous blood so that the core body temperature can remain constant in cold water, and it cools the arterial blood so that less heat is lost when the arterial blood reaches the tip of the extremity.

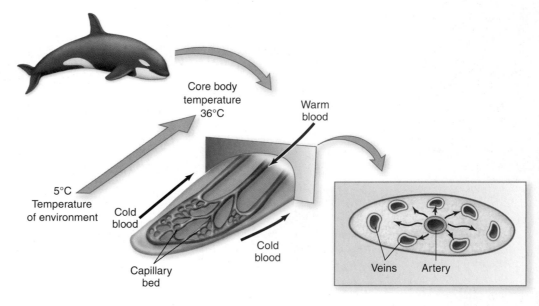

Core body temperature 36°C

Warm blood

5°C Temperature of environment

Cold blood

Cold blood

Capillary bed

Veins   Artery

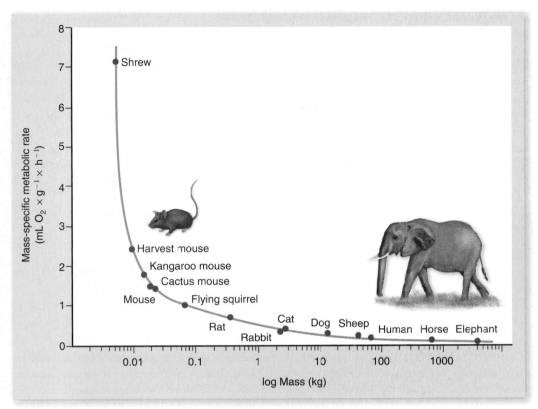

figure 50.4

RELATIONSHIP BETWEEN
BODY MASS AND METABOLIC
RATE IN MAMMALS. Smaller
animals have a much higher
metabolic rate per unit body mass
relative to larger animals. In the
figure, mass-specific metabolic
rate (expressed as O_2 consumption
per unit mass) is plotted against
body mass. Note that the body
mass axis is a logarithmic scale.

inquiry

? *What does this graph predict
about the different challenges
faced by smaller versus larger
mammals in hot and cold
environments?*

lating blood vessels increases the amount of blood flowing to the surface, which in turn increases thermal heat exchange and dissipation of heat. In contrast, constriction of blood vessels decreases the amount of blood flowing to the surface and decreases thermal heat exchange, limiting the amount of heat lost due to conduction.

When ambient temperatures rise, many endotherms take advantage of evaporative cooling in the form of sweating or panting. Sweating is found in some mammals, including humans, and involves the active extrusion of water from sweat glands onto the surface of the body. As the water evaporates, it cools the skin, and this cooling can be transferred internally by capillaries near the surface of the skin. Panting is a similar adaptation used by some mammals and birds that takes advantage of respiratory surfaces for evaporative cooling. For evaporative cooling to be effective, the animal must be able to tolerate the loss of water.

The advantage of endothermy is that it allows sustained high-energy activity. The tradeoff for endotherms is that the high metabolic rate has a corresponding cost in requiring relatively constant and high rates of energy intake (food).

Body size and insulation

Size is one important characteristic affecting animal physiology. Changes in body mass have a large effect on metabolic rate. Smaller animals consume much more energy per unit body mass than larger animals. This relationship is summarized in the "mouse to elephant" curve that shows the nonproportionality of metabolic rate versus size of mammals (figure 50.4).

For small animals with a high metabolic rate, surface area is also large relative to their volume. In a cold environment, this can be disastrous as they cannot produce enough internal heat to balance conductive loss through their large surface

area. Thus, small endotherms in cold environments require significant insulation to maintain their body temperature. The amount of insulation can also vary seasonally and geographically with thicker coats in the north and in winter.

Conversely, large animals in hot environments have the opposite problem: Although their metabolic rate is relatively low, they still produce a large amount of heat with much less relative surface area to dissipate this heat by conduction. Thus large endotherms in hot environments usually have little insulation, and will use behavior to lose heat such as elephants flapping their ears to increase convective heat loss.

Thermogenesis

When temperatures fall below a critical lower threshold, normal endothermic responses are not sufficient to warm an animal. In this case, the animal resorts to **thermogenesis,** or the use of normal energy metabolism to produce heat. Thermogenesis takes two forms: shivering and nonshivering thermogenesis.

In nonshivering thermogenesis, fat metabolism is altered to produce heat instead of ATP. Nonshivering thermogenesis takes place throughout the body, but in some mammals, special stores of fat called brown fat are utilized specifically for this purpose. This brown fat is stored in small deposits in the neck and between the shoulders. This fat is highly vascularized, allowing efficient transfer of heat away from the site of production.

Shivering thermogenesis uses muscles to generate heat without producing useful work. It occurs in some insects, such as the earlier example of a butterfly warming its flight muscles, and in endothermic vertebrates. Shivering involves the use of antagonistic muscles to produce little net generation of movement, but hydrolysis of ATP, generating heat.

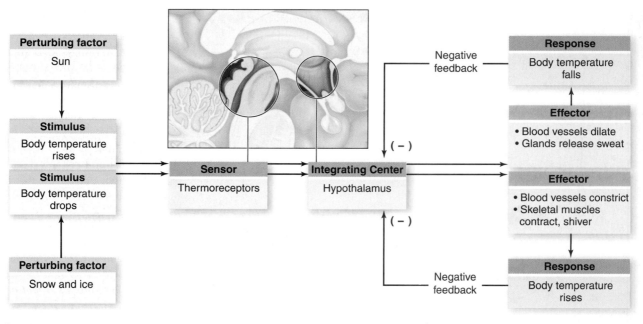

figure 50.5

THE CONTROL OF BODY TEMPERATURE BY THE HYPOTHALAMUS. Central thermoreceptors in the brain and interior of the abdomen sense changes in core temperature. These thermoreceptors synapse with neurons on the hypothalamus, which acts as an integrating center. The hypothalamus then controls effectors such as blood vessels and sweat glands via sympathetic nerves. The hypothalamus also causes the release of hormones that stimulate the thyroid to produce thyroxin, which modulates metabolism.

Mammalian thermoregulation is controlled by the hypothalamus

Mammals that maintain a relatively constant core temperature need an overall control system (summarized in figure 50.5). The system functions much like the heating/cooling system in your house that has a thermostat connected to a furnace to produce heat and an air conditioner to remove heat. Such a system will maintain the temperature of your house about a set point by alternately heating or cooling as necessary.

When the temperature of your blood exceeds 37°C (98.6°F), neurons in the hypothalamus detect the temperature change (see chapters 44 and 46). This leads to stimulation of the *heat-losing center* in the hypothalamus. Sympathetic nerves from this area cause a dilation of peripheral blood vessels, bringing more blood to the surface to dissipate heat. Other sympathetic nerves stimulate the production of sweat, leading to evaporative cooling. Production of hormones that stimulate metabolism is also inhibited.

When your temperature falls below 37°C, an antagonistic set of effects are produced by the hypothalamus. This is under control of the *heat-promoting center*, which has sympathetic nerves that constrict blood vessels to reduce heat transfer, and inhibit sweating to prevent evaporative cooling. The adrenal medulla is stimulated to produce epinephrine, and the anterior pituitary to produce TSH, both of which stimulate metabolism. In the case of TSH, this is indirect as it stimulates the thyroid to produce thyroxin, which stimulates metabolism (see chapter 46). A combination of epinephrine and sympathetic nerve stimulation of fat tissue can induce thermogenesis to produce more internal heat. Again, as tem-

perature rises, negative feedback to the hypothalamus reduces the heat-producing response.

Fever

Substances that cause a rise in temperature are called **pyrogens,** and they produce the state we call **fever.** Fever is a result of resetting the body's normal set point to a higher temperature. A number of gram-negative bacteria have components in their cell walls called endotoxins that act as pyrogens. Substances produced by circulating white blood cells act as pyrogens as well. Pyrogens act on the hypothalamus to increase the set point.

The adaptive value of fever seems to be that increased temperature can inhibit the growth of bacteria. Evidence for this comes from the observation that some ectotherms respond to pyrogens as well. When desert iguanas were injected with pyrogen-producing bacteria, they spent more time in the Sun, producing an elevated temperature: They induced fever behaviorally!

These observations have led to a reevaluation of fever as a state that should be treated medically. Fever is a normal response to infection, and treatment to reduce fever may be working against this natural defense system. Extremely high fevers, however, can be dangerous, inducing symptoms ranging from seizures to hallucinations.

Torpor

Endotherms can also reduce both metabolic rate and body temperature to produce a state of dormancy called **torpor.** Torpor allows an animal to reduce the need for food intake by reducing metabolism. Some birds, such as the hummingbird, allow their body temperature to drop as much as 25°C at night. This strategy is found in smaller endotherms; larger mammals have too large a mass to allow rapid cooling.

Hibernation is an extreme state in which deep torpor lasts for several weeks or even several months. In this case, the animal's temperature may drop as much as 20°C below its normal set point for an extended period of time. The animals that practice hibernation seem to be in the midrange of size; smaller endotherms quickly consume more energy than they can easily store, even by reducing their metabolic rate.

Very large mammals do not appear to hibernate. It was long thought that bears hibernate, but in reality their temperature is reduced only a few degrees. They instead undergo a prolonged winter sleep. With their large thermal mass and low rate of heat loss, they do not seem to require the additional energy savings of hibernation.

Body heat is equal to heat produced plus heat transferred. Heat is transferred by conduction, convection, radiation, and evaporation. Organisms that generate heat and can maintain a temperature above ambient levels are called endotherms. Organisms that conform to their surroundings are called ectotherms. Both types can regulate temperature, but ectotherms use mainly behavior. Mammals maintain a consistent body temperature through regulation by the hypothalamus. Two negative feedback loops act to raise or lower temperature as needed.

50.2 Osmolarity and Osmotic Balance

Water in a multicellular animal's body is distributed between the intracellular and extracellular compartments (figure 50.6). To maintain osmotic balance, the extracellular compartment of an animal's body (including its blood plasma) must be able to take water from the environment or to excrete excess water into the environment. Inorganic ions must also be exchanged between the extracellular body fluids and the external environment to maintain homeostasis. Exchanges of water and electrolytes between the body and the external environment occur across specialized epithelial cells and, in most vertebrates, through a filtration process in the kidneys.

Most vertebrates maintain homeostasis in regard to both the total solute concentration of their extracellular fluids and the concentration of specific inorganic ions. Sodium

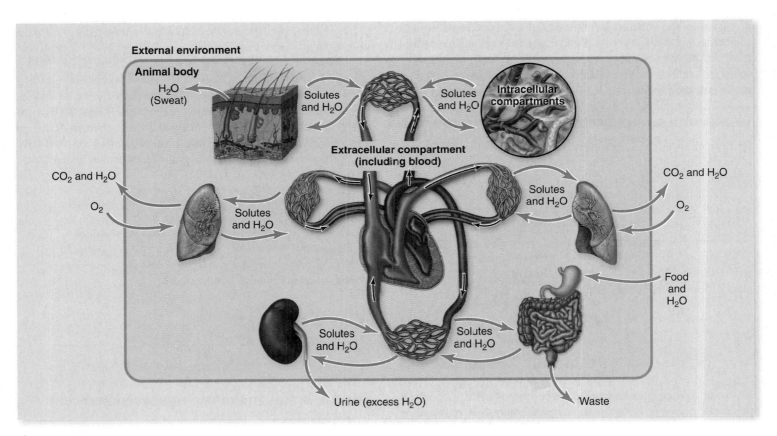

figure 50.6

THE INTERACTION BETWEEN INTRACELLULAR AND EXTRACELLULAR COMPARTMENTS OF THE BODY AND THE EXTERNAL ENVIRONMENT. Water can be taken in from the environment or lost to the environment. Exchanges of water and solutes between the extracellular fluids of the body and the environment occur across transport epithelia, and water and solutes can be filtered out of the blood by the kidneys. Overall, the amount of water and solutes that enters and leaves the body must be balanced in order to maintain homeostasis.

chapter 50 temperature, osmotic regulation, and the urinary system **1023**

(Na$^+$) is the major cation in extracellular fluids, and chloride (Cl$^-$) is the major anion. The divalent cations, calcium (Ca^{2+}) and magnesium (Mg^{2+}), the monovalent cation K$^+$, as well as other ions, also have important functions and are maintained at constant levels.

Osmotic pressure is a measure of concentration difference

You learned in chapter 5 that **osmosis** is the diffusion of water across a semipermeable membrane. Osmosis always occurs from a more dilute solution (with a lower solute concentration) to a less dilute solution (with a higher solute concentration). The **osmotic pressure** of a solution is a measure of its tendency to take in water by osmosis. This is the amount of pressure needed to balance the pressure created by the movement of water.

A solution with a higher concentration of solute exerts more osmotic pressure. This is measured as the **osmolarity** of a solution, the number of osmotically active moles of solute per liter of solution. Notice that osmolarity can differ from molar concentration if a substance dissociates in solution into more than one osmotically active particle. For example, a 1 (M) solution of sucrose is also 1 osmolar (Osm), but a 1 M solution of NaCl is 2 Osm as it dissociates into two osmotically active ions.

The **tonicity** of a solution is a measure of the ability of the solution to change the volume of a cell by osmosis. An animal cell placed in a **hypertonic** solution loses water to the surrounding solution and shrinks. In contrast, an animal cell placed in a **hypotonic** solution gains water and expands. A cell in an **isotonic** solution shows no net water movement. In medical care, isotonic solutions such as normal saline and 5% dextrose are used to bathe exposed tissues and are given as intravenous fluids.

Osmoconformers live in marine environments

In most marine invertebrates, the osmolarity of their body fluids is the same as that of seawater (although the concentrations of particular solutes, such as Mg^{2+}, are not equal). Because the extracellular fluids are isotonic to seawater, no osmotic gradient exists, and there is no tendency for water to leave or enter the body. Such organisms are termed **osmoconformers**—they are in osmotic equilibrium with their environment.

Among the vertebrates, only the primitive hagfish are strict osmoconformers. The sharks and their relatives in the class Chondrichthyes (cartilaginous fish) are also isotonic to

seawater, even though their blood level of NaCl is lower than that of seawater; the difference in total osmolarity is made up by retaining urea, as described later on.

Osmoregulators control their osmolarity internally

All other vertebrates are **osmoregulators**—that is, animals that maintain a relatively constant blood osmolarity despite the different concentration in the surrounding environment. The maintenance of a relatively constant body-fluid osmolarity has permitted vertebrates to exploit a wide variety of ecological niches. Achieving this constancy, however, requires continuous regulation.

Freshwater vertebrates have a much higher solute concentration in their body fluids than that of the surrounding water. In other words, they are hypertonic to their environment. Because of their cells' higher osmotic pressure, water tends to enter their bodies. Consequently, they have adapted to prevent water from entering their bodies as much as possible and to eliminate the excess water that does enter. In addition, freshwater vertebrates tend to lose inorganic ions to their environment and so must actively transport these ions back into their bodies.

In contrast, most marine vertebrates are hypotonic to their environment; their body fluids have only about one-third the osmolarity of the surrounding seawater. These animals are therefore in danger of losing water by osmosis, and adaptations have evolved to help them retain water to prevent dehydration. They do this by drinking seawater and eliminating the excess ions through kidneys and gills.

The body fluids of terrestrial vertebrates clearly have a higher concentration of water than does the air surrounding them. Therefore, they tend to lose water to the air by evaporation from the skin and lungs. All reptiles, birds, and mammals, as well as amphibians during the time when they live on land, face this problem. Urinary/osmoregulatory systems have evolved in these vertebrates that help them retain water.

> Marine invertebrates are osmoconformers; their body fluids are isotonic to their environment. Most vertebrates are osmoregulators; their body fluids are either hypertonic or hypotonic to their environment. Physiological mechanisms help most vertebrates maintain a relatively constant blood osmolarity and ion concentrations.

Osmoregulatory Organs

A variety of mechanisms have evolved in animals to cope with problems of water balance. In many animals, the removal of water or salts from the body is coupled with the removal of metabolic wastes through the excretory system. Single-celled protists employ contractile vacuoles for this purpose, as do sponges. Other multicellular animals have a system of excretory tubules (little tubes) that expel fluid and wastes from the body. In addition, more elaborate systems can be found in invertebrates. In vertebrates, the urinary system is highly complex.

Invertebrates make use of specialized cells and tubules

In flatworms, tubules called **protonephridia** branch throughout the body into bulblike **flame cells** (figure 50.7). Although these simple excretory structures open to the outside of the body, they do not open to the inside; rather, the movement of cilia within the flame cells must draw in fluid from the body. Water and metabolites are then reabsorbed, and the substances to be excreted are expelled through excretory pores.

Other invertebrates have a system of tubules that open both to the inside and to the outside of the body. In the earthworm, these tubules are known as **nephridia** (orange structures in figure 50.8). The nephridia obtain fluid from the body cavity through a process of filtration into funnel-shaped structures called *nephrostomes*. The term *filtration* is used because the fluid is formed under pressure and passes through small openings, so that molecules larger than a certain size are excluded. This filtered fluid is isotonic to the fluid in the coelom, but as it passes through the tubules of the nephridia, NaCl is removed by active transport processes.

The general term for transport out of the tubule and into the surrounding body fluids is **reabsorption.** Because salt is reabsorbed from the filtrate, the urine excreted is more dilute than the body fluids—that is, the urine is hypotonic. The kidneys of mollusks and the excretory organs of crustaceans (called *antennal glands*) also produce urine by filtration and reclaim certain ions by reabsorption.

Insects have a unique osmoregulatory system

The excretory organs in insects are the **Malpighian tubules** (figure 50.9), extensions of the digestive tract that branch off anterior to the hindgut. Urine is not formed by filtration in these tubules because there is no pressure difference between the blood in the body cavity and the tubule. Instead, waste molecules and potassium (K^+) ions are secreted into the tubules by active transport.

Secretion is the opposite of reabsorption—ions or molecules are transported from the body fluid into the tubule. The secretion of K^+ creates an osmotic gradient that causes water to enter the tubules by osmosis from the body's open circulatory

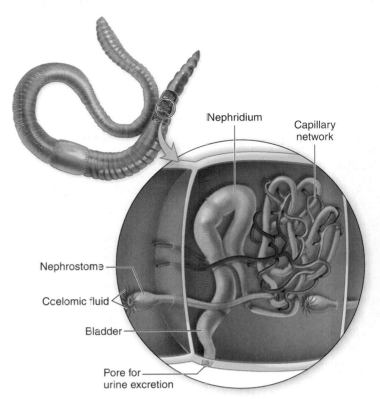

figure 50.8

THE NEPHRIDIA OF ANNELIDS. Most invertebrates, such as the annelid shown here, have nephridia (*orange*). These consist of tubules that receive a filtrate of coelomic fluid, which enters the funnel-like nephrostomes. Salt can be reabsorbed from these tubules, and the fluid that remains, urine, is released from pores into the external environment.

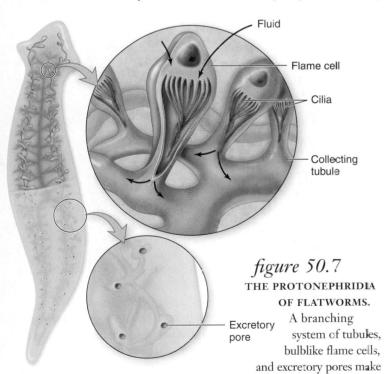

figure 50.7

THE PROTONEPHRIDIA OF FLATWORMS. A branching system of tubules, bulblike flame cells, and excretory pores make up the protonephridia of flatworms. Cilia inside the flame cells draw in fluids from the body by their beating action. Substances are then expelled through pores that open to the outside of the body.

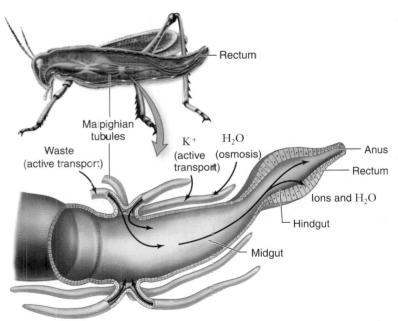

figure 50.9

THE MALPIGHIAN TUBULES OF INSECTS. The Malpighian tubules of insects are extensions of the digestive tract that collect water and wastes from the body's circulatory system. K^+ is secreted into these tubules, drawing water with it osmotically. Much of this water (*arrows*) is reabsorbed across the wall of the hindgut.

system. Most of the water and K⁺ is then reabsorbed into the circulatory system through the epithelium of the hindgut, leaving only small molecules and waste products to be excreted from the rectum along with feces. Malpighian tubules thus provide a very efficient means of water conservation.

The vertebrate kidney filters and then reabsorbs

The **kidneys** of vertebrates, unlike the Malpighian tubules of insects, create a tubular fluid by filtering the blood under pressure. In addition to waste products and water, the filtrate contains many small molecules, including glucose, amino acids, and vitamins, that are of value to the animal. These molecules and most of the water are reabsorbed from the tubules into the blood, while wastes remain in the filtrate. Additional wastes may be secreted by the tubules and added to the filtrate, and the final waste product, urine, is eliminated from the body.

It may seem odd that the vertebrate kidney should filter out almost everything from blood plasma (except proteins, which are too large to be filtered) and then spend energy to take back or reabsorb what the body needs. But selective reabsorption provides great flexibility. Various vertebrate groups have evolved the ability to reabsorb molecules that are especially valuable in particular habitats. This flexibility is a key factor underlying the successful colonization of many diverse environments by the vertebrates. In the rest of this chapter, we focus on the vertebrate kidney and its elimination of waste materials, notably nitrogen compounds.

> Many invertebrates filter fluid into a system of tubules and then reabsorb ions and water, leaving waste products for excretion. Insects create an excretory fluid by secreting K⁺ and waste products into tubules, which draws water osmotically. The vertebrate kidney produces a filtrate that enters tubules and is modified to become urine.

50.4 Evolution of the Vertebrate Kidney

The kidney is a complex organ made up of thousands of repeating units called **nephrons,** each with the structure of a loop that penetrates deep into the medulla of the kidney (shown schematically in figure 50.10). Blood pressure forces the fluid in blood out of a ball of capillaries called the *glomerulus* into *Bowman's capsule,* the beginning of the tubule system. This process filters the blood forming the tubular filtrate that can then be modified by the rest of the nephron. The glomerulus retains blood cells, proteins, and other useful large molecules in the blood, but it allows the water, and the small molecules and wastes dissolved in it, to pass through and into the tubule system of the nephron. As the filtered fluid passes through the nephron tube, useful nutrients and ions are reabsorbed from it by both active and passive transport mechanisms, leaving the water and metabolic wastes behind in a fluid urine. (The details of this process are described in a later section.)

Although the same basic design has been retained in all vertebrate kidneys, a few modifications have occurred. Because the original glomerular filtrate is isotonic to blood, all vertebrates can produce a urine that is isotonic to blood by reabsorbing ions and water in equal proportions. Or, they can produce a urine that is hypotonic to blood—more dilute than the blood—by reabsorbing relatively less water. Only birds and mammals can reabsorb enough water from their glomerular filtrate to produce a urine that is hypertonic to blood— more concentrated than the blood—by reabsorbing relatively more water.

Freshwater fishes must retain electrolytes and keep water out

Kidneys are thought to have evolved among the freshwater teleosts, or bony fishes. Because the body fluids of a freshwater fish are hypertonic with respect to the surrounding water, these

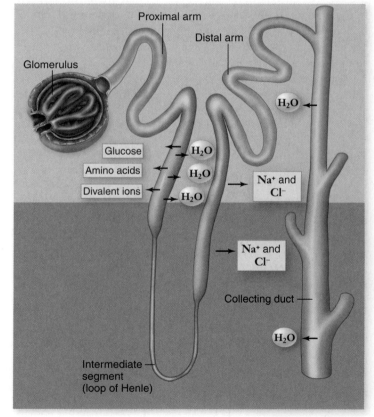

figure 50.10

ORGANIZATION OF THE VERTEBRATE NEPHRON. The nephron tubule is a basic design that has been retained in the kidneys of vertebrates. Sugars, amino acids, water, important monovalent ions, and divalent ions are reabsorbed in the proximal arm; water and monovalent ions such as Na⁺ and Cl⁻ are reabsorbed in the loop of Henle; varying amounts of water and monovalent ions (Na⁺ and Cl⁻) can be reabsorbed in the distal arm and the collecting duct, depending on hormonal influences.

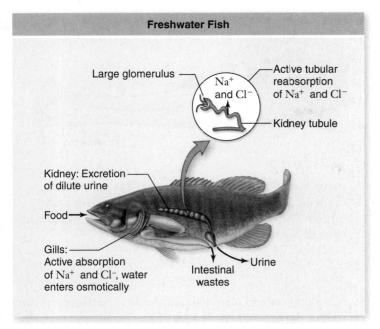

Freshwater Fish

Large glomerulus

Na^+ and Cl^-

Active tubular reabsorption of Na^+ and Cl^-

Kidney tubule

Kidney: Excretion of dilute urine

Food

Gills: Active absorption of Na^+ and Cl^-, water enters osmotically

Intestinal wastes

Urine

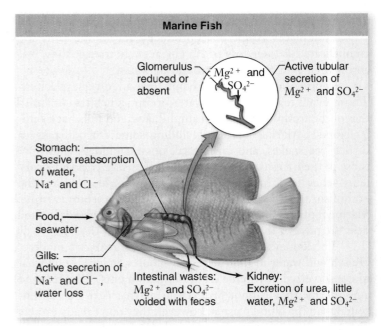

Marine Fish

Glomerulus reduced or absent

Mg^{2+} and SO_4^{2-}

Active tubular secretion of Mg^{2+} and SO_4^{2-}

Stomach: Passive reabsorption of water, Na^+ and Cl^-

Food, seawater

Gills: Active secretion of Na^+ and Cl^-, water loss

Intestinal wastes: Mg^{2+} and SO_4^{2-} voided with feces

Kidney: Excretion of urea, little water, Mg^{2+} and SO_4^{2-}

figure 50.11

FRESHWATER AND MARINE TELEOSTS FACE DIFFERENT OSMOTIC PROBLEMS. Whereas the freshwater teleost is hypertonic to its environment, the marine teleost is hypotonic to seawater. To compensate for its tendency to take in water and lose ions, a freshwater fish excretes dilute urine, avoids drinking water, and reabsorbs ions across the nephron tubules. To compensate for its osmotic loss of water, the marine teleost drinks seawater and eliminates the excess ions through active transport across epithelia in the gills and kidneys.

animals face two serious problems: (1) Water tends to enter the body from the environment; and (2) solutes tend to leave the body and enter the environment.

Freshwater fish address the first problem by *not* drinking water and by excreting a large volume of dilute urine, which is hypotonic to their body fluids. They address the second problem by reabsorbing ions across the nephron tubules, from the glomerular filtrate back into the blood. In addition, they actively transport ions across their gill surfaces from the surrounding water into the blood (figure 50.11, *left*).

Marine bony fishes must excrete electrolytes and keep water in

Although most groups of animals seem to have evolved first in the sea, marine bony fish (teleosts) probably evolved from freshwater ancestors. They faced significant new problems in making the transition to the sea because their body fluids were, and are, hypotonic to seawater. Consequently, water tends to leave their bodies by osmosis across their gills, and they also lose water in their urine. To compensate for this continuous water loss, marine fish drink large amounts of seawater (figure 50.11, *right*).

Many of the divalent cations (principally, Ca^{2+} and Mg^{2+}) in the seawater that a marine fish drinks remain in the digestive tract and are eliminated through the anus. Some, however, are absorbed into the blood, as are the monovalent ions K^+, Na^+, and Cl^-. Most of the monovalent ions are actively transported out of the blood across the gill surfaces, while the divalent ions that enter the blood are secreted into the nephron tubules and excreted in the urine. In these two ways, marine bony fish elim-

inate the ions they get from the seawater they drink. The urine they excrete is isotonic to their body fluids. It is more concentrated than the urine of freshwater fish, but not as concentrated as that of birds and mammals.

Cartilaginous fishes pump out electrolytes and retain urea

The elasmobranchs, including sharks and rays, are by far the most common subclass in the class Chondrichthyes (cartilaginous fish). Elasmobranchs have solved the osmotic problem posed by their seawater environment in a different way. Instead of having body fluids that are hypotonic to seawater, so that they have to continuously drink seawater and actively pump out ions, the elasmobranchs reabsorb urea from the nephron tubules and maintain a blood urea concentration that is 100 times higher than that of mammals.

The added urea makes elasmobranchs' blood approximately isotonic to the surrounding sea. Because no net water movement occurs between isotonic solutions, water loss is therefore prevented. As a result, these fishes do not need to drink seawater for osmotic balance, and their kidneys and gills do not have to remove large amounts of ions from their bodies. The enzymes and tissues of the cartilaginous fish have evolved to tolerate the high urea concentrations.

Amphibians and reptiles have osmotic adaptations to their environments

The first terrestrial vertebrates were the amphibians, and the amphibian kidney is identical to that of freshwater fish. This is not surprising because amphibians spend a significant portion

chapter 50 temperature, osmotic regulation, and the urinary system **1027** of their time in fresh water, and when on land, they generally stay in wet places. Amphibians produce a very dilute urine and compensate for their loss of Na+ by actively transporting Na+ across their skin from the surrounding water.

Reptiles, on the other hand, live in diverse habitats. Those living mainly in fresh water occupy a habitat similar to that of the freshwater fish and amphibians, and thus have similar kidneys. Marine reptiles, including some crocodilians, sea turtles, sea snakes, and one lizard, possess kidneys similar to those of their freshwater relatives, but they face opposite problems—they tend to lose water and take in salts. Like marine bony fish, they drink seawater and excrete an isotonic urine. Marine reptiles eliminate excess salt through salt glands located near the nose or the eye.

The kidneys of terrestrial reptiles also reabsorb much of the salt and water in their nephron tubules, helping somewhat to conserve blood volume in dry environments. Like fish and amphibians, they cannot produce urine that is more concentrated than the blood plasma; however, they don't really excrete urine, but instead empty it into a cloaca (the common exit of the digestive and urinary tracts), where additional water can be reabsorbed and the wastes excreted with the feces.

Mammals and birds are able to excrete concentrated urine and retain water

Mammals and birds are the only vertebrates able to produce urine with a higher osmotic concentration than their body fluids. These vertebrates are therefore able to excrete their waste products in a small volume of water, so that more water can be retained in the body.

Human kidneys can produce urine that is as much as 4.2 times as concentrated as blood plasma, but the kidneys of some other mammals are even more efficient at conserving water. For example, camels, gerbils, and pocket mice of the genus *Perognathus* can excrete urine 8, 14, and 22 times as concentrated as their blood plasma, respectively. The kidneys of kangaroo rats (genus *Dipodomys*) are so efficient that they never have to drink water; they can obtain all the water they need from their food and from water produced in aerobic cellular respiration.

The production of hypertonic urine is accomplished by the *loop of Henle* portion of the nephron (figures 50.10 and

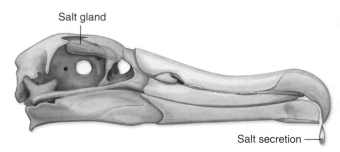

Salt gland

Salt secretion

figure 50.12

HOW MARINE BIRDS COPE WITH EXCESS SALT. Marine birds drink seawater and then excrete the salt through salt glands. The extremely salty fluid excreted by these glands can then dribble down the beak.

50.16), found only in mammals and birds. The degree of concentration depends on the length of the loop; most mammals have some nephrons with short loops and other nephrons with much longer loops. Birds, however, have relatively few or no nephrons with long loops, so they cannot produce urine that is as concentrated as that of mammals. At most, they can only reabsorb enough water to produce a urine that is about twice the concentration of their blood. Marine birds solve the problem of water loss by drinking salt water and then excreting the excess salt from salt glands near the eyes (figure 50.12).

The moderately hypertonic urine of a bird is delivered to its cloaca, along with the fecal material from its digestive tract. If needed, additional water can be absorbed across the wall of the cloaca to produce a semisolid white paste or pellet, which is excreted.

> The kidneys of freshwater fishes must excrete copious amounts of very dilute urine; marine bony fishes drink seawater and excrete an isotonic urine. The basic design and function of the nephron of freshwater fishes have been retained in the terrestrial vertebrates. Modifications, particularly the presence of a loop of Henle, allow mammals and birds to reabsorb more water and produce a urine hypertonic to their body fluids.

50.5 Nitrogenous Wastes: Ammonia, Urea, and Uric Acid

Amino acids and nucleic acids are nitrogen-containing molecules. When animals catabolize these molecules for energy or convert them into carbohydrates or lipids, they produce nitrogen-containing by-products called **nitrogenous wastes** (figure 50.13) that must be eliminated from the body.

Ammonia is toxic and must be quickly removed

The first step in the metabolism of amino acids and nucleic acids is the deamination, that is, removal of the amino (—NH₂) group, and its combination with H+ to form **ammonia** (NH₃) in the liver. Am-

monia is quite toxic to cells and therefore is safe only in very dilute concentrations. The excretion of ammonia is not a problem for the bony fishes and amphibian tadpoles, which eliminate most of it by diffusion through the gills and less by excretion in very dilute urine.

Urea and uric acid are less toxic but have different solubilities

In elasmobranchs, adult amphibians, and mammals, the nitrogenous wastes are eliminated in the far less toxic form of **urea.** Urea is water-soluble and so can be excreted in large amounts

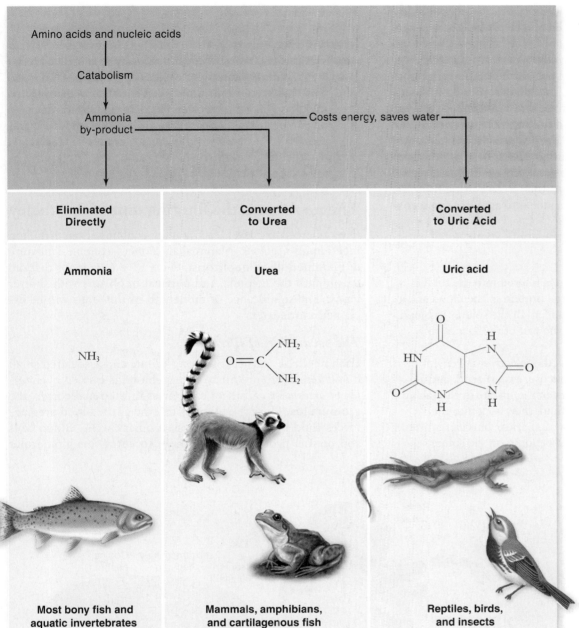

figure 50.13

NITROGENOUS WASTES. When amino acids and nucleic acids are metabolized, the immediate by-product is ammonia, which is quite toxic but can be eliminated through the gills of teleost fish. Mammals convert ammonia into urea, which is less toxic. Birds and terrestrial reptiles convert it instead into uric acid, which is insoluble in water. Production of uric acid is the most energetically expensive of the three but also saves the most water.

in the urine. It is carried in the bloodstream from its place of synthesis in the liver to the kidneys where it is excreted.

Reptiles, birds, and insects excrete nitrogenous wastes in the form of **uric acid,** which is only slightly soluble in water. As a result of its low solubility, uric acid precipitates and thus can be excreted using very little water. Uric acid forms the pasty white material in bird droppings called *guano*. It costs the animal energy to synthesize uric acid, but this is offset by the conservation of water.

The ability to synthesize uric acid in these groups of animals is also important because their eggs are encased within shells, and nitrogenous wastes build up as the embryo grows within the egg. The formation of uric acid, although a lengthy process that requires considerable energy, produces a compound that crystallizes and precipitates. As a solid precipitate, uric acid is unable to affect the embryo's development even though it is still inside the egg.

Mammals also produce some uric acid, but it is a waste product of the degradation of purine nucleotides, not of amino acids. Most mammals have an enzyme called *uricase*, which converts uric acid into a more soluble derivative, **allantoin.** Only humans, apes, and the Dalmatian dog lack this enzyme, and they must excrete the uric acid. In humans, excessive accumulation of uric acid in the joints produces a condition known as *gout*.

The metabolic breakdown of amino acids and nucleic acids produces ammonia as a by-product. Ammonia is excreted by bony fishes and gilled amphibians, but other vertebrates convert nitrogenous wastes into urea and uric acid, which are less toxic nitrogenous wastes.

The Mammalian Kidney

In humans, the kidneys are fist-sized organs located in the lower back. Each kidney receives blood from a renal artery, and from this blood, urine is produced. Urine drains from each kidney through a **ureter,** which carries the urine to a **urinary bladder.** From the bladder, urine is passed out of the body through the **urethra** (figure 50.14).

Within the kidney, the mouth of the ureter flares open to form a funnel-like structure, the *renal pelvis.* The renal pelvis, in turn, has cup-shaped extensions that receive urine from the renal tissue. The renal tissue is divided into an outer **renal cortex** and an inner **renal medulla.**

The kidney has three basic functions summarized in figure 50.15:

- *Filtration:* Fluid in the blood is filtered into the tubule system, leaving cells and large protein in the blood and a filtrate composed of water and all of the solutes in blood. This filtrate is modified by the rest of the kidney to produce urine for excretion.
- *Reabsorption:* Reabsorption is the selective movement of important solutes such as glucose, amino acids, and a variety of inorganic ions, out of the filtrate in the tubule system to the extracellular fluid, then back into the bloodstream via peritubular capillaries. This can utilize active or passive processes depending on the solute.

Water is also reabsorbed, and this can be controlled to control the amount of water loss.

- *Secretion:* Secretion is the movement of substances from the blood into the extracellular fluid, then into the filtrate in the tubule system. Unlike reabsorption, which preserves substances in the body, this adds to what will be expelled from the body and can be used to remove toxic substances.

The nephron is the filtering unit of the kidney

On a microscopic level, each kidney contains about 1 million functioning *nephrons.* Mammalian kidneys contain a mixture of **juxtamedullary nephrons,** which have long loops that dip deeply into the medulla, and **cortical nephrons** with shorter loops. The significance of the length of the loops will be explained a little later.

The production of filtrate

Each nephron consists of a long tubule and associated small blood vessels (figure 50.16). First, blood is carried by an *afferent arteriole* to a tuft of capillaries in the renal cortex, the **glomerulus.** Here the blood is filtered as the blood pressure forces fluid through the porous capillary walls. Blood cells and plasma proteins are too large to enter this glomerular

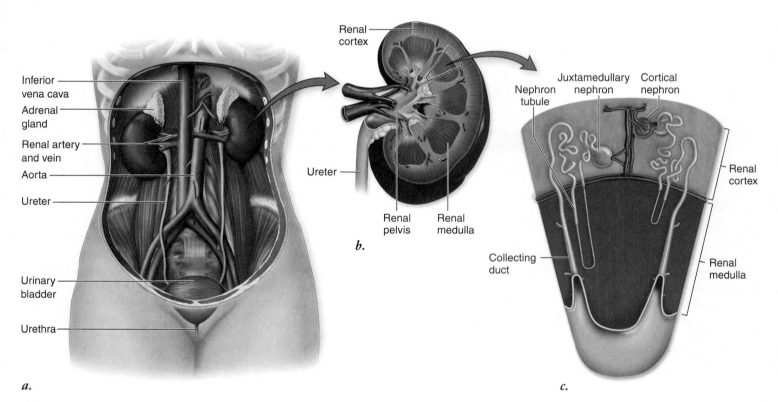

a.

c.

figure 50.14

THE HUMAN RENAL SYSTEM. *a.* The positions of the organs of the urinary system. *b.* A sectioned kidney, revealing the internal structure. *c.* The position of nephrons in the mammalian kidney. Cortical nephrons are located predominantly in the renal cortex; juxtamedullary nephrons have long loops that extend deep into the renal medulla.

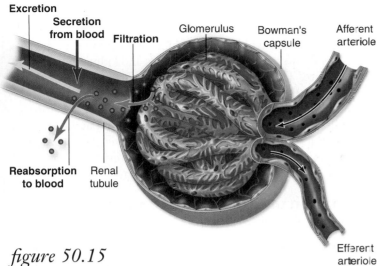

figure 50.15

FOUR FUNCTIONS OF THE KIDNEY. Molecules enter the urine by *filtration* out of the glomerulus and by *secretion* into the tubules from surrounding peritubular capillaries. Molecules that entered the filtrate can be returned to the blood by *reabsorption* from the tubules into surrounding peritubular capillaries. The fluid exiting the kidney is eliminated from the body by *excretion* through the tubule to a ureter and then to the bladder.

filtrate, but large amounts of the plasma, consisting of water and dissolved molecules, leave the vascular system at this step. The filtrate immediately enters the first region of the nephron tubules. This region, **Bowman's capsule,** envelops the glomerulus much as a large, soft balloon surrounds your hand if you press your fist into it. The capsule has slit openings so that the glomerular filtrate can enter the system of nephron tubules.

Blood components that were not filtered out of the glomerulus drain into an *efferent arteriole,* which then empties into a second bed of capillaries called **peritubular capillaries** that surround the tubules. This is only one of several locations in the body where two capillary beds occur in series. In juxtamedullary nephrons, efferent arteriole and peritubular capillaries also feed the **vasa recta** capillaries that surround the loop of Henle. As described later, the peritubular capillaries are needed for the processes of reabsorption and secretion.

After the filtrate enters Bowman's capsule, it goes into a portion of the nephron called the **proximal convoluted**

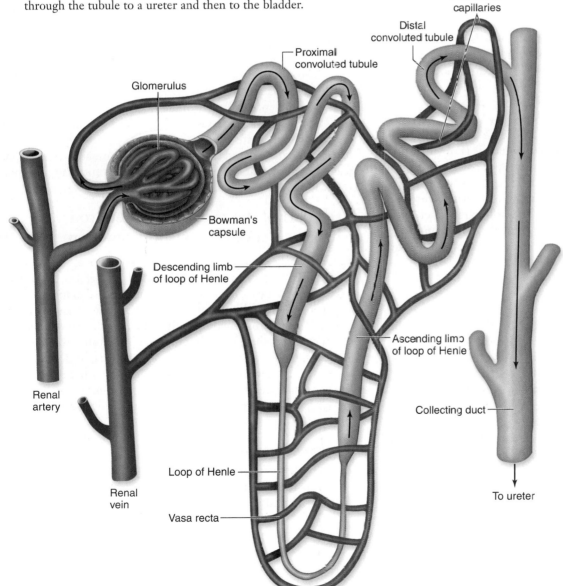

figure 50.16

A NEPHRON IN A MAMMALIAN KIDNEY. The nephron tubule is surrounded by peritubular capillaries in the cortex, and their vasa recta extensions surround the loop of Henle in the medulla. This capillary bed carries away molecules and ions that are reabsorbed from the filtrate.

tubule, located in the cortex. In a cortical nephron, the fluid then flows through the **loop of Henle** that dips only minimally into the medulla before ascending back into the cortex. In juxtamedullary nephrons, the loop of Henle extends much deeper into the medulla before ascending back up into the cortex. More water can be reabsorbed from juxtamedullary nephrons than from cortical nephrons. The fluid then moves deeper into the medulla and back up again into the cortex in a loop of Henle. As mentioned earlier, only the kidneys of mammals and birds have loops of Henle, and this is why only birds and mammals have the ability to concentrate their urine.

Collection of urine

After leaving the loop, the fluid is delivered to a **distal convoluted tubule** in the cortex that next drains into a **collecting duct.** The collecting duct again descends into the medulla, where it merges with other collecting ducts to empty its contents, now called urine, into the renal pelvis.

Water, some nutrients, and some ions are reabsorbed; other molecules are secreted

Most of the water and dissolved solutes that enter the glomerular filtrate must be returned to the blood by reabsorption, or the animal would literally urinate to death. In a human, for example, approximately 2000 L of blood passes through the kidneys each day, and 180 L of water leaves the blood and enters the glomerular filtrate.

Water

Because humans have a total blood volume of only about 5 L and produce only 1–2 L of urine per day, it is obvious that each liter of blood is filtered many times per day, and most of the filtered water is reabsorbed. Water is reabsorbed from the filtrate by the proximal convoluted tubule, as it passes through the descending loop of Henle and the collecting duct. The selective reabsorption in the collecting duct is driven by an osmotic gradient produced by the loop of Henle, as is described shortly.

Glucose and other nutrients

The reabsorption of glucose, amino acids, and many other molecules needed by the body is driven by active transport and secondary active transport (cotransport) carriers. As in all carrier-mediated transport, a maximum rate of transport is reached whenever the carriers are saturated (see chapter 5).

In the case of the renal glucose carriers in the proximal convoluted tubule, saturation occurs when the concentration of glucose in the blood (and thus in the glomerular filtrate) is about 180 mg/100 mL of blood. If a person has a blood glucose concentration in excess of this amount, as happens in untreated diabetes mellitus, the glucose remaining in the filtrate is expelled in the urine. Indeed, the presence of glucose in the urine is diagnostic of diabetes mellitus.

Secretion of wastes

The secretion of foreign molecules and particular waste products of the body involves the transport of these molecules across the membranes of the blood capillaries and kidney tubules into the filtrate. This process is similar to reabsorption, but it proceeds in the opposite direction.

Some secreted molecules are eliminated in the urine so rapidly that they may be cleared from the blood in a single pass through the kidneys. This rapid elimination explains why penicillin, which is secreted by the nephrons, must be administered in very high doses and several times per day.

Excretion of toxins and excess ions maintains homeostasis

A major function of the kidney is the elimination of a variety of potentially harmful substances that animals eat and drink. In addition, urine contains nitrogenous wastes, described earlier, that are products of the catabolism of amino acids and nucleic acids. Urine may also contain excess K^+, H^+, and other ions that are removed from the blood.

Urine's generally high H^+ concentration (pH 5–7) helps maintain the acid–base balance of the blood within a narrow range (pH 7.35–7.45). Moreover, the excretion of water in urine contributes to the maintenance of blood volume and pressure (chapter 49); the larger the volume of urine excreted, the lower the blood volume.

The purpose of kidney function is therefore homeostasis; the kidneys are critically involved in maintaining the constancy of the internal environment. When disease interferes with kidney function, it causes a rise in the blood concentration of nitrogenous waste products, disturbances in electrolyte and acid–base balance, and a failure in blood pressure regulation. Such potentially fatal changes highlight the central importance of the kidneys in normal body physiology.

Each part of the mammalian nephron performs a specific transport function

As previously described, approximately 180 L of isotonic glomerular filtrate enters the Bowman's capsules of human kidneys each day. After passing through the remainder of the nephron tubules, this volume of fluid would be lost as urine if it were not reabsorbed back into the blood. It is clearly impossible to produce this much urine, yet water is only able to pass through a cell membrane by osmosis, and osmosis is not possible between two isotonic solutions. Therefore, some mechanism is needed to create an osmotic gradient between the glomerular filtrate and the blood, to allow reabsorption of water.

Proximal convoluted tubule

Virtually all the nutrient molecules in the filtrate are reabsorbed back into the systemic blood by the proximal convoluted tubule. In addition, approximately two-thirds of the NaCl and water filtered into Bowman's capsule is immediately reabsorbed across the walls of the proximal convoluted tubule.

This reabsorption is driven by the active transport of Na^+ out of the filtrate and into surrounding peritubular capillaries. Cl^- follows Na^+ passively because of electrical attraction, and water follows them both because of osmosis. Because NaCl and water are removed from the filtrate in proportionate amounts, the filtrate that remains in the tubule is still isotonic to the blood plasma.

Although only one-third of the initial volume of filtrate remains in the nephron tubule after the initial reabsorption of NaCl and water, it still represents a large volume (60 L out of the original 180 L of filtrate). Obviously, no animal can afford to excrete that much urine, so most of this water must also be reabsorbed. It is reabsorbed primarily across the wall of the collecting duct.

Loop of Henle

The function of the loop of Henle is to create a gradient of increasing osmolarity from the cortex to the medulla. This allows water to be reabsorbed by osmosis in the collecting duct as it runs down into the medulla past the loop of Henle. The descending and ascending limbs of the loop of Henle differ structurally and in their permeability to ions and water. This produces a gradient of increasing osmolarity from cortex to medulla (figure 50.17). The structure of the loop also forms another example of a countercurrent system, this time acting to increase the osmolarity of interstitial fluid. To understand the functioning of the loop of Henle, it is easiest to start in the ascending limb:

1. The entire ascending limb is impermeable to water. The thick portion of the ascending limb actively transports Na^+ out of the tubule, with Cl^- passively following. The thin ascending limb is permeable to both Na^+ and Cl^-, which move out by diffusion.
2. The descending limb is thin and permeable to water but not to NaCl. Because of the Na^+ and Cl^- lost by the ascending limb, the osmolarity of the interstitial fluid is higher than in the descending limb, and water moves out of the descending limb by osmosis. This also increases the osmolarity of the fluid in the tubule such that as it turns at the bottom, it will lose NaCl by diffusion in the thin ascending loop as described earlier.
3. The loss of water from the descending limb multiplies the concentration that can be achieved at each level of the loop through the active extrusion of Na^+ (with Cl^- following passively) by the ascending limb. The longer the loop of Henle, the longer the region of interaction between the descending and ascending limbs, and the greater the total concentration that can be achieved. In a human kidney, the concentration of filtrate entering the loop is 300 milliosmolar (mOsm), and this concentration is multiplied to more than 1200 mOsm at the bottom of the longest loops of Henle in the renal medulla.

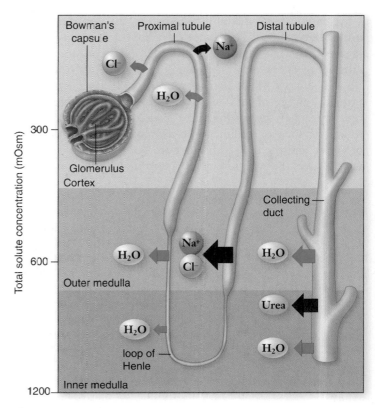

figure 50.17

THE REABSORPTION OF SALT AND WATER IN THE MAMMALIAN KIDNEY. Active transport of Na^+ out of the proximal tubules is followed by the passive movement of Cl^- and water. Active extrusion of NaCl from the ascending limb of the loop of Henle creates the osmotic gradient required for the reabsorption of water from the descending limb of the loop of Henle and the collecting duct. The two limbs of the loop form a countercurrent multiplier system that increases the osmotic gradient. The changes in osmolarity from the cortex to the medulla are indicated to the left of the figure.

4. The NaCl pumped out of the ascending limb of the loop is reabsorbed from the surrounding interstitial fluid into the loops of the *vasa recta*, so that NaCl can diffuse from the blood leaving the medulla to the blood entering the medulla. Thus, the vasa recta also functions in a countercurrent exchange, similar to that described for oxygen in the countercurrent flow of water and blood in the gills of fish (see chapter 44). In the case of the vasa recta, this prevents the flow of blood through the capillaries from destroying the osmotic gradient established by the loop of Henle. Thus, blood can be supplied to this region of the kidney without affecting the ability of the collecting duct to selectively reabsorb water.

Because fluid flows in opposite directions in the two limbs of the loop, the action of the loop of Henle in creating a hypertonic renal medulla is known as the **countercurrent multiplier system.** The osmotic gradient that is established is greater than what would be produced by just active transport of salts out of the tubule system.

The high solute concentration of the renal medulla is primarily the result of NaCl accumulation by the countercurrent multiplier system, but urea also contributes to the total osmolarity of the medulla. The descending limb of the loop of Henle and the collecting duct are both permeable to urea, which leaves these regions of the nephron by diffusion.

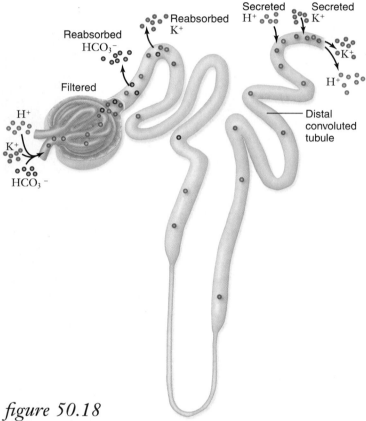

figure 50.18

CONTROLLING SALT BALANCE. The nephron controls the amounts of K^+, H^+, and HCO_3^- excreted in the urine. K^+ is completely reabsorbed in the proximal tubule and then secreted in hormonally regulated amounts into the distal tubule. HCO_3^- is filtered but normally completely reabsorbed. H^+ is filtered and also secreted into the distal tubule, so that the final urine has an acidic pH.

Distal convoluted tubule and collecting duct

Because NaCl was pumped out of the ascending limb, the filtrate that arrives at the distal convoluted tubule and enters the collecting duct in the renal cortex is hypotonic (with a concentration of only 100 mOsm). The collecting duct carrying this dilute fluid now plunges into the medulla. As a result of the hypertonic interstitial fluid of the renal medulla, a strong osmotic gradient pulls water out of the collecting duct and into the surrounding blood vessels.

The osmotic gradient is normally constant, but the permeability of the distal convoluted tubule and the collecting duct to water is adjusted by a hormone, *antidiuretic hormone* (ADH), mentioned in chapters 46 and 49. When an animal needs to conserve water, the posterior pituitary gland secretes more ADH, and this hormone increases the number of water channels in the plasma membranes of the collecting duct cells. This increases the permeability of the collecting ducts to water so that more water is reabsorbed and less is excreted in the urine. The animal thus excretes a hypertonic urine.

In addition to regulating water balance, the kidneys regulate the balance of electrolytes in the blood by reabsorption and secretion. For example, the kidneys reabsorb K^+ in the proximal tubule and then secrete an amount of K^+ needed to maintain homeostasis into the distal convoluted tubule (figure 50.18). The kidneys also maintain acid–base balance by excreting H^+ into the urine and reabsorbing HCO_3^-.

The reabsorption of NaCl in the distal convoluted tubule and collecting duct depends on the needs of the body and is under the control of the hormone *aldosterone.* Both ADH and aldosterone influence the distal convoluted tubule and collecting duct, although aldosterone is more significant in terms of NaCl. We present more about hormonal control of excretion in the next section.

> The mammalian kidney is divided into a cortex and medulla and contains about a million functional units called nephrons. The nephron tubules receive a blood filtrate from the glomeruli and modify this filtrate to produce urine, which empties into the renal pelvis and is expelled from the kidney through the ureter.
>
> The loop of Henle creates a hypertonic renal medulla as a result of the active extrusion of NaCl from the ascending limb and the interaction with the descending limb. The hypertonic medulla then draws water osmotically from the distal convoluted tubule and collecting duct, which are permeable to water under the influence of antidiuretic hormone.

50.7 Hormonal Control of Osmoregulatory Functions

In mammals and birds, the amount of water excreted in the urine, and thus the concentration of the urine, varies according to the changing needs of the body. Acting through the mechanisms described next, the kidneys excrete a hypertonic urine when the body needs to conserve water. If an animal drinks excess water, the kidneys excrete a hypotonic urine.

As a result, the volume of blood, the blood pressure, and the osmolarity of blood plasma are maintained relatively constant by the kidneys, no matter how much water you drink. The kidneys also regulate the plasma K^+ and Na^+ concentrations and blood pH within very narrow limits. These homeostatic functions of the kidneys are coordinated primarily by hormones.

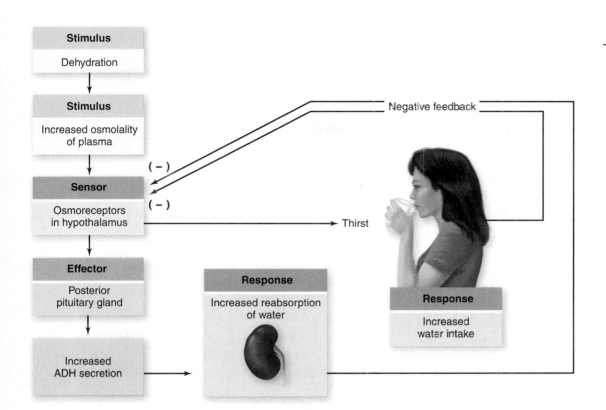

figure 50.19

ANTIDIURETIC HORMONE STIMULATES THE REABSORPTION OF WATER BY THE KIDNEYS. This action completes a negative feedback loop and helps to maintain homeostasis of blood volume and osmolarity.

Antidiuretic hormone causes water to be conserved

Antidiuretic hormone (ADH) is produced by the hypothalamus and secreted by the posterior pituitary gland. The primary stimulus for ADH secretion is an increase in the osmolarity of the blood plasma. When a person is dehydrated or eats salty food, the osmolarity of plasma increases. Osmoreceptors in the hypothalamus respond to the elevated blood osmolarity by sending increasing action potentials to the integration center (also in the hypothalamus). This, in turn, triggers a sensation of thirst and an increase in the secretion of ADH (figure 50.19).

ADH causes the walls of the distal convoluted tubules and collecting ducts in the kidney to become more permeable to water. Water channels called aquaporins (chapter 5) are contained within the membranes of intracellular vesicles in the epithelium of the distal convoluted tubules and collecting ducts; ADH stimulates the fusion of the vesicle membrane with the plasma membrane, similar to the process of exocytosis. The aquaporins are now in place and allow water to flow out of the tubules and ducts in response to the hypertonic condition of the renal medulla. This water is reabsorbed into the bloodstream.

When secretion of ADH is reduced, the plasma membrane pinches in to form new vesicles that contain aquaporins, so that the plasma membrane becomes less permeable to water. More water is then excreted in urine.

Under conditions of maximal ADH secretion, a person excretes only 600 mL of highly concentrated urine per day. A person who lacks ADH due to pituitary damage has the disorder known as *diabetes insipidus* and constantly excretes a large volume of dilute urine. Such a person is in danger of becoming severely dehydrated and succumbing to dangerously low blood pressure.

Aldosterone and atrial natriuretic hormone control sodium ion concentration

Sodium ions are the major solute in the blood plasma. When the blood concentration of Na^+ falls, therefore, the blood osmolarity also falls. This drop in osmolarity inhibits ADH secretion, causing more water to remain in the collecting duct for excretion in the urine. As a result, the blood volume and blood pressure decrease.

A decrease in extracellular Na^+ also causes more water to be drawn into cells by osmosis, partially offsetting the drop in plasma osmolarity, but further decreasing blood volume and blood pressure. If Na^+ deprivation is severe, the blood volume may fall so low that blood pressure is insufficient to sustain life. For this reason, salt is necessary for life. Many animals have a "salt hunger" and actively seek salt, such as deer do at "salt licks."

A drop in blood Na^+ concentration is normally compensated for by the kidneys under the influence of the hormone **aldosterone,** which is secreted by the adrenal cortex. Aldosterone stimulates the distal convoluted tubules and collecting ducts to reabsorb Na^+, decreasing the excretion of Na^+ in the urine. Indeed, under conditions of maximal aldosterone secretion, Na^+ may be completely absent from the urine. The reabsorption of Na^+ is followed by reabsorption of Cl^- and by water, so aldosterone has the net effect of promoting the retention of both salt and water. It thereby helps to maintain blood volume, osmolarity, and pressure.

The secretion of aldosterone in response to a decreased blood level of Na^+ is indirect. Because a fall in blood Na^+ is accompanied by decreased blood volume, the flow of blood past a group of cells called the juxtaglomerular apparatus is reduced. The juxtaglomerular apparatus is located in the region of the

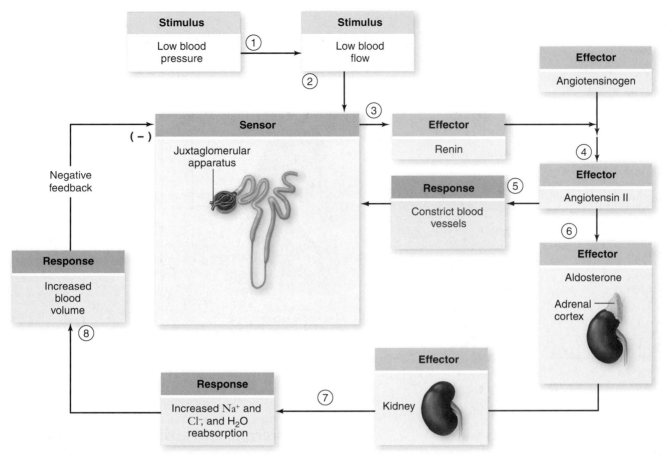

figure 50.20

A LOWERING OF BLOOD VOLUME ACTIVATES THE RENIN–ANGIOTENSIN–ALDOSTERONE SYSTEM. (*1*) Low blood volume and a decrease in blood Na$^+$ levels reduce blood pressure. (*2*) Reduced blood flow past the juxtaglomerular apparatus triggers (*3*) the release of renin into the blood, which catalyzes the production of angiotensin I from angiotensinogen. (*4*) Angiotensin I converts into a more active form, angiotensin II. (*5*) Angiotensin II stimulates blood vessel constriction and (*6*) the release of aldosterone from the adrenal cortex. (*7*) Aldosterone stimulates the reabsorption of Na$^+$ in the distal convoluted tubules. Increased Na$^+$ reabsorption is followed by the reabsorption of Cl$^-$ and water. (*8*) This increases blood volume. An increase in blood volume may also trigger the release of an atrial natriuretic hormone, which inhibits the release of aldosterone. These two systems work together to maintain homeostasis.

kidney between the distal convoluted tubule and the afferent arteriole (figure 50.20).

When blood flow is reduced, the juxtaglomerular apparatus responds by secreting the enzyme *renin* into the blood. Renin catalyzes the production of the polypeptide angiotensin I from the protein angiotensinogen. Angiotensin I is then converted by another enzyme into angiotensin II, which stimulates blood vessels to constrict and the adrenal cortex to secrete aldosterone. Thus, homeostasis of blood volume and pressure can be maintained by the activation of this renin–angiotensin–aldosterone system. In addition to stimulating Na$^+$ reabsorption, aldosterone also promotes the secretion of K$^+$ into the distal convoluted tubules and collecting ducts. Consequently, aldosterone lowers the blood K$^+$ concentration, helping to maintain constant blood K$^+$ levels in the face of changing amounts of K$^+$ in the diet. People who lack the ability to produce aldosterone will die if untreated because of the excessive loss of salt and water in the urine and the buildup of K$^+$ in the blood.

The action of aldosterone in promoting salt and water retention is opposed by another hormone, **atrial natriuretic hormone,** mentioned in chapter 49. This hormone is secreted by the right atrium of the heart in response to an increased blood volume, which stretches the atrium. Under these conditions, aldosterone secretion from the adrenal cortex decreases, and atrial natriuretic hormone secretion increases, thus promoting the excretion of salt and water in the urine and lowering the blood volume.

ADH stimulates the insertion of water channels into the cells of the distal convoluted tubule and collecting duct, making them more permeable to water. Thus, ADH stimulates the reabsorption of water and the excretion of a hypertonic urine. Aldosterone promotes the reabsorption of Na$^+$, Cl$^-$, and water across the distal convoluted tubule and collecting duct, as well as the secretion of K$^+$ into the tubules. ANH decreases Na$^+$ and Cl$^-$ reabsorption.

50.1 Regulating Body Temperature

Temperature is one of the most important aspects of the environment that organisms must contend with.

- The Q_{10} is the ratio of reaction rates at two temperatures 10°C apart. For chemical reactions Q_{10} is about 2.

- Body temperature is determined by internal factors such as metabolism and external factors that affect heat transfer.

- Heat is transferred by four mechanisms: radiation, conduction, convection, and evaporation (figure 50.1).

- Transfer of heat depends on surface-to-volume ratio, temperature differential, and specific heat conductance.

- Organisms are classified by their heat source: endotherms generate their own heat, and ectotherms rely on the environment.

- Ecotherms have low metabolic rates and regulate their body temperature by changes in behavior (figures 50.2 and 50.3).

- Endotherms can regulate body temperature by changes in metabolic rate, peripheral blood flow, and sweating or panting.

- Nonshivering thermogenesis generates heat by metabolism of fat, and shivering thermogenesis uses muscular contractions.

- Mammalian thermoregulation is controlled by the hypothalamus (figure 50.5).

- Pyrogens stimulate fever, a rise in the body temperature set point.

- Torpor is a reduction in metabolic rate that allows reduced food intake. Over a longer period of time this is called hibernation.

50.2 Osmolarity and Osmotic Balance

To maintain osmotic balance an animal's body must be able to take water up from or excrete excess water to the environment, and regulate ions.

- Osmotic pressure is a measure of a solution's propensity to take in water by osmosis.

- Osmolarity is defined as moles of solute per liter of solution.

- Cells in a hypertonic solution lose water, cells in a hypotonic solution gain water, and cells in an isotonic solution do not change size.

- Osmoconformers are in osmotic equilibrium with their environment, osmoregulators control osmolarity of body fluids.

- Freshwater vertebrates are hypertonic to their environment, and marine vertebrates are hypotonic to the environment.

50.3 Osmoregulatory Organs

Most organisms, including one-celled organisms, have evolved mechanisms to cope with problems of water balance.

- Flatworms use tubular protonephridia connected to flame cells that draw fluid from the body (figure 50.7).

- Other invertebrates have nephridia open to both the inside and outside of the body where body fluids are filtered then salts reabsorbed before excretion (figure 50.8).

- Insects have Malpighian tubules through which uric acid and wastes are secreted into the excretory organ and water and salts are reabsorbed before excretion (figure 50.9).

- Vertebrates have kidneys that produce urine by filtration, secretion, and selective reabsorption.

50.4 Evolution of the Vertebrate Kidney

The kidney is a complex organ made up of thousands of units called nephrons that regulate body fluids.

- The nephron is composed of a glomerulus, Bowman's capsule, proximal convoluted tubule, intermediate loop of Henle, distal convoluted tubule, and collecting duct (figure 50.10).

- Freshwater fishes are hypertonic to the environment. They excrete large quantities of water and reabsorb ions (figure 50.11).

- Saltwater fish are hypotonic to their environment. They drink large amounts of water, ions are excreted or actively transported out through the gills, and urine is isotonic to the blood.

- Cartilaginous fishes are isotonic to their environment because they retain urea, actively pump out electrolytes, and produce isotonic urine.

- Freshwater amphibians and reptiles act like freshwater fishes, and the kidneys of marine reptiles function like marine bony fishes.

- Terrestrial reptiles can't produce hypertonic urine.

- Mammals and birds can excrete hypertonic urine and retain water (figure 50.12).

50.5 Nitrogenous Wastes: Ammonia, Urea, and Uric Acid

When animals catabolize amino acids and nucleic acids they produce toxic nitrogenous wastes (figure 50.13).

- Ammonia is very toxic, and much water is used to remove it quickly from the body or it is eliminated through the gills of fish.

- Urea is less toxic than ammonia and requires less water, but requires energy to manufacture it

- Uric acid is the least toxic, requires the least amount of water to eliminate, and is the most energetically expensive to make.

50.6 The Mammalian Kidney

Mammalian kidneys process 180 L of filtrate each day and produce urine excreted through the urethra (figure 50.14).

- The kidney has three basic functions: filtration, reabsorption, and secretion (figure 50.15).

- Filtration occurs in the glomerulus, removing everything from the blood except cells and large proteins and other molecules.

- Reabsorption is selective movement of nutrients, ions, and water, from the filtrate back into the blood.

- Secretion moves substances from the blood into the filtrate.

- Filtrate passes through the Bowman's capsule, proximal convoluted tubule, loop of Henle, distal convoluted tubule, and collecting duct (figure 50.16).

- Blood passes from the afferent arteriole to the glomerulus, afferent arteriole, the peritubular capillaries, and the vasa recta.

- The kidney also eliminates a variety of potentially harmful substances including nitrogenous wastes, excess ions, and toxins.

- The loop of Henle creates a gradient of increasing osmolarity from the renal cortex to the renal medulla.

- A longer loop of Henle produces more concentrated urine (figure 50.17).

- The kidney regulates electrolyte levels by reabsorption and secretion.

50.7 Hormonal Control of Osmoregulatory Functions

Blood volume, blood pressure, and osmolarity of blood plasma are maintained by the kidneys, coordinated by hormones (figure 50.19).

- ADH produced by the hypothalamus conserves water by increasing the permeability of the collecting duct (figure 50.19).

- Low Na^+ levels inhibit ADH secretion, and aldosterone stimulates Na^+ uptake by the distal convoluted tubule.

- Atrial natriuretic hormone antagonizes the action of aldosterone.

- The lowering of blood volume activates the renin–angiotensin–aldosterone system (figure 50.20).

SELF TEST

1. Which of the following is *not* a method used in maintaining homeostasis in the body?
   a. Behavioral changes
   b. Negative feedback loops
   c. Hormonal actions
   d. Positive feedback loops

2. Similarities between the invertebrate and vertebrate excretory systems include:
   a. filtering of body fluid.
   b. use of tubules to contain filtrate.
   c. reabsorption of ions and water.
   d. all of these.

3. Suppose that your research mentor has decided to do a project on the filtering capabilities of Malpighian tubules. Which of the following creatures will you be spending your summer studying?
   a. Ants
   b. Birds
   c. Mammals
   d. Earthworms

4. A shark's blood is isotonic to the surrounding seawater because of the reabsorption of _____ in its blood.
   a. ammonia
   b. uric acid
   c. urea
   d. NaCl

5. An important function of the excretory system is to eliminate excess nitrogen produced by metabolic processes. Which of the following organisms is most efficient at packaging nitrogen for excretion?
   a. Frog
   b. Freshwater fish
   c. Iguana
   d. Camel

6. Which of the following is a function of the kidneys?
   a. The kidneys remove harmful substances from the body.
   b. The kidneys recapture water for use by the body.
   c. The kidneys regulate the levels of salt in the blood.
   d. All of these are functions of the kidneys.

7. A viral infection that specifically interferes with the reabsorption of ions from the glomerular filtrate would have attack cells located in the—
   a. Bowman's capsule.
   b. glomerulus.
   c. renal tubules.
   d. collecting duct.

8. Humans excrete their excess nitrogenous wastes as—
   a. uric acid crystals.
   b. compounds containing protein.
   c. ammonia.
   d. urea.

9. Which of the following statements is *not* true?
   a. ADH makes the collecting duct more permeable to water.
   b. Guano contains high concentrations of uric acid.
   c. Aldosterone is produced by the hypothalamus in response to high levels of sodium ions in the blood.
   d. Uric acid is the least soluble of the nitrogenous waste products.

10. The major difference between ectotherms and endotherms in terms of maintenance of body temperature is that—
    a. ectotherms rely solely on environmental conditions to determine body temperature.
    b. endotherms are impervious to environmental conditions.
    c. ectotherms rely more heavily on behavioral adaptations than do endotherms.
    d. endotherms are larger than ectotherms.

11. An osmoregulator would maintain its internal fluids at a concentration that is _____ relative to its surroundings.
    a. isotonic
    b. hypertonic
    c. hypotonic
    d. either hypertonic, hypotonic, or isotonic

12. If humans are endotherms, why do conditions like hypothermia (low body temperature) and hyperthermia (high body temperature) occur?
    a. Mechanisms that control body heat are inefficient and allow large fluctuations in temperature.
    b. Body heat is the sum of heat produced by metabolic reactions and the heat transferred from the environment.
    c. Metabolic reaction rates may change due to illness or activity level.
    d. Both (b) and (c).

13. You and your study partner want to draw the pathway that controls the reabsorption of sodium ion when blood pressure falls. Which of the following is the correct sequence of events?
    1. Aldosterone is released.
    2. Kidney tubules reabsorb Na$^+$.
    3. Renin is released.
    4. Juxtaglomerular apparatus (JGA) recognizes a drop in blood pressure.
    5. Angiotensin II is produced.
    a. 1, 3, 5, 2, 4
    b. 4, 2, 3, 1, 5
    c. 4, 3, 5, 2, 1
    d. 2, 4, 3, 1, 5

CHALLENGE QUESTIONS

1. Indicate the areas of the nephron that the following hormones target, and describe when and how the hormones elicit their actions.
   a. Antidiuretic hormone
   b. Aldosterone
   c. Atrial natriuretic hormone

2. John's doctor is concerned that John's kidneys may not be functioning properly due to a circulatory condition. The doctor wants to determine if the blood volume that is flowing through the kidneys (called renal blood flow rate) is within normal range. Calculate what would be a "normal" renal blood flow rate based on the following information:
   John weighs 90 kg. Assume a normal total blood volume is 80 mL/kg of body weight, and a normal heart pumps the total blood volume through the heart once per minute (cardiac output). Also assume that the normal renal blood flow rate is 21% of cardiac output.

3. Why can an elephant in a zoo in a temperate climate go outside on a cold day while a smaller mammal like a mouse would stay inside?

introduction

WHEN YOU CONSIDER HOW ANIMALS defend themselves, it is natural to think of turtles and armadillos with their obvious external armor. However, armor offers little protection against the greatest dangers vertebrates face—microorganisms and viruses. We live in a world awash with organisms too tiny to see with the naked eye, and no vertebrate could long withstand their onslaught unprotected. We survive because we have evolved a variety of very effective defenses. However, our defenses are far from perfect. Some 40 million people died from influenza in 1918–1919, and more than a million people will die of malaria this year. Attempts to improve our defenses against infectious diseases are being actively researched.

Just as a computer network has multiple levels of protection against external attack, vertebrate animals have several levels of defense from invasion by microorganisms.

1. **The Integumentary System.** The outermost layer of the vertebrate body is the first barrier to invasion by microbes. This includes the skin (**integument**) as well as mucous membranes in the respiratory, digestive, and urogenital tracts. Various chemicals and proteins produced by cells within the epithelial surfaces also help deter colonization by infectious agents, referred to collectively as **pathogens.**

2. **Nonspecific (Innate) Immunity.** If the first line of defense is penetrated, such as when a cut or other injury breaks the skin, the body mounts a cellular and chemical defense, using a battery of cells that are part of the **nonspecific,** or **innate, immune system.** Additional chemicals and proteins are produced that aid in the destruction of microbes. These defenses act very rapidly after the onset of infection.

3. **Specific Immunity.** Finally, the body is also guarded by cells that are part of the **specific immune system.** This system can specifically identify and remove any foreign invading microorganisms or viruses that were not eliminated by components of the innate immune system. One kind of immune cell aggressively attacks and kills any cell identified as foreign, including any host cells infected by viruses; another cell type marks the foreign cell or virus for elimination by the nonspecific immune cells.

In this chapter, we discuss details of all three lines of defense, beginning in this section with the integumentary system.

The skin is a barrier to infection

The skin is the largest organ of the body, accounting for 15% of an adult human's total weight. The integument not only defends the body by providing a nearly impenetrable barrier, but also reinforces this defense with chemical weapons on the surface. Oil and sweat glands give the skin's surface a pH of 3–5, acidic enough to inhibit the growth of many pathogenic microorganisms. Sweat also contains the enzyme lysozyme, which digests bacterial cell walls.

The skin is also home to many **normal flora,** nonpathogenic bacteria or fungi that are well adapted to the skin conditions in different regions of the body. Pathogenic bacteria that might attempt to colonize the skin generally are unable to compete with the normal flora.

In addition to defending the body against invasion by viruses and microorganisms, the skin also prevents excessive loss of water to the air through evaporation.

The **epidermis** of skin is approximately 10–30 cells thick, about as thick as this page. The outer layer, called the *stratum corneum*, contains cells that are continuously abraded, injured, and worn by friction and stress during the body's many activities. Cells are shed continuously from the stratum corneum and are replaced by new cells produced in the innermost layer of the epidermis, the *stratum basale*, which contains some of the most actively dividing cells in the vertebrate body. The cells formed in this layer migrate upward and enter a broad, intermediate *stratum spinosum* layer. As they move upward, they form the protein keratin, which makes skin tough and water resistant.

These new cells eventually arrive at the stratum corneum, where they normally remain for about a month before they are shed and replaced by newer cells from below. Psoriasis, which afflicts some 4 million Americans, is a chronic skin disorder in which epidermal cells are replaced every 3–4 days, about eight times faster than normal. The **dermis** of skin contains connective tissue and is 15–40 times thicker than the epidermis. It provides structural support for the epidermis and a matrix for the many blood vessels, nerve endings, muscles, and other structures situated within skin.

The layer of subcutaneous tissue below the dermis contains primarily adipose (fat) cells. These cells act as shock absorbers and provide insulation, conserving body heat. Subcutaneous tissue varies greatly in thickness in different parts of the body. It is nonexistent in the eyelids, is a half-centimeter thick or more on the soles of the feet, and may be much thicker in other areas of the body, such as the buttocks and thighs.

Mucosal epithelial surfaces also prevent entry of pathogens

In addition to the skin, three other potential routes of entry by microorganisms and viruses must be guarded: the digestive tract, the respiratory tract, and the urogenital tract. Recall that each of these tracts opens to the external environment. Each of these tracts is lined by epithelial cells, which are continuously replaced, as are those of the skin.

A layer of mucus, secreted by specialized cells scattered between the epithelial cells, covers all these epithelial surfaces. Pathogens are frequently trapped within this mucus layer and are eliminated by mechanisms specific to the particular tract.

Microbes are present in food, but many are killed by saliva (which contains lysozyme), by the very acidic environment of the stomach, and by digestive enzymes in the intestine. Additionally, the gastrointestinal tract is home to a vast array of nonpathogenic normal flora, whose presence inhibits the growth of pathogenic competitors. These nonpathogenic organisms not only outcompete pathogens, but they also may secrete substances that kill harmful agents.

Microorganisms present in inhaled air are trapped by the mucus within the smaller bronchi and bronchioles before they can reach the warm, moist lungs, which would provide ideal breeding grounds for them. The epithelial cells lining these passages have cilia that continually sweep the mucus toward the glottis. There the mucus can be swallowed, carrying potential invaders out of the lungs and into the digestive tract. One of the pitfalls of smoking is that nicotine paralyzes the cilia of the respiratory system so that this natural cleaning of the air passages does not take place.

Vaginal secretions are sticky and acidic, and they also promote the growth of normal flora; all of these characteristics help prevent foreign invasion. In both males and females, acidic urine continually washes potential pathogens from the urinary tract. In addition to these physical and chemical barriers to pathogen invasion, the body also uses defense mechanisms such as vomiting, diarrhea, coughing, and sneezing to expel potential pathogens.

> The integument forms the first line of body defense. Epithelial cells, covered by a mucus layer in many body locations, provide a physical barrier to the pathogens. The mucus layer is often swept away and eliminated, along with trapped microorganisms and other debris.

51.2 Nonspecific Immunity: The Second Line of Defense

When surface defenses fail, the body uses a host of nonspecific cellular and chemical devices to defend itself. We refer to this as the second line of defense. These devices all have one property in common: they respond to any microbial infection without needing to determine the invader's identity, and they do so quite rapidly.

Several types of cells nonspecifically eliminate invading pathogens

Among the most important nonspecific defenses are some types of **leukocytes**, or white blood cells, that circulate through the body and nonspecifically attack pathogens within tissues. Three basic kinds of defending leukocytes have been identified, and each kills invading microorganisms differently.

Macrophages

Macrophages ("big eaters") are large, irregularly shaped cells that kill microorganisms by ingesting them through phagocytosis, much as an amoeba engulfs a food particle (figure 51.1). Once within the macrophage, the pathogen is inside a membrane-bound phagosome, which fuses with a lysosome. Fusion activates lysosomal enzymes that kill and digest the microorganism.

Additionally, large quantities of oxygen-containing free radicals are frequently produced within the phagosome; these free radicals are very reactive and degrade the pathogen.

In addition to bacteria, macrophages also engulf viruses, cellular debris, and dust particles in the lungs. Macrophages roam continuously in the extracellular fluid bathing tissues, and their phagocytic actions supplement those of the specialized phagocytic cells that are part of the structure of organs such as the liver, spleen, and bone marrow. In response to an infection, monocytes, undifferentiated macrophages found in the blood, squeeze through the endothelial cells of capillaries to enter the connective tissues. There, at the site of the infection, the monocytes mature into active, phagocytic macrophages.

Neutrophils

Neutrophils are the most abundant circulating leukocytes, accounting for 50–70% of the peripheral blood leukocytes. They are the first type of cell to appear at the site of tissue damage or infection. Like macrophages, they squeeze between capillary endothelial cells to enter infected tissues, where they ingest a variety of pathogens by phagocytosis. Their mechanism of pathogen destruction is similar to that of macrophages except that they produce an even greater range of reactive oxygen radicals.

Natural killer cells

Natural killer (NK) cells do not attack invading microbes directly. Instead, they kill cells of the body that have been infected with viruses. They kill not by phagocytosis, but rather by inducing apoptosis (programmed cell death) of the target cell (see chapter 19). Proteins called *perforins*, released from the NK cells, insert into the membrane of the target cell, polymerizing into a pore. Other NK produced proteins, *granzymes*, then enter the pores created by the perforin molecules and activate proteins known as *caspases*, found within the target cells, which

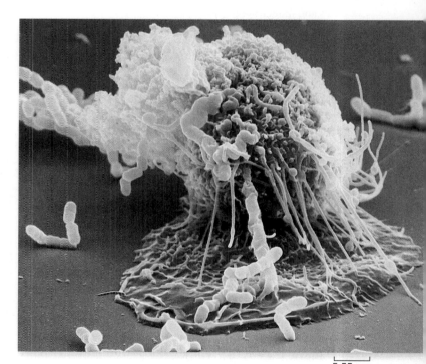

5.55 µm

figure 51.1

A MACROPHAGE IN ACTION. In this scanning electron micrograph, a macrophage is "fishing" with long, sticky cytoplasmic extensions. Bacterial cells that come in contact with the extensions are drawn toward the macrophage and engulfed.

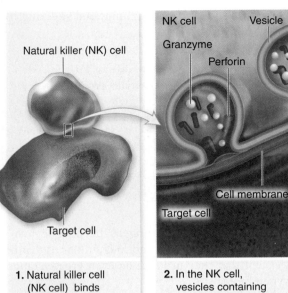

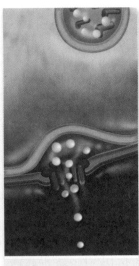

1. Natural killer cell (NK cell) binds tightly to target cell.

2. In the NK cell, vesicles containing perforin molecules and granzymes move to the plasma membrane and release their contents by exocytosis into the space between the two cells.

3. The perforin molecules polymerize in the plasma membrane of the target cell forming pores in the membrane.

4. Granzymes pass through the pores and activate caspase enzymes that induce apoptosis in the target cell.

5. During apoptosis, the target cell is broken down into membrane enclosed vesicles containing the cellular contents. Macrophages phagocytose these vesicles to prevent release of their contents into the tissues.

figure 51.2

HOW NATURAL KILLER CELLS ELIMINATE TARGET CELLS. Natural killer cells kill virally infected cells by programmed cell death, or apoptosis. This is accomplished by secreting proteins that form pores in the cell to be killed, along with proteins that diffuse through these pores and induce apoptosis.

inquiry

What would happen if an NK cell killed a virally infected target cell by simply causing the cell to burst, releasing all the cell contents into the tissues?

in turn induce apoptosis (figure 51.2). Macrophages ingest the resulting membrane-bound vesicular cell debris.

NK cells also attack tumor cells, often before the tumor cells have had a chance to divide sufficiently to be detectable as a tumor. The vigilant surveillance by natural killer cells is one of the body's most potent defenses against cancer. Thus, these cells are often said to play a role in **immune surveillance.**

The inflammatory response is a nonspecific response to infection or tissue injury

The inflammatory response involves several systems of the body, and it may be either localized or systemic. An acute response is one that generally starts rapidly but lasts for only a relatively short while.

Certain infected or injured cells release chemical alarm signals—most notably histamine, along with prostaglandins and bradykinin (see chapter 46). These chemicals promote the dilation of local blood vessels, which increases the flow of blood to the site and causes the area to become red and warm, two of the hallmark signs of inflammation.

These chemicals also increase the permeability of capillaries in the area, producing the third hallmark sign of inflammation, the edema (tissue swelling) often associated with

infection. Swelling puts pressure on nerve endings in the region and this, in combination with the release of other mediators, leads to pain and potential loss of function, the final two hallmark signs of inflammation.

Increased capillary permeability initially promotes the migration of phagocytic neutrophils from the blood to the extracellular fluid bathing the tissues, where the neutrophils can ingest and degrade pathogens; the pus associated with some infections is a mixture of dead or dying pathogens, tissue cells, and neutrophils. The neutrophils also secrete signaling molecules that attract monocytes several hours later; as the monocytes differentiate into macrophages, they too engulf pathogens and the remains of the dead cells (figure 51.3).

The inflammatory response is accompanied by an **acute-phase response.** One manifestation of this response is an elevation of body temperature, or fever (see chapter 50). Macrophages that encounter invading microbes release a cytokine, a regulatory protein called **interleukin-1 (IL-1),** which is carried by the blood to the brain. IL-1 causes neurons in the hypothalamus to raise the body's temperature several degrees above the normal value of 37°C (98.6°F). This increase in body temperature promotes the activity of phagocytic cells and impedes the growth of some microorganisms.

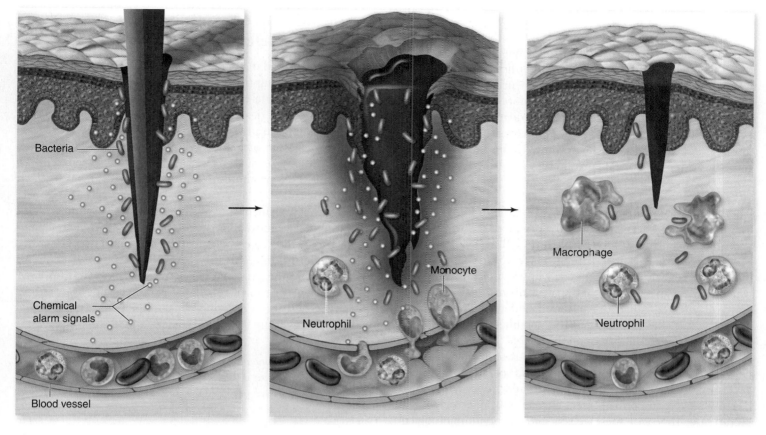

figure 51.3

THE EVENTS IN LOCAL INFLAMMATION. When invading pathogens have penetrated an epithelial surface, chemical alarm signals such as histamine and prostaglandins, released from damaged cells, cause nearby blood vessels to dilate and increase in permeability. Increased blood flow causes swelling and promotes the accumulation of phagocytic cells, specifically neutrophils followed by macrophages, which attack and engulf invading the pathogens.

Fever contributes to the body's defense by stimulating phagocytosis and causing the liver and spleen to store iron. This storage reduces blood levels of iron, which bacteria need in large amounts to grow. Very high fevers are hazardous, however, because excessive heat may denature critical enzymes. In general, temperatures greater than 39.4°C (103°F) are considered dangerous for humans, and those greater than 40.6°C (105°F) are often fatal.

A group of proteins collectively referred to as *acute-phase proteins* are also released from cells of the liver during an inflammatory response, sometimes at levels 1000-fold above the normal serum concentration. These proteins bind to a variety of microorganisms and promote their ingestion by phagocytic cells—the neutrophils and macrophages.

Complement proteins and interferons aid in the elimination of pathogens

The cellular defenses of vertebrates are enhanced by a very effective chemical defense called the **complement system.** This system consists of approximately 30 different proteins that circulate freely in the blood plasma, generally in an inactive form, and that can enter the tissues during an inflammatory response. **Interferons** are another class of protein that participates in system defense.

The complement system and MAC

When the complement proteins encounter many pathogens, a cascade of activation occurs, with the result that some of these proteins aggregate to form a **membrane attack complex (MAC)** that inserts itself into the pathogen's plasma membrane (or the lipid membrane around an enveloped virus), forming a pore.

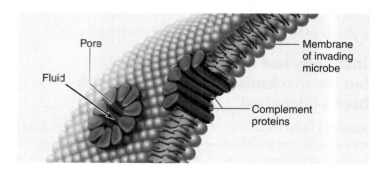

Extracellular fluid enters the pathogen through this pore, causing the pathogen to swell and burst. Activation of the complement proteins is also triggered in a specific fashion by the binding of antibodies, which are secreted by B lymphocytes, to invading pathogens, as we describe in a later section.

Other complement proteins, particularly one known as *C3b*, may coat the surface of invading pathogens. Phagocytic neutrophils and macrophages, which have receptors for C3b, may thus be "directed" to bind to the pathogens, promoting their phagocytosis and destruction. This is a particularly efficient way of eliminating those pathogens that do not have an outer lipid membrane into which a MAC may insert. Some complement proteins stimulate the release of histamine and other mediators by cells known as mast cells and basophils, promoting the dilation and increased permeability of capillaries; other complement proteins attract more phagocytes, especially neutrophils, to the area of infection through the more-permeable blood vessels.

Interferons

Another class of proteins that play a key role in body defense are interferons. There are three major categories of interferons: alpha, beta, and gamma (IFN-α, IFN-β, IFN-γ).

Almost all cells in the body make IFN-α and IFN-β. These polypeptides are synthesized upon viral infection of a cell and act as messengers that protect normal cells in the vicinity of infected cells from becoming infected. Although viruses are still able to penetrate the neighboring cells, IFN-α and IFN-β induce the degradation of RNA and block protein production in these cells. While this leads to the death of the cells, it also prevents virus production and spread.

IFN-γ is produced only by particular leukocytes called T lymphocytes (described later) and natural killer cells. The secretion of IFN-γ by these cells is part of the immunological defense against infection and cancer.

Many different organisms exhibit nonspecific defenses

All vertebrates and many invertebrates possess phagocytic cells that attack invading pathogens. Perhaps these phagocytic cells evolved from single-celled eukaryotes, such as amoebae, that phagocytose bacteria as a source of food.

In simple animals such as sponges, which lack either a circulatory system or a body cavity, the phagocytic cells roam within the mesohyl. Binding of antimicrobial peptides to pathogens often promotes ingestion of the pathogens by these phagocytic cells.

The complement system is considered to be another evolutionarily ancient form of innate immunity. Proteins homologous to several of the key complement proteins have been described in echinoderms, and expression of these proteins has been shown to increase in the presence of bacteria. Homologous proteins, which may play a similar role in promoting phagocytosis of bacteria, have also been described in the horseshoe crab (*Limulus polyphemus*), a distant relative of spiders, and in some insects.

Nonspecific immune responses are a second line of defense against infection. Phagocytic cells and apoptosis-inducing cells eliminate many pathogens that penetrate the epithelium. A variety of chemicals and proteins, including complement proteins and inteferons, further promote inflammation and the elimination of pathogens.

51.3 The Specific Immune Response: The Third Line of Defense

Few of us pass through childhood without contracting a variety of infectious illnesses. Prior to the advent of an effective vaccine in about 1991, most children contracted chicken pox before reaching their teens. Chicken pox and some other such diseases were considered diseases of childhood because most people, once recovered, never experienced them again. They developed immunity to the chicken pox-causing *varicella zoster* virus and maintained this immunity as long as their immune systems remained intact. Similarly, immunization today with a nonpathogenic form of *varicella* virus can also confer protection. This immunity is produced by specific immune defense mechanisms.

Immunity had long been observed, but the mechanisms have only recently been understood

Societies have known for over 2000 years that an individual who experiences an infectious disease is often protected against a subsequent occurrence of the same disease. The scientific study of immunity, however, did not begin until 1796, when an English country doctor, Edward Jenner, carried out an experiment to protect people again smallpox.

Jenner and the smallpox virus

Smallpox, caused by the *variola* virus, was a common and deadly disease in the 1700s and earlier centuries. As with chicken pox, those who survived smallpox rarely caught the disease again,

and people had been known to deliberately infect themselves through inoculation hoping to survive a mild case and become immune. Jenner observed, however, that milkmaids who had caught a much milder form of "the pox" called cowpox (presumably from cows) rarely experienced smallpox.

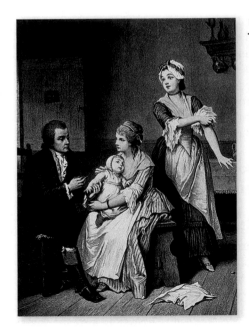

figure 51.4

THE BIRTH OF IMMUNOLOGY. This painting shows Edward Jenner inoculating patients with cowpox in the 1790s and thus protecting them from smallpox.

Jenner set out to test the idea that cowpox could confer protection against smallpox. He innoculated a healthy child with fluid from a cowpox vesicle and later deliberately infected him with fluid from a smallpox vesicle; as he had predicted, the child did not become ill. (Jenner's experiment would be considered unethical today.) Subsequently, many people were protected from smallpox by immunization with fluid from cowpox vesicles, a much less risky proposition (figure 51.4).

We now know that smallpox and cowpox are caused by two different viruses that have similar surfaces. Jenner's patients who were injected with the cowpox virus mounted a defense that was also effective against a later infection of the smallpox virus.

Jenner's procedure of injecting a harmless agent to confer resistance to a dangerous one is called **vaccination**. Modern attempts to develop resistance to malaria, herpes, and other diseases often involve delivering antigens via a harmless *vaccinia* virus related to the cowpox virus (chapter 17).

Pasteur and avian cholera

Many years passed before anyone learned how exposure to an infectious agent can confer resistance to a disease. A key step toward answering this question was taken more than a half-century later by the famous French scientist Louis Pasteur. Pasteur was studying avian cholera, a form that infects birds. He isolated a culture of bacteria from diseased chickens that would produce the disease if injected into healthy birds.

It is reported that before departing on a two-week vacation, he accidentally left his bacterial culture out on a shelf. When he returned, he injected this old culture into healthy birds and found that it had apparently been weakened; the injected birds became only slightly ill and then recovered. Surprisingly, however, those birds did not get sick when subsequently infected with fresh cholera bacteria that did produce the disease in control chickens. Clearly, something about the bacteria could elicit immunity, as long as the bacteria did not kill the animals first. We now know that molecules protruding from the surfaces of the bacteria evoked active immunity in the chickens.

Antigens stimulate specific immune responses

An **antigen** is a molecule that provokes a specific immune response. The most effective antigens are large, complex molecules such as proteins. The greater their "foreignness," or put another way, their phylogenetic distance from the host, the greater will be the immune response they elicit.

Antigens may be components of a microorganism or a virus, but they may also be proteins or glycoproteins on the surface of transfused red blood cells or on transplanted tissue. They may also be components of foods or pollens. A large antigen is likely to have many different parts, known as **antigenic determinants**, or **epitopes** (figure 51.5), each of which can stimulate a distinct immune response.

Lymphocytes carry out the specific immune responses

The four distinguishing characteristics of the specific, or adaptive, immune system are:

1. specificity of recognition of antigen,

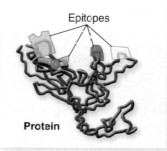
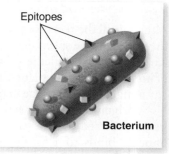

a.   *b.*

figure 51.5

MANY DIFFERENT EPITOPES ARE EXHIBITED BY ANY ONE ANTIGEN. *a.* A single protein may have many different antigenic determinants called epitopes, each of which can stimulate a distinct immune response. *b.* A pathogen such as a bacterium has many proteins on its surface, and there are likely to be multiple copies of each. Note that the protein and bacterium are not drawn to scale with respect to each other.

2. the wide diversity of antigens that can be specifically recognized,
3. memory, such that the immune system responds more quickly to an antigen it has previously encountered than to one it is meeting for the first time, and
4. the ability to distinguish self-antigens from nonself.

Particular leukocytes called **lymphocytes** have receptor proteins on their surfaces that recognize specific epitopes on an antigen and direct an immune response against either the antigen or the cell that carries the antigen (figure 51.6).

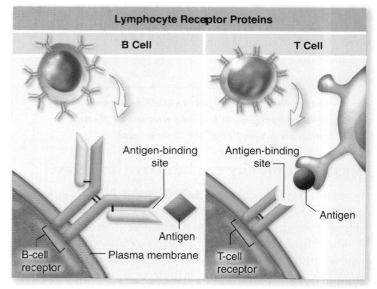

figure 51.6

B- AND T-CELL RECEPTORS BIND ANTIGENS. B-cell receptors are immunoglobulin (Ig) molecules with a characteristic Y-shaped structure. Every B cell has a single kind of Ig on its surface that binds to a single antigenic determinant. T-cell receptors are simpler than Ig molecules, but also bind to specific antigenic determinants. T cells only bind to antigens bound to another cell.

Lymphocytes and antigen recognition

Although all the receptor proteins on any one lymphocyte have the same epitope specificity, it is rare that any two lymphocytes have identical specificities. This feature produces the diversity of immune responses that ensures that at least some epitopes of any antigen that might be encountered are recognized.

A lymphocyte that has never before encountered antigen is referred to as a *naive lymphocyte*. When a naive lymphocyte binds to a foreign antigen, this activates the cell, causing it to divide producing a clone of cells with identical antigen specificity, a process called **clonal selection.** Some of these cells respond immediately to the antigen, and others become memory cells, which can remain in our bodies for years and perhaps for the remainder of our lives. Memory cells are easily and rapidly activated on subsequent encounters with the same antigen.

B cells

Lymphocytes called **B lymphocytes,** or **B cells,** respond to antigens by secreting proteins called **antibodies** or **immunoglobulins (Ig).** Antigen recognition occurs when an antigen binds to immunoglobulins on the B cell's membrane. Binding to antigen, in conjunction with other signals to be described later, initiates a signaling pathway that leads to the B cell secreting antibodies specific for the epitope recognized by the antibody in the B-cell membrane. This B-cell–mediated response producing secreted antibodies is called **humoral immunity.**

T cells

Other lymphocytes, called **T lymphocytes,** or **T cells,** do not secrete antibodies but instead regulate the immune responses of other cells or directly attack the cells that carry the specific antigens. These cells participate in the other arm of specific immunity called **cell-mediated immunity.** Both cell-mediated and humoral immunity processes are described in detail in later sections.

inquiry

Jenner used cowpox virus to elicit an immune response against smallpox. What does this tell us about the antigenic properties of the two viruses?

Specific immunity can be active or passive

Immunity can be acquired in different ways. First, an individual can gain immunity on infection by a pathogen and perhaps developing the disease it causes. Alternatively, an individual can be immunized with portions of a pathogen or with a less virulent form of the pathogen. Both of these situations result in *active immunity*, associated with the activation of specific lymphocytes and the generation of memory cells by the individual. Second, an individual can gain immunity by obtaining antibodies from another individual. This happened to you before you were born, as some antibodies made by your mother were transferred to your body across the placenta. Immunity gained in this way is called *passive immunity*, and it does not result in the generation of memory cells. The immunity is only effective as long as the antibodies remain in your body. Like any other proteins, they will degrade in time.

Hematopoiesis gives rise to the cells of the immune system

All the cells that are found in the blood are derived from the division and differentiation of hematopoietic stem cells, a process called **hematopoiesis** (see chapter 49). Embryologically, these stem cells are initially found in the yolk sac, then migrate to the fetal liver and spleen and finally to the bone marrow. Stem cells give rise to lymphoid progenitors and myeloid progenitors. A lymphoid progenitor, in turn, gives rise to both the B and T lymphocytes as well as to natural killer cells. A myeloid progenitor gives rise to all the other cells of the immune system as well as to erythrocytes and platelets (figure 49.16).

Although the lymphocytes are responsible for the specific third line of defense against pathogens, all the other leukocytes illustrated in figure 49.16 play supporting roles in this specific response or are important in the nonspecific second line of defense. **Monocytes** give rise to the macrophages, and these along with the **neutrophils** are phagocytic cells. **Eosinophils** are important in the elimination of helminths (flatworms; chapter 33), either through secretion of digestive enzymes through perforin pores into the helminths or occasionally by phagocytosis. They also play a role in exacerbating chronic inflammatory diseases such as asthma or inflammatory bowel disease.

Basophils and **mast cells** are not phagocytic but rather secrete inflammatory mediators such as histamine and prostaglandin in response to the binding of complement proteins during the elimination of pathogens. These cells, and mast cells in particular, are also activated during an allergic response, and the inflammatory mediators they release cause the symptoms of allergy. **Dendritic cells** are important in the activation of T cells, as will be described further. The roles of these cells are summarized in table 51.1.

The immune system is supported by two classes of organs

The organs of the immune system consist of the **primary lymphoid organs,** the bone marrow and the thymus, as well as the **secondary lymphoid organs,** the lymph nodes, spleen, and mucosal-associated lymphoid tissue, or MALT (figure 51.7).

The primary lymphoid organs

The **bone marrow** is the site of maturation of B cells. After hematopoiesis gives rise to the most immature B cells, progenitor B cells, these cells complete their maturation in the bone marrow. It is here that DNA rearrangements of the immunoglobulin genes, to be discussed later, dictate the specificity of each B cell. Every B cell has about 10^5 Ig molecules on its surface, all with identical specificity of epitope binding and all different from cell to cell.

Any lymphocytes that are likely to bind to self-antigens undergo apoptosis (figure 51.8*a*). The remainder are released to circulate in the blood and lymph and pass through the secondary lymphoid organs, where they may encounter antigen.

After their origin in the bone marrow, progenitor T cells migrate to the **thymus,** a primary lymphoid organ located just above the heart. The thymus is very large in infants; it starts to shrink in the teenage years, which it continues to do throughout life.

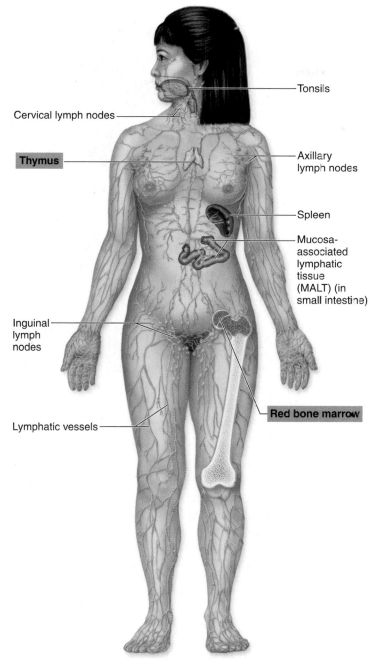

Tonsils

Cervical lymph nodes

Thymus

Axillary lymph nodes

Spleen

Mucosa-associated lymphatic tissue (MALT) (in small intestine)

Inguinal lymph nodes

Red bone marrow

Lymphatic vessels

figure 51.7

ORGANS OF THE SPECIFIC IMMUNE SYSTEM. There are two types of immune system organs, primary lymphoid organs (red boxes), in which B and T lymphocytes mature and acquire their specific receptors, and secondary lymphoid organs (labeled in black) in which antigen is collected and through which the mature naive lymphocytes circulate in order to meet and be stimulated by antigen.

The antigen receptor on T cells is designated the **T-cell receptor,** or **TCR.** The TCR is produced by gene rearrangements as T cells mature in the thymus, similar to those that occur for Ig genes of progenitor B cells. Thus, T cells may express about 10^5 identical TCRs per T cell, all likely to be different from one T cell to the next.

A B cell recognizes an epitope of an intact antigen that may or may not be a protein. In contrast, a T cell recognizes only a peptide fragment of a protein antigen, and this peptide fragment

TABLE 51.1	Cells of the Immune System	
Cell Type		**Function**
Helper T cell		Specifically recognizes foreign peptides on antigen-presenting cells, inducing the release of cytokines that activate B cells or macrophages
Cytotoxic T cell		Specifically recognizes and kills "altered-self" cells: virally infected, or tumor cells
B cell		Binds specific soluble or particulate antigens with its membrane-bound antibody; also serves as an antigen-presenting cell to T_H cells
Plasma cell		Activated B cell that is a biochemical factory devoted to the secretion of antibodies directed against specific antigens
Natural killer cell		Rapidly and nonspecifically recognizes and kills virally infected or tumor cells
Monocyte		Precursor of macrophage; located in blood
Macrophage		Phagocytic tissue cell that is a component of the body's first cellular line of defense; also serves as an antigen-presenting cell to T_H cells
Neutrophil		A phagocytic cell that is a component of the body's first cellular line of defense; found in the blood in large numbers until attracted to tissues during inflammation
Eosinophil		Important to the elimination of parasites and involved in chronic inflammatory diseases
Basophil		Circulating cell that releases mediators such as histamine that promote inflammation
Mast cell		Located primarily under mucosal surfaces and releases mediators such as histamine that promote inflammation; triggered both during inflammatory and allergic responses
Dendritic cell		Important antigen-presenting cell to naive T_H cells; also helps in the activation of naive T_C cells

figure 51.8

SELECTION AGAINST SELF-REACTIVE LYMPHOCYTES IN PRIMARY LYMPHOID ORGANS.
After B and T lymphocytes acquire their specific receptors, self-reactive cells are eliminated by apoptosis. ***a.*** If neighboring Igs on the surface of a maturing B cell are crosslinked by binding to an epitope on a bone marrow stromal cell, then that cell will undergo apoptosis. The small percentage (10%) of B cells whose Igs do not recognize stromal cell epitopes will be released from the bone marrow. ***b.*** If TCRs on a maturing T cell bind too tightly to self-MHC/self-peptide complexes on dendritic cells in the thymus, then that T cell will undergo apoptosis. Cells that do not bind the MHC complexes at all are also eliminated. Only the very small percentage (2-5%) of maturing T cells that can bind MHC peptide complexes with intermediate affinity will be released from the thymus. These cells will bind self-MCH/foreign peptide complexes with high affinity.

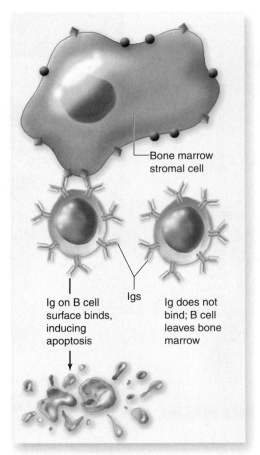

a.

Bone marrow stromal cell

Ig on B cell surface binds, inducing apoptosis

Igs

Ig does not bind; B cell leaves bone marrow

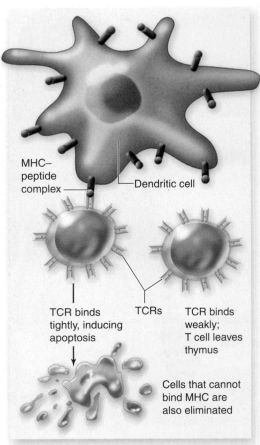

b.

MHC–peptide complex

Dendritic cell

TCR binds tightly, inducing apoptosis

TCRs

TCR binds weakly; T cell leaves thymus

Cells that cannot bind MHC are also eliminated

---

must be bound to one of a series of self-proteins that are present on the surface of almost all of the body's cells. These proteins are encoded by genes in the **major histocompatibility complex,** or **MHC.** The MHC is discussed in detail in subsequent sections.

During selection in the thymus, T cells are exposed to many thymic cells, all expressing self-MHC proteins with bound self-peptides on their surfaces. If a T cell's TCRs bind too strongly to these self-MHC protein complexes, then that T cell is "self-reactive" and undergoes apoptosis (figure 51.8*b*). Conversely, if the T cell's TCR does not bind MHC complexes at all, it is also eliminated. Only about 5% of the progenitor T cells that enter the thymus pass this rigorous two-step selection and avoid apoptosis.

The secondary lymphoid organs

The locations of the secondary lymphoid organs promote the filtering of antigens that enter any part of an individual's body. Bacteria attached to a thorn stuck in the skin, for example, enter the lymph that bathes the tissues. Lymph is eventually returned to the blood circulation through a series of vessels referred to as the lymphatics (see chapter 49). On its way, the lymph is filtered in the thousands of lymph nodes, which are located at the junction of lymphatic vessels (see figure 51.7).

The many mature but naive B and T lymphocytes that have entered a lymph node on exiting the primary lymphoid organs, or the memory cells that are located here, become activated on meeting with antigen. Antibodies secreted on activation of B cells in the lymph nodes, as well as the clonal progeny of the activated B and T cells, then leave the lymph node and enter the blood circulation when the lymph is returned to blood vessels near the heart.

Lymphocytes responding to antigens in a lymph node may pass out of capillaries supplying blood to the lymph node and enter the node's tissues. This is the cause of the "swollen glands" that sometimes accompany infection. The local lymph nodes enlarge due to the vast influx of lymphocytes.

Some antigens are found primarily in the blood, or in the blood as well as in the tissues. One example is the bacterium *Neisseria meningitidis,* a cause of a potentially fatal meningitis (infection of the meninges of the brain; chapter 44). Immune responses to such antigens occur in the **spleen.**

The splenic artery carries blood to the spleen where it then subdivides into arterioles. Antigens released into the ground tissue of the spleen are recognized by B and T cells present in the white pulp, regions of the spleen immediately surrounding the arterioles. Lymphocytes in the white pulp may be activated, as in the lymph node. Antibodies along with some of the activated lymphocytes exit via the splenic vein.

The final important secondary lymphoid organ is the **mucosal-associated lymphoid tissue (MALT),** which includes the tonsils, the appendix, and a large number of follicles located in the connective tissue under mucosal surfaces. These follicles are composed of lymphocytes, primarily B cells but also some T cells, and some macrophages. Any antigens that pass through the mucosa will immediately encounter lymphocytes in these follicles and their entry further into the body may be stopped at this point.

If invading organisms manage to evade the nonspecific defenses of mucosal surfaces as well as the specific responses of the lymphocytes in the MALT, then they still face a further chance of being stopped by responses in the other secondary lymphoid organs.

Two forms of adaptive immunity have evolved

Adaptive immunity, involving the ability to distinguish between self and nonself, was long thought to have evolved once in vertebrates. The type of adaptive immunity described in this chapter first arose in the cartilaginous fish (see chapter 35). These vertebrates evolved some 450 MYA.

Sharks and rays possess a thymus and a spleen, as well as a rather diffuse MALT. These animals mount cell-mediated responses with T cells bearing TCRs, and humoral responses with B cells that secrete Ig. Bone marrow in which hematopoiesis occurs appeared first in amphibians, although its exact role appears to vary in different species. Lymph nodes appeared first in birds, and their immune system differs little from that of mammals.

Recently, a second form of adaptive immunity has been described in jawless fish. This system does not involve B and T cells with their characteristic receptors. Instead, lymphocytes have receptor proteins composed of variable repeats rich in the amino acid leucine. These proteins appear to function much like Ig, but with a completely different protein architecture. The number of different receptor proteins produced by this system appears similar to the number of potential Ig. The generation of diversity in the two systems appears to have some similarity as the different lymphocyte receptors in jawless fish are also assembled by DNA rearrangements. The genetic makeup of the genes involved and the mechanism of these rearrangements is currently unknown.

It is not clear whether this newly described form of adaptive immunity was present in the ancestor to all chordates, or if it evolved in the lineage that gave rise to jawless fish. Given the differences in the two systems, it is likely that they represent independent events. If this other form of adaptive immunity was present in the ancestor to all chordates, some vestige may remain in modern vertebrates, including humans.

> All types of blood cells and platelets are produced in the bone marrow. Lymphocytes must acquire their specific receptors and undergo selection for self-reactivity in primary lymphoid organs. These mature but naive lymphocytes circulate to secondary lymphoid organs, where antigens are collected and where the lymphocytes may recognize and respond to foreign antigens that have evaded nonspecific defenses.

51.4 T Cells: Cell-Mediated Immunity

T cells may be characterized as either **cytotoxic T cells (T_C)** or **helper T cells (T_H)**. These cells can also be identified based on cell surface markers. T_C cells have CD8 protein on their cell surface, making them CD8$^+$ cells. T_H cells have CD4 protein on their cell surface, making them CD4$^+$ cells.

To be activated, both of these T cell types must recognize peptide fragments bound to MHC proteins, but the two cell types may be distinguished by (a) recognition of different classes of MHC proteins, which have distinct cell distributions, and (b) differing roles of the T cells after they are activated.

The MHC carries self and nonself information

As discussed earlier, the surfaces of most vertebrate cells exhibit glycoproteins encoded by the MHC. In humans, the name given to the proteins encoded by the MHC complex is **human leukocyte antigens (HLAs)**. The genes encoding the MHC proteins are highly polymorphic (have many alleles). For example, the HLA proteins are specified by genes that are the most polymorphic known, with nearly 500 alleles detected for some of the proteins. Only rarely will two individuals have the same combination of alleles, and the HLAs are thus different for each individual, much as fingerprints are.

MHC proteins on the tissue cells serve as self markers that enable an individual's immune system, specifically its T cells, to distinguish its own cells from foreign cells, an ability called *self versus nonself recognition*.

There are two classes of MHC proteins. **MHC class I proteins** are present on every nucleated cell of the body. **MHC class II proteins,** however, are found only on **antigen-presenting cells** (in addition to MHC class I); these cells include macrophages, B cells, and dendritic cells (table 51.2). T_C cells respond to peptides bound to MHC class I proteins, and T_H cells respond to peptides bound to MHC class II proteins.

Most of the time, the peptides bound to MHC proteins are derived from self-proteins from the individual's own cells. For this reason, it is important that T cells undergo selection in the thymus so that those that bind too strongly to peptides of self-proteins on self-MHC are eliminated. In this way, T cells normally are activated only outside the primary lymphoid organs in which they mature, when they encounter peptides of foreign proteins on self-MHC—for example, in the case of viral infection or cancer.

TABLE 51.2	Lymphocyte Recognition of Antigen			
	Recognize epitopes of soluble or particulate antigen	Recognize peptides bound to self-MHC proteins	Class of MHC proteins recognized	Cell types on which recognized MHC is expressed
B cells	yes	no	none	NA
T_H (CD4$^+$) cells	no	yes	class II	antigen-presenting cells: dendritic cells, B cells, and macrophages
T_C (CD8$^+$) cells	no	yes	class I	all nucleated cells

Cytotoxic T cells eliminate virally infected cells and tumor cells

Activated cytotoxic T cells recognize "altered-self" cells, particularly those that are virally infected or tumor cells. The TCRs of cytotoxic T lymphocytes recognize peptides of endogenous antigens bound to MHC class I proteins. Peptides of endogenous antigens are generated in a cell's cytosol, and then are pumped by special transport proteins into the rough endoplasmic reticulum where they become bound to MHC class I proteins. These proteins continue on their way through the endomembrane system to the cell surface.

An endogenous antigen may be a self-protein, or it may be a viral protein produced within a virally infected cell or an unusual protein produced by a cancerous cell. T_C cells respond only to the peptides of these unusual proteins bound to self-MHC class I. T-cell activation occurs in a secondary lymphoid organ, as described earlier. In a lymph node, for example, T cells encounter antigen-presenting cells. Dendritic cells in particular often present antigens that activate T_C cells.

Because not all viruses can infect dendritic cells, the dendritic cells must ingest viruses or tumor cells and then, through a mechanism referred to as cross-presentation, place the viral or tumor peptides on MHC class I proteins. Binding of the T_C cell through its TCR and its CD8 site to the dendritic cell induces clonal expansion of the T_C cell, generating many activated T_C cells as well as memory T_C cells (figure 51.9). The activated T_C cells then circulate around the body where they bind to "target" host cells expressing the same combination of foreign peptide on self-MHC class I (figure 51.10).

Apoptosis of the target cell is induced in a very similar fashion to that used by NK cells; a T_C cell secretes perforin monomers that create pores in the target's membrane; granzymes enter and activate caspases, which in turn cause apoptosis of the target.

Helper T cells secrete proteins that direct immune responses

Activated helper T cells, T_H cells, secrete low-molecular-weight proteins known as **cytokines.** A vast array of cytokines is known, many but not all of which are secreted by T_H cells. These cytokines bind to specific receptors on the membranes of many other cells, particularly but not exclusively those of the immune system. On binding, they initiate signaling cascades in these cells that promote their activation or differentiation.

Because cytokines are quite potent, they are generally secreted at very low concentrations so that, with a few exceptions, they bind only to nearby cells. IL-1 is an exception in that it travels to the hypothalamus to induce the fever response. Different subsets of T_H cells secrete cytokines specific for different cell receptors, so it is largely the T_H cells and the cytokines they secrete that determine whether an immune response will be humoral or cell-mediated in nature.

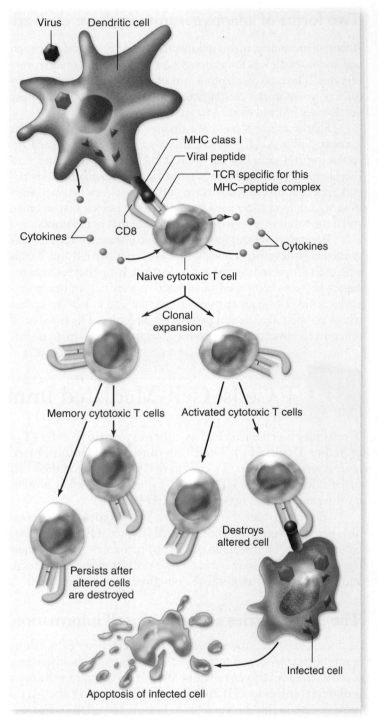

figure 51.9

CYTOTOXIC T CELLS INDUCE APOPTOSIS OF "ALTERED-SELF" CELLS. Naive cytotoxic T cells are initially activated on TCR recognition of foreign peptide displayed on self-MHC class I proteins on dendritic cells in a secondary lymphoid organ. Activation results in clonal expansion and differentiation into memory cells and activated cells. Activated progeny of the T_C cell can induce apoptosis of any cell in the periphery (outside the secondary lymphoid organ) that displays the same self-MHC class I–peptide combination on its surface. This will most likely be a virally infected cell or a tumor cell.

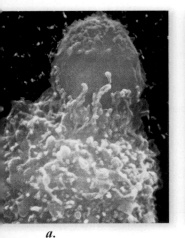
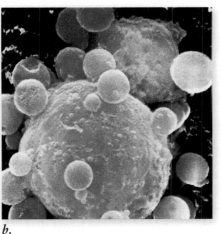

a.　　　　　b.

figure 51.10

CYTOTOXIC T CELLS DESTROY TUMOR CELLS. *a.* The cytotoxic T cell (*orange*) comes into contact with a tumor cell (*purple*). *b.* The T cell recognizes that the tumor cell is "altered-self" and induces the apoptosis of the tumor cell.

T_H cells respond to exogenous antigen that has been brought into an antigen-presenting cell. Macrophages or dendritic cells acquire these antigens by phagocytosis or endocytosis, and B cells gain them through receptor-mediated endocytosis. Once inside these cells, the antigen is gradually degraded in increasingly acidic endosomes/lysosomes. Peptides of the antigen join with MHC class II proteins in certain of these endosomes, and the MHC class II–peptide complexes are then transported to and displayed on the cell surface of the antigen-presenting cell. T_H cells encounter these cells within the secondary lymphoid organs and bind to the complexes. The CD4 protein of the T_H cells additionally bind to conserved regions of MHC class II.

A naive T_H cell expresses a protein called CD28 that must bind to a protein called B7 if that T cell is to be activated. B7 is found only on antigen-presenting cells and is at highest levels on dendritic cells. This requirement ensures that T_H cells are activated only when needed; this careful regulation is necessary due to the potency of the cytokines these cells release.

As with T_C cells, an activated T_H cell gives rise to a clone of T_H cells including both effector T_H cells and memory T_H cells, with identical TCR specificity. Most of the effector cells will leave the lymphoid organ and circulate around the body.

T cells are the primary cells that mediate transplant rejection

When T cells encounter the non-self MHC–peptide complexes present on transplanted tissue, such as a kidney, the TCRs on many of the T cells can weakly bind to these complexes. This is simply a case of cross-reactivity: The structure of a nonself MHC–peptide complex sufficiently resembles that of the self-MHC–foreign-peptide complex. The result is that the T cell binds to the foreign tissue cell.

Although the interactions between TCRs and nonself MHC–peptide complexes are relatively weak, many interactions occur between any one T cell and any one transplanted cell because a high density of MHC proteins is present on the surface of all cells. This induces activation of the T cells and attack of the foreign tissues.

Because of the genetic basis of MHC proteins, the more closely two individuals are related, the less their MHC proteins vary, and thus the more likely they will be to tolerate each other's tissues. As a result, relatives are often sought for kidney transplants, and HLA typing is done to find matching alleles.

A variety of drugs are used to suppress immune system rejection of a transplant; most individuals with a non-MHC-matched transplant continue to take some of these drugs for the remainder of their lives. One very effective drug is cyclosporin, which blocks the activation of lymphocytes.

Many cells of the immune system release cytokines

Many cells in addition to T_H cells release cytokines, always in a carefully regulated fashion. For example, macrophages that have been activated by phagocytosis of antigen or by the binding of certain bacterial products release cytokines such as interleukin-12 (IL-12) that can, in turn, bind to T_H cells to increase their level of activation. Macrophages may also release other cytokines, such as tumor necrosis factor-α (TNF-α), that bind to blood vessels to induce a local or even systemic increase in vascular permeability.

> T cells respond to peptides of foreign antigens displayed on self-MHC proteins. Activated T_C cells induce apoptosis of altered-self cells, those that are virally infected or tumor cells, whereas T_H cells secrete cytokines that promote either cell-mediated or humoral immune responses.

51.5 B Cells: Humoral Immunity and Antibody Production

The B-cell receptors for antigen are the immunoglobulin molecules present as integral proteins in the plasma membrane. As noted earlier, each B cell exhibits about 10^5 immunoglobulin molecules of identical specificity for a particular epitope of an antigen. Naive B cells in secondary lymph organs encounter antigens. When immunoglobulin molecules on a B cell bind to a specific epitope on an antigen, and the B cell receives additional required signals, particularly cytokines secreted by

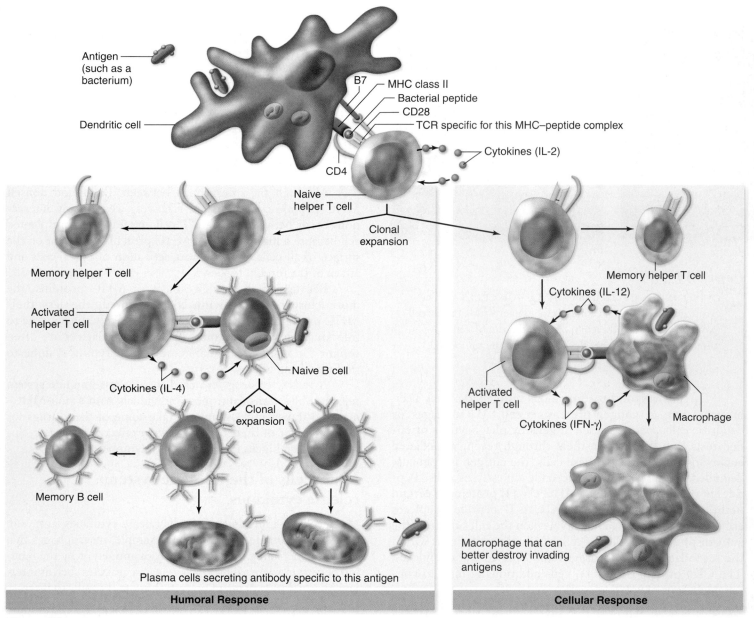

figure 51.11

HELPER T CELLS SECRETE CYTOKINES PROMOTING EITHER CELL-MEDIATED OR HUMORAL IMMUNE RESPONSES. Naive helper T cells are initially activated by TCR binding to foreign peptide displayed on self-MHC class II proteins on dendritic cells. Activation results in clonal expansion and differentiation into memory cells and activated cells. T_H cells promote the humoral response when they recognize the same antigen displayed by a B cell. Cytokines such as interleukin-4 (IL-4) released from the T_H cell will activate the B cell, producing memory cells and plasma that secrete antibodies against the antigen. T_H cells also secrete interferon-γ (IFN-γ), which stimulates cells involved in the cellular response such as the macrophage shown here. Macrophages secrete other cytokines that stimulate T_H cells.

T_H cells, then that B cell becomes activated, proliferating into plasma cells and memory cells (figure 51.11).

Each plasma B cell is a miniature factory producing soluble antibodies of the same specificity as the membrane-bound antibodies of the parent B cell. These antibodies enter the lymph and blood circulation as well as the extracellular fluid, and they bind to the appropriate epitopes of antigen encountered anywhere in the body. Any one antigen may present a variety of epitopes, so that different B cells might recognize different epitopes of a single antigen.

Once immunoglobulins coat an antigen, many other cells and processes may be activated to eliminate the antigen. The immunity to avian cholera that Pasteur observed in his chickens resulted from such antibodies and from the continued presence of the progeny of the B cells that produced them.

Immunoglobulin structure reveals variable regions and constant regions

Each immunoglobulin molecule consists of two identical short polypeptides called **light chains** and two identical longer polypeptides called **heavy chains** (figure 51.12). The four chains in an immunoglobulin molecule are held together by disulfide bonds, forming a Y-shaped molecule (figure 51.12*a*). Each "arm" of the molecule is referred to as an Fab region, and the "stem" is the Fc region (figure 51.12*b*).

Antibody specificity: The variable region

Comparison of the amino acid sequences of many different immunoglobulin molecules has demonstrated that the specificity of immunoglobulins for antigen epitopes resides in the amino-terminal half of each of the Fab regions. This half of the Fab has an amino acid sequence that varies from one immunoglobulin to the next, and is thus designated the **variable region.** Both the light chain and the heavy chain have a variable region.

The amino acid sequence of the remainder of the immunoglobulin is relatively constant from one immunoglobulin to the next and is thus designated the **constant region** (figure 51.12*c*). Both light and heavy chains also exhibit constant regions. Careful

analysis shows that light-chain constant regions of mammalian immunoglobulins consist of two different sequences, designated κ (kappa) and λ (lambda), of apparently equivalent function. The heavy-chain constant regions consist of five different sequences, named μ (mu), δ (delta), γ (gamma), α (alpha), and ε (epsilon). Each of these heavy chains, when bound to either type of light chain, gives rise to a particular class of immunoglobulin, IgM, IgD, IgG, IgA, and IgE, respectively.

Binding of antibody with antigen

The variable regions of the heavy and light chains fold together to form a sort of cleft, the **antigen-binding site** (see figure 51.12). The size and shape of the antigen-binding site, as well as which amino acids line its surface, determine the specificity of each immunoglobulin for an antigen epitope.

Because each immunoglobulin is composed of two identical halves, each immunoglobulin can bind two identical epitopes, although not generally on the same antigen because of steric (shape) constraints. This ability to bind

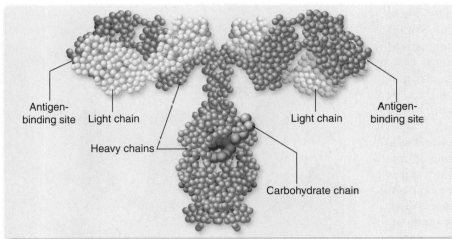

Antigen-binding site   Light chain   Light chain   Antigen-binding site

Heavy chains

Carbohydrate chain

a.

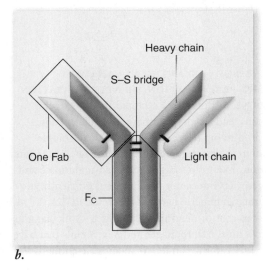

Heavy chain

S–S bridge

One Fab

Fₒ

Light chain

b.

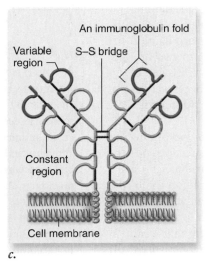

An immunoglobulin fold

Variable region   S–S bridge

Constant region

Cell membrane

c.

figure 51.12

**THE STRUCTURE OF AN IMMUNOGLOBULIN MOLECULE. *a.* In this molecular model of an immunoglobulin (Ig) molecule, each amino acid in the protein chains is represented by a small sphere. The molecule consists of two heavy chains (*brown*) and two light chains (*yellow*). The four chains wind about one another to form a Y shape, with two identical antigen-binding sites at the arms of the Y, the Fab regions, and a stem region, or Fc region. The two Fab regions are joined to the Fc region by a flexible hinge. *b.* A more schematic depiction of an Ig molecule showing heavy chains (*brown*) and light chains (*yellow*) as rods. The two identical halves of the molecule are joined by disulfide bonds (*red*) as are the heavy and light chains of each half. *c.* Ig molecule shown as a membrane protein. This depiction highlights the domain structure of heavy and light chains. Each chain is arranged into a series of domains, each of about 110 amino acids, which include a motif called an immunoglobulin fold. These immunoglobulin fold domains are represented as loops. The globular structure of these domains is maintained by disulfide bonds (*red*). The amino-terminal half of each Fab is a variable region (*blue*) that binds to an epitope and the remainder of the molecule is the constant region.

with two epitopes allows the formation of antigen–antibody complexes (figure 51.13*a*).

Function of antibody classes: The constant region

Although the specificity of each immunoglobulin is determined by its variable region, the function of the immunoglobulin depends on its class, as determined by the heavy-chain constant region, and particularly the F_C portion of the constant region.

Many cells have F_C receptors that can bind the F_C region of a particular class of immunoglobulin. Therefore, when an immunoglobulin binds to an antigen through its antigen-binding site, another cell, such as a phagocytic cell, may be brought close to the antigen by binding to the F_C region of the immunoglobulin (figure 51.13*c*). This binding of antigen–antibody complex to F_C receptors can also activate these cells. In this way, specific immunoglobulins can promote the interaction of nonspecific cells with the antigen, generally resulting in the elimination of the antigen.

The five classes of immunoglobulins have different functions

The five classes of antibodies are based on the sequence and structure of the constant regions of their heavy chains. These five classes have different functions in the protection of an individual. Characteristics of the different classes are summarized in table 51.3 and are described in the following sections.

Keep in mind that antibodies don't kill invading pathogens directly; rather, they cause destruction of pathogens by targeting them for attack by other, nonspecific cells or by activating the complement system.

IgM is a receptor on the surface of all mature, naive B cells and is the first type of antibody to be secreted during an immune response. Although IgM in the membrane of a B cell is monomeric in form, it is secreted as a pentamer (five units) of about 900,000 kDa. Its large size restricts it to the circulation, but its pentameric form means that it very efficiently promotes agglutination of larger antigens (figure 51.13*a*) and precipitation of soluble antigens (figure 51.13*b*). IgM bound to an antigen also activates a complement protein cascade, triggered by the binding of certain complement proteins to the exposed F_C ends.

IgD is also present, along with IgM, on mature naive B cells. The B cells can be activated by cross-linking of two IgD molecules, although under normal circumstances this class of immunoglobulin is not secreted by the cells. On B-cell activation, IgD is no longer displayed on the cell surface. Other roles for IgD remain elusive.

IgG is the major form of antibody in the blood plasma and in most tissues, making up about 75% of plasma antibodies. It is the most common form of antibody produced in a secondary immune response (any response triggered on a subsequent exposure to an antigen). IgG can bind to an antigen in such quantity that the antigen—a virus, bacterium, or bacterially derived toxin—is said to be neutralized, meaning that it can no longer bind to the host. Macrophages and neutrophils have F_C receptors that bind to IgGs bound to antigens, and in this

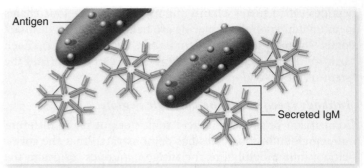

a.

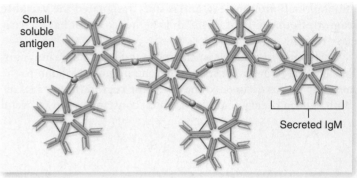

b.

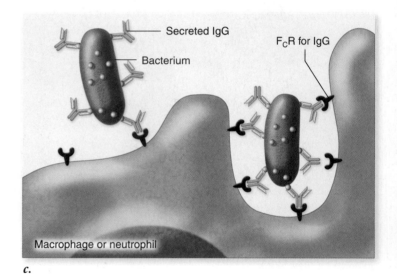

c.

figure 51.13

BINDING OF ANTIBODY TO ANTIGENS CAN CAUSE AGGLUTINATION, PRECIPITATION, OR NEUTRALIZATION OF THE ANTIGENS. *a.* Binding of secreted IgM to larger particulate antigens leads to the clumping, or agglutination, of the antigens. *b.* Binding of secreted IgM to small soluble antigens can lead to their precipitation. Secreted IgG, due to its high concentration, 75% of plasma Ig, can also agglutinate and precipitate antigens. IgG does not precipitate as efficiently as IgM because of the pentameric nature of secreted IgM.
c. Secreted IgG can coat or neutralize an antigen by blocking its ability to bind to a host. Macrophages and neutrophils that have F_C receptors for IgG can thus attach to an antigen–antibody complex, which they will then phagocytose and destroy.

way IgG binding or coating of antigens facilitates their elimination by phagocytosis (figure 51.13c). IgG is also important in providing passive immunity to a fetus; it readily crosses the placenta from the mother. Finally, IgG can also activate complement, although not as efficiently as IgM, leading to pathogen elimination.

IgA is the major form of antibody in external secretions, such as saliva, tears, and the mucus that coats the gastrointestinal tract, bronchi, and genitourinary tract. IgA plays a major role in protection of these surfaces; it is usually secreted as a dimer. The many plasma cells present in the MALT, under the mucosal surfaces, secrete IgA that crosses the epithelial cells to the lumen of these tracts; here it can bind and neutralize antigens. Additionally, any pathogen that passes through a mucosal surface becomes bound to IgA because it is secreted by cells in follicles under that surface. The bound IgA crosses the epithelial cells into the lumen, taking the pathogen with it. The pathogen can then be eliminated by nonspecific defenses. IgA

also provides passive immunity to a nursing infant since it is present in mother's milk.

IgE is present at very low concentration in the plasma. On secretion, most becomes bound to mast cells and basophils that recognize the F_C portion of IgE. As described later, binding of certain normally harmless antigens to IgE molecules bound to mast cells and basophils produces the symptoms of allergy, such as the runny nose and itchy eyes of hay fever. IgE is also often secreted in response to an infection by helminth worms. In this instance, secreted IgE binds to epitopes on the worms, and is then recognized by F_C receptors on eosinophils. The eosinophils generally kill the worms by secreting digestive enzymes through perforin pores into the worms.

Immunoglobulin diversity is generated through DNA rearrangement

The vertebrate immune system is capable of recognizing as foreign virtually any nonself molecule presented to it. It is estimated that human or mouse B cells can generate antibodies with over 10^{10} different antigen-binding sites. Although an individual probably does not have antibodies specific to all epitopes of an antigen, it is fairly certain that antibodies will recognize some of the epitopes, which is all that is required to generate an effective immune response. How do vertebrates generate such diversity of antigen recognition?

The answer lies in the unusual genetics of the variable region. This region in each chain of an immunoglobulin is not encoded by one single stretch of DNA but rather is assembled by joining two or three separate DNA segments together to produce the variable region. This process is called **DNA rearrangement** and is similar to the crossing over that occurs during meiosis (see chapter 11) with two key differences: It occurs between loci on the same chromosome and it is site-specific.

DNA rearrangement occurs as a progenitor B cell matures in the bone marrow. After DNA rearrangement, RNA transcription produces an mRNA that can be translated into either a heavy- or a light-chain immunoglobulin polypeptide, depending on the locus transcribed.

Cells contain homologous pairs of chromosomes, but DNA rearrangement occurs for the heavy-chain and light-chain loci on only one homologue, a process referred to as *allelic exclusion*. Thus, each B cell makes immunoglobulins of only one specificity.

Variable region DNA rearrangements

Sequencing of human immunoglobulin heavy-chain gene loci from several different individuals shows that the locus contains a cluster of approximately 50 sequential DNA segments, termed *V segments*, followed by a cluster of approximately 30 smaller segments, *D segments*, and finally by another cluster of 6 smaller segments, *J segments*. Each V segment is approximately the same size as any other, but they are all of different nucleotide sequence and thus encode different amino acids; the situation is similar for the D and the J segments.

The first DNA rearrangement during B-cell maturation is a site-specific recombination event joining one of the D

TABLE 51.3	Five Classes of Immunoglobulins	
Class		**Function**
IgM	Pentamer	First antibody secreted during the primary immune response; promotes agglutination and precipitation reactions and activates complement
IgD	Monomer	Present only on surfaces of B cells; serves as antigen receptor
IgG	Monomer	Major antibody secreted during the secondary response; neutralizes antigens and promotes their phagocytosis and activates complement
IgA	Dimer	Most abundant form of antibody in body secretions; high density of IgA-secreting plasma cells in the MALT
IgE	Monomer	F_C binds to mast cells and basophils; allergen binding to V regions promotes the release of mediators, which triggers allergic reactions

segments to one of the J segments (figure 51.14). Recombination between two sites on the same chromosome results in the deletion of the intervening DNA, which is subsequently degraded.

This is followed by another site-specific recombination joining a V segment to the rearranged DJ, with the deletion of all the intervening DNA. Which V, which D, and which J are chosen by any cell appears to be completely random.

Because of the many combinations of V, D, and J that can be formed, one can calculate the generation of about 9000 different heavy-chain variable-region sequences. A similar situation occurs for light-chain variable region formation, except that each light-chain variable region is encoded by only a V segment and a J segment.

Other processes contribute even further to the diversity of variable region sequence. As the DNA segments are joined to each other, a few nucleotides may be added to or deleted from the end of each segment, and this is generally followed by somewhat imprecise joining of the segments to each other, resulting in a shift of the reading frame. Finally, a B cell may end up expressing any heavy-chain variable region with any light-chain variable region during its maturation. Taking all these processes into account has allowed the estimate of more than 10^{10} possible variable regions.

Transcription and translation

After the DNA rearrangements that encode the variable region are complete, pre-mRNA transcripts are formed with 5′ ends that begin at the rearranged variable region-encoding segments and continue through constant region exons. More specifically, heavy-chain-encoding pre-mRNA transcripts start at the rearranged

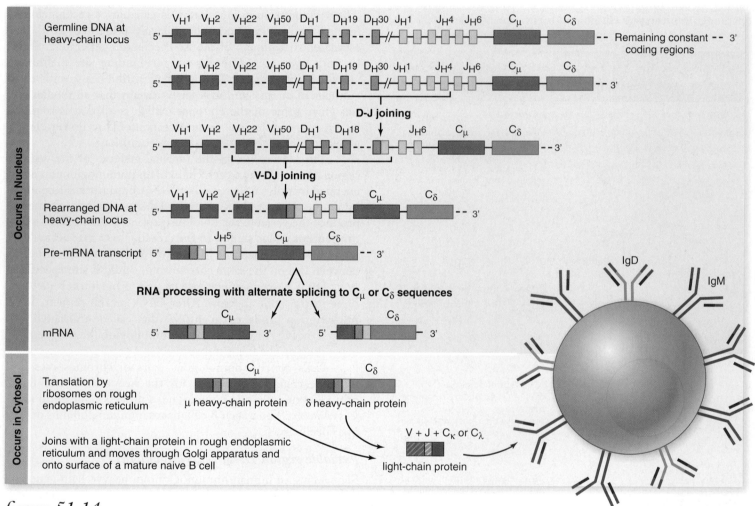

figure 51.14

IMMUNOGLOBULIN MOLECULES ARE ENCODED BY REARRANGEMENT OF SEGMENTS OF DNA. An immunoglobulin protein is encoded by different segments of DNA, a V (variable) segment, a D (diversity) segment, a J (joining) segment, and a constant region. These segments are joined by a precise sequence of DNA rearrangements during maturation in the bone marrow. This process first joins a D segment to a J segment, then this combined DJ segment to a V segment. Other cells will select other V, D, and J segments, contributing to the vast diversity of specific immune responses. Pre-mRNA transcripts start at the rearranged VDJ and continue through constant region exons. The extra downstream J regions are removed during preRNA splicing, which joins the variable region to either a μ or a δ constant region. These transcripts are translated by ribosomes on the rough endoplasmic reticulum to produce heavy-chain polypeptides that join with light chains (encoded by a V, a J, and a C) in the rough endoplasmic reticulum. These proteins are transported to the cell surface, resulting in expression by a mature naive B cell of both IgM (μ constant region) and IgD (δ constant region), both with the same variable region and thus the same antigen specificity.

VDJ and continue through exons encoding μ and δ constant regions (see figure 51.14).

Alternate splicing of these RNA transcripts removes any extra J segments that remain 3′ to the rearranged VDJ, as well as either δ or μ sequences, resulting in transcripts that all encode the same variable region but either μ or δ constant region exons, respectively. Translation results in a μ or δ heavy-chain polypeptide, which associates with a light-chain polypeptide in the rough endoplasmic reticulum. Thus, the mature naive B cell expresses both IgM and IgD on its surface, both having the same antigen-binding specificity (see figure 51.14).

T-cell receptors

At this point, we must briefly return to T-cell receptors and examine their similarity to the immunoglobulins of B cells. The structure of a TCR is essentially like a single Fab region of an immunoglobulin molecule (figure 51.15).

The TCR is a dimeric (two-chain) protein, with about 95% of TCRs composed of an α chain and a β chain. The amino terminal halves of the two polypeptides are the variable domains that bind to self-MHC plus peptide, and the membrane-proximal halves are the constant domains of each polypeptide. The TCR variable-region gene loci also contain multiple DNA segments—V, D, and J, or only V and J—that are joined by the same enzymes and in a similar fashion to the immunoglobulin gene segments. This similarity in structure and DNA rearrangements produces similar diversity of TCRs as immunoglobulins.

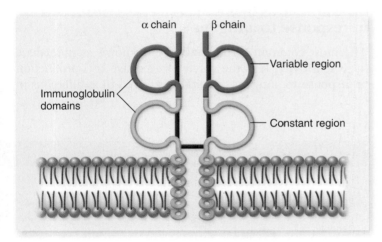

figure 51.15

THE STRUCTURE OF A TCR IS SIMILAR TO AN IMMUNOGLOBULIN FAB. TCRs are composed of two chains, generally α and β, joined by a disulfide bond (*red*), and each composed of two immunoglobulin domains as in an Ig Fab. The amino terminal domain of each chain is the variable region that binds to an MHC–peptide complex, and the membrane proximal domain is the constant region. Unlike Igs, TCRs are not secreted.

inquiry

? *What does the common structure and mechanism of formation of Igs and TCRs suggest about the evolution of B and T lymphocytes and these proteins?*

The secondary response to an antigen is more effective than the primary response

When a particular antigen enters the body, it must, by chance, encounter naive lymphocytes with the appropriate receptors to provoke an immune response. The first time a pathogen invades the body only a few B or T cells may exist with receptors able to recognize the pathogen's epitopes, or for infected or otherwise abnormal cells, foreign peptides bound to self-MHC. Thus, in this first encounter, a person develops symptoms of illness because only a few cells are present that can mount an immune response. Clonal expansion of T and B cells occurs, as well as secretion of IgM antibodies (see figure 51.16).

Because a clone of many memory cells develops during the primary response, the next time the body is invaded by the same pathogen, the immune system is ready. Memory cells are more rapidly activated than are naive lymphocytes, so a secondary immune response both initiates and peaks much more rapidly than a primary response. Further clonal expansion takes place, along with the secretion of large amounts of antibodies that are generally of the IgG class, although IgA and IgE are also possible (see figure 51.16). The class of immunoglobulin produced is dictated by the identity of the cytokines, derived from activated T_H memory cells, that bind to the B cells during their secondary response.

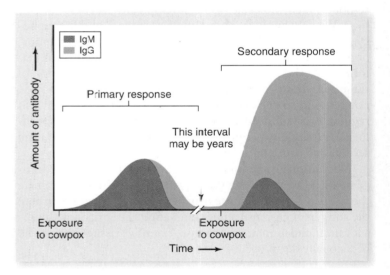

figure 51.16

THE DEVELOPMENT OF ACTIVE IMMUNITY. Immunity to smallpox in Jenner's patients occurred because their inoculation with cowpox stimulated the development of lymphocyte clones, including memory cells, with receptors that could bind not only to cowpox but also to smallpox antigens. A second exposure stimulates the memory cells to produce large amounts of antibody of the same specificity and much more rapidly than during the primary immunization. The first antibodies produced during the primary response are IgM in class (*red*) although IgG (*blue*) is secreted near the end of the primary response. The majority of the antibody secreted during a secondary response is IgG, although IgA could be secreted if the antigen has activated B cells in the MALT, or in some circumstances, such as allergies, IgE is secreted.

It is advantageous for an individual to produce immuno-globulins of different classes during an immune response because each class has a different function. During a second exposure to the same antigen, while memory cells are activated and secrete isotypes other than IgM, other naive B cells also recognize the antigen for the first time, become activated, and secrete IgM.

Memory cells can survive for several decades, which is why people rarely contract chicken pox a second time after they have had it once or been vaccinated against it. The vaccine triggers a primary response, so that if the actual pathogen is encountered later, a large and rapid secondary response occurs and stops the infection before disease symptoms are even detected. The viruses causing childhood diseases have surface antigens that change little from year to year, so the same antibody is effective for decades.

Antibodies produced by B cells have variable regions by which they recognize and bind antigen. The variable regions are encoded by the joining of distinct DNA segments, which provides diversity of antigen recognition. Each antibody also has one of five kinds of constant region that determines its function. Antibody binding to an antigen may thus lead to its elimination in different ways. Vaccination artificially presents an antigen to elicit the primary immune response; when a pathogen carrying this antigen is encountered later, the secondary response eliminates it quickly.

51.6 Autoimmunity and Hypersensitivity

Sometimes the immune system is the cause of disease rather than the cure. Inappropriate responses to self-antigens may occur, as well as inappropriate or greatly heightened responses to foreign antigens, which, in turn, causes tissue damage.

A mature animal's immune system normally does not respond to that animal's own tissue. This acceptance of self cells is known as **immunological tolerance.** The immune system of a fetus undergoes the process of tolerance to lose the ability to respond to self-molecules as its development proceeds.

We now know that not all self-reactive T and B lymphocytes undergo apoptosis during selection in primary lymphoid organs. Normal healthy individuals are known to possess mature, potentially self-reactive lymphocytes. The activity of these cells, however, is regulated or suppressed so that they do not respond to the self-antigens they encounter. When this regulation or suppression breaks down, then humoral or cell-mediated responses can occur against self-antigens, causing serious and sometimes fatal disease.

Additionally, an immune response against a foreign antigen may be a greater one than is actually required to eliminate the antigen, or the response may be seemingly inappropriate to the antigen. Thus, instead of eliminating the antigen with only a localized inflammatory response, extensive tissue damage and occasionally death occurs.

Autoimmune diseases result from immune system attack on the body's own tissues

Autoimmune diseases are produced by the failure of immunological tolerance. Autoreactive T cells become activated, and autoreactive B cells produce autoantibodies, causing inflammation and organ damage. There are over 40 known or suspected autoimmune diseases, and they affect 5–7% of the population. For reasons that are not understood, two-thirds of the people with autoimmune diseases are women.

Autoimmune diseases can result from a variety of mechanisms. For example, the self-antigen may normally be hidden from the immune system; if later it is exposed, the immune system may treat it as foreign. This happens, for example, when a protein normally trapped in the thyroid follicles triggers autoimmune destruction of the thyroid (Hashimoto thyroiditis). It also occurs in sympathetic opthalmia in which antigens are released from the eye.

Because the immune attack triggers inflammation, and inflammation causes tissue and organ damage, the immune system must be suppressed to alleviate the symptoms of autoimmune diseases. Immune suppression is generally accomplished by administering corticosteroids and nonsteroidal antiinflammatory drugs, including aspirin.

Allergies are caused by IgE secretion in response to antigens

The most common form of allergy is known as immediate hypersensitivity. It is the result of excessive IgE production in response to antigens, generally referred to as allergens in

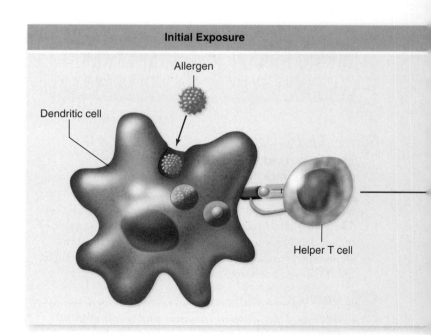

Initial Exposure

Allergen

Dendritic cell

Helper T cell

this context. Allergens that provoke immediate hypersensitivity include various foods, the venom in insect stings, molds, animal danders, and pollen grains. The most common allergy of this type is seasonal hay fever, which may be provoked by ragweed (*Ambrosia* spp.) or other pollen. Allergies have earned the designation "immediate" because a response to an allergen occurs within seconds or minutes.

The first time or even the first few times that one encounters an allergen, the allergen binds to and activates B cells, which start to secrete allergen-specific immunoglobulins. Activated T_H cells release cytokines such as IL-4, which bind to the B cells and dictate that the antibodies secreted should be IgE. The B cells rapidly switch from the more common IgG secretion to IgE secretion.

Unlike IgG, IgE rapidly binds to mast cells and basophils. When the individual is again exposed to the same allergen, the allergen now specifically binds to the exposed variable regions of identical IgE molecules attached to mast cells and basophils. This binding triggers these cells to secrete histamine, prostaglandins, and other chemical mediators, which produce the symptoms of allergy (figure 51.17).

In **systemic anaphylaxis,** the allergic reaction is severe and potentially life-threatening because of the rapid inflammatory response and release of chemical mediators. The individual experiences a tremendous drop in blood pressure; swelling of the epiglottis can block the trachea, and bronchial constriction can prevent the exit of air from the lungs. This combination of effects is referred to as **anaphylactic shock.** Death can result within 20–30 min without prompt medical treatment.

Fortunately, most people with allergies only experience **local anaphylaxis,** such as the itchy welts from hives, or the respiratory constriction of mild asthma. Diarrhea from response to food allergens is another form of local anaphylaxis.

Allergies have traditionally been treated with antihistamines that prevent histamine released by mast cells from binding to its receptor. More recently, a variety of drugs have been developed that block the activation of mast cells and basophils, so that they do not release their mediators. Hyposensitization treatment is another alternative; it consists of the injection, over several months, of an increasing concentration of the allergens to which one is allergic. In some individuals, particularly those with allergic rhinitis (runny nose and eyes) or asthma, this treatment seems to cause a preferential secretion of IgG rather than IgE, and allergy symptoms diminish over time.

Delayed-type hypersensitivity is mediated by T_H cells and macrophages

Delayed-type hypersensitivity, which is mediated by T_H cells and macrophages, produces symptoms within about 48 hr after a second exposure to an antigen. (A first exposure causes a much slower primary response, such as described earlier, frequently without the manifestation of any symptoms.) A form of delayed-type hypersensitivity is contact dermatitis, caused by such varied materials as poison ivy, nickel in jewelry, and some cosmetics. After contact with poison ivy, oils that enter the skin complex with skin proteins, causing the proteins to appear foreign. A delayed-type hypersensitivity response requires that antigen entering the body travel to a secondary lymphoid organ,

figure 51.17

AN ALLERGIC RESPONSE. On initial exposure to an allergen, B cells are activated to secrete antibodies of the IgE class. These antibodies are in very low levels in the plasma but rapidly bind to F_C receptors on mast cells or basophils. On subsequent exposure to allergen, the allergen cross-links variable regions of two neighboring IgEs with the same epitope specificity on mast cells and basophils. This induces the cells to release histamine and other mediators of inflammation that will cause the symptoms of allergy.

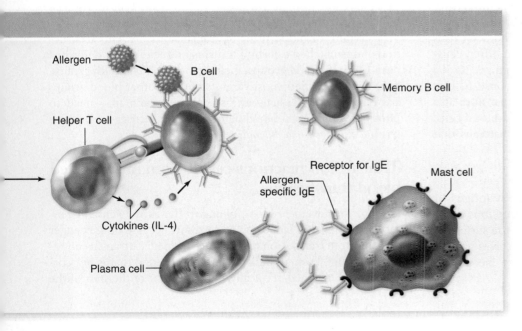

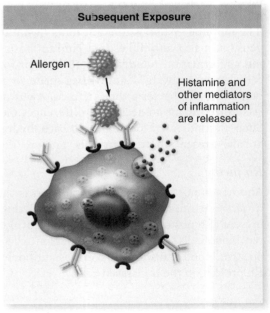

generally lymph nodes, where T_H cells can be activated. These activated T_H cells then recirculate around the body, and on encountering macrophages that have ingested the antigen, they release cytokines that activate the macrophages. This induces the macrophages to release other cytokines, and in the case of poison ivy, itchy welts on the skin erupt. The time required for the activation of T_H cells and then of macrophages is the reason for the "delayed" response to the antigens.

Autoimmunity, allergies (immediate hypersensitivity), and delayed hypersensitivity are all examples of inappropriate or heightened immune responses. Autoimmunity results from a loss of self-tolerance. Allergies are associated with a rapid response from mast cells when allergen binds to IgE on their membrane. Delayed hypersensitivity to pathogens or irritants such as poison ivy is mediated by T_H cells and macrophages.

51.7 Antibodies in Medical Treatment and Diagnosis

The vertebrate immune system can have a range of effects on medical treatment of disease. As two examples, we discuss blood type and its effect on transfusion, as well as the use of monoclonal antibodies for diagnosis and treatment.

Blood type indicates the antigens present on an individual's red blood cells

A person's blood type is determined by certain antigens found on the red blood cell surface. These antigens are clinically important because they must be matched between donors and recipients during blood transfusions.

ABO groups

In chapter 12, you learned about the genetic basis of the human *ABO blood groups*. To review, three alleles are possible at a single gene locus, namely I^A, I^B, and i. The enzyme produced by these alleles adds or does not add a sugar molecule to a protein found in the membrane of red blood cells. The I^A allele adds galactosamine, I^B adds galactose, and i does not add any sugar. Each individual has two alleles, so a person may be blood type A ($I^A I^A$ or $I^A i$), type B ($I^B I^B$ or $I^B i$), type AB ($I^A I^B$), or type O ($i\,i$).

The protein–sugar complex on the surface of the red blood cells acts as an antigen, and these antigens differ with regard to the sugar present (or absent, in the case of type O). The immune system is tolerant to its own red blood cell antigens but makes antibodies that bind to those that differ, causing agglutination (clumping) and lysis of foreign red blood cells. Apparently, IgM antibodies made in response to carbohydrates on bacteria that are part of our normal flora also recognize the monosaccharide differences on red blood cells. Such antibodies are not made to carbohydrate patterns that are also present on our own cells.

Rh factor

Another important blood-borne antigen is the Rh antigen or *Rh factor*. Again, as described in chapter 12, this protein is either present (Rh positive) or absent (Rh negative) on the surface of red blood cells. An Rh-negative person who receives an Rh-positive blood transfusion produces antibodies to the foreign Rh protein on the transfused cells.

An additional complication occurs when Rh-negative mothers carry Rh-positive fetuses, which may result in the infant exhibiting hemolytic disease of newborns (HDN; chapter 12). A first child is usually not harmed; however, at the time of the first birth, the Rh-negative mother's immune system may be exposed to fetal blood. As a result, the woman may become sensitized and produce antibodies and memory B cells against the Rh antigen. If any exposure to fetal blood occurs during a subsequent pregnancy, IgG antibodies, secreted on activation of these memory cells, can cross the placenta and cause destruction of the red blood cells of the fetus.

Blood is typed by agglutination

Blood typing is done by taking advantage of the circulating IgM antibodies, which are produced against foreign blood antigens but not against self. If type A blood is mixed with serum from a person with type B or type O blood, the anti-A antibodies in the serum cause the type A red blood cells to agglutinate. This does not happen if type A blood is mixed with serum from another type A individual or from a type AB individual.

Similarly, if serum from an Rh-negative individual is added to red blood cells, agglutination of the red blood cells indicates that they came from an Rh-positive individual. This individual's blood would not be an appropriate match for transfusion.

Typing of blood prior to transfusions prevents destruction of mismatched cells by a transfusion recipient, as described next. Over 20 blood groups, including the ABO and Rh groups, have been identified; most variants of these other blood groups are rare, but individuals at risk for mismatch may need to "stockpile" their own blood before elective surgery—a practice termed *autologous blood donation*.

Transfusion reactions result from mismatched blood transfusions

Prior to the advent of blood typing in the early twentieth century, transfusion of blood was a last resort because of the danger of death from transfusion reaction. An immediate transfusion reaction occurs when an individual receives blood that is not correctly ABO matched. Typically, within 5–8 hr of the start of the transfusion, tremendous intravascular hemolysis (rupture) of the transfused red blood cells is detected. This rupture is a result of IgM binding to foreign antigens and activating the complement system. The result is the formation of MACs in the red blood cell membranes and rapid osmotic lysis of the cells.

The hemoglobin released from the red blood cells is converted to a molecule called *bilirubin*, which is particularly toxic to cells and can cause severe organ damage, especially to the kidneys. The major treatment in such situations is to discontinue the transfusion immediately and to administer large amounts of intravenous fluids to "wash" the bilirubin from the body.

Monoclonal antibodies are a valuable tool for diagnosis and treatment

Antibodies to a known antigen may be obtained by chemically purifying an antigen and then injecting it into a laboratory animal (vertebrate). Periodic bleeding of the animal after a few immunizations allows the isolation of serum antibodies against the antigen. But because an antigen typically has many different epitopes, the antibodies obtained by this method are *polyclonal*, that is, they are secreted by B-cell clones with many different specificities. Their polyclonal nature decreases their sensitivity to any one particular epitope, and it may result in some degree of cross-reactivity with closely related epitopes of different antigens.

Monoclonal antibodies, by contrast, exhibit specificity for one epitope only. In the preparation of monoclonal antibodies, an animal, generally a mouse, is immunized several times with an antigen and is subsequently killed. B lymphocytes, many of which should now be specific for epitopes of the antigen, are collected from the animal's spleen. These B cells would soon die, but utilizing a technique first described in 1975, they are fused with cancerous multiple myeloma cells. These myeloma cells have all the characteristics of plasma cells, except for the secretion of immunoglobulins—but more importantly, they are immortal, meaning they will divide indefinitely. The outcome of a B-cell/myeloma cell fusion is a **clonal hybrid,** or **hybridoma,** that can divide indefinitely and that continues to secrete a large quantity of identical, monoclonal antibodies of the specificity produced by a single B cell (figure 51.18).

Monoclonal antibodies and diagnostic testing

The availability of large quantities of pure monoclonal antibodies has allowed the development of much more sensitive clinical laboratory tests. Some pregnancy tests, for example, use a monoclonal antibody produced against the hormone human chorionic gonadotrophin (HCG, secreted early in pregnancy). The test uses HCG-coated latex particles that are exposed to a urine sample and anti-HCG antibody. If the urine contains HCG, it will block binding of the antibody to the HCG-coated particles and prevent their agglutination, indicating pregnancy based on the presence of HCG (figure 51.19).

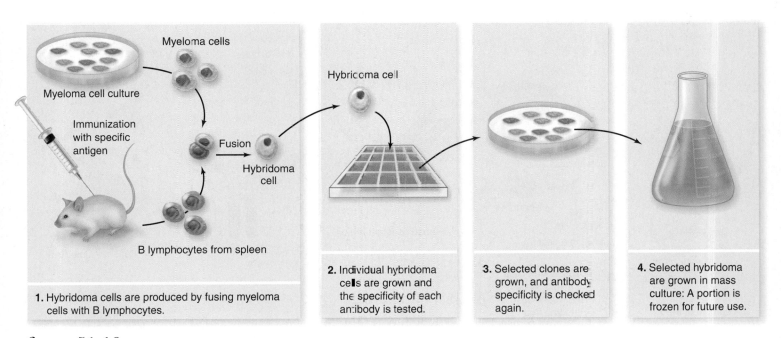

figure 51.18

THE PRODUCTION OF MONOCLONAL ANTIBODIES. These antibodies are of a single specificity and are produced by "hybridoma" cells. These result from the fusion of a B cell, specific for a particular antigenic determinant, with a myeloma cell, a B-cell tumor, that no longer secretes Ig but provides immortality to the fusion. After hybridoma production, the antibody produced by each hybridoma is tested to see whether it produces specific antibodies against the desired antigen. Selected hybridomas are grown in mass culture for antibody production and are frozen for future use.

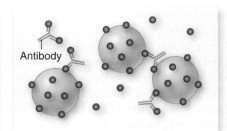
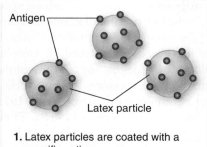
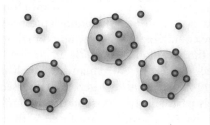

1. Latex particles are coated with a specific antigen.

2. A sample to be tested for presence of the same antigen is added.

3. An antibody against the antigen is added. Concentration of antigen in the sample alters the degree of agglutination of particles.

figure 51.19

USING MONOCLONAL ANTIBODIES TO DETECT AN ANTIGEN. Many clinical tests, such as pregnancy testing, use monoclonal antibodies. A specific antigen is attached to latex beads that are mixed with the test sample and a monoclonal antibody specific for the antigen. If no antigen is present in the sample, the antibody will cause agglutination of the beads. If the sample contains the antigen, it will bind to the antibodies and prevent agglutination of the beads by the antibody.

 inquiry

How would a high level of HCG present in a urine sample be indicated in this agglutination test?

Acquired immune deficiency syndrome (AIDS) is characterized in part by destruction of T_H cells. The progression of this disease can be monitored by examining the reactivity of a patient's leukocytes with a monoclonal antibody against CD4, a marker of T_H cells, to track a decrease in the number of these cells.

Monoclonal antibodies and cancer treatment
Scientists utilize a variety of techniques to kill tumor cells in individuals who develop cancer. One of these techniques is to create monoclonal antibodies specific to tumor cells isolated from a patient, attach a toxin to the antibodies, and then inject this **immunotoxin** into the patient. The variable

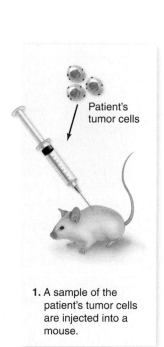
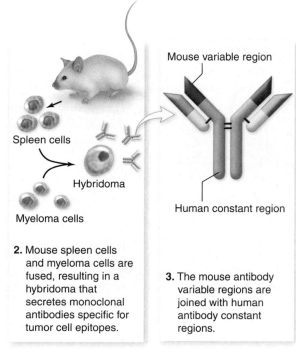
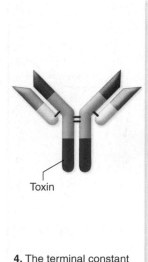
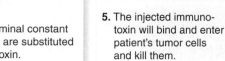

1. A sample of the patient's tumor cells are injected into a mouse.

2. Mouse spleen cells and myeloma cells are fused, resulting in a hybridoma that secretes monoclonal antibodies specific for tumor cell epitopes.

3. The mouse antibody variable regions are joined with human antibody constant regions.

4. The terminal constant regions are substituted with a toxin.

5. The injected immunotoxin will bind and enter patient's tumor cells and kill them.

Mouse variable region

Human constant region

Toxin

Patient's tumor cell

figure 51.20

IMMUNOTOXIN PRODUCTION. Tumor-specific immunotoxins may be formed by combining mouse Ig variable regions with a partial human Ig constant region and a toxin that inhibits protein synthesis. The lack of a complete F_C region prevents the immunotoxin from binding to inappropriate cells bearing F_C receptors.

regions of the antibody direct its binding to only the tumor cells, and the attached immunotoxin conjugate then enters the tumor cells by receptor-mediated endocytosis. The toxin then kills the cells.

Antibodies to human tumor cells are best raised in another species, such as mice—but they could stimulate an immune response in the person they are designed to help. Through molecular biology techniques, immunotoxins can be designed such that the mouse immunoglobulin variable regions that recognize the tumors are joined to human constant regions. The toxin, in turn, is attached to the human constant region (figure 51.20). These antibodies are less likely to elicit an immune system response. Clinical trials in patients with such cancers

as leukemia and lymphoma have shown promise, and further experiments are designed to improve the safety and efficacy of the immunotoxins.

> The presence in the plasma of blood group antibodies for blood types different from the individual's own type made blood transfusion risky and often fatal in the past. Blood typing has improved the safety of blood transfusions and has allowed prompt treatment of Rh-positive babies born to Rh-negative mothers. The development of hybridoma technology allows the production of large quantities of monoclonal antibodies that are used in both diagnostic tests and in the elimination of tumors.

51.8 Pathogens That Evade the Immune System

For any pathogen to establish itself in a host and to cause a productive infection in which the pathogen successfully reproduces, the pathogen must evade both the nonspecific and specific immune systems. Many pathogens can alter the structure of their surface antigens so that they are no longer recognized. In such a case, the immune response selects for mutated pathogens, which survive and continue to cause infection. Other pathogens have simply evolved ways to evade intracellular destruction. Infection by still other pathogens can actually cause the death of cells of the immune system.

Many pathogens change surface antigens to avoid immune system detection

Influenza virus is perhaps the most universally known example of an organism or virus altering its surface antigens and thus avoiding immune system recognition and destruction. Because of this tendency of the virus to change, yearly immunizations against influenza virus are recommended.

The two viral proteins expressed on the influenza virus' envelope, as well as on the surface of cells infected by influenza virus, are hemaglutinin (HA) and neuraminidase (NA). Because this virus has an RNA genome, it is replicated by a viral RNA polymerase that lacks proofreading ability. As a result, mutations are likely to accumulate over time, including point mutations to the HA and NA genes. This is referred to as **antigen drift.**

Even more dramatically but less frequently, the HA and NA proteins may also undergo **antigen shift,** referring to the sudden appearance of a new subtype of influenza virus in which the expressed HA or NA proteins (or both) are completely different. Such a change makes the population particularly susceptible to infection. Immunization with a new vaccine created every year using the most recently isolated strains of the virus attempts to establish immunity in the population prior to infection by these new strains.

Antigen shifting and the resulting lack of immunity is the reason for the recent interest in "bird flu." The subtype

of influenza that causes bird flu is characterized as H5N1, a primarily avian form of influenza to which people have no immunity. There is no evidence as of this writing, however, that the H5N1 virus can be contracted by people except through contact with infected birds.

Many other pathogens can alter or shift their surface antigens in order to avoid immune system destruction. As another example, every year, more than 1 million people, the vast majority of whom are African children under the age of 5, die from malaria. This disease, as described in chapter 29, is caused by the protozoan parasite *Plasmodium* and contracted when humans are bitten by an infected *Anopheles* mosquito. These protozoans have several life-cycle stages and are hidden from the immune system alternately within host hepatocytes or red blood cells. In addition, they can alter some of the proteins expressed during certain life cycle stages. Continued use of certain anti-*Plasmodium* drugs has also selected for the emergence of multidrug-resistant organisms. Work is ongoing to develop a vaccine that would induce effective immunity to specific life cycle stages and that would thus promote immune system elimination of the *Plasmodium*.

inquiry

Why were we able to eliminate smallpox virus using a vaccine but cannot eliminate influenza?

Many mechanisms have evolved in bacteria to evade immune system attack

Salmonella typhimurium, a common cause of food poisoning, can alternate between expression of two different flagellar proteins, so that antibodies made to one protein do not recognize the other protein and therefore cannot be used to promote phagocytosis of the bacteria.

Mycobacterium tuberculosis bacteria, once phagocytosed into macrophages, inhibit fusion of the phagosome with lysosomes. These organisms can then multiply quite successfully within the macrophages.

Other bacteria that invade mucosal surfaces, such as *Neisseria meningitidis* or *Neisseria gonorrhoeae*, secrete proteases that degrade the IgA antibodies that protect the mucosal surface. External capsules on many particularly pathogenic strains of bacteria block binding of the phagocytosis-inducing complement protein C3b, slowing the phagocytosis response. Because bacteria utilizing any of these mechanisms are better able to survive, the immune response acts as selective pressure favoring the evolution of such mechanisms.

HIV infection kills T_H cells and causes immunosuppression

One mechanism for defeating vertebrate defenses is to attack the adaptive immune system itself. CD4$^+$ T_H cells play a central role in the activity of the immune system: The cytokines they secrete directly or indirectly affect the activity of all other cells of the immune system.

HIV, human immunodeficiency virus, mounts a direct attack on T_H cells (see chapter 27). It binds to the CD4 proteins present on these cells and utilizes these proteins to promote endocytosis into the cells. (The virus infects monocytes as well because they too express CD4.) HIV-infected cells die only after releasing replicated viruses that infect other CD4$^+$ cells (figure 51.21). Over time, the number of T_H cells in an infected individual decreases.

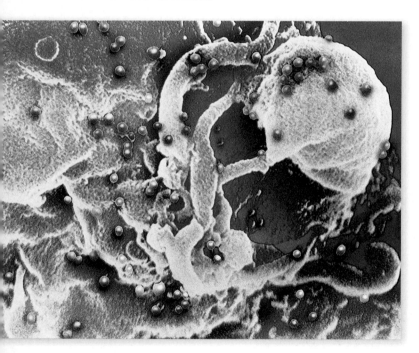

figure 51.21

HIV, THE VIRUS THAT CAUSES AIDS. Viruses released from infected CD4$^+$ T_H cells soon spread to neighboring T_H cells, infecting them in turn. The individual viruses, colored red in this scanning electron micrograph, are extremely small; over 200 million would fit on the period at the end of this sentence.

The individual is considered to have **AIDS** when his or her T_H cell levels have dropped dramatically, leading to the individual experiencing an increase in infections due to opportunistic organisms as well as other diseases.

The progression of HIV infection

The immune system initially controls an HIV infection by the production of antibodies to the virus and by the elimination of virally infected cells by cytotoxic T cells. For a period of time, the level of HIV in the serum does not increase beyond a steady state, and the number of T_H cells does not significantly decrease.

As the virus reproduces in the T_H cells, it rapidly kills some of them, but many others continue to divide on antigen stimulation. Eventually, however, HIV kills T_H cells more rapidly than they can proliferate. HIV-encoded proteins also cause a decrease in the expression of MHC class I on the infected cells, so that these cells are less likely to be recognized and killed by T_C cells.

Finally, because HIV is a retrovirus (see chapter 27), it can integrate itself into the genome of infected cells and hide in a "latent" form. When any of these cells divides, the HIV genome is present in the progeny cells and thus these progeny can start to produce HIV at any point in time

The combined effect of these responses to HIV infection is to wipe out the human immune system. With no defense against infection, any of a variety of otherwise commonplace infections may prove fatal. Death by cancer also becomes far more likely. In fact, AIDS was first recognized as a disease when a cluster of previously healthy young men all died of *Pneumocystis jiroveci*–induced pneumonia, a disease generally of the immunosuppressed, or of Kaposi's sarcoma, a rare form of cancer.

The human effect of HIV

Although HIV became a human disease only recently, AIDS is already clearly one of the most serious diseases in human history. It is estimated that between 35 and 42 million people are living with AIDS, with the greatest number in sub-Saharan Africa (about 25 million), followed by south and southeast Asia (6.5 million).

By the end of 2004, almost 514,000 people had died of AIDS in the United States alone. The fatality rate of AIDS is close to 100%. No patient exhibiting the symptoms of AIDS has been known to survive more than a few years without treatment. The disease is not highly contagious, however; it is transmitted from one individual to another through the transfer of internal body fluids, typically semen and blood.

Pathogens have evolved a variety of ways by which to avoid actions of the immune system. Many pathogens, such as influenza virus, can alter epitopes on their surfaces and thus thwart specific immune recognition. Other pathogens, such as HIV, infect and destroy T_H cells, simply disabling the immune system.

51.1 The Integumentary System: The First Line of Defense

The integument defends the body with a nearly impenetrable barrier reinforced with chemical weapons on the surface.

- Oil and sweat glands give the skin's surface a pH of 3–5, which inhibits growth of many pathogenic microorganisms.
- Sweat also contains the enzyme lysozyme.
- The skin is home to many common flora, nonpathogenic bacteria and fungi that out-compete pathogenic bacteria.
- The epithelia of the digestive, respiratory, and urogenital tracts contain cells that produce mucus to trap microorganisms.

51.2 Nonspecific Immunity: The Second Line of Defense

The body uses a host of nonspecific cellular and chemical defenses that respond to any microbial infection.

- Three leukocyte types are involved in nonspecific immunity: macrophages, neutrophils, and natural killer (NK) cells (figure 51.2).
- Natural killer cells kill by forming pores with perforins, and releasing granzymes that enter the pores induce apoptosis.
- The inflammatory response involves histamines that increase blood flow and permeability of capillaries (figure 51.3).
- The acute-phase response leads to fever, which promotes phagocytic activity and impedes growth of microbes.
- Complement forms pores in invading cells, coats pathogen with C3b proteins, and targets cells for destruction (figure 51.4).
- Interferons prevent viral infections by inducing RNA degradation.

51.3 The Specific Immune Response: The Third Line of Defense

Jenner's work on cowpox and Pasteur's work on avian cholera led to the idea that prior exposure to a disease prevented subsequent infections.

- Antigens provoke a specific immune response (figure 51.5).
- Lymphocytes have receptors on their surface that recognize antigen and direct a specific immune response (figure 51.6).
- When a naive lymphocyte binds to a foreign antigen it divides, producing a clone of activated cells and memory cells (clonal selection).
- Humoral immunity is the production of Ig by B cells.
- Cell-mediated immunity involves T cells that regulate the immune responses of other cells or directly attack cells.
- Immunity can be active, due to pathogen infection, or passive, antibodies from another source entering the body.
- Immune cells are derived from hematopoietic stem cells (table 51.1).
- The organs of the immune system include the primary lymphoid organs and secondary lymphoid organs (figure 51.8).
- B cells recognize any epitope, T cells recognize peptide/MHC complex.
- Lymphocytes likely to bind to self-antigens are destroyed before the remaining cells are released (figure 51.8).

51.4 T Cells: Cell-Mediated Immunity

T cells are identified by surface markers, their ability to recognize MHC markers, and roles after activation (table 51.2).

- T_C eliminate infected cells, and T_H secrete cytokines.
- The MHC is composed of surface glycoproteins, and the proteins encoded by the MHCs are called human leukocyte antigens (HLAs).
- MHC class I proteins are present on every nucleated cell, MHC class II proteins are found only on antigen-presenting cells.
- T_C cells recognize virally infected cells and tumor cells. They destroy cells in a fashion similar to NK cells (figure 51.9).
- T_H cells secrete cytokines in response to exogenous antigens and promote both cell-mediated and humoral immune responses.
- T_H cells respond to foreign antigens brought by antigen-presenting cells or B cells.
- Activated T_C and T_H produce effector and memory cells.
- T cells are the primary cells that mediate transplant rejection.

51.5 B Cells: Humoral Immunity and Antibody Production

Humoral immunity begins when naive B cells in secondary lymphoid organs encounter an antigen and ends with Ig production.

- B cells are activated by membrane Ig molecules binding to a specific epitope on an antigen and receiving required signals.
- Activated B cells produce plasma and memory cells.
- Immunoglobulins consists of two light chain and two longer heavy chain polypeptides (figure 51.12).
- The antigen-binding site is a cleft in the Fab region of an Ig.
- Antibodies can agglutinate, precipitate, or neutralize antigens (figure 51.13).
- The five classes of Ig have different functions (table 51.3).
- Ig diversity is generated by DNA rearrangements (figure 51.14).
- TCR are similar to a single Fab region of an Ig. Their diversity is also caused by DNA rearrangements (figure 51.15).
- The second exposure to a pathogen results in a rapid secondary immune response due to memory cells (figure 51.16).

51.6 Autoimmunity and Hypersensitivity

The immune system can be the cause of disease due to inappropriate responses to self-antigens or heightened responses to foreign antigens.

- The acceptance of self cells is called immunological tolerance.
- Autoimmunity is caused by a failure of immunological tolerance.
- Immediate hypersensitivity is caused by excessive IgE. Allergen binding to IgE triggers the release of histamine (figure 51.17).
- Anaphylaxis is a severe response caused by rapid inflammation and release of chemical mediators.
- Delayed-type hypersensitivity is mediated by T_H cells and macrophage. Symptoms appear about 48 hr after second exposure.

51.7 Antibodies in Medical Treatment and Diagnosis

Immune molecules are used in diagnosis and treatment of diseases.

- Antigens can be used to match tissues between donors and recipients in transplants.
- Monoclonal antibodies exhibit specificity for one epitope and can be used in a variety of tests and to make immunotoxins.

51.8 Pathogen That Evade the Immune System

To cause infection, pathogens must evade specific and nonspecific immune systems.

- Many pathogens exhibit antigen drift and antigen shift, changing their surface antigens to avoid immune detection.
- Some bacteria have evolved mechanisms to avoid immune system attack by inhibiting normal cellular processes.
- HIV kills T_H cells and causes immunosuppression.

SELF TEST

1. You start a new job in a research lab. The lab protocols state that you should check your hands for any breaks in the skin before handling infectious agents. This is because the epidermis fights microbial infections by—
   a. making the surface of the skin acidic.
   b. excreting lysozyme to attack bacteria.
   c. producing mucus to trap microorganisms.
   d. all of these.

2. Cells that target and kill body cells infected by viruses are—
   a. macrophages.
   b. natural killer cells.
   c. monocytes.
   d. neutrophils.

3. Structures on invading cells recognized by the immune system are known as—
   a. antigens.
   b. interleukins.
   c. antibodies.
   d. lymphocytes.

4. Which one of the following acts as the "alarm signal" to activate the body's immune system by stimulating helper T cells?
   a. B cells
   b. Interleukin-1
   c. Complement
   d. Histamines

5. Cytotoxic T cells are called into action by the—
   a. presence of histamine.
   b. presence of interleukin-1.
   c. presence of interleukin-2.
   d. interferon.

6. The primary response to invading microorganisms involves _____ , and the secondary response involves _____ .
   a. IgG/IgA
   b. IgM/IgE
   c. IgE/histamines
   d. IgM/IgG

7. How is your body able to detect millions of different antigens?
   a. The few hundred immunoglobulin genes can be rearranged or can undergo mutations to form millions of antibody molecules.
   b. There are millions of different antibody genes.
   c. The few hundred immunoglobulin genes undergo antigen shifting.
   d. Each B cell has a different set of immunoglobulin genes, and so the activation of different B cells produces different antibodies.

8. If you have type AB blood, which of the following results would be expected?
   a. Your blood agglutinates with anti-A antibodies only.
   b. Your blood agglutinates with anti-B antibodies only.
   c. Your blood agglutinates with both anti-A and anti-B antibodies.
   d. Your blood would not agglutinate with either anti-A or anti-B antibodies.

9. The HIV virus is particularly dangerous because it attacks—
   a. cells with the CD4$^+$ coreceptor.
   b. helper T cells.
   c. 60–80% of circulating T cells in the body.
   d. all of these.

10. Diseases in which the person's immune system no longer recognizes its own MHC proteins are called—
    a. allergies.
    b. autoimmune diseases.
    c. immediate hypersensitivity.
    d. delayed hypersensitivity.

11. Suppose that you get a paper cut while studying. Which of the following responses will occur last?
    a. Injured epidermal cells release histamine.
    b. Bacteria enter the cut.
    c. Helper T cells are activated.
    d. Macrophages engulf bacteria.

12. If you wanted to cure allergies by bioengineering an antibody that would bind and disable the antibody responsible for allergic reactions, which of the following would you target?
    a. IgG
    b. IgA
    c. IgE
    d. IgD

13. Why don't we become immune to influenza viruses?
    a. Because they attack only the helper T cells, thereby suppressing the immune system.
    b. Because they alter their surface proteins and thus avoid recognition.
    c. Because they don't actually generate an immune response, the "flu" is actually an inflammatory response.
    d. Because they are too small to serve as antigens.

14. Suppose that a new disease is discovered that suppresses the immune system. Which of the following would indicate that the disease specifically affects the B cells rather than the helper or cytotoxic T cells?
    a. A decrease in the production of interleukin-2
    b. A decrease in interferon production
    c. A decrease in the number of plasma cells
    d. A decrease in the production of interleukin-1

15. If you wanted to design an artificial cell that could safely carry drugs inside the body. Which of the following molecules would you need to mimic to deter the immune system?
    a. MHC-1
    b. Interleukin-1
    c. Antigen
    d. Complement

CHALLENGE QUESTIONS

1. Suppose you take a job in the marketing department of a cosmetic company. Always seeking a competitive advantage, the vice president of marketing has decided to advertise the new skin lotion as having immune-enhancing effects. The lotion is produced from secretions of a plant that releases extremely watery, alkaline fluid. Explain how you will market this product as an immune-enhancer.

2. Your new kitten scratches your roommate. Her skin is reddened and feels warm and sore to the touch, she thinks she has contracted some kind of fatal infection. In order to deflect her anger (she is definitely not a cat person!), you try telling her about the activities of the nonspecific defense system. Explain what is actually happening to her skin.

3. Some people claim that they never catch colds. How could you show that this is due to a difference in receptors on their respective cell surfaces?

chapter 52

The Reproductive System

introduction

BIRD SONG IN THE SPRING, insects chirping outside the window, frogs croaking in swamps, and wolves howling in a frozen northern forest are all sounds of evolution's essential act, reproduction. These distinctive noises, as well as the bright coloration of some animals, such as the tropical golden toads in the figure, function to attract mates. Few subjects pervade our everyday thinking more than sex, and few urges are more insistent. This chapter deals with sex and reproduction among the vertebrates, including humans.

concept outline

52.1 Animal Reproductive Strategies

- *Some species have developed novel reproductive methods*
- *Sex determination in mammals occurs in the embryo*

52.2 Vertebrate Fertilization and Development

- *Internal fertilization has led to three strategies for development of offspring*
- *Most fishes and amphibians have external fertilization*
- *Reptiles and birds have internal fertilization and lay eggs*
- *Mammals generally do not lay eggs, but give birth to their young*

52.3 Structure and Function of the Human Male Reproductive System

- *Sperm cells are produced by the millions*

- *Male accessory sex organs aid in sperm delivery*
- *Hormones regulate male reproductive function*

52.4 Structure and Function of the Human Female Reproductive System

- *Usually only one egg is produced per menstrual cycle*
- *Female accessory sex organs receive sperm and provide nourishment and protection to the embryo*

52.5 Contraception and Infertility Treatments

- *Contraception is aimed at preventing fertilization or implantation*
- *Infertility occurs in both males and females*
- *Treatment of infertility often involves assisted reproductive technologies*

Animal Reproductive Strategies

Most animals, including humans, reproduce sexually. As described in chapter 11, sexual reproduction requires a specialized form of cell division, meiosis, to produce haploid **gametes** each with a single complete set of chromosomes. These gametes, including **sperm** and **eggs** (or *ova*; singular *ovum*), are united by fertilization to restore the diploid complement of chromosomes. The diploid fertilized egg, or **zygote**, develops by mitotic division into a new multicellular organism.

Bacteria, archaea, protists, and multicellular animals including cnidarians and tunicates, as well as many of the more complex animals, reproduce asexually. In **asexual reproduction**, genetically identical cells are produced from a single parent cell through mitosis. In single-celled organisms, an individual organism divides, a process called **fission**, and then each part becomes a separate but identical organism. Cnidarians commonly reproduce by **budding**, whereby a part of the parent's body becomes separated from the rest and differentiates into a new individual (figure 52.1). The new individual may become an independent animal or may remain attached to the parent, forming a colony.

Some species have developed novel reproductive methods

Another form of asexual reproduction, **parthenogenesis**, is common in many species of arthropods. In parthenogenesis, females produce offspring from unfertilized eggs. Some species are exclusively parthenogenic (and all female), whereas others switch between sexual reproduction and parthenogenesis, producing progeny that are both diploid and haploid, respectively. In honeybees, for example, a queen bee mates only once and stores the sperm. She then can control the release of the sperm. If no sperm are released, the eggs develop parthenogenetically into haploid drones, which are males. If sperm are allowed to fertilize the eggs, the fertilized eggs develop into diploid worker bees, which are female. When fertilized eggs are exposed to the appropriate hormone, they will also develop into another queen.

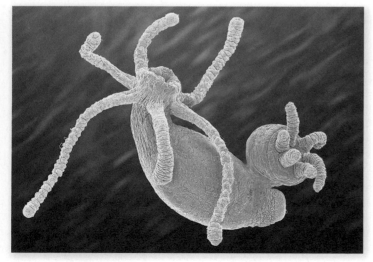

figure 52.1

CNIDARIAN BUDDING. This cnidarian reproduces asexually by budding. The new individual can be seen in the lower right of the micrograph.

a.   *b.*

figure 52.2

HERMAPHRODITISM AND PROTOGYNY. *a.* The hamlet bass (genus *Hypoplectrus*) is a deep-sea fish that is a hermaphrodite—both male and female at the same time. In the course of a single pair-mating, one fish may switch sexual roles as many as four times, alternately offering eggs to be fertilized and fertilizing its partner's eggs. Here, the fish acting as a male curves around its motionless partner, fertilizing the upward-floating eggs. *b.* The bluehead wrasse, *Thalassoma bifasciatium*, is protogynous—females sometimes turn into males. Here, a large male, or sex-changed female, is seen among females, which are typically much smaller.

In 1958, the Russian biologist Ilya Darevsky reported one of the first cases of unusual modes of reproduction among vertebrates. He observed that some populations of small lizards of the genus *Lacerta* were exclusively female, and he suggested that these lizards could lay eggs that were viable even if they were not fertilized. In other words, they were capable of asexual reproduction in the absence of sperm, a type of parthenogenesis. Further work has shown that parthenogenesis also occurs in populations of other lizard genera.

Another variation in reproductive strategies is **hermaphroditism**, in which one individual has both testes and ovaries, and so can produce both sperm and eggs (figure 52.2*a*). A tapeworm is hermaphroditic and can fertilize itself, a useful strategy because it is unlikely to encounter another tapeworm. Most hermaphroditic animals, however, require another individual in order to reproduce. Two earthworms, for example, are required for reproduction—each functions as both male and female during copulation, and each leaves the encounter with fertilized eggs.

Some deep-sea fish are hermaphrodites. Numerous fish genera include species whose individuals can change their sex, a process called *sequential hermaphroditism*. Among coral reef fish, for example, both **protogyny** ("first female," a change from female to male) and **protandry** ("first male," a change from male to female) occur. In fish that practice protogyny (figure 52.2*b*),

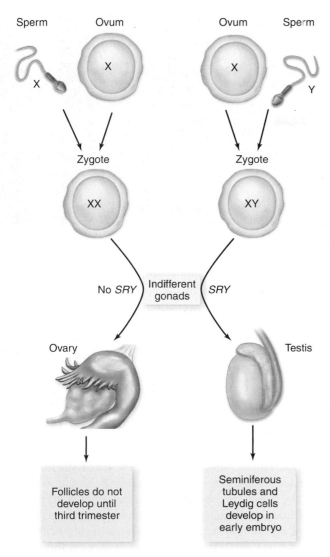

figure 52.3

SEX DETERMINATION IN MAMMALS. The sex-determining region of the mammalian Y chromosome is designated *SRY*. Testes are formed when the Y chromosome and *SRY* are present; ovaries are formed when they are absent.

Labels in figure:
Sperm — X
Ovum — X
Ovum — X
Sperm — Y
Zygote — XX
Zygote — XY
No *SRY* | Indifferent gonads | *SRY*
Ovary
Testis
Follicles do not develop until third trimester
Seminiferous tubules and Leydig cells develop in early embryo

the sex change appears to be under social control. These fish commonly live in large groups, or schools, where successful reproduction is typically limited to one or a few large, dominant males. If those males are removed, the largest female rapidly changes sex and becomes a dominant male.

Sex determination in mammals occurs in the embryo

Among the fish just described and in some species of reptiles, environmental changes can cause changes in the sex of the animal. In mammals, the sex is determined early in embryonic development. The reproductive systems of human males and females appear similar for the first 40 days after conception. During this time, the cells that will give rise to ova or sperm migrate from the yolk sac to the embryonic gonads, which have the potential to become either ovaries in females or testes in males (figure 52.3). For this reason, the embryonic gonads are said to be "indifferent."

If the embryo is a male, it has a Y chromosome with a gene whose product converts the indifferent gonads into testes. In females, which lack a Y chromosome, this gene and the protein it encodes are absent, and the gonads become ovaries. An important gene involved in sex determination is known as *SRY* (for "sex determining region of the Y chromosome"). The *SRY* gene appears to have been highly conserved during the evolution of different vertebrate groups.

Once testes have formed in the embryo, the testes secrete testosterone and other hormones that promote the development of the male external genitalia and accessory reproductive organs.

If the embryo lacks the *SRY* gene, the embryo develops female external genitalia and accessory organs. In other words, all mammalian embryos will develop into females unless a functional *SRY* gene is present.

Sexual reproduction is most common among animals, but many reproduce asexually by fission, budding, or parthenogenesis. Sexual reproduction involves the fusion of gametes derived from different individuals of a species, although some hermaphroditic species can self-fertilize.

Vertebrate Fertilization and Development

Vertebrate sexual reproduction evolved in the ocean before vertebrates colonized the land. The females of most species of marine bony fish produce eggs in batches and release them into the water. The males generally release their sperm into the water containing the eggs, where the union of the free gametes occurs. This process is known as **external fertilization.**

Although seawater is not a hostile environment for gametes, it does cause the gametes to disperse rapidly, so their release by females and males must be almost simultaneous. Thus, most marine fish restrict the release of their eggs and sperm to a few brief and well-defined periods. Some reproduce just once a year, while others do so more frequently. The ocean has few

seasonal clues that organisms can use as signals for synchronizing reproduction, but one all-pervasive signal is the cycle of the Moon. Once each month, the Moon approaches closer to the Earth than usual, and when it does, its increased gravitational attraction causes somewhat higher tides. Many marine organisms sense the tidal changes and link the production and release of their gametes to the lunar cycle.

Once vertebrates began living on land, they encountered a new danger—desiccation, a problem that can be especially severe for the small and vulnerable gametes. On land, the gametes could not simply be released near each other because they would soon dry up and perish. Consequently, intense selective pressure

figure 52.4

VIVIPAROUS FISH CARRY LIVE, MOBILE YOUNG WITHIN THEIR BODIES. The young complete their development within the body of the mother and are then released as small but competent adults. Here, a lemon shark has just given birth to a young shark, which is still attached by the umbilical cord.

resulted in the evolution of **internal fertilization** in terrestrial vertebrates (as well as some groups of fish)—that is, the introduction of male gametes directly into the female reproductive tract. By this means, fertilization still occurs in a nondesiccating environment, even when the adult animals are fully terrestrial.

Internal fertilization has led to three strategies for development of offspring

The vertebrates that practice internal fertilization exhibit three strategies for embryonic and fetal development, namely *oviparity*, *ovoviviparity*, and *viviparity*.

1. **Oviparity** is found in some bony fish, most reptiles, some cartilaginous fish, some amphibians, a few mammals, and all birds. The eggs, after being fertilized internally, are deposited outside the mother's body to complete their development.
2. **Ovoviviparity** is found in some bony fish (including mollies, guppies, and mosquito fish), some cartilaginous fish, and many reptiles. The fertilized eggs are retained within the mother to complete their development, but the embryos still obtain all of their nourishment from the egg yolk. The young are fully developed when they are hatched and released from the mother.
3. **Viviparity** is found in most cartilaginous fish, some amphibians, a few reptiles, and almost all mammals. The young develop within the mother and obtain nourishment directly from their mother's blood through an *umbilical cord*, rather than from the egg yolk (figure 52.4).

Most fishes and amphibians have external fertilization

Most fishes and amphibians, unlike other vertebrates, reproduce by means of external fertilization, although internal fertilization takes place in some groups of fishes.

Fishes

Fertilization in most species of bony fish (teleosts) is external, and the eggs contain only enough yolk to sustain the developing embryo for a short time. After the initial supply of yolk has been exhausted, the young fish must seek its food from the waters around it. Development is speedy, and the young that sur-

vive mature rapidly. Although thousands of eggs are fertilized in a single mating, many of the resulting individuals succumb to microbial infection or predation, and few grow to maturity.

In marked contrast to the bony fish, fertilization in most cartilaginous fish is internal. The male introduces sperm into the female through a modified pelvic fin. Development of the young in these vertebrates is generally viviparous.

Amphibians

The life cycle of amphibians is still tied to the water. Fertilization is external in most amphibians, just as it is in most species of bony fish. Gametes from both males and females are released through the cloaca. Among the frogs and toads, the male grasps the female and discharges fluid containing the sperm onto the eggs as the female releases them into the water (figure 52.5).

Although the eggs of most amphibians develop in the water, there are some interesting exceptions (figure 52.6). In two species of frogs, for example, the eggs develop in the vocal sacs

figure 52.5

THE EGGS OF FROGS ARE FERTILIZED EXTERNALLY. When frogs mate, the clasp of the male induces the female to release a large mass of mature eggs, over which the male discharges his sperm.

 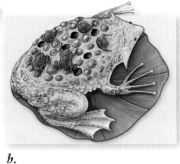  

a.   b.   c.   d.

figure 52.6

DIFFERENT WAYS YOUNG DEVELOP IN FROGS. *a.* In the poison arrow frog, the male carries the tadpoles on his back. *b.* In the female Surinam frog, froglets develop from eggs in special brooding pouches on the back. *c.* In the South American pygmy marsupial frog, the female carries the developing larvae in a pouch on her back. *d.* Tadpoles of the Darwin's frog develop into froglets in the vocal pouch of the male and emerge from the mouth.

and stomach of the parents, and the young frogs leave through their parents' mouths.

The process of development in most amphibians is divided into embryonic, larval, and adult stages. The embryo develops within the egg, obtaining nutrients from the yolk. After hatching from the egg the aquatic larva, commonly called a tadpole, then functions as a free-swimming, food-gathering machine, often for a considerable period of time. Some tadpoles grow in a matter of weeks from creatures no bigger than the tip of a pencil into individuals as big as a goldfish. When the larva has grown to a sufficient size, it undergoes a developmental transition, or metamorphosis, into the terrestrial adult form.

Reptiles and birds have internal fertilization and lay eggs

Most reptiles and all birds are oviparous. After the eggs are fertilized internally, they are deposited outside the mother's body to complete their development.

Reptiles

Like most vertebrates that fertilize internally, male reptiles utilize a tubular organ, the penis, to inject sperm into the female in a process called copulation (figure 52.7).

Most oviparous reptiles lay eggs and then abandon them. These eggs are surrounded by a leathery shell that is deposited as the egg passes through the oviduct, the part of the female reproductive tract leading from the ovary. Other species of reptiles are ovoviviparous, forming eggs that develop into embryos within the body of the mother, and a few species are viviparous.

Birds

All birds practice internal fertilization, though most male birds lack a penis. However, in most birds (including swans, geese, and ostriches), the male cloaca extends to function as a penis.

As the egg passes along the oviduct, glands secrete albumin proteins (the egg white) and the hard, calcareous shell that distinguishes bird eggs from reptilian eggs. Although modern reptiles are ectotherms, birds are endotherms; therefore, most birds incubate their eggs after laying them to keep them warm (figure 52.8). The young that hatch from the eggs of most bird species are unable to survive unaided because their development

figure 52.7

THE INTRODUCTION OF SPERM BY THE MALE INTO THE FEMALE'S BODY. Reptiles such as these tortoises were the first terrestrial vertebrates to develop this form of reproduction, called copulation, which is particularly suited to a terrestrial environment.

figure 52.8

CRESTED PENGUINS INCUBATING THEIR EGG. This nesting pair is changing the parental guard in a stylized ritual.

is still incomplete. These young birds are fed and nurtured by their parents, and they grow to maturity gradually.

The shelled eggs of reptiles and birds constitute one of the most important adaptations of these vertebrates to life on land. As described in chapter 35, these eggs are known as *amniotic eggs* because the embryo develops within a fluid-filled cavity surrounded by a membrane called the *amnion*. Other extraembryonic membranes in amniotic eggs include the *chorion*, which lines the inside of the eggshell, the *yolk sac*, and the *allantois*. Together, these extraembryonic membranes in combination with the external calcareous shell help form a desiccation-resistant egg that can be laid in dry places. In contrast, the eggs of fish and amphibians contain only one extraembryonic membrane, the yolk sac, and need to be deposited in an aquatic habitat.

The viviparous mammals, including humans, also have extraembryonic membranes, as described in the following chapter.

Mammals generally do not lay eggs, but give birth to their young

Some mammals are seasonal breeders, reproducing only once a year, while others have more frequent reproductive cycles. Among the latter, the females generally undergo the reproductive cycles, and the males are more constant in their reproductive capability.

Female reproductive cycles

Cycling in females involves the periodic release of a mature ovum from the ovary in a process known as *ovulation*. Most female mammals are "in heat," or sexually receptive to males, only around the time of ovulation. This period of sexual receptivity is called **estrus,** and the reproductive cycle is therefore called an **estrous cycle.** Reproductive cycles continue in females until they become pregnant.

In the estrous cycle of most mammals, changes in the secretion by the anterior pituitary gland of follicle-stimulating hormone (FSH) and luteinizing hormone (LH) cause changes in egg cell development and hormone secretion in the ovaries (chapter 46). Humans and apes have **menstrual cycles** that are similar to the estrous cycles of other mammals in their pattern of hormone secretion and ovulation. Unlike mammals with estrous cycles, however, human and ape females bleed when they shed the inner lining of their uterus, a process called **menstruation,** and they may engage in copulation at any time during the cycle.

Rabbits and cats differ from most other mammals in that they are induced ovulators. Instead of ovulating in a cyclic fashion regardless of sexual activity, the females ovulate only after copulation, as a result of a reflex stimulation of LH secretion. This makes these animals extremely fertile.

Monotremes, marsupials, and placental mammals

The most primitive mammals, the **monotremes** (consisting solely of the duck-billed platypus and the echidna), are oviparous, like the reptiles from which they evolved. They incubate their eggs in a nest (figure 52.9*a*) or specialized pouch, and the young hatchlings obtain milk from their mother's mammary glands by licking her skin (because monotremes lack nipples). All other mammals are viviparous, and are divided into two subcategories based on how they nourish their young.

The **marsupials,** a group that includes opossums and kangaroos, give birth to fetuses that are incompletely developed. The fetuses complete their development in a pouch of their mother's skin, where they can obtain nourishment from nipples of the mammary glands (figure 52.9*b*).

The **placental mammals** (figure 52.9*c*) retain their young for a much longer period of development within the mother's uterus. The fetuses are nourished by a structure known as the placenta, which is derived from both an extraembryonic membrane (the chorion) and the mother's uterine lining. Because the fetal and maternal blood vessels are in very close proximity in the placenta, the fetus can obtain nutrients by diffusion from the mother's blood. The functioning of the placenta is discussed in more detail in chapter 53.

Fertilization is external in most fishes and amphibians but internal in most other vertebrates. Most reptiles and all birds are oviparous, laying amniotic eggs that are protected by watertight membranes from desiccation. Birds, being endotherms, must keep the eggs warm by incubation. The large majority of mammals are viviparous. Mammals may breed seasonally, or if not, the females undergo estrous cycles in most mammals and menstrual cycles in humans and apes. Fertilization can only occur during these cycles. Monotremes, marsupials, and placental mammals differ in details of where the fetus develops.

figure 52.9

REPRODUCTION IN MAMMALS. *a.* Monotremes, such as the duck-billed platypus shown here, lay eggs in a nest. *b.* Marsupials, such as this kangaroo, give birth to small fetuses that complete their development in a pouch. *c.* In placental mammals, such as this doe nursing her fawn, the young remain inside the mother's uterus for a longer period of time and are born relatively more developed.

a.

b.

c.

Structure and Function of the Human Male Reproductive System

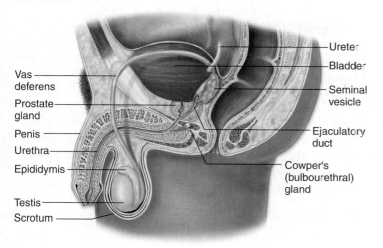

figure 52.10

ORGANIZATION OF THE HUMAN MALE REPRODUCTIVE SYSTEM. The penis and scrotum are the external genitalia, the testes are the gonads, and the other organs are accessory sex organs, aiding the production and ejaculation of semen.

The structures of the human male reproductive system, typical of male mammals, are illustrated in figure 52.10. When testes form in the human embryo, they develop **seminiferous tubules,** the sites of sperm production, beginning around 43–50 days after conception. At about 9–10 weeks, the **Leydig cells,** located in the interstitial tissue between the seminiferous tubules, begin to secrete testosterone (the major male sex hormone, or androgen). Testosterone secretion during embryonic development converts indifferent structures into the male external genitalia, the *penis* and the *scrotum*, the latter being a sac that contains the testes. In the absence of testosterone, these structures develop into the female external genitalia. Testosterone is also responsible at puberty for male secondary sex characteristics, such as development of the beard, a deeper voice, and body hair.

In an adult, each testis is composed primarily of the highly convoluted seminiferous tubules (figure 52.11, *left*). Although the testes are actually formed within the abdominal cavity, shortly before birth they descend through an opening called the inguinal canal into the scrotum, which suspends them outside the abdominal

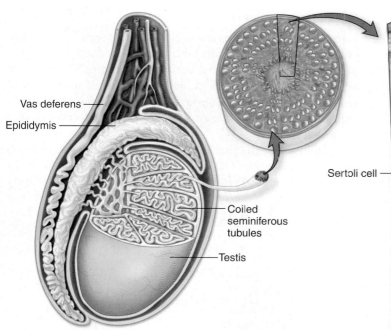

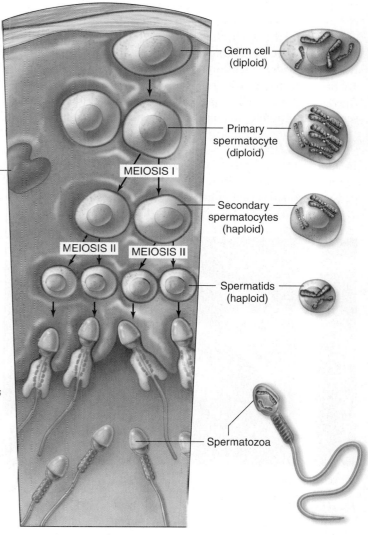

figure 52.11

THE TESTIS AND SPERMATOGENESIS. Spermatogenesis occurs in the seminiferous tubules, shown on the left. Enlargements show the radial arrangement of meiotic cells within the tubule, then the process of meiosis and differentiation to produce spermatozoa. Sertoli cells are nongerminal cells within the walls of the seminiferous tubules that assist spermatogenesis. Events begin on the outside of the tubule progressing inwards to release mature spermatozoa into the tubule. The first meiotic division separates homologous chromosomes, forming two haploid secondary spermatocytes. The second meiotic division separates sister chromatids to form four haploid spermatids, which are converted into spermatozoa.

cavity. The scrotum maintains the testes at around 34°C, slightly lower than the core body temperature (37°C). This lower temperature is required for normal sperm development in humans.

Sperm cells are produced by the millions

The wall of the seminiferous tubule consists of **spermatogonia,** or *germ cells*, and supporting **Sertoli cells.** The germ cells near the outer surface of the seminiferous tubule are diploid and are the only cells that will undergo meiosis to produce gametes (see chapter 11). The developing gamete cells, located closer to the lumen of the tubule, are haploid.

Cell divisions leading to sperm

A spermatogonium cell divides by mitosis to produce two diploid cells. One of these two cells then undergoes meiotic division to produce four haploid cells that will become sperm while the other remains as a spermatogonium. In that way, the male never runs out of spermatogonia to produce sperm. Adult males produce an average of 100–200 million sperm each day and can continue to do so throughout most of the rest of their lives.

The diploid daughter cell that begins meiosis is called a **primary spermatocyte.** In humans it has 23 pairs of homologous chromosomes (46 chromosomes total), and each chromosome is duplicated, with two chromatids. The first meiotic division separates the homologous chromosome pairs, producing two haploid **secondary spermatocytes.** However, each chromosome still consists of two duplicate chromatids.

Each of these cells then undergoes the second meiotic division to separate the chromatids and produce two haploid cells, the **spermatids.** Therefore, a total of four haploid spermatids are produced from each primary spermatocyte (see figure 52.11, *right*). All of these cells constitute the germinal epithelium of the seminiferous tubules because they "germinate" the gametes.

Supporting tissues

In addition to the germinal epithelium, the walls of the seminiferous tubules contain nongerminal cells such as the Sertoli cells mentioned earlier. These cells nurse the developing sperm and secrete products required for spermatogenesis. They also help convert the spermatids into **spermatozoa (sperm)** by engulfing their extra cytoplasm.

Sperm structure

Spermatozoa are relatively simple cells, consisting of a head, body, and flagellum (tail) (figure 52.12). The head encloses a compact nucleus and is capped by a vesicle called an **acrosome,** which is derived from the Golgi complex. The acrosome contains enzymes that aid in the penetration of the protective layers surrounding the egg. The body and tail provide a propulsive mechanism: Within the tail is a flagellum, and inside the body are a centriole, which acts as a basal body for the flagellum, and mitochondria, which generate the energy needed for flagellar movement.

Male accessory sex organs aid in sperm delivery

After the sperm are produced within the seminiferous tubules, they are delivered into a long, coiled tube called the **epididymis.** The sperm are not motile when they arrive in the epididymis, and they must remain there for at least 18 hours before their motility develops. From the epididymis, the sperm enter another long tube, the **vas deferens,** which passes into the abdominal cavity via the inguinal canal.

Semen production

Semen is a complex mixture of fluids and sperm. The vas deferens from each testis joins with one of the ducts from a pair of glands called the **seminal vesicles** (see figure 52.10), which produce a fructose-rich fluid, which constitutes about 60% of semen volume. From this point, the vas deferens continues as the ejaculatory duct and enters the prostate gland at the base of the urinary bladder.

In humans, the **prostate gland** is about the size of a golf ball and is spongy in texture. It contributes up to 30% of the bulk of the semen. Within the prostate gland, the ejaculatory duct merges with the urethra from the urinary bladder. The

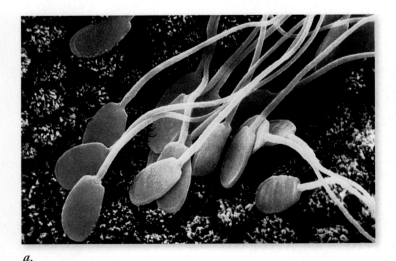

a.

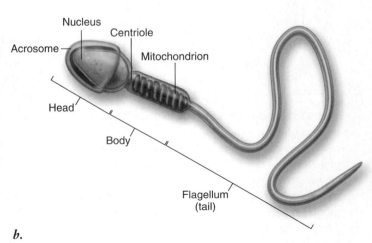

b.

figure 52.12

HUMAN SPERM. *a.* A scanning electron micrograph with sperm colored yellow. *b.* A diagram of the main components of a sperm cell.

urethra carries the semen out of the body through the tip of the penis. A pair of pea-sized **bulbourethral glands** add secretions to make up the last 10% of semen, also secreting a fluid that lines the urethra and lubricates the tip of the penis prior to coitus (sexual intercourse).

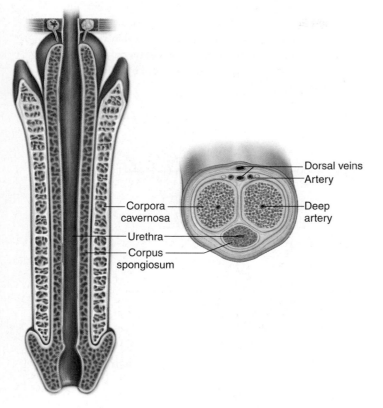

figure 52.13

A PENIS IN CROSS-SECTION (*RIGHT*) AND LONGITUDINAL SECTION (*LEFT*). During erection, tissues of the corpora cavernosa fill with blood enlarging the penis. The corpus spongiosum prevents compression of the urethra during erection.

Structure of the penis and erection

In addition to the urethra, the penis has two columns of erectile tissue, the corpora cavernosa, along its dorsal side and one column, the corpus spongiosum, along the ventral side (figure 52.13). Penile erection is produced by neurons in the parasympathetic division of the autonomic nervous system, which release nitric oxide (NO), causing arterioles in the penis to dilate. The erectile tissue becomes turgid as it is engorged with blood. This increased pressure in the erectile tissue compresses the veins, so blood flows into the penis but cannot flow out.

Some mammals, such as the walrus, have a bone in the penis that contributes to its stiffness during erection, but humans do not.

Ejaculation

The result of erection and continued sexual stimulation is **ejaculation,** the ejection from the penis of about 2–5 mL of semen containing an average of 300 million sperm. Successful fertilization requires such a high sperm count because the odds against any one sperm cell completing the journey to the egg and fertilizing it are extraordinarily high, and the acrosomes of many sperm need to interact with the egg before a single sperm can penetrate the egg (fertilization is described in chapter 53). Males with fewer than 20 million sperm per milliliter are generally considered sterile. Despite their large numbers, sperm constitute only about 1% of the volume of the semen ejaculated.

Hormones regulate male reproductive function

As you saw in chapter 46, the anterior pituitary gland secretes two gonadotropic hormones: follicle-stimulating hormone (FSH) and luteinizing hormone (LH). Although these hormones are named for their actions in the female, they are also involved in regulating male reproductive function (table 52.1). In males, FSH stimulates the Sertoli cells to facilitate

TABLE 52.1	Mammalian Reproductive Hormones
MALE	
Follicle-stimulating hormone (FSH)	Stimulates spermatogenesis via Sertoli cells
Luteinizing hormone (LH)	Stimulates secretion of testosterone by Leydig cells
Testosterone	Stimulates development and maintenance of male secondary sexual characteristics, accessory sex organs, and spermatogenesis
FEMALE	
Follicle-stimulating hormone (FSH)	Stimulates growth of ovarian follicles and secretion of estradiol
Luteinizing hormone (LH)	Stimulates ovulation, conversion of ovarian follicles into corpus luteum, and secretion of estradiol and progesterone by corpus luteum
Estradiol (estrogen)	Stimulates development and maintenance of female secondary sexual characteristics; prompts monthly preparation of uterus for pregnancy
Progesterone	Completes preparation of uterus for pregnancy; helps maintain female secondary sexual characteristics
Oxytocin	Stimulates contraction of uterus and milk-ejection reflex
Prolactin	Stimulates milk production

sperm development, and LH stimulates the Leydig cells to secrete testosterone.

The principle of negative feedback inhibition applies to the control of FSH and LH secretion (figure 52.14). The hypothalamic hormone gonadotropin-releasing hormone (GnRH) stimulates the anterior pituitary gland to secrete both FSH and LH. FSH causes the Sertoli cells to release a peptide hormone called inhibin, which specifically inhibits FSH secretion. Similarly, LH stimulates testosterone secretion, and testosterone feeds back to inhibit the release of LH, both directly at the anterior pituitary gland and indirectly by reducing GnRH release from the hypothalamus.

The importance of negative feedback inhibition can be demonstrated by removing the testes; in the absence of testosterone and inhibin, the secretion of FSH and LH from the anterior pituitary is greatly increased.

An adult male produces sperm continuously by meiotic division of germ cells lining the seminiferous tubules. Semen consists of sperm from the testes and fluid contributed by the seminal vesicles and prostate gland. Production of sperm and secretion of testosterone from the testes are controlled by FSH and LH from the anterior pituitary.

figure 52.14

HORMONAL INTERACTIONS BETWEEN THE TESTES AND ANTERIOR PITUITARY. The hypothalamus secretes GnRH, which stimulates the anterior pituitary to produce LH and FSH. LH stimulates the Leydig cells to secrete testosterone, which is involved in development and maintenance of secondary sexual characteristics, and stimulates spermatogenesis. FSH stimulates the Sertoli cells of the seminiferous tubules, which facilitate spermatogenesis. FSH also stimulates Sertoli cells to secrete inhibin. Testosterone and inhibin exert negative feedback inhibition on the secretion of LH and FSH, respectively.

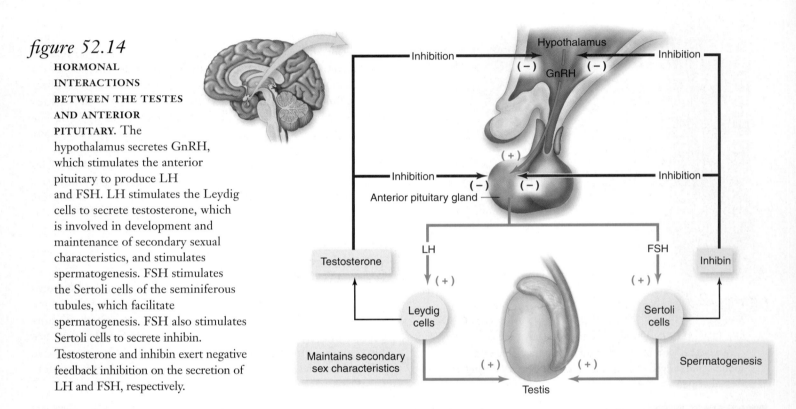

52.4 Structure and Function of the Human Female Reproductive System

The structures of the reproductive system in a human female are shown in figure 52.15. In contrast to the testes, the ovaries develop much more slowly. In the absence of testosterone, the female embryo develops a **clitoris** and **labia majora** from the same embryonic structures that produce a penis and a scrotum in males. Thus, the clitoris and penis, and the labia majora and scrotum, are said to be homologous structures. The clitoris, like the penis, contains corpora cavernosa and is therefore erectile.

The ovaries contain microscopic structures called **ovarian follicles,** which each contain a potential egg cell called a **primary oocyte** and smaller **granulosa cells.**

At puberty, the granulosa cells begin to secrete the major female sex hormone, estradiol (also called estrogen), triggering

menarche, the onset of menstrual cycling. Estradiol also stimulates the formation of the female secondary sexual characteristics, including breast development and the production of pubic hair. In addition, estradiol and another steroid hormone, progesterone, help maintain the female accessory sex organs: the fallopian tubes, uterus, and vagina.

Usually only one egg is produced per menstrual cycle

At birth, a female's ovaries contain about 1 million follicles, each containing a **primary oocyte** that has begun meiosis but is arrested in prophase of the first meiotic division. Some of these primary oocyte-containing follicles are stimulated to develop

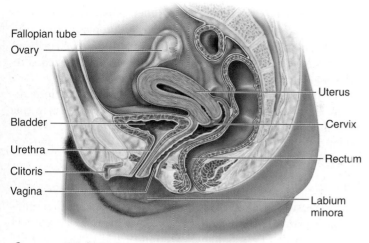

figure 52.15

ORGANIZATION OF THE HUMAN FEMALE REPRODUCTIVE SYSTEM. The ovaries are the gonads, the Fallopian tubes receive the ovulated ova, and the uterus is the womb, the site of development of an embryo if the egg cell becomes fertilized.

during each cycle. The human menstrual cycle lasts approximately one month (28 days on the average) and can be divided in terms of ovarian activity into a follicular phase and luteal phase, with the two phases separated by the event of ovulation (figure 52.16).

Follicular phase

During each **follicular phase,** several follicles in the ovaries are stimulated to grow under FSH stimulation, but only one achieves full maturity as a **tertiary,** or **Graafian, follicle** by ovulation. This follicle forms a thin-walled blister on the surface of the ovary. The uterus is lined with a simple columnar epithelial membrane called the endometrium, and during the follicular phase, estradiol causes growth of the endometrium. This phase is therefore also known as the **proliferative phase** of the endometrium (see figure 52.16).

The primary oocyte within the Graafian follicle completes the first meiotic division during the follicular phase. Instead of forming two equally large daughter cells, however, it produces one large daughter cell, the **secondary oocyte** (figure 52.17),

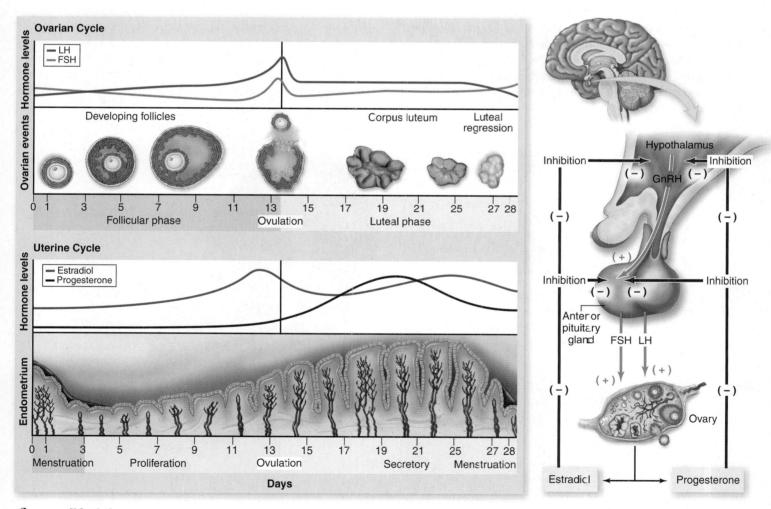

figure 52.16

THE HUMAN MENSTRUAL CYCLE. *Left:* Hormone levels during the cycle are correlated with ovulation and the growth of the endometrial lining of the uterus. Growth and thickening of the endometrium is stimulated by estradiol during the proliferative phase. Estradiol and progesterone maintain and regulate the endometrium during the secretory phase. Decline in the levels of these two hormones triggers menstruation. *Right:* Production of estradiol and progesterone by the anterior pituitary is controlled by negative feedback.

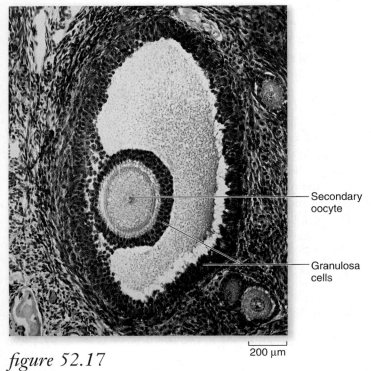

figure 52.17

A MATURE GRAAFIAN FOLLICLE IN A CAT OVARY. Note the ring of granulosa cells that surrounds the secondary oocyte. This ring will remain around the egg cell when it is ovulated, and sperm must tunnel through the ring in order to reach the plasma membrane of the secondary oocyte.

and one tiny daughter cell, called a **polar body.** Thus, the secondary oocyte acquires almost all of the cytoplasm from the primary oocyte (unequal cytokinesis), increasing its chances of sustaining the early embryo should the oocyte be fertilized. The polar body, on the other hand, disintegrates.

The secondary oocyte then begins the second meiotic division, but its progress is arrested at metaphase II. It is in this form that the potential egg cell is discharged from the ovary at ovulation, and it does not complete the second meiotic division unless it becomes fertilized in the Fallopian tube.

Ovulation

The increasing level of estradiol in the blood during the follicular phase stimulates the anterior pituitary gland to secrete LH about midcycle. This sudden secretion of LH causes the fully developed Graafian follicle to burst in the process of ovulation, releasing its secondary oocyte.

The released oocyte enters the abdominal cavity near the fimbriae, the feathery projections surrounding the opening to the Fallopian tube. The ciliated epithelial cells lining the Fallopian tube draw in the oocyte and propel it through the Fallopian tube toward the uterus.

If it is not fertilized, the oocyte disintegrates within a day following ovulation. If it is fertilized, the stimulus of fertilization allows it to complete the second meiotic division, forming a fully mature ovum and a second polar body (figure 52.18). Fusion of the nuclei from the ovum and the sperm produces a diploid zygote. Fertilization normally occurs in

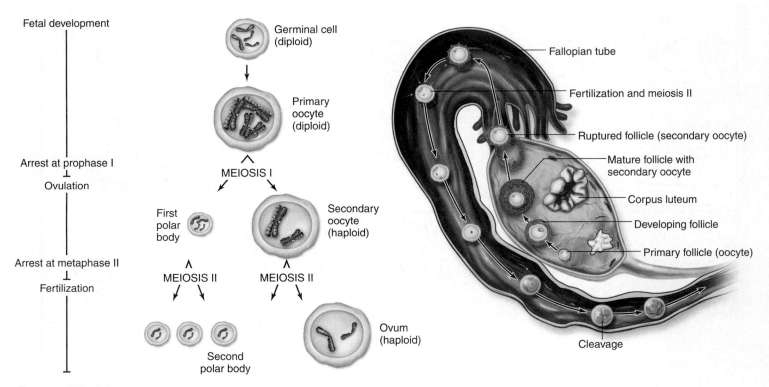

figure 52.18

THE MEIOTIC EVENTS OF OOGENESIS IN HUMANS. A primary oocyte is diploid. At the completion of the first meiotic division, one division product is eliminated as a polar body, while the other, the secondary oocyte, is released during ovulation. The secondary oocyte does not complete the second meiotic division until after fertilization; that division yields a second polar body and a single haploid egg, or ovum. Fusion of the haploid egg nucleus with a haploid sperm nucleus produces a diploid zygote.

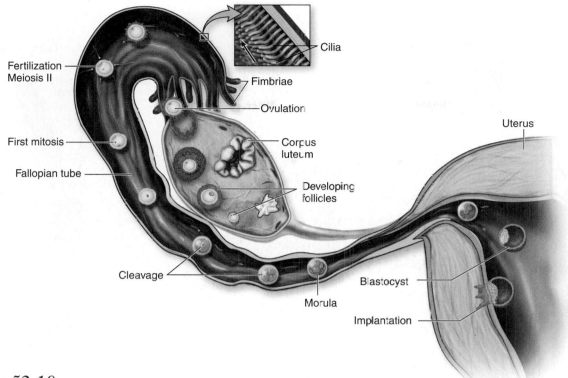

figure 52.19

THE JOURNEY OF AN EGG. Produced within a follicle and released at ovulation, the secondary oocyte is swept into a Fallopian tube and carried along by waves of ciliary motion in the tube walls. Sperm journeying upward from the vagina penetrate the secondary oocyte, meiosis is completed and fertilization of the resulting ovum occurs within the Fallopian tube. The resulting zygote undergoes several mitotic divisions while still in the tube. By the time it enters the uterus, it is a hollow sphere of cells called a blastocyst. The blastocyst implants within the wall of the uterus, where it continues its development. (The egg and its subsequent stages have been enlarged for clarification.)

the upper one-third of the Fallopian tube, and in humans the zygote takes approximately 3 days to reach the uterus and then another 2–3 days to implant in the endometrium (figure 52.19).

Luteal phase

After ovulation, LH stimulation completes the development of the Graafian follicle into a structure called the **corpus luteum.** For this reason, the second half of the menstrual cycle is referred to as the **luteal phase.** The corpus luteum secretes both estradiol and another steroid hormone, progesterone. The high blood levels of estradiol and progesterone during the luteal phase now exert negative feedback inhibition of FSH and LH secretion by the anterior pituitary gland (see figure 52.16). This inhibition during the luteal phase is in contrast to the stimulation exerted by estradiol on LH secretion at midcycle, which caused ovulation. The inhibitory effect of estradiol and progesterone after ovulation acts as a natural contraceptive mechanism, preventing both the development of additional follicles and continued ovulation.

During the luteal phase of the cycle, the combination of estradiol and progesterone cause the endometrium to become more vascular, glandular, and enriched with glycogen deposits. Because of the endometrium's glandular appearance and function, this portion of the cycle is known as the **secretory phase** of the endometrium. These changes prepare the uterine lining for embryo implantation.

In the absence of fertilization, the corpus luteum degenerates due to the decreasing levels of LH and FSH, toward the end of the luteal phase. Estradiol and progesterone, which the corpus luteum produces, inhibit the secretion of LH, the hormone needed for its survival. The disappearance of the corpus luteum results in an abrupt decline in the blood concentration of estradiol and progesterone at the end of the luteal phase, causing the built-up endometrium to be sloughed off with accompanying bleeding. This is menstruation, and the portion of the cycle in which it occurs is known as the **menstrual phase** of the endometrium.

If the ovulated oocyte is fertilized, however, regression of the corpus luteum and subsequent menstruation are averted by the tiny embryo. It does this by secreting **human chorionic gonadotropin (hCG),** an LH-like hormone produced by the chorionic membrane of the embryo. By maintaining the corpus luteum, hCG keeps the levels of estradiol and progesterone high and thereby prevents menstruation, which would terminate the pregnancy. Because hCG comes from the embryonic chorion and not from the mother, it is the hormone tested for in all pregnancy tests.

Mammals with estrous cycles

Menstruation is absent in mammals with an estrous cycle. Although such mammals do cyclically shed cells from the endometrium, they don't bleed in the process. The estrous cycle is divided into four phases: proestrus, estrus, metestrus, and diestrus, which

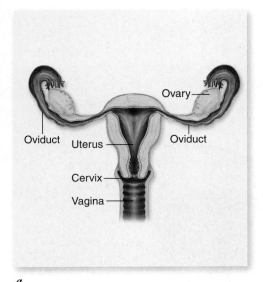
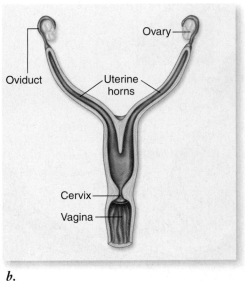
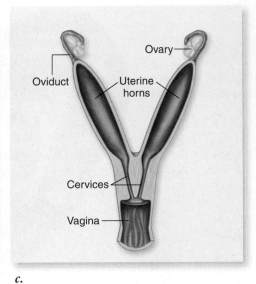

a.   b.   c.

figure 52.20

A COMPARISON OF MAMMALIAN UTERUSES. *a.* Humans and other primates; (*b*) cats, dogs, and cows; and (*c*) rats, mice, and rabbits.

correspond to the proliferative, midcycle, secretory, and menstrual phases of the endometrium in the menstrual cycle.

Female accessory sex organs receive sperm and provide nourishment and protection to the embryo

The **Fallopian tubes** (also called uterine tubes or oviducts) transport ova from the ovaries to the uterus. In humans, the **uterus** is a muscular, pear-shaped organ that narrows to form a neck, the **cervix,** which leads to the **vagina** (figure 52.20*a*).

The entrance to the vagina is initially covered by a membrane called the *hymen.* This will eventually be disrupted by vigorous activity or actual sexual intercourse. In the latter case, this can make the first experience painful when the hymen is ruptured.

During sexual arousal, the labia minora, clitoris, and vagina all become engorged with blood, much like the male erectile response. The clitoris has many sensory nerve endings and is one of the most sensitive and responsive areas for female arousal. During sexual arousal, glands located near the vaginal opening called Bartholin's glands, secrete a lubricating fluid that facilitates penetration by the penis. Ejaculation by the male introduces sperm cells that must then make the long swim out of the vagina and up the Fallopian tubes to encounter a secondary oocyte for fertilization to occur.

Mammals other than primates have more complex female reproductive tracts, in which part of the uterus divides to form uterine "horns," each of which leads to an oviduct (figure 52.20*b, c*). Cats, dogs, and cows, for example, have one cervix but two uterine horns separated by a septum, or wall. Marsupials, such as opossums, carry the split even further, with two unconnected uterine horns, two cervices, and two vaginas. A male marsupial has a forked penis that can enter both vaginas simultaneously.

During each menstrual cycle, several ovarian follicles develop under FSH stimulation, culminating in the ovulation of one follicle under LH stimulation. During the follicular and luteal phases, the hormones secreted by the ovaries stimulate the development of the endometrium, so an embryo can implant there if fertilization occurs. A secondary oocyte is released from an ovary at ovulation, and it only completes meiosis if it is fertilized.

52.5   Contraception and Infertility Treatments

In most vertebrates, copulation is associated solely with reproduction. Reflexive behavior that is deeply ingrained in the female limits sexual receptivity to those periods of the sexual cycle when she is fertile. In humans and a few species of apes, the female can be sexually receptive throughout her reproductive cycle, and this extended receptivity to sexual intercourse serves a second important function—it reinforces pair-bonding, the emotional relationship between two individuals.

Sexual intercourse may be a necessary and important part of humans' emotional lives—and yet not all couples desire to initiate a pregnancy every time they engage in sex. Throughout history, people and cultures have attempted to control reproduction without avoiding sexual intercourse. The prevention of pregnancy or giving birth is known as **birth control.** Physiologically, pregnancy begins not at fertilization but approximately a week later with successful implantation. Methods of birth control that act prior to implantation are usually termed **contraception.**

In contrast, some couples desire to have children, but find for a variety of reasons that pregnancy is not occurring—a condition termed **infertility.** Technologies have also been developed to assist these couples in having children.

Contraception is aimed at preventing fertilization or implantation

A variety of approaches, differing in effectiveness and in their acceptability to different couples, religions, and cultures, are commonly taken to prevent pregnancy (figure 52.21 and table 52.2).

Abstinence

The most reliable way to avoid pregnancy is **abstinence,** or not to have sexual intercourse at all. Of all the methods of contraception, this is the most certain. It is also the most limiting and the most difficult method to sustain. The drive to engage in sexual intercourse is compelling, and many unwanted pregnancies result when a couple who desire each other and are attempting to adhere to abstinence fail in the attempt.

Sperm blockage

If sperm cannot reach the uterus, fertilization cannot occur. One way to prevent the delivery of sperm is to encase the penis within a thin sheath, or **condom.** Some males do not favor the use of condoms, which tend to decrease males' sensory pleasure during intercourse. In principle, this method is easy to apply and foolproof, but in practice it has a failure rate of 3–15% because of incorrect or inconsistent use or condom failure. Nevertheless, condom use is the most commonly employed form of contraception in the United States. Condoms are also widely used to prevent the transmission of AIDS and other sexually transmitted diseases (STDs). Over a billion condoms are sold in the United States each year.

A second way to prevent the entry of sperm into the uterus is to place a cover over the cervix. The cover may be a relatively tight-fitting **cervical cap,** which is worn for days at a time, or a rubber dome called a **diaphragm,** which is inserted before intercourse. Because the dimensions of individual cervices vary, a cervical cap or diaphragm must be initially fitted by a physician. Failure rates average 4–25% for diaphragms. Failure rates for cervical caps are somewhat lower.

Sperm destruction

A third general approach to pregnancy prevention is to eliminate the sperm after ejaculation. This can be achieved in principle by washing out the vagina immediately after intercourse, before the sperm have a chance to enter the uterus. Such a procedure is called a **douche.** The douche method is difficult to

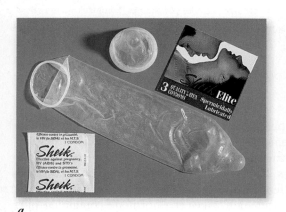
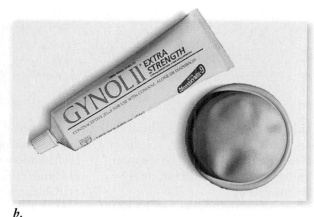

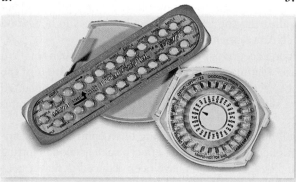
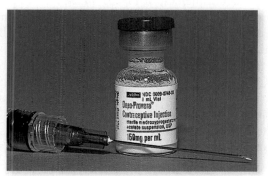

a.

b.

c.

d.

figure 52.21

FOUR COMMON METHODS OF BIRTH CONTROL. *a.* Condom; (*b*) diaphragm and spermicidal jelly; (*c*) oral contraceptives; (*d*) medroxyprogesterone acetate (Depo-Provera).

TABLE 52.2 — Methods of Birth Control

Device	Action	Failure Rate*	Advantages	Disadvantages
Oral contraceptive	Hormones (progesterone analogue alone or in combination with other hormones) primarily prevent ovulation	1–5, depending on type	Convenient; highly effective; provides significant noncontraceptive health benefits such as protection against ovarian and endometrial cancers	Must be taken regularly; possible minor side effects, which new formulations have reduced; not for women with cardiovascular risks (mostly smokers over age 35)
Condom	Thin sheath for penis collects semen; "female condoms" sheath vaginal walls	3–15	Easy to use, effective, inexpensive, protects against some sexually transmitted diseases	Requires male cooperation, may diminish spontaneity, may deteriorate on the shelf
Diaphragm	Soft rubber cup covers entrance to uterus; prevents sperm from reaching egg, holds spermicide	4–25	No dangerous side effects; reliable if used properly; provides some protection against sexually transmitted diseases and cervical cancer	Requires careful fitting, some inconvenience associated with insertion and removal; may be dislodged during intercourse
Intrauterine device (IUD)	Small plastic or metal device placed in the uterus; prevents implantation; some contain copper, others release hormones	1–5	Convenient, highly effective; infrequent replacement	Can cause excess menstrual bleeding and pain; risk of perforation, infection, expulsion, pelvic inflammatory disease, and infertility; not recommended for those who eventually intend to conceive or are not monogamous; dangerous in pregnancy
Cervical cap	Miniature diaphragm covers cervix closely, prevents sperm from reaching egg, holds spermicide	Probably similar to that of diaphragm	No dangerous side effects; fairly effective; can remain in place longer than diaphragm	Problems with fitting and insertion; comes in limited number of sizes
Foams, creams, jellies, vaginal suppositories	Chemical spermicides inserted in vagina before intercourse prevent sperm from entering uterus	10–25	Can be used by anyone who is not allergic; protect against some sexually transmitted diseases; no known side effects	Relatively unreliable; sometimes messy; must be used 5–10 minutes before each act of intercourse
Implant (levonorgestrel; Norplant)	Capsules surgically implanted under skin slowly release hormone that blocks ovulation	0.03	Very safe, convenient, and effective; very long-lasting (5 years); may have nonreproductive health benefits like those of oral contraceptives	Irregular or absent periods; minor surgical procedure needed for insertion and removal; some scarring may occur
Injectable contraceptive (medroxyprogesterone; Depo-Provera)	Injection every 3 months of a hormone that is slowly released and prevents ovulation	1	Convenient and highly effective; no serious side effects other than occasional heavy menstrual bleeding	Animal studies suggest it may cause cancer, though new studies in humans are mostly encouraging; occasional heavy menstrual bleeding

*Failure rate is expressed as pregnancies per 100 actual users per year.

Source: Data from American College of Obstetricians and Gynecologists: Contraception, Patient Education Pamphlet No. AP005. ACOG, Washington, D.C., 1990.

apply well, because it involves a rapid dash to the bathroom immediately after ejaculation and a very thorough washing. Douching can, in fact, increase the possibility of conception by forcing sperm farther up into the vagina and uterus, thereby accounting for its high failure rate (40%).

Alternatively, sperm delivered to the vagina can be destroyed there with spermicidal agents, jellies, or foams. These treatments generally require application immediately before intercourse. Their failure rates vary from 10–25%. The use of a spermicide with a condom or diaphragm increases the effectiveness over each method used independently.

Prevention of ovulation

Since about 1960, a widespread form of contraception in the United States has been the daily ingestion of **birth control pills,** or **oral contraceptives,** by women. These pills contain analogues of progesterone, sometimes in combination with estrogens. As described earlier, progesterone and estradiol act by negative feedback to inhibit the secretion of FSH and LH dur-

ing the luteal phase of the ovarian cycle, thereby preventing follicle development and ovulation. They also cause a buildup of the endometrium. The hormones in birth control pills have the same effects. Because the pills block ovulation, no ovum is available to be fertilized.

A woman generally takes the hormone-containing pills for 3 weeks; during the fourth week, she takes pills without hormones, allowing the levels of those hormones in her blood to fall, which causes menstruation.

Oral contraceptives provide a very effective means of birth control, with a failure rate of only 1–5%. In a variation of the oral contraceptive, hormone-containing capsules are implanted beneath the skin. These implanted capsules have failure rates below 1%.

A small number of women using birth control pills or implants experience undesirable side effects, such as blood clotting and nausea. These side effects have been reduced in newer generations of birth control pills, which contain less estrogen and different analogues of progesterone. Moreover, these new oral contraceptives provide a number of benefits, including reduced risks of endometrial and ovarian cancer, cardiovascular disease, and osteoporosis (for older women). However, they may increase the risk of developing breast cancer and cervical cancer.

The risks involved with birth control pills increase in women who smoke and increase greatly in women over 35 who smoke. The current consensus is that, for many women, the health benefits of oral contraceptives outweigh their risks, although a physician must help each woman determine the relative risks and benefits.

Prevention of embryo implantation

The insertion of an **intrauterine device (IUD),** such as a coil or other irregularly shaped object, is an effective means of contraception because the irritation it produces prevents the implantation of an embryo. IUDs have a failure rate of only 1–5%. Their high degree of effectiveness probably reflects their convenience; once they are inserted, they can be forgotten. The great disadvantage of this method is that almost a third of the women who attempt to use IUDs experience cramps, pain, and sometimes bleeding and therefore must discontinue using them. There is also a risk of uterine infection with insertion of the IUD.

Another method of preventing embryo implantation is the "morning-after pill," or Plan B, which contains 50 times the dose of estrogen present in birth control pills. The pill works by temporarily stopping ovum development, by preventing fertilization, or by stopping the implantation of a fertilized ovum. Its failure rate is 1–10%.

Many women are uneasy about taking such high hormone doses because side effects can be severe. This pill is not recommended as a regular method of pregnancy prevention, but rather as a method of emergency contraception.

Sterilization

Sterilization is usually accomplished by the surgical removal of portions of the tubes that transport the gametes from the gonads (figure 52.22). It is an almost 100% effective means of contraception. Sterilization may be performed on either males or females, preventing sperm from entering the semen in males and preventing an ovulated oocyte from reaching the uterus in females.

In males, sterilization involves a vasectomy, the removal and tying off of a portion of the vas deferens from each testis. In females, the comparable operation, called tubal ligation, involves the removal of a section of each Fallopian tube and tying off the tube. In very rare cases, it is possible for the tubes to grow back together, restoring fertility. This is more common in vasectomy but does occur in both at a very low level. This accounts for the less than 100% effectiveness statistically.

Infertility occurs in both males and females

Infertility is defined as the inability to conceive after 12 months of contraception-free sexual intercourse. In about 40% of cases, the failure to conceive is due to problems on the male side with about 45% due to problems on the female side, leaving another 15% unexplained (idiopathic infertility). Given these background statistics, it is clear that we still have a lot to learn about human fertility, despite a significant amount of study.

Female infertility

Infertility in females can occur due to a failure at any stage from the production of an oocyte, to the implantation of the zygote. The most common problems arise from failure to ovulate, and from some kind of mechanical blockage preventing either fertilization or implantation.

The leading cause of infertility worldwide is pelvic inflammatory disease (PID). This can be caused by infection with a number of different bacteria that all lead to blockage of the Fallopian tubes. This blockage then causes problems in sperm passage, and of transfer of fertilized eggs to the uterus.

Endometriosis, the presence of ectopic endometrial tissue, can lead to infertility by a mechanism similar to PID. The body responds to the ectopic tissue by trying to wall it off with

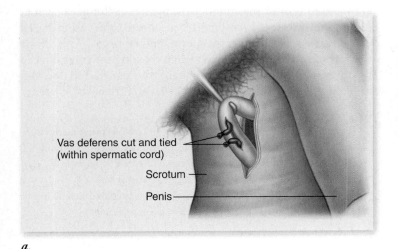

a.

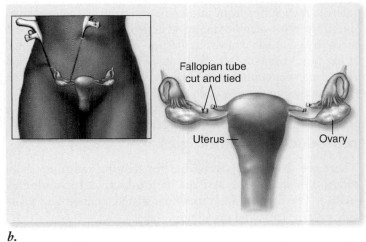

b.

figure 52.22

BIRTH CONTROL THROUGH STERILIZATION. *a.* Vasectomy; *(b)* tubal ligation.

scar tissue. The buildup of scar tissue can then prevent the transfer of eggs to the uterus.

Another common cause of infertility in females is age, or premature ovarian failure (POF). Fertility declines significantly in females with age, and the incidence of some genetic abnormalities caused by nondisjunction of chromosomes increases (see chapter 13). If a women under 40 years of age has a diminished supply of eggs, this is considered diagnostic of POF.

Disruption of the normal hormonal control of ovulation discussed earlier is also a common cause of infertility in females. Decreased levels of GnRH will disrupt ovulation, a condition referred to as hypogonadotropic hypogonadism. This can arise from damage to the hypothalamus or pituitary, or by any disorder that affects normal levels of hypothalamic hormones. For example, diabetes, thyroid disease, and excessive adrenal androgen production all affect hormonal feedback to the hypothalamus and can disrupt its normal function, leading to decreased levels of GnRH and infertility. Excessive exercise and anorexia can also lead to reduced GnRH levels and produce infertility.

Hormonal imbalances can occur during the luteal phase as well. Inadequate levels of progesterone during the luteal phase reduce the thickening of the uterine wall. If the uterine wall is inadequately prepared, implantation may not occur or can lead to an increased likelihood of spontaneous abortion.

Male infertility

Infertility in males can be due to a reduced number, viability, or motility of sperm in the ejaculate. These can be due to a variety of factors from infection to hormonal imbalances. Analysis on the male side is easier as sperm collection is noninvasive. Sperm can be easily analyzed for number, viability, morphology, and motility.

Infertility can arise from autoimmunity to sperm, leading to sperm loss, as well as due to abnormalities of all of the glands that contribute to the production of semen. Damage to the vas deferens or to the seminiferous tubules can also result in infertility. Anything that disrupts the maturation process of sperm can result in possible infertility.

After all possible causes have been ruled out, up to 5% of infertile men suffer from idiopathic infertility, or unexplained infertility. This may be due to genetic causes as the numbers seem to be similar worldwide despite different environments. It has been estimated in studies of *Drosophila* that up to 1500 recessive genes contribute to male fertility. Work is ongoing to examine the human genome for evidence of similar genes.

Treatment of infertility often involves assisted reproductive technologies

There are two basic possibilities for treating infertility: hormonal treatment and **assisted reproductive technologies (ART)**. The number and variety of assisted technologies available today is large and growing.

Hormone treatment

In the case of female infertility due to ovulatory defects, treatment is designed to produce high levels of FSH and LH at a single point during the normal menstrual cycle. Given the complexity of the hormonal control of the cycle, it is not surprising that this can be achieved in a number of ways. The most common drug currently used is clomiphine (Clomid), which is a competitive inhibitor of the estrogen receptor. This interferes with the negative feedback loop controlling estradiol production by the ovaries and consequently increases FSH and LH levels. If this is not successful, gonadatropins can be injected to stimulate ovulation.

Assistive reproductive technology

The simplest method to assist reproduction is to use artificial insemination, a process by which sperm are introduced into the female reproductive tract artificially. This is widely used in reproduction of domestic animals and is also used in humans. This has also been extended in cases of infertility in which both sperm and egg are introduced artificially by a technique called *gametic intrafallopian transfer*, or *GIFT*.

The birth of the first "test tube baby" in 1978 was heralded as the beginning of a new age of reproductive technology. Even the early pioneers may not have envisioned how far this technology would proceed. The basic technique of external fertilization is called *in vitro fertilization* (*IVF*) and transfer of the developing embryo is called simply *embryo transfer* (*ET*). When the sperm are unable to successfully fertilize an egg in vitro, they can be directly injected into an egg by *intracytoplasmic sperm injection* (*ICSI*).

One of the downsides of much of this assisted technology is multiple births. This is due to the common practice of transferring more than one embryo to ensure that at least one implants and develops normally. With advances in understanding of human development, it is possible to monitor early embryo growth to select the "best" embryos for transfer and to therefore transfer fewer embryos to reduce multiple births.

It is also possible to freeze sperm, eggs, and even human embryos to reduce the number of invasive techniques such as harvesting oocytes. Live births have been achieved using all combinations of frozen eggs, sperm, and embryos. This allows the transfer of a single embryo while freezing others produced by in vitro fertilization. If the first embryo transferred does not implant, then the others can be thawed and transferred later.

Pregnancy can be prevented by a variety of contraception methods, including abstinence, barrier contraceptives, hormonal inhibition, and sterilization surgery. Infertility can be treated by hormonal manipulation to induce ovulation or by the use of assisted reproductive technologies. These assisted technologies include in vitro fertilization and intracytoplasmic sperm injection.

52.1 Animal Reproduction Strategies

Although most animals reproduce sexually, this is not the only way to produce progeny.

- Sexual reproduction involves the production of haploid gametes—egg and sperm—by meiosis that join during fertilization to produce a diploid zygote.

- Asexual reproduction produces offspring with the same genes as the parent cell.

- Single-celled bacteria reproduce by fission, and cnidarians reproduce by budding, in which a part becomes separated and develops into a new identical individual.

- In parthenogenesis the females produce offspring from unfertilized eggs. Parthenogenesis is common in arthropods and some populations of small lizards.

- Hermaphroditism is a variant sexual reproductive strategy in which one individual has both testes and ovaries but does not necessarily fertilize itself.

- Hermaphroditism can occur simultaneously or sequentially.

- In some animals sex determination is environmentally controlled, but human sex determination is genetically controlled.

- If an embryo has a Y chromosome, specifically the *SRY* gene, it will develop into a male. If not, the embryo will develop into a female (figure 52.3).

52.2 Vertebrate Fertilization and Development

Because of the threat of desiccation, internal fertilization is common on land, but external fertilization is common in aquatic organisms.

- Internal fertilization has lead to three forms of development: oviparity, ovoviviparity, and viviparity.

- Most fishes and amphibians reproduce by means of external fertilization, and reptiles and birds have internal fertilization and are oviparous.

- The embryos of reptiles and birds develop in a fluid-filled cavity surrounded by the amnion and extraembryonic membranes as the shell helps prevent desiccation.

- Mammals are viviparous and reproduce at various times of the year.

- Most mammals have an estrus cycle, but primates have a menstrual cycle.

- Mammals are divided into three reproductive categories: oviparous monotremes, marsupials, and placental mammals.

52.3 Structure and Function of the Human Male Reproductive System

The male reproductive system begins with the production of testosterone and sperm and ends in the ejaculation of semen (figure 52.10).

- Haploid sperm are produced by meiosis of spermatogonia and assisted by Sertoli cells (figure 52.11).

- Sperm have three parts: a head with an acrosome, a body containing mitochondria, and a flagellar tail.

- Sperm complete their development in the epididymis before being transported through the vas deferens.

- Semen is a complex mixture of sperm and fluids from the seminal vesicles, prostate gland, and bulbourethral glands.

- Testosterone is produced by the Leydig cells and is responsible for development of male secondary sex characteristics and sperm production.

- The penis contains the urethra for transport of sperm and urine, two columns of erectile tissue (the corpora cavernosa and the corpora spongiosum), blood vessels, and nerves (figure 52.13).

- Ejaculation is the ejection of semen from the penis by smooth muscle contractions of the vas deferens and urethra.

- Male reproductive function is controlled by hormones and negative feedback loops (figure 52.14, and table 52.1).

52.4 Structure and Function of the Human Female Reproductive System

The female reproductive system is more complex than that of the male, and egg production occurs more slowly (figure 52.15).

- If testosterone is not present, the female embryo develops a clitoris and labial lips that have the same embryonic origin and are homologous to the male genitalia.

- At birth, the ovaries contain about a million ovarian follicles that each contain an egg cell and smaller granulosa cells that produce estrogen.

- FSH stimulates follicular development, which in turn produces estrogen, LH stimulates ovulation and corpus luteum development, which in turn produces progesterone and more estrogen. Estrogen and progesterone are necessary for development and maintenance of the uterine lining (figure 52.16).

- The menstrual cycle includes the coordination of the ovarian cycle and the uterine cycle.

- The ovarian cycle has three phases: follicular phase, ovulation, and luteal phase.

- The uterine cycle has three stages that mirror the ovarian cycle: menstruation, proliferation, and secretion.

- The primary oocyte is arrested in the first meiotic division, and one will complete meiosis I each month. The resultant secondary oocyte begins the second meiotic division and arrests until the egg is fertilized (figure 52.18).

- The secondary oocyte is released from the Graafian follicle during ovulation and is swept up by currents produced by the fimbriae and transported down the Fallopian tubes to the uterus.

- If fertilized, by the time the zygote leaves the Fallopian tube a blastocyst is formed and it implants in the wall of the uterus.

- If fertilization and implantation do not occur, the production of hormones declines, causing the built-up endometrium to be sloughed off during menstruation.

- If implantation occurs the embryo produces human chorionic gonadotropin (hCG), which maintains the corpus luteum and prevents menstruation until the placenta takes over.

- Females may have accessory organs that receive sperm and are involved in sexual response (figure 52.20).

52.5 Contraception and Infertility Treatments

Although intercourse is important in human pair-bonding, not every couple wishes to initiate pregnancy every time they engage in sex. Other couples, however, may desire to get pregnant and cannot (table 52.2).

- Pregnancy can be avoided by abstinence, blocking sperm from reaching the ovum, destroying sperm after ejaculation, preventing ovulation and embryo implantation, or sterilization.

- Female infertility ranges from failure of oocyte production to zygote implantation. Causes include hormonal imbalance, lack of ovulation, blockage of the Fallopian tubes, and age.

- Male infertility is usually due to a reduction in sperm number, viability, motility, hormonal imbalance, or damage to the vas deferens or seminiferous tubules.

- Hormonal treatment and numerous assisted reproduction technologies can be used to treat infertility in many instances.

SELF TEST

1. You have discovered a new organism living in tide pools at your favorite beach. Every so often, one of the creature's appendages will break off and gradually grow into a whole new organism, identical to the first. This is an example of—
   a. sexual reproduction.
   b. fission.
   c. budding.
   d. parthenogenesis.

2. If you decided that the organism you discovered in question 1 used parthenogenesis, what would you also know about this species?
   a. It is asexual.
   b. All the individuals are female.
   c. Each individual develops from an unfertilized egg.
   d. All of these would be true.

3. Sequential hermaphroditism differs from general hermaphroditism in that sequential hermaphrodites—
   a. contain the reproductive structures of both sexes.
   b. may change sexes as the result of social stimuli.
   c. change species as they mature.
   d. start as male and change to female.

4. Which of the following terms describes your first stage as a diploid organism?
   a. Sperm
   b. Egg
   c. Gamete
   d. Zygote

5. In which of the following groups of mammals would a mother not produce milk to feed her offspring?
   a. Monotremes
   b. Marsupials
   c. Placentals
   d. All of these produce milk

6. The major difference between an estrous cycle and a menstrual cycle is that—
   a. sexual receptivity occurs only around ovulation in the estrous cycle, but it can occur during any time of the menstrual cycle.
   b. estrous cycles occur in reptiles, but menstrual cycles occur in mammals.
   c. estrous cycles are determined by FSH, but menstrual cycles are determined by LH.
   d. estrous cycles occur monthly, but menstrual cycles occur sporadically.

7. Which of the following structures is the site of spermatogenesis?
   a. Prostate
   b. Bulbourethral gland
   c. Urethra
   d. Seminiferous tubule

8. Which of the following is a major difference between spermatogenesis and oogenesis?
   a. Spermatogenesis involves meiosis, and oogenesis involves mitosis.
   b. Spermatogenesis is continuous, but oogenesis is variable.
   c. Spermatogenesis produces fewer gametes per precursor cell than oogenesis.
   d. All of these are significant differences between oogenesis and spermatogenesis.

9. FSH and LH are produced by—
   a. the ovaries.
   b. the testes.
   c. the anterior pituitary.
   d. the adrenal glands.

10. Gametogenesis requires the conclusion of meiosis II. When does this occur in females?
    a. During fetal development
    b. At the onset of puberty
    c. After fertilization
    d. After implantation

11. Mutations that affect proteins in the acrosome would impede which of the following functions?
    a. Fertilization
    b. Locomotion
    c. Meiosis
    d. Semen production

12. In humans, fertilization occurs in the _____ , and implantation of the zygote occurs in the _____ .
    a. seminiferous tubules, uterus
    b. vagina, oviduct
    c. oviduct, uterus
    d. urethra, uterus

13. Infertility—
    a. occurs in females only.
    b. is always related to ovulation.
    c. is not known to be the result of sexually transmitted infections.
    d. All of these are false statements.

14. An animal that is oviparous reproduces by—
    a. giving birth to free-living young.
    b. producing internally fertilized eggs that develop externally.
    c. producing eggs that are fertilized externally.
    d. incubating eggs internally while the fetus develops.

15. The testicles of male mammals are suspended in the scrotum because—
    a. the optimum temperature for sperm production is less than the normal core body temperature of the organism.
    b. the optimum temperature for sperm production is higher than the normal core body temperature of the organism.
    c. there is not enough room in the pelvic area for the testicles to be housed internally.
    d. it is easier for the body to expel sperm during ejaculation.

CHALLENGE QUESTIONS

1. Suppose that the *SRY* gene mutated such that a male embryo could not produce functional protein. What kinds of changes would you expect to see in the embryo?

2. Why do you think that amphibians and many fish have external fertilization, whereas lizards, birds, and mammals rely on internal fertilization?

3. How are the functions of FSH and LH similar in male and female mammals? How do they differ?

4. You are interested in developing a contraceptive that blocks hCG receptors. Will it work? Why or why not?

5. Why are all parthenogenic parents female?

Animal Development

introduction

SEXUAL REPRODUCTION in all but a few animals unites two haploid gametes to form a single diploid cell called a **zygote.** The zygote develops by a process of cell division and differentiation into a complex multicellular organism, composed of many different tissues and organs, as the picture illustrates. At the same time, a group of cells that constitute the **germ line** are set aside to enable the developing organism to engage in sexual reproduction as an adult. In this chapter, we focus on the stages that all coelomate animals pass through during embryogenesis: fertilization, cleavage, gastrulation, and organogenesis (table 53.1). Development is a dynamic process, and so the boundaries between these stages are somewhat artificial. Although differences can be found in the details, developmental genes and cellular pathways have been greatly conserved, and they create similar structures in different organisms.

concept outline

53.1 Fertilization

- *A sperm must penetrate to the plasma membrane of the egg for membrane fusion to occur*
- *Membrane fusion activates the egg*
- *The fusion of nuclei restores the diploid state*

53.2 Cleavage and the Blastula Stage

- *The blastula is a hollow mass of cells*
- *Cleavage patterns are highly diverse and distinctive*
- *Blastomeres may or may not be committed to developmental paths*

53.3 Gastrulation

- *Gastrulation produces the three germ layers*
- *Gastrulation patterns also vary according to the amount of yolk*
- *Extraembryonic membranes are an adaptation to life on dry land*

53.4 Organogenesis

- *Changes in gene expression lead to cell determination*
- *Development of selected systems in* Drosophila melanogaster *illustrates organogenesis*
- *In vertebrates, organogenesis begins with neurulation and somitogenesis*
- *Migratory neural crest cells differentiate into many cell types*
- *Neural crest derivatives are important in vertebrate evolution*

53.5 Vertebrate Axis Formation

- *The Spemann organizer determines dorsal–ventral axis*
- *Maternally encoded dorsal determinants activate Wnt signaling*
- *Signaling molecules from the Spemann organizer inhibit ventral development*
- *Evidence indicates that organizers are present in all vertebrates*
- *Induction can be primary or secondary*

53.6 Human Development

- *During the first trimester, the zygote undergoes rapid development and differentiation*
- *During the second trimester, the basic body plan develops further*
- *During the third trimester, organs mature to the point at which the baby can survive outside the womb*
- *Critical changes in hormones bring on birth*
- *Nursing of young is a distinguishing feature of mammals*
- *Postnatal development in humans continues for years*

TABLE 53.1	Stages of Animal Development (Using a Mammal as an Example)	
Fertilization	The haploid male and female gametes fuse to form a diploid zygote.	
Cleavage	The zygote rapidly divides into many cells, with no overall increase in size. In many animals, these divisions affect future development because different cells receive different portions of the egg cytoplasm and, hence, different cytoplasmic determinants. Cleavage ends with formation of a blastula (called a blastocyst in mammals), which varies in structure among animal embryos.	 Blastocyst
Gastrulation	The cells of the embryo move, forming the three primary germ layers—ectoderm, mesoderm, and endoderm.	 Ectoderm Mesoderm Endoderm
Organogenesis	Cells from the three primary germ layers interact in various ways to produce the organs of the body. In chordates, organogenesis begins with formation of the notochord and the hollow dorsal nerve cord in the process of neurulation.	Neural groove Notochord Neural crest — Neural tube Notochord

53.1 Fertilization

In all sexually reproducing animals, the first step in development is the union of male and female gametes, a process called *fertilization*. As you learned in the preceding chapter, fertilization is typically external in aquatic animals. In contrast, internal fertilization is used by most terrestrial animals to provide a nondessicating environment for the gametes.

One physical challenge of sexual reproduction is for gametes to get together. Many elaborate strategies have evolved to enhance the likelihood of such encounters. For example, most marine invertebrates release hundreds of millions of eggs and sperm into the surrounding sea water on spawning; others use lunar cycles to time gamete release. Elaborate courtship behaviors are typical of many animals that utilize internal fertilization (see chapter 54). Fertilization itself consists of three events: sperm penetration and membrane fusion, egg activation, and fusion of nuclei.

A sperm must penetrate to the plasma membrane of the egg for membrane fusion to occur

Embryonic development begins with the fusion of the sperm and egg plasma membranes. But the unfertilized egg presents a challenge to this process, since it is enveloped by one or more protective coats. These protective coats include the **chorion** of insect eggs, the **jelly layer** and **vitelline envelope** of sea urchin and frog eggs, and the **zona pellucida** of mammalian eggs. Mammalian oocytes are also surrounded by a layer of supporting granulosa cells (figure 53.1). Thus, the first challenge of fertilization is that sperm have to penetrate these external layers to reach the plasma membrane of the egg.

A saclike organelle named the **acrosome** is positioned between the plasma membrane and the nucleus of the sperm head. The acrosome contains digestive enzymes, which are released by the process of exocytosis when a sperm reaches the outer layers of the egg. These enzymes create a hole in the protective layers, enabling the sperm to tunnel its way through to the egg's plasma membrane.

In sea urchin sperm, actin monomers assemble into cytoskeletal filaments just under the plasma membrane to create a long narrow process—the **acrosomal process.** The acrosomal process extends through the vitelline envelope to the egg's plasma membrane, and the sperm nucleus then passes through the acrosomal process to enter the egg.

In mice, an acrosomal process is not formed, and the entire sperm head burrows through the zona pellucida to the egg. Membrane fusion of the sperm and egg will then allow the sperm nucleus to pass directly into the egg cytoplasm. In many species, egg cytoplasm bulges out at membrane fusion to engulf the head of the sperm (figure 53.2).

Membrane fusion activates the egg

After ovulation, the egg remains in a quiescent state until fusion of the sperm and egg membranes triggers reactivation of the egg's metabolism. In most species, there is a dramatic

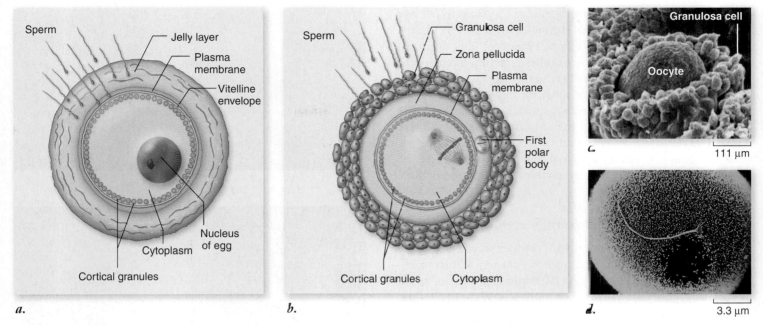

figure 53.1

ANIMAL REPRODUCTIVE CELLS. *a.* The structure of a sea urchin egg at fertilization. The relative sizes of the sperm and egg are represented in the diagram. *b.* A mammalian sperm must penetrate a layer of granulosa cells and then a glycoprotein layer called the zona pellucida before it reaches the oocyte membrane. The scanning electron micrographs show *(c)* a human oocyte surrounded by numerous granulosa cells and *(d)* a human sperm on an egg.

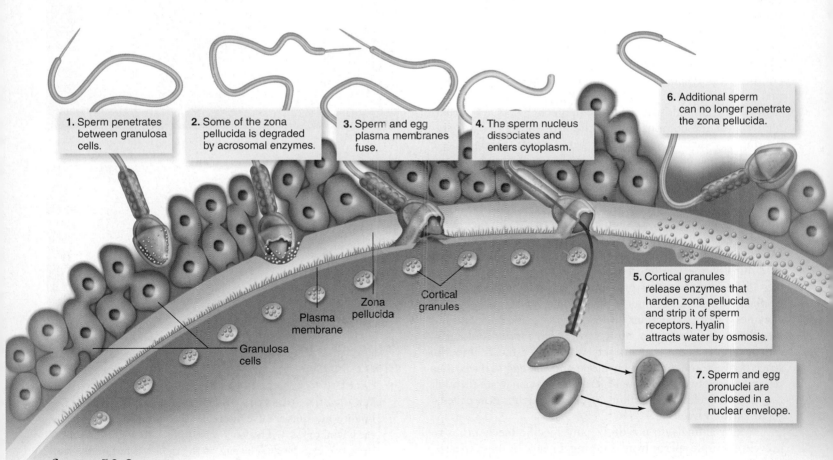

1. Sperm penetrates between granulosa cells.

2. Some of the zona pellucida is degraded by acrosomal enzymes.

3. Sperm and egg plasma membranes fuse.

4. The sperm nucleus dissociates and enters cytoplasm.

5. Cortical granules release enzymes that harden zona pellucida and strip it of sperm receptors. Hyalin attracts water by osmosis.

6. Additional sperm can no longer penetrate the zona pellucida.

7. Sperm and egg pronuclei are enclosed in a nuclear envelope.

figure 53.2

SPERM PENETRATION AND FUSION. The sperm must penetrate the outer layers around the egg before fusion of sperm and egg plasma membranes can occur. Fusion activates the egg and leads to a series of events that prevent polyspermy.

increase in the levels of free intracellular Ca^{2+} ions in the egg shortly after the sperm makes contact with the egg's plasma membrane. This increase is due to release of Ca^{2+} from internal, membrane-bounded organelles, starting at the point of sperm entry and traversing across the egg.

Scientists have been able to watch this wave of Ca^{2+} release by preloading unfertilized eggs with a dye that fluoresces when bound to free Ca^{2+}, and then fertilizing the eggs (figure 53.3). The released Ca^{2+} act as second messengers in the cytoplasm of the egg, to initiate a host of changes in protein activity. These many events initiated by membrane fusion are collectively called **egg activation.**

Blocking of additional fertilization events

Because large numbers of sperm are released during spawning or ejaculation, many more than one sperm is likely to reach, and try to fertilize, a single egg. Multiple fertilization would result in a zygote that has three or more sets of chromosomes, a condition known as polyploidy. Polyploidy is incompatible with animal development, although it is frequently found in plants. As a result, an early response to sperm fusion in many animal eggs is to prevent fusion of additional sperm—in other words, to initiate a block to *polyspermy*.

In sea urchins, membrane contact by the first sperm results in a rapid, transient change in membrane potential of the egg, which prevents other sperm from fusing to the egg's plasma membrane. The importance of this event was shown by experiments where sea urchin eggs are fertilized in low-sodium, artificial seawater. The change in membrane potential is mostly due to an influx of Na^+, so fertilization in low-sodium water prevents this. Under these conditions polyspermy is much more frequent than in normal seawater.

Many animals use additional mechanisms to permanently alter the composition of the exterior egg coats, preventing any further sperm from penetrating through these layers. In sea urchins and mammals, specialized vesicles called **cortical granules,** located just beneath the plasma membrane of the egg, release their contents by exocytosis into the space between the plasma membrane and the vitelline envelope or zona pellucida, respectively. In each case, cortical granule enzymes remove critical sperm receptors from the outer coat of the egg.

Finally, the vitelline envelopes in many sea urchin species "lift off" the surfaces of the eggs via the combined action of different cortical granule enzymes and hyalin release. The enzymes digest connections between the vitelline envelope and the plasma membrane to allow separation. *Hyalin* is a sugar-rich macromolecule that attracts water by osmosis into the space between the vitelline envelope and the egg surface, thus separating the two. Additional sperm cannot penetrate through the hardened, elevated vitelline envelope, which is now called a **fertilization envelope.**

Many animals do not utilize any specific mechanisms to prevent multiple sperm from entering an egg. In these species, all but one of the sperm nuclei is degraded or subsequently extruded from the egg to prevent polyploidy.

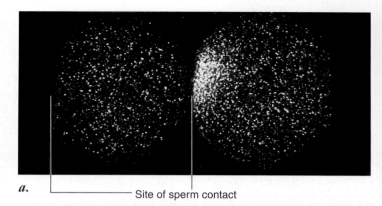

a. — Site of sperm contact

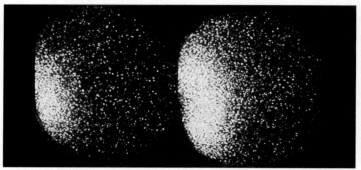

b.

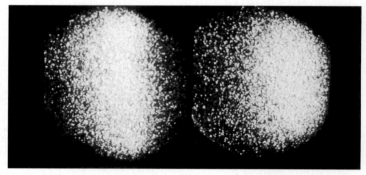

c.

d.

figure 53.3

CALCIUM IONS ARE RELEASED IN WAVE ACROSS TWO SEA URCHIN EGGS FOLLOWING SPERM CONTACT. The bright white dots are dye molecules that fluoresce when they are bound to Ca^{2+}. The Ca^{2+} wave moves from left to right in these two eggs *(a–d)*. The egg on the right was fertilized a few seconds before the egg on the left. The wave takes about 30 sec to cross the entire egg.

Primary Oocyte	First Metaphase of Meiosis	Second Metaphase of Meiosis	Meiosis Complete

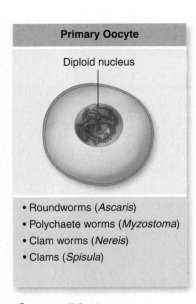
Diploid nucleus

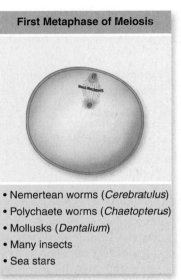

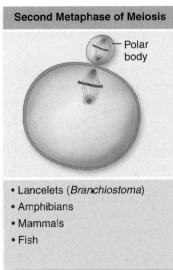
Polar body

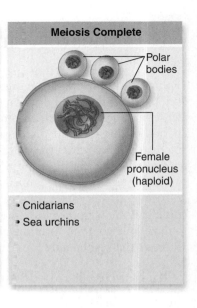
Polar bodies

Female pronucleus (haploid)

- Roundworms (*Ascaris*)
- Polychaete worms (*Myzostoma*)
- Clam worms (*Nereis*)
- Clams (*Spisula*)

- Nemertean worms (*Cerebratulus*)
- Polychaete worms (*Chaetopterus*)
- Mollusks (*Dentalium*)
- Many insects
- Sea stars

- Lancelets (*Branchiostoma*)
- Amphibians
- Mammals
- Fish

- Cnidarians
- Sea urchins

figure 53.4

STAGE OF EGG MATURATION AT TIME OF SPERM BINDING IN REPRESENTATIVE ANIMALS.

Other effects of sperm penetration

In addition to the previously mentioned surface changes, sperm penetration can have three other effects on the egg. First, in many animals, the nucleus of the unfertilized egg is not yet haploid because it had not entered or completed meiosis prior to ovulation (figure 53.4). Fusion of the sperm plasma membrane then triggers the eggs of these animals to complete meiosis. In mammals, a single large egg with a haploid nucleus and one or more small polar bodies, which contain the other nuclei, are produced (see chapter 52).

Second, sperm penetration in many animals triggers movements of the egg cytoplasm. In chapter 19, we discussed the cytoplasmic rearrangements of newly fertilized tunicate eggs, which result in the asymmetrical localization of pigment granules that determine muscle development. In amphibian embryos, the point of sperm entry is the focal point of cytoplasmic movements in the egg, and these movements ultimately establish the bilateral symmetry of the developing animal.

In some frogs, for example, sperm penetration causes an outer pigmented cap of egg cytoplasm to rotate toward the point of entry, uncovering a gray crescent of interior cytoplasm opposite the point of penetration (figure 53.5). The position of this gray crescent determines the orientation of the first cell division. A line drawn between the point of sperm entry and the gray crescent would bisect the right and left halves of the future adult.

Third, activation is characterized by a sharp increase in protein synthesis and an increase in metabolic activity in general. Experiments demonstrate that the burst of protein synthesis in an activated egg uses mRNAs that were deposited into the cytoplasm of the egg during oogenesis.

In some animals, it is possible to artificially activate an egg without the entry of a sperm, simply by pricking the egg membrane. An egg that is activated in this way may go on to develop parthenogenetically. A few kinds of amphibians, fish, and reptiles rely entirely on parthenogenetic reproduction in nature, as we mentioned in chapter 52.

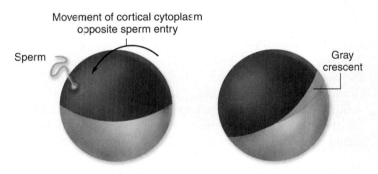
Movement of cortical cytoplasm opposite sperm entry

Sperm

Gray crescent

figure 53.5

GRAY CRESCENT FORMATION IN FROG EGGS. The gray crescent forms on the side of the egg opposite the point of penetration by the sperm.

The fusion of nuclei restores the diploid state

In the third and final stage of fertilization, the haploid sperm nucleus fuses with the haploid egg nucleus to form the diploid nucleus of the zygote. The process involves migration of the two nuclei toward each other along a microtubule-based aster. A centriole that enters the egg cell with the sperm nucleus organizes the microtubule array, which is made from stored tubulin proteins in the egg's cytoplasm.

Following sperm penetration through external layers, membrane fusion with the egg initiates a complex series of developmental events. These include egg activation, blocks to polyspermy, and major rearrangements of cytoplasm. Polyspermy is blocked by changes in membrane polarity and alterations to the surface of the egg. The final stage of fertilization is fusion of the egg and sperm nuclei to create a diploid zygote.

53.2 Cleavage and the Blastula Stage

Following fertilization, the second major event in animal development is the rapid division of the zygote into a larger and larger number of smaller and smaller cells (see table 53.1). This period of division, called **cleavage,** is not accompanied by an increase in the overall size of the embryo. Each individual cell in the resulting tightly packed mass of cells is referred to as a **blastomere.** In many animals, the two ends of the egg and subsequent embryo are traditionally referred to as the **animal pole** and the **vegetal pole.** In general, the blastomeres of the animal pole go on to form the external tissues of the body, and those of the vegetal pole form the internal tissues.

The blastula is a hollow mass of cells

In many animal embryos, the outermost blastomeres in the ball of cells produced during cleavage become joined to one another by tight junctions, belts of protein that encircle a cell and weld it to its neighbors (see chapter 9). These tight junctions create a seal that isolates the interior of the cell mass from the surrounding medium.

Subsequently, cells in the interior of the mass begin to pump Na$^+$ from their cytoplasm into the spaces between cells. The resulting osmotic gradient causes water to be drawn into the center of the embryo, enlarging the intercellular spaces. Eventually, the spaces coalesce to form a single large cavity within the embryo. The resulting hollow ball of cells is called a **blastula** (or **blastocyst** in mammals), and the fluid-filled cavity within the blastula is known as the **blastocoel** (see table 53.1).

Cleavage patterns are highly diverse and distinctive

Cleavage divisions are quite rapid in most species, and chapter 19 provides an overview of the conserved set of proteins that control the cell cycle in animal embryos. Cleavage patterns are quite diverse, and there are about as many ways to divide up the cytoplasm of an animal egg during cleavage as there are phyla of animals! Nonetheless, we can make some generalizations.

First, the relative amount of nutritive yolk in the egg is the characteristic that most affects the cleavage pattern of an animal embryo (figure 53.6). Vertebrates exhibit a variety of reproductive strategies involving different patterns of yolk utilization.

Cleavage in insects

Insects have yolk-rich eggs, and in chapter 19 we discussed the *syncytial blastoderm* of insects, in which multiple mitotic divisions of the nucleus occur in the absence of cytokinesis. Because there are no membranes separating the early embryonic nuclei of insects, gradients of diffusible proteins termed **morphogens** within the egg's cytoplasm can directly and differentially affect the activity of these embryonic nuclei, and thus the pattern of the early embryo. The nuclei eventually migrate to the periphery of the egg, where cell membranes form around each nucleus. The resulting **cellular blastoderm** of an insect has a single layer of cells surrounding a central mass of yolk (see figure 19.12 and table 53.2).

Cleavage of eggs with moderate or little yolk

In eggs that contain moderate to little yolk, cleavage occurs throughout the whole egg, a pattern called **holoblastic cleavage** (figure 53.7). This pattern of cleavage is characteristic of invertebrates such as mollusks, annelids, echinoderms and tunicates, and also of amphibians and mammals (described shortly).

In sea urchins, holoblastic cleavage results in the formation of a symmetrical blastula composed of a single layer of cells of approximately equal size surrounding a spherical blastocoel. In contrast, amphibian eggs contain much more cytoplasmic yolk in the vegetal hemisphere than in the animal hemisphere. Because yolk-rich regions divide much more slowly than areas with little yolk, horizontal cleavage furrows

Sea Urchin	Frog	Chicken

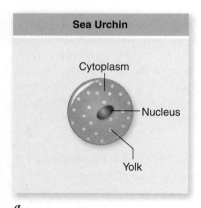
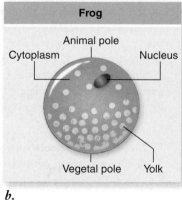
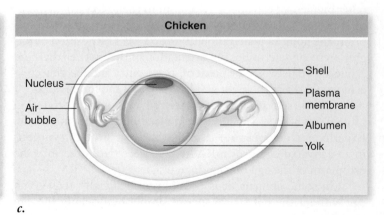

a.   *b.*   *c.*

figure 53.6

YOLK DISTRIBUTION IN THREE KINDS OF EGGS. *a.* In a sea urchin egg, the cytoplasm contains a small amount of evenly distributed yolk and a centrally located nucleus. *b.* In a frog egg, there is much more yolk, and the nucleus is displaced toward one pole. *c.* Bird eggs are complex, with the nucleus contained in a small disc of cytoplasm that sits on top of a large, central yolk mass.

TABLE 53.2 — The Major Cleavage Patterns of Animal Embryos

HOLOBLASTIC (COMPLETE) CLEAVAGE

Isolecithal (Sparse, evenly distributed yolk)

Radial cleavage — Echinoderms	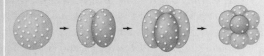
Spiral cleavage — Annelids, Mollusks, Flatworms	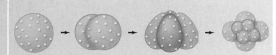
Rotational cleavage — Mammals, Nematodes	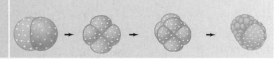

Mesolecithal (Moderate vegetal yolk disposition)

Displaced radial cleavage — Amphibians	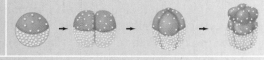

MEROBLASTIC (INCOMPLETE) CLEAVAGE

Telolecithal (Dense yolk throughout most of cell)

Discoidal cleavage — Fish, Reptiles, Birds	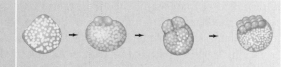

Centrolecithal (Yolk in center of egg)

Syncytial cleavage — Most insects	

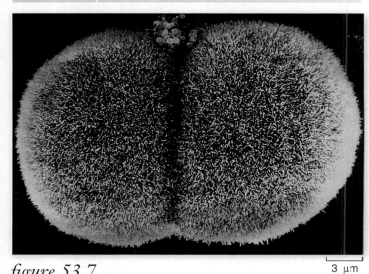

figure 53.7

HOLOBLASTIC CLEAVAGE. In this type of cleavage, which is characteristic of eggs with relatively small amounts of yolk, cell division occurs throughout the entire egg.

3 μm

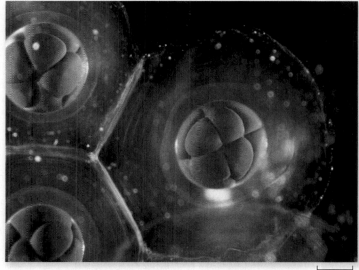

a.

333.3 μm

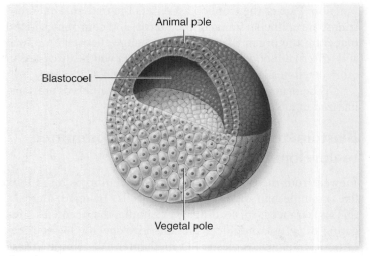

b.

Animal pole
Blastocoel
Vegetal pole

figure 53.8

FROG CLEAVAGE AND BLASTULA FORMATION. a. The closest cells in this photo (those near the animal pole) divide faster and are smaller than those near the vegetal pole (below cells of the animal pole). **b.** A cross-section of a frog blastula, showing an eccentric blastocoel, larger yolk-filled cells at the vegetal pole, and smaller cells with little yolk at the animal pole.

are displaced toward the animal pole (figure 53.8a). Thus, holoblastic cleavage in frog eggs results in an asymmetrical blastula, with a displaced blastocoel. The blastula consists of large cells containing a lot of yolk at the vegetal pole, and smaller, more numerous cells containing little yolk at the animal pole (figure 53.8b).

Cleavage of eggs with large amounts of yolk

The eggs of reptiles, birds, and some fishes are composed almost entirely of yolk, with a small amount of clear cytoplasm concentrated at one pole called the **blastodisc.** Cleavage in these eggs is restricted to the blastodisc. The yolk is essentially an inert mass. This type of cleavage pattern is called

meroblastic cleavage (figure 53.9). The resulting embryo is not spherical, but rather has the form of a thin cap perched on the yolk.

Cleavage in mammals

Mammalian eggs contain very little yolk; however, mammalian embryogenesis has many similarities to development of their reptilian and avian relatives.

Because cleavage is not impeded by yolk in mammalian eggs, it is holoblastic, forming a structure called a **blastocyst,** in which a single layer of cells surrounds a central fluid-filled blastocoel. In addition, an **inner cell mass (ICM)** is located at one pole of the blastocoel cavity (figure 53.10). The ICM is similar to the blastodisc of reptiles and birds, and it goes on to form the developing embryo.

The outer layer of cells, called the **trophoblast,** is similar to the cells that form the membranes underlying the tough outer shell of the reptilian egg. These cells have changed during the course of mammalian evolution to carry out a very different function: Part of the trophoblast enters the maternal endometrium (the epithelial lining of the uterus) and contributes to the **placenta,** the organ that permits exchanges between the fetal and maternal blood supplies. The placenta will be discussed in more detail in a later section.

The major cleavage patterns of animal embryos are summarized in table 53.2.

Blastomeres may or may not be committed to developmental paths

Viewed from the outside, cleavage-stage embryos often look like a simple ball or disc of similar cells. In many animals, this appearance is misleading; for example, the unequal segregation of cytoplasmic determinants into specific blastomeres of tunicate embryos, described in chapter 19, commits those cells to different developmental paths. The experimental de-

figure 53.10

THE EMBRYOS OF MAMMALS AND BIRDS ARE MORE SIMILAR THAN THEY SEEM. A mammalian blastula (*left*), called a blastocyst, is composed of a sphere of cells, the trophoblast, surrounding a cavity, the blastocoel, and an inner cell mass (ICM). An avian (bird) blastula consists of a cap of cells, the blastodisc, resting atop a large yolk mass (*right*). The blastodisc will form an upper and a lower layer with a compressed blastocoel in between.

struction or removal of these committed cells results in embryos deficient in the tissues that would have developed from those cells.

In contrast, mammals exhibit highly regulative development, in which early blastomeres do not appear to be committed to a particular fate. For example, if a blastomere is removed from an early eight-cell stage human embryo (as done in the process of preimplantation genetic diagnosis), the remaining seven cells of the embryo will "regulate" and develop into a complete individual if implanted into the uterus of a woman. Similarly, embryos that are split into two (either naturally or experimentally) form identical twins. It therefore appears that inheritance of maternally encoded determinants is not an important mechanism in mammalian development, and body form is determined primarily by cell–cell interactions.

The earliest patterning events in mammalian embryos occur during the preimplantation stages that lead to formation of the blastocyst. At the eight-cell stage, the outer surfaces of many mammalian blastomeres flatten against each other in a process called **compaction,** which serves to polarize the blastomeres. The polarized blastomeres then undergo asymmetrical cell divisions. Cell lineage studies have shown that cells that are in the interior of the embryo most often become ICM cells of the mammalian blastocyst, whereas cells on the exterior of the embryo usually become trophoblast cells.

> A series of rapid cell divisions called cleavage transforms the zygote into a hollow ball of cells, the blastula, in animals. The cleavage pattern is influenced by the amount of yolk and its distribution in the egg. Eggs with little yolk cleave completely (holoblastic cleavage); eggs with a large amount of yolk cannot cleave completely (meroblastic cleavage).

figure 53.9

MEROBLASTIC CLEAVAGE. Only a portion of the egg actively divides to form a mass of cells in this type of cleavage, which occurs in eggs with relatively large amounts of yolk.

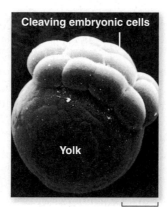

Cleaving embryonic cells

Yolk

25 µm

53.3 Gastrulation

In a complex series of cell shape changes and cell movements, the cells of the blastula rearrange themselves to form the basic body plan of the embryo. This process, called **gastrulation,** forms the three primary germ layers and converts the blastula into a bilaterally symmetrical embryo with a central progenitor gut and visible anterior–posterior and dorsal–ventral axes.

TABLE 53.3	Developmental Fates of the Primary Germ Layers in Vertebrates	
Ectoderm	Epidermis of skin, nervous system, sense organs	
Mesoderm	Skeleton, muscles, blood vessels, heart, blood, gonads, kidneys, dermis of skin	
Endoderm	Lining of digestive and respiratory tracts, liver, pancreas, thymus, thyroid	

Gastrulation produces the three germ layers

Gastrulation creates the three primary **germ layers:** endoderm, ectoderm, and mesoderm. The cells in each germ layer have very different developmental fates. The cells that move into the embryo to form the tube of the primitive gut are **endoderm;** they give rise to the lining of the gut and its derivatives (pancreas, lungs, liver, etc.). The cells that remain on the exterior are **ectoderm,** and their derivatives include the epidermis on the outside of the body and the nervous system. The cells that move into the space between the endoderm and ectoderm are **mesoderm;** they eventually form the notochord, bones, blood vessels, connective tissues, muscles and internal organs such as the kidneys and gonads (table 53.3).

Cells move during gastrulation using a variety of cell shape changes. Some cells use broad, actin-filled extensions called *lamellipodia* to crawl over neighboring cells. Other cells send out narrow extensions called *filopodia*, which are used to "feel out" the surfaces of other cells or the extracellular matrix. Once a satisfactory attachment is made, the filopodia retract to pull the cell forward. Contractions of actin filament bundles are responsible for many of these cell shape changes. Cells that are tightly attached to one another via desmosomes or adherens junctions will move as cell sheets.

In embryos with little yolk and a hollow blastula, the cell sheet at the vegetal pole of the blastula **invaginates** (dents inward) to form the primitive gut tube. In embryos with large yolky cells that are hard to move, sheets of smaller cells will **involute** (roll inward) from the surface of the blastula and move over the basal surfaces of the outer cells. Other cells break away from cell sheets and migrate as individual cells during **ingression.**

Avian and mammalian gastrulation begins with **delamination,** in which one sheet of cells splits into two sheets. Each migrating cell possesses particular cell surface glycoproteins, which adhere to specific molecules on the surfaces of other cells or in the extracellular matrix. Changes in cell adhesiveness, as described in chapter 19, are key events in gastrulation. The extracellular matrix protein fibronectin and the corresponding integrin receptors of cells are essential molecules of gastrulation in many animals.

Gastrulation patterns also vary according to the amount of yolk

Just as in cleavage patterns, yolk quantity also affects the types of cell movements that occur during gastrulation. Here, we examine gastrulation in four representative classes of embryos with differing quantities of yolk.

Gastrulation in sea urchins

Echinoderms such as sea urchins develop from relatively yolk-poor eggs and form hollow, symmetrical blastulas. Gastrulation begins when cells at the vegetal surface of the blastula change their shape to form a flattened **vegetal plate.** In an example of ingression, a subset of cells in the vegetal plate breaks away from the blastula wall and moves into the blastocoel cavity. These **primary mesenchyme cells** are future mesoderm cells, and they use *filopodia* to migrate through the blastocoel cavity (figure 53.11). Eventually, they become localized in the ventrolateral corners of the blastocoel, where they form the larval skeleton.

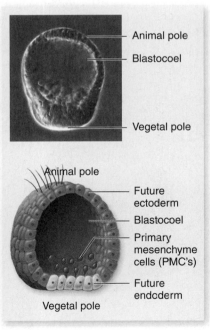

a.

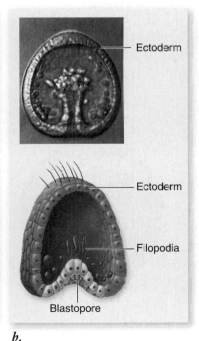

b.

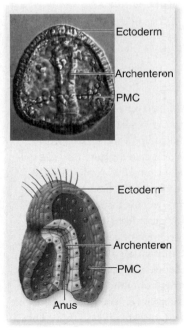

c.

figure 53.11

GASTRULATION IN A SEA URCHIN. *a.* Gastrulation begins with formation of the vegetal plate and ingression of primary mesenchyme cells (prospective mesoderm cells) into the blastocoel cavity. *b.* The endoderm is then formed by invagination of the remaining vegetal plate cells and extension of a cellular tube to produce the primitive gut, or archenteron. *c.* Cells that remain on the surface form the ectoderm.

The remaining cells of the vegetal plate then invaginate into the blastocoel to form the endoderm layer, creating a structure that looks something like an indented tennis ball. Eventually, the inward-moving tube of cells contacts the opposite side of the gastrula and stops moving. The hollow structure resulting from the invagination is called the **archenteron,** and it is the progenitor of the digestive tube. The opening of the archenteron, the future anus, is known as the **blastopore.** A secondary opening will develop at the point where the archenteron contacts the opposite side of the gastrula, forming the mouth (see figure 53.11). Animals in which the anus develops first and the mouth second are termed *deuterostomes*, as was discussed in chapter 32.

Gastrulation in frogs

The blastula of an amphibian has an asymmetrical yolk distribution, and the yolk-laden cells of the vegetal pole are less numerous but much larger than the yolk-free cells of the animal pole. Consequently, gastrulation is more complex than it is in sea urchins. In frogs, a layer of surface cells first invaginates to form a small, crescent-shaped slit, which initiates formation of the blastopore. Next, cells from the animal pole involute over the dorsal lip of the blastopore (see figure 53.12*a*), which forms at the same location as the gray crescent of the fertilized egg (see figure 53.5).

The involuting cell layer eventually presses against the inner surface of the opposite side of the embryo, eliminating the blastocoel and producing an archenteron with a blastopore. In this case, however, the blastopore is filled with yolk-rich cells, forming the **yolk plug** (figure 53.12*b, c*). The outer layer of cells resulting from these movements is the ectoderm, and the inner layer is the endoderm. Other cells that involute over the dorsal lip and ventral lip (the two lips of the blastopore that are separated by the yolk plug) migrate between the ectoderm and endoderm to form the third germ layer, the mesoderm (figure 53.12*c–e*).

Gastrulation in birds

At the end of cleavage in a bird or reptile, the developing embryo is a small cap of cells called the **blastoderm,** which sits on top of the large ball of yolk (figure 53.13*a*). As a result, gastrulation proceeds somewhat differently,

In birds, the blastoderm first separates into two layers, and a blastocoel cavity forms between them (figure 53.13*b*). The deep, internal layer of the bilayered blastoderm gives rise to extraembryonic tissues only (described later on), whereas all cells of the embryo proper are derived from the upper layer of cells. Thus, the upper layer of the blastoderm gives rise to all three germ layers.

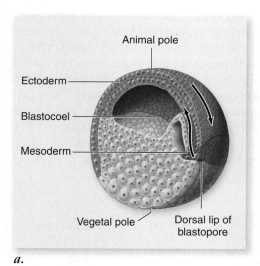

a.

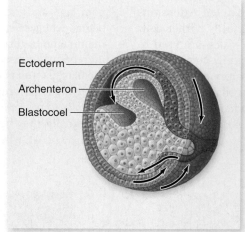

b.

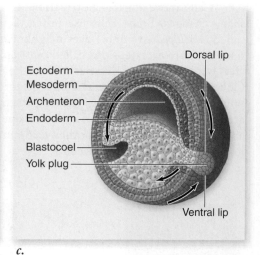

c.

figure 53.12

FROG GASTRULATION. *a.* A layer of cells from the animal pole moves toward the vegetal pole, ultimately involuting through the dorsal lip of the blastopore. *b.* Cells in the dorsal lip zone then involute into the hollow interior, or blastocoel, eventually pressing against the far wall. The three primary germ tissues (ectoderm, mesoderm, and endoderm) become distinguished. Ectoderm is shown in *blue*, mesoderm in *red*, and endoderm in *yellow*. *c.* The movement of cells through the blastopore creates a new internal cavity, the archenteron, which displaces the blastocoel. *d.* Organogenesis begins when the neural plate forms from dorsal ectoderm to begin the process of neurulation. *e.* The neural plate will next form a neural groove and then a neural tube. The cells of the neural ectoderm are shown in *purple*.

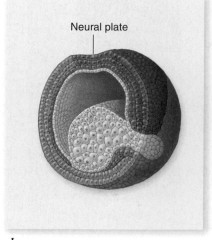

d.

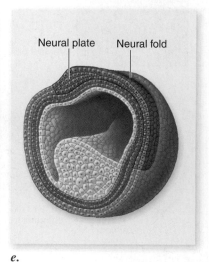

e.

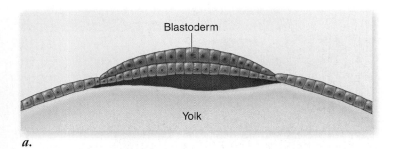

a.

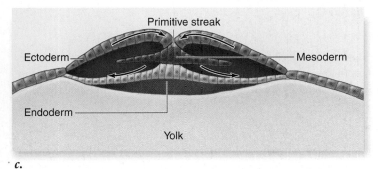

b.

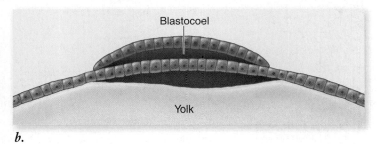

c.

figure 53.13

GASTRULATION IN BIRDS. *a.* The avian blastula is made up of a disc of cells sitting atop the large yolk mass. *b.* Gastrulation commences with the delamination of the blastoderm into two layers. All three germ layers are derived from the upper layer of the blastoderm. *c.* Cells that migrate through primitive streak into the interior of the embryo are future endoderm or mesoderm cells. Cells that remain in the upper layer form the ectoderm.

Some of the surface cells begin moving to the midline, where they break away from the surface sheet of cells and ingress into the blastocoel cavity. A furrow along the longitudinal midline marks the site of this ingression (figure 53.13*c*). This furrow, analogous to an elongated blastopore, is called the **primitive streak.** Some cells migrate through the primitive streak and across the blastocoel cavity to displace cells in the lower layer. These deep-migrating cells form the endoderm. Other cells that move through the primitive streak migrate laterally into intermediate regions and form a new layer—the mesoderm. Cells that remain on the surface and do not enter the primitive streak form the ectoderm.

Gastrulation in mammals

Mammalian gastrulation proceeds much the same way as it does in birds. In both types of animals, the embryo develops from a flattened collection of cells—the blastoderm in birds or the inner cell mass in mammals. Although the blastoderm of a bird is flattened because it is pressed against a mass of yolk, the inner cell mass of a mammal is flat despite the absence of a yolk mass.

In mammals, the placenta has made yolk dispensable; the embryo obtains nutrients from its mother following implantation into the uterine wall. However, the embryo still gastrulates as though it were sitting on top of a ball of yolk.

In mammals, a primitive streak forms and cell movements through the primitive streak give rise to the three primary germ layers, much the same as in birds (figure 53.14). Similarly, mammalian embryos envelop their "missing" yolk by forming a yolk sac from extraembryonic cells that migrate away from the lower layer of the blastoderm and line the blastocoel cavity.

Extraembryonic membranes are an adaptation to life on dry land

As an adaptation to terrestrial life, the embryos of reptiles, birds, and mammals develop within a fluid-filled **amniotic membrane** or **amnion** (chapter 35). The amniotic membrane and several other membranes form from embryonic cells, but they

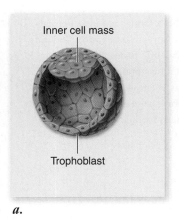

a.

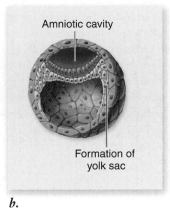

b.

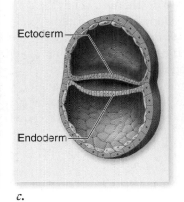

c.

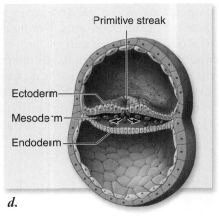

d.

figure 53.14

MAMMALIAN GASTRULATION. *a.* Cross section of the mammalian blastocyst at the end of cleavage. *b.* The amniotic cavity forms between the inner cell mass (ICM) and the pole of the embryo. Meanwhile, the ICM flattens and delaminates into two layers that will become ectoderm and endoderm. *b.* and *c.* Cells of the lower layer migrate out to line the blastocoel cavity to form the yolk sac. *d.* A primitive streak forms the ectoderm layer, and cells destined to become mesoderm migrate into the interior, similar to gastrulation in birds.

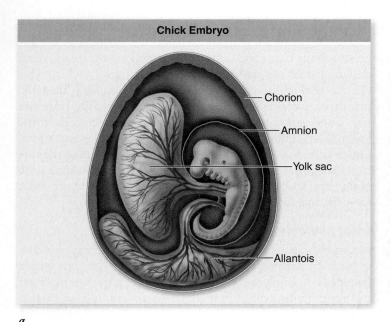

Chick Embryo

Chorion

Amnion

Yolk sac

Allantois

a.

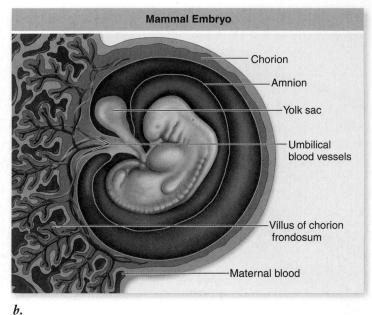

Mammal Embryo

Chorion

Amnion

Yolk sac

Umbilical
blood vessels

Villus of chorion
frondosum

Maternal blood

b.

figure 53.15

THE EXTRAEMBRYONIC MEMBRANES. The extraembryonic membranes in (*a*) a chick embryo and (*b*) a mammalian embryo share some of the same characteristics. However, in the chick, the allantois continues to grow until it eventually unites with the chorion just under the eggshell, where it is involved in gas exchange. In the mammalian embryo, the allantois contributes blood vessels to the developing umbilical cord.

are located outside of the body of the embryo. For this reason, they are known as **extraembryonic membranes.** The extraembryonic membranes include the amnion, chorion, yolk sac, and allantois.

In birds, the amnion and chorion arise from two folds that grow to completely surround the embryo (figure 53.15*a*). The amnion is the inner membrane that surrounds the embryo and suspends it in *amniotic fluid*, thereby mimicking the aquatic environments of fish and amphibian embryos. The chorion is located next to the eggshell and is separated from the other membranes by a cavity, the *extraembryonic coelom*.

The **yolk sac** plays a critical role in the nutrition of bird and reptile embryos; it is also present in mammals, although it does not nourish the embryo. The **allantois** is derived as an outpouching of the gut and serves to store the uric acid excreted in the urine of birds. During development, the allantois of a bird embryo expands to form a sac that eventually fuses with the overlying chorion, just under the eggshell. The fusion of the allantois and chorion form a functioning unit, the chorioallantoic membrane, in which embryonic blood vessels, carried in the allantois, are brought close to the porous eggshell for gas exchange. The chorioallantoic membrane is thus the respiratory membrane of a bird embryo.

In mammals, the trophoblast cells of the blastocyst implant into the endometrial lining of the mother's uterus and become the chorionic membrane (figure 53.15*b*). The part of the chorion in contact with endometrial tissue contributes to the placenta. The other part of the placenta is composed of modified endometrial tissue of the mother's uterus, as is described in more detail in a later section. The allantois in mammals contributes blood vessels to the structure that will become the umbilical cord, so that fetal blood can be delivered to the placenta for gas exchange.

Gastrulation produces the three germ layers from which all body tissues arise. In sea urchins, gastrulation involves formation of the endoderm by invagination of the blastula, and mesodermal cells form from other surface cells. In vertebrates with moderate to extensive amounts of yolk, gastrulation proceeds by the movement of surface cells through a blastopore or primitive streak, respectively. Mammalian gastrulation is similar to gastrulation in birds. The extraembryonic membranes of amniote species (reptiles, birds, and mammals) include the yolk sac, amnion, chorion, and allantois.

53.4 Organogenesis

Gastrulation establishes the basic body plan and creates the three primary germ layers of animal embryos. The stage is now set for **organogenesis**—the formation of the organs in their proper locations—which occurs by interactions of cells within and between the three germ layers. Thus, organogenesis follows rapidly on the heels of gastrulation, and in many animals begins before gastrulation is complete. Over the course of subsequent development, tissues develop into organs and animal embryos assume their unique body form (see table 53.1).

Changes in gene expression lead to cell determination

All of the cells in an animal's body, with the exception of a few specialized ones that have lost their nuclei, have the same complement of genetic information. Despite the fact that all of its cells are genetically identical, an adult animal contains dozens to hundreds of cell types, each expressing some unique aspect of the total genetic information for that individual. The information for other cell types is not lost, but most cells within a developing organism progressively lose the capacity to express ever-larger portions of their genomes. What factors determine which genes are to be expressed in a particular cell?

To a large degree, a cell's location in the developing embryo determines its fate. By changing a cell's location, an experimenter can often alter its developmental destiny, as mentioned in chapter 19. But this is only true up to a certain point in the cell's development. At some stage, every cell's ultimate fate becomes fixed, a process referred to as **cell determination.**

A cell's fate can be established by inheritance of cytoplasmic determinants or by interactions with neighboring cells. The process by which a cell or group of cells instructs neighboring cells to adopt a particular fate is called **induction.** If a nonporous barrier, such as a layer of cellophane, is imposed between the inducer and the target tissue, no induction takes place. In contrast, a porous filter, through which proteins can pass, does permit induction to occur.

In these experiments, researchers concluded that the inducing cells secrete a paracrine signal molecule that binds to the cells of the target tissue. Such signal molecules are capable of producing changes in the patterns of gene transcription in the target cells. You will learn more about the origin of embryonic induction a little later in this chapter.

Development of selected systems in *Drosophila melanogaster* illustrates organogenesis

In chapter 19, you saw how the creation of morphogen gradients in a fruit fly embryo leads to hierarchies of gene expression that direct cell fate decisions along both the anterior–posterior and dorsal–ventral axes. These two axes form a coordinate system to specify the position of tissues and organs within the *Drosophila* embryo. In this section we look at development of three different organs: salivary glands, the heart, and the tracheae of the respiratory system.

Salivary gland development

The fruit fly larva is a mobile eating machine, and thus it has very active salivary glands. The primordia of the salivary glands develop as simple tubular invaginations of ectodermal cells on the ventral surface of the third head segment.

Salivary glands develop only from an anterior strip of cells that express the *sex combs reduced* (*scr*) gene. No salivary glands form in *scr*-deficient embryos, whereas experimental expansion of *scr* expression along the anterior–posterior axis results in

the formation of additional salivary gland primordia along the length of the embryo.

The *scr* gene is one of the homeotic genes in the Antennapedia complex, which encode transcription factors that bind to DNA via their homeodomains to regulate gene expression (chapter 19). One downstream target of the *scr* gene is the *fork head* (*fkh*) gene, which has Scr binding sites in its enhancer. The *fkh* gene is required for secretory cell development in salivary gland rudiments, and it encodes a transcription factor that directly activates expression of salivary gland-specific genes. Thus, action of the *scr* gene activates *fkh* expression at the proper anterior location for salivary gland formation.

The inhibitory action of a dorsally expressed protein, Decapentaplegic (Dpp), determines the ventral position of the salivary glands. Activation of the Dpp-signaling pathway represses salivary gland specification in neighboring cells. This restricts development of salivary gland rudiments to their specific ventral patch of ectoderm cells (figure 53.16). In mutant embryos deficient for Dpp or any of the downstream Dpp-signaling proteins, salivary gland rudiments are not restricted to this ventral patch, and they form from the entire ectoderm of the third segment.

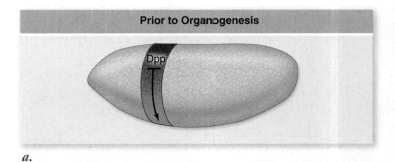

Prior to Organogenesis

Dpp

a.

During Organogenesis

Salivary gland

Labium

b.

figure 53.16

SALIVARY GLAND FORMATION IN *Drosophila*. Prospective salivary gland cells are determined by the intersection of the anterior–posterior and dorsal–ventral axes. *a.* Prior to organogenesis, the *sex combs reduced* (*scr*) gene is expressed in an anterior band of cells (shaded *blue*). At the same time, Decapentaplegic protein (Dpp) is released by cells on the dorsal side of the embryo, forming a gradient in the dorsal–ventral direction. Dpp specifies dorsal cell fates and inhibits formation of salivary gland rudiments. *b.* During organogenesis, the salivary glands develop in areas where Scr is expressed but Dpp is absent. Each salivary gland rudiment forms as a ventral invagination of the surface ectoderm on either side of the third head segment (the labium).

Heart development

The heart is a mesoderm-derived structure in all animals, and it is the first organ to become functional during embryonic development. The dorsal vessel is the heart-equivalent structure in *Drosophila melanogaster*. The homeobox-containing gene *tinman* is expressed in the prospective heart mesoderm and in the developing dorsal vessel, and its activity is required for dorsal vessel development in *Drosophila* (figure 53.17).

Dorsal vessel development in *Drosophila* is also dependent on two other types of transcription factors (known as GATA and T-box factors). In an illuminating case of evolutionary conservation, scientists have discovered gene families similar to each of these three *Drosophila* genes in vertebrates. Moreover, members of these gene families play important roles in vertebrate heart specification.

This evolutionary conservation includes not just the structure of these genes, but their function as well. Researchers have discovered that specification of cardiac mesoderm is subject to inductive signals from adjoining germ layers in both *Drosophila* and vertebrates. In vertebrates, the heart develops in an internal location, and the inductive signals come from the underlying anterior endoderm. In *Drosophila*, the dorsal vessel forms in a more superficial location, and the signals come from the overlying ectoderm.

Despite the different sources, the signals that regulate the expression of these three key types of transcription factors are themselves conserved between *Drosophila* and vertebrates. Given the critical and conserved circulatory function of the heart, it is perhaps not surprising that similar gene families mediate the specification of heart mesoderm in both *Drosophila* and vertebrates.

Tracheae: Branching morphogenesis

As you learned in chapters 34 and 49, insects exchange gases via a branching system of finer and finer tubes called **tracheae.** The repeated branching of simple epithelial tubes that leads to formation of the tracheal system is an example of **branching morphogenesis.**

Mutations in the *branchless* gene in *Drosophila* result in embryos with greatly reduced tracheal systems. The *branchless* gene encodes a member of the large family of **fibroblast growth factors (FGF),** which bind to receptor tyrosine kinase proteins (see chapter 9) to stimulate proliferation of target cells. In another interesting case of evolutionary conservation, the mammalian FGF homologue of the *branchless* gene is required for branching morphogenesis that creates the alveolar passageways in the mammalian lung.

In both animals, loose clusters of mesenchymal cells adjacent to distal regions of the epithelial tube secrete FGF. The FGF binds to a specific FGF receptor in the membrane of the epithelial cells, stimulating them to proliferate and to grow out into a new tube bud.

In vertebrates, organogenesis begins with neurulation and somitogenesis

The process of organogenesis in vertebrates begins with the formation of two morphological features found only in chordates: the **notochord** and the hollow **dorsal nerve cord** (see chapter 35). The development of the dorsal nerve cord is called **neurulation.**

Development of the neural tube

The notochord forms from mesoderm and is first visible soon after gastrulation is complete. It is a flexible rod located along the dorsal midline in the embryos of all chordates, although its function as a supporting structure is supplanted by the subsequent development of the vertebral column in the vertebrates. After the notochord has been laid down, the region of dorsal ectodermal cells situated above the notochord begins to thicken to form the **neural plate.**

The thickening is produced by the elongation of the dorsal ectoderm cells. Those cells then assume a wedge shape because of contracting bundles of actin filaments at their apical end. This change in shape causes the neural tissue to roll up into a **neural groove** running down the long axis of the embryo. The edges of the neural groove then move toward each other and fuse, creating a long hollow cylinder, the **neural tube** (figure 53.18). The neural tube eventually pinches off from the surface ectoderm to end up beneath the surface of the embryo's back. Regional changes, which are under control of the *Hox* gene complexes (see chapter 19), then occur in the neural tube as it differentiates into the spinal cord and brain.

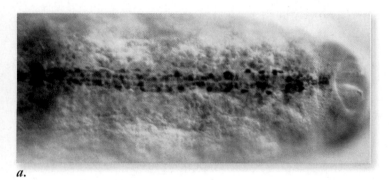

a.

b.

figure 53.17

A GENE CONTROLLING HEART FORMATION IN *Drosophila*. Called *tinman*, this gene is responsible for formation of the dorsal vessel (heart-equivalent). *a.* The brown dye shows expression of *tinman* in a normal embryo, in which the dorsal vessel develops along the center of the embryo. *b.* The dorsal vessel is missing in *tinman* mutant embryos.

inquiry

Why do you suppose that geneticists named this gene tinman?

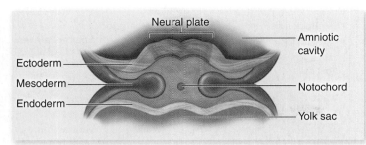

a.

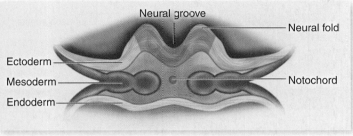

b.

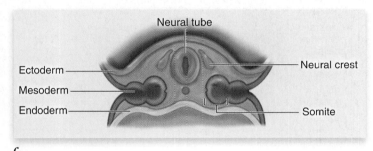

c.

figure 53.18

MAMMALIAN NEURAL TUBE FORMATION. *a.* The neural plate forms from ectoderm above the notochord. *b.* The cells of the neural plate fold together to form the neural groove. *c.* The neural groove eventually closes to form a hollow tube called the neural tube, which will become the brain and spinal cord. As the tube closes, some of the cells from the dorsal margin of the neural tube differentiate into the neural crest, migratory cells that form a variety of structure and are characteristic of vertebrates.

Generation of somites

While the neural tube is forming from dorsal ectoderm, the rest of the basic architecture of the body is being rapidly established by changes in the mesoderm. The sheets of mesoderm on either side of the developing notochord separate into a series of rounded regions called **somitomeres.** The somitomeres then separate into segmented blocks called **somites** (see figure 53.18). The mesoderm in the head region does not separate into discrete somites but remains connected as somitomeres, which form the skeletal muscles of the face, jaws, and throat.

Somites form in an anterior–posterior wave with a regular periodicity, which can be easily timed—for example, by using a vital dye, which marks cells without killing them, to mark each somite as it forms in a chick embryo. Cells at the presumptive boundary regions in the presomitic mesoderm instruct cells anterior to them to condense and separate into somites at certain times (for example, every 90 min in a chick embryo).

This "clock" appears to be regulated by contact-mediated cell signaling between neighboring cells.

Somites themselves are transient embryonic structures, and soon after their formation, cells disperse and start differentiating along different pathways to ultimately form the skeleton, skeletal musculature, and associated connective tissues. The total number of somites formed is species-specific; for example, chickens form 50 somites, whereas some species of snakes form as many as 400 somites.

Some body organs, including the kidneys, adrenal glands, and gonads, develop within a strip of mesoderm that runs lateral to each row of somites. The remainder of the mesoderm, which is most ventrally located, moves out and around the endoderm and eventually surrounds it completely. As a result of this movement, the mesoderm becomes separated into two layers. The outer layer is associated with the inner body wall, and the inner layer is associated with the outer lining of the gut tube. Between these two layers of mesoderm is the **coelom** (see chapter 32), which becomes the body cavity of the adult. Figure 53.19 shows the major mesoderm lineages of amniote embryos.

Migratory neural crest cells differentiate into many cell types

Neurulation occurs in all chordates, and the process in the simple lancelet, a nonvertebrate chordate, is much the same as it is in a human. However, neurulation is accompanied by an additional step in vertebrates. Just before the neural groove closes to form the neural tube, its edges pinch off, forming a small cluster of cells—the **neural crest**—between the roof of the neural tube and the surface ectoderm (figure 53.18c).

In another example of extensive cell movements during animal development, the neural crest cells then migrate away

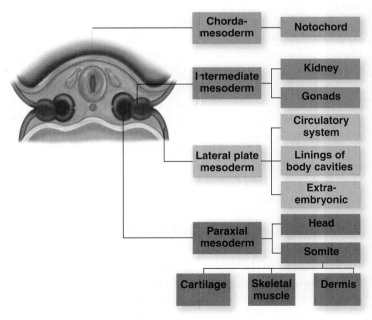

figure 53.19

MESODERM-DERIVED STRUCTURES OF BIRDS AND MAMMALS.

from the neural tube to colonize many different regions of the developing embryo. The appearance of the neural crest was a key event in the evolution of the vertebrates because neural crest cells, after reaching their final destinations, ultimately develop into many structures characteristic of the vertebrate body.

The differentiation of neural crest cells depends on their migration pathway and final location. Neural crest cells migrate along one of three pathways in the embryo. Cranial neural crest cells are anterior cells that migrate into the head and neck; trunk neural crest cells migrate along one of two different pathways (to be described shortly). Each population of neural crest cells develops into a variety of cell types.

Cranial neural crest cells' migration

Cranial neural crest cells contribute significantly to development of the skeletal and connective tissues of the face and skull, as well as differentiating into nerve and glial cells of the nervous system, and melanocyte pigment cells. Changes in the placement of cranial neural crest cells during development have led to the evolution of the great complexity and variety of vertebrate heads.

There are two waves of cranial neural crest cell migration. The first produces both dorsal and ventral structures, and the second produces only dorsal structures and makes much less cartilage and bone. Transplantation experiments indicate that the developmental potential of the cells in these two waves is identical. The differences in cell fate are due to the environment the migrating cells encounter and not due to prior determination of cell fate.

Trunk neural crest cells: Ventral pathway

Neural crest cells located in more posterior positions have very different developmental fates depending on their migration pathway. The first trunk neural crest cells that migrate away from the neural tube pass through the anterior half of each adjoining somite to ventral locations (figure 53.20a).

Some of these cells form the sensory neurons of the dorsal root ganglia, which send out projections to connect the periphery of the animal with the spinal cord (see chapter 44). Others become specialized as Schwann cells, which insulate nerve fibers to facilitate the rapid conduction of impulses along peripheral nerves. Still others form nerves of the autonomic ganglia, which regulate the activity of internal organs, and endocrine cells of the adrenal medulla (figure 53.20b). The chemical similarity of the hormone epinephrine and the neurotransmitter norepinephrine, released by sympathetic neurons of the autonomic nervous systems, may result because both adrenal medullary cells and sympathetic neurons derive from the neural crest.

Trunk neural crest cells: Lateral pathway

The second group of trunk neural crest cells migrate away from the neural tube in the space just under the surface ectoderm, to occupy this space around the entire body of the embryo. There, they will differentiate into the pigment cells of the skin (figure 53.20a, b). Mutations in genes that affect the survival and migration of neural crest cells lead to white spotting in the skin on ventral surfaces, as well as internal problems in other neural crest-derived tissues (figure 53.20c).

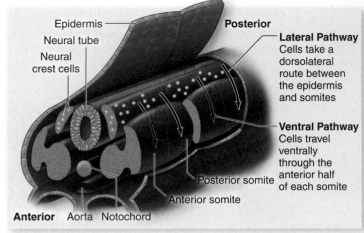

a.

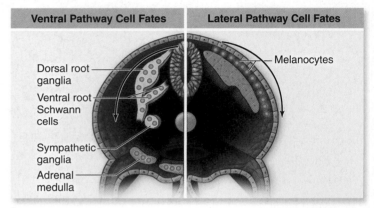

b.

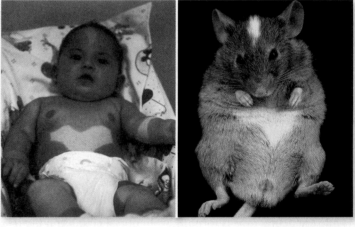

c.

figure 53.20

MIGRATION PATHWAYS AND CELL FATES OF TRUNK NEURAL CREST CELLS. *a.* The first wave of trunk neural crest cells migrates ventrally through the anterior half of each somite, whereas the second wave of cells leaves dorsally and migrates through the space between the epidermis and the somites. *b.* Ventral pathway neural crest cells differentiate into a variety of specialized cell types, but lateral pathway cells develop into the melanocytes (pigment cells) of the skin. *c.* A mutation in a gene that promotes survival of neural crest cells in all mammals leads to white spotting on the bellies and foreheads of both human babies and mice! Each individual is heterozygous for this mutation and thus has only half as much of the survival factor as unaffected individuals.

Because the fate of a neural crest cell is dictated by its migration pathway, many studies have been done to identify the molecules that control the migration pathways of neural crest cells. Cell adhesion molecules on cell surfaces and in the extracellular matrix are expected to play prominent roles. For example, prospective neural crest cells down-regulate the expression of *N*-cadherin on their surfaces, which enables them to break away from the neural tube. Then, soon after leaving the neural tube, integrin receptors appear on the surfaces of neural crest cells, allowing them to interact with proteins in the extracellular matrix pathways along which they will migrate.

Neural crest derivatives are important in vertebrate evolution

Primitive chordates such as lancelets are filter feeders, using the rapid beating of cilia to draw water into their mouths, which then exits through slits in their pharynx. These pharyngeal slits evolved into the vertebrate gill chamber, a structure that provides a greatly improved means of gas exchange. Thus, evolution of the gill chamber was certainly a key event in the transition from filter feeding to active predation, which requires a much higher metabolic rate.

In the development of the gill chamber, some of the cranial neural crest cells form cartilaginous bars between the embryonic pharyngeal slits. Other cranial neural crest cells induce portions of the mesoderm to form muscles along the cartilage, and still others form neurons that carry impulses between the central nervous system and these muscles.

Many of the unique vertebrate adaptations that contribute to their varied ecological roles involve structures that arise from neural crest cells. The vertebrates became fast-swimming predators with much higher metabolic rates. This accelerated metabolism permitted a greater level of activity than was possible among the more primitive chordates. Other evolutionary changes associated with the derivatives of the neural crest provided better detection of prey, a greatly improved ability to orient spatially during prey capture, and the means to respond quickly to sensory information. The evolution of the neural crest and of the structures derived from it were thus crucial steps in the evolution of the vertebrates (figure 53.21).

The genetic control of organogenesis in vertebrates and invertebrates relies on conserved families of cell-signaling molecules and transcription factors. The control of heart development in *Drosophila* and mammals uses some of the same proteins. The process of neurulation forms the basic nervous system in vertebrates. Neural crest cells arise from the neural tube and migrate to many sites to form a variety of cell types. The evolution of the neural crest led to the appearance of many vertebrate-specific adaptations.

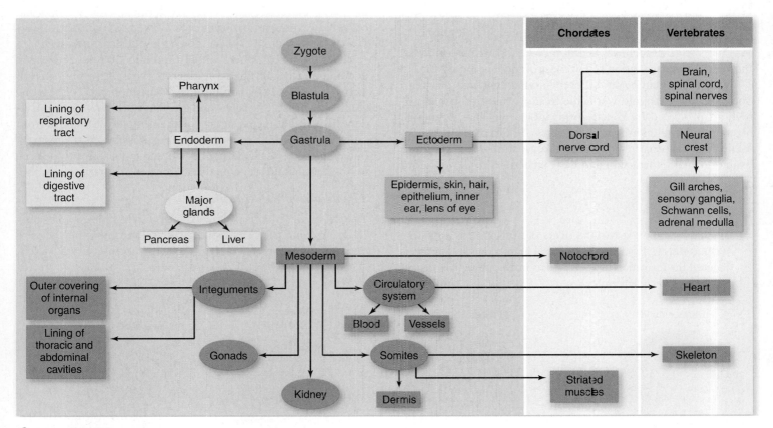

figure 53.21

GERM-LAYER DERIVATION OF THE MAJOR TISSUE TYPES IN ANIMALS. The three germ layers that form during gastrulation give rise to all the organs and tissues in the body, but the neural crest cells that form from ectodermal tissue give rise to structures that are prevalent in vertebrates, such as gill arches and bones of the face and skull.

chapter 53 *animal development* **1103**

In animal development, the relative position of cells in particular germ layers determines, to a large extent, the organs that develop from them. In *Drosophila*, you have seen that formation of morphogen gradients in the syncytial blastoderm establishes the anterior–posterior and dorsal–ventral axes of the embryo. The *Hox* gene complexes in vertebrates function similarly to the homeotic genes of *Drosophila* to specify the position of organs along the anterior–posterior axis. But how is cell fate selection along the dorsal–ventral axis accomplished in vertebrate embryos? Put another way, how do cells of the dorsal ectoderm "know" they are above the mesoderm-derived notochord, and thus fated to develop into the neural tube? The solution to this puzzle is one of the outstanding accomplishments of experimental embryology.

The Spemann organizer determines dorsal–ventral axis

The renowned German biologist Hans Spemann and his student Hilde Mangold solved this puzzle early in the twentieth century. Normally, cells derived from the dorsal lip of the blastopore of a gastrulating amphibian embryo give rise to the notochord. Spemann and Mangold removed cells of the dorsal lip from one embryo and transplanted them to a different location on another embryo (figure 53.22). The new location corresponded to that of the animal's future belly. They found that some of the embryos developed two notochords: a normal dorsal one, and a second one along the belly. Moreover, a complete set of dorsal axial structures (e.g., notochord, neural tube, and somites) formed at the ventral transplantation site in most of these embryos.

By using genetically different donor and host blastulas, Spemann and Mangold were able to show that the second notochord produced by transplanting dorsal lip cells contained host cells as well as transplanted ones. The transplanted dorsal lip cells had thus acted as **organizers,** stimulating cells that would normally form skin and belly structures to develop into dorsal axial structures. The belly cells must clearly contain the genetic information for dorsal axial developmental program, but they do not express it in the normal course of their development. Signals from the transplanted dorsal lip cells, however, must have caused them to do so.

How the organizer works

An organizer is a cluster of cells that release diffusible signal molecules, which convey positional information to other cells. As seen earlier, organizers can have a profound influence on the development of surrounding tissues. Working as signal beacons, they inform surrounding cells of their distance from the organizer. The closer a particular cell is to an organizer, the higher the concentration of the signal molecule (morphogen) it experiences (figure 53.23). Organizers and the diffusible morphogens that they release are thought to be part of a widespread mechanism for determining relative position and cell fates during vertebrate development.

The action of morphogens

The action of morphogens can be studied by using isolated portions of the blastula. The blastula can be bisected into an animal half (the animal cap) and vegetal half (the vegetal cap). If animal caps are removed from a frog blastula and cultured alone, they will form only ectoderm-derived epidermal cells. Similarly, cultured vegetal caps will form only endodermal cells. However, if animal caps are cultured combined with vegetal caps, the animal caps will form mesodermal structures.

The molecules involved in this induction have not been unambiguously identified. Members of the transforming growth factor beta (TGF-β) family have been implicated. These include activin, and *Xenopus* nodal-related proteins (Xnrs). Evidence for the inducing action of these molecules ranges from indirect: the timing and pattern of expression correlates with inducing tissue, to depleting developing embryos of these proteins with specific reagents that block gene expression.

The origin of the organizer

The problem remains of how cells of the frog blastopore's dorsal lip become the Spemann organizer and how they acquire their ability to specify cell fate along the dorsal–ventral axis. In

figure 53.22

SPEMANN AND MANGOLD'S DORSAL LIP TRANSPLANT EXPERIMENT. Tissue from the dorsal lip of a donor embryo induced the formation of a second axis in the future belly region of a second, recipient embryo.

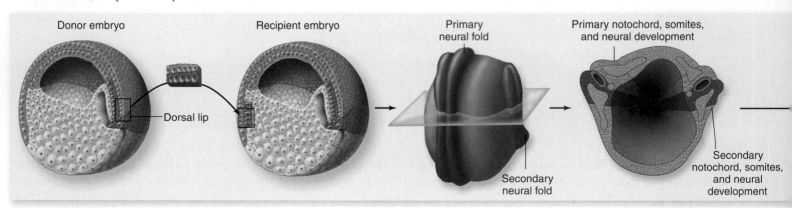

Donor embryo — Dorsal lip

Recipient embryo

Primary neural fold

Secondary neural fold

Primary notochord, somites, and neural development

Secondary notochord, somites, and neural development

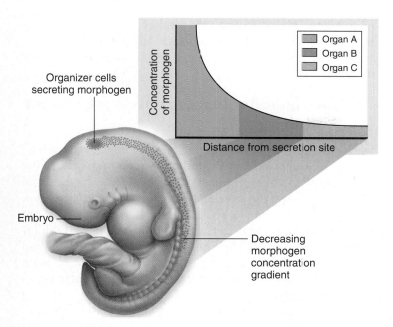

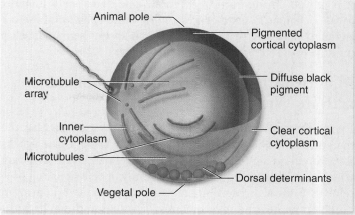

a.

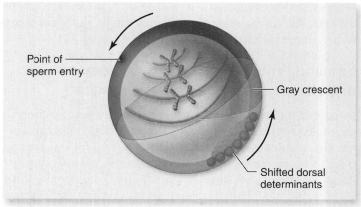

b.

figure 53.23

AN ORGANIZER CREATES A MORPHOGEN GRADIENT. As a morphogen diffuses from the organizer site, it becomes less concentrated. Different concentrations of the morphogen stimulate the development of different organs.

frogs, as in fruit flies, this process starts during oogenesis in the mother. At that time, maternally encoded dorsal determinants are put into the developing oocyte, one of which accumulates at the vegetal pole of the unfertilized egg. At fertilization, cytoplasmic rearrangements cause this determinant to shift to the future dorsal side of the egg.

First, a signal from the point of sperm entry initiates the assembly of a microtubule array, which enables the egg's plasma membrane and the underlying cortical cytoplasm to rotate over the surface of the deeper cytoplasm. This physical rotation shifts this maternally encoded dorsal determinant to the opposite side of the egg from the point of sperm entry (figure 53.24a, b). In some frogs, a gray crescent forms opposite the sperm entry point, as mentioned earlier, and this crescent marks the future site of the dorsal lip.

Cells that form in this area during cleavage (called the Nieuwkoop center for the scientist who did the previously mentioned animal cap studies) receive the dorsal determinants

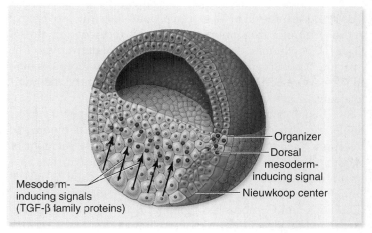

c.

figure 53.24

CREATION OF THE SPEMANN ORGANIZER. *a.* Dorsal determinants are localized at the vegetal pole of the unfertilized frog egg. At fertilization, a microtubule array forms at the site of sperm entry. These microtubules organize parallel microtubules to line the vegetal half of the egg between the cortex and cytoplasm. *b.* The cortical cytoplasm and dorsal determinants ride on this parallel array of microtubules, shifting to a site opposite sperm entry. *c.* Cells that inherit these shifted dorsal determinants form the Nieuwkoop center, which releases diffusible signaling molecules that specify the cells in the overlying dorsal marginal zone to become the organizer. The organizer forms at the area of the gray crescent, visible following the cytoplasmic rearrangements at fertilization.

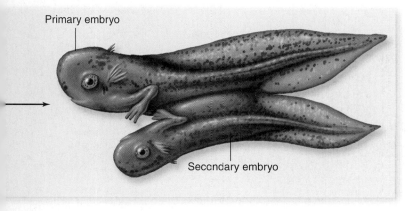

that moved during cortical rotation. The dorsal determinants cause a change in gene expression in these cells, producing a signaling molecule that induces the cells above them to develop into the dorsal lip of the blastopore (figure 53.24c).

Maternally encoded dorsal determinants activate Wnt signaling

Experiments carried out over the last 10 years suggest that the maternally encoded dorsal determinants in *Xenopus* are mRNAs for proteins that function in the intracellular **Wnt** signaling pathway. *Wnt* genes encode a large family of cell-signaling proteins that affect the development of a number of structures in both vertebrates and invertebrates. Turning on the Wnt pathway in the dorsal vegetal cells of the Nieuwkoop center leads ultimately to activation of a transcription factor, which moves into the nucleus to activate the expression of genes necessary for organizer specification.

Signaling molecules from the Spemann organizer inhibit ventral development

It has taken decades to establish the identity and function of the molecules that are synthesized by cells of the Spemann organizer to subsequently specify dorsal mesoderm cell fates in frogs. A surprising finding of recent experiments indicates that dorsal lip cells do not directly *activate* dorsal development. Instead, dorsal mesoderm development is a result of the *inhibition* of ventral development.

A protein called **bone morphogenetic protein 4 (BMP4)** is expressed in all marginal zone cells (the prospective mesoderm) of a frog embryo. Cells with receptors for BMP4 have the potential to develop into mesodermal derivatives. The specific mesodermal fate depends on how many receptors bind BMP4: More BMP4 binding induces a more ventral mesodermal fate.

The organizer functions by secreting a host of *inhibitory* molecules that can bind to BMP4 and prevent its binding to receptor. Such molecules are referred to as BMP4 antagonists. Up to 13 different proteins have been identified in the Spemann organizer, most of which appear to function as BMP4 antagonists. These include the proteins Noggin, Chordin, Dickkopf, and Cerebrus. Noggin and BMP4 are also involved in toe and finger joint formation, so humans homozygous for a *Noggin* mutation have fused joints.

Thus, the gradient of *inhibitory* molecules that emanates from the Spemann organizer leads to a declining level of BMP4 *function* in the ventral-to-dorsal direction. Cells farthest from the organizer bind the highest levels of BMP4 and differentiate into ventral mesoderm structures such as blood and connective tissues. Cells that are midway from the organizer bind intermediate amounts of BMP4, differentiate into intermediate mesoderm, and form organs such as the kidneys and gonads. BMP4 binding is completely inhibited by the high levels antagonists in the organizer itself. Thus, these cells adopt the most dorsal of mesoderm fates and develop into somites. The influence of the organizer also extends to ectoderm as inhibition of BMP4 in ectoderm leads to formation of neural tissue instead of epidermis (figure 53.25).

Evidence indicates that organizers are present in all vertebrates

In chicks, a group of cells at the anterior limit of the primitive streak called **Hensen's node** functions similarly to the dorsal lip of the blastopore: Hensen's node induces a second axis when transplanted to another area of a chick embryo. Recent studies have shown that cells of Hensen's node act like the Spemann organizer, secreting molecules that inhibit ventral development. These molecules are the same as found in frog embryos. Therefore, these experiments once again illustrate the evolutionary conservation of particular genes in animal development.

In addition, notochord signaling acts to pattern the neural tube. The notochord produces the signaling molecule sonic hedgehog (Shh), which is related to a signaling molecule in *Drosophila* called hedgehog. Signaling by Shh specifies ventral cell fate with dose-related effects similar to those described for TGF-β family proteins discussed earlier. In this way, induction by the notochord causes somites to form vertebrae, ribs, muscle, and skin depending on the levels of Shh cells are exposed to.

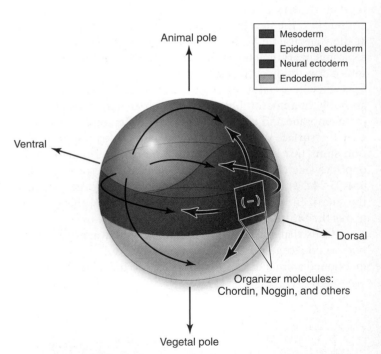

figure 53.25

FUNCTION OF THE SPEMANN ORGANIZER. The organizer is a hotbed of secreted molecules that bind to and antagonize the action of BMP4, a morphogen that at high levels specifies ventral mesoderm cell fates.

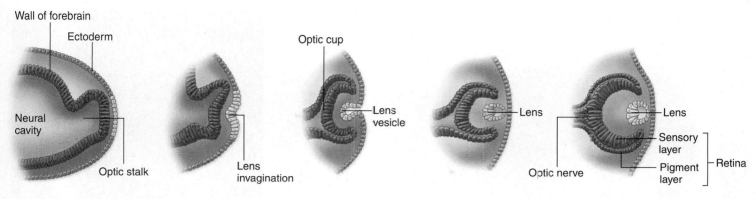

figure 53.26

DEVELOPMENT OF THE VERTEBRATE EYE BY INDUCTION. An extension of the optic stalk grows until it contacts the surface ectoderm, where it induces a section of the ectoderm to pinch off and form the lens. Other structures of the eye develop from the optic stalk, with lens cells reciprocally inducing the formation of photoreceptors in the optic cup.

Induction can be primary or secondary

The process of induction that Spemann initially discovered appears to be a fundamental mode of development in vertebrates. Inductions between the three primary germ layers—ectoderm, mesoderm, and endoderm—are referred to as **primary inductions.** The differentiation of the central nervous system during neurulation by the interaction of dorsal ectoderm and dorsal mesoderm to form the neural tube is an example of primary induction.

Inductions between tissues that have already been specified to develop along a particular developmental pathway are called **secondary inductions.** An example of secondary induction is the development of the lens of the vertebrate eye. The eye develops as an extension of the forebrain, a stalk that grows outward until it comes into contact with the surface ectoderm (figure 53.26). At a point directly above the growing stalk, a layer of the surface ectoderm pinches off, forming a transparent lens. The formation of lens from the surface ectoderm requires induction by the underlying neural ectoderm.

This was shown by transplantation experiments performed by Spemann. When the optic stalks of the two eyes have just started to project from the brain prior to lens formation, one of the budding stalks can be removed and transplanted underneath surface ectoderm in a region that would normally develop into the epidermis of the skin (such as that of the belly). When this is done, a lens forms from belly ectoderm cells in the region above where the budding stalk was transplanted. This lens forms due to inductive signals from the underlying optic stalk.

The dorsal lip of the blastopore and Henson's node of amniote embryos play equivalent roles in axis formation in vertebrates. By inhibiting BMP4, the organizer induces ectoderm to form neural tissue and mesoderm to form dorsal mesoderm. The notochord acts to pattern both the neural tube and somites with Shh. Primary inductions between germ layers lead to development of the vertebrate nervous system, whereas secondary inductions result in formation of structures such as the lens of the eye.

53.6 Human Development

Human development from fertilization to birth takes an average of 266 days, or about 9 months. This time is commonly divided into three periods called **trimesters.** We describe here the development of the embryo as it takes place during these trimesters. Later, we summarize the process of birth, nursing of the infant, and postnatal development of infants.

During the first trimester, the zygote undergoes rapid development and differentiation

About 30 hr after fertilization, the zygote undergoes its first cleavage; the second cleavage occurs about 30 hr after that. By the time the embryo reaches the uterus, 6–7 days after

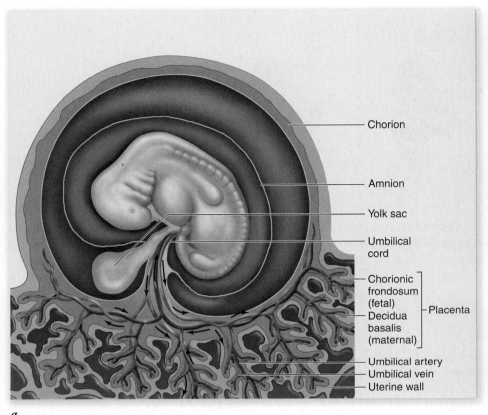

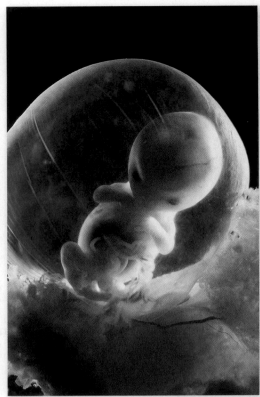

a.   *b.*

figure 53.27

STRUCTURE OF THE PLACENTA. *a.* The placenta contains a fetal component, the chorionic frondosum, and a maternal component, the decidua basalis. Deoxygenated fetal blood from the umbilical arteries (shown in *blue*) enters the placenta, where it picks up oxygen and nutrients from the mother's blood. Oxygenated fetal blood returns in the umbilical vein (shown in *red*) to the fetus. *b.* Note that the 7-week embryo is surrounded by a fluid-filled amniotic sac.

fertilization, it has differentiated into a blastocyst. As mentioned earlier, the blastocyst consists of an inner cell mass, which will become the body of the embryo, and a surrounding layer of trophoblast cells (see figure 53.10).

The trophoblast cells of the blastocyst digest their way into the endometrial lining of the uterus in the process known as **implantation.** The blastocyst begins to grow rapidly and initiates the formation of the amnion and the chorion.

Development in the first month

During the second week after fertilization, the developing chorion and the endometrial tissues of the mother engage to form the placenta (figure 53.27). Within the placenta, the mother's blood and the blood of the embryo come into close proximity but do not mix. Gases are exchanged, however, and the placenta provides nourishment for the embryo, detoxifies certain molecules that may pass into the embryonic circulation, and secretes hormones. Certain substances, such as alcohol, drugs, and antibiotics, are not stopped by the placenta and pass from the mother's bloodstream into the embryo.

One of the hormones released by the placenta is human chorionic gonadotropin (hCG), which was discussed in chapter 52. This hormone is secreted by the trophoblast cells even before they become the chorion, and it is the hormone assayed in pregnancy tests. Human chorionic gonadotropin maintains the mother's corpus luteum. The corpus luteum, in turn, continues to secrete estradiol and progesterone, thereby preventing menstruation and further ovulations.

Gastrulation also takes place in the second week after fertilization, and the three germ layers are formed. Neurulation occurs in the third week. The first somites appear, which give rise to the muscles, vertebrae, and connective tissues. By the end of the third week, over a dozen somites are evident, and the blood vessels and gut have begun to develop. At this point, the embryo is about 2 mm long.

Organogenesis begins during the fourth week (figure 53.28*a*). The eyes form. The tubular heart develops its four chambers and starts to pulsate rhythmically, as it will for the rest of the individual's life. At 70 beats per minute, the heart is destined to beat more than 2.5 billion times during a lifetime of 70 years. Over 30 pairs of somites are visible by the end of

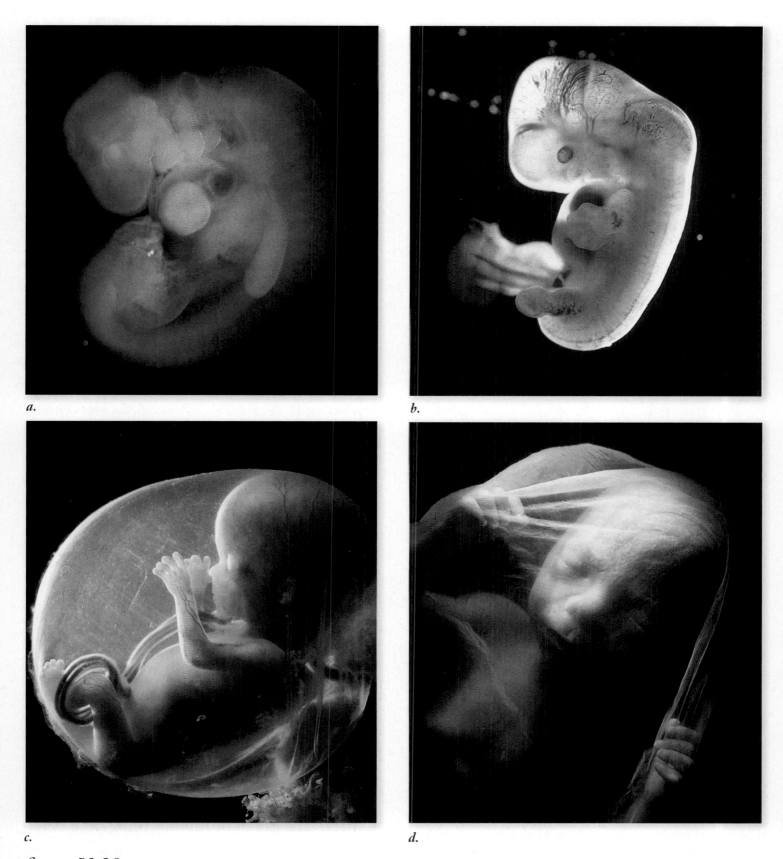

a.

b.

c.

d.

figure 53.28

THE DEVELOPING HUMAN. *a.* 4 weeks, (*b*) 7 weeks, (*c*) 3 months, and (*d*) 4 months.

the fourth week, and the arm and leg buds have begun to form. The embryo has increased in length to about 5 mm. Although the developmental scenario is now far advanced, many women are still unaware they are pregnant at this stage. Most spontaneous abortions (miscarriages), which frequently occur in the case of a defective embryo, occur during this period.

The second month

Organogenesis continues during the second month (figure 53.28b). The miniature limbs of the embryo assume their adult shapes. The arms, legs, knees, elbows, fingers, and toes can all be seen—as well as a short bony tail. The bones of the embryonic tail, an evolutionary reminder of our past, later fuse to form the coccyx.

Within the abdominal cavity, the major organs, including the liver, pancreas, and gallbladder, become evident. By the end of the second month, the embryo has grown to about 25 mm in length, weighs about 1 g, and begins to look distinctly human. The ninth week marks the transition from embryo to fetus. At

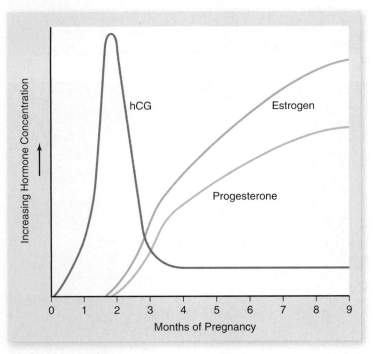

figure 53.29

HORMONAL SECRETION BY THE PLACENTA. The placenta secretes human chorionic gonadotropin (hCG), which peaks in the second month and then declines. After 5 weeks, it secretes increasing amounts of estrogen and progesterone.

 inquiry

The high levels of estradiol and progesterone secreted by the placenta prevent ovulation and thus formation of any additional embryos during pregnancy. What would be the expected effect of these high hormone levels in the absence of pregnancy?

this time, all of the major organs of the body have been established in their proper locations.

The third month

The nervous system develops during the third month, and the arms and legs start to move (see figure 53.28c). The embryo begins to show facial expressions and carries out primitive reflexes such as the startle reflex and sucking.

At around 10 weeks, the secretion of hCG by the placenta declines, and the corpus luteum regresses as a result. However, menstruation does not occur because the placenta itself secretes estradiol and progesterone (figure 53.29).

The high levels of estradiol and progesterone in the blood during pregnancy continue to inhibit the release of FSH and LH, thereby preventing ovulation. They also help maintain the uterus and eventually prepare it for labor and delivery, and they stimulate the development of the mammary glands in preparation for lactation after delivery.

During the second trimester, the basic body plan develops further

Bones actively enlarge during the fourth month (see figure 53.28d), and by the end of the month, the mother can feel the baby kicking. By the end of the fifth month, the rapid heartbeat of the fetus can be heard with a stethoscope, although it can also be detected as early as 10 weeks with a fetal monitor.

Growth begins in earnest in the sixth month; by the end of that month, the fetus weighs 600 g (1.3 lb) and is over 300 mm (1 ft) long. Most of its prebirth growth is still to come, however. The fetus cannot yet survive outside the uterus without special medical intervention.

During the third trimester, organs mature to the point at which the baby can survive outside the womb

The third trimester is predominantly a period of growth and maturation of organs. The weight of the fetus doubles several times, but this increase in bulk is not the only kind of growth that occurs. Most of the major nerve tracts in the brain, as well as many new neurons (nerve cells), are formed during this period. Neurological growth is far from complete when birth takes place, however. If the fetus remained in the uterus until its neurological development was complete, it would grow too large for safe delivery through the pelvis. Instead, the infant is born as soon as the probability of its survival is high, and its brain continues to develop and produce new neurons for months after birth.

Critical changes in hormones bring on birth

In some mammals, changing hormone levels in the developing fetus initiate the process of birth. The fetuses of these mammals have an extra layer of cells in their adrenal cortex, which secrete corticosteroids that induce the uterus of the mother to manufacture prostaglandins. Prostaglandins trigger powerful contractions of the uterine smooth muscles.

The adrenal glands of human fetuses lack this extra layer of cells, and human birth does not seem to be initiated by this mechanism. The mother's uterus releases prostaglandins, possibly as a result of the high levels of estradiol secreted by the placenta. Estradiol also stimulates the uterus to produce more oxytocin receptors, and as a result, the uterus becomes increasingly sensitive to oxytocin.

Prostaglandins begin the uterine contractions, but then sensory feedback from the uterus stimulates the release of oxytocin from the mother's posterior pituitary gland. Working together, oxytocin and prostaglandins further stimulate uterine contractions, forcing the fetus downward (figure 53.30). This positive feedback mechanism accelerates during labor. Initially, only a few contractions occur each hour, but the rate eventually increases to one contraction every 2–3 min. Finally, strong contractions, aided by the mother's voluntary pushing, expel the fetus, which is now a newborn baby, or *neonate*.

After birth, continuing uterine contractions expel the placenta and associated membranes, collectively called the *afterbirth*. The umbilical cord is still attached to the baby, and

to free the newborn, a doctor or midwife clamps and cuts the cord. Blood clotting and contraction of muscles in the cord prevent excessive bleeding.

Nursing of young is a distinguishing feature of mammals

Milk production, or **lactation,** occurs in the alveoli of mammary glands when they are stimulated by the anterior pituitary hormone prolactin. Milk from the alveoli is secreted into a series of alveolar ducts, which are surrounded by smooth muscle and lead to the nipple.

During pregnancy, high levels of progesterone stimulate the development of the mammary alveoli, and high levels of estradiol stimulate the development of the alveolar ducts. However, estradiol blocks the actions of prolactin on the mammary glands, and it inhibits prolactin secretion by promoting the release of prolactin-inhibiting hormone from the hypothalamus. During pregnancy, therefore, the mammary glands are prepared for, but prevented from, lactating.

When the placenta is discharged after birth, the concentrations of estradiol and progesterone in the mother's blood decline rapidly. This decline allows the anterior pituitary gland to secrete prolactin, which stimulates the mammary alveoli to produce milk. Sensory impulses associated with the baby's suckling trigger the posterior pituitary gland to release oxytocin. Oxytocin stimulates contraction of the smooth muscle surrounding the alveolar ducts, thus causing milk to be ejected by the breast. This pathway is known as the *milk let-down reflex,* and it is found in other mammals as well. The secretion of oxytocin during lactation also causes some uterine contractions, as it did during labor. These contractions help restore the tone of uterine muscles in mothers who are breast-feeding.

The first milk produced after birth is a yellowish fluid called **colostrum,** which is both nutritious and rich in maternal antibodies. Milk synthesis begins about 3 days following the birth and is referred to as the milk "coming in." Many mothers nurse for a year or longer. When nursing stops, the accumulation of milk in the breasts signals the brain to stop secreting prolactin, and milk production ceases.

Postnatal development in humans continues for years

Growth of the infant continues rapidly after birth. Babies typically double their birth weight within 2 months. Because different organs grow at different rates and cease growing at different times, the body proportions of infants are different from those of adults. The head, for example, is disproportionately large in newborns, but after birth it grows more slowly than the rest of the body. Such a pattern of growth, in which different components grow at different rates, is referred to as **allometric growth.**

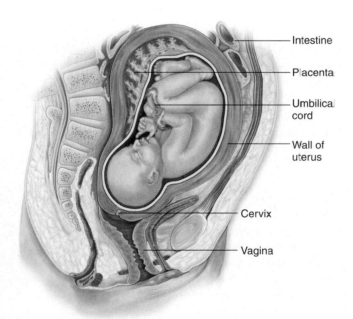

figure 53.30

POSITION OF THE FETUS JUST BEFORE BIRTH. A developing fetus causes major changes in a woman's anatomy. The stomach and intestines are pushed far up, and considerable discomfort often results from pressure on the lower back. In a natural delivery, the fetus exits through the cervix, which must dilate (expand) considerably to permit passage.

Labels in figure: Intestine, Placenta, Umbilical cord, Wall of uterus, Cervix, Vagina

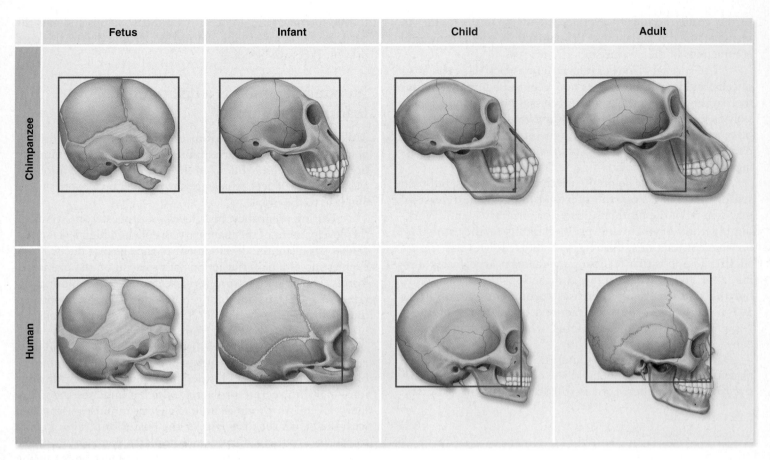

	Fetus	Infant	Child	Adult
Chimpanzee				
Human				

figure 53.31

ALLOMETRIC GROWTH. In young chimpanzees, the jaw grows at a faster rate than the rest of the head. As a result, the shape of the adult head differs greatly from its shape as a newborn. In humans, the difference in growth between the jaw and the rest of the head is much smaller, and the adult head shape is similar to that of the newborn.

In most mammals, brain growth is mainly a fetal phenomenon. In chimpanzees, for instance, the brain and the cerebral portion of the skull grow very little after birth, whereas the bones of the jaw continue to grow. As a result, the head of an adult chimpanzee looks very different from that of a fetal or infant chimpanzee (figure 53.31). In human infants, by contrast, the brain and cerebral skull grow at the same rate as the jaw. Therefore, the jaw–skull proportions do not change after birth, and the head of a human adult looks very similar to that of a human fetus or infant.

The fact that the human brain continues to grow significantly for the first few years of postnatal life means that adequate nutrition and a safe environment are particularly crucial during this period for the full development of a person's intellectual potential.

The critical stages of human development take place quite early in gestation, during the first trimester; the subsequent 6 months are essentially a period of growth and maturation. The growth of the brain is not yet complete, however, by the end of the third trimester, and must be completed postnatally. Hormones in the bloodstream of the mother maintain the nutritive uterine environment for the developing fetus; changes in hormone secretion and levels stimulate birth and lactation.

53.1 Fertilization

In all sexually reproducing organisms, the first step is fertilization—the union of male and female gametes (table 53.1).

■ Fertilization occurs in three stages: sperm penetration and membrane fusion, egg activation, and fusion of nuclei.

■ For a sperm to fertilize an egg it must penetrate the external layers to reach the plasma membrane (figure 53.1).

■ The acrosome contains digestive enzymes that enable the sperm to tunnel its way through the external layers to the ovum.

■ Fusion of the plasma membranes of the egg and sperm allow the sperm nucleus to pass directly into the egg cytoplasm.

■ Fusion of membranes triggers egg activation by the release of calcium, which initiates changes in the egg (figure 53.2).

■ Blocks to polyspermy include changes in membrane potential and altering of the external coat of the egg.

■ Sperm penetration has other effects on the egg: meiosis is completed; cytoplasmic rearrangements occur and protein synthesis increases sharply (figures 53.4, and 53.5).

■ Fertilization is complete when the haploid sperm nucleus fuses with the haploid egg nucleus.

53.2 Cleavage and Blastula Stage

The second major event in animal development is the rapid division of the zygote resulting in a large number of small cells.

■ Cleavage is a periodic series of cell divisions that does not increase the size of the embryo but produces smaller cells called blastomeres.

■ The outermost blastomeres join by tight junctions, and Na^+ pumped into the intracellular spaces creates an osmotic gradient, bringing in water and creating a hollow ball of cells called a blastula.

■ Cleavage patterns are diverse and are influenced by the amount of yolk (table 53.2).

■ Cleavage in mammals is holoblastic, but the hollow blastocyst contains an ICM similar to the blastoderm found in reptiles and birds (figure 53.10).

■ Blastomeres may or may not be committed to a developmental path.

53.3 Gastrulation

Gastrulation establishes the basic body plan and creates the three primary germ layers of animal embryos.

■ Gastrulation produces the three germ layers: endoderm, ectoderm, and mesoderm (table 53.3).

■ Cells move during gastrulation using a variety of cell shape changes, resulting in invagination or other movements such as involution, ingression, and delamination.

■ Cell movement in gastrulation also varies with the amount of yolk.

■ In yolk-poor eggs gastrulation begins with the ingression of cells from the flattened vegetal plate to form the primitive gut, or archenteron.

■ In frogs, a layer of cells migrates toward the vegetal pole, ultimately involuting through the dorsal lip of the blastopore and eventually encasing the yolk (figure 53.12).

■ In birds, the blastoderm delaminates to form a blastocoel, and surface cells migrate through the primitive streak (figure 53.13).

■ Mammalian gastrulation is similar to birds (figure 53.14).

■ As an adaptation to life on dry land, amniotic species develop several extraembryonic membranes: the yolk sac, amnion, chorion and allantois, which nourish and protect the developing embryo (figure 53.15).

53.4 Organogenesis

Organogenesis is the formation of organs in their proper locations by interactions of cells within and between the three germ layers.

■ To a large degree, a cell's location in the developing embryo determines its fate—a process called cell determination.

■ A cell's fate can be established by inheritance of cytoplasmic determinants from the mother and by interactions of cells, called induction (figures 53.16 and 53.17).

■ In vertebrates, organogenesis begins with neurulation and somitogenesis (figures 53.18–53.20).

■ Neural crest derivatives are important in vertebrate evolution and have led to their many ecological roles (figure 53.21).

53.5 Vertebrate Axis Formation

Experimental embryology determined how cell fate selection occurs in the dorsal–ventral axis.

■ The dorsal–ventral axis is determined by the Spemann organizer. Organizers are a cluster of cells that produce gradients of diffusible signal molecules, conveying positional information to other cells (figures 53.23 and 53 24).

■ Axis formation begins with maternally encoded dorsal determinants, probably mRNA, that are transported 180° away from the point of sperm entry.

■ Morphogens can either activate or inhibit development along a certain path. The Spemann organizer induces formation of the dorsum by inhibiting ventral development (figure 53.25).

■ Primary induction occurs between the three germ layers and secondary induction occurs between tissues that are already specified.

53.6 Human Development

Human development takes an average of 266 days and is divided into three 3-month periods called trimesters.

■ Trophoblast cells digest their way into the endometrial tissues of the uterus and implant.

■ During the first trimester the zygote undergoes rapid development and differentiation.

■ During the second week the chorion of the embryo and the endometrial tissues of the woman form the placenta (figure 53.27).

■ Gastrulation occurs in the second week, and by the third week neurulation and somitogenesis occur, and blood vessels appear.

■ Organogenesis begins during the fourth week, and the arms and legs begin to appear.

■ During the second month organogenesis continues, and the appendages and digits form along with a tail.

■ The eighth week marks the transition from embryo to fetus, when all the major organs are in position.

■ During the third month the nervous system develops.

■ During the second trimester growth continues, and the basic body plan develops further.

■ The third trimester is a period of growth and organ maturation such that the fetus can survive outside the womb.

■ Birth is initiated by secretions of corticosteroids from the fetal adrenal cortex that induce prostaglandins, which cause contractions.

■ Nursing involves a neuroendocrine reflex, causing the release of oxytocin and the milk let-down response.

■ Postnatal development continues with different organs growing at different rates—called allometric growth (figure 53.31).

SELF TEST

1. Your cousin just had twins. She tells you that twinning occurs when two sperm fertilize the same egg. You reply that—
   a. Yes, she is right, that is the most common source of twinning.
   b. No, only one sperm survives passage through the uterine cervix, so two sperm are never present at fertilization.
   c. No, cortical granules are used to prevent additional sperm penetration.
   d. No, twinning occurs when unfertilized eggs divide spontaneously and thus is parthenogenic in nature.

2. Prior to fertilization, the mammalian egg completes the second meiotic division.
   a. This statement is true; meiosis II is actually completed before ovulation.
   b. This statement is false because eggs don't undergo meiosis; they are descended by cloning from a haploid germ cell line.
   c. This statement is false because mammalian eggs don't begin meiosis until after fertilization.
   d. This statement is false because mammalian eggs do not release the second polar body until the sperm has entered the egg.

3. Which of the following events occur immediately after fertilization?
   a. Egg activation
   b. Polyspermy defense
   c. Cytoplasm changes
   d. All of these occur after fertilization.

4. Suppose that a burst of electromagnetic radiation were to strike the blastomeres of only the animal pole. Which of the following would be most likely to occur?
   a. A change or mutation relevant to the epidermis or skin
   b. A switching of the internal organs so that reverse orientation (left/right) occurs along the midline of the body
   c. The migration of the nervous system to form outside of the body
   d. Failure of the reproductive system to develop

5. Which of the following plays the greatest role in determining how cytoplasmic division occurs during cleavage?
   a. Number of chromosomes
   b. Amount of yolk
   c. Orientation of the vegetal pole
   d. Sex of the zygote

6. Gastrulation is a critical event during development. Why?
   a. Gastrulation converts a hollow ball of cells into a bilaterally symmetrical structure.
   b. Gastrulation causes the formation of a primitive digestive tract.
   c. Gastrulation causes the blastula to develop a dorsal–ventral axis.
   d. All of these are significant events that occur during gastrulation.

7. Gastrulation in a mammal would be most similar to gastrulation in—
   a. gecko.
   b. tuna.
   c. eagle.
   d. no other species; mammalian gastrulation is unique.

8. Somites—
   a. begin forming at the tail end of the embryo and then move forward in a wavelike fashion.
   b. are derived from endoderm.
   c. develop into only one type of tissue per somite.
   d. may vary in number from one species to the next.

9. Of the following processes, which occurs last?
   a. Cleavage
   b. Neurulation
   c. Gastrulation
   d. Fertilization

10. Which of the following would qualify as a secondary induction?
    a. The formation of the lens of the eye due to induction by the neural ectoderm
    b. Differentiation during neurulation by the dorsal ectoderm and mesoderm
    c. Both of these
    d. Neither of these

11. Your Aunt Ida thinks that babies can stimulate the onset of their own labor. You tell her that—
    a. No, among mammals the onset of labor has been most closely linked to a change in the phases of the Moon.
    b. No, it is the mother's circadian clock that determines the onset of labor.
    c. Yes, body weight determines the onset of labor.
    d. Yes, changes in fetal hormone levels can determine the onset of labor.

12. Drug or alcohol exposure during which of the following stages is most likely to have a profound effect on the neural development of the fetus?
    a. Preimplantation
    b. First trimester
    c. Second trimester
    d. Third trimester

13. Axis formation in amniotic embryos can be affected by—
    a. mutation of the dorsal lip of the blastopore.
    b. mutation of the primitive streak structure.
    c. both of these.
    d. neither of these.

14. How would a given cell know whether it was to become skin or muscle tissue?
    a. The level of Sonic hedgehog molecules it is exposed to.
    b. The size of nearby cells.
    c. The number of chromosomes within its nucleus.
    d. The age of the embryo when the cell differentiates.

CHALLENGE QUESTIONS

1. Suppose you discover a new species whose development mechanisms have not been documented before. How could you determine at what stage the cell fate is determined?

2. You look up from your studying to see your dog, Fifi, acting silly again. Using this as a teachable moment, compare and contrast the homeoboxes in your dog and the fruit fly she just ate.

3. Why doesn't a woman menstruate while she is pregnant?

4. Spemann and Mangold were able to demonstrate that some cells act as "organizers" during development. What types of cells did they use? How did they determine that these cells were organizers?

chapter **54**

Behavioral Biology

introduction

ORGANISMS INTERACT WITH their environment in many ways. To understand these interactions, we need to appreciate the internal factors that shape the way an animal behaves, as well as aspects of the external environment that affect individual organisms. In this chapter, we explore the mechanisms that determine an animal's behavior and examine the field of behavioral ecology, which investigates how natural selection has molded behavior through evolutionary time.

Approaches to the Study of Behavior

Behavior can be defined as the way an animal responds to stimuli in its environment. A stimulus might be as simple as detection of the presence of food in the environment. In this sense, a bacterial cell "behaves" by moving toward higher concentrations of a sugar in its surrounding medium. This behavior is very simple and is well suited to the life of bacteria, allowing these organisms to live and reproduce.

As animals evolved, they occupied different environments and faced diverse problems that affected their survival and reproduction. Their nervous systems and behavior concomitantly became more complex. Nervous systems perceive and process information and trigger adaptive motor responses, which we see as patterns of behavior.

Behavior's two components are its immediate cause and its evolutionary origin

We can talk about animal behavior in two different ways. First, we might ask *how* it all works—that is, how the animal's senses, nerve networks, or internal state provide a physiological basis for the behavior. The "how" of behavior is a question about *proximate causation.* To analyze the proximate cause of behavior, we might measure hormone levels or record the impulse activity of neurons in the animal. For example, a male songbird may sing during the breeding season because of an increased level of the steroid sex hormone testosterone, which binds to receptors in the brain and triggers the production of song; this explanation would describe the proximate cause of the male bird's song.

Why a behavior evolved—an investigation of its adaptive value—is a question concerning *ultimate causation.* To study the ultimate cause of a behavior, we attempt to determine how it influenced the animal's survival or reproductive success. A male bird sings to defend a territory from other males and to attract a female with which to reproduce; this is the ultimate, or evolutionary, explanation for the male's vocalization.

The study of behavior has had a long history of controversy. One source of controversy has been the question of whether behavior is determined more by an individual's genes or by its learning and experience. In other words, is behavior the result of nature (instinct) or nurture (experience)? In the past, this question was considered an either/or proposition, but we now know that instinct and experience both play significant roles, often interacting in complex ways to produce the final behavior.

Innate behavior does not require learning

Early research in the field of animal behavior focused on behavioral patterns that are always exhibited by members of a species in response to a particular stimulus; that is, they appeared to be instinctive, or **innate, behaviors.** Because behavior is often stereotyped (appearing in the same way in different individuals of a species), these early researchers argued that it must be based on preset paths in the nervous system. In their view, these paths are structured from genetic blueprints and cause animals to show essentially the same behavior from the first time it is produced throughout their lives.

These researchers based their opinions on behaviors such as egg retrieval by geese. Geese incubate their eggs in a nest. If a goose notices that an egg has been knocked out of the nest, it will extend its neck toward the egg, get up, and roll the egg back into the nest with a side-to-side motion of its neck while the egg is tucked beneath its bill (figure 54.1). Even if the egg is removed during retrieval, the goose completes the behavior, as if driven by a program released by the initial sight of the egg outside the nest.

Egg retrieval behavior is triggered by a **sign stimulus,** the appearance of an egg out of the nest. A component of the goose's nervous system, the **innate releasing mechanism,** provides the neural instructions for the motor program, or **fixed action pattern,** involved in egg retrieval. More generally, the sign stimulus is a signal in the environment that triggers a behavior. The innate releasing mechanism is the sensory mechanism that detects the signal, and the fixed action pattern is the stereotyped act.

One interesting aspect of sign stimuli is that they are often not very specific; in some situations, a wide variety of objects will trigger a fixed action pattern. For example, geese will attempt to roll baseballs and even beer cans back into their nests. Moreover, once the objects are in the nest, the goose recognizes that they are not eggs and removes them!

A similar example is provided by male stickleback fish. During the breeding season, males develop bright red coloration on their undersides. Territorial males react aggressively to the approach of other males, performing an aggressive dis-

figure 54.1

INNATE EGG-ROLLING RESPONSE IN GEESE. The series of movements used by a goose to retrieve an egg is a fixed action pattern. Once it detects the sign stimulus (in this case, an egg outside the nest), the goose goes through the entire set of movements: It will extend its neck toward the egg, get up, and roll the egg back into the nest with a side-to-side motion of its neck while the egg is tucked beneath its bill.

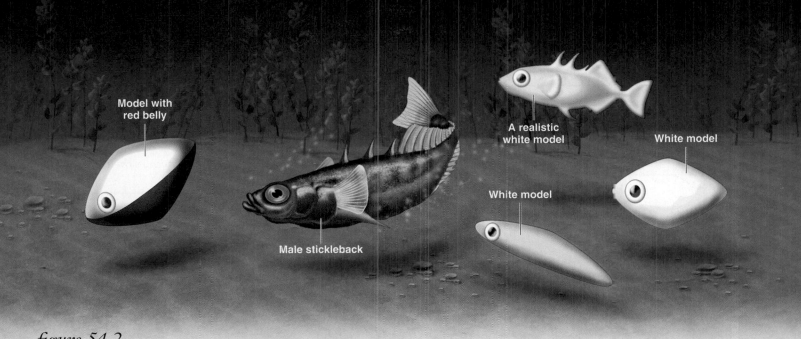

figure 54.2

SIGN STIMULUS IN STICKLEBACK FISH. Male sticklebacks are very territorial and aggressively defend their territories against other males. Territorial males have red bellies; as a result, red is the sign stimulus that elicits aggressive behavior. In laboratory studies (in which fish are usually presented only one model at a time), the presence of a red stripe determines whether the male reacts aggressively.

play and even attacking. When researcher Niko Tinbergen observed a male stickleback in a laboratory aquarium displaying aggressively when a red mail truck passed by the window, he realized that the red coloration was the sign stimulus (figure 54.2). Subsequent experiments revealed that males would respond to many unfishlike models as long as the models had a red stripe. Tinbergen won the 1973 Nobel Prize in medicine or physiology for this work.

This phenomenon is taken one step further by what are termed **supernormal stimuli.** Given a choice between two sign stimuli, one of normal size and the other much larger, many animals will respond to the larger of the two. Thus, given a choice of a normal goose egg and one the size of a volleyball, a goose will attempt to roll the bigger one back to the nest.

Why supernormal stimuli exist is not always clear. One aspect to keep in mind, however, is that in many cases, supernormal stimuli do not occur in nature. Thus, geese may prefer eggs the size of volleyballs, but they never encounter eggs of that size. It may be that geese have evolved to respond to the larger object so that they will attend to eggs, rather than to smaller, circular rocks. As a result, natural selection may have favored the evolution of a preference for larger objects. This general response may lead to unexpected outcomes in experiments, but probably doesn't often lead to maladaptive behavior in nature.

Early research in animal behavior emphasized innate behaviors that are the result of preset pathways in the nervous system and thus are likely to be genetically controlled.

54.2 Behavioral Genetics

The study of behavioral genetics deals with those components of behavior that are hereditary—governed by genes and passed from one generation to the next. A wide variety of data, from artificial selection experiments to modern molecular genetics, indicates that behavioral differences among individuals often result from genetic differences.

Rats can be artificially selected for learning capacity

A famous experiment in the 1940s studied the ability of rats to find their way through a maze consisting of many blind alleys and only one exit, where a reward of food awaited them.

Some rats quickly learned to zip through the maze to the food, making few incorrect turns, but other rats took much longer to learn the correct path.

Researchers bred the fast learners with one another to establish a "maze-bright" colony, and the slow learners with one another to establish a "maze-dull" colony. Offspring in each colony were then tested to see how quickly they learned the maze. The offspring of maze-bright rats learned even more quickly than their parents had, and the offspring of maze-dull parents were even poorer at maze learning. Repeating this selection for several generations led to two behaviorally distinct types of rat with very different maze-learning abilities (figure 54.3).

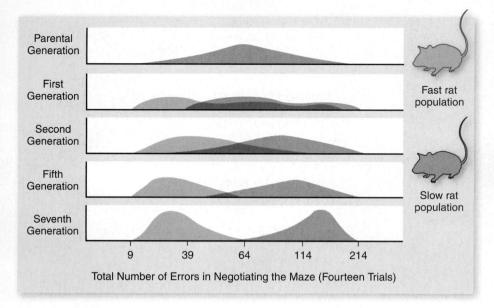

figure 54.3

THE GENETICS OF LEARNING. The fastest rats (those that made the fewest errors) in the parental population were used to establish one population (*green*), and the slowest rats to establish the other (*red*). In subsequent generations, the same procedure was used to select for fast (*green*) and slow (*red*) rats.

inquiry

? *What would happen if, after the seventh generation, rats were randomly assigned mates regardless of their ability to learn the maze?*

Clearly, the ability to learn the maze was to some degree hereditary. Furthermore, those genes appeared to be specific to this behavior, because the two groups of rats did not differ in their ability to perform other behavioral tasks, such as running a completely different kind of maze. This research demonstrates how a study can reveal that behavior has a heritable component.

Human twin studies reveal similarities independent of environment

The role of genetics can also be seen in humans by comparing the behavior of identical twins. Identical twins are, as their name implies, genetically identical, and most sets of identical twins are raised in the same environment, so it is not possible to determine whether similarities in behavior result from their genetic similarity or from environmental experiences shared as they grew up (the classic nature-versus-nurture debate). In some cases, however, twins have been separated at birth and raised in different families.

A recent study of 50 such sets of twins revealed many similarities in personality, temperament, and even leisure-time activities, even though the twins had often been raised in very different environments. These similarities indicate that genetics plays a role in determining behavior even in humans, although the relative importance of genetics versus environment is still hotly debated.

Some behaviors appear to be controlled by a single gene

The maze-learning and identical twins studies just described suggest that genes play a role in behavior, but recent research has provided much greater detail on the genetic basis of behavior. In both *Drosophila* and mice, many mutations have been associated with particular behavioral abnormalities.

In fruit flies, for example, individuals that possess alternative alleles for a single gene differ greatly in their feeding behavior as larvae: Larvae with one allele move around a great deal as they eat, whereas individuals with the alternative allele move hardly at all. A wide variety of mutations at other genes are now known in *Drosophila* that affect almost every aspect of courtship behavior.

The ways in which genetic differences affect behavior have been worked out for several mouse genes. For example, some mice with one mutation have trouble remembering information learned two days earlier about where objects are located. This difference appears to result because the mutant mice do not produce the enzyme α-calcium-calmodulin-dependent kinase II, which plays an important role in the functioning of the hippocampus, a part of the brain important for spatial learning (described in chapter 46)

Modern molecular biology techniques allow the role of genetics in behavior to be investigated with ever-greater precision. For example, male mice genetically engineered to lack the ability to synthesize nitric oxide, a brain neurotransmitter, show increased aggressive behavior.

A particularly fascinating breakthrough occurred in 1996, when scientists discovered a new gene, *fosB*, that seems to determine whether female mice nurture their young in particular ways. Females with both *fosB* alleles disabled initially investigate their newborn babies, but then ignore them, in stark contrast to the caring and protective maternal behavior displayed by normal females (figure 54.4).

The cause of this inattentiveness appears to result from a chain reaction. When mothers of new babies initially inspect them, information from their auditory, olfactory, and tactile senses is transmitted to the hypothalamus, where *fosB* alleles are activated. The *fosB* alleles produce a particular protein, which in turn activates other enzymes and genes that affect the neural circuitry within the hypothalamus. These modifications within the brain cause the female to react maternally toward her offspring. In contrast, in mothers lacking the *fosB* alleles, this reaction is stopped midway. No protein is activated, the brain's neural circuitry is not rewired, and maternal behavior does not result.

Another fascinating example of the genetic basis of behavior concerns two species of North American rodents: the prairie and montane voles. These closely related species differ in their social behavior: Male and female prairie voles form monogamous pair bonds and work together to raise their young, whereas montane voles mate and go their separate ways.

The differences between these species have been intensively studied. The act of mating leads to the release of the neuropeptides vasopressin and oxytocin in both vole species (as well as in many other mammal species). How the voles respond to the release of these peptides, however, differs dramatically. Injection of either of these peptides into prairie voles leads to pair bonding even in the absence of mating. Conversely, injecting a chemical that blocks the action of these neuropeptides causes prairie voles not to form pair bonds after mating. By contrast, montane voles are unaffected by either of these manipulations.

These different responses have been traced to interspecific differences in brain structure (figure 54.5). The prairie vole has many receptors for these peptides in a particular part of the brain, the nucleus accumbens, which seems to be involved in the expression of pair-bonding behavior. By contrast, few such

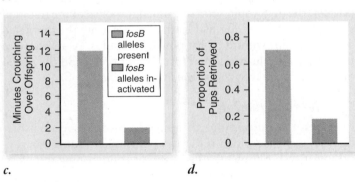

c. *d.*

figure 54.4

GENETICALLY CAUSED DEFECT IN MATERNAL CARE. *a.* In mice, normal mothers take very good care of their offspring, retrieving them if they move away and crouching over them. *b.* Mothers with the mutant *fosB* allele perform neither of these behaviors, leaving their pups exposed. *c.* Amount of time female mice were observed crouching in a nursing posture over offspring. *d.* Proportion of pups retrieved when they were experimentally moved.

 inquiry

*Why does the lack of **fosB** alleles lead to maternal inattentiveness?*

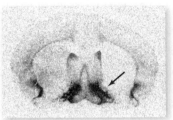

a. Prairie vole   *b.* Montane vole

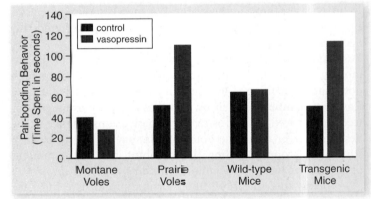

c.

figure 54.5

GENETIC BASIS OF DIFFERENCES IN PAIR-BONDING BEHAVIOR IN TWO RODENT SPECIES. *a.* and *b.* The prairie (*Microtus ochrogaster*) and montane (*M. montanus*) voles differ in the distribution of one type of vasopressin receptor in the brain. *c.* Transgenic mice created with the prairie-vole version of the receptor genes respond to injections of vasopression by exhibiting heightened levels of pair-bonding behavior in 5 min trials compared with their response to a control injection. By contrast, normal wild-type mice show no increase in such behaviors.

receptors occur in the same brain region in the montane vole. In laboratory experiments with prairie voles, blocking these receptors tends to prevent pair-bonding, whereas stimulating them leads to pair-bonding behavior.

Recently, the genetic basis for these differences has been uncovered. Scientists have identified the gene that codes for the peptide receptors and have discovered that a difference exists in the DNA structure between the species. To test the hypothesis that this genetic difference was responsible for the differences in behavior, scientists created transgenic mice with the prairie vole version of the gene, and

sure enough, when injected with vasopressin, the transgenic mice exhibited pair-bonding behavior very similar to that of prairie voles, whereas normal mice showed no response (see figure 54.5).

> The genetic basis of behavior is supported by artificial selection experiments, studies of identical twins, and studies on the behavior of genetic mutants. Recent developments in molecular biology have led to the discovery of specific genes that control behavior.

54.3 Learning

Many of the behavioral patterns displayed by animals are not solely the result of instinct. In many cases, animals alter their behavior as a result of previous experiences, a process termed **learning.** The role of learning was first studied intensively in laboratory rodents, but now researchers investigate the learning processes and capabilities of a wide range of organisms.

Habituation occurs when organisms respond less to a stimulus over time

The simplest type of learning does not require an animal to form an association between two stimuli or between a stimulus and a response. One form of *nonassociative learning* is **habituation,** which can be defined as a decrease in response to a repeated stimulus that has no positive or negative consequences. In many cases, the stimulus evokes a strong response when it is first encountered, but the magnitude of the response gradually declines with repeated exposure.

As one example, young birds see many types of objects moving overhead. At first, they may respond by crouching down and remaining still. Some of the objects, such as falling leaves or members of their own species flying by, are seen very frequently and have no positive or negative consequence to the nestlings. Over time, the young birds may habituate to such stimuli and stop responding. Thus, habituation can be thought of as learning not to respond to a stimulus.

Being able to ignore unimportant stimuli is critical for an animal confronting a barrage of stimuli in a complex environment; animals that could not do so would fail to focus their attention on important activities, such as finding food and avoiding predators, and probably would leave few offspring in the next generation.

Associative learning links stimulus with response

A change in behavior that involves an association between two stimuli or between a stimulus and a response is termed **associative learning** (figure 54.6). The behavior is modified, or *condi-*

tioned, through the association. This form of learning is more complex than habituation. The two major types of associative learning are classical conditioning and operant conditioning; they differ in the way the associations are established.

Classical conditioning

In **classical conditioning,** the paired presentation of two different kinds of stimuli causes the animal to form an association between the stimuli. Classical conditioning is also called **pavlovian conditioning,** after Russian psychologist Ivan Pavlov, who first described it.

Pavlov presented meat powder, an *unconditioned stimulus,* to a dog and noted that the dog responded by salivating, an *unconditioned response.* If an unrelated stimulus, such as the ringing of a bell, repeatedly was presented at the same time as the meat powder, the dog would soon salivate in response to the sound of the bell alone. The dog had learned to associate the unrelated sound stimulus with the meat powder stimulus. Its response to the sound stimulus was, therefore, conditioned, and the sound of the bell is referred to as a *conditioned stimulus.*

Operant conditioning

In **operant conditioning,** an animal learns to associate its behavioral response with a reward or punishment. American psychologist B. F. Skinner studied operant conditioning in rats by placing them in an apparatus that came to be called a "Skinner box." As the rat explored the box, it would occasionally press a lever by accident, causing a pellet of food to appear. At first, the rat would ignore the lever, eat the food pellet, and continue to move about. Soon, however, it learned to associate pressing the lever (the behavioral response) with obtaining food (the reward). When it was hungry, it would spend all its time pressing the lever. This sort of trial-and-error learning is of major importance to most vertebrates.

Comparative psychologists used to believe that any two stimuli could be linked in classical conditioning and that animals could be conditioned to perform any learnable behavior in response to any stimulus by operant conditioning. As you will see in the following discussion, this view has changed. Today, researchers think that instinct guides learning by determining what type of information can be learned through conditioning.

Instinct and learning

It is now clear that some animals have innate predispositions toward forming certain associations. For example, if a rat is offered a food pellet at the same time it is exposed to X-rays (which later produce nausea), the rat remembers the taste of the food pellet but not its size, and in the future will avoid food with that taste, but will readily eat pellets of the same size if they have a different taste.

Similarly, pigeons can learn to associate food with colors, but not with sounds. In contrast, they can associate danger with sounds, but not with colors.

These examples of learning preparedness demonstrate that what an animal can learn is biologically influenced—that is, learning is possible only within the boundaries set by instinct. Innate programs have evolved because they underscore adaptive responses. In nature, food that is toxic to a rat is likely to have a particular taste; thus, it is adaptive to be able to associate a taste with a feeling of sickness that may develop hours later. The seed a pigeon eats may have a distinctive color that the pigeon can see, but it makes no sound the pigeon can hear.

An animal's ecology is key to understanding its mental capabilities. Some species of birds, such as Clark's nutcracker, feed on seeds. When seeds are abundant, these birds store seeds in buried caches so they will have food during the winter. Thousands of seed caches may be buried and then later recovered, sometimes as many as nine months later. One would expect the birds to have an extraordinary spatial memory, and this is indeed what has been found (figure 54.7). Clark's nutcracker, and other seed-hoarding birds, have an unusually large hippocampus, the center for memory storage in the brain.

figure 54.7

THE CLARK'S NUTCRACKER HAS AN EXTRAORDINARY MEMORY. A Clark's nutcracker, *Nucifraga columbiana*, can remember the locations of up to 2000 seed caches months after hiding them. After conducting experiments, scientists have concluded that the birds use features of the landscape and other surrounding objects as spatial references to memorize the locations of the caches.

Habituation is a simple form of learning that does not require an association between stimuli and responses. In contrast, associative learning (classical and operant conditioning) involves the formation of an association between two stimuli or between a behavior and a response.

a. b. c.

figure 54.6

LEARNING WHAT IS EDIBLE. Associative learning is involved in predator–prey interactions. ***a.*** A naive toad is offered a bumblebee as food. ***b.*** The toad is stung, and (***c***) subsequently avoids feeding on bumblebees or any other insects having black-and-yellow coloration. The toad has associated the appearance of the insect with pain and modifies its behavior.

The Development of Behavior

Behavioral biologists now recognize that behavior has both genetic and learned components. Thus far in this chapter, we have discussed the influence of genes and learning separately. But as you will see, these factors interact during development to shape behavior.

Parent–offspring interactions influence cognition and behavior

As an animal matures, it may form social attachments to other individuals or develop preferences that will influence behavior later in life. This process, called **imprinting,** is sometimes considered a type of learning.

In *filial imprinting*, social attachments form between parents and offspring. For example, young birds of some species begin to follow their mother within a few hours after hatching, and their following response results in a bond between mother and young. However, the young birds' initial experience determines how this imprint is established. The German behaviorist Konrad Lorenz showed that geese will follow the first object they see after hatching and direct their social behavior toward that object. Lorenz raised geese from eggs, and when he offered himself as a model for imprinting, the goslings treated him as if he were their parent, following him dutifully (figure 54.8). The success of imprinting is highest during a critical period (roughly 13 to 16 hours after hatching in geese).

Several studies demonstrate that the social interactions that occur between parents and offspring are key to the normal development of behavior. The psychologist Harry Harlow gave orphaned rhesus monkey infants the opportunity to form social

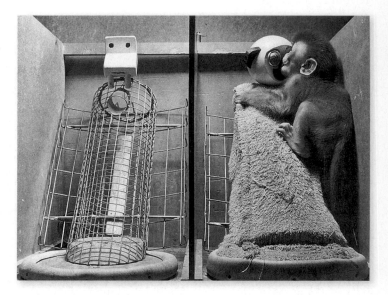

figure 54.9

CHOICE TRIAL ON INFANT MONKEYS. Given a choice between a wire frame that provided food and a similar frame covered with cloth and given a monkey-like head, orphaned rhesus monkeys chose the monkey-like figure over the food.

attachments with two surrogate "mothers," one made of soft cloth covering a wire frame and the other made only of wire (figure 54.9). The infants chose to spend time with the cloth mother, even if only the wire mother provided food, indicating that texture and tactile contact, rather than provision of food, may be among the key qualities in a mother that promote infant social attachment. Other studies show that if infant monkeys are deprived of normal social contact, their development is abnormal. Greater degrees of deprivation lead to greater abnormalities in social behavior during childhood and adulthood. Studies of orphaned human infants similarly suggest that a constant "mother figure" is required for normal growth and psychological development.

Recent research has revealed a biological need for the stimulation that occurs during parent–offspring interactions early in life. Female rats lick their pups after birth, and this stimulation inhibits the release of a hormonelike chemical that can block normal growth. Pups that receive normal tactile stimulation also have more brain receptors for glucocorticoid hormones, longer-lived brain neurons, and a greater tolerance for stress. Premature human infants who are massaged gain weight rapidly. These studies indicate that the need for normal social interaction is based in the brain, and that touch and other aspects of contact between parents and offspring are important for physical as well as behavioral development.

Sexual imprinting is a process in which an individual learns to direct its sexual behavior toward members of its own species. **Cross-fostering** studies, in which individuals of one species are raised by parents of another species, reveal that this

figure 54.8

AN UNLIKELY PARENT. The eager goslings follow Konrad Lorenz as if he were their mother. He is the first object they saw when they hatched, and they have used him as a model for imprinting. Lorenz won the 1973 Nobel Prize in medicine or physiology for this work.

form of imprinting also occurs early in life. In most species of birds, these studies have shown that a fostered bird will attempt to mate with members of its foster species when it is sexually mature.

Instinct and learning may interact as behavior develops

Mature male white-crowned sparrows sing a species-specific courtship song during mating season. Young male birds acquire the song by a combination of instinct and learning.

In one experiment, researchers reared male birds in sound-proof incubators equipped with speakers and microphones. In this way, they could control what a bird heard as it matured and then record the song it produced as an adult. White-crowned sparrows that heard no song at all during development, or that heard only the song of a different species, the song sparrow, sang a poorly developed song as adults (figure 54.10). But birds that heard the song of their own species, or that heard the songs of *both* the white-crowned sparrow and the song sparrow, sang a fully developed, white-crowned sparrow song as adults.

These results suggest that these birds have a **genetic template,** or innate program, that guides them to learn the appropriate song. During a critical period in development, the template will accept the correct song as a model. Thus, song acquisition depends on learning, but only the song of the correct species can be learned; the genetic template for learning is selective.

But learning plays a prominent role as well. If a young white-crowned sparrow becomes deaf after it hears its species'

figure 54.11

BROOD PARASITE. Cuckoos lay their eggs in the nests of other species of birds. Because the young cuckoos (large bird to the right) are raised by a different species (such as this meadow pipit, smaller bird to the left), they have no opportunity to learn the cuckoo song; the cuckoo song they later sing is innate.

song during the critical period, it will sing a poorly developed song as an adult. Therefore, the bird must "practice" listening to himself sing, matching what he hears to the model his template has accepted.

Although this explanation of song development stood unchallenged for many years, recent research has shown that white-crowned sparrow males *can* learn another species' song under certain conditions. If a live male strawberry finch is placed in a cage next to a young male sparrow, the young sparrow will learn to sing the strawberry finch's song. This finding indicates that social stimuli—in this case, being able to see the other bird—may be more effective than a tape-recorded song in altering the innate program that guides song development.

The males of some bird species have no opportunity to hear the song of their own species. In such cases, it appears that the males instinctively "know" their own species' song. For example, cuckoos are brood parasites; females lay their eggs in the nest of another species of bird, and the young that hatch are reared by the foster parents (figure 54.11). When the cuckoos become adults, they sing the song of their own species rather than that of their foster parents. Because male brood parasites would most likely hear the song of their host species during development, it is adaptive for them to ignore such "incorrect" stimuli. They hear no adult males of their own species singing, so no correct song models are available. In these species, natural selection has produced a completely genetically guided song.

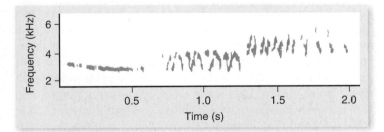

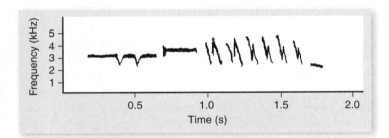

a.

b.

figure 54.10

SONG DEVELOPMENT IN BIRDS. *a.* The sonograms of songs produced by male white-crowned sparrows (*Zonotrichia leucophrys*) that had been exposed to their own species' song during development are different from (*b*) those of male sparrows that heard no song during rearing. This difference indicates that the genetic program itself is insufficient to produce a normal song.

Interactions that occur during sensitive phases of development are critical to normal behavioral development. Physical contact plays an important role in growth and in the development of psychological well-being.

Animal Cognition

The degree to which animals "think" is a subject of lively dispute. Many of us have observed the behavior of a pet cat or dog that would suggest the animal had a degree of reasoning ability or was capable of thinking. For many decades, however, students of animal behavior flatly rejected the notion that nonhuman animals can think. In fact, behaviorist Lloyd Morgan stated in the late nineteenth century that one should never assume a behavior represents conscious thought if there is any other explanation that precludes the assumption of consciousness. The prevailing approach was to treat animals as though they responded to the environment through instinctive behaviors and simple, innately programmed learning.

In recent years, serious attention has been given to the topic of animal awareness. The central question is whether animals show **cognitive behavior**—that is, do they process information and respond in a manner that suggests thinking (figure 54.12)?

What kinds of behavior would demonstrate cognition? A number of cases are suggestive of cognitive capabilities:

- Some birds in urban areas in the midtwentieth century, when milk delivery to homes was common, learned to remove the foil caps from nonhomogenized milk bottles to get at the cream beneath; other birds learned the behavior through observation;
- Japanese macaques learned to wash sand off potatoes and to float grain to separate it from sand;
- Chimpanzees have been observed to pull the leaves off a tree branch and then stick the branch into the en-

figure 54.13

PROBLEM SOLVING BY A CHIMPANZEE. Unable to get the bananas by jumping, the chimpanzee devises a solution.

trance of a termite nest; when termites climb onto the branch, the chimpanzees remove the branch from the mound and eat the termites.

Only a few experiments have tested the thinking ability of nonhuman animals. Some of these studies suggest that animals may deliberately give false information. Currently, researchers are trying to determine if some primates deceive others to manipulate the behavior of the other members of their troop. Many anecdotal accounts appear to support the idea that deception occurs in some nonhuman primate species, such as baboons and chimpanzees, but it has been difficult to devise field-based experiments to test this idea. Much of this type of research on animal cognition is in its infancy, but it is sure to grow and to raise controversy. In any case, nothing is to be gained by dogmatically denying the possibility of animal consciousness.

Some behaviors, particularly those involving problem solving by animals, are hard to explain in any way other than as a result of some sort of mental process. For example, in a series of classic experiments conducted in the 1920s, a chimpanzee was left in a room with bananas hanging from the ceiling out of reach. Also in the room were several boxes, each lying on the floor. After some unsuccessful attempts to jump up and grab the bananas, the chimpanzee suddenly looked at the boxes and then immediately proceeded to move them underneath the bananas, stack one on top of another, and climb up to claim its prize (figure 54.13).

Recent studies have found that animals other than primates also show evidence of cognition. Ravens have always been considered among the most intelligent of birds. A recent experiment using hand-reared ravens that lived in an outdoor aviary provided evidence of reasoning ability. A piece of meat was placed on the end of a string and hung from a branch in the aviary. The birds liked to eat meat, but had never seen string before and were unable to get at the meat. After sev-

a.   *b.*

figure 54.12

ANIMAL THINKING? *a.* This chimpanzee is stripping the leaves from a twig, which it will then use to probe a termite nest. This behavior strongly suggests that the chimpanzee is consciously planning ahead, with full knowledge of what it intends to do. *b.* This sea otter is using a rock as an "anvil," against which it bashes a clam to break it open. A sea otter will often keep a favorite rock for a long time, as though it has a clear idea of its future use of the rock. Behaviors such as these suggest that animals have cognitive abilities.

eral hours, during which time the birds periodically looked at the meat but did nothing else, one bird flew to the branch, reached down, grabbed the string with its beak, pulled it up, and placed the string under its foot. It then reached down and grabbed another piece of the string, repeating this action over and over, each time bringing the meat closer (figure 54.14). Eventually, the raven brought the meat within reach and grasped it. Three of the other five ravens also figured out how to get the meat. Presented with a novel problem, the raven had devised a solution.

> **Research on the cognitive behavior of animals is in its infancy, but some examples are compelling.**

figure 54.14

PROBLEM SOLVING BY A RAVEN. Confronted with a problem it has never previously faced, the raven figures out how to get the meat at the end of the string by repeatedly pulling up a bit of string and stepping on it.

54.6 Orientation and Migratory Behavior

Some animals engage in a number of seemingly goal-oriented movements. They may, for example, travel to and from a nest or a watering hole. To do so, they must orient themselves by tracking stimuli in the environment, a process called **orientation.** Animals with a homing instinct, such as pigeons, recognize complex environmental clues to return to their home, often from great distances.

Movement toward or away from a stimulus is called a **taxis.** The attraction of flying insects to outdoor lights is an example of positive phototaxis. Insects that avoid light, such as the common cockroach, exhibit negative phototaxis. Other stimuli may be used as orienting cues. For example, trout orient themselves in a stream so as to face against the current.

Not all responses involve a specific orientation, however. Some animals just become more or less active when stimulus intensity increases; such responses are called **kineses.**

Migration often involves populations moving large distances

Long-range, two-way movements are known as **migrations.** Each fall, ducks, geese, and many other birds migrate south along flyways from Canada across the United States, heading as far as South America, only to return each spring.

Monarch butterflies also migrate each fall from central and eastern North America to several small, geographically isolated areas of coniferous forest in the mountains of central Mexico (figure 54.15). Each August, the butterflies begin a flight southward to their overwintering sites. At the end of winter, the monarchs begin the return flight to their summer breeding ranges. What is amazing about the migration of the monarch, however, is that two to five generations may be produced as the butterflies fly north. The butterflies that migrate in the autumn to the precisely located overwintering grounds in Mexico have never been there before.

Recent geographic range expansions by some migrating birds have revealed how migratory patterns change. When colonies of bobolinks became established in the western United States, far from their normal range in the Midwest and East, they did

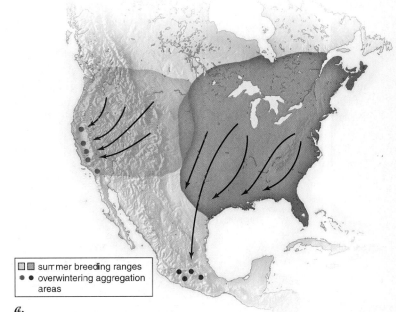

☐☐ summer breeding ranges
● ● overwintering aggregation areas

a.

b.　　　　*c.*

figure 54.15

MIGRATION OF MONARCH BUTTERFLIES (*Danaus plexippus*). *a.* Monarchs from western North America overwinter in areas of mild climate along the Pacific coast. Those from the eastern United States and southeastern Canada migrate to Mexico, a journey of over 3000 kilometers. *b.* Monarch butterflies arrive at the remote fir forests of the overwintering grounds in Mexico, where they (*c*) form aggregations on the tree trunks.

not migrate directly to their winter range in South America. Instead, they migrated east to their ancestral range, and then south along the original flyway (figure 54.16). Rather than changing the original migration pattern, they simply added a new segment. Scientists continue to study the western bobolinks to learn whether, in time, a more efficient migration path will evolve or whether the birds will always follow their ancestral course.

Migrating animals must be capable of orientation and navigation

Biologists have studied migration with great interest, and we now have a good understanding of how these feats of navigation are achieved. It is important to understand the distinction between orientation, the ability to follow a bearing, and **navigation,** the ability to set or adjust a bearing, and then follow it. The former is analogous to using a compass, while the latter is like using a compass in conjunction with a map. Experiments on starlings indicate that inexperienced birds migrate by orientation, but older birds that have migrated previously use true navigation (figure 54.17).

Birds and other animals navigate by looking at the Sun and the stars. The indigo bunting, which flies during the day and uses the Sun as a guide, compensates for the movement of the Sun in the sky as the day progresses by reference to the North Star, which does not move in the sky. Buntings also use

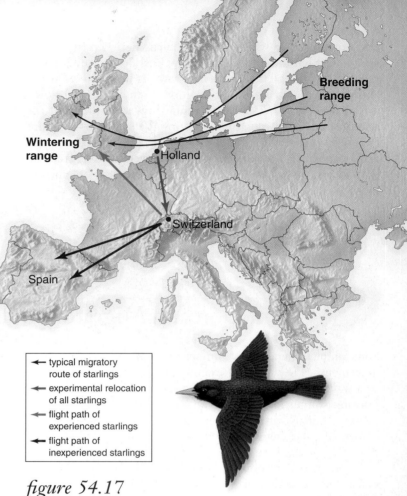

← typical migratory route of starlings

← experimental relocation of all starlings

← flight path of experienced starlings

← flight path of inexperienced starlings

figure 54.17

MIGRATORY BEHAVIOR OF STARLINGS (*Sturnus vulgaris*). The navigational abilities of inexperienced birds differ from those of adults that have made the migratory journey before. Starlings were captured in Holland, halfway along their full migratory route from Baltic breeding grounds to wintering grounds in the British Isles; these birds were transported to Switzerland and released. Experienced older birds compensated for the displacement and flew toward the normal wintering grounds (*blue arrow*). Inexperienced young birds kept flying in the same direction, on a course that took them toward Spain (*red arrows*). These observations imply that inexperienced birds fly by orientation, while experienced birds learn true navigation.

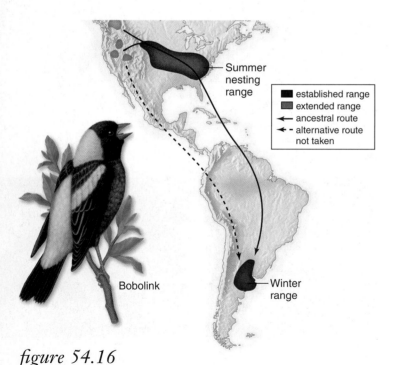

established range
extended range
← ancestral route
← - alternative route not taken

figure 54.16

BIRDS ON THE MOVE. The summer range of bobolinks (*Dolichonyx oryzivorus*) recently extended to the far western United States from their more established range in the Midwest. When birds in these newly established populations migrate to South America in the winter, they do not fly directly to the winter range; instead, they fly to the Midwest first and then use the ancestral flyway, going considerably farther than if they flew directly to their winter range.

the positions of the constellations and the position of the North Star in the night sky, cues they learn as young birds.

Many migrating birds also have the ability to detect Earth's magnetic field and to orient themselves with respect to it. In a closed indoor cage, they will attempt to move in the correct geographic direction, even though there are no visible external cues. However, the placement of a magnet near the cage can alter the direction in which the birds attempt to move. Researchers have found magnetite, a magnetized iron ore, in the heads of some birds, but have not been able to identify the sensory receptors birds employ to detect magnetic fields.

The first migration of a bird appears to be innately guided by both celestial cues (the birds fly mainly at night) and Earth's magnetic field. When the two cues are experimentally manipulated to give conflicting directions, the information provided by the stars seems to override the magnetic information. Recent studies, however, indicate that celestial cues indicate the general direction for migration, whereas magnetic cues indicate the specific migratory path (perhaps a turn the bird must make midroute).

We know relatively little about how other migrating animals navigate. For instance, green sea turtles (*Chelonia mydas*) migrate from Brazil halfway across the Atlantic Ocean to Ascension Island, where the females lay their eggs. How do they find this tiny island in the middle of the ocean, which they haven't seen for perhaps 30 years? How do the young that hatch on the island know how to find their way to Brazil? Researchers still have few answers to these questions.

Many animals migrate in predictable ways, navigating by looking at the Sun and stars, and in some cases by detecting magnetic fields.

54.7 Animal Communication

Communication among members of the same species as well as between species can play a key role in many behaviors. Much of the research in animal behavior is devoted to analyzing the nature of communication signals, determining how they are perceived, and identifying their ecological roles and evolutionary origins. Communication is particularly important in reproduction and in social interactions in a population. Communication can take many forms, including visual, acoustic, tactile, and chemical signals.

Successful reproduction depends on appropriate signals and responses

Species recognition

During courtship, animals produce signals to communicate with potential mates and with other members of their own sex. A **stimulus–response chain** sometimes occurs, in which the behavior of one individual in turn releases a behavior by another individual (figure 54.18).

Courtship signals are often species-specific, limiting communication to members of the same species and thus playing a key role in reproductive isolation (described in chapter 22). The flashes of fireflies (which are actually beetles) are an example of a species-specific signal. Females recognize conspecific males by their flash pattern (figure 54.19), and males recognize conspecific females by their flash response. This series of reciprocal responses provides a continuous "check" on the species identity of potential mates.

Long-distance communication

Chemical signals also mediate interactions between males and females. **Pheromones,** chemical messengers used for communication between individuals of the same species, serve as

figure 54.18

A STIMULUS–RESPONSE CHAIN. Stickleback courtship involves a sequence of behaviors leading to the fertilization of eggs.

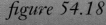

1. Female gives head-up display to male

2. Male swims zigzag to female and then leads her to nest

3. Male shows female entrance to nest

4. Female enters nest and spawns while male stimulates tail

5. Male enters nest and fertilizes eggs

sex attractants, among other functions, in many animals. Female silk moths (*Bombyx mori*) produce a sex pheromone called *bombykol* in a gland associated with the reproductive system. Neurophysiological studies show that the male's antennae contain numerous sensory receptors specific for bombykol. These receptors are extraordinarily sensitive; in some moth species, males can detect extremely low concentrations of bombykol and locate females from as far as 7 km away.

Many insects, amphibians, and birds produce species-specific acoustic signals to attract mates. Bullfrog males call by inflating and discharging air from their vocal sacs, located beneath the lower jaw. Females can distinguish a conspecific male's call from the call of other frogs that may be in the same habitat and calling at the same time. As mentioned earlier, male birds sing to advertise their presence and to attract females. In many species, variations in the males' songs identify individual males in a population. In these species, the song is individually specific as well as species-specific.

Courtship behaviors are a major factor in sexual selection, which we discuss later in this chapter.

Communication facilitates group living

Many insects, fish, birds, and mammals live in social groups in which information is communicated between group members. For example, some individuals in mammalian societies serve as "guards." When a predator appears, the guards give an alarm call, and group members respond by seeking shelter (figure 54.20). Social insects such as ants and honeybees produce alarm pheromones that trigger attack behavior. Ants also deposit trail pheromones between the nest and a food source to lead other colony members to food. Honeybees have an extremely complex dance language that directs hivemates to rich nectar sources.

The dance language of the honeybee

The European honeybee lives in hives consisting of 30,000 to 40,000 individuals whose behaviors are integrated into a complex colony. Worker bees may forage miles from the hive, collecting nectar and pollen from a variety of plants and switching between plant species on the basis of how energetically rewarding their food is.

The food sources used by bees tend to occur in patches, and each patch offers much more food than a single bee can transport to the hive. A colony is able to exploit the resources of a patch because of the behavior of scout bees, which locate patches and communicate their location to hivemates through a dance language. Over many years, Nobel laureate Karl von Frisch (who shared the 1973 prize in physiology or medicine with Tinbergen and Lorenz) was able to unravel the details of this communication system.

After a successful scout bee returns to the hive, she performs a remarkable behavior pattern called a *waggle dance* on a vertical comb (figure 54.21). The path of the bee during the dance resembles a figure-eight. On the straight part of the path, the bee vibrates or waggles her abdomen while producing bursts of sound. The bee may stop periodically to give her hivemates a sample of the nectar carried back to the hive in her crop. As she dances, she is followed closely by other bees, which soon appear as foragers at the new food source.

Von Frisch and his colleagues claimed that the other bees use information in the waggle dance to locate the food source. According to their explanation, the scout bee indicates

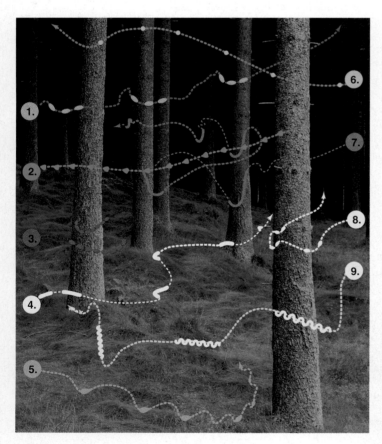

figure 54.19

FIREFLY FIREWORKS. The bioluminescent displays of these lampyrid beetles are species-specific and serve as behavioral mechanisms of reproductive isolation. Each number represents the flash pattern of a male of a different species.

figure 54.20

ALARM CALLING BY A PRAIRIE DOG (*Cynomys ludovicianus*). When a prairie dog sees a predator, it stands on its hind legs and gives an alarm call, which causes other prairie dogs to rapidly return to their burrows.

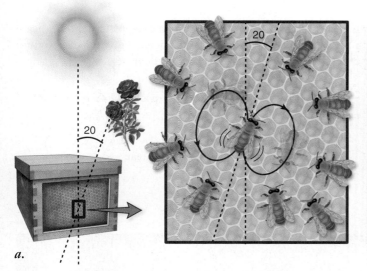

a.

b.

figure 54.21

THE WAGGLE DANCE OF HONEYBEES (*Apis mellifera*). *a.* The angle between the food source, the nest, and the Sun is represented by a dancing bee as the angle between the straight part of the dance and vertical. The food is 20° to the right of the Sun, and the straight part of the bee's dance on the hive is 20° to the right of vertical. *b.* A scout bee dances on a comb in the hive.

the *direction* of the food source by representing the angle between the food source, the hive, and the Sun as the deviation from vertical of the straight part of the dance performed on the hive wall (that is, if the bee moved straight, then the food source would be in the direction of the Sun, but if the food were at a 30° angle relative to the Sun's position, then the bee would move upward at a 30° angle from vertical) (figure 54.21*a*). The distance to the food source is indicated by the duration of the dance.

Adrian Wenner, a scientist at the University of California, challenged von Frisch's explanation. Wenner maintained that flower odor was the most important cue leading bees to arrive at a new food source. A heated controversy ensued as the two groups of researchers published articles supporting their positions.

Such controversies can be very beneficial, because they often generate innovative experiments. In this case, the "dance language controversy" was resolved (in the minds of most scientists) in the mid-1970s by the creative research of James L. Gould. Gould devised an experiment in which hive members were tricked into misinterpreting the directions given by the scout bee's dance. As a result, Gould was able to manipulate where the hive members would go if they were using visual signals. If the bees were using odor as the cue, they would have appeared at the food source regardless, but instead they appeared exactly where Gould had predicted. This result confirmed von Frisch's ideas.

Recently, researchers have extended the study of the honeybee dance language by building robot bees whose dances can be completely controlled. Their dances are programmed by a computer and perfectly reproduce the natural honeybee dance—the robots even stop to give food samples. The use of robot bees has allowed scientists to determine precisely which cues direct hivemates to food sources.

Primate language

Some primates have a "vocabulary" that allows individuals to communicate the identity of specific predators. Different vocalizations of African vervet monkeys, for example, indicate eagles, leopards, or snakes (figure 54.22). Chimpanzees and gorillas can learn to recognize a large number of symbols and use them to communicate abstract concepts.

The complexity of human language would at first appear to defy biological explanation, but closer examination suggests that the differences are in fact superficial—all languages share many basic structural similarities. All of the roughly 3000 languages draw from the same set of 40 consonant sounds (English uses two dozen of them), and any human can learn them. Researchers believe these similarities reflect the way our brains

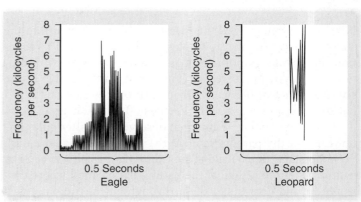

a.

b.

figure 54.22

PRIMATE SEMANTICS. Vervet monkeys, *Cercopithecus aethiops*, give different alarm calls (*a*) when troop members sight an eagle, leopard, or snake. *b.* Each distinctive call elicits a different and adaptive escape behavior.

handle abstract information, a genetically determined characteristic of all humans.

Although language is the primary channel of human communication, odor and other nonverbal signals (such as "body language") may also convey information. However, it is difficult to determine the relative importance of these other communication channels in humans.

Signals vary in their degree of specificity

Different signals provide different levels of information about the sender. The **level of specificity** relates to the function of the signal. Many courtship signals are species-specific to help animals avoid making errors in mating that would produce nonviable hybrids or otherwise waste reproductive effort.

But not all signals have this type of specificity. Many mammals mark their territories with pheromones that signal individual identity, encoded as a blend of a number of chemicals. Members of the same species can detect this chemical signal, but so can many other animals, who are thus informed of the marking animal's presence. Other signals, such as the alarm calls of birds, are anonymous, conveying no information about the identity of the sender. These signals may permit communication about the presence of a predator common to several bird species.

Communication also plays a part in interspecies relationships, which are described in detail in chapter 56. Fish with parasites adopt a specific posture in the presence of "cleaner fish" that indicates they are ready to be cleaned (figure 54.23). In a similar vein, some animals send signals to predators. White-tailed deer, for example, raise their tails to display their prominent white undersides while running away from a preda-

tor. These "pursuit-deterrent" signals are presumed to indicate to the predator that it has already been seen and thus should not waste its time trying to catch the deer.

Animal communications serve many purposes and are transmitted in many ways.

The study of animal communication involves analysis of the specificity of signals, their information content, and the methods used to produce and receive them.

figure 54.23

CLEANER FISH. This grouper has entered the cleaner fish's "station" and adopted a posture that allows the cleaner fish to enter the mouth and gills and feed on attached parasites.

54.8 Behavioral Ecology

Niko Tinbergen divided the investigation of behavior into the study of its development, its physiological basis, and its function, including evolutionary significance. One type of evolutionary analysis pioneered by Tinbergen was the study of the **survival value** of behavior. That is, how does an animal's behavior allow it to stay alive or keep its offspring alive?

As one example, Tinbergen observed that after gull nestlings hatch, the parents remove the eggshells from the nest. To understand this behavior, he camouflaged chicken eggs by painting them to resemble the natural background where they would lie and distributed them throughout the area in which the gulls were nesting (figure 54.24). He placed broken eggshells next to some of the eggs, and as a control, he left other camouflaged eggs alone without eggshells.

He then noted which eggs were found more easily by crows. Because the crows could use the white interior of a broken eggshell as a cue, they ate more of the camouflaged eggs that were near eggshells. Tinbergen concluded that eggshell

removal behavior is adaptive: It reduces predation and thus increases the offspring's chances of survival.

Tinbergen is credited with being one of the founders of **behavioral ecology,** the study of how natural selection shapes behavior. This branch of ecology examines the **adaptive significance** of behavior, or how behavior may increase survival and reproduction. Current research in behavioral ecology focuses on how behavior contributes to an animal's reproductive success, or fitness. As we saw in section 54.2, differences in behavior among individuals often result from genetic differences. Therefore, natural selection operating on behavior has the potential to produce evolutionary change.

Consequently, the field of behavioral ecology is concerned with two questions. First, is behavior adaptive? Although it is tempting to assume that the behavior produced by individuals must in some way represent an adaptive response to the environment, this need not be the case. As you saw in chapter 20, traits can appear for many reasons other than

Foraging behavior can directly influence individual fitness

The best way to introduce behavioral ecology is by examining one well-defined behavior in detail. Although many behaviors might be chosen, we will focus on foraging behavior.

For many animals, food comes in a variety of sizes. Larger foods may contain more energy but may be harder to capture and less abundant. In addition, animals may forage for some types of food that are farther away than other types. For these animals foraging involves a trade-off between a food's energy content and the cost of obtaining it. The *net energy* (in calories or joules) gained by feeding on prey of each size is simply the energy content of the prey minus the energy costs of pursuing and handling it. According to **optimal foraging theory,** natural selection favors individuals whose foraging behavior is as energetically efficient as possible. In other words, animals tend to feed on prey that maximize their net energy intake per unit of foraging time.

A number of studies have demonstrated that foragers do preferentially utilize prey that maximize the energy return. Shore crabs, for example, tend to feed primarily on intermediate-sized mussels, which provide the greatest energy return; larger mussels yield more energy, but also take considerably more energy to crack open (figure 54.25).

This optimal foraging approach makes two assumptions. First, natural selection will only favor behavior that maximizes energy acquisition if the increased energy reserves lead to increases in reproductive success. In both Colombian ground squirrels and captive zebra finches, a direct relationship exists between net energy intake and the number of offspring raised;

figure 54.24

THE ADAPTIVE VALUE OF EGG COLORATION. Niko Tinbergen, a winner of the 1973 Nobel Prize in Physiology or Medicine, painted chicken eggs to resemble the mottled brown camouflage of gull eggs. The eggs were used to test the hypothesis that camouflaged eggs are more difficult for predators to find and thus increase the young's chances of survival.

natural selection, such as genetic drift, gene flow, or the correlated consequences of selection on other traits. Moreover, traits may be present in a population because they evolved as adaptations in the past, but are no longer useful. These possibilities hold true for behavioral traits as much as for any other kind of trait.

If a trait is adaptive, the next question is: How is it adaptive? Although the ultimate criterion is reproductive success, behavioral ecologists are interested in *how* a trait can lead to greater reproductive success. Does a behavior enhance energy intake, thus increasing the number of offspring produced? Does it increase mating success? Does it decrease the chance of predation? The job of a behavioral ecologist is to determine the effect of a behavioral trait—for example, foraging efficiency—on each of these activities and then to discover whether increases translate into increased fitness.

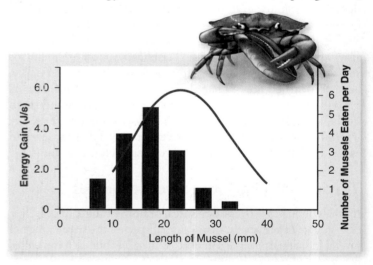

figure 54.25

OPTIMAL DIET. The shore crab selects a diet of energetically profitable prey. The curve describes the net energy gain (equal to energy gained minus energy expended) derived from feeding on different sizes of mussels. The bar graph shows the numbers of mussels of each size in the diet. Shore crabs tend to feed on those mussels that provide the most energy.

inquiry

What factors might be responsible for the slight difference in peak prey length relative to the length optimal for maximum energy gain?

similarly, the reproductive success of orb-weaving spiders is related to how much food they can capture.

Animals have other needs besides energy, however, and sometimes these needs conflict. One obvious need is the avoidance of predators: Often, the behavior that maximizes energy intake is not the one that minimizes predation risk. In this case, the behavior that maximizes fitness often may reflect a trade-off between obtaining the most energy at the least risk of being eaten. Not surprisingly, many studies have shown that a wide variety of animal species alter their foraging behavior—becoming less active, spending more time watching for predators, or staying nearer to cover—when predators are present.

Still another need is finding mates: Males of many species, for example, will greatly reduce their feeding rate to enhance their ability to attract and defend females.

Even within foraging behavior itself, trade-offs must be made, because maximizing energy is not the sole goal of foraging; particular nutrients are needed as well. Moose, for example, will feed on energy-poor aquatic vegetation to ensure that they get an adequate supply of calcium.

The second assumption of optimal foraging theory is that optimal behavior has evolved by natural selection. As described in earlier chapters, natural selection can lead to evolutionary change only when differences among individuals have a genetic basis. A few studies have investigated whether differences in an individual's ability to maximize energy intake are the result of genetic differences. One such study found that female zebra finches that were particularly successful in maximizing net energy intake tended to have similarly successful offspring. Because young birds were removed from their mothers before they were able to leave the nest, this similarity indicated that foraging behavior probably has a large innate component.

Differences in foraging behavior among individuals may also be a function of age. Inexperienced yellow-eyed juncos *Junco phaeontus* (a small North American bird), for example, have not learned how to handle large prey items efficiently. As a result, the energy costs of eating such prey are higher than the benefits, and these birds tend to focus on smaller prey. Only when the birds are older and more experienced do they learn to easily dispatch these prey, which are then included in the diet.

Territorial behavior secures resources

Animals often move over a large area, or **home range,** during their daily course of activity. In many animal species, the home range of several individuals overlaps in time or in space, but each individual defends a portion of its home range and uses it and its resources exclusively. This behavior is called **territoriality** (figure 54.26).

The critical aspect of territorial behavior is defense against intrusion by other individuals. Territories are defended by displays advertising that the territories are occupied, and by overt aggression. A bird sings from its perch within a territory to prevent takeover by a neighboring bird. If an intruder is not deterred by the song, the territory owner may attack and try to drive it away. But territorial defense has its costs. Singing is energetically expensive, and attacks can lead to injury. In addition, advertisement through song or visual display can reveal a bird's position to a predator.

Why does an animal bear the costs of territorial defense? An economic approach can be useful in answering this question. Al-

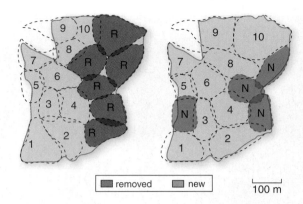

figure 54.26

COMPETITION FOR SPACE. Territory size in birds is adjusted according to the number of competitors. When six pairs of great tits (*Parus major*) were removed from their territories (indicated by R in the left figure), their territories were taken over by other birds in the area and by four new pairs (indicated by N in the right figure). Numbers correspond to the birds present before and after.

though there are costs to defending a territory, there are also benefits; these benefits may take the form of increased food intake, exclusive access to mates, or access to refuges from predators.

Studies of nectar-feeding birds such as hummingbirds and sunbirds provide an example (figure 54.27). A bird benefits from having the exclusive use of a patch of flowers because it can efficiently harvest the nectar the flowers produce. To maintain exclusive use, however, the bird must actively defend the patch. The benefits of exclusive use outweigh the costs of defense only under certain conditions.

Sunbirds, for example, expend 3000 Calories per hour chasing intruders from a territory. Whether the benefit of de-

figure 54.27

THE BENEFIT OF TERRITORIALITY. Sunbirds (on the left), found in Africa and ecologically similar to New World hummingbirds (on the right), protect their food source by attacking other sunbirds that approach flowers in their territory.

c.

figure 54.28

PRODUCTS OF SEXUAL SELECTION.
Attracting mates with long feathers is common in bird species such as (*a*) the African paradise whydah, *Vidua paradisaea*, and (*b*) the peacock, *Pavo cristatus*, which show pronounced sexual dimorphism. *c.* Female peahens prefer to mate with males having greater numbers of eyespots in their tail feathers.

inquiry

? *Why do females prefer males with more spots?*

a.   *b.*

fending a territory will exceed this cost depends on the amount of nectar in the flowers and how efficiently the bird can collect it. When flowers are very scarce or nectar levels are very low, a nectar-feeding bird may not gain enough energy to balance the energy used in defense. Under this circumstance, it is not advantageous to be territorial. Similarly, when flowers are very abundant, a bird can efficiently meet its daily energy requirements without behaving territorially and adding the costs of defense. So from an energetic standpoint, defending abundant resources isn't worth the cost, either. Territoriality therefore only occurs at intermediate levels of flower availability and nectar production, when the benefits of defense outweigh the costs.

In many species, exclusive access to females is a more important determinant of territory size for males than is food availability. In some lizards, for example, males maintain enormous territories during the breeding season. These territories, which encompass the territories of several females, are much larger than would be re-

quired to supply enough food, and they are defended vigorously. In the nonbreeding season, by contrast, male territory size decreases dramatically, as does aggressive territorial behavior.

Behavioral ecology is the study of how natural selection shapes behavior.

Natural selection may favor the evolution of foraging behaviors that maximize the amount of energy gained per unit time spent foraging. Animals that acquire energy efficiency during foraging may increase their fitness, but other considerations, such as avoiding predators, are also important in determining reproductive success.

An economic approach can be used to explain the evolution and ecology of behaviors such as territoriality. This approach assumes that animals that gain more energy from a behavior than they expend will have an advantage in survival and reproduction.

54.9 Reproductive Strategies and Sexual Selection

During the breeding season, animals make several important "decisions" concerning their choice of mates, how many mates to have, and how much time and energy to devote to rearing offspring. These decisions are all aspects of an animal's **reproductive strategy,** a set of behaviors that presumably have evolved to maximize reproductive success.

Reproductive strategies have evolved partly in response to the energy costs of reproduction. They also appear to be responses to the way food resources, nest sites, and members of the opposite sex are spatially distributed in the environment.

The sexes often have different reproductive strategies

Males and females usually differ in their reproductive strategies. Darwin was the first to observe that females often do not simply mate with the first male they encounter, but instead seem to evaluate a male's quality and then decide whether to mate. Peahens prefer to mate with peacocks that have more spots in their long tail feathers (figure 54.28*b,c*). Similarly, female frogs prefer to mate with males having more complex calls. This behavior, called **mate choice,** has since been described in many invertebrate and vertebrate species.

Males engage in mate choice much less frequently than females do. Why should this be? Many of the differences in reproductive strategies between the sexes can be understood by comparing the parental investment made by males and females. **Parental investment** refers to the contributions each sex makes in producing and rearing offspring; it is, in effect, an estimate of the energy expended by males and females in each reproductive event.

Numerous studies have shown that females in general have a higher parental investment. One reason is that eggs are much larger than sperm—195,000 times larger in humans! Eggs contain proteins and lipids in the yolk and other nutrients for the developing embryo, but sperm are little more than mobile DNA packages. In some groups of animals, females are responsible for gestation and lactation, costly reproductive functions only they can carry out.

The consequence of such great disparities in reproductive investment is that the sexes face very different selective pressures. Because any single reproductive event is relatively cheap for males, they can best increase their fitness by mating with as many females as possible—male fitness is rarely limited by the amount of sperm they can produce. By contrast, each reproductive event for females is much more costly, and the number of eggs that can be produced often does limit reproductive success. For this reason, a female has an incentive to be choosy, trying to pick the male that can provide the greatest benefit to her offspring.

These conclusions hold only when female reproductive investment is much greater than that of males. In species with biparental care, males may contribute equally to the cost of raising young; in this case, the degree of mate choice should be equal between the sexes.

In some cases, male investment exceeds that of females. For example, male Mormon crickets transfer a protein-containing packet (called a spermatophore) to females during mating. Almost 30% of a male's body weight is made up by the spermatophore, which provides nutrition for the female and helps her develop her eggs. As we might expect, in this case it is the females that compete with one another for access to males, which are the choosy sex. Indeed, males are quite selective, favoring heavier females. The selective advantage of this strategy results because heavier females have more eggs; thus, males that choose larger females leave more offspring (figure 54.29).

Males take care of the eggs and developing young in many species, including seahorses and a number of bird and insect species. In these species, as with Mormon crickets, males are often choosy, and females must compete for mates.

Parental investment explains why one sex in a species may be choosier than the other. In the following section we will consider the factors that contribute to mate choice.

Sexual selection occurs in many different ways

As discussed in chapter 20, the reproductive success of an individual is determined by a number of factors: how long the individual lives, how frequently it mates, and how many offspring it produces per mating. The second of these factors, competition for mating opportunities, has been termed **sexual selection.** Some people consider sexual selection to be distinctive from natural selection, but others see it as a subset of natural selection, just one of the many factors affecting an organism's fitness.

Sexual selection involves both **intrasexual selection,** or interactions between members of one sex ("the power to conquer other males in battle," as Darwin put it), and **intersexual selection,** which is another name for mate choice ("the power to charm"). Sexual selection leads to the evolution of structures used in combat with other males, such as a deer's antlers and a ram's horns, as well as ornaments used to "persuade" members of the opposite sex to mate, such as long tail feathers and bright plumage (see figure 54.28a,b). These traits are called *secondary sexual characteristics.*

Intrasexual selection

In many species, individuals of one sex—usually males—compete with one another for the opportunity to mate. These competitions may take place over ownership of a territory in which females reside or direct access to the females themselves. The latter case is exemplified by many species, such as the impala (*Aepyceros melampus*), in which females travel in large groups with a single male that gets exclusive access to mate with the females. The male strives vigorously to defend this access against other males.

In mating systems such as these, a few males engage in an inordinate number of matings, and most males do not mate at all. In elephant seals, males control territories on the breeding beaches, and a few dominant males do most of the breeding (figure 54.30). On one beach, for example, eight males impregnated 348 females, while the remaining males mated rarely, if at all.

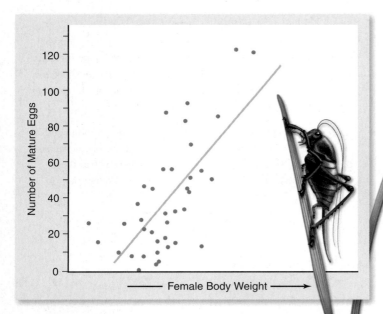

figure 54.29

THE ADVANTAGE OF MALE MATE CHOICE. Male Mormon crickets (*Anabrus simplex*) choose heavier females as mates, and larger females have more eggs. Thus, male mate selection increases fitness.

inquiry

? *Is there a benefit to females for mating with large males?*

figure 54.30

FEMALE DEFENSE POLYGYNY IN NORTHERN ELEPHANT
SEALS (*Mirounga angustirostris*). Male elephant seals
fight with one another for possession of territories. Only the
largest males can hold territories, which contain many females.

For this reason, selection strongly favors any trait that
confers greater ability to outcompete other males. In many
cases, larger males are able to dominate smaller ones. As a result, in many territorial species, males are considerably larger
than females, for the simple reason that the largest males have
been the ones that mate. Such differences between the sexes are
referred to as **sexual dimorphism**. In other species, structures
used for fighting, such as horns, antlers, and large canine teeth,
have evolved in males. These traits are often sexually dimor-
phic also, and they may have evolved because of the advantage
they give in intrasexual conflicts.

Sometimes competition occurs not between the males
themselves, but between their sperm, a phenomenon called
sperm competition. In species in which females mate with
multiple males, many features have evolved to maximize
sperm success: The testes are often large to produce many
sperm per mating, and the sperm themselves are also often
larger and swim more rapidly, enhancing the likelihood of
fertilizing an egg.

Intersexual selection

As described in the preceding section, many species engage in mate
choice rather than in random mating. Usually the gender having
the greater parental investment is the one that does the choosing, and in many species, including many birds and mammals, the
female does the choosing. A number of secondary characteristics,
such as bright colors, loud or elaborate songs, or particular display
behaviors, have evolved in the males of such species.

Direct Benefits of Mate Choice In some cases, the benefits
of mate choice are obvious. In many species of birds and mammals and in some species of other types of animals, males help
raise the offspring. In these cases, females would benefit by
choosing the male that can provide the best care—the better
the parent, the more offspring she is likely to rear.

In other species, males provide no care, but maintain territories that provide food, nesting sites, and predator refuges.
In such species, females that choose males with the best territories will maximize their reproductive success.

Indirect Benefits of Mate Choice In many species, however,
males provide no direct benefits of any kind to females. In such
cases, it is not intuitively obvious what females have to gain by
being "choosy." Moreover, what could be the possible benefit of
choosing a male with an extremely long tail or a complex song?

A number of theories have been proposed to explain the
evolution of such preferences. One idea is that females choose the
male that is the healthiest or oldest. Large males, for example, have
probably been successful at living long, acquiring a lot of food, and
resisting parasites and disease. In other species, features other than
size may indicate a male's condition. In guppies and some birds,
the brightness of a male's color reflects the quality of his diet and
overall health. Females may gain two benefits from mating with
the healthiest males. First, healthy males are less likely to be carrying diseases, which might be transmitted to the female during
mating. Second, to the extent that the males' success in living long
and prospering is the result of genetic makeup, the female will be
ensuring that her offspring receive good genes from their father.

Several experimental studies in fish and moths have examined whether female mate choice leads to greater reproductive success. In these experiments, females in one group were
allowed to choose males, whereas males were randomly mated
to a different group of females. Offspring of females that chose
their mates were more vigorous and survived better than offspring from females given no choice, which suggests that females preferred males with a better genetic makeup.

A variant of this theory goes one step further. In some cases,
females prefer mates with traits that appear to be detrimental to
survival (see figure 54.28). The long tail of the peacock is a hin-
drance in flying and makes males more vulnerable to predators.
Why should females prefer males with such traits? The **handicap hypothesis** states that only genetically superior mates can
survive with such a handicap. By choosing a male with the largest
handicap, the female is ensuring that her offspring will receive
these quality genes. Of course, the male offspring will also inherit the genes for the handicap. For this reason, evolutionary
biologists are still debating the merit of this hypothesis.

figure 54.31

MALE TÚNGARA FROG (*Physalaemus pustulosus*) CALLING.
Female frogs of several species in the genus *Physalaemus* prefer
males that include a "chuck" in their call. However, only males
of the Túngaru frog produce such calls.

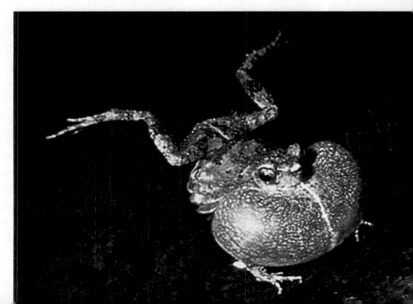

Alternative Theories About the Evolution of Mate Choice Other courtship displays appear to have evolved from a predisposition in the female's sensory system toward a certain type of stimulus. For example, females may be better able to detect particular colors or sounds at a certain frequency. **Sensory exploitation** involves the evolution in males of an attractive signal that "exploits" these preexisting biases—for example, if females are particularly adept at detecting red objects, then red coloration will often evolve in males.

As another example, consider the vocalizations of the Túngara frog (figure 54.31). Unlike related species, males include a short burst of harmonic sound, termed a "chuck," at the end of their calls. Recent research suggests that not only are females of this species particularly attracted to calls of this sort, but so are females of related species, even though males of these species do not produce "chucks." Why this preference evolved is unknown, but male Túngara frogs clearly are taking advantage of it.

A great variety of other hypotheses have been proposed to explain the evolution of mating preferences. Many of these hypotheses may be correct in some circumstances, but none seems capable of explaining all of the variation in mating behavior in the animal world. This is an area of vibrant research, with new discoveries appearing regularly.

Mating systems reflect adaptations for reproductive success

An animal's reproductive strategies have evolved partly in response to the energy costs of reproduction and the way food resources, nest sites, and members of the opposite sex are spatially distributed in the environment.

The number of individuals with which an animal mates during the breeding season varies throughout the animal kingdom. **Mating systems** include *monogamy* (one male mates with one female), *polygyny* (one male mates with more than one female; see figure 54.30), and *polyandry* (one female mates with more than one male). Like mate choice, mating systems have evolved to maximize reproductive fitness.

Much research has shown that mating systems are strongly influenced by ecology. A male may defend a territory, for example, that holds nest sites or food sources sufficient for more than one female. If territories vary in quality or quantity of resources, a female's fitness is maximized if she mates with a male holding a high-quality territory. Such a male may already have a mate, but it is still more advantageous for the female to breed with that male than with an unmated male holding a low-quality territory. In this case, natural selection would favor the evolution of polygyny.

Mating systems are also constrained by the needs of offspring. If the presence of both parents is necessary for young to be reared successfully, then monogamy may be favored. Generally this is the case for birds, in which over 90% of all species are monogamous. A male may either remain with his mate and provide care for the offspring or desert that mate to search for others; both strategies may increase his fitness. The strategy that natural selection will favor depends on the requirement for male assistance in feeding or defending the offspring. In some species, offspring are **altricial**—they require prolonged and extensive care. In these species, the need for care by two parents reduces the tendency for the male to desert his mate and seek other matings. In species where the young are **precocial** (requiring little parental care), males may be more likely to be polygynous.

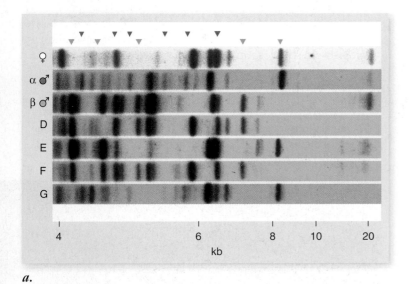

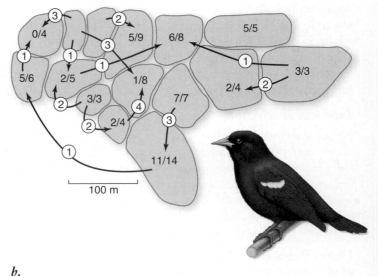

a.   *b.*

figure 54.32

THE STUDY OF PATERNITY. *a.* A DNA fingerprinting gel from the dunnock (*Prunella modularis*). The bands represent fragments of DNA of different lengths. The four nestlings (D–G) were in the nest of the female. By comparing the bands present in the two males and the female, we can determine which male fathered which offspring. The triangles point to the bands that are diagnostic for one male and not the other. In this case, the β-male fathered three (D, E, F, but not G) of the four offspring. *b.* Results of a DNA fingerprinting study in red-winged blackbirds (*Agelaius phoeniceus*). Fractions indicate the proportion of offspring fathered by the male in whose territory the nest occurred. Arrows indicate how many offspring were fathered by particular males outside of each territory. Nests on some territories were not sampled.

Although polygyny is much more common, polyandrous systems—in which one female mates with several males—are known in a variety of animals. For example, in spotted sandpipers (*Actitus macularia*), males take care of all incubation and parenting, and females mate and leave eggs with two or more males.

Extra-pair copulations

In recent years, researchers have uncovered many unexpected aspects of animal mating. Some of these discoveries have resulted from the application of new technologies, whereas others have come from detailed and intensive field studies.

Chapter 15 describes how DNA fingerprinting can be used to identify blood samples. Another common use of this technology is to establish paternity. Using DNA fingerprinting, behavioral ecologists can precisely quantify the reproductive success of individual males and thus assess how successful their particular reproductive strategies have been (figure 54.32*a*).

In one classic study of red-winged blackbirds (figure 54.32*b*), researchers established that half of all nests contained at least one hatchling fertilized by a male other than the territory owner; overall, 20% of the offspring were the result of such **extra-pair copulations (EPCs).**

Studies such as this have established that EPCs are much more pervasive in the bird world than originally suspected. Even in some species that were believed to be monogamous on the basis of behavioral observations, the incidence of offspring being fathered by a male other than the female's mate is sometimes surprisingly high.

What is the evolutionary advantage of EPCs? For males, the answer is obvious: increased reproductive success. For females, it is less clear, because in most cases, it does not result in an increased number of offspring. One possibility is that females tend to mate with genetically superior individuals even if already paired with a male, thus enhancing the genes passed on to their offspring. Another possibility is that females can increase the amount of help they get in raising their offspring. This is exactly what happens in a common English bird, the dunnock. Females mate not only with the territory owner, but also with subordinate males that hang around the edge of the territory. If these subordinates mate a sufficient number of times with a female, they will help raise her young, presumably because they may have fathered some of these young.

Alternative mating strategies

Natural selection has led to the evolution of many ways of increasing reproductive success. For example, in many species of fish, there are two genetic classes of males. One group is large and defends territories to obtain matings. The other group is small and adopts a completely different strategy. These males do not maintain territories, but loiter at the edge of the territories of large males. Just at the end of a male's courtship, when the female is laying her eggs and the territorial male is depositing sperm, the smaller male darts in and releases its own sperm into the water, thus fertilizing some of the eggs. If this strategy is successful, natural selection will favor the evolution of these two different male reproductive strategies.

Similar patterns are seen in other organisms. In some dung beetles, territorial males have large horns that they use to guard the chambers in which females reside, whereas genetically small males don't have horns. Instead, the smaller males dig side tunnels and attempt to intercept the female inside her chamber. In Isopods, there are three genetic size classes. The medium-sized males pass for females and enter a large male's territory in this way; the smallest class are so tiny, they are able to sneak in completely undetected.

This is just a glimpse of the rich diversity in mating systems that have evolved. The bottom line is: If there is a way of increasing reproductive success, natural selection will favor its evolution.

The relative parental investment of the sexes determines reproductive strategies. The sex that invests more tends to be choosier with respect to mate choice.

In some species, members of one sex compete with one another for access to members of the other sex. In other species, members of one sex choose the members of the other sex with whom they will mate. Many different factors may affect the evolution of mate choice.

54.10 Altruism and Group Living

Altruism—the performance of an action that benefits another individual at a cost to the actor—occurs in many guises in the animal world. In many bird species, for example, parents are assisted in raising their young by other birds, which are called "helpers at the nest." In species of both mammals and birds, individuals that spy a predator will give an alarm call, alerting other members of their group, even though such an act would seem to call the predator's attention to the caller. Finally, lionesses with cubs will allow all cubs in the pride to nurse, including cubs of other females.

The existence of altruism has long perplexed evolutionary biologists. If altruism imposes a cost to an individual, how could an allele for altruism be favored by natural selection? One would expect such alleles to be at a disadvantage, and thus their frequency in the gene pool should decrease through time.

A number of explanations have been put forward to explain the evolution of altruism. One suggestion often heard on television documentaries is that such traits evolve for the good of the species. The problem with such explanations is that natural selection operates on individuals within species, not on species themselves. Thus, natural selection will not favor alleles that lead an individual to act in ways that benefit others only at a cost to itself; it is even possible for traits to evolve that are detrimental to the species as a whole, as long as they benefit the individual.

In some cases, selection can operate on groups of individuals, but such **group selection** is rare. For example, if an allele for cannibalism evolved within a population, individuals with that allele would be favored because they would have more to eat; however, the group might eventually eat itself to extinction, and the allele would be removed from the species. Thus, selection *among* groups would lead to a decrease in the allele's frequency in the species, even though selection *within* a particular group favored that allele. Although group selection can occur, the necessary conditions are rarely met in nature. In most cases, consequently, the "good of the species" cannot explain the evolution of altruistic traits.

Another possibility is that seemingly altruistic acts aren't altruistic after all. For example, helpers at the nest are often young and gain valuable parenting experience by assisting established breeders, which may give them an advantage when they breed. Moreover, by hanging around an area, such individuals may inherit the territory when the established breeders die. Similarly, alarm calls may actually benefit callers by causing other animals to panic. In the ensuing confusion, the caller may be able to slip off undetected. Detailed field studies in recent years have demonstrated that some acts truly are altruistic, but others are not.

Reciprocity may explain some altruism

One explanation of altruism proposes that individuals may form "partnerships" in which mutual exchanges of altruistic acts occur because they benefit both participants. In the evolution of such **reciprocal altruism,** "cheaters" (nonreciprocators) are discriminated against and are cut off from receiving future aid. According to this hypothesis, if the altruistic act is relatively inexpensive, the small benefit a cheater receives by not reciprocating is far outweighed by the potential cost of not receiving future aid. Under these conditions, cheating behavior should be eliminated by selection.

Vampire bats roost in hollow trees, caves, and mines in groups of 8 to 12 individuals (figure 54.33). Because these bats have a high metabolic rate, individuals that have not fed recently may die. Bats that have found a host imbibe a great deal of blood, so giving up a small amount to keep a roostmate from starvation presents no great energy cost to the donor. Vampire bats tend to share blood with past reciprocators. If an individual fails to give blood to a bat from which it received blood in the past, it will be excluded from future bloodsharing.

Kin selection proposes a direct genetic advantage to altruism

The most influential explanation for the origin of altruism was presented by William D. Hamilton in 1964. It is perhaps best introduced by quoting a passing remark made in a pub in 1932 by the great population geneticist J. B. S. Haldane. Haldane said that he would willingly lay down his life for *two brothers* or *eight first cousins.*

Evolutionarily speaking, Haldane's statement makes sense, because for each allele Haldane received from his parents, his brothers each had a 50% chance of receiving the same allele (figure 53.34). Consequently, it is statistically expected that two of his brothers would pass on as many of Haldane's particular combination of alleles to the next generation as Haldane himself would. Similarly, Haldane and a first cousin would share an eighth of their alleles (see figure

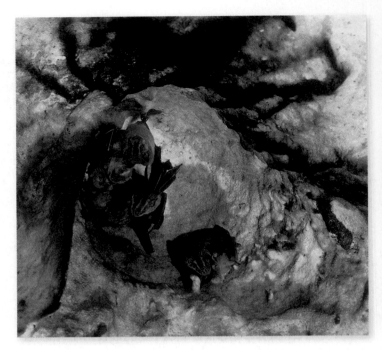

figure 54.33

TRUTH IS STRANGER THAN FICTION: RECIPROCAL ALTRUISM IN VAMPIRE BATS (*Desmodus rotundus*). Vampire bats do feed on the blood of large mammals, but they don't transform into people and sleep in coffins. Vampires live in groups and share blood meals. They remember which bats have provided them with blood in the past, and are more likely to share with those bats that have shared with them previously.

54.34). Their parents, who are siblings, would each share half their alleles, and each of their children would receive half of these, of which half on the average would be in common: $\frac{1}{2} \times \frac{1}{2} \times \frac{1}{2} = \frac{1}{8}$. Eight first cousins would therefore pass on as many of those alleles to the next generation as Haldane himself would. Hamilton saw Haldane's point clearly: Natural selection will favor any strategy that increases the net flow of an individual's alleles to the next generation.

Hamilton showed that by directing aid toward close genetic relatives, an altruist may increase the reproductive success of its relatives enough to compensate for the reduction in its own fitness. Because the altruist's behavior increases the propagation of alleles in relatives, it will be favored by natural selection. Selection that favors altruism directed toward relatives is called **kin selection.** Although the behaviors being favored are cooperative, the genes are actually "behaving selfishly," because they encourage the organism to support copies of themselves in other individuals. In other words, if an individual has a dominant allele that causes altruism, any action that increases the frequency of this allele in future generations will be favored, even if that action is detrimental to the particular individual taking the action.

Hamilton's kin selection model predicts that altruism is likely to be directed toward close relatives. The more closely related two individuals are, the greater the potential genetic payoff. This relationship is described by **Hamilton's rule,** which states that altruistic acts are favored when $rb > c$. In this expression, b and c are the benefits and costs of the altruistic

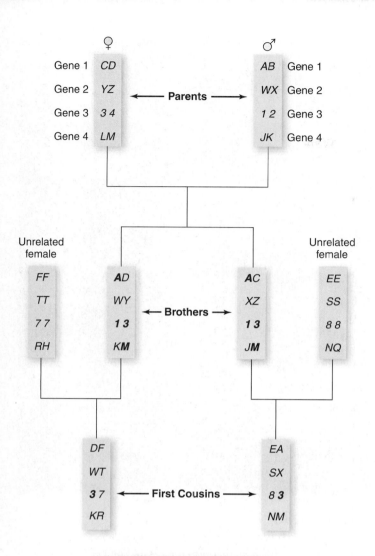

figure 54.34

HYPOTHETICAL EXAMPLE OF GENETIC RELATIONSHIPS.
On average, full siblings share half of their alleles. By contrast, cousins only share one-eighth of their alleles on average. Each letter and number represents a different allele.

Average Genetic Relatedness	
Parent-offspring	= 1/2
Full siblings	= 1/2
Half-siblings	= 1/4
1st cousins	= 1/8

act, respectively, and *r* is the coefficient of relatedness, the proportion of alleles shared by two individuals through common descent. For example, an individual should be willing to have one less child ($c = 1$) if such actions allow a half-sibling, which shares one-quarter of its genes ($r = 0.25$), to have five or more additional offspring ($b = 5$).

Examples of kin selection

Many examples of kin selection are known from the animal world. For example, Belding's ground squirrels (*Spermophilus belding*) give alarm calls when they spot a predator such as a coyote or a badger. Such predators may attack a calling squirrel, so giving a signal places the caller at risk.

The social unit of a ground squirrel colony consists of a female and her daughters, sisters, aunts, and nieces. When males mature, they disperse long distances from where they are born, so adult males in the colony are not genetically related to the females. By marking all squirrels in a colony with an individual dye pattern on their fur and by recording which individuals gave calls and the social circumstances of their calling, researchers found that females who have relatives living nearby are more likely to give alarm calls than females with no kin nearby. Males tend to call much less frequently, as would be expected because they are not related to most colony members.

Another example of kin selection comes from a bird called the white-fronted bee-eater which lives along rivers in Africa in colonies of 100 to 200 birds (figure 54.35). In contrast to ground squirrels, the male bee-eaters usually remain in the colony in which they were born, and the females disperse to join new colonies. Many bee-eaters do not raise their own offspring, but rather help others. Most helpers are young birds, but older birds whose nesting attempts have failed may also be helpers. The presence of a single helper, on average, doubles the number of offspring that survive. Two lines of evidence support the idea that kin selection is important in determining helping behavior in this species. First, helpers are normally males, which are usually related to other birds in the colony, and not females, which are not related. Second, when birds have the choice of helping different parents, they almost invariably choose the parents to which they are most closely related.

figure 54.35

KIN SELECTION IN THE WHITE-FRONTED BEE-EATER
Merops bullockoides. Bee-eaters are small insectivorous birds that live in Africa in large colonies. Bee-eaters often help others raise their young; helpers usually choose to help close relatives.

Haplodiploidy and Hymenopteran social evolution

Probably the most famous application of kin selection theory has been to social insects. A hive of honeybees consists of a single queen, who is the sole egg-layer, and up to 50,000 of her offspring, nearly all of whom are female workers with nonfunctional ovaries (figure 54.36). In addition to this reproductive division of labor, honeybees exhibit cooperative care of the brood and overlap of generations, such that queens live alongside their offspring. These are the hallmarks of a **eusocial** system.

The evolutionary origin of eusociality was long a mystery. How could natural selection favor the evolution of sterile workers that provided no offspring? Hamilton explained the origin of eusociality in hymenopterans (bees, wasps, and ants) with his kin selection model. In these insects, males are haploid and females are diploid. This unusual system of sex determination, called *haplodiploidy*, leads to an unusual situation. If the queen is fertilized by a single male, then all female offspring will inherit exactly the same alleles from their father (because he is haploid and has only one copy of each allele). These female offspring will also share among themselves, on average, half of the alleles they get from the queen. Consequently, each female offspring will share, on average, 75% of her alleles with each sister (to verify this, rework figure 54.34, but allow the father to only have one allele for each gene).

By contrast, should a female offspring have offspring of her own, she would share only half of her alleles with these offspring (the other half would come from their father). Thus, because of this close genetic relatedness, *workers propagate more of their own alleles by giving up their own reproduction to assist their mother in rearing their sisters, some of whom will be new queens and start new colonies and reproduce.*

In this way, the unusual haplodiploid system may have set the stage for the evolution of eusociality in hymenopterans, and indeed, such systems have evolved as many as 12 or more separate times in the Hymenoptera. One wrinkle in this theory, however, is that eusocial systems have evolved in several other groups, including thrips, termites, and naked mole rats. Although thrips are also haplodiploid, both termites and naked mole rats are not. Thus, although haplodiploidy may have facilitated the evolution of eusociality, it is not a necessary prerequisite.

figure 54.36

REPRODUCTIVE DIVISION OF LABOR IN HONEYBEES. The queen (shown here with a red spot painted on her thorax) is the sole egg-layer. Her daughters are sterile workers.

Many factors could be responsible for the evolution of altruistic behaviors. Individuals may benefit directly if altruistic acts are reciprocated; kin selection explains how alleles for altruism can increase in frequency if altruistic acts are directed toward relatives. Kin selection is a potent force favoring, in some situations, the evolution of altruism and even complex social systems.

54.11 The Evolution of Social Systems

Organisms as diverse as prokaryotes, cnidarians, insects, fish, birds, prairie dogs, lions, whales, and chimpanzees exist in social groups. To encompass the wide variety of social phenomena, we can broadly define **a society** as a group of organisms of the same species that are organized in a cooperative manner.

Why have individuals in some species given up a solitary existence to become members of a group? We have just seen that one explanation is kin selection: Groups may be composed of close relatives. In other cases, individuals may benefit directly from social living. For example, a bird that joins a flock may receive greater protection from predators. As flock size increases, the risk of predation decreases because there are more individuals to scan the environment for predators (figure 54.37).

A member of a flock may also increase its feeding success if it can acquire information from other flock members about the location of new, rich food sources. In some predators, hunting in groups can increase success and allow the group to tackle prey too large for any one individual.

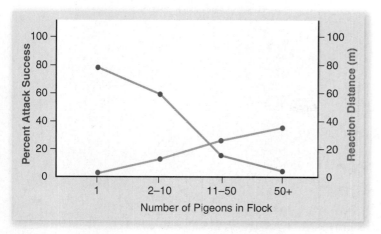

figure 54.37

FLOCKING BEHAVIOR DECREASES PREDATION. When more pigeons are present in the flock, they can detect hawks at greater distances, thus allowing more time for the pigeons to escape. As a result, as the size of a pigeon flock increases, hawks are less successful at capturing pigeons.

inquiry

Would living in a flock affect the time available for foraging by pigeons?

Insect societies include individuals specialized for different tasks

In insects, sociality has chiefly evolved in two orders, the Hymenoptera (ants, bees, and wasps) and the Isoptera (termites), although a few other insect groups include social species. As just discussed, a number of types of insects have evolved eusocial systems. These social insect colonies are composed of different **castes,** groups of individuals that differ in size and morphology and perform different tasks, such as workers and soldiers (figure 54.38).

Honeybees

In honeybees, the queen maintains her dominance in the hive by secreting a pheromone, called "queen substance," that suppresses development of the ovaries in other females, turning them into sterile workers. Drones (male bees) are produced only for purposes of mating. When the colony grows larger in the spring, some members do not receive a sufficient quantity of queen substance and begin to develop into queens, and the colony begins preparations for swarming.

Workers make several new queen chambers, in which new queens begin to develop. Scout workers look for a new nest site and communicate its location to the colony. The old queen and a swarm of female workers then move to the new site. Left behind, a new queen emerges, kills the other potential queens, flies out to mate, and returns to assume "rule" of the hive.

Leaf-cutter ants

Leaf-cutter ants provide another fascinating example of the remarkable lifestyles of social insects. Leaf-cutters live in colonies of up to several million individuals, growing crops of fungi

beneath the ground. Their moundlike nests are underground "cities" covering more than 100 m², with hundreds of entrances and chambers as deep as 5 m beneath the ground. Recent molecular studies suggest that ants have been cultivating these fungi for more than 50 million years.

The division of labor among the worker ants is related to their size. Every day, workers travel along trails from the nest to a tree or a bush, cut its leaves into small pieces, and carry

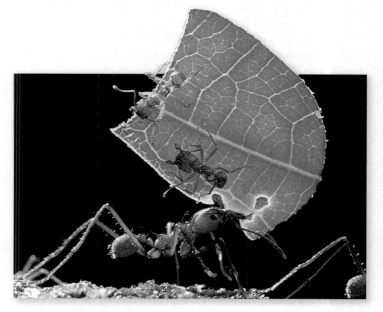

figure 54.38

CASTES OF ANTS. These leaf-cutter ants are members of different castes. The large ant is a worker carrying leaves to the nest, whereas the smaller ants are protecting the worker from attack.

the pieces back to the nest (see figure 54.38). Smaller workers chew the leaf fragments into a mulch, which they spread like a carpet in the underground fungus chambers. Even smaller workers implant fungal hyphae in the mulch. Soon a luxuriant garden of fungi is growing.

While other workers weed out undesirable kinds of fungi, nurse ants carry the larvae of the nest to choice spots in the garden, where the larvae graze. Some of these larvae grow into reproductive queens that will disperse from the parent nest and start new colonies, repeating the cycle.

Vertebrate societies come in many forms and structures

In contrast to the highly structured and integrated insect societies and their remarkable forms of altruism, vertebrate social groups are usually less rigidly organized and cohesive. It seems paradoxical that vertebrates, which have larger brains and are capable of more complex behaviors, are generally less altruistic than insects. Nevertheless, in some complex vertebrate social systems, individuals may be exhibiting both reciprocity and kin-selected altruism. Vertebrate societies also generally display more conflict and aggression among group members than do insect societies. Conflict in vertebrate societies generally centers on access to food and mates.

Like insect societies, vertebrate societies have particular types of organization. Each social group of vertebrates has a certain size, stability of members, number of breeding males and females, and type of mating system. Behavioral ecologists have learned that the way a group is organized is influenced most often by ecological factors such as food type and predation. For example, meerkats take turns watching for predators while other group members forage for food (figure 54.39).

African weaver birds, which construct nests from vegetation, provide an excellent example of the relationship between ecology and social organization. Their roughly 90 species can be divided according to the type of social group they form. One set of species lives in the forest and builds camouflaged, solitary nests. Males and females are monogamous; they forage for insects to feed their young. The second group of species nests in colonies in trees on the savanna. They are polygynous and feed in flocks on seeds.

The feeding and nesting habits of these two sets of species are correlated with their mating systems. In the forest, insects are hard to find, and both parents must cooperate in feeding the young. The camouflaged nests do not call the attention of predators to their brood. On the open savanna, building a hidden nest is not an option. Rather, savanna-dwelling weaver birds protect their young from predators by nesting in trees, which are not very abundant. This shortage of safe nest sites means that birds must nest together in colonies. Because seeds occur abundantly, a female can acquire all the food needed to rear young without a male's help. The male, free from the duties of parenting, spends his time courting many females—a polygynous mating system.

One exception to the general rule that vertebrate societies are not organized like those of insects is the naked mole rat (*Heterocephalus glaber*), a small, hairless rodent that lives in and near East Africa. Unlike other kinds of mole rats, which

figure 54.39

FORAGING AND PREDATOR AVOIDANCE. A meerkat sentinel on duty. Meerkats, *Suricata suricata*, are a species of highly social mongoose living in the semiarid sands of the Kalahari Desert in southern Africa. This meerkat is taking its turn to act as a lookout for predators. Under the security of its vigilance, the other group members can focus their attention on foraging.

live alone or in small family groups, naked mole rats form large underground colonies with a far-ranging system of tunnels and a central nesting area. It is not unusual for a colony to contain 80 individuals.

Naked mole rats feed on bulbs, roots, and tubers, which they locate by constant tunneling. As in insect societies, there is a division of labor among the colony members, with some mole rats working as tunnelers while others perform different tasks, depending on the size of their bodies. Large mole rats defend the colony and dig tunnels.

Naked mole rat colonies have a reproductive division of labor similar to the one normally associated with the eusocial insects. All of the breeding is done by a single female, or "queen," who has one or two male consorts. The workers, consisting of both sexes, keep the tunnels clear and forage for food.

Eusocial insects exhibit an advanced social structure that includes reproductive division of labor and workers with different tasks. Social behavior in vertebrates is often characterized by kin-selected altruism. Altruistic behavior is involved in cooperative breeding in birds and alarm calling in mammals.

54.1 Approaches to the Study of Behavior

Behavior is defined as the way an animal responds to stimuli in its environment.

■ Proximate causation, the "how" of behavior, studies the immediate cause of behavior.

■ Ultimate causation, the "why" of behavior, studies the evolutionary significance and how it influences survival and reproduction.

■ Innate, or instinctive, behavior is a response to an environmental trigger that does not require learning.

54.2 Behavioral Genetics

Behavioral genetics deals with those components of behavior governed by genes that are inherited.

54.3 Learning

Learning is a process in which behavior is altered as a result of prior experience.

■ Habituation, a form of nonassociative learning, is defined as a decrease in response to repeated nonessential stimuli.

■ Associative learning is a change in behavior by association of two stimuli or by a behavior and a response.

■ Classical conditioning, also called Pavlovian conditioning, occurs when two stimuli are associated with each other.

■ Operant conditioning occurs when an animal associates a behavior with reward or punishment.

54.4 The Development of Behavior

Behavior has both genetic and learned components.

■ Imprinting is a form of learning in which a young animal forms a social attachment to other individuals or develops preferences that will influence behavior later in life.

■ Animals may have an innate genetic template that guides their learning as behavior develops.

54.5 Animal Cognition

Some animals may exhibit cognitive behavior, that is, they can solve problems.

54.6 Orientation and Migratory Behavior

Some animals engage in apparent goal-oriented movements.

■ Orientation is the mechanism by which animals move by tracking environmental stimuli.

■ Migrations are long-range, two-way movements of animals using orientation or navigation (or both).

■ In birds celestial cues determine the direction of migration, whereas magnetic cues determine a specific migratory path.

■ Inexperienced animals migrate by orientation, and experienced animals migrate by navigation.

54.7 Animal Communication

Communication between and among species is key to many behaviors.

■ Successful reproduction depends on the progressive exchange of appropriate signals and responses.

■ Courtship signals are usually species-specific and limit communication to members of the same species.

■ Communication includes visual displays, sounds, electrical signals, and pheromones.

■ Communication among individuals of the same species promotes group living.

54.8 Behavioral Ecology

Behavioral ecology is the study of how natural selection influences behavior that increases survival and reproduction.

■ Natural selection favors optimal foraging strategies in which energy acquisition (cost) is minimized and reproductive success (benefit) is maximized.

■ Some animals are territorial. Territorial defense has its costs, and territoriality may only occur when the benefits outweigh the costs.

54.9 Reproductive Strategies and Sexual Selection

Reproductive strategy is a set of behaviors that maximize reproductive success. They include mate choice, how many mates to have, and parental investment in care of the offspring.

■ Reproductive strategies are responses to the spatial distribution of food resources, nest sites, and members of the opposite sex.

■ Relative parental investment of the sexes affects reproductive strategies.

■ Intrasexual selection involves members of the same sex for the chance to mate. Sperm competition is also a form of intrasexual selection.

■ Intersexual selection refers to a choice of mate by members of the opposite sex.

■ Direct benefits of mate choice refer to advantages such as territory quality or degree of parental care that differ between possible mates.

■ Indirect benefits of mate choice include factors such as the genetic quality of the mate, which can be passed on, producing higher quality offspring.

■ Mating systems include monogamy, polygyny, and polyandry, and they are influenced by ecology and constrained by the needs of the offspring.

54.10 Altruism and Group Living

Altruism refers to the action of an individual that benefits the fitness of another individual or individuals.

■ Seemingly altruistic acts may not be altruistic because the helpers benefit by learning valuable experiences, inheriting territories, or increasing the ability to escape from a predator.

■ Kin selection increases the reproductive success of relatives and frequency of alleles shared by kin.

■ Hamilton's rule states that altruistic acts are favored when the product of the benefits of altruism and the coefficient of relatedness is greater than the costs of the altruistic act.

54.11 The Evolution of Social Systems

A social system is a group of organisms of the same species that are organized in a cooperative manner.

■ Individuals benefit from social living, and the benefits increase with the number of organisms.

■ Social insects are composed of different castes that each have specialized tasks.

■ Vertebrate social systems are less rigidly organized and cohesive and are influenced by food availability and predation.

SELF TEST

1. A sign stimulus, innate releasing mechanism, fixed action pattern, and supernormal stimuli
   a. are mechanisms associated with behaviors that are learned.
   b. are components of behaviors that are innate.
   c. involve behaviors that cannot be explained in terms of ultimate causation.
   d. involve behaviors that are not subject to natural selection.

2. In operant conditioning
   a. an animal learns that a particular behavior leads to a reward or punishment.
   b. an animal associates an unconditioned stimulus with a conditioned response.
   c. learning is unnecessary.
   d. habituation is required for an appropriate response.

3. The study of song development in sparrows showed that
   a. the acquisition of a species-specific song is innate.
   b. there are two components to this behavior: a genetic template and learning.
   c. song acquisition is an example of associative learning.
   d. All of these are correct.

4. The difference between following a set of driving directions given to you by somebody on the street (for example "... take a right at the next light, go four blocks and turn left ...") and using a map to find your destination is the difference between
   a. navigation and orientation, respectively.
   b. learning and migration, respectively.
   c. orientation and navigation, respectively.
   d. is why birds are not capable of orientation.

5. One of most important differences between signaling in courtship and that in territorial marking is that
   a. the former is always species-specific.
   b. the latter is always species-specific.
   c. both always involve a stimulus–response chain.
   d. the former is always initiated by males.

6. Behavioral ecology assumes
   a. that all behavioral traits are innate.
   b. learning is the dominant determinate of behavior.
   c. behavioral traits are subject to natural selection.
   d. behavioral traits do not affect fitness.

7. According to optimal foraging theory
   a. individuals minimize energy intake per unit of time.
   b. energy content of a food item is the only determinant of a forager's food choice.
   c. time taken to capture a food item is the only determinant of a forager's food choice.
   d. a higher energy item might be less valuable than a lower energy item if it takes too much time to capture the larger item.

8. The showy tail feathers of a male peacock evolved because they
   a. improve reproductive success.
   b. improve male survival.
   c. reduce survival and reproductive success.
   d. None of the above.

9. From the perspective of females, extra-pair copulations (epc)
   a. are always disadvantageous.
   b. always provide indirect benefits such as good genes.
   c. might be beneficial if the benefit gained from an epc mate outweighs the cost.
   d. can only occur if the epc male has a "handicap" trait.

10. In the haplodiploidy system of sex determination, males are
    a. haploid.
    b. diploid.
    c. sterile.
    d. not present because bees exist as single-sex populations.

11. According to kin selection, saving the life of your _____ would do the least for increasing your inclusive fitness.
    a. mother
    b. brother
    c. sister-in-law
    d. niece

12. Altruism
    a. is only possible with reciprocity.
    b. is only possible with kin selection.
    c. cannot be explained given the way natural selection operates.
    d. will only occur when the fitness benefit of a given act is greater than the fitness cost.

CHALLENGE QUESTIONS

1. Refer to figure 54.25. Data on size of mussels eaten by shore crabs suggests they eat sizes smaller than expected by an optimal foraging model. Suggest a hypothesis for why and describe an experiment to test your hypothesis.

2. Refer to figure 54.26. Six pairs of birds were removed but only four pairs moved in. Where did the new pairs come from? Additionally, it appears that many of the birds that were not removed expanded their territories and that the new residents ended up with smaller territories than the pairs they replaced. Explain.

3. Refer to figure 54.28. Peahens prefer to mate with peacocks that have more eyespots in their tail feathers (that is, longer tail feathers). It has also been suggested that the longer the tail feathers, the more impaired the flight of the males. One possible hypothesis to explain such a preference by females is that the males with the longest tail feathers experience the most severe handicap, and if they can nevertheless survive, it reflects their "vigor." Suggest some studies that would allow you to test this idea. Your description should include the kinds of traits that you would measure and why.

4. An altruistic act is defined as one that benefits another individual at a cost to the actor. There are two theories about how to explain how such behavior evolves: reciprocity and kin selection. In the context of natural selection, is an altruistic act "costly" to an individual who performs it?

5. What are the potential consequences for evolution of prey for a predator such as a frog associating black and yellow coloration with the pain of a bee sting?

introduction

ECOLOGY, THE STUDY OF HOW organisms relate to one another and to their environments, is a complex and fascinating area of biology that has important implications for each of us. In our exploration of ecological principles, we first consider how organisms respond to the abiotic environment in which they occur and how these responses affect the properties of populations, emphasizing population dynamics. In chapter 56, we discuss communities of coexisting species and the interactions that occur among them. In subsequent chapters, we discuss the functioning of entire ecosystems and of the biosphere, concluding with a consideration of the problems facing our planet and our fellow species.

55.1 The Environmental Challenge

The nature of the physical environment in large measure determines which organisms live in a particular climate or region. Key elements of the environment include:

Temperature. Most organisms are adapted to live within a relatively narrow range of temperatures and will not thrive if temperatures are colder or warmer. The growing season of plants, for example, is importantly influenced by temperature.

Water. All organisms require water. On land, water is often scarce, so patterns of rainfall have a major influence on life.

Sunlight. Almost all ecosystems rely on energy captured by photosynthesis; the availability of sunlight influences the amount of life an ecosystem can support, particularly below the surface in marine communities.

Soil. The physical consistency, pH, and mineral composition of the soil often severely limit plant growth, particularly the availability of nitrogen and phosphorus.

An individual encountering environmental variation may maintain a "steady-state" internal environment, a condition known as **homeostasis.** Many animals and plants actively employ physiological, morphological, or behavioral mechanisms to maintain homeostasis. The beetle in figure 55.1 is using a behavioral mechanism to cope with drastic changes in water availability. Other animals and plants are known as **conformers** because they conform to the environment in which they find themselves, their bodies adopting the temperature, salinity, and other physical aspects of their surroundings.

Responses to environmental variation can be seen over both the short and the long term. In the short term, spanning periods of a few minutes to an individual's lifetime, organisms have a variety of ways of coping with environmental change. Over longer periods, natural selection can operate to make a population better adapted to the environment.

figure 55.1

MEETING THE CHALLENGE OF OBTAINING MOISTURE. On the dry sand dunes of the Namib Desert in southwestern Africa, the fog-basking beetle *Onymacris unguicularis* collects moisture from the fog by holding its abdomen up at the crest of a dune to gather condensed water; water condenses as droplets and trickles down to the beetle's mouth.

TABLE 55.1	Physiological Changes at High Elevation
Increased rate of breathing	
Increased erythrocyte production, raising the amount of hemoglobin in the blood	
Decreased binding capacity of hemoglobin, increasing the rate at which oxygen is unloaded in body tissues	
Increased density of mitochondria, capillaries, and muscle myoglobin	

Organisms are capable of responding to environmental changes that occur during their lifetime

During the course of a day, a season, or a lifetime, an individual organism must cope with a range of living conditions. They do so through the physiological, morphological, and behavioral abilities they possess. These abilities are a product of natural selection acting in a particular environmental setting over time, which explains why an individual organism that is moved to a different environmental may not survive.

Physiological responses

Many organisms are able to adapt to environmental change by making physiological adjustments. For example, you sweat when it is hot, increasing evaporative heat loss and thus preventing overheating. Similarly, people who visit high altitudes may initially experience altitude sickness—the symptoms of which include heart palpitations, nausea, fatigue, headache, mental impairment, and in serious cases, pulmonary edema—because of the lower atmospheric pressure and consequent lower oxygen availability in the air. After several days, however, the same people usually feel fine, because a number of physiological changes have increased the delivery of oxygen to their body tissues (table 55.1).

Some insects avoid freezing in the winter by adding glycerol "antifreeze" to their blood; others tolerate freezing by converting much of their glycogen reserves into alcohols that protect their cell membranes from freeze damage.

Morphological capabilities

Animals that maintain a constant internal temperature (endotherms) in a cold environment have adaptations that tend to minimize energy expenditure. For example, many mammals grow thicker coats during the winter, their fur acting as insulation to retain body heat. In general, the thicker the fur, the greater the insulation (figure 55.2). Thus, a wolf's fur is about three times thicker in winter than in summer and insulates more than twice as well.

Behavioral responses

Many animals deal with variation in the environment by moving from one patch of habitat to another, avoiding areas that are unsuitable. The tropical lizard in figure 55.3 manages to

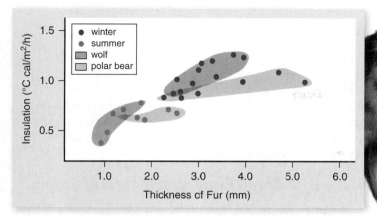

figure 55.2

MORPHOLOGICAL ADAPTATION. Fur thickness in North American mammals has a major effect on the degree of insulation the fur provides.

maintain a fairly uniform body temperature in an open habitat by basking in patches of sunlight and then retreating to the shade when it becomes too hot. By contrast, in shaded forests, the same lizard does not have the opportunity to regulate its body temperature through behavioral means. Thus, it becomes a conformer and adopts the temperature of its surroundings.

Behavioral adaptations can be extreme. Spadefoot toads (genus *Scaphiophus*), which live in the deserts of North America, can burrow nearly a meter below the surface and remain there for as long as nine months of each year, their metabolic rates greatly reduced as they live on fat reserves. When moist, cool conditions return, the toads emerge and breed. The young toads mature rapidly and burrow underground.

Natural selection leads to evolutionary adaptation to environmental conditions

The ability of an individual to alter its physiology, morphology, or behavior is itself an evolutionary adaptation, the result of natural selection. The results of natural selection can also be detected by comparing closely related species that live in different environments. In such cases, species often have evolved striking adaptations to the particular environment in which they live.

For example, animals that live in different climates show many differences. Mammals from colder climates tend to have shorter ears and limbs—a phenomenon termed *Allen's rule*—which reduces the surface area across which animals lose heat. Lizards that live in different climates exhibit physiological adaptations for coping with life at different temperatures. Desert lizards are unaffected by high temperatures that would kill a lizard from northern Europe, but the northern lizards are capable of running, capturing prey, and digesting food at cooler temperatures at which desert lizards would be completely immobilized.

Many species also exhibit adaptations to living in areas where water is scarce. Everyone knows of the camel and other desert animals that can go extended

periods without drinking water. Another example of desert adaptation is seen in frogs. Most frogs have moist skins through which water permeates readily. Such animals could not survive in arid climates because they would rapidly dehydrate and die. However, some frogs have solved this problem by evolving a greatly reduced rate of water loss through the skin. One species, for example, secretes a waxy substance from specialized glands that waterproofs its skin and reduces rates of water loss by 95%.

Adaptation to different environments can also be studied experimentally. For example, when strains of *E. coli* were grown at high temperatures (42°C), the speed at which the bacteria utilized resources improved through time. After 2000 generations, this ability increased 30% over what it had been when the experiment started. The mechanism by which efficiency of resource use was increased is still unknown and is the focus of current research.

Organisms use a variety of physiological, morphological, and behavioral mechanisms to adjust to environmental variation. Over time, species evolve adaptations to living in different environments.

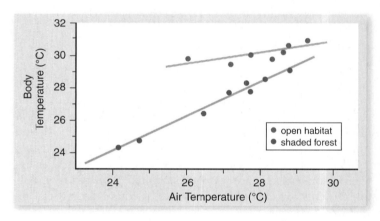

figure 55.3

BEHAVIORAL ADAPTATION. In open habitats, the Puerto Rican crested lizard (*Anolis cristatellus*) maintains a relatively constant temperature by seeking out and basking in patches of sunlight; as a result, it can maintain a relatively high temperature even when the air is cool. In contrast, in shaded forests, this behavior is not possible, and body temperature conforms to that of its surroundings.

inquiry

When given the opportunity, lizards regulate their body temperature to maintain a temperature optimal for physiological functioning. Would lizards in open habitats exhibit different escape behaviors from lizards in shaded forest?

Populations: Groups of Individuals of the Same Species in One Place

Organisms live as members of **populations,** groups of individuals that occur together at one place and time. In the rest of this chapter, we consider the properties of populations, focusing on elements that influence whether a population will grow or shrink, and at what rate. The explosive growth of the world's human population in the last few centuries provides a focus for our inquiry.

The term *population* can be defined narrowly or broadly. This flexibility allows us to speak in similar terms of the world's human population, the population of protists in the gut of a termite, or the population of deer that inhabit a forest. Sometimes the boundaries defining a population are sharp, such as the edge of an isolated mountain lake for trout, and sometimes they are fuzzier, as when individual deer readily move back and forth between two forests separated by a cornfield.

Three characteristics of population ecology are particularly important: population range, the area throughout which a population occurs; the pattern of spacing of individuals within that range; and how the population changes in size through time.

A population's geographic distribution is termed its range

No population, not even one composed of humans, occurs in all habitats throughout the world. Most species, in fact, have relatively limited geographic ranges, and the range of some species is miniscule. For example, the Devil's Hole pupfish lives in a single spring in southern Nevada (figure 55.4), and the Socorro isopod (*Thermosphaeroma thermophilus*) is known from a single spring system in New Mexico. At the other extreme, some species are widely distributed. The common dolphin (*Delphinus delphis*), for example, is found throughout all the world's oceans.

As discussed earlier, organisms must be adapted for the environment in which they occur. Polar bears are exquisitely adapted to survive the cold of the Arctic, but you won't find them in the tropical rain forest. Certain prokaryotes can live in the near-boiling waters of Yellowstone's geysers, but they do not occur in cooler streams nearby. Each population has its own requirements—temperature, humidity, certain types of

food, and a host of other factors—that determine where it can live and reproduce and where it can't. In addition, in places that are otherwise suitable, the presence of predators, competitors, or parasites may prevent a population from occupying an area, a topic we will take up in chapter 56.

Ranges undergo expansion and contraction

Population ranges are not static but change through time. These changes occur for two reasons. In some cases, the environment changes. As the glaciers retreated at the end of the last ice age, approximately 10,000 years ago, many North American plant and animal populations expanded northward. At the same time, as climates warmed, species experienced shifts in the elevation at which they could live (figure 55.5).

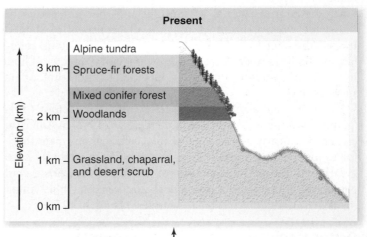

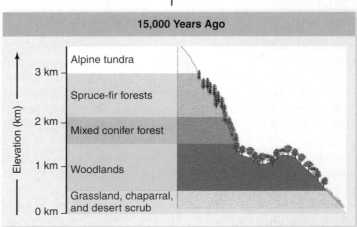

figure 55.4

THE DEVIL'S HOLE PUPFISH (*Cyprinodon diabolis*). This fish has the smallest range of any vertebrate species in the world.

figure 55.5

ALTITUDE SHIFTS IN ALTITUDINAL DISTRIBUTIONS OF TREES IN THE MOUNTAINS OF SOUTHWESTERN NORTH AMERICA. During the glacial period 15,000 years ago, conditions were cooler than they are now. As the climate has warmed, tree species that require colder temperatures have shifted their range upward in altitude so that they live in the climatic conditions to which they are adapted.

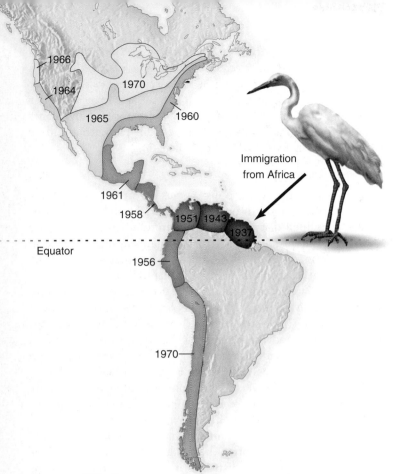

figure 55.6

RANGE EXPANSION OF THE CATTLE EGRET (*Bubulcus ibis*).
The cattle egret—so named because it follows cattle and other hoofed animals, catching any insects or small vertebrates it disturbs—first arrived in South America in the late 1800s. Since the 1930s, the range expansion of this species has been well documented, as it has moved northward into much of North America, as well as southward along the western side of the Andes to near the southern tip of South America.

In addition, populations can expand their ranges when they are able to circumvent inhospitable habitat to colonize suitable, previously unoccupied areas. For example, the cattle egret is native to Africa. Some time in the late 1800s, these birds appeared in northern South America, having made the nearly 3500-km transatlantic crossing, perhaps aided by strong winds. Since then, they have steadily expanded their range and now can be found throughout most of the United States (figure 55.6).

The human effect

By altering the environment, humans have allowed some species, such as coyotes, to expand their ranges and move into areas they previously did not occupy. Moreover, humans have served as an agent of dispersal for many species. Some of these transplants have been widely successful, as is discussed in greater detail in chapter 59. For example, 100 starlings were introduced into New York City in 1896 in a misguided attempt to establish every species of bird mentioned by Shakespeare. Their population steadily spread so that by 1980, they occurred throughout the United States.

Similar stories could be told for countless plants and animals, and the list increases every year. Unfortunately, the success of these invaders often comes at the expense of native species.

Dispersal mechanisms

Dispersal to new areas can occur in many ways. Lizards have colonized many distant islands, as one example, probably due to individuals or their eggs floating or drifting on vegetation. Bats are often the only mammals on distant islands because they can fly to them.

Seeds of plants are designed to disperse in many ways (figure 55.7). Some seeds are aerodynamically designed to be blown long distances by the wind. Others have structures that stick to the fur or feathers of animals, so that they are carried long distances before falling to the ground. Still others are enclosed in fleshy fruits. These seeds can pass through the digestive systems of mammals or birds and then germinate where they are defecated. Finally, seeds of mistletoes (*Arceuthobium*) are violently propelled from the base of the fruit in an explosive discharge. Although the probability of long-distance dispersal events leading to successful establishment of new populations is low, over millions of years, many such dispersals have occurred.

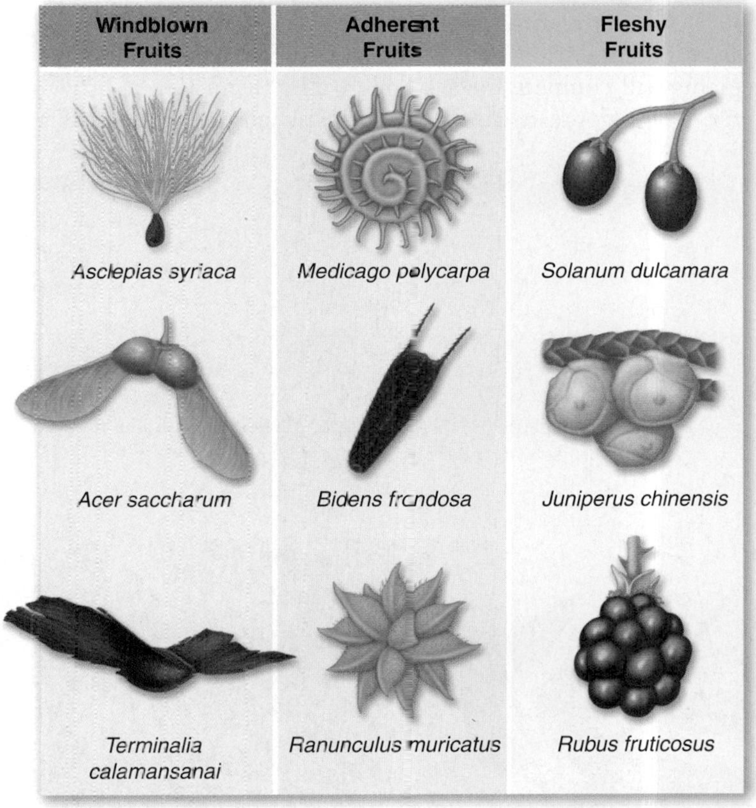

Windblown Fruits	Adherent Fruits	Fleshy Fruits
Asclepias syriaca	*Medicago polycarpa*	*Solanum dulcamara*
Acer saccharum	*Bidens frondosa*	*Juniperus chinensis*
Terminalia calamansanai	*Ranunculus muricatus*	*Rubus fruticosus*

figure 55.7

SOME OF THE MANY ADAPTATIONS OF SEEDS. Seeds have evolved a number of different means of facilitating dispersal from their maternal plant. Some seeds can be transported great distances by the wind, whereas seeds enclosed in adherent or fleshy fruits can be transported by animals.

Individuals in populations exhibit different spacing patterns

Another key characteristic of population structure is the way in which individuals of a population are distributed. They may be randomly spaced, uniformly spaced, or clumped (figure 55.8a).

Random spacing

Random spacing of individuals within populations occurs when they do not interact strongly with one another and when they are not affected by nonuniform aspects of their environment. Random distributions are not common in nature. Some species of trees, however, appear to exhibit random distributions in Panamanian rain forests (figure 55.8b).

Uniform spacing

Uniform spacing within a population may often, but not always, result from competition for resources. This spacing is accomplished, however, in many different ways

In animals, uniform spacing often results from behavioral interactions, as described in chapter 54. In many species, individuals of one or both sexes defend a territory from which other individuals are excluded. These territories provide the owner with exclusive access to resources such as food, water, hiding refuges, or mates, and tend to space individuals evenly across the habitat. Even in nonterritorial species, individuals often maintain a defended space into which other animals are not allowed to intrude.

Among plants, uniform spacing is also a common result of competition for resources. Closely spaced individual plants will compete for available sunlight, nutrients, or water. These contests can be direct, as when one plant casts a shadow over another, or indirect, as when two plants compete by extracting nutrients or water from a shared area. In addition, some plants, such as the creosote bush, produce chemicals in the surrounding soil that are toxic to other members of their species. In all of these cases, only plants that are spaced an adequate distance from each other will be able to coexist, leading to uniform spacing.

Clumped spacing

Individuals clump into groups or clusters in response to uneven distribution of resources in their immediate environments. Clumped distributions are common in nature because individual animals, plants, and microorganisms tend to occur in habitats defined by soil type, moisture, or other aspects of the environment to which they are best adapted.

Social interactions also can lead to clumped distributions. Many species live and move around in large groups, which go by a variety of names (for example, flock, herd, pride). These groupings can provide many advantages, including increased awareness of and defense against predators, decreased energy cost of moving through air and water, and access to the knowledge of all group members.

At a broader scale, populations are often most densely populated in the interior of their range and less densely distributed toward the edges. Such patterns usually result from the manner in which the environment changes in different areas.

Populations are often best adapted to the conditions in the interior of their distribution. As environmental conditions change, individuals are less well adapted, and thus densities decrease. Ultimately, the point is reached at which individuals cannot persist at all; this marks the edge of a population's range.

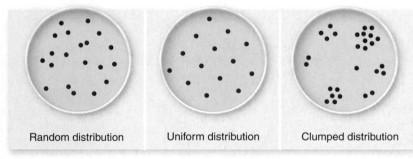

Random distribution Uniform distribution Clumped distribution

a.

figure 55.8

POPULATION DISPERSION. The different patterns of dispersion are exhibited by (*a*) different arrangements of bacterial colonies and (*b*) three different species of trees from the same locality in Panama.

Source: Data from Elizabeth Losos, Center for Tropical Forest Science, Smithsonian Tropical Research Institute.

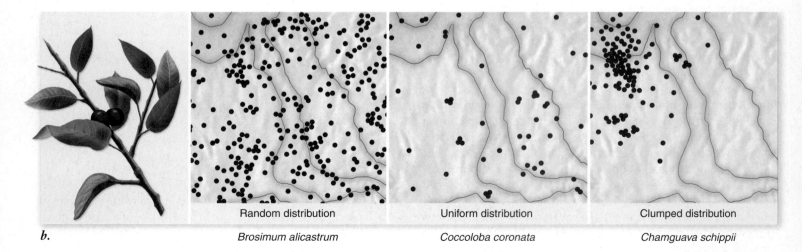

b.

Random distribution Uniform distribution Clumped distribution

Brosimum alicastrum *Coccoloba coronata* *Chamguava schippii*

A metapopulation comprises distinct populations that may exchange members

Species often exist as a network of distinct populations that interact with one another by exchanging individuals. Such networks, termed **metapopulations,** usually occur in areas in which suitable habitat is patchily distributed and is separated by intervening stretches of unsuitable habitat.

Dispersal and habitat occupancy

The degree to which populations within a metapopulation interact depends on the amount of dispersal; this interaction is often not symmetrical: Populations increasing in size tend to send out many dispersers, whereas populations at low levels tend to receive more immigrants than they send off. In addition, relatively isolated populations tend to receive relatively few arrivals.

Not all suitable habitats within a metapopulation's area may be occupied at any one time. For a number of reasons, some individual populations may become extinct, perhaps as a result of an epidemic disease, a catastrophic fire, or the loss of genetic variation following a population bottleneck (see chapter 59). Dispersal from other populations, however, may eventually recolonize such areas. In some cases, the number of habitats occupied in a metapopulation may represent an equilibrium in which the rate of extinction of existing populations is balanced by the rate of colonization of empty habitats.

Source–sink metapopulations

A species may also exhibit a metapopulation structure in areas in which some habitats are suitable for long-term population maintenance, but others are not. In these situations, termed **source–sink metapopulations,** the populations in the better areas (the sources) continually send out dispersers that bolster the populations in the poorer habitats (the sinks). In the absence of such continual replenishment, sink populations would have a negative growth rate and would eventually become extinct.

Metapopulations of butterflies have been studied particularly intensively. In one study, researchers sampled populations of the Glanville fritillary butterfly at 1600 meadows in southwestern Finland (figure 55.9). On average, every year, 200 populations became extinct, but 114 empty meadows were colonized. A variety of factors seemed to increase the likelihood of a population's extinction, including small population size, isolation from sources of immigrants, low resource availability (as indicated by the number of flowers on a meadow), and lack of genetic variation within the population.

The researchers attribute the greater number of extinctions than colonizations to a string of very dry summers. Because none of the populations is large enough to survive on its own, continued survival of the species in southwestern Finland would appear to require the continued existence of a metapopulation network in which new populations are continually created and existing populations are supplemented by immigrants. Continued bad weather thus may doom the species, at least in this part of its range.

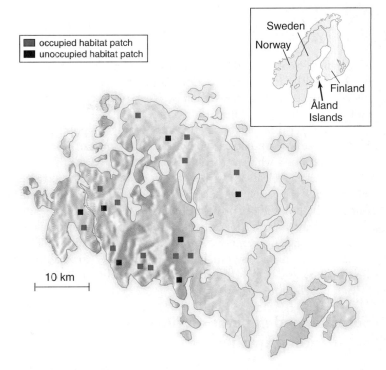

figure 55.9

METAPOPULATIONS OF BUTTERFLIES. The Glanville fritillary butterfly (*Melitaea cinxia*) occurs in metapopulations in southwestern Finland on the Åland Islands. None of the populations is large enough to survive for long on its own, but continual immigration of individuals from other populations allows some populations to survive. In addition, continual establishment of new populations tends to offset extinction of established populations, although in recent years, extinctions have outnumbered colonizations.

Metapopulations, where they occur, can have two important implications for the range of a species. First, through continuous colonization of empty patches, metapopulations prevent long-term extinction. If no such dispersal existed, then each population might eventually perish, leading to disappearance of the species from the entire area. Moreover, in source–sink metapopulations, the species occupies a larger area than it otherwise might, including marginal areas that could not support a population without a continual influx of immigrants. For these reasons, the study of metapopulations has become very important in conservation biology as natural habitats become increasingly fragmented.

A population is a group of individuals of the same species existing together in an area. Its range, the area a population occupies, changes over time.

The distribution of individuals within a population can be random, uniform, or clumped and is determined in part by the availability of resources.

Across broader areas, individuals may occur in populations that are loosely interconnected, termed metapopulations.

55.3 Population Demography and Dynamics

The dynamics of a population—how it changes through time—are affected by many factors. One important factor is the age distribution of individuals—that is, what proportion of individuals are adults, juveniles, and young.

Demography is the quantitative study of populations. How the size of a population changes through time can be studied at two levels: as a whole or broken down into parts. At the most inclusive level, we can study the whole population to determine whether it is increasing, decreasing, or remaining constant. Put simply, populations grow if births outnumber deaths and shrink if deaths outnumber births. Understanding these trends is often easier, however, if we break the population into smaller units composed of individuals of the same age (for example, 1-year-olds) and study the factors affecting birth and death rates for each unit separately.

Sex ratio and generation time affect population growth rates

Population growth can be influenced by the population's **sex ratio.** The number of births in a population is usually directly related to the number of females; births may not be as closely related to the number of males in species in which a single male can mate with several females. In many species, males compete for the opportunity to mate with females, as you learned in the preceding chapter; consequently, a few males have many matings, and many males do not mate at all. In such species, a female-biased sex ratio would not affect population growth rates; reduction in the number of males simply changes the identities of the reproductive males without reducing the number of births. By contrast, among monogamous species, pairs may form long-lasting reproductive relationships, and a reduction in the number of males can then directly reduce the number of births.

Generation time is the average interval between the birth of an individual and the birth of its offspring, and it can also affect population growth rates. Species differ greatly in generation time. Differences in body size can explain much of this variation—mice go through approximately 100 generations during the course of one elephant generation (figure 55.10). But small size does not always mean short generation time. Newts, for example, are smaller than mice, but have considerably longer generation times.

Everything else being equal, populations with short generations can increase in size more quickly than populations with long generations. Conversely, because generation time and life span are usually closely correlated, populations with short generation times may also diminish in size more rapidly if birthrates suddenly decrease.

Age structure is determined by the numbers of individuals in different age groups

A group of individuals of the same age is referred to as a **cohort.** In most species, the probability that an individual will reproduce or die varies through its life span. As a result, within a population, every cohort has a characteristic birthrate, or **fecundity,** defined as the number of offspring produced in a stan-

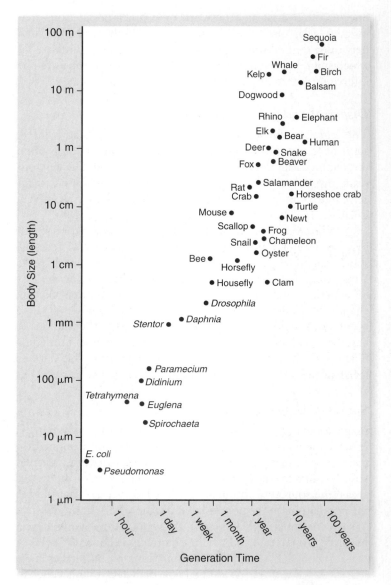

figure 55.10

THE RELATIONSHIP BETWEEN BODY SIZE AND GENERATION TIME. In general, larger organisms have longer generation times, although there are exceptions.

 inquiry

If resources became more abundant, would you expect smaller or larger species to increase in population size more quickly?

dard time (for example, per year), and death rate, or **mortality,** the number of individuals that die in that period.

The relative number of individuals in each cohort defines a population's **age structure.** Because different cohorts have different fecundity and death rates, age structure has a critical influence on a population's growth rate. Populations with a large proportion of young individuals, for example, tend to grow rapidly because an increasing proportion of their individuals are reproductive. Human populations in many developing countries are an example, as will be discussed later in

Age (in 3-month intervals)	Number Alive at Beginning of Time Interval	Proportion of Cohort Alive at Beginning of Time Interval (survivorship)	Deaths During Time Interval	Mortality Rate During Time Interval	Seeds Produced During Time Interval	Seeds Produced per Surviving Individual (fecundity)	Seeds Produced per Member of Cohort (fecundity × survivorship)
TABLE 55.2							
0	843	1.000	121	0.143	0	0.00	0.00
1	722	0.857	195	0.271	303	0.42	0.36
2	527	0.625	211	0.400	622	1.18	0.74
3	316	0.375	172	0.544	430	1.36	0.51
4	144	0.171	90	0.626	210	1.46	0.25
5	54	0.064	39	0.722	60	1.11	0.07
6	15	0.018	12	0.800	30	2.00	0.04
7	3	0.004	3	1.000	10	3.33	0.01
8	0	0.000	—		Total = 1665		Total = 1.98

TABLE 55.2 — Life Table Cohort of the Grass *Poa annua*, Containing 843 Seedlings

this chapter. Conversely, if a large proportion of a population is relatively old, populations may decline. This phenomenon now characterizes Japan and some countries in Europe.

Life tables show probability of survival and reproduction through a cohort's life span

To assess how populations in nature are changing, ecologists use a **life table,** which tabulates the fate of a cohort from birth until death, showing the number of offspring produced and the number of individuals that die each year. Table 55.2 shows an example of a life table analysis from a study of the meadow grass *Poa annua*. This study follows the fate of 843 individuals through time, charting how many survive in each interval and how many offspring each survivor produces.

In table 55.2, the first column indicates the age of the cohort (that is, the number of 3-month intervals from the start of the study). The second and third columns indicate the number of survivors and the proportion of the original cohort still alive at the beginning of that interval. The fifth column presents the **mortality rate,** the proportion of individuals that started that interval alive but died by the end of it. The seventh column indicates the average number of seeds produced by each surviving individual in that interval, and the last column shows the number of seeds produced relative to the size of the original cohort.

Much can be learned by examining life tables. In the case of *P. annua*, we see that both the probability of dying and the number of offspring produced per surviving individual steadily increases with age. By adding up the numbers in the last column, we get the total number of offspring produced per individual in the initial cohort. This number is almost 2, which means that for every original member of the cohort, on average two new individuals have been produced. A figure of 1.0 would be the break-even number, the point at which the population was neither growing nor shrinking. In this case, the population appears to be growing rapidly.

In most cases, life table analysis is more complicated than this. First, except for organisms with short life spans, it

is difficult to track the fate of a cohort until the death of the last individual. An alternative approach is to construct a cross-sectional study, examining the fate of cohorts of different ages in a single period. In addition, many factors—such as offspring reproducing before all members of their parents' cohort have died—complicate the interpretation of whether populations are growing or shrinking.

Survivorship curves demonstrate how survival probability changes with age

The percentage of an original population that survives to a given age is called its **survivorship.** One way to express some aspects of the age distribution of populations is through a *survivorship curve*. Examples of different survivorship curves

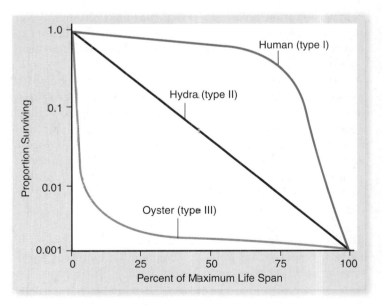

figure 55.11

SURVIVORSHIP CURVES. By convention, survival (the vertical axis) is plotted on a log scale. Humans have a type I life cycle, hydra (an animal related to jellyfish) type II, and oysters type III.

chapter 55 population ecology **1153**

are shown in figure 55.11. Oysters produce vast numbers of offspring, only a few of which live to reproduce. However, once they become established and grow into reproductive individuals, their mortality rate is extremely low (type III survivorship curve). Note that in this type of curve, survival and mortality rates are inversely related. Thus, the rapid decrease in the proportion of oysters surviving indicates that few individuals survive, thus producing a high mortality rate. In contrast, the relatively flat line at older ages indicates high survival and low mortality.

In hydra, animals related to jellyfish, individuals are equally likely to die at any age. The result is a straight survivorship curve (type II).

Finally, mortality rates in humans, as in many other animals and in protists, rise steeply later in life (type I survivorship curve).

Of course, these descriptions are just generalizations, and many organisms show more complicated patterns. Examination of the data for *P. annua*, for example, reveals that it is most similar to a type II survivorship curve (figure 55.12).

> **The growth rate of a population is a sensitive function of its age structure. The age structure of a population and the manner in which mortality and birthrates vary among different age cohorts determine whether a population will increase or decrease in size.**

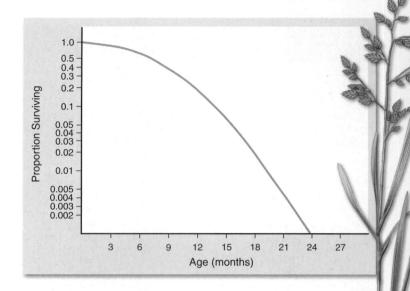

figure 55.12

SURVIVORSHIP CURVE FOR A COHORT OF THE MEADOW GRASS *POA ANNUA*. After several months of age, mortality increases at a constant rate through time.

inquiry

? *Suppose you wanted to keep meadow grass in your room as a houseplant. Suppose, too, that you wanted to buy an individual plant that was likely to live as long as possible. What age plant would you buy? How might the shape of the survivorship curve affect your answer?*

55.4 Life History and the Cost of Reproduction

Natural selection favors traits that maximize the number of surviving offspring left in the next generation. Two factors affect this quantity: how long an individual lives, and how many young it produces each year.

Why doesn't every organism reproduce immediately after its own birth, produce large families of offspring, care for them intensively, and perform these functions repeatedly throughout a long life, while outcompeting others, escaping predators, and capturing food with ease? The answer is that no one organism can do all of this, simply because not enough resources are available. Consequently, organisms allocate resources either to current reproduction or to increasing their prospects of surviving and reproducing at later life stages.

The complete life cycle of an organism constitutes its **life history.** All life histories involve significant trade-offs. Because resources are limited, a change that increases reproduction may decrease survival and reduce future reproduction. As one example, a Douglas fir tree that produces more cones increases its current reproductive success—but it also grows more slowly. Because the number of cones produced is a function of how large a tree is, this diminished growth will decrease the number of cones it can produce in the fu-

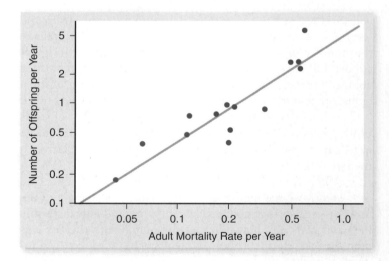

figure 55.13

REPRODUCTION HAS A PRICE. Data from many bird species indicate that increased fecundity in birds correlates with higher mortality, ranging from the albatross (lowest) to the sparrow (highest). Birds that raise more offspring per year have a higher probability of dying during that year.

ture. Similarly, birds that have more offspring each year have a higher probability of dying during that year or of producing smaller clutches the following year (figure 55.13). Conversely, individuals that delay reproduction may grow faster and larger, enhancing future reproduction.

In one elegant experiment, researchers changed the number of eggs in the nests of a bird, the collared flycatcher (figure 55.14). Birds whose clutch size (the number of eggs produced in one breeding event) was decreased expended less energy raising their young and thus were able to lay more eggs the next year, whereas those given more eggs worked harder and consequently produced fewer eggs the following year. Ecologists refer to the reduction in future reproductive potential resulting from current reproductive efforts as the **cost of reproduction.**

Natural selection will favor the life history that maximizes lifetime reproductive success. When the cost of reproduction is low, individuals should produce as many offspring as possible because there is little cost. Low costs of reproduction may occur when resources are abundant and may also be relatively low when overall mortality rates are high. In the latter case, individuals may be unlikely to survive to the next breeding season anyway, so the incremental effect of increased reproductive efforts may have little effect on future survival.

Alternatively, when costs of reproduction are high, lifetime reproductive success may be maximized by deferring or minimizing current reproduction to enhance growth and survival rates. This situation may occur when costs of reproduction significantly affect the ability of an individual to survive, or decrease the number of offspring that can be produced in the future.

Investment per offspring is geared to maximize fitness

In terms of natural selection, the number of offspring produced is not as important as how many of those offspring themselves survive to reproduce. Assuming that the amount of energy to be invested in offspring is limited, a balance must be reached between the number of offspring produced and the size of each offspring (figure 55.15). This trade-off has been experimentally demonstrated in the side-blotched lizard, which normally lays, on average, between four and five eggs at a time. When some of the eggs are removed surgically early in the reproductive cycle, the female lizard produces only one to three eggs, but supplies each of these eggs with greater amounts of yolk, producing eggs and, subsequently, hatchlings that are much larger than normal (figure 55.16). Alternatively, by removing yolk from eggs, scientists have demonstrated that smaller young would be produced.

In the side-blotched lizard and many other species, the size of offspring critically affects their survival prospects—larger

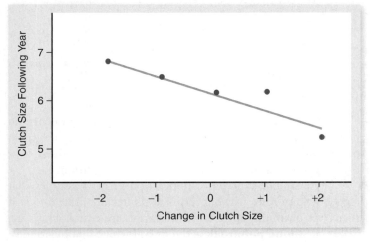

figure 55.14

REPRODUCTIVE EVENTS PER LIFETIME. Adding eggs to nests of collared flycatchers (*Ficedula albicollis*), which increases the reproductive efforts of the female rearing the young, decreases clutch size the following year; removing eggs from the nest increases the next year's clutch size. This experiment demonstrates the trade-off between current reproductive effort and future reproductive success.

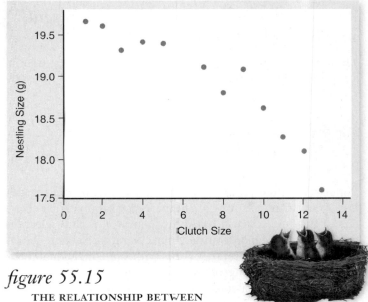

figure 55.15

THE RELATIONSHIP BETWEEN CLUTCH SIZE AND OFFSPRING SIZE. In great tits (*Parus major*), the size of the nestlings is inversely related to the number of eggs laid. The more mouths they have to feed, the less the parents can provide to any one nestling.

inquiry

? *Would natural selection favor producing many small young or a few large ones?*

chapter 55 population ecology **1155**

figure 55.16

VARIATION IN BABY SIDE-BLOTCHED LIZARDS (*Uta stansburiana*) SIZE PRODUCED BY EXPERIMENTAL MANIPULATIONS. In clutches in which some developing eggs were surgically removed, the remaining offspring were larger (*center*) than lizards produced in control clutches in which all the eggs were allowed to develop (*right*). In experiments in which some of the yolk was removed from the eggs, smaller lizards hatched (*left*).

offspring have a greater chance of survival. Producing many offspring with little chance of survival might not be the best strategy, but producing only a single, extraordinarily robust offspring also would not maximize the number of surviving offspring. Rather, an intermediate situation, in which several fairly large offspring are produced, should maximize the number of surviving offspring.

Reproductive events per lifetime represent an additional trade-off

The trade-off between age and fecundity plays a key role in many life histories. Annual plants and most insects focus all their reproductive resources on a single large event and then die. This life history adaptation is called **semelparity.** Organisms that produce offspring several times over many seasons exhibit a life history adaptation called **iteroparity.**

Species that reproduce yearly must avoid overtaxing themselves in any one reproductive episode so that they will be able to survive and reproduce in the future. Semelparity, or "big bang" reproduction, is usually found in short-lived species that have a low probability of staying alive between broods, such as plants growing in harsh climates. Semelparity is also favored when fecundity entails large reproductive cost, exemplified by Pacific salmon migrating upriver to their spawning grounds. In these species, rather than investing some resources in an unlikely bid to survive until the next breeding season, individuals put all their resources into one reproductive event.

Age at first reproduction correlates with life span

Among mammals and many other animals, longer-lived species put off reproduction longer than short-lived species, relative to expected life span. The advantage of delayed reproduction is that juveniles gain experience before expending the high costs of reproduction. In long-lived animals, this advantage outweighs the energy that is invested in survival and growth rather than reproduction.

In shorter-lived animals, on the other hand, time is of the essence; thus, quick reproduction is more critical than juvenile training, and reproduction tends to occur earlier.

> Life history adaptations involve many tradeoffs between reproductive cost and investment in survival. Different kinds of animals and plants employ quite different approaches.

55.5 Population Growth and Environmental Limits

Populations often remain at a relatively constant size, regardless of how many offspring are born. As you saw in earlier chapters of this text, Darwin based his theory of natural selection partly on this seeming contradiction. Natural selection occurs because of checks on reproduction, with some individuals producing fewer surviving offspring than others. To understand populations, we must consider how they grow and what factors in nature limit population growth.

The exponential growth model applies to populations with no growth limits

The rate of population increase, r, is defined as the difference between the birthrate, b, and the death rate, d, corrected for movement of individuals in or out of the population (e, rate of movement out of the area; i, rate of movement into the area). Thus,

$$r = (b - d) + (i - e)$$

Movements of individuals can have a major influence on population growth rates. For example, the increase in human population in the United States during the closing decades of the twentieth century was mostly due to immigration.

The simplest model of population growth assumes that a population grows without limits at its maximal rate and also that rates of immigration and emigration are equal. This rate, called the **biotic potential,** is the rate at which a population of a given species will increase when no limits are placed on its rate of growth. In mathematical terms, this is defined by the following formula:

$$\frac{dN}{dt} = r_i N$$

where N is the number of individuals in the population, dn/dt is the rate of change in its numbers over time, and r_i is the intrinsic rate of natural increase for that population—its innate capacity for growth.

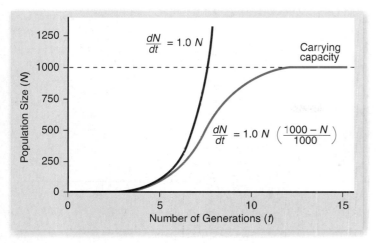

figure 55.17

TWO MODELS OF POPULATION GROWTH. The red line illustrates the exponential growth model for a population with an r of 1.0. The blue line illustrates the logistic growth model in a population with $r = 1.0$ and $K = 1000$ individuals. At first, logistic growth accelerates exponentially; then, as resources become limiting, the death rate increases and growth slows. Growth ceases when the death rate equals the birthrate. The carrying capacity (K) ultimately depends on the resources available in the environment.

The biotic potential of any population is exponential (red line in figure 55.17). Even when the *rate* of increase remains constant, the actual *number* of individuals accelerates rapidly as the size of the population grows. The result of unchecked exponential growth is a population explosion.

A single pair of houseflies, laying 120 eggs per generation, could produce more than 5 trillion descendants in a year. In 10 years, their descendants would form a swarm more than 2 m thick over the entire surface of the Earth! In practice, such patterns of unrestrained growth prevail only for short periods, usually when an organism reaches a new habitat with abundant resources. Natural examples include dandelions arriving in the fields, lawns, and meadows of North America from Europe for the first time; algae colonizing a newly formed pond; or cats introduced to an island with many birds, but previously lacking predators.

Carrying capacity

No matter how rapidly populations grow, they eventually reach a limit imposed by shortages of important environmental factors, such as space, light, water, or nutrients. A population ultimately may stabilize at a certain size, called the *carrying capacity* of the particular place where it lives. The **carrying capacity,** symbolized by K, is the maximum number of individuals that the environment can support.

The logistic growth model applies to populations that approach their carrying capacity

As a population approaches its carrying capacity, its rate of growth slows greatly, because fewer resources remain for each new individual to use. The growth curve of such a population, which is always limited by one or more factors in the

environment, can be approximated by the following logistic growth equation:

$$\frac{dN}{dt} = rN\left(\frac{K - N}{K}\right)$$

In this model of population growth, the growth rate of the population (dN/dt) is equal to its intrinsic rate of natural increase (r multiplied by N, the number of individuals present at any one time), adjusted for the amount of resources available. The adjustment is made by multiplying rN by the fraction of K, the carrying capacity, still unused [$(K - N)/K$]. As N increases, the fraction of resources by which r is multiplied becomes smaller and smaller, and the rate of increase of the population declines.

Graphically, if you plot N versus t (time), you obtain a **sigmoidal growth curve** characteristic of many biological populations. The curve is called 'sigmoidal" because its shape has a double curve like the letter S. As the size of a population stabilizes at the carrying capacity, its rate of growth slows down, eventually coming to a halt (blue line in figure 55.17).

In mathematical terms, as N approaches K, the *rate* of population growth (dN/dt) begins to slow, reaching 0 when $N = K$ (figure 55.18). Conversely, if the population size exceeds the carrying capacity, then $K - N$ will be negative, and the population will experience a negative growth rate. As the population size then declines toward the carrying capacity, the magnitude of this negative growth rate will decrease until it reaches 0 when $N = K$.

Notice that the population will tend to move toward the carrying capacity regardless of whether it is initially above or below it. For this reason, logistic growth tends to return a population to the same size. In this sense, such populations are considered to be in equilibrium because they would be expected to be at or near the carrying capacity at most times.

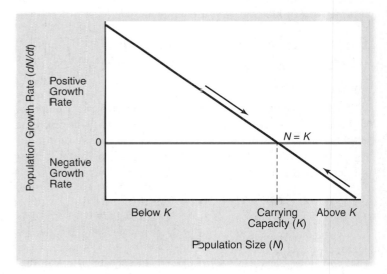

figure 55.18

RELATIONSHIP BETWEEN POPULATION GROWTH RATE AND POPULATION SIZE. Populations far from the carrying capacity (K) will have high growth rates—positive if the population is below K, and negative if it is above K. As the population approaches K, growth rates approach zero.

inquiry

? Why does the growth rate converge on zero?

figure 55.19

MANY POPULATIONS EXHIBIT LOGISTIC GROWTH. *a.* A fur seal (*Callorhinus ursinus*) population on St. Paul Island, Alaska. *b.* Two laboratory populations of the cladoceran *Bosmina longirostris*. Note that the populations first exceeded the carrying capacity, before decreasing to a size that was then maintained.

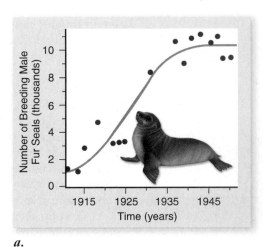

a.

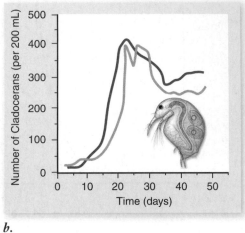

b.

In many cases, real populations display trends corresponding to a logistic growth curve. This is true not only in the laboratory, but also in natural populations (figure 55.19*a*). In some cases, however, the fit is not perfect (figure 55.19*b*), and as we shall see shortly, many populations exhibit other patterns.

> The size at which a population stabilizes in a particular place is defined as the carrying capacity of that place for that species. Populations often grow to the carrying capacity of their environment.

Factors that Regulate Populations

A number of factors may affect population size through time. Some of these factors depend on population size and are therefore termed *density-dependent*. Other factors, such as natural disasters, affect populations regardless of size; these factors are termed *density-independent*. Many populations exhibit cyclic fluctuations in size that may result from complex interactions of factors.

Density-dependent effects occur when reproduction and survival are affected by population size

The reason population growth rates are affected by population size is that many important processes have **density-dependent effects.** That is, as population size increases, either reproductive rates decline or mortality rates increase, or both, a phenomenon termed *negative feedback* (figure 55.20).

Populations can be regulated in many different ways. When populations approach their carrying capacity, competition for resources can be severe, leading both to a decreased birthrate and an increased risk of death (figure 55.21). In addition, predators often focus their attention on a particularly common prey species, which also results in increasing rates of mortality as populations increase. High population densities can also lead to an accumulation of toxic wastes in the environment.

Behavioral changes may also affect population growth rates. Some species of rodents, for example, become antisocial, fighting more, breeding less, and generally acting stressed-out. These behavioral changes result from hormonal actions, but their ultimate cause is not yet clear; most likely, they have evolved as adaptive responses to situations in which resources are scarce. In addition, in crowded populations, the population growth rate may decrease because of an increased rate of emigration of individuals attempting to find better conditions elsewhere (figure 55.22).

figure 55.20

DENSITY-DEPENDENT POPULATION REGULATION. Density-dependent factors can affect birthrates, death rates, or both.

inquiry

? *Why might birthrates be density-dependent?*

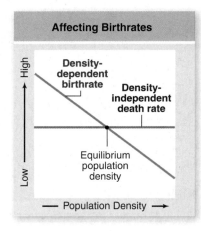

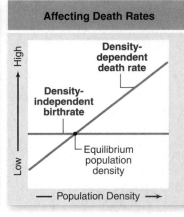

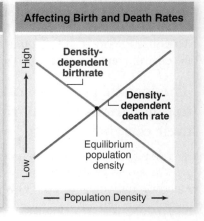

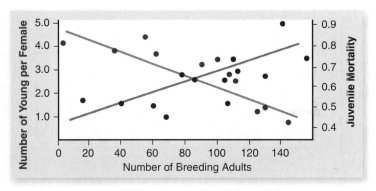

figure 55.21

DENSITY DEPENDENCE IN THE SONG SPARROW (*Melospiza melodia*) ON MANDARTE ISLAND. Reproductive success decreases and mortality rates increase as population size increases.

inquiry

What would happen if researchers supplemented the food available to the birds?

However, not all density-dependent factors are negatively related to population size. In some cases, growth rates increase with population size. This phenomenon is referred to as the **Allee effect** (after Warder Allee, who first described it), and is an example of *positive feedback*. The Allee effect can take several forms. Most obviously, in populations that are too sparsely distributed, individuals may have difficulty finding mates. Moreover, some species may rely on large groups to deter predators or to provide the necessary stimulation for breeding activities.

Density-independent effects include environmental disruptions and catastrophes

Growth rates in populations sometimes do not correspond to the logistic growth equation. In many cases, such patterns result because growth is under the control of **density-independent effects.**

figure 55.22

DENSITY-DEPENDENT EFFECTS. Migratory locusts, *Locusta migratoria*, are a legendary plague of large areas of Africa and Eurasia. At high population densities, the locusts have different hormonal and physical characteristics and take off as a swarm.

In other words, the rate of growth of a population at any instant is limited by something unrelated to the size of the population.

A variety of factors may affect populations in a density-independent manner. Most of these are aspects of the external environment, such as extremely cold winters, droughts, storms, or volcanic eruptions. Individuals often are affected by these occurrences regardless of the size of the population.

Populations in areas where such events occur relatively frequently will display erratic growth patterns in which the populations increase rapidly when conditions are benign, but exhibit large reductions whenever the environment turns hostile (figure 55.23). Needless to say, such populations do not produce the sigmoidal growth curves characteristic of the logistic equation.

Population cycles may reflect complex interactions

In some populations, density-dependent effects lead not to an equilibrium population size but to cyclic patterns of increase and decrease. For example, ecologists have studied cycles in hare populations since the 1820s. They have found that the North American snowshoe hare (*Lepus americanus*) follows a "10-year cycle" (in reality, the cycle varies from 8 to 11 years). Hare population numbers fall 10-fold to 30-fold in a typical cycle, and 100-fold changes can occur (figure 55.24). Two factors appear to be generating the cycle: food plants and predators.

Food plants. The preferred foods of snowshoe hares are willow and birch twigs. As hare density increases,

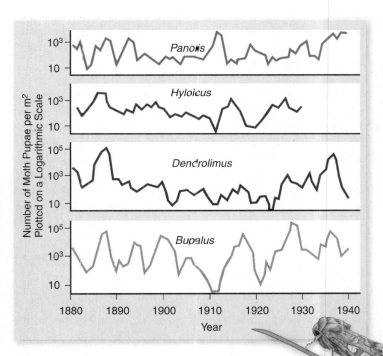

figure 55.23

FLUCTUATIONS IN THE NUMBER OF PUPAE OF FOUR MOTH SPECIES IN GERMANY. The population fluctuations suggest that density-independent factors are regulating population size. The concordance in trends through time suggests that the same factors are regulating population size in all four species.

the quantity of these twigs decreases, forcing the hares to feed on high-fiber (low-quality) food. Lower birthrates, low juvenile survivorship, and low growth rates follow. The hares also spend more time searching for food, an activity that increases their exposure to predation. The result is a precipitous decline in willow and birch twig abundance, and a corresponding fall in hare abundance. It takes 2 to 3 years for the quantity of mature twigs to recover.

Predators. A key predator of the snowshoe hare is the Canada lynx. The Canada lynx shows a "10-year" cycle of abundance that seems remarkably entrained to the hare abundance cycle (see figure 55.24). As hare numbers increase, lynx numbers do too, rising in response to the increased availability of lynx food. When hare numbers fall, so do lynx numbers, their food supply depleted.

Which factor is responsible for the predator–prey oscillations? Do increasing numbers of hares lead to overharvesting of plants (a hare–plant cycle), or do increasing numbers of lynx lead to overharvesting of hares (a hare–lynx cycle)? Field experiments carried out by C. Krebs and coworkers in 1992 provide an answer.

In Canada's Yukon, Krebs set up experimental plots that contained hare populations. If food is added (no food shortage effect) and predators are excluded (no predator effect) in an experimental area, hare numbers increase 10-fold and stay there—the cycle is lost. However, the cycle is retained if either of the factors is allowed to operate alone: exclude predators but don't add food (food shortage effect alone), or add food in the presence of predators (predator effect alone). Thus, both factors can affect the cycle, which in practice seems to be generated by the interaction between the two.

Population cycles traditionally have been considered to occur rarely. However, a recent review of nearly 700 long-term (25 years or more) studies of trends within populations found that cycles were not uncommon; nearly 30% of the studies—including birds, mammals, fish, and crustaceans—provided evidence of some cyclic pattern in population size through time, although most of these cycles are nowhere near as dramatic in amplitude as the snowshoe hare and lynx cycles. In some cases, such as that of the snowshoe hare and lynx, density-dependent factors may be involved, whereas in other cases, density-independent factors, such as cyclic climatic patterns, may be responsible.

Resource availability affects life history adaptations

As you have seen, some species usually maintain stable population sizes near the carrying capacity, whereas in other species population sizes fluctuate markedly and are often far below carrying capacity. The selective factors affecting such species differ markedly. Populations near their carrying capacity may face stiff competition for limited resources; by contrast, populations far below carrying capacity have access to abundant resources.

We have already described the consequences of such differences. When resources are limited, the cost of repro-

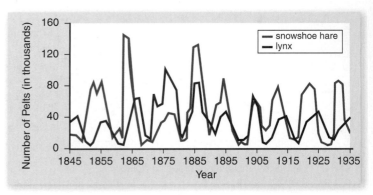

figure 55.24

LINKED POPULATION CYCLES OF THE SNOWSHOE HARE (*Lepus americanus*) AND THE NORTHERN LYNX (*Lynx canadensis*). These data are based on records of fur returns from trappers in the Hudson Bay region of Canada. The lynx population carefully tracks that of the snowshoe hare, but lags behind it slightly.

inquiry

? *Suppose experimenters artificially kept the hare population at a high and constant level; what would happen to the lynx population? Conversely, if experimenters artificially kept the lynx population at a high and constant level, what would happen to the hare population?*

duction often will be very high. Consequently, selection will favor individuals that can compete effectively and utilize resources efficiently. Such adaptations often come at the cost of lowered reproductive rates. Such populations are termed **K-selected** because they are adapted to thrive when the population is near its carrying capacity (*K*). Table 55.3 lists some of the typical features of *K*-selected populations. Examples of *K*-selected species include coconut palms, whooping cranes, whales, and humans.

TABLE 55.3	**r-Selected and K-Selected Life History Adaptations**	
Adaptation	**r-Selected Populations**	**K-Selected Populations**
Age at first reproduction	Early	Late
Life span	Short	Long
Maturation time	Short	Long
Mortality rate	Often high	Usually low
Number of offspring produced per reproductive episode	Many	Few
Number of reproductions per lifetime	Few	Many
Parental care	None	Often extensive
Size of offspring or eggs	Small	Large

By contrast, in populations far below the carrying capacity, resources may be abundant. Costs of reproduction will be low, and selection will favor those individuals that can produce the maximum number of offspring. Selection here favors individuals with the highest reproductive rates; such populations are termed **r-selected.** Examples of organisms displaying r-selected life history adaptations include dandelions, aphids, mice, and cockroaches.

Most natural populations show life history adaptations that exist along a continuum ranging from completely r-selected traits to completely K-selected traits. Although these tendencies hold true as generalities, few populations are purely r- or K-selected and show all of the traits listed in table 55.3. These attributes should be treated as generalities, with the recognition that many exceptions exist.

> Density-dependent effects are caused by factors that come into play particularly when the population size is larger; density-independent effects result from factors that operate regardless of population size. Some life history adaptations favor near-exponential growth; others favor the more competitive logistic growth. Most natural populations exhibit a combination of the two.

55.7 Human Population Growth

Humans exhibit many K-selected life history traits, including small brood size, late reproduction, and a high degree of parental care. These life history traits evolved during the early history of hominids, when the limited resources available from the environment controlled population size. Throughout most of human history, our populations have been regulated by food availability, disease, and predators. Although unusual disturbances, including floods, plagues, and droughts, no doubt affected the pattern of human population growth, the overall size of the human population grew slowly during our early history.

Two thousand years ago, perhaps 130 million people populated the Earth. It took a thousand years for that number to double, and it was 1650 before it had doubled again, to about 500 million. In other words, for over 16 centuries, the human population was characterized by very slow growth. In this respect, human populations resembled many other species with predominantly K-selected life history adaptations.

Human populations have grown exponentially

Starting in the early 1700s, changes in technology gave humans more control over their food supply, enabled them to develop superior weapons to ward off predators, and led to the development of cures for many diseases. At the same time, improvements in shelter and storage capabilities made humans less vulnerable to climatic uncertainties. These changes allowed humans to expand the carrying capacity of the habitats in which they lived and thus to escape the confines of logistic growth and re-enter the exponential phase of the sigmoidal growth curve.

Responding to the lack of environmental constraints, the human population has grown explosively over the last 300 years. Although the birthrate has remained unchanged at about 30 per 1000 per year over this period, the death rate has fallen dramatically, from 20 per 1000 per year to its present level of 13 per 1000 per year. The difference between birth and death rates meant that the population grew as much as 2% per year, although the rate has now declined to 1.2% per year.

A 1.2% annual growth rate may not seem large, but it has produced a current human population of 6.5 billion people (figure 55.25). At this growth rate, 78 million people would be

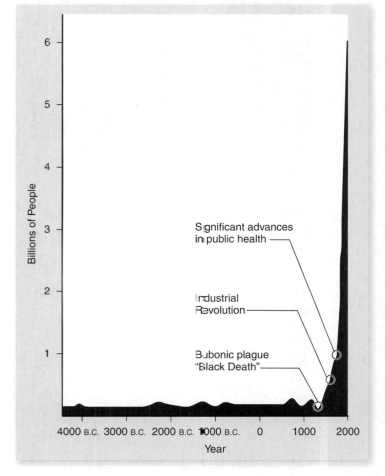

figure 55.25

HISTORY OF HUMAN POPULATION SIZE. Temporary increases in death rate, even a severe one such as that occurring during the Black Death of the 1300s, have little lasting effect. Explosive growth began with the Industrial Revolution in the 1800s, which produced a significant, long-term lowering of the death rate. The current world population is 6.5 billion, and at the present rate, it will double in 58 years.

 inquiry

Based on what we have learned about population growth, what do you predict will happen to human population size?

chapter 55 population ecology **1161**

added to the world population in the next year, and the human population would double in 58 years. Both the current human population level and the projected growth rate have potentially grave consequences for our future.

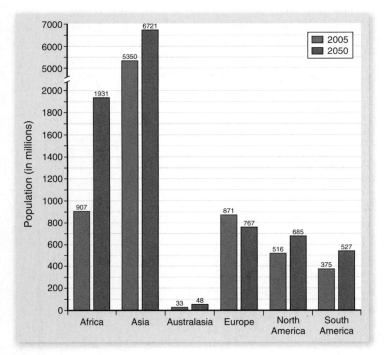

figure 55.26

PROJECTED POPULATION GROWTH IN 2050. Developed countries are predicted to grow little; almost all of the population increase will occur in less-developed countries.

Population pyramids show birth and death trends

Although the human population as a whole continues to grow rapidly at the beginning of the twenty-first century, this growth is not occurring uniformly over the planet. Rather, most of the population growth is occuring in Africa, Asia, and Latin America (figure 55.26). By contrast, populations are actually decreasing in some countries in Europe.

The rate at which a population can be expected to grow in the future can be assessed graphically by means of a **population pyramid,** a bar graph displaying the numbers of people in each age category (figure 55.27). Males are conventionally shown to the left of the vertical age axis, females to the right. A human population pyramid thus displays the age composition of a population by sex. In most human population pyramids, the number of older females is disproportionately large compared with the number of older males, because females in most regions have a longer life expectancy than males.

Viewing such a pyramid, we can predict demographic trends in births and deaths. In general, a rectangular pyramid is characteristic of countries whose populations are stable, neither growing nor shrinking. A triangular pyramid is characteristic of a country that will exhibit rapid future growth because most of its population has not yet entered the childbearing years. Inverted triangles are characteristic of populations that are shrinking, usually as a result of sharply declining birthrates.

Examples of population pyramids for Sweden and Kenya in 2005 are shown in figure 55.27. The two countries exhibit very different age distributions. The nearly rectangular population pyramid for Sweden indicates that its population is not expanding because birthrates have decreased and average life span has increased. The very triangular pyramid of Kenya, by

figure 55.27

POPULATION PYRAMIDS FROM 2005. Population pyramids are graphed according to a population's age distribution. Kenya's pyramid has a broad base because of the great number of individuals below childbearing age. When the young people begin to bear children, the population will experience rapid growth. The Swedish pyramid exhibits a slight bulge among middle-aged Swedes, the result of the "baby boom" that occurred in the middle of the twentieth century, and many postreproductive individuals resulting from Sweden's long average life span.

inquiry

What will the population distributions look like in 20 years?

TABLE 55.4	A Comparison of 2005 Population Data in Developed and Developing Countries		
	United States (highly developed)	Brazil (moderately developed)	Ethiopia (poorly developed)
Fertility rate	2.1	1.9	5.3
Doubling time at current rate (years)	75	65	29
Infant mortality rate (per 1000 births)	6.5	30	95
Life expectancy at birth (years)	78	72	49
Per capita GNP (U.S. $)	$40,100	$8100	$800
Population < 15 years old (%)	21	26	44

contrast, results from relatively high birthrates and shorter average life spans, which can lead to explosive future growth. The difference is most apparent when we consider that only 17% of Sweden's population is less than 15 years old, compared with nearly half of all Kenyans. Moreover, the fertility rate (offspring per woman) in Sweden is 1.6; in Kenya, it is 4.4. As a result, Kenya's population could double in less than 35 years, whereas Sweden's will remain stable.

Humanity's future growth is uncertain

Earth's rapidly growing human population constitutes perhaps the greatest challenge to the future of the **biosphere,** the world's interacting community of living things. Humanity is adding 78 million people a year to its population—over a million every 5 days, 150 every minute! In more rapidly growing countries, the resulting population increase is staggering (table 55.4). India, for example, had a population of 1.05 billion in 2002; by 2050, its population likely will exceed 1.6 billion.

A key element in the world's population growth is its uneven distribution among countries. Of the billion people added to the world's population in the 1990s, 90% live in developing countries (figure 55.28). The fraction of the world's population that lives in industrialized countries is therefore diminishing. In 1950, fully one-third of the world's population lived in industrialized countries; by 1996, that proportion had fallen to one-quarter; and in 2020, the proportion will have fallen to one-sixth. In the future, the world's population growth will be centered in the parts of the world least equipped to deal with the pressures of rapid growth.

Rapid population growth in developing countries has had the harsh consequence of increasing the gap between rich and poor. Today, the 19% of the world's population that lives in the industrialized world have a per capita income of $22,060, while 81% of the world's population lives in developing countries and has a per capita income of only $3,580. Furthermore, of the people in the developing world, about one-quarter of the population gets by on $1 per day. Eighty percent of all the energy used today is consumed by the industrialized world, but only 20% is used by developing countries.

No one knows whether the world can sustain today's population of 6.5 billion people, much less the far greater numbers expected in the future. As chapter 58 outlines, the world

ecosystem is already under considerable stress. We cannot reasonably expect to expand its carrying capacity indefinitely, and indeed we already seem to be stretching the limits.

Despite using an estimated 45% of the total biological productivity of Earth's landmasses and more than one-half of all renewable sources of fresh water, between one-fourth and one-eighth of all people in the world are malnourished. Moreover, as anticipated by Thomas Malthus in his famous 1798 work, *Essay on the Principle of Population*, death rates are beginning to rise in some areas. In sub-Saharan Africa, for example, population projections for the year 2025 have been scaled back from 1.33 billion to 1.05 billion (21%) because of the effect of AIDS. Similar decreases are projected for Russia as a result of higher death rates due to disease.

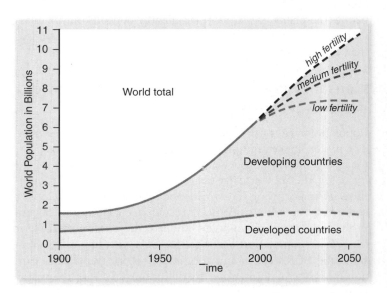

figure 55.28

DISTRIBUTION OF POPULATION GROWTH. Most of the worldwide increase in population since 1950 has occurred in developing countries. The age structures of developing countries indicate that this trend will increase in the near future. World population in 2050 likely will be between 7.3 and 10.7 billion, according to a recent United Nations study. Depending on fertility rates, the population at that time will either be increasing rapidly or slightly, or in the best case, declining slightly.

If we are to avoid catastrophic increases in the death rate, birthrates must fall dramatically. Faced with this grim dichotomy, significant efforts are underway worldwide to lower birthrates.

The population growth rate has declined

The world population growth rate is declining, from a high of 2.0% in the period 1965–1970 to 1.2% in 2005. Nonetheless, because of the larger population, this amounts to an increase of 78 million people per year to the world population, compared with 53 million per year in the 1960s.

The United Nations attributes the growth rate decline to increased family planning efforts and the increased economic power and social status of women. The United States has led the world in funding family planning programs abroad, but some groups oppose spending money on international family planning. The opposition believes that money is better spent on improving education and the economy in other countries, leading to an increased awareness and lowered fertility rates. The U.N. certainly supports the improvement of education programs in developing countries, but interestingly, it has reported increased education levels *following* a decrease in family size as a result of family planning.

Most countries are devoting considerable attention to slowing the growth rate of their populations, and there are genuine signs of progress. For example, from 1984 to 2005, family planning programs in Kenya succeeded in reducing the fertility rate from 8.0 to 5.0 children per couple, thus lowering the population growth rate from 4.0% per year to 2.6% per year. Because of these efforts, the global population may stabilize at about 8.9 billion people by the middle of the current century. How many people the planet can support sustainably depends on the quality of life that we want to achieve; there are already more people than can be sustainably supported with current technologies.

Consumption in the developed world further depletes resources

Population size is not the only factor that determines resource use; per capita consumption is also important. In this respect, we in the industrialized world need to pay more attention to lessening the impact each of us makes because, even though the vast majority of the world's population is in developing countries, the vast majority of resource consumption occurs in the industrialized countries. Indeed, the wealthiest 20% of the world's population accounts for 86% of the world's consumption of resources and produces 53% of the world's carbon dioxide emissions, whereas the poorest 20% of the world is responsible for only 1.3% of consumption and 3% of carbon dioxide emissions. Looked at another way, in terms of resource use, a child born today in the industrialized world will consume many more resources over the course of his or her life than a child born in the developing world.

One way of quantifying this disparity is by calculating what has been termed the **ecological footprint,** which is the amount of productive land required to support an individual at the standard of living of a particular population through the course of his or her life. This figure estimates the acreage used for the production of food (both plant and animal), forest products, and housing, as well as the area of forest required to

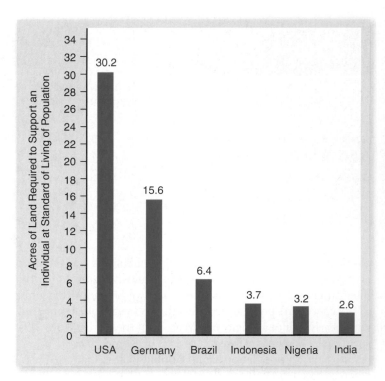

figure 55.29

ECOLOGICAL FOOTPRINTS OF INDIVIDUALS IN DIFFERENT COUNTRIES. An ecological footprint calculates how much land is required to support a person through his or her life, including the acreage used for production of food, forest products, and housing, in addition to the forest required to absorb the carbon dioxide produced by the combustion of fossil fuels.

 inquiry

Which is a more important cause of resource depletion, overpopulation or overconsumption?

absorb carbon dioxide produced by the combustion of fossil fuels. As figure 55.29 illustrates, the ecological footprint of an individual in the United States is more than 10 times greater than that of someone in India.

Based on these measurements, researchers have calculated that resource use by humans is now one-third greater than the amount that nature can sustainably replace. Moreover, consumption is increasing rapidly in parts of the developing world; if all humans lived at the standard of living in the industrialized world, two additional planet Earths would be needed.

Building a sustainable world is the most important task facing humanity's future. The quality of life available to our children will depend to a large extent on our success in limiting both population growth and the amount of per capita resource consumption.

In 2005, the global human population of 6.5 billion people was growing at a rate of approximately 1.2% annually. At that rate, the population would double in 58 years. Growth rates, however, are declining, but consumption per capita in the developed world is also a significant drain on resources.

55.1 The Environmental Challenge

The physical environment primarily determines which organisms live in a particular climate or region.

- Key environmental factors include temperature, water, sunlight, and soil type.

- Organisms can maintain homeostasis by regulating their internal environment independent of the external environment or they can conform to environmental changes.

- Organisms adapt to the environment by evolving physiological, morphological, or behavioral changes that make them better adapted to the environment in which they occur.

55.2 Populations: Groups of Single Species in One Place

Populations are defined as groups of individuals that occur at one place and time.

- Most populations have limited geographic ranges.

- Population ranges vary over time because populations expand and contract as the environment changes.

- Within a population individuals are distributed randomly, uniformly, or are clumped (figure 55.8).

- Metapopulations are networks of loosely connected populations that exchange individuals. The degree of exchange is highest when populations are large and more connected.

- Metapopulations prevent long-term extinction and lead to utilization of marginal areas.

55.3 Population Demography and Dynamics

Demography is the statistical study of the factors that affect population change over time.

- Sex ratio affects population growth; the more females the greater the number of offspring and the faster a population will grow.

- The shorter the generation time, the faster a population will grow.

- The age structure of a population affects population growth because every age cohort has a characteristic fecundity and death rate.

- Survivorship curves demonstrate how survival and mortality changes with age (figure 55.11).

55.4 Life History and the Cost of Reproduction (figure 55.13)

Natural selection favors a life history that maximizes lifetime reproductive success.

- The total number of surviving offspring is influenced by how long an individual lives and how many young are produced each year.

- Life tables summarize the probability of survival and reproduction throughout a cohort's life span.

- The cost of reproduction is low when resources are abundant or when mortality rates are high.

- When the cost of reproduction is high, reproductive success of a population can be maximized by deferring reproduction and enhancing parental survival or producing few large-sized young that have a greater chance of survival (figure 55.15).

- Organisms can be semelparous, having one reproductive event, or iteroparous, having multiple reproductive events during their life.

- Short-lived organisms reproduce early, and long-lived organisms reproduce later in life.

55.5 Population Growth and Environmental Limits

To understand populations, scientists need to know how populations grow and what factors limit their growth.

- The rate of population increase, r, is defined as the difference between birthrate, b, and death rate, d.

- Migration into and from a population needs to be considered when studying population growth.

- Biotic potential refers to population growth without limits and not influenced by migrations.

- Exponential growth is expressed as $dN/dt = r_iN$ (figure 55.17).

- Logistic growth is exhibited when a population reaches its carrying capacity, that is, resources become limiting. Logistic growth is expressed as $dN/dt = r_rN[(K - N)/K]$.

55.6 Factors that Regulate Populations

Natural factors and population density affect population size.

- Density-dependent factors depend on population size and usually involve negative feedback. To stabilize a population size birthrates must decline, death rates must increase or both.

- Density-independent factors are not related to population size and include environmental disruptions and catastrophes that result in mortality.

- In some populations size is cyclical because of environmental factors such as food supply and predation pressure.

- Resource availability affects reproductive strategies. If food is limiting, natural selection favors K-selection; if food is not limiting, r-selection is favored.

- The Allee effect occurs when populations become so small that they have lower rates of reproduction and survival of individuals.

55.7 Human Population Growth

Humans exhibit many K-selected traits, including small brood size, late reproduction, and high degree of parental care. Environmental disturbances can influence human population growth.

- Human populations have expanded environmental carrying capacity by the use of technology and other innovations resulting in an exponential growth rate.

- Population pyramids, a bar graph of people in various age categories, allow us to predict demographic trends in births and deaths (figure 55.27).

- The future of human population growth is uncertain in part due to the uneven distribution of the population in the biosphere.

- Two factors affect the utilization of resources and the future of human populations: population size and per capita consumption.

- Resource use by humans is one-third greater than the amount that nature can sustainably replace.

- A sustainable future requires limits to population growth and per capita resource consumption.

SELF TEST

1. When individuals respond physiologically to an environmental change
   a. natural selection has occurred.
   b. differences among individuals could lead to evolution by natural selection.
   c. they cannot respond simultaneously through changes in behavior and morphology.
   d. None of the above.

2. Geographic ranges of populations
   a. were static until human disturbance led to extinction and introductions.
   b. are never affected by the distribution and abundance of predators.
   c. do not respond to long-term climatic changes.
   d. None of the above.

3. Source–sink metapopulations are distinct from other types of metapopulations because
   a. exchange of individuals only occurs in the former.
   b. populations with negative growth rates are a part of the former.
   c. populations never go extinct in the former.
   d. all populations eventually go extinct in the former.

4. I would expect the potential for social interactions among individuals to be maximized when individuals
   a. are randomly distributed in their environment.
   b. are uniformly distributed in their environment.
   c. have a clumped distribution in their environment.
   d. are non-randomly distributed in their environment.

5. When ecologists talk about the cost of reproduction they mean
   a. the reduction in future reproductive output as a consequence of current reproduction.
   b. the amount of calories it takes for all the activity used in successful reproduction.
   c. the amount of calories contained in eggs or offspring.
   d. None of the above.

6. A life history trade-off between clutch size and offspring size
   a. means that as clutch size increases, offspring size increases.
   b. means that as clutch size increases, offspring size decreases.
   c. means that as clutch size increases, adult size increases.
   d. means that as clutch size increases, adult size decreases.

7. The difference between exponential and logistic growth rates is
   a. exponential growth depends on birth and death rates and logistic does not.
   b. in logistic growth, emigration and immigration are unimportant.
   c. that both are affected by density, but logistic growth is slower.
   d. that only logistic growth reflects density-dependent effects on births or deaths.

8. Humans are an example of an organism with a type I survivorship curve. This means
   a. mortality rates are highest for younger individuals.
   b. mortality rates are highest for older individuals.
   c. mortality rates are constant over the life span of individuals.
   d. the population growth rate is high.

9. According to the Population Reference Bureau (2002), the worldwide intrinsic rate of human population growth (r) is currently 1.3%. In the United States, $r = 0.6\%$. How will the U.S. population change relative to the world population?
   a. The world population will grow, while the population of the United States will decline.
   b. The world population will grow, while the population of the United States will remain the same.
   c. Both the world and the U.S. populations will grow, but the world population will grow more rapidly.
   d. The world population will decline, while the U.S. population will increase.

10. The logistic population growth model, $dN/dt = rN[(K - N)/K]$, describes a population's growth when an upper limit to growth is assumed. As N approaches (numerically) the value of K
    a. dN/d increases rapidly.
    b. dN/d approaches 0.
    c. dN/d increases slowly.
    d. the population becomes threatened by extinction.

11. Which of the following is an example of a density-dependent effect on population growth?
    a. An extremely cold winter
    b. A tornado
    c. An extremely hot summer in which cool burrow retreats are fewer than number of individuals in the population
    d. A drought

CHALLENGE QUESTIONS

1. Refer to figure 55.9. What are the implications for evolutionary divergence among populations that are part of a metapopulation versus populations that are independent of other populations?

2. Refer to figure 55.14. Given a trade-off between current reproductive effort and future reproductive success (the so-called cost of reproduction), would you expect old individuals to have the same "optimal" reproductive effort as young individuals?

3. Refer to figure 55.15. Because the number of offspring that a parent can produce often trades off with the size of individual offspring, many circumstances lead to an intermediate number and size of offspring being favored. If the size of an offspring was completely unrelated to the quality of that offspring (its chances of surviving until it reaches reproductive age), would you expect parents to fall on the left or right side of the x-axis (clutch size)? Explain.

4. Refer to figure 55.27. Would increasing the mean generation time have the same kind of effect on population growth rate as reducing the number of children that an individual female has over her lifetime? Which effect would have a bigger influence on population growth rate? Explain.

 **Do you need additional review?** *Visit www.ravenbiology.com for practice quizzes, animations, videos, and activities designed to help you master the material in this chapter.*

Community Ecology

introduction

ALL THE ORGANISMS THAT LIVE together in a place are members of a community. The myriad of species that inhabit a tropical rain forest are a community. Indeed, every inhabited place on Earth supports its own particular array of organisms. Over time, the different species that live together have made many complex adjustments to community living, evolving together and forging relationships that give the community its character and stability. Both competition and cooperation have played key roles; in this chapter, we look at these and other factors in community ecology.

concept outline

Biological Communities: Species Living Together

Almost any place on Earth is occupied by species, sometimes by many of them, as in the rain forests of the Amazon, and sometimes by only a few, as in the near-boiling waters of Yellowstone's geysers (where a number of microbial species live). The term **community** refers to the species that occur at any particular locality (figure 56.1). Communities can be characterized either by their constituent species or by their properties, such as **species richness** (the number of species present) or **primary productivity** (the amount of energy produced).

Interactions among community members govern many ecological and evolutionary processes. These interactions, such as predation and mutualism, affect the population biology of particular species—whether a population increases or decreases in abundance, for example—as well as the ways in which energy and nutrients cycle through the ecosystem. Moreover, the community context in many ways affects the patterns of natural selection faced by a species, and thus the evolutionary course it takes.

Scientists study biological communities in many ways, ranging from detailed observations to elaborate, large-scale experiments. In some cases, studies focus on the entire community, whereas in other cases only a subset of species that are likely to interact with one another are studied. Although scientists sometimes refer to such subsets as communities (for example, the "spider community"), the term **assemblage** is more appropriate to connote that the species included are only a portion of those present within the entire community.

Communities have been viewed in different ways

Two views exist on the structure and functioning of communities. The **individualistic concept** of communities, first championed by H. A. Gleason of the University of Chicago early in the twentieth century, holds that a community is nothing more than an aggregation of species that happen to occur together at one place.

By contrast, the **holistic concept** of communities, which can be traced to the work of F. E. Clements, also about a century ago, views communities as an integrated unit. In this sense, the community could be viewed as a superorganism whose constituent species have coevolved to the extent that they function as part of a greater whole, just as the kidneys, heart, and lungs all function together within an animal's body. In this view, then, a community would amount to more than the sum of its parts.

These two views make differing predictions about the integrity of communities across space and time. If, as the individualistic view implies, communities are nothing more than a combination of species that occur together, then moving geographically across the landscape or back through time, we would not expect to see the same community. That is, species should appear and disappear independently, as a function of each species' own unique ecological requirements. By contrast, if a community is an integrated whole, then we would make the opposite prediction: Communities should stay the same through space or time, until being replaced by completely different communities when environmental differences are sufficiently great.

Communities change over time

Most ecologists today favor the individualistic concept. For the most part, species seem to respond independently to changing environmental conditions. As a result, community composition changes gradually across landscapes as some species appear and become more abundant, while others decrease in abundance and eventually disappear.

figure 56.1

AN AFRICAN SAVANNA COMMUNITY. A community consists of all the species—plants, animals, fungi, protists, and prokaryotes—that occur at a locality, in this case Etosha National Park in Namibia.

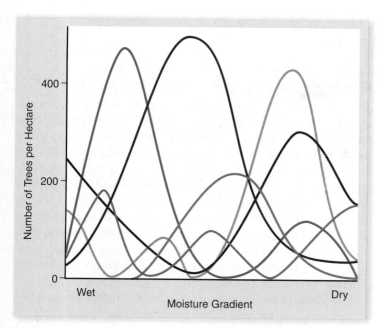

figure 56.2

ABUNDANCE OF TREE SPECIES ALONG A MOISTURE GRADIENT IN THE SANTA CATALINA MOUNTAINS OF SOUTHEASTERN ARIZONA. Each line represents the abundance of a different tree species. The species' patterns of abundance are independent of one another. Thus, community composition changes continually along the gradient.

 inquiry

Why do species exhibit different patterns of response to change in moisture?

A famous example of this pattern is the abundance of tree species in the Santa Catalina Mountains of Arizona along a geographic gradient running from very dry to very moist. Figure 56.2 shows that species can change abundance in patterns that are for the most part independent of one another. As a result, tree communities at different localities in these mountains fall on a continuum, one merging into the next, rather than representing discretely different sets of species.

Similar patterns through time are seen in paleontological studies. For example, a very good fossil record exists for the trees and small mammals that occurred in North America over the past 20,000 years. Examination of prehistoric communities shows little similarity to those that occur today. Many species that occur together today were never found together in the past. Conversely, species that used to occur in the same communities often do not overlap in their geographic ranges today. These findings suggest that as climate has changed during the waxing and waning of the Ice Ages, species have responded independently, rather than shifting their distributions together, as would be expected if the community were an integrated unit.

Nonetheless, in some cases the abundance of species in a community does change geographically in a synchronous pattern. Often, this occurs at **ecotones**, places where the environment changes abruptly. For example, in the western United States, certain patches of habitat have serpentine soils. This soil differs from normal soil in many ways—for example, high concentrations of nickel, chromium, and iron; low concentrations of copper and calcium. Comparison of the plant species that occur on different soils shows that distinct communities exist on each type, with an abrupt transition from one to the other over a short distance (figure 56.3). Similar transitions are seen wherever greatly different habitats come into contact, such as at the interface between terrestrial and aquatic habitats or where grassland and forest meet.

A community comprises all species that occur at one site. In most cases, community members vary independently of one another in abundance across space and through time.

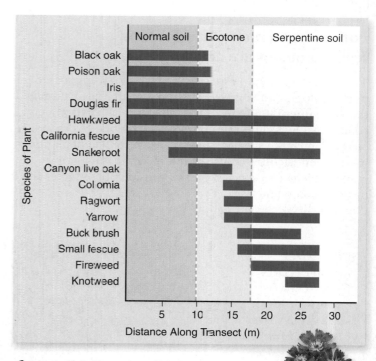

figure 56.3

CHANGE IN COMMUNITY COMPOSITION ACROSS AN ECOTONE. The plant assemblages on normal and serpentine soils are greatly different, and the transition from one community to another occurs over a short distance.

 inquiry

Why is there a sharp transition between the two community types?

The Ecological Niche Concept

Each organism in a community confronts the challenge of survival in a different way. The **niche** an organism occupies is the total of all the ways it uses the resources of its environment. A niche may be described in terms of space utilization, food consumption, temperature range, appropriate conditions for mating, requirements for moisture, and other factors.

Sometimes species are not able to occupy their entire niche because of the presence or absence of other species. Species can interact with one another in a number of ways, and these interactions can either have positive or negative effects. One type of interaction is **interspecific competition,** which occurs when two species attempt to use the same resource and there is not enough of the resource to satisfy both. Physical interactions over access to resources—such as fighting to defend a territory or displacing an individual from a particular location—are referred to as **interference competition;** consuming the same resources is called **exploitative competition.**

Fundamental niches are potential; realized niches are actual

The entire niche that a species is capable of using, based on its physiological tolerance limits and resource needs, is called the **fundamental niche.** The actual set of environmental conditions, including the presence or absence of other species, in which the species can establish a stable population is its **realized niche.** Because of interspecific interactions, the realized niche of a species may be considerably smaller than its fundamental niche.

Competition between species for niche occupancy

In a classic study, J. H. Connell of the University of California, Santa Barbara, investigated competitive interactions between two species of barnacles that grow together on rocks along the coast of Scotland. Of the two species Connell studied, *Chthamalus stellatus* lives in shallower water, where tidal action often exposes it to air, and *Semibalanus balanoides* (called *Balanus balanoides* prior to 1995) lives lower down, where it is rarely exposed to the atmosphere (figure 56.4). In these areas, space is at a premium. In the deeper zone, *S. balanoides* could always outcompete *C. stellatus* by crowding it off the rocks, undercutting it, and replacing it even where it had begun to grow, an example of interference competition.

When Connell removed *S. balanoides* from the area, however, *C. stellatus* was easily able to occupy the deeper zone, indicating that no physiological or other general obstacles prevented it from becoming established there. In contrast, *S. balanoides* could not survive in the shallow-water habitats where *C. stellatus* normally occurs; it does not have the physiological adaptations to warmer temperatures that allow *C. stellatus* to occupy this zone. Thus, the fundamental niche of *C. stellatus* includes both shallow and deeper zones, but its realized niche is much narrower because *C. stellatus* can be outcompeted by *S. balanoides* in parts of its fundamental niche. By contrast, the realized and fundamental niches of *S. balanoides* appear to be identical.

Other causes of niche restriction

Processes other than competition can also restrict the realized niche of a species. For example, the plant St. John's wort (*Hypericum perforatum*) was introduced and became widespread in

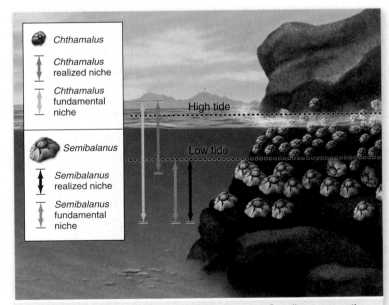

figure 56.4

COMPETITION AMONG TWO SPECIES OF BARNACLES. The fundamental niche of *Chthamalus stellatus* includes both deep and shallow zones, but *Semibalanus balanoides* forces *C. stellatus* out of the part of its fundamental niche that overlaps the realized niche of *Semibalanus*.

Chthamalus

Chthamalus realized niche

Chthamalus fundamental niche

Semibalanus

Semibalanus realized niche

Semibalanus fundamental niche

High tide

Low tide

S.balanoides and *C.stellatus* competing

C.stellatus fundamental and realized niches are identical when *S.balanoides* is removed.

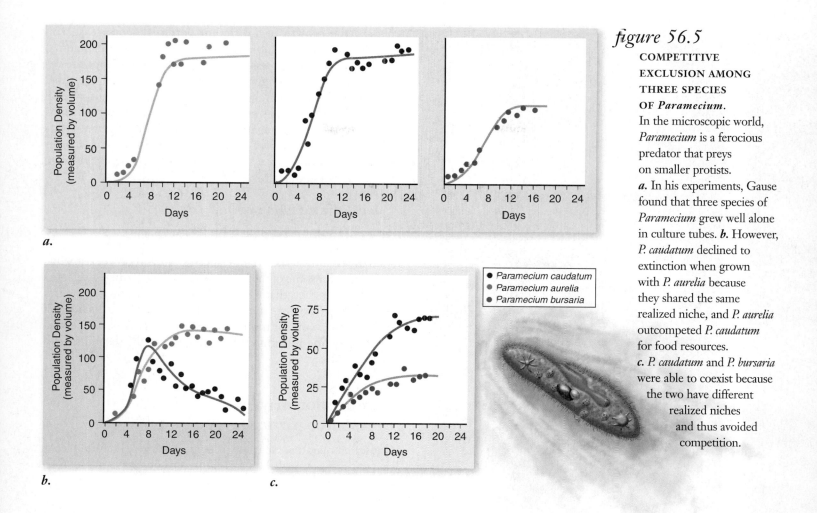

figure 56.5

COMPETITIVE EXCLUSION AMONG THREE SPECIES OF *Paramecium*. In the microscopic world, *Paramecium* is a ferocious predator that preys on smaller protists. *a.* In his experiments, Gause found that three species of *Paramecium* grew well alone in culture tubes. *b.* However, *P. caudatum* declined to extinction when grown with *P. aurelia* because they shared the same realized niche, and *P. aurelia* outcompeted *P. caudatum* for food resources. *c. P. caudatum* and *P. bursaria* were able to coexist because the two have different realized niches and thus avoided competition.

open rangeland habitats in California until a specialized beetle was introduced to control it. Population size of the plant quickly decreased, and it is now only found in shady sites where the beetle cannot thrive. In this case, the presence of a predator limits the realized niche of a plant.

In some cases, the absence of another species leads to a smaller realized niche. Many North American plants depend on insects for pollination; indeed, the value of insect pollination for American agriculture has been estimated at greater than $2 billion per year. However, pollinator populations are currently declining for a variety of reasons. Conservationists are concerned that if these insects disappear from some habitats, the niche of many plant species will decrease or even disappear entirely. In this case, then, the absence—rather than the presence—of another species will be the cause of a relatively small realized niche.

Competitive exclusion can occur when species compete for limited resources

In classic experiments carried out in 1934 and 1935, Russian ecologist G. F. Gause studied competition among three species of *Paramecium*, a tiny protist. Each of the three species grew well in culture tubes by themselves, preying on bacteria and yeasts that fed on oatmeal suspended in the culture fluid (figure 56.5*a*). However, when Gause grew *P. aurelia* together with *P. caudatum*

in the same culture tube, the numbers of *P. caudatum* always declined to extinction, leaving *P. aurelia* the only survivor (figure 56.5*b*). Why did this happen? Gause found that *P. aurelia* could grow six times faster than its competitor *P. caudatum* because it was able to better utilize the limited available resources, an example of exploitative competition.

From experiments such as this, Gause formulated what is now called the principle of **competitive exclusion**. This principle states that if two species are competing for a limited resource, the species that uses the resource more efficiently will eventually eliminate the other locally. In other words, no two species with the same niche can coexist when resources are limiting.

Niche overlap and coexistence

In a revealing experiment, Gause challenged *Paramecium caudatum*—the defeated species in his earlier experiments—with a third species, *P. bursaria*. Because he expected these two species to also compete for the limited bacterial food supply, Gause thought one would win out, as had happened in his previous experiments. But that's not what happened. Instead, both species survived in the culture tubes, dividing the food resources.

The explanation for the species' coexistence is simple. In the upper part of the culture tubes, where the oxygen concentration and bacterial density were high, *P. caudatum* dominated because it was better able to feed on bacteria. In the lower part of the tubes, however, the lower oxygen concentration

chapter 56 community ecology **1171**

favored the growth of a different potential food, yeast, and *P. bursaria* was better able to eat this food. The fundamental niche of each species was the whole culture tube, but the realized niche of each species was only a portion of the tube. Because the realized niches of the two species did not overlap too much, both species were able to survive. However, competition did have a negative effect on the participants (figure 56.5c). When grown without a competitor, both species reached densities three times greater than when they were grown with a competitor.

Competitive exclusion refined

Gause's principle of competitive exclusion can be restated as: No two species can occupy the same niche *indefinitely* when resources are limiting. Certainly species can and do coexist while competing for some of the same resources. Nevertheless, Gause's hypothesis predicts that when two species coexist on a long-term basis, either resources must not be limited or their niches will always differ in one or more features; otherwise, one species will outcompete the other, and the extinction of the second species will inevitably result.

Competition may lead to resource partitioning

Gause's competitive exclusion principle has a very important consequence: If competition for a limited resource is intense, then either one species will drive the other to extinction, or natural selection will reduce the competition between them.

When the late Princeton ecologist Robert MacArthur studied five species of warblers, small insect-eating forest songbirds, he discovered that they all appeared to be competing for the same resources. But when he studied them more carefully, he found that each species actually fed in a different part of spruce trees and so ate different subsets of insects. One species fed on insects near the tips of branches, a second within the dense foliage, a third on the lower branches, a fourth high on the trees, and a fifth at the very apex of the trees. Thus, each species of warbler had evolved so as to utilize a different portion of the spruce tree resource. They had *subdivided the niche* to avoid direct competition with one another. This niche subdivision is termed **resource partitioning**.

Resource partitioning is often seen in similar species that occupy the same geographic area. Such sympatric species often avoid competition by living in different portions of the habitat or by using different food or other resources (figure 56.6). This pattern of resource partitioning is thought to result from the process of natural selection causing initially similar species to diverge in resource use to reduce competitive pressures.

Whether such evolutionary divergence occurs can be investigated by comparing species whose ranges only partially overlap. Where the two species occur together, they often tend to exhibit greater differences in morphology (the form and structure of an organism) and resource use than do allopatric populations of the same species that do not occur with the other species. Called **character displacement**, the differences

figure 56.6

RESOURCE PARTITIONING AMONG SYMPATRIC LIZARD SPECIES. Species of *Anolis* lizards on Caribbean islands partition their habitats in a variety of ways. **a.** Some species occupy leaves and branches in the canopy of trees, (**b**) others use twigs on the periphery, and (**c**) still others are found at the base of the trunk. In addition, (**d**) some use grassy areas in the open. When two species occupy the same part of the tree, they either utilize different-sized insects as food or partition the thermal microhabitat; for example, one might only be found in the shade, whereas the other would only bask in the sun.

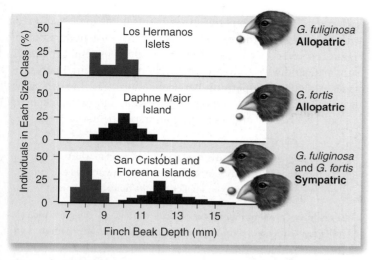

figure 56.7

CHARACTER DISPLACEMENT IN DARWIN'S FINCHES. These two species of finches (genus *Geospiza*) have beaks of similar size when allopatric, but different size when sympatric.

evident between sympatric species are thought to have been favored by natural selection as a means of partitioning resources and thus reducing competition.

As an example, the two Darwin's finches in figure 56.7 have bills of similar size where the finches are allopatric (that is, each living on an island where the other does not occur). On islands where they are sympatric (that is, occur together), the two species have evolved beaks of different sizes, one adapted to larger seeds and the other to smaller ones. Character displacement such as this may play an important role in adaptive radiation, leading new species to adapt to different parts of the environment, as discussed in chapter 22.

Detecting interspecific competition can be difficult

It is not simple to determine when two species are competing. The fact that two species use the same resources need not imply competition if that resource is not in limited supply. Even if the population sizes of two species are negatively correlated, such that where one species has a large population, the other species has a small population and vice versa, the two species may not be competing for the same limiting resource. Instead, the two species might be independently responding to the same feature of the environment—perhaps one species thrives best in warm conditions and the other where it's cool.

Experimental studies of competition

Some of the best evidence for the existence of competition comes from experimental field studies. By setting up experiments in which two species occur either alone or together, scientists can determine whether the presence of one species has a negative effect on a population of the second species.

For example, a variety of seed-eating rodents occur in North American deserts. In 1988, researchers set up a series of 50-m × 50-m enclosures to investigate the effect of kangaroo

rats on smaller, seed-eating rodents. Kangaroo rats were removed from half of the enclosures, but not from the others. The walls of all of the enclosures had holes that allowed rodents to come and go, but in the plots in which the kangaroo rats had been removed, the holes were too small to allow the kangaroo rats to reenter.

Over the course of the next 3 years, the researchers monitored the number of the smaller rodents present in the plots. As figure 56.8 illustrates, the number of other rodents was substantially higher in the absence of kangaroo rats, indicating that kangaroo rats compete with the other rodents and limit their population sizes.

A great number of similar experiments have indicated that interspecific competition occurs between many species of plants and animals. The effects of competition can be seen in aspects of population biology other than population size, such as behavior and individual growth rates. For example, two species of *Anolis* lizards occur on the Caribbean island of St. Maarten. When one of the species, *A. gingivinus*, is placed in 12-m × 12-m enclosures without the other species, individual lizards grow faster and perch lower than do lizards of the same species when placed in enclosures in which *A. pogus*, a species normally found near the ground, is also present.

Limitations of experimental studies

Although experimental studies can be a powerful means of understanding interactions between coexisting species, they have their limitations.

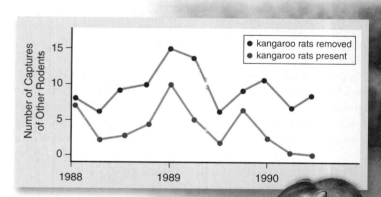

figure 56.8

DETECTING INTERSPECIFIC COMPETITION. This experiment tests how removal of kangaroo rats affects the population size of other rodents. Immediately after kangaroo rats were removed, the number of other rodents increased relative to the enclosures that still contained kangaroo rats. Notice that population sizes (as estimated by number of captures) changed in synchrony in the two treatments, probably reflecting changes in the weather.

inquiry

Why are there more individuals of other rodent species when kangaroo rats are excluded?

First, care is necessary in interpreting the results of field experiments. Negative effects of one species on another do not automatically indicate the existence of competition. For example, many similarly sized fish have a negative effect on one another, but it results not from competition, but from the fact that adults of each species prey on juveniles of the other species.

In addition, the presence of one species may attract predators or parasites, which then also prey on the second species. In this case, even if the two species are not competing, the second species may have a lower population size in the presence of the first species due to predators or parasites. Indeed, we can't rule out this possibility with the results of the kangaroo rat exclusion study just mentioned, although the close proximity of the enclosures (they were adjacent) would suggest that the same predators and parasites were present in all of them. Thus, experimental studies are most effective when combined with detailed examination of the ecological mechanisms causing the observed effect of one species on another.

Second, experimental studies are not always feasible. For example, the coyote population has increased in the United States in recent years concurrently with the decline of the grey wolf. Is this trend an indication that the species compete? Because of the size of the animals and the large geographic areas occupied by each individual, manipulative experiments involving fenced areas with only one or both species—with each experimental treatment replicated several times for statistical analysis—are not practical. Similarly, studies of slow-growing trees might require many centuries to detect competition between adult trees. In such cases, detailed studies of the ecological requirements of each species are our best bet for understanding interspecific interactions.

A niche may be defined as the total number of ways in which an organism utilizes its environment. Interspecific interactions may cause a species' realized niche to be smaller than its fundamental niche.

If resources are limiting, no two species can occupy the same niche indefinitely without competition driving one to local extinction.

Sympatric species often partition available resources, reducing competition between them.

Experimental studies can provide strong tests of the hypothesis that interspecific competition occurs, but such studies have limitations. Detailed ecological studies are important regardless of whether experiments are conducted.

56.3 Predator–Prey Relationships

Predation is the consuming of one organism by another. In this sense, predation includes everything from a leopard capturing and eating an antelope, to a deer grazing on spring grass.

When experimental populations are set up under simple laboratory conditions, as illustrated in figure 56.9 with the predatory protist *Didinium* and its prey *Paramecium*, the predator often exterminates its prey and then becomes extinct itself, having nothing left to eat. If refuges are provided for the *Paramecium*, however, its population drops to low levels but not to extinction. Low prey population levels then provide inadequate food for the *Didinium*, causing the predator population to decrease. When this occurs, the prey population can recover.

Predation strongly influences prey populations

In nature, predators often have large effects on prey populations. As the previous example indicates, however, the interaction is a two-way street: prey can also affect the dynamics of predator populations. The outcomes of such interactions are complex and depend on a variety of factors.

Prey population explosions and crashes

Some of the most dramatic examples of the interconnection between predators and their prey involve situations in which humans have either added or eliminated predators from an area. For example, the elimination of large carnivores from much of the eastern United States has led to population explosions of white-tailed deer, which strip the habitat of all edible plant life

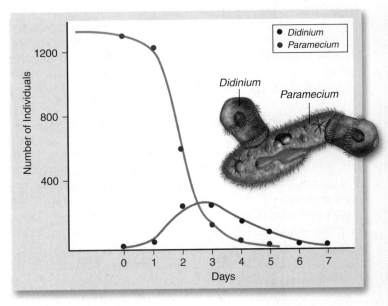

figure 56.9

PREDATOR–PREY IN THE MICROSCOPIC WORLD. When the predatory *Didinium* is added to a *Paramecium* population, the numbers of *Didinium* initially rise, while the numbers of *Paramecium* steadily fall. When the *Paramecium* population is depleted, however, the *Didinium* individuals also die.

inquiry

? *Can you think of any ways this experiment could be changed so that* **Paramecium** *might not go extinct?*

within their reach. Similarly, when sea otters were hunted to near extinction on the western coast of the United States, populations of sea urchins, a principal prey item of the otters, exploded.

Conversely, the introduction of rats, dogs, and cats to many islands around the world has led to the decimation of native fauna. Populations of Galápagos tortoises on several islands are endangered by introduced rats, dogs, and cats, which eat the eggs and the young tortoises. Similarly, in New Zealand, several species of birds and reptiles have been eradicated by rat predation and now only occur on a few offshore islands that the rats have not reached. On Stephens Island, near New Zealand, every individual of the now-extinct Stephens Island wren was killed by a single lighthouse keeper's cat.

A classic example of the role predation can play in a community involves the introduction of prickly pear cactus to Australia in the nineteenth century. In the absence of predators, the cactus spread rapidly, so that by 1925 it occupied 12 million hectares of rangeland in an impenetrable morass of spines that made cattle ranching difficult. To control the cactus, a predator from its natural habitat in Argentina, the moth *Cactoblastis cactorum*, was introduced, beginning in 1926. By 1940, cactus populations had been greatly reduced and it now generally occurs in small populations.

Predation and coevolution

Predation provides strong selective pressures on prey populations. Any feature that would decrease the probability of capture should be strongly favored. In turn, the evolution of such features causes natural selection to favor counteradaptations in predator populations. The process by which these adaptations are selected in lockstep fashion in both predator and prey is termed **coevolution**. A coevolutionary race may ensue in which predators and prey are constantly evolving better defenses and better means of circumventing these defenses. In the sections that follow, you'll learn more about these defenses and responses.

Plant adaptations defend against herbivores

Plants have evolved many mechanisms to defend themselves from herbivores. The most obvious are morphological defenses: Thorns, spines, and prickles play an important role in discouraging browsers, and plant hairs, especially those that have a glandular, sticky tip, deter insect herbivores. Some plants, such as grasses, deposit silica in their leaves, both strengthening and protecting themselves. If enough silica is present, these plants are simply too tough to eat.

Chemical defenses

As significant as morphological adaptations are, the chemical defenses that occur so widely in plants are even more widespread. Plants exhibit some amazing chemical adaptations to combat herbivores. For example, recent work demonstrates that when attacked by caterpillars, wild tobacco plants emit a chemical into the air that attracts a species of bug that feeds on that caterpillar (discussed in greater detail in chapter 40).

The best known and perhaps most important of the chemical defenses of plants against herbivores are **secondary chemical compounds**. These chemicals are distinguished from primary compounds, which are the components of a ma-jor metabolic pathway, such as respiration. Many plants, and apparently many algae as well, contain structurally diverse secondary compounds that are either toxic to most herbivores or disturb their metabolism greatly, preventing, for example, the normal development of larval insects. Consequently, most herbivores tend to avoid the plants that possess these compounds.

The mustard family (Brassicaceae) produces a group of chemicals known as mustard oils. These substances give the pungent aromas and tastes to plants such as mustard, cabbage, watercress, radish, and horseradish. The flavors we enjoy indicate the presence of chemicals that are toxic to many groups of insects. Similarly, plants of the milkweed family (Asclepiadaceae) and the related dogbane family (Apocynaceae) produce a milky sap that deters herbivores from eating them. In addition, these plants usually contain cardiac glycosides molecules that can produce drastic deleterious effects on the heart function of vertebrates.

The coevolutionary response of herbivores

Certain groups of herbivores are associated with each family or group of plants protected by a particular kind of secondary compound. These herbivores are able to feed on these plants without harm, often as their exclusive food source.

For example, cabbage butterfly caterpillars (subfamily Pierinae) feed almost exclusively on plants of the mustard and caper families, as well as on a few other small families of plants that also contain mustard oils (figure 56.10). Similarly, caterpillars of monarch butterflies and their relatives (subfamily Danainae) feed on plants of the milkweed and dogbane families. How do these animals manage to avoid the chemical defenses of the plants, and what are the evolutionary precursors and ecological consequences of such patterns of specialization?

a.

figure 56.10

INSECT HERBIVORES WELL SUITED *b.*
TO THEIR PLANT HOSTS. *a.* The green caterpillars of the cabbage white butterfly, *Pieris rapae*, are camouflaged on the leaves of cabbage and other plants on which they feed. Although mustard oils protect these plants against most herbivores, the cabbage white butterfly caterpillars are able to break down the mustard oil compounds. *b.* An adult cabbage white butterfly.

a.

b.

figure 56.11

A BLUE JAY LEARNS NOT TO EAT MONARCH BUTTERFLIES.
a. This cage-reared jay had never seen a monarch butterfly
before it tried eating one. *b.* The same jay regurgitated the
butterfly a few minutes later. This bird will probably avoid
trying to capture all orange-and-black insects in the future.

We can offer a potential explanation for the evolution
of these particular patterns. Once the ability to manufacture
mustard oils evolved in the ancestors of the caper and mustard
families, the plants were protected for a time against most or
all herbivores that were feeding on other plants in their area.
At some point, certain groups of insects—for example, the
cabbage butterflies—evolved the ability to break down mustard oils and thus feed on these plants without harming themselves. Having developed this new capability, the butterflies
were able to use a new resource without competing with other
herbivores for it. As we saw in chapter 22, exposure to an un-
derutilized resource often leads to evolutionary diversification
and adaptive radiation.

Animal adaptations defend against predators

Some animals that feed on plants rich in secondary compounds receive an extra benefit. For example, when the caterpillars of monarch butterflies feed on plants of the milkweed family, they do not
break down the cardiac glycosides that protect these plants from
herbivores. Instead, the caterpillars concentrate and store the cardiac glycosides in fat bodies; they then pass them through the chrysalis
stage to the adult and even to the eggs of the next generation.

The incorporation of cardiac glycosides protects all stages of
the monarch life cycle from predators. A bird that eats a monarch
butterfly quickly regurgitates it (figure 56.11) and in the future
avoids the conspicuous orange-and-black pattern that characterizes the adult monarch. Some bird species have evolved the ability
to tolerate the protective chemicals; these birds eat the monarchs.

Chemical defenses

Animals also manufacture and use a startling array of defensive substances. Bees, wasps, predatory bugs, scorpions, spiders, and many
other arthropods use chemicals to defend themselves and to kill
their own prey. In addition, various chemical defenses have evolved
among many marine invertebrates, as well as a variety of vertebrates, including frogs, snakes, lizards, fishes, and some birds.

The poison-dart frogs of the family Dendrobatidae produce toxic alkaloids in the mucus that covers their brightly
colored skin; these alkaloids are deadly to animals that try to
eat the frogs (figure 56.12). Some of these toxins are so powerful that a few micrograms will kill a person if injected into
the bloodstream. More than 200 different alkaloids have been
isolated from these frogs, and some are playing important roles
in neuromuscular research. Similarly intensive investigations of
marine animals, venomous reptiles, algae, and flowering plants
are underway in search of new drugs to fight cancer and other
diseases, or to use as sources of antibiotics.

figure 56.12

VERTEBRATE CHEMICAL DEFENSES. Frogs of the family
Dendrobatidae, abundant in the forests of Central and South
America, are extremely poisonous to vertebrates; 80 different
toxic alkaloids have been identified from different species in
this genus. Dendrobatids advertise their toxicity with bright
coloration. As a result of either instinct or learning, predators
avoid such brightly colored species that might otherwise be
suitable prey.

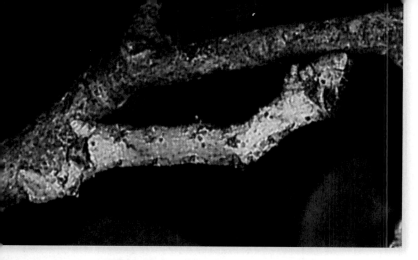

figure 56.13

CRYPTIC COLORATION AND FORM. An inchworm caterpillar (*Nacophora quernaria*) (hanging from the upper twig) closely resembles a twig.

Defensive coloration

Many insects that feed on milkweed plants are brightly colored; they advertise their poisonous nature using an ecological strategy known as **warning coloration.**

Showy coloration is characteristic of animals that use poisons and stings to repel predators, while organisms that lack specific chemical defenses are seldom brightly colored. In fact, many have cryptic coloration—color that blends with the surroundings and thus hides the individual from predators (figure 56.13). Camouflaged animals usually do not live together in groups because a predator that discovers one individual gains a valuable clue to the presence of others.

Mimicry allows one species to capitalize on defensive strategies of another

During the course of their evolution, many species have come to resemble distasteful ones that exhibit warning coloration. The mimic gains an advantage by looking like the distasteful model. Two types of mimicry have been identified: Batesian mimicry and Müllerian mimicry.

Batesian mimicry

Batesian mimicry is named for Henry Bates, the British naturalist who first brought this type of mimicry to general attention in 1857. In his journeys to the Amazon region of South America, Bates discovered many instances of palatable insects that resembled brightly colored, distasteful species. He reasoned that the mimics would be avoided by predators, who would be fooled by the disguise into thinking the mimic was the distasteful species.

Many of the best-known examples of Batesian mimicry occur among butterflies and moths. Predators of these insects must use visual cues to hunt for their prey; otherwise, similar color patterns would not matter to potential predators. Increasing evidence indicates that Batesian mimicry can involve nonvisual cues, such as olfaction, although such examples are less obvious to humans.

The kinds of butterflies that provide the models in Batesian mimicry are, not surprisingly, members of groups whose caterpillars feed on only one or a few closely related plant families. The plant families on which they feed are strongly protected by toxic chemicals. The model butterflies incorporate the poisonous molecules from these plants into their bodies. The mimic butterflies, in contrast, belong to groups in which the feeding habits of the caterpillars are not so restricted. As caterpillars, these butterflies feed on a number of different plant families that are unprotected by toxic chemicals.

One often-studied mimic among North American butterflies is the tiger swallowtail, whose range occurs throughout the eastern United States and into Canada (figure 56.14*a*). In areas in which the poisonous pipevine swallowtail occurs, female tiger swallowtails are polymorphic and one color form is extremely similar in appearance to the pipevine swallowtail.

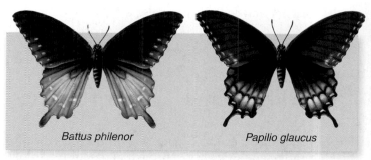

Battus philenor *Papilio glaucus*

a. **Batesian mimicry:** Pipevine swallowtail butterfly (*Battus philenor*) is poisonous; Tiger swallowtail (*Papilio glaucus*) is a palatable mimic.

Heliconius erato *Heliconius melpomene*

Heliconius sapho *Heliconius cydno*

b. **Müllerian mimicry:** Two pairs of mimics; all are distasteful.

figure 56.14

MIMICRY. *a.* Batesian mimicry. Pipevine swallowtail butterflies, *Battus philenor*, are protected from birds and other predators by the poisonous compounds they derive from the food they eat as caterpillars and store in their bodies. Adult pipevine swallowtails advertise their poisonous nature with warning coloration. Tiger swallowtails, *Pepilio glaucus*, are Batesian mimics of the poisonous pipevine swallowtail and are not chemically protected. *b.* Pairs of Müllerian mimics. *Heliconius erato* and *H. melpomene* are sympatric, and *H. sapho* and *H. cydno* are sympatric. All of these butterflies are distasteful. They have evolved similar coloration patterns in sympatry to minimize predation; predators need only learn one pattern to avoid.

The caterpillars of the tiger swallowtail feed on a variety of trees, including tulip, aspen, and cherry, and neither caterpillars nor adults are distasteful to birds. Interestingly, the Batesian mimicry seen in the adult tiger swallowtail butterfly does not extend to the caterpillars: Tiger swallowtail caterpillars are camouflaged on leaves, resembling bird droppings, while the pipevine swallowtail's distasteful caterpillars are very conspicuous.

Müllerian mimicry

Another kind of mimicry, **Müllerian mimicry,** was named for the German biologist Fritz Müller, who first described it in 1878. In Müllerian mimicry, several unrelated but protected animal species come to resemble one another (figure 55.14*b*). If animals that resemble one another are all poisonous or dangerous, they gain an advantage because a predator will learn more quickly to avoid them. In some cases, predator populations even evolve an innate avoidance of species; such evolution may occur more quickly when multiple dangerous prey look alike.

In both Batesian and Müllerian mimicry, mimic and model must not only look alike but also act alike. For example, the members of several families of insects that closely resemble wasps behave surprisingly like the wasps they mimic, flying often and actively from place to place.

Predation can have substantial effects on prey populations. As a result, prey species often evolve defensive adaptations.

Many plants are protected from most herbivores by secondary compounds. Once the members of a particular herbivore group evolve the ability to feed on one of these types of plants, these herbivores gain access to a new resource, which they can exploit without competition from other herbivores.

Animals defend themselves against predators with chemical defenses, warning coloration, and camouflage.

In Batesian mimicry, unprotected species resemble others that are distasteful. Both species exhibit warning coloration. In Müllerian mimicry, two or more unrelated but protected species resemble one another, thus achieving a kind of group defense.

56.4 The Many Types of Species Interactions

The plants, animals, protists, fungi, and prokaryotes that live together in communities have changed and adjusted to one another continually over millions of years. We have already discussed competition and predation, but other types of ecological interactions commonly occur. For example, many features of flowering plants have evolved in relation to the dispersal of the plant's gametes by animals (figure 56.15). These animals, in turn, have evolved a number of special traits that enable them to obtain food or other resources efficiently from the plants they visit, often from their flowers. While doing so, the animals pick up pollen, which they may deposit on the next plant they visit, or seeds, which may be left elsewhere in the environment, sometimes a great distance from the parent plant.

Symbiosis involves long-term interactions

In **symbiosis,** two or more kinds of organisms interact in often elaborate and more-or-less permanent relationships. All symbiotic relationships carry the potential for coevolution between the organisms involved, and in many instances the results of this coevolution are fascinatingly complex.

Examples of symbiosis include lichens, which are associations of certain fungi with green algae or cyanobacteria. Another important example are mycorrhizae, associations between fungi and the roots of most kinds of plants. The fungi expedite the plant's absorption of certain nutrients, and the plants in turn provide the fungi with carbohydrates (both mycorrhizae and lichens are discussed in greater detail in chapter 29). Similarly, root nodules that occur in legumes and certain other kinds of plants contain bacteria that fix atmospheric nitrogen and make it available to their host plants.

figure 56.15

POLLINATION BY A BAT. Many flowers have coevolved with other species to facilitate pollen transfer. Insects are widely known as pollinators, but they're not the only ones: birds, bats, and even small marsupials and lizards serve as pollinators for some species. Notice the cargo of pollen on the bat's snout.

In the tropics, leaf-cutter ants are often so abundant that they can remove a quarter or more of the total leaf surface of the plants in a given area (see figure 29.17). They do not eat these leaves directly; rather, they take them to underground nests, where they chew them up and inoculate them with the spores of particular fungi. These fungi are cultivated by the ants and brought from one specially prepared bed to another, where they grow and reproduce. In turn, the fungi constitute the primary food of the ants and their larvae. The relationship between leaf-cutter ants and these fungi is an excellent example of symbiosis. Recent phylogenetic studies using DNA and assuming a molecular clock (see chapter 23) suggest that these symbioses are ancient, perhaps originating more than 50 million years ago.

The major kinds of symbiotic relationships include (1) **commensalism,** in which one species benefits while the other neither benefits nor is harmed; (2) **mutualism,** in which both participating species benefit; and (3) **parasitism,** in which one species benefits but the other is harmed. Parasitism can also be viewed as a form of predation, although the organism that is preyed on does not necessarily die.

Commensalism benefits one species and is neutral to the other

In commensalism, one species benefits while the other is neither hurt nor helped by the interaction. In nature, individuals of one species are often physically attached to members of another. For example, epiphytes are plants that grow on the branches of other plants. In general, the host plant is unharmed, while the epiphyte that grows on it benefits. An example is Spanish moss, which hangs on trees in the southern United States. This plant and other members of its genus, which is in the pineapple family, grow on trees to gain access to sunlight; they generally do not harm the trees (figure 56.16).

figure 56.17

COMMENSALISM, MUTUALISM, OR PARASITISM? In this symbiotic relationship, oxpeckers definitely receive a benefit in the form of nutrition from the ticks and other parasites they pick off their host (in this case, an impala, *Aepyceros melampus*). But the effect on the host is not always clear. If the ticks are harmful, their removal benefits the host, and the relationship is mutually beneficial. If the oxpeckers also pick at scabs, causing blood loss and possible infection, the relationship may be parasitic. If the hosts are unharmed by either the ticks or the oxpeckers, the relationship may be an example of commensalism.

Similarly, various marine animals, such as barnacles, grow on other, often actively moving sea animals, such as whales, and thus are carried passively from place to place. These "passengers" presumably gain more protection from predation than they would if they were fixed in one place, and they also reach new sources of food. The increased water circulation that these animals receive as their host moves around may also be of great importance, particularly if the passengers are filter feeders. Unless the number of these passengers gets too large, the host species is usually unaffected.

When commensalism may not be commensalism

One of the best-known examples of symbiosis involves the relationships between certain small tropical fishes (clownfish) and sea anemones, shown in the first figure of this chapter. The fish have evolved the ability to live among the stinging tentacles of sea anemones, even though these tentacles would quickly paralyze other fishes that touched them. The clownfish feed on food particles left from the meals of the host anemone, remaining uninjured under remarkable circumstances.

On land, an analogous relationship exists between birds called oxpeckers and grazing animals such as cattle or antelopes (figure 56.17). The birds spend most of their time clinging to the

figure 56.16

AN EXAMPLE OF COMMENSALISM. Spanish moss (*Tillandsia usneoides*) benefits from using trees as a substrate, but the trees generally are not affected positively or negatively.

animals, picking off parasites and other insects, carrying out their entire life cycles in close association with the host animals.

No clear-cut boundary exists between commensalism and mutualism; in each of these instances, it is difficult to be certain whether the second partner receives a benefit or not. It may be advantageous to the sea anemone to have particles of food removed from its tentacles because it may then be better able to catch other prey. Similarly, while often thought of as commensalism, the association of grazing mammals and gleaning birds is actually an example of mutualism. The mammal benefits by having parasites and other insects removed from its body, but the birds also benefit by gaining a dependable source of food.

On the other hand, commensalism can easily transform itself into parasitism. Oxpeckers are also known to pick not only parasites, but also scabs off their grazing hosts. Once the scab is picked, the birds drink the blood that flows from the wound. Occasionally, the cumulative effect of persistent attacks can greatly weaken the herbivore, particularly when conditions are not favorable, such as during droughts.

Mutualism benefits both species

Mutualism is a symbiotic relationship between organisms in which both species benefit. Mutualistic relationships are of fundamental importance in determining the structure of biological communities.

Mutualism and coevolution

Some of the most spectacular examples of mutualism occur among flowering plants and their animal visitors, including insects, birds, and bats. During the course of flowering-plant evolution, the characteristics of flowers evolved in relation to the characteristics of the animals that visit them for food and, in the process, spread their pollen from individual to individual. At the same time, characteristics of the animals have changed, increasing their specialization for obtaining food or other substances from particular kinds of flowers.

Another example of mutualism involves ants and aphids. Aphids are small insects that suck fluids from the phloem of living plants with their piercing mouthparts. They extract a certain amount of the sucrose and other nutrients from this fluid, but they excrete much of it in an altered form through their anus. Certain ants have taken advantage of this—in effect, domesticating the aphids. The ants carry the aphids to new plants, where they come into contact with new sources of food, and then consume as food the "honeydew" that the aphids excrete.

Ants and acacias: A prime example of mutualism

A particularly striking example of mutualism involves ants and certain Latin American tree species of the genus *Acacia*. In these species, certain leaf parts, called stipules, are modified as paired, hollow thorns. The thorns are inhabited by stinging ants of the genus *Pseudomyrmex*, which do not nest anywhere else (figure 56.18). Like all thorns that occur on plants, the acacia thorns serve to deter herbivores.

At the tip of the leaflets of these acacias are unique, protein-rich bodies called Beltian bodies, named after the nineteenth-century British naturalist Thomas Belt. Beltian bodies do not occur in species of *Acacia* that are not inhabited by ants, and their role is clear: they serve as a primary food for the ants. In addition, the plants secrete nectar from glands near

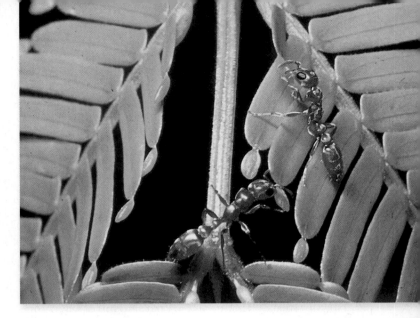

figure 56.18

MUTUALISM: ANTS AND ACACIAS. Ants of the genus *Pseudomyrmex* live within the hollow thorns of certain species of acacia trees in Latin America. The nectaries at the bases of the leaves and the Beltian bodies at the ends of the leaflets provide food for the ants. The ants, in turn, supply the acacias with organic nutrients and protect the acacias from herbivores and shading from other plants.

the bases of their leaves. The ants consume this nectar as well, feeding it and the Beltian bodies to their larvae.

Obviously, this association is beneficial to the ants, and one can readily see why they inhabit acacias of this group. The ants and their larvae are protected within the swollen thorns, and the trees provide a balanced diet, including the sugar-rich nectar and the protein-rich Beltian bodies. What, if anything, do the ants do for the plants?

Whenever any herbivore lands on the branches or leaves of an acacia inhabited by ants, the ants, which continually patrol the acacia's branches, immediately attack and devour the herbivore. The ants that live in the acacias also help their hosts compete with other plants by cutting away any encroaching branches that touch the acacia in which they are living. They create, in effect, a tunnel of light through which the acacia can grow, even in the lush tropical rain forests of lowland Central America. In fact, when an ant colony is experimentally removed from a tree, the acacia is unable to compete successfully in this habitat. Finally, the ants bring organic material into their nests. The parts they do not consume, together with their excretions, provide the acacias with an abundant source of nitrogen.

When mutualism may not be mutualism

As with commensalism, however, things are not always as they seem. Ant–acacia associations also occur in Africa; in Kenya, several species of acacia ants occur, but only a single species is found on any one tree. One species, *Crematogaster nigriceps*, is competitively inferior to two of the other species. To prevent invasion by these other ant species, *C. nigriceps* prunes the branches of the acacia, preventing it from coming into contact with branches of other trees, which would serve as a bridge for invaders.

Although this behavior is beneficial to the ant, it is detrimental to the tree because it destroys the tissue from which flowers are produced, essentially sterilizing the tree. In this case, what initially evolved as a mutualistic interaction has instead become a parasitic one.

Parasitism benefits one species at the expense of another

Parasitism may be regarded as a special form of symbiosis in which the parasite usually is much smaller than the prey and remains closely associated with it. Parasitism is harmful to the prey organism and beneficial to the parasite. In many cases, the parasite kills its host, and thus the ecological effects of parasitism can be similar to those of predation. Although in the past parasitism was usually studied mostly in terms of its effects on individuals and the populations in which they live, in recent years researchers have realized that parasitism can be an important factor affecting the structure of communities.

External parasites

Parasites that feed on the exterior surface of an organism are external parasites, or **ectoparasites** (figure 56.19). Many instances of external parasitism are known in both plants and animals. **Parasitoids** are insects that lay eggs on living hosts. This behavior is common among wasps, whose larvae feed on the body of the unfortunate host, often killing it.

Internal parasites

Parasites that live within the body of their hosts, termed **endoparasites,** occur in many different phyla of animals and protists. Internal parasitism is generally marked by

Infected ant

figure 56.20

PARASITIC MANIPULATION OF HOST BEHAVIOR. Due to a parasite in its brain, an ant climbs to the top of a grass blade, where it may be eaten by a grazing herbivore, thus passing the parasite from insect to mammal.

much more extreme specialization than external parasitism, as shown by the many protist and invertebrate parasites that infect humans.

The more closely the life of the parasite is linked with that of its host, the more its morphology and behavior are likely to have been modified during the course of its evolution (the same is true of symbiotic relationships of all sorts). Conditions within the body of an organism are different from those encountered outside and are apt to be much more constant. Consequently, the structure of an internal parasite is often simplified, and unnecessary armaments and structures are lost as it evolves (for example, see descriptions of tapeworms in chapter 33).

Parasites and host behaviors

Many parasites have complex life cycles that require several different hosts for growth to adulthood and reproduction. Recent research has revealed the remarkable adaptations of certain parasites that alter the behavior of the host and thus facilitate transmission from one host to the next. For example, many parasites cause their hosts to behave in ways that make them more vulnerable to their predators; when the host is ingested, the parasite is able to infect the predator.

One of the most famous examples involves a parasitic flatworm, *Dicrocoelium dendriticum*, which lives in ants as an intermediate host, but reaches adulthood in large herbivorous mammals such as cattle and deer. Transmission from an ant to a cow might seem difficult, because cows do not normally eat insects. The flatworm, however, has evolved a remarkable adaptation. When an ant is infected, one of the flatworms migrates into the brain and causes the ant to climb to the top of vegetation and lock its mandibles onto a grass blade at the end of the day, just when herbivores are grazing (figure 56.20). The result is that the ant is eaten along with the grass, leading to infection of the grazer.

figure 56.19

AN EXTERNAL PARASITE. The yellow vines are the flowering plant dodder (*Cuscuta*), a parasite that has lost its chlorophyll and its leaves in the course of its evolution. Because it is heterotrophic (unable to manufacture its own food), dodder obtains its food from the host plants it grows on.

Ecological processes have interactive effects

We have seen the different ways in which species can interact with one another. In nature, however, more than one type of interaction often occurs at the same time. In many cases, the outcome of one type of interaction is modified or even reversed when another type of interaction is also occurring.

Predation reduces competition

When resources are limiting, a superior competitor can eliminate other species from a community through competitive exclusion. However, predators can prevent or greatly reduce exclusion by lowering the numbers of individuals of competing species.

A given predator may often feed on two, three, or more kinds of plants or animals in a given community. The predator's choice depends partly on the relative abundance of the prey options. In other words, a predator may feed on species A when it is abundant and then switch to species B when A is rare. Similarly, a given prey species may become a primary source of food for increasing numbers of species as it becomes more abundant. In this way, superior competitors may be prevented from competitively excluding other species.

Such patterns are often characteristic of communities in marine intertidal habitats. For example, in preying selectively on bivalves, sea stars prevent bivalves from monopolizing a habitat, opening up space for many other organisms (figure 56.21). When sea stars are removed from a habitat, species diversity falls precipitously, and the seafloor community comes to be dominated by a few species of bivalves.

Predation tends to reduce competition in natural communities, so it is usually a mistake to attempt to eliminate a major predator, such as wolves or mountain lions, from a community. The result may be a decrease in biological diversity.

Parasitism may counter competition

Parasites may affect sympatric species differently and thus influence the outcome of interspecific interactions. One classic experiment investigated interactions between two sympatric flour beetles, *Tribolium castaneum* and *T. confusum*, with and without a parasite, *Adelina*. In the absence of the parasite, *T. castaneum* is dominant, and *T. confusum* normally becomes extinct. When the parasite is present, however, the outcome is reversed, and *T. castaneum* perishes.

Similar effects of parasites in natural systems have been observed in many species. For example, in the *Anolis* lizards of St. Maarten mentioned previously, the competitively inferior species is resistant to lizard malaria (a form of the disease related to human malaria), whereas the other species is highly susceptible. Only in areas in which the malaria parasite occurs are the two species capable of coexisting.

Indirect effects

In some cases, species may not directly interact, yet the presence of one species may affect a second by way of interactions with a third. Such effects are termed **indirect effects.**

The desert rodents described earlier in the experiment with kangaroo rats eat seeds, and so do the ants in their community; thus, we might expect them to compete with each other. But when all rodents were completely removed from large experimental enclosures and not allowed back in (unlike the previous experiment, no holes were placed in the enclosure walls), ant populations first increased but then declined (figure 56.22).

The initial increase was the expected result of removing a competitor; why did it reverse? The answer reveals the intricacies of natural ecosystems. Rodents prefer large seeds, whereas ants prefer smaller ones. Furthermore, in this system, plants with large seeds are competitively superior to plants with small seeds. The removal of rodents therefore led to an increase in the number of plants with large seeds, which reduced the number of small seeds available to ants, which in turn led to a decline in ant populations. In summary, the effect of rodents on ants is complicated: a direct, negative effect of resource competition and an indirect, positive effect mediated by plant competition.

figure 56.21

PREDATION REDUCES COMPETITION. *a.* In a controlled experiment in a coastal ecosystem, Robert Paine of the University of Washington removed a key predator, sea stars (*Pisaster*). *b.* In response, fiercely competitive mussels, a type of bivalve mollusk, exploded in population growth, effectively crowding out seven other indigenous species.

a.

b.

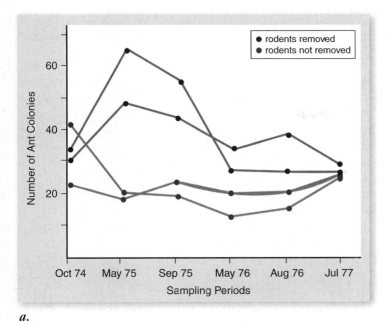

a.

b.

(−) Indirect positive effect

figure 56.22

DIRECT AND INDIRECT EFFECTS IN AN ECOLOGICAL COMMUNITY. *a.* In the enclosures in which kangaroo rats had been removed, ants initially increased in population size relative to the ants in the control enclosures, but then these ant populations declined. *b.* Rodents and ants both eat seeds, so the presence of rodents has a direct negative effect on ants, and vice versa. However, the presence of rodents has a negative effect on large seeds. In turn, the number of plants with large seeds has a negative effect on plants that produce small seeds. Hence, the presence of rodents should increase the number of small seeds. In turn, the number of small seeds has a positive effect on ant populations. Thus, indirectly, the presence of rodents has a positive effect on ant population size.

inquiry

Why do ant populations increase and then decrease in the absence of rodents?

Keystone species have major effects on communities

Species whose effects on the composition of communities are greater than one might expect based on their abundance are termed **keystone species.** Predators, such as the sea star described earlier, can often serve as keystone species by preventing one species from outcompeting others, thus maintaining high levels of species richness in a community.

A wide variety of other types of keystone species also exist. Some species manipulate the environment in ways that create new habitats for others. Beavers, for example, change running streams into small impoundments, altering the flow of water and flooding areas (figure 56.23). Similarly, alligators excavate deep holes at the bottoms of lakes. In times of drought, these holes are the only areas where water remains, thus allowing aquatic species that otherwise would perish to persist until the drought ends and the lake refills.

In symbiosis, two or more species interact closely, with at least one species benefitting.

Commensalism is a type of symbiosis in which one species benefits and the other is not affected positively or negatively.

Mutualism involves interactions between species that are mutually beneficial.

Parasitism is a form of symbiosis that is beneficial to the parasite, but harmful to the host.

Many different processes are likely to occur simultaneously within communities. Only by understanding how these processes interact will we be able to understand how communities function.

---

figure 56.23

EXAMPLE OF A KEYSTONE SPECIES. Beavers, by constructing dams and transforming flowing streams into ponds, create new habitats for many plant and animal species.

Ecological Succession, Disturbance, and Species Richness

Even when the climate of an area remains stable year after year, communities have a tendency to change from simple to complex in a process known as **succession.** This process is familiar to anyone who has seen a vacant lot or cleared woods slowly become occupied by an increasing number of species.

Succession produces a change in species composition

If a wooded area is cleared or burned and left alone, plants will slowly reclaim the area. Eventually, all traces of the clearing will disappear, and the area will again be woods. This kind of succession, which occurs in areas where an existing community has been disturbed but organisms still remain, is called **secondary succession.**

In contrast, **primary succession** occurs on bare, lifeless substrate, such as rocks, or in open water, where organisms gradually move into an area and change its nature. Primary succession occurs in lakes left behind and on land exposed after the retreat of glaciers, and on volcanic islands that rise above the sea (figure 56.24).

Primary succession on glacial moraines provides an example (see figure 56.24). On the bare, mineral-poor ground exposed when glaciers recede, soil pH is basic as a result of carbonates in the rocks, and nitrogen levels are low. Lichens are the first vegetation able to grow under such conditions. Acidic secretions from the lichens help break down the substrate and reduce the pH, as well as adding to the accumulation of soil. Mosses then colonize these pockets of soil, eventually building up enough nutrients in the soil for alder shrubs to take hold. Over a hundred years, the alders, which have symbiotic bacteria that fix atmospheric nitrogen (described in chapter 26), increase soil nitrogen levels, and their acidic leaves further lower soil pH. At this point, spruce are able to become established, and they eventually crowd out the alders and form a dense spruce forest.

In a similar example, an *oligotrophic* lake—one poor in nutrients—may gradually, by the accumulation of organic matter, become *eutrophic*—rich in nutrients. As this occurs, the composition of communities will change, first increasing in species richness and then declining.

Why succession happens

Succession happens because species alter the habitat and the resources available in it in ways that favor other species. Three dynamic concepts are of critical importance in the process: tolerance, facilitation, and inhibition.

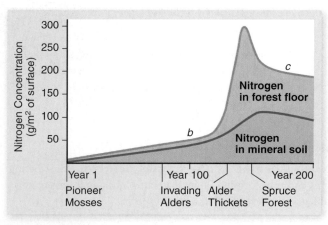

figure 56.24

PRIMARY SUCCESSION AT ALASKA'S GLACIER BAY. *a.* Initially, the glacial moraine at Glacier Bay, Alaska had little soil nitrogen *b.* The first invaders of these exposed sites are pioneer moss species with nitrogen-fixing, mutualistic microbes. *c.* Within 20 years, young alder shrubs take hold. Rapidly fixing nitrogen, they soon form dense thickets. *d.* As soil nitrogen levels rise, spruce crowd out the mature alders, forming a forest.

a.

b.

c.

d.

1. **Tolerance.** Early successional stages are characterized by weedy, *r*-selected species that are tolerant of the harsh, abiotic conditions in barren areas (the preceding chapter discussed *r*-selected and *K*-selected species).

2. **Facilitation.** The weedy early successional stages introduce local changes in the habitat that favor other, less weedy species. Thus, the mosses in the Glacier Bay succession convert nitrogen to a form that allows alders to invade (see figure 56.24). The alders in turn lower soil pH as their fallen leaves decompose, and spruce and hemlock, which require acidic soil, are able to invade.

3. **Inhibition.** Sometimes the changes in the habitat caused by one species, while favoring other species, also inhibit the growth of the original species that caused the changes. Alders, for example, do not grow as well in acidic soil as the spruce and hemlock that replace them.

Over the course of succession, the number of species typically increases as the environment becomes more hospitable. In some cases, however, as ecosystems mature, more *K*-selected species replace *r*-selected ones, and superior competitors force out other species, leading ultimately to a decline in species richness.

Succession in animal communities

The species of animals present in a community also change through time in a successional pattern. As the vegetation changes during succession, habitat disappears for some species and appears for others.

A particularly striking example occurred on the Krakatau islands, which were devastated by an enormous volcanic eruption in 1883. Initially composed of nothing but barren ash-fields, the three islands of the group experienced rapid successional change as vegetation became reestablished. A few blades of grass appeared the next year, and within 15 years the coastal vegetation was well established and the interior was covered with dense grasslands. By 1930, the islands were almost entirely forested (figure 56.25).

The fauna of Krakatau changed in synchrony with the vegetation. Nine months after the eruption, the only animal found was a single spider, but by 1908, 200 animal species were found in a 3-day exploration. For the most part, the first animals were grassland inhabitants, but as trees became established, some of these early colonists, such as the zebra dove and the long-tailed shrike (a type of predatory bird), disappeared and were replaced by forest-inhabiting species, such as fruit bats and fruit-eating birds.

Although patterns of succession of animal species have typically been caused by vegetational succession, changes in the composition of the animal community in turn have affected plant occurrences. In particular, many plant species that are animal-dispersed or pollinated could not colonize Krakatau until their dispersers or pollinators had become established. For example, fruit bats were slow to colonize Krakatau, and until they appeared, few bat-dispersed plant species were present.

a.

b.

figure 56.25

SUCCESSION AFTER A VOLCANIC ERUPTION. A major volcanic explosion in 1883 on the island of Krakatau destroyed all life on the island. *a.* This photo shows a later, much less destructive eruption of the volcano. *b.* Krakatau, forested and populated by animals.

Disturbances can play an important role in structuring communities

Traditionally, many ecologists considered biological communities to be in a state of equilibrium, a stable condition that resisted change and fairly quickly returned to its original state if disturbed by humans or natural events. Such stability was usually attributed to the process of interspecific competition.

In recent years, this viewpoint has been reevaluated. Increasingly, scientists are recognizing that communities are constantly changing as a result of climatic changes, species invasions, and disturbance events. As a result, many ecologists now invoke nonequilibrium models that emphasize change, rather than stability. A particular focus of ecological research concerns the role that disturbances play in determining the structure of communities.

Disturbances can be widespread or local. Severe disturbances, such as forest fires, drought, and floods, may affect large areas. Animals may also cause severe disruptions. Gypsy moths can devastate a forest by consuming all of the leaves on its trees. Unregulated deer populations may grow explosively, the deer overgrazing and so destroying the forest in which they live. On the other hand, local disturbances may affect only a small area, as when a tree falls in a forest or an animal digs a hole and uproots vegetation.

Intermediate disturbance hypothesis

In some cases, disturbance may act to increase the species richness of an area. According to the **intermediate disturbance hypothesis,** communities experiencing moderate amounts of disturbance will have higher levels of species richness than communities experiencing either little or great amounts of disturbance.

Two factors could account for this pattern. First, in communities where moderate amounts of disturbance occur, patches of habitat will exist at different successional stages. Within the area as a whole, then, species diversity will be greatest because the full range of species—those characteristic of all stages of succession—will be present. For example, a pattern of intermittent episodic disturbance that produces gaps in the rain forest (as when a tree falls) allows invasion of the gap by other species (figure 56.26). Eventually, the species inhabiting the gap will go through a successional sequence, one tree replacing another, until a canopy tree species comes again to occupy the gap. But if there are many gaps of different ages in the forest, many different species will be coexisting, some in young gaps and others in older ones.

Second, moderate levels of disturbance may prevent communities from reaching the final stages of succession, in which a few dominant competitors eliminate most of the other species. In contrast, too much disturbance might leave the community continually in the earliest stages of succession, when species richness is relatively low.

Ecologists are increasingly realizing that disturbance is common, rather than exceptional, in many communities. As a result, the idea that communities inexorably move along

figure 56.26

INTERMEDIATE DISTURBANCE. A single fallen tree created a small light gap in the tropical rain forest of Panama. Such gaps play a key role in maintaining the high species diversity of the rain forest. In this case, a sunlight-loving plant is able to sprout up among the dense foliage of trees in the forest.

a successional trajectory culminating in the development of a predictable end-state or "climax" community is no longer widely accepted. Rather, predicting the state of a community in the future may be difficult because the unpredictable occurrence of disturbances will often counter successional changes. Understanding the role that disturbances play in structuring communities is currently an important area of investigation in ecology.

Communities change through time by a process termed succession. Species richnes tends to increase through time, although ultimately it may decline.

Disturbance is often common in ecological communities. In some cases, intermediate levels of disturbance may maximize species richness.

56.1 Biological Communities: Species Living Together

A community is a group of different species that occupy a given location.

- The term *assemblage* denotes a subset of a larger community.

- Communities are characterized by their principal species, species richness, or primary productivity.

- The individualistic concept of a community holds that a community is a random assemblage of species that happen to occur in a given place.

- The holistic concept of a community views a community as an integrated unit composed of species that work together as part of a functional whole.

- Species respond independently to environmental conditions, and community composition gradually changes over space and time.

- Ecotones are transitional zones between two distinct environments where environmental conditions rapidly change and the species from each environment overlap (figure 56.3).

56.2 The Ecological Niche Concept

A niche is the total of all the ways a species uses environmental resources.

- A fundamental, or hypothetical, niche describes how a species is capable of using resources if there are no intervening factors.

- A realized niche is the actual environmental conditions that allow the establishment of a stable population.

- Niches are realized because of competition, predation, and diseases.

- The principle of competitive exclusion states that if two species simultaneously occupy the same area and compete for the same resource, one species will be eliminated if the resource is limited (figure 56.5b).

- Resource partitioning occurs in ecologically similar species that occupy the same geographic area to reduce competition pressure (figure 56.7).

56.3 Predator–Prey Relationships

Predation is the consumption of one species by another.

- Predation exerts strong influences on prey populations.

- There are strong selection pressures on prey to avoid predation and vice versa, resulting in the coevolution of the two.

- Plant and animal prey evolve chemical defenses to protect themselves against predation.

- Warning coloration occurs in prey organisms that use poisons or stings to repel predators.

- Batesian mimicry occurs when a palatable species looks like a distasteful species.

- Müllerian mimicry occurs when two distasteful species look similar.

56.4 The Many Types of Species Interactions

Symbiosis occurs when two or more species interact in more-or-less permanent associations.

- Commensalism benefits one species and does not help or hurt the other species.

- Mutualism is a relationship in which both species benefit.

- Parasitism is a relationship in which the prey or host is harmed; often fatally.

- In nature more than one interaction usually occurs at the same time.

- Indirect effects can occur when one member of two interacting species is affected by a third species.

- Keystone species can maintain a more diverse community by reducing competition between species or altering the environment to create new habitats.

56.5 Ecological Succession, Disturbance, and Species Richness

Communities have a tendency to change over time in a process called succession.

- Succession occurs as species alter their habitat and resources, to thus facilitating other species.

- Primary succession begins with a barren, lifeless substrate (figure 56.24b).

- Secondary succession occurs after an existing community is disturbed.

- Successional dynamics include tolerance, facilitation, and inhibition.

- As succession occurs species richness increases so as to become stable or decline in time if *K*-selected species out-compete other species.

- Animal species succession changes with plant succession, and, in turn, animal communities may affect plant succession.

- Community composition changes as a result of local and global disturbances that "reset" succession.

- Intermediate levels of disturbance may maximize species richness.

SELF TEST

1. Studies that demonstrate that species living in an ecological community change independently of one another in space and time
   a. support the individualistic concept of ecological communities.
   b. support the holistic concept of ecological communities.
   c. suggest species interactions are the sole determinant of which species coexist in a community.
   d. None of the above.
2. If two species have very similar realized niches and are forced to coexist and share a limiting resource indefinitely,
   a. both species would be expected to coexist.
   b. both species would be expected to go extinct.
   c. the species that uses the limiting resource most efficiently should drive the other species extinct.
   d. both species would be expected to become more similar to one another.
3. According to the idea of coevolution between predator and prey, when a prey species evolves a novel defense against a predator
   a. the predator is expected to always go extinct.
   b. the prey population should increase irreversibly out of control of the predator.
   c. the predator population should increase.
   d. evolution of a predator response should be favored by natural selection.
4. In order for mimicry to be effective in protecting a species from predation, it must
   a. occur in a palatable species that looks like a distasteful species.
   b. have cryptic coloration.
   c. occur such that mimics look and act like models.
   d. occur in only poisonous or dangerous species.
5. Which of the following is an example of commensalism?
   a. A tapeworm living in the gut of its host
   b. A clownfish living among the tentacles of a sea anemone
   c. An acacia tree and acacia ants
   d. Bees feeding on nectar from a flower
6. Oxpeckers' eating
   a. noninjurious insects off mammals is an example of commensalism.
   b. injurious ectoparasites off mammals is an example of mutualism.
   c. scabs so they can feed on a mammal's blood is an example of parasitism.
   d. All of the above.
7. A species whose effect on the composition of a community is greater than expected based on its abundance can be called a
   a. predator.
   b. primary succession species.
   c. secondary succession species.
   d. keystone species.
8. When a predator preferentially eats the superior competitor in a pair of competing species
   a. the inferior competitor is more likely to go extinct.
   b. the superior competitor is more likely to persist.
   c. coexistence of the competing species is more likely.
   d. None of the above.
9. Keystone species
   a. always tend to increase species diversity.
   b. always tend to decrease species diversity.
   c. are always predators.
   d. None of the above.
10. Species that are the first colonists in a habitat undergoing primary succession
    a. are usually the fiercest competitors.
    b. help maintain their habitat constant so their persistence is ensured.
    c. may change their habitat in a way that favors the invasion of other species.
    d. must first be successful secondary succession specialists.
11. Species diversity of very early and late successional stages
    a. is expected to be low.
    b. is expected to be high.
    c. is always at equilibrium.
    d. depends only on the physical characteristics of the habitat.
12. The fundamental niche of an organism
    a. is always more restricted than its realized niche.
    b. is usually less restricted than its realized niche.
    c. takes into account actual environmental conditions and the presence of other species.
    d. None of the above.
13. Resource partitioning
    a. reduces niche overlap.
    b. increases competition.
    c. increases niche overlap.
    d. All of the above.
14. Lichen growing on the surface of rocks provides an example of
    a. facilitation.
    b. tolerance.
    c. inhibition.
    d. secondary succession.

CHALLENGE QUESTIONS

1. Competition is traditionally indicated by documenting the effect of one species on the population of another. Are there alternative ways to study the potential effects of competition on organisms that are impractical to study with experimental manipulations because they are too big or live too long?
2. Refer to figure 56.9. If the single prey species of *Paramecium* was replaced by several different potential prey species that varied in their palatability or ease of subduing by the predator (leading to different levels of preference by the predator) what would you expect the dynamics of the system to look like (that is, would the system be more or less likely to go to extinction?)?
3. Refer to figure 56.22. Are there alternative hypotheses that might explain the increase followed by the decrease in ant colony numbers subsequent to rodent removal in the experiment described in figure 56.22? If so, how would you test the mechanism hypothesized in the figure?
4. Refer to figure 56.7. Examine the pattern of beak size distributions of two species of finches on the Galápagos Islands. One hypothesis that can be drawn from this pattern is that character displacement has taken place. Are there other hypotheses? If so, how would you test them?

chapter 57

Dynamics of Ecosystems

introduction

THE EARTH IS A RELATIVELY CLOSED system with respect to chemicals. It is an open system in terms of energy, however, because it receives energy at visible and near-visible wavelengths from the Sun and steadily emits thermal energy to outer space in the form of infrared radiation. The organisms in ecosystems interact in complex ways as they participate in the the cycling of chemicals and as they capture and expend energy. All organisms, including humans, depend on the specialized abilities of other organisms—plants, algae, animals, fungi, and prokaryotes—to acquire the essentials of life, as explained in this chapter. In chapters 57 and 58, we consider the many different types of ecosystems that constitute the biosphere and discuss the threats to the biosphere and the species it contains.

concept outline

57.1 Biogeochemical Cycles

An **ecosystem** includes all the organisms that live in a particular place, plus the abiotic (nonliving) environment in which they live—and with which they interact—at that location. Ecosystems are intrinsically dynamic in a number of ways, including their processing of matter and energy. We start with matter.

The atomic constituents of matter cycle within ecosystems

During the biological processing of matter, the atoms of which it is composed, such as the atoms of carbon or oxygen, maintain their integrity even as they are assembled into new compounds and the compounds are later broken down. The Earth has an essentially fixed number of each of the types of atoms of biological importance, and the atoms are recycled.

Each organism assembles its body from atoms that previously were in the soil, the atmosphere, other parts of the abiotic environment, or other organisms. When the organism dies, its atoms are released unaltered to be used by other organisms or returned to the abiotic environment. Because of the cycling of the atomic constituents of matter, your body is likely during your life to contain a carbon or oxygen atom that once was part of Julius Caesar's body or Cleopatra's.

The atoms of the various chemical elements are said to move through ecosystems in **biogeochemical cycles,** a term emphasizing that the cycles of chemical elements involve not only biological organisms and processes, but also geological (abiotic) systems and processes. Biogeochemical cycles include processes that occur on many spatial scales, from cellular to planetary, and they also include processes that occur on multiple timescales, from seconds (biochemical reactions) to millennia (weathering of rocks).

Biogeochemical cycles usually cross the boundaries of ecosystems to some extent, rather than being self-contained within individual ecosystems. For example, one ecosystem might import or export carbon to others.

In this section, we consider the cycles of some major elements along with the compound water. We also present an example of biogeochemical cycles in a forest ecosystem.

Carbon, the basis of organic compounds, cycles through most ecosystems

Carbon is a major constituent of the bodies of organisms because carbon atoms help form the framework of all organic compounds (see chapter 3); almost 20% of the weight of the human body is carbon. From the viewpoint of the day-to-day dynamics of ecosystems, carbon dioxide (CO_2) is the most significant carbon-containing compound in the abiotic environments of organisms. It makes up 0.03% of the volume of the atmosphere, meaning the atmosphere contains about 750 billion metric tons of carbon. In aquatic ecosystems, CO_2 reacts spontaneously with the water to form bicarbonate ions (HCO_3^-).

The basic carbon cycle

The carbon cycle is straightforward, as shown in figure 57.1. In terrestrial ecosystems, plants and other photosynthetic organisms take in CO_2 from the atmosphere and use it in

figure 57.1

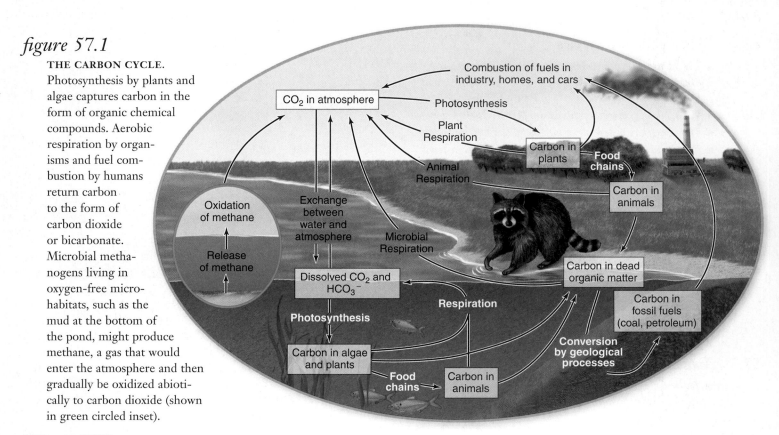

THE CARBON CYCLE. Photosynthesis by plants and algae captures carbon in the form of organic chemical compounds. Aerobic respiration by organisms and fuel combustion by humans return carbon to the form of carbon dioxide or bicarbonate. Microbial methanogens living in oxygen-free microhabitats, such as the mud at the bottom of the pond, might produce methane, a gas that would enter the atmosphere and then gradually be oxidized abiotically to carbon dioxide (shown in green circled inset).

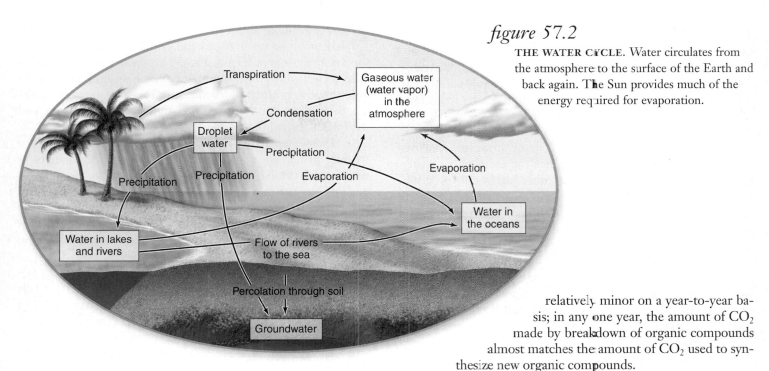

figure 57.2

THE WATER CYCLE. Water circulates from the atmosphere to the surface of the Earth and back again. The Sun provides much of the energy required for evaporation.

photosynthesis to synthesize the carbon-containing organic compounds of which they are composed (see chapter 8). The process is sometimes called **carbon fixation;** fixation refers to metabolic reactions that make nongaseous compounds from gaseous ones.

Animals eat the photosynthetic organisms and build their own tissues by making use of the carbon atoms in the organic compounds they ingest. Both the photosynthetic organisms and the animals obtain energy during their lives by breaking down some of the organic compounds available to them, through aerobic cellular respiration (see chapter 7). When they do this, they produce CO_2. Decaying organisms also produce CO_2. Carbon atoms returned to the form of CO_2 are available once more to be used in photosynthesis to synthesize new organic compounds.

In aquatic ecosystems, the carbon cycle is fundamentally similar, except that inorganic carbon is present in the water not only as dissolved CO_2, but also as HCO_3^- ions, both of which act as sources of carbon for photosynthesis by algae and aquatic plants.

Methane producers

Microbes that break down organic compounds by anaerobic cellular respiration (see chapter 7) provide an additional dimension to the global carbon cycle. Methanogens, for example, are microbes that produce methane (CH_4) instead of CO_2. One major source of CH_4 is wetland ecosystems, where methanogens live in the oxygen-free sediments. Methane that enters the atmosphere is oxidized abiotically to CO_2, but CH_4 that remains isolated from oxygen can persist for great lengths of time.

The rise of atmospheric carbon dioxide

Another dimension of the global carbon cycle is that over long periods of time, the cycle may proceed faster in one direction than the other. These differences in rate have ordinarily been relatively minor on a year-to-year basis; in any one year, the amount of CO_2 made by breakdown of organic compounds almost matches the amount of CO_2 used to synthesize new organic compounds.

Small mismatches, however, can have large consequences if continued for many years. The Earth's present reserves of coal were built up over geologic time. Organic compounds such as cellulose accumulated by being synthesized faster than they were broken down, and then they were transformed by geological processes into the fossil fuels. Most scientists believe that the world's petroleum reserves were created in the same way.

Human burning of fossil fuels today is creating large contemporary imbalances in the carbon cycle. Carbon that took millions of years to accumulate in the reserves of fossil fuels is being rapidly returned to the atmosphere, driving the concentration of CO_2 in the atmosphere upward year by year and helping to spur fears of global warming (see chapter 58).

The availability of water is fundamental to terrestrial ecosystems

The water cycle, seen in figure 57.2, is probably the most familiar of all biogeochemical cycles. All life depends on the presence of water; even organisms that can survive without water in resting states require water to regain activity. The bodies of most organisms consist mainly of water. The adult human body, for example, is about 60% water by weight. The amount of water available in an ecosystem often determines the nature and abundance of the organisms present, as illustrated by the difference between forests and deserts (see chapter 58).

Each type of biogeochemical cycle has distinctive features. A distinctive feature of the water cycle is that water is a compound, not an element, and thus it can be synthesized and broken down. It is synthesized during aerobic cellular respiration (see chapter 7) and chemically split during photosynthesis (see chapter 8). The rates of these processes are ordinarily about equal, and therefore a relatively constant amount of water cycles through the biosphere.

The basic water cycle

One key part of the water cycle is that liquid water from the Earth's surface evaporates into the atmosphere. The change of water from a liquid to a gas requires a considerable addition of thermal energy, explaining why evaporation occurs more rapidly when solar radiation beats down on a surface.

Evaporation occurs directly from the surfaces of oceans, lakes, and rivers. In terrestrial ecosystems, however, approximately 90% of the water that reaches the atmosphere passes through plants. Trees, grasses, and other plants take up water from soil via their roots, and then the water evaporates from their leaves and other surfaces through a process called transpiration (see chapter 38).

Evaporated water exists in the atmosphere as a gas, just like any other atmospheric gas. The water can condense back into liquid form, however, mostly because of cooling of the air. Condensation of gaseous water (water vapor) into droplets or crystals causes the formation of clouds, and if the droplets or crystals are large enough, they fall to the surface of the Earth as precipitation (rain or snow).

Groundwater

Less obvious than surface water, which we see in rivers and lakes, is water under ground—termed **groundwater.** Groundwater occurs in **aquifers,** which are permeable, underground layers of rock, sand, and gravel that are often saturated with water. Groundwater is the most important reservoir of water on land in many parts of the world, representing over 95% of all fresh water in the United States, for example.

Goundwater consists of two subparts. The upper layers of the groundwater constitute the water table, which is unconfined in the sense that it flows into streams and is partly accessible to the roots of plants. The lower, confined layers of the groundwater are generally out of reach to streams and plants, but can be tapped by wells. Groundwater is recharged by water that percolates downward from above, such as from precipitation. Water in an aquifer flows much more slowly than surface water, anywhere from a few millimeters to a meter or so per day.

In the United States, groundwater provides about 25% of the water used by humans for all purposes, and it supplies about 50% of the population with drinking water. In the Great Plains states, the deep Ogallala Aquifer is tapped extensively as a water source for agricultural and domestic needs. The aquifer is being depleted faster than it is recharged—a local imbalance in the water cycle—posing an ominous threat to the agricultural production of the area. Similar threats exist in many of the drier portions of the globe.

Changes in ecosystems brought about by changes in the water cycle

Water is so crucial for life that changes in the supply of water to an ecosystem can radically alter the nature of the ecosystem. Such changes have occurred often during the geological history of the Earth.

Consider, for example, the ecosystem of the Serengeti Plain in Tanzania, famous for its seemingly endless grasslands occupied by vast herds of antelopes and other grazing animals. The semiarid grasslands of today's Serengeti were rain forests 25 MYA. Starting at about that time, mountains such as Mount Kilimanjaro rose up between the rain forests and the Indian Ocean, their source of moisture. The presence of the mountains forced winds from the Indian Ocean upward, cooling the air and causing much of its moisture to precipitate before the air reached the rain forests. The land became much drier, and the forests turned to grasslands.

Today, human activities can alter the water cycle so profoundly that major changes occur in ecosystems. Changes in rain forests caused by deforestation provide an example. In healthy tropical rain forests, more than 90% of the moisture that falls as rain is taken up by plants and returned to the air by transpiration. The plants, in a very real sense, create their own rain: The moisture they return to the atmosphere falls back on the forests.

When human populations cut down or burn the rain forests in an area, the local water cycle is broken. Water that falls as rain thereafter drains away in rivers instead of rising to form clouds and fall again on the forests. Just such a transformation is occurring today in many tropical rain forests (figure 57.3). Large areas in Brazil, for example, were transformed in the twentieth century from lush tropical forest to semiarid desert, depriving many unique plant and animal species of their native habitat.

The nitrogen cycle depends on nitrogen fixation by microbes

Nitrogen is a component of all proteins and nucleic acids and is required in substantial amounts by all organisms; proteins are 16% nitrogen by weight. In many ecosystems, nitrogen is the chemical element in shortest supply relative to the needs of organisms. A paradox is that the atmosphere is 78% nitrogen by volume.

figure 57.3

DEFORESTATION DISRUPTS THE LOCAL WATER CYCLE. Tropical deforestation can have severe consequences, as shown by the extensive erosion in this deforested area in the Amazon region of Brazil.

Nitrogen availability

How can nitrogen be in short supply if the atmosphere is so rich with it? The answer is that the nitrogen in the atmosphere is in its elemental form—molecules of nitrogen gas (N_2)—and the vast majority of organisms, including all plants and animals, have no way to use nitrogen in this chemical form.

For animals, the ultimate source of nitrogen is nitrogen-containing organic compounds synthesized by plants or by algae or other microbes. Herbivorous animals, for example, eat plant or algal proteins and use the nitrogen-containing amino acids in them to synthesize their own proteins.

Plants and algae use a number of simple nitrogen-containing compounds as their sources of nitrogen to synthesize proteins and other nitrogen-containing organic compounds in their tissues. Two commonly used nitrogen sources are ammonia (NH_3) and nitrate ions (NO_3^-). As described in chapter 39, certain prokaryotic microbes can synthesize ammonia and nitrate from N_2 in the atmosphere, thereby constituting a part of the nitrogen cycle that makes atmospheric nitrogen accessible to plants and algae (figure 57.4). Other prokaryotes turn NH_3 and NO_3^- into N_2, making the nitrogen inaccessible. The balance of the activities of these two sets of microbes determines the accessibility of nitrogen to plants and algae.

Microbial nitrogen fixation, nitrification, and denitrification

The synthesis of nitrogen containing compounds from N_2 is known as **nitrogen fixation.** The first step in this process is the synthesis of NH_3 from N_2, and biochemists sometimes use the term "nitrogen fixation" to refer specifically to this step. After NH_3 has been synthesized, other prokaryotic microbes oxidize part of it to form NO_3^-, a process called **nitrification.**

Certain genera of prokaryotes have the ability to accomplish nitrogen fixation using a system of enzymes known as the nitrogenase complex (the *nif* gene complex; chapter 28). Most of the microbes are free-living, but on land some are found in symbiotic relationships with the roots of legumes (plants of the pea family, Fabaceae), alders, myrtles, and other plants.

Additional prokaryotic microbes (including both bacteria and archaea) are able to convert the nitrogen in NO_3^- into N_2 (or other nitrogen gases such as N_2O), a process termed **denitrification.** Ammonia can be subjected to denitrification indirectly by being converted first to NO_3^- and then to N_2.

Nitrogenous wastes and fertilizer use

Most animals, when they break down proteins in their metabolism, excrete the nitrogen from the proteins as NH_3. Humans and other mammals excrete nitrogen as urea in their urine (see chapter 50); a number of types of microbes convert the urea to NH_3. The NH_3 from animal excretion can be picked up by plants and algae as a source of nitrogen.

Human populations are radically altering the global nitrogen cycle by the use of fertilizers on lawns and agricultural fields. The fertilizers contain forms of fixed nitrogen that crops can use, such as ammonium (NH_4) salts manufactured industrially from atmospheric N_2. Partly because of the production of fertilizers, humans have already doubled the rate of transfer of N_2 in usable forms into soils and waters.

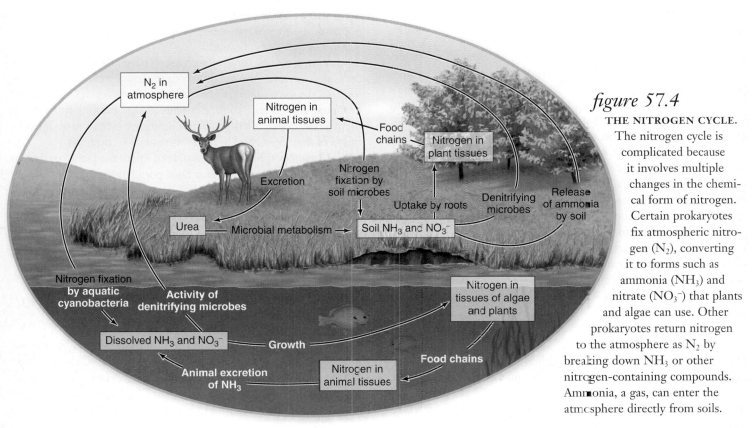

figure 57.4

THE NITROGEN CYCLE. The nitrogen cycle is complicated because it involves multiple changes in the chemical form of nitrogen. Certain prokaryotes fix atmospheric nitrogen (N_2), converting it to forms such as ammonia (NH_3) and nitrate (NO_3^-) that plants and algae can use. Other prokaryotes return nitrogen to the atmosphere as N_2 by breaking down NH_3 or other nitrogen-containing compounds. Ammonia, a gas, can enter the atmosphere directly from soils.

figure 57.5

PHOSPHORUS CYCLE. In contrast to carbon, water, and nitrogen, phosphorus occurs only in the liquid and solid states and thus does not enter the atmosphere.

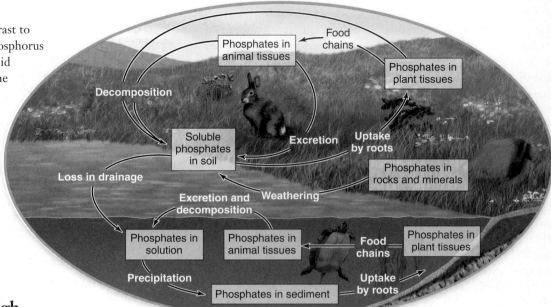

Phosphorus cycles through terrestrial and aquatic ecosystems, but not the atmosphere

Phosphorus is required in substantial quantities by all organisms; it occurs in nucleic acids, membrane phospholipids, and other essential compounds, such as adenine triphosphate (ATP).

Unlike carbon, water, and nitrogen, phosphorus has no significant gaseous form and does not cycle through the atmosphere (figure 57.5). In this respect, the phosphorus cycle exemplifies the sorts of cycles also exhibited by calcium, silicon, and many other mineral elements. Another feature that greatly simplifies the phosphorus cycle compared with the nitrogen cycle is that phosphorus exists in ecosystems in just a single oxidation state, phosphate (PO_4^{3-}).

Phosphate availability

Plants and algae use free inorganic PO_4^{3-} in the soil or water for synthesizing their phosphorus-containing organic compounds. Animals then tap the phosphorus in plant or algal tissue compounds to build their own phosphorus compounds. When organisms die, decay microbes—in a process called PO_4^{3-} remineralization—break up the organic compounds in their bodies, releasing phosphorus as inorganic PO_4^{3-} that plants and algae again can use.

The phosphorus cycle includes critical abiotic chemical and physical processes. Free PO_4^{3-} exist in soil in only low concentrations both because they combine with other soil constituents to form insoluble compounds and because they tend to be washed away by streams and rivers. Weathering of many sorts of rocks releases new PO_4^{3-} into terrestrial systems, but then rivers carry the PO_4^{3-} into the ocean basins. There is a large one-way flux of PO_4^{3-} from terrestrial rocks to deep-sea sediments.

Phosphates as fertilizers

Human activities have greatly modified the global phosphorus cycle since the advent of crop fertilization. Fertilizers are typically designed to provide PO_4^{3-} because crops might otherwise be short of it; the PO_4^{3-} in fertilizers is typically derived from crushed phosphate-rich rocks and bones. Detergents are another potential culprit in adding PO_4^{3-} to ecosystems, but laws now mandate low-phosphate detergents in much of the world.

Limiting nutrients in ecosystems are those in short supply relative to need

A chain is only as strong as its weakest link. For the plants and algae in an ecosystem to grow—and to thereby provide food for animals—they need many different chemical elements. The simplest theory is that in any particular ecosystem, one element will be in shortest supply relative to the needs for it by the plants and algae. That element is the **limiting nutrient**—the weak link—in the ecosystem.

The cycle of a limiting nutrient is particularly important because it determines the rate at which the nutrient is made available for use. We gave the nitrogen and phosphorus cycles close attention precisely because those elements are the limiting nutrients in many ecosystems. Nitrogen is the limiting nutrient in about two-thirds of the oceans and in many terrestrial ecosystems.

Oceanographers have discovered in just the last 15 years that iron is the limiting nutrient for algal populations (phytoplankton) in about one-third of the world's oceans. In these waters, wind-borne soil dust seems often to be the chief source of iron. When wind brings in iron-rich dust, algal populations proliferate, provided the iron is in a usable chemical form. In this way, sand storms in the Sahara Desert, by increasing the dust in global winds, can increase algal productivity in Pacific waters (figure 57.6).

Biogeochemical cycling in a forest ecosystem has been studied experimentally

An ongoing series of studies at the Hubbard Brook Experimental Forest in New Hampshire has yielded much of the available information about the cycling of nutrients in forest ecosystems. Hubbard Brook is the central stream of a large watershed that drains the hillsides of a mountain range covered with temperate

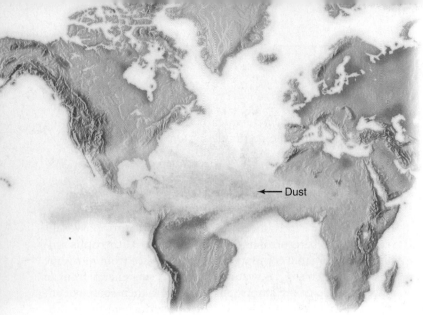

figure 57.6

ONE WORLD. Every year, millions of metric tons of iron-rich dust is carried westward by the trade winds from the Sahara Desert and neighboring Sahel area. A working hypothesis of many oceanographers is that this dust fertilizes parts of the ocean, including parts of the Pacific Ocean, where iron is the limiting nutrient. Land use practices in Africa, which are increasing the size of the north African desert, can thus affect ecosystems on the other side of the globe.

deciduous forest. Multiple tributary streams carry water off the hillsides into Hubbard Brook.

Six tributary streams, each draining a particular valley, were equipped with measurement devices when the study was started. All of the water that flowed out of each valley had to pass through the measurement system, where the flow of water and concentrations of nutrients was quantified.

According to the measurements the researchers made, the undisturbed forests around Hubbard Brook are efficient at retaining nutrients. In a year, only small quantities of nu-

trients enter a valley from outside, doing so mostly as a result of precipitation. The quantities carried out in stream waters are small also. When we say "small," we mean the influxes and outfluxes represent just minor fractions of the total amounts of nutrients in the system—about 1% in the case of calcium, for example.

In 1965 and 1966, the investigators felled all the trees and cleared all shrubs in one of the six valleys and prevented regrowth (figure 57.7a). The effects were dramatic. The amount of water running out of that valley increased by 40%, indicating that water previously taken up by vegetation and evaporated into the atmosphere was now running off. The amounts of a number of nutrients running out of the system also greatly increased. For example, the rate of loss of calcium increased nine-fold. Phosphorus, on the other hand, did not increase in the stream water; it apparently was locked up in insoluble compounds in the soil.

The change in the status of nitrogen in the disturbed valley was especially striking (figure 57.7b). The undisturbed forest in this valley had been accumulating NO_3^- at a rate of about 5 kg per hectare per year, but the deforested ecosystem lost NO_3^- at a rate of about 53 kg per hectare per year. The NO_3^- concentration in the stream water rapidly increased. The fertility of the valley decreased dramatically, while the run-off of nitrate generated massive algal blooms downstream, and the danger of downstream flooding greatly increased.

This experiment is particularly instructive at the start of the twenty-first century because forested land continues to be cleared worldwide (see chapter 58).

> The atoms of which matter is composed cycle between the tissues of organisms and the abiotic environment, such as the atmosphere or soil. This cycle is termed the biogeochemical cycle. Individual elements such as carbon, nitrogen, and phosphorus cycle in known ways, as does water, which is critical to ecosystems. The activities of growing human populations are affecting many of the rates, such as the rate at which carbon dioxide is added to the atmosphere.

a.   b.

figure 57.7

THE HUBBARD BROOK EXPERIMENT. *a.* A 38-acre watershed was completely deforested, and the runoff monitored for several years. *b.* Deforestation greatly increased the loss of nutrients in runoff water from the ecosystem. The orange curve shows the nitrate concentration in the runoff water from the deforested watershed; the green curve shows the nitrate concentration in runoff water from an undisturbed neighboring watershed.

The dynamic nature of ecosystems includes the processing of energy as well as that of matter. Energy, however, follows very different principles than does matter. Energy is never recycled. Instead, radiant energy from the Sun that reaches the Earth makes a one-way pass through our planet's ecosystems before being converted to heat and radiated back into space, signifying that the Earth is an open system for energy.

Energy can neither be created nor destroyed, but changes form

Why is energy so different from matter? A key part of the answer is that energy exists in several different forms, such as light, chemical-bond energy, motion, and heat. Although energy is neither created nor destroyed in the biosphere (the First Law of Thermodynamics), it frequently changes form.

A second key point is that organisms cannot convert heat to any of the other forms of energy. Thus, if organisms convert some chemical-bond or light energy to heat, the conversion is one-way; they cannot cycle that energy back into its original form.

Living organisms can use many forms of energy, but not heat

To understand why the Earth must function as an open system with regard to energy, two additional principles need to be recognized. The first is that organisms can use only certain forms of energy. For animals to live, they must have energy specifically as chemical-bond energy, which they acquire from their foods. Plants must have energy as light. Neither animals nor plants (nor any other organisms) can use heat as a source of energy to stay alive.

The second principle is that whenever organisms use chemical-bond or light energy, some of it is converted to heat; the Second Law of Thermodynamics states that a partial conversion to heat is inevitable. Put another way, animals and plants require chemical-bond energy and light to stay alive, but as they use these forms of energy, they convert them to heat, which they cannot use to stay alive and which they cannot cycle back into the original forms.

Fortunately for organisms, the Earth functions as an open system for energy. Light arrives every day from the Sun. Plants and other photosynthetic organisms use the newly arrived light to synthesize organic compounds and stay alive. Animals then eat the photosynthetic organisms, making use of the chemical-bond energy in their organic molecules to stay alive. Light and chemical-bond energy are partially converted to heat at every step. In fact, the light and chemical-bond energy are ultimately converted completely to heat. The heat leaves the Earth by being radiated into outer space at invisible, infrared wavelengths of the electromagnetic spectrum. For life to continue, new light energy is always required.

The Earth's incoming and outgoing flows of radiant energy must be equal for global temperature to stay constant. One concern is that human activities are changing the composition of the atmosphere in ways that impede the outgoing flow—the so-called *greenhouse effect*, which is described in the following chapter. Heat may be accumulating on Earth, causing global warming (see chapter 58).

Energy flows through trophic levels of ecosystems

In chapter 7, we introduced the concepts of autotrophs ("self-feeders") and heterotrophs ("fed by others"). **Autotrophs** synthesize the organic compounds of their bodies from inorganic precursors such as CO_2, water, and NO_3^- using energy from an abiotic source. Some autotrophs use light as their source of energy and therefore are **photoautotrophs;** they are the photosynthetic organisms, including plants, algae, and cyanobacteria. Other autotrophs are **chemoautotrophs** and obtain energy by means of inorganic oxidation reactions, such as the microbes that use hydrogen sulfide available at deep water vents (see chapter 58). All chemoautotrophs are prokaryotic. The photoautotrophs are of greatest importance in most ecosystems, and we focus on them in the remainder of this chapter.

Heterotrophs are organisms that cannot synthesize organic compounds from inorganic precursors, but instead live by taking in organic compounds that other organisms have made. They obtain the energy they need to live by breaking up some of the organic compounds available to them, thereby liberating chemical-bond energy for metabolic use (see chapter 7). Animals, fungi, and many microbes are heterotrophs.

When living in their native environments, species are often organized into chains that eat each other sequentially. For example, a species of insect might eat plants, and then a species of shrew might eat the insect, and a species of hawk might eat the shrew. Food passes through the four species in the sequence: plants → insect → shrew → hawk. A sequence of species like this is termed a **food chain.**

In a whole ecosystem, many species play similar roles; there is typically not just a single species in each role. For example, the animals that eat plants might include not just a single insect species, but perhaps 30 species of insects, plus perhaps 10 species of mammals. To organize this complexity, ecologists recognize a limited number of feeding levels called **trophic levels** (figure 57.8).

Definitions of trophic levels

The first trophic level in an ecosystem, called the **primary producers,** consists of all the autotrophs in the system. The other trophic levels consist of the heterotrophs, the **consumers.** All the heterotrophs that feed directly on the primary producers are placed together in a trophic level called the **herbivores.** In turn, the heterotrophs that feed on the herbivores (eating them or being parasitic on them) are collectively termed **primary carnivores,** and those that feed on the primary carnivores are called **secondary carnivores.**

Advanced studies of ecosystems need to take into account that organisms often do not line up in simple linear sequences in terms of what they eat; some animals, for ex-

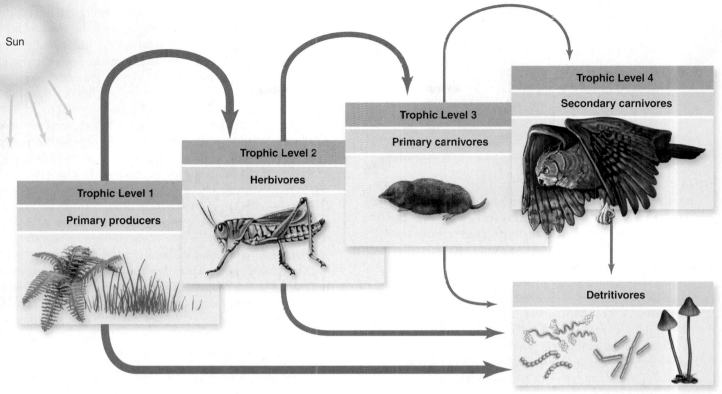

figure 57.8

TROPHIC LEVELS WITHIN AN ECOSYSTEM. Primary producers such as plants obtain their energy directly from the Sun, placing them in trophic level 1. Animals that eat plants, such as plant-eating insects, are herbivores and are in trophic level 2. Animals that eat the herbivores, such as shrews, are primary carnivores and are in trophic level 3. Animals that eat the primary carnivores, such as owls, are secondary carnivores in trophic level 4. Each trophic level, although illustrated here by a particular species, consists of all the species in the ecosystem that function in a similar way in terms of what they eat. The organisms in the detritivore trophic level consume dead organic matter they obtain from all the other trophic levels.

ample, eat both primary producers and other animals. A linear sequence of trophic levels is a useful organizing principle for many purposes, however.

An additional consumer level is the **detritivore** trophic level. Detritivores differ from the organisms in the other trophic levels in that they feed on the remains of already-dead organisms; **detritus** is dead organic matter. A subcategory of detritivores is the **decomposers,** which are mostly microbes and other minute organisms that live on and break up dead organic matter.

Concepts to describe trophic levels

Trophic levels consist of whole populations of organisms. For example, the primary producer trophic level consists of the whole populations of all the autotrophic species in an ecosystem. Ecologists have developed a special set of terms to refer to the properties of populations and trophic levels.

The **productivity** of a trophic level is the rate at which the organisms in the trophic level collectively synthesize new organic matter (new tissue substance). **Primary productivity** is the productivity of the primary producers. An important complexity in analyzing the primary producers is that not only do they synthesize new organic matter by photosynthesis, but they also break down some of the organic matter to release energy by means of aerobic cellular respiration (see chapter 8). The

respiration of the primary producers, in this context, is the rate at which they break down organic compounds. **Gross primary productivity (GPP)** is simply the raw rate at which the primary producers synthesize new organic matter; **net primary productivity (NPP)** is the GPP less the respiration of the primary producers. The NPP represents the organic matter available for herbivores to use as food.

The productivity of a heterotroph trophic level is termed **secondary productivity.** For instance, the rate that new organic matter is made by means of individual growth and reproduction in all the herbivores in an ecosystem is the secondary productivity of the herbivore trophic level. Each heterotroph trophic level has its own secondary productivity.

As you learn about ecological principles, one of your top goals should be to distinguish between *dynamic* and *static* properties of populations. Productivity is a dynamic property, and it is always expressed as a rate because it goes on and on, rather than having a meaning independent of the passage of time.

The chief static property of a population or trophic level is the amount of organic matter present at a particular time, termed the **standing crop biomass** of the population or trophic level, or often simply the *standing crop* or the *biomass.* Imagine taking a snapshot of a trophic level. The organic matter present in the snapshot would be the standing crop biomass at that moment.

17% growth

33% cellular respiration

50% feces

figure 57.9

THE FATE OF INGESTED CHEMICAL-BOND ENERGY: WHY ALL THE ENERGY INGESTED BY A HETEROTROPH IS NOT AVAILABLE TO THE NEXT TROPHIC LEVEL. A heterotroph such as this herbivorous insect assimilates only a fraction of the chemical-bond energy it ingests. In this example, 50% is not assimilated and is egested in feces; this egested chemical-bond energy cannot be used by the primary carnivores. A third (33%) of the ingested energy is used to fuel cellular respiration and thus is converted to heat, which cannot be used by the primary carnivores. Only 17% of the ingested energy is converted into insect biomass through growth and can serve as food for the next trophic level, but not even that percentage is certain to be used in that way because some of the insects will die before they are eaten.

How trophic levels process energy

The fraction of incoming solar radiant energy that the primary producers capture is small. Averaged over the course of a year, something around 1% of the solar energy impinging on forests or oceans is captured. Investigators sometimes observe far lower levels, but also see percentages as high as 5% under some conditions. The solar energy not captured as chemical-bond energy through photosynthesis is immediately converted to heat.

The primary producers, as noted before, carry out respiration in which they break down some of the organic compounds in their bodies to release chemical-bond energy. They use a portion of this chemical-bond energy to make ATP, which they in turn use to power various energy-requiring processes. Ultimately, the chemical-bond energy they release by respiration turns to heat.

Remember that organisms cannot use heat to stay alive. As a result, whenever energy changes form to become heat, it loses much or all of its usefulness for organisms as a fuel source. What we have seen so far is that about 99% of the solar energy impinging on an ecosystem turns to heat because it fails to be used by photosynthesis. Then some of the energy captured by photosynthesis also becomes heat because of respiration by the primary producers. All the heterotrophs in an ecosystem must live on the chemical-bond energy that is left.

An example of energy loss between trophic levels

As chemical-bond energy is passed from one heterotroph trophic level to the next, a great deal of the energy is diverted all along the way. This principle has dramatic consequences. It means

figure 57.10

THE FLOW OF ENERGY THROUGH AN ECOSYSTEM. Blue arrows represent the flow of energy that enters the ecosystem as light and is then passed along as chemical-bond energy to successive trophic levels. At each step energy is diverted, meaning that the chemical-bond energy available to each trophic level is less than that available to the preceding trophic level. Red arrows represent diversions of energy into heat. Tan arrows represent diversions of energy into feces and other organic materials useful only to the detritivores.

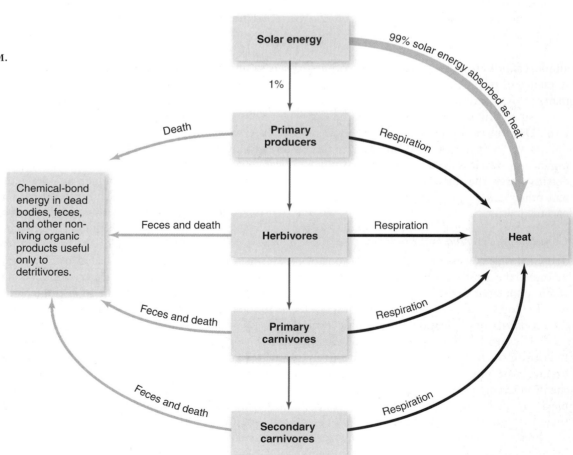

that, over any particular period of time, the amount of chemical-bond energy available to primary carnivores is far less than that available to herbivores, and the amount available to secondary carnivores is far less than that available to primary carnivores.

Why does the amount of chemical-bond energy decrease as energy is passed from one trophic level to the next? Consider the use of energy by the herbivore trophic level as an example (figure 57.9). After a herbivore such as a leaf-eating insect ingests some food, it produces feces. The chemical-bond energy in the compounds in the feces is not passed along to the primary carnivore trophic level. The chemical-bond energy of the food that is assimilated by the herbivore is used for a number of functions. Part of the assimilated energy is liberated by cellular respiration to be used for tissue repair, body movements, and other such functions. The energy used in these ways turns to heat and is not passed along to the carnivore trophic level. Some chemical-bond energy is built into the tissues of the herbivore and can serve as food for a carnivore. However, some herbivore individuals die of disease or accident rather than being eaten by predators.

In the end, of course, some of the initial chemical-bond energy acquired from the leaf is built into the tissues of herbivore individuals that are eaten by primary carnivores. Much of the initial chemical-bond energy, however, is diverted into heat, feces, and the bodies of herbivore individuals that carnivores do not get to eat. The same scenario is repeated at each step in a series of trophic levels (figure 57.10).

Ecologists figure as a rule of thumb that the amount of chemical-bond energy available to a trophic level over a period of time is about 10% of that available to the preceeding level over the same period of time. In some instances the percentage is higher, even as high as 30%.

Heat as the final energy product

Essentially all of the chemical-bond energy captured by photosynthesis in an ecosystem eventually becomes heat as the chemical-bond energy is used by various trophic levels. To see this important point, recognize that when the detritivores in the ecosystem metabolize all the dead bodies, feces, and other materials made available to them, they produce heat just like the other trophic levels do.

Productive ecosystems

Ecosystems vary considerably in their NPP. Wetlands and tropical rain forests are examples of particularly productive ecosystems (figure 57.11); in them, the NPP, measured as dry weight of new organic matter produced, is often around 2000 g/m²/year. By contrast, the corresponding figures for some other types of ecosystems are 1200–1300 for temperate forests, 900 for savanna, and 90 for deserts. (These general ecosystem types, termed *biomes*, are described in the following chapter.)

The number of trophic levels is limited by energy availability

The rate at which chemical-bond energy is made available to organisms in different trophic levels decreases exponentially as energy makes its way from primary producers to herbivores and then to various levels of carnivores. To envision this

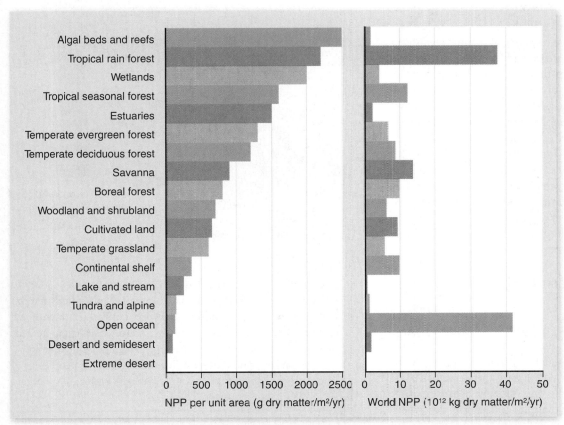

figure 57.11

ECOSYSTEM PRODUCTIVITY PER YEAR. The first column of data shows the average net primary productivity (NPP) per square meter per year. The second column of data factors in the area covered by the ecosystem type; it is the product of the productivity per square meter per year times the number of square meters occupied by the ecosystem type worldwide. Note that an ecosystem type that is very productive on a square-meter basis may not contribute much to global productivity if it is an uncommon type, such as wetlands. On the other hand, a very widespread ecosystem type, such as the open ocean, can contribute greatly to global productivity even if its productivity per square meter is low.

Source: Data in: Begon, M., J.L. Harper, and C. R. Townsend, *Ecology* 3/e, Blackwell Science, 1996, page 715. Original source: Whittaker, R. H. *Communities and Ecosystems*, 2/e, Macmillan, London, 1975.

critical point, assume for simplicity that the primary producers in an ecosystem gain 1000 units of chemical-bond energy over a period of time. If the energy input to each trophic level is 10% of the input to the preceding level, then the input of chemical-bond energy to the herbivore trophic level is 100 units, to the primary carnivores, 10 units, and to the secondary carnivores, 1 unit over the same period of time.

Limits on top carnivores

The exponential decline of chemical-bond energy in a trophic chain limits the lengths of trophic chains and the numbers of top carnivores an ecosystem can support. According to our model calculations, if an ecosystem includes secondary carnivores, only about one one-thousandth of the energy captured by photosynthesis passes all the way through the series of trophic levels to reach these animals as usable chemical-bond energy. Tertiary carnivores would receive only one ten-thousandth. This helps explain why no predators subsist on eagles or lions.

The decline of available chemical-bond energy also helps explain why the numbers of individual top-level carnivores in an ecosystem tend to be low. The whole trophic level of top carnivores receives relatively little energy, and yet such carnivores tend to be big: They have relatively large individual body sizes and great individual energy needs. Because of these two factors, the population numbers of top predators tend to be small.

The longest trophic chains probably occur in the oceans. Some tunas and other top-level ocean predators probably function as third- and fourth-level carnivores at times. The challenge of explaining such long trophic chains is obvious, but the solutions are not well understood presently.

Humans as consumers: A case study

The flow of energy in Cayuga Lake in upstate New York (figure 57.12) helps illustrate how the energetics of trophic levels can affect the human food supply. Researchers calculated from the actual properties of this ecosystem that about 150 of each 1000 calories of chemical-bond energy captured by primary producers in the lake were transferred into the bodies of herbivores. Of these calories, about 30 were transferred into the bodies of smelt, small fish that were the principal primary carnivores in the system.

If humans ate the smelt, they gained about 6 of the 1000 calories that originally entered the system. If trout ate the smelt and humans ate the trout, the humans gained only about 1.2 calories. For human populations in general, more energy is available if plants or other primary producers are eaten than if animals are eaten—and more energy is available if herbivores rather than carnivores are consumed.

Ecological pyramids illustrate the relationship of trophic levels

Imagine that the trophic levels of an ecosystem are represented as boxes stacked on top of each other. Imagine also that the width of each box is proportional to the productivity of the trophic level it represents. The stack of boxes will always have the shape of a pyramid; each box is narrower than the one under it because of the inviolable rules of energy flow. A diagram of the sort described is called a **pyramid of energy flow** or **pyramid of productivity** (figure 57.13a). It is an example of an **ecological pyramid.**

There are several types of ecological pyramids. Pyramid diagrams can be used to represent standing crop biomass or numbers of individuals, as well as productivity.

In a **pyramid of biomass,** the widths of the boxes are drawn to be proportional to standing crop biomass. Usually, trophic levels that have relatively low productivity also have relatively little biomass present at a given time. Thus, pyramids of biomass are usually upright, meaning each box is narrower than the one below it (figure 57.13b). An upright pyramid of biomass is not mandated by fundamental and inviolable rules like an upright pyramid of productivity is, however. In some ecosystems, the pyramid of biomass is **inverted,** meaning that at least one trophic level has greater biomass than the one below it (figure 57.13c).

How is it possible for the pyramid of biomass to be inverted? Consider a common sort of aquatic system in which the primary producers are single-celled algae (phytoplankton), and the herbivores are rice grain-sized animals (such as

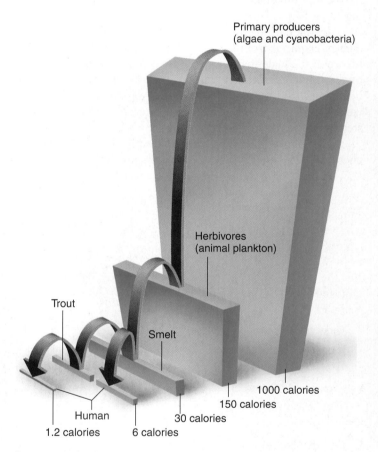

figure 57.12

FLOW OF ENERGY THROUGH THE TROPHIC LEVELS OF CAYUGA LAKE. Autotrophic plankton (algae and cyanobacteria) fix the energy of the Sun, the herbivores (animal plankton) feed on them, and both are consumed by smelt. The smelt are eaten by trout. The amount of fish flesh produced per unit time for human consumption is at least five times greater if people eat smelt rather than trout, but people typically prefer to eat trout.

inquiry

? *Why does it take so many calories of algae to support so few calories of humans?*

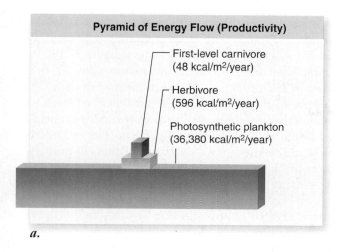

Pyramid of Energy Flow (Productivity)

First-level carnivore
(48 kcal/m²/year)

Herbivore
(596 kcal/m²/year)

Photosynthetic plankton
(36,380 kcal/m²/year)

a.

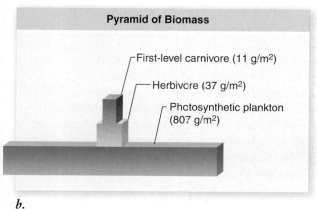

Pyramid of Biomass

First-level carnivore (11 g/m²)

Herbivore (37 g/m²)

Photosynthetic plankton
(807 g/m²)

b.

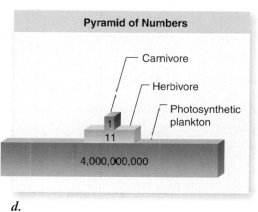

Inverted Pyramid of Biomass

Herbivorous zooplankton and
bottom fauna (21 g/m²)

Phytoplankton
(4 g/m²)

c.

Pyramid of Numbers

Carnivore

Herbivore

Photosynthetic
plankton

1
11

4,000,000,000

d.

figure 57.13

ECOLOGICAL PYRA-MIDS. In an ecological pyramid, successive trophic levels in an ecosystem are represented as stacked boxes, and the widths of the boxes represent the magnitude of an ecological property in the various trophic levels. Ecological pyramids can represent several different properties. *a.* Pyramid of energy flow (productivity). *b.* Pyramid of biomass of the ordinary type. *c.* Inverted pyramid of biomass. *d.* Pyramid of numbers.

inquiry

How can the existence of inverted pyramids of biomass be explained?

copepods) that feed directly on the algal cells. In such a system, the turnover of the algal cells is often very rapid: The cells multiply rapidly, but the animals consume them equally rapidly. In these circumstances, the algal cells never develop a large population size or large biomass. Nonetheless, because the algal cells are very productive, the ecosystem can support a substantial biomass of the animals, a biomass larger than that ever observed in the algal population.

In a **pyramid of numbers,** the widths of the boxes are proportional to the numbers of individuals present in the various trophic levels (figure 57.13*d*). Such pyramids are usually, but not always, upright.

The photosynthetic organisms in an ecosystem—the primary producers—capture about 1% of solar energy as chemical-bond energy. As the chemical-bond energy is passed through the other trophic levels, some of the energy is diverted at each step into heat, feces, and dead matter. Accordingly, the amount of chemical-bond energy made available to each trophic level per unit time is only about 10% of that available to the trophic level preceding it. This loss of chemical-bond energy limits the number of trophic levels in an ecosystem and mandates that energy pyramids for these systems be upright. This loss also means consumers can obtain more energy if they function as herbivores rather than carnivores.

57.3 Trophic-Level Interactions

The existence of food chains creates the possibility that species in any one trophic level may have effects on more than one trophic level. Primary carnivores, for example, may have effects not only on the animals they eat, but also, indirectly, on the plants or algae eaten by their prey. Conversely, increases in primary productivity may provide more food not just to herbivores, but also, indirectly, to carnivores.

The process by which effects exerted at an upper trophic level flow down to influence two or more lower levels is termed a **trophic cascade.** The effects themselves are called **top-down effects.** When an effect flows up through a trophic chain, such

as from primary producers to higher trophic levels, it is termed a **bottom-up effect.**

Top-down effects occur when changes in the top trophic level affects primary producers

The existence of top-down effects has been confirmed by controlled experiments in some types of ecosystems, particularly freshwater ones. For example, in one study, sections of a stream were enclosed with a mesh that prevented fish from entering. Brown trout—predators on invertebrates—were added to some

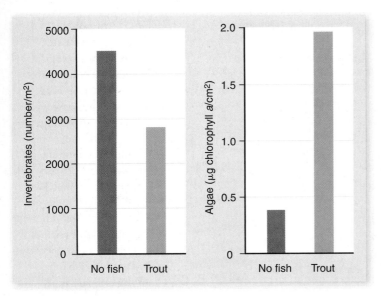

figure 57.14

TOP-DOWN EFFECTS DEMONSTRATED BY EXPERIMENT IN A SIMPLE TROPHIC CASCADE. In a New Zealand stream, enclosures with trout had fewer herbivorous invertebrates (see the left-hand panel) and more algae (see the right-hand panel) than ones without trout.

 inquiry

Why do streams with trout have more algae?

enclosures but not others. After 10 days, the numbers of invertebrates in the enclosures with trout were only two-thirds as great as the numbers in the no-fish enclosures (figure 57.14). In turn, the biomass of algae, which the invertebrates ate, was five times greater in the trout enclosures than the no-fish ones.

The logic of the trophic cascade just described leads to the expectation that if secondary carnivores are added to enclosures, they would also cause cascading effects. The secondary carnivores would be predicted to keep populations of primary carnivores in check, which would lead to a profusion of herbivores and a scarcity of primary producers.

In an experiment similar to the one just described, enclosures were created in free-flowing streams in northern California. In these streams, the principal primary carnivores were damselfly larvae (termed *nymphs*). Fish that preyed on the nymphs and on other primary carnivores were added to some enclosures but not others. In the enclosures with fish, the numbers of damselfly nymphs were reduced, leading to higher numbers of their prey, including herbivorous insects, which led in turn to a decreased biomass of algae (figure 57.15).

Trophic cascades in large-scale ecosystems are not as easy to verify by experiment as ones in stream enclosures, and the workings of such cascades are not thoroughly known. Nonetheless, certain cascades in large-scale ecosystems are recognized by most ecologists. One of the most dramatic involves sea otters, sea urchins, and kelp forests along the West Coast of North America (figure 57.16).

The otters eat the urchins, and the urchins eat young kelps, inhibiting the development of kelp forests. When the otters are abundant, the kelp forests are well developed because there are relatively few urchins in the system. But when the

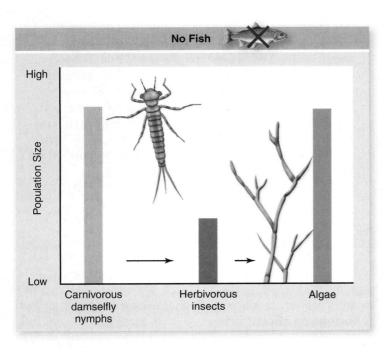

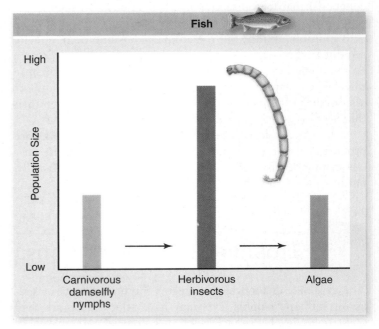

figure 57.15

TOP-DOWN EFFECTS DEMONSTRATED BY AN EXPERIMENT IN A FOUR-LEVEL TROPHIC CASCADE. Stream enclosures with large, carnivorous fish (*on right*) have fewer primary carnivores, such as damselfly nymphs, more herbivorous insects (exemplified here by the number of chironomids, a type of aquatic insect), and lower levels of algae.

inquiry

What might be the effect if snakes that prey on fish were added to the enclosures?

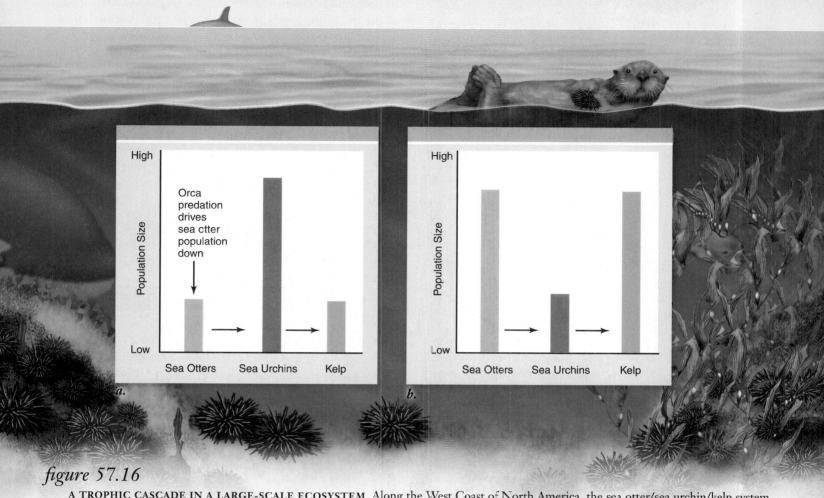

figure 57.16

A TROPHIC CASCADE IN A LARGE-SCALE ECOSYSTEM. Along the West Coast of North America, the sea otter/sea urchin/kelp system exists in two states: In the state shown in panel **a**, low populations of sea otters permit high populations of urchins, which suppress kelp populations; in the state shown in panel **b**, high populations of otters keep urchins in check, permitting profuse kelp growth. According to a recent hypothesis, a switch of orcas to preying on otters rather than other mammals is leading the ecosystem today to be mostly in the state represented on the left.

otters are sparse, the urchins are numerous and impair development of the kelp forests. Orcas (killer whales) also enter the picture because in recent years they have started to prey intensively on the otters, driving otter populations down.

Human removal of carnivores produces top-down effects

Human activities are believed to have had top-down effects in a number of ecosystems, usually by the removal of top-level carnivores. The great naturalist Aldo Leopold posited such effects long before the trophic cascade hypothesis had been scientifically articulated when he wrote in *Sand County Almanac:*

"I have lived to see state after state extirpate its wolves. I have watched the face of many a new wolfless mountain, and seen the south-facing slopes wrinkle with a maze of new deer trails. I have seen every edible bush and seedling browsed, first to anemic desuetude, and then to death. I have seen every edible tree defoliated to the height of a saddle horn."

Many similar examples exist in which the removal of predators has led to cascading effects on lower trophic levels. Large predators such as jaguars and mountain lions are absent on Barro Colorado Island, a hilltop turned into an island by the construction of the Panama Canal at the beginning of the last century. As a result, smaller predators whose populations are normally held in check—including monkeys, peccaries (a relative of the pig), coatimundis, and armadillos—have become extraordinarily abundant. These animals eat almost anything they find. Ground-nesting birds are particularly vulnerable, and many species have declined; at least 15 bird species have vanished from the island entirely.

Similarly, in the world's oceans, large predatory fish such as billfish and cod have been reduced by overfishing to an average of 10% of their previous numbers in virtually all parts of the world's oceans. In some regions, the prey of cod—such as certain shrimp and crabs—have become many times more abundant than they were before, and further cascading effects are evident at still lower trophic levels.

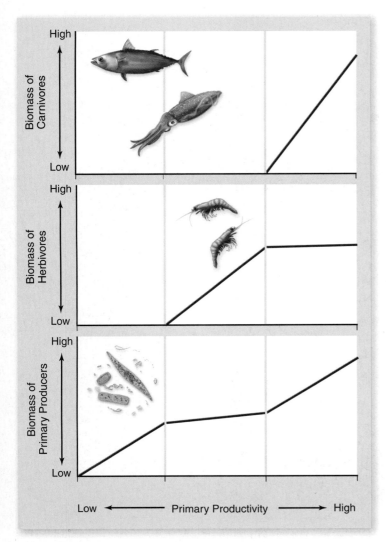

figure 57.17

A MODEL OF BOTTOM-UP EFFECTS. At low levels of primary productivity, herbivore populations cannot obtain enough food to be maintained; without herbivory, the standing crop biomass of the primary producers such as these diatoms increases as their productivity increases. Above some threshold, increases in primary productivity lead to increases in herbivore populations and herbivore biomass; the biomass of the primary producers then does not increase as primary productivity increases because the increasing productivity is cropped by the herbivores. Above another threshold, populations of primary carnivores can be sustained. As primary productivity increases above this threshold, the carnivores consume the increasing productivity of the herbivores, so the biomass of the herbivore populations remains relatively constant while the biomass of the carnivore populations increases. The biomass of the primary producers is no longer constrained by increases in the herbivore populations and thus also increases with increasing primary productivity. A key to understanding the model is to maintain a distinction between the concepts of productivity and standing crop biomass.

 inquiry

How is it possible for the biomass of the primary producers to stay relatively constant as the primary productivity increases?

Bottom-up effects occur when changes to primary producers affect higher trophic levels

In predicting bottom-up effects, ecologists must take account of the life histories of the organisms present. A model of bottom-up effects thought to apply to a number of types of ecosystems is diagrammed in figure 57.17.

According to the model, when primary productivity is low, producer populations cannot support significant herbivore populations. As primary productivity increases, herbivore

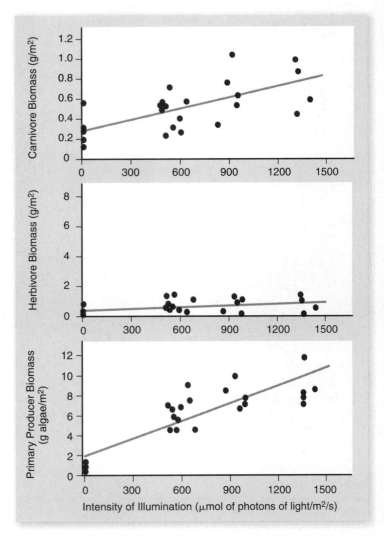

figure 57.18

AN EXPERIMENTAL STUDY OF BOTTOM-UP EFFECTS IN A RIVER ECOSYSTEM. This system, studied on the Eel River in northern California, exhibited the patterns modeled by the red graphs of figure 57.17. Increases in the intensity of illumination led to increases in primary productivity and in the biomass of the primary producers. The biomass of the carnivore populations also increased. However, herbivore biomass did not increase much with increasing primary productivity because increases in herbivore productivity were consumed by the carnivores.

 inquiry

Why is the amount of light an important determinant of carnivore biomass?

populations become a feature of the ecosystem. Increases in primary productivity are then entirely devoured by the herbivores, the populations of which increase in size while keeping the populations of primary producers from increasing.

As primary productivity becomes still higher, herbivore populations become large enough that primary carnivores can be supported. Further increases in primary productivity then does not lead to increases in herbivore populations, but rather to increases in carnivore populations.

Experimental evidence for the bottom-up effects predicted by the model was provided by a study conducted in enclosures on a river (figure 57.18). The enclosures excluded large fish (secondary carnivores). A roof was placed above each enclosure. Some roofs were clear, whereas others were tinted to various degrees, so that the enclosures differed in the amount of sunlight entering them.

The primary productivity was highest in the enclosures with clear roofs and lowest in the ones with darkly tinted roofs. As primary productivity increased in parallel with illumination, the biomass of the primary producers increased, as did the biomass of the carnivores. However, the biomass of the trophic level sandwiched in between, the herbivores, did not increase much, as predicted by the model in figure 57.17 (see red graph lines).

Because of the linked nature of food chains, populations of species at different trophic levels affect one another, and these effects can propagate both downward and upward in ecosystems. Top-down effects, or trophic cascades, are observed when changes in carnivore populations affect lower trophic levels. Bottom-up effects are observed when changes in primary productivity affect higher trophic levels.

57.4 Biodiversity and Ecosystem Stability

In the preceding chapter, we discussed *species richness*—the number of species present in a community. Ecologists have long debated the consequences of differences in species richness between one community and another. One theory is that species-rich communities are more stable—that is, more constant in composition and better able to resist disturbance. This hypothesis has been elegantly studied by David Tilman and colleagues at the University of Minnesota's Cedar Creek Natural History Area.

Species richness may increase stability: The Cedar Creek studies

Workers monitored 207 small rectangular plots of land (8–16 m²) for 11 years (figure 57.19a). In each plot, they counted the number of prairie plant species and measured the total amount of plant biomass (that is, the mass of all plants on the plot). Over the course of the study, plant species richness was related to community stability—plots with more species showed less year-to-year variation in biomass. Moreover, in two drought years, the decline in biomass was negatively related to species richness—in other words, plots with more species were less affected by drought.

In a related experiment, when seeds of other plant species were added to different plots, the ability of these species to become established was negatively related to species richness (figure 57.19b). More diverse communities, in other words, are more resistant to invasion by new species, which is another measure of community stability.

Species richness may also affect other ecosystem processes. Tilman and colleagues monitored 147 experimental plots that varied in number of species to estimate how much growth was occurring and how much nitrogen the growing plants were taking up from the soil. They found that the more species a plot

a.

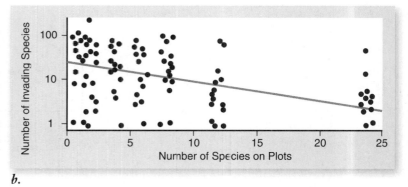

b.

figure 57.19

EFFECT OF SPECIES RICHNESS ON ECOSYSTEM STABILITY. *a.* One of the Cedar Creek experimental plots. *b.* Community stability can be assessed by looking at the effect of species richness on community invasibility. Each dot represents data from one experimental plot in the Cedar Creek experimental fields. Plots with more species are harder to invade by nonnative species.

inquiry

How could you devise an experiment on invasibility that didn't rely on species from surrounding areas?

had, the greater the nitrogen uptake and total amount of biomass produced. In his study, increased biodiversity clearly appeared to lead to greater productivity.

Laboratory studies on artificial ecosystems have provided similar results. In one elaborate study, ecosystems covering 1 m² were constructed in growth chambers that controlled temperature, light levels, air currents, and atmospheric gas concentrations. A variety of plants, insects, and other animals were introduced to construct ecosystems composed of 9, 15, or 31 species, with the lower diversity treatments containing a subset of the species in the higher diversity enclosures. As with Tilman's experiments, the amount of biomass produced was related to species richness, as was the amount of carbon dioxide consumed, another measure of the productivity of the ecosystem.

Tilman's conclusion that healthy ecosystems depend on diversity is not accepted by all ecologists, however. Critics question the validity and relevance of these biodiversity studies, arguing that the more species are added to a plot, the greater the probability that one species will be highly productive. To show that high productivity results from high species richness per se, rather than from the presence of particular highly productive species, experimental plots would have to exhibit "overyielding;" in other words, plot productivity would have to be greater than that of the single most productive species grown in isolation.

Although this point is still debated, recent work at Cedar Creek and elsewhere has provided evidence of overyielding, supporting the claim that species richness of communities enhances community productivity and stability.

Species richness is influenced by ecosystem characteristics

A number of factors are known or hypothesized to affect species richness in a community. We discussed some in chapter 56, such as loss of keystone species and moderate physical disturbance. Here we discuss three more: primary productivity, habitat heterogeneity, and climatic factors.

Primary productivity

Ecosystems differ substantially in primary productivity (see figure 57.11). Some evidence indicates that species richness is related to primary productivity, but the relationship between them is not linear. In a number of cases, for example, ecosystems with intermediate levels of productivity tend to have the greatest number of species (figure 57.20a).

Why this is so is debated. One possibility is that levels of productivity are linked with numbers of consumers. Applying this concept to plant species richness, the argument is that at low productivity, there are few herbivores, and superior competitors among the plants are able to eliminate most other plant species. In contrast, at high productivity so many herbivores are present that only the plant species most resistant to grazing survive, reducing species diversity. As a result, the greatest numbers of plant species coexist at intermediate levels of productivity and herbivory.

Habitat heterogeneity

Spatially heterogeneous abiotic environments are those that consist of many habitat types—such as soil types, for example. These heterogeneous environments can be expected to accommodate more species of plants than spatially homogeneous environments. What's more, the species richness of animals can be expected to reflect the species richness of plants present. An example of this latter effect is seen in figure 57.20b: The number of lizard species at various sites in the American Southwest mirrors the local structural diversity of the plants.

Climatic factors

The role of climatic factors is more difficult to predict. On the one hand, more species might be expected to coexist in a seasonal environment than in a constant one because a changing

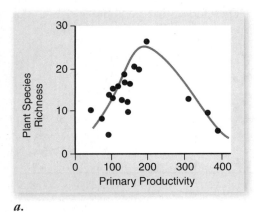

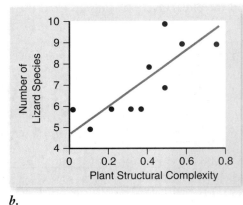

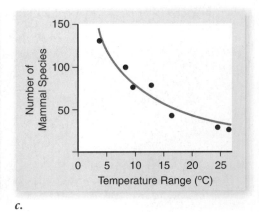

a.

b.

c.

figure 57.20

FACTORS THAT AFFECT SPECIES RICHNESS. *a. Productivity:* In plant communities of mountainous areas of South Africa, species richness of plants peaks at intermediate levels of productivity (biomass). *b. Spatial heterogeneity:* The species richness of desert lizards is positively correlated with the structural complexity of the plant cover in desert sites in the American Southwest. *c. Climate:* The species richness of mammals is inversely correlated with monthly mean temperature range along the West Coast of North America.

inquiry

? *a. Why is species richness greatest at intermediate levels of productivity? b. Why do more structurally complex areas have more species? c. Why do areas with less variation in temperature have more species?*

climate may favor different species at different times of the year. On the other hand, stable environments are able to support specialized species that would be unable to survive where conditions fluctuate. The number of mammal species at locations along the West Coast of North America is inversely correlated with the amount of local temperature variation—the wider the variation, the fewer mammalian species—supporting the latter line of argument (figure 57.20c).

Tropical regions have the highest diversity, although reasons are unclear

Since before Darwin, biologists have recognized that more different kinds of animals and plants inhabit the tropics than the temperate regions. For many types of organisms, there is a steady increase in species richness from the arctic to the tropics. Called a **species diversity cline,** this biogeographic gradient in numbers of species correlated with latitude has been reported for plants and animals, including birds (figure 57.21), mammals, and reptiles.

For the better part of a century, ecologists have puzzled over the species diversity cline from the arctic to the tropics. The difficulty has not been in forming a reasonable hypothesis of why more species exist in the tropics, but rather in sorting through these many reasonable hypotheses. Here, we consider five of the most commonly discussed suggestions.

Evolutionary age of tropical regions

Scientists have frequently proposed that the tropics have more species than temperate regions because the tropics have existed over long, uninterrupted periods of evolutionary time, whereas temperate regions have been subject to repeated glaciations. The greater age of tropical communities would have allowed complex population interactions to coevolve within them, fostering a greater variety of plants and animals.

Recent work suggests that the long-term stability of tropical communities has been greatly exaggerated, however. An examination of pollen within undisturbed soil cores reveals that during glaciations, the tropical forests contracted to a few small refuges surrounded by grassland. This suggests that the tropics have not had a continuous record of species richness over long periods of evolutionary time.

Increased productivity

A second often-advanced hypothesis is that the tropics contain more species because this part of the Earth receives more solar radiation than do temperate regions. The argument is that more solar energy, coupled to a year-round growing season, greatly increases the overall photosynthetic activity of plants in the tropics.

If we visualize the tropical forest's total resources as a pie, and its species niches as slices of the pie, we can see that a larger pie accommodates more slices. But as noted earlier, many field studies have indicated that species richness is highest at intermediate levels of productivity. Accordingly, increasing productivity would be expected to lead to lower, not higher, species richness.

Stability/constancy of conditions

Seasonal variation, though it does exist in the tropics, is generally substantially less than in temperate areas. This reduced seasonality might encourage specialization, with niches sub-

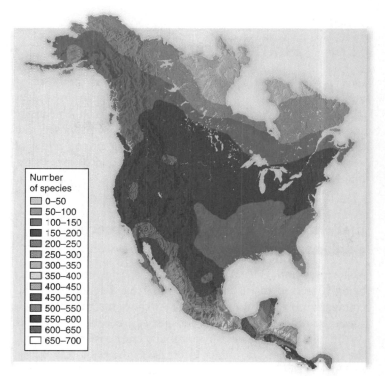

figure 57.21

A LATITUDINAL CLINE IN SPECIES RICHNESS. Among North and Central American birds, a marked increase in the number of species occurs moving toward the tropics. Fewer than 100 species are found at arctic latitudes, but more than 600 species live in southern Central America.

divided to partition resources and so avoid competition. The expected result would be a larger number of more specialized species in the tropics, which is what we see. Many field tests of this hypothesis have been carried out, and almost all support it, reporting larger numbers of narrower niches in tropical communities than in temperate areas.

Predation

Many reports indicate that predation may be more intense in the tropics. In theory, more intense predation could reduce the importance of competition, permitting greater niche overlap and thus promoting greater species richness.

Spatial heterogeneity

As noted earlier, spatial heterogeneity promotes species richness. Tropical forests, by virtue of their complexity, create a variety of microhabitats and so may foster larger numbers of species. Perhaps the long vertical column of vegetation through which light passes in a tropical forest produces a wide range of light frequencies and intensities, creating a greater variety of light environments and so promoting species diversity.

Experimental field studies support the hypothesis that species-rich communities are more stable and productive, though not all ecologists agree with this conclusion.

No one really knows why more species are present in the tropics, but a number of hypotheses have been suggested.

Island Biogeography

One of the most reliable patterns in ecology is the observation that larger islands contain more species than do smaller islands. In 1967, Robert MacArthur of Princeton University and Edward O. Wilson of Harvard University proposed that this **species-area relationship** was a result of the effect of geographic area and isolation on the likelihood of species extinction and colonization.

The equilibrium model proposes that extinction and colonization reach a balance point

MacArthur and Wilson reasoned that species are constantly being dispersed to islands, so islands have a tendency to accumulate more and more species. At the same time that new species are added, however, other species are lost by extinction. As the number of species on an initially empty island increases, the rate of colonization must decrease as the pool of potential colonizing species not already present on the island becomes depleted. At the same time, the rate of extinction should increase—the more species on an island, the greater the likelihood that any given species will perish.

As a result, at some point, the number of extinctions and colonizations should be equal, and the number of species should then remain constant. Every island of a given size, then, has a characteristic equilibrium number of species that tends to persist through time (the intersection point in figure 57.22a)—though the species composition will change as some species become extinct and new species colonize.

MacArthur and Wilson's equilibrium model proposes that island species richness is a dynamic equilibrium between colonization and extinction. Both island size and distance from the mainland would affect colonization and extinction. We would expect smaller islands to have higher rates of extinction because their population sizes would, on average, be smaller. Also, we would expect fewer colonizers to reach islands that lie farther from the mainland. Thus, small islands far from the mainland would have the fewest species; large islands near the mainland would have the most (figure 57.22b).

The predictions of this simple model bear out well in field data. Asian Pacific bird species (figure 57.22c) exhibit a positive correlation of species richness with island size, but a negative correlation of species richness with distance from the source of colonists.

The equilibrium model is still being tested

Wilson and Dan Simberloff, then a graduate student, performed initial studies in the mid-1960s on small mangrove islands in the Florida keys. These islands were censused, cleared of animal life by fumigation, and then allowed to recolonize, with censuses being performed at regular intervals. These and other such field studies have tended to support the equilibrium model.

Long-term experimental field studies, however, are suggesting that the situation is more complicated than MacArthur and Wilson envisioned. Their model predicts a high level of **species turnover** as some species perish and others arrive. But studies of island birds and spiders indicate that very little turnover occurs from year to year. Those species that do come and go, moreover, comprise a subset of species that never attain high populations. A substantial proportion of the species appear to maintain high populations and rarely go extinct.

These studies have been going on for a relatively short period of time. It is possible that over periods of centuries, the equilibrium model is a good description of what determines island species richness.

> Species richness on islands appears to be a dynamic equilibrium between colonization and extinction.

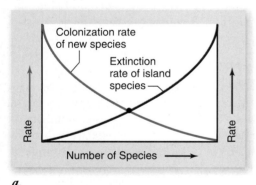

a.

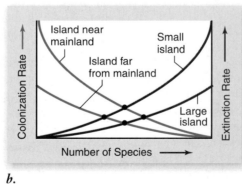

b.

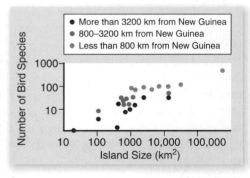

c.

figure 57.22

THE EQUILIBRIUM MODEL OF ISLAND BIOGEOGRAPHY. *a.* Island species richness reaches an equilibrium (*black dot*) when the colonization rate of new species equals the extinction rate of species on the island. ***b.*** The equilibrium shifts depending on the rate of colonization, the size of an island, and its distance to sources of colonists. Species richness is positively correlated with island size and inversely correlated with distance from the mainland. Smaller islands have higher extinction rates, shifting the equilibrium point to the left. Similarly, more distant islands have lower colonization rates, again shifting the equilibrium point leftward. ***c.*** The effect of distance from a larger island, which can be the source of colonizing species, is readily apparent. More distant islands have fewer Asian Pacific bird species compared with nearer islands of the same size.

57.1 Biogeochemical Cycles

Ecosystems are intrinsically dynamic in the way they process matter and energy.

- Essentially, the Earth has a fixed number of each type of atoms of biological importance, and these atoms are recycled in biogeochemical cycles.

- Biogeochemical cycles include processes that occur on many spatial and multiple timescales.

- Biogeochemical cycles cross the boundaries of ecosystems to some extent.

- The carbon cycle usually involves carbon dioxide, which is fixed through photosynthesis and released by respiration. In aquatic environments, the carbon is also present as bicarbonate ions, and in anoxic conditions carbon is available as methane. Human burning of fossil fuels has created an imbalance in the carbon cycle (figure 57.1).

- The water cycle is distinctive because it involves a compound that can be synthesized during cellular respiration and broken down during photosynthesis. Water enters the atmosphere via evaporation and transpiration and returns to the Earth's surface as precipitation. Much of our water, including the groundwater found in aquifers, is polluted, and human activities can not only change a supply of water but also radically alter ecosystems when they do so (figure 57.2).

- Aquifers are being depleted faster than they are recharged.

- Nitrogen is usually the chemical element in shortest supply even though the atmosphere is 78% nitrogen. Nitrogen in its elemental form cannot be used by organisms until microbes progressively convert it to ammonia by nitrogen fixation or to nitrate by nitrification. Nitrate is converted back to nitrogen by denitrification. Human activities have doubled the rate of transfer of atmospheric nitrogen by the manufacture of fertilizers (figure 57.4).

- Phosphorus, another limiting nutrient, does not cycle through the atmosphere. Phosphorus is released by weathering of rocks and it flows into the oceans where it is deposited in deep-sea sediments (figure 57.5).

- The cycle of a limiting nutrient is important because it determines the rate at which the nutrient is made available for use.

- Deforestation increases the rate at which nutrients are lost from an ecosystem (figure 57.7b).

57.2 The Flow of Energy in Ecosystems

The Earth is an open system for energy. The radiant energy from the Sun passes one way through an ecosystem before it is converted to heat and radiated back into space.

- Energy is neither created nor destroyed, but it frequently changes form. When energy is converted to heat as it passes through living organisms, it cannot be converted to other forms of usable energy.

- The Second Law of Thermodynamics states that whenever organisms use chemical-bond or light energy, some of it is inevitably converted to heat.

- Organic compounds are synthesized by autotrophs— photoautotrophs and chemoautotrophs—and utilized by heterotrophs.

- As energy passes from organism to organism, each feeding level is called a trophic level, and the sequence through progressive trophic levels is called a food chain (figure 57.8).

- The base trophic level includes the primary producers, which are autotrophs, that are eaten by herbivores which in turn are eaten by carnivores. Detritivores feed on the remains of dead organisms.

- Productivity of a trophic level is the rate at which organisms in that trophic level collectively synthesize new organic matter by growth or reproduction.

- Primary productivity by autotrophs can be expressed as gross primary production (GPP) or net primary production (NPP). NPP refers to GPP less respiration.

- The standing crop biomass is the amount of organic matter present at a particular time.

- Only about 1% of the solar energy that impinges on the Earth is captured for photosynthetic processes.

- As energy is processed through each trophic level, very little (approximately 10% of the chemical-bond energy) remains from the preceding trophic level (figure 57.10).

- NPP varies considerably among ecosystems or biomes (figure 57.11).

- The exponential decline of chemical-bond energy limits the lengths of food chains and the numbers of top carnivores that can be supported.

- Ecological pyramids based on energy flow, biomass, or numbers are usually upright. Inverted pyramids of biomass or numbers are possible if at least one trophic level has a greater biomass or more organisms than the trophic level below it (figure 57.13).

57.3 Trophic-Level Interactions

The existence of food chains creates the possibility that changes in species at any one trophic level may have multiple effects on other trophic levels.

- A trophic cascade occurs when the effects exerted at an upper trophic level affect a lower level.

- A top-down trophic effect occurs when changes in the top trophic level affect primary producers (figure 57.15).

- A bottom-up effect occurs when changes in primary productivity affects higher trophic levels (figure 57.17).

57.4 Biodiversity and Ecosystem Stability

Species richness may affect community stability, that is, a community's composition and ability to resist disturbance.

- Higher species richness has resulted in less year-to-year variation in biomass and greater resistance to drought.

- Species richness is influenced by primary production, habitat heterogeneity, and climatic factors (figure 57.20).

- Tropical regions have the highest diversity although the reasons are unclear. The higher diversity may reflect long evolutionary time, higher productivity, low seasonal variation, greater predation reducing competition, or spatial heterogeneity (figure 57.21).

57.5 Island Biogeography

The number of species in an environment appears to be a dynamic equilibrium between colonization and extinction (figure 57.22).

- Smaller islands have fewer species than large islands because of higher rates of extinction.

- Near islands have more species than far islands because of higher rates of colonization.

SELF TEST

1. Which of the statements about groundwater is *not* accurate?
   a. In the United States, groundwater provides 50% of the population with drinking water.
   b. Groundwaters are being depleted faster than they can be recharged.
   c. Groundwaters are becoming increasingly polluted.
   d. Removal of pollutants from groundwaters is easily achieved.

2. Photosynthetic organisms—
   a. fix carbon dioxide.
   b. release carbon dioxide.
   c. fix oxygen.
   d. (a) and (b).
   e. (a) and (c).

3. Nitrogen is often a limiting nutrient in many ecosystems because—
   a. there is much less nitrogen in the atmosphere than carbon.
   b. elemental nitrogen is very rapidly used by most organisms.
   c. nitrogen availability is being reduced by pollution due to fertilizer use.
   d. most organisms cannot use nitrogen in its elemental form.

4. Some bacteria have the ability to "fix" nitrogen. This means—
   a. they convert ammonia into nitrites and nitrates.
   b. they convert atmospheric nitrogen gas into biologically useful forms of nitrogen.
   c. they break down nitrogen-rich compounds and release ammonium ions.
   d. they convert nitrate into nitrogen gas.

5. Which of the following statements about the phosphorus cycle is correct?
   a. Phosphorus is fixed by plants and algae.
   b. Most phosphorus released from rocks is carried to the oceans by rivers.
   c. Animals cannot get their phosphorus from eating plants and algae.
   d. Fertilizer use has not affected the global phosphorus budget.

6. Based on results from studies at Hubbard Brook Experimental Forest, what would be the predicted effect of clearing trees from a watershed?
   a. Increase loss of water and nutrients from a watershed
   b. Decrease loss of water and nutrients from a watershed
   c. Increase availability of phosphorus
   d. Increase the availability of nitrate

7. As a general rule, how much energy is lost in the transmission of energy from one trophic level to the one immediately above it?
   a. 1%
   b. 10%
   c. 90%
   d. 50%

8. Inverted ecological pyramids of real systems usually involve—
   a. energy flow.
   b. biomass.
   c. energy flow and biomass.
   d. None of the above.

9. According to the trophic cascade hypothesis, the removal of carnivores from an ecosystem may result in—
   a. a decline in the number of herbivores and a decline in the amount of vegetation.
   b. a decline in the number of herbivores and an increase in the amount of vegetation.
   c. an increase in the number of herbivores and an increase in the amount of vegetation.
   d. an increase in the number of herbivores and a decrease in the amount of vegetation.

10. Bottom-up effects on trophic structure result from—
    a. a limitation of energy flowing to the next higher trophic level.
    b. actions of top predators on lower trophic levels.
    c. climatic disruptions on top consumers.
    d. stability of detritivores in ecosystems.

11. At Cedar Creek Natural History Area, experimental plots showed reduced numbers of invaders as species diversity of plots increased—
    a. suggesting that low species diversity increases stability of ecosystems.
    b. suggesting that ecosystem stability is a function of primary productivity only.
    c. consistent with the theory that intermediate disturbance results in the highest stability.
    d. None of the above.

12. Species diversity—
    a. increases with latitude as you move away from the equator to the arctic.
    b. decreases with latitude as you move away from the equator to the arctic.
    c. stays the same as you move away from the equator to the arctic.
    d. increases with latitude as you move north of the equator and decreases with latitude as you move south of the equator.

13. The equilibrium model of island biogeography suggests all of the following *except*—
    a. larger islands have more species than smaller islands.
    b. the species richness of an island is determined by colonization and extinction.
    c. smaller islands have lower rates of extinction.
    d. islands closer to the mainland will have higher colonization rates.

CHALLENGE QUESTIONS

1. Given that ectotherms do not utilize a large fraction of ingested food energy to maintain a high and constant body temperature (generate heat), how would you expect the food chains of systems dominated by ectothermic herbivores and carnivores to compare with systems dominated by endothermic herbivores and carnivores?

2. Given that, in general, energy input is greatest at the bottom trophic level (primary producers) and decreases with increasing transfers across trophic levels, how is it possible for many lakes to show much greater standing biomass in herbivorous zooplankton than in the phytoplankton they consume?

3. Ecologists often worry about the potential effects of the loss of species (e.g., due to pollution, habitat degradation, or other human-induced factors) on an ecosystem for reasons other than just the direct loss of the species. Using figure 57.17 explain why.

4. Explain several detailed ways in which increasing plant structural complexity could lead to greater species richness of lizards (figure 57.20*b*). Could any of these ideas be tested? How?

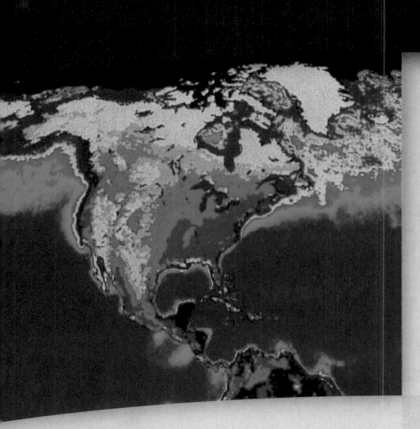

chapter 58

The Biosphere

introduction

THE BIOSPHERE INCLUDES ALL LIVING communities on
Earth, from the profusion of life in the tropical rain forests to
the planktonic communities in the world's oceans. In a very
general sense, the distribution of life on Earth reflects variations
in the world's abiotic environments, such as the variations
in temperature and availability of water from one terrestrial
environment to another. The figure on this page is a satellite
image of the Americas, based on data collected over 8 years.
The colors are keyed to the relative abundance of chlorophyll, an
indicator of the richness of biological communities. Green and
dark green areas on land are areas with high primary productivity
(such as thriving forests), whereas yellow areas include the deserts
of the Americas and the tundra of the far north, which have
lower productivity.

concept outline

58.1 Ecosystem Effects of Sun, Wind, and Water

The great global patterns of life on Earth are heavily influenced by (1) the amount of solar radiation that reaches different parts of the Earth and seasonal variations in that radiation; and (2) the patterns of global atmospheric circulation and the resulting patterns of oceanic circulation. Local characteristics, such as soil types and the altitude of the land, interact with the global patterns in Sun, winds, and water currents to determine the conditions under which life exists and thus the distributions of ecosystems.

Solar energy and the Earth's rotation affect atmospheric circulation

The Earth receives energy from the Sun at a high rate in the form of electromagnetic radiation at visible and near-visible wavelengths. Each square meter of the upper atmosphere receives about 1400 joules per second (J/sec), equivalent to the output of fourteen 100-watt (W) lightbulbs.

As the solar radiant energy passes through the atmosphere, it is modified in its intensity and wavelength composition. About half of the energy is absorbed within the atmosphere, and half reaches the Earth's surface. The gases in the atmosphere absorb some wavelengths strongly while allowing other wavelengths to pass freely through. As a result, the wavelength composition of the solar energy that reaches the Earth's surface is different from that emitted by the Sun. For example, the band of ultraviolet wavelengths known as UV-B is strongly absorbed by the ozone (O_3) in the atmosphere, and therefore this wavelength is greatly reduced by the time the solar energy reaches the Earth's surface.

How solar radiation affects climate

Some parts of the Earth's surface receive more energy from the Sun than others. These differences have a great effect on climate.

A major reason for differences in solar radiation from place to place is the fact that Earth is a sphere (figure 58.1*a*). The tropics are particularly warm because the Sun's rays arrive almost perpendicular to the surface of the Earth in regions near the equator. Closer to the poles, the angle at which the Sun's rays strike, called the *angle of incidence*, spreads the solar energy out over more of the Earth's surface, providing less energy per unit of surface area. As figure 58.2 shows, the highest annual mean temperatures occur near the equator (0° latitude).

The Earth's annual orbit around the Sun and its daily rotation on its own axis are also important in determining patterns of solar radiation and their effects on climate (figure 58.1*b*). The axis of rotation of the Earth is not perpendicular to the plane in which the earth orbits the Sun. Because the axis is tilted by approximately 23.5°, a progression of seasons occurs on all parts of the Earth, especially at latitudes far from the equator. The northern hemisphere, for example, tilts toward the Sun during some months but away during others, giving rise to summer and winter.

Global circulation patterns in the atmosphere

Hot air tends to rise relative to cooler air because the motion of molecules in the air increases as temperature increases, making it less dense. Accordingly, the intense solar heating of the Earth's surface at equatorial latitudes causes air to rise from the surface

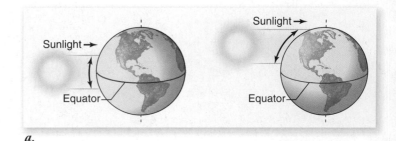

a.

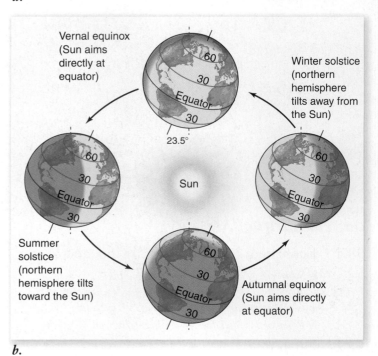

b.

figure 58.1

RELATIONSHIPS BETWEEN THE EARTH AND THE SUN ARE CRITICAL IN DETERMINING THE NATURE AND DISTRIBUTION OF LIFE ON EARTH. *a.* A beam of solar energy striking the Earth in the middle latitudes of the northern hemisphere (or the southern) spreads over a wider area of the Earth's surface than an equivalent beam striking the Earth at the equator. *b.* The fact that the Earth orbits the Sun each year has a profound effect on climate. In the northern and southern hemispheres, temperature changes in an annual cycle because the Earth's axis is not perpendicular to its orbital plane and, consequently, each hemisphere tilts toward the Sun in some months but away from the Sun in others.

to high in the atmosphere at these latitudes. This rising air is typically rich with water vapor; one reason is that the moisture-holding capacity of air increases when it is heated, and a second reason is that the intense solar radiation at the equator provides the heat needed for great quantities of water to evaporate. After the warm, moist air rises from the surface (figure 58.3), it moves away from the equator at high altitudes (above 10 km), to the north in the northern hemisphere and to the south in the southern hemisphere. To take the place of the rising air, cooler air

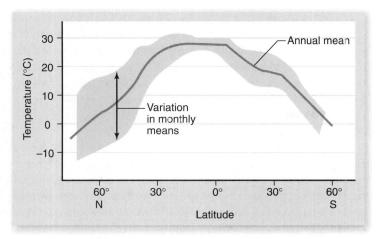

figure 58.2

ANNUAL MEAN TEMPERATURE VARIES WITH LATITUDE.
The red line represents the annual mean temperature at various
latitudes, ranging from near the North Pole at the left to near
Antarctica at the right; the equator is at 0° latitude. At each
latitude, the upper edge of the blue zone is the highest mean
monthly temperature observed in all the months of the year,
and the lower edge is the lowest mean monthly temperature.

flows toward the equator along the surface from both the north
and the south. These air movements give rise to one of the major
features of the global atmospheric circulation: air flows toward
the equator in both hemispheres at the surface, rises at the equator, and flows away from the equator at high altitudes. The exact
patterns of flow are affected by the spinning of the Earth on its
axis; we discuss this effect shortly.

For complex reasons, the air circulating up from the
equator and away at high altitudes in both hemispheres tends
to circulate back down to the surface of the Earth at about 30°
of latitude, both north and south (see figure 58.3). During the
course of this movement, the moisture content of the air changes
radically because of the changes in temperature the air undergoes. Cooling dramatically decreases air's ability to hold water
vapor. Consequently, much of the water vapor in the air rising
from the equator condenses to form clouds and rain as the air
moves upward. This rain falls in the latitudes near the equator,
latitudes that experience the greatest precipitation on Earth.

By the time the air starts to descend back to the Earth's
surface at latitudes near 30°, it is cold and thus has lost most
of its water vapor. Although the air rewarms as it descends, it
does not gain much water vapor on the way down. Many of the
greatest deserts occur at latitudes near 30° because of the steady
descent of dry air to the surface at those latitudes. The Sahara
Desert is the most dramatic example.

The air that descends at latitudes near 30° flows only
partly toward the equator after reaching the surface of the
Earth. Some of it flows toward the poles, helping to give rise
in each hemisphere to winds that blow over the Earth's surface
from 30° toward 60° latitude. At latitudes near 60° air tends to
rise from the surface toward high altitudes.

inquiry

Why is it hotter at latitudes near 0°?

The Coriolis effect

If Earth did not rotate on its axis, global air movements would
follow the simple patterns already described. Air currents—the
winds—move across a rotating surface, however. Because the
solid Earth rotates under the winds, the winds move in curved
paths across the surface, rather than straight paths. The curvature of the paths of the winds due to Earth's rotation is termed
the **Coriolis effect.**

If you were standing on the North Pole, the Earth would
appear to be rotating counterclockwise on its axis, but if you
were at the South Pole, the Earth would appear to be rotating
clockwise. This property of a rotating sphere, that its direction
of rotation is opposite when viewed from its two poles, explains
why the direction of the Coriolis effect is opposite in the two

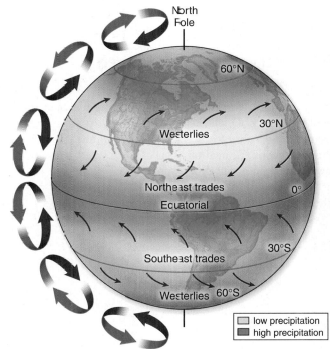

figure 58.3

GLOBAL PATTERNS OF ATMOSPHERIC CIRCULATION. The
diagram shows the patterns of air circulation that prevail on
average over weeks and months of time (on any one day the
patterns might be dramatically different from these average
patterns). Rising air that is cooled creates bands of relatively
high precipitation near the equator and at latitudes near 60°N
and 60°S. Air that has lost most of its moisture at high altitudes
tends to descend to the surface of the Earth at latitudes near
30°N and 30°S, creating bands of relatively low precipitation.
The red arrows show the winds blowing at the surface of the
Earth; the blue arrows show the direction the winds blow at
high altitude. The winds travel in curved paths relative to
the Earth's surface because the Earth is rotating on its axis
under them (the Coriolis effect). A terminological problem to
recognize is that the formal names given to winds refer to the
directions from which they come, rather than the directions
toward which they go; thus, the winds between 30° and 60°
are called Westerlies because they come out of the west.
Unfortunately, oceanographers use the opposite convention,
naming water currents for the directions in which they go.

hemispheres. In the northern hemisphere, winds always curve to the right of their direction of motion; in the southern hemisphere, they always curve to the left.

Consider the surface winds that blow from about 30° latitude toward the equator in each hemisphere. In the northern hemisphere, they curve to the right, and therefore tend to blow westward as well as toward the equator. In the southern hemisphere, they curve to the left, meaning they *also* blow westward as well as toward the equator. The result (see figure 58.3) is that winds on both sides of the equator—called the Trade Winds—blow out of the east and toward the west.

The winds between 30° and 60° follow the same principles. Those in the northern hemisphere curve to the right and thus tend to blow eastward as well as toward the pole. Those in the southern hemisphere curve to the left, also tending to blow eastward as well as toward the pole. In both hemispheres, therefore, winds between 30° and 60° blow out of the west and toward the east; these winds are called Westerlies.

Global ocean currents are largely driven by winds

The major ocean currents are driven by the winds at the surface of the Earth, which means that indirectly the currents are driven by solar energy. The radiant input of heat from the Sun sets the atmosphere in motion as already described, and then the winds set the ocean in motion.

In the north Atlantic Ocean (figure 58.4), the global wind pattern is that surface winds tend to blow out of the east and toward the west near the equator, but out of the west and toward the east at midlatitudes (between 30° and 60°). The consequence is that the surface waters of the north Atlantic Ocean tend to move in a giant closed curve—called a **gyre**—flowing from North America toward Europe at midlatitudes, then returning from Europe and Africa to North America at latitudes near the equator.

Water currents are affected by the Coriolis effect. Thus, the Coriolis effect contributes to this clockwise closed-curve motion. Water flowing across the Atlantic toward Europe at midlatitudes tends to curve to the right and enters the flow from east to west near the equator. This latter flow also tends to curve to its right and enters the flow from west to east at midlatitudes. In the south Atlantic Ocean, the same processes occur in a sort of mirror image, and similar clockwise and counterclockwise gyres occur in the north and south Pacific Ocean as well.

Regional and local differences affect terrestrial ecosystems

The environmental conditions at a particular place are affected by regional and local effects of solar radiation, air circulation, and water circulation, not just the global patterns of these processes. In this section we look at just a few examples of regional and local effects, focusing on terrestrial systems. These include rain shadows, monsoon winds, elevation, and presence of microclimate factors.

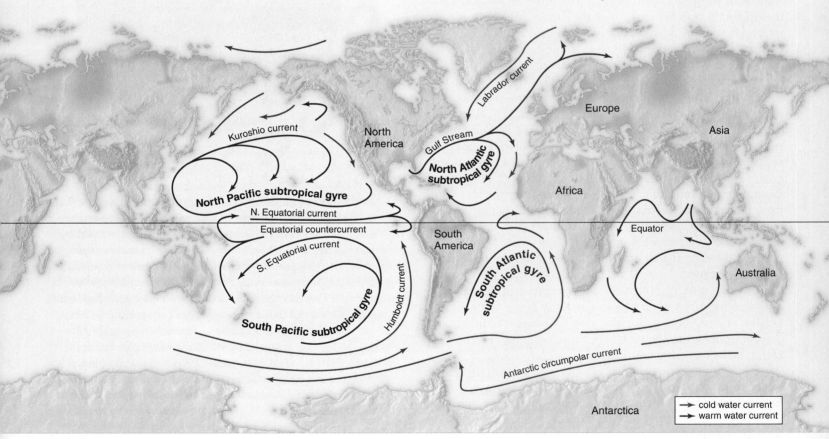

figure 58.4

OCEAN CIRCULATION. In the centers of several of the great ocean basins, surface water moves in great closed-curve patterns called gyres. These water movements affect biological productivity in the oceans and sometimes profoundly affect the climate on adjacent landmasses, as when the Gulf Stream brings warm water to the region of the British Isles.

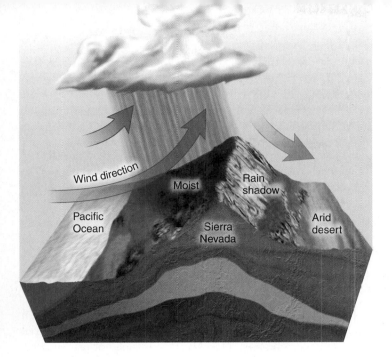

figure 58.5

THE RAIN SHADOW EFFECT EXEMPLIFIED IN CALIFORNIA.
Moisture-laden winds from the Pacific Ocean rise and are
cooled when they encounter the Sierra Nevada Mountains. As
the moisture-holding capacity of the air decreases at colder,
higher altitudes, precipitation occurs, making the seaward
facing slopes of the mountains moist; tall forests occur on
those slopes, including forests that contain the famous giant
sequoias (*Sequoiadendron giganteum*). As the air descends on
the eastern side of the mountain range, its moisture-holding
capacity increases again, and the air picks up moisture from its
surroundings. As a result, the eastern slopes of the mountains
are arid, and rain shadow deserts sometimes occur.

Rain shadows

Deserts on land sometimes occur because mountain ranges intercept moisture-laden winds from the sea. When air flowing
landward from the oceans encounters a mountain range (figure
58.5), the air rises, and its moisture-holding capacity decreases
because it becomes cooler at higher altitude, causing precipitation to fall on the mountain slopes facing the sea.

As the air—stripped of much of its moisture—then descends
on the other side of the mountain range, it remains dry even as it is
warmed, and as it is warmed its moisture-holding capacity increases,
meaning it can readily take up moisture from soils and plants.

One consequence is that the two slopes of a mountain
range often differ dramatically in how moist they are; in California, for example, the eastern slopes of the Sierra Nevada
Mountains—facing away from the Pacific Ocean—are far drier
than the western slopes. Another consequence is that a desert
may develop on the dry side, the Mojave Desert being an example. The mountains are said to produce a **rain shadow.**

Monsoons

The continent of Asia is so huge that heating and cooling of its
surface during the passage of the seasons causes massive regional
shifts in wind patterns. During summer, the surface of the Asian

landmass heats up more than the surrounding oceans, but during winter the landmass cools more than the oceans. The consequence is that winds tend to blow off the water into the interior of
the Asian continent in summer, particularly in the region of the
Indian Ocean and western tropical Pacific Ocean. These winds
reverse to flow off the continent out over the oceans in winter.
These seasonally shifting winds are called the **monsoons.** They
affect rainfall patterns, and their duration and strength can spell
the difference between food sufficiency and starvation for hundreds of millions of people in the region each year.

Elevation

Another significant regional pattern is that in mountainous regions, temperature and other conditions change with elevation.
At any given latitude, air temperature falls about 6°C for every
1000-m increase in elevation. The ecological consequences of
the change of temperature with elevation are similar to those of
the change of temperature with latitude (figure 58.6).

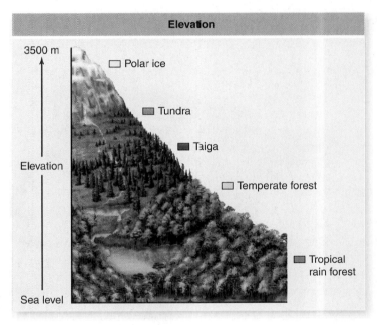

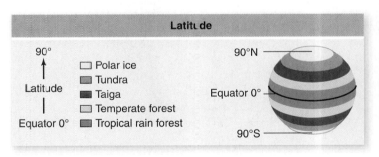

figure 58.6

**ELEVATION AFFECTS THE DISTRIBUTION OF BIOMES IN
MUCH THE SAME MANNER AS LATITUDE DOES.** Biomes
that normally occur far north of the equator at sea level also
occur in the tropics at high mountain elevations. Thus, on a tall
mountain in the tropics, one might see a sequence of biomes
like the one illustrated above. In North America, a 1000-m
increase in elevation results in a temperature drop equal to that
of an 880-km increase in latitude.

Microclimates

Conditions also vary in significant ways on very small spatial scales. For example, in a forest, a bird sitting in an open patch may experience intense solar radiation, a high air temperature, and a low humidity, even while a mouse hiding under a log 10 feet away may experience shade, a cool temperature, and air saturated with water vapor. Such highly localized sets of climatic conditions are called **microclimates.** Gardeners spread straw over newly seeded lawns to create a cool, moist microclimate suitable for seedlings.

Solar energy drives most of the Earth's dynamism. Solar heating of some global regions relative to others sets up global patterns of atmospheric circulation, which in turn set up global patterns of water circulation in the oceans. These patterns—plus seasonal changes in the solar radiation received on Earth—strongly affect the conditions (such as temperature and precipitation) that exist for living organisms in various parts of the world. Regional phenomena such as rain shadows and altitude also influence the conditions that exist for living organisms.

58.2 Earth's Biomes

Biomes are major types of ecosystems on land. Each biome has a characteristic appearance and is distributed over wide areas of land defined largely by sets of regional climatic conditions. Biomes are named according to their vegetational structures, but they also include the animals that are present.

As you might imagine from the broad definition given for biomes, there are a number of ways to classify terrestrial ecosystems into biomes. Here we recognize eight principal biomes: (1) tropical rain forest, (2) savanna, (3) desert, (4) temperate grassland, (5) temperate deciduous forest, (6) temperate evergreen forest, (7) taiga, (8) tundra.

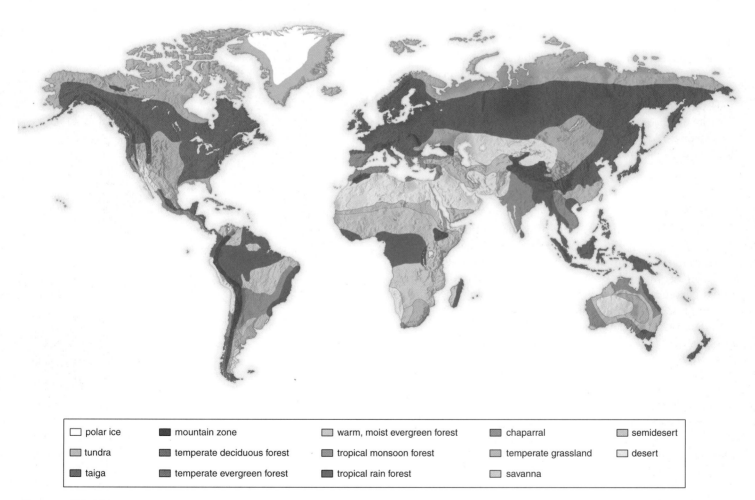

☐ polar ice	■ mountain zone	☐ warm, moist evergreen forest	■ chaparral	■ semidesert
■ tundra	■ temperate deciduous forest	☐ tropical monsoon forest	■ temperate grassland	☐ desert
■ taiga	■ temperate evergreen forest	■ tropical rain forest	☐ savanna	

figure 58.7

THE DISTRIBUTIONS OF BIOMES. Each biome is similar in vegetational structure and appearance wherever it occurs.

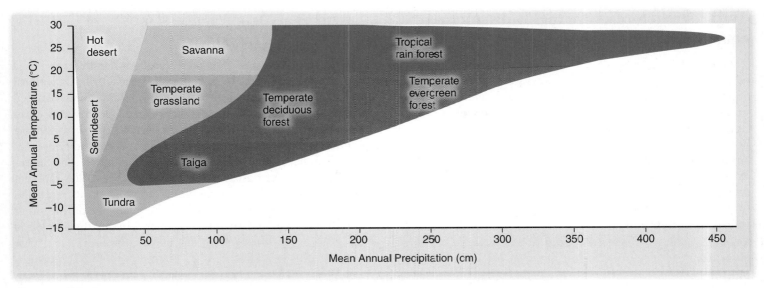

figure 58.8

PREDICTORS OF BIOME DISTRIBUTION. Temperature and precipitation are quite useful predictors of biome distribution, although other factors sometimes also play critical roles.

Six additional biomes recognized by some ecologists are: polar ice, mountain zone, chaparral, warm moist evergreen forest, tropical monsoon forest, and semidesert. Other ecologists lump these six with the eight major ones. Figure 58.7 shows the distributions of all 14 biomes.

Biomes are defined by their characteristic vegetational structures and associated climatic conditions, not by the particular species of plants present. Two regions assigned to a single biome may differ therefore in the species that dominate the landscape. Tropical rain forests around the world, for example, are all composed of tall, lushly vegetated trees—but the particular tree species that dominate a South American tropical rain forest are different from those in an Indonesian one. The similarity between such forests is a product of convergent evolution (see chapter 21).

Temperature and moisture often determine biomes

In determining which biomes are found where, two key environmental factors are temperature and moisture. As seen in figure 58.8, if you know the mean annual temperature and mean annual precipitation in a terrestrial region, you often can predict the biome that dominates. Temperature and moisture affect ecosystems in a number of ways. One reason they are so influential is that primary productivity is strongly correlated with them, as described in the preceding chapter (figure 58.9).

Different places that are similar in mean annual temperature and precipitation sometimes support different biomes, indicating that temperature and moisture are not the only factors that can be important. Soil structure and mineral composition (see chapter 39) are among the other factors that can be influential. The biome that is present may also depend on whether the conditions of temperature and precipitation are strongly seasonal or relatively constant.

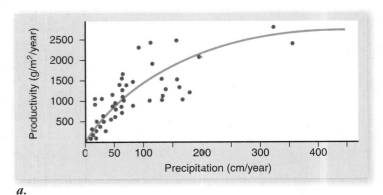

a.

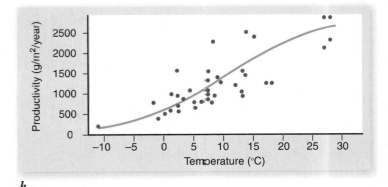

b.

figure 58.9

THE CORRELATIONS OF PRIMARY PRODUCTIVITY WITH PRECIPITATION AND TEMPERATURE. The net primary productivity of ecosystems at 52 locations around the globe correlates significantly with (*a*) mean annual precipitation and (*b*) mean annual temperature.

inquiry

? *Why might you expect primary productivity to increase with increasing precipitation and temperature?*

Tropical rain forests are highly productive equatorial systems

Tropical rain forests, which typically require 140–450 cm of rain per year, are the richest ecosystems on land (figure 58.10). They are very productive because they enjoy the advantages of both high temperature and high precipitation (see figure 58.9). They also exhibit very high biodiversity, being home to at least half of all the species of terrestrial plants and animals—over 2 million species! In a single square mile of Brazilian rain forest, there can be 1200 species of butterflies—twice the number found in all of North America. Tropical rain forests recycle nutrients rapidly, so their soils often lack great reservoirs of nutrients.

Savannas are tropical grasslands with seasonal rainfall

The **savannas** are tropical or subtropical grasslands, often dotted with widely spaced trees or shrubs (see figure 58.10). On a global scale, savannas often occur as transition ecosystems between tropical rain forests and deserts; they are characteristic of warm places where annual rainfall (50–125 cm) is too little to support rain forest, but not so little as to produce desert conditions.

Rainfall is often highly seasonal in savannas. The Serengeti ecosystem in East Africa is probably the world's most famous example of the savanna biome. In most of the Serengeti, no rain falls for many months of the year, but during other months rain is abundant. The huge herds of grazing animals in

figure 58.10

MAJOR BIOMES OF THE WORLD.

Biome	Climate	Example Location	Characteristic Flora	Characteristic Fauna
▪ Tropical Rain Forest	High temperatures year round	 Brazilian rain forest	 Plant Species	 Animal Species
▫ Savanna	Warm temperatures year round	 Serengeti	 Plant Species	 Animal Species
▫ Desert	Warm and cool temperatures	 Mojave	 Plant Species	 Animal Species
▫ Temperate Grasslands	Warm summers cool winters	 South Dakota prairie	 Plant Species	 Animal Species

the ecosystem respond to the seasonality of the rain; a number of species migrate away from permanently flowing rivers only during the months when rain falls.

Deserts are regions with little rainfall

Deserts are dry places where rain is both sparse (annual rainfall often less than 25–40 cm) and unpredictable (see figure 58.10). The unpredictability means that plants and animals cannot depend on experiencing rain even once each year. As mentioned earlier, many of the largest deserts occur at latitudes near 30°N and 30°S because of global air circulation patterns (see figure 58.3). Other deserts result from rain shadows (see figure 58.5).

Vegetation is sparse in deserts, and survival of both plants and animals depends on water conservation. Many desert organisms enter inactive stages during rainless periods. To avoid extreme temperatures, small desert vertebrates often live in deep, cool, and sometimes even somewhat moist burrows. Some emerge only at night. Among large desert animals, camels drink large quantities of water when it is available and then conserve it so well that they can survive for weeks without drinking. Oryxes (large, desert-dwelling antelopes) survive opportunistically on moisture in leaves or roots that they dig up, as well as drinking water when possible.

Temperate grasslands have rich soils

Halfway between the equator and the poles are temperate regions where rich **temperate grasslands** grow (see figure 58.10). These grasslands, also called **prairies,** once covered much of the interior of North America, and they were widespread in Eurasia and South America as well.

The roots of perennial grasses characteristically penetrate far into the soil, and grassland soils tend to be deep and

Biome	Climate	Example Location	Characteristic Flora	Characteristic Fauna
Temperate Deciduous Forest	Warm summers cool winters	Acadia National Park	Plant Species	Animal Species
Temperate Evergreen Forest	Temperate climates	Mount Hood	Plant Species	Animal Species
Taiga	Cold temperatures	Northwest Territory, Canada	Plant Species	Animal Species
Tundra	Cold temperatures	Alaska	Plant Species	Animal Species

fertile. Temperate grasslands are often highly productive when converted to agricultural use, and vast areas have been transformed in this way. In North America prior to this change in land use, huge herds of bison and pronghorn antelope inhabited the temperate grasslands. Natural temperate grasslands are one of the biomes adapted to periodic fire and therefore need fires to prosper.

Temperate deciduous forests are adapted to seasonal change

Mild but seasonal climates (warm summers and cold winters), plus plentiful rains, promote the growth of **temperate deciduous forests** in the eastern United States, eastern Canada, and Eurasia (see figure 58.10). A deciduous tree is one that drops its leaves in the winter. Deer, bears, beavers, and raccoons are familiar animals of these forests.

Temperate evergreen forests are coastal

Temperate evergreen forests occur along coastlines with temperate climates, such as in the northwest of the United States (see figure 58.10). The dominant vegetation includes trees, such as spruces, pines, and redwoods, that do not drop their leaves (thus, they are *ever green*).

Taiga is the northern forest where winters are harsh

Taiga and tundra (described next) differ from other biomes in that taiga and tundra both stretch in great unbroken circles around the entire globe (see figure 58.7). The **taiga** consists of a great band of northern forest dominated by coniferous trees (spruce, hemlock, and fir) that retain their needle-like leaves all year long (see figure 58.10).

The taiga is one of the largest biomes on Earth. The winters where taiga occurs are severely long and cold, and most of the limited precipitation falls in the summer. Many large herbivores, including elk, moose, and deer, plus carnivores such as wolves, bears, lynx, and wolverines, are characteristic of the taiga.

Tundra is a largely frozen treeless area with a short growing season

In the far north, at latitudes above the taiga but south of the polar ice, few trees grow. The landscape that occurs in this band, called **tundra,** is open, windswept, and often boggy (see figure 58.10). This enormous biome covers one-fifth of the Earth's land surface. Little rain or snow falls. **Permafrost**—soil ice that persists throughout all seasons—usually exists within a meter of the ground surface.

What trees can be found are small and mostly confined to the margins of streams and lakes. Large grazing mammals, including musk-oxen and reindeer (caribou), and carnivores such as wolves, foxes, and lynx, live in the tundra. Populations of lemmings (a small rodent native to the Arctic) rise and fall dramatically, with important consequences for the animals that prey on them.

> Major types of ecosystems called biomes can be distinguished in different climatic regions on land. These biomes are much the same wherever they are found on the Earth. Annual mean temperature and precipitation are effective predictors of biome type.

58.3 Freshwater Habitats

Fresh water covers by far the smallest percentage of the Earth's surface of the major habitats: Only 2%, compared with 27% for land and 71% for ocean. The formation of fresh water starts with the evaporation of water into the atmosphere, which essentially removes dissolved constituents, much like distillation does. When water falls back to the surface of the Earth as rain or snow, it arrives in an almost pure state, although it may have picked up biologically significant dissolved or particulate matter from the atmosphere.

Freshwater wetlands—marshes, swamps, and bogs—represent intermediate habitats between the freshwater and terrestrial realms. Wetlands are highly productive (see figure 57.11). They also play key additional roles, such as acting as water storage basins that moderate flooding.

Primary production in freshwater bodies is carried out by single-celled algae (phytoplankton) floating in the water, by algae growing as films on the bottom, and by rooted plants such as water lilies. In addition, a considerable amount of organic matter—such as dead leaves—enters some bodies of fresh water from plant communities growing on the land nearby.

Life in freshwater habitats depends on oxygen availability

The concentration of dissolved oxygen (O_2) is a major determinant of the properties of freshwater communities. Oxygen dissolves in water just like sugar or salt does. Fish and other aquatic organisms obtain the oxygen they need by taking it up from solution. The solubility of oxygen is therefore critically important.

In reality, oxygen is not very soluble in water. Consequently, even when fresh water is fully aerated and at equilibrium with the atmosphere, the amount of oxygen it contains per liter is only 5%, or less, of that in air. This means that, in terms of acquiring the oxygen they need, freshwater organisms have a far smaller margin of safety than air-breathing ones.

Oxygen is constantly added to and removed from any body of fresh water. Oxygen is added by photosynthesis and by aeration from the atmosphere, and it is removed by animals and other heterotrophs. If a lot of decaying organic matter is present in a body of water, the oxygen demand of the decay microbes can be high and have an impact on other life forms. Under conditions in which the rate of oxygen removal from water exceeds the rate of addition, the concentration of dissolved oxygen can fall so low that many aquatic animals cannot survive in it.

Lake and pond habitats change with water depth

Bodies of relatively still fresh water are called lakes if large and ponds if small. Water absorbs light passing through it, and the intensity of sunlight available for photosynthesis decreases sharply with increasing depth. In deep lakes, only water relatively near the surface receives enough light for phytoplankton to exhibit a positive net primary productivity (figure 58.11). Those waters are described as the **photic zone.**

The photic zone

The thickness of the photic zone depends on how much particulate matter is in the water. Water that is relatively free of particulate matter and clear allows light to penetrate tens of meters at sufficient intensity to support phytoplankton. Water that is thick with surface algal cells or soil from erosion may not allow light to penetrate very far before its intensity becomes too diminished for algal growth.

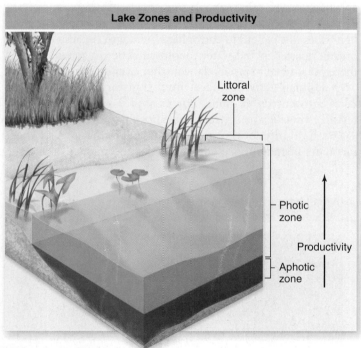

Lake Zones and Productivity

figure 58.11

LIGHT IN A LAKE. The intensity of the sunlight available for photosynthesis decreases with depth in a lake. Consequently, only some of the upper waters—termed the photic zone—receive sufficient light for the net primary productivity of phytoplankton to be positive. The depth of the photic zone depends on how cloudy the water is. The shallows at the edge of a lake are called the littoral zone. They are well-illuminated to the bottom, so rooted plants and bottom algae can thrive there.

The supply of dissolved oxygen to the deep waters of a lake can be a problem because all oxygen enters any aquatic system near its surface. In the still waters of a lake, mixing between the surface and deeper layers may not occur except occasionally. When photosynthesis produces oxygen, it adds it to the photic zone of the lake near the surface. Thermal stratification commonly affects how readily oxygen enters the deep waters from the surface waters.

Thermal stratification

Thermal stratification is characteristic of many lakes and large ponds. In summer, as shown at the bottom of figure 58.12, water warmed by the Sun forms a layer known as the *epilimnion* at the surface—because warm water is less dense

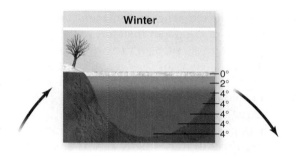

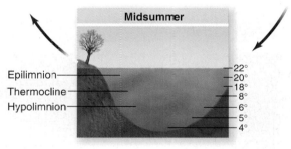

figure 58.12

THE ANNUAL CYCLE OF THERMAL STRATIFICATION IN A TEMPERATE-ZONE LAKE. During the summer (lower diagram), water warmed by the Sun (the epilimnion) floats on top of colder, denser water (the hypolimnion). The lake is also thermally stratified in winter (upper diagram) when water that is near freezing or frozen floats on top of water that is at 4°C (the temperature of greatest density for fresh water). Stratification is disrupted in the spring and fall overturns, when the lake is at an approximately uniform temperature and winds mix it from top to bottom.

than cold water and tends to float on top. Colder, denser water, called the *hypolimnion*, lies below. Between the warm and cold layers is a transitional layer, the **thermocline.** Although here we are focusing on fresh water, a similar thermal structuring of the water column occurs also in many parts of the ocean.

In a lake, thermal stratification tends to cut off the oxygen supply to the bottom waters; a consequence of the stratification is that the upper waters that receive oxygen do not mix with the bottom waters. The concentration of oxygen at the bottom may then gradually decline over time as the organisms living there use oxygen faster than it is replaced. If the rate of oxygen use is high, the bottom waters may run out of oxygen and become oxygen-free before summer is over. Oxygen-free conditions, if they occur, kill most (although not all) animals.

In autumn, the temperature of the upper waters in a stratified lake drops until it is about the same as the temperature of the deep waters. The densities of the two water layers become similar, and the tendency for them to stay apart is weakened. Winds can then force the layers to mix, a phenomenon called the **fall overturn** (see figure 58.12). High oxygen concentrations are then restored in the bottom waters.

In chapter 2, you learned about the unique properties of water. Fresh water is densest when its temperature is 4°C, and ice, at 0°, floats on top of this dense water. As a lake is cooled toward the freezing point with the onset of winter, the whole lake first reaches 4°C. Then, some water cools to an even lower temperature, and when it does, it becomes less dense and rises to the top. Further cooling of this surface water causes it to freeze into a layer of ice covering the lake. In spring, the ice melts, the surface water warms up, and again winds are able to mix the whole lake—the **spring overturn.**

Oligotrophic waters have high oxygen but low nutrient content

Bodies of fresh water that are low in algal nutrients (such as nitrate or phosphate) and low in the amount of algal material per unit of volume are termed **oligotrophic.** Such waters are often crystal clear. Oligotrophic streams and rivers tend to be high in dissolved oxygen because the movement of the flowing water aerates them; the small amount of organic matter in the water means that oxygen is used at a relatively low rate. Similarly, oligotrophic lakes and ponds tend to be high in dissolved oxygen at all depths all year because they also have a low rate of oxygen use. Because the water is relatively clear, light can penetrate the waters readily, allowing photosynthesis to occur through much of the water column, from top to bottom (figure 58.13).

Eutrophic waters are high in nutrient content and phytoplankton populations, but low in oxygen

Eutrophic bodies of water are high in algal nutrients and often populated densely with algae. They are more likely to be low in dissolved oxygen, especially in summer. In a eutrophic body of water, decay microbes often place high demands on the oxygen available because when thick populations of algae die, large amounts of organic matter are made available for decomposition. Moreover, light does not penetrate eutrophic waters well because of all the organic matter in the water; photosynthetic oxygen addition is therefore limited to just a relatively thin layer of water at the top.

Human activities have often transformed oligotrophic lakes into eutrophic ones. For example, when people over-fertilize their lawns, nitrate and phosphate from the fertilizers wash off into local water systems. Lakes that receive these nutrients become more eutrophic. A consequence is that the

Oligotrophic Lake	Eutrophic Lake

a.   *b.*

figure 58.13

OLIGOTROPHIC AND EUTROPHIC LAKES. *a.* Oligotrophic lakes are low in algal nutrients, have high levels of dissolved oxygen, and are clear. *b.* Eutrophic lakes have high levels of algal nutrients and low levels of dissolved oxygen. Light does not penetrate deeply in such lakes.

bottom waters are more likely to become oxygen-free during the summer. Many species of fish that are characteristic of oligotrophic lakes, such as trout, are very sensitive to oxygen deprivation. When lakes become eutrophic, these species of fish disappear and are replaced with species like carp that can better tolerate low oxygen concentrations. Lakes can return toward an oligotrophic state over time if people take steps to eliminate the addition of excess nitrates, phosphates, and foreign organic matter such as sewage.

Nutrients for phytoplankton and dissolved oxygen for heterotrophic organisms are major determinants of the nature of freshwater ecosystems. Thermal stratification is a major determinant of oxygen levels. Levels of nutrients, such as phosphate, determine whether a river or lake is oligotrophic or eutrophic. Levels of dissolved oxygen are partly dependent on the oligotrophic or eutrophic nature of a body of water because the more organic matter there is, the faster available oxygen tends to be used.

58.4 Marine Habitats

About 71% of the Earth's surface is covered by ocean. Near the coastlines of the continents are the **continental shelves,** where the water is not especially deep (figure 58.14); the shelves, in essence, represent the submerged edges of the continents. Worldwide, the shelves average about 80 km wide, and the depth of the water over them increases from 1 m to about 130 m as one travels from the coast toward the open ocean.

Beyond the continental shelves, the depth suddenly becomes much greater. The average depth of the open ocean is 4000–5000 m, and there are some parts—called trenches—that are far deeper, reaching 11,000 m in the Marianas Trench in the western Pacific Ocean.

In most of the ocean, the principal primary producers are phytoplankton floating in the well-lit surface waters. A revolution is currently underway in scientific understanding of the limiting nutrients for ocean phytoplankton (see chapter 57). Primary production by the phytoplankton is presently understood to be nitrogen-limited in about two-thirds of the world's ocean, but iron-limited in about one-third. The principal known iron-limited areas are the great Southern Ocean surrounding Antarctica, parts of the equatorial Pacific Ocean, and parts of the subarctic, northeast Pacific Ocean. Where the water is shallow along coastlines, primary production is carried out not just by phytoplankton but also by rooted plants such as seagrasses and by bottom-dwelling algae, including seaweeds.

The world's ocean is so vast that it includes many different types of ecosystems. Some, such as coral reefs and estuaries, are high in their net primary productivity per unit of area (see figure 57.11), but others are low in productivity per unit area. A useful way to categorize ocean ecosystems is to recognize four major types: The open oceans, continental shelf ecosystems, upwelling regions, and deep sea.

Open oceans have low primary productivity

In speaking of the **open oceans,** we mean the waters far from land (beyond the continental shelves) that are near enough to the surface to receive sunlight or to interact on a daily or weekly basis with those waters. We discuss the deep sea separately later on.

The intensity of solar illumination in the open oceans drops from being high at the surface to being essentially zero at 200 m of depth; photosynthesis is limited to this level of the ocean. However, nutrients for phytoplankton, such as nitrate, tend to be present at low concentrations in the photic zone because over eons of time in the past, ecological processes have exported nitrate and other nutrients from the upper waters to the deep waters, and no vigorous forces exist in the open ocean to return the nutrients to the sunlit waters.

Because of the low concentrations of nutrients in the photic zone, large parts of the open oceans are low in primary productivity per unit area (see figure 57.11) and aptly called a "biological desert." These parts—which correspond to the centers of the great midocean gyres (see figure 58.4)—are often

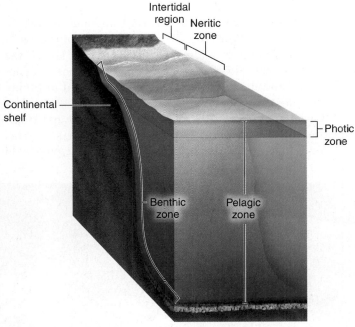

figure 58.14

BASIC CONCEPTS AND TERMINOLOGY USED IN DESCRIBING MARINE ECOSYSTEMS. The continental shelf is the submerged edge of the continent. The waters over it are termed neritic and, on a worldwide average basis, are only 130 m deep at their deepest. The region where the tides rise and fall along the shoreline is called the intertidal region. The bottom is called the benthic zone, whereas the water column in the open ocean is called the pelagic zone. The photic zone is the part of the pelagic zone in which enough light penetrates for the phytoplankton to have a positive net primary productivity. The vertical scale of this drawing is highly compressed; whereas the outer edge of the continental shelf is 130 m deep, the open ocean in fact averages 35 times deeper (4000–5000 m deep).

collectively termed the **oligotrophic ocean** (figure 58.15) in reference to their low nutrient levels and low productivity.

People fish the open oceans today for tunas and a few other species such as certain squids, and in the past they fished them for whales. Fishing in the open oceans is limited to just a few sorts of animals for two reasons. First, because of the low primary productivity per unit of area, animals tend to be thinly distributed in the open oceans. The only ones that are commercially profitable to catch are those that are individually large or tend to gather together in tight schools. Second, costs for fuel and other necessities to travel far from land are steep. All authorities agree that as we turn to the sea to help feed the burgeoning human population, we cannot expect the open ocean regions to supply great quantities of food.

Continental shelf ecosystems provide abundant resources

Many of the ecosystems on the continental shelves are relatively high in productivity per unit area. An important reason is that the waters over the shelves—termed the **neritic waters** (see figure 58.14)—tend to have relatively high concentrations of nitrate and other nutrients, averaged over the year.

Because the waters over the shelves are shallow, they have not been subject, over the eons of time, to the loss of nutrients into the deep sea, as the open oceans have. Over the shelves, nutrient-rich materials that sink hit the shallow bottom, and the nutrients they contain are stirred back into the water column by stormy weather. In addition, nutrients are continually replenished by run-off from nearby land.

Around 99% of the food people glean from the ocean comes from continental shelf ecosystems or nearby upwelling regions. The shelf ecosystems are also particularly important to humankind in other ways. Mineral resources taken from the ocean, such as petroleum, come almost exclusively from the shelves. In addition, almost all recreational uses of the ocean, from sailing to scuba diving, take place on the shelves. The shelves feature prominently in these ways because they are close to coastlines and relatively shallow.

Estuaries

Estuaries are one of the types of shelf ecosystems. An estuary is a place along a coastline, such as a bay, that is partially surrounded by land and in which fresh water from streams or rivers mixes with ocean water, creating intermediate (brackish) salinities.

Estuaries, besides being bodies of water, include intertidal marshes or swamps. An **intertidal** habitat is an area that is exposed to air at low tide but under water at high tide. The marshes of the intertidal zone are called **salt marshes.** Intertidal swamps called **mangrove swamps** (dominated by trees and bushes) occur in tropical and subtropical parts of the world.

Estuaries are a vital and highly productive ecosystem—they provide shelter and food for many aquatic animals, especially the larvae and young, that people harvest for food. Estuaries are also important to a very large number of other animal species, such as migrating birds.

Banks and coral reefs

Other types of shelf ecosystems include banks and coral reefs. **Banks** are local shallow areas on the shelves, often extremely important as fishing grounds; Georges Bank, 100 km off the shore of Massachusetts, was formerly one of the most productive and famous; much of this area has been closed to fishing since the mid-1990s because of overexploitation.

Coral reef ecosystems occur in subtropical and tropical latitudes. Their defining feature is that in them, stony corals—corals that secrete a solid, calcified type of skeleton—build

figure 58.15

MAJOR FUNCTIONAL REGIONS OF THE OCEAN. The regions classed as oligotrophic ocean (colored *dark blue*) are "biological deserts" with low productivity per unit area. Continental shelf ecosystems (*green at the edge of continents*) are typically medium to high in productivity. Upwelling regions (*yellow at the edge of continents*) are the highest in productivity per unit area and rank with the most productive of all ecosystems on Earth.

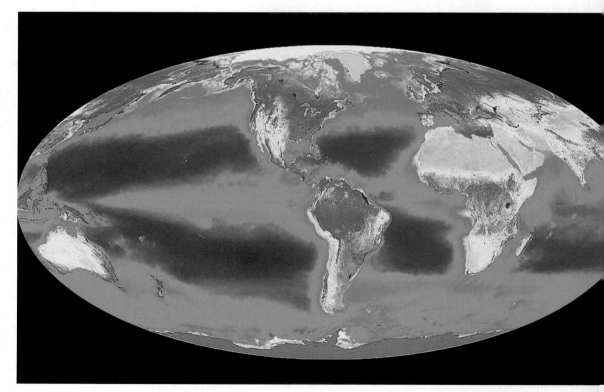

figure 58.16

A CORAL REEF ECOSYSTEM. Reef-building corals, which consist of symbioses between cnidarians and algae, construct the three-dimensional structure of the reef and carry out considerable primary production. Fish and many other kinds of animals find food and shelter, making these ecosystems among the most diverse. About 20% of all fish species occur specifically in coral reef ecosystems.

three-dimensional frameworks that form a unique habitat in which many other distinctive organisms live, including reef fish and soft corals (figure 58.16).

All the 700 or so species of reef-building corals are animal–algal symbioses; the animals are cnidarians, and dinoflagellate symbionts live within the cells of their inner cell layer (the gastrodermis). These corals depend on photosynthesis by the algal symbionts, and thus require clear waters through which sunlight can readily penetrate. Reef-building corals are threatened worldwide, as described later in this chapter.

Upwelling regions experience mixing of nutrients and oxygen

The **upwelling regions** of the ocean are localized places where deep water is drawn consistently to the surface because of the action of local forces such as local winds. The deep water is often rich in nitrate and other nutrients. Upwelling therefore steadily brings nutrients into the well-lit surface layers. Phytoplankton respond to the abundance of nutrients and light with prolific growth and reproduction. Upwelling regions have the highest primary productivity per unit area in the world's ocean.

The most famous upwelling region (see figure 58.15) is found along the coast of Peru and Ecuador, where upwelling occurs year-round. Another important upwelling region is the coastline of California, along which upwelling occurs during about half the year in the summer, explaining why swimmers find cold water at the beaches even in July and August.

Upwelling regions support prolific but vulnerable fisheries. Sardine fishing in the California upwelling region crashed a few decades ago, but previously was enormously important to the region, as Nobel Prize–winning author John Steinbeck chronicled in a number of his books, most notably *Cannery Row.*

El Niño Southern Oscillation (ENSO)

The phenomenon named **El Niño** first came to the attention of science in studies of the Peru–Ecuador upwelling region. In that region, every 2–7 years on an irregular and relatively unpredictable basis, the water along the coastline becomes profoundly warm, and simultaneously the primary productivity becomes unusually low.

Because of the low primary productivity, the ordinarily prolific fish populations weaken, and populations of seabirds and sea mammals that depend on the fish are stressed or plummet. The local people had named a mild annual warming event, which occurred around Christmas each year, "El Niño" (literally, "the child," after the Christ Child). Scientists adopted the term El Niño Southern Oscillation (ELSO) to refer to the dramatic warming events that occurred irregularly every 2–7 years.

The immediate cause of El Niño took several decades to figure out, but research ultimately showed that the cause is a weakening of the east-to-west Trade Winds in the region. The Trade Winds ordinarily blow warm surface water to the west, away from the Peru–Ecuador coast. This thins the warm surface layer of water along the coast, so that deep water—cold but highly rich in nutrients—is drawn to the surface, leading to high primary production.

Weakening of the Trade Winds allows the warm surface layer to become thicker. Upwelling continues, but under such circumstances it merely recirculates the thick warm surface layer, which is nutrient-depleted.

After these fundamentals had been discovered, researchers in the 1980s realized that the weakening of the Trade Winds is actually part of a change in wind circulation patterns that recurs irregularly every 2–7 years. One reason the Trade Winds blow east-to-west in ordinary times is that the surface waters in the western equatorial Pacific are warmer than those in the eastern equatorial Pacific; air rises from the warm western areas, creating low pressure at the surface there, and air blows out of the east into the low pressure. During an El Niño, the warmer the eastern ocean gets, the more similar it becomes to the western ocean, reducing the difference in pressure across the ocean. Thus, once the Trade Winds weaken a bit, the pressure difference that makes them blow is lessened, weakening the Trade Winds further. Warm water ordinarily kept in the west by the Trade Winds creeps progressively eastward at equatorial latitudes because of this self-reinforcing series of events. Ultimately, effects of El Niño occur across large parts of the world's weather systems, affecting sea temperatures in California, rainfall in the southwestern United States, and even systems as far distant as Africa.

One specific result is to shift the weather systems of the western Pacific Ocean 6000 km eastward. The tropical rainstorms that usually drench Indonesia and the Philippines occur when warm seawater abutting these islands causes the air above it to rise, cool, and condense its moisture into clouds. When the warm water moves east, so do the clouds, leaving the previously rainy areas in drought. Conversely, the western edge of Peru and Ecuador, which usually receives little precipitation, gets a soaking.

El Niño can wreak havoc on ecosystems. During an El Niño event, plankton can drop to one-twentieth of their normal abundance in the waters of Peru and Ecuador, and because of the drop in plankton productivity, commercial fish stocks virtually disappear (figure 58.17). In the Galápagos Islands, for example, seabird and sea lion populations crash as animals starve due to the lack of fish. By contrast, on land, the heavy rains produce a bumper crop of seeds, and land birds flourish. In Chile, similar effects on seed abundance propagate up the food chain, leading first to increased rodent populations and then to increased predator populations, a nice example of a bottom-up trophic cascade, as was discussed in chapter 57.

The deep sea is a cold, dark place with some fascinating communities

The **deep sea** is by far the single largest habitat on Earth, in the sense that it is a huge region characterized by relatively uniform conditions throughout the globe. The deep sea is seasonless, cold (2–5°C), totally dark, and under high pressure (400–500 atmospheres where the bottom is 4000–5000 m deep).

In most regions of the deep sea, food originates from photosynthesis in the sunlit waters far above. Such food takes about a month to drift down from the surface to the bottom, and along the way about 99% of it is eaten by animals living in the water column. Thus, the bottom communities receive only about 1% of the primary production and are food-poor. Nonetheless, a great many species of animals—most of them small-bodied and thinly distributed—are now known to live in the deep sea. Some of the animals are bioluminescent (figure 58.18*a*) and thereby able to communicate or attract prey by use of light.

Hydrothermal vent communities

The most astounding communities in the deep sea are the **hydrothermal vent communities.** Unlike most parts of the deep sea, these communities are thick with life (figure 58.18*b*), including large-bodied animals such as worms the size of baseball bats. The reason such a profusion of life can be supported is

a.   *b.*

figure 58.18

LIFE IN THE DEEP SEA. *a.* The luminous spot below the eye of this deep-sea fish results from the presence of a symbiotic colony of bioluminescent bacteria. Bioluminescence is a fairly common feature of mobile animals in the parts of the ocean that are so deep as to be dark. It is more common among species living part way down to the bottom than in ones living at the bottom. *b.* These large worms live along vents where hot water containing hydrogen sulfide rises through cracks in the seafloor crust. Much of the body of each worm is devoted to a colony of symbiotic sulfur-oxidizing bacteria. The worms transport sulfide and oxygen to the bacteria, which oxidize the sulfur and use the energy thereby obtained for primary production of new organic compounds, which they share with their worm hosts.

that these communities live on vigorous, local primary production rather than depending on the photic zone far above.

The hydrothermal vent communities occur at places where tectonic plates are moving apart, and seawater—circulating through porous rock—is able to come into contact with very hot rock under the seafloor. This water is heated to temperatures in excess of 350°C and, in the process, becomes rich in hydrogen sulfide.

As the water rises up out of the porous rock, free-living and symbiotic bacteria oxidize the sulfide, and from this reaction they obtain energy, which, in a manner analogous to photosynthesis, they use to synthesize their own cellular substance, grow, and reproduce. These **sulfur-oxidizing bacteria** are chemoautotrophs (see chapter 57). Animals in the communities either survive on the bacteria or eat other animals that do. The hydrothermal vent communities are among the few communities on Earth that do not depend on the Sun's energy for primary production.

The centers of the great ocean basins together constitute one of the regions of the Earth where primary productivity per unit area is low: the oligotrophic ocean, which includes the open ocean and the deep sea. Continental shelf ecosystems tend to be moderate to high in productivity; they include estuaries, salt marshes, fishing banks, and coral reefs. The highest levels of primary productivity in the oceans are found in upwelling regions, such as those along the west coasts of North and South America, where prolific but vulnerable fisheries can be found.

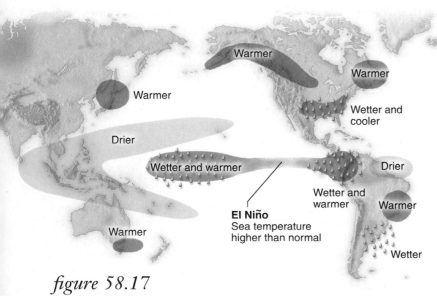

Warmer

Warmer

Warmer

Wetter and cooler

Drier

Wetter and warmer

Drier

Wetter and warmer

Warmer

El Niño
Sea temperature higher than normal

Warmer

Wetter

figure 58.17

AN EL NIÑO WINTER. This diagram shows just some of the worldwide alterations of weather that are often associated with the El Niño phenomenon.

Human Impacts on the Biosphere: Pollution and Resource Depletion

We all know that human activities can cause adverse changes in ecosystems. In discussing these, it is important to recognize that creative people can often come up with rational solutions to such problems.

An outstanding example is provided by the history of DDT in the United States. DDT is a highly effective insecticide that was sprayed widely in the decades following World War II, often on wetlands to control mosquitoes. During the years of heavy DDT use, populations of ospreys, bald eagles, and brown pelicans—all birds that catch large fish—plummeted. Ultimately, the use of DDT was connected with the demise of these birds.

Scientists established that DDT and its metabolic products became more and more concentrated in the tissues of animals as the compounds were passed along food chains (figure 58.19). Animals at the bottom of food chains accumulated relatively low concentrations in their fatty tissues. But the primary carnivores that preyed on them accumulated higher concentrations from eating great numbers, and the secondary carnivores accumulated higher concentrations yet. Top-level carnivores, such as the birds that eat large fish, were dramatically affected by the DDT. In these birds, scientists found that metabolic products of DDT disrupted the formation of eggshells. The birds laid eggs with such thin shells that the eggs often cracked before the young could hatch.

Researchers concluded that the demise of the fish-eating birds could be reversed by a rational plan to clean ecosystems of DDT, and laws were passed banning the use of DDT. Now, three decades later, populations of ospreys, eagles, and pelicans are dramatically rebounding. For some people, a major reason to study science is the opportunity to be part of success stories of this sort.

Freshwater habitats are threatened by pollution and resource use

Fresh water is not just the smallest of the major habitats, but also the most threatened. One of the simplest yet most ominous threats to fresh water is that burgeoning human populations often extract excessive amounts of water from rivers, lakes, or streams. The Colorado River, for example, is one of the greatest rivers in North America, originating with snow melt in the Rocky Mountains and flowing through Utah, Arizona, Nevada, California, and northern Mexico before emptying into the ocean. Today, water is pumped out of the river all along its way to meet the water needs of cities (even ones as distant as Los Angeles) and to irrigate crops. The river now frequently runs out of water and dries up in the desert, never reaching the sea. Worldwide, many crises in the supply of fresh water loom on the horizon.

Pollution: Point source versus diffuse

Pollution of fresh water is a global problem. **Point-source pollution** comes from an identifiable location—such as easily identified factories or other facilities that add pollutants at highly defined locations, such as an outfall pipe. Examples include sewage-treatment plants, which discharge treated effluents at specific spots on rivers, and plating factories (factories that chrome-plate car parts), which sometimes discharge water contaminated with heavy metals. Laws and technologies can readily be brought to bear to moderate point-source pollution because the exact locations and types of pollution are well defined. Great progress has in fact been made.

Diffuse pollution is exemplified by eutrophication caused by excessive run-off of nitrates and phosphates from lawn fertilization. When excessive nitrates and phosphates enter rivers and lakes, the character of the bodies of water is changed for the worse; the concentration of dissolved oxygen declines, and fish species such as carp take the place of more desirable species. The nitrates and phosphates originate on thousands of lawns spread over whole watersheds, and they often enter fresh waters at virtually countless locations. The diffuseness of this sort of pollution renders it difficult to modify by simple technical fixes. Instead, solutions often depend on public education and political action.

DDT Concentration

25 ppm in predatory birds

2 ppm in large fish

0.5 ppm in small fish

0.04 ppm in zooplankton

0.000003 ppm in water

figure 58.19

BIOLOGICAL MAGNIFICATION OF DDT CONCENTRATION. Because all the DDT an animal eats in its food tends to accumulate in its fatty tissues, DDT becomes increasingly concentrated in animals at higher levels of the food chain. The concentrations at the right are in parts per million (ppm). Before DDT was banned in the United States, bird species that eat large fish underwent drastic population declines because metabolic products of DDT made their eggshells so thin that the shells broke during incubation.

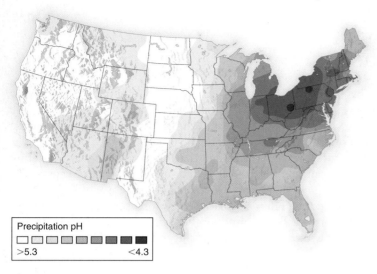

figure 58.20

pH VALUES OF RAINWATER IN THE UNITED STATES.
pH values of less than 7 represent acid conditions; the lower the values, the greater the acidity. Precipitation in parts of the United States, especially in the Northeast, is commonly more acidic than natural rainwater, which has a pH of 5.6 or higher.

Precipitation pH
>5.3   <4.3

Pollution from coal burning: Acid precipitation

A type of pollution that has properties intermediate between the point source and diffuse types is the pollution that can arise from burning of coal for power generation. Although each smokestack is a point source, there are many stacks, and the smoke and gases from these stacks spread over wide areas.

Acid precipitation is one aspect of this problem. When coal is burned, sulfur in the coal is oxidized. The sulfur oxides, unless controlled, are spewed into the atmosphere in the stack smoke, and there they combine with water vapor to produce sulfuric acid. Falling rain or snow picks up the acid and is excessively acidic when it reaches the surface of the Earth (figure 58.20).

Mercury emitted in stack smoke is a second potential problem. Burning of coal can be one of the major sources of environmental mercury, a serious public health issue because just small amounts of mercury can interfere with brain development in human fetuses and infants.

Acid precipitation and mercury pollution affect freshwater ecosystems. At pH levels below 5.0, many fish species and other aquatic animals die, unable to reproduce. Thousands of lakes and ponds around the world no longer support fish because of pH shifts induced by acid precipitation. Mercury that falls from atmospheric emissions into lakes and ponds accumulates in the tissues of food fish. In the Great Lakes region of the United States, people—especially pregnant women—are advised to eat little or no locally caught fish because of mercury.

Introductions of nonnative species of plants and animals are additional threats to fresh water (discussed in the next chapter). The increased use of recreational speedboats in shallow areas is also a problem, in part because sunlight for photosynthesis is blocked when bottom sediments are stirred up.

Terrestrial ecosystems are threatened by deforestation

Probably the single greatest problem for terrestrial habitats worldwide is deforestation by cutting or burning. There are many reasons for deforestation. In poverty-stricken countries, deforestation is often carried out diffusely by the general population; people burn wood to cook or stay warm, and they collect it from the local forests.

At the other extreme, corporations still cut large tracts of virgin forests in an industrialized fashion, often shipping the wood halfway around the world to buyers. Tropical hardwoods, such as mahogany, from Southeast Asian rain forests are shipped to the United States for use in furniture, and softwood logs are shipped from Alaska to East Asia for pulping and paper production. Forests are sometimes simply burned to open up land for farming or ranching (figure 58.21a).

Loss of habitat

The loss of forest habitat can have sad consequences. Particularly diverse sets of species depend on tropical rain forests for their habitat, for example. Thus, when rain forests are cleared, the loss of biodiversity can be extreme. Many tropical forest regions have been severely degraded, and recent estimates suggest that less than half of the world's tropical rain forests remain in pristine condition. All of the world's tropical

a.   *b.*

figure 58.21

DESTROYING THE TROPICAL RAIN FORESTS. *a.* These fires are destroying a tropical rain forest in Brazil to clear it for cattle pasture. ***b.*** The consequences of deforestation can be seen on these middle-elevation slopes in Ecuador, which once supported tropical rain forest, but now support only low-grade pastures and permit topsoil to erode into the rivers (note the color of the water, stained brown by high levels of soil erosion). This sort of picture is seen in a number of places around the world, including Madagascar and Haiti as well as Ecuador.

rain forests will be degraded or gone in about 30 years at present rates of destruction.

Besides loss of habitat, deforestation can have numerous secondary consequences, depending on local contexts. South of the Sahara Desert in Africa, in the Sahel region, deforestation has been a major contributing factor in increased desertification of this region. In the forests of the northeastern United States, as the Hubbard Brook experiment shows (see figure 57.7), deforestation can lead to both a loss of nutrients from forest soils and a simultaneous nutrient enrichment of bodies of water downstream.

Disruption of the water cycle

As discussed in chapter 57, cutting of a tropical rain forest often interrupts the local water cycle in ways that permanently or semipermanently alter the landscape. After an area of tropical rain forest is cleared, rain water often runs off the land to distant places, rather than being returned to the atmosphere immediately above by transpiration. This change may render conditions unsuitable for the rain forest trees that originally lived there. Then the poorly vegetated land—exposed and no longer stabilized by thick root systems—may be ravaged by erosion (figure 58.21b).

Acid rain and loss of topsoil

Two other problems on land are the effects of acid rain and the loss of topsoil by poor land-use practices. Acid rain affects forests as well as lakes and streams; large tracts of trees have been adversely affected by acid rain (figure 58.22). Topsoil, which consists of organic as well as mineral matter, is essential for crop productivity. In Kansas, a state that depends heavily on agriculture for its economy, topsoil is being lost at an average rate of about 1 inch per 15 to 20 years, but regeneration of an inch of topsoil requires 500 years. A net loss of topsoil is the rule worldwide.

Marine habitats are being depleted of fish and other species

Overfishing of the ocean has risen to crisis proportions in recent decades and probably represents the single greatest current problem in the ocean realm. The ocean is so huge that it has tended to be more immune than fresh water or terrestrial ecosystems to global human alteration. Nonetheless, the total world fish catch has been pushed to its maximum for over two decades, even as demand for fish has continued to rise. Fishing pressure is so excessive that 25–30% of the world's ocean fish stocks are presently rated officially as being overexploited, depleted, or in recovery; another 40–50% are rated as being maximally exploited.

Major cod fisheries in waters off of Nova Scotia, Massachusetts, and Great Britain have been closed to fishing in the past 15 years because of collapse (figure 58.23). Overfishing can have disturbing indirect effects. In impoverished parts of Africa, poaching on primates and other wild mammals in national parks increases when fish catches decline.

Aquaculture: At present only a quick fix

Production of fish by aquaculture has grown steadily in the last two decades, and it is often viewed as a straightforward solution to the fisheries problem. But the dietary protein needs of many aquacultured fish, such as salmon, are met largely with wild-caught fish. In this case, exploitation has simply shifted to different species.

In addition, current aquaculture practices often damage natural ocean ecosystems. One example is the clearing of

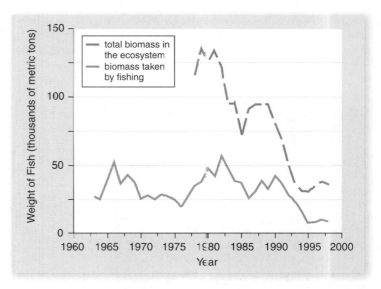

figure 58.23

THE COLLAPSE OF A FISHERY. The red line shows the biomass of cod (*Gadus morhua*) in the Georges Bank ecosystem as estimated by the U.S. National Marine Fisheries Service based on data collected by scientific sampling. The biomass declined steeply between the 1970s and 1990s because of fishing pressure. As the years passed, commercial landings of cod (*blue line*) remained fairly constant, in part because ships worked harder and harder to catch cod, until catches fell precipitously toward zero and the fishery collapsed in the mid-1990s. Regulatory agencies closed the fishery in the mid-1990s to permit the cod to recover, but even in 2005 recovery of cod was weak at best, and production from the fishery was far below historical norms.

figure 58.22

DAMAGE TO TREES BY ACID PRECIPITATION AT CLINGMAN'S DOME, TENNESSEE. Acid precipitation weakens trees and makes them more susceptible to pests and predators.

mangrove swamps along coasts to create shrimp and fish ponds, which are abandoned when their productivity declines. Research is needed to ameliorate these problems.

Pollution effects

As large as the ocean is, enough pollutants are being added that at the start of the twenty-first century, polluting materials are easily detectable on a global basis. An expedition to some of the most remote, uninhabited islands in the vast Pacific Ocean recently reported, for example, that considerable amounts of plastic could be found washed up on the beaches. Similarly, even the waters of the Arctic are laced with toxic chemicals; biopsy samples of tissue from Arctic killer whales (*Orcinus orca*) revealed extremely high levels of many chemicals, including pesticides and a flame-retardant chemical often used in carpets. Nonetheless, because of the ocean's vastness, concentrations of pollutants are not at crisis levels in the ocean at large.

Destruction of coastal ecosystems

Second to overfishing, the greatest problem in the ocean realm is deterioration of coastal ecosystems. Estuaries along coastlines are often subject to severe eutrophication; since about 1970, for example, the bottom waters of the Chesapeake Bay near Washington, DC, have become oxygen-free each summer because of the decay of excessive amounts of organic matter.

Another coastal problem is destruction of salt marshes, which (like freshwater wetlands) are often perceived as disposable. Most authorities believe that the loss of salt marshes in the twentieth century was a major contributing factor to the destruction of New Orleans by Hurricane Katrina in 2005; had the salt marshes been present at their full extent, they would have absorbed a great deal of the flooding water and buffered the city from some of the storm's violence.

Stratospheric ozone depletion has led to an ozone "hole"

The colors of the satellite photo in figure 58.24 represent different concentrations of ozone (O_3) located 20–25 km above the Earth's surface in the stratosphere. Stratospheric ozone is depleted over Antarctica (purple region in the figure) to between one-half and one-third of its historically normal concentration, a phenomenon called the **ozone hole.**

Although depletion of stratospheric ozone is most dramatic over Antarctica, it is a worldwide phenomenon. Over the United States, the ozone concentration has been reduced by about 4%, according to the U.S. Environmental Protection Agency.

Stratospheric ozone and UV-B

Stratospheric ozone is important because it absorbs ultraviolet (UV) radiation—specifically the wavelengths called **UV-B**—from incoming solar radiation. UV-B is damaging to living organisms in a number of ways; for instance, it increases risks of cataracts and skin cancer in people. Depletion of stratospheric ozone permits more UV-B to reach the Earth's surface and therefore increases the risks of UV-B damage. Every 1% drop in stratospheric ozone is estimated to lead to a 6% increase in the incidence of skin cancer, for example.

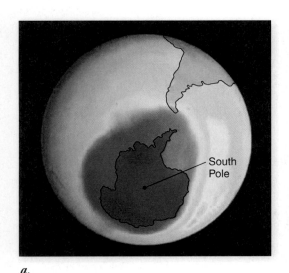

a.

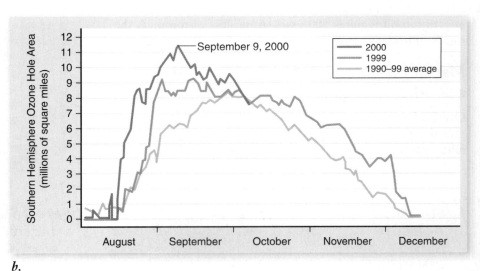

b.

figure 58.24

THE OZONE HOLE OVER ANTARCTICA. NASA satellites currently track the extent of ozone depletion in the stratosphere over Antarctica each year. Every year since about 1980, an area of profound ozone depletion, called the ozone hole, has appeared in August (early spring in the southern hemisphere) when sunlight triggers chemical reactions in cold air trapped over the South Pole during the Antarctic winter. The hole intensifies during September before tailing off as temperature rises in November–December. *a.* In 2000, the 11.4 million-square-mile hole (*purple* in the satellite image shown) covered an area larger than the United States, Canada, and Mexico combined, the largest hole ever recorded. In September 2000, the hole extended over Punta Arenas, a city of about 120,000 people in southern Chile, exposing residents to high levels of UV-B radiation. *b.* Concentrations of ozone-depleting chemical compounds in the atmosphere have probably peaked in the last few years and are expected to decline slowly over the decades ahead.

Ozone depletion and CFCs

The major cause of the depletion of stratospheric ozone is the addition of industrially produced chlorine- and bromine-containing compounds to the atmosphere. Of particular concern are chlorofluorocarbons (CFCs), used until recently as refrigerants in air conditioners and refrigerators, and in manufacturing. CFCs released into the atmosphere can ultimately liberate free chlorine atoms, which in the stratosphere catalyze the breakdown of ozone molecules (O_3) to form ordinary oxygen (O_2). Ozone is continually being made and broken down, and free chlorine atoms tilt the balance toward a faster rate of breakdown.

The extreme depletion of ozone seen in the ozone hole is, fortunately, a consequence of the unique weather conditions that exist over Antarctica. During the continuous dark of the Antarctic winter, a strong stratospheric wind, the polar-night jet, develops and, blowing around the full circumference of the Earth, isolates the stratosphere over Antarctica from the rest of the atmosphere.

The Antarctic stratosphere stays extremely cold (−80°C or lower) for many weeks as a consequence, permitting unique types of ice clouds to form. Reactions associated with the particles in these clouds lead to accumulation of diatomic chlorine, Cl_2. When sunlight returns in the early Antarctic spring, the diatomic chlorine is photochemically broken up to form free chlorine atoms in great abundance, and the ozone-depleting reactions ensue.

Phase-out of CFCs

After research revealed the causes of stratospheric ozone depletion, worldwide agreements were reached to phase out the production of CFCs and other compounds that lead to ozone depletion. Manufacture of such compounds ceased in the United States in 1996, and there is now a great deal of public awareness about the importance of using "ozone safe" alternative chemicals. The atmosphere will cleanse itself of ozone-depleting compounds only slowly because the substances are chemically stable. Nonetheless, the problem of ozone depletion is diminishing and is expected to be substantially corrected by the second half of the twenty-first century.

Fresh water is the most threatened of the major habitats on Earth because of extraction of water by human populations and pollution. Deforestation and loss of topsoil are two of the greatest problems currently faced by terrestrial ecosystems. Overfishing is the greatest problem in the ocean. Chlorine atoms from industrial CFCs catalyze loss of stratospheric ozone, exposing the Earth's surface to increased levels of harmful UV-B radiation.

58.6 Human Impacts on the Biosphere: Global Warming

By studying the Earth's history and making comparisons with other planets, scientists have determined that concentrations of gases in our atmosphere, particularly CO_2, maintain the average temperature on Earth about 25°C higher than it would be if these gases were absent. This fact emphasizes that the composition of our atmosphere is a key consideration for life on Earth as we know it. Unfortunately, human activities are now changing the composition of the atmosphere in ways that most authorities conclude will be damaging or, in the long run, disastrous.

Because of changes in atmospheric composition, the average temperature of the Earth's surface is increasing, a phenomenon called **global warming.** As you might imagine from what we said at the beginning of this chapter, changes in temperature alter global wind and water-current patterns in complicated ways. This means that as the average global temperature increases, some particular regions of the world warm to a lesser extent, whereas other regions heat up to a greater extent (figure 58.25). It also means that rainfall patterns are altered because

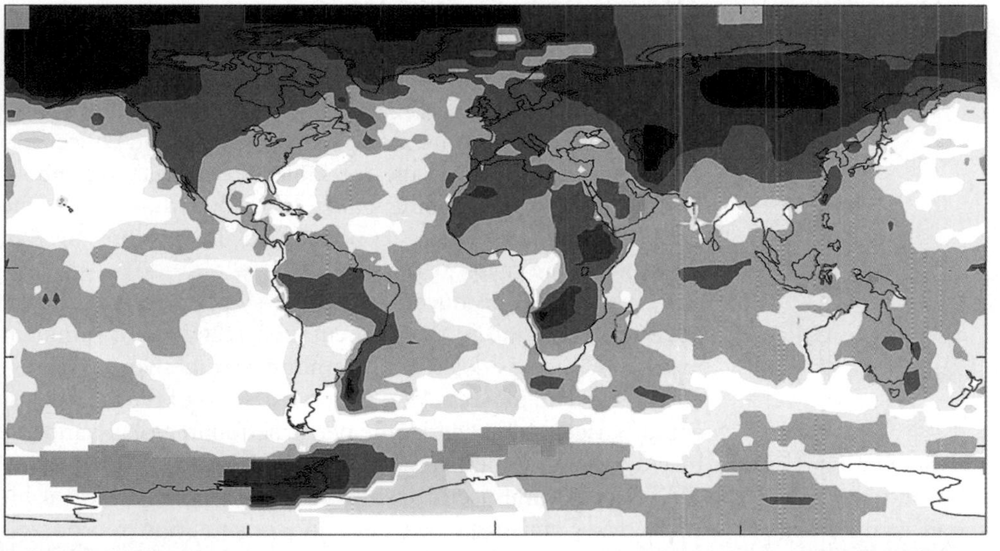

figure 58.25

GEOGRAPHIC VARIATION IN GLOBAL WARMING. 2005 was the warmest year on record, but some areas of the globe heated up more than others. Colors indicate how much warming occurred relative to the mean temperature during a reference period (1951–1980) prior to full onset of the modern greenhouse effect.

-2   -1.6   -1.2   -.8   -.4   -.2   .2   .4   .8   1.2   1.6   2.1

global precipitation patterns depend on global wind patterns. Enormous computer models are used to calculate the effects predicted in all parts of the world.

Independent computer models predict global changes

A study published late in 2005 employed four independent computer models to come up with the most reliable possible prediction for the countries in Europe. Based on the outputs of all four models, the temperature in Europe is predicted to increase by between 2°C (3.6°F) and 4°C (7.2°F) before the year 2080. Increases of temperature of this sort will be disruptive; reliable snow cover in the Swiss Alps, for example, will start 300 m higher than today.

More ominous perhaps than temperature are some of the predictions for precipitation. Although northern Europe is expected to receive more precipitation than today, all four models predict that parts of southern Europe will receive about 20% less, disrupting natural ecosystems, agriculture, and human water supplies. Some European countries may come out ahead economically, but others will come out behind, and political relationships among countries will likely change as some shift from being food exporters to the more tenuous role of requiring food imports.

Carbon dioxide is a major greenhouse gas

Carbon dioxide is the gas usually emphasized in discussing the cause of global warming (figure 58.26), although other atmospheric gases are also involved. A monitoring station on the top of the 13,700-foot (4200-m) Mauna Loa volcano

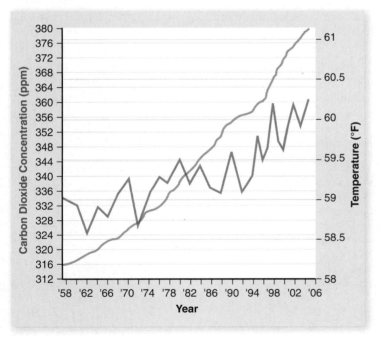

figure 58.26

THE GREENHOUSE EFFECT. The concentration of carbon dioxide in the atmosphere has increased steadily since the 1950s, as shown by the blue line. The red line shows the change in average global temperature over the same period.

on the island of Hawaii has monitored the concentration of atmospheric CO_2 since the 1950s. This station is particularly important because it is in the middle of the Pacific Ocean, far from the great continental landmasses where most people live, and it is therefore able to monitor the state of the global atmosphere without confounding influences of local events.

In 1958, the atmosphere was 0.031% CO_2. By 2004, the concentration had risen to 0.038%. All authorities agree that the cause of this steady rise in atmospheric CO_2 is the burning of coal and petroleum products by the increasing (and increasingly energy-demanding) human population.

How carbon dioxide affects temperature

The atmospheric concentration of CO_2 affects global temperature because carbon dioxide strongly absorbs electromagnetic radiant energy at some of the wavelengths that are critical for the global heat budget. As stressed in chapter 57, the Earth not only receives radiant energy from the Sun each day, but also emits radiant energy into outer space every day. The Earth's temperature will be constant only if the rates of these two processes are equal.

The incoming solar energy is at relatively short wavelengths of the electromagnetic spectrum: Wavelengths that are visible or near-visible. The outgoing energy from the Earth is at different, longer wavelengths. Carbon dioxide absorbs energy at certain of the important long-wave infrared wavelengths. This means that although carbon dioxide does not interfere with the arrival of radiant energy at short wavelengths, it retards the rate at which energy travels away from the Earth at long wavelengths into outer space.

Carbon dioxide is often called a **greenhouse gas** because its effects are analogous to those of a greenhouse. The reason that a glass greenhouse gets warm inside is that window glass is transparent to light but only slightly transparent to long-wave infrared radiation. Energy that strikes a greenhouse as light enters the greenhouse freely. Once inside, the energy is absorbed as heat and then re-radiated as long-wave infrared radiation. The infrared radiation cannot easily get out through the glass, and therefore energy accumulates inside.

Other greenhouse gases

Carbon dioxide is not the only greenhouse gas. Others include methane and nitrous oxide. The effect of any particular greenhouse gas depends on its molecular properties and concentration. For example, molecule-for-molecule, methane has about 20 times the heat-trapping effect of carbon dioxide; on the other hand, methane is less concentrated and less long-lived in the atmosphere than carbon dioxide.

Methane is produced in globally significant quantities in anaerobic soils and in the fermentation reactions of ruminant mammals, such as cows. Huge amounts of methane are presently locked up in Arctic permafrost. Melting of the permafrost could cause a sudden and large perturbation in global temperature by releasing methane rapidly.

Agricultural use of fertilizers is the largest source of nitrous oxide emissions, with energy consumption second and industrial use third.

Evidence confirms global warming is occurring

Evidence for warming can be seen in many ways. For example, on a worldwide statistical basis, ice on lakes and rivers forms later and melts sooner than it used to; on average, ice-free seasons are now 2.5 weeks longer than they were a century ago. Also, the extent of ice at the North Pole has decreased substantially, and glaciers are retreating around the world (figure 58.27).

Carbon dioxide concentrations in the year 2100 are predicted to be between 0.05 and 0.12% (somewhere in the middle being most likely). In a recent study, the U.N.-sponsored Intergovernmental Panel on Climate Change predicted that the average global temperature will rise an additional 1.4–5.8°C by 2100.

Global temperature change has affected ecosystems in the past and is doing so now

Global warming—and cooling—have occurred in the past, most recently during the ice ages and intervening warm periods. Species often responded by shifting their geographic ranges, tracking their environments. For example, a number of cold-adapted North American tree species that are now found only in the far north, or at high elevations, lived much farther south or at substantially lower elevations 10,000–20,000 years ago, when conditions were colder. Present-day

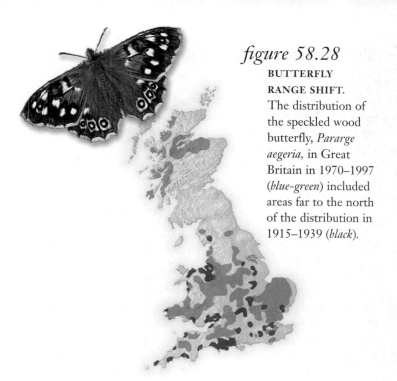

figure 58.28
BUTTERFLY RANGE SHIFT. The distribution of the speckled wood butterfly, *Pararge aegeria*, in Great Britain in 1970–1997 (*blue-green*) included areas far to the north of the distribution in 1915–1939 (*black*).

global warming is having similar effects. For example, many butterfly and bird species have shifted northward in recent decades (figure 58.28).

Many migratory birds arrive earlier at their summer breeding grounds than they did decades ago. Many insects and amphibians breed earlier in the year, and many plants flower earlier. In Australia, recent research shows that wild fruit fly populations have undergone changes in gene frequencies in the past 20 years, such that populations in cool parts of the country now genetically resemble ones in warm parts.

Reef-building corals seem to have narrow margins of safety between the sea temperatures to which they are accustomed and the maximum temperatures they can survive. Global warming seems already to be threatening some corals by inducing mass "bleaching," a disruption of the normal and necessary symbiosis between the cnidarians and algal cells.

There are reasons to think that the effects of global warming on natural ecosystems today may, overall, be more severe than those of warming events in the distant past. One concern is that the rate of warming today is rapid, and therefore evolutionary adaptations would need to occur over relatively few generations to aid the survival of species. Another concern is that natural areas no longer cover the whole landscape but often take the form of parks that are completely surrounded by cities or farms. The parks are at fixed geographic locations and in general cannot be moved. If climatic conditions in a park become unsuitable for its inhabitants, the park will cease to perform its function. Moreover, in the areas to which the park inhabitants might then find suitable climatic conditions, there will likely be no parks.

Similarly, as temperatures increase, many montane species have shifted to higher altitudes to find their preferred habitat. However, eventually they can shift no higher because they reach the mountain's peak. As the temperature continues to increase, the species' habitat disappears entirely. A number of Costa Rican frog species are thought to have become extinct for this reason.

figure 58.27
DISAPPEARING GLACIERS. Mount Kilimanjaro in Tanzania in 1970 (*top*) and 2000 (*bottom*). Note the decrease in glacier coverage over three decades.

Global warming affects human populations as well

Global warming could affect human health and welfare in a variety of ways. Some of these changes may be beneficial, but even if they are detrimental, some countries—particularly the wealthier ones—will be able to adjust. But poorer countries may not be able to make these adjustments, and some changes will require extremely costly countermeasures that even wealthy countries will be hard-pressed to afford.

Rising sea levels

During the second half of the twentieth century, sea level rose at 2–3 cm per decade. The U.S. Environmental Protection Agency predicts that sea level is likely to rise two or three times faster in the twenty-first century because of two effects of global warming: (1) the melting of polar ice and glaciers, adding water to the ocean, and (2) the increase of average ocean temperature, causing an increase in volume because water expands as it warms. Such an increase would cause increased erosion and inundation of low-lying land and coastal marshes, and other habitats would also be imperiled. As many as 200 million people would be affected by increased flooding. Should sea levels continue to rise, coastal cities and some entire islands, such as the Maldives in the Indian Ocean, would be in danger of becoming submerged.

Other climatic effects

Global warming is predicted to have a variety of effects besides increased temperatures. In particular, the frequency or severity of extreme events—such as heat waves, droughts, severe storms, and hurricanes—is expected to increase, and El Niño events, with their attendant climatic effects, may become more common.

In addition, rainfall patterns are likely to shift, and those geographic areas that are already water stressed, which are currently home to nearly 2 billion people, will likely face even graver water shortage problems in the years to come. Some evidence suggests that these effects are already evident in an increase in powerful storms, hurricanes, and the frequency of El Niño events over the past few years.

Effects on agriculture

Global warming may have both positive and negative effects on agriculture. On the positive side, warmer temperatures and increased atmospheric carbon dioxide tend to increase growth of some crops and thus may increase agricultural yields. Other crops, however, may be negatively affected. Furthermore, most crops will be affected by increased frequencies of droughts. Moreover, although crops in north temperate regions may flourish with higher temperatures, many tropical crops are already growing at their maximal temperatures, so increased temperatures may lead to reduced crop yields.

Also on the negative side, changes in rainfall patterns, temperature, pest distributions, and various other factors will require many adjustments. Such changes may come relatively easily for farmers in the developed world, but the associated costs may be devastating for those in the developing countries.

Effects on human health

Increasingly frequent storms, flooding, and drought will have adverse consequences on human health. Aside from their direct impact, such events often disrupt the fragile infrastructure of developing countries, leading to the loss of safe drinking water and other problems. As a result, epidemics of cholera and other diseases may be expected to occur more often as a consequence of these events.

In addition, as temperatures rise, areas suitable for tropical organisms will expand northward. Of particular concern are those organisms that cause human diseases. Many diseases currently limited to tropical areas may expand their range and become problematic in nontropical countries. Diseases transmitted by mosquitoes, such as malaria (see chapter 29), dengue fever, and several types of encephalitis, are examples. The distribution of mosquitoes is limited by cold; winter freezes kill many mosquitoes and their eggs. As a result, malaria only occurs in areas where temperatures are usually above 16°C, and yellow fever and dengue fever occur in areas where temperatures are normally above 10°C. (The difference results because the diseases are transmitted by different mosquito species.) Moreover, at higher temperatures, the malaria pathogen matures more rapidly.

Malaria already kills 1 million people every year; some projections suggest that the percentage of the human population at risk for malaria may increase by 33% by the end of the twenty-first century. Moreover, as predicted, malaria already appears to be on the move. By 1980, malaria had been eradicated from all of the United States except California, but in recent years it has appeared in a variety of southern, and even a few northern, states.

Dengue fever (sometimes called "breakbone fever" because of the pain it causes) is also spreading. Previously a disease restricted to the tropics and subtropics, where it infects 50–100 million people, it now occurs in the United States, southern South America, and northern Australia.

One of the most alarming aspects of these diseases is that no vaccines are available. Drug treatment is available (for malaria), but the parasites are rapidly evolving resistance and rendering the drugs ineffective. There is no drug treatment for dengue fever.

Solving the global warming problem will not be easy. It will require large reductions in the amount of CO_2 injected into the atmosphere. Some nations are taking steps to reduce their emissions, but others are not. Coordinated international effort is required to slow down the increase in global temperatures. Although the predicted effects of global warming are uncertain, most scientists agree that the impact will be severe.

Global warming caused by changes in atmospheric composition—most notably CO_2 accumulation—has the potential to alter all the major habitats on Earth. The impact on human life of global warming could also be severe, with increased violent weather events, shifts in water availability, and flooding of low-lying areas. Increased temperature could cause the range of tropical diseases to expand.

58.1 Ecosystem Effects on Sun, Wind, and Water

The global patterns of life on Earth are influenced by the amount and variation of solar radiation and patterns of atmospheric and oceanic circulation.

- As solar radiation energy passes through the atmosphere it is modified in intensity and wavelength composition.

- The amount of solar radiation reaching the Earth's surface has a great effect on climate; it is reduced as the angle of incidence increases. The season's result from changes in the Earth's position relative to the Sun (figure 58.1).

- Because of intense solar heating, hot air with its increased water content rises at the equator, then cools and loses its moisture, creating the equatorial rain forests (figure 58.3).

- As the drier cool air of the upper atmosphere moves away from the equator it descends to Earth near the 30°N and 30°S latitudes. As it returns to the equator it removes moisture from the Earth's surface and creates dry deserts. This air circulation pattern occurs again between the 30° and 60° latitudes and above the 60° latitudes.

- Winds travel in curved paths relative to the Earth's surface, called the Coriolis effect, because the Earth rotates on its axis.

- A rain shadow occurs when moisture-laden air ascends on the windward side of a mountain and loses its moisture; the now dry air then descends on the opposite side of the mountain, creating a drier environment as moisture is removed from plants and soils (figure 58.5).

- Elevation causes significant changes in temperature and moisture. For every 1000-m increase in elevation, temperature drops approximately 6°C. This temperature reduction also occurs for every 880-km increase in latitude (figure 58.6).

58.2 Earth's Biomes

Biomes are major types of ecosystems that include characteristic vegetational structures and their associated climatic conditions (figure 58.7).

- Temperature and moisture, along with soil structure and mineral content, influence primary production and define biomes (figures 58.8 and 58.9).

58.3 Freshwater Habitats

Freshwater covers only 2% of the Earth's surface; it begins with evaporation and forms by precipitation.

- The concentration of dissolved oxygen in water is the major determinant of the properties of freshwater communities. Oxygen is not very soluble in water.

- Lake and pond habitats change with water depth because of the amount of light that enters a system; light levels affect net primary production and the distribution of oxygen in the water column (figure 58.11).

- In temperate climates, thermal stratification occurs in the summer as the warmer water, the epilimnion, floats on top of the colder hypolimnion. The thermocline is the transitional layer between the epilimnion and hypolimnion. Freshwater lakes turn over twice a year because at 4°C the water is most dense and sinks (figure 58.12).

- Lakes can be categorized based on their nutrient and oxygen levels. Oligotrophic lakes have high oxygen and low nutrients, whereas eutrophic lakes are the opposite.

58.4 Marine Habitats

The oceans cover about 70% of the Earth's surface and can be more than 5000 m deep.

- The ocean is divided into several zones depending on depth, light penetration, and proximity to land or the bottom: intertidal, neritic, photic, benthic, and pelagic zones (figure 58.14).

- In oceans, phytoplankton is the major primary producer in open waters, and primary production is low because nutrients are limiting.

- The neritic waters are found over continental shelves, and they are more productive than open ocean because nutrient levels are higher (figure 58.15).

- Estuaries are another example of a shelf ecosystem. Estuaries occur where freshwater mixes with saltwater. Estuaries frequently contain intertidal zones with salt marshes or mangrove swamps.

- Other shelf ecosystems include productive banks, local shallow areas on continental shelves, and symbiotic coral reef ecosystems.

- Upwelling regions of the ocean are places where local winds bring up nutrient-rich deep waters, creating the highest rates of primary production.

- El Niño occurs when trade winds weaken and the depth of the warmer surface water increases along the coast, restricting upwellings to nutrient-poor surface waters rather than to the deeper nutrient-rich waters

- The deep sea is the single largest habitat, and it is seasonless, cold, dark, and under high pressure.

- Hydrothermal vent communities in the deep sea are located where tectonic plates are moving apart; chemoautotrophs living there obtain energy from the oxidation of sulfur.

58.5 Human Impacts on the Biosphere: Pollution and Resource Depletion

Human activities cause adverse changes in ecosystems.

- Dangerous chemicals like DDT are biomagnified as energy moves up the food chain (figure 58.19).

- Freshwater environments are not only the smallest of the major habitats but they are the most threatened by point-source and diffuse pollution, acid precipitation, and overuse (figure 58.20).

- Deforestation of terrestrial habitats leads to loss of habitat, disruption of the water cycle, acid rain, and loss of topsoil.

- Marine habitats are being depleted of fish and other species by overfishing, destruction of coastal ecosystems, and pollution (figure 58.23).

- Stratospheric ozone depletion has lead to an "ozone" hole allowing more harmful UV-B radiation to reach the Earth's surface (figure 58.24).

58.6 Human Impacts on the Biosphere: Global Warming

Global warming may cause severe impacts on the Earth and is caused by increases in the atmospheric concentration of greenhouse gases, specifically carbon dioxide.

- Carbon dioxide is a major greenhouse gas that allows solar radiation to pass through the atmosphere but prevents long-wave radiation from leaving the Earth, thereby increasing surface temperatures.

- In addition to carbon dioxide, other greenhouse gases include methane and nitrous oxide.

- In the past, fluctuating global temperatures have resulted in ice ages with intervening periods of warming.

- If temperatures change rapidly, natural selection will not occur rapidly enough to prevent many species from becoming extinct.

- Global warming will affect humans in several ways: changing sea levels, increased frequency of extreme climatic events, direct and indirect impacts (both positive and negative) on agriculture, and the expansion of tropical diseases.

SELF TEST

1. If the Earth was not tilted on its axis of rotation, the annual cycle of seasons in the northern and southern hemispheres—
   a. would be reversed.
   b. would stay the same.
   c. would be reduced.
   d. would not exist.

2. The Coriolis effect—
   a. drives the rotation of the Earth.
   b. is responsible for the relative lack of seasonality at the equator.
   c. drives global wind circulation patterns.
   d. drives global wind and ocean circulation patterns.

3. What two factors are most important in biome distribution?
   a. Temperature and latitude
   b. Rainfall and temperature
   c. Latitude and rainfall
   d. Temperature and soil type

4. In a rain shadow, air is cooled as it rises and heated as it descends, often producing a wet and dry side because the water-holding capacity of the air—
   a. is directly related to air temperature.
   b. is inversely related to air temperature.
   c. is unaffected by air temperature.
   d. produces changes in air temperature.

5. The main cause(s) driving the difference between a tropical rain forest and a temperate evergreen forest is—
   a. the amount of average annual rainfall.
   b. the average annual temperature.
   c. temperature and rainfall.
   d. none of the above.

6. Thermal stratification in a lake—
   a. is not modified by fall and spring overturn.
   b. leads to higher oxygen in deep versus surface waters.
   c. leads to higher oxygen in surface versus deep waters.
   d. is reduced when ice forms on the surface of the lake.

7. Oligotrophic lakes have—
   a. low oxygen, and high nutrient availability.
   b. high oxygen, and high nutrient availability.
   c. high oxygen, and low nutrient availability.
   d. low oxygen, and low nutrient availability.

8. Oligotrophic lakes can be turned into eutrophic lakes as a result of human activities such as—
   a. overfishing of sensitive species, which disrupts fish communities.
   b. introducing nutrients into the water, which stimulates plant and algal growth.
   c. disrupting terrestrial vegetation near the shore, which causes soil to run into the lake.
   d. spraying pesticides into the water to control aquatic insect populations.

9. Deep-sea hydrothermal vent communities—
   a. get their energy from photosynthesis in the photic zone near the surface.
   b. use bioluminescence to generate food.
   c. are built on the energy produced by the activity of chemoautotrophs that oxidize sulfur.
   d. contain only bacteria and other microorganisms.

10. Biological magnification occurs when—
    a. pollutants increase in concentration in tissues at higher trophic levels.
    b. the effect of a pollutant is magnified by chemical interactions within organisms.
    c. an organism is placed under a dissecting scope.
    d. a pollutant has a greater than expected effect once ingested by an organism.

11. Which of the following is a point source of pollution?
    a. Lawns
    b. Smokestacks of coal-fired power plants
    c. Factory effluent pipe draining into a river
    d. Acid rain

12. Statements that CO_2 is increasing compared with historical levels—
    a. are based on theory only.
    b. an based on data.
    c. are irrelevant with respect to global warming.
    d. are the basis for a theory that the Earth's global temperature is decreasing.

13. The loss of the ozone layer has serious implications for the quality of the environment because—
    a. ozone (O_3) protects organisms from ultraviolet radiation that can cause cancer.
    b. a depleted ozone layer causes rainwater to have a lower pH that kills plant life.
    c. loss of the ozone layer causes the Sun's rays to get trapped in the atmosphere and increase global temperatures.
    d. a depleted ozone layer can interact with toxic chemicals to increase their effect on organismal health.

CHALLENGE QUESTIONS

1. Discuss how figure 58.1 explains the pattern observed in figure 58.2.
2. Why are most of the Earth's deserts found at approximately 30° latitude?
3. Rain shadows are expected when prevailing winds carrying moist air intercept high mountains. What kind of rain shadow would you expect if the prevailing winds arrive on the windward side relatively dry?
4. What is the distinction between global warming and changes in average CO_2 levels?
5. If a pesticide is harmless at low concentrations (such as, DDT) and used properly, how can it become a threat to nontarget organisms?

ARIS™ **Do you need additional review?** *Visit www.ravenbiology.com for practice quizzes, animations, videos, and activities designed to help you master the material in this chapter.*

chapter 59

Conservation Biology

introduction

AMONG THE GREATEST CHALLENGES facing the biosphere is the accelerating pace of species extinctions. Not since the end of the Cretaceous period 65 million years ago have so many species become extinct in so short a time span. This challenge has led to the emergence of the discipline of conservation biology. Conservation biology is an applied science that seeks to learn how to preserve species, communities, and ecosystems. It studies the causes of declines in species richness and attempts to develop methods for preventing such declines. In this chapter, we first examine the biodiversity crisis and its importance. Then, using case histories, we identify and study factors that have played key roles in many extinctions. We finish with a review of recovery efforts at the species and community levels.

concept outline

Extinction is a fact of life. Most species—probably all—become extinct eventually. More than 99% of species known to science (most from the fossil record) are now extinct. Current rates of extinction are alarmingly high, however. Taking into account the current rapid and accelerating loss of habitat, especially in the tropics, it has been calculated that as much as 20% of the world's biodiversity may be lost by the middle of this century. In addition, many of these species may be lost before we are even aware of their existence. Scientists estimate that no more than 15% of the world's eukaryotic organisms have been discovered and given scientific names, and this proportion is probably much lower for tropical species.

These losses will affect more than poorly known groups. As many as 50,000 species of the world's total of 250,000 species of plants, 4000 of the world's 20,000 species of butterflies, and nearly 2000 of the world's 8600 species of birds could be lost during this period. Considering that the human species has been in existence for less than 200,000 years of the world's 4.5-billion-year history, and that our ancestors developed agriculture only about 10,000 years ago, this is an astonishing—and dubious—accomplishment.

Prehistoric humans were responsible for local extinctions

A great deal can be learned about current rates of extinction by studying the past. In prehistoric times, members of *Homo sapiens* wreaked havoc whenever they entered a new area. For example, at the end of the last Ice Age, approximately 12,000 years ago, the fauna of North America was composed of a diversity of large mammals similar to those living in Africa today: mammoths and mastodons, horses, camels, giant ground sloths, saber-toothed cats, and lions, among others (figure 59.1).

Shortly after humans arrived, 74% to 86% of the megafauna (that is, animals weighing more than 100 lb) became extinct. These extinctions are thought to have been caused by hunting and, indirectly, by burning and clearing of forests. (Some scientists attribute these extinctions to climate change, but that hypothesis doesn't explain why the ends of earlier Ice Ages were not associated with mass extinctions, nor does it explain why extinctions occurred primarily among larger animals, with smaller species being relatively unaffected.)

figure 59.1

NORTH AMERICA BEFORE HUMAN INHABITANTS. Animals found in North America prior to the arrival of humans included birds and large mammals, such as the ancient North American camel, saber-toothed cat, giant ground sloth, and teratorn vulture.

Around the globe, similar results have followed the arrival of humans. Forty thousand years ago, Australia was occupied by a wide variety of large animals, including marsupials similar in size and ecology to hippos and leopards, a kangaroo 9 ft tall, and a 20-ft-long monitor lizard. These all disappeared at approximately the same time as humans arrived.

Smaller islands have also been devastated. Madagascar has seen the extinction of at least 15 species of lemurs, including one the size of a gorilla; a pygmy hippopotamus; and the flightless elephant bird, *Aepyornis*, the largest bird to ever live (more than 3 m tall and weighing 450 kg). On New Zealand, 30 species of birds went extinct, including all 13 species of moas, another group of large, flightless birds. Interestingly, one continent that seems to have been spared these megafaunal extinctions is Africa. Scientists speculate that this lack of extinction in prehistoric Africa may have resulted because much of human evolution occurred in Africa. Consequently, African species had been coevolving with humans for several million years and thus had evolved counteradaptations to human predation.

Extinctions have continued in historical time

Historical extinction rates are best known for birds and mammals because these species are conspicuous—that is, relatively large and well studied. Estimates of extinction rates for other species are much rougher. The data presented in table 59.1, based on the best available evidence, show recorded extinctions from 1600 to the present. These estimates indicate that about 85 species of mammals and 113 species of birds have become extinct since the year 1600. That is about 2.1% of known mammal species and 1.3% of known birds.

The majority of extinctions have occurred in the last 150 years: one species every year during the period from 1850 to 1950, and four species per year between 1986 and 1990. This increase in the rate of extinction is the heart of the biodiversity crisis.

Unfortunately, the situation seems to be worsening. For example, the number of bird species recognized as "critically endangered" increased 8% from 1996 to 2000, and a 2002 report suggested that as many as half of Earth's plant species may be threatened with extinction. Some researchers predict that two-thirds of all vertebrate species could perish by the end of this century.

The majority of historic extinctions—though by no means all of them—have occurred on islands. For example, of the 85 species of mammals that have gone extinct in the last 400 years, 60% lived on islands. The particular vulnerability of island species probably results from a number of factors: Such species have often evolved in the absence of predators, and so have lost their ability to escape both humans and introduced predators such as rats and cats. In addition, humans have introduced competitors and diseases; malaria, for example, has devastated the bird fauna of the Hawaiian Islands. Finally, island populations are often relatively small, and thus particularly vulnerable to extinction, as we shall see later in this chapter.

In recent years, the extinction crisis has moved from islands to continents. Most species now threatened with extinction occur on continents, and these areas will bear the brunt of the extinction crisis in this century.

Some people have argued that we should not be concerned, because extinctions are a natural event and mass extinctions have occurred in the past. Indeed, mass extinctions have taken place several times over the past half-billion years (see figure 22.20). However, the current mass extinction event is notable in several respects. First, it is the only such event triggered by a single species (us). Moreover, although species diversity usually recovers after a few million years (as discussed in chapter 22), this is a long time to deny our descendants the benefits and joys of biodiversity.

In addition, it is not clear that biodiversity will rebound this time. After previous mass extinctions, new species have evolved to utilize resources newly available due to extinctions of the species that previously used them. Today, however, such resources are unlikely to be available, because humans are destroying the habitats and taking the resources for their own use.

Endemic species hotspots are especially threatened

A species found naturally in only one geographic area and no place else is said to be **endemic** to that area. The area over which an endemic species is found may be very large. For example, the black cherry tree (*Prunus serotina*) is endemic to all of temperate North America. More typically, however, endemic species occupy restricted ranges. The Komodo dragon (*Varanus komodoensis*) lives only on a few small islands in the Indonesian archipelago, and the Mauna Kea (*Argyroxiphium sandwicense*) and

TABLE 59.1	Recorded Extinctions Since 1600					
	RECORDED EXTINCTIONS				Approximate Number of Species	Percent of Taxon Extinct
Taxon	Mainland	Island	Ocean	Total		
Mammals	30	51	4	85	4,000	2.1
Birds	21	92	0	113	8,600	1.3
Reptiles	1	20	0	21	6,300	0.3
Fish	22	1	0	23	24,000	0.1
Invertebrates*	49	48	1	98	1,000,000+	0.01
Flowering plants	245	139	0	384	250,000	0.2

*Number of extinct invertebrates is probably greatly underestimated due to lack of knowledge for many species (other groups are probably underestimated to a lesser extent for the same reason).

figure 59.2

MAUNA KEA SILVERSWORD (*Argyroxiphium sandwicense*).
Many species of silverswords are endemic to very small areas.
This photo illustrates two stages in the plant's life cycle.

Haleakala silverswords (*A. s. macrocephalum*) each lives in a single volcano crater on the island of Hawaii (figure 59.3). Isolated geographic areas, such as oceanic islands, lakes, and mountain peaks, often have high percentages of endemic species, many in significant danger of extinction.

The number of endemic plant species can vary greatly from one place to another. In the United States, for example,

379 plant species are found in Texas and nowhere else, whereas New York has only one endemic plant species. California, with its varied array of habitats, including deserts, mountains, seacoast, old-growth forests, and grasslands, is home to more endemic plant species than any other state.

Species hotspots

Worldwide, notable concentrations of endemic species occur in particular regions. Conservationists have recently identified areas, termed **hotspots**, that have high endemism and are disappearing at a rapid rate. Such hotspots include Madagascar, a variety of tropical rain forests, the eastern Himalayas, areas with Mediterranean climates such as California, South Africa, and Australia, and several other climatic areas (figure 59.3 and table 59.2). Overall, 25 such hotspots have been identified, which in total contain nearly half of all the terrestrial species in the world.

Why these areas contain so many endemic species is a topic of active scientific research. Some of these hotspots occur in areas of high species diversity; for these hotspots, the explanations for high species diversity in general, such as high productivity, probably apply (see chapter 57). In addition, some hotspots occur on isolated islands, such as New Zealand, New Caledonia, and the Hawaiian Islands, where evolutionary diversification over long periods has resulted in rich biotas composed of plant and animal species found nowhere else in the world.

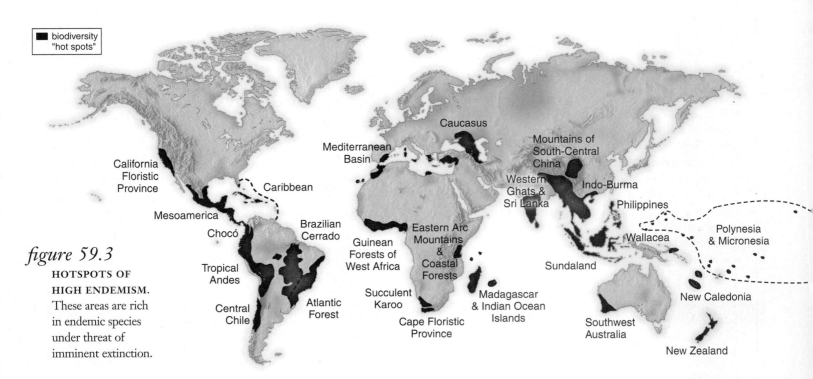

figure 59.3

HOTSPOTS OF HIGH ENDEMISM. These areas are rich in endemic species under threat of imminent extinction.

TABLE 59.2	Numbers of Endemic Species in Some Hotspot Areas			
Region	**Mammals**	**Reptiles**	**Amphibians**	**Plants**
Atlantic coastal forest (Brazil)	160	60	253	6,000
South American Chocó	60	63	210	2,250
Philippines	115	159	65	5,832
Tropical Andes	68	218	604	20,000
Southwestern Australia	7	50	24	4,331
Madagascar	84	301	187	9,704
Cape region (South Africa)	9	19	19	5,682
California Floristic Province	30	16	17	2,125
New Caledonia	6	56	0	2,551
South-central China	75	16	51	3,500

Human population growth in hotspots

Because of the great number of endemic species that hotspots contain, conserving their biological diversity must be an important component of efforts to safeguard the world's biological heritage. Or, to look at it another way, by protecting just 1.4% of the world's land surface, 44% of the world's vascular plants and 35% of its terrestrial vertebrates can be preserved.

Unfortunately, hotspots contain not only many endemic species, but also growing human populations. In 1995,

these areas contained 1.1 billion people—20% of the world's population—sometimes at high densities (figure 59.4a). More important, human populations were growing in all but one of these hotspots both because birthrates are much higher than death rates and because rates of immigration into these areas are often high. Overall, the rate of growth exceeded the global average in 19 hotspots (figure 59.4b). In some hotspots, the rate of growth is nearly twice that of the rest of the world.

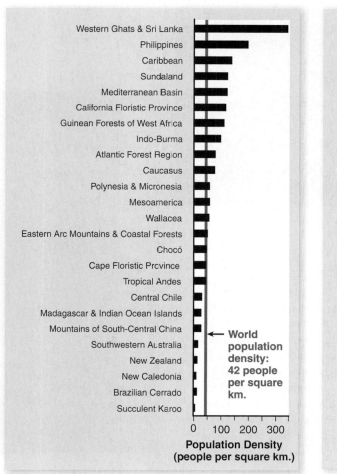

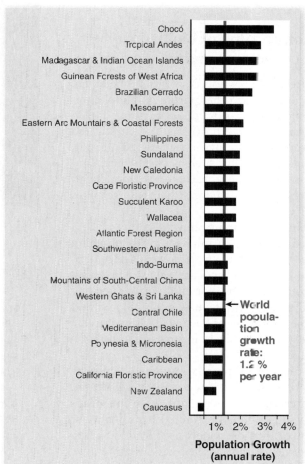

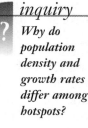

figure 59.4

HUMAN POPULATIONS IN HOTSPOTS. *a.* Human population density and (*b*) population growth rate in biodiversity hotspots.

inquiry

Why do population density and growth rates differ among hotspots?

Not surprisingly, many of these areas are experiencing high rates of habitat destruction as land is cleared for agriculture, housing, and economic development. More than 70% of the original area of each hotspot has already disappeared, and in 14 hotspots, 15% or less of the original habitat remains. In Madagascar, it is estimated that 90% of the original forest has already been lost—this on an island where 85% of the species are found nowhere else in the world. In the forests of the Atlantic coast of Brazil, the extent of deforestation is even higher: 95% of the original forest is gone.

Population pressure is not the only cause of habitat destruction in hotspots. Commercial exploitation to meet the demands of more affluent people in the developed world also plays an important role. For example, large-scale logging of tropical rain forests occurs in countries around the world to provide lumber, most of which ends up in the United States, western Europe, and Japan. Similarly, many forests in Central and South America are cleared to make way for cattle ranches that produce cheap meat for fast-food restaurants. Hotspots in more affluent countries are often at risk because they occur in areas where land has great value for real estate and commercial purposes, such as in Florida and California in the United States.

> Current estimates indicate that biodiversity is being lost at an alarming rate, and that the great majority of this loss is a result of human activities in both prehistoric and historical times.
>
> Endemic species are found in only a single region on Earth; regions with a high number of endemic species, or hotspots, are particularly threatened by human encroachment.

59.2 The Value of Biodiversity

Why should we worry about loss of biodiversity? The reason is that biodiversity has value to us in a number of ways:

- *Direct economic value* of products we obtain from species of plants, animals, and other groups;
- *Indirect economic value* of benefits produced by species without our consuming them; and
- *Ethical and aesthetic values*.

The direct economic value of biodiversity includes resources for our survival

Many species have direct value as sources of food, medicine, clothing, biomass (for energy and other purposes), and shelter. Most of the world's food crops, for example, are derived from a small number of plants that were originally domesticated from wild plants in tropical and semiarid regions. As a result, many of our most important crops, such as corn, wheat, and rice, contain relatively little genetic variation (equivalent to a founder effect; see chapter 20), whereas their wild relatives have great diversity.

In the future, genetic variation from wild strains of these species may be needed if we are to improve yields or find a way to breed resistance to new pests. In fact, recent agricultural breeding experiments have illustrated the value of conserving wild relatives of common crops. For example, by breeding commercial varieties of tomato with a small, oddly colored wild tomato species from the mountains of Peru, scientists were able to increase crop yields by 50%, while increasing both nutritional content and color.

About 70% of the world's population depends directly on wild plants as their source of medicine. In addition, about 40% of the prescription and nonprescription drugs used today have active ingredients extracted from plants or animals. Aspirin, the world's most widely used drug, was first extracted from the leaves of the tropical willow, *Salix alba*. The rosy periwinkle from Madagascar has yielded potent drugs for combating childhood leukemia (figure 59.5), and drugs effective in treating several forms of cancer and other diseases have been produced from the Pacific yew.

a.

figure 59.5

THE ROSY PERIWINKLE. *a.* Two drugs extracted from the Madagascar periwinkle (*Catharanthus roseus*), vinblastine and vincristine, effectively treat common forms of childhood leukemia, increasing chances of survival from 20% to over 95%. *b.* Cancer-fighting drugs, such as taxol, have been developed from the bark of the Pacific yew (*Taxus brevifolia*).

b.

Only recently have biologists perfected the techniques that make possible the transfer of genes from one species to another. We are just beginning to use genes obtained from other species to our advantage (see chapter 15). So-called "gene prospecting" of the genomes of plants and animals for useful genes has only begun. We have been able to examine only a minute proportion of the world's organisms to see whether any of their genes have useful properties for humans.

By conserving biodiversity, we maintain the option of finding useful benefits in the future. Unfortunately, many of the most promising species occur in habitats, such as tropical rain forests, that are being destroyed at an alarming rate.

Indirect economic value is derived from ecosystem services

Diverse biological communities are of vital importance to healthy ecosystems. They help maintain the chemical quality of natural water, buffer ecosystems against storms and drought, preserve soils and prevent loss of minerals and nutrients, moderate local and regional climate, absorb pollution, and promote the breakdown of organic wastes and the cycling of minerals.

In chapter 57, we discussed the evidence that the stability and productivity of ecosystems is related to species richness. By destroying biodiversity, we are creating conditions of instability and lessened productivity and promoting desertification, waterlogging, mineralization, and many other undesirable outcomes throughout the world.

The value of intact habitats

Economists have recently been able to compare the societal value, in monetary terms, of intact habitats compared with the value of destroying those habitats. Surprisingly, in most studies conducted so far, intact ecosystems are more valuable than the products derived by destroying them. In Thailand, as one example, coastal mangrove habitats are commonly cleared so that shrimp farms can be established. Although the shrimp produced are valuable, their value is vastly outweighed by the benefits in timber, charcoal production, offshore fisheries, and storm protection provided by the mangroves (figure 59.6a).

Similarly, intact tropical rain forest in Cameroon, West Africa, provides fruit and other forest materials. Clearing the forest for agriculture or palm plantations leads to stream-polluting erosion as well as increased flooding. Combining all the costs and benefits of the three options, maintaining intact forests has the highest economic value (figure 59.6b).

Case study: New York City watersheds

Probably the most famous example of the value of intact ecosystems is provided by the watersheds of New York City. Ninety percent of the water for the New York area's 9 million residents comes from the Catskill Mountains and the nearby headwaters of the Delaware River (figure 59.7). Water that runs off from over 4000 km² of rural, mountainous areas is collected into reservoirs and then transported by

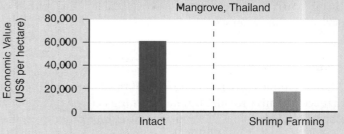

a.

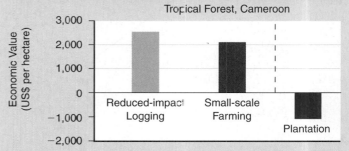

b.

figure 59.6

THE ECONOMIC VALUE OF MAINTAINING HABITATS.
a. Mangroves in Thailand are more valuable than shrimp farms.
b. Rain forests in Cameroon provide more economic benefits if they are left standing than if they are destroyed and the land used for other purposes.

 inquiry

If shrimp farms established on cleared mangrove habitats make money, how can clearing mangroves not be an economic plus?

aqueduct more than 136.8 km to New York City at a rate of 4.9 billion liters per day.

In the 1990s, New York City faced a dilemma. New federal water regulations were requiring ever cleaner water, even as development and pollution in the source areas of the water were threatening to compromise water quality. The city had two choices: either work to protect the functioning ecosystem so that it could produce clean water, or construct filtration plants to clean it on arrival. Economic analysis made the choice clear: Building the plants would cost $6 billion, with annual operating costs of $300 million, whereas spending a billion dollars over 10 years could preserve the ecosystem and maintain water purity. The decision was easy.

Economic trade-offs

These examples provide some idea of the value of the services that ecosystems provide. But maintaining ecosystems is not always more valuable than converting them to other uses. Certainly, when the United States was being settled and land was plentiful, ecosystem conversion was beneficial. Even today, habitat destruction may sometimes be economically desirable. Nonetheless, we still have only a rudimentary knowledge of the many ways intact ecosystems provide services. In many cases, it is not until they are lost that the value becomes clear, as unexpected negative effects, such as increased flooding and pollution, decreased rainfall, or vulnerability to hurricanes become apparent.

The same argument can be made for preserving particular species within ecosystems. Given how little we know about the biology of most species, particularly in the tropics, it is impossible to predict all the consequences of removing a species.

Imagine taking a parts list for an airplane and randomly changing a digit in one of the part numbers. You might change a seat cushion into a roll of toilet paper—or you might just as easily change a key bolt holding up a wing into a pencil. By removing biodiversity, we are gambling with the future of the ecosystems on which we depend and whose functioning we understand very little.

In recent years, the field of ecological economics has developed to study how the societal benefits provided by species and ecosystems can be appropriately valued. The problem is two-fold. First, until recently, we have not had a good estimate of the monetary value of services provided by ecosystems, a situation which, as you've just seen, is now changing.

The second problem, however, is that the people who gain the benefits of environmental degradation are often not the same as the people who pay the costs. For instance, in the example of the Thai mangroves, the shrimp farmers reap the financial rewards, while the local people bear the costs. The same is generally true of factories that produce air or water pollution. Environmental economists are devising ways to appropriately value and regulate the use of the environment in ways that can maximize the benefits relative to the costs for society as a whole.

Ethical and aesthetic values are based on our conscience and our consciousness

Many people believe that preserving biodiversity is an ethical issue because every species is of value in its own right, even if humans are not able to exploit or benefit from it. These people feel that along with the power to exploit and destroy other species comes responsibility: As the only organisms capable of eliminating large numbers of species and entire ecosystems, and as the only organisms capable of reflecting on what we are doing, humans should act as guardians or stewards for the diversity of life around us.

Almost no one would deny the aesthetic value of biodiversity—a wild mountain range, a beautiful flower, or a noble elephant—but how do we place a value on beauty or on the renewal many of us feel when we are in natural surroundings? Perhaps the best we can do is to consider the deep sense of loss we would feel if it no longer existed.

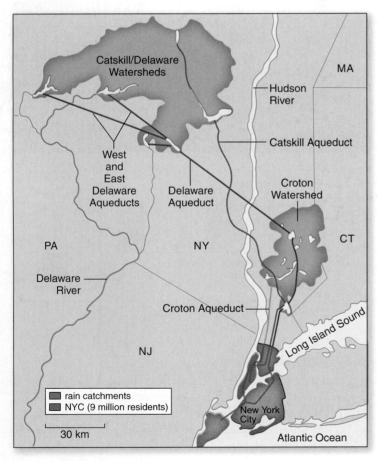

figure 59.7

NEW YORK CITY'S WATER SOURCE. New York gets its water from distant rain catchments. Preserving the ecological integrity of these areas is cheaper than building new water treatment plants.

Biodiversity is of great value in its own right, as well as for the direct economic benefits it provides, its indirect economic benefits in the form of contributions to the health of the ecosystems we depend on, and the aesthetic beauty it offers. Humans are in a unique position to make decisions about their use of the environment.

Factors Responsible for Extinction

A variety of causes, independently or in concert, are responsible for extinctions (table 59.3). Historically, overexploitation was the major cause of extinction; although it is still a factor, habitat loss is the major problem for most groups today, and introduced species rank second. Many other factors can contribute to species extinctions as well, including disruption of ecosystem interactions, pollution, loss of genetic variation, and catastrophic disturbances, either natural or human-caused.

More than one of these factors may affect a species. In fact, a chain reaction is possible in which the action of one factor predisposes a species to be more severely affected by another factor. For example, habitat destruction may lead to decreased birthrates and increased mortality rates. As a result, populations become smaller and more fragmented, making them more vulnerable to disasters such as floods or forest fires, which may eliminate populations. Also, as the habitat becomes more fragmented, populations become isolated, so that genetic interchange ceases and areas devastated by disasters are not re-colonized. Finally, as populations become very small, inbreeding increases, and genetic variation is lost through genetic drift, further decreasing population fitness. Which factor acts as the final coup de grace may be irrelevant; many factors, and the interactions between them, may have contributed to a species' eventual extinction.

Amphibians are on the decline: A case study

In 1963, herpetologist Jay Savage was hiking through pristine cloud forest in Costa Rica. Reaching a windswept ridge, he couldn't believe his eyes. Before him was a huge aggregation of breeding toads. What was so amazing was the color of the toads: bright, eye-dazzling orange, unlike anything he had ever seen before (figure 59.8).

The color of the toads was so amazing and unexpected that Savage briefly considered the possibility that his colleagues had played a practical joke, getting to the clearing before him and somehow coloring normal toads orange. Realizing that this could not be, he went on to study the toads, eventually describing a species new to science, the golden toad, *Bufo periglenes*.

For the next 24 years, large numbers of toads were seen during the breeding season each spring. Their home was legally recognized as the Monteverde Cloud Forest Reserve, a well-protected, intact, and functioning ecosystem, seemingly a successful model of conservation. Then, in 1988, few toads were seen, and in 1989, only a single male was observed. Since then, despite exhaustive efforts, no more golden toads have been found.

Despite living in a well-protected ecosystem, with no obvious threats from pollution, introduced species, overexploitation, or any other factor, the species appears to have gone extinct, right under the eyes of watchful scientists and conservationists. How could this happen?

figure 59.8

AN EXTINCT SPECIES. A breeding assemblage of the golden toad, *Bufo periglenes*, which was last seen in the wild in 1989.

TABLE 59.3	Causes of Extinctions				
	PERCENTAGE OF SPECIES INFLUENCED BY A GIVEN FACTOR*				
Group	Habitat Loss	Overexploitation	Species Introduction	Other	Unknown
EXTINCTIONS					
Mammals	19	23	20	2	36
Birds	20	11	22	2	45
Reptiles	5	32	42	0	21
Fish	35	4	30	4	48
THREATENED EXTINCTIONS					
Mammals	68	54	6	20	—
Birds	58	30	28	2	—
Reptiles	53	63	17	9	—
Fish	78	12	28	2	—

*Some species may be influenced by more than one factor; thus, some rows may exceed 100%.

Frogs in trouble

At the first World Herpetological Congress in 1989 in Canterbury, England, frog experts from around the world met to discuss conservation issues relating to frogs and toads. At this meeting, it became clear that the golden toad story was not unique. Experts reported case after case of similar losses: Frog populations that had once been abundant were now decreasing or entirely gone.

Since then, scientists have devoted a great deal of time and effort to determining whether frogs and other amphibian species truly are in trouble and, if so, why. Unfortunately, the situation appears to be even worse than originally suspected. In 2005, amphibian experts reported that 43% of all amphibian species have experienced decreases in population size, and one third of all amphibian species are threatened with extinction in countries as different as Ecuador, Venezuela, Australia, and the United States (figure 59.9).

Moreover, these numbers are probably underestimates; little information exists from many areas of the world, such as Southeast Asia and central Africa. Indeed researchers think that as many as 100 species from the island nation of Sri Lanka have recently gone extinct, news that is perhaps not surprising given that 95% of that nation's rain forests have also disappeared in recent times.

Cause for concern

Amphibian declines are worrisome for several reasons. First, many of the species—including the golden toad—have declined in pristine, well-protected habitats. If species are becoming extinct in such areas, it brings into question our ability to preserve global biological diversity.

Second, many amphibian species are particularly sensitive to the state of the environment because of their moist skin, which allows chemicals from the environment to pass into the body, and their use of aquatic habitats for larval stages, which requires unpolluted water. In other words, amphibians may be analogous to the canaries formerly used in coal mines to detect problems with air quality: If the canaries keeled over, the miners knew they had to get out.

Third, no single cause for amphibian declines is apparent. Although a single cause would be of concern, it would also suggest that a coordinated global effort could reverse the trend, as happened with chlorofluorocarbons and decreasing ozone levels (see chapter 58). However, different species are afflicted by different problems, including habitat destruction, the effects of global warming, pollution, decreased stratospheric ozone levels, parasite epidemics, and introduced species.

The implication is that the global environment is deteriorating in many different ways. Could amphibians be global "canaries," serving as indicators that the world's environment is in serious trouble?

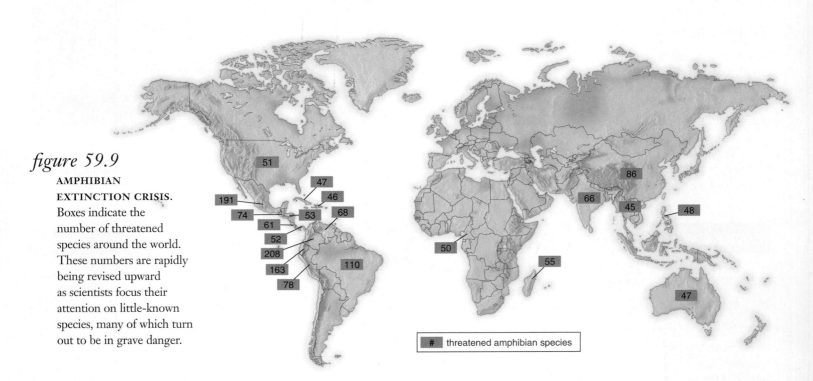

figure 59.9

AMPHIBIAN EXTINCTION CRISIS. Boxes indicate the number of threatened species around the world. These numbers are rapidly being revised upward as scientists focus their attention on little-known species, many of which turn out to be in grave danger.

threatened amphibian species

Venezuela — *Dendrobates leucomelas*

Panama — *Atelopus zeteki*

Madagascar — *Mantella aurantiaca*

Australia — *Litoria caerulea*

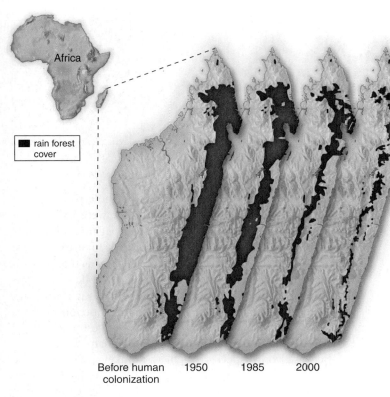

Africa

rain forest cover

Before human colonization　　1950　　1985　　2000

figure 59.10

EXTINCTION AND HABITAT DESTRUCTION. The rain forest covering the eastern coast of Madagascar, an island off the coast of East Africa, has been progressively destroyed and fragmented as the island's human population has grown. Ninety percent of the eastern coast's original forest cover is now gone. Many species have become extinct, and many others are threatened, including 16 of Madagascar's 31 primate species.

Habitat loss devastates species richness

As table 59.3 indicates, habitat loss is the most important cause of modern-day extinction. Given the tremendous amounts of ongoing destruction of all types of habitat, from rain forest to ocean floor, this should come as no surprise. Natural habitats may be adversely affected by humans in four ways:

1. destruction,
2. pollution,
3. disruption, and
4. habitat fragmentation.

Destruction of habitat

A proportion of the habitat available to a particular species may simply be destroyed. This destruction is a common occurrence in the "clear-cut" harvesting of timber, in the burning of tropical forest to produce grazing land, and in urban and industrial development. Deforestation has been, and continues to be, by far the most pervasive form of habitat disruption (figure 59.10). Many tropical forests are being cut or burned at a rate of 1% or more per year.

　　To estimate the effect of reductions in habitat available to a species, biologists often use the well-established observation that larger areas support more species (see figure 57.22). Although this relationship varies according to geographic area, type of organism, and type of area, in general a 10-fold increase in area leads to approximately a doubling in the number of species. This relationship suggests, conversely, that if the area of a habitat is reduced by 90%, so that only 10% remains, then half of all species will be lost. Evidence for this hypothesis comes from a study in Finland of extinction rates of birds on habitat islands (that is, islands of a particular type of habitat surrounded by unsuitable habitat) where the population extinction rate was found to be inversely proportional to island size (figure 59.11).

Pollution

Habitat may be degraded by pollution to the extent that some species can no longer survive there. Degradation occurs as a result of many forms of pollution, from acid rain to pesticides. Aquatic environments are particularly vulnerable; for example, many northern lakes in both Europe and North America have been essentially sterilized by acid rain (chapter 57).

Disruption

Human activities may disrupt a habitat enough to make it untenable for some species. For example, visitors to caves in Alabama and Tennessee caused significant population declines in bats over an 8-year period, some as great as 100%. When visits were fewer than one per month, less than 20% of bats were lost, but caves having more than four visits per month suffered population declines of 86% to 95%.

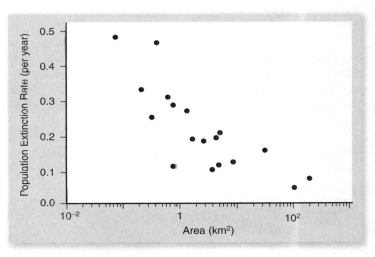

figure 59.11

EXTINCTION AND ISLAND AREA. The data present percent extinction rates for populations as a function of habitat area for birds on a series of Finnish habitat islands. Smaller islands experience far greater extinction rates.

inquiry

Why does extinction rate increase with decreasing island size?

figure 59.12

FRAGMENTATION OF WOODLAND HABITAT. From the time of settlement of Cadiz Township, Wisconsin, the forest has been progressively reduced from a nearly continuous cover to isolated woodlots covering less than 1% of the original area.

  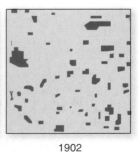 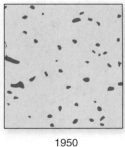

| 1831 | 1882 | 1902 | 1950 |

Habitat fragmentation

Loss of habitat by a species frequently results not only in lowered population numbers, but also in fragmentation of the population into unconnected patches (figure 59.12). A habitat also may become fragmented in nonobvious ways, as when roads and habitation intrude into forest. The effect is to carve the populations living in the habitat into a series of smaller populations, often with disastrous consequences because of the relationship between range size and extinction rate. Although detailed data are not available, fragmentation of wildlife habitat in developed temperate areas is thought to be very substantial.

As habitats become fragmented and shrink in size, the relative proportion of the habitat that occurs on the boundary, or edge, increases. **Edge effects** can significantly degrade a population's chances of survival. Changes in microclimate (such as temperature, wind, humidity) near the edge may reduce appropriate habitat for many species more than the physical fragmentation suggests. In isolated fragments of rain forest, for example, trees on the edge are exposed to direct sunlight. As a result, these trees experience hotter and drier conditions than those normally encountered in the cool, moist forest interior, leading to negative effects on their survival and growth. In one study, the biomass of trees within 100 m of the forest edge decreased by 36% in the first 17 years after fragment isolation.

Also, increasing habitat edges opens up opportunities for some parasite and predator species that are more effective at edges. As fragments decrease in size, the proportion of habitat that is distant from any edge decreases, and consequently, more and more of the habitat is within the range of these species. Habitat fragmentation is blamed for local extinctions in a wide range of species.

The impact of habitat fragmentation can be seen clearly in a major study done in Manaus, Brazil, where the rain forest was commercially logged. Landowners agreed to preserve patches of rain forest of various sizes, and censuses of these patches were taken before the logging started, while they were still part of a continuous forest. After logging, species began to disappear from the now-isolated patches (figure 59.13). First to go were the monkeys, which have large home ranges. Birds that prey on the insects flushed out by marching army ants followed, disappearing from patches too small to maintain enough army ant colonies to support them. As expected, the extinction rate was negatively related to patch size, but even the largest patches (100 hectares) lost half of their bird species in less than 15 years.

Because some species, such as monkeys, require large patches, large fragments are indispensable if we wish to preserve high levels of biodiversity. The take-home lesson is that preservation programs will need to provide suitably large habitat fragments to avoid this effect.

Case study: Songbird declines

Every year since 1966, the U.S. Fish and Wildlife Service has organized thousands of amateur ornithologists and birdwatchers in an annual bird count called the Breeding Bird Survey. In recent years, a shocking trend has emerged. While year-round residents that prosper around humans, such as robins, starlings, and blackbirds, have increased their numbers and distribution over the last 30 years, forest songbirds have declined severely. The decline has been greatest among long-distance migrants such as thrushes, orioles, tanagers, catbirds, vireos, buntings, and warblers. These birds nest in northern forests in the summer, but spend their winters in South or Central America or the Caribbean Islands.

In many areas of the eastern United States, more than three-quarters of the tropical migrant bird species have declined significantly. Rock Creek Park in Washington, D.C., for example, has lost 90% of its long-distance migrants in the past 20 years. Nationwide, American redstarts declined about 50% in the single de-

figure 59.13

A STUDY OF HABITAT FRAGMENTATION. Landowners in Manaus, Brazil, agreed to preserve patches of rain forest of different sizes to examine the effect of patch size on species extinction. Biodiversity was monitored in the isolated patches before and after logging. Fragmentation led to significant species loss within patches.

cade of the 1970s. Studies of radar images from National Weather Service stations in Texas and Louisiana indicate that only about half as many birds fly over the Gulf of Mexico each spring as did in the 1960s. This suggests a total loss of about half a billion birds.

The culprit responsible for this widespread decline appears to be habitat fragmentation and loss. Fragmentation of breeding habitat and nesting failures in the summer nesting grounds of the United States and Canada have had a major negative impact on the breeding of woodland songbirds. Many of the most threatened species are adapted to deep woods and need an area of 25 acres or more per pair to breed and raise their young. As woodlands are broken up by roads and developments, it is becoming increasingly difficult for them to find enough contiguous woods to nest successfully.

A second and perhaps even more important factor is the availability of critical winter habitat in Central and South America. Studies of the American redstart clearly indicate that birds with better winter habitat have a superior chance of successfully migrating back to their breeding grounds in the spring. In a recent study, scientists were able to determine the quality of the habitat that particular birds used during the winter by examining levels of the stable carbon isotope ^{13}C in their blood. Plants growing in the best overwintering habitats in Jamaica and Honduras (mangroves and wetland forests) have low levels of ^{13}C, and so do the redstarts that feed on the insects that live in them. Of these wet-forest birds, 65% maintained or gained weight over the winter.

By contrast, plants growing in substandard dry scrub have high levels of ^{13}C, and so do the redstarts that feed in those habitats. Scrub-dwelling birds lost up to 11% of their body mass over the winter. Now here's the key: Birds that winter in the substandard scrub leave later in the spring on the long flight to northern breeding grounds, arrive later at their summer homes, and have fewer young (figure 59.14).

The proportion of ^{13}C in birds arriving in New Hampshire breeding grounds increases as spring wears on and scrub-overwintering stragglers belatedly arrive. Thus, loss of mangrove habitat in the neotropics is having a quantifiable negative impact. As the best habitat disappears, overwintering birds fare poorly, and this leads to decreased reproduction and population declines.

Unfortunately, the Caribbean lost about 10% of its mangroves in the 1980s, and continues to lose about 1% per year. This loss of key habitat appears to be a driving force in the looming extinction of some songbirds.

Overexploitation wipes out species quickly

Species that are hunted or harvested by humans have historically been at grave risk of extinction, even when the species is initially very abundant. A century ago, the skies of North America were darkened by huge flocks of passenger pigeons, but after being hunted as free and tasty food, they were driven to extinction. The bison that used to migrate in enormous herds across the central plains of North America only narrowly escaped the same fate.

Commercial motivation for exploitation

The existence of a commercial market often leads to overexploitation of a species. The international trade in furs, for example, has severely reduced the numbers of chinchilla, vicuña, otter, and many

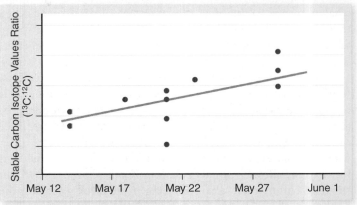

figure 59.14

THE AMERICAN REDSTART, *Setophaga ruticilla*, A MIGRATORY SONGBIRD. The numbers of this species are in serious decline. The graph presents data on the ratio of ^{13}C to ^{12}C in male redstarts arriving at summer breeding grounds. Early arrivals, which have higher reproductive success, have lower proportions of $^{13}C:^{12}C$, indicating they wintered in more favorable mangrove–wetland forest habitats.

cat species. The harvesting of commercially valuable trees provides another example: Almost all West Indies mahogany trees (*Swietenia mahogani*) have been logged, and the extensive cedar forests of Lebanon, once widespread at high elevations, now survive in only a few isolated groves.

A particularly telling example of overexploitation is the commercial harvesting of fish in the North Atlantic. During the 1980s, fishing fleets continued to harvest large amounts of cod off the coast of Newfoundland, even as the population numbers declined precipitously. By 1992, the cod population had dropped to less than 1% of its original numbers. The American and Canadian governments have closed the fishery, but no one can predict whether the fish populations will recover. The Atlantic bluefin tuna has experienced a 90% population decline in the past 10 years. The swordfish has declined even further. In both cases, the drop has led to even more intense fishing of the remaining populations.

Case study: Whales

Whales, the largest living animals that ever evolved, are rare in the world's oceans today, their numbers driven down by commercial whaling. Before the advent of cheap, high-grade oils manufactured from petroleum in the early twentieth century, oil made from whale blubber was an important commercial product in the worldwide marketplace. In addition, the fine, latticelike structure termed "baleen" used by baleen whales to filter-feed plankton from seawater was used in women's undergarments. Because a whale is such a large animal, each individual captured is of significant commercial value.

In the eighteenth century, right whales were the first to bear the brunt of commercial whaling. They were called "right" whales because they were slow and easy to capture, and they provided up to 150 barrels of blubber oil and abundant baleen, making them the right whale for a commercial whaler to hunt.

As the right whale declined, whalers turned to the gray, humpback, and bowhead whales. As their numbers declined, whalers turned to the blue, the largest of all whales, and when those were decimated, to the fin, then the Sei, and then the sperm whales. As each species of whale became the focus of commercial whaling, its numbers began a steep decline (figure 59.15).

Hunting of right whales was made illegal in 1935. By then, they had been driven to the brink of extinction, their numbers less than 5% of what they had been. Although protected ever since, their numbers have not recovered in either the North Atlantic or the North Pacific. By 1946, several other whale species faced imminent extinction, and whaling nations formed the International Whaling Commission (IWC) to regulate commercial whale hunting. Like a fox guarding the henhouse, the IWC for decades did little to limit whale harvests, and whale numbers continued to decline steeply.

Finally, in 1974, when the numbers of all but the small minke whales had been driven down, the IWC banned hunting of blue, gray, and humpback whales, and instituted partial bans on other species. The rule was violated so often, however, that the IWC in 1986 instituted a worldwide moratorium on all commercial killing

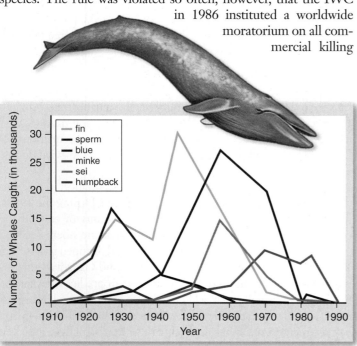

figure 59.15

WORLD CATCH OF WHALES IN THE TWENTIETH CENTURY. Each species is hunted in turn until its numbers fall so low that hunting it becomes commercially unprofitable.

 inquiry

Why might whale populations fail to recover once hunting is stopped?

of whales. Although some commercial whaling continues, often under the guise of harvesting for scientific studies, annual whale harvests have dropped dramatically in the last 20 years.

Some species appear to be recovering, but others are not. Humpback numbers have more than doubled since the early 1960s, increasing nearly 10% annually, and Pacific gray whales have fully recovered to their previous numbers of about 20,000 animals, after having been hunted to less than 1000. Right, sperm, fin, and blue whales have not recovered, and no one knows whether they ever will.

Introduced species threaten native species and habitats

Colonization, a natural process by which a species expands its geographic range, occurs in many ways: A flock of birds gets blown off course, a bird eats a fruit and defecates its seed miles away, or lowered sea levels connect two previously isolated landmasses, allowing species to freely move back and forth. Such events—particularly those leading to successful establishment of a new population—probably occur rarely, but when they do, the resulting change to natural communities can be large. The reason is that colonization brings together species with no history of interaction. Consequently, ecological interactions may be particularly strong because the species have not evolved ways of adjusting to the presence of one another, such as adaptations to avoid predation or to minimize competitive effects.

The paleontological record documents many cases in which geologic changes brought previously isolated species together, such as when the Isthmus of Panama emerged above the sea approximately 3 million years ago, connecting the previously isolated fauna and flora of North and South America. In some cases, the result has been an increase in species diversity, but in other cases, invading species have led to the extinction of natives.

Human influence on colonization

Unfortunately, what was naturally a rare process has become all too common in recent years, thanks to the actions of humans. Species introductions due to human activities occur in many ways, sometimes intentionally, but usually not. Plants and animals can be transported in the ballast of large ocean vessels; in nursery plants; as stowaways in boats, cars, and planes; as beetle larvae within wood products—even as seeds or spores in the mud stuck to the bottom of a shoe. Overall, some researchers estimate that as many as 50,000 species have been introduced into the United States.

The effects of introductions on humans have been enormous. In the United States alone, nonnative species cost the economy an estimated $140 billion per year. For example, dozens of foreign weeds in Colorado have covered more than a million acres. Just three of these species cost wheat farmers tens of millions of dollars. At the same time, leafy spurge, a plant from Europe, outcompetes native grasses, ruining rangeland for cattle at a price tag of $144 million per year.

The zebra mussel, a mollusk native to the Black Sea region, is a huge problem throughout much of the eastern and central United States, where it can attain densities as high as 700,000/m^2, clogging pipes, including those for water and power plants, and causing an estimated $3 to $5 billion damage a year (figure 59.16).

figure 59.16

ZEBRA MUSSELS (*Dreissena polymorpha*) CLOGGING A PIPE. These mussels were introduced from Europe, and are now a major problem in North American rivers.

Introduced species can also affect human health. For example, West Nile fever was probably introduced from Africa or the Middle East in the late 1990s.

The effect of species introductions on native ecosystems is equally dramatic. Islands have been particularly affected. For example, as mentioned in an earlier chapter, a single lighthouse keeper's cat wiped out an entire species, the Stephens Island wren. Rats had a devastating effect throughout the South Pacific where bird species nested on the ground and had no defense against the voracious predators to which they were evolutionarily naive. More recently, the brown tree snake, introduced to the island of Guam, essentially eliminated all species of forest birds.

In Hawaii, the problem has been slightly different: Introduced mosquitoes brought with them malaria, to which the native species had evolved no resistance. The result is that more than 100 species (greater than 70% of the native fauna) either became extinct or are now restricted to higher and cooler elevations where the mosquitoes don't occur (figure 59.17).

The effects of introduced species are not always direct, but instead may reverberate throughout an ecosystem. For example, the Argentine ant has spread through much of the southern United States, greatly reducing populations of most native ant species with which it comes in contact. The extinction of these ant species has had a dramatic negative effect on the coast horned lizard (*Phrynosoma coronatum*), which feeds on the larger native species. In their absence, the lizards have shifted to less-preferred prey species. In addition, the native ant species consume seeds, and in the process, play an important role in seed dispersal. Argentine ants, by contrast, do not eat seeds. In South Africa, where the Argentine ant has also appeared, at least one plant species has experienced decreased reproductive success due to the loss of its dispersal agent.

The most dramatic effects of introduced species, however, occur when entire ecosystems are transformed. Some plant species can completely overrun a habitat, displacing all native species and turning the area into a monoculture (that is, an area occupied by a single species). In California, the yellow star thistle now covers 4 million hectares of what was once highly productive grassland. In Hawaii, a small tree native to the Canary Islands, *Myrica faya*, has spread widely. Because it is able to fix nitrogen at high rates, it has caused a 90-fold increase in the nitrogen content of the soil, thus allowing other, nitrogen-requiring species to invade.

Efforts to combat introduced species

Once an introduced species becomes established, eradicating it is often extremely difficult, expensive, and time-consuming. Some efforts—such as the removal of goats and rabbits from certain small islands—have been successful, but many other efforts have failed. The best hope for stopping the ravages of introduced species is to prevent them from being introduced in the first place. Although easier said than done, government agencies are now working strenuously to put into place procedures that can intercept species in transit, before they have the opportunity to become established.

Case study: Lake Victoria cichlids

Lake Victoria, an immense, shallow, freshwater sea about the size of Switzerland in the heart of equatorial East Africa, used to be home to an incredibly diverse collection of over 300 species of cichlid fishes (see figure 22.14). These small, perchlike fish range from 5 to 13 cm in length, with males having endless varieties of color. Today, most of these cichlid species are threatened, endangered, or extinct.

What happened to bring about the abrupt loss of so many endemic cichlid species? In 1954, the Nile perch, a commercial

figure 59.17

THE AKIAPOLAAU (*Hemignathus munroi*) AND THE PALILA (*Loxioides bailleui*), ENDANGERED HAWAIIAN BIRDS. More than two-thirds of Hawaii's native bird species are now extinct or have been greatly reduced in population size. Bird faunas on islands around the world have experienced similar declines after human arrival.

fish with a voracious appetite, was purposely introduced on the Ugandan shore of Lake Victoria. Nile perch, which grow to over a meter in length, were to form the basis of a new fishing industry (figure 59.18). For decades, these perch did not seem to have a significant effect; over 30 years later, in 1978, Nile perch still made up less than 2% of the fish harvested from the lake.

Then something happened to cause the Nile perch population to explode and to spread rapidly through the lake, eating their way through the cichlids. By 1986, Nile perch constituted nearly 80% of the total catch of fish from the lake and the endemic cichlid species were virtually gone. Over 70% of cichlid species disappeared, including all open-water species.

So what happened to kick-start the mass extinction of the cichlids? The trigger seems to have been eutrophication. Before 1978, Lake Victoria had high oxygen levels at all depths, down to the bottom layers more than 60 m deep. However, by 1989 high inputs of nutrients from agricultural runoff and sewage from towns and villages had led to algal blooms that severely depleted oxygen levels in deeper parts of the lake. Cichlids feed on algae, and initially their population numbers are thought to have risen in response to this increase in their food supply, but unlike the conditions during similar algal blooms of the past, the Nile perch was present to take advantage of the situation. With a sudden increase in its food supply (cichlids), the numbers of Nile perch exploded, and they simply ate all available cichlids.

Since 1990, the situation has been compounded by the introduction into Lake Victoria of a floating water weed from South America, the water hyacinth *Eichhornia crassipes*. Reproducing quickly under eutrophic conditions, thick mats of water hyacinth soon covered entire bays and inlets, choking off the coastal habitats of non-open-water cichlids.

Disruption of ecosystems can cause an extinction cascade

Species often become vulnerable to extinction when their web of ecological interactions becomes seriously disrupted. Because of the many relationships linking species in an ecosystem (see chapter 57), human activities that affect one species can have ramifications throughout an ecosystem, ultimately affecting many other species.

A recent case in point involves the sea otters that live in the cold waters off Alaska and the Aleutian Islands. Otter populations have declined sharply in recent years. In a 500-mile stretch of coastline, otter numbers have dropped from 53,000 in the 1970s to an estimated 6000, a plunge of nearly 90%. Investigating this catastrophic decline, marine ecologists uncovered a chain of interactions among the species of the ocean and kelp forest ecosystems, a falling-domino series of lethal effects that illustrates the concepts of both top-down and bottom-up trophic cascades discussed in chapter 57.

Case study: Alaskan near-shore habitat

The first in a series of events leading to the sea otter's decline seems to have been the heavy commercial harvesting of whales, described earlier in this chapter. Without whales to keep their numbers in check, ocean zooplankton thrived, leading in turn to proliferation of a species of fish called pollock that feeds on the abundant zooplankton. Given this ample food supply, the pollock proved to compete very successfully with other north-

figure 59.18

NILE PERCH (*Lates niloticus*). This predatory fish, which can reach a length of 2 m and a mass of 200 kg, was introduced into Lake Victoria as a potential food source. It is responsible for the virtual extinction of hundreds of species of cichlid fishes.

ern Pacific fish, such as herring and ocean perch, so that levels of these other fish fell steeply in the 1970s.

Then the falling chain of dominoes began to accelerate. The decline in the nutritious forage fish led to an ensuing crash in Alaskan populations of sea lions and harbor seals, for which pollock did not provide sufficient nourishment. This decline may also have been hastened by orcas (also called killer whales) switching from feeding on the less-available whales to feeding on seals and sea lions. The numbers of these pinniped species have fallen precipitously since the 1970s.

When pinniped numbers crashed, some orcas, faced with a food shortage, turned to the next best thing: sea otters. In one bay where the entrance from the sea was too narrow and shallow for orcas to enter, only 12% of the sea otters have disappeared, while in a similar bay that orcas could enter easily, two-thirds of the otters disappeared in a year's time.

Without otters to eat them, the population of sea urchins exploded, eating the kelp and thus "deforesting" the kelp forests and denuding the ecosystem (figure 59.19). As a result, fish species that live in the kelp forest, such as sculpins and greenlings, are declining.

Loss of keystone species may disrupt ecosystems

As discussed in chapter 56, a keystone species is a species that exerts a greater influence on the structure and functioning of an ecosystem than might be expected solely on the basis of its abundance. The sea otters of figure 59.19 are a keystone species of the kelp forest ecosystem, and their removal can have disastrous consequences.

No hard-and-fast line allows us to clearly identify keystone species. Rather, it is a qualitative concept, a statement that indicates a species plays a particularly important role in its community. Keystone species are usually characterized by the strength of their impact on their community.

Case study: Flying foxes

The severe decline of many species of "flying foxes," a type of bat (figure 59.20), in the Old World tropics is an example of how the loss of a keystone species can dramatically affect the other species living within an ecosystem, sometimes even leading to a cascade of further extinctions.

These bats have very close relationships with important plant species on the islands of the Pacific and Indian Oceans. The family Pteropodidae contains nearly 200 species, approximately one-quarter of them in the genus *Pteropus*, and is widespread on the islands of the South Pacific, where they are the most important—and often the only—pollinators and seed dispersers.

A study in Samoa found that 80% to 100% of the seeds landing on the ground during the dry season were deposited by flying foxes, which eat the fruits and defecate the seeds, often moving them great distances in the process. Many species are entirely dependent on these bats for pollination. Some have evolved features such as night-blooming flowers that prevent any other potential pollinators from taking over the role of the fruit bats.

In Guam, where the two local species of flying fox have recently been driven extinct or nearly so, the impact on the ecosystem appears to be substantial. Botanists have found that some plant species are not fruiting or are doing so only marginally, producing fewer fruits than normal. Fruits are not being dispersed away from parent plants, so seedlings are forced to compete, usually unsuccessfully, with adult trees.

Flying foxes are being driven to extinction by human hunters who kill them for food and for sport, and by orchard farmers who consider them pests. Flying foxes are particularly vulnerable because they live in large and obvious groups of up to a million individuals. Because they move in regular and predictable patterns and can be easily tracked to their home roost, hunters can easily kill thousands at a time.

Programs aimed at preserving particular species of flying foxes are only just beginning. One particularly successful example is the program to save the Rodrigues fruit bat, *Pteropus rodricensis*, which occurs only on Rodrigues Island in the Indian

1. Whales
Overharvesting of plankton-eating whales may have caused an increase in plankton-eating pollock populations.

2. Nutritious fish
Populations of nutritious fish like ocean perch and herring declined, likely due to competition with pollock.

3. Sea lions and harbor seals
Sea lion and harbor seal populations drastically declined in Alaska, probably because the less-nutritious pollock could not sustain them.

4. Killer whales
With the decline in their prey populations of sea lions and seals, killer whales turned to a new source of food: sea otters.

5. Sea otters
Sea otter populations declined so dramatically that they disappeared in some areas.

6. Sea urchins
Usually the preferred food of sea otters, sea urchin populations now exploded and fed on kelp.

7. Kelp forests
Severely thinned by the sea urchins, the kelp beds no longer support a diversity of fish species, which may lead to a decline in populations of eagles that feed on the fish.

figure 59.19

DISRUPTION OF THE KELP FOREST ECOSYSTEM.
Overharvesting by commercial whalers altered the balance of fish in the ocean ecosystem, inducing killer whales to feed on sea otters, a keystone species of the kelp forest ecosystem.

figure 59.20

THE IMPORTANCE OF KEYSTONE SPECIES. Flying foxes, a type of fruit-eating bat, are keystone species on many Old World tropical islands. It pollinates many plants and is a key disperser of seeds. Its elimination due to hunting and habitat loss is having a devastating effect on the ecosystems of many South Pacific Islands.

Ocean near Madagascar. The population dropped from about 1000 individuals in 1955 to fewer than 100 by 1974, largely due to the loss of the fruit bat's forest habitat to farming. Since 1974, the species has been legally protected, and the forest area of the island is being increased through a tree-planting program. Eleven captive-breeding colonies have been established, and the bat population is now increasing rapidly. The combination of legal protection, habitat restoration, and captive breeding has in this instance produced a very effective preservation program.

Small populations are particularly vulnerable

Because of the factors just discussed, populations of many species are fragmented and reduced in size. Such populations are particularly prone to extinction.

Demographic factors

Small populations are vulnerable to events that decrease survival or reproduction. For example, by nature of their size, small populations are ill-equipped to withstand catastrophes, such as a flood, forest fire, or disease epidemic. One example is provided by the history of the heath hen. Although the species was once common throughout the eastern United States, hunting pressure in the eighteenth and nineteenth centuries eventually eliminated all but one population, on the island of Martha's Vineyard near Cape Cod, Massachusetts. Protected in a nature preserve, the population was increasing in number until a fire destroyed most of the preserve's habitat. The small surviving population was then ravaged the next year by an unusual congregation of predatory birds, followed shortly thereafter by a disease epidemic. The last sighting of a heath hen, a male, was in 1932.

When populations become extremely small, bad luck can spell the end. For example, the dusky seaside sparrow (figure 59.21), a now-extinct subspecies that was found on the east coast of Florida, dwindled to a population of five individuals, all of which happened to be males. In a large population, the probability that all individuals will be of one sex is infinitesimal. But in small populations, just by the luck of the draw, it is possible that 5 or 10 or even 20 consecutive births will all be individuals of one sex, and that can be enough to send a species to extinction. In addition, when populations are small, individuals may have trouble finding each other (the Allee effect discussed in chapter 55), thus leading the population into a downward spiral toward extinction.

Lack of genetic variability

Small populations face a second dilemma. Because of their low numbers, such populations are prone to the loss of genetic variation as a result of genetic drift (figure 59.22). Indeed, many small populations contain little or no genetic variability. The result of such genetic homogeneity can be catastrophic. Genetic variation is beneficial to a population both because of heterozygote advantage (see chapter 20) and because genetically variable individuals tend not to have two copies of deleterious recessive alleles. Populations lacking variation are often composed of sickly, unfit, or sterile individuals. Laboratory groups of rodents and fruit flies that are maintained at small population sizes often perish after a few generations as each generation becomes less robust and fertile than the preceding one.

Although it is difficult to demonstrate that a species has gone extinct because of lack of genetic variation, studies of both zoo and natural populations clearly reveal that more genetically variable individuals have greater fitness. Furthermore, in the longer term, populations with limited genetic variation have diminished ability to adapt to changing environments.

Interaction of demographic and genetic factors

As populations decrease in size, demographic and genetic factors combine to cause what has been termed an "extinction vortex." That is, as a population gets smaller, it becomes more vulnerable to demographic catastrophes. In turn, genetic varia-

a.

b.

figure 59.21

ALIVE NO MORE. *a.* A museum specimen of the heath hen, *Tympanuchus cupido cupido*, which went extinct in 1932. *b.* This male was one of the last dusky seaside sparrows, *Ammodramus maritimus nigrescens*.

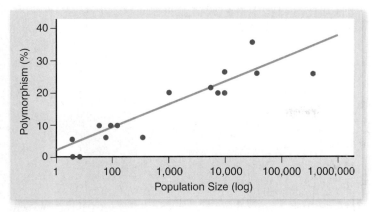

figure 59.22

LOSS OF GENETIC VARIABILITY IN SMALL POPULATIONS.
The percentage of genes that are polymorphic in isolated
populations of the tree *Halocarpus bidwillii* in the mountains of
New Zealand is a sensitive function of population size.

 inquiry

Why do small populations lose genetic variation?

tion starts to be lost, causing reproductive rates to decline and
population numbers to decline even further, and so on. Eventually, the population disappears entirely, but attributing its demise to one particular factor would be misleading.

Case study: Prairie chickens

The greater prairie chicken, a close relative of the now-extinct
heath hen, is a showy, 2-lb bird renowned for its flamboyant mating rituals (figure 59.23). Abundant in many midwestern states,
the prairie chickens in Illinois have in the past six decades undergone a population collapse.

Once, enormous numbers of birds occurred throughout the
state, but with the 1837 introduction of the steel plow, the first
that could slice through the deep, dense root systems of prairie
grasses, the Illinois prairie began to be replaced by farmland. By
the turn of the century, the prairie had all but vanished, and by
1931, the heath hen, had become locally extinct in Illinois. The
greater prairie chicken fared little better, its numbers falling to
25,000 statewide in 1933 and then to 2000 by 1962. In surrounding states with less intensive agriculture, it continued to prosper.

In 1962 and 1967, sanctuaries were established in Illinois to
attempt to preserve the greater prairie chicken. But privately owned
grasslands kept disappearing, along with their prairie chickens, and
by the 1980s the birds were extinct in Illinois except for two preserves, and even there, their numbers kept falling. By 1990, the egg
hatching rate, which at one time had averaged between 91% and
100%, had dropped to an extremely low 38%. By the mid-1990s,
the count of males had dropped to as low as six in each sanctuary.

What was wrong with the sanctuary populations? One suggestion was that because of very small population sizes and a mating
ritual whereby one male may dominate a flock, the Illinois prairie
chickens had lost so much genetic variability as to create serious
genetic problems. To test this idea, biologists at the University of
Illinois compared DNA from frozen tissue samples of birds that had
died in Illinois between 1974 and 1993, and found that in recent
years Illinois birds had indeed become genetically less diverse.

The researchers then extracted DNA from tissue in the
roots of feathers from stuffed birds collected in the 1930s from
the same population. They found that Illinois birds had lost
fully one-third of the genetic diversity of birds living in the
same place before the population collapse of the 1970s. By contrast, prairie chicken populations in other states still contained
much of the genetic variation that had disappeared from Illinois populations.

Now the stage was set to halt the Illinois prairie chicken's
race toward extinction in Illinois. Wildlife managers began to
transplant birds from genetically diverse populations of Minnesota, Kansas, and Nebraska to Illinois. Between 1992 and 1996,
a total of 518 out-of-state prairie chickens were brought in to
interbreed with the Illinois birds, and hatching rates were back
up to 94% by 1998. It looks as though the prairie chickens have
been saved from extinction in Illinois.

The key lesson to be learned is the importance of not
allowing things to go too far—not to drop down to a single
isolated population. Without the outlying genetically different
populations, the prairie chickens in Illinois could not have been
saved. When the last population of the dusky seaside sparrow
lost its last female, there was no other source of females, and
the subspecies went extinct.

Many factors are responsible for extinction. As habitats are
destroyed, remaining habitat becomes fragmented, leading
to increased risk of extinction. Over-exploitation can reduce
populations to low levels or eliminate them entirely. Introduced
species can wreak havoc on native communities through predation
or competition and can disrupt an entire ecosystem. Finally, small
populations face a variety of perils because they have lessened
ability to rebound from catastrophes and are vulnerable to loss of
genetic variation.

figure 59.23

A MATING RITUAL. The male greater prairie chicken
(*Tympanuchus cupido pinnatus*) inflates bright orange air sacs, part
of his esophagus, into balloons on each side of his head. As air
is drawn into the sacs, it creates a three-syllable low frequency
"boom-boom-boom" that can be heard for kilometers.

Approaches for Preserving Endangered Species

Once the cause of a species' endangerment is known, it becomes possible to design a recovery plan. If the cause is commercial overharvesting, regulations can be issued to lessen the impact and protect the threatened species. If the cause is habitat loss, plans can be instituted to restore the habitat. Loss of genetic variability in isolated subpopulations can be countered by transplanting individuals from genetically different populations. Populations in immediate danger of extinction can be captured, introduced into a captive-breeding program, and later reintroduced to other suitable habitat.

All of these solutions are extremely expensive. But as Bruce Babbitt, Secretary of the Interior in the Clinton administration, noted, it is much more economical to prevent "environmental trainwrecks" from occurring than to clean them up afterwards. Preserving ecosystems and monitoring species before they are threatened is the most effective means of protecting the environment and preventing extinctions.

Destroyed habitats can sometimes be restored

Conservation biology typically concerns itself with preserving populations and species in danger of decline or extinction. Conservation, however, requires that there be something left to preserve; in many situations, conservation is no longer an option. Species, and in some cases whole communities, have disappeared or been irretrievably modified. The clear-cutting of the temperate forests of Washington State leaves little behind to conserve, as does converting a piece of land into a wheat field or an asphalt parking lot. Redeeming these situations requires restoration rather than conservation.

Three quite different sorts of habitat restoration programs might be undertaken, depending on the cause of the habitat loss.

Pristine restoration

In ecosystems where all species have been effectively removed, conservationists might attempt to restore the plants and animals that are the natural inhabitants of the area, if these species can be identified. When abandoned farmland is to be restored to prairie, as in figure 59.24, how would conservationists know what to plant?

Although it is in principle possible to reestablish each of the original species in their original proportions, rebuilding a community requires knowing the identities of all the original inhabitants and the ecologies of each of the species. We rarely have this much information, so no restoration is ever truly pristine.

Removing introduced species

Sometimes the habitat has been destroyed by a single introduced species. In such a case, habitat restoration involves removing the introduced species. Restoration of the once-diverse cichlid fishes to Lake Victoria will require more than breeding and restocking the endangered species. The introduced water hyacinth and Nile perch populations will have to be brought under control or removed, and eutrophication will have to be reversed.

a.

b.

figure 59.24

HABITAT RESTORATION. The University of Wisconsin–Madison Arboretum has pioneered restoration ecology. *a.* The restoration of the prairie was at an early stage in November 1935. *b.* The prairie as it looks today. This picture was taken at approximately the same location as the 1935 photograph.

It is important to act quickly if an introduced species is to be removed. When aggressive African bees (the so-called "killer bees") were inadvertently released in Brazil, they remained confined to the local area for only one season. Now they occupy much of the western hemisphere.

Cleanup and rehabilitation

Habitats seriously degraded by chemical pollution cannot be restored until the pollution is cleaned up. The successful restoration of the Nashua River in New England is one example of how a concerted effort can succeed in restoring a heavily polluted habitat to a relatively pristine condition.

Captive breeding programs have saved some species

Recovery programs, particularly those focused on one or a few species, must sometimes involve direct intervention in natural populations to avoid an immediate threat of extinction.

Case study: The peregrine falcon

American populations of birds of prey, such as the peregrine falcon, began an abrupt decline shortly after World War II. Of the approximately 350 breeding pairs east of the Mississippi River in 1942, all had disappeared by 1960. The culprit proved to be the chemical pesticide DDT (dichlorodiphenyltrichloroethane) and related organochlorine pesticides. Birds of prey are particularly vulnerable to DDT because they feed at the top of the food chain, where DDT becomes concentrated. DDT interferes with the deposition of calcium in the bird's eggshells, causing most of the eggs to break before they are ready to hatch.

The use of DDT was banned by federal law in 1972, causing levels in the eastern United States to fall quickly. However, no peregrine falcons were left in the eastern United States to reestablish a natural population. Falcons from other parts of the country were used to establish a captive-breeding program at Cornell University in 1970, with the intent of reestablishing the peregrine falcon in the eastern United States by releasing offspring of these birds. By the end of 1986, over 850 birds had been released in 13 eastern states, producing an astonishingly strong recovery (figure 59.25).

Case study: The California condor

The number of California condors (*Gymnogyps californianus*), a large, vulturelike bird with a wingspan of nearly 3 m, has been declining gradually for

the past 200 years. By 1985, condor numbers had dropped so low that the bird was on the verge of extinction. Six of the remaining 15 wild birds disappeared in that year alone. The entire breeding population of the species consisted of the birds remaining in the wild and an additional 21 birds in captivity.

In a last-ditch attempt to save the condor from extinction, the remaining birds were captured and placed in a captive-breeding population. The breeding program was set up in zoos, with the goal of releasing offspring on a large, 5300-hectare ranch in prime condor habitat. Birds were isolated from human contact as much as possible, and closely related individuals were prevented from breeding.

By early 2004, the captive population of California condors had reached over 149 individuals. Fifty-six captive-reared condors have been released successfully in California at two sites in the mountains north of Los Angeles, after extensive prerelease training to avoid power poles and people. All of the released birds seem to be doing well, and another 53 birds released into the Grand Canyon have adapted successfully. Biologists are particularly excited by breeding activities that resulted in the first-ever offspring produced in the wild by captive-reared parents in both California and Arizona.

Case study: Yellowstone wolves

The ultimate goal of captive-breeding programs is not simply to preserve interesting species, but rather to restore ecosystems to a balanced, functional state. Yellowstone Park has been an ecosystem out of balance, due in large part to the systematic extermination of the gray wolf (*Canis lupus*) in the park early in the twentieth century. Without these predators to keep their numbers in check, herds of elk and deer expanded rapidly, damaging vegetation so that the elk themselves starve in time of scarcity.

In an attempt to restore the park's natural balance, two complete wolf packs from Canada were released into the park in 1995 and 1996. The wolves adapted well, breeding so successfully that by 2002 the park contained 16 free-ranging packs and more than 200 wolves.

While ranchers near the park have been unhappy about the return of the wolves, little damage to livestock has been noted, and the ecological equilibrium of Yellowstone Park seems well on the way to being regained. Elk are congregating in larger herds and are avoiding areas near rivers where they are vulnerable. As a result, riverside trees such as willows are increasing in number, in turn providing food for beavers, whose dams lead to the creation of ponds, a habitat type that had become rare in Yellowstone. This newly restored habitat, in turn, has led to increases in some species of birds such as the redstart that had been in decline for decades or disappeared entirely.

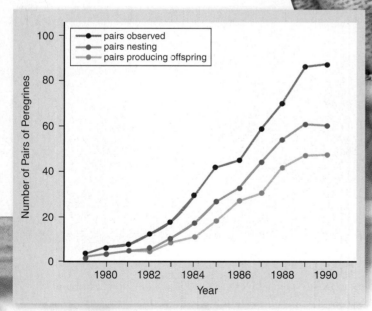

figure 59.25

SUCCESS OF CAPTIVE BREEDING. The peregrine falcon (*Falco peregrinus*) has been reestablished in the eastern United States by releasing captive-bred birds over a period of 10 years.

Efforts to preserve endangered species are as diverse as the causes of endangerment. Restoration can sometimes return an area to viability as a natural habitat. Although captive breeding is not a solution in all, or even most, cases, it has helped restore several vertebrate species.

Conservation of Ecosystems

Habitat fragmentation is one of the most pervasive enemies of biodiversity conservation efforts. As you have seen, some species require large patches of habitat to thrive, and conservation efforts that cannot provide suitable habitat of such a size are doomed to failure. Historically, conservationists strived to solve this dilemma by focusing solely on preserving as much land as possible in a pristine state in national parks and reserves. Increasingly, however, it has become apparent that the amount of land that can be preserved in such a state is limited; moreover, many areas that are not completely protected nonetheless provide suitable habitat for many species.

As a result, conservation plans are becoming multidimensional, including not only pristine areas, but also surrounding areas in which some level of human disturbance is permitted. As discussed previously, isolated patches of habitat lose species far more rapidly than large preserves do. By including these other, less pristine areas, the total amount of area available for many species is increased.

The key to managing such large tracts of land successfully over a long time is to operate them in a way compatible with local land use. For example, while no economic activity is allowed in the core pristine area, the remainder of the land may be used for nondestructive harvesting of resources. Even areas in which hunting of some species is allowed provide protection for many other species.

Corridors of dispersal are also being provided that link the pristine areas, thus effectively increasing population sizes and allowing recolonization if a population disappears in one area due to a catastrophe. Corridors can also provide protection to species that move over great distances during the course of a year. Corridors in East Africa have protected the migration routes of ungulates. In Costa Rica, a corridor linking the lowland rain forest at the La Selva Biological Station to the montane rain forest in Braulio Carrillo National Park permits the altitudinal migration of many species of birds, mammals, and butterflies (figure 59.26).

In addition to this focus on maintaining large enough reserves, in recent years conservation biologists also have recognized that the best way to preserve biodiversity is to focus on preserving intact ecosystems, rather than particular species. For this reason, attention in many cases is turning to identifying those ecosystems most in need of preservation and devising the means to protect not only the species within the ecosystem, but the functioning of the ecosystem itself. This entails making sure that reserves are not only large enough, but also that they protect elements such as watersheds so that activities outside the reserve won't threaten the ecosystem within it.

Efforts are being undertaken worldwide to preserve biodiversity in networks of reserves and other less protected areas designed to counter the influences of habitat fragmentation. Focusing on the health of entire ecosystems, rather than particular species, can often be a more effective means of preserving biodiversity.

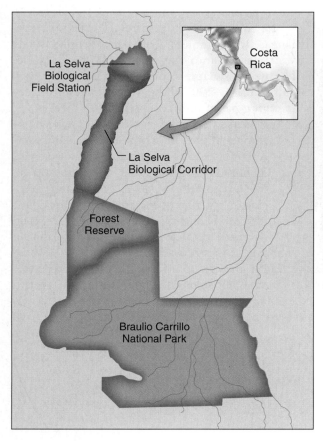

figure 59.26

CORRIDOR CONNECTING TWO RESERVES. *a.* The Organization of Tropical Studies' La Selva Biological Station in Costa Rica is connected to Braulio Carrillo National Park. *b.* The corridor allows migration of birds, mammals, butterflies, and other animals from La Selva at 35 m above sea level to mountainous habitats up to 2900 m elevation.

59.1 Overview of the Biodiversity Crisis

Multiple extinctions are well known from the fossil record, but the current rates of extinction are extremely high; many species are being lost before they can be discovered or identified.

- The majority of historical extinctions have occurred within the last 150 years and on islands.

- In recent times humans have accounted for the majority of local extinctions by overexploitation and habitat destruction.

- Agriculture, housing, and commercial purposes contribute to species losses by destroying or reducing habitat.

- Africa is the only continent where megafaunal extinctions have not occurred perhaps because of the long period of coevolution by humans and indigenous fauna.

- Endemic species are particularly vulnerable because they are found in one restricted range and areas with high rates of human population growth and immigration.

- Hotspots are areas with many endemic species exhibiting high rates of extinction.

- The current mass extinction is unique because it is the only mass extinction event triggered by one species, *Homo sapiens*, and the only one in which resources will not be widely available for evolutionary recovery afterward.

59.2 The Value of Biodiversity

Biodiversity has economic, ethical, and aesthetic values.

- The direct economic value of biodiversity comes from the products we obtain from different species and ecosystems.

- The indirect economic value of biodiversity comes from the ecosystem services they provide, such as maintaining water quality, preservation of soils and nutrients, affecting local climates, and recycling nutrients.

- Intact ecosystems are often more valuable than the products we obtain from them; however, their value is often not apparent until they are lost.

- The persons who benefit from environmental degradation are often not the same persons who pay the costs.

- Humans can and should make ethical decisions to protect the esthetic, ecological, and economic values of ecosystems.

59.3 Factors Responsible for Extinction

Many direct and indirect factors are responsible for species extinctions.

- Overexploitation and habitat loss by humans are the two most important causes of extinction.

- The susceptibility of a species to one extinction factor may increase the vulnerability of that species to other factors and hasten their demise.

- Different species are not affected by the same problems in the same ways.

- Humans reduce species richness by destroying, polluting, disrupting, and fragmenting habitats.

- As habitats become more and more fragmented, the relative proportion of the remaining habitat that occurs on the boundary or edge increases rapidly.

- As the relative proportion of edge increases, species become more exposed to parasites and predators.

- As the area of an island or fragment decreases extinction rates increase (figure 59.11).

- Natural or accidental introductions of new species will result in large and often negative changes to a community because there is no history of species interactions.

- Disruptions of ecosystem interactions may cause an extinction cascade from either the top-down or the bottom-up trophic levels.

- Loss of a keystone species may increase competition and greatly influence ecosystem structure and function.

- Small populations are particularly vulnerable to extinction because of catastrophes, lack of mates, and loss of genetic variability (figure 59.22).

59.4 Approaches for Preserving Endangered Species

Once the cause of species' endangerment is identified, it is possible to design a recovery plan that may protect them.

- Conservation biology is concerned with preserving populations and species in danger of decline or extinction.

- Restoration of damaged habitats to natural conditions is difficult because we rarely know enough about an ecosystem, its inhabitants, and their interactions.

- Restoration by removal of introduced species is very difficult and is most successful if done very soon after a new species is introduced.

- Severely polluted or damaged habitats probably cannot be restored to original conditions but may be restored to provide different environmental services.

- Recovery programs, such as captive-breeding programs, may involve direct interventions in natural populations to avoid the immediate threat of extinction.

- The goal of captive-breeding programs is to restore animal species.

59.5 Conservation of Ecosystems

Habitat fragmentation is one of the most pervasive enemies of biodiversity conservation efforts.

- The key to management of large tracts of land is to operate them in a way compatible with local needs.

- Corridors of dispersal link habitat fragments with one another and natural habitats, allowing for increased population size, genetic exchange, and recolonization.

- The best way to preserve biodiversity is to preserve intact ecosystems rather than a particular species.

SELF TEST

1. Historically, island species have tended to become extinct faster than species living on a mainland. Which of the following reasons can be used to explain this phenomenon?
   a. Island species have often evolved in the absence of predators and have no natural avoidance strategies.
   b. Humans have introduced diseases and competitors to islands, which negatively affects island populations.
   c. Island populations are usually smaller than mainland populations.
   d. All of the above.

2. Conservation hotspots are best described as
   a. areas with large numbers of endemic species that are disappearing rapidly.
   b. areas where people are particularly active supporters of biological diversity.
   c. islands that are experiencing high rates of extinction.
   d. areas where native species are being replaced with introduced species.

3 Biodiversity hotspots
   a. cover 1.4% of the world's land surface but are home to 35% of extant terrestrial vertebrate species.
   b. cover 35% of the world's land surface but are home to 1.4% of extant terrestrial vertebrate species.
   c. are found only on islands.
   d. are found only in the tropics.

4. The ability of an intact ecosystem, such as a wetland, to buffer against flooding and filter pollutants from water is a(n) _____ value of biodiversity.
   a. direct economic
   b. indirect economic
   c. ethical
   d. aesthetic

5. The economic value of indirect ecosystem services
   a. is unlikely to exceed the economic value derived from uses after ecosystem conversion.
   b. has never been carefully determined.
   c. can greatly exceed the value derived after ecosystem conversion.
   d. is entirely aesthetic.

6. The amphibian decline is best described as
   a. global disappearance of amphibian populations due to the pervasiveness of local habitat destruction.
   b. global shrinkage of amphibian populations due to global climate change.
   c. the unexplained disappearance of golden toads in Costa Rica.
   d. None of the above.

7. Habitat fragmentation can negatively affect populations
   a. by restricting gene flow among areas that were previously continuous.
   b. by increasing the relative amount of edge in suitable habitat patches.
   c. by creating patches that are too small to support a breeding population.
   d. All of the above.

8. A keystone species is one that
   a. has a higher likelihood of extinction than a nonkeystone species.
   b. exerts a strong influence on an ecosystem out of proportion to its abundance.
   c. causes other species to become extinct.
   d. has a weak influence on an ecosystem.

9. When populations are drastically reduced in size, genetic diversity and heterozygosity
   a. are likely to increase, enhancing the probability of extinction.
   b. are likely to decrease, enhancing the probability of extinction.
   c. are usually not factors that influence the probability of extinction.
   d. automatically respond in a way that protects populations from future changes.

10. A captive-breeding program followed by release to the wild
    a. is very likely, all by itself, to save a species threatened by extinction.
    b. is only likely to succeed when genetic variation of wild populations is very low.
    c. may be successful when combined with proper regulations and habitat restoration.
    d. None of the above.

11. Ninety-nine percent of all the species that ever existed have gone extinct,
    a. serving as evidence that current extinction rates are not higher than normal.
    b. but most of these losses have occurred in the last 400 years.
    c. which argues that the world just had too many species.
    d. None of the above.

12. To effectively address the biodiversity crisis, the protection of individual species
    a. must be used in concert with a principle of ecosystem management and restoration.
    b. is a sufficient management approach that merely needs to be expanded to more species.
    c. has no role to play in addressing the biodiversity crisis.
    d. usually conflicts with the principle of ecosystem management.

CHALLENGE QUESTIONS

1. If 99% of the species that ever existed are now extinct, why is there such concern over the extinction rates over the last several centuries?

2. Ecosystem conversion always has a cost and a benefit. Usually the benefit flows to a segment of society (a business or one group of people, for instance), but the costs are borne by all of society. That is what makes decisions about how and when to convert ecosystems difficult. However, is that a problem unique to conversion of ecosystems in the way we understand it today (for example, the conversion of the mangrove to a shrimp farm)? Are there other examples we can look to for guidance in how to make these decisions?

3. There is concern and evidence that amphibian populations are declining worldwide as a consequence of factors acting globally. Given that we know that species extinction is a natural process, how do we determine if there is a global decline that is different from normal species extinction?

glossary

A

ABO blood group A set of four phenotypes produced by different combinations of three alleles at a single locus; blood types are A, B, AB, and O, depending on which alleles are expressed as antigens on the red blood cell surface.

abscission In vascular plants, the dropping of leaves, flowers, fruits, or stems at the end of the growing season, as the result of the formation of a layer of specialized cells (the abscission zone) and the action of a hormone (ethylene).

absorption spectrum The relationship of absorbance vs. wavelength for a pigment molecule. This indicates which wavelengths are absorbed maximally by a pigment. For example, chlorophyll *a* absorbs most strongly in the violet-blue and red regions of the visible light spectrum.

acceptor stem The 3′ end of a tRNA molecule; the portion that amino acids become attached to during the tRNA charging reaction.

accessory pigment A secondary light-absorbing pigment used in photosynthesis, including chlorophyll *b* and the carotenoids, that complement the absorption spectrum of chlorophyll *a*.

aceolomate An animal, such as a flatworm, having a body plan that has no body cavity; the space between mesoderm and endoderm is filled with cells and organic materials.

acetyl-CoA The product of the transition reaction between glycolysis and the Krebs cycle. Pyruvate is oxidized to acetyl-CoA by NAD^+, also producing CO_2, and NADH.

achiasmate segregation The lining up and subsequent separation of homologues during meiosis I without the formation of chiasmata between homologues; found in *Drosophila* males and some other species.

acid Any substance that dissociates in water to increase the hydrogen ion (H^+) concentration and thus lower the pH.

actin One of the two major proteins that make up vertebrate muscle; the other is myosin.

action potential A transient, all-or-none reversal of the electric potential across a membrane; in neurons, an action potential initiates transmission of a nerve impulse.

action spectrum A measure of the efficiency of different wavelengths of light for photosynthesis. In plants it corresponds to the absorption spectrum of chlorophylls.

activation energy The energy that must be processed by a molecule in order for it to undergo a specific chemical reaction.

active site The region of an enzyme surface to which a specific set of substrates binds, lowering the activation energy required for a particular chemical reaction and so facilitating it.

active transport The pumping of individual ions or other molecules across a cellular membrane from a region of lower concentration to one of higher concentration (i.e., against a concentration gradient); this transport process requires energy, which is typically supplied by the expenditure of ATP.

adaptation A peculiarity of structure, physiology, or behavior that promotes the likelihood of an organism's survival and reproduction in a particular environment.

adapter protein Any of a class of proteins that acts as a link between a receptor and other proteins to initiate signal transduction.

adaptive radiation The evolution of several divergent forms from a primitive and unspecialized ancestor.

adenosine triphosphate (ATP) A nucleotide consisting of adenine, ribose sugar, and three phosphate groups; ATP is the energy currency of cellular metabolism in all organisms.

adherins junction An anchoring junction that connects the actin filaments of one cell with those of adjacent cells or with the extracellular matrix.

ATP synthase The enzyme responsible for producing ATP in oxidative phosphorylation; it uses the energy from a proton gradient to catalyze the reaction $ADP + P_i \rightarrow ATP$.

adenylyl cyclase An enzyme that produces large amounts of cAMP from ATP; the cAMP acts as a second messenger in a target cell.

adhesion The tendency of water to cling to other polar compounds due to hydrogen bonding.

adipose cells Fat cells, found in loose connective tissue, usually in large groups that form adipose tissue. Each adipose cell can store a droplet of fat (triacylglyceride).

adventitious Referring to a structure arising from an unusual place, such as stems from roots or roots from stems.

aerenchyma In plants, loose parenchymal tissue with large air spaces in it; often found in plants that grow in water.

aerobic Requiring free oxygen; any biological process that can occur in the presence of gaseous oxygen.

aerobic respiration The process that results in the complete oxidation of glucose using oxygen as the final electron acceptor. Oxygen acts as the final electron acceptor for an electron transport chain that produces a proton gradient for the chemiosmotic synthesis of ATP.

aleurone In plants, the outer layer of the endosperm in a seed; on germination, the aleurone produces α-amylase that breaks down the carbohydrates of the endosperm to nourish the embryo.

alga, pl. algae A unicellular or simple multicellular photosynthetic organism lacking multicellular sex organs.

allantois A membrane of the amniotic egg that functions in respiration and excretion in birds and reptiles and plays an important role in the development of the placenta in most mammals.

allele One of two or more alternative states of a gene.

allele frequency A measure of the occurrence of an allele in a population, expressed as proportion of the entire population, for example, an occurrence of 0.84 (84%).

allometric growth A pattern of growth in which different components grow at different rates.

allelopathy The release of a substance from the roots of one plant that block the germination of nearby seeds or inhibits the growth of a neighboring plant.

allopatric speciation The differentiation of geographically isolated populations into distinct species.

allopolyploid A polyploid organism that contains the genomes of two or more different species.

allosteric activator A substance that binds to an enzyme's allosteric site and keeps the enzyme in its active configuration.

allosteric inhibitor A noncompetitive inhibitor that binds to an enzyme's allosteric site and prevents the enzyme from changing to its active configuration.

allosteric site A part of an enzyme, away from its active site, that serves as an on/off switch for the function of the enzyme.

alpha (α) helix A form of secondary structure in proteins where the polypeptide chain is wound into a spiral due to interactions between amino and carboxyl groups in the peptide backbone.

alternation of generations A reproductive cycle in which a haploid (*n*) phase (the gametophyte), gives rise to gametes, which, after fusion to form a zygote, germinate to produce a diploid (*2n*) phase (the sporophyte). Spores produced by meiotic division from the sporophyte give rise to new gametophytes, completing the cycle.

alternative splicing In eukaryotes, the production of different mRNAs from a single primary transcript by including different sets of exons.

altruism Self-sacrifice for the benefit of others; in formal terms, the behavior that increases the fitness of the recipient while reducing the fitness of the altruistic individual.

alveolus, pl. alveoli One of many small, thin-walled air sacs within the lungs in which the bronchioles terminate.

amino acid The subunit structure from which proteins are produced, consisting of a central carbon atom with a carboxyl group (—COOH), an amino group (—NH_2), a hydrogen, and a side group (*R* group); only the side group differs from one amino acid to another.

aminoacyl-tRNA synthetase Any of a group of enzymes that attach specific amino acids to the correct tRNA during the tRNA-charging reaction. Each of the 20 amino acids has a corresponding enzyme.

amniocentesis Indirect examination of a fetus by tests on cell cultures grown from fetal cells obtained from a sample of the amniotic fluid or tests on the fluid itself.

amnion The innermost of the extraembryonic membranes; the amnion forms a fluid-filled sac around the embryo in amniotic eggs.

amniote A vertebrate that produces an egg surrounded by four membranes, one of which is the amnion; amniote groups are the reptiles, birds, and mammals.

amniotic egg An egg that is isolated and protected from the environment by a more or less impervious shell during the period of its development and that is completely self-sufficient, requiring only oxygen.

ampulla In echinoderms, a muscular sac at the base of a tube foot that contracts to extend the tube foot.

amyloplast A plant organelle called a plastid that specializes in storing starch.

anabolism The biosynthetic or constructive part of metabolism; those chemical reactions involved in biosynthesis.

anaerobic Any process that can occur without oxygen, such as anaerobic fermentation or H_2S photosynthesis.

anaerobic respiration The use of electron transport to generate a proton gradient for chemiosmotic synthesis of ATP using a final electron acceptor other than oxygen.

analogous Structures that are similar in function but different in evolutionary origin, such as the wing of a bat and the wing of a butterfly.

anaphase In mitosis and meiosis II, the stage initiated by the separation of sister chromatids, during which the daughter chromosomes move to opposite poles of the cell; in meiosis I, marked by separation of replicated homologous chromosomes.

anaphase-promoting complex (APC) A protein complex that triggers anaphase; it initiates a series of reactions that ultimately degrades cohesin, the protein complex that holds the sister chromatids together. The sister chromatids are then released and move toward opposite poles in the cell.

anchoring junction A type of cell junction that mechanically attaches the cytoskeleton of a cell to the cytoskeletons of adjacent cells or to the extracellular matrix.

androecium The floral whorl that comprises the stamens.

aneuploidy The condition in an organism whose cells have lost or gained a chromosome; Down syndrome, which results from an extra copy of human chromosome 21, is an example of aneuploidy in humans.

angiosperms The flowering plants, one of five phyla of seed plants. In angiosperms, the ovules at the time of pollination are completely enclosed by tissues.

animal pole In fish and other aquatic vertebrates with asymmetrical yolk distribution in their eggs, the hemisphere of the blastula comprising cells relatively poor in yolk.

anion A negatively charged ion.

annotation In genomics, the process of identifying and making note of "landmarks" in a DNA sequence to assist with recognition of coding and transcribed regions.

anonymous markers Genetic markers in a genome that do not cause a detectable phenotype, but that can be detected using molecular techniques.

antenna complex A complex of hundreds of pigment molecules in a photosystem that collects photons and feeds the light energy to a reaction center.

anther In angiosperm flowers, the pollen-bearing portion of a stamen.

antheridium, pl. antheridia A sperm-producing organ.

anthropoid Any member of the mammalian group consisting of monkeys, apes, and humans.

antibody A protein called immunoglobulin that is produced by lymphocytes in response to a foreign substance (antigen) and released into the bloodstream.

anticodon The three-nucleotide sequence at the end of a transfer RNA molecule that is complementary to, and base-pairs with, an amino-acid–specifying codon in messenger RNA.

antigen A foreign substance, usually a protein or polysaccharide, that stimulates an immune response.

antiporter A carrier protein in a cell's membrane that transports two molecules in opposite directions across the membrane.

anus The terminal opening of the gut; the solid residues of digestion are eliminated through the anus.

aorta (Gr. *aeirein*, to lift) The major artery of vertebrate systemic blood circulation; in mammals, carries oxygenated blood away from the heart to all regions of the body except the lungs.

apical meristem In vascular plants, the growing point at the tip of the root or stem.

apoplast route In plant roots, the pathway for movement of water and minerals that leads through cell walls and between cells.

apoptosis A process of programmed cell death, in which dying cells shrivel and shrink; used in all animal cell development to produce planned and orderly elimination of cells not destined to be present in the final tissue.

aposematic coloration An ecological strategy of some organisms that "advertise" their poisonous nature by the use of bright colors.

aquaporin A membrane channel that allows water to cross the membrane more easily than by diffusion through the membrane.

aquifers Permeable, saturated, underground layers of rock, sand, and gravel, which serve as reservoirs for groundwater.

archegonium, pl. archegonia The multicellular egg-producing organ in bryophytes and some vascular plants.

archenteron The principal cavity of a vertebrate embryo in the gastrula stage; lined with endoderm, it opens up to the outside and represents the future digestive cavity.

arteriole A smaller artery, leading from the arteries to the capillaries.

artificial selection Change in the genetic structure of populations due to selective breeding by humans.

Many domestic animal breeds and crop varieties have been produced through artificial selection.

ascomycetes A large group comprising part of the "true fungi." They are characterized by separate hyphae, asexually produced conidiospores, and sexually produced ascospores within asci.

ascus, pl. asci A specialized cell, characteristic of the ascomycetes, in which two haploid nuclei fuse to produce a zygote that divides immediately by meiosis; at maturity, an ascus contains ascospores.

asexual reproduction The process by which an individual inherits all of its chromosomes from a single parent, thus being genetically identical to that parent; cell division is by mitosis only.

A site In a ribosome, the aminoacyl site, which binds to the tRNA carrying the next amino acid to be added to a polypeptide chain.

assembly The phase of a virus's reproductive cycle during which the newly made components are assembled into viral particles.

assortative mating A type of nonrandom mating in which phenotypically similar individuals mate more frequently.

aster In animal cell mitosis, a radial array of microtubules extending from the centrioles toward the plasma membrane, possibly serving to brace the centrioles for retraction of the spindle.

atom The smallest unit of an element that contains all the characteristics of that element. Atoms are the building blocks of matter.

atrial peptide Any of a group of small polypeptide hormones that may be useful in treatment of high blood pressure and kidney failure; produced by cells in the atria of the heart.

atrioventricular (AV) node A slender connection of cardiac muscle cells that receives the heartbeat impulses from the sinoatrial node and conducts them by way of the bundle of His.

atrium An antechamber; in the heart, a thin-walled chamber that receives venous blood and passes it on to the thick-walled ventricle; in the ear, the tympanic cavity.

autonomic nervous system The involuntary neurons and ganglia of the peripheral nervous system of vertebrates; regulates the heart, glands, visceral organs, and smooth muscle.

autopolyploid A polyploid organism that contains a duplicated genome of the same species; may result from a meiotic error.

autosome Any eukaryotic chromosome that is not a sex chromosome; autosomes are present in the same number and kind in both males and females of the species.

autotroph An organism able to build all the complex organic molecules that it requires as its own food source, using only simple inorganic compounds.

auxin (Gr. *auxein*, to increase) A plant hormone that controls cell elongation, among other effects.

auxotroph A mutation, or the organism that carries it, that affects a biochemical pathway causing a nutritional requirement.

avirulent pathogen Any type of normally pathogenic organism or virus that utilizes host resources but does not cause extensive damage or death.

axil In plants, the angle between a leaf's petiole and the stem to which it is attached.

axillary bud In plants, a bud found in the axil of a stem and leaf; an axillary bud may develop into a new shoot or may become a flower.

axon A process extending out from a neuron that conducts impulses away from the cell body.

B

b6–f **complex** See cytochrome *b6–f* complex.

bacteriophage A virus that infects bacterial cells; also called a *phage*.

Barr body A deeply staining structure, seen in the interphase nucleus of a cell of an individual with more than one X chromosome, that is a condensed and inactivated X. Only one X remains active in each cell after early embryogenesis.

basal body A self-reproducing, cylindrical, cytoplasmic organelle composed of nine triplets of microtubules from which the flagella or cilia arise.

base Any substance that dissociates in water to absorb and therefore decrease the hydrogen ion (H$^+$) concentration and thus raise the pH.

base-pair A complementary pair of nucleotide bases, consisting of a purine and a pyrimidine.

basidium, pl. basidia A specialized reproductive cell of the basidiomycetes, often club-shaped, in which nuclear fusion and meiosis occur.

basophil A leukocyte containing granules that rupture and release chemicals that enhance the inflammatory response. Important in causing allergic responses.

Batesian mimicry A survival strategy in which a palatable or nontoxic organism resembles another kind of organism that is distasteful or toxic. Both species exhibit warning coloration.

B cell A type of lymphocyte that, when confronted with a suitable antigen, is capable of secreting a specific antibody protein.

behavioral ecology The study of how natural selection shapes behavior.

biennial A plant that normally requires two growing seasons to complete its life cycle. Biennials flower in the second year of their lives.

bilateral symmetry A single plane divides an organism into two structural halves that are mirror images of each other.

bile salts A solution of organic salts that is secreted by the vertebrate liver and temporarily stored in the gallbladder; emulsifies fats in the small intestine.

binary fission Asexual reproduction by division of one cell or body into two equal or nearly equal parts.

binomial distribution The distribution of phenotypes seen among the progeny of a cross in which there are only two alternative alleles.

binomial name The scientific name of a species that consists of two parts, the genus name and the specific species name, for example, *Apis mellifera*.

biochemical pathway A sequence of chemical reactions in which the product of one reaction becomes the substrate of the next reaction. The Krebs cycle is a biochemical pathway.

biodiversity The number of species and their range of behavioral, ecological, physiological, and other adaptations, in an area.

bioenergetics The analysis of how energy powers the activities of living systems.

biofilm A complex bacterial community comprising different species; plaque on teeth is a biofilm.

biogeography The study of the geographic distribution of species.

biological community All the populations of different species living together in one place; for example, all populations that inhabit a mountain meadow.

biological species concept (BSC) The concept that defines species as groups of populations that have the potential to interbreed and that are reproductively isolated from other groups.

biomass The total mass of all the living organisms in a given population, area, or other unit being measured.

biome One of the major terrestrial ecosystems, characterized by climatic and soil conditions; the largest ecological unit.

bipolar cell A specialized type of neuron connecting cone cells to ganglion cells in the visual system. Bipolar cells receive a hyperpolarized stimulus from the cone cell and then transmit a depolarization stimulus to the ganglion cell.

biramous Two-branched; describes the appendages of crustaceans.

blade The broad, expanded part of a leaf; also called the lamina.

blastocoel The central cavity of the blastula stage of vertebrate embryos.

blastodisc In the development of birds, a disclike area on the surface of a large, yolky egg that undergoes cleavage and gives rise to the embryo.

blastomere One of the cells of a blastula.

blastopore In vertebrate development, the opening that connects the archenteron cavity of a gastrula stage embryo with the outside.

blastula In vertebrates, an early embryonic stage consisting of a hollow, fluid-filled ball of cells one layer thick; a vertebrate embryo after cleavage and before gastrulation.

Bohr effect The release of oxygen by hemoglobin molecules in response to elevated ambient levels of CO$_2$.

bottleneck effect A loss of genetic variability that occurs when a population is reduced drastically in size.

Bowman's capsule In the vertebrate kidney, the bulbous unit of the nephron, which surrounds the glomerulus.

β-oxidation The oxygen-dependent reactions where 2-carbon units of fatty acids are cleaved and combined with CoA to produce acetyl-CoA, which then enters the Krebs cycle. This occurs cyclically until the entire fatty acid is oxidized.

β sheet A form of secondary structure in proteins where the polypeptide folds back on itself one or more times to form a planar structure stabilized by hydrogen bonding between amino and carboxyl groups in the peptide backbone. Also known as a β-pleated sheet.

book lung In some spiders, a unique respiratory system consisting of leaflike plates within a chamber over which gas exchange occurs.

bronchus, pl. bronchi One of a pair of respiratory tubes branching from the lower end of the trachea (windpipe) into either lung.

bud An asexually produced outgrowth that develops into a new individual. In plants, an embryonic shoot, often protected by young leaves; buds may give rise to branch shoots.

buffer A substance that resists changes in pH. It releases hydrogen ions (H$^+$) when a base is added and absorbs H$^+$ when an acid is added.

C

C$_3$ photosynthesis The main cycle of the dark reactions of photosynthesis, in which CO$_2$ binds to ribulose 1,5-bisphosphate (RuBP) to form two 3-carbon phosphoglycerate (PGA) molecules.

C$_4$ photosynthesis A process of CO$_2$ fixation in photosynthesis by which the first product is the 4-carbon oxaloacetate molecule.

cadherin One of a large group of transmembrane proteins that contain a Ca^{2+}-mediated binding between cells; these proteins are responsible for cell-to-cell adhesion between cells of the same type.

callus Undifferentiated tissue; a term used in tissue culture, grafting, and wound healing.

Calvin cycle The dark reactions of C$_3$ photosynthesis; also called the Calvin–Benson cycle.

calyx The sepals collectively; the outermost flower whorl.

Cambrian explosion The huge increase in animal diversity that occurred at the beginning of the Cambrian period.

cAMP response protein (CRP) See catabolite activator protein (CAP)

cancer The unrestrained growth and division of cells; it results from a failure of cell division control.

capillary The smallest of the blood vessels; the very thin walls of capillaries are permeable to many molecules, and exchanges between blood and the tissues occur across them; the vessels that connect arteries with veins.

capsid The outermost protein covering of a virus.

capsule In bacteria, a gelatinous layer surrounding the cell wall.

carapace (Fr. from Sp. *carapacho*, shell) Shieldlike plate covering the cephalothorax of decapod crustaceans; the dorsal part of the shell of a turtle.

carbohydrate An organic compound consisting of a chain or ring of carbon atoms to which hydrogen and oxygen atoms are attached in a ratio of approximately 2:1; having the generalized formula (CH$_2$O)$_n$; carbohydrates include sugars, starch, glycogen, and cellulose.

carbon fixation The conversion of CO$_2$ into organic compounds during photosynthesis; the first stage of the dark reactions of photosynthesis, in which carbon dioxide from the air is combined with ribulose 1,5-bisphosphate.

carotenoid Any of a group of accessory pigments found in plants; in addition to absorbing light energy, these pigments act as antioxidants, scavenging potentially damaging free radicals.

carpel A leaflike organ in angiosperms that encloses one or more ovules.

carrier protein A membrane protein that binds to a specific molecule that cannot cross the membrane and allows passage through the membrane.

carrying capacity The maximum population size that a habitat can support.

cartilage A connective tissue in skeletons of vertebrates. Cartilage forms much of the skeleton of embryos, very young vertebrates, and some adult vertebrates, such as sharks and their relatives.

Casparian strip In plants, a band that encircles the cell wall of root endodermal cells. Adjacent cells' strips connect, forming a layer through which water cannot pass; therefore, all water entering roots must pass through cell membranes and cytoplasm.

catabolism In a cell, those metabolic reactions that result in the breakdown of complex molecules into simpler compounds, often with the release of energy.

catabolite activator protein (CAP) A protein that, when bound to cAMP, can bind to DNA and activate transcription. The level of cAMP is inversely related to the level of glucose, and CAP/cAMP in *E. coli* activates the *lac* (lactose) operon. Also called *cAMP response protein (CRP)*.

catalysis The process by which chemical subunits of larger organic molecules are held and positioned by enzymes that stress their chemical bonds, leading to the disassembly of the larger molecule into its subunits, often with the release of energy.

cation A positively charged ion.

cavitation In plants and animals, the blockage of a vessel by an air bubble that breaks the cohesion of the solution in the vessel; in animals more often called embolism.

CD4⁺ cell A subtype of helper T cell that is identified by the presence of the CD4 protein on its surface. This cell type is targeted by the HIV virus that causes AIDS.

cecum In vertebrates, a blind pouch at the beginning of the large intestine.

cell cycle The repeating sequence of growth and division through which cells pass each generation.

cell determination The molecular "decision" process by which a cell becomes destined for a particular developmental pathway. This occurs before overt differentiation and can be a stepwise process.

cell plate The structure that forms at the equator of the spindle during early telophase in the dividing cells of plants and a few green algae.

cell-surface marker A glycoprotein or glycolipid on the outer surface of a cell's membrane that acts as an identifier; different cell types carry different markers.

cell-surface receptor A cell surface protein that binds a signal molecule and converts the extracellular signal into an intracellular one.

cellular blastoderm In insect embryonic development, the stage during which the nuclei of the syncitial blastoderm become separate cells through membrane formation.

cellular respiration The metabolic harvesting of energy by oxidation, ultimately dependent on molecular oxygen; carried out by the Krebs cycle and oxidative phosphorylation.

cellulose The chief constituent of the cell wall in all green plants, some algae, and a few other organisms; an insoluble complex carbohydrate formed of microfibrils of glucose molecules.

cell wall The rigid, outermost layer of the cells of plants, some protists, and most bacteria; the cell wall surrounds the plasma membrane.

central nervous system (CNS) That portion of the nervous system where most association occurs; in vertebrates, it is composed of the brain and spinal cord; in invertebrates, it usually consists of one or more cords of nervous tissue, together with their associated ganglia.

central vacuole A large, membrane-bounded sac found in plant cells that stores proteins, pigments, and waste materials, and is involved in water balance.

centriole A cytoplasmic organelle located outside the nuclear membrane, identical in structure to a basal body; found in animal cells and in the flagellated cells of other groups; divides and organizes spindle fibers during mitosis and meiosis.

centromere A visible point of constriction on a chromosome that contains repeated DNA sequences that bind specific proteins. These proteins make up the kinetochore to which microtubules attach during cell division.

cephalization The evolution of a head and brain area in the anterior end of animals; thought to be a consequence of bilateral symmetry.

cerebellum The hindbrain region of the vertebrate brain that lies above the medulla (brainstem) and behind the forebrain; it integrates information about body position and motion, coordinates muscular activities, and maintains equilibrium.

cerebral cortex The thin surface layer of neurons and glial cells covering the cerebrum; well developed only in mammals, and particularly prominent in humans. The cerebral cortex is the seat of conscious sensations and voluntary muscular activity.

cerebrum The portion of the vertebrate brain (the forebrain) that occupies the upper part of the skull, consisting of two cerebral hemispheres united by the corpus callosum. It is the primary association center of the brain. It coordinates and processes sensory input and coordinates motor responses.

chaetae Bristles of chitin on each body segment that help anchor annelid worms during locomotion.

channel protein A transmembrane protein with a hydrophilic interior that provides an aqueous channel allowing diffusion of species that cannot cross the membrane. Usually allows passage of specific ions such as K⁺, Na⁺, or Ca²⁺ across the membrane.

chaperone protein A class of enzymes that help proteins fold into the correct configuration and can refold proteins that have been misfolded or denatured.

character displacement A process in which natural selection favors individuals in a species that use resources not used by other species. This results in evolutionary change leading to species dissimilar in resource use.

character state In cladistics, one of two or more distinguishable forms of a character, such as the presence or absence of teeth in amniote vertebrates.

charging reaction The reaction by which an aminoacyl-tRNA synthetase attaches a specific amino acid to the correct tRNA using energy from ATP.

chelicera, pl. chelicerae The first pair of appendages in horseshoe crabs, sea spiders, and arachnids—the chelicerates, a group of arthropods. Chelicerae usually take the form of pincers or fangs.

chemical synapse A close association that allows chemical communication between neurons.

A chemical signal (neurotransmitter) released by the first neuron binds to receptors in the membrane of the second neurons.

chemiosmosis The mechanism by which ATP is generated in mitochondria and chloroplasts; energetic electrons excited by light (in chloroplasts) or extracted by oxidation in the Krebs cycle (in mitochondria) are used to drive proton pumps, creating a proton concentration gradient; when protons subsequently flow back across the membrane, they pass through channels that couple their movement to the synthesis of ATP.

chiasma An X-shaped figure that can be seen in the light microscope during meiosis; evidence of crossing over, where two chromatids have exchanged parts; chiasmata move to the ends of the chromosome arms as the homologues separate.

chitin A tough, resistant, nitrogen-containing polysaccharide that forms the cell walls of certain fungi, the exoskeleton of arthropods, and the epidermal cuticle of other surface structures of certain other invertebrates.

chlorophyll The primary type of light-absorbing pigment in photosynthesis. Chlorophyll *a* absorbs light in the violet-blue and the red ranges of the visible light spectrum; chlorophyll *b* is an accessory pigment to chlorophyll *a*, absorbing light in the blue and red-orange ranges. Neither pigment absorbs light in the green range, 500–600 nm.

chloroplast A cell-like organelle present in algae and plants that contains chlorophyll (and usually other pigments) and carries out photosynthesis.

choanocyte A specialized flagellated cell found in sponges; choanocytes line the body interior.

chorion The outer member of the double membrane that surrounds the embryo of reptiles, birds, and mammals; in placental mammals, it contributes to the structure of the placenta.

chorionic villi sampling A technique in which fetal cells are sampled from the chorion of the placenta rather than from the amniotic fluid; this less invasive technique can be used earlier in pregnancy than amniocentesis.

chromatid One of the two daughter strands of a duplicated chromosome that is joined by a single centromere.

chromatin The complex of DNA and proteins of which eukaryotic chromosomes are composed; chromatin is highly uncoiled and diffuse in interphase nuclei, condensing to form the visible chromosomes in prophase.

chromatin-remodeling complex A large protein complex that has been found to modify histones and DNA and that can change the structure of chromatin, moving or transferring nucleosomes.

chromosomal mutation Any mutation that affects chromosome structure.

chromosome The vehicle by which hereditary information is physically transmitted from one generation to the next; in a bacterium, the chromosome consists of a single naked circle of DNA; in eukaryotes, each chromosome consists of a single linear DNA molecule and associated proteins.

chromosomal theory of inheritance The theory stating that hereditary traits are carried on chromosomes.

cilium A short cellular projection from the surface of a eukaryotic cell, having the same internal structure of microtubules in a 9 + 2 arrangement as seen in a flagellum.

circadian rhythm An endogenous cyclical rhythm that oscillates on a daily (24-hour) basis.

circulatory system A network of vessels in coelomate animals that carries fluids to and from different areas of the body.

cisterna A small collecting vessel that pinches off from the end of a Golgi body to form a transport vesicle that moves materials through the cytoplasm.

cisternal space The inner region of a membrane-bounded structure. Usually used to describe the interior of the endoplasmic reticulum; also called the *lumen*.

cladistics A taxonomic technique used for creating hierarchies of organisms that represent true phylogenetic relationship and descent.

class A taxonomic category between phyla and orders. A class contains one or more orders, and belongs to a particular phylum.

classical conditioning The repeated presentation of a stimulus in association with a response that causes the brain to form an association between the stimulus and the response, even if they have never been associated before.

clathrin A protein located just inside the plasma membrane in eukaryotic cells, in indentations called clathrin-coated pits.

cleavage In vertebrates, a rapid series of successive cell divisions of a fertilized egg, forming a hollow sphere of cells, the blastula.

cleavage furrow The constriction that forms during cytokinesis in animal cells that is responsible for dividing the cell into two daughter cells.

climax vegetation Vegetation encountered in a self-perpetuating community of plants that has proceeded through all the stages of succession and stabilized.

cloaca In some animals, the common exit chamber from the digestive, reproductive, and urinary system; in others, the cloaca may also serve as a respiratory duct.

clone-by-clone sequencing A method of genome sequencing in which a physical map is constructed first, followed by sequencing of fragments and identifying overlap regions.

cloning Producing a cell line or culture all of whose members contain identical copies of a particular nucleotide sequence; an essential element in genetic engineering.

closed circulatory system A circulatory system in which the blood is physically separated from other body fluids.

coacervate A spherical aggregation of lipid molecules in water, held together by hydrophobic forces.

coactivator A protein that functions to link transcriptional activators to the transcription complex consisting of RNA polymerase II and general transcription factors.

cochlea In terrestrial vertebrates, a tubular cavity of the inner ear containing the essential organs for hearing.

coding strand The strand of a DNA duplex that is the same as the RNA encoded by a gene. This strand is not used as a template in transcription, it is complementary to the template.

codominance Describes a case in which two or more alleles of a gene are each dominant to other alleles but not to each other. The phenotype of a heterozygote for codominant alleles exhibit characteristics of each of the homozygous forms. For example, in human blood types, a cross between an AA individual and a BB individual yields AB individuals.

codon The basic unit of the genetic code; a sequence of three adjacent nucleotides in DNA or mRNA that codes for one amino acid.

coelom In animals, a fluid-filled body cavity that develops entirely within the mesoderm.

coenzyme A nonprotein organic molecule such as NAD that plays an accessory role in enzyme-catalyzed processes, often by acting as a donor or acceptor of electrons.

coevolution The simultaneous development of adaptations in two or more populations, species, or other categories that interact so closely that each is a strong selective force on the other.

cofactor One or more nonprotein components required by enzymes in order to function; many cofactors are metal ions, others are organic coenzymes.

cohesin A protein complex that holds sister chromatids together during cell division. The loss of cohesins at the centromere allow the anaphase movement of chromosomes.

collenchyma cell In plants, the cells that form a supporting tissue called collenchyma; often found in regions of primary growth in stems and in some leaves.

colloblast A specialized type of cell found in members of the animal phylum Ctenophora (comb jellies) that bursts on contact with zooplankton, releasing an adhesive substance to help capture this prey.

colonial flagellate hypothesis The proposal first put forth by Haeckel that metazoans descended from colonial protists; supported by the similarity of sponges to choanoflagellate protists.

commensalism A relationship in which one individual lives close to or on another and benefits, and the host is unaffected; a kind of symbiosis.

community All of the species inhabiting a common environment and interacting with one another.

companion cell A specialized parenchyma cell that is associated with each sieve-tube member in the phloem of a plant.

competitive exclusion The hypothesis that two species with identical ecological requirements cannot exist in the same locality indefinitely, and that the more efficient of the two in utilizing the available scarce resources will exclude the other; also known as Gause's principle.

competitive inhibitor An inhibitor that binds to the same active site as an enzyme's substrate, thereby competing with the substrate.

complementary Describes genetic information in which each nucleotide base has a complementary partner with which it forms a base-pair.

complementary DNA (cDNA) A DNA copy of an mRNA transcript; produced by the action of the enzyme reverse transcriptase.

complement system The chemical defense of a vertebrate body that consists of a battery of proteins that become activated by the walls of bacteria and fungi.

complete digestive system A digestive system that has both a mouth and an anus, allowing unidirectional flow of ingested food.

compound eye An organ of sight in many arthropods composed of many independent visual units called ommatidia.

concentration gradient A difference in concentration of a substance from one location to another, often across a membrane.

condensin A protein complex involved in condensation of chromosomes during mitosis and meiosis.

cone (1) In plants, the reproductive structure of a conifer. (2) In vertebrates, a type of light-sensitive neuron in the retina concerned with the perception of color and with the most acute discrimination of detail.

conidia An asexually produced fungal spore.

conjugation Temporary union of two unicellular organisms, during which genetic material is transferred from one cell to the other; occurs in bacteria, protists, and certain algae and fungi.

consensus sequence In genome sequencing, the overall sequence that is consistent with the sequences of individual fragments; computer programs are used to compare sequences and generate a consensus sequence.

conservation of synteny The preservation over evolutionary time of arrangements of DNA segments in related species.

contig A contiguous segment of DNA assembled by analyzing sequence overlaps from smaller fragments.

continuous variation Variation in a trait that occurs along a continuum, such as the trait of height in human beings; often occurs when a trait is determined by more than one gene.

contractile vacuole In protists and some animals, a clear fluid-filled vacuole that takes up water from within the cell and then contracts, releasing it to the outside through a pore in a cyclical manner; functions primarily in osmoregulation and excretion.

conus arteriosus The anteriormost chamber of the embryonic heart in vertebrate animals.

convergent evolution The independent development of similar structures in organisms that are not directly related; often found in organisms living in similar environments.

cork cambium The lateral meristem that forms the periderm, producing cork (phellem) toward the surface (outside) of the plant and phelloderm toward the inside.

cornea The transparent outer layer of the vertebrate eye.

corolla The petals, collectively; usually the conspicuously colored flower whorl.

corpus callosum The band of nerve fibers that connects the two hemispheres of the cerebrum in humans and other primates.

corpus luteum A structure that develops from a ruptured follicle in the ovary after ovulation.

cortex The outer layer of a structure; in animals, the outer, as opposed to the inner, part of an organ; in vascular plants, the primary ground tissue of a stem or root.

cotyledon A seed leaf that generally stores food in dicots or absorbs it in monocots, providing nourishment used during seed germination.

crassulacean acid metabolism (CAM) A mode of carbon dioxide fixation by which CO_2 enters open leaf stomata at night and is used in photosynthesis during the day, when stomata are closed to prevent water loss.

crista A folded extension of the inner membrane of a mitochondrion. Mitochondria contain numerous cristae.

cross-current flow In bird lungs, the latticework of capillaries arranged across the air flow, at a $90°$ angle.

crossing over In meiosis, the exchange of corresponding chromatid segments between homologous chromosomes; responsible for genetic recombination between homologous chromosomes.

ctenidia Respiratory gills of mollusks; they consist of a system of filamentous projections of the mantle that are rich in blood vessels.

cuticle A waxy or fatty, noncellular layer (formed of a substance called cutin) on the outer wall of epidermal cells.

cutin In plants, a fatty layer produced by the epidermis that forms the cuticle on the outside surface.

cyanobacteria A group of photosynthetic bacteria, sometimes called the "blue-green algae," that contain the chlorophyll pigments most abundant in plants and algae, as well as other pigments.

cyclic AMP (cAMP) A form of adenosine monophosphate (AMP) in which the atoms of the phosphate group form a ring; found in almost all organisms, cAMP functions as an intracellular second messenger that regulates a diverse array of metabolic activities.

cyclic photophosphorylation Reactions that begin with the absorption of light by reaction center chlorophyll that excites an electron. The excited electron returns to the photosystem, generating ATP by chemiosmosis in the process. This is found in the single bacterial photosystem, and can occur in plants in photosystem I.

cyclin Any of a number of proteins that are produced in synchrony with the cell cycle and combine with certain protein kinases, the cyclin-dependent kinases, at certain points during cell division.

cyclin-dependent kinase (Cdk) Any of a group of protein kinase enzymes that control progress through the cell cycle. These enzymes are only active when complexed with cyclin. The cdc2 protein, produced by the *cdc2* gene, was the first Cdk enzyme discovered.

cytochrome Any of several iron-containing protein pigments that serve as electron carriers in transport chains of photosynthesis and cellular respiration.

cytochrome *b6–f* complex A proton pump found in the thylakoid membrane. This complex uses energy from excited electrons to pump protons from the stroma into the thylakoid compartment.

cytokinesis Division of the cytoplasm of a cell after nuclear division.

cytoplasm The material within a cell, excluding the nucleus; the protoplasm.

cytoskeleton A network of protein microfilaments and microtubules within the cytoplasm of a eukaryotic cell that maintains the shape of the cell, anchors its organelles, and is involved in animal cell motility.

cytosol The fluid portion of the cytoplasm; it contains dissolved organic molecules and ions.

cytotoxic T cell A special T cell activated during cell-mediated immune response that recognizes and destroys infected body cells.

D

deamination The removal of an amino group; part of the degradation of proteins into compounds that can enter the Krebs cycle.

deductive reasoning The logical application of general principles to predict a specific result. In science, deductive reasoning is used to test the validity of general ideas.

dehydration synthesis A type of chemical reaction in which two molecules join to form one larger molecule, simultaneously splitting out a molecule of water; one molecule is stripped of a hydrogen atom, and another is stripped of a hydroxyl group (—OH), resulting in the joining of the two molecules, while the H and —OH released may combine to form a water molecule.

dehydrogenation Chemical reaction involving the loss of a hydrogen atom. This is an oxidation that combines loss of an electron with loss of a proton.

deletion A mutation in which a portion of a chromosome is lost; if too much information is lost, the deletion can be fatal.

demography The properties of the rate of growth and the age structure of populations.

denaturation The loss of the native configuration of a protein or nucleic acid as a result of excessive heat, extremes of pH, chemical modification, or changes in solvent ionic strength or polarity that disrupt hydrophobic interactions; usually accompanied by loss of biological activity.

dendrite A process extending from the cell body of a neuron, typically branched, that conducts impulses toward the cell body.

deoxyribonucleic acid (DNA) The genetic material of all organisms; composed of two complementary chains of nucleotides wound in a double helix.

dephosphorylation The removal of a phosphate group, usually by a phosphatase enzyme. Many proteins can be activated or inactivated by dephosphorylation.

depolarization The movement of ions across a plasma membrane that locally wipes out an electrical potential difference.

derived character A characteristic used in taxonomic analysis representing a departure from the primitive form.

dermal tissue In multicellular organisms, a type of tissue that forms the outer layer of the body and is in contact with the environment; it has a protective function.

desmosome A type of anchoring junction that links adjacent cells by connecting their cytoskeletons with cadherin proteins.

derepression Seen in anabolic operons where the operon that encodes the enzymes for a biochemical pathway is repressed in the presence of the end product of the pathway and derepressed in the absence of the end product. This allows production of the enzymes only when they are necessary.

determinate development A type of development in animals in which each embryonic cell has a predetermined fate in terms of what kind of tissue it will form in the adult.

deuterostome Any member of a grouping of bilaterally symmetrical animals in which the anus develops first and the mouth second; echinoderms and vertebrates are deuterostome animals.

diacylglycerol (DAG) A second messenger that is released, along with inositol-1,4,5-trisphosphate (IP_3), when phospholipase C cleaves PIP_2. DAG can have a variety of cellular effects through activation of protein kinases.

diaphragm (1) In mammals, a sheet of muscle tissue that separates the abdominal and thoracic cavities and functions in breathing. (2) A contraceptive device used to block the entrance to the uterus temporarily and thus prevent sperm from entering during sexual intercourse.

diapsid Any of a group of reptiles that have two pairs of temporal openings in the skull, one lateral and one more dorsal; one lineage of this group gave rise to dinosaurs, modern reptiles, and birds.

diastolic pressure In the measurement of human blood pressure, the minimum pressure between heartbeats (repolarization of the ventricles). *Compare with* systolic pressure.

dicer An enzyme that generates small RNA molecules in a cell by chopping up double-stranded RNAs; dicer produces miRNAs and siRNAs.

dicot Short for dicotyledon; a class of flowering plants generally characterized as having two cotyledons, net-veined leaves, and flower parts usually in fours or fives.

dideoxynucleotide A nucleotide lacking —OH groups at both the 2′ and 3′ positions; used as a chain terminator in the enzymatic sequencing of DNA.

differentiation A developmental process by which a relatively unspecialized cell undergoes a progressive change to a more specialized form or function.

diffusion The net movement of dissolved molecules or other particles from a region where they are more concentrated to a region where they are less concentrated.

dihybrid An individual heterozygous at two different loci; for example *A/a B/b*.

dihybrid cross A single genetic cross involving two different traits, such as flower color and plant height.

dikaryotic In fungi, having pairs of nuclei within each cell.

dioecious Having the male and female elements on different individuals.

diploid Having two sets of chromosomes ($2n$); in animals, twice the number characteristic of gametes; in plants, the chromosome number characteristic of the sporophyte generation; in contrast to haploid (n).

directional selection A form of selection in which selection acts to eliminate one extreme from an array of phenotypes.

disaccharide A carbohydrate formed of two simple sugar molecules bonded covalently.

disruptive selection A form of selection in which selection acts to eliminate rather than favor the intermediate type.

dissociation In proteins, the reversible separation of protein subunits from a quaternary structure without altering their tertiary structure. Also refers to the dissolving of ionic compounds in water.

disassortative mating A type of nonrandom mating in which phenotypically different individuals mate more frequently.

diurnal Active during the day.

DNA-binding motif A region found in a regulatory protein that is capable of binding to a specific base sequence in DNA; a critical part of the protein's DNA-binding domain.

DNA fingerprinting An identification technique that makes use of a variety of molecular techniques to identify differences in the DNA of individuals.

DNA gyrase A topoisomerase involved in DNA replication; it relieves the torsional strain caused by unwinding the DNA strands.

DNA library A collection of DNAs in a vector (a plasmid, phage, or artificial chromosome) that taken together represent a complex mixture of DNAs, such as the entire genome, or the cDNAs made from all of the mRNA in a specific cell type.

DNA ligase The enzyme responsible for formation of phosphodiester bonds between adjacent nucleotides in DNA.

DNA microarray An array of DNA fragments on a microscope slide or silicon chip, used in hybridization experiments with labeled mRNA or DNA to identify active and inactive genes, or the presence or absence of particular sequences.

DNA polymerase A class of enzymes that all synthesize DNA from a preexisting template. All synthesize only in the 5′-to-3′ direction, and require a primer to extend.

DNA vaccine A type of vaccine that uses DNA from a virus or bacterium that stimulates the cellular immune response.

domain (1) A distinct modular region of a protein that serves a particular function in the action of the protein, such as a regulatory domain or a DNA-binding domain. (2) In taxonomy, the level higher than kingdom. The three domains currently recognized are Bacteria, Archaea, and Eukarya.

Domain Archaea In the three-domain system of taxonomy, the group that contains only the Archaea, a highly diverse group of unicellular prokaryotes.

Domain Bacteria In the three-domain system of taxonomy, the group that contains only the Bacteria, a vast group of unicellular prokaryotes.

Domain Eukarya In the three-domain system of taxonomy, the group that contains eukaryotic organisms including protists, fungi, plants, and animals.

dominant An allele that is expressed when present in either the heterozygous or the homozygous condition.

dosage compensation A phenomenon by which the expression of genes carried on sex chromosomes is kept the same in males and females, despite a different number of sex chromosomes. In mammals, inactivation of one of the X chromosomes in female cells accomplishes dosage compensation.

double fertilization The fusion of the egg and sperm (resulting in a 2*n* fertilized egg, the zygote) and the simultaneous fusion of the second male gamete with the polar nuclei (resulting in a primary endosperm nucleus, which is often triploid, 3*n*); a unique characteristic of all angiosperms.

double helix The structure of DNA, in which two complementary polynucleotide strands coil around a common helical axis.

duodenum In vertebrates, the upper portion of the small intestine.

duplication A mutation in which a portion of a chromosome is duplicated; if the duplicated region does not lie within a gene, the duplication may have no effect.

E

ecdysis Shedding of outer, cuticular layer; molting, as in insects or crustaceans.

ecdysone Molting hormone of arthropods, which triggers when ecdysis occurs.

ecology The study of interactions of organisms with one another and with their physical environment.

ecosystem A major interacting system that includes organisms and their nonliving environment.

ecotype A locally adapted variant of an organism; differing genetically from other ecotypes.

ectoderm One of the three embryonic germ layers of early vertebrate embryos; ectoderm gives rise to the outer epithelium of the body (skin, hair, nails) and to the nerve tissue, including the sense organs, brain, and spinal cord.

ectomycorrhizae Externally developing mycorrhizae that do not penetrate the cells they surround.

ectotherms Animals such as reptiles, fish, or amphibians, whose body temperature is regulated by their behavior or by their surroundings.

electronegativity A property of atomic nuclei that refers to the affinity of the nuclei for valence electrons; a nucleus that is more electronegative has a greater pull on electrons than one that is less electronegative.

electron transport chain The passage of energetic electrons through a series of membrane-associated electron-carrier molecules to proton pumps embedded within mitochondrial or chloroplast membranes. *See* chemiosmosis.

elongation factor (Ef-Tu) In protein synthesis in *E. coli*, a factor that binds to GTP and to a charged tRNA to accomplish binding of the charged tRNA to the A site of the ribosome, so that elongation of the polypeptide chain can occur.

embryo A multicellular developmental stage that follows cell division of the zygote.

embryonic stem cell (ES cell) A stem cell derived from an early embryo that can develop into different adult tissues and give rise to an adult organism when injected into a blastocyst.

emergent properties Novel properties arising from the way in which components interact. Emergent properties often cannot be deduced solely from knowledge of the individual components.

emerging virus Any virus that originates in one organism but then passes to another; usually refers to transmission to humans.

endergonic Describes a chemical reaction in which the products contain more energy than the reactants, so that free energy must be put into the reaction from an outside source to allow it to proceed.

endocrine gland Ductless gland that secretes hormones into the extracellular spaces, from which they diffuse into the circulatory system.

endocytosis The uptake of material into cells by inclusion within an invagination of the plasma membrane; the uptake of solid material is phagocytosis, and that of dissolved material is pinocytosis.

endoderm One of the three embryonic germ layers of early vertebrate embryos, destined to give rise to the epithelium that lines internal structures and most of the digestive and respiratory tracts.

endodermis In vascular plants, a layer of cells forming the innermost layer of the cortex in roots and stems.

endomembrane system A system of connected membranous compartments found in eukaryotic cells.

endometrium The lining of the uterus in mammals; thickens in response to secretion of estrogens and progesterone and is sloughed off in menstruation.

endomycorrhizae Mycorrhizae that develop within cells

endonuclease An enzyme capable of cleaving phosphodiester bonds between nucleotides located internally in a DNA strand.

endoplasmic reticulum (ER) An internal membrane system that forms a netlike array of channels and interconnections of organelles within the cytoplasm of eukaryotic cells.

endorphin One of a group of small neuropeptides produced by the vertebrate brain; like morphine, endorphins modulate pain perception.

endosperm A storage tissue characteristic of the seeds of angiosperms, which develops from the union of a male nucleus and the polar nuclei of the embryo sac. The endosperm is digested by the growing sporophyte either before maturation of the seed or during its germination.

endospore A highly resistant, thick-walled bacterial spore that can survive harsh environmental stress, such as heat or dessication and then germinate when conditions become favorable.

endosymbiosis Theory that proposes that eukaryotic cells evolved from a symbiosis between different species of prokaryotes.

endotherm An animal capable of maintaining a constant body temperature. *See* homeotherm.

energy level A discrete level, or quantum, of energy that an electron in an atom possesses. To change energy levels, an electron must absorb or release energy.

enhancer A site of regulatory protein binding on the DNA molecule distant from the promoter and start site for a gene's transcription.

enthalpy In a chemical reaction, the energy contained in the chemical bonds of the molecule, symbolized as *H*; in a cellular reaction, the free energy is equal to the enthalpy of the reactant molecules in the reaction.

entropy A measure of the randomness or disorder of a system; a measure of how much energy in a system has become so dispersed (usually as evenly distributed heat) that it is no longer available to do work.

enzyme A protein that is capable of speeding up specific chemical reactions by lowering the required activation energy.

enzyme–substrate complex The complex formed when an enzyme binds with its substrate. This complex often has an altered configuration compared with the nonbound enzyme.

epicotyl The region just above where the cotyledons are attached.

epidermal cell In plants, a cell that collectively forms the outermost layer of the primary plant body; includes specialized cells such as trichomes and guard cells.

epidermis The outermost layers of cells; in plants, the exterior primary tissue of leaves, young stems, and roots; in vertebrates, the nonvascular external layer of skin, of ectodermal origin; in invertebrates, a single layer of ectodermal epithelium.

epididymis A sperm storage vessel; a coiled part of the sperm duct that lies near the testis.

epistasis Interaction between two nonallelic genes in which one of them modifies the phenotypic expression of the other.

epithelium In animals, a type of tissue that covers an exposed surface or lines a tube or cavity.

equilibrium A stable condition; the point at which a chemical reaction proceeds as rapidly in the reverse direction as it does in the forward direction, so that there is no further net change in the concentrations of products or reactants. In ecology, a stable condition that resists change and fairly quickly returns to its original state if disturbed by humans or natural events.

erythrocyte Red blood cell, the carrier of hemoglobin.

erythropoiesis The manufacture of blood cells in the bone marrow.

E site In a ribosome, the exit site that binds to the tRNA that carried the previous amino acid added to the polypeptide chain.

estrus The period of maximum female sexual receptivity, associated with ovulation of the egg.

ethology The study of patterns of animal behavior in nature.

euchromatin That portion of a eukaryotic chromosome that is transcribed into mRNA; contains active genes that are not tightly condensed during interphase.

eukaryote A cell characterized by membrane-bounded organelles, most notably the nucleus, and one that possesses chromosomes whose DNA is associated with proteins; an organism composed of such cells.

eutherian A placental mammal.

eutrophic Refers to a lake in which an abundant supply of minerals and organic matter exists.

evolution Genetic change in a population of organisms; in general, evolution leads to progressive change from simple to complex.

excision repair A nonspecific mechanism to repair damage to DNA during synthesis. The damaged or mismatched region is excised, and DNA polymerase replaces the region removed.

exergonic Describes a chemical reaction in which the products contain less free energy than the reactants, so that free energy is released in the reaction.

exhalant siphon In bivalve mollusks, the siphon through which outgoing water leaves the body.

exocrine gland A type of gland that releases its secretion through a duct, such as a digestive gland or a sweat gland.

exocytosis A type of bulk transport out of cells in which a vacuole fuses with the plasma membrane, discharging the vacuole's contents to the outside.

exon A segment of DNA that is both transcribed into RNA and translated into protein. *See* intron.

exonuclease An enzyme capable of cutting phosphodiester bonds between nucleotides located at an end of a DNA strand. This allows sequential removal of nucleotides from the end of DNA.

exoskeleton An external skeleton, as in arthropods.

experiment A test of one or more hypotheses. Hypotheses make contrasting predictions that can be tested experimentally in control and test experiments where a single variable is altered.

expressed sequence tag (EST) A short sequence of a cDNA that unambiguously identifies the cDNA.

expression vector A type of vector (plasmid or phage) that contains the sequences necessary to drive expression of inserted DNA in a specific cell type.

exteroceptor A receptor that is excited by stimuli from the external world.

extremophile An archaean organism that lives in extreme environments; different archaean species may live in hot springs (thermophiles), highly saline environments (halophiles), highly acidic or basic environments, or under high pressure at the bottom of oceans.

F

5′ cap In eukaryotes, a structure added to the 5′ end of an mRNA consisting of methylated GTP attached by a 5′ to 5′ bond. The cap protects this end from degradation and is involved in the initiation of translation.

facilitated diffusion Carrier-assisted diffusion of molecules across a cellular membrane through specific channels from a region of higher concentration to one of lower concentration; the process is driven by the concentration gradient and does not require cellular energy from ATP.

family A taxonomic grouping of similar species above the level of genus.

fat A molecule composed of glycerol and three fatty acid molecules.

feedback inhibition Control mechanism whereby an increase in the concentration of some molecules inhibits the synthesis of that molecule.

fermentation The enzyme-catalyzed extraction of energy from organic compounds without the involvement of oxygen.

fertilization The fusion of two haploid gamete nuclei to form a diploid zygote nucleus.

fibroblast A flat, irregularly branching cell of connective tissue that secretes structurally strong proteins into the matrix between the cells.

first filial (F₁) generation The offspring resulting from a cross between a parental generation (P); in experimental crosses, these parents usually have different phenotypes.

First Law of Thermodynamics Energy cannot be created or destroyed, but can only undergo conversion from one form to another; thus, the amount of energy in the universe is unchangeable.

fitness The genetic contribution of an individual to succeeding generations. relative fitness refers to the fitness of an individual relative to other individuals in a population.

fixed action pattern A stereotyped animal behavior response, thought by ethologists to be based on programmed neural circuits.

flagellin The protein composing bacterial flagella, which allow a cell to move through an aqueous environment.

flagellum A long, threadlike structure protruding from the surface of a cell and used in locomotion.

flame cell A specialized cell found in the network of tubules inside flatworms that assists in water regulation and some waste excretion.

flavin adenine dinucleotide (FAD, FADH₂) A cofactor that acts as a soluble (not membrane-bound) electron carrier (can be reversibly oxidized and reduced).

fluorescent in situ hybridization (FISH) A cytological method used to find specific DNA sequences on chromosomes with a specific fluorescently labeled probe.

food security Having access to sufficient, safe food to avoid malnutrition and starvation; a global human issue.

foraging behavior A collective term for the many complex, evolved behaviors that influence what an animal eats and how the food is obtained.

founder effect The effect by which rare alleles and combinations of alleles may be enhanced in new populations.

fovea A small depression in the center of the retina with a high concentration of cones; the area of sharpest vision.

frameshift mutation A mutation in which a base is added or deleted from the DNA sequence. These changes alter the reading frame downstream of the mutation.

free energy Energy available to do work.

free radical An ionized atom with one or more unpaired electrons, resulting from electrons that have been energized by ionizing radiation being ejected from the atom; free radicals react violently with other molecules, such as DNA, causing damage by mutation.

frequency-dependent selection A type of selection that depends on how frequently or infrequently a phenotype occurs in a population.

fruit In angiosperms, a mature, ripened ovary (or group of ovaries), containing the seeds.

functional genomics The study of the function of genes and their products, beyond simply ascertaining gene sequences.

functional group A molecular group attached to a hydrocarbon that confers chemical properties or reactivities. Examples include hydroxyl (—OH), carboxylic acid (—COOH) and amino groups (—NH₂).

fundamental niche Also referred to as the hypothetical niche, this is the entire niche an organism could fill if there were no other interacting factors (such as competition or predation).

G

G₀ phase The stage of the cell cycle occupied by cells that are not actively dividing.

G₁ phase The phase of the cell cycle after cytokinesis and before DNA replication called the first "gap" phase. This phase is the primary growth phase of a cell.

G₁/S checkpoint The primary control point at which a cell "decides" whether or not to divide. Also called START and the restriction point.

G₂ phase The phase of the cell cycle between DNA replication and mitosis called the second "gap" phase. During this phase, the cell prepares for mitosis.

G₂/M checkpoint The second cell-division control point, at which division can be delayed if DNA has not been properly replicated or is damaged.

gametangium, pl. gametangia A cell or organ in which gametes are formed.

gamete A haploid reproductive cell.

gametocytes Cells in the malarial sporozoite life cycle capable of giving rise to gametes when in the correct host.

gametophyte In plants, the haploid (*n*), gamete-producing generation, which alternates with the diploid (2*n*) sporophyte.

ganglion, pl. ganglia An aggregation of nerve cell bodies; in invertebrates, ganglia are the integrative centers; in vertebrates, the term is restricted to aggregations of nerve cell bodies located outside the central nervous system.

gap gene Any of certain genes in *Drosophila* development that divide the embryo into large blocks in the process of segmentation; *hunchback* is a gap gene.

gap junction A junction between adjacent animal cells that allows the passage of materials between the cells.

gastrodermis In eumetazoan animals, the layer of digestive tissue that develops from the endoderm.

gastrula In vertebrates, the embryonic stage in which the blastula with its single layer of cells turns into a three-layered embryo made up of ectoderm, mesoderm, and endoderm.

gastrulation Developmental process that converts blastula into embryo with three embryonic germ layers: endoderm, mesoderm, and ectoderm. Involves massive cell migration to convert the hollow structure into a three-layered structure.

gene The basic unit of heredity; a sequence of DNA nucleotides on a chromosome that encodes a protein, tRNA, or rRNA molecule, or regulates the transcription of such a sequence.

gene conversion Alteration of one homologous chromosome by the cell's error-detection and repair system to make it resemble the sequence on the other homologue.

gene expression The conversion of the genotype into the phenotype; the process by which DNA is transcribed into RNA, which is then translated into a protein product.

gene pool All the alleles present in a species.

gene-for-gene hypothesis A plant defense mechanism in which a specific protein encoded by a viral, bacterial, or fungal pathogen binds to a protein encoded by a plant gene and triggers a defense response in the plant.

general transcription factor Any of a group of transcription factors that are required for formation of an initiation complex by RNA polymerase II at a promoter. This allows a basal level that can be increased by the action of specific factors.

generalized transduction A form of gene transfer in prokaryotes in which any gene can be transferred between cells. This uses a lytic bacteriophage as a carrier where the virion is accidentally packaged with host DNA.

genetic counseling The process of evaluating the risk of genetic defects occurring in offspring, testing for these defects in unborn children, and providing the parents with information about these risks and conditions.

genetic drift Random fluctuation in allele frequencies over time by chance.

genetic map An abstract map that places the relative location of genes on a chromosome based on recombination frequency.

genome The entire DNA sequence of an organism.

genomic imprinting Describes an exception to Mendelian genetics in some mammals in which the phenotype caused by an allele is exhibited when the allele comes from one parent, but not from the other.

genomic library A DNA library that contains a representation of the entire genome of an organism.

genomics The study of genomes as opposed to individual genes.

genotype The genetic constitution underlying a single trait or set of traits.

genotype frequency A measure of the occurrence of a genotype in a population, expressed as a proportion of the entire population, for example, an occurrence of 0.25 (25%) for a homozygous recessive genotype.

genus, pl. genera A taxonomic group that ranks below a family and above a species.

germination The resumption of growth and development by a spore or seed.

germ layers The three cell layers formed at gastrulation of the embryo that foreshadow the future organization of tissues; the layers, from the outside inward, are the ectoderm, the mesoderm, and the endoderm.

germ-line cells During zygote development, cells that are set aside from the somatic cells and that will eventually undergo meiosis to produce gametes.

gill (1) In aquatic animals, a respiratory organ, usually a thin-walled projection from some part of the external body surface, endowed with a rich capillary bed and having a large surface area. (2) In basidiomycete fungi, the plates on the underside of the cap.

globular protein Proteins with a compact tertiary structure with hydrophobic amino acids mainly in the interior.

glomerular filtrate The fluid that passes out of the capillaries of each glomerulus.

glomerulus A cluster of capillaries enclosed by Bowman's capsule.

glucagon A vertebrate hormone produced in the pancreas that acts to initiate the breakdown of glycogen to glucose subunits.

gluconeogenesis The synthesis of glucose from noncarbohydrates (such as proteins or fats).

glucose A common six-carbon sugar ($C_6H_{12}O_6$); the most common monosaccharide in most organisms.

glucose repression In *E. coli*, the preferential use of glucose even when other sugars are present; transcription of mRNA encoding the enzymes for utilizing the other sugars does not occur.

glycocalyx A "sugar coating" on the surface of a cell resulting from the presence of polysaccharides on glycolipids and glycoproteins embedded in the outer layer of the plasma membrane.

glycogen Animal starch; a complex branched polysaccharide that serves as a food reserve in animals, bacteria, and fungi.

glycolipid Lipid molecule modified within the Golgi complex by having a short sugar chain (polysaccharide) attached.

glycolysis The anaerobic breakdown of glucose; this enzyme-catalyzed process yields two molecules of pyruvate with a net of two molecules of ATP.

glycoprotein Protein molecule modified within the Golgi complex by having a short sugar chain (polysaccharide) attached.

glyoxysome A small cellular organelle or microbody containing enzymes necessary for conversion of fats into carbohydrates.

glyphosate A biodegradable herbicide that works by inhibiting EPSP synthetase, a plant enzyme that makes aromatic amino acids; genetic engineering has allowed crop species to be created that are resistant to glyphosate.

Golgi apparatus A collection of flattened stacks of membranes (each called a Golgi body) in the cytoplasm of eukaryotic cells; functions in collection, packaging, and distribution of molecules synthesized in the cell.

G protein A protein that binds guanosine triphosphate (GTP) and assists in the function of cell-surface receptors. When the receptor binds its signal molecule, the G protein binds GTP and is activated to start a chain of events within the cell.

G protein-coupled receptor (GPCR) A receptor that acts through a heterotrimeric (three component) G protein to activate effector proteins. The effector proteins then function as enzymes to produce second messengers such as cAMP or IP₃.

gradualism The view that species change very slowly in ways that may be imperceptible from one generation to the next but that accumulate and lead to major changes over thousands or millions of years.

Gram stain Staining technique that divides bacteria into gram-negative or gram-positive based on retention of a violet dye. Differences in staining are due to cell wall construction.

granum (pl. grana) A stacked column of flattened, interconnected disks (thylakoids) that are part of the thylakoid membrane system in chloroplasts.

gravitropism Growth response to gravity in plants; formerly called geotropism.

ground meristem The primary meristem, or meristematic tissue, that gives rise to the plant body (except for the epidermis and vascular tissues).

ground tissue In plants, a type of tissue that performs many functions, including support, storage, secretion, and photosynthesis; may consist of many cell types.

growth factor Any of a number of proteins that bind to membrane receptors and initiate intracellular signaling systems that result in cell growth and division.

guard cell In plants, one of a pair of sausage-shaped cells flanking a stoma; the guard cells open and close the stomata.

guttation The exudation of liquid water from leaves due to root pressure.

gymnosperm A seed plant with seeds not enclosed in an ovary; conifers are gymnosperms.

gynoecium The aggregate of carpels in the flower of a seed plant.

H

habitat The environment of an organism; the place where it is usually found.

habituation A form of learning; a diminishing response to a repeated stimulus.

halophyte A plant that is salt-tolerant.

haplodiploidy A phenomenon occurring in certain organisms such as wasps, wherein both haploid (male) and diploid (female) individuals are encountered.

haploid Having only one set of chromosomes (*n*), in contrast to diploid (*2n*).

haplotype A region of a chromosome that is usually inherited intact, that is, it does not undergo recombination. These are identified based on analysis of SNPs.

Hardy-Weinberg equilibrium A mathematical description of the fact that allele and genotype frequencies remain constant in a random-mating population in the absence of inbreeding, selection, or other evolutionary forces; usually stated: if the frequency of allele *a* is *p* and the frequency of allele *b* is *q*, then the genotype frequencies after one generation of random mating will always be $p_2 + 2pq + q_2 = 1$.

Haversian canal Narrow channels that run parallel to the length of a bone and contain blood vessels and nerve cells.

heat A measure of the random motion of molecules; the greater the heat, the greater the motion. Heat is one form of kinetic energy.

heat of vaporization The amount of energy required to change 1 g of a substance from a liquid to a gas.

heavy metal Any of the metallic elements with high atomic numbers, such as arsenic, cadmium, lead, etc. Many heavy metals are toxic to animals even in small amounts.

helicase Any of a group of enzymes that unwind the two DNA strands in the double helix to facilitate DNA replication.

helix-turn-helix motif A common DNA-binding motif found in regulatory proteins; it consists of two α-helices linked by a nonhelical segment (the "turn").

helper T cell A class of white blood cells that initiates both the cell-mediated immune response and the humoral immune response; helper T cells are the targets of the AIDS virus (HIV).

hemoglobin A globular protein in vertebrate red blood cells and in the plasma of many invertebrates that carries oxygen and carbon dioxide.

hemopoietic stem cell The cells in bone marrow where blood cells are formed.

hermaphroditism Condition in which an organism has both male and female functional reproductive organs.

heterochromatin The portion of a eukaryotic chromosome that is not transcribed into RNA; remains condensed in interphase and stains intensely in histological preparations.

heterochrony An alteration in the timing of developmental events due to a genetic change; for example, a mutation that delays flowering in plants.

heterokaryotic In fungi, having two or more genetically distinct types of nuclei within the same mycelium.

heterosporous In vascular plants, having spores of two kinds, namely, microspores and megaspores.

heterotroph An organism that cannot derive energy from photosynthesis or inorganic chemicals, and so must feed on other plants and animals, obtaining chemical energy by degrading their organic molecules.

heterozygote advantage The situation in which individuals heterozygous for a trait have a selective advantage over those who are homozygous; an example is sickle cell anemia.

heterozygous Having two different alleles of the same gene; the term is usually applied to one or more specific loci, as in "heterozygous with respect to the *W* locus" (that is, the genotype is *W/w*).

Hfr cell An *E. coli* cell that has a high frequency of recombination due to integration of an F plasmid into its genome.

histone One of a group of relatively small, very basic polypeptides, rich in arginine and lysine, forming the core of nucleosomes around which DNA is wrapped in the first stage of chromosome condensation.

histone protein Any of eight proteins with an overall positive charge that associate in a complex. The DNA duplex coils around a core of eight histone proteins, held by its negatively charged phosphate groups, forming a nucleosome.

holoblastic cleavage Process in vertebrate embryos in which the cleavage divisions all occur at the same rate, yielding a uniform cell size in the blastula.

homeobox A sequence of 180 nucleotides located in homeotic genes that produces a 60-amino-acid peptide sequence (the homeodomain) active in transcription factors.

homeodomain motif A special class of helix-turn-helix motifs found in regulatory proteins that control development in eukaryotes.

homeosis A change in the normal spatial pattern of gene expression that can result in homeotic mutants where a wild-type structure develops in the wrong place in or on the organism.

homeostasis The maintenance of a relatively stable internal physiological environment in an organism; usually involves some form of feedback self-regulation.

homeotherm An organism, such as a bird or mammal, capable of maintaining a stable body temperature independent of the environmental temperature. *See* endotherm.

homeotic gene One of a series of "master switch" genes that determine the form of segments developing in the embryo.

hominid Any primate in the human family, Hominidae. *Homo sapiens* is the only living representative.

hominoid Collectively, hominids and apes; the monkeys and hominoids constitute the anthropoid primates.

homokaryotic In fungi, having nuclei with the same genetic makeup within a mycelium.

homologue One of a pair of chromosomes of the same kind located in a diploid cell; one copy of each pair of homologues comes from each gamete that formed the zygote.

homologous (1) Refers to similar structures that have the same evolutionary origin. (2) Refers to a pair of the same kind of chromosome in a diploid cell.

homoplasy In cladistics, a shared character state that has not been inherited from a common ancestor exhibiting that state; may result from convergent evolution or evolutionary reversal. The wings of birds and of bats, which are convergent structures, are examples.

homosporous In some plants, production of only one type of spore rather than differentiated types. *Compare with* heterosporous.

homozygous Being a homozygote, having two identical alleles of the same gene; the term is usually applied to one or more specific loci, as in "homozygous with respect to the *W* locus" (i.e., the genotype is *W/W* or *w/w*).

horizontal gene transfer (HGT) The passing of genes laterally between species; more prevalent very early in the history of life.

hormone A molecule, usually a peptide or steroid, that is produced in one part of an organism and triggers a specific cellular reaction in target tissues and organs some distance away.

host range The range of organisms that can be infected by a particular virus.

Hox **gene** A group of homeobox-containing genes that control developmental events, usually found organized into clusters of genes. These genes have been conserved in many different multicellular animals, both invertebrates and vertebrates, although the number of clusters changes in lineages, leading to four clusters in vertebrates.

humus Partly decayed organic material found in topsoil.

hybridization The mating of unlike parents.

hydration shell A "cloud" of water molecules surrounding a dissolved substance, such as sucrose or Na+ and Cl- ions.

hydrogen bond A weak association formed with hydrogen in polar covalent bonds. The partially positive hydrogen is attracted to partially negative atoms in polar covalent bonds. In water, oxygen and hydrogen in different water molecules form hydrogen bonds.

hydrolysis A reaction that breaks a bond by the addition of water. This is the reverse of dehydration, a reaction that joins molecules with the loss of water.

hydrophilic Literally translates as "water-loving" and describes substances that are soluble in water. These must be either polar or charged (ions).

hydrophobic Literally translates as "water-fearing" and describes nonpolar substances that are not soluble in water. Nonpolar molecules in water associate with each other and form droplets.

hydrophobic exclusion The tendency of nonpolar molecules to aggregate together when placed in water. Exclusion refers to the action of water in forcing these molecules together.

hydrostatic skeleton The skeleton of most soft-bodied invertebrates that have neither an internal nor an external skeleton. They use the relative incompressibility of the water within their bodies as a kind of skeleton.

hyperosmotic The condition in which a (hyperosmotic) solution has a higher osmotic concentration than that of a second solution. *Compare with* hypoosmotic.

hyperpolarization Above-normal negativity of a cell membrane during its resting potential.

hypersensitive response Plants respond to pathogens by selectively killing plant cells to block the spread of the pathogen.

hypertonic A solution with a higher concentration of solutes than the cell. A cell in a hypertonic solution tends to lose water by osmosis.

hypha, pl. hyphae A filament of a fungus or oomycete; collectively, the hyphae constitute the mycelium.

hypocotyl The region immediately below where the cotyledons are attached.

hypoosmotic The condition in which a (hypoosmotic) solution has a lower osmotic concentration than that of a second solution. *Compare with* hyperosmotic.

hypothalamus A region of the vertebrate brain just below the cerebral hemispheres, under the thalamus; a center of the autonomic nervous system, responsible for the integration and correlation of many neural and endocrine functions.

hypotonic A solution with a lower concentration of solutes than the cell. A cell in a hypotonic solution tends to take in water by osmosis.

I

icosahedron A structure consisting of 20 equilateral triangular facets; this is commonly seen in viruses and forms one kind of viral capsid.

imaginal disk One of about a dozen groups of cells set aside in the abdomen of a larval insect and committed to forming key parts of the adult insect's body.

immune response In vertebrates, a defensive reaction of the body to invasion by a foreign substance or organism. *See* antibody and B cell.

immunoglobulin An antibody.

in vitro mutagenesis The ability to create mutations at any site in a cloned gene to examine the mutations' effects on function.

inbreeding The breeding of genetically related plants or animals; inbreeding tends to increase homozygosity.

inclusive fitness Describes the sum of the number of genes directly passed on in an individual's offspring and those genes passed on indirectly by kin (other than offspring) whose existence results from the benefit of the individual's altruism.

incomplete dominance Describes a case in which two or more alleles of a gene do not display clear dominance. The phenotype of a heterozygote is intermediate between the homozygous forms. For example, crossing red-flowered with white-flowered four o'clocks yields pink heterozygotes.

independent assortment In a dihybrid cross, describes the random assortment of alleles for each of the genes. For genes on different chromosomes this results from the random orientations of different homologous pairs during metaphase I of meiosis. For genes on the same chromosome, this occurs when the two loci are far enough apart for roughly equal numbers of odd- and even-numbered multiple crossover events.

indeterminate development A type of development in animals in which the first few embryonic cells are identical daughter cells, any one of which could develop separately into a complete organism; their fate is indeterminate.

inducer exclusion Part of the mechanism of glucose repression in *E. coli* in which the presence of glucose prevents the entry of lactose such that the *lac* operon cannot be induced.

induction (1) Production of enzymes in response to a substrate; a mechanism by which binding of an inducer to a repressor allows transcription of an operon. This is seen in catabolic operons and results in production of enzymes to degrade a compound only when it is available. (2) In embryonic development, the process by which the development of a cell is influenced by interaction with an adjacent cell.

inductive reasoning The logical application of specific observations to make a generalization. In science, inductive reasoning is used to formulate testable hypotheses.

industrial melanism Phrase used to describe the evolutionary process in which initially light-colored organisms become dark as a result of natural selection.

inflammatory response A generalized nonspecific response to infection that acts to clear an infected area of infecting microbes and dead tissue cells so that tissue repair can begin.

inhalant siphon In bivalve mollusks, the siphon through which incoming water enters the body.

inheritance of acquired characteristics Also known as Lamarckism; the theory, now discounted, that individuals genetically pass on to their offspring physical and behavioral changes developed during the individuals' own lifetime.

inhibitor A substance that binds to an enzyme and decreases its activity.

initiation factor One of several proteins involved in the formation of an initiation complex in prokaryote polypeptide synthesis.

initiator tRNA A tRNA molecule involved in the beginning of translation. In prokaryotes, the initiator tRNA is charged with N-formylmethionine (tRNA^{fMet}); in eukaryotes, the tRNA is charged simply with methionine.

inorganic phosphate A phosphate molecule that is not a part of an organic molecule; inorganic phosphate groups are added and removed in the formation and breakdown of ATP and in many other cellular reactions.

insertional inactivation Destruction of a gene's function by the insertion of a transposon.

instar A larval developmental stage in insects.

integrin Any of a group of cell-surface proteins involved in adhesion of cells to substrates. Critical to migrating cells moving through the cell matrix in tissues such as connective tissue.

intercalary meristem A type of meristem that arises in stem internodes in some plants, such as corn and horsetails; responsible for elongation of the internodes.

interferon In vertebrates, a protein produced in virus-infected cells that inhibits viral multiplication.

intermembrane space The outer compartment of a mitochondrion that lies between the two membranes.

interneuron (association neuron) A nerve cell found only in the middle of the spinal cord that acts as a functional link between sensory neurons and motor neurons.

internode In plants, the region of a stem between two successive nodes.

interoceptor A receptor that senses information related to the body itself, its internal condition, and its position.

interphase The period between two mitotic or meiotic divisions in which a cell grows and its DNA replicates; includes G_1, S, and G_2 phases.

intracellular receptor A signal receptor that binds a ligand inside a cell, such as the receptors for NO, steroid hormones, vitamin D, and thyroid hormones

intron Portion of mRNA as transcribed from eukaryotic DNA that is removed by enzymes before the mature mRNA is translated into protein. *See* exon.

inversion A reversal in order of a segment of a chromosome; also, to turn inside out, as in embryogenesis of sponges or discharge of a nematocyst.

ionizing radiation High-energy radiation that is highly mutagenic, producing free radicals that react with DNA; includes X-rays and γ-rays.

isomer One of a group of molecules identical in atomic composition but differing in structural arrangement; for example, glucose and fructose.

isosmotic The condition in which the osmotic concentrations of two solutions are equal, so that no net water movement occurs between them by osmosis.

isotonic A solution having the same concentration of solutes as the cell. A cell in an isotonic solution takes in and loses the same amount of water.

isotope Different forms of the same element with the same number of protons but different numbers of neutrons.

J

jasmonic acid An organic molecule that is part of a plant's wound response; it signals the production of a proteinase inhibitor.

K

karyotype The morphology of the chromosomes of an organism as viewed with a light microscope.

keratin A tough, fibrous protein formed in epidermal tissues and modified into skin, feathers, hair, and hard structures such as horns and nails.

key innovation A newly evolved trait in a species that allows members to use resources or other aspects of the environment that were previously inaccessible.

kidney In vertebrates, the organ that filters the blood to remove nitrogenous wastes and regulates the balance of water and solutes in blood plasma.

kilocalorie Unit describing the amount of heat required to raise the temperature of a kilogram of water by 1°C; sometimes called a Calorie, equivalent to 1000 calories.

kinase cascade A series of protein kinases that phosphorylate each other in succession; a kinase cascade can amplify signals during the signal transduction process.

kinesis Changes in activity level in an animal that are dependent on stimulus intensity. *See* kinetic energy.

kinetic energy The energy of motion.

kinetochore Disk-shaped protein structure within the centromere to which the spindle fibers attach during mitosis or meiosis. *See* centromere.

kingdom The second highest commonly used taxonomic category.

kin selection Selection favoring relatives; an increase in the frequency of related individuals (kin) in a population, leading to an increase in the relative frequency in the population of those alleles shared by members of the kin group.

knockout mice Mice in which a known gene is inactivated ("knocked out") using recombinant DNA and ES cells.

Krebs cycle Another name for the citric acid cycle; also called the tricarboxylic acid (TCA) cycle.

L

labrum The upper lip of insects and crustaceans situated above or in front of the mandibles.

lac **operon** In *E. coli*, the operon containing genes that encode the enzymes to metabolize lactose.

lagging strand The DNA strand that must be synthesized discontinuously because of the 5′-to-3′ directionality of DNA polymerase during replication, and the antiparallel nature of DNA. Compare *leading strand*.

larva A developmental stage that is unlike the adult found in organisms that undergo metamorphosis. Embryos develop into larvae that produce the adult form by metamorphosis.

larynx The voice box; a cartilaginous organ that lies between the pharynx and trachea and is responsible for sound production in vertebrates.

lateral line system A sensory system encountered in fish, through which mechanoreceptors in a line down the side of the fish are sensitive to motion.

lateral meristems In vascular plants, the meristems that give rise to secondary tissue; the vascular cambium and cork cambium.

Law of Independent Assortment Mendel's second law of heredity, stating that genes located on nonhomologous chromosomes assort independently of one another.

Law of Segregation Mendel's first law of heredity, stating that alternative alleles for the same gene segregate from each other in production of gametes.

leading strand The DNA strand that can be synthesized continuously from the origin of replication. Compare *lagging strand*.

leaf primordium, pl. **primordia** A lateral outgrowth from the apical meristem that will eventually become a leaf.

lenticels Spongy areas in the cork surfaces of stem, roots, and other plant parts that allow interchange of gases between internal tissues and the atmosphere through the periderm.

leucine zipper motif A motif in regulatory proteins in which two different protein subunits associate to form a single DNA-binding site; the proteins are connected by an association between hydrophobic regions containing leucines (the "zipper").

leucoplast In plant cells, a colorless plastid in which starch grains are stored; usually found in cells not exposed to light.

leukocyte A white blood cell; a diverse array of nonhemoglobin-containing blood cells, including phagocytic macrophages and antibody-producing lymphocytes.

lichen Symbiotic association between a fungus and a photosynthetic organism such as a green alga or cyanobacterium.

ligand A signaling molecule that binds to a specific receptor protein, initiating signal transduction in cells.

light-dependent reactions In photosynthesis, the reactions in which light energy is captured and used in production of ATP and NADPH. In plants this involves the action of two linked photosystems.

light-independent reactions In photosynthesis, the reactions of the Calvin cycle in which ATP and NADPH from the light-dependent reactions are used to reduce CO_2 and produce organic compounds such as glucose. This involves the process of carbon fixation, or the conversion of inorganic carbon (CO_2) to organic carbon (ultimately carbohydrates).

lignin A highly branched polymer that makes plant cell walls more rigid; an important component of wood.

limbic system The hypothalamus, together with the network of neurons that link the hypothalamus to some areas of the cerebral cortex. Responsible for many of the most deep-seated drives and emotions of vertebrates, including pain, anger, sex, hunger, thirst, and pleasure.

linked genes Genes that are physically close together and therefore tend to segregate together; recombination occurring between linked genes can be used to produce a map of genetic distance for a chromosome.

lipase An enzyme that catalyzes the hydrolysis of fats.

lipid A nonpolar hydrophobic organic molecule that is insoluble in water (which is polar) but dissolves readily in nonpolar organic solvents; includes fats, oils, waxes, steroids, phospholipids, and carotenoids.

lipid bilayer The structure of a cellular membrane, in which two layers of phospholipids spontaneously align so that the hydrophilic head groups are exposed to water, while the hydrophobic fatty acid tails are pointed toward the center of the membrane.

lipopolysaccharide A lipid with a polysaccharide molecule attached; found in the outer membrane layer of gram-negative bacteria; the outer membrane layer protects the cell wall from antibiotic attack.

locus The position on a chromosome where a gene is located.

long interspersed element (LINE) Any of a type of large transposable element found in humans and other primates that contains all the biochemical machinery needed for transposition.

long terminal repeat (LTR) A particular type of retrotransposon that has repeated elements at its ends. These elements make up 8% of the human genome.

loop of Henle In the kidney of birds and mammals, a hairpin-shaped portion of the renal tubule in which water and salt are reabsorbed from the glomerular filtrate by diffusion.

lophophore A horseshoe-shaped crown of ciliated tentacles that surrounds the mouth of certain spiralian animals; seen in the phyla Brachiopoda and Bryozoa.

lumen A term for any bounded opening; for example, the cisternal space of the endoplasmic reticulum of eukaryotic cells, the passage through which blood flows inside a blood vessel, and the passage through which material moves inside the intestine during digestion.

luteal phase The second phase of the female reproductive cycle, during which the mature eggs are released into the fallopian tubes, a process called ovulation.

lymph In animals, a colorless fluid derived from blood by filtration through capillary walls in the tissues.

lymphatic system In animals, an open vascular system that reclaims water that has entered interstitial regions from the bloodstream (lymph); includes the lymph nodes, spleen, thymus, and tonsils.

lymphocyte A type of white blood cell. Lymphocytes are responsible for the immune response; there are two principal classes: B cells and T cells.

lymphokine A regulatory molecule that is secreted by lymphocytes. In the immune response, lymphokines secreted by helper T cells unleash the cell-mediated immune response.

lysis Disintegration of a cell by rupture of its plasma membrane.

lysogenic cycle A viral cycle in which the viral DNA becomes integrated into the host chromosome and is replicated during cell reproduction. Results in vertical rather than horizontal transmission.

lysosome A membrane-bounded vesicle containing digestive enzymes that is produced by the Golgi apparatus in eukaryotic cells.

lytic cycle A viral cycle in which the host cell is killed (lysed) by the virus after viral duplication to release viral particles.

M

macroevolution The creation of new species and the extinction of old ones.

macromolecule An extremely large biological molecule; refers specifically to proteins, nucleic acids, polysaccharides, lipids, and complexes of these.

macronutrients Inorganic chemical elements required in large amounts for plant growth, such as nitrogen, potassium, calcium, phosphorus, magnesium, and sulfur.

macrophage A large phagocytic cell that is able to engulf and digest cellular debris and invading bacteria.

madreporite A sievelike plate on the surface of echinoderms through which water enters the water–vascular system.

MADS **box gene** Any of a family of genes identified by possessing shared motifs that are the predominant homeotic genes of plants; a small number of *MADS* box genes are also found in animals.

major groove The larger of the two grooves in a DNA helix, where the paired nucleotides' hydrogen bonds are accessible; regulatory proteins can recognize and bind to regions in the major groove.

major histocompatibility complex (MHC) A set of protein cell-surface markers anchored in the plasma membrane, which the immune system uses to identify "self." All the cells of a given individual have the same "self" marker, called an MHC protein.

Malpighian tubules Blind tubules opening into the hindgut of terrestrial arthropods; they function as excretory organs.

mandibles In crustaceans, insects, and myriapods, the appendages immediately posterior to the antennae; used to seize, hold, bite, or chew food.

mantle The soft, outermost layer of the body wall in mollusks; the mantle secretes the shell.

map unit Each 1% of recombination frequency between two genetic loci; the unit is termed a centimorgan (cM) or simply a map unit (m.u.).

marsupial A mammal in which the young are born early in their development, sometimes as soon as eight days after fertilization, and are retained in a pouch.

mass extinction A relatively sudden, sharp decline in the number of species; for example, the extinction at the end of the Cretaceous period in which the dinosaurs and a variety of other organisms disappeared.

mass flow hypothesis The overall process by which materials move in the phloem of plants.

maternal inheritance A mode of uniparental inheritance from the female parent; for example, in humans mitochondria and their genomes are inherited from the mother.

matrix In mitochondria, the solution in the interior space surrounded by the cristae that contains the enzymes and other molecules involved in oxidative respiration; more generally, that part of a tissue within which an organ or process is embedded.

medusa A free-floating, often umbrella-shaped body form found in cnidarian animals, such as jellyfish.

megapascal (MPa) A unit of measure used for pressure in water potential.

megaphyll In plants, a leaf that has several to many veins connecting it to the vascular cylinder of the stem; most plants have megaphylls.

mesoglea A layer of gelatinous material found between the epidermis and gastrodermis of eumetazoans; it contains the muscles in most of these animals.

mesohyl A gelatinous, protein-rich matrix found between the choanocyte layer and the epithelial layer of the body of a sponge; various types of amoeboid cells may occur in the mesohyl.

metacercaria An encysted form of a larval liver fluke, found in muscle tissue of an infected animal; if the muscle is eaten, cysts dissolves in the digestive tract, releasing the flukes into the body of the new host.

methylation The addition of a methyl group to bases (primarily cytosine) in DNA. Cytosine methylation is correlated with DNA that is not expressed.

meiosis I The first round of cell division in meiosis; it is referred to as a "reduction division" because homologous chromosomes separate, and the daughter cells have only the haploid number of chromosomes.

meiosis II The second round of division in meiosis, during which the two haploid cells

from meiosis I undergo a mitosis-like division without DNA replication to produce four haploid daughter cells.

membrane receptor A signal receptor present as an integral protein in the cell membrane, such as GPCRs, chemically gated ion channels in neurons, and RTKs.

Mendelian ratio The characteristic dominant-to-recessive phenotypic ratios that Mendel observed in his genetics experiments. For example, the F_2 generation in a monohybrid cross shows a ratio of 3:1; the F_2 generation in a dihybrid cross shows a ratio of 9:3:3:1.

menstruation Periodic sloughing off of the blood-enriched lining of the uterus when pregnancy does not occur.

meristem Undifferentiated plant tissue from which new cells arise.

meroblastic cleavage A type of cleavage in the eggs of reptiles, birds, and some fish. Occurs only on the blastodisc.

mesoderm One of the three embryonic germ layers that form in the gastrula; gives rise to muscle, bone and other connective tissue, the peritoneum, the circulatory system, and most of the excretory and reproductive systems.

mesophyll The photosynthetic parenchyma of a leaf, located within the epidermis.

messenger RNA (mRNA) The RNA transcribed from structural genes; RNA molecules complementary to a portion of one strand of DNA, which are translated by the ribosomes to form protein.

metabolism The sum of all chemical processes occurring within a living cell or organism.

metamorphosis Process in which a marked change in form takes place during postembryonic development as, for example, from tadpole to frog.

metaphase The stage of mitosis or meiosis during which microtubules become organized into a spindle and the chromosomes come to lie in the spindle's equatorial plane.

metastasis The process by which cancer cells move from their point of origin to other locations in the body; also, a population of cancer cells in a secondary location, the result of movement from the primary tumor.

methanogens Obligate, anaerobic archaebacteria that produce methane.

microarray DNA sequences are placed on a microscope slide or chip with a robot. The microarray can then be probed with RNA from specific tissues to identify expressed DNA.

microbody A cellular organelle bounded by a single membrane and containing a variety of enzymes; generally derived from endoplasmic reticulum; includes peroxisomes and glyoxysomes.

microevolution Refers to the evolutionary process itself. Evolution within a species. Also called adaptation.

micronutrient A mineral required in only minute amounts for plant growth, such as iron, chlorine, copper, manganese, zinc, molybdenum, and boron.

microphyll In plants, a leaf that has only one vein connecting it to the vascular cylinder of the stem; the club mosses in particular have microphylls.

micropyle In the ovules of seed plants, an opening in the integuments through which the pollen tube usually enters.

micro-RNA (miRNA) A class of RNAs that are very short and only recently could be detected. *See also* small interfering RNAs (siRNAs).

microtubule In eukaryotic cells, a long, hollow protein cylinder, composed of the protein tubulin; these influence cell shape, move the chromosomes in cell division, and provide the functional internal structure of cilia and flagella.

microvillus Cytoplasmic projection from epithelial cells; microvilli greatly increase the surface area of the small intestine.

middle lamella The layer of intercellular material, rich in pectic compounds, that cements together the primary walls of adjacent plant cells.

mimicry The resemblance in form, color, or behavior of certain organisms (mimics) to other more powerful or more protected ones (models).

miracidium The ciliated first-stage larva inside the egg of the liver fluke; eggs are passed in feces, and if they reach water they may be eaten by a host snail in which they continue their life cycle.

missense mutation A base substitution mutation that results in the alteration of a single amino acid.

mitogen-activated protein (MAP) kinase Any of a class of protein kinases that activate transcription factors to alter gene expression. A mitogen is any molecule that stimulates cell division. MAP kinases are activated by kinase cascades.

mitosis Somatic cell division; nuclear division in which the duplicated chromosomes separate to form two genetically identical daughter nuclei.

molar concentration Concentration expressed as moles of a substance in 1 L of pure water.

mole The weight of a substance in grams that corresponds to the atomic masses of all the component atoms in a molecule of that substance. One mole of a compound always contains $6\ 023 \times 10^{23}$ molecules.

molecular clock method In evolutionary theory, the method in which the rate of evolution of a molecule is constant through time.

molecular cloning The isolation and amplification of a specific sequence of DNA.

monocot Short for monocotyledon; flowering plant in which the embryos have only one cotyledon, the floral parts are generally in threes, and the leaves typically are parallel-veined.

monocyte A type of leukocyte that becomes a phagocytic cell (macrophage) after moving into tissues.

monoecious A plant in which the staminate and pistillate flowers are separate, but borne on the same individual.

monomer The smallest chemical subunit of a polymer. The monosaccharide α-glucose is the monomer found in plant starch, a polysaccharide.

monophyletic In phylogenetic classification, a group that includes the most recent common ancestor of the group and all its descendants. A clade is a monophyletic group.

monosaccharide A simple sugar that cannot be decomposed into smaller sugar molecules.

monosomic Describes the condition in which a chromosome has been lost due to nondisjunction during meiosis, producing a diploid embryo with only one of these autosomes.

monotreme An egg-laying mammal.

morphogen A signal molecule produced by an embryonic organizer region that informs surrounding cells of their distance from the organizer, thus determining relative positions of cells during development.

morphogenesis The development of an organism's body form, namely its organs and anatomical features; it may involve apoptosis as well as cell division, differentiation, and changes in cell shape.

morphology The form and structure of an organism.

morula Solid ball of cells in the early stage of embryonic development.

mosaic development A pattern of embryonic development in which initial cells produced by cleavage divisions contain different developmental signals (determinants) from the egg, setting the individual cells on different developmental paths.

motif A substructure in proteins that confers function and can be found in multiple proteins. One example is the helix-turn-helix motif found in a number of proteins that is used to bind to DNA.

motor (efferent) neuron Neuron that transmits nerve impulses from the central nervous system to an effector, which is typically a muscle or gland.

M phase The phase of cell division during which chromosomes are separated. The spindle assembles, binds to the chromosomes, and moves the sister chromatids apart.

M phase-promoting factor (MPF) A Cdk enzyme active at the G_2/M checkpoint.

Müllerian mimicry A phenomenon in which two or more unrelated but protected species resemble one another, thus achieving a kind of group defense.

multidrug-resistant (MDR) strain Any bacterial strain that has become resistant to more than one antibiotic drug; MDR *Staphylococcus* strains, for example, are responsible for many infection deaths.

multienzyme complex An assembly consisting of several enzymes catalyzing different steps in a sequence of reactions. Close proximity of these related enzymes speeds the overall process, making it more efficient.

multigene family A collection of related genes on a single chromosome or on different chromosomes.

muscle fiber A long, cylindrical, multinucleated cell containing numerous myofibrils, which is capable of contraction when stimulated.

mutagen An agent that induces changes in DNA (mutations); includes physical agents that damage DNA and chemicals that alter DNA bases.

mutation A permanent change in a cell's DNA; includes changes in nucleotide sequence, alteration of gene position, gene loss or duplication, and insertion of foreign sequences.

mutualism A symbiotic association in which two (or more) organisms live together, and both members benefit.

mycelium, pl. **mycelia** In fungi, a mass of hyphae.

mycorrhiza, pl. **mycorrhizae** A symbiotic association between fungi and the roots of a plant.

myelin sheath A fatty layer surrounding the long axons of motor neurons in the peripheral nervous system of vertebrates.

myofilament A contractile microfilament, composed largely of actin and myosin, within muscle.

myosin One of the two protein components of microfilaments (the other is actin); a principal component of vertebrate muscle.

N

natural killer cell A cell that does not kill invading microbes, but rather, the cells infected by them.

natural selection The differential reproduction of genotypes; caused by factors in the environment; leads to evolutionary change.

nauplius A larval form characteristic of crustaceans.

negative control A type of control at the level of DNA transcription initiation in which the frequency of initiation is decreased; repressor proteins mediate negative control.

negative feedback A homeostatic control mechanism whereby an increase in some substance or activity inhibits the process leading to the increase; also known as feedback inhibition.

nematocyst A harpoonlike structure found in the cnidocytes of animals in the phylum Cnidaria, which includes the jellyfish among other groups; the nematocyst, when released, stings and helps capture prey.

nephridium, pl. **nephridia** In invertebrates, a tubular excretory structure.

nephrid organ A filtration system of many freshwater invertebrates in which water and waste pass from the body across the membrane into a collecting organ, from which they are expelled to the outside through a pore.

nephron Functional unit of the vertebrate kidney; one of numerous tubules involved in filtration and selective reabsorption of blood; each nephron consists of a Bowman's capsule, an enclosed glomerulus, and a long attached tubule; in humans, called a renal tubule.

nephrostome The funnel-shaped opening that leads to the nephridium, which is the excretory organ of mollusks.

nerve A group or bundle of nerve fibers (axons) with accompanying neurological cells, held together by connective tissue; located in the peripheral nervous system.

nerve cord One of the distinguishing features of chordates, running lengthwise just beneath the embryo's dorsal surface; in vertebrates, differentiates into the brain and spinal cord.

neural crest A special strip of cells that develops just before the neural groove closes over to form the neural tube in embryonic development.

neural groove The long groove formed along the long axis of the embryo by a layer of ectodermal cells.

neural tube The dorsal tube, formed from the neural plate, that differentiates into the brain and spinal cord.

neuroglia Nonconducting nerve cells that are intimately associated with neurons and appear to provide nutritional support.

neuromuscular junction The structure formed when the tips of axons contact (innervate) a muscle fiber.

neuron A nerve cell specialized for signal transmission; includes cell body, dendrites, and axon.

neurotransmitter A chemical released at the axon terminal of a neuron that travels across the synaptic cleft, binds a specific receptor on the far side, and depending on the nature of the receptor, depolarizes or hyperpolarizes a second neuron or a muscle or gland cell.

neurulation A process in early embryonic development by which a dorsal band of ectoderm thickens and rolls into the neural tube.

neutrophil An abundant type of granulocyte capable of engulfing microorganisms and other foreign particles; neutrophils comprise about 50–70% of the total number of white blood cells.

niche The role played by a particular species in its environment.

nicotinamide adenine dinucleotide (NAD) A molecule that becomes reduced (to NADH) as it carries high-energy electrons from oxidized molecules and delivers them to ATP-producing pathways in the cell.

NADH dehydrogenase An enzyme located on the inner mitochondrial membrane that catalyzes the oxidation by NAD⁺ of pyruvate to acetyl-CoA. This reaction links glycolysis and the Krebs cycle.

nitrification The oxidization of ammonia or nitrite to produce nitrate, the form of nitrogen taken up by plants; some bacteria are capable of nitrification.

nociceptor A naked dendrite that acts as a receptor in response to a pain stimulus.

nocturnal Active primarily at night.

node The part of a plant stem where one or more leaves are attached. *See* internode.

node of Ranvier A gap formed at the point where two Schwann cells meet and where the axon is in direct contact with the surrounding intercellular fluid.

nodule In plants, a specialized tissue that surrounds and houses beneficial bacteria, such as root nodules of legumes that contain nitrogen-fixing bacteria.

nonassociative learning A learned behavior that does not require an animal to form an association between two stimuli, or between a stimulus and a response.

noncompetitive inhibitor An inhibitor that binds to a location other than the active site of an enzyme, changing the enzyme's shape so that it cannot bind the substrate.

noncyclic photophosphorylation The set of light-dependent reactions of the two plant photosystems, in which excited electrons are shuttled between the two photosystems, producing a proton gradient that is used for the chemiosmotic synthesis of ATP. The electrons are used to reduce NADP to NADPH. Lost electrons are replaced by the oxidation of water producing O_2.

nondisjunction The failure of homologues or sister chromatids to separate during mitosis or meiosis, resulting in an aneuploid cell or gamete.

nonextreme archaea Archaean groups that are not extremophiles, living in more moderate environments on Earth today.

nonpolar Said of a covalent bond that involves equal sharing of electrons. Can also refer to a compound held together by nonpolar covalent bonds.

nonsense codon One of three codons (UAA, UAG, and UGA) that are not recognized by tRNAs, thus serving as "stop" signals in the mRNA message and terminating translation.

nonsense mutation A base substitution in which a codon is changed into a stop codon. The protein is truncated because of premature termination.

Northern blot A blotting technique used to identify a specific mRNA sequence in a complex mixture. *See* Southern blot.

notochord In chordates, a dorsal rod of cartilage that runs the length of the body and forms the primitive axial skeleton in the embryos of all chordates.

nucellus Tissue composing the chief pair of young ovules, in which the embryo sac develops; equivalent to a megasporangium.

nuclear envelope The bounding structure of the eukaryotic nucleus. Composed of two phospholipid bilayers with the outer one connected to the endoplasmic reticulum.

nuclear pore One of a multitude of tiny but complex openings in the nuclear envelope that allow selective passage of proteins and nucleic acids into and out of the nucleus.

nuclear receptor Intracellular receptors are found in both the cytoplasm and the nucleus. The site of action of the hormone–receptor complex is in the nucleus where they modify gene expression.

nucleic acid A nucleotide polymer; chief types are deoxyribonucleic acid (DNA), which is double-stranded, and ribonucleic acid (RNA), which is typically single-stranded.

nucleoid The area of a prokaryotic cell, usually near the center, that contains the genome in the form of DNA compacted with protein.

nucleolus In eukaryotes, the site of rRNA synthesis; a spherical body composed chiefly of rRNA in the process of being transcribed from multiple copies of rRNA genes.

nucleosome A complex consisting of a DNA duplex wound around a core of eight histone proteins.

nucleotide A single unit of nucleic acid, composed of a phosphate, a five-carbon sugar (either ribose or deoxyribose), and a purine or a pyrimidine.

nucleus In atoms, the central core, containing positively charged protons and (in all but hydrogen) electrically neutral neutrons; in eukaryotic cells, the membranous organelle that houses the chromosomal DNA; in the central nervous system, a cluster of nerve cell bodies.

nutritional mutation A mutation affecting a synthetic pathway for a vital compound, such as an amino acid or vitamin; microorganisms with a nutritional mutation must be grown on medium that supplies the missing nutrient.

O

ocellus, pl. **ocelli** A simple light receptor common among invertebrates.

octet rule Rule to describe patterns of chemical bonding in main group elements that require a total of eight electrons to complete their outer electron shell.

Okazaki fragment A short segment of DNA produced by discontinuous replication elongating in the 5′-to-3′ direction away from the replication.

olfaction The function of smelling.

ommatidium, pl. **ommatidia** The visual unit in the compound eye of arthropods; contains light-sensitive cells and a lens able to form an image.

oncogene A mutant form of a growth-regulating gene that is inappropriately "on," causing unrestrained cell growth and division.

oocyst The zygote in a sporozoan life cycle. It is surrounded by a tough cyst to prevent dehydration or other damage.

open circulatory system A circulatory system in which the blood flows into sinuses in which it mixes with body fluid and then reenters the vessels in another location.

open reading frame (ORF) A region of DNA that encodes a sequence of amino acids with no stop codons in the reading frame.

operant conditioning A learning mechanism in which the reward follows only after the correct behavioral response.

operator A regulatory site on DNA to which a repressor can bind to prevent or decrease initiation of transcription.

operculum A flat, bony, external protective covering over the gill chamber in fish.

operon A cluster of adjacent structural genes transcribed as a unit into a single mRNA molecule.

opisthosoma The posterior portion of the body of an arachnid.

oral surface The surface on which the mouth is found; used as a reference when describing the body structure of echinoderms because of their adult radial symmetry.

orbital A region around the nucleus of an atom with a high probability of containing an electron. The position of electrons can only be described by these probability distributions.

order A category of classification above the level of family and below that of class.

organ A body structure composed of several different tissues grouped in a structural and functional unit.

organelle Specialized part of a cell; literally, a small cytoplasmic organ.

orthologues Genes that reflect the conservation of a single gene found in an ancestor.

oscillating selection The situation in which selection alternately favors one phenotype at one time, and a different phenotype at a another time, for example, during drought conditions versus during wet conditions.

osculum A specialized, larger pore in sponges through which filtered water is forced to the outside of the body.

osmoconformer An animal that maintains the osmotic concentration of its body fluids at about the same level as that of the medium in which it is living.

osmosis The diffusion of water across a selectively permeable membrane (a membrane that permits the free passage of water but prevents or retards the passage of a solute); in the absence of differences in pressure or volume, the net movement of water is from the side containing a lower concentration of solute to the side containing a higher concentration.

osmotic concentration The property of a solution that takes into account all dissolved solutes in the solution; if two solutions with different osmotic concentrations are separated by a water-permeable membrane, water will move from the solution with lower osmotic concentration to the solution with higher osmotic concentration.

osmotic pressure The potential pressure developed by a solution separated from pure water by a differentially permeable membrane. The higher the solute concentration, the greater the osmotic potential of the solution; also called *osmotic potential*.

ossicle Any of a number of movable or fixed calcium-rich plates that collectively make up the endoskeleton of echinoderms.

osteoblast A bone-forming cell.

osteocyte A mature osteoblast.

outcrossing Breeding with individuals other than oneself or one's close relatives.

ovary (1) In animals, the organ in which eggs are produced. (2) In flowering plants, the enlarged basal portion of a carpel that contains the ovule(s); the ovary matures to become the fruit.

oviduct In vertebrates, the passageway through which ova (eggs) travel from the ovary to the uterus.

oviparity Refers to a type of reproduction in which the eggs are developed after leaving the body of the mother, as in reptiles.

ovoviviparity Refers to a type of reproduction in which young hatch from eggs that are retained in the mother's uterus.

ovulation In animals, the release of an egg or eggs from the ovary.

ovum, pl. **ova** The egg cell; female gamete.

oxidation Loss of an electron by an atom or molecule; in metabolism, often associated with a gain of oxygen or a loss of hydrogen.

oxidation–reduction reaction A type of paired reaction in living systems in which electrons lost from one atom (oxidation) are gained by another atom (reduction). Termed a *redox reaction* for short.

oxidative respiration Process of cellular activity in which glucose or other molecules are broken down to water and carbon dioxide with the release of energy.

oxygen debt The amount of oxygen required to convert the lactic acid generated in the muscles during exercise back into glucose.

oxytocin A hormone of the posterior pituitary gland that affects uterine contractions during childbirth and stimulates lactation.

ozone O_3, a stratospheric layer of the Earth's atmosphere responsible for filtering out ultraviolet radiation supplied by the Sun.

P

p53 **gene** The gene that produces the p53 protein that monitors DNA integrity and halts cell division if DNA damage is detected. Many types of cancer are associated with a damaged or absent *p53* gene.

pacemaker A patch of excitatory tissue in the vertebrate heart that initiates the heartbeat.

pair-rule gene Any of certain genes in *Drosophila* development controlled by the gap genes that are expressed in stripes that subdivide the embryo in the process of segmentation.

paleopolyploid An ancient polyploid organism used in analysis of polyploidy events in the study of a species' genome evolution.

palisade parenchyma In plant leaves, the columnar, chloroplast-containing parenchyma cells of the mesophyll. Also called *palisade cells.*

panspermia The hypothesis that meteors or cosmic dust may have brought significant amounts of complex organic molecules to Earth, kicking off the evolution of life.

papilla A small projection of tissue.

paracrine A type of chemical signaling between cells in which the effects are local and short-lived.

paralogues Two genes within an organism that arose from the duplication of one gene in an ancestor.

paraphyletic In phylogenetic classification, a group that includes the most recent common ancestor of the group, but not all its descendants.

parapodia One of the paired lateral processes on each side of most segments in polychaete annelids.

parasexuality In certain fungi, the fusion and segregation of heterokaryotic haploid nuclei to produce recombinant nuclei.

parasitism A living arrangement in which an organism lives on or in an organism of a different species and derives nutrients from it.

parenchyma cell The most common type of plant cell; characterized by large vacuoles, thin walls, and functional nuclei.

parthenogenesis The development of an egg without fertilization, as in aphids, bees, ants, and some lizards.

partial diploid (merodiploid) Describes an *E. coli* cell that carries an F′ plasmid with host genes. This makes the cell diploid for the genes carried by the F′ plasmid.

partial pressure The components of each individual gas—such as nitrogen, oxygen, and carbon dioxide—that together constitute the total air pressure.

passive transport The movement of substances across a cell's membrane without the expenditure of energy.

pedigree A consistent graphic representation of matings and offspring over multiple generations for a particular genetic trait, such as albinism or hemophilia.

pedipalps A pair of specialized appendages found in arachnids; in male spiders, these are specialized as copulatory organs, whereas in scorpions they are large pincers.

pelagic Free-swimming, usually in open water.

pellicle A tough, flexible covering in ciliates and euglenoids.

pentaradial symmetry The five-part radial symmetry characteristic of adult echinoderms.

peptide bond The type of bond that links amino acids together in proteins through a dehydration reaction.

peptidoglycan A component of the cell wall of bacteria, consisting of carbohydrate polymers linked by protein cross-bridges.

peptidyl transferase In translation, the enzyme responsible for catalyzing the formation of a peptide bond between each new amino acid and the previous amino acid in a growing polypeptide chain.

perianth In flowering plants, the petals and sepals taken together.

pericycle In vascular plants, one or more cell layers surrounding the vascular tissues of the root, bounded externally by the endodermis and internally by the phloem.

periderm Outer protective tissue in vascular plants that is produced by the cork cambium and functionally replaces epidermis when it is destroyed during secondary growth; the periderm includes the cork, cork cambium, and phelloderm.

peristalsis In animals, a series of alternating contracting and relaxing muscle movements along the length of a tube such as the oviduct or alimentary canal that tend to force material such as an egg cell or food through the tube.

peroxisome A microbody that plays an important role in the breakdown of highly oxidative hydrogen peroxide by catalase.

petal A flower part, usually conspicuously colored; one of the units of the corolla.

petiole The stalk of a leaf.

phage conversion The phenomenon by which DNA from a virus, incorporated into a host cell's genome, alters the host cell's function in a significant way; for example, the conversion of *Vibrio cholerae* bacteria into a pathogenic form that releases cholera toxin.

phage lambda (λ) A well-known bacteriophage that has been widely used in genetic studies and is often a vector for DNA libraries.

phagocyte Any cell that engulfs and devours microorganisms or other particles.

phagocytosis Endocytosis of a solid particle; the plasma membrane folds inward around the particle (which may be another cell) and engulfs it to form a vacuole.

pharyngeal pouches In chordates, embryonic regions that become pharyngeal slits in aquatic and marine chordates and vertebrates, but do not develop openings to the outside in terrestrial vertebrates.

pharyngeal slits One of the distinguishing features of chordates; a group of openings on each side of the anterior region that form a passageway from the pharynx and esophagus to the external environment.

pharynx A muscular structure lying posterior to the mouth in many animals; aids in propelling food into the digestive tract.

phenotype The realized expression of the genotype; the physical appearance or functional expression of a trait.

pheromone Chemical substance released by one organism that influences the behavior or physiological processes of another organism of the same species. Pheromones serve as sex attractants, as trail markers, and as alarm signals.

phloem In vascular plants, a food-conducting tissue basically composed of sieve elements, various kinds of parenchyma cells, fibers, and sclereids.

phoronid Any of a group of lophophorate invertebrates, now classified in the phylum Brachiopoda, that burrows into soft underwater substrates and secretes a chitinous tube in which it lives out its life; it extends its lophophore tentacles to feed on drifting food particles.

phosphatase Any of a number of enzymes that removes a phosphate group from a protein, reversing the action of a kinase.

phosphodiester bond The linkage between two sugars in the backbone of a nucleic acid molecule; the phosphate group connects the pentose sugars through a pair of ester bonds.

phospholipid Similar in structure to a fat, but having only two fatty acids attached to the glycerol backbone, with the third space linked to a phosphorylated molecule; contains a polar hydrophilic "head" end (phosphate group) and a nonpolar hydrophobic "tail" end (fatty acids).

phospholipid bilayer The main component of cell membranes; phospholipids naturally associate in a bilayer with hydrophobic fatty acids oriented to the inside and hydrophilic phosphate groups facing outward on both sides.

phosphorylation Chemical reaction resulting in the addition of a phosphate group to an organic molecule. Phosphorylation of ADP yields ATP. Many proteins are also activated or inactivated by phosphorylation.

photoelectric effect The ability of a beam of light to excite electrons, creating an electrical current.

photon A particle of light having a discrete amount of energy. The wave concept of light explains the different colors of the spectrum, whereas the particle concept of light explains the energy transfers during photosynthesis.

photoperiodism The tendency of biological reactions to respond to the duration and timing of day and night; a mechanism for measuring seasonal time.

photoreceptor A light-sensitive sensory cell.

photorespiration Action of the enzyme rubisco, which catalyzes the oxidization of RuBP, releasing CO_2; this reverses carbon fixation and can reduce the yield of photosynthesis.

photosystem An organized complex of chlorophyll, other pigments, and proteins that traps light energy as excited electrons. Plants have two linked photosystems in the thylakoid membrane of chloroplasts. Photosystem II passes an excited electron through an electron transport chain to photosystem I to replace an excited electron passed to NADPH. The electron lost from photosystem II is replaced by the oxidation of water.

phototropism In plants, a growth response to a light stimulus.

pH scale A scale used to measure acidity and basicity. Defined as the negative log of H^+ concentration. Ranges from 0 to 14. A value of 7 is neutral; below 7 is acidic and above 7 is basic.

phycobiloprotein A type of accessory pigment found in cyanobacteria and some algae. Complexes of phycobiloprotein are able to absorb light energy in the green range.

phycologist A scientist who studies algae.

phyllotaxy In plants, a spiral pattern of leaf arrangement on a stem in which sequential leaves are at a 137.5° angle to one another, an angle related to the golden mean.

phylogenetic species concept (PSC) The concept that defines species on the basis of their phylogenetic relationships.

phylogenetic tree A pattern of descent generated by analysis of similarities and differences among organisms. Modern gene-sequencing techniques have produced phylogenetic trees showing the evolutionary history of individual genes.

phylogeny The evolutionary history of an organism, including which species are closely related and in what order related species evolved; often represented in the form of an evolutionary tree.

phylum, pl. phyla A major category, between kingdom and class, of taxonomic classifications.

physical map A map of the DNA sequence of a chromosome or genome based on actual landmarks within the DNA.

phytochrome A plant pigment that is associated with the absorption of light; photoreceptor for red to far-red light.

phytoestrogen One of a number of secondary metabolites in some plants that are structurally and functionally similar to the animal hormone estrogen.

phytoremediation The process that uses plants to remove contamination from soil or water.

pigment A molecule that absorbs light.

pilus, pl. pili Extensions of a bacterial cell enabling it to transfer genetic materials from one individual to another or to adhere to substrates.

pinocytosis The process of fluid uptake by endocytosis in a cell.

pistil Central organ of flowers, typically consisting of ovary, style, and stigma; a pistil may consist of one or more fused carpels and is more technically and better known as the gynoecium.

pith The ground tissue occupying the center of the stem or root within the vascular cylinder.

placenta, pl. placentae (1) In flowering plants, the part of the ovary wall to which the ovules or seeds are attached. (2) In mammals, a tissue formed in part from the inner lining of the uterus and in part from other membranes, through which the embryo (later the fetus) is nourished while in the uterus and through which wastes are carried away.

plankton Free-floating, mostly microscopic, aquatic organisms.

plant receptor kinase Any of a group of plant membrane receptors that, when activated by binding ligand, have kinase enzymatic activity. These receptors phosphorylate serine or threonine, unlike RTKs in animals that phosphorylate tyrosine.

planula A ciliated, free-swimming larva produced by the medusae of cnidarian animals.

plasma The fluid of vertebrate blood; contains dissolved salts, metabolic wastes, hormones, and a variety of proteins, including antibodies and albumin; blood minus the blood cells.

plasma cell An antibody-producing cell resulting from the multiplication and differentiation of a B lymphocyte that has interacted with an antigen.

plasma membrane The membrane surrounding the cytoplasm of a cell; consists of a single phospholipid bilayer with embedded proteins.

plasmid A small fragment of extrachromosomal DNA, usually circular, that replicates independently of the main chromosome, although it may have been derived from it.

plasmodesmata In plants, cytoplasmic connections between adjacent cells.

plasmodium Stage in the life cycle of myxomycetes (plasmodial slime molds); a multinucleate mass of protoplasm surrounded by a membrane.

plasmolysis The shrinking of a plant cell in a hypertonic solution such that it pulls away from the cell wall.

plastid An organelle in the cells of photosynthetic eukaryotes that is the site of photosynthesis and, in plants and green algae, of starch storage.

platelet In mammals, a fragment of a white blood cell that circulates in the blood and functions in the formation of blood clots at sites of injury.

pleiotropy Condition in which an individual allele has more than one effect on production of the phenotype.

plesiomorphy In cladistics, another term for an ancestral character state.

plumule The epicotyl of a plant with its two young leaves.

point mutation An alteration of one nucleotide in a chromosomal DNA molecule.

polar body Minute, nonfunctioning cell produced during the meiotic divisions leading to gamete formation in vertebrates.

polar covalent bond A covalent bond in which electrons are shared unequally due to differences in electronegativity of the atoms involved. One atom has a partial negative charge and the other a partial positive charge, even though the molecule is electrically neutral overall.

polarity (1) Refers to unequal charge distribution in a molecule such as water, which has a positive region and a negative region although it is neutral overall. (2) Refers to axial differences in a developing embryo that result in anterior–posterior and dorsal–ventral axes in a bilaterally symmetrical animal.

polarize In cladistics, to determine whether character states are ancestral or derived.

pollen tube A tube formed after germination of the pollen grain; carries the male gametes into the ovule.

pollination The transfer of pollen from an anther to a stigma.

polyandry The condition in which a female mates with more than one male.

polyclonal antibody An antibody response in which an antigen elicits many different antibodies, each fitting a different portion of the antigen surface.

polygenic inheritance Describes a mode of inheritance in which more than one gene affects a trait, such as height in human beings; polygenic inheritance may produce a continuous range of phenotypic values, rather than discrete either–or values.

polygyny A mating choice in which a male mates with more than one female.

polymer A molecule composed of many similar or identical molecular subunits; starch is a polymer of glucose.

polymerase chain reaction (PCR) A process by which DNA polymerase is used to copy a sequence of interest repeatedly, making millions of copies of the same DNA.

polymorphism The presence in a population of more than one allele of a gene at a frequency greater than that of newly arising mutations.

polyp A typically sessile, cylindrical body form found in cnidarian animals, such as hydras.

polypeptide A molecule consisting of many joined amino acids; not usually as complex as a protein.

polyphyletic In phylogenetic classification, a group that does not include the most recent common ancestor of all members of the group.

polyploidy Condition in which one or more entire sets of chromosomes is added to the diploid genome.

polysaccharide A carbohydrate composed of many monosaccharide sugar subunits linked together in a long chain; examples are glycogen, starch, and cellulose.

polyunsaturated fat A fat molecule having at least two double bonds between adjacent carbons in one or more of the fatty acid chains.

population Any group of individuals, usually of a single species, occupying a given area at the same time.

population genetics The study of the properties of genes in populations.

positive control A type of control at the level of DNA transcription initiation in which the frequency of initiation is increased; activator proteins mediate positive control.

posttranscriptional control A mechanism of control over gene expression that operates after the transcription of mRNA is complete.

postzygotic isolating mechanism A type of reproductive isolation in which zygotes are produced but are unable to develop into reproducing adults; these mechanisms may range from inviability of zygotes or embryos to adults that are sterile.

potential energy Energy that is not being used, but could be; energy in a potentially usable form; often called "energy of position."

precapillary sphincter A ring of muscle that guards each capillary loop and that, when closed, blocks flow through the capillary.

pre-mRNA splicing In eukaryotes, the process by which introns are removed from the primary transcript to produce mature mRNA; pre-mRNA splicing occurs in the nucleus.

pressure potential In plants, the turgor pressure resulting from pressure against the cell wall.

prezygotic isolating mechanism A type of reproductive isolation in which the formation of a zygote is prevented; these mechanisms may range from physical separation in different habitats to gametic in which gametes are incapable of fusing.

primary endosperm nucleus In flowering plants, the result of the fusion of a sperm nucleus and the (usually) two polar nuclei.

primary growth In vascular plants, growth originating in the apical meristems of shoots and roots results in an increase in length.

primary immune response The first response of an immune system to a foreign antigen. If the system is challenged again with the same antigen, the memory cells created during the primary response will respond more quickly.

primary induction Inductions between the three primary tissue types: mesoderm and endoderm.

primary meristem Any of the three meristems produced by the apical meristem; primary meristems give rise to the dermal, vascular, and ground tissues.

primary nondisjunction Failure of chromosomes to separate properly at meiosis I.

primary phloem The cells involved in food conduction in plants.

primary plant body The part of a plant consisting of young, soft shoots and roots derived from apical meristem tissues.

primary productivity The amount of energy produced by photosynthetic organisms in a community.

primary structure The specific amino acid sequence of a protein.

primary tissues Tissues that make up the primary plant body.

primary transcript The initial mRNA molecule copied from a gene by RNA polymerase, containing a faithful copy of the entire gene, including introns as well as exons.

primary wall In plants, the wall layer deposited during the period of cell expansion.

primase The enzyme that synthesizes the RNA primers required by DNA polymerases.

primate Monkeys and apes (including humans).

primitive streak In the early embryos of birds, reptiles, and mammals, a dorsal, longitudinal strip of ectoderm and mesoderm that is equivalent to the blastopore in other forms.

primordium In plants, a bulge on the young shoot produced by the apical meristem; primordia can differentiate into leaves, other shoots, or flowers.

principle of parsimony Principle stating that scientists should favor the hypothesis that requires the fewest assumptions.

prions Infectious proteinaceous particles.

procambium In vascular plants, a primary meristematic tissue that gives rise to primary vascular tissues.

product rule *See* rule of multiplication.

proglottid A repeated body segment in tapeworms that contains both male and female reproductive organs; proglottids eventually form eggs and embryos, which leave the host's body in feces.

prokaryote A bacterium; a cell lacking a membrane-bounded nucleus or membrane-bounded organelles.

prometaphase The transitional phase between prophase and metaphase during which the spindle attaches to the kinetochores of sister chromatids.

promoter A DNA sequence that provides a recognition and attachment site for RNA polymerase to begin the process of gene transcription; it is located upstream from the transcription start site.

prophase The phase of cell division that begins when the condensed chromosomes become visible and ends when the nuclear envelope breaks down. The assembly of the spindle takes place during prophase.

proprioceptor In vertebrates, a sensory receptor that senses the body's position and movements.

prosimian Any member of the mammalian group that is a sister group to the anthropoids; prosimian means "before monkeys." Members include the lemurs, lorises, and tarsiers.

prosoma The anterior portion of the body of an arachnid, which bears all the appendages.

prostaglandins A group of modified fatty acids that function as chemical messengers.

prostate gland In male mammals, a mass of glandular tissue at the base of the urethra that secretes an alkaline fluid that has a stimulating effect on the sperm as they are released.

protease An enzyme that degrades proteins by breaking peptide bonds; in cells, proteases are often compartmentalized into vesicles such as lysosomes.

proteasome A large, cylindrical cellular organelle that degrades proteins marked with ubiquitin.

protein A chain of amino acids joined by peptide bonds.

protein kinase An enzyme that adds phosphate groups to proteins, changing their activity.

protein microarray An array of proteins on a microscope slide or silicon chip. The array may be used with a variety of probes, including antibodies, to analyze the presence or absence of specific proteins in a complex mixture.

proteome All the proteins coded for by a particular genome.

proteomics The study of the proteomes of organisms. This is related to functional genomics as the proteome is responsible for much of the function encoded by a genome.

protoderm The primary meristem that gives rise to the dermal tissue.

proton pump A protein channel in a membrane of the cell that expends energy to transport protons against a concentration gradient; involved in the chemiosmotic generation of ATP.

protooncogene A normal cellular gene that can act as an oncogene when mutated.

protostome Any member of a grouping of bilaterally symmetrical animals in which the mouth develops first and the anus second; flatworms, nematodes, mollusks, annelids, and arthropods are protostomes.

pseudocoel A body cavity located between the endoderm and mesoderm.

pseudogene A copy of a gene that is not transcribed.

pseudomurien A component of the cell wall of archaea; it is similar to peptidoglycan in structure and function but contains different components.

pseudopod A nonpermanent cytoplasmic extension of the cell body.

P site In a ribosome, the peptidyl site that binds to the tRNA attached to the growing polypeptide chain.

punctuated equilibrium A hypothesis about the mechanism of evolutionary change proposing that long periods of little or no change are punctuated by periods of rapid evolution.

Punnett square A diagrammatic way of showing the possible genotypes and phenotypes of genetic crosses.

pupa A developmental stage of some insects in which the organism is nonfeeding, immotile, and sometimes encapsulated or in a cocoon; the pupal stage occurs between the larval and adult phases.

purine The larger of the two general kinds of nucleotide base found in DNA and RNA; a nitrogenous base with a double-ring structure, such as adenine or guanine.

pyrimidine The smaller of two general kinds of nucleotide base found in DNA and RNA; a nitrogenous base with a single-ring structure, such as cytosine, thymine, or uracil.

pyruvate A three-carbon molecule that is the end product of glycolysis; each glucose molecule yields two pyruvate molecules.

Q

quantitative trait A trait that is determined by the effects of more than one gene; such a trait usually exhibits continuous variation rather than discrete either–or values.

quaternary structure The structural level of a protein composed of more than one polypeptide chain, each of which has its own tertiary structure; the individual chains are called subunits.

R

radial canal Any of five canals that connect to the ring canal of an echinoderm's water–vascular system.

radial cleavage The embryonic cleavage pattern of deuterostome animals in which cells divide parallel to and at right angles to the polar axis of the embryo.

radial symmetry A type of structural symmetry with a circular plan, such that dividing the body or structure through the midpoint in any direction yields two identical sections.

radicle The part of the plant embryo that develops into the root.

radioactive isotope An isotope that is unstable and undergoes radioactive decay, releasing energy.

radioactivity The emission of nuclear particles and rays by unstable atoms as they decay into more stable forms.

radula Rasping tongue found in most mollusks.

reaction center A transmembrane protein complex in a photosystem that receives energy from the antenna complex exciting an electron that is passed to an acceptor molecule.

reading frame The correct succession of nucleotides in triplet codons that specify amino acids on translation. The reading frame is established by the first codon in the sequence as there are no spaces in the genetic code.

realized niche The actual niche occupied by an organism when all biotic and abiotic interactions are taken into account.

receptor-mediated endocytosis Process by which specific macromolecules are transported into eukaryotic cells at clathrin-coated pits, after binding to specific cell-surface receptors.

receptor protein A highly specific cell-surface receptor embedded in a cell membrane that responds only to a specific messenger molecule.

receptor tyrosine kinase (RTK) A diverse group of membrane receptors that when activated have kinase enzymatic activity. Specifically, they phosphorylate proteins on tyrosine. Their activation can lead to diverse cellular responses.

recessive An allele that is only expressed when present in the homozygous condition, but being "hidden" by the expression of a dominant allele in the heterozygous condition.

redia A secondary, nonciliated larva produced in the sporocysts of liver flukes.

regulatory protein Any of a group of proteins that modulates the ability of RNA polymerase to bind to a promoter and begin DNA transcription.

replicon An origin of DNA replication and the DNA whose replication is controlled by this origin. In prokaryotic replication, the chromosome plus the origin consist of a single

replicon; eukaryotic chromosomes consist of multiple replicons.

replisome The macromolecular assembly of enzymes involved in DNA replication; analogous to the ribosome in protein synthesis.

reciprocal altruism Performance of an altruistic act with the expectation that the favor will be returned. A key and very controversial assumption of many theories dealing with the evolution of social behavior. *See* altruism.

reciprocal cross A genetic cross involving a single trait in which the sex of the parents is reversed; for example, if pollen from a white-flowered plant is used to fertilize a purple-flowered plant, the reciprocal cross would be pollen from a purple-flowered plant used to fertilize a white-flowered plant.

reciprocal recombination A mechanism of genetic recombination that occurs only in eukaryotic organisms, in which two chromosomes trade segments; can occur between nonhomologous chromosomes as well as the more usual exchange between homologous chromosomes in meiosis.

recombinant DNA Fragments of DNA from two different species, such as a bacterium and a mammal, spliced together in the laboratory into a single molecule.

recombination frequency The value obtained by dividing the number of recombinant progeny by the total progeny in a genetic cross. This value is converted into a percentage, and each 1% is termed a map unit.

reduction The gain of an electron by an atom, often with an associated proton.

reflex In the nervous system, a motor response subject to little associative modification; a reflex is among the simplest neural pathways, involving only a sensory neuron, sometimes (but not always) an interneuron, and one or more motor neurons.

reflex arc The nerve path in the body that leads from stimulus to reflex action.

refractory period The recovery period after membrane depolarization during which the membrane is unable to respond to additional stimulation.

reinforcement In speciation, the process by which partial reproductive isolation between populations is increased by selection against mating between members of the two populations, eventually resulting in complete reproductive isolation.

replica plating A method of transferring bacterial colonies from one plate to another to make a copy of the original plate; an impression of colonies growing on a Petri plate is made on a velvet surface, which is then used to transfer the colonies to plates containing different media, such that auxotrophs can be identified.

replication fork The Y-shaped end of a growing replication bubble in a DNA molecule undergoing replication.

repolarization Return of the ions in a nerve to their resting potential distribution following depolarization.

repression In general, control of gene expression by preventing transcription. Specifically, in bacteria such as *E. coli* this is mediated by repressor proteins. In anabolic operons, repressors bind DNA in the absence of corepressors to repress an operon.

repressor A protein that regulates DNA transcription by preventing RNA polymerase from attaching to the promoter and transcribing the structural gene. *See* operator.

reproductive isolating mechanism Any barrier that prevents genetic exchange between species.

residual volume The amount of air remaining in the lungs after the maximum amount of air has been exhaled.

resting membrane potential The charge difference (difference in electric potential) that exists across a neuron at rest (about 70 mV).

restriction endonuclease An enzyme that cleaves a DNA duplex molecule at a particular base sequence, usually within or near a palindromic sequence; also called a restriction enzyme.

restriction fragment length polymorphism (RFLP) Restriction enzymes recognize very specific DNA sequences. Alleles of the same gene or surrounding sequences may have base-pair differences, so that DNA near one allele is cut into a different-length fragment than DNA near the other allele. These different fragments separate based on size on electrophoresis gels.

retina The photosensitive layer of the vertebrate eye; contains several layers of neurons and light receptors (rods and cones); receives the image formed by the lens and transmits it to the brain via the optic nerve.

retinoblastoma susceptibility gene (Rb) A gene that, when mutated, predisposes individuals to a rare form of cancer of the retina; one of the first tumor-suppressor genes discovered.

retrovirus An RNA virus. When a retrovirus enters a cell, a viral enzyme (reverse transcriptase) transcribes viral RNA into duplex DNA, which the cell's machinery then replicates and transcribes as if it were its own.

reverse genetics An approach by which a researcher uses a cloned gene of unknown function, creates a mutation, and introduces the mutant gene back into the organism to assess the effect of the mutation.

reverse transcriptase A viral enzyme found in retroviruses that is capable of converting their RNA genome into a DNA copy.

Rh blood group A set of cell-surface markers (antigens) on the surface of red blood cells in humans and rhesus monkeys (for which it is named); although there are several alleles, they are grouped into two main types: Rh-positive and Rh-negative.

rhizome In vascular plants, a more or less horizontal underground stem; may be enlarged for storage or may function in vegetative reproduction.

rhynchocoel A true coelomic cavity in ribbonworms that serves as a hydraulic power source for extending the proboscis.

ribonucleic acid (RNA) A class of nucleic acids characterized by the presence of the sugar ribose and the pyrimidine uracil; includes mRNA, tRNA, and rRNA.

ribosomal RNA (rRNA) A class of RNA molecules found, together with characteristic proteins, in ribosomes; transcribed from the DNA of the nucleolus.

ribosome The molecular machine that carries out protein synthesis; the most complicated aggregation of proteins in a cell, also containing three different rRNA molecules.

ribosome-binding sequence (RBS) In prokaryotes a conserved sequence at the 5' end of mRNA that is complementary to the 3' end of a small subunit rRNA and helps to position the ribosome during initiation.

ribozyme An RNA molecule that can behave as an enzyme, sometimes catalyzing its own assembly; rRNA also acts as a ribozyme in the polymerization of amino acids to form protein.

ribulose 1,5-b sphosphate (RuBP) In the Calvin cycle, the five-carbon sugar to which CO_2 is attached, accomplishing carbon fixation. This reaction is catalyzed by the enzyme rubisco.

ribulose bisphosphate carboxylase/oxygenase (rubisco) The four-subunit enzyme in the chloroplast that catalyzes the carbon fixation reaction joining CO_2 to RuBP.

RNA interference A type of gene silencing in which the mRNA transcript is prevented from being translated; small interfering RNAs (siRNAs) have been found to bind to mRNA and target its degradation prior to its translation.

RNA polymerase An enzyme that catalyzes the assembly of an mRNA molecule, the sequence of which is complementary to a DNA molecule used as a template. *See* transcription.

RNA primer In DNA replication, a sequence of about 10 RNA nucleotides complementary to unwound DNA that attaches at a replication fork; the DNA polymerase uses the RNA primer as a starting point for addition of DNA nucleotides to form the new DNA strand; the RNA primer is later removed and replaced by DNA nucleotides.

RNA splicing A nuclear process by which intron sequences of a primary mRNA transcript are cut out and the exon sequences spliced together to give the correct linkages of genetic information that will be used in protein construction.

rod Light-sensitive nerve cell found in the vertebrate retina; sensitive to very dim light; responsible for "night vision."

root The usually descending axis of a plant, normally below ground, which anchors the plant and serves as the major point of entry for water and minerals.

root cap In plants, a tissue structure at the growing tips of roots that protects the root apical meristem as the root pushes through the soil; cells of the root cap are continually lost and replaced.

root hair In plants, a tubular extension from an epidermal cell located just behind the root tip; root hairs greatly increase the surface area for absorption.

root pressure In plants, pressure exerted by water in the roots in response to a solute potential in the absence of transpiration; often occurs at night. Root pressure can result in guttation, excretion of water from cells of leaves as dew.

root system In plants, the portion of the plant body that anchors the plant and absorbs ions and water.

R plasmid A resistance plasmid; a conjugative plasmid that picks up antibiotic resistance genes and can therefore transfer resistance from one bacterium to another.

rule of addition The rule stating that for two independent events, the probability of either event occurring is the sum of the individual probabilities.

rule of multiplication The rule stating that for two independent events, the probability of both events occurring is the product of the individual probabilities.

rumen An "extra stomach" in cows and related mammals wherein digestion of cellulose occurs and from which partially digested material can be ejected back into the mouth.

S

salicylic acid In plants, an organic molecule that is a long-distance signal in systemic acquired resistance.

saltatory conduction A very fast form of nerve impulse conduction in which the impulses leap from node to node over insulated portions.

saprobes Heterotrophic organisms that digest their food externally (e.g., most fungi).

sarcolemma The specialized cell membrane in a muscle cell.

sarcomere Fundamental unit of contraction in skeletal muscle; repeating bands of actin and myosin that appear between two Z lines.

sarcoplasmic reticulum The endoplasmic reticulum of a muscle cell. A sleeve of membrane that wraps around each myofilament.

satellite DNA A nontranscribed region of the chromosome with a distinctive base composition; a short nucleotide sequence repeated tandemly many thousands of times.

saturated fat A fat composed of fatty acids in which all the internal carbon atoms contain the maximum possible number of hydrogen atoms.

Schwann cells The supporting cells associated with projecting axons, along with all the other nerve cells that make up the peripheral nervous system.

sclereid In vascular plants, a sclerenchyma cell with a thick, lignified, secondary wall having many pits; not elongate like a fiber.

sclerenchyma cell Tough, thick-walled cells that strengthen plant tissues.

scolex The attachment organ at the anterior end of a tapeworm.

scrotum The pouch that contains the testes in most mammals.

scuttellum The modified cotyledon in cereal grains.

second filial (F₂) generation The offspring resulting from a cross between members of the first filial (F₁) generation.

secondary cell wall In plants, the innermost layer of the cell wall. Secondary walls have a highly organized microfibrillar structure and are often impregnated with lignin.

secondary growth In vascular plants, an increase in stem and root diameter made possible by cell division of the lateral meristems.

secondary immune response The swifter response of the body the second time it is invaded by the same pathogen because of the presence of memory cells, which quickly become antibody-producing plasma cells.

secondary induction An induction between tissues that have already differentiated.

secondary metabolite A molecule not directly involved in growth, development, or reproduction of an organism; in plants these molecules, which include nicotine, caffeine, tannins, and menthols, can discourage herbivores.

secondary plant body The part of a plant consisting of secondary tissues from lateral meristem tissues; the older trunk, branches, and roots of woody plants.

secondary structure In a protein, hydrogen-bonding interactions between —CO and —NH groups of the primary structure.

secondary tissue Any tissue formed from lateral meristems in trees and shrubs.

Second Law of Thermodynamics A statement concerning the transformation of potential energy into heat; it says that disorder (entropy) is continually increasing in the universe as energy changes occur, so disorder is more likely than order.

second messenger A small molecule or ion that carries the message from a receptor on the target cell surface into the cytoplasm.

seed bank Ungerminated seeds in the soil of an area. Regeneration of plants after events such as fire often depends on the presence of a seed bank.

seed coat In plants, the outer layers of the ovule, which become a relatively impermeable barrier to protect the dormant embryo and stored food.

segment polarity gene Any of certain genes in *Drosophila* development that are expressed in stripes that subdivide the stripes created by the pair-rule genes in the process of segmentation.

segmentation The division of the developing animal body into repeated units; segmentation allows for redundant systems and more efficient locomotion.

segmentation gene Any of the three classes of genes that control development of the segmented body plan of insects; includes the gap genes, pair-rule genes, and segment polarity genes.

segregation The process by which alternative forms of traits are expressed in offspring rather than blending each trait of the parents in the offspring.

selection The process by which some organisms leave more offspring than competing ones, and their genetic traits tend to appear in greater proportions among members of succeeding generations than the traits of those individuals that leave fewer offspring.

selectively permeable Condition in which a membrane is permeable to some substances but not to others.

self-fertilization The union of egg and sperm produced by a single hermaphroditic organism.

semen In reptiles and mammals, sperm-bearing fluid expelled from the penis during male orgasm.

semicircular canal Any of three fluid-filled canals in the inner ear that help to maintain balance.

semiconservative replication DNA replication in which each strand of the original duplex serves as the template for construction of a totally new complementary strand, so the original duplex is partially conserved in each of the two new DNA molecules.

senescent Aged, or in the process of aging.

sensory (afferent) neuron A neuron that transmits nerve impulses from a sensory receptor to the central nervous system or central ganglion.

sensory setae In insect, bristles attached to the nervous system that are sensitive mechanical

and chemical stimulation; most abundant on antennae and legs.

sepal A member of the outermost floral whorl of a flowering plant.

septation In prokaryotic cell division, the formation of a septum where new cell membrane and cell wall is formed to separate the two daughter cells.

septum, pl. septa A wall between two cavities.

sequence-tagged site (STS) A small stretch of DNA that is unique in a genome, that is, it occurs only once; useful as a physical marker on genomic maps.

seta, pl. setae (L., bristle) In an annelid, bristles of chitin that help anchor the worm during locomotion or when it is in its burrow.

severe acute respiratory syndrome (SARS) A respiratory infection with an 8% mortality rate that is caused by a coronavirus.

sex chromosome A chromosome that is related to sex; in humans, the sex chromosomes are the X and Y chromosomes.

sex-linked A trait determined by a gene carried on the X chromosome and absent on the Y chromosome.

sexual reproduction The process of producing offspring through an alternation of fertilization (producing diploid cells) and meiotic reduction in chromosome number (producing haploid cells).

sexual selection A type of differential reproduction that results from variable success in obtaining mates.

shared derived character In cladistics, character states that are shared by species and that are different from the ancestral character state.

shoot In vascular plants, the aboveground portions, such as the stem and leaves.

short interspersed element (SINE) Any of a type of retrotransposon found in humans and other primates that does not contain the biochemical machinery needed for transposition; half a million copies of a SINE element called Alu is nested in the LINEs of the human genome.

shotgun sequencing The method of DNA sequencing in which the DNA is randomly cut into small fragments, and the fragments cloned and sequenced. A computer is then used to assemble a final sequence.

sieve cell In the phloem of vascular plants, a long, slender element with relatively unspecialized sieve areas and with tapering end walls that lack sieve plates.

signal recognition particle (SRP) In eukaryotes, a cytoplasmic complex of proteins that recognizes and binds to the signal sequence of a polypeptide, and then docks with a receptor that forms a channel in the ER membrane. In this way the polypeptide is released into the lumen of the ER.

signal transduction The events that occur within a cell on receipt of a signal, ligand binding to a receptor protein. Signal transduction pathways produce the cellular response to a signaling molecule.

simple sequence repeat (SSR) A one- to three-nucleotide sequence such as CA or CCG that is repeated thousands of times.

single-nucleotide polymorphism (SNP) A site present in at least 1% of the population at which individuals differ by a single nucleotide. These can be used as genetic markers to map unknown genes or traits.

sinus A cavity or space in tissues or in bone.

sister chromatid One of two identical copies of each chromosome, still linked at the centromere, produced as the chromosomes duplicate for mitotic division; similarly, one of two identical copies of each homologous chromosome present in a tetrad at meiosis.

small interfering RNAs (siRNAs) A class of micro-RNAs that appear to be involved in control of gene transcription and that play a role in protecting cells from viral attack.

small nuclear ribonucleoprotein particles (snRNP) In eukaryotes, a complex composed of snRNA and protein that clusters together with other snRNPs to form the spliceosome, which removes introns from the primary transcript.

small nuclear RNA (snRNA) In eukaryotes, a small RNA sequence that, as part of a small nuclear ribonucleoprotein complex, facilitates recognition and excision of introns by base-pairing with the 5′ end of an intron or at a branch site of the same intron.

sodium–potassium pump Transmembrane channels engaged in the active (ATP-driven) transport of Na$^+$, exchanging them for K$^+$, where both ions are being moved against their respective concentration gradients; maintains the resting membrane potential of neurons and other cells.

solute A molecule dissolved in some solution; as a general rule, solutes dissolve only in solutions of similar polarity; for example, glucose (polar) dissolves in (forms hydrogen bonds with) water (also polar), but not in vegetable oil (nonpolar).

solute potential The amount of osmotic pressure arising from the presence of a solute or solutes in water; measure by counterbalancing the pressure until osmosis stops.

solvent The medium in which one or more solutes is dissolved.

somatic cell Any of the cells of a multicellular organism except those that are destined to form gametes (germ-line cells).

somatic cell nuclear transfer (SCNT) The transfer of the nucleus of a somatic cell into an enucleated egg cell that then undergoes development. Can be used to make ES cells and to create cloned animals.

somatic mutation A change in genetic information (mutation) occurring in one of the somatic cells of a multicellular organism, not passed from one generation to the next.

somatic nervous system In vertebrates, the neurons of the peripheral nervous system that control skeletal muscle.

somite One of the blocks, or segments, of tissue into which the mesoderm is divided during differentiation of the vertebrate embryo.

Southern blot A technique in which DNA fragments are separated by gel electrophoresis, denatured into single-stranded DNA, and then "blotted" onto a sheet of filter paper; the filter is then incubated with a labeled probe to locate DNA sequences of interest.

S phase The phase of the cell cycle during which DNA replication occurs.

specialized transduction The transfer of only a few specific genes into a bacterium, using a lysogenic bacteriophage as a carrier.

speciation The process by which new species arise, either by transformation of one species into another, or by the splitting of one ancestral species into two descendant species.

species, pl. species A kind of organism; species are designated by binomial names written in italics.

specific heat The amount of heat that must be absorbed or lost by 1 g of a substance to raise or lower its temperature 1°C.

specific transcription factor Any of a great number of transcription factors that act in a time- or tissue-dependent manner to increase DNA transcription above the basal level.

spectrin A scaffold of proteins that links plasma membrane proteins to actin filaments in the cytoplasm of red blood cells, producing their characteristic biconcave shape.

spermatid In animals, each of four haploid (n) cells that result from the meiotic divisions of a spermatocyte; each spermatid differentiates into a sperm cell.

spermatozoa The male gamete, usually smaller than the female gamete, and usually motile.

sphincter In vertebrate animals, a ring-shaped muscle capable of closing a tubular opening by constriction (e.g., between stomach and small intestine or between anus and exterior).

spicule Any of a number of minute needles of silica or calcium carbonate made in the mesohyl by some kinds of sponges as a structural component.

spindle The structure composed of microtubules radiating from the poles of the dividing cell that will ultimately guide the sister chromatids to the two poles.

spindle apparatus The assembly that carries out the separation of chromosomes during cell division; composed of microtubules (spindle fibers) and assembled during prophase at the equator of the dividing cell.

spindle checkpoint The third cell-division checkpoint, at which all chromosomes must be attached to the spindle. Passage through this checkpoint commits the cell to anaphase.

spinnerets Organs at the posterior end of a spider's abdomen that secrete a fluid protein that becomes silk.

spiracle External opening of a trachea in arthropods.

spiral cleavage The embryonic cleavage pattern of some protostome animals in which cells divide at an angle oblique to the polar axis of the embryo; a line drawn through the sequence of dividing cells forms a spiral.

spiralian A member of a group of invertebrate animals; many groups exhibit spiral cleavage. Mollusks, annelids, and flatworms are examples of spiralians.

spliceosome In eukaryotes, a complex composed of multiple snRNPs and other associated proteins that is responsible for excision of introns and joining of exons to convert the primary transcript into the mature mRNA.

spongin A tough protein made by many kinds of sponges as a structural component within the mesohyl.

spongy parenchyma A leaf tissue composed of loosely arranged, chloroplast-bearing cells. *See* palisade parenchyma.

sporangium, pl. sporangia A structure in which spores are produced.

spore A haploid reproductive cell, usually unicellular, capable of developing into an adult without fusion with another cell.

sporophyte The spore-producing, diploid ($2n$) phase in the life cycle of a plant having alternation of generations.

stabilizing selection A form of selection in which selection acts to eliminate both extremes from a range of phenotypes.

stamen The organ of a flower that produces the pollen; usually consists of anther and filament; collectively, the stamens make up the androecium.

starch An insoluble polymer of glucose; the chief food storage substance of plants.

start codon The AUG triplet, which indicates the site of the beginning of mRNA translation; this codon also codes for the amino acid methionine.

stasis A period of time during which little evolutionary change occurs.

statocyst Sensory receptor sensitive to gravity and motion.

stele The central vascular cylinder of stems and roots.

stem cell A relatively undifferentiated cell in animal tissue that can divide to produce more differentiated tissue cells.

stereoscopic vision Ability to perceive a single, three-dimensional image from the simultaneous but slightly divergent two-dimensional images delivered to the brain by each eye.

stigma (1) In angiosperm flowers, the region of a carpel that serves as a receptive surface for pollen grains. (2) Light-sensitive eyespot of some algae.

stipules Leaflike appendages that occur at the base of some flowering plant leaves or stems.

stolon A stem that grows horizontally along the ground surface and may form adventitious roots, such as runners of the strawberry plant.

stoma, pl. stomata In plants, a minute opening bordered by guard cells in the epidermis of leaves and stems; water passes out of a plant mainly through the stomata.

stop codon Any of the three codons UAA, UAG, and UGA, that indicate the point at which mRNA translation is to be terminated.

stratify To hold plant seeds at a cold temperature for a certain period of time; seeds of many plants will not germinate without exposure to cold and subsequent warming.

stratum corneum The outer layer of the epidermis of the skin of the vertebrate body.

striated muscle Skeletal voluntary muscle and cardiac muscle.

stroma In chloroplasts, the semiliquid substance that surrounds the thylakoid system and that contains the enzymes needed to assemble organic molecules from CO_2.

stromatolite A fossilized mat of ancient bacteria formed as long as 2 BYA, in which the bacterial remains individually resemble some modern-day bacteria.

style In flowers, the slender column of tissue that arises from the top of the ovary and through which the pollen tube grows.

stylet A piercing organ, usually a mouthpart, in some species of invertebrates.

suberin In plants, a fatty acid chain that forms the impermeable barrier in the Casparian strip of root endoderm.

subspecies A geographically defined population or group of populations within a single species that has distinctive characteristics.

substrate (1) The foundation to which an organism is attached. (2) A molecule on which an enzyme acts.

subunit vaccine A type of vaccine created by using a subunit of a viral protein coat to elicit an immune response; may be useful in preventing viral diseases such as hepatitis B.

succession In ecology, the slow, orderly progression of changes in community composition that takes place through time.

summation Repetitive activation of the motor neuron resulting in maximum sustained contraction of a muscle.

supercoiling The coiling in space of double-stranded DNA molecules due to torsional strain, such as occurs when the helix is unwound.

surface tension A tautness of the surface of a liquid, caused by the cohesion of the molecules of liquid. Water has an extremely high surface tension.

surface area-to-volume ratio Relationship of the surface area of a structure, such as a cell, to the volume it contains.

suspensor In gymnosperms and angiosperms, the suspensor develops from one of the first two cells of a dividing zygote; the suspensor of an angiosperm is a nutrient conduit from maternal tissue to the embryo. In gymnosperms the suspensor positions the embryo closer to stored food reserves.

swim bladder An organ encountered only in the bony fish that helps the fish regulate its buoyancy by increasing or decreasing the amount of gas in the bladder via the esophagus or a specialized network of capillaries.

swimmerets In lobsters and crayfish, appendages that occur in lines along the ventral surface of the abdomen and are used in swimming and reproduction.

symbiosis The condition in which two or more dissimilar organisms live together in close association; includes parasitism (harmful to one of the organisms), commensalism (beneficial to one, of no significance to the other), and mutualism (advantageous to both).

sympatric speciation The differentiation of populations within a common geographic area into species.

symplast route In plant roots, the pathway for movement of water and minerals within the cell cytoplasm that leads through plasmodesmata that connect cells.

symplesiomorphy In cladistics, another term for a shared ancestral character state.

symporter A carrier protein in a cell's membrane that transports two molecules or ions in the same direction across the membrane.

synapomorphy In systematics, a derived character that is shared by clade members.

synapse A junction between a neuron and another neuron or muscle cell; the two cells do not touch, the gap being bridged by neurotransmitter molecules.

synapsid Any of an early group of reptiles that had a pair of temporal openings in the skull behind the eye sockets; jaw muscles attached to these openings. Early ancestors of mammals belonged to this group.

synapsis The point-by-point alignment (pairing) of homologous chromosomes that occurs before the first meiotic division; crossing over takes place during synapsis.

synaptic cleft The space between two adjacent neurons.

synaptic vesicle A vesicle of a neurotransmitter produced by the axon terminal of a nerve. The filled vesicle migrates to the presynaptic membrane, fuses with it, and releases the neurotransmitter into the synaptic cleft.

synaptonemal complex A protein lattice that forms between two homologous chromosomes in prophase I of meiosis, holding the replicated chromosomes in precise register with each other so that base-pairs can form between nonsister chromatids for crossing over that is usually exact within a gene sequence.

syncytial blastoderm A structure composed of a single large cytoplasm containing about 4000 nuclei in embryonic development of insects such as *Drosophila*.

syngamy The process by which two haploid cells (gametes) fuse to form a diploid zygote; fertilization.

synthetic polyploidy A polyploidy organism created by crossing organisms most closely related to an ancestral species and then manipulating the offspring.

systematics The reconstruction and study of evolutionary relationships.

systemic acquired resistance (SAR) In plants, a longer-term response to a pathogen or pest attack that can last days to weeks and allow the plant to respond quickly to later attacks by a range of pathogens.

systemin In plants, an 18-amino-acid peptide that is produced by damaged or injured leaves that leads to the wound response.

systolic pressure A measurement of how hard the heart is contracting. When measured during a blood pressure reading, ventricular systole (contraction) is what is being monitored.

T

3′ poly-A tail In eukaryotes, a series of 1–200 adenine residues added to the 3′ end of an mRNA; the tail appears to enhance the stability of the mRNA by protecting it from degradation.

T box A transcription factor protein domain that has been conserved, although with differing developmental effects, in invertebrates and chordates.

tagma, pl. **tagmata** A compound body section of an arthropod resulting from embryonic fusion of two or more segments; for example, head, thorax, abdomen.

Taq polymerase A DNA polymerase isolated from the thermophilic bacterium *Thermus aquaticus* (Taq); this polymerase is functional at higher temperatures, and is used in PCR amplification of DNA.

TATA box In eukaryotes, a sequence located upstream of the transcription start site. The TATA box is one element of eukaryotic core promoters for RNA polymerase II.

taxis, pl. **taxes** An orientation movement by a (usually) simple organism in response to an environmental stimulus.

taxonomy The science of classifying living things. By agreement among taxonomists, no two organisms can have the same name, and all names are expressed in Latin.

T cell A type of lymphocyte involved in cell-mediated immunity and interactions with B cells; the "T" refers to the fact that T cells are produced in the thymus.

telencephalon The most anterior portion of the brain, including the cerebrum and associated structures.

telomerase An enzyme that synthesizes telomeres on eukaryotic chromosomes using an internal RNA template.

telomere A specialized nontranscribed structure that caps each end of a chromosome.

telophase The phase of cell division during which the spindle breaks down, the nuclear envelope of each daughter cell forms, and the chromosomes uncoil and become diffuse.

telson The tail spine of lobsters and crayfish.

temperate (lysogenic) phage A virus that is capable of incorporating its DNA into the host cell's DNA, where it remains for an indeterminate length of time and is replicated as the cell's DNA replicates.

template strand The DNA strand that is used as a template in transcription. This strand is copied to produce a complementary mRNA transcript.

tendon (Gr. *tendon*, stretch) A strap of cartilage that attaches muscle to bone.

tensile strength A measure of the cohesiveness of a substance; its resistance to being broken apart. Water in narrow plant vessels has tensile strength that helps keep the water column continuous.

tertiary structure The folded shape of a protein, produced by hydrophobic interactions with water, ionic and covalent bonding between side chains of different amino acids, and van der Waal's forces; may be changed by denaturation so that the protein becomes inactive.

testcross A mating between a phenotypically dominant individual of unknown genotype and a homozygous "tester," done to determine whether the phenotypically dominant individual is homozygous or heterozygous for the relevant gene.

testis, pl. **testes** In mammals, the sperm-producing organ.

tetanus Sustained forceful muscle contraction with no relaxation.

thalamus That part of the vertebrate forebrain just posterior to the cerebrum; governs the flow of information from all other parts of the nervous system to the cerebrum.

therapeutic cloning The use of somatic cell nuclear transfer to create stem cells from a single individual that may be reimplanted in that individual to replace damaged cells, such as in a skin graft.

thermodynamics The study of transformations of energy, using heat as the most convenient form of measurement of energy.

thigmotropism In plants, unequal growth in some structure that comes about as a result of physical contact with an object.

threshold The minimum amount of stimulus required for a nerve to fire (depolarize).

thylakoid In chloroplasts, a complex, organized internal membrane composed of flattened disks, which contain the photosystems involved in the light-dependent reactions of photosynthesis.

Ti (tumor-inducing) plasmid A plasmid found in the plant bacterium *Agrobacterium tumefaciens* that has been extensively used to introduce recombinant DNA into broadleaf plants. Recent modifications have allowed its use with cereal grains as well.

tight junction Region of actual fusion of plasma membranes between two adjacent animal cells that prevents materials from leaking through the tissue.

tissue A group of similar cells organized into a structural and functional unit.

tissue plasminogen activator (TPA) A human protein that causes blood clots to dissolve; if used within 3 hours of an ischemic stroke, TPA may prevent disability.

tissue-specific stem cell A stem cell that is capable of developing into the cells of a certain tissue, such as muscle or epithelium; these cells persist even in adults.

tissue system In plants, any of the three types of tissue; called a system because the tissue extends throughout the roots and shoots.

tissue tropism The affinity of a virus for certain cells within a multicellular host; for example, hepatitis B virus targets liver cells.

tonoplast The membrane surrounding the central vacuole in plant cells that contains water channels; helps maintain the cell's osmotic balance.

topoisomerase Any of a class of enzymes that can change the topological state of DNA to relieve torsion caused by unwinding.

torsion The process in embryonic development of gastropods by which the mantle cavity and anus move from a posterior location to the front of the body, closer to the location of the mouth.

totipotent A cell that possesses the full genetic potential of the organism.

trachea, pl. **tracheae** A tube for breathing: in terrestrial vertebrates, the windpipe that carries air between the larynx and bronchi (which leads to the lungs); in insects and some other terrestrial arthropods, a system of chitin-lined air ducts.

tracheids In plant xylem, dead cells that taper at the ends and overlap one another.

tracheole The smallest branches of the respiratory system of terrestrial arthropods; tracheoles convey air from the tracheae, which connect to the outside of the body at spiracles.

trait In genetics, a characteristic that has alternative forms, such as purple or white flower color in pea plants or different blood type in humans.

transcription The enzyme-catalyzed assembly of an RNA molecule complementary to a strand of DNA.

transcription complex The complex of RNA polymerase II plus necessary activators, coactivators, transcription factors, and other factors that are engaged in actively transcribing DNA.

transcription factor One of a set of proteins required for RNA polymerase to bind to a eukaryotic promoter region, become stabilized, and begin the transcription process.

transcription bubble The region containing the RNA polymerase, the DNA template, and the RNA transcript, so called because of the locally unwound "bubble" of DNA.

transcription unit The region of DNA between a promoter and a terminator.

transcriptome All the RNA present in a cell or tissue at a given time.

transfection The transformation of eukaryotic cells in culture.

transfer RNA (tRNA) A class of small RNAs (about 80 nucleotides) with two functional sites; at one site, an "activating enzyme" adds a specific amino acid, while the other site carries the nucleotide triplet (anticodon) specific for that amino acid.

transformation The uptake of DNA directly from the environment; a natural process in some bacterial species.

transgenic organism An organism into which a gene has been introduced without conventional breeding, that is, through genetic engineering techniques.

translation The assembly of a protein on the ribosomes, using mRNA to specify the order of amino acids.

translation repressor protein One of a number of proteins that prevent translation of mRNA by binding to the beginning of the transcript and preventing its attachment to a ribosome.

translocation (1) In plants, the long-distance transport of soluble food molecules (mostly sucrose), which occurs primarily in the sieve tubes of phloem tissue. (2) In genetics, the interchange of chromosome segments between nonhomologous chromosomes.

transmembrane domain Hydrophobic region of a transmembrane protein that anchors it in the membrane. Often composed of α-helices, but sometimes utilizing β-pleated sheets to form a barrel-shaped pore.

transmembrane route In plant roots, the pathway for movement of water and minerals that crosses the cell membrane and also the membrane of vacuoles inside the cell.

transpiration The loss of water vapor by plant parts; most transpiration occurs through the stomata.

transposable elements Segments of DNA that are able to move from one location on a chromosome to another. Also termed *transposons* or *mobile genetic elements*.

transposition Type of genetic recombination in which transposable elements (transposons) move from one site in the DNA sequence to another, apparently randomly.

transposon DNA sequence capable of transposition.

trichome In plants, a hairlike outgrowth from an epidermal cell; glandular trichomes secrete oils or other substances that deter insects.

triglyceride (triacylglycerol) An individual fat molecule, composed of a glycerol and three fatty acids.

triploid Possessing three sets of chromosomes.

trisomic Describes the condition in which an additional chromosome has been gained due to nondisjunction during meiosis, and the diploid embryo therefore has three of these autosomes. In humans, trisomic individuals may survive

if the autosome is small; Down syndrome individuals are trisomic for chromosome 21.

trochophore A specialized type of free-living larva found in lophotrochozoans.

trophic level A step in the movement of energy through an ecosystem.

trophoblast In vertebrate embryos, the outer ectodermal layer of the blastodermic vesicle; in mammals, it is part of the chorion and attaches to the uterine wall.

tropism Response to an external stimulus.

tropomyosin Low-molecular-weight protein surrounding the actin filaments of striated muscle.

troponin Complex of globular proteins positioned at intervals along the actin filament of skeletal muscle; thought to serve as a calcium-dependent "switch" in muscle contraction.

trp **operon** In *E. coli*, the operon containing genes that code for enzymes that synthesize tryptophan.

true-breeding Said of a breed or variety of organism in which offspring are uniform and consistent from one generation to the next; for example. This is due to the genotypes that determine relevant traits being homozygous.

tube foot In echinoderms, a flexible, external extension of the water–vascular system that is capable of attaching to a surface through suction.

tubulin Globular protein subunit forming the hollow cylinder of microtubules.

tumor-suppressor gene A gene that normally functions to inhibit cell division; mutated forms can lead to the unrestrained cell division of cancer, but only when both copies of the gene are mutant.

turgor pressure The internal pressure inside a plant cell, resulting from osmotic intake of water, that presses its cell membrane tightly against the cell wall, making the cell rigid. Also known as *hydrostatic pressure*.

tympanum In some groups of insects, a thin membrane associated with the tracheal air sacs that functions as a sound receptor; paired on each side of the abdomen.

U

ubiquitin A 76-amino-acid protein that virtually all eukaryotic cells attach as a marker to proteins that are to be degraded.

unequal crossing over A process by which a crossover in a small region of misalignment at synapsis causes two homologous chromosomes to exchange segments of unequal length.

uniporter A carrier protein in a cell's membrane that transports only a single type of molecule or ion.

uniramous Single-branched; describes the appendages of insects.

unsaturated fat A fat molecule in which one or more of the fatty acids contain fewer than the maximum number of hydrogens attached to their carbons.

urea An organic molecule formed in the vertebrate liver; the principal form of disposal of nitrogenous wastes by mammals.

urethra The tube carrying urine from the bladder to the exterior of mammals.

uric acid Insoluble nitrogenous waste products produced largely by reptiles, birds, and insects.

urine The liquid waste filtered from the blood by the kidney and stored in the bladder pending elimination through the urethra.

uropod One of a group of flattened appendages at the end of the abdomen of lobsters and crayfish that collectively act as a tail for a rapid burst of speed.

uterus In mammals, a chamber in which the developing embryo is contained and nurtured during pregnancy.

V

vacuole A membrane-bounded sac in the cytoplasm of some cells, used for storage or digestion purposes in different kinds of cells; plant cells often contain a large central vacuole that stores water, proteins, and waste materials.

valence electron An electron in the outermost energy level of an atom.

variable A factor that influences a process, outcome, or observation. In experiments, scientists attempt to isolate variables to test hypotheses.

vascular cambium In vascular plants, a cylindrical sheath of meristematic cells, the division of which produces secondary phloem outwardly and secondary xylem inwardly; the activity of the vascular cambium increases stem or root diameter.

vascular tissue Containing or concerning vessels that conduct fluid.

vas deferens In mammals, the tube carrying sperm from the testes to the urethra.

vasopressin A posterior pituitary hormone that regulates the kidney's retention of water.

vector In molecular biology, a plasmid, phage or artificial chromosome that allows propagation of recombinant DNA in a host cell into which it is introduced.

vegetal pole The hemisphere of the zygote comprising cells rich in yolk.

vein (1) In plants, a vascular bundle forming a part of the framework of the conducting and supporting tissue of a stem or leaf. (2) In animals, a blood vessel carrying blood from the tissues to the heart.

veliger The second larval stage of mollusks following the trochophore stage, during which the beginning of a foot, shell, and mantle can be seen.

ventricle A muscular chamber of the heart that receives blood from an atrium and pumps blood out to either the lungs or the body tissues.

vertebrate A chordate with a spinal column; in vertebrates, the notochord develops into the vertebral column composed of a series of vertebrae that enclose and protect the dorsal nerve cord.

vertical gene transfer (VGT) The passing of genes from one generation to the next within a species.

vesicle A small intracellular, membrane-bounded sac in which various substances are transported or stored.

vessel element In vascular plants, a typically elongated cell, dead at maturity, which conducts water and solutes in the xylem.

vestibular apparatus The complicated sensory apparatus of the inner ear that provides for balance and orientation of the head in vertebrates.

vestigial structure A morphological feature that has no apparent current function and is thought to be an evolutionary relic; for example, the vestigial hip bones of boa constrictors.

villus, pl. **villi** In vertebrates, one of the minute, fingerlike projections lining the small intestine that serve to increase the absorptive surface area of the intestine.

virion A single virus particle.

viroid Any of a group of small, naked RNA molecules that are capable of causing plant diseases, presumably by disrupting chromosome integrity.

virus Any of a group of complex biochemical entities consisting of genetic material wrapped in protein; viruses can reproduce only within living host cells and are thus not considered organisms.

visceral mass Internal organs in the body cavity of an animal.

vitamin An organic substance that cannot be synthesized by a particular organism but is required in small amounts for normal metabolic function.

viviparity Refers to reproduction in which eggs develop within the mother's body and young are born free-living.

voltage-gated ion channel A transmembrane pathway for an ion that is opened or closed by a change in the voltage, or charge difference, across the plasma membrane.

W

water potential The potential energy of water molecules. Regardless of the reason (e.g., gravity, pressure, concentration of solute particles) for the water potential, water moves from a region where water potential is greater to a region where water potential is lower.

water–vascular system A fluid-filled hydraulic system found only in echinoderms that provides body support and a unique type of locomotion via extensions called tube feet.

Western blot A blotting technique used to identify specific protein sequences in a complex mixture. *See* Southern blot.

wild type In genetics, the phenotype or genotype that is characteristic of the majority of individuals of a species in a natural environment.

wobble pairing Refers to flexibility in the pairing between the base at the 5′ end of a tRNA anticodon and the base at the 3′ end of an mRNA codon. This flexibility allows a single tRNA to read more than one mRNA codon.

wound response In plants, a signaling pathway initiated by leaf damage, such as being chewed by a herbivore, and lead to the production of proteinase inhibitors that give herbivores indigestion.

X

X chromosome One of two sex chromosomes; in mammals and in *Drosophila*, female individuals have two X chromosomes.

xylem In vascular plants, a specialized tissue, composed primarily of elongate, thick-walled conducting cells, which transports water and solutes through the plant body.

Y

Y chromosome One of two sex chromosomes; in mammals and in *Drosophila*, male individuals have a Y chromosome and an X chromosome; the Y determines maleness.

yolk plug A plug occurring in the blastopore of amphibians during formation of the archenteron in embryological development.

yolk sac The membrane that surrounds the yolk of an egg and connects the yolk, a rich food supply, to the embryo via blood vessels.

Z

zinc finger motif A type of DNA-binding motif in regulatory proteins that incorporates zinc atoms in its structure.

zona pellucida An outer membrane that encases a mammalian egg.

zone of cell division In plants, the part of the young root that includes the root apical meristem and the cells just posterior to it; cells in this zone divide every 12–36 hr.

zone of elongation In plants, the part of the young root that lies just posterior to the zone of cell division; cells in this zone elongate, causing the root to lengthen.

zone of maturation In plants, the part of the root that lies posterior to the zone of elongation; cells in this zone differentiate into specific cell types.

zoospore A motile spore.

zooxanthellae Symbiotic photosynthetic protists in the tissues of corals.

zygomycetes A type of fungus whose chief characteristic is the production of sexual structures called zygosporangia, which result from the fusion of two of its simple reproductive organs.

zygote The diploid (2*n*) cell resulting from the fusion of male and female gametes (fertilization).

Photographs

Chapter 1
Opener: © Soames Summerhays/Natural Visions; 1.1 (organelle): © S. Gschmeeissner/SPL/PUBLIPHOTO; (cell): © Lennart Nilsson/Albert Bonniers Förlag AB; (tissue): © Ed Reschke; (organism): © Russell Illig/Getty Images; (population): © Jeremy Woodhouse/Getty Images; (species (both)): © PhotoDisc/Volume 44/Getty Images; (community): © Steve Harper/Grant Heilman Photography; (ecosystem): © Robert & Jean Pollock; (biosphere): NASA; 1.5 (bottom left): © Huntington Library/Superstock; 1.11 (center right): © Dennis Kunkel/Phototake; (bottom right): © Karl E. Deckart/Phototake; 1.13 (1st row-left): © Alan L. Detrick/Photo Researchers Inc; (center): © David M. Dennis/Animals Animals-Earth Scenes; (right): Corbis/Volume 46; (2nd row-left): © Royalty-Free/CORBIS; (center): © Mediscan/CORBIS; (right): © PhotoDisc BS/Volume 15/Getty Images; (3rd row-left): © Royalty-Free/Corbis; (center): © Tom Brakefield/CORBIS; (right): © PhotoDisc/Volume 44/Getty Images; (4th row-left): © Corbis/Volume 64; (right): © T.E. Adams/Visuals Unlimited; (right): © Douglas P. Wilson/Frank Lane Picture Agency/CORBIS; (5th row-left): © R. Robinson/Visuals Unlimited; (right): © Kari Lounatman/Photo Researchers Inc; (6th row-left): © Dwight R. Kuhn; (right): © Alfred Pasieka/Science Photo Library/Photo Researchers Inc.

Chapter 2
Opener: © IBM; 2.1: © Veeco Digital Instruments; 2.9a: © PhotoLink/Getty Images; 2.9b: © Glen Allison/Getty Images; 2.9c: © Jeff Vanuga/CORBIS; 2.12 (bottom left): © Hermann Eisenbeiss/National Audubon Society Collection/Photo Researchers Inc.

Chapter 3
Opener: © Jacob Halaska/IndexStock; 3.9 (center right): © Asa Thoresen/Photo Researchers Inc; 3.9 (bottom right): © J. Carson/Custom Medical Stock Photo; 3.10 (center): © J.D. Litvay/Visuals Unlimited; 3.11 (bottom right): © Scott Johnson/Animals Animals/Earth Scenes; 3.12a (bottom left): © Driscoll: Youngquist & Baldeschwieler, Caltech/SPL/Photo Researchers Inc; 3.12b (bottom center): © M. Freeman/PhotoLink/Getty Images; 3.18 (top left): © Dr. Tim Evans/Photo Researchers Inc; 3.18 (top right): © PhotoDisc/Volume 6/Getty Images; 3.18 (left 2nd from top): © C.W. SCHWARTZ; 3.18 (right 2nd from top): © Royalty-Free/CORBIS; 3.18 (left bottom): © Chad Baker/Getty Images; 3.18 (right 2nd from bottom): © Lon C. Diehl/Photo Edit; 3.18 (right bottom): © DPA/NVM/The Image Works.

Chapter 4
Opener: © Dr. Gopal Murti/Photo Researchers Inc; 4.1 (1st from top): © David M. Phillips/Visuals Unlimited; (2nd from top): © Mike Abbey/Visuals Unlimited; (3rd from top): © David M. Phillips/Visuals Unlimited; (4th from top): © Mike Abbey/Visuals Unlimited; (5th from top): Dr. Torsten Wittmann/Photo Researchers Inc; (6th from top): © Med. Mic. Sciences: Cardiff Uni./Wellcome Photo Library; (7th from top): © Microworks/Phototake; (8th from top): © Stanley Flegler/Visuals Unlimited; 4.2 (top left): © Dr. Don W. Fawcett/Visuals Unlimited; 4.3 (bottom right): © Phototake; 4.4: Courtesy of E.H. Newcomb & T.D. Pugh: University of Wisconsin; 4.5 (top left): © Eye of Science/Photo Researchers Inc; 4.8 (center left): © Dr. Richard Kessel & Dr. Gene Shih/Visuals Unlimited; 4.8 (center right): © John T. Hansen, Ph.D/Phototake; 4.8d (bottom right): Courtesy of Ueli Aebi, M.E. Mueller Institute for Structural Biology, Biozentrum, University of Basel, Switzerland.Picture adapted from, Aebi, U., Cohn, J., Buhle, E.L. and Gerace, L. (1986). The Nuclear Lamina is a Meshwork of Intermediate-type Filaments. Nature (Lond.) 323: 560–564; 4.9 (bottom left): © Ed Reschke; 4.11 (bottom right): © R. Bolender & D. Fawcett/Visuals Unlimited; 4.12 (center left): © Dennis Kunkel/Phototake; 4.15 (center left): Courtesy of E.H. Newcomb & S.E. Frederick University of Wisconsin. Reprinted with permission from Science, Vol 163, 1353–1355 © 1969, American Association for the Advancement of Science; 4.16 (bottom right): © Dr. Henry Aldrich/Visuals Unlimited; 4.17 (bottom right): © Dr. Donald Fawcett & Dr. Porter/Visuals Unlimited; 4.18 (top right): © Dr. Jeremy Burgess/Photo Researchers Inc; 4.23 (both): © William Dentler, University of Kansas; 4.24 (both): © SPL/Photo Researchers Inc; 4.25 (bottom right): © BioPhoto Associates/Photo Researchers Inc.

Chapter 5
Opener: © Dr. Gopal Murti/SPL/Photo Researchers Inc; 5.1 (pg 88): © Don W. Fawcett/Photo Researchers Inc; 5.3 (top right): © Dr. Don W. Fawcett/Visuals Unlimited; 5.13 (all): © Dr. David M. Phillips/Visuals Unlimited; 5.14: © Wim van Egmond/Visuals Unlimited; 5.17a: Micrograph Courtesy of the CDC/Dr. Edwin P. Ewing, Jr; 5.17b: BCC Microimaging. Inc. Reproduced with permission; 5.17d: © The Company of Biologists Limited; 5.18b: Dr. Brigit Satir.

Chapter 6
Opener: © Robert A. Caputo/Aurora & Quanta Productions Inc; 6.3 (both): © Spencer Grant/Photo Edit; 6.10 (bottom right): Professor Emeritus Lester J. Reed, University of Texas at Austin.

Chapter 7
Opener: © Creatas/PunchStock; 7.9 (center right): Royalty-Free/CORBIS; 7.18 (bottom left): © Wolfgang Baumeister/Photo Researchers Inc; 7.18 (bottom right): National Park Service.

Chapter 8
Opener: © Royalty-Free/Corbis; 8.1 (center right): Courtesy Dr. Kenneth Miller, Brown University; 8.7 (both): © Eric Soder/Tom Stack & Associates; 8.18 (bottom right): © Dr. Jeremy Burgess/Photo Researchers Inc; 8.20 (top center): © John Shaw/Photo Researchers Inc; 8.20 (bottom center): © Joseph Nettis/National Audubon Society Collection/Photo Researchers Inc; 8.22 (top right): © Eric Soder/Tom Stack & Associates.

Chapter 9
Opener: © RMF/Scientifica/Visuals Unlimited; 9.17 (top left): Courtesy of Daniel Goodenough; 9.17 (center left): © Dr. Donald Fawcett/Visuals Unlimited; 9.17 (bottom left): © Dr. Donald Fawcett/D. Albertini/Visuals Unlimited.

Chapter 10
Opener: © Stem Jems/Photo Researchers Inc; 10.2 (both): Courtesy of William Margolin; 10.4 (bottom right): Biophoto Associates/Photo Researchers Inc; 10.6 (bottom left): CNRI/Photo Researchers Inc; 10.10 (top right): Image courtesy of S. Hauf and J-M. Peters, IMP, Vienna, Austria; 10.11 (all), 10.12; © Andrew S. Bauer, University of Oregon; 10.13 (both): © Dr. Jeremy Pickett-Heaps; 10.14 (bottom left): © Dr. David M. Phillips/Visuals Unlimited; 10.14 (bottom center): Guenter Albrecht-Buehler, Northwestern University, Chicago; 10.15 (top right): © B.A. Palevits & E.H. Newcomb/BPS/Tom Stack & Associates.

Chapter 11
Opener: © Science VU/L. Maziarski; 11.4b: Reprinted, with permission, from the Annual Review of Genetics, Volume 6 © 1972 by Annual Reviews, www.annualreviews.org; 11.8 (all): © Clare A. Hasenkamp/Biological Photo Service.

Chapter 12
Opener: © Corbis; 12.1: © Norbert Schaefer/Corbis; 12.2: © David Sieren/Visuals Unlimited; 12.3: © Leslie Holzer/Photo Researchers Inc; 12.5 (both) © Wally Eberhart/Visuals Unlimited; 12.12: From Albert & Blakeslee Corn and Man Journal of Heredity, Vol. 5, pg 511, 1914, Oxford University Press; 12.15: © DK Limited/Corbis.

Chapter 13
Opener: © Adrian T. Sumner/Photo Researchers Inc; 13.1 (both): © Cabisco/Phototake; 13.2: © Biophoto Associates/Photo Researchers Inc; 13.3: © Bettmann/Corbis; p. 241 left): PNAS 2004, Chadwick and Willard; 13.4 (top right): © Kenneth Mason; 13.11: © Jackie Lewin, Royal Free Hospital/Photo Researchers Inc; 13.13: © R. Hutchings/Photo Researchers Inc.

Chapter 14
Opener: © Vol. 29/PhotoDisc/Getty Images; 14.6 (both): From "The Double Helix," by J.D. Watson,

Atheneum Press, NY, 1968; 14.7: © Barrington Brown/Photo Researchers Inc; 14.12 (bottom right): From M. Meselson and F.W. Stahl/ Proceedings of the Nat. Acad. of Sci. 44 (1958): 671; 14.17 (both): From Biochemistry 4e by Stryer © 1995 by Lupert Stryer. Used with permission of W.H. Freeman and Company; 14.21: © Dr. Don W. Fawcett/Visuals Unlimited; 14.22: Courtesy of Dr. David Wolstenholme.

Chapter 15
Opener: © Dr. Gopal Murti/Visuals Unlimited; 15.3: © University of Missouri, Extension and Agriculture Information; 15.4: From R.C. Williams, Proc. Nat. Acad. of Sci. 74 (1977): 2313; 15.8 (top right): © Dr. Oscar Miller; 15.11b: Courtesy of Dr. Bert O'Malley, Baylor College of Medicine; 15.13 (space-filled model): Created by John Beaver using ProteinWorkshop, a product of the RCSB PDB, and built using the Molecular Biology Toolkit developed by John Moreland and Apostol Gramada (mbt.sdsc.edu). The MBT is financed by grant GM63208; 15.16: © P. Nissen, N. Ban, P.B. Moore, and T.A. Steitz (from Science 289: 920-930 [2000].

Chapter 16
Opener: © Dr. Claus Pelling; 16.10 (both): Courtesy of Dr. Harrison Echols; 16.20: Reprinted with permission from the Annual Review of Biochemistry, Volume 68 © 1999 by Annual Reviews www.annualreviews.org.

Chapter 17
Opener: © Prof. Stanley Cohen/Photo Researchers Inc; 17.2d: Courtesy of Biorad Laboratories; 17.7 (bottom right): © SSPL/The Image Works; 17.9: Courtesy of Lifecodes Corp, Stamford CT; 17.10: © Matt Meadows/Peter Arnold Inc; 17.15: R.L. Brinster, U. of Pennsylvania Sch. of Vet. Med.; 17.18: © Rob Horsch, Monsanto Company.

Chapter 18
Opener: © William C. Ray, Ohio State University; 18.2a: © B. Trask and Colleagues, Fred Hutchinson Cancer Research Center: *Nature* 409, Feb. 15, 2001, pg 953; 18.2b: © Dr. Cynthia Morton: Nature 409, Feb. 15, 2001, pg 953; 18.4: Courtesy of Celera Genomics; 18.11a,b: Photographs provided by Indra K. Vasil; 18.11c-d: With permission from Altpeter et al, Plant Cell Reports 16: 12-17, 1996, photos provided by Indra Vasil; 18.12: Courtesy of Research Collaboratory for Structural Bioinformatics; 18.13: © Royalty Free/Corbis; 18.14: © Grant Heilman/ Grant Heilman Photography.

Chapter 19
Opener: © Dwight Kuhn; 19.1 (all): © Carolina Biological Supply Company/Phototake; 19.4b: © University of Wisconsin-Madison News & Public Affairs; 19.6b: © J. Richard Whittaker, used by permission; 19.9 (bottom right): AP/WideWorld Photos; 19.13 (top left): © Steve Paddock and Sean Carroll; 19.13 (top right, center left, center right): © Jim Langeland, Steve Paddock and Sean Carroll; 19.16a: © Dr. Daniel St. Johnston/Wellcome Photo Library; 19.16b: © Schupbach, T. and van Buskirk, C.; 19.16c: from Rotha et al., 1989, courtesy of Siegfried Roth; 19.17: Courtesy of E.B. Lewis; 19.20 (both): from Boucaut et al., 1984, courtesy of J-C Boucaut; 19.22: © Christian Laforsch/Photo Researchers Inc.

Chapter 20
Opener: © Cathy & Gordon ILLG; 20.2: © Royalty-Free/Corbis.

Chapter 21
Opener: © PhotoDisc/Getty Images; 21.3 (both): © Breck P. Kent/Animals Animals/Earth Scenes; 21.8 (both): © Courtesy of Lyudmilla N. Trut, Institute of Cytology & Genetics, Siberian Dept. of the Russian Academy of Sciences; 21.11: © Kevin Schafer/Peter Arnold Inc.

Chapter 22
Opener: © Chris Johns/National Geographic/Getty Images; 22.2: © Porterfield/Chickering/Photo Researchers Inc; 22.3: © Barbara Gerlach/Visuals Unlimited; 22.5 (1): © John Shaw/Tom Stack & Associates; (2): © Rob & Ann Simpson/Visuals Unlimited; (3): © Suzanne L. Collins & Joseph T. Collins/National Audubon Society Collection/Photo Researchers; (4): © Phil A. Dotson/National Audubon Society Collection/Photo Researchers; 22.7 (left to right): © Jonathan Losos; © Chas. McRae/ Visuals Unlimited; © Jonathan Losos; 22.13a: © Jeffrey Taylor; 22.13b: © Kenneth Y. Kaneshiro, Center for Conservation Research & Training, University of Hawaii; 22.16 (left to right): © Photo New Zealand/Nick Groves; © Jim Harding/ firstlight; © Colin Harris/Light Touch Images/ Alamy; © Focus New Zealand Photo Library; 22.16: © Focus New Zealand Photo Library.

Chapter 23
Opener: © G. Mermet/Peter Arnold Inc; 23.1a: by permission of the Syndics of Cambridge University Library; 23.8a: image #5789, photo by D. Finnin/ American Museum of Natural History; 23.8b: © Roger De La Harpe/Animals Animals; 23.10 (bottom left): © Lee W. Wilcox; (bottom right): © Dr. Richard Kessel & Dr. Gene Shih/Visuals Unlimited.

Chapter 25
Opener: © Michael & Patricia Fogden/Minden Pictures; 25.4a: © Michael Persson; 25.4b: © E.R. Degginger/Photo Researchers Inc; 25.5: Dr. Anna Di Gregorio, Weill Cornell Medical College; 25.10 (bottom left): © Chuck Pefley/Getty Images; (bottom center left): © Darwin Dale/Photo Researchers Inc; (bottom center right): © Aldo Brando Peter Arnold Inc; (bottom right): © Tom E. Adams/Peter Arnold Inc; 25.11 (all): Courtesy of Walter Gehring, reprinted with permission from Induction of Ectopic Eyes by Targeted Expression of the Eyeless Gene in Drosophila, G. Halder, P. Callaerts, Walter J. Gehring, Science Vol. 267 © 24 March 1995 American Association for the Advancement of Science; 25.12 (both): © Dr. William Jeffrey.

Chapter 26
Opener: © Jeff Hunter/The Image Bank/Getty Images; 26.1: © T.E. Adams/Visuals Unlimited; pg 504 (bottom right): © NASA/Photo Researchers Inc; 26.2: © NASA/JPL-Caltech; 26.5 (top left): © Tom Walker/Riser/Getty Images; (top right): © Corbis/Volume 8; (bottom left): © Corbis/Volume 102; (bottom right): © PhotoDisc Volume 1/Getty Images; 26.14: © Sean W. Graham, UBC Botanical Garden & Centre for Plant Research, University of British Columbia.

Chapter 27
Opener: © Dr. Gopal Murti/Visuals Unlimited; 27.2: From J. H. Hogle et al, "Three Dimensional Structure of Poliovirus @ 2.9 A Resolution," Science 229: 1360, Sept. 27, 1985. © 1985 by the AAAS; 27.3a: Dept. of Biology, Biozentrum/SPL/Photo Researchers; 27.8: © Corbis/Volume 40.

Chapter 28
Opener: © Dr. David Phillips/Visuals Unlimited; 28.6, pg 544 (left to right): © SPL/Photo Researchers Inc; © Dr. R. Rachel and Prof. Dr. K. O. Stetten, University of Regensburg, Lehrstuhl fuer Mikrobiologie, Regensburg, Germany; © Andrew Syred/SPL/Photo Researchers, Inc.; © Microfield Scientific Ltd/SPL/Photo Researchers Inc; © Alfred Paseika/SPL/Photo Researchers, Inc.; pg 545 (left to right): © Science VU/S. Watson/Visuals Unlimited; © Dennis Kunkel Microscopy Inc; Prof. Dr. Hans Reichenbach, Helmholtz Centre for Infection Research, Braunschweig; 28.3 (center left to right): Dr. Gary Gaugler/Science Photo Library/Photo Researchers, Inc.; © CNRI/Photo Researchers, Inc.; © Dr. Richard Kessel & Dr. Gene Shih/Visuals Unlimited; 28.7b: © Jack Bostrack/Visuals Unlimited; 28.9b: © Julius Adler; 28.10a: © Science VU/S. W. Watson/Visuals Unlimited; 28.10b: © Norma J. Lang/Biological Photo Service; 28.11a: © Dr. Dennis Kunkel/Visuals Unlimited; 28.16: © CNRI/Photo Researchers, Inc.; 28.18: © Science/Visuals Unlimited.

Chapter 29
Opener: © Wim van Egmond/Visuals Unlimited; 29.2: © Andrew H. Knoll, Harvard University; 29.6, 29.7: © Science VU/E. White/Visuals Unlimited; 29.8a: © Andrew Syred/Photo Researchers; 29.9a: © Manfred Kage/Peter Arnold Inc; 29.9b: © Edward S. Ross; 29.10: Michael Delannoy (John Hopkins University School of Medicine Microscope Facility) and Vern B. Carruthers (University of Michigan School of Medicine); 29.12: © Dr. David M. Phillips/Visuals Unlimited; 29.14: © Michael Abbey/Visuals Unlimited; 29.15: © Prof. David J.P. Ferguson, Oxford University; 29.17: © Brian Parker/Tom Stack & Associates; 29.18: © Michele Bahr and D. J. Patterson, used under license to MBL; 29.19: © Randy Morse/Earth Scenes; 29.21: © Dennis Kunkel/Phototake; 29.22: © Andrew Syred/Photo Researchers, Inc.; 29.23 (center left): © Manfred Kage/Peter Arnold Inc; (center right): © Wim van Egmond/Visuals Unlimited; (bottom right): Runk/ Schoenberger/Grant Heilman Photography; 29.24: © William Bourland, image used under license to MBL; 29.25: © Eye of Science/Photo Researchers Inc; 29.26: Phil A. Harrington/Peter Arnold Inc; 29.27: © Manfred Kage/Peter Arnold Inc; 29.28: © Ric Ergenbright/CORBIS; 29.29: © Peter Arnold Inc/ Alamy; 29.30: © John Shaw/Tom Stack & Associates; 29.31: © Mark J. Grimson and Richard L. Blanton, Biological Sciences Electron Microscopy Laboratory, Texas Tech University.

Chapter 30
Opener: © S.J. Krasemann/Peter Arnold; 30.3: © Dr. Richard Kessel & Dr. Gene Shih/Visuals Unlimited; 30.4: © Wim van Egmond/Visuals Unlimited; 30.5: © Marevision/agefotostock; 30.6 (bottom left): © Robert Calentine/Visuals Unlimited; (bottom right): © Wim van Egmond/Visuals Unlimited; 30.7: © David Sieren/Visuals Unlimited; 30.8: © Lee Wilcox; 30.9: © Edward S. Ross; 30.11: Courtesy of

Hans Steur, The Netherlands; 30.12: © Ed Reschke/Peter Arnold; 30.13: © Kingsley R. Stern; 30.14: © Stephen P. Parker/Photo Researchers Inc; 30.15: © Mark Bowler/NHPA; 30.16 (left): © Ed Reschke; (right): © Mike Zensa/CORBIS; p. 593: © Biology Media/Photo Researchers Inc; 30.18: © Patti Murray/Earth Scenes; 30.20a: © Jim Strawser/Grant Heilman Photography; 30.20b: © Nancy Hoyt Belcher/Grant Heilman Photography; 30.20c: © Robert Gustafson/Visuals Unlimited; 30.21: David Dilcher and Ge Sun; 30.22: Courtesy of Sandra Floyd.

Chapter 31

Opener: © Ullstein-Joker/Peter Arnold Inc; 31.1a: © Dean A. Glawe/Biological Photo Service; 31.1b: © Carolina Biological Supply Company/Phototake; 31.1c: Yolande Dalpé, Agriculture and Agri-Food Canada, Spores of Glomus Intradices Schenck-Smith Glomeronmycota, In-vitro Collection–GINCO/DAOM 197198; 31.1d: © Michael & Patricia Fogden; 31.2: © Dr. Garry T. Cole/Biological Photo Service; 31.3: © Michael & Patricia Fogden/CORBIS, (inset): © Micro Discovery/CORBIS; 31.4: © Microfield Scientific Ltd/Photo Researchers Inc; 31.5a: © Carolina Biological Supply Company/Phototake; 31.5b: © L. West/Photo Researchers, Inc.; 31.7a, 31.8a: © Carolina Biological Supply Company/Phototake; 31.10: © David Scharf/Photo Researchers Inc; 31.9a: © Richard Kolar/Earth Scenes; 31.9b: © Ed Reschke/Peter Arnold Inc.; 31.11a: © Alexandra Lowry/The National Audubon Society Collection/Photo Researchers, Inc.; 31.12a: © Dr. Fred Hossler/Visuals Unlimited; 31.12b: © Eye of Science/Photo Researchers Inc; 31.13a (left): © Holt Studios International Ltd./Alamy; 31.13b (right): © B. Borrell Casal/Frank Lane Picture Agency/CORBIS; 31.14a: © Ken Wagner/Phototake; 31.14b: © Robert & Jean Pollack/Visuals Unlimited; 31.14c: © Robert Lee/Photo Researchers, Inc.; 31.15: © Ed Reschke; 31.16a: © Eye of Science/Photo Researchers, Inc.; 31.16b: © Dr. Gerald Van Dyke/Visuals Unlimited; 31.17: © Scott Camazine/Photo Researchers, Inc.; 31.18a: © Ralph Williams/USDA Forest Service; 31.18b: © Chris Mattison/Superstock; 31.18c: © USDA Forest Service Archives, USDA Forest Service, www.forestryimages.org; 31.19a: © Dayton Wild/Visuals Unlimited; 31.19b: © Manfred Kage/Peter Arnold Inc; 31.20: Courtesy of Zoology Dept/University of Canterbury, New Zealand; 31.20b (inset): Courtesy of Dr. Peter Dazak.

Chapter 32

Opener: © Corbis/Volume 53; 32.1, pg 422 (top): © Corbis/Volume 86, (2nd from top): © Corbis/Volume 65, (3rd from top): © David M. Phillips/Visuals Unlimited, (4th from top): © Royalty-Free/Corbis, pg 323 (top): © Edward S. Ross, (2nd from top): © Corbis, (3rd from top): © Cleveland P. Hickman, (4th from top): © Cabisco/Phototake, (5th from top): © Ed Reschke; 32.4: Courtesy of Dr. Igor Eeckhaut.

Chapter 33

Opener: © Denise Tackett/Tom Stack & Associates; 33.3a: © Andrew J. Martinez/Photo Researchers Inc; 33.5: © Roland Birke/Phototake; 33.6: © Ed Reschke; 33.7: © Amos Nachoum/CORBIS; 33.10: © David Wrobel/Visuals Unlimited; 33.8: © Kelvin Aitken/Peter Arnold Inc; 33.9: © Neil G. McDaniel/Photo Researchers; 33.11 (top): © Tom Adams/Visuals Unlimited; 33.12 (bottom left): © Dwight Kuhn; 33.13 (top left): © Dennis Kunkel/Phototake; 33.14 © L. Newman & A. Flowers/Photo

Researchers; 33.15: © Kjell Sandved/Butterfly Alphabet; 33.16 (top left): © Peter Funch, University of Aarhus; 33.17: © Gary D. Gaugler/Photo Researchers; 33.18: © Educational Images Ltd, Elmira NY, USA; 33.19: © T.E. Adams/Visuals Unlimited.

Chapter 34

Opener: © James H. Robinson/Animals Animals/Earth Scenes; 34.1a: © Marty Snyderman/Visuals Unlimited; 34.1b: © Alex Kerstitch/Visuals Unlimited; 34.1c: © Douglas Faulkner/Photo Researchers Inc; 34.1d: © age fotostock/SuperStock; 34.2: © A. Flowers & L. Newman/Photo Researchers Inc; 34.4: © Eye of Science/Photo Researchers Inc; 34.5b: © Kjell Sandved/Butterfly Alphabet; 34.6: © Kelvin Aitken/Peter Arnold Inc; 34.7: © Milton Rand/Tom Stack & Associates; 34.8: © PhotoDisc/Getty Images; 34.10: © AFP/Getty Images; 34.11: © Jeff Rotman/Photo Researchers Inc; 34.12: © Ken Lucas/Visuals Unlimited; 34.14: © Ronald L. Shimek; 34.15: © Fred Grassle, Woods Hole Oceanographic Institution; 34.16: © David M. Dennis/Animals Animals Enterprises; 34.17: © Pascal Goetgheluck/Photo Researchers Inc; 34.18b: © Robert Brons/Biological Photo Service; 34.19: © Fred Bavendam/Peter Arnold Inc; 34.26a: © National Geographic/Getty Images; 34.26b: © S. Camazine/K. Visscher/Photo Researchers Inc; 34.27a: © Alex Kerstich/Visuals Unlimited; 34.27b: © Edward S. Ross; 34.29: © T.E. Adams/Visuals Unlimited; 34.31: © Kjell Sandved/Butterfly Alphabet; 34.32a: © Cleveland P. Hickman; 34.32b: © Valorie Hodgson/Visuals Unlimited; 34.32c: © Gyorgy Csoka, Hungary Forest Research Institute www.forestryimages.org; 34.32d: © Kjell Sandved/Butterfly Alphabet; 34.32e: © Greg Johnston/Lonely Planet Images/Getty Images; 34.32f: © Nature's Images/Photo Researchers Inc; 34.34: © Dwight Kuhn; 34.35, 34.36: © Kjell Sandved/Butterfly Alphabet; 34.37: © Wim van Egmond/Visuals Unlimited; 34.38a: © Alex Kerstich/Visuals Unlimited; 34.38b: © Randy Morse/Tom Stack & Associates; 34.38c: © Daniel W. Gotshall/Visuals Unlimited; 34.38d: © Reinhard Dirscherl/Visuals Unlimited; 34.38e: © Jeff Rotman/Photo Researchers Inc.

Chapter 35

Opener: © Phone Ferrero J.P./Labat J.M./Peter Arnold Inc; 35.2: © Eric N. Olson, PhD/The University of Texas MD Anderson Cancer Center; 35.4a: © Rick Harbo/Marine Images; 35.5: © Heather Angel; 35.9 (top): © agefotostock/Superstock; (left): © Royalty-Free/CORBIS; (right): © Brandon Cole/www.brandoncole.com/Visuals Unlimited; 35.11: Corbis/Volume 33; 35.13a: © Federico Cabello/Superstock; 35.13b: © Raymond Tercafs-Production Services/Bruce Coleman Inc; 35.16a: © John Shaw/Tom Stack & Associates; 35.16b: © Suzanne L. Collins & Joseph T. Collins/Photo Researchers Inc; 35.16c: © Jany Sauvanet/Photo Researchers Inc; 35.22: © Paul Sereno, University of Chicago; 35.24a (left): © William J. Weber/Visuals Unlimited; (right): © Frans Lemmens/Getty Images; 35.24b: © Jonathan Losos; 35.24c (right): © Rod Planck/Tom Stack & Associates; 35.25 (right): © Rod Planck/Tom Stack & Associates; 35.25d (left): © Corbis/Volume 6; (right): © Zig Leszczynski/Animals Animals; 35.28: © Layne Kennedy/CORBIS; 35.29a: © Corbis; 35.29b: Arthur

C. Smith III/Grant Heilman Photography; 35.29c: © David Boyle/Animals Animals; 35.29d: © John Cancalosi/Peter Arnold; 35.32: © Stephen Dalton/National Audubon Society Collection/Photo Researchers; 35.33a (left): © BJ Alcock/Visuals Unlimited; (right): © Dave Watts/Alamy; 35.33b (left): © CORBIS Volume 6; (right): © W. Perry Conway/CORBIS; 35.33c (left): © Stephen J. Krasemann/DRK Photo; (right): © J. & C. Sohns/Animals Animals Enterprises; 35.34: © Alan Nelson/Animals Animals/Earth Scenes; 35.35a: © Peter Arnold Inc/Alamy; 35.35b: © Martin Harvey/Peter Arnold; 35.35c (left): © Joe McDonald/Visuals Unlimited; 35 35c (right): © Dynamic Graphics Group/IT Stock Free/Alamy; 35.38: National Museums of Kenya, Nairobi © 1985 David L. Brill; 35.40: © AP/Wide World Photos.

Chapter 36

Opener: © Susan Singer; 36.4 (left side): Dr. Robert Lyndon; (right side): © Biodisc/Visuals Unlimited; 36.6a: © Brian Sullivan/Visuals Unlimited; 36.6b, 36.6c: © EM Unit, Royal Holloway, University of London, Egham, Surrey; 36.7: © Jessica Lucas & Fred Sack; 36.8: © Andrew Syred/Science Photo Library/Photo Researchers Inc; 36.10: © Runk/Shoenberger/Grant Heilman Photography; 36.9a (both): Courtesy of Allan Lloyd; 36.11a: © Lee Wilcox; 36.11b: © George Wilder/Visuals Unlimited; 36.11c: © Lee Wilcox; 36.12 (upper): © NC Brown Center for Ultrastructure Studies, SUNY, College of Environmental Science and Forestry, Syracuse, NY; (lower): Tom Kuster/USDA; 36.13b: © Dr. Richard Kessel & Dr. Gene Shih/Visuals Unlimited; 36.14: © Biodisc/Visuals Unlimited; 36.15b: © John Schiefelbein, University of Michigan; 36.16b: © Courtesy of Dr. Philip Benfey, from Wysocka-Diller, J.W., Helariutta, Y. Fukaki, H., Malamy, J.E. and P.N. Benfey (2000) Molecular analysis of SCARECROW function reveals a radial patterning mechanism common to root and shoot development, Cell 127, 595–603; 36.17 (bottom left upper): © Carolina Biological Supply Company/Phototake; (bottom right upper): 36:17 (bottom right; upper/lower): © George Ellmore, Tufts University; (bottom left, lower): © Lee Wilcox; 36.19a: © E.R. Degginger/Photo Researchers, Inc.; 36.19b: © Peter Frischmuth/Peter Arnold Inc; 36.19c: © Walter H. Hodge/Peter Arnold Inc; 36.19d: © Gerald & Buff Corsi/Visuals Unlimited; 36.19e: © Kingsley R. Stern; 36.20: Courtesy of J.H. Troughton and L. Donaldson/Industrial Research Ltd; 36.23a,b: © Ed Reschke; 36.26: © Ed Reschke/Peter Arnold Inc; 36.27a: © Ed Reschke; 36.27b: © Biodisc/Visuals Unlimited; 36.28a: © Jerome Wexler/Visuals Unlimited; 36.28b: © Lee Wilcox; 36.28c: © Runk/Shoenberger/Grant Heilman Photography; 36.28d: © Chase Studio/Photo Researchers Inc.; 36.28e: © Charles D. Winters/Photo Researchers, Inc.; 36.28f: © Lee Wilcox; 36.29 (both): © Scott Poethig, University of Pennsylvania 36.30a (upper): © Kjell Sandved/Butterfly Alphabet; 36.30b (lower): © Pat Anderson/Visuals Unlimited; 36.31a: © Gusto/Photo Researchers; 36.31b: © Glenn M. Oliver/Visuals Unlimited; 36.31c: © Joel Arrington/Visuals Unlimited; 36.33: © Ed Reschke.

Chapter 37

Opener: © Norm Thomas/The National Audubon Society Collection/Photo Researchers Inc; 37.4: Edward

Yeung (University of Calgary) and David Meinke (Oklahoma State University); 37.6 (all): Kindly provided by Prof. Chun-ming Liu, Institute of Botany, Chinese Academy of Sciences; 37.7: Max Planck Institute for Developmental Biology; 37.8c: © Ben Scheres, University of Utrecht; 37.8d, 37.8e: Courtesy of George Stamatiou and Thomas Berleth; 37.11 (corn): © photocuisine/Corbis; 37.11 (bean): © Barry L. Runk/Grant Heilman Photography; 37.13a: © Ed Reschke/Peter Arnold Inc; 37.13b: © David Sieren/Visuals Unlimited; 37.15 (top left), (top center), (bottom center): © Kingsley R. Stern; 37.15 (bottom left): © James Richardson/Visuals Unlimited; 37.15 (top right): Courtesy of Robert A. Schisling; 37.15 (bottom right): © Charles D. Winters/Photo Researchers Inc; 37.16a: © Edward S. Ross; 37.16b, c, d: © James Castner; 37.18a: © PHONE Thiriet Claudius/Peter Arnold, Inc; 37.18b: © Holt Studios International Ltd/Alamy.

Chapter 38

Opener: © Richard Rowan's Collection Inc/Photo Researchers Inc; 38.4 (both): © Jim Strawser/Grant Heilman Photography; 38.7: © Ken Wagner/Phototake; 38.11a: (closed): © Dr. Ryder/Jason Borns/Phototake; 38.11b (open): © Dr. Ryder/Jason Borns/Phototake; 38.14: © Herve Conge/ISM/Phototake; 38.15a: © Ed Reschke; 38.15b: © Jon Bertsch/Visuals Unlimited; 38.16: © Mark Boulton/Photo Researchers Inc; 38.17a: © Andrew Syred/Photo Researchers Inc; 38.17b: © Bruce Iverson Photomicrography.

Chapter 39

Opener: © PhotoDisc/Getty Images; 39.4a: © Hulton Archive/Getty Images; 39.4b: © M.L. Bonsirven-Fontana/UNESCO; 39.5 (all): International Plant Nutrition Institute (IPNI), Norcross, GA U.S.A.; 39.7: © George Bernard/Animals Animals/Earth Scenes; 39.8 (bottom left): © Ken Wagner/Phototake; (bottom center): © Bruce Iverson; 39.10a: © Kjell Sandved/Butterfly Alphabet; 39.10b: © Runk/Schoenberger/Grant Heilman Photography; 39.10c: © Perennov Nuridsany/Photo Researchers, Inc.; 39.10d: © Barry Rice; 39.12: © Don Albert; 39.15 (both): Courtesy of Nicholas School of the Environment and Earth Sciences, Duke University; 39.17: © Greg Harvey USAF; 39.18b: © REUTERS; 39.18c: © AP/Wide World Photo.

Chapter 40

Opener: © fstop/Getty Images; 40.1: © Richard la Val/Animals Animals; 40.2, 40.3a: USDA/Agricultural Research Service; 40.3b: © USDA/Agricultural Research Service; 40.5: © Allan Morgan/Peter Arnold Inc; 40.6: © Adam Jones/Photo Researchers Inc; p. 793 (1-3): © Inga Spence/Visuals Unlimited; (4): © Heather Angel/Natural Visions; (5): © Pallava Bagla/Corbis; 40.7a: © Gilbert S. Grant/Photo Researchers Inc; 40.7b: © Lee Wilcox; 40.8: © Michael J. Doolittle/Peter Arnold Inc; 40.12: © Courtesy R.X. Latin. Reprinted with permission from Compendium of Cucurbit Diseases, 1996, American Phytopathological Society, St. Paul, MN.

Chapter 41

Opener: © Alan G. Nelson/Animals Animals; 41.3a (all):f 41.8: © Ray Evert; 41.10 (all): Jee Jung & Philip Benfey; 41.11: © Lee Wilcox; 41.13: © Frank Krahmar/Zefa/Corbis; 41.16: © Don Grall/Index Stock Imagery; 41.26: © Prof. Malcolm B. Wilkins, Botany Dept, Glasgow University; 41.28: © Robert Calentine/Visuals Unlimited; 41.29: © Runk/

Schoenberger/Grant Heilman Photography; 41.31: Amnon Lichter, The Volcani Center; 41.34a: © John Solden/Visuals Unlimited; 41.34b: Courtesy of Donald R. McCarty, from "Molecular Analysis of viviparous-1: An Abscisic Acid-Insensitive Mutant of Maize: The Plant Cell. v. 1, 523–532 © 1989 American Society of Plant Physiologists; 41.34c: © ISM/Phototake.

Chapter 42

Opener: © Heather Angel/Natural Visions; 42.3a: © Fred Habegger/Grant Heilman Photography; 42.3b: © Pat Breen, Oregon State University; 42.4: © Richard La Val/Animals Animals; 42.5a: © Mack Henley/Visuals Unlimited; 42.5a (inset): © Michael Gadomski; 42.5b: Max Planck Institute; 42.5b (inset): © Henrik Bohlenius; 42.7: © Jim Strawser/Grant Heilman Photography; 42.12 (all): © John Bowman; 42.15: © John Bishop/Visuals Unlimited; 42.16: © Paul Gier/Visuals Unlimited; 42.17 (both): Courtesy of Enrico Coen; 42.19 (both): © L. DeVos—Free University of Brussels; 42.20: © Kingsley R. Stern; 42.21: © David Cappaert, www.forestryimages.org; 42.22: © Topham/The Image Works; 42.23: © Michael Fogden; 42.24 (both): © Thomas Eisner; 42.25: © John D. Cunningham/Visuals Unlimited; 42.26: © Edward S. Ross; 42.27a: © David Sieren/Visuals Unlimited; 42.27b: © Barbara Gerlach/Visuals Unlimited; 42.30: © Jerome Wexler/Photo Researchers Inc; 42.31a: © Sinclair Stammers/Photo Researchers; 42.31 (b-d): Courtesy of Dr. Hans Ulrich Koop, from Plant Cell Reports, 17:601–604eports, 17: 601-604; 42.32a: © Lee Wilcox; 42.32b: © David Lazenby/Animals Animals/Earth Scenes.

Chapter 43

Opener: © Dr. Roger C. Wagner, Professor Emeritus of Biological Sciences, University of Delaware; Table 43.1, pg 855 (top to bottom): © Ed Reschke; © Arthur Siegelman/Visuals Unlimited; © Ed Reschke; © Gladden Willis, M.D./Visuals Unlimited; © Ed Reschke; 43.3: © J. Gross, Biozentrum/Photo Researchers Inc; 43.4: © Biophoto Associates/Photo Researchers Inc; Table 43.2, pg 856 (top to bottom): © Ed Reschke; © Gladden Willis, M.D./Visuals Unlimited; © Chuck Brown/Photo Researchers Inc; © Ed Reschke; © Kenneth Eward/Photo Researchers Inc; 43.3 (all): © Ed Reschke.

Chapter 44

Opener: Courtesy of David I. Vaney, University of Queensland Australia; 44.3: © Enrico Mugnaini/Visuals Unlimited; 44.12: © John Heuser, Washington University School of Medicine, St. Louis, MO; 44.14: © Ed Reschke; 44.25: © Dr. Marcus E. Rachle, Washington University, McDonnell Center for High Brain Function; 44.26: © Lennart Nilsson/Albert Bonniers Förlag AB; 44.29: © E.R. Lewis/Biological Photo Service.

Chapter 45

Opener: © Omikron/Photo Researchers Inc; 45.22: © Leonard L. Rue, III.

Chapter 46

Opener: © Nature's Images/Photo Researchers Inc; 46.10: © John Paul Kay/Peter Arnold Inc; 46.11: © Bettmann/Corbis; 46.16: © Robert & Linda Mitchell.

Chapter 47

Opener: © Hal Beral/Grant Heilman Photography; 47.4 (top): © Ed Reschke/Peter Arnold; (2nd row-left):

© David Scharf/Peter Arnold; (center): © Dr. Holger Jastrow; (right): © CNRI/Photo Researchers Inc; (3rd row-left): © Dr. Kessel & Dr. Kardon/Tissue & Organs/Visuals Unlimited/Getty Images; (center) © Ed Reschke/Peter Arnold; (right): © Ed Reschke; (bottom right): © Ed Reschke/Peter Arnold; 47.11 (both): © Dr. H.E. Huxley; 47.22: © Treat Davidson/Photo Researchers Inc.

Chapter 48

Opener: © John Gerlach, Animals Animals/Earth Scenes; 48.11: © Ron Boardman/Stone/Getty Images; 48.19: from O.T. Avery. C.M. McLeod & M. McCarty, "Studies on the chemical nature of the substance inducing transformation of pneumococcal types" reproduced from *The Journal of Experimental Medicine* 79 (1944): 137–158, fig 1 by copyright permission of the Rockefeller University press, reproduced by permission. Photograph made by Mr. Joseph B. Haulenbeck.

Chapter 49

Figure 49.12: © Tom Drysdale; 49.13a,b: © Ed Reschke; 49.13c: © Dr. Gladden Willis/Visuals Unlimited; 49.18: © Bruce Watkins/Animals Animals; 49.30a: © Clark Overton/Phototake; 49.30b: © Martin Rotker/Phototake; 49.31: © Kenneth Eward/BioGrafx/Photo Researchers Inc.

Chapter 50

Opener: © Ann & Steve Toon/Robert Harding World Imagery/Getty Images.

Chapter 51

Opener: AP/Wide World Photos; 51.1: © Manfred Kage/Peter Arnold Inc; 51.10 (both): © Dr. Andrejs Liepins/Photo Researchers, Inc; 51.21: © CDC/Science Source; 51.4: © Wellcome Library London.

Chapter 52

Opener: © Michael Fogden/DRK Photo; 52.1: © Dennis Kunkel Microscopy Inc; 52.2a: © Chuck Wise/Animals Animals/Earth Scenes; 52.2b: © Fred McConnaughey/The National Audubon Society Collection/Photo Researchers Inc; 52.4: © David Doubilet; 52.5: © Hans Pfletschinger/Peter Arnold Inc; 52.7: © Cleveland P. Hickman; 52.8: © Frans Lanting/Minden Pictures; 52.9a: © Jean Phllippe Varin/Jacana/Photo Researchers Inc; 52.9b: © Tom McHugh/The National Audubon Society Collection/Photo Researchers Inc; 52.9c: © Corbis Volume 86; 52.12a: © David M. Phillips/Photo Researchers Inc; 52.17: © Ed Reschke.

Chapter 53

Opener: © Lennart Nilsson/Albert Bonniers Förlag AB, A Child Is Born, Dell Publishing Company; 53.1d: © David M. Philips/Visuals Unlimited; 53.3 (all): Dr. Mathias Hafner (Mannheim University of Applied Sciences, Institute for Molecular Biology, Mannheim, Germany) and Dr. Gerald Schatten (Pittsburgh Development Center Deputy Director, Magee-Women's Research Institute Professor and Vice-Chair of Obstetrics, Gynecology & Reproductive Sciences and Professor of Cell Biology & Physiology Director, Division of Developmental and Regenerative Medicine University of Pittsburgh School of Medicine Pittsburgh, PA 15213); 53.7: © David M. Philips/Visuals Unlimited; 53.8a: © Cabisco/Phototake; 53.9: © David M. Philips/Visuals Unlimited; 53.11 (all): From "An Atlas of the Development of the Sea Urchin *Lytechinus variegatus.*" Provided by Dr. John B. Merrill (left to right) Plate 20, pg 62, #I; Plate 33, pg 93, #C; Plate 38, pg 105, #6; 53.17 (both): © Courtesy of Manfred Frasch; 53.20 (both): © Roger Fleischman, University of Kentucky; 53.27b, 53.28 (all): © Lennart Nilsson/Albert Bonniers Förlag AB, A Child Is Born, Dell Publishing Company.

Chapter 54

Opener: © K. Ammann/Bruce Coleman Inc; 54.4 (both): from J.R. Brown et al, "A defect in nurturing mice lacking . . . gene for fosB" Cell v. 86, 1996 pp. 297–308, © Cell Press; 54.5 (both): © Larry Young Emory University—Yerkes Research Center; 54.6 (all): © Lee Boltin Picture Library; 54.7: © William Grenfell/Visuals Unlimited; 54.8: © Thomas McAvoy, Life Magazine/Time, Inc.; 54.9: Harlow Primate Laboratory—University of Wisconsin-Madison; 54.11: © Roger Wilmhurst/The National Audubon Society Collection/Photo Researchers Inc; 54.12a: © Linda Koebner/Bruce Coleman; 54.12b: © Jeff Foott/Tom Stack & Associates; 54.13: © Superstock; 54.14: Courtesy of Bernd Heinrich; 54.15b: © Fred Breunner/Peter Arnold Inc; 54.15c: © George Lepp/Getty Images; 54.18: © Dwight Kuhn; 54.19: © PhotoLink/Getty Images; 54.20: © Tom Leeson; 54.21b: © Scott Camazine/Photo Researchers, Inc; 54.22a: © Gerald Cubitt; 54.23: © Corbis Volume 53; 54.24: © Nina Leen, Life Magazine/Time, Inc; 54.27a: © Peter Steyn/Getty Images; 54.27b: © Gerald C. Kelley/Photo Researchers Inc; 54.28a: © Bruce Beehler/Photo Researchers Inc; 54.28b: © B. Chudleigh/Vireo; 54.30: © Cathy & Gordon ILLG; 54.32a: Courtesy of T.A. Burke, Reprinted by permission from Nature, "Parental care and mating behavior of polyandrous dunnocks," 338: 247–251, 1989; 54.33: © Merlin B. Tuttle/Bat Conservation International; 54.35: © Heinrich Van DEN Berf/Peter Arnold Inc; 54.36: © Edward S. Ross; 54.38: © Mark Moffett/Minden Pictures; 54.39: © Nigel Dennis/National Audubon Society Collection/Photo Researchers Inc.

Chapter 55

Opener: © PhotoDisc/Volume 44/Getty Images; 55.1: © Michael Fogden/Animals Animals; 55.4: © Tom Baugh; 55.14: © Christian Kerihuel; 55.16: Courtesy of Barry Sinervo; 55.22 (bottom left): © Juan Medina/Reuters/Corbis; (bottom center): © Oxford Scientific/Photolibrary.

Chapter 56

Opener: © Corbis; 56.1: © Daryl & Sharna Balfour/Okopia/Photo Researchers Inc; 56.6 (all): © J.B. Losos; 56.10a: © Edward S. Ross; 56.10b: © Raymond Mendez/Animals Animals; 56.11 (both): © Lincoln P. Brower; 56.12: © Michael & Patricia Fogden/Corbis; 56.13: © James L. Castner; 56.15: © Merlin D. Tuttle/Bat Conservation International; 56.16: © Eastcott/Momatiuk/The Image Works; 56.17: © PhotoDisc/Volume 44/Getty Images; 56.18: © Michael Fogden/DRK Photo; 56.19: © David Moorhead; 56.21a: © F. Stuart Westmorland/Photo Researchers Inc; 56.21b: © Ann Weirtheim/Animals Animals/Earth Scenes; 56.23: © David Hosking/National Audubon Society Collection/Photo Researchers Inc; 56.24 (all): © Tom Bean; 56.25 (both): © Dani/Jeske/Earth Scenes; 56.26: © Educational Images Ltd., Elmira, NY, USA. Used by Permission.

Chapter 57

Opener: © PhotoDisc/Getty Images; 57.3: © Worldwide Picture Library/Alamy; 57.7a: © U.S. Forest Service; 57.19a: © Lane Kennedy/Corbis.

Chapter 58

Opener: NASA; 58 10, pg 1218, 1st row, (left to right): © age fotostock/SuperStock; © Luiz C. Marigo/Peter Arnold; © age fotostock/SuperStock; 2nd row, (left to right): © Mitsuaki Iwago/Minden Pictures; © Konrad Wothe/Minden Pictures; © age fotostock/SuperStock; 3rd row, (left to right): © Adam Jones/Getty Images; © age fotostock/SuperStock; 4th row, (left to right): © Jon Arnold Images/Alamy; © Jim Brandenburg/Minden Pictures; © Eastcott Momatiuk/Getty Images; p. 1219 (1st row-left to right): © SuperStock, Inc./SuperStock; © Dwight Kuhn; © Claudia Adams/Dembinsky Photo Associates; (2nd row-left to right): © Jim Lundgren/Alamy; © Craig Tuttle/Corbis; © Dennis M. Dennis/Animals Animals; (3rd row-left to right): © Stephen J. Krasemann/Photo Researchers Inc; © Joe McDonald/Corbis; © imagebroker/Alamy; (4th row-left to right): © Eastcott Momatiuk/Getty Images; © Science Faction/Getty Images; © Michio Hoshino/Minden Pictures; 58.13a: © Art Wolfe/Photo Researchers Inc; 58.13b: © Bill Banaszewski/Visuals Unlimited; 58.15: Provided by the SeaWiFS Project, NASA/Goddard Space Flight Center, and ORBIMAGE; 58.16: © Digital Vision/Picture Quest; 58.18a: © Jim Church; 58.18b: © Ralph White/Corbis; 58.21a: © Peter May/Peter Arnold Inc; 58.21b: © Frans Lanting/Minden Pictures; 58.22: © Gilbert S. Grant/National Audubon Society Collection/Photo Researchers, Inc; 58.24a: NASA; 58.25: NASA/Goddard Institute for Space Studies, New York; 58.27 (bottom left upper): © Lonnie G. Thompson, The Ohio State University; 58.27b (bottom left lower): © Dr. Bruno Messerli.

Chapter 59

Opener: Cornell Lab of Ornithology; 59.2: © John Elk; 59.3 (left to right): © Frank Krahmer/Masterfile; © Michael & Patricia Fogden/Minden Pictures; © Heather Angel/Natural Visions; © NHPA/Martin Harvey; 59.5a: © Edward S. Ross; 59.5b: © Inga Spence/Getty Images; 59.6a: © Jean-Leo Dugast/Peter Arnold Inc; 59.6b: © Oxford Scientific/PhotoLibrary; 59.8: © Michael Fogden/DRK Photo; 59.9 (left to right): © Brian Rogers/Natural Visions; © David M. Dennis/Animals Animals; © Michael Turco, 2006; © David A. Northcott/Corbis; 59.13 (right): © Randall Hyman; (left): © Dr. Morley Read/Photo Researchers Inc; 59.14: © John Gerlach/Animals Animals; 59.16: © Peter Yates/Science Photo Library; 59.17 (left): © Jack Jeffrey; (right): © Jack Jeffrey/Photo Resource Hawaii; 59.18: © Tom McHugh/Photo Researchers Inc; 59.20: © Merlin B. Tuttle/Bat Conservation International; 59.21: U.S. Fish and Wildlife Service; 59.21a: ANSP © Steven Holt/stockpix.com; 59.23: Wm. J. Weber/Visuals Unlimited; 59.24 (both): University of Wisconsin, Madison Arboretum; 59.26b: © Dani/Jeske/Earth Scenes.

Line Art and Text

Chapter 8

Figure 8.5: From Raven et al., Biology of Plants, 5e. Reprinted by permission of W.H. Freeman and Company/Worth Publishers. Figure 8.12: From Lincoln Taiz and Eduardo Zeiger, Plant Physiology, 1991 Benjamin-Cummings Publishing. Reprinted with permission of the authors.

Chapter 17

TA17.1: CALVIN AND HOBBES © 1995 Watterson. Dist. By UNIVERSAL PRESS SYNDICATE. Reprinted with permission. All rights reserved.

Chapter 18

Figure 18.8: After Nature, page 790, December 2003. Figure 18.9: C. More, K.M. Devos, Z. Wang, and M.D. Gale: "Grass, line up and form a circle," Current Biology, 1995, vol. 5, pp. 737–739.

Chapter 19

Figure 19.5: After Alberts, et al., Molecular Biology of the Cell, 3e, 1994, Garland Publishing, Inc.

Chapter 20

Figure 20.7: Data from P.A. Powers, et al., "A Multidisciplinary Approach to the Selectionist/Neutralist Controversy." Oxford Surveys in Evolutionary Biology. Oxford University Press, 1993. Figure 20.9: From R.F. Preziosi and D.J. Fairbairn, "Sexual Size Dimorphism and Selection in the Wild in the Water-strider Aquarius remigis: Lifetime Fecundity Selection on Female Total Length and Its Components," Evolution, International Journal of Organic Evolution 51:467–474, 1997. Figure 20.10: Data from M.R. MacNair in J.M. Bishops & L.M. Cook, Genetic Consequences of Man-Made Change, Academic Press, 1981, pp. 177–207. Figure 20.11: Adapted from Clark, B. "Balanced Polymorphism and the Diversity of Sympatric Species," Syst. Assoc. Publ., Vol. 4, 1962.

Chapter 21

Figure 21.2a: Data from Grant, "Natural Selection and Darwin's Finches" in Scientific American, October 1991. Figure 21.2b: Data from Grant, "Natural Selection and Darwin's Finches" in Scientific American, October 1991. Figure 21.4: Data from Grant, et al., "Parallel Rise and Fall of Melanic Peppered Moths" in Journal of Heredity, vol. 87, 1996, Oxford University Press. Figure 21.5: Data from G. Dayton and A. Roberson, Journal of Genetics, Vol. 55, p. 154, 1957.

Chapter 22

Figure 22.1: Data from R. Conant & J.T. Collins, Reptiles & Amphibians of Eastern/Central North America, 3rd edition, 1991. Houghton Mifflin Company.

Chapter 23

Figure 23.11 Adapted from Duda and Palumbi, Evolution, August 1999, Proceedings for the National Academy of Science. Figure 23.14: Adapted from Collin, Evolution, Proceedings for the National Academy of Science. Figure 23.16: Data from Hahn et al., Science, 1980.

Chapter 24

Figure 24.4: Data from Adams and Wendell, Current Opinion in Plant Biology, 2005. Figure 24.6: Data from Blane and Wolfe, The Plant Cell, 2004; Adams and Wendell, Current Opinion in Plant Biology, 2005.

Chapter 28

Figure 28.18 Data from U.S. Centers for Disease Control and Prevention, Atlanta, GA.

Chapter 37

Figure 37.3: From Ralph Quantrano, Washington University. Figure 37.9d & e: Data from The EMBO Journal, vol. 17, 1998.

Chapter 38

Figure 38.11: Data from Raven, Evert, and Eichhorn, *Biology of Plants*, 7e, W. H. Freeman and Co. Figure 38.13: Data from Raven, Evert, and Eichhorn, *Biology of Plants*, 7e, W. H. Freeman and Co.

Chapter 41

Figure 41.9: Based on data from The Arabidopsis Book. Figure 41.10: Based on data from The Arabidopsis Book.

Chapter 52

Table 52.2: Data from American College of Obstetricians and Gynecologists: Contraception, Patient Education Pamphlet No. AP005. ACOG, Washington, D.C., 1990.

Chapter 54

Figure 54.4c & d: Data from J.R. Brown et al., "A Defect in Nurturing in Mice Lacking the Immediate Early Gene for fosB" *Cell*, 1996. Figure 54.10: Reprinted from *Animal Behavior*, 51, Brood Parasite Figure by M.D. Beecher et al., Copyright © 1996 with permission from Elsevier. Figure 54.22: Reprinted from *Animal Behavior*, 36(2), Private Semantics Figure by John Alcock, p. 484, Copyright © 1988, with permission from Elsevier. Figure 54.28c: Data from M. Petrie, et al. "Peahens Prefer Peacocks with Elaborate Trains," *Animal Behavior*, 1991. Figure 54.32: Data from H.L. Gibbs et al., "Realized Reproductive Excess of Polygynous Red-Winged Blackbirds Revealed by NDA Markers," *Science*, 1990.

Chapter 55

Figure 55.5: Data from Brown & Lomolino, *Biogeography*, 3rd edition, 1998, Sinauer Associates, Inc. Figure 55.6: Data from Brown & Lomolino, *Biogeography*, 3rd edition, 1998, Sinauer Associates, Inc. After A.T. Smith, *Ecology*, 1974. Figure 55.9: Data from *Patch Occupation and Population Size of the Glanville Fritillary in the Anland Islands*, Metapopulation Research Group, Helsinki, Finland. Figure 55.19b: Data from C.E. Goulden, L.L. Henry, and A.J. Tessier, *Ecology*, 1982. Figure 55.23: From "Population Changes in German Forest Pests," by G.C. Varley in *Journal of Animal Ecology*, 18. Copyright © 1949 British Ecology Society. Reprinted by permission.

Chapter 56

Figure 56.2: Abundance of tree species along a moisture gradient in the Santa Catalina Mountains of southeastern Arizona, from *The Economy of Nature*, 4e, by Robert E. Ricklets. Copyright 1973, 1979 by Chiron Press, Inc. © 1990, 2000 by W.H. Freeman and Company. Reprinted by permission. Figure 56.3: Change in community composition across an ecotone, from *The Economy of Nature*, 4e, by Robert E. Ricklets. Copyright 1973, 1979 by Chiron Press, Inc. © 1990, 2000 by W.H. Freeman and Company. Reprinted by permission. Figure 56.5: Data from Begon et al., *Ecology*, 1996. After: W.B. Clapham, *Natural Ecosystems*, Clover, Macmillan. Figure 56.7: Data from E.J. Heske, et al., *Ecology*, 1994. Figure 56.8: Data from E.J. Heske, et al., *Ecology*, 1994.

Chapter 57

Figure 57.1: Data in: Begon, M., J.L. Harper, and C. R. Townsend, *Ecology*, 3/e, Blackwell Science 1996, page 715. Original Source: Whittaker, R. H.

Communities and Ecosystems, 2/e Macmillan, London, 1975. Figure 57.11: Data in: Begon, M., J.L. Harper, and C.R. Townsend, *Ecology*, Blackwell Science 1996, page 715. Original Source: Whittaker, R.H. *Communities and Ecosystems*, 2/e Macmillan, London, 1975. Figure 57.14: Data from Flecker, A.S. and Townsend, C.R., "Community-Wide Consequences of Trout Introduction in New Zealand Streams." In *Ecosystem Management: Selected Readings*, F.B. Samson and F.L. Knopf eds., Springer-Verlag, New York, 1996. Figure 57.18: Data from J.T. Wooten and M.E. Power, "Productivity, Consumers & the Structure of a River Food Chain," *Proceedings National Academic Science*, 1993. Figure 57.20: Data from F. Morrin, *Community Ecology*, Blackwell, 1999.

Chapter 59

Figure 59.4: From *Nature's Place: Human Population and the Future of Biological Diversity*, by Richard P. Cincotta and Robert Engelman, 2000. Reprinted by permission. Figure 59.11: Data assembled by Pima, 1991. Figure 59.14: Data from Marra, Hobson, Holmes, "Linking Winter & Summer Events," in *Science*, Dec. 1998. Figure 59.15: Data from UNEP, *Environmental Data Report*, 1993, 1994. Figure 59.22: Data from H.L. Billington, "Effects of Population Size on a Genetic Variation in a Dioecious Conifer" in *Conservation Geology*, Blackwell Scientific Publication, Inc., 1999. Figure 59.25: Data from The Peregrine Fund.

Electronic Publishing Services Inc. Illustration Team

Art Director: Kim E. Moss
Lead Illustrator: Erin Daniel
Lead Illustrator: Gwen DeCelles
Lead Illustrator: Matthew McAdams
Art Coordinator: Donna Hallera
Art Coordinator: Haydee Martinez

Art Development

Jen Christiansen: Chapters 2–7, 24, 27–29, 35, 51, 57, 58.

Martin Huber: Chapters 8–11, 14, 19, 25, 30, 43–50, 52, 53.

Eliza Jewett: Chapters 36–42.

Kim Moss: Chapters 1, 12, 13, 15–18, 20–23, 26, 31–34, 54–56, 59.

Illustrations

Philip Ashley: Figures: 32.01, 32.06, 33.19, 34.29, 58.19, 58.28

Christopher Burke: Figures: 17.07, 17.12

Raychel Ciemma: Figures: 1.07, ta4.02, 4.03, 4.05, 4.06, 4.12, 4.13, 4.14, 4.17, 4.18, 4.19, 4.20, 4.23, 4.25, ta5.02, ta5.03, ta5.04, 5.05, 5.06, 5.07, 5.09, 5.10, 5.11, 5.12, 5.17, 7.05, 7.13, 7.14, 7.15, 8.02, 8.06, 8.09, 8.14, 8.15, 9.01, 9.02, 9.03, 9.05, 9.06, 9.07, 9.10, 9.11, 9.13, 9.14, 9.15, 9.16, 9.17, 9.18, 9.19, 10.05, 11.04, 12.16, 14.09, 14.11, 14.16, 14.18, 14.19, 14.20, 14.23, 14.24, 14.25, 14.26, ta14.01, 17.13, 21.01, 21.09, 21.12, 22.09, 22.14, 36.01, 36.02, 36.09, 37.09, 37.15, 38.08, 38.11, 39.10, 42.13, 54.17, 54.18, 54.33, 55.03, 55.07, 55.19, 55.23, ta55.01, 59.19, 55.25

Erin Daniel: Figures: 21.04, 21.05, 21.06, 21.07, 21.10, 21.13, 21.14, 21.17, 21.18, 21.19, 21.20, 22.16

Laurie Decopain: Figure ta34.06

Mica Duran: Figures: 17.19, 19.11

Eli Ensor: Figures: 35.20, 35.22, 35.27

Shawn Gould: Figures: 29.01, 29.11, 29.20, 47.02, 47.21, 47.23, 53.01, 53.08, 53.10–53.14, 53.20, 53.22, 53.24, 53.25

Stephen Halker: Figure ta53.03

Joyce Lavery Hall: Figures: 20.03, 34.22, 34.25, 46.01, 46.03, 47.03b, 50.03, 50.07, 50.09, 50.10, 50.11, 50.15, 50.16, 50.17, 50.18, 50.20, 52.06, 52.11, 52.20

Jonathan Higgins: Figures: 20.09, 20.14, 20.17, 20.20, 21.14, 57.01, 57.02, 57.04, 57.05, 59.01

Kellie Holoski: Figures: 22.08, 44.11, 44.17, 44.18, 44.20, 44.21, 50.19

Martin Huber: Figures: 14.11, 15.05, 15.06, 15.07, 15.09, 15.13, 15.14, 15.15, 15.20, 15.21, 19.02, 19.08, 31.02, 31.07, 31.09, 31.16, 46.12, 46.17, 49.01, 49.03, 49.06, 49.09, 49.11, 49.17, 49.19, 49.20, 49.22, 49.28, 49.29, 49.34, 50.02, 51.02, 51.03, ta52.01, 51.05, 51.06, ta51.01, 51.08, 51.09, 51.11–51.15, 51.17, 51.18, 51.19, 51.20, 53.16, 53.26, 57.9

Sara Krause: Figures: 19.09, 24.01, 24.05, 25.03, 33.13, 34.02, 34.19, 34.20, 34.21, 35.12, 35.18, 37.17, 42.08

Tiana Litwak: Figures: 19.12, 19.14, 23.05, 23.11, 23.13, 34.34a, 35.19, 35.21, 35.22, 40.09

Jacqueline Mahannah: Figures: ta32.02, 35.14b, 35.15, 35.39, 46.08, 47.09, 47.18

Gwen DeCelles: Figures: 16.19, 16.21, 18.09, 19.05, 19.12, 19.14, 19.21, 23.11, 25.02, 25.07, 26.12, 28.11, 28.13, 29.11, 29.13, 33.12, 34.09, 35.04, 35.13, 37.08, 39.01, 41.06, 41.20, 41.29, 42.02, 48.15, 50.03, 53.06, 59.01

Matthew McAdams: Figures: 20.05, 20.06, 20.12, 22.05, 22.08, 22.15, 55.06, 55.08, 55.15

Thomas Moorman: Figures: 1.01, 6.21, 20.05, ta22.01, 54.01, 54.02, 54.18, 54.29, 55.12

Fiona Morris: Figures: 3.16, 20.05, 21.20, 54.25, 56.03, 56.04, 56.06, 56.08, 56.14, 56.20

Kim Moss: Figures: 1.12, 3.15, 3.22, 3.29, 10.05, 13.12, 15.17, 15.18, 15.19, 15.22, 16.02, 16.09, 16.12, 16.14, 17.02, 17.11, 18.03, 23.13, ta30.01, 30.23, 43.02, 45.02, 45.20, 46.08, 47.05, 47.10, 48.06, 48.15, 48.16, 49.11, 57.15, 57.16

Evelyn Pence: Figures: 4.05a, 4.06, 4.07, 4.08a, 4.10, 4.12, 4.14, 4.17, 4.18, 4.19, 4.20, 4.22, 4.26, 5.02, 8.19, 8.21, 19.06, 19.07, 22.15, CH 27, CH 28, 36.32, 37.01, 37.02, 37.03, 37.05, 41.12, 41.18, 41.19, 41.20, 41.21

Curtis Perone: Figures: 16.02, 16.07, 20.07

Michael Rothman: Figure 37.15

Tara Russo: Figures: 3.02, 3.03, 3.21, 3.24, 3.25, 3.27, 3.30, 5.13, 6.01, 6.09, 6.10, 6.12, 6.13, 6.14, 22.19

Alison Schroer: Figure: 43.02

Cameron Slayden: Figures: 35.14, 3515, 35.39

Rick Simonson: Figures: 19.13, 23.09, 25.01, 25.04, 26.16, 36.21, 42.27

Zeke Smith: Figures: ta35.04, 47.06b, 47.08, 50.12, 53.31

Tami Tolpa: Figures: 15.12, 17.02

Brook Wainwright: Figures: 16.03, 16.04, 16.05, 16.06, 20.01, 20.11, 26.08, 28.08, 28.11, 28.12

Travis Vermilye: Figures: ta43.02, ta43.03, 45.11, 45.13, 45.22

index

Boldface page numbers correspond with **boldface terms** in the text. Page numbers followed by an "f" indicate figures; page numbers followed by a "t" indicate tabular material.

A

Aardvark, 520, 520f
Aarskog-Scott syndrome, 246f
Abalone, 655
A band, 953–54, 953–54f
ABC model, of floral organ specification, 834–36, 835f
Abdominopelvic cavity, **852**, 853f
Abiotic realm, 1190
ABO blood group, 88t, 230t, 231, **232**–33, 233f, 376, 1060
Abomasum, 973, 973–74f
Abortion, spontaneous, 249, 1110
Abscisic acid, 754, **767**, 767f, 812–13, 814t, **823**–24, 824f
Abscission, **811**, 811f
  prevention of, 818
Abscission zone, 811
Absolute dating, 422
Absorption
  in digestive tract, 964, 970–72
  water and minerals in plants, 763–64, 763f
Absorption maximum, 913
Absorption spectrum, of photosynthetic pigments, 148–**49**, 148f
Abstinence, **1081**
Acacia, mutualism with ants, 795, 795f, 1180–81, 1180f
Acanthodian, 687t, 688
Acceptor stem, **290**–91, 291f
Accessory digestive organs, 965f, 970–71, 971f, 976–77
Accessory pigment, 149–50, 150f, 154–55
Accessory sex organs
  female, 1080
  male, 1074–75, 1075f
Acellular bone, 949
Acetaldehyde, 34f, 127f, 137, 137f
Acetic acid, 34f
Acetylation, of histones, 316, 316f
Acetylcholine (ACh), **879**–80, 879f, 892t, 893–94, 894f, 956, 995
Acetylcholine (ACh) receptor, 170, 875f, 893–94, 894f
Acetylcholinesterase (AChE), **880**

Acetyl-CoA
  from fat catabolism, 138, 138–39f
  oxidation in Krebs cycle, 128–31, 130f
  from protein catabolism, 138f
  from pyruvate, 124, 127–28, 127–28f
  uses of, 139
ACh. *See* Acetylcholine
achaete-scute (transcription factor), 495
AChE. *See* Acetylcholinesterase
Achiasmate segregation, **211**
Acid, 29–**30**
Acid growth hypothesis, **818**, 818f
Acid precipitation, **1228**, 1228f, 1247
Acid soil, 776
Acinar cells, 975f
Acini, **970**
Acoela, 629f, 637f, 647, 647f
Acoelomate, **625**, 625f, 629f, 636, 636f, 644–48, 645–48f
Acoelomorpha, 647
Aconitase, 130f
Acorn, 752
Acorn worm, 631t
Acquired characteristics, inheritance of, 396, 396f
Acquired immunity. *See* Active immunity
Acquired immunodeficiency syndrome. *See* AIDS
Acromegaly, **933**
Acrosomal process, **1088**
Acrosome, **1074**, 1074f, **1088**
ACTH. *See* Adrenocorticotropic hormone
Actin, 44f, 45, 45t, **77**, 187, 858, 953–55f, **954**
Actin filament, **77**–78, 77f, 80, 87, 182, 196, 389, 576. *See also* Thin myofilament
Actinobacteria, 544f
Actinomyces, 544f
Actinopoda (phylum), 577
Actinopterygii (class), 686f, 687t, **690**, 690f
Actinosphaerium, 577f
Action potential, 875–77
  all-or-none law of, 876
  falling phase of, 875–76, 876f

generation of, 875–77, 876–78f, 900f
  propagation of, 877, 877f
  rising phase of, 875–76, 876f
  undershoot of, 875–76, 876f
Action spectrum, **149**
  of chlorophyll, 148f, 149
Activation energy, **109**, 109f, 111–12
Activator, 114, **307**, **313**, 314–15f, 316
  allosteric, 114
Active immunity, 1046, 1055f
Active site, **112**, 112f
Active transport, **97**–100, 102t, 758f
Activin, 1104
Acute-phase proteins, 1043
Acute-phase response, **1042**
Adaptation, to different environments, 1147, 1147f
Adapter protein, **173**, 175
Adaptive radiation, **443**–44, 443f, 446–47, 447f
Adaptive significance, of behavior, **1130**–31
Adaptive value, of egg coloration, 1130, 1131f
Adder, 699
Adder's tongue fern, 188t
Addiction. *See* Drug addiction
Adenine, 41, 258, 258f, 819, 819f
Adenohypophysis, 921f, 922t, **928**, 934f, 937f
Adenosine deaminase deficiency, 342
Adenosine diphosphate. *See* ADP
Adenosine monophosphate. *See* AMP
Adenosine triphosphate. *See* ATP
Adenovirus, 515f, 526f
Adenylyl cyclase, **176**, 177–78f, 178–79, 928
ADH. *See* Antidiuretic hormone
Adherens junction, 181t, **182**
Adhesion, **27**, 27f, 758
ADH gene, 397
Adipose cells, **856**, 857t
Adipose tissue, **856**, 856f
ADP, **110**, 110f
Adrenal cortex, 923t, **936**, 937f, 1035
Adrenal gland, 862f, 893, 921f, **936**, 937f

Adrenal hypoplasia, 246f
Adrenaline. *See* Epinephrine
Adrenal medulla, 921f, 923t, **936**, 937f, 1102
Adrenocorticotropic hormone (ACTH), 922t, **930**, 933, 936, 937f
Adrenoleukodystrophy, 246f
Adrenomyeloneuropathy, 246f
Adult stem cells, 380, 380f
Adventitious plantlet, 846, 846f
Adventitious root, 730f, **731**, 735, 754f, 768, 828, 829f, 833f
Aerenchyma, **768**, 768–69f
Aerial root, 730f, **731**
Aerobic capacity, 958
Aerobic respiration, **121**, 127, 133f
  ATP yield from, 134–35, 134f
  evolution of, 140
  regulation of, 135, 135f
Aesthetic value, of biodiversity, 1244
Afferent arteriole, 1030
Afferent neuron. *See* Sensory neuron
Aflatoxin, 618, 618f
Africa, human migration out of, 712
African boomslang, 699
African sleeping sickness. *See* Trypanosomiasis
African violet, 736f
Afrotheria, 520
Afrovenator, 697f
Afterbirth, 1111
Agammaglobulinemia, 246f
Age, at first reproduction, 1155–56
Agent Orange, 818
Age of Amphibians, 693
Age of Mammals, 706
Age structure
  of population, **1152**–54
  population pyramids, **1162**–63, 1162f
Agglutination reaction, 1054, 1054f
  blood typing by, 1060
Aggregate fruit, 751f
Aging, telomerase and, 272
Agonist (muscle), **952**
Agnatha (superclass), 687